# BIOLOGY
# HOW
# LIFE
# WORKS

## James Morris
BRANDEIS UNIVERSITY

## Daniel Hartl
HARVARD UNIVERSITY

## Andrew Knoll
HARVARD UNIVERSITY

## Robert Lue
HARVARD UNIVERSITY

ANDREW BERRY, ANDREW BIEWENER,
BRIAN FARRELL, N. MICHELE HOLBROOK,
NAOMI PIERCE, ALAIN VIEL
HARVARD UNIVERSITY

W.H. Freeman and Company

A Macmillan Higher Education Company

PUBLISHER  Susan Winslow

LEAD DEVELOPMENTAL EDITOR  Lisa Samols

SENIOR DEVELOPMENTAL EDITOR  Susan Moran

DEVELOPMENTAL EDITOR  Erica Pantages Frost

EDITORIAL ASSISTANTS  Yassamine Ebadat, Jane Taylor

REVIEW COORDINATOR  Donna Brodman

PROJECT MANAGER  Karen Misler

ART MANAGER  Carolyn Deacy

EDITORIAL RESEARCH AND DEVELOPMENT  Shannon Howard

MARKET DEVELOPMENT MANAGER  Lindsey Veautour

ASSOCIATE DIRECTOR OF MARKETING  Debbie Clare

LEAD ASSESSMENT AUTHOR  Melissa Michael

ASSESSMENT AUTHORS/TEAM LEADERS  Mark Hens, John Merrill, Randall Phillis, Debra Pires

ART AND MEDIA DIRECTOR  Robert Lue, Harvard University

MANAGER OF DIGITAL DEVELOPMENT  Amanda Dunning

SENIOR DEVELOPMENT EDITOR FOR TEACHING & LEARNING STRATEGIES  Elaine Palucki

SENIOR MEDIA PRODUCER  Keri Fowler

PROJECT EDITOR  Robert Errera

MANUSCRIPT EDITOR  Nancy Brooks

DESIGN  MGMT. design

SENIOR ILLUSTRATION COORDINATOR  Bill Page

ILLUSTRATIONS  Imagineering

CREATIVE DIRECTOR  Mark Mykytiuk, Imagineering

ART DIRECTOR  Diana Blume

LAYOUT ARTIST  Tom Carling, Carling Design Inc.

PHOTO EDITOR  Christine Buese

PHOTO RESEARCHERS  Jacquelin Wong and Deborah Anderson

PRODUCTION MANAGER  Paul Rohloff

COMPOSITION  MPS Limited

PRINTING AND BINDING  Quad Graphics–Versailles

Library of Congress Control Number: 2012948406

ISBN-13: 978-1-4292-1870-2

ISBN-10: 1-4292-1870-3

International Edition

ISBN-13: 978-1-4641-5601-4

ISBN-10: 1-4641-5601-8

Printed in the United States of America

Second printing

Macmillan

W. H. Freeman and Company

41 Madison Avenue

New York, NY 10010

Houndmills, Basingstoke RG21 6XS, England

www.macmillanhighereducation.com/international

www.whfreeman.com

*To all who are curious
about life and how it works.*

# ABOUT THE AUTHORS

**James R. Morris** is Associate Professor in the Biology Department at Brandeis University. He teaches introductory biology, as well as a wide variety of courses for majors and non-majors in evolution, genetics, genomics, and anatomy. In addition, he teaches a first-year seminar focusing on Darwin's *On the Origin of Species*. He is the recipient of numerous teaching awards. His research focuses on the rapidly growing field of epigenetics, using the fruit fly *Drosophila melanogaster* as a model organism. He currently pursues this research with undergraduates in order to give them the opportunity to do genuine, laboratory-based research early in their scientific careers. Dr. Morris received a PhD in genetics from Harvard University and an MD from Harvard Medical School. He was a Junior Fellow in the Society of Fellows at Harvard University, has given talks to the public on current science at the Museum of Science in Boston, and works on promoting public understanding of personal genetics and genomics.

**Daniel L. Hartl** is Higgins Professor of Biology in the Department of Organismic and Evolutionary Biology at Harvard University. He has taught highly popular courses in genetics and evolution at both the introductory and advanced levels. His lab studies molecular evolutionary genetics and population genetics and genomics. Dr. Hartl is the recipient of the Samuel Weiner Outstanding Scholar Award and the Medal of the Stazione Zoologica Anton Dohrn, Naples. He is a member of the National Academy of Sciences and the American Academy of Arts and Sciences. He has served as President of the Genetics Society of America and President of the Society for Molecular Biology and Evolution. Dr. Hartl's PhD was awarded by the University of Wisconsin, and he did postdoctoral studies at the University of California, Berkeley. Before joining the Harvard faculty, he served on the faculties of the University of Minnesota, Purdue University, and Washington University Medical School. In addition to publishing more than 350 scientific articles, Dr. Hartl has authored or coauthored 30 books.

**Andrew H. Knoll** is Fisher Professor of Natural History in the Department of Organismic and Evolutionary Biology at Harvard University. He is also Professor of Earth and Planetary Sciences. Dr. Knoll teaches introductory courses in both departments. His research focuses on the early evolution of life, Precambrian environmental history, and the interconnections between the two. He has also worked extensively on the early evolution of animals, mass extinction, and plant evolution. He currently serves on the science team for NASA's mission to Mars. Dr. Knoll received the Phi Beta Kappa Book Award in Science for *Life on a Young Planet*. Other honors include the Paleontological Society Medal and Wollaston Medal of the Geological Society, London. He is a member of the National Academy of Sciences, the American Academy of Arts and Sciences, and the American Philosophical Society. He received his PhD from Harvard University and then taught at Oberlin College before returning to Harvard.

**Robert A. Lue** is Professor in the Department of Molecular and Cellular Biology and Director of Life Science Education at Harvard University. He regularly teaches courses in Harvard's first-year Life Sciences program and upper-level courses in cell biology. He has a long-standing commitment to interdisciplinary teaching and research, and chaired the faculty committee that developed an integrated science course to serve science majors and premedical students. Dr. Lue has also developed award-winning multimedia, including the animation "The Inner Life of the Cell." He has coauthored undergraduate biology textbooks and chaired education conferences on college biology for the National Academies and the National Science Foundation and on diversity in science for the Howard Hughes Medical Institute and the National Institutes of Health. He also founded and directs a Harvard Life Sciences outreach program that serves over 50 high schools. He received his PhD from Harvard University.

**Andrew Berry** is Lecturer in the Department of Organismic and Evolutionary Biology and an undergraduate advisor in the Life Sciences at Harvard University. He teaches courses in Harvard's first-year Life Sciences program, as well as courses on evolution and Darwin. His research interests are in evolutionary biology and the history of science. He has coauthored two books: *Infinite Tropics*, a collection of the writings of Alfred Russel Wallace, and *DNA: The Secret of Life*, which is part history, part exploration of the controversies swirling around DNA-based technology.

**Andrew A. Biewener** is Charles P. Lyman Professor of Biology in the Department of Organismic and Evolutionary Biology at Harvard University and Director of the Concord Field Station. He teaches both introductory and advanced courses in anatomy, physiology, and biomechanics. His research focuses on the comparative biomechanics and neuromuscular control of mammalian and avian locomotion, with relevance to biorobotics. He is currently Deputy Editor-in-Chief for the *Journal of Experimental Biology*. He also served as President of the American Society of Biomechanics.

**Brian D. Farrell** is Professor of Biology in the Department of Organismic and Evolutionary Biology at Harvard University and Curator of Entomology in the Museum of Comparative Zoology. He has collaborated with Los Niños de Leonardo y Meredith in the Dominican Republic to teach children about native insects, and participates in an All Taxa Biodiversity Inventory of the Boston Harbor Islands National Recreation Area. His research focuses on the interplay of adaptation and historical contingency in species diversification, particularly of beetles. In 2011–2012, Dr. Farrell was a Fulbright Scholar to the Universidad Autonoma de Santo Domingo.

**N. Michele Holbrook** is Charles Bullard Professor of Forestry in the Department of Organismic and Evolutionary Biology at Harvard University. She teaches an introductory course on biodiversity as well as advanced courses in plant biology. She studies the physics and physiology of vascular transport in plants with the goal of understanding how constraints on the movement of water and solutes between soil and leaves influences ecological and evolutionary processes.

**Naomi E. Pierce** is Hessel Professor of Biology in the Department of Organismic and Evolutionary Biology at Harvard University and Curator of Lepidoptera in the Museum of Comparative Zoology. She studies and teaches animal behavior and behavioral ecology. Her lab focuses on the ecology of species interactions, such as insect–host plant associations, and on the life-history evolution and systematics of Lepidoptera. She has also been involved in reconstructing the evolutionary tree of life of insects such as ants, bees, and butterflies.

**Alain Viel** is Director of Undergraduate Research and Senior Lecturer in the Department of Molecular and Cellular Biology at Harvard University. He teaches research-based courses as well as courses in molecular biology and biochemistry. He is a founding member of BioVisions, a collaboration between scientists, teaching faculty, students, and multimedia professionals that focuses on science visualization. Dr. Viel worked with his colleague and *Biology: How Life Works* coauthor Robert Lue on the animation "The Inner Life of the Cell."

## *Biology: How Life Works* Assessment Authors

**Melissa Michael**, Lead Assessment Author, is Director for Core Curriculum and Assistant Director for Undergraduate Instruction for the School of Molecular and Cellular Biology at the University of Illinois at Urbana-Champaign. A cell biologist, she primarily focuses on the continuing development of the School's undergraduate curricula. She is currently engaged in several projects aimed at improving instruction and assessment at the course and program levels. She continues to work in several different arenas to improve undergraduate biology education.

**Mark Hens** is Associate Professor of Biology at the University of North Carolina at Greensboro. He has taught introductory biology at this institution since 1996. He is the director of his department's Introductory Biology Program and is chair of the university's General Education Council. In these administrative capacities, he leads efforts on his campus to establish learning objectives and develop assessment tools. Dr. Hens is a National Academies Education Mentor at the National Academies/HHMI Summer Institute for Undergraduate Education in Biology. He also serves on the advisory board of an NSF-funded project focused on the assessment of student learning in college science curricula.

**John Merrill** is Director of the Biological Sciences Program in the College of Natural Science at Michigan State University. This program administers the core biology course sequence required for all science majors. In recent years, Dr. Merrill has focused his research on teaching and learning. With the support of several NSF grants, he is exploring innovative classroom interventions coupled to enhanced assessment. A particularly active area is the use of computers to analyze student's written responses to conceptual assessment questions, with the goal of making it easier to use open-response questions in large-enrollment classes.

**Randall Phillis** is Associate Professor of Biology at the University of Massachusetts Amherst. He has taught in the majors introductory biology course at this institution for 19 years and is a National Academies Education Mentor in the Life Sciences. With help from the PEW Center for Academic Transformation (1999), he has been instrumental in transforming the introductory biology course to an active learning format that makes use of classroom communication systems. He also participates in an NSF-funded project to design model-based reasoning assessment tools for use in class and on exams. These tools are being designed to develop and evaluate student scientific reasoning skills, with a focus on topics in introductory biology.

**Debra Pires** is an Academic Administrator at the University of California, Los Angeles. She teaches the introductory courses in the Life Sciences Core Curriculum. Her research focuses on creating assessment tools to evaluate how well students understand concepts taught in the introductory courses and how well they retain those concepts during their time at UCLA. Student Learning Outcome (SLO)-centered assessments have become a major component of the introductory curriculum, and workshops with faculty in two departments have begun to help instructors develop rubrics and assessment strategies that are aligned with the goals of the long-term study.

# BRIEF CONTENTS

Dear Students and Instructors,

We wrote this book in recognition of recent and exciting changes in biology, education, and technology. There was a time when introductory biology could cover all of biology over the course of a single year. This is no longer possible. The amount of scientific information has grown exponentially, necessitating that we, as teachers, rethink the role of introductory biology and the resources that support it. One goal remains paramount: to help students think like biologists. To think like a biologist means understanding key concepts that span all of biology. It means being able to communicate in the shared language of biologists. It means recognizing the powerful ability of evolution to explain both the unity and diversity of life. It means thinking about how biological research can help solve some of the world's most pressing issues, from cancer to infectious diseases to biodiversity loss to climate change.

We have also noticed a change in the way biological problems are approached. We now have a "parts list" of genes and proteins for how life works, and many scientists today are focused on how the parts work together. As a result, we can no longer divide information into discrete topics. To prepare students for science as it is currently practiced, we must integrate concepts from different areas of biology as well as from other scientific disciplines.

What is particularly exciting for us as teachers is that the remarkable changes in the science of biology are paralleled by a new appreciation for and understanding of how students learn. There is now good evidence that teaching students only by lecturing does not lead to mastery of core concepts. Lecturing alone does not help develop the scientific skills and habits of mind that students need to become successful scientists and health-care workers or thoughtful, scientifically informed citizens. Students learn most effectively when they are actively involved in their learning and construct their own knowledge through a combination of lectures, problem solving, hands-on experiences, and collaborative work.

At the same time, technology is transforming how and where students access information. The Internet provides all kinds of information at a click. There is no need for a modern textbook to be a reference book. What, then, is the role of a textbook? A textbook needs to be selective, help students see connections between seemingly disparate topics, and make the material engaging and relevant. Technology is also making possible new and unprecedented ways to visualize biological processes and provide interactive ways for students to learn.

To support 21st-century student learning and instructor teaching, we feel that it is time to rethink what takes place both in and out of the introductory biology classroom and to reimagine the resources that can best support these efforts. *Biology: How Life Works* provides an integrated set of resources to engage students, encourage critical thinking, help students make connections, and provide a framework for further studies.

Sincerely,
The *Biology: How Life Works* author team

## Rethinking Biology

For the *Biology: How Life Works* team, it has been an exciting experience to reimagine what a resource for today's students should look like. We began with the question, "What do we want students to understand and apply at the end of the course?" Once we decided where we wanted students to end up, we started asking other questions: What content should be covered? How should it be organized? How should it be delivered—in the narrative, a figure, an animation? What questions should we ask students to gauge their understanding? What questions should we ask to help students work toward that understanding? Through this process of questioning and answering, one point became clear—we'd have to start from scratch and build *Biology: How Life Works* from the ground up.

## Rethinking biology means rethinking the text, visual program, and assessment.

Rethinking biology required rethinking the way a textbook is created. Ordinarily, textbooks are developed by first writing chapters, then making decisions about art and images, and finally, once the book is complete, assembling a test bank and ancillary media. This process dramatically limits integration across resources and reduces art, media, and assessments to supplements rather than essential resources for student learning.

*Biology: How Life Works* is the first project to develop three pillars of learning—the text, visual program, and assessment—at the same time. These three pillars are all tied to the same set of core concepts, share a common language, and use the same visual palette. In this way, the visual program and assessments are integral parts of student learning, rather than accessories to the text. In addition, every concept is conveyed and explored in multiple ways, allowing for authentic learning.

## *Rethinking biology means rethinking*
# THE TEXT

*Biology: How Life Works* includes a text that is uniquely *integrated*, *selective*, and *thematic*.

### INTEGRATED

Textbooks commonly present biology as a series of minimally related chapters. This style of presentation lends itself to memorization and offers students little guidance on how concepts connect to one another and to the bigger picture. *Biology: How Life Works* moves away from this model and toward an integrated approach. Across the book, we present chemistry in context, and cover structure and function together. We introduce the flow of information in a cell where it makes the most conceptual sense and use cases as a framework for connecting and assimilating information.

### SELECTIVE

With the ever-increasing scope of biology, it is unrealistic to expect the majors course or a textbook to cover everything. From the start, we envisioned *Biology: How Life Works* not as a reference book for all of biology, but as a resource focused on foundational concepts, terms, and experiments. We explain fundamental topics carefully, with an appropriate amount of supporting detail. This allows students to more easily identify, understand, and apply critical concepts. In this way, students will leave an introductory biology class with a framework on which to build.

### THEMATIC

We wrote *Biology: How Life Works* with six themes in mind. Deciding on these themes in advance helped us make decisions about which concepts to include and how to organize them. Introduced in Chapter 1 and revisited throughout the text, the themes provide a framework that helps students see biology as a set of connected concepts. In particular, we emphasize the theme of evolution for its ability to explain and predict so many patterns in biology.

1. We learn how life works by applying the scientific method, which involves making observations, generating hypotheses, and testing hypotheses through experiment and observation.
2. Life works according to fundamental principles of chemistry and physics. All organisms share a limited number of molecules and chemical processes.
3. All of the chemical and physical functions of life are packaged within cells. Multicellular organisms function by the differentiation and coordinated operation of many cells.
4. Both the features that organisms share and those that set them apart are explained by evolution. Variation exists within as well as between species.
5. Organisms interact in nature, with basic features of anatomy, physiology, and behavior shaping the ecological systems that sustain life.
6. Humans have emerged as major agents in ecology and evolution. Our future welfare depends, in part, on improving our knowledge of how life works.

> I think the figures are much simpler for beginning students to understand. I found all of the figures in my chapter to be clear, concise, and distilled into the most relevant facts necessary to illustrate the concepts.
>
> DALE CASAMATTA, INSTRUCTOR
> UNIVERSITY OF NORTH FLORIDA

> The figures in this chapter [Chapter 22: Species and Speciation] are particularly good. These are not the only great figures, but they are among my favorites for how they demonstrate a concept so easily that might otherwise require a couple of paragraphs of text.
>
> MATT BREWER, INSTRUCTOR
> GEORGIA STATE UNIVERSITY

> My initial reaction was "Wow." It [animation on gene expression] helped me visualise the spatial relationships associated with information flow at the cellular level, and I think it is thus likely to really help undergraduates.
>
> DAVE KUBIEN, INSTRUCTOR
> UNIVERSITY OF NEW BRUNSWICK

## *Rethinking biology means rethinking*
# THE VISUAL PROGRAM

The *Biology: How Life Works* visual program—all art and visual media—is an *integrated, engaging, visual framework* for understanding and connecting concepts.

### INTEGRATED

Just as we decided on a consistent vocabulary to explain concepts and processes, we also created a consistent "visual language." This means that across *Biology: How Life Works*—whether students are looking at a figure in the book, watching an animation, or interacting with a simulation—they see a consistent use of color, shapes, and design. Having a coherent, integrated visual program allows students to recognize concepts they have already encountered, and assimilate new information.

### ENGAGING

We have a shared goal with instructors—we want students to *want* to learn more. Cognitive science tells us visualization is tremendously effective at triggering student interest and passion. With that in mind, the *Biology: How Life Works* team committed to designing and developing a visual program with the same attention and care that goes into text development. Every image—still and in motion—engages students by being vibrant, clear, and approachable. The result is a visual environment that pulls students in, deepens their interest, and helps them see a world of biological processes.

### VISUAL FRAMEWORK

Scientists often build a contextual picture, or visual framework, in their mind upon which they hang facts and connect ideas. To help students think like biologists, our visual program deliberately provides this type of framework. Individual figures present foundational concepts; Visual Synthesis figures tie together multiple concepts across multiple chapters; animations bring these figures to life, allowing students to explore concepts in space and time; and simulations have  students interact with the concepts. Collectively, this visual framework gives students a way to contextualize information—to move seamlessly back and forth between the big picture and the details, from the basic to the complex.

## *Rethinking biology means rethinking*
# ASSESSMENT

*Biology: How Life Works* represents a groundbreaking departure from traditional assessment materials. Written by leaders in science education in collaboration with the authors, our assessments are fully *integrated* with the text and visual program, span the *full range of Bloom's Taxonomy,* and are properly *aligned.*

### INTEGRATED

If assessment is important, it cannot be ancillary—it must be integral to the learning process. Each time an instructor asks a student to engage with *Biology: How Life Works*—whether it is reading a chapter, watching an animation, or working through an experiment—the opportunity to assess that experience exists. This unprecedented level of integration results from developing assessments alongside the text and visual program. This parallel development also allows for consistency of language and focus, helping students connect what they have read with the questions they are being asked to answer.

### RANGE

For the first time, instructors in majors biology have access to a set of thoughtfully developed, peer-reviewed formative and summative assessments. Most test banks consist entirely of low-order, recall questions. Consequently, instructors wanting to teach and test at a range of levels are left with the time-consuming task of creating their own higher-order questions. The *Biology: How Life Works* assessment team has done the "heavy lifting" of question writing for instructors—providing assessments that span everything from recall to synthesis. They are designed to be used in a range of settings (pre-class, in-class, post-class, and exam) and come in a variety of formats (multiple choice, multiple true/false, free response). In addition to questions, our assessments include in-class activities and interactive, online exercises.

### ALIGNED

Traditionally, assessment questions are provided as test banks, organized by chapter but not as a series designed to work together. We believe questions aren't just for testing—they are for teaching. To help instructors assess students in more meaningful ways, we wrote and organized sets of questions we call *progressions*. Progressions include formative and summative assessment—reading comprehension questions, in-class activities, post-class assignments, and exam questions—all properly aligned with the text's core concepts. Used in sequence, questions within a progression provide a connected learning path for students and a suggested teaching path for instructors.

# AUTHORING *BIOLOGY: HOW LIFE WORKS*

*Biology: How Life Works* is authored by a team of nationally recognized scientists and educators.

James Morris, Lead Author, Brandeis University
Daniel Hartl, Lead Author, Harvard University
Andrew Knoll, Lead Author, Harvard University
Robert Lue, Lead Author and Visual Program Director, Harvard University
Andrew Berry, Author, Harvard University
Andrew Biewener, Author, Harvard University
Brian Farrell, Author, Harvard University
N. Michele Holbrook, Author, Harvard University
Naomi Pierce, Author, Harvard University
Alain Viel, Author, Harvard University

Melissa Michael, Lead Assessment Author, University of Illinois at Urbana-Champaign
Mark Hens, Assessment Author, University of North Carolina at Greensboro
John Merrill, Assessment Author, Michigan State University
Randall Phillis, Assessment Author, University of Massachusetts Amherst
Debra Pires, Assessment Author, University of California, Los Angeles

The mission of *Biology: How Life Works* is not to cover all of biology, but to provide a proper introduction. This selective approach requires a large team of educators who are experts in their respective fields. Because the authors possess an intimate knowledge of their fields, they were able to step back and consider where their fields have been, what questions are currently being asked, and where they are heading. It is only with a *collective* perspective that *Biology: How Life Works* is able to present contemporary biology in a complete but selective manner.

The multi-author approach was also necessary for the amount of writing, design, development, and collaboration this project required. With the text, visual program, and assessments all being developed in parallel, *Biology: How Life Works* needed more than one author. Additionally, the level of integration, creativity, and new ideas present in the three pillars of *Biology: How Life Works* was largely achieved by having a community of authors who were able to dialogue, discuss, and debate.

At the same time, *Biology: How Life Works* has a single voice. This is possible because every chapter, image, animation, and assessment flowed through one author, James Morris. In this way, it benefits from multi-author expertise while telling a single, cohesive story.

In sum, the size of the author team is purposeful and directly supports the goals of the project.

# KEY FEATURES SUPPORTING THE VISION

## TOPIC COVERAGE AND ORDER:

The goal of a single narrative of biology is supported by the Table of Contents. a select number of important and purposeful changes. Below is an annotated table of

Chapter 1 introduces evolution as a major theme of the book before discussing microevolution in Chapter 4 as a foundation for later discussions of conservation of metabolic pathways and enzyme structure (Chapters 6–8) and genetic and phenotypic variation (Chapters 14 and 15). After the chapters on the mechanisms of evolution (Chapters 21–24), we discuss diversity of all organisms in terms of adaptations and comparative features, culminating in ecology as the ultimate illustration of evolution in action.

The first set of chapters emphasizes three key aspects of a cell—information, homeostasis, and energy. Chemistry is taught in the context of biological processes, highlighting the principle that structure determines function.

I found [Chapter 2: The Molecules of Life] quite readable and it makes the chemistry seem much more relevant to organisms from the beginning.
UDO SAVALLI, INSTRUCTOR
ARIZONA STATE UNIVERSITY WEST

The genetics chapters start with genomes, move to mutation and genetic variation, and then consider inheritance to provide a modern, molecular look at genetic variation and how traits are transmitted.

# AND STORY OF BIOLOGY: HOW LIFE WORKS

The book's chapters and sections are arranged in a familiar way to be used in a range of introductory biology courses, with contents, highlighting these changes and the reasons behind them.

[Chapter 25: Cycling Carbon] does a great job connecting the "halves"; providing examples in the text and asking questions along the way forces the student to give some thought to the topic. . . . Well done.
STEPHEN TRUMBLE
BAYLOR UNIVERSITY

We present the carbon cycle as a bridge between the molecular and organismal parts of the book, integrating ecology and diversity.

The plant chapters combine content in novel ways to better integrate content and provide context for biological processes, such as plant reproduction with the timing of reproductive events.

Diversity follows physiology in order to provide a basis for understanding the groupings of organisms.

■ *Biology: How Life Works* includes chapters that don't traditionally appear in introductory biology texts, one in almost every major subject area. These novel chapters represent shifts toward a more modern conception of certain topics in biology:
- The basis of complex traits (Chapter 18)
- Human evolution (Chapter 24)
- Cycling carbon (Chapter 25)
- Multicellularity (Chapter 28)
- Plant defenses (Chapter 32)
- Human impact on the environment (Chapter 48)

## CONNECTED LEARNING PATH

**CHAPTER 21**

## EVOLUTION

### How Genotypes and Phen
Change over Time

**CORE CONCEPTS**

**21.1** Genetic variation is the result of differences in DNA sequences.

**21.2** Information about allele frequencies is key to understand patterns of genetic variation.

**21.3** Evolution is a change in the

Variation is a fact of nature: A walk down any street reveals h
cies is. Skin color and hair color, for example, vary from person
publication in 1859 of Charles Darwin's *On the Origin of Species*
view all the variation we see in humans and other species as b...
ant. According to the traditional view at the time, not only we
created by God in their modern forms, but, because the Creat...
in mind for each one, they were fixed and unchanging. Depart...
this divinely ordained type were therefore ignored.

Since Darwin, however, we have appreciated that a species...
a type. Rather, a species consists of a range of variants: In o...
may be tall, short, dark-skinned, fair-skinned, and so on. Varia...
ingredient of Darwin's theory because **natural selection dep**...
success—in terms of surviving and reproducing—of variants...
we view variation: Before Darwin, variation was irrelevant, so...

The authors of *Biology: How Life Works* rethought the standard pedagogical tools like chapter summaries and end-of-chapter questions. Each element purposefully relates to the same set of core concepts and functions as a connected learning path for students.

Core Concepts listed at the beginning of each chapter map to the chapter's numbered sections and give students a preview of what ideas they should know by the time they finish reading. We built each chapter around core concepts in order to focus on the most important and overarching ideas in any one topic.

The breakdown of a molecule of glucose by fermentation yields only two molecules of ATP. The energetic gain is relatively small compared to the total yield of aerobic respiration because the end products, lactic acid and ethanol, still contain a large amount of chemical energy in their bonds and are not fully oxidized. The modest yield explains why organisms that produce ATP by fermentation must consume a large quantity of fuel molecules to power the cell.

→ **Quick Check 6** Breadmaking involves ethanol fermentation and requires yeast, sugar, flour, and water. Why are yeast and sugar needed?

Throughout the chapter, Quick Checks ask students to pause and make sure they're ready to move on to the next idea. These brief questions pointedly ask the student to demonstrate a full understanding of a tough topic before reading on.

### Core Concepts Summary

**4.1 PROTEINS ARE LINEAR POLYMERS OF AMINO ACIDS THAT FORM THREE-DIMENSIONAL STRUCTURES WITH SPECIFIC FUNCTIONS.**

An amino acid consists of an α carbon connected by covalent bonds to an amino group, a carboxyl group, a hydrogen atom, and a side chain or R group. page 4-2

There are 20 common amino acids that differ in their side chains. page 4-2

Amino acids are connected by peptide bonds to form proteins. page 4-3

The primary structure of a protein is its amino acid sequence.

tRNA
mRNA

Ami...
tRN...

The...
lett...
it w...

The...
spec...

Tran...
tera...

4.3 PRO...
SEL...

Students revisit the core concepts in the Core Concepts Summary. Here, they can remind themselves of the big-picture ideas from the beginning of the chapter and see them in the context of key supporting ideas.

### Self-Assessment

1. Describe two ways in which Mendel's experimental approach differed from others of the time.
2. Distinguish among gene, allele, genotype, and phenotype.
3. Name and describe Mendel's two laws.
4. Explain the relationship between Mendel's two laws and the mechanics of meiosis.
5. Explain how you can predict the genotypes and phenotypes of offspring if you know the genotypes of the parents.
6. Describe an instance when a testcross would be useful.
7. Define the multiplication and addition rules, and explain how these rules can help you predict the outcome of a cross between

Students test their understanding of core concepts with Self-Assessments. These assessments, along with the online answer guides, allow students to see for themselves whether or not they have understood the major ideas of the chapter.

Learning doesn't end when the printed chapter ends. A wealth of media organized by core concept, including dynamic, zoomable Visual Synthesis figures, animations, and simulations, reinforce what students have read. Assessments give students opportunities to apply what they have learned and hone problem-solving skills.

**4.1.4**
The function of a protein is dependent upon the shape into which the chain of amino acids folds. Many noncovalent interactions are responsible for maintaining the protein's shape. Assume you have isolated a protein from an organism in its proper shape, and you

**4.1.5**
The interactions between amino acids are major factors in determining the shape of a protein. These interactions can be affected by the environment surrounding a protein. Which of the following would have an effect on the shape of a protein?

A. the temperature of the environment
B. the pH of the environment
C. whether the other molecules in the environment are predominantly hydrophilic o
hydrophobic
D. the concentrations of ions present in the environment
E. All of the above.

| Core Concept | LEVEL | Connections | Guidance |
|---|---|---|---|
| 4.1 | LOC, although a thorough explanation of the free response questions would be HOC (comprehension) for many (most?) first-year biology students. | Figures 4.2, 4.4, 2.11 | Pre-instruction, post-class or exa... |
| | | | MCQ Small group discu... |
| | | | This question pr... good review of... concentration o... |

Instructors can use and assign material from a set of questions we call progressions. Progressions offer questions designed as a sequence to use before, during, and after lecture. Progressions help instructors align the questions they ask on exams to the in-class activities and homework questions they have students engage in before an exam.

## VISUAL SYNTHESIS FIGURES

Just as we integrate topics in biology to create a single narrative, it's our goal to extend integration to the media by creating a seamless path from the book to all online materials. Visual Synthesis figures represent one way that we integrate the text and media. In the book, a Visual Synthesis figure is a two-page spread and serves as a visual summary integrating concepts across multiple chapters. These figures help students look beyond chapter divisions and see how concepts combine to tell a single story. Students continue their exploration online, where they can interact with a dynamic Visual Synthesis figure. Zooming in and out, students can explore both the big picture and the details, building a framework for how concepts connect and relate. The Visual Synthesis figure also functions as a launch pad to other resources, like animations, simulations, and assessment questions.

The course I teach strives to integrate concepts across multiple levels of biological complexity and Visual Synthesis would help with this goal.
BRETTON KENT, INSTRUCTOR
UNIVERSITY OF MARYLAND

Visual Synthesis does a great job of summarizing the content of [Chapter 42: Reproduction] in terms of human development. It appears to be very comprehensive including the concepts of fertilization, cleavage (including both maternal and fetal transcription controls) all the way through cell differentiation and tissue development.
CYNTHIA LITTLEJOHN, INSTRUCTOR
UNIVERSITY OF SOUTHERN MISSISSIPPI

The zoomable Visual Synthesis image is an incredibly powerful visual aid for students, allowing instructors to bridge the gap between the details and the larger concept.
ROBERT MAXWELL, INSTRUCTOR
GEORGIA STATE UNIVERSITY

The zoomable digital media is an excellent tool for students and instructors! The interactive capabilities and ability to view content on different levels of complexity are unlike anything I have seen from other texts.
JEANELLE MORGAN, INSTRUCTOR
GAINESVILLE STATE COLLEGE

## ANIMATIONS AND SIMULATIONS

Biological processes are not static—they are dynamic, always in motion. We use animations as an engaging and revealing way to bring biological processes to life, place them in context, and give students an intuitive sense of how these processes work.

Some concepts are best learned by doing, rather than by reading or watching. *Biology: How Life Works* simulations allow students to explore biological processes directly and problem solve by doing. Each simulation asks students to work within or break a system, and assessment questions lead students through the system so they can work through a core concept.

The animations and simulations are available independently but are also integrated into the online, zoomable *Visual Synthesis* figures. In exploring the *Visual Synthesis* figure, students can watch animations, answer assessment questions, and interact with simulations to guide them through the illustration.

> I would like to incorporate more animations, virtual labs, and real-life scenarios relating to the concepts we are learning in lecture. It is incredibly helpful when these resources are created by the textbook publisher, so they correspond directly to the material I am covering.
>
> ASHLEY SPRING
> BREVARD COMMUNITY COLLEGE

> Students have trouble visualizing cellular activities, so we constantly look for tools to enhance their understanding.
>
> MARGARET OTT
> TYLER JUNIOR COLLEGE

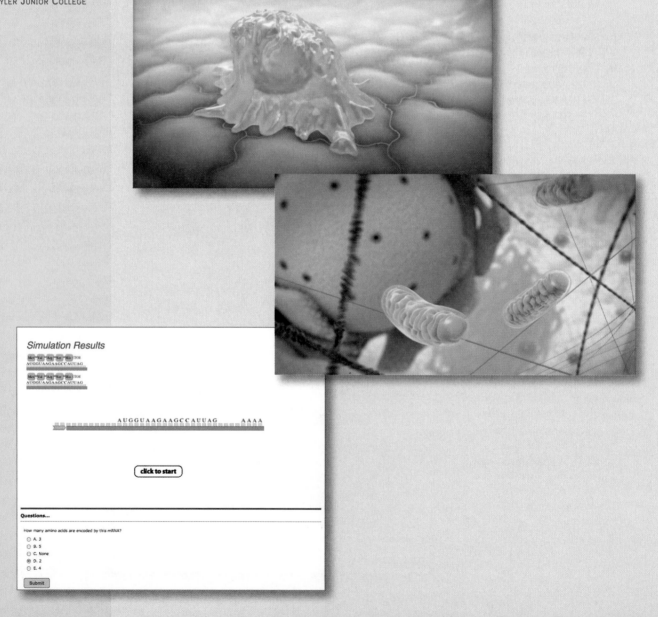

## CASES AND CASE ACTIVITIES

Biology is best understood when presented using real and engaging examples as a framework for putting information together. One of the ways we provide this framework is through cases. Each case begins with a two-page essay, providing background and raising questions about the case's subject. Those questions are answered in subsequent chapters, reinforcing information students have already learned and providing a framework for learning new information. For example, our case about malaria is introduced before the set of chapters on evolution and is revisited in each of these chapters where it serves to reinforce important evolutionary concepts.

The *Case Activities* tie chapter core concepts and cases together, showing students again how what they are learning is relevant to their lives and to a larger picture of biology. We created several activities, varying in length and complexity, for each case. Activities can be assigned as homework or used in class, giving instructors the ability to choose what kind of activity best fits their course needs.

The connection of the case study to the two sections in the chapter is a great way to maintain reader engagement and show the connection of material being presented to real-life problems and solutions.
STEPHEN JURIS, INSTRUCTOR
CENTRAL MICHIGAN UNIVERSITY

This [Cancer: When Good Cells Go Bad] case study is perfect for this group of chapters because it covers cell division, uncontrolled cell growth, cell communication, genetic control, and transformation. In addition, this topic is relevant to the health of the students taking this course.
SHANNON MCQUAIG, INSTRUCTOR
ST. PETERSBURG COLLEGE

I really like this [Cancer: When Good Cells Go Bad] case, including the integration of vaccinations and some public policy questions. This will help students see the real-world application of what they are learning.
JOHN KAUWE, INSTRUCTOR
BRIGHAM YOUNG UNIVERSITY

CASE 1

### The First Cell

LIFE'S ORIGINS

Deep underground, in Mexico's Cueva de Villa Luz, the cave walls drip with slime. The rocky surfaces are teeming with colonies of mucus-producing bacteria. No sunlight reaches these organisms far beneath Earth's surface. Instead, the bacteria survive by capturing energy from hydrogen sulfide gas within the cave. As a by-product of that reaction, the microbes produce sulfuric acid. The stalactite-like slime formations oozing from the cave walls—dubbed "snottites" by researchers—are as corrosive as battery acid.

Snottites might be stomach turning, but they're intriguing, too. Called "extremophiles" because they live in places where humans and most other animals cannot survive, such microorganisms may tell us something about life when Earth was young.

From cave-dwelling bacteria to 100-ton blue whales, the diversity of life on Earth is astounding. Yet all of the planet's organisms, living and extinct, exist on branches of the same family tree. Snottites, swordfish, humans, hydrangeas—all evolved from one single common ancestor.

When, where, and how life originated are some of the biggest questions in biology. Chemical evidence from 3.5-billion-year-old rocks in Australia suggests that biologically driven carbon and sulfur cycles existed at the time those rocks were formed. In the eons since, the first primitive life-forms have evolved into the 100 million or so organisms thought to populate the planet today.

How did the first living cell arise? Before scientists can hope to tackle such a question, they must agree on a definition of life. That's not necessarily as straightforward as it sounds. In our modern world, the features that separate life from non-life are relatively easy to discern. But Earth's first organisms were almost certainly much less complicated than even the simplest bacteria alive today. And before

those first truly living things appeared, molecular systems presumably existed that hovered somewhere between the domains of the living and the nonliving.

All cells require an archive of information, a membrane to maintain the inside of the cell different from the outside, and the ability to harness energy from the environment. In modern organisms, that archive is in the form of DNA, the double-stranded molecule that contains the instructions needed for cells to grow, differentiate, and reproduce. Without that molecular machinery, life as we know it would not exist.

DNA is critical, and it's also complex. Among the organisms alive today, the smallest known genome belongs to the bacterium *Carsonella ruddi*. Even that relatively small genome contains nearly 160,000 DNA base pairs. How could such sophisticated molecular systems have arisen by chance?

The likely answer to that question is step by step. Laboratory experiments have shown how precursors to nucleic acids might have come together under the chemical conditions present on the young Earth. It's exceedingly unlikely that a molecule as complex as DNA was the first archive of information employed by the very first living cells. As you'll see in the chapters that follow, scientists have gathered evidence that hints at what the earliest nucleic acid molecules might have looked like.

While some kind of information archive was necessary for life to unfold, life requires more than a collection of nucleic acids replicating in a warm primordial pond. Living things must have some barrier that separates them from their

> All cells require an archive of information, a membrane to maintain the inside of the cell different from the outside, and the ability to harness energy from the environment.

Snottites deep underground in a Mexican cave. These stalactite-like slime formations are produced by bacteria that break down hydrogen sulfide gas as a source of energy.

environment. The cells of all living things, single-celled organisms or multicellular creatures, are each encased in a cell membrane.

Once again, scientists can only guess at how the first cell membranes came about. But research shows the molecules that make up modern membranes possess some interesting properties that may have led them to arise spontaneously. At first, the membranes were probably quite simple—straightforward (but leaky) barriers that kept the contents of early cells separated from the world at large. Over time, as chance variations arose, those membranes that provided a better barrier and that provided molecular gates were subject to an early example of the same natural selection process that is still happening today. A third essential characteristic of living things is the

as metal ions have been shown to play a role in key cellular reactions. Billions of years after the first cells arose, some of those metal ions—such as iron-sulfur minerals—still play a critical role in cells.

That's one reason researchers are so keen to study creatures like the sulfur-hungry snottites in the Cueva de Villa Luz. When life arose, the planet's atmosphere contained no oxygen—humans couldn't survive in such a world. By

---

? **CASE 1 QUESTIONS**

*Answers to Case 1 questions can be found in Chapters 2–8.*

1. How did the molecules of life form? *See page 2-17.*
2. What was the first nucleic acid molecule, and how did it arise?
3. How did the genetic code originate? *See page 4-14.*
4. How did the first cell membranes form? *See page 5-3.*
5. What naturally occurring elements might have spurred the first
6. What were the earliest energy-harnessing reactions? *See page 7*
7. How did early cells meet their energy requirements? *See page 7*

---

**SECTION 2.6** HOW DID THE MOLECULES OF LIFE FORM?

? **CASE 1 The First Cell: Life's Origins**

### 2.6 HOW DID THE MOLECULES OF LIFE FORM?

In Chapter 1, we considered the similarities and differences between living and non-living things. Four billion years ago, however, the differences may not have been so pronounced. Scientists believe that life originated early in our planet's history by a set of chemical processes that, through time, produced organisms that could be distinguished from their non-living surroundings. How can we think scientifically about one of biology's deepest and most difficult problems? We can't reconstruct key events from the geological record—the study of fossils tells us that life already existed when Earth's oldest surviving fossil-containing rocks were deposited. The alternative is to approach life's origins experimentally, asking whether chemical reactions likely to have taken place on the early Earth can generate the molecules of life. It is important to note that even the simplest living organisms living today are far more complicated than our earliest ancestors. No one suggests that cells as we know them emerged directly from primordial chemical reactions. Rather, the quest is to discover simple molecular systems able to replicate themselves, while subject to natural selection.

A key starting point is the observation, introduced earlier in this chapter, that the principal macromolecules found in organisms are themselves made of simpler molecules joined together. Thus, if we want to understand how proteins might have emerged on the early Earth, we should begin with the synthesis of amino acids, and if we are interested in nucleic acids, we should focus on nucleotides.

**The building blocks of life can be generated in the laboratory.**
Research into the origins of life was catapulted into the experimental age in 1953, with an elegant experiment carried out by Stanley Miller, then a graduate student in the laboratory

I liked the way that experiments were shown in *How Life Works,* particularly Fig 8.7. It summarized three experiments in a way that students could understand them.

DAVID BYRES
FLORIDA STATE COLLEGE JACKSONVILLE

This was a great figure [Fig 13.1]. It achieves the goal, easy to immediately grasp, an appropriate experiment to highlight because it explains how the technique is done conceptually without any of the annoying details. It is perfect.

VICTORIA CORBIN, INSTRUCTOR
UNIVERSITY OF KANSAS

I like the background portion of those figures in . . . *How Life Works;* they help understand the rationale behind the experiment or the technique that was used.

PETER KOURTEV, INSTRUCTOR
CENTRAL MICHIGAN UNIVERSITY

## HOW DO WE KNOW? FIGURES AND ACTIVITIES

Biology is not just about what we know, but also *how we know what we know.* An important skill that introductory students must develop is asking questions and forming hypotheses like scientists. Therefore, we introduce key experiments in the narrative flow of the text and then visually unpack the steps of scientific inquiry in How Do We Know? figures. These figures emphasize that science is a process of asking and answering questions and convey an authentic sense of scientific inquiry.

The exploration of *How Do We Know?* continues with online activities that emphasize using the scientific method. A *How Do We Know?* activity begins with a tutorial that reviews the experimental design of the text figure, followed by questions that ask students to apply what they learned to new experiments, as well as predict how different variables shape the outcome of the experiment. Our goal is to ensure that students understand the concepts covered in the *How Do We Know?* figures, and that they are able to apply the concepts to new situations or experiments.

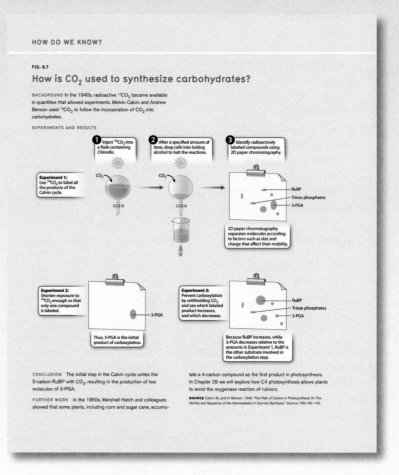

In keeping with our modern approach, we are taking full advantage of the online environment and making **Biology: How Life Works** dynamic. We will continually update and add to our collection of online assessments, visual media, and activities based on feedback, our continued interaction with the teaching community, and the needs of students and instructors.

# ACKNOWLEDGMENTS

A book like this one is a team effort, and we have had the good fortune of working with a remarkable, caring, and dedicated team. First and foremost, we would like to thank the thousands of students we have collectively taught. Their curiosity, intelligence, and enthusiasm have been sources of motivation for all of us. Our teachers and mentors have likewise provided us with models of patience, persistence, and inquisitiveness that we bring into our own teaching and research. They encourage us to be lifelong learners, teachers, and scholars.

We all feel very lucky to be a partner with W. H. Freeman and Company. From the start, they have embraced our project, giving us the space and room to carve out something unique, while at the same time providing guidance, support, and input from the broader community of instructors and students.

Susan Winslow, our publisher, deserves special thanks for her ability to expertly juggle all aspects of this first edition, keeping a watchful eye on important trends in education and science, carefully listening to what we wanted to do, and providing a gentle nudge from time to time.

This book represents a genuine, rewarding, and unique partnership with a team of assessment leaders and authors. We especially thank and acknowledge the contributions of lead assessment author Melissa Michael and assessment team leaders Mark Hens, John Merrill, Randall Phillis, and Debra Pires. We all learned a great deal about pedagogy and how to use questions as tools for teaching, not just testing.

We also especially thank Mark Hens of the University of North Carolina at Greensboro for his contribution to several chapters, and Kirsten Weir, an award-winning science writer and editor, who helped us to set the tone, write, and put together the case introductions.

Lead developmental editor Lisa Samols has just the right touch—she has the ability to listen as well as offer intelligent suggestions; she is serious with a touch of humor; she is quiet but persistent. Senior developmental editor Susan Moran has an eye for detail and the uncanny ability to read the manuscript like a student. Developmental editor Erica Pantages Frost brought intelligence and thoughtfulness to her edits.

Karen Misler kept us all on schedule in a clear and firm, but always understanding and compassionate, way. Thanks also to Dusty Friedman, who helped us start this project on schedule. Carolyn Deacy expertly managed the art program, keeping all the various components coordinated and offering intelligent advice, and Bill Page coordinated our initial relationship with Imagineering.

Shannon Howard, who focused on editorial research and development, and Lindsey Veautour, our market development manager, are remarkable for their energy and enthusiasm, and their creativity in ways to reach out to instructors and students.

We thank Robert Errera for coordinating the move from manuscript to the page. We also thank Diana Blume, our art director, and Tom Carling, the layout artist. Together, they managed the look and feel of the book, coming up with creative solutions for page layout.

Imagineering under the patient and intelligent guidance of Mark Mykytiuk provided creative, insightful art to complement, support, and reinforce the text. Christine Buese, our photo editor, and Jacquelin Wong and Deborah Anderson, our photo researchers, provided us with a steady stream of stunning photos, and never gave up on hard-to-find shots.

On the digital media front, we thank Amanda Dunning for her editorial insight into making pedagogically useful media tools, and Keri Fowler for managing and coordinating the media and websites. They both took on this project with dedication, persistence, enthusiasm, and attention to detail that we deeply appreciate.

We are very grateful to Debbie Clare for her work in marketing, Elaine Palucki for her insight into teaching and learning strategies, Donna Brodman for coordinating the many reviewers, and Yassamine Ebadat and Jane Taylor for their consistent and tireless support.

We are also extremely grateful for all of the hard work and expertise of the sales representatives, regional managers, and regional sales specialists. We have enjoyed meeting and working with this dedicated sales staff, who are the ones that ultimately put the book in the hands of instructors for the first time.

We would also like to acknowledge Charles Smith and Neil Patterson for their early support of the author team and helping to foster the idea that would eventually become *How Life Works*, and special thanks to Charles for many spectacular photos that appear throughout the book.

Countless reviewers made invaluable contributions to this book and deserve special thanks. From catching mistakes to suggesting new and innovative ways to organize the content, they provided substantial input to the book. They brought their collective years of teaching to the project, and their suggestions are tangible in every chapter.

And none of this would have been possible without the support, inspiration, and encouragement of our families.

# Contributors

*Thank you to all the instructors who worked in collaboration with the authors and assessment authors to write* Biology: How Life Works *assessments, activities, and exercises.*

Allison Alvarado, University of California, Los Angeles

Peter Armbruster, Georgetown University

Zane Barlow-Coleman, formerly of University of Massachusetts Amherst

James Bottesch, Brevard Community College

Jessamina Blum, Yale University

Jere Boudell, Clayton State University

David Bos, Purdue University

Laura Ciaccia West, Yale University*

Laura DiCaprio, Ohio University

Tod Duncan, University of Colorado Denver

Cindy Giffen, University of Wisconsin Madison

Paul Greenwood, Colby College

Stanley Guffey, University of Tennessee, Knoxville

Alison Hill, Duke University

Meg Horton, University of North Carolina at Greensboro

Kerry Kilburn, Old Dominion University

Jo Kurdziel, University of Michigan

David Lampe, Duquesne University

Brenda Leady, University of Toledo

Sara Marlatt, Yale University*

Kelly McLaughlin, Tufts University

Brad Mehrtens, University of Illinois at Urbana-Champaign

Nancy Morvillo, Florida Southern College

Jennifer Nauen, University of Delaware

Kavita Oommen, Georgia State University

Patricia Phelps, Austin Community College

Melissa Reedy, University of Illinois at Urbana-Champaign

Lindsay Rush, Yale University*

Sukanya Subramanian, Collin College

Michelle Withers, West Virginia University

*Graduate student, Yale University Scientific Teaching Fellow

# Reviewers, Class Testers, and Focus Group Participants

*Thank you to all the instructors who reviewed and/or class tested chapters, art, assessment questions, and other* Biology: How Life Works *materials.*

Thomas Abbott, University of Connecticut

Tamarah Adair, Baylor University

Sandra Adams, Montclair State University

Jonathon Akin, University of Connecticut

Eddie Alford, Arizona State University

Chris Allen, College of the Mainland

Sylvester Allred, Northern Arizona University

Shivanthi Anandan, Drexel University

Andrew Andres, University of Nevada, Las Vegas

Michael Angilletta, Arizona State University

Jonathan Armbruster, Auburn University

Jessica Armenta, Lone Star College System

Brian Ashburner, University of Toledo

Andrea Aspbury, Texas State University

Nevin Aspinwall, Saint Louis University

Felicitas Avendano, Grand View University

Yael Avissar, Rhode Island College

Ricardo Azpiroz, Richland College

Jessica Baack, Southwestern Illinois College

Charles Baer, University of Florida

Brian Bagatto, University of Akron

Alan L. Baker, University of New Hampshire

Ellen Baker, Santa Monica College

Mitchell Balish, Miami University

Teri Balser, University of Florida

Paul Bates, University of Minnesota, Duluth

Michel Baudry, University of Southern California

Jerome Baudry, University of Tennessee, Knoxville

Mike Beach, Southern Polytechnic State University

Andrew Beall, University of North Florida

Gregory Beaulieu, University of Victoria

John Bell, Brigham Young University

Michael Bell, Richland College

Rebecca Bellone, University of Tampa

Anne Bergey, Truman State University

Laura Bermingham, University of Vermont

Aimee Bernard, University of Colorado, Denver

Annalisa Berta, San Diego State University

Joydeep Bhattacharjee, University of Louisiana, Monroe

Arlene Billock, University of Louisiana, Lafayette

Daniel Blackburn, Trinity College

Mark Blackmore, Valdosta State University

Justin Blau, New York University

Andrew Blaustein, Oregon State University

Mary Bober, Santa Monica College

Robert Bohanan, University of Wisconsin, Madison

Jim Bonacum, University of Illinois at Springfield

Laurie Bonneau, Trinity College

David Bos, Purdue University

James Bottesch, Brevard Community College

Jere Boudell, Clayton State University

Nancy Boury, Iowa State University

Matthew Brewer, Georgia State University

Mirjana Brockett, Georgia Institute of Technology

Andrew Brower, Middle Tennessee State University

Heather Bruns, Ball State University

Jill Buettner, Richland College

Stephen Burnett, Clayton State University

Steve Bush, Coastal Carolina University

David Byres, Florida State College at Jacksonville

James Campanella, Montclair State University

Darlene Campbell, Cornell University

Jennifer Campbell, North Carolina State University

John Campbell, Northwest College

David Canning, Murray State University

Richard Cardullo, University of California, Riverside

Sara Carlson, University of Akron

Jeff Carmichael, University of North Dakota

Dale Casamatta, University of North Florida

Anne Casper, Eastern Michigan University

David Champlin, University of Southern Maine

Rebekah Chapman, Georgia State University

Samantha Chapman, Villanova University

Mark Chappell, University of California, Riverside

P. Bryant Chase, Florida State University

Young Cho, Eastern New Mexico University

Tim Christensen, East Carolina University

Steven Clark, University of Michigan

Ethan Clotfelter, Amherst College

Catharina Coenen, Allegheny College

Mary Colavito, Santa Monica College

Craig Coleman, Brigham Young University

Alex Collier, Armstrong Atlantic State University

Sharon Collinge, University of Colorado, Boulder

Jay Comeaux, McNeese State University

Reid Compton, University of Maryland

Ronald Cooper, University of California, Los Angeles

Victoria Corbin, University of Kansas

Asaph Cousins, Washington State University

Will Crampton, University of Central Florida

Kathryn Craven, Armstrong Atlantic State University

Scott Crousillac, Louisiana State University

Kelly Cude, College of the Canyons

Stanley Cunningham, Arizona State University

Karen Curto, University of Pittsburgh

Bruce Cushing, The University of Akron

Rebekka Darner, University of Florida

James Dawson, Pittsburg State University

Elizabeth De Stasio, Lawrence University

Jennifer Dechaine, Central Washington University

James Demastes, University of Northern Iowa

D. Michael Denbow, Virginia Polytechnic Institute and State University Joseph Dent, McGill University

Terry Derting, Murray State University

Jean DeSaix, University of North Carolina at Chapel Hill

Donald Deters, Bowling Green State University

Hudson DeYoe, The University of Texas, Pan American

Leif Deyrup, University of the Cumberlands

Laura DiCaprio, Ohio University

Jesse Dillon, California State University, Long Beach

Frank Dirrigl, The University of Texas, Pan American

Kevin Dixon, Florida State University

Elaine Dodge Lynch, Memorial University of Newfoundland

Hartmut Doebel, George Washington University

Jennifer Doll, Loyola University, Chicago

Logan Donaldson, York University

Blaise Dondji, Central Washington University

Christine Donmoyer, Allegheny College

James Dooley, Adelphi University

Jennifer Doudna, University of California, Berkeley

John DuBois, Middle Tennessee State University

Richard Duhrkopf, Baylor University

Kamal Dulai, University of California, Merced

Arthur Dunham, University of Pennsylvania

Mary Durant, Lone Star College System

Roland Dute, Auburn University

Andy Dyer, University of South Carolina, Aiken

William Edwards, Niagara University

John Elder, Valdosta State University

William Eldred, Boston University

David Eldridge, Baylor University

Inge Eley, Hudson Valley Community College

Lisa Elfring, University of Arizona

Richard Elinson, Duquesne University

Kurt Elliott, Northwest Vista College

Miles Engell, North Carolina State University

Susan Erster, Stony Brook University

Joseph Esdin, University of California, Los Angeles

Jean Everett, College of Charleston

Brent Ewers, University of Wyoming

Melanie Fierro, Florida State College at Jacksonville

Michael Fine, Virginia Commonwealth University

Jonathan Fingerut, St. Joseph's University

Ryan Fisher, Salem State University

David Fitch, New York University

Paul Fitzgerald, Northern Virginia Community College

Jason Flores, University of North Carolina at Charlotte

Matthias Foellmer, Adelphi University

Barbara Frase, Bradley University

Caitlin Gabor, Texas State University

Michael Gaines, University of Miami

Jane Gallagher, The City College of New York, The City University of New York

Kathryn Gardner, Boston University

J. Yvette Gardner, Clayton State University

Gillian Gass, Dalhousie University

Jason Gee, East Carolina University

Topher Gee, University of North Carolina at Charlotte

Vaughn Gehle, Southwest Minnesota State University

Tom Gehring, Central Michigan University

John Geiser, Western Michigan University

Alex Georgakilas, East Carolina University

Peter Germroth, Hillsborough Community College

Arundhati Ghosh, University of Pittsburgh

Carol Gibbons Kroeker, University of Calgary

Phil Gibson, University of Oklahoma

Cindee Giffen, University of Wisconsin, Madison

Matthew Gilg, University of North Florida

Sharon Gillies, University of the Fraser Valley

Leonard Ginsberg, Western Michigan University

Florence Gleason, University of Minnesota

Russ Goddard, Valdosta State University

Miriam Golbert, College of the Canyons

Jessica Goldstein, Barnard College, Columbia University

Steven Gorsich, Central Michigan University

Sandra Grebe, Lone Star College System

Robert Greene, Niagara University

Ann Grens, Indiana University, South Bend

Theresa Grove, Valdosta State University

Stan Guffey, The University of Tennessee

Nancy Guild, University of Colorado, Boulder

Lonnie Guralnick, Roger Williams University

Laura Hake, Boston College

Kimberly Hammond, University of California, Riverside

Paul Hapeman, University of Florida

Luke Harmon, University of Idaho

Sally Harmych, University of Toledo

Jacob Harney, University of Hartford

Sherry Harrel, Eastern Kentucky University

Dale Harrington, Caldwell Community College and Technical Institute

J. Scott Harrison, Georgia Southern University

Diane Hartman, Baylor University

Mary Haskins, Rockhurst University

Bernard Hauser, University of Florida

David Haymer, University of Hawaii

David Hearn, Towson University

Marshal Hedin, San Diego State University

Paul Heideman, College of William and Mary

Gary Heisermann, Salem State University

Brian Helmuth, University of South Carolina

Christopher Herlihy, Middle Tennessee State University

Albert Herrera, University of Southern California

Brad Hersh, Allegheny College

David Hicks, The University of Texas at Brownsville

Karen Hicks, Kenyon College

Alison Hill, Duke University

Kendra Hill, South Dakota State University

Jay Hodgson, Armstrong Atlantic State University

John Hoffman, Arcadia University

Jill Holliday, University of Florida

Sara Hoot, University of Wisconsin, Milwaukee

Margaret Horton, University of North Carolina at Greensboro

Lynne Houck, Oregon State University

Kelly Howe, University of New Mexico

William Huddleston, University of Calgary

Jodi Huggenvik, Southern Illinois University

Melissa Hughes, College of Charleston

Randy Hunt, Indiana University Southeast

Tony Huntley, Saddleback College

Brian Hyatt, Bethel College

Jeba Inbarasu, Metropolitan Community College

Colin Jackson, The University of Mississippi

Eric Jellen, Brigham Young University

Dianne Jennings, Virginia Commonwealth University

Scott Johnson, Wake Technical Community College

Mark Johnston, Dalhousie University

Susan Jorstad, University of Arizona

Stephen Juris, Central Michigan University

Julie Kang, University of Northern Iowa

Jonghoon Kang, Valdosta State University

George Karleskint, St. Louis Community College at Meramec

David Karowe, Western Michigan University

Judy Kaufman, Monroe Community College

Nancy Kaufmann, University of Pittsburgh

John Kauwe, Brigham Young University

Elena Keeling, California Polytechnic State University

Jill Keeney, Juniata College

Tamara Kelly, York University

Chris Kennedy, Simon Fraser University

Bretton Kent, University of Maryland

Jake Kerby, University of South Dakota

Jeffrey Kiggins, Monroe Community College

Scott Kight, Montclair State University

Stephen Kilpatrick, University of Pittsburgh, Johnstown

Kelly Kissane, University of Nevada, Reno

David Kittlesen, University of Virginia

Jennifer Kneafsey, Tulsa Community College

Jennifer Knight, University of Colorado, Boulder

Ross Koning, Eastern Connecticut State University

David Kooyman, Brigham Young University

Olga Kopp, Utah Valley University

Anna Koshy, Houston Community College

Todd Kostman, University of Wisconsin, Oshkosh

Peter Kourtev, Central Michigan University

William Kroll, Loyola University, Chicago

Dave Kubien, University of New Brunswick

Allen Kurta, Eastern Michigan University

Ellen Lamb, University of North Carolina at Greensboro

Troy Ladine, East Texas Baptist University

David Lampe, Duquesne University

Evan Lampert, Gainesville State College

James Langeland, Kalamazoo College

John Latto, University of California, Santa Barbara

Brenda Leady, University of Toledo

Jennifer Leavey, Georgia Institute of Technology

Hugh Lefcort, Gonzaga University

Brenda Leicht, University of Iowa

Craig Lending, The College at Brockport, The State University of New York

Nathan Lents, John Jay College of Criminal Justice, The City University of New York

Michael Leonardo, Coe College

Army Lester, Kennesaw State University

Cynthia Littlejohn, University of Southern Mississippi

Zhiming Liu, Eastern New Mexico University

Jonathan Lochamy, Georgia Perimeter College

Suzanne Long, Monroe Community College

Julia Loreth, University of North Carolina at Greensboro

Jennifer Louten, Southern Polytechnic State University

Janet Loxterman, Idaho State University

Ford Lux, Metropolitan State College of Denver

Jose-Luis Machado, Swarthmore College

C. Smoot Major, University of South Alabama

Charles Mallery, University of Miami

Mark Maloney, Spelman College

Carroll Mann, Florida State College at Jacksonville

Carol Mapes, Kutztown University of Pennsylvania

Nilo Marin, Broward College

Diane Marshall, University of New Mexico

Heather Masonjones, University of Tampa

Scott Mateer, Armstrong Atlantic State University

Luciano Matzkin, The University of Alabama in Huntsville

Robert Maxwell, Georgia State University

Meghan May, Towson University

Douglas Meikle, Miami University

Michael McGinnis, Spelman College

Kathleen McGuire, San Diego State University

Maureen McHale, Truman State University

Shannon McQuaig, St. Petersburg College

Susan McRae, East Carolina University

Lori McRae, University of Tampa

Mark Meade, Jacksonville State University

Brad Mehrtens, University of Illinois at Urbana-Champaign

Michael Meighan, University of California, Berkeley

Richard Merritt, Houston Community College

Jennifer Metzler, Ball State University

James Mickle, North Carolina State University

Brian Miller, Middle Tennessee State University

Allison Miller, Saint Louis University

Yuko Miyamoto, Elon University

Ivona Mladenovic, Simon Fraser University

Marcie Moehnke, Baylor University

Chad Montgomery, Truman State University

Jennifer Mook, Gainesville State College

Daniel Moon, University of North Florida

Jamie Moon, University of North Florida

Jeanelle Morgan, Gainesville State College

David Morgan, University of West Georgia

Julie Morris, Armstrong Atlantic State University

Becky Morrow, Duquesne University

Mark Mort, University of Kansas

Nancy Morvillo, Florida Southern College

Anthony Moss, Auburn University

Mario Mota, University of Central Florida

Alexander Motten, Duke University

Tim Mulkey, Indiana State University

John Mull, Weber State University

Michael Muller, University of Illinois at Chicago

Beth Mullin, University of Tennessee, Knoxville

Paul Narguizian, California State University, Los Angeles

Jennifer Nauen, University of Delaware

Paul Nealen, Indiana University of Pennsylvania

Diana Nemergut, University of Colorado, Boulder

Kathryn Nette, Cuyamaca College

Jacalyn Newman, University of Pittsburgh

James Nienow, Valdosta State University

Alexey Nikitin, Grand Valley State University

Tanya Noel, York University

Fran Norflus, Clayton State University

Celia Norman, Arapahoe Community College

Eric Norstrom, DePaul University

Jorge Obeso, Miami Dade College

Kavita Oommen, Georgia State University

David Oppenheimer, University of Florida

Joseph Orkwiszewski, Villanova University

Rebecca Orr, Collin College

Don Padgett, Bridgewater State College

Joanna Padolina, Virginia Commonwealth University

One Pagan, West Chester University

Kathleen Page, Bucknell University

Daniel Papaj, University of Arizona

Pamela Pape-Lindstrom, Everett Community College

Bruce Patterson, University of Arizona, Tucson

Shelley Penrod, Lone Star College System

Roger Persell, Hunter College, The City University of New York

John Peters, College of Charleston

Chris Petrie, Brevard Community College

Patricia Phelps, Austin Community College

Steven Phelps, The University of Texas at Austin

Kristin Picardo, St. John Fisher College

Aaron Pierce, Nicholls State University

Debra Pires, University of California, Los Angeles

Thomas Pitzer, Florida International University

Nicola Plowes, Arizona State University

Crima Pogge, City College of San Francisco

Darren Pollock, Eastern New Mexico University

Kenneth Pruitt, The University of Texas at Brownsville

Sonja Pyott, University of North Carolina at Wilmington

Rajinder Ranu, Colorado State University

Philip Rea, University of Pennsylvania

Amy Reber, Georgia State University

Ahnya Redman, West Virginia University

Melissa Reedy, University of Illinois at Urbana-Champaign

Brian Ring, Valdosta State University

David Rintoul, Kansas State University

Michael Rischbieter, Presbyterian College

Laurel Roberts, University of Pittsburgh

George Robinson, The University at Albany, The State University of New York

Peggy Rolfsen, Cincinnati State Technical and Community College

Mike Rosenzweig, Virginia Polytechnic Institute and State University

Doug Rouse, University of Wisconsin, Madison

Yelena Rudayeva, Palm Beach State College

Ann Rushing, Baylor University

Shereen Sabet, La Sierra University

Rebecca Safran, University of Colorado

Peter Sakaris, Southern Polytechnic State University

Thomas Sasek, University of Louisiana, Monroe

Udo Savalli, Arizona State University

H. Jochen Schenk, California State University, Fullerton

Gregory Schmaltz, University of the Fraser Valley

Jean Schmidt, University of Pittsburgh

Andrew Schnabel, Indiana University, South Bend

Roxann Schroeder, Humboldt State University

David Schultz, University of Missouri, Columbia

Andrea Schwarzbach, The University of Texas at Brownsville

Erik Scully, Towson University

Robert Seagull, Hofstra University

Pramila Sen, Houston Community College

Alice Sessions, Austin Community College

Vijay Setaluri, University of Wisconsin

Jyotsna Sharma, The University of Texas at San Antonio

Elizabeth Sharpe-Aparicio, Blinn College

Patty Shields, University Of Maryland

Cara Shillington, Eastern Michigan University

James Shinkle, Trinity University

Rebecca Shipe, University of California, Los Angeles

Marcia Shofner, University of Maryland

Laurie Shornick, Saint Louis University

Jill Sible, Virginia Polytechnic Institute and State University

Allison Silveus, Tarrant County College

Kristin Simokat, University of Idaho

Sue Simon-Westendorf, Ohio University

Sedonia Sipes, Southern Illinois University, Carbondale

John Skillman, California State University, San Bernardino

Marek Sliwinski, University of Northern Iowa

Felisa Smith, University of New Mexico

John Sollinger, Southern Oregon University

Scott Solomon, Rice University

Morvarid Soltani-Bejnood, The University of Tennessee

Vladimir Spiegelman, University of Wisconsin, Madison

Chrissy Spencer, Georgia Institute of Technology

Kathryn Spilios, Boston University

Ashley Spring, Brevard Community College

Bruce Stallsmith, The University of Alabama in Huntsville

Jennifer Stanford, Drexel University

Barbara Stegenga, University of North Carolina, Chapel Hill

Patricia Steinke, San Jacinto College, Central Campus

Asha Stephens, College of the Mainland

Robert Steven, University of Toledo

Eric Strauss, University of Wisconsin, La Crosse

Sukanya Subramanian, Collin College

Mark Sugalski, Southern Polytechnic State University

Brad Swanson, Central Michigan University

Ken Sweat, Arizona State University

David Tam, University of North Texas

Ignatius Tan, New York University

William Taylor, University of Toledo

Christine Terry, Lynchburg College

Sharon Thoma, University of Wisconsin, Madison

Pamela Thomas, University of Central Florida

Carol Thornber, University of Rhode Island

Patrick Thorpe, Grand Valley State University

Briana Timmerman, University of South Carolina

Chris Todd, University of Saskatchewan

Gail Tompkins, Wake Technical Community College

Martin Tracey, Florida International University

Randall Tracy, Worcester State University

James Traniello, Boston University

Bibit Traut, City College of San Francisco

Terry Trier, Grand Valley State University

Stephen Trumble, Baylor University

Jan Trybula, The State University of New York at Potsdam

Alexa Tullis, University of Puget Sound

Marsha Turell, Houston Community College

Mary Tyler, University of Maine

Marcel van Tuinen, University of North Carolina at Wilmington

Dirk Vanderklein, Montclair State University

Jorge Vasquez-Kool, Wake Technical Community College

William Velhagen, New York University

Dennis Venema, Trinity Western University

Laura Vogel, North Carolina State University

Jyoti Wagle, Houston Community College

Jeff Walker, University of Southern Maine

Gary Walker, Appalachian State University

Andrea Ward, Adelphi University

Fred Wasserman, Boston University

Elizabeth Waters, San Diego State University

Douglas Watson, The University of Alabama at Birmingham

Matthew Weand, Southern Polytechnic State University

Michael Weber, Carleton University

Cindy Wedig, The University of Texas, Pan American

Brad Wetherbee, University of Rhode Island

Debbie Wheeler, University of the Fraser Valley

Clay White, Lone Star College System

Lisa Whitenack, Allegheny College

Maggie Whitson, Northern Kentucky University

Stacey Wild East, Tennessee State University

Herbert Wildey, Arizona State University and Phoenix College

David Wilkes, Indiana University, South Bend

Lisa Williams, Northern Virginia Community College

Elizabeth Willott, University of Arizona

Mark Wilson, Humboldt State University

Ken Wilson, University of Saskatchewan

Bob Winning, Eastern Michigan University

Candace Winstead, California Polytechnic State University

Robert Wise, University of Wisconsin, Oshkosh

D. Reid Wiseman, College of Charleston

MaryJo Witz, Monroe Community College

David Wolfe, American River College

Kevin Woo, University of Central Florida

Denise Woodward, Penn State

Shawn Wright, Central New Mexico Community College

Grace Wyngaard, James Madison University

Aimee Wyrick, Pacific Union College

Joanna Wysocka-Diller, Auburn University

Ken Yasukawa, Beloit College

John Yoder, The University of Alabama

Kelly Young, California State University, Long Beach

James Yount, Brevard Community College

Min Zhong, Auburn University

# CONTENTS

CHAPTER 3    **NUCLEIC ACIDS AND THE ENCODING OF BIOLOGICAL INFORMATION** 3-1

CHAPTER 4    **TRANSLATION AND PROTEIN STRUCTURE** 4-1

CHAPTER 7  **CELLULAR RESPIRATION**
Harvesting Energy from Carbohydrates and Other Fuel Molecules    7-1

CHAPTER 8  **PHOTOSYNTHESIS**
Using Sunlight to Build Carbohydrates    8-1

CHAPTER 9   **CELL COMMUNICATION**    9-1

CHAPTER 10   **CELL FORM AND FUNCTION**
Cytoskeleton, Cellular Junctions, and
Extracellular Matrix    10-1

CHAPTER 11 **CELL DIVISION**
Variations, Regulation, and Cancer 11-1

CHAPTER 12 **DNA REPLICATION AND MANIPULATION** 12-1

CHAPTER 13    **GENOMES**    13-1

CHAPTER 14    **MUTATION AND DNA REPAIR**    14-1

---

CHAPTER 17 **BEYOND MENDEL**
Sex Chromosomes, Linkage, and Organelles 17-1

CHAPTER 18 **THE GENETIC AND ENVIRONMENTAL BASIS OF COMPLEX TRAITS** 18-1

## CHAPTER 19  GENETIC AND EPIGENETIC REGULATION

## CHAPTER 20  GENES AND DEVELOPMENT

CHAPTER 23  **EVOLUTIONARY PATTERNS**

CHAPTER 24  **HUMAN ORIGINS AND EVOLUTION** 24-1

PART 2 FROM ORGANISMS TO THE ENVIRONMENT

CHAPTER 25 CYCLING CARBON 25-1

CHAPTER 26 BACTERIA AND ARCHAEA 26-1

---

## CHAPTER 27 EUKARYOTIC CELLS
Origins and Diversity 27-1

---

## CHAPTER 28 BEING MULTICELLULAR 28-1

CHAPTER 29  **PLANT STRUCTURE AND FUNCTION**
Moving Photosynthesis onto Land                         29-1

CHAPTER 30  **PLANT REPRODUCTION**
Finding Mates and Dispersing Young                      30-1

CHAPTER 31   **PLANT GROWTH AND DEVELOPMENT**
Building the Plant Body    31-1

CHAPTER 32   **PLANT DEFENSE**
Keeping the World Green    32-1

CHAPTER 33 **PLANT DIVERSITY** 33-1

CHAPTER 34 **FUNGI**
Structure, Function, and Diversity 34-1

CHAPTER 35 **ANIMAL NERVOUS SYSTEMS** 35-1

CHAPTER 36 **ANIMAL SENSORY SYSTEMS AND BRAIN FUNCTION** 36-1

CHAPTER 37 **ANIMAL MOVEMENT**
Muscles and Skeletons 37-1

CHAPTER 38 **ANIMAL ENDOCRINE SYSTEMS** 38-1

## CHAPTER 39 ANIMAL CARDIOVASCULAR AND RESPIRATORY SYSTEMS — 39-1

## CHAPTER 40 ANIMAL METABOLISM, NUTRITION, AND DIGESTION — 40-1

CHAPTER 43 **ANIMAL IMMUNE SYSTEMS** 43-1

CHAPTER 44 **ANIMAL DIVERSITY** 44-1

---

CHAPTER 45 **ANIMAL BEHAVIOR** 45-1

---

CHAPTER 46 **POPULATION ECOLOGY** 46-1

CHAPTER 47 **SPECIES INTERACTIONS,
COMMUNITIES, AND ECOSYSTEMS** 47-1

CHAPTER 48 **THE ANTHROPOCENE**
Humans as a Planetary Force 48-1

# PART 1

# FROM CELLS TO ORGANISMS

"Nothing in biology makes sense except in the light of evolution."

—THEODOSIUS DOBZHANSKY

## CHAPTER 1

# LIFE

## Chemical, Cellular, and Evolutionary Foundations

### Core Concepts

**1.1** The scientific method is a deliberate way of asking and answering questions about the natural world.

**1.2** Life works according to fundamental principles of chemistry and physics.

**1.3** The fundamental unit of life is the cell.

**1.4** Evolution explains the features that organisms share and those that set them apart.

**1.5** Organisms interact with one another and with their physical environment, shaping ecological systems that sustain life.

**1.6** In the 21st century, humans have become major agents in ecology and evolution.

Every day, remarkable things happen within and all around you. Strolling through a local market, you come across a bin full of crisp apples, pick one up, and take a bite. Underlying this unremarkable occurrence is a remarkable series of events. Your eyes sense the apple from a distance, and nerves carry that information to your brain, permitting identification. Biologists call this cognition, an area of biological study. Stimulated by the apple and recognizing it as ripe and tasty, your brain transmits impulses through nerves to your muscles. How we respond to external cues motivates behavior, another biological discipline. Grabbing the apple requires the coordinated activities of dozens of muscles that move your arm and hand to a precise spot. These movements are described by biomechanics, yet another area of biological research. And, as you bite down on the apple, glands in your mouth secrete saliva, starting to convert energy stored in the apple as sugar into energy that you will use to fuel your own activities. Physiology, like biomechanics, lies at the heart of biological function.

The study of cognition, behavior, biomechanics, and physiology are all ways of approaching **biology,** the science of how life works. **Biologists,** scientists who study life, have come to understand a great deal about these and other processes at levels that run from molecular mechanisms within the cell, through the integrated actions of many cells within an organ or body, to the interactions among different organisms in nature. We don't know everything about how life works—in fact, it seems as if every discovery raises new questions. But biology provides us with an organized way of understanding ourselves and the world around us.

Why study biology? The example of eating an apple, which we will follow through this chapter, was deliberately chosen because it is an everyday occurrence that we ordinarily wouldn't think twice about. The scope of modern biology, however, is vast, raising questions that can fire our imaginations, affect our health, and influence our future. How, for example, will our understanding of the human genome change the way that we fight cancer? How do bacteria in our digestive system help determine health and well-being? Will expected increases in the temperature and acidity of seawater doom coral reefs? Is there, or has there ever been, life on Mars? And, to echo the great storyteller Rudyard Kipling, why do leopards have spots, and tigers stripes?

We can describe six grand themes that connect and unite the many dimensions of life science, from molecules to the biosphere. These six themes are stated as Core Concepts and are introduced in the following sections. Throughout the book, these themes will be visited again and again. We view them as the keys to understanding the many details in subsequent chapters and relating them to one another. Our

**FIG. 1.1** **A hummingbird visiting a flower.** This simple observation leads to questions: Why do hummingbirds pay so much attention to flowers? Why do they hover near red flowers?

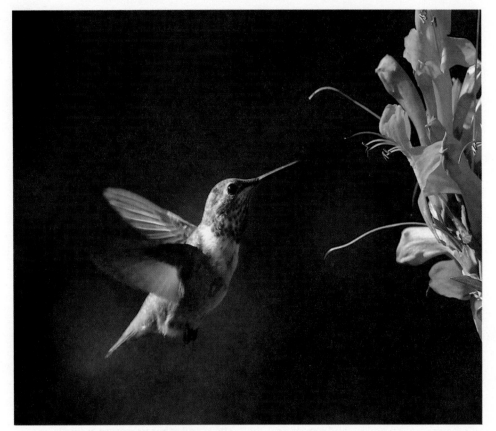

### Observation allows us to draw tentative explanations called hypotheses.

Observations allow us to ask focused questions about nature. Let's say you observe a hummingbird like the one pictured in **Fig. 1.1** hovering near a red flower, occasionally dipping its long beak into the bloom. What motivates this behavior? Is the bird feeding on some substance within the flower? Is it drawn to the flower by its vivid color? What benefit, if any, does the flower derive from this busy bird?

Observations such as these, and the questions they raise, allow us to propose tentative explanations, or **hypotheses.** We might, for example, hypothesize that the hummingbird is carrying pollen from one flower to the next, facilitating reproduction in the plant. Or we might hypothesize that nectar produced deep within the flower provides nutrition for the hummingbird—that the hummingbird's actions reflect the need to take in food. Both hypotheses provide a reasonable explanation of the behavior we observed, but they may or may not be correct. To find out, we have to test them.

hope is that by the time you finish this book, you will have an understanding of how life works, from the molecular machines inside cells and the metabolic pathways that cycle carbon through the biosphere to the process of evolution, which has shaped the living world that surrounds (and includes) us. You will, we hope, see the connections among these different ways of understanding life, and come away with a greater understanding of how scientists think about and ask questions about the natural world. How, in fact, do we know what we think we know about life? And we hope you will develop a basis for making informed decisions about your career and the actions you take as a citizen.

## 1.1 THE SCIENTIFIC METHOD

How do we go about trying to understand the vastness and complexity of nature? For most scientists, studies of the natural world involve the complementary processes of observation and experimentation. **Observation** is the act of viewing the world around us. **Experimentation** is a disciplined and controlled way of asking and answering questions about the world in an unbiased manner.

Piecing together individual observations to construct a working hypothesis is beautifully illustrated in Charles Darwin's classic book, *On the Origin of Species*, published in 1859. In his text, Darwin discussed a wide range of observations, from pigeon breeding to fossils and from embryology to the unusual animals and plants found on islands. Darwin noted the success of animal breeders in selecting specific individuals for reproduction and thereby generating new breeds for agriculture or show. He appreciated that selective breeding is successful only if specific features of the animals can be passed from one generation to the next by inheritance. Reading economic treatises by the English clergyman Thomas Malthus, he understood that limiting environmental resources could select among the variety of different individuals in populations in much the way that breeders do among cows or pigeons. Gathering all these seemingly disparate pieces of information, he argued that life has evolved over time by means of natural selection. Since its formulation, Darwin's initial hypothesis has been tested by experiments, many thousands of them. Our knowledge of many biological phenomena, ranging from biodiversity to the way the human brain is wired, depends on direct observation followed by careful inferences that lead to models of how things work.

## A hypothesis makes predictions that can be tested by observation and experiments.

Not just any idea qualifies as a hypothesis. Two features set hypotheses apart from other ways of attacking problems. First, a good hypothesis makes **predictions** about observations not yet made or experiments not yet run. Second, because hypotheses make predictions, we can **test** them. That is, we can devise an experiment to see whether the predictions made by the hypothesis actually occur, or we can go into the field to try to make further observations predicted by the hypothesis. A hypothesis, then, is a statement about nature that can be tested by experiments or by new observations. Hypotheses are testable because even as they suggest an explanation for observations made previously, they make predictions about observations yet to be made.

Returning to the hummingbird and flower, we can test the hypothesis that the bird is transporting pollen from one flower to the next, enabling the plant to reproduce. Observation provides one type of test: If we catch and examine the bird just after it visits a flower, do we find pollen stuck to its beak or feathers? If so, our hypothesis survives the test. Note, however, that we haven't proved the case. Pollen might be stuck on the bird for a different reason—perhaps it provides food for the hummingbird. However, if the birds *didn't* carry pollen from flower to flower, we would reject the hypothesis that they facilitate pollination. In other words, a single observation or experiment can lead us to reject a hypothesis, or it can support the hypothesis, but it cannot prove that a hypothesis is correct. To move forward, then, we might make a second set of observations. Does pollen that adheres to the hummingbird rub off when the bird visits a second flower of the same species? If so, we have stronger support for our hypothesis.

We might also use observations to test a more general hypothesis about birds and flowers. Does red color generally attract birds and so facilitate pollination in a wide range of flowers? To answer this question, we might catalog the pollination of many red flowers and ask whether they are pollinated mainly by birds. Or we might go the opposite direction and catalog the flowers visited by many different birds—are they more likely to be red than chance alone might predict?

Finally, we can test the hypothesis that the birds visit the flowers primarily to obtain food, spreading pollen as a side effect of their feeding behavior. We can measure the amount of nectar in the flower before and after the bird visits and calculate how much energy has been assimilated by the bird during its visit. Continued observations over the course of the day will tell us whether the birds gain the nutrition they need by eating nectar, and whether the birds have other sources of food.

In many cases, experiments provide the most powerful tests of hypotheses because the scientist can ensure that conditions are tightly controlled. We might test whether hummingbirds

facilitate pollination by surrounding the flowers with a mesh that allows small insects access to the plant but keeps hummingbirds away. Will the flowers be pollinated? Experimental or observational tests may support the initial hypothesis, in which case the hypothesis becomes less tentative and more certain, or the results might refute the hypothesis, in which case the scientist may discard it for another explanation or amend it to account for the new information.

→   **Quick Check 1**  Mice that live in sand dunes commonly have light tan fur. Develop a hypothesis to explain this coloration. How can you test your hypothesis?

Using observations to generate a hypothesis and then making predictions based on that hypothesis that can be tested experimentally are the first two steps in the **scientific method**, which is outlined in **Fig. 1.2.** The scientific method is a deliberate and careful way of asking questions about the unknown. We make observations, collect field or laboratory samples, and design and carry out experiments to make sense of things we initially do not understand. The scientific method has proved to be spectacularly successful in helping us to understand the world around us.

**FIG. 1.2**     The scientific method.

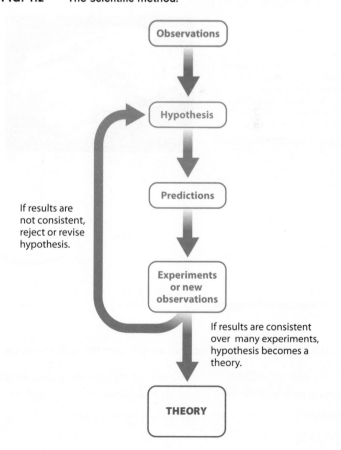

If results are not consistent, reject or revise hypothesis.

If results are consistent over many experiments, hypothesis becomes a theory.

**FIG. 1.3**

# What caused the extinction of the dinosaurs?

**BACKGROUND** Dinosaurs were diverse and ecologically important for nearly 150 million years, but became extinct about 65 million years ago.

**OBSERVATION**

Iridium, common in meteorites, was discovered in rock layers corresponding to the time of extinction.

**HYPOTHESIS** The impact of a large meteorite disrupted communities on land and in the sea, causing the extinction of the dinosaurs and many other species.

**PREDICTIONS** Independent evidence of a meteor impact should be found in rock layers corresponding to the time of the extinction, and be rare or absent in older and younger beds.

**FURTHER OBSERVATIONS**

Quartz crystals that form only at high temperature and pressure—conditions met by giant meteors as they crash into the Earth—occur abundantly in rock layers dated to the time of the extinction.

By 1990, geologists had located the "smoking gun"— a crater of just the right age and size in the Yucatán Peninsula of Mexico.

**CONCLUSION** A giant meteor struck the Earth 65 million years ago, causing the extinction of the dinosaurs and other species.

**FOLLOW-UP WORK** Researchers have documented additional episodes of mass extinction, but the event that eliminated the dinosaurs appears to be unique in its association with a meteorite impact.

**SOURCE** Alvarez, W. 1998. *T. rex and the Crater of Doom*. New York: Vintage Press.

To emphasize the power of the scientific method, we turn to a famous riddle drawn from the fossil record (**Fig. 1.3**). Since the nineteenth century, paleontologists have known that before mammals expanded to their current ecological importance, other large animals dominated Earth. Dinosaurs evolved about 210 million years ago and disappeared 65 million years ago, along with many other species of plants, animals, and microscopic organisms. In many cases, the skeletons and shells of these creatures were buried in sediment and became fossilized. Layers of sedimentary rock therefore record the history of Earth. Working in Italy, the American geologist Walter Alvarez collected samples from the precise point in the rock layers that corresponds to the time of the extinction 65 million years ago. Careful chemical analysis showed that rocks at this level are unusually enriched in the element iridium. Iridium is rare in most rocks on continents and the seafloor, but is relatively common in rocks that fall from space—that is, in meteorites. From these observations, Alvarez and his colleagues developed a remarkable hypothesis: 65 million years ago, a large (11-km diameter) meteor slammed into Earth, and in the resulting environmental havoc, dinosaurs and many other species became extinct. This hypothesis makes specific predictions, described in Fig. 1.3, which turned out to be supported by further observations. Thus, observational tests support the hypothesis that 150 million years of dinosaur evolution were undone in a moment.

→ **Quick Check 2** Devise a test for the hypothesis that cigarette smoke causes lung cancer.

### General explanations of natural phenomena supported by many experiments and observations are called theories.

As already noted, a hypothesis may initially be tentative. Commonly, in fact, it will provide only one of several possible ways of explaining existing data. With repeated observation and experimentation, however, a good hypothesis gathers strength, and we have more and more confidence in it. When a number of related hypotheses survive repeated testing and so come to be accepted as good bases for explaining what we see in nature, scientists articulate a broader explanation that accounts for all of the hypotheses and the results of their tests. We call this statement a **theory,** a general explanation of the world supported by a large body of experiments and observations (see Fig. 1.2).

Note that scientists use the word "theory" in a very particular way. In general conversation, "theory" is often synonymous with "hypothesis," "idea," or "hunch"—"I've got a theory about that." But in a scientific context, the word "theory" has a specific meaning. Only if hypotheses have withstood testing to the point where they provide a general explanation for many observations and experimental results do scientists speak in terms of theories. Just as a good hypothesis makes testable predictions, a good theory both generates good hypotheses and predicts their outcomes. Thus, scientists talk about the theory of gravity—

a set of hypotheses you test every day by walking down the street or dropping a fork. Similarly, the theory of evolution is not one explanation among many for the unity and diversity of life. It is a set of hypotheses that has been tested for more than a century and shown to be an extraordinarily powerful means of explaining biological observations that range from amino acid sequences of proteins to the diversity of ants in a rain forest. In fact, as we discuss repeatedly in this book, evolution is the single most important theory in all of biology. It provides the most general and powerful explanation of how life works.

## 1.2 CHEMICAL AND PHYSICAL PRINCIPLES

We stated earlier that biology is the study of life. But what exactly *is* life? As simple as this question seems, it is frustratingly difficult to answer. We all recognize life when we see it, but coming up with a definition is harder than it first appears.

Living organisms are clearly different from nonliving things. But just how different is an organism from the rock shown in **Fig. 1.4**? On one level, the comparison is easy: The rock is much simpler than any living organism we can think of. It has far fewer

**FIG. 1.4** **A climber scaling a rock.** Living organisms like this climber contain chemicals that are found in rocks, but only living organisms reproduce in a manner that allows for evolution over time.

components, and it is largely static, with no apparent response to environmental change on time scales that are readily tracked.

In contrast, even an organism as relatively simple as a bacterium contains many hundreds of different chemical compounds organized in a complex manner. The bacterium is also dynamic in that it changes continuously, especially in response to the environment. Organisms reproduce, which minerals do not. And organisms do something else that rocks and minerals don't: They evolve. Indeed, the molecular biologist Gerald Joyce has defined life as a chemical system capable of undergoing Darwinian evolution.

From these simple comparisons we can highlight four key characteristics of living organisms: (1) complexity, with precise spatial organization on several scales; (2) the ability to change in response to the environment; (3) the ability to reproduce; and (4) the capacity to evolve. Nevertheless, the living and nonliving worlds share an important attribute: Both are subject to the basic laws of chemistry and physics.

### The living and nonliving worlds share the same chemical foundations and obey the same physical laws.

The chemical elements found in rocks and other nonliving things are no different from those found in living organisms. In other words, all the elements that make up living things can be found in the nonliving environment—there is nothing special about our chemical components when taken individually. That said, the *relative* abundances of elements in organisms differ greatly from those in the nonliving world. In the universe as a whole, hydrogen and helium make up more than 99% of known matter, while Earth's crust contains mostly oxygen and silicon, with significant amounts of aluminum, iron, and calcium (**Fig. 1.5a**). In organisms, by contrast, oxygen, carbon, and hydrogen are by far the most abundant elements (**Fig. 1.5b**). As discussed more fully in Chapter 2, carbon provides the backbone of life's chemistry. The particular properties of carbon make possible a wide diversity of molecules that, in turn, support a wide range of functions and activities within cells.

All living organisms are subject to the physical laws of the universe. Physics helps us to understand how animals move and why trees don't fall over; it explains how redwoods conduct water upward through their trunks and how oxygen gets into the cells that line your lungs. Indeed, two laws of thermodynamics, both of which describe how energy is transformed in any system, determine how living organisms are able to do work and maintain their spatial organization.

The **first law of thermodynamics** states that energy can neither be created nor destroyed; it can only be transformed from one form into another. In other words, the total energy in the universe is constant, but the form that energy takes can change. Living organisms are energy transformers. They acquire energy from the environment and transform it into a chemical form

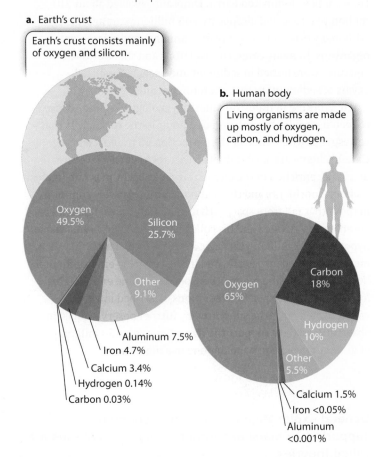

**FIG. 1.5**    **Composition of (a) the Earth's crust and (b) the human body.**  The Earth beneath our feet is made up of the same elements found in our feet, but in strikingly different proportions.

**a.** Earth's crust

Earth's crust consists mainly of oxygen and silicon.

Oxygen 49.5%
Silicon 25.7%
Other 9.1%
Aluminum 7.5%
Iron 4.7%
Calcium 3.4%
Hydrogen 0.14%
Carbon 0.03%

**b.** Human body

Living organisms are made up mostly of oxygen, carbon, and hydrogen.

Carbon 18%
Oxygen 65%
Hydrogen 10%
Other 5.5%
Calcium 1.5%
Iron <0.05%
Aluminum <0.001%

that cells can use. All organisms obtain energy from the sun or from chemical compounds. As they harness this energy, some is used to do work—such as moving, reproducing, and building cellular components—and the rest is dissipated as heat. The energy that is used to do work plus the heat that is generated is the total amount of energy, which is the same as the input energy (**Fig. 1.6**). In other words, the total amount of energy remains constant.

The **second law of thermodynamics** states that the degree of disorder in the universe tends to increase. To understand order and disorder in this context, think about a box full of marbles distributed more or less randomly; if you want to line up all the red ones or blue ones in a row, you have to do work. That is, you have to add energy. In this case, the addition of energy increases the order of the system, or, put another way, decreases its disorder. Physicists quantify the amount of disorder in a system, describing it as the **entropy** of the system.

Living organisms are highly organized. As with lining up marbles in a row, energy is needed to maintain this organization.

**FIG. 1.6    Energy transformation and the first law of thermodynamics.** The first law states that the total amount of energy in any system remains the same. Organisms transform energy from one form to another, but the total energy in any system is constant.

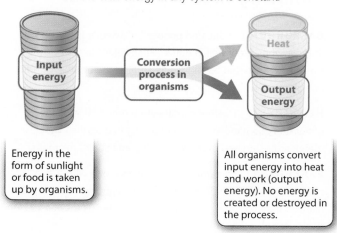

Energy in the form of sunlight or food is taken up by organisms.

All organisms convert input energy into heat and work (output energy). No energy is created or destroyed in the process.

**FIG. 1.7    Energy transformation and the second law of thermodynamics.** The second law states that the disorder in any system tends to increase. Entropy can decrease locally (inside a cell, for example) because the heat released increases disorder in the environment.

As monomers combine into polymers, the entropy inside the cell decreases, requiring energy.

Some of the energy is released as heat, increasing the entropy (motion of molecules) in the environment. The total entropy of the cell plus the environment increases.

Given the tendency toward greater disorder, the high level of organization of even a single cell would appear to violate the second law. But it does not. The key is that a cell is not an isolated system and therefore cannot be considered on its own; it exists in an environment. So we need to take into account the whole system, the cell plus the surrounding environment. As energy is harnessed by cells, only some is used to do work; the rest is dissipated as heat (Fig. 1.6). That is, conversions of energy from one form to another are never 100% efficient. Heat is a form of energy, so the total amount of energy is conserved, as dictated by the first law. In addition, heat corresponds to the motion of small molecules—the greater the heat, the greater the motion, and the greater the disorder. Therefore, the release of heat as organisms harness energy means that the total entropy for the combination of the cell and its surroundings increases, in keeping with the second law (**Fig. 1.7**).

### The scientific method shows that living organisms come from other living organisms.

Life is made up of chemical components that also occur in the nonliving environment and obey the same laws of chemistry and physics. Can life spontaneously arise from these nonliving materials? We all know that living organisms come from other living organisms, but it is worth asking *how* we know this. Direct observation can be misleading here. For example, raw meat, if left out on a plate, will rot and become infested with maggots (fly larvae). It might seem as though the maggots appear spontaneously. In fact, the question of where maggots come from was a matter of vigorous debate for centuries, until application of the scientific method settled the issue. In the 1600s, the

Italian physician and naturalist Francesco Redi hypothesized that maggots (and hence flies) in rotting meat come only from other flies that laid their eggs in the meat.

To test his hypothesis, Redi set up an experiment in which he placed meat in three glass jars (**Fig. 1.8**). One jar was left open, a second was covered with gauze, and the third was sealed with a cap. The jars were left in a room with flies. Note that in this experiment, the three jars were subject to the same conditions—the only difference was the opening of the jar. The open jar allowed for the passage of flies and air; the jar with the gauze allowed for the passage of air but not flies; and the sealed jar did not allow air or flies to enter. Over time, Redi observed that maggots appeared only on the meat in the open jar. No maggots appeared in the other two jars, which did not allow access to the meat by flies. These observations supported Redi's hypothesis that flies come from other flies, and did not provide support for the alternative hypothesis that maggots and flies arise spontaneously from meat.

Redi demonstrated that living organisms come from other organisms, but some argued that his conclusion might apply only to larger organisms—microscopic life might be another matter entirely. It was not until the nineteenth century that the French

**FIG. 1.8**

# Can living organisms arise from nonliving matter?

**BACKGROUND** Until the 1600s, many people believed that rotting meat spontaneously generates maggots (fly larvae).

**HYPOTHESIS** Francesco Redi hypothesized that maggots come only from flies and are not spontaneously generated.

**EXPERIMENT** Redi used three jars containing meat. One jar was left open; one was covered with gauze; one was sealed with a cap.

**RESULTS**

| Open jar | Gauze-covered jar | Sealed jar |
|---|---|---|
| Maggots appeared in the open jar. | No maggots appeared in the gauze-covered jar. | No maggots appeared in the sealed jar. |

**CONCLUSION** The presence of maggots in the open jar and the absence of maggots in the gauze-covered and sealed jars supported the hypothesis that maggots come from flies, and allowed Redi to reject the hypothesis that maggots are spontaneously generated.

**FOLLOW-UP WORK** Redi's experiment argued against the idea of spontaneous generation for insects. However, it was unclear whether his results could be extended to microbes. Applying the scientific method, Louis Pasteur used a similar approach about 200 years later to investigate this question (see Fig. 1.9).

chemist and biologist Louis Pasteur tested the hypothesis that microorganisms can arise by spontaneous generation (**Fig. 1.9**).

Pasteur filled two glass flasks with heat-sterilized broth—one with a straight vertical neck and the other with a curved swan neck. As in Redi's experiments, there was only one variable, in this case the shape of the neck of the flask. The straight-neck flask allowed airborne dust particles carrying microbes to fall into the sterile broth, while the swan-neck flask prevented dust from getting inside. Over time, Pasteur observed that microbes grew in the broth inside the straight-neck flask but not in the

**FIG. 1.9**

# Can microscopic life arise from nonliving matter?

**BACKGROUND** Educated people in Pasteur's time knew that microbes grow well in nutrient-rich liquids like broth. It was also known that boiling would sterilize the broth, killing the microbes.

**HYPOTHESIS** Pasteur hypothesized that if microbes were generated spontaneously from nonliving matter, they should reappear in sterilized broth without the addition of microbes.

**EXPERIMENT** Pasteur used two flasks, one with a straight neck and one with a swan neck. The straight-neck flask allowed dust particles with microbes to enter. The swan-neck flask did not.

swan-neck flask. From these observations, Pasteur rejected the hypothesis that microbes arise spontaneously from sterile broth. Instead, exposure to microbes carried on airborne dust particles is necessary for microbial growth.

Redi's and Pasteur's experiments demonstrated that living organisms come from other living organisms and are not generated spontaneously from chemical components. But this raises the question of how life arose in the first place. If life comes from life, where did the first living organisms come from? Although today all organisms are produced by parental organisms, early in Earth's history this was probably not the case. Scientists hypothesize that life initially emerged from chemical compounds about 4 billion years ago. That is, chemical systems capable of evolution arose from chemical reactions that took place on the early Earth. We'll return to the great question of the origin of life in Case 1: The First Cell and in Chapters 2 through 8.

## 1.3 THE CELL

To understand the universal features of life, we need to know what constitutes the minimal unit that can be considered alive. The **cell** is the simplest entity that can exist as an independent unit of life. Every known living organism is either a single cell or an ensemble of many cells (**Fig. 1.10**). Most bacteria (like those in Pasteur's experiment), yeasts, and the tiny algae that float in oceans and ponds spend their lives as single cells. In contrast, multicellular plants and animals contain billions to trillions of cells that function in a coordinated fashion.

**RESULTS**

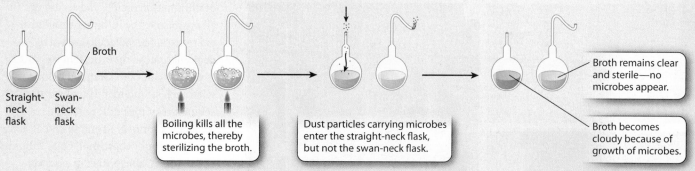

Straight-neck flask · Swan-neck flask · Broth

Boiling kills all the microbes, thereby sterilizing the broth.

Dust particles carrying microbes enter the straight-neck flask, but not the swan-neck flask.

Broth remains clear and sterile—no microbes appear.

Broth becomes cloudy because of growth of microbes.

**CONCLUSION** The presence of microbes in the straight-neck flask and the absence of microbes in the swan-neck flask supported the hypothesis that microbes come from other microbes and are not spontaneously generated.

**DISCUSSION** Redi's and Pasteur's research illustrate classic attributes of well-designed experiments. Multiple treatments are set up, and nearly all conditions are the same in them all—they are constant, and therefore cannot be the cause of different outcomes of the experiment. One key feature—the **variable**—is changed by the experimenter from one treatment to the next. This is a place to look for explanations of different experimental outcomes.

**FIG. 1.10 Unicellular and multicellular organisms.** All living organisms are made up of cells. (a) Bacteria; (b) brewer's yeast; (c) algae; (d) cheetahs; (e) humans

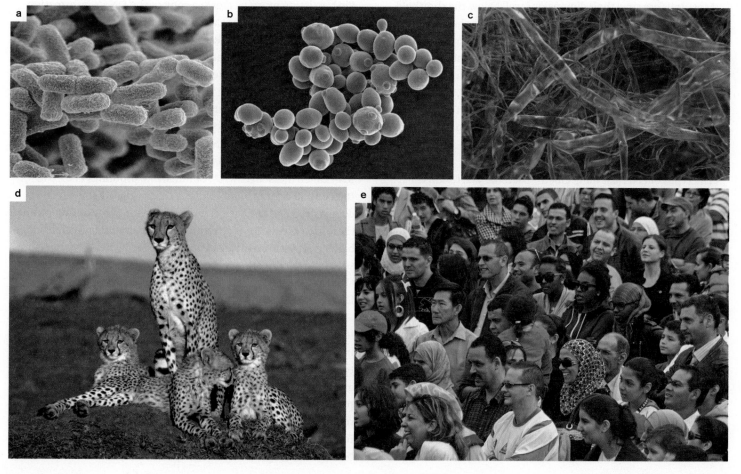

**FIG. 1.11** **Cell diversity.** Cells vary greatly in size and shape. (a) Skin cells; (b) nerve cells; (c) ostrich egg.

Most cells are tiny, their dimensions well below the threshold of detection by the naked eye **(Fig. 1.11).** The cells that make up the layers of your skin (Fig. 1.11a) average about 100 microns or 0.1 mm in diameter, which means that about 10 would fit in a row across the period at the end of this sentence. Many bacteria are less than a micron long. Certain specialized cells, however, can be quite large. Some nerve cells in humans, like the ones pictured in Fig. 1.11b, extend slender projections known as axons for distances as great as a meter, and the cannonball-size egg of an ostrich in Fig. 1.11c is a single giant cell.

The types of cells just mentioned—bacteria, yeast, skin cells, nerve cells, and an egg—seem very different, but all are organized along broadly similar lines. In particular, they all contain a stable blueprint of information in molecular form; they have a discrete boundary that separates the interior of the cell from its external environment; and they have the ability to harness materials and energy from the environment.

### Nucleic acids store and transmit information needed for growth, function, and reproduction.

The first essential feature of a cell is its ability to store and transmit information. To accomplish this, cells require a stable archive of information that encodes and helps determine their physical attributes. Just as the construction and maintenance of a house requires a blueprint that defines the walls, plumbing, and electrical wiring, organisms require an accessible and reliable archive of information that helps determine their structure and metabolic

activities. Another hallmark of life is the ability to reproduce. To reproduce, cells must be able to copy their archive of information rapidly and accurately. In all organisms, the information archive is a remarkable molecule known as deoxyribonucleic acid or **DNA** (**Fig. 1.12**).

DNA is a double-stranded helix, with each strand made up of varying sequences of four different kinds of molecules connected end to end. It is the arrangement of these molecular subunits that makes DNA special; in essence, they provide a four-letter alphabet that encodes cellular information. Notably, the information encoded in DNA directs the formation of **proteins,** the key structural and functional molecules that do the work of the cell. Virtually every aspect of the cell's existence—its internal architecture, its shape, its ability to move, and its various chemical reactions—depends on proteins.

How does the information stored in DNA direct the synthesis of proteins? First, existing proteins create a copy of the DNA's information in the form of a closely related molecule called ribonucleic acid, or **RNA.** The synthesis of RNA from a DNA template is called **transcription,** a term that describes the copying of information from one form into another. Specialized molecular structures within the cell then "read" the RNA molecule to determine what building blocks to use to create a protein. This process, called **translation,** converts information stored in the language of nucleic acids to information in the language of proteins.

The pathway from DNA to RNA (specifically to a form of RNA called messenger RNA, or mRNA) to protein is known as the **central dogma** of molecular biology (**Fig. 1.13**). The central dogma describes the basic flow of information in a cell and, while there are exceptions, it constitutes a fundamental principle in biology. As proteins are ultimately encoded by DNA, we can define specific stretches or segments of DNA according to the proteins that they encode. This is the simplest definition of a **gene:** the DNA sequence that corresponds to a specific protein product.

DNA has another remarkable feature. In addition to storing information, it is easily copied, or **replicated,** allowing genetic information to be passed from cell to cell or from one organism to its progeny. Each organism's DNA archive can be stably and reliably passed from generation to generation in large part because of its double-stranded helical structure. During replication, each strand of the double helix serves as a template for a new strand. Replication is necessarily precise and accurate because mistakes introduced into the cell's information archive

**FIG. 1.12** **A molecule of DNA.** DNA is a double helix made up of varying sequences of four different subunits.

may be lethal to the cell. That said, errors in DNA can and do occur during the process of replication, and environmental insults can damage DNA. Such changes are known as **mutations;** they can spell death for the cell, or they can lead to the variations that underlie the diversity of life and the process of evolution.

→ **Quick Check 3** How does the central dogma help us to understand how mutations in DNA can result in disease?

**Membranes define cells and spaces within cells.**
The second essential feature of all cells is a **plasma membrane** that separates the living material within the cell from the nonliving environment around it (**Fig. 1.14**). This boundary between inside and outside does not mean that cells are closed systems independent of the environment. On the contrary, there is an active and dynamic interplay between cells and their surroundings that is mediated by the plasma membrane. All cells require

**FIG. 1.13** **The central dogma of molecular biology, defining the flow of information in all living organisms from DNA to RNA to protein.**

sustained contributions from their surroundings, both nutrients and the building blocks required to manufacture macromolecules. They also release waste products into the environment. As discussed more fully in Chapter 5, the plasma membrane controls the movement of materials into and out of the cell.

In addition to the plasma membrane, many cells have internal membranes that divide the cell into discrete compartments, each specialized for a particular function. A notable example is the **nucleus,** which houses the cell's DNA. Like the plasma membrane, the nuclear membrane selectively controls movement of molecules into and out of it. As a result, the nucleus occupies a discrete space within the cell, separate from the space outside of the nucleus, called the **cytoplasm.**

Not all cells have a nucleus. In fact, cells can be grouped into two broad classes depending on whether or not they have a nucleus. Cells without a nucleus are **prokaryotes,** and cells with a nucleus are **eukaryotes.**

The first cells were prokaryotes. They evolved about 4 billion years ago, and their descendants include familiar bacteria that are around today. They are found wherever life can persist. Some prokaryotes live in peaceful coexistence with humans, inhabiting our gut and aiding digestion. Others cause disease—salmonella, tuberculosis, and cholera are familiar examples. Their success depends in part on their small size, their ability to reproduce rapidly, and their ability to obtain energy and nutrients from diverse sources. Most prokaryotes live as single-celled organisms, but some have simple multicellular forms.

**FIG. 1.14** **The plasma membrane.** The plasma membrane surrounds every cell and controls the exchange of material with the environment.

Transmission electron micrograph of a cell

Eukaryotes evolved much later, roughly 2 billion years ago, from one lineage among a diversity of prokaryotes that existed at the time. They include familiar groups such as animals, plants, and fungi, along with a wide diversity of single-celled microorganisms called protists. Eukaryotic organisms exist as single cells like yeast or as multicellular organisms like humans. In multicellular organisms, cells may specialize to perform different functions. For example, in humans, muscle cells contract; red blood cells carry oxygen to tissues; and skin cells provide an external barrier.

The terms "prokaryotes" and "eukaryotes" are useful in drawing attention to a fundamental distinction between these two groups of cells. However, today, biologists recognize three domains of life—**Bacteria, Archaea,** and **Eukarya** (Chapters 26 and 27). Bacteria and Archaea both lack a nucleus and are therefore prokaryotes, while Eukarya are eukaryotes. In spite of the similar cellular organization of Bacteria and Archaea, the Eukarya and Archaea are more closely related to each other than either is to Bacteria. Archaea are single-celled microorganisms, many of which flourish under seemingly hostile conditions, such as the hot springs of Yellowstone National Park.

### Metabolism converts energy from the environment into a form that can be used by cells.

A third key feature of cells is the ability to harness energy from the environment. Let's go back to our example of eating an apple. The apple contains sugars, which store energy in their chemical bonds. By breaking down sugar, our cells harness this energy and convert it into a form that can be used to do the work of the cell. Energy from the food we eat allows us to grow, move, communicate, and do all the other things that we do.

Organisms acquire energy from just two sources—the sun and chemical compounds. Regardless of their source of energy, all organisms use chemical reactions to break down molecules, releasing energy in the process and storing it in a chemical form called **ATP** (adenosine triphosphate) that cells can use. The term **metabolism** describes chemical reactions that cells use to convert energy from one form to another and to build and break down molecules. These reactions are required to sustain life.

Many metabolic reactions are highly conserved between organisms, suggesting that they evolved early in the history of life and have been maintained for billions of years because of their fundamental importance to cellular biochemistry.

### A virus is genetic material in need of a cell.

It's worth taking a moment to consider viruses. A virus is an agent that infects cells. It is smaller and simpler than cells. Why, then, aren't viruses the smallest unit of life? We just considered three essential features of cells—the capacity to store and transmit information, a membrane that selectively controls movement in and out, and the ability to harness energy from the environment. Viruses have a stable archive of genetic information, which can be RNA or DNA, and the genetic material of viruses is surrounded by a protein coat and sometimes a lipid envelope. But viruses cannot harness energy from the environment. Therefore, on their own viruses cannot read and use the information contained in their genetic material, nor can they regulate the passage of substances across their protein coats or lipid envelopes the way that cells do. To replicate, they require a cell.

A virus infects a cell by binding to the cell surface, inserting its genetic material into the cell, and, in most cases, using the cellular machinery to replicate its own genetic material, synthesize proteins, and produce more viruses. In this way, it is often said that a virus "hijacks" a cell. The infected cell may produce more viruses, sometimes by lysis, or breakage, of the cell, and the new viruses can then infect more cells. In some cases, the genetic material of the virus integrates into the DNA of the host cell.

We discuss viruses many times throughout the book, as they are interesting in themselves and infect all kinds of cells. Each species of Bacteria, Archaea, and Eukarya is susceptible to many types of virus that are specialized to attack its cells. Several hundred types of virus are known to infect humans, and the catalog is still incomplete. Useful tools in biological research, viruses have provided a model system for many problems in biology, including how genes are turned on and off and how cancer develops.

## 1.4 EVOLUTION

The themes introduced in the last two sections stress life's unity: Cells form the basic unit of all life; DNA, RNA, and proteins carry out the molecular functions of all cells; and metabolic reactions build and break down macromolecules. We need only look around us, however, to recognize that for all its unity, life displays a remarkable degree of diversity. We don't really know how many species share our planet, but reasonable estimates run to 10 million or more. Both the unity and the diversity of life are explained by the process of **evolution,** change over time.

### Variation in populations provides the raw material for evolution, or change over time.

Described in detail, evolution by **natural selection** calls on complex mathematical formulations, but at heart its main principles are simple, indeed unavoidable. When there is variation within a population of organisms, and when that variation can be inherited (that is, when it can be passed from one generation to the next), the variants best suited for growth and reproduction in a given environment will contribute disproportionately to the next generation. As Darwin recognized, farmers have used this principle for thousands of years to select for crops with high yield or improved resistance to drought and disease. It is how people around the world have developed breeds of dog ranging from terriers to huskies

**FIG. 1.15** **Artificial selection.** Selection over many centuries has resulted in remarkable variations among dogs. Charles Darwin called this "selection under domestication" and noted that it resembles selection that occurs in nature.

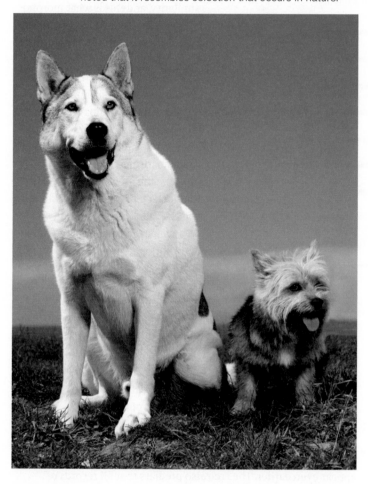

(**Fig. 1.15**). And it is why antibiotic resistance is on the rise in many disease-causing microorganisms. Life has been shaped by evolution since its origin, and the capacity for Darwinian evolution may be life's most fundamental property.

The apples in the bin from which you made your choice didn't all look alike. Had you picked your apple in an orchard, you would have seen that different apples on the same tree looked different— some smaller, some greener, some misshapen, a few damaged by worms. Such variation is so commonplace that we scarcely pay attention to it. Variation is observed among individuals in virtually every species of organism. Variation that can be inherited provides the raw material on which evolution acts.

The causes of variation among individuals within a species are usually grouped into two broad categories. Variation among individuals is sometimes due to differences in the environment; this is called **environmental variation.** Among apples on the same tree, some may have good exposure to sunlight, some may be hidden in the shade, some were lucky enough to escape the female codling moth, whose egg develops into a caterpillar

that eats its way into the fruit. These are all examples of environmental variation.

The other main cause of variation among individuals is differences in the genes that are transmitted from parents to their offspring; this is known as **genetic variation.** Genes are composed of DNA and contain coded information that instructs the cell to produce specific types of RNA and protein. Differences among individuals' DNA lead to differences among the individuals' RNA and proteins, which affect the molecular functions of the cell and ultimately can lead to physical differences that we can observe. Genetic differences among apples produce varieties whose mature fruits differ in taste and color, such as the green Granny Smith, the yellow Golden Delicious, and the scarlet Red Delicious.

But even on a single tree, each apple contains seeds that are genetically distinct, because the apple tree is a sexual organism. Bees carry pollen from the flowers of one tree and deposit it in the flowers of another, enabling the sperm inside pollen grains to fertilize egg cells within that single flower. All the seeds on an apple tree contain shared genes from one parent, the tree on which they developed. But they contain distinct sets of genes contributed by sperm transported in pollen from other trees. In all sexual organisms, fertilization produces unique combinations of genes, which explains in part why sisters and brothers with the same parents can be so different from one another.

Genetic variation arises ultimately from mutations. A few mutations arise in each generation as a result of errors in DNA replication and environmental factors such as UV radiation. Most genetic variation, however, is due to mutations that occurred in previous generations. To put a human face on this, consider that lung cancer can reflect either environmental insult from cigarette smoking, for example, or an inherited genetic susceptibility.

In nature, most mutations that are harmful to growth and reproduction die out after a handful of generations. Those that are neither harmful nor beneficial can persist for hundreds or thousands of generations. And those that are beneficial to growth and reproduction can gradually become incorporated into the genetic makeup of every individual in the species. That is how evolution works: The genetic makeup of a population changes over time.

### Evolution predicts a nested pattern of relatedness among species, depicted as a tree.

Evolutionary theory predicts that new species arise by the divergence of populations through time from a common ancestor. As a result, closely related species are likely to resemble each other more closely than they do more distantly related species. You know this to be true from common experience. All of us recognize the similarity between a chimpanzee's body and our own (**Fig. 1.16a**), and biologists have long known that we share more features with chimpanzees than we do with any other species.

**FIG. 1.16    Phylogenetic relationships among primates.**  (a) Humans share many features with chimpanzees. (b) Humans and chimpanzees, in turn, share more features with gorillas than they do with other species, and so on down through the evolutionary tree of primates. Treelike patterns of nested similarities are the predicted result of evolution.

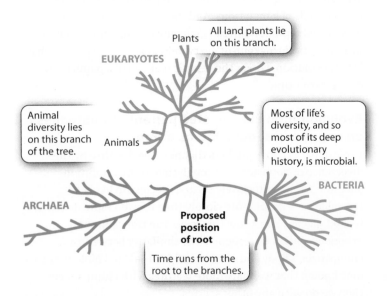

Humans and chimpanzees, in turn, share more features with gorillas than they do with any other species. And humans, chimpanzees, and gorillas share more features with orangutans than they do with any other species. And so on. We can continue to include a widening diversity of species, successively adding monkeys, lemurs, and other primates, to construct a set of evolutionary relationships that can be depicted as a tree (**Fig. 1.16b**). In this tree, the tips or branches on the right represent different groups of organisms, nodes (where lines split) represent

**FIG. 1.17    The tree of life.**

the most recent common ancestor, and time runs from left to right.

Evolutionary theory predicts that primates show a nested pattern of similarity, and this is what morphological and molecular observations reveal. We can continue to add other mammals, and then other vertebrate animals, to our comparison, in the process generating a pattern of evolutionary relationships that forms a larger tree, with the primates confined to one limb. And using molecular comparisons among species, we can generate still larger trees, ones that include plants as well as animals and the full diversity of microscopic organisms. Biologists call the full set of evolutionary relationships among all organisms the **tree of life.** This tree, illustrated in **Fig. 1.17,** has three major branches representing the three domains mentioned earlier and is made up mostly of microorganisms. The last common ancestor of all living organisms, which form a root to the tree, is thought to lie between the branch leading to Bacteria and the branch leading to Archaea and Eukarya. The plants and animals so conspicuous in our daily existence make up only two branches on the eukaryotic limb of the tree.

The tree of life makes predictions for the order of appearance of different life-forms in the fossil record. For example, primates should show up before humans, and the first mammals should appear even earlier. The tree also predicts that all records of animal life should be preceded by a long interval of microbial evolution. As we will see in subsequent chapters, these predictions are met by the geologic record.

Shared features, then, sometimes imply inheritance from a common ancestor. In practice, molecular comparisons and fossils show that the close similarity between humans and chimpanzees reflects descent from a common ancestor that lived about 6 million years ago. Their differences reflect what Darwin called "descent with modification"—evolutionary changes that have accumulated over time since the two lineages split. For example, the flat face of humans, our small teeth, and our upright posture all evolved within our ancestors after they diverged from the ancestors of chimpanzees. As discussed in Chapter 24, fossils document the accumulation of uniquely human features over the past 6 million years. At a broader scale, the fundamental features of molecular function and cell organization that are shared by all organisms reflect inheritance from a common ancestor that lived billions of years ago. And the nested patterns of similarity and differences that characterize the many branches on the tree of life have formed through the continuing action of evolution *since* the time of our earliest ancestors.

Nearly four decades ago, the geneticist Theodosius Dobzhansky wrote, "Nothing in biology makes sense except in the light of evolution." For this reason, evolution permeates discussions throughout this book, whether we are explaining the molecular biology of cells, how organisms function and reproduce, how species interact in nature, or the remarkable biological diversity of our planet.

**Evolution can be studied by means of experiments.** Both the patterns of similarity among living organisms and the succession of fossils in the geologic record fit the predictions of evolutionary theory. Can we actually capture evolutionary processes in action? One way to accomplish this is in the laboratory. Bacteria are ideal for these experiments because they reproduce rapidly and can form laboratory populations with millions of individuals. Large population size means that mutations will form in nearly every generation, even though the probability that any individual cell will acquire a mutation is small. (In contrast to bacteria, think about trying evolutionary experiments on elephants!)

One such experiment is illustrated in **Fig. 1.18.** The microbiologists Santiago Elena and Richard Lenski grew populations of the common intestinal bacterium *Escherichia coli* in a liquid medium, with the organic acid succinate as the only source of food. In general, *E. coli* cells can feed on succinate poorly

**HOW DO WE KNOW?**

**FIG. 1.18**

# Can evolution be demonstrated in the laboratory?

**BACKGROUND** *Escherichia coli* is an intestinal bacterium commonly used in the laboratory. It grows poorly in liquid media where the only source of food is succinate. Santiago Elena and Richard Lenski wondered whether *E. coli* grown for 20,000 generations in succinate would evolve in ways that improved their ability to metabolize this compound.

**HYPOTHESIS** Any bacterium with a random mutation that happens to increase its ability to utilize succinate will reproduce at a faster rate than other bacteria in the population. Over time, such a mutant will increase in frequency relative to other types of bacteria, thereby demonstrating evolution in a bacterial population.

**EXPERIMENT** Cells of *E. coli* can be frozen in liquid nitrogen, which keeps them in a sort of suspended animation in which no biological processes take place, but the cells survive. At the beginning of the experiment, the researchers froze a large number of samples of the starting bacteria ("ancestral"). As the experiment progressed, they took samples of the bacterial populations at intervals ("later") and grew them together with a thawed sample of the starting bacteria in succinate. They then compared the rate of growth of the ancestral bacteria to that of the bacteria taken at later time points.

**RESULTS** At each time interval, the cells from later time points grew more rapidly than the ancestral cells when the two populations were grown together in succinate.

Over time, the bacteria grown in succinate showed more and more improvement in growth compared to the starting bacteria.

Improvement in growth of later strains compared to ancestral strains increases as the number of generations increases.

**CONCLUSION** Evolution occurred in the population: the bacteria evolved an improved ability to metabolize succinate.

**FOLLOW-UP WORK** Other experimental studies have shown that bacteria are able to evolve adaptations to a wide variety of environmental conditions.

**SOURCE** Elena, S. F., and R. E. Lenski. 2003. "Evolution Experiments with Microorganisms." *Nature Review Genetics* 4:457–469.

if at all, leading the researchers to hypothesize that any bacterium with a mutation that increased its ability to utilize succinate would reproduce at a faster rate than other bacteria in the population. The key questions were whether bacteria from later generations differed from those of earlier generations—that is, did evolution occur in the population?—and in particular whether the bacteria had evolved an improved ability to metabolize succinate.

In fact, the bacteria did evolve an improved ability to use succinate, which is demonstrated by the results shown in Fig. 1.18. This experiment illustrates several key points. It shows how experiments can be used to test hypotheses and, of course, it shows evolution in action. Not only did the experiments demonstrate evolutionary adaptation of *E. coli* populations to a specific laboratory environment, but follow-up studies of their DNA identified differences in genetic makeup that resulted in the observed changes. Many experiments of this general type have been carried out, applying the scientific method to demonstrate how bacteria adapt through mutation to any number of imposed environments.

Experiments in laboratory evolution help us to understand how life works, and they have an immensely important practical side as well. They allow biologists to develop new and beneficial strains of microorganisms that, for example, remove toxins from lakes and rivers and show how some of our worst pathogens develop resistance to drugs designed to eliminate them.

→ **Quick Check 4** How might the heavy-handed use of antibiotics result in the increase of antibiotic-resistant cells in bacterial populations?

## 1.5 ECOLOGICAL SYSTEMS

Watching a movie, we don't need a narrator to tell us whether the action is set in a tropical rain forest, the African savanna, or Arctic tundra. The plants in the scene tell us. Plants can be linked closely with environment—palm trees with the tropics, for example, or cacti with the desert—because the environmental distributions of different species reflect the sum of their biological features. Palms cannot tolerate freezing and so are confined to warmer environments; the cactus can store water in its tissues, enabling it to withstand prolonged drought. But the geographic ranges of palms and cacti reflect more than just their physical tolerances. They are also strongly influenced by interactions with other species, including other plants that compete for the limited resources available for growth, animals that feed on them or spread their pollen, and microorganisms that infest their tissues. **Ecology** is the study of how organisms interact with one another and with their physical environment in nature.

### Basic features of anatomy, physiology, and behavior shape ecological systems.

Apple trees, we noted, reproduce sexually. How is this accomplished, given that the trees have no moving parts? The answer is that bees carry pollen from one flower to the next, enabling sperm carried within the pollen grain to fuse with an egg cell protected within the flower (**Fig. 1.19**). The process is much like the one discussed at the beginning of the chapter in which hummingbirds carry pollen from flower to flower. In both examples, the plants complete their life cycles by exploiting close interactions with animal species. Birds and bees visit flowers, attracted, as we are, by their color and odor. Neither visits flowers with the intent to pollinate; they come in expectation of a meal. As the birds and bees nestle into flowers to collect nutritious nectar, pollen rubs off on their bodies.

But if animals help apples to fertilize their eggs, how do apple trees disperse from one site to another? Again, the plants rely on animals—the apple's flesh attracts animals that eat the fruit, seeds and all, and spread the seeds through defecation. Humans tend to be finicky—we carefully eat only the sweet outer flesh, discarding the seedy core. Nonetheless, even as humans have selected apples for their quality of fruit, we have dispersed apple trees far beyond their natural range.

The many ways that organisms interact in nature reflect their basic functional requirements. If a plant is to grow and reproduce, for example, it must have access to light, carbon dioxide, water, and basic nutrients. All plants have these

**FIG. 1.19** **Bee pollination.** Many plants reproduce sexually by exploiting the behavior of animals. Here a honey bee pollinates an apple flower.

**FIG. 1.20**  **Ecological relationships.** (a) The trees in a rain forest all require water and nutrients. Neighboring plants compete for the limited supplies of these materials. (b) Plants provide food for many animals. Leaf-cutter ants cut slices of leaves from tropical plants and transport them to their nests, where the leaves grow fungi that the ants eat.

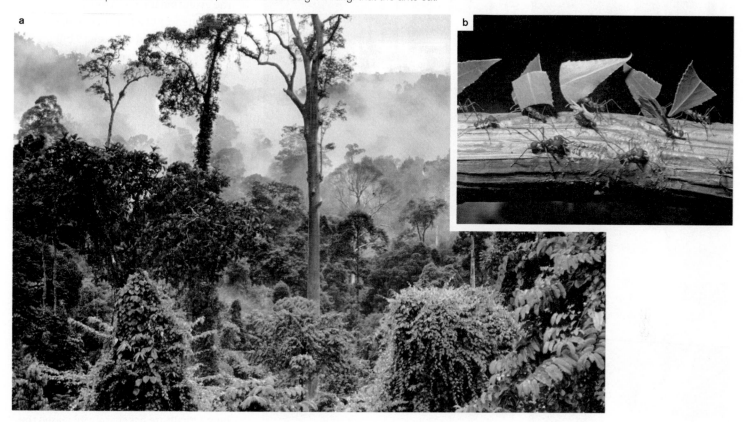

requirements, so one plant's success in gathering nutrients may mean that fewer nutrients are available for its neighbor—the plants may compete for limited resources needed for growth (**Fig. 1.20a**). Animals in the same area require organic molecules for nutrition, and some of them may eat the plant's leaves or bark (**Fig. 1.20b**). Predation also benefits some organisms at the expense of others. In combination, the physical tolerances of organisms and the ways that organisms interact with one another determine the structure and diversity of communities.

In short, ecological relationships reflect the biomechanical, physiological, and behavioral traits of organisms in nature. Form and function, in turn, arise from molecular processes within cells, governed by the expression of genes.

### Ecological interactions play an important role in evolution.

G. Evelyn Hutchinson, one of the founders of modern ecology, wrote a book called *The Ecological Theater and the Evolutionary Play* (1965). This wonderful title succinctly captures a key feature of biological relationships. It suggests that ecological communities provide the "stage" on which the "play" of evolution takes place.

For example, plants often compete for resources. As a result, natural selection may favor plants that have more efficient uptake of nutrients or water. If animals eat plants, natural selection may favor animals with greater jaw strength or more efficient extraction of nutrients in the digestive system. In turn, plants that avoid predation by synthesizing toxic compounds in their leaves may gain the upper hand. In each case, interactions between organisms lead to the evolution of particular traits.

To take another example from mammal–plant interactions: Selection for fleshy fruits improved the dispersal of apples because it increased the attractiveness of these seed-bearing structures to hungry mammals. Tiny yeasts make a meal on sugary fruits as well, in the process producing alcohol that deters potential competitors for the food. Already in prehistoric times, humans had learned to harness this physiological capability of yeasts, and for this reason the total abundance and distribution of *Vitis vinifera*, the wine grape, has increased dramatically through time.

**FIG. 1.21**    Species that have benefited from human activity, including (a) corn, (b) rats, and (c) cockroaches.

## 1.6 THE HUMAN FOOTPRINT

The story of life has a cast of millions, with humans playing only one of many roles in an epic 4 billion years in the making. Our own species, *Homo sapiens*, has existed for only the most recent 1/200th of 1% of life's history, yet there are compelling reasons to pay special attention to ourselves. We want to understand how our own bodies work and how humans came to be: Curiosity about ourselves is after all a deeply human trait.

We also want to understand how biology can help us to conquer disease and improve human welfare. Epidemics have decimated human populations throughout history. The Black Death—bubonic plague caused by the bacterium *Yersinia pestis*—is estimated to have killed half the population of Europe in the fourteenth century. Casualties from the flu pandemic of 1918 exceeded those of World War I. Even King Tut, we now know, suffered from malaria in his Egyptian palace. Throughout this book, we discuss how basic biological principles are helping scientists to prevent and cure the great diseases that have

persisted since antiquity, as well as modern ones such as AIDS.

Furthermore, we need to understand the evolutionary and ecological consequences of a human population that has expanded to exceed 7 billion. All species affect the world around them, and we are no different in this regard. Indeed, in the 21st century human activities have taken on special importance because our numbers and technological abilities make our footprint on Earth's ecology so large. Human activities now emit more carbon dioxide than do volcanoes, through industrial processes we convert more atmospheric nitrogen to ammonia than nature does, and we commandeer, either directly or indirectly, as much as 25% of all photosynthetic production on land. To chart our environmental future, we need to understand our role in the Earth system as a whole.

And, as we have become major players in ecology, humans have become important agents of evolution. As our population has expanded, some species have expanded along with us. We've seen how agriculture has sharply increased the abundance and distribution of grapes, and the same is true for corn, cows, and apple trees. At the same time, we have inadvertently helped other species to expand—the crowded and not always clean environments of cities provide excellent habitats for cockroaches and rats (**Fig. 1.21**).

Other species, however, are in decline, their populations reduced by hunting and fishing, changes in land use, and other human activities. When Europeans first arrived on the Indian Ocean island of Mauritius, large flightless birds called dodos were plentiful (**Fig. 1.22**). Within a century, the dodo was gone. Early Europeans in midcontinental North America were greeted by vast populations of passenger pigeons, more than a million in a single flock. By the early 20th century, the species was extinct, a victim of hunting and habitat change through expanding agriculture. Other organisms both great (the Bali tiger) and small (the dusky seaside sparrow) have become extinct in recent decades. Still others are imperiled by human activities; the magnificent white rhinoceros of Africa evolved over millions of years, but today is threatened by both habitat destruction and poaching. Whether any rhinos will exist at the end of this century will depend almost entirely on decisions we make today.

**FIG. 1.22** **Extinct and endangered species.** Humans have caused many organisms to become extinct, such as the (a) dodo, (b) passenger pigeon, (c) Bali tiger, and (d) dusky seaside sparrow. Burgeoning human populations have diminished the ranges of many others, including the (e) white rhinoceros.

Throughout this book, we return to the practical issues of life science. How can we use the principles of biology to improve human welfare, and how can we live our lives in ways that control our impact on the world around us? The answers to the questions critically depend on understanding biology in an *integrated* fashion. While it is tempting to consider molecules, cells, organisms, and ecosystems as separate entities, they are inseparable in nature. To tackle biological problems, whether building an artificial cell, stopping the spread of infectious diseases such as HIV or malaria, feeding a growing population, or preserving endangered habitats and species, we need an integrated perspective. In decisively important ways, our future welfare depends on improving our knowledge of how life works.

# Core Concepts Summary

**1.1 THE SCIENTIFIC METHOD IS A DELIBERATE WAY OF ASKING AND ANSWERING QUESTIONS ABOUT THE NATURAL WORLD.**

Observations are used to generate a hypothesis, a tentative explanation that makes predictions that can be tested.  page 1-2

On the basis of a hypothesis, scientists design experiments and make additional observations that test the hypothesis.  page 1-3

If a hypothesis is supported through continued observation and experiments over long periods of time, it is elevated to a theory, a sound and broad explanation of some aspect of the world.  page 1-5

**1.2 LIFE WORKS ACCORDING TO FUNDAMENTAL PRINCIPLES OF CHEMISTRY AND PHYSICS.**

Living organisms and nonliving objects are made up of the same chemicals and obey the same physical laws.  page 1-6

Experiments by Redi in the 1600s and Pasteur in the 1800s demonstrated that organisms come from other organisms and are not spontaneously generated.  page 1-7

Life originated on Earth about 4 billion years ago and most likely arose from nonliving matter.  page 1-8

**1.3 THE FUNDAMENTAL UNIT OF LIFE IS THE CELL.**

The cell is the simplest biological entity that can exist independently.  page 1-8

Information in a cell is stored in the form of the nucleic acid DNA. page 1-10

The central dogma describes the usual flow of information in a cell, from DNA to RNA to protein.  page 1-10

The plasma membrane is the boundary that separates the cell from its environment.  page 1-11

Cells with a nucleus are eukaryotes; cells without a nucleus are prokaryotes.  page 1-11

Metabolism is the set of chemical reactions that are used by cells to build and break down macromolecules and to harness energy. page 1-12

**1.4 BOTH THE FEATURES THAT ORGANISMS SHARE AND THOSE THAT SET THEM APART ARE EXPLAINED BY EVOLUTION.**

When there is variation within a population of organisms, and when that variation can be inherited, the variants best able to grow and reproduce in a particular environment will contribute disproportionately to the next generation, leading to a change in the population over time, or evolution.  page 1-12

Variation can be genetic or environmental. The ultimate source of genetic variation is mutation.  page 1-12

Organisms show a nested pattern of similarity, with humans more similar to primates than other organisms, primates more similar to mammals, mammals more similar to vertebrates, and so on. page 1-13

Evolution can be demonstrated by laboratory experiments. page 1-15

**1.5 ORGANISMS INTERACT WITH ONE ANOTHER AND WITH THEIR PHYSICAL ENVIRONMENT, SHAPING ECOLOGICAL SYSTEMS THAT SUSTAIN LIFE.**

Ecology is the study of how organisms interact with one another and with their physical environment in nature.  page 1-16

These interactions are driven in part by the anatomy, physiology, and behavior of organisms.  page 1-16

**1.6 IN THE 21ST CENTURY, HUMANS HAVE BECOME MAJOR AGENTS IN ECOLOGY AND EVOLUTION.**

Humans have existed for only the most recent 1/200th of 1% of life's 4-billion-year history.  page 1-18

In spite of our recent arrival, our growing numbers are leaving a large ecological and evolutionary footprint.  page 1-18

Solving biological problems requires an integrated understanding of life, with contributions from all the fields of biology, including molecular biology, cell biology, genetics, organismal biology, and ecology, as well as from chemistry, physics, and engineering. page 1-19

# Self-Assessment

1. Describe the steps in the scientific method.

2. Differentiate among a guess, a hypothesis, and a theory.

3. State the first and second laws of thermodynamics and describe how they relate to living organisms.

4. Describe what it means to say that a cell is life's functional unit.

5. Explain how we know that living organisms come from other living organisms.

6. Explain how evolution accounts for both the unity and diversity of life.

7. List features of organisms that shape ecological systems.

8. Name three ways that humans have affected life on Earth.

9. Take a step back and describe the six themes that are discussed in this chapter.

# The First Cell

## LIFE'S ORIGINS

Deep underground, in Mexico's Cueva de Villa Luz, the cave walls drip with slime. The rocky surfaces are teeming with colonies of mucus-producing bacteria. No sunlight reaches these organisms far beneath Earth's surface. Instead, the bacteria survive by capturing energy from hydrogen sulfide gas within the cave. As a by-product of that reaction, the microbes produce sulfuric acid. The stalactite-like slime formations oozing from the cave walls—dubbed "snottites" by researchers—are as corrosive as battery acid.

Snottites might be stomach turning, but they're intriguing, too. Called "extremophiles" because they live in places where humans and most other animals cannot survive, such microorganisms may tell us something about life when Earth was young.

From cave-dwelling bacteria to 100-ton blue whales, the diversity of life on Earth is astounding. Yet all of the planet's organisms, living and extinct, exist on branches of the same family tree. Snottites, swordfish, humans, hydrangeas—all evolved from one single common ancestor.

When, where, and how life originated are some of the biggest questions in biology. Chemical evidence from 3.5-billion-year-old rocks in Australia suggests that biologically driven carbon and sulfur cycles existed at the time those rocks were formed. In the eons since, the first primitive life-forms have evolved into the 100 million or so organisms thought to populate the planet today.

How did the first living cell arise? Before scientists can hope to tackle such a question, they must agree on a definition of life. That's not necessarily as straightforward as it sounds. In our modern world, the features that separate life from non-life are relatively easy to discern. But Earth's first organisms were almost certainly much less complicated than even the simplest bacteria alive today. And before

those first truly living things appeared, molecular systems presumably existed that hovered somewhere between the domains of the living and the nonliving.

All cells require an archive of information, a membrane to maintain the inside of the cell different from the outside, and the ability to harness energy from the environment. In modern organisms, that archive is in the form of DNA, the double-stranded molecule that contains the instructions needed for cells to grow, differentiate, and reproduce. Without that molecular machinery, life as we know it would not exist.

DNA is critical, and it's also complex. Among the organisms alive today, the smallest known genome belongs to the bacterium *Carsonella rudii*. Even that relatively small genome contains nearly 160,000 DNA base pairs. How could such sophisticated molecular systems have arisen by chance?

The likely answer to that question is step by step. Laboratory experiments have shown how precursors to nucleic acids might have come together under the chemical conditions present on the young Earth. It's exceedingly unlikely that a molecule as complex as DNA was the first archive of information employed by the very first living cells. As you'll see in the chapters that follow, scientists have gathered evidence that hints at what the earliest nucleic acid molecules might have looked like.

While some kind of information archive was necessary for life to unfold, life requires more than a collection of nucleic acids replicating in a warm primordial pond. Living things must have some barrier that separates them from their

> All cells require an archive of information, a membrane to maintain the inside of the cell different from the outside, and the ability to harness energy from the environment.

**Snottites deep underground in a Mexican cave.** These stalactite-like slime formations are produced by bacteria that break down hydrogen sulfide gas as a source of energy.

environment. The cells of all living things, single-celled organisms or multicellular creatures, are each encased in a cell membrane.

Once again, scientists can only guess at how the first cell membranes came about. But research shows the molecules that make up modern membranes possess some interesting properties that may have led them to arise spontaneously. At first, the membranes were probably quite simple—straightforward (but leaky) barriers that kept the contents of early cells separated from the world at large. Over time, as chance variations arose, those membranes that provided a better barrier and that provided molecular gates were subject to an early example of the same natural selection process that is still happening today.

A third essential characteristic of living things is the ability to harness energy from the environment. Here, too, it's feasible that a series of natural chemical processes led to entities that could achieve this feat. Simple reactions may have produced molecular by-products that enabled more complex reactions down the road. Ultimately, that collection of reactions—combined with an archive of information and enclosed in some kind of primitive membrane—evolved into individual units that could breathe, grow, reproduce, and evolve.

Such a series of events may sound unlikely. However, some scientists argue that given the chemicals present on early Earth, it was likely—if not inevitable—that they would come together in such a way that life would emerge. Indeed, relatively simple, naturally occurring materials such as metal ions have been shown to play a role in key cellular reactions. Billions of years after the first cells arose, some of those metal ions—such as iron-sulfur minerals—still play a critical role in cells.

That's one reason researchers are so keen to study creatures like the sulfur-hungry snottites in the Cueva de Villa Luz. When life arose, the planet's atmosphere contained no oxygen—humans couldn't survive in such a world. By studying modern extremophiles—including bacteria that thrive in caves and near superheated hydrothermal vents at the bottom of the sea—scientists may uncover clues about how Earth's first cells came together and functioned.

Did life arise just once? Or could it have started up and died out several times before it finally got a foothold? If, given Earth's early chemistry, life here was inevitable, could it have arisen elsewhere in the universe? The study of life's origins involves many more questions than answers—and not just for biologists. The mystery of life spans the fields of biology, chemistry, physics, and planetary science. Though the questions are vast, our understanding of life's origins is likely to come about the same way life itself arose: step by step.

A hydrothermal vent. Some scientists think that this type of vent provided a favorable environment for chemical reactions that led to the origin of life.

## ? CASE 1 QUESTIONS

*Answers to Case 1 questions can be found in Chapters 2–8.*

1. **How did the molecules of life form?** *See page 2-17.*
2. **What was the first nucleic acid molecule, and how did it arise?** *See page 3-10.*
3. **How did the genetic code originate?** *See page 4-14.*
4. **How did the first cell membranes form?** *See page 5-3.*
5. **What naturally occurring elements might have spurred the first reactions that led to life?** *See page 6-15.*
6. **What were the earliest energy-harnessing reactions?** *See page 7-10.*
7. **How did early cells meet their energy requirements?** *See page 7-16 and page 8-15.*

## CHAPTER 2

# THE MOLECULES OF LIFE

## Core Concepts

**2.1** The atom is the fundamental unit of matter.

**2.2** Atoms can combine to form molecules linked by chemical bonds.

**2.3** Water is abundant and essential for life.

**2.4** Carbon is the backbone of organic molecules.

**2.5** Organic molecules include proteins, nucleic acids, carbohydrates, and lipids, each of which is built from simpler units.

**2.6** Life likely originated on Earth by a set of chemical reactions that gave rise to the molecules of life.

When biologists speak of diversity, they commonly point to the 2 million or so species named and described to date, or to the 10–100 million living species thought to exist in total. Life's diversity can also be found at a very different level of observation: in the molecules that make up each and every cell. Life depends critically on many essential functions, including establishing a boundary to separate cells from their surroundings, storing and transmitting genetic information, and harnessing energy from the environment. These functions ultimately depend on the chemical characteristics of the molecules that make up organisms.

In spite of the diversity of molecules and functions, the chemistry of life is based on just a few types of molecule, which in turn are made up of just a few elements. Of the 100 or so chemical elements, only about a dozen are found in more than trace amounts in living organisms. These elements interact with one another in only a limited number of ways. So, the question arises: How is diversity generated from a limited suite of chemicals and interactions? The answer lies in some basic features of chemistry.

## 2.1 PROPERTIES OF ATOMS

Since antiquity, it has been accepted that the materials of nature are made up of a small number of fundamental substances combined in various ways. Aristotle called these substances **elements,** and recognized four: earth, air, fire, and water. From the seventeenth century through the end of the nineteenth, elements were defined as pure substances that could not be broken down further by the methods of chemistry. In time, it was recognized that each element contains only one type of **atom,** the basic unit of matter. By 1850, about 60 elements were known, including such common ones as oxygen, copper, gold, and sodium. Today, 118 elements are known, of which 94 occur naturally and 24 have been created artificially in the laboratory. Elements are often indicated by a chemical symbol, which consists of a one- or two-letter abbreviation of the name of the element. For example, carbon is represented by C, hydrogen by H, and helium by He.

### Atoms consist of protons, neutrons, and electrons.

Elements are composed of atoms. The atom contains a dense central **nucleus** made up of positively charged particles called **protons** and electrically neutral particles called **neutrons.** A third type of particle, the negatively charged **electron,** moves around the

**FIG. 2.1**     **A carbon atom.** Each carbon atom has six protons, six neutrons, and six electrons. The net charge of any atom is neutral because there are as many electrons as protons.

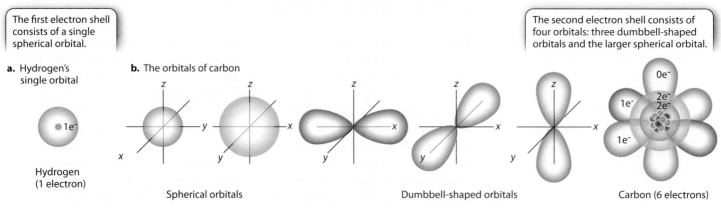

Nucleus
(6 protons + 6 neutrons)

e⁻   Electron

+   Proton

Neutron

Carbon atom

nucleus at some distance from it. Carbon, for example, typically has six protons, six neutrons, and six electrons (**Fig. 2.1**).

The number of protons, or the atomic number, specifies the atom as a particular element; an atom with one proton is hydrogen, for example, and an atom with six protons is carbon. The atomic number is sometimes indicated as a left subscript to the chemical symbol, for example $_6$C. Here, the "6" represents the number of protons in a carbon atom.

Together, the protons and neutrons determine the **atomic mass,** the mass of the atom. Each proton and neutron, by definition, has a mass of 1, whereas an electron has negligible mass. The number of neutrons in atoms of a particular element can vary, changing its mass. **Isotopes** are atoms of the same element that have different numbers of neutrons. For example, carbon has three isotopes: About 99% of carbon atoms have six neutrons and six protons, for an atomic mass of 12; about 1% has seven neutrons and six protons, for an atomic mass of 13; and only a very small fraction has eight neutrons and six protons, for an atomic mass of 14. The atomic mass is sometimes indicated as a left superscript to the chemical symbol. For instance, $^{12}$C is the isotope of carbon with six neutrons and six protons.

Typically, an atom has the same number of protons and electrons. Because a carbon atom has six protons and six electrons, the positive and negative charges cancel each other out and the

**FIG. 2.2**     **Electron orbitals and energy levels (shells) for hydrogen and carbon.** The orbital of an electron can be visualized as a cloud of points that are more dense where the electron is more likely to be. The hydrogen atom contains a single orbital, in a single energy level (a and c). The carbon atom has five orbitals, one in the first energy level and four in the second energy level (b and c).

The first electron shell consists of a single spherical orbital.

The second electron shell consists of four orbitals: three dumbbell-shaped orbitals and the larger spherical orbital.

**a.** Hydrogen's single orbital

1e⁻

Hydrogen
(1 electron)

**b.** The orbitals of carbon

Spherical orbitals

Dumbbell-shaped orbitals

0e⁻
1e⁻
2e⁻
2e⁻
1e⁻

Carbon (6 electrons)

**c.** Energy levels of hydrogen and carbon

Hydrogen

Carbon

In this simplified diagram, the electron energy levels (shells) are depicted as circles and the electrons that occupy them as dots.

carbon atom is electrically neutral. Some chemical processes cause an atom to either gain or lose electrons. An atom that has lost an electron is positively charged, and one that has gained an electron is negatively charged. Electrically charged atoms are called **ions.** The charge of an ion is specified in the right superscript position next to the chemical symbol. Thus, $H^+$ indicates a hydrogen ion that has lost an electron and is positively charged.

### Electrons occupy regions of space called orbitals.

Electrons move around the nucleus, but not in the simplified way shown in Fig. 2.1. The exact path that an electron takes is unknown, but it is possible to identify a region in space, called an **orbital,** where an electron is present most of the time. For example, **Fig. 2.2a** shows the orbital for hydrogen, which is simply a sphere occupied by a single electron. Most of the time, the electron is found within the space defined by the sphere, although its exact location at any instant is unpredictable.

Orbitals have certain properties. The maximum number of electrons in any orbital is two. Most atoms have more than two electrons and so have several orbitals positioned at different distances from the nucleus. These orbitals differ in size and shape. Electrons in orbitals close to the nucleus have less energy than do electrons in orbitals farther away, so electrons fill up orbitals close to the nucleus before occupying those farther away. Several orbitals can exist at a given energy level, or **shell.** The first shell consists of the spherical orbital shown in Fig. 2.2a.

**Fig. 2.2b** shows electron orbitals for carbon. Of carbon's six electrons, two occupy the small spherical orbital representing the lowest energy level. The remaining four are distributed among four possible orbitals at the next highest energy level: One of these four orbitals is a sphere (larger in diameter than that at the lowest energy level) and three are dumbbell-shaped. In carbon, the outermost spherical orbital has two electrons, two of the dumbbell-shaped orbitals have one electron each, and one of the dumbbell-shaped orbitals is empty. Because a full orbital contains two electrons, it would take a total of four additional electrons to completely fill all of the orbitals at this energy level. Therefore,

after the first shell, the maximum number of electrons per energy level is eight. **Fig. 2.2c** shows that the highest energy level, or shell, of carbon, represented by the outermost circle, contains four electrons, and that of hydrogen contains one electron.

→ **Quick Check 1** In the early 1900s, Ernest Rutherford produced a beam of very small negative particles and directed it at a thin piece of gold foil just a few atoms thick. Most of the particles passed through the foil without changing their path; very rarely, a particle was deflected. What conclusions can you draw from this experiment about the structure of an atom?

### Elements have recurring, or periodic, chemical properties.

The chemical elements are often arranged in a tabular form known as the **periodic table of the elements**, shown in **Fig. 2.3** and generally credited to the nineteenth-century Russian chemist Dmitri Mendeleev. The table provides a way to organize all the chemical elements in terms of their chemical properties.

In the periodic table, the elements are indicated by their chemical symbols and arranged in order of increasing atomic number. For example, the second row of the periodic table begins with lithium (Li) with 3 protons and ends with neon (Ne) with 10 protons.

Elements in a given horizontal row all have the same number of shells, and so also have the same number and types of orbitals.

**FIG. 2.3** **The periodic table of the elements.** Elements are arranged by increasing number of protons, the atomic number. The elements in any column share similar chemical properties.

| Abundance in cells |
| --- |
| High · Low · Trace · None |

| 1 H | | | | | | | | | | | | | | | | | 2 He |
| --- | --- | --- | --- | --- | --- | --- | --- | --- | --- | --- | --- | --- | --- | --- | --- | --- | --- |
| 3 Li | 4 Be | | | | | | | | | | | 5 B | 6 C | 7 N | 8 O | 9 F | 10 Ne |
| 11 Na | 12 Mg | | | | | | | | | | | 13 Al | 14 Si | 15 P | 16 S | 17 Cl | 18 Ar |
| 19 K | 20 Ca | 21 Sc | 22 Ti | 23 V | 24 Cr | 25 Mn | 26 Fe | 27 Co | 28 Ni | 29 Cu | 30 Zn | 31 Ga | 32 Ge | 33 As | 34 Se | 35 Br | 36 Kr |
| 37 Rb | 38 Sr | 39 Y | 40 Zr | 41 Nb | 42 Mo | 43 Tc | 44 Ru | 45 Rh | 46 Pd | 47 Ag | 48 Cd | 49 In | 50 Sn | 51 Sb | 52 Te | 53 I | 54 Xe |
| 55 Cs | 56 Ba | 57-71 La-Lu | 72 Hf | 73 Ta | 74 W | 75 Re | 76 Os | 77 Ir | 78 Pt | 79 Au | 80 Hg | 81 Tl | 82 Pb | 83 Bi | 84 Po | 85 At | 86 Rn |
| 87 Fr | 88 Ra | 89-103 Ac-Lr | 104 Rf | 105 Db | 106 Sg | 107 Bh | 108 Hs | 109 Mt | 110 Ds | 111 Rg | 112 Cn | 113 Uut | 114 Uuq | 115 Uup | 116 Uuh | 117 Uus | 118 Uuo |

| 57 La | 58 Ce | 59 Pr | 60 Nd | 61 Pm | 62 Sm | 63 Eu | 64 Gd | 65 Tb | 66 Dy | 67 Ho | 68 Er | 69 Tm | 70 Yb | 71 Lu |
| --- | --- | --- | --- | --- | --- | --- | --- | --- | --- | --- | --- | --- | --- | --- |
| 89 Ac | 90 Th | 91 Pa | 92 U | 93 Np | 94 Pu | 95 Am | 96 Cm | 97 Bk | 98 Cf | 99 Es | 100 Fm | 101 Md | 102 No | 103 Lr |

**FIG. 2.4    Energy levels (shells) of row 2 of the periodic table.** The complete complement of electrons in the outer shell of this row of elements is 8.

Across the row, electrons are added until the outer shell contains its complete complement of eight electrons.

| Lithium | Beryllium | Boron | Carbon | Nitrogen | Oxygen | Fluorine | Neon |

Across a row, therefore, electrons fill the shell until a full complement of electrons is reached on the right-hand side of the table. **Fig. 2.4** shows the filling of the shells for the second row of the periodic table. The elements in a vertical column are called a group or family. Members of a group all have the same number of electrons in their outermost shell. For example, carbon (C) and lead (Pb) both have four electrons in their outermost shell. The number of electrons in the outermost shell determines in large part how elements behave and interact with other elements, as we will see in the next section.

## 2.2 MOLECULES AND CHEMICAL BONDS

Atoms can combine with other atoms to form **molecules,** which are substances made up of two or more atoms. When two atoms form a molecule, the individual atoms interact in what is called a **chemical bond,** a form of attraction between atoms that holds them together. The ability of atoms to form molecules in part explains why just a few types of element can make many different molecules and perform diverse functions in a cell. There

are many different ways in which atoms can interact with one another, and therefore many different types of chemical bond, as we describe in this section.

### A covalent bond results when two atoms share electrons.

The ability of atoms to combine with other atoms is determined in large part by the electrons farthest from the nucleus—those in the outermost orbitals of an atom. These electrons are called **valence electrons,** and they are at the highest energy level. In many cases, when atoms combine with other atoms to form a molecule, the atoms share valence electrons with each other. Specifically, when the outermost orbitals of two atoms come into proximity, two atomic orbitals each containing one electron merge into a single orbital containing a full complement of two electrons. The merged orbital is called a **molecular orbital,** and each shared pair of electrons constitutes a **covalent bond** that holds the atoms together.

Hydrogen gas ($H_2$), illustrated in **Fig. 2.5,** is one of the simplest molecules. Note that the right subscript position is used to indicate the number of atoms in a molecule. Each hydrogen atom has a single electron in a spherical orbital. When the atoms join into a molecule, the two orbitals merge into a single molecular orbital containing two electrons that are shared by the hydrogen atoms. A covalent bond between atoms is denoted by a single line connecting the two chemical symbols for the atoms, as shown in the structural formula in Fig. 2.5.

→  **Quick Check 2** From their positions in the periodic table (see Fig. 2.3), can you predict how many lithium atoms and hydrogen atoms combine to form a molecule?

Molecules tend to be most stable when they share enough electrons to completely occupy the outermost energy level or shell. This simple rule of thumb is known as the octet rule and applies to many, but not all, elements. For example, as shown in **Fig. 2.6,** one carbon atom (C, with four valence electrons) combines with four hydrogen atoms (H, with one valence electron each) to form $CH_4$ (methane); nitrogen (N, with five valence electrons) combines with three H atoms to form $NH_3$

**FIG. 2.5    A covalent bond.** A covalent bond is formed when two atoms share a pair of electrons in a molecular orbital.

Chemical formula    $H_2$
Structural formula   $H-H$

There is one electron in the colored area representing an orbital.

Molecular orbital

H    +    H    →    H    H

Hydrogen atom (1 electron)    Hydrogen atom (1 electron)    Hydrogen gas (2 electrons)

Two atoms of hydrogen combine to form hydrogen gas by sharing electrons in a molecular orbital.

**FIG. 2.6    Four molecules.** Atoms tend to combine in such a way as to complete the complement of electrons in the outer shell.

The electrons in an atom's outermost shell are the valence electrons.

Hydrogen gas, H₂          Methane, CH₄          Ammonia, NH₃          Water, H₂O

(ammonia); and oxygen (O, with six valence electrons) combines with two H atoms to form H₂O (water). Interestingly, the elements of the same column in the next row behave similarly. This is just one example of the recurring, or periodic, behavior of the elements.

### A polar covalent bond is characterized by unequal sharing of electrons.

In hydrogen gas, the electrons are shared equally by the two hydrogen atoms. In many bonds, however, the electrons are not shared equally by the two atoms. A notable example is provided by the bonds in a water molecule (H₂O), which consists of two hydrogen atoms each covalently bound to a single oxygen atom (**Fig. 2.7a**).

In a molecule of water, the electrons are not shared equally between the hydrogen and oxygen atoms; rather, the electrons are more likely to be located near the oxygen atom. Unequal sharing of electrons results from a difference in the ability of the atoms to attract electrons, a property known as **electronegativity.** Electronegativity tends to increase across a row in the periodic table; as the number of protons across a row increases, electrons are held more tightly to the nucleus.

Therefore, oxygen is more electronegative than hydrogen and attracts electrons more than does hydrogen. In a molecule of water, oxygen has a slight negative charge, while the two hydrogen atoms have a slight positive charge (**Fig. 2.7b**). When electrons are shared unequally between the two atoms, the resulting interaction is described as a **polar covalent bond.**

### A hydrogen bond is an interaction of a hydrogen atom and an electronegative atom.

Because the oxygen and hydrogen atoms have slight charges, water molecules orient themselves to minimize the repulsion of like charges so that positive charges are near negative charges. A **hydrogen bond** results when a hydrogen atom covalently bound to an electronegative atom (such as oxygen or nitrogen) interacts with an electronegative atom of another molecule. In the case of water, a hydrogen atom covalently bound to an oxygen atom is attracted to and interacts with an oxygen atom of another water molecule. The result of many such interactions is a kind of molecular network stabilized by hydrogen bonds. Typically, a hydrogen bond is depicted by a dotted line, as in **Fig. 2.8.**

**FIG. 2.7    Polar covalent bond.** Polar covalent bonds do not share the electrons equally.

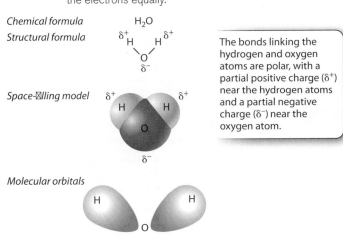

*Chemical formula*    H₂O
*Structural formula*

*Space-filling model*

*Molecular orbitals*

The bonds linking the hydrogen and oxygen atoms are polar, with a partial positive charge (δ⁺) near the hydrogen atoms and a partial negative charge (δ⁻) near the oxygen atom.

**FIG. 2.8    Hydrogen bonds in liquid water.** Because of thermal motion, hydrogen bonds in water are continually breaking and reforming between different pairs of molecules.

Hydrogen bonds

Hydrogen bonds form when the partial positive charge of hydrogen atoms are attracted to the partial negative charge of oxygen atoms.

Hydrogen bonds are much weaker than covalent bonds, but it is hydrogen bonding that gives water many of its unusual properties, which are described in the next section. In addition, many weak hydrogen bonds can help stabilize biological molecules, as in the case of nucleic acids and proteins.

## An ionic bond forms between oppositely charged ions.

In water, the difference in electronegativity between the oxygen and the hydrogen atoms leads to unequal sharing of electrons. In more extreme cases, when an atom of very high electronegativity is paired with an atom of very low electronegativity, the difference in electronegativity is so great that the electronegative atom "steals" the electron from its less electronegative partner. In this case, the atom with the extra electron has a negative charge and is a negative ion. The atom that has lost an electron has a positive charge and is a positive ion. The two ions are not covalently bound, but they associate with each other because of the attraction of opposite charges in what is called an **ionic bond.** An example of a compound formed by the attraction of a positive ion and a negative ion is table salt, or sodium chloride (NaCl) (**Fig. 2.9a**).

When sodium chloride is placed in water, the salt dissolves to form sodium ions (written as "$Na^+$") that have lost an electron and so are positively charged, and chloride ions ($Cl^-$) that have gained an electron and are negatively charged. In solution, the two ions are pulled apart and become surrounded by water molecules: The negatively charged ends of water molecules are attracted to the positively charged sodium ion and the positively charged ends of other water molecules are attracted to the negatively charged chloride ion (**Fig. 2.9b**). Only as the water evaporates do the concentrations of $Na^+$ and $Cl^-$ increase to the point where the ions join and precipitate as salt crystals.

## A chemical reaction involves breaking and forming chemical bonds.

The chemical bonds that link atoms in molecules can change in a **chemical reaction,** a process by which given molecules, called **reactants**, are transformed into different molecules, called **products.** During a chemical reaction, atoms keep their identity but change their chemical bonds.

For example, two molecules of hydrogen gas ($2H_2$) and one molecule of oxygen ($O_2$) can react to form two molecules

**FIG. 2.9    An ionic bond.** (a) Sodium chloride (salt) is formed by the attraction of two ions. (b) In solution, the ions are surrounded by water molecules.

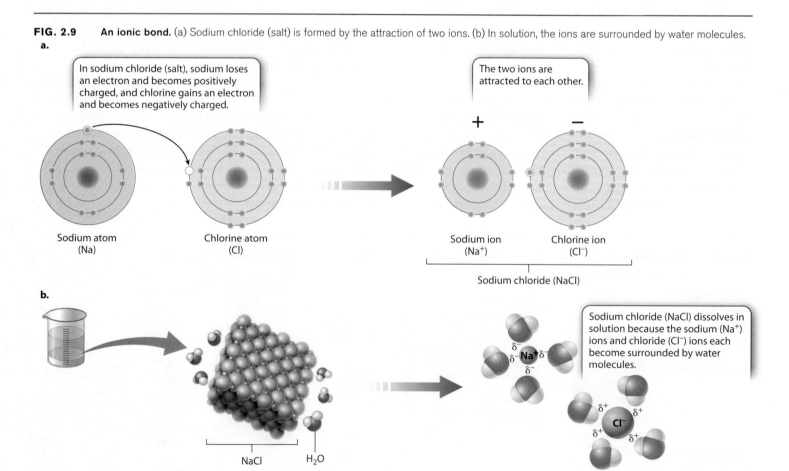

of water ($2H_2O$), as shown in **Fig. 2.10.** In this reaction, the numbers of each type of atom are conserved, but their arrangement is different in the reactants and the products. Specifically, the H–H bond in hydrogen gas and the O=O bond in oxygen are broken. At the same time, each oxygen atom forms new covalent bonds with two hydrogen atoms, forming two molecules of water. In fact, this reaction is the origin of the name "hydrogen," which literally means "water former." The reaction releases a good deal of energy; it was used in the main engine of the Space Shuttle.

In biological systems, chemical reactions provide a way to build and break down molecules for use by the cell, as well as to harness energy, which can be stored in chemical bonds (Chapter 6).

## 2.3 WATER: THE MEDIUM OF LIFE

On Earth, all life depends on water. Indeed, life originated in water, and the availability of water strongly influences the environmental distributions of different species. Furthermore, water is the single most abundant molecule in all cells, so water is the medium in which the molecules of life interact. In the late 1990s, the National Aeronautic and Space Administration (NASA) announced that the search for extraterrestrial life would guide continuing exploration of the solar system and beyond. NASA's operational strategy was simple: Follow the water. NASA's logic was straightforward: Within our solar system, Earth stands out both for its abundance of water and the life it supports. What makes water so special as the medium of life?

### Water is a polar molecule.

As we saw earlier, water molecules have polar covalent bonds, characterized by an uneven distribution of electrons. A molecule like water that has regions of positive and negative charge is called a **polar** molecule. Molecules, or even different regions of the same molecule, fall into two general classes, depending on how they interact with water: **hydrophilic** ("water loving") and **hydrophobic** ("water fearing").

Hydrophilic compounds, like water itself, are polar; they dissolve readily in water. That is, water is a good **solvent,** capable of dissolving many substances. Think of what happens when you stir a teaspoon of sugar into water: The sugar seems to disappear as the sugar dissolves. What is happening is that the sugar molecules are dispersing through the water and becoming separated from one another. Sugar is in solution in the watery, or **aqueous,** environment.

By contrast, hydrophobic compounds are nonpolar and arrange themselves to minimize their contact with water. For example, when oil and water are mixed, oil molecules organize themselves into droplets that limit the oil–water interface.

**FIG. 2.10** **A chemical reaction.** During a chemical reaction, atoms retain their identity, but their connections change as bonds are broken and new bonds are formed.

$2H_2$ + $O_2$ ⟶ $2H_2O$

Hydrogen gas    Oxygen    Water

H—H    O=O ⟶

H—H

Reactants    Products

This **hydrophobic effect,** in which polar molecules like water exclude nonpolar ones, drives such biological processes as the formation of cell membranes (Chapter 3) and the folding of proteins (Chapter 5).

### pH is a measure of the concentration of protons in solution.

A small proportion of the molecules in water exist as protons ($H^+$) and hydroxide ions ($OH^-$). The pH of a solution measures the proton concentration ($[H^+]$) by the following formula:

$$pH = -\log [H^+]$$

The pH of a solution can range from 0 to 14. Since the pH scale is logarithmic, a difference of one pH unit corresponds to a tenfold difference in hydrogen ion concentration. A solution is neutral (pH = 7) when the concentrations of protons ($H^+$) and hydroxide ions ($OH^-$) are equal. When the concentration of protons is higher than that of hydroxide ions, then the pH is lower than 7 and the solution is **acidic.** When the concentration of protons is lower than that of hydroxide ions, then the pH is higher than 7, and the solution is **basic.** An acid therefore is a molecule that releases a proton ($H^+$), and a base is a molecule that accepts a proton in aqueous solution.

Pure water has a pH of 7—that is, it is neutral, with an equal concentration of protons and hydroxide ions. The pH of most cells is approximately 7 and is tightly regulated, as most chemical reactions can be carried out only in a narrow pH range. Certain cellular compartments, however, have a much lower pH. The pH of blood is slightly basic, with a pH around 7.4. This value is sometimes referred to in medicine as physiological pH, as it can change in response to certain diseases. Freshwater lakes, ponds, and rivers tend to be slightly acidic because of dissolved carbon dioxide from the air, which forms carbonic acid in water.

**FIG. 2.11** **Liquid water and ice.** Hydrogen bonds create a dense structure in water (a), and a highly ordered, less dense, crystalline structure in ice (b).

**a.** Liquid water

**b.** Ice

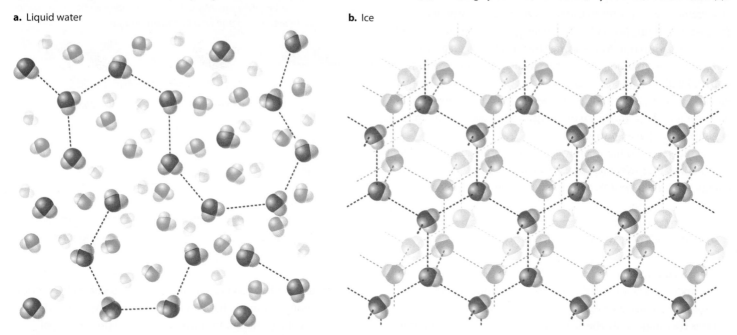

## Hydrogen bonds give water many unusual properties.

Water is also characterized by extensive hydrogen bonding, as we have seen. Hydrogen bonds influence the structure of both liquid water (**Fig. 2.11a**) and ice (**Fig. 2.11b**). When water freezes, most water molecules become hydrogen bonded to four other water molecules, forming an open lattice-like, crystalline structure we call ice. As the temperature increases and the ice melts, some of the hydrogen bonds are destabilized. This allows the water molecules to pack more closely, and it is the reason why liquid water is more dense than solid water. As a result, ice floats on water, and ponds and lakes freeze from the top down, and therefore do not freeze completely. This special property allows fish and aquatic plants to survive winter in the cold water under the layer of ice.

→ **Quick Check 3** Why do containers of water, milk, soda, or other liquids sometimes burst when frozen?

Hydrogen bonds also make water molecules cohesive, meaning that they tend to stick to one another. A consequence of cohesion is high surface tension, a measure of the difficulty of breaking the surface of a liquid. Water cohesion and surface tension contribute to water movement in plants. As water evaporates from leaves, water is pulled upward, sometimes as high as 300 feet above the ground in giant sequoia and coast redwood trees, which are among the tallest trees on Earth.

The hydrogen bonds of water also influence how water responds to heating. Molecules are in constant motion, and this motion increases as the temperature increases. When water is heated, however, the increased motion first breaks hydrogen bonds, and only afterward leads to a temperature increase. The need to break hydrogen bonds first means that water resists temperature changes more than do other substances, a property that is important for living organisms on a variety of scales. In the cell, water resists temperature variations that would otherwise result from numerous biochemical reactions. On a global scale, the oceans minimize temperature fluctuations, stabilizing the temperature on Earth in a range compatible with life.

In short, water is clearly the medium of life on Earth, but is this because water is uniquely suited for life, or is it because life on Earth has adapted through time to a watery environment? We don't know the answer, but probably both explanations are partly true. Chemists have proposed that under conditions of high pressure and temperature, other small molecules, among them ammonia ($NH_3$) and some simple carbon-containing molecules, might display similar characteristics friendly to life. However, under the conditions of pressure and temperature that exist on Earth, water is the only molecule uniquely suited to life. Water is a truly remarkable substance, and life on Earth would not be possible without it. Leonardo da Vinci once wrote that water is the driving force of all nature. Without a doubt, water is the driving force of all biology.

## 2.4 CARBON: LIFE'S CHEMICAL BACKBONE

Hydrogen and helium are far and away the most abundant elements in the universe. In contrast, the solid Earth is dominated by silicon, oxygen, aluminum, iron, and calcium (Chapter 1). In other words, Earth is not a typical sample of the universe. Nor, as it turns out, is the cell a typical sample of the solid Earth. **Fig. 2.12** shows the relative abundance by mass of chemical elements present in human cells after all the water has been removed. Note that just four elements—carbon (C), oxygen (O), hydrogen (H), and nitrogen (N)—constitute 94% of the total dry mass, and that the most abundant element is carbon. The elemental composition of human cells is typical of all cells. Human life, and all life as we know it, is based on carbon. Carbon molecules play such an important role in living organisms that carbon-containing molecules have a special name—they are called **organic molecules.** Their central role in life implies that there must be something very special about carbon, and there is. Carbon has the ability to combine with many other elements to form a wide variety of molecules, each specialized for the functions it carries out in the cell.

### Carbon atoms form four covalent bonds.

One of the special properties of carbon is that, in forming molecular orbitals, a carbon atom behaves as if it had four unpaired electrons. This behavior occurs because one of the electrons in the outermost sphere moves into the empty dumbbell-shaped orbital. In this process, the single large spherical orbital and three dumbbell-shaped orbitals change shape, becoming four equivalent hybrid orbitals each with one electron.

**Fig. 2.13** shows the molecular orbitals that result when one atom of carbon combines with four atoms of hydrogen to form the gas methane ($CH_4$). Each of the four valence electrons of carbon shares a new molecular orbital with the electron of one of the hydrogen atoms. These bonds can rotate freely about their axis. Furthermore, because of the shape of the orbitals, the carbon atom lies at the center of a three-dimensional structure called a tetrahedron, and the four molecular orbitals point toward the four corners of this structure. The ability of carbon to form four covalent bonds, the spatial orientation of these bonds in the form of a tetrahedron, and the ability of each bond to rotate freely all contribute importantly to the structural diversity of carbon-based molecules.

### Carbon-based molecules are structurally and functionally diverse.

Carbon has other properties that contribute to its ability to form a diversity of molecules. For example, carbon atoms can link with each other by covalent bonds to form long chains. These chains can be branched, or two carbons at the ends of the chain or within the chain can link to form a ring structure.

**FIG. 2.12** Approximate proportions by dry mass of chemical elements found in human cells.

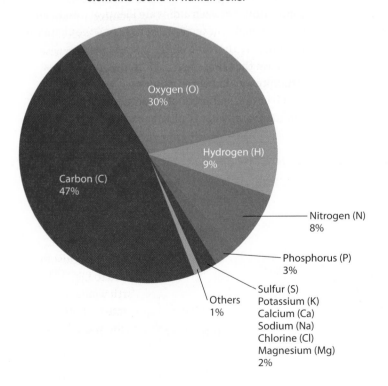

**FIG. 2.13** **A carbon atom with four covalent bonds.** One carbon atom combines with four hydrogen atoms to form methane.

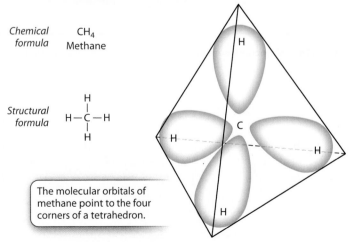

**FIG. 2.14** **Diverse carbon-containing molecules.** (a) The molecule ethane contains one C–C bond. (b) A linear chain of carbon atoms and a ring structure that also contains oxygen contain multiple C—C bonds.

**a.**

Chemical formula: $C_2H_6$ Ethane

Structural formula

C–C bond

**b.**

Chemical formula: $C_7H_{16}$     $C_5H_{10}O$

Structural formula

Simplified structure

These structures are shown in simplified form, where a carbon atom is at the end of a line or the angle where two lines join; the hydrogen atoms that share any otherwise unshared electrons of each carbon atom are assumed but not shown.

**FIG. 2.15** **Molecules containing double bonds between carbon atoms.**

Chemical formula: $C_2H_4$     $C_7H_{14}$     $C_6H_6$

Structural formula

**FIG. 2.16** **The isomers isoleucine and leucine.** Isoleucine and leucine are isomers: Their chemical formulas are the same, but their structures differ.

Chemical formula: $C_6H_{13}O_2N_1$ Isoleucine     $C_6H_{13}O_2N_1$ Leucine

Structural formula

Among the simplest chains is ethane, shown in **Fig. 2.14a,** in which two carbon atoms become connected by a covalent bond. In this case, the orbitals of unpaired electrons in two carbon atoms merge to create the covalent bond. Some more complex examples of carbon-containing molecules, and an abbreviated form, are shown in **Fig. 2.14b.**

Two adjacent carbon atoms can also share two pairs of electrons, forming a **double bond,** as shown in **Fig. 2.15.** Note that each carbon atom has exactly four covalent bonds, but in this case two are shared between adjacent carbon atoms. The double bond is shorter than a single bond and is not free to rotate so that all of the covalent bonds in the carbon atoms connected by a double bond are in the same geometrical plane. As with single bonds, double bonds can be found in chains of atoms or cyclical structures. A double bond between atoms is represented by a double line connecting the two chemical symbols for the atoms.

While the types of atoms making up a molecule help characterize the molecule, the spatial arrangement of atoms is also important. For example, 6 carbon atoms, 13 hydrogen atoms, 2 oxygen atoms, and 1 nitrogen atom can join covalently in many different arrangements to produce molecules with different structures. Two of these many arrangements are shown in **Fig. 2.16.** Note that some of the connections between atoms are identical (black) and some are different (green), even though the chemical formulas are the same ($C_6H_{13}O_2N_1$). Molecules that have the same chemical formula but different structures are known as **isomers.**

We have seen that carbon-containing molecules can adopt a wide range of arrangements, a versatility that helps us to understand how a limited number of elements can create an astonishing variety of molecules. We might ask whether carbon is *uniquely* versatile. Put another way, if we ever discover life on a distant planet, will it be carbon based, like us? Silicon, just below carbon in the periodic table (see Fig. 2.3), is the one other element that is both reasonably abundant and characterized by four atomic orbitals with one electron each. Some scientists have speculated that silicon might provide an alternative to carbon as a chemical basis for life. However, silicon readily binds oxygen. On Earth, nearly all of the silicon atoms found in molecules are covalently bound to oxygen. Studies of Mars and of meteorites show that silicon is tightly bound to oxygen throughout our solar system, and that is likely to be true everywhere we might explore. There are about 1000 different silicate minerals on Earth, but this diversity pales before the millions of known carbon-based molecules. If we ever discover life beyond Earth, very likely its chemistry will be based on carbon.

## 2.5 ORGANIC MOLECULES

The chemistry of carbon creates diverse geometries, which in turn lead to molecules with different structures and functions. These molecules perform essential tasks of the cell, including the establishment of a boundary to separate the inside of the cell from the environment, the storage and transmission of genetic information, and the capture, storage, and utilization of energy from the environment. These processes depend on just a few classes of carbon-based molecules. **Proteins** provide structural support and act as catalysts that facilitate chemical reactions. **Nucleic acids** encode and transmit genetic information. **Carbohydrates** provide a source of energy and make up the cell wall in bacteria, plants, and algae. **Lipids** make up cell membranes, store energy, and act as signaling molecules.

These molecules are all relatively large, and most are **polymers,** complex molecules made up of repeated simpler units connected by covalent bonds. Proteins are polymers of **amino acids,** nucleic acids are made up of **nucleotides,** and carbohydrates such as starch are built from simple **sugars.** Lipids are a bit different, as we will see—they are defined by a property rather than by their chemical structure. The lipid membranes that define cell boundaries consist of **fatty acids** bonded to other organic molecules.

Building macromolecules from simple, repeating units provides a means of generating virtually limitless chemical diversity. Indeed, in macromolecules, the building blocks of polymers play a role much like that of the letters in words. In written language, a change in the content or order of letters changes the meaning of the word (or renders it meaningless). For example, by reordering the letters of the word "SILENT" you

can write "LISTEN," a word with a different meaning. Similarly, rearranging the building blocks that make up macromolecules provides an important way to make a large number of diverse macromolecules.

In the following section, we focus on the building blocks of these four key molecules of life, reserving a discussion of the structure and function of the macromolecules for later chapters.

### Proteins are composed of amino acids.

Proteins do much of the cell's work. Some function as catalysts that accelerate the rates of chemical reactions (in which case they are called **enzymes**), and some act as structural components necessary for cell shape and movement. Your body contains many thousands of proteins that perform a wide range of functions. Since proteins consist of amino acids linked covalently to form a chain, we need to examine the chemical features of amino acids to understand the diversity and versatility of proteins.

The general structure of an amino acid is shown in **Fig. 2.17a.** Each amino acid contains a central carbon atom, called the **α (alpha) carbon,** covalently linked to four groups: a **carboxyl group** (COOH; red), an **amino group** ($NH_2$; blue), a hydrogen atom (H), and an **R group,** or **side chain,** (green) that differs from one amino acid to the next. The identity of each amino acid is determined by the structure and composition of the side chain. The side chain of the amino acid glycine is simply H, for example, and that of alanine is $CH_3$. In most amino acids, the α-carbon is covalently linked to four different groups. Glycine is the exception, since its R group is a hydrogen atom.

Amino acids are linked in a chain to form a protein (**Fig. 2.17b**). The carbon atom in the carboxyl group of one amino acid is

**FIG. 2.17  Amino acids and peptide bonds** (a) An amino acid contains four groups attached to a central carbon. (b) Peptide bonds link amino acids to form a protein.

joined to the nitrogen atom in the amino group of the next by a covalent linkage called a **peptide bond.** In Fig. 2.17b, the chain of amino acids includes four amino acids, and the peptide bonds are indicated in red. The formation of a peptide bond involves the loss of a water molecule since in order to form a C–N bond, the carbon atom must release an oxygen atom and the nitrogen must release two hydrogen atoms. These can then combine to form a water molecule ($H_2O$). The loss of a water molecule is a consistent feature in the linking of subunits to form polymers such as nucleic acids and complex carbohydrates.

Cellular proteins are composed of 20 amino acids, which can be classified according to the chemical properties of their side chains. The particular sequence, or order, in which amino acids are present in a protein determines how it folds into its three-dimensional structure. The three-dimensional structure, in turn, determines the protein's function. In Chapter 4, we will examine how the sequence of amino acids in a particular protein is specified and discuss how proteins fold into their three-dimensional configuration.

### Nucleic acids encode genetic information in their nucleotide sequence.

Nucleic acids are examples of informational molecules—that is, they are large molecules that carry information in the sequence of nucleotides that make them up. This molecular information is much like the information carried by the letters in an alphabet, but in the case of nucleic acids, the information is in chemical form.

The nucleic acid **deoxyribonucleic acid (DNA)** is the genetic material in all organisms. It is transmitted from parents to offspring, and it contains the information needed to specify the amino acid sequence of all the proteins synthesized in an organism. The nucleic acid **ribonucleic acid (RNA)** has multiple functions; it is a key player in protein synthesis and the regulation of gene expression.

DNA and RNA are long molecules consisting of nucleotides bonded covalently one to the next. Nucleotides, in turn, are composed of three components: a 5-carbon sugar, a nitrogen-containing compound called a **base,** and one or more phosphate groups (**Fig. 2.18**). The sugar in RNA is ribose, and the sugar in DNA is deoxyribose. The sugars differ in that ribose has a hydroxyl (OH) group on the second carbon (designated the 2′ carbon), while deoxyribose has a hydrogen atom at this position (hence, *deoxy*ribose). (By convention, the carbons in the sugar are numbered with primes—1′, 2′, etc.—to distinguish them from carbons in the base—1, 2, etc.)

The bases are built from nitrogen-containing rings and are of two types. The **pyrimidine** bases (**Fig. 2.19a**) have a single ring and include **thymine (T), cytosine (C),** and **uracil (U).** The **purine** bases (**Fig. 2.19b**) have a double-ring structure and include **adenine (A)** and **guanine (G).** DNA contains the bases A, T, G, and C, and RNA contains the bases A, U, G, and C. Just as the order of amino acids provides the unique information carried in proteins, so, too, does the sequence of nucleotides determine the information in DNA and RNA molecules.

**FIG. 2.18** **A ribonucleotide and a deoxyribonucleotide, the units of RNA and DNA.**

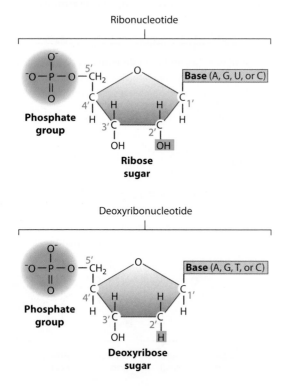

**FIG. 2.19** **Pyrimidine bases (a) and purine bases (b).** Pyrimidines have a single-ring structure, and purines have a double-ring structure.

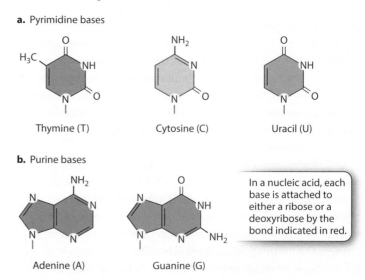

**FIG. 2.20    The phosphodiester bond.** Phosphodiester bonds link successive deoxyribonucleotides, forming the backbone of the DNA strand.

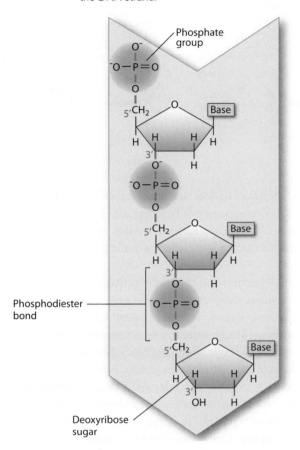

occur along the molecule. Successive nucleotides along a DNA strand can occur in *any* order, and hence a long molecule could contain any of an immense number of possible nucleotide sequences. This is one reason why DNA is an efficient carrier of genetic information. In Chapter 3, we consider the structure and function of DNA and RNA in greater detail.

### Complex carbohydrates are made up of simple sugars.

Many of us, when we feel tired, reach for a candy bar for a quick energy boost. The quick energy in a candy bar comes from sugars, which are quickly broken down to release energy. Sugars belong to a class of molecules called **carbohydrates,** distinctive molecules composed of C, H, and O atoms, usually in the ratio 1:2:1. Carbohydrates provide a principal source of energy for metabolism.

The simplest carbohydrates are sugars (also called **saccharides**). Simple sugars are linear or, far more commonly, cyclic molecules containing five or six carbon atoms. All six-carbon sugars have the same chemical formula ($C_6H_{12}O_6$) and differ only in configuration. Glucose (the product of photosynthesis), galactose (found in dairy products), and fructose (a commercial sweetener) are examples; they share the same

**FIG. 2.21    The structure of DNA.** (a) DNA is most commonly in the form of a double helix, with the sugar and phosphate groups forming the backbone and the bases oriented inward. (b) The bases are complementary, with A pairing with T and G pairing with C. (c) Base pairing results from hydrogen bonds.

In DNA and RNA, each adjacent pair of nucleotides is connected by a **phosphodiester bond,** which forms when a phosphate group in one nucleotide is covalently joined to the sugar unit in another nucleotide (**Fig. 2.20**). As in the formation of a peptide bond, the formation of a phosphodiester bond involves the loss of a water molecule.

DNA in cells usually consists of two strands of nucleotides twisted around each other in the form of a **double helix** (**Fig. 2.21a**). The sugar–phosphate backbones of the strands wrap like a ribbon around the outside of the double helix, and the bases are pointed inward. The bases form specific purine–pyrimidine pairs that are said to be **complementary:** Where one strand carries an A, the other carries a T; and where one strand carries a G, the other carries a C (**Fig. 2.21b**). Base pairing results from hydrogen bonding between the bases (**Fig. 2.21c**).

Genetic information in DNA is contained in the sequence, or order, in which successive nucleotides

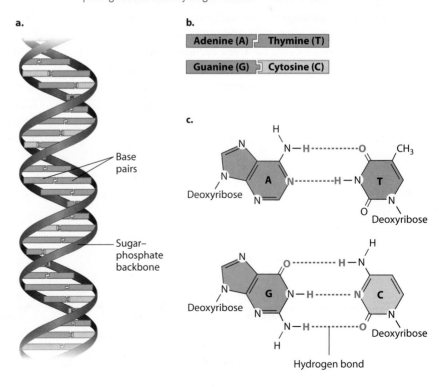

**FIG. 2.22** Structural formulas for some 6-carbon aldoses and ketoses.

Glucose
(an aldose)

Galactose
(an aldose)

Fructose
(a ketose)

**FIG. 2.23** Formation of the cyclic form of glucose.

Linear glucose

Cyclic glucose

formula ($C_6H_{12}O_6$) but differ in the arrangement of their atoms (**Fig. 2.22**).

→ **Quick Check 4** Take a close look at Fig. 2.22. How is glucose different from galactose?

A simple sugar is also called a **monosaccharide** (*mono* means "one"), and two simple sugars linked together by a covalent bond is called a disaccharide (*di* means "two"). Sucrose ($C_{12}H_{22}O_{11}$), or table sugar, is a disaccharide that combines one molecule each of glucose and fructose. As noted above, nucleotides contain a 5-carbon sugar, either ribose (in RNA) or deoyxribose (in DNA). Simple sugars combine in many ways to form polymers called **polysaccharides** (*poly* means "many") that provide long-term energy storage (starch and glycogen) or structural support (cellulose in plant cell walls). Long, branched chains of monosaccharides are called **complex carbohydrates.**

Let's take a closer look at monosaccharides, the simplest sugars. Monosaccharides are unbranched carbon chains with either an aldehyde (HC=O) or a ketone (C=O) group (Fig. 2.22). Monosaccharides with an aldehyde group are called **aldoses** and those with a ketone group are known as **ketoses.** In

both types of monosaccharides, the other carbons each carry one hydroxyl (OH) group and one hydrogen (H) atom. When the linear structure of a monosaccharide is written with the aldehyde or ketone group at the top, the carbons are numbered from top to bottom.

Virtually all of the monosaccharides in cells are in ring form (**Fig. 2.23**), not linear structures. To form a ring, the carbon in the aldehyde or ketone group forms a covalent bond with the oxygen of a hydroxyl group carried by another carbon in the same molecule. For example, cyclic glucose is formed when the oxygen atom of the hydroxyl group on carbon 5 forms a covalent bond with carbon 1, which is part of an aldehyde group. The cyclic structure is approximately flat, and you can visualize it perpendicular to the plane of the paper with the covalent bonds indicated by the thick lines in the foreground. The groups attached to any carbon therefore project either above or below the ring. When the ring is formed, the aldehyde oxygen becomes a hydroxyl group.

Monosaccharides, especially 6-carbon sugars, are the building blocks of complex carbohydrates. Monosaccharides are attached to each other by covalent bonds called **glycosidic bonds** (**Fig. 2.24**). As with peptide bonds, the formation of glycosidic bonds involves the loss of a water molecule. A glycosidic bond is formed between carbon 1 of one monosaccharide and a hydroxyl

**FIG. 2.24 Glycosidic bonds.** Glycosidic bonds link glucose monomers together to form the polysaccharide starch.

Glycosidic bond

group carried by a carbon atom in a different monosaccharide molecule.

Carbohydrate diversity stems in part from the monosaccharides that make them up, similar to the way that protein and nucleic acid diversity stems from the sequence of their subunits. Some complex carbohydrates are composed of a single type of monosaccharide, while others are a mix of different kinds of monosaccharide. Starch, for example, is a sugar storage molecule in plants composed completely of glucose molecules, whereas pectin, a component of the cell wall, contains up to five different monosaccharides.

## Lipids are hydrophobic molecules.

Proteins, nucleic acids, and carbohydrates all are polymers made up of smaller, repeating units with a defined structure. Lipids are different. Instead of being defined by a chemical structure, they share a particular property: Lipids are all hydrophobic. Because they share a property and not a structure, lipids are a chemically diverse group of molecules. They include familiar fats that make up part of our diet, components of cell membranes, and signaling molecules. Let's briefly consider each in turn.

**Triacylglycerol** is an example of a lipid that is used for energy storage. It is the major component of animal fat and vegetable oil. A triacylglycerol molecule is made up of three fatty acids joined to **glycerol** (**Fig. 2.25**). A fatty acid is a long chain of carbons attached to a carboxyl group (COOH) at one end (Figs. 2.25a and 2.25b). Glycerol is a 3-carbon molecule with OH groups attached to each carbon (Fig. 2.25c).

Fatty acids differ in the length (that is, in the number of carbons) of their hydrocarbon chain. Most fatty acids in cells contain an even number of carbons because they are synthesized by the stepwise addition of 2-carbon units. Some fatty acids have one or more carbon–carbon double bonds; these double bonds can differ in number and location. Fatty acids that do not contain double bonds are described as **saturated.** Because there are no double bonds, the maximum number of hydrogen atoms is attached to each carbon atom, so all of the carbons are

**FIG. 2.25**   Triacylglycerol and its components.

**a.** Palmitic acid (fatty acid)

**b.** Palmitoleic acid (fatty acid)

An unsaturated fatty acid contains one or more double bonds.

**c.** Glycerol

**d.** Triacylglycerol

Fats are formed by the addition of three fatty acid chains to glycerol.

said to be "saturated" with hydrogen atoms (Fig. 2.25a). Fatty acids that contain carbon–carbon double bonds are **unsaturated** (Fig. 2.25b). The chains of saturated fatty acids are straight, while the chains of unsaturated fatty acids have a kink at each double bond.

Triacylglycerols can contain different types of fatty acids attached to the glycerol backbone. They are all extremely hydrophobic and, therefore, triacylglycerols form oil droplets inside the cell. Triacylglycerols are an efficient form of energy storage because by excluding water molecules a large number can be packed into a small volume.

The hydrocarbon chains of fatty acids do not contain polar covalent bonds like those in a water molecule. Their electrons are distributed uniformly over the whole molecule and thus these molecules are uncharged. However, the constant motion

**FIG. 2.26    Van der Waals forces.** Transient asymmetry in the distribution of electrons along fatty acid chains leads to asymmetry in neighboring molecules, resulting in weak electrostatic attractions.

der Waals bonds between the chains also increases, and this in turn increases the melting temperature. Kinks introduced by double bonds reduce the tightness of the molecular packing and lead to fewer intermolecular interactions and a lower melting temperature. Therefore, an unsaturated fatty acid has a lower melting point than a saturated fatty acid of the same length. Animal fats such as butter are composed of triacylglycerols with saturated fatty acids and are solid at room temperature, whereas plant fats and fish oils are composed of triacylglycerols with unsaturated fatty acids and are liquid at room temperature.

**Steroids** such as cholesterol are a second type of lipid (**Fig. 2.27**). Like other steroids, cholesterol has a core composed of 20 carbon atoms bonded to form four fused rings, and it is hydrophobic. Cholesterol is a component of animal cell membranes (Chapter 5) and serves as a precursor for the synthesis of steroid hormones such as estrogen and testerone (Chapter 38).

**Phospholipids** are a third type of lipid and a major component of the cell membrane. Whereas triacylglycerol is made up of glycerol attached to three fatty acids, most phospholipids are made up of glycerol attached to two fatty acids and a third molecule that contains a phosphate group (**Fig. 2.28a**). The phosphate "head" group is hydrophilic, while the fatty acid "tails" are hydrophobic. As a result, phospholipids have hydrophobic and hydrophilic groups in the same molecule, giving them an interesting property when placed in water: They form a variety of structures all of which limit the exposure of the hydrophobic tails to water. One important structure is a **bilayer,** a two-layered structure with the hydrophilic heads pointing outward toward the aqueous environment and the hydrophobic tails oriented inward, away from water (**Fig. 2.28b**). The formation of lipid bilayers is discussed more fully in the next chapter.

of electrons leads to regions of slight positive and negative charges (**Fig. 2.26**). These charges in turn attract or repel electrons in neighboring molecules, setting up areas of positive and negative charge in those molecules as well. The temporarily polarized molecules weakly bind to one another because of the attraction of opposite charges. These interactions are known as **van der Waals forces.** The van der Waals forces are weaker than hydrogen bonds, but many of them acting together help to stabilize molecules.

Because of van der Waals forces, the melting points of fatty acids depend on their length and level of saturation. As the length of the hydrocarbon chains increases, the number of van

**FIG. 2.27    The chemical structure of cholesterol.**

**FIG. 2.28** **Phospholipids.** (a) Most phospholipids are made up of glycerol attached to two fatty acids and a phosphate-containing head group. (b) Phospholipids form a bilayer in aqueous solution.

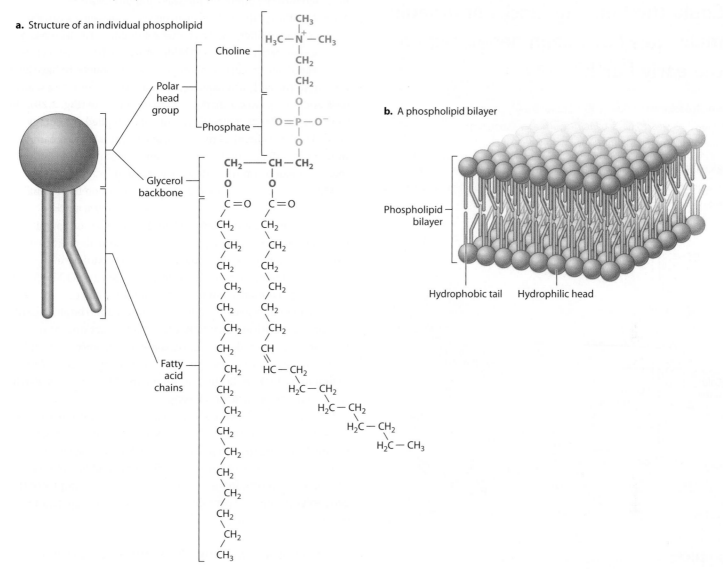

**a.** Structure of an individual phospholipid

**b.** A phospholipid bilayer

Phospholipid bilayer

Hydrophobic tail    Hydrophilic head

? **CASE 1 The First Cell: Life's Origins**

## 2.6 HOW DID THE MOLECULES OF LIFE FORM?

In Chapter 1, we considered the similarities and differences between living and non-living things. Four billion years ago, however, the differences may not have been so pronounced. Scientists believe that life originated early in our planet's history by a set of chemical processes that, through time, produced organisms that could be distinguished from their non-living surroundings. How can we think scientifically about one of biology's deepest and most difficult problems? We can't reconstruct key events from the geological record—the study of fossils tells us that life already existed when

Earth's oldest surviving fossil-containing rocks were deposited. The alternative is to approach life's origins experimentally, asking whether chemical reactions likely to have taken place on the early Earth can generate the molecules of life. It is important to note that even the simplest living organisms living today are far more complicated than our earliest ancestors. No one suggests that cells as we know them emerged directly from primordial chemical reactions. Rather, the quest is to discover simple molecular systems able to replicate themselves, while subject to natural selection.

A key starting point is the observation, introduced earlier in this chapter, that the principal macromolecules found in organisms are themselves made of simpler molecules joined together. Thus, if we want to understand how proteins might have emerged on the early Earth, we should begin with the synthesis of amino acids, and if we are interested in nucleic acids, we should focus on nucleotides.

**FIG. 2.29**

# Could the building blocks of organic molecules have been generated on the early Earth?

**BACKGROUND** In the 1950s, Earth's early atmosphere was widely believed to have been rich in water vapor, methane, ammonia, and hydrogen gas, with no free oxygen.

**EXPERIMENT** Stanley Miller built an apparatus, shown below, designed to simulate Earth's early atmosphere. Then he passed a spark through the mixture to simulate lightning.

**RESULTS** As the experiment proceeded, reddish material accumulated on the walls of the flask. Analysis showed that the brown matter included a number of amino acids.

**CONCLUSION** Amino acids can be generated in conditions that mimic those of the early Earth.

**FOLLOW-UP WORK** Recent analysis of the original extracts, saved by Miller, shows that the experiment actually produced about 20 different amino acids, not all of them found in organisms.

SOURCE Miller, S. L. 1953. "Production of Amino Acids Under Possible Primitive Earth Conditions." *Science* 117:528–529.

## The building blocks of life can be generated in the laboratory.

Research into the origins of life was catapulted into the experimental age in 1953, with an elegant experiment carried out by Stanley Miller, then a graduate student in the laboratory of Nobel laureate Harold Urey. Miller started with gases such as water vapor, methane, and hydrogen gas, all thought to have been present in the early atmosphere. He put these gases into a sealed flask and then passed a spark through the mixture (**Fig. 2.29**). On the primitive Earth, lightning might have supplied the energy needed to drive chemical reactions, and the spark was meant to simulate its effects. Analysis of the contents of the flask showed that a number of amino acids were generated.

Miller and others conducted many variations on his original experiment, all with similar results. Today, many scientists doubt that the early atmosphere had the composition found in Miller's experimental apparatus, but amino acids and other biologically important molecules can form in a variety of simulated atmospheric compositions. If oxygen gas ($O_2$) is absent and hydrogen is more common in the mixture than carbon, the addition of energy generates diverse amino acids. The absence of oxygen gas is critical, since Miller-Urey-type reactions cannot run to completion in modern air or seawater. Here, however, geology supports the experiments: Chemical analyses of Earth's oldest sedimentary rocks indicate that, for its first 2 billion years, Earth's surface contained little or no oxygen.

Later experiments have shown that other chemical reactions can generate simple sugars, the bases found in nucleotides, and the lipids needed to form primitive membranes. Independent evidence that simple chemistry can form the building blocks of life comes from certain meteorites, which provide samples of the early solar system and contain diverse amino acids, lipids, and other organic compounds.

## Experiments show how life's building blocks can form macromolecules.

From the preceding discussion, we have seen that life's simple building blocks can be generated under conditions likely to have been present on the early Earth—but can these simple units be stitched together to form the polymeric molecules of life? Once again, careful experiments have shown how polymers could have formed in the conditions of the early Earth. Clay minerals that form from volcanic rocks can bind nucleotides on their surfaces (**Fig. 20.30a**). The clays provide a surface that places the nucleotides in proximity to one another, making it possible for them to join to form chains, or simple strands of nucleic acid.

In a classic experiment, biochemist Leslie Orgel placed a short nucleic acid sequence into a reaction vessel and then added individual chemically modified nucleotides. The nucleotides spontaneously joined into a polymer, forming the sequence complementary to the nucleic acid already present (**Fig. 20.30b**).

Such experiments show that nucleic acids can be synthesized experimentally from nucleotide building blocks, but until recently the synthesis of nucleotides themselves presented a formidable problem for research on the origins of life. Many tried to generate nucleotides from their sugar, base, and phosphate constituents, but no one succeeded until 2009. That year, John Sutherland and his colleagues showed that nucleotides can be synthesized under conditions thought to be like those on the young Earth. These chemists showed how simple organic molecules likely to have formed in abundance on the early Earth react in the presence of phosphate molecules, yielding the long-sought nucleotides.

Such humble beginnings eventually gave rise to the abundant diversity of life we see around us, described memorably by Charles Darwin in the final paragraph of *The Origin of Species*:

**FIG. 2.30** **Spontaneous polymerization of nucleotides.** (a) Clays may have played an important role in the origin of life by providing surfaces for nucleotides to form nucleic acids. (b) Addition of a short RNA molecule, the template strand, to a flask containing modified nucleotides results in the formation of a complementary RNA strand.

a.

b.

It is interesting to contemplate an entangled bank, clothed with many plants of many kinds, with birds singing on the bushes, with various insects flitting about, and with worms crawling through the damp earth, and to reflect that these elaborately constructed forms, so different from each other, and dependent on each other in so complex a manner, have all been produced by laws acting around us. . . . There is grandeur in this view of life . . . from so simple a beginning endless forms most beautiful and most wonderful have been, and are being, evolved.

# Core Concepts Summary

## 2.1 THE ATOM IS THE FUNDAMENTAL UNIT OF MATTER.

Atoms consist of positively charged protons and electrically neutral neutrons in the nucleus, and negatively charged electrons darting around the nucleus. page 2-1

The number of protons determines the identity of an atom. page 2-2

Protons and neutrons together determine the mass of an atom. page 2-2

Protons and electrons determine the charge of an atom. page 2-3

Negatively charged electrons travel around the nucleus in regions called orbitals. page 2-3

The periodic table of the elements reflects a regular and repeating pattern in the chemical behavior of elements. page 2-3

## 2.2 ATOMS CAN COMBINE TO FORM MOLECULES LINKED BY CHEMICAL BONDS.

Electrons that occupy the outermost energy level (shell) of an atom (valence electrons) determine its ability to combine with other atoms to form molecules. page 2-4

A covalent bond results from the sharing of electrons between atoms to form molecular orbitals. page 2-4

A polar covalent bond results when two atoms do not share electrons equally as a result of a difference in the ability of the atoms to attract electrons, a property called electronegativity. page 2-5

A hydrogen bond results when a hydrogen atom covalently bonded to an electronegative atom interacts with an electronegative atom of another molecule.  page 2-5

An ionic bond results from the attraction of oppositely charged ions.  page 2-6

### 2.3  WATER IS ABUNDANT AND ESSENTIAL FOR LIFE.

Water is a polar molecule because shared electrons are distributed asymmetrically between the oxygen and hydrogen atoms.  page 2-7

Hydrophilic molecules dissolve readily in water, while hydrophobic molecules in water tend to associate with one another, minimizing their contact with water.  page 2-7

The pH of an aqueous solution is a measure of the acidity of the solution.  page 2-7

Water forms hydrogen bonds, which help explain its high cohesion, surface tension, and resistance to rapid temperature change.  page 2-8

### 2.4  CARBON IS THE BACKBONE OF ORGANIC MOLECULES.

A carbon atom can form up to four covalent bonds with other atoms.  page 2-9

The geometry of these covalent bonds helps explain the structural and functional diversity of organic molecules.  page 2-9

### 2.5  ORGANIC MOLECULES INCLUDE PROTEINS, NUCLEIC ACIDS, CARBOHYDRATES, AND LIPIDS, EACH OF WHICH IS BUILT FROM SIMPLER UNITS.

Amino acids are linked by covalent bonds to form proteins.  page 2-11

An amino acid consists of a carbon atom (the $\alpha$-carbon) attached to a carboxyl group, an amino group, a hydrogen atom, and a side chain.  page 2-11

The side chain determines the properties of each amino acid.  page 2-11

Nucleotides assemble to form nucleic acids, which store and transmit genetic information.  page 2-12

Nucleotides are composed of a 5-carbon sugar, a nitrogen-containing base, and a phosphate group.  page 2-12

Nucleotides in DNA incorporate the sugar ribose, and nucleotides in RNA incorporate the sugar deoxyribose.  page 2-12

The bases are pyrimidines (thymine, cytosine, and uracil) and purines (adenine and guanine).  page 2-12

Sugars are carbohydrates, distinctive molecules composed of C, H, and O atoms, usually in the ratio 1:2:1, that are a source of energy.  page 2-13

Monosaccharides assemble to form disaccharides or longer polymers called complex carbohydrates.  page 2-14

Lipids are hydrophobic.  page 2-15

Triacylglycerols store energy and are made up of glycerol and fatty acids.  page 2-15

Fatty acids consist of a linear hydrocarbon chain of variable length with a carboxyl group at one end.  page 2-15

Fatty acids are either saturated (no carbon–carbon double bonds) or unsaturated (one or more carbon–carbon double bonds).  page 2-15

The tight packing of fatty acids in lipids is the result of van der Waals forces, a type of weak, noncovalent bond.  page 2-16

### 2.6  LIFE LIKELY ORIGINATED ON EARTH BY A SET OF CHEMICAL REACTIONS THAT GAVE RISE TO THE MOLECULES OF LIFE.

In 1953, Stanley Miller and Harold Urey demonstrated that amino acids can be generated in the laboratory in conditions that mimic those of the early Earth.  page 2-18

Other experiments have shown that sugars, bases, and lipids can be generated in a similar way.  page 2-18

Once the building blocks were synthesized, they could join together in the presence of clay minerals to form polymers.  page 2-18

## Self-Assessment

1.  Name and describe the components of an atom.

2.  Explain how the periodic table of the elements is organized.

3.  Differentiate among covalent, polar covalent, hydrogen, and ionic bonds.

4.  List three features of water that make it conducive to life.

5.  Name the four most common elements in organic molecules.

6.  List features of carbon that allow it to form diverse structures.

7.  List essential functions of proteins, nucleic acids, carbohydrates, and lipids.

8.  Describe how diversity is achieved in polymers.

9.  Sketch the basic structures of amino acids, nucleotides, monosaccharides, and fatty acids.

10.  State and defend current hypotheses about how life originated on Earth.

Do you understand the chapter's Core Concepts? Log in to check your answers to the Self-Assessment questions, then practice what you've learned and reinforce this chapter's concepts by working through the problems and multimedia tutorials provided there.

📖 www.biologyhowlifeworks.com

## CHAPTER 3

# NUCLEIC ACIDS AND THE ENCODING OF BIOLOGICAL INFORMATION

## Core Concepts

**3.1** Deoxyribonucleic acid (DNA) stores and transmits genetic information.

**3.2** DNA is a polymer of nucleotides and forms a double helix.

**3.3** Transcription is the process by which RNA is synthesized from a DNA template.

**3.4** The primary transcript is processed to become messenger RNA (mRNA).

So much in biology depends on shape. Take your hand, for example. You can pick up a pin, text on a cellphone, and touch your pinky to your thumb. These activities are made possible by the coordinated movement of dozens of bones, muscles, nerves, and blood vessels that give shape to your hand. The functional abilities of your hand emerge from its structure. A causal connection between structure and function exists in many molecules, too. Proteins are a good example. Composed of linear strings of 20 amino acids, each protein folds into a specific three-dimensional shape due to chemical interactions between the amino acids along the chain. The three-dimensional structure of the protein determines its functional properties, such as what other molecules it can bind with, and enables the protein to carry out its job in the cell.

Another notable example is the macromolecule **deoxyribonucleic acid (DNA)**, a linear polymer of four subunits. DNA molecules from all cells and organisms have a very similar three-dimensional structure, reflecting their shared ancestry. This structure, called a **double helix,** is composed of two strands coiled around each other to form a sort of spiral staircase. The banisters of the spiral staircase are formed by the linear backbone of the paired strands, and the steps are formed by the pairing of the subunits at the same level in each strand.

The spiral-staircase structure is common to all cellular DNA molecules, and its structure gave immediate clues to its function. First, DNA *stores* information. Some of the information in DNA encodes for proteins that provide structure and do much of the work of the cell. Information in DNA is called **genetic information,** and it is organized in the form of **genes,** as textual information is organized in the form of words. Genes can differ from one individual to the next, even within a single species. Differences in genes can affect the shape of the hand, for example, yielding long or short fingers or extra or missing fingers. As we will see, it is the order of individual subunits (bases) of DNA that accounts for differences in genes.

Second, DNA plays a special role in *transmitting* genetic information from one generation to the next. The transmission of genetic information from parents to their offspring enables species of organisms to maintain their identity through time. The genetic information in DNA guides the development of the offspring, ensuring that parental apple trees give rise to apple seedlings and parental geese give rise to goslings. As we will see, determining the double-helical structure of DNA provided one of the first hints of how genetic information could be faithfully copied from cell to cell, and from one generation to the next.

In this chapter, we examine the structure of DNA in more detail and show how its structure is well suited to its biological function as the carrier and transmitter of genetic information.

## 3.1 MAJOR BIOLOGICAL FUNCTIONS OF DNA

DNA is the molecule by which hereditary information is transmitted from generation to generation. Today the role of DNA is well known, but at one time hardly any biologist would have bet on it. Any poll of biologists before about 1950 would have shown overwhelming support for proteins as life's information molecule. Compared to the seemingly monotonous, featureless structure of DNA, the three-dimensional structures of proteins are highly diverse. Proteins carry out most of the essential activities in the life of a cell, and so it seemed logical to assume that they would play a key role in heredity, too. But while proteins do play a role in heredity, they play a supporting role in looking after the DNA—rather like the way worker bees are essential in maintaining the queen bee, who alone is able to reproduce.

### DNA can transfer biological characteristics from one organism to another.

The first experiments to demonstrate that molecules can transfer genetic information from one organism to another were carried out in 1928 by Frederick Griffith, working with the bacterium *Streptococcus pneumoniae*. This organism causes a variety of infections in humans and a deadly form of pneumonia in mice. In the original experiments, illustrated in **Fig. 3.1,**

## HOW DO WE KNOW?

### FIG. 3.1

# What is the nature of the genetic material?

**BACKGROUND** In the 1920s, it was not clear what biological molecule carries genetic information. Fred Neufeld, a German microbiologist, identified several strains of the bacterium *Streptococcus pneumoniae*, one of which was virulent and caused death when injected into mice (Fig. 3.1a), and another which was nonvirulent and did not cause illness when injected into mice (Fig. 3.1b).

**a.**
Virulent bacteria

Mouse dies of pneumonia.

**b.**
Nonvirulent bacteria

Mouse remains healthy.

**EXPERIMENT** Frederick Griffith was also a microbiologist interested in bacterial virulence. He made a puzzling observation. He noted that nonvirulent bacteria do not cause mice to get sick (Fig. 3.1b) and killed virulent bacteria do not cause mice to get sick (Fig. 3.1c), but when the two were mixed, the injected mice got sick and died (Fig. 3.1d). Furthermore, when he isolated bacteria from the dead mice, they had the appearance of the virulent strain, even though he had injected nonvirulent bacteria.

**RESULTS**

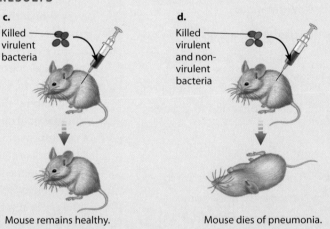

**c.**
Killed virulent bacteria

Mouse remains healthy.

**d.**
Killed virulent and non-virulent bacteria

Mouse dies of pneumonia.

**CONCLUSION** One strain of bacteria (nonvirulent) can be transformed into another (virulent) by an unknown molecule from the virulent cells. In other words, the unknown molecule carries information that causes virulence.

**FOLLOW-UP WORK** Griffith's experiments were followed up by many researchers, most notably Oswald Avery, Colin MacLeod, and Maclyn McCarty, who identified DNA as the molecule responsible for transforming bacteria from one strain to the other (see Fig. 3.2). In addition, the process in which DNA is taken up by cells, called transformation, is now a common technique used in molecular biology.

**SOURCE:** Griffith, F. 1928. "The Significance of Pneumococcal Types." *Journal of Hygiene* 27:113–159.

Griffith studied two strains of the bacterium, one a virulent strain that caused pneumonia and death in injected mice and the other a mutant, nonvirulent strain that allowed injected mice to survive. When the debris of dead virulent cells was mixed with nonvirulent cells, some of the nonvirulent cells became virulent. Griffith concluded that some type of molecule in the debris carried the genetic information for virulence, but he did not identify the molecule.

Experiments carried out in 1944 by Oswald Avery, Colin MacLeod, and Maclyn McCarty showed that the molecule responsible for the conversion, or **transformation,** of nonvirulent cells into virulent cells is DNA (**Fig. 3.2**). These collaborators made a highly purified preparation from the debris of virulent cells, and they found that the preparation could carry out transformation. When enzymes were used to destroy any trace of protein or RNA in the preparation, the transforming ability of the preparation remained. But when an enzyme that destroys DNA was used, the transforming ability was lost.

These experiments, along with others, established that DNA is the genetic material. Today, transformation is widely used in biological research and in the genetic modification of agricultural plants and animals (Chapter 12).

### DNA molecules are copied in the process of replication.

DNA can serve as the genetic material because it is unique among cellular molecules in being able to specify exact copies of itself. This copying process, known as **replication,** allows the genetic information from one DNA molecule to be copied into that of another DNA molecule. Faithful replication is critical in that it allows DNA to pass genetic information from cell to cell and from parent to offspring. The copying must reproduce the sequence of subunits almost exactly because most changes that occur are harmful to the cell or organism.

A rare error that can take place in DNA replication is known as a **mutation,** which is a change in the genetic information in DNA. A mutation in DNA causes the genetic difference between virulent and nonvirulent *Streptococcus pneumoniae*. While most mutations are harmful, rare favorable mutations are essential in the process of evolution because they allow populations of organisms to change through time and adapt to their environment.

### Genetic information flows from DNA to RNA to protein.

Biologists often say that genetic information in DNA directs the activities in a cell or guides the development of an organism, but these effects of DNA are indirect. Most of the key molecules in cells and development are proteins, including the enzymes that convert energy into usable forms and the proteins that provide structural support for the cell. DNA acts indirectly by specifying the sequence of amino acid subunits of which each protein is

**FIG. 3.2**

# What is the nature of the genetic material?

**BACKGROUND** Oswald Avery, Colin MacLeod, and Maclyn McCarty also studied virulence in pneumococcal bacteria. They recognized the significance of Griffith's experiments (see Fig. 3.1) and wanted to identify the molecule responsible for transforming nonvirulent bacteria into virulent ones.

**EXPERIMENT** Avery, MacLeod, and McCarty extracted DNA from virulent bacteria, which allowed them to perform a series of tests and control variables. To identify what caused transformation, they treated the extract, which contained DNA as well as trace amounts of RNA and protein, with enzymes that destroyed one of the three molecules. Their hypothesis was that transformation would not occur if they destroyed the molecule responsible for it.

**RESULTS**

Virulent bacteria (killed)

DNA is extracted from heat-killed virulent cells, along with trace amounts of RNA and protein.

RNase    Protease    DNase

Nonvirulent bacteria    Nonvirulent bacteria    Nonvirulent bacteria    Nonvirulent bacteria

Virulent and nonvirulent bacteria    Virulent and nonvirulent bacteria    Virulent and nonvirulent bacteria    Nonvirulent bacteria only

The untreated extract can transform nonvirulent cells into virulent cells.

Only extracts treated with the enzyme that destroys DNA were unable to transform nonvirulent bacteria.

**CONCLUSION** DNA is the molecule responsible for transforming nonvirulent bacteria into virulent bacteria. This experiment provided a key piece of evidence that DNA is the genetic material.

**FOLLOW-UP WORK** These experiments were followed up by Alfred Hershey and Martha Chase, who used a different system to confirm that DNA is the genetic material.

**SOURCE:** Avery, O., C. MacLeod, and M. McCarty. 1944. "Studies on the Chemical Nature of the Substance Inducing Transformation of Pneumococcal Types." *Journal of Experimental Medicine.* 79:137–158.

composed, and this sequence in turn determines the three-dimensional structure of the protein, its chemical properties, and its biological activities.

In specifying the amino acid sequence of proteins, DNA acts through an intermediary molecule known as **ribonucleic acid (RNA),** another type of linear polymer. As we saw in Chapter 1, the flow of information from DNA to RNA to protein has come to be known as the **central dogma** of molecular biology **(Fig. 3.3).** The central dogma states that genetic information can be transferred from DNA to RNA to protein. Through the years, some exceptions to this "dogma" have been discovered, including the transfer of genetic information from RNA to DNA (as in HIV, which causes AIDS), from RNA to RNA (as in replication of the genetic material of influenza virus), and even from protein to protein (in the unusual case of prions). Nevertheless, the central dogma still conveys the basic idea that in most cases the flow of information is from DNA to RNA to protein.

The first step in this process is **transcription,** in which the genetic information in a molecule of DNA is used as a **template**, or pattern, to generate a molecule of RNA. The term "transcription" is used because it emphasizes that both molecules use the same language of nucleic acids. Transcription is the first step in **gene expression,** which is the production of a functional gene product. The second step in the readout of genetic information is **translation,** in which a molecule of RNA is used as a code for the sequence of amino acids in a protein. The term "translation" is used to indicate a change of languages, from nucleotides that make up nucleic acids to amino acids that make up proteins.

The processes of transcription and translation are regulated, meaning that they do not occur at all times in all cells. Genes are expressed, or turned "on," only at certain times and places, and not expressed, or turned "off," at other times and places. In multicellular organisms, for instance, cells are specialized for certain functions, and these different functions depend on which genes are on and which genes are off in specific cells. Muscle cells express genes that encode for proteins involved in muscle contraction, but these genes are not expressed in skin cells or liver cells, for example. Similarly, during development of a multicellular organism, genes may be required at certain times, but not at others. In this case, the timing of expression is carefully controlled.

In prokaryotes, transcription and translation occur in the cytoplasm, but in eukaryotes, the two processes are separated from each other, with transcription occurring in the nucleus and translation in the cytoplasm. The separation of transcription and translation in time and space in eukaryotic cells allows for additional levels of gene regulation that are not possible in prokaryotic cells. In spite of this and other differences in the details of transcription and translation between prokaryotes and eukaryotes, the processes are sufficiently similar that they must have evolved early in the history of life.

**FIG. 3.3** The central dogma of molecular biology, which defines the usual flow of information in a cell from DNA to RNA to protein.

Proteins provide structure and carry out many essential activities in a cell.

## 3.2 CHEMICAL COMPOSITION AND STRUCTURE OF DNA

By about 1950, some biologists were convinced by the work of Griffith, Avery, and others that DNA is the genetic material. To serve as the genetic material, DNA would have to be able to replicate itself, undergo rare mutations, replicate mutant forms as faithfully as the original forms, and direct the synthesis of other macromolecules in the cell. How could one molecule do all this? Part of the answer emerged in 1953, when James D. Watson and Francis H. C. Crick announced a description of the three-dimensional structure of DNA. This discovery marked a turning point in modern biology. The discovery of the structure of DNA opened the door to understanding how genetic information is stored, faithfully replicated, and altered by rare mutations.

**A DNA strand consists of subunits called nucleotides.**
In the 50 years since the publication of Watson and Crick's paper, we have all become familiar with the iconic double helix of DNA. The elegant shape of the twisting strands relies on the structure of DNA's subunits, called **nucleotides.** As we saw in Chapter 2, nucleotides consist of three components: a 5-carbon **sugar,** a **base,** and one or more **phosphate groups (Fig. 3.4).** Each component plays an important role in DNA structure. The 5-carbon sugars and phosphate groups form the backbone of the molecule, with each sugar linked to the phosphate group of the

**FIG. 3.4** Nucleotide structure.

**FIG. 3.5** Bases normally found in DNA.

Adenine (A)    Guanine (G)

Thymine (T)    Cytosine (C)

neighboring nucleotide. The bases sticking out from the sugar give each nucleotide its chemical identity. Each strand of DNA consists of an enormous number of nucleotides linked one to the next.

DNA had been discovered 85 years before its three-dimensional structure was determined, and in the meantime a great deal had been learned about its chemistry, specifically the chemistry of nucleotides. Fig. 3.4 illustrates a nucleotide. In the figure, the 5-carbon sugar is indicated by the pentagon, in which each vertex represents the position of a carbon atom. By convention the carbon atoms of the sugar ring are numbered with primes (1′, 2′, and so forth, read as "one prime," "two prime" and so forth). Technically, the sugar in DNA is 2′-deoxyribose because the chemical group projecting downward from the 2′ carbon is a hydrogen atom (–H) rather than a hydroxyl group (–OH), but for our purposes the term **deoxyribose** will suffice.

Note in Fig. 3.4 that the phosphate group attached to the 5′ carbon has negative charges on two of its oxygen atoms. These charges are present because at cellular pH (around 7), the free hydroxyl groups attached to the phosphorus atom are ionized by the loss of a proton, and hence are negatively charged. It is these negative charges that make DNA a mild acid, which you will recall from Chapter 2 is a molecule that tends to lose protons to the aqueous environment.

Each base is attached to the 1′ carbon of the sugar and projects above the sugar ring. A nucleotide normally contains one of four kinds of bases, denoted A, G, T, and C (**Fig. 3.5**). Two of the bases are double-ring structures known as **purines;** these are the bases **adenine (A)** and **guanine (G),** shown across the top of the figure. The other two bases are single-ring structures known as **pyrimidines;** these are the bases **thymine (T)** and **cytosine (C),** shown across the bottom.

The combination of sugar and base is known as a **nucleoside,** which is shown in simplified form in **Fig. 3.6**. A nucleoside with one or more phosphate groups constitutes a nucleotide. When a nucleoside has one phosphate group attached to the 5′ carbon, it is called a nucleoside 5′-monophosphate, in which the 5′ indicates where a phosphate group is attached to the ring. The nucleotide shown in Fig. 3.4 is therefore a nucleoside 5′-monophosphate. When one of the free hydroxyl

groups attached to the phosphorus atom is in turn attached to another phosphate group, the nucleotide becomes a nucleoside 5′-diphosphate, and if yet another phosphate group is attached to the second, it becomes a nucleoside 5′-triphosphate (Fig. 3.6). The nucleoside triphosphates are particularly important because, as we will see later in this chapter, they are the molecules that are used to form nucleotide polymers, such as DNA and RNA. In addition, nucleoside triphosphates have other functions in the cell, notably as carriers of chemical energy in the form of ATP and GTP.

**FIG. 3.6** **Nucleoside and nucleoside phosphates.** A nucleoside is a sugar attached to a base, and nucleotides are nucleosides with one, two, or three phosphate groups.

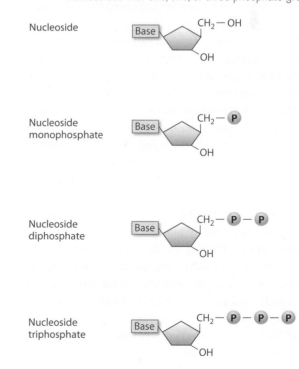

Nucleoside

Nucleoside monophosphate

Nucleoside diphosphate

Nucleoside triphosphate

## DNA is a linear polymer of nucleotides linked by phosphodiester bonds.

Not only were the nucleotide building blocks of DNA known before the structure was discovered, it was also known how they were linked into a polymer. The chemical linkages between nucleotides in DNA are shown in **Fig. 3.7.** The characteristic covalent bond that connects one nucleotide to the next is indicated by the vertical red lines that connect the 3′ carbon of one nucleotide to the 5′ carbon of the next nucleotide in line through the 5′-phosphate group. This C–O–P–O–C linkage is known as a **phosphodiester bond,** which in DNA is a relatively stable bond that can withstand stress like heat and substantial changes in pH that would break weaker bonds. The succession of phosphodiester bonds traces the backbone of the DNA strand.

The phosphodiester linkages in a DNA strand give it **polarity,** which means that one end differs from the other. In Fig. 3.7, the nucleotide at the top has a free 5′ phosphate, and is known as the **5′ end** of the molecule. The nucleotide at the bottom has a free 3′ hydroxyl and is known as the **3′ end**. The DNA strand in Fig. 3.7 has the sequence of bases AGCT from top to bottom, but because of strand polarity we need to specify which end is which. For this strand of DNA, we could say that the base sequence is 5′-AGCT-3′ or equivalently 3′-TCGA-5′. When a base sequence is stated without specifying the 5′ end, by convention the end at the left is the 5′ end. Hence, we could say the sequence in Fig. 3.7 is AGCT, which means 5′-AGCT-3′.

## Cellular DNA molecules take the form of a double helix.

To the knowledge of the chemical makeup of the nucleotides and their linkages in a DNA strand, Watson and Crick added results from earlier physical studies indicating that DNA is a long molecule. They also relied on important information from X-ray diffraction studies by Rosalind Franklin implying that DNA molecules form a helix with a simple repeating structure. Analysis of the pattern of X-rays diffracted from a crystal of a molecule can indicate the arrangement of atoms in the molecules.

With these critical pieces of information in hand, Watson and Crick set out to build a model of DNA that could account for the results of all previous chemical and physical experiments, using sheet metal cutouts of the bases and wire ties for the sugar–phosphate backbone. After many false starts and much disappointment, they finally found a structure that worked. They realized immediately that they had made one of the most important discoveries in all of biology, and that day, February 28, 1953, they lunched at the Eagle, a pub across the street from their laboratory, where Crick loudly pronounced, "We have discovered the secret of life." The Eagle is still there in Cambridge, England, and sports on its wall a commemorative plaque marking the table where the two ate.

Why all the fuss (and why the Nobel Prize 9 years later)? First, let's look at the structure, and you will see that the

**FIG. 3.7** Nucleotides linked by phosphodiester bonds to form a DNA strand.

structure itself tells you how DNA carries and transmits genetic information. The Watson-Crick structure, now often called the double helix, is shown in **Fig. 3.8.** Fig. 3.8a is a space-filling model, in which each atom is represented as a color-coded sphere. The big surprise of the structure is that it consists of two DNA strands like that in Fig. 3.7, each wrapped around the other in the form of a helix coiling to the right, with the sugar–phosphate backbones winding around the outside of the molecule and the bases pointing inward. In the double helix, there are 10 base pairs per complete turn, and the diameter of the molecule is 2 nm, a measurement that is hard to relate to everyday objects, but it might help to know that the cross section of a bundle of 100,000 DNA molecules would be about the size of the period at the end of this sentence. The outside contours of the twisted strands form an uneven pair of grooves, called the **major groove** and the

**minor groove.** These grooves are important because proteins that interact with DNA often recognize a particular sequence of bases by making contact with the bases via the major or minor groove or both.

Importantly, the individual DNA strands in the double helix are **antiparallel,** which means that they run in opposite directions. That is, the 3′ end of one strand is opposite the 5′ end of the other. In Fig. 3.8a, the strand that starts at the bottom left and coils upward begins with the 3′ end and terminates at the top with the 5′ end, whereas its partner strand begins with its 5′ end at the bottom and terminates with the 3′ end at the top.

Fig. 3.8b shows a different depiction of double-stranded DNA, called a ribbon model, which clearly shows the sugar–phosphate backbones winding around the outside with the bases paired between the strands. The ribbon model of the structure closely resembles a spiral staircase, with the backbones forming the banisters and the base pairs the steps. If the amount of DNA in

the human genome (3 billion base pairs) were scaled to the size of a real spiral staircase, it would reach from Earth to the moon.

Note that, as shown in Fig. 3.8b, an A in one strand pairs only with a T in the other, and G pairs only with C. Each base pair contains a purine and a pyrimidine. This precise pairing maintains the structure of the double helix, since pairing two purines would cause the backbones to bulge, and pairing two pyrimidines would cause them to narrow. The pairing of one purine with one pyrimidine preserves the distance between the backbones along the length of the entire molecule.

Because they form specific pairs, the bases A and T are said to be **complementary,** as are the bases G and C. Why is it that A pairs only with T, and G only with C? **Fig. 3.9** illustrates the answer. The specificity of base pairing is brought about by hydrogen bonds that form between A and T (two hydrogen bonds) and between G and C (three hydrogen bonds). A hydrogen bond in DNA is formed when an electronegative atom (O or N) in one

---

**FIG. 3.8**    **Structure of DNA.** The DNA double helix can be shown with (a) the atoms as solid spheres or (b) the backbones as ribbons.

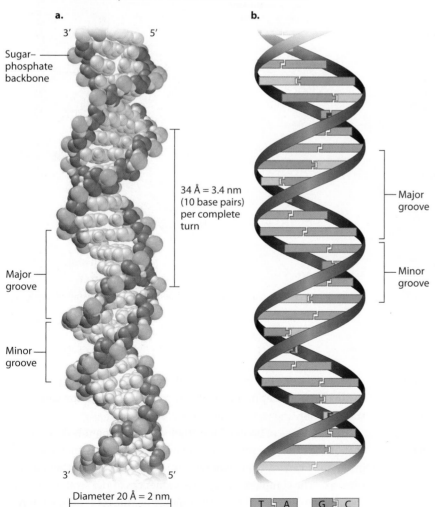

a.

Sugar–phosphate backbone

3′        5′

34 Å = 3.4 nm (10 base pairs) per complete turn

Major groove

Minor groove

3′        5′

Diameter 20 Å = 2 nm

b.

Major groove

Minor groove

T  A    G  C

---

**FIG. 3.9**    **Base pairing.** Adenine pairs with thymine, and guanine pairs with cytosine. These base pairs differ in the number of hydrogen bonds.

A and T are held together by two hydrogen bonds.

Deoxyribose

**Adenine (A)**        **Thymine (T)**

G and C are held together by three hydrogen bonds.

Deoxyribose

**Guanine (G)**        **Cytosine (C)**

base shares a proton (H) with another electronegative atom in the base across the way. Hydrogen bonds are relatively weak bonds, typically 5% to 10% of the strength of covalent bonds, and can be disrupted by high pH or heat. However, in total they contribute to the stability of the DNA double helix.

An almost equally important factor contributing to the stability of the double helix is the interactions between bases in the same strand (**Fig. 3.10**). This stabilizing force is known as **base stacking,** and it occurs because the nonpolar, flat surfaces of the bases tend to group together away from water molecules, and hence stack on top of one another as tightly as possible.

### The three-dimensional structure of DNA gave important clues about its functions.

The double helix is a good example of how chemical structure and biological function come together. The structure of the molecule itself immediately suggested how genetic information can be stored in DNA. One of the most important features of DNA structure is that there is no restriction on the sequence of bases along a DNA strand. The lack of sequence constraint suggested that the genetic information in DNA could be encoded in the sequence of bases along the DNA, much as textual information in a book is stored in a sequence of letters of the alphabet. With any of four possible bases at each nucleotide site, the information-carrying capacity of a DNA molecule is unimaginable. The number of possible base sequences of a DNA molecule only 133 nucleotides in length is equal to the estimated number of electrons, protons, and neutrons in the entire universe! This is the secret of how DNA can carry the genetic information for so many different types of organisms, and how variation in DNA sequence even within a single species can underlie genetic differences among individuals.

Watson and Crick's model still left many questions unanswered in regard to how the information is read out

**FIG. 3.10** **Interactions stabilizing the double helix.** Hydrogen bonds between the bases in opposite strands and base stacking of bases within a strand contribute to the stability of the DNA double helix.

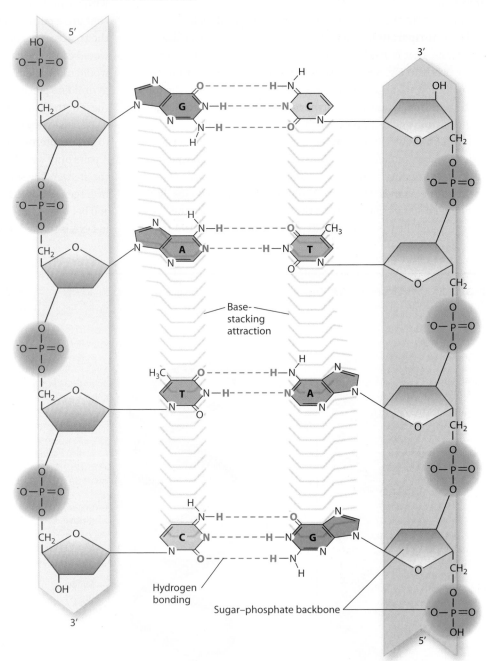

and what it does, but in the years following it became clear that the major processes in the readout of genetic information were transcription and translation (see Fig. 3.3).

The sequence of bases along either strand completely determines that of the other because wherever one strand carries an A, the other must carry a T, and wherever one carries a G, the other must carry a C. The complementary base sequences of the strands means that, in any double-stranded DNA molecule, the total number of A bases must equal that of T, and the total number of G bases must equal that of C. These equalities are often written in terms of the percent (%) of each base in double-stranded DNA. In these terms, the base-pairing rules imply that %A = %T and %G = %C. Interestingly, these equalities were known and described by the American biochemist Erwin Chargaff before the double helix was discovered, and they provided important clues to the structure of DNA. It was the double helix that finally showed why they are observed.

→ **Quick Check 1** The letter R is conventionally used to represent any purine base (A or G) and Y to represent any pyrimidine base (T or C). In double-stranded DNA, what is the relation between %R and %Y?

The complementary, double-stranded nature of the double helix also suggested a mechanism by which DNA replication could take place. In fact, in their 1953 paper describing the structure of DNA, Watson and Crick wrote, "It has not escaped our notice that the specific pairing we have postulated immediately suggests a possible copying mechanism for the genetic material."

A simplified outline of DNA replication is shown in **Fig. 3.11.** In brief, the two strands of a parental double helix unwind, and as they do each of the parental strands serves as a pattern, or template, for the synthesis of a complementary daughter strand. When the process is complete, there are two molecules, each of which is identical in sequence to the original molecule, except possibly for rare errors (mutations) that cause one base pair to be replaced with another. Although the process of replication as depicted in Fig. 3.11 looks exceedingly simple and straightforward, there are technical details that make the actual process more complex. These are discussed in Chapter 12.

### Cellular DNA is coiled and packaged with proteins.

The DNA molecules inside cells are highly convoluted. They have to be because DNA molecules in cells have a length far

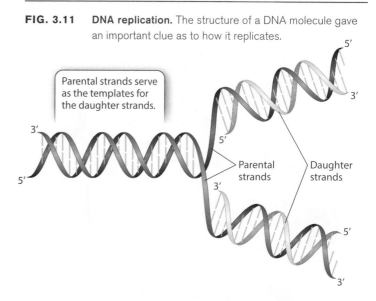

**FIG. 3.11 DNA replication.** The structure of a DNA molecule gave an important clue as to how it replicates.

Parental strands serve as the templates for the daughter strands.

Parental strands

Daughter strands

greater than the diameter of the cell itself. The DNA of a bacterium known as *Mycoplasma*, for example, if stretched to its full linear extent, would be about 1000 times longer than the diameter of the bacterial cell. Many of the double-stranded DNA molecules in prokaryotic cells are circular and form **supercoils** in which the circular molecule coils upon itself, much like what happens to a rubber band when you twist it between your thumb and forefinger (**Fig. 3.12**). Supercoiling is caused by enzymes called **topoisomerases** that cleave, partially unwind, and reattach a DNA strand, which puts strain on the DNA double helix. Supercoils then relieve the strain and help to preserve the 10 base pairs per turn in the double helix.

In eukaryotic cells, most DNA molecules in the nucleus are linear, and each individual molecule forms one **chromosome.** There is a packaging problem here, too, which you can appreciate by considering that the length of the DNA molecule contained in a single human chromosome is roughly 6000 times greater than the average diameter of the cell nucleus

**FIG. 3.12 Supercoils.** A highly twisted rubber band forms coils of coils (supercoils), much as a circular DNA molecule does when it contains too many base pairs per helical turn.

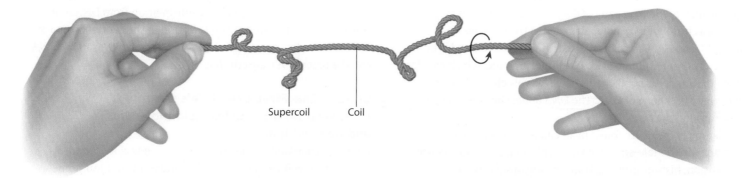

Supercoil    Coil

**FIG. 3.13**   **Chromatin.**  (a) Double-stranded DNA is usually packaged with histone proteins into nucleosomes forming a chromatin fiber. The diameter of the fiber differs according to how closely the nucleosomes are packaged. (b) A nucleosome consists of a segment of DNA wrapped around eight histone proteins.

nucleosomes with the DNA of any other organism. The reason for this ability is that these proteins are **evolutionarily conserved,** which means that they are very similar in sequence from one organism to the next. Conserved DNA, RNA, or protein sequences indicate that they serve an essential function and therefore have not changed very much over long stretches of evolutionary time. The more distantly related two organisms are that share conserved sequences, the more highly conserved the sequence is. Among the histone proteins, histones H3 and H4 are the most conserved; for example, 98% of the amino acids in histone H4 are identical in organisms as diverse as domesticated cattle and the garden pea. Histones are very highly conserved, reflecting their key role in packaging DNA.

## 3.3 RETRIEVAL OF GENETIC INFORMATION STORED IN DNA: TRANSCRIPTION

Although the three-dimensional structure of DNA gave important clues about how DNA stores and transmits information, it left open many questions about how the genetic information in DNA is read out to control cellular processes. In 1953, when the double helix was discovered, virtually nothing was known about these processes. Within a few years, however, evidence was already accumulating that DNA carries the genetic information for proteins, and that proteins are synthesized on particles in the cytoplasm called **ribosomes.** But ribosomes contain no DNA! There must therefore be an intermediary molecule by which the genetic information is transferred from the DNA to the ribosome, and some researchers began to suspect that this intermediary was another type of nucleic acid called ribonucleic acid (RNA).

This hypothesis was supported by a clever experiment carried out in 1961 by Sydney Brenner, François Jacob, and Matthew Meselson. They used the virus T2, which infects cells of the bacterium *Escherichia coli* and hijacks the cellular machinery to produce viral proteins. The researchers showed that, shortly after infection and before viral proteins are made, the infected cells produce a burst of RNA molecules that correspond to the viral DNA. This finding and others suggested that RNA is used to retrieve the genetic information stored in DNA for use in protein synthesis. The transfer of genetic information from DNA to RNA constitutes the key step of transcription in the central dogma of molecular biology (see Fig. 3.2). In this section, we examine RNA and the process of transcription.

**?** **CASE 1 The First Cell: Life's Origins**
**What was the first nucleic acid molecule, and how did it arise?**

RNA is a remarkable molecule. Like DNA, it is able to store information in its sequence of nucleotides. In addition, some

(**Fig. 3.13**). Double-stranded DNA molecules in eukaryotes are usually packaged with proteins called histones into a **30-nm chromatin fiber** (Fig. 3.13a). In regions of the nucleus in which transcription is actively taking place, the 30-nm fiber is usually relaxed into a **10-nm chromatin fiber** (Fig. 3.13a), which is about 5 times the diameter of the double helix. Chromatin fibers of either dimension are often referred to simply as **chromatin.**

The 10-nm chromatin fiber is composed of a beadlike repeating unit known as a **nucleosome** (Fig. 3.13b). The core particle of each nucleosome is a flattened sphere consisting of approximately 150 base pairs of double-stranded DNA wrapped around a cluster of two molecules each of the histones H2A, H2B, H3, and H4. Each of these **histone** proteins is rich in the positively charged amino acids lysine and arginine, which enables them to form ionic bonds with the negatively charged sugar–phosphate backbone of DNA. The DNA double helix is wrapped twice around the core particle.

Histone proteins are found in all eukaryotes, and they interact with double-stranded DNA without regard to sequence. In addition, histone proteins from any organism can form

RNA molecules can actually act as enzymes that facilitate chemical reactions. Because RNA has properties of both DNA (information storage) and proteins (enzymes), many scientists think that RNA, not DNA, was the original information-storage molecule in the earliest forms of life on Earth. This idea, sometimes called the **RNA world hypothesis,** is supported by other evidence as well. Notably, as we will see, RNA is used in key cellular processes, including DNA replication, transcription, and translation. Many scientists believe that this involvement is a remnant of a time when RNA played a more central role in life's fundamental processes.

In addition, ingenious experiments show how RNA could have evolved the ability to catalyze a simple reaction. A strand of RNA was synthesized in the laboratory and then replicated many times to produce a large population of identical RNA molecules. Next, the RNA was exposed to a chemical that induced random changes in the identity of some of the nucleotides in these molecules. These random changes were mutations that created a population of diverse RNA molecules, much in the way that mutation builds genetic variation in cells.

Next, all of these RNA molecules were placed into a container, and those RNA variants that successfully catalyzed a simple reaction—cleaving a strand of RNA, for example, or joining two strands together—were isolated, and the cycle was repeated. In each round of the experiment, the RNA molecules that functioned best were retained, replicated, subjected to treatments that induced additional mutations, and then tested for the ability to catalyze the same reaction. With each generation, the RNA catalyzed the reaction more efficiently, and after only a few dozen rounds of the procedure, very efficient RNA catalysts had evolved. Experiments such as this one suggest that RNA molecules can evolve over time and act as catalysts. Therefore, many scientists believe that RNA, with its dual functions of information storage and catalysis, was a key molecule in the very first forms of life.

If RNA played a key role in the origin of life, why do cells now use DNA for information storage and proteins to carry out other cellular processes? RNA is much less stable than DNA, and proteins are more versatile, so a plausible explanation is that life evolved from an RNA-based world to one in which DNA, RNA, and proteins are specialized for different functions.

RNA is a polymer of nucleotides linked by phosphodiester bonds, similar to those in DNA (see Fig. 3.7). Each RNA strand therefore has a polarity determined by which end of the chain carries the 3′ hydroxyl (–OH) and which end carries the 5′ phosphate. There are a number of important differences that distinguish RNA from DNA, however (**Fig. 3.14**). First, the sugar in RNA is **ribose,** which carries a hydroxyl group on the 2′ carbon (Fig. 3.14a). Hydroxyls are reactive functional groups, so the additional hydroxyl group on ribose in part explains why RNA is a less stable molecule than DNA. Second, the base **uracil** found

in RNA replaces thymine found in DNA (Fig. 3.14b). The groups that participate in hydrogen bonding (highlighted in pink in Fig. 3.14b) are identical so that uracil pairs with adenine (U–A) just as thymine pairs with adenine (T–A). Third, while the 5′ end of a DNA strand is typically a monophosphate, the 5′ end of an RNA molecule is typically a triphosphate.

Two other features that distinguish RNA from DNA are physical rather than chemical. One is that RNA molecules are usually much shorter than DNA molecules. A typical RNA molecule used in protein synthesis consists of a few thousand nucleotides, whereas a typical DNA molecule consists of millions or tens of millions of nucleotides. The other major distinction is that most RNA molecules in the cell are single stranded, whereas DNA molecules, as we saw, are double stranded. Single-stranded RNA molecules often form complex three-dimensional structures by folding back upon themselves, which enhances their stability.

## In transcription, DNA is used as a template to make complementary RNA.

Conceptually, the process of transcription is straightforward. As a region of the DNA duplex unwinds, one strand is used as a template, or pattern, for the synthesis of an **RNA transcript** that is complementary in sequence to the template according to the

**FIG. 3.14** **RNA.** RNA differs from DNA in that (a) RNA contains the sugar ribose rather than deoxyribose and (b) the base uracil rather than thymine.

base-pairing rules, except that the transcript contains U (uracil) where the template has an A (**Fig. 3.15**). The transcript is produced by polymerization of ribonucleoside triphosphates. The enzyme that carries out the polymerization is known as **RNA polymerase,** which acts by adding successive nucleotides to the 3′ end of the growing transcript (Fig. 3.15). Only the template strand of DNA is transcribed. Its partner, called the **nontemplate strand,** is not transcribed.

It is important to keep in mind the direction of growth of the RNA transcript and the direction that the DNA template is read. All nucleic acids are synthesized by addition of nucleotides to the 3′ end. That is, they grow in a 5′-to-3′ direction, also described simply as the 3′ direction. Because of the antiparallel nature of the DNA–RNA duplex, the DNA template that is being transcribed runs in the opposite direction, from 3′ to 5′ (Fig. 3.15).

→   **Quick Check 2**  A segment of one strand of a double-stranded DNA molecule has the sequence 5′-ACTTTCAGCGAT-3′. What is the sequence of an RNA molecule synthesized from this DNA template?

### Transcription starts at a promoter and ends at a terminator.

A long DNA molecule typically contains thousands of genes, most of them coding for proteins or RNA molecules with specialized functions, and hence thousands of different transcripts are produced. For example, the DNA molecule in the bacterium *E. coli* has about 4 million base pairs and produces about 4000 transcripts, most of which code for proteins. A typical map of a small part of a long DNA molecule is shown in **Fig. 3.16**. Each green segment indicates the position where a transcription is initiated, and each purple segment indicates the position where it ends.

The green segments are **promoters,** regions of typically a few hundred base pairs where RNA polymerase and associated proteins bind to the DNA duplex. Many eukaryotic and archaeal promoters contain a sequence similar to 5′-TATAAA-3′, which is known as a **TATA box** because the TATA sequence is usually present. The first nucleotide to be transcribed is usually positioned about 25 base pairs from the TATA box, and transcription takes place as the RNA polymerase moves along the template strand in the 3′-to-5′ direction.

Transcription continues until the RNA polymerase encounters a sequence known as a **terminator** (shown in purple in Fig. 3.16). Transcription stops at the terminator, and the transcript is released. Transcripts from a DNA duplex may be produced from either strand, depending on the orientation of the promoter. As shown in Fig. 3.16, promoters oriented in one direction result in transcription from right to left (top strand), whereas those

**FIG. 3.15**  **Template and nontemplate strands.** When the DNA double helix unwinds for transcription, usually only one strand is transcribed.

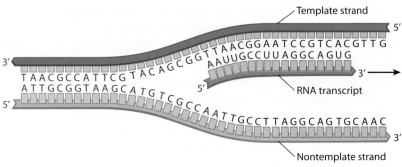

oriented in the reverse direction result in transcription from left to right (bottom strand). Orientation matters, as noted earlier, because transcription can proceed only by successive addition of nucleotides to the 3′ end of the transcript.

Transcription does not take place indiscriminately from promoters but is a regulated process. For genes called **housekeeping genes,** whose products are needed at all times in all cells, transcription takes place continually. But most genes are transcribed only at certain times, under certain conditions, or in certain cell types. In *E. coli,* for example, the genes that encode proteins needed to utilize the sugar lactose (milk sugar) are transcribed only when lactose is present in the environment. For such genes, regulation of transcription often depends on whether the RNA polymerase and associated proteins are able to bind with the promoter.

In bacteria, promoter recognition is mediated by a protein called **sigma factor,** which associates with RNA polymerase

**FIG. 3.16**  **Transcription along a stretch of DNA.** A DNA molecule usually contains many genes that are transcribed individually and at different times, often from opposite strands.

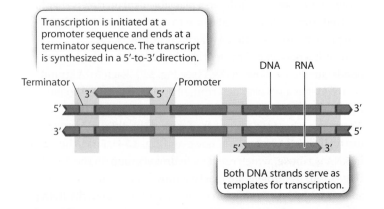

and facilitates its binding to specific promoters. One primary sigma factor is used for transcription of housekeeping genes and many others, but there are other sigma factors for genes whose expression is needed under special environmental conditions such as lack of nutrients or excess heat.

Promoter recognition in eukaryotes is considerably more complicated. Transcription requires the combined action of at least six proteins known as **general transcription factors** that assemble at the promoter of a gene. The general transcription factors attract, or recruit, the RNA polymerase and its associated proteins to the site (**Fig. 3.17**). Cells have several different types of RNA polymerase enzymes, but in both prokaryotes and eukaryotes all protein-coding genes are transcribed by just one of them. In eukaryotes, the RNA polymerase complex responsible for transcription of protein-coding genes is called **Pol II.**

Assembly of the general transcription factors with the RNA polymerase complex is necessary for transcription to occur, but not sufficient. Also needed is the presence of one or more types of **transcriptional activator protein,** each of which binds to a specific DNA sequence known as an **enhancer** (Fig. 3.17). The transcriptional activator proteins recruit a **mediator complex** of proteins, which in turn interacts with the Pol II complex, and transcription begins. The initiation of transcription of any gene therefore depends on the availability of the transcriptional activator proteins that bind with the enhancers controlling the expression of the gene.

### RNA polymerase adds successive nucleotides to the 3' end of the transcript.

Transcription actually takes place in a sort of bubble in which the strands of the DNA duplex are separated and the growing end of

**FIG. 3.17**  The eukaryotic transcription complex composed of many different proteins.

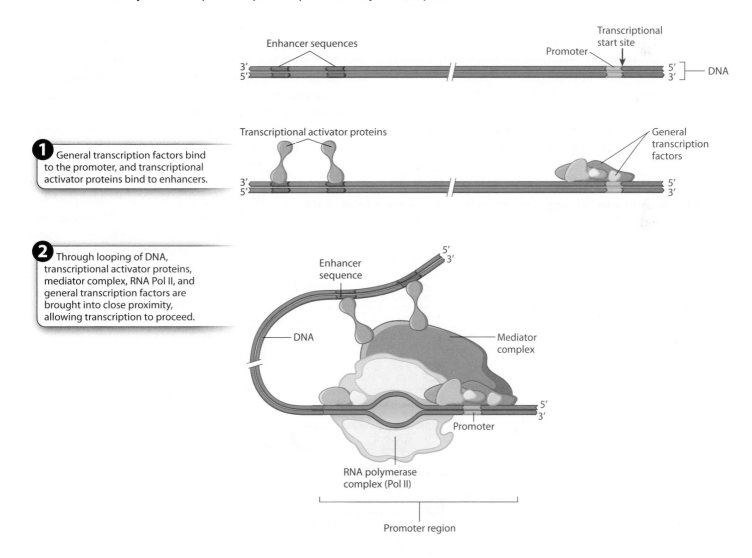

**FIG. 3.18   Transcription bubble.** Within the polymerase, the two strands of DNA separate and the growing RNA strand forms a duplex with the DNA template.

the RNA transcript is paired with the template strand, creating an RNA–DNA duplex (**Fig. 3.18**). In bacteria, the total length of the transcription bubble is about 14 base pairs, and the length of the RNA–DNA duplex in the bubble is about 8 base pairs.

Details of the polymerization reaction are shown in **Fig. 3.19.** The incoming ribonucleoside triphosphate, shown at the bottom right, is accepted by the RNA polymerase only if it undergoes proper

base pairing with the base in the template DNA strand. In Fig. 3.19, there is a proper match because U pairs with A. At this point, the RNA polymerase orients the oxygen in the hydroxyl group at the 3' end of the growing strand into a position from which it can attack the innermost phosphate of the triphosphate, competing for the covalent bond. The bond connecting the innermost phosphate to the next is a **high-energy phosphate bond,** which when cleaved provides the energy to drive the reaction that creates the phosphodiester bond attaching the incoming nucleotide to the 3' end of growing chain. The term "high-energy" here refers to the amount of energy released when the phosphate bond is broken that can be used to drive other chemical reactions.

The polymerization reaction releases a phosphate–phosphate group (pyrophosphate), shown at the lower right in Fig. 3.19, which also has a high-energy phosphate bond that is cleaved by another enzyme. Cleavage of the pyrophosphate molecule makes the polymerization reaction irreversible, and the next ribonucleoside triphosphate that complements the template is brought into line.

**FIG. 3.19   The polymerization reaction that allows the RNA transcript to be elongated.**

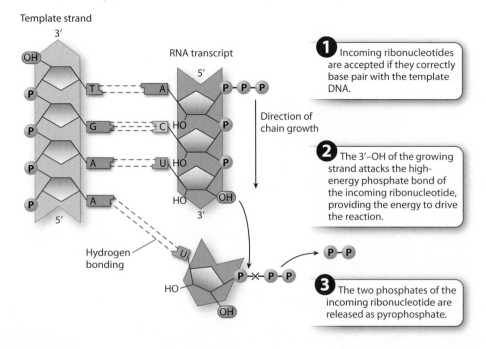

Template strand

RNA transcript

**1** Incoming ribonucleotides are accepted if they correctly base pair with the template DNA.

Direction of chain growth

**2** The 3'–OH of the growing strand attacks the high-energy phosphate bond of the incoming ribonucleotide, providing the energy to drive the reaction.

Hydrogen bonding

**3** The two phosphates of the incoming ribonucleotide are released as pyrophosphate.

→ **Quick Check 3** What is the consequence in terms of elongation for a growing RNA transcript after incorporating an abnormal nucleotide with a 3' H rather than a 3' OH? How about a 2' H rather than a 2' OH?

**The RNA polymerase complex is a molecular machine that opens, transcribes, and closes duplex DNA.**
Transcription does not take place spontaneously. It requires template DNA, a supply of ribonucleoside triphosphates, and RNA polymerase, a large multiprotein complex in which transcription occurs. To illustrate RNA polymerase in action, we consider here bacterial RNA polymerase because of its relative simplicity.

The transcription bubble forms and transcription takes place in the polymerase (**Fig. 3.20**). RNA polymerase contains structural features that separate the DNA strands, allow an RNA–DNA duplex to form, elongate the transcript nucleotide by nucleotide, release the finished transcript, and restore the original DNA double helix. Fig. 3.20 shows how structure and function come together in the bacterial RNA polymerase.

DNA enters the polymerase through a channel and is threaded out another channel via a structure that splits the strands apart, forming a bubble, so that ribonucleotides can assemble on the template strand. Another part of the enzyme forces the growing RNA out a different channel. As the DNA travels through the RNA polymerase, the RNA transcript is elongated at the active site of the enzyme as successive ribonucleotides are added to its 3′ end.

RNA polymerase is a remarkable molecular machine capable of adding thousands of nucleotides to a transcript before dissociating from the template. It is also very accurate, with only about 1 incorrect nucleotide incorporated per 10,000 nucleotides.

## 3.4 FATE OF THE RNA PRIMARY TRANSCRIPT

The RNA transcript that comes off the template DNA strand is known as the **primary transcript,** and it contains the genetic information of the gene that was transcribed. For protein-coding genes, this means that the primary transcript includes the information needed to direct the ribosome to produce the protein corresponding to the gene (Chapter 4). The RNA molecule that combines with the ribosome to direct protein synthesis is known as the **messenger RNA (mRNA)** because it serves to carry the genetic "message" (information) from the DNA to the ribosome. As we will see in this section, there is a major difference between prokaryotes and eukaryotes in the manner in which the primary transcript relates to the mRNA. We will also see that some genes do not code for proteins, but for RNA molecules that have functions of their own.

**Messenger RNA carries information for the synthesis of a specific protein.**
In prokaryotes, the relation between the primary transcript and the mRNA is as simple as can be: The primary transcript *is* the mRNA. Even as the 3′ end of the primary transcript is still being synthesized, ribosomes bind with special sequences near its 5′ end and begin the process of protein synthesis (**Fig. 3.21**). This intimate connection between transcription and translation can take place because prokaryotes have no nuclear envelope to spatially separate transcription from translation; the two processes are coupled, which means that they are connected in space and time.

**FIG. 3.20** **The RNA polymerase complex in prokaryotes.** This molecular machine has channels for DNA input and output, nucleotide input and RNA output, and features that disrupt the DNA double helix, stabilize the RNA–DNA duplex, and allow the DNA double helix to re-form.

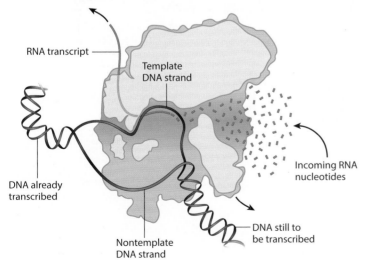

**FIG. 3.21** **Fate of the primary transcript for protein-coding genes in prokaryotes.**

Primary transcripts for protein-coding genes in prokaryotes have another feature not shared with those in eukaryotes: They often contain the genetic information for the synthesis of two or more different proteins, usually proteins that code for successive steps in the biochemical reactions that produce small molecules needed for growth, or successive steps needed to break down a small molecule used for nutrients or energy. Molecules of mRNA that code for multiple proteins are known as **polycistronic mRNA** because the term "cistron" was once widely used to refer to a protein-coding sequence in a gene.

## Primary transcripts in eukaryotes undergo several types of chemical modification.

In eukaryotes, the nuclear envelope is a barrier between the processes of transcription and translation. Transcription takes place in the nucleus, and translation in the cytoplasm. The separation allows for a complex chemical modification of the primary transcript, known as **RNA processing,** which converts the primary transcript into the finished mRNA, which can then be translated by the ribosome.

RNA processing consists of three principal types of chemical modification, illustrated in **Fig. 3.22.** First, the 5′ end of the primary transcript is modified by the addition of a special nucleotide attached in an unusual linkage. This addition is called the **5′ cap,** and it consists of a modified nucleotide called 7-methylguanosine. An enzyme attaches the cap to the 5′ end of the primary transcript essentially backwards: in a normal linkage between two nucleotides, the phosphodiester bridge forms between the 5′ carbon of one and the 3′-OH group of the next, but here the cap is linked to the RNA transcript by a triphosphate bridge between the 5′ carbons of both ribose sugars (**Fig. 3.23**). The 5′ cap is essential for translation because in eukaryotes the ribosome recognizes an mRNA by its 5′ cap. Without the cap, the ribosome would not attach the mRNA and translation would not occur.

The second major modification of eukaryotic primary transcripts is **polyadenylation,** the addition of a string of about 250 consecutive A-bearing ribonucleotides to the 3′ end, forming a **poly(A) tail** (see Fig. 3.22). Polyadenylation plays an important role in transcription termination as well as in the export of the mRNA to the cytoplasm of the cell. In addition, both the 5′ cap and poly(A) tail help to stabilize the RNA transcript. Single-stranded nucleic

**FIG. 3.22**   Fate of the primary transcript for protein-coding genes in eukaryotes.

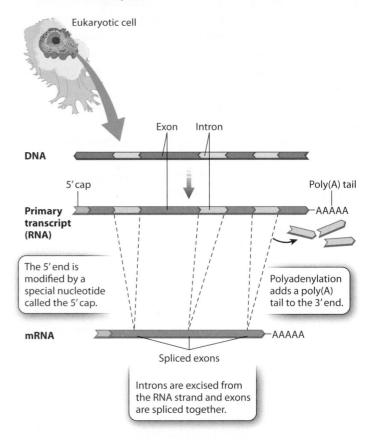

The 5′ end is modified by a special nucleotide called the 5′ cap.

Polyadenylation adds a poly(A) tail to the 3′ end.

Introns are excised from the RNA strand and exons are spliced together.

**FIG. 3.23**   Structure of the 5′ cap on eukaryotic messenger RNA.

The 5′ cap on eukaryotic messenger RNA consists of 7-methylguanosine attached in a 5′-to-5′ phosphate linkage.

acids can be unstable and are even susceptible to enzymes that break them down. In eukaryotes, the 5′ cap and poly(A) tail protect the two ends of the transcript and increase the stability of the RNA transcript until it is translated in the cytoplasm.

A third modification of the primary transcript is the excision of certain sequences, which are known as **introns,** from the transcript (see Fig. 3.22). What's left intact are sequences known as **exons.** The process of intron removal is known as **RNA splicing,** which is catalyzed by a complex of RNA and protein known as the **spliceosome.** The mechanism of splicing is outlined in **Fig. 3.24.** In the first step, specific sequences near the ends of the intron (Fig. 3.24a) undergo base pairing with RNA molecules in the spliceosome and are brought into close proximity (Fig. 3.24b). The spliceosome enables a reaction that cuts one end of the intron and connects it to a nucleotide

near the other, forming a loop and tail called a **lariat** (Fig. 3.24c). In the next step, the exon on one end of the intron is brought close to the exon at the other end (Fig. 3.24c). These exons are joined and the introns are released (Fig. 3.24d). The lariat making up the intron is quickly broken down into its constituent nucleotides.

About 90% of all human genes contain at least one intron. Although most genes contain 6 to 9 introns, the largest number is 147, found in a muscle gene. Most introns are just a few thousand nucleotides in length, but about 10% are longer than 10,000 nucleotides. The presence of multiple introns in most genes allows for a process known as **alternative splicing,** in which primary transcripts from the same gene can be spliced in different ways to yield different mRNAs and therefore different protein products (**Fig. 3.25**). More than 80% of human genes are alternatively spliced. In most cases, the alternatively spliced forms differ in whether a particular exon is or is not removed from the primary transcript along with its flanking introns.

→ **Quick Check 4** When a region of DNA that contains the genetic information for a protein is isolated from a bacterial cell and inserted into a eukaryotic cell in a proper position between a promoter and a terminator, the resulting cell usually produces the correct protein. But when the experiment is done in the reverse direction (eukaryotic DNA into a bacterial cell), the correct protein is often not produced. Can you suggest an explanation?

---

**FIG. 3.24   RNA splicing.**

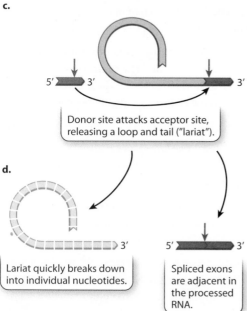

**a.**
Primary transcript — Exon   Intron

**b.**
Donor splice site   5′   Branch site   Acceptor splice site   3′

Branch site attacks donor site, forming a loop.

**c.**
5′ 3′   3′

Donor site attacks acceptor site, releasing a loop and tail ("lariat").

**d.**
3′   5′ 3′

Lariat quickly breaks down into individual nucleotides.

Spliced exons are adjacent in the processed RNA.

---

**FIG. 3.25   Alternative splicing.** A single primary transcript can be spliced in different ways .

**Gene**
Intron  Exon

**Primary transcription**
E1   E2   E3   E4

Alternative splicing

**mRNA**
E3
E1  E2  E4

E4
E1  E2  E3

**Protein A**          **Protein B**

## Some RNA transcripts are processed differently from protein-coding transcripts and have functions of their own.

Not all primary transcripts are processed into mRNA because not all genes code for proteins. Some RNA transcripts have functions of their own, and many of these transcripts are produced by RNA polymerases other than Pol II. These primary transcripts undergo different types of RNA processing, and their processed forms include such important noncoding RNA types as:

- **Ribosomal RNA** (**rRNA**), found in all ribosomes that aid in translation.

- **Transfer RNA** (**tRNA**) that carries individual amino acids for use in translation.

- **Small nuclear RNA** (**snRNA**), found in eukaryotes and involved in splicing, polyadenylation, and other processes in the nucleus.

- Small, regulatory RNA molecules that can inhibit translation (**microRNA** or **miRNA**) or cause destruction of an RNA transcript (**small interfering RNA** or **siRNA**).

By far, the most abundant transcripts in mammalian cells are those for ribosomal RNA and transfer RNA. In a typical mammalian cell, about 80% of all of the RNA consists of ribosomal RNA, and another approximately 10% consists of transfer RNA. Why are these types of RNA so abundant? The answer is that they are needed in large amounts to synthesize the proteins encoded in the messenger RNA. The roles of mRNA, tRNA, and rRNA in protein synthesis are discussed in the next chapter.

# Core Concepts Summary

## 3.1 DEOXYRIBONUCLEIC ACID (DNA) STORES AND TRANSMITS GENETIC INFORMATION.

Experiments carried out by Griffith in 1928 demonstrated that bacteria can transmit genetic information from one strain to another.  page 3-2

Experiments performed by Avery, MacLeod, and McCarty in 1944 showed that DNA is the molecule that transmits genetic information.  page 3-3

DNA is copied in the process of replication.  page 3-3

Ribonucleic acid (RNA) is synthesized from a DNA template. page 3-4

The central dogma of molecular biology states that the usual flow of genetic information is from DNA to RNA to protein. DNA is transcribed to RNA, and RNA is translated to protein.  page 3-4

## 3.2 DNA IS A POLYMER OF NUCLEOTIDES AND FORMS A DOUBLE HELIX.

A nucleotide consists of a 5-carbon sugar, a phosphate group, and a base.  page 3-4

The four bases of DNA are adenine, guanine, cytosine, and thymine.  page 3-5

Successive nucleotides are linked by phosphodiester bonds to form a linear DNA molecule.  page 3-6

DNA strands have polarity, with a 5′-phosphate group at one end and a 3′-hydroxyl group at the other end.  page 3-6

Cellular DNA molecules consist of a helical spiral of two paired, antiparallel strands called a double helix.  page 3-6

In a DNA double helix, A pairs with T, and G pairs with C.  page 3-7

The structure of DNA relates to its function. Information is coded in the sequence of bases, and the structure suggests a mechanism for replication, in which each parental strand serves as a template for a daughter strand.  page 3-8

Cellular DNA is packaged with evolutionary conserved proteins called histones. Histones form a nucleosome core particle around which DNA is wrapped to form a 10-nm and a 30-nm chromatin fiber.  page 3-9

## 3.3 TRANSCRIPTION IS THE PROCESS BY WHICH RNA IS SYNTHESIZED FROM A DNA TEMPLATE.

RNA, like DNA, is a polymer of nucleotides linked by phosphodiester bonds.  page 3-10

Some types of RNA can store genetic information and other types can catalyze chemical reactions. These characteristics have led to the RNA world hypothesis, the idea that RNA played a critical role in the early evolution of life on Earth.  page 3-11

Unlike DNA, RNA incorporates the sugar ribose instead of deoxyribose, and the base uracil instead of thymine.  page 3-11

RNA is synthesized from one of the two strands of DNA, called the template strand.  page 3-11

RNA is synthesized by RNA polymerase in a 5′-to-3′ direction, starting at a promoter and ending at a terminator in the DNA template.  page 3-12

RNA polymerase separates the two DNA strands, allows an RNA–DNA duplex to form, elongates the transcript, releases the transcript, and restores the DNA duplex.  page 3-13

## 3.4 THE PRIMARY TRANSCRIPT IS PROCESSED TO BECOME MESSENGER RNA (mRNA).

In prokaryotes, the primary transcript is immediately translated into protein.  page 3-15

In eukaryotes, transcription and translation are separated in time and space, with transcription first occurring in the nucleus and then translation in the cytoplasm. page 3-16

In eukaryotes, there are three major types of modification to the primary transcript—the addition of a 5′ cap, polyadenylation, and splicing. page 3-16

The 5′ cap is a 7-methylguanosine added to the 5′ end of the transcript. page 3-16

The poly(A) tail is a stretch of adenines added to the 3′ end of the transcript. page 3-16

Splicing is the excision of introns from the transcript, bringing exons together. page 3-16

Alternative splicing is a process in which primary transcripts from the same gene are spliced in different ways to yield different protein products. page 3-17

Some RNAs, called noncoding RNAs, do not code for proteins, but instead have functions of their own. page 3-18

## Self-Assessment

1. Explain how the function of DNA is attributable to and dependent on its structural features.

2. Describe how DNA molecules are replicated.

3. Explain how the sequence of just four nucleotide monomers found in DNA can encode the enormous amount of genetic information stored in the chromosomes of living organisms.

4. Draw a nucleosome, indicating the positions of DNA and proteins.

5. Describe the usual flow of genetic information in a cell.

6. Name two differences between the structure of DNA and RNA.

7. Describe how a molecule of RNA is synthesized using a DNA molecule as a template.

8. Explain the relationship between RNA structure and function.

9. Name and describe three mechanisms of RNA processing in eukaryotes, and explain their importance to the cell.

10. List three types of noncoding RNA and describe their functions.

Do you understand the chapter's Core Concepts? Log in to check your answers to the Self-Assessment questions, then practice what you've learned and reinforce this chapter's concepts by working through the problems and multimedia tutorials provided there.

**www.biologyhowlifeworks.com**

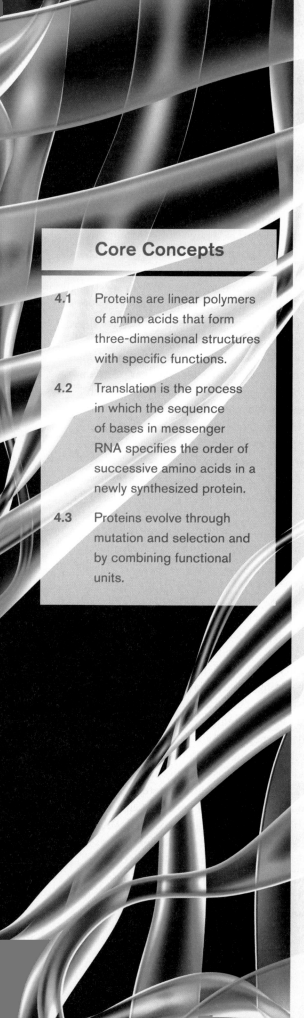

## CHAPTER 4

# TRANSLATION AND PROTEIN STRUCTURE

## Core Concepts

**4.1** Proteins are linear polymers of amino acids that form three-dimensional structures with specific functions.

**4.2** Translation is the process in which the sequence of bases in messenger RNA specifies the order of successive amino acids in a newly synthesized protein.

**4.3** Proteins evolve through mutation and selection and by combining functional units.

Hardly anything happens in the life of a cell that does not require proteins. They are the most versatile of macromolecules, each with its own inbuilt ability to carry out a cellular function. Some proteins aggregate to form relatively stiff filaments that help define the cell's shape and hold organelles in position. Others span the cell membrane and form channels or pores through which ions and small molecules can move. Many others are enzymes that catalyze the thousands of chemical reactions needed to maintain life. Still others are signaling proteins that enable cells to coordinate their internal activities or to communicate with other cells.

Want to see some proteins? Look at the white of an egg. Apart from the 90% or so that is water, most of what you see is protein. The predominant type of protein is ovalbumin. Easy to obtain in large quantities, ovalbumin was one of the first proteins studied by the scientific method (Chapter 1). In the 1830s, ovalbumin was shown to consist largely of carbon, hydrogen, nitrogen, and oxygen. Each molecule of ovalbumin was estimated to contain at least 400 carbon atoms. Leading chemists of the time scoffed at this number, believing that no organic molecule could possibly be so large. Little did they know: The number of carbon atoms in ovalbumin is actually closer to 2000 than to 400!

The reason that proteins can be such large organic molecules began to become clear only about 100 years ago, when scientists hypothesized that proteins are polymers (large molecules made up of repeated subunits). Now we know that proteins are linear polymers of any combination of 20 amino acids, each of which differs from the others in its chemical characteristics. In size, ovalbumin is actually an average protein, consisting of a chain of 385 amino acids.

In Chapter 3, we discussed how genetic information flows from the sequence of bases of DNA into the sequence of bases in a transcript of RNA. For protein-coding genes, the transcript is processed into messenger RNA (mRNA). In this chapter, we examine how protein polymers are assembled by ribosomes by means of a template of messenger RNA and transfer RNAs. We also discuss how the amino acid sequences of proteins help determine their three-dimensional structures and diverse chemical activities, as well as how proteins change through evolutionary time.

## 4.1 MOLECULAR STRUCTURE OF PROTEINS

If you think of a protein as analogous to a word in the English language, then the amino acids are like letters. The comparison is not altogether fanciful, as there are about as many amino acids in proteins as letters in the alphabet, and the order of both amino acids and letters is important. For example, the word PROTEIN has the same letters as POINTER, but the two words have completely different meanings. Similarly, the exact order of amino acids in a protein makes a big difference because that order determines the protein's shape and function.

**FIG. 4.1**   **Structure of an amino acid.** The central carbon atom is attached to an amino group, a carboxyl group, an R group or side chain, and a hydrogen atom.

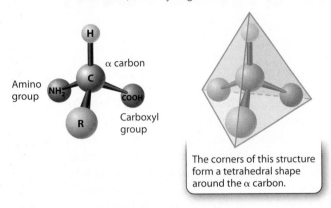

The corners of this structure form a tetrahedral shape around the α carbon.

## Amino acids differ in their side chains.

The general structure of an amino acid was discussed in Chapter 2 and is shown again in **Fig. 4.1.** It consists of a central carbon atom, called the **α (alpha) carbon,** connected by covalent bonds to four different chemical groups: an **amino group** (–NH₂ shown in dark blue), a **carboxyl group** (–COOH shown in brown), a hydrogen atom (–H shown in light blue), and a **side chain** or **R group** (shown in green). The four covalent bonds from the α carbon are at equal angles. As a result, an amino acid forms a tetrahedron, a sort of pyramid with four triangular faces (Fig. 4.1).

The R groups of the amino acids are also known as side chains, and they differ from one amino acid to the next. They are what make the "letters" of the amino acid "alphabet" distinct from one another. Just as letters differ in their shapes and sounds—vowels like E, I, and O and hard consonants like B, P, and T—amino acids differ in their chemical and physical properties.

The chemical structures of the 20 amino acids commonly found in proteins are shown in **Fig. 4.2.** The side chains (shown in green) are chemically diverse and are grouped according to their properties, with a particular emphasis on whether they are hydrophobic or hydrophilic, or have special characteristics that might affect a protein's structure. Within these broad categories are additional groupings based on whether they are nonpolar or polar, basic or acidic. These properties strongly influence how a polypeptide folds, and hence the three-dimensional shape of the protein.

For example, as we saw in Chapter 2, hydrophobic molecules do not readily interact with water and do not easily form hydrogen bonds. Water molecules in the cell therefore tend to form hydrogen bonds with each other instead of with hydrophobic side chains. The aggregation of hydrophobic side chains is also stabilized by weak van der Waals forces (Chapter 2), in which asymmetries in electron distribution create temporary

charges in the interacting molecules, which are then attracted to each other. The tendency for hydrophilic water molecules to interact with each other and for hydrophobic molecules to interact with each other is the very same tendency that leads to the formation of oil droplets in water. This is also the reason why most hydrophobic amino acids tend to be buried in the interior of folded proteins, where they are kept away from water.

Amino acids with polar side chains have a permanent charge separation, in which one end of the side chain is slightly more negatively charged than the other. As we saw in Chapter 2, polar molecules are hydrophilic, and they tend to form hydrogen bonds with each other or with water molecules.

The side groups of the basic and acidic amino acids are strongly polar and are also hydrophilic. The side groups of basic amino acids tend to be positively charged at intracellular pH, and those of the acidic amino acids tend to be negatively charged. For this reason, basic and acidic side chains are usually located on the outside surface of the folded molecule. The charged groups can also form ionic bonds (negative with positive) with each other and with other charged molecules in the environment. This ability to bind another molecule of opposite charge is one important way that proteins can associate with each other or with other macromolecules such as DNA.

The properties of several amino acids are noteworthy because of their effect on protein structure. These amino acids include glycine, proline, and cysteine. Glycine is different from the other amino acids because its R group is hydrogen, exactly like the hydrogen on the other side, and therefore it is not asymmetric. All of the other amino acids have four different groups attached to the α carbon and are asymmetric. In addition, glycine is nonpolar and small enough to tuck into spaces where other R groups would not fit. The small size of glycine's R group also allows for freer rotation around the C–N bond since its R group does not get in the way of the R groups of neighboring amino acids. Thus, glycine increases the flexibility of the polypeptide backbone, which can be important in the folding of the protein.

Proline is also distinctive, but for a different reason. Note how its R group is linked back to the amino group. This linkage creates a kink or bend in the polypeptide chain and restricts rotation of the C–N bond, thereby imposing constraints on protein folding in its vicinity, an effect the very opposite of glycine's.

Cysteine makes a special contribution to protein folding through its –SH group. When two cysteine side chains in the same or different polypeptides come into proximity, they can react to form an S–S disulfide bond, which covalently joins the side chains. Such disulfide bonds form cross-bridges that can connect different parts of the same protein or even different proteins. This property contributes to the overall structure of single proteins or combinations of proteins.

**FIG. 4.2**    Structures of the 20 amino acids commonly found in proteins.

### Successive amino acids in proteins are connected by peptide bonds.

Amino acids are linked together to form proteins. **Fig. 4.3** shows how amino acids in a protein are bonded together. The bond formed between the two amino acids is a **peptide bond,** shown in red in Fig. 4.3. In forming the peptide bond, the carboxyl group of one amino acid reacts with the amino group of the next amino acid in line, and a molecule of water is released. Note that in the resulting molecule, the R groups of each amino acid point in different directions.

The C=O group in the peptide bond is known as a carbonyl group, and the N–H group is an amide group. Note in Fig. 4.3 that these two groups are on either side of the peptide bond. This arrangement results in the delocalization of electrons, with the effect that the peptide bond has some of the characteristics of a double bond. The peptide bond is shorter than a single bond, for example, and it is not free to rotate like a single bond. The other bonds are free to rotate around their central axes.

Polymers of amino acids ranging from as few as two to many hundreds share a chemical feature common to individual amino

**FIG. 4.3**    **Formation of a peptide bond.**  A peptide bond forms between the carboxyl group of one amino acid and the amino group of another.

**FIG. 4.4**    **Levels of protein structure.**  The primary structure of an amino acid determines its shape and function.

The primary structure is the sequence of amino acids.

The secondary structure results from interactions of nearby amino acids.

The tertiary structure is the three-dimensional shape of a protein.

The quaternary structure results from interactions of polypeptide subunits.

acids: namely, that the ends are chemically distinct from each other. One end, shown at the left in Fig. 4.3, has a free amino group; this is the **amino end** of the peptide. The other end has a free carboxyl group, which constitutes the **carboxyl end** of the molecule. More generally, a polymer of amino acids connected by peptide bonds is known as a **polypeptide.** Typical polypeptides produced in cells consist of a few hundred amino acids. In human cells, the shortest polypeptides are about 100 amino acids in length; the longest is the muscle protein titin, with 34,350 amino acids. The term **protein** is often used as a synonym for polypeptide, especially when the polypeptide chain has folded into a stable, three-dimensional conformation. Amino acids that are incorporated into a protein are often referred to as amino acid **residues.**

## The sequence of amino acids dictates protein folding, which determines function.

Up to this point, we have considered the sequence of amino acids that make up a protein. This is the first of several levels of protein structure, illustrated in **Fig. 4.4.** The sequence of amino acids in a protein is its **primary structure.** The sequence of amino acids ultimately determines how a protein folds. Interactions between stretches of amino acids in a protein form local **secondary structures.** Longer-range interactions between these secondary structures in turn support the overall three-dimensional shape of the protein, which is its **tertiary structure.** Finally, some proteins are made up of several individual polypeptides that interact with each other, and the resulting ensemble is the **quaternary structure.**

Proteins have a remarkably wide range of functions in the cell, from serving as structural elements to communicating with the external environment to accelerating the rate of chemical reactions. No matter what the function of a protein is, the ability to carry out this function depends on the three-dimensional shape of the protein. When fully folded, some proteins contain pockets with positively or negatively charged side chains at just the right positions to trap small molecules; others have surfaces that can bind another protein or a sequence of nucleotides in DNA or RNA; some form rigid rods for structural support; and still others keep their hydrophobic side chains away from water molecules by inserting into the cell membrane.

The sequence of amino acids in a protein (its primary structure) is usually represented by a series of three-letter or one-letter abbreviations for the amino acids (abbreviations for the 20 common amino acids are given in Fig. 4.2). By convention, the amino acids in a protein are listed in order from left to right, starting at the amino end and proceeding to the carboxyl end. The amino and the carboxyl ends are different, so the order matters. Just as TIPS is not the same word as

SPIT, the sequence Thr–Ile–Pro–Ser is not the same peptide as Ser–Pro–Ile–Thr.

→ **Quick Check 1**  A mutation leads to a change in one amino acid in a protein. The result is that the protein no longer functions properly. How is this possible?

## Secondary structures result from hydrogen bonding in the polypeptide backbone.

Hydrogen bonds can form between the carbonyl group in one peptide bond and the amide group in another, thus allowing localized regions of the polypeptide chain to fold. This localized folding is a major contributor to the secondary structure of the protein. In the early 1950s, American structural biologists Linus Pauling and Robert Corey used a technique known as X-ray crystallography to study the structure of proteins. This technique was pioneered by British biochemists Dorothy Crowfoot Hodgkin, Max Perutz, and John Kendrew, among others **(Fig. 4.5).** Pauling and Corey studied crystals of highly purified proteins and discovered that two types of secondary structure are found in many different proteins. These are the **α helix** and the **β (beta) sheet.** Both these secondary structures are stabilized by hydrogen bonding along the peptide backbone.

## HOW DO WE KNOW?

**FIG. 4.5**

# What are the shapes of proteins?

**BACKGROUND**  The three-dimensional shapes of proteins can be determined by X-ray crystallography. One of the pioneers in this field was Dorothy Crowfoot Hodgkin, who used this technique to define the structures of cholesterol, vitamin B$_{12}$, penicillin, and insulin. She was awarded the Nobel Prize in Chemistry in 1964 for her early work. Max Perutz and John Kendrew shared the Nobel Prize in Chemistry in 1962 for defining the structures of myoglobin and hemoglobin using this method.

**METHOD**  X-ray crystallography can be used to determine the shape of proteins, as well as other types of molecules. The first step, which can be challenging, is to make a crystal of the protein. Then X-rays are aimed at the crystal while it is rotated. Some X-rays pass through the crystal, while others are scattered in different directions. A film or other detector records the pattern as a series of spots, which is known as a diffraction pattern. The locations and intensities of these spots can be used to infer the position and arrangement of the atoms in the molecule.

**RESULTS**  The X-ray diffraction pattern for hemoglobin looks like this:

From this two-dimensional pattern, researchers can use mathematical methods to determine the three-dimensional shape of the protein.

**FOLLOW-UP WORK**  Linus Pauling and Robert Corey used X-ray crystallography to determine two types of secondary structures commonly found in proteins—the α helix and the β sheet. Today, this technique is a common method for determining the shape of proteins.

**SOURCES**  Crowfoot, D. 1935. "X-Ray Single Crystal Photographs of Insulin." *Nature* 135:591–592. Kendrew, J.C., G. Bodo, H. M. Dintzis, et al. 1958. "A Three-Dimensional Model of the Myoglobin Molecule Obtained by X-Ray Analysis." *Nature* 181:662–666.

**FIG. 4.6**     **An α helix.** Hydrogen bonding between carbonyl and amide groups in the backbone stabilize the helix.

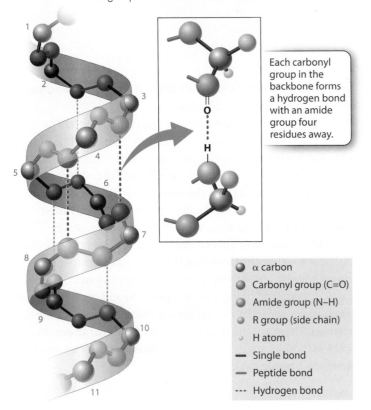

Each carbonyl group in the backbone forms a hydrogen bond with an amide group four residues away.

- α carbon
- Carbonyl group (C=O)
- Amide group (N–H)
- R group (side chain)
- H atom
— Single bond
— Peptide bond
--- Hydrogen bond

In the α helix shown in **Fig. 4.6**, the polypeptide backbone is twisted tightly in a right-handed coil with 3.6 amino acids per complete turn. The helix is stabilized by hydrogen bonds that form between each amino acid's carbonyl group and the amide group four residues ahead in the sequence, as indicated

by the dashed lines in Fig. 4.6. Note that the side chains project outward from the α helix. The chemical properties of the projecting side chains largely determine where the α helix is positioned in the folded protein, and how it might interact with other molecules.

The other secondary structure that Pauling and Corey found is the β sheet, depicted in **Fig. 4.7.** In a β sheet, the polypeptide folds back and forth on itself, forming a pleated sheet that is stabilized by hydrogen bonds between carbonyl groups in one chain and amide groups in the other chain across the way (dashed lines). The side chains (shown in green) project alternately above and below the plane of the β sheet. β sheets typically consist of 4 to 10 polypeptide chains aligned side by side, with the amides in each chain hydrogen-bonded to the carbonyls on either side (except for those at the edges).

β sheets are typically denoted by broad arrows, where the direction of the arrow runs from the amino end of the polypeptide segment to the carboxyl end. In Fig. 4.7, the arrows run in opposite directions, and the polypeptide chains are said to be antiparallel. β sheets can also be formed by hydrogen bonding between polypeptide chains that are parallel (pointing in the same direction). However, the antiparallel configuration is more stable because the carbonyl and amide groups are more favorably aligned for hydrogen bonding.

### Tertiary structures result from interactions between amino acid side chains.

The **tertiary structure** of a protein is the three-dimensional conformation of a single polypeptide chain, usually made up of several secondary structure elements. The shape of a protein is defined largely by interactions between the amino acid side chains. By contrast, the formation of secondary structures

**FIG. 4.7**     **A β sheet.** Hydrogen bonds between neighboring strands stabilize the structure.

Adjacent strands can run in the same direction (parallel), or in opposite directions (antiparallel), as shown here.

Hydrogen bonds form between carbonyl groups in one polypeptide and amide groups in a different part of the polypeptide.

- α carbon
- Carbonyl group (C=O)
- Amide group (N–H)
- R group (side chain)
- H atom
— Single bond
— Peptide bond
--- Hydrogen bond

**FIG. 4.8** **Three ways of showing the structure of the protein tubulin:** (a) ball-and-stick model; (b) ribbon model; (c) space-filling model.

a.

b.

c.

discussed earlier relies on interactions in the polypeptide backbone and is relatively independent of the side chains. Tertiary structure is determined by the spatial distribution of hydrophilic and hydrophobic side chains along the molecule, as well as by different types of chemical bonds and interactions (ionic, hydrogen, and van der Waals) that form between various side chains. The amino acids whose side chains form bonds with each other may be far apart in the polypeptide chain, but can end up near each other in the folded protein. Hence, the tertiary structure usually includes loops or turns in the backbone that allow these side chains to sit near each other in space and for bonds to form.

The three-dimensional shapes of proteins can be illustrated in different ways, as shown in **Fig. 4.8.** A ball-and-stick model (Fig. 4.8a) draws attention to the atoms in the amino acid chain. A ribbon model (Fig. 4.8b) emphasizes secondary structures, with α helices depicted as twisted ribbons and β sheets as broad arrows. Finally, a space-filling model (Fig. 4.8c) shows the overall shape and contour of the folded protein.

Remember that the folding of a polypeptide chain is determined by the sequence of amino acids. The primary structure determines the secondary and tertiary structures. Furthermore, tertiary structure determines function because it is the three-dimensional shape of the molecule—the contours and distribution of charges on the outside of the molecule and the presence of pockets that might bind with smaller molecules on the inside—that enables the protein to serve as structural support, membrane channel, enzyme, or signaling molecule. **Fig. 4.9** shows the tertiary structure of a bacterial protein that contains a pocket in the center in which certain side groups can form hydrogen bonds with a specific small molecule and hold it in place.

The principle that structure determines function can be demonstrated by many observations. For example, most proteins can be unfolded, or **denatured,** by chemical treatment or high temperature, and under these conditions they lose their functional activity; but when the chemicals are removed or the

temperature reduced, the proteins refold and their functional activity returns. Similarly, mutant proteins containing an amino acid that prevents proper folding are often devoid of their functional activity.

## Polypeptide subunits can come together to form quaternary structures.

Although many proteins are complete and fully functional as a single polypeptide chain with a tertiary structure, there are many other proteins that are composed of two or more polypeptide chains or subunits, each of which has a tertiary structure, and that also come together to form a higher-order quaternary structure. In the case of a multi-subunit protein, the activity of the complex depends on the quaternary structure formed by the combination of the various tertiary structures.

**FIG. 4.9** **Tertiary structure determines function.** This bacterial protein has a cavity that can bind with a small molecule (shown as a ball-and-stick model in the center).

**FIG. 4.10**    **Quaternary structure.** Polypeptide units of proteins may be identical, as in an enzyme from HIV (a), or different, as in hemoglobin (b).

**a.**
                 **b.**

This enzyme consists of two identical polypeptide subunits, shown in light green and dark green.

Hemoglobin is made up of four subunits: two copies of the polypeptide depicted in magenta and two copies of the polypeptide depicted in blue.

The polypeptide subunits may be identical or different (**Fig. 4.10**). Fig. 4.10a shows an example of a protein produced by HIV that consists of two identical polypeptide subunits. By contrast, many proteins, such as hemoglobin shown in Fig. 410b, are composed of different subunits. In either case, the subunits can influence each other in subtle ways and influence their function. For example, the hemoglobin in red blood cells that carries oxygen has four subunits. When one of these binds oxygen, a slight change in its structure is transmitted to the other subunits, making it easier for them to take up oxygen. In this way, oxygen transport from the lungs to the tissues is improved.

### Chaperones help some proteins fold properly.

The amino acid sequence (primary structure) of a protein determines how it forms its secondary, tertiary, and quaternary structures. For about 75% of proteins, the folding process takes place within milliseconds as the molecule is synthesized. Some proteins fold more slowly, however, and for these molecules folding is a dangerous business. The longer these polypeptides remain in a denatured (unfolded) state, the longer their hydrophobic groups are exposed to other macromolecules in the crowded cytoplasm. The hydrophobic effect, along with van der Waals interactions, tends to bring the exposed hydrophobic groups together, and their inappropriate aggregation may prevent proper folding. Correctly folded proteins can sometimes unfold because of elevated temperature, for example, and in the denatured state they are subject to the same risks of aggregation.

Cells have evolved proteins called **chaperones** that help protect slow-folding or denatured proteins until they can attain their proper three-dimensional structure. Chaperones bind with hydrophobic groups and nonpolar side chains to shield them from inappropriate aggregation, and in repeated cycles of binding and release they give the polypeptide time to find its correct shape.

## 4.2 TRANSLATION: HOW PROTEINS ARE SYNTHESIZED

The three-dimensional structure of a protein determines what it can do and how it works, and the immense diversity in the tertiary and quaternary structures among proteins explains their wide range of functions in cellular processes. Yet it is the sequence of amino acids along a polypeptide chain—its primary structure—that governs how the molecule folds into a stable three-dimensional configuration. How is the sequence of amino acids specified? It is specified by the sequence of nucleotides in the DNA, in coded form. The decoding of the information takes place according to the central dogma of molecular biology, which defines information flow in a cell from DNA to RNA to protein (**Fig. 4.11**). The key steps are known as **transcription** and **translation.** In transcription, the sequence of bases along part of a DNA strand is used as a template in the synthesis of a complementary sequence of bases in a molecule of RNA, as described in Chapter 3. In translation, the sequence of bases in an RNA molecule known as **messenger RNA** (**mRNA**) is used to specify the order in which successive amino acids are added to a newly synthesized polypeptide chain.

### Translation uses many molecules found in all cells.

Translation requires many components. Well over 100 genes encode components needed for translation, some of which

**FIG. 4.11**    **The central dogma, showing how information flows from DNA to RNA to protein.** Note the large number of cellular components required for translation.

are shown in Fig. 4.11. What are these needed components? First, the cell needs **ribosomes,** which are complex structures of RNA and protein. Ribosomes bind with mRNA, and it is on ribosomes that translation takes place. In prokaryotes, translation occurs as soon as the mRNA comes off the DNA template. In eukaryotes, the processes of transcription and translation are physically separated: Transcription takes place in the nucleus, and translation takes place on ribosomes in the cytoplasm.

In both eukaryotes and prokaryotes, the ribosome consists of a small subunit and a large subunit, each composed of both RNA and protein. The sizes of the subunits are given in Svedberg units (S), a measure of size and shape. Eukaryotic ribosomes are larger than prokaryotic ribosomes. As indicated in **Fig. 4.12**, the large subunit of the ribosome includes three binding sites for molecules of **transfer RNA (tRNA),** which are called the **A (aminoacyl)** site, the **P (peptidyl)** site, and the **E (exit) site.**

A major role of the ribosome is to ensure that, when the mRNA is in place on the ribosome, the sequence in the mRNA coding for amino acids is read in successive, non-overlapping groups of three nucleotides, much as you would read the sentence

THEBIGBOYSAWTHEBADMANRUN

Each non-overlapping group of three adjacent nucleotides (like THE or BIG or BOY, for example) constitutes a **codon,** and each codon in the mRNA codes for a single amino acid in the polypeptide chain.

In the example above, it is clear that the sentence begins with THE. However, in a long linear mRNA molecule, the ribosome could begin at any of three possible positions. These are known as **reading frames.** As an analogy, if the letters THE were the start codon for reading text, then we would know immediately how to read

ZWTHEBIGBOYSAWTHEBADMANRUN

However, without knowing the first codon of this string of letters, we could find three ways to break the sentence into three-letter words:

ZWT HEB IGB OYS AWT HEB ADM ANR UN
Z WTH EBI GBO YSA WTH EBA DMA NRU N
ZW THE BIG BOY SAW THE BAD MAN RUN

Obviously, only one of these frames is correct. The same is true for mRNAs.

While the ribosome establishes the correct reading frame for the codons, the actual translation of each codon in the mRNA into one amino acid in the polypeptide is carried out by means of transfer RNA (tRNA). Transfer RNAs are small RNA molecules of 70 to 90 nucleotides (**Fig. 4.13**). Each has a characteristic

**FIG. 4.13** Transfer RNA structure depicted in (a) a cloverleaf configuration and (b) a more realistic three-dimensional structure.

**FIG. 4.12** Composition and simplified structure of ribosomes in prokaryotes and eukaryotes.

The large subunit includes three binding sites for tRNAs.

| | Eukaryote | Prokaryote |
|---|---|---|
| **Large subunit** | | |
| Size | 60S | 50S |
| RNAs | 5S (120) 5.8S (160) 28S (4700) | 5S (120) 23S (2900) |
| Proteins | ≈49 | ≈34 |
| **Small subunit** | | |
| Size | 40S | 30S |
| RNAs | 18S (1900) | 16S (1540) |
| Proteins | ≈33 | ≈21 |
| **Total size** | 80S (MW 4.2 million) | 70S (MW 2.5 million) |

self-pairing structure that can be drawn as a cloverleaf, as shown in Fig. 4.13a, but the actual structure is more like that in Fig. 4.13b. In Fig. 4.13a, the bases found in all tRNA molecules are indicated by the letters; the Greek letter Ψ (psi) stands for pseudouracil, a slightly modified form of uracil. Three bases in the anticodon loop make up the anticodon; these are the three nucleotides that undergo base pairing with the corresponding codon.

Each tRNA has the nucleotide sequence CCA at its 3′ end (Fig. 4.13), and the 3′ hydroxyl of the A is the attachment site for the amino acid corresponding to the anticodon. Enzymes called **aminoacyl tRNA synthetases** connect specific amino acids to specific tRNA molecules (**Fig. 4.14**). Therefore, they are directly responsible for actually translating the codon sequence in a nucleic acid to a specific amino acid in a polypeptide chain. Most organisms have one aminoacyl tRNA synthetase for each amino acid. The enzyme binds to multiple sites on any tRNA that has an anticodon corresponding to the amino acid, and it catalyzes formation of the covalent bond between the amino acid and tRNA. A tRNA that has no amino acid attached is said to be uncharged, and one with its amino acid attached is said to be charged. Amino acid tRNA synthetases are very accurate and attach the wrong amino acid far less often than 1 time in 10,000. This accuracy is greater than that of subsequent steps in translation.

Although the specificity for attaching an amino acid to the correct tRNA is a property of aminoacyl tRNA synthetase, the specificity of DNA–RNA and codon–anticodon interactions result from base pairing. **Fig. 4.15** shows the relationships for one codon in double-stranded DNA, in the corresponding mRNA, and in the codon–anticodon pairing between the mRNA and the tRNA. Note that the first (5′) base in the codon in mRNA pairs with the last (3′) base in the anticodon because, as noted in Chapter 3, nucleic acid strands that undergo base pairing must be antiparallel.

### The genetic code shows the correspondence between codons and amino acids.

Fig. 4.15 shows how the codon AUG specifies the amino acid methionine (Met) by base pairing with the anticodon of a charged tRNA, denoted tRNA^Met. Most codons specify an amino acid according to a **genetic code.** This code is sometimes called the "standard" genetic code because, while it is used by almost all cells, some minor differences are found in a few organisms as well as in mitochondria.

The codon that initiates the process of translation is AUG, which specifies Met. The polypeptide is synthesized from the

**FIG. 4.14   Function of aminoacyl tRNA synthetase enzymes.** Aminoacyl tRNA synthetases attach specific amino acids to tRNAs and are therefore responsible for translating the codon sequence into an amino acid sequence in a protein.

Each aminoacyl tRNA synthetase binds to one uncharged tRNA and its corresponding amino acid.

There is a specific enzyme for each amino acid.

The enzyme attaches the amino acid to the 3′ end of the tRNA.

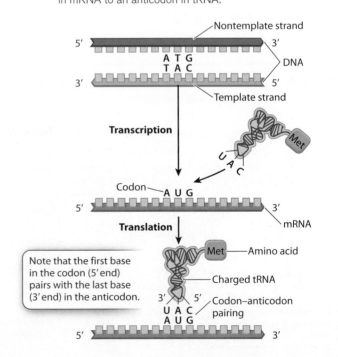

**FIG. 4.15   Role of base pairing.** Base pairing determines the relationship between a three-base triplet in double-stranded DNA to a codon in mRNA, and between a codon in mRNA to an anticodon in tRNA.

Note that the first base in the codon (5′ end) pairs with the last base (3′ end) in the anticodon.

amino end to the carboxyl end, and so Met forms the amino end of any polypeptide being synthesized; however, in many cases the Met is cleaved off by an enzyme after synthesis is complete. The AUG codon is also used to specify the incorporation of Met at internal sites within the polypeptide chain.

As is apparent in Fig. 4.15, the AUG codon that initiated translation is preceded by a region in the mRNA that is not translated. The position of the initiator AUG codon in the mRNA establishes the reading frame that determines how the downstream codons (those following the AUG) are to be read.

Once the initial Met creates the amino end of a new polypeptide chain, the downstream codons are read one by one in non-overlapping groups of three bases. At each step,

the ribosome binds to a tRNA with an anticodon that can base pair with the codon, and the amino acid on that tRNA is attached to the growing chain to become the new carboxyl end of the polypeptide chain. This process continues until one of three "stop" codons is encountered: UAA, UAG, or UGA. (The stop codons are also called termination codons or sometimes nonsense codons.) At this point, the polypeptide is finished and released into the cytosol.

The standard genetic code was deciphered in the 1960s by a combination of techniques, but among the most ingenious was the development by American biochemist Har Gobind Khorana and his colleagues of chemical methods for making synthetic RNAs of known sequence. This experiment is illustrated in **Fig. 4.16.**

## HOW DO WE KNOW?

**FIG. 4.16**

# How was the genetic code deciphered?

**BACKGROUND** The genetic code is the correspondence between three-letter nucleotide codons in RNA and amino acids in a protein. American biochemist Har Gobind Khorana performed key experiments that helped to crack the code.

**METHOD** Khorana and his group made RNAs of known sequence. They then added these synthetic RNAs to a solution containing all of the other components needed for translation. By adjusting the concentration of magnesium and other factors, the researchers could get the ribosome to initiate synthesis with any codon, even if not AUG.

**EXPERIMENT 1 AND RESULTS** When a synthetic poly(U) was used as the mRNA, the resulting polypeptide was polyphenylalanine (Phe–Phe–Phe...):

**a.**

**CONCLUSION** The codon UUU corresponds to Phe. The poly(U) mRNA can be translated in three possible reading frames, depending on which U is the 5′ end of the start codon, but in each of them, all the codons are UUU.

**EXPERIMENT 2 AND RESULTS** When a synthetic mRNA with alternating U and C was used, the resulting polypeptide had alternating serine (Ser) and leucine (Leu):

**b.**

**CONCLUSION** Here again there are three reading frames, but each of them has alternating UCU and CUC codons. The researchers could not deduce from this result whether UCU corresponds to Ser and CUC to Leu or the other way around; the correct assignment came from experiments using other synthetic mRNA molecules.

**EXPERIMENT 3 AND RESULTS** When a synthetic mRNA with repeating UCA was used, three different polypeptides were produced—polyserine (Ser), polyhistidine (His), and polyisoleucine (Ile).

**c.**

**CONCLUSION** The results do not reveal which of the three reading frames corresponds to which amino acid, but this was sorted out by studies of other synthetic polymers.

**SOURCE** Khorana, H. G. 1972. "Nucleic Acid Synthesis in the Study of the Genetic Code." In *Nobel Lectures, Physiology or Medicine 1963–1970.* Amsterdam: Elsevier.

The standard genetic code shown in **Table 4.1** has 20 amino acids specified by 64 codons. Many amino acids are therefore specified by more than one codon, and hence the genetic code is redundant, or degenerate. The redundancy has strong patterns, however:

- The redundancy results almost exclusively from the third codon position.

- When an amino acid is specified by two codons, they differ either in whether the third position is a U or a C (both pyrimidine bases), or in whether the third position is an A or a G (both purine bases).

- When an amino acid is specified by four codons, the identity of the third codon position does not matter; it could be U, C, A, or G.

The chemical basis of these patterns results from two features of translation. First, in many tRNA anticodons the 5′ base that pairs with the 3′ (third) base in the codon is chemically modified into a form that can pair with two or more bases at the third position in the codon. Second, in the ribosome, there is less-than-perfect alignment between the third position of the codon and the base that pairs with it in the anticodon, so the requirements for base pairing are somewhat relaxed; this feature of the codon–anticodon interaction is referred to as wobble.

→  **Quick Check 2**  What polypeptide sequences would you expect to result from a synthetic mRNA with the repeating sequence 5′-UUUGGGUUUGGGUUUGGG-3′?

## Translation consists of initiation, elongation, and termination.

Translation is usually divided into three separate processes. The first is **initiation,** in which the initiator AUG codon is recognized and Met is established as the first amino acid in the new polypeptide chain. The second process is **elongation,** in which successive amino acids are added one by one to the growing chain. And the third process is **termination,** in which the addition of amino acids stops and the completed polypeptide chain is released from the ribosome.

Initiation of translation (**Fig. 4.17**) requires a number of protein **initiation factors** that bind to the mRNA. In eukaryotes, one group of initiation factors binds to the 5′ cap that is added to the mRNA during processing (Fig. 4.17a). These recruit a small subunit of the ribosome, and other initiation factors bring up a transfer RNA charged with methionine (Met). The initiation complex then moves along the mRNA until it encounters the first AUG triplet. The position of this AUG establishes the translational reading frame.

When the first AUG codon is encountered, a large ribosomal subunit joins the complex, the initiation factors are released, and the next tRNA is ready to join the ribosome (Fig. 4.17b). Note in Fig. 4.17b that the tRNA^Met binds with the P (peptidyl) site in the ribosome and that the next tRNA in line comes in at the A (aminoacyl) site. Once the new tRNA is in place, a coupled reaction takes place in which the bond connecting the Met to its tRNA is transferred to the amino group of the next amino acid in line as the first peptide bond is formed (Fig. 4.17c). The new peptide is now attached to the tRNA in the A site. Formation of the peptide bond requires multiple proteins in the large subunit, but an RNA in the large subunit is the actual catalyst. The ribosome then shifts one codon to the right (Fig. 4.17d), which moves the uncharged tRNA^Met to the E site and the peptide-bearing tRNA to the P site, freeing the A site for the next charged tRNA in line to come in.

Once translation has been initiated, it proceeds step by step through the

**TABLE 4.1  The standard genetic code.**

| First position (5′ end) | Second position | | | | Third position (3′ end) |
|---|---|---|---|---|---|
| | U | C | A | G | |
| U | UUU Phe ⎤ F<br>UUC Phe ⎦<br>UUA Leu ⎤ L<br>UUG Leu ⎦ | UCU Ser ⎤<br>UCC Ser ⎥ S<br>UCA Ser ⎥<br>UCG Ser ⎦ | UAU Tyr ⎤ Y<br>UAC Tyr ⎦<br>UAA Stop<br>UAG Stop | UGU Cys ⎤ C<br>UGC Cys ⎦<br>UGA Stop<br>UGG Trp  W | U<br>C<br>A<br>G |
| C | CUU Leu ⎤<br>CUC Leu ⎥ L<br>CUA Leu ⎥<br>CUG Leu ⎦ | CCU Pro ⎤<br>CCC Pro ⎥ P<br>CCA Pro ⎥<br>CCG Pro ⎦ | CAU His ⎤ H<br>CAC His ⎦<br>CAA Gln ⎤ Q<br>CAG Gln ⎦ | CGU Arg ⎤<br>CGC Arg ⎥ R<br>CGA Arg ⎥<br>CGG Arg ⎦ | U<br>C<br>A<br>G |
| A | AUU Ile ⎤<br>AUC Ile ⎥ L<br>AUA Ile ⎦<br>AUG Met  M | ACU Thr ⎤<br>ACC Thr ⎥ T<br>ACA Thr ⎥<br>ACG Thr ⎦ | AAU Asn ⎤ N<br>AAC Asn ⎦<br>AAA Lys ⎤ K<br>AAG Lys ⎦ | AGU Ser ⎤ S<br>AGC Ser ⎦<br>AGA Arg ⎤ R<br>AGG Arg ⎦ | U<br>C<br>A<br>G |
| G | GUU Val ⎤<br>GUC Val ⎥ V<br>GUA Val ⎥<br>GUG Val ⎦ | GCU Ala ⎤<br>GCC Ala ⎥ A<br>GCA Ala ⎥<br>GCG Ala ⎦ | GAU Asp ⎤ D<br>GAC Asp ⎦<br>GAA Glu ⎤ E<br>GAG Glu ⎦ | GGU Gly ⎤<br>GGC Gly ⎥ G<br>GGA Gly ⎥<br>GGG Gly ⎦ | U<br>C<br>A<br>G |

▨ Nonpolar    ▨ Polar    ▨ Basic    ▨ Acidic    ▨ Stop codon

**FIG. 4.17   Initiation of translation.** Initiation factors bind the small subunit of the ribosome, which scans until it reaches an AUG codon, where the large subunit of the ribosome joins the complex.

**a.**

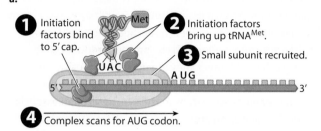

**1** Initiation factors bind to 5' cap.

**2** Initiation factors bring up tRNA^Met.

**3** Small subunit recruited.

**4** Complex scans for AUG codon.

**b.**

**1** Initiation factors released.

**2** Large subunit joins complex.

**3** Next tRNA binds with the A site.

**c.**

Met detaches from tRNA^Met.

Peptide bond

**d.**

Next tRNA binds with the A site.

Ribosome shifts one codon.

**FIG. 4.18   Elongation of the polypeptide chain.** Note that the growing polypeptide chain remains attached to a tRNA throughout the process.

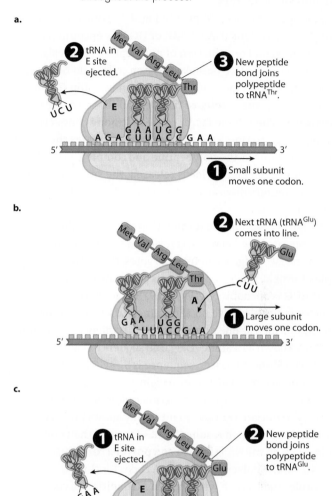

**a.**

**2** tRNA in E site ejected.

**3** New peptide bond joins polypeptide to tRNA^Thr.

**1** Small subunit moves one codon.

**b.**

**2** Next tRNA (tRNA^Glu) comes into line.

**1** Large subunit moves one codon.

**c.**

**1** tRNA in E site ejected.

**2** New peptide bond joins polypeptide to tRNA^Glu.

**3** Small subunit moves one codon.

process of elongation (**Fig. 4.18**). Fig. 4.18a shows the situation immediately after the formation of a new peptide bond that attaches the growing polypeptide chain to the amino acid linked to the tRNA in the A site (in this case tRNA^Thr); the small ribosomal subunit has moved one codon farther along the mRNA, and the tRNA in the E (exit) site is being ejected. Fig. 4.18b shows the situation an instant later, when the large ribosomal subunit has moved downstream one codon, emptying the A site by shifting the tRNA^Thr to the P site in the middle and the tRNA^Leu to the E site. Note that, while there are three tRNA binding sites on the ribosome, only two of them are occupied at any one time. Fig. 4.18c shows the situation slightly later; it is almost identical to that in Fig. 4.18a, except that the polypeptide is one amino acid longer and the ribosome has moved one codon farther to the right. Ribosome movement along the mRNA and formation of the peptide bonds require energy, which is obtained by breaking the high-energy bonds of the molecule GTP bound with proteins called **elongation factors.**

The elongation process shown in Fig. 4.18 continues until one of the stop codons (UAA, UAG, or UGA) is encountered; these codons signal termination of polypeptide synthesis. Termination takes place because the stop codons do not have corresponding tRNA molecules. Rather, when a stop codon is encountered, a protein **release factor** binds to the A site of the ribosome. The release factor causes the bond connecting the polypeptide to the tRNA to break, which creates the carboxyl terminus of the polypeptide and completes the chain. Once the finished polypeptide is released, the small and large ribosomal subunits disassociate from the mRNA and from each other.

Although elongation and termination are very similar in prokaryotes and in eukaryotes, translation initiation differs between the two (**Fig. 4.19**). In eukaryotes, the initiation complex forms at the 5′ cap and scans along the mRNA until the first AUG is encountered (Fig. 4.19a). In prokaryotes, the mRNA molecules have no 5′ cap. Instead, the initiation complex is formed at one or more internal sequences present in the mRNA known as a Shine–Dalgarno sequence (Figure 4.19b). In *E. coli*, the Shine–Dalgarno sequence is 5′-AGGAGGU-3′, and it is followed by an AUG codon eight nucleotides farther downstream that serves as an initiation codon for translation. The ability to initiate translation internally allows prokaryotic mRNAs to contain open reading frames for more than one protein. Such an mRNA is known as a **polycistronic mRNA.** In Fig. 4.19b, the polycistronic mRNA codes for three different polypeptide chains, each with its own AUG initiation codon preceded eight nucleotides upstream by its own Shine–Dalgarno sequence. Each Shine–Dalgarno sequence can serve as

an initiation site for translation, and so all three polypeptides can be translated.

A polycistronic mRNA results from transcription of a group of functionally related genes located in tandem along the DNA and transcribed as a single unit from one promoter. This type of gene organization is known as an **operon.** Prokaryotes have many of their genes organized into operons because the production of a polycistronic mRNA allows all the protein products to be expressed together whenever they are needed. Typically, the genes organized into operons are those whose products are needed either for successive steps in the synthesis of an essential small molecule, such as an amino acid, or else for successive steps in the breakdown of a source of energy, such as a complex carbohydrate.

→   **Quick Check 3**  Bacterial DNA containing an operon encoding three enzymes is introduced into chromosomal DNA in yeast (a eukaryote) in such a way that it is properly flanked by a promoter and a transcriptional terminator. The bacterial DNA is transcribed and the RNA correctly processed, but only the protein nearest the promoter is produced. Can you suggest why?

? **CASE 1 The First Cell: Life's Origins**
**How did the genetic code originate?**
During transcription and translation, proteins and nucleic acids work together to convert the information stored in DNA into proteins. If we think about how such a system might have originated, however, we immediately confront a chicken-and-egg problem: Cells need nucleic acids to make proteins, but proteins are required to make nucleic acids. Which came first? In Chapter 2, we discussed the special features that make RNA an attractive candidate for both information storage and catalysis in early life. Early in evolutionary history, then, proteins had to be added to the mix. No one fully understands

---

**FIG. 4.19**  **Initiation in eukaryotes and in prokaryotes.**  (a) In eukaryotes, translation is initiated only at the 5′ cap. (b) In prokaryotes, initiation takes place at any Shine–Dalgarno sequence; this mechanism allows a single mRNA to include coding sequences for multiple polypeptides.

**a.** Monocistronic mRNA

**b.** Polycistronic mRNA

how they were incorporated, but researchers are looking closely at tRNA, the molecule that completes the "translating" step of translation.

In modern cells, tRNA shuttles amino acids to the ribosome, but an innovative hypothesis suggests that in early life tRNA-like molecules might have served a different function. This proposal holds that the early precursors of the ribosome were RNA molecules that facilitated the replication of other RNAs, not proteins. In this version of an RNA world, precursors to tRNA would have shuttled nucleotides to growing RNA strands. Researchers hypothesize that tRNAs bound to amino acids may have acted as simple catalysts, facilitating more accurate RNA synthesis. Through time, amino acids brought into close proximity in the process of building RNA molecules might have polymerized to form peptide chains. From there, natural selection would favor the formation of peptides that enhanced replication of RNA molecules, bringing proteins into the chemistry of life.

All of the steps in gene expression, including transcription and translation, are summarized in **Fig. 4.20** on the next page.

## 4.3 PROTEIN EVOLUTION AND THE ORIGIN OF NEW PROTEINS

The amino acid sequences of more than a million proteins are known, and the particular three-dimensional structure has been determined for each of more than 10,000 proteins. While few of the sequences and structures are identical, many are sufficiently similar that the proteins can be grouped into about 25,000 **protein families** that are structurally and functionally related. Why are there not more types of proteins? The number of possible sequences is unimaginably large. For example, for a polypeptide of only 62 amino acids, there are $20^{62}$ possible sequences (because each of the 62 positions could be occupied by any of the 20 amino acids). The number $20^{62}$ equals approximately $10^{80}$; this number is also the estimated total number of electrons,

protons, and neutrons in the entire universe! So why are there so few protein families? The most likely answer is that the chance that any random sequence of amino acids would fold into a stable configuration and carry out some useful function in the cell is very close to zero.

**Most proteins are composed of modular folding domains.** If functional proteins are so unlikely, how could life have evolved? The answer is that the earliest proteins were probably much shorter than modern proteins and needed only a trace of function. Only as proteins evolved through billions of years did they become progressively longer and more specialized in their functions. Many protein families that exist today exhibit small regions of three-dimensional structure in which the protein folding is similar. These regions range in length from 25 to 100 or more amino acids. A region of a protein that folds in a similar way relatively independently of the rest of the protein is known as a **folding domain.**

Several examples of folding domains are illustrated in **Fig. 4.21.** Many folding domains are functional units in themselves. The folding domain in Fig. 4.21a is a globin fold composed of multiple α helices that have a hydrophobic core and a hydrophilic exterior. The globin fold is characteristic of hemoglobin and other oxygen-carrying proteins. Fig. 4.21b is a Rossman fold, which in many enzymes binds a nicotinamide adenine dinucleotide (NAD) used in oxidation reactions. The TIM barrel (Fig. 4.21c) is named after the enzyme triose phosphate isomerase in which it is a prominent feature. The TIM barrel consists of alternating α helices and parallel β sheets connected by loops. In many enzymes with a TIM barrel, the active site is formed by the loops at the carboxyl ends of the sheets. Finally, Fig. 4.21d is a β barrel formed from antiparallel β sheets. β barrel structures occur in proteins in some types of bacteria, usually in proteins that span the cell membrane, where the β barrel provides a channel that binds hydrophobic molecules.

The number of known folding domains is only about 2500, which is far fewer than the number of protein families. The

---

**FIG. 4.21**   **Examples of folding domains:**   (a) globin fold; (b) Rossman fold; (c) TIM barrel; (d) β barrel.

**a.**

**b.**

**c.**

**d.**

# VISUAL SYNTHESIS  Gene Expression

FIG. 4.20

**Integrating concepts from Chapters 3 and 4**

## Transcription

During transcription, RNA polymerase reads a DNA sequence and produces a complementary, antiparallel strand of RNA.

DNA → RNA

## RNA processing

In eukaryotes, the primary RNA transcript is modified in the nucleus, creating mature mRNA and noncoding RNA.

→ mRNA

Noncoding RNAs do not encode proteins and include tRNA and rRNA. These RNAs are transcribed from DNA found in the nucleolus.

Nucleolus

tRNA

DNA

Gene

During mRNA processing, introns are excised from the strand while exons are spliced together. The mRNA receives a 5′ cap and a poly(A) tail, then travels out of the nucleus.

RNA polymerase

Primary transcript (RNA)

5′ cap

Spliceosome

Exon     Intron     — AAAAA

— AAAAA

Intron lariat

— AAAAA

— AAAAA

RNA transcription begins at a promoter region and continues until a terminator sequence is encountered.

RNA transcript

5′

Promoter region

3′

5′

RNA–DNA duplex

Template DNA

Terminator sequence

5′

3′

3′

Gene

Polymerase movement

*Nucleus*

4-16

**Translation**
Ribosomes translate mRNA in the cytosol, producing polypeptide chains. The resulting proteins carry out vital cellular functions.

mRNA

Protein

Growing polypeptide chain

Proteins have many different roles in the cell. They provide structural support, act as enzymes that facilitate chemical reactions, and are involved in cell signaling and communication.

Met
Leu
Arg
Thr
Ser
Pro
Glu
Glu
Asp
Val
Phe
Ala
Met
Gly
Val
Arg
Thr

During elongation, the appropriate tRNA anticodon matches with the available codon on the mRNA, bringing the next amino acid to the polypeptide chain.

tRNA

U C U

E site

P site

A site

Ribosome

A G A A C C G A A

U G G

5'

3'

Codon

mRNA

Glu

C U U

Incoming tRNA charged with amino acid

Anticodon

Ribosome moves along mRNA in a 5' to 3' direction.

Growing actin filament

Actin protein

A ribosome attaches to the mRNA and scans the transcript for an AUG start codon. The ribosome initiates translation at that point and synthesizes a polypeptide chain based on the sequence of the mRNA until it reaches a stop codon.

Ribosomes

*Cytosol*

*Extracellular fluid*

reason for the discrepancy is that different protein families contain different combinations of folding domains. Modern protein families are composed of different combinations of a number of folding domains, each of which contributes some structural or functional feature of the protein. Different types of protein folds occur again and again in different contexts and combinations. The earliest proteins may have been little more than single folding domains that could aggregate to form more complex functional units. As life evolved, the proteins became longer by joining the DNA coding for the individual folding units together into a single molecule.

For example, human tissue plasminogen activator, a protein that is used in treating strokes and heart attacks because it dissolves blood clots, contains domains shared with cell-surface receptors, a domain shared with cellular growth factors, and a domain that folds into large loops facilitating protein–protein interactions. Hence, novel proteins do not always evolve from random combinations of amino acids; instead, they often evolve by combining already functional folding domains into novel combinations.

## Amino acid sequences evolve through mutation and selection.

Another important reason that complex proteins can evolve seemingly against long odds is that evolution proceeds stepwise through the processes of mutation and selection. A **mutation** is a change in the sequence of a gene. The process of mutation is discussed in Chapter 14, but for now all you need to know is that mutations affecting proteins occur at random in regard to their effects on protein function. In protein-coding genes, some mutations may affect the amino acid sequence; others might change the level of protein expression or the time in development or type of cell in which the protein is produced. Here, we will consider only those mutations that change the amino acid sequence.

By way of analogy, we can use a word game that we call change-one-letter. The object of the game is to change an ordinary English word into another meaningful English word by changing exactly one letter. Consider the word GONE. To illustrate "mutations" of the word that are random with respect to function (that is, random with respect to whether the change will yield a meaningful new word), we wrote a computer program that would choose one letter in GONE at random and replace it with a different random letter. The first 26 "mutants" of GONE are:

| | | | |
|---|---|---|---|
| UONE | GNNE | GONJ | GOZE |
| GONH | GOLE | GFNE | XONE |
| NONE | GKNE | GJNE | DONE |
| GCNE | GONB | GOIE | GGNE |
| GONI | GFNE | GPNE | GENE |
| BONE | GOWE | OONE | GYNE |

Most of the mutant words are gibberish, corresponding to the biological reality that most random amino acid replacements impair protein function to a greater or lesser extent. On the other hand, some mutant proteins function just as well as the original, and a precious few change function. In the word-game analogy, the mutants that can persist correspond to meaningful words, those words shown in red.

In a population of organisms, random mutations are retained or eliminated through the process of **selection** among individuals on the basis of their ability to survive and reproduce. This process is considered in greater detail in Chapter 21, but the principle is straightforward. Most mutations that impair protein function will be eliminated because, to the extent that the function of the protein contributes to survival and reproduction, the individuals carrying these mutations will leave fewer offspring than others. Mutations that do not impair function may remain in the population for long periods because their carriers survive and reproduce in normal numbers; a mutation of this type has no tendency to either increase or decrease in frequency over time. In contrast, individuals that carry the occasional mutation that improves protein function will reproduce more successfully than others. Because of the enhanced reproduction, the mutant gene encoding the improved protein will gradually increase in frequency and spread throughout the entire population.

In the word game, any of the mutants in red may persist in the population, but suppose that one of them, GENE for example, is actually superior to GONE (considered more euphonious, perhaps). Then GENE will gradually displace GONE, and eventually GONE will be gone. In a similar way that one meaningful word may replace another, one amino acid sequence may be replaced with a different one in the course of evolution.

A real-world example that mirrors the word game is found in the evolution of resistance of the malaria parasite to the drug pyrimethamine. This drug inhibits an enzyme known as dihydrofolate reductase, which the parasite needs to survive and reproduce inside red blood cells. Resistance to pyrimethamine is known to have evolved through a stepwise sequence of four amino acid replacements. In the first replacement, serine (S) at the 108th amino acid in the polypeptide sequence (position 108) was replaced with asparagine (N); then cysteine (C) at position 59 was replaced with arginine (R); asparagine (N) at position 51 was then replaced with isoleucine (I); finally isoleucine (I) at position 164 was replaced with leucine (L). If we list the amino acids according to their single-letter abbreviation in the order of their occurrence in the protein, the evolution of resistance followed this pathway:

NCSI → NCNI → NRNI → IRNI → IRNL

where the mutant amino acids are shown in red. Each successive amino acid replacement increased the level of resistance, so that a greater concentration of drug was needed to treat the disease. The quadruple mutant IRNL is resistant to such high levels that the drug is no longer useful.

Depicted according to stepwise amino acid replacements, the analogy between the evolution of pyrimethamine resistance and the change-one-letter game is clear. It should also be clear from our earlier discussion that hundreds of other mutations causing amino acid replacements in the enzyme must have

occurred in the parasite during the course of evolution, but only these amino acid changes occurring in this order persisted and increased in frequency because they conferred greater survival and reproduction of the parasite under treatment with the drug.

→ **Quick Check 4** What do you think happened to the mutations that decreased survival or reproduction of the parasites?

## Core Concepts Summary

**4.1 PROTEINS ARE LINEAR POLYMERS OF AMINO ACIDS THAT FORM THREE-DIMENSIONAL STRUCTURES WITH SPECIFIC FUNCTIONS.**

An amino acid consists of an α carbon connected by covalent bonds to an amino group, a carboxyl group, a hydrogen atom, and a side chain or R group. page 4-2

There are 20 common amino acids that differ in their side chains. page 4-2

Amino acids are connected by peptide bonds to form proteins. page 4-3

The primary structure of a protein is its amino acid sequence. The primary structure determines how a protein folds, which in turn determines how it functions. page 4-4

The secondary structure of a protein results from the interactions of nearby amino acids. Examples include the α helix and β sheet. page 4-5

The tertiary structure of a protein is its three-dimensional shape, which results from long-range interactions of amino acid side chains. page 4-6

Some proteins are made up of several polypeptide subunits; this group of subunits is the protein's quaternary structure. page 4-7

Chaperones help some proteins fold properly. page 4-8

**4.2 TRANSLATION IS THE PROCESS BY WHICH THE SEQUENCE OF BASES IN MESSENGER RNA SPECIFIES THE ORDER OF SUCCESSIVE AMINO ACIDS IN A NEWLY SYNTHESIZED PROTEIN.**

Translation requires many cellular components, including ribosomes, tRNAs, and proteins. page 4-8

mRNAs have three possible reading frames comprised of three-nucleotide codons. page 4-9

tRNAs have an anticodon that base pairs with the codon in the mRNA and carries a specific amino acid. page 4-10

Aminoacyl tRNA synthetases attach specific amino acids to tRNAs. page 4-10

The genetic code defines the relationship between the three-letter codons of nucleic acids and their corresponding amino acids. It was deciphered using synthetic RNA molecules. page 4-10

The genetic code is redundant in that many amino acids are specified by more than one codon. page 4-12

Translation consists of three steps: initiation, elongation, and termination. page 4-12

**4.3 PROTEINS EVOLVE THROUGH MUTATION AND SELECTION AND BY COMBINING FUNCTIONAL UNITS.**

Protein families are groups of proteins that are structurally and functionally related. page 4-15

There are far fewer protein families than the total number of possible proteins because the probability that a random sequence of amino acids will fold properly to carry out a specific function is very small. page 4-15

A region of a protein that folds in a particular way and that carries out a specific function is called a folding domain. page 4-15

Proteins evolve by combining different folding domains. page 4-18

Proteins also evolve by changes in amino acid sequence, which occurs by mutation and selection. page 4-18

## Self-Assessment

1. Draw one of the 20 amino acids and label the amino group, the carboxyl group, the side chain (R group), and the α carbon.

2. Name four major groups of amino acids, categorized by the properties of their side chains. Explain how the chemical properties of each group affect protein shape.

3.  Describe the importance to proteins of peptide bonds, hydrogen bonds, ionic bonds, disulfide bridges, and noncovalent interactions (e.g. van der Waals forces and the hydrophobic effect).

4.  Explain how the order of amino acids determines the way in which a protein folds.

5.  Explain the relationship between protein folding and protein function.

6.  Describe the relationship between codons of mRNA, anticodons of tRNA, and amino acids.

7.  Describe the process by which ribosomes synthesize polypeptides.

8.  Name and describe two ways that proteins can acquire new functions in the course of evolution.

Do you understand the chapter's Core Concepts? Log in to check your answers to the Self-Assessment questions, then practice what you've learned and reinforce this chapter's concepts by working through the problems and multimedia tutorials provided there.

　 **www.biologyhowlifeworks.com**

## CHAPTER 5

# ORGANIZING PRINCIPLES

## Lipids, Membranes, and Cell Compartments

"With the discovery of the cell, biologists found their atom." So stated François Jacob, the French biologist who shared the Nobel Prize in Physiology or Medicine in 1965. Just as the atom is the smallest, most basic unit of matter, the cell is the smallest, most basic unit of living organisms. All organisms, from single-celled algae to complex multicellular organisms like humans, are made up of cells. Therefore, essential properties of life, including growth, reproduction, and metabolism, must be understood in terms of cell structure and function.

Cells were first seen sometime around 1665, when the English scientist Robert Hooke built a microscope that he used to observe thin sections of dried cork tissue derived from plants. In these sections, Hooke observed arrays of small cavities and named them "cells" (**Fig. 5.1**). Although Hooke was probably looking at empty cell walls that make up the plant vascular system rather than at living cells, his observations nevertheless led to the concept that cells are the fundamental units of life. Later in the seventeenth century, the Dutch microbiologist Anton van Leeuwenhoek greatly improved the magnifying power of microscope lenses, enabling him to see and describe unicellular organisms, including bacteria, protozoa, and algae. Today, modern microscopy is providing unprecedented detail and a deeper understanding of the inner architecture of cells.

Cells differ in size and shape, but they share many features. This similarity in the microscopic organization of all living organisms led to the development, in the middle of the nineteenth century, of one of the pillars of modern biology: the **cell theory**. Based on the work and ideas of Matthias Schlieden, Theodor Schwann, Rudolf Virchow, and others, the cell theory states that all organisms are made up of cells, that the cell is the fundamental unit of life, and that cells come from preexisting cells. There is no life without cells, and the cell is the smallest unit of life.

This chapter focuses on cells and their internal organization. We pay particular attention to the key role that membranes play in separating a cell from the external environment and defining structural and functional spaces within cells.

## 5.1 STRUCTURE OF CELL MEMBRANES

Cells are defined by membranes. After all, membranes physically separate cells from their external environment. In addition, membranes define spaces within many cells that allow them to carry out their diverse functions. Lipids are the main component of cell membranes. Proteins are often embedded in or associated with the membrane. Carbohydrates can also be found in cell membranes, usually attached to lipids (glycolipids) and proteins (glycoproteins).

**FIG. 5.1**    **The first observation of cells.** Robert Hooke used a simple microscope to observe small chambers in a sample of cork tissue that he described as "cells."

Drawing by Hooke                Cork tissue

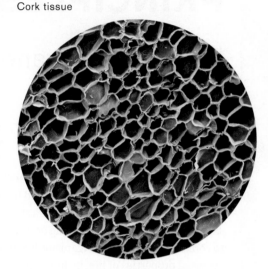

## Cell membranes are composed of two layers of lipids.

The major types of lipid found in cell membranes are phospholipids, introduced in Chapter 2. Phospholipids have both hydrophilic ("water-loving") and hydrophobic ("water-fearing") regions in a single molecule (**Fig. 5.2**). The phosphate "head" group is hydrophilic because it is polar, enabling it to form hydrogen bonds with water. By contrast, the two long fatty acid "tails" are hydrophobic because they are nonpolar and do not form hydrogen bonds with water. Molecules with both hydrophilic and hydrophobic regions are termed **amphipathic**.

In an aqueous environment, amphipathic molecules such as phospholipids behave in an interesting way. They spontaneously arrange themselves into various structures that have the polar head groups on the outside interacting with water and the nonpolar tail groups grouped together on the inside away from water. This arrangement results from the tendency of polar molecules like water to exclude nonpolar molecules or nonpolar groups within molecules. The shape of the structure is determined by the bulkiness of the head group relative to that of the hydrophobic tails. For example, lipids with bulky heads and a single hydrophobic fatty acid tail are wedge-shaped and pack into spherical structures called **micelles** (**Fig. 5.3a**). By contrast, lipids with less bulky head groups and two hydrophobic tails form a **bilayer** (**Fig. 5.3b**). A lipid bilayer is a two-layered structure organized in such a way that the hydrophilic portion of the lipid faces out toward the aqueous environment and the hydrophobic portion faces in toward the hydrophobic portions of other lipids and away from the aqueous environment.

The resulting bilayers form closed structures with an inner space since free edges would expose the hydrophobic chains to the aqueous environment. This organization in part explains why bilayers are effective cell membranes. It also explains why

**FIG. 5.2**    **Phospholipid structure.** Phospholipids, the major component of cell membranes, are termed amphipathic because they have both hydrophilic and hydrophobic domains.

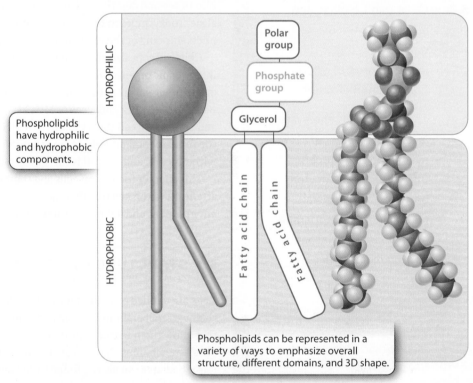

**FIG. 5.3** **Lipid structures.** Lipids can form micelles (a), bilayers (b), or liposomes (c) when placed in water.

**a.** Micelle

**b.** Bilayer

Polar group (hydrophilic)

Nonpolar group (hydrophobic)

**c.** Liposome

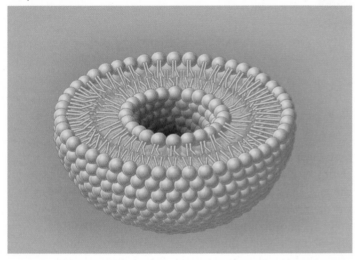

membranes are "self-healing." Small tears in a membrane are rapidly sealed by a spontaneous rearrangement of the lipids surrounding the damaged region, due to the tendency of water to exclude nonpolar molecules.

**? CASE 1 The First Cell: Life's Origins**
**How did the first cell membranes form?**
We have seen that the phospholipid bilayer forms spontaneously in water, arranged in such a way that the hydrophobic fatty acids do not interact with water and the polar head groups do interact with water. The bilayer structure forms spontaneously as long as the concentration of free phospholipids is high enough and the pH of the solution is similar to that of a cell. The pH is important because it ensures that the head groups are in their ionized (charged) form and thus suitably hydrophilic. Thus, if phospholipids are added to a test tube of water at neutral pH, they spontaneously form enclosed bilayer structures called **liposomes** (**Fig. 5.3c**). An enzyme is not required to catalyze

any chemical reaction. The membrane enclosure forms by a process of self-assembly that depends only on the chemical properties of the lipids. As the liposomes form, they may capture macromolecules present in solution.

Such a process may have been at work in the early evolution of life on Earth. Experiments show that liposomes can form and re-form in environments like tidal flats where wetting and drying occur repeatedly. The liposomes can even grow, incorporating more and more lipids from the environment. In addition, the liposomes can incorporate nucleic acids and other molecules into their interiors. Depending on their chemical composition, early membranes might have been either leaky or almost impervious to the molecules of life. Over time, they evolved in such a way as to allow at least limited molecular traffic between the environment and cell interior. At some point, new lipids no longer had to be incorporated from the environment. Instead, proteins guided lipid synthesis within the cell, although how this switch to synthesis happened remains uncertain.

All evidence suggests that membranes formed originally by straightforward physical processes, but that their composition and function evolved over time. François Jacob once said that evolution works more like a tinkerer than an engineer, modifying already existing materials rather than designing systems from scratch. It seems that the evolution of membranes is no exception to this pattern.

### Cell membranes are dynamic.

Lipids freely associate with one another because of extensive van der Waals forces between their fatty acid tails (Chapter 2). These weak interactions are easily broken and re-formed, so lipid molecules are able to move within the plane of the membrane. For example, a single phospholipid can move across the entire length of a bacterial cell in less than a second. Lipids can also rapidly rotate around their vertical axis, and individual fatty acid chains are able to flex, or bend. As a result, membranes are dynamic, forming and re-forming continually during the lifetime of a cell.

Because membrane lipids are able to move in the plane of the membrane, the membrane is said to be **fluid.** The degree of membrane fluidity depends on which types of lipid make up the membrane. In a single layer of the lipid bilayer, most of the van der Waals and hydrophobic interactions occur between the hydrophobic tails of lipids. The strength of these interactions depends on the length of the fatty acid tails and the presence of double bonds between neighboring carbon atoms. The longer the fatty acid tails, the more surface is available to participate in van der Waals interactions. The tighter packing that results tends to reduce lipid mobility. Likewise, saturated fatty acid tails, which have no double bonds, are straight and tightly packed—again reducing

**FIG. 5.4**     **Saturated and unsaturated fatty acids in phospholipids.** The composition of cell membranes affects the tightness of packing.

**a.** Stearic acid, $CH_3(CH_2)_{16}COOH$

**b.** Oleic acid, $CH_3(CH_2)_7CH = CH(CH_2)_7COOH$

Saturated fatty acid chains lack double bonds, resulting in phospholipids with a straight structure that favors tight packing.

Unsaturated fatty acids have one or more double bonds that introduce kinks in the phospholipids, reducing the tightness of packing.

mobility (**Fig. 5.4a**). The double bonds in unsaturated fatty acids introduce kinks in the fatty acid tails, reducing the tightness of packing and enhancing lipid mobility in the membrane (**Fig. 5.4b**).

→   **Quick Check 1**  Animal fats tend to be solid at room temperature, whereas plant and fish oils tend to be liquid. Both contain fatty acids. Can you predict which type of fat contains saturated fatty acids, and which type contains unsaturated fatty acids?

In addition to phospholipids, cell membranes often contain other types of lipid, and these can also influence membrane fluidity. For example, **cholesterol** is a major component of animal cell membranes, representing about 30% by mass of the membrane lipids. Like phospholipids, cholesterol is amphipathic, with both hydrophilic and hydrophobic groups in the same molecule. In the case of cholesterol, the hydrophilic region is simply a hydroxyl group (OH) and the hydrophobic region consists of four interconnected carbon rings with an attached hydrocarbon chain (**Fig. 5.5**). This structure allows cholesterol to insert into the lipid bilayer. Specifically, the hydrophilic head group of cholesterol interacts with the hydrophilic head group of phospholipids, while the ring structure participates in van der Waals interactions with the fatty acid chains.

Cholesterol increases or decreases membrane fluidity depending on temperature. At normal temperatures typically found in a cell, the interaction of the rigid ring structure of cholesterol with the phospholipid fatty acid tails reduces the mobility of the phospholipids and hence the fluidity of the membrane. However, at low temperatures, cholesterol prevents phospholipids from packing tightly with other phospholipids and therefore increases membrane fluidity. Thus, cholesterol helps maintain a consistent state of membrane fluidity by preventing dramatic transitions from a fluid to solid state.

For many decades, it was thought that the various types of lipid found in the membrane were randomly distributed throughout the bilayer. However, more recent studies show that specific types of lipid sometimes assemble into defined patches called **lipid rafts.** For example, sphingolipids (a type of lipid) cluster in discrete patches in the membrane. Furthermore, cholesterol and other membrane components such as proteins appear to accumulate in some of these regions. Thus, membranes are not always a uniform fluid bilayer, but instead contain regions with discrete components.

Although lipids are free to move in the plane of the membrane, the spontaneous transfer of a lipid between layers of the bilayer, known as lipid flip-flop, is very rare. This is not surprising, since flip-flop requires the hydrophilic head group to pass through the hydrophobic interior of the membrane. As a result, there is little exchange of components between

**FIG. 5.5** **Cholesterol in the lipid bilayer.** Cholesterol molecules embedded in the lipid bilayer affect the fluidity of the membrane.

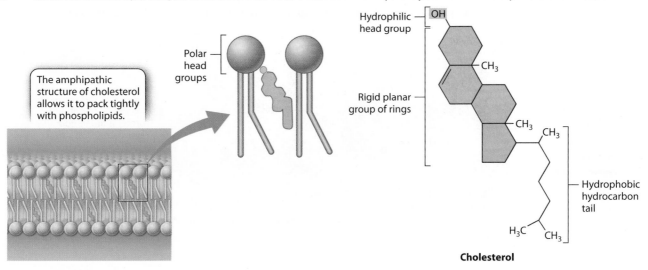

Cholesterol

the two layers of the membrane, which in turn allows the two layers to differ in composition. In fact, in many membranes, different types of lipid are present primarily in one layer or the other.

## Proteins associate with cell membranes in different ways.

Most membranes contain proteins as well as lipids. For example, the cell membrane of a red blood cell contains a large variety of proteins, representing as much as 50% of the membrane by mass. These membrane proteins serve different functions (**Fig. 5.6**). Some act as **transporters,** moving ions or other molecules across the membrane. Transporters include **channels** that allow movement of molecules through them, and **carriers** that facilitate movement. Other membrane proteins act as **receptors** that allow the cell to receive signals from the environment. Still others are **enzymes** that catalyze chemical reactions or **anchors** that attach to other proteins that help to maintain cell structure and shape.

These various kinds of membrane protein can be classified into two groups depending on how they associate with the

**FIG. 5.6** **Functions of membrane proteins.**

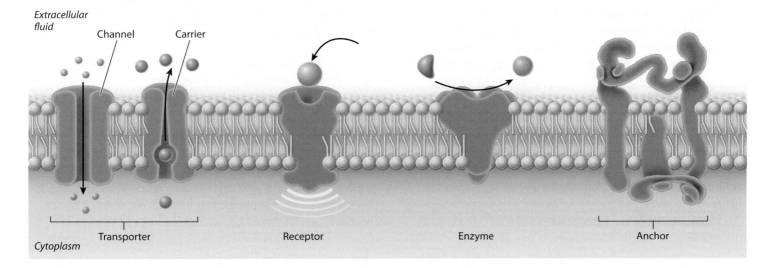

**FIG. 5.7**     **Integral and peripheral membrane proteins.** Integral membrane proteins are permanently associated with the membrane. Peripheral membrane proteins are temporarily associated with one of the two lipid bilayers or with an integral membrane protein.

membrane (**Fig. 5.7**). **Integral membrane proteins** are permanently associated with cell membranes and cannot be separated from the membrane experimentally without destroying the membrane itself. **Peripheral membrane proteins** are temporarily associated with the lipid bilayer or with integral membrane proteins through weak noncovalent interactions. They are easily separated from the membrane by simple experimental procedures that leave the structure of the membrane intact.

Most integral membrane proteins are **transmembrane proteins** that span the entire lipid bilayer, as shown in Fig. 5.7. These proteins are composed of three regions: two hydrophilic regions, one protruding from each face of the membrane, and a connecting hydrophobic region that spans the membrane. The hydrophilic region on the internal side of the membrane often interacts with other proteins in the cytoplasm of the cell, whereas the hydrophilic region on the external side can interact with signaling molecules.

Peripheral membrane proteins may be associated with either the internal or external side of the membrane (Fig. 5.7). These proteins interact either with the polar heads of lipids or with integral membrane proteins via weak noncovalent interactions such as hydrogen bonds. Peripheral membrane proteins are only transiently associated with the membrane and can play a role in transmitting information received from external signals. Other peripheral membrane proteins limit the ability of transmembrane proteins to move within the membrane and assist proteins to cluster in lipid rafts.

Proteins, like lipids, are free to move in the membrane. How do we know this? The mobility of proteins in the cell membrane can be demonstrated using an elegant experimental technique called fluorescence recovery after photobleaching, or FRAP (**Fig. 5.8**).

In this technique, proteins embedded in the cell membrane are labeled with fluorescent dye molecules. Labeling an entire membrane creates a fluorescent cell that can be visualized using a fluorescent microscope. A laser is then used to bleach the fluorescent dye molecules in a small area of the cell membrane, leaving a nonfluorescent spot on the surface of the cell. If proteins in the cell membrane were not capable of movement, the bleached area would remain nonfluorescent. However, over time fluorescence appears in the bleached area, telling us that fluorescent proteins that were not bleached moved into the bleached area.

The idea that lipids and proteins coexist in the membrane, and that both are able to move in the plane of the membrane, led American biologists S. Jonathan Singer and Garth Nicolson to propose the **fluid mosaic model** in 1972. According to this model, the lipid bilayer is a fluid structure that allows molecules to move laterally within the membrane, and is a mosaic (a mixture) of two types of molecules, lipids and proteins.

## 5.2 THE PLASMA MEMBRANE AND CELL WALL

All cells are enclosed by a **plasma membrane,** also called the cell membrane. The plasma membrane is a fundamental, defining feature of all cells. It is the boundary that defines the space of the cell, separating its internal contents from the surrounding environment. But the plasma membrane is not simply a passive boundary or wall. Instead, it serves an active and important function. The environment outside the cell is changing all the time. In contrast, the internal environment of a cell operates within a persistent and narrow window of conditions, such as pH

**FIG. 5.8**

# Do proteins move in the plane of the membrane?

**BACKGROUND** Fluorescent recovery after photobleaching (FRAP) is a technique used to measure mobility of molecules in the plane of the membrane. A fluorescent dye is attached to proteins embedded in the cell membrane in a process called labeling. A laser is then used to bleach a small area of the membrane.

**HYPOTHESIS** If membrane components such as proteins move in the plane of the membrane, the bleached spot should become fluorescent over time as unbleached fluorescent molecules move into the bleached area. If membrane components do not move, the bleached spot should remain intact.

**EXPERIMENT AND RESULTS**

Fluorescent molecules

Laser beam

Bleached area

The membrane is labeled with fluorescent molecules, resulting in a uniform fluorescence over the entire membrane.

Photobleaching causes a bleached area where there is no fluorescence.

Fluorescence gradually returns to the bleached area because of the movement of fluorescent proteins into the area.

Fluorescence (in bleached area)

Before bleaching

Laser bleaching

Recovery

Time

**CONCLUSION** The gradual recovery of fluorescence in the bleached area indicates that proteins move in the plane of the membrane.

**SOURCE** Peters, R., J. Peters, K.H. Tews, and W. Bähr. 1974. "A Microfluorimetric Study of Translational Diffusion in Erythrocyte Membranes." *Biochim Biophys Acta* 367:282–294.

**FIG. 5.9**   Diffusion, the movement of molecules due to random motion. Net movement of molecules only results when there are concentration differences.

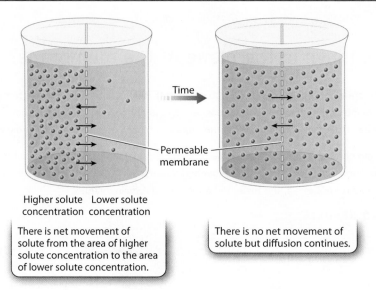

Time

Permeable membrane

Higher solute concentration    Lower solute concentration

There is net movement of solute from the area of higher solute concentration to the area of lower solute concentration.

There is no net movement of solute but diffusion continues.

or salt concentration. It is the plasma membrane that actively maintains intracellular conditions compatible with life.

In addition, the cells of many different groups of organisms have a **cell wall** external to the plasma membrane. The cell wall plays an important role in maintaining the shape and internal composition of these cells. In this section, we consider the functions performed by these two key structures.

### The plasma membrane maintains homeostasis.

The active maintenance of a constant environment is known as **homeostasis,** and it is a critical attribute of cells and of life itself. Chemical reactions and protein folding, for example, are carried out efficiently only within a narrow range of conditions. How does the plasma membrane maintain homeostasis? The answer is that it acts as a **selective barrier.** This means that the plasma membrane lets some molecules in and out freely; it lets others in and out only under certain conditions; and it prevents still other molecules from passing through at all.

The membrane's ability to act as a selective barrier is the result of the combination of lipids and embedded proteins that make it up. The hydrophobic lipid bilayer prevents ions as well as charged or polar molecules from diffusing freely across the plasma membrane. Furthermore, many macromolecules such as proteins and polysaccharides are too large to cross the plasma membrane on their own. At the same time, gases, other lipids, and small polar molecules freely move across the lipid bilayer. Protein transporters in the membrane allow the export and import of molecules, including certain ions and nutrients, that cannot cross the cell membrane on their own.

The identity and abundance of these membrane-associated proteins vary among cell types, reflecting the specific functions of different cells. For example, cells in your gut contain membrane transporters that specialize in the uptake of glucose, while nerve cells have different types of ion channel that are involved in electrical signaling.

### Passive transport involves diffusion.

The simplest form of movement into and out of cells is passive transport. Passive transport works by **diffusion,** which is the random movement of molecules. When there is a concentration difference (that is, a concentration gradient) in the distribution of a molecule, with areas of higher and lower concentrations, diffusion results in net movement of the molecule from an area of higher concentration to an area of lower concentration (**Fig. 5.9**). However, diffusion occurs even in the absence of concentration differences, due to the random motion of molecules, but in this case there is no net movement of the substance.

Some molecules diffuse freely across the plasma membrane as a result of differences in concentrations between the inside and outside of a cell. Oxygen and carbon dioxide, for example, move into and out of the cell in this way. Certain hydrophobic molecules, such as triacylglycerols (Chapter 2), are also able to diffuse through the cell membrane, which is not surprising since the lipid bilayer is likewise hydrophobic.

Some molecules that cannot move across the lipid bilayer directly can move passively toward a region of lower concentration through protein channels or carriers. When a molecule moves by diffusion through a membrane protein and bypasses the lipid bilayer, the process is called **facilitated diffusion.** Diffusion and facilitated diffusion both result from the random motion of molecules, and net movement of the substance occurs when there are concentration differences (**Fig. 5.10**). In the case of facilitated diffusion, the molecule moves through a membrane protein channel or carrier, whereas in the case of simple diffusion, the molecule moves directly through the lipid bilayer.

**FIG. 5.10**  **Simple diffusion and facilitated diffusion through the cell membrane.**

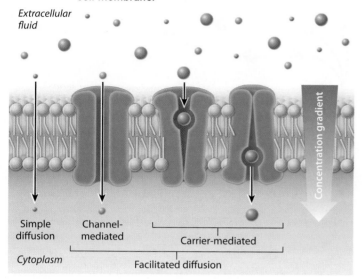

Extracellular fluid

Concentration gradient

Simple diffusion

Channel-mediated

Carrier-mediated

Facilitated diffusion

Cytoplasm

Although some membrane proteins provide a channel or opening between the inside and outside of the cell, most are carriers selective for specific molecules or are even "gated," opening and closing only in response to the binding of specific molecules or some other signal. These membrane proteins exist in two shapes or conformations, one that is open to one side of the cell, and another that is open to the other side of the cell. Binding of the transported molecule induces a conformational change in the membrane protein, allowing the molecule to be transported across the lipid bilayer, as shown on the right in Fig. 5.10.

Up to this point, we have focused our attention on the movement of molecules (solutes) in water (solvent). We can take a different perspective and focus instead on water movement. Water itself also moves into and out of cells by passive transport.

Although the plasma membrane is hydrophobic, water molecules are small enough to move passively through the membrane to a limited extent by simple diffusion. In addition, many cells have specific protein channels, known as **aquaporins,** which allow water to flow through the plasma membrane more readily by facilitated diffusion.

The diffusion of water is known as **osmosis.** As in any form of diffusion, water moves from regions of higher *water* concentration to regions of lower *water* concentration (**Fig. 5.11**). Because water is a solvent within which nutrients such as glucose or ions such as sodium or potassium are dissolved, water concentration drops as solute concentration rises. Therefore, it is sometimes easier to think about water moving from regions of lower *solute* concentration to regions of higher *solute* concentration. Either way, the direction of water movement is the same.

→   **Quick Check 2**  A container is divided into two compartments by a membrane that is fully permeable to water and small ions. Water is added to one side of the membrane (side A), and a 5% solution of sodium chloride (NaCl) is added to the other (side B). In which direction will water molecules move? In which direction will sodium and chloride ions move? When the concentration is equal on both sides, will diffusion stop?

**Primary active transport uses the energy of ATP.**
Passive transport works to the cell's advantage only if the concentration gradient is in the right direction, from higher on the outside to lower on the inside for nutrients that the cell needs to take in, and from higher on the inside and lower on the outside for wastes that the cell needs to export. However, many of the molecules that cells require are in low concentration in the environment. Although some of these molecules can be synthesized by the cell, others need to be taken up from the environment and concentrated inside the cell. In other

**FIG. 5.11**  **Osmosis.** Osmosis is the diffusion of water across a semipermeable membrane from an area of lower solute concentration to an area of higher solute concentration.

Solute

Osmosis  Water

Semipermeable membrane

Time

Higher solute concentration, lower water concentration

Lower solute concentration, higher water concentration

Membrane allows passage of water but not solute.

Equal solute and water concentration

**FIG. 5.12    Primary active transport.** The sodium-potassium pump is a membrane protein that uses the energy stored in ATP to move sodium and potassium ions against their concentration gradients.

words, cells have to move these substances from areas of lower concentration to areas of higher concentration. The "uphill" movement of substances against a concentration gradient, called **active transport,** requires energy. The transport of many kinds of molecules across membranes requires energy, either directly or indirectly. In fact, most of the energy used by a cell goes into keeping the inside of the cell different from the outside, a function carried out by proteins in the plasma membrane.

During active transport, cells move substances through transport proteins embedded in the cell membrane. Some of these proteins act as pumps, using energy directly to move a substance into or out of a cell. A good example is the sodium-potassium pump (**Fig. 5.12**). Within cells, sodium is kept at concentrations much lower than in the exterior environment; the opposite is true of potassium. Therefore, both sodium and potassium have to be moved against a concentration gradient. The sodium-potassium pump actively moves sodium out of the cell and potassium into the cell. This movement of ions takes energy, which comes from the chemical energy stored in ATP. Active transport that uses energy directly in this manner is called **primary active transport.** Note that the sodium and potassium ions move in opposite directions. Protein transporters that work in this way are sometimes referred to as antiporters. Other transporters move two molecules in the same direction. These transport proteins are referred to as symporters or cotransporters.

### Secondary active transport is driven by an electrochemical gradient.

Active transport can also work in another way. Because small ions cannot cross the lipid bilayer, many cells use a transport protein

to build up the concentration of a small ion on one side of the membrane. The resulting concentration gradient stores energy that can be harnessed to drive the movement of other substances across the membrane against their concentration gradient.

For example, some cells actively pump protons (H⁺) across the cell membrane using ATP (**Fig. 5.13a**). As a result, the concentration of protons is higher on one side of the membrane and lower on the other side. In other words, the pump generates a concentration gradient, also called a chemical gradient because the entity forming the gradient is a chemical (**Fig. 5.13b**). We have already seen that concentration differences favor the movement of protons back to the other side of the membrane. However, the lipid bilayer blocks the movement of protons to the other side and therefore stores energy, just like a dam or battery.

In addition to the chemical gradient, another force favors the movement of protons back across the membrane: a difference in charge. Because protons carry a positive charge, the side of the membrane with more protons is more positive than the other side. This difference in charge is called an electrical gradient. Protons (and other ions) move from areas of like charge to areas of unlike charge, driven by an electrical gradient. Together, the charge and chemical gradients are known as an **electrochemical gradient** (Fig. 5.13b).

If protons are then allowed to pass through the cell membrane by a transport protein, they will move down their electrochemical gradient toward the region of lower proton concentration. These transport proteins can use the movement of protons to drive the movement of other molecules against their concentration gradient (**Fig. 5.13c**). The movement of protons is always from regions of higher to lower concentration, while the movement of the coupled molecule is from regions of lower to higher

**FIG. 5.13    Secondary active transport.** Protons are pumped across a membrane by primary active transport (a), resulting in an electrochemical gradient (b), which drives the movement of another molecule against its concentration gradient (c).

**a.**

Extracellular fluid

Proton

ADP

ATP

Cytoplasm

Protons are pumped across membrane by primary active transport.

**b.**

Proton concentration gradient

Electrical gradient

The proton pump generates an electrochemical gradient, with a higher concentration of protons outside the cell and a lower concentration of protons inside the cell.

**c.**

An antiporter uses the proton electrochemical gradient to move a different molecule out of the cell against its concentration gradient.

concentration. Because the movement of the coupled molecule is driven by the movement of protons and not by ATP directly, this form of transport is called **secondary active transport.** Secondary active transport uses the energy of an electrochemical gradient to drive the movement of molecules; by contrast, primary active transport uses the energy of ATP directly.

The use of an electrochemical gradient as a temporary energy source is a common cellular strategy. For example, cells use the sodium electrochemical gradient generated by the sodium-potassium pump to transport glucose and amino acids into cells. In addition, cells use the proton electrochemical gradient to move other molecules and, as we discuss below and in Chapter 7, to synthesize ATP.

### Many cells maintain size and composition using active transport.

Many cells use active transport to maintain their size and composition. Consider human red blood cells placed in a variety of different solutions (**Fig. 5.14**). If a red blood cell is placed in a hypertonic solution (one with a higher solute concentration than that inside the cell), water leaves the cell by osmosis and the cell shrinks. By contrast, if a red blood cell is placed in a

hypotonic solution (one with a lower solute concentration than that inside the cell), water moves into the cell by osmosis and the cell lyses, or bursts. Animal cells solve the problem of water movement in part by maintaining the intracellular fluid isotonic (that is, at the same solute concentration) as the extracellular fluid. Cells use the active transport of ions to maintain equal concentrations inside and out, and the sodium-potassium pump plays an important role in keeping the inside of the cell isotonic with the extracellular fluid.

→    **Quick Check 3** In the absence of the sodium-potassium pump, the extracellular solution becomes hypotonic relative to the inside of the cell. Poisons such as the snake venom ouabain can interfere with the action of the sodium-potassium pump. What are the consequences for the cell?

Human red blood cells avoid shrinking or bursting by maintaining an intracellular environment isotonic with the extracellular environment, the blood. But what about a single-celled organism, like *Paramecium*, swimming in a freshwater lake? In this case, the extracellular environment is hypotonic

**FIG. 5.14    Changes in red blood cell shape due to osmosis.** Red blood cells shrink, swell, or burst because of net water movement (osmosis) driven by differences in solute concentration between the inside and the outside of the cell.

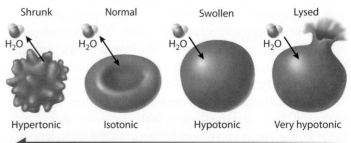

Shrunk

Normal

Swollen

Lysed

H₂O

H₂O

H₂O

H₂O

Hypertonic

Isotonic

Hypotonic

Very hypotonic

Solute concentration in extracellular space

**FIG. 5.15    The plant cell wall and vacuoles.** The plant cell wall is a rigid structure that maintains the shape of the cell. The pressure exerted by water absorbed by vacuoles also provides support.

Vacuoles exert hydrostatic pressure on the cell walls and provide rigidity to the plant cell.

Dehydration collapses vacuoles, relieves the hydrostatic pressure, and results in wilting.

compared to the concentration in the cell's interior. As a result, *Paracemium* faces the risk of bursting caused by water moving in by osmosis. *Paramecium* and some other single-celled organisms contain **contractile vacuoles** that solve this problem. Contractile vacuoles are compartments that take up excess water from inside the cell and then, by contraction, expel it into the external environment. The mechanism by which water moves into the contractile vacuole differs depending on the organism. The contractile vacuole of some organisms takes in water through aquaporins, while other organisms have proton pumps that first move protons into the contractile vacuole, with water following by osmosis.

### The cell wall provides another means of maintaining cell shape.

Most organisms do not contain contractile vacuoles. How do organisms such as plants, fungi, and bacteria maintain cell size and composition? A key feature of these organisms is the cell wall, a rigid structure that surrounds the plasma membrane (**Fig. 5.15**). The cell wall plays a critical role in the maintenance of cell shape and volume. When Hooke looked at cork through his microscope, what he saw was not living cells but the remains of cell walls.

The cell wall provides structural support and protection to the cell. Because the cell wall is fairly rigid and resists expansion, it allows pressure to build up in a cell when it absorbs water. The force exerted by water pressing against an object results in hydrostatic pressure, or **turgor pressure.** The pressure exerted by water provides structural support for many organisms that is similar in function to the support provided by the skeletons or shells of animals. In addition, many plant cells have another structure, called the **vacuole** (distinct from the contractile vacuole), that also absorbs water and contributes to turgor pressure (Fig. 5.15). It is therefore easy to understand why plants wilt when dehydrated—the loss of water from the vacuoles reduces turgor pressure and the tissue can no longer maintain rigidity.

The cell wall is made up of many different components, including carbohydrates and proteins. The specific components differ depending on the organism. The plant cell wall is composed of polysaccharides, one of the best known of which is cellulose, a polymer of the sugar glucose. Cellulose is the most abundant biological material in nature. Many types of algae have cell walls made up of cellulose, as in plants. Fungi have cell walls made of chitin, another polymer based on sugars. In bacteria, the cell wall is made up primarily of peptidoglycan, a mixture of amino acids and sugars. Animal cells do not have walls.

## 5.3 THE INTERNAL ORGANIZATION OF CELLS

In addition to the plasma membrane, many cells contain membrane-bound regions within which specific functions are carried out. Such a cell can be compared to a large factory with many rooms and different departments. Each department has a specific function and internal organization that contribute to the overall "life" of the factory. In this section, we give an overview of two broad classes of cells that can be distinguished by the presence or absence of these membrane-enclosed compartments.

### Eukaryotes and prokaryotes differ in internal organization.

All cells have a plasma membrane and contain genetic material. In some cells, the genetic material is housed in a membrane-bound space called the **nucleus.** Cells can be divided into two classes based on the presence or absence of a nucleus (Chapter 1). **Prokaryotes,** including bacteria and archaea, lack a nucleus; **eukaryotes,** including plants, animals, fungi, and protists, have a nucleus (**Fig. 5.16**).

Although the two groups are defined by the presence or absence of a nucleus, they differ in many other aspects as well. For example, we saw in Chapter 3 that promoter recognition during transcription is different in prokaryotes and in eukaryotes. In addition, there are differences in the specific types of lipid that make up their cell membranes. In mammals,

---

**FIG. 5.16  Prokaryotic and eukaryotic cells.** Prokaryotic cells lack a nucleus and extensive internal compartmentalization. Eukaryotic cells have a nucleus and extensive internal compartmentalization.

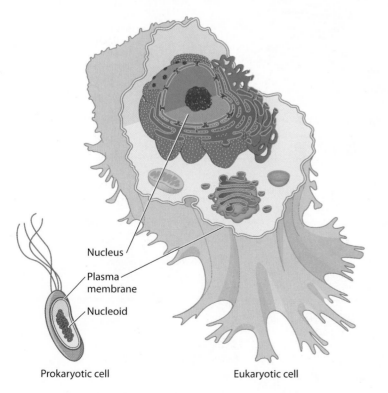

Nucleus

Plasma membrane

Nucleoid

Prokaryotic cell          Eukaryotic cell

as we have seen, cholesterol is present in cell membranes. Cholesterol belongs to a group of chemical compounds known as sterols, which are molecules containing a hydroxyl group attached to a four-ringed structure. In eukaryotes other than mammals, diverse sterols are synthesized and present in cell membranes. Most prokaryotes do not synthesize sterols, but some synthesize compounds called hopanoids. These five-ringed structures are thought to serve a function similar to that of cholesterol in mammalian cell membranes. Additional differences between prokaryotes and eukaryotes are discussed below.

### Prokaryotic cells lack a nucleus and extensive internal compartmentalization.

Prokaryotes do not have a nucleus—that is, there is no physical barrier separating the genetic material from the rest of the cell. Instead, the DNA is concentrated in a discrete region of the cell interior known as the **nucleoid.** Bacteria often contain additional small circular molecules of DNA known as **plasmids** that carry a small number of genes. Plasmids are commonly transferred between bacteria through threadlike, hollow structures known as **pili** (singular, pilus), which extend from one cell to another. Genes for antibiotic resistance are commonly transferred in this way, which accounts for the quick spread of antibiotic resistance among bacterial populations.

Although the absence of a nucleus is a defining feature of prokaryotes, other features also stand out. For example, prokaryotes are small, just 1–2 micrometers (1/1000 of a meter) in diameter or smaller. By contrast, eukaryotic cells are commonly much larger, on the order of 10 times larger in diameter and 1000 times larger in volume. The small size of prokaryotic cells means that they have a relatively high ratio of surface area to volume, which makes sense for an organism that absorbs nutrients from the environment. In other words, there is a large amount of membrane surface area for absorption relative to the volume of the cell that it serves. In addition, most prokaryotes lack the extensive internal organization characteristic of eukaryotes.

### Eukaryotic cells have a nucleus and specialized internal structures.

Eukaryotes are defined by the presence of a nucleus, which houses the vast majority of the cell's DNA. The nuclear membrane allows for more complex regulation of gene expression than is possible in prokaryotic cells. In eukaryotes, DNA is transcribed to RNA inside the nucleus, but the RNA molecules carrying the genetic message travel from inside to outside the nucleus, where they instruct the synthesis of proteins.

Eukaryotes have a remarkable internal array of membranes. These membranes define compartments, called **organelles,** that divide the cell contents into smaller spaces specialized for different functions. **Fig. 5.17a** shows a typical animal cell (a macrophage) with various organelles. The **endoplasmic reticulum (ER)** is involved in the synthesis of proteins and lipids. The **Golgi apparatus** modifies proteins and lipids produced by the ER and acts as a sorting station as they move to their final destinations. **Lysosomes** contain enzymes that break down macromolecules such as proteins, nucleic acids, lipids, and complex carbohydrates. **Mitochondria** (singular, mitochondrion) are specialized organelles that harness energy for the cell. Many

cell membranes that define these organelles are also associated with a protein scaffold called the **cytoskeleton** that helps cells to maintain their shape and serves as a network of tracks for the movement of substances within cells. Some cells even move with the help of the cytoskeleton.

**Fig. 5.17b** shows a typical plant cell. In addition to the organelles described above, plant cells have a cell wall outside the plasma membrane, vacuoles specialized for water uptake, and **chloroplasts** that convert energy of sunlight into chemical energy.

**FIG. 5.17    An animal cell and a plant cell.** Animal and plant cells have many cell components in common.

**a.** Typical features of an animal cell

The **nucleus** is the storehouse for the cell's genetic information and the site for RNA synthesis.

The **endoplasmic reticulum** is involved in both protein and lipid synthesis.

**Lysosomes** degrade macromolecules.

The **Golgi apparatus** modifies and sorts proteins and lipids as they move to their final destinations in or out of the cell.

**Mitochondria** produce most of the ATP that serves as the energy currency of the cell.

The **plasma membrane** is composed of phospholipids and proteins, and regulates the passage of materials into and out of the cell.

The **cytoskeleton** is a network of protein filaments and other associated proteins that provide the cell with an internal structural framework.

The entire contents of a cell other than the nucleus make up the **cytoplasm.** The region of the cell inside the plasma membrane but outside the organelles is referred to as the **cytosol.** This is the jelly-like internal environment surrounding the organelles.

In the next two sections, we consider these organelles in more detail, focusing on the role of membranes in forming distinct compartments within the cell. The cytoskeleton is discussed in Chapter 10.

## 5.4 THE ENDOMEMBRANE SYSTEM

In eukaryotes, the total surface area of intracellular membranes is about tenfold greater than that of the plasma membrane. This high ratio of internal membrane area to plasma membrane area underscores the significant degree to which a eukaryotic cell is divided into internal compartments. Many of the organelles inside cells are not distinct, isolated entities,

**b.** Typical features of a plant cell

The plant **cell wall** is a rigid barrier composed of polysaccharides.

**Vacuoles** are water-filled organelles that contribute to the structural rigidity of plants by maintaining turgor pressure against cell walls.

Plasma membrane

Mitochondria

**Chloroplasts** enable plant cells to harness the energy of sunlight to synthesize sugars.

Nucleus

Endoplasmic reticulum

Golgi apparatus

Cytoskeleton

**FIG. 5.18** **The endomembrane system.** The endomembrane system is a series of membrane-bound internal compartments in eukaryotic cells.

Nuclear envelope

Endoplasmic reticulum

Golgi apparatus

Vesicle

Lysosome

Plasma membrane

Exocytosis

Endocytosis

fusing of a vesicle between these organelles. Similarly, a molecule associated with the ER membrane can move to the Golgi membrane or the plasma membrane by vesicle transport. However, molecules in the cytosol are in a different physical space, separated by membranes of the endomembrane system. This physical separation allows specific functions to take place within the spaces defined by the membranes and also within the membrane itself.

In spite of forming a continuous and interconnected system, the various compartments have unique properties and maintain distinct identities determined by which lipids and proteins are present in their membranes. Recall that membranes are not fixed but dynamic, able to change their lipid and protein composition over time.

As we have seen, vesicles can bud off and fuse with components of the endomembrane system, creating a set of interconnected spaces. They can even fuse with the plasma membrane. This process, called **exocytosis,** provides a way for a vesicle to empty its contents to the extracellular space or to deliver proteins embedded in the vesicle membrane to the plasma membrane (Fig. 5.18). The process also works in reverse: A vesicle can bud off

but instead communicate with one another. In fact, the membranes of these organelles are either physically connected by membrane "bridges" or communicate by the budding off and fusing of **vesicles,** small membrane-enclosed sacs that transport substances. In total, these membranes make up the **endomembrane system.** The endomembrane system includes the nuclear envelope, endoplasmic reticulum, Golgi apparatus, lysosomes, the plasma membrane, and the vesicles that move between them (**Fig. 5.18**).

Extensive internal membranes are not common in prokaryotic cells. However, photosynthetic bacteria do have internal membranes that are specialized for harnessing light energy (Chapters 8 and 26).

**The endomembrane system compartmentalizes the cell.**
Because of the selective permeability of cell membranes, the endomembrane divides the interior of a cell into two distinct "worlds," one inside the spaces defined by these membranes and one outside these spaces. Note that a molecule within the interior space of the ER can stay in the ER or end up within the Golgi apparatus or even outside the cell by the budding off and

**FIG. 5.19** **A surface view of the nuclear envelope.** The nucleus is surrounded by a double membrane and houses the cell's DNA.

Nucleus

Nuclear pores

1 µm

The nuclear envelope is perforated by membrane protein openings called nuclear pores. Small molecules and ions can passively diffuse through the pores, but large proteins and RNA require active transport.

from the plasma membrane, bringing material from outside the cell into a vesicle, which can then fuse with other organelles. This process is called **endocytosis.** Together, exocytosis and endocytosis provide a way to move material into and out of cells without passing through the cell membrane.

### The nucleus houses the genome and is the site of RNA synthesis.

The **nucleus** stores DNA, the genetic material that encodes the information necessary for all the activities and structures of the cell. The **nuclear envelope** defines the boundary of the nucleus (**Fig. 5.19**). It consists of two membranes, the inner and outer membranes, and each is a lipid bilayer with associated proteins.

These two membranes are continuous with each other at protein openings called **nuclear pores.** These pores act as gateways that allow molecules to move into and out of the nucleus, and thus are essential for the nucleus to communicate with the rest of the cell. In fact, the transfer of information

encoded by DNA depends on the movement of RNA molecules out of the nucleus, while the control of how and when this information is expressed depends on the movement of proteins into the nucleus. The nuclear envelope with its associated protein pores regulates which molecules move into and out of the nucleus.

### The endoplasmic reticulum is involved in protein and lipid synthesis.

The outer membrane of the nuclear envelope is physically continuous with the endoplasmic reticulum (ER), an organelle bounded by a single membrane (**Fig. 5.20**). The ER is a conspicuous feature of many eukaryotic cells, accounting in some cases for as much as half of the total amount of membrane. The ER produces and transports many of the lipids and proteins used inside and outside the cell. It is the site of production of most of the lipids that make up the various cell membranes. In addition, transmembrane proteins and proteins destined for the Golgi apparatus, lysosomes, or export out of the cell are synthesized in the ER.

---

**FIG. 5.20    The endoplasmic reticulum (ER).** The ER is a major site for lipid and protein synthesis. (Proteins are also synthesized in the cytoplasm.)

The rough ER is associated with ribosomes. Many proteins, including those that are destined for secretion, are synthesized by ribosomes associated with the rough ER.

The smooth ER lacks ribosomes and is the primary site of lipid synthesis.

Smooth endoplasmic reticulum

Rough endoplasmic reticulum

Ribosome bound to ER

Protein

Ribosome free in cytoplasm

Protein

Unlike the nucleus, which is a single rounded structure in the cell, the ER consists of a complex network of interconnected tubules and flattened sacs. The interior of the ER is continuous throughout and is called the **lumen.** As shown in Fig. 5.20, the ER has an almost mazelike appearance when sliced and viewed in cross section. The ER membrane is extensively convoluted, allowing a large amount of membrane surface area to fit within the cell. The amount of ER membrane in a cell varies among cells of different functions. For example, mucus-producing cells of the gut synthesize a large amount of protein for export and thus have an extensive ER, as do the cells of the pancreas that produce the protein insulin. In cells that do not secrete large quantities of protein, the ER can be quite small.

When viewed through an electron microscope, ER membranes have two different appearances (Fig. 5.20). In most cells, the majority of ER membranes have small, rounded particles associated with them that are exposed to the cytosol. This portion of the ER is referred to as **rough endoplasmic reticulum (RER).** The particles bound to the cytosolic face of the RER are **ribosomes.** Ribosomes are the sites of protein synthesis, in which amino acids are assembled into polypeptides guided by the information stored in mRNA (Chapter 4). Ribosomes can be free in the cytosol or associated with the ER membrane.

In most cells, there is a small amount of ER membrane that lacks ribosomes and is consequently called **smooth endoplasmic reticulum (SER)** (Fig. 5.20). Portions of the smooth ER membrane actively bud off to produce vesicles that are free to move in the cytosol. Since each vesicle is formed from a patch of ER membrane and encloses a portion of the ER lumen, vesicles are an effective means of moving proteins that are either embedded in the ER membrane or free floating inside.

SER is also the site of fatty acid and phospholipid biosynthesis. Thus, this type of ER predominates in cells specialized for the production of lipids. For example, cells that synthesize steroid hormones have a well-developed SER that produces large quantities of cholesterol and contains enzymes that convert cholesterol into steroid hormones.

## The Golgi apparatus modifies and sorts proteins and lipids.

Although it is not physically continuous with the ER, the Golgi apparatus is often the next stop for vesicles that bud off the ER. These vesicles carry lipids and proteins that are either located in the vesicle interior or embedded in their membranes. The movement of these vesicles from the ER to the Golgi apparatus and then to the rest of the cell is part of a biosynthetic pathway in which lipids and proteins are successively modified and delivered to their final destinations. The Golgi apparatus has three primary roles: (1) It further modifies proteins and lipids produced by the ER; (2) it acts as a sorting station as they move to their final destinations; and (3) it is the site where most of the cell's carbohydrates are synthesized.

Under the microscope, the Golgi apparatus looks like a series of flattened membrane sacs, which are called **cisternae** (**Fig. 5.21**). The cisternae, which are stacked, are surrounded by many small vesicles. These vesicles transport proteins from the ER to the Golgi apparatus, between the various cisternae, and between the Golgi apparatus and the plasma membrane or other organelles. Vesicles are therefore the primary means

**FIG. 5.21**   **The Golgi apparatus.** The Golgi apparatus sorts proteins and lipids to other organelles, the plasma membrane, or the cell exterior.

Vesicle

Cisternae

Golgi apparatus

The Golgi apparatus receives proteins and lipids from the ER and sorts them to other organelles or the plasma membrane. These proteins are transported in vesicles.

Soluble protein

Transmembrane protein

Carbohydrate

by which proteins and lipids move through the Golgi apparatus to their final destinations.

Enzymes within the Golgi apparatus chemically modify proteins and lipids as they pass through it. These modifications are sequential since each region of the Golgi apparatus contains a different set of enzymes that catalyze specific reactions. As a result, there is a general movement of vesicles from the ER through the Golgi apparatus and then to their final destinations.

An example of a chemical modification that occurs predominantly in the Golgi apparatus is glycosylation, in which sugars are covalently linked to lipids or specific amino acids of proteins. As these lipids and proteins move through the Golgi apparatus, sugars are added and trimmed in a sequential fashion. Glycoproteins are important components of the eukaryotic cell surface because their attached sugars can protect the protein from enzyme digestion by blocking access to the peptide chain. As a result, glycoproteins form a relatively flexible and protective coating over the plasma membrane. The distinctive shapes that sugars contribute to glycoproteins and glycolipids also allow cell surface components to be recognized specifically by other cells and molecules in the external environment. For example, human blood types (A, B, AB, and O) are distinguished by the particular sugars that are linked to proteins and lipids on the surface of red blood cells.

While traffic usually travels from the ER to the Golgi apparatus, there is also a small amount of traffic moving in the reverse direction, from the Golgi apparatus to the ER. This reverse pathway is important to retrieve ER or Golgi resident proteins that are accidentally moved forward, and to recycle membrane components.

### Lysosomes degrade macromolecules.

The ability of the Golgi apparatus to sort and dispatch proteins to particular destinations is dramatically illustrated by lysosomes. **Lysosomes** are specialized vesicles derived from the Golgi apparatus that degrade damaged or unneeded macromolecules. They contain a variety of enzymes that break down macromolecules such as proteins, nucleic acids, lipids, and complex carbohydrates (**Fig. 5.22**). Macromolecules destined for degradation are packaged by the Golgi apparatus into vesicles. The vesicles then fuse with lysosomes, delivering their contents to the lysosome interior.

**FIG. 5.22** **Lysosomes.** Lysosomes are vesicles that degrade macromolecules.

Enzymes

Lysosomes

The Golgi apparatus delivers enzymes that break down macromolecules to a lysosome. A vesicle with macromolecules also merges with the lysosome.

Proton pump

Protons

Transporter

Broken-down macromolecules

Proton pumps help maintain an acidic pH, while broken-down macromolecules are transported out of the lysosome by transporters.

The formation of lysosomes also illustrates the ability of the Golgi apparatus to sort key proteins. The enzymes inside of lysosomes are synthesized in the RER, sorted in the Golgi apparatus, and then packaged into lysosomes. In addition, the Golgi apparatus sorts and delivers specialized proteins that become embedded in lysosome cell membranes. These include proton pumps that keep the internal environment at an acidic pH of around 5, the optimum pH for the activity of the enzymes inside. Other proteins in the lysosomal membranes transport the breakdown products of macromolecules, such as amino acids and simple sugars, across the membrane to the cytosol for use by the cell.

The function of lysosomes underscores the importance of having separate compartments within the cell bounded by selectively permeable membranes. Many of a cell's enzymes and proteins would unfold and degrade if the entire cell were at a pH of 5, and lysosomal enzymes cannot function in the normal cellular environment, which has a pH of about 7. By restricting the activity of these enzymes to the lysosome, proteins and organelles in the cytosol are protected from degradation.

## Protein sorting directs proteins to their proper location in or out of the cell.

As we have seen, eukaryotic cells have many compartments, and different proteins function in different places, such as enzymes that act in lysosomes or transmembrane proteins embedded in the plasma membrane. **Protein sorting** is the process by which proteins end up where they need to be to perform their function. Protein sorting directs proteins to the cytosol, the lumen of organelles, the membranes of the endomembrane system, or even out of the cell entirely. Let's consider in more detail how proteins produced from free ribosomes and those produced from membrane-bound ribosomes on the RER are sorted to their final destinations in the cell.

Proteins produced on free ribosomes start off in the cytosol and are sorted to their final destination after translation. These proteins are often directed to their proper cellular compartments by means of particular amino acid sequences called **signal sequences.** As shown in **Fig. 5.23,** there are several types of signal sequences that direct proteins to different cellular compartments. Proteins with no signal sequence remain in the cytosol (Fig. 5.23a). Most proteins with a signal sequence at their amino ends are targeted to mitochondria or chloroplasts (Fig. 5.23b). Proteins targeted to the nucleus usually have signal sequences located internally (Fig. 5.23c). The signal sequence for the nucleus, also called a **nuclear localization signal,** enables proteins to move through pores in the nuclear envelope.

Proteins produced by ribosomes on the RER end up within the lumen of the endomembrane system or embedded in its membrane, or they may be secreted out of the cell. These proteins are sorted as they are translated: Proteins destined for the lumen or for secretion are fed into the ER lumen as they are synthesized,

whereas the proteins destined for membranes are inserted in the ER membrane as they are synthesized (**Fig. 5.24**).

The polypeptide chains of proteins can move into the lumen or membrane because of the presence of an amino-terminal signal sequence shown in **Fig. 5.25**. This signal sequence is recognized

---

**FIG. 5.23** **Signal sequences on proteins synthesized by free ribosomes.** Most proteins with no signal peptide remain in the cytosol (a). Other signal sequences direct proteins to mitochondria and chloroplasts (b) or to the nucleus (c).

**a.** No signal peptide

To cytosol

**b.** Amino-terminal signal

To chloroplast
To mitochondrion

**c.** Internal signal

To nucleus

---

**FIG. 5.24** Pathways for proteins destined to be secreted (left) or transported to the plasma membrane (right).

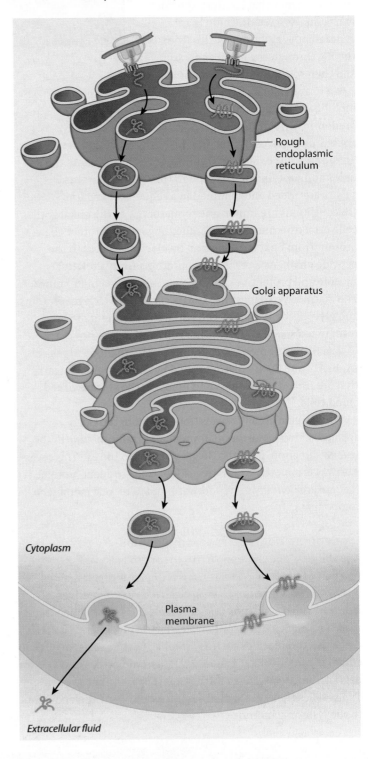

Rough endoplasmic reticulum

Golgi apparatus

Cytoplasm

Plasma membrane

Extracellular fluid

**FIG. 5.25** **Interaction of a signal sequence, signal recognition particle (SRP), and SRP receptor.** Binding of a signal recognition particle (SRP) halts translation, followed by docking of the ribosome on the ER membrane, release of the SRP, and continuation of translation.

**a.** The signal recognition particle (SRP) binds to a signal sequence in the amino-terminal end of the growing polypeptide and halts translation.

**b.** The SRP binds to the SRP receptor on the ER membrane.

**c.** The SRP receptor brings the ribosome to a transmembrane channel, the SRP dissociates, protein synthesis resumes, and the growing polypeptide chain is threaded through the channel.

**d.** The protein ends up in the lumen of the ER, where it may remain, be transported to the lumen of another organelle, or be secreted out of the cell.

almost immediately after synthesis by an RNA–protein complex known as a **signal-recognition particle (SRP).** The SRP binds to both the signal sequence and the ribosome and brings about a pause in translation (Fig. 5.25a). The SRP then binds with a receptor on the RER (Fig. 5.25b). The SRP dissociates and translation continues. The SRP receptor brings the ribosome to a channel through the hydrophobic membrane of the RER. Each growing polypeptide chain is threaded through the channel (Fig. 5.25c). A specific protease cleaves the signal sequence as it emerges in the lumen of the ER. The finished protein ends up in the lumen of the RER (Fig. 5.25d).

Whether the polypeptide chain ends up in the membrane or in the lumen depends on whether or not the polypeptide chain has a signal-anchor sequence in its interior. Proteins destined for the lumen of the ER, the Golgi apparatus, a lysosome, or the exterior of the cell have no signal-anchor sequence. These chains are fed completely through the channel into the lumen. Then, chaperone proteins assist with protein folding. Some proteins are retained in the ER, while others are transported in

vesicles to the Golgi apparatus. Some of these proteins end up being secreted by exocytosis, as shown on the left in Fig. 5.24.

Proteins destined to be embedded in membranes (shown on the right in Fig. 5.24) do contain signal-anchor sequences (**Fig. 5.26**). The growing polypeptide chain is transported across the

**FIG. 5.26** **Targeting of a transmembrane protein by means of a hydrophobic signal-anchor sequence.** Proteins with a signal-anchor sequence end up embedded in the membrane.

**a.** Proteins with signal-anchor sequences are threaded through a channel in the ER membrane until the signal-anchor sequence is encountered.

**b.** The ER channel releases the protein into the membrane.

**c.** When translation is completed, the protein remains in the membrane.

channel in the ER membrane until the signal-anchor sequence is encountered (Fig. 5.26a). In the case of transmembrane proteins, the signal-anchor sequence is hydrophobic and can become so intimately associated with the membrane that it exits the channel and diffuses laterally in the lipid bilayer (Fig. 5.26b). At this point, the ribosome disassociates from the channel while translation continues (Fig. 5.26c). When translation is completed and the polypeptide is released, the carboxyl end of the chain remains on the cytosolic side of the ER membrane, the amino end is in the ER lumen, and the membrane-bound region between them anchors both sides to the membrane. Transmembrane proteins such as these may stay in the membrane of the ER or end up in other internal membranes or the plasma membrane, where they serve as channels, pumps, receptors, or enzymes.

## 5.5 MITOCHONDRIA AND CHLOROPLASTS

The membranes of two organelles, **mitochondria** and **chloroplasts,** are not part of the endomembrane system. Both of these organelles are specialized to harness energy for the cell. Interestingly, they are both semi-autonomous organelles that grow and multiply independently of the other membrane-bound compartments, and they contain their own genomes. As we discuss in Chapter 27, the similarities between mitochondrial and chloroplast DNA and the DNA of certain bacteria have led scientists to conclude that these organelles originated as bacteria that were captured by a eukaryotic cell and, over time, evolved to their current function.

### Mitochondria provide the eukaryotic cell with most of its usable energy.

Mitochondria are organelles that harness energy from chemical compounds like sugars and convert it into ATP, which serves as the universal energy currency of the cell. ATP is able to drive the many chemical reactions in the cell. Mitochondria are present in nearly all eukaryotic cells.

Mitochondria are rod-shaped organelles with an outer membrane and a highly convoluted inner membrane whose folds project into the interior (**Fig. 5.27**). The inner mitochondrial membrane is where a proton electrochemical gradient is generated, and the energy stored in the gradient is used to synthesize ATP for use by the cell. In the process of breaking down sugar and synthesizing ATP, oxygen is consumed and carbon

**FIG. 5.27    Mitochondria.** Mitochondria synthesize most of the ATP required to meet the cell's energy needs.

Mitochondria have a double membrane, consisting of an inner and outer membrane and an aqueous compartment in between.

Outer membrane
Inner membrane

**FIG. 5.28    Chloroplasts.** Chloroplasts capture energy from sunlight and use it to synthesize sugars.

Chloroplasts are surrounded by a double membrane like mitochondria and in addition have a third membrane in the interior, the thylakoid membrane.

Outer membrane
Inner membrane
Thylakoid membrane

dioxide is released. Does this process sound familiar? It also describes your own breathing, or respiration. Mitochondria are the site of cellular respiration, and the oxygen that you take in with each breath is used by mitochondria to produce ATP. Cellular respiration is discussed in greater detail in Chapter 7.

### Chloroplasts capture energy from sunlight.

Both animal and plant cells have mitochondria to provide them with life-sustaining ATP.  In addition, plant cells and green algae have organelles called chloroplasts that capture the energy of sunlight to synthesize simple sugars (**Fig. 5.28**). This process, called **photosynthesis,** results in the release of oxygen as a waste product. Like the nucleus and mitochondria, chloroplasts are surrounded by a double membrane. They also have a separate internal membrane-bound compartment called the **thylakoid.** The thylakoid membrane contains specialized light-collecting molecules called pigments, of which **chlorophyll** is the most common.

Chlorophyll plays a key role in the organelle's ability to capture energy from sunlight. The green color of chlorophyll explains why so many plants have green leaves. Using the light energy collected by pigments, enzymes present in the cytoplasm use carbon dioxide as a carbon source to produce carbohydrates. Photosynthesis is discussed in greater detail in Chapter 8.

## Core Concepts Summary

**5.1 CELL MEMBRANES ARE COMPOSED OF LIPIDS, PROTEINS, AND CARBOHYDRATES.**

Phospholipids have both hydrophilic and hydrophobic regions. As a result, they spontaneously form structures such as micelles and bilayers when placed in an aqueous environment.  page 5-2

Membranes are fluid, meaning that membrane components are able to move laterally in the plane of the membrane.  page 5-3

Membrane fluidity is influenced by length of fatty acid chains, presence of carbon–carbon double bonds in fatty acid chains, and amount of cholesterol.  page 5-4

Many membranes also contain proteins that span the membrane (transmembrane proteins) or are temporarily associated with one layer of the lipid bilayer (peripheral proteins).  page 5-6

**5.2  THE PLASMA MEMBRANE IS A SELECTIVE BARRIER THAT CONTROLS THE MOVEMENT OF MOLECULES BETWEEN THE INSIDE AND OUTSIDE OF THE CELL.**

Selective permeability results from the combination of lipids and proteins that make up cell membranes.  page 5-8

Passive transport works by diffusion, the random movement of molecules. Net movement of molecules occurs from regions of higher concentration to regions of lower concentration.  page 5-8

Passive transport can occur directly through the plasma membrane (simple diffusion) or be aided by protein channels (facilitated diffusion).  page 5-8

Active transport moves molecules from regions of lower concentration to regions of higher concentration and requires energy.  page 5-10

Primary active transport uses energy stored in ATP; secondary active transport uses the energy stored in an electrochemical gradient.  page 5-10

Cells maintain size and composition by protein pumps that actively move ions in and out of the cell, contractile vacuoles, and the cell wall.  page 5-11

**5.3  CELLS CAN BE CLASSIFIED AS PROKARYOTES OR EUKARYOTES; THESE DIFFER IN THE DEGREE OF INTERNAL COMPARTMENTALIZATION.**

Prokaryotic cells lack a nucleus and other internal membrane-enclosed compartments.  page 5-13

Prokaryotic cells include bacteria and archaea and are much smaller than eukaryotes.  page 5-13

Eukaryotic cells have a nucleus and other internal compartments called organelles.  page 5-13

Eukaryotes include animals, plants, fungi, and protists.  page 5-13

**5.4  THE ENDOMEMBRANE SYSTEM IS AN INTERCONNECTED SYSTEM OF MEMBRANES THAT INCLUDES THE NUCLEAR ENVELOPE, ENDOPLASMIC RETICULUM, GOLGI APPARATUS, LYSOSOMES, VESICLES, AND PLASMA MEMBRANE.**

The nucleus, which is enclosed by a double membrane called the nuclear envelope, houses the genome.  page 5-17

The endoplasmic reticulum is continuous with the outer nuclear envelope and manufactures proteins and lipids for use by the cell or for export out of the cell.  page 5-17

The Golgi apparatus communicates with the ER via transport vesicles. It receives proteins and lipids from the endoplasmic reticulum and directs them to their final destinations.  page 5-18

Lysosomes break down macromolecules like proteins to simpler compounds that can be used by the cell.  page 5-19

Protein sorting directs proteins to their final destinations in or out of the cell.  page 5-20

Proteins synthesized on free ribosomes are sorted after translation and proteins synthesized on ribosomes associated with the endoplasmic reticulum are sorted during translation.  page 5-20

Proteins synthesized on free ribososomes are sorted by means of an amino acid sequence known as a signal sequence.  page 5-21

Proteins synthesized on ribosomes associated with the endoplasmic reticulum have a signal sequence that is recognized by a signal recognition particle (SRP).  page 5-21

**5.5  MITOCHONDRIA AND CHLOROPLASTS ARE ORGANELLES INVOLVED IN HARNESSING ENERGY; THEY ARE LIKELY EVOLVED FROM FREE-LIVING PROKARYOTES.**

Mitochondria harness energy from chemical compounds for use by both animal and plant cells.  page 5-22

Chloroplasts harness the energy of sunlight to build sugars.  page 5-23

## Self-Assessment

1.  Describe how lipids with hydrophilic and hydrophobic regions behave in an aqueous environment.

2.  Describe two ways in which proteins associate with membranes.

3.  Describe an experiment that demonstrates that proteins move in membranes.

4.  Name three parameters that need to be stably maintained inside a cell.

5.  Explain the role of lipids and proteins in maintaining the selective permeability of membranes.

6.  Distinguish between passive and active transport mechanisms across cell membranes.

7.  Describe three different ways in which cells maintain size and composition.

8.  Compare the organization, degree of compartmentalization, and size of prokaryotic and eukaryotic cells.

9.  Name the major organelles in eukaryotic cells and describe their functions.

10. Explain how a protein ends up in the cytosol, in the plasma membrane, or secreted from the cell.

Do you understand the chapter's Core Concepts? Log in to check your answers to the Self-Assessment questions, then practice what you've learned and reinforce this chapter's concepts by working through the problems and multimedia tutorials provided there.

📖 **www.biologyhowlifeworks.com**

## CHAPTER 6

# MAKING LIFE WORK

## Capturing and Using Energy

## Core Concepts

**6.1** Metabolism is the set of biochemical reactions that transforms biomolecules and transfers energy.

**6.2** The energy of a system is its capacity to do work.

**6.3** The laws of thermodynamics govern energy flow in biological systems.

**6.4** Chemical reactions are subject to the laws of thermodynamics.

**6.5** The rate of biochemical reactions is controlled by protein catalysts called enzymes.

We have seen that cells require a way to encode and transmit information and a membrane to separate inside from out. The third requirement of a cell is energy, which cells need to do work. Among other things, they must grow and divide, move, change shape, pump ions in and out of the cell, transport vesicles, and synthesize macromolecules such as DNA, RNA, proteins, and complex carbohydrates. All these activities are considered work, and they require energy.

We are all familiar with different forms of energy—the sun and wind provide a source of energy, as do fossil fuels such as oil and natural gas. We have learned to harness the energy from these sources and convert it into other forms, such as electricity, to provide needed power to our homes and cities.

Cells are faced with similar challenges. They must harness energy from the environment and convert it to a form that allows them to do work necessary to sustain life. Cells harness energy from the sun and from chemical compounds, including carbohydrates, lipids, and proteins. Although the source of energy may differ among cells, all cells convert energy to a form that can be easily used for cellular processes. All cells use energy in the form of a molecule called **adenosine triphosphate,** or **ATP.**

ATP is often called the universal "currency" of cellular energy to indicate that ATP provides energy in a form that all cells can readily use to perform the work of the cell. Although the term "currency" is a useful shorthand, keep in mind an important distinction between actual currency, such as a dollar bill, and ATP. A dollar bill *represents* a certain value but does not in fact have any value in itself. By contrast, ATP does not represent energy; it actually contains energy in its chemical bonds. Nevertheless, the analogy is useful because ATP, like a dollar bill, engages in a broad range of energy "transactions" in the cell, as we discuss below.

In this chapter, we consider energy in the context of cells. What exactly is energy? What principles govern its flow in biological systems? And how do cells make use of it?

## 6.1 AN OVERVIEW OF METABOLISM

When considering a cell's use of energy, it is also useful to consider the cell's sources of carbon, because carbon is the backbone of the organic molecules that make up cells and because cells often use carbon-based compounds as a source of energy to carry out their functions. How organisms obtain the energy and carbon needed for growth and other vital functions is so fundamental that it is sometimes used to provide a metabolic classification of life (**Fig. 6.1**). Simply put, organisms have two ways of harvesting energy from their environment and two sources of carbon. Together, this means that there are four principal ways in which organisms acquire the energy and materials needed to grow, function, and reproduce.

**FIG. 6.1**     **A metabolic classification of organisms.** Organisms can be classified according to their energy and carbon sources.

## Organisms can be classified according to their energy and carbon sources.

Organisms have two ways of harvesting energy from their environment: They can obtain energy either from the sun or from chemical compounds (Fig. 6.1). Organisms that capture energy from sunlight are called **phototrophs.** Plants are the most familiar example. Plants use the energy of sunlight to convert carbon dioxide and water into sugar and oxygen (Chapter 8). Sugars, such as glucose, contain energy in their chemical bonds that is used to synthesize ATP, which in turn can power the work of the cell.

Other organisms derive their energy directly from organic molecules such as glucose. These organisms are called **chemotrophs,** and animals are familiar examples. Animals ingest other organisms, obtaining organic molecules such as glucose that they break down in the presence of oxygen to produce carbon dioxide and water. In this process, the energy in the organic molecule is converted to energy carried in the bonds of ATP (Chapter 7).

In drawing this distinction between plants and animals, we have to be careful: Although sunlight provides the energy that

plants use to synthesize glucose, plants still power most cellular processes by breaking down the sugar they make, just as animals do. And although chemotrophs harness energy from organic molecules, the energy in these organic molecules is generally derived from the sun. Bearing these two points in mind, we use the terms "phototroph" and "chemotroph" because they call our attention to the flow of energy from the sun to organisms and then from one organism to the next (discussed more fully in Chapter 25).

Organisms can also be considered in terms of their source of carbon (Fig. 6.1). Some organisms are able to convert carbon dioxide (an inorganic form of carbon) into glucose (an organic form of carbon). These organisms are **autotrophs,** or "self feeders," because they make their own organic sources of carbon. Plants again are an example, so plants are both phototrophs and autotrophs, or photoautotrophs.

Other organisms do not have the ability to convert carbon dioxide into organic forms of carbon. Instead, they obtain their carbon from organic molecules synthesized by other organisms (called preformed organic molecules)—that is, they eat other organisms or molecules derived from other organisms. Such organisms are **heterotrophs,** or "other feeders," as they rely on other organisms for their organic source of carbon. Animals obtain their carbon in this way, and so animals are both chemotrophs and heterotrophs, or chemoheterotrophs.

As we move away from the familiar examples of plants and animals and begin in Part 2 to explore the diversity of life, we will see that not all organisms fit into these two categories (Fig. 6.1). Some microorganisms gain energy from sunlight, but obtain their carbon from preformed organic molecules; such organisms are called photoheterotrophs. Other microorganisms extract energy from inorganic sources but build their own organic molecules; these organisms are termed chemoautotrophs. They are often found in extreme environments, such as deep-sea vents, where sunlight is absent and inorganic compounds such as hydrogen sulfide are plentiful.

### Metabolism is the set of chemical reactions that sustain life.

The building and breaking down of sugars such as glucose, and the harnessing and release of energy in the process, are driven by chemical reactions in the cell. The term **metabolism** encompasses the entire set of these chemical reactions that convert molecules into other molecules and transfer energy in living organisms. These chemical reactions are occurring all of the time in your cells. Many of these reactions are linked, in that the products of one are the reactants of the next, forming long pathways and intersecting networks.

---

**FIG. 6.2** **Catabolism and anabolism.** The energy harvested as ATP during the breakdown of molecules in catabolism can be used to synthesize molecules in anabolism.

Metabolism is divided into two branches: **Catabolism** is the set of chemical reactions that break down molecules into smaller units and, in the process, produce ATP, and **anabolism** is the set of chemical reactions that build molecules from smaller units and require an input of energy, usually in the form of ATP (**Fig. 6.2**). For example, carbohydrates can be broken down, or catabolized, into sugars, fats into fatty acids and glycerol, and proteins into amino acids. These initial products can be broken down further to release energy stored in their chemical bonds. The synthesis of macromolecules such as sugars and proteins, by contrast, is anabolic.

---

## 6.2 ENERGY

The energy of a system has a formal definition: It is the system's capacity to do work. For a cell, this work involves processes we discussed in earlier chapters, such as synthesizing DNA, RNA, and proteins, pumping substances into and out of the cell, and moving vesicles between various compartments of a cell. The sun is a familiar source of energy, and we recognize energy in movement. On a cellular level, energy is held in chemical bonds, such as those of ATP, glucose, and lipids. How do these molecules contain energy in their chemical bonds? In this section, we discuss different sources of energy.

### Kinetic and potential energy are two forms of energy.

Light, electricity, wind, and fossil fuels are all different sources of energy, and they can be classified as one of two types: kinetic energy or potential energy (**Fig. 6.3**). **Kinetic energy** is the energy of motion, and it is perhaps the most familiar form of energy. Moving objects possess kinetic energy because they perform work, resulting in the movement of themselves and the surrounding matter. For example, a ball bouncing down a set of stairs possesses kinetic energy. Kinetic energy

---

**FIG. 6.3** **Potential and kinetic energy.** Some of the high potential energy of a ball at the top of a set of stairs is transformed to kinetic energy as the ball rolls down the stairs.

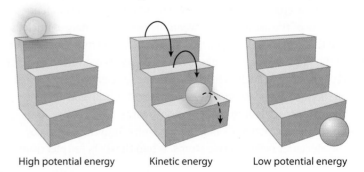

High potential energy     Kinetic energy     Low potential energy

is associated with any kind of movement, such as a person running or a muscle contracting. Similarly, light is associated with the movement of photons, electricity with the movement of electrons, and thermal energy (perceived as heat) with the movement of molecules, so these, too, are forms of kinetic energy.

Energy is not always associated with motion. An immobile object can still possess a form of energy called **potential energy,** or stored energy. Potential energy depends on the structure of the object or its position relative to its surroundings, and it is released by a change in the object's structure or position. For example, the potential energy of a ball is higher at the top of a flight of stairs than at the bottom (Fig. 6.3). If it were not blocked by the floor, it would move from the position of higher potential energy (the top of the stairs) to the position of lower potential energy (the bottom of the stairs) because of the force of gravity. Similarly, an electrochemical gradient of molecules across a cell membrane is a form of potential energy. Given the chance, the molecules move down their concentration and electrical gradients (Chapter 5).

Energy can be converted from one form to another. The ball at the top of the stairs has a certain amount of potential energy because of its position. As it rolls down the stairs, this potential energy is converted to kinetic energy associated with movement of the ball and the surrounding air. When the ball reaches the bottom of the stairs, the remaining energy is stored as potential energy. Conversely, it takes an input of energy to move the ball back to the top of the stairs, and this input of energy is stored as potential energy.

### Chemical energy is a form of potential energy.

We obtain the energy we need from the food we eat, which contains chemical energy, a form of potential energy. How do food molecules contain potential energy? Recall from Chapter 2 that atoms consist of a nucleus with positively charged protons and electrically neutral neutrons, and negatively charged electrons occupying orbitals at various distances from the nucleus. Like the ball on the stairs, electrons have potential energy based on their position. The farther away an electron is from the nucleus, the more potential energy it has (**Fig. 6.4**). Given the chance, an electron will move closer to the nucleus because of the attraction between the positively charged nucleus and the negatively charged electron. When electrons move closer to the nucleus, some of this potential energy is converted to other types of energy, such as heat or light energy. Conversely, it takes an input of energy to move electrons farther from the nucleus, and this energy is stored as potential energy.

Recall, too, that a covalent bond results from the sharing of the electrons between two atoms. When these shared electrons

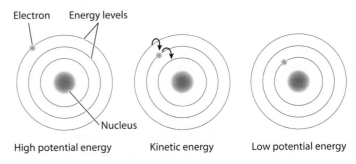

**FIG. 6.4** **Potential and kinetic energy of electrons.** Some of the high potential energy of an electron farther from the nucleus is converted to the kinetic energy of light or heat as the electron moves closer to the nucleus.

are far from both nuclei, they have a large amount of potential energy; when close, they have less potential energy. In this way, molecules carry potential energy in their chemical bonds. Organic molecules contain many covalent bonds, including carbon–carbon (C–C) bonds and carbon–hydrogen (C–H) bonds. In these bonds, the electrons are far from the two atoms, so they store a large amount of chemical potential energy. Organic molecules such as carbohydrates, lipids, and proteins are rich sources of chemical energy and therefore are sometimes called fuel molecules.

### ATP is the cell's energy currency.

The chemical energy carried in carbohydrates, lipids, and proteins is harnessed by cells to do work. But cells do not use this energy all at once. Instead, through a series of chemical reactions described in Chapter 7, they package this energy into a chemical form that is readily accessible to the cell. One form of chemical energy is adenosine triphosphate, or ATP, shown in **Fig. 6.5.** The chemical energy in the bonds of ATP is used in turn to drive many cellular processes, such as muscle contraction, cell movement, and membrane pumps. In this way, ATP serves as a "go-between," acting as an intermediary between fuel molecules that store a large amount of potential energy in their bonds and the activities of the cell that require an input of energy.

ATP is composed of adenosine, which is made up of the base adenine and the five-carbon sugar ribose, attached to triphosphate, or three phosphate groups (Fig. 6.5). Its chemical relatives are ADP (adenosine diphosphate) and AMP (adenosine monophosphate), with two phosphate groups and one phosphate group, respectively. The use of ATP as an energy source in nearly all cells reflects its use early in the evolution of life on Earth.

The chemical energy of ATP is held in the bonds connecting the phosphate groups. At physiological pH, these phosphate groups are negatively charged and have a tendency to repel each other.

**FIG. 6.5** The structure of ATP.

The chemical bonds connecting the phosphate groups therefore store potential energy that is released when the bonds are broken. In turn, the released energy can power the work of the cell.

## 6.3 THE LAWS OF THERMODYNAMICS

Energy from the sun can be converted by plants and other phototrophs into chemical potential energy held in the bonds of molecules. Chemical reactions can transfer this chemical energy between different molecules. And energy from these molecules can be used by the cell to do work. In all these instances, energy is subject to the laws of thermodynamics. In fact, all processes in a cell, from diffusion to osmosis to the pumping of ions to cell movement, are subject to these laws. There are four laws of thermodynamics, two of which are particularly relevant to biological processes.

### The first law of thermodynamics: Energy is conserved.

The **first law of thermodynamics** is the law of conservation of energy. It states that the universe contains a constant amount of energy. Therefore, energy is neither created nor destroyed. New energy is never formed, and energy is never lost. Instead, energy simply changes from one form to another. For example, kinetic energy can change to potential energy and vice versa, but the total amount of energy always remains the same (**Fig. 6.6**).

**FIG. 6.6** The first law of thermodynamics. When energy is transformed from one form to another, the total amount of energy remains the same.

In the simple example of a ball rolling down a set of stairs (see Fig. 6.3), the potential energy stored at the top of the stairs is converted to kinetic energy associated with the movement of the ball and the surrounding air. The total amount of energy in this energy transformation is constant—that is, the difference in potential energy at the top and the bottom of the stairs is equal to the total amount of kinetic energy associated with the movement of the ball and air molecules.

Similarly, when an electron moves from a higher energy level, or shell, to a lower one, it emits light or heat (see Fig. 6.4). The difference in potential energy between the two energy levels is equal to the light or heat energy emitted, consistent with the first law.

### The second law of thermodynamics: Disorder tends to increase.

When energy changes form, the *total amount of energy* remains constant. However, in going from one form of energy to another, the *energy available to do work* decreases (**Fig. 6.7**). How can the

**FIG. 6.7** The second law of thermodynamics. Because the amount of disorder increases when energy is transformed from one form to another, some of the energy available to do work is lost.

total energy remain the same, but the energy available to do work decrease? The answer is that some of the energy is available to do work, and some is not available to do work. So, energy transfers are never 100% efficient, since the amount of energy available to do work decreases every time energy changes form.

The energy that is not available to do work takes the form of an increase in disorder. In other words, there is a universal price to pay in transforming energy from one form to another, which takes the form of an increase in disorder. For example, when kinetic energy is changed into potential energy or vice versa, the amount of disorder always increases. This principle is summarized by the **second law of thermodynamics,** which states that the transformation of energy is associated with an increase in disorder of the universe. The degree of disorder is called **entropy.**

In chemical reactions, most of the entropy increase occurs through the transformation of various forms of energy into thermal energy, which we experience as heat. Thermal energy is a form of kinetic energy corresponding to the random motion of molecules, and results in a given temperature. The higher the temperature, the more rapidly the molecules move, and the higher the disorder.

Consider the contraction of a muscle, a form of kinetic energy associated with the shortening of muscle cells. The contraction of muscle is powered by chemical potential energy in fuel molecules. Some of this potential energy is transformed into kinetic energy (movement), and the rest is dissipated as thermal energy (which is why your muscles warm as you exercise). The amount of chemical potential energy is equal to the amount of kinetic energy plus the amount of thermal energy, consistent with the first law of thermodynamics. In addition, because of the entropy "tax," not all of the chemical energy stored in molecules is converted to kinetic energy, as thermal energy flowing as heat is a necessary by-product of the process.

In living organisms, catabolic reactions result in an increase of entropy as a single ordered biomolecule is broken down into several smaller ones with more freedom to move around, consistent with the second law. Anabolic reactions, by contrast, might seem to decrease entropy because they use individual building blocks to synthesize more ordered biomolecules such as proteins or nucleic acids. Do these anabolic reactions violate the second law of thermodynamics? No, because the second law of thermodynamics always applies to the universe as a whole, not to a chemical reaction in isolation. A local decrease in entropy is always accompanied by an even higher increase in the entropy of the surroundings. The production of heat in a chemical reaction increases entropy because heat is associated with the motion of molecules. The combination of the decrease in entropy associated with the building of a macromolecule and

the increase in entropy associated with heat always results in a net increase in entropy.

The key point here is that the maintenance of the high degree of function and organization of a single cell or a multicellular organism requires a constant input of energy, either from the sun or from the energy stored in chemical compounds. This energy allows molecules to be built and work to be carried out, but also leads to greater disorder in the surroundings.

→   **Quick Check 1**  Cold air is more ordered (that is, it has less entropy) than hot air. The second law of thermodynamics states that entropy always increases. Do air conditioners violate this law?

## 6.4 CHEMICAL REACTIONS

Living organisms build and break down molecules, pump ions across membranes, move, and in general function largely through chemical reactions. As we have seen, organisms break down food molecules, storing energy in the bonds of ATP, which then powers chemical reactions that sustain life. Chemical reactions are therefore central to life processes. We describe here several features of a chemical reaction.

**A chemical reaction occurs when molecules interact.**
As we saw in Chapter 2, a chemical reaction is the process by which molecules, called reactants, are transformed into other molecules, called products. During a chemical reaction, atoms keep their identity, but the bonds linking the atoms change. For example, carbon dioxide ($CO_2$) and water ($H_2O$) can react to produce carbonic acid ($CH_2O_3$), as shown here and in **Fig. 6.8:**

$$CO_2 + H_2O \rightarrow H_2CO_3$$

---

**FIG. 6.8**   **A chemical reaction.** Atoms retain their identity during a chemical reaction as bonds are broken and new bonds formed to yield new molecules.

A single bond connecting carbon and oxygen in $CO_2$ and a single bond connecting hydrogen and oxygen in $H_2O$ are broken.

A covalent bond is formed between the carbon of $CO_2$ and the oxygen of $H_2O$, and between one oxygen of $CO_2$ and one hydrogen of $H_2O$.

New covalent bonds

$CO_2$         +         $H_2O$                                     $H_2CO_3$

Carbon dioxide       Water                                     Carbonic acid

The preceding reaction occurs commonly. For example, in an aqueous (watery) environment, carbonic acid exists as bicarbonate ions ($HCO_3^-$) and protons ($H^+$). This reaction occurs in the oceans, and explains why the oceans are becoming more acidic with increasing carbon dioxide levels in the atmosphere. In addition, when cells break down glucose in organisms, they produce carbon dioxide. In animals, this carbon dioxide diffuses out of cells and into the blood. Rather than remaining dissolved in solution, most of the carbon dioxide is converted into carbonic acid according to the preceding reaction.

Most chemical reactions in cells are reversible; the products can react to form the reactants. For example, carbon dioxide and water react to form carbonic acid, and carbonic acid can also dissociate to produce carbon dioxide and water. This reverse reaction also occurs in the blood, allowing carbon dioxide to be removed from the lungs or gills (Chapter 39). The reversibility of the reaction is indicated by a double arrow:

$$CO_2 + H_2O \rightleftharpoons H_2CO_3$$

The way the reaction is written defines the forward and reverse reactions: A forward reaction proceeds from left to right and the reactants are placed on the left side of the arrow; a reverse reaction proceeds from right to left and the reactants are placed on the right side of the arrow.

The direction of a reversible reaction can be influenced by the concentrations of reactants and products. For example, increasing the concentration of the reactants or decreasing the concentration of the products favors the forward reaction. This effect explains how many reactions in metabolic pathways proceed: The products of many reactions are quickly consumed by the next reaction, helping to drive the first reaction forward.

### Chemical reactions are subject to the laws of thermodynamics.

Chemical reactions, like all processes in the cell, are subject to the laws of thermodynamics. We have seen that cells use chemical reactions to perform much of the work of the cell. The amount of energy available to do work is called **Gibbs free energy ($G$)**. In a chemical reaction, we can compare the free energy of the reactants and products to determine whether there is energy available to do work. The difference between two values is denoted by the Greek letter delta ($\Delta$). In this case, $\Delta G$ is the free energy of the products minus the free energy of the reactants (**Fig. 6.9**). If the products of a reaction have more free energy than the reactants, then $\Delta G$ is positive and a net input of energy is required to drive the reaction forward (Fig. 6.9a). By contrast, if the products of a reaction have less free energy than the reactants, $\Delta G$ is negative and energy is released and available to do work (Fig. 6.9b).

Reactions with a negative $\Delta G$ that release energy and proceed spontaneously are called **exergonic**, and reactions with a positive

$\Delta G$ that require an input of energy and are not spontaneous are called **endergonic**. Note that the term "spontaneous" does not imply instantaneous or even rapid. "Spontaneous" in this context means that a reaction releases energy; "non-spontaneous" means that a reaction requires a sustained input of energy.

Recall that the total amount of energy is equal to the energy available to do work plus the energy that is not available to do work because of the increase in disorder. We can write this relationship as an equation, in which the total amount of energy is **enthalpy ($H$)**, the energy available to do work is Gibbs free energy ($G$), and the degree of disorder is **entropy ($S$)** multiplied by the **absolute temperature ($T$)** (measured in degrees Kelvin), since temperature influences the movement of molecules (and hence the degree of disorder). Therefore,

Total amount of energy ($H$) =
energy available to do work ($G$) + energy lost to entropy ($TS$)

We are interested in the energy available for a cell to do work, or $G$. Therefore, we express $G$ in terms of the other two parameters, $H$ and $TS$, as follows:

$$G = H - TS$$

In a chemical reaction, we can compare the total energy and entropy of the reactants with the total energy and entropy of the products to see if there is energy available to do work. As a result, we get

$$\Delta G = \Delta H - T\Delta S$$

This equation is a useful way to see if a chemical reaction takes place, what direction it proceeds in, and whether net energy is required or released. Let's take a step back and make sure it makes sense intuitively. The value of $\Delta G$ depends on *both* the change in enthalpy and the change in disorder. Catabolic reactions are those in which the products have less chemical energy in their bonds than in those of the reactants and in which the products

---

**FIG. 6.9** **Endergonic and exergonic reactions.** An endergonic reaction requires an input of energy; an exergonic reaction releases energy.

**a.** Endergonic reaction (Non-spontaneous)

**b.** Exergonic reaction (Spontaneous)

**FIG. 6.10** **Energy in catabolism and anabolism.** Catabolic reactions have a negative $\Delta G$ and release energy often in the form of ATP, whereas anabolic reactions have a positive $\Delta G$ and require an input of energy often in the form of ATP.

in a positive value of $\Delta G$ and requires a net input of energy.

There are also cases in which the change in enthalpy and the change in entropy are both positive or both negative. In these cases, the absolute value of these parameters determines whether $\Delta G$ is positive or negative and therefore whether a reaction is spontaneous or not.

→ **Quick Check 2** How does increasing the temperature affect the change in free energy ($\Delta G$) of a chemical reaction?

### The hydrolysis of ATP releases energy.

Let's apply these concepts to a specific chemical reaction. Earlier, we introduced ATP, the molecule that drives many cellular processes using the chemical potential energy in its chemical bonds. ATP reacts with water to lose a phosphate group ($HPO_4^{2-}$), forming ADP and inorganic phosphate ($P_i$) (**Fig. 6.11**):

$$ATP + H_2O \rightarrow ADP + P_i$$

This is an example of a hydrolysis reaction, a chemical reaction in which a water molecule is split into a proton ($H^+$) and a hydroxyl group ($OH^-$). Hydrolysis reactions often break down polymers into their subunits, and in the process one product gains a proton and the other gains a hydroxyl group.

The reaction of ATP with water is an exergonic reaction that releases energy. This negative free energy difference between the products and reactants can be explained by referring to the formula we derived in the last section. Recall that the phosphate groups of ATP are negatively charged at physiological pH and repel each other. ATP has three phosphate groups, and ADP has two. Therefore, ADP contains less chemical potential energy in its bonds than does ATP, resulting in a negative value of $\Delta H$. In addition, a single molecule of ATP is broken down into two molecules, ADP and $P_i$. Therefore, the reaction is also associated with an increase in entropy, or a positive value of $\Delta S$. Since $\Delta G = \Delta H - T\Delta S$, $\Delta G$ is negative and the reaction is a spontaneous

are more disordered than the reactants. In other words, such reactions have a negative value of $\Delta H$ and a positive value of $\Delta S$, and therefore a negative value of $\Delta G$ and proceed spontaneously (**Fig. 6.10**). The reverse is also true: Increasing chemical potential energy and decreasing disorder, as in the synthesis of proteins from individual amino acids and other anabolic reactions, results

**FIG. 6.11** **ATP hydrolysis.** ATP hydrolysis is an exergonic reaction that releases free energy.

one that releases free energy. The release of free energy during the hydrolysis of ATP allows ATP to drive chemical reactions and other processes that require a net input of energy, as we will see next.

## Non-spontaneous reactions are often coupled to spontaneous reactions.

If the conversion of reactant A into product B is spontaneous, the reverse reaction leading to the formation of B from A is not. The $\Delta G$'s for the forward and reverse reactions have the same absolute value but opposite signs. You might expect that the direction of the reaction would always be from A to B. However, in a living organism, not all chemical reactions are spontaneous. Anabolic reactions are a good example; they require an input of energy to drive them in the right direction. This raises the question: What drives non-spontaneous reactions? The coupling of a non-spontaneous reaction to a spontaneous one can drive a non-spontaneous reaction, as long as the net $\Delta G$ of the two reactions is negative.

Let's consider the following two reactions:

(1) A → B      $\Delta G_1 > 0$ (energy consumed)
(2) B → C      $\Delta G_2 < 0$ (energy released)

Only reaction 2 is spontaneous since $\Delta G_2 < 0$, but if the sum of $\Delta G_1$ and $\Delta G_2$ is negative, then the overall reaction A →→ C is spontaneous and the transformation of A into B occurs despite a positive $\Delta G$. This type of **energetic coupling,** in which a spontaneous reaction drives a non-spontaneous one, provides the thermodynamic driving force of a non-spontaneous biochemical reaction.

In a cell, the most common reaction involving energetic coupling is the hydrolysis of ATP. This reaction can drive an endergonic reaction that requires an input of energy if the two reactions are coupled together (**Fig. 6.12a**). Coupled reactions involving ATP share a phosphate group, which is transferred between ATP and other molecules in the reaction.

Following ATP hydrolysis, the cell needs to replenish its ATP so that it can carry out additional

**FIG. 6.12   Energetic coupling.** A spontaneous (exergonic) reaction drives a non-spontaneous (endergonic) reaction. (a) The hydrolysis of ATP drives the formation of glucose 6-phosphate from glucose. (b) The hydrolysis of phosphoenolpyruvate drives the synthesis of ATP.

> The coupled reaction proceeds because $\Delta G$ is less than 0 and $P_i$ is shared between the two reactions.

**a.**

| | | | | | | | |
|---|---|---|---|---|---|---|---|
| ATP | + | $H_2O$ | ⟶ | ADP | + $P_i$ | $\Delta G_1 = -7$ kcal/mol | **Exergonic reaction** |
| Glucose | + | $P_i$ | ⟶ | Glucose 6-phosphate | + $H_2O$ | $\Delta G_2 = +3.3$ kcal/mol | **Endergonic reaction** |
| Glucose | + | ATP | ⟶ | Glucose 6-phosphate | + ADP | $\Delta G = -3.7$ kcal/mol | **Coupled reaction** |

**b.**

| | | | | | | | |
|---|---|---|---|---|---|---|---|
| Phosphoenolpyruvate | + | $H_2O$ | ⟶ | Pyruvate | + $P_i$ | $\Delta G_1 = -14.8$ kcal/mol | **Exergonic reaction** |
| ADP | + | $P_i$ | ⟶ | ATP | + $H_2O$ | $\Delta G_2 = +7$ kcal/mol | **Endergonic reaction** |
| Phosphoenolpyruvate | + | ADP | ⟶ | Pyruvate | + ATP | $\Delta G = -7.8$ kcal/mol | **Coupled reaction** |

chemical reactions. The synthesis of ATP from ADP and $P_i$ is an endergonic reaction with a positive $\Delta G$, requiring an input of energy. In some cases, exergonic reactions can drive the synthesis of ATP by energetic coupling (**Fig. 6.12b**). The sum of the $\Delta G$'s of the two reactions is negative, allowing the overall reaction to proceed.

Phosphorylated molecules, like ATP, can be hydrolyzed, releasing free energy. Hydrolysis reactions can be ranked by their free energy differences. **Fig. 6.13** shows that ATP hydrolysis has an intermediate free energy difference compared to the free energy difference for hydrolysis of other common phosphorylated molecules. Those reactions that have a $\Delta G$ more negative than that of ATP hydrolysis give a phosphate group to ADP, and those reactions that have a $\Delta G$ less negative than that of ATP hydrolysis

**FIG. 6.13   $\Delta G$ of common hydrolysis reactions in a cell.** ATP hydrolysis has an intermediate $\Delta G$.

$\Delta G$ of hydrolysis reaction (in kcal/mol)

−14.8 — Phosphoenolpyruvate + $H_2O$ ⟶ Pyruvate + $P_i$

−7 — ATP + $H_2O$ ⟶ ADP + $P_i$

> $\Delta G$ for ATP hydrolysis is intermediate compared to $\Delta G$ of hydrolysis of common phosphorylated molecules, allowing ATP to drive reactions as well as be replenished.

−3.3 — Glucose 6-phosphate + $H_2O$ ⟶ Glucose + $P_i$

receive a phosphate group from ATP by energetic coupling. Therefore, ADP is an energy acceptor and ATP an energy provider. The ATP–ADP system is at the core of energetic coupling between catabolic and anabolic reactions. In this way, ATP continually turns over in a cell.

## 6.5 ENZYMES AND THE RATE OF CHEMICAL REACTIONS

Up to this point, we have focused on the spontaneity and direction of chemical reactions. Now we address their rate. As mentioned earlier, a spontaneous reaction is not necessarily a fast one. For example, the breakdown of glucose into carbon dioxide and water is spontaneous with a negative $\Delta G$, but the rate of the reaction is close to zero and the breakdown of glucose is imperceptible. However, glucose is readily broken down inside cells all the time. How is this possible? The answer is that chemical reactions in a cell are accelerated by chemical catalysts.

The rate of a chemical reaction is defined as the amount of product formed (or reactant consumed) per unit of time. Catalysts are substances that increase the rate of chemical reactions without themselves being consumed. In biological systems, the catalysts are usually proteins called **enzymes,** although, as we saw in Chapter 2, some RNA molecules have catalytic activity as well. Enzymes can increase the rate of chemical reactions dramatically. Moreover, because they are highly specific, acting only on certain reactants and catalyzing only some reactions, enzymes play a critical role in determining which chemical reactions take place from all the possible reactions that could occur in a cell. In this section, we discuss how enzymes increase the rate of chemical reactions and how that ability gives them a central role in metabolism.

### Enzymes reduce the activation energy of a chemical reaction.

Earlier, we saw that exergonic reactions release free energy and endergonic reactions require free energy. Nevertheless, all chemical reactions require an input of energy to proceed, even exergonic reactions that release energy. For this type of reaction, the energy released is more than the initial input of energy, so there is a net release of energy.

Why is an input of energy needed for all chemical reactions? As a chemical reaction proceeds, existing chemical bonds break and new ones form. For an extremely brief period of time, a compound is formed in which the old bonds are breaking and the new ones are forming. This intermediate stage between reactants and products is called the **transition state.** It is highly unstable and therefore has a large amount of free energy.

In all chemical reactions, reactants adopt at least one transition state before their conversion into products. The graph in **Fig. 6.14** represents the free energy levels of the reactant, transition state, and product. This reaction is spontaneous since the free energy of the reactant is higher than the free energy of the product and $\Delta G$ is negative. However, the highest free energy value corresponds to the transition state.

To reach the transition state, the reactant must absorb energy from its surroundings (the uphill portion of the curve in Fig. 6.14). As a result, all chemical reactions, even spontaneous ones that release energy, require an input of energy that we can think of as an "energy barrier." The energy input necessary to reach the transition state is called the **activation energy ($E_A$).** Once the transition state is reached, the reaction proceeds, products are formed, and energy is released into the surroundings (the downhill portion of the curve in Fig. 6.14). There is an inverse correlation between the rate of a reaction and the height of the energy barrier: the lower the energy barrier, the faster the reaction; the higher the barrier, the slower the reaction.

In chemical reactions that take place in the lab, heat is a common source of energy used to overcome the energy barrier. In living organisms, reactions are accelerated by the action of enzymes. However, enzymes do not act by supplying heat. Instead, they reduce the activation energy by stabilizing the transition state and decreasing its free energy (the red line in Fig. 6.14). As the activation energy decreases, the speed of the reaction increases.

Although an enzyme accelerates a reaction by reducing the activation energy, the difference in free energy between reactants and products ($\Delta G$) does not change. In other words, an enzyme

**FIG. 6.14** **An enzyme-catalyzed reaction.** An enzyme accelerates a reaction by lowering the activation energy, $E_A$.

changes the path of the reaction between reactants and products, but not the starting or end point (Fig. 6.14). Consider the breakdown of glucose into carbon dioxide and water. The $\Delta G$ of the reaction is the same whether it proceeds by combustion or by the action of multiple enzymes in a metabolic pathway in a cell.

→ **Quick Check 3** Which of the following do enzymes change? $\Delta G$; reaction rate; types of products generated; activation energy; the laws of thermodynamics.

## Enzymes form a complex with reactants and products.

An important characteristic of enzymes is that, as catalysts, they participate in a chemical reaction but are not themselves consumed in the process. In other words, they emerge unchanged from a chemical reaction, ready to catalyze the same reaction again. How do enzymes increase the reaction rate without being consumed? The answer is that enzymes form a complex with the reactants and products.

In an uncatalyzed reaction, a reactant, often called a **substrate (S)**, is converted to a product (P):

$$S \rightleftharpoons P$$

In the presence of an enzyme (E), the substrate first forms a complex with the enzyme (enzyme–substrate, or ES). While still part of the complex, the substrate is converted to product (enzyme–product, or EP). Finally, the complex dissociates, releasing the enzyme and product. Therefore, a reaction catalyzed by an enzyme can be described as

$$S + E \rightleftharpoons ES \rightleftharpoons EP \rightleftharpoons E + P$$

The formation of this complex is critical for accelerating the rate of a chemical reaction. Recall from Chapter 4 that proteins adopt three-dimensional shapes and that the shape of a protein is linked to its function. Enzymes are folded into three-dimensional shapes that bring particular amino acids into proximity to form an **active site.** The active site of the enzyme is the portion of the enzyme that binds substrate and converts it to product (**Fig. 6.15**). In the active site, the enzyme and substrate form both transient covalent bonds and/or weak noncovalent interactions. Together, these interactions stabilize the transition state and decrease the activation energy.

Enzymes also reduce the energy of activation by positioning two substrates to react. The formation of the enzyme–substrate complex promotes the reaction between two substrates by aligning their reactive chemical groups and limiting their motion relative to each other.

The size of the active site is extremely small compared to the size of the enzyme. If only a small fraction of the enzyme is necessary for the catalysis of a reaction, why are enzymes so large? Of the many amino acids that form the active site, only

**FIG. 6.15** **The active site of the enzyme catalase, shown schematically and as a surface model.** Substrate binds and is converted to product at the active site.

Schematic model of an enzyme

Molecular model of catalase

a few actively contribute to catalysis. Each of these amino acids has to occupy a very specific spatial position to align with the correct reactive group on the substrate. If the few essential amino acids were part of a short peptide, the alignment of chemical groups between the peptide and the substrate would be difficult or even impossible because the length of the bonds and the bond angles in the peptide would constrain its three-dimensional structure. In fact, in many cases the catalytic amino acids are spaced far apart in the primary structure of the enzyme, but brought close together in the formation of the active site by protein folding (**Fig. 6.16**). In other words, the large size of many enzymes is required at least in part to bring the catalytic amino acids into very specific positions in the active site of the folded enzyme.

The formation of a complex between enzymes and substrates

**FIG. 6.16** **Formation of the active site of an enzyme by protein folding.**

Unfolded enzyme

Amino acids

COOH

Protein folding brings specific amino acids close to each other to form the active site.

The amino acids that form the active site are often far apart in the linear sequence of the unfolded enzyme.

Folded enzyme

COOH

Bound substrate

NH₂

NH₂

can be demonstrated experimentally, as illustrated in **Fig. 6.17** and **Fig. 6.18**, using two different techniques to answer the same question.

### Enzymes are highly specific.

Enzymes are remarkably specific both for the substrate and the reaction that is catalyzed. In general, enzymes recognize either a unique substrate or a class of substrates that share common chemical structures. In addition, enzymes catalyze only one reaction or a very limited number of reactions.

For example, the enzyme succinate dehydrogenase acts only on succinate, whereas the enzyme β- (beta-) galactosidase is specific for a class of molecules. The enzyme β-galactosidase catalyzes the cleavage of the glycosidic bond that links galactose to glucose in the disaccharide lactose, as well as any glycosidic bond that links galactose to one of several types of

**HOW DO WE KNOW?**

**FIG. 6.17**

# Do enzymes form complexes with substrates?

**BACKGROUND** The idea that enzymes form complexes with substrates to catalyze a chemical reaction was first proposed in 1888 by the Swedish chemist Svante Arrhenius. One of the earliest experimental demonstrations in support of this idea was made by American chemist Kurt Stern. He studied an enzyme called catalase, which is very abundant in animal and plant tissues. Catalase converts hydrogen peroxide ($H_2O_2$), which is toxic to cells, to water ($H_2O$) and oxygen ($O_2$):

$$2H_2O_2 \quad \rightarrow \quad 2H_2O \quad + \quad O_2$$

  Hydrogen peroxide          Water            Oxygen

Catalase is very efficient, able to convert hydrogen peroxide to water and oxygen very rapidly.

**HYPOTHESIS** Stern hypothesized that catalase forms a complex with hydrogen peroxide.

**METHOD** Stern used spectral analysis to test his hypothesis. He shined a white light on the reaction as it took place. Different molecules absorb different wavelengths of light. The maximum absorption of a wavelength by each different molecule is called the absorption peak. By analyzing the patterns of absorption peaks as the reaction proceeded, Stern could determine whether or not an intermediate complex was formed.

**RESULTS** A schematic representation of the absorption peaks is shown here. Note the appearance of a new absorption peak in panel II, consistent with the formation of an enzyme–substrate complex.

Each molecule gives a unique pattern of peaks, which represent absorption of a given wavelength of light.

I — Free enzyme

II — Enzyme–substrate complex

III — Product and free enzyme

IV — Free enzyme

Wavelength (μm)

**CONCLUSION** The simplest interpretation of the data is that catalase forms a complex with hydrogen peroxide during its conversion to water and oxygen, providing support for the initial hypothesis.

**FOLLOW-UP WORK** As Stern wrote, "It remains to be seen to which extent the findings of this study apply to enzyme action in general." In other words, the next steps were to see if other enzymes also work by forming a complex with their substrates.

**SOURCE** Stern, K. G. 1936. "On the Mechanism of Enzyme Action: A Study of the Decomposition of Monoethyl Hydrogen Peroxide by Catalase and of an Intermediate Enzyme–Substrate Compound." *J. Biol. Chem.* 114:473–494.

**FIG. 6.18**

# Do enzymes form complexes with substrates?

**BACKGROUND** The enzyme β-galactosidase catalyzes the cleavage of the glycosidic bond that links galactose to glucose in the disaccharide lactose. Lactose belongs to a family of molecules called β-galactosides since it contains a galactose unit attached to the rest of the molecule by a glycosidic bond. In a related compound called β-thiogalactoside, the oxygen atom in the glycosidic bond is replaced by sulfur. The enzyme β-galactosidase binds β-thiogalactoside but cannot cleave or release it.

β-galactoside        β-thiogalactoside

**METHOD** A container is separated into two compartments by a semipermeable membrane. The membrane is permeable to β-galactoside and β-thiogalactoside, but not permeable to the enzyme.

Semipermeable membrane

**EXPERIMENT 1** Radioactively labeled β-thiogalactoside (S) is added to compartment 1 and the movement of S is followed by measuring the level of radioactivity in the two compartments.

**RESULT** Over time, the level of radioactivity is the same in the two compartments.

Substrate        Time

**EXPERIMENT 2** Radioactively labeled β-thiogalactoside (S) is added to compartment 1, enzyme (E) is added to compartment 2, and the movement of S is followed by measuring the level of radioactivity in the two compartments.

**RESULT** Over time, the level of radioactivity is greater in compartment 2 than in compartment 1.

Enzyme        Time

**CONCLUSION** These results can be explained if the substrate diffuses from compartment 1 to compartment 2, forms a complex with the enzyme, and is not released. In other words, E and S form a complex.

**SOURCE** Adapted from Doherty, D. G., and F. Vaslow. 1952. "Thermodynamic Study of an Enzyme–Substrate Complex of Chymotrypsin." *J. Am. Chem. Soc.* 74:931–936.

molecule. In this case, the enzyme does not recognize the whole substrate, but a particular structural motif within it, and very small differences in the structure of this motif affect the activity of the enzyme. For example, β-galactosidase cleaves β-galactoside but does not cleave α- (alpha-) galactoside, which differs from β-galactoside only in the orientation of the glycosidic bond. Hence, the enzyme is able to discriminate between two identical bonds with different orientations within a molecule. The specificity of enzymes can be attributed to the structure of their active sites. The enzyme active site interacts only with substrates having a precise three-dimensional structure.

## Enzyme activity can be influenced by inhibitors and activators.

The activity of enzymes can be influenced by **inhibitors** and **activators.** Inhibitors decrease the activity of enzymes, whereas activators increase the activity of enzymes. Enzyme inhibitors are quite common. They are synthesized naturally by many plants and animals as a defense against predators. Similarly, pesticides and herbicides often target enzymes to inactivate them. Many drugs used in medicine are enzyme inhibitors, including drugs used to treat infections as well as cancer and other diseases. Given the importance of chemical reactions and the role of enzymes in metabolism, it is not surprising that enzyme inhibitors have such widespread applications.

Inhibitors bind either to the active site of the enzyme or elsewhere, but in both cases they hinder the enzyme's ability to accelerate the conversion of reactants to products. Inhibitors fall into two classes. **Irreversible inhibitors** usually form covalent bonds with enzymes and irreversibly inactivate them. **Reversible inhibitors** form weak bonds with enzymes and easily dissociate from them.

Reversible inhibitors can act competitively or non-competitively with enzymes (**Fig. 6.19**). **Competitive inhibitors** bind to the active site of the enzyme and prevent the binding of the substrate (Fig. 6.19a). In other words, they compete with substrate for the active site of the enzyme. Competitive inhibitors are often structurally similar to the substrate. By binding to the active site of the enzyme, competitive inhibitors reduce the affinity of the enzyme for the substrate. Therefore, competitive inhibitors can be overcome by increasing the concentration of substrate.

**Non-competitive inhibitors** usually have a structure very different from that of the substrate and bind to the enzyme at a site different from the active site (Fig. 6.19b). When the inhibitor is bound, the enzyme is still able to bind the substrate and there is no competition between the binding of the substrate and the inhibitor. Therefore, the affinity of the enzyme for its substrate is unchanged by a non-competitive inhibitor. However, a non-competitive inhibitor slows down the reaction by altering the shape of the enzyme and reducing its activity.

## Allosteric enzymes in the cell are regulated by activators and inhibitors.

Enzyme activators and inhibitors are sometimes important in the normal operation of a cell—for example, to regulate a metabolic pathway. Consider the synthesis of isoleucine from threonine, a pathway found in some bacteria. This conversion requires five reactions, each catalyzed by a different enzyme (**Fig. 6.20**). Once the bacterium has enough isoleucine for its needs, it would be a waste of energy to continue synthesizing the amino acid. To shut down the pathway once it is no longer needed, the cell relies on an enzyme inhibitor. The inhibitor is isoleucine, the final product of the five reactions. Isoleucine binds to the first enzyme in the pathway, threonine dehydratase, at a site distinct from the active site. The binding of isoleucine changes the shape of the enzyme and in this way inhibits its function. This is an example of **negative feedback**, in which the final product inhibits the first step of the reaction.

Threonine dehydratase is an example of an **allosteric enzyme.** Allosteric enzymes bind to activators and inhibitors

---

**FIG. 6.19**    **(a) Competitive and (b) non-competitive inhibition.** Both types of inhibitor reduce the activity of an enzyme by binding to it reversibly, but competitive inhibitors bind the active site and non-competitive inhibitors bind a different site.

**a.** Competitive inhibition

In competitive inhibition, the enzyme binds either its substrate or its inhibitor.

**b.** Non-competitive inhibition

In non-competitive inhibition, the binding sites of the substrate and inhibitor are different. The binding of a non-competitive inhibitor causes a decrease in the rate at which the enzyme converts the substrate to product.

---

**FIG. 6.20**    **The regulation of threonine dehydratase, an allosteric enzyme.** The enzyme is inhibited by the final product of this synthetic pathway and activated by the initial substrate.

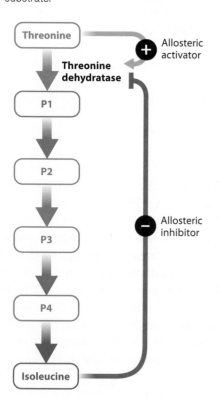

at sites that are different from the active site, resulting in a change in the shape and activity of the enzyme. Threonine dehydratase is allosterically inhibited by isoleucine, as we have seen. In addition, it is allosterically activated by the first substrate in the series of reactions, threonine (Fig. 6.20). Threonine binds to the enzyme and changes its conformation in a way that increases its activity. At a low concentration of threonine, the rate of the reaction is very slow. As the concentration of threonine increases, the activity of the enzyme increases. At a particular threshold, a small increase in threonine concentration results in a large increase in reaction rate. Finally, when there is excess substrate, the reaction rate slows down. In effect, the allosteric enzyme adjusts the rate of a reaction depending on substrate concentration.

In the next two chapters, we examine more closely two key metabolic processes, cellular respiration and photosynthesis. Both of these processes require many chemical reactions acting in a coordinated fashion. Allosteric enzymes catalyze key reactions in metabolism and are usually found at or near the start of a metabolic pathway or at the crossroads between two metabolic pathways. Allosteric enzymes are one way that the cell coordinates the activity of multiple metabolic pathways.

**? CASE 1  The First Cell: Life's Origins**
**What naturally occurring elements might have spurred the first reactions that led to life?**
Many enzymes are not just made up of protein, but also contain metal ions. These ions are one type of **cofactor,** a substance that

associates with an enzyme and plays a key role in its function. Metallic cofactors, especially iron, magnesium, manganese, cobalt, copper, zinc, and molybdenum, bind to diverse proteins, including those used in DNA synthesis, nitrogen metabolism, and the transport of electrons for cellular respiration and photosynthesis, processes that are discussed in the next two chapters. With this in mind, scientists have asked whether metal ions might, by themselves, catalyze chemical reactions thought to have played a role in the origin of life. They do. For example, magnesium and zinc ions added to solutions can accelerate the linking of nucleotides to form RNA and DNA molecules.

Enzymes that contain iron and sulfur clustered together are particularly important in the transport of electrons within cells. Iron–sulfur minerals, especially pyrite (or fool's gold, $FeS_2$), form commonly in mid-ocean hydrothermal vent systems and other environments where oxygen is absent. It has been proposed that reactions now carried out in cells by iron–sulfur proteins reflect chemical reactions that took place spontaneously on the early Earth. The idea that your cells preserve an evolutionary memory of ancient hydrothermal environments may seem like science fiction, but it finds support in laboratory experiments. For example, the reaction of $H_2S$ and FeS to form pyrite has been shown to catalyze a number of plausibly pre-biotic chemical reactions, including the formation of pyruvate (a key intermediate in energy metabolism discussed in Chapter 7). Thus, the metals in enzymes help connect the chemistry of life to the chemistry of Earth.

## Core Concepts Summary

**6.1 METABOLISM IS THE SET OF BIOCHEMICAL REACTIONS THAT TRANSFORMS BIOMOLECULES AND TRANSFERS ENERGY.**

Organisms can be grouped according to their source of energy: Phototrophs acquire energy from sunlight and chemotrophs obtain energy from chemical compounds. page 6-2

Organisms can also be grouped according to the source of carbon they use to build organic molecules: Heterotrophs obtain carbon from organic molecules, and autotrophs obtain carbon from inorganic carbon, such as carbon dioxide. page 6-2

Catabolism is the set of reactions that break down molecules and release energy, and anabolism is the set of reactions that build molecules and require energy. page 6-3

**6.2 THE ENERGY OF A SYSTEM IS ITS CAPACITY TO DO WORK.**

Kinetic energy is the energy of motion. page 6-3

Potential energy is stored energy and depends on the structure of an object or its position relative to its surroundings. page 6-4

Chemical energy is a form of potential energy held in the bonds of molecules. page 6-4

**6.3 THE LAWS OF THERMODYNAMICS GOVERN ENERGY FLOW IN BIOLOGICAL SYSTEMS.**

The first law of thermodynamics states that energy cannot be created or destroyed. page 6-5

The second law of thermodynamics states that there is an increase in disorder or entropy in the universe over time. page 6-5

**6.4 CHEMICAL REACTIONS ARE SUBJECT TO THE LAWS OF THERMODYNAMICS.**

In a chemical reaction, atoms themselves do not change, but the bonds linking the atoms change to form new molecules. page 6-6

Many chemical reactions are reversible. page 6-7

The direction of reversible reactions is influenced by the concentration of reactants and products. page 6-7

Gibbs free energy (G) is the amount of energy available to do work. page 6-7

Three thermodynamic parameters define a chemical reaction: Gibbs free energy (G), enthalpy (H), and entropy (S), and the change of free energy in a chemical reaction is described by $\Delta G = \Delta H - T\Delta S$. page 6-7

Exergonic reactions are spontaneous ($\Delta G < 0$) and release energy. page 6-7

Endergonic reactions are non-spontaneous ($\Delta G > 0$) and require energy. page 6-7

The hydrolysis of ATP is an exergonic reaction that is used to drive many endergonic reactions in a cell. page 6-8

In living systems, non-spontaneous reactions are often coupled to spontaneous ones. page 6-9

## 6.5 THE RATE OF BIOCHEMICAL REACTIONS IS CONTROLLED BY PROTEIN CATALYSTS CALLED ENZYMES.

Enzymes reduce the free energy level of the transition state between reactants and products, thereby reducing the energy input, or activation energy, required for a chemical reaction to proceed. page 6-10

During catalysis, the substrate and product form a complex with the enzyme. Transient covalent bonds and/or weak noncovalent interactions stabilize the complex. page 6-11

The size of the active site of an enzyme is small compared to the size of the enzyme as a whole and the active site amino acids occupy a very specific spatial arrangement. page 6-11

An enzyme is very specific toward its substrate and the types of reactions it catalyzes. page 6-12

Inhibitors reduce the activity of enzymes. They can act irreversibly or reversibly. page 6-13

Activators increase the activity of enzymes. page 6-13

Allosteric enzymes bind activators and inhibitors at sites other than the active site, resulting in a change in their shape and activity. Allosteric enzymes are often found at or near the start of a metabolic pathway or at the crossroads of multiple pathways. page 6-14

# Self-Assessment

1. Describe four ways in which organisms obtain energy and carbon from the environment.

2. Distinguish between catabolism and anabolism.

3. Define energy; describe the two forms that it takes and provide an example of each.

4. Explain how molecules contain chemical energy in their bonds.

5. Draw the structure of ATP, indicating the bonds that are broken during hydrolysis.

6. Name and describe the first and second laws of thermodynamics.

7. Given a set of thermodynamic parameters, indicate whether a chemical reaction is spontaneous or not.

8. Describe how the hydrolysis of ATP can drive non-spontaneous reactions in a cell.

9. Give three characteristics of enzymes.

10. Explain how protein folding allows for enzyme specificity.

Do you understand the chapter's Core Concepts? Log in to check your answers to the Self-Assessment questions, then practice what you've learned and reinforce this chapter's concepts by working through the problems and multimedia tutorials provided there.

www.biologyhowlifeworks.com

## CHAPTER 7

# CELLULAR RESPIRATION

## Harvesting Energy from Carbohydrates and Other Fuel Molecules

The ability to harness energy from the environment is a key attribute of life. We have seen that energy is needed for all kinds of tasks—among them cell movement and division, muscle contraction, growth and development, and the synthesis of macromolecules. Organic molecules such as carbohydrates, lipids, and proteins are good sources of energy. Some organisms, like humans and other heterotrophs, obtain organic molecules by consuming them in their diet. Others, like plants and other autotrophs, synthesize these molecules on their own, as we discuss more fully in the next chapter. Regardless of how these molecules are obtained, nearly all organisms—animals, plants, fungi, and microbes—break them down in the process of **cellular respiration,** releasing energy that can be used to do the work of the cell. Cellular respiration is a series of chemical reactions that convert the energy stored in fuel molecules into a chemical form that can be readily used by cells.

It is tempting to think that organic molecules are converted to energy in this process, but this is not the case. Recall from Chapter 6 that the first law of thermodynamics (the law of conservation of energy) states that energy cannot be created or destroyed. Biological processes, like all processes, are subject to the laws of thermodynamics. As a result, the process of cellular respiration converts the chemical potential energy stored in organic molecules to a form of chemical potential energy that is useful to cells: **adenosine triphosphate,** or **ATP.** ATP is the universal energy currency for all cells (Chapter 6).

It is also easy to forget that organisms other than animals, such as plants, use cellular respiration. If plants use sunlight as a source of energy, why would they need cellular respiration? As we will see in the next chapter, plants use the energy of sunlight to make carbohydrates. Plants then break down these carbohydrates in the process of cellular respiration to produce ATP.

In this chapter, we discuss the breakdown, storage, and mobilization of sugars such as glucose, the synthesis of ATP, and the coordination and regulation of metabolic pathways that supply the energy needs of a cell.

---

## 7.1 AN OVERVIEW OF CELLULAR RESPIRATION

In the last chapter, we saw that catabolism describes the set of chemical reactions that break down molecules into smaller units. In the process, these reactions release chemical energy and store it in molecules of ATP. Anabolism, by contrast, is the set of chemical reactions that build molecules from smaller units. Anabolic reactions require an input of energy, usually in the form of ATP.

Cellular respiration is one of the major sets of catabolic reactions in a cell. During cellular respiration, fuel molecules such as glucose, fatty acids, and proteins are catabolized into smaller units, releasing the energy stored in their chemical bonds to power the work of the cell.

## Cellular respiration occurs in four stages.

Cellular respiration is a series of catabolic reactions that convert the energy stored in food molecules, such as glucose, into ATP. It can occur in the presence of oxygen (termed aerobic respiration) or in the absence of oxygen (termed anaerobic respiration). Most organisms are capable of aerobic respiration; some bacteria respire anaerobically (Chapter 26). Here we focus on aerobic respiration. Oxygen is consumed in aerobic respiration, and carbon dioxide and water are produced. The process occurs in four stages (**Fig. 7.1**).

In stage 1, glucose, fatty acids, or amino acids are partially broken down and a modest amount of energy is released. In this chapter, we focus on the breakdown of glucose to make pyruvate, a process known as **glycolysis.**

In stage 2, pyruvate is converted to another molecule called acetyl-coenzyme A (acetyl-CoA) and carbon dioxide is produced.

Stage 3 is the **citric acid cycle.** During this stage, acetyl-CoA is broken down and more carbon dioxide is released.

In stages 1–3, chemical energy is transferred to two types of energy-storing molecules: ATP and **electron carriers.** ATP was discussed in Chapter 6. Electron carriers are molecules that store and transfer energy in the form of "high-energy" or "excited" electrons, discussed more fully below.

Stage 4 is **oxidative phosphorylation.** In this series of reactions, electron carriers generated in stages 1–3 donate their high-energy electrons to an **electron transport chain** (also called a respiratory chain). Electron transport chains transfer electrons along a series of membrane-associated proteins to a final electron acceptor and harness the energy of the electrons to produce a large amount of ATP. In aerobic respiration, oxygen is the final electron acceptor, so oxygen is consumed and water is produced in the process.

In eukaryotes, glycolysis takes place in the cytoplasm, and the citric acid cycle and oxidative phosphorylation take place in mitochondria. The electron transport chain is made up of proteins associated with the inner mitochondrial membrane (Chapter 5). In some bacteria, these reactions take place in the cytoplasm, and the electron transport chain is located in the plasma membrane. It is thought that mitochondria were once free-living bacteria but are now unable to live outside their host eukaryotic cell. This idea, known as the endosymbiotic theory, is discussed more fully in Chapter 27.

Before turning to the details of the four stages of cellular respiration, we take a moment to consider how energy is stored and used in cells. In particular, we focus on why carbohydrates and lipids are such good sources of energy.

**FIG. 7.1**    **Cellular respiration.** In most organisms, cellular respiration consumes oxygen and produces carbon dioxide, water, and ATP.

*Stage 1*

*Glycolysis*

**Stages 1 & 2:** Fuel molecules are partially broken down, producing ATP and electron carriers.

*Stage 2*

*Acetyl-CoA synthesis*

*Stage 3*

*Citric acid cycle*

**Stage 3:** Fuel molecules are fully broken down, producing ATP and electron carriers.

*Stage 4*

*Oxidative phosphorylation*

**Stage 4:** Electron carriers donate electrons to the electron transport chain, leading to the synthesis of ATP.

## Cellular respiration involves a series of redox reactions.

You know from everyday experience that carbohydrates and lipids are good sources of energy. The key to understanding how energy is stored in these molecules is to understand **oxidation–reduction reactions** ("redox reactions" for short). In biological systems, oxidation–reduction reactions are often used to store or release chemical energy.

Oxidation is the loss of electrons, and reduction is the gain of electrons. The loss and gain of electrons always occur together in a single oxidation–reduction reaction: Electrons are transferred from one molecule to another so that one molecule loses electrons and one molecule gains those electrons.

A familiar example of an oxidation–reduction reaction is the reaction of iron and oxygen to make iron oxide—that is, rust. The overall reaction can be written as follows:

$$4Fe + 3O_2 \rightarrow 2Fe_2O_3$$

We can also write this reaction as two half-reactions in order to follow the electrons:

$$4Fe \rightarrow 2Fe_2^{3+} + 12e^-$$
$$3O_2 + 12e^- \rightarrow 2O_3^{2-}$$

As you can see from these half-reactions, iron loses electrons and is therefore oxidized, while oxygen gains electrons and is reduced.

$$
\begin{array}{c}
\overset{\text{Oxidation}}{\overbrace{\phantom{aaaaaaaaaaaaa}}} \\
4Fe \rightarrow 2Fe_2^{3+} + 12e^- \\
3O_2 + 12e^- \rightarrow 2O_3^{2-} \\
\underset{\text{Reduction}}{\underbrace{\phantom{aaaaaaaaaaaaa}}}
\end{array}
$$

The electrons lost by iron are gained by oxygen, so the original reaction can be written as follows:

$$
\begin{array}{c}
\overset{\text{Oxidation}}{\overbrace{\phantom{aaaaaaaaaaaaa}}} \\
4Fe + 3O_2 \rightarrow 2Fe_2O_3 \\
\underset{\text{Reduction}}{\underbrace{\phantom{aaaaaaaaaaaaa}}}
\end{array}
$$

In many biological systems, electrons are not completely transferred between molecules. Instead, there is a change in electron density around an atom. In this case, oxidation is a decrease in electron density and reduction is an increase in electron density. Another way to think about these reactions is that electrons are partially lost or gained. The breakdown of the sugar glucose provides an example. The overall reaction can be written as follows:

$$C_6H_{12}O_6 + 6O_2 \rightarrow 6CO_2 + 6H_2O + \text{energy}$$
$$\text{Glucose} \quad \text{Oxygen} \quad \underset{\text{dioxide}}{\text{Carbon}} \quad \text{Water}$$

This reaction is an example of an oxidation–reduction reaction in which glucose is oxidized to carbon dioxide and at the same time oxygen is reduced to water:

$$
\begin{array}{c}
\overset{\text{Oxidation}}{\overbrace{\phantom{aaaaaaaaaaaaa}}} \\
C_6H_{12}O_6 + 6O_2 \rightarrow 6CO_2 + 6H_2O + \text{energy} \\
\underset{\text{Reduction}}{\underbrace{\phantom{aaaaaaaaaaaaa}}}
\end{array}
$$

To understand why this is an oxidation–reduction reaction, consider where the electrons are in the reactants and products. Let's first compare glucose and carbon dioxide. The carbon atoms of glucose are bound to other carbon atoms, hydrogen atoms, and oxygen atoms. In the case of C—C and C—H covalent bonds, electrons are shared about equally between the two atoms (**Fig. 7.2**). By contrast, in carbon dioxide, they are not shared equally. The oxygen atom is more electronegative than the carbon atom, so the electrons that are shared between carbon and oxygen spend more time near the oxygen atom (Fig. 7.2a). As a result,

---

**FIG. 7.2    Electron sharing in covalent bonds.** Electrons are more likely to be found near the more electronegative atom in a bond.

**a.** Carbon atoms are oxidized in the conversion of glucose to carbon dioxide.

**b.** Oxygen atoms are reduced in the conversion from oxygen to water.

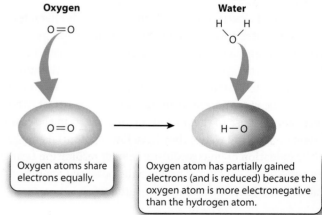

in the preceding reaction, carbon has partially lost electrons to oxygen. We say that the carbon atoms in glucose are oxidized, or simply that glucose is oxidized.

Let's consider the reduction reaction by comparing oxygen gas and water (Fig. 7.2b). In oxygen gas, electrons are shared equally between two oxygen atoms. In water, the electrons that are shared between hydrogen and oxygen spend more time near oxygen because oxygen is more electronegative than hydrogen. As a result, in the preceding reaction, the electron density around the oxygen atom has increased, so oxygen has partially gained electrons and is reduced.

Because oxygen gains electrons, it is called an **electron acceptor.** It also oxidizes glucose, so it can be referred to as an **oxidizing agent.** Glucose, conversely, is an **electron donor** and therefore a **reducing agent.**

Oxidation and reduction reactions are defined by the loss or gain of electrons. But note that in the oxidation of glucose, electrons always travel with hydrogen atoms. Therefore, the movement of hydrogen atoms can be used as an aid to recognize which molecules become oxidized and which ones become reduced. For example, we can see that the carbon atoms of glucose lose hydrogen atoms in the conversion to carbon dioxide; therefore, glucose loses electrons and is oxidized. In the same reaction, oxygen gains hydrogen atoms in becoming water; therefore, it gains electrons and is reduced. Using hydrogen atoms as a marker for electrons is often a convenient way to follow redox reactions that occur in biological systems. However, the movement of electrons is key, as they are the entities that carry and transfer energy in redox reactions.

## Chemical energy is stored in reduced molecules such as carbohydrates and lipids.

We can now address the question of why reduced molecules, such as carbohydrates and lipids, have so much energy. The answer has to do with how the atoms of these molecules share electrons.

The shared electrons of C—C and C—H bonds characteristic of carbohydrates and lipids have high potential energy. The high potential energy results from the fact that the electrons are, on average, far from the nucleus of the atoms. As discussed in Chapter 6, the farther an electron is from the nucleus, the more potential energy it has. By contrast, when electrons are close to the nucleus of an atom, as in the case of carbon dioxide and water, they have lower potential energy. So the oxidation of glucose during cellular respiration results in a large release of energy because the potential energy of the reactants is higher than that of the products.

The reaction for the oxidation of glucose helps us focus on the starting reactants, final products, and release of energy. However, it misses the many intermediate steps that take place as the cell oxidizes glucose. Tossing a match into the gas tank of a car would release a tremendous amount of energy in the form of an explosion, but this energy would not be used productively. Similarly, if all the energy stored in glucose were released at once, most of it would be released as heat and the cell would not be able to harness it to do work. The energy in organic molecules is released gradually in a series of chemical reactions (**Fig. 7.3**). By oxidizing glucose slowly and in a controlled manner, the chemical energy stored in glucose can be harnessed in the chemical bonds of other molecules, such as ATP and electron carriers. ATP carries energy in its chemical bonds (Chapter 6). Electron carriers contain chemical energy in the form of high-energy electrons, which can be harnessed to produce ATP.

## Electron carriers transport high-energy electrons.

As we just saw, many of the chemical reactions involved in cellular respiration are redox reactions that store energy in the form of electron carriers. Two important electron carriers in cells are the molecules nicotinamide adenine dinucleotide ($NAD^+$/NADH) and flavin adenine dinucleotide (FADH/ $FADH_2$). These electron carriers exist in two forms—an oxidized form ($NAD^+$ and FAD) and a reduced form (NADH and $FADH_2$). The oxidized form accepts electrons in many of the reactions of cellular respiration and becomes reduced. The reduced form has high potential energy. The electrons it gains in redox reactions are then donated to the electron transport chain. The electron transport chain transfers these electrons to oxygen, the final electron acceptor, releasing energy used to synthesize ATP.

Electron transport chains are used in respiration to extract energy from fuel molecules such as glucose, as well as in photosynthesis to extract energy from sunlight (Chapter 8).

## ATP is generated by substrate-level phosphorylation and oxidative phosphorylation.

The chemical energy stored in a molecule of glucose is used to produce ATP in two different ways during cellular respiration. In some reactions, a phosphorylated organic molecule directly transfers a phosphate group to ADP. The free energy difference ($\Delta G$) of hydrolysis of these phosphorylated molecules is more negative than that of ATP hydrolysis and so can drive the synthesis of ATP by energetic coupling (Chapter 6). This way of generating ATP is called **substrate-level phosphorylation** because a phosphate group is transferred to ADP from an organic

**FIG. 7.3** **Energy is released gradually in cellular respiration.** (a) Glucose is oxidized slowly through many reactions. (b) Many of these reactions release energy in the form of ATP and electron carriers.

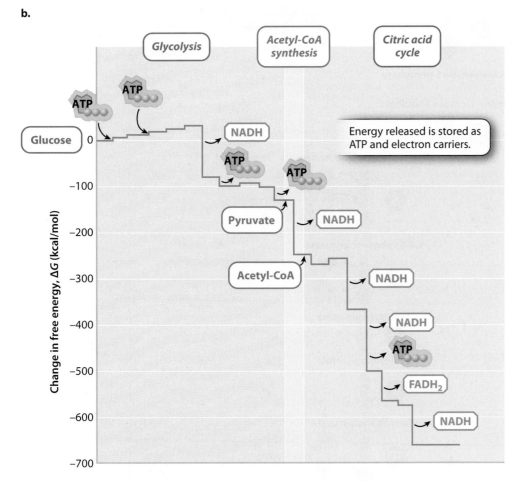

molecule, which acts as a phosphate donor or substrate. Substrate-level phosphorylation produces only a small amount of the total ATP generated in the process of cellular respiration.

Most of the ATP generated in cellular respiration is produced in an entirely different manner—by oxidative phosphorylation. ATP is generated indirectly in the reactions of oxidative phosphorylation, through the reduction of electron carriers, the transfer of high-energy electrons from electron carriers to the electron transport chain, and the subsequent synthesis of ATP from ADP and inorganic phosphate ($P_i$).

→ **Quick Check 1** For each of the following pairs of molecules, indicate which member of the pair has higher and which has lower levels of chemical energy: ADP/ATP; $NAD^+$/NADH; FAD/$FADH_2$; $CO_2$/$C_6H_{12}O_6$; oxidized molecules/corresponding reduced molecules.

## 7.2 GLYCOLYSIS: THE SPLITTING OF SUGAR

Glucose is the most common fuel molecule in animals, plants, and microbes. It is the starting molecule for glycolysis, which results in the partial oxidation of glucose and the synthesis of a relatively small amount of ATP and electron carriers. Glycolysis literally means "splitting sugar," an apt name because in glycolysis a 6-carbon sugar (glucose) is split in two, yielding two 3-carbon molecules. The process is anaerobic because oxygen is not consumed. Glycolysis evolved very early in the evolution of life, when oxygen was not present in Earth's atmosphere.

**FIG. 7.4** **Glycolysis**. Glucose is partially oxidized to pyruvate, with the net production of 2 ATP and 2 NADH.

**TABLE 7.1    Approximate Total ATP Yield in Cellular Respiration**

| PATHWAY | SUBSTRATE-LEVEL PHOSPHORYLATION | OXIDATIVE PHOSPHORYLATION | TOTAL ATP |
|---|---|---|---|
| Glycolysis (Glucose → 2 pyruvate) | 2 ATP | 2 NADH = 5 ATP | 7 |
| Acetyl-CoA Synthesis (2 Pyruvate → 2 acetyl-CoA) | 0 ATP | 2 NADH = 5 ATP | 5 |
| Citric acid cycle (2 turns, 1 for each acetyl-CoA) | 2 ATP | 6 NADH = 15 ATP $\phantom{xx}$ 2 FADH$_2$ = 3 ATP | 20 |
| Total | 4 ATP | 28 ATP | 32 |

It occurs in nearly all living organisms and is therefore probably the most widespread metabolic pathway among organisms.

### Glycolysis is the partial breakdown of glucose.

Glycolysis begins with a molecule of glucose and produces two 3-carbon molecules of pyruvate and a net total of two molecules of ATP and two molecules of the electron carrier NADH. ATP is produced directly by the addition of a phosphate group to ADP, or substrate-level phosphorylation.

Glycolysis is a series of 10 chemical reactions (**Fig. 7.4**). These reactions can be divided into three phases. The first phase (reactions 1–3) prepares glucose for the next two phases by the addition of two phosphate groups to glucose, producing fructose 1,6-bisphosphate. This phase requires an input of energy. To supply that energy and the phosphate groups, two molecules of ATP are hydrolyzed per molecule of glucose. In other words, the first phase of glycolysis is an endergonic process. The phosphorylation of glucose has two important consequences. While glucose enters and exits cells through specific membrane transporters, phosphorylated glucose is trapped inside the cell. In addition, the presence of two negatively charged phosphate groups in proximity destabilizes the molecule so that it can be broken apart in the second phase of glycolysis.

The second phase (reactions 4 and 5) is the cleavage phase. In reaction 4, the 6-carbon molecule fructose 1,6-bisphosphate is split into two 3-carbon molecules, glyceraldehyde 3-phosphate and its isomer dihydroxyacetone phosphate. Then, in reaction 5, dihydroxyacetone phosphate is converted to glyceraldehyde 3-phosphate. For each molecule of glucose entering glycolysis, two 3-carbon glyceraldehyde 3-phosphate molecules enter the third phase.

Reactions 6–10 are the third and final phase of glycolysis. This phase ends with the production of two molecules of pyruvate. It is sometimes called the payoff phase because ATP

and the electron carrier NADH are produced. Later, NADH will contribute to the synthesis of ATP during oxidative phosphorylation.

In summary, glycolysis begins with a single molecule of glucose (six carbons) and produces two molecules of pyruvate (three carbons each). These reactions yield four molecules of ATP and two molecules of NADH. However, two ATP molecules are consumed during the initial phase of glycolysis, resulting in a net gain of two ATP molecules and two molecules of NADH (Fig. 7.4; **Table 7.1**). The overall chemical equation for glycolysis is as follows:

Glucose + 2 NAD$^+$ + 2 ADP + 2 P$_i$ →
$$2 \text{ pyruvate} + 2 \text{ ATP} + 2 \text{ NADH} + 2\text{H}^+ + 2\text{H}_2\text{O}$$

→ **Quick Check 2** At the end of glycolysis, but before the subsequent steps in cellular respiration, which molecules contain the energy held in the original glucose molecule?

## 7.3 ACETYL-CoA SYNTHESIS

Glycolysis occurs in almost all living organisms, but it does not generate very much energy in the form of ATP. The end product, pyruvate, still contains a good deal of chemical potential energy in its bonds. In the presence of oxygen, pyruvate can be further broken down to release more energy, first to acetyl-CoA and then even further in a series of reactions known as the citric acid cycle. The synthesis of acetyl-CoA is a key step that links glycolysis to the citric acid cycle. In eukaryotes, this is the first step that takes place inside the mitochondria.

### The oxidation of pyruvate connects glycolysis to the citric acid cycle.

The end product of glycolysis is pyruvate, which can be transported into mitochondria. Mitochondria are rod-shaped organelles surrounded by a double membrane (**Fig. 7.5**; Chapter 5). The inner and outer mitochondrial membranes are not close to each other in all areas because the inner membrane has folds that project inward. These membranes define two spaces. The space between the inner and outer membranes is called the **intermembrane space,** and

the space enclosed by the inner membrane is called the mitochondrial **matrix.**

Pyruvate is transported into the mitochondrial matrix, where it is converted into acetyl-CoA (**Fig. 7.6**). First, part of the pyruvate molecule is oxidized and splits off to form carbon dioxide, the most oxidized (and therefore the least energetic) form of carbon. The electrons lost in this process are donated to $NAD^+$, which is reduced to NADH. The remaining part of the pyruvate molecule—an acetyl group ($CH_3OC—$)—still contains a large amount of potential energy that can be harnessed. It is transferred to coenzyme A (CoA), a molecule that carries the acetyl group to the next set of reactions. All these reactions are catalyzed by a group of enzymes called the pyruvate dehydrogenase complex.

Overall, the synthesis of one molecule of acetyl-CoA from pyruvate results in the formation of one molecule of carbon dioxide and one molecule of NADH. Recall, however, that a single molecule of glucose forms two molecules of pyruvate during glycolysis. Therefore, two molecules of carbon dioxide, two molecules of NADH, and two molecules of acetyl-CoA are produced from a single starting glucose molecule in this stage of cellular respiration (Table 7.1). Acetyl-CoA is the substrate of the first step in the citric acid cycle.

**FIG. 7.5**    **Mitochondrial membranes and compartments.**

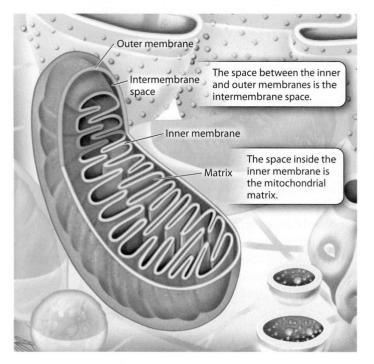

## 7.4 THE CITRIC ACID CYCLE

The citric acid cycle is the step in cellular respiration in which fuel molecules are completely oxidized. It is also called the Krebs cycle and the tricarboxylic acid (TCA) cycle. During the citric acid cycle, the chemical energy in the bonds of acetyl-CoA is transferred to ATP by substrate-level phosphorylation and to the electron carriers NADH and $FADH_2$. In this way, the citric acid cycle supplies high-energy electrons to the electron transport chain, leading to the production of much more energy in the form of ATP than is obtained by glycolysis alone.

### The citric acid cycle produces ATP and electron carriers.

Like the synthesis of acetyl-CoA, the citric acid cycle takes place in the mitochondrial matrix. It is composed of eight reactions and is called a cycle because the starting molecule, oxaloacetate, is regenerated at the

**FIG. 7.6**    **The formation of acetyl-CoA, the first substrate in the citric acid cycle.** Pyruvate is oxidized to acetyl-CoA in the mitochondrial matrix.

**FIG. 7.7** **The citric acid cycle.** The acetyl group of acetyl-CoA is completely oxidized, with the net production of one ATP, three NADH, and one $FADH_2$.

end (**Fig. 7.7**). In the first reaction, a 4-carbon molecule of oxaloacetate reacts with the 2-carbon acetyl group of acetyl-CoA to form a 6-carbon molecule of citrate. The last reaction of the cycle regenerates a molecule of oxaloacetate, allowing the cycle to continue.

The citric acid cycle results in the complete oxidation of the acetyl group of acetyl-CoA. Since the first reaction creates a molecule with six carbons and the last reaction regenerates a 4-carbon molecule, two carbons are eliminated during the cycle. These carbons are released as carbon dioxide in reactions 3 and 4. Along with the release of carbon dioxide from pyruvate in its conversion to acetyl-CoA, these reactions are the sources of the $CO_2$ released during cellular respiration and therefore the sources

of the $CO_2$ that we exhale when we breathe.

The oxidation of carbon atoms to produce $CO_2$ in reactions 3 and 4 is coupled with the reduction of the electron carrier NAD+ to NADH (Fig. 7.7). In this way, energy released in the oxidation reactions is transferred to NADH. Additional electron carriers are produced in reactions 6 and 8, which are also redox reactions. In fact, the citric acid cycle produces a large quantity of electron carriers: three molecules of NADH and one molecule of $FADH_2$ per turn of the cycle. These electron carriers donate high-energy electrons in the next stage of cellular respiration to power the synthesis of ATP.

Reaction 5 is a substrate-level phosphorylation reaction that generates a molecule of GTP (Fig. 7.7). GTP can transfer its terminal phosphate to a molecule of ADP to form ATP. Reaction 5 is the only substrate-level phosphorylation in the citric acid cycle.

Overall, two molecules of acetyl-CoA produced from a single molecule of glucose yield two molecules of ATP, six molecules of NADH, and two molecules of $FADH_2$ in the citric acid cycle (Table 7.1).

→  **Quick Check 3**  At the end of the citric acid cycle, but before the subsequent steps of cellular respiration, which molecules contain the energy held in the original glucose molecule?

**?** **CASE 1  The First Cell: Life's Origins**
**What were the earliest energy-harnessing reactions?**
Some bacteria run the citric acid cycle in reverse, incorporating carbon dioxide into organic molecules instead of liberating it. Running the citric acid cycle in reverse requires energy, which is supplied by sunlight (Chapter 8) or chemical reactions (Chapter 26).

Why would an organism run the citric acid cycle in reverse? The answer is that the intermediates generated step by step as the cycle turns provide the building blocks for synthesizing the cell's key biomolecules. This is true whether the cycle is run in the reverse or forward direction (**Fig. 7.8**). Pyruvate, for example, is the starting point for the synthesis of sugars and the amino

acid alanine; acetate is the starting point for the synthesis of the cell's lipids; oxaloacetate is modified to form different amino acids and pyrimidine bases; and α- (alpha-) ketoglutarate is modified to form other amino acids. For organisms that run the cycle in the forward direction, the citric acid cycle is used to generate both energy-storing molecules (ATP and electron carriers) and intermediates in the synthesis of other molecules. For organisms that run the cycle in reverse, it is used to generate intermediates in the synthesis of other molecules and also to reduce carbon dioxide and incorporate carbon into organic molecules.

The centrality of the citric acid cycle to both the synthesis of biomolecules and to meeting the energy requirements of cells suggests to some biologists that this cycle evolved early, appearing in some of the first cells to feature metabolism. This early appearance of the citric acid cycle, in turn, implies that the great variety of biosynthetic and energy-yielding pathways found in modern cells evolved through the extension and modification of this deeply rooted cycle. As always in evolution, new cellular capabilities arose by the modification of preexisting capabilities. Complex networks of highly specific pathways evolved from a simpler, more general set of reactions.

---

## 7.5 THE ELECTRON TRANSPORT CHAIN AND OXIDATIVE PHOSPHORYLATION

The complete oxidation of glucose during glycolysis and the citric acid cycle results in the production of two kinds of electron carriers: NADH and $FADH_2$. We are now going to see how the energy stored in these electron carriers is used to synthesize ATP.

The energy in these electron carriers is released in a series of redox reactions that occur as electrons pass through a chain of protein complexes in the inner mitochondrial membrane to the final electron acceptor, oxygen, which is reduced to water. The energy in these electrons is not converted directly into the chemical energy of ATP, however. Instead, the passage of electrons is coupled to the transfer of protons ($H^+$) across the inner mitochondrial membrane, creating a concentration and charge gradient (Chapter 5). This electrochemical gradient provides a source of potential energy that is then used to drive the synthesis of ATP.

We next explore the properties of the electron transport chain, the proton gradient, and the synthesis of ATP.

### The electron transport chain transfers electrons and pumps protons.

Electrons are not directly transported from NADH and $FADH_2$ to oxygen. Instead, they are transported along a series of four large protein complexes that form the electron transport chain (complexes I to IV). These are shown in **Fig. 7.9**. These membrane proteins are embedded in the mitochondrial inner membrane (see Fig. 7.5). The inner mitochondrial membrane contains one of the highest concentrations of proteins found in eukaryotic membranes.

---

**FIG. 7.8**    **Molecules produced by the citric acid cycle.** Many organic molecules can be synthesized from citric acid cycle intermediates.

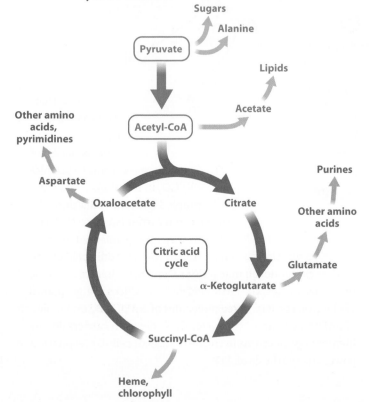

**FIG. 7.9    The electron transport chain.** (a) The electron transport chain consists of four complexes (I to IV) in the inner mitochondrial membrane. (b) Electrons flow from electron carriers to oxygen, the final electron acceptor. (c) The proton gradient formed from the electron transport chain has potential energy that is used to synthesize ATP.

**a.** The electron transport chain in cellular respiration

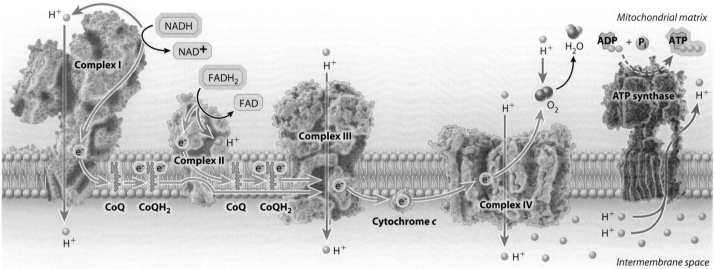

**b.** Electron transport

Complexes I and II harvest electrons from NADH and FADH$_2$.

Cytochrome *c* moves to complex IV where oxygen is reduced to form water.

Coenzyme Q transports electrons from complexes I and II to complex III and is reduced to CoQH$_2$.

**c.** Proton transport and ATP synthesis

ATP synthase uses the electrochemical proton gradient to drive the synthesis of ATP.

The transport of electrons in complexes I, III, and IV is coupled with the transport of protons across the inner membrane, from the mitochondrial matrix to the intermembrane space.

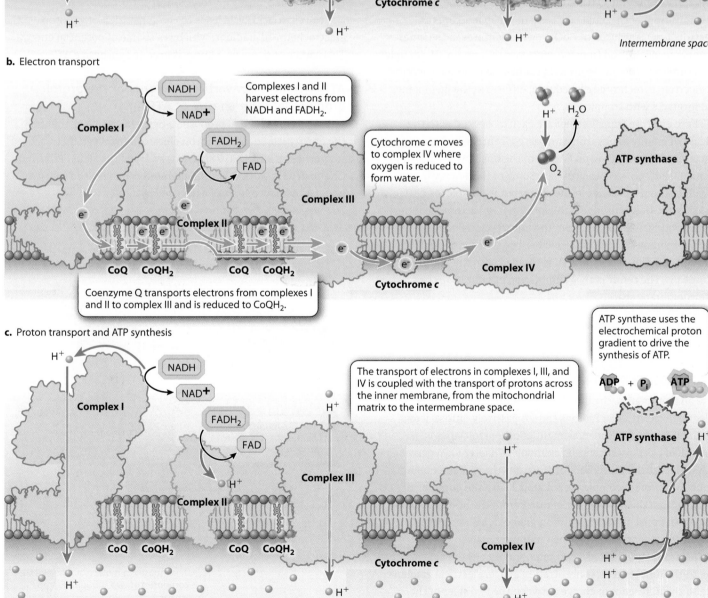

Electrons enter the electron transport chain via either complex I or II. Electrons donated by NADH enter through complex I, and electrons donated by FADH$_2$ enter through complex II (complex II is the same enzyme that catalyzes step 6 in the citric acid cycle). These electrons are transported through either complex I or II to complex III and then through complex IV.

Within each protein complex of the electron transport chain, electrons are passed from electron donors to electron acceptors. Each donor and acceptor is a redox couple, consisting of an oxidized and a reduced form of a molecule. The electron transport chain contains many of these redox couples. As electrons are passed from donors to acceptors, the energy of the electrons is reduced. Each electron acceptor therefore binds electrons more strongly than the previous one in the chain. Oxygen is the final electron acceptor, and it has the highest affinity for electrons. When oxygen accepts an electron, it is reduced to water.

Electrons also must be transported between the four complexes (Fig. 7.9). **Coenzyme Q** (CoQ), also called ubiquinone, accepts electrons from both complexes I and II. When it accepts an electron, CoQ is reduced to CoQH$_2$, which diffuses in the inner membrane, docks, and transfers electrons to complex III. Complex III in turn transfers electrons to **cytochrome c**. When it accepts an electron, cytochrome *c* is reduced, diffuses in the membrane, and interacts with complex IV.

These electron transfer steps are each associated with the release of energy as electrons are passed from the high-energy electron carriers NADH and FADH$_2$ to the final low-energy (high-affinity) electron acceptor, oxygen. Some of this energy is used to reduce the next carrier in the chain, but in complexes I, III, and IV some of it is used to pump protons (H$^+$) across the inner mitochondrial membrane, from the mitochondrial matrix to the intermembrane space (Fig. 7.9). Thus, the transfer of electrons through complexes I, III, and IV is coupled with the pumping of protons. The result is an accumulation of protons in the intermembrane space.

→   **Quick Check 4** Animals breathe in air that contains more oxygen than the air they breathe out. Where is oxygen consumed?

### The proton gradient is a source of potential energy.

Like all membranes, the inner mitochondrial membrane is selectively permeable: Protons cannot passively diffuse across this membrane, and the movement of other molecules is controlled by transporters and channels (Chapter 5). We have just seen that the movement of electrons through membrane-embedded protein complexes is coupled with the pumping of protons from the mitochondrial matrix into the intermembrane space. The consequence is a proton gradient, a difference in proton concentration across the inner membrane.

The proton gradient has two components: a chemical gradient due to the difference in concentration and an electrical gradient due to the difference in charge between the two sides of the

membrane. To reflect the dual contribution of the concentration gradient and the electrical gradient, the proton gradient is also called an electrochemical gradient.

The proton gradient is a source of potential energy, as discussed in Chapters 5 and 6. It stores energy much in the same way that a battery or a dam does. Through the actions of the electron transport chain, protons have a high concentration in the intermembrane space and a low concentration in the mitochondrial matrix. As a result, there is a tendency for protons to diffuse back to the mitochondrial matrix, driven by a difference in concentration and charge on the two sides of the membrane. This movement, however, is blocked by the membrane, so the gradient holds potential energy. That energy can be harnessed if a pathway is opened through the membrane because, as we will see shortly, the resulting movement of the protons through the membrane can be used to perform work.

In sum, the oxidation of the electron carriers NADH and FADH$_2$ formed during glycolysis, acetyl-CoA synthesis, and the citric acid cycle leads to the generation of a proton electrochemical gradient, which is a source of potential energy. This source of potential energy is used to synthesize ATP.

### ATP synthase converts the energy of the proton gradient into the energy of ATP.

In 1961, Peter Mitchell proposed a hypothesis to explain how the energy stored in the proton electrochemical gradient is used to synthesize ATP. In 1978, he was awarded the Nobel Prize in Chemistry for work that fundamentally changed the way we understand how energy is harnessed by a cell.

---

**FIG. 7.10   ATP synthase.** ATP synthase drives the synthesis of ATP by means of an electrochemical proton gradient.

**2** The F$_1$ subunit then uses this rotational energy to catalyze the synthesis of ATP.

**1** The F$_0$ subunit forms a channel that rotates as protons pass through it.

ATP synthase is a molecular machine that is composed of two subunits, F$_0$ and F$_1$.

According to Mitchell's hypothesis, the gradient of protons provides a source of potential energy that is converted into chemical energy stored in ATP. First, for the potential energy of the proton gradient to be released, there must be an opening in the membrane for the protons to flow through. Mitchell suggested that protons in the intermembrane space diffuse down their electrical and concentration gradients through a transmembrane protein channel into the mitochondrial matrix. Second, the movement of protons through the enzyme must be coupled with the synthesis of ATP. This coupling is made possible by **ATP synthase,** a remarkable enzyme composed of two distinct subunits called $F_0$ and $F_1$ (**Fig. 7.10**). $F_0$ forms the channel in the inner mitochondrial membrane through which protons flow; $F_1$ is the catalytic unit that synthesizes ATP. Proton flow through the channel ($F_0$) makes it possible for the enzyme ($F_1$) to synthesize ATP.

Proton flow through the $F_0$ channel causes it to rotate, converting the energy of the proton gradient into mechanical rotational energy, a form of kinetic energy. The rotation of the $F_0$ subunit leads to rotation of the $F_1$ subunit in the mitochondrial matrix (Fig. 7.10). The rotation of the $F_1$ subunit in turn causes conformational changes that allow it to catalyze the synthesis of ATP from ADP and $P_i$. In this way, mechanical rotational energy is converted into the chemical energy of ATP.

Direct experimental evidence for Mitchell's hypothesis did not come for over a decade. One of the key experiments that provided support for Mitchell's hypothesis is illustrated in **Fig. 7.11**.

## HOW DO WE KNOW?

**FIG. 7.11**

# Can a proton gradient drive the synthesis of ATP?

**BACKGROUND** Peter Mitchell's hypothesis that a proton gradient can drive the synthesis of ATP was proposed before experimental evidence supported it and was therefore met with skepticism. In the 1970s, biochemist Efraim Racker and his collaborator Walther Stoeckenius tested the hypothesis.

**EXPERIMENT** Racker and Stoeckenius built an artificial system consisting of a membrane, a bacterial proton pump activated by light, and ATP synthase.

They measured the concentration of protons in the external medium and the amount of ATP produced in the presence and absence of light.

**RESULTS** In the presence of light, the concentration of protons increased inside the vesicles, suggesting that protons were taken up by the vesicles.

In the dark, the concentration of protons returned to the starting level. ATP was generated in the light, but not in the dark.

| Condition | Relative level of ATP |
|-----------|----------------------|
| Light | 594 |
| Dark | 23 |

**INTERPRETATION** In the presence of light, the proton pump was activated and protons were pumped to one side of the membrane, leading to the formation of a proton gradient. The proton gradient, in turn, powered synthesis of ATP via ATP synthase.

**CONCLUSION** A membrane, proton gradient, and ATP synthase are sufficient to synthesize ATP. This result provided experimental evidence for Mitchell's hypothesis.

**SOURCES** Mitchell, P. 1961. "Coupling of Phosphorylation to Electron and Hydrogen Transfer by a Chemiosmotic Type of Mechanism." *Nature* 191:144–148; Racker, E., and W. Stoeckenius. 1974. "Reconstitution of Purple Membrane Vesicles Catalyzing Light-Driven Proton Uptake and Adenosine Triphosphate Formation." *J. Biol. Chem.* 249:662–663.

**FIG. 7.12    The flow of energy in cellular respiration.** A single glucose molecule yields 32 ATP molecules.

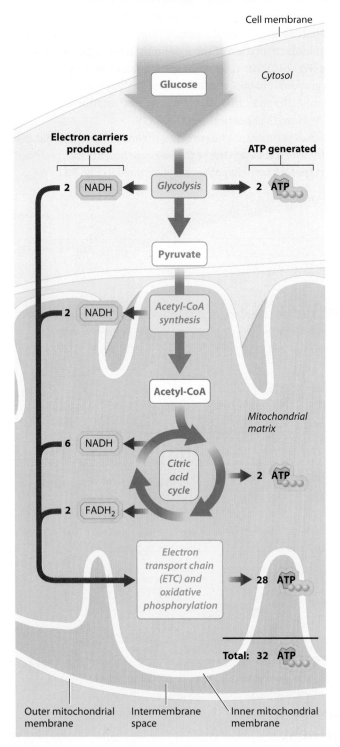

→  **Quick Check 5** Uncoupling proteins are proteins spanning the inner mitochondrial membrane that allow protons to pass through the membrane and bypass the channel of ATP synthase. Describe the consequences to the proton gradient and ATP production.

Approximately 2.5 molecules of ATP are produced for each NADH that donates electrons to the chain and 1.5 molecules of ATP for each FADH$_2$. Therefore, overall, the complete oxidation of glucose yields about 32 molecules of ATP from glycolysis, acetyl-CoA synthesis, the citric acid cycle, and oxidative phosphorylation (Table 7.1). The energy held in the bonds of glucose is now stored in a molecule that can be readily used by cells.

It is worth taking a moment to follow the flow of energy in cellular respiration, illustrated in its full form in **Fig. 7.12**. We began with glucose and noted that it held chemical potential energy in its covalent bonds. This energy is released slowly in a series of reactions and captured in chemical form. Some of these reactions generate ATP directly by substrate-level phosphorylation. Others are redox reactions that transfer energy to the electron carriers NADH and FADH$_2$. These electron carriers donate electrons to the electron transport chain, which uses the energy stored in the electron carriers to pump protons across the inner membrane of the mitochondria. In other words, the energy of the electron carriers is transformed into energy stored in a proton electrochemical gradient. ATP synthase then converts the energy of the proton gradient to rotational energy, which drives the synthesis of ATP. The cell now has a form of energy that it can use in many ways to perform work.

## 7.6 ANAEROBIC METABOLISM AND THE EVOLUTION OF CELLULAR RESPIRATION

Up to this point, we have followed a single metabolic path: the breakdown of glucose in the presence of oxygen to produce carbon dioxide and water. However, metabolic pathways more often resemble intersecting roads rather than a single, linear path. We saw this earlier in the discussion of the citric acid cycle, where intermediates in the cycle often feed into other metabolic pathways.

One of the major forks in the metabolic road occurs at pyruvate, the end product of glycolysis (section 10.2). When oxygen is present, it is converted to acetyl-CoA, which then enters the citric acid cycle, resulting in the production of ATP and electron carriers to fuel the electron transport chain, as we saw.

When oxygen is not present, however, pyruvate is metabolized along a number of different pathways. These pathways occur in many living organisms today and played an important role in the early evolution of life on Earth.

## Fermentation extracts energy from glucose in the absence of oxygen.

Pyruvate, the end product of glycolysis, is a molecule shared by many metabolic pathways and it therefore has many possible fates in the cell. In the presence of oxygen, most of the pyruvate is used to synthesize acetyl-CoA, which in turn fuels the citric acid cycle. In the absence of oxygen, pyruvate can be broken down by **fermentation,** which does not rely on oxygen or a similar outside electron acceptor. Fermentation is accomplished through a wide variety of metabolic pathways that extract energy from fuel molecules such as glucose. Fermentation pathways are important for anaerobic organisms that live without oxygen, as well as some organisms such as yeast that favor fermentation over oxidative phosphorylation even in the presence of oxygen. It is also sometimes used in aerobic organisms when oxygen cannot be delivered fast enough to meet the cell's metabolic needs, as in exercising muscle.

Recall that during glycolysis, glucose is oxidized to form pyruvate, and $NAD^+$ is reduced to form NADH. For glycolysis to continue, NADH must be oxidized to $NAD^+$. The regeneration of $NAD^+$ is important because without it glycolysis would grind to a halt. In the presence of oxygen, $NAD^+$ is regenerated when NADH donates its electrons to the electron transport chain. In the absence of oxygen during fermentation, NADH is oxidized to $NAD^+$ when pyruvate or a derivative of pyruvate is reduced.

There are many fermentation pathways, especially in bacteria. Two of the major ones are **lactic acid fermentation** and **ethanol fermentation** (**Fig. 7.13**). Lactic acid fermentation occurs in animals and bacteria. During lactic acid fermentation, electrons from NADH are transferred to pyruvate to produce lactic acid and $NAD^+$ (Fig. 7.13a). The overall chemical reaction is written as follows:

$$\text{Glucose} + 2\,\text{ADP} + 2\,\text{P}_i \rightarrow 2\,\text{lactic acid} + 2\,\text{ATP} + 2\text{H}_2\text{O}$$

Ethanol fermentation occurs in plants and fungi. During ethanol fermentation, pyruvate releases carbon dioxide to form acetaldehyde, and electrons from NADH are transferred to acetaldehyde to produce ethanol and $NAD^+$ (Fig. 7.13b). The overall chemical reaction is written as follows:

$$\text{Glucose} + 2\,\text{ADP} + 2\,\text{P}_i \rightarrow 2\,\text{ethanol} + 2\text{CO}_2 + 2\,\text{ATP} + 2\text{H}_2\text{O}$$

In both fermentation pathways, NADH is oxidized to $NAD^+$. However, NADH and $NAD^+$ do not appear in the overall chemical equations because there is no net production or loss of either molecule. $NAD^+$ molecules that are reduced during glycolysis are oxidized when lactic acid or ethanol is formed.

**FIG. 7.13    Lactic acid and ethanol fermentation pathways.**

**a.** Lactic acid fermentation

**b.** Ethanol fermentation

The breakdown of a molecule of glucose by fermentation yields only two molecules of ATP. The energetic gain is relatively small compared to the total yield of aerobic respiration because the end products, lactic acid and ethanol, still contain a large amount of chemical energy in their bonds and are not fully oxidized. The modest yield explains why organisms that produce ATP by fermentation must consume a large quantity of fuel molecules to power the cell.

→   **Quick Check 6** Breadmaking involves ethanol fermentation and typically uses yeast, sugar, flour, and water. Why are yeast and sugar used?

## ❓ CASE 1  The First Cell: Life's Origins
### How did early cells meet their energy requirements?
The four stages of cellular respiration lead to the full oxidation of glucose, resulting in the release of a large amount of energy stored in its chemical bonds. The first stage, glycolysis, results in only the partial oxidation of glucose, so just some of the energy held in its chemical bonds is released. Nearly all organisms are capable of partially breaking down glucose, suggesting that glycolysis evolved very early in the history of life.

Life first evolved about 4 billion years ago in the absence of atmospheric oxygen. The earliest organisms probably used one of the fermentation pathways to generate the ATP necessary to power cellular processes because fermentation does not require atmospheric oxygen. This process occurs in the cytoplasm and does not require proteins embedded in specialized membranes.

As we have seen, cellular respiration involves an electron transport chain, composed of proteins embedded in a membrane and capable of transferring high-energy electrons from one protein to the next and pumping protons. The resulting proton gradient powers the synthesis of ATP. Like fermentation, cellular respiration can occur in the absence of oxygen, but in that case molecules other than oxygen, such as sulfate and nitrate, are the final electron acceptor (Chapter 26). This form of respiration is known as anaerobic respiration and occurs in some present-day bacteria. The electron transport chain in these bacteria is located in the plasma membrane, not in an internal membrane.

How might such a system have evolved? An intriguing possibility is that early prokaryotes evolved pumps to drive protons out of the cell in response to an increasingly acidic environment (**Fig. 7.14**). Some pumps might have used the energy of ATP to pump protons, while others used electron transport proteins to pump protons (Fig. 7.14a). At some point, proton pumps powered by electron transport might have become efficient enough that the protons could pass back

through the ATP-driven pumps, running them in reverse to synthesize ATP (Fig. 7.14b).

Organisms capable of producing oxygen, the cyanobacteria, did not evolve until about 2.5 billion years ago, maybe earlier. The evolution of this new form of life introduced oxygen into Earth's atmosphere. This dramatic change led to the evolution of new life-forms with new possibilities for extracting energy from fuel molecules such as glucose. Aerobic respiration, in which oxygen serves as the final electron acceptor in the electron transport chain, generates much more energy than does anaerobic respiration or fermentation.

**FIG. 7.14**    **The possible evolution of the electron transport chain and oxidative phosphorylation.**

a.

Early cells evolved mechanisms to pump protons out of the cell, powered by ATP and electron transport.

b.

Eventually, electron-transport-powered pumps became efficient enough to run the ATP-driven pump in reverse.

The evolution of cellular respiration illustrates that evolution often works in a stepwise fashion, building on what is already present. In this case, aerobic respiration picked up where anaerobic respiration left off, making it possible to harness more energy from organic molecules to power the work of the cell.

## 7.7 METABOLIC INTEGRATION

In this chapter, we have focused on the breakdown of glucose. What happens if there is more glucose than is needed by the cell? As well as glucose, you probably consume diverse carbohydrates, lipids, and proteins. How are these broken down? And how are these various metabolic pathways coordinated so that the intracellular level of ATP is maintained in a narrow range? In this final section, we consider how the cell responds to these challenges.

### Excess glucose is stored as glycogen in animals and starch in plants.

Glucose is a readily available form of energy in organisms, but it is not always broken down immediately. Excess glucose can be stored in cells and then mobilized—that is, broken down—when necessary. Glucose can be stored in two major forms: as **glycogen** in animals and **starch** in plants (**Fig. 7.15**). Both these molecules are large branched polymers of glucose.

Carbohydrates that are consumed by animals are broken down into simple sugars and circulate in the blood. The level of glucose in the blood is tightly regulated. When the blood glucose level is high, as it is after a meal, glucose molecules that are not consumed by glycolysis are linked together to form glycogen in liver and muscle. Glycogen stored in muscle is used to provide ATP for muscle contraction. By contrast, the liver does not store glycogen primarily for its own use, but is a central glycogen storehouse for the whole body, able to release glucose into the bloodstream when it is needed elsewhere. Glycogen provides a source of glucose 6-phosphate to feed glycolysis when the level of blood glucose is low. Glucose molecules located at the end of glycogen chains can be cleaved one by one, and they are released in the form of

**FIG. 7.15**  **Storage forms of glucose.** (a) Glycogen is a storage form of glucose in animal cells and (b) starch is a storage form of glucose in plant cells.

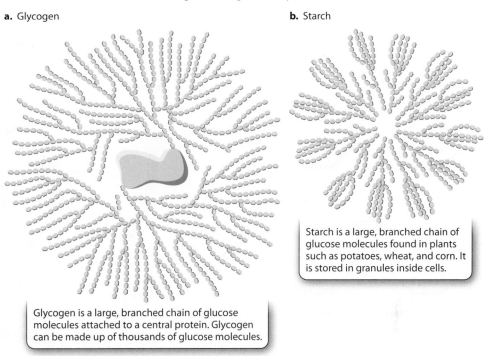

**a.** Glycogen

**b.** Starch

Starch is a large, branched chain of glucose molecules found in plants such as potatoes, wheat, and corn. It is stored in granules inside cells.

Glycogen is a large, branched chain of glucose molecules attached to a central protein. Glycogen can be made up of thousands of glucose molecules.

glucose 1-phosphate. Glucose 1-phosphate is then converted into glucose 6-phosphate, an intermediate in glycolysis (see Fig. 7.4). One glucose molecule cleaved off a glycogen chain produces three and not two molecules of ATP by glycolysis because the ATP-consuming step 1 of glycolysis is bypassed.

### Sugars other than glucose contribute to glycolysis.

The carbohydrates in your diet are digested to produce a variety of sugars (**Fig. 7.16**). Some of these are disaccharides (maltose, lactose, and sucrose) with two sugar units; others are monosaccharides (fructose, mannose, and galactose) with a single sugar unit. The disaccharides are hydrolyzed into monosaccharides, which are transported into cells.

The hydrolysis of some disaccharides produces glucose molecules that directly enter glycolysis. What happens to other monosaccharides? They, too, enter glycolysis, although not as glucose. Instead, they are converted into intermediates of glycolysis that come later in the pathway. For example, fructose is produced by the hydrolysis of sucrose (table sugar) and receives a phosphate group to form either fructose 6-phosphate or fructose 1-phosphate. In the liver, fructose 1-phosphate is cleaved and converted into glyceraldehyde 3-phosphate, which enters glycolysis at reaction 6 (see Fig. 7.4).

**FIG. 7.16** Common sugars in your diet.

## Fatty acids and proteins are useful sources of energy.

In addition to carbohydrates, lipids are also a good source of energy. We know this from common experience. Butter, oils, ice cream, and the like all contain lipids and are high in calories, which are units of energy. We can also infer that lipids are a good source of energy from their chemical structure. Recall from Chapter 2 that a type of fat called triacylglycerol is composed of three fatty acid molecules bound to a glycerol backbone. These fatty acid molecules are rich in carbon–carbon and carbon–hydrogen bonds, which, as we saw earlier, carry chemical energy.

Following a meal, the small intestine very quickly absorbs triacylglycerols, which are then transported by the bloodstream and either consumed or stored in fat (adipose) tissue. Triacylglycerols are broken down inside cells to glycerol and fatty acids. Then, the fatty acids themselves are shortened by a series of reactions that sequentially remove two carbon units from their ends (**Fig. 7.17**). This process is called β- (**beta-**) **oxidation.** It does not produce ATP, but releases a large number of NADH and FADH₂ molecules that

**FIG. 7.17** β-oxidation of fatty acids. The fatty acids from lipids are broken down to produce NADH and FADH₂ as well as acetyl-CoA.

provide high-energy electrons for the synthesis of ATP by oxidative phosphorylation. In addition, the end product of the reaction is acetyl-CoA, which feeds the citric acid cycle and leads to the production of an even larger quantity of electron carriers.

The oxidation of fatty acids produces a large amount of ATP. For example, the complete oxidation of a molecule of palmitic acid, a fatty acid containing 16 carbons, yields about 106 molecules of ATP. By contrast, glycolysis yields just 2 molecules of ATP, and the complete oxidation of a glucose molecule produces about 32 molecules of ATP (Table 7.1). Fatty acids therefore are a useful and efficient source of energy, but they cannot be used by all tissues of the body. Notably, the brain and red blood cells depend primarily on glucose for energy.

Proteins, like fatty acids, are a source of chemical energy that can be broken down, if necessary, to power the cell. Proteins are typically first broken down to amino acids, some of which can then enter glycolysis and others the citric acid cycle.

## The intracellular level of ATP is a key regulator of cellular respiration.

ATP is the key end product of cellular respiration, holding in its bonds energy that can be used for all kinds of cellular processes. ATP is constantly being turned over in a cell, broken down to ADP and $P_i$ to supply the cell's energy needs, and re-synthesized by fermentation and cellular respiration. The level of ATP inside a cell can therefore be an indicator of how much energy a cell has available. When ATP levels are high, the cell has a high amount of free energy and is poised to carry out cellular processes. In this case, pathways that generate ATP are slowed, or down-regulated. By contrast, when ATP levels are low, the cell activates, or up-regulates, pathways that lead to ATP synthesis. Other intermediates of cellular respiration, such as NADH, have a similar effect in that high $NAD^+$ levels stimulate cellular respiration, whereas high NADH levels inhibit it (**Fig. 7.18**).

How is this kind of coordinated response of the cell possible? The cell uses several mechanisms, one of which is the regulation of enzymes that control key steps of the pathway. One of these key reactions is reaction 3 of glycolysis—the conversion of fructose 6-phosphate to fructose 1,6-bisphosphate (see Fig. 7.4). This is a key step in glycolysis because it is highly endergonic and irreversible. As a result, it is considered a "committed" step and is subject to tight control. This reaction is catalyzed by the enzyme phosphofructokinase-1 (PFK-1), which can be thought of as a metabolic valve that regulates the rate of glycolysis.

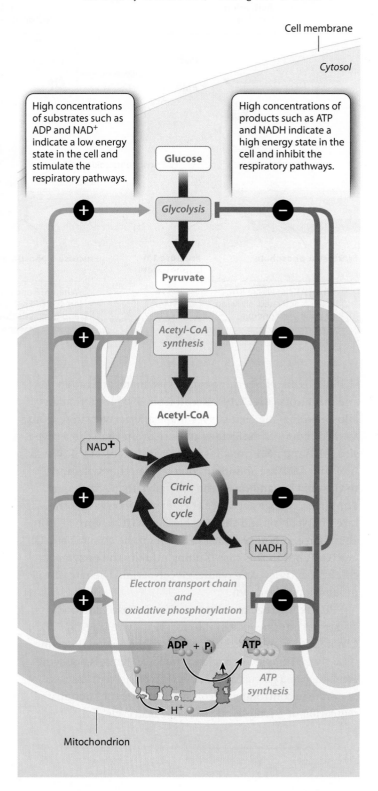

**FIG. 7.18** **Regulation of cellular respiration.** Cellular respiration is inhibited by its products, including ATP and NADH, and activated by its substrates, including ADP and $NAD^+$.

**FIG. 7.19.** **Regulation of PFK-1.** The regulation of the glycolytic enzyme phosphofructokinase-1 (PFK-1) is an example of integrated metabolic control.

## Exercise requires several types of fuel molecules and the coordination of metabolic pathways.

In the last two chapters, we considered what energy is and how it is harnessed by cells. Let's apply the concepts we discussed to a familiar example: exercise. Exercise such as running, walking, and swimming is a form of kinetic energy, powered by ATP in muscle cells. Where does this ATP come from?

Muscle cells, like all cells, do not contain a lot of ATP, and stored ATP is depleted by exercise in a matter of seconds. As a result, muscle cells rely on fuel molecules to generate ATP. For a short sprint or a burst of activity, muscle can convert stored glycogen to glucose, and then break down glucose anaerobically to pyruvate and lactic acid by lactic acid fermentation. This pathway is rapid, but it does not generate a lot of ATP. In addition, it is limited by the production of lactic acid, which lowers the pH of the blood.

PFK-1 is an allosteric enzyme with many activators and inhibitors (**Fig. 7.19**). Recall from Chapter 6 that an allosteric enzyme changes its shape and activity in response to the binding of molecules at a site other than the active site. ADP and AMP are allosteric activators of PFK-1. When ADP and AMP are abundant, one or the other will bind to the enzyme and cause the enzyme's shape to change. The shape change activates the enzyme, increasing the rate of glycolysis and the synthesis of ATP. When ATP is in abundance, it binds to the same site on the enzyme as ADP and AMP, but in this case binding inhibits the enzyme's catalytic activity. As a result, glycolysis and the rate of ATP production slow down.

PFK-1 is also regulated by one of its downstream products, citrate, an intermediate in the citric acid cycle (see Fig. 7.7). Citrate acts as an allosteric inhibitor of the enzyme, slowing its activity. High levels of citrate indicate that it is not being consumed and glucose breakdown can be slowed. The role of citrate in controlling glycolysis illustrates the coordinated regulation of glycolysis and the citric acid cycle.

For longer, more sustained exercise, other metabolic pathways come into play. Muscle cells contain many mitochondria, which produce ATP by aerobic respiration. The energy yield of aerobic respiration is much greater than that of fermentation, but the process is slower. This slower production of ATP by aerobic respiration in part explains why runners cannot maintain the pace of a sprint for longer runs.

For even longer exercise, liver glycogen supplements muscle glycogen: The liver releases glucose into the blood that is taken up by muscle cells and oxidized to produce ATP. In addition, fatty acids are released from adipose tissue and taken up by muscle cells, where they are broken down by β-oxidation. β-oxidation yields even more ATP than does the complete oxidation of glucose, but the process is again slower. Storage forms of energy molecules, such as fatty acids and glycogen, contain large reservoirs of energy, but are slower to mobilize. Thus, exercise takes coordination between different cells, tissues, and metabolic pathways to ensure adequate ATP to meet the needs of working muscle.

# Core Concepts Summary

**7.1 CELLULAR RESPIRATION IS A SERIES OF CATABOLIC REACTIONS THAT CONVERT THE ENERGY STORED IN FUEL MOLECULES INTO ATP.**

Cellular respiration is a four-stage process that includes (1) glycolysis; (2) pyruvate oxidation and acetyl-CoA synthesis; (3) the citric acid cycle; and (4) oxidative phosphorylation. page 7-2

The complete oxidation of glucose in the presence of oxygen to make carbon dioxide and water is an oxidation–reduction reaction. page 7-3

In oxidation–reduction reactions, electrons are transferred from one molecule to another. Oxidation is the loss of electrons, and reduction is the gain of electrons. page 7-3

The chemical bonds of sugar molecules like glucose have high potential energy because the electrons that are shared by neighboring atoms are on average far from the nucleus. page 7-4

The chemical energy in glucose and other fuel molecules is released slowly in a series of chemical reactions that produce energy-storing molecules, including ATP and the electron carriers NADH and $FADH_2$. page 7-4

Electron carriers transfer electrons to an electron transport chain, which harnesses the energy of these electrons to generate ATP. page 7-4

ATP is generated by substrate-level phosphorylation and oxidative phosphorylation in cellular respiration. page 7-4

**7.2 GLYCOLYSIS IS THE PARTIAL OXIDATION OF GLUCOSE AND RESULTS IN THE PRODUCTION OF PYRUVATE, A SMALL AMOUNT OF ATP, AND HIGH-ENERGY ELECTRON CARRIERS.**

Glycolysis takes place in the cytoplasm. page 7-2

Glycolysis is a series of 10 reactions in which glucose is converted to pyruvate. page 7-7

Glycolysis involves preparatory, cleavage, and payoff phases. page 7-7

For each molecule of glucose consumed during glycolysis, a net gain of two molecules of ATP and two molecules of NADH is produced. page 7-7

The synthesis of ATP in glycolysis results from the direct transfer of a phosphate group to ADP, a process called substrate-level phosphorylation. page 7-7

**7.3 PYRUVATE IS OXIDIZED TO ACETYL-CoA, CONNECTING GLYCOLYSIS TO THE CITRIC ACID CYCLE.**

The conversion of pyruvate to acetyl-CoA results in the production of one molecule of NADH and one molecule of carbon dioxide. page 7-7

Acetyl-CoA synthesis occurs in the mitochondrial matrix. page 7-7

Acetyl-CoA is the first substrate in the citric acid cycle. page 7-8

**7.4 THE CITRIC ACID CYCLE RESULTS IN THE COMPLETE OXIDATION OF FUEL MOLECULES AND THE GENERATION OF ATP AND HIGH-ENERGY ELECTRON CARRIERS.**

The citric acid cycle takes place in the mitochondrial matrix. page 7-8

The acetyl group of acetyl-CoA combines with oxaloacetate, followed by a series of decarboxylation and redox reactions that regenerate oxaloacetate. page 7-9

A complete turn of the citric acid cycle results in the production of one molecule of GTP (which is converted to ATP), three molecules of NADH, and one molecule of $FADH_2$. page 7-9

Citric acid cycle intermediates are starting points for the synthesis of many different organic molecules. page 7-10

**7.5 THE ELECTRON TRANSPORT CHAIN TRANSFERS ELECTRONS FROM ELECTRON CARRIERS TO OXYGEN, USING THE ENERGY TO PUMP PROTONS AND SYNTHESIZE ATP BY OXIDATIVE PHOSPHORYLATION.**

NADH and $FADH_2$ donate electrons to the electron transport chain. page 7-10

In the electron transport chain, electrons move from one redox carrier to the next. page 7-10

The electron transport chain is made up of four complexes. Complexes I and II accept electrons from NADH and $FADH_2$ respectively. The electrons are transferred from these two complexes to coenzyme Q. page 7-10

Reduced coenzyme Q transfers electrons to complex III and cytochrome *c* transfers electrons to complex IV. Complex IV contains oxygen, the final electron acceptor. page 7-12

The transfer of electrons through the electron transport chain is coupled with the movement of protons across the inner mitochondrial membrane into the intermembrane space. page 7-12

The buildup of protons in the intermembrane space results in a proton electrochemical gradient, which stores potential energy. page 7-12

The movement of protons back into the mitochondrial matrix through the $F_0$ subunit of ATP synthase is coupled with the formation of ATP, a reaction catalyzed by the $F_1$ subunit of ATP synthase. page 7-13

**7.6 GLUCOSE CAN BE BROKEN DOWN IN THE ABSENCE OF OXYGEN BY FERMENTATION, PRODUCING A MODEST AMOUNT OF ENERGY IN THE FORM OF ATP.**

Pyruvate, the end product of glycolysis, is processed differently in the presence and the absence of oxygen. page 7-14

In the absence of oxygen, pyruvate enters one of several fermentation pathways. page 7-15

In lactic acid fermentation, pyruvate is reduced to lactic acid. page 7-15

In ethanol fermentation, pyruvate is converted to acetaldehyde, which is reduced to ethanol. page 7-15

During fermentation, NADH is oxidized to $NAD^+$, allowing glycolysis to proceed. page 7-15

Glycolysis and fermentation are ancient biochemical pathways and were likely used in the common ancestor of all organisms living today. page 7-16

## 7.7 METABOLIC PATHWAYS ARE INTEGRATED, ALLOWING CONTROL OF THE ENERGY LEVEL OF CELLS.

Excess glucose is polymerized and stored in molecules called glycogen (in animals) and starch (in plants). page 7-17

Other monosaccharides derived from the digestion of dietary carbohydrates are converted into intermediates of glycolysis. page 7-17

Fatty acids contained in triacylglycerols are an important form of energy storage in cells. The breakdown of fatty acids is called β-oxidation. page 7-18

Phosphofructokinase-1 controls a key step in glycolysis. It has many allosteric activators, including ADP and AMP, and allosteric inhibitors, including ATP and citrate. page 7-20

The ATP in muscle cells used to power exercise is generated by lactic acid fermentation, aerobic respiration, and β-oxidation. page 7-20

## Self-Assessment

1. Name and describe the four major stages of cellular respiration.

2. Explain what an oxidation–reduction reaction is and why the breakdown of glucose in the presence of oxygen to produce carbon dioxide and water is an example of an oxidation–reduction reaction.

3. Describe two ways in which ATP is generated in cellular respiration.

4. Write the overall chemical equation for glycolysis, noting the starting and ending products and highlighting the energy-storing molecules that are produced.

5. Describe two different metabolic pathways that pyruvate can enter.

6. Name the products of the citric acid cycle.

7. Describe how the movement of electrons along the electron transport chain leads to the generation of a proton gradient.

8. Describe how a proton gradient generates ATP.

9. Explain how muscle tissue generates ATP during short-term and long-term exercise.

Do you understand the chapter's Core Concepts? Log in to check your answers to the Self-Assessment questions, then practice what you've learned and reinforce this chapter's concepts by working through the problems and multimedia tutorials provided there.

www.biologyhowlifeworks.com

## CHAPTER 8

# PHOTOSYNTHESIS

## Using Sunlight to Build Carbohydrates

## Core Concepts

Walk through a forest and you will be struck, literally if you aren't careful, by the substantial nature of trees. Where does the material to construct these massive organisms come from? Because trees grow upward from a firm base in the ground, a reasonable first guess is the soil. In the first recorded experiment on this topic, the Flemish chemist and physiologist Jan Baptist van Helmont (1580–1644) found that the initial 300 pounds of soil into which he had planted a small willow tree decreased by only 2 ounces over a 5-year period. During this same period, the tree gained more than 150 pounds. Van Helmont concluded that water must be responsible for the tree's growth. He was, in fact, half right: A tree is roughly half liquid water. But what he missed entirely is that the other half of his tree had been created out of thin air.

The process that allowed Van Helmont's tree to increase in mass using substances pulled from the air is called **photosynthesis.** Photosynthesis is a biochemical process for building carbohydrates from sunlight and carbon dioxide ($CO_2$) taken from the air. These carbohydrates are used as both structural components of the plant and as a source of energy used to produce ATP.

## 8.1 THE NATURAL HISTORY OF PHOTOSYNTHESIS

Carbohydrates have more energy stored in their chemical bonds than is contained in the bonds of the $CO_2$ molecules from which they are synthesized during photosynthesis. Therefore, to build carbohydrates using $CO_2$ requires an input of energy. In photosynthesis, this energy comes from sunlight.

### Photosynthesis is a redox reaction.

Energy is added to molecules during carbohydrate synthesis through the transfer of high-energy electrons. It is this addition of energy and electrons that allows the incoming $CO_2$ molecules to form the higher-energy bonds found in a carbohydrate molecule. In Chapter 7, we saw that **reduction** reactions are reactions in which a molecule acquires electrons and gains energy, whereas **oxidation** reactions are reactions in which a molecule loses electrons and releases energy. During photosynthesis, $CO_2$ molecules are reduced to form higher-energy carbohydrate molecules (**Fig. 8.1**).

Where do the electrons used to reduce $CO_2$ come from? These electrons can only come from the oxidation of other molecules, illustrating once again that reduction–oxidation (or redox) reactions always come in pairs. In photosynthesis carried out by plants and many algae, the ultimate **electron donor** is water. However, as we will see in Chapter 26, photosynthetic bacteria can use a variety of other electron donors. The oxidation of water results in the production of electrons, protons, and $O_2$. Thus, oxygen is formed in photosynthesis as a by-product of water's role as a source of electrons. We can demonstrate that water is the source of the oxygen released during photosynthesis using isotopes, molecules that can be distinguished on the basis of their molecular mass (**Fig. 8.2**).

Overall, then, the equation for photosynthesis can be described as follows:

$$\overset{\text{Oxidation}}{CO_2 + H_2O \longrightarrow C_6H_{12}O_6 + O_2}_{\text{Reduction}}$$

The oxidation of water is linked with the reduction of $CO_2$ through a series of redox reactions in which electrons are passed from one compound to another. This series of reactions constitutes the **photosynthetic electron transport chain.** The process begins with the absorption of light by protein–pigment complexes known as **photosystems.** Photosystems use absorbed light energy to drive redox reactions and thereby set the photosynthetic electron transport chain in motion. In turn, the movement of electrons through this transport chain is used to drive the synthesis of ATP and NADPH. And finally, ATP and NADPH are the

**FIG. 8.1**   **Overview of photosynthesis.**   In photosynthesis, energy from sunlight is used to reduce $CO_2$ to form carbohydrates.

**FIG. 8.2**

# Does the oxygen released by photosynthesis come from $H_2O$ or $CO_2$?

**BACKGROUND**   The reactants in photosynthesis are water and carbon dioxide. Therefore, it is unclear where the oxygen that is produced in the reaction comes from.

**METHOD**   Most of the oxygen in the atmosphere is $^{16}O$, with 8 protons and 8 neutrons. A small amount (0.2%) is $^{18}O$, with 8 protons and 10 neutrons. $^{16}O$ and $^{18}O$ are stable isotopes. The relative abundance of molecules containing $^{16}O$ versus $^{18}O$ can be measured using a mass spectrometer. $H_2O$ and $CO_2$ containing a high percentage of $^{18}O$ can be used to determine whether the oxygen produced in photosynthesis comes from water or carbon dioxide.

**EXPERIMENT**

**RESULTS**

**CONCLUSION**   The ratio of $^{18}O_2/^{16}O_2$ only increases when water contains $^{18}O$, but not when carbon dioxide contains $^{18}O$. This finding indicates that the oxygen produced in photosynthesis comes from water, not carbon dioxide.

**FOLLOW-UP WORK**   Carbon also has several isotopes, and their measurement has been used to determine the source of increased $CO_2$ in the atmosphere today (Chapter 25).

**SOURCE** Adapted from Ruben, S., M. Randall, M. Kamen, and J. L. Hyde. 1941. "Heavy Oxygen (O$^{18}$) as a Tracer in the Study of Photosynthesis." *Journal of the American Chemical Society.* 63:877–879.

energy sources needed to synthesize carbohydrates using $CO_2$ in a process called the **Calvin cycle** (see Fig. 8.1).

Each year, photosynthesis removes more than 100 billion metric tons of carbon from the atmosphere, while incorporating more than 100 terawatts (a terawatt is $10^{12}$ watts) of solar energy into chemical bonds. To put these quantities in perspective, the energy captured by photosynthesis in a year is 10 times the energy of all the oil and coal used worldwide in 2006. Fossil fuels are themselves the legacy of ancient photosynthesis: Oil has its origin in the bodies of marine phytoplankton and the organisms that graze on them, while coal represents the geologic remains of terrestrial (land) plants. Thus, one motivation to understand photosynthesis is the sheer magnitude and importance of this process for life on Earth. Before exploring the details of how photosynthesis actually occurs, let's look at what types of organisms are photosynthetic, where they live, how they have evolved to be photosynthetic, and what structural components are needed to allow cells to capture energy in this remarkable way.

→ **Quick Check 1** If you want to produce carbohydrates containing the heavy oxygen ($^{18}O$) isotope, should you water your plants with $H_2^{18}O$ or inject $C^{18}O_2$ into the air?

## Photosynthesis is widely distributed.

Photosynthesis occurs among prokaryotic as well as eukaryotic organisms, on land as well as in the sea. Approximately 50% of global photosynthesis is carried out by terrestrial organisms, with the other half taking place in the ocean. The majority of photosynthetic organisms in marine environments are unicellular. About half of oceanic photosynthesis is carried out by phytoplankton (single-celled marine eukaryotes), while the other half is carried out by cyanobacteria (photosynthetic bacteria). On land, photosynthesis is dominated by multicellular plants.

Photosynthesis takes place almost everywhere sunlight is available to serve as a source of energy. In the ocean, photosynthesis occurs in the surface layer about 100 m deep, called the **photic zone,** through which enough sunlight penetrates to enable photosynthesis. On land, photosynthesis occurs most readily in environments that are both moist and warm. However, photosynthetic organisms have evolved adaptations that allow them to tolerate a wide range of environmental conditions, like those illustrated in **Fig. 8.3.** In very dry regions, a combination of photosynthetic bacteria and unicellular algae forms an easily disturbed layer on the

**FIG. 8.3** **Photosynthesis in extreme environments.** (a) Desert crust in Utah formed by photosynthetic bacteria and algae. (b) A hot spring in Yellowstone National Park. The yellow color is due to photosynthetic bacteria. (c) The surface of a permanent snow pack. The red color is due to photosynthetic algae.

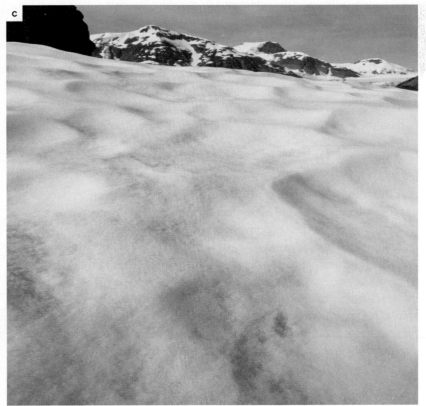

surface of the soil known as desert crust. Photosynthetic bacteria are also found in the hot springs of Yellowstone National Park at temperatures up to 75°C. At the other extreme, unicellular eukaryotic algae can grow on the surfaces of glaciers, causing the surface of the snow to appear red. In section 8.4, we discuss why photosynthetic organisms growing in such extreme environments often appear red or yellow rather than green.

### The evolutionary history of photosynthesis includes both horizontal gene transfer and endosymbiosis.

The most ancient forms of photosynthesis have simple photosynthetic electron transport chains with only a single photosystem. However, a single photosystem cannot capture enough energy to pull electrons from (oxidize) water and then use them to reduce $CO_2$. Thus, photosynthetic organisms with a single photosystem must use more easily oxidized compounds, such as $H_2S$, as electron donors. These organisms are limited to environments where the electron-donor molecules are abundant. Because these organisms do not use water as an electron donor, they do not produce $O_2$ during photosynthesis (Chapter 26).

A major event in the history of life was the evolution of photosynthetic electron transport chains that could bridge the energy difference between the oxidation of water and the reduction of $CO_2$. Water is abundant, but it is extremely difficult to pull electrons from water. The first organisms to accomplish this feat were the cyanobacteria. These photosynthetic bacteria incorporated two different photosystems into a single photosynthetic electron transport chain. Having two photosystems allows the energy level of the electrons moving through the photosynthetic electron chain to be increased in two steps, much as a series of locks serves to raise the elevation of boats traveling through a canal. The addition of light energy in the first step allows electrons to be pulled from a water molecule, while the addition of light energy in the second step raises the energy level of these electrons so that they can be used to reduce $CO_2$. Stripping electrons from water results in the release of oxygen. Indeed, all the oxygen in Earth's atmosphere results from photosynthesis by organisms containing two photosystems.

Each of the two photosystems present in cyanobacteria has a high degree of similarity to the single photosystem found in other lineages of photosynthetic bacteria. The structural similarity is so great as to make it highly unlikely that the two photosystems in cyanobacteria evolved independently. An alternative hypothesis is that the genetic material associated with one photosystem was transferred to a bacterium with

the other photosystem, resulting in a single bacterium with the genetic material to produce both types of photosystems (**Fig. 8.4**). The transfer of genetic material between organisms that are not parent and offspring is called **horizontal gene transfer,** and it is common among prokaryotes, where it provides an important means of producing genetic diversity. The mechanisms of horizontal gene transfer are discussed more fully in Chapter 26.

Photosynthesis is hypothesized to have gained a foothold among eukaryotic organisms when a free-living cyanobacterium was engulfed by a eukaryotic cell (Fig. 8.4). Over time, the engulfed cyanobacterium lost its ability to survive outside of its host cell and evolved into the chloroplast, the organelle in eukaryotic cells that carries out photosynthesis. The process in which one cell takes up residence inside of another cell is called endosymbiosis. Therefore, the idea that chloroplasts and mitochondria (Chapter 7) arose in this way is called the endosymbiotic hypothesis. It is discussed in Chapter 27.

**FIG. 8.4**    **The evolutionary history of photosynthesis.** Bacteria with a single photosystem obtained two photosystems by horizontal gene transfer and later, by endosymbiosis, became the chloroplasts of eukaryotic cells.

## The photosynthetic electron transport chain takes place on specialized membranes.

The ability to absorb sunlight and convert it into chemical forms allows photosynthetic organisms to produce their own food. However, high-energy electrons moving through the photosynthetic electron transport chain have the potential to damage the cell. To prevent damage, the movement of electrons through the photosynthetic electron transport chain is constrained: The electrons follow a defined network of pathways through large protein complexes embedded in specialized "photosynthetic" membranes. These protein complexes provide a scaffold along which the pigments and other elements of the photosynthetic electron transport chain are arrayed. In fact, the movement of electrons through the photosynthetic electron transport chain is not controlled by the specificity of individual enzymes. Instead, it depends upon the spatial arrangement of the sites where the redox reactions take place within and between these major protein complexes.

Chloroplasts are enclosed by a double membrane. Filling much of the center of the chloroplast is a third, highly folded, membrane known as the **thylakoid membrane** (**Fig. 8.5**). The photosynthetic electron transport chain is located on the thylakoid membrane.

The name "thylakoid" is derived from *thylakois,* the Greek word for "sac." Thylakoid membranes form structures that resemble flattened sacs, which are grouped into structures called **grana** (singular, granum) that look like stacks of interlinked pancakes. Grana are connected to one another by membrane bridges so that the thylakoid forms a single continuous membrane. Thus, while it may look as though the individual grana are independent from one another, in fact the thylakoid membrane encloses a common fluid-filled interior compartment called the **lumen.** The region surrounding the thylakoid membrane is called the **stroma.** Carbohydrate synthesis takes place in the stroma, whereas sunlight is captured and transformed via the photosynthetic electron transport chain on the thylakoid membrane.

The outer membrane of the chloroplast double membrane is thought to have originated from the plasma membrane of the ancestral eukaryotic cell, which surrounded the ancestral cyanobacterium as it was being engulfed. The inner chloroplast membrane is thought to correspond to the plasma membrane of the ancestral free-living cyanobacterium. The thylakoid membrane then corresponds to the internal photosynthetic membrane found in cyanobacteria. Finally, the stroma corresponds to the cytoplasm of the ancestral cyanobacterium.

A photosynthetic cell can have more than 100 chloroplasts. Photosynthetic cells also contain mitochondria. Although

**FIG. 8.5** **Chloroplast structure.** Chloroplasts contain highly folded thylakoid membranes.

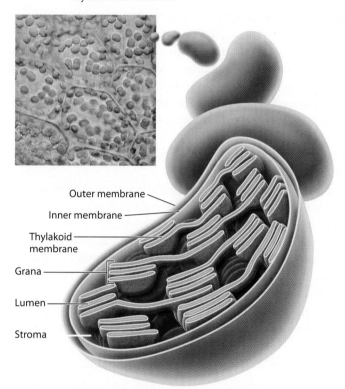

- Outer membrane
- Inner membrane
- Thylakoid membrane
- Grana
- Lumen
- Stroma

photosynthetic organisms are correctly described as autotrophs because they can form carbohydrates from $CO_2$, they also require mitochondrial respiration, in which carbohydrates are broken down to generate ATP to meet the cell's energy requirements (Chapter 7). Cellular respiration is therefore one of several features that heterotrophic organisms like ourselves share with photosynthetic organisms.

We now consider the underlying biochemistry of photosynthesis. Though in this chapter we focus on photosynthesis as carried out by eukaryotic cells, there is a remarkable diversity in the way photosynthesis is carried out in bacteria. We discuss this diversity in Chapter 26. Because carbohydrates are the major product of photosynthesis, we first examine the Calvin cycle, the biochemical pathway used in photosynthesis to synthesize carbohydrates from $CO_2$. Once we understand what energy forms are needed to drive this autotrophic pathway, we turn our attention to how energy is captured from sunlight. Finally, we examine some of the challenges of coordinating these two metabolic stages and consider ways in which photosynthetic organisms cope with the inherent biochemical challenges of photosynthesis.

## 8.2 THE CALVIN CYCLE

The Calvin cycle consists of 15 chemical reactions that synthesize carbohydrates from $CO_2$. These reactions can be grouped into three main steps: (1) **carboxylation,** in which $CO_2$ absorbed from the air is added to a 5-carbon molecule; (2) **reduction,** in which energy and electrons are transferred to the molecules formed from carboxylation; and (3) **regeneration** of the 5-carbon molecule needed for carboxylation (**Fig. 8.6**).

### The incorporation of $CO_2$ is catalyzed by the enzyme rubisco.

The first step of the Calvin cycle is a carboxylation reaction (Fig. 8.6). Specifically, $CO_2$ is added to a 5-carbon sugar called **ribulose-1,5-bisphosphate** (**RuBP**). This step is catalyzed by the enzyme **ribulose bisphosphate carboxylase oxygenase,** or **rubisco** for short. Although many enzymes are active in the Calvin cycle, rubisco has a disproportionate influence over the functioning of the entire pathway.

Compared to other enzymes, rubisco is surprisingly slow. As a result, photosynthetic cells must produce large amounts of this single enzyme. In fact, rubisco is the most abundant protein on Earth and constitutes as much as half of the total protein found in a leaf. As the principal gatekeeper through which $CO_2$ becomes incorporated into carbohydrates, rubisco plays a key role in the processes that permit life on Earth.

Before rubisco can act as a carboxylase, RuBP and $CO_2$ must diffuse into its active site. Once the active site is occupied, carboxylation proceeds spontaneously. The product is a 6-carbon compound that immediately breaks into two molecules of **3-phosphoglycerate** (**3-PGA**). These 3-carbon molecules are the first stable products of the Calvin cycle. However, if $O_2$ instead of $CO_2$ diffuses into the active site, the reaction can still proceed, although $O_2$ is added to RuBP in place of $CO_2$. An enzyme like rubisco that adds $O_2$ to a substrate molecule is called an oxygenase, explaining part of this enzyme's long name. The addition of $O_2$ by rubisco creates a major challenge for photosynthesis, a topic we discuss in section 8.4.

### NADPH is the reducing agent of the Calvin cycle.

Rubisco is responsible for the addition of the carbon atoms needed for the formation of carbohydrates, but by itself rubisco does not increase the amount of energy stored within the

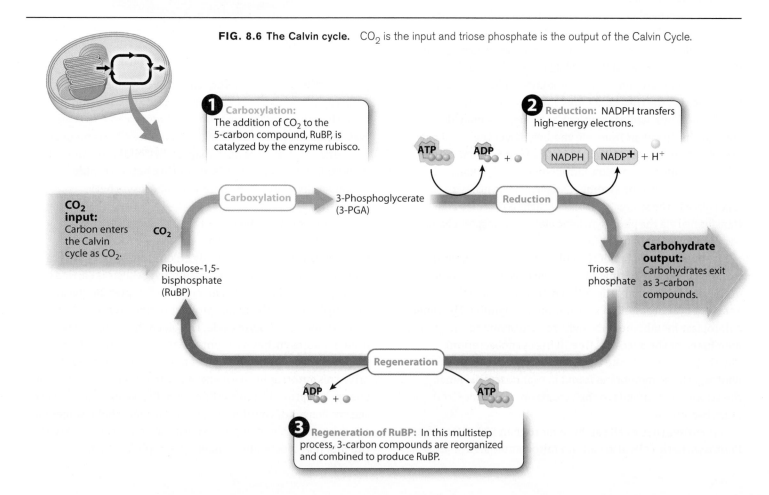

**FIG. 8.6 The Calvin cycle.** $CO_2$ is the input and triose phosphate is the output of the Calvin Cycle.

newly formed bonds. For this energy increase to take place, the carbon compounds formed by rubisco must be reduced. **Nicotinamide adenine dinucleotide phosphate** (**NADPH**) is the reducing agent used in the Calvin cycle. NADPH transfers the energy and electrons that allow carbohydrates to be synthesized from $CO_2$, the most oxidized of all carbon compounds (Fig. 8.6).

NADPH is produced by the photosynthetic electron transport chain. Like all components of the Calvin cycle, NADPH can move freely within the stroma of the chloroplast. Although NADPH is a powerful reducing agent, energy and electrons are transferred from NADPH only under the catalysis of a specific enzyme, thus providing a high degree of control over the fate of these high-energy electrons. In the Calvin cycle, the reduction of 3-PGA involves two steps: (1) ATP is used to phosphorylate 3-PGA, and (2) NADPH transfers two high-energy electrons to the phosphorylated compound. Because two molecules of 3-PGA are formed each time rubisco catalyzes the incorporation of one molecule of $CO_2$, two ATP and two NADPH are required for each molecule of $CO_2$ incorporated by rubisco. NADPH provides most of the energy incorporated in the bonds of the carbohydrate molecules produced by the Calvin cycle. Nevertheless, ATP plays an essential role in preparing 3-PGA for the addition of energy and electrons from NADPH.

These energy transfer steps result in the formation of glyceraldehyde 3-phosphate (GAP), which is reversibly interconverted to dihydroxyacetone phosphate (DHAP) by the enzyme triose phosphate isomerase. Together GAP and DHAP constitute a pool of 3-carbon carbohydrate molecules known as **triose phosphates.** Triose phosphates are the true products of the Calvin cycle because they are the molecules exported from the chloroplast. Larger sugars, such as glucose and sucrose, are assembled from triose phosphates in the cytoplasm. However, if every triose phosphate molecule produced by the Calvin cycle were exported from the chloroplast, RuBP could not be regenerated and the Calvin cycle would grind to a halt. Thus, for every six triose phosphate molecules that are produced, only one can be withdrawn from the Calvin cycle.

### The regeneration of RuBP requires ATP.

Of the 15 chemical reactions that make up the Calvin cycle, 12 are directly involved in the last step, the regeneration of RuBP (Fig. 8.6). The large number of reactions involved in this step reflects the degree of reshuffling of carbon atoms needed to produce three 5-carbon RuBP molecules from five 3-carbon triose phosphate molecules. ATP is required for the regeneration of RuBP, raising the Calvin cycle's total energy requirements to two molecules of NADPH and three molecules of ATP for each molecule of $CO_2$ incorporated by rubisco. Thus, NADPH and ATP play distinct roles

in the formation of carbohydrates from $CO_2$. NADPH provides energy that is incorporated into the higher-energy bonds that characterize carbohydrate molecules, whereas ATP provides energy that is needed to regenerate RuBP.

The Calvin cycle does not utilize sunlight directly. For this reason this pathway is sometimes referred to as the light-independent or even the "dark" reactions of photosynthesis. However, this pathway cannot operate without the energy input provided by a steady supply of NADPH and ATP. Both are supplied by the photosynthetic electron transport chain, in which light is captured and transformed into chemical energy. Thus, under natural conditions, photosynthesis, including the Calvin cycle, occurs only in the light.

→ **Quick Check 2** The Calvin cycle requires both ATP and NADPH. Which of these provides the major input of energy needed to synthesize carbohydrates?

### The steps of the Calvin cycle were determined using radioactive $CO_2$.

In a series of experiments conducted between 1948 and 1954 (**Fig. 8.7**), the American chemist Melvin Calvin and colleagues supplied radioactively labeled $CO_2$ ($^{14}CO_2$) to the unicellular green alga *Chlorella* and then plunged the cells into boiling alcohol, thus halting all enzymatic reactions. By examining the compounds that became radioactively labeled, Calvin and his colleagues were able to determine the identity of the carbon compounds produced in photosynthesis (Experiment 1 in Fig. 8.7).

Figuring out the chemical reactions that connected these labeled compounds, however, required both ingenuity and hard work. For example, by using a very short exposure to $^{14}CO_2$, Calvin and colleagues determined that the incorporation of $CO_2$ results in the production of 3-PGA (Experiment 2 in Fig. 8.7).

To determine how 3-PGA is formed, they supplied radioactive $^{14}CO_2$ to label the products, but then cut off the supply of $^{14}CO_2$ to block the carboxylation reaction. In this experiment, the amount of RuBP increased relative to the amount seen in the first experiment. They concluded that the first step was the addition of $CO_2$ to RuBP (Experiment 3 in Fig. 8.7).

### Carbohydrates are stored in the form of starch.

The Calvin cycle is capable of producing more carbohydrates than the cell needs or, in a multicellular organism, more than the cell is able to export. If carbohydrates accumulated in the cell, they would upset the osmotic balance, causing water to enter the cell. Instead, excess carbohydrates are converted to starch, a storage form of carbohydrates discussed in Chapter 2. Because starch molecules are not soluble, they provide a means of carbohydrate storage that

FIG. 8.7

# How is CO$_2$ used to synthesize carbohydrates?

**BACKGROUND** In the 1940s, radioactive $^{14}CO_2$ became available in quantities that allowed experiments. Melvin Calvin and Andrew Benson used $^{14}CO_2$ to follow the incorporation of $CO_2$ into carbohydrates.

**EXPERIMENTS AND RESULTS**

**1** Inject $^{14}CO_2$ into a flask containing *Chlorella*.

**2** After a specified amount of time, drop cells into boiling alcohol to halt the reactions.

**3** Identify radioactively labeled compounds using 2D paper chromatography.

**Experiment 1:** Use $^{14}CO_2$ to label all the products of the Calvin cycle.

$^{14}CO_2$

RuBP
Triose phosphates
3-PGA

2D paper chromatography separates molecules according to factors such as size and charge that affect their mobility.

**Experiment 2:** Shorten exposure to $^{14}CO_2$ enough so that only one compound is labeled.

3-PGA

Thus, 3-PGA is the initial product of carboxylation.

**Experiment 3:** Prevent carboxylation by withholding $CO_2$ and see which labeled product increases, and which decreases.

RuBP
Triose phosphates
3-PGA

Because RuBP increases, while 3-PGA decreases relative to the amounts in Experiment 1, RuBP is the other substrate involved in the carboxylation step.

**CONCLUSION** The initial step in the Calvin cycle unites the 5-carbon RuBP with $CO_2$, resulting in the production of two molecules of 3-PGA.

**FOLLOW-UP WORK** In the 1950s, Marshall Hatch and colleagues showed that some plants, including corn and sugar cane, accumulate a 4-carbon compound as the first product in photosynthesis. In Chapter 29, we explore how C4 photosynthesis allows plants to avoid the oxygenase reaction of rubisco.

**SOURCE** Calvin, M., and H. Benson. 1949. "The Path of Carbon in Photosynthesis IV: The Identity and Sequence of the Intermediates in Sucrose Synthesis." *Science* 109:140–142.

**FIG. 8.8**    A chloroplast containing starch granules, shown here in yellow.

does not affect the cell's osmotic balance. The formation of starch during the day provides photosynthetic cells with a source of carbohydrates that they can use during the night (**Fig. 8.8**).

## 8.3 CAPTURING SUNLIGHT INTO CHEMICAL FORMS

To use sunlight to power the Calvin cycle, the cell must be able to convert light energy into both NADPH and ATP. NADPH is the reducing agent used to synthesize carbohydrates from $CO_2$, whereas ATP is required for the regeneration of RuBP. The production of both molecules is the task of the photosynthetic electron transport chain. NADPH is produced by the linear flow of electrons through the photosynthetic electron transport chain from water to $NADP^+$. To produce ATP, the photosynthetic electron transport chain functions as a proton pump, leading to the accumulation of protons in the thylakoid lumen. The resulting proton gradient is used to drive the synthesis of ATP, just as it drives the synthesis of ATP in mitochondrial respiration (Chapter 7).

**Photosystems use light energy to set the photosynthetic electron transport chain in motion.**
The function of a photosystem is to convert absorbed light energy into the movement of electrons, making it a key element in the photosynthetic electron transport chain. To understand how a photosystem works, we need to know a little about light. Light is a type of electromagnetic radiation, as are radio waves, X-rays, and other forms of radiation. Electromagnetic radiation is energy in the form of waves; the type of electromagnetic energy depends on the wavelength. **Visible light** is the portion of the electromagnetic spectrum apparent to our eyes, and it includes the range of wavelengths used in photosynthesis. The wavelengths of visible light are from 400 nm to 700 nm.

Pigments are molecules that absorb some wavelengths of visible light (**Fig. 8.9**). Pigments look colored because they reflect light enriched in the wavelengths that they do not absorb. Physically, a photosystem is a complex of proteins and pigments that is embedded in the thylakoid membrane. **Chlorophyll** is the major photosynthetic pigment; it appears green because it is poor at absorbing green wavelengths. The chlorophyll molecule consists of a large, light-absorbing "head" containing a magnesium atom at its center and a long hydrocarbon "tail" that allows the pigment to be anchored in the lipid membrane.

The thylakoid membrane also contains other pigments, most notably the orange-yellow carotenoids. Carotenoids are able to absorb light from regions of the visible spectrum that are poorly absorbed by chlorophyll. Thus, the presence of these **accessory pigments** allows photosynthetic cells to absorb a broader range of visible light than would be possible with just chlorophyll alone. As we will see in section 8.4, carotenoids play an important role in protecting the photosynthetic electron transport chain from damage.

Absorption of visible light by a chlorophyll molecule results in one of its electrons being elevated to a higher energy state

**FIG. 8.9**    **Light wavelengths absorbed by leaves.** The electromagnetic spectrum below highlights the range of visible wavelengths used in photosynthesis; the graph shows the extent to which these wavelengths are absorbed by pigments in an intact leaf.

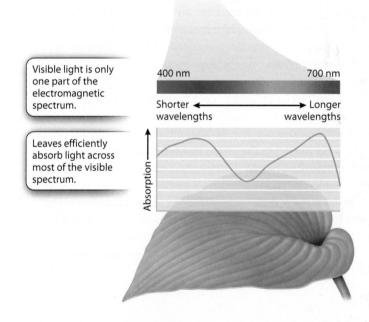

**FIG. 8.10    Absorption of light energy by chlorophyll.** Absorption of light energy by (a) an isolated chlorophyll molecule in the lab and (b) a chlorophyll molecule in a plant cell.

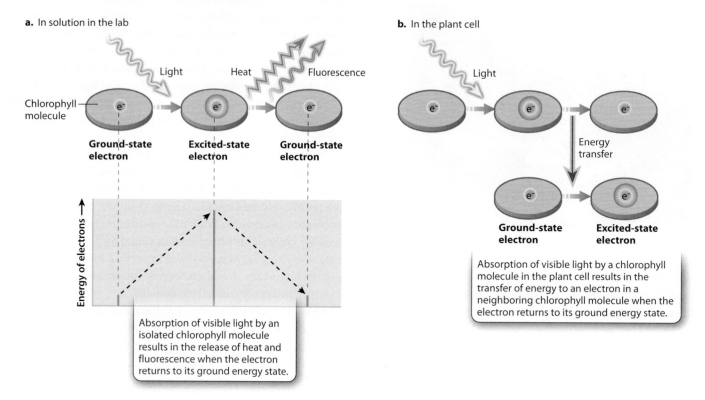

**a.** In solution in the lab

Light    Heat    Fluorescence

Chlorophyll molecule

**Ground-state electron**    **Excited-state electron**    **Ground-state electron**

Energy of electrons →

Absorption of visible light by an isolated chlorophyll molecule results in the release of heat and fluorescence when the electron returns to its ground energy state.

**b.** In the plant cell

Light

Energy transfer

**Ground-state electron**    **Excited-state electron**

Absorption of visible light by a chlorophyll molecule in the plant cell results in the transfer of energy to an electron in a neighboring chlorophyll molecule when the electron returns to its ground energy state.

(**Fig. 8.10**). For chlorophyll molecules that have been isolated in the laboratory, this absorbed light energy is rapidly released, allowing the electron to return to its initial "ground" energy state (Fig. 8.10a). Most of the energy (>95%) is converted into heat; a small amount is reemitted as light (fluorescence). By contrast, for chlorophyll molecules situated within the photosystems of the thylakoid, something entirely different occurs: The absorbed light energy is transferred to another chlorophyll molecule and then on to another (Fig. 8.10b).

Most of the chlorophyll molecules in the thylakoid membrane function as an antenna: Absorbed light energy is transferred from one chlorophyll molecule to another until it is finally transferred to a specially configured pair of chlorophyll molecules known as the **reaction center.** As we will see, the reaction center is where light energy is converted into electron transport. This division of labor among chlorophyll molecules was discovered in the 1940s in a series of experiments by the American biophysicists Robert Emerson and William Arnold, who showed that only a small fraction of chlorophyll molecules are directly involved in electron transport (**Fig. 8.11**). We now know that several hundred antenna chlorophyll molecules are associated with each reaction center. The antenna chlorophylls allow the photosynthetic electron transport chain to operate efficiently. Without the antenna to gather light energy, reaction centers would sit idle much of the time, even in bright sunlight.

The chlorophyll molecules that make up the antenna are precisely spaced so that when a chlorophyll molecule absorbs light it transfers energy, but not electrons, to an adjacent chlorophyll molecule (see Fig. 8.10b). A good analogy for this form of energy transfer is the transfer of vibrational energy between two tuning forks held close together. The transfer of energy between antenna chlorophyll molecules is highly efficient, so little energy is lost as heat, in contrast to what happens with an isolated chlorophyll molecule. Light energy absorbed by the antenna is passed from one chlorophyll molecule to the next until eventually it is transferred to the reaction center, as shown in **Fig. 8.12a.**

The reaction center chlorophylls have a configuration distinct from that of the antenna chlorophylls. As a result, reaction center chlorophylls are able to transfer both absorbed light energy *and* an associated high-energy electron to an adjacent molecule that acts as an electron acceptor. When the transfer takes place, the reaction center becomes oxidized and the adjacent electron-acceptor molecule is reduced. The result is the conversion of light energy into a chemical form. This electron transfer sets in motion

**FIG. 8.11**

# Do chlorophyll molecules operate on their own or in groups?

**BACKGROUND** By about 1915, scientists knew that chlorophyll was the pigment responsible for absorbing light energy in photosynthesis. However, it was unclear how these pigments contributed to the reduction of $CO_2$. The American physiologists Robert Emerson and William Arnold set out to determine the nature of the "photochemical unit" by quantifying how many chlorophyll molecules were needed to incorporate one $CO_2$ molecule into carbohydrate.

**EXPERIMENT** Emerson and Arnold exposed flasks of the green alga *Chlorella* to flashes of light of such short duration ($10^{-5}$ s) that each chlorophyll molecule could be "excited" only once, and the time between flashes was long enough to allow the reactions resulting from this light energy to run to completion. In step 1, they recorded the maximum rate of $CO_2$ uptake by increasing the intensity of the light flashes until the rate could not go any higher. In step 2, they determined the concentration of chlorophyll present in their solution of cells. They then compared the maximum rate of $CO_2$ uptake to the number of chlorophyll molecules.

**RESULTS**

**CONCLUSION** Because the maximum rate of $CO_2$ uptake per flash was much smaller than the amount of chlorophyll in their flask, Emerson and Arnold concluded that each photochemical unit contains many chlorophyll molecules.

**FOLLOW-UP WORK** Emerson and Arnold's work was followed by studies that demonstrated that the photosynthetic electron-transport chain contains two photosystems (photosynthetic units) arranged in series.

**SOURCE** Emerson, R., and W. Arnold. 1932. "The Photochemical Reaction in Photosynthesis." *Journal of General Physiology.* 16:191–205.

---

**FIG. 8.12**   **The reaction center.** (a) Antenna chlorophylls deliver absorbed light energy to the reaction center, where energy and electrons are transported to the electron transport chain. (b) After the reaction center has lost an electron, it is reduced by gaining an electron, so it is ready to absorb additional light energy.

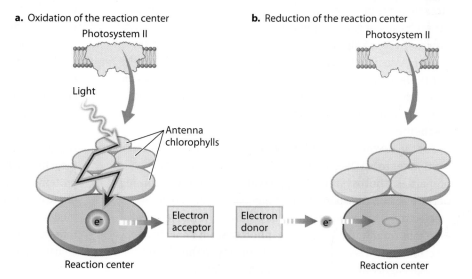

the light-driven chain of redox reactions that constitute the photosynthetic electron transport chain.

Once the reaction center has lost an electron, it is no longer able to absorb light. From a visual perspective, it has been bleached. Thus, for the photosystem to continue contributing electrons to the photosynthetic electron transport chain, another electron must be delivered to take the place of the one that has entered the transport chain (**Fig. 8.12b**). As we will see below, these replacement electrons ultimately come from water.

→   **Quick Check 3**  How do antenna chlorophylls differ from reaction center chlorophylls?

### The photosynthetic electron transport chain connects two photosystems.

Experiments conducted in the 1940s led to the surprising finding that photosynthetic cells use light energy at only half the maximum efficiency predicted. We now know that the reason that the efficiency is lower than predicted is that the photosynthetic electron transport chain includes not one but two photosystems arranged in series. Two photosystems are necessary to provide enough energy to pull electrons from water and then use them to reduce $NADP^+$.

If you follow the flow of electrons from water, through both photosystems, and on to $NADP^+$, as shown in **Fig. 8.13**, you can see a large increase in the energy level of the electrons as they pass through each of the two photosystems. You can also see that every other step along the photosynthetic electron transport chain is associated with a small decrease in the energy level of the electrons. Because the overall energy trajectory has an up-down configuration resembling a "Z," the photosynthetic electron transport chain is sometimes referred to as the **Z scheme.**

For the two photosystems to work together to move electrons from water to NADPH, they must have distinct chemical properties. **Photosystem II** supplies electrons to the beginning of the electron transport chain. When photosystem II loses an electron (that is, when it is itself oxidized), it is able to pull electrons from water. In contrast, **photosystem I** energizes electrons with a second input of light energy so they have enough energy to reduce $NADP^+$. The key point here is that photosystem I when oxidized is not a sufficiently strong oxidant

**FIG. 8.13**  **The Z scheme.** The use of water as an electron donor requires input of light energy at two places in the photosynthetic electron transport chain.

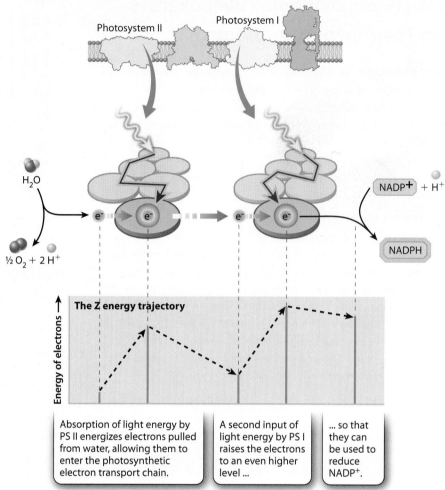

to split water, whereas photosystem II cannot produce electrons with enough energy to form NADPH.

The major protein complexes of the photosynthetic electron transport chain include the two photosystems as well as the **cytochrome $b_6 f$ complex,** through which electrons pass between photosystem II and photosystem I (**Figure 8.14**). Small, relatively mobile compounds convey electrons between these protein complexes. Plastoquinone, a lipid-soluble mobile compound, carries electrons from photosystem II to the cytochrome $b_6 f$ complex, while plastocyanin, a water-soluble protein, carries electrons from the cytochrome $b_6 f$ complex to photosystem I by diffusing through the thylakoid lumen. The modular nature of the photosynthetic electron transport chain, interconnected by diffusible elements, enables this pathway to adjust in response to changes in the availability of light.

Water, as the electron donor, sits at one end of the photosynthetic electron transport chain, whereas $NADP^+$, the

**FIG. 8.14    The photosynthetic electron transport chain.** (a) An overview of the production of NADPH and ATP. (b) The linear flow of electrons from $H_2O$ to NADPH. (c) The use of a proton electrochemical gradient to synthesize ATP.

**a.** The production of NADPH and ATP by photosynthesis

**b.** Electron transport and the synthesis of NADPH

**c.** Proton transport and ATP synthesis

electron acceptor that transports high-energy electrons to the Calvin cycle, sits at the other. NADPH is formed when high-energy electrons are passed from photosystem I to a membrane-associated protein called ferredoxin (Fig. 8.14b). The enzyme ferredoxin–NADP$^+$ reductase then catalyzes the formation of NADPH by transferring two high-energy electrons to NADP$^+$ as well as a proton from the surrounding solution:

$$NADP^+ + 2e^- + H^+ \longrightarrow NADPH$$

→ **Quick Check 4** Why are two photosystems needed if H$_2$O is used as an electron donor?

## The accumulation of protons in the thylakoid lumen drives the synthesis of ATP.

So far, we have considered only how the photosynthetic electron transport chain leads to the formation of NADPH. However, we know that the Calvin cycle also requires ATP. In chloroplasts, as in mitochondria, ATP is synthesized by ATP synthase, a transmembrane protein powered by the movement of protons across the membrane. In chloroplasts, the ATP synthase is oriented such that the movement of protons from the thylakoid lumen to the stroma results in the synthesis of ATP, as shown in Fig. 8.14c.

How do protons accumulate in the thylakoid lumen? Two features of the photosynthetic electron transport chain are responsible for the build-up of protons in the thylakoid lumen. First, the oxidation of water releases protons and O$_2$ into the lumen (Fig. 8.14c). Second, the cytochrome-$b_6 f$ complex, the protein complex situated between photosystem II and photosystem I, functions as a proton pump. As electrons transit through the cytochrome-$b_6 f$ complex, some of the energy that is released by each redox reaction is used to drive protons from the stroma side of the thylakoid membrane to the lumen (Fig. 8.14c).

These two mechanisms are quite powerful. When the photosynthetic electron transport chain is operating at full capacity, the concentration of protons in the lumen can be more than 1000 times greater than that in the stroma (equivalent to a difference of 3 pH units). This accumulation of protons on one side of the thylakoid membrane can then be used to power the synthesis of ATP. If offered a pathway through the membrane, protons diffuse from this region of higher concentration and net positive charge to the region of lower concentration and net negative charge in the stroma, releasing energy. The pathway most readily available is through a channel in the ATP synthase. When protons pass through this enzyme, some of the energy that is released is used to drive the synthesis of ATP (Chapter 7).

## Cyclic electron transport increases the production of ATP.

Sunlight varies in intensity throughout the day. When light levels are low, light is shared evenly between the two photosystems. However, as light levels increase, the light energy absorbed begins to overwhelm the capacity of the Calvin cycle to make use of NADPH. The danger is that when there is no NADP$^+$ returning from the Calvin cycle to act as the terminal electron acceptor for the photosynthetic electron transport chain, the high-energy electrons can damage the cell.

To prevent this from happening, electrons are shunted into an alternative pathway that increases the production of ATP while decreasing the production of NADPH. Electrons from photosystem I are redirected from ferredoxin back into the electron transport chain (**Fig. 8.15**). These electrons reenter the photosynthetic electron transport chain upstream of the cytochrome-$b_6 f$ proton pump. Because these electrons

**FIG. 8.15  Cyclic electron transport.** When light levels are high, some electrons are diverted from NADPH synthesis back into the electron transport chain, entering upstream of the cytochrome-$b_6 f$ proton pump.

eventually return to photosystem I, this alternative pathway is referred to as **cyclic electron transport,** in contrast to the linear movement of electrons from water to NADPH.

Cyclic electron transport leads to the production of ATP because as the high-energy electrons from ferredoxin pass through the cytochrome-$b_6f$ complex, additional protons are pumped into the lumen. As a result, there are more protons in the lumen that can be used to drive the synthesis of ATP. In addition, a subset of the photosystem II antennae migrate through the thylakoid membrane to become associated with photosystem I. This redistribution of light-harvesting capacity increases the rate of cyclic electron transport relative to linear electron transport, further enhancing the production of ATP.

**The spatial organization of the thylakoid membrane contributes to its functioning.**

We are now ready to understand how the major components of the photosynthetic electron transport chain are positioned on the thylakoid membrane in a manner consistent with their roles in photosynthesis. To do this, we must revisit the peculiar shape of this photosynthetic membrane, shown in Fig. 8.5. Recall that, despite its convoluted arrangement, the thylakoid membrane forms a single large sac, with the lumen on the inside and the stroma on the outside. The convoluted arrangement of interlinked granal stacks increases the total surface area of the thylakoid that can be accommodated within a single chloroplast. Nevertheless, the tight packing of the granal stacks means that only membranes on the outside of each granal stack and the links between grana are in direct contact with the stroma. The inner folds of the granal stacks have no direct contact with the stroma.

The components of the photosynthetic electron transport chain are distributed in such a way as to make tight packing possible, increasing surface area without sacrificing function. Photosystem I and the ATP synthase are concentrated on the outer regions of the granal stacks (**Fig. 8.16**). From there they can readily supply NADPH and ATP to the stroma, where these molecules are used to power the Calvin cycle. In contrast, photosystem II is located primarily on the closely packed inner regions of the thylakoid membrane that have little contact with the stroma. The cytochrome-$b_6f$ complex, which is located between the two photosystems, has a relatively even distribution.

**FIG. 8.16**    Distribution of PS I, PS II, and ATP synthase in the thylakoid membrane.

PS I and **ATP synthase** occur primarily where they have good contact with the stroma.

PS II predominates in the granal stacks.

Thylakoid lumen

Thylakoid membrane (cut)

PS II    PS I    Cytochrome $b_6f$ complex    ATP synthase

This distribution of components to different regions is made possible by the presence of the mobile compounds that carry electrons between the major components of the photosynthetic electron transport chain. In particular, the diffusion of the protein plastocyanin through the lumen allows photosystem II and photosystem I to be located in different regions of the thylakoid membrane. The close packing of thylakoid membrane regions helps explain why the thylakoid membrane and the inner mitochondrial membrane look so different, even though they both support electron transport chains.

**? CASE 1  The First Cell: Life's Origins**
**How did early cells meet their energy requirements?**
Photosynthesis is a highly complex process, involving the coordination of many reactions and enzymes. This complexity is the result of several billion years of evolution. The first organisms that were able to draw upon sunlight as an energy source would have been much simpler.

How simple? The earliest reaction centers may have used light energy to drive the movement of electrons from a soluble inorganic electron donor in the surrounding medium to an electron-acceptor molecule within the cell. Reduced iron, $Fe^{2+}$, is thought to have been abundant in the early ocean, and therefore could have served as the first electron donor. Alternatively, the first forms of light-driven electron transport may have been coupled to the net movement of protons across the membrane, allowing for the synthesis of ATP.

Similarly, it is unlikely that these first photosynthetic organisms employed chlorophyll, a complex molecule whose biosynthetic pathway contains at least 17 enzymatic steps. Because some of the intermediate compounds leading to chlorophyll are themselves capable of absorbing light energy, perhaps each of these now-intermediate compounds was, at one time, a functional end product used by an early photosynthetic organism. Selection for chemical variants with increased efficiency or the ability to absorb new portions of the visible spectrum may then have led to the further elaboration of this pathway. Selection would have eventually resulted in the chlorophyll pigments that are used by photosynthetic organisms today.

## 8.4 CHALLENGES TO PHOTOSYNTHETIC EFFICIENCY

The efficient functioning of photosynthesis faces two major challenges. The first is that if more energy is generated than the Calvin cycle can use, excess light energy can damage the cell. The second challenge stems from rubisco's ability to catalyze the addition of either carbon dioxide or oxygen to RuBP. The addition of oxygen instead of carbon dioxide can substantially reduce the amount of carbohydrate produced.

### Excess light energy can cause damage.

Photosynthesis is an inherently dangerous enterprise. Unless the photosynthetic reactions are carefully controlled, molecules will be formed that can damage cells through the indiscriminate oxidization of lipids, proteins, and nucleic acids (**Fig. 8.17**).

Under normal conditions, the redox reactions that make up the photosynthetic electron transport chain do not allow either absorbed light energy or the resulting high-energy electrons to stray. However, when $NADP^+$ is in short supply, either the absorbed light energy or the energy and the associated electron can be transferred to $O_2$, resulting in the formation of highly reactive forms of oxygen known collectively as **reactive oxygen species** (Fig. 8.17a).

$NADP^+$ is returned to the photosynthetic electron transport chain by the Calvin cycle's consumption of NADPH. Thus, any factor that causes the rate of $NADP^+$ regeneration to fall behind the rate of light-driven electron transport can potentially lead to damage. Such an imbalance is likely to occur, for example, in the middle of the day when light intensity is highest. Photosynthetic cells could prevent excess light energy by synthesizing sufficient quantities of Calvin cycle enzymes to process all the energy

absorbed by chlorophyll even at full sunlight. This strategy, however, would be energetically expensive because light levels vary dramatically over the course of the day. When light levels are low, such as in the morning and late afternoon, Calvin cycle enzymes would sit idle.

The rate at which the Calvin cycle can utilize NADPH is also influenced by a number of factors that are independent of light intensity. For example, cold temperatures cause the enzymes of the Calvin cycle to function more slowly, but they have little impact on the absorption of light energy by the photosynthetic electron transport chain. On a cold, sunny day, more light energy is absorbed than can be used by the Calvin cycle. Other factors that depress the rate at which the Calvin cycle can function include shortages of nitrogen, which reduces protein levels overall, and of $CO_2$.

Photosynthetic organisms employ two major lines of defense to deal with the stresses that occur when the Calvin cycle cannot keep up with light harvesting (Fig. 8.17b). First among these are chemicals that detoxify reactive oxygen species. Ascorbate (vitamin C), beta-carotene, and other antioxidants are able to neutralize reactive oxygen species. These compounds exist in high concentration in chloroplasts. Some of these antioxidant molecules are brightly colored, like the red pigments found in algae that live on snow shown in Fig. 8.3c. The presence of antioxidant compounds is one of the many reasons that eating photosynthetic tissues is good for your health.

A second line of defense is to prevent reactive oxygen species from forming in the first place. **Xanthophylls** are yellow-orange pigments that slow the formation of reactive oxygen species by reducing excess light energy. These pigments accept absorbed light energy directly from chlorophyll and then convert this energy to heat (Fig. 8.17b). Photosynthetic organisms that live in extreme environments often appear brown or yellow because of high levels of xanthophyll pigments, as seen in Figs. 8.3a and 8.3b. Plants that lack xanthophylls grow poorly when exposed to moderate light levels and die in full sunlight.

Converting absorbed light energy into heat is beneficial at high light levels, but at low light levels it would decrease the production of carbohydrates. Therefore, this ability is switched on only when the photosynthetic electron transport chain is working at high capacity. The creation of a strong proton gradient across the thylakoid membrane at high light levels activates the enzyme that converts inactive xanthophyll molecules into their light-absorbing form.

Xanthophyll pigments are predominantly associated with photosystem II. When activated by high light, these light-absorbing pigments reduce linear electron transport, but

**FIG. 8.17  Defenses against reactive oxygen species.** (a) Reactive oxygen species are generated when excess light energy and electrons are transferred to oxygen. (b) Defenses against the reactive oxygen species include antioxidants that neutralize reactive oxygen species and xanthophylls that convert excess light energy into heat.

**a.** The problem: Harmful reactive oxygen species are generated.

When not enough $NADP^+$ is available, the high-energy electrons will reduce $O_2$ instead of $NADP^+$.

**b.** The solution: Xanthophylls and antioxidants reduce the amount of reactive oxygen species.

Antioxidants detoxify reactive oxygen species.

$$O_2 \quad \text{Reactive oxygen species} + \text{antioxidants} \longrightarrow H_2O$$

Xanthophylls convert excess light energy to heat, reducing the rate at which energy and electrons enter the electron transport chain.

allow cyclic electron transport to continue. Light absorbed by photosystem I thus remains available to power the synthesis of ATP, which can then be used to help repair cellular components that have been damaged by reactive oxygen species.

**Photorespiration leads to a net loss of energy and carbon.**
A major challenge to photosynthetic efficiency is the fact that rubisco can add $O_2$ to RuBP instead of $CO_2$. The idea that an enzyme could make a "mistake" may seem surprising, yet, from

the point of view of a very large enzyme, $CO_2$ and $O_2$ have a lot in common: They are similar in size, have similar chemical bonds, and are both uncharged. Moreover, $O_2$ is approximately 500 times more abundant in the atmosphere than $CO_2$.

When rubisco adds $O_2$ instead of $CO_2$ to RuBP, the result is one molecule with three carbon atoms (3-PGA) and one molecule with only two carbon atoms (2-phosphoglycolate). The production of 2-phosphoglycolate creates a serious problem because this molecule cannot be utilized by the Calvin cycle either to produce triose phosphate or to regenerate RuBP.

A metabolic pathway to recycle 2-phosphoglycolate is present in photosynthetic cells. However, this pathway is not able to return all of the carbon atoms in 2-phosphoglycolate to the Calvin cycle; some are released as $CO_2$. Because the overall effect of oxygenation is a release of $CO_2$ in the presence of light, this process is referred to as **photorespiration** (**Fig. 8.18**). However, unlike respiration, which produces ATP, photorespiration actually consumes ATP. In photorespiration, ATP is used to drive the reactions that convert a portion of the carbon atoms in 2-phosphoglycolate into 3-PGA, which can reenter the Calvin cycle. Thus, photorespiration represents a net energy drain on two accounts: First, it results in the oxidation and loss, in the form of $CO_2$, of carbon atoms that had previously been incorporated and reduced by the Calvin cycle, and second, it consumes ATP.

Rubisco plays a key role in photosynthesis, and yet it is an enzyme that makes mistakes, "confusing" $O_2$ and $CO_2$. What accounts for its evolutionary success? A partial answer is that rubisco first evolved long before oxygen appeared in Earth's atmosphere. Still, why would photorespiration persist in the face of what must be strong evolutionary pressure to reduce or eliminate the unwanted oxygenation reaction?

The difficulty is that for rubisco to favor the addition of $CO_2$ over $O_2$ requires a high degree of selectivity, and the price of selectivity is speed. The better rubisco is at discriminating between $CO_2$ and $O_2$, the slower its catalytic rate. Nowhere is this trade-off more evident than in land plants, whose photosynthetic cells acquire $CO_2$ from an $O_2$-rich and $CO_2$-poor atmosphere. The rubiscos of land plants are highly selective, but incredibly slow, with catalytic rates on the order of three reactions per second. To put this rate in perspective, it is not uncommon for metabolic enzymes to achieve a catalytic rate of a hundred thousand reactions per second.

This trade-off between selectivity and speed is a key constraint for photosynthetic organisms. For land plants, rubisco's low catalytic rate means that photosynthetic cells must produce huge amounts of this enzyme; as much as 50% of the total protein within a leaf is rubisco. At the same time, the low concentration of $CO_2$ in the atmosphere (0.0385%) relative to $O_2$ (21%) means that as much as one-quarter of the reduced carbon formed in photosynthesis can be lost via photorespiration.

→ **Quick Check 5**  Why does rubisco have such a low catalytic rate (that is, why is it so slow)?

---

**FIG. 8.18**  **Photorespiration.**  Carbon and energy are lost when rubisco acts as an oxygenase in the process called photorespiration.

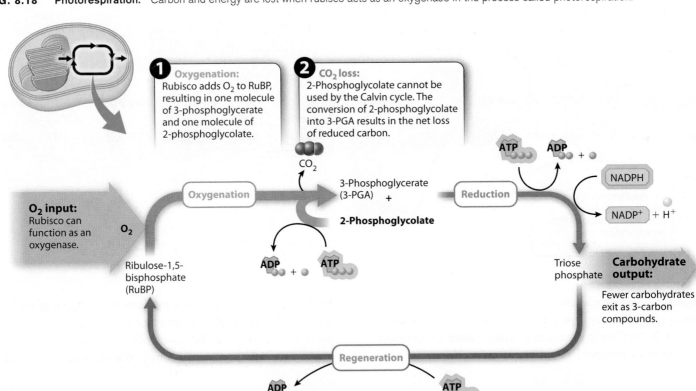

## Photosynthesis captures just a small percentage of incoming solar energy.

Typically, only 1% to 2% of the sun's energy that lands on a leaf ends up in carbohydrates. Does this mean that photosynthesis is incredibly wasteful? Or is this process, the product of billions of years of evolution, surprisingly efficient? This is not an idle question. The effects of rising $CO_2$ concentrations on Earth's climate, the search for a renewable, carbon-neutral fuel to power our transportation sector, and the agricultural demands of our skyrocketing human population all point to photosynthesis as a metabolic process relevant to solving some of our most pressing global issues.

Photosynthetic efficiency is typically calculated relative to the total energy output of the sun (**Fig. 8.19**). However, only visible light has the appropriate energy levels to produce the high-energy electrons required by the photosynthetic electron transport chain. Most of the sun's output (~60%) is not absorbed by chlorophyll and thus cannot be used in photosynthesis. In addition, leaves are not perfect at absorbing visible light—about 8% is either reflected or passes through the leaf. Finally, even under optimal conditions, not all of the light energy absorbed by chlorophyll can be transferred to the reaction center and instead is given off as heat (also ~8%). As we have seen, when light levels are high, excess light is actively converted into heat by xanthophyll pigments.

The photosynthetic electron transport chain therefore captures at most about 24% of the sun's usable energy arriving at the surface of a leaf (100% − 60% − 8% − 8% = 24%). While this number may appear low, it is on a par with the number of high-performance photovoltaic cells in solar panels, which convert sunlight into electricity. This comparison is even more impressive when you consider that photosynthetic organisms must build and maintain all their biochemical machinery. However, energy is lost at another step as well. The incorporation of $CO_2$ into carbohydrates results in considerable loss in free energy, equivalent to ~20% of the total incoming solar radiation. Much of this loss in free energy is due to photorespiration.

In total, therefore, the maximum energy conversion efficiency of photosynthesis is calculated to be around 4% (24% − 20%). Efficiencies achieved by real plants growing in nature, however, are typically much lower, on the order of 1% to 2%. In Chapter 29, we explore the many factors that can constrain the

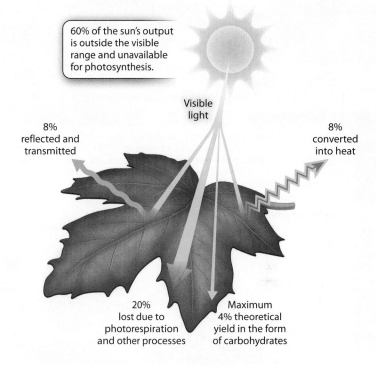

**FIG. 8.19 Photosynthetic efficiency.** Maximum photosynthetic efficiency is theoretically around 4% of incoming solar energy, but actual yields are closer to 1% to 2%.

60% of the sun's output is outside the visible range and unavailable for photosynthesis.

Visible light

8% reflected and transmitted

8% converted into heat

20% lost due to photorespiration and other processes

Maximum 4% theoretical yield in the form of carbohydrates

photosynthetic output of land plants, and see how some plants have evolved ways to minimize losses in productivity due to drought and photorespiration.

Cellular respiration and photosynthesis are complementary metabolic processes. Cellular respiration breaks down carbohydrates in the presence of oxygen to supply the energy needs of the cell, producing carbon dioxide and water as byproducts, while photosynthesis uses carbon dioxide and water in the presence of sunlight to build carbohydrates, releasing oxygen as a byproduct. We summarize the two processes in **Fig. 8.20**.

→ **Quick Check 6** In what ways is photorespiration similar to cellular respiration ( Chapter 7) and in what ways does it differ?

# Harnessing Energy: Photosynthesis and Cellular Respiration

**Integrating concepts from Chapters 6–8**

## Photosynthesis

Energy captured from sunlight is stored in carbohydrates.

*Photosynthesis* requires light energy, $CO_2$, and $H_2O$, and produces carbohydrates and $O_2$. During *light dependent reactions*, an electron in PS II is excited by light energy and is captured in NADPH. An electron from a water molecule replaces the excited PS II electron captured in NADPH. The accumulation of protons in the lumen is used to drive the synthesis of ATP.

Carbon dioxide is reduced by NADPH and, using the energy provided by ATP, is incorporated into a carbohydrate molecule in the *Calvin cycle*.

### Electron movement

*Photosynthetic electron transport chain*

*Calvin cycle*

**Inputs:**

**Outputs:**

## Aerobic respiration

Energy released from the oxidation of carbohydrates is used to make ATP.

*Aerobic respiration* requires carbohydrates and $O_2$ and produces $CO_2$, $H_2O$, and ATP. *Glycolysis* splits glucose into two molecules of pyruvate and begins the harvest of high energy electrons.

The *citric acid cycle* completes the harvest of high energy electrons and transfers them to NADH and $FADH_2$.

The energy released from the electrons as they are passed down the *electron transport chain* maintains the proton gradient needed by ATP synthase.

NADH — NADH — NADH

$FADH_2$

$NAD^+$     $NAD^+$     FAD     $NAD^+$

Carbohydrate → Pyruvate → Acetyl-CoA

$CO_2$

NADH     $FADH_2$     $NAD^+$     FAD

Mitochondrial matrix

$H^+$     $H_2O$

$O_2$

ADP     ATP

Complex I     Complex II     Complex III     Complex IV     ATP synthase

Intermembrane space

*Glycolysis*     *Citric acid cycle*     *Electron transport chain*

**Inputs:**     ADP     ADP     $O_2$     $H^+$     ADP

Carbohydrate → Pyruvate → Acetyl-CoA → ⟳ → $FADH_2$ / NADH → Complex I → II → III → IV → $e^-$ →

**Outputs:**     ATP     $CO_2$     $CO_2$     ATP     $H_2O$     ATP

# Core Concepts Summary

## 8.1 PHOTOSYNTHESIS IS THE MAJOR PATHWAY BY WHICH ENERGY AND CARBON ARE INCORPORATED INTO CARBOHYDRATES.

In photosynthesis, water is oxidized to form oxygen, and carbon dioxide is reduced to form carbohydrates. page 8-1

Photosynthesis consists of two sets of reactions: (1) the Calvin cycle, in which carbon dioxide is reduced to form carbohydrates, and (2) light-harvesting reactions, in which ATP and NADPH are generated to drive the Calvin cycle. page 8-2

The evolution of photosynthesis involved horizontal gene transfer in bacteria and endosymbiosis in eukaryotes. page 8-4

In eukaryotes, photosynthesis takes place in chloroplasts: The Calvin cycle occurs in the stroma, and the light-harvesting reactions take place on the thylakoid membrane. page 8-5

## 8.2 THE CALVIN CYCLE IS A THREE-STEP PROCESS THAT RESULTS IN THE INCORPORATION OF CARBON DIOXIDE INTO CARBOHYDRATES.

The three steps of the Calvin cycle are (1) carboxylation; (2) reduction; and (3) regeneration. page 8-6

The first step is the addition of $CO_2$ to the 5-carbon sugar RuBP. This step is catalyzed by the enzyme rubisco, considered the most abundant protein on Earth. The resulting 6-carbon compound immediately breaks down into two 3-carbon compounds. page 8-6

The second step is phosphorylation of the 3-carbon compounds by ATP followed by reduction by NADPH to produce 3-carbon triose phosphate molecules that are exported from the chloroplast to the cytosol, where they are used to build larger sugars. page 8-7

The third step is the regeneration of RuBP from five 3-carbon molecules. page 8-7

Starch formation provides chloroplasts with an osmotically inactive way of storing carbohydrates. page 8-7

## 8.3 THE LIGHT-HARVESTING REACTIONS USE SUNLIGHT TO PRODUCE ATP AND NADPH REQUIRED BY THE CALVIN CYCLE.

Visible light is absorbed by chlorophyll. The absorbed light energy can be released as heat, reemitted as light (fluorescence), or transferred to an adjacent chlorophyll molecule. Special chlorophyll molecules in the reaction center transfer both energy and electrons, thus initiating the photosynthetic electron transport chain. page 8-9

The electron transport chain consists of multisubunit protein complexes and diffusible compounds. Water is the electron donor and $NADP^+$ is the final electron acceptor. page 8-9

Antenna chlorophylls transfer absorbed light energy to the reaction center. page 8-10

Reaction centers are located within pigment–protein complexes known as photosystems. page 8-10

The linear transport of electrons from water to NADPH requires the energy input of two photosystems. page 8-12

Photosystem II pulls electrons from water, resulting in the production of oxygen and protons on the lumen side of the membrane. Photosystem I passes electrons to $NADP^+$, producing NADPH for use in the Calvin cycle. page 8-12

The buildup of protons drives ATP synthase to produce ATP on the stroma side of the membrane, where it is used by the Calvin cycle. page 8-14

## 8.4 PHOTOSYNTHESIS FACES SEVERAL CHALLENGES TO ITS EFFICIENCY.

An imbalance between the light-harvesting reactions and the Calvin cycle can lead to the formation of reactive oxygen species. page 8-16

Protection from excess light energy includes antioxidant molecules that neutralize reactive oxygen species and xanthophyll pigments that dissipate excess light energy as heat. page 8-16

Rubisco can act catalytically on oxygen as well as on carbon dioxide. When it acts as an oxygenase, there is a loss of energy and reduced carbon from the Calvin cycle. page 8-18

Rubisco has evolved to favor carbon dioxide over oxygen, but the cost of this selectivity is reduced speed. page 8-18

The synthesis of carbohydrates via the Calvin cycle results in significant energy losses, largely due to photorespiration. page 8-18

The maximum theoretical efficiency of photosynthesis is approximately 4% of total incident solar energy. page 8-19

# Self-Assessment

1. Write the overall photosynthetic reaction and identify which molecules are oxidized and which molecules are reduced.

2. Describe how photosynthesis evolved in prokaryotes and in eukaryotes.

3.  Explain the functions of the Calvin cycle and the light harvesting reactions in photosynthesis.

4.  Name the major inputs and outputs of the Calvin cycle.

5.  Describe the three major steps in the Calvin cycle and the role of the key enzyme rubisco.

6.  Explain why photosynthetic electron transport requires two photosystems.

7.  Describe the role of cyclic electron transport.

8.  List two strategies that plants use to limit the formation and effects of reactive oxygen species.

9.  Explain the trade-off that rubisco faces in terms of selectivity and enzymatic speed.

10. Estimate the overall efficiency of photosynthesis and where in the pathway energy is dissipated.

Do you understand the chapter's Core Concepts? Log in to check your answers to the Self-Assessment questions, then practice what you've learned and reinforce this chapter's concepts by working through the problems and multimedia tutorials provided there.

 www.biologyhowlifeworks.com

# Cancer

## WHEN GOOD CELLS GO BAD

Imagine a simple vaccine that could prevent about 500,000 cases of cancer worldwide every year. You'd think such a discovery would be hailed as a miracle. Not quite. A vaccine to prevent cervical cancer—which affects about half a million women and kills as many as 275,000 annually—has been available since 2006. But in the United States, the vaccine has stirred unanticipated controversy.

Cancer is uncontrolled cell division. Cell division is a normal process that occurs during development of a multicellular organism and subsequently as part of the maintenance and repair of adult tissues. It is carefully regulated so that it occurs only at the right time and place. But sometimes, this careful regulation can be disrupted. When the normal checks on cell division become derailed, cancer can result.

Most cancers are caused by inherited or acquired mutations, but nearly all cases of cervical cancer are caused by a virus called human papillomavirus (HPV). There are hundreds of strains of HPV. Some of these strains cause minor problems such as common warts and plantar warts in men and women. Other strains are sexually transmitted. Some can cause genital warts in men and women but aren't associated with cancer. However, a handful of "high-risk" HPV strains are strongly tied to cancer of the cervix. More than 99% of cervical cancer cases are believed to arise from HPV infections.

HPV infects epithelial cells, a type of cell that lines the body cavities and covers the outer surface of the body. Once inside a cell, the virus hijacks the cellular machinery to produce new viruses. The high-risk strains of HPV take it one step further: Those viruses that aren't fought off by the immune system can permanently integrate their own eight-gene DNA sequence into the host cell's DNA.

Once integrated into the DNA of human epithelial cells, two viral genes, E6 and E7, ramp up their activity and produce two proteins. These viral proteins inhibit the activity of key tumor suppressor genes. Tumor suppressor genes code for proteins that keep cell division in check by slowing down cell division, repairing DNA replication errors, or instructing defective cells to die. When tumor suppressors are prevented from doing these jobs, cancer can grow.

HPV strikes in two ways. The viral E6 protein inhibits a protein called p53, an important tumor suppressor, which in healthy cells controls responses to cellular stresses such as DNA damage. However, when bound by E6, p53 becomes essentially inactive. Meanwhile, the viral E7 protein inhibits a protein called Rb, which normally blocks transcription factors that promote cell division. Without Rb and p53 to put the brakes on cell division, cervical cells divide uncontrollably.

As they grow and multiply, the abnormal cells push through the basal lamina, a thin layer that separates the epithelial cells lining the cervix from the connective tissue beneath. Untreated, invasive cancer can spread to other organs. Once cancer travels beyond its primary location, the situation is grim. Most cancers cannot be cured after they have spread.

The U.S. Food and Drug Administration (FDA) has approved two vaccines for HPV, both of which protect against the high-risk strains responsible for cervical cancer. Medical groups such as the American Academy of Pediatrics and governmental organizations such as the Centers for Disease Control and Prevention (CDC) recommend

> When the normal checks on cell division become derailed, cancer can result.

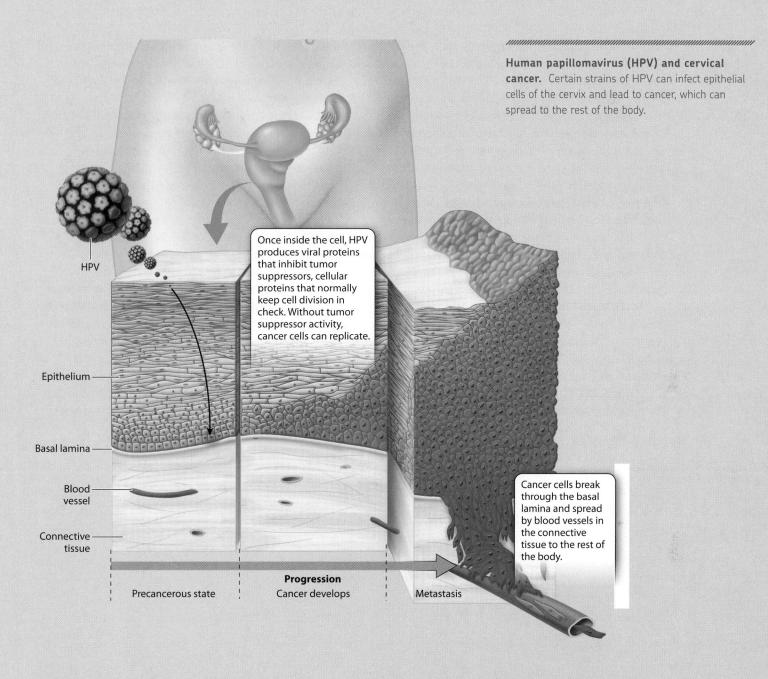

**Human papillomavirus (HPV) and cervical cancer.** Certain strains of HPV can infect epithelial cells of the cervix and lead to cancer, which can spread to the rest of the body.

HPV

Once inside the cell, HPV produces viral proteins that inhibit tumor suppressors, cellular proteins that normally keep cell division in check. Without tumor suppressor activity, cancer cells can replicate.

Epithelium

Basal lamina

Blood vessel

Connective tissue

Cancer cells break through the basal lamina and spread by blood vessels in the connective tissue to the rest of the body.

**Progression**

Precancerous state | Cancer develops | Metastasis

vaccinating girls at age 11 or 12, before they become sexually active. In addition, recent CDC guidelines also recommend vaccinating boys because they can transmit the virus to women when they become sexually active.

On the basis of these recommendations, Virginia and Washington, D.C., require that middle-school girls receive the vaccine, and many other states are considering similar legislation. But there are critics of these mandates. Some say the government shouldn't be making medical decisions. Others argue that the vaccine condones sexual activity among young girls and boys. Still others have expressed concern over the vaccine's safety.

Studies have found the HPV vaccine to be safe, a conclusion backed by a 2010 report from the Institute of Medicine, an independent nonprofit organization that advises the U.S government on issues of health. In spite of these findings, the HPV vaccine has not been widely accepted. The CDC found that by 2010, a full 5 years after the vaccine was introduced, only 32% of teenage girls had been vaccinated.

Proponents of HPV vaccination say those numbers should cause concern. HPV is common: A 2007 study found that nearly 27% of American women ages 14–59 were infected with the virus. Among

20- to 24-year-olds, the infection rate was nearly 45%. Certainly, not everyone who contracts HPV will develop cancer. Most people manage to clear the virus from their bodies within 2 years of infection. But in some cases, the virus hangs on and cancer results.

Because cancer is often difficult to treat, especially in its later stages, many researchers focus on early detection or prevention. Early detection of cervical cancer can be made from routinely taken Pap smears, in which cells from the cervix are collected and observed under a microscope. Cervical cancer is also an obvious target for prevention, since it is caused by a virus that can be averted with a conventional vaccine. However, most cancers are not caused by viruses, and developing vaccines to prevent those cancers is a trickier proposition—but researchers are pushing ahead.

At the Mayo Clinic in Rochester, Minnesota, researchers are working to design vaccines to prevent breast and ovarian cancers from recurring in women who have been treated for these diseases. The researchers have zeroed in on proteins on the surface of cancerous cells. Cells communicate with one another by releasing signaling molecules that are picked up by receptor molecules on another cell's surface, much

the way a radio antenna picks up a radio signal sent across a distance. Cellular communication is critical for a functioning organism. Sometimes, though, the signaling process goes awry.

The Mayo Clinic team is focusing on two cell surface receptors that, when malfunctioning, lead to cancer. One, Her2/neu, promotes the growth of aggressive breast cancer cells. The other, folate receptor α (alpha) protein, is frequently overexpressed in breast and ovarian tumors. The researchers hope to train patients' immune systems to generate antibodies that recognize these proteins, and then destroy the cancerous cells. The approach successfully prevented tumors in mice. Now the researchers are testing the vaccines in humans.

Even if these new therapies turn out to be successful, cancer researchers have much more work to do. Cancer is not one disease but many, and most of them are caused by a complex interplay of genetics and environmental factors. Any number of things can go wrong as cells communicate, grow, and divide. But as researchers learn more about the cellular processes involved, they can step in to prevent or treat cancer. As the HPV vaccine shows, there is significant progress to be made.

## ? CASE 2 QUESTIONS

1. How do cell signaling errors lead to cancer? *See page 9-15.*
2. How can doctors test for the spread of cancer? *See page 10-5.*
3. How do cancer cells spread throughout the body? *See page 10-18.*
4. What genes are involved in cancer? *See page 11-17.*

## CHAPTER 9

# CELL COMMUNICATION

## Core Concepts

**9.1** Cells communicate primarily by sending and receiving chemical signals.

**9.2** Cell signaling can be classified according to the distance between the signaling and responding cells.

**9.3** Signaling molecules bind to and activate specific cell-surface and intra-cellular receptors.

**9.4** Signals are transmitted across the plasma membrane, are often amplified in the cytosol leading to a cellular response, and are eventually terminated.

Up to this point, we have considered how life works by looking mainly at what happens inside individual cells. We have seen how a cell uses the information stored in genes to synthesize the proteins necessary to carry out their functions. We have also seen how the plasma membrane actively keeps the environment inside the cell different from that outside it. Finally, we have explored how a cell harvests and uses energy from the environment. Most cells, however, do not exist in isolation. Instead, they exist in sometimes simple, but more often highly complex, communities of cells.

Successful communities have several features in common. First, individuals within a community communicate with one another. Cells respond to information provided by neighboring cells or to information provided by the surrounding environment. Second, individuals or groups of individuals within a community have particular skills that enable them to perform specific tasks essential to the well-being of the community. Cells in multicellular organisms have specialized functions: Muscle cells are specialized for contraction, red blood cells for transporting oxygen, and skin cells for establishing a physical barrier. Third, members of a community stick together—literally, in the case of cells. Most cells in multicellular organisms are physically attached to other cells or to other materials in their surroundings. The physical associations among some cells are very strong. In other instances, the connections between cells are transient and far less elaborate.

In the next three chapters, we consider communities of cells. In Chapter 9, we look at how cells communicate with one another. In Chapter 10, we consider cell shape and structure, which are both important for the ability of cells to perform their specific tasks, and we examine how cells physically interact with one another. Finally, in Chapter 11, we examine the means by which new members are added to cellular communities through cell division.

## 9.1 PRINCIPLES OF CELL COMMUNICATION

As members of our own communities, we often base our actions on information we receive from others in the group, responding very differently to a friendly handshake or affectionate hug than to a hard shove or violent blow. We also glean information from inanimate things around us: We respond differently when we walk barefoot on hot pavement than when we take a seat in a comfortable chair. Similarly, the activities of virtually all cells are influenced by their physical surroundings, including interactions with other members of the community of cells. Cells receive large amounts of information from their surroundings, from numerous sources and in different ways. Often, information is conveyed from one cell to another in the form of small molecules secreted into the environment. Other information is provided by physical contact with neighboring cells. Still other information is provided by the meshwork of proteins and polysaccharides that underlies or surrounds most cells in multicellular organisms (Chapter 10).

In this section, we take a look at general principles of cell communication, including how cells send and receive signals and how a cell responds after it receives a signal. These basic principles apply to all living organisms, both prokaryotes and eukaryotes, and both unicellular and multicellular organisms.

## Cells communicate using chemical signals that bind to specific receptors.

An example from prokaryotes illustrates some important principles underlying cell communication. *Streptococcus pneumoniae* (also known as pneumococcus) is a disease-causing bacterium that is associated not only with pneumonia but also with meningitis and some kinds of arthritis. Many bacteria, including pneumococcus, are able to take up DNA from the environment and incorporate it into their own genome. By this means, individual bacterial cells can share genes with advantageous properties, including the ability to resist antibiotic drugs.

In the 1960s, it was observed that the rate of DNA uptake by pneumococcal cells increased sharply once the bacterial population reached a certain density. Scientists concluded that the bacteria were able to coordinate DNA uptake across the population so that uptake occurred only at the appropriate time. How is it possible that these bacteria "know" how many other bacteria are present and are able to communicate this information to one another?

In the 1990s, scientists discovered a peptide consisting of 17 amino acids that is continuously synthesized and released by pneumococcal cells. Not long after, a receptor for this peptide was discovered on the surface of the pneumococcal cells. The binding of this peptide to its receptor causes a bacterium to express the genes required for DNA uptake. When the bacteria are at low density, the concentration of the peptide is too low to activate the genes required for DNA uptake (**Fig. 9.1a**). As the population density increases, so does the concentration of the peptide, until it reaches a level high enough to cause the cells to turn on the genes necessary for DNA uptake (**Fig. 9.1b**).

This simple example of DNA uptake by pneumococcus illustrates the four essential elements involved in communication between all cells, whether prokaryotic or, as shown in **Fig. 9.2**, eukaryotic. These elements are a **signaling cell**, a **signaling molecule**, a **receptor molecule**, and a **responding cell.** The signaling cell is the source of the signaling molecule, which binds to a receptor molecule on or in the responding cell. In the case of pneumococcus, the bacterial cells are both the signaling cells and the responding cells. In other cases, the signaling and responding cells may be different cells. Nevertheless, the general idea that cells are able to communicate by sending a signaling molecule that binds to a receptor on a responding cell is universal among both prokaryotes and eukaryotes. These four players

**FIG. 9.1**    **Communication among bacterial cells.**

**a.** Low cell density

At low population density, the concentration of the signaling peptide is too low to stimulate DNA uptake by bacteria.

Receptor

Bacterial cell

Signaling molecule

**b.** High cell density

At high population density, the concentration of the signaling peptide is high enough to stimulate the DNA-uptake response.

DNA molecule

interact in very much the same way in diverse types of cellular communication.

## Signaling involves receptor activation, signal transduction, response, and termination.

What happens after a signaling molecule binds to a receptor on a responding cell? On binding the signal, the receptor is turned on, or **activated** (**Fig. 9.3**). Once activated, the receptor transmits the message through the cytoplasm, often by intracellular signaling

**FIG. 9.2**    **Four components required for cellular communication: a signaling cell, a signaling molecule, a receptor, and a responding cell.**

Signaling molecule      Receptor

The signaling cell releases signaling molecules.

The responding cell has receptor proteins that bind to the signaling molecule.

## FIG. 9.3 Steps in cell signaling: receptor activation, signal transduction, response, and termination.

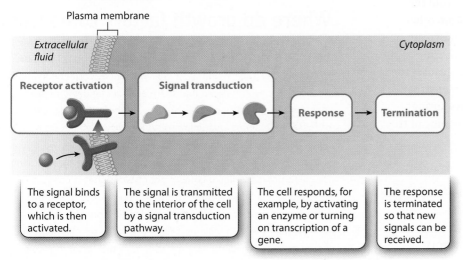

Plasma membrane

*Extracellular fluid* / *Cytoplasm*

**Receptor activation** | **Signal transduction** | **Response** | **Termination**

| The signal binds to a receptor, which is then activated. | The signal is transmitted to the interior of the cell by a signal transduction pathway. | The cell responds, for example, by activating an enzyme or turning on transcription of a gene. | The response is terminated so that new signals can be received. |

series of steps, which have been evolutionarily conserved over long periods of time.

## 9.2 TYPES OF CELL SIGNALING

In prokaryotes and unicellular eukaryotes, cell communication is communication between individuals. In complex multicellular eukaryotes, cell signaling involves communication between cells that make up an individual. The same principles apply in both instances, but there are differences to be considered. In multicellular organisms, the distance between communicating cells varies considerably. In addition, many cells in multicellular organisms are physically attached to one another, in which case the signaling molecule is not released from the signaling cell. In this section, we explore communication over long and short distances, as well as communication among cells that are physically associated with one another.

pathways or cascades, in a process called **signal transduction.** The message is carried from outside the cell into the cytosol or nucleus. Along the way, the initial signal is often amplified, so that a small signal can have a large effect on the responding cell.

Next, there is a cellular **response,** which can take different forms depending on the nature of the signal and the type of responding cell. For example, signaling pathways can activate enzymes or turn on genes that cause the cell to divide, change shape, or signal other cells. Finally, the signal is terminated. **Termination** allows the cell to respond to new signals.

In the case of pneumococcal cells, the peptide-signaling molecule binds to a receptor on the cell surface. When enough receptors are bound by the signaling molecule, the message is relayed by signal-transduction pathways to the nucleoid in the cytosol, where genes are turned on that express proteins involved in DNA uptake from the environment. Eventually, when the density of bacteria is low, the initiating signal falls below a critical threshold and gene expression is turned back off. This example is relatively simple. Remarkably, however, other signaling pathways in a wide range of organisms involve the same four players and a common

### Endocrine signaling acts over long distances.

Signaling molecules released by a cell may have to travel great distances in the body, usually via the circulatory system, to reach their target responding cells. Signaling by means of molecules that travel through the bloodstream is called **endocrine signaling** (**Fig. 9.4a** and Chapter 38).

**FIG. 9.4** **Types of cell communication.** Cell communication can be classified according to the distance between the signaling and responding cells.

**a.** Endocrine signaling

Signaling molecule

Signal travels through the circulatory system.

Receptor

**b.** Paracrine signaling

**c.** Autocrine signaling

**d.** Juxtacrine signaling

Familiar examples of endocrine-signaling molecules include the mammalian steroid hormones estradiol (an estrogen) and testosterone (an androgen). These hormones travel from the ovaries and the testes, respectively (although there are other minor sources of these hormones), through the bloodstream, to target cells in various tissues throughout the body. The increased amount of estrogen in girls during puberty causes the development of breast tissue and the beginning of menstrual cycles. The increased amount of testosterone in boys during puberty causes the growth of muscle cells, the deepening of the voice, and the growth of facial hair (Chapter 42).

→ **Quick Check 1** If a hormone is released into the bloodstream and therefore comes into contact with many cells, what determines which cells in the body respond to the hormone?

## Paracrine and autocrine signaling act over short distances.

In contrast to endocrine signaling, **paracrine signaling** (**Fig. 9.4b**) involves two cells that are close together. In this case, a signaling molecule needs to diffuse only a short distance to the nearest neighboring cell in order to bind its receptor and deliver its message. Signaling molecules in paracrine communication travel distances of up to 20 cell diameters, or a few hundred micrometers. In some cases, signaling molecules may be secreted by a cell and then bind to receptors on the very same cell, as we saw in the case of pneumococcal cells. Such cases, where signaling cell and responding cell are one and the same, are examples of **autocrine signaling** (**Fig. 9.4c**). Paracrine and autocrine signaling are especially important to multicellular organisms during the development of the embryo (Chapter 20).

In paracrine signaling, the signal is usually a small, soluble molecule, such as a **growth factor.** The first growth factor discovered was found by scientists who were attempting to understand how to maintain cultures of cells in the laboratory for use in their research. For decades, medical research has worked with cells maintained in culture. Initially, these cultured cells had limited use because they failed to divide outside the body unless they were supplied with unidentified factors from mammalian blood serum. In 1974, American scientists Nancy Kohler and Allan Lipton discovered that these factors are secreted by platelets (**Fig. 9.5**). Consequently, the first growth factor was named "platelet-derived growth factor," or PDGF. We now know of scores of proteins secreted by cells that function as growth factors, and in most cases their effects are confined to neighboring cells. These effects include the stimulation of cell growth, cell division, and changes in gene expression.

**FIG. 9.5**

# Where do growth factors come from?

**BACKGROUND** Cells can be grown outside the body in culture. However, they survive and grow well only under certain conditions. Researchers hypothesized that there are substances that are required for growth of cells in culture, but the nature and source of these substances was unknown. A key insight came from the observation that chicken cells grew much better if they were cultured in the presence of blood serum rather than blood plasma. Blood serum is the liquid component of blood that is collected after blood has been allowed to clot. Blood plasma is also the liquid component of blood, but it is collected from blood that has not clotted. American biologists Nancy Kohler and Allan Lipton were interested in identifying the source of the factor in blood serum that allows cells to survive in culture.

**HYPOTHESIS** Since Kohler and Lipton knew that clotting depends on the release of substances from platelets, they hypothesized that a growth-promoting factor was activated or introduced into the blood by platelets during the clotting reaction.

**EXPERIMENT 1** The investigators first confirmed earlier observations using cells called fibroblasts, which they took from mice. They cultured two sets of fibroblasts in small plastic dishes. To one of the cultures they added serum; to the other culture they added plasma. Then they monitored the rate of cell division in both culture dishes for nine days.

**RESULTS 1** They observed that the rate of cell division in the fibroblasts cultured in serum was far greater than that of the cells cultured in plasma, as expected (Fig. 9.5a).

Growth factors secreted by cells in an embryo work over short distances to influence the kind of cells their neighbors will become. In this way, they help shape the structure of the adult's tissues, organs, and limbs. For example, in developing vertebrates, paracrine signaling by the growth factor Sonic Hedgehog (yes, it's named after a video game character) ensures that the motor neurons in your spinal cord are located properly, that the bones of your vertebral column form correctly, and that your thumb and pinky fingers are on the correct sides of your hands.

A specialized form of paracrine signaling in many organisms is the communication between neurons (nerve cells), or between

**a.**

**EXPERIMENT 2** To see if the factor is released directly from platelets, they prepared a solution of proteins made from purified platelets, added these proteins to cultured fibroblasts, and measured cell growth.

**RESULTS 2** They found that the solution of platelet proteins also caused an increase in the growth of fibroblasts compared to the growth of fibroblasts in plasma (Fig. 9.5b).

**b.**

**CONCLUSION** Kohler and Lipton concluded that the growth-promoting factor is a protein that is released by platelets on activation and clot formation, and therefore normally present in serum.

**FOLLOW-UP WORK** Subsequent work over the next few years by these and other scientists resulted in the purification and characterization of the growth factor we know today as platelet-derived growth factor, or PDGF.

**SOURCE** Kohler, N., and A. Lipton. 1974. "Platelets as a Source of Fibroblast Growth-Promoting Activity." *Experimental Cell Research* 87:297–301.

neurons and muscle cells (Chapter 35). When neurotransmitters, a type of signaling molecule, are released from a neuron, they diffuse across a very small space, called a synapse, between the signaling cell and the responding cell. If the adjacent cell is a neuron, it often responds by transmitting the nerve impulse further and then releasing additional neurotransmitters. If the responding cell is a muscle cell, it often responds by contracting.

**Juxtacrine signaling depends on direct cell–cell contact.**
In some cases, a cell communicates with another cell through direct physical contact, without a chemical signal that

diffuses or circulates through an external medium. This type of communication is called contact-dependent signaling, or **juxtacrine signaling** (**Fig. 9.4d**). It requires that the two communicating cells be in physical contact with each other. A transmembrane protein on the surface of one cell acts as the signaling molecule, and a transmembrane protein on the surface of an adjacent cell acts as the receptor.

As an example of juxtacrine signaling, let's look at the development of the central nervous system of vertebrate animals. In the brain and spinal cord, neurons transmit information in the form of electrical signals from one part of the body to another. The neurons in the central nervous system are greatly outnumbered by supporting cells, called glial cells, that nourish and insulate the neurons. Both the neurons and the glial cells start out as similar cells in the embryo, but some of these undifferentiated cells become neurons and many more become glial cells.

For reasons that are not entirely clear, the amount of a transmembrane protein called Delta dramatically increases on the surface of some of these undifferentiated cells. Delta binds to a transmembrane protein called Notch on the surface of adjacent cells. In this case, the signaling cell is the cell with elevated levels of Delta protein. The Delta protein in turn is the signaling molecule, and Notch is its receptor. The signal Delta sends to the adjacent, undifferentiated cell through Notch is a message to become a glial cell and not a neuron. Because one signaling cell sends this same message to all the cells it touches, it is easy to understand how there can be so many more glial cells than neurons in the central nervous system.

As you can see from these examples, the ways in which cells of a developing embryo communicate with one another and the way neurons signal muscle cells or other neurons are based on the same principles at work in DNA uptake by pneumococcal cells in response to a peptide signal. All these communications are based on small-molecule signals that are exchanged between cells. These signaling molecules are the language of cellular communication.

## 9.3 RECEPTORS AND RECEPTOR ACTIVATION

Receptors are proteins that receive and interpret information carried by signaling molecules. Regardless of the distance

between communicating cells, a message is received by a cell when the signaling molecule binds noncovalently to its appropriate and usually highly specific receptor protein. For this reason, the signaling molecule is often referred to as a **ligand** (from the Latin *ligare*, which means "to bind"). The signaling molecule binds to a specific location on the receptor protein called the **ligand-binding site.**

Almost without exception, the binding of a signaling molecule to the ligand-binding site of a receptor causes a conformational change in the receptor. We say that the conformational change "activates" the receptor because it is through this change that the receptor passes the message from the signaling molecule to the interior of the cell. In many ways, this change in receptor shape is similar to the change that occurs when a substrate binds to the active site of an enzyme (Chapter 6). The conformational change in the receptor ultimately triggers various chemical reactions or other changes in the cytosol, and is therefore a crucial step in the reception and interpretation of communications from other cells.

### Receptors can be on the cell surface or in the interior of the cell.

The location of a particular receptor in a cell depends largely on whether the signaling molecule is polar or nonpolar (**Fig. 9.6**). Many signaling molecules, such as the growth factors we just encountered, are small, polar proteins that cannot pass through the hydrophobic core of the plasma membrane. The receptor proteins for these signals are on the outside surface of the responding cell (Fig. 9.6a).

Receptor proteins for growth factors and other polar ligands are transmembrane proteins with an extracellular

---

**FIG. 9.6** **Cell-surface and intracellular receptors.** (a) Cell-surface receptors interact with polar signaling molecules that cannot cross the plasma membrane. (b) Intracellular receptors interact with nonpolar signaling molecules that can cross the plasma membrane.

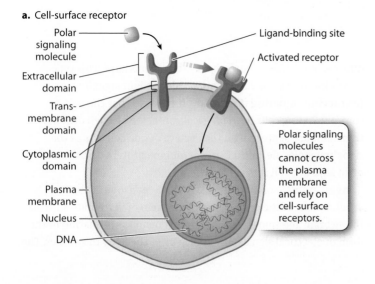

**a.** Cell-surface receptor

Polar signaling molecule
Ligand-binding site
Activated receptor
Extracellular domain
Trans-membrane domain
Cytoplasmic domain
Plasma membrane
Nucleus
DNA

Polar signaling molecules cannot cross the plasma membrane and rely on cell-surface receptors.

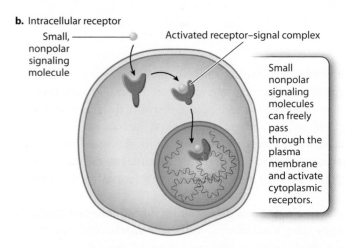

**b.** Intracellular receptor

Small, nonpolar signaling molecule
Activated receptor–signal complex

Small nonpolar signaling molecules can freely pass through the plasma membrane and activate cytoplasmic receptors.

domain, a short transmembrane domain, and a cytoplasmic domain. When a signaling molecule binds to the ligand-binding site in the extracellular domain, the entire molecule, including the cytoplasmic domain of the receptor, undergoes a conformational change, and as a result the molecule is activated. In this way, the receptor acts as a bridge between the inside and outside of the responding cell that carries the message of the hydrophilic signal across the hydrophobic core of the plasma membrane. The type of receptor determines which signaling pathway is activated and ultimately determines how the cell responds to the signaling molecule. We discuss several examples of signaling pathways in section 9.4.

Some nonpolar signaling molecules, such as the steroid hormones involved in endocrine signaling, don't need a receptor on the cell surface in order to relay information to the interior of the cell. Since steroids are derived from cholesterol, they are hydrophobic and pass easily through the hydrophobic core of the phospholipid bilayer and into the target cell. Once inside, steroid hormones bind to receptor proteins located in the cytosol or in the nucleus to form receptor–steroid complexes (Fig. 9.6b). Steroid–receptor complexes formed in the cytosol enter the nucleus, where they act as transcriptional regulators to control the expression of specific genes. Steroid receptors located in the nucleus are often already bound to DNA and need only to bind their steroid counterpart to become active.

The importance of steroid hormones and their receptors cannot be overstated. Sex hormones, glucocorticoids (which raise blood glucose levels), and ecdysone (involved in insect molting) are examples of steroid hormones. However, since the majority of information received by cells is transmitted across the plasma membrane through transmembrane receptors, we focus our attention on the sequence of events that takes place when receptors on the surface of cells bind their ligands.

### There are three major types of cell-surface receptor, which act like molecular switches.

As we saw earlier, receptor activation occurs after a signaling molecule binds to the ligand-binding site on its receptor. Many receptors, in fact, act as binary molecular switches, existing in only two alternative states, either "on" or "off," like a light switch without a dimmer (**Fig. 9.7a**). Receptors behave similarly to a light switch. When bound to their signaling molecule, the molecular switch is turned on. When the signaling molecule is no longer bound, the switch is turned off.

There are thousands of different receptor proteins on the surface of any given cell. Most of them can be placed into one of three groups according to the way they are activated. The first type of cell-surface receptor is called a **G protein-coupled receptor** (**Fig. 9.7b**). As its name suggests, this type of receptor couples to, or associates with, **G proteins,** proteins that bind to

**FIG. 9.7** **Three types of cell-surface receptor.** G protein-coupled receptors (b), receptor kinases (c), and ligand-gated ion channels (d) can be either "off" (not activated) or "on" (activated), and in this way they act like a light switch (a).

**a.** A light switch

Inactive          Active

**b.** G protein-coupled receptor

Ligand
Receptor    G protein
GDP
Inactive

GTP
Active
Signaling pathways
Response

**c.** Receptor kinase

Phosphate group
Response

**d.** Ligand-gated ion channel

Ions
Response

the guanine nucleotides GTP and GDP. When a G protein is bound to GTP, it is active, and when it is bound to GDP, it is inactive. When a ligand binds to a G protein-coupled receptor, the receptor binds to and activates the G protein by replacing GDP with GTP. As long as the G protein is bound to GTP, it is in the "on" position and the signal is propagated.

→   **Quick Check 2** Is the term "G protein" just a shorter name for a G protein-coupled receptor?

The second type of receptor is an enzyme called a **receptor kinase** (**Fig. 9.7c**). A kinase is an enzyme that adds a phosphate group to another molecule in a process called phosphorylation. **Phosphatases** have the opposite effect and remove a phosphate group, a process called dephosphorylation. Usually (but not always), when a protein is phosphorylated by a kinase, it becomes active and is switched on. When the protein is dephosphorylated by a phosphatase, it is switched off and becomes inactive.

Receptors in the third group, the **ligand-gated ion channels,** alter the flow of ions across the plasma membrane when bound by their ligand (**Fig. 9.7d**). Recall from Chapter 5 that channel proteins help ions and other molecules diffuse into and out of the cell by providing a hydrophilic pathway through the hydrophobic core of the phospholipid bilayer. Most of the time, the channels are closed to prevent the free movement of ions across the membrane. However, when a signaling molecule binds to the extracellular ligand-binding site of the channel protein, the channel undergoes a conformational change that opens it and allows ions to flow in and out. The channel remains open as long as the signaling molecule remains bound. This type of signaling is especially important for neurons and muscle cells, since their primary functions depend on a rapid change in ion transport across the plasma membrane.

## 9.4 SIGNAL TRANSDUCTION, RESPONSE, AND TERMINATION

In this section, we focus on what happens after a signaling molecule binds to its receptor and flips a molecular switch. Although the molecular switches are different, the subsequent steps are similar. Following the binding of a ligand and receptor activation, most signaling pathways involve signal transduction and amplification, a cellular response, and termination of the signal. We examine the signaling pathways activated by G protein-coupled receptors, receptor kinases, and ligand-gated ion channels. We also consider the role of signal transduction in cancer and how different signaling pathways intersect with one another.

### Signals transmitted by G protein-coupled receptors are amplified and regulated at several steps.

Signaling by G protein-coupled receptors is widespread, and its effects are diverse. The receptors are highly conserved evolutionarily and are found in virtually every eukaryotic organism. In humans, more than 900 different G protein-coupled receptors have been found. Signaling through these receptors makes possible our senses of sight, smell, and taste. They also enable the cells that express them to respond to a number of different hormones. Despite this great variability in the nature of the signals, the signal transduction events triggered by these receptors are remarkably similar.

**FIG. 9.8     Activation of a G protein through a G protein-coupled receptor.** A G protein-coupled receptor is activated when it binds a signaling molecule, which leads to exchange of GDP for GTP, separation of the α subunit of the G protein, and downstream effects.

When the α subunit is bound to GDP, the three subunits are joined together and the G protein is inactive.

When the G protein is bound by an activated receptor, the GDP bound to the α subunit is replaced by GTP, which activates the α subunit to bind to target proteins.

Activated α subunit binds to and activates target protein.

As we saw in section 9.3, G protein-coupled receptors are transmembrane proteins that bind signaling molecules. These receptors are positioned in proximity to G proteins, which are located on the cytoplasmic side of the plasma membrane. When a signaling molecule binds to the receptor, the cytoplasmic domain of the receptor binds to and activates the nearby G protein by exchanging GDP with GTP. The activated G protein goes on to activate additional proteins in the signaling pathway.

G proteins are composed of three subunits called the α (alpha), β (beta), and γ (gamma) subunits (**Fig. 9.8**). The α subunit is the part of the G protein that binds to either GDP or GTP. When the α subunit is bound to GDP, the three subunits are joined and the G protein is inactive. When the G protein is bound by an activated receptor, the GDP bound to the α subunit is replaced by GTP, which causes the α subunit to separate from the β and γ subunits. The isolated GTP-bound α subunit is now active and able to bind to specific target proteins in the cell, activating them in turn.

Let's consider an example. You're startled or you're scared, and you experience that strange feeling in the pit of the stomach and rapid heartbeat that we've all felt at one time or another. This effect, which accompanies fright or other forms of stress, is mediated by adrenaline, a hormone released by the adrenal glands that binds to a G protein-coupled receptor. When adrenaline binds to its G protein-coupled receptor on cardiac muscle cells, GDP in the G protein is replaced by GTP and activates the G protein. The GTP-bound α subunit of the activated G protein then binds to and activates an enzyme in the cell membrane called adenylyl cyclase. Adenylyl cyclase converts the nucleotide ATP into cyclic AMP, abbreviated as cAMP. This molecule is one of several intermediate, cytosolic signaling molecules known as **second messengers** (adrenaline being the first messenger). Once formed, cAMP binds to and activates another enzyme, a kinase called protein kinase A, or PKA. The phosphorylation of specific proteins in the heart muscle by PKA causes the rate of contraction to increase (**Fig. 9.9**). As long as adrenaline is bound to its receptor, the heart rate remains high, resulting in increased blood flow to the brain and skeletal muscles needed to deal with the stress.

A little adrenaline goes a long way as a result of signal amplification. A single receptor bound to adrenaline can activate several individual G protein molecules and continues to do so until the adrenaline leaves the binding site of the receptor. Every one of the activated G protein α subunits binds to an adenylyl cyclase molecule and, as long as the α subunit remains bound, the enzyme continues to generate cAMP. Every cAMP molecule binds to an available protein kinase A molecule, and likewise, as long as cAMP remains bound, the kinase continues to phosphorylate its target proteins. These sequential molecular changes in the cytosol amplify the signal so that a very small amount of signaling molecule has a large effect on a responding cell.

In this example, the adrenaline signal is amplified in three places: (1) Each receptor activates multiple G proteins; (2) each molecule of adenylyl cyclase produces large amounts of the second messenger cAMP; and (3) each protein kinase A

---

**FIG. 9.9** **Adrenaline signaling in heart muscle.** Adrenaline binds to a G protein-coupled receptor, leading to production of the second messenger cAMP, activation of protein kinase A, and increased heart rate.

molecule activated by cAMP activates multiple protein targets by phosphorylation (**Fig. 9.10**).

After a good scare, we eventually calm down and our heartbeat returns to normal. This change means that the signal initiated by adrenaline has been terminated. Let's look at how this happens (**Fig. 9.11**). First, most ligands, including adrenaline, do not bind

to their receptors permanently. The amount of time a signaling molecule remains bound to its receptor depends on how tightly the receptor holds on to it, a property called **binding affinity.** Once adrenaline leaves the receptor, the receptor reverts to its inactive conformation and no longer activates adjacent G proteins. If the concentration of adrenaline in the system is high enough, then as

**FIG. 9.10    Amplification of G protein signaling.** Signaling through G protein-coupled receptors is amplified at several places, so that a small amount of signal can produce a large response in the cell.

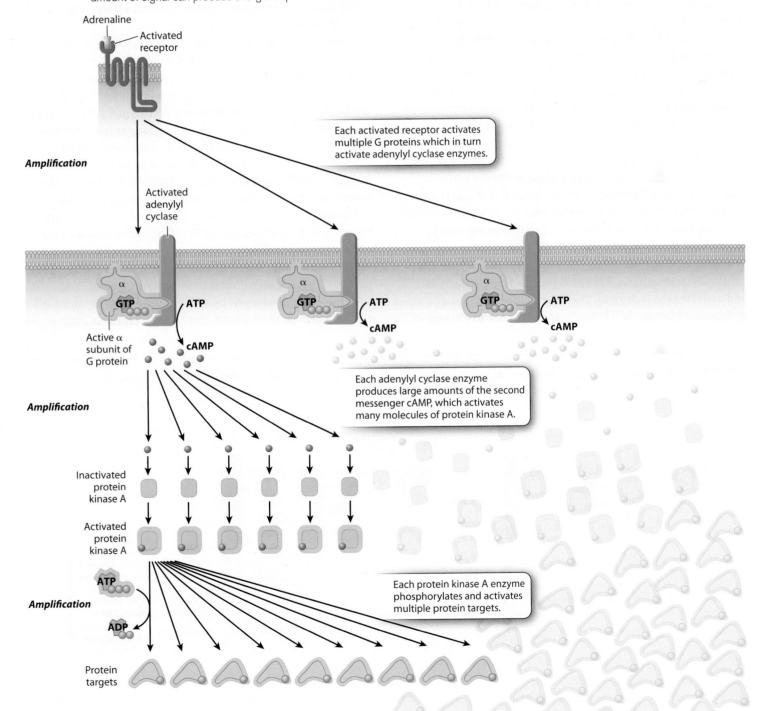

Amplification

Each activated receptor activates multiple G proteins which in turn activate adenylyl cyclase enzymes.

Amplification

Each adenylyl cyclase enzyme produces large amounts of the second messenger cAMP, which activates many molecules of protein kinase A.

Amplification

Each protein kinase A enzyme phosphorylates and activates multiple protein targets.

**FIG. 9.11 Termination of a G protein signal.** G protein-coupled signaling is terminated at several places, allowing the cell to respond to new signals.

soon as one adrenaline molecule leaves the receptor, another takes its place and reactivates the receptor. However, once the concentration of adrenaline returns to normal low levels, the receptors are no longer continually occupied and they become inactive.

Even once a receptor has been turned off, a signal will continue to be transmitted unless all the other components of the signaling pathway are also inactivated. In our example of the adrenaline-signaling pathway, a second place where the signal is terminated is at the G protein itself (Fig. 9.11). As we saw earlier, the α subunit is able to bind to and activate other proteins only when it is bound to GTP. G proteins have enzymatic activity, and they are able to catalyze the hydrolysis of GTP to GDP and inorganic phosphate. This means that an active, GTP-bound α subunit in the "on" position automatically turns itself "off" by converting the GTP to GDP. In fact, the α subunit converts GTP to GDP almost as soon as a molecule of GTP binds to it. Thus, a G protein is able to activate adenylyl cyclase, and adenylyl cyclase is able to make cAMP only during the very short time it takes the α subunit to convert its bound GTP to GDP, a timescale on the order of seconds. Without an active receptor to generate more active G protein α subunits, transmission of the signal quickly comes to a halt.

Farther down the pathway, enzymes shut down other components. An enzyme converts the second messenger cAMP to AMP, which inactivates protein kinase A. Phosphatases remove the phosphate groups, inactivating proteins targeted by protein kinase A (Fig. 9.11). In fact, most signaling events counteract kinases and phosphatases at one or more points in the pathway as a means of increasing or decreasing the response of the cell to the signal.

→ **Quick Check 3** Name four ways in which the adrenaline signal to the heart is terminated.

## Receptor kinases phosphorylate each other and activate intracellular signaling pathways.

Like the communication that takes place through G protein-coupled receptors, signaling through receptor kinases causes cells to respond in many ways. During embryonic development, receptor kinase signaling is responsible for the formation and elongation of structures called limb buds that eventually become our arms and legs. As adults, insulin signaling through its receptor kinase enables virtually every

cell in our body to transport glucose across the plasma membrane into the cytosol. When we cut a finger, platelet-derived growth factor is released from platelets in the blood and binds to its receptor kinase on the surface of cells at the site of the wound, where it triggers the cell division necessary to repair the damaged tissue.

Signaling through receptor kinases takes place in most eukaryotic organisms, and the structure and function of these receptors has been remarkably conserved evolutionarily. A receptor kinase called Kit has been studied in a number of animal models. In vertebrates, signaling through the Kit receptor kinase is important for the production of pigment in skin, feathers, scales, and hair. The conserved function of this receptor is obvious in individuals with mutations in the *kit* gene, as

**FIG. 9.12    Mutations in the Kit receptor kinase.** Similar patterns of incomplete pigmentation are present in mammals, reptiles, birds, and fish with this mutated receptor.

shown in **Fig. 9.12**. (By convention, the name of a protein, like Kit, is capitalized and in roman type. The name of the gene that encodes the protein, like *kit*, is italicized and all lower case.) As you can see from Fig. 9.12, mammals, reptiles, birds, and fish with a mutation in the *kit* gene have a remarkably similar appearance.

Let's take a look at how receptor kinases work. The cytoplasmic portion of these receptors is a kinase, that is, an enzyme that phosphorylates other molecules. When a signaling

molecule binds to the extracellular portion of the receptor, a conformational change causes the receptor to partner up with another receptor kinase bound to another molecule of the same ligand in a process called **dimerization.** Dimerization activates the cytoplasmic kinase domains of the paired receptors, causing them to phosphorylate each other at multiple sites on their cytoplasmic tails (**Fig. 9.13**). The addition of these phosphate groups provides places on the receptor where other proteins bind and become active.

**FIG. 9.13    Receptor kinase activation and signaling.** Receptor kinases bind signaling molecules, dimerize, phosphorylate each other, and activate intracellular signal molecules.

Signaling through receptor kinases follows the same basic sequence of events that we saw in signaling though G protein-coupled receptors. Binding of the ligand activates the receptor, which causes a set of cytoplasmic proteins to become sequentially activated, and, since many of these proteins are enzymes, the signal is amplified. And as we saw earlier in the G protein-coupled pathway, intracellular signaling proteins in receptor kinase pathways are also activated by GTP binding and phosphorylation, and are inactivated by GTP hydrolysis to GDP and dephosphorylation by phosphatases.

Now let's look at an example of signaling through a receptor kinase pathway. Think about the last time you got a paper cut. The cut likely bled for a minute or two, and then the bleeding stopped. If the cut was small, it probably healed completely in a week or so. PDGF has a lot to do with getting the healing process started. When platelets in the blood encounter damaged tissue, they release a number of proteins, including PDGF. PDGF is the signaling molecule that binds to PDGF-specific receptor kinases on the surface of cells at the site of a wound, causing the receptors to dimerize and become active. The activated receptors then recruit and activate additional signaling proteins present in the cytosol.

One of these cytoplasmic signaling proteins is **Ras,** which is very similar to the α subunit of G proteins. In the absence of a signal, Ras is bound to GDP and is inactive. However, when Ras binds to an activated receptor kinase, it releases GDP and binds GTP to become active. Activated GTP-bound Ras triggers the activation of a series of kinases that together are called the mitogen-activated protein kinase pathway, or, for short, the **MAP kinase pathway** (**Fig. 9.14**). The final activated kinase in the series enters the nucleus, where it phosphorylates its target proteins. Some of these proteins include regulators of transcription that switch on genes needed for the cell division so that your paper cut can heal.

The signals received by receptor kinases are amplified as the signal is passed from kinase to kinase. Each phosphorylated kinase in the series activates multiple molecules of the downstream kinase, each of which in turn activates many molecules of another kinase still farther downstream. In this way, a very small amount of signaling molecule in the environment (PDGF, in our example) can cause a large-scale response in the cell.

Receptor kinase signaling is also reversed or terminated by the same basic mechanisms that are at work in G protein-coupled receptor pathways. For example, protein phosphatases play important roles in the inactivation of receptor kinases and the other enzymes of the MAP kinase pathway. Furthermore, Ras is a GTPase, just like the G protein α subunit. Shortly after Ras binds to GTP and becomes active, Ras converts GTP to GDP, making it inactive. Without an active receptor kinase to generate more active Ras, activation of MAP kinase pathway components stops.

**FIG. 9.14** **The MAP kinase pathway.** Some receptor kinases signal through Ras, which in turn activates the MAP kinase pathway, leading to changes in gene expression.

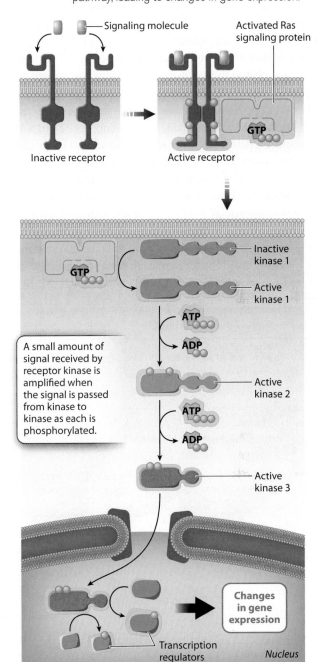

### Ligand-gated ion channels alter the movement of ions across the plasma membrane.

Let's turn our attention now to the third group of receptors, the ligand-gated ion channels. Instead of activating cytosolic enzymes, these receptors function as ion channels to regulate the permeability of the cell membrane to ions. Nevertheless, the same principles of cell communication apply here as well: A signal binds to a receptor, which is activated and causes a cellular response until the signal is terminated.

There are many examples of ligand-gated ion channels, each specific for one or more different ions. Ligand-gated ion channels are particularly important in neurons and muscle cells. The plasma membranes of neurons and muscle cells are often referred to as "excitable membranes" because of the dramatic changes that take place in the membrane and in the entire cell when ions, especially sodium ions (Na+), are permitted to flow into the cell. At rest, almost all of the Na+ channels of neurons and muscle cells are closed, and in fact, these cells expend large amounts of ATP to pump Na+ out of the cell. As a result, the concentration of Na+ inside the cell is very low compared to its concentration outside the cell.

The difference in Na+ concentration also contributes to a difference in electrical charge across the plasma membrane, called the **membrane potential.** At rest, the inside of the cell has a negative charge relative to the outside. This charge difference is in part the result of the large amount of positively charged sodium ions outside the cell and the high concentration of negatively charged ions and proteins in the cytosol. When a signaling molecule released by a neuron binds a ligand-gated Na+ channel, the channel opens, allowing Na+ to move down its concentration gradient into the cell. The influx causes a drastic decline in the charge difference between the outside and inside of the cell. In turn, this change in membrane potential causes the cell to respond: The neuron responds by sending a nerve impulse, and the muscle cell responds by contracting.

Let's look more closely at the signal that sets off a muscle contraction (**Fig. 9.15**). When a neuron sends a signal to a muscle cell, it causes the muscle cell to shorten, and when many muscle cells shorten, the muscle as a whole contracts. The signaling molecule released from the neuron is a neurotransmitter called acetylcholine, which diffuses across the small space called the neuromuscular junction between the neuron and the muscle cell, until it encounters and binds to its receptor on the muscle cell. The acetylcholine receptor is a ligand-gated ion channel, so when acetylcholine binds to the extracellular ligand binding site, the conformational change that takes place opens the channel, and Na+ ions rush into the muscle cell. The sudden influx of Na+ ions changes the membrane potential, activating the contractile machinery of the cell and causing it to contract.

**FIG. 9.15    Ligand-gated ion channels.** Ligand-gated ion channels open in response to a bound signaling molecule.

This sort of cell signaling needs a mechanism for shutting off just as the other receptor pathways do. In the case of ligand-gated channels, there are several places where regulation takes place. An important point of regulation is the length of time the ligand remains in the binding site of the receptor, since the occupation of the binding site causes the channel to remain open. A second point of regulation is provided by an enzyme called acetylcholinesterase, which is present in the neuromuscular junction and terminates the acetylcholine signal by rapidly breaking it down. When the binding site is unoccupied, the channel reverts to the closed conformation. The flow of $Na^+$ ions into the cells ceases and the membrane potential returns to its resting state.

This signaling pathway is the target of a deadly poison known as curare. Native hunters of South America isolate curare from indigenous plants and rub it on arrowheads to kill their prey. Curare blocks the acetylcholine receptor, thereby relaxing voluntary muscles, including the diaphragm, which supports normal breathing. The result is death by asphyxiation. Interestingly, during the twentieth century, curare and its derivatives were used as a component of anesthesia to relax muscles during surgery. Today, anesthesiologists routinely use much safer and more effective blockers of the acetylcholine receptor to ease the manipulation of muscles and tissues during surgery.

**? CASE 2 Cancer: When Good Cells Go Bad**
**How do cell signaling errors lead to cancer?**
Many cancers arise when something goes wrong with the way a cell responds to a signal that leads to cell division or, in some cases, when a cell behaves as if it has received a signal for cell division when in fact it hasn't. Problems with cell signaling that can lead to cancer can take place at just about every step in the cell-signaling process. In some rare cases, a tumor may form as the result of the overproduction of a signaling molecule or the production of an altered form of a signaling molecule. More frequently, the source of the problem is in the receptor. For example, individuals with some forms of cancer have from 10 to 100 times the normal number of receptors for a signaling molecule called epidermal growth factor (EGF), making the cell more responsive to the normally low levels of EGF in the environment. Under normal conditions, binding of EGF to its receptor leads to the controlled division of cells. But in these cancer patients, the heightened response of the signaling pathway leads to abnormal gene expression and excess cell division.

Individuals with other cancers, including certain breast cancers, have mutant EGF receptors that activate intracellular pathways even when EGF isn't present. One of these mutant receptors, called Her2/neu, is the target of antibody therapy. Antibodies are proteins synthesized by cells of the immune system that target foreign substances in the body for destruction. Pharmaceutical companies are able to produce antibodies that specifically target the Her2/neu receptor. When these antibodies are used as a form of chemotherapy, they bind to the cancer cells with Her2/neu on their surfaces, targeting them for destruction by the immune system.

Farther down the pathway, mutant forms of the Ras protein are often present in cancers. One especially harmful mutation prevents Ras from converting its bound GTP to GDP. The protein remains locked in the active, GTP-bound state, causing the sustained activation of the MAP kinase pathway. More than 30% of all human cancers involve abnormal Ras activity as a result of one or more mutations in the *ras* gene.

**Signaling pathways can intersect with one another in a cell.**
In this chapter, we have focused on individual signaling pathways to emphasize the general principles of communication between cells. In each case, we saw that a cell releases or has a cell-surface signaling molecule, which binds to and activates a cell-surface or intracellular receptor. Binding of the signaling molecule to the receptor induces a conformational change in the receptor, causing it to become active. The activated receptor in turn causes changes in the interior of the cell, frequently turning on a signal-transduction cascade that is amplified, leading to a cellular response. Cellular responses can vary depending on the nature of the signal and type of cell. They include turning genes on or off, changes in cellular metabolism, and cell division. Eventually, the signal and response terminate.

Focusing on each pathway one at a time allowed us to understand how each pathway operates. But in the context of an organism, cell signaling can be quite complex. There are scores upon scores of different signaling molecules, most of which have their own specific receptor. The receptors may or may not be present on or in a particular cell, and if a certain receptor is present, the number of copies of that receptor may vary widely.

In addition, the cellular response to a particular signaling molecule can vary from one kind of cell to another. For example, in mammalian cells the MAP kinase pathway promotes cell division, whereas in yeast cells it triggers sexual reproduction. The various effects that activating the MAP kinase pathway can have on the cell depends in part on the different sets of proteins that are found in different cell types.

It is also important to realize that different signaling molecules can bind to a single cell and activate several signaling pathways simultaneously, so that the final response of a cell depends on how the pathways intersect with one another. The integration of different signals gives cells a wide range of possible responses to their environment. Receiving two different signals may enhance a particular response, such as cell growth; or one signal may inhibit a component of the signaling pathway triggered by the other signal, resulting in a weakened response. Thus, cells are able to use a similar set of signaling cascades to communicate with other cells and respond to the environment in myriad different ways.

For example, recent studies have shown that enzymes in the MAP kinase pathway can be inhibited by active PKA. Recall that PKA is activated by elevated cAMP levels associated with the G protein-coupled receptor pathway. The regulation of the MAP kinase pathway by PKA therefore illustrates how one signaling pathway can inhibit another.

Researchers have made use of this molecular cross talk. Many patients with breast cancer have elevated MAP kinase activity in their tumor cells. When human breast cancer cells are genetically modified to express an activated G protein $\alpha$ subunit, their growth after transplantation into mice is significantly inhibited. In addition, in cell culture, elevated cAMP levels block cells from responding to growth factors that signal through Ras to the MAP kinase pathway. As we understand more about how different signaling pathways integrate their outputs in specific cell types, we will stand a better chance of tailoring therapies to inhibit the pathway components relevant to a particular cancer.

## Core Concepts Summary

**9.1 CELLS COMMUNICATE PRIMARILY BY SENDING AND RECEIVING CHEMICAL SIGNALS.**

There are four essential players in communication between two cells: a signaling cell, a signaling molecule, a receptor, and a responding cell. page 9-2

The signaling molecule binds to its receptor on the responding cell, leading to receptor activation, signal transduction and amplification, response, and eventually termination of the signal and response. page 9-2

**9.2 CELL SIGNALING CAN BE CLASSIFIED ACCORDING TO THE DISTANCE BETWEEN THE SIGNALING AND RESPONDING CELLS.**

Endocrine signaling takes place over long distances and relies on the circulatory system for delivery of signaling molecules. page 9-3

Paracrine signaling takes place over short distances between neighboring cells. page 9-4

Autocrine signaling occurs when a cell signals itself. page 9-4

Juxtacrine communication depends on physical contact between cells. page 9-5

**9.3 SIGNALING MOLECULES BIND TO AND ACTIVATE SPECIFIC CELL-SURFACE AND INTRACELLULAR RECEPTORS.**

A signaling molecule, or ligand, binds to the ligand-binding site of the receptor. Binding causes the receptor to undergo a conformational change that activates the receptor. page 9-6

Receptors for polar signaling molecules, including growth factors, are located on the plasma membrane. page 9-6

Receptors for nonpolar signaling molecules, such as steroid hormones, are located in the cytosol or in the nucleus. page 9-7

There are three major types of cell-surface receptor: G protein-coupled receptors, receptor kinases, and ligand-gated ion channels. All act as molecular switches. page 9-7

G protein-coupled receptors associate with G proteins, which are active when bound to GTP and inactive when bound to GDP. page 9-8

Receptor kinases are activated by phosphorylation and inactivated by dephosphorylation. page 9-8

Ligand-gated ion channels open in response to a signal, allowing the movement of ions across the plasma membrane. page 9-8

**9.4 SIGNALS ARE TRANSMITTED ACROSS THE PLASMA MEMBRANE AND ARE OFTEN AMPLIFIED IN THE CYTOSOL, LEADING TO A CELLULAR RESPONSE FOLLOWED BY TERMINATION.**

G proteins are associated with G protein-coupled receptors that bind signaling molecules. page 9-8

G proteins are composed of three subunits, denoted $\alpha$, $\beta$, and $\gamma$. When a G protein encounters an activated receptor, the $\alpha$ subunit exchanges GDP for GTP, dissociates from the $\beta$ and $\gamma$ subunits, and becomes active. page 9-9

Second messengers like cAMP amplify the signal in the cytosol (brought to the cell by the extracellular first messenger). page 9-9

Intracellular, cytosolic signals are short-lived before they are terminated. page 9-11

Ligand binding to a receptor kinase causes it to dimerize with another receptor bound to the same ligand. The two receptor kinases become active when they phosphorylate each other's cytoplasmic domains, allowing activation of intracellular signaling pathways. page 9-12

The phosphorylated receptors are bound by other proteins, which cause the subsequent activation of other cytosolic signaling molecules, such as Ras. page 9-13

When Ras is activated, the GDP to which it is bound is released and is replaced by GTP. The active, GTP-bound Ras temporarily binds to and activates the first in a series of kinases.  page 9-13

Neurons communicate with other neurons and with muscle cells using ligand-gated ion channels, which lead to movement of ions across the membrane and a change in the membrane charge or potential.  page 9-14

Cancer can be caused by malfunction of any step in the signaling processes involved in cell division.  page 9-15

Mutations in receptor kinases and Ras are frequently associated with human cancers.  page 9-15

Several signaling pathways can take place simultaneously in a single cell, and the cellular response depends on the integration of several signals.  page 9-15

## Self-Assessment

1.  Name the four players in cell communication.

2.  Name the steps that are often involved in cell communication.

3.  Describe one way in which endocrine-, paracrine-, autocrine-, and juxtacrine-signaling pathways are similar to one another and one way they differ from one another.

4.  Explain how cells respond to external signals even when those signals cannot enter the cell.

5.  Explain how signals can be specific in targeting only some cells, even if they are released into the bloodstream and come into contact with many cells.

6.  List several possible types of responses a cell might have to a signaling molecule.

7.  Compare and contrast cell-surface receptors and intracellular receptors in terms of their location, chemical properties of their signaling molecules, and types of responses mediated.

8.  List several ways in which a signal may be amplified in an intracellular signaling pathway.

9.  Describe ways in which the response of a cell to a signal can be terminated.

Do you understand the chapter's Core Concepts? Log in to check your answers to the Self-Assessment questions, then practice what you've learned and reinforce this chapter's concepts by working through the problems and multimedia tutorials provided there.

  www.biologyhowlifeworks.com

## CHAPTER 10

# CELL FORM AND FUNCTION

## Cytoskeleton, Cellular Junctions, and Extracellular Matrix

## Core Concepts

**10.1** Tissues and organs are communities of cells that perform a specific function.

**10.2** The cytoskeleton is composed of microtubules, microfilaments, and intermediate filaments that help to maintain cell shape.

**10.3** The cytoskeleton interacts with motor proteins to permit the movement of cells and substances within cells.

**10.4** Cells adhere to other cells and the extracellular matrix by means of cell adhesion molecules and junctional complexes.

**10.5** The extracellular matrix provides structural support and informational cues.

A recurring theme in this book, and in all of biology, is that form and function are inseparably linked. This is true at every level of structural organization, from molecules to organelles to cells and to organisms themselves. We saw in Chapter 4 that the function of proteins depends on their shape. In Chapters 7 and 8, we saw that the function of organelles like mitochondria and chloroplasts depends on the large internal surface areas of the inner membrane of mitochondria and the thylakoid membrane of chloroplasts.

The functions of different cell types are also reflected in their shape and internal structural features (**Fig. 10.1**). Consider a typical red blood cell. It is shaped like a disk that is slightly indented in the middle, and it lacks a nucleus and other organelles (Fig. 10.1a). This unusual shape and internal organization allow it to be remarkably flexible as it carries oxygen in the bloodstream. Because it is able to deform readily, it can pass through blood vessels with diameters smaller than that of the red blood cell itself. Cells in the liver (Fig. 10.1b) that synthesize proteins and glycogen look very different from muscle cells (Fig. 10.1c) that contract to exert force. A neuron (Fig. 10.1d), with its long and extensively branched extensions that communicate with other cells, is structurally nothing like a cell lining the intestine (Fig. 10.1e) that absorbs nutrients.

As we saw in Chapter 9, cells often exist in communities, forming tissues and organs in multicellular organisms. Again, form and function are intimately linked. Think of the branching structure and large surface area of the mammalian lung, which is well adapted for gas exchange, or the muscular heart, adapted for pumping blood through large circulatory systems.

In this chapter, we look at what determines cell shape. We also examine the different ways cells adhere to one another to build tissues and organs, and see how this adhesion differs depending on the function of the cellular community. We also learn about the physical environment outside cells, which is synthesized by cells themselves and influences their behavior.

## 10.1 TISSUES AND ORGANS

At the beginning of Chapter 9, we considered the features that enable a group of individuals to function as a community. One of these is the presence of members able to perform specialized tasks. In addition, community members stick together and respond to their surrounding environment. Now, think of our own tissues and organs—a heart, a stomach, a kidney, a brain—all are communities of cells that work together to perform highly specific and important tasks in response to cues from the environment.

**FIG. 10.1   Diverse cell types.** Cells differ in shape and are well adapted for their various functions.

The shape of a red blood cell maximizes its surface area for gas exchange and allows it to deform as it passes through the circulatory system.

Hepatocytes (liver cells) contain large amounts of rough ER needed for protein synthesis.

Long, multinucleated muscle cells are packed with actin and myosin needed for contraction.

Long extensions of the plasma membrane allow neurons to communicate with other cells.

Microvilli on intestinal epithelial cells increase their absorptive surface area.

## Tissues and organs are communities of cells.

A biological **tissue** is a collection of cells that work together to perform a specific function. Two or more tissues often combine and function together as an **organ,** such as a heart or lung. Animals and plants have tissues and organs that allow them to carry out the various processes necessary to sustain them. In animals, for example, four types of tissue—epithelial, connective, nervous, and muscle—combine to make up all the organs of the body. Plants are made up of just three types of tissue: dermal (the outer protective layer), ground (the bulk of the plant body), and vascular (the channels that transport nutrients and water throughout the plant).

Tissues and organs have distinctive shapes that reflect how they work and what they do. In the same way, the different cell types that make up these organs have distinctive shapes based on what they do in the organ. In animals, the shape of cells is determined and maintained by structural protein networks in the cytoplasm called the **cytoskeleton.** The shape and structural integrity of tissues and organs depend on the ability of cells to adhere to one another. In turn, the adhesion of cells to one another depends on cell adhesion molecules and other cytosolic proteins that assemble into structures called **cellular junctions.** Equally important to a strong, properly shaped tissue or organ is the ability of cells to adhere to a meshwork of proteins and polysaccharides outside the cell called the **extracellular matrix.** In plants, the extracellular matrix provides both a mechanism for cells to adhere to one another and structural support that maintains their shape.

## The structure of skin relates to its function.

To start our investigation of the cytoskeleton, cellular junctions, and the extracellular matrix, let's consider a community of cells very familiar to all of us—our own skin (**Fig. 10.2**). The structure of mammalian skin is clearly tied to its function. Skin has two main layers. The outer layer, the **epidermis,** serves as a water-resistant, protective barrier. The layer beneath the epidermis is the **dermis.** This layer of the skin supports the epidermis, both physically and by supplying it with nutrients. It also provides a cushion surrounding the body.

As you can see in Fig. 10.2, the epidermis is several cell layers thick. Cells arranged in one or more layers are called epithelial cells and together make up a type of animal tissue called **epithelial tissue.** Epithelial tissue covers the outside of the body and lines many internal structures, such as the digestive tract and vertebrate blood vessels. The epidermal layer of skin is primarily composed of epithelial cells called keratinocytes. The epidermis also contains melanocytes that produce the pigment that gives skin its coloration. Different kinds of epithelial cells are specialized to carry out different functions, such as protection, secretion, and absorption.

Keratinocytes in the epidermis are specialized to protect underlying tissues and organs. They are able to perform this function in part because of their elaborate system of cytoskeletal filaments. These filaments are often connected to the cellular

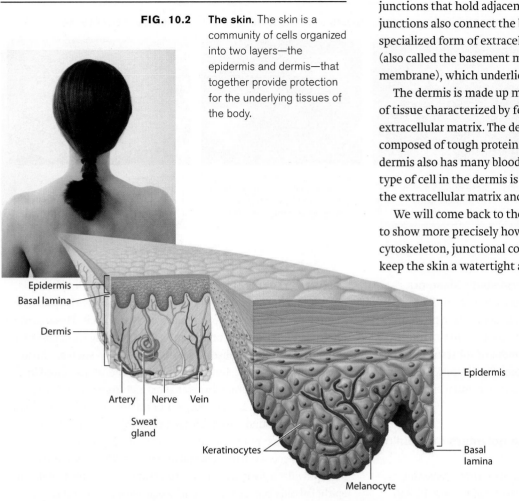

**FIG. 10.2** **The skin.** The skin is a community of cells organized into two layers—the epidermis and dermis—that together provide protection for the underlying tissues of the body.

Epidermis
Basal lamina
Dermis
Artery Nerve Vein
Sweat gland
Keratinocytes
Epidermis
Basal lamina
Melanocyte

junctions that hold adjacent keratinocytes together. Cellular junctions also connect the bottom layer of keratinocytes to a specialized form of extracellular matrix called the **basal lamina** (also called the basement membrane, although it is not in fact a membrane), which underlies and supports all epithelial tissues.

The dermis is made up mostly of **connective tissue,** a type of tissue characterized by few cells and substantial amounts of extracellular matrix. The dermis is strong and flexible because it is composed of tough protein fibers of the extracellular matrix. The dermis also has many blood vessels and nerve endings. The main type of cell in the dermis is the fibroblast. Fibroblasts synthesize the extracellular matrix and repair wounds.

We will come back to the skin several times in this chapter to show more precisely how these multiple connections among cytoskeleton, junctional complexes, and the extracellular matrix keep the skin a watertight and strong protective barrier.

## 10.2 THE CYTOSKELETON

Just as the bones of vertebrate skeletons provide internal support for the body, the protein fibers of the cytoskeleton provide internal support for cells (**Fig. 10.3**). All eukaryotic cells have at least two cytoskeletal elements, **microtubules** and **microfilaments.**

**FIG. 10.3** **Three types of cytoskeletal element.** (a) Microtubules; (b) microfilaments; and (c) intermediate filaments.

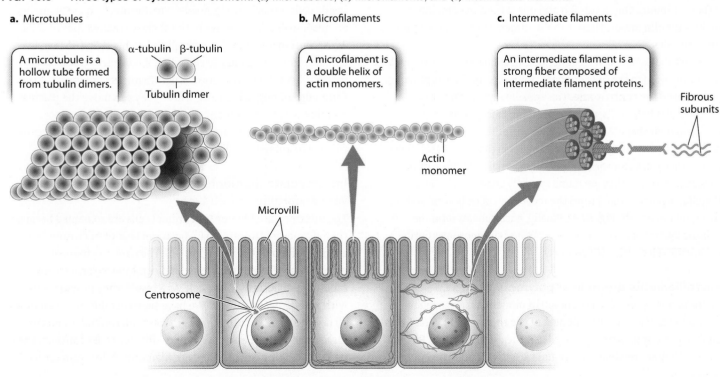

**a.** Microtubules

A microtubule is a hollow tube formed from tubulin dimers.

α-tubulin β-tubulin

Tubulin dimer

**b.** Microfilaments

A microfilament is a double helix of actin monomers.

Actin monomer

**c.** Intermediate filaments

An intermediate filament is a strong fiber composed of intermediate filament proteins.

Fibrous subunits

Microvilli

Centrosome

**FIG. 10.4    Cilia and flagella in diverse cell types.** Cilia and flagella, composed of microtubules, move cells or propel substances by them.

Coordinated beating of the cilia that cover the paramecium moves the cell through its environment.

The cilia in these human airway epithelial cells propel mucus containing debris out of the lungs.

These unicellular algae are propelled by two flagella.

Some cells of multicellular organisms, including sperm cells, swim by movement of a flagellum.

Animal cells have a third element, **intermediate filaments.** All three of these cytoskeletal elements are long chains, or polymers, made up of protein subunits. In addition to providing structural support, microtubules and microfilaments enable the movement of many cells and the movement of substances within cells. We first consider the structural role of the cytoskeleton and return to its role in cell motility and intracellular transport in section 10.3.

### Microtubules are hollow, tubelike polymers of tubulin dimers.

Microtubules, which are hollow tubelike structures, have the largest diameter of the three cytoskeletal elements (Fig. 10.3a). They are polymers of protein dimers. Each protein dimer is made up of two slightly different **tubulin** proteins, called α (alpha) and β (beta) tubulin. One α tubulin and one β tubulin combine to make a tubulin dimer. These tubulin dimers are assembled into microtubules.

Microtubules have diverse functions. In animal cells, microtubules radiate outward to the cell periphery from a microtubule organizing center called the **centrosome.** This spokelike arrangement of microtubules helps cells withstand compression and thereby maintain their shape. Many organelles are tethered to microtubules, which guide the arrangement of organelles in the cell. Microtubules also provide tracks for the transport of material from one part of the cell to another. They are found specially arranged in **cilia** and **flagella,** organelles that propel the movement of cells or substances surrounding the cell (**Fig. 10.4**). Finally, microtubules form the **spindle apparatus** that separates replicated chromosomes during eukaryotic cell division (Chapter 11).

### Microfilaments are helical polymers of actin.

Microfilaments are polymers of **actin** monomers, arranged to form a helix. They are the thinnest of the three cytoskeletal fibers and are present in various locations in the cytoplasm (see Fig. 10.3b). They are relatively short and extensively branched in the cell cortex, the area of the cytoplasm just beneath the plasma membrane. At the cortex, microfilaments reinforce the plasma membrane and organize proteins associated with it. These cortical microfilaments are also important in maintaining the shape of a cell. For example, in absorptive epithelial cells such as those in the small intestine, bundles of microfilaments are found in microvilli, hairlike projections that extend from the surface of the cell (**Fig. 10.5**). Longer bundles of microfilaments form a band that extends around the circumference of epithelial cells. This band is attached to a type of cell junction called an adherens junction, which connects a cell to its neighbors (section 10.4). As a result, the band provides a great deal of structural support not only to individual epithelial cells, but also to the entire epithelial layer of cells.

In addition to providing structural support to the cell, microfilaments are important for the transport of materials inside cells, especially plant cells. Furthermore, microfilaments are responsible for changes in the shape of many types of cells. An exceptionally dramatic example of such a shape change is the shortening of a muscle cell when it contracts, discussed more fully in the next section. Microfilaments are found at the front edge of migrating cells, where they can move the plasma membrane forward. Finally, microfilaments, arranged in a structure called the contractile ring separate the daughter cells at the end of animal cell division (Chapter 11).

### Intermediate filaments are polymers of proteins that vary according to cell type.

The intermediate filaments of animal cells are so named because their diameter is intermediate between that of microtubules and microfilaments (see Fig. 10.3c). They are polymers of intermediate filament proteins that combine to form strong, cable-like structures in the cell. As a result, they provide cells with mechanical strength. We have seen that different cell types all use the same tubulin dimers to form microtubules and the same actin monomers to form microfilaments. By contrast, the proteins making up intermediate filaments differ from one cell

**FIG. 10.5** **Microfilaments in intestinal microvilli.** Actin microfilaments help to maintain the structure of microvilli in the intestine.

Circumferential band of microfilament bundles

Microvillus

Plasma membrane

Cross-linking actin-binding proteins

Actin microfilaments

Intestinal epithelial cells

type to another. For example, in epithelial cells, these protein subunits are keratins; in fibroblasts, they are vimentins; and in neurons, they are neurofilaments. Some intermediate filaments, called lamins, are even found inside the nucleus, where they provide support for the nuclear envelope (**Fig. 10.6**). In fact, there are well over 100 different kinds of intermediate filaments.

Once assembled, many intermediate filaments become attached at the cytoplasmic side of cellular junctions called desmosomes (**Fig. 10.7**; section 10.4). Like the association of microfilaments with adherens junctions, the anchoring of intermediate filaments to desmosomes provides strong support for the cells. In the case of epithelial cells, this anchoring results in structural continuity from one cell to another that greatly strengthens the entire epithelial tissue, enhancing the ability of epithelial cell layers to withstand physical stress. This is especially important in a wide range of tissues that are regularly subject to such stress, including the skin and the lining of the intestine.

Genetic defects that disrupt the intermediate filament network can have severe consequences. For example, individuals with epidermolysis bullosa, a rare genetic disease, have defective keratin genes inherited from their parents. Intermediate filaments do not polymerize properly in these individuals, weakening connections between the layers of cells that make up the epidermis. As a consequence, the outer layers can detach, resulting in extremely fragile skin that blisters in response to the slightest trauma (Fig. 10.7). The sensitivity to physical stress is so extreme that infants with epidermolysis bullosa often suffer significant damage to the skin during childbirth. Therefore, caesarean section is sometimes recommended in cases where the disease is diagnosed during pregnancy.

**? CASE 2 Cancer: When Good Cells Go Bad**
**How can doctors test for the spread of cancer?**
Cancer is characterized by uncontrolled cell division. Many types of cancer result in the formation of a mass of cells called a tumor. In most cases, the tumor itself is not the problem. The real concern is that tumor cells will metastasize, or spread to other parts of the body, where they can affect the function of distant organs. One way physicians check for metastasis is to look for intermediate filament proteins in places where they don't belong. For example, the presence of certain keratins in the bloodstream is an indicator of metastasis.

Physicians may also examine lymph nodes for signs that a cancer has spread. Lymph nodes are part of the lymphatic system,

**FIG. 10.6** **Intermediate filaments in the cytoplasm and the nucleus of a cell.**

Intermediate filaments composed of keratins in the cytoplasm are stained red.

Intermediate filaments composed of lamins in the nucleus are stained blue.

**FIG. 10.7 Intermediate filaments in the epidermis of the skin.** Intermediate filaments bind to cellular junctions called desmosomes, forming a strong, interconnected network. Epidermolysis bullosa is a blistering disease of the skin caused by a defect in intermediate filaments.

Epidermis

Keratinocytes

Basal lamina

Melanocyte

Intermediate filaments

Desmosome

The strutlike intermediate filaments linked to desmosomes in keratinocytes provide a great deal of structural support to the skin.

Defects in intermediate filaments make the skin far less resistant to physical stress.

which collects extracellular fluid from tissues in the body and returns it to the circulatory system. Each part of the body drains into a specific set of lymph nodes. So if a patient is diagnosed with a melanoma, a cancer of melanocytes, a surgeon can remove the lymph node that drains the area of the body where the cancer is. A pathologist then examines the cells of the lymph node for the presence of melanocyte keratins. Since the cells of the lymph node do not normally contain melanocyte keratins, the presence of these intermediate filament proteins in the lymph node helps doctors determine how advanced the cancer is and thus design the best course of treatment. This lymph node test is useful in detecting other kinds of cancer as well, including cancers of the breast and prostate.

### Microtubules and microfilaments are dynamic structures.

When we think of the parts of a skeleton, we are inclined to think of our bones. Unlike our bones, which undergo remodeling but do not change rapidly, microtubules and microfilaments are highly dynamic structures. They become longer by the addition of subunits to their ends, and shrink by the loss of subunits. The growth and shrinkage of microtubules and microfilaments are influenced by many factors, including the concentration of free tubulin and actin subunits and the activity of regulatory proteins that attach to the cytoskeleton.

The rate at which protein subunits are added depends on the concentrations of tubulin and actin in that region of the cell. At high concentrations of subunits, microtubules and microfilaments may grow at both ends, although these subunits are added more quickly to one end than to the other. The faster-growing end is called the plus end and the slower-growing end is called the minus end (**Fig. 10.8**). The minus ends of microtubules in animal cells are positioned at the organizing center of the centrosome, and the plus ends project outward toward the plasma membrane.

In addition, microtubules have an important property not shared by the other cytoskeletal elements: Their

**FIG. 10.8    Different rates of growth at the plus and minus ends of microtubules and microfilaments.** The plus end grows quickly and the minus end grows slowly.

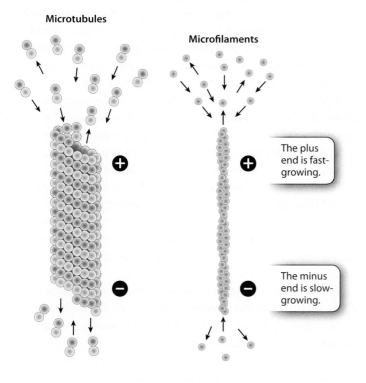

plus ends undergo seemingly random cycles of rapid shrinkage (depolymerization) followed by slower growth (polymerization). These cycles of shrinkage and growth are called **dynamic instability** (**Fig. 10.9**). The dramatic shrinkage is often called microtubule catastrophe and takes place because the plus end of a microtubule is structurally unstable. Once depolymerization has occurred, tubulin dimers are added to reassemble the microtubule.

Dynamic instability might seem like an undesirable feature for a component of the cytoskeleton, but it is actually very important for many functions of microtubules. For example, it allows microtubules to explore the space of the cell by growing into new areas and then shrinking back. This ability is especially important in the process of dividing chromosomes between the two daughter cells during cell division. In Chapter 11, we will see that the segregation of chromosomes requires that each chromosome be attached to two microtubules. The way that microtubules encounter chromosomes fast enough to form these attachments is by random exploration of the cytoplasm, which is driven by dynamic instability.

### The cytoskeleton is an ancient feature of cells.

Actin and tubulin are found in all eukaryotic cells, and evolutionarily they are highly conserved. The amino acid sequences of yeast tubulin and human tubulin are 75% identical. Similar comparisons of actin from amoebas and animals show that they are 80% identical after close to a billion years of evolution. In fact, a mixture of yeast and human actin monomers forms hybrid microfilaments able to function normally in the cell.

Not long ago, it was believed that cytoskeletal proteins were present only in eukaryotic cells. However, a number of studies have shown that many prokaryotes also have a system of proteins similar in structure to the cytoskeletal elements of eukaryotic cells and are involved in similar processes, including the separation of daughter cells during cell division. Interestingly, at least one of these prokaryotic cytoskeleton-like proteins is expressed in chloroplasts and mitochondria of some eukaryotic cells. The presence of this protein in these organelles lends support to the theory that chloroplasts and mitochondria were once independent prokaryotic cells that developed a symbiotic

**FIG. 10.9    Dynamic instability.** The plus ends of microtubules undergo cycles of rapid shrinkage followed by slow growth.

relationship with ancestral eukaryotic cells. This idea is called the endosymbiotic theory and we discuss it in more detail in Chapter 27.

## 10.3 CELLULAR MOVEMENT

In addition to providing structure and support, some cytoskeletal elements also allow cellular movement. We can think of cellular movement in three different ways. First, there is the movement of the cell itself, and many cells are capable of moving great distances on their own. The second kind of movement is the change in shape of a cell, as when a contracting muscle cell shortens. Changes in cell shape are also common in embryos and are an essential part of early development. The third kind of cellular movement is the movement of molecules and organelles within a cell. For example, melanocytes in the epidermis transport their packets of melanin along their branching dendrites to deliver them to neighboring keratinocytes (see Fig. 10.2). Similarly, in neurons, neurotransmitters are transported from the cell body to the end of the axon, where they are stored until they are needed. In this section, we discuss the molecular basis for these kinds of cellular movement.

### Motor proteins associate with microtubules and microfilaments to cause movement.

A motor is a device that imparts motion. We saw that microtubules and microfilaments have some capacity to move by polymerization and depolymerization, but on their own their capacity for movement is fairly limited. However, when joined by small accessory proteins called **motor proteins,** microtubules and microfilaments are capable of causing amazing movements.

First let's consider how these motor proteins contribute to a change in cell shape. Motor proteins cause muscle contraction by moving the actin microfilaments inside muscle cells. The cytoplasm of muscle cells shown in Fig. 10.1c is packed with actin microfilaments that are anchored to the ends of the cell. **Myosin,** a motor protein found in muscle cells, binds to actin and undergoes a conformational change. As a result, the actin microfilaments slide relative to myosin, causing the cell to shorten, or contract (**Fig. 10.10**). The energy required for the conformational change in myosin and the movement of actin microfilaments comes from the energy stored in ATP and is discussed in Chapter 37.

In other cells, myosin attached to various types of cellular cargo, such as transport vesicles, works by a similar mechanism to move materials from one part of the cell to another, using microfilaments as tracks (**Fig. 10.11a**).

**FIG. 10.10 Muscle contraction.** The shortening of a muscle cell is driven by interactions between the motor protein myosin and actin microfilaments.

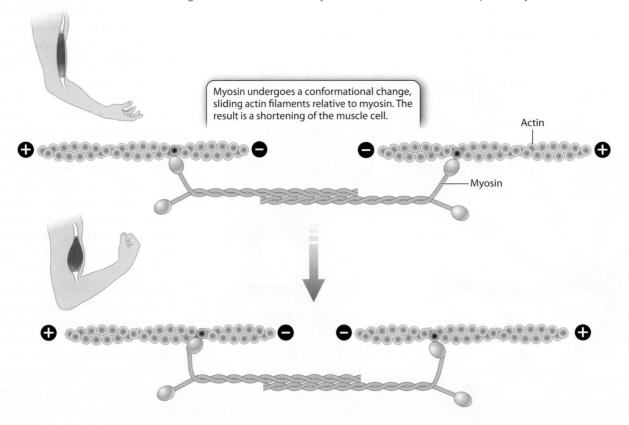

Myosin undergoes a conformational change, sliding actin filaments relative to myosin. The result is a shortening of the muscle cell.

Actin

Myosin

**FIG. 10.11   Intracellular transport.** The motor protein myosin interacts with microfilaments and the motor protein kinesin interacts with microtubules to move vesicles in the cell.

**a.** Microfilaments

**b.** Microtubules

Myosin and kinesin carry cargo toward the plus ends of microfilaments and microtubules, respectively.

**FIG. 10.12   Color change in zebrafish embryos driven by motor proteins kinesin and dynein.** Melanin granules are redistributed along microtubules in the melanophores of the skin.

**Night (dark)**      **Day (light)**

In the dark, melanin granules are dispersed outward by kinesin, causing the embryo to be darkly colored.

In the light, melanin granules are aggregated toward the center by dynein, causing the embryo to be lightly colored.

Microtubules also function as tracks for transport within the cell. Transport along microtubules takes place in a similar fashion to transport along microfilaments, except that the two motor proteins that transport cargo are **kinesin** and **dynein.** Kinesin is similar in structure to myosin and transports cargo toward the plus end of microtubules at the periphery of the cell (**Fig. 10.11b**). By contrast, dynein carries its load away from the plasma membrane toward the minus end. As with myosin, movement along microtubules by kinesin and dynein is driven by conformational changes in the motor proteins and is powered by energy harvested from ATP.

Let's look at an especially striking example of this system at work in the specialized skin cells called melanophores present in some vertebrates. Melanophores are similar to the melanocytes in our skin, but rather than hand off their melanin to other cells, as in humans, melanophores keep their pigment granules and move them around the cell in response to hormones or neuronal signals. This redistribution of melanin within the cell allows animals such as fish or amphibian embryos to change color. For example, at night the melanin granules in the skin of a zebrafish embryo are dispersed throughout the melanophores, making it darkly colored. As morning comes and the day brightens, the pigment granules are aggregated at the center of the cell around the centrosome, causing the embryo's color to lighten (**Fig. 10.12**). The melanin granules

in the melanophores move back and forth along microtubules, transported by kinesin and dynein. Kinesin moves the granules out toward the plus end of the microtubule during dispersal, and dynein moves them back toward the minus end during aggregation. The daytime and nighttime camouflage provided by this mechanism of color change keeps young, developing organisms from being spotted by hungry predators lurking below.

→   **Quick Check 1**  Would a defect in dynein or kinesin cause a zebrafish embryo to remain darkly colored after daybreak?

## Organelles with special arrangements of microtubules propel cells through the environment.

Many single-celled eukaryotic organisms that live in aquatic environments propel themselves through the water by means of the motion of short cilia or long flagella (see Fig. 10.4). These organelles are fiberlike extensions of the plasma membrane that have microtubules extending their entire length. Some cells of multicellular organisms also have cilia or flagella. For example, the sperm cells of algae, some plants, and many animals are propelled by one or more flagella. Epithelial cells in a number of animal tissues, such as the lining of the trachea and the upper respiratory tract, have cilia that move substances along the surface of the cell layer.

The microtubules in cilia and flagella are distributed in a characteristic "9 + 2" arrangement, that is, nine pairs of microtubules are located around the periphery of these organelles and two microtubules are at the center (**Fig. 10.13**). The outer microtubules are connected to the center pair by cross-linking proteins and to their neighbors by dynein molecules. Energy harvested by the hydrolysis of ATP powers the motion of cilia and flagella. Dynein undergoes a conformational change that causes the pairs of microtubules to slide past each other. The sliding of the microtubules results in a whiplike motion in the case of flagella and an oarlike rowing motion in the case of cilia.

## Actin polymerization moves cells forward.

Both single-celled amoebas foraging for food on the bottom of a pond and mammalian white blood cells chasing down foreign bacteria rely on actin polymerization to get from one place to another. In fact, many cells move through their environment by crawling across a substrate or by squeezing between other cells and burrowing through connective tissues, rather than by using cilia or flagella. This type of movement relies on microfilaments. Commonly, new microfilaments are assembled and extended at one end of the cell and existing microfilaments are pulled

Microtubule doublets

Dynein arms

Plasma membrane

**FIG. 10.13  Cilia with a characteristic "9 + 2" arrangement of microtubules.** Cilia have 9 microtubule doublets around the periphery and 2 microtubules in the center.

A cross section through an individual cilium reveals the 9 + 2 arrangement of microtubules.

Dynein arms

Cross-linking proteins

Central microtubules

Microtubule doublets

An electron micrograph shows cilia of this paramecium with microtubules at their cores.

**FIG. 10.14   Cell migration.** Cells move by extending a lamellipodium at the leading edge and contracting microfilaments at the trailing edge.

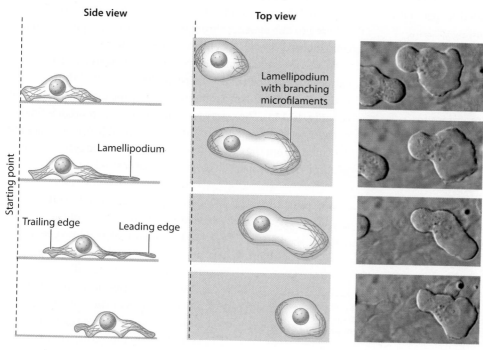

Side view        Top view

Starting point

Lamellipodium with branching microfilaments

Lamellipodium

Trailing edge        Leading edge

together at the other end (**Fig. 10.14**). The polymerization of actin into a microfilament exerts a considerable amount of force, enough to push the plasma membrane out into a thin, sheetlike structure called a lamellipodium, from the Greek for "thin-layered foot." New points of adhesion are established between the lamellipodium and the substrate.

Meanwhile, bundles of actin filaments at the other end of the cell contract by the interactions of myosin and actin, squeezing the cytoplasm and its contents forward toward the lamellipodium. The edge of the cell where the lamellipodium forms is called the leading edge, and the end where contraction takes place is called the trailing edge. Repeated cycles of actin polymerization at the leading edge and microfilament contraction at the trailing edge propel the cell forward. Cells are able to migrate considerable distances by this method.

Cells do not have permanent "front" and "back" ends. Instead, the locations of the leading and trailing edges can change in an instant. This flexibility allows a migrating cell to change its direction of movement, toward a source of nutrients, for example. The formation of a lamellipodium is triggered when a cell-surface receptor binds to a nutrient molecule or other substance, so the lamellipodium forms in the region of the plasma membrane where the signal is strongest. In this way, cells can follow the concentration gradient of a nutrient to its source.

## 10.4   CELL ADHESION

The cell, as we know, is the fundamental unit of living organisms (Chapter 1). Some conservative estimates place the number of cells in an adult human being at between 50 and 75 trillion, whereas others place the number at well above 100 trillion. Whichever estimate is more accurate, humans, as well as all complex multicellular organisms, are made up of a *lot* of cells! What then keeps us (or any other plant or animal) from slumping into a pile of cells? And what keeps cells organized into tissues, and tissues into organs?

Multicellular organisms are made up of much more than cells. The cells in most tissues and organs are connected to a complex meshwork of proteins and polysaccharides known as the extracellular matrix. The architecture and structural integrity of a particular tissue or organ depends on these attachments between cells and the extracellular matrix. In this section, we explore the means by which cells are connected to one another and to proteins in the extracellular environment.

Cells are attached to one another and to the extracellular matrix by cell-surface proteins called **cell adhesion molecules.** The regions in the plasma membrane where cells make contact with and adhere to other cells or the extracellular matrix are called **cellular junctions.** Together, cell adhesion molecules, cellular junctions, and the extracellular matrix keep us (and all multicellular organisms) intact. The importance of cell adhesion is reflected by the fact that between 5% and 10% of the genes in the human genome are involved in attaching cells to one another or to the extracellular matrix.

### Cell adhesion molecules allow cells to attach to other cells and to the extracellular matrix.

In 1907, American embryologist H. V. Wilson discovered that if he pressed a live sponge through fine cloth he could break up the sponge into individual cells. Then if he swirled the cells together, they would coalesce back into a group resembling a sponge. If he swirled together the cells from sponges of two different species, he observed that the cells sorted themselves out—that is, cells from one species of sponge associated only with cells from that same species (**Fig. 10.15a**). Fifty years later, German-born embryologist Johannes Holtfreter observed that if he took neuronal cells and skin cells from an amphibian embryo and treated them the same way that Wilson had treated sponge cells, the embryonic cells would sort themselves according to tissue type (**Fig. 10.15b**).

Cells can sort themselves because of the presence of various cell adhesion molecules. While a number of cell adhesion molecules are now known, the **cadherins** (for *calcium-dependent adherence proteins*) are especially important in the adhesion of cells to other cells. There are many different kinds of cadherins, and a given cadherin may bind only to another cadherin of the same type. This property explains

Holtfreter's observations of the cells from amphibian embryos. E-cadherin (for "epidermal cadherin") is present on the surface of embryonic epidermal cells, and N-cadherin (for "neural cadherin") is present on neuronal cells. The epidermal cells adhered to one another through E-cadherin, and the neuronal cells adhered to each other through N-cadherin.

Cadherins are transmembrane proteins (Chapter 5). The extracellular domain of a cadherin molecule binds to the extracellular domain of a cadherin of the same type on an adjacent cell (**Fig. 10.16a**). The cytoplasmic portion of the protein is linked to the internal cytoskeleton. This arrangement provides structural continuity from the cytoskeleton of one cell to the cytoskeleton of another, increasing the strength of tissues and organs.

In addition to being stably connected to other cells, cells also attach to proteins of the extracellular matrix. This type of attachment provides structural reinforcement, especially to tissues under physical stress like the epidermis of the skin and the lining of the digestive tract. The cell adhesion molecules that enable cells to adhere to the extracellular matrix are called **integrins.** Like the cadherins, integrins are transmembrane proteins, and their cytoplasmic domain is linked to the cytoskeleton (**Fig. 10.16b**). Also like the cadherins, integrins are of many different types, each with a specificity for a different extracellular matrix protein. Integrins are present on the surface of virtually every animal cell. Though integrins have not been observed in plants, integrin-like proteins are present on the surface of some plant cells as well. We return to integrins and how they interact with the extracellular matrix in section 10.5.

## Adherens junctions and desmosomes connect adjacent animal cells and are anchored to the cytoskeleton.

Cadherins are not distributed randomly in the plasma membrane but are located in adherens junctions

---

**FIG. 10.15  Cell type-specific cell adhesion.** Experiments showed that cells from (a) sponges and from (b) amphibian embryos are in each case able to adhere to one another in a specific manner.

**a.** Sponge

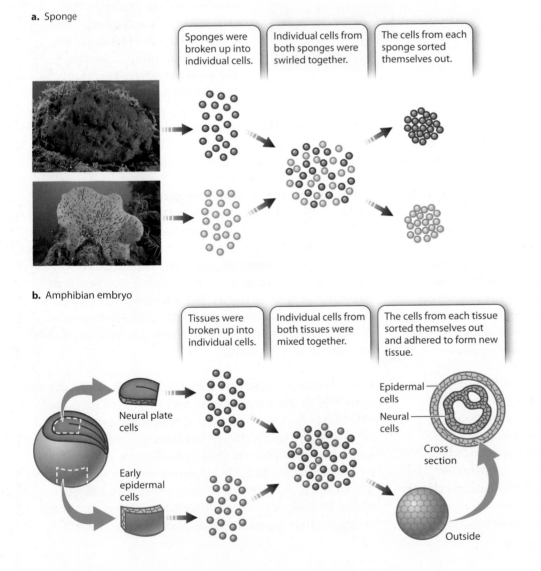

Sponges were broken up into individual cells.

Individual cells from both sponges were swirled together.

The cells from each sponge sorted themselves out.

**b.** Amphibian embryo

Neural plate cells

Early epidermal cells

Tissues were broken up into individual cells.

Individual cells from both tissues were mixed together.

The cells from each tissue sorted themselves out and adhered to form new tissue.

Epidermal cells

Neural cells

Cross section

Outside

**FIG. 10.16 Cell adhesion by (a) cadherins and (b) integrins.** Cadherins are transmembrane proteins that connect cells to other cells; integrins are transmembrane proteins that connect cells to the extracellular matrix.

**a.** Cadherins

**b.** Integrins

The structural support provided by desmosomes and their associated network of intermediate filaments is crucial to the function of several organs, such as the skin and heart. Mutations in several of the genes for desmosomal proteins are responsible for a number of serious diseases. In these instances, intercellular connections are profoundly weakened. The results range from serious skin conditions due to fragile epidermis, as we saw with epidermolysis bullosa, to a high risk of early heart failure due to severe weakening of the heart muscle.

→ **Quick Check 2** Adherens junctions and desmosomes both attach cells to other cells and are made up of cadherins. How, then, are they different?

## Tight junctions prevent the movement of substances through the space between animal cells.

Epithelial cells, which form layers or sheets that cover or line other tissues and organs, define the boundaries of many of these tissues, including the digestive tract, respiratory tract, and outer layer of the skin. Like any effective boundary, a layer of epithelial cells must limit or control the passage of material across it. Adherens junctions and desmosomes provide strong connections between cells, but they do not prevent the free passage of materials through the spaces between the cells they connect. Junctional complexes called **tight junctions** (Fig. 10.17) establish a seal between cells so that the only way a substance can travel from one side of a sheet of epithelial cells to the other is by moving *through* the cells by means of one of the cellular transport mechanisms discussed in Chapter 5.

A tight junction is a band of interconnected strands of integral membrane proteins that, like adherens junctions, encircles the epithelial cell. It also binds to tight-junction proteins on adjacent cells. Unlike adherens junctions, however, tight junctions are not connected to the cytoskeleton. Their function is to prevent passage of materials in between cells, not to anchor the cells together.

Cells that have tight junctions have two distinct sides because the tight junction divides the plasma membrane into two distinct regions (Fig. 10.17). The portion of the plasma

and desmosomes. These junctional complexes anchor cells to one another and are reinforced by the cytoskeleton (**Fig. 10.17**).

In our earlier discussion of microfilaments (section 10.2), we saw that a long bundle of actin microfilaments forms a band that extends around the circumference of epithelial cells, such as epithelial cells that line the intestine. This band of actin is attached to the plasma membrane by cadherins in a beltlike junctional complex called an **adherens junction** (Fig. 10.17). The cadherins in the adherens junction of one cell attach to the cadherins in the adherens junctions of adjacent cells. This arrangement establishes a physical connection among the actin cytoskeletons of all cells present in an epithelial layer of cells.

**Desmosomes** are buttonlike points of adhesion that hold the plasma membrane of adjacent cells together (Fig. 10.17). Cadherins are at work here, too, strengthening the connection between cells in a manner similar to adherens junctions. The cadherins in the desmosome of one cell bind to the cadherins in the desmosomes of adjacent cells. The cytoplasmic domain of these cadherins is linked to intermediate filaments in the cytoskeleton. This second type of physical connection among neighboring cells greatly enhances the structural integrity of epithelial cell layers.

**FIG. 10.17 Junctional complexes.** Junctional complexes connect cells to other cells or to the basal lamina and are reinforced by the cytoskeleton.

**TABLE 10.1** Interactions among cellular junctions, cell adhesion molecules, and cytoskeleton elements.

| DIFFERENT TYPES OF CELLULAR JUNCTION | USE SPECIFIC CELL ADHESION MOLECULES | TO ANCHOR THEMSELVES TO THE CYTOSKELETON |
|---|---|---|
| Adherens junctions | Cadherins | Microfilaments |
| Desmosomes | Cadherins | Intermediate filaments |
| Hemidesmosomes | Integrins | Intermediate filaments |

membrane in contact with the lumen, or the inside of any tubelike structure like the gut, is called the apical membrane. The apical membrane defines the "top" side of the cell. The rest of the plasma membrane is the basolateral membrane, which defines the bottom ("baso") and sides ("lateral") of the cell. These two regions of the plasma membrane are structurally different because the tight junction prevents lipids and proteins in the membrane on one side of the junction from diffusing to the other side. As a result, the apical and basolateral membranes of a cell are likely to have different integral membrane proteins, which causes them to be functionally different as well. In the small intestine, for example, glucose is transported from the lumen into intestinal epithelial cells by transport proteins on the apical side of the cells, and is transported out of the cells into the circulation by facilitated diffusion through glucose transporters restricted to the basolateral sides of the cells.

From this discussion, we see that cellular junctions and cytoskeletal elements interact to create stable communities of cells in the form of tissues and organs. These interactions are specific, with certain junctions associated with specific components of the cytoskeletal network. These interactions are summarized in **Table 10.1.**

### Gap junctions and plasmodesmata allow the passage of substances from one cell to another.

Not all junctions are involved in anchoring cells in place or sealing off one side of a layer of cells from the other. The **gap junctions** (Fig. 10.17) of animal cells and **plasmodesmata** of plant cells are connections between the plasma membranes of adjacent cells that permit materials to pass directly from the cytoplasm of one cell to the cytoplasm of another. Gap junctions are formed when a set of integral membrane proteins arranged in a ring connects to a similar ring of proteins in the membrane of another cell. Ions and signaling molecules pass through these junctions, allowing cells to communicate. In the heart, for example, ions pass though gap junctions connecting cardiac muscle cells. This rapid electrical communication allows the muscle cells to beat in a well-timed, coordinated rhythm.

Plasmodesmata (the singular form is "plasmodesma") are passages through the cell walls of adjacent plant cells. They are similar to gap junctions in that they allow cells to exchange ions

and small molecules directly, but the similarity ends there. In plasmodesmata, the plasma membranes of the two connected cells are actually continuous. The size of the opening is considerably larger than that of gap junctions, large enough for cells to transfer RNA molecules and proteins, a capability that is especially important during plant embryonic development. Plasmodesmata allow plant cells to send signals to one another despite being enclosed within rigid cell walls.

→ **Quick Check 3** Which type(s) of junction prevent substances from moving through the space between cells? Which one(s) permit the exchange of cytoplasmic materials?

## 10.5 THE EXTRACELLULAR MATRIX

So far, we have looked at how the cytoskeleton maintains the shape of cells. We have also seen how the stable association of animal cells with one another and with the extracellular matrix is made possible by cellular junctions, and that these junctions are reinforced by the cytoskeleton. As important as the cytoskeleton and cellular junctions are to the structure of cells and tissues, it is the extracellular matrix that provides the molecular framework that ultimately determines the structural architecture of plants and animals.

The extracellular matrix is synthesized, secreted, and modified by many different kinds of cell. It is an insoluble meshwork composed of proteins and polysaccharides. There are many different forms of extracellular matrix, which differ in the amount, type, and organization of the proteins and polysaccharides that make them up.

In plants, the extracellular matrix takes the form of the cell wall, which provides the support needed by individual cells and results in turgor pressure (Chapter 5). Collectively, the interconnected cell walls of a plant act as an internal framework to support the entire organism. In animals, the extracellular matrix is present in most tissues and is especially abundant in connective tissue, where it provides support and protection, and in the basal lamina found underneath all epithelial tissue. In both plants and animals, the extracellular matrix not only contributes structural support but also provides informational cues that determine the activity of the cells that are in contact with it.

### The extracellular matrix of plants is the cell wall.

The paper we write on, the cotton fibers in the clothes we wear, the wood in the chairs we sit on are, in fact, the extracellular matrix of plants. In plants, the extracellular matrix forms the cell wall, and the main component of the plant cell wall is the polysaccharide cellulose discussed in Chapter 2. Its presence in the cell wall of every plant makes cellulose the most widespread organic macromolecule on Earth.

The plant cell wall represents possibly one of the most complex examples of an extracellular matrix. It is certainly one of the most diverse in the functions it performs. Cell walls maintain the shape and turgor pressure of plant cells, allow plant cells to grow, and act as a barrier that prevents foreign materials and pathogens from reaching the plasma membrane. In many plants, cell walls collectively serve as a skeletal support structure for the entire plant.

The plant cell wall is composed of as many as three layers: the outermost middle lamella, the primary cell wall, and, closest to the plasma membrane, the secondary cell wall (**Fig. 10.18**). The middle lamella is made first, and is synthesized during the late stages of cell division. It is made of a gluelike complex carbohydrate, and it is the main mechanism by which plant cells adhere to one another. The primary cell wall is formed next and consists mainly of cellulose fibers, but it also contains a number of other molecules, including pectin and several proteins. The primary cell wall is laid down while the cells are still growing. It is assembled by enzymes on the

surface of the cell and remains thin and flexible. Once cell growth has stopped, the secondary cell wall is constructed in many, but not all, plant cells. It also is made largely of cellulose fibers but in addition contains a substance called lignin. Lignin hardens the cell wall and makes it water resistant. In woody plants, the cell wall can be up to 25% lignin. The rigid secondary cell wall permits the growth of woody plants to tremendous heights. Giant sequoia trees grow to more than 300 feet and are supported entirely by the lignin-reinforced cellulose fibers of the interconnected cell walls.

For a plant cell to grow, the cell wall must accommodate an increase in cell volume and surface area. Even though the matrix of cellulose and other polysaccharides of the primary cell wall is relatively flexible compared to the lignin-containing secondary cell wall, it is still very strong. When a plant cell grows, additional cell wall components must be synthesized to expand the area of the wall. Unlike the extracellular matrix components that are secreted by animal cells, the cellulose polymer is assembled outside the cell, on the extracellular surface of the plasma membrane. Both the glucose monomers that form the polymer and the enzymes that attach them are delivered to the cell surface by arrays of microtubules. Here is yet another example of how the cytoskeleton plays an indispensable role in regulating the shape of cells.

**FIG. 10.18** **The three layers of the plant cell wall: middle lamella, primary cell wall, and secondary cell wall.** The major component of the plant cell wall is cellulose, a polymer of glucose.

## The extracellular matrix is abundant in connective tissues of animals.

The extracellular matrix of animals, like that of plants, is secreted by cells and is a mixture of proteins and polysaccharides. The animal extracellular matrix is composed of large fibrous proteins, including collagen, elastin, and laminin, which impart tremendous tensile strength. These fibrous proteins are embedded in a gel-like polysaccharide matrix. The matrix is negatively charged, attracting positively charged ions and water molecules that provide protection against compression and other physical stress.

The extracellular matrix can be found in abundance in animal connective tissue (**Fig. 10.19**). Connective tissue underlies all epithelial tissues, as we have seen. For example, the dermis of the skin is connective tissue, providing support and nutrients to the overlying epidermis. A number of cell types, including the fibroblasts that synthesize most of the extracellular matrix proteins, reside in this tissue. Connective tissue is unusual compared to other tissue types in that it is dominated by the extracellular matrix and has a low cell density. Consequently, the extracellular matrix determines the properties of different types of connective tissue. Other more specialized types of animal connective tissue include bone, cartilage, and tendon.

Collagen is the most abundant protein in the extracellular matrix of animals, and in fact is the most abundant animal protein on the planet. There are more than 20 different forms of collagen, and in humans collagen accounts for almost a quarter of the protein present in the body. Over 90% of this collagen is type I collagen, which is present in the dermis of your skin, where it provides strong, durable support for the overlying epidermis. The tendons that connect your muscles to bones and the ligaments that connect your bones to other bones are able to withstand the physical stress placed on them because they are made up primarily of collagen.

Collagen's strength is related to its structure. Like a rope or a cable, collagen is composed of intertwined fibers that make it much stronger than if it were a single fiber of the same diameter. A collagen molecule consists of three polypeptides wound around one another in a triple helix. A bundle of collagen molecules forms a fibril, and the fibrils are assembled into fibers (**Fig. 10.20**). Once multiple collagen fibers are assembled into a ligament or tendon, the final structure is incredibly strong.

**FIG. 10.19  Animal connective tissue.** Connective tissue is composed of protein fibers in a gel-like polysaccharide matrix.

Cells in the connective tissue of animals reside in a complex meshwork of extracellular matrix fibers.

Elastin fibers

Collagen fibers

Fibroblast nuclei

Elastin fibers

Collagen fibers

Fibroblast cell

Fibroblast nucleus

Gel-like matrix

**FIG. 10.20  Collagen.** Type I collagen molecules are arranged in a triple helix and are grouped into bundles called fibrils, which in turn are grouped into bundles called fibers, resulting in tremendous strength like a steel cable.

Collagen molecules (triple helices)

Collagen fibrils

Collagen fiber

Collagen molecules are bundled like steel wires in a cable.

## The basal lamina is a special form of extracellular matrix.

The basal lamina is a specialized layer of extracellular matrix that is present beneath all epithelial tissues, including the lining of the digestive tract, epidermis of the skin, and endothelial cells that line the blood vessels of vertebrates (**Fig. 10.21**). The role of the basal lamina is to provide structural foundation for these epithelial tissues. The basal lamina is made of several proteins, including a special type of collagen. The triple-helical structure of collagen provides flexible support to the epithelial sheet and also provides a scaffold on which other proteins are assembled.

Epithelial cells are firmly anchored to the basal lamina by a version of the desmosome called a **hemidesmosome** (see Fig. 10.17). Integrins are the prominent cell adhesion molecules in hemidesomsomes. Their extracellular domains bind to the extracellular matrix proteins in the basal lamina, and their cytoplasmic domains are linked to intermediate filaments of the cytoskeleton. The intermediate filaments connected to hemidesomsomes are linked to those connected to the desmosomes in the lateral membrane. The result is a firmly anchored and tremendously reinforced layer of cells.

A migrating cell must be able to attach to extracellular matrix proteins in order to move forward. These contacts are also made by means of integrins. Cell migration in animals is normal and necessary during embryonic development (Chapter 42), wound

---

**FIG. 10.21   The basal lamina.** The basal lamina, found beneath epithelial tissue, is a specialized form of extracellular matrix.

Basal lamina

healing, and the immune response to infection (Chapter 43). However, inappropriate cell migration can have devastating consequences, as we discuss now.

**? CASE 2   Cancer: When Good Cells Go Bad**
**How do cancer cells spread throughout the body?**

Nonmalignant, or benign, tumors are encapsulated masses of cells that divide continuously because regulation of cell division has gone awry (Chapter 11). As the tumor grows, its border pushes outward against adjacent tissues. Benign tumors are rarely life threatening unless the tumor interferes with the function of a vital organ.

Malignant tumors are different. They contain some cells that can metastasize, that is, break away from the main tumor and colonize distant sites in the body. Metastatic tumor cells have an enhanced ability to adhere to extracellular matrix proteins, especially those in the basal lamina. This is significant because for a cell to metastasize, it must enter and leave the bloodstream through capillaries. Since all blood vessels, including capillaries, have a basal lamina, a metastatic tumor cell needs to cross a basal lamina at least twice—once on the way into the bloodstream and again on the way out (**Fig. 10.22**). Since cells attach to basal lamina proteins by means of integrins, many studies have compared the integrins in metastatic and non-metastatic cells in the search for potential targets for treatment.

In some types of cancer, the number of specific integrins on the cell surface is an indicator of metastatic potential. Melanoma provides an example. A specific type of integrin is present in high amounts on metastatic melanoma cells but is absent on non-metastatic cells from the same tumor. In laboratory tests, blocking these integrins eliminates the melanoma cell's ability to cross an artificial basal lamina. Drugs targeting this integrin protein are currently in clinical trials.

## Extracellular matrix proteins influence cell shape and gene expression.

The extracellular matrix is not an inert, three-dimensional environment in which cells live; instead, it plays an active role. Cells continue to interact with the extracellular matrix long after they have synthesized it or moved into it, and these interactions can have profound effects on the cell. Biologists have observed the results of these interactions in experiments conducted with cells grown in culture in the laboratory. For example, fibroblasts that secrete components of the extracellular matrix are in turn influenced by the extracellular matrix. Fibroblasts cultured on a two-dimensional surface coated with extracellular matrix proteins attach to the matrix and flatten as they maximize their adhesion to the matrix. By contrast, the same cells cultured in a three-dimensional gel of extracellular matrix look and behave like the spindle-shaped, highly migratory fibroblasts present in

**FIG. 10.22 Metastatic cancer cells.** Some cancer cells spread from the original site of cancer formation to the bloodstream and then to distant organs of the body.

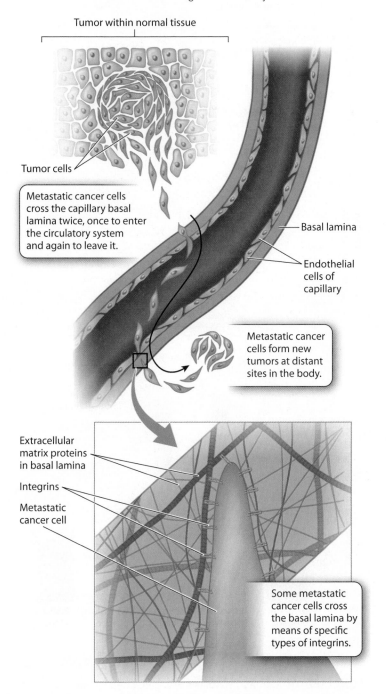

Tumor within normal tissue

Tumor cells

Metastatic cancer cells cross the capillary basal lamina twice, once to enter the circulatory system and again to leave it.

Basal lamina

Endothelial cells of capillary

Metastatic cancer cells form new tumors at distant sites in the body.

Extracellular matrix proteins in basal lamina

Integrins

Metastatic cancer cell

Some metastatic cancer cells cross the basal lamina by means of specific types of integrins.

**FIG. 10.23 Cell shape determined by the structure of the extracellular matrix.** Fibroblasts adopt different shapes depending on whether they are grown on a two-dimensional or a three-dimensional matrix.

Fibroblasts grown on a two-dimensional matrix attach and become flattened.

On glass

Fibroblasts grown in a three-dimensional matrix are spindle shaped and look as they do in vivo.

In suspension

**FIG. 10.24 Cell shape determined by composition of the extracellular matrix.** Neurons adopt different shapes depending on whether or not they are cultured with the extracellular matrix protein laminin.

Neurons maintained in the absence of laminin attach to the surface but do not take on the appearance of a nerve cell.

Neurons cultured on a surface coated with laminin develop axon-like extensions of the plasma membrane.

living connective tissue (**Fig. 10.23**). So the structure of the extracellular matrix can influence the shape of cells.

A second example is provided by nerve cells. When these cells are grown in culture on a plastic surface, they attach to the surface of the dish, but do not take on a neuron-like shape.

However, when these cells are grown on the same surface coated with the extracellular matrix protein laminin, they develop long extensions that resemble the axons and dendrites of normal nerve cells (**Fig. 10.24**). In this case, the composition of the extracellular matrix can affect the shape of cells.

## FIG. 10.25

# Can extracellular matrix proteins influence gene expression?

**BACKGROUND** The adhesion of cells to the extracellular matrix is required for cell division, DNA synthesis, and proper cell shape. Research by Iranian-American cell biologist Mina Bissell and colleagues indicated that cellular interaction with extracellular matrix proteins influences gene expression. Bissell discovered that mammary cells expressed and secreted high levels of the milk protein β-casein when grown in a three-dimensional collagen matrix but not in a two-dimensional collagen matrix. American cell biologist Joan Caron followed up these studies using liver cells called hepatocytes.

**HYPOTHESIS** Caron hypothesized that a specific protein in the extracellular matrix is necessary for the expression of the protein albumin from hepatocytes grown in culture. Albumin is a major product of liver cells.

**EXPERIMENT** Caron cultured hepatocytes on a thin layer of type I collagen, which does not induce albumin synthesis. Next, she added a mixture of several different extracellular matrix proteins to the culture and looked for changes in albumin gene expression and protein secretion into the media. She then tested individual extracellular matrix proteins from the mixture to see which one was responsible for the increase in albumin gene expression.

**RESULTS** Caron found that when she cultured cells on collagen with a combination of three extracellular matrix proteins—laminin, type IV collagen, and heparin sulfate proteoglycan (HSPG)—the cells synthesized albumin mRNA and secreted albumin protein for several weeks, but if she cultured the cells on collagen alone, they did not (top and middle graphs). In addition, when she tested individual extracellular matrix proteins, she found that laminin, but not any of the other proteins, caused an increase in albumin gene expression (bottom graph).

**CONCLUSION** Caron's hypothesis was supported by the experiments. A specific extracellular matrix protein, laminin, influences the expression of albumin by hepatocytes.

**FOLLOW-UP WORK** Bissell continued her work with mammary cells and found that the expression of the β-casein gene was also increased by laminin in the same way as the albumin gene in hepatocytes.

**SOURCES** Lee, E. Y., W. H. Lee, C. S. Kaetzel, G. Parry, and M. J. Bissell. 1985. "Interaction of Mouse Mammary Epithelial Cells with Collagen Substrata: Regulation of Casein Gene Expression and Secretion." *PNAS* 82:1419–1423; Caron, J. M. 1990. "Induction of Albumin Gene Transcription in Hepatocytes by Extracellular Matrix Proteins." *Molecular and Cellular Biology* 10:1239–1243.

In addition to influencing cell shape, the structure and composition of the extracellular matrix can influence gene expression of the cells that are grown in it. Mouse mammary epithelial cells grown in culture on plain glass coated with collagen remain alive and apparently healthy, and even make and secrete a minimal amount of milk proteins, including β-casein, when stimulated by the milk-inducing hormone prolactin. However, if the mammary epithelial cells are grown in three-dimensional collagen gels, they synthesize and secrete up to 10 times more β-casein.

→ **Quick Check 4** Do you think cadherins or integrins are responsible for the dependence of mammary cells on the extracellular matrix for their ability to produce milk proteins? Why?

Two related experiments that further explore the importance of extracellular matrix proteins in the regulation of gene expression are described in **Fig. 10.25**. Taken together, these experiments demonstrate that there is a dynamic interplay between the extracellular matrix and the cells that synthesize it.

## Core Concepts Summary

**10.1 TISSUES AND ORGANS ARE COMMUNITIES OF CELLS THAT PERFORM A SPECIFIC FUNCTION.**

A tissue is a collection of cells that work together to perform a specific function. page 10-2

Two or more tissues often work together to form an organ. page 10-2

Cytoskeletal structures determine the shape of the cell. page 10-2

Cell junctions connect cells to one another and to the extracellular matrix, a meshwork of proteins and polysaccharides outside the cell. page 10-2

**10.2 THE CYTOSKELETON IS COMPOSED OF MICROTUBULES, MICROFILAMENTS, AND INTERMEDIATE FILAMENTS THAT HELP TO MAINTAIN CELL SHAPE.**

All eukaryotic cells have microtubules and microfilaments. Animal cells also have intermediate filaments. page 10-3

Microtubules, microfilaments, and intermediate filaments are polymers of protein subunits. page 10-4

Microtubules provide structural support and maintain cytoplasmic organization. They are hollow polymers of tubulin dimers. page 10-4

Microfilaments also provide structural support necessary to maintain the shape of a cell. They are polymers of actin monomers. page 10-4

Microtubules and microfilaments are dynamic structures and can assemble and disassemble rapidly. page 10-4

Intermediate filaments are polymers of proteins that differ depending on cell type. They do not exhibit the dynamic properties of microtubules and microfilaments and provide stable structural support for many types of cells. page 10-5

Microtubules and microfilaments have a plus end and a minus end, determined by the rate at which subunits are added. page 10-6

Microtubules go through rounds of growth and rapid shrinkage called dynamic instability. page 10-7

The cytoskeleton is an ancient feature of cells. Some prokaryotic cells have protein polymer-like systems that function similarly to microtubules and microfilaments. page 10-7

**10.3 THE CYTOSKELETON INTERACTS WITH MOTOR PROTEINS TO PERMIT THE MOVEMENT OF CELLS AND SUBSTANCES WITHIN CELLS.**

Motor proteins associated with microtubules and microfilaments enable intracellular transport and cell movement. Myosin binds to actin in microfilaments. Dynein and kinesin bind to tubulin to facilitate microtubule-dependent movement. page 10-8

Microfilaments are involved in intracellular vesicle transport and enable many cells to move by crawling across a surface or through a tissue. page 10-8

Microtubules provide intracellular tracks along which cargo is transported in the cytoplasm. page 10-9

Flagella and cilia are organelles that contain microtubules that propel cells through aqueous environments. page 10-10

**10.4 CELLS ADHERE TO OTHER CELLS AND THE EXTRACELLULAR MATRIX BY MEANS OF CELL ADHESION MOLECULES AND JUNCTIONAL COMPLEXES.**

Cell adhesion molecules allow cells to attach to other cells and to extracellular matrix proteins. page 10-11

Cadherins are transmembrane proteins that allow cells to adhere to other cells. Integrins are transmembrane proteins that facilitate cell adhesion to the extracellular matrix. page 10-12

Adherens junctions form a continuous band around the circumference of a cell. Desmosomes are buttonlike points of adhesion between cells. page 10-13

Cadherins in adherens junctions are anchored to the microfilaments of the cytoskeleton. Cadherins in desmosomes are anchored to intermediate filaments. page 10-13

Tight junctions prevent the passage of substances through the space between cells. They also divide the plasma membrane into apical and basolateral regions. page 10-13

Gap junctions and plasmodesmata allow the passage of substances from the cytoplasm of one cell to that of its neighbor. page 10-15

## 10.5 THE EXTRACELLULAR MATRIX PROVIDES STRUCTURAL SUPPORT AND INFORMATIONAL CUES.

The extracellular matrix is an insoluble meshwork of proteins and polysaccharides secreted by the cells it surrounds. It provides structural support to cells, tissues, and organs. page 10-15

In plants, the extracellular matrix is found in the cell wall, and the main component of the plant cell wall is the polysaccharide cellulose. page 10-16

In animals, the extracellular matrix is found in abundance in connective tissue. page 10-17

Collagen is the primary component of connective tissues in animals and is exceptionally strong. page 10-17

A highly specialized extracellular matrix called the basal lamina is present under all epithelial cell layers. Cells are anchored to the basal lamina by integrins in hemidesmosomes. page 10-18

In addition to providing structural support for cells, the extracellular matrix can influence cell shape, movement, and gene expression. page 10-18

# Self-Assessment

1. Name three types of filaments that make up the cytoskeleton, including the subunits of each, and state which cytoskeletal element is not present in plant cells.

2. Explain how microtubules and microfilaments are dynamic structures.

3. Describe the functions of the three major motor proteins that are required for cellular movements involving the cytoskeleton.

4. Compare the cytoskeletal dynamics in cells that move by beating cilia or flagella with cells that move by crawling.

5. Identify the cytoskeletal component linked to adherens junctions, desmosomes, and hemidesmosomes.

6. Compare and contrast the structural features and functional roles of cadherins and integrins.

7. Name examples of the extracellular matrix in plants and animals.

8. Explain the functional relationships among the basal lamina, integrins, and intermediate filaments in epithelial tissues.

9. Predict the effect of inadequate or inappropriate adhesion of a cell to the extracellular matrix on gene expression in the cell.

Do you understand the chapter's Core Concepts? Log in to check your answers to the Self-Assessment questions, then practice what you've learned and reinforce this chapter's concepts by working through the problems and multimedia tutorials provided there.

☐ **www.biologyhowlifeworks.com**

## CHAPTER 11

# CELL DIVISION

## Variations, Regulation, and Cancer

Cells come from preexisting cells. This is one of the fundamental principles of biology and a key component of the cell theory, which was introduced in Chapter 1. **Cell division** is the process by which cells make more cells. In multicellular organisms, which begin life as a single cell, cell division produces the millions, billions, or in the case of humans, trillions of cells that make up the fully developed organism. Even after a multicellular organism has achieved its adult size, cell division continues. In plants, cell division is essential for continued growth. In many animals, cell division replaces worn-out blood cells, skin cells, and cells that line much of the digestive tract. And if you fall and scrape your knee, the cells at the site of the wound begin dividing to replace the damaged cells and heal the scrape.

Cell division is also important in reproduction. In bacteria, for example, a new generation is created when the parent cell divides and forms two daughter cells. This type of reproduction is called **asexual reproduction.** When a single bacterium reproduces by cell division, it first makes identical copies of its genetic material so that each of the two daughter cells has the same genetic material as the parent cell. There are many connotations to the word "sex," but strictly speaking, sex is the combining of genetic material from two individuals. Since the genetic material in the bacterial daughter cells comes from just one parent, reproduction occurred without sex, or asexually. In principle, the daughter cells are genetically identical to the parental cell, but because DNA replication is not completely error free, the daughter cells may carry small genetic differences or mutations compared to the parent cell.

By contrast, **sexual reproduction** results in offspring that receive genetic material from two parents. Half the genetic material is supplied by the female parent and is present in the egg; the other half is contributed by the male's sperm. Eggs and sperm are specialized cells called **gametes.** A female gamete and a male gamete merge during fertilization to form a new organism (Chapter 42). A unique feature of gametes is that they contain half the number of chromosomes as the other cells in the parent organism. So when fertilization occurs, the combination of genetic material from the egg and the sperm results in a new organism with the same number of chromosomes as the parents. The production of gametes comes about by a form of cell division that results in daughter cells with half the number of chromosomes as the parent cell. As we will see, the products of this cell division are *not* genetically identical to the parent.

What determines when cells divide and, importantly, when they should not? And what determines which cells divide? To answer these questions, we must understand the process of cell division and how it is controlled. This discussion will lay the groundwork for exploring how cancer results from a loss of control of cell division.

## 11.1 CELL DIVISION

Cell division is the process by which a single cell becomes two daughter cells. While this process may seem simple, successful cell division must satisfy several important requirements. First, the two daughter cells must each receive the full complement of genetic material (DNA) present in the single parent cell. Second, the parent cell must be large enough to divide in two and still contribute sufficient cytoplasmic components such as proteins, lipids, and other macromolecules to each daughter cell. Satisfying these requirements means that key cellular components must be duplicated before cell division takes place. This duplication of material is achieved in a series of steps that constitute the life cycle of every cell. When you think of a life cycle, you might think of various stages beginning with birth and ending with death. In the case of a single cell, the life cycle begins and ends with cell division.

In this section, we explore the different mechanisms by which prokaryotic and eukaryotic cells divide. Prokaryotic cells divide by **binary fission.** When eukaryotic cells divide, they first divide the nucleus during **mitosis,** then divide the cytoplasm into two daughter cells during **cytokinesis.** As we discuss, it is likely that mitosis evolved from binary fission.

### Prokaryotic cells reproduce by binary fission.

The majority of prokaryotic cells, namely bacteria and archaeons, divide by binary fission. In this form of cell division, a cell replicates its DNA, increases in size, and divides into two daughter cells. Each daughter cell receives one copy of the replicated parental DNA. The molecular mechanisms that drive binary fission have been studied most extensively in bacteria. The process of binary fission is similar in archaeons, as well as in chloroplasts and mitochondria, organelles within plant and animal cells which evolved from free-living prokaryotic cells (Chapters 5 and 27).

Let's consider the process of binary fission in the intestinal bacterium *Escherichia coli* (**Fig. 11.1**). The circular genome of *E. coli* is attached by proteins to the inside of the plasma membrane. DNA replication is initiated at a specific location on the circular DNA molecule and proceeds in opposite directions around the circle. The result is two DNA molecules, both attached to the plasma membrane. The cell then grows. As the cell elongates, the two DNA attachment sites move apart. When the cell is about twice its original size and the DNA molecules are well separated, a constriction forms at the midpoint of the cell. Eventually, new membrane and cell wall are synthesized at the site of the constriction, dividing the single cell into two. The result is two daughter cells, each having the same genetic material as the parent cell.

Like most cellular processes, binary fission requires the coordination of many components in both time and space. Recent research has identified several genes whose products play a key role in bacterial cell division. One of these genes, called *FtsZ*, has been especially well studied. The protein it encodes forms a ring at the site of constriction where the new cell wall forms between the two daughter cells. *FtsZ* is present in the genomes of diverse bacteria and archaeons, suggesting that it plays a fundamental role in prokaryotic cell division. Interestingly, it appears to be evolutionarily related to tubulin, which you will recall from Chapter 10 makes up the dynamic microtubules found in eukaryotic cells that are important in intracellular transport, cell movement, and cell division.

→ **Quick Check 1** What do you predict would be the consequence of a mutation in *FtsZ* that disrupts the function of the protein it encodes?

**FIG. 11.1** **Binary fission.** Cell division in bacteria and archaeons occurs by binary fission.

1 The circular bacterial DNA molecule is attached by proteins to the inner membrane (red).

— Protein
— DNA

2 DNA replication begins at a specific location and proceeds bidirectionally around the circle.

3 The newly synthesized DNA molecule is also attached to the inner membrane, near the attachment site of the initial molecule.

— Newly synthesized DNA

4 As replication proceeds, the cell elongates symmetrically around the midpoint, separating the DNA attachment sites.

5 Cell division begins with the synthesis of new membrane and wall material at the midpoint.

6 Continued synthesis completes the constriction and separates the daughter cells.

## Eukaryotic cells reproduce by mitotic cell division.

The basic steps of binary fission that we just saw—replication of DNA, segregation of replicated DNA to daughter cells, and division of one cell into two—occur in all forms of cell division. However, cell division in eukaryotes is more complicated than cell division in prokaryotes. Instead of the single, relatively small, circular DNA molecule that is the genome of prokaryotic cells, the genome of eukaryotic cells is typically much larger and organized into one or more linear chromosomes, each of which needs to be replicated and separated into daughter cells. And whereas the DNA of prokaryotes is attached to the inside of the plasma membrane, allowing separation of replicated DNA into daughter cells by cell growth, the DNA of eukaryotes is located in the nucleus. As a result, eukaryotic cell division requires first the breakdown and then the re-formation of the nuclear envelope, as well as mechanisms other than cell growth to separate replicated DNA. As we saw in Chapter 10 and discuss in more detail in section 11.2, chromosomes of dividing eukaryotic cells attach to the mitotic spindle, which separates them into daughter cells.

Interestingly, some unicellular eukaryotes exhibit forms of cell division that have characteristics of binary fission and mitosis. For example, dinoflagellates, like all eukaryotes, have a nucleus and linear chromosomes. However, unlike most eukaryotes, the nuclear envelope does not break down but stays intact during cell division. Furthermore, the replicated DNA is attached to the nuclear envelope. The nucleus then grows and divides in a manner reminiscent of binary fission. These and observations of intermediate forms of cell division in other organisms strongly suggest that mitosis evolved from binary fission.

## The cell cycle describes the life cycle of a eukaryotic cell.

Cell division in eukaryotic cells proceeds through a number of steps that make up the **cell cycle** (**Fig. 11.2**). The cell cycle consists of two distinct stages: **M phase** and **interphase.** During M phase, the parent cell divides into two daughter cells. M phase consists of two different events: (1) mitosis, the separation of the chromosomes into two nuclei, and (2) cytokinesis, the division of the cell itself into two separate cells. Usually, these two processes go hand in hand, with cytokinesis usually beginning even before mitosis is complete. In most mammalian cells, M phase lasts about an hour.

The second stage of the cell cycle, called interphase, is the time between two successive M phases (Fig. 11.2). This stage lasts about 10 to 14 hours in most actively multiplying mammalian cells. For many years, it was thought that the relatively long period of interphase is uneventful. Today we know that during this stage the cell makes many preparations for division. These preparations include replication of the DNA in the nucleus so that each daughter cell receives a copy of the genome, and an increase in

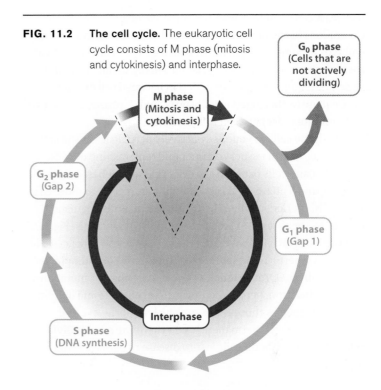

**FIG. 11.2** **The cell cycle.** The eukaryotic cell cycle consists of M phase (mitosis and cytokinesis) and interphase.

cell size so that each daughter cell receives sufficient amounts of cytoplasmic and membrane components to survive on its own.

Interphase can be divided into three phases, as shown in Fig. 11.2. Among the many preparations that the cell must make during interphase, one particularly important task is the replication of the entire DNA content of the nucleus. Since replication involves the synthesis of DNA, this stage is called **S phase** ("S" standing for "synthesis").

In most cells, S phase does not immediately precede or follow mitosis but is separated from it by two gap phases: $G_1$ **phase** between the end of M phase and the start of S phase, and $G_2$ **phase** between the end of S phase and the start of M phase. Many essential processes occur during both "gap" phases, despite the name. For example, both the size and protein content of the cell increase in preparation for division. In addition, the $G_1$ phase is a time when specific regulatory proteins are made and activated. Once active, the regulatory proteins, many of which are kinases, then promote the activity of enzymes that synthesize DNA. Thus, $G_1$ is a time of preparation for S-phase DNA synthesis, and $G_2$ is a time of preparation for M-phase mitosis and cytokinesis.

The time it takes for cells to pass through the cell cycle depends on the type of cell and the organism's stage of development. Actively dividing cells in some human tissues such as the intestine and skin require frequent replenishing. Cells in these tissues usually need about 12 hours to complete the cell cycle. Most other actively dividing cells in your body take about 24 hours to complete the cycle. A unicellular eukaryote like yeast can complete an entire cell cycle in just 90 minutes. Champions in the race through the cell cycle are the cells of the embryos of some frog species, in which early cell divisions divide the cytoplasm of the large egg cell into many smaller cells and so do not require any growth period in between cell divisions. Consequently, there are virtually no $G_1$ and $G_2$ phases and as little as 30 minutes pass between cell divisions.

Not all the cells in your body actively participate in the cell cycle, since not all tissues require rapid replenishing of cells. Instead, many cells pause in the cell cycle somewhere between M phase and S phase for periods ranging from days to more than a year. This period is called the **$G_0$ phase** and is distinguished from $G_1$ by the absence of preparations for DNA synthesis (Fig. 11.2). Although cells in $G_0$ have exited the cell cycle, they are active in other ways, performing specialized functions. For example, liver cells remain in $G_0$ for as much as a year, carrying out metabolism and detoxification, before reentering the cell cycle and dividing. Other cells such as those that form the lens of the eye and nerve cells enter $G_0$ permanently and therefore are nondividing cells. Thus, many brain cells lost to disease or damage cannot be replaced.

## 11.2 MITOTIC CELL DIVISION

Mitotic cell division (mitosis followed by cytokinesis) is the normal mode of asexual reproduction in unicellular eukaryotes, and it is the means by which an organism's cells, tissues, and organs develop and are maintained in multicellular eukaryotes. During mitosis and cytokinesis, the parental cell's DNA is divided and passed on to two daughter cells. This process is continuous, but is divided into discrete steps marked by dramatic changes in the cytoskeleton and in the packaging and movement of the chromosomes.

### The DNA of eukaryotic cells is organized as chromosomes.

One of the key challenges faced by a dividing eukaryotic cell is ensuring that both daughter cells receive an equal and complete set of chromosomes. The length of DNA contained in the nucleus of an average eukaryotic cell is on the order of 1 to 2 meters, well beyond the diameter of a cell. The DNA therefore needs to be condensed to fit into the nucleus, and then further condensed during cell division so that it does not become tangled as it segregates into daughter cells.

In eukaryotic cells, DNA is organized with histones and other proteins into chromatin, which can be looped and packaged to form the structures we know as chromosomes (Chapter 3). One of the earliest events in mitosis is the condensing of chromosomes from long, thin, threadlike structures typical of interphase to short, dense forms that are identifiable under the microscope during M phase.

**FIG. 11.3** A human karyotype showing 22 pairs of chromosomes plus 2 sex chromosomes, or 46 chromosomes in total.

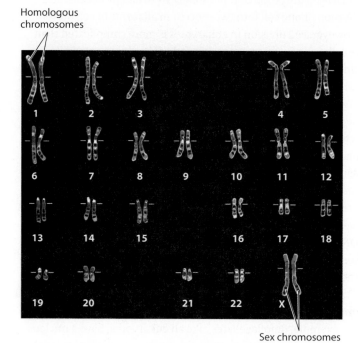

Every species is characterized by a specific number of chromosomes, and each chromosome contains a single molecule of DNA carrying a specific set of genes. When chromosomes condense and become visible during mitosis, they adopt characteristic shapes and sizes that allow each chromosome to be identified by its appearance in the microscope. The portrait formed by the number and shapes of chromosomes representative of a species is called its **karyotype.** Most of the cells in the human body, with the exception of the gametes, contain 46 chromosomes (**Fig. 11.3**). In contrast, cells from horses have 64 chromosomes, and cells from corn have 20.

**FIG. 11.4** Homologous chromosomes and sister chromatids. Sister chromatids result from the duplication of chromosomes. They are held together at the centromere.

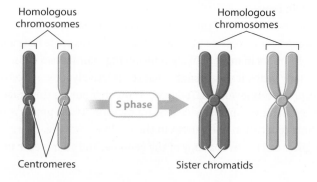

In a normal human karyotype, the 46 chromosomes can be arranged into 23 pairs, 22 pairs of **homologous chromosomes** numbered 1 to 22 from the longest to the shortest chromosome and 1 pair of **sex chromosomes** (Fig. 11.3). Each pair of homologous chromosomes represents two of the same type of chromosome (both carrying the same set of genes), one of which was received from the mother and the other from the father. The sex chromosomes are the X and Y chromosomes. Individuals with two X chromosomes are female, and those with an X and a Y chromosome are male.

The number of complete sets of chromosomes in a cell is known as ploidy. A cell with one complete set of chromosomes is **haploid,** and a cell with two complete sets of chromosomes is **diploid.** Some organisms, such as plants, can have four or sometimes more complete sets of chromosomes. Such cells are polyploid.

In order for cell division to proceed normally, every chromosome in the parent cell must be duplicated so that each daughter cell receives a full set of chromosomes. This duplication occurs during S phase. Even though the DNA in each chromosome duplicates, the two identical copies, called **sister chromatids,** do not separate. They stay side by side, physically held together at a constriction called the **centromere.** At the beginning of mitosis, the nucleus of a human cell contains 46 chromosomes, each of which is a pair of identical sister chromatids linked together at the centromere (**Fig. 11.4**). Thus, to count chromosomes, it is simply a matter of counting centromeres.

→ **Quick Check 2** Which DNA sequences are more alike: a pair of sister chromatids or a pair of homologous chromosomes?

## Prophase: Chromosomes condense and become visible.

Mitosis takes place in five stages, each of which is easily identified by events that can be observed in the microscope (**Fig. 11.5**). When you look in a microscope at a cell in interphase, specific chromosomes cannot be distinguished because they are long and thin. As the cell moves from $G_2$ phase to the start of mitosis, the chromosomes condense and become visible in the nucleus. The first stage of mitosis is known

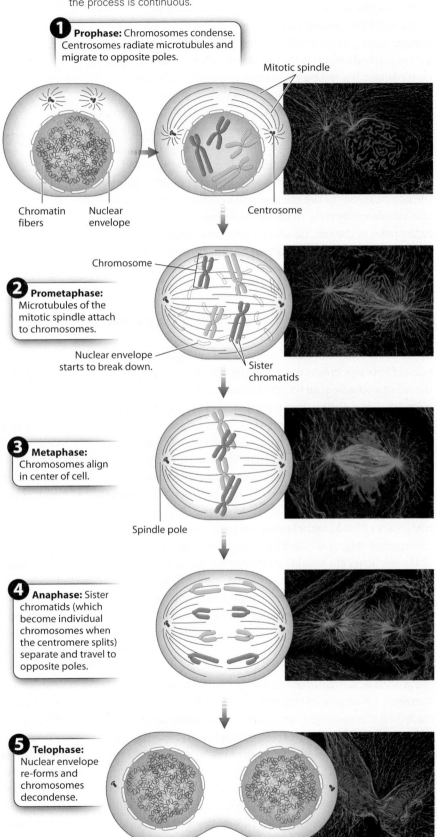

**FIG. 11.5    Mitosis, or nuclear division.** Mitosis can be divided into separate steps, but the process is continuous.

**1 Prophase:** Chromosomes condense. Centrosomes radiate microtubules and migrate to opposite poles.

Mitotic spindle

Chromatin fibers

Nuclear envelope

Centrosome

**2 Prometaphase:** Microtubules of the mitotic spindle attach to chromosomes.

Chromosome

Nuclear envelope starts to break down.

Sister chromatids

**3 Metaphase:** Chromosomes align in center of cell.

Spindle pole

**4 Anaphase:** Sister chromatids (which become individual chromosomes when the centromere splits) separate and travel to opposite poles.

**5 Telophase:** Nuclear envelope re-forms and chromosomes decondense.

as **prophase** and is characterized by the appearance of visible chromosomes.

Outside the nucleus, in the cytosol, the cell begins to assemble the **mitotic spindle,** a structure made up predominantly of microtubules that pull the chromosomes into separate daughter cells. Recall from Chapter 10 that the **centrosome** is a compact structure that is the microtubule organizing center for animal cells. Plant cells also have microtubule-based mitotic spindles, but they lack centrosomes. As part of the preparation for mitosis during S phase in animal cells, the centrosome duplicates and each one begins to migrate around the nucleus, the two ultimately halting at opposite poles in the cell. The final locations of the centrosomes define the opposite ends of the cell that will eventually be separated into two daughter cells. As the centrosomes make their way to the poles of the cell, tubulin dimers assemble around them, forming microtubules that radiate from each centrosome. These radiating filaments form the mitotic spindle and later serve as the guide wires for chromosome movement.

### Prometaphase: Chromosomes attach to the mitotic spindle.

In the next stage of mitosis, known as **prometaphase,** the nuclear envelope breaks down and the microtubules of the mitotic spindle attach to chromosomes (Fig. 11.5). The microtubules radiating from the centrosome poles grow and shrink as they explore the region of the cell where the nucleus once was. This process of growing and shrinking depends on the dynamic instability of microtubules, discussed in Chapter 10.

As the ends of the microtubules encounter chromosomes, they attach to the chromosomes at their centromeres. Associated with the centromere of each chromosome are two protein complexes called **kinetochores,** one located on each side of the constriction (**Fig. 11.6**). Each kinetochore is associated with one of the two sister chromatids and forms the site of attachment for a single microtubule. This arrangement ensures that each sister chromatid is attached to a microtubule radiating from one of the poles of the cell. The symmetrical tethering of each chromosome to the two poles of the cell is essential for proper chromosome segregation.

### Metaphase: Chromosomes align as a result of dynamic changes in the mitotic spindle.

Once each chromosome is attached to both spindle poles, the microtubules of the mitotic spindle lengthen or shorten to move the chromosomes into position in the middle of the cell. There the chromosomes are lined up in a single plane that is roughly equidistant from both of the spindle poles. This stage of mitosis, when the chromosomes are aligned in the middle of the dividing cell, is called **metaphase** (see Fig. 11.5). It is one of the most visually distinctive stages under the microscope.

**FIG. 11.6**    **Kinetochores.** Kinetochores, the sites of spindle microtubule attachment, are located on both sides of the centromere.

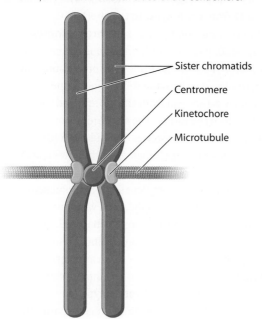

Sister chromatids
Centromere
Kinetochore
Microtubule

### Anaphase: Sister chromatids fully separate.

In the next stage of mitosis, called **anaphase,** the sister chromatids separate (see Fig. 11.5). The centromere holding a pair of sister chromatids together divides, allowing the two sister chromatids to separate from each other. After separation, each chromatid is considered to be a full-fledged chromosome. The microtubules attached to the kinetochores gradually shorten, pulling the newly separated chromosomes to the opposite spindle poles of the cell.

This remarkable event is the reason for the equal segregation of chromosomes between the two daughter cells. In chromosome duplication during S phase in a human cell, each of the 46 chromosomes is duplicated to yield 46 pairs of identical sister chromatids. Thus, when the chromatids are separated at anaphase, an identical set of 46 chromosomes arrives at each spindle pole, the complete genetic material for one of the daughter cells.

### Telophase: Nuclear envelopes re-form around newly segregated chromosomes.

Once a complete set of chromosomes arrives at a spindle pole, the chromosomes have entered the area that will form the cytosol of a new daughter cell. This event marks the beginning of **telophase**, during which cytosolic changes occur in preparation for the cell's division into two new cells (see Fig. 11.5). The microtubules of the mitotic spindle break down and disappear, while nuclear envelopes begin to re-form around each set of chromosomes, creating two new nuclei. As the nuclei become increasingly division; only when they are mutated do they have the potential to cause cancer. Today, we know of scores of proto-oncogenes, most of which were identified through the study of cancer-causing viruses in chickens, mice, and cats.

Oncogenes also play a major role in human cancers. Most human cancers are not caused by viruses. Instead, human proto-oncogenes can be mutated into cancer-causing oncogenes by environmental agents such as chemical pollutants. A well-studied environmental factor linked to cancer is the set of aromatic amines present in the smoke of a burning cigarette. Aromatic amines are organic chemicals that can enter cells and associate with DNA, leading to damage or errors during DNA replication. Such changes result in mutations in DNA that can in some cases convert a normal proto-oncogene into an oncogene.

Not all proto-oncogenes encode protein kinases like Src. Nearly every protein that performs a key step in a signaling cascade that promotes cell division can be the product of a proto-oncogene. In other words, normal proto-oncogenes encoding growth factors, cell-surface receptors, G proteins, and protein kinases can be mutated to become oncogenes. Since the outcome of cell division depends on a complex and interconnected cascade of protein activities, any player in this scheme that does not function properly can lead to uncontrolled cell division.

Let's consider an example of a proto-oncogene that codes for a cell-surface receptor. In Chapter 9, we discussed platelet-derived growth factor, or PDGF, which promotes cell division by binding to and dimerizing a receptor kinase in the membrane of the target cell. This action leads to the downstream activation of several signaling pathways, including the MAP kinase pathway and the promotion of cell division. In one type of leukemia (a cancer of blood cells), a mutation in the PDGF receptor gene results in an altered receptor that is missing the extracellular portion needed to bind the growth factor. Instead, the receptor dimerizes on its own, independent of PDGF binding. As a consequence, the receptor is always turned on.

Just like an overactive Src kinase, the overactive PDGF receptor turns on too many target proteins over too long a time period, leading to uncontrolled proliferation of blood cells. In both cases, mutations that change proto-oncogenes to oncogenes activate the protein inappropriately. That is, the mutant oncogene has more activity than its proto-oncogene counterpart.

**FIG. 11.20** **Multiple-mutation model for the development of cancer.** Cells that become cancerous typically carry mutations in several different genes.

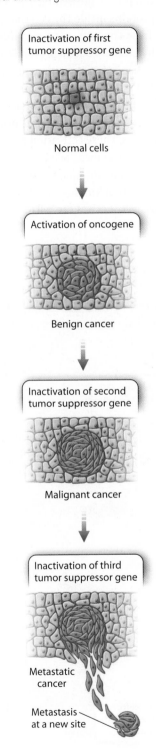

Inactivation of first tumor suppressor gene

Normal cells

Activation of oncogene

Benign cancer

Inactivation of second tumor suppressor gene

Malignant cancer

Inactivation of third tumor suppressor gene

Metastatic cancer

Metastasis at a new site

facilitating the identification of genes involved in cancer. Their investigation therefore provided a major breakthrough in our understanding of cancer.

As discussed in Chapter 9, growth factors activate several types of signaling proteins inside the cell that promote cell division. The gene from the Rous sarcoma virus that promotes uncontrolled cell division encodes a protein kinase similar to the cellular kinases that function as signaling proteins. This viral gene was named *v-src*, for *viral-src* (pronounced "sarc" and short for "sarcoma," the type of cancer it causes).

Protein kinases are enzymes that activate other proteins by adding phosphate groups. These activation events are usually counterbalanced by other enzymes, called phosphatases, that remove phosphate groups. The viral protein encoded by *v-src* is overactive compared to cellular protein kinases. When the *v-src* gene is introduced into a host cell, it drives the cell to divide uncontrollably.

The *v-src* gene is one of several examples of an **oncogene,** or cancer-causing gene, found in viruses. A real surprise was the discovery that the *v-src* oncogene is found not just in the Rous sarcoma virus. It is in fact an altered version of a gene normally found in the host animal cell, known as *c-src* (*cellular-src*). The *c-src* gene plays a role during embryonic development and is involved in the control of cell division.

## Proto-oncogenes are genes that when mutated may cause cancer.

The realization that the *v-src* oncogene has a normal counterpart in the host cell was an important step toward determining the cellular genes that participate in cell growth and division. These normal cellular genes are called **proto-oncogenes** because they are related to viral oncogenes but do not themselves cause cancer. Proto-oncogenes are normal genes important in cell

The extract is injected into a healthy chicken.

The chicken develops a tumor at the site of injection.

The filtered cell-free extract is injected into a healthy chicken.

The chicken develops a tumor at the site of injection.

extract through a filter to remove all cells, including cancer cells and bacterial cells, and then injected this cell-free extract into a healthy chicken.

**RESULTS** In both cases, healthy chickens injected with the extract from the sarcoma developed cancer at the site of injection. Microscopic examination of the cancer showed it to be the same type of cancer as the original one.

**CONCLUSIONS AND FOLLOW-UP WORK** Rous concluded, contrary to his hypothesis, that a small agent—a virus or chemical—is capable of causing cancer. Later experiments confirmed that the cause of the cancer is a virus. This result was surprising, controversial, and dismissed at the time. A second cancer-causing virus was not found until the 1930s. Although most cancers are not caused by viruses, work with cancer-causing viruses helped to identify cellular genes that, when mutant, can lead to cancer. In 1966, Rous shared the Nobel Prize in Physiology or Medicine for his discovery.

**SOURCES** Rous, P. 1910. "A Transmissable Avian Neoplasm (Sarcoma of the Common Fowl)." *J. Exp. Med.* 12:696–705; Rous, P. 1911. "Transmission of a malignant new growth by means of a cell-free filtrate." *JAMA* 56:198.

**? CASE 2 Cancer: When Good Cells Go Bad**

## 11.5 WHAT GENES ARE INVOLVED IN CANCER?

As we have just seen, cells have evolved multiple regulatory mechanisms to ensure that they divide only at the appropriate time and place, and that cell division proceeds without error. What happens when cell division is not properly regulated? One possibility is that a cell divides rapidly, leading to **cancer**, a group of diseases characterized by uncontrolled cell division. When cells inappropriately activate mechanisms that promote cell division or lose the normal checks that prevent cell division, they can divide rapidly.

In this section, we explore the various ways in which cell cycle control can fail and lead to cancer. We begin with a discussion of a cancer caused by a virus. Although most cancers are not caused by viruses, the study of cancers caused by viruses helped us to understand how cancers develop.

### Oncogenes promote cancer.

Our understanding of cancer is based partly on early observations of cancers in animals. In the first decade of the twentieth century, Peyton Rous studied cancers called sarcomas in chickens (**Fig. 11.19**). His work and that of others led to the discovery of the first virus known to cause cancer in animals, named the Rous sarcoma virus.

Viruses are assemblages of protein surrounding a core of either RNA or DNA. They multiply by infecting cells and using the biochemical machinery of their host to synthesize the proteins encoded in their genome and to make more copies of themselves. Viruses typically carry only a handful of genes,

**HOW DO WE KNOW?**

**FIG. 11.19**

# Can a virus cause cancer?

**BACKGROUND** In the early 1900s, little was known about the cause of cancer or the nature of viruses. Peyton Rous, an American pathologist, studied a form of cancer called a sarcoma in chickens. First, he moved a cancer tumor from a diseased chicken to a healthy chicken, and found that the cancer could be transplanted. Then, he tried to isolate the factor that causes the cancer in chickens. He made an extract of the tumor, filtered it to remove all the cells, and injected the extract into a healthy chicken to see if it could induce cancer.

**HYPOTHESIS** Experiments in other organisms, such as mice, rats, and dogs, showed that an extract free of cells from a tumor does not cause cancer. Therefore, Rous hypothesized that a cell-free extract of the chicken sarcoma would not induce cancer in healthy chickens.

**EXPERIMENT** Rous took a sample of the sarcoma from a chicken, ground it up, suspended it in solution, and centrifuged it to remove the debris. In one experiment, he injected this extract into a healthy chicken. In a second experiment, he passed the

A sample of sarcoma is obtained from a chicken.

The tumor is ground up, suspended, and centrifuged to create an extract.

The extract is passed through a filter to remove cells and bacteria.

**FIG. 11.17 Cell cycle checkpoints.** Cell cycle checkpoints monitor key steps in the cell cycle.

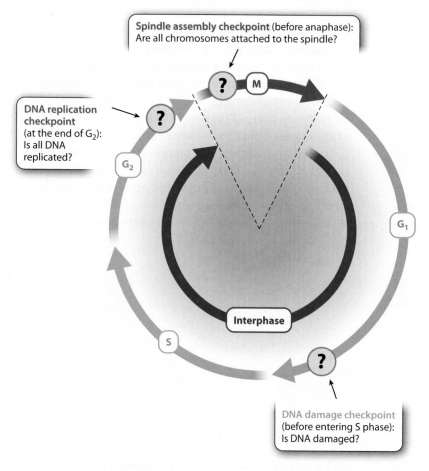

**Spindle assembly checkpoint** (before anaphase): Are all chromosomes attached to the spindle?

**DNA replication checkpoint** (at the end of $G_2$): Is all DNA replicated?

Interphase

**DNA damage checkpoint** (before entering S phase): Is DNA damaged?

DNA. If the cell progresses through mitosis with DNA damage, the damage might be inherited by the daughter cells or the chromosomes might not segregate normally. Some checkpoints delay progression through the cell cycle until DNA damage is repaired. An important one occurs in late $G_1$. This DNA damage checkpoint depends on several regulatory proteins, some of which recognize damaged DNA while others arrest cell cycle progression before S phase (Fig. 11.17).

When DNA is damaged by radiation, a specific protein kinase is activated that phosphorylates a protein called **p53.** Normally, p53 is exported from the nucleus and therefore is at very low levels in the nucleus. Phosphorylation of p53 prevents its export and thereby increases its levels in the nucleus. As p53 levels rise, p53 activates the transcription of a gene that expresses a CDK inhibitor protein. This inhibitor binds to and blocks the activity of the $G_1$/S cyclin–CDK complex (**Fig. 11.18**). In so doing, p53 arrests the cell at the $G_1$/S transition, giving the cell time to repair the damaged DNA.

→ **Quick Check 7** Can you think of two ways in which the function of p53 can be disrupted?

**FIG. 11.18 DNA damage checkpoint mediated by p53.**

*Genes that inhibit the cell cycle are OFF.*

*Genes that inhibit the cell cycle are OFF.*

*Genes that inhibit the cell cycle are ON.*

Nucleus

p53

Cytoplasm

Normally p53 in the nucleus is not phosphorylated and is rapidly exported from the nucleus and degraded.

DNA damage

P

p53

DNA damage activates protein kinases that phosphorylate p53, blocking the export and degradation process.

Gene turned ON

As p53 levels in the nucleus increase, p53 acts as a transcription factor to turn on genes that inhibit the cell cycle.

Inhibiting the cell cycle gives the cell time to repair the damaged DNA.

## Different cyclin–CDK complexes regulate each stage of the cell cycle.

In mammals, there are several different cyclins and CDKs that act at specific steps of the cell cycle. Three steps in particular are subject to cyclin–CDK regulation in all eukaryotes (**Fig. 11.16**).

The $G_1$/S cyclin–CDK complex, which is active at the end of the $G_1$ phase, is necessary for the cell to enter S phase. For example, this cyclin–CDK complex activates a protein that promotes the expression of histone proteins needed for packaging the newly replicated DNA strands.

The S cyclin–CDK complex is involved in the initiation of DNA synthesis. It activates protein complexes involved in DNA replication that contain enzymes necessary for DNA synthesis.

Once replication has begun at a particular place on the DNA, S cyclin–CDK activity prevents the replication complex from reassembling at the same place and re-replicating the same DNA sequence. Synthesizing the same sequence repeatedly would be dangerous because cells with too much or not enough DNA could die or become cancerous.

The M cyclin–CDK complex initiates multiple events associated with mitosis. For example, M cyclin–CDK phosphorylates structural proteins in the nucleus, triggering the breakdown of the nuclear envelope in prophase. M cyclin–CDK also phosphorylates proteins that regulate the assembly of tubulin into microtubules, promoting the formation of the mitotic spindle.

**FIG. 11.16** Three cyclin–CDK complexes active at different times in the cell cycle.

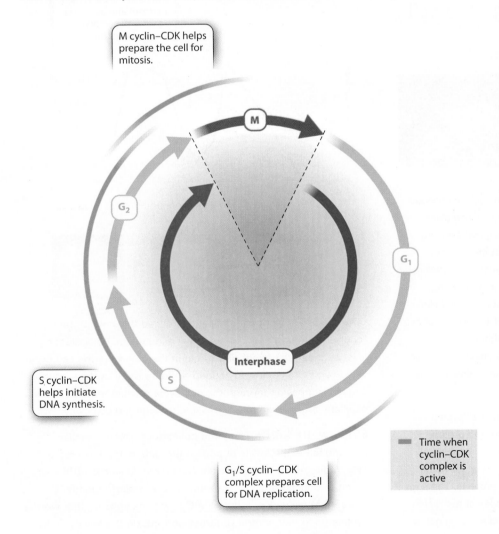

M cyclin–CDK helps prepare the cell for mitosis.

S cyclin–CDK helps initiate DNA synthesis.

$G_1$/S cyclin–CDK complex prepares cell for DNA replication.

Interphase

M

$G_2$

$G_1$

S

Time when cyclin–CDK complex is active

## Cell cycle progression requires successful passage through multiple checkpoints.

The regulatory system represented by cyclins and CDKs gives the cell opportunities to halt the cell cycle should something go wrong. In other words, if the preparations for the next stage of the cell cycle are incomplete or if there is some kind of damage, regulatory mechanisms block the cyclin–CDK activity required for the next step, pausing cell division until preparations are complete or the damage is repaired. Each regulatory mechanism is called a **checkpoint.**

Cells have many cell cycle checkpoints. Three major checkpoints that have been well studied are illustrated in **Fig. 11.17.** The presence of unreplicated DNA arrests the cell at the end $G_2$ before the cell enters mitosis, and abnormalities in chromosome attachment to the spindle arrest the cell in early mitosis. By way of illustration, we focus on a key checkpoint that responds to the presence of damaged DNA that occurs at the end of $G_1$.

DNA can be damaged by environmental insults such as ultraviolet radiation or chemical agents. Typically, damage takes the form of double-stranded breaks in the proteins that phosphorylate other proteins (Chapter 9). The timing of kinase activity is delayed slightly relative to the appearance of the cyclical proteins.

These and many other observations led to the following view of cell cycle control: Proteins are synthesized that activate the kinases. These regulatory proteins are called **cyclins** because their levels rise and fall with each turn of the cell cycle. Once activated by cyclins, the kinases phosphorylate target proteins involved in promoting cell division (**Fig. 11.14**). These kinases, **cyclin-dependent kinases,** or **CDKs,** are always present within the cell but are active only when bound to the appropriate cyclin. It is the kinase activity of the cyclin–CDK complexes that triggers the required cell cycle events. Therefore, the cyclical change in cyclin–CDK activity depends on the cyclical levels of the cyclins.

These cell cycle control proteins, including cyclins and CDKs, are widely conserved across eukaryotes, reflecting their fundamental role in controlling cell cycle progression, and have been extensively studied in yeast, sea urchins, mice, and humans (**Fig. 11.15**).

---

## HOW DO WE KNOW?

**FIG. 11.15**

# How is progression through the cell cycle controlled?

**BACKGROUND** The cell cycle is characterized by cyclical changes in many components of the cell: Chromosomes condense and decondense, the mitotic spindle forms and breaks down, the nuclear envelope

breaks down and re-forms. How these regular and cyclical changes are controlled is the subject of active research. First, clues came from studies in yeast in which mutations in certain genes blocked progression through the cell cycle, suggesting that these genes encode proteins that play a role in cell cycle progression. Additional clues came from studies in embryos of the sea urchin *Arbacia punctulata*. These embryos are large and divide rapidly by mitosis, making them a good model system for the study of the control of cell division. In the early 1980s, it was known that inhibition of protein synthesis blocks key steps of cell division in sea urchins. It was also known that an enzyme called MPF, or M-phase promotion factor, is important for the transition from $G_2$ to M phase.

**EXPERIMENT** To better understand events in the cell cycle, English biochemist Tim Hunt and colleagues measured protein levels in sea urchin embryos as they divide rapidly by mitosis. He added radioactive methionine (an amino acid) to eggs, which became incorporated into any newly synthesized proteins. The eggs were then fertilized and allowed to develop. Samples of the rapidly dividing embryos were taken every 10 minutes and run on a gel to visualize the levels of different proteins.

**RESULTS** Most protein bands became darker as cell division proceeded, indicating more and more protein synthesis. However, the level of one protein band oscillated, increasing in intensity and then decreasing with each cell cycle, as shown in the graph.

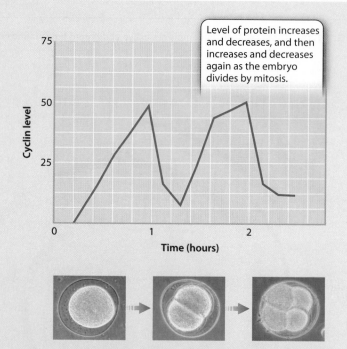

Level of protein increases and decreases, and then increases and decreases again as the embryo divides by mitosis.

**CONCLUSION** Hunt and colleagues called this new protein "cyclin." Although they did not know its function, its fluctuating level suggested it might play a role in the control of the cell cycle. However, more work was needed to figure out if cyclins actually cause progression through the cell cycle or whether their levels oscillate in response to progression through the cell cycle.

**FOLLOW-UP WORK** Hunt and colleagues found a similar rise and fall in the levels of certain proteins in the sea urchin *Lytechinus pictus* and surf clam *Spisula solidissima*. MPF was found to consist of a cyclin protein and a cyclin-dependent kinase that play a key role in the $G_2$–M transition. In other words, cyclins do in fact control progression through the cell cycle by their interactions with cyclin-dependent kinases. Hunt shared the Nobel Prize in Physiology or Medicine in 2001 for his work on cyclins.

**SOURCE** T. Evans, E. T. Rosenthal, J. Youngblom, D. Distel, and T. Hunt. 1983. "Cyclin: A Protein Specified by Maternal mRNA in Sea Urchin Eggs That Is Destroyed at Each Cleavage Division." *Cell* 33:389–396.

During fertilization, these gametes fuse to form a single cell called a **zygote.** The zygote is diploid, having two complete sets of chromosomes, one from each parent. Therefore, fertilization restores the original chromosome number.

As we discuss further in Chapters 16 and 42, sexual reproduction plays a key role in increasing genetic diversity. Genetic diversity results from meiotic cell division (the cells that are produced are each genetically different from one another), and from fertilization (different gametes are combined to produce a new, unique individual). The increase in genetic diversity made possible by sexual reproduction allows organisms to evolve and adapt more quickly to their environment than is possible with asexual reproduction.

In the life cycle of multicellular animals like humans, then, the diploid organism produces single-celled gametes that fuse to make a diploid zygote. In this case, the only haploid cells in the life cycle are the gametes, and the products of meiotic cell division do not undergo mitotic cell division but instead fuse to become a diploid zygote. However, there are a number of life cycles in other organisms that differ in the timing of meiotic cell division and fertilization, discussed more fully in Chapter 27. Some organisms, like most fungi, are haploid. These haploid cells can fuse to produce a diploid zygote, but this cell immediately undergoes meiotic cell division to produce haploid cells, so that the only diploid cell in the life cycle is the zygote (Chapter 34). Other organisms, like plants, have both multicellular haploid and diploid phases (Chapter 30). In this case, meiotic cell division produces haploid spores that divide by mitotic cell division to produce a multicellular haploid phase, and subsequently haploid cells fuse to form a diploid zygote that also divides by mitotic cell division to produce a multicellular diploid phase.

## 11.4 REGULATION OF THE CELL CYCLE

Both mitotic and meiotic cell division must occur only at certain times and places. Mitotic cell division, for example, occurs during growth of a multicellular organism, wound healing, or in the maintenance of actively dividing tissues such as the skin or lining of the intestine. Similarly, meiotic cell division occurs only at certain times during development. Cell division cannot occur all the time because uncontrolled cell proliferation is dangerous—in fact, it is a hallmark of cancer. Even for unicellular organisms, cell division needs to be regulated so that it takes place only when conditions are favorable—for example, when enough nutrients are present in the environment.

Cells do not proceed to cell division until they are ready. Has all of the DNA been accurately replicated during S phase? Has the cell grown to a sufficient size to support division into viable daughter cells? If these and other preparations have not been accomplished, the cell halts its progression through the cell cycle. The cell has regulatory mechanisms for spotting faulty or incomplete preparations and arresting cell division. When these mechanisms fail, the resulting daughter cells are abnormal, and either may not

survive or may run the risk of dividing inappropriately. In this section, we consider how cells control their passage through the cell cycle. Our focus is on control of mitotic cell division. However, many of the same factors also regulate meiotic cell division, a reminder of the close evolutionary connection between these two forms of cell division.

### Protein phosphorylation controls passage through the cell cycle.

Early animal embryos, such as those of frogs and sea urchins, are useful models for studying cell cycle control because they are large and undergo many rapid mitotic cell divisions following fertilization. During these rapid cell divisions, mitosis and S phase alternate with virtually no $G_1$ and $G_2$ phases in between. Studies of animal embryos revealed two interesting patterns. First, as the cells undergo this rapid series of divisions, several proteins appear and disappear in a cyclical fashion. Researchers interpreted this observation to mean that they might play a role in the control of the progression through the cell cycle. Second, several enzymes become active and inactive in cycles. These enzymes are kinases,

**FIG. 11.14 Cyclins and cyclin-dependent kinases (CDKs).** Cyclins and CDKs control progression through the cell cycle.

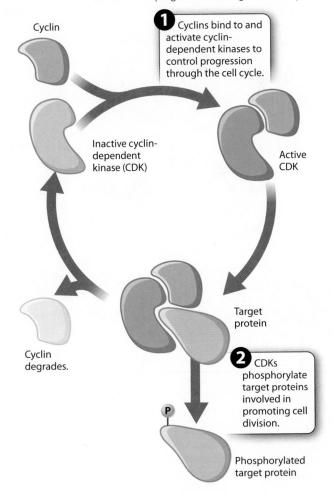

Cyclin

**1** Cyclins bind to and activate cyclin-dependent kinases to control progression through the cell cycle.

Inactive cyclin-dependent kinase (CDK)

Active CDK

Cyclin degrades.

Target protein

**2** CDKs phosphorylate target proteins involved in promoting cell division.

P

Phosphorylated target protein

mammals (**Fig. 11.13a**), the cytoplasm is divided very unequally in both meiotic divisions. Most of the cytoplasm is retained in one meiotic product, a very large cell called the oocyte, which can develop into the functional egg cell, and the other meiotic products receive only small amounts of cytoplasm. These smaller cells are called **polar bodies.** In male mammals (**Fig. 11.13b**), the cytoplasm divides about equally in both meiotic divisions, and each of the resulting meiotic products goes on to form a functional sperm. During the development of the sperm, most of the cytoplasm is eliminated, and what is left is essentially a nucleus in the sperm head equipped with a long whiplike flagellum to help propel it toward the egg.

## Meiosis is the basis of sexual reproduction.

Sexual reproduction involves two processes: meiotic cell division and fertilization. Meiotic cell division, as we just saw, produces cells with half the number of chromosomes present in the parent cell. In multicellular animals, the products of meiotic cell division are gametes: An egg cell is a gamete and a sperm cell is a gamete. Each gamete is haploid, containing a single set of chromosomes. In humans, meiosis takes place in the ovaries of the female and the testes of the male, and each resulting gamete contains 23 chromosomes, including one each of the 22 numbered chromosomes plus either an X or a Y chromosome.

**FIG. 11.13** **Cytoplasmic division in (a) females and (b) males.** In females, cytoplasmic division results in one oocyte and three polar bodies, and in males it results in four sperm cells.

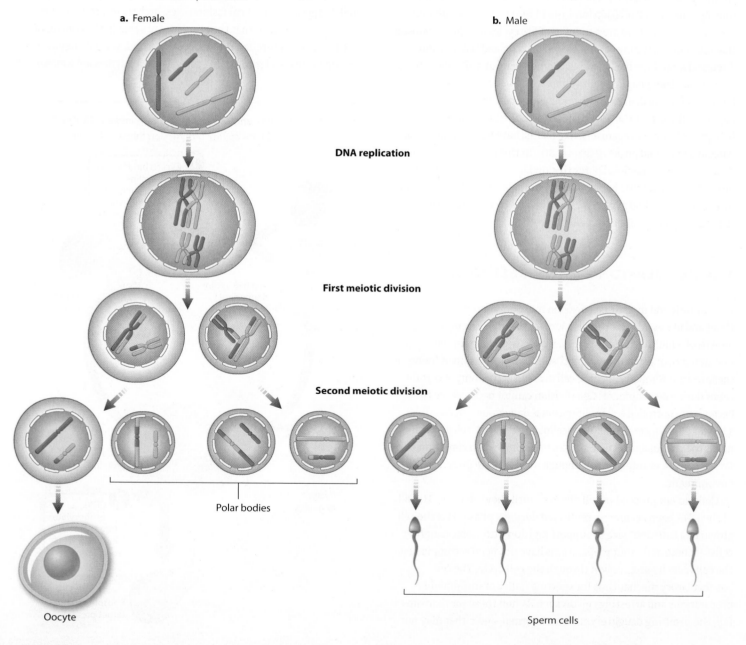

a. Female      b. Male

DNA replication

First meiotic division

Second meiotic division

Polar bodies

Oocyte      Sperm cells

**FIG. 11.12**  Comparison of mitosis and meiosis.

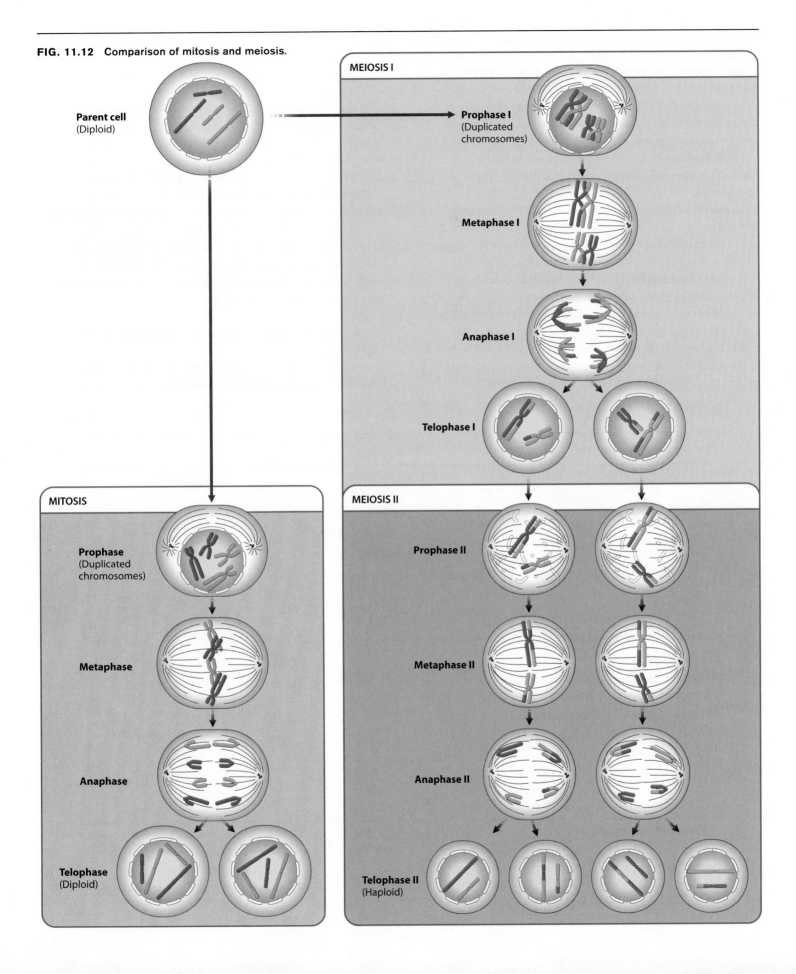

division is in many respects like a conventional mitotic division, except that the nuclei in prophase II have the haploid number of chromosomes (23 in humans), not the diploid number. In prophase II, the chromosomes recondense to their maximum extent. Toward the end of prophase II, the nuclear envelope disappears (in those species in which it has formed), and the spindle begins to be set up.

In **metaphase II,** the chromosomes line up so that their centromeres lie on an imaginary plane cutting across the spindle.

In **anaphase II,** the centromere of each chromosome splits. The separated chromatids, now each regarded as a full-fledged chromosome, are pulled toward opposite poles of the spindle. In this sense, anaphase II resembles anaphase of mitosis.

Finally, in **telophase II,** the chromosomes uncoil and become diffuse, a nuclear envelope re-forms around each set of chromosomes, and the cytoplasm divides by cytokinesis. Each resulting nucleus after telophase II has the haploid number of chromosomes. Because cells in meiosis II have the same number of chromosomes at the beginning and at the end, meiosis II is often called the **equational division.**

→ **Quick Check 6** The genetic constitution of each cell after telophase II is different from the others. What two processes during meiosis result in these differences?

A comparison of mitosis and meiosis gives us hints about how meiosis might have evolved (**Table 11.1** and **Fig. 11.12**). During meiosis I, maternal and paternal homologs separate from each other, whereas during meiosis II, sister chromatids separate from each other, similar to mitosis. The similarity of meiosis II and mitosis suggests that meiosis likely evolved from mitosis. Mitosis occurs in all eukaryotes and was certainly present in the common ancestor of all living eukaryotes. Meiosis is present in most, but not all, eukaryotes. Because the steps of meiosis are the same in all eukaryotes, meiosis is thought to have evolved in the common ancestor of all eukaryotes and has been subsequently lost in some lineages.

### Division of the cytoplasm often differs between the sexes.

In multicellular organisms, division of the cytoplasm in meiotic cell division differs between the sexes. In female

**TABLE 11.1**  Comparison of Mitosis and Meiosis

|  | MITOSIS | MEIOSIS |
|---|---|---|
| Function | Asexual reproduction in unicellular eukaryotes<br>Development in multicellular eukaryotes<br>Tissue regeneration and repair in multicellular eukaryotes | Sexual reproduction<br>Production of gametes and spores |
| Organisms | All eukaryotes | Most eukaryotes |
| Number of rounds of DNA synthesis | 1 | 1 |
| Number of cell divisions | 1 | 2 |
| Number of daughter cells | 2 | 4 |
| Chromosome complement of daughter cell compared to parent cell | Same | Half |
| Pairing of homologous chromosomes | No | Meiosis I—Yes<br>Meiosis II—No |
| Crossing over | No | Meiosis I—Yes<br>Meiosis II—No |
| Separation of homologous chromosomes | No | Meiosis I—Yes<br>Meiosis II—No |
| Centromere division | Yes | Meiosis I—No<br>Meiosis II—Yes |
| Separation of sister chromatids | Yes | Meiosis I—No<br>Meiosis II—Yes |

At the beginning of **anaphase I,** the two homologous chromosomes of each bivalent separate as they are pulled in opposite directions. The key feature of anaphase I is that the centromeres do not split, and the two chromatids that make up each chromosome remain together. Anaphase I is thus very different from anaphase of mitosis, in which the centromeres split and each pair of chromatids becomes separated.

The end of anaphase I coincides with the arrival of the chromosomes at the poles of each spindle. Only one of the two homologous chromosomes goes to each pole, so in human cells there are 23 chromosomes at each pole at the end of meiosis I. Each of these chromosomes consists of two chromatids attached to a single centromere. Meiosis I is sometimes called the **reductional division,** since it reduces the number of chromosomes by half.

In **telophase I,** the chromosomes may uncoil slightly, a nuclear envelope briefly reappears, and in many species (including humans) the cytoplasm divides, producing two separate cells. The chromosomes do not completely decondense, however, and so telophase I blends into prophase II. Importantly, there is no DNA synthesis between the two meiotic divisions.

→   **Quick Check 5** List three ways in which meiosis I differs from mitosis.

## The second meiotic division resembles mitosis.

Now let's turn to the second meiotic division, meiosis II, shown in **Fig. 11.11.** Starting with **prophase II,** the second meiotic

---

**FIG. 11.11   Meiosis II.** Meiosis II is the equational division: The number of chromosomes stays the same.

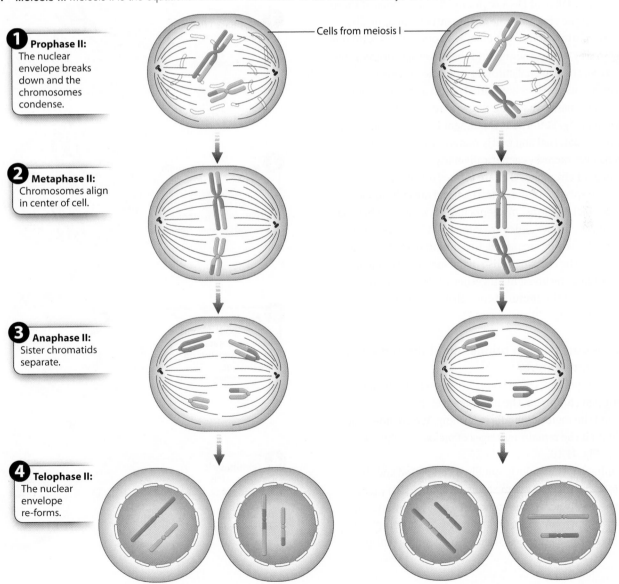

Cells from meiosis I

**1 Prophase II:** The nuclear envelope breaks down and the chromosomes condense.

**2 Metaphase II:** Chromosomes align in center of cell.

**3 Anaphase II:** Sister chromatids separate.

**4 Telophase II:** The nuclear envelope re-forms.

**FIG. 11.9** **Chiasmata.** Crossing over between non-sister chromatids results in recombinant chromatids.

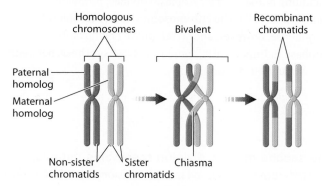

Through the process of crossing over, meiosis allows homologous chromosomes of maternal origin and paternal origin to undergo an exchange of DNA segments. The positions of these exchanges along the chromosome are essentially random, and therefore each chromosome that emerges from meiosis is unique, containing some DNA segments from the maternal chromosome and others from the paternal chromosome. The process is very precise: No nucleotides are gained or lost as homologous chromosomes exchange material. Note the results of crossing over as shown in Fig. 11.9: The recombinant chromatids are those that carry partly paternal and partly maternal segments. In this way, crossing over increases genetic diversity.

The number of chiasmata that are formed during meiosis depends on the species. In humans, the usual range is 50–60 chiasmata per meiosis. Most bivalents have at least one chiasma. Even the X and Y chromosomes are joined by a chiasma in the small region where they are paired. In addition to exchanging genetic material, the chiasmata also play a mechanical role in meiosis by holding the bivalents together while they become properly oriented in the center of the cell during metaphase, the stage we turn to next.

### The first meiotic division brings about the reduction in chromosome number.

At the end of prophase I, the chromosomes are fully condensed and have formed chiasmata, the nuclear envelope has begun to disappear, and the meiotic spindle is forming. We are now ready to move through the remaining stages of meiosis I, which are illustrated in **Fig. 11.10.**

In **metaphase I,** the meiotic spindle is completed and the bivalents move so that they come to lie on an imaginary plane cutting transversely across the spindle. Each bivalent lines up so that its two centromeres lie on opposite sides of this plane, pointing toward opposite poles of the spindle. Importantly,

the orientation of these bivalents is random with respect to each other. For some, the maternal homolog is attached to the spindle radiating from a given pole and the paternal homolog is attached to the spindle originating from the other pole. For others, the orientation is reversed. As a result, when the homologous chromosomes separate from each other in the next step (anaphase I), a complete set of chromosomes moves toward each pole, and that chromosome set is a random mix of maternal and paternal homologs. The random alignment of chromosomes on the spindle in metaphase I further increases genetic diversity in the products of meiosis.

**FIG. 11.10** **Meiosis I.** Meiosis I is the reductional division: The number of chromosomes is halved.

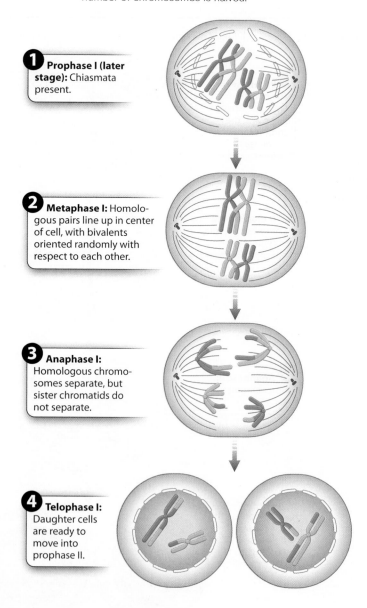

**1** **Prophase I (later stage):** Chiasmata present.

**2** **Metaphase I:** Homologous pairs line up in center of cell, with bivalents oriented randomly with respect to each other.

**3** **Anaphase I:** Homologous chromosomes separate, but sister chromatids do not separate.

**4** **Telophase I:** Daughter cells are ready to move into prophase II.

section, we consider the steps by which meiosis occurs, its role in sexual reproduction, and how it likely evolved.

## Pairing of homologous chromosomes is unique to meiosis.

Like mitotic cell division, meiotic cell division follows one round of DNA synthesis, but, unlike mitotic cell division, meiotic cell division consists of two successive cell divisions. The two cell divisions are called **meiosis I** and **meiosis II,** and they occur one after the other. Each cell division results in two cells, so that by the end of meiotic cell division a single parent cell has produced four daughter cells.

Meiosis I begins with **prophase I,** illustrated in **Fig. 11.8.** The beginning of prophase I marks the earliest visible manifestation of chromosome condensation. The chromosomes first appear as long, thin threads present throughout the nucleus. By this time, DNA replication has already taken place, so each chromosome has become two sister chromatids held together at the centromere.

What happens next is an event of enormous importance and is unique to meiosis. The homologous chromosomes pair with each other, coming together to lie side by side, gene for gene, in a process known as **synapsis.** Even the X and Y chromosomes pair, but only at the tip where their DNA sequences are nearly identical. Because one of each pair of homologs is maternal in origin and the other is paternal in origin, chromosome pairing provides an opportunity for the maternal and paternal chromosomes to exchange genetic information, as described in the next section.

Because each homologous chromosome is a pair of sister chromatids attached to a single centromere, a pair of synapsed chromosomes creates a four-stranded structure: two pairs of sister chromatids aligned along their length. The whole unit is called a **bivalent,** and the chromatids attached to different centromeres are called **non-sister chromatids,** in contrast to sister chromatids, which result from replication of a single chromosome. The important distinction between sister chromatids and non-sister chromatids is that sister chromatids, as replicas of the same chromosome, are genetically identical; non-sister chromatids, which are replicas of different chromosomes, are genetically similar but not identical.

→ **Quick Check 4** In a human cell at the end of prophase I, how many chromatids, centromeres, and bivalents are present?

## Crossing over between DNA molecules results in exchange of genetic material.

Within the bivalents are crosslike structures, each called a **chiasma** (from the Greek meaning a "cross piece"; the plural is "chiasmata") (**Fig. 11.9**). Each chiasma is a visible manifestation of a **crossover,** the physical breakage and reunion between non-sister chromatids.

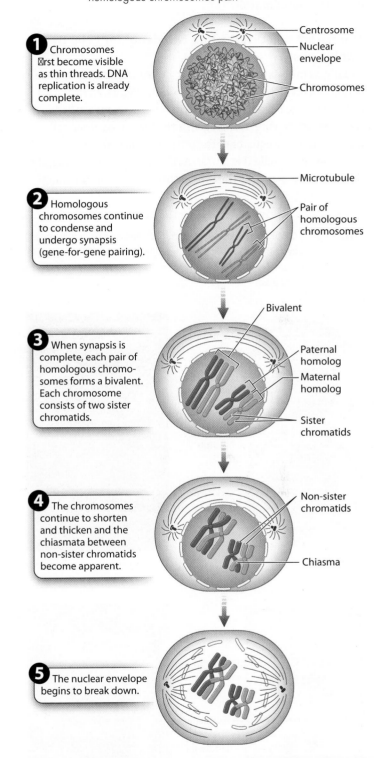

**FIG. 11.8** **Prophase I of meiosis.** Chromosomes condense and homologous chromosomes pair.

**1** Chromosomes first become visible as thin threads. DNA replication is already complete.
- Centrosome
- Nuclear envelope
- Chromosomes

**2** Homologous chromosomes continue to condense and undergo synapsis (gene-for-gene pairing).
- Microtubule
- Pair of homologous chromosomes

**3** When synapsis is complete, each pair of homologous chromosomes forms a bivalent. Each chromosome consists of two sister chromatids.
- Bivalent
- Paternal homolog
- Maternal homolog
- Sister chromatids

**4** The chromosomes continue to shorten and thicken and the chiasmata between non-sister chromatids become apparent.
- Non-sister chromatids
- Chiasma

**5** The nuclear envelope begins to break down.

distinct in the cell, the chromosomes contained within them decondense, becoming less visible in the microscope. This stage marks the end of mitosis.

### The parent cell divides into two daughter cells by cytokinesis.

Usually, as mitosis is nearing its end, cytokinesis begins and the parent cell divides into two daughter cells (**Fig. 11.7**). In animal cells, this stage begins when a ring of actin filaments, called the **contractile ring**, forms against the inner face of the cell membrane at the equator of the cell perpendicular to the axis of what was the spindle (Fig. 11.7a). As if pulled by a drawstring, the ring contracts, pinching the cytoplasm of the cell and dividing

it in two. This process is similar to what occurs in binary fission, though in the case of binary fission, the process is driven by FtsZ protein, a homolog of tubulin, not by actin as in eukaryotic cells. The constriction of the contractile ring is driven by motor proteins that slide bundles of actin filaments in opposite directions. Successful division results in two daughter cells, each with its own nucleus. The daughter cells are now free to enter $G_1$ phase and start the process anew.

For the most part, mitosis is similar in animal and plant cells, but cytokinesis is different (Fig. 11.7b). Since plant cells have a cell wall, the division of the cell is achieved by constructing a new cell wall. During telophase, dividing plant cells form a structure called the **phragmoplast** in the middle of the cell. The phragmoplast consists of overlapping microtubules that guide vesicles containing cell wall components to the middle of the cell. During late anaphase and telophase, these vesicles fuse to form a new cell wall, called the **cell plate,** in the middle of the dividing cell. Once this developing cell wall is large enough, it fuses with the original cell wall at the perimeter of the cell. Cytokinesis is then complete and the plant cell has divided into two daughter cells.

→ **Quick Check 3** What would be the consequence if a cell underwent mitosis but not cytokinesis?

**FIG. 11.7 Cytokinesis, or cytoplasmic division.** (a) In animal cells, cytokinesis involves a contractile ring made of actin. (b) In plant cells, it involves the growth of a new cell wall called a cell plate.

**a.**

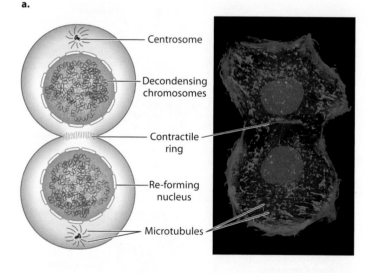

- Centrosome
- Decondensing chromosomes
- Contractile ring
- Re-forming nucleus
- Microtubules

**b.**

- Decondensing chromosomes
- Cell plate
- Re-forming nucleus

## 11.3 MEIOTIC CELL DIVISION

As we have seen, mitotic cell division is important in the development of a multicellular organism and in the maintenance and repair of tissues and organs. Mitotic cell division is also the basis of asexual reproduction in unicellular eukaryotes. Sexual reproduction involves the fusion of gametes during fertilization to form a new organism. A new organism produced by sexual reproduction has the same number of chromosomes as its parents because the egg and sperm each have half the number of chromosomes as the diploid parents.

Gametes are produced by **meiotic cell division,** a form of cell division that includes two rounds of nuclear division. Meiotic cell division makes sexual reproduction possible. There are several major differences between meiotic cell division and mitotic cell division. First, meiotic cell division results in four daughter cells instead of two. Second, each of the four daughter cells contains half the number of chromosomes as the parent cell. (The word "meiosis" is from the Greek for "diminish" or "lessen.") Third, the four daughter cells are each genetically unique. In other words, they are genetically different from each other and from the parental cell.

In multicellular animals, the cells produced by meiosis are the haploid eggs and sperm that fuse in sexual reproduction. In other organisms, such as fungi, the products are spores, and in some unicellular eukaryotes, the products are new organisms. In this

## Tumor suppressors block specific steps in the development of cancer.

From our discussion so far, you might think that oncogenes are solely responsible for the rampant cell growth that leads to cancer. This is true in some cases such as *v-src* in chickens, but more often something else must also go wrong before cancer can occur. Cells have safeguards that must be overcome before the controls on cell division are totally removed. These safeguards are a family of genes collectively called **tumor suppressors** because they encode proteins whose normal activities inhibit cell division.

Tumor suppressors prevent cell division by acting in opposition to proto-oncogenes. Therefore, whether a cell divides or not depends on the activities of both proto-oncogenes and tumor suppressors: Proto-oncogenes must be turned on and tumor suppressors must be turned off. Given the importance of controlling cell division, it is not surprising that cells have two counterbalancing systems that must be in agreement before division takes place.

We have already introduced a well-known example of a tumor suppressor, namely *p53*, which is defective in a majority of human cancers. As discussed earlier in this chapter and in Case 2: Cancer, the p53 protein normally halts cell division in response to DNA damage. When p53 activity is missing from the cell, excessive cell division is promoted.

→ **Quick Check 8** How do oncogenes differ from tumor suppressor genes?

## Most cancers require the accumulation of multiple mutations.

Most human cancers require more than the overactivation of one oncogene or the inactivation of a single tumor suppressor. Given the multitude of different tumor suppressor proteins that are produced in the cell, it is likely that one will compensate for even the complete loss of another. Cells have evolved a redundant arrangement of these control mechanisms to ensure that cell division is properly regulated.

When several different cell cycle regulators fail, leading to both the overactivation of oncogenes and the loss of tumor suppressor activity, it is likely that cancer will develop. The cancer may be benign, which means that it is relatively slow growing and does not invade the surrounding tissue, or it may be malignant, which means that it grows rapidly and invades surrounding tissues. In many cases of malignant colon cancer, for example, tumor cells contain at least one overactive oncogene and several inactive tumor suppressor genes (**Fig. 11.20**). The gradual accumulation of these mutations over a period of years can be correlated with the stepwise progression of the cancer from a benign form to full malignancy.

Taken together, we can now define some of the key characteristics that make a cell cancerous. Uncontrolled cell division is certainly important, but given the communities of cells and extracellular matrix we have discussed over the last few chapters, additional characteristics should also be considered. In 2000, American biologists Douglas Hanahan and Robert Weinberg wrote a paper called "Hallmarks of Cancer," in which they highlighted key features of cancer cells. These include the ability to divide on their own in the absence of growth signals; resistance to signals that inhibit cell division and cell death; the ability to invade local and distant tissues (metastasis); and the production of signals to promote new blood vessel growth for nutrients to support cell division. Considered in this light, a cancer cell is one that no longer plays by the rules of a normal cellular community. Cancer therefore serves to remind us of the normal controls and processes that are required to allow cells to exist in a community. These processes, which work together, are summarized in **Fig. 11.21** on the following page.

# Cellular Communities

**Integrating concepts from Chapters 9–11**

## Cell division

Mitotic cell division is the process by which a single cell divides into two cells that, except for rare mutations, are genetically identical to the parent cell.

**1 Prophase:** Chromosomes condense. Centrosomes radiate microtubules and migrate to opposite poles.

**2 Metaphase:** Chromosomes align in center of cell.

**3 Anaphase:** Sister chromatids (individual chromosomes when the centromere splits) separate and travel to opposite poles.

**4 Telophase and cytokinesis:** Nuclear envelope re-forms, chromosomes condense, and the cytoplasm divides.

M phase (Mitosis and cytokinesis)

G₂ phase (Gap 2)

G₁ phase (Gap 1)

Interphase

S phase (DNA synthesis)

Epidermis

Dermis

## Cell signaling

Cells communicate with each other and respond to their environment using signaling molecules that bind to receptors, leading to responses inside the cell.

Termination

## Cell adhesion

Tissues are communities of cells that are held together and communicate by various types of cellular junctions. Other junctions connect cells with protein networks in the extracellular environment.

Response

Signal transduction and amplification

Receptor activation

Signaling molecules

Tight junction

Adherens junction

Desmosome

Hemidesmosome

Gap junction

Basal lamina

Connective tissue

Blood vessel

# Core Concepts Summary

## 11.1 DURING CELL DIVISION, A SINGLE PARENTAL CELL DIVIDES INTO TWO DAUGHTER CELLS.

Prokaryotic cells divide by binary fission, in which a cell replicates its DNA, segregates its DNA, and divides into two cells. page 11-2

Eukaryotic cells divide by mitosis (nuclear division) and cytokinesis (cytoplasmic division). Together, mitosis and cytokinesis are known as mitotic cell division. page 11-3

M phase (mitosis and cytokinesis) alternates with interphase, which consists of $G_1$, S (synthesis), and $G_2$ phases. These four stages together constitute the cell cycle. page 11-3

Cells that do not need to divide exit the cell cycle and are in $G_0$. page 11-4

## 11.2 MITOTIC CELL DIVISION IS THE BASIS OF ASEXUAL REPRODUCTION IN UNICELLULAR EUKARYOTES AND THE PROCESS BY WHICH CELLS DIVIDE IN MULTICELLULAR EUKARYOTES.

DNA in a eukaryotic cell is packaged as linear chromosomes. page 11-4

Humans have 46 chromosomes: 22 pairs of homologous chromosomes and one pair of sex chromosomes. Each parent contributes one complete set of 23 chromosomes at fertilization. page 11-5

During S phase, chromosomes replicate, resulting in the formation of sister chromatids held together at the centromere. page 11-5

Mitosis involves five steps following DNA replication: (1) prophase—the chromosomes condense and become visible under the light microscope; (2) prometaphase— the microtubule spindles attach to the centromeres; (3) metaphase—the chromosomes line up in the middle of the cell; (4) anaphase—the centromeres divide and the chromosomes move to opposite poles; and (5) telophase—the nuclear envelope re-forms and chromosomes decondense. page 11-5

Mitosis is followed by cytokinesis, in which one cell divides into two. In animals, a contractile ring of actin pinches the cell in two. In plants, a new cell wall, called the cell plate, is synthesized between the daughter cells. page 11-7

## 11.3 MEIOTIC CELL DIVISION IS ESSENTIAL FOR SEXUAL REPRODUCTION, THE PRODUCTION OF OFFSPRING THAT COMBINES GENETIC MATERIAL FROM TWO PARENTS.

Sexual reproduction involves meiosis and fertilization, both of which are important in increasing genetic diversity. page 11-7

Meiotic cell division is a form of cell division that reduces the number of chromosomes by half to produce haploid gametes or spores that have one copy of each chromosome. page 11-7

Fertilization involves the fusion of haploid gametes to produce a diploid cell. page 11-7

Meiosis consists of two successive cell divisions: The first is reductional (the chromosome number is halved), and the second is equational (the chromosome number stays the same). Each division consists of prophase, metaphase, anaphase and telophase. page 11-8

In meiosis I, homologous chromosomes pair and exchange genetic material at chiasmata, or regions of crossing over. In contrast to mitosis, centromeres do not divide and sister chromatids do not separate. page 11-8

Genetic diversity is generated by crossing over and random alignment and subsequent segregation of maternal and paternal homologs on the metaphase plate in meiosis I. page 11-9

Meiosis II is similar to mitosis, in which chromosomes align on the metaphase plate, centromeres divide, and sister chromatids separate from each other. page 11-10

The similarity of meiosis II and mitosis suggests that meiosis evolved from mitosis. page 11-11

The division of the cytoplasm differs between the sexes: Male meiotic cell division produces four functional sperm cells, whereas female meiotic cell division produces a single functional egg cell and three polar bodies. page 11-11

## 11.4 THE CELL CYCLE IS REGULATED SO THAT CELL DIVISION OCCURS ONLY AT APPROPRIATE TIMES.

Levels of proteins called cyclins increase and decrease during the cell cycle. page 11-15

Cyclins form complexes with cyclin-dependent kinases (CDKs), activating the CDKs to phosphorylate target proteins involved in cell division. page 11-15

Different cyclin–CDK complexes control progression through the cell cycle at key steps, including $G_1$/S phase, S phase, and M phase. page 11-16

These complexes can be inhibited at key checkpoints, halting progression through the cell cycle if something is not right. page 11-16

## 11.5 CANCER IS UNCONTROLLED CELL DIVISION THAT USUALLY RESULTS FROM MUTATIONS IN GENES THAT CONTROL CELL DIVISION.

Cancers can be caused by certain viruses carrying oncogenes that promote uncontrolled cell division, although most cancers are not caused by viruses. page 11-17

Viral oncogenes have cellular counterparts called proto-oncogenes that play normal roles in cell growth and division and that, when mutated, can cause cancer. page 11-19

Oncogenes and proto-oncogenes often encode proteins involved in signaling pathways that promote cell division. page 11-19

The normal cellular role of tumor suppressor proteins, such as p53, is to block cell division and inhibit cancers. page 11-19

Cancers usually result from several mutations in proto-oncogenes and tumor suppressor genes that have accumulated over time within the same cell. page 11-20

## Self-Assessment

1. Compare and contrast the ways in which prokaryotic cells and eukaryotic cells divide.

2. Describe three situations in which mitotic cell division occurs.

3. Name the five steps of mitosis, and draw the changes in the structure and position of the chromosomes at each step.

4. Describe how chromosomes behave in meiosis. Be able to state when chromosomes are duplicated (forming sister chromatids) and when they are not duplicated.

5. Compare and contrast mitotic cell division and meiotic cell division in terms of number of products, number of cell divisions, and processes unique to each.

6. Name two ways in which meiotic cell division creates genetic diversity, and explain how each occurs.

7. Explain how cytokinesis differs between animal and plant cells.

8. Describe the roles of cyclins and cyclin–dependent kinases in the cell cycle.

9. Give three examples of checkpoints that the cell monitors before proceeding through the cell cycle.

10. Describe the differences among an oncogene, a proto-oncogene, and a tumor suppressor gene.

Do you understand the chapter's Core Concepts? Log in to check your answers to the Self-Assessment questions, then practice what you've learned and reinforce this chapter's concepts by working through the problems and multimedia tutorials provided there.

**www.biologyhowlifeworks.com**

# You, From A to T

## YOUR PERSONAL GENOME

**Answers in the genetic code.** Claudia Gilmore found that her genome contains a mutation that can lead to breast cancer.

As an American Studies major at Georgetown University, Claudia Gilmore had plenty of experience taking exams. But at the age of 21, she faced an altogether different kind of test—and no amount of studying could have prepared her for the result.

Gilmore had decided to be tested for a mutation in a gene known as *BRCA1*. Specific mutations in the *BRCA1* and *BRCA2* genes are associated with an increased risk of breast and ovarian cancers.

Gilmore's grandmother had battled breast cancer and later passed away from ovarian cancer. Before she died, she tested positive for the *BRCA1* mutation. Her son, Gilmore's father, was tested and discovered he had inherited the mutation as well. After talking with a genetic counselor to help her understand the implications of the test, Gilmore gave a sample of her own blood and crossed her fingers.

"I had a fifty-fifty chance of inheriting the mutation. I knew there was a great possibility it would be a part of my future," she says. "But I was 21, I was healthy. A part of me also thought this could never really happen to me."

> **Mutations that increase the risk of developing a particular disease are called risk factors.**

Unfortunately, it could. Two weeks after her blood was drawn, she learned that she, too, carried the mutated gene.

Genetic testing is becoming increasingly common—in some cases, even routine. In 2003, after 13 years of painstaking work, scientists published the first draft of the complete human genome. The human genetic code contains about 3 billion base pairs, or structural units of DNA. In the years that followed, much attention has been placed on understanding the genetic differences between individuals.

In reality, there isn't one single human genome. Everyone on Earth (with the exception of identical twins) has his or her own unique genetic sequence. Your personal genome is the blueprint that codes for your hair color, the length of your nose, and your susceptibility to certain diseases. On average, the genomes of two people are 99.9% identical, meaning that they differ at about 3 million sites.

Oftentimes, those individual differences have no impact on health. In some cases, however, a particular genetic signature is associated with disease. Sometimes a gene mutation makes a given illness inevitable. Certain mutations in a gene called *HTT*, for instance, always result in Huntington's disease, a degenerative brain condition that usually appears in middle age.

The link between mutations and disease isn't always so clear-cut, however. Most genetic diseases are complex in origin, and may require multiple genetic mutations, as well as other nongenetic factors, for the disease to develop. Mutations that increase the risk of developing a particular disease are called risk factors.

Certain mutations in the *BRCA1* and *BRCA2* genes are known risk factors for breast and ovarian cancers, for instance. But not everyone with these mutations develops

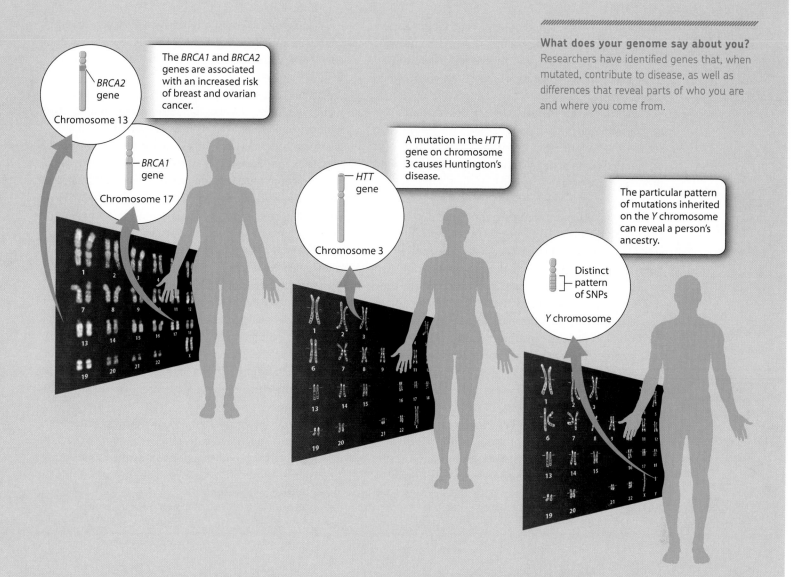

**What does your genome say about you?** Researchers have identified genes that, when mutated, contribute to disease, as well as differences that reveal parts of who you are and where you come from.

The *BRCA1* and *BRCA2* genes are associated with an increased risk of breast and ovarian cancer.

*BRCA2* gene

Chromosome 13

*BRCA1* gene

Chromosome 17

A mutation in the *HTT* gene on chromosome 3 causes Huntington's disease.

*HTT* gene

Chromosome 3

The particular pattern of mutations inherited on the *Y* chromosome can reveal a person's ancestry.

Distinct pattern of SNPs

*Y* chromosome

cancer. According to the National Cancer Institute, about 60% of women with a harmful *BRCA* mutation will develop breast cancer in her lifetime, compared to about 12% of women in the general population. And 15% to 40% of women with a *BRCA* mutation will be diagnosed with ovarian cancer, versus just 1.4% of women without that genetic signature.

The *BRCA* mutations are just some of the thousands of harmful genetic changes that geneticists have identified so far. Other common mutations have been shown to elevate one's risk of developing heart disease, diabetes, various cancers, and numerous other common illnesses. Often, these mutations involve changes to just a single base pair of DNA. These common single-letter mutations are known as single nucleotide polymorphisms, or SNPs.

Genome-sequencing technology has improved significantly over the last decade, making it easier and less expensive to scan an individual's DNA for potentially harmful SNPs. A number of companies now offer genetic tests directly to the public. Unlike tests such as the *BRCA* blood test that Gilmore was given, these direct-to-consumer (DTC) tests are offered to customers without any involvement from a medical professional, at a cost of a few hundred dollars.

Some of these tests aren't related to health at all. Your personal genome contains many unique features—from the shape of your fingernails to the shade of your skin—that don't impact your health, but make you the person you are. Some DTC testing services aim to tell customers about their heritage, by screening genes to identify mutations that are more common in certain geographical regions or among members of certain ethnic groups. One such company offers genetic tests to African-Americans to determine from what part of the African continent their ancestors originated.

Other popular DTC tests inspect DNA samples for SNPs associated with certain diseases and physical traits—everything from Parkinson's disease and age-related macular degeneration to earwax type and propensity for baldness.

Advocates of the tests say the technology puts the power of genetic information in the hands of consumers. Critics, on the other hand, argue that the information provided by DTC tests isn't always very meaningful. Some SNPs might raise the risk of an already-rare disease by just 2% or 3%, for example. In many cases, the precise link between mutation and disease is still being sorted out. Also, without input from a genetic counselor or medical professional, consumers may not know how to interpret the information revealed by the tests. The American Medical Association has recommended that a doctor always be involved when any genetic testing is performed.

Moreover, knowing your genetic risk factors isn't the whole story. When it comes to your health and well-being, the environment also plays a significant role. Someone might override a genetic predisposition for skin cancer by using sunscreen faithfully everyday. On the other hand, a person might have a relatively low genetic risk for type 2 diabetes, but still boost the odds of developing the disease by eating a poor diet and getting little physical exercise.

Claudia Gilmore is especially careful to exercise regularly and eat a healthy diet. Still, she can only control her environment to a degree. She knew that, given her genetic status, her risk of breast cancer remained high. She made the extraordinary decision to eliminate that risk by undergoing a mastectomy at the age of 23. It wasn't an easy decision, she says, but she feels privileged to have been able to take proactive steps to protect her health. "I'll be a 'previvor' instead of a survivor," she says.

For now, it's still too costly to sequence every individual's entire genome. But each year, many more genetic tests hit the market. Already, doctors are beginning to design medical treatments based on a patient's personal genome. People with a certain genetic profile, for example, are less likely than others to benefit from statins, medications prescribed to lower cholesterol. Doctors are also choosing which cancer drugs to prescribe based on the unique genetic signatures of patients and their tumors.

We've only just entered the era of personal genomics. While there's much left to decipher, it's clear that each of our individual genomes contains a wealth of biological knowledge. And, as Gilmore says, "I've always been taught that knowledge is power."

## ❓ CASE 3 QUESTIONS

*Answers to Case 3 questions can be found in Chapters 12-20.*

1. What new technologies will be required to sequence your personal genome? *See page 12-16.*
2. Why sequence your personal genome? *See page 13-4.*
3. What can your personal genome tell you about your genetic risk factors? *See page 14-4.*
4. How can genetic risk factors be detected? *See page 15-9.*
5. How do genetic tests identify disease risk factors? *See page 16-17.*
6. How can the Y chromosome be used to trace ancestry? *See page 17-14.*
7. How can mitochondrial DNA be used to trace ancestry? *See page 17-16.*
8. Can personalized medicine lead to effective treatments of common diseases? *See page 18-12.*
9. How do lifestyle choices affect expression of your personal genome? *See page 19-10.*
10. Can cells with your personal genome be reprogrammed for new therapies? *See page 20-5.*

## CHAPTER 12

# DNA REPLICATION AND MANIPULATION

## Core Concepts

**12.1** In DNA replication, a single parental molecule of DNA produces two daughter molecules.

**12.2** The replication of linear chromosomal DNA requires mechanisms that ensure efficient and complete replication.

**12.3** Techniques for manipulating DNA follow from the basics of DNA structure and replication.

**12.4** Recombinant DNA technology combines DNA from two or more organisms.

One of the overarching themes of biology discussed in Chapter 1 is that the functional unit of life is the cell, a theme that rests on the fundamental concept that all cells come from preexisting cells. In Chapter 11, we considered the mechanics of cell division and how it is regulated. Prokaryotic cells multiply by binary fission, whereas eukaryotic cells multiply by mitosis and cytokinesis. These processes ensure that cellular reproduction results in daughter cells that are like the parental cell except for any mutations that may have taken place during the cell cycle. In other words, in cell division, like begets like.

The two main processes of cell division—binary fission for prokaryotes and mitosis followed by cytokinesis for eukaryotes—differ primarily because of the way the DNA is organized in the two types of cell. Most prokaryotes have just a single circular DNA molecule, whereas eukaryotes have multiple linear DNA molecules that are typically very long and packaged into chromosomes. The eukaryotic chromosomes must condense during mitosis into compact structures that can be maneuvered properly into the daughter cells by the elaborate spindle apparatus. Concealed in the details that make the two processes different is the fact that in *all* dividing cells a double-stranded DNA molecule gives rise to two double-stranded daughter DNA molecules that are identical to each other except for rare mutations.

The process of duplicating a DNA molecule is called **DNA replication,** and it occurs in virtually the same way in all organisms, reflecting its evolution very early in life's history. The process is conceptually simple—the parental strands separate and new partner strands are made—but the molecular details are more complicated. Once scientists understood the molecular mechanisms of DNA replication, they could devise improved methods for manipulating and studying DNA. An understanding of DNA replication is therefore fundamental not only to understanding how cells and organisms produce offspring like themselves, but also to understanding some of the key experimental methods in modern biology.

## 12.1 DNA REPLICATION

You may recall from Chapter 3 that double-stranded DNA consists of a pair of deoxyribonucleotide polymers wound around each other in antiparallel helical coils in such a way that, across the center of the double helix, a purine base (A or G) in one strand is paired with a pyrimidine base (T or C, respectively) in the other strand. To say that the strands are antiparallel means that the 3' hydroxyl of any deoxyribose sugar in the backbone of one strand is opposite the 5' phosphate group of the corresponding deoxyribose sugar in the backbone across the way. These key elements of DNA structure are the only essential pieces of information needed to understand the mechanism of DNA replication.

**During DNA replication, the parental strands separate and new partners are made.**

When Watson and Crick published their paper describing the structure of DNA, they also coyly laid claim to another discovery: "It has not escaped our notice that the specific pairing we have postulated [A with T, and G with C] immediately suggests a copying mechanism for the genetic material." The copying mechanism they had in mind is exquisitely simple. The two strands of the parental duplex molecule separate (**Fig. 12.1**), and each individual parental strand serves as a model, or **template strand,** for the synthesis of a **daughter strand.** As each daughter strand is synthesized, the order of the bases in the template strand determines the order of the complementary bases added to the daughter strand. For example, the sequence 5'-ATGC-3' in the template strand specifies the sequence 3'-TACG-5' in the daughter strand because A pairs with T and G pairs with C. (The designations 3' and 5' convey the antiparallel orientation of the strands.)

A key prediction of the model shown in Fig. 12.1 is **semiconservative replication.** That is, after replication, each new DNA duplex will consist of one strand that was originally present in the parental duplex and one newly synthesized strand. An alternative model is conservative replication, which proposes that the original DNA duplex remains intact and the daughter DNA duplex is completely new. The correct model cannot be determined without knowing just how DNA is replicated. If there were a way to distinguish newly synthesized daughter DNA strands ("new strands") from previously synthesized parental strands ("old strands"), the products of replication could be observed and the mode of replication determined.

The American molecular biologists Matthew S. Meselson and Franklin W. Stahl carried out an experiment to determine how DNA replicates. This experiment, described in **Fig. 12.2,** has been called "the most beautiful experiment in biology" because it so elegantly demonstrated the scientific method (Chapter 1) of hypothesis, prediction, and experimental test. They distinguished "old" from "new" DNA strands by labeling them with different isotopes of nitrogen, either the normal form of nitrogen $^{14}$N or a heavier form with an extra neutron, denoted $^{15}$N.

Meselson and Stahl's finding that DNA replicates semiconservatively (Fig. 12.2) also predicted the results when cells are allowed to undergo two rounds of replication in a medium containing only light $^{14}$N nitrogen. The heavy strand and light strand each serve as templates for a new light daughter strand. The result is that half of the DNA molecules will have one heavy old strand and one light new strand and an intermediate

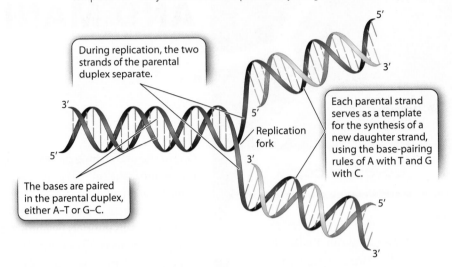

**FIG. 12.1** **DNA replication.** During DNA replication, each parental strand serves as a template for the synthesis of a complementary daughter strand.

During replication, the two strands of the parental duplex separate.

Each parental strand serves as a template for the synthesis of a new daughter strand, using the base-pairing rules of A with T and G with C.

Replication fork

The bases are paired in the parental duplex, either A–T or G–C.

**HOW DO WE KNOW?**

**FIG. 12.2**

# How is DNA replicated?

**BACKGROUND** Watson and Crick's discovery of the structure of DNA in 1953 suggested a mechanism by which DNA is replicated. Experimental evidence came with research by American molecular biologists Matthew Meselson and Franklin Stahl in 1958.

**HYPOTHESIS** DNA replicates in a semiconservative manner, meaning that each new DNA molecule consists of one parental strand and one newly synthesized strand.

**ALTERNATIVE HYPOTHESIS** DNA replicates in a conservative manner, meaning that one DNA molecule consists of two parental strands, and the other consists of two newly synthesized strands.

**METHOD** Meselson and Stahl distinguished parental strands ("old") from newly synthesized strands ("new") using two isotopes of nitrogen atoms. Old strands were labeled with a heavy form of nitrogen with an extra neutron ($^{15}$N), and new strands were labeled with the normal, lighter form of nitrogen ($^{14}$N).

**EXPERIMENT** The researchers first grew bacterial cells on medium containing only the heavy $^{15}$N form of nitrogen. As the cells grew, $^{15}$N was incorporated into the DNA bases, resulting, after several generations, in DNA containing only $^{15}$N. They then transferred the cells into medium containing only light $^{14}$N nitrogen. After one round of replication in this medium, cell replication was halted. The researchers could not observe the DNA directly, but instead they measured the density of the DNA by spinning it in a high-speed centrifuge in tubes containing a solution of cesium chloride.

density, and half of the DNA molecules will have one light old strand and one light new strand and a low density. This is precisely what they observed (Fig. 12.2).

→ **Quick Check 1** Suppose Meselson and Stahl had done their experiment the other way around, starting with cells fully labeled with $^{14}$N light DNA and then transferring them to medium containing only $^{15}$N heavy DNA. What density of DNA molecule would you predict after one and two rounds of replication?

Important as it was in demonstrating semiconservative replication in bacteria, the Meselson–Stahl experiment left open the possibility that DNA replication in eukaryotes might be different. It was only some years after the Meselson–Stahl experiment that methods for labeling DNA with fluorescent nucleotides were developed. These methods allowed researchers to visualize entire strands of eukaryotic DNA and follow each strand through replication. **Fig. 12.3** shows a human

**FIG. 12.3** **Eukaryotic DNA replication.** Further evidence that DNA replication is semiconservative came from observing the uptake of fluorescent nucleotides into chromosomal DNA.

Half-labeled    Fully labeled

After two rounds of DNA replication in a labeled medium, one daughter molecule is half-labeled, and the other is fully labeled (compare with Figure 12.2).

**1** At the beginning of the experiment, both strands are labeled with heavy nitrogen ($^{15}$N). The molecules form a band with the density of "heavy" DNA.

1.722 — Density of DNA (gm/cm$^3$)

**2** After one round of replication in the absence of heavy nitrogen, the parental strand still contains the heavy nitrogen ($^{15}$N), but the daughter strand contains light nitrogen ($^{14}$N). The daughter DNA molecules form a band with an intermediate density.

1.715

**3** After two rounds of replication, half of the duplex DNA molecules have one strand with $^{15}$N and one with $^{14}$N, and the other half have both strands with $^{14}$N. The daughter DNA molecules therefore form two bands, one intermediate in density and the other light.

1.708
1.715

**PREDICTION** If DNA replicated conservatively, half of the DNA in the cells should be composed of two heavy parental strands containing $^{15}$N, and half should be composed of two light daughter strands containing $^{14}$N. If DNA replicated semiconservatively, the daughter DNA molecules should each consist of one heavy strand and one light strand.

**RESULTS** When the fully $^{15}$N-labeled parental DNA was spun, it concentrated in a single heavy band with a density of 1.722 gm/cm$^3$. After one round of replication, the DNA formed a band at a density of 1.715 gm/cm$^3$, which is the density expected of a duplex molecule containing one heavy ($^{15}$N-labeled) strand and one light ($^{14}$N-labeled) strand. DNA composed only of $^{14}$N would have a density of 1.708 gm/cm$^3$. After two rounds of replication, half the molecules exhibited a density of 1.715 gm/cm$^3$, indicating a duplex molecule with one heavy strand and one light strand, and the other half exhibited a density of 1.708 gm/cm$^3$, indicating a duplex molecule containing two light strands.

**CONCLUSION** DNA replicates semiconservatively, supporting the first hypothesis.

**SOURCE** Meselson, M., and F. W. Stahl. 1958. "The Replication of DNA in *Escherichia coli.*" *PNAS* 44:671–82.

chromosome with unlabeled DNA that subsequently underwent two rounds of replication in medium containing a fluorescent nucleotide. The chromosome was photographed at metaphase of mitosis, after chromosome duplication but before the separation of the chromatids into the daughter cells. Notice that one chromatid contains hybrid DNA with one labeled strand and one unlabeled strand, which fluoresces faintly (light); the other chromatid contains two strands of labeled DNA, which fluoresces strongly (dark). This result is conceptually the same as what was seen by Meselson and Stahl after two rounds of replication and exactly as predicted by the semiconservative replication model. The result also demonstrates that each eukaryotic chromosome contains a single DNA molecule that runs continuously all along its length.

### New DNA strands grow by the addition of nucleotides to the 3′ end.

Although replication is semiconservative, as predicted by the model in Fig. 12.1, the model alone does not tell us the details of replication. For example, the model implies that both daughter strands should grow in length by the addition of nucleotides near the site where the parental strands separate, a site called the **replication fork.** As more and more parental DNA is unwound and the replication fork moves forward (to the left in Fig. 12.1), both new strands would also grow in the direction of replication fork movement. But it turns out that this scenario is impossible.

We have seen that the two DNA strands in a double helix run in an antiparallel fashion: One of the template strands (the bottom one in Fig. 12.1) has a left-to-right 5′-to-3′ orientation, whereas the other template strand (the top one in Fig. 12.1) has a left-to-right 3′-to-5′ orientation. Therefore, the new daughter strands will also have opposite orientations, so that near the replication fork the daughter strand in the bottom duplex terminates in a 3′ hydroxyl, whereas that in the top duplex terminates in a 5′ phosphate. There's the rub: The strand that terminates in the 5′ phosphate cannot grow in the direction of the replication fork because new DNA strands can grow only by the addition of successive nucleotides to the 3′ end. That is, DNA always grows in the 5′-to-3′ direction.

DNA polymerization occurs only in the 5′-to-3′ direction because of the chemistry of DNA synthesis, discussed in Chapter 3. The building blocks of DNA (Chapter 2) are nucleotides, each consisting of a deoxyribose sugar with three phosphate groups attached to the 5′ carbon, a nitrogenous base (A, T, G, or C) attached to the 1′ carbon, and a free hydroxyl (OH) group attached to the 3′ carbon. DNA polymerization occurs when the 3′ hydroxyl at the growing end of the polynucleotide chain attacks the triphosphate at the 5′ end of an incoming nucleotide (**Fig. 12.4**). Each of these incoming nucleotides is a triphosphate (three phosphate groups attached to the 5′ carbon of the deoxyribose).

**FIG. 12.4**    DNA synthesis by nucleotide addition to the 3′ end of a growing DNA strand.

When the incoming nucleotide triphosphate is added to the growing DNA strand, the outermost two phosphates (called pyrophosphate) are cleaved off and diffuse away. The energy released by cleaving off the pyrophosphate is what drives the polymerization reaction that attaches the incoming nucleotide to the growing strand.

The polymerization reaction is catalyzed by **DNA polymerase,** an enzyme that is a critical component of a large protein complex that carries out DNA replication. DNA polymerases exist in all organisms and are highly conserved, meaning that they vary little from one species to another because they carry out an essential function. A cell typically contains several different DNA polymerase enzymes, each specialized for a particular situation. But all DNA polymerases share the same basic function in that they synthesize a new DNA strand from an existing template. Most, but not all, also correct mistakes in replication, as we will see. DNA polymerases have many practical applications in the laboratory, which we will discuss later in this chapter.

### In replicating DNA, one daughter strand is synthesized continuously and the other in a series of short pieces.

Because a new DNA strand can be elongated only at the 3′ end, the two daughter strands are synthesized in quite different ways (**Fig. 12.5**). The daughter strand shown at the bottom of Fig. 12.5 has its 3′ end pointed toward the replication fork, so that as the parental double helix unwinds, this daughter strand can be synthesized as one long, continuous polymer. This daughter strand is called the **leading strand.**

The situation is different for the daughter strand shown at the top in Fig. 12.5. Its 5′ end is pointed toward the replication fork, but the strand cannot grow in that direction. Instead, as the replication fork unwinds, it forms a stretch of single-stranded DNA of a few hundred to a few thousand nucleotides, depending on the species. Then a new daughter strand is initiated with its 5′ end near the replication fork, and this strand is elongated at the 3′ end as usual. The result is that the daughter strand shown at the top in Fig. 12.5 is actually synthesized in short, discontinuous pieces. As the parental double helix unwinds, a new piece is initiated at intervals, and each new piece is elongated at its 3′ end until it reaches the piece in front of it. This daughter strand is called the

**lagging strand.** The short pieces in the lagging strand are sometimes called **Okazaki fragments** after their discoverer, Japanese molecular biologist Reiji Okazaki.

The presence of leading and lagging strands during DNA replication is a consequence of the antiparallel nature of the two strands in a DNA double helix, and the fact that DNA polymerase can synthesize DNA in only one direction.

## A small stretch of RNA is needed to begin synthesis of a new DNA strand.

Each new DNA strand must begin with a short stretch of RNA that serves as a **primer,** or starter, for DNA synthesis. The primer is needed because the DNA polymerase complex cannot begin a new strand on its own; it can only elongate the end of an existing piece of DNA or RNA. The primer is made by an RNA polymerase called **RNA primase,** which synthesizes a short piece of RNA complementary to the DNA template

**FIG. 12.5 Leading and lagging strand synthesis.** In DNA replication, one daughter strand (the leading strand) is synthesized continuously and the other (the lagging strand) is synthesized in smaller pieces.

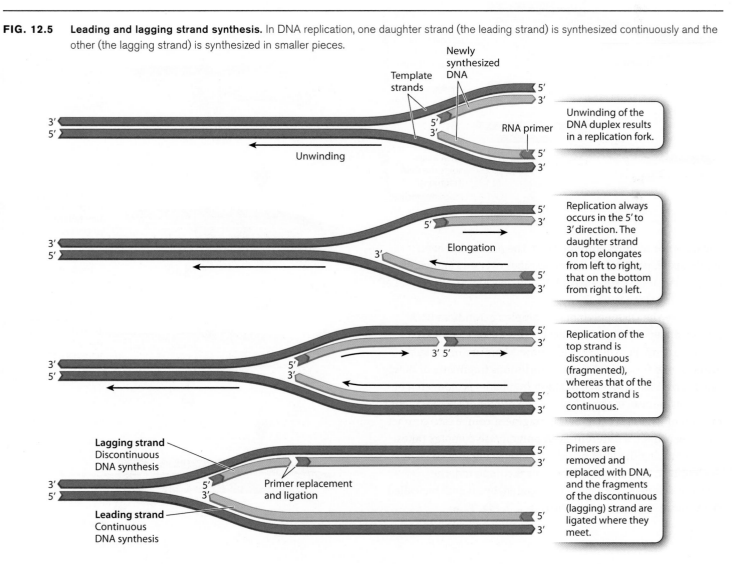

**FIG. 12.6** Synthesis and removal of RNA primers.

RNA primase lays down an RNA primer.

DNA polymerase extends the RNA primer.

A different DNA polymerase removes the primer and replaces it with DNA.

DNA ligase joins the DNA fragments.

and does not require a primer. Once the primer has been synthesized, the DNA polymerase takes over and elongates the primer, adding successive DNA nucleotides to the 3′ end of the growing strand.

Because the DNA polymerase complex extends an RNA primer, all new DNA strands have a short stretch of RNA at their 5′ end. For the lagging strand, there are many such primers, one for each of the discontinuous fragments of newly synthesized DNA. As each of these fragments is elongated by DNA polymerase, it grows toward the primer of the fragment in front of it. When the growing fragment comes into contact with the primer, a different DNA polymerase complex takes over, removing the RNA primer, and replacing it with DNA nucleotides. When the replacement is completed, the adjacent fragments are joined, or ligated, by an enzyme called **DNA ligase.** This process is illustrated in **Fig. 12.6**.

## DNA polymerase is self-correcting because of its proofreading function.

Most DNA polymerases can correct their own errors in a process called **proofreading,** which is a separate enzymatic activity from strand-elongation (synthesis) (**Fig. 12.7**).

When each new nucleotide comes into line in preparation for attachment to the growing DNA strand, the nucleotide is

**FIG. 12.7** Proofreading, the process by which an incorrect nucleotide is removed immediately after it is incorporated.

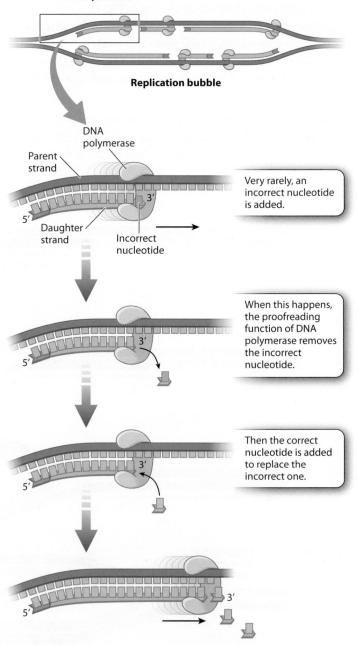

**Replication bubble**

Very rarely, an incorrect nucleotide is added.

When this happens, the proofreading function of DNA polymerase removes the incorrect nucleotide.

Then the correct nucleotide is added to replace the incorrect one.

temporarily held in place by hydrogen bonds that form between the base in the new nucleotide and the base across the way in the template strand. The strand being synthesized and the template strand therefore have complementary bases—A paired with T, or G paired with C. However, on rare occasions, improper hydrogen bonds form, with the result that an incorrect nucleotide is attached to the new DNA strand. DNA polymerase can correct errors because it detects mispairing between the template and the most recently added nucleotide. Mispairing between a base in the parental strand and a newly added base in the daughter strand activates a DNA-cleavage function of DNA polymerase that removes the incorrect nucleotide, and inserts the correct one in its place.

Mutations resulting from errors in nucleotide incorporation still occur, but proofreading reduces their number. In the bacterium *E. coli*, for example, about 99% of the incorrect nucleotides that are incorporated during replication are removed and repaired by the proofreading function of DNA polymerase. Those that slip past proofreading and other repair systems (Chapter 14) lead to mutations, which are then faithfully copied and passed on to daughter cells. Some of these mutations may be harmful, but others are neutral and a rare few may be beneficial. These mutations are the ultimate source of genetic variation that we see among individuals of the same species and among species, as we explore in Chapter 15.

## Many proteins participate in DNA replication.

We have seen that DNA replication requires the DNA polymerase complex that elongates each strand at the 3′ end, the RNA primase that makes the primer, the DNA polymerase complex that replaces the RNA primer with DNA, and the DNA ligase that joins the DNA fragments in the lagging strand. Many other proteins are also involved (**Fig. 12.8**). One of these, **helicase,** unwinds the parental double helix at the replication fork. **Single-stranded binding proteins** then bind the resulting single-stranded regions of DNA to prevent the template strands from coming back together. **Topoisomerase II** works upstream from the replication fork to relieve the stress on the double helix that results from its unwinding at the replication fork.

Although many evolutionarily conserved proteins are required for DNA replication, the underlying process is quite simple and the same in all organisms. The two strands of the parental DNA duplex separate, and each serves as a template for the synthesis of a daughter strand according to the base-pairing rules of A–T and G–C, with each successive nucleotide being added to the 3′ end of the growing strand.

**FIG. 12.8** Major proteins involved in DNA replication.

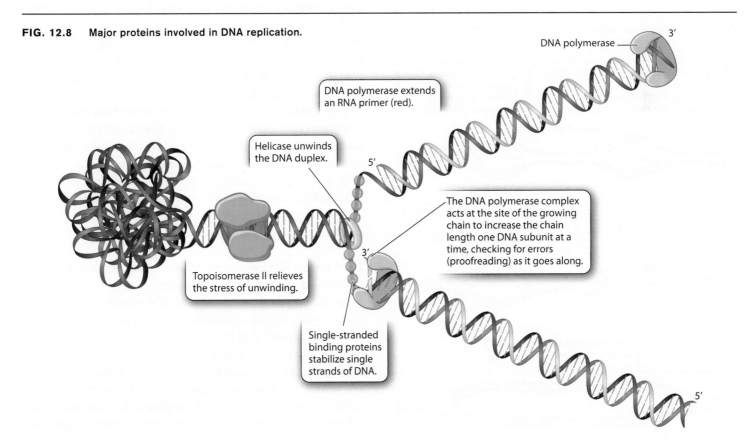

DNA polymerase extends an RNA primer (red).

Helicase unwinds the DNA duplex.

DNA polymerase

Topoisomerase II relieves the stress of unwinding.

The DNA polymerase complex acts at the site of the growing chain to increase the chain length one DNA subunit at a time, checking for errors (proofreading) as it goes along.

Single-stranded binding proteins stabilize single strands of DNA.

## 12.2 REPLICATION OF CHROMOSOMES

The steps involved in DNA replication are universal, suggesting that they evolved in the common ancestor of all living organisms. In addition, they are the same whether they occur in a test tube or in a cell, or whether the segment of DNA being replicated is short or long. However, the replication of an entire linear chromosome poses particular challenges. Here, we consider two such challenges encountered by cells in replicating their chromosomes: how replication starts, and how it ends.

### Replication of DNA in chromosomes starts at many places almost simultaneously.

DNA replication is relatively slow. In eukaryotes, it occurs at a rate of about 50 nucleotides per second. At this rate, replication from end to end of the DNA molecule in the largest human chromosome would take almost two months. In fact, it takes only a few hours. This fast pace is possible because, in a long DNA molecule, replication begins almost simultaneously at many places. Each point at which DNA synthesis is initiated is called an **origin of replication.** The opening of the double helix at each origin of replication forms a **replication bubble** with a replication fork

**FIG. 12.9    Origins of replication.** The replication of chromosomal DNA is initiated at many different places on the chromosome.

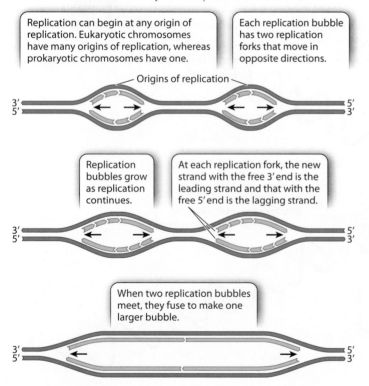

**FIG. 12.10    Replication of a circular bacterial chromosome.**

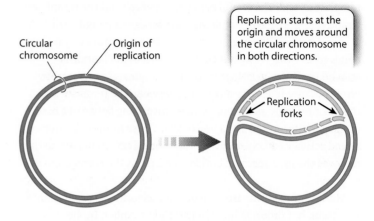

on each side, each with a leading strand and a lagging strand (**Fig. 12.9**). DNA synthesis takes place at each replication fork, and as the replication forks move in opposite directions the replication bubble increases in size. When two replication bubbles meet, they fuse to form one larger replication bubble.

Note in Fig. 12.9 that within a single replication bubble, the same daughter strand is the leading strand at one replication fork and the lagging strand at the other replication fork. This situation results from the fact that the replication forks in each replication bubble move away from each other. When two replication bubbles fuse and the leading strand from one meets the lagging strand from the other, the ends of the strands that meet are joined by DNA ligase, just as happens when the discontinuous fragments within the lagging strand meet (see Fig. 12.6).

Some DNA molecules, including most of the DNA molecules in bacterial cells and the DNA in mitochondria and chloroplasts (Chapter 3) are small circles, not long linear molecules. Such circular DNA molecules typically have only one origin of replication (**Fig. 12.10**). Replication takes place at both replication forks, and the replication forks proceed in opposite directions around the circle until they meet and fuse on the opposite side, completing one round of replication.

### Telomerase restores tips of linear chromosomes shortened during DNA replication.

A circular DNA molecule can be replicated completely because it has no ends, and the replication forks can move completely around the circle (Fig. 12.10). Linear DNA molecules have ends, however, and at each round of DNA replication the ends become slightly shorter. The reason for the shortening is illustrated in **Fig. 12.11.** Recall that each fragment of newly synthesized DNA starts with an RNA primer. On the leading strand, the only primer required is at the origin of replication when synthesis begins. The leading strand is elongated in the same direction as the moving replication fork and is able to replicate the template strand all the way to the end. But on the lagging strand, which grows away from the replication fork, many primers are required, and the

**FIG. 12.11 Shortening of linear DNA at the ends.** Because an RNA primer cannot be synthesized precisely at the 3′ end of a DNA strand, its partner strand becomes a little shorter in each round of replication.

Ends of template DNA strands

← Replication fork

The leading strand replicates the whole template strand.

Lagging strand

Leading strand

The last RNA primer on the lagging strand sits near the end of the template strand.

Template strand

3′
5′

RNA primer

Lagging strand

Unreplicated template DNA

The RNA primer is removed, and a section of template DNA remains unreplicated.

3′
5′

In the next round of replication, the shortened template results in a shorter chromosome.

Daughter strand of next generation

3′
5′

Shorter template strand of next generation

If this pattern were allowed to persist, the chromosomes would be severely shortened after several generations.

last RNA primer is synthesized on the end of the template strand about 100 nucleotides from the 3′ end of the template. When DNA replication is complete, this new daughter DNA strand (light blue in Fig. 12.11) is therefore missing about 100 base pairs from the tip. When the daughter strand is itself replicated, the newly synthesized strand must terminate at the shortened end of the template strand, and so the new duplex molecule is shortened by about 100 base pairs from the original parental molecule. The strand shortening in each round of DNA replication is a problem because without some mechanism to restore the tips, the DNA in the chromosome would eventually be nibbled away to nothing.

**Fig. 12.12** illustrates the mechanism that eukaryotic organisms have evolved to solve the problem of shortened ends. Each end of a eukaryotic chromosome is capped by a repeating sequence called the **telomere,** shown in purple in Fig. 12.12. In human chromosomes, the telomere consists of the sequence 5′-TTAGGG-3′ repeated over and over again in about 1500–3000 copies. The telomere is slightly shortened in each round of DNA replication, as shown in Fig. 12.11, but before the next round of replication the shortened end is restored by an enzyme known as **telomerase,** which replaces the lost telomere repeats. Because there are no genes in the telomere, the slight shortening and subsequent restoration that take place have no harmful consequences.

Telomerase activity differs from one cell type to the next. It is fully active in **germ cells,** which produce sperm or eggs, and also in **stem cells,** which are undifferentiated cells that can undergo an unlimited number of mitotic divisions and can differentiate into any of a large number of specialized cell types. Stem cells are found in embryos, where they differentiate into all the various cell types. Stem cells are also found in some tissues of the body after embryonic development, where they replenish cells with a high rate of turnover, such as blood and intestinal cells, and play a role in tissue repair.

In contrast to the high activity of telomerase in germ cells and stem cells, telomerase is almost inactive in most cells in the adult body, the **somatic cells.** In these cells, the telomeres are actually shortened by about 100 base pairs in each mitotic division. Telomere shortening limits the number of mitotic divisions that the cells can undergo because human cells stop dividing when their chromosomes have telomeres with fewer than about 100 copies of the telomere repeat. Adult somatic cells can therefore undergo only about 50 mitotic divisions until the telomeres are so short that the cells stop dividing.

Many biologists believe that the limit on the number of cell divisions explains in part why our tissues become less youthful and wounds heal more slowly with age. The telomere hypothesis of aging is still controversial, but increasing evidence suggests that it is one of several factors that lead to aging. The flip side of the coin is observed in cancer cells, in which telomerase is reactivated and helps support the uncontrolled growth and division of abnormal cells.

**FIG. 12.12  Telomerase.** Telomerase prevents successive shortening of chromosomes.

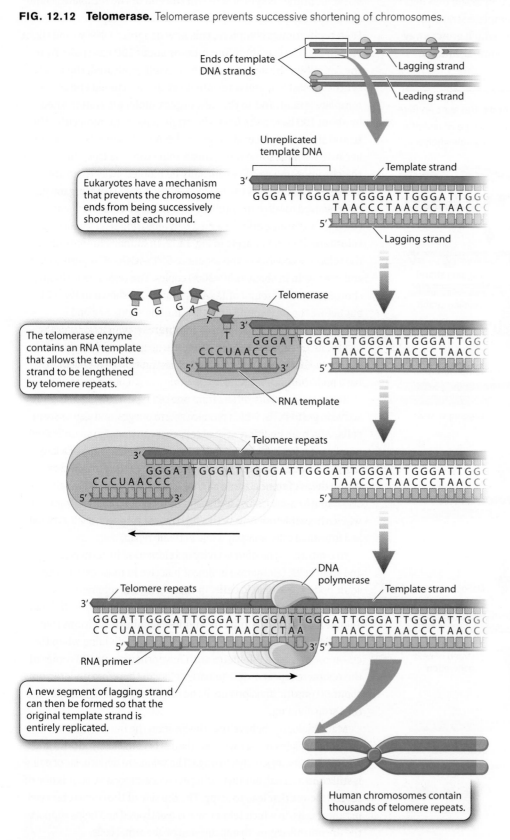

Ends of template
DNA strands

Lagging strand

Leading strand

Unreplicated
template DNA

Template strand

Eukaryotes have a mechanism that prevents the chromosome ends from being successively shortened at each round.

3′
GGGATTGGGATTGGGATTGGGATTGGG
TAACCCTAACCCTAACCC
5′

Lagging strand

Telomerase

The telomerase enzyme contains an RNA template that allows the template strand to be lengthened by telomere repeats.

3′
GGGATTGGGATTGGGATTGGGATTGGG
CCCUAACCC              TAACCCTAACCCTAACCC
5′                                5′

RNA template

Telomere repeats

3′
GGGATTGGGATTGGGATTGGGATTGGGATTGGGATTGGG
CCCUAACCC                              TAACCCTAACCC
5′

DNA
polymerase

Telomere repeats                    Template strand

3′
GGGATTGGGATTGGGATTGGGATTGGGGATTGGGATTGGGATTGGG
CCCUAACCCTAACCCTAACCCTAA      TAACCCTAACCCTAACCC
5′                                3′ 5′

RNA primer

A new segment of lagging strand can then be formed so that the original template strand is entirely replicated.

Human chromosomes contain thousands of telomere repeats.

## 12.3 ISOLATION, IDENTIFICATION, AND SEQUENCING OF DNA FRAGMENTS

Watson and Crick's discovery of the structure of DNA and knowledge of the mechanism of replication allowed biologists not only to understand some of life's central processes, but also to create tools to study how life works. Biologists often need to isolate, identify, and determine the nucleotide sequence of particular DNA fragments. Such procedures can determine whether a genetic risk factor for diabetes has been inherited, whether blood at a crime scene matches that of a suspect, whether a variety of rice or wheat carries a genetic factor for insect resistance, and how closely two species of organisms are related. Many of the experimental procedures for the isolation, identification, and sequencing of DNA are based on knowledge of DNA structure and the physical properties of DNA. Others make use of the principles of DNA replication. In this section, we discuss how particular fragments of double-stranded DNA can be produced, how DNA fragments of different sizes can be physically separated, and how the nucleotide sequence of a piece of DNA can be determined.

**The polymerase chain reaction selectively amplifies regions of DNA.**

DNA that exists in the nuclei of your cells is present in just one copy per cell (in the case of the X or Y chromosome in males) or two copies per cell (in the case of the two X chromosomes in females and  chromosomes other than the sex chromosomes). In the laboratory, it is very difficult to manipulate or visualize a sample containing just one

**RESULTS** The computer program the group had written to assemble the fragments worked. The researchers were able to sequence the entire *Drosophila* genome by piecing together the fragments according to their overlaps. In the sentence analogy, the fragments (Fig. 13.1a) can be assembled into the complete sentence (Fig. 13.1b) by matching the overlaps between the fragments.

**a.**

1 "It has not escaped our no
2 specific pairing we have post
3 suggests a plausible copying m
4 d our notice that the specifi
5 ve postulated immediately suggests
6 pying mechanism for the genetic material."

The sequences of DNA fragments are analogous to sentence fragments in the English language.

**b.**

1 − 4 − 2 − 5 − 3 − 6

"It has not escaped our no
d our notice that the specifi
specific pairing we have post
ve postulated immediately suggests
suggests a plausible copying m
pying mechanism for the genetic material."

The sentence fragments can be assembled in the correct order according to their overlaps and the original complete sentence reconstructed.

**CONCLUSION** The hypothesis was supported: Celera Genomics could determine the entire genomic sequence of an organism by sequencing small, random fragments and piecing them together at their overlapping ends.

**FOLLOW-UP WORK** Today, the computer assembly method is routinely used to determine genome sequences. This method is also used to infer the genome sequences of hundreds of bacterial species simultaneously—for example, in bacterial communities sampled from seawater or from the human gut.

**SOURCE** Adams, M. D., et al. 2000. "The Genome Sequence of *Drosophila melanogaster.*" *Science* 287:2185–2195.

---

**FIG. 13.2 Principal types of sequence repeats found in eukaryotic genomes (not drawn to scale).** Repeats often pose problems in DNA sequencing.

**a.**

Tandem repeats

**b.**

Dispersed repeats

**c.**

Simple-sequence repeats

...ATATATATATATATATATATATATATATAT...
...TATATATATATATATATATATATATATATA...

The difficulty with long repeated sequences is that they typically are much longer than the short sequences obtained by automated sequencing. As a result, the repeat may not be detected at all. And if the repeat is detected, there is no easy way of knowing the number of copies of the repeat, that is, whether the segment includes two, three, four, or any number of copies of the repeat. Sometimes, researchers can use the ends of repeats, where the fragments overlap with an adjacent, nonrepeating sequence, as a guide to the position and number of repeats.

In another type of repeat, the repeating sequence is short, even as short as two nucleotides, such as AT, repeated over and over again in a stretch of DNA (Fig. 13.2c). Short repeating sequences of this kind are troublesome for sequencing machines because any single-stranded fragment consisting of alternating AT can fold back upon itself to form a double-stranded structure in which A is paired with T. Such structures are more stable than the unfolded single-stranded structures and are not easily sequenced.

→ **Quick Check 2** Let's say that a stretch of repeated AT is successfully sequenced. From what you know of the difficulties of sequencing long repeated sequences, what other problems might you encounter in assembling these fragments?

**?** **CASE 3 You, From A To T: Your Personal Genome**
**Why sequence your personal genome?**
The Human Genome Project, which began in 1990, had the goal of sequencing the human genome as well as the genomes of certain key organisms used as models in genetic research. The model organisms chosen are the mainstays of laboratory biology— a species each of bacteria, yeast, nematode worm, fruit fly, and mouse. By 2003, the genome sequences of these model organisms had been completed, as well as the human genome. By then, the cost of sequencing had become so low and the sequence output so high that many more genomes were sequenced than originally planned. Large-scale genome sequencing is still done today, and soon you could choose to have your personal genome sequenced.

Why sequence more genomes? And if the human genome is sequenced, why sequence yours? As we saw in Case 3: You, From A to T, there is really no such thing as *the* human genome, any more than there is *the* fruit fly genome or *the* mouse genome. With the exception of identical twins, every person's genome is unique, the product of a fusion of a unique egg with a unique sperm. The sequence that is called "the human genome" is actually a composite of sequences from different individuals. This sequence is nevertheless useful because most of us share the same genes and regulatory regions, organized the same way on chromosomes.

Detailed knowledge of your own personal genome can be valuable. Our individual DNA sequences differ at millions of nucleotide sites from one person to the next. Some of these differences account in part for the physical differences we see among us; others have the potential to predict susceptibility to disease and response to medication. For Claudia Gilmore, knowledge of the sequence of her *BRCA1* gene had a significant impact on her life.

Determining these differences is a step toward **personalized medicine,** in which an individual's genome sequence, by revealing his or her disease susceptibilities and drug sensitivities, allows treatments to be tailored to the individual. There may come a time, perhaps within your lifetime, when personal genome sequencing becomes part of routine medical testing. Information about a patient's genome will bring benefits but also raises ethical concerns and poses risks to confidentiality and insurability.

## 13.2 GENOME ANNOTATION

A goal of biology is to identify all the component macromolecules in biological systems and to understand their individual functions and the ways in which they interact—their cellular organization, metabolism, growth, adaptation, reproduction, and other aspects of how life works. This research has practical applications: Increased understanding of the molecular and cellular basis of disease, for example, can lead to improved diagnosis and treatment.

The value of genome sequencing in identifying macromolecules is that the genome sequence contains, in coded form, the nucleotide sequence of all RNA molecules transcribed from the DNA as well as the amino acid sequence of all proteins. There is a catch, however. A genome sequence is merely an extremely long list of A's, T's, G's, and C's that represent the order in which nucleotides occur along the DNA in one strand of the double helix. (Because A in one strand is paired with T in the other and G is paired with C, knowing the sequence of one strand specifies the other.) The catch is that in multicellular organisms, not all of the DNA is transcribed into RNA, and not all of the RNA that is transcribed is translated into protein. Therefore, genome sequencing is just the first step in understanding the function of any particular DNA sequence. Following genome sequencing, the next step is to identify the locations and functions of the various types of sequence present in the genome.

### Genome annotation identifies various types of sequence.

Genomes contain many different types of sequence, among them protein-coding genes. Protein-coding genes are themselves composed of different regions, including regulatory elements that specify when and where an RNA transcript will be produced, noncoding introns that are removed from the RNA transcript during RNA processing (Chapter 3), and protein-coding exons that contain the codons that specify the amino acid sequence of a polypeptide chain (Chapter 4). Genomes also contain coding sequences for RNAs that are not translated into protein (noncoding RNAs), such as ribosomal RNA, transfer RNA, and other types of small RNA molecule. Finally, while much of the DNA in the genomes of multicellular organisms is transcribed at least in some cell types, the functions of these transcripts in metabolism, physiology, development, or behavior are unknown.

**Genome annotation** is the process by which researchers identify the various types of sequence present in genomes. Genome annotation is essentially an exercise in adding commentary to a genome sequence that identifies which types of sequence are present and where they are located. It can be thought of as a form of pattern recognition, where the patterns are regularities in sequence that are characteristic of protein-coding genes or other types of sequence.

An example of genome annotation is shown in **Fig. 13.3.** Genes present in one copy per genome are indicated in orange. Most of these single-copy genes are protein-coding genes. The annotation of a single-copy gene typically specifies any nearby regulatory regions that control transcription, the intron–exon boundaries in the gene, and any known or predicted alternative forms in which the introns and exons are spliced (Chapter 3). Each single-copy gene is given a unique name and its protein product identified. Note in Fig. 13.3 that single-copy genes can differ in size from

**FIG. 13.3** **Genome annotation.** Given the DNA sequence of a genome, researchers can pinpoint locations of various types of sequence.

Noncoding RNA
Single-copy gene
Dispersed repeat
Tandem repeat
Simple-sequence repeat

Double-stranded DNA

one gene to the next. The annotations in Fig. 13.3 also specify the locations of sequences that encode RNAs that are not translated into proteins, as well as various types of repeated sequence.

Small genomes such as that of HIV and other viruses can be annotated by hand, but for large genomes like the human genome, computers are essential. In the human genome, some protein-coding genes extend for more than a million nucleotides. A comparison helps convey the challenge of genome annotation. Roughly speaking, if the sequence of the approximately 3 billion nucleotides in a human egg or sperm was printed in normal-sized type, the length of the ribbon would stretch 4000 miles (6440 kilometers), about the distance from Fairbanks, Alaska, to Miami, Florida. By contrast, for the approximately 10,000 nucleotides in the HIV genome, the ribbon would extend a mere 70 feet (21 meters).

Genome annotation is an ongoing process because, as macromolecules and their functions and interactions become better understood, the annotations to the genome must be updated. A sequence that is annotated as nonfunctional today may be found to have a function tomorrow. For this reason, the annotation of certain genomes—including the human genome—will certainly continue to change.

## Genome annotation includes searching for sequence motifs.

Because genome annotation is essentially pattern recognition, it begins with the identification of a **sequence motif,** a telltale sequence that indicates what type of sequence it is (**Fig. 13.4**). This sequence pattern can be found in the DNA itself, or in the RNA sequence inferred from the DNA sequence. Once identified, sequence motifs are typically confirmed by experimental methods.

An example of a sequence motif is an **open reading frame (ORF)** (Fig. 13.4a). The motif for an open reading frame is a long string of codons for amino acids with no stop codon. The presence of such a motif by itself is enough to annotate the DNA segment as potentially protein coding.

Fig. 13.4b shows another type of sequence motif, this one also present in a hypothetical RNA transcript inferred from the DNA sequence. The nucleotide sequence in one part of the RNA is complementary to that in another part, so the single-stranded molecule is able to fold back on itself and undergo base pairing to form a hairpin-shaped structure. Such hairpin structures are characteristic of certain types of RNA that function in gene regulation (Chapter 19). The DNA from which this RNA is transcribed would have complementary sequences on either end as well.

**FIG. 13.4** **Some common sequence motifs useful in genome annotation.** (a) An open reading frame; (b) a noncoding RNA molecule; (c) transcription factor binding sites.

**a.**

RNA from a protein-coding region contains an open reading frame consisting of triplets of nucleotides that can specify amino acids.

**b.**

Regions of some RNA molecules, such as those in transfer RNA, form hairpin structures in which the molecule folds back on itself.

**c.**

DNA sequences that bind transcription factors are often short sequences present in multiple copies near a protein-coding gene.

Some sequence motifs are detected directly in the double-stranded DNA. Fig. 13.4c shows two copies of a short sequence that are known binding sites for DNA-binding proteins called transcription factors (Chapter 3), whose binding to DNA initiates transcription. Transcription factor binding sites are often present in multiple copies and in either strand of the DNA. Sometimes they are located near the region of a gene where transcription is initiated because the transcription factor helps determine when the gene will be transcribed. However, they can also be located far upstream of the gene, downstream of the gene, or in introns, and so their identification is difficult.

### Comparison of genomic DNA with messenger RNA reveals the intron–exon structure of genes.

In annotating an entire genome, researchers typically make use of information beyond the genome sequence itself. This information may include sequences of messenger RNA molecules that are isolated from various tissues or from various stages of development of the organism. Recall from Chapter 3 that messenger RNA (mRNA) molecules undergo processing and are therefore usually simpler than the DNA sequences from which they are transcribed—for example, introns are removed and exons are spliced together. The resulting mature mRNA therefore contains a long sequence of codons uninterrupted by a stop codon—in other words, an open reading frame. The open reading frame in an mRNA is the region that is actually translated into protein on the ribosome.

One aspect of genome annotation is the determination of which portions of the genome sequence correspond to sequences in mRNA transcripts. An example is shown in **Fig. 13.5**, which compares the DNA and mRNA for the β (beta) chain of hemoglobin, the oxygen-carrying protein in red blood cells. Note that the genomic DNA contains some sequences present in the mRNA, which correspond to exons, and some sequences that are not present in the mRNA, which correspond to introns. Comparison of mRNA with genomic DNA therefore reveals the intron–exon structure of protein-coding genes. In fact, introns were first discovered by comparing β-globin mRNA with genomic DNA.

### An annotated genome summarizes knowledge, guides research, and reveals evolutionary relationships among organisms.

Genome annotation, which aims to identify all the functional and repeat sequences present in the genome, is an imperfect science. Even in a well-annotated genome, some protein-coding sequences or other important features may be overlooked, and occasionally the annotation of a sequence motif is incorrect.

Because researchers often have to rely on sequence motifs alone and not experimental data, their descriptions may be vague. For example, a common annotation in large genomes is "hypothetical protein." In some cases, such as the genome of the malaria parasite, this type of annotation accounts for about 50% of the possible protein-coding genes. There is no hint of what a hypothetical protein may do or even whether it is actually produced, since it is determined solely by the presence of an ORF in the genomic sequence, and not by the presence of actual mRNA or protein. Other annotations might be "DNA-binding protein," "possible hairpin RNA," or "tyrosine kinase"—with no additional detail. In short, although some genome annotations summarize experimentally verified facts, many others are hypotheses and guides to future research.

Genome sequences contain information about ancestry and evolution, and so comparisons among genomes can reveal how different species are related. For example, the sequence of the human genome is significantly more similar to that of the chimpanzee than to that of the gorilla, indicating a more recent common ancestry of humans and chimpanzees (Chapter 1).

Analysis of the similarities and differences in protein-coding genes and other types of sequence in the genomes of different species is an area of study called **comparative genomics.** Such studies help us understand how genes and genomes evolve. They can also guide genome annotation because the sequences of important functional elements are often very similar among genomes of different organisms. Sequences that are similar in different organisms are said to be **conserved.** A sequence motif that is conserved is likely to be important even if its function is unknown, since it has changed very little over evolutionary time.

### The HIV genome illustrates the utility of genome annotation and comparison.

HIV and related viruses provide an example of how genomes are annotated and how comparisons among genome sequences can reveal evolutionary relationships. A **virus** is a small infectious agent that contains a nucleic acid genome packaged inside a protein coat called a capsid. In some viruses, a lipid envelope surrounds the capsid. Viruses can bind to surface receptor molecules on cells of the host organism, enabling the viral genome to enter the cell. Infection of a host cell is essential to viral reproduction because viruses use cellular ATP and hijack cellular machinery to replicate, transcribe, and translate their genome in order to make more viruses.

**FIG. 13.5**  Identification of exons and introns by comparison of genomic DNA with mRNA sequence.

**FIG. 13.6** Evolutionary relationships among viruses related to HIV.

Whereas the genome of all cells consists of double-stranded DNA, the genomes of viruses may be double-stranded DNA, single-stranded DNA, double-stranded RNA, or single-stranded RNA. The genome of HIV is single-stranded RNA. The sequence of the HIV genome identifies it as a retrovirus that replicates via a DNA intermediate that can be incorporated into the host genome. More narrowly, the sequence of the HIV genome groups it among the mammalian lentiviruses, so named because of the long lag between the initial time of infection and the appearance of symptoms (*lenti-* means "slow").

**Fig. 13.6** shows the evolutionary relationships among a sample of lentiviruses, grouped according to the similarity of their genome sequences. The evolutionary tree shows that closely related viruses have closely related hosts. For example, cat lentiviruses are more closely related to one another than they are to cow lentiviruses. This observation implies that the genomes of the viruses evolve along with the genomes of their hosts. A second feature shown by the evolutionary tree is that human HIV originated from at least two separate simian viruses that switched hosts from simians (most likely chimpanzees) to humans.

The annotated sequence of the HIV genome tells us a lot about the biology of this virus (**Fig. 13.7**). Many details are left out here, but the main point is to show the functional elements of HIV in the form of an annotated genome. The open reading frame denoted *gag* encodes protein components of the capsid, *pol* encodes proteins needed for reverse transcription of the viral RNA into DNA and incorporation into the host genome, and *env* encodes proteins that are embedded in the lipid envelope. The annotation in Fig. 13.7 also includes the genes *tat* and *rev,* encoding proteins essential for the HIV life cycle, as well as the genes *vif, vpr, vpu,* and *nef,* which encode accessory proteins that enhance virulence in organisms. Identification of the genes necessary to complete the HIV cycle is the first step to finding drugs that can interfere with the cycle and prevent infection.

→ **Quick Check 3** Fig. 13.6 shows that closely related lentiviruses have closely related host organisms. Given this pattern, how many different kinds of organisms would you expect one lentivirus to be capable of infecting?

## 13.3 GENE NUMBER, GENOME SIZE, AND ORGANISMAL COMPLEXITY

Before whole-genome sequencing, molecular biology and evolution research tended to focus on single genes. Researchers studied how individual genes are turned on and off, and how a gene in one organism is related to a gene in another organism. Now, with the availability of genome sequences from multiple species, it is possible to make comparisons across full genomes. Some of the results have been striking.

### Gene number is not a good predictor of biological complexity.

The complete genome sequences of many organisms allow us to make comparisons among them. One of the surprising results is the relatively low number of genes in humans relative to the

**FIG. 13.7** The annotated genome of HIV.

**TABLE 13.1    Gene Numbers of Several Organisms.**
Humans have more genes than fruit flies and nematode worms, but not as many as might be expected on the basis of complexity.

| COMMON NAME | SPECIES NAME | APPROXIMATE NUMBER OF GENES |
|---|---|---|
| Mustard plant | *Arabidopsis thaliana* | 27,000 |
| Human | *Homo sapiens* | 25,000 |
| Nematode worm | *Caenorhabditis elegans* | 20,000 |
| Fruit fly | *Drosophila melanogaster* | 14,000 |
| Baker's yeast | *Saccharomyces cerevisiae* | 6000 |
| Gut bacterium | *Escherichia coli* | 4000 |

number in other species sequenced (**Table 13.1**). Because humans are so much more complex in cell number and type, and in behavior, it was expected that we would have many more genes than we do.

Humans have about 25,000 genes. Surprisingly, the mustard plant *Arabidopsis*, a common weed and model organism in plant biology, has 27,000, about the same number. Perhaps the most sobering comparison is with the nematode worm *Caenorhabditis elegans*, another model organism, which has only 959 cells (we have an estimated 100 trillion), of which 302 are nerve cells that form the worm's brain (our brain has 100 billion nerve cells). Despite having 100 million times as many cells, we have roughly the same number of genes as the worm. How can we account for the disconnect between levels of complexity and gene number?

One hypothesis is that human cells are able to do many more things with the genes they have. A number of mechanisms are likely to be important. The expression of our relatively few genes can be regulated in many subtle ways, causing different gene products to be made in different amounts in different cells at different times (Chapters 19 and 20). This differential gene expression contributes to the large number of different cell types. Human proteins interact with one another so that, even though there are relatively few actual proteins, they are capable of combining in many different ways to perform different functions. And we know that a single gene may yield multiple proteins, either because different exons are spliced together to make different proteins (Chapter 3) or because the proteins undergo biochemical changes after they have been translated.

Overall, the discovery of the relatively low human gene number poses many tantalizing evolutionary questions and suggests that major evolutionary changes can be accomplished not only by the acquisition of whole new genes, but also by modifying existing genes and their regulation in subtle ways.

## Viruses, bacteria, and archaeons have small, compact genomes.

As well as comparing numbers of genes, we can also compare sizes of genomes in different organisms. Before making such comparisons, we need to understand how genome size is measured. Genomes are measured in numbers of base pairs, and the yardsticks of genome size are a thousand base pairs (a kilobase, kb), a million base pairs (a megabase, Mb), and a billion base pairs (a gigabase, Gb).

Most viral genomes range in size from 3 kb to 300 kb, but a few are very large. The largest viral genome, found in a virus that infects the amoeba *Acanthamoeba polyphaga*, is 1.2 Mb. This viral genome contains almost 1000 protein-coding genes, including some for sugar, lipid, and amino acid metabolism not found in any other viruses.

The largest viral genome is twice as large as that of the bacterium *Mycoplasma genitalium*. At 580 kb and encoding only 471 genes, the genome of *M. genitalium* is the smallest known among free-living bacteria, those capable of living entirely on their own. The complete sequence of small bacterial genomes has allowed researchers to define the smallest, or minimal, genome (and therefore the minimal set of proteins) necessary to sustain life. Current findings suggest that the small *M. genitalium* is about two times larger than the minimal genome size thought to be necessary to encode all the functions essential to life.

The genomes of bacteria and archaeons are information dense, meaning that most of the genome has a defined function. Roughly speaking, 90% or more of their genomes consist of

**TABLE 13.2    Genome Sizes of Several Organisms**
Genome size varies tremendously among eukaryotic organisms, and there is no correlation between genome size and the complexity of an organism.

| COMMON NAME | SPECIES NAME | APPROXIMATE GENOME SIZE (MB) |
|---|---|---|
| Fruit fly | *Drosophila melanogaster* | 180 |
| Fugu fish | *Fugu rubripes* | 400 |
| Boa constrictor | *Boa constrictor* | 2100 |
| Human | *Homo sapiens* | 3100 |
| Locust | *Schistocerca gregaria* | 9300 |
| Onion | *Allium cepa* | 18,000 |
| Newt | *Amphiuma means* | 84,000 |
| Lungfish | *Protopterus aethiopicus* | 140,000 |
| Fern | *Ophioglossum petiolatum* | 160,000 |
| Amoeba | *Amoeba dubia* | 670,000 |

protein-coding genes. Bacterial genomes range in size from 0.5 to 10 Mb. The bigger genomes have more genes, allowing these bacteria to synthesize small molecules that other bacteria have to scrounge for, or to use chemical energy in the covalent bonds of substances that other bacteria cannot. Archaeons, whose genomes range in size from 0.5 to 5.7 Mb, have similar capabilities.

## Among eukaryotes, there is no relationship between genome size and organismal complexity.

In eukaryotes, just as the number of genes does not correlate well with organismal complexity, the size of the genome is unrelated to the metabolic, developmental, and behavioral complexity of the organism (**Table 13.2**). The range of genome sizes is huge, even among similar organisms (**Fig. 13.8**). The largest eukaryotic genome exceeds the size of the smallest by a factor of more than 500,000—and both the smallest and the largest are found among protozoa. The range among flowering plants (angiosperms) is about three orders of magnitude, and the range among animals is about seven orders of magnitude. One species of lungfish has a genome size more than 45 times larger than the human genome (Table 13.2). Clearly, there is no relationship between the size of the genome and the complexity of the organism.

The disconnect between genome size and organismal complexity is called the **C-value paradox.** The C-value is the amount of DNA in a reproductive cell, and the "paradox" is the apparent contradiction between genome size and organismal complexity, and hence the difficulty of predicting one from the other.

If organismal complexity is not a good predictor of genome size, what is? In eukaryotic organisms, large genomes can differ from small ones for a number of reasons. One reason

is **polyploidy,** or having more than two sets of chromosomes in the genome. Polyploidy is especially prominent in many groups of plants. Humans have two sets of 23 chromosomes, giving us 46 chromosomes in total. But the polyploid bread wheat *Triticum aestivum,* for example, has six sets of seven chromosomes.

Polyploidy, which is widespread, has played an important role in plant evolution. Many agricultural crops are polyploid, including wheat, potato, olive, banana, sugarcane, and coffee. Among flowering plants, it is estimated that 30% to 80% of existing species have polyploidy in their evolutionary history, either because of the duplication of the complete set of chromosomes in a single species, or because of hybridization, or crossing, between related species followed by duplication

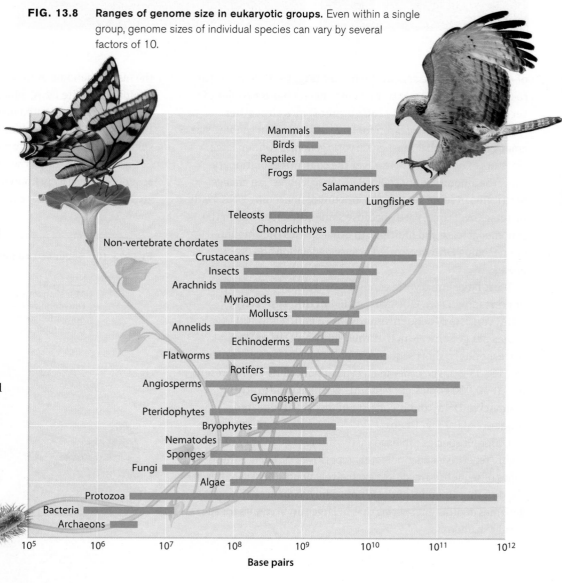

**FIG. 13.8** **Ranges of genome size in eukaryotic groups.** Even within a single group, genome sizes of individual species can vary by several factors of 10.

**FIG. 13.9**  **Polyploidy in plants.** This species was formed by hybridization between two different species, and it contains a full set of chromosomes from each parent (colored yellow and green).

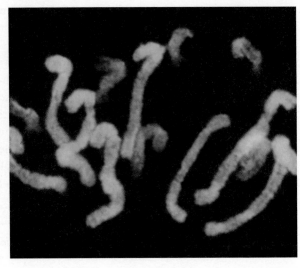

of the chromosome sets in the hybrid (**Fig. 13.9**). Some ferns take polyploidy to an extreme: One species has 84 copies of a set of 15 chromosomes—1260 chromosomes altogether.

Another important reason for large differences in genome size among species is that most eukaryotic genomes contain large amounts of noncoding DNA. Over the course of evolutionary time, the amount of noncoding DNA in a genome can change drastically. In some species, the amount of noncoding DNA vastly increases over time, partly because some sequences within the genome increase and multiply. In other species the amount of noncoding DNA decreases because of deletion and other processes.

Most large eukaryotic genomes contain two main types of noncoding DNA. One type, called **highly repetitive DNA,** consists of sequences present in more than 100,000 copies per genome. Much of the highly repetitive DNA corresponds to short sequence repeats such as we saw earlier (see Fig. 13.2c). The other main type of noncoding DNA, present in 100–10,000 copies per genome, is called **moderately repetitive DNA,** and it consists of dispersed repeated sequences (see Fig. 13.2b).

Different species can have vastly different quantities of highly repetitive and moderately repetitive DNA. Since these types of sequence are almost exclusively noncoding DNA, it is the differing amounts of these noncoding sequences among the genomes of different species that in large part accounts for the C-value paradox.

→  **Quick Check 4**  Given our knowledge of genome sizes in different organisms, would you predict that humans or amoebas (a single-celled eukaryote) have a larger genome?

**About half of the human genome consists of repetitive DNA and transposable elements.**

Complete genome sequencing has allowed the different types of noncoding DNA to be specified more precisely in a variety of organisms. It came as a great surprise to learn that in the human genome only about 2.5% of the genome actually codes for proteins. The other 97.5% includes sequences we have encountered earlier, including sequences that specify noncoding RNA, repetitive DNA, and noncoding regions of genes, such as introns.

In **Fig. 13.10**, we see the principal repetitive sequences in the human genome. Among the highly repetitive sequences is α (alpha) satellite DNA, which consists of tandem copies of a 171-bp sequence repeated near each centromere an average of 18,000 times. The α satellite DNA is essential for attachment of spindle fibers to the centromeres during cell division (Chapter 11).

Fig. 13.10 also shows the proportions of several types of sequence collectively known as **transposable elements,** which are DNA sequences that can replicate and insert themselves

**FIG. 13.10**  **Sequence composition of the human genome.**

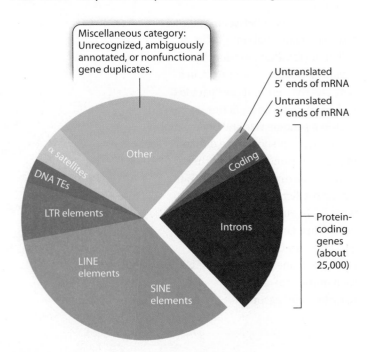

into new positions in the genome. As a result, they have the potential to increase their copy number in the genome over time. Transposable elements are sometimes referred to as "selfish" DNA because it seems that their only function is to duplicate themselves and proliferate in the genome, making them the ultimate parasite.

Transposable elements make up about 45% of the DNA in the human genome. They can be classified into two groups based on the way they replicate. One class consists of **DNA transposable elements (DNA TEs),** which replicate and transpose via DNA replication and repair. The other class consists of elements that transpose by means of an RNA intermediate. Among these are **LTR elements,** which are characterized by long repeated sequences, called long terminal repeats (LTRs), at their ends. Also in this class are two types that are distinguished by their length: the **LINEs** (long interspersed nuclear elements), of about 1000 base pairs, and the **SINEs** (short interspersed nuclear elements), with about 300 base pairs.

## 13.4 ORGANIZATION OF GENOMES

The genomes of all organisms are large relative to the size of the cell. For example, if the circular genome of the intestinal bacterium *Escherichia coli* were fully extended, its length would be 200 times greater than the diameter of the cell itself. The fully extended length of DNA in human chromosome 1, our longest chromosome, would be 10,000 times greater than the diameter of the average human cell. There is consequently a need to package an enormous length of DNA into a form that will fit inside the cell while still allowing the DNA to replicate and carry out its coding functions. The mechanism of packaging differs substantially in bacteria, archaeons, and eukaryotes. We focus here on bacteria and eukaryotes, primarily because less is known about how DNA is packaged in archaeons.

### Bacterial cells package their DNA as a nucleoid composed of many loops.

Bacterial genomes are circular, and the DNA double helix is underwound, which means that it makes fewer turns in going around the circle than would allow every base in one strand to pair with its partner base in the other strand. Underwinding is caused by an enzyme, **topoisomerase II,** that breaks the double helix, rotates the ends, and then seals the break. Underwinding creates strain on the DNA molecule, which is relieved by the formation of **supercoils,** in which the DNA molecule coils on itself. Supercoiling allows all the base pairs to form, even though the molecule is underwound. (You can make your own supercoil by stretching and twirling the ends of a rubber band, then relaxing the stress slightly to allow the twisted part to form coils around itself.) Supercoils that result from underwinding

are called negative supercoils, and those that result from overwinding are positive supercoils. In most organisms, DNA is negatively supercoiled.

In bacteria, the supercoils of DNA form a structure with multiple loops called a **nucleoid** (**Fig. 13.11**). The supercoil loops are bound together by proteins. In *E. coli*, the nucleoid has about 100 loops, each containing about 50 kb of DNA. In addition to the protein binding that forms the loops, the negative supercoiling of the DNA serves to compress the molecule into a compact volume.

The supercoils in the loops of the nucleoid can be relaxed by introducing nicks, or interruptions, in the DNA backbones by means of enzymes or chemicals. The nicks allow the strands to rotate around each another to remove the supercoils, and the

**FIG. 13.11** **Bacterial nucleoid.** The circular bacterial chromosome twists on itself to form supercoils, which are anchored by proteins.

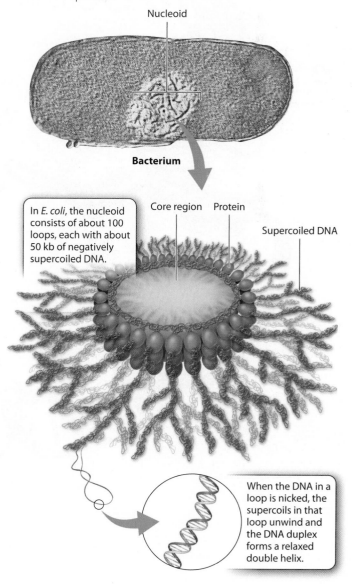

Nucleoid

**Bacterium**

In *E. coli*, the nucleoid consists of about 100 loops, each with about 50 kb of negatively supercoiled DNA.

Core region    Protein

Supercoiled DNA

When the DNA in a loop is nicked, the supercoils in that loop unwind and the DNA duplex forms a relaxed double helix.

**FIG. 13.12**  Levels of chromosome condensation.

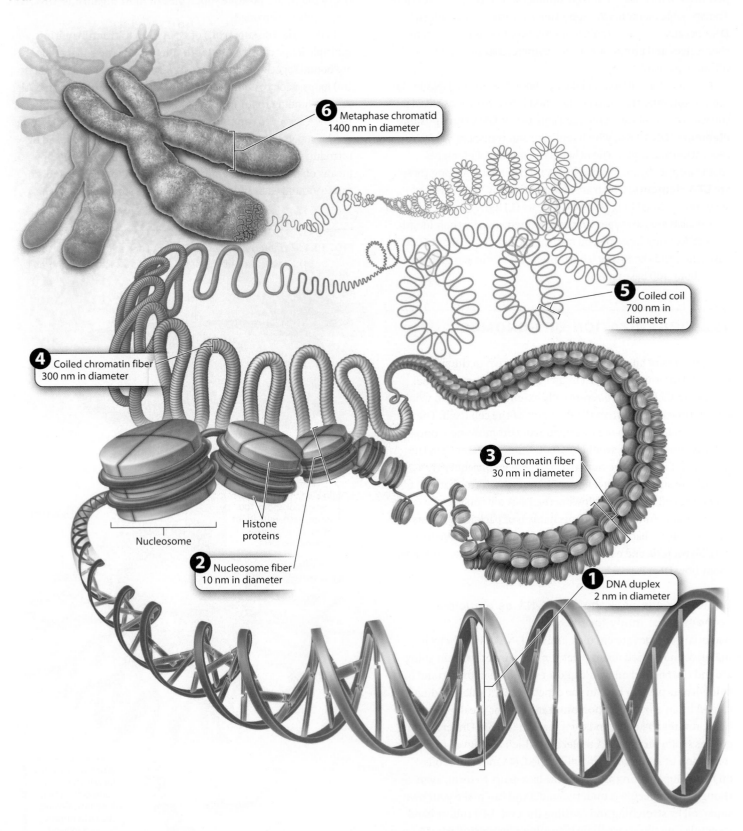

**6** Metaphase chromatid 1400 nm in diameter

**5** Coiled coil 700 nm in diameter

**4** Coiled chromatin fiber 300 nm in diameter

**3** Chromatin fiber 30 nm in diameter

Nucleosome

Histone proteins

**2** Nucleosome fiber 10 nm in diameter

**1** DNA duplex 2 nm in diameter

DNA molecule relaxes into the standard double helix. Any nick will relieve the supercoils in a loop, but only in the loop that is nicked. This means that each loop is supercoiled independently of the others (because of the proteins that hold the loops together).

## Eukaryotic cells package their DNA as one molecule per chromosome.

In eukaryotic cells, DNA in the nucleus is packaged differently from that in bacteria (**Fig. 13.12**). Eukaryotic DNA is first wrapped around a group of histone proteins called a **nucleosome** (Chapter 3). Each nucleosome consists of two molecules, and each molecule consists of four different histone proteins. The histone proteins are rich in the amino acids lysine and arginine, whose positive charges neutralize the negative charges of the phosphates along the backbone of each DNA strand. This first level of packaging of the DNA is sometimes referred to as "beads on a string," with the nucleosomes the beads, and the DNA the string.

The nucleosomes and associated DNA are then coiled to form a structure called the **30-nm chromatin fiber** (Fig. 13.12). When the chromosomes in the nucleus condense in preparation for cell division, each chromosome becomes progressively shorter and thicker as the 30-nm fiber coils in a manner that is still not fully understood. The progressive coiling constitutes **chromosome condensation,** an active, energy-consuming process requiring the participation of several types of proteins.

Greater detail of the structure of a fully condensed chromosome is revealed when the histones are chemically removed (**Fig. 13.13**). Without histones to coil around, the DNA spreads out in loops around a supporting protein structure called the chromosome **scaffold.** Each loop of relaxed DNA is 30 to 90 kb long and anchored to the scaffold at its base. Before removal of the histones, the loops are compact and supercoiled. Each human chromosome contains 2000–8000 such loops, depending on its size.

Despite intriguing similarities between the nucleoid model in Fig. 13.11 and the chromosome scaffold model in Fig. 13.13, the structures evolved independently and make use of different types of protein to bind the DNA and to form the folded structure of DNA and protein. Furthermore, the similarities are misleading, in part because the size of the eukaryotic chromosome is vastly greater than the size of the bacterial nucleoid. To appreciate the difference in scale, keep in mind that the volume of a fully condensed human *chromosome* is five times larger than the volume of a bacterial *cell*.

## The human genome consists of 22 pairs of chromosomes and two sex chromosomes.

As emphasized in Chapter 11, the orderly process of meiosis is possible because chromosomes occur in pairs. The pairs usually match in size, general appearance, and position of the centromere, but there are exceptions, such as the X and Y sex chromosomes. The pairs of chromosomes that match in size and appearance are called **homologous chromosomes**. The members of each pair of homologous chromosomes have the same genes arranged in the same order along their length. If the DNA duplexes in each pair of homologs were denatured, each DNA strand could form a duplex with its complementary strand from the other homolog. There would be some differences in DNA sequence due to genetic variation, but not so many differences as to prevent DNA hybridization.

Chromosome painting illustrates the nearly identical nature of the DNA molecules in each pair of homologs. In this technique, individual chromosomes are isolated from cells in metaphase of mitosis. Metaphase of mitosis is the easiest stage in which to isolate chromosomes because of the availability of chemicals that prevent the spindle from forming. These chemicals block the cell cycle at metaphase, so cells progress to metaphase and then stop.

Once the chromosomes have been isolated, the DNA from each chromosome is fragmented, denatured, and labeled with a unique combination of fluorescent dyes. Under fluorescent light, the dyes give the DNA in each type of chromosome a different color. The fluorescently labeled DNA fragments are then mixed and hybridized to intact metaphase chromosomes from another cell. Each labeled fragment hybridizes to its complementary sequence in the metaphase chromosomes, "painting" each metaphase chromosome with dye-labeled DNA fragments.

**FIG. 13.13** (a) A chromosome with histones and (b) a chromosome depleted of histones, showing the underlying scaffold.

DNA with histones

Relaxed DNA after histones removed

a

b

Chromosome

Scaffold

**FIG. 13.14   A chromosome paint of chromosomes from a human male.** (a) Condensed chromosomes are "painted" with fluorescent dyes. (b) Chromosomes are arranged in the standard form of a karyotype. The colors in (a) and (b) do not match exactly because the "paints" are different.

A chromosome paint of the chromosomes in a human male is shown in **Fig. 13.14**. Fig. 13.14a shows the chromosomes in the random orientation in which they were found in the metaphase cell, and Fig. 13.14b shows them arranged in a standard form called a **karyotype** (Chapter 11). To make a karyotype, the images of the homologous chromosomes are arranged in pairs from longest to shortest, with the sex chromosomes placed at the lower right. In this case, the sex chromosomes are XY (one X chromosome and one Y chromosome), indicating that the individual is male, whereas in a female the sex chromosomes would be XX (a pair of X chromosomes). Including the sex chromosomes, humans have 23 pairs of chromosomes.

An important observation from the chromosome paint shown in Fig. 13.14 is that the two members of each pair of homologous chromosomes show the same pattern of fluorescent color. This means that a particular labeled DNA fragment only hybridized with the two homologs of one chromosome. Hence, the DNA in each pair of homologous chromosomes is different from that in any other pair of homologous chromosomes.

Higher resolution of human chromosomes can be obtained by the use of stains that bind preferentially to certain chromosomal regions and produce a visible pattern of **bands,** or crosswise striations, in the chromosomes. One such stain is the Giemsa stain; a Giemsa-stained karyotype is shown in **Fig. 13.15**. Note that each chromosome has a unique pattern of bands and that homologous chromosomes can readily be identified by their identical banding patterns. Procedures using the Giemsa stain yield about 300 bands that are used as landmarks for describing the location of genes along the chromosome.

Every species of eukaryote has its characteristic number of chromosomes. The number differs from one species to the next, with little relation between chromosome number and genome size. Despite variation in number, the rule that chromosomes come in pairs holds up pretty well, with the exception of the sex chromosomes. Polyploids, too, are an exception, but even in polyploids the number of copies of each homologous chromosome is usually an even number, so there are pairs of homologs after all.

The occurrence of chromosomes in pairs allows eukaryotes to reproduce sexually. When reproductive cells are formed during meiotic cell division, each cell receives one and only one copy of each of the pairs of homologous chromosomes (Chapter 11). When reproductive cells from two individuals fuse to form an offspring cell, the chromosome number characteristic of the species is reconstituted.

## Organelle DNA forms nucleoids that differ from those in bacteria.

Most eukaryotic cells contain mitochondria, and many contain chloroplasts. Each type of organelle has its own DNA, meaning that eukaryotic cells have multiple genomes. Each has a **nuclear genome** consisting of the DNA in the chromosomes. Cells with mitochondria also have a **mitochondrial genome**, and those with chloroplasts also have a **chloroplast genome**.

Because the genome organization and mechanisms of protein synthesis in these organelles resemble those of bacteria, most biologists subscribe to the theory that the organelles originated as free-living bacterial cells that were engulfed by primitive eukaryotic cells billions of years ago (Chapter 27). In Chapters 7

**FIG. 13.15   Giemsa bands in a karyotype of a human male.** Each dashed line marks the position of the centromere.

and 8, we saw that the likely ancestor of mitochondria resembled a group of today's non-photosynthetic bacteria (a group that includes *E. coli*), and the likely ancestor of chloroplasts resembled today's photosynthetic cyanobacteria. In both cases, the DNA of the organelles became smaller during the course of evolution because most of the genes were transferred to the DNA in the nucleus. If the products of these transferred genes are needed in the organelles, they are synthesized in the cytoplasm and targeted for entry into the organelles by signal peptides (Chapter 5).

DNA in mitochondria and chloroplasts is often but not always circular, and there are usually multiple copies per organelle. Among animals, the size of a mitochondrial DNA molecule ranges from 14 kb to 18 kb. Mitochondrial DNA in plants is generally much larger, up to 100 kb. Plant chloroplast DNA is more uniform in size than mitochondrial DNA, with a range of 130 to 200 kb.

On the basis of size alone, some kind of packaging of organelle DNA is necessary. For example, the mitochondrial DNA in human cells is a circular molecule of 16 kb. Fully extended, it would have a circumference about as large as that of the mitochodrion itself. From their bacterial origin, you might expect organelle DNA to be packaged as a nucleoid rather than as a chromosome, and indeed it is. But the structures of the nucleoid in mitochondria and chloroplasts differ from each other, and also differ from those in free-living bacteria.

## 13.5 VIRUSES AND VIRAL GENOMES

The first genome sequenced was not the human genome, nor did it come from any of the three domains of life—Bacteria, Archaea, or Eukarya. It was a viral genome. As we discussed, viral genomes are small and compact, so they are easier to sequence than the larger genomes of cellular organisms.

We mostly think of viruses as causing disease, and indeed they do. Familiar examples include influenza ("flu"), polio, and HIV. In some cases, they cause cancer (Case 2: Cancer). In fact, the term "virus" comes from the Latin for "poison." But viruses play other roles as well. Some viruses transfer genetic material from one cell to another. This process is called horizontal gene transfer to distinguish it from parent-to-offspring (vertical) gene transfer. Horizontal gene transfer has played a major role in the evolution of bacterial and archaeal genomes, as well as in the origin and spread of antibiotic-resistance genes (Chapter 26). Molecular biologists have learned to make use of this ability of viruses to deliver genes into cells. Viruses also play vital roles in ecosystems: Scientists estimate that 1 to 100 million viruses can be found in a teaspoon of seawater. They are not only abundant, but also have important functions, infecting and limiting the number of bacteria and other microorganisms in soil and water.

In this section, we focus on viral diversity with an emphasis on viral genomes. Thousands of viruses have been described in detail, and probably millions more have yet to be discovered.

### The host range of a virus is determined by viral and host surface proteins.

Recall that a virus consists simply of nucleic acid, a protein coat called a **capsid**, and sometimes a lipid **envelope** (Chapter 1). Viruses can reproduce only by infecting living cells and subverting cellular metabolism and protein synthesis to produce more viruses. Most viruses are tiny, some hardly larger than a eukaryotic ribosome, 25–30 nm in diameter. Roughly speaking, the average size of a virus, relative to that of the host cell it infects, may be compared to the size of an average person relative to that of a commercial airliner.

All known cells and organisms are susceptible to viral infection, including bacteria, archaeons, and eukaryotes. A cell in which viral reproduction occurs is called a **host cell**. Some viruses kill the host cell; others do not. Although viruses can infect all types of living organism, a given virus can infect only certain species or types of cell. At one extreme, a virus can infect just a single species. Smallpox, which infects only humans, is a good example. In this case, we say that the virus has a narrow host range. For other viruses, the host range can be broad. For example, rabies infects many different types of mammal, including squirrels, dogs, and humans. Similarly, tobacco mosaic virus, a plant virus, infects more than 100 different species of plants (**Fig. 13.16**). No matter how broad the host range, however, plant viruses cannot infect bacteria or animals, for example, and viruses that infect bacteria cannot infect plants or animals.

Host specificity relates to the way that viruses gain entry into cells. Proteins on the surface of the capsid or envelope (if present) bind to proteins on the surface of host cells. These proteins interact in a specific manner, so the presence of the host protein on the cell surface determines which host cells a virus can infect. For example, a protein on the surface of the HIV envelope called gp120 (encoded by the *env* gene; see Fig. 13.7) binds to a protein called CD4 on the surface of certain immune cells (Chapter 43), so HIV infects these cells but not other human cells.

Following attachment, the virus gains entry into the cell, where it can replicate by using the host cell machinery to make new viruses. In some cases, the viral genome integrates into the host cell genome (Chapter 19).

The host range of a virus can change because of mutation and other mechanisms (Chapter 43). For example, avian, or bird, flu is a type of influenza virus that was once restricted primarily to birds but now can infect humans. The first reported case in humans was in 1996, and the disease has since spread widely. Similarly, HIV once infected nonhuman primates, but its host range expanded in the twentieth century to include humans. Canine distemper virus, which infects dogs, expanded its host range in the early 1990s, leading to infection and death of lions in Tanzania.

**FIG. 13.16** A plant infected by tobacco mosaic virus.

## Viruses can be classified by their genomes.

The genomes of viruses are diverse. Some genomes are composed of RNA and others of DNA. Some are single stranded, others are double stranded, and still others have both single- and double-stranded regions. Some are circular and others are made up of a single piece or multiple linear pieces of DNA (called linear and segmented genomes, respectively).

Unlike forms of cellular life, there is no evidence that all viruses share a single common ancestor. Different types of virus may have evolved independently more than once. Since classification of viruses based on evolutionary relatedness is not possible, other criteria are necessary. One of the most useful classifications is based on the type of nucleic acid the virus contains and how the messenger RNA, which produces viral proteins, is synthesized (**Fig. 13.17**). The classification is called the Baltimore system after David Baltimore, who shared the Nobel Prize in Physiology or Medicine in 1975 for the discovery of the enzyme reverse transcriptase.

According to the Baltimore system, there are seven major groups of viruses, designated I–VII, as shown in Fig. 13.17.

These groups are largely based on whether their nucleic acid is double-stranded DNA, double-stranded RNA, partially double-stranded and partially single-stranded, or single-stranded RNA or DNA with a positive (+) or negative (−) sense. The sense of a nucleic acid molecule is positive if its sequence is the same as the sequence of the mRNA that is used for protein synthesis, and negative if it is the complementary sequence. For example, a (+)RNA strand has the same nucleotide sequence as the mRNA, whereas a (−)RNA strand has the complementary sequence. Similarly, a (+)DNA strand has the same sequence as the mRNA, and a (−)DNA strand has the complementary sequence (except that U in RNA is replaced with T in DNA). Because mRNA is synthesized from a DNA template, it is the (−)DNA strand that is used for mRNA synthesis (Chapter 3).

Two groups synthesize mRNA by the enzyme reverse transcriptase, and therefore are placed into their own groups (VI and VII). Reverse transcriptase synthesizes DNA from an RNA template (Fig. 13.17), and therefore reverses the usual flow of genetic information from DNA to RNA (see Fig. 4.20).

**FIG. 13.17** **The Baltimore system of virus classification.** This system classifies viruses by type of nucleic acid and the way mRNA is produced.

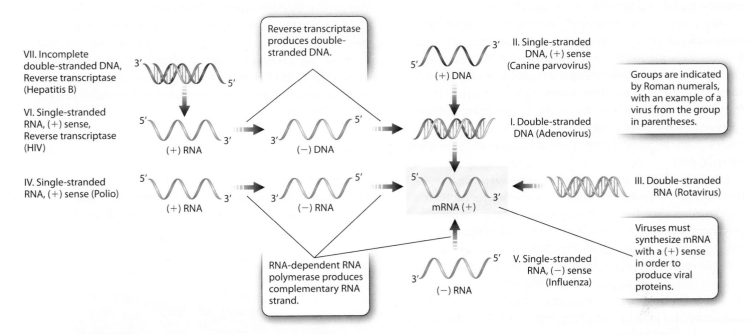

As we saw earlier, genome size varies greatly among different viruses. RNA viral genomes tend to be smaller than DNA viral genomes. Most eukaryotic viruses that have RNA genomes of one type or another, and most plant viruses including tobacco mosaic virus are in group IV. Among bacterial and archaeal viruses, most genomes consist of double-stranded DNA.

## Viruses have diverse sizes and shapes.

Viruses show a wide variety of shapes and sizes. Three examples are shown in **Fig. 13.18**. The T4 virus (Fig. 13.18a) infects cells of the bacterium *Escherichia coli*. Viruses that infect bacterial cells are often called **bacteriophages**, which literally means "bacteria eaters." The T4 bacteriophage has a complex structure that includes a head composed of protein surrounding a molecule

**FIG. 13.18** **Virus structure.** Viruses come in a variety of shapes, including (a) the head-and-tail structure of a bacteriophage, (b) the helical shape of tobacco mosaic virus, and (c) the icosahedral shape of adenovirus.

of double-stranded DNA, a tail, and tail fibers. In infecting a host cell, the T4 tail fibers attach to the surface, and the DNA and some proteins are injected into the cell through the tail.

Most viruses are not structurally so complex. Consider the tobacco mosaic virus, which infects plants (see Fig. 13.16). It has a helical shape formed by the arrangement of protein subunits entwined with a molecule of single-stranded RNA (Fig. 13.18b). Tobacco mosaic virus was the first virus to be discovered, revealed in experiments showing that the infectious agent causing brown spots and discoloration of tobacco leaves was so small that it could pass through the pores of filters that could trap even the smallest bacterial cells.

Many viruses have an approximately spherical shape formed from polygons of protein subunits that come together at their edges to form a polyhedral capsid. Among the most common polyhedral shapes is an icosahedron, which has 20 identical triangular faces. The example in Fig. 13.18c is adenovirus, a common cause of upper respiratory infections in humans. Many viruses that infect eukaryotic cells, such as adenovirus, are surrounded by a glycoprotein envelope composed of a lipid bilayer with embedded proteins and glycoproteins that recognize and attach to host cell receptors.

**Viruses are capable of self-assembly.**

The nucleic acid contained in a virus particle contains the genetic information needed to specify all the structural components of the virus. Progeny particles are formed according to a principle of **molecular self-assembly:** When the viral components are present in the proper relative amounts and under the right conditions, the components interact spontaneously to assemble themselves into the mature virus particle.

Tobacco mosaic virus illustrates the process of self-assembly. In the earliest stages, the coat-protein monomers assemble into two circular layers forming a cylindrical disk. This disk binds with the RNA, and the combined structure forms the substrate for polymerization of all the other protein monomers into a helical filament that incorporates the rest of the RNA as the filament grows (**Fig. 13.19**). The mature virus particle consists of 2130 protein monomers and 1 single-stranded RNA molecule of 6400 ribonucleotides.

**FIG. 13.19  Tobacco mosaic virus.** The protein and RNA components assemble themselves inside a cell.

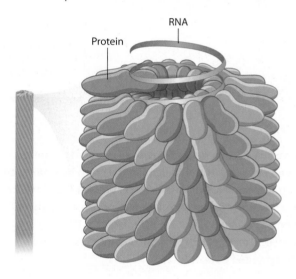

# Core Concepts Summary

**13.1 A GENOME IS ALL THE GENETIC MATERIAL OF AN ORGANISM TRANSMITTED FROM PARENTS TO OFFSPRING, AND ITS SEQUENCE IS THE ORDER OF BASES ALONG A DNA MOLECULE.**

The sequence of an organism's genome can be determined by breaking up the genome into small fragments, sequencing these fragments, and then putting the sequences together at their overlaps. page 13-2

Sequences that are repeated in the genome can make sequence assembly difficult. page 13-2

**13.2 RESEARCHERS ANNOTATE GENOME SEQUENCES TO IDENTIFY GENES AND OTHER FUNCTIONAL ELEMENTS.**

Genome annotation is the process by which the types and locations of the different types of sequences, such as protein-coding genes, are identified. page 13-4

Genome annotation sometimes involves scanning the DNA sequence for characteristic sequence motifs. page 13-5

Comparison of DNA sequences with messenger RNA sequences reveals the intron–exon structure of protein-coding genes. page 13-6

By comparing annotated genomes of different organisms, we can gain insight into their ancestry and evolution. page 13-6

The annotated HIV genome shows it is a retrovirus, and it contains the genes *gag, pol,* and *env.* page 13-7

**13.3 THE NUMBER OF GENES IN A GENOME AND THE SIZE OF A GENOME DO NOT CORRELATE WELL WITH THE COMPLEXITY OF AN ORGANISM.**

The C-value paradox describes the observation that the size of a genome (measured by its C-value) does not correlate with an organism's complexity. page 13-9

In eukaryotes, the C-value paradox can be explained by differences in the amount of noncoding DNA. Some genomes have a lot of noncoding DNA; others do not. page 13-10

### 13.4 THE ORDERLY PACKAGING OF DNA ALLOWS IT TO CARRY OUT ITS FUNCTIONS AND FIT INSIDE THE CELL.

Bacteria package their circular DNA in a structure called a nucleoid. page 13-11

Eukaryotic cells package their DNA into linear chromosomes. page 13-13

DNA in eukaryotes is wound around groups of histone proteins called nucleosomes, which in turn coil to form higher-order structures, such as the 30-nm fiber. page 13-13

Diploid organisms have two copies of each chromosome, and these two copies are called homologous chromosomes. page 13-13

Humans have 23 pairs of chromosomes, including the X and Y sex-chromosome pair. Females are XX, and males are XY. page 13-14

The genomes of mitochondria and chloroplasts are organized into nucleoids that resemble, but are distinct from, those of bacteria. page 13-14

### 13.5 VIRUSES HAVE DIVERSE GENOMES, BUT ALL REQUIRE A HOST CELL TO REPLICATE.

Viruses can infect all types of organism, but a given virus can infect only some types of cell. page 13-15

The host range of a virus is determined by proteins on viral and host cells. page 13-15

Viruses can be classified by the Baltimore system, which defines seven groups on the basis of type of nucleic acid and the way the mRNA is synthesized. page 13-16

Viruses have diverse shapes, including head-and-tail, helical, and icosahedral. page 13-17

Viruses are capable of self-assembly under the right conditions. page 13-18

## Self-Assessment

1. Describe a method for determining the complete genome sequence of an organism.

2. Describe several types of sequence present in many genomes.

3. Explain the purpose of genome annotation.

4. Describe how the comparison of genomic DNA to messenger RNA can identify the exons and introns in a gene.

5. Explain how comparing the sequences of two genomes can help to infer evolutionary relationships.

6. Define the C-value paradox and explain why it is a paradox.

7. Compare and contrast the mechanisms by which bacterial cells and eukaryotic cells package their DNA.

8. Define "homologous chromosomes" and describe a technique that you could use to show their similarity.

9. Describe the steps necessary to synthesize mRNA from each of the following: double-stranded DNA, single-stranded (+)DNA, single-stranded (−)DNA, single-stranded (+)RNA, and single-stranded (−)RNA.

Do you understand the chapter's Core Concepts? Log in to check your answers to the Self-Assessment questions, then practice what you've learned and reinforce this chapter's concepts by working through the problems and multimedia tutorials provided there.

**www.biologyhowlifeworks.com**

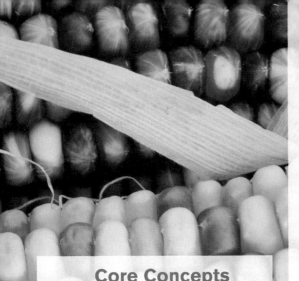

## CHAPTER 14

# MUTATION AND DNA REPAIR

## Core Concepts

**14.1** Mutations are very rare for any given nucleotide and occur randomly without regard to the needs of an organism.

**14.2** Small-scale mutations include point mutations, insertions and deletions, and movement of transposable elements.

**14.3** Chromosomal mutations involve large regions of one or more chromosomes.

**14.4** DNA can be damaged by mutagens, but most DNA damage is repaired.

The sequencing of the human genome was a great step forward. However, as we saw in the previous chapter, it is a simplification to speak of *the* human genome. Virtually all species have abundant amounts of genetic variation, that is, different nucleotides at the same site. Hence the accepted human genome sequence is actually a composite, displaying the most common nucleotide at most sites.

Differences among genomes arise from **mutations.** Any heritable change in the genetic material is a mutation, and by "heritable" we mean that the mutation is stable and therefore passed on through cell division (meiotic cell division, mitotic cell division, or binary fission). The process by which mutations occur is fundamental in biology because mutation is the ultimate source of genetic variation, which accounts in part for the physical differences we see among individuals, such as differences in hair color, eye color, and height. And, on a much larger scale, genetic variation results in the diversity of organisms on this planet, from bacteria to blue whales.

There are many different types of mutation, from small changes affecting a single base to larger alterations, such as the duplication or deletion of a segment of a chromosome, that affect one or more chromosomes. In this chapter, we examine some of the basic principles of mutation: the different types of mutation, how and when they occur, and how they are repaired. In Chapter 15, we look at common types of mutation, or genetic variation, present in populations, and, in Chapter 16, at how this variation is inherited from one generation to the next.

## 14.1 THE RATE AND NATURE OF MUTATIONS

Most mutations are **spontaneous,** occurring by chance in the absence of any assignable cause. They also occur randomly, unconnected to an organism's needs—it makes no difference whether or not a given mutation would benefit the organism. Whether a favorable mutation does or does not occur is purely a matter of chance. This key principle, that mutations are spontaneous and random, as well as other basic features of mutation, are discussed in this section.

### For individual nucleotides, mutation is a rare event.

For any individual nucleotide, mutation is a very rare event, and the rate of mutation varies among organisms. **Fig. 14.1** compares the probability of the occurrence of a new mutation in a given base pair in a single round of replication in different organisms. As seen in the graph, the mutation rates for different organisms range across almost eight orders of magnitude.

The highest rates of mutation per nucleotide per replication are found among RNA viruses and retroviruses, including HIV. Lower rates occur in DNA viruses, and even lower rates in unicellular organisms such as bacteria and yeast. The rates of mutation

**FIG. 14.1** **Mutation rate.** The rate of mutation per nucleotide per replication varies among organisms.

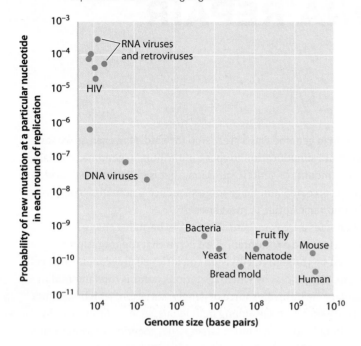

mutations that occur in **germ cells** (the reproductive cells and the cells that give rise to them) and mutations that occur in **somatic cells** (the other cells of the body). In mammals, the rate of mutation per nucleotide per replication is greater in somatic cells than in germ-line cells.

## Across the genome as a whole, mutation is common.

So far, we have considered the mutation rate for any given nucleotide, which varies from organism to organism, but in all cases is relatively small. For a particular cell, the overall mutation rate depends on the number of times DNA replication occurs and the size of the genome. In multicellular organisms, there are sometimes many cycles of DNA replication and cell division that take place in the germ line before meiosis and the formation of reproductive cells. In a human male at age 30, for example, about 400 cycles of DNA replication and cell division take place before meiosis. Therefore, the overall rate of mutation per nucleotide in sperm is 400 times greater than the rate per single cycle of replication plotted in Fig. 14.1. Furthermore, some genomes are much larger than others (Chapter 13), and across a genome as a whole, there will be a greater number of mutations per genome in large genomes than in small ones.

**Fig. 14.2** shows the average number of mutations for a genome in an organismal generation for the same organisms

per nucleotide per DNA replication are nearly the same for all multicellular animals, including mice and humans.

It is unclear why the rates vary over such a large range. RNA viruses and retroviruses might be expected to have a relatively high rate of mutation, first because the backbone of an RNA strand is more prone to spontaneous breakage than that of a DNA strand, and second because the replication of these genomes lacks a proofreading function. For the other genomes, the rate of mutation per nucleotide per DNA replication might simply reflect the limits of proofreading and other repair mechanisms. Some mutations must inevitably occur because of chemical errors such as base mispairing and DNA-strand breakage.

For the cellular organisms plotted in Fig. 14.1, the average probability of a new mutation per DNA replication is about $10^{-10}$, which means that, on average, only 1 nucleotide in every 10 billion is mistakenly substituted for another.

But averaging conceals many details. First, certain nucleotides are especially prone to mutation and can exhibit rates of mutation that are greater than the average by a factor of 10 or more. Sites in the genome that are especially mutable are called **hotspots.** Second, in some multicellular organisms, the rates of mutation differ between the sexes. In humans, for example, the rate of mutation is substantially greater in males than in females. Finally, the rates for the multicellular animals plotted in Fig. 14.1 depend on the type of cell: A distinction must be made between

**FIG. 14.2** **The average number of new mutations per genome per organismal generation.**

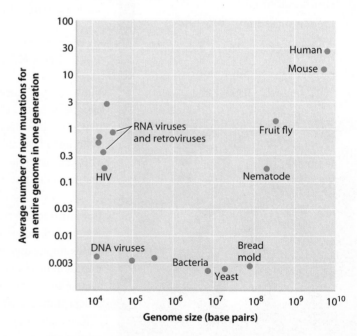

as shown in Fig. 14.1, but in this case taking into account the number of germ-line cell divisions as well as genome size. The picture is now quite different. For the DNA viruses, bacteria, and fungi (the bread mold), the average number of new mutations per genome per generation is remarkably similar, about 0.003. This number indicates that, in bacteria and fungi, roughly one cell division in every eight yields a daughter cell with a new mutation somewhere in the genome.

For RNA viruses and retroviruses, the average number of new mutations per genome reflects the high rate of mutation per nucleotide seen in Fig. 14.1. Among multicellular animals, the average number of new mutations per genome reflects the effects of multiple germ-line cell divisions per generation of organisms as well as larger genome size. It is these mutations, accumulated through many generations and shuffled by recombination and independent assortment (Chapter 11), that account for the genetic diversity of most populations and the genetic uniqueness of each individual.

Such a large number of new mutations as occurs in the human genome would be intolerable in organisms with a high density of protein-coding genes, such as bacteria or fungi. It is tolerable in humans and other mammals only because a mere 2.5% of the genome codes for protein (Chapter 13). The vast majority of new mutations therefore occur in noncoding DNA and are neutral or very nearly neutral in their effects.

**Only germ-line mutations are transmitted to progeny.**
Which is more important—the rate of mutation per cycle of DNA replication or the rate of mutation per organismal generation? That depends on context. Mutations can take place in any type of cell. Those that occur in eggs and sperm and the cells that give rise to these reproductive cells are called **germ-line mutations,** and those in nonreproductive cells are called **somatic mutations.** This distinction is important because somatic mutations affect only the individual in which they occur. In other words, they are not transmitted to future generations. In contrast, germ-line mutations are transmitted to future generations because they occur in reproductive cells.

For germ-line mutations, it is the rate of mutation per organismal generation that matters most. Germ-line mutations are important to the evolutionary process because, through transmission between generations, they may eventually come to be present in many individuals descended from the original carrier.

For somatic mutations, the mutation rate that matters is the rate of mutation per cycle of DNA replication. Although somatic mutations are not transmitted to future generations,

**FIG. 14.3** **Somatic mutation.** Somatic mutations in the Japanese morning glory (*Ipomoea nil*) in cell lineages that differ in their ability to make purple pigment cause sectors of different pigmentation in the flower.

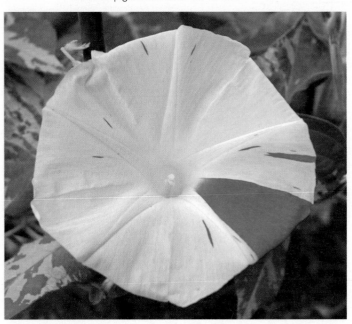

they are transmitted to progeny cells in mitotic cell divisions (Chapter 11). Hence, a somatic mutation affects not only the cell in which it occurs, but also all the cells that descend from it. The areas of different color or pattern that appear in "sectored" flowers, valued as ornamental plants (**Fig. 14.3**), are usually due to somatic mutations in flower-color genes. A mutation in a flower-color gene occurs in one cell, and as it replicates during development of the flower, all its descendants in the cell lineage—the generations of cells that originate from a single ancestral cell—carry that mutation, producing a sector with altered coloration.

Most cancers result from mutations in somatic cells (Chapter 11). In some cases, the mutation increases the activity of a gene that promotes cell growth and division, while in other cases, it decreases the activity of a gene that restrains cell growth and division. In either event, the mutant cell and its descendants escape from one of the normal control processes. Fortunately, a single somatic mutation is usually not sufficient to cause cancer— usually two or three or more mutations in different genes are required to derail control of normal cell division so extensively that cancer results.

To cause cancer, the mutations must all be present in the same cell lineage, which means that they all must have occurred

**FIG. 14.4    Three somatic mutations implicated in the origin of invasive colon cancer.** These mutations must occur in the same cell lineage for cancer to develop.

Normal cells

Mutation in APC gene

Normal colon tissue

Cells mutant in APC gene

Mutation in Ras gene

Small benign polyp

Cells mutant in APC and Ras genes

Mutation in p53 gene

Large benign polyp

Cells mutant in APC, Ras, and p53 genes

Malignant colon cancer

The accumulation of three successive mutations in a lineage of colon cells results in malignant colon cancer.

in the same original cell. **Fig. 14.4** shows three key mutations that have been implicated in the origin of invasive colon cancer: *p53, Ras,* and *APC.* Each mutation occurs randomly, but if, by chance, a mutation in the *Ras* gene occurs in a cell that is derived from one in which the *APC* gene has been mutated, that cell's progeny forms a polyp. Another chance mutation in the same cell line, which now carries mutations in both the *APC* and *Ras* genes, could lead to malignant cancer. Normally, the occurrence of multiple mutations in a single cell lineage is rare, but in people exposed to chemicals that cause mutations or who carry mutations in DNA repair processes, multiple mutations in a single cell lineage are more likely to occur and so the risk of cancer is increased.

? **CASE 3   You, From A To T: Your Personal Genome**
**What can your personal genome tell you about your genetic risk factors?**

Cancer is usually due to a series of mutations that occur sequentially in a single lineage of somatic cells, as illustrated for colon cancer in Fig. 14.4. Each type of cancer is caused by its own particular sequence of mutations, although some genes are implicated in several different types of cancer. An example is the *p53* gene, the product of which detects DNA damage and slows the cell cycle to allow time for DNA repair (Chapter 11). Mutations in *p53* are one step in the mutational progression of many different types of cancer, including colon cancer and breast cancer.

In most individuals with cancer, all the sequential mutations that cause the cancer are spontaneous mutations that take place in somatic cells. They are not transmitted through the germ line, and so there is little or no increased risk of cancer in the offspring. In some families, however, there is a germ-line mutation in one of the genes implicated in cancer that is transmitted from parents to their children. In any child who inherits the mutation, all cells in the body contain the defective gene, and hence the cells already have taken one of the mutational steps that lead to cancer. The effect of such a germ-line mutation is therefore to reduce the number of additional mutations that would otherwise be necessary to produce cancer cells.

Any mutation that increases the risk of disease in an individual is known as a **genetic risk factor** for that disease. For colon cancer, the major genetic risk factors are mutations in *APC, Ras,* and *p53.* For breast cancer, the major genetic risk factors are mutations in the *BRCA1* and *BRCA2* genes (Case 3: You, From A to T). A risk factor does not cause the disease, but it makes the disease more likely to occur. For the genes implicated in colon cancer, each is a risk factor because, when a mutation is present, only two more mutations are needed for the cancer to form, whereas in its absence three mutations are required. Because a cell lineage is much more likely to undergo two mutations than three, a mutation in each of the genes implicated in colon cancer is a risk factor that substantially increases the likelihood of the disease.

The DNA sequence of each of our personal genomes can reveal the genetic risk factors that each of us carries, not only for cancer but for many other diseases as well. Not all genetic risk factors are known for all diseases, and a great deal of current research aims to identify new ones. But many genetic risk factors are already known for a large number of common diseases, including high blood pressure, diabetes, inflammatory bowel disease, age-related macular degeneration, Alzheimer's disease, and many forms of cancer. Therefore, your personal genome can be of great value in identifying diseases for which you carry risk factors, as was the case for Claudia Gilmore (Case 3: You, From A to T).

Our personal genomes can identify only *genetic* risk factors, however. In many cases, disease risk is substantially increased by environmental risk factors as well, especially lifestyle choices. While there are genetic risk factors for lung cancer, for example, the single biggest environmental risk factor is cigarette smoking. For skin cancer, the greatest environmental risk factor is exposure to the damaging ultraviolet rays in sunlight. For heart disease, it is smoking, lack of physical activity, and obesity. For diabetes, it is an unhealthy diet.

For breast cancer, the environmental risk factors include certain forms of hormone therapy, lack of physical activity, and alcohol. While we may not be able to do much about the genetic risk factors for any of these conditions, knowing that we have them may make us more careful about the lifestyle choices that we make.

### Mutations are random with regard to an organism's needs.

How do mutations arise? Consider the following example: If an antibiotic is added to a liquid culture of bacterial cells that are growing and dividing, most of the cells are killed, but a few survivors continue to grow and divide. These survivors are found to contain mutations that confer resistance to the antibiotic. This simple observation raises a profound question. Does this experiment reveal the presence of individual bacteria with mutations that had arisen spontaneously and were already present? Or do the antibiotic-resistant mutants arise in response to the presence of the antibiotic?

These alternative hypotheses had deep implications for all of biology because they suggested two very different ways in which mutations might arise. The first suggested that mutations occur without regard to the needs of an organism. According to this hypothesis, the presence of the antibiotic in the experiment with bacterial cells would not direct or induce antibiotic resistance in the cells, but instead would allow the small number of preexisting antibiotic-resistant mutants to flourish at the expense of the antibiotic-sensitive cells. The second hypothesis suggested that there is some sort of feedback between the needs of an organism and the process of mutation, and the environment directs specific mutations that would be beneficial to the organism.

To distinguish between these two hypotheses, in 1952 Joshua and Esther Lederberg carried out a now-famous experiment, described in **Fig. 14.5.** Bacterial cells were grown and formed colonies on agar plates in the absence of antibiotic. Then, using a technique they invented called replica plating, the Lederbergs transferred these colonies to new plates containing antibiotic. Only bacteria that were resistant to the antibiotic grew on the new plates. Because replica plating preserved the arrangement of the colonies, the Lederbergs were able to go back to the original plate and identify the colony that produced the antibiotic-resistant colony on the replica plate. From that original colony, they were then able to isolate a pure culture of antibiotic-resistant bacteria.

Replica plating allowed the Lederbergs to isolate a pure culture of antibiotic-resistant bacteria, even though the original bacteria actually never were exposed to antibiotic. This result supported the first hypothesis: Mutations occur randomly, and without regard to the needs of the organisms. The role of the environment is not to create specific mutations, but instead to select them. The principle the Lederbergs demonstrated is true of all organisms so far examined.

→   **Quick Check 1** If mutations occur at random with respect to an organism's needs, then how does a species become more adapted to its environment over time?

## 14.2 SMALL-SCALE MUTATIONS

At the molecular level, a mutation is a change in the nucleotide sequence of a genome. Such changes can be small, affecting one or a few bases, or large, affecting entire chromosomes. We begin by considering the origin and effects of small-scale changes to the DNA sequence. While mutation provides the raw material that allows evolution to take place, it can play this role only because of an important feature of living systems: Once a mutation has taken place in a gene, the mutant genome is replicated as faithfully as the nonmutant genome.

### Point mutations are changes in a single nucleotide.

Most DNA damage or errors in replication are immediately removed or corrected by specialized enzymes in the cell. (Some examples of DNA repair will be discussed in section 14.4.) We have already seen one example of DNA repair, the proofreading function of DNA polymerase, which acts to remove an incorrect nucleotide from the 3′ end of growing DNA strand (see Fig. 12.7). Damage that is corrected is not regarded as a mutation because the DNA sequence is immediately restored to its original state.

If a change in DNA is to become stable and subsequently inherited through mitotic or meiotic cell divisions, it must escape correction by the DNA repair systems. The example in

FIG. 14.5

# Do mutations occur randomly, or are they directed by the environment?

**BACKGROUND** Researchers have long observed that beneficial mutations tend to persist in environments where they are useful— in the presence of antibiotic, bacteria become antibiotic resistant; in the presence of insecticides, insects become insecticide resistant.

**HYPOTHESES** These observations lead to two hypotheses about how a mutation, such as one that confers antibiotic resistance on bacteria, might arise. The first suggests that mutations occur randomly in bacterial populations and over time become more common in the population in the presence of antibiotic (which destroys those bacteria without the mutation). In other words, they occur randomly with respect to the needs of an organism. The second hypothesis suggests that the environment, in this case the application of antibiotic, induces or directs antibiotic resistance.

**METHOD** To distinguish between these two hypotheses, Joshua and Esther Lederberg developed replica plating. In this technique, bacteria are grown on agar plates, where they form colonies (Fig. 14.5a). The cells in any one colony result from the division of a single original cell, and thus they constitute a group of cells that are genetically identical except for rare mutations that occur in the course of growth and division. Then a disk of sterilized velvet is pressed onto the plate. Cells from each colony stick to the velvet disk (in mirror image, but the relative positions of the colonies are preserved). The disk is then pressed onto the surface of a fresh plate, transferring to the new plate a few cells that originate from each colony on the first agar plate, in their initial positions.

**EXPERIMENT** First, the Lederbergs grew bacterial colonies on medium without antibiotic, called a nonselective medium because all cells are able to grow and form colonies on it. Then, by replica plating, they transferred some cells from each colony to a plate containing antibiotic, so only antibiotic-resistant cells could multiply and form colonies. (Medium containing antibiotic is a selective medium because it "selects" for a particular attribute or element, in this case antibiotic-resistant cells.) Because replica plating preserves the arrangement of the colonies, the location of an antibiotic-resistant colony on the selective medium reveals the location of its parental colony on the nonselective plate (Fig. 14.5b). Finally, the Lederbergs were able to go back to the original colony and isolate a pure culture of antibiotic-resistant bacteria (Fig. 14.5c).

**CONCLUSION** The Lederbergs' replica-plating experiments demonstrated that antibiotic-resistant mutants can arise in the absence of antibiotic because at no time in the experiments did the cells on nonselective medium come into contact with the antibiotic. Only the successive generations of daughter cells carried over to selective medium by replica plating were exposed to the antibiotic. Nevertheless, by their procedure the Lederbergs were able to isolate pure colonies of antibiotic-resistant cells.

**FOLLOW-UP WORK** These results have been extended to other types of mutation and other organisms, suggesting that mutations are random and not directed by the environment.

**SOURCE** Lederberg, J., and E. M. Lederberg. 1952. "Replica Plating and Indirect Selection of Bacterial Mutants." *Journal of Bacteriology* 63:399–406.

a.

Agar plate 1

Agar plate 2

Incubation to allow growth of colonies

b.

By chance, this colony contains a few mutant cells.

The sterile velvet template picks up both mutant and nonmutant cells.

Nonselective medium

Incubation to allow growth of colonies

Only the mutant cells grow on selective medium; the position of the colony tells you which colony on the nonselective medium contains the mutant cells.

Selective medium

c.

Mutant colony identified on original plate

Nonselective medium

Culture diluted and cells spread on selective medium

Pure culture of antibiotic-resistant bacteria

**FIG. 14.6** **A point mutation.** A point mutation is a change in a single nucleotide.

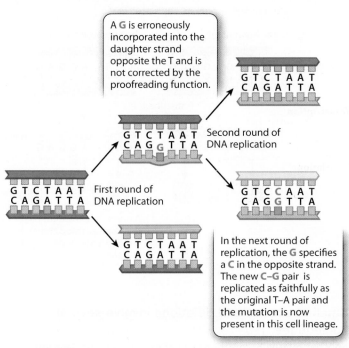

A **G** is erroneously incorporated into the daughter strand opposite the T and is not corrected by the proofreading function.

G T C T A A T
C A G A T T A

G T C T A A T
C A G **G** T T A

Second round of DNA replication

G T C T A A T
C A G A T T A

First round of DNA replication

G T C C A A T
C A G G T T A

G T C T A A T
C A G A T T A

In the next round of replication, the **G** specifies a **C** in the opposite strand. The new **C–G** pair is replicated as faithfully as the original T–A pair and the mutation is now present in this cell lineage.

**Fig. 14.6** shows how a mutation incorporated during replication can become a permanent change to the genome. The figure shows the replication of a DNA molecule in which a T–A base pair temporarily becomes a mismatched T–G base pair because DNA polymerase misincorporates a G into the daughter strand opposite the T in the template strand. Ordinarily, this G would be removed immediately after incorporation by the proofreading function of DNA polymerase and replaced with an A. In this example, we assume that the proofreading function failed to catch the mistake. This leaves the T–G mismatch in the double-stranded DNA, and at the next replication the G in the new template strand specifies a C in the daughter strand, with the result that the daughter DNA duplex has a perfectly matched C–G base pair. From this point forward, the DNA molecule containing the mutant C–G base pair will replicate as faithfully as the original molecule

bearing the nonmutant T–A base pair. A mutation in which one base pair (in this example, T–A) is replaced by a different base pair (in this example, C–G) is called a **nucleotide substitution** or **point mutation.** This is the most frequent type of mutation.

The effect of a point mutation depends in part on where in the genome it occurs. In many multicellular eukaryotes, including humans, the vast majority of DNA in the genome does not code for protein or RNA (Chapter 13). Most of the sequences in noncoding DNA have no known function, which may explain why many mutations in noncoding DNA have no detectable effects on the organism.

On the other hand, mutations in coding sequences do have predictable consequences in an organism. **Fig. 14.7** shows an example in the human DNA sequence coding for the amino acid chain of β- (beta-) globin, a subunit of the protein hemoglobin, which carries oxygen in red blood cells. Fig. 14.7a shows a small part of the nonmutant DNA sequence, transcription and translation of which results in incorporation of the amino acids Pro-Glu-Glu in the β-globin polypeptide.

Fig. 14.7b shows an example of a harmless nucleotide substitution in this coding sequence. The mutation consists of the substitution of an A–T base pair for the normal G–C base pair. In the mRNA, the mutation changes the normal GAG codon into the mutant GAA codon. But GAG and GAA both code for the same amino acid, glutamic acid (Glu). In other words, they are synonymous codons, and so the resulting amino acid sequences are the same: Pro-Glu-Glu. Such mutations are called **synonymous (silent) mutations.** This example is typical in that the synonymous codons differ at their third position (the 3′ end of the codon).

**FIG. 14.7** **Synonymous and nonsynonymous mutations.** Synonymous mutations do not change the amino acid sequence. Nonsynonymous mutations change the amino acid sequence.

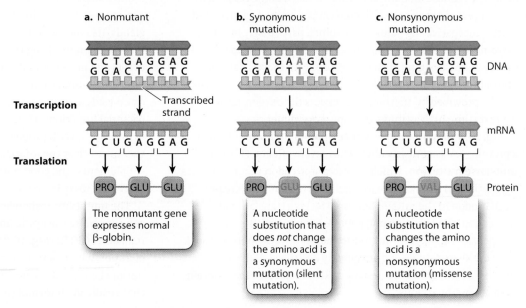

**a.** Nonmutant

C C T G A G G A G
G G A C T C C T C

Transcription

Transcribed strand

C C U G A G G A G

Translation

PRO — GLU — GLU

The nonmutant gene expresses normal β-globin.

**b.** Synonymous mutation

C C T G A A G A G
G G A C T T C T C

C C U G A A G A G

PRO — GLU — GLU

A nucleotide substitution that does *not* change the amino acid is a synonymous mutation (silent mutation).

**c.** Nonsynonymous mutation

C C T G T G G A G
G G A C A C C T C

DNA

C C U G U G G A G

mRNA

PRO — VAL — GLU

Protein

A nucleotide substitution that changes the amino acid is a nonsynonymous mutation (missense mutation).

A quick look at the genetic code (see Table 4.1) shows that most amino acids can be specified by synonymous codons, and that in most cases the synonymous codons differ in the identity of the nucleotide at the third position.

On the other hand, a point mutation in coding sequences can sometimes have a drastic effect on an organism. Fig. 14.7c shows such an example. In this case, a point mutation substitutes the first A–T base pair for a T–A base pair. The result is a change in the mRNA from GAG, which specifies Glu (glutamic acid), into GUG, which specifies Val (valine). The resulting protein therefore contains the amino acids Pro–Val–Glu instead of Pro–Glu–Glu. In other words, the nucleotide substitution results in an **amino acid replacement.** Point mutations that cause amino acid replacements are called **nonsynonymous (missense) mutations.**

Since the complete β-globin chain consists of 146 amino acids, it may seem that a change in only one amino acid would have little effect. But the change in even a single amino acid can affect the three-dimensional structure of a protein, and therefore change its ability to function. Individuals who inherit two copies of the mutant β-globin gene that specifies the Glu-to-Val replacement have a disease known as **sickle-cell anemia.** In this condition, the hemoglobin molecules tend to crystallize when exposed to lower-than-normal levels of oxygen. The crystallization of hemoglobin causes the cell to collapse from its normal ellipsoidal shape into the shape of a half-moon, or "sickle." In this form, the red blood cell is unable to carry the normal amount of oxygen. More important, the sickled cells tend to block tiny capillary vessels, interrupting the blood supply to vital tissues and organs.

Children with sickle-cell anemia tire easily, may be slow in their physical development, and tend to be susceptible to infections as a result of their general weakened condition brought on by the reduced amount of normal hemoglobin. Occasionally, those affected experience sickle-cell crises marked by fever and severe, incapacitating pains in the joints, particularly in the extremities, and in the chest, back, and abdomen. The pains are caused by the clogging of the blood supply to these vital areas. The painful episodes may last from hours to days to weeks and may be provoked by anything that lowers the oxygen supply: overexertion, high altitude, or respiratory ailments.

**Fig. 14.8** shows a third way that a point mutation can affect a protein, one that nearly always has severe effects. This is a **nonsense mutation,** which creates a stop codon that terminates translation. In the example in Fig. 14.8, the mutation creates a UAG codon in the mRNA. Because UAG is a translational stop codon, the resulting polypeptide terminates after Pro. Polypeptides that are truncated are nearly always nonfunctional. They are also unstable and are quickly destroyed. Eukaryotic cells have mechanisms to destroy mRNA molecules that contain premature stop codons.

**FIG. 14.8    A nonsense mutation.** A nonsense mutation can change an amino acid to a stop codon, resulting in a shortened and unstable protein.

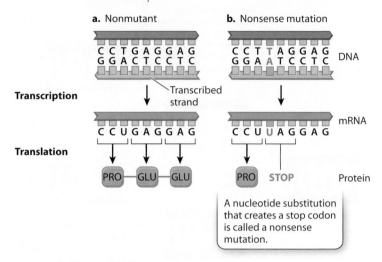

A nucleotide substitution that creates a stop codon is called a nonsense mutation.

## Small insertions and deletions involve several nucleotides.

Another relatively common type of mutation is the deletion or insertion of a small number of nucleotides. In noncoding DNA, such mutations have little or no effect. In protein-coding regions, their effects depend on their size. A small deletion or insertion that is an exact multiple of three nucleotides results in a polypeptide with as many fewer (in the case of a deletion) or more (in the case of an insertion) amino acids as there are codons deleted or inserted. Thus, a deletion of three nucleotides eliminates one amino acid, and an insertion of six adds two amino acids.

The effects of a deletion of three nucleotides can be seen in cystic fibrosis. This disease is characterized by the production of abnormal secretions in the lungs, liver, and pancreas and other glands. Its chief symptoms are recurrent respiratory infections, malnutrition resulting from incomplete digestion and absorption of fats and proteins, and liver disease. Patients with cystic fibrosis have an accumulation of thick, sticky mucus in their lungs, which often leads to respiratory complications, including recurrent bacterial infections. Untreated, 95% of affected children die before age 5. With proper medical care, including regular physical therapy to clear the lungs, antibiotics, pancreatic enzyme supplements, and good nutrition, the average life expectancy is currently 30 to 40 years.

The mutations responsible for cystic fibrosis are in the gene encoding the cystic fibrosis transmembrane conductance regulator (CFTR) (**Fig. 14.9**). The CFTR protein is a chloride channel, which acts as a transporter to pump chloride ions out of the cell. Malfunction of CFTR causes ion imbalances that result in abnormal secretions from the many cell types in

**FIG. 14.9** CFTR, the cystic fibrosis transmembrane conductance regulator.

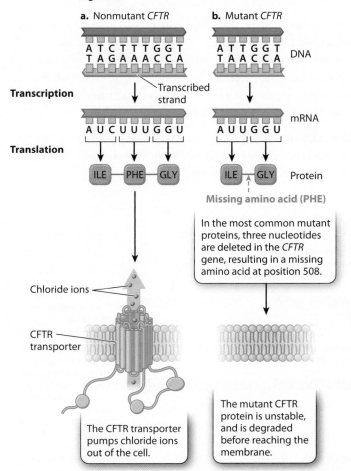

In the most common mutant proteins, three nucleotides are deleted in the *CFTR* gene, resulting in a missing amino acid at position 508.

The CFTR transporter pumps chloride ions out of the cell.

The mutant CFTR protein is unstable, and is degraded before reaching the membrane.

which the *CFTR* gene is expressed. About 70% of the mutations associated with cystic fibrosis have a specific mutation known as Δ508 (*delta 508*), which is a deletion of three nucleotides that eliminates a phenylalanine normally present at position 508 in the protein. The missing amino acid results in a CFTR protein that does not fold properly and is degraded before reaching the membrane. Some researchers hope that future therapy will include drugs that stabilize the mutant protein or even treatments that repair the defective gene in affected cells.

Small deletions or insertions that are not exact multiples of 3 can cause changes in amino acid sequence because they do not insert or delete entire codons. The effect of such a mutation can be appreciated by seeing how deletion of a single letter turns a perfectly sensible sentence of three-letter words into gibberish. Consider the sentence:

THE BIG BOY SAW THE CAT EAT THE BUG

If the red E is deleted, the new reading frame for three-letter words is as follows:

THB IGB OYS AWT HEC ATE ATT HEB UG

The result is unintelligible. Similarly, an insertion of a single

nucleotide causes a one-nucleotide shift in the reading frame of the mRNA, and it changes all codons following the site of insertion. For this reason, such mutations are called **frameshift mutations.**

**Fig. 14.10** shows the consequences of a frameshift mutation in the β-globin gene. The normal sequence in Fig. 14.10a corresponds to amino acids 5–10. The frameshift mutation in Fig. 14.10b is caused by the insertion of a C–G base pair. The mRNA transcript of the DNA therefore also has a single-base insertion. When this mRNA is translated, the one-nucleotide shift in the reading frame results in an amino acid sequence that has no resemblance to the original protein. All amino acids downstream of the site of insertion are changed, resulting in loss of protein function.

→ **Quick Check 2** The coding sequence of genes that specify proteins with the same function in related species sometimes include an insertion or deletion of contiguous nucleotides. The number of nucleotides that are inserted or deleted is almost always an exact multiple of 3. Why is this expected?

**FIG. 14.10 A frameshift mutation.** A frameshift mutation changes the translational reading frame.

An insertion or deletion that is not an exact multiple of three nucleotides changes the reading frame of translation. Such a mutation is called a frameshift mutation.

**FIG. 14.11**

# What causes sectoring in corn kernels?

**BACKGROUND** In the late 1940s, Barbara McClintock discovered what are now called transposable elements, DNA sequences that can move from one position to another in the genome. She studied corn (*Zea mays*). Wild-type corn has purple kernels, resulting from expression of purple anthocyanin pigment (Fig. 14.11a). A mutant with yellow kernels results from lack of purple anthocyanin pigment. McClintock noticed that streaks of purple pigmentation could be seen in many yellow kernels (Fig. 14.11b). This observation indicated that the mutation causing yellow color was unstable and could revert to the normal purple color.

**HYPOTHESIS** McClintock hypothesized that the yellow mutant color resulted from a transposable element, which she called *Dissociator* (*Ds*), jumping into a site near or in the anthocyanin gene and disrupting its function. She attributed the purple streaks to cell lineages in which the transposable element had jumped out again, restoring the anthocyanin gene.

**EXPERIMENT AND RESULTS** By a series of genetic crosses, McClintock showed that the genetic instability of *Ds* was due to something on another chromosome that she called *Activator* (*Ac*). She set up crosses in which she could track the *Ac*-bearing chromosome. She observed that in the presence of *Ac*, mutant yellow kernels reverted to normal purple, resulting in purple sectors in an otherwise

yellow kernel. From this observation, she inferred that the *Ds* element had jumped out of the anthocyanin gene, restoring its function. She also demonstrated that restoration of the original purple color was associated with mutations elsewhere in the genome. From this observation, she inferred that the *Ds* element had integrated elsewhere in the genome, where it disrupted the function of another gene.

**CONCLUSION** McClintock's conclusion is illustrated in Fig. 14.11c: Transposable elements can be excised from their original position in the genome and inserted into another position.

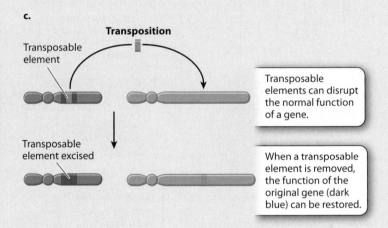

**FOLLOW-UP WORK** McClintock won the Nobel Prize in Physiology or Medicine in 1983. Later experiments showed that *Ds* is a transposable element that lacks a functional gene for transposase, the protein needed for the element to move, and *Ac* is a transposable element that encodes transposase. Presence of *Ac* produces active transposase that allows *Ds* to move. Much additional work showed that there are many different types of transposable element and that they are ubiquitous among organisms.

**SOURCE** McClintock, B. 1950. "The Origin and Behavior of Mutable Loci in Maize." *Proceedings of the National Academy of Sciences of the USA.* 36:344–355.

## Some mutations are due to the insertion of a transposable element.

An important source of new mutations in many organisms is the insertion of movable DNA sequences into or near a gene. Such movable DNA sequences are called **transposable elements.** As we saw in Chapter 13, the genomes of virtually all organisms contain several types of transposable element, each present in multiple copies per genome.

Transposable elements were discovered by American geneticist Barbara McClintock in the 1940s (**Fig. 14.11**).

She studied corn (maize) because genetic changes that affect pigment formation can be observed directly in the kernels. The normal color of maize kernels is purple. Each kernel consists of many cells, and if no mutations affecting pigmentation occur in a kernel, it will be uniformly purple. The kernel pigments are synthesized by enzymes in a metabolic pathway, and any of these enzymes can be rendered nonfunctional if a transposable element jumps into the gene that codes for them. Since her work, we have learned that most transposable elements are segments of DNA a thousand or more base pairs long. When such a large piece of DNA inserts into a gene, it can interfere with transcription, cause errors in RNA processing, or disrupt the open reading frame. The result in the case of maize is that the cell is unable to produce pigment, and so the kernels will be yellow.

McClintock observed that, while the kernels in the mutant corn were mostly yellow, most of them had purple sectors. Each colored sector consists of a lineage of progeny cells from a single progenitor cell in which pigment synthesis had been restored. McClintock realized that, just as the inability of the kernel cells to produce pigment was caused by a transposable element jumping into a gene, restoration of that ability could be caused by the transposable element jumping out again. Her hypothesis that transposable elements are responsible for the pigment mutations was confirmed when she found that in cells where pigment had been restored, mutations affecting other genes had also occurred. She deduced that these other mutations were due to a transposable element jumping out of a pigment gene and into a different gene in the same cell.

The movement or **transposition** of transposable elements, including those McClintock studied, requires specialized enzymes, different for each type of element, which are often encoded in the sequence of the element itself. These enzymes recognize specific sequences in the ends of the transposable element, cut it out of the DNA molecule it is in, and insert it somewhere else in the genome. In some cases, the excised transposable element leaves a copy of itself in its original position, resulting in multiple identical sequences throughout the genome as the transposable element jumps around. In other cases, the transposable element is removed entirely, leading to restoration of gene function, and the sectoring shown in Fig. 14.11b. Unstable mutations due to transposable elements can occur in almost any organism, including the Japanese morning glory, whose sectored flowers are shown in Fig. 14.3.

## 14.3 CHROMOSOMAL MUTATIONS

Whereas some mutations involve only one or a few nucleotides, some affect larger regions extending over thousands or millions of nucleotides, and a few have effects on chromosome structure that are large enough to be visible through a microscope. In this section, we briefly consider some of the major chromosome abnormalities. The most important are those in which the mutation affects the **centromere,** the site associated with attachment of the spindle fibers that move the chromosome during cell division (Chapter 11). An abnormal chromosome without a centromere or one with two centromeres is usually lost within a few cell divisions because it cannot be directed properly into the daughter cells during cell division.

### Duplications and deletions result in gain or loss of DNA.

Among the most common chromosomal abnormalities are those in which a segment of the chromosome is either present in two copies or is missing altogether (**Fig. 14.12**). A chromosome in which a region is present twice instead of once is said to contain a **duplication** (Fig. 14.12a). Although large duplications that include hundreds or thousands of genes are usually harmful and quickly eliminated from the population, small duplications including only one or a few genes can be maintained over many generations. Usually, duplication of a region of the genome is less harmful than deletion of the same region.

An example of a **deletion,** in which a region of the chromosome is missing, is shown in Fig. 14.12b. A deletion can result from an error in replication or from the joining of breaks in a chromosome that occur on either side of the deleted region. Even though a deletion may eliminate a gene that is essential for survival, the deletion can persist in the population because chromosomes usually occur in homologous pairs. If one member of a homologous pair has a deletion of an essential gene but the gene is present in the other member of the pair, that one copy of the gene is often sufficient for survival and reproduction. In these cases, the deletion can be transmitted from generation to generation and persist harmlessly, as long as the chromosome is present along with a normal chromosome.

But some deletions decrease the chance of survival or reproduction of an organism even when the homologous chromosome is normal. In general, the larger the deletion, the smaller the chance of survival. In the fruit fly *Drosophila,* individuals with deletions of more than 100–150 genes rarely survive even when the homologous chromosome is normal. The interpretation of the reduced survival is that organisms are sensitive to the **dosage,** or number of copies, of each gene. Normal embryonic development requires that genes be present in a particular dosage. Although small deviations from normal gene dosage can be tolerated, as indicated by the survival of individuals containing small duplications or deletions, the cumulative effects of large deviations from normal gene dosage are incompatible with life. It is usually not the total number of

**FIG. 14.12 Duplication and deletion.** (a) In a duplication, a segment of chromosome is repeated. (b) In a deletion, a segment of chromosome is missing.

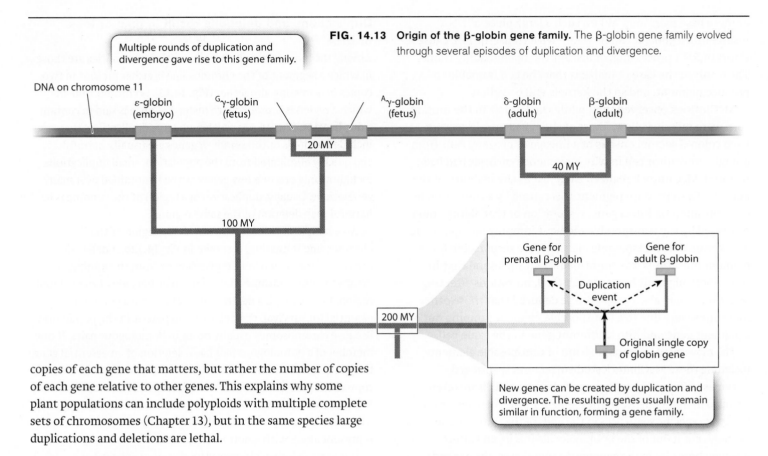

**FIG. 14.13 Origin of the β-globin gene family.** The β-globin gene family evolved through several episodes of duplication and divergence.

Multiple rounds of duplication and divergence gave rise to this gene family.

DNA on chromosome 11

ε-globin (embryo)

<sup>G</sup>γ-globin (fetus)

<sup>A</sup>γ-globin (fetus)

δ-globin (adult)

β-globin (adult)

20 MY

40 MY

100 MY

200 MY

Gene for prenatal β-globin

Gene for adult β-globin

Duplication event

Original single copy of globin gene

New genes can be created by duplication and divergence. The resulting genes usually remain similar in function, forming a gene family.

copies of each gene that matters, but rather the number of copies of each gene relative to other genes. This explains why some plant populations can include polyploids with multiple complete sets of chromosomes (Chapter 13), but in the same species large duplications and deletions are lethal.

### Gene families arise from gene duplication and evolutionary divergence.

Small duplications play an important role in the origin of new genes in the course of evolution. In most cases, when a gene is duplicated, one of the copies is free to change without causing harm to the organism because the other copy continues to carry out the normal function of the gene. Occasionally, a mutation in the "extra" copy of the gene may result in a beneficial effect on survival or reproduction, and gradually a new gene is formed from the duplicate. These new genes usually have a function similar to that of the original gene.

This process of creating new genes from duplicates of old ones is known as **duplication and divergence.** Multiple rounds of duplication and divergence can give rise to a group of genes with related functions known as a **gene family**. The largest gene family in the human genome has about 400 genes and encodes for proteins that detect odors. These proteins are structurally very similar, but differ in the region that binds small odor molecules. It is the diversity of the odorant binding sites that allows us to identify so many different smells.

Most gene families are not as large and diverse as that for odor detection. **Fig. 14.13** shows the evolutionary origin of the family of globin genes, which in humans are spread across about 50 kb of chromosome 11. The globin gene family consists of five different genes that are expressed at various times during

development (embryo, fetus, or adult). The two γ- (gamma-) globin polypeptides are nearly identical in amino acid sequence but expressed at different levels in the fetus, whereas the adult δ- and β-polypeptides differ somewhat more and are expressed at very different levels. The sequences are all similar enough to imply that the genes arose through duplication and divergence.

Globin sequences have changed through evolutionary time at a relatively constant rate, and so the fraction of differences between any two sequences is proportional to the time since they were created by duplication. The relative constancy of rates of evolutionary change in a DNA nucleotide sequence or a protein amino acid sequence is known as a **molecular clock,** and it allows molecular differences among genomes to be correlated with the fossil record and dated accordingly (Chapter 21). For example, the earliest duplication event in the tree in Fig. 14.13, which produced distinct genes for fetal and adult hemoglobin, took place at about the same time (200 million years ago) that the common ancestor of today's placental mammals became a distinct species from the common ancestor of today's marsupial mammals.

### An inversion has a chromosomal region reversed in orientation.

Chromosomes in which the normal order of a block of genes is reversed contain an **inversion** (**Fig. 14.14**). An inversion is typically produced when the region between two breaks in

**FIG. 14.14** **Inversion.** A chromosome with an inversion has a segment of chromosome present in reverse orientation.

Normal chromosome

Inversion

**FIG. 14.15** **Reciprocal translocation.** A reciprocal translocation results from an interchange of parts between nonhomologous chromosomes.

Chromosome breaks

Reciprocal translocation

a chromosome is flipped in orientation before the breaks are repaired. Especially in large genomes, the breaks are likely to occur in noncoding DNA rather than within a gene. Whereas large inversions can cause problems in meiosis, small inversions are common in many populations and play an important role in chromosome evolution. The accumulation of inversions over evolutionary time explains in part why the order of genes along a chromosome can differ even among closely related species.

### A reciprocal translocation joins segments from nonhomologous chromosomes.

A **reciprocal translocation** (**Fig. 14.15**) is said to have occurred when two different (nonhomologous) chromosomes undergo an exchange of parts. In the formation of a reciprocal translocation, both chromosomes are broken and the terminal segments are

exchanged before the breaks are repaired. In large genomes, the breaks are likely to occur in noncoding DNA, so the breaks themselves do not usually disrupt gene function.

Since reciprocal translocations change only the arrangement of genes and not their number, most reciprocal translocations do not affect the survival of organisms. Proper gene dosage requires the presence of both parts of the reciprocal translocation, however, as well as one copy of each of the normal homologous chromosomes. Problems can arise in meiosis because both chromosomes involved in the reciprocal translocation may not move together into the same daughter cells, resulting in gametes with only one part of the reciprocal translocation. This inequality does upset gene dosage, because these gametes have extra copies of genes in one of the chromosomes and are missing copies of genes in the other. In the next chapter, we will see that such abnormalities are observed in a significant number of human embryos.

## 14.4 DNA DAMAGE AND REPAIR

DNA is a fragile molecule, prone to damage of many different kinds. Every day, in each cell in our bodies, the DNA is damaged in some way at tens of thousands of places along the molecule. Fortunately, cells have evolved mechanisms for repairing various types of damage and restoring the DNA to its original condition. In this section, we examine some of the major types of DNA damage and some of the ways in which it is repaired.

### DNA damage can affect both DNA backbone and bases.

Most mutations are spontaneous, occurring without an assignable cause. However, mutations can also be induced by radiation or chemicals. **Mutagens** are agents that increase the probability of mutation. The presence of a mutagen can increase the probability of mutation by a factor of 100 or more.

Some of the most important types of DNA damage induced by mutagens are illustrated in **Fig. 14.16**, which shows a highly

**FIG. 14.16** **Major types of DNA damage.** Most types of DNA damage can be repaired by specialized enzymes.

Single-stranded break in DNA backbone

Cross-linked thymine bases

Missing base

Bulky side group attached to a base

Double-stranded break in DNA backbone

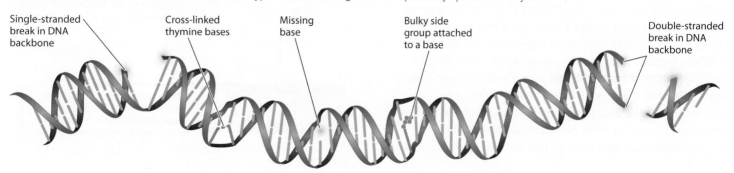

mutated DNA molecule. Some types of damage affect DNA structure. These include breaks in the sugar–phosphate backbone, one of the main mutagenic effects of X-rays. Breaks can occur in just one strand of the DNA or both. Ultraviolet light can cause cross-links between adjacent pyrimidine bases, especially thymine, resulting in the formation of thymine dimers.

Yet another type of structural damage is loss of a base from one of the deoxyribose sugars, resulting in a gap in one strand where no base is present. Spontaneous loss of a purine base is one of the most common types of DNA damage, occurring at the rate of about 13,000 purines lost per human cell per day. The rate increases with age, and it can also be increased by exposure to oxidizing agents such as household bleach or hydrogen peroxide.

Other types of damage affect the bases themselves. Bases that are chemically damaged tend to mispair. Some bases are damaged spontaneously when reaction with a water molecule replaces an amino group ($-NH_2$) with an atom of oxygen ($=O$). Some naturally occurring molecules mimic bases and can be incorporated into DNA and cause nucleotide substitutions. Caffeine mimics a purine base, for example (although to be mutagenic, the amount of caffeine required is far more than any normal person could possibly consume).

Chemicals that are highly reactive tend to be mutagenic, often because they add bulky side groups to the bases that hinder proper base pairing. The main environmental source of such chemicals is tobacco smoke. Other chemicals can perturb the DNA replication complex and cause the insertion or deletion of one or occasionally several nucleotides.

→  **Quick Check 3** Earlier, we described the Lederbergs' experiment, which demonstrated that mutations are not directed by the environment. But mutagens, which are environmental, can lead to mutations. What's the difference?

### Most DNA damage is corrected by specialized repair enzymes.

Cells contain many specialized DNA-repair enzymes that correct specific kinds of damage, and in this section we examine a few examples. Perhaps the simplest is the repair of breaks in the sugar–phosphate backbone, which are sealed by **DNA ligase,** an enzyme that can repair the break by using the energy in ATP to join the 3' hydroxyl of one end to the 5' phosphate of the other end. Most organisms have multiple different types of DNA ligase, some of which participate in DNA replication (Chapter 12) and others in DNA repair. One type of ligase seals single-stranded breaks in DNA, and a different type seals double-stranded breaks. Double-stranded breaks often result in point mutations or chromosomal rearrangements because they are less likely to be repaired than

single-stranded breaks. In addition to their importance in DNA replication and repair, ligases are an important tool in research in molecular biology because they allow DNA molecules from different sources to be joined to produce recombinant DNA (Chapter 12).

As we saw earlier in this chapter, mispairing of bases during DNA replication leads to the incorporation of incorrect nucleotides and potentially to nucleotide substitutions. Although 99% of the mispaired bases are corrected immediately by the proofreading function of DNA polymerase, another mechanism also scans the DNA for mismatches to afford a second level of quality control. This process is called **mismatch repair** (**Fig. 14.17**). In mismatch repair, one of the strand backbones in the vicinity of the mismatch is cleaved and then degraded nucleotide by nucleotide to a point at a distance past the site of the mismatch. Then the gap is filled by new synthesis, which corrects the mismatch.

**FIG. 14.17  Mismatch repair.** In mismatch repair, the segment of a DNA strand containing the mismatch is removed and then resynthesized.

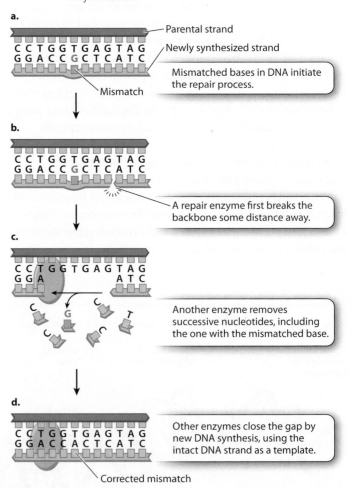

**a.**
Parental strand
Newly synthesized strand
C C T G G T G A G T A G
G G A C C G C T C A T C
Mismatched bases in DNA initiate the repair process.
Mismatch

**b.**
C C T G G T G A G T A G
G G A C C G C T C A T C
A repair enzyme first breaks the backbone some distance away.

**c.**
C C T G G T G A G T A G
G G A          A T C
Another enzyme removes successive nucleotides, including the one with the mismatched base.

**d.**
C C T G G T G A G T A G
G G A C C A C T C A T C
Other enzymes close the gap by new DNA synthesis, using the intact DNA strand as a template.
Corrected mismatch

**FIG. 14.18 Base excision repair.** In base excision repair, an improper base in DNA and its deoxyribose sugar are both removed, and the resulting gap is then repaired.

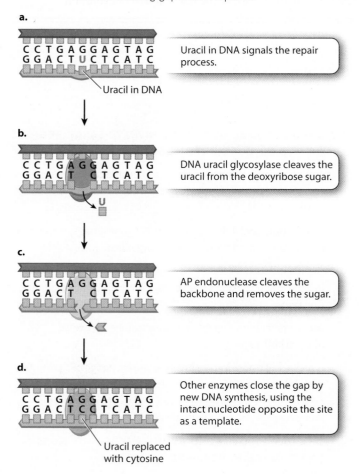

**FIG. 14.19 Nucleotide excision repair.** In nucleotide excision repair, a damaged segment of a DNA strand is removed and resynthesized using the undamaged partner strand as a template.

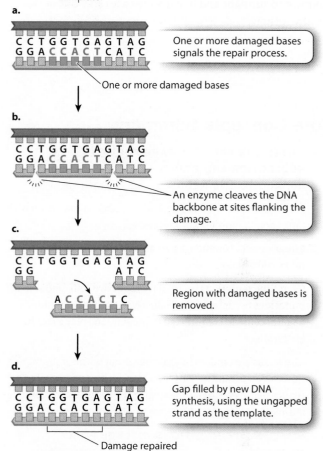

**Fig. 14.18** shows an example of a more specialized repair system called **base excision repair.** Fig. 14.18a depicts a molecule in which uracil was mistakenly incorporated into a new DNA strand instead of cytosine. Recall that uracil is normally found in mRNA, but not in DNA. Most organisms have an enzyme called DNA uracil glycosylase that cleaves the uracil from its deoxyribose sugar, resulting in a sugar in the DNA backbone lacking a base (Fig. 14.18b). Such sites are recognized by an enzyme called AP endonuclease (Fig. 14.18c), which cleaves the backbone on both sides of the sugar that lacks a base. This cleavage results in a one-nucleotide gap, which is filled by new synthesis using the intact DNA strand as a template. The result is that uracil is removed and replaced with cytosine. Fig. 14.18 shows DNA repair systems that work in tandem, in this case DNA uracyl glycosylase acting with the AP endonuclease. Other types of glycosylases remove abnormal or damaged bases from

the DNA, and these enzymes are also coupled with the AP endonuclease.

**Nucleotide excision repair,** which is similar in some ways to mismatch repair, is illustrated in **Fig. 14.19.** The major difference is that mismatch repair operates on single mispaired bases, whereas nucleotide excision repair recognizes multiple mismatched or damaged bases across a region. The region excised in mismatch repair is small, but that in nucleotide excision repair can extend for thousands of nucleotides. Nucleotide excision repair is also used to remove nucleotides with some types of bulky side group, as well as thymine dimers resulting from ultraviolet light.

The importance of DNA repair mechanisms is illustrated by considering what happens when they do not function properly. Xeroderma pigmentosum (XP) is a disease in which nucleotide excision repair is defective. People with XP are exquisitely sensitive to the UV radiation in sunlight. Because of the defect in nucleotide excision repair, damage to DNA resulting from UV light is not corrected, leading to the accumulation of mutations in skin cells. The result is high rates of skin cancer. People with XP must minimize their exposure to sunlight and in extreme cases stay out of the sun altogether.

When DNA repair mechanisms function properly, they reduce the rate of mutation to a level that is compatible with life. But they are not perfect, and they do not catch all the mistakes. The result is genetic variation, the raw material of evolution.

## Core Concepts Summary

**14.1 MUTATIONS ARE VERY RARE FOR ANY GIVEN NUCLEOTIDE AND OCCUR RANDOMLY WITHOUT REGARD TO THE NEEDS OF AN ORGANISM.**

For an individual nucleotide, the frequency of mutation is very rare, though it differs among species. page 14-1

Mutations are common across an entire genome and across many cell divisions. page 14-2

Germ-line mutations occur in reproductive cells, and somatic mutations occur in nonreproductive cells. Only mutations in germ cells are transmitted to progeny and play a role in evolution. page 14-3

Mutations are random in that they occur without regard to the needs of an organism. This principle can be demonstrated by replica plating bacterial colonies in the absence and presence of antibiotic. page 14-5

**14.2 SMALL-SCALE MUTATIONS INCLUDE POINT MUTATIONS, INSERTIONS AND DELETIONS, AND MOVEMENT OF TRANSPOSABLE ELEMENTS.**

A point mutation, or nucleotide substitution, is a change of one base for another. page 14-5

The effect of a point mutation depends on where it occurs. If it occurs in noncoding DNA, it will likely have no effect on an organism. If it occurs in a protein-coding gene, it can result in a change of the amino acid sequence (nonsynonymous mutation), no change of the amino acid sequence (synonymous mutation), or the introduction of a stop codon (nonsense mutation). page 14-7

Small insertions or deletions in DNA add or remove one base or a few contiguous bases. Their effect depends on where in the genome they occur and on their size. An insertion of a single nucleotide in a protein-coding gene results in a frameshift mutation, in which all the codons downstream of the insertion or deletion are changed. page 14-8

Transposable elements are DNA sequences that can jump from one place in a genome to another. They can affect the expression of a gene if they insert into or near a gene. page 14-10

**14.3 CHROMOSOMAL MUTATIONS INVOLVE LARGE REGIONS OF ONE OR MORE CHROMOSOMES.**

A duplication is a region of the chromosome that is present two times, and a deletion is the loss of part of a chromosome. page 14-11

Duplications and deletions both affect gene dosage, which can have important effects on a cell or organism. page 14-11

Gene duplication followed by evolutionary divergence results in gene families made up of genes with related but not identical functions. page 14-12

An inversion is a segment of a chromosome in reverse orientation. page 14-12

A reciprocal translocation involves the exchange of a part of one chromosome with another. Reciprocal translocations do not affect gene dosage, but can lead to problems during meiosis. page 14-13

**14.4 DNA CAN BE DAMAGED BY MUTAGENS, BUT MOST DNA DAMAGE IS REPAIRED.**

Some types of damage alter the structure of DNA, such as single-stranded or double-stranded breaks, cross-linked thymine dimers, or missing bases. page 14-13

Other types of DNA damage, such as changes in the side groups that form hydrogen bonds or the addition of side groups that interfere with base pairing, affect the bases themselves. page 14-14

Common mutagens are X-rays, which cause single-stranded and double-stranded breaks, ultraviolet light, which creates thymine dimers, and reactive chemicals in cigarette smoke, which damage the bases. page 14-14

Many specialized DNA repair enzymes can correct DNA damage. page 14-14

DNA ligase seals breaks in DNA and is an important tool for molecular biologists. page 14-14

Mismatch repair provides a backup mechanism for mistakes not caught by the proofreading function of DNA polymerase. page 14-14

Base excision repair corrects individual nucleotides and involves several DNA repair enzymes working together.  page 14-14

Nucleotide excision repair functions similarly to mismatch repair but excises longer stretches of damaged nucleotides.  page 14-15

## Self-Assessment

1.  Describe different effects a mutation can have on an organism.

2.  Explain the difference between the mutation rate for a given nucleotide and the mutation rate for a given cell.

3.  Explain what it means to say that mutations are random.

4.  Describe several types of small-scale mutation and their likely effect on an organism.

5.  Explain why the location of a small-scale mutation in the genome can make a difference in the phenotype.

6.  Describe the differences among silent, missense, nonsense, and frameshift mutations.

7.  Describe several types of large-scale mutation and their likely effect on an organism.

8.  Explain how a gene family, such as the odorant receptor gene family, is thought to have evolved.

9.  Name several common mutagens and their effect on DNA.

10.  Describe a DNA repair mechanism.

Do you understand the chapter's Core Concepts? Log in to check your answers to the Self-Assessment questions, then practice what you've learned and reinforce this chapter's concepts by working through the problems and multimedia tutorials provided there.

📖  **www.biologyhowlifeworks.com**

## CHAPTER 15

# GENETIC VARIATION

## Core Concepts

**15.1** Genetic variation describes common genetic differences (polymorphisms) among the individuals in a population at any given time.

**15.2** Human genetic variation can be detected by DNA typing, which can uniquely identify each individual.

**15.3** Two common types of genetic variation are single-nucleotide polymorphisms (SNPs) and copy-number variation (CNV).

**15.4** Chromosomal variants can also occur but are usually harmful.

In the last chapter, we saw how mutations arise and what common types of mutation occur. What is the fate of all these mutations? Some, as we have seen, are corrected by DNA repair mechanisms. In this case, the DNA sequence is not changed. Some mutations escape these repair mechanisms and are replicated as faithfully as the original DNA sequence. If such mutations occur in somatic (body) cells, they can be passed on in the individual's cells through mitotic cell divisions, but they will not be passed on to progeny. If they occur in the germ line, they can be passed on to progeny. Over time, through evolution (Chapter 21), these mutations may increase or decrease in frequency. Therefore, if we look at any present-day population of organisms, such as the human population, we will find that it harbors lots of genetic differences, all of which result from mutations that occurred sometime in the past. **Genetic variation** refers to genetic differences that exist among individuals in a population at a particular point in time.

In the human population, you can observe the effects of common genetic differences or mutations by looking at the people around you. They differ in height, weight, facial features, skin color, eye color, hair color, hair texture, and in many other ways. These traits differ in part because of genetic variation, and in part because of the environment. Weight is affected by diet, for example, and skin color by exposure to sunlight.

In this chapter, we consider examples of major types of DNA and chromosomal variation in populations today and examine their consequences for the organism. We also describe key molecular techniques that allow genetic variation to be studied directly in DNA molecules. The emphasis here is primarily, but not exclusively, on human populations.

## 15.1 GENOTYPE AND PHENOTYPE

We tend to think of mutations as something negative or harmful; the term "mutant" ordinarily connotes something unpleasantly abnormal. But because mutations result in genetic variation among individuals and organisms, we are in fact all mutants, different from one another genetically because of mutations, that is, differences in our DNA. Whereas some mutations are harmful, some have no effect on an organism, and some indeed are beneficial. Without mutations, evolution would not be possible. Mutations generate the occasional favorable variants that allow organisms to evolve and become adapted to their environment over time (Chapter 21).

### Genotype is the genetic makeup of a cell or organism; phenotype is its observed characteristics.

The genetic makeup of a cell or organism constitutes its **genotype.** A population with a gene pool that has many variants in many different genes will consist of organisms with numerous different genotypes. For example, any two human genomes are likely to differ at about 3 million nucleotide sites. Most humans have the same genes as one another, organized in the same way along the same set of chromosomes. We differ primarily from one person to the next in a very small fraction of nucleotides.

Mutations, as we have seen, are the ultimate source of differences among genotypes. If we consider any present-day population of organisms, we find that some mutations are very common. Geneticists use the term **polymorphism** to refer to any genetic difference among individuals that is sufficiently common that it would almost certainly be present in a group of 50 randomly chosen individuals. For example, if many people have an A–T pair at a particular site in the genome, but many others have a G–C pair at the same site, this difference is a polymorphism. Of course, the polymorphism is the result of a mutation. All individuals once had the same genotype at this site, but a mutation in one or a few individuals made different genotypes possible. In this example, the mutation that made either A–T or G–C possible occurred sometime in the past, and is now commonly found in the population.

**Phenotype** is an individual's observable characteristics or traits, such as a person's height, weight, eye color, and so forth. The phenotype may be visible, as in these characteristics, or may be seen in the development, physiology, or behavior of a cell or organism. For example, color blindness and lactose intolerance are phenotypes. The phenotype results in part from the genotype: A genotype with a mutated gene for an enzyme that would normally metabolize lactose can lead to the phenotype of lactose intolerance. However, the environment also commonly plays an important role, so it is most accurate to say that a phenotype results from an interaction between the genotype and environment. These genotype–environment interactions are discussed in Chapter 18.

### The effect of a genotype often depends on several factors.

Let's examine how genotype influences phenotype with an example of a genetic polymorphism that we introduced in the previous chapter. **Fig. 15.1** shows genetic variation in the gene for β- (beta-) globin, one of the subunits of hemoglobin that carries oxygen in red blood cells. Three forms of the β-globin gene are shown in the figure: A, S, and C. These three forms of the gene are relatively common in certain African populations and in people of African descent (the mutations originated in Africa). The different forms of any gene are called **alleles,** and they correspond to different DNA sequences (polymorphisms) in the genes. In this case, the most common allele is the A allele, which has a GAG codon in the position indicated. This codon translates to glutamic acid in the resulting polypeptide.

The allele denoted S in Fig. 15.1 is associated with sickle-cell anemia (Chapter 14). In this allele, the GAG codon seen in the A allele is instead a GTG codon, with the result that the glutamic acid in the protein is replaced with valine (Val). The third allele in Fig. 15.1 is the C allele. It has a variation in exactly the same codon as the S allele, but in this case the change is from the normal GAG to AAG, which results in glutamic acid being replaced with lysine (Lys). Although in each case only one amino acid of the β-globin protein is affected, this change can have a dramatic effect on the function of the protein, since the amino acid sequence determines how a protein folds, and protein folding in turn determines the protein's function (Chapter 4).

An individual who inherits an allele of the same type from each parent is said to be **homozygous**. For the hemoglobin A, S, and C alleles, there are three possible homozygous genotypes, AA, SS, and CC. (The first letter or symbol indicates the allele inherited from one parent, and the second indicates the allele inherited from the other parent. By convention, alleles and genotypes are designated by italic letters.)

By contrast, individuals who inherit different types of alleles from their parents are **heterozygous.** For the hemoglobin A, S, and C alleles, there are three possible heterozygous genotypes: AS, AC, and SC. Note that while each individual can have only two alleles of a gene, many more alleles can exist in an entire population. In the hemoglobin example, a person can have two identical alleles (homozygous) or two different alleles (heterozygous), but there are three different alleles available in the population to mix and match in each individual. How those alleles are inherited is explored in Chapter 16.

**FIG. 15.1  Three alleles of the gene encoding β-globin, a subunit of hemoglobin.** Adult hemoglobin is made up of 2 α-globin and 2 β-globin subunits.

the same positions in the homologous chromosomes inherited from the mother and the father. With five alleles there are 5 possible homozygous genotypes and 10 possible heterozygous genotypes. These can be visualized using PCR, in which primers are used to amplify a specific region of DNA and the resulting fragments are separated by size on a gel and then visualized with a dye. The pattern of bands expected from the DNA in each genotype is shown in Fig. 15.5b. Note that each genotype yields a distinct pattern of bands, and so these patterns can be used to identify the genotype of any individual for a particular region of DNA.

With as few as 10 equally frequent alleles, the chance that any two individuals would have the same genotype by chance alone is about 1 in 50, and with 20 equally frequent alleles the chance of identity is about 1 in 200. These numbers suggest why DNA typing can yield a sort of genetic fingerprint that can uniquely match DNA samples.

The lane labeled "E" at the far right in Fig. 15.5b shows the DNA bands for this same polymorphism from biological material collected at the scene of a crime. In this case, the bands in lane E imply that the source of the evidence is an individual of

genotype $A_2/A_4$ because this is the only genotype with bands with vertical positions in the gel that match those of the evidence sample exactly. A single polymorphism of the type shown in Fig. 15.5 is not enough to say for sure that two pieces of DNA come from the same person, but because the human genome has such polymorphisms scattered throughout, many multiple-allele polymorphisms can be examined. If six or eight of these polymorphisms match between two samples, it is very likely that the samples come from the same individual (or from identical twins). On the other hand, if any of the polymorphisms fails to match band for band, then it is almost certain that the samples come from different individuals (barring such technical mishaps as samples whose DNA is contaminated, degraded, or mislabeled).

One of the advantages of DNA typing is that a large amount of DNA is not needed. Even minuscule amounts of DNA can be typed, so tiny spots of blood, semen, or saliva are sufficient samples. A cotton swab wiped across the inside of your cheek will contain enough DNA for typing; so will a discarded paper cup or a cigarette butt. One enterprising researcher in England used PCR to type the dog feces on his lawn and matched it to the DNA from hair he obtained from neighborhood dogs, much to the chagrin of the guilty dog's owner.

→  **Quick Check 3** In typing DNA from a sample found at a crime scene, how can a DNA mismatch prove that a suspect is not the source of that sample, whereas a DNA match does not necessarily prove that a suspect is the source?

### Some polymorphisms add or remove restriction sites in the DNA.

Researchers can use restriction enzymes to detect another type of polymorphism commonly present in genomes and useful in DNA typing. Sometimes, because of common polymorphisms or sequence differences, a restriction site will be present along a stretch of DNA in some chromosomes but not in others.

**Fig. 15.6** shows an example. Here, a restriction enzyme cleaves DNA near the sequence GAGGAG. The three sites shown are in and flanking the β-globin gene from the sickle-cell example we

---

**FIG. 15.6** Restriction fragment length polymorphism between the *A* and *S* alleles of the β-globin gene.

**a.**

A restriction enzyme cleaves both DNA samples at all of the sites it recognizes, producing fragments of different lengths. Following electrophoresis, the gels show the position of the fragments to which the probe hybridizes.

**b.**

Number and position of labeled bands indicate the genotype.

explored earlier in the chapter. The restriction site in the middle includes the GAG that is mutated to GTG in the *S* allele (see Fig. 15.1). Note that whereas the DNA of the *A* allele is cleaved at three sites, the DNA of the *S* allele is cleaved at only two sites because the *S* mutation changes the GAGGAG restriction site into GTGGAG.

These differences can be visualized by a Southern blot (Chapter 12), in which DNA is isolated from different individuals, cut with restriction enzymes, separated by size on a gel, transferred to a blot, and hybridized to a probe to detect the bands of interest. In this case, we would use a restriction enzyme that recognizes the GAGGAG sequence. The *A* allele produces two fragments: a short fragment that moves rapidly through the gel and ends up near the bottom, and a longer fragment that ends up near the top. Since the probe hybridizes only where indicated in Fig. 15.6a, the longer restriction fragment produced by digestion of the *A* allele is not detected. The band produced by the *S* allele is larger because there is no restriction site in the middle. This larger fragment moves more slowly in gel electrophoresis and ends up forming a band near the top of the gel. This kind of polymorphism is called a **restriction fragment length polymorphism**, or **RFLP,** because the length of the restriction fragments is polymorphic—that is, different—in the two alleles.

As seen in the restriction fragment length polymorphism in Fig. 15.6b, DNA from homozygous *AA* individuals yields only the small fragment, and that from homozygous *SS* individuals yields only the large fragment. DNA from heterozygous *AS* individuals yields *both* the small fragment and the large fragment. Therefore, each of the three genotypes *AA*, *AS*, and *SS* yields a unique pattern of bands, allowing each of the possible genotypes to be identified.

How many restriction fragment length polymorphisms are there in the human genome? As in the case of VNTRs, the human genome is littered with restriction sites, so there are numerous RFLPs in the human population, making it another good individual "fingerprint."

→  **Quick Check 4** VNTRs and RFLPs both result in restriction fragments of different lengths. How, then, are they different?

## 15.3 GENOMEWIDE STUDIES OF GENETIC VARIATION

Restriction enzymes are extremely useful for some purposes, but their utility as a tool for studying genomewide genetic variation is limited because they reveal only sequence variation in restriction sites or differences in the distances between adjacent restriction sites. In this section, we examine

methods by which genetic differences in any individual nucleotide can be studied.

### Single-nucleotide polymorphisms (SNPs) are single base changes in the genome.

One of the most common types of genetic variation is a difference in a nucleotide at a specific site. A **single-nucleotide polymorphism (SNP)** is a site in the genome where there is a base-pair difference among chromosomes that is common enough to be present in a random sample of 50 diploid individuals. The differences among the hemoglobin alleles are SNPs. The *A*, *S*, and *C* alleles differ from each other at just one nucleotide site (see Fig. 15.1).

Eye color is also associated with a SNP. The blue-eyed phenotype results from reduced expression of a gene called *OCA2*, which encodes a membrane protein involved in the transport of small molecules including the amino acid tyrosine, which is a precursor of the melanin pigment associated with brown eyes. Although more than a dozen alleles of *OCA2* are known that have an amino acid replacement in the protein, none of these results in blue eyes.

The SNP nucleotide implicated in blue eyes is a C–G base pair in one allele. The other common variant is a T–A, which is not associated with blue eyes (**Fig. 15.7**). In one study, the homozygous *C–G* genotype was found in 94% of 183 individuals with blue eyes and in only 2% of 176 individuals with brown eyes. This strong association is quite unexpected because the SNP associated with blue eyes is a noncoding SNP present in an intron of a neighboring gene. The proposed mechanism for the association is that the *C–G* allele makes the adjacent *OCA2* gene less accessible to transcription factors, thereby reducing the amount of the transporter protein and consequently the production of melanin

**FIG. 15.7**   **Single-nucleotide polymorphisms (SNPs).** A SNP in the region neighboring *OCA2* is strongly associated with blue eyes.

in the iris. The proposed mechanism may be right or wrong; this example emphasizes that scientists have much to learn about the possible phenotypic effects of noncoding DNA.

The *C–G* versus *T–A* SNP is typical of most SNPs in that only two of the four possible base pairs are present in the population at any appreciable frequency.

→ **Quick Check 5** What's the difference between a point mutation and a SNP?

## ? CASE 3 You, From A To T: Your Personal Genome
### How can genetic risk factors be detected?

SNPs result from point mutations, the most frequent type of mutation. A point mutation is the substitution of one base pair for another in double-stranded DNA (Chapter 14). Each of us is genetically unique partly because of the abundance of SNPs in the human genome. There are approximately 3 million SNPs that distinguish any one human genome from any other. SNPs are abundant in the genomes of most species; some are thought to be the main source of evolutionary innovation, and others are major contributors to inherited disease. Practically speaking, SNPs are important because they can be used to detect the presence of a genetic risk factor for a disease before the onset of the disease. For example, the ability to detect the β-globin SNPs shown in Fig. 15.1 allows prenatal identification of the β-globin genotype of a fetus.

While there is great interest in developing ultrafast DNA sequencing machines that can determine anyone's personal genome quickly at relatively low cost, between any two genomes 99.9% of the nucleotides are identical. An alternative is to focus on genotyping just SNPs. As many as one million SNPs at different positions in the genome can be genotyped simultaneously, and the genotyping can be carried out on thousands or tens of thousands of individuals. Such massive genotyping allows any SNP associated with a disease to be identified, which is especially important for complex diseases affected by many different genetic risk factors (Chapter 18).

What does it mean to say that a SNP is associated with a disease? It means that individuals carrying one of the SNP alleles are more likely to develop the disease than those carrying the other allele. The increased risk depends on the disease and can differ from one SNP to the next. Sickle-cell anemia affords an example at one extreme of the spectrum of effects. In this case, the *T–A* base pair in the *S* allele of the β-globin gene is the SNP that results in the amino acid replacement of glutamic acid with valine in the protein. Because *SS* individuals always have sickle-cell anemia, it would be fair to say that homozygous *S* "causes" sickle-cell anemia.

But except for inherited diseases due to single mutant genes, which are usually rare, the vast majority of SNPs implicated in disease increase the risk only moderately, typically by 10% to 50% as compared with individuals lacking the risk factor. We then say that the SNP is "associated" with the disease, since the SNP alone does not cause the disease but only increases the risk. For heart disease, diabetes, and some other diseases, many SNPS at different places in the genome, as well as environmental risk factors, can be associated with the disease. Usually, genetic and environmental risk factors act cumulatively: the more you have, the greater the risk.

As emphasized in the case of Claudia Gilmore's genome, certain SNPs in the *BRCA1* and *BRCA2* genes are associated with an increased risk of breast and ovarian cancers. Women who carry a mutation in either of these genes can minimize their risk by frequent mammograms and other tests. Both genes are large—*BRCA1* codes for a protein of 1863 amino acids and *BRCA2* for one of 3418 amino acids—and many different mutations in these genes can predispose to breast or ovarian cancer. In certain high-risk populations, however, such as Ashkenazi Jews, only a few mutations predominate, and the SNPs associated with these mutations can be detected easily and efficiently.

### SNPs can be detected by DNA microarrays.

A widely used method for SNP genotyping is shown in **Fig. 15.8.** The technique involves a **DNA microarray,** which consists of a waferlike supporting surface about the size of a postage stamp to which are attached millions of different **oligonucleotides,** short, single-stranded DNA molecules of known sequence. Microarrays have many applications in biology, one of which is the detection of SNPs. For each SNP, researchers design oligonucleotides that contain a base in the middle that is complementary to one or the other of the SNP alleles. In the example shown in Fig. 15.8a, the bases on the oligonucleotide are G and C, which complement one SNP allele, and A and T, which complement the other. The flanking bases are complementary to those in both SNP alleles, which are the same. Although for clarity each oligonucleotide is shown as a single strand in Fig. 15.8a, there are actually millions of identical copies of each oligonucleotide deposited at each spot on the microarray.

Once a microarray with oligonucleotides complementary to both SNP alleles has been constructed, researchers add single-stranded DNA taken from an individual. Recall that DNA hybridization involves two single-stranded DNA molecules forming a double-stranded DNA molecule (Chapter 12). Under suitable conditions of hybridization, DNA fragments from individuals homozygous for the *C–G* allele will hybridize only with the oligonucleotides that contain G or C (Fig. 15.8b) and those from individuals homozygous for the *T–A* allele will hybridize with the oligonucleotides that contain A or T (Fig. 15.8d). The DNA fragments from heterozygous *C–G/T–A* individuals will hybridize with all the complementary oligonucleotides (Fig. 15.8c). The DNA fragments are usually labeled with a fluorescent dye,

**FIG. 15.8    SNP genotyping.** SNPs can be genotyped by observing their pattern of hybridization with oligonucleotides on a microarray.

**a.** Genotyping microarray

**G, C, A, T:**
Nucleotides that match one strand of one allele of the SNP

**X, Y:** Deliberate mismatches adjacent to SNP

Oligonucleotides attached to a microarray

**b.** DNA from a homozygous C-G/C-G individual

Pattern of fluorescence: **CG/CG**

**c.** DNA from a heterozygous C-G/T-A individual

Pattern of fluorescence: **CG/TA**

**d.** DNA from a homozygous T-A/T-A individual

Pattern of fluorescence: **TA/TA**

and any fragment that hybridizes with an oligonucleotide will fluoresce at that site.

The patterns of fluorescence are shown beneath the microarrays in Fig. 15.8. The key point is that each SNP genotype yields a unique pattern of fluorescence. DNA microarrays also include oligonucleotides that serve as experimental controls (important in any experiment) to ensure that the correct SNP has been identified. These contain deliberate mismatches adjacent to the site of the SNP. The control oligonucleotides in Fig. 15.8 are denoted X and Y; and they are expected not to show any fluorescence.

## Copy-number variation constitutes a significant proportion of genetic variation.

Genotyping microarrays are useful for detecting **copy-number variation (CNV),** the differences among individuals in the number of copies of a region of the genome. In contrast to the short repeats characteristic of VNTRs, the regions involved in CNVs are large and may include one or more complete genes. An example is shown in **Fig. 15.9.** In this case, a region of the genome that is normally present in only one copy per chromosome (Fig. 15.9a) may in some chromosomes be duplicated (Fig. 15.9b) or deleted (Fig. 15.9c). As in the figure, the multiple copies of the CNV region are usually adjacent to one another along the chromosome.

One of the surprises that emerged from sequencing the human genome was that CNV is quite common in the human population. Any two individuals' genomes differ in copy number at about five different regions, each with an average length of 200 to 300 kb. Across the genome as a whole, about 10% to 15% of the genome is subject to copy-number variation.

**FIG. 15.9    Copy-number variation in a region of a chromosome.**

**a.**

One copy of region

**b.**

Two copies of region (duplication)

**c.**

No copies (deletion)

In copy-number variation, the size of the duplicated or deleted region can include one or more complete genes.

## FIG. 15.10 Copy-number variation. Copy numbers of *AMY1* vary between societies with high-starch and with low-starch diets.

In societies with low-starch diets the average *AMY1* copy number per chromosome 1 is about 2.5, whereas in those with high-starch diets it is about 3.5.

Some CNVs occur in noncoding regions, but others consist of genes that are present in multiple tandem copies along the chromosome. An example is the human gene *AMY1* for the salivary-gland enzyme amylase, which aids in the digestion of starch. This gene is located in chromosome 1, and the *AMY1* copy number differs from one chromosome 1 to the next. **Fig. 15.10** shows the distribution of *AMY1* copy number along chromosome 1 in two groups of people: societies with a long history of a high-starch diet, and societies with a long history of a low-starch diet. There is a clear tendency for chromosomes from the latter group to have fewer copies of *AMY1* than those from the former group. On average, a chromosome 1 from the low-starch group has a copy number of about

2.5, whereas one from the high-starch group has an average of about 3.5. Because each individual has two copies of chromosome 1, this difference means that the average individual in the low-starch group has about 5 copies of *AMY1*, whereas the average individual in the high-starch group has about 7 copies. A plausible hypothesis is that extra copies were selected in groups with a high-starch diet because of the advantage extra copies conferred in digesting starch.

The detection of CNV is based on the relative intensity of hybridization of spots in a DNA microarray. Recall that each tiny spot on a DNA microarray holds millions of identical copies of a particular oligonucleotide sequence. In this case, the researcher chooses oligonucleotides in regions of the genome of interest, where CNV is likely. The greater the number of copies of a sequence in the genome, the more hybridization will occur when total genomic DNA is hybridized with complementary oligonucleotides on the microarray. Extra copies of a region result in greater intensity of fluorescence, and missing copies result in decreased fluorescence compared to the level of fluorescence observed for single-copy regions (**Fig. 15.11**).

## 15.4 GENETIC VARIATION IN CHROMOSOMES

Copy-number variation usually involves only one or a small number of genes, and the size of the duplicated or deleted region is physically so small that the differences are undetectable with conventional microscopy. In the human genome, some common variants involve large regions of the chromosome and are big enough to be visible through a microscope. Most of these variations involve regions of the genome with an extremely low density of genes, such as regions around the centromere or the long arm of the *Y* chromosome.

It may come as a surprise to learn that major chromosomal differences occur quite commonly. They are infrequently observed, however, because most of them are lethal. Nevertheless, a few major chromosomal differences are found in the general population. Some are common enough that their phenotypic effects are familiar. Others are less common, and a few have no major phenotypic effects at all. In this section, we focus on the types of difference that are observed, what causes them, and their phenotypic effects.

### Nondisjunction in meiosis results in extra or missing chromosomes.

**Nondisjunction** is the failure of a pair of chromosomes to separate during anaphase of cell division (Chapter 11). The result is that one daughter cell receives an extra copy of one chromosome, and the other daughter cell receives no copy of that chromosome. Nondisjunction can take place in mitosis, and it leads to cell lineages with extra or missing chromosomes—which is often observed in cancer cells.

## FIG. 15.11 CNV genotyping. Copy-number variation can be detected by the intensity of hybridization on a DNA microarray.

Duplicated DNA sequence (CNV)

The intensity of fluorescence is proportional to the number of hybridized oligonucleotides, which in turn is proportional to the number of copies of the sequence in the chromosome.

Position on chromosome

**FIG. 15.12 Nondisjunction.** (a) Chromosomes separate evenly into gametes in normal meiotic cell division. (b) Homologous chromosomes fail to separate in first-division nondisjunction. (c) Sister chromatids fail to separate in second-division nondisjunction.

**a.** Normal meiosis

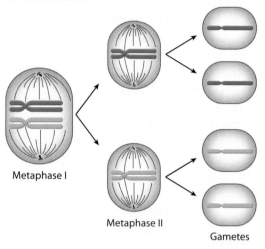

Metaphase I

Metaphase II

Gametes

**b.** First-division nondisjunction

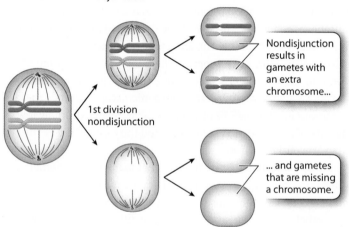

1st division nondisjunction

Nondisjunction results in gametes with an extra chromosome...

... and gametes that are missing a chromosome.

**c.** Second-division nondisjunction

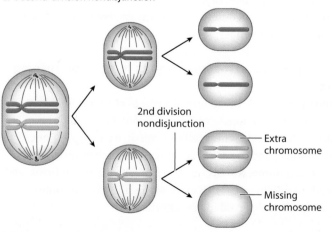

2nd division nondisjunction

Extra chromosome

Missing chromosome

When nondisjunction takes place in meiosis, the result is reproductive cells (gametes) that have either an extra or a missing chromosome. **Fig. 15.12** illustrates nondisjunction in meiosis. Fig. 15.12a shows key stages in a normal meiosis, in which chromosome separation (disjunction) takes place in both meiotic divisions, with the result that each gamete receives one and only one copy of each chromosome.

When failure of chromosome separation occurs, it can take place in either of the two meiotic divisions. Both types of nondisjunction result in gametes with an extra chromosome or a missing chromosome, but there is an important difference. In **first-division nondisjunction** (Fig. 15.12b), the homologous chromosomes fail to separate and so the chromosomes that move together are not genetically identical. On the other hand, in **second-division nondisjunction** (Fig. 15.12c), the sister chromatids of a chromosome fail to separate and therefore the two copies of the chromosome that end up in the same gamete are genetically identical. (In this discussion, we are ignoring the process of crossing over discussed in Chapter 11.)

**Some human disorders result from nondisjunction.**
Approximately 1 out of 155 live-born children (0.6%) has a major chromosome abnormality of some kind. About half of these have a major abnormality in chromosome structure, such as an inversion or translocation (Chapter 14). The others have a chromosome that is present in an extra copy or a chromosome that is missing because of nondisjunction—a parent has produced an egg or a sperm with an extra or a missing chromosome (**Fig. 15.13**).

Perhaps the most familiar of the conditions resulting from nondisjunction is **Down syndrome,** which results from the

**FIG. 15.13 Incidence of extra and missing chromosomes in human live births.**

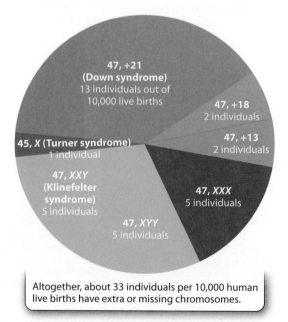

47, +21
(Down syndrome)
13 individuals out of
10,000 live births

47, +18
2 individuals

47, +13
2 individuals

45, X (Turner syndrome)
1 individual

47, XXY
(Klinefelter
syndrome)
5 individuals

47, XXX
5 individuals

47, XYY
5 individuals

Altogether, about 33 individuals per 10,000 human live births have extra or missing chromosomes.

**FIG. 15.14**

# What is the genetic basis of Down syndrome?

**BACKGROUND** The detailed images of cells and chromosomes available today are produced by microscopic and imaging techniques not available during most of the history of genetics and cell biology. Techniques for observing human chromosomes were originally so unreliable that the correct number of human chromosomes was firmly established only in the mid-1950s. Mistakes were common, such as counting two nearby chromosomes as one or by including chromosomes in the count of one nucleus when they actually belonged to a nearby nucleus. Despite these limitations, in 1959 French scientists Jérôme Lejeune, Marthe Gautier, and Raymond Turpin decided to investigate whether any common birth abnormalities were associated with chromosomal abnormalities. They chose Down syndrome, one of the most common birth abnormalities, because its genetic basis was an enigma.

**HYPOTHESIS** Down syndrome is associated with extra or missing chromosomes.

**EXPERIMENT** The chromosomes of five boys and four girls with Down syndrome were studied. A small patch of skin was taken from each child and placed in growth medium. In this medium, fibroblast cells, which synthesize the extracellular matrix that provides the structural framework of the skin, readily undergo DNA synthesis and divide. After a period allowing cellular growth and division, the fibroblast cells were prepared in a way that allowed the researchers to count the number of chromosomes under a microscope.

**RESULTS** Chromosomes were counted in 103 cells, yielding counts of 46 chromosomes (10 cells), 47 chromosomes (84 cells), and 48 chromosomes (9 cells). For each cell examined, the researchers noted whether the count was "doubtful" or "absolutely certain." The researchers marked cells as "doubtful" if the chromosomes or nuclei were not well separated from one another. They encountered 57 cells of which they felt "absolutely certain," and in each of these they counted 47 chromosomes. The extra chromosome was always one of the smallest chromosomes in the set.

**CONCLUSION** The researchers concluded that Down syndrome is the result of a chromosomal abnormality, an extra copy of one of the smallest chromosomes. The discovery helped resolve many of the questions surrounding the genetics of Down syndrome.

**FOLLOW-UP WORK** Improved techniques including chromosome banding (shown in Fig. 13.15) later indicated that the extra chromosome was chromosome 21. The finding stimulated many other chromosome studies. Today it is clear that, except for the sex chromosomes, the vast majority of fetuses with extra or missing chromosomes undergo spontaneous abortion. Furthermore, the Human Genome Project has indicated that genes on chromosome 21 have direct roles in many of the signs and symptoms of Down syndrome.

**SOURCE** J. Lejeune, Gautier, M., and Turpin, R. 1959. "Premier exemple d'aberration autosomique humaine." *Comptes rendus des séances de l'Académie des Sciences.* 248:1721–1722. Translated from the French in *Landmarks in Medical Genetics*, edited by P. S. Harper. Oxford, New York, 2004.

presence of an extra copy of chromosome 21. Down syndrome is also known as **trisomy 21** because affected individuals have three copies of chromosome 21. The extra chromosome can be seen in a chromosome layout called a **karyotype.** French researchers Jérôme Lejeune, Marthe Gautier, and Raymond Turpin first saw the extra chromosome when they developed a technique to visualize a patient's chromosomes (**Fig. 15.14**). The shorthand for the condition is 47,+21, meaning that the individual has 47 chromosomes with an extra copy of chromosome 21.

The presence of an extra chromosome 21 leads to a set of traits that, taken together, constitute a syndrome. These traits presumably result from an increase in dosage of the genes and other elements present on chromosome 21, although the specific genes or genetic elements have not been identified. Children with Down syndrome are mentally disabled to varying degrees; most are in the mild to moderate range and with special education and support they can acquire basic communication, self-help, and social skills. People with Down syndrome are short because of delayed maturation of the skeletal system; their muscle tone is low; and they have a characteristic facial appearance. About 40% of children with Down syndrome have major heart defects, and life expectancy is currently about 55 years. Among adults with Down syndrome, the risk of dementia is about 15% to 50%, which is about five times greater than the risk in the general population.

Although Down syndrome affects approximately 1 in 750 live-born children overall, there is a pronounced effect of mother's age on the risk of Down syndrome. For mothers of ages 45–50, the risk of having a baby with Down syndrome is approximately 50 times greater than it is for mothers of ages 15–20. The increased risk is the reason why many physicians recommend tests to determine abnormalities in the fetus for pregnant women over the age of 35.

Two other trisomies of autosomes (that is, chromosomes other than the X and Y chromosomes), are sometimes found in live births. These are trisomy 13 and trisomy 18. Both are much rarer than Down syndrome and also much more severe. In both cases, the developmental abnormalities are so profound that newborns with these conditions usually do not survive the first year of life. Other trisomies occur, but they are not compatible with life.

## Extra sex chromosomes have fewer effects than extra autosomes.

Extra or missing **sex chromosomes** (X or Y) are also quite common. In most mammals, females have two X chromosomes (XX), and males have one X and one Y chromosome (XY). The presence of the Y chromosome, not the number of X chromosomes, leads to male development. The female karyotype 47, XXX (47 chromosomes, including three X chromosomes) and the male karyotype 47, XYY (47 chromosomes, including one X and two Y chromosomes) are found among healthy females and males. These persons are in the normal range of physical development and mental capability and usually their extra sex chromosome remains undiscovered until their chromosomes are examined for some other reason.

The reason that 47, XYY males show no detectable phenotypic effects is because of the unusual nature of the Y chromosome.

The Y contains only a few functional genes other than the gene that stimulates the embryo to take the male developmental pathway. The absence of detectable phenotypic effects in 47, XXX females has a completely different explanation that has to do with the manner in which the activity of genes in the X chromosome is regulated. In the cells of female mammals, all X chromosomes except one are inactivated and gene expression is largely repressed. (This process, called X-inactivation, is discussed in Chapter 20.) Because of X-inactivation, a 47, XXX female has one active X chromosome per cell, the same number as in a normal 46, XX female.

→   **Quick Check 6**  A male baby is born with the sex-chromosome constitution *XYY*. Both parents have normal sex chromosomes (*XY* in the father, *XX* in the mother). In which meiotic division of which parent did the nondisjunction take place that produced the *XYY* baby?

Two other sex-chromosomal abnormalities do have phenotypic effects, especially on sexual development. The more frequent is 47, XXY (47 chromosomes, including two X chromosomes and one Y chromosome). Individuals with this karyotype are male because of the presence of the Y chromosome, but they have a distinctive group of symptoms known as **Klinefelter syndrome (Fig. 15.15a)**. Characteristics of the syndrome are very small testes but normal penis and scrotum. Growth and physical development are, for the most part, normal, although affected individuals tend to be tall for their age. About half of the individuals have some degree of mental impairment. When a person with Klinefelter syndrome reaches puberty, the testes fail to enlarge, the voice remains high pitched, pubic and facial hair remains sparse, and there is some enlargement of the breasts. Affected males do not produce sperm and hence are sterile; up to 10% of men who seek aid at infertility clinics turn out to be XXY.

Much rarer than all the other sex-chromosome abnormalities is 45, X (45 chromosomes, with just one X chromosome), the karyotype associated with characteristics known as **Turner syndrome (Fig. 15.15b)**. Affected females are short and often have a distinctive webbing of the skin between the neck and shoulders. There is no sexual maturation. The external sex organs remain immature, the breasts fail to develop, and pubic hair fails to grow. Internally, the ovaries are small or absent, and menstruation does not occur. Although the mental abilities of affected females are very nearly normal, they have specific defects in spatial abilities and arithmetical skills.

Earlier, we saw that in XX females one X chromosome undergoes inactivation. Why, then, do individuals with just one X chromosome show any symptoms at all? The explanation is that some genes on the inactive X chromosome in XX females are not completely inactivated. Evidently, for normal development to take place, both copies of some genes that escape complete inactivation must be expressed.

45, *X* is one of the rarest karyotypes seen in live-born babies. The reason is that more than 99% of the fetuses that are 45, *X* undergo spontaneous abortion—the subject of the next section.

## Nondisjunction is a major cause of spontaneous abortion.

Trisomies and the sex-chromosome abnormalities discussed in the previous section account for most of the simpler chromosomal abnormalities that are found among babies born alive. However, these chromosomal abnormalities represent only a minority of those that actually occur. Most fertilized eggs with chromosome abnormalities fail to complete embryonic development. At some time during pregnancy— in some cases very early, in other cases relatively late—the chromosomally abnormal embryo or fetus undergoes spontaneous abortion.

The relative proportions of some of the major chromosomal abnormalities in spontaneous abortion are shown in **Fig. 15.16**. The bars in red represent recognized pregnancies

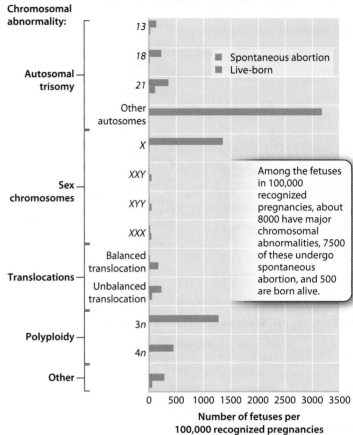

**FIG. 15.16** Chromosomal abnormalities in spontaneous abortion.

Among the fetuses in 100,000 recognized pregnancies, about 8000 have major chromosomal abnormalities, 7500 of these undergo spontaneous abortion, and 500 are born alive.

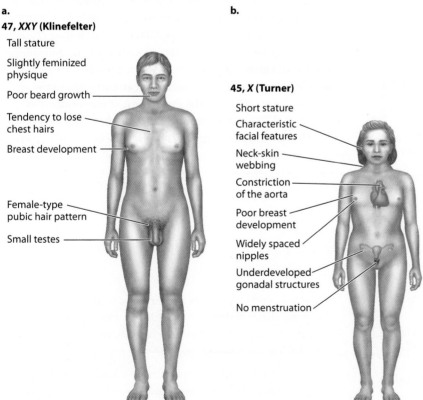

**FIG. 15.15** Symptoms associated with (a) Klinefelter syndrome and (b) Turner syndrome.

**a.**

**47, *XXY* (Klinefelter)**

Tall stature

Slightly feminized physique

Poor beard growth

Tendency to lose chest hairs

Breast development

Female-type pubic hair pattern

Small testes

**b.**

**45, *X* (Turner)**

Short stature

Characteristic facial features

Neck-skin webbing

Constriction of the aorta

Poor breast development

Widely spaced nipples

Underdeveloped gonadal structures

No menstruation

that terminate in spontaneous abortion, and those in blue represent those in which the fetus develops to term and is born alive. Note the large number of autosomal trisomies, none of which (with the exception of trisomies 13, 18, and 21) permit live births. Even among these, about 75% of fetuses with trisomy 21 undergo spontaneous abortion, and the proportions are even greater for trisomies 13 and 18. The 45, *X* karyotype is also very frequent among spontaneous abortions.

A surprisingly large number of fetuses that undergo spontaneous abortion are **triploid** (with three complete sets of chromosomes, 69 altogether) or **tetraploid** (four complete sets, 92 altogether). These karyotypes usually result from a defective spindle apparatus and failure of cell division in anaphase. When this occurs in meiosis, the result is a diploid gamete and a triploid fertilized egg, and when it occurs

**FIG. 15.17  Formation of polyploid organisms.** (a) A triploid organism can result from failure of division in meiosis. (b) A tetraploid organism can result from failure of cell division in mitosis.

**a.** Formation of triploid organisms

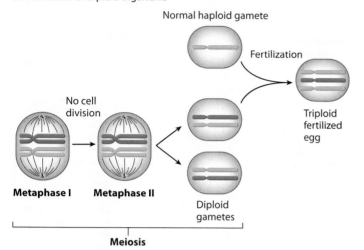

**b.** Formation of tetraploid organisms

**FIG. 15.18  (a) Balanced and (b) unbalanced translocations.**

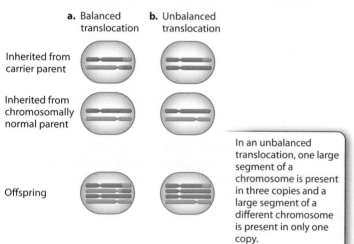

In an unbalanced translocation, one large segment of a chromosome is present in three copies and a large segment of a different chromosome is present in only one copy.

in the first mitosis in a normal fertilized egg, the result is a tetraploid (**Fig. 15.17**). Many spontaneous abortions result from an **unbalanced translocation** (Chapter 14), in which only part of a reciprocal translocation (along with one of the nontranslocated chromosomes) is inherited from one of the parents (**Fig. 15.18**).

Altogether, about 15% of all recognized pregnancies terminate with spontaneous abortion of the fetus, and roughly half of these are due to major chromosomal abnormalities. This number tells only part of the story because embryos with a missing autosome are not found among spontaneously aborted fetuses. These must occur at least as frequently as those with an extra autosome because both are created by the same event of nondisjunction (see Fig. 15.12). The explanation seems to be that in fertilized eggs with a missing autosome the abortion occurs shortly after fertilization, and most cases are not recognized.

## Core Concepts Summary

### 15.1 GENETIC VARIATION DESCRIBES COMMON GENETIC DIFFERENCES (POLYMORPHISMS) AMONG THE INDIVIDUALS IN A POPULATION AT ANY GIVEN TIME.

The genotype of an organism is its genetic makeup. page 15-1

The phenotype of an organism is any observable characteristic of that organism, such as its appearance, physiology, or behavior. The phenotype results from a complex interplay of the genotype and environment. page 15-2

A person's genotype can be homozygous, in which he or she has two alleles of the same type (one from each parent), or heterozygous, in which he or she has inherited different types of alleles from each parent. page 15-2

Genetic variation can be neutral (no effect on survival or reproduction), harmful (associated with decreased survival or reproduction), or beneficial (associated with increased survival or reproduction). page 15-3

## 15.2 HUMAN GENETIC VARIATION CAN BE DETECTED BY DNA TYPING, WHICH CAN UNIQUELY IDENTIFY EACH INDIVIDUAL.

DNA typing analyzes genetic polymorphisms at multiple genes with multiple alleles. Because there are so many possible genotypes in the population, this procedure can uniquely identify an individual with high probability. page 15-6

A variable number tandem repeat (VNTR) results from differences in the number of small-sequence repeats in a given area of the genome. page 15-6

A restriction fragment length polymorphism (RFLP) results from small changes in the DNA sequence between chromosomes, such as point mutations, that create or destroy restriction sites. page 15-8

## 15.3 TWO COMMON TYPES OF GENETIC POLYMORPHISM ARE SINGLE-NUCLEOTIDE POLYMORPHISMS (SNPs) AND COPY-NUMBER VARIATION (CNV).

The DNA in any two human genomes differs at about 3 million individual nucleotide sites; these are called SNPs. page 15-8

SNP genotyping allows many SNPs to be identified simultaneously in many individuals. It is carried out by the hybridization of genomic DNA onto a microarray containing millions of different DNA fragments of known sequence corresponding to places in the genome where SNPs are present. page 15-9

CNV is a difference in the number of copies of a particular DNA sequence among chromosomes. CNVs are detected in a similar way to SNPs, except that the readout is signal intensity, which depends on number of copies. page 15-10

## 15.4 CHROMOSOMAL VARIANTS CAN ALSO OCCUR BUT ARE USUALLY HARMFUL.

Nondisjunction is the failure of a pair of chromosomes to separate during anaphase of cell division. It can occur in mitosis or meiosis and results in daughter cells with extra or missing chromosomes. page 15-11

Nondisjunction in meiosis leads to gametes with extra or missing chromosomes. page 15-12

Most human fetuses with extra or missing chromosomes spontaneously abort, but some are viable and exhibit characteristic syndromes. page 15-13

Trisomy 21 (Down syndrome) is usually characterized by three copies of chromosome 21. page 15-13

Extra or missing sex chromosomes (X or Y) are common and result in syndromes such as Klinefelter syndrome (47, XXY) and Turner syndrome (45, X). page 15-14

## Self-Assessment

1. Differentiate between a genotype and a phenotype.

2. Describe common effects of mutations on an organism, and provide an example of each.

3. Explain why it is sometimes an oversimplification to consider a mutation harmful, beneficial, or neutral.

4. Describe two types of genetic polymorphism that are useful in DNA typing.

5. Define the term "SNP" and explain why researchers are interested in detecting SNPs.

6. Explain how SNPs and CNVs are detected in a population of individuals.

7. Diagram how nondisjunction in meiosis I or II can result in extra or missing chromosomes in reproductive cells (gametes).

8. Describe the consequences of an extra copy of chromosome 21 (Down syndrome).

Do you understand the chapter's Core Concepts? Log in to check your answers to the Self-Assessment questions, then practice what you've learned and reinforce this chapter's concepts by working through the problems and multimedia tutorials provided there.

**www.biologyhowlifeworks.com**

# MENDELIAN INHERITANCE

## Core Concepts

From differences in appearances to differences in the ways our bodies work, we have focused so far on how genetic variation leads to our individuality. Examples of genetic variation in the human population range from harmless curiosities like the genetic difference in taste receptors that determine whether or not broccoli is perceived as having an unpleasantly bitter taste, to mutant forms of genes resulting in serious diseases like sickle-cell anemia or emphysema.

This chapter focuses on how that genetic variation is inherited. **Transmission genetics** deals with the manner in which genetic differences among individuals are passed from generation to generation. We are all aware of genetics. We know that children resemble their parents and that there are sometimes uncanny similarities among even distant relatives. But some patterns are more difficult to discern. Traits such as eye color, nose shape, or risk for a particular disease may be passed down faithfully generation after generation, but sometimes they are not, and sometimes they appear and disappear in seemingly random ways.

As a modern science, transmission genetics began with pea-breeding experiments carried out by the monk Gregor Mendel in the 1860s. However, even before then, people understood enough about inheritance that they were able to select crops and livestock with particular characteristics.

## 16.1 EARLY THEORIES OF INHERITANCE

Thousands of years before Mendel, many societies carried out practical plant and animal breeding. The ancient practices were based on experience rather than on a full understanding of the rules of genetic transmission, but they were nevertheless highly successful. In Mesoamerica, for example, Native Americans chose corn (maize) plants for cultivation that had the biggest ears and softest kernels. Over many generations of such selection, their cultivated corn came to have less and less physical resemblance to its wild ancestral species. Similarly, people in the Eurasian steppes selected their horses for a docile temperament suitable for riding or for hitching to carts or sleds. Practical breeding of this kind provided the crops and livestock from which most of our modern domesticated animals and plants derive.

### Early theories of heredity predicted the transmission of acquired characteristics.

The first written speculations about mechanisms of heredity were made by the ancient Greeks. Hippocrates (460–377 BCE), considered the founder of Western medicine, proposed that each part of the body in a sexually mature adult produces a substance that collects in the reproductive organs and that determines the inherited characteristics of the offspring. An implication of this theory is that any **trait,** or characteristic, of an individual can be transmitted from parent to offspring. Even traits that are acquired during the lifetime of an individual, such as muscle strength or bodily injury, were thought to be heritable because of the substance supposedly passed from each body part to the reproductive organs.

The theory that acquired characteristics can be inherited was invoked to explain such traits as the webbed feet of ducks, which were thought to result from many successive generations in which adult ducks stretched the skin between their toes while swimming and passed this trait to offspring. A few decades later, however, Aristotle (384–322 BCE) emphasized several observations that the theory of inheritance of acquired characteristics cannot account for:

- Traits such as hair color can be inherited, but it is difficult to see how hair—a nonliving tissue—could send substances to the reproductive organs.

- Traits that are not yet present in an individual can be transmitted to the offspring. For example, a father and his adult son can both be bald, even though the son was born before the father became bald.

- Parts of the body that are lost as a result of surgery or accident are not missing in the offspring.

From these and other observations, Aristotle concluded that the process of heredity transmits only the *potential* for producing traits present in the parents, and not the traits themselves. Nevertheless, Hippocrates' theory influenced biology until well into the 1800s. It was incorporated into an early theory of evolution proposed by the French biologist Jean-Baptiste Lamarck around 1800, and was also accepted by Charles Darwin, who developed an alternative theory—the theory of evolution by natural selection (Chapter 21).

### Belief in blending inheritance discouraged studies of hereditary transmission.

Darwin also subscribed to the now-discredited model of **blending inheritance,** in which traits in the offspring resemble the average of those in the parents. For example, the offspring of plants with blue flowers and those with red flowers should have purple flowers. While traits of offspring are sometimes the average of those of the parents (think of certain cases of human height, for example), the idea of blending inheritance—which implies the blending of the genetic material—as a general rule presents problems. For example, it cannot explain the reappearance of a trait several generations after it apparently "disappeared" in a family, such as red hair or blue eyes.

Another difficulty with the concept of blending inheritance is that variation will be lost over time. Consider an example where black rabbits are rare and white rabbits are common (**Fig. 16.1**).

**FIG. 16.1    Blending inheritance.** This model predicts loss of variation over time and blending of genetic material, which is not observed.

Black rabbits, being rare, are most likely to mate with the much more common white rabbits. If blending inheritance occurs, the result will be gray rabbits. These will also mate with white rabbits, producing even lighter gray rabbits. Over time, the population will end up being all white, or very close to white, and there will be less variation. The only way that black rabbits will be present is if they are re-introduced into the population through mutation or migration. Inheritance will tend to be a homogenizing force, producing in each generation a blend of the original phenotypes. But we know from common experience that variation in most populations is plentiful.

It is ironic that Darwin believed in blending inheritance because this mechanism of inheritance is incompatible with his theory of evolution by means of natural selection. The incompatibility was pointed out by some of Darwin's contemporaries, and Darwin himself recognized it as a serious problem. The problem with blending inheritance is that rare variants, such as black rabbits in the above example, will have no opportunity to increase in frequency even if they survive and reproduce more than white rabbits, since they gradually disappear over time.

Although Darwin was convinced that he was right about natural selection, he was never able to reconcile his theory with the concept of blending inheritance. Unknown to him, the solution had already been discovered in experiments carried out by Gregor Mendel (1822–1884), an Augustinian monk in a monastery in the city of Brno in what is now the Czech Republic. Mendel's key discovery was this: It is not *traits* that are transmitted in inheritance—it is *genes* that are transmitted.

## 16.2 THE FOUNDATIONS OF MODERN TRANSMISSION GENETICS

Mendel's scientific fame rests on the pea-breeding experiments that he carried out in the years 1856 to 1864 (**Fig. 16.2**), in which he demonstrated the basic principles of transmission genetics. His experiments were so simple and presented with such clarity that they still serve as examples of the scientific method at its best (Chapter 1).

**Mendel's experimental organism was the garden pea.** For his experimental material, Mendel used several strains of ordinary garden peas (*Pisum sativum*) that he obtained from a local seed supplier. His experimental approach was similar to that of a few botanists of the eighteenth and nineteenth centuries who studied the results of **hybridization**, or interbreeding between two different varieties or species of an organism. Where Mendel differed from his predecessors was in paying close attention to a small number of easily classified traits with contrasting characteristics. For example, where one strain of peas

had yellow seeds, another had green seeds; and where one had round seeds, the other had wrinkled seeds (**Fig. 16.3**). Altogether Mendel studied seven physical features expressed in contrasting fashion among the strains: seed color, seed shape, pod color, pod shape, flower color, flower position, and plant height.

The expression of each trait in each strain was **true breeding,** which means that the physical appearance of the offspring in each successive generation is identical to the previous one. In other words, plants of the strain with yellow seeds produced only

**FIG. 16.2** Gregor Mendel and his experimental organism, the pea plant.

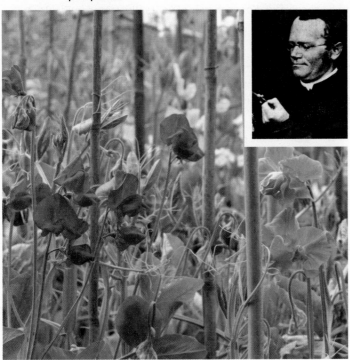

**FIG. 16.3** **Contrasting traits.** Mendel focused on seven contrasting traits.

| | Dominant | Recessive |
|---|---|---|
| **a.** Color of seeds (yellow or green) | | |
| **b.** Shape of seeds (round or wrinkled) | | |
| **c.** Color of pod (green or yellow) | | |
| **d.** Shape of pod (smooth or indented) | | |
| **e.** Color of flower (purple or white) | | |
| **f.** Position of flowers (along stem or at tip) | | |
| **g.** Plant height (tall or dwarfed) | | |

yellow seeds, and those of the strain with green seeds produced only green seeds. Likewise, plants of the strain with round seeds produced only round seeds, and those of the strain with wrinkled seeds produced only wrinkled seeds.

The objective of Mendel's experiments was simple. By means of crosses between the true-breeding strains and crosses among their progeny, Mendel hoped to determine whether there are statistical patterns in the occurrence of the contrasting traits, such as yellow seeds or green seeds. If such patterns could be found, he would seek to devise a hypothesis to explain them and then use his hypothesis to predict the outcome of further crosses.

In designing his experiments, Mendel departed from other plant hybridizers of the time in three important ways:

1. Mendel studied true breeding strains, unlike many other plant hybridizers who used complex and poorly defined material.

2. Mendel focused on one trait, or a small number of traits, at a time, with characteristics that were easily contrasted among the true-breeding strains. Other plant hybridizers crossed strains differing in many traits and tried to follow all the traits at once. This resulted in amazingly complex inheritance, and no underlying patterns could be discerned.

3. Mendel counted the progeny of his crosses, looking for statistical patterns in the offspring of his crosses. Others typically noted only whether offspring with a particular characteristic were present or absent, but did not keep track of and count all the progeny of a particular cross.

## In crosses, one of the traits was dominant in the offspring.

Mendel began his studies with crosses between true-breeding strains that differed in a single contrasting characteristic. Crossing peas is not as easy as it may sound. Pea flowers include both female and male reproductive organs enclosed together within petals (**Fig. 16.4**). Because of this arrangement, pea plants usually fertilize themselves (self-fertilize). Performing crosses between different plants is a tedious and painstaking process. First, the flower of the designated female parent must be opened at an early stage, and the immature anthers (plant male reproductive organs) clipped off and discarded so the plant cannot self-fertilize. Then mature pollen from the designated male parent must be collected and deposited on the stigma (female reproductive organ) of the

female parent, from where it travels to the reproductive cells in the ovule at the base of the flower. And finally, a cloth bag must be tied around the female flower to prevent stray pollen from entering. No wonder Mendel complained that his eyes hurt!

For each of the seven pairs of contrasting traits, true-breeding strains differing in the trait were crossed. **Fig. 16.5** illustrates a typical result, in this case for yellow versus green seeds. In these kinds of crosses, the parental generation is referred to as the **$P_1$ generation,** and the first offspring, or filial, generation, is referred to as the **$F_1$ generation.** In the cross of $P_1$ yellow $\times$ $P_1$ green, Mendel observed that all the $F_1$ progeny were yellow. This result was shown to be independent of the seed color, yellow or green, of the pollen donor because both of the crosses shown below yielded progeny plants with yellow seeds:

Crosses like these, in which the expressions of the trait in the female and male parents are interchanged, are known as **reciprocal crosses.** Mendel showed that reciprocal crosses yielded the same result for each of his seven pairs of contrasting traits.

Moreover, for each pair of contrasting traits, only one of the traits appeared in t he $F_1$ generation. For these simple crosses, the trait that appears in the $F_1$ generation is said to be **dominant,** and the contrasting trait that does not appear is said to be **recessive.** For each pair of traits illustrated in Fig. 16.3, the dominant trait is shown on the left, and the recessive on the

---

**FIG. 16.4    Crossing pea plants.** Crossing plants in a way that isolated either the ovules or the pollen controlled what genetic material each parent contributed to the offspring.

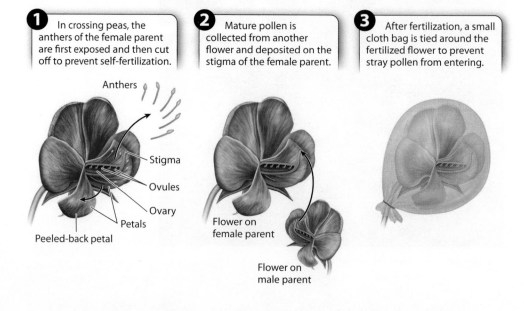

**1** In crossing peas, the anthers of the female parent are first exposed and then cut off to prevent self-fertilization.

**2** Mature pollen is collected from another flower and deposited on the stigma of the female parent.

**3** After fertilization, a small cloth bag is tied around the fertilized flower to prevent stray pollen from entering.

**FIG. 16.5** **The first-generation hybrid (F₁).** A cross between two of Mendel's true-breeding plants (the parental, or P₁, generation) yielded first-generation hybrids displaying the dominant trait.

Plant grown from true-breeding strain with yellow seeds

Plant grown from true-breeding strain with green seeds

×

True-breeding strains that are crossed constitute the P₁ or parental generation.

**P₁ generation**

The trait that appears in the F₁ generation (in this case yellow seeds) is dominant, and the other trait (green seeds) is recessive.

**F₁ generation**

right. Thus, yellow seed is dominant to green seed, round seed is dominant to wrinkled seed, and so forth. Mendel explained these findings by supposing that there is a hereditary factor for yellow seeds and a different hereditary factor for green seeds, likewise a hereditary factor for round seeds and a different one for wrinkled seeds, and so on. We now know that the hereditary factors that result in contrasting traits are different forms of a gene that affect the trait. The different forms of a gene are called **alleles.** The particular combination of alleles present in an individual constitutes its **genotype,** and the expression of the trait in the individual constitutes its **phenotype.**

In the cross between true-breeding plants with yellow seeds and true-breeding plants with green seeds, the genotype of each F₁ seed includes an allele for yellow from the yellow parent and an allele for green from the green parent. The phenotype of each F₁ seed is nevertheless yellow, because yellow is dominant to green.

The molecular basis of dominance is the fact that, in diploid organisms, only one copy of most genes is needed to carry out

the normal function of the gene. For example, yellow pea seeds result from an enzyme that breaks down green chlorophyll, allowing yellow pigments to show through. Green seeds result from a mutation in this gene that inactivates the enzyme, and so the green chlorophyll is retained. In an F₁ hybrid, such as that in Fig. 16.5, which receives a nonmutant gene from one parent and a mutant gene from the other, the seeds are yellow because one copy of the nonmutant gene produces enough of the enzyme to break down the chlorophyll to yield a yellow seed.

## 16.3 SEGREGATION: MENDEL'S KEY DISCOVERY

Mendel's most important discovery was that the F₁ progeny of a cross between plants with different traits did *not* breed true. In the **F₂ generation,** produced by allowing the F₁ flowers to undergo self-fertilization, the recessive trait reappeared (**Fig. 16.6**). Not only did the recessive trait reappear, it reappeared in a definite numerical proportion. Among a large number of F₂ progeny,

**FIG. 16.6** **Self-fertilization of the F₁ hybrid, resulting in seeds of the F₂ generation.** The recessive trait appeared again in the F₂ generation.

Seeds from F₁ plants produced from a cross of true-breeding yellow-seed and green-seed plants are yellow because yellow is dominant and green is recessive in seed color.

**F₁ generation**

Peas are normally self-fertilizing, and so if they are left alone, the pollen produced in each flower fertilizes the ovules.

Because fertilization takes place at random, any individual pod can have a ratio of dominant:recessive that deviates from 3:1.

**F₂ generation**
(ratio 3 yellow seeds:1 green seed)

**TABLE 16.1** Observed F$_2$ Ratios in Mendel's Experiments

| TRAIT | DOMINANT TRAIT | RECESSIVE TRAIT | RATIO |
|---|---|---|---|
| Seed color | 6,022 | 2,001 | 3.01:1 |
| Seed shape | 5,474 | 1,850 | 2.96:1 |
| Pod color | 428 | 152 | 2.82:1 |
| Pod shape | 882 | 299 | 2.95:1 |
| Flower color | 705 | 224 | 3.15:1 |
| Flower position | 651 | 207 | 3.14:1 |
| Plant height | 787 | 277 | 2.84:1 |

Mendel found that the dominant:recessive ratio was very close to 3:1. The results he observed among the F$_2$ progeny for each of the seven pairs of traits are given in **Table 16.1.** Across experiments for all seven traits, the ratio of dominant:recessive F$_2$ offspring was 14,949:5,010. Although there is variation from one experiment to the next, the overall ratio of 14,949:5,010 equals 2.98:1, which is a very close approximation to 3:1.

### Genes come in pairs that segregate in the formation of reproductive cells.

The explanation for the 3:1 ratio in the F$_2$ and Mendel's observations in general can be summarized with reference to **Fig. 16.7:**

---

**FIG. 16.7** The principle of segregation.

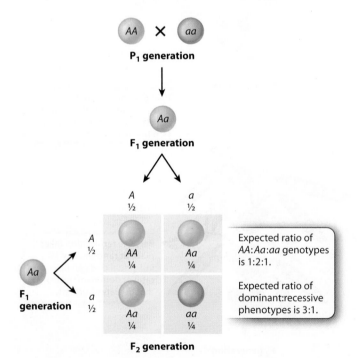

**F$_2$ generation**

1. Except for cells involved in reproduction, each cell of a pea plant contains two alleles of each gene. In each true-breeding strain constituting the P$_1$ generation, the two alleles are identical. In Fig. 16.7, we designate the allele associated with yellow seeds as *A* and that associated with green seeds as *a*. The genotype of the true-breeding strain with yellow seeds can therefore be written as *AA*, and that of the true-breeding strain with green seeds as *aa*. The *AA* and *aa* genotypes are said to be **homozygous,** which means that both alleles present in an individual are the same.

→ **Quick Check 1** What are the genotypes and phenotypes for Mendel's true-breeding parent plants?

2. Each reproductive cell, or **gamete,** contains only one allele of each gene. In this case, a gamete can contain the *A* allele or the *a* allele, but not both.

3. In the formation of gametes, the two members of a gene pair **segregate** (or separate) equally into gametes, so that half the gametes get one allele and half get the other allele. This separation of alleles into different gametes defines the **principle of segregation.** In the case of homozygous plants (such as *AA* or *aa*), all the gametes from an individual are the same. That is, the homozygous *AA* strain with yellow seeds produces gametes containing the *A* allele, and the homozygous *aa* strain with green seeds produces gametes containing the *a* allele.

4. The fertilized egg cell, called the **zygote,** is formed from the random union of two gametes, one from each parent. For the cross *AA* × *aa*, the zygote is an F$_1$ hybrid formed from the union of an *A*-bearing gamete with an *a*-bearing gamete. Each F$_1$ hybrid therefore has the genotype *Aa* (Fig. 16.7). The *Aa* genotype is **heterozygous,** which means that the two alleles for a given gene are different. The F$_1$ seeds are yellow because yellow is dominant to green. Note that each F$_1$ progeny contains an *a* allele because its genotype is *Aa*, but the phenotype of the seed is yellow and indistinguishable from that of an *AA* genotype.

→ **Quick Check 2** Is it possible for two individuals to have the same phenotype but different genotypes? The same genotype, but different phenotypes? How?

5. When the F$_1$ progeny (genotype *Aa*) form gametes, by the principle of segregation the *A* and *a* alleles again separate, so that half the gametes contain only the *A* allele and the other half contain only the *a* allele (Fig. 16.7).

6. In the formation of the F$_2$ generation, the gametes from the F$_1$ parents again combine at random. The consequences of random union of gametes can be worked out by means of

a checkerboard of the sort shown in Fig. 16.7. This kind of square is called a **Punnett square** after its inventor, Reginald Punnett, a British geneticist.

Across the top of a Punnett square are the gametes from one parent, each with its respective frequency; and down the side are the gametes from the other parent, again each with its respective frequency. These frequencies are the probabilities of each type of gamete occurring. In the case illustrated in Fig. 16.7, both parents are $F_1$ hybrids with the genotype $Aa$, and because the $A$ and $a$ alleles undergo segregation in the formation of gametes, half of the gametes from each parent are $A$ and half are $a$.

Inside the Punnett square, each box contains the genotype of a possible zygote formed by random union of the gametes. The genotype of the zygote in each cell corresponds to the alleles in the gametes in the corresponding row and column, and the relative frequency of each zygote is obtained by multiplication of the gametic frequencies in the corresponding row and column. For example, the zygote in the top left corner in the Punnett square in Fig. 16.7 is formed by the union of an $A$-bearing gamete from the top and an $A$-bearing gamete from the side. The zygote genotype is therefore $AA$, and because each of the gametic types has a relative frequency of $\frac{1}{2}$, the relative frequency of the $AA$ zygote is $\frac{1}{2} \times \frac{1}{2} = \frac{1}{4}$. It is the multiplication of the gametic frequencies that corresponds to the random union of gametes.

For the cross in Fig. 16.7, the boxes of the Punnett square correspond to all of the possible individuals in the $F_2$ generation. The expected *genotypes* of the progeny are therefore $\frac{1}{4}$ $AA$, $\frac{1}{2}$ $Aa$, and $\frac{1}{4}$ $aa$ (or 1:2:1). When there is dominance, however, as there is in this case, the $AA$ and $Aa$ genotypes have the same phenotype, and so the ratio of dominant:recessive *phenotypes* is 3:1. The Punnett square in Fig. 16.7 illustrates the biological basis of the 3:1 ratio of phenotypes that Mendel observed in the $F_2$ generation. Nevertheless, the underlying ratio of $AA:Aa:aa$ genotypes is 1:2:1.

→ **Quick Check 3** What are the expected progeny (genotypes and phenotypes) from a cross of an $AA$ plant with an $Aa$ plant?

### The principle of segregation was tested by predicting the outcome of crosses.

The model of segregation of gene pairs (alleles) and their random combination in the formation of a zygote depicted in Fig. 16.7 was Mendel's hypothesis, an explanation he put forward to explain an observed result. But without further experiments, he could not know if his hypothesis was right or wrong. The true test of a hypothesis is whether it can predict the results of experiments that have not yet been carried out (Chapter 1). If the predictions are correct, one's confidence in the hypothesis is strengthened. Mendel appreciated this intuitively, even though the scientific method as understood today had not been formalized.

The Punnett square in Fig. 16.7 makes two strong predictions. The first is that the seeds in the $F_2$ generation showing the recessive green phenotype should be homozygous $aa$. If the green $F_2$ seeds have the genotype $aa$, then they should breed true. That is, when the seeds are grown into mature plants and self-fertilization is allowed to take place, the self-fertilized $aa$ plants should produce only green seeds ($aa$). This prediction was confirmed by examining seeds actually produced by plants grown from the $F_2$ green seeds.

The second prediction from the Punnett square in Fig. 16.7 is more complex. It has to do with the seeds in the $F_2$ generation that show the dominant yellow phenotype. Note that although these seeds have the same phenotype, they have two different genotypes ($AA$ and $Aa$). Among just the yellow seeds, $\frac{1}{3}$ should have the genotype $AA$ and $\frac{2}{3}$ should have the genotype $Aa$, for a ratio of 1 $AA$:2 $Aa$. (The proportions are $\frac{1}{3}$:$\frac{2}{3}$ because we are considering *only* the seeds that are yellow.) The $AA$ and $Aa$ genotypes can be distinguished by the types of seeds they produce when self-fertilized. The $AA$ plants produce only seeds with the dominant yellow phenotype (that is, they are true breeding), whereas the $Aa$ plants yield dominant yellow and recessive green seeds in the ratio 3:1. Mendel did such experiments, and the prediction turned out to be correct. His data confirming the 1:2 ratio of $AA:Aa$ among $F_2$ individuals with the dominant phenotype are shown in **Table 16.2**.

### A testcross is a mating to an individual with the homozygous recessive genotype.

A more direct test of segregation is to cross the $F_1$ progeny with the true-breeding recessive strain instead of allowing them to self-fertilize. Any cross of an unknown genotype with a homozygous recessive genotype is known as a **testcross.**

The $F_1$ progeny show the dominant yellow seed phenotype. A yellow seed phenotype can result from either of two possible genotypes, $Aa$ or $AA$. A testcross with plants of the homozygous

---

**TABLE 16.2** Genetic ratios from self-fertilization of plants showing the dominant phenotype

| TRAIT | HOMOZYGOUS DOMINANT | HETERO-ZYGOUS | RATIO |
|---|---|---|---|
| Yellow seeds | 166 | 353 | 0.94:2 |
| Round seeds | 193 | 372 | 1.04:2 |
| Green pods | 40 | 60 | 1.33:2 |
| Smooth pods | 29 | 71 | 0.82:2 |
| Purple flowers | 36 | 64 | 1.13:2 |
| Flowers along stem | 33 | 67 | 0.99:2 |
| Tall plants | 28 | 72 | 0.78:2 |

recessive genotype (*aa*) can distinguish between these two possibilities (**Fig. 16.8**). The testcross is highly informative because the homozygous recessive parent produces only gametes carrying the recessive allele (*a*). In Fig. 16.8, the gametes from the homozygous recessive parent are shown across the top, and they consist only of *a*-bearing gametes. Down the side are gametes from the individual with the unknown genotype.

Let's first consider what happens in a testcross to an *Aa* individual (Fig. 16.8a). An *Aa* individual produces both *A*-bearing gametes and *a*-bearing gametes. Inside the Punnett square are the predicted offspring from the testcross, which consist of ½ *Aa* zygotes, which yield yellow seeds, and ½ *aa* zygotes, which yield green seeds. By contrast, an *AA* individual produces only *A*-bearing gametes, so all the zygotes have the *Aa* genotype, which yields yellow seeds (Fig. 16.8b). In other words, the testcross gives different results depending on whether the parent is heterozygous (*Aa*) or homozygous (*AA*).

Note that in a testcross the *phenotypes* of the progeny reveal the *alleles* present in the gametes from the tested parent. An *Aa* individual yields ½ *Aa* (yellow seeds) and ½ *aa* (green seeds), since the *Aa* parent produces ½ *A*-bearing and ½ *a*-bearing gametes. An *AA* individual yields only *Aa* (yellow seeds), since the *AA* parent produces only *A*-bearing gametes. These results are a direct demonstration of the principle of segregation, since the ratio of the phenotypes of progeny reflect the equal segregation of alleles into gametes. Some of Mendel's testcross data indicating 1:1 segregation in heterozygous genotypes are shown in **Table 16.3.**

### Segregation of alleles reflects the separation of chromosomes in meiosis.

The principles of transmission genetics have a physical basis in the process of meiosis (Chapter 11). During meiosis I, maternal and paternal chromosomes (homologous chromosomes) align on the metaphase plate. Then, during anaphase I, the homologous

**FIG. 16.8**   **A testcross.** A cross with a homozygous recessive individual reveals the genotype of the other parent.

chromosomes separate, and each chromosome goes to a different pole. Because gene pairs are carried on homologous chromosomes, segregation of alleles observed by Mendel corresponds to the separation of chromosomes that takes place in anaphase I.

**Fig. 16.9** illustrates the separation of a pair of homologous chromosomes in anaphase I. In the configuration shown, the copies of the *A* allele (dark blue) separate from the copies of the *a* allele (light blue) in anaphase I. The separation of chromosomes is the physical basis of the segregation of alleles.

### Dominance is not universally observed.

Many traits do not show complete dominance such as Mendel observed with pea plants. Instead, they show **incomplete dominance,** in which the phenotype of the heterozygous

**FIG. 16.9**   **Segregation of alleles of a single gene.** Homologous chromosomes separate during meiosis, leading to segregation of alleles.

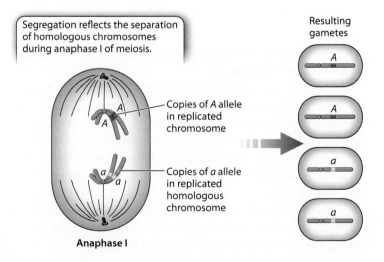

**TABLE 16.3**   **Phenotype of progeny from testcrosses of heterozygotes**

| TRAIT | DOMINANT TRAIT | RECESSIVE TRAIT | RATIO |
|---|---|---|---|
| Seed color | 196 | 189 | 1.04:1 |
| Seed shape | 193 | 192 | 1.01:1 |
| Flower color | 85 | 81 | 1.05:1 |
| Plant height | 87 | 79 | 1.10:1 |

genotype is intermediate between those of the homozygous genotypes. In such cases, the result of segregation can be observed directly because each genotype has a distinct phenotype. An example is flower color in the snapdragon (*Antirrhinum majus*), in which the homozygous genotypes have red ($C^R C^R$) or white ($C^W C^W$) flowers, and the heterozygous genotype $C^R C^W$ has pink flowers (**Fig. 16.10**). In notating incomplete dominance, we use superscripts to indicate the alleles, rather than upper-case and lower-case letters, because neither allele is dominant to the other. A cross of homozygous $C^R C^R$ and $C^W C^W$ strains results in hybrid $F_1$ progeny that are pink ($C^R C^W$), and when these are crossed the resulting $F_2$ generation consists of ¼ red ($C^R C^R$), ½ pink ($C^R C^W$), and ¼ white ($C^W C^W$).

**FIG. 16.10 Incomplete dominance for flower color in snapdragons.** In incomplete dominance, an intermediate phenotype is seen.

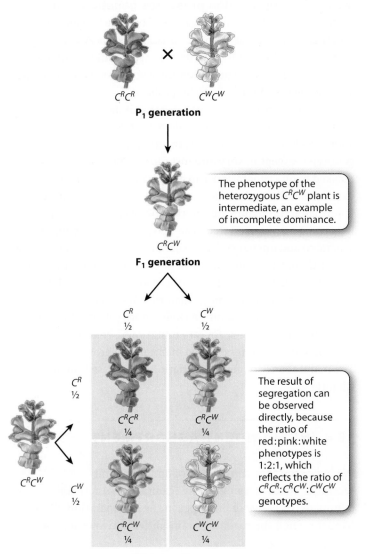

$C^R C^R$      $C^W C^W$

**P₁ generation**

The phenotype of the heterozygous $C^R C^W$ plant is intermediate, an example of incomplete dominance.

$C^R C^W$

**F₁ generation**

$C^R$ ½      $C^W$ ½

$C^R$ ½

$C^R C^R$ ¼      $C^R C^W$ ¼

$C^R C^W$

$C^W$ ½

$C^R C^W$ ¼      $C^W C^W$ ¼

The result of segregation can be observed directly, because the ratio of red:pink:white phenotypes is 1:2:1, which reflects the ratio of $C^R C^R : C^R C^W : C^W C^W$ genotypes.

**F₂ generation**

Note that in this case the genotype ratio and the phenotype ratio are both 1:2:1, since each genotype has a distinct phenotype. Such a direct demonstration of segregation makes one wonder whether Mendel's work might have been appreciated more readily if his traits had shown incomplete dominance!

### The principles of transmission genetics are statistical and stated in terms of probabilities.

The element of chance in fertilization implies that the genotype of any particular progeny cannot be determined in advance. However, one can deduce the likelihood, or **probability,** that a specified genotype will occur. The probability of occurrence of a genotype must always lie between 0 and 1; a probability of 0 means that the genotype cannot occur, and a probability of 1 means that the occurrence of the genotype is certain. For example, in the cross $Aa \times AA$, no offspring can have the genotype $aa$, so in this mating the probability of $aa$ is 0. Similarly, in the mating $AA \times aa$, all offspring must have the genotype $Aa$, so in this mating the probability of $Aa$ is 1.

In many cases, the probability of a particular genotype is neither 0 nor 1, but some intermediate value. For one gene, the probabilities for a single individual can be deduced from the parental genotypes in the mating and the principle of segregation. For example, the probability of producing a homozygous recessive individual from the cross $Aa \times Aa$ is ¼ (see Fig. 16.7), and that from the cross $Aa \times aa$ is ½ (see Fig. 16.8a).

The probabilities for a single individual can also be inferred from observed data because the overall proportions of two (or more) genotypes among a large number of observations approximates the probability of each of the genotypes for a single observation. For example, in Mendel's $F_2$ data (see Table 16.1), the overall ratio of dominant:recessive is 2.98:1, or very nearly 3:1. This result implies that the probability that an individual $F_2$ plant has the homozygous recessive phenotype is very close to ¼, which is the value inferred from the principle of segregation.

Sometimes it becomes necessary to combine the probabilities of two or more possible outcomes of a cross, and in such cases either of two rules may be helpful.

1. **Addition rule.** This principle applies when the possible outcomes being considered cannot occur simultaneously. For example, suppose that a single offspring is chosen at random from the progeny of the mating $Aa \times Aa$, and we wish to know the probability that the offspring is either $AA$ or $Aa$. The key words here are "either" and "or." Each of these outcomes is possible, but both cannot occur simultaneously in a single individual; the outcomes are mutually exclusive. When the possibilities are mutually exclusive, the addition rule states that the probability of either event occurring is given by the sum of their individual probabilities. In this example, the chosen offspring could either have genotype $AA$ (with probability ¼,

according to Fig. 16.7), or the offspring could have genotype *Aa* (with probability ½). Therefore, the probability that the chosen individual has either the *AA* or the *Aa* genotype is given by ¼ + ½ = ¾ (see Fig. 16.7). Alternatively, the ¾ could be interpreted to mean that, among a large number of offspring from the mating *Aa* × *Aa*, the proportion exhibiting the dominant phenotype will be very close to ¾. This interpretation is verified by the data in Table 16.1.

2. **Multiplication rule**. This principle applies when outcomes can occur simultaneously, and the occurrence of one has no effect upon the likelihood of the other. Events that do not influence one another are independent, and the multiplication rule states that the probability of two independent events occurring together is the product of their respective probabilities. This rule is widely used to determine the probabilities of successive offspring of a cross, because each event of fertilization is independent of any other. For example, in the mating *Aa* × *Aa*, one may wish to determine the probability that, among four peas in a pod, the one nearest the stem is green and the others yellow. Here, the word "and" is a simple indicator that the multiplication rule should be used. Because each seed results from an independent fertilization, this probability is given by the product of the probability that the seed nearest the stem is *aa* and the probability that each of the other seeds is either *AA* or *Aa*, and hence the probability is ¼ × ¾ × ¾ × ¾ = 27/256, as shown for the top pod in **Fig. 16.11.**

---

**FIG. 16.11** Application of the multiplication and addition rules.

(¼) (¾) (¾) (¾) = ²⁷/₂₅₆

Probability that the seed closest to the stem is green *and* the others are yellow is determined by the multiplication rule.

(¾) (¼) (¾) (¾) = ²⁷/₂₅₆

(¾) (¾) (¼) (¾) = ²⁷/₂₅₆

(¾) (¾) (¾) (¼) = ²⁷/₂₅₆

———

¹⁰⁸/₂₅₆ = 42%

Probability that *any* seed is green and the others are yellow is determined by first using the multiplication rule to determine the probability that a particular seed is green, then the addition rule to determine that any seed is green.

The addition and multiplication rules are very powerful when used in combination. Consider the following question: In the mating *Aa* × *Aa*, what is the probability that, among four seeds in a pod, exactly one is green? We have already seen in Fig. 16.11 that the multiplication rule gives the probability of the seed nearest the stem being green as 27/256. As illustrated in Fig. 16.11, there are only four possible ways in which exactly one seed can be green, each of which has a probability of 27/256, and these outcomes are mutually exclusive. Therefore, by the addition rule, the probability of there being exactly one green and three yellow seeds in a pod, occurring in any order, is given by 27/256 + 27/256 + 27/256 + 27/256 = 108/256, or approximately 42%.

→ **Quick Check 4** What is the probability that *any* two peas are green and two are yellow in a pea pod with exactly four seeds?

**Mendelian segregation preserves genetic variation.**
As noted earlier, Darwin was befuddled because blending inheritance would make genetic variation disappear so rapidly that evolution by means of natural selection could not occur. Although Darwin was completely unaware of Mendel's findings, segregation was the answer to his problem.

The importance of segregation is that it demonstrates that the alleles encoding a trait do not alter or influence one another when they are present together in a heterozygous genotype (except in very rare instances). The recessive trait, masked in one generation, can appear in the next, looking exactly as it did in the true-breeding strains. Mendel fully appreciated the significance of this discovery. In one of his letters, he emphasized that "the two parental traits appear, separated and unchanged, and there is nothing to indicate that one of them has either inherited or taken over anything from the other." In other words, no hint of any sort of blending between the parental genetic material takes place. Because the individual genes maintain their identity down through the generations (except for rare mutations), genetic variation in a population also tends to be maintained through time. The maintenance of genetic variation is discussed further in Chapter 21.

## 16.4 INDEPENDENT ASSORTMENT

We have seen that segregation of dominant and recessive alleles of a single gene results in a 3:1 ratio of dominant:recessive phenotypes in the $F_2$ generation. What happens when the parental strains differ in two traits, for example when a strain having yellow and wrinkled seeds is crossed with a strain having green and round seeds? The results of these kinds of experiments constitute Mendel's second key discovery, the **principle of independent assortment.** This principle states that segregation of one set of alleles of a gene pair is independent of the segregation of another set of alleles of a different gene pair. That is, different gene pairs assort (segregate) independently of one another.

### Independent assortment is observed when genes segregate independently of one another.

In the cross between a strain with yellow and wrinkled seeds and a strain with green and round seeds, the phenotype of the $F_1$ seeds is easily predicted. Because yellow is dominant to green, and round is dominant to wrinkled, the $F_1$ seeds are expected to be yellow and round, and in fact they are (**Fig. 16.12**). When these seeds are grown and the $F_1$ plants are allowed to undergo self-fertilization, the result is as shown in Fig. 16.12. Among 639 seeds from this cross, Mendel observed the following:

| | |
|---|---|
| yellow round | 367 |
| green round | 122 |
| yellow wrinkled | 113 |
| green wrinkled | 37 |

The ratio of these phenotypes is $9.9:3.3:3.1:1.0$, which Mendel realized is close to $9:3:3:1$. The latter ratio is that expected if the $A$ and $a$ alleles for seed color undergo segregation and form gametes independently of the $B$ and $b$ alleles for seed shape. For seed color alone we expect a ratio of ¾ yellow:¼ green, and for seed shape alone we expect a ratio of ¾ round:¼ wrinkled. If the traits are independent, then we can use the multiplication rule to predict the outcomes for both traits:

| | | |
|---|---|---|
| yellow round | $(¾) \times (¾)$ | = 9/16 |
| green round | $(¼) \times (¾)$ | = 3/16 |
| yellow wrinkled | $(¾) \times (¼)$ | = 3/16 |
| green wrinkled | $(¼) \times (¼)$ | = 1/16 |

Note that $9/16:3/16:3/16:1/16$ is equivalent to $9:3:3:1$.

The underlying reason for the $9:3:3:1$ ratio of phenotypes in the $F_2$ generation is that the alleles for yellow versus green and those for round versus wrinkled are assorted into gametes independently of each other. In other words, the hereditary transmission of either gene has no effect on the hereditary transmission of the other. A Punnett square depicting

**FIG. 16.12** **Mendel's crosses with two traits.** Plants homozygous for two traits produce offspring with a phenotypic ratio of 9:3:3:1.

Plant grown from true-breeding strain with yellow and wrinkled seeds

Plant grown from true-breeding strain with green and round seeds

**P₁ generation**

(genotype *AA bb*)        (genotype *aa BB*)

Because of dominance, the seeds in the $F_1$ generation are yellow and round.

**F₁ generation**
(genotype *Aa Bb*)

The expected ratio of the four types of seeds is
9 yellow, round
3 green, round
3 yellow, wrinkled
1 green, wrinkled

**F₂ generation**

**FIG. 16.13  Independent assortment of two gene pairs.** The Punnett square reveals the underlying phenotypic ratio of 9:3:3:1.

There are 9 possible genotypes and 4 possible phenotypes. The ratio of phenotypes is 9:3:3:1.

independent assortment is shown in **Fig. 16.13.** The *A* and *a* alleles segregate equally into gametes as ½ *A* : ½ *a*, and likewise the *B* and *b* alleles segregate equally into gametes as ½ *B* : ½ *b*. The result of independent assortment is that the four possible gametic types are produced in equal proportions:

*AB* gametes   (½) × (½) = ¼

*Ab* gametes   (½) × (½) = ¼

*aB* gametes   (½) × (½) = ¼

*ab* gametes   (½) × (½) = ¼

As the Punnett square in Fig. 16.13 shows, random union of these gametic types produces the expected ratio of 9 yellow round, 3 green round, 3 yellow wrinkled, and 1 green wrinkled. **Fig. 16.14** summarizes how Mendel's experiments led him to formulate his two laws.

### Independent assortment reflects the random alignment of chromosomes in meiosis.

Independent assortment of genes on different chromosomes results from the mechanics of meiosis (Chapter 11), in which different pairs of homologous chromosomes align randomly on the metaphase plate. For some pairs of chromosomes, the maternal chromosome goes toward one pole, and the paternal chromosome goes to the other pole, but for other pairs, just the

**FIG. 16.14**

# How are simple traits inherited?

**BACKGROUND** The experiments of Gregor Mendel, carried out in the years 1856–1863, are among the most important in all of biology.

**EXPERIMENT** Mendel set out to improve upon previous research in heredity. He writes that "among all the numerous experiments made, not one has been carried out to such an extent and in such a way as to make it possible to determine the number of different forms under which the offspring of the hybrids appear, or to arrange these forms with certainty according to their separate generations, or definitely to ascertain their statistical relations." By studying simple traits across several generations of crosses, Mendel observed how these traits were inherited.

**RESULTS** The "statistical relations" became very clear. Crosses between plants that were hybrids of a single trait displayed two phenotypes in a ratio of 3:1. Crosses between plants that were hybrids of two traits displayed four different phenotypes in a ratio of 9:3:3:1.

**Two traits (seed color and shape)**

AAbb ✕ aaBB

↓

AaBb

> Alleles of one gene (A and a) segregate independently of alleles of another gene (B and b).

| | A B ¼ | A b ¼ | a B ¼ | a b ¼ |
|---|---|---|---|---|
| **A B ¼** | AA BB 1/16 | AA Bb 1/16 | Aa BB 1/16 | Aa Bb 1/16 |
| **A b ¼** | AA Bb 1/16 | AA bb 1/16 | Aa Bb 1/16 | Aa bb 1/16 |
| **a B ¼** | Aa BB 1/16 | Aa Bb 1/16 | aa BB 1/16 | aa Bb 1/16 |
| **a b ¼** | Aa Bb 1/16 | Aa bb 1/16 | aa Bb 1/16 | aa bb 1/16 |

AaBb

> 9:3:3:1 ratio of phenotypes

**CONCLUSION** From observing these ratios among several different traits, Mendel made two key conclusions, now called Mendel's laws:

1. The principle of segregation states that individuals inherit two copies (alleles) of each gene, one from the mother and one from the father, and when the individual forms reproductive cells, the two copies separate (segregate) equally in the eggs or sperm.

2. The principle of independent assortment states that the two copies of each gene segregate into gametes independently of the two copies of another gene.

**FOLLOW-UP WORK** Mendel's work was ignored during his lifetime, and its importance was not recognized until 1900, 16 years after his death. The rediscovery marks the beginning of the modern science of genetics.

**SOURCE** Mendel's paper in English is available at http://www.mendelweb.org/Mendel.html.

opposite occurs. Because the alignment is random, gene pairs on different chromosomes assort independently of one another.

**Fig. 16.15** illustrates two possible alignments that are equally likely. In one alignment, the B allele (dark red) goes to the same pole as the A allele (dark blue), and in the other alignment it is the b allele (light red) that goes in the same direction as the A allele. The first type of alignment results in a 1:1 ratio of $AB:ab$ gametes, and the second type of alignment results in a 1:1 ratio of $Ab:aB$ gametes. Because the two orientations are equally likely, the overall ratio of $AB:ab:Ab:aB$ from a large number of cells undergoing meiosis is expected to be 1:1:1:1. This is the principle of independent assortment for genes located in different chromosomes.

Not all genes undergo independent assortment. For example, genes that are sufficiently close together in one chromosome do not assort independently of one another. Genes in the same chromosome that fail to show independent assortment are said to be **linked.** Genes that are genetically linked are discussed in Chapter 17.

**FIG. 16.15 Independent assortment of genes in different chromosomes.** Chromosomes are sorted into daughter cells randomly during meiosis, resulting in independent assortment of genes.

> Independent assortment of genes in different chromosomes reflects the fact that nonhomologous chromosomes can orient in either of two ways that are equally likely.

Anaphase I          Anaphase I

↓                   ↓

Resulting gametes    Resulting gametes

## Phenotypic ratios can be modified by interactions between genes.

The 9:3:3:1 ratio results from independent assortment of two genes when one allele of each gene is dominant and when the two genes affect different traits. However, even with complete dominance, the ratio of phenotypes may be different if the two genes affect the same trait. In this case, the genes can interact to affect the phenotypic expression of the genotypes, resulting in a modification of the expected ratio. Genes that modify the phenotypic expression of other genes are said to show **epistasis.**

There are many types of epistasis leading to different modifications of the 9:3:3:1 ratio. **Fig. 16.16** shows one example, in which the ratio is modified to 13:3. This ratio occurs in the $F_2$ generation of a cross between White Leghorn and White Wyandotte chickens. Both breeds are white, but for different genetic reasons. There are two genes involved in pigment production, each with two alleles. The $C$ gene encodes a protein that affects coloration in feathers. This gene has two alleles, the dominant allele $C$, which produces pigment, and the recessive allele $c$, which does not produce pigment. A different gene, $I$, codes for an inhibitor protein. The dominant allele, $I$, inhibits the expression of $C$, whereas the recessive allele, $i$, does not inhibit the expression of $C$. Therefore, the White Leghorn (genotype $CC\ II$) is white because the dominant allele $I$ inhibits the pigment in the feathers, and the White Wyandotte (genotype $cc\ ii$) is white because the recessive allele $c$ does not produce feather pigment to begin with. The $F_1$ generation has genotype $Cc\ Ii$ and is also white. With independent assortment, only the three $C–ii$ offspring have colored feathers in the $F_2$ generation (the dash indicates that the second allele could be either $C$ or $c$), while the rest have white feathers, and so the ratio of white:colored is 13:3.

→ **Quick Check 5** Why do the $F_1$ chickens with genotype $Cc\ Ii$ have white feathers?

**FIG. 16.16    Epistasis, the interaction of genes affecting the same trait.** Epistasis can modify the 9:3:3:1 ratio of phenotypes, in this example to 13:3.

The White Leghorn is white because the inhibitor allele *I* blocks expression of the pigment allele *C*.

The White Wyandotte is white because the pigment allele *c* does not produce pigment.

P₁ generation

$CC\ II$ × $cc\ ii$

F₁ generation

$Cc\ Ii$ × $Cc\ Ii$

F₂ generation

Genotypes of the form *C–ii* have colored feathers, whereas all other genotypes have white feathers. The result is an $F_2$ ratio of white:colored of 13:3, which is a modified form of the expected 9:3:3:1.

**Alleles**

| | |
|---|---|
| C | Pigment |
| c | No pigment |
| I | Inhibitor |
| i | No inhibitor |

## 16.5 PATTERNS OF INHERITANCE OBSERVED IN FAMILY HISTORIES

Segregation of alleles takes place in human meiosis just as it does in peas and most other sexual organisms. The results are not so easily observed as in peas, however, for several reasons. First, humans do not choose their mating partners for the convenience of biologists, and so experimental crosses are not possible. Second, the number of children in human families is relatively small, and so the Mendelian ratios are often obscured by random fluctuations due to chance. Nevertheless, in many cases the occurrence of segregation can be observed even in human families. The patterns are often easiest to spot for inherited traits that are rare because in those cases the mutant alleles occur only in affected individuals and their close relatives.

In studying human families, the record of the ancestral relationships among individuals is summarized in a diagram of

**FIG. 16.17 Some conventions used in drawing human pedigrees.**

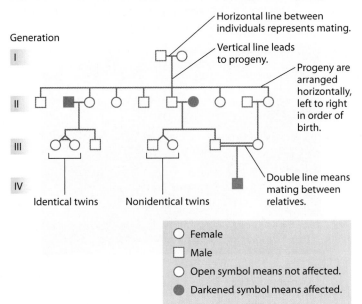

right in order of birth. The same patterns of dominance and recessiveness that Mendel observed in his pea plants can be seen in some pedigrees. Over the next few sections, we'll see how to recognize telltale features of a pedigree that reveal the genotypic nature of a trait.

### Dominant traits appear in every generation.

The pedigree shown in **Fig. 16.18** is for a rare dominant trait, brachydactyly, in which the middle long bone in the fingers fails to grow and therefore the fingers remain very short. This pedigree, published in 1905, was the first demonstration of dominant Mendelian inheritance in humans. This particular form of brachydactyly results from a mutation in a gene whose normal product is a protein involved in cartilage formation, which is necessary for bone growth.

These are the features of the pedigree that immediately suggest dominant inheritance:

1  Affected individuals are equally likely to be females or males.

2  Most matings that produce affected offspring have only one affected parent. This occurs because the brachydactyly trait is rare, and therefore a mating between two affected individuals is extremely unlikely.

3.  Among matings in which one parent is affected, approximately half the offspring are affected.

If a dominant trait is rare, then affected individuals will almost always be heterozygous (*Aa*) and not homozygous (*AA*). In a mating in which one parent is heterozygous for the dominant

family history called a **pedigree.** Some typical symbols used in pedigrees are shown in **Fig. 16.17.** Females are represented as circles and males as squares. If the pedigree is following the transmission of a disease from one generation to the next, unaffected individuals are indicated by open symbols and affected individuals by darkened symbols. A mating is represented as a horizontal line connecting the partners, and the children of the partners are arranged below them, left to

**FIG. 16.18 Pedigree of a trait caused by a dominant allele.** This pedigree shows the inheritance of shortened fingers associated with a form of brachydactyly (inset).

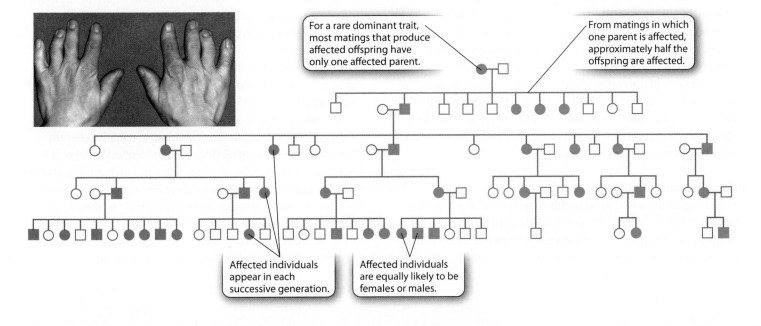

gene (*Aa*) and the other is homozygous recessive (*aa*), half the offspring are expected to be heterozygous (*Aa*) and the other half homozygous recessive (*aa*).

## Recessive traits skip generations.

Recessive inheritance shows a pedigree pattern very different from that of dominant inheritance. The pedigree shown in **Fig. 16.19** pertains to albinism, in which the amount of melanin pigment in the skin, hair, and eyes is reduced. In most populations, the frequency of albinism is about 1 in 36,000, but it has a much higher frequency—about 1 in 200—among the Hopi and several other Native American Indian tribes of the Southwest. (It is not unusual for genetic diseases to have elevated frequencies among certain isolated populations.) This type of albinism is due to a mutation in the gene *OCA2*, which encodes a membrane transporter protein. As noted in Chapter 15, another type of mutation affecting expression of this same gene is associated with blue eyes. As shown in this pedigree, double lines represent matings between relatives, and in both cases shown here the mating is between first cousins.

These are the principal pedigree characteristics of recessive traits:

1. The trait may skip one or more generations.

2. Affected individuals are equally likely to be females or males.

3. Affected individuals may have unaffected parents, as in the offspring of the second mating in the second generation in Fig. 16.19. For a recessive trait that is sufficiently rare, virtually all affected individuals have unaffected parents.

4. Affected individuals often result from mating between relatives, typically first cousins.

Recessive inheritance has these characteristics because recessive genes can be transmitted from generation to generation without manifesting the recessive phenotype. In order for an affected individual to occur, the recessive gene must be inherited from both parents. Mating between relatives often allows rare recessive alleles to become homozygous because an ancestor that is shared between the relatives may carry the gene (*Aa*). The recessive gene in the common ancestor can be transmitted to both parents, making them each a carrier of the gene as well. If both parents are unaffected carriers of the allele, they both have the genotype *Aa*, and therefore ¼ of their offspring are expected to be homozygous *aa* and affected.

## Many genes have multiple alleles.

The examples discussed so far in this chapter involve genes with only two alleles, such as *A* for yellow seeds and *a* for green seeds, or the allele of human *OCA2* for normal skin pigmentation and the mutant *OCA2* allele for albinism. Because a gene consists

**FIG. 16.19  Pedigree of a trait caused by a recessive allele.** The photograph is of a group of Hopi males, three with albinism, which is caused by a recessive allele.

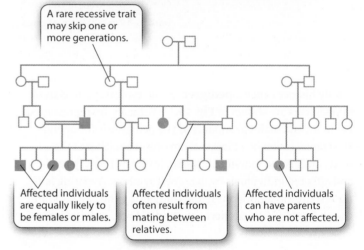

A rare recessive trait may skip one or more generations.

Affected individuals are equally likely to be females or males.

Affected individuals often result from mating between relatives.

Affected individuals can have parents who are not affected.

of a sequence of nucleotides, however, any nucleotide or set of nucleotides in the gene can undergo mutation. Each of the mutant forms that exists in a population constitutes a different allele, and hence a population of organisms may contain many different alleles of the same gene, which are called **multiple alleles.** As noted in Chapter 15, some genes have so many alleles that they can be used for individual identification by means of DNA typing.

In considering genetic diseases, multiple alleles are often grouped into categories such as "mutant" and "normal." But there are often many different "mutant" and many different "normal" alleles in a population. For the "mutant" alleles, the DNA sequences are different from one another, but none of them produces a functional protein product. For the "normal" alleles, the DNA sequences are also different, but they all are able to produce functional protein.

For example, more than 400 different recessive alleles that cause phenylketonuria (PKU) have been discovered across the world. PKU is a moderate to severe form of mental retardation caused by mutations in the gene encoding the enzyme phenylalanine hydroxylase. Children affected with PKU are unable to break down the excess phenylalanine present in a normal diet, and the buildup impairs the development of neurons in the brain. About 1 in 10,000 newborns inherits two mutant alleles, which could be two copies of the same mutant allele or two different mutant alleles, and is affected.

The "normal" form of the gene encoding phenylalanine hydroxylase also exists in the form of multiple alleles, each of which differs from the others, but nevertheless encodes a functional form of the enzyme.

→ **Quick Check 6** How is it possible that there are multiple different alleles in a population and yet any individual can have only two alleles?

## Incomplete penetrance and variable expression can obscure inheritance patterns.

Many traits with single-gene inheritance demonstrate complications that can obscure the expected patterns in pedigrees. Chief among these are traits with **incomplete penetrance,** which means that individuals with a genotype corresponding to a trait do not actually show the phenotype, either because of environmental effects or because of interactions with other genes. Penetrance is the proportion of individuals with a particular genotype that show the expected trait. If the penetrance is less than 100%, then the trait shows reduced, or incomplete, penetrance. Familial cancers often show incomplete penetrance. For example, retinoblastoma caused by certain mutations in the *Rb* gene and breast cancer caused by certain mutations in the *BRCA1* and *BRCA2* genes are incompletely penetrant. That is, some people who inherit a mutation in these genes that predisposes to cancer do not in fact develop cancer. Similarly, some mutations in the *apoliprotein E* (*APOE*) gene increase the risk of developing Alzheimer's disease. However, not everyone who inherits these forms of the gene develops Alzheimer's disease, so these alleles are also incompletely penetrant.

Another common complication in human pedigrees is **variable expressivity,** which means that a particular phenotype is expressed with a different degree of severity in different individuals. Don't confuse variable expressivity with incomplete penetrance. With variable expressivity, the trait is always expressed, though the severity varies; with incomplete penetrance, the trait is either expressed or it is not. Variation among individuals in the expression of a trait can result from the action of other genes, from effects of the environment, or both. An example is provided by deficiency of the enzyme alpha-1 antitrypsin ($\alpha$1AT) discussed in Chapter 15, which is associated

with loss of lung elasticity and emphysema. Among individuals with emphysema due to $\alpha$1AT deficiency, the severity of the symptoms varies dramatically from one patient to the next. In this case, tobacco smoking is an environmental factor that increases the severity of the disease.

Incomplete penetrance and variable expressivity both provide examples where a given genotype does not always produce the same phenotype, since the expression of genes is often influenced by other genes, the environment, or a combination of the two.

**?** **CASE 3 You, From A To T: Your Personal Genome How do genetic tests identify disease risk factors?** Your personal genome, as well as that of every human being, contains a unique combination of alleles of thousands of different genes. Most of these have no detectable effects on health or longevity, but many are risk factors for genetic diseases. Molecular studies have discovered particular alleles of genes associated with a large number of such conditions, and the presence of these alleles can be tested. More than a thousand genetic tests have already been deployed, and many more are actively being developed. A **genetic test** is a method of identifying the genotype of an individual. The tests may be carried out on entire populations or restricted to high-risk individuals.

The benefits of genetic testing can be appreciated by an example. Screening of newborns for phenylketonuria identifies babies with high blood levels of phenylalanine. In the absence of treatment, 95% of such newborns will progress to moderate or severe mental retardation, whereas virtually all those placed on a special diet with a controlled amount of phenylalanine will have mental function within the normal range. For recessive conditions like phenylketonuria, tests can be carried out on people with affected relatives to identify the heterozygous genotypes. Testing can also identify genetic risk factors for disease, and carriers can take additional precautions. For example, individuals with $\alpha$1AT deficiency can prolong and improve the quality of their lives by not smoking tobacco, women with genetic risk factors *BRCA1* and *BRCA2* for breast cancer can have frequent mammograms, and those with the *TCF7L2* risk factor for type 2 diabetes can decrease their risk by lifestyle choices that include weight control and exercise.

While there are many potential benefits to genetic testing, there are also some perils. One major concern is maintaining the privacy of those who choose to be tested. With medical records increasingly going online, who will have access to your test results, and how will this information be used? Could your test results be used to deny you health or life insurance because you have a higher than average risk of some medical condition? Or could an employer who got hold of your genetic test results decide to reassign you to another job, or even eliminate your position because of your genetic predispositions? There are some safeguards designed to protect you from such discrimination. The Genetic Information Nondiscrimination Act (GINA) was signed into law in 2008 and forbids the use of genetic information in decisions concerning employment and health insurance. The protection provided by GINA will, it is hoped, allow for the responsible and productive use of genetic information.

There is also increasing concern about the reliability and accuracy of genetic tests, especially direct-to-consumer (DTC) genetic tests. DTC tests can be purchased directly without the intervention of medical professionals. Since the consumer sends a biological sample and DTC tests are carried out by the provider, the tests are not regarded as medical devices and so are unregulated. One problem is that some DTC tests are based on flimsy and unconfirmed evidence connecting a gene with a disease. Another is that the link between genotype and risk may be exaggerated for marketing purposes. Yet another is lack of information on quality control in the DTC laboratories. Finally, consumer misinterpretation may regard genotype as destiny, at one extreme descending into depression and despair, and at the other using a low-risk genotype to justify an unhealthy lifestyle.

## Core Concepts Summary

### 16.1 THE EARLIEST THEORIES OF HEREDITY INCORRECTLY ASSUMED THE INHERITANCE OF ACQUIRED CHARACTERISTICS AND BLENDING OF PARENTAL TRAITS IN THE OFFSPRING.

The inheritance of acquired characteristics suggests that traits that develop during the lifetime of an individual can be passed on to offspring. With rare exceptions, this mode of inheritance does not occur. page 16-1

Blending inheritance is the incorrect hypothesis that characteristics in the parents are averaged in the offspring. This model predicts the blending of genetic material, which does not occur. Different forms of a gene maintain their separate identities even when present together in the same individual. page 16-2

### 16.2 MODERN TRANSMISSION GENETICS BEGAN WITH GREGOR MENDEL, WHO USED THE GARDEN PEA AS HIS EXPERIMENTAL ORGANISM AND STUDIED TRAITS WITH CONTRASTING CHARACTERISTICS.

Mendel started his experiments with true-breeding plants, ones whose progeny are identical to their parents. He followed just one or two traits at a time, allowing him to discern simple patterns, and he counted all the progeny of his crosses. page 16-3

In crosses of one true-breeding plant with a particular trait and another true-breeding plant with a contrasting trait, just one of the two characteristics appeared in the offspring. The trait that appeared in this generation is dominant, and the trait that is not seen is recessive. page 16-4

Mendel explained this result by hypothesizing that there is a hereditary factor for each trait (now called a gene), that each pea plant carries two copies of the gene for each trait, and that one of two different forms of the gene (alleles) is dominant to the other one. page 16-4

### 16.3 MENDEL'S FIRST KEY DISCOVERY WAS THE PRINCIPLE OF SEGREGATION, WHICH STATES THAT MEMBERS OF A GENE PAIR SEPARATE EQUALLY INTO GAMETES.

When Mendel allowed the progeny of the first cross to self-fertilize, he observed a 3:1 ratio of the dominant and recessive traits among the progeny. page 16-6

Mendel reasoned that the parent in this generation must have two different forms of the same gene ($A$ and $a$) and that these alleles segregate from each other to form gametes with each gamete getting $A$ or $a$ but not both. When the gametes combine at random, they produce progeny in the genotypic ratio 1 $AA$:2 $Aa$:1 $aa$, which yields a phenotypic ratio of 3:1 because $A$ is dominant to $a$. This idea became known as the principle of segregation. page 16-6

The principle of segregation reflects the separation of homologous chromosomes that occurs in anaphase I of meiosis. page 16-8

Some traits show incomplete dominance, in which the phenotype of the heterozygous genotype is intermediate between those of the two homozygous genotypes. page 16-8

The expected frequencies of progeny of crosses can be predicted using the addition rule, which states that when two possibilities are mutually exclusive, the probability of either event occurring is the sum of their individual probabilities, and the multiplication rule, which states that when two possibilities occur independently, the probability of both events occurring is the product of the probabilities of each of the two events. page 16-9

## 16.4 MENDEL'S SECOND KEY FINDING WAS THE PRINCIPLE OF INDEPENDENT ASSORTMENT, WHICH STATES THAT DIFFERENT GENE PAIRS SEGREGATE INDEPENDENTLY OF ONE ANOTHER.

In a cross with two traits, each with two contrasting characteristics, the two traits behave independently of each other. This idea became known as the principle of independent assortment. For example, in a self-cross of a double heterozygote, the phenotypic ratio of the progeny is 9:3:3:1, reflecting the independent assortment of two 3:1 ratios. page 16-10

Independent assortment results from chromosome behavior during meiosis, specifically from the random orientation of different chromosomes on the meiotic spindle. page 16-12

In some cases, genes interact with each other, modifying the expected ratios in crosses. Epistasis is a form of gene interaction in which one gene affects the expression of another. page 16-14

## 16.5 THE PATTERNS OF INHERITANCE THAT MENDEL OBSERVED IN PEAS CAN ALSO BE SEEN IN HUMANS.

In a human pedigree, females are represented by circles and males as squares; affected individuals are shown as filled symbols and unaffected individuals as open symbols; and horizontal lines denote matings. page 16-14

Dominant traits appear in every generation and affect males and females equally. page 16-15

Recessive traits skip one or more generations and affect males and females equally. Affected individuals often result from matings between close relatives. page 16-16

Although a given individual has only two alleles of each gene, there can be many alleles of a particular gene in the population as a whole. page 16-16

Interpreting pedigrees can be difficult because of incomplete penetrance and variable expressivity. Penetrance is the percentage of individuals with a particular genotype who show the expected phenotype, and expressivity is the degree to which a genotype is exhibited. page 16-17

Genetic testing enables the genotype of an individual to be determined for one or more genes. Such tests carry both benefits and risks. page 16-17

## Self-Assessment

1. Describe two ways in which Mendel's experimental approach differed from others of the time.

2. Distinguish among gene, allele, genotype, and phenotype.

3. Name and describe Mendel's two laws.

4. Explain the relationship between Mendel's two laws and the mechanics of meiosis.

5. Explain how you can predict the genotypes and phenotypes of offspring if you know the genotypes of the parents.

6. Describe an instance when a testcross would be useful.

7. Define the multiplication and addition rules, and explain how these rules can help you predict the outcome of a cross between parents with known genotypes.

8. Construct a human pedigree for a dominant and a recessive trait and explain the patterns of inheritance.

9. Discuss the benefits and risks of genetic testing and personal genomics.

Do you understand the chapter's Core Concepts? Log in to check your answers to the Self-Assessment questions, then practice what you've learned and reinforce this chapter's concepts by working through the problems and multimedia tutorials provided there.

📖 www.biologyhowlifeworks.com

## CHAPTER 17

# BEYOND MENDEL

## Sex Chromosomes, Linkage, and Organelles

### Core Concepts

**17.1** The *X* and *Y* chromosomes of some animals, including humans, determine sex and are inherited differently from the autosomes.

**17.2** *X*-linked genes, which show a crisscross inheritance pattern, provided the first evidence that genes are present in chromosomes.

**17.3** Genetic linkage occurs when two genes are sufficiently close together in the same chromosome that the combination of alleles present in the chromosome tends to remain together in inheritance.

**17.4** Most *Y*-linked genes are passed from father to son.

**17.5** Mitochondria and chloroplast DNA follow their own inheritance pattern.

Mendel's principles of segregation and independent assortment are the foundation of transmission genetics (Chapter 16). For traits such as pea color and shape, these principles predict simple progeny phenotypic ratios in self-crosses of heterozygous genotypes, such as 3:1 for one gene or 9:3:3:1 for two genes that undergo independent assortment. The principle of segregation also defines the inheritance patterns expected in human pedigrees for traits due to recessive or dominant mutations, as we saw in the case of albinism and brachydactyly.

Some traits, such as color blindness in humans and white eyes in fruit flies, do not show the inheritance patterns predicted by Mendel. In this chapter, we consider some examples in which the expected ratios and pedigree patterns are modified by principles Mendel knew nothing about. One example pertains to genes in the *X* and *Y* sex chromosomes, which are transmitted differently from other chromosomes. Although the patterns of inheritance of genes in the sex chromosomes do not follow Mendel's laws, those patterns can easily be understood by following the movement of chromosomes in which they are present.

Another important exception to Mendel's laws has to do with genes that are located close to each other in the same chromosome. Although each gene pair undergoes segregation in the normal manner, independent assortment of different genes does not take place. Finally, the laws of Mendelian segregation and independent assortment do not apply to genes in the DNA molecules in mitochondria and chloroplasts. All of these cases involve genetic transmission with patterns of inheritance beyond those discovered by Mendel.

## 17.1 THE *X* AND *Y* SEX CHROMOSOMES

Among Mendel's discoveries was that reciprocal crosses, in which the sexes of the parents associated with each genotype are reversed, yield the same types of progeny (Chapter 16). This principle is true of most genes, but an important exception applies to genes located in the *X* and *Y* chromosomes, which are associated with the sex of an individual. Mendel did not have to deal with this complication because peas and most other plants do not have sex chromosomes. In organisms that do have them, such as humans, specific rules govern the transmission of their genes.

**In many animals, sex is genetically determined and associated with chromosomal differences.**

Most chromosomes come in pairs that match in shape and size. The members of each pair are known as homologous chromosomes because they have the same arrangement of genes along their length (Chapter 11). One member of each pair of homologous chromosomes is inherited from the mother and the other from the father. In many animal species, however, the sex of an individual is determined by a special pair of unmatched chromosomes known as the **sex chromosomes,** which are usually designated as the **X chromosome** and the **Y chromosome.** Chromosomes other than the sex chromosomes are known as **autosomes.**

In humans, a normal female has two copies of the X chromosome (a sex-chromosome constitution denoted XX), and a normal male has one X chromosome and one Y chromosome (XY). The sizes of the human X and Y chromosomes are very different (**Fig. 17.1**). The X chromosome DNA molecule is more than 150 Mb long, while the Y chromosome is only about 50 Mb long. Except for a small region near each tip, the gene contents of the X and Y chromosomes are different from each other. The X chromosome, which includes more than 1000 genes, has a gene density similar to that of most autosomes. The vast majority of these genes have no counterpart in the Y chromosome. In contrast to the X chromosome, the Y chromosome contains only about 50 genes.

The regions of homology between the X and Y chromosomes consist of about 2.7 Mb of DNA near the tip of the short arm and about 0.3 Mb of DNA near the tip of the long arm. These regions of homology allow the chromosomes to pair during meiosis (Chapter 11). In most cells undergoing meiosis, a crossover

(physical breakage, exchange of parts, and rejoining of the DNA molecules) occurs in the larger of these regions. The crossover allows the chromosomes to align correctly at metaphase I and to separate from each other at anaphase I.

The relative size and gene content of the X and Y chromosomes differ greatly among species. In some species of mosquitoes, the X and Y chromosomes are virtually identical in size and shape, and the regions of homology include almost the entire chromosome. This situation is unusual, however. In most species, the Y chromosome contains many fewer genes than the X chromosome. Some species, like grasshoppers, have no Y chromosome: Females in these species have two X chromosomes, and males have only one X chromosome. The total number of chromosomes in grasshoppers therefore differs between females and males.

## Segregation of the sex chromosomes predicts a 1:1 ratio of females to males.

It is ironic that Mendel did not interpret sex as an inherited trait. If he had, he might have realized that sex itself provides one of the most convincing demonstrations of segregation. In human males, segregation of the X chromosome from the Y chromosome during anaphase I of meiosis results in half the sperm bearing an X chromosome and the other half bearing a Y chromosome (**Fig. 17.2**).

**FIG. 17.2  Inheritance of the X and Y chromosomes.** Segregation of the sex chromosomes in meiosis and random fertilization result in a 1:1 ratio of female:male embryos.

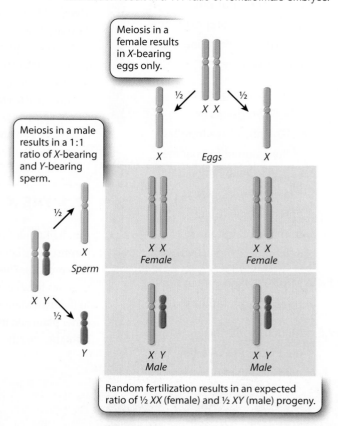

**FIG. 17.1  Human sex chromosomes.** The human X and Y sex chromosomes differ in size and number of genes.

Meiosis in human females results in eggs that each contain one X chromosome. With random fertilization, as shown in Fig. 17.2, half of the fertilized eggs are expected to be chromosomally XX (and therefore female) and half are expected to be chromosomally XY (and therefore male). This 1:1 ratio of female:male is characteristic of Mendelian segregation.

The expected ratio of 1:1 of females:males refers to the sex ratio at the time of conception. This is the primary sex ratio, and it is not easily observed because the earliest stages of fertilization and development are inaccessible to large-scale study. What can be observed is the sex ratio at birth, called the secondary sex ratio, which differs among populations but usually shows a slight excess of males. In the United States, for example, the secondary sex ratio is approximately 100 females:105 males. The explanation seems to be that, for reasons that are not entirely understood, females are slightly less likely to survive from conception to birth. On the other hand, males are slightly less likely to survive from birth to reproductive maturity, and so, at the age of reproductive maturity, the sex ratio is very nearly 1:1. Male mortality continues to be greater than that of females throughout life, so that by age 85 and over, the sex ratio is about 100 females:50 males.

The sex of each birth appears to be random relative to previous births—that is, there do not seem to be tendencies for some families to have boys or for other ones to have girls. Many people are surprised by this, for they know of one or more large families consisting mostly of boys or mostly of girls. But the occasional family with children of predominantly one sex would be expected simply by chance. When one occurs, its unusual sex distribution commands attention out of proportion to the actual numbers of such families. In short, families with unusual sex distributions are not more frequent than would be expected by chance.

## 17.2 INHERITANCE OF GENES IN THE X CHROMOSOME

Genes in the X chromosome are called **X-linked genes**. These genes have a unique pattern of inheritance first discovered by Thomas Hunt Morgan in 1910. Morgan's pioneering studies of genetics of the fruit fly *Drosophila melanogaster* helped bring Mendelian genetics into the modern era. The discovery of X-linked inheritance was not only important in itself, but also provided the first experimental evidence that chromosomes contain genes.

### *X*-linked inheritance was discovered through studies of male fruit flies with white eyes.

Morgan's discovery of X-linked genes began when he noticed a white-eyed male in a bottle of fruit flies in which all the others had normal, or wild-type, red eyes. (The most common phenotype in a population is often called the **wild type**.) This

was the first mutant he discovered, and finding it was a lucky break. As we saw in Chapter 16, most mutations are recessive, which means that when they occur their effect on the organism (the phenotype) is not observed because of the presence of the nonmutant gene in the homologous chromosome. Recall that the nonmutant form of a recessive mutant gene is dominant, and that the different forms of the gene are alleles.

Morgan's initial crosses are outlined in **Fig. 17.3**. In the first generation, he crossed the white-eyed male with a wild-type red-eyed female. All of the progeny fruit flies had red eyes, as you would expect from any recessive mutation. Morgan then carried out brother–sister matings, and he found that the phenotype of white eyes reappeared among the progeny. This result, too, was expected. However, there was a surprise: Morgan observed that the white-eye phenotype was associated with the sex of the fly. In the second generation, all the white-eyed fruit flies were male, and the white-eyed males appeared along with red-eyed males in a ratio of 1:1. No females with white eyes were observed; all the females had red eyes.

**FIG. 17.3  Morgan's white-eyed fly.** Morgan's discovery of *X*-linked genes derived from his crosses of a white-eyed male in 1910.

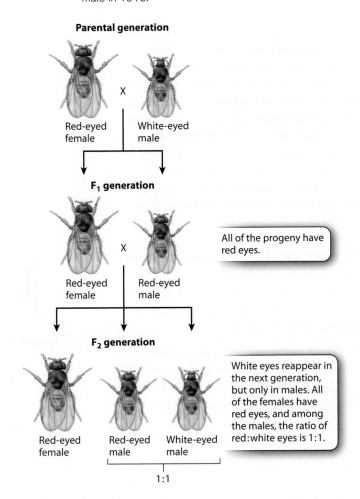

**Parental generation**

Red-eyed female  X  White-eyed male

**F₁ generation**

Red-eyed female  X  Red-eyed male

All of the progeny have red eyes.

**F₂ generation**

Red-eyed female    Red-eyed male    White-eyed male

White eyes reappear in the next generation, but only in males. All of the females have red eyes, and among the males, the ratio of red:white eyes is 1:1.

1:1

## Genes in the X chromosome exhibit a "crisscross" inheritance pattern.

When Morgan did his crosses with the white-eyed male, the X chromosome had only recently been discovered by microscopic examination of the chromosomes in male and female grasshoppers. Morgan was the first to understand that the pattern of inheritance of the X chromosome would be different from that of the autosomes, and he proposed the hypothesis that the white-eyed phenotype was due to a mutation in a gene in the X chromosome. This hypothesis could explain the pattern of inheritance shown in Fig. 17.3.

The key features of X-linked inheritance are shown in **Fig. 17.4.** In *Drosophila*, as in humans, females are XX and males are XY. Fig. 17.4a shows an XY male in which the X chromosome contains a recessive mutation. Because the Y chromosome does not carry an allele of this gene, the recessive mutation will be reflected in the male's phenotype—in this case, white eyes.

The Punnett square in Fig. 17.4a explains why all the offspring had red eyes when Morgan crossed the white-eyed male with a red-eyed female. During meiosis in the male, the mutant X chromosome segregates from the Y chromosome, and each type of sperm, X or Y, combines with a normal X-bearing egg. The result is that the female progeny are heterozygous. They have only one copy of the mutant allele, and because the mutant allele is recessive, the heterozygous females do not express the white-eye trait. The male progeny are also red-eyed because they receive their X chromosome from their wild-type red-eyed mother.

The Punnett square in Fig. 17.4a illustrates two important principles governing the inheritance of X-linked genes:

1. The XX offspring indicate that a male transmits his X chromosome only to his daughters. In this case, the X chromosome transmitted by the male carries the white-eye mutation.

**FIG. 17.4   X-linkage.** X-linked recessive alleles (red) are expressed in males because males have only one X chromosome.

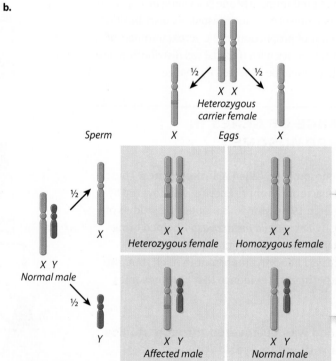

**a.**

Most genes in the X chromosome have no counterparts in the Y chromosome. Therefore, a recessive mutation in an X-linked gene is expressed in males.

Homozygous female

Sperm          X          Eggs          X

X Y
Affected male

X X
Heterozygous female

X X
Heterozygous female

X Y
Normal male

X Y
Normal male

A male with an X-linked recessive trait will have heterozygous (carrier) daughters and normal sons.

**b.**

Heterozygous carrier female

Sperm          X          Eggs          X

X Y
Normal male

X X
Heterozygous female

X X
Homozygous female

X Y
Affected male

X Y
Normal male

Among progeny from a heterozygous carrier female, half of the daughters are expected to be heterozygous carriers and half the sons are expected to be affected.

2. The XY offspring indicate that a male inherits his X chromosome from his mother. In this case, the X chromosome transmitted by the mother carries the nonmutant allele of the gene.

These principles underlie a pattern often referred to as **crisscross inheritance:** An X chromosome present in a male in one generation must be transmitted to a female in the next generation, and in the generation after that can be transmitted back to a male. Therefore, an X chromosome can "crisscross," or alternate, between the sexes in successive generations.

The cross illustrated in Fig. 17.4b shows another important feature of X-linked inheritance, one that explains the results of Morgan's $F_1$ cross. In this case, the mother is a heterozygous female. During meiosis, the mutant allele and the nonmutant allele undergo segregation. Half of the resulting eggs contain an X chromosome with the mutant allele, and half contain an X chromosome with the nonmutant allele. These combine at random with either X-bearing sperm or Y-bearing sperm. Among the female progeny in the next generation, half are heterozygous for the mutant allele and the other half are homozygous for the

normal allele. Thus, none of the females exhibits the white-eye trait because the mutation is recessive. Among the male progeny, half receive the mutant allele and have white eyes, whereas the other half receive the nonmutant allele and have red eyes. The expected progeny from the cross in Fig. 17.4b therefore consist of all red-eyed females, and there is a 1:1 ratio of red-eyed to white-eyed males, which is what Morgan observed in the $F_2$ generation of his crosses, as shown in Fig. 17.3.

Knowing the patterns revealed by the Punnett squares, we can now assign genotypes to Morgan's original crosses (**Fig. 17.5**). In Fig. 17.5, the white-eyed male that Morgan used in his first crosses are given the genotype of $w^- Y$, and the red-eyed female has a genotype of $w^+ w^+$. The symbol "$w^-$" stands for the recessive white-eye mutation in one X chromosome, and the symbol "$w^+$" stands for the dominant nonmutant allele (red eyes) in the other X chromosome. The symbol "Y" stands for the Y chromosome, and it is important to remember that the Y chromosome does not contain an allele of the white-eye gene.

In the cross between a wild-type red-eyed female and a white-eyed male, illustrated in Fig. 17.5a, the male offspring

**FIG. 17.5  Genotypes of Morgan's white-eyed fruit fly crosses.**

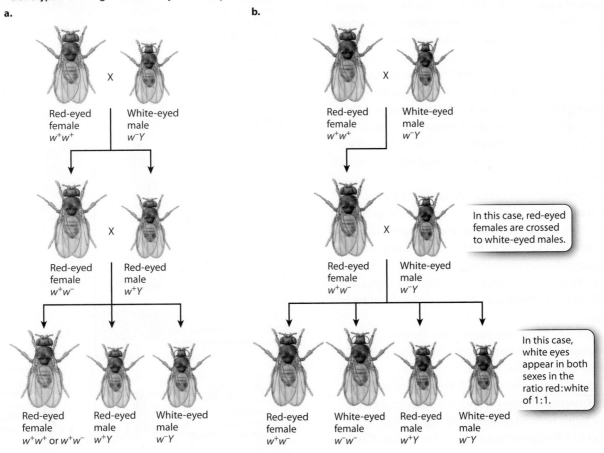

have red eyes because they receive their X chromosome from their mother. The female offspring from this cross receive one of their X chromosomes from their father, and hence they are heterozygous, $w^+w^-$. When the male and female progeny are mated together, their offspring consist of all red-eyed females (half of which are heterozygous) and a 1:1 ratio of red-eyed to white-eyed males, exactly as Morgan had observed.

The hypothesis of X-linkage not only explained the original data, but it also predicted the results of other crosses. One important test is outlined in Fig. 17.5b. Here the original cross is the same as that in Fig. 17.5a, but instead of mating the hybrid females to their brothers, they are mated to white-eyed males. The prediction is that there should be a 1:1 ratio of red-eyed to white-eyed females as well as a 1:1 ratio of red-eyed to white-eyed males. Again, these were the results observed. By the results of these crosses and others, Morgan demonstrated that the pattern of inheritance of the white-eye mutation parallels the pattern of inheritance of the X chromosome.

### *X*-linkage provided the first experimental evidence that genes are in chromosomes.

Morgan's original experiments indicated that the white-eye mutation showed a pattern of inheritance like that expected of the X chromosome. However, it was one of Morgan's students who showed experimentally that the white-eye mutation was actually a physical part of the X chromosome. Today it seems obvious that genes are in chromosomes because we know that genes consist of DNA and that DNA in the nucleus is found in chromosomes. But in 1916, when Calvin B. Bridges, who had joined Morgan's laboratory as a freshman, was working on his PhD research under Morgan's direction, neither the chemical nature of the gene nor the chemical composition of chromosomes was known.

In one set of experiments, Bridges crossed white-eyed females with red-eyed males (**Fig. 17.6**). Usually, the progeny consisted of red-eyed females and white-eyed males (Fig. 17.6a). This is the result expected when the X chromosomes in the mother separate normally at anaphase I in meiosis because all the daughters receive a $w^+$-bearing X chromosome from their father and all the sons receive a $w^-$-bearing X chromosome from their mother.

But Bridges noted a few rare exceptions. He saw that about 1 offspring in 2000 from the cross was "exceptional"— either a female with white eyes or a male with red eyes. The exceptional females were fertile, and the exceptional males were sterile. To explain these exceptional progeny, Bridges proposed the hypothesis diagrammed in Fig. 17.6b: The X chromosomes in a female occasionally fail to separate in anaphase I in meiosis, and both X chromosomes go to the same spindle pole. Recall from Chapter 15 that chromosomes sometimes fail to separate normally in meiosis, a process

**FIG. 17.6    Nondisjunction as evidence that genes are present in chromosomes.** (a) Normal chromosome segregation yields expected progeny. (b) Nondisjunction yields exceptional progeny.

**a.** Normal chromosome separation

**b.** Nondisjunction

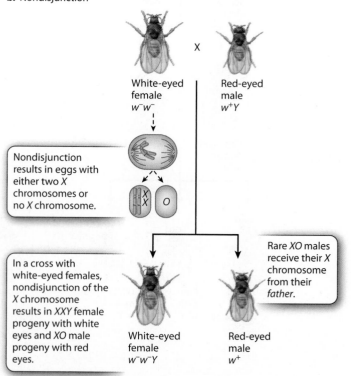

known as **nondisjunction.** Nondisjunction of X chromosomes results in eggs containing either two X chromosomes or no X chromosome. Figure 17.6b shows the implications for eye color in the progeny if the hypothesis is correct. The exceptional white-eyed females would contain two X chromosomes plus a Y chromosome (genotype $w^-/w^-/Y$), and the exceptional red-eyed males would contain a single X chromosome and no Y chromosome (genotype $w^+$).

Bridges's hypothesis for the exceptional progeny in Fig. 17.6b was bold, as it assumed that *Drosophila* males could develop in the absence of a Y chromosome (XO embryos yielding sterile males, where "O" indicates absence of a chromosome), and that females could develop in the presence of a Y chromosome (XXY embryos yielding fertile females). The hypothesis was accurate as well as bold. Microscopic examination of the chromosomes in the exceptional fruit flies confirmed that the exceptional white-eyed females had XXY sex chromosomes and that the exceptional sterile red-eyed males had an X but no Y. Because fruit flies with three X chromosomes (XXX) or no X chromosome (OY) were never observed, Bridges concluded that embryos with these chromosomal constitutions are unable to survive. He also showed that nondisjunction can take place in males as well as in females. From the phenotypes of these exceptional fruit flies and their chromosome constitutions, Bridges concluded that the white-eye gene (and by implication any other X-linked gene) is physically present within the X chromosome.

Bridges's demonstration that genes are present in chromosomes was also the first experimental evidence of nondisjunction. *Drosophila* differ from humans in that the Y chromosome is necessary for male fertility but not for male development. As we will see later in this chapter, a gene in the Y chromosome itself is the trigger for male development in humans and other mammals, and so for these organisms, the Y chromosome is needed both for male development and male fertility. Nondisjunction occasionally takes place in meiosis in humans as well as in fruit flies. When nondisjunction takes place in the human sex chromosomes, it results in chromosomal constitutions such as 47, XXY and 47, XYY males as well as 47, XXX and 45, XO females. Nondisjunction of autosomes can also occur, resulting in fetuses that have extra copies or missing copies of entire chromosomes. The consequences of nondisjunction of human chromosomes were examined in Chapter 15.

## Genes in the *X* chromosome show characteristic patterns in human pedigrees.

The features of X-linked inheritance can be seen in human pedigrees for traits due to an X-linked recessive mutation. These are illustrated in **Fig. 17.7** for red–green color blindness, a condition that affects about 1 in 20 males. An individual with red–green color blindness will have difficulty seeing the number in the colored dots in Fig. 17.7. The key features of X-linked inheritance, noted in the pedigree, are listed here:

1. Affected individuals are almost always males because males need only one copy of the mutant gene to be affected, whereas females need two copies to be affected.

2. Affected males have unaffected sons because males transmit their X chromosome only to their daughters.

3. A female whose father is affected can have affected sons because such a female must be a heterozygous carrier of the recessive mutant allele.

An additional feature worth mentioning is that the sisters of an affected male each have a 50% chance of being a heterozygous carrier because when a brother is affected the mother must be heterozygous for the recessive allele.

→ **Quick Check 1** Is it possible for an unaffected female to have female offspring with red–green color blindness?

**FIG. 17.7** **Inheritance of an *X*-linked recessive mutation.** This pedigree shows the inheritance of red–green color blindness, an X-linked recessive mutation.

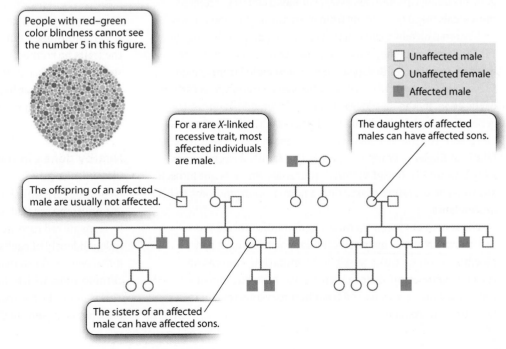

People with red–green color blindness cannot see the number 5 in this figure.

☐ Unaffected male
○ Unaffected female
■ Affected male

For a rare X-linked recessive trait, most affected individuals are male.

The daughters of affected males can have affected sons.

The offspring of an affected male are usually not affected.

The sisters of an affected male can have affected sons.

**FIG. 17.8** *X*-linked hemophilia in European royalty.

A pedigree for one of the most famous examples of human *X*-linked inheritance is shown in **Fig. 17.8.** The trait is **hemophilia,** which results from a recessive mutation in a gene encoding a protein necessary for blood clotting. Affected individuals bleed excessively from even minor cuts and bruises, and internal bleeding can cause excruciating pain. Affecting about 1 in 7000 males, hemophilia is famous because of its presence in many members of European royalty descended from Queen Victoria of England (1819–1901), who was a heterozygous carrier of the gene. By the marriages of her carrier granddaughters, the gene was introduced into the royal houses of Germany, Russia, and Spain. The mutant allele is not present in the present royal family of England, however, because this family descends from King Edward VII, one of Victoria's four sons, who was not himself affected and therefore passed only a normal *X* chromosome to his descendants.

The source of Queen Victoria's hemophilia mutation is not known. None of her ancestors is reported to have had a bleeding disorder. Quite possibly the mutation was present for a few generations before Victoria was born but remained hidden because it was passed from heterozygous female to heterozygous female.

## 17.3 GENETIC LINKAGE AND RECOMBINATION

Mendel was fortunate not only because peas do not possess sex chromosomes, but also because the genes that influenced the traits he studied, such as round/wrinkled and green/yellow seeds, were on separate chromosomes or far apart on the same chromosome. What happens when genes are close to each other in the same chromosome? We explore the answer to this question in this section.

### Nearby genes in the same chromosome show linkage.

Genes that are sufficiently close together in the same chromosome are said to be **linked.** That is, they tend to be transmitted together in inheritance and do not assort independently of each other as Mendel observed. Note that *linked genes* refer to two genes that are close together in the same chromosome, which may be an autosome or sex chromosome. This is not to be confused with an *X-linked gene*, which is simply one that is present in the *X* chromosome.

Linkage was discovered in *Drosophila* by Alfred H. Sturtevant, another of Morgan's students. Once again, genes in the *X* chromosome played a key role in the discovery, as seen in **Fig. 17.9**. Sturtevant worked with male fruit flies that have an *X* chromosome carrying two recessive mutations. One is in the *white* gene (*w*) discussed earlier, which when nonmutant results in fruit flies with red eyes and when mutant results in fruit flies with white eyes. The other recessive mutation is in a gene called *crossveinless* (*cv*), which when nonmutant results in fruit flies with tiny crossveins in the wings and when mutant results in the absence of these crossveins.

Sturtevant crossed this doubly mutant male with a female carrying the nonmutant forms of the genes ($w^+$ and $cv^+$) in both *X* chromosomes. He saw that the offspring consist of phenotypically wild-type females that are heterozygous for both genes, and phenotypically wild-type males. When these are crossed with each other, the female $F_2$ progeny do not tell us anything because they are all wild type; each female receives the $w^+cv^+$ *X* chromosome from her father and therefore has red eyes and normal crossveins. In the male $F_2$ progeny, the situation is different: Each male progeny receives its *X* chromosome from the mother and its *Y* chromosome from the father, and so the phenotype of each male immediately reveals the genetic constitution of the *X* chromosome that the male inherited from the mother.

As shown in Fig. 17.9, the male $F_2$ progeny consist of four types:

| Genotype of $F_2$ progeny | Number of fruit flies |
|---|---|
| $w^+cv^+$/Y (red eyes, normal crossveins) | 357 |
| $w^-cv^-$/Y (white eyes, missing crossveins) | 341 |
| $w^+cv^-$/Y (red eyes, missing crossveins) | 52 |
| $w^-cv^+$/Y (white eyes, normal crossveins) | 45 |

Although all four possible classes of maternal gametes are observed in the male progeny, they do not appear in the ratio 1:1:1:1 expected when gametes contain two independently assorting genes (Chapter 16). The lack of independent assortment means that the genes show linkage.

The male progeny fall into two groups. One group, represented by fewer numbers of progeny, consists of $w^+cv^-$ and $w^-cv^+$ combinations of alleles. These are called **recombinants**,

**FIG. 17.9** **Linkage of the *white* (*w*) and *crossveinless* (*cv*) genes in the *X* chromosome.** Note that genes are written according to their order along the chromosome, with a line between homologous chromosomes, in this case between the *X* and *Y* chromosomes.

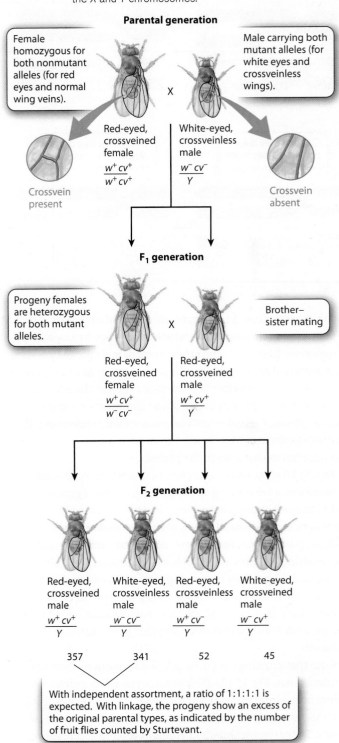

**FIG. 17.10  Linkage and recombination.** Crossing over during meiosis results in recombination between linked genes.

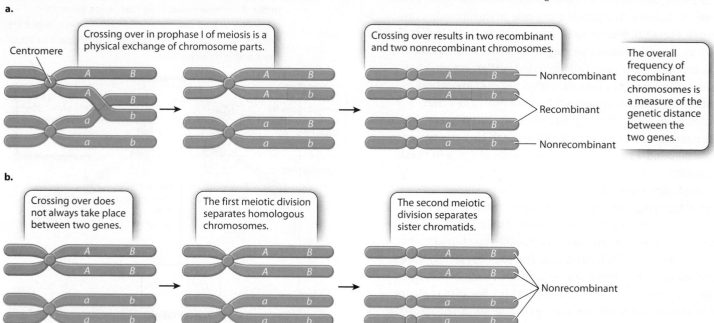

and they result from a **crossover,** the physical exchange of parts of homologous chromosomes, which takes place in prophase I of meiosis (Chapter 11). The other group of male progeny derives from maternal gametes containing either $w^+cv^+$ or $w^-cv^-$. These gametes each carry a chromosome that contains the alleles in the same combination as in one of the chromosomes in the mother. The $w^+cv^+$ and $w^-cv^-$ combinations are therefore called **nonrecombinants** (that is, the alleles are present in the same combination as that seen in the parent).

**Fig. 17.10** shows how recombinant and nonrecombinant chromosomes arise, using the hypothetical linked genes $A$ and $B$. In a cell undergoing meiosis in which a crossover takes place between the linked genes, the allele combinations are broken up in the chromatids involved in the exchange, and the resulting gametes are $AB$, $Ab$, $aB$, and $ab$. In Fig. 17.10a, the $Ab$ and $aB$ gametes have undergone recombination: The allele combinations that were originally together in each chromosome ($A$ with $B$ and $a$ with $b$) have recombined ($A$ with $b$ and $a$ with $B$) in the chromosomes that participated in the crossover.

Note that crossing over does not just result in recombinant chromosomes. Fig. 17.10a shows that, even when a crossover occurs in the interval between the genes, two of the resulting chromosomes contain the nonrecombinant configuration of alleles. These nonrecombinant configurations occur because crossing over occurs at the four-strand stage of meiosis (when

each homologous chromosome is a pair of sister chromatids), but only two of the four strands (one sister chromatid from each homologous chromosome) are included in any crossover.

Fig. 17.10b shows a second way in which nonrecombinant chromosomes originate. When genes are linked, some cells undergoing meiosis do not undergo a crossover in the interval between the genes. In this case, the $AB$ combination of alleles in one chromosome and the $ab$ combination of alleles in the homologous chromosome both remain intact, and the gametes produced are $AB$ and $ab$.

### Recombination frequency is a measure of the distance between linked genes.

When two genes are on separate chromosomes, a ratio of 1:1:1:1 is expected for the nonrecombinant (parental) and recombinant (nonparental) gametic types (Chapter 16). For two genes present in the same chromosome, we can consider two extreme situations. If they are located very far apart from each other, crossing over will certainly occur, and there will be a 1:1:1:1 ratio of nonrecombinant and recombinant gametes (as shown in Fig. 17.10a). At the other extreme, if two genes are so close together that crossing over never takes place between them, we would expect only nonrecombinant chromosomes (Fig. 17.10b).

What happens in between these extremes? In these cases, the nonrecombinant chromosomes in the offspring are more numerous than the recombinant chromosomes, as we saw with *w* and *cv* (see Fig. 17.9). The frequency of recombinants depends on the distance between the genes. The closer, or more tightly linked, that two genes are to each other, the smaller the frequency of recombinants among the progeny. The distance between the genes is important because whether or not a crossover occurs between the genes is a matter of chance, and the closer the genes are along the chromosome, the less likely it is that a crossover will take place in the interval between them. Because the formation of recombinant chromosomes requires a crossover between the genes, genes that are close together show less recombination than genes that are far apart. In fact, the proportion of recombinant chromosomes observed among the total, which is called the **frequency of recombination,** is a convenient measure of distance between the genes along the chromosome.

In the example with the genes *w* and *cv*, the total number of chromosomes observed among the progeny is $357 + 341 + 52 + 45 = 795$, and the number of recombinant chromosomes is $52 + 45 = 97$. The frequency of recombination between *w* and *cv* is therefore $97/795 = 0.122$, or 12.2%, and this serves as a measure of distance between the genes. In studies of genetic linkage, the distance between genes is not measured directly by physical distance between them, but rather by the frequency of recombination.

The frequency of recombination between any two genes ranges from 0% (when crossing over between the genes never takes place) to 50% (when the genes are so far apart that crossing over between the genes always takes place). Genes that are linked have a recombination frequency somewhere between 0% and 50%.

→ **Quick Check 2** Why is the upper limit of recombination 50% rather than 100%?

Recombination plays an important role in creating new combinations of alleles in each generation and in ensuring the genetic uniqueness of each individual. If there were no recombination (that is, if all the alleles in each chromosome were completely linked), any individual human would be able to produce only $2^{23} = 8.4$ million types of reproductive cells. While this is a large number, the average number of sperm per ejaculate is much larger—approximately 350 million. Because recombination does occur, and because the crossovers resulting in recombination can occur at any of thousands of different positions in the genome, each of the 350 million sperm is virtually certain to carry a different combination of alleles.

## Genetic mapping assigns a location to each gene along a chromosome.

With the exception of a few regions, such as the area near the centromere, the likelihood of a crossover occurring somewhere between two points on a chromosome is approximately proportional to the length of the interval between the points. Therefore, the frequency of recombination can be used as a measure of the physical distance between genes. These distances are used in the construction of a **genetic map,** which is a diagram showing the relative position of genes along a chromosome. The maps are drawn using a scale in which one unit of distance (called a **map unit**) is the distance between genes resulting in 1% recombination. Thus, in a *Drosophila* genetic map containing the genes *w* and *cv*, the distance between the genes is 12.2 map units.

Genetic maps are built up step by step as new genes are discovered that are genetically linked with known genes, as shown in **Fig. 17.11.** Across distances that are less than about 15 map units, the map distances are approximately additive, which means that the distances between adjacent genes can be added to get the distance between the genes at the ends. For example, in Fig. 17.11, there are two genes between *w* and *cv*. The map distance between *w* and the next gene, *ec*, is 4.0, the distance between *ec* and the next gene, *rb*, is 2.0, and the distance between *rb* and *cv* is 6.2. The map distance between *w* and *cv* is therefore $4.0 + 2.0 + 6.2 = 12.2$ map units, and hence the expected frequency of recombination between these genes is 12.2%, which is the value observed.

However, for two genes that are farther apart than about 15 map units, the observed recombination frequency is somewhat smaller than the sum of the map distances between the genes. The reason is that, with greater distances, two or more crossovers between the genes may occur in the same chromosome, and thus an exchange produced by one crossover may be reversed by another crossover farther along the way.

→ **Quick Check 3** For two genes that show independent assortment, what is the frequency of recombination?

## Genetic risk factors for disease can be localized by genetic mapping.

The discovery of abundant genetic variation in DNA sequences in human populations, such as single-nucleotide polymorphisms (Chapter 15), made it possible to study genetic linkage in the human genome. At first, the focus was on finding mutations that cause disease, such as a mutation that causes cystic fibrosis (Chapter 14). The method was to study large families extending over three or more generations in which the disease was present and then to identify the genotypes of each of the individuals for thousands of genetic markers (previously discovered DNA polymorphisms) throughout the genome. The goal was to find genetic markers that showed a statistical association with the disease gene, which would indicate genetic

**FIG. 17.11**

# Can recombination be used to construct a genetic map of a chromosome?

**BACKGROUND** In 1910, Thomas Hunt Morgan discovered *X*-linkage by studying the white-eye mutation in *Drosophila*. Soon other *X*-linked mutations were found. Alfred H. Sturtevant, Morgan's student, decided to test whether mutant genes in the same *X* chromosome were inherited together, that is, linked. He found that genes in the *X* chromosome were linked, but not completely, and that different pairs of genes showed great differences in their linkage. Some genes showed almost no recombination, whereas others underwent so much recombination that they showed independent assortment.

**HYPOTHESIS** Sturtevant hypothesized that recombination was due to crossing over between the genes, and that genes farther apart in the chromosome would show more recombination.

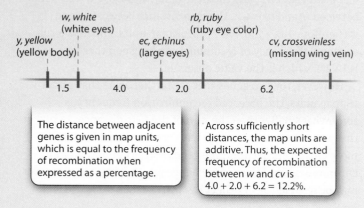

The distance between adjacent genes is given in map units, which is equal to the frequency of recombination when expressed as a percentage.

Across sufficiently short distances, the map units are additive. Thus, the expected frequency of recombination between *w* and *cv* is 4.0 + 2.0 + 6.2 = 12.2%.

**EXPERIMENT** Taking this idea a step further, Sturtevant reasoned that if one knew the frequency of recombination between genes *a* and *b*, between *b* and *c*, and between *a* and *c*, then one should be able to deduce the order of genes along the chromosome. He also predicted that, if the order of genes were known to be *a–b–c*, then if the genes were close enough, the frequency of recombination between *a* and *c* should equal the sum of the frequencies between *a* and *b* and that between *b* and *c*.

**RESULTS** Sturtevant studied the frequencies of recombination between many pairs of genes along the *X* chromosome, including some of those shown in the accompanying illustration.

**CONCLUSION** The results confirmed Sturtevant's hypothesis and showed that genes could be arranged in the form of a genetic map, depicting their linear order along the chromosome, with the distance between any pair of genes proportional to the frequency of recombination between them. Across sufficiently short regions, the frequencies of recombination are additive.

**FOLLOW-UP WORK** Genetic mapping remains a cornerstone of genetic analysis, showing which chromosome contains a mutant gene and where along the chromosome the gene is located. The method helped to identify the genes responsible for many single-gene inherited disorders, including Huntington's disease, cystic fibrosis, and muscular dystrophy.

**SOURCE** Sturtevant, A. H. 1913. "The Linear Arrangement of Six Sex-Linked Factors in *Drosophila*, as Shown by Their Mode of Association." *Journal of Experimental Zoology* 14: 43–59.

linkage and reveal the approximate location of the disease gene along the chromosome.

**Fig. 17.12** illustrates the underlying concept. It assumes 100 chromosomes observed among different individuals in a pedigree, of which 50 carry a mutant allele of a gene and 50 carry a nonmutant allele. These chromosomes are tested for a marker of known location, in this case a single-nucleotide polymorphism (SNP) in which one of the nucleotide pairs in the DNA is a G–C base pair in some chromosomes and A–T in others. In Fig. 17.12a, there is clearly an association between the disease gene and the SNP. Almost all of the chromosomes that carry the mutant allele show the G–C nucleotide pair, whereas almost all of the chromosomes that carry the nonmutant allele show the A–T nucleotide pair. The two chromosomes in which

the mutant and nonmutant genes are associated with the other SNPs can be attributed to recombination.

The association in Fig. 17.12a may be contrasted with the pattern in Fig. 17.12b, in which there is no association. In this case, each of the alleles of the disease gene is equally likely to carry either form of the SNP. The failure to find an association means that the SNP is not closely linked to the disease gene, and may actually be in a different chromosome. In actual studies, an association is almost never observed, but the lucky find of an association helps identify the location of the disease gene in the genetic map. Using such association methods, hundreds of important disease genes have been located by genetic mapping. Once the location of the disease gene is known, the identity and normal function of the gene can be determined.

**FIG. 17.12 Genetic mapping.** SNPs associated with a mutant gene show where that gene is located in the genetic map.

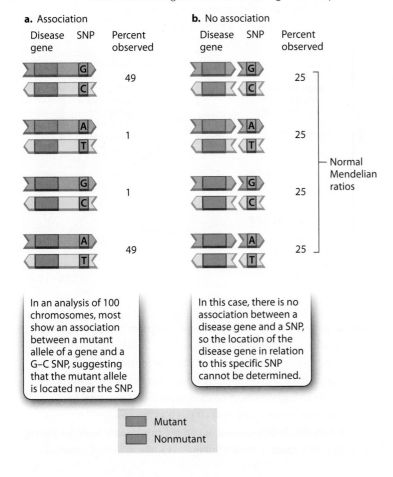

**a. Association**

Disease SNP    Percent
gene           observed

49

1

1

49

In an analysis of 100 chromosomes, most show an association between a mutant allele of a gene and a G–C SNP, suggesting that the mutant allele is located near the SNP.

**b. No association**

Disease SNP    Percent
gene           observed

25

25            Normal
              Mendelian
25            ratios

25

In this case, there is no association between a disease gene and a SNP, so the location of the disease gene in relation to this specific SNP cannot be determined.

Mutant
Nonmutant

## 17.4 INHERITANCE OF GENES IN THE Y CHROMOSOME

Like the X chromosome, the Y chromosome exhibits a particular pattern of inheritance because of its association with the male sex. In humans and other mammals, all embryos initially develop immature internal sexual structures of both females and males. SRY, a gene in the Y chromosome, encodes a protein that is the trigger for male development. ("SRY" stands for "sex-determining region in the Y chromosome.") In the presence of SRY, male structures complete their development and female structures degenerate. In the absence of SRY, male embryonic structures degenerate and female structures complete their

development. The SRY gene is therefore the male-determining gene in humans and other mammals.

Because they are linked to SRY, most genes in the Y chromosome show a distinctive pattern of inheritance different from that Mendel observed. In this case, they are passed from father to son.

### Y-linked genes are transmitted from father to son to grandson.

Genes that are present in the unique region of the Y chromosome (the part that cannot cross over with the X) are known as **Y-linked genes,** of which there are not many. As well as the SRY male-determining gene, they include a number of genes in which mutations are associated with impaired fertility and low sperm count.

The pedigree characteristics of Y-linked inheritance are striking (**Fig. 17.13**):

1. Only males are affected with the trait.

2. Females never inherit or transmit the trait, regardless of how many affected male relatives they have.

3. All sons of affected males are also affected.

Because the Y chromosome is always transmitted from father to son (and never transmitted to daughters), a trait determined by a Y-linked gene will occur in fathers, sons, grandsons, and so forth. Traits resulting from Y-linked genes cannot be present in females nor can they be transmitted by females. However, other than maleness itself and some types of impaired fertility, no physical traits are known that follow a strict Y-linked pattern of inheritance. This observation emphasizes the extremely low density of functional genes in the Y chromosome.

**FIG. 17.13 Inheritance of Y-linked traits.** The pedigree pattern is that of father to son to grandson to great-grandson, and so forth.

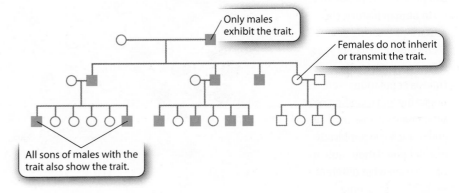

Only males exhibit the trait.

Females do not inherit or transmit the trait.

All sons of males with the trait also show the trait.

**? CASE 3  You, From A to T: Your Personal Genome**
**How can the *Y* chromosome be used to**
**trace ancestry?**

The example of Claudia Gilmore illustrates how, through tests of her personal genome with regard to the *BRCA1* and *BRCA2* mutations, she became aware of her elevated risk of breast cancer. Your personal genome not only can tell you about your genetic risk factors for disease, but it also contains important information about your genetic ancestry. For example, your male ancestors can be traced through your *Y* chromosome. The regions at the tips in which the *X* and *Y* chromosomes share homology is only about 6% of the entire length of the *Y* chromosome. This means that 94% of the *Y* chromosome consists of sequences in which genetic linkage is complete because it does not pair with another chromosome and does not undergo crossing over.

Because of this complete linkage, each hereditary lineage of *Y* chromosomes is separate from every other lineage. As mutations occur along the *Y* chromosome, they are completely linked to any past mutations that may be present and also completely linked to any future mutations that may take place. The mutations therefore accumulate, and this allows the evolutionary history of a set of sequences to be reconstructed.

**Fig. 17.14** shows an evolutionary tree based on the accumulation of mutations at a set of nucleotide sites along the *Y* chromosome. Because the nucleotide sites are completely linked, the nucleotides shown need not be adjacent but can be anywhere along the *Y*. Each unique combination of nucleotides constitutes a *Y*-chromosome **haplotype,** or haploid genotype. In the figure, the most ancient *Y* chromosomes are at the top, and the accumulation of new mutations as the generations proceed results in the successive creation of new haplotypes. Each *Y*-chromosome lineage may leave some nonmutant descendants as well as some mutant descendants, and hence any or all of the sequences shown may coexist in a present-day population.

In human history, the mutations creating new *Y*-chromosome haplotypes were occurring at the same time as populations were migrating and founding new settlements across the globe, and so each geographically distinct population came to have a somewhat different set of *Y*-chromosome haplotypes. The differences

**FIG. 17.14  *Y*-chromosome haplotypes.** Because *Y* chromosomes in a lineage conform to an evolutionary tree, the ancestry of a male's *Y* chromosome can be traced.

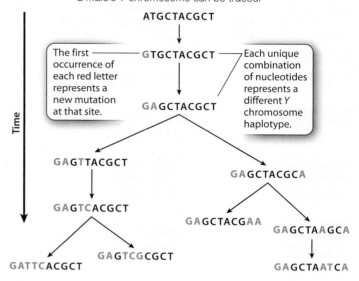

among populations are offset to some extent by migration among populations, which mixes the geographical locations of various haplotypes. Nevertheless, the fact that the mutations accumulate through time and are completely linked allows the evolutionary history of the haplotypes to be reconstructed. It also enables *Y* chromosomes to be traced to their likely ethnic origin.

The worldwide distribution of real *Y*-chromosome lineages among human populations is shown in **Fig. 17.15.** The different colors in

**FIG. 17.15  Geographical distribution of *Y*-chromosome haplotypes.** The evolutionary trees of *Y*-chromosome haplotypes reflect the origin and movement of different *Y* chromosomes over time.

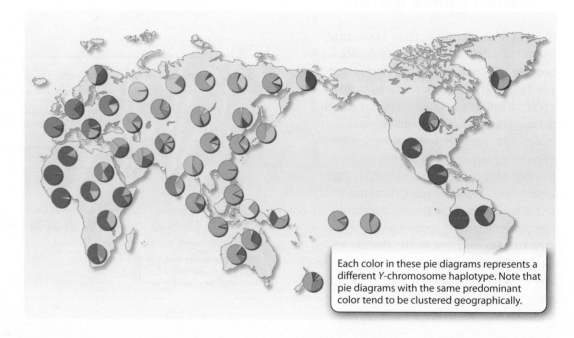

Each color in these pie diagrams represents a different *Y*-chromosome haplotype. Note that pie diagrams with the same predominant color tend to be clustered geographically.

the pie diagrams represent different mutations. Neighboring populations tend to have more closely related Y chromosomes than more distant populations. Four major clusters of Y-chromosome lineages can be recognized in Fig. 17.15. One is concentrated in Africa, another in Southeast Asia and Australia, a third in Europe and central and western Asia, and the fourth in North and South America. These clusters correspond roughly with the spread of human settlements around the globe inferred from archaeological evidence.

The implication of Fig. 17.15 is that the haplotype of your Y chromosome (if you have a Y chromosome) contains genetic information about its origin. And you can learn what this information is from genetic testing companies that sell direct-to-consumer (DTC) services. Their tests are not regarded as medical devices and so are unregulated, and quality control is sometimes uncertain. Nevertheless, you can send saliva or other biological samples to a DTC provider, which will (for a fee) test your Y chromosome and send you a report that details its possible origin. Of course, because of recombination and independent assortment of genes in other chromosomes, the ethnic origin of your Y chromosome may have little or nothing to do with the ethnic origins of genes in any of your other chromosomes.

## 17.5 INHERITANCE OF MITOCHONDRIAL AND CHLOROPLAST DNA

Sex chromosomes and linked genes are not the only genes that are inherited in ways Mendel did not describe. Genes in mitochondria and chloroplasts also show distinct inheritance patterns. Mitochondria and chloroplasts are ancient organelles of eukaryotic cells originally acquired by the engulfing of prokaryotic cells (Chapter 5). Mitochondria generate most of the ATP that cells use for their chemical energy, and they also function in the cell cycle, programmed cell death, and other important processes. Chloroplasts are found only in plant cells and eukaryotic algae. These organelles contain chlorophyll, a green pigment that absorbs light and, in the process of photosynthesis (Chapter 8), produces the sugars that are essential for growth of the plant or algal cells and of the organisms that eat them. Mitochondria and chloroplasts have their own genomes, so genes present in these genomes move with the organelle during cell division.

### Mitochondrial and chloroplast genomes often show uniparental inheritance.

During sexual reproduction, cellular organelles do not show the regular, highly choreographed movements that chromosomes undergo during Mendelian segregation. These organelles are partitioned to the gametes along with other cytoplasmic components, and therefore their mode of inheritance depends on how the gametes are formed, how much cytoplasm is included in the gametes, and the fate of the cytoplasm in each parental gamete after fertilization.

Considering the great diversity in the details of reproduction in different groups of organisms, it is not surprising that there is a diversity of types of inheritance of cytoplasmic organelles. The three most important types are:

- **Maternal inheritance,** in which the organelles in the offspring cells derive from those in the mother.

- **Paternal inheritance,** in which the organelles in the offspring cells derive from those in the father.

- **Biparental inheritance,** in which the organelles in the offspring cells derive from those in both parents.

In most organisms, either maternal inheritance or paternal inheritance predominates, but sometimes there is variation from one offspring to the next. For example, transmission of the chloroplasts ranges from strictly paternal in the giant redwood *Sequoia,* to strictly maternal in the sunflower *Helianthus,* to either maternal or paternal (or less frequently biparental) in the fern *Scolopendrium,* to mostly maternal but sometimes paternal or biparental in the snapdragon *Antirrhinum.*

There is likewise great diversity among organisms in the inheritance of mitochondrial DNA. Most animals show maternal transmission of the mitochondria, as would be expected from their large, cytoplasm-rich eggs and the small, cytoplasm-poor sperm. Among other organisms, there is again much variation, including maternal transmission of mitochondria in flowering plants and paternal transmission in the green alga *Chlamydomonas.*

### Maternal inheritance is characteristic of mitochondrial diseases.

In humans and other mammals, mitochondria show strictly maternal inheritance—the mitochondria in the offspring cells derive from those in the mother. **Fig. 17.16** shows the characteristic pedigree patterns of a trait due to maternal inheritance:

1. Both males and females can show the trait.

2. All offspring from an affected female show the trait.

3. Males never transmit the trait to their offspring.

The pedigree in Fig. 17.16 follows a mitochondrial disease known as MERRF syndrome (the acronym stands for "myoclonic epilepsy with ragged red fibers"). As its name suggests, the syndrome is characterized by epilepsy, a neurological disease characterized by seizures, as well as by the accumulation of abnormal mitochondria in muscle fibers. This extremely rare disease is associated with a single point mutation in a mitochondrial gene involved in protein synthesis that affects oxidative phosphorylation (Chapter 7).

More than 40 different diseases show these pedigree characteristics. They all result from mutations in the mitochondrial DNA, but the tissues and organs affected as well as the severity differs from one to the next. All of the mutations

**FIG. 17.16 Maternal inheritance.** Human mitochondrial DNA is transmitted from a mother to all of her offspring.

Inherited mitochondrial diseases are often associated with muscle weakness reflecting deficient production of ATP. The red patches in the microscopic image result from clumps of defective mitochondria in muscle fibers observed in one form of epilepsy due to mutation in mitochondrial DNA.

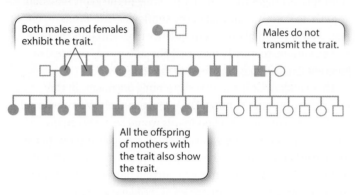

Both males and females exhibit the trait.

Males do not transmit the trait.

All the offspring of mothers with the trait also show the trait.

affect energy production in one way or another. In some cases, disease results directly from lack of adequate amounts of ATP, in other cases from intermediates in energy production that are toxic to the cell or that damage the mitochondrial DNA.

**? CASE 3 You, From A to T: Your Personal Genome**
**How can mitochondrial DNA be used to trace ancestry?**
Sequencing of mitochondrial DNA reveals mitochondrial haplotypes analogous to those in the Y chromosome. As in the Y chromosome, mutations in mitochondrial DNA accumulate through time, and each mitochondrial haplotype remains intact through successive generations because recombination between mitochondrial genomes is rare. This means that mitochondrial DNA can be used to trace ancestry and population history, much as described previously for the Y chromosome. The mitochondrial DNA is actually more informative than Y-chromosomal DNA because it shows substantially more genetic variation.

The use of mitochondrial DNA to trace human origins and migration is discussed in greater detail in Chapter 24. As with the Y chromosome, your personal mitochondrial genome includes information about its origin. If you want to learn about your mitochondrial DNA, you can send a tissue sample to any of a number of direct-to-consumer genotyping services that will, for a fee, analyze the DNA and provide you with a report.

## Core Concepts Summary

**17.1 THE X AND Y CHROMOSOMES OF SOME ANIMALS, INCLUDING HUMANS, DETERMINE SEX AND ARE INHERITED DIFFERENTLY FROM THE AUTOSOMES.**

In humans and other mammals, XX individuals are female and XY individuals are male. page 17-2

The human X and Y chromosomes are different lengths and contain different genes, except for small regions of homology that allow the two chromosomes to pair in meiosis. page 17-2

Segregation of the X and Y chromosomes during male meiosis results in half of the sperm receiving an X chromosome and half a Y chromosome so that random union of gametes predicts a 1:1 female:male sex ratio at the time of fertilization. page 17-2

**17.2 X-LINKED GENES, WHICH SHOW A CRISSCROSS INHERITANCE PATTERN, PROVIDED THE FIRST EVIDENCE THAT GENES ARE PRESENT IN CHROMOSOMES.**

Morgan studied a mutation in the fruit fly *Drosophila melanogaster* that resulted in fruit flies with white eyes rather than normal red eyes. In this species, as in mammals, females are XX and males are XY. page 17-3

In a cross of a normal red-eyed female with a mutant white-eyed male, all of the male and female progeny had red eyes. When brothers and sisters of this cross were mated with each other, all of the females had red eyes, but males were red-eyed and white-eyed in a 1:1 ratio. page 17-3

This pattern of inheritance is observed because the gene Morgan studied, called *white*, is located in the X chromosome. The nonmutant $w^+$ allele is dominant to the mutant $w^-$ allele, and the gene is present only in the X chromosome and not in the Y chromosome. page 17-4

X-linked genes show a crisscross inheritance pattern, in which the X chromosome with the mutant gene that is present in males in one generation is present in females in the next generation. page 17-5

Bridges observed rare fruit flies that did not follow the usual pattern for X-linked inheritance and inferred that these exceptional fruit flies resulted from nondisjunction, or failure of homologous chromosomes to segregate, in male or female meiosis. His observations provided evidence that genes are carried in chromosomes. page 17-6

In humans, X-linked inheritance shows a pattern in which affected individuals are almost always males, affected males have unaffected sons, and a female whose father is affected can have affected sons. page 17-7

## 17.3 GENETIC LINKAGE OCCURS WHEN TWO GENES ARE SUFFICIENTLY CLOSE TOGETHER IN THE SAME CHROMOSOME THAT THE COMBINATION OF ALLELES PRESENT IN THE CHROMOSOME TENDS TO REMAIN TOGETHER IN INHERITANCE.

Genes that are close together in the same chromosome are linked and do not undergo independent assortment. page 17-8

Recombinant chromosomes result from crossing over between genes on the same chromosome and show a nonparental combination of alleles. page 17-9

Nonrecombinant chromosomes have the same configuration of alleles as one of the parental chromosomes. page 17-10

In genetic mapping, the observed proportion of recombinant chromosomes is the frequency of recombination and can be used as a measure of distance along a chromosome. A recombination frequency of 1% is one map unit. page 17-10

Gene linkage and mapping are used to identify the locations of disease genes in the human genome. page 17-11

## 17.4 MOST Y-LINKED GENES ARE PASSED FROM FATHER TO SON.

In humans and other mammals, the Y chromosome contains a gene called *SRY* that results in male development. page 17-13

In Y-linked inheritance, only males are affected and all sons of an affected male are affected. Females are never affected and do not transmit the trait. page 17-13

Most Y-linked genes show complete linkage, which allows their evolutionary history to be traced. page 17-14

## 17.5 MITOCHONDRIA AND CHLOROPLAST DNA FOLLOW THEIR OWN INHERITANCE PATTERN.

Mitochondria and chloroplasts have their own genomes, which reflect their evolutionary history as free-living prokaryotes. page 17-15

Mitochondria in humans and other mammals show maternal inheritance, in which individuals inherit their mitochondrial DNA from their mother. page 17-15

Because mitochondrial DNA does not undergo recombination and is maternally inherited, it can be used to trace human ancestry and migration. page 17-16

## Self-Assessment

1. Explain how the human X and Y chromosomes can pair during meiosis even though they are of different lengths and most of their genes are different.

2. Describe the biological basis for the 1:1 ratio of males and females at conception in mammals.

3. For a recessive X-linked mutation, such as color blindness, draw and explain its pattern of inheritance through a set of crosses.

4. Explain why linked genes do not exhibit independent assortment.

5. Describe how recombination frequency can be used to build a genetic map.

6. Describe the pattern of inheritance expected from a Y-linked gene in a human pedigree.

7. Describe the pattern of inheritance expected from a gene present in mitochondrial DNA in a human pedigree.

8. Explain how Y-chromosome and mitochondrial DNA data can be used to trace ancestry.

Do you understand the chapter's Core Concepts? Log in to check your answers to the Self-Assessment questions, then practice what you've learned and reinforce this chapter's concepts by working through the problems and multimedia tutorials provided there.

📖 **www.biologyhowlifeworks.com**

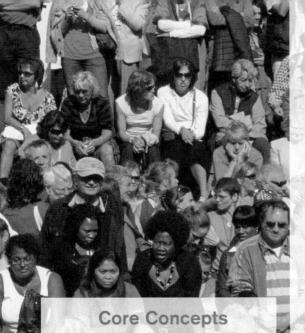

# THE GENETIC AND ENVIRONMENTAL BASIS OF COMPLEX TRAITS

## Core Concepts

**18.1** Complex traits are those influenced both by the action of many genes and by environmental factors.

**18.2** Genetic effects on complex traits are reflected in resemblance between relatives.

**18.3** Twin studies help separate the effects of genotype and environment on variation in a trait.

**18.4** Many common diseases and birth defects are affected by multiple genetic and environmental risk factors.

Biologists initially had a hard time accepting the principles of Mendelian inheritance because they seem so at odds with everyday observations. Common and easily observed traits like height, weight, hair color, and skin color give no evidence of segregation in pedigrees, and simple phenotypic ratios like 3:1 or 9:3:3:1 are not observed for them. The lack of these characteristic ratios raised serious doubt whether Mendel's principles are valid for common traits—some biologists concluded that they apply only to seemingly trivial traits like round and wrinkled seeds in peas.

At about the time that Mendel was studying inheritance in garden peas, the biologist Francis Galton, a friend and cousin of Charles Darwin, was studying common traits including human height. From studies of height and other common traits in parents and their offspring, Galton discovered general principles in the inheritance of such traits. For example, parents who are tall tend to have offspring who are taller than average but not as tall as themselves. Galton's principles for the inheritance of common traits did not invoke genes, segregation, independent assortment, or other features of Mendelian inheritance, but they did describe the observations. Not only were genes—what Mendel called "hereditary factors"—thought to be unnecessary in Galton's theory of inheritance, but also many biologists thought that Galton's theories and Mendel's were incompatible.

This, it turned out, was not the case. The traits that Mendel studied are now called **single-gene traits** because each one is determined by variation at a single gene and the traits are for the most part not influenced by the environment. By focusing on single-gene traits, Mendel was able to infer underlying mechanisms of inheritance based on physical factors we now call genes. By contrast, **complex traits,** such as human height, are influenced by multiple genes as well as by the environment. As a result, their inheritance patterns are more difficult to follow and simple phenotypic ratios are not observed.

In many ways, complex traits are more important than single-gene Mendelian traits. One reason is their prevalence—complex traits are found everywhere and include most of the traits we can see around us. By contrast, there are relatively few examples of single-gene traits. Complex traits are also important in human health and disease. In the most common disorders—among them heart disease, diabetes, and cancer—single-gene Mendelian inheritance is seldom found. It is therefore important to understand the inheritance of complex traits and common disorders, which are the subjects of this chapter. We begin by describing some of the features of complex traits, and then show how these principles are not only compatible with, but in fact predicted by, Mendelian inheritance. Finally, we describe how modern molecular genetics and genomics have allowed the identification of genes affecting complex traits.

## 18.1 HEREDITY AND ENVIRONMENT

Complex traits are important not only in humans, but also in agricultural plants and animals. We will examine human height in some detail because it has been widely studied, but equally well known complex traits are number of eggs laid by hens, milk production in dairy cows, and yield per acre of grain (**Fig. 18.1**).

Many common human diseases, including high blood pressure, obesity, diabetes, and depression, are complex traits (**Fig. 18.2**). High blood pressure, for example, affects about one-third of the U.S. population, and obesity another third. Type 2 diabetes affects around 8% of the U.S. population, and an estimated 15% will suffer at least one episode of severe depression in the course of a lifetime. Taken together, about 200 million Americans—two-thirds of the entire population—suffer from one or more of these common disorders. None of these traits shows single-gene Mendelian inheritance.

In many complex traits, the phenotype of an individual is determined by measurement: Human height is measured in inches, milk yield by the gallon, grain yield by the bushel, egg production by the number of eggs, blood pressure by millimeters of mercury, and blood sugar by millimoles per liter. Because the phenotype of complex traits such as these is measured along a continuum with only small intervals between similar individuals, complex traits like these are often called **quantitative traits.** By contrast, single-gene traits often appear in one of two or more different phenotypes, such as round versus wrinkled seeds, or green versus yellow seeds.

### Complex traits are affected by the environment.

Expression of complex traits is notoriously susceptible to lifestyle choices and other environmental factors. Inadequate nutrition is linked to slow growth rate and short stature in adults. Salt intake is associated with an increased likelihood of high blood pressure and is therefore an **environmental risk factor** for this common disorder. An environmental risk factor is a characteristic in a person's surroundings that increases the likelihood of developing a particular disease. A junk-food diet high in fat and carbohydrates is an environmental risk factor for obesity and diabetes.

Environmental effects are also important in agriculture. Farmers are well aware that adequate nutrition is essential to normal growth of chicks, lambs, piglets, and calves, and that continued high-quality feed is necessary for high egg production and milk yield when the animals reach adulthood. In crop plants like grains, adequate soil moisture and nutrients are necessary for sustained high yields.

Environmental factors not only affect the average phenotype for complex traits, but they also affect the variation in phenotype from one individual to the next. For example, most fields of corn you see planted along the roadside come from seeds that are genetically identical to one another, yet there is phenotypic variation in complex traits like plant height (**Fig. 18.3**). Because the plants are genetically identical, the differences in phenotype result from differences in the environment. Some parts of the field may receive more sunlight than others, and some parts may have better water drainage. No matter how uniform an environment may seem, there are always minor differences from one area to the next, and these differences can result in variation in complex traits.

Environmental effects on complex traits in animals are evident in true-breeding, totally homozygous strains produced by many generations of brother–sister matings, like those used by Mendel in his experiments with pea plants. Such true-breeding, homozygous strains are called **inbred lines,** and they are often used for research. Even though all animals in any inbred line are genetically identical, and are caged in the same facility and fed the same food, there is variation from one animal to the next. In one study of cholesterol levels in an inbred line of mice, for example, average serum cholesterol was 120 mg/dl (milligrams per deciliter), but the range was 60–180 mg/dl. Because the mice are

---

**FIG. 18.1   Examples of complex traits.** (a) Human height, (b) egg number, (c) milk production, and (d) grain yield.

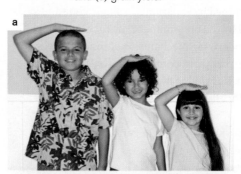

**FIG. 18.2  Examples of human diseases that are complex traits.** (a) High blood pressure, (b) obesity, (c) diabetes, and (d) depression.

genetically identical, the variation in serum cholesterol resulted from variation in the environment.

## Complex traits are affected by multiple genes.

Most complex traits are affected by many genes, in contrast to Mendel's traits, which are primarily affected only by one. Therefore, in complex traits the familiar phenotypic ratios such as 3:1 that Mendel observed when he crossed two true-breeding strains with contrasting characters are not observed. For complex

**FIG. 18.3  Phenotypic variation due to variation in the environment.** Genetically identical corn plants vary in height because of variation in exposure to sun and in soil composition at different locations in the same field.

traits, the effects of individual genes are obscured by variation in phenotype that is due to multiple genes affecting the trait and also due to the environment. The number of genes affecting complex traits is usually so large that different genotypes can have very similar phenotypes, which also makes it difficult to see the effects of individual genes on a trait.

In a few traits, however, the effects of the environment are minor and the number of genes is small, and in these cases the genetic basis of the trait can be analyzed. A classic example, studied by Herman Nilsson-Ehle about a century ago, concerns the color of seed casing in wheat (the "seed color"), which ranges from nearly white to dark red (**Fig. 18.4**). His experiment demonstrated that complex traits are subject to the same laws that Mendel worked out for single-gene traits, but that the inheritance patterns are more difficult to see because of the number of genes involved.

In studying seed color in true-breeding varieties and their first-generation and second-generation hybrids, Nilsson-Ehle realized that the relative frequencies of different shades of red color could be explained by the effects of three genes that undergo independent assortment, as illustrated by the Punnett square shown in Fig. 18.4. Each of the genes has two alleles, designated by combinations of upper-case and lower-case letters. In the development of seed color, each upper-case allele in a genotype intensifies the red coloration in an additive fashion. Altogether there are seven possible phenotypes, ranging from a phenotype of 0 (nearly colorless, genotype *aa bb cc*) to 6 (dark red, genotype *AA BB CC*).

In the cross shown at the top of Fig. 18.4, the parental genotypes are both *Aa Bb Cc*, and so they each have a phenotype of 3, an intermediate seed color on the 0-to-6 color scale. If the three genes show independent assortment (Chapter 16), then the distribution of phenotypes expected in progeny from the cross *Aa Bb Cc × Aa Bb Cc* is as shown in the Punnett square, in which the numbers indicate the number of upper-case alleles in each class of offspring. The bar graph below the Punnett square shows the distribution of the seed-color phenotypes and the probability

**FIG. 18.4** **Multiple genes contributing to a complex trait.** Three unlinked genes, each with two alleles, influence the intensity of red coloration of the seed casing in wheat.

Two plants heterozygous for each of three unlinked genes affecting the color of seed casing are crossed.

$Aa\ Bb\ Cc \times Aa\ Bb\ Cc$

The genes act cumulatively to determine the intensity of red coloration; in this example, each upper-case allele in the genotype adds more red to the phenotype.

Female gametes

Male gametes

The bar graph of color phenotypes among the progeny approximates a bell-shaped curve.

Relative probability of phenotype

Intensity of red coloration

of each of the possible phenotypes, which is approximated reasonably well by the bell-shaped curve known as the **normal distribution.** The phenotypes of many complex traits, including human height, conform to the normal distribution. Nelson-Ehle's results conformed to a normal distribution, consistent with the hypothesis of three unlinked genes each with two alleles influencing seed color in wheat.

When differences in phenotype due to the environment can be ignored, the genetic variation affecting complex traits can be detected more easily. And when studying inbred lines, differences in phenotype due to genotype can be ignored because all individuals have the same genotype, and the effects of environment can be observed. In most cases, however, genetic variation among individuals and environmental variation among individuals are both present, and it is difficult to quantify how much variation in phenotype is due to genetics and how much due to environment.

It's important to point out that complex traits are not really more "complex" than any other biological trait. The term is used merely to imply that both genetic factors and environmental factors contribute to variation in phenotype among individuals. As there are environmental factors that affect complex traits such as height, so there are genetic factors that affect height (as we discuss below). Similarly, just as an environmental risk factor increases the likelihood of a common disease, so does a genetic risk factor predispose an individual to the condition. For example, the human gene *ApoE* encodes a protein that helps transport fat and cholesterol. Certain alleles of *ApoE* are associated with high levels of cholesterol, and one particular allele is a genetic risk factor for Alzheimer's disease, the most common serious form of age-related loss of cognitive ability.

### The relative importance of genes and environment can be determined by differences among individuals.

For any one individual, it is impossible to specify the relative roles of genes and environment in the expression of a complex trait. For example, in an individual 66 inches tall, it would be meaningless to attribute 33 inches of height to parentage (genes) and 33 inches of height to nutrition (environment). This kind of partitioning makes no sense because genes and environment act together so intimately in each individual that their effects are inseparable. To attempt to separate them for any individual would be like asking to what extent it is breathing or oxygen that keeps us alive. Both are important, and both must take place together if we are to stay alive.

It is nevertheless possible to separate genes and environment in regard to their effects on the *differences*, or variation, among individuals within a particular population. For some traits, the variation seen among individuals is due largely if not exclusively to differences in the environment. For other traits, the variation is due mainly to genetic differences. In the case of human height, roughly 80% of the variation among individuals of the same sex is due to genetic differences, and the remaining 20% to differences in their environment, primarily differences in nutrition during their years of growth.

### Genetic and environmental effects can interact in unpredictable ways.

One of the features of complex traits is that genetic and environmental effects may interact, often in unpredictable ways. Consider the example shown in **Fig. 18.5.** In this experiment, two strains of corn were each grown in a series of soils in which the amount of nitrogen had been enriched by the growth of legumes. The yield of strain 1 varies little across these different environments. The yield of strain 2, however, increases dramatically with soil nitrogen. Each of the lines is known as a **norm of reaction,** which for any genotype graphically depicts how the environment (shown on the x-axis) affects phenotype (shown on the y-axis) across a range of environments. Note that in one case in Fig. 18.5, the environment has little to no effect on the phenotype (the norm of reaction is nearly flat), but in the other it has a very noticeable effect. Therefore, it is impossible to predict the phenotype of a given genotype without knowing what the environmental conditions are and how the phenotype changes in response to variation in the environment.

Such variation in the effects of the environment on different genotypes is known as **genotype-by-environment interaction.** This type of interaction is important because it implies that the effect of a genotype cannot be specified without knowing the environment, and the other way around. A further implication is that there may be no genotype that is the "best" across a broad range of environments, and likewise

**FIG. 18.6** **Genotype-by-environment interaction for obesity in mice.** When fed a normal diet, both inbred lines grow to be about the same size. When fed a high-fat diet, mouse A becomes obese but mouse B does not.

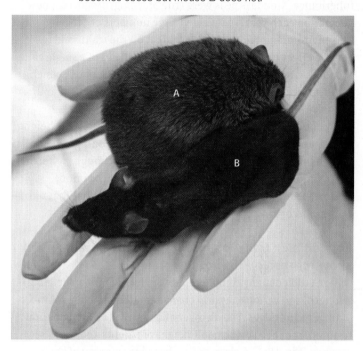

no environment that is "best" for all genotypes. For complex traits, phenotype depends on both genotype *and* environment. For the strains of corn in Fig. 18.5, for example, which strain is "better" depends on the soil nitrogen. With little nitrogen enrichment, strain 1 is better; at intermediate values, both strains yield about the same; and with high nitrogen enrichment, strain 2 is better. In this case, knowing the norms of reaction, and recognizing the magnitude and direction of genotype-by-environment interaction, allows each farmer to use the strains likely to perform best under the available conditions of cultivation.

Genotype-by-environment interaction implies that the interplay between genes and the environment is difficult to predict. An example of genotype-by-environment interaction affecting obesity is shown in **Fig. 18.6.** The animals shown are adults of two inbred lines of mice. When fed a normal diet, both inbred lines grow to be about the same size. When fed a high-fat diet, however, the inbred line A becomes obese, whereas inbred line B does not. Hence, it is not genotype alone that causes obesity in line A because with a normal diet, line A does not become obese. Nor is it a high-fat diet alone that causes obesity because the high-fat diet does not cause obesity in line B. Rather, the obesity in line A results from a genotype-by-environment interaction (which in this case is a genotype-by-diet interaction).

**FIG. 18.5** **Genotype-by-environment interaction for grain yield in corn.** The effect of soil nitrogen on grain yield is minimal in strain 1, but dramatic in strain 2.

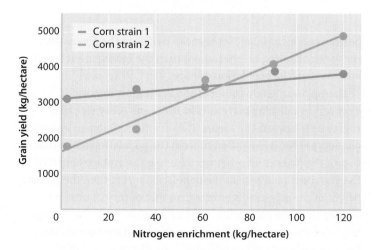

## 18.2 RESEMBLANCE AMONG RELATIVES

Mendel had many advantages over Galton in his studies of inheritance. Mendel's peas were true breeding, produced a new generation each year, and yielded large numbers of progeny. The environment had a negligible effect on the traits, and the genetic effect on each trait was due to the alleles of a single gene. Mendel's crosses yielded simple ratios such as 3:1 or 1:1, which could be interpreted in terms of segregation of dominant and recessive alleles.

By contrast, Galton studied variation in such traits as height in humans. Humans are obviously not true breeding and have few offspring. Galton had one big advantage, though: Whereas most simple Mendelian traits are relatively uncommon, Galton's traits are readily observed in everyday life. What did Galton discover?

**For complex traits, offspring resemble parents but show regression toward the mean.**

Galton studied many complex traits, including human height, strength, and various physical characteristics, including number of fingerprint ridges. The discovery he regarded as fundamental resulted from his data on adult height of parents and their progeny (**Fig. 18.7**). Galton noted that each category of parent (tall parents or short parents) produced a range of progeny forming a distribution with its own mean. He also noted that the mean height of the progeny of tall parents was taller than the average height in the entire population but shorter than that of the parents. Finally, he noted that the mean height of the progeny of short parents was shorter than the population average but taller than that of the parents.

The bar graph in Fig. 18.7a shows the distribution of height among the progeny of the tallest parents, whose average height is 72 inches. The mean height of the offspring is 71 inches, which is greater than the mean height of the whole population of the study (68.25 inches) but less than that of the parents. The bar graph in Fig. 18.7b is the distribution of height of progeny of the shortest parents, who averaged 66 inches. In this case, the mean height of the progeny is 67 inches, which is less than the mean height of the population but greater than that of the parents. Note, however, that *some* of the offspring of tall parents are taller than their parents, and likewise *some* of the offspring of short parents are shorter than their parents. It is only *on average* that the height of the offspring is less extreme than that of the parents.

Galton's analysis led him to the trend shown in **Fig. 18.8**. The graph plots the average height of parents and the average height of their progeny across the whole range of heights observed. Consider two cases. If the mean parent height is 72 inches (above the mean height in the population as a whole), the mean offspring height is less than the mean parent height but more than the population average. By contrast, if the mean parent

**FIG. 18.7**    Galton's data showing distribution of height of offspring of (a) the tallest parents and (b) the shortest parents.

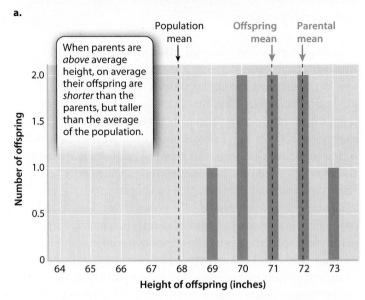

a.

> When parents are *above* average height, on average their offspring are *shorter* than the parents, but taller than the average of the population.

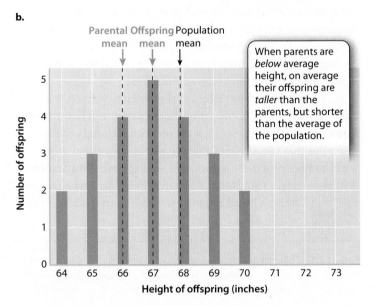

b.

> When parents are *below* average height, on average their offspring are *taller* than the parents, but shorter than the average of the population.

height is 66 inches (below the mean height in the population), the mean offspring height is more than the mean parent height but less than the population average.

Galton regarded this observation as his most important discovery, publishing his results in 1886, and today we call it **regression toward the mean.** The offspring exhibit an average phenotype that is less different from the population mean than that of the parents. In other words, when the average height of the parents is *smaller* than the population mean, then the average height of the offspring is *greater* than that of the parents (but smaller than the population mean). Likewise, when the average height of the parents is *greater* than the population mean, then the average height of the offspring is *smaller* than that of the parents (but greater than the population mean).

**FIG. 18.8 Regression toward the mean.** Galton's data on the average adult height of parents and that of their offspring show that offspring mean height (blue line) falls between the parental mean (dashed red line) and the population mean (dashed black line).

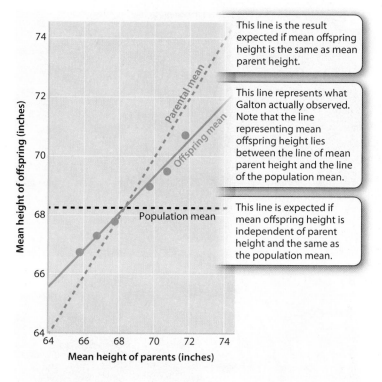

This line is the result expected if mean offspring height is the same as mean parent height.

This line represents what Galton actually observed. Note that the line representing mean offspring height lies between the line of mean parent height and the line of the population mean.

This line is expected if mean offspring height is independent of parent height and the same as the population mean.

Regression toward the mean is observed for two reasons. The first is that during meiotic cell division (Chapter 11), segregation and recombination break up combinations of genes that result in extreme phenotypes, such as very tall or very short, that are present in the parents. The second reason is that the phenotype of the parents results not only from genes but also from the environment. Environmental effects are not inherited, so to the extent the parents' phenotypes result from environmental effects, the same phenotypes will not necessarily appear in the offspring.

→ **Quick Check 1** Does regression toward the mean imply that the human population is getting shorter over time?

### Heritability is the proportion of the total variation due to genetic differences among individuals.

How much of the difference in height among individuals is due to genetic differences, and how much is due to environmental differences? The slope of the line that relates the average phenotype of parents to the average phenotype of their offspring (the blue line in the graph in Fig. 18.8) can answer this question because it provides a measure of the **heritability** of the trait. The heritability of a trait in a population of organisms is the proportion of the total variation in the trait that is due to genetic differences among individuals. For a complex trait, the heritability determines how closely the mean of the progeny resembles that of the parents.

When heritability is 100%, all of the variation among individuals in the population is due to differing genotypes, and in this case the mean phenotype of the offspring is identical to that of the parents. If height had a heritability of 100%, then the blue line in Fig. 18.8 would be the same as the line representing the parental mean (the dashed red line). In contrast, when the heritability of a trait is 0%, all of the variation in the trait in a population is due to differences in environment, and in this case the mean phenotype of offspring would have no resemblance to that of their parents. In Fig. 18.8, if the heritability of height were 0%, then the blue line would be superimposed on the horizontal line depicting the population mean (the dashed black line).

The term "heritability" is often misinterpreted. The problem is that, in non-scientific contexts, the word means "the capability of being inherited or being passed by inheritance." This definition suggests that heritability has something to do with the inheritance of a trait. But "heritability" as used for complex traits means no such thing. It refers only to the *variation* in a trait among individuals, and specifically to the proportion of the variation among individuals in a population due to differences in genotype.

Hence, a heritability of 100% does not imply that the environment cannot affect a trait. As emphasized earlier, the environment is always important, just as oxygen is important to life. What a heritability of 100% means is that *variation* in the environment does not contribute to differences among individuals in a specific population. For example, if genetically different strains of corn are grown in a greenhouse and subject to identical environmental conditions, then differences in height have to be due to genetic differences, and the heritability of the trait would be 100% (**Fig. 18.9a**). Similarly, a heritability of 0% does not imply that genotype cannot affect the trait. A heritability of 0% merely means that differences in genotype do not contribute to the *variation* in

**FIG. 18.9 Heritability, the proportion of the total phenotypic variation that is the result of genetic differences.** Heritability can vary, even for a particular trait like corn height, so it is defined only for a particular population in a particular set of environmental conditions.

**a.** Genetically different strains of corn

All of these corn plants are grown in the same environment, so differences in height are due to genetic differences.

Heritability of height of corn = 100%

**b.** Genetically identical strains of corn

All of these corn plants are genetically identical, so differences in height are due to environmental differences.

Heritability of height of corn = 0%

the trait among individuals in a specific population. If genetically identical strains of corn are grown in different environments, then differences in height must be due to the environment, and heritability would be 0% (**Fig. 18.9b**).

The magnitude of heritability therefore is not an intrinsic property of a trait. For corn height, as we have seen, the heritability in one case was 100% and in the second 0%. Heritability applies only to the trait in a particular population across the range of environments that exist at a specific time. Similarly, the heritability depicted by the slope of the blue line in Fig. 18.8 applies only to the population studied (205 pairs of British parents and their 930 adult offspring, in the late nineteenth century), and may be larger or smaller in different populations at different times. In particular, the magnitude of the heritability cannot specify how much of the difference in average phenotype *between* two populations is due to genotype and how much due to environment.

If the heritability of a trait can change depending on the population and the conditions being studied, why is it a useful measure? Heritability is important in evolution, particularly in studies of artificial selection, a type of selective breeding in which only certain chosen individuals are allowed to reproduce (Chapter 21). Practiced over many generations, artificial selection can result in considerable changes in morphology or behavior or almost any trait that is selected. The large differences among breeds of pigeons and other domesticated animals prompted Charles Darwin to point to artificial selection as an example of what natural selection could achieve. Heritability is important because this quantity determines how rapidly a population can be changed by artificial selection. A trait with a high heritability responds rapidly to selection, whereas a trait with a low heritability responds slowly or not at all.

→  **Quick Check 2** Many people are surprised to learn that, while each individual's fingerprints are unique, the total number of fingerprint ridges is highly heritable, about 90% heritability in many populations. What does high heritability of this trait mean?

## 18.3 TWIN STUDIES

Galton pioneered studies of twins as a way to separate the effects of genotype and environment in phenotype differences among individuals. Depending on whether they arise from one or two egg cells, twins can be **identical (monozygotic twins)** or **fraternal (dizygotic twins).** Identical twins arise from a single fertilized egg (the zygote), which, after several rounds of cell division, separates into two distinct but genetically identical embryos. Strikingly similar in overall appearance (**Fig. 18.10a**), identical twins have stimulated the imagination since antiquity, inspiring stories by the Roman playwright Plautus, William Shakespeare (himself the father of twins), Alexander Dumas, Mark Twain, and many others.

**FIG. 18.10  Identical twins.** (a) Identical (monozygotic) twins arise from a single fertilized egg and are genetically identical. (b) Fraternal (dizygotic) twins arise from two different fertilized eggs and are no more closely related than other pairs of siblings.

In contrast, fraternal twins result when two separate eggs, produced by a double ovulation, are fertilized by two different sperm. Whereas identical twins are genetically identical, fraternal twins are only as closely related as any other pair of siblings (**Fig. 18.10b**).

### Identical twins share the same genotype.

The utility of twins in the study of complex traits derives from the genetic identity of identical twins. Whereas differences between fraternal twins with regard to any trait may arise because of genetic or environmental factors (or genotype-by-environment interactions), differences between identical twins must be due only to environmental factors because the twins are genetically identical. Consequently, if variation in a trait has an important genetic component, then identical twins will be more similar to each other than fraternal twins are similar to each other. But if the genetic contribution to variation is negligible, then identical twins will not be more similar to each other than fraternal twins are to each other.

One criticism that arises in twin studies—and it is a serious one—is that the environment of identical twins is often more similar than that of fraternal twins. Within the uterus, identical twins frequently share the same embryonic membranes and placenta, whereas fraternal twins rarely do. After birth, the striking resemblance in facial appearance, body shape, and other features between identical twins often leads their parents, teachers, and friends to treat them more similarly than if they were fraternal twins. Resemblances between identical twins due to similarities in their environments will mistakenly be attributed to their genetic identity. This difficulty can be minimized in part by studying identical twins who were separated from each other shortly after birth and reared in different environments, but identical twins reared apart are rare.

Ongoing studies at the University of Minnesota begun about 30 years ago united dozens of twins reared apart and found some astonishing coincidences—separated twins wearing the same color of clothes, sporting the same number of rings on the same fingers, giving their dogs the same name, sending their friends the same birthday cards. But some skepticism of such coincidences may be in order. If you sit down with anybody— even a complete stranger—of your age, sex, ethnic group, and socioeconomic background and compare your life histories, likes, dislikes, hobbies, sports, and personality traits, the two of you are likely to discover many quirky similarities. This observation does not invalidate twin studies but suggests that they should be interpreted with caution.

### Twin studies help separate the effects of genes and environment in differences among individuals.

For complex traits, such as depression or diabetes, the extent to which twins are alike is measured according to the **concordance** of the trait, which is defined as the percentage of cases in which both members of a pair of twins show the trait when it is known that at least one member shows it. The relative importance of genetic and environmental factors in causing differences in phenotype can be estimated by comparing the concordance of identical and fraternal twins. A significantly higher concordance rate for a given trait for identical twins compared to that for fraternal twins suggests that the trait has a strong genetic component. Similar concordance rates for identical and fraternal twins suggest that the genetic component is less important.

In the study of concordance, twin pairs in which neither member is affected contribute no information, and these twin pairs must be removed from the analysis. The concordance is based solely on twin pairs in which one or both members express the trait. To take a concrete example, let us symbolize each member of a twin pair as a closed square (■) if the member shows the trait and as an open square (□) if the member does not show the trait. Then, in any sample of identical twins, there are three possibilities, illustrated with data for adult-onset diabetes:

■ ■ 36 identical twin pairs, both showing adult-onset diabetes

■ □ 40 identical twin pairs, only one showing adult-onset diabetes

□ □ 336 identical twin pairs, neither showing adult-onset diabetes

In this example, the 336 twin pairs who do not show the trait provide no information. The concordance is based only on the first two types of twin pairs, among which in 36 pairs both members show the trait and in 40 pairs only one shows the trait. In this case the concordance among identical twins is $36/(36 + 40) = 47\%$. Among fraternal twins, by contrast, the concordance for adult-onset diabetes is only 10%. The difference in concordance between identical and fraternal twins (47% for identical twins versus 10% for fraternal twins) implies that differences in the risk of adult-onset diabetes have an important genetic component. Furthermore, the observation that the concordance for identical twins is much less than 100% implies that environmental factors (which we now know to include diet and exercise) are also important in the risk of adult-onset diabetes.

**Table 18.1** shows identical and fraternal twin concordance rates for a number of other disorders. Just as with diabetes, the marked difference in concordance rates between identical and fraternal twins for high blood pressure, asthma, rheumatoid arthritis, and epilepsy suggests that all these disorders have an important genetic component. In addition, the fact that the concordance is not 100% even for identical twins implies that both genes and environment—"nature" *and* "nurture"—play important roles in the differences in risk among people.

**TABLE 18.1**  Concordance Rates for Identical and Fraternal Twins for Several Diseases

| DISORDER | CONCORDANCE IN IDENTICAL TWINS (%) | CONCORDANCE IN FRATERNAL TWINS (%) |
|---|---|---|
| High blood pressure | 25 | 7 |
| Asthma | 47 | 24 |
| Rheumatoid arthritis | 34 | 7 |
| Epilepsy | 37 | 10 |
| Cancer at any site | 16 | 13 |
| Acute infection leading to death | 8 | 9 |

FIG. 18.11

# What is the relative importance of genes and the environment for complex traits?

**BACKGROUND** Twin studies remain important for assessing the relative importance of "nature" (genotype) and "nurture" (environment) in determining variation among individuals for complex traits. The idea of using twins to distinguish nature from nurture is usually attributed to Francis Galton because of an article he wrote about twins in 1875. In fact, the modern twin study does not trace to Galton but to Curtis Merriman in the United States and Hermann Siemens in Germany, who independently hit upon the idea in 1924. In Galton's time, it was not even known that there are genetically two kinds of twins.

**EXPERIMENT** The rationale of a twin study is to compare identical twins with same-sex fraternal twins. In principle, identical twins differ only because of environment, whereas fraternal twins differ because of genotype as well as environment. The extent to which fraternal twins differ more from each other than identical twins differ from each other measures the effects of genotype. Some twin studies focus on twins reared apart in order to compensate for shared environmental influences that may be stronger for identical twins than for fraternal twins.

**RESULTS** The bar graph shows the results of typical twin studies for various traits. The concordance between twins is the fraction of twin pairs in which both twins show the trait among all those pairs in which at least one twin shows the trait. Roughly speaking, the difference in the concordance between identical twins and fraternal twins is a measure of the relative importance of genotype. In the data shown here, for example, autism and clinical depression both show strong genetic influences, whereas female alcoholism shows almost no genetic influence.

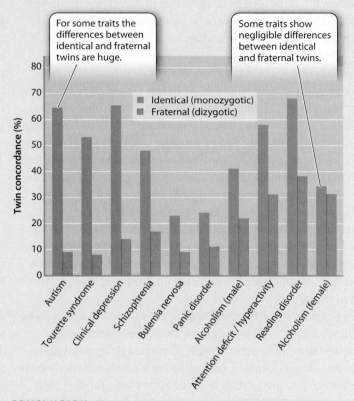

For some traits the differences between identical and fraternal twins are huge.

Some traits show negligible differences between identical and fraternal twins.

**CONCLUSION** The examples shown here indicate very large differences in the importance of genotype versus environment among complex traits. These results are typical of most complex traits.

**FOLLOW-UP WORK** Twin studies must be interpreted in light of other studies comparing complex traits among individuals with various degrees of genetic relatedness. On the whole, twin studies of complex traits have yielded results that are consistent with other available evidence.

**SOURCES** Rende, R. D., R. Plomin, S. G. Vandenberg. 1990. "Who Discovered the Twin Method?" *Behavior Genetics* 20:277–285; McGue, M., and T. J. Bouchard, Jr. 1998. "Genetic and Environmental Influences on Human Behavioral Differences." *Annual Review of Neuroscience* 21:1–24.

Contrast these results with the data for cancer and death from acute infection in Table 18.1. Note that in these cases the concordance rates for identical twins are very similar to those for fraternal twins—there is no significant genetic component to variation in risk, at least insofar as can be determined from twin studies. For these conditions, the environment plays an important role. (There are some inherited cancer syndromes, but these are so rare relative to the overall incidence of cancer that they have a negligible effect on twin concordance.)

**Fig. 18.11** shows the identical and fraternal concordance rates for a number of common behavioral disorders. For some, such as schizophrenia, depression, and autism, the difference between identical and fraternal twins is very great, again implying an important role for genetic factors. For other traits, such as alcoholism in females, the difference is negligible, suggesting an important role for environmental and social factors. This example also illustrates that complex traits can have quite different risk factors and outcomes in different sexes.

## 18.4 COMPLEX TRAITS IN HEALTH AND DISEASE

Much evidence beyond twin studies implies an important role for particular alleles as risk factors for diabetes, high blood pressure, asthma, rheumatoid arthritis, epilepsy, schizophrenia, clinical

**FIG. 18.12** Incidence of the most common birth anomalies.

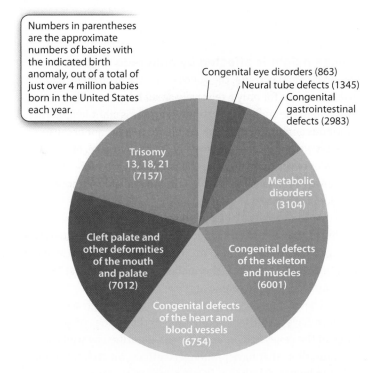

Numbers in parentheses are the approximate numbers of babies with the indicated birth anomaly, out of a total of just over 4 million babies born in the United States each year.

Congenital eye disorders (863)
Neural tube defects (1345)
Congenital gastrointestinal defects (2983)
Trisomy 13, 18, 21 (7157)
Metabolic disorders (3104)
Cleft palate and other deformities of the mouth and palate (7012)
Congenital defects of the skeleton and muscles (6001)
Congenital defects of the heart and blood vessels (6754)

depression, autism, and many other conditions. These are all complex traits, affected by genotype, environment, and genotype-by-environment interactions. Even the most common birth defects are affected by multiple genetic risk factors. The most common birth anomalies and the numbers of affected babies born in the United States each year—collectively more than 35,000—are indicated in **Fig. 18.12.** About 20% of these anomalies are due to an extra chromosome, and trisomy 21 (Down syndrome) is by far the most common. Another roughly 10% are due to defects in metabolism. Many of these—among them diseases such as cystic fibrosis, sickle-cell anemia, and phenylketonuria (the inability to break down the amino acid phenylalanine)—are rare simple Mendelian disorders. But the majority of birth anomalies are affected by both genetic and environmental risk factors.

**Most common diseases and birth defects are affected by many genes that each have relatively small effects.**
Because the most common birth anomalies as well as childhood and adult disorders are complex traits, biologists are keenly interested in identifying the genes that contribute to differences in risk, understanding what these genes do, and translating this knowledge into prevention or treatment. They have therefore begun to apply modern molecular methods—genome sequencing (Chapter 13), genome annotation (Chapter 14), and genotyping (Chapter 15)—to complex traits.

The identification of genes affecting complex traits is not only a major goal of much current research in human genetics, but also in domesticated animals and plants. Gene identification is

also important in model organisms used in research, especially the laboratory mouse. Much is already known about these model organisms and they are easily manipulated, making it possible to discover the molecular mechanisms by which genes affecting complex traits exert their effects.

The study of complex traits has already reached a stage at which patterns are beginning to emerge. Many of these patterns are exemplified by the chromosome map in **Fig. 18.13,** which shows the location of genes in the human genome that affect cholesterol levels. The first observation is that many genes contribute to cholesterol levels in humans—specifically,

**FIG. 18.13** Genes affecting the level of serum cholesterol in the human genome.

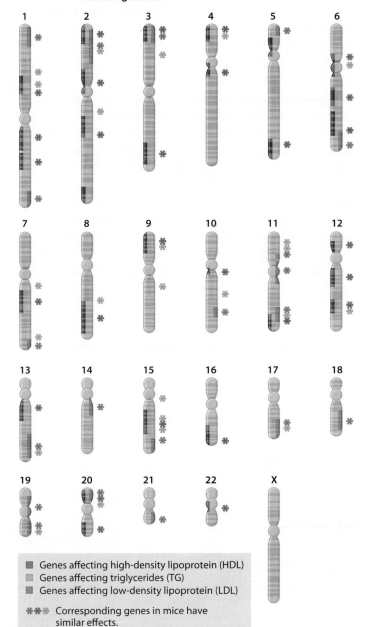

■ Genes affecting high-density lipoprotein (HDL)
■ Genes affecting triglycerides (TG)
■ Genes affecting low-density lipoprotein (LDL)

✻✻✻ Corresponding genes in mice have similar effects.

more than 50 genes contribute to the level of HDL (high-density lipoproteins, or "good cholesterol"), LDL (low-density lipoproteins, or "bad cholesterol"), or triglycerides. A second observation is that many of these genes affect two or even all three of the types of molecule. (A single gene that has multiple effects is said to show **pleiotropy.**) Third, many of the genes show epistasis, that is, they interact with one another (Chapter 16). Fourth, many of the genes occur in clusters, being physically close together in the same chromosome. Clustering of genes reflects the fact that many genes arose through the process of duplication of a single gene followed by divergence over time, generating a family of genes near one another with related functions (Chapter 14). Finally, the functions of all of the genes that contribute to a complex trait are not known. Some of the genes in Fig. 18.13 affect serum cholesterol through known metabolic pathways, but many others work through unknown pathways. Many of the human genes shown have counterparts in the mouse genome that have similar effects on serum cholesterol, so these genes are open to direct experimental investigation.

A principle that is not evident in Fig. 18.13 is that the effects of the genes on cholesterol levels are very unequal. Some have relatively large effects, while others have small ones. The distribution of the magnitude of gene effects for complex traits resembles that shown in **Fig. 18.14.** The axes are labeled only in relative terms, because the actual numbers and effect sizes differ from one trait to the next. However, the main point is that for the majority of genes that contribute to a complex trait, the magnitude of their individual effects is typically quite small. The magnitude of genetic effects also often differs between the sexes, which helps explain why complex traits so often differ in prevalence or severity between males and females.

---

**FIG. 18.14** Relationship between number of genes and their effect on complex traits.

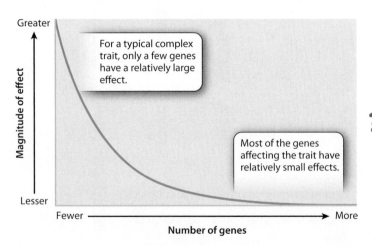

→ **Quick Check 3** When genes for complex traits have effects that are distributed as shown in Fig. 18.14, which are easier to identify: those that are numerous with small effects, or those that are few with large effects?

## Human height is affected by hundreds of genes.

For most complex traits, the genes that have been identified to date account for only a relatively small fraction of the total variation in the trait. One extreme example is adult height. An enormous amount of data is available for height, not because height has been studied extensively for its own sake, but because in studies of common diseases the height of each individual is recorded, and hence data on height are available without additional effort or expense.

In one analysis, results from 46 separate studies were combined. These studies included about 185,000 individuals who were genotyped for common nucleotide variants (SNPs) at about 3 million nucleotide sites. The analysis identified at least 180 genes affecting height. Some of these genes are known to affect skeletal development, growth hormones, or other growth factors, but most have no obvious connection to the biology of growth. Some of the genes had previously been identified by studies of rare mutations that have pronounced effects on skeletal growth, either in human families or in laboratory mice. An unexpected finding was that a few genes affecting height are known to be associated with bone mineral density, obesity, and rheumatoid arthritis. These and many other of the genes identified might affect height indirectly.

The 180 genes for height identified among the 185,000 individuals account for only about 12% of the genetic variation in height. The authors calculate that studies of 500,000 individuals would reveal approximately 520 additional genes with effects as large as or larger than those of the 180 genes identified. They also estimate that these 700 genes in total would still account for only 20% of the genetic variation in height, suggesting that many more genes of still smaller effect contribute to variation in height. And this analysis does not address at all the effects of the environment on human height, nor genotype-by-environment interactions, which likely play an important role as well. Complex traits therefore require sophisticated studies to tease apart all the genetic and environmental factors that play a role.

**? CASE 3 You, From A to T: Your Personal Genome**
**Can personalized medicine lead to effective treatments of common diseases?**
The multiple genetic and environmental factors affecting complex traits imply that different people can have the same disease for different reasons. For instance, one person might develop breast cancer because of a mutation in the *BRCA1* or *BRCA2* genes, while

another might develop breast cancer because of other genetic risk factors or even environmental ones. Similarly, one person might develop emphysema because of a mutation in the gene that encodes the enzyme alpha-1 antitrypsin ($\alpha$1AT) (Chapter 15), and another might have emphysema as a result of cigarette smoking. Other examples where the same disease can be the result of different underlying genetic or environmental factors include elevated cholesterol levels, high blood pressure, and depression. Because the underlying genetic basis for the same disease may be different in different patients, some patients respond well to certain drugs and others do not.

The traditional strategy for treating diseases is to use the same medicine for the same disease. However, because we now know that the same disease may have different causes, another possibility has emerged: to identify each patient's genotype for each of the relevant genes, and then to match the treatment to the genetic risk factors in each patient. The approach is known as **personalized medicine.** Personalized medicine matches the treatment to the patient, not the disease.

Personalized medicine not only aims to identify ahead of time medicine that will work effectively, but also to avoid medicines that may lead to harmful side effects, even death. Advertisements and enclosures with prescription drugs enumerate long lists of side effects that you may get if you take a particular medicine. At present, it is difficult to know in advance who will get one or more of these side effects and who won't—that is, the side effects of taking medicine are themselves complex traits and the result of many underlying genetic and environmental factors. If we could identify ahead of time those patients who will respond negatively to a particular medicine and those who won't, we could minimize potentially harmful effects of medicines on certain individuals.

Someday, it may be possible to determine each patient's genome sequence quickly, reliably, and cheaply. At present, personalized medicine is restricted to studies of a few key genes known to have important effects on treatment outcomes. In the treatment of asthma, for example, some of the differences in response to albuterol inhalation have been traced to genetic variation in the gene *ADRB2*, encoding the $\beta$-(beta-)2-adrenergic receptor. Similarly, certain drugs used in the treatment of Alzheimer's disease are less effective in women with a particular *apolipoprotein E* (*APOE*) genotype than in other classes of patients. Also, more than half of the cases of muscle weakness occurring as an adverse effect of drug treatment used to control high cholesterol can be traced to genetic variation in the gene *SLCO1B1*, which encodes a liver transport protein.

Other factors in addition to genes are involved in these disorders, but genes play an important role. Knowledge of a patient's genotype can help guide treatment. These are only a few of many examples, but they demonstrate the substantial potential benefits of personalized medicine.

## Core Concepts Summary

**18.1 COMPLEX TRAITS ARE THOSE INFLUENCED BOTH BY THE ACTION OF MANY GENES AND BY ENVIRONMENTAL FACTORS.**

Complex traits that are measured on a continuous scale, like human height, are called quantitative traits. page 18-2

It is usually difficult to assess the relative roles of genes and the environment ("nature" vs. "nurture") in the production of a given trait in an individual, but it is reasonable to consider the relative roles of genetic and environmental variation in accounting for differences among individuals for a given trait. page 18-2

The relative importance of genes and environment in causing differences in phenotype among individuals differs among traits. For some traits (like height), genetic differences are the more important source of variation, whereas for others (such as cancer), environmental differences can be the more important. page 18-4

Genetic and environmental factors can interact in unpredictable ways, resulting in genotype-by-environment interactions. page 18-5

**18.2 GENETIC EFFECTS ON COMPLEX TRAITS ARE REFLECTED IN RESEMBLANCE BETWEEN RELATIVES.**

In an analysis of heights of parents and offspring, Galton observed regression toward the mean, in which the offspring exhibit an average phenotype that is less different from the population mean than that of the parents. page 18-6

"Heritability" refers to the proportion of the total variation in a trait that can be attributed to genetic differences among individuals. page 18-7

The heritability of the same trait can differ among populations because of differences in genotype or environment. page 18-8

**18.3 TWIN STUDIES HELP SEPARATE THE EFFECTS OF GENO-TYPE AND ENVIRONMENT ON VARIATION IN A TRAIT.**

Monozygotic, or identical, twins result from the fertilization of a single egg and are genetically identical. page 18-8

Dizygotic, or fraternal, twins result from the fertilization of two eggs and are genetically related to each other in the same way that other siblings are related to each other. page 18-8

Concordance is the percentage of cases in which both members of a pair of twins show the trait when it is known that at least one member shows it. page 18-9

Comparisons of concordance rates of identical twins and concordance rates of fraternal twins can help to determine to what extent variation in a particular trait has a genetic component. page 18-9

**18.4 MANY COMMON DISEASES AND BIRTH DEFECTS ARE AFFECTED BY MULTIPLE GENETIC AND ENVIRONMENTAL RISK FACTORS.**

Complex traits are often influenced by many genes with multiple, interacting, and unequal effects. page 18-12

Hundreds of genes affect human height. page 18-12

Personalized medicine tailors treatment to an individual's genetic makeup. page 18-12

## Self-Assessment

1. Give three examples of a complex trait.

2. Explain why complex traits are often called quantitative traits.

3. Name several factors that influence variation in complex traits.

4. Explain why it does not make sense to try to separate the effects of genes ("nature") and the environment ("nurture") in a single individual.

5. Explain how you would go about determining the relative importance of genes and the environment for a particular trait.

6. Graph a trait, like human height, with height on the x-axis and number of individuals on the y-axis, and describe the shape of the resulting graph.

7. Explain why the effect of a genotype on a phenotype cannot always be determined without knowing what the environment is, and why the effect of a particular environment on a phenotype cannot always be determined without knowing the underlying genotype.

8. Define what is meant by "regression toward the mean."

9. Define the heritability of a trait and explain why it depends on the population being studied.

10. Define concordance and explain how twin studies can be used to investigate the importance of genetic and environmental factors in the expression of a trait.

11. For a typical complex trait, describe the relationship between the number of genes affecting the trait and the magnitude of their effects on the trait.

12. Explain what personalized medicine is and how it relates to complex traits such as human diseases.

Do you understand the chapter's Core Concepts? Log in to check your answers to the Self-Assessment questions, then practice what you've learned and reinforce this chapter's concepts by working through the problems and multimedia tutorials provided there.

　**www.biologyhowlifeworks.com**

## CHAPTER 19

# GENETIC AND EPIGENETIC REGULATION

## Core Concepts

In the discussion of complex traits in Chapter 18, we emphasized the principle that complex traits are not determined by single genes with simple Mendelian inheritance. The traits we usually encounter in ourselves or in others, such as height and weight, diabetes and high blood pressure, are influenced by multiple genes that interact with one another and with environmental factors such as diet and exercise. The number of genes influencing complex traits can be large—for example, hundreds of genes contribute to adult height. The activities of these genes must be coordinated in time (for instance, at a particular point in development) and place (for instance, in a particular type of tissue). For example, growth to normal adult height requires that the production of growth hormone be coordinated in time and place with the production of growth-hormone receptor.

In Chapters 3 and 4, we outlined the basic steps of information flow in a cell, focusing on transcription and translation. In these processes, DNA is a template for the production of messenger RNA (mRNA), which directs the synthesis of a protein. For these processes to work in a living organism to produce the traits that we see, they must be coordinated so that genes are only **expressed,** or turned on, in the right place and time, and in the right amount. In a multicellular organism, for example, certain genes are expressed in some cells but not in others. Muscle actin and myosin are turned on in muscle cells but not in liver or kidney cells. And even for single-celled organisms like bacteria, certain genes are expressed only in response to environmental signals, such as the availability of nutrients, as we discuss in section 19.3.

**Gene regulation** encompasses the ways in which cells control gene expression. It can be thought of as the *where? when?* and *how much?* of gene expression. Where (in which cells) are genes turned on? When (during development or in response to changes in the environment) are they turned on? How much gene product is made? This chapter provides an overview of the most important ways in which gene expression is regulated.

One of the important principles we discuss is that gene regulation can occur at almost any step in the path from DNA to mRNA to protein—at the level of the chromosome itself, by controlling transcription or translation, and, perhaps surprisingly, even after the protein product is made. Each of these steps or levels of gene expression may be subject to regulation. In other words, each successive event that takes place in the expression of a gene is a potential control point for gene expression.

We begin by discussing gene regulation in eukaryotes, and then turn to regulation in prokaryotes and viruses. All life, from the simplest to the most complex, requires gene regulation. Gene regulation relies on similar processes in all organisms because all life shares common ancestry. Nevertheless, certain features of eukaryotes—the packaging of DNA into chromosomes, mRNA processing, and the separation in space of transcription and translation—provide additional levels of gene regulation in eukaryotes that are not possible in prokaryotes.

## 19.1 CHROMATIN TO MESSENGER RNA IN EUKARYOTES

Gene regulation in multicellular eukaryotes leads to cell specialization: Different types of cell express different genes. The human body contains about 200 major cell types, and although for the most part they share the same genome, they look and function differently from one another because each type of cell expresses different sets of genes. For example, the insulin needed to regulate sugar levels in the blood is produced only by small patches of cells in the pancreas. Every cell in the body contains the genes that would lead to the production of insulin, but only in these patches of pancreatic cells are they expressed. In this section, we take a look at gene regulation as it occurs in eukaryotic cells, focusing on regulation at the level of DNA, chromatin, and mRNA.

### Gene expression can be influenced by chemical modification of DNA or histones.

**Fig. 19.1** shows the major places where gene regulation in eukaryotes can take place. The first level of control is at the chromosome, even before transcription takes place. The manner in which DNA is packaged in the nucleus in eukaryotes provides an important opportunity for regulating gene expression.

In many eukaryotic organisms, gene expression is affected by chemical modification of certain bases in the DNA (**Fig. 19.2**), the most common of which is the addition of a methyl group to the base cytosine. Methylation often occurs in cytosine bases that are adjacent to guanosine bases on a DNA strand. Such pairs of nucleotides are abbreviated "CpG" (the "p" represents the phosphate in the backbone of the DNA strand between the two nucleotides). In mammalian protein-coding genes, CpG sites are often clustered in small regions located in or near the promoter of the gene, the region where RNA polymerase and associated proteins bind to the DNA to initiate transcription. Such a cluster of CpG sites is known as a **CpG island.**

Some CpG sites in the genome are methylated and some are not. Methylated CpG sites are rarely seen in CpG islands near active genes (Fig. 19.2a), whereas transcriptional repression of a gene is often accompanied by heavy methylation of a nearby CpG island (Fig. 19.2b). The methylation state of a CpG island can change over time or in response to environmental cues, providing a way to turn genes on or off. Cells sometimes heavily methylate CpG islands of transposable elements or viral DNA sequences that are integrated into the genome, thus preventing the expression of genes in viruses and transposable elements (Chapter 14). In cancer cells, CpG island methylation often takes place, repressing genes that could restrict the cells' growth.

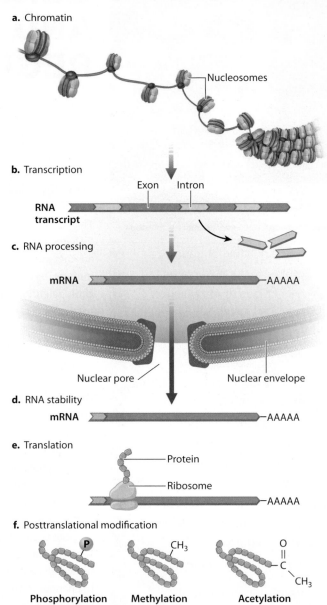

**FIG. 19.1**　Levels of gene expression regulation.

**a.** Chromatin

Nucleosomes

**b.** Transcription

Exon　Intron

RNA transcript

**c.** RNA processing

mRNA ▸ AAAAA

Nuclear pore　Nuclear envelope

**d.** RNA stability

mRNA ▸ AAAAA

**e.** Translation

Protein

Ribosome

AAAAA

**f.** Posttranslational modification

P

Phosphorylation

CH₃

Methylation

O
‖
C
CH₃

Acetylation

DNA in eukaryotes is packaged as **chromatin,** a complex of DNA, RNA, and proteins that gives chromosomes their structure. Chromatin includes a thread of nucleosome particles in which about 150 base pairs of DNA are wrapped around each octamer of histone proteins. (Greater detail in the structure of chromatin and nucleosomes is depicted in Fig. 3.13.) When chromatin is in its coiled state, the DNA is not accessible to the proteins that carry out transcription. The chromatin must unravel to allow space for transcriptional enzymes and proteins to work. This is accomplished through **chromatin remodeling,** in which the nucleosomes are repositioned to expose different stretches of

**FIG. 19.2** Methylation states of CpG islands in or near the promoter of a gene.

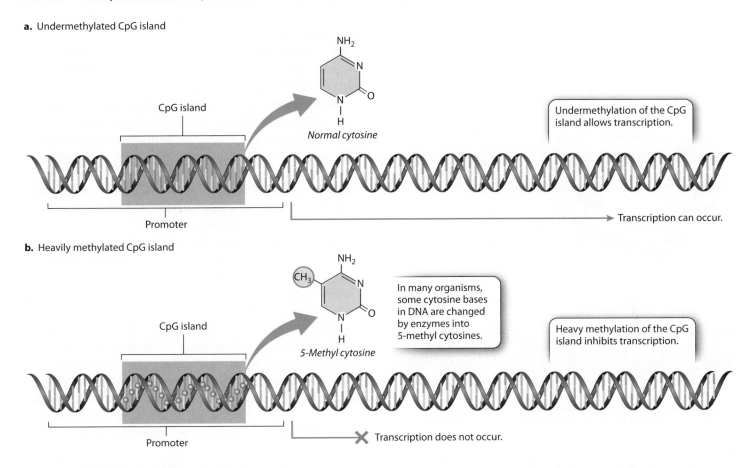

**a.** Undermethylated CpG island

Normal cytosine

Undermethylation of the CpG island allows transcription.

CpG island

Promoter

Transcription can occur.

**b.** Heavily methylated CpG island

5-Methyl cytosine

In many organisms, some cytosine bases in DNA are changed by enzymes into 5-methyl cytosines.

Heavy methylation of the CpG island inhibits transcription.

CpG island

Promoter

Transcription does not occur.

DNA to the nuclear environment. Chromatin remodeling is another level at which gene expression can be controlled.

One way in which chromatin is remodeled is by chemical modification of the histones around which DNA is wound (**Fig. 19.3**). Modification usually occurs on **histone tails**, strings of amino acids that protrude from the histone proteins in the nucleosome (Fig. 19.3a). Individual amino acids in the tails can be modified by the addition (or later removal) of different chemical groups, including methyl groups (—$CH_3$) and acetyl groups (—$COCH_3$). Most often, methylation or acetylation occurs on the lysine residues of the histone tails.

The pattern of modifications of the histone tails is thought to constitute a **histone code** that affects chromatin structure and gene transcription (Fig. 19.3b). For example, methylation of lysine with a single methyl group and lysine acetylation are often associated with the activation of transcription (+ in Fig. 19.3b), whereas methylation of lysine with three methyl groups is often associated with repression of transcription (− in Fig. 19.3b). Modification of histones takes place at key times in development to ensure that the proper genes are turned on or off, as well as in response to environmental cues.

Together, these modifications of bases, changes to histones, and alterations in chromatin structure are often termed **epigenetic,** from the Greek *epi-* ("over and above," "in addition to") and *genetic* ("inherited"). That is, epigenetic mechanisms of gene regulation typically involve changes not to the DNA sequence itself but to the manner in which DNA is packaged. Epigenetic modifications can in some cases affect gene expression. They can be inherited through cell divisions, just as genes are, but are often reversible and responsive to changes in the environment.

### Gene expression can be regulated at the level of an entire chromosome.

A striking example of an epigenetic form of gene regulation is the manner in which mammals equalize the expression of X-linked genes in XX females and XY males. For most genes, there is a direct relation between the number of copies of the gene (the gene dosage) and the level of expression of the gene. An increase in gene dosage increases the level of expression because each copy of the gene is regulated independently of

**FIG. 19.3    Histone modifications.** (a) Typical modifications of the amino acid lysine observed in the histone tails of nucleosomes include addition of a methyl group (Me), addition of three methyl groups (Me₃), and acetylation (Ac). (b) Modification of specific lysine amino acids is associated with increases (+) or decreases (−) in transcription, and constitutes a histone code.

**a.**

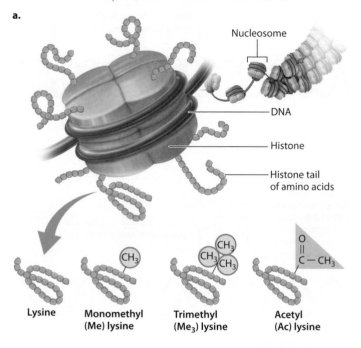

Nucleosome

DNA

Histone

Histone tail of amino acids

| Lysine | Monomethyl (Me) lysine | Trimethyl (Me₃) lysine | Acetyl (Ac) lysine |

**b.**

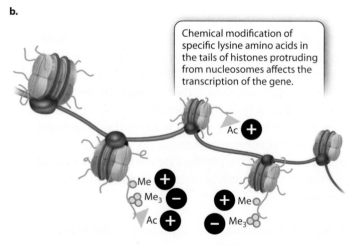

Chemical modification of specific lysine amino acids in the tails of histones protruding from nucleosomes affects the transcription of the gene.

other copies. For example, as we saw in Chapter 15, the presence of an extra copy of most human chromosomes results in spontaneous abortion because of the increase in the expression of the genes in that chromosome.

XX females and XY males have different numbers of X chromosomes. For genes contained in the X chromosome, the dosage of genes is twice as great in females as it is in males. However, the level of expression of X-linked genes is about the same in both sexes. These observations imply that the regulation of X-chromosomal genes is different in females and in males. The differential regulation is called **dosage compensation.**

Different species have evolved different mechanisms of dosage compensation. In *Drosophila* fruit flies, males double the transcription of the single X chromosome to achieve equal expression compared to the two X chromosomes in females. In *Caenorhabditis* nematode worms, females decrease the transcription of both X chromosomes in females to one-half the level of the single X chromosome in males. In mammals, including humans, dosage compensation occurs through the inactivation of one X chromosome in each cell in females. This process, now known as **X-inactivation,** was first proposed by Mary F. Lyon in the early 1960s.

Soon after a fertilized egg with two X chromosomes implants in the mother's uterine wall, one X chromosome is selected at random and inactivated (**Fig. 19.4**). In Fig. 19.4a, the inactive X chromosome is shown in gray. The inactive state persists through cell division, so in each cell lineage, the same X chromosome that was originally inactivated remains inactive. The result is that a normal female is a mosaic, or patchwork, of tissue. In some patches, the genes on the maternal X are expressed (and the paternal X is inactivated), whereas in other patches, the genes on the paternal X are expressed (and the maternal X is inactivated). The term "inactive X" is a slight exaggeration since a substantial number of genes are still transcribed, although usually at a low level.

As one argument for her X-inactivation hypothesis, Lyon called attention to calico cats, which are nearly always female (Fig. 19.4b). In calico cats, the orange or black fur colors are due to different alleles of a single gene in the X chromosome. In a heterozygous female, X-inactivation predicts discrete patches of orange and black, and this is exactly what is observed. (The white patches on a calico cat are due to an autosomal gene.)

How does X-inactivation work? Some of the details are still unknown, but the main features of the process are shown in **Fig. 19.5.** A key player is a small region in the X chromosome called the X-chromosome inactivation center (*XIC*), which contains a gene called *Xist* (X-inactivation specific transcript). The *Xist* gene is normally transcribed at a very low level, and the RNA is unstable, but in an X chromosome about to become inactive, *Xist* transcription markedly increases. The transcript undergoes RNA splicing, but it does not encode a protein. *Xist* RNA is therefore an example of a noncoding RNA, introduced in Chapter 3. Instead of being translated, the processed *Xist* RNA coats the *XIC* region, and as it accumulates, the coating spreads outward from the *XIC* until the entire chromosome is coated with *Xist* RNA. The presence of *Xist* RNA along the chromosome recruits factors that promote DNA methylation, histone modification, and other changes associated with transcriptional silencing.

**FIG. 19.4** ***X*-inactivation in female mammals.** *X*-inactivation equalizes the expression of most genes in the *X* chromosome between *XX* females and *XY* males.

**a.**

Maternal *X* chromosome — Paternal *X* chromosome

Fertilized egg

Early divisions

Early divisions

Inactivated *X* chromosome

Random *X*-inactivation in the embryo occurs at about the time of implantation in the uterine wall.

In each cell lineage, the inactivated *X* remains inactivated.

**b.**

The mosaic black and orange colors of a calico cat result from *X*-chromosome inactivation in a heterozygous female.

## Transcription is a key control point in gene expression.

While access to DNA and appropriate histone modifications are necessary for transcription, they are not sufficient. The molecular machinery that actually carries out transcription is also required once the template DNA is made accessible through chromatin remodeling and histone modification. The mechanisms that regulate whether or not transcription occurs are known collectively as **transcriptional regulation** (see Fig. 19.1b).

Transcriptional regulation in eukaryotic cells requires the coordinated action of many proteins that interact with one another and with DNA sequences near the gene. Let's first review the basic process of transcription (Chapter 3). An important group of proteins are the **general transcription factors.** These proteins bind to the gene's promoter, which is the region of a gene that recruits factors necessary to start transcription. The transcription factors are brought there by one of the proteins that binds to a short sequence called the TATA box, which is usually situated 25–30 nucleotides upstream of the nucleotide site where transcription begins. Once bound to the promoter, the transcription factors recruit the components of the **RNA polymerase complex,** which synthesizes the RNA transcript complementary to the template strand of DNA.

Where in the many steps of transcription initiation does regulation occur? The first place where transcription can be

**FIG. 19.5** **The role of *Xist* in *X*-inactivation.** *Xist* is noncoding RNA that is expressed from the inactive *X* chromosome and coats the entire chromosome, leading to inactivation of most of the genes on the chromosome.

The *Xist* gene is transcribed and spliced, and *Xist* noncoding RNA binds with the *X*-chromosome inactivation center (*XIC*).

Transcription of *Xist* continues, and the *X* chromosome becomes coated with *Xist* RNA.

Eventually the entire chromosome becomes coated with *Xist* RNA.

Presence of *Xist* RNA triggers DNA methylation and other changes associated with reduced transcriptional activity.

**FIG. 19.6**    Protein–DNA and protein–protein interactions in the eukaryotic transcription complex.

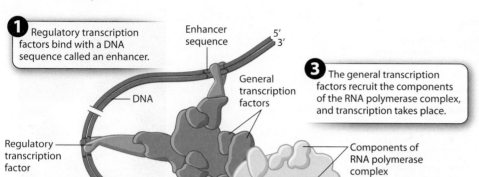

**RNA processing is also important in gene regulation.**

A great deal happens in the nucleus after transcription takes place. The initial transcript, called the **primary transcript,** undergoes several types of modification, collectively called **RNA processing** (Chapter 3), that includes the addition of a nucleotide cap to the 5′ end and a string of tens to hundreds of adenosine nucleotides to the 3′ end to form the poly(A) tail. These modifications are necessary for the RNA molecule to be transported to the cytoplasm and recognized by the translational machinery, and they also help to determine how long the RNA will persist before being degraded. RNA processing is therefore an important point where gene

regulated is in the recruitment of the general transcription factors and components of the RNA polymerase complex (**Fig. 19.6**). Recruitment of these elements is controlled by proteins called **regulatory transcription factors.** Transcription does not occur if the regulatory transcription factors do not recruit the components of the transcription complex to the gene. Each regulatory transcription factor has two binding sites, one of which binds with a particular sequence in the DNA in or near a gene known as an **enhancer** (Fig. 19.6). A second binding site on the regulatory transcription factor recruits one or more general transcription factors to the promoter region. The general transcription factors then recruit the RNA polymerase complex, and transcription can begin (Chapter 3).

Hundreds of different regulatory transcription factors control the transcription of thousands of genes. A typical gene has several different types of enhancer sequence, each with its own regulatory transcription factor or set of transcription factors. Transcription takes place only when all the regulatory transcription factors are present and work together, as shown in Fig. 19.6. Transcription of a gene therefore depends on the presence of a particular combination of enhancers and their regulatory transcription factors, a type of regulation called combinatorial control.

→    **Quick Check 1**  The idea that the expression of some genes is controlled by the products of other genes was originally criticized on the ground that, if *n* genes were to be controlled, then another *n* genes would be needed to control them, and then another *n* genes would be needed to control the controllers, and so on and on. How does combinatorial control help refute this criticism?

regulation can occur (see Fig. 19.1c).

In eukaryotes, the primary transcript of many protein-coding genes is far longer than the messenger RNA ultimately used in protein synthesis. The long primary transcript consists of regions that are retained in the messenger RNA (the **exons**) interspersed with regions that are excised and degraded (the **introns**). The introns are excised during **RNA splicing** (Chapter 3). The exons are joined together in their original linear order to form the processed messenger RNA.

RNA splicing provides an opportunity for regulating gene expression because the same primary transcript can be spliced in different ways to yield different proteins in a process called **alternative splicing.** This process takes place because what the spliceosome—the splicing machinery—recognizes as an exon in some primary transcripts it recognizes as part of an intron in other primary transcripts. The alternative-splice forms may be produced in the same cells or in different types of cell. Alternative splicing accounts in part for the observation that we produce many more proteins than our total number of genes. By some estimates, over 90% of human genes undergo alternative splicing.

**Fig. 19.**7 shows the primary transcript of a gene encoding an insulin receptor found in humans and other mammals. During RNA splicing in liver cells, exon 11 is included in the messenger RNA, and the insulin receptor produced from this messenger RNA has low affinity for insulin. In contrast, in cells of skeletal muscle, the 36 nucleotides of exon 11 are spliced out of the primary transcript along with the flanking introns. The resulting protein is 12 amino acids shorter, and

**FIG. 19.7** **Alternative splicing of a mammalian insulin-receptor transcript.** Alternative splicing generates different processed mRNAs and different proteins from the same primary transcript.

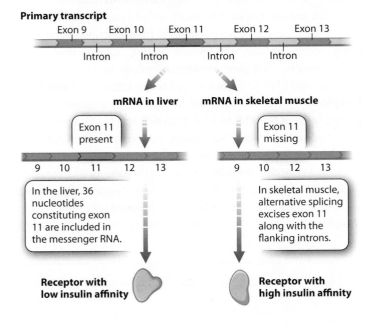

**Primary transcript**

Exon 9    Exon 10    Exon 11    Exon 12    Exon 13

Intron    Intron    Intron    Intron

**mRNA in liver**    **mRNA in skeletal muscle**

Exon 11 present    Exon 11 missing

9  10  11  12  13    9  10  12  13

In the liver, 36 nucleotides constituting exon 11 are included in the messenger RNA.    In skeletal muscle, alternative splicing excises exon 11 along with the flanking introns.

**Receptor with low insulin affinity**    **Receptor with high insulin affinity**

this form of the insulin receptor has high affinity for insulin. The different forms of the protein are important: The higher sensitivity of muscle cells to insulin enables them to absorb enough glucose to fulfill their energy needs.

Some RNA molecules can become a substrate for enzymes that modify particular bases in the RNA, thereby changing its

**FIG. 19.8** **RNA editing.** RNA editing results in chemical modifications to the bases in mRNA, which can lead to changes in the amino acid sequence of the protein.

**a.**

Deamination of adenosine produces inosine.

Adenosine    Inosine

**b.**

Deamination of cytosine produces uracil.

Cytosine    Uracil

sequence and what it codes for. This process is known as **RNA editing** (**Fig. 19.8**). One type of editing enzyme (Fig. 19.8a) removes the amino group (—NH$_2$) from adenosine and converts it to inosine, a base that in translation functions like guanosine. Another enzyme (Fig. 19.8b) removes the amino group from cytosine and converts it to uracil. In the human genome, hundreds if not thousands of transcripts undergo RNA editing. In many cases, not all copies of the transcript are edited, and some copies may be edited more extensively than others. The result is that transcripts from the same gene can produce multiple types of proteins even in a single cell.

Transcripts from the same gene may undergo different editing in different cell types. An example of tissue-specific RNA editing is shown in **Fig. 19.9.** The mRNA fragments show part of the coding sequence for apolipoprotein B. The unedited mRNA in the liver (Fig. 19.9a) is translated into a protein that transports cholesterol in the blood. In contrast, RNA editing of the message occurs in the intestine (Fig. 19.9b). The cytosine nucleotide in codon 2153 is edited to uracil. The edited codon is UAA, which is a stop codon. Translation therefore terminates at this point, releasing a protein only about half as long as the liver form. This shorter form of the protein helps the cells of the intestine absorb lipids from the foods we eat.

**FIG. 19.9** **Tissue-specific RNA editing of the human apolipoprotein B transcript.** Different RNA editing in the (a) liver and (b) intestine result in proteins with different functions.

**a.** Processed apolipoprotein mRNA in liver

5′ – · · · U A U A U G A U A C A A U U U G A U C A G · · · – 3′

Liver (C not edited)

5′ – · · · U A U A U G A U A C A A U U U G A U C A G · · · – 3′

Tyr  Met  Ile  Gln  Phe  Asp  Gln

In the liver, the unedited mRNA encodes a protein of 4563 amino acids that transports cholesterol in the blood.

**b.** Processed apolipoprotein mRNA in intestine

5′ – · · · U A U A U G A U A C A A U U U G A U C A G · · · – 3′

Intestine (C edited to U)

5′ – · · · U A U A U G A U A U A A U U U G A U C A G · · · – 3′

Tyr  Met  Ile    Termination codon

In the intestine, the edited mRNA encodes a protein of 2152 amino acids that absorbs lipids from foods.

## 19.2 MESSENGER RNA TO PHENOTYPE IN EUKARYOTES

In eukaryotes, a processed mRNA must exit the nucleus before the translation step of gene expression can occur. The mRNA migrates to the cytoplasm through one of a few thousand nuclear pores, large protein complexes that traverse the nuclear envelope and regulate the inflow and outflow of macromolecules. Once the mRNA is in the cytoplasm, there are multiple opportunities for gene regulation at the levels of mRNA stability, translation, and protein activity (see Figs. 19.1d–19.1f).

### Small regulatory RNAs inhibit translation or promote RNA degradation.

Regulatory RNA molecules known as **small regulatory RNAs** are among the most exciting recent discoveries in gene regulation. They often work by binding to transcripts and blocking translation. Small regulatory RNAs are of exceptional interest to biologists and drug researchers because their small size allows easy synthesis in the laboratory, and researchers can design their sequences to target transcripts of interest.

One type of small regulatory RNA, **microRNA (miRNA)**, starts out just like the RNA transcribed from protein-coding genes, using the same RNA polymerase for transcription and going through the same processes of capping, splicing, and polyadenylation. But there the similarity ends. In miRNA, the processed RNA folds back upon itself to form one or more **hairpin** structures, or stems-and-loops, stabilized by base pairing in the stem (**Fig. 19.10a**). Enzymes specifically recognize processed and folded miRNAs and cleave the stems from the hairpin, then further cleave the stem into small double-stranded fragments typically about 20–25 nucleotide pairs long.

One strand from each RNA fragment is incorporated into a protein complex known as **RISC (RNA-induced silencing complex)**. The small, single-stranded RNA (the miRNA) targets

**FIG. 19.10 MicroRNA (miRNA) and small interfering RNA (siRNA).** (a) miRNAs are small noncoding RNAs that inhibit translation. (b) siRNAs are small noncoding RNAs that degrade mRNA.

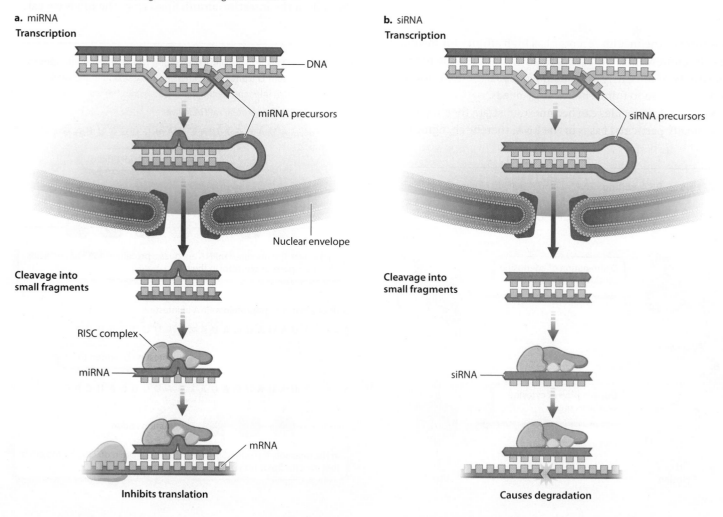

**a.** miRNA

Transcription

DNA

miRNA precursors

Nuclear envelope

Cleavage into small fragments

RISC complex

miRNA

mRNA

**Inhibits translation**

**b.** siRNA

Transcription

siRNA precursors

Cleavage into small fragments

siRNA

**Causes degradation**

the RISC complex to specific mRNA molecules by base pairing with short regions on the target mRNA (Fig. 19.10a). The miRNA in the RISC complex does not base pair exactly with the target mRNA, but has enough sequence in common that the two strands anneal. Once bound to the target mRNA, the RISC complex inhibits translation. Human chromosomes are thought to encode about 1000 miRNAs, each of which can inhibit translation of tens or hundreds of mRNA molecules. Half or more of human proteins may have their synthesis regulated in part by miRNA.

A second type of small regulatory RNA is known as **small interfering RNA (siRNA)**. Transcription and processing of siRNA and miRNA are virtually identical, including incorporation into a RISC complex. But unlike miRNAs, which have a few mismatches in their double-stranded region, siRNA molecules have strands that match exactly. RISC complexes containing siRNA cleave the targeted RNA, instead of just binding to the target mRNA (**Fig. 19.10b**). The cleavage exposes unprotected 5′ and 3′ ends of the single-stranded RNA, which are vulnerable to attack by other nucleases. RISC complexes containing siRNA can lead to degradation of RNA transcripts in the nucleus at any stage during transcription, RNA processing, or export from the nucleus. Regulation by siRNA is widespread in eukaryotes, and is thought to have evolved originally as a defense against viruses and transposable elements.

→  **Quick Check 2** How do small regulatory RNAs differ from messenger RNA?

## Translational regulation controls the rate, timing, and location of protein synthesis.

Translation of mRNA into protein provides another level of control of gene expression. **Fig. 19.11** shows the structure of a hypothetical mRNA molecule in a eukaryotic cell and highlights some of the features that help regulate its translation (Chapter 4). Not all mRNA molecules have all the features shown, but all mRNA molecules have a 5′ cap, a 5′ untranslated region (5′ UTR), an open reading frame (ORF) containing the codons that determine the amino acid sequence of the protein, a 3′ untranslated region (3′ UTR), and a poly(A) tail. The 5′ UTR and the 3′ UTR may contain regions that bind with proteins.

**FIG. 19.11  Some features of mRNA that affect gene expression, including the 5′ cap, 5′ UTR, 3′ UTR, and poly(A) tail.**

Sequences in the 5′ untranslated region (5′ UTR) can affect translation initiation or serve as targets for protein binding to transport the mRNA to particular locations in the cell.

5′ cap   5′ UTR   AUG

Cap                                          ORF

The 5′ cap structure is the major site of translational initiation.

3′ UTR                  Poly(A) tail

Translation is often regulated by sequences in the 3′ untranslated region (3′ UTR), sometimes by binding with specific proteins.

The poly(A) tail is essential for translation.

In some cases, these proteins interact with molecular motors that transport the mRNA to particular regions of the cell. In other cases, the proteins are localized in the cell and repress translation of the mRNAs to which they are bound. By either transport or repression, these proteins cause the mRNA to be translated only in certain places in the cell.

The cap structure is one of the main recognition signals for translation initiation, which requires the coordinated action of about 25 proteins. These are present in most cells in limiting amounts, and so at any one time while some mRNAs from a gene are being translated, other mRNAs transcribed from the same gene may not have a translation initiation complex assembled. Upon formation, the initiation complex moves along the 5′ UTR, scanning for an AUG codon (the initiation codon) to allow the complete ribosome to assemble and begin translation. Although translation initiation is the principal mode of translational regulation, not all mRNA molecules are equally accessible to translation. Among the key variables are the secondary (folded) structure of the 5′ UTR, the distance from the 5′ cap to the AUG initiation codon, and the sequences flanking the AUG initiation codon.

The 3′ UTR and the poly(A) tail are also important in translation initiation. The mRNA forms a loop during initiation and the 3′ end is brought into proximity with the 5′ end. Because of their proximity to the start site for translation, sequences at the 3′ end help regulate translation. In fact, most mRNA sequences that regulate translation are present in the 3′ UTR.

## Protein structure and chemical modification modulate protein effects on phenotype.

Once translation is completed, the resulting protein can alter the phenotype of the cell or organism by affecting metabolism, signaling, or cell structure. After translation, proteins are modified in multiple ways that regulate their structure and function. Collectively, these processes are called **posttranslational modification** (see Fig. 19.1e). Regulation at this level is essential because some proteins are downright dangerous. For example, proteases such as the digestive enzyme trypsin must be kept inactive until secreted out of the cell. If they were not, their activity would kill the cell itself. These types of protein are often controlled by being translated in inactive forms that are made active by modification after secretion.

Folding and acquiring stability are key control points for some proteins (Chapter 4). While many proteins fold properly as they come off the ribosome, others require help from other proteins, called chaperones, which act as folding facilitators. Correct folding is important because improperly folded proteins may form aggregates that are destructive to cell function. Many diseases are associated with protein aggregates, including Alzheimer's disease, Huntington's disease, and mad cow disease.

Posttranslational modification also helps regulate protein activity. Many proteins are modified by the addition of one or more sugar molecules to the side chains of some amino acids. This modification can alter the protein's folding and stability, or target the molecule to particular cellular compartments. Reversible addition of a phosphate group to the side groups of amino acids such as serine, threonine, or tyrosine is a key regulator of protein activity (Chapter 9). Introduction of the negatively charged phosphate group alters the conformation of the protein, in some cases switching it from an inactive state to an active state and in other cases the reverse. Because the function of a protein molecule results from its shape and charge (Chapter 4), a change in protein conformation affects protein function.

Marking proteins for enzymatic destruction by the addition of chemical groups after translation is also important in controlling their activity. We have already seen how the destruction of successive waves of cyclin proteins helps move the cell through its division cycle (Chapter 11).

**?** **CASE 3 You, From A to T: Your Personal Genome**
**How do lifestyle choices affect expression of your personal genome?**

What does gene regulation have to do with *you*? If you examine Fig. 19.1 as a whole and consider the DNA sequence shown as your personal genome, the situation looks pretty grim. You might be led to believe that genes dictate everything, and that biology is destiny. But if you focus on the lower levels of regulation in Fig. 19.1, a different picture emerges. The picture is different because much of the regulation that occurs after transcription (posttranslational modifications, regulation of translation, regulation of RNA

stability) is determined by the physiological state of your cells, which in turn is strongly influenced by your lifestyle choices. For example, your cells can synthesize 12 of the amino acids, but if any of these is present in sufficient amounts in your diet, it is absorbed during digestion and not synthesized. The essential amino acid you ingest blocks the synthetic pathway through feedback effects.

The effect of an intervention—genetic or environmental—at any given level can affect regulatory processes at both higher and lower levels. This cascade of regulatory effects in both directions can occur because the expression of any gene is regulated at multiple levels, and because there is much feedback and signaling back and forth between nucleus and cytoplasm. It is because of these feedback and signaling mechanisms that the effects of lifestyle choices can be propagated *up* the regulatory hierarchy. For example, it has been shown that dietary intake of fats and cholesterol affects not only the activity of enzymes directly involved in the metabolism of fats and cholesterol, but also the levels of transcription of the genes encoding these enzymes by affecting the activity of their regulatory transcription factors. Similarly, lifestyles that combine balanced diets with exercise and stress relief have been shown to increase transcription of genes whose products prevent cellular dysfunction and to decrease transcription of genes whose products promote disease, including cancer.

So far, we have been talking primarily about complex traits of the type discussed in Chapter 18, which are affected by multiple genes and by multiple environmental factors as well as by genotype-by-environment interaction. For example, there are both genetic and environmental risk factors for breast and ovarian cancers, as we have seen. Simple Mendelian traits caused by mutations in single genes, such as cystic fibrosis and alpha-1 antitrypsin ($\alpha$1AT) deficiency (Chapter 17), are less responsive to lifestyle choices. But even in these cases, lifestyle does matter—for example, people with $\alpha$1AT deficiency should not smoke tobacco and should avoid environments with low air quality.

## 19.3 TRANSCRIPTIONAL REGULATION IN PROKARYOTES

The central message of Fig. 19.1 is that the regulation of gene expression occurs by means of a hierarchy of regulatory mechanisms acting at different levels (and usually at multiple levels) from DNA to protein. Gene regulation in prokaryotes is simpler than gene regulation in eukaryotes since DNA is not packaged into chromosomes, mRNA is not processed, and transcription and translation are not separated by a nuclear envelope. In prokaryotes, expression of a protein-coding gene entails transcription of the gene into messenger RNA and translation of the messenger RNA into protein. Each of these levels of gene expression is subject to regulation.

Because gene regulation is simpler in prokaryotes than gene regulation in eukaryotes, prokaryotes have served as model organisms for understanding how genes are turned on and off. In this section, we consider in more detail how gene expression is regulated at the level of transcription in bacteria and in viruses that infect bacteria. We focus on two well-studied systems: (1) the regulation of genes in the intestinal bacterium *Escherichia coli* that allows proteins needed to utilize the sugar lactose to be produced only when lactose is present in the environment and only when it is the best nutrient available, and (2) the regulation of genes in a virus that infects *E. coli* that control whether the virus integrates its DNA into the bacterial host or lyses (breaks open) the cell. In both cases, specific genes are turned on and off in response to environmental conditions.

### Transcriptional regulation can be positive or negative.

Transcription can be positively or negatively regulated. In **positive regulation,** a regulatory molecule (usually a protein) must bind to the DNA at a site near the gene in order for transcription to take place. In **negative regulation,** a regulatory molecule (again, usually a protein) must bind to the DNA at a site near the gene in order for transcription to be prevented.

**Fig. 19.12** illustrates positive regulation. The main players are DNA, the RNA polymerase complex, and a regulatory protein called a transcriptional **activator.** As shown in Fig. 19.12a, when the activator protein is present in a state that can interact with its binding site in the DNA, the RNA polymerase complex is recruited to the promoter of the gene and transcription takes place. When the activator is not present, or not able to bind with the DNA (Fig. 19.12b), transcription does not occur. The binding site for the activator may be upstream of the promoter, as shown in the figure, downstream of the promoter, or even overlap the promoter.

Sometimes, the activator protein combines with a small molecule in the cell and undergoes a change in shape that alters its binding affinity for DNA. The change in shape is an example of an **allosteric effect** (Chapter 6). In some cases, combining with the small molecule allows the activator to bind with DNA, and the presence of the small molecule in the cell results in transcription of the gene. Genes subject to this type of positive control typically encode proteins needed only when the small molecule is present in the cell. For example, in *E. coli*, the genes for degradation of the sugar arabinose are regulated by an activator that binds to the sugar. When arabinose is present, the genes are transcribed, but when arabinose is absent, the genes are not transcribed. For some other genes, activator proteins can bind DNA only when a small molecule is absent from the cell. These genes typically encode proteins needed for synthesis of the small molecule. In *E. coli*, the genes for synthesis of the amino acid cysteine are regulated in this fashion.

**Fig. 19.13** illustrates negative regulation. In this case, the DNA in its native state can recruit the RNA polymerase complex, and

---

**FIG. 19.12 Role of the activator in positive transcriptional regulation in prokaryotes.** When an activator protein binds to DNA, it promotes transcription of a gene.

**a.**

> In positive transcriptional regulation, RNA polymerase can bind to the promoter only if an activator protein binds to a site near the promoter.

Activator protein · · · RNA polymerase complex

Activator binding site · · · Polymerase binding site (promoter)

→ Transcription can occur.

**b.**

→ Transcription does not occur.

> If the activator does not bind to the DNA, the RNA polymerase cannot bind and transcription does not occur.

---

**FIG. 19.13 Role of the repressor in negative transcriptional regulation in prokaryotes.** When a repressor protein binds to DNA, it prevents transcription of a gene.

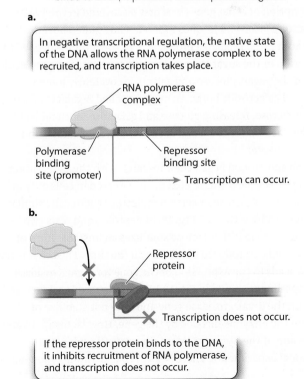

**a.**

> In negative transcriptional regulation, the native state of the DNA allows the RNA polymerase complex to be recruited, and transcription takes place.

RNA polymerase complex

Polymerase binding site (promoter) · · · Repressor binding site

→ Transcription can occur.

**b.**

Repressor protein

→ Transcription does not occur.

> If the repressor protein binds to the DNA, it inhibits recruitment of RNA polymerase, and transcription does not occur.

transcription takes place at a constant rate unless something turns it off (Fig. 19.13a). What turns it off is binding with a protein called a **repressor** (Fig. 19.13b). Again, the binding site for the repressor can be upstream from the promoter, downstream, or overlapping.

As with positive control, the active state of the repressor is often determined by an allosteric interaction with a small molecule. A small molecule that prevents binding by the repressor is called an **inducer.** In the next section, we will look closely at the regulation of genes for the breakdown of the sugar lactose in *E. coli,* and we will see that lactose binds to and inhibits the repressor and is therefore an inducer of these genes. In other cases, the small molecule changes the conformation of the repressor so that it can bind with the repressor binding site. Genes regulated in this way are often needed for synthesis of the small molecule. In *E. coli*, for example, the genes for the synthesis of the amino acid tryptophan are negatively regulated by tryptophan. When tryptophan is present in sufficient amounts, it binds with a regulatory protein to form the functional repressor, and transcription of the genes does not occur. When the level of tryptophan drops too low to form the repressor, transcription of the genes is initiated.

→ **Quick Check 3** An activator and inducer both activate gene expression. What's the difference between an activator and an inducer?

## Lactose utilization in *E. coli* is the pioneering example of transcriptional regulation.

The principle that the product of one gene can regulate transcription of other genes was first discovered in the 1960s by François Jacob and Jacques Monod, who studied how the bacterium *E. coli* regulates production of the proteins needed for utilization of the sugar lactose. Lactose consists of one molecule each of the sugars glucose and galactose covalently joined by a β- (beta-) galactoside bond. An enzyme called β-galactosidase cleaves lactose, releasing glucose and galactose. Both molecules can then be broken down and used as a source of carbon and energy (Chapter 7).

Jacob and Monod began their research with the observation that active β-galactosidase enzyme is observed in cells only in the presence of lactose or certain molecules chemically similar to lactose. Why is this so? **Fig. 19.14** describes and tests two hypotheses. One is that lactose stabilizes an unstable form of β-galactosidase produced by all cells all the time. The other is that lactose leads to the expression of the gene for β-galactosidase. The experiment shown in Fig. 19.14 demonstrated that lactose turns on the β-galactosidase gene and does not stabilize or activate the enzyme encoded by the gene. How lactose activates expression of the β-galactosidase gene was the subject of additional experiments. These Nobel Prize–winning follow-up

**HOW DO WE KNOW?**

**FIG. 19.14**

# How does lactose lead to the production of active β-galactosidase enzyme?

**BACKGROUND** Active β-galactosidase enzyme is observed only in *E. coli* cells that are growing in the presence of lactose.

**HYPOTHESES** One hypothesis is that the enzyme is always being produced, but is produced in an unstable form that breaks down rapidly in the absence of lactose. A second hypothesis is that the enzyme is stable, but is produced only in the presence of lactose.

**EXPERIMENT** François Jacob and Jacques Monod exposed a culture of growing cells to lactose and later removed it. They measured the amount of β-galactosidase present in the culture during the experiment.

**RESULTS** Almost immediately upon addition of lactose, β-galactosidase began to accumulate, and its amount steadily increased. When lactose was removed, the enzyme did not disappear immediately (as would be the case if it were unstable). Instead, the amount of enzyme remained the same as when lactose was present. This result is expected only if β-galactosidase is a stable enzyme that is synthesized when lactose is added and stops being synthesized when lactose is removed.

**CONCLUSION** Synthesis of β-galactosidase is turned on when lactose is added and turned off when lactose is removed, in support of the second hypothesis.

**FOLLOW-UP WORK** These results stimulated pioneering experiments that ultimately led to the discovery of the lactose operon and the mechanism of transcriptional regulation by a repressor protein.

**SOURCE** Monod, J. 1965. "From Enzymatic Adaption to Allosteric Transitions." Nobel Prize lecture. http://nobelprize.org/nobel_prizes/medicine/laureates/1965/monod-lecture.htm.

**FIG. 19.15** **Structural and regulatory elements of the lactose operon.** The *lac* operon consists of the promoter, the operator, and all the structural genes that are transcribed into a single mRNA called a polycistronic mRNA.

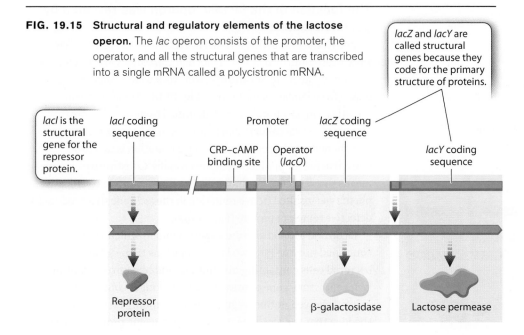

*lacZ* and *lacY* are called structural genes because they code for the primary structure of proteins.

*lacI* is the structural gene for the repressor protein.

*lacI* coding sequence

Promoter

*lacZ* coding sequence

CRP–cAMP binding site

Operator (*lacO*)

*lacY* coding sequence

Repressor protein

β-galactosidase

Lactose permease

studies of Jacob and Monod gave the first demonstrated example of transcriptional gene regulation.

To understand how the genes for lactose utilization are regulated, you need to know the main players, and these are shown in **Fig. 19.15.** The overall situation looks quite complicated, but when you take it apart and look at the individual pieces, it is in fact quite simple. Let us start with the two coding sequences on the right:

- *lacZ* is the gene (coding sequence) for β-galactosidase, which cleaves the lactose molecule into its glucose and galactose constituents.

- *lacY* is the gene (coding sequence) for lactose permease, which transports lactose from the external medium into the cell.

These genes are called **structural genes** because they code for the sequence of amino acids making up the primary structure of each protein. Bacteria that contain mutations in the *lacZ* gene or the *lacY* gene or both cannot utilize lactose as a source of energy. Without a functional product from *lacY*, lactose cannot enter the cell, and without a functional product from *lacZ*, lactose cannot be cleaved into its component sugars. Thus, β-galactosidase and permease are essential for the utilization of lactose for cell growth.

Regulation of the *lacZ* and *lacY* structural genes is controlled by the product of another structural gene, called *lacI*, which encodes a repressor protein. Located between *lacI* and *lacZ* are a series of regulatory sequences in the DNA that include a promoter, whose function is to recruit the RNA polymerase complex and initiate transcription, and an **operator** (*lacO*), which is the binding site for the repressor protein. Another regulatory region is a binding site for a protein called CRP, which is discussed later.

The kind of gene organization depicted in Fig. 19.15 is common in bacteria. Typically, a group of functionally related genes are located next to one another along the bacterial DNA, and when they are transcribed they are transcribed together into a single molecule of messenger RNA. Such an mRNA is called a **polycistronic RNA** ("cistron" is an old term for "coding sequence"). In Fig. 19.15, the polycistronic RNA includes the coding sequences for β-galactosidase and lactose permease. The region of DNA consisting of the promoter, the operator, and the coding sequence for the structural genes is called an **operon.**

Operons are found in bacteria and archaeons, whose cells can translate polycistronic mRNA molecules correctly because their ribosomes can initiate translation anywhere along an mRNA that contains a proper ribosome-binding site (Chapter 4). In a polycistronic mRNA, each of the coding sequences is preceded by a ribosome-binding site, so translation can be initiated there.

## The repressor protein binds with the operator and prevents transcription, but not in the presence of lactose.

The lactose operon is negatively regulated by the repressor protein—that is, the structural genes of the lactose operon are always expressed unless the operon is turned off by a regulatory molecule, in this case the repressor. What the operon looks like in the absence of lactose is shown in **Fig. 19.16.** The *lacI* gene,

**FIG. 19.16** **The lactose operon in the repressed state in the absence of lactose.**

RNA polymerase complex

Promoter

*lacI* coding sequence

Operator (*lacO*)

In the absence of lactose, the repressor protein binds to the operator and prevents transcription from taking place.

Repressor protein

**FIG. 19.17  The lactose operon in the induced state in the presence of lactose.**

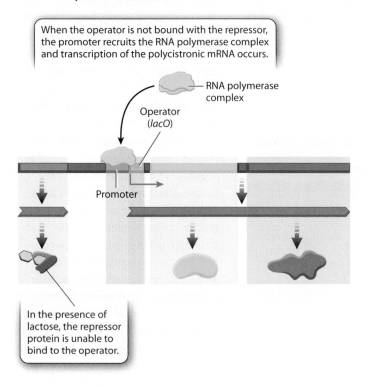

When the operator is not bound with the repressor, the promoter recruits the RNA polymerase complex and transcription of the polycistronic mRNA occurs.

RNA polymerase complex

Operator (*lacO*)

Promoter

In the presence of lactose, the repressor protein is unable to bind to the operator.

encoding the repressor protein, is expressed constantly at a low level. The repressor protein binds with the operator (*lacO*), the RNA polymerase complex is not recruited, and transcription does not take place.

The configuration of the lactose operon in the presence of lactose is shown in **Fig. 19.17.** When lactose is present in the external medium, the repressor protein is unable to bind to the operator, RNA polymerase is recruited, and transcription occurs. In other words, lactose acts as an inducer since it prevents binding of the repressor protein. The inducer is not actually lactose itself, but rather an isomer of lactose called allolactose, which differs in the way the sugars are linked with each other. Allolactose is produced in small amounts whenever lactose is present in the cell.

The binding of the inducer with the repressor protein results in an allosteric change in repressor structure that inhibits the protein's ability to bind to the operator. The absence of repressor from the operator allows the RNA polymerase complex to be recruited to the promoter, and the polycistronic mRNA is produced. The resulting lactose permease allows the lactose to be transported into the cell on a large scale, and β-galactosidase cleaves the molecules to allow the constituents to be used as a source of energy and carbon. The lactose operon is therefore an example of negative regulation by the repressor, whose function is modulated by an inducer.

## The function of the lactose operon was revealed by genetic studies.

Although the interactions shown in Fig. 19.16 and Fig. 19.17 have since been confirmed by direct biochemical studies, the inferences about how the repressor and operator work were originally drawn from studies of mutations (**Fig. 19.18**). As part of their investigation, Jacob and Monod identified bacterial mutants that expressed β-galactosidase and permease in the absence of lactose. The phenotype of a cell carrying such a mutation is said to be **constitutive** for production of the proteins. Constitutive expression means that it occurs continuously. The most common constitutive phenotype resulted from a mutation in the *lacI* gene that produced a defective repressor protein (Fig. 19.18a).

Jacob and Monod also did experiments in which *E. coli* contained not one but two lactose operons (Fig. 19.18b). In bacterial cells containing one mutant and one normal copy of *lacI* (repressor), gene expression was no longer constitutive but instead showed normal regulation. This finding is consistent with the idea that the *lacI* gene produces a diffusible protein since the normal repressor is able to bind to and repress transcription from both operons, not just the one that it is physically linked to.

A much less common class of constitutive mutants identified the operator (Fig. 19.18c). Genetic studies of the mutations in these cells showed that they were not located in the *lacI* gene that encodes the repressor but, rather, closer to the coding sequence of *lacZ*. The genetic element in which the mutations occurred was called the lactose *operator* (*lacO*), and the constitutive mutations were designated *lacO*^c ("c" for "constitutive"). When two different lactose operons, one normal and one with *lacO*^c, were in the same cell, the operon carrying *lacO*^c was transcribed constitutively, even in the presence of normal repressor, because the repressor was unable to bind to the mutant operator site (Fig. 19.18d). "To explain this effect," Jacob and Monod wrote, "it seems necessary to invoke a new type of genetic entity, called an 'operator,' which would be: (a) adjacent to the group of genes and would control their activity; and (b) would be sensitive to the repressor produced by a particular regulatory gene."

→  **Quick Check 4** Predict the consequence of a mutation in *lacI* (repressor gene) that produces repressor protein that is able to bind to the operator, but not able to bind allolactose.

## The lactose operon is also positively regulated by CRP–cAMP.

Fig. 19.16 and Fig. 19.17 show how the ability of the repressor to bind with either the operator (in the absence of lactose) or with the inducer (in the presence of lactose) provides a simple and elegant way for the bacterial cell to transcribe the genes needed for lactose utilization only in the presence of lactose. The elucidation of these interactions was as far as the research tools used by Jacob and Monod could take them. Since the original

**FIG. 19.18  Lactose operon regulatory mutants.** Jacob and Monod found two classes of constitutive mutants, one affecting the repressor (a) and the other the operator (c). In cells containing both a mutant and a nonmutant lactose operon, those with the repressor mutant became regulated (b), and those with the operator mutant remained constitutive (d).

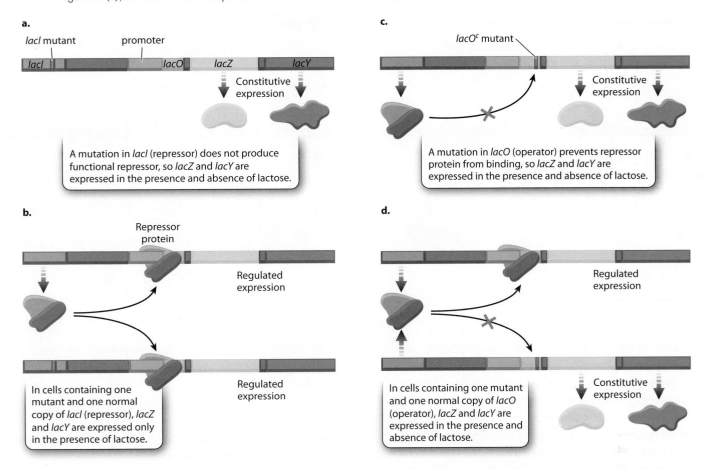

experiments, the lactose operon has been studied in much greater detail and additional levels of regulation have been discovered.

One of these additional levels involves the CRP binding site shown in Fig. 19.15, which in **Fig. 19.19** is occupied by a protein called the CRP–cAMP complex. The CRP–cAMP complex is a positive regulator of the lactose operon, which you will recall is a protein that activates gene expression upon binding to DNA. "CRP" stands for "cAMP receptor protein." The role of CRP–cAMP is to provide another level of control of transcription that is more sensitive to the nutritional needs of the cell than the level of control provided by the presence or absence of lactose. *E. coli* can utilize many kinds of molecules as sources of energy. When more than one type of energy source is available in the environment, certain sources are used before others. For example, glucose is preferred to lactose, and lactose is preferred to glycerol. The CRP–cAMP complex helps regulate which compounds are utilized.

The concentration of the small molecule cAMP in the cell is a signal about the nutritional state of the cell. In the absence of glucose, cAMP levels are high, and cAMP binds to CRP, changing the shape of CRP so that it can bind DNA and activate

transcription. In this way, cAMP is an allosteric activator of CRP, acting as described earlier. If lactose is present, the repressor can't bind and the lactose operon is induced (Fig. 19.19a). If lactose is not present, the repressor is bound and the lactose operon is not transcribed even in the presence of the cAMP–CRP complex.

In the presence of glucose, cAMP levels are low, and the cAMP–CRP complex does not bind the lactose operon. As a result, even in the presence of lactose, the lactose operon is not transcribed to high levels (Fig. 19.19b). In this way, *E. coli* preferentially utilizes glucose when both glucose and lactose are present, and utilizes lactose only when glucose is depleted.

### Transcriptional regulation determines the outcome of infection by a bacterial virus.

Transcriptional regulation has also been well studied in viruses. Bacterial cells are susceptible to infection by a variety of viruses known as **bacteriophages** ("bacteriophage" literally means "bacteria-eater," and is often shortened to just "phage"). Among bacteriophages is a type that can undergo either of two fates when infecting a cell. The best known example is bacteriophage λ (lambda), which infects cells of *E. coli*. The possible fates of

**FIG. 19.19 The CRP–cAMP complex, a positive regulator of the lactose operon.** (a) In the absence of glucose, cAMP levels are high and the CRP–cAMP complex binds to a site near the promoter, where it activates transcription. (b) In the presence of glucose, cAMP levels are low and the CRP–cAMP complex does not bind, so transcription is not induced to high levels, even in the presence of lactose.

**a.** Low glucose, high cAMP

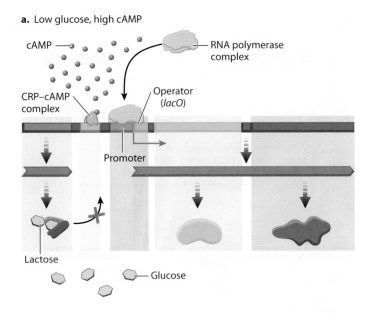

**b.** High glucose, low cAMP

λ infection are illustrated in **Fig. 19.20.** Upon infection, the linear DNA of the phage genome is injected into the bacterial cell, and almost immediately the ends of the molecule join to form a circle. In normal cells growing in nutrient medium, the usual outcome of infection is the **lytic pathway,** shown at the left in Fig. 19.20. In the lytic pathway, the virus hijacks the cellular machinery to replicate the viral genome and produce viral proteins. After about an hour, the infected cell undergoes **lysis** and the cell bursts open to release a hundred or more progeny phage capable of infecting other bacterial cells.

The alternative to the lytic pathway is **lysogeny,** shown at the right in Fig. 19.20. In lysogeny, the bacteriophage DNA and the bacterial DNA undergo a process of recombination at a specific site in both molecules, which results in a slightly enlarged bacterial DNA molecule that now includes the bacteriophage DNA. Lysogeny often takes place in cells growing in poor conditions. The relative sizes of the DNA molecules in Fig. 19.20 are not to scale. In reality, the length of the bacteriophage DNA is only about 1% of that of the bacterial DNA. When the bacteriophage genome is integrated in lysogeny, the only bacteriophage gene transcribed and translated is one that represses the transcription of other phage genes, preventing entry into the lytic pathway. The bacteriophage DNA is replicated along with the bacterial DNA and transmitted to the bacterial progeny when the cell divides. Under stress, such as exposure to ultraviolet light, recombination is reversed, freeing the bacterial genome and initiating the lytic pathway.

At the molecular level, the choice between the lytic and lysogenic pathways is determined by the positive and negative regulatory effects of a small number of bacteriophage

**FIG. 19.20 Alternative outcomes of infection by bacteriophage λ.** The bacteriophage can enter either the lytic or the lysogenic pathway.

**FIG. 19.21** Transcriptional regulation of *cI* and *cro* genes, which determine the lytic pathway versus lysogeny.

proteins produced soon after infection. **Fig. 19.21** shows the small region of the bacteriophage DNA in which the key interactions take place. Almost immediately after infection and circularization of the bacteriophage DNA, transcription takes place from the promoters $P_L$ and $P_R$. The transcript from $P_R$ encodes the proteins cro and cII. The cro protein represses transcription from another promoter $P_M$, which prevents transcription and synthesis of the protein cI. In normal cells growing in nutrient medium, proteases present in the bacterial cell degrade cII and prevent its accumulation. With cro protein preventing *cI* expression and cII protein unable to accumulate, transcription of bacteriophage genes in the lytic pathway takes place, including those genes needed for bacteriophage DNA replication, those encoding proteins in the bacteriophage head and tail, and, finally, those needed for lysis.

Alternatively, in bacterial cells growing in poor conditions, reduced protease activity allows cII protein to accumulate. When cII protein reaches a high enough level, it stimulates transcription from the promoter $P_E$. The transcript from $P_E$ includes the coding sequence for cI protein, and the cI protein has three functions:

- It binds with the operator $O_R$ and prevents further expression of *cro* and *cII*.

- It stimulates transcription of its own coding sequence from the promoter $P_M$, establishing a positive feedback loop that keeps the level of cI protein high.

- It binds with the operator $O_L$ and prevents further transcription from $P_L$.

The result is that cI production shuts down transcription of all bacteriophage genes except its own gene, and this is the regulatory state that produces lysogeny. (The protein needed for recombination between the bacteriophage DNA and the bacterial DNA is produced by transcription from the $P_L$ promoter before it is shut down by cI.) When cells that have undergone lysogeny are exposed to ultraviolet light or certain other stresses, the cI protein is degraded. In this case, cro and cII are produced again, and the lytic pathway follows.

Regulation of the lytic and lysogenic pathways works to the advantage of bacteriophage λ, but the process is not as simple and elegant as something an engineer might design. That is because biological systems are not engineered, they evolve. Regulatory mechanisms are built up over time by the selection of successive mutations. Each evolutionary step refines the regulation in such a way as to be better adapted to the environment than it was before. Each successive step occurs only because it increases survival and reproduction.

We summarize these two alternative outcomes of infection by a bacterial virus, and the earlier discussions of viral diversity, replication, host range, and effects on a cell in Fig. 19.22 on the following page.

# Virus: A Genome in Need of a Cell

**Integrating concepts from Chapters 11, 13, and 19**

## Diversity

Viruses come in many different shapes and viral genomes include a diverse array of types and structures of nucleic acids.

**Bacteriophage λ (dsDNA)**  **Adenovirus (dsDNA)**  **Picobirnavirus (dsRNA)**  **Rabies (-ssRNA)**  **Tobacco mosaic virus (+ssRNA)**  **HIV (+ssRNA)**

Mature assembled viruses are released

Fragmented bacterial DNA

Replicated viral components

**Lysis**

*Bacterial cell*

Mature λ phage

Circular DNA

**Lysogeny**

Viral DNA integrated into bacterial DNA

## Transcriptional regulation

When bacterial cells are grown in favorable conditions, bacterial proteases degrade a key viral protein and genes are transcribed that lead to the bursting of the bacterial cell (lysis). When bacterial cells are grown in poor conditions, the viral protein accumulates and genes are transcribed that allow the viral genome to remain integrated in host genome (lysogeny).

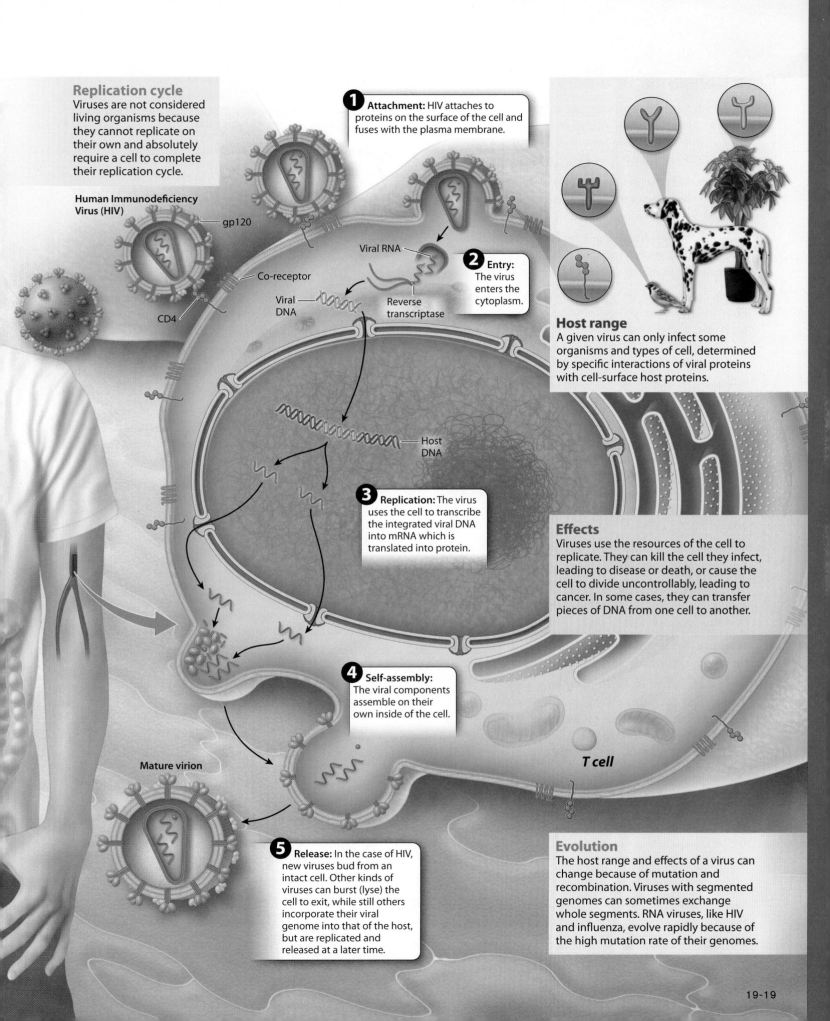

## Replication cycle
Viruses are not considered living organisms because they cannot replicate on their own and absolutely require a cell to complete their replication cycle.

**Human Immunodeficiency Virus (HIV)**

gp120

Co-receptor

CD4

**1 Attachment:** HIV attaches to proteins on the surface of the cell and fuses with the plasma membrane.

Viral RNA

**2 Entry:** The virus enters the cytoplasm.

Viral DNA

Reverse transcriptase

Host DNA

**3 Replication:** The virus uses the cell to transcribe the integrated viral DNA into mRNA which is translated into protein.

**4 Self-assembly:** The viral components assemble on their own inside of the cell.

Mature virion

**5 Release:** In the case of HIV, new viruses bud from an intact cell. Other kinds of viruses can burst (lyse) the cell to exit, while still others incorporate their viral genome into that of the host, but are replicated and released at a later time.

*T cell*

## Host range
A given virus can only infect some organisms and types of cell, determined by specific interactions of viral proteins with cell-surface host proteins.

## Effects
Viruses use the resources of the cell to replicate. They can kill the cell they infect, leading to disease or death, or cause the cell to divide uncontrollably, leading to cancer. In some cases, they can transfer pieces of DNA from one cell to another.

## Evolution
The host range and effects of a virus can change because of mutation and recombination. Viruses with segmented genomes can sometimes exchange whole segments. RNA viruses, like HIV and influenza, evolve rapidly because of the high mutation rate of their genomes.

# Core Concepts Summary

**19.1 THE REGULATION OF GENE EXPRESSION IN EUKARYOTES TAKES PLACE AT MANY LEVELS, INCLUDING DNA PACKAGING IN CHROMOSOMES, TRANSCRIPTION, AND RNA PROCESSING.**

Gene expression involves the turning on or turning off of a gene. Gene regulation determines where, when, and how much gene product is made. page 19-1

Regulation at the level of chromatin involves chemical modifications of DNA and histones that make a gene accessible or inaccessible to the transcriptional machinery. page 19-2

Dosage compensation is the process by which the expression of X-linked genes is equalized in XX individuals (who have two copies of each X chromosome and hence two copies of each X-linked gene) and XY individuals (who have only one copy of each X chromosome and X-linked gene). page 19-4

One mechanism of dosage compensation involves the inactivation of one of the two X chromosomes in females; this mechanism, known as X-inactivation, is observed in mammals. page 19-4

X-inactivation occurs by the transcription of a noncoding RNA known as *Xist*, which binds to the X-chromosome inactivation center (*XIC*) and coats the entire X chromosome, leading to DNA and histone modifications and transcriptional repression. page 19-4

Transcriptional regulation controls whether or not transcription of a gene occurs. page 19-5

Transcription can be regulated by regulatory transcription factors that bind to specific DNA sequences near genes known as enhancers. page 19-6

Further levels of regulation after a gene is transcribed to mRNA include RNA processing, splicing, and editing. page 19-6

**19.2 AFTER AN mRNA IS TRANSCRIBED AND EXPORTED TO THE CYTOPLASM IN EUKARYOTES, GENE EXPRESSION CAN BE REGULATED AT THE LEVEL OF mRNA STABILITY, TRANSLATION, AND POSTTRANSLATIONAL MODIFICATION OF PROTEINS.**

Small regulatory RNAs, especially microRNA (miRNA) and small interfering RNA (siRNA), affect gene expression through their effects on translation (miRNA) or RNA stability (siRNA). page 19-8

Translational regulation controls the rate, timing, and location of protein synthesis. page 19-9

Translational regulation is determined by many features of an mRNA molecule, including the 5' and 3' UTR, the cap, and the poly(A) tail. page 19-9

Posttranslational modification comes into play after a protein is synthesized, and includes chemical modification of side groups of amino acids, affecting the structure and activity of a protein. page 19-10

Gene regulation is influenced by both genetic and environmental factors. page 19-10

**19.3 TRANSCRIPTIONAL REGULATION IS ILLUSTRATED IN BACTERIA BY THE CONTROL OF THE PRODUCTION OF PROTEINS NEEDED FOR THE UTILIZATION OF LACTOSE, AND IN VIRUSES BY THE CONTROL OF THE LYTIC AND LYSOGENIC PATHWAYS.**

Transcriptional regulation can be positive, in which a gene is usually off and is turned on in response to the binding to DNA of a regulatory protein called an activator, or negative, in which a gene is usually on and is turned off in response to the binding to DNA of a regulatory protein called a repressor. page 19-11

Jacob and Monod studied the lactose operon in *E. coli* as a model for bacterial gene regulation. page 19-12

When lactose is added to culture of bacteria, the genes for the uptake of lactose (permease, encoded by *lacY*) and cleavage of lactose (β-galactosidase, encoded by *lacZ*) are expressed. page 19-12

The lactose operon is negatively regulated by the repressor protein (encoded by *lacI*), which binds to DNA sequences known as the operator. page 19-13

When lactose is added to the medium, it induces an allosteric change in the repressor protein, preventing it from binding to the operator and allowing transcription of *lacY* and *lacZ*. In this way, lactose acts as an inducer of the lactose operon. page 19-14

An additional level of regulation of the lactose operon is provided by the CRP–cAMP complex, a positive activator of transcription. page 19-14

In infection of *E. coli* cells by bacteriophage λ, predominance of cro protein results in the lytic pathway, whereas predominance of the cI protein results in the lysogenic pathway. page 19-15

# Self-Assessment

1. Distinguish between gene expression and gene regulation.

2. Explain what is meant by "different levels" of gene regulation and give some examples.

3. Give a few examples of how DNA bases or chromatin can be modified to regulate gene expression.

4. Explain how X-inactivation in female mammals results in patchy coat color in calico cats.

5. Explain how one protein-coding gene can code for more than one polypeptide chain.

6.  Name and describe three ways in which gene expression can be influenced after mRNA is processed and leaves the nucleus.

7.  Diagram the lactose operon in *E. coli* with the proper order of the elements *lacI*, *lacO*, *lacY*, and *lacZ*, and explain how expression is controlled in the presence and absence of lactose.

8.  Describe the role of the CRP–cAMP complex in positive regulation of the lactose operon in *E. coli*.

9.  Describe what is meant by lysis and lysogeny, and explain how gene regulation controls these two pathways.

Do you understand the chapter's Core Concepts? Log in to check your answers to the Self-Assessment questions, then practice what you've learned and reinforce this chapter's concepts by working through the problems and multimedia tutorials provided there.

📖 **www.biologyhowlifeworks.com**

## CHAPTER 20

# GENES AND DEVELOPMENT

## Core Concepts

**20.1** In the development of humans and other animals, stem cells become progressively more restricted in their possible pathways of cellular differentiation.

**20.2** The genetic control of development is a hierarchy in which genes are deployed in groups that in turn regulate the next set of genes.

**20.3** Many proteins that play key roles in development are evolutionarily conserved but can have dramatically different effects in different organisms.

**20.4** Combinatorial control is a developmental strategy in which cellular differentiation depends on the particular combination of transcription factors present in a cell.

**20.5** Ligand–receptor interactions activate signal transduction pathways that converge on transcription factors and other genes that determine cell fate.

Altogether, the human body contains about 200 different types of cell, all of which derive from a single cell, the zygote. In animals, some cells become muscle cells, others nerve cells, and still others connective tissue. Almost all of these cells have exactly the same genome: They differ not in their content of genes, but instead in the groups of genes that are expressed or repressed. In other words, these cell types differ as a result of gene regulation, discussed in the previous chapter.

Gene regulation is especially important in multicellular organisms because it underlies **development,** the process in which a fertilized egg undergoes multiple rounds of cell division to become an embryo with specialized tissues and organs. During development, cells undergo changes in gene expression as genes are turned on and off at specific times and places. Gene regulation causes cells to become progressively more specialized, a process known as **differentiation.**

In this chapter, we focus on the general principles by which genes control development. We will see that, as cells differentiate along one pathway, they progressively lose their ability to differentiate along other pathways. Yet gene expression can sometimes be reprogrammed to reopen pathways of differentiation that had previously been shut off, a process that has important implications for therapeutic replacement of diseased or damaged tissue. We will also see that, while some of the key molecular mechanisms of development are used over and over again in different organisms, they have evolved to yield such differences in shape and form as to conceal the underlying similarity in mechanism.

## 20.1 GENETIC PROGRAMS OF DEVELOPMENT

Genetic programs and computer programs have a lot in common. Computer code is written as a linear string of letters, which corresponds to the sequence of nucleotides in genomic DNA. Once a computer program is initiated, it automatically runs and performs its coded task. Small mistakes in the code, analogous to mutations, can have big consequences and even cause the program to crash.

The analogy between genetic programs and computer programs has an important limitation. Computer programs, designed by humans, are consciously written, whereas genetic programs evolve. The genetically encoded developmental programs of all living organisms emerged over hundreds of millions of years through mutation and natural selection. These developmental programs changed gradually through time, persisting only if they produced organisms that could successfully survive and reproduce in the existing environment. Here, we explore the genetic program of development— that is, the genetic instructions that lead from a single fertilized egg to a complex multicellular organism.

## The fertilized egg is a totipotent cell.

In all sexually reproducing organisms, the fertilized egg is special because of its developmental potential. The fertilized egg is said to be **totipotent,** which means that it can give rise to a complete organism. In mammals, the egg also forms the membranes that surround and support the developing embryo (Chapter 42).

Development begins with **fertilization,** which in human females normally occurs in one of the two fallopian tubes stretching from the upper corners of the uterus to the ovaries on either side, where the tubes flare out with tiny fingerlike projections to catch the egg released at ovulation. The fertilized egg, or **zygote,** travels down the fallopian tube toward the uterus, undergoing mitotic cell division as it moves along. One cell becomes two, two become four, four become eight, eight become sixteen, and so on, with all the cells contained within the egg's outer membrane (**Fig. 20.1**). This clump of cells,

called the **morula,** reaches the uterus about 4 or 5 days after fertilization.

These early cell divisions are different from mitotic cell division later in life because the cells do not grow between divisions; they merely replicate their chromosomes and divide again. The result is that the cytoplasm of the egg is partitioned into smaller and smaller packages, the new cells all bunched together inside the gelatinous envelope that covers the developing embryo.

Cell division continues in the morula until there are a few thousand cells. The cells then begin to move in relation to one another, pushing against and expanding the membrane that encloses them and rearranging themselves to form a hollow sphere called a **blastocyst** (Fig. 20.1). In one region of the inner wall of the blastocyst, there is a group of cells known as the **inner cell mass,** from which the body of the embryo develops. The wall of the blastocyst forms several membranes that envelop and support the developing embryo. Once the blastocyst forms, 5 or 6 days after fertilization, it implants in the uterine wall. From implantation onward, the cells in the blastocyst approximately double their size with each cell division. This is the first great trial of the embryo, which can no longer draw on the cytoplasm in the egg produced by the mother. To survive, the cells of the embryo must now manufacture their own cytoplasm.

Once implanted in the uterine wall, the multiplying cells of the inner cell mass reorganize to form a **gastrula.** It is at this stage that the three **germ layers** are established (Fig. 20.1). Germ layers are sheets of cells that include the ectoderm, mesoderm, and endoderm and that differentiate further into specialized cells. Those formed from the **ectoderm** include epithelial cells and pigment cells in the skin and nerve cells in the brain; cells from the **mesoderm** include cells that make up the inner layer of the skin, muscle cells, and red blood cells; and cells formed from the **endoderm** include cells of the lining of the digestive tract and lung, as well as liver cells and pancreas cells.

## Cellular differentiation increasingly restricts alternative fates.

At each successive stage in development, as the cells become differentiated they lose the potential to develop into any kind of cell. The fertilized egg is totipotent because it can differentiate into both the inner cell mass and supporting membranes, and eventually into an entire organism. The cells of the inner cell mass, called embryonic stem cells, are **pluripotent** because they are able to give rise to any of the three germ layers, and therefore to any cell of the body. However, pluripotent cells cannot on their own give rise to an entire organism, as a totipotent cell can. Cells further along in differentiation are **multipotent;** these cells can form a limited number of types of specialized cell. Totipotent, pluripotent, and multipotent cells are all considered **stem cells,** cells that are capable of differentiating into different cell types.

**FIG. 20.1** **Early development of a human embryo.** The zygote is a totipotent cell because its daughter cells can develop into any cell type and eventually into a complete organism.

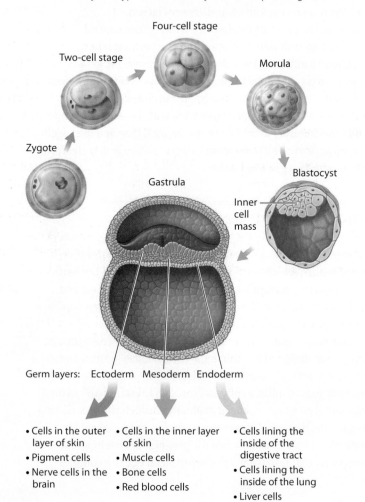

Four-cell stage

Two-cell stage

Morula

Zygote

Blastocyst

Gastrula

Inner cell mass

Germ layers:   Ectoderm   Mesoderm   Endoderm

- Cells in the outer layer of skin
- Pigment cells
- Nerve cells in the brain

- Cells in the inner layer of skin
- Muscle cells
- Bone cells
- Red blood cells

- Cells lining the inside of the digestive tract
- Cells lining the inside of the lung
- Liver cells
- Pancreas cells

→ **Quick Check 1** From what you know about embryonic development, do you think that a cell from the inner cell mass or one from the ectoderm has more developmental potential?

Why do differentiating cells increasingly lose their developmental potential? One hypothesis focuses on gene regulation. When cells become committed to a particular developmental pathway, genes no longer needed are turned off (that is, repressed) and are difficult to turn on again. Another hypothesis is genome reduction: As cells become differentiated, they delete the DNA for genes they no longer need.

These hypotheses can be distinguished by an experiment in which differentiated cells are reprogrammed to mimic earlier states. If loss of developmental potential is due to gene regulation, then cells could be reprogrammed to become pluripotent or multipotent. If loss of developmental potential is due to genome reduction, then differentiated cells could not be reprogrammed to become pluripotent or multipotent.

British developmental biologist John Gurdon carried out such experiments in the early 1960s (**Fig. 20.2**). Gurdon used a procedure called **nuclear transfer,** in which a hollow glass needle is used to insert the nucleus of a cell into the cytoplasm

## HOW DO WE KNOW?

**FIG. 20.2**

# How do stem cells lose their ability to differentiate into any cell type?

① The nucleus of an egg cell is inactivated with ultraviolet light.

Unfertilized egg

**716 experiments:** Development terminated prior to tadpole stage.

*Xenopus laevis* tadpole

② The nucleus from an intestinal cell of a tadpole is injected in to the egg cell.

**10 experiments:** Development to tadpole stage occurred.

**BACKGROUND** During differentiation, cells become progressively more specialized and restricted in their fates. Early studies left it unclear whether cell differentiation occurs because of changes in gene expression, or whether cell differentiation results from loss of genes.

**HYPOTHESES** One hypothesis is that differentiation occurs as a result of changes in gene expression. A second hypothesis is that differentiation occurs as a result of genome reduction, in which genes that are not needed are deleted.

**EXPERIMENT** John Gurdon carried out experiments in the amphibian *Xenopus laevis* to test these hypotheses. He transferred nuclei from differentiated cells into unfertilized eggs whose nuclei had been inactivated with ultraviolet light. If differentiation is due to changes in gene expression, then the differentiated nucleus should be able to reprogram itself in the egg cytoplasm and differentiate again into all the cells of a tadpole. If differentiation is accompanied by loss of genes, then differentiation is irreversible and development will not proceed.

**RESULTS** The experiment was carried out 726 times. In 716 cases, development did not occur; in 10, development proceeded normally.

**CONCLUSION** Although the experiment succeeded in only 10 of 726 attempts, it showed that the nucleus of an intestinal cell and the cytoplasm of the unfertilized egg are able to support complete development of a normal animal. This result allows us to reject the hypothesis that differentiation occurs by the loss of genes. Differentiated cells must contain a complete genome. The first hypothesis—that cells become differentiated as a result of changes in gene expression—was supported. But, because of the small number of successes in reprogramming, additional experiments were needed to validate the conclusions.

**FOLLOW-UP WORK** This work was controversial. Some critics argued that the successful experiments resulted from a small number of undifferentiated cells present in intestinal epithelium. Others accepted the conclusion but expressed misgivings about possible applications to humans. Later experiments that succeeded in cloning mammals from fully differentiated cells confirmed the original conclusion.

**SOURCE** Gurdon, J. B. 1962. "The Developmental Capacity of Nuclei Taken from Intestinal Epithelium Cells of Feeding tadpoles." *Journal of Embryology & Experimental Morphology* 10: 622–640.

**FIG. 20.3**    Results of nuclear transfer of differentiated cells.

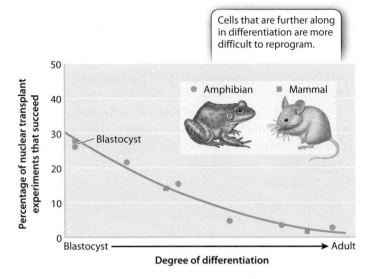

of an egg whose own nucleus has been destroyed or removed. Previous nuclear transfer experiments had been carried out in the leopard frog, *Rana pipiens*. Whereas nuclei from pluripotent or multipotent cells could often be reprogrammed to develop into normal tadpoles, attempts with nuclei from fully differentiated cells failed.

Gurdon tried the experiments in a different organism, the clawed toad *Xenopus laevis*, and demonstrated that nuclei from fully differentiated intestinal cells could be reprogrammed to support normal development of the tadpole (Fig. 20.2). Only 10 of 726 experiments succeeded, but this was sufficient

to show that intestinal cell nuclei still contained a complete *Xenopus* genome. In other words, his findings supported the first hypothesis—all of the same genes are present in intestinal cells as in early embryonic cells, but some of the genes are turned off, or repressed, during development.

**Fig. 20.3** summarizes the results of many nuclear transfer experiments carried out in mammals and amphibians. The percentage of reprogramming experiments that fail increases as cells differentiate. The best chance of success is to use pluripotent nuclei from cells in the blastocyst (or its amphibian equivalent, the blastula). However, even some experiments using nuclei from fully differentiated cells have been successful.

When nuclear transfer succeeds, the result is a **clone**—an individual that carries an exact copy of the nuclear genome of another individual. In this case, the new individual shares the same genome as that of the individual from which the donor nucleus was obtained. (The mitochondrial DNA is not from the nuclear donor, but from the donor of the egg cytoplasm.) The first mammalian clone was a lamb called Dolly (**Fig. 20.4a**), born in 1996. She was produced from the transfer of the nucleus of a cell in the mammary gland of a sheep, and was the only successful birth among 277 nuclear transfers. Successful cloning in sheep soon led to cloning in cattle, pigs, and goats.

The first household pet to be cloned was a kitten named CopyCat (**Fig. 20.4b**), born in 2001 and derived from a differentiated ovarian cell. CopyCat was the only success among 87 tries. As shown in Fig. 20.4b, the cat from which the donor nucleus was obtained was a calico, but CopyCat herself was not, even though the two cats are clones of each other. The reason for their different appearance has to do with X-inactivation, discussed in Chapter 19. Recall

**FIG. 20.4**    Celebrity clones and their genetic mothers. (a) Dolly. (b) CopyCat.

that the mottled orange and black calico pattern results from random inactivation of one of the two X chromosomes during development. The lack of a calico pattern in CopyCat implies that the X chromosomes in the transferred nucleus did not "reset" as they do in normal embryos. Instead, the inactive X in the donor nucleus remained inactive in all the cells in the clone. Hence, while CopyCat and her mother share the same nuclear genome, the genes were not expressed in the same way because of irreversible epigenetic regulation in the donor nucleus.

→  **Quick Check 2** *X*-inactivation results in two clones of cells differing in the genes expressed. Can you think of other reasons why two genetically identical individuals might look different from each other?

**?** **CASE 3  You, From A to T: Your Personal Genome**
**Can cells with your personal genome be reprogrammed for new therapies?**
Stem cells play a prominent role in **regenerative medicine,** which aims to use the natural processes of cell growth and development to replace diseased or damaged tissues. Stem cells are already used in bone marrow transplantation and may someday be used to treat Parkinson's disease, Alzheimer's disease, heart failure, certain types of diabetes, severe burns and wounds, and spinal cord injury.

At first it seemed as though the use of embryonic stem cells gave the greatest promise for regenerative medicine because of their pluripotency. This approach proved ethically controversial because obtaining embryonic stem cells requires the destruction of human blastocysts. A major breakthrough took place in 2006 when Japanese scientists demonstrated that adult cells can be reprogrammed by activation of just a handful of genes, most of them encoding transcription factors or chromatin proteins. The reprogrammed cells were pluripotent and were therefore called **induced pluripotent cells (iPS cells).**

The success rate was only about one iPS cell per thousand, and the genetic engineering technique required the use of viruses that can sometimes cause cancer. Nevertheless, the result was regarded as spectacular. Other researchers soon found other genes that could be used for reprogramming adult cells into pluripotent or multipotent stem cells, and still other investigators developed virus-free methods for delivering the genes. In recent years, researchers have even discovered small organic molecules that can reprogram adult cells.

This kind of reprogramming opens the door to personalized stem cell therapies. The goal is to create stem cells derived from the adult cells of the individual patient. Since these cells contain the patient's own genome, problems with tissue rejection are minimized or eliminated (Chapter 43). There remains much to learn before therapeutic use of induced stem cells becomes routine. Researchers will face challenges such as increasing the efficiency of reprogramming, verifying that reprogramming is complete, making sure that the reprogrammed cells are not prone to cancer, and demonstrating that the reprogrammed cells differentiate as they should. Nevertheless, researchers hope that someday soon your own cells containing your personal genome could be reprogrammed to restore cells or organs damaged by disease or accident.

## 20.2 HIERARCHICAL CONTROL OF DEVELOPMENT

During development of a complex multicellular organism, many genes are activated and repressed at different times, thus restricting cell fates. One of the key principles of development is that genes expressed early in an organism's development control the activation of other groups of genes that act later in development. Gene regulation during development is therefore **hierarchical** in the sense that genes expressed at each stage in the process control the expression of genes that act later.

*Drosophila* **development proceeds through egg, larval, and adult stages.**
The fruit fly *Drosophila melanogaster* has played a prominent role in our understanding of the genetic control of early development, and in particular the hierarchical control of development. Researchers have isolated and analyzed a large number of mutant genes that lead to a variety of defects at different stages in development. These studies have revealed many of the key genes and processes in development, which are the focus of the following sections.

The major events in *Drosophila* development are illustrated in **Fig. 20.5**. DNA replication and nuclear division begin soon after the egg and sperm nuclei fuse (Fig. 20.5a). Unlike in mammalian development, the early nuclear divisions in the *Drosophila* embryo occur without cell division, and therefore the embryo consists of a single cell with many nuclei in the center (Fig. 20.5b). When there are roughly 5000 nuclei, they migrate to the periphery (Fig. 20.5c), where each nucleus becomes enclosed in its own cell membrane, and together they form the **cellular blastoderm** (Fig. 20.5d).

Then begins the process of **gastrulation,** in which the cells of the blastoderm migrate inward, creating layers of cells within the embryo. As in humans and most other animals (section 20.1 and Chapter 42), gastrulation forms the three germ layers (ectoderm, mesoderm, and endoderm) that differentiate into different types of cell. A *Drosophila* embryo during gastrulation is shown in Fig. 20.5e. At this stage, the embryo already shows an organization into discrete parts or segments, the formation of which is known

**FIG. 20.5**　**Life cycle of the fruit fly *Drosophila melanogaster.*** The life cycle begins with (a) a fertilized egg, followed by (b–e) a developing embryo, (f–h) larval stages, (i) pupa, and (j) adult.

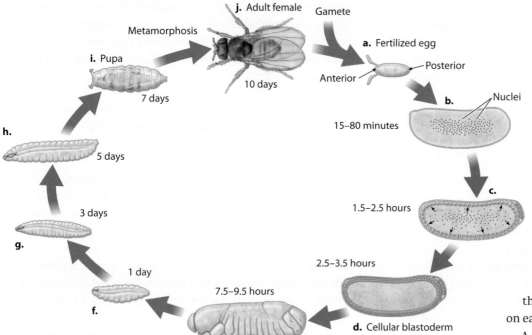

*Images not drawn to scale.

as **segmentation.** There are three cephalic segments, C1–C3 (the term "cephalic" refers to the head); three thoracic segments, T1–T3 (the thorax is the middle region of an insect); and eight abdominal segments (A1–A8). Each of these segments has a different fate in development.

About one day after fertilization, the embryo hatches from the egg as a larva (Fig. 20.5f). Over the next eight days, the larva grows and replaces its rigid outer shell, or cuticle, twice (Fig. 20.5g and h). After a week of further growth, the cuticle forms a casing—called the pupa—in which the larva is immobilized (Fig. 20.5i), and where it undergoes dramatic developmental changes known as metamorphosis that give rise to the adult fruit fly.

### The egg is a highly polarized cell.

How genes control development in *Drosophila* was inferred from systematic studies of mutants by Christiane Nusslein-Volhard and Eric F. Wieschaus, work for which they were awarded the 1995 Nobel Prize in Physiology or Medicine. One of their findings was that development starts even before a zygote is formed, in the maturation of the **oocyte,** the unfertilized egg cell produced by the mother. The oocyte, which matures under control of the mother's genes, is nonetheless important for normal embryonic development. This finding applies not only to insects like *Drosophila,* but also to many multicellular animals.

Among the striking mutants Nusslein-Volhard and Wieschaus generated and investigated were ones that significantly affected

early development (**Fig. 20.6**). These defects in very early development can be easily seen by the time the mutants reach the larval stage. In one class of mutants, called *bicoid,* larvae are missing segments at the anterior end. In another class of mutants, called *nanos,* larvae are missing segments at the posterior end. The mutant larvae are grossly abnormal and do not survive. Nusslein-Volhard and Wieschaus were able to identify each segment that was missing based on each segment's distinctive pattern of hairlike projections. In Fig. 20.6, the patterns are shown as dark shapes on each segment.

A distinguishing feature of *bicoid* and *nanos* mutants is that the abnormalities in the embryo depend on the genotype of the mother, not the genotype of the embryo. The reason the genotype of the mother can affect the phenotype of the developing embryo is that successful development requires a functioning oocyte. In

**FIG. 20.6**　**Normal and mutant *Drosophila* larvae.** (a) Nonmutant larva have anterior, middle, and posterior segments. (b) *Bicoid* mutant larva lack anterior segments. (c) *Nanos* mutant larva lack posterior structures.

**a.** Nonmutant larva

**b.** Larva from *bicoid* mutant

**FIG. 20.7** **Gradients of *bicoid* and *nanos* mRNA and protein in the developing embryo.** (a) The mRNA and protein for *bicoid* are localized in the anterior end of the egg. (b) The mRNA and protein for *nanos* are localized in the posterior end of the egg.

**a.** Distribution of *bicoid* mRNA and protein in the egg

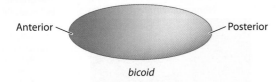

*bicoid*

**b.** Distribution of *nanos* mRNA and protein in the egg

*nanos*

*Drosophila* and many other organisms, the composition of the egg includes macromolecules (such as RNA and protein) synthesized by cells in the mother and transported into the egg. If the mother carries mutations in genes involved in the development of the oocyte, the offspring can be abnormal. Genes such as *bicoid* and *nanos* that are expressed by the mother but affect the phenotype of the offspring (in this case the developing embryo and larvae) are called **maternal-effect genes.**

A normal *Drosophila* oocyte is highly polarized, meaning that one end is distinctly different from the other. For example, there are gradients of macromolecules that define the anterior–posterior axis of the embryo as well as the dorsal–ventral axis. The best known of these gradients are those of messenger RNAs that correspond to the maternal-effect genes *bicoid* and *nanos*. **Fig. 20.7a** shows the gradient of *bicoid* mRNA across the oocyte. The bulk of *bicoid* mRNA comes from the mother, not from the oocyte, and is localized in the anterior of the egg by proteins that attach them to the cytoskeleton.

The mRNA corresponding to the maternal-effect gene *nanos* is also present in a gradient, but its maximum is at the posterior end (**Fig. 20.7b**). Like *bicoid*, mRNA for *nanos* is synthesized by the mother's cells and then imported into the oocyte. After fertilization, the zygote produces Bicoid and Nanos proteins,

and they have concentration gradients resembling those of the mRNAs in the oocyte (Fig. 20.7).

The anterior–posterior axis set up by the gradients of Bicoid and Nanos proteins is reinforced by gradients of two transcription factors called Caudal and Hunchback (**Fig. 20.8**). Like the mRNAs for Bicoid and Nanos, the mRNAs for Caudal and Hunchback are transcribed from the mother's genome and transported into the egg. As shown in Fig. 20.8a, the mRNAs for *caudal* and *hunchback* are spread uniformly in the cytoplasm of the fertilized egg. However, the mRNAs are not translated uniformly in the egg. Bicoid protein represses translation of *caudal*, and Nanos protein represses translation of *hunchback* (Fig. 20.8b). Caudal protein is therefore concentrated at the posterior end and Hunchback protein is concentrated at the anterior end. The expression of Caudal and Hunchback illustrates gene regulation at the level of translation, discussed in Chapter 19. Bicoid protein is also a transcription factor that promotes transcription of the *hunchback* gene from zygotic nuclei, which reinforces the localization of Hunchback protein at the anterior end.

The Hunchback and Caudal gradients set the stage for the subsequent steps in development. The Hunchback transcription factor targets genes needed for the development of anterior structures like eyes and antennae, and Caudal

**FIG. 20.8** **Caudal and Hunchback gradients in the developing embryo.** (a) mRNA levels of *hunchback* and *caudal* are uniform across the embryo. (b) Hunchback and Caudal protein levels are localized to the anterior and posterior ends of the embryo, respectively, because Bicoid and Nanos control the translation of *hunchback* and *caudal* mRNA.

**a.**

The gradient of *bicoid* mRNA peaks near the anterior end of the oocyte.

The gradient of *nanos* mRNA peaks near the posterior end of the oocyte.

**b.**

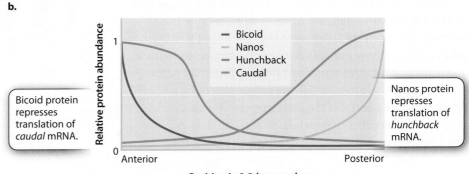

Bicoid protein represses translation of *caudal* mRNA.

Nanos protein represses translation of *hunchback* mRNA.

targets genes needed for the development of posterior structures like legs. In this way, genes expressed early in development influence the expression of genes important in later development. Because the products of the *bicoid* and *nanos* mRNA are the ones initially responsible for organizing the anterior and posterior ends of the embryo, respectively, mothers that are mutant for *bicoid* have larvae that lack anterior structures, and mothers that are mutant for *nanos* have larvae that lack posterior structures.

→   **Quick Check 3**  Do you think development would happen normally if the mother has normal *bicoid* function, but the embryo does not?

## Development proceeds by progressive regionalization and specification.

Nusslein-Volhard and Wieschaus also discovered mutants in three other classes of developmental genes. Analysis of these mutants showed that genes controlling development are turned on in groups, and that each successive group acts to refine and narrow the pattern of differentiation generated by previous groups. This, too, is a general principle of development in many multicellular organisms.

The anterior–posterior gradient set up by the maternal-effect genes is first narrowed by genes called gap genes (**Fig. 20.9**), each of which is expressed in a broad region of the embryo. The name "gap gene" derives from the phenotype of mutant embryos, which are missing groups of adjoining segments, leaving a gap in the pattern of segments. Fig. 20.9 shows the expression pattern of the gap gene *Krüppel*, which is expressed in the middle region of the embryo. Mutants of *Krüppel* lack some larval thoracic and abdominal segments. *Krüppel* is expressed in the pattern shown in Fig. 20.9 because it is under the control of the transcription factor Hunchback, which in this embryo is stained in green. High concentrations of Hunchback repress *Krüppel* transcription entirely, and low concentrations fail to induce *Krüppel* transcription. Because Hunchback is present in an anterior–posterior gradient (see Fig. 20.8), the pattern of Hunchback expression means that *Krüppel* is transcribed only in the middle region of the embryo where Hunchback is present but not too abundant.

The next level of the regulatory hierarchy is that of pair-rule genes (**Fig. 20.10**), which receive their name because their mutants lack alternate larval segments. The example in Fig. 20.10 is *hairy*, whose mutants lack the odd-numbered thoracic segments and the even-numbered abdominal segments. The pair-rule genes help to establish the uniqueness of each of seven broad stripes across the anterior–posterior axis.

The pair-rule genes in turn help to regulate the next level in the segmentation hierarchy, which consists of

**FIG. 20.9**     **Normal gap-gene expression pattern and mutant phenotype.**

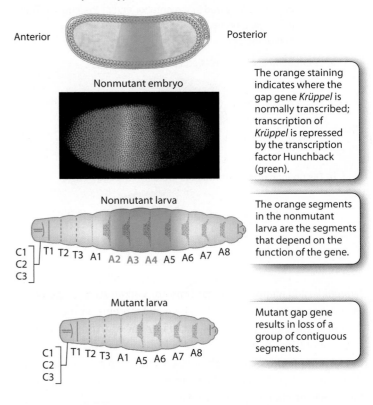

Anterior                              Posterior

Nonmutant embryo

The orange staining indicates where the gap gene *Krüppel* is normally transcribed; transcription of *Krüppel* is repressed by the transcription factor Hunchback (green).

Nonmutant larva

The orange segments in the nonmutant larva are the segments that depend on the function of the gene.

C1
C2   T1 T2 T3  A1  A2 A3 A4 A5 A6 A7  A8
C3

Mutant larva

Mutant gap gene results in loss of a group of contiguous segments.

C1
C2   T1 T2 T3  A1  A5 A6 A7  A8
C3

**FIG. 20.10**    **Normal pair-rule gene expression pattern and mutant phenotype.**

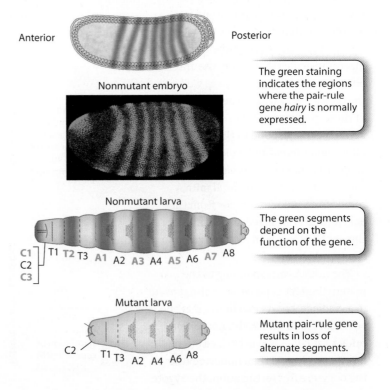

Anterior                              Posterior

The green staining indicates the regions where the pair-rule gene *hairy* is normally expressed.

Nonmutant embryo

Nonmutant larva

The green segments depend on the function of the gene.

C1   T1 T2 T3  A1  A2 A3 A4 A5 A6 A7  A8
C2
C3

Mutant larva

Mutant pair-rule gene results in loss of alternate segments.

C2    T1 T3  A2 A4 A6  A8

**FIG. 20.11 Normal segment-polarity gene expression pattern and mutant phenotype.**

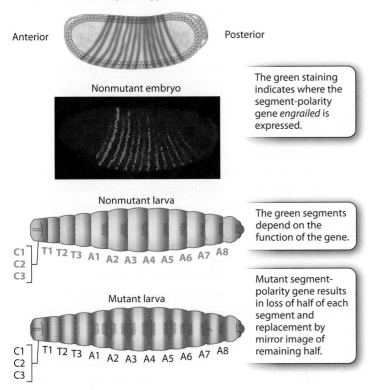

Anterior

Posterior

Nonmutant embryo

The green staining indicates where the segment-polarity gene *engrailed* is expressed.

Nonmutant larva

The green segments depend on the function of the gene.

C1 C2 C3  T1 T2 T3 A1 A2 A3 A4 A5 A6 A7 A8

Mutant larva

Mutant segment-polarity gene results in loss of half of each segment and replacement by mirror image of remaining half.

C1 C2 C3  T1 T2 T3 A1 A2 A3 A4 A5 A6 A7 A8

segment-polarity genes (**Fig. 20.11**). The segment-polarity genes refine the 7-striped pattern still further into a 14-striped pattern. Each of the 14 stripes has distinct anterior and posterior ends determined by the segment-polarity genes. Embryos with mutations in segment-polarity genes lose this anterior–posterior differentiation, with the result that the anterior and the posterior halves of each segment are mirror images. The example in Fig. 20.11 is *engrailed*, which eliminates the posterior pattern element in each stripe and replaces it with a mirror image of the anterior pattern element.

→ **Quick Check 4** Do you think that the pattern of segment-polarity gene expression would be normal if one or more gap genes are not expressed properly?

## Homeotic genes determine where different body parts develop in the organism.

Together, the segment-polarity genes and certain genes expressed earlier in the hierarchy control the pattern of expression of another set of genes called **homeotic genes**. Originally discovered in *Drosophila*, homeotic genes encode some of the most important transcription factors in animal development. A homeotic gene is a gene that specifies the identity of a body part or segment during embryonic development. For example, homeotic genes instruct the three thoracic segments (T1, T2, and T3) to each develop a set of legs, and the second thoracic segment (T2) to develop wings.

Two classic examples of the consequences of mutations in homeotic genes in *Drosophila* are shown in **Fig. 20.12**. Fig. 20.12a shows what happens when the homeotic gene *Antennapedia*, which specifies the development of the leg, is inappropriately expressed in anterior segments. The mutation causes legs to grow where antennae usually would. Similarly, a mutation in the homeotic gene *Bithorax* results in the transformation of thoracic segment 3 (T3) into thoracic

**FIG. 20.12 Homeotic genes and segment identity.** (a) Normal antennae are transformed into legs in an *Antennapedia* mutant. (b) Normal structures in the third thoracic segment are transformed into wings in a *Bithorax* mutant.

Normal

Mutant

**a.** *Antennapedia*

**b.** *Bithorax*

Haltere

**FIG. 20.13  Tissues controlled by homeotic genes.** Homeotic genes specify the fate of clumps of tissue in the *Drosophila* larva.

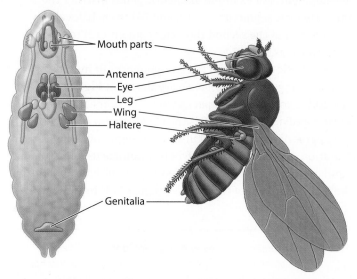

- Mouth parts
- Antenna
- Eye
- Leg
- Wing
- Haltere
- Genitalia

segment 2 (T2), so that the fruit fly has two T2 segments in a row. As shown in the Fig. 20.12b, the result is a fruit fly with two complete sets of wings.

In *Drosophila*, the adult body parts like legs, antennae, and wings are formed from organized collections of tissue located throughout the larval body (**Fig. 20.13**). The development of these tissues and their metamorphosis into the adult body parts are regulated by the homeotic genes. First expressed at about the same time as the segment-polarity genes (see Fig. 20.11), the homeotic genes continue to be expressed even after the genes that regulate early development have shut down. Their continuing activity is due to the presence of chromatin remodeling proteins that keep the chromatin physically accessible to the transcription complex (Chapter 19).

Homeotic genes encode transcription factors. The DNA-binding domain in the homeotic proteins is a sequence of 60 amino acids called a **homeodomain,** whose sequences are very similar from one homeotic protein to the next. The homeodomain is specified by a DNA sequence within the homeotic genes called the **homeobox** (the name is the origin of the term **Hox gene,** which is often used as a synonym for "homeotic gene").

The *Drosophila* genome contains eight *Hox* genes comprising two distinct clusters, the *Antennapedia* complex and the *Bithorax* complex (**Fig. 20.14**). The genes are arranged along the chromosome in the same order as their products function in anterior–posterior segments along the embryo. In addition, the timing of their expression corresponds to their order along the chromosome and location of expression, with genes that

are expressed closer to the anterior end turned on earlier than genes that are expressed closer to the posterior end. The correlation among linear order along the chromosome, anterior–posterior position in the embryo, and timing of expression is observed in *Hox* clusters in almost all organisms studied.

Because the amino acid sequences of the homeodomains of *Hox* gene products are very similar from one organism to the next, *Hox* gene clusters have been identified in a wide variety of animals with bilateral symmetry (organisms in which both sides of the midline are mirror images), from insects to mammals. Comparison of the number and types of *Hox* genes in different species supports the hypothesis that the ancestral *Hox* gene cluster had an organization very similar to what we now see in most organisms with *Hox* gene clusters. In its evolutionary history, the vertebrate genome underwent two whole-genome duplications; hence, vertebrates have four copies of the *Hox* gene cluster (**Fig. 20.15**).

Unlike the *Hox* genes in *Drosophila*, mammalian *Hox* genes do not specify limbs but are important in the embryonic development of structures that become parts of the hindbrain, spinal cord, and vertebral column (Fig. 20.15). As in *Drosophila*, the genes in each cluster are expressed according to their linear order along the chromosome, which coincides with the linear order of regions the genes affect in the embryo. Each gene helps to specify the identity of the region in which it is expressed. Many of the genes in the mammalian *Hox* clusters

**FIG. 20.14  Organization of the Hox gene clusters in Drosophila and the body parts that they affect.** The order of genes along the chromosome corresponds to their position along the anterior–posterior axis in the developing embryo.

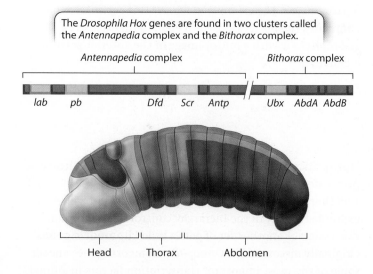

The *Drosophila Hox* genes are found in two clusters called the *Antennapedia* complex and the *Bithorax* complex.

*Antennapedia* complex          *Bithorax* complex

lab    pb          Dfd  Scr  Antp        Ubx  AbdA  AbdB

Head        Thorax        Abdomen

**FIG. 20.15** Organization and content of *Hox* gene clusters in the mouse and the regions of the embryo in which they are expressed.

**a.**

Inferred *Hox* gene organization in unidentified common ancestor.

Gene duplications

The mammalian genome contains four copies of *Hox* gene clusters. The genes figure prominently in the regional differentiation of parts of the brain and vertebral column.

**b.**

HoxA — a-1 a-2 a-3 a-4 a-5 a-6 a-7 a-9 a-10 a-11 a-13

HoxB — b-1 b-2 b-3 b-4 b-5 b-6 b-7 b-8 b-9

HoxC — c-4 c-5 c-6 c-8 c-9 c-10 c-11 c-12 c-13

HoxD — d-1 d-3 d-4 d-8 d-9 d-10 d-11 d-12 d-13

Gene deletions

Anterior

Posterior

have redundant or overlapping functions so that learning exactly what each gene does continues to be a research challenge.

## 20.3 EVOLUTIONARY CONSERVATION OF KEY TRANSCRIPTION FACTORS IN DEVELOPMENT

As we have seen, *Hox* genes and the proteins they encode are very similar in a wide range of organisms. As a result, they can be identified by their similarity in DNA or amino acid sequence. Molecules that are similar in sequence among distantly related organisms are said to be evolutionarily conserved. Their similarity in sequence suggests that they were present in the most recent common ancestor and have changed very little over time because they carry out a vital function. Many transcription factors important in development are evolutionarily conserved. An impressive example is found in the development of animal eyes.

### Animals have evolved a wide variety of eyes.
Animal eyes show an amazing diversity in their development and anatomy (**Fig. 20.16**). Among the simplest eyes are those of the planarian flatworm (Fig. 20.16a), which are only pit-shaped cells containing light-sensitive photoreceptors. The planarian eye has no lens to focus the light, but the animal can perceive differences in light intensity. The incorporation of a spherical lens, as in some jellyfish (Fig. 20.16b), improves the image.

Among the most complex eyes are the camera-type eyes of the squid (Fig. 20.16c) and human (Fig. 20.16d), which have a single

**FIG. 20.16** Diversity of animal eyes.

**a.** Planaria

**b.** Jellyfish

**c.** Squid

**d.** Human

**e.** House fly

**f.** Trilobite

lens to focus light onto a light-sensitive tissue, the retina (Chapter 36). Though the single-lens eyes of squids and vertebrates are similar in external appearance, they are vastly different in their development, anatomy, and physiology.

Some organisms, such as the house fly (Fig. 20.16e) and other insects, have compound eyes, that is, eyes consisting of hundreds of small lenses arranged on a convex surface pointing in slightly different directions. This arrangement of lenses allows a wide viewing angle and detection of rapid movement.

In the history of life, eyes are very ancient. By the time of the extraordinary diversification of animals 542 million years ago known as the Cambrian explosion, organisms such as trilobites (Fig. 20.16f) already had well-formed compound eyes. These differed greatly from the eyes in modern insects, however. For example, trilobite eyes had hard mineral lenses composed of calcium carbonate (the world's first safety glasses, so to speak). Altogether, about 96% of living animal species have true eyes that produce an image, as opposed to simply being able to detect differences in light intensity. Despite this common feature, the extensive diversity among the eyes of different organisms encouraged evolutionary biologists in the belief that the ability to perceive light may have evolved independently about 40 to 60 times.

### Pax6 is a master regulator of eye development.

The diversity of animal eyes suggests that they may have evolved independently in different organisms. However, more recent evidence suggests an alternative hypothesis—that they evolved once, very early in evolution and subsequently diverged over time. One argument against the idea of multiple independent origins for light perception is the observation that the light-sensitive molecule in all light-detecting cells is the same, a derivative of vitamin A complexed with a protein called opsin (Chapter 36). The presence of the same light-sensitive molecule in diverse eyes argues that it may have been present in the common ancestor of all animals with eyes and has been retained over time.

Another argument against multiple independent origins of eyes came from studies of eye development. Researchers identified *eyeless,* a gene in the fruit fly *Drosophila.* As its name implies, the phenotype of *eyeless* mutants is abnormal eye development (**Fig. 20.17a**). When the protein product of the *eyeless* gene was identified, it was found to be a transcription factor called Pax6. Mutant forms of a *Pax6* gene were already known to cause small eyes in the mouse (**Fig. 20.17b**) and aniridia (absence of the iris) in humans.

The mutations in the *Pax6* gene that cause the development defects in the eye in fruit flies, mice, and humans are **loss-of-function mutations.** A loss-of-function mutation is just that: a mutation that inactivates the normal function of a gene. In this case, loss-of-function mutations in *Pax6* make a defective version of the Pax6 transcription factor that is not able to carry out its function.

The strong conservation of Pax6 in *Drosophila* and mouse and the similarity in their mutant phenotypes led Swiss developmental

**FIG. 20.17 Effect of Pax6 mutations on eye development.** (a) Normal and *eyeless* mutant in *Drosophila;* (b) normal and *small eye* mutant in mouse.

| Normal | Mutant |
|---|---|

**a.** *Drosophila*

**b.** Mouse

biologist Walter Gehring to hypothesize that Pax6 might be a master regulator of eye development. In other words, he hypothesized that Pax6 binds to regulatory regions of a set of genes that turns on a developmental program that induces eye development. In theory, this means that Pax6 can induce the development of an eye in any tissue in which it is expressed.

To test this hypothesis, Gehring and collaborators genetically engineered fruit flies that would produce the normal Pax6 transcription factor in the antenna, where it is not normally expressed (**Fig. 20.18a**). Any mutation in which a gene is expressed in the wrong place or at the wrong time is known as a **gain-of-function mutation.** (We saw an example of gain-of-function mutations when the *Hox* genes were expressed in the wrong segments of the fruit fly, producing legs where antennae normally develop and thus resulting in four-winged fruit flies.) For genes that control a developmental pathway, loss-of-function mutations and gain-of-function mutations often have opposite effects on phenotype.

→   **Quick Check 5**   For genes that control pathways of development, loss-of-function mutations are usually recessive whereas gain-of-function mutations are usually dominant. Can you suggest a reason why?

Since loss-of-function mutations in *Pax6* result in an eyeless phenotype, a gain-of-function mutation should result in eyes developing in whatever tissue *Pax6* is expressed. The result is shown in **Fig. 20.18b**. In the gain-of-function mutation, the antenna developed into a miniature compound eye, and electrical recordings demonstrated that some of these antennal eyes were

**FIG. 20.18 Pax6, a master switch controlling eye development.**
(a) Normal fly antenna; (b) eye tissue induced by expression of fruit fly *Pax6* in the antenna; (c) eye tissue induced by expression of mouse *Pax6* in the antenna.

Normal antenna

Antennal eye
induced by
*Drosophila Pax6* gene

Antennal eye
induced by mouse
*Pax6* gene

functional. The researchers also created other gain-of-function mutations that led to eyes on the legs, wings, and other tissues, which the *New York Times* publicized in an article headlined "With New Fly, Science Outdoes Hollywood."

Gehring and his group then went one step further. They took the *Pax6* gene from mice and expressed it in fruit flies to see whether the mouse *Pax6* gene is similar enough to the fruit fly version of the gene that it could induce eye development in the fruit fly. Specifically, they created transgenic fruit flies that expressed the *Pax6* gene from mice in the fruit fly antenna. The result is shown in **Fig. 20.18c**. The mouse gene induced a miniature eye in the fly. Note, however, that the *Pax6* gene from mice induced the development of a compound eye of *Drosophila*, not the single-lens eye of mouse.

The ability of mouse Pax6 to make an eye in fruit flies suggests that mouse and fruit fly *Pax6* are not only similar in DNA and amino acid sequence, but also similar in function, and indeed act as a master switch that can turn on a developmental program leading to the formation of an eye. But these observations lead to another question: Why did the mouse *Pax6* gene produce fruit fly eyes instead of mouse eyes? The answer is that the fruit fly genome does not include the genes needed to make mouse eyes. Mouse Pax6 protein induces fly eyes because of the **downstream genes** affected by Pax6, those that function later in the process of eye development. Transcription factors like Pax6 interact with their target genes by binding with short DNA sequences adjacent to the gene, usually at the 5′ end, called **cis-regulatory elements.** Some transcription factors act as repressors that prevent transcription of the target gene, and others serve as activators by recruiting the transcriptional machinery to the target gene (Chapter 19). A transcription factor can even repress some of its target genes and activate others.

In the fruit fly, Pax6 binds to cis-regulatory elements in many genes, turning some genes on and others off. The products

of these downstream genes in turn affect the expression of further downstream genes. The total number of genes that are downstream of Pax6 and that are needed for eye development is estimated at about 2000. Most are not direct targets of Pax6 but are activated indirectly through other transcription factors downstream of Pax6. When mouse Pax6 is expressed in fruit flies, it is similar enough in sequence to activate the genes involved in fruit fly eye development, so it makes sense that mouse Pax6 leads to the development of a fruit fly eye, not a mouse eye.

One scenario of how Pax6 became a master switch for eye development in a wide range of organisms, but produces a diversity of eyes in these organisms, is that Pax6 evolved early in the history of life as a transcription factor able to bind to and regulate genes involved in early eye development. Over time, different genes in different organisms acquired new Pax6-binding cis-regulatory elements by mutation, and if these were beneficial they persisted. The downstream genes that are targets of Pax6 therefore are different in different organisms, but they share two features—they are regulated by Pax6 and they are involved in eye development. So, the early steps are conserved, but the later ones are not.

These studies of Pax6 argue that the genetic pathway for eye development evolved early and is shared among a wide range of animals, even though the eyes that are produced are quite diverse.

## 20.4 COMBINATORIAL CONTROL IN DEVELOPMENT

Most cis-regulatory elements are located near one or more binding sites for transcription factors, some of which are activators of transcription and others repressors of transcription. The rate of transcription of any gene in any type of cell is therefore determined by the combination of transcription factors that are present in the cell and by the relative balance of activators and repressors. Regulation of gene transcription according to the mix of transcription factors in the cell is known as **combinatorial control.** Combinatorial control of transcription is another general principle often seen in all multicellular organisms at many stages of development. Here, we discuss flower development as an example of this general principle.

### Floral differentiation is a model for plant development.
The plant *Arabidopsis thaliana*, a weed commonly called mouse-ear cress, is a model organism for developmental studies and presents a clear case of combinatorial control. As in all plants, *Arabidopsis* has regions of undifferentiated cells, called meristems, where growth can take place (Chapter 31). Meristem cells are similar to stem cells in animals. They are the growing points where shoots, roots, and flowers are formed. In floral meristems of *Arabidopsis,* the flowers develop from a pattern of four concentric circles, or whorls, of cells.

**FIG. 20.19** Development of the *Arabidopsis* floral organs from concentric whorls of cells in the floral meristem.

**a.**

■ Whorl 4 ⇒ Carpel
■ Whorl 3 ⇒ Stamen
■ Whorl 2 ⇒ Petal
■ Whorl 1 ⇒ Sepal

Cells in the floral meristem are organized in a pattern of four concentric whorls (1–4).

**b.**

Each whorl of cells in the floral meristem gives rise to a different organ of the mature flower.

From the outermost whorl (whorl 1) of cells to the innermost whorl (whorl 4), the floral organs are formed as shown in **Fig. 20.19**:

• Cells in whorl 1 form the green sepals, which are modified leaves forming a protective sheath around the petals in the flower bud that unfold like petals when the flower opens.

• Cells in whorl 2 form the petals.

• Cells in whorl 3 form the stamens, the male sexual structures in which pollen is produced. (The small yellow granules in Fig. 20.19b are pollen grains.)

• Cells in whorl 4 form the carpels, the female reproductive structures that receive pollen and contain the ovaries.

### The identity of the floral organs is determined by combinatorial control.

The genetic control of flower development in *Arabidopsis* was discovered by the analysis of mutant plants, an investigation similar in principle to the way that genetic control of development in fruit flies was determined. The plant mutants fell into three classes showing characteristic floral abnormalities (**Fig. 20.20**). Mutations in the gene *APETALA-2* inactivate the gene's function and affect whorls 1 and 2 (Fig. 20.20b), those in *APETALA-3* or *PISTILLATA* affect whorls 2 and 3 (Fig. 20.20c), and those in *AGAMOUS* affect whorls 3 and 4 (Fig. 20.20d). (There are different standards for naming genes and proteins in different organisms. In *Arabidopsis*, wild-type alleles are written in capital letters with italics, and wild-type proteins in capital letters without italics. Mutant alleles and proteins follow the same rules, but are written in lower-case letters.)

The observation that the mutant phenotypes affect development of pairs of adjacent whorls already hints at combinatorial control, which researchers incorporated into a model of floral development called the **ABC model.** The model invokes three activities arbitrarily called A, B, and C. These activities represent the function of a protein or proteins that is hypothesized to be present in cells of each whorl as shown in **Fig. 20.21**. The A activity is present in whorls 1 and 2, B in whorls 2 and 3, and C in whorls 3 and 4. According to the ABC model, activity of A alone results in the formation of sepals, A and B together result in petals, B and C together result in stamens, and C alone results in carpels. Like any good hypothesis (Chapter 1), this one makes specific predictions. Mutants lacking A will have defects

**FIG. 20.20 Phenotypes of the normal and floral mutants of *Arabidopsis*.**

a. 1 2 3 4
In a nonmutant flower, whorl 1 makes sepals, whorl 2 makes petals, whorl 3 makes stamens, whorl 4 makes carpels.

b. 1 2 3 4
In the mutant *APETALA-2*, whorl 1 makes carpels, whorl 2 makes stamens, whorl 3 makes stamens, whorl 4 makes carpels.

c. 1 2 3 4
In the mutant *APETALA-3* or the mutant *PISTILLATA*, whorl 1 makes sepals, whorl 2 makes sepals, whorl 3 makes carpels, whorl 4 makes carpels.

d. 1 2 3 4
In the mutant *AGAMOUS*, whorl 1 makes sepals, whorl 2 makes petals, whorl 3 makes petals, whorl 4 makes sepals.

**FIG. 20.21 The ABC model for flower development in *Arabidopsis*.**

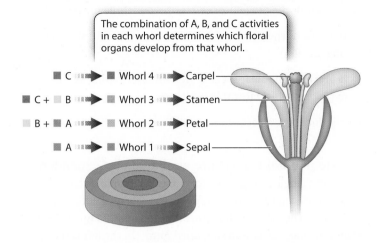

The combination of A, B, and C activities in each whorl determines which floral organs develop from that whorl.

- C ➡ Whorl 4 ➡ Carpel
- C + B ➡ Whorl 3 ➡ Stamen
- B + A ➡ Whorl 2 ➡ Petal
- A ➡ Whorl 1 ➡ Sepal

in whorls 1 and 2, mutants lacking B will have defects in whorls 2 and 3, and mutants lacking C will have defects in whorls 3 and 4.

Matching these predictions with the mutants in Fig. 20.20 yields the following correspondence:

- The A activity is encoded by the gene *APETALA-2*.

- The B activity is encoded jointly by the genes *APETALA-3* and *PISTILLATA*.

- The C activity is encoded by the gene *AGAMOUS*.

To account for the mutant phenotypes in Fig. 20.20, you need only to postulate that when A is absent, C is expressed in whorls 1 and 2 in addition to its normal expression in whorls 3 and 4; and when C is absent, A expression expands into whorls 3 and 4 in addition to its normal expression in whorls 1 and 2. The domains of expression expand in this way because, in addition to its role in activating certain downstream genes, the product of *APETALA-2* represses the expression of *AGAMOUS* in whorls 1 and 2. As a result, *AGAMOUS* is normally not expressed in whorls 1 and 2. Similarly, in addition to its role in activating other genes, the product of *AGAMOUS* represses the expression of *APETALA-2* in whorls 3 and 4, so *APETALA-2* is normally not expressed in whorls 3 and 4.

→ **Quick Check 6** Predict the phenotype of a flower in which *APETALA-2* (activity A) and *AGAMOUS* (activity C) are expressed normally, but *APETALA-3* and *PISTILLATA* (activity B) are expressed in all four whorls.

The identification of the products of these genes demonstrates combinatorial control and explains the mutant phenotypes. All of the genes encode transcription factors, which are called AP2, AP3, PI, and AG, corresponding to the four genes. The transcription factors bind to cis-regulatory elements of genes that encode proteins necessary for each whorl's development, similar to what we saw earlier for Pax6 and eye development. AP3 and PI are both necessary for the B activity

because the proteins form a heterodimer, a protein made up of two different subunits.

Although other transcription factors associated with activities D and E that also contribute to flower development were discovered later, the original ABC model is still valid and serves as an elegant example of combinatorial control. As in the case of the *Pax6* gene in the development of animals' eyes, evolution of the cis-regulatory sequences in downstream genes has resulted in a wide variety of floral morphologies in different plant lineages, some of which are shown in **Fig. 20.22.**

## 20.5 CELL SIGNALING IN DEVELOPMENT

As we have seen in the discussion of stem cells, differentiated cells can be reprogrammed by the action of only a few key genes or small organic molecules. In some cases, the processes that push differentiation in a forward direction are also quite simple. An important example is **signal transduction,** in which an extracellular molecule acts as a signal to activate a membrane protein that in turn activates molecules inside the cell that control differentiation (Chapter 9). The signaling molecule is called the **ligand** and the membrane protein that it activates is called the **receptor.** The following example shows how a simple receptor–ligand pair can

**FIG. 20.22 Floral diversity.** (a) Colorado blue columbine (*Aquilegia coaerulea*); (b) slipper orchid (*Paphiopedilum holdenii*); (c) ginger flower (*Smithatris supraneanae*); (d) garden rose (hybrid).

have profound effects not only on the differentiation of the cell that carries the receptor but also on its neighbors.

## A signaling molecule can cause multiple responses in the cell.

Pioneering experiments on signal transduction in development were carried out in the soil nematode *Caenorhabditis elegans* (**Fig. 20.23**). This worm is a favored model organism for developmental studies because of its small size (the adult is about 1 millimeter long), short generation time (2½ days from egg to sexually mature adult), and stereotyped process of development. (A stereotyped process is one that is always the same.) The adult animal consists of exactly 959 somatic (body) cells, and the cells' patterns of division, migration, differentiation, and morphogenesis are identical in each individual *C. elegans*. The adult is typically a hermaphrodite that produces both male and female reproductive cells. The eggs are fertilized internally by sperm, and, after undergoing five or six cell divisions, the eggs are laid through an organ known as the vulva (Fig. 20.23).

Vulva development is relatively simple and therefore amenable to experimental studies. The entire structure arises from only three cells that undergo either of two types of differentiation and divide to produce a vulva consisting of exactly 22 cells. Cells resulting from type 1 differentiation form the opening of the vulva, and cells resulting from type 2 differentiation form a supporting structure around the vulval opening. In each of the two types of differentiation, different genes are activated or repressed, resulting in differentiated cells with different functions.

How each of the three original cells, called progenitor cells, differentiates is determined by its position in the developing worm (**Fig. 20.24**). The fate of the progenitor cells is determined by their proximity to another cell, called the anchor cell (AC in Fig. 20.24a), which secretes a protein called epidermal growth factor (EGF) that binds to and activates a transmembrane EGF

receptor (Chapter 9). The progenitor cell closest to the anchor cell receives the most amount of signal, and upon activation of its EGF receptor the progenitor cell carries out three functions:

- It activates the genes for differentiation into a type 1 cell (Fig. 20.24a)

- It prevents the adjacent cells from differentiating as type 1 (a process called **lateral inhibition**).

- It induces the adjacent cells to differentiate as type 2 cells.

## Developmental signals are amplified and expanded.

How can a single ligand–receptor pair cause so many changes in gene expression that it determines the pathway of differentiation not only of the cell itself but also of its neighbors? **Fig. 20.25** shows the mechanisms in simplified form. On the left is the progenitor cell nearest the anchor cell, which receives the strongest EGF signal. Activation of the EGF receptor by the ligand initiates a process of signal transduction in the cytoplasm, in which the signal is transmitted from one protein to the next by means of phosphorylation, which amplifies the signal at each

---

**FIG. 20.23  The nematode worm *Caenorhabditis elegans*.**

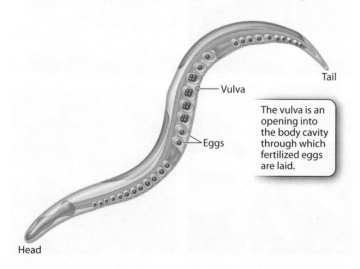

The vulva is an opening into the body cavity through which fertilized eggs are laid.

---

**FIG. 20.24  Vulval development in *Caenorhabditis elegans*.** The cells of the vulva differentiate in response to molecular signals from other cells.

**a.**

The anchor cell (AC) induces vulval development.

The EGF ligand is a protein signal that induces type 1 differentiation in the nearest cell.

The type 1 cell sends out signals to the adjacent cells that prevent type 1 differentiation (lateral inhibition) and that promote type 2 differentiation.

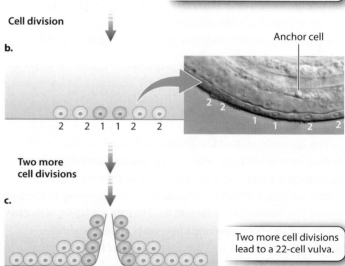

Two more cell divisions lead to a 22-cell vulva.

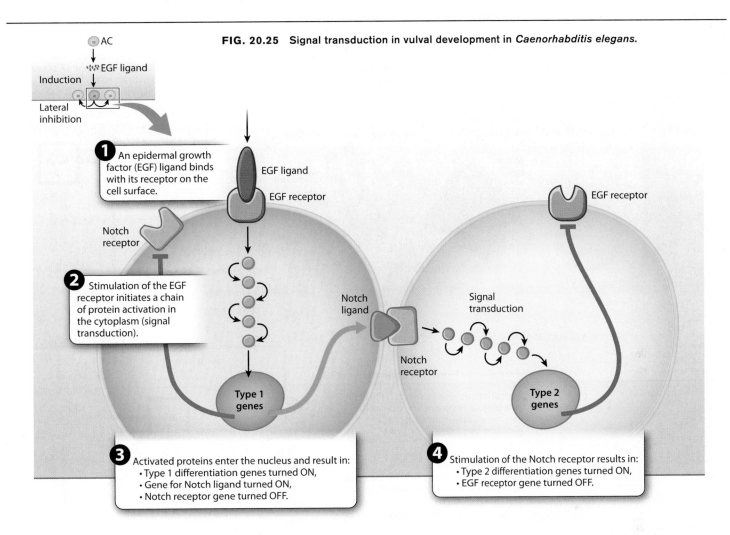

**FIG. 20.25 Signal transduction in vulval development in *Caenorhabditis elegans*.**

**1** An epidermal growth factor (EGF) ligand binds with its receptor on the cell surface.

**2** Stimulation of the EGF receptor initiates a chain of protein activation in the cytoplasm (signal transduction).

**3** Activated proteins enter the nucleus and result in:
- Type 1 differentiation genes turned ON,
- Gene for Notch ligand turned ON,
- Notch receptor gene turned OFF.

**4** Stimulation of the Notch receptor results in:
- Type 2 differentiation genes turned ON,
- EGF receptor gene turned OFF.

stage (Chapter 9). The result of signal transduction is that a set of transcription factors is activated.

In the nucleus, the transcription factors activate transcription of genes for type 1 differentiation. The transcription factors also activate transcription of genes that produce another type of protein ligand, called Notch. The Notch ligand is a transmembrane protein that activates Notch receptors in the neighboring cells. Activation of the Notch receptor in these cells activates another signal transduction cascade, which results in transcription of the genes for type 2 differentiation. The cascade started by the binding of Notch also activates transcription of other genes whose products inhibit the EGF receptor. Inhibiting the EGF receptor in type 2 cells prevents EGF from eliciting a type 1 response in the type 2 cell. In addition, at the same time that the Type 1 cell produces the Notch ligand, it produces proteins that inhibit its own Notch receptors, and this prevents Notch from initiating a type 2 response in the type 1 cell.

While vulva development in nematodes is a fairly simple example of the importance of ligand–receptor signaling in development, EGF and Notch ligands and their receptors are found in virtually all animals. They are among dozens of ligand–receptor pairs that have evolved as signaling mechanisms to regulate processes in cellular metabolism and development.

Humans have several distinct but related gene families of EGF and Notch ligands and receptors. Human EGF is important in cell survival, proliferation, and differentiation. EGF functions in the differentiation and repair of multiple types of cells in the skin; it is present in all body fluids and helps regulate rapid metabolic responses to changing conditions. Human Notch ligands are involved in development of the nervous and immune systems as well as heart, pancreas, and bone. Abnormalities in EGF or Notch signaling are associated with many different types of cancer.

Therefore, just as we saw in our discussion of eye and flower development, the molecular players involved in development are often evolutionarily conserved across a wide range of organisms. This is true even of genes that we typically associate with disease, such as *BRCA1* and *BRCA2* and their link with cancer. These genes not only play a role in cell cycle control in the adult, but also in early development in many organisms. In fact, although heterozygous mutations in these genes predispose to breast and ovarian cancers in humans, homozygous mutations are lethal in early embryonic stages. We focus on the *BRCA1* gene to summarize key concepts about DNA replication, mutation, genetic variation, inheritance, gene regulation, and development in **Fig. 20.26**.

# Genetic Variation and Inheritance

**Integrating concepts from Chapters 12–20**

## BRCA1 Genetics

The *BRCA1* gene on chromosome 17 encodes a protein that repairs double-strand breaks in DNA that result from DNA replication or environmental agents. The *BRCA1* gene is present in all cells, but is expressed, or turned on, in rapidly dividing cells, including breast and ovarian cells. If *BRCA1* does not function properly due to a mutation, damaged DNA is not repaired, which in turn increases the risk for certain kinds of cancers, particularly breast and ovarian cancers in women.

**Genetic variation**:
Everyone carries two copies of the *BRCA1* gene on chromosome 17. Some individuals have an allele of the gene that increases the risk of breast and ovarian cancer. In some populations, such as Ashkenazi Jews, the frequency of specific mutations in the gene is high.

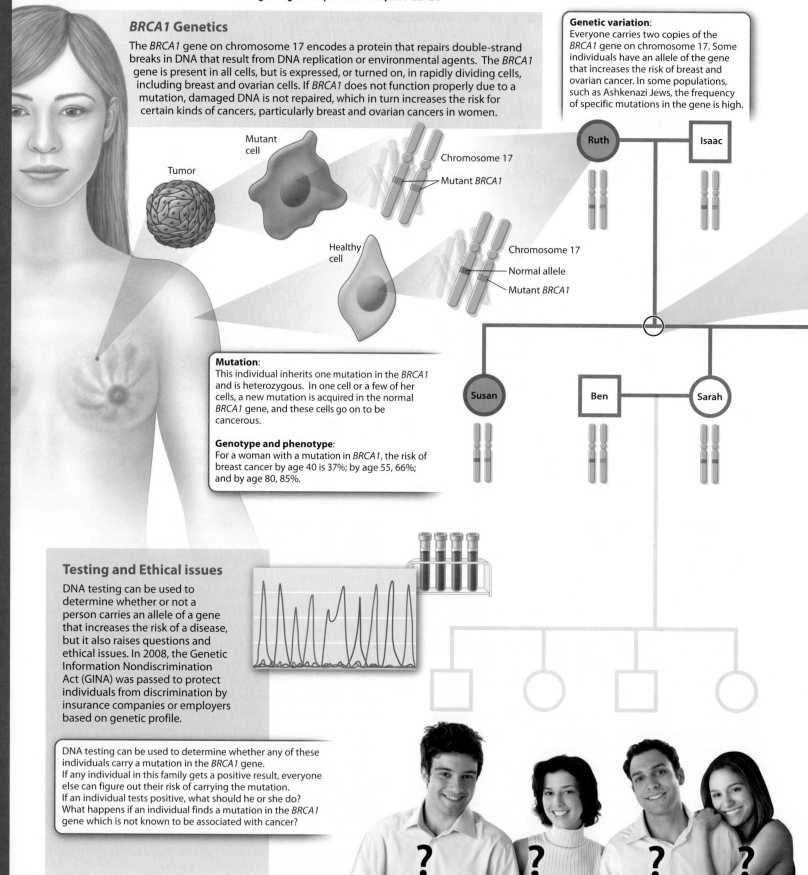

Mutant cell

Tumor

Chromosome 17

Mutant *BRCA1*

Healthy cell

Chromosome 17

Normal allele

Mutant *BRCA1*

**Mutation**:
This individual inherits one mutation in the *BRCA1* and is heterozygous. In one cell or a few of her cells, a new mutation is acquired in the normal *BRCA1* gene, and these cells go on to be cancerous.

**Genotype and phenotype**:
For a woman with a mutation in *BRCA1*, the risk of breast cancer by age 40 is 37%; by age 55, 66%; and by age 80, 85%.

Ruth    Isaac

Susan    Ben    Sarah

## Testing and Ethical issues

DNA testing can be used to determine whether or not a person carries an allele of a gene that increases the risk of a disease, but it also raises questions and ethical issues. In 2008, the Genetic Information Nondiscrimination Act (GINA) was passed to protect individuals from discrimination by insurance companies or employers based on genetic profile.

DNA testing can be used to determine whether any of these individuals carry a mutation in the *BRCA1* gene.
If any individual in this family gets a positive result, everyone else can figure out their risk of carrying the mutation.
If an individual tests positive, what should he or she do?
What happens if an individual finds a mutation in the *BRCA1* gene which is not known to be associated with cancer?

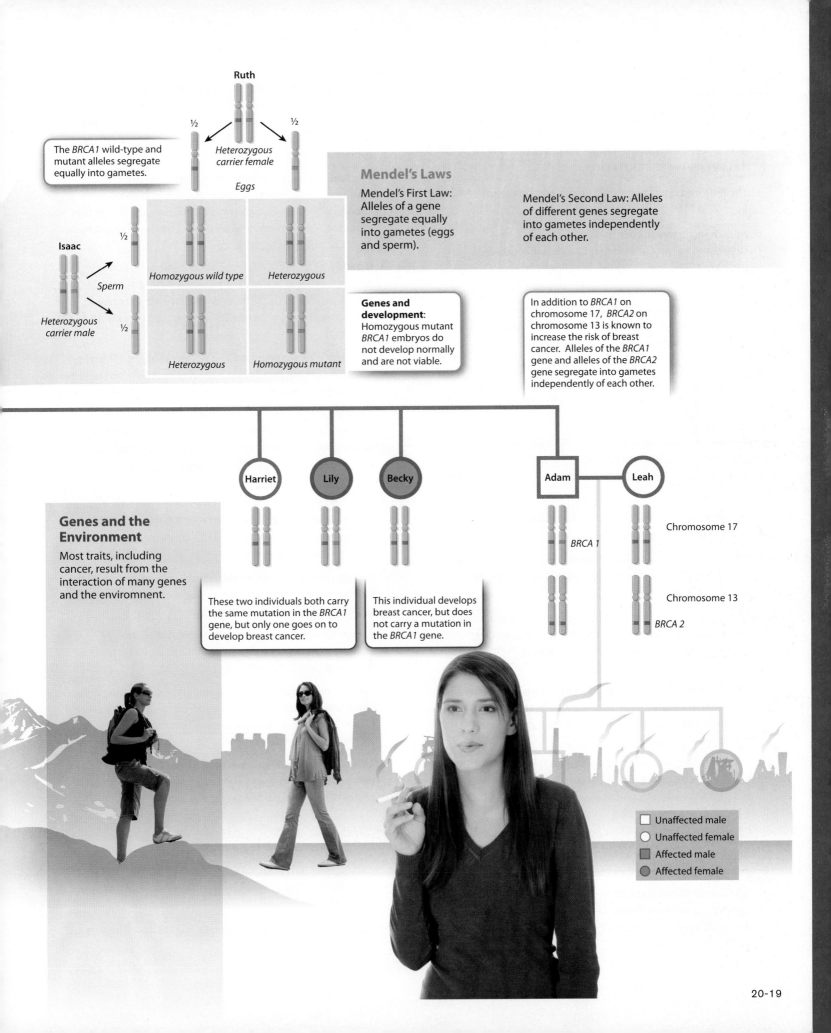

**Ruth**

½    ½

The *BRCA1* wild-type and mutant alleles segregate equally into gametes.

*Heterozygous carrier female*

*Eggs*

**Mendel's Laws**

Mendel's First Law: Alleles of a gene segregate equally into gametes (eggs and sperm).

Mendel's Second Law: Alleles of different genes segregate into gametes independently of each other.

**Isaac**

½

*Homozygous wild type*    *Heterozygous*

*Sperm*

½

*Heterozygous carrier male*

*Heterozygous*    *Homozygous mutant*

**Genes and development**: Homozygous mutant *BRCA1* embryos do not develop normally and are not viable.

In addition to *BRCA1* on chromosome 17, *BRCA2* on chromosome 13 is known to increase the risk of breast cancer. Alleles of the *BRCA1* gene and alleles of the *BRCA2* gene segregate into gametes independently of each other.

**Harriet**    **Lily**    **Becky**    **Adam**    **Leah**

**Genes and the Environment**

Most traits, including cancer, result from the interaction of many genes and the enviromnent.

These two individuals both carry the same mutation in the *BRCA1* gene, but only one goes on to develop breast cancer.

This individual develops breast cancer, but does not carry a mutation in the *BRCA1* gene.

*BRCA 1*

Chromosome 17

Chromosome 13

*BRCA 2*

☐ Unaffected male
○ Unaffected female
▨ Affected male
⬤ Affected female

20-19

# Core Concepts Summary

**20.1 IN THE DEVELOPMENT OF HUMANS AND OTHER ANIMALS, STEM CELLS BECOME PROGRESSIVELY MORE RESTRICTED IN THEIR POSSIBLE PATHWAYS OF CELLULAR DIFFERENTIATION.**

The fertilized egg can give rise to a complete organism. page 20-2

At each successive stage in development, cells lose developmental potential as they become differentiated. page 20-2

Embryonic stem cells can give rise to any of the three germ layers, those further along in differentiation can form only a limited number of specialized cell types, and those still further along can form only one cell type. page 20-2

Stem cells play a prominent role in regenerative medicine, which uses stem cells—in some cases reprogrammed cells from the patient's own body—to replace diseased or damaged tissues. page 20-5

**20.2 THE GENETIC CONTROL OF DEVELOPMENT IS A HIERARCHY IN WHICH GENES ARE DEPLOYED IN GROUPS THAT IN TURN REGULATE THE NEXT SET OF GENES.**

Hierarchical gene control can be seen in fruit fly (*Drosophila*) development. page 20-5

The oocyte of a fruit fly is highly polarized, with gradients of mRNA established by the mother that set up anterior–posterior and dorsal–ventral axes. page 20-6

These gradients in turn affect the expression of segmentation genes in the zygote, including the gap, pair-rule, and segment-polarity genes, which define specific regions in the developing embryo. page 20-8

The segmentation genes direct the expression of homeotic genes, key transcription factors that specify the identity of each segment of the fly and that are conserved in animal development. page 20-9

**20.3 MANY PROTEINS THAT PLAY KEY ROLES IN DEVELOPMENT ARE EVOLUTIONARILY CONSERVED BUT CAN HAVE DRAMATICALLY DIFFERENT EFFECTS IN DIFFERENT ORGANISMS.**

Many proteins important in development are similar in sequence from one organism to the next. They are said to be evolutionarily conserved. page 20-11

The downstream targets, or cis-regulatory elements, of homeotic genes are different in different animals, allowing homeotic genes to activate different developmental pathways in different organisms. page 20-12

Although animals exhibit an enormous diversity in eye morphology, the observation that the proteins involved in light perception are evolutionarily conserved suggests that the ability to perceive light may have evolved once, early in the evolution of animals. page 20-12

The Pax6 transcription factor is a master regulator of eye development. Loss-of-function mutations in *Pax6* result in abnormalities in eye development, whereas gain-of-function mutations result in eye development in tissues in which eyes do not normally form. page 20-12

**20.4 COMBINATORIAL CONTROL IS A DEVELOPMENTAL STRATEGY IN WHICH THE PATHWAY OF CELLULAR DIFFERENTIATION DEPENDS ON THE PARTICULAR COMBINATION OF TRANSCRIPTION FACTORS PRESENT IN A CELL.**

By analyzing mutants that affect flower development in the plant *Arabidopsis*, researchers were able to determine the genes involved in normal flower development. page 20-13

The ABC model of flower development invokes three activities (A, B, and C) present in circular regions (whorls) of the developing flower, with the specific combination of factors determining the developmental pathway in each whorl. page 20-14

**20.5 LIGAND–RECEPTOR INTERACTIONS ACTIVATE SIGNAL TRANSDUCTION PATHWAYS THAT CONVERGE ON TRANSCRIPTION FACTORS AND OTHER GENES THAT DETERMINE CELL FATE.**

Cell signaling involves a ligand, an extracellular molecule that acts as a signal to activate a membrane receptor protein, which in turn activates molecules inside the cell. page 20-15

Activation of a receptor sets off a pathway of signal transduction, in which a series of proteins in the cytoplasm become sequentially activated. page 20-16

Because signal transduction can amplify and expand a developmental signal, a single ligand–receptor pair can cause major changes in gene expression and ultimately determine the pathway of differentiation. page 20-16

An example of signal transduction in development is differentiation of the nematode vulva, which is determined by means of an EGF ligand and receptor. page 20-16

## Self-Assessment

1. Distinguish among totipotent, pluripotent, and multipotent stem cells, and give an example of where you would find each type of cell.

2. Explain how an individual's own cells might be used in stem cell therapy.

3. Draw a diagram to illustrate how overlapping domains of gap-gene expression along the anterior–posterior axis of a *Drosophila* embryo can create a series of segments, each expressing a unique combination of gap genes.

4. Explain the role of homeotic genes in *Drosophila* development, and describe one phenotype in *Drosophila* resulting from a mutation in a homeotic gene that leads to development of a body part in an inappropriate segment.

5. Explain why some genes, particularly those involved in the early stages of development, are evolutionarily conserved.

6. Define combinatorial control in the context of the ABC model of floral development.

7. Diagram a pathway of signal transduction including a ligand, receptor, and ultimately a transcription factor that activates a gene that inhibits the receptor.

Do you understand the chapter's Core Concepts? Log in to check your answers to the Self-Assessment questions, then practice what you've learned and reinforce this chapter's concepts by working through the problems and multimedia tutorials provided there.

⌨ **www.biologyhowlifeworks.com**

# Malaria

### CO-EVOLUTION OF
### HUMANS AND A PARASITE

**1** As it feeds, the mosquito injects *Plasmodium* parasites along with its saliva into the host. The parasite migrates through the bloodstream to the host's liver.

Most people would say the world has more than enough mosquitoes, but in 2010 scientists at the University of Arizona conjured up a new variety. The high-tech bloodsuckers were genetically engineered to resist *Plasmodium*, the single-celled eukaryote that causes malaria.

Normally, the parasite grows in the mosquito's gut and is spread to humans by the insect's bite. By altering a single gene in the mosquito's genome, the researchers had made the insects immune to the malaria parasite. The accomplishment is a noteworthy advance, the latest in a long line of efforts to stop *Plasmodium* in its tracks.

Malaria is one of the most devastating diseases on the planet. The World Health Organization estimates that 500 million people contract malaria annually, primarily in tropical regions. The disease is thought to claim about a million lives each year. Of those deaths, 85% to 90% occur in sub-Saharan Africa, mostly among children under 5.

> **As humans have tried a succession of weapons to defeat *P. falciparum*, the parasite has evolved, thwarting their efforts.**

Four species of *Plasmodium* can cause malaria in humans. One of them, *P. falciparum*, is particularly dangerous, accounting for the vast majority of malaria fatalities. Over thousands of years, this parasite and its human host have played a deadly game of tug-of-war. As humans have tried a succession of weapons to defeat *P. falciparum*, the parasite has evolved, thwarting their efforts. And in turn, the tiny organism has helped to shape human evolution.

*Plasmodium* is a wily and complicated parasite, requiring both humans and mosquitoes to complete its life cycle. The part of the cycle in humans begins with a single bite from an infected mosquito. As the insect draws blood, it releases *Plasmodium*-laden saliva into the bloodstream. Once inside their human host, the *Plasmodium* parasites invade the liver cells. There they undergo cell division for several days, their numbers increasing. Eventually, they infect red blood cells, where they continue to grow and multiply. The mature parasites burst from the red blood cells at regular intervals, triggering malaria's telltale cycle of fever and chills.

Some of the freed *Plasmodium* parasites go on to infect new red blood cells; others divide to form gametocytes that travel through the victim's blood vessels. When another mosquito bites the infected individual, it takes up the gametocytes with its blood meal. Inside the insect, the parasite completes its life cycle. The gametocytes fuse to form zygotes. Those zygotes bore into the mosquito's stomach, where they form oocysts that give rise to a new generation of parasites. When the infected mosquito sets out to feed, the cycle begins again.

The battle between malaria and humankind has raged through the ages. Scientists have recovered *Plasmodium* DNA from the bodies of 3500-year-old Egyptian mummies —evidence that those ancient humans were infected with the malaria parasite. The close connection between humans, mosquitoes, and the malaria parasite almost certainly extends back much further.

In fact, people who hail from regions where malaria is endemic are more likely than others to have certain genetic signatures that offer some degree of protection from the parasite. That indicates that *Plasmodium* has been exerting evolutionary pressure on humankind for quite some time.

Meanwhile, we've done our best to fight back. For centuries, humans fought the infection with quinine, a chemical found in the bark of the South American cinchona tree. In the 1940s, scientists developed a more sophisticated drug based on the cinchona compound. That drug, chloroquine, was effective, inexpensive, easy to administer, and caused few side effects. As a result, it was widely used, and in the late 1950s, *P. falciparum* began showing signs of resistance to chloroquine. Within 20 years, resistance had spread to Africa, and today most strains of *P. falciparum* have evolved resistance to the once-potent medication.

Just as bacteria develop resistance to antibiotics, *Plasmodium* evolves resistance to the antiparasitic drugs designed to fight it. In poor, rural areas where malaria is prevalent, people often can't follow the recommended protocols for antimalarial treatment. Sick individuals may be able to afford only a few pills rather than the

The life cycle of *Plasmodium falciparum*.

Liver

Malaria parasite

Liver cell

**2** *Plasmodium* incubates in liver cells, maturing into a form that can invade red blood cells.

**5** Gametocytes combine to form zygotes in the mosquito gut. These mature into oocysts, and spread thousands of parasites through the mosquito.

**6** This mosquito then leaves and infects another human host, beginning the cycle again.

Blood vessel

**3** The parasite continues to grow and reproduce, bursting from red blood cells to infect other red blood cells.

**4** Some parasites develop into male and female gametocytes, which infect another mosquito when it takes a blood meal from the host.

Red blood cells

Malaria parasites

Mature gametocytes

Gamete

Zygote

Malaria parasites

Oocyst

full recommended dose. The strength and quality of those pills may be questionable, and the drugs are often taken without oversight from a medical professional.

Unfortunately, inadequate use of the drugs fuels resistance. When the pills are altered or the course of treatment is abbreviated, not all *Plasmodium* parasites are wiped out. Those that survive in the presence of the drug are likely to evolve resistance to the drug. Because the resistant parasites have a survival advantage, the genes for drug resistance spread quickly through the population.

Since chloroquine resistance emerged, pharmaceutical researchers have developed a variety of new medications to prevent or treat malaria. Most, however, are far too expensive for people in the poverty-stricken regions where malaria is rampant. And just as was the case with chloroquine, almost as quickly as new drugs are developed, *Plasmodium* begins evolving resistance.

One of the more recent weapons added to the drug arsenal is artemisinin, a compound derived from the Chinese wormwood tree. It has turned out to be an effective and relatively inexpensive way to treat malaria infections. However, pockets of artemisinin resistance have already been uncovered in Southeast Asia. Public health workers now recommend that artemisinin be given in combination with other drugs. Treating infected patients with multiple drugs is more likely to wipe out *Plasmodium* in their bodies, reducing the chances that more drug-resistant strains will emerge.

Given the challenges of developing practical drugs to prevent or treat malaria, some scientists have turned their attention to other approaches. One goal is to produce a malaria vaccine. The parasite's complex life cycle makes that a complicated endeavor, though. Several vaccines are now in various stages of testing, and some show promise.

But researchers expect it will be years before a safe, effective vaccine for malaria could be available.

Other researchers are focusing their efforts on the mosquitoes that carry the parasite, rather than on *Plasmodium* itself. The genetically engineered insects created by the team in Arizona are a promising step in that direction.

The Arizona researchers set out to alter a cellular signaling gene that plays a role in the mosquito's life cycle. Mosquitoes normally live 2 to 3 weeks, and *Plasmodium* takes about 2 weeks to mature in the mosquito's gut. The researchers hoped to create mosquitoes that would die prematurely, before the parasite is mature. The genetic modification worked as planned. The engineered mosquitoes' life-spans were shortened by 18% to 20%. The genetic tweak also had a surprising side effect. The altered gene completely blocked the development of *Plasmodium* in the mosquitoes' guts. The engineered mosquitoes are incapable of spreading malaria to humans, regardless of how long they live.

While the finding was a laboratory success, it will be much harder to translate the results to the real world. To create malaria-free mosquitoes in the wild, scientists would have to release the genetically modified mosquitoes and hope that their altered gene spreads through the wild mosquito population. But that gene would be passed on only if it gave the insects a distinct evolutionary advantage. The engineered mosquitoes may be malaria free, but so far they're no fitter than their wild counterparts.

It's clear that slashing malaria rates will not be an easy task. Despite decades of research, insecticide-treated bed nets are still the best method for preventing the disease. For millennia, the malaria parasite has managed to withstand our efforts to squelch it, yet science continues to push the boundaries. Who will emerge the victor? Stay tuned.

**? CASE 4 QUESTIONS**

*Answers to Case 4 questions can be found in Chapters 21–24.*

1. **What genetic differences have made some individuals more and some less susceptible to malaria?** *See page 21-10.*
2. **How did malaria come to infect humans?** *See page 22-11.*
3. **What human genes are under selection for resistance to malaria?** *See page 24-15.*

## CHAPTER 21

# EVOLUTION

## How Genotypes and Phenotypes Change over Time

## Core Concepts

Variation is a fact of nature. A walk down any street reveals how variable our species is: Skin color and hair color, for example, vary from person to person. Until the publication in 1859 of Charles Darwin's *On the Origin of Species*, scientists tended to view all the variation we see in humans and other species as biologically unimportant. According to the traditional view at the time, not only were species individually created by God in their modern forms, but, because the Creator had a specific design in mind for each one, they were fixed and unchanging. Departures or variations from this divinely ordained type were therefore ignored.

Since Darwin, however, we have appreciated that a species does not conform to a type. Rather, a species consists of a range of variants. In our own species, people may be tall, short, dark-skinned, fair-skinned, and so on. Furthermore, variation is an essential ingredient of Darwin's theory because **natural selection** depends on the differential success—in terms of surviving and reproducing—of variants. Darwin changed how we view variation. Before Darwin, variation was irrelevant, something to be ignored, but after him it was recognized as the key to the evolutionary process.

---

## 21.1 GENETIC VARIATION

Variation is a major feature of the natural world. We humans are particularly good at noticing phenotypic variation among individuals of our own species. As we discussed in Chapter 16, a phenotype is an observable trait, such as human height or wing color in butterflies. Two factors contribute to phenotype: an individual's genotype, the set of alleles possessed by the individual at relevant genetic loci, and the environment in which the individual lives. We can take the environment out of the equation by looking directly at genotypic differences through sequencing DNA regions in multiple individuals. We now explore genetic variation directly, in terms of differences at the DNA sequence level.

**Population genetics is the study of patterns of genetic variation.**
Remarkably, in spite of a high degree of phenotypic variation, humans actually rank low in terms of overall genetic variation compared to other species. Specifically, if we ask how many DNA bases differ from one individual to another within the same species, we find that the fruit fly *Drosophila melanogaster* is about 10 times more genetically variable than we are. Even one of the most seemingly uniform species on

**FIG. 21.1** **Genetic diversity in Adelie penguins.** Adelie penguins are uniform in appearance but are actually more genetically diverse than humans.

the planet, the Adelie penguin seen in **Fig. 21.1,** is two to three times more genetically variable than we are.

As we discuss in Chapter 22, a **species** consists of individuals that can exchange genetic material through interbreeding. From a genetic perspective, a species is therefore a group of individuals capable, through reproduction, of sharing alleles with one another. Individuals represent different combinations of alleles drawn from a single **gene pool,** that is, all the alleles present in all individuals in a species. The human gene pool includes alleles that cause differences in skin color, hair type, eye color, and so on. Each one of us has a different set of those alleles—alleles that cause brown hair and brown eyes, for example, or black hair and blue eyes—drawn from that gene pool.

Population genetics is the study of genetic variation in natural **populations,** which are interbreeding groups of organisms of the same species living in the same geographical area. What factors determine levels of variation in a population and in a species? Why are humans genetically less variable than penguins? What factors affect the distribution of particular variations? Population genetics addresses detailed questions about patterns of variation. And small differences, given enough time, can lead to the major differences we see among organisms today.

## Mutation and recombination are the two sources of genetic variation.

Genetic variation has two sources: mutation and recombination. Mutation generates new variation and recombination shuffles mutations to create new combinations of mutations. In both cases, new alleles are formed, as shown in **Fig. 21.2.**

Mutations can be **somatic,** occurring in the body's tissues, or **germ-line,** occurring in the reproductive cells and therefore passed on to the next generation. From an evolutionary viewpoint, we are primarily interested in germ-line mutations. A somatic mutation affects only the cells descended from the one cell in which the mutation originally arose, and thus affects only that one individual. However, a germ-line mutation appears in every cell of an individual derived from the fertilization involving the mutation-bearing gamete, and thus in its descendants.

## Mutations can be harmful, neutral, or beneficial.

Organisms, even the very simplest, are exquisitely complicated. In a large multicellular species such as humans, the level of organizational complexity is staggering. Consider the extraordinary process of development whereby, starting from a single cell at

**FIG. 21.2** The formation of new alleles through mutation and recombination.

Recombination shuffles mutations to produce new sequences.

conception, an adult human being develops that consists of a few hundred trillion interacting, cooperating cells (Chapter 20). What is the likely impact of a random change—a mutation—on a complex piece of biological machinery? Random changes to the working parts of complex, organized systems are almost always **deleterious** (harmful). Imagine making a random change to the engine of a car, a relatively simple machine compared to complex biological machines. Pull out a hose here or put in a steel rod there, and you will almost certainly have a problem with your car.

Many mutations, however, have little or no effect on the organism. These mutations are neither deleterious nor advantageous because they occur in regions of the genome that are not functionally important. Such mutations are termed **neutral.**

Occasionally, a mutation occurs that is actually beneficial in terms of survival or reproduction. For example, a gene encoding hemoglobin might by chance acquire a mutation that allows it to take up and deliver oxygen to tissues more easily. Similarly, a gene encoding a particular enzyme might be mutated in such a way that it is more efficient at catalyzing a chemical reaction. Mutations like these are considered **advantageous** if they improve their carriers' chances of survival or reproduction. Advantageous mutations, as we will see, can increase in frequency in a population until eventually they are carried by every member of a species. These mutations are the ones that result in a species that is **adapted** to its environment—better able to survive and reproduce in that environment.

## 21.2 MEASURING GENETIC VARIATION

Mutations, whether harmful, neutral, or advantageous, are sources of genetic variation. The goal of population genetics is to make inferences about the evolutionary process from patterns of genetic variation in nature. The raw information for this comes from the rates of occurrence of alleles in populations, or **allele frequencies.**

### To understand patterns of genetic variation, we require information about allele frequencies.

The allele frequency of an allele $x$ is simply the number of $x$'s present in the population divided by the total number of alleles. Consider, for example, pea color in Mendel's pea plants. In Chapter 16, we discussed how pea color (yellow or green) results from variation at a single gene. Two alleles of this gene are the dominant $A$ (yellow) allele and the recessive $a$ (green) allele. $AA$ homozygotes and $Aa$ heterozygotes produce yellow peas, and $aa$ homozygotes produce green peas. Imagine that in a population every pea plant produces green peas, meaning that only one allele, $a$, is present: The allele frequency of $a$ is 100%, whereas the allele frequency of $A$ is 0%. When a population exhibits only one allele at a particular gene, we say that the population is **fixed** for that allele.

Now consider another population of 100 pea plants with genotype frequencies of 50% $aa$, 25% $Aa$, and 25% $AA$. (A genotype frequency is the proportion in a population of each genotype at a particular gene or set of genes.) These genotype frequencies give us 50 green-pea pea plants ($aa$), 25 yellow-pea heterozygotes ($Aa$), and 25 yellow-pea homozygotes ($AA$). What is the allele frequency of $a$ in this population? Each of the 50 $aa$ homozygotes has two $a$ alleles and each of the 25 heterozygotes has one $a$ allele. Of course, there are no $a$ alleles in $AA$ homozygotes. The total number of $a$ alleles is thus $(2 \times 50) + 25 = 125$. To determine the allele frequency of $a$, we divide the number of $a$ alleles by the total number of alleles in the population, 200 (because each pea plant is diploid, meaning that it has two alleles): $\frac{125}{200}$ = 62.5%. Because we are dealing with only two alleles in this example, the allele frequency of A is 100% - 62.5% = 37.5%

Thus, the allele frequencies of $A$ and $a$ provide a measure of genetic variation at one gene in a given population. In this example, we were given the genotype frequencies, and from this information we determined the allele frequencies. But how are genotype and allele frequencies measured? We consider three ways to measure genotype and allele frequencies in populations: observable traits, gel electrophoresis, and DNA sequencing.

→ **Quick Check 1** In the example of pea color in Mendel's pea plants, we worked with numbers of plants of each genotype to calculate allele frequencies. Can you figure out how to determine allele frequencies directly from genotype frequencies, even when we have no information on the number of individuals being studied?

### Early population geneticists relied on observable traits to measure variation.

It would be a simple matter to measure genetic variation in a population if we could use observable traits. Then we could simply count the individuals displaying variant forms of a trait and have a measure of the variation of that trait's gene. However, as we saw in Chapter 18, this approach can work only rarely for two important reasons. First, many traits are encoded by a large number of genes. In these cases, it is difficult, if not impossible, to make direct inferences from a phenotype to the underlying genotype. Even apparently straightforward traits often prove to have a complicated genetic basis. For instance, human skin color is determined by at least six different genes. Second, the phenotype is a product of both the genotype and the environment.

Until the 1960s, there was only one workable solution: to limit population genetics to the study of phenotypes that are encoded by a single gene. As these are few, the number of genes that population geneticists could study was extremely small. Human blood groups, including the ABO system, provided an early example of a trait encoded by a single gene with multiple alleles. At this gene, there are three alleles in the population—*A*, *B*, and *O*—and therefore six possible genotypes, which result in four different phenotypes (**Table 21.1**).

Other instances in which phenotypic variation can be readily correlated with genotype include certain markings in invertebrates. For example, the coloring of the two-spot ladybug *Adalia bipunctata* is controlled by a single gene (**Fig. 21.3**). However, the genetic basis of most traits is not so simple.

### Gel electrophoresis facilitates the detection of genetic variation.

Single-gene variation became much easier to detect in the 1960s with the application of gel electrophoresis. In Chapter 12, we saw how gel electrophoresis separates segments of DNA according to their size. Before DNA technologies were developed, the same basic process was applied to proteins to separate them according

**TABLE 21.1    The ABO blood system.**

| PHENOTYPE | GENOTYPE |
|---|---|
| A | *AA* or *AO* |
| B | *BB* or *BO* |
| AB | *AB* |
| O | *OO* |

to their electrical charge and their size. In gel electrophoresis, the proteins being studied migrate through a gel when an electrical charge is applied, creating an electrical field. The rate at which the proteins move from one end of the gel to the other is determined by their charge and their size. Proteins with more negatively charged amino acids migrate more rapidly toward the positively charged end, and vice versa.

Early studies of protein electrophoresis focused on enzymes that catalyze reactions that can be induced to produce a dye when the substrate for the enzyme is added. If we add some of the substrate, we can see the locations of the proteins in the gel. **Fig. 21.4** shows this sort of experiment. Material from different individuals is loaded in each lane of the gel. One individual, a homozygote for a particular protein sequence, produces a single band on the gel. A heterozygote for two differently charged alleles produces two bands. So the bands in the gel can provide a visual picture of genetic variation.

### DNA sequencing is the gold standard for measuring genetic variation.

Protein gel electrophoresis was a significant leap forward in our ability to detect genetic variation, but even this technique had significant limitations. We could only study enzymes because we needed to be able to stain specifically for enzyme activity, and we could only detect mutations that resulted in amino acid substitutions that changed a protein's mobility in the gel. Only with DNA sequencing did we finally have an unambiguous means of detecting all genetic variation in a stretch of DNA, whether in a coding region or not. The variations studied by modern population geneticists are differences in DNA sequence, such as a *T* rather than a *G* at a specified nucleotide position in a particular gene.

Calculating allele frequencies, then, simply involves collecting a population sample and counting the number of occurrences of a given mutation. Take the *A*-or-*G* mutation at a specific nucleotide position in

**FIG. 21.3    A genetic difference in color in the two-spot ladybug, *Adalia bipunctata*, that results from variation in a single gene.**

**FIG. 21.4**

# How is genetic variation measured?

**BACKGROUND** The introduction of protein gel electrophoresis in 1966 gave researchers the opportunity to identify differences in amino acid sequence in proteins both among individuals and, in the case of heterozygotes, within individuals. Proteins with different amino acid sequences run at different rates through a gel in an electric field. Often, a single amino acid difference is enough to affect the mobility of a protein in a gel.

**METHOD** Starting with crude tissue—the whole body of a fruit fly, or a blood sample from a human—we load the material on a gel, and turn on the current. The rate at which a protein migrates depends on its size and charge, both of which may be affected by its amino acid sequence. To visualize the protein at the end of the gel run, we use a biochemical indicator that produces a stain when the protein of interest is active. The result is a series of bands on the gel.

Test material may include blood samples or ground up samples of organisms.

Human    Blood sample    or    *Drosophila*    The body of the fly is ground up.

The samples are loaded into separate wells on the gel, and the proteins in each sample migrate toward the positive electrode according to their charge and size.

**RESULTS** The genotypes of eight individuals for a gene with two alleles are analyzed. Four are allele 1 homozygotes; two are allele 2 homozygotes; and two are heterozygotes. Note that the heterozygotes do not stain as strongly on the gel because each band has half the intensity of the single band in the homozygote. We can measure the allele frequencies simply by counting the alleles. Each homozygote has two of the same allele, and each heterozygote has one of each.

The gel is stained with a biochemical agent to see how far the proteins have migrated.

Heterozygote

Homozygote for allele 1    Homozygote for allele 2

Total number of alleles in the population = 8 × 2 = 16

Number of allele 1 in the population = 2 × (number of allele 1 homozygotes) + (number of heterozygotes) = 8 + 2 = 10

Frequency of allele 1 = $\frac{10}{16}$ = $\frac{5}{8}$

Number of allele 2 in the population = 2 × (number of allele 2 homozygotes) + (number of heterozygotes) = 4 + 2 = 6

Frequency of allele 2 = $\frac{6}{16}$ = $\frac{3}{8}$

Note that the two allele frequencies add to 1.

**CONCLUSION** We now have a profile of genetic variation at this gene for these individuals. Population genetics involves comparing data such as these with data collected from other populations to determine the forces shaping patterns of genetic variation.

**FOLLOW-UP WORK** This technique is seldom used these days because it is easy now to recover much more detailed genetic information about genetic variation from DNA sequencing.

**SOURCE** Lewontin, R. C., and J. L. Hubby. 1966. "A Molecular Approach to the Study of Genic heterozygosity in natural populations. II. Amount of variation and degree of heterozygosity in natural populations of *Drosophila pseudoobscura.*" *Genetics* 54:595–609.

the fruit fly *Alcohol dehydrogenase (Adh)* gene. We can sequence the gene from 50 individual flies. We will then have 100 gene sequences from these diploid individuals. We find 70 sequences have an *A* and 30 have a *G* at a given position. Therefore, the allele frequency of *A* is $\frac{70}{100} = 0.7$ and the allele frequency of *G* is 0.3. In general, in a sample of *n* diploid individuals, the allele frequency is the number of occurrences of that allele divided by twice the number of individuals.

→   **Quick Check 2** Population genetics data have become ever more resolved over time, from phenotypes that are determined by a single gene, to gel electrophoresis that looks at variation among genes that encode for enzymes, to analysis of the DNA sequence. What is the next logical step?

## 21.3 EVOLUTION AND THE HARDY– WEINBERG EQUILIBRIUM

Determining allele frequencies gives us information about genetic variation. Following that variation over time is key to understanding the genetic basis of evolution.

### Evolution is a change in allele or genotype frequency over time.

At the genetic level, evolution is simply a change in the frequency of an allele or a genotype from one generation to the next. For example, if there are 200 copies of an allele that causes blue eye color in a population in generation 1 and there are 300 copies of that allele in a population of the same size in generation 2, evolution has occurred. In principle, evolution may occur without allele frequencies changing. For instance, even if, in our fruit fly example, the *A/G* allele frequencies stay the same from one generation to the next, the frequencies of the different genotypes (that is, of *AA*, *AG*, and *GG*) may change. This would be evolution *without* allele frequency change.

Evolution is therefore a change in the genetic makeup of a population over time. Note an important and often misunderstood aspect of this definition: *populations* evolve, not *individuals*. Note, too, that this definition does not specify a mechanism for this change. As we will see, many mechanisms can cause allele or genotype frequencies to change. Regardless of which mechanisms are involved, any change in allele frequencies, genotype frequencies, or both constitutes evolution.

### The Hardy–Weinberg equilibrium describes situations in which allele and genotype frequencies do not change.

Allele and genotype frequencies change over time only if specific mechanisms act on the population. This principle was demonstrated independently in 1908 by the English mathematician G. H. Hardy and the German physician Wilhelm Weinberg, and has become known as the **Hardy–Weinberg equilibrium.** In essence, the Hardy–Weinberg equilibrium is the situation in which evolution does not occur.

The Hardy–Weinberg equilibrium specifies the relationship between allele frequencies and genotype frequencies when a number of key conditions are met. When these conditions are met—and the Hardy–Weinberg equilibrium holds—we can conclude that evolutionary mechanisms are not acting on the gene in the population we are studying. In many ways, then, the Hardy–Weinberg equilibrium is most interesting when we find instances in which allele or genotype frequencies *depart* from expectations. This finding implies that one or more of the conditions are *not* met and that evolutionary mechanisms are at work.

These are the required conditions for the Hardy–Weinberg equilibrium:

1. **There can be no differences in the survival and reproductive success of individuals**. Given two alleles, *A* and *a*, imagine an extreme case in which *a*, a recessive mutation, is lethal. All *aa* individuals die. Therefore, in every generation, there is a selective elimination of *a* alleles, meaning that the frequency of *a* will gradually decline (and the frequency of *A* correspondingly increase) over the generations. As we discuss below, we call this differential success of alleles **selection.**

2. **Populations must not be added to or subtracted from by migration**. Imagine a second population adjacent to the one we used in the preceding example in which all the alleles are *A* and all individuals have the genotype *AA*. Recall that when a population has only one allele of a particular gene, it is fixed for that allele. Now imagine that there is a sudden influx of individuals from the first population into the second. The frequency of *A* in the second population will change in proportion to the number of immigrants.

3. **There can be no mutation.** If *A* alleles mutate into *a* alleles, and vice versa, then again we will see changes in the allele frequencies over the generations. In general, because mutation is so rare, it is not a serious consideration on the timescales studied by population geneticists. However, as we have seen, mutation is the ultimate source of genetic variation, so it is a key input on which other evolutionary mechanisms act.

4. **The population must be sufficiently large to prevent sampling errors**. Chance events are more likely in small samples than in large ones. Campus-wide, a college's sex ratio may be close to 50:50, but in a small class of 8 individuals it is not improbable that we would have 6 women and 2 men (a 75:25 ratio). Sample size, in the form of population size, also affects the Hardy–Weinberg equilibrium such that it technically holds only for infinitely large populations. A change in the frequency of an allele due to the random effects of small population size is called **genetic drift**.

5. **Individuals must mate at random.** For the Hardy–Weinberg equilibrium to hold, mate choice must be made without regard to genotype. For example, an *AA* homozygote when offered a choice of mate from among *AA*, *Aa*, or *aa* individuals should choose at random. In contrast, **non-random mating** occurs when individuals do not mate randomly. For example, *AA* homozygotes might preferentially mate with other *AA* homozygotes. Non-random mating affects genotype frequencies from generation to generation, but does not affect allele frequencies.

### The Hardy–Weinberg equilibrium translates allele frequencies into genotype frequencies.

Now that we have established the conditions required for a population to be in Hardy–Weinberg equilibrium, let us explore the idea in detail. In the example we looked at earlier, we know the frequency of the two alleles, one with *A* and the other with *G* in the *Adh* gene in *Drosophila*. What are the genotype frequencies? That is, how many *AA* homozygotes, *AG* heterozygotes, and *GG* homozygotes do we see? The Hardy–Weinberg equilibrium predicts genotype frequencies from allele frequencies.

The logic is simple. Random mating is the equivalent of putting all the population's gametes into a single pot and drawing out pairs of them at random to form a zygote. We therefore put in our 70 *A* alleles and 30 *G* alleles and pick pairs at random. What is the probability of picking an *AA* homozygote (that is, what is the probability of picking an *A* allele followed by another *A* allele)? The probability of picking an *A* allele is its frequency in the population, so the probability of picking the first *A* is 0.7. What is the probability of picking the second *A*? Also 0.7. What then is the probability of picking an *A* followed by another *A*? It is the product of the two probabilities: 0.7 × 0.7 = 0.49. Thus, the frequency of an *AA* genotype is 0.49. We take an identical approach to determine the genotype frequency for the *GG* genotype: Its frequency is 0.3 × 0.3, or 0.09.

What about the frequency of the heterozygote, *AG*? This is the probability of drawing *G* followed by *A*, or *A* followed by *G*. There are thus two separate ways in which we can generate the heterozygote. Its frequency is therefore (0.7 × 0.3) + (0.3 × 0.7) = 0.42.

We can generalize these calculations algebraically by substituting letters for the numbers we have computed to derive the Hardy–Weinberg equilibrium. If the allele frequency of one allele, *A*, is *p*, and the other, *G*, is *q*, then

$p + q = 1$ (because there are no other alleles at this site).

| Genotypes | *AA* | *AG* | *GG* |
|---|---|---|---|
| Frequencies | $p^2$ | $2pq$ | $q^2$ |

Not only do these relationships predict genotype frequency from allele frequencies, but they work in reverse, too. **Fig. 21.5** shows how Hardy–Weinberg values are calculated from allele frequencies and, in the graph on the right, how allele and genotype frequencies are related. Note that we can use the graph to determine both allele frequencies for a given genotype frequency (to take a simple case, a 0.5 frequency of heterozygotes, *Aa*, implies that both *p* and *q* are 0.5) and genotype frequencies for a specified allele frequency.

We can do this mathematically as well. Knowing the genotype frequency of *AA*, for example, permits us to calculate allele frequencies: if, as in our *Adh* example, $p^2$ is 0.49 (that is, 49% of the population has genotype *AA*), then *p*, the allele frequency of *A*, is $\sqrt{0.49} = 0.7$. Because $p + q = 1$, then *q*, the allele frequency of *a*, is $1 - 0.7 = 0.3$.

Note that these relationships hold only if the Hardy–Weinberg conditions are met. If not, then allele frequencies can only be deduced by tallying alleles from genotype frequencies as described earlier in section 21.2.

---

**FIG. 21.5  Hardy-Weinberg Relation.** The Hardy-Weinberg relation predicts genotype frequencies from allele frequencies, and vice versa.

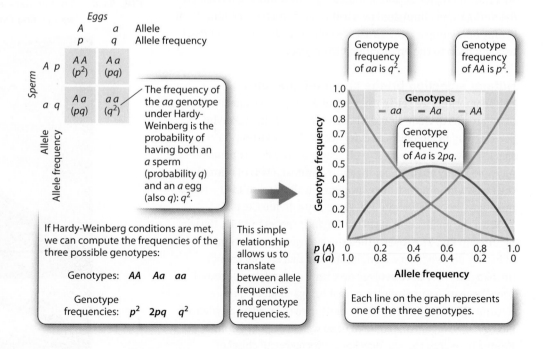

## The Hardy–Weinberg equilibrium is the starting point for population genetic analysis.

Recall our definition of evolution: a change in allele or genotype frequency from one generation to the next. Given this definition, it might seem odd to be discussing factors necessary for allele frequencies to stay the same. The Hardy–Weinberg equilibrium not only provides a means of converting between allele and genotype frequencies, but, critically, it also serves as an indicator that something interesting is happening in a population when it is not upheld.

If we find a population whose allele or genotype frequencies are not in Hardy–Weinberg equilibrium, we can infer that evolution has occurred. We can then consider, for the gene under study, whether the population is subject to selection, migration, mutation (unlikely because of its rarity), genetic drift, or non-random mating. These are the primary mechanisms of evolution. The Hardy–Weinberg equilibrium gives us a baseline from which to explore the evolutionary processes affecting populations. We will start by considering one of the most important evolutionary mechanisms: natural selection.

→   **Quick Check 3**. When we find a population whose allele frequencies are *not* in Hardy–Weinberg equilibrium, what can and can't we conclude about that population?

## 21.4 NATURAL SELECTION

Natural selection results in allele frequencies changing from generation to generation according to the allele's impact on the survival and reproduction of individuals. New mutations that are deleterious and eliminated by natural selection have no long-term evolutionary impact; ones that are beneficial, however, can result in adaptation to the environment over time.

### Natural selection brings about adaptations.

The adaptations we see in the natural world—the exquisite fit of organisms to their environment—were typically taken by pre-Darwinian biologists as evidence of God's existence. Each species, they argued, was so well adapted—the desert plant so physiologically adept at coping with minimal levels of rainfall and the fast-swimming fish so hydrodynamically streamlined—that it must have been deliberately designed by a divine Creator.

With the publication of *On the Origin of Species* in 1859, Darwin, pictured in **Fig. 21.6,** overturned the biological convention of his day on two fronts. First, he showed that species are not unchanging; they have evolved over time. Second, he suggested a mechanism, natural selection, that brings about adaptation. Natural selection was a brilliant solution to the central problem of biology: how organisms come to fit so well in their environments. From where does the woodpecker get its powerful chisel of a bill? And the hummingbird its long delicate bill for probing

the nectar stores in flowers? Darwin showed how a simple mechanism, without foresight or intentionality, could result in the extraordinary range of adaptations that all of life is testimony to.

For 20 years after originally conceiving the essence of his theory, Darwin collected supporting evidence. In 1858, however, he was spurred to begin writing *On the Origin of Species* by a letter from a little-known naturalist collecting specimens in what is today Indonesia. By a remarkable coincidence, Alfred Russel Wallace, shown in **Fig. 21.7,** had also developed the theory of evolution by natural selection. Aware that Darwin was interested in the problem but having no idea that Darwin was working on the same theory, Wallace wrote to Darwin in 1858 to see what he thought of his idea.

Suddenly, Darwin was confronted with the prospect of losing his claim on the theory that he had been quietly nurturing for the previous 20 years. But all was not lost. Darwin's colleagues arranged for the publication of a joint paper by Wallace and Darwin in 1858. This was done without consulting Wallace, who nonetheless never resented Darwin and afterward was careful to insist that the idea rightly belonged to the older man. It was Darwin's publication of *On the Origin of Species* in 1859 that brought both evolution and its underlying mechanism of adaptation, natural selection, to public attention. Wallace is only fleetingly mentioned in Darwin's great work, and Darwin is now the name, not Wallace, associated with the discovery.

Both Darwin and Wallace recognized their debt to the writings of a British clergyman, Thomas Malthus. In his *Essay on the Principle of Population*, first published in 1798, Malthus pointed out that natural populations have the potential to increase

**FIG. 21.6**   **Charles Darwin, photographed at about the time he was writing *On the Origin of Species*.**

**FIG. 21.7  Alfred Russel Wallace, photographed in Singapore during his expedition to Southeast Asia.**

geometrically, meaning that there is potentially an accelerating increase in population size. Imagine that human couples can have just four children (two males and two females), meaning that the population doubles every generation. Starting with a single couple, by the twentieth generation, the population will have grown to over a million—1,048,576 to be precise. Note that this growth assumes a rather modest rate of increase (twofold). Consider how much faster population size would ramp up in an insect that lays hundreds of eggs in its lifetime.

However, this geometric expansion of populations does not occur. In fact, population sizes are typically reasonably stable from generation to generation. This is because the resources upon which populations are dependent—food, water, places to live—are limited. This imbalance suggests that in each generation many fail to survive or reproduce; there simply is not enough food and other resources to go around. This implies in turn that individuals within a population must compete for these resources.

Which individuals will win the competition? Darwin and Wallace suggested that those that are best adapted would most likely survive and leave more offspring. Genetic variation among individuals results in some being more likely to survive and reproduce. As a result, the next generation will be enriched for these same advantageous alleles. Darwin used the term "natural selection" for the filtering process that acts against deleterious alleles and in favor of advantageous ones.

Competitive advantage is a function of how well an organism is adapted to its environment. A desert plant that is more efficient

at minimizing water loss than another plant is better adapted to the desert environment. An organism that is better adapted to its environment is more fit. By fit, we do not mean physically fit, or even optimally adapted. **Fitness,** in this context, is a measure of the extent to which the individual's genotype is represented in the next generation. We say that the first plant's fitness is higher than the second's if it leaves more surviving offspring. Natural selection then acts over generations to increase the overall fitness of a population. A plant population newly arrived in a desert may be poorly adapted to its environment, but, over time, alleles that minimize water loss will increase under natural selection, resulting in a population that is better adapted to the desert.

Such changes in populations take time. Borrowing from the geologists of his day, Darwin recognized that time is a critical ingredient of his theory. Geologists had put forward a view of Earth's history that argued that large geological changes—like the carving out of the Grand Canyon—can be explained by simple day-to-day processes—in this case, erosion—operating over vast timescales. Darwin applied this worldview to biology. He recognized that small changes, like subtle shifts in the frequencies of alleles, could add up to major changes given long enough time periods. What might seem to us to be a trivial change over the short term can, over the long term, result in substantial differences among populations.

### The Modern Synthesis is a marriage between Mendelian genetics and Darwinian evolution.

Darwinian evolution involves the change over time of the genetic composition of populations and is thus a genetic theory. Although Mendel published his genetic studies of pea plants in 1866, not long after *The Origin*, Darwin never saw them, so a key component of the theory was missing. However, instead of clarifying the role of genetic processes underlying natural selection, the rediscovery of Mendel's work in 1900 unexpectedly provoked a major controversy among evolutionary biologists. Some argued that Mendel's discoveries did not apply to most genetic variation because the traits studied by Mendel were **discrete,** meaning that they had clear alternative states, such as peas that were either yellow or green. Most of the variation we see in natural populations, in contrast, is **continuous,** meaning that variation occurs across a spectrum. Human height, for example, does not come in discrete classes. People are not either 5 feet tall or 6 feet tall and no height in between. Instead, they may be any height within a certain range.

How could Mendel's discrete traits account for the continuous variation seen in natural populations? This question was answered by the English theoretician Ronald Fisher, who realized that, instead of a single gene contributing to a trait like human height, there could be several genes that contribute to the trait. He argued that extending Mendel's theory to include multiple genes per trait could account for patterns of continuous variation that we see all around us.

**FIG. 21.8** **The effect of a single base-pair mutation.** (a) Normal red blood cells look different from (b) sickled ones, whose shape has been distorted by hemoglobin molecules with the sickle variant.

different perspective. Instead of following individual mutations, we can look at changes over time of a particular trait of an organism. For example we might track the evolution of height in a population, despite not knowing the specifics of the genetic basis of height differences. When we look at natural selection

from this perspective, we see three types of patterns: **stabilizing, directional,** and **disruptive,** illustrated in **Fig. 21.9.**

Stabilizing selection maintains the status quo and acts against extremes. A good example is provided by human birth weight, a trait affected by many factors, including many fetal

**FIG. 21.9** **Stabilizing, directional, and disruptive selection.** Selection can change the distribution of phenotypes (and therefore genotypes) in three different ways.

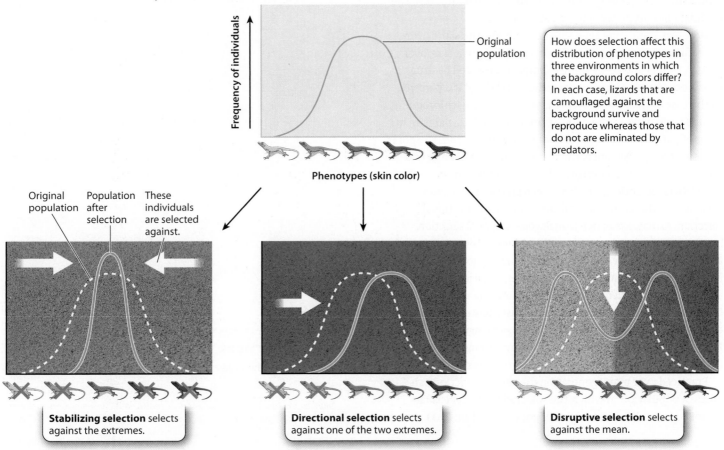

How does selection affect this distribution of phenotypes in three environments in which the background colors differ? In each case, lizards that are camouflaged against the background survive and reproduce whereas those that do not are eliminated by predators.

Frequency of individuals

Original population

Phenotypes (skin color)

Original population

Population after selection

These individuals are selected against.

**Stabilizing selection** selects against the extremes.

**Directional selection** selects against one of the two extremes.

**Disruptive selection** selects against the mean.

21-12

**FIG. 21.10 Stabilizing selection.** Stabilizing selection on human birth weight results in selection against babies that are either too small or too large.

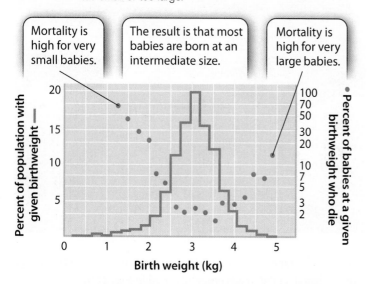

Mortality is high for very small babies.

The result is that most babies are born at an intermediate size.

Mortality is high for very large babies.

genes (**Fig. 21.10**). If a baby is too small, then its chances of survival after birth are low. However, if it is too big, there may be complications that endanger both mother and baby during delivery. Thus, the optimum birth weight is between these two extremes. In this case, natural selection acts against the extremes. The vast majority of natural selection is of this kind as deleterious mutations that cause a departure from the optimal phenotype are selected against.

Whereas stabilizing selection keeps a trait the same over time, **directional selection** leads to a change in a trait over time. A timely and concerning example is provided by drug resistance in many viruses, bacteria, and the malaria parasite. In the case of malaria, widespread use of the drug chloroquine has led to changes in the population of the malaria parasite in favor of parasites that are resistant to this drug. From the point of view of drug resistance, the population has moved in one direction since the 1960s toward increased resistance.

**Artificial selection,** which has been practiced by humans since at least the dawn of agriculture, is a form of directional selection. Artificial selection is analogous to natural selection, but the competitive element is removed. Successful genotypes are selected by the breeder, not through competition. Because it can be carefully controlled by the breeder, artificial selection is astonishingly efficient at generating genetic change. Dogs are basically domesticated wolves that, over just a few millennia of careful breeding, have produced morphologies ranging from the tiny Pekingese to the vast Great Dane and behaviors ranging from the herding abilities of Border Collies to the retrieval skills of a Labrador.

Practiced over many generations, artificial selection can create a population in which the selected phenotype is far removed from that of the starting population. **Fig. 21.11** shows

HOW DO WE KNOW?

**FIG. 21.11**

# How far can artificial selection be taken?

**BACKGROUND** Begun in the 1890s and continuing to this day at the University of Illinois, this experiment is one of the longest-running biological experiments in history.

**HYPOTHESIS** Researchers hypothesized that there would be a limit to the extent to which the population could respond to continued directional selection, that there would be a point at which there was no longer a response to selection.

**EXPERIMENT** Corn was artificially selected for either high oil content or low oil content. Every generation, kernels showed a range of oil levels, but only the 12 kernels with the highest or the lowest oil content were used for the next generation.

**RESULTS** In the line selected for high oil content, the percentage of oil more than quadrupled, from about 5% to more than 20%. In the line selected for low oil content, the oil content fell so close to zero that it could no longer be measured accurately, and the selection was terminated. Both selected lines are completely outside the range of any phenotype observed at the beginning of the experiment.

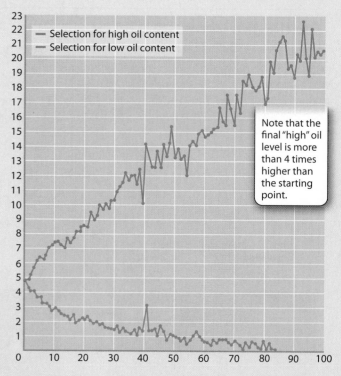

Note that the final "high" oil level is more than 4 times higher than the starting point.

**FOLLOW-UP WORK** Genetic analysis of the selected lines indicates that the differences in oil content are due to the effects of at least 50 genes.

SOURCE Moose, S. P., J. W. Dudley, and T. R. Rocheford. 2004. "Maize Selection Passes the Century Mark: A Unique Resource for 21st Century Genomics." *Trends in Plant Science* 9:358–364.

the result of long-continued artificial selection for the oil content in kernels of corn.

A third mode of selection, known as **disruptive selection,** operates in favor of extremes and against intermediate forms. For example, we may see disruptive selection operating on the size of seed-eating birds' bills, with small (for delicate seeds) and large (for chunky seeds) bills favored over intermediate sizes that perform poorly with both delicate and chunky seeds. We explore this mechanism in more detail in the next chapter, as it can lead to the evolution of new species.

### Sexual selection increases an individual's reproductive success.

Initially, Darwin was puzzled by features of organisms that seemed to reduce an individual's chances of survival. In a letter a few months after the publication of *The Origin,* he wrote, "The sight of a feather in a peacock's tail, whenever I gaze at it, makes me sick!" The tail is metabolically expensive to produce; it is an advertisement to potential predators; and it is a massive encumbrance in any attempt to escape a predator. How could such a feature evolve under natural selection?

In his 1871 book, *The Descent of Man,* Darwin introduced a solution to this problem. Natural selection is indeed acting to reduce the showiness and size of the peacock's tail, but another form of selection, **sexual selection,** is acting in the opposite direction. Sexual selection promotes traits that increase an individual's access to reproductive opportunities. Peahens are thought to select their mates on the basis of the showiness and size of a male's tail. In the absence of sexual selection, natural selection would act to minimize the size of the peacock's tail. Presumably, the peacocks' tails we see are a compromise, a trade-off between these conflicting demands of reproduction and survival (Chapter 45).

## 21.5 MIGRATION, MUTATION, AND GENETIC DRIFT

Selection is evolution's major driving force, enriching each new generation for the mutations that best fit organisms to their environments. However, as we have seen from the discussion of the Hardy–Weinberg equilibrium, it is not the only evolutionary mechanism. There are other forces that can cause allele frequencies to change. These are migration or gene flow, mutation, and the random effects of finite population size (that is, genetic drift). Like natural selection, these mechanisms can cause allele frequencies to change. Unlike natural selection, they do not lead to adaptations. Therefore, they are often considered non-adaptive evolutionary mechanisms.

### Migration reduces genetic variation between populations.

**Migration** is the movement of individuals from one population to another, resulting in **gene flow,** the movement of alleles from one population to another. It is relatively simple to see how movements of individuals and alleles can lead to changes in allele frequencies. Consider two isolated island populations of rabbits, one white, the other black. Now imagine that the isolation breaks down—a bridge is built between the islands—and migration occurs. Over time, black alleles enter the white population and vice versa, and the allele frequencies of the two populations gradually become the same.

The consequence of migration is therefore the homogenizing of populations, making them more similar to each other and reducing genetic differences between them. Because populations are often adapted to their particular local conditions (think of dark-skinned humans in regions of high sunlight versus fair-skinned humans in regions of low sunlight), migration may be worse than merely non-adaptive—it may be maladaptive, in that it causes a decrease in a population's average fitness. Fair-skinned people arriving in an equatorial region are, for example, at risk of sunburn and skin cancer.

### Mutation increases genetic variation.

As we saw earlier in this chapter, mutation is a rare event. This means that it is generally not important as an evolutionary mechanism that leads allele frequencies to change. However, as we have also seen, it is the source of new alleles and the raw material on which the other forces act. Without mutation, there would be no genetic variation, and therefore no evolution.

### Genetic drift is particularly important in small populations.

Genetic drift is the random change in allele frequencies from generation to generation. By "random" change, we mean that frequencies can either go up or down simply by chance. An extreme case is a population **bottleneck,** which occurs when a population falls to just a few individuals. Consider a rare allele, $A$, with frequency of $\frac{1}{1000}$. Now imagine that habitat destruction reduces the population to just one pair of individuals, one of which is carrying $A$. The frequency of $A$ in this new population is $\frac{1}{4}$ because each individual has two alleles, giving a total of four alleles. In other words, the bottleneck resulted in a massive change in allele frequencies. Another kind of bottleneck, called a **founder event,** may be important in the establishment of new populations, for example when just a few individuals arrive to colonize an island.

Earlier, we considered the fate of beneficial and harmful mutations under the influence of natural selection. What about neutral mutations? Natural selection, by definition, does not

**FIG. 21.12 Genetic drift.** The fate of neutral mutations is governed by genetic drift, the effect of which is more extreme in small populations than in large populations.

**a.** Population size = 4

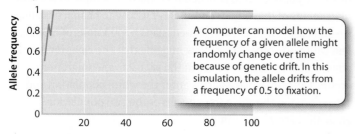

A computer can model how the frequency of a given allele might randomly change over time because of genetic drift. In this simulation, the allele drifts from a frequency of 0.5 to fixation.

**b.** Population size = 4

Each line represents a different simulation of the effect of drift on a population of 4 individuals. All simulations result in fixation or extinction of the allele in relatively few generations.

**c.** Population size = 40

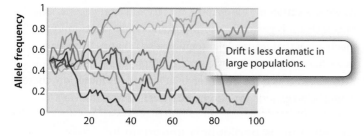

Drift is less dramatic in large populations.

**d.** Population size = 400

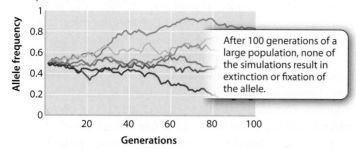

After 100 generations of a large population, none of the simulations result in extinction or fixation of the allele.

Generations

govern the fate of neutral mutations. So what happens to them? Consider a neutral mutation, *m*, which has no effect on the fitness of its carrier. At first, it is in just a single heterozygous individual. What happens if that individual fails to reproduce (for reasons unrelated to *m*)? In this case, *m* will be lost from the population, but not by natural selection (which does not

discriminate against *m*). Alternatively, the *m*-bearing individual might by chance leave many offspring (again for reasons unrelated to *m*), in which case the frequency of *m* will increase. In principle, it is possible over a long period of time for *m* to take over the population. At the end of the process, every member of the population is homozygous *mm*.

Like natural selection, genetic drift leads to allele frequency changes and therefore to evolution. Unlike natural selection, however, it does not lead to adaptations, since the alleles whose frequencies are changing do not affect an individual's ability to survive or reproduce.

The impact of genetic drift depends on population size (**Fig. 21.12**). If *m* arises in a very small population, its frequency will change rapidly, as shown in Figs. 21.12a and 21.12b. Imagine *m* arising in a population of just six individuals (or three pairs). Its initial frequency is 1 in 12, or about 8% (there are a total of 12 alleles because each individual is diploid). If, by chance, one pair fails to breed and the other two (including the one who is an *Mm* heterozygote) each produce three offspring, and all three of the *Mm* individual's offspring happen to inherit the *m* allele, then the frequency of *m* will increase to 3 in 12 (25%) in a single generation. In effect, genetic drift is equivalent to a sampling process. In a small sample, extreme departures from the expected outcome are common. Toss a coin 5 times, and you might well end up with zero heads.

On the other hand, if the population is large, as in Figs. 21.12c and 21.12d, then shifts in allele frequency from generation to generation are much smaller, typically less than 1%. A large population is analogous to a large sample size, in which we tend not to see marked departures from expectation. Toss a coin 1000 times, and you will end up with approximately 500 heads. In a small sample of coin tosses, we are much more likely to see marked departures from our 50:50 expectation than in a large sample. The same is true of genetic drift. It is likely to be much more significant in small populations than in large ones.

→ **Quick Check 4:** Why, of all the evolutionary forces, is selection the only one that can result in adaptation?

## 21.6 MOLECULAR EVOLUTION

How do DNA sequence differences arise among species? Imagine starting with two pairs of identical twins, one pair male and the other female. Now we place one member of each pair together on either side of a mountain range (**Fig. 21.13**). Let's assume the mountain range completely isolates the couples on each side. What, in genetic terms, will happen over time? The original pairs will found populations on each side of the mountain range. The genetic starting point is, in each case,

**FIG. 21.13  Genetic divergence in isolated populations.**

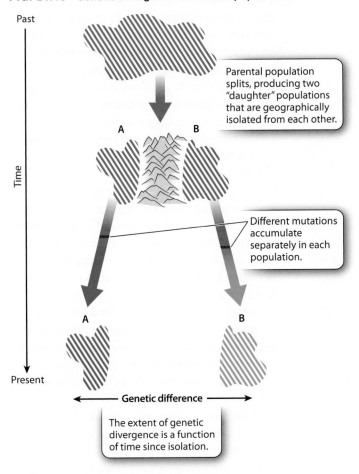

Past

Time

Parental population splits, producing two "daughter" populations that are geographically isolated from each other.

A          B

Different mutations accumulate separately in each population.

A                    B

Present

◄——— Genetic difference ———►

The extent of genetic divergence is a function of time since isolation.

exactly the same, but, as the generations tick by, differences will accumulate between the two populations. Over time, mutations will occur in one population that will not have arisen in the other population, and vice versa. Because mutations occur at a low rate and are scattered randomly throughout the genome, the chance that both populations acquire the same mutations is practically zero.

A mutation in either population ultimately suffers one of three fates: It goes to fixation (either through genetic drift or through positive selection); it is maintained at intermediate frequencies (by balancing selection); or it is eliminated (either through natural selection or genetic drift). Different mutations will be fixed in each population. When we come back thousands of generations later and sequence the DNA of our original identical individuals' descendants, we will find that many differences have accumulated. The populations have diverged genetically. What we are seeing is evidence of **molecular evolution.**

Species are the biological equivalents of islands because they, too, are isolated. They are *genetically* isolated because, by definition, members of one species cannot exchange genetic material with

members of another (Chapter 22). The amount of time that two species have been isolated from each other is the time since their most recent common ancestor. Thus, humans and chimpanzees, whose most recent common ancestor lived about 6-7 million years ago, have been isolated from each other for about 6-7 million years. Mutations arose and were fixed in the human lineage over that period; mutations, usually different ones, also arose and were fixed in the chimpanzee lineage over the same period. The result is the genetic difference between humans and chimpanzees. If we sequence a particular gene for both species, we find a number of sequence differences.

### The extent of sequence difference between species is a function of the time since the species diverged.

The extent of genetic difference, or genetic divergence, between two species is a function of the time they have been genetically isolated from each other. The longer they have been apart, the greater the opportunity for mutation and fixation to occur in each population. This correlation between the time two species have been evolutionarily separated and the amount of genetic divergence between them is known as the **molecular clock.**

For a clock to function properly, it not only needs to keep time, but it also needs to be set. We set the clock using dates from the fossil record. For example, in a 1967 study, Vince Sarich and Allan Wilson deduced from fossils that the lineages that gave rise to the Old and New World monkeys separated about 30 million years ago. Finding that the amount of genetic divergence between humans and chimpanzees was about one-fifth of that between Old and New World monkeys, they concluded that humans and chimpanzees had been separated one-fifth as long, or about 6 million years. Although this is the generally accepted number today, Sarich and Wilson's result was revolutionary at the time, when it was thought that the two species were separated for as long as 25 million years.

### The rate of the molecular clock varies.

Molecular clocks can be useful for dating evolutionary events like the separation of humans and chimpanzees. However, because the rates of molecular clocks vary from gene to gene, clock data should be interpreted cautiously. These rate differences can be attributed largely to differences in intensity of negative selection (which results in the elimination of harmful mutations) among different genes. The slowest molecular clock on record belongs to the histone genes, which encode the proteins around which DNA is wrapped to form chromatin (Chapter 3). These proteins are exceptionally similar in all organisms; only 2 amino acids (in a chain of about 100) distinguish plant and animal histones. Plants and animals last shared a common ancestor more than 1 billion years ago, which means, because each evolutionary

**FIG. 21.14  The molecular clock.** Different genes evolve at different rates because of differences in the intensity of negative selection.

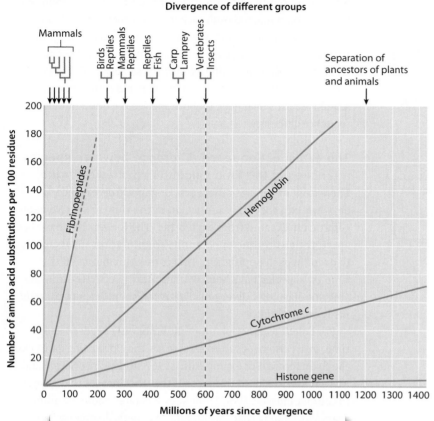

Since vertebrates and insects diverged 600 mya, the amino acid sequences of cytochrome *c* have acquired 25 differences. Amino acid sequences of hemoglobin have acquired many more differences in the same time period.

lineage is separate, that there have been at least 2 billion years of evolution since they were in genetic contact. And yet, in all that time, the histones have hardly changed at all. Almost any amino acid change fatally disrupts the histone protein, preventing it from carrying out its proper function. Negative selection has thus been extremely effective in eliminating just about every amino acid-changing histone mutation over 2 billion years of evolution. The histone molecular clock is breathtakingly slow.

Other proteins are less subject to such rigorous nagative selection. Occasional mutations may therefore become fixed, either through drift (if they are neutral) or selection (if beneficial). The extreme case of a fast molecular clock is that derived from a **pseudogene,** a gene that is no longer functional. Because *all* mutations in a pseudogene are by definition neutral—there is no function for a mutation to disrupt, so a mutation is neither deleterious nor beneficial—we expect to see a pseudogene's molecular clock tick at a very fast rate. In the histone genes, virtually all mutations are selected against, constraining the rate of evolution; in pseudogenes, none are. **Fig. 21.14** vividly shows the varying rates of the molecular clock for different genes.

In the next chapter, we will see how genetic divergence between populations can lead to the evolution of new species.

# Core Concepts Summary

## 21.1 GENETIC VARIATION IS THE RESULT OF DIFFERENCES IN DNA SEQUENCES.

Visible differences among members of a species (phenotypic variation) are the result of differences at the DNA level (genetic variation) as well as the influence of the environment. page 21-1

Mutation, a change in DNA sequence, is the ultimate source of genetic variation. This variation is reconfigured by recombination. page 21-2

Mutations can be somatic (in body tissues) or germ-line (in gametes), but germ-line mutations are the only ones that can be passed on to the next generation. page 21-2

When a mutation occurs in a gene, it creates a new allele. page 21-2

## 21.2 INFORMATION ON ALLELE FREQUENCIES IS KEY TO UNDERSTANDING PATTERNS OF GENETIC VARIATION.

An allele frequency is the number of occurrences of a particular allele divided by the total number of occurrences of all alleles of that gene in a population. page 21-3

In the past, population geneticists relied on observable traits determined by a single gene to measure genetic variation, but this method is not generally useful as most traits are influenced by many genes and by the environment. page 21-4

Allele frequencies can be measured by protein gel electrophoresis. page 21-4

DNA sequencing is now the standard technique for measuring allele frequencies. page 21-4

## 21.3 EVOLUTION IS A CHANGE IN THE FREQUENCY OF ALLELES OR GENOTYPES OVER TIME.

Allele frequencies change in response to selection, migration, mutation, and genetic drift. These are the primary mechanisms of evolution. page 21-6

The Hardy–Weinberg equilibrium describes situations in which allele frequencies do not change. In this way, it provides a null hypothesis against which to test whether or not evolution is occurring in a population. page 21-6

The Hardy–Weinberg equilibrium makes five assumptions. These assumptions are no selection, no migration, no mutation, no sampling error due to small population size, and random mating. page 21-6

The Hardy–Weinberg equilibrium allows allele frequencies and genotype frequencies to be calculated from each other. page 21-7

## 21.4 NATURAL SELECTION LEADS TO ADAPTATION, WHICH ENHANCES THE FIT BETWEEN AN ORGANISM AND ITS ENVIRONMENT.

Independently conceived by Charles Darwin and Alfred Russel Wallace, natural selection is the differential reproductive success of genetic variants. page 21-8

Under natural selection, a harmful allele is eliminated over time, and a beneficial one increases in frequency. Natural selection does not affect the frequency of neutral mutations. page 21-10

Natural selection can maintain alleles at intermediate frequencies by balancing selection. page 21-10

Changes in phenotype show that natural selection can be stabilizing, directional, or disruptive. page 21-11

In artificial selection, a form of directional selection, a breeder governs the selection process. page 21-12

Sexual selection involves the evolution of traits that increase an individual's access to members of the opposite sex. page 21-13

## 21.5 MIGRATION, MUTATION, AND GENETIC DRIFT ARE NON-ADAPTIVE MECHANISMS OF EVOLUTION.

Migration involves the movement of alleles between populations (gene flow) and tends to have a homogenizing effect. page 21-13

Mutation is the ultimate source of variation, but it also can change allele frequencies on its own. page 21-13

Genetic drift is a kind of sampling error, which is more marked in small populations. page 21-13

## 21.6 MOLECULAR EVOLUTION LOOKS AT CHANGES AT THE LEVEL OF DNA OR AMINO ACID SEQUENCES.

The extent of sequence difference between two species is a function of the time they have been genetically isolated from each other. page 21-15

Correlation between sequence differences among species and time since common ancestry of those species is known as the molecular clock. page 21-15

The rate of the molecular clock varies among genes because some genes are more selectively constrained than others. page 21-15

## Self-Assessment

1. Differentiate between a phenotype and a genotype.

2. Name three types of mutations in terms of their effect on an organism.

3. Define genetic variation.

4. Describe three ways to measure genetic variation, and state which one is used today.

5. Given a set of genotype frequencies, calculate allele frequencies.

6. Define evolution.

7. Describe what happens to allele and genotype frequencies under the Hardy–Weinberg equilibrium.

8. Name and describe the five assumptions of the Hardy–Weinberg equilibrium.

9. Given a set of allele frequencies, calculate genotype frequencies if the population is in Hardy–Weinberg equilibrium.

10. Name and describe the five mechanisms of evolution.

11. Define natural selection and indicate how it is different from other mechanisms of evolution.

12. Explain how a molecular clock can be used to determine the time of divergence of two species.

Do you understand the chapter's Core Concepts? Log in to check your answers to the Self-Assessment questions, then practice what you've learned and reinforce this chapter's concepts by working through the problems and multimedia tutorials provided there.

**www.biologyhowlifeworks.com**

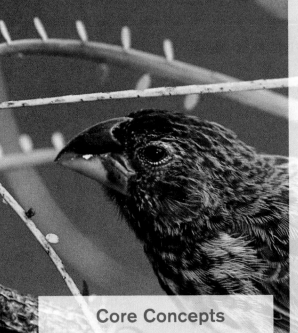

## CHAPTER 22

# SPECIES AND SPECIATION

## Core Concepts

**22.1** Species are reproductively isolated from one another.

**22.2** Reproductive isolation is caused by barriers to reproduction before or after egg fertilization.

**22.3** Speciation underlies the diversity of life on Earth.

**22.4** Speciation can occur with or without natural selection.

Imagine for a moment a world without **speciation,** the process that produces new and distinct forms of life. Life would have originated and natural selection would have done its job of winnowing advantageous mutations from disadvantageous ones, but the planet would be inhabited by a single kind of generally adapted organism. Instead of the staggering biological diversity we see around us—current estimates are that between 10 and 100 million species call Earth home—there would be just a single life-form. From a biological perspective, the planet would be a decidedly dull place. Evolution is as much about adaptation, the result of natural selection, as it is about speciation, the engine that generates that breathtaking biodiversity.

## 22.1 THE BIOLOGICAL SPECIES CONCEPT

The definition of **species** has been a long-standing problem in biology. Many biologists respond to the problem in the same way that Darwin himself did. In *On the Origin of Species*, Darwin wrote, "No one definition has as yet satisfied all naturalists; yet every naturalist knows vaguely what he means when he speaks of a species." The difficulty of defining species has come to be called the species problem.

Here is the problem in a nutshell: The species, as an evolutionary unit, must by definition be fluid and capable of changing, giving rise through evolution to new species. The whole point of the Darwinian revolution is that species are not fixed. How, then, can we define something that changes over time, and, by the process of speciation, even leads to the origin of two species from one?

### The species is the fundamental evolutionary unit.

We can plainly see biodiversity, but are what we call "species" real biological entities? Or was the term coined by biologists to simplify their description of the natural world? If so, species would then be convenient categories that derive not from nature but from our desire to categorize what we see around us. To test whether or not species are real, we can examine the natural world, measure some characteristic of different living organisms we see, and then plot these measurements on a graph. **Fig. 22.1** shows such a plot, graphing antenna length and wing length of three different types of butterfly. Note that the dots, representing individual organisms, fall into non-overlapping clusters. Each cluster is a species, and the fact that the clusters are distinct implies that species are biologically real. The distances we see between the dots within a cluster reflect variation from one individual to the next within a species (Chapter 21). Humans are highly variable, but overall we are more similar to one another than to our most

**FIG. 22.1    Species clusters on the basis of two characteristics.**

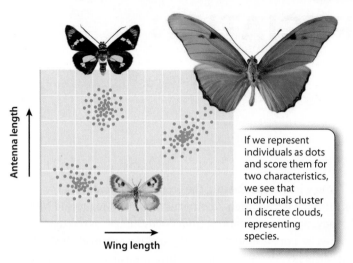

If we represent individuals as dots and score them for two characteristics, we see that individuals cluster in discrete clouds, representing species.

Antenna length

Wing length

humanlike relative, the chimpanzee. We form a messy cluster, but that cluster does not overlap with the chimpanzee cluster.

The species, then, is a fundamental biological unit. Taxonomists group similar species into genera (singular, genus), and similar genera into families, and so forth on the basis of shared ancestry. The lines between these higher taxonomic categories, however, are arbitrary; their composition depends on the judgment of taxonomists. If, in their view, two species are enough alike, they are placed in the same genus. If they are sufficiently different, they are placed in separate genera.

Whether or not two individuals are members of the same species, however, is not a matter of judgment, but rather a reflection of their ability (or inability) to exchange genetic material by producing fertile offspring. Furthermore, the species is the ultimate evolutionary unit. Think back to Chapter 21 and consider a new advantageous mutationthat appears initially in a single individual. That individual and its offspring inheriting the mutation will have a competitive advantage over other members of the population, and the mutation will increase in frequency until it reaches 100%. Migration among populations causes the mutation to spread further until all individuals within the species have it. The mutation spreads *within* the species but cannot spread beyond it. Therefore, the species is the key evolutionary unit. It is species that become extinct, and it is species that, through genetic divergence, give rise to new species.

### Reproductive isolation is the key to the biological species concept (BSC).

Although biologists took until the middle part of the twentieth century to reach any kind of agreement on the definition of species, Alfred Russel Wallace, who discovered natural selection independently of Darwin, published a definition in 1865 that is almost identical to the one we use today:

> Species are merely those strongly marked races or local forms which when in contact do not intermix, and when inhabiting distinct areas are generally believed to have had a separate origin, and to be incapable of producing a fertile hybrid offspring.

Wallace's definition anticipates the key elements of the most widely used and generally accepted definition of a species, the **biological species concept (BSC).** The BSC was described by the great evolutionary biologist Ernst Mayr (1904–2005) as follows:

> Species are groups of actually or potentially interbreeding populations that are reproductively isolated from other such groups.

Let us look at this definition closely. At the heart of it is the idea of reproductive compatibility. Members of the same species are capable of producing offspring together, whereas members of different species are incapable of producing offspring together. In other words, members of different species are reproductively isolated from one another. As Wallace and Mayr both realized, however, reproductive compatibility in fact entails more than just the ability to produce offspring. The offspring must be fertile and therefore capable of passing their genes on to their own offspring. For example, although a horse and a donkey, two different species, can mate to produce a mule, the mule is infertile. If, like a mule, a hybrid offspring is infertile, then it is a genetic dead end.

Note, too, the "actual or potential" part of Mayr's definition. An Asian elephant living on the island of Sri Lanka and one living in nearby India are considered members of the same species, *Elephas maximus,* even though Sri Lankan and Indian Asian elephants never have a chance to mate with each other in nature because they are geographically separated (**Fig. 22.2**).

### The BSC is more useful in theory than in practice.

Species, then, are real biological entities and the BSC is the most useful way to define them. However, the BSC has shortcomings, the most important of which is that it can be difficult to apply. Imagine you are on a field expedition in a rainforest and you find two insects that look reasonably alike. Are they members of the same species or not? To use the BSC, you would need to test whether or not they are capable of producing fertile offspring. However, in practice, you probably will not have the time and resources to perform such a test. And even if you do, it is possible that members of the same species will not reproduce when you place them together in a petri dish in your field camp.

**FIG. 22.2** Sri Lankan and Indian Asian elephants.

The conditions may be too unnatural or too threatening for the insects to behave in a normal way. And even if they do mate, you may not have time to determine whether or not the offspring are fertile. Wallace recognized this practical limitation of a reproduction-based definition of species, pointing out that in practice, "the test of hybridity cannot be applied in more than one case in ten thousand."

Thus, we should consider the BSC as a valid framework for thinking about species, but one that is difficult to test. On a day-to-day basis, therefore, biologists often use a rule of thumb called the **morphospecies concept.** Stated simply, the morphospecies concept holds that members of the same species usually look alike, which is why natural history guidebooks are useful. For example, the shape, size, and coloration of the bald eagle make it easy to determine whether or not a bird observed in the wild is a member of the bald eagle species. Here we are returning to those biological clusters of Fig. 22.1 in which members of a species fall close to one another within a single, discrete group.

Although the morphospecies concept is a useful and generally applicable rule of thumb, it is not an infallible way to identify species. Members of a species may not always look alike, but instead show different phenotypes called polymorphisms (literally, "many forms")—for example, color difference in some species of birds. In Chapter 21, we saw how minor genetic differences can result in two individuals within a species having very different appearances. Sometimes, males of a species may look different from females—think of the showy peacock, which differs dramatically from the relatively drab peahen. Young can be very different from old. Caterpillars that mature into butterflies are a striking example.

Whereas members of a species may look quite different from one another, the opposite can also be true: Members of different species can look quite similar. Some species of butterfly, for example, appear identical but have chromosomal differences that can be observed only with the aid of a microscope (**Fig. 22.3**). These chromosomal differences prevent the formation of interspecies hybrids, so those butterflies meet the BSC criterion—they are different species—but they do not meet the morphospecies criterion because they look so much alike.

## The BSC does not apply to asexual or extinct organisms.

In addition to the difficulties of putting the BSC into practice, a second problem is that it overlooks some organisms. For example, because it is based on the sexual exchange of genetic information, the BSC cannot apply to species that reproduce asexually, as do many bacteria. Although some asexual species do occasionally exchange genetic information in a process

**FIG. 22.3** Three species of *Agrodiaetus* butterflies. These similar-appearing butterflies are identifiable only by differences in their chromosome numbers.

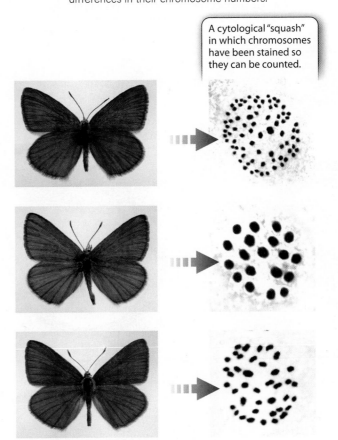

A cytological "squash" in which chromosomes have been stained so they can be counted.

known as conjugation (Chapter 26), true asexual organisms do not fit the BSC.

In addition, because it depends on reproduction, the BSC obviously cannot be applied to species that have become extinct and are known only through the fossil record, like different "species" of trilobites and dinosaurs.

## Ring species and hybridization complicate the BSC.

An unusual but interesting geographic pattern shown by **ring species** highlights another shortcoming of the BSC. Here we find that some populations within a species are reproductively isolated from each other, but others are not.

For example, populations of the greenish warbler, *Phylloscopus trochiloides,* are distributed in a large geographic loop around the Himalayas, as shown in **Fig. 22.4**. In Russia to the north, members of two neighboring populations do not interbreed.

Therefore, according to the BSC, they are different species. However, the more western of the two Russian populations is capable of reproducing with the population to the south of it, and that population is capable of reproducing with the population to the east of it, and so on. Eventually, the loop of genetically exchanging populations comes all the way round to the more eastern of the two Russian populations. Thus, though members of the two Russian populations cannot exchange genes directly, they can do so indirectly, with the genetic material passing through many intermediate populations. This situation is more complicated than anything predicted by the BSC. The two Russian populations are reproductively isolated from each other but they are not genetically isolated from each other because of gene flow around the ring.

We also see a complicated situation in some groups of closely related species of plants. Despite apparently being good morphospecies that can be distinguished by appearance alone, many different species of willows (*Salix*), oaks (*Quercus*), and dandelions (*Taraxacum*) are still capable of exchanging genes with other species in their genera through **hybridization,** or interbreeding, between species. By the BSC, these different forms should be considered one large species because they are able to reproduce and produce fertile offspring. However, because they maintain their distinct appearances, natural selection must work against the hybrid offspring. This unusual phenomenon seems to occur mainly in plants.

## Ecology and evolution can extend the BSC.

Because of these problems, much effort has been put into modifying and improving the BSC and many alternative definitions of species have been suggested. In general, these efforts highlight how difficult it is to make all species fit easily into one definition. The natural world truly defies neat categorization!

Several useful ideas, however, have come out of this literature. First is the notion that a species can sometimes be characterized by its **ecological niche,** which, as we will discuss further in Chapter 47, is a complete description of the role the species plays in its environment—its habitat requirements, its nutritional and water needs, and the like. It

---

**FIG. 22.4** **Ring species.** Ring species contain populations that are reproductively isolated from each other but can still exchange genetic material via other linking populations.

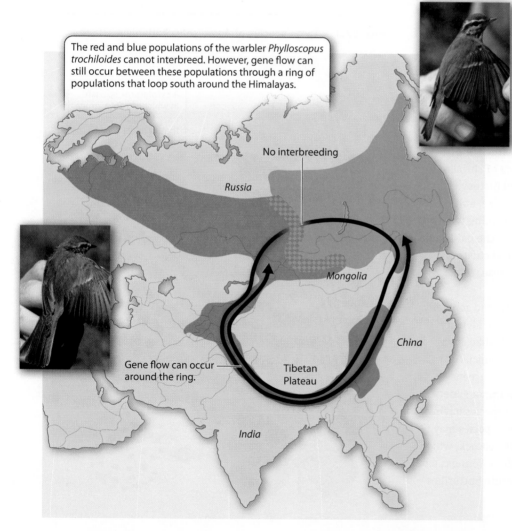

The red and blue populations of the warbler *Phylloscopus trochiloides* cannot interbreed. However, gene flow can still occur between these populations through a ring of populations that loop south around the Himalayas.

No interbreeding

*Russia*

*Mongolia*

*China*

Gene flow can occur around the ring.

Tibetan Plateau

*India*

turns out that it is impossible for two species to coexist in the same location if their niches are too similar because competition between them for resources will inevitably lead to the extinction of one of them. This observation has given rise to the **ecological species concept (ESC),** the idea that there is a one-to-one correspondence between a species and its niche. Thus, we can determine whether or not asexual bacterial lineages are distinct species on the basis of differences or similarities in their ecological requirements. If two lineages have very different nutritional needs, for example, we can infer on ecological grounds that they are separate species.

Second is the **evolutionary species concept (EvSC),** which is the idea that members of a species all share a common ancestry and a common fate. It is, after all, species rather than individuals that become extinct. The EvSC requires that all members of a species are descended from a single common ancestor. It does not specify, however, on what scale this idea should be applied. All mammals derive from a single common ancestor that lived about 200 million years ago, but there are thousands of what we recognize as species of mammals that have evolved since that long-ago common ancestor. But, under a strict application of the EvSC, would we consider all mammals to be a single species? Similarly, siblings and cousins are all descended from a common ancestor, a grandmother, but that is surely not sufficient grounds to classify more distant relatives as a distinct species. The EvSC can be useful when thinking about asexual species, but, given the arbitrariness of the decisions involved in assessing whether or not the descendants of a single ancestor warrant the term "species," its utility is distinctly limited.

It is worth bearing these ecological and evolutionary considerations in mind when discussing and thinking about species. Although the BSC remains our most useful definition of species, the ESC and EvSC broaden and generalize the concept. For example, in the case of a normally asexual species that is difficult to study because conjugation is so infrequent, we can jointly apply the ESC and EvSC. We can use the ESC to loosely define the species in terms of its ecological characteristics (for example, its nutritional requirements), and we can refine that definition by using genetic analyses to determine whether the group is indeed a species by the EvSC's standard (that is, that all its members derive from a single common ancestor).

Despite the shortcomings of the BSC and the usefulness of alternative ideas, we stress that the BSC is the most constructive way to think about species. In particular, by focusing on reproductive isolation—the inability to produce viable, fertile offspring—the BSC gives us a means of studying and understanding speciation, the process by which two populations, originally members of the same species, become distinct.

→ **Quick Check 1** Why haven't we been able to come up with a single, comprehensive, and agreed-upon species concept?

## 22.2 REPRODUCTIVE ISOLATION

Factors that cause reproductive isolation are generally divided into two categories, depending on when they act. **Pre-zygotic** isolating factors act before the fertilization of an egg, and **post-zygotic** factors come into play after fertilization. In other words, pre-zygotic factors prevent fertilization from taking place, whereas post-zygotic factors result in the failure of the fertilized egg to develop into a fertile individual.

### Pre-zygotic isolating factors occur before egg fertilization.

Most species are reproductively isolated by pre-zygotic isolating factors. Among animals, species are often **behaviorally isolated,** meaning that individuals only mate with other individuals based on specific courtship rituals, songs, and other kinds of behaviors. Chimpanzees may be our closest relative—and therefore the species we are most likely to confuse with our own—but a chimpanzee of the appropriate sex, however attractive to a chimpanzee of the opposite sex, fails to provoke even the faintest reproductive impulse in a human. In this case, the pre-zygotic reproductive isolation of humans and chimpanzees is behavioral.

Behavior does not play a role in plants, but pre-zygotic factors can still be important in reproductive isolation. Pre-zygotic isolation in plants can take the form of incompatibility between the incoming pollen and the receiving flower so fertilization fails to take place. We see similar forms of isolation between members of marine species, such as abalone, which simply discharge their gametes into the water. In these cases, membrane-associated proteins on the surface of sperm interact specifically with membrane-associated proteins on the surface of eggs of the same species but not with those of different species. These specific interactions ensure that a sperm from one abalone species, *Haliotis rufescens*, fertilizes only an egg of its own species and not an egg from *H. corrugata*, a closely related species.

In some animals, especially insects, incompatibility arises earlier in the reproductive process. The genitalia of males of the fruit fly *Drosophila melanogaster* are configured in such a way that they fit only with the genitalia of females of the same species. Attempts by males of *D. melanogaster* to copulate with females of another species of fruit fly, *D. virilis*, are thwarted by mechanical incompatibility.

In all three of these cases—plants, marine invertebrates, and insects—mating with members of one species (and reproductive isolation from other species) is promoted by **lock and key** systems that require both components, whether physical or biochemical, to match for a successful interaction to take place.

Both plants and animals may also be pre-zygotically isolated in time (**temporal separation**) and space (**ecological separation**). For example, closely related plant species may flower at different times, so there is no chance that the pollen of one will come into contact with the flowers of the other. Similarly, members of a nocturnal animal species simply will not encounter members of a closely related species that are active only during the day.

Closely related species that are separated by vast distances or physical features of the landscape obviously have no opportunity to exchange gametes, but spatial separation may also be more subtle. For example, the two Japanese species of ladybug beetle shown in **Fig. 22.5** can be found living side by side in the same field, but they feed on different plants. Because their life cycles are so intimately associated with their host plants (adults even mate on their host plants), these two species never breed with each other. This ecological separation is what causes their pre-zygotic isolation.

### Post-zygotic isolating factors occur after egg fertilization.

Post-zygotic isolating factors involve mechanisms that come into play after fertilization of the egg. Typically, they involve some kind of **genetic incompatibility.** One example, which we saw earlier in Fig. 22.3 and will explore later in the chapter, is the case of two organisms with different numbers of chromosomes.

In some instances, the effect can be extreme. For example, the zygote may fail to develop after fertilization because the two parental genomes are sufficiently different to prevent normal development. In others, the effect is less obvious. Some matings between different species produce perfectly viable adults, as in the case of the horse–donkey hybrid, the mule. As we have seen, though, all is not well with the mule from an evolutionary perspective. The horse and donkey genomes are different enough to cause the mule to be sterile. As a general rule, the more closely related—and therefore genetically similar (Chapter 21)—a pair of species, the less extreme the genetic incompatibility between their genomes.

## 22.3 SPECIATION

Recognizing that species are groups of individuals that are reproductively isolated from other such groups, we are now in a position to recast the key question: How does speciation occur? Instead of asking, "How do new species arise?" we can ask, "How does reproductive isolation arise between populations?"

### Speciation is a by-product of the genetic divergence of separated populations.

The key to speciation is the fundamental evolutionary process of genetic divergence between genetically separated populations. As we saw in Chapter 21, if a single population is split into two populations that are unable to interbreed, different mutations will appear by chance in the two populations. Like all mutations, these will be subject to genetic drift or natural selection, resulting over time in the genetic divergence of the two populations. Two separate populations that are initially identical will, over long periods of time, gradually become distinct as different mutations are introduced and propagated in each population. At some stage in the course of divergence, changes occur in one population that lead to its members being reproductively isolated from members of the other population (**Fig. 22.6**). It is this process that results in speciation.

Speciation—the development of reproductive isolation between populations—is, therefore, just a by-product of the genetic divergence of separated populations.

As Fig. 22.6 shows, speciation is typically a gradual process. If we try to cross members of two populations that have genetically diverged but not diverged far enough for full reproductive isolation (that is, speciation) to have arisen, we may find that the populations are **partially reproductively isolated.** They are not yet separate species, but the genetic differences between them are extensive enough that the hybrid offspring they produce have reduced fertility or viability compared to offspring produced by crosses between individuals within each population.

### Allopatric speciation is speciation that results from the geographical separation of populations.

Reproductive isolation between two populations ultimately results in two species. Because this process requires genetic isolation between the diverging populations and because geography is

---

**FIG. 22.5 Ecological separation.** The ladybugs (a) *Henosepilachna yasutomii* and (b) *H. niponica* are reproductively separated from each other because they feed and mate on different host plants.

a.

b.

**FIG. 22.6  Speciation.** Speciation occurs when two populations that are genetically diverging become reproductively isolated from each other.

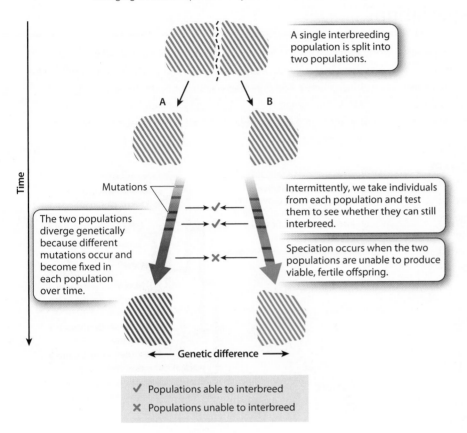

A single interbreeding population is split into two populations.

Mutations

The two populations diverge genetically because different mutations occur and become fixed in each population over time.

Intermittently, we take individuals from each population and test them to see whether they can still interbreed.

Speciation occurs when the two populations are unable to produce viable, fertile offspring.

Time

← Genetic difference →

✓ Populations able to interbreed
✗ Populations unable to interbreed

name. For example, Sri Lankan Asian elephants, subspecies *Elephas maximus maximus,* are generally larger and darker than Indian ones, subspecies *Elephas maximus indicus* (see Fig. 22.2).

How do populations become allopatric? There are two basic ways. The first is by **dispersal,** in which some individuals colonize a distant place, such as an island, far from the main source population. The second is by **vicariance,** in which a geographic barrier arises within a single population, separating it into two or more isolated populations. For example, when sea levels rose at the end of the most recent ice age, new islands formed along the coastline as the low-lying land around them was flooded. The populations on those new islands suddenly found themselves isolated from other populations of their species. This kind of island formation is a vicariance event.

Regardless of how the allopatric populations came about—whether through dispersal or vicariance—the outcome is the same. The two separated populations will diverge genetically until speciation occurs.

Often, vicariance-derived speciation events are the easiest to study because we can date the time at which the populations were separated if we know when the vicariance event occurred. One such event whose history is well known is the formation of the Isthmus of Panama between Central and South America, shown in **Fig. 22.7.** This event took place about 3.5 million years ago. As a result, populations of marine organisms in the western Caribbean and eastern Pacific that had formerly been able to interbreed freely were separated from each other. After a period of time, the result was the formation of many distinct species, each of whose closest relatives are on the other side of the isthmus.

Dispersal is important in a specific kind of allopatric speciation known as **peripatric speciation.** In this model, a few individuals from a **mainland population** (the central population of a species) disperse to a new location remote from the original population and evolve separately. This may be an intentional act of dispersal, or it could be an accident brought about by, for example, an unusual storm that blows migrating birds off their normal route. The result is a distant, isolated **island population.** "Island" in this case may refer to a true island—like Hawaii—or may simply refer to a patch of habitat on the mainland that is appropriate for the species but is geographically remote from the initial mainland population's habitat area. For a species adapted to life on mountaintops, a new island might be another previously

the easiest way to ensure genetic isolation, models of speciation focus on geography.

The process usually begins with the creation of **allopatric** (literally, "different place") populations, populations that are geographically separated from each other. Clearly, physically separating two populations is the simplest way to ensure that they become genetically isolated from each other. It is therefore not surprising that most speciation is thought to be allopatric. Note, however, that the size and ecology of a species must be considered in deciding whether or not two populations are truly allopatric. Two populations of salamanders separated by a large highway may be allopatric, as salamanders may not be able to cross the highway, whereas such a barrier would obviously not suffice to separate populations of birds. To assess, then, whether two populations are truly allopatric, it is necessary to take more than just location into account.

Because genetic divergence is typically gradual, we often find allopatric populations that have yet to evolve even partial reproductive isolation but which have accumulated a few population-specific traits. This genetic distinctness is sometimes recognized by taxonomists, who deem each geographic form a **subspecies** by adding a further designation after its species

**FIG. 22.7**

# Can a vicariance event cause speciation?

**BACKGROUND** 3.5 million years ago, the Isthmus of Panama was not completely formed. Several marine corridors remained open, allowing interbreeding between marine populations in the Caribbean and the eastern Pacific. Subsequently, the gaps in the isthmus were plugged, separating the Caribbean and eastern Pacific populations and preventing gene flow between them.

**HYPOTHESIS** Patterns of speciation will reflect the impact of the vicariance event resulting from the closing of the direct marine connections between the Pacific and the Caribbean. Specifically, researchers predicted that each ancestor species (from the time before the formation of the isthmus) split into two "daughter" species, one in the Caribbean and the other in the Pacific. The closest relative of each current Pacific species, then, is predicted to be a Caribbean species (and vice versa).

Interbreeding between Eastern Pacific and Caribbean populations of *Alpheus* was possible by the corridors that existed before the final formation of the Isthmus of Panama.

**3 million years ago**

Caribbean population

Land
Shallow water

Eastern Pacific population

Interbreeding between Eastern Pacific and Caribbean populations is no longer possible because of the geographic barrier.

**Today**

Nicaragua

Caribbean population

Costa Rica

Caribbean Sea

Panama

Eastern Pacific population

Pacific Ocean

Colombia

**EXPERIMENT** This study focused on 17 species of snapping shrimp in the genus *Alpheus*, a group that is distributed on both sides of the isthmus. The first step was to sequence the same segment of DNA from each species. The next step was to compare those sequences in order to reconstruct the phylogenetic relationships among the species.

The closing of marine corridors resulted over time in the speciation of *Alpheus* into Eastern Pacific and Caribbean species.

Pacific 6'
Pacific 6
Caribbean 6
Pacific 3
Caribbean 3
Caribbean 3'
Pacific 5
Caribbean 5
Caribbean 4
Pacific 4
Pacific 1
Caribbean 1
Pacific 2
Caribbean 2
Pacific 7'
Pacific 7
Caribbean 7

**RESULTS** The phylogeny reveals that species show a distinctly paired pattern of relatedness: The closest relative of each species is one from the other side of the isthmus.

**CONCLUSION** That we see these consistent Pacific/Caribbean sister species pairings strongly supports the hypothesis that the vicariant event caused by the formation of the isthmus has driven speciation in *Alpheus*. Each Pacific/Caribbean pairing is derived from a single ancestral species (indicated by a red dot) whose continuous distribution between the Caribbean and eastern Pacific was disrupted by the formation of the isthmus. Here we see striking evidence of the role of vicariance in speciation.

**FOLLOW-UP WORK** Speciation is about more than just genetic differences between isolated populations, as shown here. Knowlton and colleagues also tested the different species for reproductive isolation and found that there were high levels of isolation between Caribbean/Pacific pairs. Given that we know that each species pair has been separated for at least 3 million years, we can then calibrate the rate at which new species are produced.

**SOURCE** Knowlton, N., L. A. Weigt, L. A. Solórzano, D. K. Mills, and E. Bermingham. 1993. "Divergence in Proteins, Mitochondrial DNA, and Reproductive Compatibility Across the Isthmus of Panama." *Science* 260 (5114):1629–1632.

uninhabited mountaintop. For a rainforest tree species, that new island might be a patch of lowland forest on the far side of a range of mountains that separates it from its mainland forest population.

The island population is classically small and often in an environment that is slightly different from that of the mainland population. The peripatric speciation model suggests that change accumulates faster in these peripheral isolates than in the large mainland populations, both because genetic drift is more pronounced in smaller populations than in larger ones and because the environment may differ between the mainland and island in a way that results in natural selection driving differences between the two populations. These mechanisms cause genetic divergence of the island population from the mainland one, ultimately leading to speciation.

It is possible to glimpse peripatric differentiation in action (**Fig. 22.8**). Studies of a kingfisher, *Tanysiptera galatea*, in New Guinea and nearby islands show it under way. There are eight recognized subspecies of *T. galatea*, three on mainland New Guinea (where they exist in large populations separated by mountain ranges) and five on nearby islands (where, because the islands are small, the populations are correspondingly small). The mainland subspecies are still quite similar to one another, but the island subspecies are much more distinct, suggesting that genetic divergence is occurring faster in the small island populations. If we wait long enough, these subspecies will probably diverge into new species.

Because dating such dispersal events is tricky—newcomers on an island tend not to leave a record of when they arrived—we can some times use vicariance information to study the timing of peripatric speciation. For instance, we know that the oldest of the Galápagos Islands were formed 4-5 million years ago by volcanic action. Some time early in the history of the Galápagos, individuals of a small South American finch species arrived there. Conditions on the Galápagos are very different from those on the South American continent, where the mainland population of this ancestral finch lived, and so the isolated island population evolved to become distinct from its mainland ancestor and eventually became a new species. The finches' subsequent dispersal among the other islands of the Galápagos has promoted further peripatric speciation (**Fig. 22.9**). The result was the evolution of 13 different species of finches, known today as Darwin's finches.

The Galápagos finches and their frenzy of speciation illustrate the important evolutionary idea of **adaptive radiation,** a bout of unusually rapid evolutionary diversification in which natural selection accelerates the rates of both speciation and adaptation. Adaptive radiation occurs when there are many

---

**FIG. 22.8** Peripatric speciation in action among populations of New Guinea kingfishers.

Mainland subspecies of *Tanysiptera galatea* are similar to one another, but island subspecies are more distinct, suggesting faster genetic divergence on island populations.

The eight sub-species on New Guinea and nearby islands are marked in different colors.

Australia

New Guinea

**FIG. 22.9** **Adaptive radiation in Darwin's finches.** Today, there are 13 different species of finch on the Galápagos Islands.

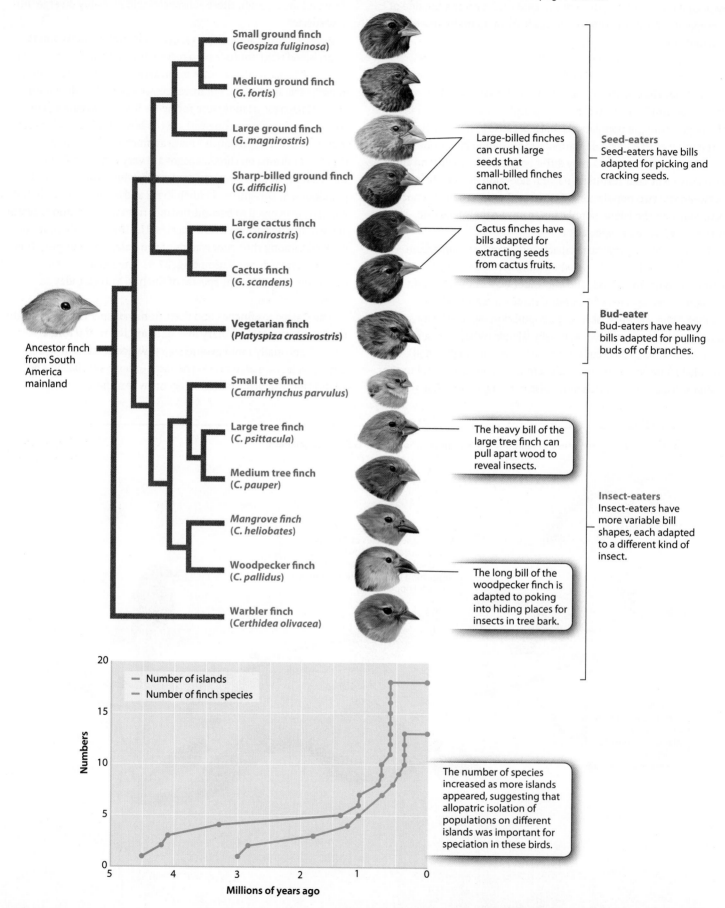

Small ground finch
(*Geospiza fuliginosa*)

Medium ground finch
(*G. fortis*)

Large ground finch
(*G. magnirostris*)

Sharp-billed ground finch
(*G. difficilis*)

Large cactus finch
(*G. conirostris*)

Cactus finch
(*G. scandens*)

**Vegetarian finch**
(***Platyspiza crassirostris***)

Small tree finch
(*Camarhynchus parvulus*)

Large tree finch
(*C. psittacula*)

Medium tree finch
(*C. pauper*)

*Mangrove finch*
(*C. heliobates*)

Woodpecker finch
(*C. pallidus*)

Warbler finch
(*Certhidea olivacea*)

Ancestor finch
from South
America
mainland

Large-billed finches can crush large seeds that small-billed finches cannot.

Cactus finches have bills adapted for extracting seeds from cactus fruits.

The heavy bill of the large tree finch can pull apart wood to reveal insects.

The long bill of the woodpecker finch is adapted to poking into hiding places for insects in tree bark.

**Seed-eaters**
Seed-eaters have bills adapted for picking and cracking seeds.

**Bud-eater**
Bud-eaters have heavy bills adapted for pulling buds off of branches.

**Insect-eaters**
Insect-eaters have more variable bill shapes, each adapted to a different kind of insect.

— Number of islands
— Number of finch species

Numbers

Millions of years ago

The number of species increased as more islands appeared, suggesting that allopatric isolation of populations on different islands was important for speciation in these birds.

ecological opportunities available for exploitation. Consider the ancestral finch immigrants arriving on the Galápagos. A wealth of ecological opportunities was open and available. Until the arrival of the colonizing finches, there were no birds on the islands to eat the plant seeds, or to eat the insects on the plants, and so on. Suppose that the ancestral finches fed specifically on medium-sized seeds on the South American mainland (that is, they were medium-seed specialists, with bills that are the right size for handling medium seeds). On the mainland, they were constrained to that size of seed because any attempt to eat larger or smaller ones brought them into conflict with other species—a large-seed specialist and a small-seed specialist—that already used these resources. In effect, stabilizing selection (Chapter 21) was operating on the mainland to eliminate the extremes of the bill-size spectrum in the medium-seed specialists. A somewhat smaller bill, which is good for eating slightly smaller seeds, was selected against because its possessors competed poorly against a small-billed species that is a small-seed specialist. A slightly larger bill was similarly disadvantageous because its possessors ended up competing for large seeds with a species that is a large seed specialist.

When the medium-billed immigrant finches first arrived on the Galápagos, however, no such competition existed. Therefore, natural selection promoted the formation of new species of small- and large-seed specialists from the original medium-seed-eating ancestral stock. With the elimination of the competition-mediated stabilizing selection that kept the medium-billed finches medium billed on the mainland, selection actually operated in the opposite direction, favoring the large and small extremes of the bill-size spectrum because these individuals could take advantage of the abundance of unused resources, small and large seeds. It is this combination of emptiness—the availability of ecological opportunity—and the potential for allopatric speciation that results in adaptive radiation.

→ **Quick Check 2.** Why do we see so many wonderful examples of adaptive radiation on mid-ocean volcanic archipelagos like the Galápagos?

## Co-speciation is speciation that occurs in response to speciation in another species.

As we have seen, physical separation is often a critical ingredient in speciation. Two populations that are not fully separated from each other—that is, there is gene flow between them—will typically not diverge from each other genetically, because genetic exchange homogenizes them. This is why most speciation is allopatric. Allopatric speciation brings to mind populations separated from each other by stretches of ocean or deserts or mountain ranges. However, separation can be just as complete even in the absence of geographic barriers. Consider an organism that parasitizes a single host species. Suppose that the host undergoes speciation, producing two daughter species. The original parasite population will also be split into two populations, one for each host species. Thus, the two new parasite populations are physically separated from each other and will diverge genetically, ultimately undergoing speciation. This divergence results in a pattern of coordinated host–parasite speciation called **co-speciation,** a process in which two groups of organisms speciate in response to each other and at the same time.

Phylogenetic analysis of lineages of parasites and their hosts that undergo co-speciation reveals trees that are similar for each group. Each time a branching event—that is, speciation—has occurred in one lineage, a corresponding branching event has occurred in the other (**Fig. 22.10**).

**? CASE 4 Malaria: Co-Evolution of Humans and a Parasite**
**How did malaria come to infect humans?**
Now let's look at a human parasite, *Plasmodium falciparum*, the causative agent of malaria. It had been suggested that *P. falciparum*'s closest relative is another *Plasmodium* species, *P. reichenowi*, found in chimpanzees, our closest living relative. Maybe *P. falciparum* and *P. reichenowi* were the products of co-speciation. When the ancestral population split millions of years ago to give rise to human and chimpanzee lineages, that population's parasitic *Plasmodium* population could also have

**FIG. 22.10 Co-speciation.** Parasites and their hosts often evolve together, and the result is similar phylogenies.

Dashed lines represent host-parasite pairs.

Chewing lice (parasites) | Pocket gophers (host)

*Geomydoecus chapini* - - - - - - - - - - *Orthogeomys hispidus*
*Geomydoecus setzeri* - - - - - - - - - - *Orthogeomys underwoodi*
*Geomydoecus panamensis* - - - - - - - - - - *Orthogeomys cavator*
*Geomydoecus cherriei* - - - - - - - - - - *Orthogeomys cherriei*
*Geomydoecus costaricensis* - - - - - - - - - - *Orthogeomys heterodus*

been split, ultimately yielding *P. falciparum* and *P. reichenowi*.

Recent studies, however, have disproved this hypothesis. We now know that *P. falciparum* was introduced to humans relatively recently from gorillas. Why doesn't the evolutionary history of *Plasmodium* follow the classical host–parasite co-speciation pattern? We know that this history is complex and is still being unraveled. However, we also know that the mosquito-borne phase of its life cycle facilitates transfer to new hosts. Malaria parasites are thus not as inextricably tied to their hosts as the pocket gopher lice in Fig 22.10. Presumably, it was just such a mosquito-mediated event that resulted in the introduction of *P. falciparum* into an ancestral human population in Africa from gorillas.

### Can sympatric populations—those not geographically separated—undergo speciation?

Can speciation occur *without* complete physical separation of populations? Evolutionary biologists are still exploring this question. Recall how separated populations inevitably diverge genetically over time (see Fig. 22.6). If a mutation arises in population A after it has separated from population B, that mutation is present only in population A and may eventually become fixed (100% frequency) in that population, either through natural selection, if it is advantageous, or through drift, if it is neutral. Once the mutation is fixed in population A, it represents a genetic difference between populations A and B. Repeated independent fixations of different mutations in the two populations result over time in the genetic divergence of separated populations.

Now imagine that populations A and B are not completely separated and there is some gene flow between them. The mutation that arose in population A can, in principle, appear in population B. The new mutation may indeed have become fixed in population A, but a migrant from population A to population B may introduce the mutation to population B as well. Gene flow effectively negates the genetic divergence of populations. If there is gene flow, a pair of populations may change over time, but they do so together. How, then, can speciation occur if gene flow exists? The term we use to describe populations that are in the same geographic location is **sympatric** (literally, "same place"). So we can rephrase the question in technical terms: Can speciation occur sympatrically?

One school of thought insists that sympatric speciation cannot occur and that all speciation is allopatric. If speciation can and does occur

sympatrically, one point is clear: Natural selection must act strongly to counteract the homogenizing effect of gene flow. Consider two sympatric populations of finch-like birds, represented in the graph in **Fig. 22.11**. One population begins to specialize on small seeds and the other on large seeds. If the two populations freely interbreed, no genetic differences between the two will occur and speciation will not take place. Now suppose that the offspring produced by the pairing of a big-seed specialist with a small-seed specialist is an individual best adapted to eat medium seeds, and there are no medium-sized seeds available in the environment. Natural selection will act against the hybrids, which will starve to death because there are no medium-sized seeds for them to eat and they are not well adapted to compete with the big- or small-seed specialists. Natural selection would then, in effect, eliminate the products of gene flow. So, although gene flow is occurring, it does not affect the divergence of the two populations because the hybrid individuals do not survive to reproduce. As discussed in Chapter 21, this form of natural selection, which operates against the middle of a spectrum of variation, is called disruptive selection.

Is there evidence for sympatric speciation? It turns out to be difficult to find evidence of sympatric speciation in nature, though it may not be especially rare in plants, as we will see. We might find two very closely related species in the same

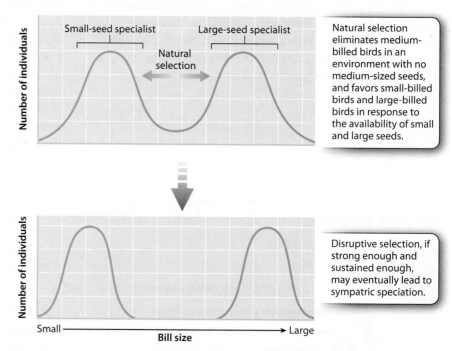

**FIG. 22.11  Sympatric speciation through disruptive selection.** Natural selection eliminates individuals in the middle of the spectrum.

location and argue that they must have arisen through sympatric speciation. However, there is an alternative explanation. One species could have arisen elsewhere by, for example, peripatric speciation and subsequently moved into the environment of the other one. In other words, the speciation occurred in the past in allopatry, and the two species are only currently sympatric because of migration after speciation occurred.

However, recent studies of fish in isolated lakes have provided strong, if not definite, evidence of sympatric speciation. In the volcanic mountains of Cameroon in West Africa, there are several small crater lakes inhabited by a diverse array of fish. In each case, all the fish species in a single lake can be shown to be one another's closest relatives. Presumably, all the species in a lake are descended from a single common ancestor, the fish species that first arrived in the lake. Each lake is small, so fish populations within them share a single habitat, implying that the speciation events that must have occurred to generate the current species diversity were sympatric.

This demonstration that sympatric speciation has in all probability occurred shows, at the least, that sympatric speciation is possible. We must recognize that we do not yet know just how much of all speciation is sympatric and how much is allopatric. **Fig. 22.12** summarizes modes of speciation based on geography.

→   **Quick Check 3** There are hundreds of species of cichlid fish in Lake Victoria in Africa. Some scientists argue that they evolved sympatrically, but recent studies of the lake suggest that it periodically dried out, leaving a series of small ponds. Why is this observation relevant to evaluating the hypothesis that these species arose by sympatric speciation?

## Speciation can occur instantaneously.

Although speciation is typically a lengthy process, it can occasionally occur in a single generation, making it sympatric by definition. Typically cases of such **instantaneous speciation** are caused by hybridization between two species in which the offspring are reproductively isolated from both parents.

For example, hybridization in the past between two sunflower species, *Helianthus annuus* and *H. petiolaris* (the ancestor of the cultivated sunflower) has apparently given rise to three new sunflower species, *H. anomalus, H. paradoxus,* and *H. deserticola*

**FIG. 22.12  Modes of speciation.**

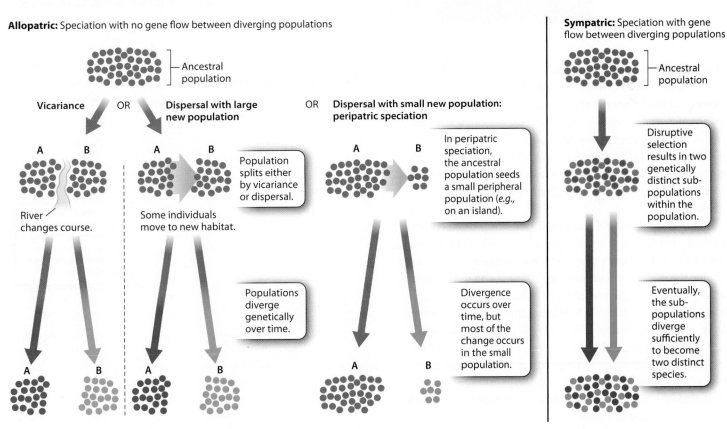

**FIG. 22.13 Speciation through hybridization**. In sunflowers, *Helianthus anomalus* (bottom panel) is the product of natural hybridization between *H. annuus* (top left) and *H. petiolaris* (top right).

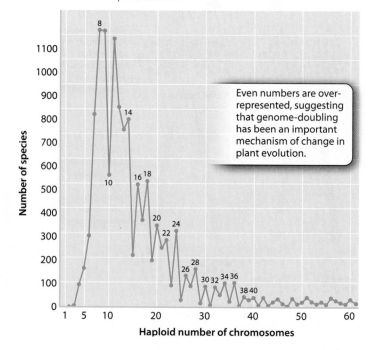

*Helianthus annuus*

*Helianthus petiolaris*

×

*Helianthus anomalus*

(**Fig. 22.13**). *H. petiolaris* and *H. annuus* have probably formed innumerable hybrids in nature, virtually all of them inviable. However, a few—the ones with a workable genetic complement— are the ones that survived to yield these three daughter species. In this case, each one of these new species has acquired a different mix of parental chromosomes. It is this species-specific chromosome complement that makes all three distinct and reproductively isolated from the parent species and from one another.

In many cases of hybridization, chromosome numbers may change. Two diploid parent species with 5 pairs of chromosomes, for a total of 10 chromosomes each, may produce a hybrid with double the number of chromosomes (that is, the hybrid inherits a full paired set of chromosomes from each parental species)—a total of 20. In this case, the hybrid has four genomes rather than the diploid number of two. We call such a double diploid a **tetraploid.** In general, animals cannot sustain this kind of expansion in chromosome complement, but plants are more likely to do so. As a result, the formation of new species through **polyploidy**—multiple chromosome sets (Chapter 13)—has been relatively common in plants.

Polyploids may be allopolyploids, meaning that they are produced from hybridization of two different species.

For example, related species of *Chrysanthemum* appear, on the basis of their chromosome numbers, to be allopolyploids. Alternatively, polyploids may be autopolyploids, meaning that they are derived from an unusual reproductive event between members of a single species. In this case, through an error of meiosis in one or both parents, the chromosome number is again increased. However, in contrast to the situation with allopolyploids, all the chromosomes derive from the single parental species. For example, *Anemone rivularis*, a plant in the buttercup family, has 16 chromosomes, and its close relative *A. quinquefolia* has 32. *A. quinquefolia* appears to be an autopolyploid derived from the joining of *A. rivularis* gametes with two full sets of chromosomes each.

So rampant is speciation by polyploidy in plants that it affects the pattern of chromosome numbers across all plants. In **Fig. 22.14**, we see the haploid chromosome numbers

**FIG. 22.14 Plant chromosome numbers.** Plant chromosome numbers suggest that polyploidy has played an important role in plant evolution.

Even numbers are over-represented, suggesting that genome-doubling has been an important mechanism of change in plant evolution.

Number of species

Haploid number of chromosomes

of thousands of plant species plotted on a graph. Note that, as the numbers get higher, even numbers tend to predominate, suggesting that the doubling of total chromosome complement (a form of polyploidy that will always result in an even number of chromosomes) is an important factor in plant evolution.

## 22.4 SPECIATION AND SELECTION

The association of both the origin of species and natural selection with Charles Darwin may make it seem that one cannot occur without the other. However, speciation can occur in the presence or absence of natural selection, and natural selection does not always lead to speciation.

### Speciation can occur with or without natural selection.

Natural selection may or may not play a role in speciation. The genetic divergence of two populations can be entirely due to genetic drift, for example, with no role for natural selection.

We have also seen two ways in which natural selection can be involved in speciation. First, sympatric speciation requires some form of disruptive natural selection, as when hybrid offspring are competitively inferior. Second, allopatric speciation (and adaptive radiation) may be facilitated by natural selection. For example, when a peripheral population is in a new environment, natural selection will act to promote its adaptation to the new conditions, accelerating in the process the rate of genetic divergence between it and its parent population.

### Natural selection can enhance reproductive isolation.

There is a third way in which natural selection may play an important role in speciation. Recall the two hypothetical bird populations, one with large bills and the other with small bills, whose medium-billed hybrid offspring are at a disadvantage because of a lack of medium-sized seeds in their environment.

If a large-billed individual cannot distinguish between large- and small-billed individuals as potential mates (that is, there is a lack of pre-zygotic isolation between them), it will frequently make the "wrong" mate choice, picking a short-billed individual and paying a considerable evolutionary cost of producing poorly adapted hybrid offspring.

Now imagine a new mutation in the large-billed population that permits individuals carrying it to distinguish between the two groups of birds and to mate only with other large-billed individuals. Such a mutation would spread under natural selection because it would prevent the wasted reproductive effort of producing disadvantaged hybrid offspring. This is an example of **reinforcement of reproductive isolation,** or **reinforcement** for short. Reinforcement is the process by which diverging populations undergo natural selection in favor of enhanced pre-zygotic isolation to prevent the production of less fit hybrid offspring. In this case, enhanced pre-zygotic isolation takes the form of mating discrimination, that is, an increased ability to recognize and mate with members of one's own population.

The best evidence in support of reinforcement comes from a study of related fruit-fly species living either in allopatry (geographically separated) or sympatry (in the same place, without geographical barriers). Sympatric species evolve pre-zygotic isolating mechanisms more rapidly than allopatric species. Why would this be? When the two populations are geographically separated, the formation of less fit hybrids is impossible since the two populations do not interbreed, so a mutation that increases mating discrimination between the two groups would not be favored by natural selection. In sympatry, however, where the production of less fit hybrids is a day-to-day problem, such a mutation provides a fitness benefit, so natural selection will favor its spread in the population and reinforce reproductive isolation between the two groups.

**Fig. 22.15** summarizes the evolutionary mechanisms that lead to speciation.

## Genetic Drift

Neutral mutations increase and decrease in frequency in the population because of random changes over time.

**1** Different mutations arise in each population. Some (black) are deleterious and are eliminated by selection. Some (red) are advantageous and swept to fixation (100% frequency) by selection. Some (blue) are neutral, so their fate is governed by genetic drift, and they ultimately drift either to extinction or fixation.

**Parent population**

Past

**Change in pigment**

A neutral mutation that briefly drifts up in frequency before drifting to extinction

1.0
0.6
0.4
0.2
0.0

Generations

**Daughter population A**

Generations
— 1
— 2
— 3
— 4

**Change in pigment**

A neutral mutation that drifts to fixation

1.0
0.6
0.4
0.2
0.0

Generations

Time

A fixed difference occurs when the frequency of an allele reaches 100%. Here, a fixed difference is shown when an entire row is red or blue.

Because neutral mutations, like these that change fish coloration, do not affect fitness, their frequencies vary from generation to generation by chance alone. In small populations, these shifts in frequency can be extreme. If this process runs for long enough, neutral mutations eventually drift to extinction (0%) or fixation (100%).

1 fixed difference between the populations

There are two fixed differences between the populations. The fish in population B are all larger than the fish in population A, and they lack the distinct pigmentation of the fish in population A.

2 fixed differences between the populations

**Present**

## Natural Selection

Mutations that increase the fitness of individuals become more common, and those that decrease the fitness of individuals become less common over time.

**A deleterious mutation** decreases the fitness of the organism.

**A neutral mutation** does not affect the fitness of the organism.

**A beneficial mutation** increases the fitness of an organism.

### Less streamlined fin shape

A deleterious mutation that is eliminated by natural selection

**Daughter population B**

Generations
1
2
3
4

### Change in pigment

A neutral mutation that briefly drifts up in frequency before drifting to extinction

1.0
0.6
0.4
0.2
0.0

**Generations**

### Allopatric speciation

When two populations are separated by a geographic boundary, they diverge genetically over time and eventually become reproductively isolated.

### Larger mass

A beneficial mutation driven to fixation by natural selection

1.0
0.6
0.4
0.2
0.0

**Generations**

**2** Mutations may be fixed by selection or drift in both populations. Different mutations are fixed in each population. Through time the populations diverge genetically.

Fish with the mutation are larger than fish without it. Because large size is advantageous (a larger fish is less likely to be eaten by a predator), larger fish are more likely to survive and reproduce than smaller fish. The frequency of the mutation in the population therefore increases over time by natural selection. This process continues until the mutation is fixed and all the fish in the population are large.

**3** Eventually, the two populations become so genetically divergent through the accumulation of fixed differences that mating between individuals of different populations fails to produce viable, fertile offspring. Speciation has occurred.

# Core Concepts Summary

## 22.1 SPECIES ARE REPRODUCTIVELY ISOLATED FROM ONE ANOTHER.

The best working definition of a species is the biological species concept (BSC). page 22-2

The BSC states that species are groups of actually or potentially interbreeding populations that are reproductively isolated from other such groups. page 22-2

We cannot apply the reproduction-based BSC to asexual or extinct organisms. page 22-3

Ring species and hybridization further demonstrate that the BSC is not a comprehensive or watertight definition of species. page 22-4

The BSC is nevertheless especially useful because it emphasizes reproductive isolation. page 22-4

## 22.2 REPRODUCTIVE ISOLATION IS CAUSED BY BARRIERS TO REPRODUCTION BEFORE OR AFTER EGG FERTILIZATION.

Reproductive barriers can be pre-zygotic, occurring before egg fertilization. page 22-5

Pre-zygotic isolation may be behavioral, mediated by differences in the timing of reproduction or by ecological differences, or brought about by mechanical or biochemical reproductive incompatibilities. page 22-5

Reproductive barriers can be post-zygotic, occurring after egg fertilization. page 22-6

In post-zygotic isolation, mating occurs but genetic incompatibilities prevent the development of a viable, fertile offspring. page 22-6

## 22.3 SPECIATION UNDERLIES THE DIVERSITY OF LIFE ON EARTH.

Speciation is typically a by-product of genetic divergence that occurs as a result of the fixation of different mutations in two populations that are not regularly exchanging genes. page 22-6

If divergence continues long enough, chance differences will arise that result in reproductive barriers between the two populations. page 22-6

Most speciation is thought to be allopatric, involving two geographically separated populations. page 22-6

Geographic separation may be caused by dispersal, resulting in the establishment of a new and distant population, or by vicariance, in which the range of a species is split by a change in the environment. page 22-7

A special case of allopatric speciation by dispersal is peripatric speciation, in which the new population is small and outside the species' original range. page 22-7

Adaptive radiation, in which speciation occurs rapidly to generate a variety of ecologically diverse forms, is best documented on oceanic islands after the arrival of a single ancestral species. Darwin's finches on the Galápagos are a good example of adaptive radiation. page 22-9

Co-speciation occurs when one species undergoes speciation in response to speciation in another. In parasites and their hosts, for example, co-speciation can result in host and parasite phylogenies that have the same branching patterns. page 22-11

Speciation may be sympatric, meaning that there is no geographic separation between the diverging populations. For this type of speciation to occur, natural selection for two or more different types within the population must act so strongly that it overcomes the homogenizing effect of gene flow. page 22-12

## 22.4 SPECIATION CAN OCCUR WITH OR WITHOUT NATURAL SELECTION.

Separated populations can diverge as a result of genetic drift, with no role for natural selection. page 22-15

Natural selection may act on mutations that allow individuals to identify and mate with individuals that are more like themselves. This process is called reinforcement of reproductive isolation. page 22-15

# Self-Assessment

1. Define the term "species."

2. Given a group of organisms, describe how you would test whether they all belong to one species or whether they belong to two separate species.

3. Name two types of organisms that do not fit easily into the biological species concept.

4. Explain how ecological and evolutionary considerations can help inform whether or not a group of organisms represents a single species.

5. Name four reproductive barriers and indicate whether each is pre- or post-zygotic.

6. Describe how genetic divergence and reproductive isolation are related to each other.

7. Differentiate between allopatric and sympatric speciation, and state which is thought to be more common.

8. Differentiate between allopatric speciation by dispersal and by vicariance, and give one example of each.

9. Describe how genetic drift can result in speciation.

10. Describe how natural selection can result in speciation.

Do you understand the chapter's Core Concepts? Log in to check your answers to the Self-Assessment questions, then practice what you've learned and reinforce this chapter's concepts by working through the problems and multimedia tutorials provided there.

www.biologyhowlifeworks.com

## CHAPTER 23

# EVOLUTIONARY PATTERNS

## Phylogeny and Fossils

## Core Concepts

**23.1** A phylogenetic tree is a reasoned hypothesis of the evolutionary relationships of organisms.

**23.2** A phylogenetic tree is built on the basis of shared derived characters.

**23.3** The fossil record provides a direct glimpse of evolutionary history.

**23.4** Phylogeny and fossils provide independent and corroborating evidence of evolution.

Nature displays, all around us, nested patterns of similarity among species. For example, as noted in Chapter 1, humans are more similar to chimpanzees than either humans or chimpanzees are to monkeys. Humans, chimpanzees, and monkeys, in turn, are more similar to one another than any of them is to a mouse. And humans, chimpanzees, monkeys, and mice are more similar to one another than any of them is to a catfish. This pattern of nested similarity was recognized more than 200 years ago and used by the great Swedish naturalist Carolus Linnaeus to classify biological diversity. A century later, Charles Darwin recognized this pattern as the expected outcome of a process of "descent with modification," or evolution.

Evolution produces two distinct but related patterns, both evident in nature. First is the nested pattern of similarities found among species on present-day Earth. The second is the historical pattern of evolution recorded by fossils. Life, in its simplest form, originated more than 3.5 billion years ago. Today, an estimated 10 million species inhabit the planet. Short of inventing a time machine, how can we reconstruct those 3.5 billion years of evolutionary history in order to understand the extraordinary events that have ultimately resulted in the biological diversity we see around us today? These two great patterns provide the answer.

Darwin recognized that the species he observed were the modified descendants of earlier ones. Distinct populations of an ancestral species separate and diverge through time, again and again, giving rise to multiple descendant species. The result is the pattern of nested similarities observed in nature (see Fig. 1.17). Just as a tree sprouts new branches on old ones and adds rings to a thickening trunk, the pattern of nested similarities among species strongly indicates a process of descent with modification and the accumulation of change. This history of descent with branching is called **phylogeny,** and is much like the genealogy that records our own family histories.

The evolutionary changes inferred from the patterns of relatedness among present-day species make predictions about the historical pattern of evolution we should see in the fossil record. For example, groups with features that we infer to have evolved earlier than others should appear earlier in time as fossils. Paleontological research reveals that the history of life is indeed laid out in the chronological order predicted on the basis of comparative biology. How do we reconstruct life's two great evolutionary patterns, and how do they compare?

## 23.1 READING A PHYLOGENETIC TREE

Chapter 22 introduced the concept of speciation, the set of processes by which physically, physiologically, or ecologically isolated subsets of a population diverge from one another to the point where they can no longer produce fertile offspring. As

**FIG. 23.1**   **The relationship between speciation and a phylogenetic tree.** The phylogenetic tree on the right depicts the evolutionary relationships that result from the two successive speciation events diagrammed on the left.

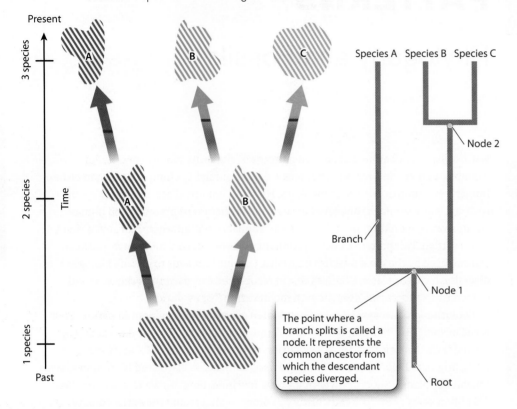

The point where a branch splits is called a node. It represents the common ancestor from which the descendant species diverged.

member of the Order Chiroptera. All the rest is understood (or can be looked up in a reference).

Phylogenetics, on the other hand, aims to discover the pattern of evolutionary relatedness among groups of species or other groups by comparing their anatomical or molecular features, and to depict these relationships as a **phylogenetic tree.** A phylogenetic tree is a hypothesis about the evolutionary history, or phylogeny, of the species. Phylogenetic trees are hypotheses because they represent the best model, or explanation, of the relatedness of organisms on the basis of all the existing data. However, as with any model or hypothesis, new data may require adjustments to the pattern of the tree. Many phylogenetic trees explore the relatedness of particular groups of individuals, populations, or species. We may, for example, want to understand how wheat is related to other, non-commercial grasses, or how pathogenic populations of *Escherichia coli* relate to more benign strains of the bacterium. At a much larger scale, universal similarities of molecular biology indicate that all living organisms are descended from a single common ancestor. This means that we can attempt to build a phylogenetic tree for all species, commonly referred to as the tree of life. In Part 2 of this book, we will make use of the tree of life and many smaller-scale phylogenetic trees to understand our planet's biological diversity.

**Fig. 23.2** shows a phylogenetic tree for vertebrate animals. The informal name at the end of each branch represents a group of organisms, many of them familiar. We will sometimes find it useful to refer to groups of species this way (for example, "frogs," or the "Class Anura") rather than name all the individual species or list the characteristics they have in common. It is important, however, to remember that such named groups represent a number of member species. If, for example, we were able to zoom in on the branch labeled "Frogs," we would see that it consists of many smaller branches, each representing a distinct species of frog.

This tree provides information about evolutionary relationships among vertebrates. For example, it proposes that the closest living relatives of birds are crocodiles and alligators. The tree also proposes that the closest relatives of all tetrapod (four-legged) vertebrates are lungfish. Phylogenetic trees are built from careful analyses of the morphological and molecular attributes of the species or other groups under study.

illustrated in **Fig. 23.1,** speciation can be thought of as a process of branching. Now consider what happens as this process occurs over and over in a lineage through time. As species proliferate, their evolutionary relationships to one another unfold in a treelike pattern, with present-day species as the tips of branches and their last common ancestors indicated by the point (called a **node**) from which they branched off.

## Phylogenetic trees provide hypotheses of evolutionary relationships.

Phylogenetics is one of two related disciplines in systematics, the study of evolutionary and genetic relationships among organisms. The other is taxonomy, the classification of organisms.

The aim of taxonomy is to recognize and name groups of individuals as species, and, subsequently, to group closely related species into the more inclusive taxonomic group of the genus, and so on up through the taxonomic ranks—species, genus, order, class, phylum, kingdom, domain. Taxonomy, then, provides us with a hierarchical classification of species in more and more inclusive groups, giving us a convenient way to communicate information about the features each group possesses. So, if we want to tell someone about a small animal we have seen with fur, mammary glands, and extended finger bones that permit it to fly, we can give them this long description, or we can just say we saw a bat (or, a

**FIG. 23.2** **A phylogeny of vertebrate animals.** The branching order constitutes a hypothesis of evolutionary relationships within the group.

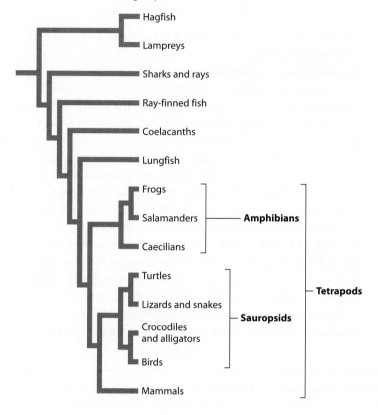

a frog or salamander, but lungfish are more closely related to frogs and salamanders than they are to other fish because lungfish share a common ancestor with frogs and salamanders that was more recent than their common ancestor with other fish (Fig. 23.2).

Groups that are more closely related to each other than either of them is to any other group, like lungfish and tetrapods, are called **sister groups**. Simply put, phylogenetic hypotheses amount to determining sister-group relationships because the simplest phylogenetic question we can ask is which two of any three species (or other groups) are more closely related to each other than either is to the third. In this light, we can see that a phylogenetic tree is simply a set of sister-group relationships in which each addition of a species beyond the original three entails finding its sister group in the tree.

Closeness of relationship is then determined by looking to see how recently two groups share a common ancestor. Shared ancestry is indicated by a node, or branch point, on a phylogenetic tree. An important aspect of a node is that it can be rotated without changing the evolutionary relationships of the groups. **Fig. 23.3,** for example, shows four phylogenetic trees depicting evolutionary relationships among birds, crocodiles and alligators, and lizards and snakes. In all four trees, birds are a sister group to crocodiles and alligators because birds, crocodiles, and alligators share a common ancestor not shared by lizards and snakes. Information about evolutionary relationships therefore lies in the order of nodes over time, not the order of groups along the tips.

→ **Quick Check 1** Does either of these two phylogenetic trees indicate that humans are more closely related to lizards than to mice?

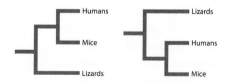

## The search for sister groups lies at the heart of phylogenetics.

Two species, or groups of species, are considered to be closest relatives if they share a common ancestor not shared by any other species or group. In Fig. 23.2, for example, we can see that frogs are more closely related to salamanders than to any other group of organisms because frogs and salamanders share a common ancestor not shared by any other group. Similarly, lungfish are more closely related to tetrapods than to any other group. A lungfish may look more like a fish than it does

**FIG. 23.3** **Depicting sister groups.** The four trees illustrate the same set of sister-group relationships.

The four trees are equivalent because nodes can be rotated without changing evolutionary relationships.

## A monophyletic group consists of a common ancestor and all its descendants.

So far in our discussion, we have used the word "group" as a kind of summary word to mean all the species in some taxonomic entity under discussion. A more technical word is **taxon** (plural, taxa), with taxonomy providing a formal means of naming groups. In recent decades, biologists have worked to integrate evolutionary history with taxonomic classification. The resulting classification emphasizes groups that are **monophyletic,** meaning that all members share a single common ancestor not shared with any other species or group of species. In Fig. 23.2, the tetrapods are monophyletic because they all share a common ancestor not shared by any other taxa. Similarly, amphibians are monophyletic.

In contrast, consider the group of animals traditionally recognized as reptiles, which includes turtles, snakes, lizards, crocodiles, and alligators (see Fig. 23.2). The group "reptiles" excludes birds, although they share a common ancestor with the included animals. Such a group is considered **paraphyletic.** A paraphyletic group includes some, but not all, of the descendants of a common ancestor. Early zoologists separated birds from reptiles because they are so distinctive, but feathers and other unique features of birds do not tell us about their relationship to other vertebrates. In fact, many features of skeletal anatomy and DNA sequence strongly support the placement of birds as a sister group to the crocodiles and alligators.

There is a simple way to distinguish between monophyletic and paraphyletic groups, illustrated in **Fig. 23.4**. If in order to separate a group from the rest of the phylogenetic tree you need only to make one cut, the group is monophyletic. If you need at least two cuts to separate the group from the tree, it is paraphyletic.

Groupings that do not include the last common ancestor of all members are called **polyphyletic.** For example, clustering bats and birds together as flying tetrapods results in a polyphyletic group (Fig. 23.4).

Determining what sets of organisms are considered monophyletic groups is a main goal of phylogenetics because monophyletic groups include *all* descendants of a common ancestor and *only* the descendants of that common ancestor. This means that monophyletic groups alone show the evolutionary path a given group has taken since its origin. Omitting some members of a group, as in the case of reptiles and other paraphyletic groups, can provide a misleading sense of evolutionary history. By using

monophyletic groups in taxonomic classification, we effectively convey our knowledge of their evolutionary history.

→  **Quick Check 2** Look at Fig. 23.4. Are fish a monophyletic group?

## Taxonomic classifications are information storage and retrieval systems.

The nested pattern of similarities among species has been recognized by naturalists for centuries. As we saw in Chapter 22, closely related species group together on a single branch of the tree of life. In the vocabulary of formal classification, closely related species are grouped into a **genus** (plural, genera). Closely related genera, in turn, belong to a larger, more inclusive branch of the tree—they are classified as a **family.** Closely related families, in turn, form an **order,** orders form a **class,** classes form a **phylum** (plural, phyla), and phyla form a **kingdom,** each more inclusive taxonomic level occupying a successively larger limb on the tree (**Fig. 23.5**). Biologists today commonly refer to the three largest limbs of the entire tree of life as **domains** (the Eukarya, or eukaryotes; Bacteria; and Archaea, or archaeons, introduced in Chapter 1 and discussed more fully in Chapters 26 and 27).

The ranks of classification form a nested hierarchy, but the boundaries of ranks above the species level are arbitrary in that there

---

**FIG. 23.4   Monophyletic, paraphyletic, and polyphyletic groups.** Only monophyletic groups reflect evolutionary relationships because only they include all the descendants of a common ancestor.

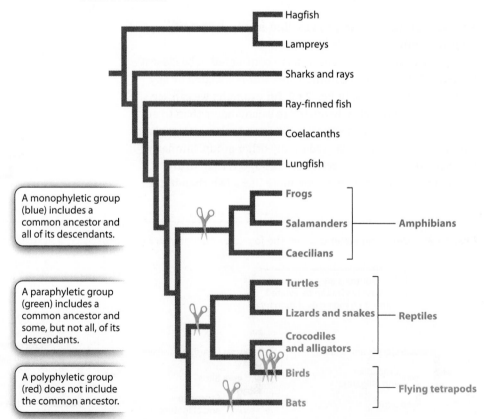

A monophyletic group (blue) includes a common ancestor and all of its descendants.

A paraphyletic group (green) includes a common ancestor and some, but not all, of its descendants.

A polyphyletic group (red) does not include the common ancestor.

**FIG. 23.5** **Classification.** Classification reflects our understanding of phylogenetic relationships, and the taxonomic hierarchy reflects the order of branching.

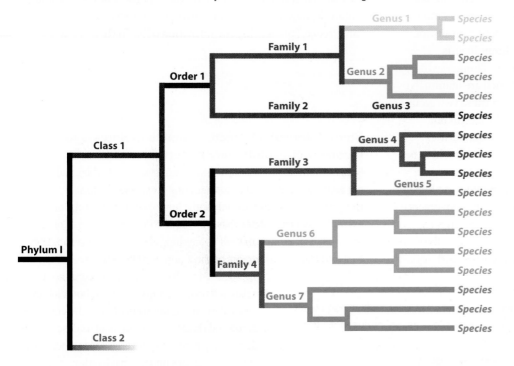

## Homology is similarity by common descent.

Phylogenetic trees are inferred by comparison of character states shared among different groups of organisms. **Characters** are the anatomical, physiological, or molecular features that make up organisms. To be useful for phylogenetic reconstruction, they must vary among but not within species and have a genetic basis. In general, characters have several observed conditions, called **character states.** In the simplest case, a character can be present or absent—lungs are present in tetrapods and lungfish, but absent in other vertebrate animals. Commonly, however, multiple character states are apparent—petals are a character of flowers, for example, and their arrangement can be considered a character state. Flowers can have many petals arranged in a helical pattern, many petals arranged in a whorl, few petals arranged in a whorl, or few petals fused into a tube. All species contain some character states that are shared with other members of their group, some that are shared with members of other groups, and some that are unique.

Character states can be similar for one of two reasons: The character state was present in the common ancestor of the two groups and retained over time (common ancestry), or the character state independently evolved in the two groups as an adaptation to similar environments (convergent evolution). Consider two examples. In the vertebrate phylogeny shown in **Fig. 23.6**, mammals and birds both

is nothing particular about a group that makes it, for example, a class rather than an order. For this reason, it is not necessarily true that orders or classes of, for example, birds and ferns are equivalent in any meaningful way. In contrast, sister groups are equivalent to one another in several ways—notably, in that they diverged from a single ancestor at a single point in time. Therefore, if one branch is a sister group that contains 500 species and the other has 6, the branches have experienced different rates of speciation, extinction, or both since they diverged. In Fig 23.5, Families 1 and 2 are sister groups but Family 1 has five species while Family 2 has just one.

## 23.2 BUILDING A PHYLOGENETIC TREE

Up to this point, we have focused on how to interpret a phylogenetic tree, a diagram that depicts the evolutionary history of organisms. But how do we infer their evolutionary history from a group of organisms? That is, how do we actually construct a phylogenetic tree? Biologists use characteristics of organisms to figure out their relationships. Similarities among organisms are particularly important in that similarities sometimes suggest shared ancestry. However, a key principle of constructing trees is that only *some* similarities are actually useful. Others can in fact be misleading.

**FIG. 23.6** **Homology and analogy.** A homology is a similarity that results from shared ancestry, while an analogy is a similarity that results from convergent evolution.

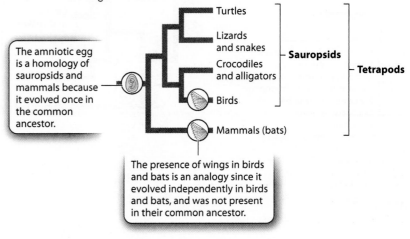

produce amniotic eggs. Because amniotic eggs occur only in groups descended from the common ancestor at the node connecting the mammal and sauropsid branches of the tree, we reason that birds and mammals each inherited this character from a common ancestor in which the trait first evolved. Characters that are similar because of descent from a common ancestor are said to be **homologous.**

Not all similarities arise in this way, however. Think of wings, a character exhibited by both birds and bats (which are mammals). Much evidence supports the view that wings in these two groups do not reflect descent from a common, winged ancestor but rather evolved independently in the two groups. Similarities due to independent adaptation by different species are said to be **analogous.** They are the result of convergent evolution.

Innumerable convergences less dramatic than wings are known. In some, we even understand the genetic basis of the convergence. For example, echolocation has evolved in bats and in dolphins, which are also mammals. Prestin is a protein in the hair cells of ears that is involved in hearing ultrasonic frequencies. Both bats

and dolphins have evolved parallel changes in the *Prestin* genes, apparently convergent adaptations for echolocation. Similarly, unrelated fish that live in freezing water at the poles, Arctic and Antarctic, have evolved similar glycoproteins that act as molecular "antifreeze," preventing the formation of ice in their tissues.

→    **Quick Check 3** Fish and dolphins have many traits in common, including a streamlined body and fins. Are these traits homologous or analogous?

## Shared derived characters enable biologists to reconstruct evolutionary history.

Because homologies result from shared ancestry, only they, and not analogies, are useful in constructing phylogenetic trees. However, it turns out that only some homologies are useful. For example, character states that are unique to a given species or other monophyletic group can't tell us anything about its sister group. They evolved after the divergence of the group from its sister group and so can be used to characterize a group but not to relate it to other groups. Similarly, homologies formed in the common ancestor of the entire group and therefore present in all its descendants do not help to identify sister-group relationships among the groups under consideration.

What we need to develop hypotheses of evolutionary relationship are homologies shared by some, but not all, of the members of the group under consideration. These are shared derived characters and are called **synapomorphies.** A derived character state is an evolutionary innovation (for example, the change from five toes to a single toe - the hoof - in the ancestor of horses and donkeys). When such a novelty arises in the common ancestor of two taxa, it is shared by both (thus the hoof is a synapomorphy defining horses and donkeys as sister groups).

In **Fig. 23.7**, we indicate the major synapomorphies that have helped us construct the phylogeny of vertebrates. For example, the lung is a character present in lungfish and tetrapods, but absent in other vertebrates. Thus, the presence of lungs provides one piece of evidence that lungfish are the sister group of tetrapods. Phylogenetic reconstruction on the basis of synapomorphies is called **cladistics.**

## The simplest tree is often favored among multiple possible trees.
To show how synapomorphies help us chart out evolutionary relationships, let's consider

**FIG. 23.7    Synapomorphies, or shared derived characters.** Homologies that are present in some, but not all, members of a group help us to construct phylogenetic trees.

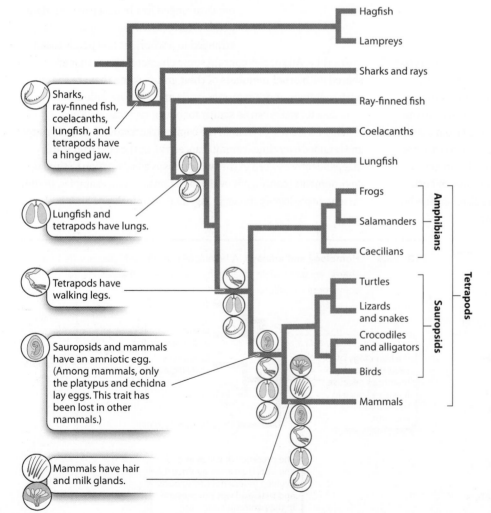

**FIG. 23.8** **Constructing a phylogenetic tree from shared derived traits.** The strongest hypothesis of evolutionary relationships overall is the tree with the fewest number of changes because it minimizes the total number of independent origins of character states.

**a.** Character states

**b.** Possible phylogenetic trees

This phylogeny requires four changes.

This phylogeny requires five changes.

This phylogeny requires six changes.

**c.** Phylogeny of Hexapoda (insects)

the simple example in **Fig. 23.8**. We begin with four species of animals in a group we wish to study, labeled "A" through "D," that we will call our ingroup; we believe them to be closely related to each other. For comparison, we have a species that we believe is outside this ingroup—that is, it falls on an earlier branch of the tree—and so is called an outgroup (labeled "OG" in Fig. 23.8). Each species in the ingroup and outgroup has a different combination of characters, such as leg number, presence of wings, and whether

development of young to adult is direct or goes through a pupal stage (Fig. 23.8a).

We are interested in the relationships among species A–D and so focus on potential synapomorphies—character states shared by some but not all species within the group. For example, only C and D have pupae. This character suggests that C and D are more closely related to each other than either is to the other species. The alternatives are that C and D each evolved pupae independently or that pupae were present in the common ancestor of A–D but were lost in A and B.

How do we choose among the alternatives? Comparison with the outgroup shows that it does not form pupae, supporting the hypothesis that pupal development evolved within the ingroup. In practice, biologists examine multiple characters and choose the phylogenetic hypothesis that best fits all of the data.

How do we determine "best fit"? The phylogenetic tree shown on the left in Fig. 23.8b considers four characters and their various character states and reflects the sister-group relationship between C and D proposed earlier. This tree requires exactly four character-state changes during the evolution of these species: reduction of leg pairs to three in the group ABCD, wings in the group BCD, and pupae in group CD, plus a change of form of the mandible, or jaw, in species D.

Now consider the middle tree in Fig. 23.8b. It groups A and B together, and so differs from the tree on the left in requiring either a loss of wings in A, or an additional origin of wings in B, independent of that in the ancestor of CD—five changes in all. The tree on the right groups A and B together, and requires two extra steps for a total of six steps. No tree that we can construct from species A–D requires fewer than four evolutionary changes, so the left-hand version in Fig. 23.8b is the best available hypothesis of evolutionary relatedness. In fact, this is the phylogeny for a sample of species from the largest group of animals on Earth, the Hexapoda—or insects (Fig. 23.8c).

In general, trees with fewer character changes are preferred to ones that require more because they provide the simplest explanation of the data. This approach is an example of **parsimony,** that is, choosing the simpler of two or more hypotheses to account for a given set of observations. In systematics, parsimony suggests counting character changes on a phylogenetic tree to find the simplest tree for the data (the one with the fewest number of changes). Each change corresponds to a mutation (or mutations) in an ancestral species, and the more changes or steps we propose, the more independent mutations we must also hypothesize.

Note also that it isn't necessary to make decisions in advance about which characters are homologies and which are analogies. We can construct all possible trees and then choose the one requiring the fewest evolutionary changes. This is a simple matter for the example in Fig. 23.8, because four species can be arranged into only 15 possible different trees. As the number of groups increases, however, the number of possible trees connecting them increases as well, and dramatically so. There are 105 trees for 5 groups and 945 trees for 6 groups, and there are nearly 2 million possible trees for 10. For 50 groups the possibilities balloon to $3 \times 10^{76}$! Clearly, computers are required to sort through all the possibilities.

As is true for all hypotheses, phylogenetic hypotheses can be strongly supported or weakly supported. Biologists use statistical methods to evaluate a given phylogenetic hypothesis.

Available character data may not strongly favor any hypothesis. When support for a specific branching pattern is weak, biologists commonly depict the relationships as unresolved and show multiple groups diverging from one node, rather than just two, as we have seen. Such branching patterns are not meant to suggest that multiple species diverged simultaneously, but instead to indicate that we lack the data to choose unequivocally among several different hypotheses of relationships. In Part 2 of this book, we will encounter unresolved branches in a number of groups. These shouldn't be read as admissions of defeat, but instead as problems awaiting resolution and opportunities for future research.

## Molecular data complement comparative morphology in reconstructing phylogenetic history.

Trees can be built using anatomical features as characters, but increasingly, tree construction relies on molecular data. The amino acids at particular positions in the primary structure of a protein can be used as characters, as can the nucleotides at specific positions along a strand of DNA.

From genealogy to phylogeny, tracing mutations in DNA or RNA sequences has revolutionized the reconstruction of historical genetic connections. Whether we are tracing the paternity of the children of Sally Hemings, mistress of Thomas Jefferson, identifying the origin of a recent cholera epidemic in Haiti, or placing baleen whales near the Hippopotamus family in a phylogenetic tree of mammals, molecular data give insights through their highly detailed information.

Much of what we know already about molecular evolution comes from phylogenetic comparisons previously established by comparative morphology, the study of structures and forms in different taxa. It is important to recognize that there is nothing about molecular data that provides a better record of history than does comparative morphology; molecular data simply provide more details. Indeed, for microbes and viruses, there is very little morphology available, so molecular information is critical for phylogenetic reconstruction in these microscopic taxa. Once a gene or other stretch of DNA or RNA with suitable levels of variation is identified in two or more species, sequences are obtained and aligned to identify homologous nucleotide sites. Analyses of this kind have typically involved comparisons of sequences of about 1000 nucleotides from one or more genes. Increasingly, though, the availability of whole-genome sequences is changing the way we do molecular phylogenetics: Rather than comparing the sequences of a few genes, we compare the sequences of entire genomes.

The process of using molecular data is conceptually similar to the process described earlier for morphological data. Through comparison to an outgroup, we can identify derived and ancestral molecular characters (whether DNA nucleotides or amino acids) and generate the phylogeny on the basis of synapomorphies as before.

**FIG. 23.9** Phylogenetic trees of DNA sequences based on (b) synapomorphies and (c) distance.

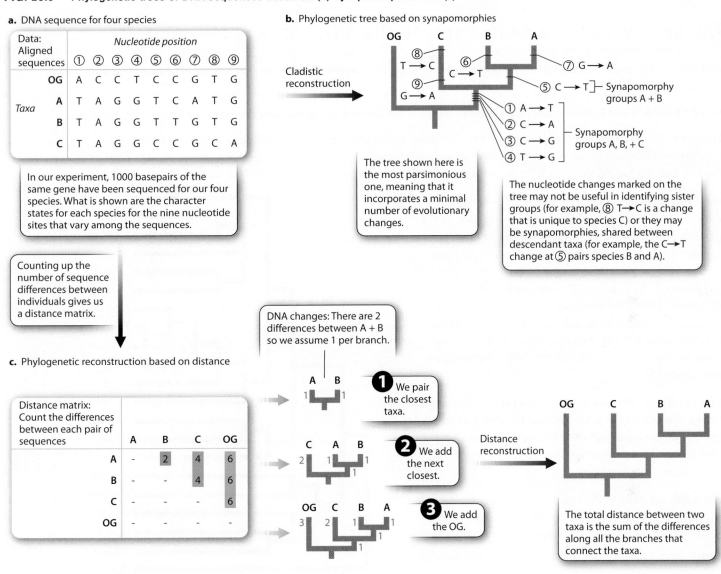

**a.** DNA sequence for four species

In our experiment, 1000 basepairs of the same gene have been sequenced for our four species. What is shown are the character states for each species for the nine nucleotide sites that vary among the sequences.

Counting up the number of sequence differences between individuals gives us a distance matrix.

**b.** Phylogenetic tree based on synapomorphies

Cladistic reconstruction

The tree shown here is the most parsimonious one, meaning that it incorporates a minimal number of evolutionary changes.

The nucleotide changes marked on the tree may not be useful in identifying sister groups (for example, ⑧ T→C is a change that is unique to species C) or they may be synapomorphies, shared between descendant taxa (for example, the C→T change at ⑤ pairs species B and A).

**c.** Phylogenetic reconstruction based on distance

DNA changes: There are 2 differences between A + B so we assume 1 per branch.

**1** We pair the closest taxa.

**2** We add the next closest.

**3** We add the OG.

Distance reconstruction

The total distance between two taxa is the sum of the differences along all the branches that connect the taxa.

An alternative method of reconstruction is based on distance rather than synapomorphies. Here the premise is simple: The descendants of a recent common ancestor will have had relatively little time to evolve differences, whereas the descendants of an ancient common ancestor have had a lot of time to evolve differences. Thus similarity (or low distance) indicates the recency of common ancestry.

Underpinning this approach is the assumption that the rate of evolution is constant (otherwise, a pair of taxa with a recent common ancestor could be more different than expected because of an unusually fast rate of evolution). This rate-constancy assumption is less likely to be violated when we are using molecular data than when we are using morphological data. Recall from Chapter 21 that the molecular clock is based on the observation of constant accumulation of genetic divergence

through time. **Fig. 23.9** shows a simple DNA sequence dataset which we can analyze either on the basis of synapomorphies (Fig. 23.9b) or on the basis of distance (Fig. 23.9c). Note that both give the same result in this case.

Molecular data are often combined with morphological data in analyses, and each can also serve as an independent assessment of the other. Not surprisingly, results from analyses of each kind of data are commonly compatible, at least for plants and animals rich in morphological characters.

Today, the single largest library of taxonomic information is the National Institutes of Health's genetic data storage facility, called GenBank. As of this writing, GenBank gives users access to more than 100 billion observations (mostly nucleotides) collected under more than 400,000 taxonomic names. A growing web resource is Encyclopedia of Life, which is gathering additional

**FIG. 23.10**

# Did an HIV-positive dentist spread the AIDS virus to his patients?

**BACKGROUND** In the late 1980s, several patients of a Florida dentist contracted AIDS. Molecular analysis showed that the doctor was HIV-positive.

**HYPOTHESIS** It was hypothesized that the patients acquired HIV during dental procedures carried out by the infected dentist.

**METHOD** Two HIV samples (denoted 1 and 2 in the figure) were obtained from several people, including the dentist (Dentist 1 and Dentist 2), several of his patients (Patients A through G), and other HIV-positive individuals chosen at random from the local population (LP). In addition, a strain of HIV from Africa (HIVELI) was included in the analysis.

**RESULTS** Biologists constructed a phylogeny based on the nucleotide sequence of a rapidly evolving gene in the genome of HIV. Because the gene evolves so quickly, its mutations preserve a record of evolutionary relatedness on a very fine scale. HIV in some of the infected patients — patients A, B, C, E, and G — were more similar to the dentist's HIV than they were to samples from other infected individuals. Some patients' sequences, however, did not cluster with the dentist's, suggesting that these patients, D and F, had acquired their HIV infections from other sources.

**CONCLUSION** HIV phylogeny makes it highly likely that the dentist infected several of his patients. The details of how the patients were infected remain unknown, but rigidly observed safety practices make it unlikely that such a tragedy could occur again.

**FOLLOW-UP WORK** Phylogenies based on molecular sequence characters are now routinely used to study the origin and spread of infectious diseases, such as swine flu.

**SOURCE** Hillis, D. M., J. P. Huelsenbeck, and C. W. Cunningham. 1994. "Application and Accuracy of Molecular Phylogenies." *Science* 264:671–677.

biological information about species, including ecology, geographic distributions, photographs, and sounds in pages for individual species that are easy to navigate. Another web-based resource, the Tree of Life, provides information on phylogenetic trees for many groups of organisms.

## Phylogenetic trees can help solve practical problems.

The sequence of changes on a tree from its base to its tips documents evolutionary changes that have accumulated through time. Trees suggest which lineages are older than others, and which traits came first and which followed later. Proper phylogenetic placement thus reveals a great deal about evolutionary history, and it can have practical consequences as well. For example, oomycetes, microorganisms responsible for potato blight and

other important diseases of food crops, were long thought to be fungi because they look a lot like some fungal species. The discovery, using molecular characters, that oomycetes belong to a very different group of eukaryotic organisms, has opened up new possibilities for understanding and controlling these plant pathogens. Similarly, in 2006, researchers used DNA sequences to identify the Malaysian parent population of the lime swallowtails that had become an invasive species in the Dominican Republic.

Phylogenetics solved a famous case in which an HIV-positive dentist in Florida was accused of infecting his patients (**Fig. 23.10**). Nucleotide sequence in the genome of HIV evolves so rapidly that biologists can build phylogenetic trees that trace the spread of specific strains from one individual to the next. Phylogenetic study of HIV present in samples from several infected patients, the dentist, and other individuals provided evidence that the dentist had, indeed, infected his patients.

Similarly, phylogenetic studies of influenza virus strains show their origins and subsequent movements among geographic regions and individual patients. Today, there is a growing effort to use particular DNA sequences as a kind of fingerprint or barcode for tracking biological material. Such information could quickly identify samples of shipments as meat from endangered species, or track newly emerging pests. The Consortium for the Barcode of Life has already accumulated on its website species-specific DNA barcodes for more than 100,000 species. Phylogenetic evidence provides a powerful tool for evolutionary analysis and is useful across timescales ranging from months to the entire history of life, from the rise of epidemics to the origins of metabolic diversity. Few other breakthroughs in science have had as broad an impact as that of phylogenetic systematics and the new sources of molecular data that have carried it so far.

## 23.3 THE FOSSIL RECORD

Phylogenies based on living organisms provide hypotheses about evolutionary pattern. Branches toward the base of the tree occurred earlier than those near the top, and characters change and accumulate along the path from the root to the tips. Fossils provide direct documentation of ancient life, and so, in combination, fossils and phylogenies provide strong complementary insights into evolutionary history.

### Fossils provide unique information.

Fossils can and do provide evidence for phylogenetic hypotheses, showing, for example, that organisms that branch early in phylogenies appear early in the geologic record. But the fossil record does more than this. First, fossils enable us to calibrate phylogenies in terms of time. It is one thing to infer that mammals diverged from the common ancestor of birds, crocodiles, lizards, and turtles before crocodiles and birds diverged from a common ancestor (see Fig. 23.7), but another matter to state that birds and crocodiles diverged about 220 million years ago, whereas the lineage represented today by mammals branched from other vertebrates about 100 million years earlier. As we saw in Chapter 21, estimates of divergence time can be made using molecular sequence data, but all such estimates must be calibrated using fossils.

The evolutionary relationship between birds and crocodiles highlights a second kind of information provided by fossils. Not only do fossils record past life, they also provide our only record of extinct species. The phylogeny in Fig. 23.7 contains a great deal of information, but it is silent about dinosaurs. Fossils demonstrate that dinosaurs once roamed Earth, and details of skeletal structure place birds among the dinosaurs in the vertebrate tree. Indeed, some remarkable fossils from China show that the dinosaurs most closely related to birds had feathers (**Fig. 23.11**).

**FIG. 23.11** *Microraptor gui*, **a remarkable fossil discovered in approximately 125-million-year-old rocks from China.** The structure of its skeleton identifies *M. gui* as a dinosaur, yet it had feathers on its arms, tail, and legs.

**FIG. 23.12 The Grand Canyon.** Erosion has exposed layers of sedimentary rock that record Earth history.

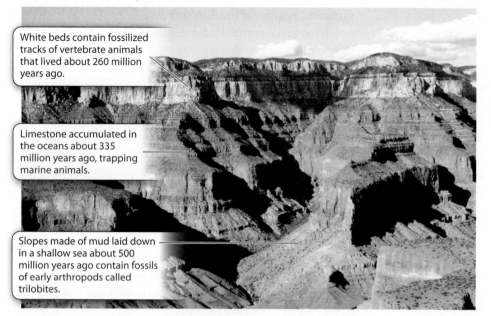

White beds contain fossilized tracks of vertebrate animals that lived about 260 million years ago.

Limestone accumulated in the oceans about 335 million years ago, trapping marine animals.

Slopes made of mud laid down in a shallow sea about 500 million years ago contain fossils of early arthropods called trilobites.

record of marine life is more complete than that for land-dwelling creatures because marine habitats are more likely than those on land to be places where sediments accumulate and become rock. Thus, trees and elks living high in the Rocky Mountains have a low probability of fossilization, whereas clams and corals on the shallow seafloor are commonly buried and become fossils.

Biological factors contribute to the incompleteness of the fossil record. Most fossils preserve the hard parts of organisms, those features that resist decay after death. For animals, this usually means mineralized skeletons. Clams and snails that secrete shells of calcium carbonate have excellent fossil records. More than 80% of the clam species found today along California's coast also occur as fossils in sediments deposited during the past million years. In contrast, nematodes, tiny worms that may be the most abundant animals on Earth, have no mineralized skeletons and almost no fossil record. The wood and pollen of plants, which are made in part of decay-resistant organic compounds, enter the fossil record far more commonly than

A third, and also unique, contribution of fossils is that they place evolutionary events in the context of Earth's dynamic environmental history. Again, dinosaurs illustrate the point. As discussed in Chapter 1, geologic evidence from several continents suggests that a large meteorite triggered drastic changes in the global environment 65 million years ago, leading to the extinction of most of the dinosaurs. In fact, at half a dozen moments in the past, large environmental disturbances sharply decreased Earth's biological diversity. These events, called mass extinctions, have played a major role in shaping the course of evolution.

**Fossils provide a selective record of past life.**
**Fossils** are the remains of once-living organisms, preserved through time in sedimentary rocks. The feathered dinosaur shown in Fig. 23.11 is an example of one kind of fossil. If we wish to use fossils to complement phylogenies based on modern organisms, we must understand how fossils form and how the processes of formation govern what is and is not preserved.

For all its merits, the fossil record should not be thought of as a complete dictionary of everything that ever walked, crawled, or swam across our planet's surface. Fossilization requires burial, as when a clam dies on the seafloor and is quickly covered by sand, or a leaf falls to the forest floor and ensuing floods cover it in mud. Through time, accumulating sediments harden into sedimentary rocks such as those exposed so dramatically in the walls of the Grand Canyon (**Fig. 23.12**). Without burial, the remains of organisms are eventually recycled by biological and physical processes, and no fossil forms. In general, the fossil

**FIG. 23.13 Trace fossils.** These footprints in 150-million-year-old rocks record both the structure and behavior of the dinosaurs that made them.

**FIG. 23.14** A fossil of *Opabinia regalis*, an early relative of the arthropods, from the 505-million-year-old Burgess Shale.

do flowers. And, among unicellular organisms, the skeleton-forming diatoms, radiolarians, and foraminiferans have exceptionally good fossil records, whereas most amoebas are unrepresented.

Together, then, the properties of organisms (do they make skeletons or other features that resist decay after death?) and environment (did the organisms live in a place where burial was likely?) determine the probability that an ancient species will be represented in the fossil record. Fortunately, for species that make mineralized skeletons and live on the shallow seafloor, the fossil record is very good, preserving a detailed history of evolution through time.

Organisms that lack hard parts can leave a fossil record in two other, distinctive ways. Many animals leave tracks and trails as they move about or burrow into sediments. These **trace fossils,** from dinosaur tracks to the feeding trails of snails and trilobites, preserve a record of both anatomy and behavior (**Fig. 23.13**).

Organisms can also contribute **molecular fossils** to the rocks. Most biomolecules decay quickly after death. Proteins and DNA, for example, generally break down before they can be preserved, although, remarkably, a sizable fraction of the Neanderthal genome has been pieced together from DNA in 40,000-year-old bones. Other molecules, especially lipids like cholesterol, are more resistant to decomposition. Sterols, bacterial

lipids, and some pigment molecules can accumulate in sedimentary rocks, documenting organisms that rarely form conventional fossils, especially bacteria and single-celled eukaryotes.

Rarely, unusual conditions preserve fossils of unexpected quality, including animals without shells, delicate flowers or mushrooms, fragile seaweeds or bacteria, even the embryos of plants and animals. For example, 505 million years ago, during the Cambrian Period, a sedimentary rock formation called the **Burgess Shale** accumulated on a relatively deep seafloor covering what is now British Columbia. Waters just above the basin floor contained little or no oxygen, so that when mud swept into the basin, entombed animals were sealed off from scavengers, disruptive burrowing activity, and even bacterial decay. For this reason, Burgess rocks preserve a remarkable sampling of marine life during the initial diversification of animals (**Fig. 23.14**).

The **Messel Shale** formed more recently—about 50 million years ago—in a lake in what is now Germany. Release of toxic gases from deep within the Messel Lake suffocated local animals, and their carcasses settled into oxygen-poor muds on the lakefloor. Fish, birds, mammals, and reptiles are preserved as complete and articulated skeletons, and mammals retain impressions of fur and color patterning (**Fig. 23.15**). Plants were also beautifully preserved, as were insects, some with a striking iridescence still intact. Messel rocks provide a truly outstanding snapshot of life on land as the age of mammals began.

**FIG. 23.15** **The Messel Shale, Germany.** Even the furry tail of this extinct squirrel-like mammal is preserved in this remarkable 50-million-year-old fossil.

In general, the fossil record preserves some aspects of biological history well and others poorly. Fossils provide a good sense of how the forms, functions, and diversity of skeletonized animals have changed over the past 500 million years. The same is true for land plants and unicellular organisms that form mineralized skeletons. These fossils shed light on major patterns of morphological evolution and diversity change through time; their geographic distributions record the movements of continents over millions of years; and the radiations and extinctions they document show how life responds to environmental change, both gradual and catastrophic.

## Geological data indicate the age and environmental setting of fossils.

How do we know the ages of fossils? Beginning in the nineteenth century, geologists recognized that groups of

**FIG. 23.16** The geologic timescale, showing major events in the history of life on Earth.

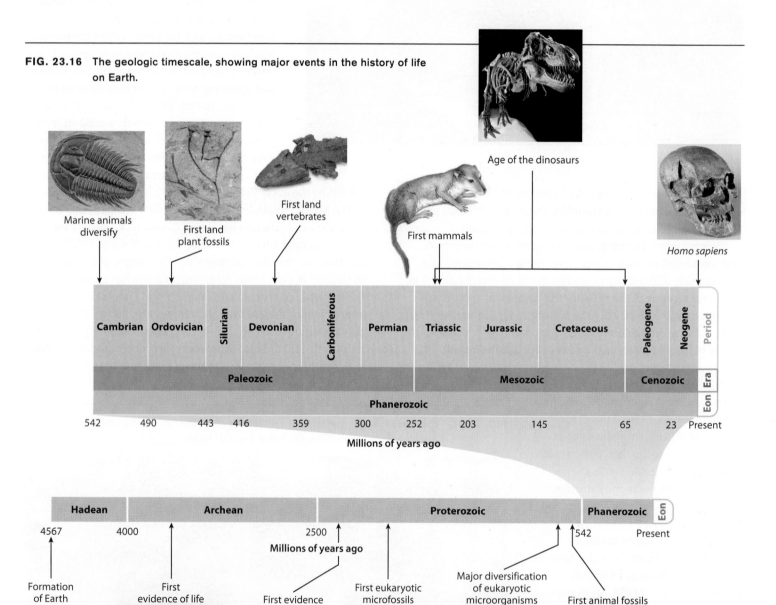

fossils change systematically from the bottom of a sedimentary rock formation to its top. As more of Earth's surface was mapped and studied, it became clear that certain fossils always occur in layers that lie beneath (and so are older than) layers that contain other species. From these patterns, geologists concluded that fossils mark time in Earth history. At first, geologists didn't know why fossils changed from one bed to the next, but after Darwin the reason became apparent: Fossils record the evolution of life on Earth. They eventually mapped out the **geologic timescale**, the series of time divisions that mark Earth's long history (**Fig. 23.16**).

The layers of fossils in sedimentary rocks can tell us that some rocks are older than others, but they cannot by themselves provide an absolute age. Calibration of the timescale became possible with the discovery of radioactive decay. In Chapter 2, we discussed isotopes, variants of an element that differ from one another in the number of neutrons they contain. Many isotopes are unstable and spontaneously break down to form other, more stable isotopes. In the laboratory, scientists can measure how fast unstable isotopes decay. Then, by measuring the amounts of the unstable isotope and its stable daughter inside a mineral, they can determine when the mineral formed.

Archaeologists commonly use the radioactive decay of the isotope carbon-14, or $^{14}C$, to date wood and bone. As shown in **Fig. 23.17**, cosmic rays continually generate $^{14}C$ in the atmosphere. Through photosynthesis, carbon dioxide that contains $^{14}C$ is incorporated into wood, and animals also incorporate small amounts of $^{14}C$ into their tissues when they eat plant material. After death, the unstable $^{14}C$ in these materials begins to break down, losing an electron to form $^{14}N$, a stable isotope of nitrogen. Laboratory measurements indicate that half of the $^{14}C$ in a given sample will decay to nitrogen in 5730 years, a period called its **half-life** (Fig. 23.17). Armed with this information, scientists can measure the amount of $^{14}C$ in an archaeological sample and, by comparing it to the amount of $^{14}C$ in a sample of known age—annual rings in trees, for example, or yearly growth coral skeletons—determine the age of the sample.

Because its half-life is so short (by geological standards), $^{14}C$ is useful only in dating materials younger than 50,000 to 60,000 years. Beyond that, there is too little $^{14}C$ left to measure accurately. Older geological materials are commonly dated using the radioactive decay of uranium (U) to lead (Pb): $^{238}U$ breaks down to $^{206}Pb$ with a half-life of 4.47 billion years, and $^{235}U$ decays to $^{207}Pb$ with a half-life of 704 million years. The calibration of the geologic timescale is based mostly on radiometric ages of volcanic ash beds found within sedimentary rock that contain key fossils, as well as volcanic rocks that intrude into (and so are younger than) layers of rock containing fossils. In turn, the age of fossils provide calibration points for phylogenies.

**FIG. 23.17** $^{14}C$ decay. Scientists can determine the age of relatively young materials such as wood and bone from the amount of $^{14}C$ they contain.

In the atmosphere, neutrons generated by cosmic radiation can collide with $^{14}N$ to create $^{14}C$.

$^{14}C$ that ends up in carbon dioxide can be incorporated in plants through photosynthesis. $^{14}C$ is next incorporated into animals when they eat the plants.

After the plant or animal dies and is buried, its $^{14}C$ decays to the more stable $^{14}N$.

Half of the $^{14}C$ in a fossil turns to $^{14}N$ in 5730 years. In another 5730 years, half of the remaining $^{14}C$ decays to $^{14}N$, and so on.

**FIG. 23.18** **Pangaea, 290 million years ago.** Plate tectonics has shaped and reshaped Earth's geography through time. The white near the South Pole is glacial ice.

The sedimentary rocks that contain fossils also preserve, encrypted in their physical features and chemical composition, information about the environment in which they formed. Sandstone beds, for example, may have rippled surfaces, like the ripples produced by currents that we see today in the sand of a seashore or a lake margin. Enrichment in pyrite ($FeS_2$) or other iron minerals may signal oxygen depletion.

We might think our moment in geologic time is representative of the Earth as it has always existed, but nothing could be further from the truth. In the location and sizes of its continents, ocean chemistry, and atmospheric composition, the Earth we experience is unlike any previous state of the planet. Today, for example, the continents are distributed widely over the planet's surface, but 290 million years ago they were clustered in a supercontinent called Pangaea (**Fig. 23.18**). Oxygen gas permeates most surface environments of Earth today, but 3 billion years ago, there was no $O_2$ anywhere. And, just 20,000 years ago, 2 km of glacial ice stood where Boston lies today. Sedimentary rocks record the changing state of Earth's surface over billions of years and show that life and environment have changed together through time, each influencing the other.

### Fossils can contain unique combinations of characters.

Phylogenies hypothesize impressive morphological and physiological shifts through time—amphibians from fish, for

example, or land plants from green algae. Do fossils capture a record of these transitions as they took place?

Let's begin with an example introduced earlier in this chapter. Phylogenies based on living organisms generally place birds as the sister group to crocodiles and alligators, but birds and crocodiles are decidedly different from each other in structure—birds have wings, feathers, toothless bills, and a number of other skeletal features distinct from those of crocodiles. In 1861, just 2 years after publication of *On the Origin of Species*, German quarry workers discovered a remarkable fossil that remains paleontology's most famous example of a transitional form. *Archaeopteryx lithographica*, now known from 11 specimens splayed for all time in fine-grained limestone, lived 150 million years ago. Its skeleton shares many characters with dromaeosaurs, a group of small, agile dinosaurs, but several features—its pelvis, its braincase, and, especially, its winglike forearms—are distinctly birdlike. Spectacularly, the fossils preserve evidence of feathers. *Archaeopteryx* clearly suggests a close relationship between birds and dinosaurs, and phylogenetic reconstructions that include information from fossils show that many of the characters found today in birds accumulated through time in their dinosaur ancestors. And, as noted earlier, even feathers first evolved in dinosaurs (**Fig. 23.19**).

*Tiktaalik roseae* and other skeletons in rocks deposited 375 to 362 million years ago record an earlier but equally fundamental transition: the colonization of land by vertebrates. Phylogenies

**FIG. 23.19 Dinosaurs and birds.** A number of dinosaur fossils link birds phylogenetically to their closest living relatives, the crocodiles. The fossil at the bottom is *Archaeopteryx*.

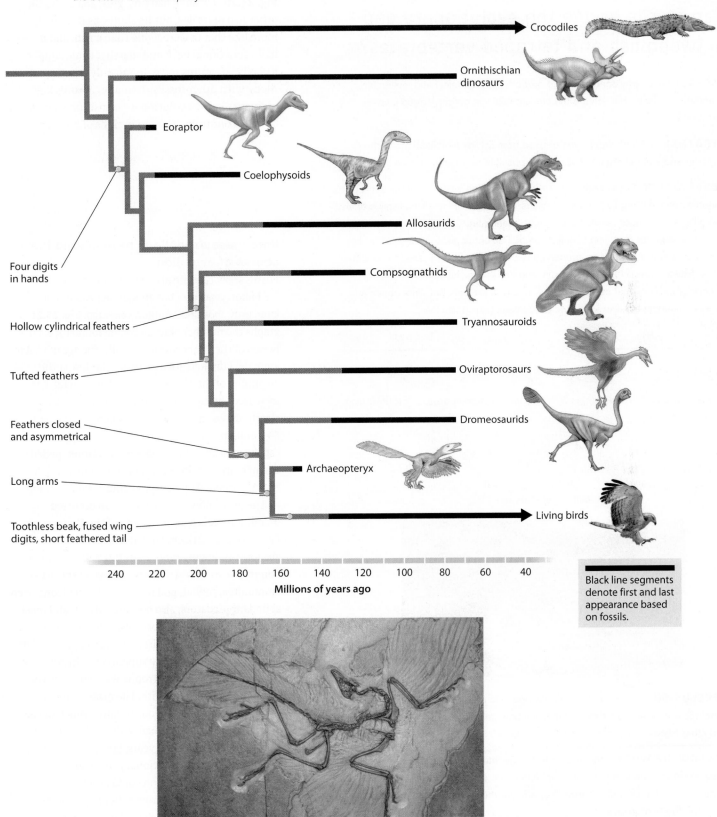

**FIG. 23.20**

# Can fossils bridge the evolutionary gap between fish and tetrapod vertebrates?

**BACKGROUND** Phylogenies based on both morphological and molecular characters indicate that fish are the closest relatives of four-legged land vertebrates.

**HYPOTHESIS** Land vertebrates evolved from fish by modifications of the skeleton and internal organs that made it possible for them to live on land.

**OBSERVATION** Fossil skeletons 390 to 360 million years old show a mix of features seen in living fish and amphibians. Older fossils have fins, fishlike heads, and gills, and younger fossils have weight-bearing legs, skulls with jaws able to grab prey, and ribs that help ventilate lungs. Paleontologists predicted that key intermediate fossils would be preserved in 380–370-million-year-old rocks. In 2004, Edward Daeschler, Neil Shubin, and Farish Jenkins discovered *Tiktaalik*, a remarkable fossil that has fins, scales, and gills like fish, but wrist bones and fingers, an amphibian-like skull, and a true neck (which fish lack).

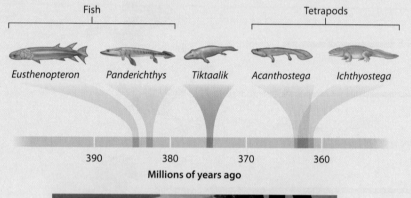

| Fish | | | Tetrapods | |
| --- | --- | --- | --- | --- |
| *Eusthenopteron* | *Panderichthys* | *Tiktaalik* | *Acanthostega* | *Ichthyostega* |

| 390 | 380 | 370 | 360 |

**Millions of years ago**

**CONCLUSION** Fossils confirm the phylogenetic prediction that tetrapod vertebrates evolved from fish by the developmental modification of limbs, skulls, and other features.

**FOLLOW-UP WORK** Research into the genetics of vertebrate development shows that the limbs of fish and amphibians are shaped by similar patterns of gene expression, providing further support for the phylogenetic connection between the two groups.

**SOURCE** Daeschler, E. B., N. H. Shubin, and F. A. Jenkins, Jr. 2006. "A Devonian Tetrapod-like Fish and the Evolution of the Tetrapod Body Plan." *Nature* 440:757–763.

show that all land vertebrates, from amphibians to mammals, are descended from fish. As seen in **Fig. 23.20,** *Tiktaalik* had fins, gills, and scales like other fish of its day, but its skull was flattened, more like that of a crocodile than a fish, and it had a functional neck and ribs that could support its body—features today found only in tetrapods. Along with other fossils, *Tiktaalik* captures key moments in the evolutionary transition from water to land, confirming the predictions of phylogeny.

→ **Quick Check 4** You have just found a novel vertebrate skeleton in 200-million-year-old rocks. How would you integrate this new fossil species into the phylogenetic tree depicted in Fig. 23.7?

## Rare mass extinctions have altered the course of evolution.

Fossils show that life originated more than 3.5 billion years ago, but animals appeared much later, only about 600 million years ago. **Fig. 23.21** graphs the number of fossil genera found for each period of the Phanerozoic (literally, the age of visible animals) Eon. Recorded diversity was low at the beginning of the Phanerozoic and, in general, has increased through time. Nonetheless, the graph clearly shows that at several moments in the past 542 million years, recorded diversity dropped catastrophically. These **mass extinctions** spelled the end of many previously important groups of species, but they opened up new possibilities for evolution.

The best-known mass extinction occurred 65 million years ago, at the end of the Cretaceous Period (Fig. 23.21). On land, dinosaurs disappeared abruptly, following more than 150 million years of dominance in terrestrial ecosystems. In the oceans, ammonites, cephalopod mollusks that had long been abundant predators, also became extinct, and most skeleton-forming microorganisms in the oceans disappeared as well. As discussed in Chapter 1, a large body of geologic evidence supports the hypothesis that this biological catastrophe was caused by the impact of a giant meteorite. The mass extinction at the end of the Cretaceous removed much of the biological diversity that had built up through genetic changes over millions of years, but it had another, equally important evolutionary consequence. Survivors of the extinction could proliferate with little competition from other populations. Mammals, our own branch of the animal tree, diversified across continents once dinosaurs had disappeared from the landscape.

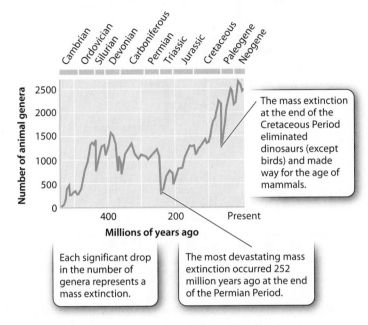

**FIG. 23.21** **Mass extinctions documented by compilations of fossil diversity through time.** Though they eliminate much of the life found on Earth at the time, mass extinctions allow new groups to proliferate and diversify.

The mass extinction at the end of the Cretaceous Period eliminated dinosaurs (except birds) and made way for the age of mammals.

Each significant drop in the number of genera represents a mass extinction.

The most devastating mass extinction occurred 252 million years ago at the end of the Permian Period.

An even larger mass extinction took place much earlier at the end of the Permian Period, 252 million years ago (Fig. 23.21). More than 90% of all genera recorded in late Permian oceans disappeared, especially animals like corals and sponges with heavy skeletons. There is no compelling evidence for meteorite impact, so geologists have hypothesized that this mass extinction resulted from the catastrophic effects of massive volcanic eruptions.

At the end of the Permian Period, most continents were gathered into the supercontinent Pangaea (see Fig. 23.18), and so a huge ocean covered more than half the Earth. Levels of oxygen in the deep waters of this ocean fell, the result of sluggish circulation and warm seawater temperatures. Then, a massive outpouring of ash and lava erupted across what is now Siberia. Enormous emissions of carbon dioxide and methane from the volcanoes caused global warming (so even less oxygen reached deep oceans) and ocean acidification (so it was difficult for animals and algae to secrete calcium carbonate skeletons).

The three-way insult of lack of oxygen, ocean acidification, and global warming doomed many species on land and in the seas. Seascapes dominated by corals and shelled invertebrates called brachiopods disappeared. After the extinction of these organisms, new populations of mollusks, new forms of coral, and other animals we see today rose to prominence. Again, major environmental disruption removed diversity that had slowly accumulated through many millions of years but also opened up new evolutionary possibilities for the survivors.

Five great mass extinctions occurred over the past 500 million years, but only the two at the end of the Permian and Cretaceous Periods so thoroughly changed the course of evolution. At present, we live at a time of accelerating ocean acidification and global warming. Will future geologists recognize our evolutionary moment as a time of mass extinction? We discuss this possibility in Chapter 48.

## 23.4 COMPARING EVOLUTION'S TWO GREAT PATTERNS

The diversity of life we see today is the result of evolutionary processes playing out over geologic time. Evolutionary process can be studied by experiment, both in the field and in the laboratory, but evolutionary history is another matter. There is no experiment we can do to determine why the dinosaurs became extinct—we cannot rerun the events of 65 million years ago, this time without the meteorite impact. The history of life must be reconstructed from evolution's two great patterns: the nested similarity observed in the forms and macromolecular sequences of living organisms, and the direct historical archive of the fossil record.

**Phylogeny and fossils complement each other.**
The great advantage of reconstructing evolutionary history from living organisms is that we can use a full range of features—skeletal morphology, cell structure, DNA sequence—to generate phylogenetic hypotheses. The disadvantage of using comparative biology is that we lack evidence of extinct species, the time dimension, and the environmental context. This, of course, is where the fossil record comes into play. Fossil evidence has strengths and limitations that complement the evolutionary information in the living organisms.

We can use phylogenetic methods based on DNA sequences to infer that birds and crocodiles are closely related, but only fossils can show that the evolutionary link between birds and crocodiles runs through dinosaurs. And only the geologic record can show that mass extinction removed the dinosaurs, paving the way for the emergence of modern mammals. Paleontologists and biologists work together to understand evolutionary history. Biology provides a functional and phylogenetic framework for the interpretation of fossils, and fossils provide a record of life's history in the context of continual planetary change.

**Agreement between phylogenies and the fossil record provides strong evidence of evolution.**
Phylogenies based on morphological or molecular comparisons of living organisms make hypotheses about the timing of evolutionary changes through Earth history. We humans are a case in point. As discussed in more detail in Chapter 24, comparisons of DNA sequences suggest that chimpanzees are our closest living relatives. This is hardly surprising, as simple

observation shows that chimpanzees and humans share many features. However, among other differences, chimpanzees have smaller stature, long arms that facilitate knuckle-walking (the arms help support the body as the chimpanzee moves forward), a more prominent snout, and larger teeth. Fossils painstakingly unearthed over the past century show that the morphological features that mark us as human accumulated through a series of speciation events. Six to seven million years ago, when the human and chimpanzee lineages first split, the two newly diverged taxa looked much more alike than humans and chimpanzees do today.

The agreement of comparative biology and the fossil record can be seen at all scales of observation. Humans form one tip of a larger branch that contains all members of the primate family. Primates, in turn, are nested within a larger branch occupied by mammals, and mammals nest within a still larger branch containing all vertebrate animals, which include fish. This arrangement predicts that the earliest fossil fish should be older than the earliest fossil mammals, the earliest fossil mammals should be older than the earliest primates, and the earliest primates older than the earliest humans. This is precisely what the fossil record shows (**Fig. 23.22**).

The agreement between fossils and phylogenies drawn from living organisms can be seen again and again when we examine different branches of the tree of life or, for that matter, the tree as a whole. All phylogenies indicate that microorganisms diverged early in evolutionary history, and mammals, flowering plants, and other large complex organisms diverged more recently. The tree's shape implies that diversity has accumulated through time, beginning with simple organisms and later adding complex macroscopic forms.

The geologic record shows the same pattern. For nearly 3 billion years of Earth history, microorganisms dominate the fossil record, with the earliest animals appearing about 600 million years ago, the earliest vertebrate animals 520 million years ago, the earliest tetrapod vertebrates about 360 million years ago, the first mammals

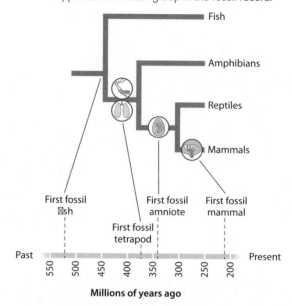

**FIG. 23.22  Phylogeny and fossils of vertebrates.** The branching order of the phylogeny corresponds to the order of appearance of each group in the fossil record.

210 million years ago, the first primates perhaps 55 million years ago, the oldest fossils of our own species a mere 200,000 years ago. Similarly, if we focus on photosynthetic organisms, we find a record of photosynthetic bacteria beginning at least 3500 million years ago, algae 1200 million years ago, simple land plants 470 million years ago, seed plants 370 million years ago, flowering plants about 140 million years ago, and the earliest grasses 70 million years ago.

In Part 2, we explore the evolutionary history of life in some detail. Here, it is sufficient to draw the key general conclusion: The fact that comparative biology and fossils, two complementary but independent approaches to reconstructing the evolutionary past, yield the same history is powerful evidence of evolution.

## Core Concepts Summary

**23.1 A PHYLOGENETIC TREE IS A REASONED HYPOTHESIS OF THE EVOLUTIONARY RELATIONSHIPS OF ORGANISMS.**

The nested pattern of similarities seen among organisms is a result of descent with modification and can be represented as a phylogenetic tree. page 23-1

The order of branches on a phylogenetic tree indicates the sequence of events in time. page 23-2

Sister groups are more closely related to one another than they are to any other group. page 23-3

A node is a branching point on a tree, and it can be rotated without changing evolutionary relationships. page 23-3

A monophyletic group includes all the descendants of a common ancestor, and it is considered a natural grouping of organisms based on shared ancestry. page 23-4

A paraphyletic group includes some, but not all, of the descendants of a common ancestor. page 23-4

A polyphyletic group includes organisms from distinct lineages based on shared characters, but it does not include a common ancestor. page 23-4

Organisms are classified into domain, kingdom, phylum, class, order, family, genus, and species. page 23-4

**23.2 A PHYLOGENETIC TREE IS BUILT ON THE BASIS OF SHARED DERIVED CHARACTERS.**

Characters, or traits, existing in different states are used to build phylogenetic trees. page 23-5

Homologies are similarities based on shared ancestry, while analogies are similarities based on independent adaptations. page 23-6

Homologies can be ancestral, unique to a particular group, or present in some, but not all, of the descendants of a common ancestor (shared derived characters). page 23-6

Only shared derived characters, or synapomorphies, are useful in constructing a phylogenetic tree. page 23-6

Molecular data provide a wealth of characters that complement other types of characters in building phylogenetic trees. page 23-8

Phylogenetic trees can be used to understand evolutionary relationships of organisms and solve practical problems, such as how viruses evolve over time. page 23-10

**23.3 THE FOSSIL RECORD PROVIDES A DIRECT GLIMPSE OF EVOLUTIONARY HISTORY.**

Fossils are the remains of organisms preserved in sedimentary rocks. page 23-11

The fossil record is imperfect because fossilization requires burial in sediment, sediments accumulate episodically and discontinuously, and fossils typically preserve only the hard parts of organisms. page 23-12

Radioactive decay provides a means of dating rocks. page 23-14

*Archaeopteryx* and *Tiktaalik* are two fossil organisms that document, respectively, the bird–dinosaur transition and the fish–tetrapod transition. page 23-16

The history of life is characterized by rare mass extinctions, including the extinction at the end of the Cretaceous Period 65 million years ago in which the dinosaurs (other than birds) became extinct, and the extinction at the end of the Permian Period 252 million years ago, the largest documented mass extinction in the history of Earth. page 23-18

**23.4 PHYLOGENY AND FOSSILS PROVIDE INDEPENDENT AND CORROBORATING EVIDENCE OF EVOLUTION.**

Phylogeny makes use of living organisms, and the fossil record supplies a record of species that no longer exist, absolute dates, and environmental context. page 23-19

Data from phylogeny and fossils are often in agreement, providing strong evidence for evolution. page 23-19

## Self-Assessment

1. Give two examples of a nested pattern of similarity among organisms.

2. With reference to a phylogenetic tree, show how a nested pattern of similarity is the necessary result of evolution.

3. Distinguish among monophyletic, paraphyletic, and polyphyletic groups, and give an example of each.

4. List the levels of classification, from the least inclusive (species) to the most inclusive (domain).

5. Describe two traits that are homologous and two that are analogous.

6. Name a type of homology that is useful in building phylogenetic trees and explain why this kind of homology, and not others, is useful.

7. Name three types of fossil.

8. Explain why there are gaps in the fossil record.

9. Explain how the fossil record can be used to determine both the relative and the absolute timescales of past events.

10. Describe the significance of *Archeopteryx* and *Tiktaalik*.

11. Describe how mass extinctions have shaped the ecological landscape.

Do you understand the chapter's Core Concepts? Log in to check your answers to the Self-Assessment questions, then practice what you've learned and reinforce this chapter's concepts by working through the problems and multimedia tutorials provided there.

**www.biologyhowlifeworks.com**

**FIG. 24.1** The primate family tree, showing the evolutionary relationships of prosimians, monkeys, and apes.

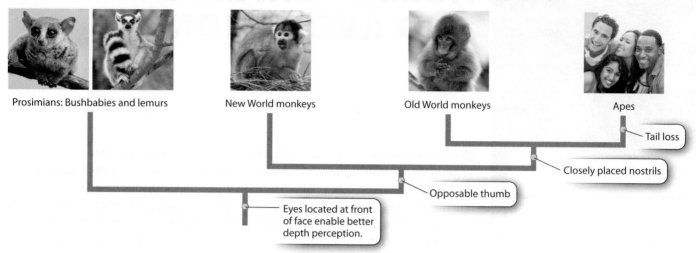

Prosimians: Bushbabies and lemurs

New World monkeys

Old World monkeys

Apes

Tail loss

Closely placed nostrils

Opposable thumb

Eyes located at front of face enable better depth perception.

from the one that gave rise to humans. Lemurs, which today are confined to the island of Madagascar, are thus only distantly related to humans.

Monkeys underwent independent bouts of evolutionary change in the Americas and in Africa and Eurasia, so the family tree is split along geographic lines into New and Old World monkeys. Though both groups have evolved similar habits, there are basic distinctions. There are differences between the teeth of the two groups, and in New World monkeys the nostrils tend to be widely spaced, whereas in Old World species they are closer together.

One line of Old World monkeys gave rise to the apes, which lack a tail and show more sophisticated behaviors than other monkeys. The apes are split into two groups, the lesser and the great apes (**Fig. 24.2**). Lesser apes include the fourteen species of gibbon,

**FIG. 24.2** The family tree of the lesser and great apes. The apes consist of two major groups, the lesser apes and great apes. The great apes group includes humans.

**Gibbon (14 species in 4 genera)**

**Orangutan**

**Gorilla**

**Bonobo**

**Chimpanzee**

**Human**

Delayed puberty

Earlobes

Robust canine teeth

Prosimians: Bushbabies and lemurs

New World monkeys

Old World monkeys

**Apes**

The characters listed in the ancestral lineages certainly do not represent all the most important evolutionary changes that occurred in these ancestors. Rather, they are a sample of the changes that occurred at each stage.

▪ Lesser apes

▪ Great apes

all of which are found in Southeast Asia. The great apes include the orangutan, gorilla, chimpanzee and, strictly speaking, humans. Taxonomists classify all the descendants of a specified common ancestor as belonging to a monophyletic group (Chapter 23). Thus, humans, blue whales, and hedgehogs are all mammals because all three are descended from the first mammal, the original common ancestor of all mammals. Because we trace our ancestry to the common ancestor of orangutans, gorillas, and chimpanzees, we, too, are a member of the monophyletic great ape group.

### Molecular analysis reveals that our lineage split from the chimpanzee lineage about 5-7 million years ago.

Which great ape is most closely related to humans? That is, which is our sister group? Traditional approaches of reconstructing evolutionary history by comparing anatomical features failed to distinguish between two candidates, gorillas and chimpanzees. The implication is that humans, gorillas, and chimpanzees split into separate lineages at about the same time. It was only with the introduction of molecular methods of assessing evolutionary relationships—through the comparison of DNA and protein amino acid sequences from the different species—that we had the answer. Our closest relative is the chimpanzee (Fig. 24.2)—or, more accurately, the chimpanzees, plural, because there are two closely related chimpanzee species, the smaller of which is often called the bonobo.

Just how closely are humans and chimpanzees related? To answer this question, we need to know the timing of the evolutionary split that led along one fork to chimpanzees and along the other to us. As we saw in Chapter 21, DNA sequence differences accumulate between isolated populations or species, and they do so at a more or less constant rate. As a result, the extent of the sequence difference between two species is a good indication of the amount of time they have been separate lineages, that is, the amount of time since their last common ancestor.

The first thorough comparison of DNA molecules between humans and chimpanzees was carried out before the advent of DNA sequencing methods by Mary-Claire King and Allan Wilson at the University of California at Berkeley (**Fig. 24.3**). One of their methods to measure molecular differences between species relied on DNA–DNA hybridization (Chapter 12). Two complementary strands of DNA in a double helix can be separated by heating the sample.

**FIG. 24.3**

# How closely related are humans and chimpanzees?

**BACKGROUND** Phylogenetic analysis based on anatomical characteristics had established that chimpanzees are closely related to humans. Mary-Claire King and Allan Wilson used molecular techniques to determine *how* closely related the two species are.

**HYPOTHESIS** Despite marked anatomical and behavioral differences between the two species, the genetic distance between the two is small, implying a relatively recent common ancestor.

**METHOD** When two complementary strands of DNA are heated, the hydrogen bonds pairing the two helices are broken at around 95°C and the double helix denatures, or separates (Chapter 12). Two complementary strands with a few mismatches separate at a temperature slightly lower than 95°C because fewer hydrogen bonds hold the helix together. Many more mismatches between the two sequences results in an even lower denaturation temperature. Using hybrid DNA double helices with one strand contributed by each species—humans and chimpanzees, in this case—and determining their denaturation temperature, King and Wilson could infer the genetic distance (the extent of genetic divergence) between the two species.

| Perfect complementarity: 95°C denaturation | Some mismatch: 93°C denaturation | More mismatch: 91°C denaturation |
| --- | --- | --- |
| T C C G T T A A G C<br>A G G C A A T T C G | T C C T T T C A G C<br>A G G C A A T T C G | T C A T T T C A G C<br>A G G C A A T T T G |

**RESULTS** King and Wilson found that human–chimpanzee DNA molecules separated at a temperature approximately 1°C less than the temperature at which human–human DNA molecules separate. This difference could be calibrated on the basis of studies of other species whose genetic distances were known from other methods. The DNA of humans differs from that of chimpanzees by about 1%.

**CONCLUSION AND INTERPRETATION** King and Wilson noted the discrepancy between the extent of genetic divergence (small) and the extent of anatomical and behavioral divergence (large) between humans and chimpanzees. They suggested that one way in which relatively little genetic change could produce extensive phenotypic change is through differences in gene regulation (Chapter 19). A small genetic change in a control region responsible for switching a gene on and off might have major consequences for the organism.

**FOLLOW-UP WORK** The sequences of the chimpanzee and human genomes allow us to compare the two sequences directly, and these data confirm King and Wilson's observations.

**SOURCE** King, M. C., and A. C. Wilson. 1975. "Evolution at Two Levels in Humans and Chimpanzees." *Science* 188:107–116.

If the two strands are not perfectly complementary, as is the case if there is a basepair mismatch (for example, a G paired with a T rather than a C), less heat is required to separate the strands.

King and Wilson used this fact to examine the differences between a human strand and the corresponding chimpanzee strand. They inferred the extent of DNA sequence divergence from the melting temperature and made a striking discovery: Human and chimpanzee DNA differ in sequence by just 1%.

King and Wilson's conclusions have been confirmed by subsequent research. Now that we have both the human and chimpanzee genome sequences in hand, we can literally count the differences between the two sequences: The genomes differ by about 1%. The overall amount of difference between human and chimpanzee genomes goes up somewhat if you take into account segments of the genomes that are absent in one or other of the species (that is, sequences where DNA has been inserted or deleted in one lineage or the other).

Because the amount of sequence difference is correlated with the length of time the two species have been isolated, we can convert the divergence results into an estimate of the timing of the split between the human and chimpanzee lineages. That split occurred about 5–7 million years ago. All the extraordinary characteristics that set our species apart from the rest of the natural world—those attributes that are ours and ours alone—arose in just 5–7 million years.

→ **Quick Check 1** Did humans evolve from chimpanzees? Explain.

### The fossil record gives us direct information about our evolutionary history.

Molecular analysis is a powerful tool for comparing species and populations within species. It allows us to compare humans and chimpanzees and look at differences among groups of humans or groups of chimpanzees. But we are limited in what we can study because we need samples from living individuals. We can study the end products of evolution but not the intermediate stages, which have long since disappeared and cannot provide us with DNA samples. For a full picture of human evolution, we must turn to fossils (Chapter 23).

For the first several million years of human evolution, all the fossils from the human lineage are found in Africa. This is not surprising. Charles Darwin himself noted that it was likely that the human lineage originated there, as humans' two closest relatives, chimpanzees and gorillas, live only in Africa. The fossil material varies in quality, and a great deal of ingenuity is often required to reconstruct the appearance and attributes of an individual from fragmentary fossil material. It's hard to determine whether two fossil specimens with slight differences belong to the same or different species. And, of course, interbreeding—the criterion that defines a species (Chapter 22)—cannot be applied to fossils. As a result, experts disagree over the details of the human fossil record. However, several robust general conclusions may be drawn.

The members of all the different species in the lineage leading to humans are called **hominins.** Let's start with the earliest known member of the hominin lineage, *Sahelanthropus tchadensis* (**Fig. 24.4**). Discovered in Chad in 2002, the skull of *S. tchadensis* combines both modern (human) and ancestral features. *S. tchadensis,* which has been dated to about 7 million years ago and has a chimpanzee-sized brain but hominin-type brow ridges, probably lived shortly after the split between the hominin and chimpanzee lineages, although this conclusion is controversial.

An important early hominin, dating from about 4.4 million years ago, is a specimen of *Ardipithecus ramidus* from

**FIG. 24.4** **The skull of *Sahelanthropus tchadensis.*** This skull, which is about 7 million years old and was found in Chad, has both human and chimpanzee features.

**FIG. 24.5** Lucy, a specimen of ***Australopithecus afarensis.*** Lucy was fully bipedal, establishing that the ability to walk upright evolved at least 3.2 million years ago.

Ethiopia. This individual, known as **Ardi,** was capable of walking upright, using two legs on the ground but all four limbs in the trees. An unusually complete early hominin fossil, **Lucy,** found in 1974 at Hadar, Ethiopia, represents the next step in the evolution of hominin gait: She was fully **bipedal,** habitually walking upright (**Fig. 24.5**). Her name comes from the Beatles' song "Lucy in the Sky with Diamonds," which was playing in the paleontologists' field camp when she was unearthed. This fossil dates from around 3.2 million years ago. Lucy was a member of the species *Australopithecus afarensis* and was much smaller than modern humans, less than 4 feet tall, and had a considerably smaller brain (even when the difference in body size is taken into consideration). In many ways, then, Lucy was similar to the common ancestor of the human and chimpanzee lineages, except that she was bipedal.

Remarkably, the hominin lineage produced many different species in Africa, so that there were at times as many as three species living at the same time. **Fig. 24.6** shows the main hominin species and the longevity of each in the fossil record, along with suggested evolutionary relationships among the species. Note that all hominins have a common ancestor, but not all of these lineages lead to modern humans, producing instead other branches of the hominin tree that ultimately went extinct.

A number of trends can be seen when we look over the entire record. Body size increased, and most striking is the increase in size of the cranium and therefore, by inference, of the brain, as shown in **Fig. 24.**7.

The fossil record indicates that about 2 million years ago, the hominin lineage ventured out of Africa. The first hominin that left Africa is sometimes called *Homo ergaster* and sometimes *Homo erectus.* The confusion stems from the controversies that surround the naming of fossil species. Some researchers contend that *H. ergaster* is merely an early form of *H. erectus.* We designate this first hominin *Homo ergaster* and a later, descendant species, *H. erectus* (see Fig. 24.6). The naming details, however, are relatively unimportant. What matters is that some hominins first left Africa about 2 million years ago. Fossils have been found throughout Eurasia, though not in Australia or the Americas. It was from relatives of *Homo ergaster* that our species, *Homo sapiens,* derived.

Another species closely related to *Homo sapiens* was *Homo neanderthalensis,* whose fossils appear in Europe and the Middle

**FIG. 24.6** **Hominin lineages.** Three main lineages, *Ardipithecus, Australopithecus,* and *Homo* existed at various and sometimes overlapping times in history. The *Homo* lineage led to modern humans, *Homo sapiens.*

East. Thicker boned than us, and with flatter heads that contained brains about the same size as, or slightly larger than, ours, **Neanderthals** first appeared in the fossil record around 600,000 years ago and disappeared around 30,000 years ago. As we will see, genetic analysis suggests that this disappearance was perhaps not as complete as the fossil record suggests.

Another species that probably ultimately derived from the original *H. ergaster* emigration from Africa became extinct only about 12,000 years ago. This was *H. floresiensis,* known popularly as the Hobbit. *H. floresiensis* is peculiar: Limited to the Indonesian island of Flores, adults were only just over 3 feet tall. Some have suggested that *H. floresiensis* is not a genuinely distinct species, but, rather, is an aberrant *H. sapiens.* Plenty of evidence, however, suggests that *H. floresiensis* is a distinct species derived from an archaic *Homo* species, probably *H. erectus.* Mammals often evolve small body size on islands because of the limited availability of food.

**FIG. 24.7** Increase in brain size (as inferred from fossil cranium volume) over 3 million years.

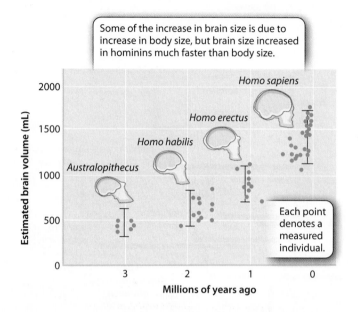

Some of the increase in brain size is due to increase in body size, but brain size increased in hominins much faster than body size.

*Homo sapiens*

*Homo erectus*

*Homo habilis*

*Australopithecus*

Each point denotes a measured individual.

Estimated brain volume (mL)

Millions of years ago

## 24.2 AFRICAN ORIGINS

For a long time, it was argued that modern humans, *Homo sapiens*, derive from the *Homo ergaster* populations that spread around the world starting about 2 million years ago. This idea is called the **multiregional hypothesis** of human origins because it implies that different *Homo ergaster* populations throughout the Old World evolved in parallel, with some limited gene flow among them, to each produce modern *H. sapiens* populations. In short, modern human traits evolved convergently in multiple populations (Chapter 23). This idea was overturned in 1987 by another study from Allan Wilson's laboratory, which instead suggested that modern humans arose much more recently from *Homo ergaster* descendants (sometimes called *Homo heidelbergensis;* see Fig 24.6) in Africa, about 200,000 years ago. This newer idea is the **out-of-Africa hypothesis** of human origins.

→ **Quick Check 2** What do the multiregional and out-of-Africa hypotheses predict about the age of the common ancestor of all humans living today?

### Studies of mitochondrial DNA reveal that modern humans evolved in Africa.

To test these two hypotheses about human origins, Rebecca Cann, a student in Allan Wilson's laboratory, chose to analyze DNA sequences to reconstruct the human family tree. Specifically, she studied sequences of a segment of mitochondrial DNA from people living around the world (**Fig. 24.8**).

**FIG. 24.8**

# When and where did the most recent common ancestor of all living humans live?

**BACKGROUND** With the availability of DNA sequence analysis tools, it became possible to investigate human prehistory by comparing the DNA of living people from different populations.

**HYPOTHESIS** The multiregional hypothesis of human origins suggests that our most recent common ancestor was living at the time *Homo ergaster* populations first left Africa, about 2 million years ago.

**METHOD** Rebecca Cann compared mitochondrial DNA (mtDNA) sequences from 147 people from around the world. This approach required a substantial amount of mtDNA from each individual, which she acquired by collecting placentas from women after childbirth. Instead of sequencing the mtDNA, Cann inferred differences among sequences by digesting the mtDNA with different restriction enzymes, each of which cuts DNA at a specific sequence (Chapter 12). If the sequence is present, the enzyme cuts. If any base in the recognition site of the restriction enzyme has changed in an individual, the enzyme does not cut. The resulting fragments were then separated using gel electrophoresis. By using 12 different enzymes, Cann was able to assay a reasonable proportion of all the mtDNA sequence variation present in the sample.

Here, we see a sample dataset for four people and a chimpanzee. There are seven varying restriction sites:

Restriction site number

1 2 3 4 5 6 7

mtDNA (Japan)

mtDNA (S. Africa)

mtDNA (E. Africa)

mtDNA (Native America)

mtDNA (Chimpanzee)

✂ Presence of a restriction site at a specified location

**ANALYSIS** The data were converted into a family tree by using shared derived characters, described in Chapter 23. By mapping the pattern of changes in this way, the phylogenetic relationships of the mtDNA sequences can be reconstructed. For example, the derived "cut" state at restriction site 1 shared by the East African, Japanese, and Native American sequences implies that the three groups had a common ancestor in which the mutation that created the new restriction site occurred.

The chimpanzee is the phylogenetic outlier because its common ancestor with the rest of the group is the oldest node on the tree. We can thus treat it as an outgroup and assume that its sequence represents the ancestral state.

Most recent common ancestor of the four human sequences analyzed.

Both Japanese and Native American sequences share the gain of restriction site 5, implying that the site gain occurred in their common ancestor.

**RESULTS AND CONCLUSION** A simplified version of Cann's phylogenetic tree based on mtDNA is shown below. Modern humans arose relatively recently in Africa, and their most common ancestor lived about 200,000 years ago. The multiregional hypothesis is rejected. Also, the data revealed that all non-Africans derive from within the African family tree, implying that *H. sapiens* left Africa relatively recently. The Cann study gave rise to the out-of-Africa theory of human origins.

The common ancestor of all living humans

**FOLLOW-UP WORK** This study set the stage for an explosion in genetic studies of human prehistory that used the same approach of comparing sequences in order to identify migration patterns. Today, with molecular tools vastly more powerful than those available to Cann, studies of the distribution of genetic variation are giving us a detailed pattern, even on relatively local scales, of demographic events.

**SOURCE** Cann, R. L., M. Stoneking, and A. C. Wilson. 1987. "Mitochondrial DNA and Human Evolution." *Nature* 325:31–36.

**Mitochondrial DNA** (**mtDNA**) is a small circle of DNA, about 17,000 base pairs long, found in every mitochondrion (Chapter 17). Cann chose to study mtDNA for several reasons. Although a typical cell contains a single nucleus with just two copies of nuclear DNA, each cell has many mitochondria, each carrying multiple copies of mtDNA. Thus, mtDNA is much more abundant than nuclear DNA and therefore easier to extract. Most important, however, is its mode of inheritance. All your mtDNA is inherited from your mother in the egg she produces because sperm do not contribute mitochondria to the zygote. This means that there is no opportunity for genetic recombination between different mtDNA molecules, so the only way in which sequence variation can arise is through mutation. In nuclear DNA, by contrast, differences between two sequences can be introduced through both mutation and recombination. Recombination obscures genealogical relationships because it mixes segments of DNA with different evolutionary histories.

Mitochondrial DNA sequences offered Cann a clear advantage. With sequence information from the mtDNA of 147 people from around the world, Cann reconstructed the human family tree (Fig. 24.8). The tree contained two major surprises. First, the two deepest branches—the ones that come off the tree earliest in time—are African. All non-Africans are branches off the African tree. The implication is that *Homo sapiens* evolved in Africa and only afterward did populations migrate out of Africa and become established elsewhere. This finding contradicted the expectations of the multiregional theory, which predicted that *Homo sapiens* evolved independently in different locations throughout the Old World.

Second, the tree is remarkably shallow, that is, even the most distantly related modern humans have a relatively recent common ancestor. By calibrating the rate at which mutations occur in mtDNA, Cann was able to estimate the time back to the common ancestor of all modern humans as about 200,000 years. Subsequent analyses have somewhat refined this estimate, but the key message has not changed: The data contradict the multiregional theory, which predicted a number closer to 2 million years.

Two hundred thousand years ago is 10 times more recent than 2 million years ago, so this change represents a revolution in our understanding of human prehistory. But surely it is risky to make such major claims on the basis of a single study. Maybe there's something peculiar about mtDNA, and it doesn't offer an accurate picture of the human past. To find independent evidence for a recent origin of modern humans in Africa, another dataset was required. One such dataset comes from studies of the Y chromosome, another segment of human DNA that does not undergo recombination (Chapter 17). When an approach similar to Cann's was used to reconstruct the human family tree using Y chromosome DNA sequences, the result was completely in agreement with the mtDNA result: The human family is young, and it arose in Africa.

## Neanderthals disappear from the fossil record as modern humans appear but have contributed to the modern human gene pool.

If we exclude *H. floresiensis*, the last of the nonmodern humans were the Neanderthals (**Fig. 24.9**), who lived in Europe and Western Asia until about 30,000 years ago. What happened? Were they eliminated by the first population of *Homo sapiens* to arrive in Europe (a group known as **Cro-Magnon** for the site in France from which specimens were first described)? Or did they interbreed with the Cro-Magnons, and, if so, does that mean that Neanderthals should be included among our direct ancestors? The controversy surrounding this question flares up from time to time when paleontologists find specimens that have anatomical features intermediate between the heavyset Neanderthal form and the lighter human one.

In 1997, we thought we had a clear answer. Matthias Krings working in Svante Pääbo's lab in Germany extracted intact stretches of mtDNA from 30,000-year-old Neanderthal material. If Neanderthals had interbred with our ancestors, we would expect to see evidence of their genetic input in the form of Neanderthal mtDNA sequences in the modern human gene pool. However, the

---

**FIG. 24.9**  (a) *Homo sapiens* and (b) *Homo neanderthalensis.* Neanderthals disappeared from Europe about 30,000 years ago when *H. sapiens*, called Cro-Magnon after the site at which they were discovered, first appeared in Europe.

**a.** Cro-Magnon (modern)

**b.** Neanderthal

Neanderthal brains were slightly larger than ours, but their skulls were shaped differently. The Neanderthal forehead was much less pronounced than ours, and the skull in general lower.

Neanderthal sequence was strikingly different from that of modern humans. That we see nothing even close to the Neanderthal mtDNA sequence in modern humans strongly suggested that Neanderthals did not interbreed with our ancestors (**Fig. 24.10a**).

This conclusion, however, has been reversed following remarkable technological advances that permitted Pääbo and others to sequence the entire Neanderthal genome (rather than just its mtDNA). Careful population genetic analysis revealed that our ancestors did interbreed with Neanderthals, and that 1% to 4% of the genome of every non-African is Neanderthal-derived. A scenario emerged: As populations of our ancestors headed from Africa into the Middle East, they encountered Neanderthals, and it was then, before modern humans had spread out across the world, that interbreeding took place (**Fig. 24.10b**).

How can we reconcile these two apparently contradictory results? The mtDNA study indicates there was no genetic input from Neanderthals in modern humans, but the whole-genome study suggests that in fact there was. One possible solution lies in the difference in patterns of transmission between mtDNA and genomic material (Chapter 17). Recall that mtDNA is maternally inherited because sperm do not contribute mitochondrial material. Your mtDNA, whether you are male or female, is derived solely from your mother. One possibility is that the Neanderthal mtDNA lineage has been lost through genetic drift (Chapter 21). An alternative is that the discrepancy in the ancient DNA stems from a sex-based difference in interbreeding. If Neanderthal females did not interbreed with our ancestors, Neanderthal mtDNA would not have entered the modern human gene pool. If only male Neanderthals interbred with our ancestors, we would expect exactly the pattern we observe: no Neanderthal mtDNA in modern humans, but Neanderthal genomic DNA in the modern human gene pool.

It is likely that our interpretation of these events will always be speculative. The basic result, however, forces us to reimagine the ancestry of a large portion of the human population. Every non-African on the planet is part Neanderthal, albeit a very small part.

→ **Quick Check 3** Explain how genetic data suggest that only male Neanderthals interbred with our ancestors.

---

**FIG. 24.10 Evidence for and against early interbreeding between Neanderthals and early humans.** Studies using only mtDNA (a) did not support the idea of interbreeding, but later studies using whole genomes (b) did.

**a.** mtDNA phylogeny for Neanderthals and modern humans

Because Neanderthal mtDNA does not fall within the human family tree (red), we initially believed that Neanderthals and our ancestors did not interbreed.

However, studies of the full Neanderthal genome revealed that there was interbreeding, probably in the Middle East as *H. sapiens* first emerged from Africa about 60,000 years ago.

— Initial *H. sapiens* out-of-Africa migration
— Expansion of *H. sapiens* populations after interbreeding with Neanderthals

---

## 24.3 DISTINCT FEATURES OF OUR SPECIES

Many extraordinary changes in anatomy and behavior occurred in the 5–7 million years since our lineage split from the lineage that gave rise to the chimpanzees. Fossils tell us a great deal about those changes, especially when high-quality material such as Lucy or Ardi is available, but in general this is an area in which fossils are hard to come by and there is a lot of speculation. Speculation is especially common when we try to explain the reasons behind the evolution of a particular trait. Why, for example, did language evolve? It is easy enough to think of a scenario in which natural selection favors some ability to communicate—maybe language arose to facilitate group hunting. There are plenty of plausible ideas on the subject, but, in most cases, no evidence, so it is impossible to distinguish among competing hypotheses. We can, however, be confident that the events that produced language occurred in Africa, and, through paleontological studies of past environments, we can conclude that humans evolved in an environment similar in many ways to today's East African savanna.

### Bipedalism was a key innovation.

The shift from walking on four legs to walking on two was probably one of the first changes in our lineage. Many primates are partially bipedal—chimpanzees, for example, may be observed knuckle-walking—but ours is the only species that is wholly bipedal. As we've seen, Lucy, dating from about 3.2 million years ago, was already bipedal, and 4.4 million years ago, Ardi was partially bipedal. We can further refine our estimate of when full bipedalism arose from the evidence of a trace fossil. A set of 3.5-million-year-old fossil footprints discovered in Laetoli, Tanzania, reveal a truly upright posture.

Becoming bipedal is not simply a matter of standing up on hind legs. The change required substantial shifts in a number of basic anatomical characteristics, described in **Fig. 24.11.**

Why did hominins become bipedal? Maybe it gave them access to berries and nuts located high on a bush. Maybe it allowed them to scan the vicinity for predators. Maybe it made long-distance travel easier. And, very important, bipedalism freed up our ancestors' hands. No longer did they need their hands for locomotion, so for the first time there arose the possibility that specialized hand function could evolve. Most primates have some kind of opposable thumb, but the human version is much more refined. The human thumb has three muscles that are not present in the thumb of chimpanzees, and these allow much finer motor control of the thumb. Tool use, present but crude in chimpanzees, can be much more subtle and sophisticated with a human hand.

Bipedalism also made it possible to carry material over long distances. Hominins could then set up complex foraging strategies whereby some individuals supply others with resources. Exactly when sophisticated tool use arose in our ancestors is controversial, but it is indisputable that bipedalism contributed. Similarly, the ability to manipulate food with the hands, and to carry material using the hands rather than the mouth, likely permitted the evolution of the human jaw—indeed, the entire facial structure—in such a way that language became a possibility.

### Adult humans share many features with juvenile chimpanzees.

King and Wilson's 1975 discovery that the DNA of humans and chimpanzees are 99% identical has recently been confirmed by DNA sequence comparison of the human and chimpanzee genomes. It turns out that the two genomes are extraordinarily similar: All our genes are also present in the chimpanzee and their sequences are extremely similar, suggesting that the functions of the proteins they code for are the same. If we are so similar, how can we account for the extensive differences between the two species?

In their original paper, King and Wilson suggested that much of the most significant evolution along the hominin lineage came about through changes in the regulation of genes (Chapter 19). A small change—one that causes a gene to be transcribed at a different stage of development, for example—can have a major effect. In other words, King and Wilson introduced a model whereby small changes in the software could have a major impact even though the basic hardware is the same.

One of the gene-regulatory pathways that changed may be responsible for human **neoteny**, the long-term evolutionary process in which the timing of development is altered so that a sexually mature organism still retains the physical characteristics of the juvenile form. As early as 1836, French naturalist Étienne Geoffroy Saint-Hilaire noted that a young orangutan on exhibit in

**FIG. 24.11 The shift from four to two legs.** Anatomical changes were required for the shift, especially in the structure of the skull, spine, legs, and feet.

The foramen magnum, the hole in the base of the skull through which the spinal cord extends, is repositioned so that the human skull is balanced directly on top of the vertebral column.

The human spine is approximately S-shaped, ensuring that the weight is directly over the pelvis.

The pelvis is extensively reconfigured for an upright posture, and the internal organs can sit within it as though it were a kind of basin. The legs are longer, enabling long stride length and therefore efficient locomotion. The anatomy of the legs is also altered – the legs are directly under the body.

The human foot is narrower and has a much more developed heel and a larger big toe, features that contribute to a springier foot.

**FIG. 24.12 A juvenile and an adult orangutan.** Humans look more like juvenile orangutans than adult orangutans, suggesting that humans may be neotenous great apes.

Paris looked considerably more like a human than the adult of its own species (**Fig. 24.12**).

Several human attributes support this model, including our large heads (and correspondingly large brains), a feature of juvenile great apes. A second human attribute is our lack of hair. The juveniles of other great apes are not as hairless as humans, but they're considerably less hairy than adult apes. A third attribute is the position of the foramen magnum at the base of the skull. In primate development, the foramen magnum starts off in the position it occupies in the adult human and then, in nonhuman great apes, migrates toward the back of the skull. Adult humans have retained the juvenile great ape foramen magnum position. Finally, it has been suggested that our mentality, with its questioning and playfulness, is equivalent in many ways to that of a juvenile ape rather than to that of the comparatively inflexible adult ape.

In keeping with King and Wilson's idea, this shift in development could conceivably be achieved with relatively little genetic change. All that it might take would be a few changes to the regulatory switches that control the timing of development.

## Humans have large brains relative to body size.

In mammals, brain size is typically correlated with body size. Humans are relatively large-bodied mammals, but our brains are large for even our body size (**Fig. 24.13**). It is our large brains that have allowed our species' success, extraordinary technological achievements, and at times destructive dominion over the planet.

What factors promoted the evolution of the large human brain? Again, speculation is common, and it is unlikely we will ever have a definitive answer. Because a large brain is metabolically expensive to produce and to maintain, we can conclude that natural selection must have acted in favor of large brains. What are the selective factors? Here are some possibilities:

- *Tool use.* Bipedalism permitted the evolution of manual dexterity, which in turn requires a complex nervous organization if delicate hands are to be useful.

- *Social living.* Groups require coordination, and coordination requires some form of communication and the means of integrating and acting upon the information conveyed. One scenario, for example, sees group hunting as critical in the evolution of the brain: Natural selection favored those individuals who cooperated best as they pursued large prey.

- *Language.* Did the evolution of language drive the evolution of large brains? Or did language arise as a result of having large brains? Again, we will probably never know, but it is tempting to speculate that the brain and our extraordinary powers of communication evolved in concert.

Probably, as with bipedalism, there was no single factor but a mix of elements that worked together to result in the evolution of a large brain size. We tend to focus on brain size because it is a convenient stand-in for mental power and because we can measure it in the fossil record by making the reasonable assumption that the volume of a fossil's cranium reflects the size of its brain. However,

**FIG. 24.13 Brain size plotted against body size for different species.** Humans have large brains for their body size.

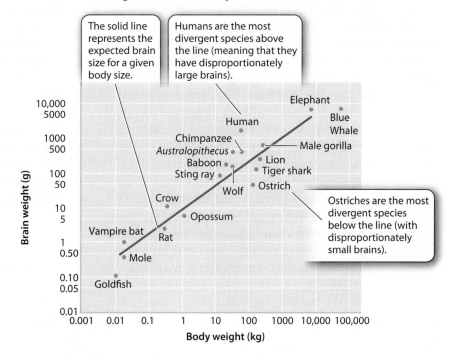

it should be emphasized that our minds are not the products solely of larger brains. Rather, the reorganization of existing structures and pathways is the more important product of the evolution of the brain. Not only was the brain expanding over those 7 million years of human evolution, it was also being rewired.

The human brain evolved through natural selection. What we have today is a learning machine capable of generating many more skills and abilities than just those that directly enhance evolutionary fitness. For example, let's assume that the brain evolved to make group hunting more efficient. The result is that the brain does indeed enhance our ability to hunt together, but there are numerous by-products of this brain that take its abilities beyond just hunting. Playing the piano, for instance, has nothing to do with group hunting, and yet the brain allows us to do it. A brain evolved for a mundane task like group hunting lies at the heart of much that is wonderful about humanity.

### The human and chimpanzee genomes help us identify genes that make us human.

The key genetic differences between humans and chimpanzees must lie in the approximately 1% of DNA sequence that differs between the two genomes. What do we see when we compare human and chimpanzee genomes?

A gene that has attracted a lot of interest is *FOXP2*, a member of a large family of evolutionarily conserved genes that encode transcription factors that play important roles in development. Individuals with mutations in *FOXP2* often have difficulty with speech and language. Interestingly, other animals whose *FOXP2* has been knocked out also have communication impairments. Songbirds are less capable of learning new songs, and the ultrasonic "songs" of mice are disrupted. In addition to its effects on the brain, *FOXP2* plays a role in the development of many tissues.

Studies of the gene's amino acid sequence reveal a pattern of extreme conservation. The sequences in mice and chimpanzees, whose most recent common ancestor lived about 75 million years ago, differ by a single amino acid (**Fig. 24.14**). However, two amino acids are present in humans but absent in chimpanzees. Studies of Neanderthal DNA have shown that Neanderthals possessed the modern human version of *FOXP2*. The presence of the same version of the gene in both species suggests that the differences arose in the hominin line before the split between our species and Neanderthals, which occurred at least 600,000 years ago.

*FOXP2* gives us a glimpse of the genetic architecture of the traits that are likely to be important in the determination of "humanness." Those two differences in amino acid sequence are intriguing, but they do not a human make. Substitute the human *FOXP2* gene into a mouse and you will not get a talking mouse. The critical genetic differences are many, subtle, and interacting.

→ **Quick Check 4** The *FOXP2* gene is sometimes called the "language gene." Why is this name inaccurate?

**FIG. 24.14** **A family tree of mammals showing amino acid changes in *FOXP2*.** The gene is highly conserved, but two changes are seen in the Neanderthals and modern humans.

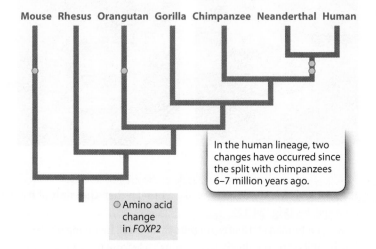

There is only a single amino change in *FOXP2* between mice and primates, although their most recent common ancestor lived about 75 million years ago.

Mouse  Rhesus  Orangutan  Gorilla  Chimpanzee  Neanderthal  Human

In the human lineage, two changes have occurred since the split with chimpanzees 6–7 million years ago.

○ Amino acid change in *FOXP2*

## 24.4 HUMAN GENETIC VARIATION

So far we have treated humans as all alike, and in many ways, of course, we are. But there are also many differences from one person to the next. Those differences ultimately have two sources: genetic variation and differences in environment (Chapter 18). A person may be born with dark skin, or a person born with pale skin may acquire darker skin—a tan—in response to exposure to sun.

The differences we see from one person to the next are deceptive. Despite appearances, ours is not the most genetically variable species on Earth. While it is certainly true that everyone alive today (except for identical twins) is genetically unique, our species is actually rather low in overall amounts of genetic variation. Modern estimates based on comparisons of many human DNA sequences indicate that, on average, about 1 in every 1000 base pairs differs among individuals (that is, our level of DNA variation is 0.1%). That's about 10 times less genetic variation than in fruit flies (which nevertheless all look the same to us) and about two to three times less than in Adélie penguins, which look strikingly similar to one another (see Fig. 21.1).

Why, then, are we all so phenotypically different if there is so little genetic variation in our species? Given the large size of our genome, a level of variation of 0.1% translates into a great many genetic differences. Our genome consists of approximately 3 billion base pairs, so 0.1% variation means that 3 million bp differ between any two people chosen at random. Many of those differences are in noncoding DNA, but some fall in regions of DNA that encode proteins and therefore influence the phenotype, so there is a fairly large reservoir of genetic variation in humans. When those mutations are reshuffled by recombination, we get the vast array of genetic combinations present in the human population.

## The prehistory of our species has had an impact on the distribution of genetic variation.

The reasons for our species' relative lack of genetic variation compared to other species lie in prehistory, and factors affecting the geographical distribution of that variation also lie in the past. Studies of the human family tree initiated by Rebecca Cann's original mtDNA analysis are giving a detailed picture of how our ancestors colonized the planet and how that process affected the distribution of genetic variation across populations today. Detailed analyses of different populations, often using mtDNA or Y chromosomes, allow us to reconstruct the history of human population movements.

As we have seen, *Homo sapiens* arose in Africa. Perhaps 60,000 years ago, populations started to venture out through the Horn of Africa and into the Middle East (**Fig. 24.15**). The first phase of

colonization took our ancestors through Asia and into Australia by about 50,000 years ago. Not until about 15,000 years ago did the first modern humans cross from Siberia to North America to populate the New World.

Other colonizations were even later. Despite its closeness to the African mainland, the first humans arrived in Madagascar only about 2000 years ago, and the colonists came from Southeast Asia, not Africa. Madagascar populations to this day bear the genetic imprint of this surprising Asian input. The Pacific Islands were among the last habitable places on Earth to be colonized during the Polynesians' extraordinary seaborne odyssey from Samoa, which began about 2000 years ago. Hawaii was colonized about 1500 years ago, and New Zealand only 1000 years ago.

By evolutionary standards, the beginning of the spread of modern humans out of Africa about 60,000 years ago is very recent. There has therefore been relatively little time for differences to accumulate among regional populations, and most of the variation in human populations today arose in ancestral populations before any humans left Africa. When we compare levels of variation in a contemporary African population to that in a non-African population, like Europeans or Asians, we find there is more variation in the African population. This is because the individuals that left Africa 60,000 years ago to found populations in the Middle East and beyond were a relatively small sample of

**FIG. 24.15 Human migratory routes.** Tracking the spread of mitochondrial DNA mutations around the globe allows us to reconstruct the colonization history of our species.

Humans arrived late in the New World. The precise dates are disputed, but the first humans reached North America about 15,000 years ago.

After *Homo sapiens* evolved in Africa, groups moved out, beginning the colonization of the rest of the planet.

→ Migratory route 1 (60,000–45,000 years ago)

→ Migratory route 2 (30,000–20,000 years ago)

→ Migratory route 3 (15,000?–10,000 years ago)

→ Migratory route 4 (<10,000 years ago)

the total amount of genetic variation then present in the human population. Non-African populations therefore began with less genetic variation, and that initial lower variation is reflected in their genetic profiles today.

### The recent spread of modern humans means that there are few genetic differences between groups.

Because the out-of-Africa migration was so recent, the genetic differences we see among geographical groups—sometimes called races—are minor. This fact is highly counterintuitive. We see many superficial differences between an African and a Caucasian, such as skin color, facial form, and hair type, and assume that these superficial differences must reflect extensive genetic differences. This assumption made sense when the standard theory about the origin of modern humans was the multiregional one. If European and African populations really had been geographically isolated from each other for as long as 2 million years, then we would expect significant genetic differences between the populations.

We expect isolated populations to diverge genetically over time as different mutations occur in each population. The longer two populations have been isolated from each other, the more genetic differences between them we expect to see (Chapter 21). Isolation lasting 2 million years implies that the differences are extensive, but isolation of 60,000 years suggests they are relatively few. Patterns of genetic variation among different human populations support the hypothesis that human populations dispersed as recently as 60,000 years ago. What we see when we look at genetic markers—variable A's, T's, G's, and C's in human DNA—is that there is indeed very little genetic differentiation by what sometimes is called race.

In short, there's a disconnect: Different groups may *look* very different, but, from a genetic perspective, they're not very different at all. Statistical analyses have shown that approximately 85% of the total amount of genetic variation in humans occurs within a population (for example, Swedes); 8% occurs between populations within races (for example, in Caucasians, between Swedes and Italians); and the remaining 7% occurs between races. The characteristics we use when we assess an individual's ethnicity, such as skin color, eye type, and hair form, are encoded by genetic variants that lie in that 7%. If Earth were threatened with destruction and only one population—Swedes, for example—survived, 85% of the total amount of human genetic variation that exists today would still be present in that Swedish population.

### Some human differences have likely arisen by natural selection.

We can conclude that the genetic variants affecting traits we can easily see are overrepresented in the 7% of human genetic variation that sorts by race. If we look at other genetic variants, ones that don't affect traits that we can see, there is no racial pattern. An African is as likely to have a particular base-pair mutation in a gene as a European. So why are visible traits so markedly different among races and others are not? Given the short amount of time (by evolutionary standards) since all *Homo sapiens* were in Africa, it is likely that the differences we see between groups are the product of selection.

People with dark skin tend to live in lower latitudes with high levels of solar radiation, and people with light skin tend to live in higher latitudes with low levels of solar radiation. It is likely that natural selection is responsible for the physical differences between these populations. Assuming that the ancestors of non-African populations were relatively dark-skinned, what selective factors can account for the loss of pigmentation?

A likely factor is an essential vitamin, vitamin D, which is particularly important in childhood because it is needed for the production of bone. A deficiency of vitamin D can result in the skeletal malformation known as rickets. The body can synthesize vitamin D, but the process requires ultraviolet radiation. Heavily pigmented skin limits the entry of UV radiation into cells and so limits the production of vitamin D. This does not present a difficulty in parts of the world where there is plenty of sunlight, but it can be problematical in regions of low sunlight. Presumably, natural selection favored lighter skin in the ancestors of Eurasian populations because lighter skin favored the production of the vitamin.

Some aspects of body shape and size may also have been influenced by natural selection. In hot climates, where dissipating body heat is a priority, a tall and skinny body form has evolved. Exemplified by East African Masai, this body type maximizes the ratio of surface area to volume and thus aids heat loss. In colder climates, by contrast, selection has favored a more robust, stockier body form, as exemplified by the Inuit, who have a low ratio of surface area to volume that promotes the retention of heat (**Fig. 24.16**). In these two cases, these are plausible explanations of body form. We should bear in mind, however, that simple one-size-fits-all explanations of human difference are almost always simplistic. Our species is complex and diverse—and often defies generalizations.

Attempts have been made to identify the adaptive value of obvious visual differences between races, such as facial features. It's possible that natural selection played a role in the evolution of these differences, but an alternative explanation, one originally suggested by Charles Darwin, is more compelling: sexual selection.

As we have seen, there is an apparent mismatch between the extent of difference among groups in visible characters, such as facial features, and the overall level of genetic difference between human groups. Sexual selection can account for this mismatch because it operates solely on characteristics that can readily be seen, such as the peacock's tail. As we learn more about the genetic underpinnings of the traits in question, we will be able to investigate directly the factors responsible for the differences we see among groups. Sexual selection is discussed in Chapter 45.

**FIG. 24.16 Evolutionary responses of body shape to climate.** (a) Heat-adapted Masai in Kenya and (b) cold-adapted Inuit in Greenland.

? **CASE 4 Malaria: Co-Evolution of Humans and a Parasite**
**What human genes are under selection for resistance to malaria?**

We see evidence of regional genetic variation in response to local challenges, especially those posed by disease. Malaria, for example, is largely limited to warm climates because it is transmitted by a species of mosquito that can survive only in these regions. Historically, the disease has been devastating in Africa and the Mediterranean. As we saw in Chapter 21, the sickle allele of hemoglobin, *S*, has evolved to be present at high frequencies in these regions because in heterozygotes it confers some protection against the disease. But in homozygotes, the *S* allele is highly detrimental because it causes sickle-cell anemia. Homozygotes for the allele encoding normal hemoglobin are also at a disadvantage because they are entirely unprotected from the parasite.

The *S* allele is beneficial only in the presence of malaria. If there is no malaria in an area, the *S* allele is disadvantageous, so natural selection presumably acted rapidly to eliminate it in the ancestors of Europeans when they arrived in malaria-free regions. The continued high frequency of the *S* allele in Africans, some Mediterranean populations, and in populations descended recently from Africans (such as African-Americans) is, however, a reflection of the response of natural selection to a regional disease.

The hemoglobin genes are not the only genes that are under selection for resistance to malaria. *Glucose-6-phosphate dehydrogenase (G6PD)*, a gene involved in glucose metabolism, is one of several other genes implicated. People who are heterozygotes for a mutation in the *G6PD* gene—and therefore have a G6PD enzyme deficiency—can develop severe anemia when they eat certain foods (most notably fava beans; hence, the condition is called favism). People who are heterozygotes for a mutation in the *G6PD* gene, however, also have increased resistance to malaria, apparently because they are better at clearing infected red blood cells from their bloodstream. In areas where malaria is common, the advantage of malaria resistance offsets the disadvantage of favism.

Detailed evolutionary analysis of mutations in *G6PD* shows that favism has arisen multiple times, each time selectively favored because of its role in the body's response to the malaria parasite. As expected, favism, like sickle-cell anemia, is mainly a feature of populations in malarial areas or of populations whose evolutionary roots lie in these areas.

## 24.5 CULTURE, LANGUAGE, AND CONSCIOUSNESS

Culture is generally defined as a body of learned behavior that is socially transmitted among individuals and passed down from one generation to the next. Culture has permitted us to transcend our biological limits. To take a simple example, clothing and ingeniously constructed shelters have enabled us to live in extraordinarily inhospitable parts of the planet, like the Arctic (**Fig. 24.17**). The capacity to innovate coupled to the ability to transmit culture is the key to the success of humans.

### Culture changes rapidly.

Culture, of course, changes over time. In many ways, cultural change is responsible for our species' extraordinary achievements. Genetic evolution is slow because it involves mutation followed by changes in allele frequencies that take place over many generations. Cultural change, on the other hand, can occur much more rapidly. Ten years ago, nobody had heard of smartphones, but today, millions of people own them. Or think of the speed at which a change in clothing style—a shift from flared to straight leg jeans, for example—spreads through a population. Today's human population as a whole is genetically almost identical to the population when your grandparents were young. But think of the cultural changes that have occurred in the 50 years or so between your grandparents' youth and your own.

Despite this clear contrast between biological evolution and cultural change, we should not necessarily think of the two processes as independent of each other. Sometimes cultural change drives biological evolution.

A good example of the interaction between cultural change and biological evolution is the evolution of lactose tolerance in populations for which domesticated animals became an important source of dairy product. Most humans are lactose intolerant. Lactose, a sugar, is a major component of mammalian milk, including human breast milk. We have an enzyme, lactase, that breaks down lactose in the gut, but, typically, the enzyme is produced only in the first years of life, when we are breast-feeding. Once a child is weaned, lactase production is turned off. Lactase, however, is clearly a useful enzyme to have if there is a major dairy component to your diet.

Archaeological and genetic analyses indicate that cattle were domesticated probably three separate times in the past 10,000 years: in the Middle East, in East Africa, and in the Indus Valley. In at least two of these cases, there has been subsequent human biological evolution in favor of lactose tolerance, that is, continued lactase production throughout life. Analysis of the gene region involved in switching lactase production on and off has revealed mutations in European lactose-tolerant people that are different from those in African lactose-tolerant people, implying independent, convergent evolution of this trait in the two populations. Furthermore, we see evidence that these changes have evolved very recently, implying that they arose as a response to the domestication of cattle. Here, we see the interaction between cultural change and biological evolution. Biological evolution of continued lactase production has resulted from the change in a cultural practice, namely cattle domestication.

### Is culture uniquely human?

Even nonprimates are known to be capable of learning from another member of their own species—culture is not uniquely human. Bird songs, for instance, may vary regionally because juveniles learn the song from local adults. In a famous case, shown in **Fig. 24.18a**, small birds called blue tits (*Parus caeruleus*) in Britain learned from each other how to peck through the aluminum caps of milk bottles left on doorsteps by milkmen to reach the rich cream at the top of the milk. Presumably, one individual discovered accidentally how to peck through a milk bottle cap and the others then imitated it.

Nor is teaching, one way in which culture is transmitted, limited to humans. In teaching, one individual tailors the information available to another in order to facilitate learning. Teaching and imitation together make learning highly efficient. Adult meerkats teach young meerkats how to handle prey by giving them the opportunity to interact with live prey (**Fig. 24.18b**).

The most sophisticated example of nonhuman culture is shown by chimpanzees. Detailed studies of several geographically isolated populations have revealed 39 culturally transmitted behaviors, such as ways of using tools to catch insects, that are specific to a particular population (**Fig. 24.18c**). West African Chimpanzees in Guinea and the Ivory Coast both use tools to help them harvest the same species of army ants for food (and avoid the painful defensive bites of the ants), but each population has its own distinct way of doing this. Chimpanzees, then, are like us in having regional variation in culture.

Despite the cultural achievements of other species, it is clear that culture

**FIG. 24.17  The power of culture.** Inventions (such as clothing) have allowed our species to expand its geographic range.

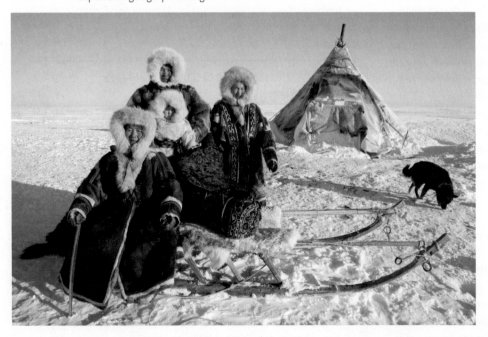

**FIG. 24.18** **Nonhuman culture.** (a) English blue tits steal cream from a milk bottle on the doorstep. (b) Adult meerkats teach their young how to handle their prey. (c) Chimpanzees in different populations have devised different ways of using tools to hunt insects.

in *Homo sapiens* is very far removed from anything seen in the natural world. We are amazingly adept at imitation, learning, and acquiring culture. Teaching, learning, and cultural transmission all benefit from another extraordinary human attribute, language.

## Is language uniquely human?

The short answer is again no, language is not unique to humans. The waggle dance of a worker bee on its return to the hive communicates information on the direction, distance, and nature of a food resource (Chapter 45). Vervet monkeys use warning vocalizations to specify the identity of a potential predator: They have one call for "leopard," for example, and another for "snake." However, most animal languages are limited. Even if a chimpanzee wished to communicate conversationally as we do, it would not be able to because its larynx is not capable of such subtle vocalizations.

Attempts have therefore been made to explore the linguistic capabilities of chimpanzees by teaching individual chimpanzees sign language or other forms of visual communication (**Fig. 24.19**). In a number of famous experiments, chimpanzees were able to acquire extensive vocabularies of signs and to construct elementary sentences. It is difficult, however, to evaluate the meaning of this research because the animals' achievements are entirely the product of the training program. Perhaps the limitation is most clearly seen when we try to teach an animal to use metaphor. Time, for example, is often described using a spatial metaphor. An

event in a time line on the left-hand side of a page is understood to come before an event on the right in time. Not even the best-trained chimpanzees can grasp this concept.

Grammar provides a set of rules that allow the combination of words into a virtually infinite array of meanings. Noam Chomsky, father of modern linguistics, has pointed out that all human languages are basically similar from a grammatical viewpoint. Humans have what he has called a "universal grammar" that would lead a visiting linguist from another planet to conclude that all Earth's languages are dialects of the same basic language. That universal grammar is in some way hard wired into the human

**FIG. 24.19** **Dr. Susan Savage Rumbaugh with Panbanisha, a female bonobo who learned to communicate using sign language.** Chimpanzees and bonobos are able to learn and use sign language to express words and simple sentences.

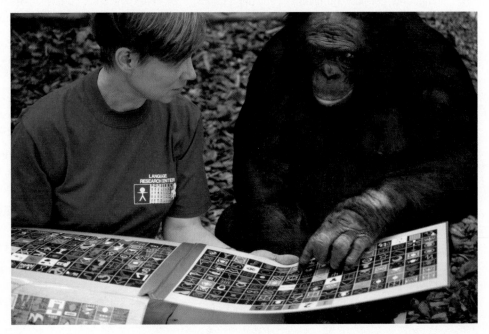

brain in such a way that every human infant spontaneously strives to acquire language. The specific attributes of the language depend on the baby's environment—a baby in France will learn French, and one in Japan will learn Japanese—but the basic process is similar in every case. Chimpanzees, and the entire natural world, lack the drive toward the acquisition of a grammatical language.

## Is consciousness uniquely human?

Descartes famously wrote, "Cogito, ergo sum"—"I think, therefore I am." Can we legitimately rewrite his statement to declare, "Animals think, therefore they are"? With the growth of the animal rights movement, particularly in reference to the treatment of animals in factory farms, this question is of more than academic interest. We now have many examples of animal thinking from a range of species, including, not surprisingly, chimpanzees and gorillas.

Perhaps more remarkable are the examples that come from animal species that are not closely related to us. In experiments carried out by Alex Kacelnik in Oxford, England, a pair of New Caledonian Crows was presented with two pieces of wire, one straight and the other hooked, and offered a food reward that could be obtained only by using the hooked wire. One member of the pair, the male, disregarded the experiment and flew off with the hooked wire. The female, however, having discovered that she could not get the food reward with the straight wire, went to some considerable trouble to bend a hook into the straight wire. She succeeded in getting the food. It is difficult to deny that the crow thought about the problem and was able to solve it, perhaps in the same way as we would. Definitions of consciousness are contested, but, as with language and culture, it seems clear that other species are capable of some form of conscious thought.

The evolutionary biologist Theodosius Dobzhansky once said, "All species are unique, but humans are uniquest." Our "uniquest" status is not derived from having attributes absent in other species, but from the extent to which those attributes are developed in us. Human language, culture, and consciousness are extraordinary products of our extraordinary brains. Nevertheless, as Darwin taught us and as we should never forget, we are fully a part of the natural world.

## Core Concepts Summary

**24.1 ANATOMICAL, MOLECULAR, AND FOSSIL EVIDENCE SHOWS THAT THE HUMAN LINEAGE BRANCHES OFF THE GREAT APES TREE.**

Anatomical features indicate that primates are a monophyletic group that includes prosimians, monkeys, and apes. The apes in turn include the lesser and great apes. The great apes include orangutans, gorillas, chimpanzees, and humans. page 24-1

Analysis of sequence differences between humans and our closest relatives, chimpanzees, indicate that our lineage split from chimpanzees 5-7 million years ago. page 24-3

Lucy, an unusually complete specimen of *Australopithecus afarensis,* demonstrates that our ancestors were bipedal by about 3.2 million years ago. page 24-4

Human lineage fossils occur only in Africa until about 2 million years ago, when *Homo ergaster* migrated out of Africa to colonize the Old World. page 24-5

**24.2 PHYLOGENETIC ANALYSIS OF MITOCHONDRIAL DNA AND THE *Y* CHROMOSOME SHOWS THAT OUR SPECIES AROSE IN AFRICA.**

Studies of mitochondrial DNA (mtDNA) suggest that the time back to the common ancestor of modern humans is about 200,000 years, implying that modern humans (*Homo sapiens*) arose in Africa (the out-of-Africa theory). page 24-6

The mtDNA out-of-Africa pattern is supported by Y-chromosome analysis, which also shows a recent African origin of modern humans. page 24-8

Analysis of Neanderthal DNA from 30,000-year-old material indicates that, as the ancestors of non-African humans emigrated from Africa, they interbred with the Neanderthals. page 24-8

Our species originated in Africa and subsequently colonized the rest of the planet, starting about 60,000 years ago. page 24-9

**24.3 DURING THE 5–7 MILLION YEARS SINCE THE MOST RECENT COMMON ANCESTOR OF HUMANS AND CHIMPANZEES, OUR LINEAGE ACQUIRED A NUMBER OF DISTINCTIVE FEATURES.**

The development of bipedalism involved a wholesale restructuring of anatomy. page 24-9

Neoteny is the acquisition of sexual maturity in an otherwise juvenile state; humans are neotenous, exhibiting many traits as adults that chimpanzees exhibit as juveniles. page 24-10

There are many possible selective factors that explain the evolution of our large brain, including tool use, social living, and language. page 24-11

FOXP2, a transcription factor involved in brain development, may be important in language, as mutations in the gene that encodes FOXP2 are implicated in speech pathologies. page 24-12

## 24.4 HUMAN HISTORY HAS HAD AN IMPORTANT IMPACT ON PATTERNS OF GENETIC VARIATION IN OUR SPECIES.

Because our ancestors left Africa very recently in evolutionary terms, there has been little chance for genetic differences to accumulate among geographically separated populations. page 24-13

Humans have very little genetic variation, with only about 1 in every 1000 base pairs varying among individuals. page 24-14

Most of the variation segregates within populations. As much as 85% of the total amount of genetic variation in humans can be found within a single population. Only about 7% of human genetic variation segregates between races. page 24-14

Some racial differences, such as skin color and resistance to malaria, have probably arisen via natural selection. page 24-14

Other racial differences have probably arisen via sexual selection. page 24-15

## 24.5 CULTURE IS A POTENT FORCE FOR CHANGE IN MODERN HUMANS.

Cultural evolution and biological evolution may interact, as in the case of the evolution of lactose tolerance in regions where cattle were domesticated. page 24-16

Other animals possess simple versions of culture, language, and even consciousness, but the capabilities of our species in all three are truly exceptional. page 24-16

# Self-Assessment

1. Describe what evidence suggests that chimpanzees are the closest living relatives of humans.

2. Explain the out-of-Africa theory of human origins and how studies of mitochondrial DNA and the Y chromosome support it.

3. List three anatomical differences between chimpanzees and humans, and explain how these changes facilitated walking upright.

4. Given the high genetic similarity of humans and chimpanzees, how can we account for the differences we see between the two species?

5. Describe three possible selective factors underlying the evolution of large brains in our ancestors.

6. Explain how differences among different human populations arose by natural and sexual selection.

7. Provide arguments for and against the idea that culture, language, and consciousness are uniquely human.

Do you understand the chapter's Core Concepts? Log in to check your answers to the Self-Assessment questions, then practice what you've learned and reinforce this chapter's concepts by working through the problems and multimedia tutorials provided there.

**www.biologyhowlifeworks.com**

# PART 2

# FROM ORGANISMS TO THE ENVIRONMENT

"From so simple a beginning endless forms most beautiful and most wonderful have been, and are being, evolved."

—CHARLES DARWIN

## CHAPTER 25

# CYCLING CARBON

## Core Concepts

**25.1** Photosynthesis and respiration are the key biochemical pathways for the biological, or short-term, carbon cycle.

**25.2** Physical processes govern the long-term carbon cycle.

**25.3** The carbon cycle can help us understand ecological interactions and the evolution of biological diversity.

In 1958, American chemist Charles Keeling began a novel program to monitor the atmosphere. At the time he began his study, scientists had only a vague notion of how carbon dioxide ($CO_2$) behaves in air. Many thought that $CO_2$ might vary unpredictably from time to time, and from place to place. Keeling decided to find out if in fact it did. From five towers set high on Mauna Loa in Hawaii, 3400 m above sea level, he sampled the atmosphere every hour and measured its composition with an infrared gas analyzer. Within the first few years of his project, a pattern of seasonal oscillation became apparent: $CO_2$ concentration in the air reached its annual high point in spring and then declined by about six parts per million (ppm, by volume) to a minimum in early fall (**Fig. 25.1**). Such regular variation was unexpected: What could be causing it?

Perhaps the pattern was local. Airplane traffic to Hawaii peaked in winter, and maybe that was affecting the air around Mauna Loa. As measurements continued through a number of years, however, the pattern persisted, even as trends in air traffic changed. Moreover, monitoring stations in many other parts of the globe gave similar results. Knowing that the Mauna Loa measurements are representative of the atmosphere as a whole allows us to appreciate what an annual variation of 6 ppm really means. With calendar-like regularity, approximately 47 billion metric tons of $CO_2$ were entering and leaving the atmosphere annually—that's 13 billion metric tons of the element carbon (the rest of the mass is oxygen). To explain such a pattern, we must consider processes that affect the planet as a whole.

A second pattern is evident in Fig. 25.1. Summertime removal of $CO_2$ from the atmosphere does not quite balance winter increase, with the result that the amount of $CO_2$ in the atmosphere on any given date is greater than it was a year earlier. Atmospheric $CO_2$ levels have therefore increased steadily through the period of monitoring. For example, mean annual $CO_2$ levels, about 315 ppm in 1958, reached 397 ppm by May 2012. This is an increase of more than 25%, and there is no indication that the rise will slow or stop any time soon. What causes this increase? What might be its consequences?

Recognizing the processes reflected in the Keeling curve provides a crucial first step toward understanding the **carbon cycle,** the intricately linked network of biological and physical processes that shuttles carbon among rocks, soil, oceans, air, and organisms. The carbon cycle, in turn, provides a glimpse of the interactions that underpin ecology and promote biological diversity. It focuses our attention on the ways that physical and biological processes interact to determine the properties of environments. And it provides a basis for assessing the role that humans play in our environmental present and future.

## 25.1 THE SHORT-TERM CARBON CYCLE

What causes atmospheric carbon dioxide to vary through the year and over a timescale of decades? To answer this question, we need to ask what processes introduce $CO_2$ into the atmosphere and what processes remove it. $CO_2$ is added to the atmosphere by (1) geologic inputs, mainly from volcanoes and mid-ocean ridges; (2) biological inputs, especially respiration; and (3) human activities, including deforestation and the burning of fossil fuels. Processes that remove $CO_2$ from the atmosphere include (1) geologic removal, especially by the chemical weathering of rocks, and (2) biological

**FIG. 25.1** **The Keeling curve.** The Keeling curve provides a record of atmospheric carbon dioxide concentrations over half a century, measured from an observatory at Mauna Loa, Hawaii.

CO₂ concentrations regularly cycle up and down over the course of a year.

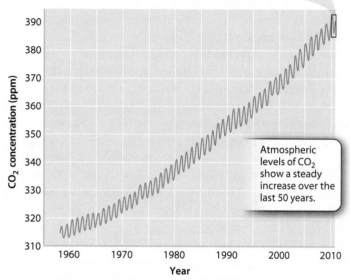

Atmospheric levels of CO₂ show a steady increase over the last 50 years.

removal, mainly through photosynthesis. As we will see, geologic processes play key roles in the carbon cycle on timescales of centuries and longer. Only biology can account for the annual rhythm of the Keeling curve.

### Photosynthesis and respiration are key processes in short-term carbon cycling.

To grow, all organisms require both carbon and energy. As we saw in Chapter 8, photosynthetic organisms convert energy from the sun into ATP, which they use to reduce $CO_2$ to sugar. As photosynthesis pulls $CO_2$ out of the atmosphere (or water, for aquatic organisms), the carbon is transferred to carbohydrates, and oxygen is given off as a by-product. The summary chemical reaction is shown here:

$$6CO_2 + 6H_2O \rightarrow C_6H_{12}O_6 + 6O_2$$

It is estimated that photosynthetic organisms remove 190 billion metric tons of carbon (in nearly 700 billion metric tons of $CO_2$) from the atmosphere each year, about 60% by land plants and the rest by phytoplankton and seaweeds in the oceans. This annual photosynthetic removal of $CO_2$ equals about 25% of the total $CO_2$ in the atmosphere. In the absence of processes that

restore $CO_2$ to the atmosphere, high rates of photosynthesis would use up the atmospheric reservoir of carbon dioxide in just a few years. But the Keeling curve tells us that atmospheric $CO_2$ is *not* decreasing rapidly—it is rising. Moreover, the geologic record indicates that photosynthesis has fueled ecosystems for billions of years without causing $CO_2$ depletion.

What processes return $CO_2$ to the atmosphere, balancing photosynthetic removal? In Chapter 7, we discussed aerobic respiration. Humans and many other organisms gain both the energy and carbon needed for growth from organic molecules (that is, by consuming food). As explained in Chapter 7, aerobic respiration uses oxygen to oxidize organic molecules to carbon dioxide, converting chemical energy in the organic compounds to ATP for use in cellular processes. Aerobic respiration can be summarized by this chemical equation:

$$C_6H_{12}O_6 + 6O_2 \rightarrow 6CO_2 + 6H_2O$$

Plants respire, too, converting the chemical energy in carbohydrates into useful work. On the present-day Earth, the amount of $CO_2$ returned annually to the atmosphere by aerobic respiration and related processes is about equal to the amount that plants and algae remove by photosynthesis. That is why photosynthesis does not deplete the $CO_2$ in air.

As we discussed in Chapter 8, the equations for oxygenic photosynthesis and aerobic respiration mirror each other. Photosynthesis continually converts $CO_2$ to organic molecules, using energy from the sun, and respiration harvests the chemical energy stored in organic molecules, releasing $CO_2$ and water in the process. Photosynthesis and aerobic respiration are complementary metabolic pathways: Each uses the products of the other and generates a new supply of substrates for its complement. The result is a cycle (**Fig. 25.2**).

### The regular oscillation of CO₂ reflects the seasonality of photosynthesis in the Northern Hemisphere.

Now that we have identified the major processes in the short-term carbon cycle, we can examine what makes atmospheric $CO_2$ level oscillate seasonally. Except in a narrow band around the equator, photosynthesis is seasonal, with higher rates in the summer and lower rates in the winter. Respiration, in contrast, remains more or less constant through the year.

**FIG. 25.2** **The short-term carbon cycle.** The complementary metabolic processes of photosynthesis and respiration drive the short-term cycling of carbon through the biosphere.

Carbon cycles by means of the complementary processes of photosynthesis and respiration.

Even cursory examination shows that land on Earth is distributed asymmetrically, with more land—and hence more plants—in the Northern Hemisphere than in the Southern Hemisphere. For this reason, global atmospheric $CO_2$ declines through the northern summer, when the ratio of photosynthesis to respiration is highest, and then increases through fall and winter, when the ratio is reversed. The result is the seasonal oscillation of atmospheric $CO_2$ levels documented by Keeling.

### Human activities play an important role in the modern carbon cycle.

Now let us ask about the second major pattern evident in the Keeling curve. Although $CO_2$ levels oscillate on an annual basis, the overall pattern is one of sustained increase. During the 1960s, the observed increase was less than 1 ppm each year. In the first decade of the new millennium, it has been closer to 2 ppm. The current level of atmospheric $CO_2$ is 25% higher than it was 50 years ago. Why has $CO_2$ input into the atmosphere outstripped removal for the past half century? Should we consider this pattern unusual?

Answering the second question helps us to address the first. Conclusions about whether it is unusual for atmospheric $CO_2$ to increase 25% in 50 years can be answered only if we can compare the Keeling curve to longer records of atmospheric $CO_2$ levels. Before Keeling, no one systematically collected air samples. Fortunately, however, nature did it for us. When glacial ice forms in Antarctica and Greenland, it traps tiny bubbles of air. Each year, snowfall gives rise to a new layer of ice, and as layers accumulate through time their bubbles preserve a history of the atmosphere. **Fig. 25.3** shows the amount of $CO_2$ in air bubbles trapped in

---

## HOW DO WE KNOW?

**FIG. 25.3**

# How much $CO_2$ was in the atmosphere 1000 years ago?

**BACKGROUND** Direct measurements of the atmosphere show that $CO_2$ has increased by 25% over the past 50 years. To know whether or not such a change is unusual, we need to know $CO_2$ levels over a longer time interval.

**OBSERVATION** Scientists recognized that the snow that accumulates on glaciers to form new layers of ice is initially full of small spaces that are in contact with the air. Through time, as snow changes to ice, these spaces become sealed off from the surrounding environment and form bubbles, preserving a minute sample of air at the time of ice formation.

**MEASUREMENTS** Physical layering and chemical indicators that vary on an annual basis enabled scientists to assign ages to ice samples taken from cores drilled through the Law Dome, a large ice dome in Antarctica. The youngest ice samples overlap in age with the first part of the Keeling data, providing a check that air samples trapped in ice have $CO_2$ concentrations similar to those in air measured directly over the same interval.

Ice core drilled from a glacier in Antarctica

Viewed under the microscope, glacial ice contains small bubbles that trap samples of air.

- Mauna Loa (direct measurements of $CO_2$ levels)
- Law Dome (ice core measurements of $CO_2$ levels)

The fact that ice core data agree with direct measurements where both are available gives confidence in the $CO_2$ curve reconstructed from ice cores.

Measurements of air samples in glacial ice from the Law Dome, Antarctica, show that atmospheric $CO_2$ varied little between 1000 and 1800.

**CONCLUSIONS** Before the Industrial Revolution, atmospheric $CO_2$ levels had varied little over 1000 years, generally falling between 270 and 280 ppm. From this, scientists have concluded that current changes in atmospheric $CO_2$ are unusual on the timescale of the past millennium.

**SOURCE** Etheridge, D. M., L. P. Steele, R. L. Langenfelds, R. J. Francey, J.-M. Barnola, and V. I. Morgan. 1996. "Natural and Anthropogenic Changes in Atmospheric $CO_2$ over the Last 1000 Years from Air in Antarctic Ice and Firn." *Journal of Geophysical Research* 101:4115–4128.

Antarctic ice that has accumulated over the past 1000 years. Ice samples show that $CO_2$ levels actually began to increase slowly in the 1800s, the time of major transformations in mining, manufacturing, and transportation we call the Industrial Revolution. Before that, however, atmospheric $CO_2$ levels had varied little since the Middle Ages, staying at 270–280 ppm for centuries on end. On a 1000-year timescale, then, big changes in $CO_2$ abundance happened only once, over the last 200 years. Atmospheric $CO_2$ inputs and outputs were approximately in balance until the Industrial Revolution.

Fig. 25.3 shows that there is a **correlation** between the increasing $CO_2$ content of the atmosphere and human activities. A correlation simply indicates that two events or processes occur together. Can we actually demonstrate that humans have played a role in recent atmospheric change? By itself, correlation does not establish **causation,** a relationship in which one event leads to another. We might propose instead that recent $CO_2$ increase reflects natural processes and only matches the period of industrialization by coincidence.

We can test the hypothesis that human activities have contributed to the increases in atmospheric $CO_2$ measured at Mauna Loa and in ice cores by making careful measurements of a chemical detail: the isotopic composition of atmospheric $CO_2$.

→ **Quick Check 1** How does the graph in Fig. 25.3 suggest that human activities have influenced carbon dioxide levels in the atmosphere? What might be a plausible alternative hypothesis?

### Carbon isotopes show that much of the $CO_2$ added to air over the past half century comes from burning fossil fuels.

In Chapter 2, we saw that many elements have several isotopes, atoms of the element that vary in atomic mass because they have different numbers of neutrons. Carbon has three isotopes: $^{12}C$ (with six neutrons, about 99% of all carbon atoms); $^{13}C$ (with seven neutrons, most of the remaining 1%); and the rare $^{14}C$ (with eight neutrons, about one part per trillion of atmospheric carbon). Beginning more than 40 years ago, the Austrian-born chemist Hans Suess measured the relative abundances of the three isotopes of carbon in atmospheric $CO_2$, a program that continues today. He observed that the proportion of $^{13}C$ in atmospheric $CO_2$ has declined as the total amount of $CO_2$ has increased (**Fig. 25.4**). This subtle change in isotopic composition occurred because the carbon dioxide being added to the atmosphere has less $^{13}C$ than the $CO_2$ already in the air.

How does this result influence the hypotheses we developed to account for increasing atmospheric $CO_2$ levels? The various candidate sources of increasing carbon dioxide have different proportions of $^{13}C$ and $^{12}C$. Potential $CO_2$ contributors such as volcanic gases and inorganic carbon dissolved in the oceans do

**FIG. 25.4**

# What is the major source of the $CO_2$ that has accumulated in Earth's atmosphere over the past two centuries?

**BACKGROUND** Chemical analyses show that $CO_2$ levels in the current atmosphere are about 40% higher than they were at the time of the American Revolution. This rise coincides with major advances in manufacturing and transportation, which are driven by the burning of fossil fuels.

**HYPOTHESIS** The burning of fossil fuels, central to the emergence of modern industrial societies, has been the principal source of the measured increase of $CO_2$ in the atmosphere.

**EXPERIMENT AND RESULTS** Hans Suess measured the abundance of the two stable isotopes of carbon, $^{13}C$ and $^{12}C$, in air samples and demonstrated that the ratio of $^{13}C$ to $^{12}C$ in atmospheric $CO_2$ has decreased over the second half of the twentieth century. Additional measurements of coral skeletons and wood show that the ratio of $^{13}C$ to $^{12}C$ in air has been decreasing over the entire period during which atmospheric $CO_2$ levels have been increasing.

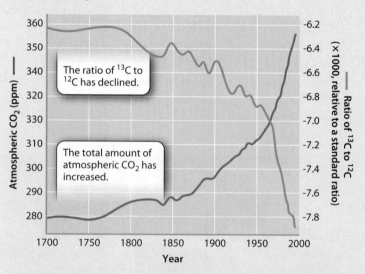

**DISCUSSION** The ratio of $^{13}C$ to $^{12}C$ in $CO_2$ added to the atmosphere over the past 200 years is lower than that in $CO_2$ already in the air. The $^{13}C:^{12}C$ ratio of $CO_2$ emitted by volcanoes is too high to account for the data, as is that of $CO_2$ released from the oceans. In contrast, organic matter formed by photosynthesis has just the right isotopic composition to account for the isotopic measurements.

**FOLLOW-UP WORK** Further studies, using $^{14}C$, the third, unstable isotope of C, also show a strong decrease in the proportional abundance of $^{14}C$ in atmospheric $CO_2$ over time. While the burning of vegetation to clear land for agriculture results in $CO_2$ with the right ratio of $^{13}C$ to $^{12}C$, modern vegetation has far too much $^{14}C$ to account for observed changes in the $^{14}C$ of air. Only fossil fuels have the right ratios of all three carbon isotopes—$^{12}C$, $^{13}C$, and $^{14}C$—to account for the pattern of isotopic change in atmospheric $CO_2$ measured by Suess and others. In fact, $^{14}C$ data from the high Rocky Mountains indicate that if fossil fuels were the principal source of added $CO_2$, carbon dioxide levels should have increased by a little less than 2 ppm per year between 2004 and 2008—which is very close to the increase actually measured.

### Air measurements at Niwot Ridge, Colorado

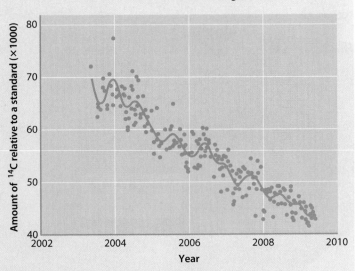

**CONCLUSION** Fossil fuel burning by industrialized societies has been, and continues to be, a principal source of carbon dioxide build-up in Earth's atmosphere.

**SOURCES** Revelle, R., and H. E. Suess. 1957. "Carbon dioxide exchange between atmosphere and ocean and the question of an increase of atmospheric $CO_2$ during the past decades." *Tellus* 9:18–27; Turnbull, J. C., S. J. Lehman, J. B. Miller, R. J. Sparks, J. R. Southon, P. P. Tans. 2007. "A new high precision $^{14}CO_2$ time series for North American continental air." *Journal of Geophysical Research-Atmospheres* 112, no. D11. Article Number: D11310. doi: 10.1029/2006JD008184.

not have the right isotopic composition to explain the observed change. Organic carbon in living organisms and in soil has the right $^{13}C$ composition. Therefore, processes that cause a net conversion of organic matter to carbon dioxide must be adding $CO_2$ to the atmosphere.

Here's where the other isotope of carbon, $^{14}C$, comes in. Measurements also show that the sources of added carbon dioxide are depleted in $^{14}C$ relative to $CO_2$ already in the air. Modern organic matter contains too much $^{14}C$ to account for the observed pattern, but ancient organic matter—the coal, petroleum, and natural gas burned as fossil fuels—is isotopically just right. Chemical analyses, therefore, support the hypothesis that human activities cycle carbon in amounts high enough to affect the chemical composition of the atmosphere.

That atmospheric $CO_2$ is increasing is not a hypothesis; it is a measurement. That the burning of fossil fuels plays an important role in this increase is also unambiguously documented by chemical analyses. Current debate focuses not on these observations, but instead on the more difficult problem of understanding how increasing $CO_2$ may affect climate. We discuss this debate in Chapter 48. For now, however, we can use some relevant numbers to reveal another important aspect of the carbon cycle.

As shown in the top graph in **Fig. 25.5**, humans add about 8 billion metric tons of carbon to the atmosphere each year, nearly

**FIG. 25.5** **Carbon dioxide sources and sinks.** Much of the carbon dioxide introduced by humans into the atmosphere has not stayed there. About half of it has accumulated as dissolved inorganic carbon in the oceans or as new biomass on the continents.

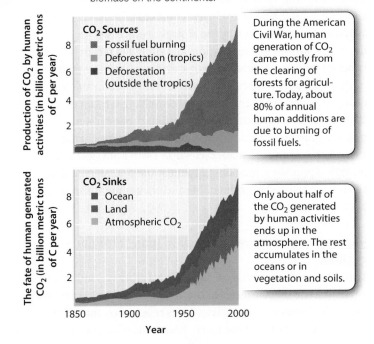

all of it as carbon dioxide. Much of this material comes from power plants, airplanes, ships, and automobiles, which oxidize ancient organic matter to $CO_2$ on a grand scale. Further important inputs come from the conversion of forests to agriculture or pastureland. Clearing and burning forest trees also oxidizes organic carbon to $CO_2$.

Interestingly, if we compare our best estimates of nonrespiratory human $CO_2$ generation since 1958 with the measured increase in atmospheric $CO_2$, we find that about half of the $CO_2$ released by human activities *did not* end up in the air. Where did it go? This has been a major question for years, and it now appears that much of the "missing carbon" has been stored as inorganic carbon in the form of $CO_2$, bicarbonate ($HCO_3^-$), and carbonate ($CO_3^{2-}$) ions dissolved in the oceans. The bottom graph in Fig. 25.5 shows an estimate of the various places where human-produced $CO_2$ ends up. These observations indicate that Earth's carbon cycle is more complicated than a simple exchange of carbon between photosynthesis and respiration. There are additional storehouses of carbon and additional processes that move carbon from one store to another, topics we discuss next.

## 25.2 THE LONG-TERM CARBON CYCLE

Up to this point, we have considered only the short-term carbon cycle: exchanges over days, years, and decades driven by the biological processes of photosynthesis and respiration, and altered in recent times by human activities. Geologic evidence, however, tells us that on longer timescales, carbon dioxide levels in air have changed dramatically. These longer-term changes mean that we must consider additional contributions to the carbon cycle: physical processes, including volcanism and climate change. Indeed, the complete carbon cycle links Earth's physical and biological processes, providing a foundation for understanding the interconnected histories of life and environment through our planet's long history.

### Reservoirs and fluxes are key in long-term carbon cycling.

To understand how biological and physical processes interact to govern carbon dioxide levels in the air, we must first look at how carbon is distributed among its various **reservoirs,** the places where carbon resides in the Earth. Reservoirs of carbon include organisms, the atmosphere, soil, the oceans, and sedimentary

**FIG. 25.6**  **Movement of carbon between reservoirs.** The numbers for each reservoir are gigatons (1000 million metric tons) of carbon, and those for fluxes are gigatons of carbon per year.

rocks, as shown in the purple boxes in **Fig. 25.6.** How much carbon is stored in each of these reservoirs?

If we add up all the carbon contained in the total mass of organisms living on land, that is, in their **biomass,** that amount of carbon is just a bit smaller than the amount of carbon stored as $CO_2$ in the atmosphere. Soil, by comparison, stores as much carbon as do land organisms and the

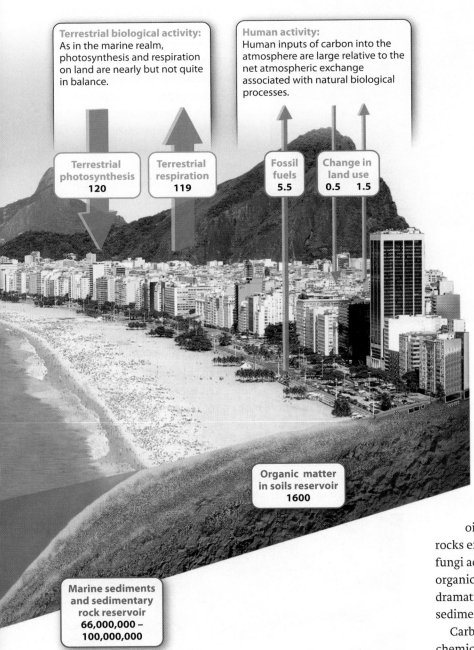

**Terrestrial biological activity:**
As in the marine realm, photosynthesis and respiration on land are nearly but not quite in balance.

**Human activity:**
Human inputs of carbon into the atmosphere are large relative to the net atmospheric exchange associated with natural biological processes.

Terrestrial photosynthesis 120

Terrestrial respiration 119

Fossil fuels 5.5

Change in land use 0.5  1.5

Organic matter in soils reservoir 1600

Marine sediments and sedimentary rock reservoir 66,000,000 – 100,000,000

The biggest carbon reservoir of all, however, lies beneath our feet. Calcium carbonate minerals ($CaCO_3$), which form limestone, and organic matter preserved in sedimentary rocks dwarf all other carbon reservoirs combined by three orders of magnitude. Coal, petroleum, and natural gas make up only a small part of the organic matter stored in sedimentary rocks, but as we have seen, they play an important role in the modern carbon cycle.

**Fluxes** are the rates at which carbon flows from one reservoir to another. The sensitivity of different reservoirs to change depends on the relative sizes of the reservoir and of the movement of material into and out of it. When fluxes are large relative to the size of the reservoir, reservoir size can change rapidly. As mentioned earlier, the amount of carbon stored as $CO_2$ in air is not that much larger than the annual fluxes into and out of the atmosphere. For this reason, atmospheric $CO_2$ abundance can be influenced by a number of processes at work in the carbon cycle.

## Physical processes add and remove $CO_2$ from the atmosphere.

Physical processes, like biological ones, are capable of adding and removing $CO_2$ from the atmosphere. A key process is volcanism: Volcanoes and mid-ocean ridges release an estimated 0.1 billion metric tons of carbon (as $CO_2$) into the atmosphere each year. More $CO_2$, about 0.05 billion metric tons of carbon per year, is released by the slow oxidation of coal, oil, and other ancient organic material in sedimentary rocks exposed at the Earth's surface. In nature, bacteria and fungi accomplish most of this oxidation by respiring old organic molecules. Burning fossil fuel accelerates this process dramatically, increasing a hundredfold the rate at which sedimentary organic carbon is oxidized.

Carbon dioxide is removed from the atmosphere by chemical reactions between air and exposed rocks, a process called chemical weathering. How can air and rock interact? To understand how weathering fits into the carbon cycle, we must first understand that carbon dioxide reacts with rainwater to form carbonic acid ($H_2CO_3$). This acid slowly reacts with rock-forming minerals, generating (among other substances) calcium and bicarbonate ($HCO_3^-$) ions that are transported by rivers to the oceans. Within the oceans, the calcium and bicarbonate ions react with each other to form calcium carbonate ($CaCO_3$) minerals that accumulate on the seafloor, forming limestone. In essence, carbon moves from

atmosphere combined, mostly as slowly decaying organic compounds. In the oceans, the amount of carbon contained in living organisms is actually very small. Much more resides as inorganic carbon dissolved in the water—most of it in the deep oceans as $CO_2$, bicarbonate, and carbonate ions. Perhaps fortunately for us, the oceans have sopped up some of the $CO_2$ generated by human activities.

FIG. 25.7    Clams and corals with skeletons made up of calcium carbonate.

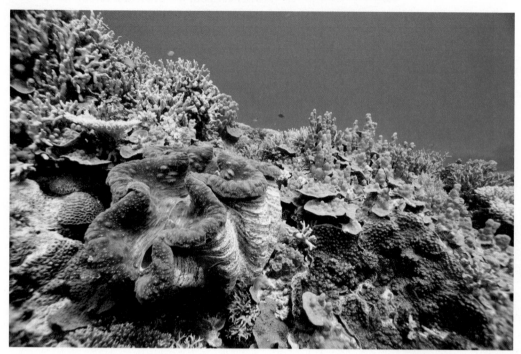

the atmosphere to the sediments, where it can be stored for millions of years. The two linked processes of chemical weathering and mineral precipitation in the oceans can be summarized by a simple chemical reaction (here, $CaSiO_3$ stands in for rock-forming minerals as a whole):

$$CaSiO_3 + CO_2 \rightarrow CaCO_3 + SiO_2$$

In present-day oceans, the carbonate minerals that accumulate on the seafloor are precipitated predominantly as skeletons and shells. Clams, corals, and many other organisms form beautiful and functional shells of $CaCO_3$ (**Fig. 25.7**). Silica ($SiO_2$) also precipitates from the ocean water largely as skeletons, especially as tiny but exquisite shells fashioned by single-celled algae called diatoms (**Fig. 25.8**). Skeleton formation is an example of **biomineralization**, the precipitation of minerals by organisms. Biomineralization provides yet another link between Earth and life, and one that has varied through evolutionary history.

The annual carbon fluxes associated with geologic processes are small relative to the yearly inputs and outputs from photosynthesis and respiration. As we see in Fig. 25.6, however, photosynthesis and respiration more or less cancel each other out. In the long-term carbon cycle, it is the net result of these rapid biological processes—photosynthesis input minus respiratory output—that matters. Is $CO_2$ produced or consumed when all biological processes are considered together?

Today, the amount of carbon incorporated into organic matter by photosynthesis is actually a bit larger than the amount returned to the atmosphere by respiration. The difference amounts to about 0.05 billion metric tons per year. This number, you will note, is on the same order as volcanic and weathering fluxes. The "excess" organic carbon produced by photosynthesis is deposited in sediments as they accumulate on the seafloor, sustaining a small but constant leak of material from the short-term biological components of the carbon cycle to long-term geologic reservoirs.

One final set of geological processes completes the long-term carbon cycle. The Earth's crust is constantly in motion, propelled by heat within the underlying mantle. **Plate tectonics** is the name given to this dynamic movement of our planet's outer layer. New crust forms at spreading centers, the places where molten rock ascends from the

FIG. 25.8    Silica ($SiO_2$) skeletons of diatoms captured by an electron microscope. Most $SiO_2$ leaves the ocean as diatom skeletons.

**FIG. 25.9  Geological processes that drive the long-term carbon cycle.** Numbers are estimates, in gigatons, of annual fluxes of carbon.

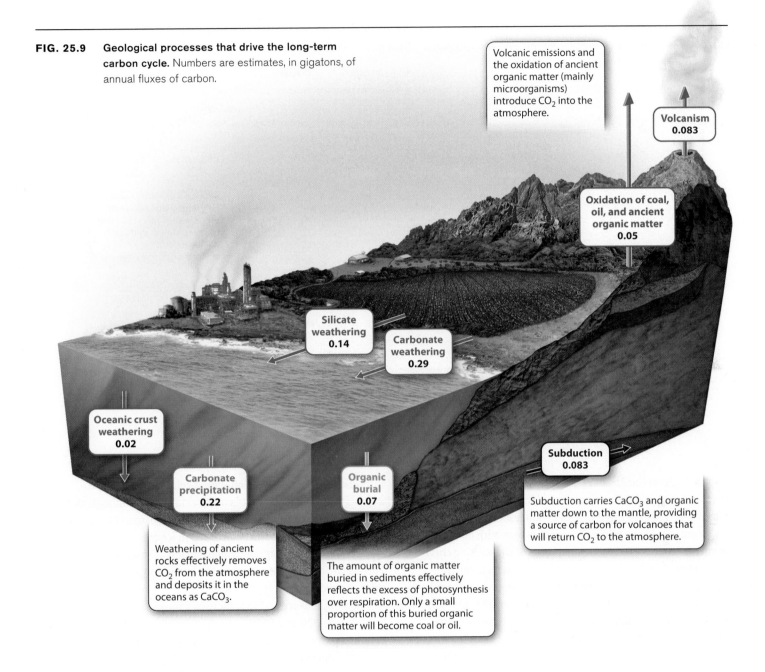

Volcanic emissions and the oxidation of ancient organic matter (mainly microorganisms) introduce $CO_2$ into the atmosphere.

**Volcanism 0.083**

**Oxidation of coal, oil, and ancient organic matter 0.05**

**Silicate weathering 0.14**

**Carbonate weathering 0.29**

**Oceanic crust weathering 0.02**

**Carbonate precipitation 0.22**

**Organic burial 0.07**

**Subduction 0.083**

Subduction carries $CaCO_3$ and organic matter down to the mantle, providing a source of carbon for volcanoes that will return $CO_2$ to the atmosphere.

Weathering of ancient rocks effectively removes $CO_2$ from the atmosphere and deposits it in the oceans as $CaCO_3$.

The amount of organic matter buried in sediments effectively reflects the excess of photosynthesis over respiration. Only a small proportion of this buried organic matter will become coal or oil.

mantle. Old crust is destroyed in subduction zones, where one slab of crust slides beneath another, returning material to the mantle. Plate tectonics provides geology's best explanation for the formation of mountains and ocean basins, and it plays an important role in the long-term carbon cycle. The crust that descends into subduction zones carries with it sediments, including carbonate minerals and sedimentary organic matter. Subduction removes carbon from Earth's surface, but this carbon will be recycled to the surface as carbon dioxide emitted from volcanoes and mid-ocean ridges. **Fig. 25.9** shows the physical processes at work in Earth's long-term carbon cycle.

→  **Quick Check 2**  If plate tectonics form a chain of high mountains, would you expect atmospheric $CO_2$ to increase or decrease?

### Records of atmospheric composition over 400,000 years show periodic shifts in $CO_2$ content.

Earlier in this chapter, we discussed evidence that before the Industrial Revolution, atmospheric $CO_2$ levels had not changed appreciably for 1000 years or more. Longer-term records, however, show that the carbon dioxide levels in air can change substantially through time.

**FIG. 25.10 Atmospheric CO₂ content for the past 400,000 years.** These measurements were recorded from air bubbles trapped in glacial ice at Vostok, Antarctica.

For the past 400,000 years, temperatures have followed a pattern similar to that of CO₂ levels.

Cold glacial and warm interglacial periods correlate with minimum and maximum atmospheric CO₂ levels, respectively.

At Vostok, high on the Antarctic ice sheet, glacial ice records more than 400,000 years of environmental history (**Fig. 25.10**). As shown in the top graph of Fig. 25.10, the youngest samples show about 285 ppm CO₂ in the atmosphere, consistent with direct measurements of air, including the first years of the Keeling curve. Notice, however, that 20,000 years ago, CO₂ levels were much lower—about 180 ppm. In fact, the Vostok ice core in its entirety shows that atmospheric CO₂ has oscillated between 285 ppm and 180 ppm for at least 400,000 years. On long timescales, therefore, the natural variations in the carbon cycle can be large.

The bottom graph in Fig. 25.10 shows an estimate of surface temperature obtained by chemical analysis of oxygen isotopes in ice from the same glacier. Over the last 400,000 years, Antarctic temperature has oscillated between peaks of a few degrees warmer than today's temperature and temperatures as much as 6° to 8° C colder than present. Interestingly, the temperature and carbon dioxide curves closely parallel each other. Carbon dioxide is known to be an effective **greenhouse gas,** meaning that it allows incoming solar radiation to reach Earth's surface but traps heat that is reemitted from land and sea. Therefore, it is not surprising that temperature and atmospheric CO₂ levels show the parallel history documented in the figure.

The two curves correlate closely with one further phenomenon, the periodic growth and decay of continental ice sheets. Large glaciers expanded in the Northern and Southern hemispheres a few million years ago, ushering in an ice age. Today, we live in an interglacial interval, when climate is relatively mild, but 20,000 years ago thick sheets of ice extended far enough away from the poles to cover the present site of Boston (**Fig. 25.11**). The repeated climatic shifts recorded in ice cores reflect periodic variations in the amount and distribution of solar radiation on Earth's surface, which are caused by oscillating changes in Earth's orbit around the sun.

The temperature and CO₂ increases recorded by Vostok ice between 20,000 and 10,000 years ago coincide with the last great retreat of continental ice sheets. What processes might explain how atmospheric CO₂ could increase by 100 ppm in just a few thousand years, as glaciers began to retreat? Can the short-term carbon cycle processes of photosynthesis and respiration account for this much carbon? Certainly, the amount of forests on the Earth's surface has varied through the past 500,000 years as ice sheets grew and decayed, but forests expand as glaciers shrink, so changes in forests cannot account for a pattern of *increasing* atmospheric CO₂ with the retreat of glaciers. Volcanism and weathering also fail to account for the observed pattern. There is no evidence that volcanic activity has waxed and waned in a pattern that could explain observed CO₂ variations. And rates of weathering, which remove CO₂ from the atmosphere, should increase as temperature rises, but carbon dioxide levels have actually increased. Something else must be going on.

Scientists continue to debate why atmospheric CO₂ oscillates in parallel with glacial expansion and retreat. Proposed

**FIG. 25.11 Earth's most recent ice age.** Twenty thousand years ago, glacial ice covered much of North America and Europe, as well as mountain ranges such as the Andes and Himalayas. Sea ice expanded markedly in both northern and southern oceans.

mechanisms suggest interactions involving the ocean and its large reservoir of inorganic carbon. For example, it has been hypothesized that during glacial advances, the circulation of carbon-rich deep-ocean waters back to the sea surface slows, causing more inorganic carbon to accumulate in the deep sea. With glacial retreat, the oceans circulate more vigorously, returning $CO_2$ to the surface and then to the atmosphere. Whatever the explanation, the historical record of the past 400,000 years shows that climate can and does change without any input from humans, something we must take into account when considering our climatic future (Chapter 48).

## Variations in atmospheric $CO_2$ over hundreds of millions of years reflect plate tectonics and evolution.

The atmospheric record trapped in glacial ice extends backward less than 1,000,000 years. To estimate atmospheric $CO_2$ levels for earlier intervals of Earth history, we must rely on computer models and measurements of chemical or paleontological features of ancient rocks that are thought to reflect atmospheric $CO_2$ at their time of formation. One substitute for direct measurements of ancient carbon dioxide levels is the anatomy of fossil leaves: Experiments show that stomata, the small pores on leaf surfaces (Chapter 29), decrease in density as atmospheric $CO_2$ levels increase. The chemistry of ancient soils is also thought to reflect atmospheric history. Such indirect observations and analyses come with a large degree of uncertainty, but most Earth scientists accept at least the broad pattern of atmospheric history shown in **Fig. 25.12.**

For the past 30 million years or so, atmospheric $CO_2$ levels have probably not exceeded those recorded today at Mauna Loa (Fig. 25.12). Before that, higher levels of $CO_2$—perhaps 4 to 6 times the 1958 level—characterized the Mesozoic Era (252 to 65 million years ago), the age of dinosaurs. Levels more like today's

levels existed earlier, during much of the late Paleozoic Era (350–252 million years ago), another time of extensive glaciation. And even before that, about 500 million years ago, early in the Paleozoic Era, atmospheric $CO_2$ may have reached values as high as 15 to 20 times the present-day level.

Not even variations in ocean circulation can account for the scale of these variations. On timescales of millions of years, atmospheric $CO_2$ and, hence, climate are determined by changes in the rate of organic carbon burial in sediments, continental weathering of rocks uplifted into mountains, and volcanic gas release. All these processes reflect the action of Earth's great physical engine: plate tectonics. Seafloor formation and destruction together influence rates of volcanic gas release, and mountain formation strongly influences rates of weathering and, therefore, the fluxes of sediments that bury organic carbon beneath the seafloor.

But life plays a role, even here. For example, the large drop in atmospheric $CO_2$ suggested for the mid-Paleozoic Era, 400–350 million years ago, is thought to reflect the evolution of a new player in the carbon cycle: woody plants. The evolution of trees

**FIG. 25.12 Atmospheric $CO_2$ content for the past 542 million years.** Models and indirect estimates from geological measurements suggest that $CO_2$ levels have varied substantially through Earth's history.

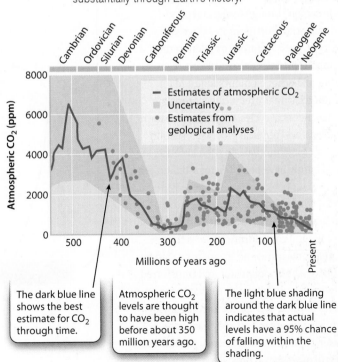

increased the size of the carbon reservoir on land and ushered in an important new mechanism for removing carbon from the air and ultimately to sedimentary rocks. That mechanism was the burial of plant material on land, forming peat and, eventually, coal.

## 25.3 THE CARBON CYCLE, ECOLOGY, AND EVOLUTION

As you sit quietly by a woodland pond, the carbon cycle is at work all around you. In the forest, plants generate carbohydrates as they photosynthesize, and algae and photosynthetic bacteria are doing the same in the pond. These organisms are **primary producers,** generating organic compounds that will provide food for other organisms in the local environment.

An insect grazes on leaves, and ducks feed on both insects and algae. Ducks and insects obtain the carbon they need for growth and reproduction from the foods they eat, and they also gain energy by respiring food molecules. Such organisms are **consumers,** and they include all the animals in both pond and forest. Beneath our feet, similar interactions play out on a microscopic scale, especially among the fungi, protists, and bacteria that recycle the organic carbon in soil back to $CO_2$.

The complementary metabolic processes of photosynthesis and respiration cycle carbon through forest and pond communities. Furthermore, as we will see in the next chapter, complementary metabolic processes also cycle nitrogen, sulfur, and other elements required for life. By continually recycling materials, biogeochemical cycles, which involve both biological and physical processes, sustain life over long intervals. In their absence, life could hardly have persisted for 4 billion years.

### Food webs trace the movement of carbon through communities.

In biological communities across the planet, carbon and energy are packaged by photosynthetic organisms and transferred to heterotrophs (Chapter 6) that graze on plants or photosynthetic microorganisms. Carnivorous organisms may eat these grazers, and they, in turn, may provide food for still other carnivores. Eventually, having

passed from one consumer to another, the carbon originally fixed by photosynthesis is returned to the atmosphere by the respiration of fungi and other **decomposers** that break down dead tissues.

The linear transfer of carbon from one organism to another is called a **food chain.** Because most heterotrophs within a community can consume or be consumed by a number of other species, biologists often prefer to speak of **food webs,** a term that provides a better sense of the complexity of interactions within the carbon cycle (**Fig. 25.13**). Food webs define the interactions among organisms in ponds, forests, and many other habitats. Simply put, the carbon cycle underpins the ecological structure of biological communities (Chapter 47).

**FIG. 25.13   A food web.** Food webs trace the cycling of carbon through communities.

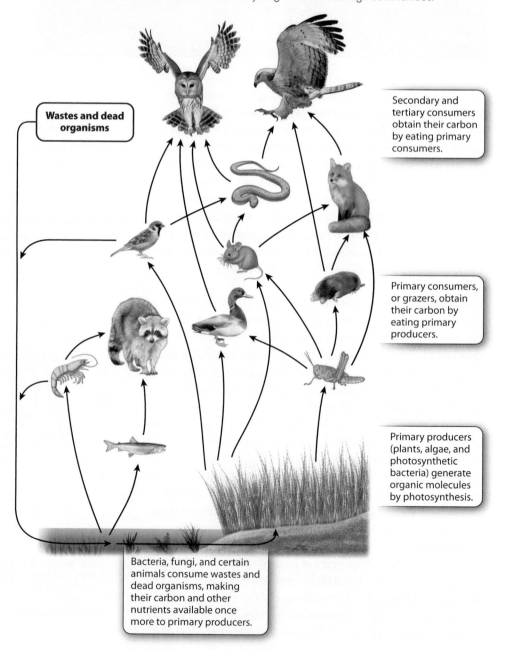

**Wastes and dead organisms**

Secondary and tertiary consumers obtain their carbon by eating primary consumers.

Primary consumers, or grazers, obtain their carbon by eating primary producers.

Primary producers (plants, algae, and photosynthetic bacteria) generate organic molecules by photosynthesis.

Bacteria, fungi, and certain animals consume wastes and dead organisms, making their carbon and other nutrients available once more to primary producers.

## Trophic pyramids trace the flow of energy through communities.

Throughout this chapter and earlier ones, we have emphasized that photosynthesis and respiration transfer energy as they cycle carbon. Carbon is removed from the environment by photosynthetic organisms and passed from one consumer to another, eventually returning to the environment by respiration carried out by consumers and decomposers. At the same time as they remove carbon from the atmosphere, photosynthetic organisms also harvest energy from sunlight and convert it into chemical energy that is stored in the bonds of organic molecules. Consumers store some of the energy in their food as fat, starch, or other organic molecules, but they use much of their food energy to do work or generate heat. The energy used for work or dissipated as heat cannot be captured again by photosynthetic organisms. Whereas carbon cycles through communities, energy transfer is directional.

Within a biological community, the various producers and consumers are commonly grouped into functionally similar sets called trophic groups, from *trophé*, the Greek word for "nourishment." Diagrams showing the amount of energy available at each level to feed the next are called **trophic pyramids** (**Fig. 25.14**). These diagrams have a pyramidal shape because the biomass of primary producers generally much larger than the biomass of primary consumers, and the biomass of primary consumers is, in turn, much greater than that of secondary consumers, and so on. The reason for this is that energy transfer from level to level is inefficient. Because of wastes, work, and heat dissipation, only about 10% to 15% of the energy available in biomass at one level gets incorporated into biomass at the next level. Like food webs, trophic pyramids relate community structure to the fundamental biological processes at work in the carbon cycle.

→ **Quick Check 3** Why are antelopes more abundant than lions in the African savanna?

## The diversity of photosynthetic organisms reflects adaptation to a wide range of environments.

An estimated 500,000 species of photosynthetic organisms fuel the carbon cycle in all but a few deep-sea and subterranean environments. In terms of carbon and energy metabolism, all these species do pretty much the same thing. Why, then, is there such a diversity of photosynthetic organisms? Why don't just a few species dominate the world's photosynthesis?

The example of the pond and forest helps to make the basis of photosynthetic diversity clear. In the forest community, the leaves of several different tree species form a photosynthetic canopy above the forest floor. Below, shrubs grow, making use of light not absorbed by the leaves above them. And below the shrubs are grasses, herbs, ferns, and mosses that can grow in the reduced light levels of the forest floor. In the adjoining pond, a few species of aquatic plants line the water's edge, but beyond that algae and photosynthetic bacteria dominate photosynthesis, with some species anchored to the pond bottom and others floating in the water column.

On the largest scale, then, the reason for photosynthetic diversity is obvious—cyanobacteria, algae, and land plants all perform comparable metabolic functions in different habitats. This kind of specialization for different environments also applies on a finer scale, although it is complicated by grazing and environmental disturbances such as fire and landslides (Chapter 47). In the forests of New England, the tree species that thrive in wet bottomlands differ from those found on well-drained hillsides. And seasonally dry woodlands in southern California support yet another set of plant species. In general, plants of varying size, shape, and physiology inhabit physically and biologically distinct environments, and the same is true of photosynthetic organisms in water.

Thus, the immense diversity of photosynthetic organisms found today reflects not so much evolutionary variations in the biochemistry of photosynthesis (although some of that occurs; see Chapter 29) as structural and physiological adaptations. These

**FIG. 25.14 Trophic pyramids.** Trophic pyramids show the transfer of energy through communities.

| Organism | Trophic levels | Available energy units |
|---|---|---|
| Herring | 4 | 3.4 |
| Carnivorous zooplankton | 3 | 22.5 |
| Herbivorous zooplankton | 2 | 150 |
| Phytoplankton | 1 | 1000 |

Coastal regions
15% efficiency

| Organism | Trophic levels | Available energy units |
|---|---|---|
| Tuna | 5 | 0.1 |
| Carnivorous fish | 4 | 1 |
| Carnivorous zooplankton | 3 | 10 |
| Herbivorous zooplankton | 2 | 100 |
| Phytoplankton | 1 | 1000 |

Open ocean
10% efficiency

Because energy is lost at each step of the pyramid, the amount of biomass present at any one level is only about 10% to 15% of that at the preceding level.

**FIG. 25.15  Microbial communities in hot springs in New Zealand.** This environment and others that few eukaryotes tolerate provide a glimpse of how the carbon cycle worked before algae, amoebas, plants, and animals evolved.

## The carbon cycle provides a framework for understanding life's evolutionary history.

The world did not always support the biological diversity we see today. Indeed, 2 billion years ago, the carbon cycle was anchored by photosynthetic bacteria and microbial heterotrophs because there were no plants or animals (**Fig. 25.15**). Subsequently, algae gained an ecological foothold as nutrients such as nitrate became more widely available, and single-celled eukaryotic heterotrophs expanded their reach by gathering food in ways not possible for bacteria—that is, by ingesting cells and other types of particulate food. With the evolution of multicellularity, animals and seaweeds added further complexity to carbon cycling in the sea. Eventually, some aquatic green algae evolved the capacity to live on land, and animals soon followed, building unprecedented levels of diversity on our planet.

The history of life is one of accumulating variety and complexity, based mostly on the great biological components of the carbon cycle: photosynthesis and respiration. Indeed, the carbon cycle did more than provide a framework for accumulating

adaptations allow the effective gathering of light, nutrients, and—critical to life on land—water, in widely varying local environments. Natural selection, acting on local populations, links the diversity of photosynthetic organisms to the carbon cycle.

## The diversity of respiring organisms reflects many sources of food.

About 10 million species help to cycle carbon through respiration. These include the plants, algae, and bacteria that generate carbohydrates by photosynthesis, as well as legions of animals, fungi, and microorganisms that obtain both carbon and energy from preexisting organic compounds.

Heterotrophic bacteria, amoebas, and humans may use essentially the same biochemical pathway to respire organic molecules, but they differ markedly in how they feed and, therefore, what they can eat. Bacteria (and also fungi) absorb molecules from their environment, but amoebas and many other eukaryotic microorganisms can capture and ingest cells— they are capable of predation. Animals capture prey, as well, but commonly feed on organisms far too large for an amoeba to eat. As photosynthetic organisms have adapted structurally and physiologically to local environments across the globe, heterotrophs have adapted by means of locomotion, mouth and limb specialization, perception, and behavior to obtain their food.

**FIG. 25.16  Oxygen levels in Earth's atmosphere during the history of life, based on analysis of oxygen-sensitive minerals in sedimentary rocks.**

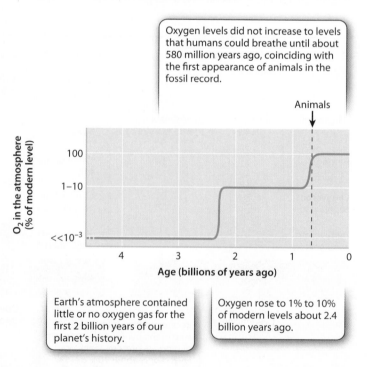

Oxygen levels did not increase to levels that humans could breathe until about 580 million years ago, coinciding with the first appearance of animals in the fossil record.

Earth's atmosphere contained little or no oxygen gas for the first 2 billion years of our planet's history.

Oxygen rose to 1% to 10% of modern levels about 2.4 billion years ago.

diversity; it changed the very nature of Earth surface environments in ways that enabled new types of organism to evolve.

When life began, the atmosphere and oceans contained little or no oxygen gas. How did our present environment come to be? The key lies in the formulas for photosynthesis and respiration, shown in Fig. 25.2. Photosynthesis and respiration not only cycle carbon, they cycle oxygen (and water) as well. When photosynthesis and respiration are in balance, the oxygen generated by photosynthesis is completely consumed by respiration. But when some of the organic carbon generated by photosynthesis is buried in sediments, oxygen production can exceed oxygen consumption. In other words, because

sedimentary organic matter burial can break the tight coupling depicted in Fig. 25.2, it can facilitate an increase in $O_2$ levels that may accumulate in the atmosphere and oceans. The geologic history of oxygen shown in **Fig. 25.16** results from the interactions through time between the carbon (and sulfur) cycle with plate tectonics. Our oxygen-rich world is the result.

In the chapters that follow, we examine life's diversity in terms of function, phylogeny, and ecology, but don't be fooled into thinking that these are discrete topics: They are inextricably intertwined. The remarkable diversity of life on Earth reflects the ecology and evolution of populations that interact to cycle carbon through the biosphere.

## Core Concepts Summary

**25.1 PHOTOSYNTHESIS AND RESPIRATION ARE THE KEY BIOCHEMICAL PATHWAYS FOR THE BIOLOGICAL, OR SHORT-TERM, CARBON CYCLE.**

Keeling began a program to monitor carbon dioxide levels in the atmosphere in 1958. Two patterns emerged: (1) carbon dioxide levels oscillate on a yearly basis, and (2) carbon dioxide levels have increased steadily to the present day since measurements began. page 25-1

The first pattern, seasonal oscillation, can be explained by the imbalance between photosynthesis and respiration in the summer and the winter months. page 25-2

The second pattern, of steady increase, can be extended further back in time by measuring the composition of air trapped in arctic gas bubbles. These measurements indicate that carbon dioxide levels in the atmosphere remained steady for more than a thousand years, but then started to increase in the mid-1800s, the time of the Industrial Revolution. page 25-3

The ratio of different isotopes of carbon in the atmosphere indicates that most of the carbon added to the atmosphere in recent decades comes from human activities, particularly the burning of fossil fuels. page 25-4

**25.2 PHYSICAL PROCESSES GOVERN THE LONG-TERM CARBON CYCLE.**

Physical processes, which act at slower rates and over longer time scales, are the major drivers of the long-term carbon cycle. page 25-6

In following carbon around its cycle, it is important to consider where carbon is stored (reservoirs) and rates at which it is added or removed from these stores (fluxes). page 25-6

Important reservoirs of carbon include organisms, the atmosphere, soils, the oceans, and sedimentary

rocks. Sedimentary rocks are by far the largest carbon reservoir. page 25-7

Physical processes that remove $CO_2$ from the atmosphere include the weathering of rocks, and those that add $CO_2$ to the atmosphere include volcanoes and the oxidation of ancient organic matter in sedimentary rocks. page 25-7

The short- and long-term carbon cycles are linked by the slow leakage of organic matter from biological communities to sediments accumulating on the seafloor. page 25-8

Over the last 400,000 years, $CO_2$ levels in the atmosphere have gone through periodic shifts that reflect repeated cycles of glaciation. page 25-10

$CO_2$ levels in the atmosphere 500 million years ago may have been as much as 15 to 20 times higher than present-day levels. Over timespans of hundreds of millions of years, movements of Earth's plates and the evolution of new life-forms are major players in the carbon cycle. page 25-11

**25.3 THE CARBON CYCLE CAN HELP US UNDERSTAND ECOLOGICAL INTERACTIONS AND THE EVOLUTION OF BIOLOGICAL DIVERSITY.**

Photosynthetic and other autotrophic organisms are primary producers, converting carbon dioxide into organic molecules. page 25-12

Heterotrophic organisms that eat primary producers are called primary consumers. Other heterotrophs that prey on these grazers are called secondary consumers. page 25-12

Interacting networks of species, linked by predator–prey interactions, are called food webs, and they are an important feature of the carbon cycle. page 25-12

Communities contain a much greater biomass of producers than consumers because the movement of energy through communities is inefficient. page 25-13

Photosynthetic organisms are well adapted to obtaining sunlight and nutrients in different environments. page 25-13

Respiring organisms are adapted to obtain different sources and sizes of food. page 25-14

The evolution of life on Earth can be considered a long history during which organisms have evolved different ways of obtaining energy and carbon from their environment. page 25-14

The interactions between biological and geologic processes at work in the carbon cycle have not only produced our current biological diversity. They are also responsible for Earth's oxygen-rich atmosphere and oceans. page 25-15

## Self-Assessment

1. Name two processes that drive the short-term carbon cycle.

2. Draw and explain the curve representing changing atmospheric levels of $CO_2$ over the course of a year.

3. Draw and explain the curve representing changing atmospheric levels of $CO_2$ over the past 150 years.

4. Explain one way by which the source of the increase in $CO_2$ in the atmosphere can be determined.

5. Name processes that influence the long-term carbon cycle.

6. Draw and explain the curve representing changing atmospheric levels of $CO_2$ over the last 400,000 years.

7. Trace the flow of carbon and energy through biological communities.

Do you understand the chapter's Core Concepts? Log in to check your answers to the Self-Assessment questions, then practice what you've learned and reinforce this chapter's concepts by working through the problems and multimedia tutorials provided there.

🖳 **www.biologyhowlifeworks.com**

# The Human Microbiome

## DIVERSITY WITHIN

*Clostridium difficile* doesn't grab many headlines. Perhaps it should. This bacterium is thought to be the most common cause of hospital-acquired infections in the United States. It sickens about 165,000 U.S. hospital patients each year, causing fevers, chills, diarrhea, and severe intestinal pain—and adds more than $1 billion annually to the nation's health care costs. Altogether, the infection kills about 28,000 people every year, according to the Centers for Disease Control and Prevention.

Now a promising (if stomach-turning) new treatment is gaining traction: fecal transplants. The vast majority of *C. difficile* cases occur after patients have taken antibiotics for other infections. The drugs kill the beneficial bacteria normally found in a healthy gut, allowing *C. difficile* to proliferate. Now some pioneering researchers have begun to replace the missing microbes by transplanting fecal material from healthy donors into patients sick with *C. difficile*. Some doctors have reported that the procedure alleviates symptoms in more than 80% of patients.

While transplanting fecal bacteria may not yet be mainstream medicine, there is no question that our gut, as well as our skin, mouth, nose, and urogenital tract, is home to a diversity of bacterial species. No one has yet identified the many microbes living on and within our bodies, but estimates suggest that each of us contains 10 times as many bacterial cells as human cells. The relationship between "us and them" is a type of symbiosis, in which both humans and bacteria benefit from the close association. Indeed, the microbiome, as our assemblage of bacterial houseguests is known, affects human health in many important ways.

Some of the most mysterious, and most intriguing, members of our microbiome live in our gut. Bacteria living in our intestines help us break down food and synthesize vitamins. On the other hand, abnormal communities of microbes in the gut have been linked to Crohn's disease, colon cancer, and nonalcoholic fatty liver disease.

Your own personal gut flora is a unique collection of species—even identical twins can have surprisingly different microbiomes. This variability has made it difficult for scientists to decipher exactly what role individual species play in health and in disease. The sheer number of bacteria in your gut only compounds the problem. One international team of scientists recently reported that each human gut contains at least 160 different bacterial species.

Despite these challenges, researchers have made strides toward understanding the unique bacterial communities within our intestinal tract. In 2011, scientists compared all the DNA sequences from gut microbes of people living in Europe, America, and Japan. They found that each individual fell into one of three enterotypes, or "gut types." Most people had a bacterial mix dominated by organisms from the genus *Ruminococcus*. But others had gut types dominated by the genus *Bacteroides* or the genus *Prevotella*. The three different gut compositions appear to produce different vitamins and may even make a person more, or less, susceptible to certain diseases.

> No one has yet identified the many microbes living on and within our bodies, but estimates suggest that each of us contains 10 times as many bacterial cells as human cells.

Surprisingly, the researchers reported that enterotype was unrelated to gender, body mass index, or nationality. So what determines your gut type? A follow-up study suggested that enterotypes were associated with long-term dietary patterns. People who eat a diet high in animal protein and saturated fats are more likely to contain a gut community dominated by *Bacteroides*. People who stick to

**The human microbiome.** Each part of the body has different groups of bacteria.

Hair

Nostril

Oral cavity

Esophagus

Skin

Stomach
*H. pylori* (+)

Stomach
*H. pylori* (−)

Colon

Urogenital
tract

Actinobacteria
Firmicutes
Proteobacteria
Bacteroidetes
Cyanobacteria
Fusobacteria

a high-carbohydrate diet rich in plant-based nutrition are more likely to host an abundance of *Prevotella*.

The researchers don't know yet what a person's enterotype means for his or her health. But scientists are eager to explore the link between microbiome and physical well-being. Meanwhile, various researchers are exploring the microbiome of other organisms—specifically, cows.

As it happens, there are good reasons to peer into a cow's gut. The cow's rumen (one of the four stomach chambers) contains microbes that break down the cellulose in plant material. Without them, cattle wouldn't be able to extract adequate nutrition from the grass

they graze. Those microbes are of particular interest to scientists at the Department of Energy (DOE), who are interested in converting material containing cellulose, such as switchgrass, into biofuels.

As researchers look for more sustainable replacements for fossil fuels, many have pinned their hopes on biofuels. But the enzymes currently used to break down cellulose aren't efficient or cost effective enough to produce biofuels on an industrial scale. That's where the cows come in. DOE scientists recently analyzed the microbial genes present in the cow rumen. They discovered nearly 28,000 genes for proteins that metabolize carbohydrates, and at least 50 new proteins that may help break down cellulose in ways that will produce biofuels more efficiently and more cheaply.

Mammals such as cows and humans are hardly the only organisms that have symbiotic relationships with bacteria. All living animals have their own microbiomes. Even tiny single-celled eukaryotes harbor still-tinier bacterial guests.

Symbiotic bacteria provide a variety of benefits to their hosts. Consider the bobtail squid, *Euprymna scolopes*, which possesses a specialized organ to house bacteria. The bacteria, *Vibrio fischeri*, are luminescent, that is, capable of generating light. The squid use their *Vibrio*-filled light organs like spotlights, projecting light downward to match light from the surface as a way to camouflage themselves from would-be predators below.

In many cases, animals and their bacterial symbionts are so intimately connected that it's hard to separate one from the other. Many human gut bacteria are poorly studied because scientists haven't been able to grow them outside the body. But some species have taken their intimate bacterial partnership a step further.

Aphids are small insects that feed on plant sap. They rely on the bacteria *Buchnera aphidicola* to synthesize certain amino acids for them that aren't available from their plant-based diet. The partnership between aphids and *Buchnera* dates back as far as 250 million to 150 million years. Over time, the organisms have become completely dependent on one another for survival. The *Buchnera* bacteria actually live inside the aphid's cells. In some ways, the bacteria resemble organelles rather than independent organisms. Intracellular symbionts may not be so unusual. In fact, mitochondria and chloroplasts—key organelles of eukaryotic cells—are thought to have originated from symbiotic bacteria. Over time, the progenitor bacteria became so intertwined with their host cells that they developed into an integral part of the cells themselves. Studying simple sapsucking aphids, it turns out, may offer insights into the evolutionary history of all eukaryotic cells.

Clearly, there are many good reasons to learn more about our bacterial comrades. From evolution to energy to human health, our microbiomes hold great promise—and great mystery. We've evolved hand in hand with these bacteria for millions of years. Let's hope it takes less time to uncover their secrets.

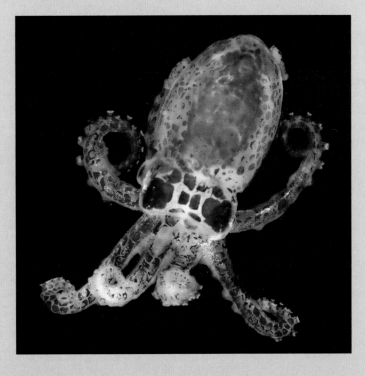

**The bobtail squid and its luminescent bacteria.** This squid harbors bacteria that emit light in specialized organs.

## ❓ CASE 5 QUESTIONS

*Answers to Case 5 questions can be found in Chapters 26 and 27.*

1. How do intestinal bacteria influence human health? *See page 26-22.*
2. What role did symbiosis play in the origin of chloroplasts? *See page 27-4.*
3. What role did symbiosis play in the origin of mitochondria? *See page 27-6.*
4. How did the eukaryotic cell originate? *See page 27-7.*

# BACTERIA AND ARCHAEA

## Core Concepts

In Chapter 25, we outlined the carbon cycle as it operates in the oxygen-rich environments of our daily existence. But can carbon cycle through the deep waters of the Black Sea, within black muds beneath swamps and marshlands, or in other habitats where oxygen is limited or absent? Can biological communities survive without oxygen? The answer, emphatically, is yes. Indeed, life existed on Earth for more than a billion years before oxygen-rich habitats first appeared on our planet.

Exploration of seemingly alien oxygen-poor environments on today's Earth reveals further aspects of the carbon cycle. These habitats are populated by organisms that neither produce nor consume oxygen, yet are able to cycle carbon. The organisms responsible for this expanded cycle share one fundamental feature: They have prokaryotic cell organization. Prokaryotes, which include bacteria and archaeons, lack a nucleus (Chapter 5). We'll look at the carbon cycle again later in this chapter, but first we discuss **Bacteria** and **Archaea,** the prokaryotic domains of microscopic organisms so critical to the cycling of carbon and other elements essential to life. We examine their diversity and how they have evolved through time.

## 26.1 TWO PROKARYOTIC DOMAINS

Chapter 5 outlined the two distinct ways that cells are organized internally. Eukaryotic cells, like those that make up our bodies, have a membrane-bound nucleus and organelles that form separate compartments for many cell functions. Prokaryotic cells have a simpler organization. No membrane surrounds the prokaryotic cell's DNA, and there is little in the way of cell compartments. Prokaryotic cell organization is an ancestral character for life as a whole—that is, it is a feature that was present in the last common ancestor of all organisms alive today. The group defined traditionally as the prokaryotes, however, is paraphyletic in that it excludes some descendants of the last common ancestor of all living things, namely eukaryotic organisms (Chapter 23).

Bacteria and Archaea are the two great domains characterized by prokaryotic cell structure. These organisms are present almost everywhere on Earth's surface. What they lack in complexity of cell structure, these tiny cells more than make up for in their dazzling metabolic diversity. As will become clear over the course of this chapter, Bacteria and Archaea underpin the efficient operation of ecosystems on our planet.

### The bacterial cell is small but powerful.

Because of their small size and deceptively simple cell organization, bacteria were long dismissed as primitive organisms, distinguished mostly by the eukaryotic features they lack: They have no membrane-bounded nuclei, no energy-producing organelles, no sex. This point of view turns out to be more than a little misleading. Bacteria are the diverse and remarkably successful products of nearly 4 billion years of evolution. On present-day Earth, bacterial cells outnumber eukaryotic cells by several orders of magnitude.

**FIG. 26.1    A bacterial cell.** The cells of Bacteria (and Archaea) do not have a membrane-bound nucleus or other organelles.

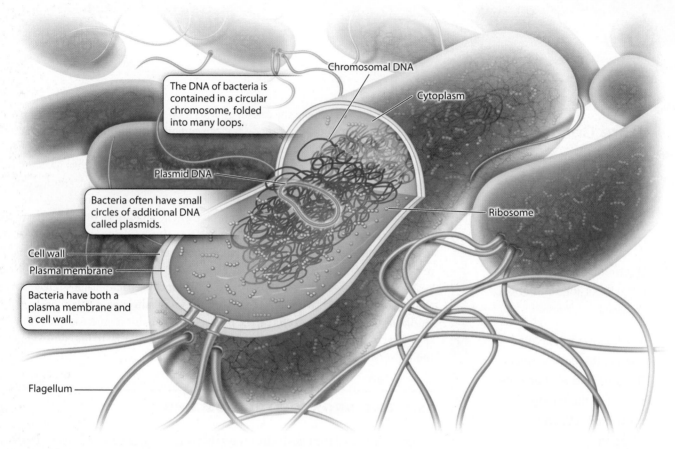

Chromosomal DNA

The DNA of bacteria is contained in a circular chromosome, folded into many loops.

Cytoplasm

Plasmid DNA

Bacteria often have small circles of additional DNA called plasmids.

Cell wall

Plasma membrane

Ribosome

Bacteria have both a plasma membrane and a cell wall.

Flagellum

**Fig. 26.1** illustrates the bacterial cell, which was briefly introduced in Chapter 5. The cell's DNA is present in a single circular chromosome, unlike the multiple linear chromosomes characteristic of eukaryotic cells. Many bacteria carry additional DNA in the form of **plasmids,** small circles of DNA that replicate independently of the cell's circular chromosome. In general, plasmid DNA is not essential for the cell's survival, but it may contain genes that have adaptive value under specific environmental conditions. No nuclear membrane separates DNA from the surrounding cytoplasm, and so transcribed mRNA is immediately translated into proteins by ribosomes.

Bacteria lack the membrane-bounded organelles found in eukaryotic cells. Instead, cell processes such as metabolism are carried out by proteins that float freely in the cytoplasm or are embedded in the plasma membrane. A few bacteria, notably the photosynthetic bacteria, contain internal membranes. The light reactions of photosynthetic bacteria take place in association with membranes distributed within the cytoplasm.

Structural support is provided by a **cell wall** made of **peptidoglycan,** a complex polymer of sugars and amino acids (Chapter 10). Some bacteria have thick walls made up of multiple peptidoglycan layers, while others have thin walls surrounded by an outer layer of lipids. For many years, it was believed that

bacteria lacked the cytoskeletal framework that organizes cytoplasm in eukaryotic cells. However, careful studies now show that bacteria do have an internal scaffolding of proteins that plays an important role in determining the shape, polarity, and other spatial properties of bacterial cells.

### Diffusion limits cell size in bacteria.

Most bacterial cells are tiny: The smallest are only 200–300 nanometers (nm) in diameter, and relatively few are more than 1–2 micrometers (μm) long. Why are bacteria so small? The answer has to do with **diffusion,** the random motion of molecules, a critical process introduced in Chapter 5. When there is a concentration difference, diffusion results in net movement of molecules from a region of higher concentration to a region of lower concentration. If you could watch the movement of any particular molecule, you would see it is random, sometimes moving in one direction and sometimes in another. On average, however, more molecules move from a region with a higher concentration of the molecule to a region with a lower concentration of the molecule than in the other direction. Net movement stops only when the two regions achieve equal concentration, but diffusion continues. Photosynthetic bacteria gain the carbon dioxide they need by the diffusion of $CO_2$ from

**FIG. 26.2** Cell shape and size in Bacteria and Archaea.

**a.** *Streptococcus*, strings of spheroidal or coccoidal bacteria

**b.** *E. coli*, bacterial rods

**c.** *Haloquadratum walsbyi*, a square archaeon that lives in salt ponds

**d.** *Streptomyces*, helical bacteria that produce antibiotics

**e.** A myxobacterium, a bacterium in which cells aggregate to form fruiting bodies

small cell's interior. The surface area of a spherical cell—the area available for taking up molecules from the environment—increases as the square of the radius. However, the cell's volume—the amount of cytoplasm that is supported by diffusion—increases as the cube of the radius. Therefore, as cell size increases, it becomes harder to supply the cell with the materials needed for growth. For this reason, most bacterial cells are small spheres, rods, spirals, or filaments that facilitate diffusion of molecules into cell interiors (**Fig. 26.2**).

A few exceptional bacteria exceed 100 μm in maximum dimension. For example, *Thiomargarita namibiensis*, a bacterium that lives in oxygen-poor sediments off the coast of southwestern Africa, has a total volume about 100,000,000 times larger than that of *Escherichia coli* (**Fig. 26.3**). But, in one sense, *T. namibiensis* cheats: 98% of its volume is taken up by a large vacuole, so the metabolically active cytoplasm is restricted to a thin film around the cell's periphery. Thus, the distance through which nutrients move by diffusion is only a few miocrometers, as in many other bacteria.

Some bacteria are multicellular, forming simple filaments or sheets of cells. More unusual are myxobacteria, which aggregate to form multicellular reproductive structures that are composed of several distinct cell types (see Fig. 26.2e).

**FIG. 26.3** *Thiomargarita namibiensis*, **the largest known bacterial cell.** The volume of the cell is mostly taken up by a vacuole, so the active cytoplasm is only a few micrometers thick.

the environment into the cell, and that is also how respiring bacteria take in small organic molecules and oxygen.

Diffusion explains why bacterial cells tend to be small. A small cell has more surface area in proportion to its volume, and so the interior parts of a small cell are closer to the surrounding environment than those of a larger cell. As a consequence, slowly diffusing molecules do not have to travel far to reach every part of a

## Horizontal gene transfer promotes genetic diversity in bacteria.

Bacterial genomes are generally smaller than those of eukaryotes, in part because bacteria lack the large stretches of noncoding DNA characteristic of eukaryotic chromosomes (Chapter 13). The streamlining of the bacterial genome confers certain benefits. For example, bacteria can reproduce rapidly when nutrients required for growth are available in the local environment. Bacteria replicate their DNA from only one or a small number of initiation sites, so genome size can influence rate of reproduction.

Bacteria do not undergo meiotic cell division and cell fusion. That is, they lack the sexual processes characteristic of eukaryotic organisms (Chapter 11). Despite this, bacterial populations display remarkable genetic diversity. For example, different strains of *Pseudomonas aeruginosa*, a common disease-causing organism, may differ in genome size by nearly a factor of 2 (3.7 million base pairs compared to 7.1 million base pairs), yet these strains are quite similar in function.

How does this variation arise? In eukaryotic organisms, genes generally pass from parent to offspring. Bacteria also inherit most of their genes from parental cells. However, bacterial cells can also obtain new genes from distant relatives. Called **horizontal gene transfer,** this process is a major source of genetic diversity in bacteria.

How does DNA move from one bacterial cell into another independently of cell division? Some bacteria synthesize thin strands of membrane-bound cytoplasm called pili that connect them to other cells (**Fig. 26.4a**). A pilus provides a migration route for the direct cell-to-cell transfer of DNA. This process, called **conjugation,** commonly transfers plasmids from one cell to another, spreading novel genes throughout a population. Genes that confer resistance to antibiotics are a well-studied example of horizontal gene transfer by conjugation. Biologists interested in genetic engineering use conjugation in the laboratory to introduce genes into cells. But conjugation is just the beginning of genetic exchange in bacteria.

Genes can be transferred from one cell to another without any direct bridge between

**FIG. 26.4    Horizontal gene transfer in bacteria.** DNA shown in red originates from the donor cell. DNA shown in blue is that of the recipient cell.

**a.** DNA transfer by conjugation

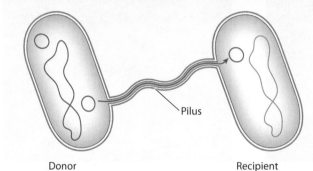

In conjugation, DNA (usually a plasmid) from a donor cell is transferred through a pilus into the recipient cell.

Pilus

Donor          Recipient

**b.** DNA transfer by transformation

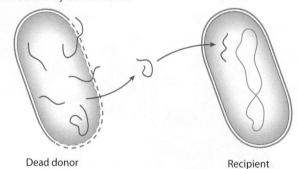

In transformation, DNA released into the environment by dead cells is taken up by a recipient cell.

Dead donor          Recipient

**c.** DNA transfer by transduction

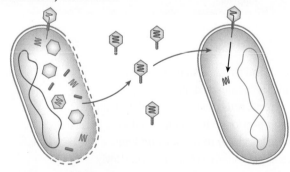

In transduction, DNA is transferred from a donor to a recipient cell by a virus.

Virus-infected donor          Recipient

the cells. DNA released to the environment by cell breakdown can be taken up by other cells, a process called **transformation** (**Fig. 26.4b**). Transformation was revealed when experiments showed that harmless strains of the bacteria causing pneumonia could be transformed into virulent strains by exposure to media containing dead cells of disease-causing strains (Chapter 3). Scientists reasoned that the transformation occurred because some substance was being taken up by the living bacteria. The "transforming substance" was later shown to be DNA.

Viruses provide a third mechanism of horizontal gene transfer. Recall from Chapter 19 that viruses in bacterial cells sometimes integrate their DNA into the host bacterial DNA and persist within the cells as they grow and divide. Before leaving the cell to infect others, the viral DNA removes itself from the bacterial genome. This excision is not always precise, and sometimes additional genetic material from the bacterial host is incorporated into the virus. Viruses released from their host cell go on to infect others, bringing host-derived genes with them. Horizontal gene transfer by means of viruses is called **transduction** (**Fig. 26.4c**). It is common in nature and is also widely used in the laboratory to to introduce novel genes into bacteria for medical research.

Horizontal gene transfer allows bacterial cells to gain beneficial genes from organisms distributed throughout the bacterial domain and beyond. Indeed, bacteria everywhere are constantly reshaping their genomes. Bacteria, therefore, are not the poor cousins of eukaryotes, unable to generate genetic diversity by sexual recombination. Instead, bacteria have highly efficient mechanisms for adding and subtracting genes that permit them to evolve and adapt rapidly to local conditions. Perhaps the most widely discussed and worrisome manifestation of horizontal gene transfer is the rapid spread of antibiotic resistance among bacteria (Chapter 48).

→ **Quick Check 1** In eukaryotes, sexual reproduction is the main process by which new gene combinations are generated. How do bacteria generate new gene combinations in the absence of sexual reproduction?

## The Archaea form a second prokaryotic domain.

For many years, all prokaryotic cells were classified as bacteria. However, in 1977, George Fox and Carl Woese published a revolutionary hypothesis. From comparisons of RNA molecules from the small subunits of ribosomes (and later, comparisons of the genes for these RNAs), Fox

**FIG. 26.5** **A three-branched tree of life, based on sequence comparisons of the genes for the small subunit of ribosomal RNA.** Archaea and Bacteria form distinct branches, although they inherited prokaryotic cell organization from a common ancestor.

and Woese argued that prokaryotic organisms actually fall into two distinct groups, as different from each other as either is from eukaryotes. If this is true, the tree of life must have three great branches: the Bacteria, the Eukarya, and the limb recognized by Fox and Woese, called the Archaea (**Fig. 26.5**). Research since the late 1970s confirms this hypothesis. Like bacterial cells, archaeal cells are prokaryotic—they have no membrane-bound nucleus and their genes are arrayed along a single circular chromosome. Moreover, cell size in Archaea is limited by diffusion, and genetic diversity is promoted by horizontal gene transfer, much as in Bacteria.

In detail, however, the Archaea are quite distinctive (**Table 26.1**). Their membranes are made from lipids different from the fatty acids found in bacterial and eukaryotic membranes.

**TABLE 26.1** Principal Differences among Archaea, Bacteria, and Eukarya

| CHARACTERISTIC | ARCHAEA | BACTERIA | EUKARYA |
|---|---|---|---|
| Cell contains a nucleus and other membrane-bound organelles | No | No | Yes |
| DNA occurs in a circular form* | Yes | Yes | No |
| Ribosome size | 70S | 70S | 80S |
| Membrane lipids ester-linked† | No | Yes | Yes |
| Photosynthesis with chlorophyll | No | Yes | Yes |
| Capable of growth at temperatures greater than 80°C | Yes | Yes | No |
| Histone proteins present in cell | Yes | No | Yes |
| Operons present in DNA | Yes | Yes | No |
| Introns present in most genes | No | No | Yes |
| Capable of methanogenesis | Yes | No | No |
| Sensitive to the antibiotics chloramphenicol, kanamycin, and streptomycin | No | Yes | No |
| Capable of nitrogen fixation | Yes | Yes | No |
| Capable of chemoautotrophy | Yes | Yes | No |

*Eukaryote DNA is linear. †Archaea membrane lipids are ether-linked.

Archaea also show a diversity of molecules in their cell walls, but none has the peptidoglycan characteristic of Bacteria or the cellulose and chitin found in most eukaryotic cell walls.

Archaea differ from Bacteria in another intriguing way. DNA transcription in archaeons employs RNA polymerase and ribosomes more similar to those of eukaryotes than to bacteria (Chapter 3). Moreover, many of the antibiotics that target protein synthesis in bacteria are ineffective against archaeons, suggesting fundamental differences in translation as well.

Many of the microorganisms first identified as Archaea have unusual physiological properties or inhabit extreme environments. For example, some archaeons live in acid mine water at pH 1 (or less!). Others live in water salty enough to precipitate NaCl (**Fig. 26.6**), or in deep-sea hydrothermal vents where temperatures can exceed 100°C. We now know, however, that many archaeons live under less extreme conditions in soils, lakes, and the sea. Indeed, they may be the most abundant organisms throughout much of the ocean.

**FIG. 26.6  Archaeons in San Francisco Bay.** The ponds are flooded and then evaporated to precipitate table salt. Salt-loving archaeons, made conspicuous by their purple pigment, thrive here.

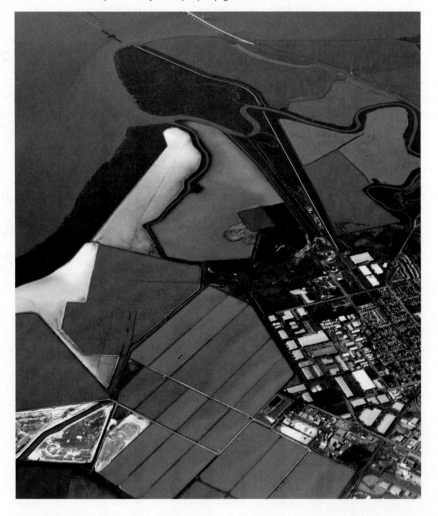

→  **Quick Check 2**  Why were Archaea originally thought to be simply unusual forms of Bacteria? What lines of evidence showed this domain to form a distinct branch on the tree of life?

## 26.2 AN EXPANDED CARBON CYCLE

In our discussion of energy metabolism in Chapters 7 and 8 and of the carbon cycle in Chapter 25, we emphasized oxidation–reduction (redox) chemistry. In redox reactions, a pair of molecules reacts, one molecule becoming more oxidized and the other becoming more reduced. If we follow the transfer of electrons in a redox reaction, we see that the molecule that is reduced gains electrons, whereas the molecule that is oxidized loses electrons. Reduction therefore requires a source of electrons, or an electron donor, and oxidation requires a sink for electrons, or an electron acceptor.

In photosynthesis, as we discussed in Chapter 8, carbon dioxide ($CO_2$) is reduced to carbohydrates, and water is oxidized to oxygen gas ($O_2$). Water is the electron donor needed to reduce $CO_2$, generating $O_2$ as a by-product. In respiration, as described in Chapter 7, organic molecules are oxidized to $CO_2$, and $O_2$ is reduced to water. In this case, $O_2$ serves as the electron acceptor needed to oxidize organic molecules.

In all known photosynthetic eukaryotes, the photosynthetic reaction is **oxygenic,** or oxygen producing. There is a good reason for this. Water occurs nearly everywhere and, in the oxygen-rich environments where most eukaryotic organisms thrive, no other molecule is available to donate electrons. Similarly, essentially all respiration in eukaryotic cells is **aerobic,** or oxygen utilizing. Again, this makes good chemical sense because oxidation of carbohydrates using oxygen generates far more energy than can be obtained using other oxidants. Nonetheless, where oxygen is limited or absent, other electron donors are available for photosynthesis, and other electron acceptors can be used for respiration. The organisms that can use these alternative electron donors and electron acceptors are Bacteria and Archaea.

We've already noted that Bacteria and Archaea differ from Eukarya in cell structure, biochemistry, and genome organization, but it is the distinctive ways in which Bacteria and Archaea gather carbon and harvest energy that make them indispensable parts of nature. Prokaryotic organisms are thus the real foundation of the carbon cycle, capable of building and dismantling organic molecules throughout the full breadth of habitats on Earth.

**FIG. 26.7** **Microbial mats.** Mats, like the example in (a) from Western Australia, cover tidal flats and lagoons where salinity or other environmental factors inhibit animals and seaweeds. As shown by the example from Cape Cod in (b), microbial mats have sharp vertical gradients of light, oxygen, and other chemical compounds, and so support nearly all the energy metabolisms found in the expanded carbon cycle (c).

a.

b.

The blue-green pigment identifies cyanobacteria, the oxygenic photosynthetic bacteria that dominate the well-lit, oxygen-rich mat surface.

The purple pigment marks purple bacteria, anoxygenic photosynthetic bacteria that live where there is light but no oxygen gas.

In the black subsurface layer, anaerobic respiration and fermentation support microbial populations.

c.

### Many photosynthetic bacteria do not produce oxygen.

In saline bays along the west coast of Australia and many other low-lying tropical coastlines, carpets of deep blue-green cover tidal flats along restricted lagoons (**Fig. 26.7a**). Called microbial mats, these carpets are densely packed communities of (mostly) bacteria and archaeons that thrive where animals and seaweeds cannot grow. The top surface of a microbial mat is dominated by cyanobacteria, bacteria that photosynthesize like plants and algae, using water as an electron donor and releasing oxygen gas (**Fig. 26.7b**). It is the photosynthetic pigments in the cyanobacteria that impart the blue-green color to microbial mats. The cyanobacteria (which synthesize their own organic molecules) and other, heterotrophic bacteria (which get their carbon from preformed organic molecules) respire aerobically, using $O_2$ to oxidize organic molecules to $CO_2$. Thus, at the mat surface, prokaryotic microorganisms carry out a biological carbon cycle much like that for plants and animals described in Chapter 25.

What happens beneath the mat surface is a different story. Mats generate a great deal of organic matter, and aerobic respiration of this material consumes $O_2$ rapidly. In the mat interior below the level at which cyanobacteria supply oxygen during the day, $O_2$ is supplied only slowly by diffusion, with the result that $O_2$ becomes depleted within a few millimeters of the mat surface. Light is also greatly reduced beneath the mat surface, but in many mats, light actually penetrates a few millimeters deeper than oxygen (**Fig. 26.7c**). What organisms live where light is present and oxygen absent?

If we label $CO_2$ with the radioisotope [14]C, we find that carbon dioxide continues to be reduced to carbohydrates within this subsurface zone of the mat. In fact, the newly generated carbohydrates, identified by the [14]C they have incorporated, occur in cells with bright purple or green pigments (**Fig. 26.7b**). These are **anoxygenic** photosynthetic bacteria, photosynthetic microorganisms that harvest light energy to drive the synthesis of carbohydrates, but which do not gain electrons from water and so do not generate oxygen gas in the process.

Anoxygenic photosynthetic bacteria absorb sunlight using **bacteriochlorophyll,** a light-harvesting pigment closely related to the chlorophyll found in plants, algae, and cyanobacteria. **Fig. 26.8a** shows the chemical structure of one form of bacteriochlorophyll, called bacteriochlorophyll $a$, and one form of chlorophyll, called chlorophyll $a$. Note the similarity in their overall structure. The two molecules absorb light of similar wavelengths as well, though small differences in their chemical structures account for small differences in which wavelengths they absorb more strongly (**Fig. 26.8b**).

Anoxygenic photosynthesis differs from oxygenic photosynthesis in another important way: It employs only a single photosystem. As discussed in Chapter 8, the amount of energy that can be captured this way is not sufficient to pull electrons from (that is, to oxidize) water and then use those electrons to reduce $CO_2$, and so oxygenic photosynthesis has two photosystems in series. With only a single

**FIG. 26.8** **Structures and absorption spectra of bacteriochlorophyll *a* and chlorophyll *a*.** Note that bacteria absorb wavelengths of light that are outside the visible spectrum.

a.

Bacteriochlorophyll *a*

Chlorophyll *a*

The differences in structure (shown in red) account for differences in wavelengths absorbed by the two molecules.

b.

acceptor for the oxidation of those molecules. Does heterotrophy simply cease below the level where $O_2$ disappears?

Not at all. Careful measurements show that organic molecules can be oxidized in the absence of $O_2$. Oxygen is not unique in its capacity to accept electrons in redox reactions involving organic molecules, but where $O_2$ is present it is used preferentially for respiration. This is because more energy is released by the oxidation of organic compounds by oxygen gas than by any alternative oxidant. In habitats where $O_2$ is absent, oxidized forms of nitrogen ($NO_3^-$), sulfur ($SO_4^{2-}$), manganese ($Mn^{4+}$), iron ($Fe^{3+}$), and even arsenic ($AsO_4^{3-}$) are used as electron acceptors in cellular respiration.

**Fermentation** provides an alternative to cellular respiration as a way of extracting energy from organic molecules. Whereas cellular respiration is the full oxidation of carbon compounds to $CO_2$, fermentation is the partial oxidation of carbon compounds to molecules that are less oxidized than $CO_2$ (Chapter 7). Fermentation has one advantage over cellular respiration in that it does not require an external electron acceptor, such as $O_2$. On the other hand, fermentation yields only a modest amount of energy.

Fermentation plays an important role in oxygen-poor environments that are rich in organic matter, such as landfills and the digestive tracts of animals. Here, the breakdown of organic molecules typically requires more than one organism, each able to metabolize different intermediates. These cooperating groups of fermenters enable the breakdown of substances that could not be metabolized by any one organism alone. Groups of fermenters thus form what is essentially an ecological solution to the metabolically difficult problem of gaining energy from complex organic molecules. Fermentation is widespread in Bacteria and Archaea, but except for yeasts, which are world-class fermenters, fermentation is of only minor importance in the energy metabolism of most Eukarya.

Therefore, where $O_2$ is present, many different kinds of organisms participate in the carbon cycle. Photosynthetic plants, algae, and cyanobacteria all transform $CO_2$ into organic molecules, and animals, fungi, single-celled eukaryotes, bacteria, and archaeons return $CO_2$ to the environment by aerobic respiration. In Chapter 25, we noted that this biological carbon cycle leaks a bit. In other words, some of the organic carbon produced by photosynthesis escapes aerobic respiration and accumulates in oxygen-depleted waters or sediments. Only prokaryotic heterotrophs can complete the recycling of organic carbon in oxygen-poor environments. Thus, in carrying out anaerobic respiration or a diversity of fermentation reactions, prokaryotes play a major role in the carbon cycle. The key point is that prokaryotic organisms are required to sustain Earth's carbon cycle, while eukaryotic organisms are optional.

photosystem, bacteria that carry out anoxygenic photosynthesis use electron donors that donate their electrons more easily than water. As a result, these microorganisms do not release $O_2$ as a by-product. Electron donors in anoxygenic photosynthesis include hydrogen sulfide ($H_2S$), hydrogen gas ($H_2$), ferrous iron ($Fe^{2+}$), and even the arsenic compound arsenite ($AsO_3^{3-}$). Where $O_2$ is present, these compounds themselves become oxidized to other molecules and so are not available for photosynthesis. Therefore, microorganisms that conduct anoxygenic photosynthesis are restricted to sunlit habitats where $O_2$ is absent.

### Many bacteria respire without oxygen.

In microbial mat communities, heterotrophic microorganisms in the upper, oxygen-rich portion of the mat respire aerobically, just as we do, gaining carbon by taking in organic molecules from their surroundings and energy by using oxygen as an electron

### Photoheterotrophs obtain energy from light but obtain carbon from preformed organic molecules.

In Chapter 6, we saw that organisms have two sources of energy: the sun (phototrophs) and chemical compounds (chemotrophs). Organisms also have two sources of the carbon needed for growth and other functions: inorganic molecules like $CO_2$ (autotrophs) and organic molecules like glucose (heterotrophs). Taken together, this means that there are four ways that organisms acquire the energy and carbon they need (see Fig. 6.1).

Up to this point, we have considered only two of these ways. Plants, algae, and cyanobacteria are photoautotrophs, gaining energy from the sunlight and carbon from $CO_2$, whereas animals, fungi, and many prokaryotes are chemoheterotrophs, gaining energy and carbon from organic molecules taken up from the environment. Among the Bacteria and Archaea, additional metabolisms are possible. Some microorganisms use the energy from sunlight to make ATP, just as plants do, but rather than reducing $CO_2$ to make their own organic molecules, they rely on organic molecules obtained from the environment as the source of carbon for growth and other vital functions. These organisms are known as **photoheterotrophs.**

Photoheterotrophy can be advantageous in environments rich in dissolved organic compounds. It allows organisms to use all their absorbed light energy to make ATP while directing all absorbed organic molecules toward growth and reproduction. Examples of organisms capable of photoheterotrophy include heliobacteria and most green nonsulfur bacteria.

## Chemoautotrophy is a uniquely prokaryotic metabolism.

On the deep seafloor, animals are relatively uncommon, their abundance limited by the slow descent of organic matter from surface oceans. Locally, however, where hydrothermal springs punctuate the deep seafloor, animal populations can be remarkably dense (**Fig. 26.9**). Why are animals so abundant around hydrothermal vents?

A clue comes from microbial mats. In mats where oxygen penetrates more deeply than light, $CO_2$ continues to get reduced into organic molecules along the base of the zone containing oxygen. What is happening within the mat?

The organisms that incorporate $CO_2$ into organic molecules deep within microbial mats and the organisms that provide food for abundant animals around deep-sea hydrothermal vents employ similar strategies for obtaining energy and carbon. Like plants and other autotrophs, they gain carbon by reducing $CO_2$ to form carbohydrates. However, they obtain the energy to fuel this process not from sunlight but from chemical reactions. Hence, these microorganisms are called **chemoautotrophs.**

Our discussions of respiration emphasized that energy released by the oxidation of organic matter is used to generate ATP (Chapter 7).

Chemoautotrophic microorganisms also use chemical reactions to generate ATP, but they use inorganic molecules present in their local environment. Chemoautotrophic prokaryotes use the oxidation of molecules such as $H_2$, $H_2S$, and even $Fe^{2+}$ to generate the ATP and reducing power required to incorporate $CO_2$ into organic molecules. In chemoautotrophic metabolism, the electron acceptor is most commonly $O_2$ or nitrate ($NO_3^-$).

Because chemoautotrophy requires access to both oxidized and reduced molecules, chemoautotrophs tend to live along the interface between oxygen-rich and oxygen-poor environments. Microbial mats commonly have a sharp boundary between oxygen-rich and oxygen-poor layers, and at mid-ocean hydrothermal ridges, chemoautotophic prokaryotes (and the animals that harvest them) thrive where reduced gases released from Earth's interior meet oxygen-rich seawater. The ability to use inorganic sources of chemical energy is widespread among Bacteria and Archaea, but entirely absent in Eukarya.

→ **Quick Check 3** How could the biological carbon cycle have worked on the primitive Earth, where oxygen gas was essentially absent from the atmosphere and oceans?

**FIG. 26.9 Deep-sea hydrothermal vents.** Chemoautotrophs fuel the carbon cycle in these environments.

## 26.3 OTHER BIOGEOCHEMICAL CYCLES

Let's think further about prokaryotic metabolism. Anoxygenic photosynthetic bacteria commonly use reduced sulfur compounds as electron donors for photosynthesis (instead of water), and bacterial heterotrophs often use oxidized sulfur compounds as electron acceptors in anaerobic respiration (instead of oxygen gas). By reducing oxidized sulfur compounds in the course of anaerobic respiration, microorganisms generate the reduced sulfur compounds required for anoxygenic photosynthesis. The reverse is also true: Oxidation of sulfur compounds during anoxygenic photosynthesis generates the oxidized sulfur molecules used in anaerobic respiration.

Once again, we see complementary metabolic pathways. The expanded carbon cycle of Bacteria and Archaea also promotes the cycling of sulfur, nitrogen, and other biologically important elements.

### Bacteria and archaeons dominate Earth's sulfur cycle.

Our bodies are mostly carbon, oxygen, and hydrogen. We also contain about 0.2% sulfur by weight. Sulfur is a component of the amino acids cysteine and methionine and therefore is present in many proteins (Chapter 2). Sulfur is also present in iron–sulfur clusters that are key components of enzymes fundamental to electron transfer. It occurs, as well, in some membranes and is found in a number of vitamins and other cofactors, including coenzyme A, thiamine (vitamin $B_1$), and biotin. Where do we get the sulfur we need? From the food we eat—protein in the steak or fish on our dinner plate provides a rich source of sulfur. Where do cows and fish get the sulfur they need? Like us, from the food they eat. Food chains don't go on forever, so somewhere in the system, some organisms must take up inorganic sulfur from the environment. **Primary producers,** organisms that reduce $CO_2$ to form carbohydrates, fill this role.

On land, plants take up sulfate ($SO_4^{2-}$) ions from the soil and reduce them within their cells to hydrogen sulfide ($H_2S$) that can be incorporated into cysteine and other biomolecules (**Fig. 26.10**). This process is called **assimilation.** Algae and photosynthetic bacteria do much the same thing in lakes, rivers, and oceans. Why don't primary producers take up $H_2S$ directly? There are two answers to this question. First, $H_2S$ is rapidly oxidized in the presence of oxygen and so does not occur in environments where oxygenic photosynthesis is common. Second, $H_2S$ is generally toxic to eukaryotic organisms, so plants and algae do not thrive where it is abundant. (The $H_2S$ produced within eukaryotic cells has a short lifetime and is restricted to intracellular sites distant from those that are vulnerable to $H_2S$.)

We've now seen half the sulfur cycle, the conversion of sulfate to $H_2S$ within cells. How did sulfate molecules get into the soil in the first place, and what happens to biological sulfur

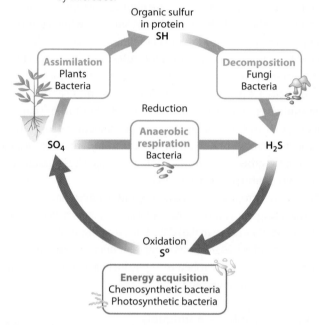

**FIG. 26.10  The sulfur cycle.** Plants take up sulfur as sulfate ions for incorporation into proteins and other compounds, and fungi release sulfur during decomposition. However, the major role in cycling sulfur through the biosphere is played by microbes.

compounds when cells die? After death, fungi and bacteria decompose cells, returning carbon, sulfur, and other compounds to the environment. Reduced sulfur compounds released from decomposing cells are then oxidized by bacteria and archaeons, completing the cycle (Fig. 26.10).

How do those reduced sulfur compounds get oxidized? There are two pathways, both of which involve microorganisms (Fig. 26.10). Chemoautotrophic bacteria react $H_2S$ with $O_2$ to obtain energy, producing sulfate in the process. And in anoxygenic photosynthesis, bacteria use electrons from $H_2S$ to form carbohydrates from $CO_2$, again generating sulfate as a by-product. The sulfate produced by these processes is consumed by heterotrophic bacteria living in oxygen-free environments. Sulfate rather than oxygen is used as the final electron acceptor in respiration and is reduced to $H_2S$. In fact, most of the sulfur cycled biologically is used to drive energy metabolism in oxygen-poor environments, not to build proteins.

Note that eukaryotes use neither $H_2S$ for photo- (or chemo-) synthesis nor sulfate in respiration. Thus, the biological sulfur cycle is completed by bacteria and archaeons alone.

### The nitrogen cycle is also driven by bacteria and archaeons.

The model provided by the sulfur cycle can be extended to other biologically important elements. Nitrogen is the fourth most abundant element in the human body, contributing 4% by weight

**FIG. 26.11 The nitrogen cycle.** Bacteria and archaeons make possible nitrogen fixation, nitrification and denitrification, as well as anammox.

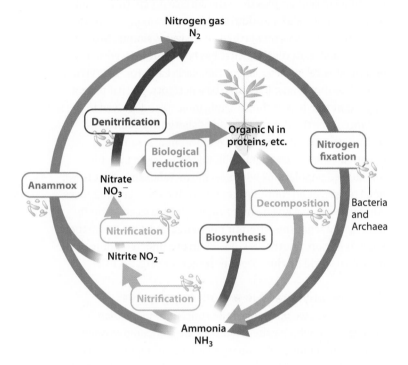

by the decomposition of dead cells can be taken up by primary producers, but commonly it is oxidized to nitrate ($NO_3^-$) by chemoautotrophic bacteria, a process called **nitrification** (see Fig. 26.11). This process works efficiently in both soil and water, and so nitrate is the most common form of nitrogen available to life in both environments.

Anaerobic respiration completes the cycle by returning $N_2$ to the atmosphere. Some bacteria can use nitrate as an electron acceptor in respiration, a process known as **denitrification** (see Fig. 26.11). Because denitrification is so efficient at returning nitrogen to the atmosphere as nitrogen gas, nitrogen fixation must continually make nitrogen available for organisms. Without nitrogen fixation, the nitrogen cycle would slowly wind down, and the productivity of Earth's biota—that is, of all its organisms—would wind down with it. This is why nitrogen fixation is such an important process.

Until the 1990s, the nitrogen cycle was described— completely, it was thought—in terms of the metabolic reactions just discussed. Then, a new and unexpected form of nitrogen metabolism was discovered in the oceans. A number of prokaryotic microorganisms function as chemoautotrophs by reacting ammonium ion ($NH_4^+$) with nitrite ($NO_2^-$) to gain energy. The reaction is shown here:

$$NH_4^+ + NO_2^- \rightarrow N_2 + 2H_2O$$

**FIG. 26.12 Roots of a soybean plant.** The nodules contain bacteria capable of reducing nitrogen gas to ammonia.

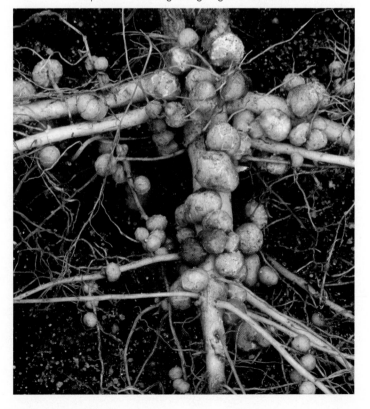

as a component of proteins, nucleic acids, and other compounds. As was true of sulfur, primary producers incorporate inorganic forms of nitrogen into biomolecules, and we get the nitrogen we need from our food. As we'll see, however, there is a twist.

Let's walk through the biological nitrogen cycle (**Fig. 26.11**). Carbon is a minor component of the atmosphere and sulfur only a trace constituent, but the air we breathe consists mostly (79%) of nitrogen gas ($N_2$). Indeed, most of the nitrogen at Earth's surface resides in the atmosphere. Although nitrogen is plentiful, the nitrogen available to primary producers in many environments is limited. Why should this be? The simple answer is that plants cannot make use of nitrogen gas and neither can algae. Only certain bacteria and archaeons can reduce $N_2$ to ammonia ($NH_3$), a form of nitrogen that can be incorporated into biomolecules.

The process of converting $N_2$ into a biologically useful form such as ammonia is called **nitrogen fixation,** one of the most important reactions ever evolved by organisms. Farmers plant soybeans to regenerate nitrogen nutrients in soils, but it is not the soybean that fixes nitrogen. Rather, soybean roots have nodules that harbor nitrogen-fixing bacteria (**Fig. 26.12**). That is, plants like soybean have entered into an intimate partnership with nitrogen-fixing bacteria to obtain the biologically useful forms of nitrogen needed for growth.

The nitrogen cycle, then, begins with nitrogen fixation, and it continues through many additional metabolisms, most of them confined to bacteria and archaeons. Ammonia released

Note that this is a redox reaction, like other reactions involved in energy metabolism. In this case, the ammonium ion is oxidized to nitrogen gas, and nitrite is reduced to nitrogen gas, with water produced as a by-product. The reaction, called **anammox** (**an**aerobic **amm**onia **ox**idation), provides only a small amount of energy, but it supports many bacteria and archaeons in oxygen-depleted parts of the ocean, providing a major route by which fixed nitrogen returns to the atmosphere as $N_2$ (see Fig. 26.11).

The biological nitrogen cycle, then, begins and ends with prokaryotic metabolisms. Bacteria and archaeons fix $N_2$ into a biologically available form, nitrifying bacteria oxidize ammonia to nitrate, and still other prokaryotes gain energy from denitrification and anammox, returning $N_2$ to the air. Unlike the carbon and sulfur cycles, the nitrogen cycle does not rely at any point on elements stored as minerals in rocks and sediments. Most of Earth's nitrogen was released to the atmosphere as gas early in our planet's history, and so movement into sediments and out of volcanoes contributes relatively little to the cycle. Humans, however, contribute in important ways, most notably through the industrial synthesis of nitrate fertilizer (Chapter 48).

→ **Quick Check 4** How would the nitrogen cycle operate in the absence of bacteria and archaeons?

## 26.4 THE DIVERSITY OF BACTERIA

Traditionally, bacterial taxonomy was a laborious process. Microbiologists grew bacteria in beakers or on petri dishes, carefully separating and enriching individual populations until only one type of bacterium—that is, a pure culture—was present. Species were identified on the basis of details of cell structure and division observed under the microscope, metabolic capabilities established through experiments and reactions to chemical stains or antibiotics. By the 1980s, a few thousand species had been described in this way.

With molecular sequencing technologies, however, bacterial species are now characterized by DNA sequences, especially the genes for the small subunit of ribosomal RNA (rRNA). A novel application of sequencing technology paved the way for modern studies of bacterial diversity and ecology. Microbiologists reasoned that the same procedures used to extract, amplify, and sequence genes from pure cultures in the lab could be used to study bacteria in nature. That is, extracting and reading gene sequences directly from soil or seawater could reveal the diversity of the microbial communities in these environments.

The results were stunning. Most bacteria in nature—perhaps 99% or more—are species that have not been cultured in the lab. Indeed, species well characterized by studies of pure cultures commonly represent only minor components of natural communities. To give just one illustration, a bacterium called

SAR11 was originally identified on the basis of gene sequences amplified from seawater. SAR11 caught the attention of microbiologists because it makes up about a third of all sequences identified in the samples from the surface ocean. How could we be ignorant of a bacterium that must be among the most abundant organisms on Earth? The answer is simple: No one had cultured it. Recently, SAR11 has been cultured. Formally described as *Pelagibacter ubique* (*ubique* is Latin for "everywhere"), this bacterium is now known to be a heterotroph that plays an important role in recycling organic molecules dissolved in the sea.

More recently, the application of shotgun sequencing techniques (Chapter 13) to the environmental sampling of genomes has allowed microbiologists to begin matching uncultured populations with specific metabolisms. That is, we need no longer be content to know from molecular sequences what is living in a given environment. Now we can begin to understand what the different microorganisms are doing there. For this reason, another revolution in our understanding of the microbial world lies just around the corner (**Fig. 26.13**).

### Bacterial phylogeny is a work in progress.

In Chapter 23, we introduced the concept of molecular phylogeny, in which gene sequences are compared to draw conclusions about evolutionary relatedness among populations in nature. Gene sequencing made it possible to test traditional bacterial groupings based on form and physiology. Some traditionally defined groups—for example, the cyanobacteria—passed the test and stand today as a well-defined bacterial group. Many, however, did not.

Molecular sequence comparisons bring enormous new possibilities to the study of prokaryotic diversity and evolution, but they introduce new problems as well. One set of problems arises because major groups of Bacteria and Archaea separated from one another billions of years ago. Inherited sequence similarities can be masked by continuing molecular evolution within groups. That is, specific nucleotides in gene sequences may change not once but multiple times through a long evolutionary history, erasing molecular features that were once the same because of inheritance from a common ancestor. Also, in some groups rates of sequence evolution appear to be higher than in others. As a result, these groups have especially divergent sequences and so may misleadingly fall toward the bottoms of phylogenetic trees.

Another problem arises because of horizontal gene transfer. Phylogenies may falsely group distantly related bacteria because they contain genes passed on by conjugation, transformation, or transduction. Statistical analyses of complete bacterial genomes suggest that at least 85% of all genes in bacterial genomes have been transferred horizontally at least once. Such apparently rampant gene exchange might well doom attempts to build a bacterial tree of life, and some biologists argue that bacterial

**FIG. 26.13**

# How many kinds of bacterium live in the oceans?

**BACKGROUND** The oceans are full of bacteria—by some estimates, the sea harbors more than $10^{29}$ bacterial cells. How diverse are these bacterial communities?

**METHOD 1** Different types of bacterium can be recognized by the unique nucleotide sequences of their genes. In fact, individual types of bacterium can be identified on the basis of nucleotide sequence in a relatively small region of the gene for the small subunit of ribosomal RNA (rRNA). This molecular "barcode" can be sequenced quickly and accurately for samples of DNA drawn from the environment.

Using this strategy, scientists associated with the International Census of Marine Life obtained samples of seawater and seafloor sediment from deep-ocean environments. In the laboratory, the rRNA sequences were amplified by PCR, separated by gel electrophoresis, and sequenced, thus providing a library of tags that document bacterial diversity.

**METHOD 2** In another set of experiments, scientists collected samples of seawater from the Sargasso Sea, and from these they amplified whole genomes—more than a billion nucleotides' worth. They used small-subunit rRNA and other genes to characterize species richness, and also characterized a large assortment of additional sequence data to understand genetic and physiological diversity.

**RESULTS** Both surveys found that the bacterial diversity of marine environments is much higher than had been thought. The barcode survey (method 1) found as many as 20,000 distinct types of bacteria in a single liter of seawater and estimated that as many as 5–10 million different microbes may live in the world's oceans. Some of these bacteria are abundant and widespread, but most are rare. The genomic survey (method 2) of the Sargasso Sea found more than 1800 distinct bacteria in sea-surface samples and, remarkably, identified more than 1.2 million previously unknown genes within these genomes.

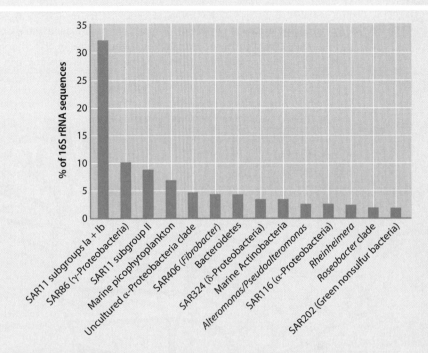

Relative abundance of different bacterial groups in samples from the Sargasso Sea, based on percent representation of small subunit rRNA sequences. Note the abundance of SAR11, only recently characterized.

Computer-aided comparison of sequences from the world's oceans shows that about 75% of all sequences collected are similar to other sequences in the library of sequence tags. However, 25% are distinctive, forming a rare but enormously diverse pool of bacterial types.

**CONCLUSION** The bacterial diversity of the oceans is huge and still largely unexplored. Both barcode and genomic surveys are continuing, and promise a more nearly complete accounting of marine diversity. Together, these surveys will help us understand how marine bacteria function and how they sort into communities.

**SOURCES** Venter, J. C., et al. 2004. "Environmental Genome Shotgun Sequencing of the Sargasso Sea." *Science* 304:66–74; Sogin, M. L., et al. 2006. "Microbial Diversity in the Deep Sea and the Underexplored 'Rare Biosphere'" *Proceedings of the National Academy of Sciences USA* 103:12115–12120.

**FIG. 26.14** **Bacterial and archaeal phylogeny.** Some scientists argue that the evolution of prokaryotes should not be viewed as a tree but as a series of branches that diverge and then come back together, representing genes that diverge from one another but then come to reside together in new organisms because of horizontal gene transfer.

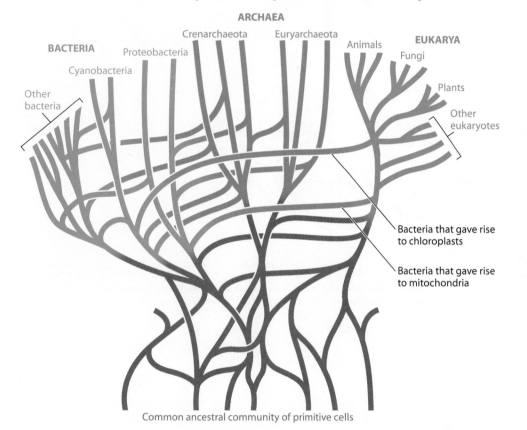

Common ancestral community of primitive cells

evolution is more properly depicted as a complex bush with many connecting branches (**Fig. 26.14**).

Other microbiologists are more optimistic. Small-subunit rRNA genes, for example, show little evidence of horizontal transfer and so may faithfully record the evolutionary history of the organisms in which they occur. Nonetheless, we need to be cautious when using bacterial trees to study the evolution of specific metabolic or physiological characteristics. Nitrogen fixation, photosystems, drug tolerance—such features can and do depend on genes known to jump from branch to branch on the tree.

At present, molecular signatures have been extremely useful in identifying major groups of bacteria, but less clear in establishing branching order among these groups. About 50 major groups of bacteria (each one very roughly equivalent to a eukaryotic phylum) have been recognized. Nearly half of these groups are not represented in pure cultures. Rather, they have been identified only by the cloning of genes from environmental

samples. To support our discussion of bacterial diversity, we present a phylogeny based on comparisons of whole genomes in **Fig. 26.15**. As you examine this figure, remember that microbiologists continue to debate branching order on the bacterial tree.

### What, if anything, is a bacterial species?

What do we mean when we talk about bacterial species? As we saw in Chapter 22, Ernst Mayr defined species as "groups of interbreeding natural populations that are reproductively isolated from other such groups." This widely accepted definition is called the biological species concept (Chapter 22), but a moment's reflection will show that it can't be applied to the most abundant organisms on Earth—Bacteria and Archaea. In the absence of sexual reproduction, how do we think about species in the prokaryotic domains?

Some microbiologists argue that we should abandon the concept of species when talking about bacteria, especially since distantly related bacteria can exchange genes. Nonetheless, nature is full of microbial populations with well-defined features of form and function.

In the age of molecular sequence comparisons, bacterial species have sometimes been defined operationally: Two populations belong to the same species if the gene sequences of their small-subunit rRNA are more than 97% identical. It would be useful, however, to conceive of bacterial species in terms of population-level processes, much as Ernst Mayr did for eukaryotic organisms. In plants, animals, and single-celled eukaryotes, interbreeding ensures that populations share the same pool of genes, providing a counterforce to mutations and local selection pressures that promote divergence.

A similar process exists in bacteria. In early research on the population genetics of bacteria, biologists observed that genetic diversity in laboratory cultures gradually increases through time and then rapidly decreases with the emergence of a successful variant that outcompetes the rest. The episodic loss of diversity is called **periodic selection.** Some biologists argue that this process provides a means of recognizing bacterial species: Populations subject to the same episodes of periodic selection belong to a single species. This is a useful way to define bacterial species, but other definitions are possible, and as yet there is no consensus among microbiologists.

### Proteobacteria are the most diverse bacteria.

**Proteobacteria** are far and away the most diverse of all bacterial groups (Fig. 26.15). Defined largely by similarities in gene sequences of rRNA, Proteobacteria include many of the organisms that populate our expanded carbon cycle and other biogeochemical cycles. For example, this phylum includes anoxygenic photosynthetic bacteria and chemoautotrophs that oxidize $NH_3$, $H_2S$, and $Fe^{2+}$, as well as bacteria able to respire using $SO_4^{2-}$, $NO_3^-$, or $Fe^{3+}$.

Many Proteobacteria have evolved intimate ecological relationships with eukaryotic organisms. Some of these are mutually beneficial, such as the nitrogen-fixing bacteria in soybean root nodules. Other Proteobacteria, however, are our worst pathogens: the rickettsias that cause typhus and spotted fever,

vibrios that trigger cholera, salmonellas responsible for typhoid fever and food poisoning, and *Pseudomonas aeruginosa* strains that infect burn victims and others whose resistance is reduced. *Escherichia coli*, famous as a laboratory model organism (see Fig. 26.4b), is usually a harmless presence in our intestinal tract, but pathogenic strains can cause serious diarrhea and even death.

### The gram-positive bacteria include organisms that cause and cure disease.

In 1884, the Danish scientist Hans Christian Gram developed a diagnostic dye that, after washing, is retained by bacteria with thick peptidoglycan walls, but not by those with thin walls. Molecular sequence comparisons confirm that most **gram-positive bacteria** (those that retain the dye) form a well-defined branch of the bacterial tree (Fig. 26.15). One species in this group, *Bacillus subtilis*, is well known to biologists because it has been the focus of many studies about basic molecular function in bacteria. All of us harbor *Streptococcus* species in our mouths, and many of us have suffered from strep throat, produced by streptococcal pathogens. More serious infections, including some forms of meningitis, are also associated with streptococci. *Staphylococcus* bacteria are commonly found on human skin, but only occasionally do we suffer from maladies such as staph infection, toxic shock syndrome, or food poisoning induced by staphylococcal bacteria.

In contrast to these pathogens, gram-positive bacteria called streptomycetes have proved invaluable to medicine because of a remarkable property. More than half of the species in this group secrete compounds that kill other bacteria and fungi in their vicinity. Streptomycetes have provided us with tetracycline, streptomycin, erythromycin, and numerous other antibiotics that combat infectious disease.

### Photosynthesis is widely distributed on the bacterial tree.

As noted earlier, molecular studies confirm that all bacteria capable of oxygenic photosynthesis form a single branch, the **cyanobacteria.** Cyanobacteria can be found in

**FIG. 26.15 Evolutionary relationships among bacteria, inferred from data from complete genomes.** Photosynthetic groups are shown in green (the Acidobacteria have one known photoheterotrophic member). Other groups discussed in the text are shown in red.

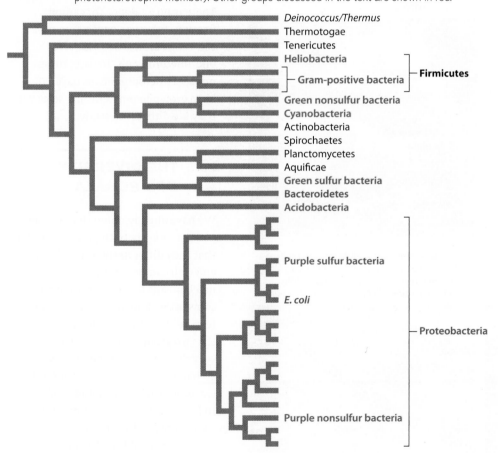

environments that range from deserts to the open ocean. Their diverse species include unicellular rods and spheroidal cells, as well as multicellular balls and filaments (**Fig. 26.16**). Some filamentous cyanobacteria can even form several different cell types, including specialized cells for nitrogen fixation and resting cells that provide protection when the local environment does not favor growth.

In contrast to the monophyly of cyanobacteria, molecular sequence comparisons show that *anoxygenic* photosynthesis is distributed widely on the bacterial tree (see Fig. 26.15). Think of the purple layer beneath the surface of many microbial mats (see Fig. 26.7). The vivid color comes from the light-sensitive pigments of the aptly named purple bacteria, found within the Proteobacteria. Purple bacteria are capable of photoautotrophic growth using bacteriochlorophyll and a single photosystem. Most, however, show evidence of metabolic diversity and can grow heterotrophically in the absence of light or appropriate electron donors. The green layer beneath the surface of some microbial mats reflects the pigments of another group of photosynthetic bacteria, called green sulfur bacteria because they commonly gain electrons from hydrogen sulfide and deposit elemental sulfur on their walls. Unlike most groups of photosynthetic bacteria, the green sulfur bacteria are highly intolerant of oxygen gas. Oddly, a species of green sulfur bacteria has been found in hydrothermal rift vents 2 km beneath the surface of the ocean. Light doesn't penetrate to these depths, so if these organisms harvest light, it must come from incandescent lavas as they erupt.

Light-harvesting bacteria occur on several other branches of the bacterial tree—the photoheterotrophic heliobacteria, for example, and the green nonsulfur bacteria often found in freshwater hot springs. Despite a century of research, our knowledge of photosynthesis within the bacterial tree remains incomplete, as illustrated by the discovery in 2007 of unique light-harvesting bacteria in Yellowstone National Park. Genomic surveys of hot springs turned up evidence of a previously unknown bacteriochlorophyll-containing microorganism, and gene sequencing established that this organism belongs to the phylum Acidobacteria. There is no evidence for carbon fixation, however, suggesting that the Yellowstone bacterium is a photoheterotroph. To date, these unusual microorganisms have not been grown in pure culture, so as yet they are difficult to study.

## 26.5 THE DIVERSITY OF ARCHAEA

We have already noted that Archaea and Bacteria are both prokaryotic, but that they differ in their membrane and wall composition, molecular mechanisms for transcription and translation, and many other biological features (see Table 26.1). As a result, the two groups tend to thrive in very different environments. Many archaeons tolerate environmental extremes such as heat and acidity. In fact, for many environmental conditions, archaeons define the known limits of life. What do these disparate environments have in common? One proposal is that

**FIG. 26.16 Diversity of form among cyanobacteria.**

**a.** *Aphanothece*, small single cells

**b.** *Chroococcus*, larger cells bound into a colony by a common extracellular envelope made of polysaccharides

**c.** *Oscillatoria*, filaments with no cell differentiation

**d.** *Anabaena*, strings of undifferentiated cells at either end of the filament, flanking two elongated resting cells, and in the middle, a heterocyst specialized for nitrogen fixation

**FIG. 26.17** Phylogeny of Archaea, showing the evolution of temperature preference, salt tolerance, acid tolerance, and production of methane within the domain.

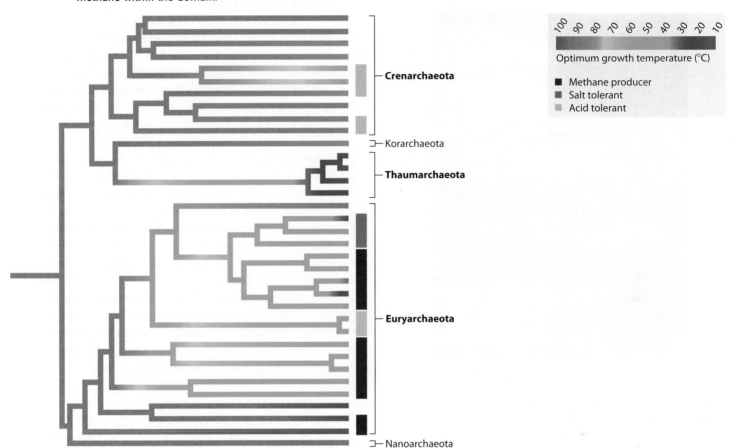

archaeons thrive in environments where the energy available to fuel cell activities is limited—where there is no light, or a limited abundance of fuels and oxidants for chemoautotrophy or respiration, or both. That is, archaeons commonly live where energy resources are too low to support bacteria and eukaryotes.

Bacteria and Archaea diverged from their last common ancestor more than 3 billion years ago. Just as Bacteria have evolved a phylogenetic diversity that defies easy categorization, Archaea also include many distinct types of organism. Many microbiologists recognize three major divisions of Archaea, called the **Crenarchaeota**, **Euryarchaeota,** and **Thaumarchaeota,** but a few poorly known groups may indicate further major branches within the domain (for example, the Korarchaeota and Nanoarchaeota noted in **Fig. 26.17**).

### The archaeal tree has anaerobic, hyperthermophilic organisms near its base.

Archaeons found near the base of the Crenarchaeota and Euryarchaeota branches exhibit a remarkable ability to grow and reproduce at temperatures of 80°C or more (Fig. 26.17).

Called **hyperthermophiles,** these organisms do not simply tolerate high temperature, they require it. At present, the world record for high temperature growth is held by *Methanopyrus kandleri,* which has been shown to grow at 122°C among hydrogen- and $CO_2$-rich hydrothermal vents on the seafloor beneath the Gulf of California. Most of these heat-loving archaeons are anaerobic, living as anaerobic respirers or as chemoautotrophs that obtain energy from the oxidation of hydrogen gas.

Why do anaerobic hyperthermophiles sit on the lowest branches of the archaeal tree? Many biologists believe that these features characterized the first Archaea and therefore imply that the Archaea evolved in hot environments with no oxygen. As we discussed in Chapter 23, geologists agree that oxygen gas was not present on the early Earth, consistent with the phylogenetic inference that early Archaea were anaerobic. A hot environment for early Archaea can be understood in two ways. Either the entire ocean was hot—for which there is little evidence—or Archaea first evolved in hot springs where local supplies of hydrogen and metals enabled them to live as chemoautotrophs. Because some versions of the bacterial tree also show hyperthermophilic

**FIG. 26.18 Headwaters of the Rio Tinto, southwestern Spain.** Archaeons thrive in these highly acidic waters, pH 1–2.

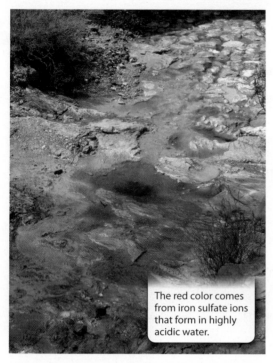

The red color comes from iron sulfate ions that form in highly acidic water.

species on their lowest branches, it has been suggested that the last common ancestor of *all* living organisms lived in a hot environment.

→   **Quick Check 5** If early branches on the bacterial and archaeal trees are dominated by hyperthermophilic microorganisms, does this mean that the early oceans were very hot?

### The Crenarchaeota and Euryarchaeota both include acid-loving microorganisms.

In many parts of the globe where humans mine coal or metal ores, abandoned mines fill up with very acidic water. Called acid mine drainage, this water has been acidified through the oxidation of pyrite ($FeS_2$) and other sulfide minerals to produce sulfuric acid ($H_2SO_4$). The pH of these waters is commonly around 1 to 2, and can actually fall below 0. To us, acid mine drainage is an environmental catastrophe, but a number of archaeons thrive in it (**Fig. 26.18**). Modified protein and membrane structures enable them to tolerate their harsh surroundings, making it possible to use sulfur compounds in the environment for both chemoautrophic and heterotrophic growth. What is the limit to growth at low pH? *Picrophilus torridus*, a euryarchaeote archaeon, grows optimally at 60°C and in environments with a pH of 0.7—the pH of battery acid. It is capable of growth in environments down to pH 0—

a record, as far as we know. In fact, *Picrophilus torridus* requires an acidic environment. It will not grow in acid rainwater because the pH is too high.

### Euryarchaeote archaeons include heat-loving, methane-producing, and salt-loving microorganisms.

Some euryarchaeote archaeons, like *Picrophilus torridus*, are thermophilic—heat-loving. Many other euryarchaeotes, uniquely among organisms, generate natural gas, or methane ($CH_4$), as a by-product of fermentation or chemoautotrophic metabolism. Methanogenic archaeons play an important role in the carbon cycle, recycling organic molecules back to simple gases in environments where the supplies of electron acceptors for respiration (oxygen gas, sulfate ion, ferric iron) are limited. Methanogenic archaeons are prominent in peats and lake bottoms where sulfate levels are low and bacterial sulfate reduction is therefore limited. Most methanogenic organisms live in soil or sediments, but some live in the rumens (a specialized chamber in the gut) of cows, helping to maximize the energy yield from ingested grass. Methane doesn't last long in the atmosphere—it is quickly oxidized to carbon dioxide. However, the steady supply of methane generated by archaeons contributes significantly to the greenhouse properties of the atmosphere. For this reason, biologists are interested in how methanogenic archaeons will respond to 21st-century climate change.

A related group of euryarchaeote archaeons illustrates the limits of tolerance to another environmental condition—salinity, or saltiness. Halophilic (salt-loving) archaeons are photoheterotrophs that use the protein bacteriorhodopsin to absorb energy from sunlight (see Fig. 26.6). They live in waters that are salty enough to precipitate table salt (NaCl). As is correspondingly true of extreme acidophiles, these extreme halophiles require high salt conditions and cannot live in dilute water.

### Thaumarchaeota may be the most abundant cells in the oceans.

Another surprise of environmental sequencing has been the discovery that the immense volumes of ocean beneath the surface waters contain vast numbers of archaeons that have not yet been cultured in the lab. In the deep ocean, these organisms may be nearly as abundant as the combined number of all bacteria on and below the surface. Recently segregated as the Thaumarchaeota, these archaeons are chemoautotrophs, deriving energy from the oxidation of ammonia (**Fig. 26.19**). Thus, like the discovery of SAR11, the discovery of marine thaumarchaeotes makes it clear that before the still-emerging age of environmental genetics and genomics, we simply didn't have a very good idea of which microorganisms lived where and did what in the oceans. Fortunately, that situation is changing rapidly.

We also influence our gut biota by ingesting antibiotics. Although prescribed for the control of pathogens, most antibiotics kill a wide range of bacteria and so can change the balance of our intestinal microbiota. Clinical studies show that the incidence of inflammatory bowel disease, a disabling inflammation of the colon, increases in humans who have been treated with antibiotics.

To understand the relationships between the human microbiome and human health, we need to know what constitutes a healthy gut biota, and this means studying the metabolism, ecology, and population genetics of the bacteria in our bodies. We need to know which bacteria have evolved to take advantage of the environments provided by the human digestive system, and which are "just passing through." Future visits to the doctor may include routine genetic fingerprinting of our bacterial biota, as well as treatments designed to keep our microbiome—and, so, ourselves—healthy.

## Core Concepts Summary

**26.1 THE TREE OF LIFE HAS THREE MAIN BRANCHES, CALLED DOMAINS: EUKARYA, BACTERIA, AND ARCHAEA.**

Prokaryotic cells are cells that lack a nucleus; they include Bacteria and Archaea. page 26-1

Bacteria are small and lack membrane-bound organelles. page 26-1

The bacterial genome is circular. Some bacteria also carry smaller circles of DNA called plasmids. page 26-2

Diffusion limits size in bacterial cells. page 26-2

Bacteria can obtain DNA by horizontal gene transfer from organisms that may be distantly related. page 26-4

Like Bacteria, Archaea lack a nucleus, but form a second prokaryotic domain distinct from Bacteria. page 26-5

Some archaeons are extremophiles, living in extreme environments characterized by low pH, high salt, or high temperatures, but others live in less extreme environments like the upper ocean or soil. page 26-6

**26.2 BACTERIA AND ARCHAEA ARE NOTABLE FOR THEIR METABOLIC DIVERSITY.**

Bacteria are capable of oxygenic photosynthesis, using water as a source of electrons and producing oxygen as a by-product, and of anoxygenic photosynthesis, using electron donors other than water, such as $H_2S$, $H_2$, and $Fe^{2+}$. page 26-7

Some bacteria and archaeons are capable of anaerobic respiration, in which $NO_3^-$, $SO_4^{2-}$, $Mn^{4+}$, and $Fe^{3+}$ serve as the electron acceptor instead of oxygen gas. page 26-8

Many bacteria and archaeons obtain energy from fermentation, which involves the partial oxidation of organic molecules and the production of ATP by substrate-level phosphorylation. page 26-8

Many photosynthetic bacteria are photoheterotrophs, obtaining energy from sunlight and using preformed organic compounds as a source of carbon instead of $CO_2$. page 26-9

Chemoautotrophy, in which chemical energy is used to convert $CO_2$ to organic molecules, is unique to Bacteria and Archaea. page 26-9

**26.3 IN ADDITION TO THEIR KEY ROLES IN THE CARBON CYCLE, BACTERIA AND ARCHAEA ARE CRITICAL TO THE BIOLOGICAL CYCLING OF SULFUR AND NITROGEN.**

Plants and algae can take up sulfur and incorporate it into proteins, but bacteria and archaeons dominate the sulfur cycle by means of oxidation and reduction reactions that are linked to the carbon cycle. page 26-10

Bacteria and archaeons can to reduce nitrogen gas to ammonia in a process called nitrogen fixation. page 26-11

The nitrogen cycle also involves oxidation and reduction reactions that are linked to the carbon cycle. page 26-11

**26.4 THE EXTENT OF BACTERIAL DIVERSITY WAS RECOGNIZED ONLY WHEN SEQUENCING TECHNOLOGIES COULD BE APPLIED TO NON-CULTURABLE BACTERIA.**

Traditionally, bacterial groups were recognized by morphology, physiology, and the ability to take up specific stains in culture. page 26-12

Direct sequencing of ribosomal RNA genes from organisms in soil and seawater samples revealed new groups of bacteria. page 26-12

Proteobacteria are the most diverse group of bacteria and are involved in many of the biogeochemical cycles that are linked to the carbon cycle. page 26-15

Gram-positive bacteria include important disease-causing strains as well as species that are principal sources of antibiotics. page 26-15

Photosynthetic bacteria are not limited to a single branch of the bacterial tree.  page 26-15

## 26.5 THE DIVERSITY OF ARCHAEA HAS ONLY RECENTLY BEEN RECOGNIZED.

Archaeons tend to thrive where energy available for growth is limited.  page 26-16

Archaea are divided into three major groups, the Crenarchaeota, Thaumarchaeota, and Euryarchaeota.  page 26-17

Archaeons at the base of the Crenarchaeota and Euryarchaeota are hyperthermophiles, meaning that they grow at high temperatures.  page 26-17

A number of archaeons grow in highly acidic waters, such as those associated with acid mine drainage.  page 26-18

Some archaeons generate methane, and others tolerate high-salt conditions.  page 26-18

Thaumarchaeotes may be the most abundant cells in the oceans.  page 26-18

## 26.6 THE EARLIEST FORMS OF LIFE ON EARTH WERE BACTERIA AND ARCHAEA.

Evidence for the early history of life on Earth comes from microfossils, fossilized structures called stromatolites, and the isotopic composition of rocks and organic matter.  page 26-20

Fossils indicate that life on Earth originated more than 3.5 billion years ago.  page 26-20

The early atmosphere and ocean contained little or no free oxygen.  page 26-20

Oxygen began to accumulate in the atmosphere and oceans about 2.4 billion years ago as a result of the success of cyanobacteria utilizing oxygenic photosynthesis.  page 26-21

Prokaryotic metabolisms were not only essential in the early history of the Earth, but are also vital today, as many forms of life depend on biogeochemical cycles and metabolisms unique to Bacteria and Archaea.  page 26-21

Most animals, including humans, live in intimate association with bacteria, which in turn affect our health.  page 26-22

## Self-Assessment

1. Name and describe the three domains of life.

2. Describe shared and contrasting features of bacterial and archaeal cells.

3. Explain how prokaryotic cells obtain nutrients and how this process puts constraints on their size.

4. Describe how surface area and volume change with size.

5. Explain how photosynthesis can occur without the production of oxygen, and how respiration can occur without requiring oxygen.

6. Describe the roles of bacteria and archaeons in the sulfur and nitrogen cycles.

7. Explain how horizontal gene transfer complicates our understanding of evolutionary relationships among bacteria and archaeons.

8. Name and describe three major groups of Bacteria.

9. Name and describe three major groups of Archaea.

10. State the age of the Earth and the time when life is thought to have first originated.

Do you understand the chapter's Core Concepts? Log in to check your answers to the Self-Assessment questions, then practice what you've learned and reinforce this chapter's concepts by working through the problems and multimedia tutorials provided there.

⊡ www.biologyhowlifeworks.com

## CHAPTER 27

# EUKARYOTIC CELLS
## Origins and Diversity

## Core Concepts

**27.1** Eukaryotic cells are defined by the presence of a nucleus, but other features, particularly a dynamic cytoskeleton and membranes, explain their success in diversifying.

**27.2** The endosymbiotic hypothesis proposes that the chloroplasts and mitochondria of eukaryotic cells were originally free-living bacteria that were incorporated into a host cell.

**27.3** Eukaryotes have historically been divided into four kingdoms—animals, plants, fungi, and protists—but are now divided into at least seven superkingdoms.

**27.4** The fossil record extends our understanding of eukaryotic diversity by providing perspectives on the timing and environmental context of eukaryotic evolution.

Eukaryotic cells have a nucleus—that's their defining characteristic. As we will see, however, it is quite another set of features that distinguishes eukaryotes in terms of function and the capacity for evolutionary innovation. Eukaryotic diversity—the animals, plants, fungi, and protists we are familiar with—does not reflect exceptional metabolic versatility. In fact, compared to Bacteria and Archaea, the Eukarya display only a limited range of metabolisms. For example, oxygenic photosynthesis is widespread among eukaryotes, but anoxygenic photosynthesis is unknown. Furthermore, aerobic respiration is present almost everywhere, but anaerobic respiration has been reported in only a handful of species. Chemoautotrophy has not been documented in eukaryotes at all. Many eukaryotes can ferment organic substrates—yeasts are particularly important in this regard—but most use fermentation only as a supplementary metabolism and lack the diversity of reactions known among prokaryotic organisms. The reasons for eukaryotic success lie elsewhere, in their dazzling diversity of shapes and sizes.

## 27.1 THE EUKARYOTIC CELL: A REVIEW

Chapter 5 introduced the fundamental features of eukaryotic cells. Here, we revisit those attributes (**Fig. 27.1**) and consider their consequences for function and diversity. How do eukaryotic cells differ from bacteria and archaeons, and how do these differences explain the roles that eukaryotes play in modern ecosystems?

### Internal protein scaffolding and dynamic membranes organize the eukaryotic cell.

All cells require a mechanism to maintain spatial order in the cytoplasm. As we saw in Chapter 26, bacteria and archaeons rely primarily on walls that support the cell from the outside, along with a framework of proteins within the cytoplasm. Eukaryotes also use an internal scaffolding of proteins, mostly filaments of actin and microtubules, to organize the cell. This **cytoskeleton** (Chapter 10), found in all eukaryotic cells, differs in one key property from the rigid protein framework of bacteria: It can be remodeled quickly, enabling cells to change shape.

This characteristic provides some eukaryotes with new possibilities for movement, and it permits them to engulf particles, including other cells, something that prokaryotic cells simply cannot do. Packaged in membrane-lined vesicles, the particles can be transported and processed inside the cell by endocytosis. The same process also works in reverse, and it is then called exocytosis. In exocytosis, vacuoles package newly formed molecules or cytoplasmic waste in vesicles and move them to the cell surface for excretion (Chapter 5). These vesicles and the molecules they carry are

**FIG. 27.1    A eukaryotic cell.** Eukaryotic cells have a nucleus and extensive internal compartmentalization.

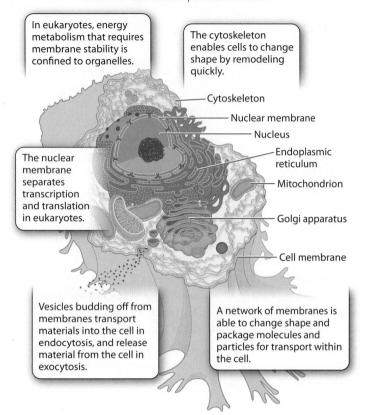

In eukaryotes, energy metabolism that requires membrane stability is confined to organelles.

The cytoskeleton enables cells to change shape by remodeling quickly.

— Cytoskeleton

— Nuclear membrane

— Nucleus

— Endoplasmic reticulum

— Mitochondrion

— Golgi apparatus

— Cell membrane

The nuclear membrane separates transcription and translation in eukaryotes.

Vesicles budding off from membranes transport materials into the cell in endocytosis, and release material from the cell in exocytosis.

A network of membranes is able to change shape and package molecules and particles for transport within the cell.

transported through the cytoplasm by means of molecular motors associated with the cytoskeleton. In this way, molecules move through the cell at speeds much greater than diffusion would allow. A major consequence is that eukaryotic cells can be much larger than most bacteria.

Dynamic cytoskeletons require dynamic membranes, and eukaryotes maintain within their cells a remarkable network of membranes called the endomembrane system (Chapter 5). This network includes the nuclear envelope, an assembly of membranes that runs through the cytoplasm called the endoplasmic reticulum (ER) and Golgi apparatus, and a cell or plasma membrane that surrounds the cytoplasm. Moreover, small pieces of membrane can bud off to form tiny vesicles that package molecular signals, nutrients, and wastes within the cell.

All membranes of the endomembrane system are interconnected, either directly or by the movement of vesicles. Many are also capable of changing shape rapidly. In fact, the different membranes are interchangeable in the sense that material originally added to the endoplasmic reticulum may in time be transferred to the cell membrane or nuclear membrane. Biologists like to say that the membranes of eukaryotic cells are in dynamic continuity. Membrane stability, required for energy metabolism, is confined to the mitochondria and chloroplasts.

## In eukaryotic cells, energy metabolism is localized in mitochondria and chloroplasts.

We have already noted that, relative to prokaryotic organisms, eukaryotes are fairly limited in the ways they obtain carbon and energy. Moreover, the metabolic processes that power eukaryotic cells take place only in specific organelles—aerobic respiration in the mitochondrion (Chapter 7) and photosynthesis in the chloroplast (Chapter 8). Only limited anaerobic processing of food molecules takes place within the cytoplasm.

Because energy metabolism is confined to these organelles, eukaryotes free the rest of the cell for other, complementary functions. Unlike bacteria, which absorb individual molecules for respiration, single-celled eukaryotic heterotrophs can engulf particulate food, including other cells (**Fig. 27.2**). The cells engulf food particles and package them inside a membrane vesicle, which is then transported into the cytoplasm in a process called phagocytosis. Within the cytoplasm, enzymes break down the particles into molecules that can be processed by the mitochondria. Animals can ingest larger foodstuffs, including other animals and plants. In consequence, eukaryotes can exploit sources of food not readily available to bacterial heterotrophs, which feed on individual molecules. This ability opens up a great new ecological possibility—predation—increasing the complexity of interactions among organisms.

The functional flexibility of eukaryotic cells also allows photosynthetic eukaryotes to interact with their environment in ways that photosynthetic bacteria cannot. Unicellular algae (which are eukaryotes) can move effectively through the water column vertically as well as horizontally and therefore can seek and exploit local patches of nutrients. Diatoms, discussed shortly, go one step further. Large internal vacuoles allow them to store nutrients when they are plentiful for later use. Plants, of course, can capture sunlight many meters above the ground, and so have a tremendous advantage on land, as long as their leaves can obtain water and nutrients from the soil in which the plants are rooted.

Taken together, the dynamic cytoskeleton and membrane system, and the related compartmentalization of basic energy metabolism, go a long way toward explaining why eukaryotic cells have so many shapes and prokaryotes have so few.

## The organization of the eukaryotic genome also helps explain eukaryotic diversity.

The prokaryotic genome has evolved in such a way that it can replicate quickly. Bacteria and Archaea absorb available nutrients quickly, so both rapid deployment of metabolic proteins and reproduction are key to ecological and evolutionary success. As the majority of a prokaryote's DNA is arrayed in a single circular chromosome, speed of replication allows for speed of reproduction. As a result, selection favors those strains of Bacteria and Archaea that retain only the genetic material vital to the organism.

**FIG. 27.2** **Phagocytosis.** The flexibility of the eukaryotic cytoskeleton and membrane system enables eukaryotic cells to engulf food particles, including other cells. This sequence shows an amoeba engulfing a yeast cell.

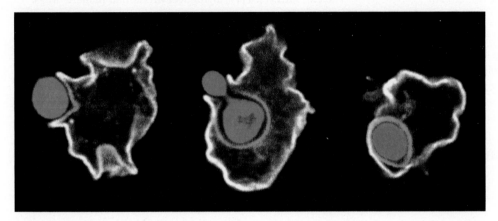

Eukaryotes have multiple linear chromosomes and can begin replication from many sites on each one, enabling them to replicate multiple strands of DNA simultaneously and rapidly (Chapter 12). This ability relieves the evolutionary pressure for streamlining, allowing eukaryotic genomes to build up large amounts of DNA that do not code for proteins. Most of this additional DNA was originally considered to have no function and, indeed, was called "junk DNA." That view, however, has lost ground in recent years, as the complete nature of the genome has come to be better understood (Chapter 13). At least some of the DNA that does not code for proteins has functions in gene regulation (Chapter 19). This regulatory DNA gives eukaryotes the fine control of gene expression required for both multicellular development and complex life cycles, two major features of eukaryotic diversity.

The evolutionary success of eukaryotes really rests on a combination of features. The innovations of dynamic cytoskeletal and membrane systems gave eukaryotes the structure required for larger cells with complex shapes and the functional ability to ingest other cells. Thus, early unicellular eukaryotes did not gain a foothold in microbial ecosystems by outcompeting bacteria and archaeons. Instead, they succeeded by evolving novel functions. Along with the capacity to remodel cell shape, eukaryotes evolved complex patterns of gene regulation, which in turn enabled unicellular eukaryotes to evolve complex life cycles and multicellular eukaryotes to generate multiple, interacting cell types during growth and development. These abilities opened up still more possibilities for novel functions, which we explore in this and later chapters.

→ **Quick Check 1** How did the evolutionary expansion of eukaryotic organisms change the way carbon is cycled through biological communities?

## Sex promotes genetic diversity in eukaryotes and gives rise to distinctive life cycles.

In Chapter 26, we saw how Bacteria and Archaea generate genetic diversity by means of horizontal gene transfer. Horizontal gene transfer has been documented in eukaryotic species, but it is relatively uncommon. How then do eukaryotes build and maintain genetic diversity within populations? The answer is sex.

Sexual reproduction involves the formation of **gametes** by meiosis and their fusion by fertilization (Chapters 11 and 42). Sex promotes genetic variation in two simple ways. First, meiotic cell division results in daughter cells (gametes or spores) that are genetically unique—that is, the combinations of alleles in each gamete are different from each other and from the parental cell—as a result of recombination and independent assortment. Second, in fertilization, new combinations of genes are brought together by the fusion of gametes. Interestingly, a few eukaryotic groups have lost the capacity for sexual reproduction. In the best-studied of these eukaryotes, tiny animals called bdelloid rotifers, genetic diversity is actually high, maintained by high rates of horizontal gene transfer. Apparently, horizontal gene transfer and sexual reproduction are alternative and, to a large extent, mutually exclusive ways of maintaining genetic diversity within populations.

As we saw in Chapter 11, meiotic cell division results in cells with one set of chromosomes. Such cells are **haploid.** Sexual fusion brings two haploid ($1n$) cells together to produce a **diploid** ($2n$) cell that has two sets of chromosomes. The life cycle of sexually reproducing eukaryotes, then, necessarily alternates between haploid and diploid states.

Many single-celled eukaryotes normally exist in the haploid stage and reproduce asexually by mitotic cell division (**Fig. 27.3a**). An example is provided by the green alga *Chlamydomonas.* Typically, starvation or other environmental stress can cause two cells to fuse, forming a diploid cell, or **zygote.** The zygote formed by these single-celled eukaryotes commonly functions as a resting cell. It covers itself with a protective wall and then lies dormant until environmental conditions improve. In time, further signals from the environment induce meiotic cell division, resulting in four genetically distinct haploid cells that emerge from their protective coating to complete the life cycle.

As shown in **Fig. 27.3b**, some single-celled eukaryotes normally exist as diploid cells. For example, in diatoms, single-celled eukaryotes commonly found in lakes, soils, and the oceans, cells are mostly diploid and reproduce asexually by mitotic cell division to make more diploid cells. Because of the constraints of their mineralized skeletons, diatoms become smaller with each asexual division. Once a critical size is reached, meiotic cell division is triggered, producing haploid gametes that fuse to regenerate the diploid state as a round, thick-walled cell.

---

**FIG. 27.3** **Eukaryotic life cycles.** (a and b) The life cycles of single-celled eukaryotes differ in the proportion of time spent as haploid (1*n*) versus diploid (2*n*) cells. (c) The life cycles of animals have many mitotic divisions between formation of the zygote and meiosis. (d) Vascular plants have two multicellular phases.

**a.** Unicellular eukaryote with prominent haploid phase

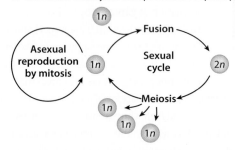

**b.** Unicellular eukaryote with prominent diploid phase

**c.** Animal

**d.** Vascular plant

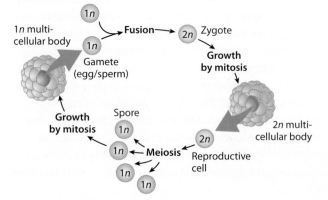

This cell grows and then germinates to form actively growing, skeletonized cells. In diatoms, then, short-lived gametes constitute the only haploid phase of the life cycle.

The two life cycles just introduced are similar in many ways. Both involve haploid cells that fuse to form diploid cells, and diploid cells that undergo meiotic cell division to generate haploid cells. Both life cycles also commonly include cells capable of persisting in a protected form when the environment becomes stressful. Why some single-celled eukaryotes usually occur as haploid cells and others generally occur as diploid cells remains unknown—a good question for continuing research.

Sexual reproduction has never been observed in some eukaryotes, but most species appear to be capable of sex, even if they reproduce asexually most of the time. Variations on the eukaryotic life cycle can be complex, especially in parasitic microorganisms that have multiple animal hosts. All, however, have the same fundamental components as the two life cycles described here.

In succeeding chapters, we discuss multicellular organisms in detail. Here we note only that animals, plants, and other complex multicellular organisms have life cycles with the same features as those just discussed. The big difference is that in animals, the zygote divides many times to form a multicellular diploid body before a small subset of cells within the body undergoes meiotic cell division to form haploid gametes (eggs and sperm). During fertilization, the egg and sperm combine sexually to form a zygote (**Fig. 27.3c**). In animals, as in diatoms, the only haploid phase of the life cycle is the gamete. As we will see in Chapter 32, plants have two multicellular phases in their life cycle, one haploid and one diploid (**Fig. 27.3d**). Like single-celled eukaryotes, many plants and animals can reproduce asexually. Humans and other mammals are unusual in that we have no capacity for asexual reproduction.

---

## 27.2 EUKARYOTIC ORIGINS

In the preceding section, we reviewed the basic features of eukaryotic cells. The distinctive cellular organization of Eukarya differs markedly from the simpler cells of Bacteria and Archaea. How did eukaryotes come to form such a remarkably diverse branch of the tree of life? In Case 5: The Human Microbiome, we saw that symbioses are common, even in humans. Is the relationship between humans and bacteria unusual, or are symbioses so fundamental that they played key roles in the evolution of the eukaryotic cell?

**? CASE 5 The Human Microbiome: Diversity Within**
**What role did symbiosis play in the origin of chloroplasts?**

The chloroplasts found in plant cells closely resemble certain photosynthetic bacteria, specifically cyanobacteria. The molecular workings of photosynthesis are nearly identical in the two, and the way that internal membranes organize the photosynthetic machinery of cyanobacteria closely resembles the way that the thylakoid membranes organize the photosynthetic machinery of chloroplasts (Chapter 8). The Russian botanist Konstantin Sergeevich Merezhkovsky recognized this similarity more than a century ago. He also understood that corals and some other organisms harbor algae within their tissues as symbionts that aid the growth of their host. A **symbiont** is an organism that lives in closely evolved association with another species. This association, called **symbiosis,** is discussed in Chapter 47.

Putting these two observations together, Merezhkovsky came up with a radical hypothesis. Chloroplasts, he argued, originated as symbiotic cyanobacteria that through time became permanently incorporated into their hosts. Merezhkhovsky's hypothesis of chloroplast origin by **endosymbiosis** (a symbiosis in which one partner lives within the other) was difficult to test with the tools available in the early twentieth century, and his idea was dismissed, more neglected than disproved, by most biologists.

In 1967, American biologist Lynn Margulis resurrected the endosymbiotic hypothesis, supporting her arguments with new types of data made possible by the then-emerging techniques of cell and molecular biology (**Fig. 27.4**). Transmission electron microscopy showed that structural similarities between chloroplasts and cyanobacteria extend to the submicrometer level, for example in

## HOW DO WE KNOW?

### FIG. 27.4

## What is the evolutionary origin of chloroplasts?

**HYPOTHESIS** Chloroplasts evolved from cyanobacteria living as endosymbionts within a eukaryotic cell.

**OBSERVATION** Chloroplasts and cyanobacteria have closely similar internal membranes that organize the light reactions of photosynthesis (Fig. 27.4a).

**EXPERIMENT** Like cyanobacteria, chloroplasts have DNA organized in a single circular chromosome. Phylogenies based on molecular sequence comparisons place chloroplasts among the cyanobacteria (Fig. 27.4b).

Transmission electron microscopy shows that cyanobacteria (top), red algal chloroplasts (middle), and plant chloroplasts (bottom) have very similar internal membranes that organize the light reactions of photosynthesis.

**CONCLUSION** Molecular and electron microscope data support the hypothesis that chloroplasts originated as endosymbiotic cyanobacteria.

**SOURCE** Giovannoni, S. J., S. Turner, G. J. Olsen, S. Barns, D. J. Lane, and N. R. Pace. 1988. "Evolutionary Relationships Among Cyanobacteria and Green Chloroplasts." *Journal of Bacteriology* 170:3584–3592.

the organization of the photosynthetic membranes (Fig. 27.4a). It also became clear that the chloroplasts in red and green algae (and in land plants) are separated from the cytoplasm that surrounds them by two membranes. This is expected if a cyanobacterial cell had been engulfed by a eukaryotic cell. The inner membrane corresponds to the cell membrane of the cyanobacterium, and the outer one is part of the engulfing cell's plasma membrane. In addition, the biochemistry of photosynthesis was found to be essentially the same in cyanobacteria and chloroplasts. Both use two linked photosystems, a common mechanism for extracting electrons from water, and the same reactions to reduce carbon dioxide ($CO_2$) into organic matter (Chapter 8).

Such observations kindled renewed interest in the endosymbiotic hypothesis, but the decisive tests were made possible by another, and unexpected, discovery. Chloroplasts, it turns out, have their own DNA, organized into a single circular chromosome, like that of bacteria. Armed with this knowledge, investigators could use the tools of molecular sequence comparison (Chapter 23). The sequences of nucleotides in chloroplast genes closely match those of cyanobacterial genes but are strikingly different from sequences of genes in the nuclei of photosynthetic eukaryotes (Fig. 27.4b). This finding provides strong support for the hypothesis of Merezkhovsky and Margulis. Chloroplasts are indeed the descendants of symbiotic cyanobacteria that lived within eukaryotic cells.

Eukaryotes were able to acquire photosynthesis because they could engulf and retain cyanobacterial cells. But engulfing another microorganism was just the beginning. The cyanobacteria most closely related to chloroplasts have 2000 to 3000 genes. In contrast, among photosynthetic eukaryotes, chloroplast gene numbers vary from just 60 to 200. Where did the rest of the cyanobacterial genome go? Some genes may simply have been lost through evolution; unneeded chloroplast genes may have been lost if similar nuclear genes could supply chloroplast requirements. Many, however, were transported to the nucleus when chloroplasts broke or by hitching a ride with viruses. The nuclear genome of the flowering plant *Arabidopsis* contains several thousand genes of cyanobacterial origin. Some code for proteins destined for use within the chloroplast.

Although foreign to mammals and other vertebrate animals, symbiosis between a heterotrophic host and photosynthetic partner is, in fact, common throughout the eukaryotic domain. Reef corals, for example, harbor photosynthetic cells that live symbiotically within their tissues. Some corals have even lost the capacity to capture food from surrounding waters. *Tridacna*, the giant clam of tropical Pacific waters, obtains its nutrition from symbiotic algae that live within its tissues.

Until recently, most biologists agreed that chloroplasts originated from endosymbiotic cyanobacteria only once, in a common ancestor of the green algae and red algae. Now,

**FIG. 27.5** **A second example of chloroplast endosymbiosis.** The photosynthetic amoeba *Paulinella chromatophora* acquired photosynthesis by endosymbiosis independently of and more recently than the endosymbiosis that gave rise to chloroplasts in green plants and other eukaryotes.

remarkably, a second case of cyanobacterial endosymbiosis has come to light. *Paulinella chromatophora* is a photosynthetic amoeba (**Fig. 27.5**). Its chloroplast genes show that its chloroplast originated from a branch of the cyanobacteria different from and arising long after the one that gave rise to chloroplasts in other photosynthetic eukaryotes. About a third to a half of the ancestral genome remains in *P. chromatophora* chloroplasts, suggesting that this organism reflects chloroplast evolution still in progress.

**? CASE 5 The Human Microbiome: Diversity Within What role did symbiosis play in the origin of mitochondria?**

Like chloroplasts, mitochondria closely resemble free-living bacteria in organization and biochemistry. And like chloroplasts, mitochondria contain DNA that confirms their close phylogenetic relationship to a form of bacteria—in this case proteobacteria (Chapter 26). Like chloroplasts, then, mitochondria originated as endosymbiotic bacteria.

Mitochondria are also like chloroplasts in having a small genome. Indeed, the mitochondrial genome is dramatically reduced compared to the ancestral proteobacterial genome, in most eukaryotes containing only a handful of functioning genes. Human mitochondria, for example, code for just 13 proteins and 24 RNAs. Once again, many genes from the original bacterial endosymbiont migrated to the nucleus, where they still reside.

Most eukaryotic cells contain mitochondria, but a few single-celled eukaryotic organisms found in oxygen-free environments do not. Biologists earlier hypothesized that these eukaryotes evolved

**FIG. 27.6** Transmission electron microscope images comparing the mitochondrion within the aerobic eukaryotic microorganism *Euplotes* (left) with a hydrogen-producing mitochondrion adapted for anaerobic metabolism in the closely related genus *Nictotherus* (right). The organelle in *Nictotherus* is halfway between a normal mitochondrion and a hydrogenosome, supporting the view that these two organelles are descended from a common ancestor.

before the endosymbiotic event that established mitochondria in cells having a nucleus, but that proposal turns out to be wrong. We know this because of the propensity for genes to migrate from the endosymbiont to the host's nucleus. Every mitochondria-free eukaryote examined to date has relic mitochondrial genes in its nuclear genome, providing support for a second hypothesis, that eukaryotic cells without mitochondria had them once but have lost them. In fact, many eukaryotes that lack mitochondria contain small organelles called hydrogenosomes that generate ATP by anaerobic processes (**Fig. 27.6**). These organelles have little or no DNA, but genes of mitochondrial origin in the cells' nuclei code for proteins that function in the hydrogenosome. Thus, hydrogenosomes appear to be highly altered mitochondria adapted to life in oxygen-poor environments. Other anaerobic eukaryotes contain still smaller organelles called mitosomes that also appear to be remnant mitochondria.

## CASE 5 The Human Microbiome: Diversity Within
### How did the eukaryotic cell originate?

If mitochondria originated as proteobacteria and chloroplasts are descended from cyanobacteria, where does the rest of the eukaryotic cell come from? Analysis of the nuclear genome alone provides no clear picture because genes of bacterial, archaeal, and purely eukaryotic origin are all present. As discussed, many nuclear genes originated with the mitochondria and chloroplasts acquired from specific bacteria. However, genes from other groups of bacteria also reside in the eukaryotic nucleus,

recording multiple episodes of horizontal gene transfer through evolutionary history. In contrast, some genes are present only in eukaryotes and apparently evolved after the domain originated. Still others are clearly related to the genes of Archaea, including the genes that govern DNA transcription and translation.

Two starkly different hypotheses have been proposed to explain this mix of genes. Some biologists believe that the host for mitochondrion-producing endosymbiosis was itself a true eukaryotic cell, with nucleus, cytoskeleton, and endomembrane system, but only limited ability to derive energy from organic molecules (**Fig. 27.7**). In this view, nuclear genes in Eukarya resemble those of Archaea because the primordial eukaryotic host cell was closely related to Archaea. Others, however, argue that no eukaryotic cell existed before there were mitochondria. Instead, they propose that the eukaryotic cell as a whole began as a symbiotic association between a proteobacterium and an archaeon (Fig. 27.7). The proteobacterium became the mitochondrion and provided many genes to the nuclear genome. The archaeon provided other genes, including those used to transcribe DNA and translate it into proteins.

Biologists continue to debate these alternatives. Both hypotheses address the hybrid nature of the eukaryotic genome, but neither explains the origins and evolution of the nucleus, linear chromosomes, the eukaryotic cytoskeleton, or a cytoplasm subdivided by ever-changing membranes. There is no consensus on the question of eukaryotic origins—it is one of biology's deepest unanswered questions, awaiting novel observations by a

**FIG. 27.7**   **Two hypotheses for the origin of the eukaryotic cell.** Both hypotheses lead to the evolution of a mitochondrion-bearing protist.

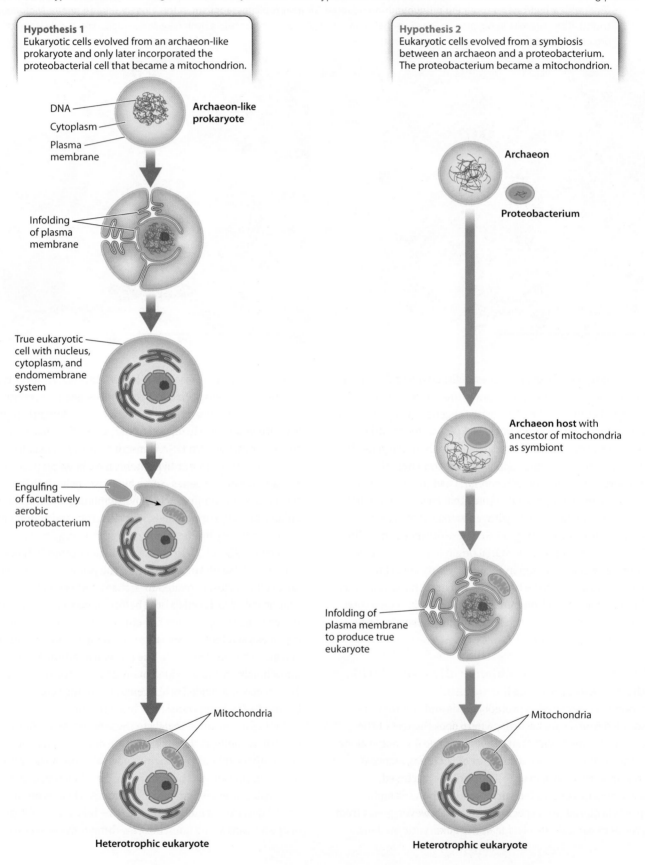

**Hypothesis 1**
Eukaryotic cells evolved from an archaeon-like prokaryote and only later incorporated the proteobacterial cell that became a mitochondrion.

DNA
Cytoplasm
Plasma membrane

**Archaeon-like prokaryote**

Infolding of plasma membrane

True eukaryotic cell with nucleus, cytoplasm, and endomembrane system

Engulfing of facultatively aerobic proteobacterium

Mitochondria

**Heterotrophic eukaryote**

**Hypothesis 2**
Eukaryotic cells evolved from a symbiosis between an archaeon and a proteobacterium. The proteobacterium became a mitochondrion.

**Archaeon**

**Proteobacterium**

**Archaeon host** with ancestor of mitochondria as symbiont

Infolding of plasma membrane to produce true eukaryote

Mitochondria

**Heterotrophic eukaryote**

**FIG. 27.8** Scanning electron microscope image of bacterial cells on the surface of a single-celled eukaryote found in oxygen-depleted sediments in the Santa Barbara Basin, off the coast of California. The bacteria are hypothesized to metabolize the $H_2S$ in this environment, thereby protecting the eukaryote that they enclose.

2 µm

new generation of biologists. Once a dynamic cytoskeleton became coupled to a flexible membrane system, however, the evolutionary possibilities of eukaryotic form were established.

### In the oceans, many single-celled eukaryotes harbor symbiotic bacteria.

The evolution of the eukaryotic cell is one of intimate associations between formerly free-living organisms. An unexpected symbiosis was recently discovered in the Santa Barbara Basin, a local depression in the seafloor off the coast of southern California. Sediments accumulating in this basin contain little oxygen but large amounts of hydrogen sulfide ($H_2S$) generated by anaerobic bacteria within the sediments. We might predict that eukaryotic cells would be uncommon in these sediments. Not only is oxygen scarce, but sulfide actively inhibits respiration in mitochondria. It came as a surprise, then, that samples of sediment from the Santa Barbara Basin contain large populations of single-celled eukaryotes.

Some of these cells have hydrogenosomes, mitochondria altered by evolution to generate energy where oxygen is absent. Others, however, appear to thrive by supporting populations of symbiotic bacteria on or within their cells. In one case, in which rod-shaped bacteria cover the surface of their eukaryotic host (**Fig. 27.8**), research has shown that the bacteria metabolize sulfide in the local environment, thereby protecting their host from this toxic substance. Scientists hypothesize that the bacteria benefit from this association as well because they get

a free ride through the sediments, enabling them to maximize chemoautotrophic growth by remaining near the boundary between waters that contain oxygen and those rich in sulfide.

The diversity of these eukaryotic–bacterial symbioses is remarkable—and poorly studied. What has been learned to date, however, shows that single-celled eukaryotes have evolved numerous symbiotic relationships with chemoautotrophic bacteria, associations that feed and protect the eukaryotes, enabling them to colonize habitats where most eukaryotes cannot live. Clearly, then, the types of symbioses that led to mitochondria and chloroplasts on the early Earth were not rare events, but basic associations between cells that continue to evolve today.

## 27.3 EUKARYOTIC DIVERSITY

Historically, Domain **Eukarya** was divided into four kingdoms: plants, animals, fungi, and protists. Plants, animals, and fungi received special consideration for the obvious reason that they include large and relatively easily studied species. All remaining eukaryotes were grouped together as **protists,** which were defined as organisms having a nucleus but lacking other features specific to plants, animals, or fungi. The term "protist" is frowned on by some biologists because it doesn't refer to a monophyletic grouping of species—that is, an ancestral form and all its descendants. Nonetheless, as an informal term that draws attention to a group of organisms sharing particular characteristics, "protist" usefully describes the diverse world of microscopic eukaryotes and seaweeds.

Two other terms scorned by systematists but embraced by ecologists are "algae" and "protozoa." **Algae** are photosynthetic protists. They may be microscopic single-celled organisms or the highly visible, multicelled organisms we call seaweed. **Protozoa** are heterotrophic protists. These are almost exclusively single-celled organisms. "Algae" and "protozoa" in fact are simple and useful terms that have little phylogenetic meaning but which convey a great deal of information about the structure and function of these organisms.

Protist cells exhibit remarkable diversity. Some have cell walls, while others do not. Some make skeletons of silica ($SiO_2$) or calcium carbonate ($CaCO_3$), and others are naked or live within **tests,** or "houses," constructed exclusively of organic molecules. Some are photosynthetic, while others are heterotrophic. Some move by beating flagella, and others, like *Amoeba*, can extend fingers of cytoplasm, called pseudopodia, to move and capture food. Most are aerobic, but there are anaerobic protists as well. Surprisingly, most of these distinctive features have evolved multiple times and so do not define phylogenetically coherent groups. As a result, our understanding of evolutionary pattern in Eukarya had to wait for the molecular age.

**FIG. 27.9**    **The eukaryotic tree of life showing major groups.**
Branches with photosynthetic species (shown in green)
are distributed widely. Additional branches, including
animals, foraminiferans, radiolarians, and ciliates, contain
species that harbor photosynthetic symbionts.

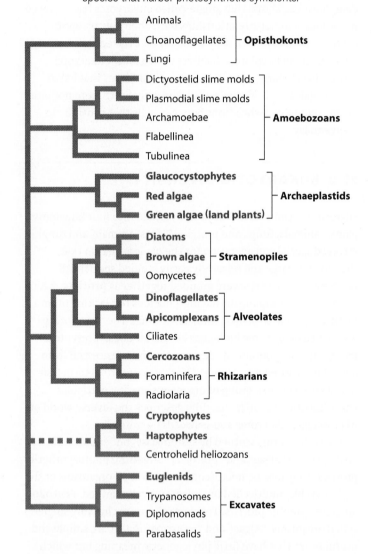

**Fig. 27.9** shows a phylogenetic tree of eukaryotes as biologists
currently understand it. As in the case of Bacteria, it is easier
to identify major groups of Eukarya than to establish their
evolutionary relationships to one another. Molecular sequence
comparisons consistently recognize seven major groups,
commonly called **superkingdoms,** noted in the figure. Animals
fall within one of these superkingdoms, the Opisthokonta,
whereas plants fall within another, the Archaeplastida. Most,
but not all, of the divisions into these major eukaryotic branches
are strongly supported by data from genes and genomes. A few

groups, notably the cryptophytes, haptophytes, and centrohelid
heliozoans, seem to wander from position to position from one
phylogenetic analysis to the next, sometimes moving together,
sometimes separately. Whether the well-resolved superkingdoms
constitute the only major limbs on the eukaryotic tree is less
certain. Many microscopic protists remain unstudied, and we may
yet be surprised by what they ultimately show.

In the following sections, we introduce the major features of
eukaryotic diversity. The species numbers we present indicate
known taxonomic diversity, and we emphasize that most
protistan species remain unknown. Many biologists estimate that
only about 1 in 10 protist species has been described so far. As this
is also the case for animals and fungi (but arguably not for land
plants because they are all visible to the naked eye), the relative
numbers of species within different superkingdoms may reflect
something like the actual distribution of biodiversity among
eukaryotes.

## Our own group, the opisthokonts, is the most diverse eukaryotic superkingdom.

Over the past 250 years, biologists have described about
1.8 million species. Of these, 75% or so fall within the
Opisthokonta (**Fig. 27.10**), a name derived from Greek words
meaning "posterior pole," calling attention to the fact that
cell movement within this group is propelled by a single
flagellum attached to the posterior end of the cell. Not all cells
in this superkingdom move around. In humans, for example,
the flagellum is limited mostly to sperm. Opisthokonts are
heterotrophic, although some species harbor photosynthetic
symbionts. Animals are the most diverse and conspicuous
opisthokonts. More than 1.3 million animal species, mostly insects
and their relatives, have been described. Fungi are also diverse,
with more than 75,000 described species, including the visually
arresting mushrooms (Fig. 27.10a). Here, however, we focus on
opisthokont protists, somewhat poorly known microorganisms
that hold clues to the origins of complex multicellularity in this
superkingdom (Chapter 28).

It isn't easy to find morphological characters that unite all
opisthokonts. We face this difficulty partly because animals and
fungi have diverged so strikingly from what must have been
the ancestral condition of the group. However, as molecular
sequence comparisons began to reshape our understanding
of eukaryotic phylogeny, it soon became clear that fungi and
animals are closely related.

Even more closely related to the animals are
**choanoflagellates,** a group of mostly unicellular protists
characterized by a ring of microvilli, fingerlike projections that
form a collar around the cell's single flagellum (Fig. 27.10b). About

**FIG. 27.10 Opisthokonts.** This superkingdom includes animals and fungi, as well as several protistan groups, including choanoflagellates, the closest protistan relatives of animals.

**a.**

**b.** Choanoflagellates

150 choanoflagellate species have been described from marine and freshwater environments, where they prey on bacteria. As early as 1841, the close similarity between choanoflagellates and the collared feeding cells of sponges suggested that these minute organisms might be our closest protistan relatives. This view gained widespread popularity with the discovery in 1880 of *Proterospongia haeckeli*, a choanoflagellate that lives in colonies of cells joined by adhesive proteins.

Molecular sequence comparisons now confirm the close relationship between animals and choanoflagellates. For example, the complete genome of the choanoflagellate *Monosiga brevocollis* shows that a number of signaling molecules known to play a role in animal development are present in our choanoflagellate relatives, although their function in these single-celled organisms remains largely unknown. More generally, many genes once thought to be unique features of animals have now been identified in choanoflagellates, underscoring the deep evolutionary roots of animal biology.

Microsporidia form another group of single-celled opisthokonts. Microsporidia are parasites that live inside animal cells. Only their spores survive in the external environment, where they await the opportunity to infect a host and complete

their life cycle. Microsporidia infect all animal phyla, so the approximately 1000 known species probably represent only the tip of the iceberg. More than a dozen species have been isolated from human intestinal tissues. Microsporidia infections can be particularly devastating in AIDS patients. Beyond their importance in human health, microsporidians have attracted the attention of biologists because of features they *lack*: Microsporidian cells have no aerobically respiring mitochondria, no Golgi apparatus, and no flagella. Furthermore, they have a highly reduced metabolism and among the smallest genomes of any known eukaryote. The cellular simplicity of microsporidians does not mean that they are early-evolved organisms, however. Molecular sequencing studies show that microsporidians are the descendants of more complex organisms. Their simplicity then is an adaptation for life as an intracellular parasite. It is now widely accepted that microsporidians are closely related to the fungi.

Other protistan opisthokonts have been identified in recent years. These are mostly parasites of aquatic animals, but like the choanoflagellates described earlier, their biology is beginning to illuminate the evolutionary path to complex multicellularity in animals and fungi.

**FIG. 27.11 Amoebozoans.** This group includes protists with an amoeboid stage in their life cycle.

**a.**

- Opisthokonts
- **Amoebozoans**
- Archaeplastids
- Stramenopiles
- Alveolates
- Rhizarians
- Excavates

**b.** *Amoeba proteus*, with clearly visible pseudopodia

200 μm

**c.** *Entamoeba histolytica*, the infectious agent in amoebic dysentery

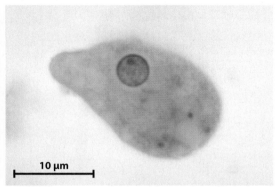

10 μm

## Amoebozoans include slime molds that produce multicellular structures.

As their name implies, the superkingdom **Amoebozoa** (**Fig. 27.11**) is a group of eukaryotes with amoeba-like cells that move and gather food by means of pseudopodia (illustrated by *Amoeba proteus* in Fig. 27.11b). More than 1000 amoebozoan species have been described. In some amoebozoan species, large numbers of cells aggregate to form multicellular structures as part of their life cycle.

Although there are amoeba-like cells that are not members of the Amoebozoa superkingdom, all members of this group are basically amoeboid in cellular organization, at least at some stage of their life cycle. Amoebozoans play an important role in soils as predators on other microorganisms. Some amoebozoans also influence human health. *Entamoeba histolytica* (Fig. 27.11c), an anaerobic protist that causes amoebic dysentery, is responsible for 50,000–100,000 deaths every year.

Beyond considerations of human health, the amoebozoans of greatest biological interest are the slime molds. In plasmodial slime molds (**Fig. 27.12**), haploid cells fuse to form zygotes that subsequently undergo repeated rounds of mitosis but not cell division to form colorful, often lacy structures visible to the naked eye (Fig. 27.12a). These structures, called plasmodia, are **coenocytic,** which means they contain many nuclei within one giant cell. The plasmodia can move and so seek out and feed on the bacteria and small fungi commonly found on bark or plant litter on the forest floor. Triggered by poorly understood environmental signals, plasmodia eventually differentiate to form stalked structures called sporangia that can be 1 to 2 mm high (Fig. 27.12b). Within the mature sporangium, cell walls form around the many nuclei, producing discrete cells that undergo meiosis, generating haploid spores that disperse into the environment. Germination of these spores begins the life cycle anew.

Cellular slime molds (**Fig. 27.13**) spend most of their life cycle as solitary amoeboid cells feeding on bacteria in the soil.

**FIG. 27.12 Plasmodial slime molds.** (a) Plasmodia, such as this pretzel slime mold, are coenocytic structures containing many nuclei. (b) Plasmodia generate sporangia, stalked structures that produce spores for dispersal.

a

10 mm

b

1 mm

Starvation, however, causes cells to produce the chemical signal cyclic AMP (Chapter 9), which induces as many as 100,000 cells to aggregate into a large multicellular slug-like form (Fig. 27.13a). This "slug" can migrate by a coordinated movement of its cells governed by actin and myosin, the same proteins that control muscle movement in animals. Migrating "slugs" can forage for food, leaving behind a trail of mucilage as they move—that's why they're called slime molds. Like plasmodial slime molds, the "slugs" of cellular slime molds eventually differentiate to form stalk-borne sporangia that produce the spores responsible for dispersal of the organism and renewal of the life cycle (Fig. 27.13b).

Slime molds are not true intermediates in the evolution of multicellularity—that is, they are not closely related to animals, fungi, or other groups in which complex multicellularity evolved. Nonetheless, the cellular slime mold *Dictyostelium* has been extensively studied by biologists as a model organism because it has much to teach us about fundamental eukaryotic processes of cell signaling and differentiation, cellular motility, and mitosis.

## Archaeplastids are photosynthetic organisms, including land plants.

Next to the opisthokonts, the most conspicuous and diverse eukaryotes fall within the superkingdom **Archaeplastida,** another major branch of the eukaryotic tree (**Fig. 27.14a**). This is where we find the land plants, whose 300,000 described species dominate eukaryotic biomass on this planet (Chapter 33). The archaeplastids contain three major phylogenetic groups, and with the exception of a few parasitic and saprophytic (living on dead matter) species, all are photosynthetic. This indicates that the last common ancestor of living archaeplastids was photosynthetic. In fact, archaeplastids are the direct descendants of the protist that first evolved chloroplasts from endosymbiotic cyanobacteria. Some molecular analyses also place the group that includes cryptophyte and haptophyte algae within this branch, but this remains a topic of continuing research.

The least conspicuous group of archaeplastids are the glaucocystophytes (**Fig. 27.14b**), a small group of single-celled algae found in freshwater ponds and lakes. We might skip over them entirely, except for one illuminating observation: Glaucocystophyte chloroplasts appear to retain more features of the ancestral cyanobacterial endosymbiont than any

**FIG. 27.13 Cellular slime molds.** (a) Starvation causes amoeboid feeding cells of cellular slime molds to aggregate into a multicellular "slug" that can migrate along surfaces, leaving behind a trail of slime. (b) "Slugs" differentiate to form stalked sporangia that produce spores.

other algae. Their chloroplasts have walls of peptidoglycan, the same molecule found in cyanobacterial walls, and their photosynthetic pigments include biliproteins, also found in cyanobacteria.

More abundant and diverse is the second major archaeplastid group, the red algae (**Fig. 27.14c**). About 5000 species are known, mostly from marine environments. Most are multicellular and some have differentiated tissues and complex morphology. Like other archaeplastids, most also have walls made of cellulose. The principal photosynthetic pigments in red algae are a form of chlorophyll called chlorophyll *a* and the biliproteins also found in glaucocystophytes and most cyanobacteria.

Red algae can be conspicuous as seaweeds on the shallow seafloor. One subgroup of red algae, known as the coralline algae, secretes skeletons of calcium carbonate ($CaCO_3$) that deter predators and help the algae withstand the energy of waves in shallow marine environments. If you examine the margin of coral reefs, where waves crash into the reef, you will often see that it is coralline algae and not the corals themselves that resist breaking waves and so enable reefs to expand upward into wave-swept environments. While most common and diverse in near-shore environments, red algae can also grow where light penetrates most deeply in the ocean. Their photosynthetic pigments enable them to take advantage of the short (blue) wavelengths of light.

Red algae are important as food in certain cultures, for example, laver in the British Isles (particularly in Wales, in cakes similar to oatcakes) and nori in Japan. Without knowing it, you often ingest carageenan, a polysaccharide synthesized by red algae. Carageenan molecules stabilize mixtures of biomolecules

**FIG. 27.14  Archaeplastids.** (a) Archaeplastids are a photosynthetic group descended from the protist that acquired photosynthesis from an endosymbioic cyanobacterium. They include the (b) glaucocystophytes, (c) red algae, and (d) green algae.

a.

Opisthokonts

Amoebozoans

**Archaeplastids**

Stramenopiles

Alveolates

Rhizarians

Excavates

◄ Most diverse groups

☆ Groups with complex multicellularity

■ Groups containing photosynthetic organisms

Glaucocystophytes

Red algae ☆

Chlorophytes

Flagellated unicells

**Green algae**

Streptophyte algae

Viridoplantae

Land plants ☆

b. Glaucocystophyte

2 µm

c. Red algae

100 µm

d. Green algae

10 µm

and promote gel formation, and so they are used in products ranging from ice cream to toothpaste. Agar, another red algal polysaccharide, is found universally in microbiology laboratories because the gel it forms at room temperature provides a nearly ideal substrate for bacterial growth.

The land plants and their algal relatives form Viridoplantae, the third branch of Archaeplastida. In this chapter, we are concerned with the protistan members of this group, the green algae (**Fig. 27.14d**). Green algae include a diverse assortment of species (about 10,000 have been described) that live in marine and, especially, freshwater environments. Green algae range from tiny single-celled flagellates to meter-scale seaweeds (**Fig. 27.15**). They all are united by two features: the presence of chlorophyll *a* and chlorophyll *b* in chloroplasts that have two membranes, and a unique attachment for flagella.

It is thought that green algae originated as small flagellated cells in seawater, and today most species that branch early on the green algal tree live as photosynthetic cells in the water. These species are called phytoplankton, and one of them is illustrated in Fig. 27.15a. The larger diversity of green algae, however, is found on two branches that diverged from these flagellated ancestors. One branch, the chlorophytes, radiated mostly in the sea and includes common seaweeds such as sea lettuce, which is seen globally on seashores as flotsam. The chlorophyte branch also includes relatively large and complex seaweeds that can form structures up to tens of centimeters long in tropical and temperate oceans (Fig. 27.15e). Moreover, chlorophyte green algae have played a major role in laboratory studies of photosynthesis (*Chlamydomonas,* Fig. 27.15a) and multicellularity (*Volvox,* Fig. 27.15d).

**FIG. 27.15 Diversity of green algae.** (a) *Chlamydomonas reinhardtii*, a tiny flagellated unicell widely used in laboratory research on photosynthesis; (b) *Spirogyra*, a freshwater green alga with distinctive helical chloroplasts; (c) a desmid, among the most diverse and widespread of all green algae in freshwater; (d) *Volvox*, a simple multicellular organism; and (e) *Acetabularia*, a macroscopic (but single-celled!) green alga found in coastal marine waters in tropical climates.

For humans, however, the more important branch of the green algal tree is the one that diversified on land. Called streptophytes, the species on this branch show a progression of form from unicells at the base through simple cell clusters and filaments on intermediate branches (see Figs. 27.15b and 21.15c), to complex multicellular algae that form the closest relatives of land plants (Chapter 33). As was true of choanoflagellates and animals, green algae display features of cell biology and genomics that tie them unequivocally to land plants.

→   **Quick Check 2** Do plants and animals have a common multicellular ancestor?

## Other photosynthetic organisms occur in the stramenopiles and alveolates.

The superkingdom **Stramenopila** forms a diverse and distinctive branch of the eukaryotic tree (see Fig. 27.9). The group includes unicellular organisms and giant kelps, algae and protozoa,

free-living cells and parasites. All have an unusual flagellum that bears two rows of stiff hairs (**Fig. 27.16**). Most also have a second, smooth flagellum.

**FIG. 27.16 Stramenopiles.** This electron micrograph of a unicellular stramenopile shows a flagellum with two rows of stiff hairs. Like this cell, most stramenopiles also have a second, smooth flagellum.

**FIG. 27.17** Kelps, complex multicellular brown algae that can form "forests" tens of meters high in the ocean.

Despite the presence of heterotrophic species, most stramenopiles are photosynthetic. There are about a dozen groups of stramenopile algae, of which the diatoms and the brown algae deserve special mention. Brown seaweeds are common along rocky shorelines across the world, and in the Sargasso Sea, huge masses of ropy brown algae (aptly named *Sargassum*) float at the surface. Easily the most impressive brown algae are the kelps, giant seaweeds that form forests above the seafloor (**Fig. 27.17**). Kelps not only develop functionally distinct tissues, but have evolved distinct organs specialized for attachment, photosynthesis, flotation, and reproduction.

The most diverse stramenopiles, however, are the diatoms (**Fig. 27.18**). Recognized by their distinctive skeletons of silica, diatoms account for half of all primary production in the sea (and, so, nearly a quarter of all photosynthesis on Earth). The 10,000 known diatom species thrive in environments that range from wet soil to the open ocean.

Among heterotrophic stramenopiles, the oomycetes, or water molds, are particularly interesting. Originally classified as fungi, these multicellular pathogens cause a number of plant diseases, including the potato blight that devastated Ireland during the 1840s (Chapter 33). Other species infect animals, including humans.

Like stramenopiles, the superkingdom **Alveolata** is a diverse group. It includes dinoflagellates (**Fig. 27.19a**), less than half of which contain chloroplasts and are photosynthetic, and ciliates (**Fig. 27.19b**), heterotrophic protists that have two nuclei in each cell and numerous short flagella called cilia. The group is united by the presence of cortical alveoli, small vesicles packed beneath the cell surface that, in some species at least, store calcium ions for use by the cell (**Fig. 27.19c**). Dinoflagellates are important as photosynthesizers in many parts of the oceans, and they affect humans when they form red tides, large and toxic blooms in coastal waters. Red tides occur when nutrients are supplied to coastal waters in large amounts. This may occur for natural reasons, but it commonly reflects human activities such as sewage seeps or runoff from highly fertilized croplands.

Scientists have described nearly 4000 dinoflagellate and 8000 ciliate species, but it may be the third principal group of alveolates, the Apicomplexans, that attracts most scientific attention. The reason is simple: The apicomplexan *Plasmodium falciparum* and related species cause malaria, a lethal disease that sickened millions and killed more than 600,000 people in 2011 (Case 4: Malaria). Although *Plasmodium* cells are heterotrophic, they contain vestigial chloroplasts that suggest evolutionary loss of photosynthesis. Recently, this hypothesis has been confirmed by the discovery in Sydney Harbour, Australia, of photosynthetic cells closely related to parasitic *Plasmodium*.

### Photosynthesis spread through eukaryotes by repeated endosymbioses involving eukaryotic algae.

A curious feature of the eukaryotic tree is that photosynthetic eukaryotes are distributed widely but discontinuously among its branches (see Fig. 27.9). How can we explain this pattern?

Two hypotheses have been suggested: Either photosynthesis was established early in eukaryotic evolution and was subsequently lost in some lineages, or eukaryotes acquired photosynthesis multiple times by repeated episodes of endosymbiosis. The second hypothesis, it turns out, is correct. The twist is that most of the symbionts involved in the spread of photosynthesis across eukaryotic phylogeny were photosynthetic eukaryotes, not cyanobacteria.

Once more, DNA sequence comparisons provide the decisive data. We can construct phylogenetic trees for photosynthetic

**FIG. 27.18 Diatoms.** Diatoms are the most diverse stramenopiles, and among the most diverse of all protists. Their shapes, outlined by their tiny skeletons made of silica ($SiO_2$) can be (a) like a pill-box, (b) elongated, or (c) twisted around the long axis of the cell. Some diatom species form colonies, like those shown in (d).

**FIG. 27.19 Alveolates.** (a) Dinoflagellates have cell walls made of interlocking plates of cellulose. (b) A ciliate has many short flagella, or cilia, that line the cell. (c) The arrows point to alveoli, flattened vesicles just beneath the surface of the cell that give alveolates their name.

eukaryotes using chloroplast genes and then repeat the exercise using nuclear genes. **Fig. 27.20a** shows a phylogenetic tree for chloroplasts. Notice that this tree differs substantially from the branching order in trees based on comparison of nucleotide sequences for nuclear genes, such as the tree shown in **Fig. 27.20b**. The differences between the chloroplast and nuclear trees favor the hypothesis that photosynthesis spread through the eukaryotes by means of repeated symbiotic events.

Let's look more closely at the chloroplast phylogeny in Fig. 27.20a. The lower part of the phylogeny shows that chloroplasts form a monophyletic grouping nested within the cyanobacteria. This is a principal line of evidence supporting the endosymbiotic origin of chloroplasts. Moving upward into the chloroplast branch, if we look only at the branches shown in black, we see the phylogeny indicated by nuclear genes for the archaeplastids—glaucocystophytes are sister to red algae and green algae. But notice the branches in green that sprout from within the green algae, the chlorarachniophytes and the euglenids. Both of these groups have chloroplasts that contain the same pigments as green algae (chlorophyll *a* and chlorophyll *b*), and molecular sequence comparison clearly indicates that their chloroplasts are closely related to the chloroplasts of green algae. Nuclear genes, however, place these groups far from the archaeplastids. Chlorarachniophytes are found within the superkingdom Rhizaria, and euglenids fall among the superkingdom Excavata. How do we explain this discordance? The answer is that chlorarachniophytes gained photosynthesis by establishing green algal cells (containing chloroplasts derived from an earlier symbiotic assimilation of cyanobacteria) as endosymbionts. Euglenids did the same thing, independently.

FIG. 27.20

# How did photosynthesis spread through the Eukarya?

**BACKGROUND** Many different branches on the eukaryotic tree include photosynthetic species, commonly interspersed with non-photosynthetic lineages. Did eukaryotes gain photosynthesis once, by the endosymbiotic incorporation of a cyanobacterium, and then lose this capacity multiple times? Or, is the history of photosynthetic endosymbioses more complicated, involving multiple events of symbiont capture and transformation?

**a.** Phylogenetic relationships based on chloroplast genes

**b.** Phylogenetic relationships based on nuclear genes

**HYPOTHESIS 1** Photosynthesis was established early in eukaryotic evolution and was subsequently lost in some lineages.

**HYPOTHESIS 2** Eukaryotes acquired photosynthesis multiple times by repeated episodes of endosymbiosis.

**PREDICTIONS** The two hypotheses make different predictions. If photosynthesis were acquired in a single common ancestor of all living eukaryotes and then lost in some lineages, we would expect chloroplast and nuclear gene phylogenies to show the same pattern of branching. If photosynthesis were acquired multiple independent times, we would expect different chloroplast and nuclear phylogenies to show different patterns.

**EXPERIMENT** Chloroplasts have DNA, so scientists developed a phylogenetic hypothesis based on molecular sequence comparison of chloroplast genes (Fig. 27.20a). This was then compared with a second phylogeny, for all eukaryotes, based on molecular sequence comparison of nuclear genes (Fig. 27.20b).

**RESULTS** Molecular sequence comparisons show that evolutionary relationships among chloroplasts do not mirror those based on nuclear genes. The chlorarachniophyte algae and photosynthetic euglenids (shown in green) fall in a monophyletic group with the green algae when chloroplast DNA is analyzed (Fig. 27.20a). However, these groups lie far from green algae in the phylogeny based on nuclear genes (Fig. 27.20b). Similarly, brown algae, diatoms, most photosynthetic dinoflagellates, and cryptophyte algae (shown in red) form a monophyletic group with the red algae in the phylogeny based on chloroplast DNA (Fig. 27.20a), but analysis of nuclear genes places them in different groups (Fig. 27.20b).

**CONCLUSION** The chloroplast and nuclear phylogenies show different patterns of branching, supporting the hypothesis that photosynthesis spread through the Eukarya by means of multiple eukaryotic endosymbionts.

**SOURCE** Hackett, J. D., H.S. Yoon, N. J. Butterfield, M. J. Sanderson, and D. Bhattacharya. 2007. "Plastid Endosymbiosis: Sources and Timing of the Major Events." In *Evolution of Primary Producers in the Sea*, edited by P. G. Falkowski and A. H. Knoll, 109–132. Boston: Elsevier Academic Press.

Now focus on the branch dominated by red algae. Once again, the branches shown in red indicate photosynthetic eukaryotes whose nuclear genes place them far away from red algae on phylogenetic trees. Thus, groups like the photosynthetic stramenopiles (diatoms, kelps, and others), photosynthetic dinoflagaellates, and a few others also appear to have arisen by endosymbiosis involving another photosynthetic endosymbiont that was eukaryotic—in this case, a red algal cell.

Several features of cell biology support this view. We noted earlier that chloroplasts in green and red algae are bounded by two membranes, but most of the algae shown in red and green in Fig. 27.20 have four membranes around their chloroplasts. The presence of four membranes is expected under the hypothesis of eukaryotic endosymbionts: The inner two membranes are those that bounded the chloroplast in the green or red algal endosymbiont, the third membrane is derived from the outer cell membrane of the symbiont, and the fourth, outer membrane is part of the cell membrane of the host (**Fig. 27.21**).

Chlorarachniophyte (and, on the red branch, cryptophyte) algae also retain another feature that supports the eukaryotic origin of their chloroplasts: a small organelle called a nucleomorph. The nucleomorph in chlorarachniophyte algae is, in fact, the remnant nucleus of a green algal symbiont. It retains only a few genes, but those show close phylogenetic affinity with green algae. Nucleomorph genes in cryptophyte algae similarly provide evidence for a red algal symbiont.

How many endosymbiotic events are required to account for the diversity of photosynthetic eukaryotes? There was the primary endosymbiosis that gave rise to green and red algae, two further events that established chloroplasts derived from green algae in euglenids and chlorarachniophytes, and at least one to three events that established chloroplasts derived from red algae in cryptophytes, haptophytes, and stramenopiles.

The dinoflagellates add a remarkable coda to this accounting. Like stramenopiles, most dinoflagellates have chloroplasts derived from red algae. Three dinoflagellate groups, however, have chloroplasts that are clearly derived from green, cryptophyte, and haptophyte algae. Include *Paulinella* and at least 8 to 10 endosymbiotic events are required to explain eukaryotic photosynthesis. Eukaryotes never evolved oxygenic photosynthesis— only the cyanobacteria accomplished that. Instead, eukaryotes appropriated the biochemistry of cyanobacteria and spread it throughout the domain by the horizontal transfer of entire cells.

→   **Quick Check 3**  How do the molecular sequences of genes in chloroplasts show how photosynthesis spread throughout the eukaryotic domain?

## The first branch of the eukaryotic tree may separate animals and slime molds from plants and diatoms.

Molecular sequence comparisons have succeeded in dividing eukaryotic species into the seven superkingdoms, but determining relationships among these major branches has proved difficult. This uncertainty is apparent in Fig. 27.9, which shows seven major branches of eukaryotic organisms but, for the most part, does not suggest how they relate to one another. Biologists call this an unrooted tree because we cannot go from the tips of the finest branches (present-day organisms) and trace backward through successive branch points to arrive at the "root"—the last common ancestor of the total group.

---

**FIG. 27.21**  Successive endosymbiotic events that established photosynthesis in chlorarachniophyte algae.

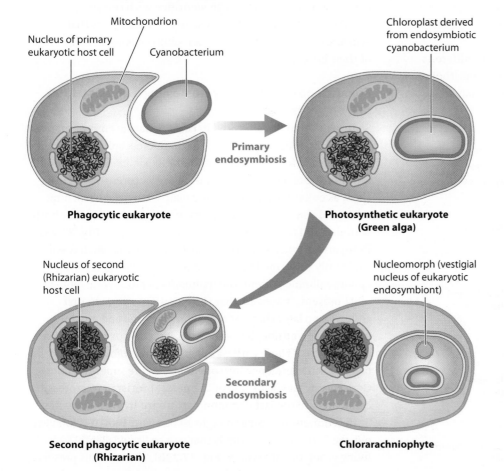

Nucleus of primary eukaryotic host cell

Mitochondrion

Cyanobacterium

Chloroplast derived from endosymbiotic cyanobacterium

**Primary endosymbiosis**

**Phagocytic eukaryote**

**Photosynthetic eukaryote (Green alga)**

Nucleus of second (Rhizarian) eukaryotic host cell

Nucleomorph (vestigial nucleus of eukaryotic endosymbiont)

**Secondary endosymbiosis**

**Second phagocytic eukaryote (Rhizarian)**

**Chlorarachniophyte**

As discussed in Chapter 23, biologists use synapomorphies, features shared by some but not all members of a group through descent from a common ancestor, to recognize sister groups, or closest evolutionary relatives. Skull shape, for example, unites birds and crocodiles, while separating them from other tetrapod vertebrates. Can we find characters that will allow us to identify sister-group relationships among the eukaryotic superkingdoms? It isn't easy, because the groups diverged more than a billion years ago and have accumulated so many differences in their cellular features that synapomorphies among superkingdoms are hard to identify. This is true of molecular sequences as well as features of cell shape and internal structure.

A leading hypothesis for sorting the superkingdoms reflects a seemingly minor feature of gene organization, but one that has so far been successful in dividing known eukaryotes into two groups. In opisthokonts and amoebozoans, genes that code for the enzymes dihydrofolate reductase and thymidylate synthase do not lie next to each other on a single chromosome. In all other eukaryotes, the two genes are fused together along one chromosome. As the two genes are also separate in bacteria, this observation suggests that the gene fusion observed in many eukaryotes is a synapomorphy that unites most eukaryotes and separates them from the branch we share with fungi and amoebozoans. Other lines of evidence support this interpretation, particularly the observation that opisthokont and amoebozoan cells have only one flagellum, but other eukaryotes (if they have them at all) usually have two or more.

This being the case, we can draw the tree in two different ways, as shown in **Fig. 27.22**. The hypothesis that opisthokonts and amoebozoans are sisters and that these two groups

together are sister to all other eukaryotes (the top tree) is consistent with the data on gene location and flagella number. But so is an alternative hypothesis—that the opisthokonts are sister to all other eukaryotes, with gene fusion occurring near the base of the limb that contains archaeplastids, stramenopiles, alveolates, rhizarians, and excavates (the bottom tree). At present, we lack data to choose between these hypotheses, but the resolution of branching pattern in eukaryotic phylogeny will help us understand which features were present in the last common ancestor of living eukaryotes and how the immense diversity of eukaryotic organisms evolved from those humble beginnings.

## 27.4 THE FOSSIL RECORD OF PROTISTS

Not all algae and protozoa fossilize easily, but those that do provide a striking record of evolution that both predates and parallels the better-known fossil record of animals. Fossilized protists do not contain DNA, nor do they preserve the features of cell biology that define eukaryotes, such as nuclear membranes and organelles. Thus, for a fossil to be identified as eukaryotic, it must preserve other features of morphology found today in Eukarya, but not in Bacteria or Archaea. For this reason, fossils that can be identified with certainty as eukaryotic are limited to those left by eukaryotes that synthesized distinctive and preservable cell walls at some stage of their life cycle.

### Fossils show that eukaryotes existed at least 1800 million years ago.

Sedimentary rocks deposited in coastal marine environments 1800–1400 million years ago contain microscopic fossils (called **microfossils** for short) up to about 300 μm in diameter. Many of these have complicated wall structures that identify them as eukaryotic. Interlocking plates, long and branching arms, and complex internal layering have been imaged by both scanning and transmission electron microscopes (**Fig. 27.23a**). Comparison with living organisms suggests that such fossils could only be formed by organisms with a cytoskeleton and endomembrane system, the hallmarks of eukaryotic biology.

At present, it appears that the eukaryotic domain first appeared no later than 1800 million years ago and possibly much earlier, although the record is difficult to trace in older rocks. None of these earliest eukaryotic microfossils can be placed with confidence into one of the living superkingdoms. Conceivably, at least some of the oldest protistan microfossils could predate the last common ancestor of living eukaryotes.

In sedimentary rocks thought to be about 1200 million years old, we find the oldest fossils that can be linked clearly to a living group of eukaryotes (**Fig. 27.23b**). These fossils preserve features found today only in the red algae. We can therefore

**FIG. 27.22** Two possible phylogenies for eukaryotes based on number of flagella and whether the genes for dihydrofolate reductase and thymidylate synthase are close or far apart.

**FIG. 27.23 Eukaryotic fossils in Precambrian sedimentary rocks.** (a) Simple protist in 1500-million-year-old rocks from Australia. (b) Simple multicellular red algae in 1200-million-year-old rocks from Canada. (c) Amoebozoan test in 750-million-year-old rocks from the Grand Canyon, Arizona.

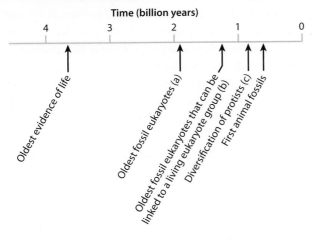

appears to have declined while oxygen increased. It may be, then, that eukaryotic organisms diversified as the oceans and atmosphere began to draw closer to their modern states.

## Protists have continued to diversify during the age of animals.

Modern protistan diversity took shape alongside evolving animals during the Phanerozoic Eon, 542 million years ago to the present. For example, foraminiferans and radiolarians, marine protozoans with shells of calcium carbonate and silica, respectively, appear for the first time in rocks deposited near the beginning of the Phanerozoic Eon and have continued to diversify since that time (**Fig. 27.24**). Some algal seaweeds also evolved skeletons of calcium carbonate at this time, which diversified along with animals in Phanerozoic oceans. Why did algae and protozoans evolve mineralized skeletons? Presumably to keep from being eaten.

Marine protists underwent a new round of radiation during the Mesozoic Era, 252–65 million years ago. The most significant changes occurred among photosynthesizers. During the preceding Paleozoic Era, green algae and cyanobacteria were the primary photosynthesizers in the oceans, but as the Mesozoic Era began,

**FIG. 27.24** (a) Foraminifera and (b) Radiolaria, showing skeletons of calcium carbonate and silica, respectively.

conclude that by 1200 million years ago Archaeplastida had already diverged from other eukaryotic superkingdoms, photosynthesis had become established in eukaryotes, and simple multicellularity had already evolved within the domain.

Much more eukaryotic diversity is recorded in rocks 800–700 million years old, including green algae and the remains of test-forming amoebozoans (**Fig. 27.23c**). Molecular fossils further indicate the presence of ciliates and dinoflagellates in the emerging eukaryotic world and show that algae were expanding to become major photosynthesizers in the oceans. Until about 800 million years ago, subsurface waters of the oceans were commonly poorly supplied with oxygen and were sometimes rich in sulfide, as well. After this time, sulfide

**FIG. 27.25** **Coccolithophorids.** (a) Coccolithophorid algae, like *Emiliania huxleyi*, form tiny scales of calcium carbonate. (b) A satellite image captures a coccolithophorid bloom off the coast of Newfoundland, visible as light blue streamers of calcium carbonate in the otherwise dark blue ocean.

new groups of photosynthetic dinoflagellates and coccolithophorid algae appeared, followed by diatoms. Today, these three groups dominate photosynthesis in many parts of the sea.

Within the oceans, diatom and coccolithophorid evolution had another important consequence. The silica skeletons of diatoms are now the principal means by which $SiO_2$ is transferred from seawater to sediments. Similarly, the $CaCO_3$ scales made by coccolithophorids and the skeletons of planktonic foraminifera are a primary source of carbonate sediments on the seafloor (**Fig. 27.25**). In this way, the ongoing evolution of protists in the age of animals has continued to transform marine ecosystems.

Today, microscopic eukaryotes dominate photosynthetic production in coastal oceans and play a major role in biogeochemical cycles. And for thousands of protists, plants and animals are simply habitats, ripe for colonization—we need look no further than the many eukaryotic vectors of human disease. The key conclusion is that a remarkable, and still largely undocumented, diversity of eukaryotic organisms has evolved through time to form an integral part of ecosystems across our planet, shaping in ways both positive and negative the world in which we live.

# Core Concepts Summary

**27.1 EUKARYOTIC CELLS ARE DEFINED BY THE PRESENCE OF A NUCLEUS, BUT OTHER FEATURES, PARTICULARLY A DYNAMIC CYTOSKELETON AND MEMBRANES, EXPLAIN THEIR SUCCESS IN DIVERSIFYING.**

Eukaryotic cells have a network of proteins inside the cell that allow them to change shape, move, and transfer substances in and out of the cell. page 27-1

Eukaryotic cells have dynamic membranes that facilitate movement and feeding. page 27-2

Eukaryotic cells compartmentalize their machinery for energy metabolism into mitochondria and chloroplasts, freeing the rest of the cell to interact with the environment in novel ways not available to prokaryotes. page 27-2

The eukaryotic genome is larger than that of prokaryotes, allowing new mechanisms of gene regulation. page 27-3

Eukaryotes reproduce sexually, which promotes genetic diversity, but cannot on its own help us understand their great diversity. page 27-3

**27.2 THE ENDOSYMBIOTIC HYPOTHESIS PROPOSES THAT THE CHLOROPLASTS AND MITOCHONDRIA OF EUKARYOTIC CELLS WERE ORIGINALLY FREE-LIVING BACTERIA THAT WERE INCORPORATED INTO A HOST CELL.**

The endosymbiotic hypothesis is based on physical, biochemical, and genetic similarities between chloroplasts and cyanobacteria, and between mitochondria and proteobacteria. page 27-5

Chloroplasts and mitochondria have their own genomes, but their genomes are small relative to free-living bacteria, to which they are closely related, mostly because of gene migration to the nuclear genome. page 27-6

Symbiosis between a heterotrophic host and photosynthetic partner is common throughout the eukaryotic domain; reef-forming corals are an example. page 27-6

Photosynthesis spread through the Eukarya by means of multiple independent events involving a protozoan host and a eukaryotic endosymbiont. page 27-6

Most eukaryotic cells have mitochondria, but a few do not. Evidence suggests that cells lacking mitochondria once had them but lost them. page 27-6

The eukaryotic nuclear genome contains genes unique to Eukarya, but also genes related to Bacteria and Archaea, suggesting either that the ancestor of the modern eukaryotic cell was a primitive eukaryote or an archaeon that engulfed a bacterium. page 27-7

**27.3 EUKARYOTES HAVE HISTORICALLY BEEN DIVIDED INTO FOUR KINGDOMS—ANIMALS, PLANTS, FUNGI, AND PROTISTS—BUT ARE NOW DIVIDED INTO AT LEAST SEVEN SUPERKINGDOMS.**

The terms "protist," "algae," and "protozoa" are useful, but do not describe monophyletic groups. page 27-9

Protists are eukaryotes that do not have features of animals, plants, and fungi; algae are photosynthetic protists; and protozoa are heterotrophic protists. page 27-9

The seven superkingdoms of eukaryotes are Opisthokonta, Amoebozoa, Archaeplastida, Stramenopila, Alveolata, Rhizaria, and Excavata. page 27-10

Opisthokonts are the most diverse eukaryotic superkingdom. They include animals and fungi, as well as choanoflagellates (our closest protistan relatives). page 27-10

Amoebozoans produce multicellular structures by the aggregation of amoeba-like cells, and include organisms that cause human disease and those important in biological research. page 27-12

The archaeplastids are photosynthetic, include land plants, and are divided into three major groups. page 27-13

Algae occur within the stramenopiles and several other groups. page 27-15

The earliest branch of the eukaryotic tree may separate Opisthokonta and Amoebozoa from the other superkingdoms, although this is still debated. page 27-19

**27.4 THE FOSSIL RECORD EXTENDS OUR UNDERSTANDING OF EUKARYOTIC DIVERSITY BY PROVIDING PERSPECTIVES ON THE TIMING AND ENVIRONMENTAL CONTEXT OF EUKARYOTIC EVOLUTION.**

Fossils in sedimentary rocks as old as 1800 million years have unmistakable signs of eukaryotic cells, including complicated wall structures. page 27-20

The earliest fossil eukaryotes that can be placed into one of the present-day superkingdoms are 1200 million years old and belong to the red algae. page 27-20

Eukaryotic fossils diversified around 800 million years ago, coinciding with an increase in oxygen and a decline in sulfide in the atmosphere and oceans. page 27-21

Protists continue to diversify to the present. Green algae and cyanobacteria dominated primary production in earlier oceans, but since about 200 million years ago dinoflagellates, coccolithophorids, and diatoms have become the primary photosynthetic organisms in the oceans. page 27-21

Protists have evolved to take advantage of the environments provided by animal and plants; these include many protists that cause disease in humans. page 27-22

## Self-Assessment

1. List key features that distinguish a eukaryotic cell from a prokaryotic cell.

2. Describe the forms of energy metabolism found in eukaryotes.

3. Describe the origin and evolution of the chloroplast and mitochondrion.

4. Present two hypotheses for the origin of the eukaryotic cell.

5. Name the seven superkingdoms of eukaryotes and an organism in each one.

6. State when the eukaryotic cell is first thought to have evolved and the evidence that supports this date.

Do you understand the chapter's Core Concepts? Log in to check your answers to the Self-Assessment questions, then practice what you've learned and reinforce this chapter's concepts by working through the problems and multimedia tutorials provided there.

**www.biologyhowlifeworks.com**

## CHAPTER 28

# BEING MULTICELLULAR

## Core Concepts

28.1 Complex multicellularity arose several times in evolution.

28.2 In complex multicellular organisms, bulk transport circumvents the limitations of diffusion.

28.3 Complex multicellularity depends on cell adhesion, communication, and a genetic program for development.

28.4 Plants and animals evolved multicellularity independently of each other and solved similar problems with different sets of genes.

28.5 The evolution of large and complex multicellular organisms, which required abundant oxygen, is recorded in the fossil record.

In the human body, several trillion cells work in close coordination, enabling us to sense our surroundings and respond to them, move, eat, grow, and reproduce. Humans are an example of a **complex multicellular organism.** Our cells are specialized for specific functions, so our body as a whole can perform a broad range of tasks, but individual cells for the most part cannot. We require trillions of cells to accomplish the same broad range of biological activities that can be performed by single-celled protists and bacteria.

Complex multicellular organisms are relative latecomers in the history of life, arising more than 3 billion years after microorganisms first evolved. In subsequent chapters, we discuss the biology of three groups that have evolved complex multicellularity: animals, land plants, and fungi. First, however, we need to ask, how did complex multicellular organisms arise from simple precursors, and what new ways of life were made possible by their evolution? As we will see, complex multicellularity depended on several key evolutionary innovations. Each step built on the previous one, leading to the evolution of organisms with tissues and organs able to transport oxygen, signaling molecules, and nutrients rapidly over long distances.

## 28.1 THE PHYLOGENETIC DISTRIBUTION OF MULTICELLULAR ORGANISMS

As we discussed in Chapter 26, most prokaryotic organisms are composed of a single cell, although some form simple filaments or live in colonies. A few types of bacterium, notably some cyanobacteria, develop several different cell types. No bacteria, however, develop macroscopic bodies with distinct types of tissues. Only eukaryotes have evolved those.

### Simple multicellularity is widespread among eukaryotes.

A recent survey of eukaryotic organisms recognized 119 major groups within the superkingdoms discussed in Chapter 27. Of these, 83 contain only single-celled organisms, predominantly cells that engulf other microorganisms or ingest small organic particles, photosynthetic cells that live suspended in the water column, or parasitic cells that live within other organisms. Each of the 36 remaining branches exhibits some cases of **simple multicellularity,** mostly in the form of filaments, hollow balls, or sheets of little-differentiated cells (**Fig. 28.1**).

Simple multicellular eukaryotes share several properties. In many, adhesive molecules cause adjacent cells to stick together, but there is relatively little communication or transfer of resources between cells and little differentiation of specialized cell types. Most or all of the cells in simple multicellular organisms retain a full range of functions, including reproduction, and so the penalty paid by the organism for individual cell death is usually small. Importantly, in simple multicellular organisms, every cell is in direct contact with the external environment, at least during phases of the life cycle when the cells must acquire nutrients.

**FIG. 28.1**   **Simple multicellularity, in which cells show little differentiation and remain in close contact with the external environment.** Many groups of eukaryotic organisms display simple multicellularity: (a) *Uroglena*, a stramenopile alga found commonly in lakes; (b) *Epistylis,* a stalked ciliate protozoan that lives attached to the surfaces of fish and crabs; and (c) *Prasiola*, a bladelike green alga only a single cell thick, found on tree trunks and rock surfaces.

Most simple multicellular species are forms of algae, although stalklike colonies of particle feeders have evolved at least three times, and simple filamentous fungi absorb organic molecules as sources of carbon and energy. In addition, four eukaryotic groups have achieved simple multicellularity by a different route, aggregating during just one stage of the life cycle. Slime molds are the best-known example (Chapter 27).

Six other groups (two algal and three protozoan, plus a group of fungi) have evolved what is called **coenocytic** organization. In these organisms, the nucleus divides multiple times, but the nuclei are not partitioned into individual cells. The result is large cells—sometimes even visible to the naked eye—with many nuclei. The green algae *Codium* and *Caulerpa* are common

examples of coenocytic organisms found along the shorelines of temperate and tropical seas, respectively (**Fig. 28.2**). Coenocytic rhizarians on the deep seafloor can be 10 to 20 cm long. There is no evidence that any of these groups evolved from truly multicellular ancestors, nor have any given rise to complex multicellular descendants.

Why did simple multicellular organisms evolve? One selective advantage is that multicellularity helps organisms to avoid protozoan predators. In an illuminating experiment, single-celled green algae were grown in the presence of a protistan predator. Within 10 to 20 generations, most of the algae were living in eight-cell colonies that were essentially invulnerable to predator attack. Another advantage is that multicellular organisms may

**FIG. 28.2**   **Coenocytic organization, in which nuclei divide repeatedly but cytokinesis does not occur.** Two different groups of green algae have evolved large size as coenocytic organisms: (a) *Codium*, found abundantly along temperate coastlines; and (b) *Caulerpa*, common in shallow tropical seas and aquaria.

**FIG. 28.3 Complex multicellular organisms.** Complex multicellularity evolved independently in (a) red algae, (b) brown algae, (c) land plants, (d) animals, and (e) fungi (at least twice).

be able to maintain their position on a surface or in the water column better than their single-celled relatives. Seaweeds, for example, live anchored to the seafloor in places where light and nutrients support growth. And in colonial heterotrophs such as the stalked ciliate *Epistylis* (see Fig. 28.1b), the coordinated beating of flagella assists feeding by directing currents of food-laden water toward the cells. As we will see, complex multicellular organisms evolved from simple multicellular ancestors, but most taxonomic groups containing simple multicellular organisms never gave rise to complex descendants.

## Complex multicellularity evolved several times.

Complex multicellular organisms mark the biological world of our daily existence. Plants, animals, and, if you look a little more closely, mushrooms and complex seaweeds define our perceptions of landscapes and the coastal ocean (**Fig. 28.3**). Plants and animals differ in many ways, but they have some fundamental features in common. Both have highly developed molecular mechanisms for adhesion between cells, and they have specialized structures that allow cells to communicate with one another. Plants and animals also both display complex patterns

of cellular and tissue differentiation, guided by networks of regulatory genes. All complex multicellular organisms exhibit these properties. In general, in complex multicellular organisms only a small subset of all cells contributes to reproduction. Cell or tissue loss can be lethal for the entire organism. Notably, complex multicellular organisms have a three-dimensional organization, so only some cells are in direct contact with the environment.

The presence of both exterior and interior cells is an important characteristic of complex multicellular organisms. Because some cells are buried within tissues, relatively far from the exterior of the organism, they do not have direct access to nutrients. Therefore, interior cells cannot grow as fast as surface cells unless there is a mechanism to transfer resources from one cell to another. Also, interior cells do not receive signals directly from the environment, even though all cells must be able to respond to environmental signals if the organism is to grow, reproduce, and survive. Complex multicellular organisms, therefore, require mechanisms for transferring environmental signals received by cells at the body's surface to interior cells, where genes will be activated or repressed in response. That process has a familiar ring to it. Development in complex multicellular organisms can be

**FIG. 28.4** The seven superkingdoms on the eukaryotic tree of life, showing that complex multicellular organisms (found on branches marked by blue) evolved at least six times.

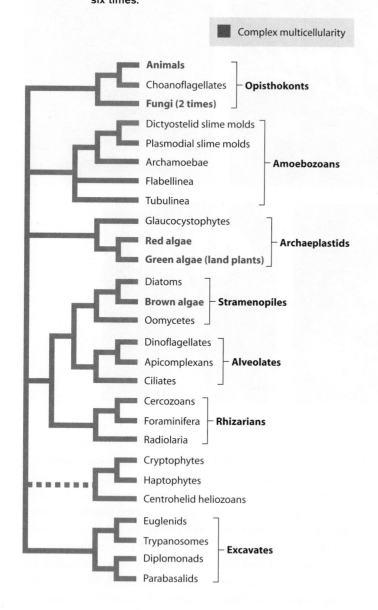

## 28.2 DIFFUSION VS. BULK TRANSPORT

How does oxygen get from the air in your lungs into your bloodstream? How does atmospheric carbon dioxide get into leaves? How does ammonia get from seawater into the cells of planktonic algae? The answer to all three questions is the same: by **diffusion.** But oxygen absorbed by your lungs doesn't reach your toes by diffusion alone—it is transported actively, and in bulk, by blood pumped through your circulatory system. Bulk transport of oxygen, nutrients, and signaling molecules, at rates and across distances far larger than can be achieved by diffusion alone, lies at the functional heart of complex multicellularity.

### Diffusion is effective only over short distances.

Diffusion is the random motion of molecules, with net movement occurring when there are areas of higher and lower concentration of the molecules (Chapter 5). Because diffusion supplies key molecules for metabolism, it exerts a strong constraint on the size, shape, and function of cells. Diffusion is effective only over small distances and so it strictly limits the size and shape of bacterial cells, as discussed in Chapter 26. Diffusion also constrains how eukaryotic organisms function.

Oxygen provides a good example. Most eukaryotes require oxygen for respiration. If a cell or tissue must rely on diffusion for its oxygen supply, its thickness is limited by the concentration difference in oxygen between the cell or tissue and its environment. The concentration difference is dependent on the amount of oxygen in the environment and the rate at which oxygen is used for respiration inside the organism. In shallow water that is in direct contact with the atmosphere, diffusion permits animals to reach thicknesses of about 1 mm to 1 cm, depending on their metabolic rate. Of course, a quick swim along the seacoast will reveal many animals much larger than this. And, obviously, *you* are larger than this. How do you and other large animals get enough oxygen to all of your cells to enable you to survive?

### Animals achieve large size by circumventing limits imposed by diffusion.

Sponges can reach overall dimensions of a meter or more, but they actually consist of only a few types of cell that line a dense network of pores and canals and so remain in close contact with circulating seawater (**Fig. 28.5a**). The large size of a sponge is therefore achieved without placing metabolically active cells at any great distance from their environment. Likewise, in jellyfish, active metabolism is confined to thin tissues that line the body. Essentially, a large flat surface is folded up to produce a three-dimensional structure (**Fig. 28.5b**). The jellyfish's bell-shaped body is often thicker than the metabolically active

defined as increasing or decreasing gene expression in response to molecular signals from surrounding cells (Chapter 20).

Complex multicellularity evolved at least six separate times in different eukaryotic groups (**Fig. 28.4**). Obviously, complex multicellularity characterizes animals, but it also evolved at least twice in the fungi, once in the green algal lineage that gave rise to land plants, once in the red algae, and once in the brown algae, producing the giant kelps that form forests in the sea.

→ **Quick Check 1** How do simple multicellular organisms differ from complex multicellular organisms?

**FIG. 28.5** **Circumventing limits imposed by diffusion.** (a) Sponges can attain a large size because the many canals in their bodies ensure that all cells are in close proximity to the environment. (b) Jellyfish also have thin layers of metabolically active tissue, but their familiar bell can be relatively thick because it is packed with metabolically inert molecules (the mesoglea, or "jelly").

tissue, but its massive interior is filled by materials that are not metabolically active. This material constitutes the mesoglea, the jellyfish's "jelly." The mesoglea provides structural support but does not require much oxygen.

But what about us? How does the human body circumvent the constraints of diffusion? Our lungs gather the oxygen we need for respiration, but the lung is a prime example of diffusion in action, not a means of avoiding it. Because lung tissues have a very high ratio of surface area to volume, oxygen can diffuse efficiently from the air you breathe into lung tissue (Chapter 39). A great deal of oxygen can be taken in this way, but how does it get from the lungs to our brains or toes? The distances are far too large for diffusion to be effective. The answer is that oxygen binds to molecules of hemoglobin in red blood cells and is carried through the bloodstream to distant sites of respiration. We circumvent diffusion by actively transporting oxygen through our bodies. More generally, the circumvention of diffusion is what makes complex multicellularity possible.

## Complex multicellular organisms have structures specialized for bulk transport.

We have seen how humans circumvent the limitations of oxygen diffusion. Many other animals similarly rely on the active circulation of fluids to transport oxygen and other essential molecules, including food and molecular signals, across distances far larger than those that could be traversed by diffusion alone. Indeed, without a mechanism like bulk transport, animals could not have achieved the range of size, shape, and function familiar to us.

**Bulk transport** is the means by which molecules move through organisms at rates beyond those possible by diffusion across a concentration gradient (**Fig. 28.6**). In animals, the active pumping of blood through the circulatory system supplies oxygen to tissues that may be more than a meter distant from the lungs (Fig. 28.6a). Bulk transport carries the organic molecules required for respiration large distances from the intestinal cells that absorb these molecules from the

**FIG. 28.6** **Bulk transport.** (a) The circulatory system in animals and (b) the vascular system in plants allow these organisms to get around the size limits of diffusion.

**a.** Animal circulatory system

**b.** Plant vascular system

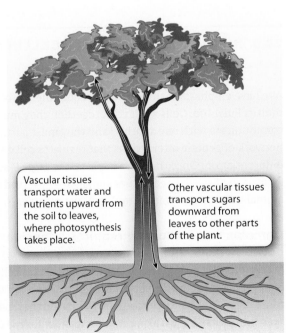

The circulatory system transports oxygen and nutrients to tissues and carbon dioxide and other wastes away from tissues.

Vascular tissues transport water and nutrients upward from the soil to leaves, where photosynthesis takes place.

Other vascular tissues transport sugars downward from leaves to other parts of the plant.

digestive tract. (Again, the molecules are transported actively in the bloodstream.) Endocrine signaling molecules such as hormones (Chapter 9) also move rapidly through the body by means of the blood and other fluids. Bulk transport is key to complex multicellularity in humans and most other animals.

Complex organisms other than animals also rely on bulk transport. A redwood tree must transport water upward from its roots to leaves that may be 100 meters above the soil. If plants relied on diffusion to transport water, they would be only a few millimeters tall. How, then, do they move water? Plants move water by bulk transport through a system of specialized tissues powered by the evaporation of water from leaf surfaces (Fig. 28.6b; Chapter 29). Vascular plants also have specialized tissues for the transport of nutrients and signaling molecules upward and downward through roots, stems, and leaves. Thus, like animals, plants rely on specialized tissues to transport essential materials over long distances at rates many times faster than could be accomplished by diffusion alone.

The circumvention of diffusion can be seen in other groups characterized by complex multicellularity. Fungi transport nutrients through networks of filaments that may be meters long, relying on osmosis to pump materials from sites of absorption to sites of metabolism (Chapter 34). The giant kelps have an internal network of tubular cells that transports molecules through a body that can be tens of meters long. In general, when some cells within an organism are buried within tissues, far from the external environment, bulk transport is required to supply those cells with molecules needed for metabolism.

→   **Quick Check 2** How do mechanisms for bulk transport enable organisms to achieve large size?

## 28.3 HOW TO BUILD A MULTICELLULAR ORGANISM

We have emphasized three general requirements for complex multicellular life: Cells must stick together; they must communicate with one another; and they must participate in a network of genetic interactions that regulates cell division and differentiation. Once these are in place, the stage is set for the formation of specialized tissues and organs observed in plants, animals, and other complex multicellular organisms.

### Complex multicellularity requires adhesion between cells.

If a fertilized egg is to develop into a complex multicellular organism, it must divide many times, and the cells produced from those divisions must stick together. Moreover, they must retain a specific relationship with one another in order for the developing organism to function. Chapter 10 introduced the molecular mechanisms that hold multicellular organisms together. We

begin by reviewing those mechanisms briefly before placing them in evolutionary context.

In animals, proliferating cells commonly become organized into sheets of cells called epithelia that line the inside and cover the outside of the body. Within epithelia, adjacent cells adhere to one another by means of transmembrane proteins, especially cadherins, that form molecular bonds between cells. Epithelial cells also secrete an extracellular matrix made of proteins and glycoproteins and attach themselves to this matrix by means of other transmembrane proteins called integrins. Cadherins, integrins, and additional transmembrane proteins thus provide the molecular mechanisms for adhesion in animals (Chapter 10).

Cadherins and integrins do not occur in other complex multicellular organisms, but all such organisms require some means of keeping their cells stuck together. How do they do it? Plants also synthesize adhesive molecules that bind cells into tissues, but in this case the molecules are polysaccharides called pectins (pectins are what make fruit jelly jell). Molecules that control adhesion between adjacent cells are less well known in other complex multicellular organisms, but the general picture is clear. Without adhesion there can be no complex multicellularity, and different groups of eukaryotes have employed distinct types of molecular "glue" for this purpose.

## HOW DO WE KNOW?

**FIG. 28.7**

# How do bacteria influence the life cycles of choanoflagellates?

**BACKGROUND** Choanoflagellates are the closest protistan relatives of animals, so an understanding of cell adhesion, cell differentiation, and multicellularity in these organisms stands to illuminate early events in the evolution of animal multicellularity. Nicole King and her colleagues cultured the choanoflagellate species *Salpingoeca rosetta* in the laboratory. High-speed videos revealed that these choanoflagellates differentiate into several distinct cell types in the course of their life cycle. Actively swimming cells can be fast swimmers or slow swimmers that differ in shape and ornamentation (Fig. 28.7a). When swimming cells make contact with a solid surface, they can differentiate a stalk that tethers them to the substrate (Fig. 28.7b). Simple multicellular colonies in the shape of rosettes (Fig. 28.7c) were also seen when the choanoflagellates were cultured in the presence of many bacteria, including *Algoriphagus machipongonesis*.

## How did animal cell adhesion develop?

The genome of the choanoflagellate *Monosiga brevicollis* provides fascinating insight into the evolution of adhesion molecules. Choanoflagellates, the closest protistan relatives of animals, are unicellular microorganisms. Therefore, it came as a surprise that the genes of *M. brevicollis* code for many of the same protein families that promote cell adhesion in animals. The choanoflagellate genome contains genes for both cadherin and integrin proteins. Clearly, these proteins are not supporting epithelia in *M. brevicollis*, so what are they doing?

One approach to an answer came from the observation that cells stick not only to one another but also to their substrates. Proteins that originally evolved to promote adhesion to sand grains or a rock surface may have been co-opted by evolution for cell–cell adhesion. It is also possible that adhesion molecules originally facilitated the capture of bacterial cells, improving predation by causing prey to adhere to the predator.

An intriguing clue to this problem comes from careful laboratory studies of choanoflagellates (**Fig. 28.7**). Although choanoflagellates are fundamentally unicellular, simple multicellular structures can be induced in a number of species. Multicellularity is induced by molecular signals, and, unexpectedly, the source of these signals is a bacterium, the preferred prey of the choanoflagellates. When they detect their food bacteria, the choanoflagellates form a novel multicellular structure. The function of adhesion molecules in these choanoflagellates has not yet been determined, but these observations support the hypothesis that they facilitate predation.

To date, cadherins have been found only in choanoflagellates and animals, but proteins of the integrin complex extend even deeper into eukaryotic phylogeny, occurring in single-celled protists that branch near the base of the opisthokont superkingdom (Chapter 27). Such a phylogenetic distribution provides strong support for the general hypothesis that cell adhesion in animals resulted from the redeployment of protein families that evolved to perform other functions before animals diverged from their closest protistan relatives.

## Complex multicellularity requires communication between cells.

In complex multicellular organisms, it is not sufficient for cells to stick together. They must also be able to communicate. As discussed in Chapter 9, cells communicate by molecular signals. A signaling molecule (generally a protein) synthesized by one cell binds with a receptor molecule (also a protein) on the surface of a second cell, essentially flipping a molecular switch that activates or represses gene expression in the receptor cell's nucleus. Plants

*Salpingoeca rosetta* choanoflagellates can exist as single cells that swim through their watery environment (a); they can be stalked cells that live attached to surfaces (b); or, in the presence of specific bacteria, they can form simple multicellular structures (c).

**HYPOTHESIS** The bacterium *Algoriphagus machipongonesis* induces the choanoflagellate to form rosette-like colonies.

**EXPERIMENT** The scientists grew the chonaoflagellates in a medium containing only *A. machipongonesis*.

**RESULTS** The multicellular state formed (Fig. 28.7c).

**CONCLUSION** Choanoflagellates differentiate several cell types upon cues from the environment. In particular, molecular signals from prey bacteria can induce a simple form of multicellularity. The direct ancestors of animals may have gained simple multicellularity by a similar route.

**FOLLOW-UP WORK** Research is continuing to investigate the molecular mechanisms of cell differentiation, cell adhesion, and signal reception in these distant relatives of ours.

**SOURCE** Dayel, M., R. A. Alegado, S. R. Fairclough, T. C. Levine, S. A. Nichols, K. McDonald, and N. King. 2011. "Cell Differentiation and Morphogenesis in the Colony-Forming Choano-flagellate *Salpingoeca rosetta*." *Developmental Biology* 357:73–82.

and animals both have signaling pathways based on receptor kinases, and, as in the case of adhesion molecules, these signaling pathways, or components of the pathways, have also been identified in their close protistan relatives.

Evidently, many of the signaling pathways used for communication between cells in complex multicellular organisms first evolved in single-celled eukaryotes. Again we can ask, what function did molecular signals and receptors have in the protistan ancestors of complex organisms?

All cells have transmembrane receptors that respond to signals from the environment. In some cases, the signal is supplied by food organisms (such as the bacteria that induce simple multicellularity in choanoflagellates), and in some cases the cells sense nutrients, temperature, or oxygen level. Single-celled eukaryotes also communicate with other cells within the same species, not least to ensure that two cells can find each other to fuse in sexual reproduction. Signaling between two cells within an animal's body can be seen as a variation on this more general theme of cellular responses to other cells and the physical environment.

While all eukaryotic cells have molecular mechanisms for communication between cells, complex multicellular organisms have distinct cellular pathways for the movement of molecules from one cell to another. For example, animals more complex than sponges have **gap junctions,** protein channels that allow ions and signaling molecules to move from one cell into another (**Fig. 28.8**). Gap junctions not only help cells to communicate with their neighbors, they allow targeted communication between a cell and only one or a few of the cells that surround it (Chapter 10).

Complex photosynthetic organisms, in contrast, have an intrinsic barrier to intercellular communication: the cell wall. Careful observation of plant cell walls using the electron microscope reveals that these walls have tiny holes through which thin strands of cytoplasm extend from one cell to the next (**Fig. 28.9**). Called **plasmodesmata,** these intercellular strands are lined by extensions of the cell membrane and contain tubules connecting the endomembrane systems of the two cells. Like gap junctions, plasmodesmata permit signaling molecules to pass between cells in a targeted fashion.

Other complex multicellular organisms also have channels that connect adjacent cells. As similar channels do not occur in most other eukaryotic organisms, they appear to represent an important step in the evolution of complex multicellularity.

### Complex multicellularity requires a genetic program for coordinated growth and cell differentiation.

All of the cells in your body derive from a single fertilized egg, and most of those cells contain the same genes. Yet your body contains about 200 distinct cell types precisely arranged in a variety of tissues and organs. How can two cells with the same genes become different cell types? The answer lies in development, the system of gene regulation that guides growth from zygote to adult (Chapter 20).

Development is the product of molecular communication between cells. Cells have different fates depending on which genes are switched off or on by the molecular signals they receive. Different signals induce the reorganization of cytoskeletons and alter the production of proteins. As a result, a stem cell may become

---

**FIG. 28.8    Gap junctions.** Gap junctions facilitate the targeted transport of ions and molecules between adjacent cells in animals.

Gap junction

Gap junctions are intercellular connections made up of a ring of proteins.

Plasma membrane

Intercellular space

**FIG. 28.9 Plasmodesmata.** Plasmodesmata facilitate the movement of ions and molecules between cells in land plants and in complex red and brown algae.

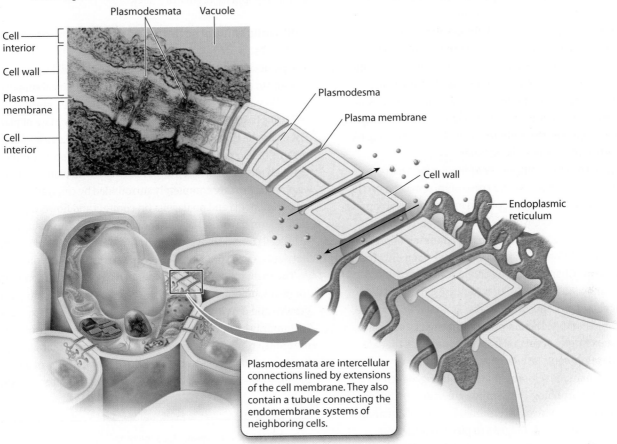

Plasmodesmata are intercellular connections lined by extensions of the cell membrane. They also contain a tubule connecting the endomembrane systems of neighboring cells.

an epithelial cell, or a muscle cell, or a neuron. This observation leads to another question: What causes the same gene to be turned on in one cell and off in another? The ultimate answer is that two cells in the same developing organism can be exposed to very different environments.

When we think about development as a process of programmed cell division and differentiation, the link to unicellular ancestors becomes clearer. Many biological innovations accompanied the evolution of complex multicellularity, but, as noted in Chapter 27, the differentiation of distinct cell types is not one of them. Many unicellular organisms have life cycles in which different cell types alternate in time, depending on environmental conditions. For example, if we experimentally starve dinoflagellate cells, two cells will undergo sexual fusion to form morphologically and physiologically distinct resting cells protected by thick walls. When food becomes available again, the cells undergo meiotic cell division to form new feeding cells. Many other single-celled eukaryotes differentiate resting cells in response to environmental cues, especially deprivation of nutrients or oxygen.

The innovation of complex multicellularity was to differentiate cells in *space* instead of time. In a three-dimensional multicellular organism, only surface cells are in direct contact with the outside environment. Interior cells are exposed to a different physical and chemical environment because the availability of nutrients, oxygen, and light decreases with increasing depth within tissues. In effect, there is a gradient of environmental signals within multicellular organisms. It is, therefore, possible that in the earliest organisms with three-dimensional multicellularity, a nutrient or oxygen gradient triggered differentiation of interior cells, causing cell differentiation of oxygen- or nutrient-starved interior cells much as these conditions induce the expression and repression of specific genes in their single-celled relatives. With increasing genetic control of cellular responses to signaling gradients, the seeds of complex development were sown.

A fascinating example supporting this hypothesis has been reported from research on green algae. The simple multicellular organism *Volvox* (see Fig. 27.15) has two types of cells, vegetative cells that photosynthesize and control movement of the organism, and reproductive cells. Cell differentiation in *Volvox* is regulated by a gene also involved in cell remodeling during the life cycle of *Volvox*'s single-celled relative *Chlamydomonas*, supporting the hypothesis that the spatial differentiation of cells in multicellular organisms began with the redeployment of genes that regulate cell differentiation in single-celled ancestors.

Bulk transport, discussed earlier with regard to the transport of nutrients and water within complex multicellular organisms, also affects developmental signaling. It increases the distance over which signals can travel, circumventing the limitations of diffusion and making possible spatially specific patterns of differentiation along the path of signal transport. For example, in animals the endocrine system releases hormones directly into the bloodstream, enabling them to affect cells far from those within which they formed (Chapter 38). Thus, the sex hormones estrogen and testosterone are synthesized in reproductive organs, but regulate development throughout the body, contributing to the differences between males and females.

The genome of the choanoflagellate *Monosiga brevicollis*, discussed earlier, has been a treasure trove of information on the antiquity of signaling molecules deployed in animal development. In addition to proteins that govern cell adhesion and epithelial cohesion in animals, *M. brevicollis* expresses a number of proteins active in animal cell differentiation. For example, signaling based on receptor kinases (Chapter 9), long thought to be restricted to animals, occurs in *M. brevicollis*. In other cases, individual domains of signaling proteins occur in the *M. brevicollis* genome, but not the complex multidomain proteins formed by animals. This is true, for example, of several protein complexes that are important in development. Similarly, molecules that play an important role in plant development are also being identified in the genomes of morphologically simple green algae. The key point is that genome sequences interpreted in light of eukaryotic phylogeny are now enabling biologists to piece together the patterns of gene evolution that accompanied the evolution of morphological complexity and diversity in plants and animals.

## 28.4 VARIATIONS ON A THEME: PLANTS VS. ANIMALS

Plants and animals are the best-known examples of complex multicellular organisms. The study of phylogenetic relationships makes it clear that complex multicellularity evolved independently in the two groups. In other words, they do not share a common ancestor that was multicellular. We can see the results of these separate evolutionary events by examining cell adhesion, communication, and development in both plants and animals.

Both plants and animals have evolved sophisticated systems that cause adjacent cells to adhere to each other and that promote the targeted movement of signaling molecules between cells. Plant and animal mechanisms, however, must differ, because plant cells have cell walls and animal cells do not. Likewise, plants and animals have evolved similar genetic logic to govern development, but use mostly distinct sets of genes. In both plants and animals, many proteins switch other genes on or off, so that the spatial organization of multicellular organisms arises from networks of interacting genes and their protein products. Ancestral plants and animals simply recruited distinct families of genes to populate regulatory networks.

### Cell walls shape patterns of growth and development in plants.

The plant cell wall (Chapter 5), made of cellulose, imparts structural support to cells and, in fact, provides the mechanical support that allows plants to stand erect (Chapter 29). The presence of cell walls has largely determined the evolutionary fate of plants. For example, because their cells cannot engulf particles or absorb organic molecules, most plants gain carbon and energy only through photosynthesis. Moreover, because all plant cells except eggs and sperm are completely surrounded by cell walls, they have no pseudopodia and no flagella (in conifers and flowering plants, even sperm have lost flagella; see Chapter 30). This being the case, plant cells cannot move, either during development or to obtain nutrients, evade predators, or escape stressful conditions.

The inability of plant cells to move has major consequences for development. At the level of the cell, the entire program of growth and development involves cell division, cell expansion (commonly by developing large vacuoles in cell interiors), and cell differentiation. The mechanical consequence is that plant growth is confined to **meristems** (**Fig. 28.10**), populations of actively dividing cells at the tips of stems and roots (Chapter 31). More

**FIG. 28.10  A meristem.** Because of their walls, plant cells can't move. For this reason, roots grow by cell division of the undifferentiated cells in the meristem region at the root tip.

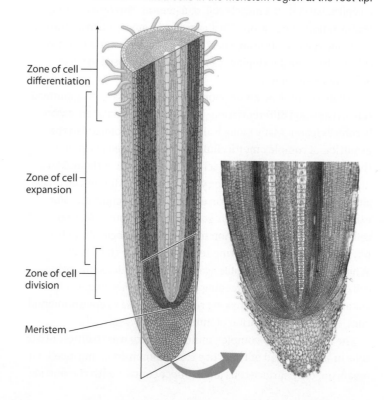

Zone of cell differentiation

Zone of cell expansion

Zone of cell division

Meristem

or less permanently undifferentiated cells in meristem regions repeatedly undergo mitosis. A few millimeters from the region of active cell division in a stem or root meristem, cells stop dividing and begin to expand. Within another millimeter or so, this activity is curtailed as well. There, individual cells respond to signaling molecules that induce differentiation, forming the mature cells that will function in photosynthesis, storage, bulk transport, or mechanical support. In general, then, plant development involves cell division, adhesion among the cells that form from meristems, adsorption of water and nutrients supplied by adjacent cells, and differentiation into distinct cell types that govern the function of the plant as a whole.

Because cell walls render plants immobile, plants have evolved mechanisms to transport water and nutrients from the soil to leaves that may be tens of meters distant without the use of any moving parts or even the expenditure of ATP. And because plants are anchored in place, they are unable to move in response to unfavorable growth conditions. For this reason, plants have intricate systems that feed information from the environment to their meristems. Heat, drought, floods, and fire can all leave their mark in altered patterns of growth. In addition, plants can't flee predators, so they have evolved mechanical structures (hairs and spines) and poisons to keep from being eaten. Indeed, it has been suggested that, in terms of responding to the environment, growth plays a role in plants similar to that played by behavior in animals.

The details of plant growth, development, reproduction, and function are presented in Chapters 29–33. Here, the point to bear in mind is that the properties of a pine tree or a rice plant depend fundamentally on the basic features of cell adhesion, intercellular communication, and regulatory network to guide development, all carried out under the constraints imposed by cell walls.

### Animal cells can move relative to one another.

Having outlined the growth and development of plants, we can appreciate anew the remarkable ballet that characterizes development in animal embryos (**Fig. 28.11;** Chapter 42). Fertilized eggs undergo several rounds of mitosis to form a ball of undifferentiated cells called a **blastula.** Then something happens that has no parallel in plant development: Unconstrained by cell walls, animal cells can move relative to one another. Blastula cells migrate, becoming reorganized into a hollow ball that folds inward at one location to form a layered structure called the **gastrula.**

Gastrula formation brings new populations of cells into direct contact with one another, inducing patterns of molecular signaling and gene regulation that begin the long process of growth and tissue specification. As cells proliferate, molecular signals are expressed in some cells and diffuse into others, generating what amounts to a three-dimensional molecular map of the organism that guides cell differentiation. Gradients in signaling molecules define top, bottom, front, back, left,

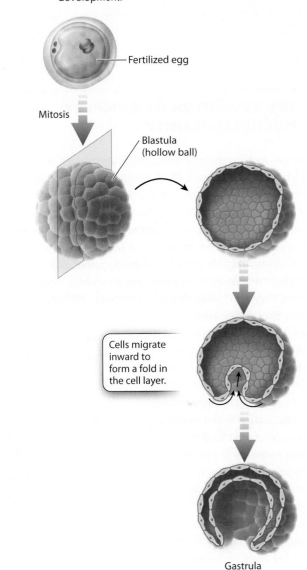

**FIG. 28.11  Gastrulation, the movement of cells during embryogenesis that transforms a blastula to a gastrula.** Unlike plant cells, animal cells move during development.

Fertilized egg

Mitosis

Blastula (hollow ball)

Cells migrate inward to form a fold in the cell layer.

Gastrula

and right. Plants do much the same thing, but because animal cells can move during development, animal embryos are not restricted to growth only from localized regions like meristems. Cell division and tissue differentiation occur throughout the developing animal body.

Because they are not constrained by cell walls, animals can form organs with moving parts—muscles that power active transport of food and fluids and allow movement. Thus, animals have possibilities for function that are far different from those of plants. If the environment offers a challenge, like drought or predators, animals can respond by changing their behavior. For example, they can move to a new location when threatened or stressed, as is discussed more fully in Chapter 37.

Plants and animals, then, display contrasting patterns of development and function that reflect both their independent origins from different groups of protists and the constraints imposed by cell walls in plants.

→    **Quick Check 3** How do plants and animals differ in the ways their cells adhere, communicate, and differentiate during development?

## 28.5 THE EVOLUTION OF COMPLEX MULTICELLULARITY

The gulf between an amoeba and a lobster seems vast, but research over the past decade increasingly shows how this apparent gap was bridged through a series of evolutionary innovations. Throughout this chapter, we have emphasized that complex multicellular organisms require mechanisms for cell adhesion, modes of communication between cells, and a genetic program to guide growth and development. Not only were all three required for the evolution of complex multicellular organisms, they also had to be acquired in a specific order. If the products of cell division don't stick together in a pattern that has some usefulness in function, there can be no complex multicellularity. Adhesion, however, is not sufficient. It must be followed by mechanisms for communication between cells. Moreover, cells must be able to send molecular messages in a targeted or spatially specific pattern, facilitated in animals by gap junctions and in plants by plasmodesmata.

With these requirements in place, evolution would favor the increase and diversification of the genes that regulate growth and development, making possible more complex morphology and anatomy. The biological stage was finally set for the functional key to complex multicellularity: the differentiation of tissues and organs that govern the bulk transport of fluids, nutrients, signaling molecules, and oxygen through increasingly large and complex bodies. This differentiation freed organisms from the tight constraints imposed by diffusion. When all these features are placed onto phylogenies, they show both the predicted order of acquisition and the tremendous evolutionary consequences of complex multicellularity (**Fig. 28.12**).

### Complex multicellularity appeared in the oceans 575 to 555 million years ago.

In Chapter 23, we noted that phylogenies based on living organisms make predictions about the fossil record. Characters and groups associated with lower branches in phylogenies should appear as earlier fossils than those associated with later branches. Although fossils don't preserve molecular features, the record supports the phylogenetic pattern shown in Fig. 28.12.

Single-celled protists are found in sedimentary rocks as old as 1800 million years, and a number of simple multicellular forms have been discovered that are nearly as old (see Fig. 27.23). Nonetheless, the oldest fossils that record complex multicellularity occur in rocks that were deposited only 575 to 555 million years ago. **Fig. 28.13a** shows one of the oldest known animal fossils, a frondlike form preserved in 575-million-year-old rocks from Newfoundland, near the eastern tip of Canada. This fossil and many others found in rocks of this age are enigmatic. There is complex morphology to be sure, but it is difficult

**FIG. 28.12   The evolution of complex multicellularity.** Phylogenetic relationships of complex (a) animals and (b) plants, with their close evolutionary relatives, show similar patterns of character accumulation and its consequences for biological diversity. Estimated species numbers are shown.

\* Only species on the branch leading to land plants
† Limited development of transport tissues in some mosses

**FIG. 28.13** **The earliest fossils of animals.** (a) The oldest known macroscopic animal fossil is a frondlike form made up of tubular structures, preserved in 575-million-year-old rocks from Newfoundland. (b) The oldest known animal fossil showing bilateral symmetry is a mollusk-like form preserved in 560–555-million-year-old rocks from northern Russia. (c) Some of the oldest tracks and trails of mobile animals are common in rocks 555 million years old and younger.

to understand these forms by comparing them to the body plans of living animals. The Newfoundland fossils show no evidence of head or tail, no limbs, and no opening that might have functioned as a mouth. They appear to be very simple organisms that gained both carbon and oxygen by diffusion. These fossils are long and wide, but they are not thick. In fact, it is likely that active metabolism occurred only along the surfaces of these structures.

Such animals lie near the base of the animal tree (Chapter 44), and probably would have developed under the guidance of regulatory genes similar to those present in modern sponges or, perhaps, jellyfish. By 560–555 million years ago, more complex animals, with distinct head and tail, top and bottom, left and right enter the record (**Fig. 28.13b**). At the same time, we begin to see tracks and trails made by animals whose muscles enabled them to move across and through surface sediments (**Fig. 28.13c**), an indirect but complementary record of increasing animal complexity. These, then, are the first known occurrences of the types of animal that dominate ecosystems today, both in the sea and on land. These animals must have possessed a diverse array of genes to guide development, something akin to those seen today in insects or snails.

Clearly, by 575–555 million years ago, the animal tree was beginning to branch. This, however, raises a new question: Why did complex multicellularity appear so much later than simple multicellular organisms?

**Oxygen is necessary for complex multicellular life.**
Earlier in the chapter, we discussed the key biological requirements for complex multicellularity. There is a critical environmental requirement as well: the presence of oxygen. Only the oxidation of organic molecules by $O_2$ provides sufficient energy to support biological communities that include large and active predators like wolves and lions. Not only do other potential electron acceptors for respiration, like sulfate and ferric iron, occur in much lower abundances, they are not gases and so do not accumulate in air. And only oxygen that is present in concentrations approaching those of the present day is able to diffuse into the interior cells of large, active organisms. On our planet—and probably, on all planets—no other oxidant is both sufficiently abundant and has the oxidizing power needed to support the biology of large complex organisms.

On the present-day Earth, lake or ocean bottoms with oxygen levels below about 10% of surface levels support few animals, and those that do tend to be tiny. Large, active, and diverse animals can live only in oxygen-rich environments. The clear restrictions on animal form and diversity observed along an environmental gradient of oxygen availability have important implications for the distribution of large, active, and diverse animals through time. We noted in Chapter 26 that oxygen first began to accumulate in the atmosphere and surface oceans about 2400 million years ago, but the chemistry of sedimentary rocks deposited after that time tells us that oxygen

**FIG. 28.14  Oxygen and the evolution of complex multicellularity.**

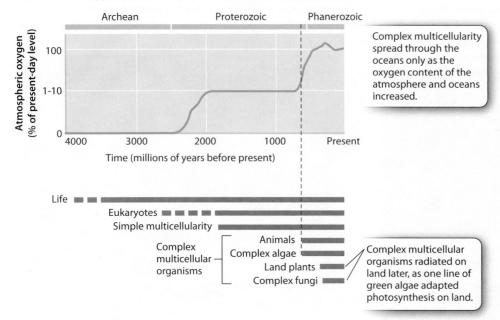

Complex multicellularity spread through the oceans only as the oxygen content of the atmosphere and oceans increased.

Complex multicellular organisms radiated on land later, as one line of green algae adapted photosynthesis on land.

lacked the mechanical strength to hold their position along the shoreline.

The key point is that complex multicellular organisms didn't succeed by doing the same things as simpler eukaryotes. In an oxygen-rich world, complex multicellularity spread through the oceans because it opened up new and unprecedented evolutionary possibilities.

### Land plants evolved from green algae that could carry out photosynthesis on land.

Complex multicellular land plants originated about 460 million years ago, well after marine animals and algae. In some ways, the pattern of evolution of land plants parallels that of animals. Ancestral green algae evolved molecular means for cell–cell adhesion and communication between cells, and the algae that form the sister group to land plants share with them the innovation of plasmodesmata. Then, as they developed larger and more complex three-dimensional bodies, plant ancestors evolved a succession of genes and their protein products to guide cell division, expansion, and differentiation. Land plants, however, had a different challenge from that of animals. The diversification of plants across the land surface requires that photosynthesis, evolved earlier by aquatic organisms, be carried out within tissues bathed by air. It also requires that nutrients and water be absorbed from the soil and transported throughout the plants rather than simply taken in by diffusion from the surrounding water.

levels remained low for a very long time. Our modern world of abundant oxygen came to exist only 580–560 million years ago, about the time when fossils first record complex multicellular organisms (**Fig. 28.14**).

Scientists continue to debate the causes of this environmental transformation, but the correspondence in time between oxygen enrichment and the first appearance of complex animals (and algae) suggests that more oxygen permitted greater size. Greater size in turn created opportunities for tissue differentiation, leading to the bulk transport of nutrients and signaling molecules. The evolution of bulk transport in turn permitted still larger size, setting up a positive feedback that eventually resulted in the complex marine organisms we see today.

With morphological complexity came new functions, including predation on other animals. Protozoan predators capture other microorganisms, but do so one cell at a time. In contrast, animals that obtain food by filtering seawater gather cells by the thousands, and larger animals can eat smaller ones, opening up endless possibilities for evolutionary specialization and, therefore, diversification. Among photosynthetic organisms, multicellular red and green algae were able to establish populations in wave-swept coastlines and other environments where simpler organisms

**FIG. 28.15  Evolution of complex fungi.** (a) This treelike fossil, discovered in Saudi Arabia, is actually a giant fungus that towered over early land plants 375 million years ago. (b) This image shows the interior of the fungus, which consists of tubes that transported nutrients through the large body.

Fossils indicate that by 400 million years ago, plants with specialized tissues for bulk transport of water and nutrients had begun to spread across the continents. As plant biomass built up on land, it created opportunities for additional types of complex multicellular organism. In particular, the fungi, which decompose organic matter, radiated by taking advantage of this resource (**Fig. 28.15**), evolving complex multicellularity in two distinct lineages (Chapter 34).

**Regulatory genes played an important role in the evolution of complex multicellular organisms.**
Complex multicellular organisms account for a large proportion of all eukaryotic species, and the evolutionary consequences of complex multicellularity are seen clearly when we compare the diversity of complex multicellular groups with their simpler phylogenetic sisters. Choanoflagellates, for example, number about 150 species, simple animals such as sponges and jellyfish about 20,000, and complex animals with circulatory systems that circumvent diffusion perhaps as many as 10 million. Plants and their relatives show a comparable pattern. There are about 6000 species of green algae on the branch that includes land plants, but about 400,000 species of anatomically complex land plants capable of bulk transport. Similarly, complex fungi and red algae include more species by an order of magnitude than their simpler relatives.

How did this immense diversity arise? At one level, the answer is functional and ecological. Complex multicellular organisms can perform a range of functions that simpler organisms cannot, and so they have evolved many specific types of interaction with other organisms and the physical environment. This explanation, however, prompts another question: What is the genetic basis for the bewildering range of sizes and shapes displayed by complex multicellular organisms?

The answer has to do with the network of genes that guide development. Since the 1950s, biologists have learned a remarkable amount about this genetic network and its host of interacting genes. Many of the genes involved in development are what we might consider middle managers: A molecular signal induces the expression of a gene, and the protein product, in turn, prompts the expression or repression of another gene. It is the complex interplay of these genetic switches—turning specific genes on in one cell and off in another, depending on where those cells occur in the developing body—that results in crabs with large pincers or butterflies with patterned wings (**Fig. 28.16**).

It makes sense that if regulatory genes guide development of the body in each complex multicellular

**FIG. 28.16**

# What controls color pattern in butterfly wings?

**BACKGROUND** Butterfly wings commonly show a striking pattern of color, including circular features known as eyespots. Eyespots have adaptive value, for example in deterring predation by birds.

**HYPOTHESIS** The expression of regulatory genes during wing development governs eyespot formation on wing surfaces.

**EXPERIMENT** Paul Brakefield and his colleagues mapped the expression of a regulatory gene called *Distalless* in developing butterfly wings. *Distalless* genes were fused to genes for green fluorescent protein (GFP), a protein that glows vivid green when illuminated under blue light. GFP would then be expressed along with *Distalless*, allowing the spatial pattern of *Distalless* expression in developing wings to be followed by GFP visualization.

**RESULTS** The expression pattern of *Distalless* in developing wings closely resembles the pattern of eyespots on the wing. Moreover, mutations that change the spatial pattern of *Distalless* expression result in different eyespot patterns.

The five pairs show the pattern of *Distalless* gene expression and eyespot development in wild-type (top left) and mutant butterfly wings. High levels of *Distalless* expression (green dots and ovals) define the centers of eyespots.

**CONCLUSION** Regulatory genes play an important role in butterfly wing coloration, and mutations in these genes can account for differences in wing coloration among species.

**SOURCE** Brakefield, P. M., J. Gates, D. Keys, F. Kesbeke, P. J. Wijngaarden, A. Monteiro, V. French, and S. B. Carroll. 1996. "Development, Plasticity and Evolution of Butterfly Eyespot Patterns." *Nature* 384: 236–242.

species, then mutations in regulatory genes may account for many of the differences we observe among different species. Our growing understanding of how developmental genes underpin evolutionary change has given rise to a whole new field of biology called evolutionary-developmental biology, or **evo-devo** for short.

In evo-devo research, scientists compare the genetic programs for growth and development in species found on different branches of phylogenetic trees. In addition to wanting to discover the molecular mechanisms that guide development in individual species, these scientists aim to understand the genetic differences associated with differences in form and function among species. Evo-devo is an exciting and rapidly expanding field of research because it is helping to illuminate the long-suspected relationship between the development of individuals and patterns of evolutionary relatedness among species. Continuing research promises to shed new light on the similarities and differences among plants, animals, and other complex multicellular organisms.

# Core Concepts Summary

## 28.1 COMPLEX MULTICELLULARITY AROSE SEVERAL TIMES IN EVOLUTION.

Bacteria are unicellular or form simple multicellular structures. page 28-1

Eukaryotes are unicellular or multicellular and exhibit both simple and complex multicellularity. page 28-1

Simple multicellularity involves the adhesion of cells with little cell differentiation. Complex multicellularity involves cell adhesion, cell signaling, and differentiation and specialization among cells. page 28-2

Complex multicellular organisms have evolved independently at least six times: in animals, vascular plants, red algae, brown algae (the kelps), and at least twice in the fungi. page 28-3

## 28.2 IN COMPLEX MULTICELLULAR ORGANISMS, BULK TRANSPORT OF MOLECULES CIRCUMVENTS THE LIMITATIONS OF DIFFUSION.

Diffusion is the random motion of molecules, with net movement occurring from regions of higher to regions of lower concentration. It generally acts over small distances, placing limits on the size of multicellular organisms. page 28-4

Bulk transport is an active process that allows multicellular organisms to nourish cells located far from the external environment, thereby circumventing the constraints imposed by diffusion. page 28-5

## 28.3 COMPLEX MULTICELLULARITY DEPENDS ON CELL ADHESION, COMMUNICATION, AND A GENETIC PROGRAM FOR DEVELOPMENT.

Animal and plant cells are organized into tissues characterized by specific molecular attachments between cells. page 28-6

Choanoflagellates express some of the same proteins that permit cell adhesion in animals, even though choanoflagellates are unicellular. Experiments suggest that choanoflagellates use these proteins to capture bacteria, not for cell adhesion. page 28-7

Gap junctions in animals and plasmodesmata in plants allow cells to communicate with each other in a targeted fashion. page 28-7

Complex multicellularity involves the genetic programming of cells so that they differentiate in space. page 28-8

A number of gene families known to play developmental roles in multicellular organisms also occur in their unicellular relatives, where at least some of them play a role in life cycle differentiation. page 28-9

## 28.4 PLANTS AND ANIMALS EVOLVED MULTICELLULARITY INDEPENDENTLY OF EACH OTHER AND SOLVED SIMILAR PROBLEMS WITH DIFFERENT SETS OF GENES.

The cell wall characteristic of plant cells provides structural and mechanical support, but does not allow plant cells to move. page 28-10

The cell wall of plants has led to distinct solutions to the problems of cell adhesion, cell communication, and development. For example, plants grow by the activity of meristems, populations of actively dividing cells at the tips of stems and roots. page 28-10

Animal cells do not have cell walls, allowing cell movement that is not possible in plants. For example, during animal development, embryos undergo gastrulation, a process in which cells migrate inward to form a layered structure called a gastrula. page 28-11

28.5 THE EVOLUTION OF LARGE AND COMPLEX MULTICELLULAR ORGANISMS, WHICH REQUIRED ABUNDANT OXYGEN, IS RECORDED IN THE FOSSIL RECORD.

Oxygen may be required for the evolution of complex multicellularity because of its chemical properties and its abundance. page 28-13

The evolution of complex multicellularity in the fossil record correlates with increases in atmospheric oxygen that occurred 580-560 million years ago. page 28-14

Complex multicellular organisms evolved later on land, as ancestral plants evolved the capacity to photosynthesize surrounded by air rather than water. page 28-14

Evolutionary-developmental biology, or evo-devo, is a field of research that looks at both individual development and evolutionary patterns in an attempt to understand the developmental changes that allowed organisms to diversify and adapt to changing environments. page 28-15

## Self-Assessment

1. Describe differences between simple and complex multicellularity.

2. Describe the phylogenetic distribution of complex multicellularity.

3. Explain how diffusion limits the size of organisms.

4. Explain how multicellular organisms get around the size limits imposed by diffusion.

5. Describe a key difference between multicellular plants and animals.

6. Describe environmental changes recorded by the sedimentary rocks that contain the oldest fossils of large active animals.

Do you understand the chapter's Core Concepts? Log in to check your answers to the Self-Assessment questions, then practice what you've learned and reinforce this chapter's concepts by working through the problems and multimedia tutorials provided there.

⊡ www.biologyhowlifeworks.com

# Agriculture

## FEEDING A GROWING POPULATION

You may have noticed that while your grocery store or supermarket has several types of apple for sale, there is usually only one type of banana. In fact, it's possible that every banana you have ever eaten has been the same variety. Now the popular yellow fruit may be in trouble.

Years ago, the Western world favored a banana variety known as the Gros Michel. In the first part of the twentieth century, a devastating fungal infection called Panama disease wiped out Gros Michel plantations around the world. In the 1950s, banana growers turned instead to the Cavendish, a variety that displayed some natural resistance to Panama disease. Half a century later, the Cavendish remains the top-selling banana in supermarkets in North America and beyond.

Yet the reign of the Cavendish may be coming to an end. For years, banana growers have battled black sigatoka, a fungal infection that can cause losses of 50% or more of the banana yield in infected regions. Making matters worse, Panama disease is once again posing a threat. A strain of the disease has emerged against which the Cavendish plants have no resistance. The disease has spread across Asia and Australia, and experts fear it's only a matter of time before it reaches prime banana-growing regions in Latin America.

**Agriculture has transformed our planet and our species; without it, modern civilization would not be possible.**

For those who enjoy bananas sliced into their breakfast cereal, a loss of bananas would be annoying. For the millions of people in tropical countries who depend on bananas and related plantains for daily sustenance, the destruction of those crops would be much more serious. Agriculture has transformed our planet and our species; without it, modern civilization would not be possible. But as the case of the banana shows, there are challenges to overcome if we're to continue feeding a growing population.

Cultivated bananas are vulnerable to infection in a way that wild bananas are not. The bananas that we eat are sterile; they do not produce seeds. This means they must be propagated vegetatively by replanting cuttings taken from parent plants. As a result, banana plantations contain virtually no genetic diversity. As Panama disease and black sigatoka are proving, that lack of diversity can have disastrous effects.

As sterile clones, bananas are particularly vulnerable to disease outbreaks. Other agricultural crops face similar threats. For example, a new, virulent strain of yellow wheat rust fungus emerged in the Middle East in 2010. In its first year, the disease wiped out as much as half of Syria's wheat crop and has since spread to several other countries in the region. Meanwhile, crop scientists have warned that Ug99, a new, virulent strain of an even more damaging fungal wheat pathogen that first surfaced in eastern Africa, could devastate global wheat crops as it spreads.

Wheat was one of the first crops that our ancestors cultivated when agriculture emerged 10,000 years ago. The early farmers learned to choose wheat plants that were easiest to harvest and whose seeds germinated without delay. Over time, wheat evolved by this process of artificial selection. Those earliest farmers set in motion a long chain of genetic modifications that ultimately led to the high-yield wheat varieties grown today.

In the wild, natural selection favored wheat whose ears were protected by tough barbs that broke apart into individual units for efficient dispersal across the landscape. Farmers, on the other hand, preferred wheat plants whose ears didn't separate, and so were easier to harvest. Such features would interfere with the plant's ability to self-seed in the wild. By selecting for traits that would be disadvantageous in nature, humans created a plant dependent on humans for its survival. The co-dependency between domestic plants and people is a hallmark of agriculture.

In addition to selecting desirable traits in their crop plants, early agriculturalists altered the environment to be favorable for those plants. They provided water, chose planting sites to ensure optimal sunlight, and removed nearby plants that would compete with the crops for vital resources such as water and nutrients. By changing the community structure and the availability of resources to support growth and reproduction, humans allowed domesticated crops to thrive.

By the 1940s, scientists had developed industrially fixed nitrogenous fertilizers that significantly boosted plant growth. They had also learned enough about genetics and DNA to begin applying that knowledge to plant breeding. These advances led to an amazingly productive period during the 1960s and 1970s now known as the Green Revolution. During that time, plant yields ballooned. Food production actually outstripped population growth through the second half of the twentieth century.

Today, we are entering a new phase of agriculture, with genetic engineering being added to the traditional approaches of crossbreeding and artificial selection. Most of the corn and soy products now consumed in the United States come from plants that have been genetically engineered to resist certain pests and herbicides. Some of the genes employed come from different species altogether. For example, corn and cotton plants are often engineered to contain a gene from the bacterium *Bacillus thuringiensis* (*Bt*) that confers pest resistance.

At a glance, it would seem that we've successfully taken domesticated species under our control, and indeed, we've made great strides in increasing yields to feed an ever-growing population. But evolution never takes a break. Crop pests and pathogens continue to evolve ways to evade our savviest plant breeders.

The nature of modern agriculture only compounds the problem. Most modern farms consist largely of monocultures, single types of crop each grown over a large area. Growing only a single crop at a time makes it much easier to mechanize planting and harvesting. In a given field of corn or wheat, therefore, the individual plants are often genetically quite similar to one another. With so little genetic variation, each plant is vulnerable to pathogens in exactly the same way. This means that a pathogen that can overcome one plant's natural defenses has the potential to wipe out the entire field.

Lately, many consumers are taking a hard look at these problems and at the source of the food on their plates. Critics of modern agriculture argue that monocultures, with their heavy reliance on fungicides and pesticides, are unsustainable and unhealthy. Proponents of genetic modification and high-tech agriculture argue that we will need every tool to meet food-production demands for the expanding human population. Sometime in 2011, the human population surpassed 7 billion people. The population continues to grow, expanding at its fastest rate at any time in human history. By 2100, experts predict, our planet will be home to more than 9 billion people. One point that both sides agree on is that unless agricultural productivity continues to increase, more and more land will be needed to produce food.

In many parts of the world, agricultural productivity is limited by the availability of water or nutrients or both. Global climate change is predicted to cause regional droughts and flooding that may further weaken the global food supply. There will be no easy solutions as we move forward into civilization's next phase of agriculture. One thing that is certain, however, is that understanding how plants grow, reproduce, and protect themselves from pests will help us survive in an increasingly crowded world.

## ? CASE 6 QUESTIONS

*Answers to Case 6 questions can be found in Chapters 29–34.*

1. How has nitrogen availability influenced agricultural productivity? *See page 29-18.*
2. What is the basis for the spectacular increases in the yield of cereal grains during the Green Revolution? *See page 30-21.*
3. What is the developmental basis for the shorter stems of high-yielding rice and wheat? *See page 31-9.*
4. Can modifying plants genetically protect crops from herbivores and pathogens? *See page 32-17.*
5. What can be done to protect the genetic diversity of crop species? *See page 33-20.*
6. How do fungi threaten global wheat production? *See page 34-19.*

## CHAPTER 29

# PLANT STRUCTURE AND FUNCTION

## Moving Photosynthesis onto Land

## Core Concepts

**29.1** The evolution of land plants from aquatic ancestors introduced a major challenge for photosynthesis: acquiring $CO_2$ without losing excessive amounts of water.

**29.2** Leaves have a waxy cuticle that retards water loss but inhibits the diffusion of $CO_2$, and pores, called stomata, that regulate $CO_2$ gain and water loss.

**29.3** Xylem allows vascular plants to replace water evaporated from leaves with water pulled from the soil.

**29.4** Phloem transports carbohydrates that support the growth and respiration of non-photosynthetic organs such as stems and roots.

**29.5** Roots expend energy to obtain nutrients from the soil.

In the ocean, where life first evolved, organisms are bathed in seawater that mirrors the osmotic concentration of their cells. In contrast, on land, organisms are surrounded by air and are at risk of drying out. Excessive water loss, or **desiccation,** is a constant challenge for life on land, especially for photosynthetic organisms, which expose large surface areas to the air in order to obtain sunlight and carbon dioxide ($CO_2$). In Chapter 27, we saw that land plants evolved from aquatic green algae. What structural and functional innovations enabled plants to colonize the land so successfully?

The story begins about 470 million years ago, when a lineage of green algae began the transition from water to land. There are few fossils to show us what the first land plants looked like. We know they were small, at most only a few centimeters tall. These early colonists resembled their algal ancestors; they had no means of obtaining water from the soil and, at best, only a limited capacity to restrict water loss from cells. Therefore, their water content and their ability to carry out photosynthesis would have fluctuated wildly.

Today, descendants of those first land plants dominate terrestrial habitats, having evolved the ability to draw water from the soil and limit water loss from their leaves. These are the **vascular plants,** and their evolution transformed the physical and biological environment on land. Vascular plants can carry out photosynthesis even when the soil surface is dry and can sustain the hydration of photosynthetic leaves elevated as much as 100 meters into the air. In many ways, the capacity to control the gain and loss of water made possible the extraordinary evolutionary success of vascular plants.

## 29.1 PLANT STRUCTURE AND FUNCTION: AN EVOLUTIONARY PERSPECTIVE

Phylogeny provides a road map of land plant evolution. We will reserve detailed discussion of plant diversity until we have discussed how plants function, and we begin now with a quick sketch of evolutionary relationships among land plants. **Fig. 29.1** shows that land plants form a monophyletic group descended from green algae. The land surface, in turn, is covered by two groups of plants: the **bryophytes** (which include the mosses, along with the less-familiar liverworts and hornworts) and the vascular plants. For many years, botanists treated both bryophytes and vascular plants as monophyletic, but phylogenetic analysis based on molecular sequences makes it clear that bryophytes are a paraphyletic grouping, including some of but not all the descendants of a common ancestor. Note in Fig. 29.1 that the descendants of the last common ancestor of bryophytes also include vascular plants.

Vascular plants make up more than 95% of all land plant species found today, and they similarly dominate the photosynthetic output of terrestrial environments. The phylogenetic tree shows that vascular plants form a monophyletic group, with four main subgroups, three of which are probably familiar to you. **Ferns and horsetails** form a monophyletic group, as do plants that reproduce by seeds. Molecular sequence comparisons suggest that living seed plants can be subdivided into two monophyletic groups, gymnosperms and angiosperms. The **gymnosperms** include pine trees and other conifers. The **angiosperms,** or flowering plants, include oak trees, grasses, and sunflowers.

The fourth major vascular plant group, the **lycophytes,** may be less familiar. This group is particularly interesting to scientists because it forms the sister group to all other vascular plants. Keep Fig. 29.1 in mind as you read the next several chapters because, while the major groups of vascular plants share many features, they have each evolved distinctive variations in growth, anatomy, and reproduction that help to explain their present-day diversity and distribution.

The evolution of vascular plants was a game-changing event in the history of life. To appreciate why this is so, we have to understand the "game" played by the lineages that preceded the vascular plants. In particular, how are the bryophytes (the three groups of nonvascular land plants) able to survive on land? Bryophytes do not produce roots and so are totally dependent upon surface water to keep their photosynthetic cells hydrated. As the environment dries, so do they.

All cells require a high degree of hydration to function. As bryophytes dry out, they lose the ability to carry out photosynthesis. However, many bryophytes exhibit **desiccation tolerance,** a suite of biochemical traits that allows their cells to survive extreme dehydration without damage to membranes or macromolecules (**Fig. 29.2**). Bryophytes can then resume photosynthesis when surface water is once again available. In

**FIG. 29.2** **Desiccation tolerance in bryophytes.** *Syntrichia princeps,* a moss, can withstand complete desiccation (right) but regains photosynthetic capacity rapidly when rewetted (left).

contrast to bryophytes, vascular plants maintain the hydration of their photosynthetic cells by drawing upon water in the soil. As a result, vascular plants can photosynthesize across a wider range of environments and conditions than can bryophytes.

Bryophytes provide insights into how the first plants may have dealt with the challenge of carrying out photosynthesis on land. The ability to survive desiccation is likely to have been critical. But we must avoid the temptation to think of living bryophytes as representative of the ancestral state. Bryophyte diversity has been shaped by their long coexistence with vascular plants. Indeed, bryophytes thrive today only in places where roots would not provide a significant advantage—on rocks and the trunks of trees, for example, and in regions too dry or too cold for roots to penetrate the soil.

We will return to bryophytes when we examine plant reproduction (Chapter 30) and diversity (Chapter 33). For now, however, let's focus on the vascular plants.

## 29.2 THE LEAF: ACQUIRING CO₂ WHILE AVOIDING DESICCATION

**Fig. 29.3** shows the general features of a vascular plant. Aboveground, we see three major types of organ—leaves, stems, and reproductive organs—which collectively form the **shoot. Roots,** which are generally hidden from view in the ground, make

**FIG. 29.1** **The phylogenetic tree of plants.** Plants can be divided into two groups: the vascular plants, which actively control their hydration, and the bryophytes, which do not.

Sister group of green algae

Liverworts

Mosses

Hornworts

Lycophytes

Ferns and horsetails

Gymnosperms

Angiosperms

Land plants

Green algae
Bryophytes
Vascular plants

**FIG. 29.3** Major organs of a vascular plant.

up the fourth major organ system. We can answer the question of how vascular plants photosynthesize on land by examining how leaves, stems, and roots work together.

The **leaf** is the principal site of photosynthesis in vascular plants. A cross section shows the leaf's three major tissues: sheets of cells called the **epidermis** that line the leaf's upper and lower

surfaces; loosely packed photosynthetic cells that make up the **mesophyll** (literally, "middle leaf"); and **veins,** the system of vascular conduits that connects the leaf to the rest of the plant (**Fig. 29.4**). We discuss how water and carbohydrates are transported through the vascular system in sections 29.3 and 29.4. In this section, we explore how the structure of leaves allows vascular plants to photosynthesize in air without drying out.

## $CO_2$ uptake results in water loss.

Water is the resource that most often limits a plant's ability to grow and function. Water availability is the single most important factor limiting agricultural yields, and rainfall patterns have a major impact on the structure and productivity of natural ecosystems (Chapter 47). Water has such a significant effect on the growth and functioning of plants because plants use large amounts of water. Why do plants require so much water?

The answer is *not* because water is consumed in supplying electrons for photosynthesis (Chapter 8). That process accounts for less than 1% of the water required by vascular plants. Instead, most of a plant's need for water arises as a consequence of $CO_2$ uptake in air.

As we saw in Chapter 25, $CO_2$ is a minor constituent of air, about 400 parts per million at present. This low concentration limits the rate at which $CO_2$ can diffuse into the leaf. Therefore, how fast a plant can take up $CO_2$ depends in large part on the degree to which leaves can expose their photosynthetic cells to the surrounding air. If we look inside a leaf, we see that each mesophyll cell is largely

**FIG. 29.4** **Leaf structure.** The uptake of $CO_2$ by diffusion is accompanied by a much larger outward diffusion of water vapor.

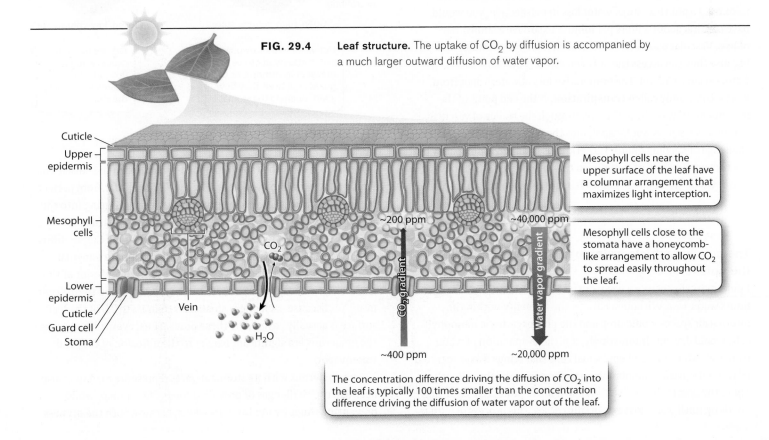

surrounded by air. The mesophyll cells obtain the $CO_2$ that they need for photosynthesis directly from these air spaces. If the air spaces within the leaf were completely sealed off, photosynthesis would quickly run out of $CO_2$ and come to a halt. But they're not sealed off—the leaf's air spaces are connected to the air surrounding the leaf by pores in the epidermis. As photosynthesis draws down the concentration of $CO_2$ molecules within the leaf's air spaces relative to the concentration of $CO_2$ in the outside air, the difference in concentration between the inside and the outside of the leaf causes $CO_2$ to diffuse into the leaf, replenishing the supply of $CO_2$ for photosynthesis.

The ability of leaves to draw in $CO_2$ comes at a price, however: If $CO_2$ can diffuse into the leaf, water vapor can diffuse out (Fig. 29.4). Furthermore, the rate at which water vapor diffuses out of a leaf is much greater than the rate at which $CO_2$ diffuses inward. Molecules diffuse from regions of higher concentration to regions of lower concentration, and the rate of diffusion is proportional to the difference in concentration. On a sunny summer day, the difference in water vapor concentration between the air spaces within a leaf and the air outside can be more than 100 times larger than the difference in concentration of $CO_2$. Add the fact that water is lighter than $CO_2$, and so diffuses 1.6 times faster *for the same concentration gradient*, and it becomes clear why, on a sunny summer day, several hundred water molecules are lost for every molecule of $CO_2$ acquired for photosynthesis.

Because of this difference, the water costs associated with photosynthesis in air are large. A sunflower leaf, for example, can lose an amount equal to its total water content in as little as 20 minutes. To put this rate of water loss in perspective, you would have to drink about 2 liters per minute to survive a similar rate of loss. Vascular plants can sustain such high rates of water loss because they can access the only consistently available source of water on land: the soil. The evaporative loss of water vapor from leaves, commonly called **transpiration,** is the end point of the movement of water from the soil, through the bodies of vascular plants, and then, as water vapor, into the atmosphere. Therefore, the challenge of keeping photosynthetic cells hydrated is met partly by the continual supply of water from the soil and partly by limiting the rate at which water already in leaf tissues is lost to the atmosphere.

→   **Quick Check 1**  Why do plants transpire?

### The cuticle restricts water loss from leaves but inhibits the uptake of $CO_2$.

Epidermal cells secrete a waxy **cuticle** on their outer surface that limits water loss. Without a cuticle, the humidity within the internal air spaces would drop and the photosynthetic mesophyll cells would dry out. Interestingly, it is the evaporation of water from the exterior face of epidermal cells that brings waxes secreted by the cell's plasma membrane up to the surface of the cell wall. There the wax molecules self-organize into a protective layer that builds up until water loss essentially ceases. This process of cuticle

**FIG. 29.5**   **Stomata.** (a) A scanning electron microscope (SEM) image of stomata on a leaf surface; (b) guard cells of a stoma in open and closed positions.

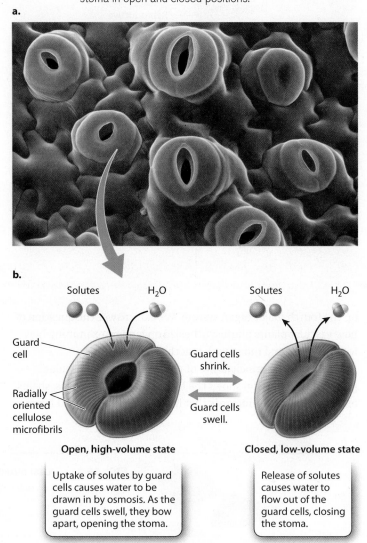

formation ensures the rapid repair of any damage to the cuticle that would allow water to evaporate. The mechanism is far from perfect, however, because the cuticle prevents $CO_2$ from diffusing into the leaf even as it restricts water vapor from diffusing out of it.

As noted above, small pores in the epidermis allow $CO_2$ to diffuse into the leaf (**Fig. 29.5a**). These pores, called **stomata** (singular, stoma), regulate the diffusion of gases between the interior of the leaf and the atmosphere. Stomata can be numerous, up to 1000 per mm². Yet because each one is small, less than 1% to 2% of the leaf surface is actually covered by these pores. Thus, even with stomata, the epidermis is a significant barrier to the diffusion of $CO_2$ and water vapor.

The epidermis with its stomatal pores represents a compromise between the challenges of providing food ($CO_2$ uptake) while preventing thirst (water loss). However, because both the dryness of the air and the wetness of the soil are variable, leaves must be able to alter the porosity of the leaf epidermis to maintain a balance between the rates of water loss to the atmosphere and water delivery from the soil.

## Stomata allow leaves to regulate water loss and carbon gain.

Stomata are more than just holes in the leaf epidermis; they are hydromechanical valves that can open and close. Each stoma consists of two **guard cells** surrounding a central pore. The guard cells can shrink or swell, changing the size of the pore between the guard cells (**Fig. 29.5b**). How does this valve system work?

Cellulose microfibrils in the guard cell walls are oriented radially, that is, wrapped around the cell. The radially oriented microfibrils make it easier for swelling guard cells to expand in length rather than in girth. Because the guard cells are firmly connected at their ends, this increase in length causes them to bow apart, opening the stoma. Conversely, a decrease in cell volume causes the pore to close.

Guard cells control their volume by altering the concentration of solutes, for example the ions $K^+$ and $Cl^-$, in their cytoplasm. An increase in solute concentration causes water to flow into the cell by osmosis, whereas a decrease in solutes causes water to flow out of the cell. Recall from Chapter 5 that osmosis is the diffusion of water across a semipermeable membrane, such as the cell's plasma membrane, from a region of higher water concentration to a region of lower water concentration. An increase in the concentration of solutes in guard cells is accompanied by a corresponding decrease in the concentration of water molecules. Thus, as guard cells add solutes, water diffuses into the cells, causing them to swell.

Stomata close by the same process, only in reverse. A decrease in solute concentration results in a higher concentration of water molecules inside a guard cell relative to the solution in the surrounding cell wall space, and water diffuses out of the cell, causing it to shrink.

The accumulation of solutes inside guard cells requires an expenditure of energy in the form of ATP, which drives the uptake of solutes across the plasma membrane (Chapter 5). Closing, in contrast, does not require direct energy expenditure. Thus, the default or resting state of guard cells is "closed"; this is the state that conserves the hydration of the leaf. In angiosperms (the flowering plants), $K^+$ and $Cl^-$ ions are transferred between guard cells and an adjacent epidermal cell, so an increase in the volume of one cell type is accommodated by a decrease in the other. This reciprocal change in cell size allows the stomatal pore to open much wider than it could if the guard cells operated on their own (**Fig. 29.6**). It takes approximately 10 minutes for angiosperm stomata to go

from fully closed to fully open. During this time, the volume of each guard cell can increase as much as threefold.

To function effectively, stomata must open and close in response to both $CO_2$ demand and water loss. Stomata are thus key sites for the processing of physiological information. Light stimulates stomatal opening, while high levels of $CO_2$ inside the leaf (a signal that $CO_2$ is being supplied faster than photosynthesis can take it up) cause stomata to close. When the soil can't supply enough water to keep up with evaporation from the leaf, guard cells become dehydrated and shrink in volume, closing the stomata. Guard cell volume also changes in response to signaling molecules. For example, abscisic acid, a hormone produced during drought, causes stomata to close. This signal allows leaves to close their stomata early enough to prevent dehydration, rather than responding only when water has already been lost. Clearly, then, stomata are a key innovation in the evolution of land plants. Not surprisingly, fossils indicate that stomata evolved early in the colonization of land.

Before leaving the topic of how stomata affect rates of water loss from leaves, we pause to consider the following question: Is there any evidence that plants control the temperature of their leaves by regulating transpiration? The answer, perhaps surprisingly, is no. Although our own experience leads us to think of evaporation (for example, sweating) as a means of preventing overheating, plants transpire when their leaves are too hot *or* too cold. When leaves are above their optimal temperature for photosynthesis, cooling through transpiration is beneficial. But when temperatures are

**FIG. 29.6 Opening and closing of guard cells in a fern and an angiosperm.** (a) Guard cells in ferns shrink and swell independently of other cells, limiting the size of the stomatal pore; (b) in angiosperms, changes in guard cell volume are matched by reciprocal changes in the adjacent epidermal cells.

**a.** *Nephrolepis exaltata* (fern)

Stoma closed

Epidermal cell   Guard cell

Stoma open

Guard cell
Stoma (open)

Top view

**b.** *Tradescantia virginiana* (angiosperm)

Stoma closed

Stoma open

Top view

below the optimum, transpiration results in leaf temperatures that are even less favorable. In addition, the hot environments in which evaporative cooling would be beneficial often coincide with low availability of water in the soil. For this reason, plants in arid regions tend to rely on mechanisms other than evaporative cooling, such as reflective hairs and waxes and small leaf size, to prevent their leaves from overheating.

## CAM plants use nocturnal CO₂ storage to avoid water loss during the day.

In most plants, water loss and photosynthesis both peak at the same time, when the sun is high overhead. What if a plant could open its stomata to capture $CO_2$ at night, when cool air limits rates of evaporation?

In fact, a number of plants have evolved precisely this mechanism, called **crassulacean acid metabolism,** or **CAM,** to help balance $CO_2$ gain and water loss (**Fig. 29.7**). CAM (named after the Crassulaceae family of plants, which uses this pathway), provides a system for overnight storage, converting $CO_2$ into a form that will not diffuse away. The storage form of $CO_2$ is produced by the activity of the enzyme PEP carboxylase, which combines a dissolved form of $CO_2$ (bicarbonate ion, $HCO_3^-$) with a 3-carbon compound called phosphoenol pyruvate (PEP). The resulting product is a 4-carbon organic acid that is stored in the cell's vacuole.

When the sun comes up the next morning, the stomata close, conserving water. At the same time, the 4-carbon organic acids are retrieved from the vacuole and decarboxylated—that is, their $CO_2$ is released into the cytoplasm. Because the stomata are now closed, this newly released $CO_2$ does not escape from the leaf. Instead, it diffuses into the chloroplast, where it can become incorporated into carbohydrates by the Calvin cycle. As we saw in Chapter 8, the Calvin cycle requires a continual supply of energy in the form of ATP and NADPH and so can operate only in the light.

The buildup of organic acid during the night results in a marked decrease in the pH of vacuoles. In 1815, the British naturalist Benjamin Heyne wrote that some of the plants in his garden in India tasted "as acid as sorrel" early in the morning, but by afternoon the acid taste had disappeared.

The nocturnal opening of stomata exhibited by CAM plants greatly increases the amount of $CO_2$ gain per unit of water loss. Indeed, the $CO_2 : H_2O$ exchange ratio for CAM plants is in the range of 1 : 50, nearly 10 times higher than that of plants that open their stomata during the day. However, CAM has a drawback. The rates of photosynthetic carbohydrate production by CAM plants tend to be low because ATP is needed to drive the uptake of organic acids into the vacuole and because only so much organic acid can accumulate in the vacuole. CAM is therefore most common in habitats such as deserts, where water conservation is crucial, and among **epiphytes,** plants that grow high in the canopy of other plants, without contact with the soil.

Because CAM uses enzymes (for example, PEP carboxylase) and compartments (for example, vacuoles) that exist in all plant cells, it is not surprising that it has evolved multiple times. CAM occurs in all four groups of vascular plants shown in Fig. 29.1. It is most widespread in the angiosperms, occurring in 5% to 10% of species. Vanilla orchids and Spanish moss, two well-known epiphytes, as well as many cacti, exhibit CAM.

**FIG. 29.7    CAM photosynthesis.** CAM plants store $CO_2$ at night in the vacuole; leaves can then photosynthesize during the day without opening their stomata.

A CAM plant: Organ pipe cactus (*Stenocereus thurberi*) in the Arizona desert.

## $C_4$ plants suppress photorespiration.

Photorespiration adds another wrinkle to the challenge of acquiring $CO_2$ while minimizing water loss. Recall from Chapter 8 that either $CO_2$ or $O_2$ can be a substrate for rubisco, the key enzyme in the Calvin cycle. $CO_2$ as a substrate leads to the production of carbohydrates through photosynthesis; $O_2$ as a substrate results in a net loss of energy and the release of $CO_2$, a process called photorespiration. (Photorespiration is similar to aerobic respiration only in the sense that it uses $O_2$ and releases $CO_2$. However, plants do not gain energy; they lose it.)

Photorespiration presents a significant challenge for land plants because air contains approximately 21% $O_2$ but less than 0.04% $CO_2$. Although rubisco reacts more readily with $CO_2$ than with $O_2$, the sheer abundance of $O_2$ means that $O_2$ is a substrate for rubisco some of the time. At moderate leaf temperatures, $O_2$ is the substrate instead of $CO_2$ as often as 1 time out of 4. At higher temperatures, $O_2$ is even more likely to be the substrate for rubisco, and the rates of photorespiration are yet higher. The reason is that higher temperatures decrease both rubisco's affinity for $CO_2$ relative to $O_2$ and the solubility of $CO_2$ relative to $O_2$ in water.

Some plants have evolved a way to reduce the energy and carbon losses associated with photorespiration. These are the **$C_4$ plants,** which suppress photorespiration by increasing the concentration of $CO_2$ in the immediate vicinity of rubisco. $C_4$ plants take their name from the fact that they, like CAM plants, use PEP carboxylase to produce 4-carbon organic acids that subsequently supply the Calvin cycle with $CO_2$. The Calvin cycle produces 3-carbon compounds (Chapter 8), and plants that do not use 4-carbon organic acids to supply the Calvin cycle with $CO_2$ are thus referred to as **$C_3$ plants.**

Both CAM and $C_4$ plants produce 4-carbon organic acids as the entry point for photosynthesis. However, in CAM plants, $CO_2$ capture and the Calvin cycle take place *at different times*; in $C_4$ plants, they take place *in different cells*. It is the spatial separation between $CO_2$ capture and the Calvin cycle that allows $C_4$ plants to suppress photorespiration.

$C_4$ plants intitially capture $CO_2$ in mesophyll cells by means of PEP carboxylase, which combines a dissolved form of $CO_2$ (bicarbonate ion, $HCO_3^-$) with the 3-carbon compound PEP. This produces 4-carbon organic acids that diffuse through plasmodesmata into the **bundle sheath** (**Fig. 29.8**), a cylinder of cells that surrounds each vein. Once inside bundle-sheath

**FIG. 29.8** **$C_4$ photosynthesis.** $C_4$ plants suppress photorespiration by concentrating $CO_2$ in bundle-sheath cells.

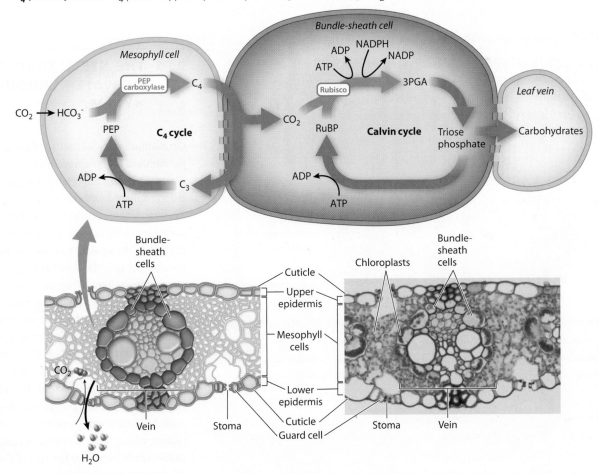

cells, the 4-carbon compounds are decarboxylated, releasing $CO_2$ that is then incorporated into carbohydrates through the Calvin cycle (Chapter 8). The $C_4$ cycle is completed as the 3-carbon molecules generated during decarboxylation diffuse back to the mesophyll, where ATP is used to re-form PEP. Because the $C_4$ cycle operates much faster than the Calvin cycle, the concentration of $CO_2$ within bundle-sheath cells builds up, reaching levels as much as five times higher than in the air surrounding the leaf. The increase in the concentration of $CO_2$ in bundle-sheath cells makes it much less likely that rubisco will use $O_2$ as a substrate.

$C_4$ plants have high rates of photosynthesis because they do not suffer the losses in energy and reduced carbon associated with photorespiration (**Fig. 29.9**). In addition, $C_4$ plants often have a more favorable $CO_2 : H_2O$ exchange ratio than plants that lack the ability to concentrate $CO_2$. However, $C_4$ photosynthesis has a greater energy requirement than conventional ($C_3$) photosynthesis, as ATP must be used to regenerate PEP in the $C_4$ cycle. Thus, $C_4$ photosynthesis confers an advantage in hot, sunny environments where rates of photorespiration would otherwise be high. $C_4$ photosynthesis has evolved as many as 20 times, but is most common among tropical grasses and plants of open habitats with warm temperatures. $C_4$ plants include a number of important crops, especially maize (corn), sugarcane, and sorghum.

→　**Quick Check 2**　Why does the formation of 4-carbon organic acids increase the efficiency of water use in CAM plants but suppress photorespiration in $C_4$ plants?

## HOW DO WE KNOW?

**FIG. 29.9**

# How do we know that $C_4$ photosynthesis suppresses photorespiration?

**BACKGROUND** Studies using radioactively labeled $CO_2$ showed that some species initially incorporate $CO_2$ into 4-carbon compounds instead of the 3-carbon compounds that are the first products in the Calvin cycle. These $C_4$ plants also have high rates of photosynthesis. Is this a new, more efficient photosynthetic pathway? Or do $C_4$ plants have high rates of photosynthesis because they are able to avoid the carbon and energy losses associated with photorespiration?

**HYPOTHESIS** $C_4$ plants do not exhibit photorespiration.

**METHOD** "Air tests," as these experiments were first called, compared rates of photosynthesis in normal air (21% $O_2$) and in an experimental gas mixture in which the concentration of $O_2$ is only 1%. When the concentration of $O_2$ is low, rubisco has a low probability of using $O_2$ (instead of $CO_2$) as a substrate, and thus photorespiration does not occur.

**RESULTS**

**CONCLUSION** Photosynthesis in $C_4$ plants is not affected by differences in $O_2$ concentration, indicating that significant photorespiration is not occurring in these plants. In contrast, the photosynthetic rate of the $C_3$ plants increased in the low $O_2$ environment, indicating that photorespiration depresses rates of photosynthesis in 21% $O_2$.

**FOLLOW-UP WORK** The higher photosynthetic efficiency of $C_4$ photosynthesis has prompted efforts, so far unsuccessful, to incorporate this pathway into $C_3$ crops such as rice.

SOURCE Bjorkman, O., and J. Berry. 1973. "High-Efficiency Photosynthesis." *Scientific American* 229:80–93.

## 29.3 THE STEM: TRANSPORT OF WATER THROUGH XYLEM

On a summer day, a tree can transport many hundreds of liters of water from the soil to its leaves. This is an impressive feat given that it is accomplished without any moving parts. Even more remarkable, trees and other plants transport water without any direct expenditure of energy. The upward transport of water is possible because the structure of vascular plants allows them to use the evaporation of water from leaves to pull water from the soil.

**Fig. 29.10** shows the cross section of a sunflower stem. Like a leaf, it has a surface layer of epidermal cells. This layer encloses thin-walled **parenchyma** cells in the interior. Notice that the stem also contains differentiated tissues that lie in a ring near the outside of the stem. These are the vascular tissues, which form a continuous pathway that extends from near the tips of the roots, through the stem, and into the network of veins within leaves. The outer tissue, called **phloem,** transports carbohydrates from leaves to the rest of the plant body. The inner tissue, called **xylem,** transports water and nutrients from the roots to the leaves.

### Xylem provides a low-resistance pathway for the movement of water.

Water travels with relative ease through xylem because of the structure of the water-transporting cells within this tissue. As they develop, these cells become greatly elongated. When they complete their growth, they secrete an additional wall layer that is very thick and which contains lignin, a chemical compound that increases mechanical strength. Most remarkable is the final stage of development, when the nucleus and cytoplasm are lost, leaving behind only the cell walls.

These thick walls form a hollow conduit through which water can flow (**Fig. 29.11**). Water enters and exits xylem conduits through **pits,** circular or ovoid regions in the walls where the lignified cell wall layer is not produced. Instead, pits contain only the thin, water-permeable cell wall that surrounded each cell as it grew. As we will see, pits play an important role in xylem because they allow the passage of water, but not air, from one conduit to another.

Xylem conduits can be formed from a single cell or multiple cells stacked to form a hollow tube. Unicellular conduits are called **tracheids** (Fig. 29.11a), and multicellular conduits are called **vessels** (Fig. 29.11b). Because tracheids are the product of a single cell, they are typically 4 to 40 μm in diameter and no more than 2 to 3 cm long. Vessels, which are made up of individual cells called **vessel elements,** can be much wider and longer. Vessel diameters range from 5 to 500 μm, and lengths can be up to several meters.

**FIG. 29.11  Xylem structure.** Xylem conduits are formed by hollow cells with stiff walls and no cell content or membrane. (a) Single-celled tracheids and (b) multicellular vessels are interconnected by pits.

**a.** Tracheids

**b.** Multicellular vessels

Pits

Thick, lignified walls

Vessel element

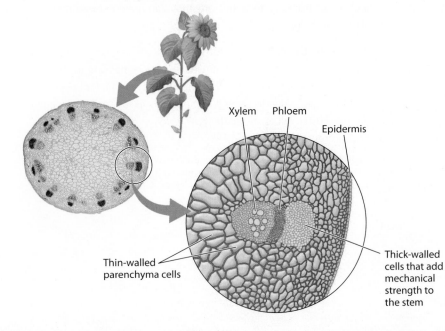

**FIG. 29.10  A cross section of the stem of a sunflower showing the two vascular tissues, xylem and phloem.**

Xylem  Phloem

Epidermis

Thin-walled parenchyma cells

Thick-walled cells that add mechanical strength to the stem

Water enters a tracheid through pits, travels upward through the conduit interior, and then flows outward through other pits into an adjacent, partially overlapping, tracheid. Water also enters and exits a vessel through pits. In contrast to tracheids, however, once the water is inside the vessel, little or nothing blocks the flow of water from one cell to the next. That is because during the development of a xylem vessel, the end walls of the vessel elements are digested away, allowing water to flow along the entire length of the vessel without having to cross any pits. At the end of a vessel, however, the water must flow through pits if it is to enter an adjacent vessel and thereby continue its journey from the soil to the leaves.

The rate at which water moves through xylem is a function of both the number of conduits and their size. Conduit length determines how often water must flow across pits, which exert a significant resistance to flow. Conduit width also has a strong effect on the rate of water transport. Like the flow through pipes, water flow through xylem conduits is greater when the conduits are wider. The flow is proportional to the radius of the conduit raised to the fourth power, so doubling the radius increases flow sixteenfold. Because vessels are both longer and wider than tracheids, plants with vessels achieve greater rates of water transport than is possible through tracheids alone. Tracheids are the most common conduits in lycophytes, ferns and horsetails, and gymnosperms, whereas vessels are the principal conduit in angiosperms.

**Water is pulled through xylem by an evaporative pump.**
If you cut a plant's roots off under water, the leaves continue to transpire for some time. The persistence of transpiration when roots are absent demonstrates that the driving force for water

transport is not generated in the roots, but instead comes from the leaves. In essence, water is pulled through xylem from above rather than being pushed from below.

The forces pulling water upward through the plant are large. Not only must these forces be able to lift water against gravity, they must also be able to pull water from the soil, which becomes increasingly difficult as the soil dries. In addition, they must be able to overcome the physical resistance associated with moving water through the plant itself. To replace water lost by transpiration with water pulled from the soil, the leaves must exert forces that are many times greater than the suctions that we can generate with a vacuum pump (**Fig. 29.12**). How do leaves exert this force?

## HOW DO WE KNOW?

**FIG. 29.12**

# How large are the forces that allow leaves to pull water from the soil?

**BACKGROUND** The idea that water is pulled through the plant by forces generated in leaves was first suggested in 1895. However, without a way to measure these forces, there was no way to know how large they were.

**HYPOTHESIS** In 1912, German physiologist Otto Renner hypothesized that leaves are able to exert stronger suctions than could be generated with a vacuum pump.

**EXPERIMENT** Renner measured the rate at which water flowed from a reservoir into the cut tip of a branch of a transpiring plant. He next cut off about 10 cm of the branch and attached a vacuum pump in place of the transpiring plant. He then compared the flow rate through the branch when attached to the transpiring plant with the flow rate generated by the vacuum pump.

**RESULTS** Renner found that the flow rates generated by transpiring plants were two to nine times greater than the flow rates generated by the vacuum pump.

**CONCLUSION** A transpiring plant pulls water through a branch faster than a vacuum pump. Therefore, the pulling force generated by a transpiring plant must be greater than that of a vacuum pump. Because vacuum pumps create suctions by reducing the pressure in the air, the maximum suction that a vacuum pump can generate is 1 atmosphere. Thus, the maximum height that one can lift water using a vacuum pump is 10 m (33 feet). The forces that pull water through plants have no comparable limit because they are generated within the hydrated cell wall. This is why plants can pull water from dry soils and why they can grow to more than 100 m in height.

**FOLLOW-UP WORK** In 1965, Per Fredrick Scholander developed a pressurization technique that uses reverse osmosis to measure the magnitude of the suctions exerted by leaves of intact plants. He used this technique to show that the leaves at the top of a Douglas fir tree 100 m tall exert greater pulling forces than do leaves closer to the ground.

**SOURCES** Renner, O. 1925. "Zum Nachweis negativer Drucke im Gefässwasser bewurzeler Holzgewachse." *Flora* 118/119:402-408.; Scholander, P. F., H. T. Hammel, E. D. Bradstreet, and E. A. Hemmingson. 1965. "Sap Pressure in Vascular Plants." *Science* 148:339–346.

**FIG. 29.13    Xylem transport from roots to leaves.** Water is pulled through the plant by forces generated in the leaves, preventing them from drying out.

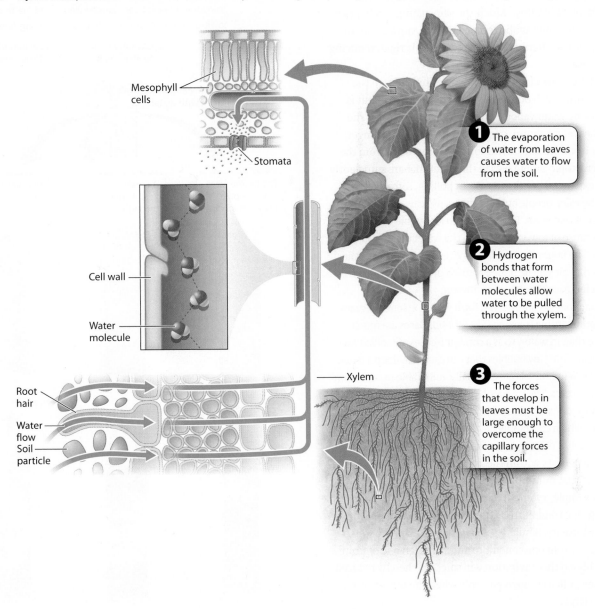

Mesophyll cells

Stomata

Cell wall

Water molecule

Root hair

Water flow

Soil particle

Xylem

**1** The evaporation of water from leaves causes water to flow from the soil.

**2** Hydrogen bonds that form between water molecules allow water to be pulled through the xylem.

**3** The forces that develop in leaves must be large enough to overcome the capillary forces in the soil.

When stomata are open, water evaporates from the walls of cells lining the intercellular air spaces of leaves. The partial dehydration of the cell walls creates a force that pulls water towards the sites of evaporation. One hypothesis for how this force is generated is that water is pulled by capillary action into spaces between the cellulose microfibrils in the cell wall, the same reason that water is drawn into a sponge. A second hypothesis is that the pectin gel in which the cellulose microfibrils are embedded causes water to flow into the partially dehydrated cell walls by osmosis. Osmotic forces can be generated in cell walls because the negatively charged pectin network restricts the diffusion of positively charged ions, much as the plasma membrane maintains a high concentration of solutes in the cytoplasm.

Once generated by the evaporation of water from leaves, this force is transmitted through the xylem, beginning in the leaf veins, then down through the stem, and out through the roots to the soil (**Fig. 29.13**). Water can be pulled through xylem because of the strong hydrogen bonds that form between water molecules. This mechanism of water transport only works, however, if there is a continuous column of water in the xylem that extends from the roots to the leaves.

### The structure of xylem conduits reduces the risks of collapse and cavitation.

The fact that water is pulled from the soil means that xylem conduits must be structured in such a way as to minimize two distinct risks (**Fig. 29.14**). The first is the danger of collapse (Fig. 29.14a). If you suck too hard on a drinking straw, it will collapse inward, blocking flow. Much the same thing can happen in xylem. Although in metabolic terms, lignin is more costly to produce than cellulose, lignin makes conduit walls rigid, reducing the risk of collapse.

The second danger associated with pulling water through the xylem is **cavitation**, which occurs when the water in a conduit is abruptly replaced by water vapor. Because cavitation disrupts the continuity of the water column, cavitated conduits can no longer transport water from the soil. Cavitation in xylem results from the presence of microscopic gas bubbles in the water that are sufficiently large that they expand under the tensile, or pulling, force exerted by the leaves. The microscopic bubbles that cause cavitation can be formed in either of two ways. The first is if an air bubble is pulled through a pit because of lower pressure in the water compared to the air (Fig. 29.14b). The larger the tensile forces exerted by leaves, the more likely it is that air will be pulled across a pit. Thus, higher rates of transpiration increase the risk of cavitation.

The second way that gas bubbles can form within xylem is if gases come out of solution during freezing (Fig. 29.14c). Gases are much less soluble in ice than in water, so as a conduit freezes, dissolved gas molecules aggregate into tiny bubbles that can cause cavitation when the conduit thaws. Wide vessels are especially vulnerable to cavitation at freezing temperatures. The susceptibility of wide vessels to cavitation partly explains why boreal (that is, subarctic) forests contain few angiosperms, which commonly have large vessel diameters.

Once cavitation has occurred, the liquid phase will not re-form as long as tension persists within the xylem. Thus, xylem is organized in ways that minimize the likelihood and impact of cavitation. For example, xylem consists of many conduits in parallel, so the loss of any one conduit to cavitation does not result in a major loss of transport capacity. Similarly, as water flows from the soil to the leaves, it passes from one conduit to another, each one of finite length. The likelihood that cavitation will spread is thereby reduced because air can be pulled through pits only when the tensile forces in the xylem are large.

→   **Quick Check 3** How is the statement that water is transported through the xylem without requiring any input in energy by the plant both correct and incorrect?

## 29.4 THE STEM: TRANSPORT OF CARBOHYDRATES THROUGH PHLOEM

Much of the body of a vascular plant is devoted to the uptake and transport of the water required by leaves rather than to photosynthesis. In contrast, all cells of the algal ancestors of plants would have been capable of photosynthesis. Roots and stems contribute indirectly to photosynthesis but produce little or no carbohydrate themselves. Thus, although vascular plants

**FIG. 29.14   Collapse and cavitation.** The forces exerted by transpiring leaves can cause xylem conduits to collapse inward or to cavitate, in both cases blocking flow.

**a.** Collapse

**b.** Cavitation due to an air leak

**c.** Cavitation due to freeze and thaw

Xylem conduit filled with water

The tensile force of water pulls the walls inward.

Air leaks into xylem conduit through a pit.

Bubbles form as water freezes.

Pits

Thaw

Collapse

Cavitation

Cavitation

Collapsed xylem

Air-filled xylem

Air-filled xylem

are photosynthetic, a large part of their body must be supplied internally with food.

## Carbohydrates are pushed through phloem by an osmotic pump.

Phloem transport takes place through multicellular **sieve tubes,** which are composed of highly modified cells called **sieve elements** that are connected end to end (**Fig. 29.15**). During

---

**FIG. 29.15** Sieve tubes and associated companion cells.

development, sieve elements lose much of their intracellular structure, including the nucleus and the vacuole. At maturity, sieve elements retain an intact plasma membrane that encloses a modified cytoplasm containing only smooth endoplasmic reticulum and a small number of organelles, including mitochondria. Cellular functions such as protein synthesis are carried out by an adjacent **companion cell,** to which the sieve element is connected by numerous plasmodesmata. Sieve elements are linked by **sieve plates,** which are modified end walls with large (1 to 1.5 μm diameter) pores. The plasma membrane of adjacent sieve elements is continuous through each of these pores, so each multicellular sieve tube can be considered a single cytoplasm-filled compartment. **Phloem sap** is the sugar-rich solution that flows through both the lumen of the sieve tubes and the sieve plate pores.

Phloem transports carbohydrates as sucrose (glucose plus fructose) or larger sugars, assembled from monosaccharides in the cytoplasm. Phloem also transports amino acids, inorganic forms of nitrogen, and ions including K⁺, which are present in much lower concentrations. Finally, phloem transports informational molecules such as hormones, protein signals, and even RNA. Thus, phloem forms a multicellular highway for the movement of raw materials and signaling molecules across the entire length of the plant.

Phloem transports its molecular cargo from **source** to **sink** (**Fig. 29.16**). In plants, sources are the regions that produce

---

**FIG. 29.16** **Phloem transport from source to sink.** Sources produce carbohydrates or supply them from storage, and sinks require carbohydrates for growth and respiration.

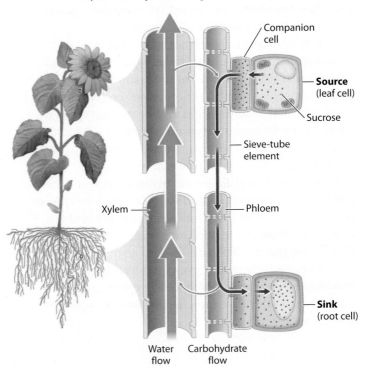

or store carbohydrates. For example, leaves are sources because they produce carbohydrates by photosynthesis, and potato tubers (the part we eat) are sources because they can supply stored carbohydrates to the rest of the plant body. Sinks are any portion of the plant that needs carbohydrates to fuel growth and respiration—examples are roots, young leaves, and developing fruits. Whereas the direction of xylem transport is always up toward the leaves, the direction of phloem transport can be either up or down, depending on where the source and sink are relative to each other.

How does phloem transport sugars from source to sink? In some plants, active transporters powered by ATP move sucrose into phloem. The buildup of sugar concentration causes water to be drawn into the phloem by osmosis. Because the cell walls of the sieve tube resist being stretched outward to accommodate the influx of water, they press back (inward) against the cytoplasm. Their resistance to stretching increases the turgor pressure (Chapter 5) at the source end of a sieve tube. At sinks, sugars are transported out of the phloem into surrounding cells. This withdrawal of sugars causes water to leave the sieve tube, again by osmosis, reducing turgor pressure at the sink end. It is the difference in turgor pressure that drives the movement of phloem sap from source to sink.

The water that exits the phloem can be used locally to support the enlargement of sink cells or it can be carried back to the leaves in the xylem. Thus, some of the water in the phloem sap is recirculated in the xylem (Fig. 29.16). The volume of water that moves through the phloem, however, is tiny compared to the amount that must be transported through the xylem to replace water lost by transpiration. Therefore, the number and size of xylem conduits greatly exceeds the number and size of sieve tubes.

Phloem transport is sometimes referred to as "translocation," a term that pairs well with "transpiration." Yet in almost every way, phloem and xylem are a study in contrasts. In phloem, the plant generates the gradient that drives transport, whereas water moving through xylem is driven by the difference in hydration between soil and air. Furthermore, water is pulled upward through xylem conduits, while transport through phloem is more of a push.

These fundamental differences in function explain the cytological differences between xylem and phloem: Phloem conduits retain an intact plasma membrane and modified cytoplasm, whereas xylem conduits retain only their cell walls. What xylem and phloem have in common is that both are essential in enabling vascular plants to carry out photosynthesis on land. Moreover, like xylem, phloem is subject to risks that arise from the way flow through sieve tubes is generated.

Damaged sieve tubes are at risk of having their contents leak out, pushed out by high turgor pressures in the phloem. Damage is an ever-present danger because the sugar content of phloem makes it an attractive target for insects. Cell damage activates sealing mechanisms that repair breaks in sieve tubes. In some respects, these mechanisms are comparable to the formation of blood clots in humans, except that phloem can seal itself much more rapidly, typically in less than a second.

→   **Quick Check 4**  How is phloem able to transport carbohydrates from the shoot to the roots, as well as from the roots to the shoot (although not at the same time)?

### Phloem feeds both the plant and the rhizosphere.

All the cells in a plant's body contain mitochondria since all cells need a constant supply of ATP. Typically, about 50% of the carbohydrates produced by photosynthesis in one day are converted back to $CO_2$ by respiration within 24 hours.

Carbohydrates that are not immediately consumed in respiration can be used as raw materials for growth, or they can be stored for later use. Carbohydrates stored within roots and stems, or in tubers (specialized storage organs such as potatoes), can support new growth in the spring or following a period of drought. Stored reserves can also be used to repair mechanical damage or replace leaves consumed by insects or grazing mammals. Like their green algal ancestors, vascular plants store carbohydrates primarily as starch, a large molecule that is not soluble and so does not affect the osmotic balance of cells.

What determines how carbohydrates become distributed within the plant? Where phloem sap ends up is determined by the sinks. Phloem transport to reproductive organs appears to have priority over movement to stems and leaves, and these have priority over movement to roots. In Chapter 31, we discuss the role that hormones play in controlling the growth and development of plants and how these hormones may influence the ability of different sinks to compete successfully for resources delivered by the phloem.

Phloem also supplies carbohydrates to organisms outside the plant. A fraction of the carbohydrates transported to the roots spills out into the **rhizosphere,** the soil layer that surrounds actively growing roots. This supply of carbohydrates stimulates the growth of soil microbes. As a result, the densities of microbial organisms near roots are much greater than in the rest of the soil. These soil bacteria are important decomposers of soil organic matter, which is rich in nutrients such as nitrogen and phosphorus. Thus, the release of carbohydrates by roots into the soil is thought to enhance the ability of roots to acquire nutrients from the soil.

## 29.5 THE ROOT: UPTAKE OF WATER AND NUTRIENTS FROM THE SOIL

As anyone who has dug a hole in a forest or a field can attest, plants make a lot of roots (**Fig. 29.17**). Laid end to end, the roots of a single corn plant would extend over 600 km. A major reason for such extensive investment belowground is that, with the important exception of $CO_2$, everything that a vascular plant needs to build and sustain its body enters through its roots. This includes, of course, water, as well as nutrients such as nitrogen, potassium, and phosphorus.

Roots are an evolutionary innovation of the vascular plants; there were no roots when the ancestors of plants first moved onto land. The first roots would have encountered a land surface much less hospitable than today's soils. To a large extent soils have a biological origin. The processes by which roots obtain nutrients from the soil accelerate rates of chemical weathering, stimulate the growth of microbes, and slow erosion, all of which contribute to the development of soils.

### Nutrient uptake by roots is highly selective.

Xylem moves both water and nutrients (for example, nitrogen and phosphorus) from the root to the rest of the plant. Xylem, however, is never in direct contact with the soil. The living cells that surround the xylem allow roots to exert a high degree of selectivity over the movement of nutrients from the soil. By taking up only the nutrients they need, roots can concentrate certain nutrients; they can also prevent the entry of substances that could cause damage.

**FIG. 29.17 Roots and root hairs.** Extensive branching of roots (a) and the presence of root hairs (b) create a large surface area in contact with the soil.

**FIG. 29.18 Water movement through endodermal cells.** As water carrying nutrients flows from the soil to the xylem of roots, it must pass through the plasma membranes of endodermal cells, which act as a selective barrier.

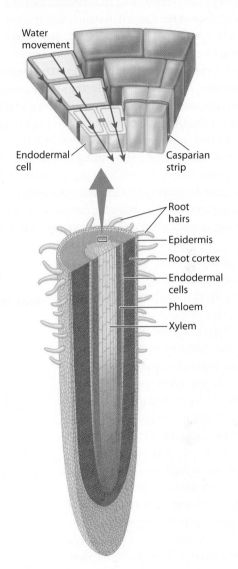

Like leaves and stems, roots have an outer cell layer of **epidermis** (**Fig. 29.18**). Epidermal cells in active areas of the root produce slender outgrowths known as **root hairs.** Root hairs require little material to build, but they greatly increase the surface area of the root available for absorbing nutrients. Inside the epidermis is the **cortex,** composed of parenchyma cells. Depending on species, the cortex can vary in thickness from a single cell to many cells. Xylem and phloem are located at the center of the root and are surrounded by the **endodermis,** a layer of cells that acts as a gatekeeper controlling the movement of nutrients into the xylem.

Water flowing across the epidermis and the root cortex can travel through the cells, crossing plasma membranes, and can also flow through the relatively porous walls that surround each cell. At the endodermis, flow through the cell wall is blocked by the **Casparian strip,** a thin band of hydrophobic material that encircles each cell. Water can gain access to the xylem only by passing through the cytoplasm of the endodermal cells.

The significance of the Casparian strip is that it allows the root to control which materials enter the xylem. How does it do that? In order for solutes to pass through the plasma membrane into the cytoplasm, they must be transported by ATP-powered proteins in the membranes of endodermal cells. These transporters move only certain compounds through the endodermis and actively exclude others. Thus, the solute composition of xylem sap—the fluid moving through the xylem—does not mirror the solute composition of water in the soil. By using the energy of ATP to control the movement of materials across the plasma membrane of endodermal cells, roots can concentrate nutrients that are needed in large amounts (for example, nitrogen) and exclude others that can damage plant cells (for example, sodium).

### Nutrient uptake requires energy.

Roots expend a great deal of energy acquiring nutrients from the soil. We have already seen that energy is required to move nutrients across the cell membranes of the endodermis on their journey to the xylem. Energy is also required to make specific nutrients in the soil accessible to the root. And still more energy is required to support symbioses with soil microorganisms that take up nutrients and transfer them to plants. The mitochondria of root cells must, therefore, respire at high rates to provide energy needed for nutrient uptake. Their high respiration rates explain why plants are harmed by flooding: Waterlogged soils slow the diffusion of oxygen, limiting respiration.

The most abundant elements in the plant body—carbon, hydrogen, and oxygen—are obtained primarily from $CO_2$ and $H_2O$. All the other elements in the plant's body come from the soil. To understand how roots acquire nutrients, let's first look at how nutrients move within the soil.

Nutrients move through soil by diffusing through films of water that surround individual soil particles. In addition, when transpiration rates are high, the flow of water through the soil helps to convey nutrients toward the root. The uptake of nutrients by roots decreases their concentration at the root surface, creating a region of lower concentration toward

which nutrients diffuse from the soil. However, because diffusion transports compounds efficiently only over short distances, each root is able to acquire nutrients only from a small volume of soil. This limitation explains why plants produce so many roots and so many root hairs: Both structures increase the volume of soil from which a root can obtain nutrients. It also explains why roots elongate more or less continuously, exploiting new regions of soil to obtain nutrients required for growth.

Nutrients vary dramatically in the ease and speed with which they move through the soil. For example, nitrate ($NO_3^-$) and sulfate ($SO_4^-$) are extremely mobile. In contrast, interactions with clay minerals keep zinc and inorganic phosphate ($P_i$) attached to mineral surfaces. One way roots gain access to highly immobile nutrients is by releasing protons or compounds that make the environment more acidic. Nutrients stuck to soil particles dissolve as water becomes acidic and diffuse to the root.

Soil microorganisms—both fungi and bacteria—are more adept than plants at obtaining nutrients. Fungi (Chapter 34) are particularly good at mobilizing phosphorus, and some bacteria can

**FIG. 29.19  Mycorrhizae.** These fungus–root associations allow plants to use carbohydrates to help gain nutrients from the soil.

Ectomycorrhizae

Fungal strands extending from the root

Epidermis
Root cortex
Endoderm
Phloem
Xylem
Fungal sheath

Fungal cells surround but do not penetrate root cells. Carbon and nutrients are exchanged through the plasma membrane.

Endomycorrhizae

Fungal cells
Arbuscule

Fungal cells penetrate inside root cells, enhancing carbon and nutrient exchange.

convert gaseous nitrogen ($N_2$) into a chemical form that can be incorporated into proteins and nucleic acids (Chapter 26). What plants have at their disposal is a renewable source of carbohydrates. Using these carbohydrates, plants feed symbiotic fungi and bacteria, in effect exchanging carbon for phosphorus or nitrogen.

### Mycorrhizae enhance the uptake of phosphorus.

Plants are able to access soil nutrients that their roots can't reach by forming partnerships with fungi. **Mycorrhizae** are symbioses between roots and fungi that enhance nutrient uptake. Mycorrhizae are particularly important in enhancing the uptake of phosphorus, one of the most immobile nutrients in soil. The fungi absorb phosphate ions and transport some of them to root tissues; the fungi receive carbohydrates produced in leaves and transported to roots by the phloem.

Why are fungi so effective at obtaining phosphorus from the soil? Roots are thin and root hairs thinner, but the cells produced by fungi are thinner still. Elongate fungal cells can extend more than 10 cm from the root surface, greatly increasing the absorptive surface area in contact with the soil by acting more or less like "super root hairs." Mycorrhizae aid in nutrient uptake because solutes can be transported through the fungal cytoplasm much faster than they can diffuse through the soil. Furthermore, because fungal cells are so thin, they can penetrate small spaces between soil particles and thus access nutrients that would otherwise be unavailable. Finally, mycorrhizal fungi secrete enzymes that release phosphorus from soil organic matter and enhance the mobility of inorganic phosphorus in the soil.

Mycorrhizae are of two main types (**Fig. 29.19**). **Ectomycorrhizae** produce a thick sheath of fungal cells that surrounds the root tip. Some of these cells penetrate the walls of root cells, producing a dense network that surrounds individual root cells. The root and fungal cells are then able to exchange carbon and nutrients through their plasma membranes. **Endomycorrhizae** do not form structures that are visible on the outside of the root. Instead, they form highly branched structures, called arbuscules, that protrude into the interior of root cells. Despite this invasion, the plasma membranes of both cell types remain intact. The structure of the arbuscules increases the contact area between the two cell types, enhancing carbon and nutrient exchange.

Mycorrhizal associations are costly to the plant, consuming between 4% and 20% of its total carbohydrate production. When phosphorus is readily available from the soil, the percentage of mycorrhizal roots decreases, indicating some level of control by the plant. Nevertheless, mutualistic associations between roots and fungi occur in more than 85% of plant species. Mycorrhizal roots are thus the norm rather than an exception. Fossils from more than 400 million years ago suggest that interactions between plants and fungi may be ancient and perhaps helped plants gain a foothold on land.

### Symbiotic nitrogen-fixing bacteria supply nitrogen to both plants and ecosystems.

Nitrogen is the nutrient that plants must acquire in greatest abundance from the soil, and in many ecosystems nitrogen is the nutrient that most limits growth. Nitrogen itself is not rare; 79% of Earth's atmosphere is nitrogen gas ($N_2$). However, as discussed in Chapter 26, plants and other eukaryotic organisms cannot use $N_2$ directly. Only certain bacteria and archaeons have the metabolic ability to convert $N_2$ into a biologically useful compound, a process known as nitrogen fixation. Nitrogen-fixing bacteria transform $N_2$ into forms such as ammonia ($NH_3$) that can be used to build proteins, nucleic acids, and other compounds. Some of these usable forms of nitrogen are released into the soil, making them available for use by plants and other organisms.

Although many nitrogen-fixing bacteria live freely in the soil, some live in plant roots. When the level of available soil nitrogen is low, some plants form symbiotic relationships with nitrogen-fixing bacteria (**Fig. 29.20**). These interactions are highly species

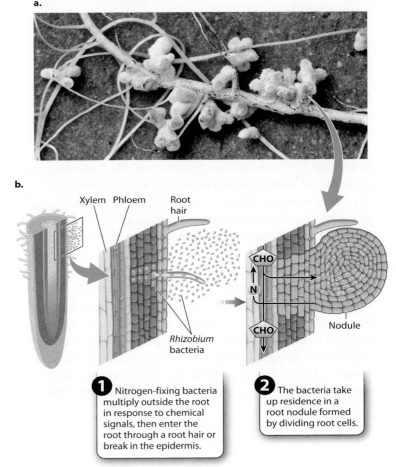

**FIG. 29.20  Symbiosis between roots and nitrogen-fixing bacteria.** (a) Symbiotic root nodules on alfalfa; (b) the formation of a root nodule.

**a.**

**b.**

Xylem  Phloem  Root hair

*Rhizobium* bacteria

CHO

N

CHO

Nodule

**1** Nitrogen-fixing bacteria multiply outside the root in response to chemical signals, then enter the root through a root hair or break in the epidermis.

**2** The bacteria take up residence in a root nodule formed by dividing root cells.

specific. In response to chemical signals transmitted between the root and bacteria, bacteria multiply within the rhizosphere. At the same time, root cells begin dividing, leading to the formation of a **root nodule.** Nitrogen-fixing bacteria enter the nodule either through root hairs or through breaks in the epidermis. Even though the bacteria appear to have taken up residence within the cytoplasm of nodule cells, in fact each one is contained within a membrane-bound vesicle produced by the host plant cell.

The plant supplies its nitrogen-fixing partners with carbohydrates transported by the phloem and carries away the products of nitrogen fixation in the xylem. As in mycorrhizal symbioses, the carbon cost to the plant is high: It is estimated that as much as 25% of these plants' total carbohydrate production goes to nitrogen-fixing bacteria.

Relatively few plant species have evolved symbiotic associations with nitrogen-fixing bacteria, but those that do—especially members of the legume (bean) family—provide an important way in which biologically available nitrogen enters into ecosystems. In addition, nitrogen fixation in legume roots helps to sustain fertility in many agricultural systems.

**?** CASE 6 Agriculture: Feeding a Growing Population
**How has nitrogen availability influenced agricultural productivity?**

Harvesting crops removes nutrients—including nitrogen—from an ecosystem. These nutrients must be returned if the land is to to be agriculturally productive.  In the early days of agriculture, the practice of shifting cultivation solved the problem by leaving fields unplanted for long periods between crops. During these periods, nitrogen-fixing bacteria replenished soil nitrogen, either in symbiotic association with roots or as free-living bacteria in the soil.

As populations grew, fields could not be left unplanted for long periods. Nutrients removed during harvest had to be replaced or the fertility of fields would decline. Organic fertilizers such as crop residues, manure, and even human feces provide a way to recyle other nutrients removed during harvest, but unless they are brought in from off the farm, they add little or no new nitrogen to the system.

Crop rotation has been widely practiced throughout the history of agriculture to resupply the soil with nitrogen. Every few years, the field is planted with legumes harboring nitrogen-fixing bacteria. Typically, the legume rotation is not harvested, but instead is treated as "green manure" that is allowed to decay and replenish soil nitrogen. During the eighteenth and nineteenth centuries, legume rotation contributed to the higher crop yields needed to support increasing populations. Therefore, along with

the mechanization of agriculture, nitrogen-fixing symbioses contributed to the social conditions that fueled the Industrial Revolution.

By the second half of the nineteenth century, more abundant and more concentrated supplies of nitrogen fertilizers were needed to promote crop growth to feed growing populations. Mining of guano and sodium nitrate mineral deposits, both found in arid regions such as the northern deserts of Chile, helped supply the demand for the nitrogen needed to increase crop production. With time, however, the reserves of these natural nitrogen sources declined and a mechanism was needed to obtain nitrogen fertilizer from the vast storehouse of nitrogen in the atmosphere.

An industrial method of fixing nitrogen was developed by German chemist Fritz Haber in the years 1908–1911 and was subsequently adapted for industrial production by Karl Bosch. Their method allows humans to achieve what, until then, had been an entirely prokaryotic activity. Like nitrogen-fixing bacteria, the Haber–Bosch process requires large inputs of energy, but unlike bacteria, it occurs at high temperatures and pressures. The prospect of an essentially unlimited supply of nitrogen in a form that plants could make use of has proved irresistible. Industrial fertilizers produced by the Haber–Bosch process now account for more than 99% of fertilizer use. Industrial fixation of nitrogen today is approximately equal to the entire rate of natural fixation of nitrogen, which is largely due to bacteria, although a small amount of nitrogen is fixed by lightning.

As we discuss in Chapter 48, human use of fertilizer is altering the nitrogen cycle on a massive scale, posing a threat to biodiversity and the stability of natural ecosystems. Yet abundant nitrogen fertilizers are a cornerstone of the Green Revolution. The high-yielding varieties of wheat and rice introduced in the mid-twentieth century achieve their high levels of growth and grain production only when abundantly supplied with nitrogen. The fact that plants such as legumes (including beans, soybeans, alfalfa, and clover) form symbiotic relationships with nitrogen-fixing bacteria suggests a possible way around this problem. If it were possible to engineer nitrogen fixation into crops such as wheat and rice, either by inducing them to form symbiotic relationships with nitrogen-fixing bacteria or by transferring the genes that would enable them to fix nitrogen on their own, would that eliminate the need for nitrogen fertilizers? The answer is yes—but with a caveat. As we have seen, nitrogen-fixation requires a large amount of energy, diverting resources that could otherwise be used to support growth and reproduction.

# Core Concepts Summary

**29.1 THE EVOLUTION OF LAND PLANTS FROM AQUATIC ANCESTORS INTRODUCED A MAJOR CHALLENGE FOR PHOTOSYNTHESIS: ACQUIRING $CO_2$ WITHOUT LOSING EXCESSIVE AMOUNTS OF WATER.**

Early branching land plant lineages, commonly grouped as bryophytes, balance $CO_2$ gain and water loss passively, photosynthesizing in wet conditions and tolerating desiccation in dry ones. page 29-2

Vascular plants actively regulate the hydration of their photosynthetic cells. page 29-5

**29.2 LEAVES HAVE A WAXY CUTICLE THAT RETARDS WATER LOSS BUT INHIBITS THE DIFFUSION OF $CO_2$, AND PORES, CALLED STOMATA, THAT REGULATE $CO_2$ GAIN AND WATER LOSS.**

The low concentration of $CO_2$ in the atmosphere forces plants to expose their photosynthetic cells directly to the air. The outward diffusion of water vapor leads to a massive loss of water. page 29-3

The waxy cuticle on the outside of the epidermis slows rates of water loss from leaves but also slows the diffusion of $CO_2$ into leaves. page 29-4

Stomata are pores in the epidermis that open and close, allowing $CO_2$ to enter into the leaf but also allowing water vapor to diffuse out of the leaf. page 29-5

CAM plants capture $CO_2$ at night when evaporative rates are low. During the day, they close their stomata and use this stored $CO_2$ to supply the Calvin cycle during the day, resulting in increases in the exchange ratio of $CO_2$ and $H_2O$. page 29-6

$C_4$ plants suppress photorespiration by concentrating $CO_2$ in bundle-sheath cells. The buildup of $CO_2$ in bundle-sheath cells results from the rapid production of $C_4$ compounds in the mesophyll cells, which then diffuse into the bundle-sheath cells and release $CO_2$ that can be used in the Calvin cycle. page 29-7

**29.3 XYLEM ALLOWS VASCULAR PLANTS TO REPLACE WATER EVAPORATED FROM LEAVES WITH WATER PULLED FROM THE SOIL.**

Water flows through xylem conduits from the soil to the leaves. page 29-9

Xylem is formed from cells that lose all their cell contents as they mature. page 29-9

Xylem conduits have thick, lignified cell walls. Water flows into xylem conduits across small thin-walled regions called pits. page 29-9

Xylem conduits are of two types: tracheids and vessels. Tracheids are unicellular xylem conduits; vessels are formed from many cells. page 29-9

The driving force for water movement in xylem is generated by the evaporation of water from leaves. Because of the strong hydrogen bonds between water molecules, water can be pulled from the soil and transported through xylem. page 29-11

Pulling water through xylem creates the risk of mechanical failure, either by the collapse inward of conduit walls or by cavitation in which a gas bubble expands to fill the entire conduit. page 29-12

**29.4 PHLOEM TRANSPORTS CARBOHYDRATES AND SUPPORTS THE GROWTH AND RESPIRATION OF NON-PHOTOSYNTHETIC ORGANS SUCH AS STEMS AND ROOTS.**

Phloem transports carbohydrates to the non-photosynthetic portions of the plant for use in growth and respiration. page 29-13

Phloem transport occurs through sieve tubes, which are formed from elongate cells that are connected end to end by sieve plates. Sieve plates contain large pores that allow phloem sap to flow from one sieve tube cell to another. page 29-13

The active loading of solutes brings water into sieve tubes by osmosis, increasing the turgor pressure. At sites of use, the removal of solutes leads to an outflow of water and a drop in turgor pressure. page 29-14

The difference in turgor pressure drives the movement of phloem sap from source to sink. page 29-14

Carbohydrates exuded from roots provide carbon and energy sources for the rhizosphere microbial community. page 29-14

**29.5 ROOTS EXPEND ENERGY TO OBTAIN NUTRIENTS FROM THE SOIL.**

The endodermis allows roots to control which materials enter the xylem. The uptake of solutes by roots is therefore highly selective. page 29-16

Symbiotic associations with fungi allow plants to exchange carbohydrates for assistance in obtaining phosphorus from the soil. page 29-17

Plants supply carbohydrates to symbiotic nitrogen-fixing bacteria in exchange for nitrogen. page 29-17

Nitrogen fixation results in a net increase in ecosystem nitrogen. page 29-18

Agricultural productivity is closely linked to nitrogen supply. page 29-18

## Self-Assessment

1.  Explain why vascular plants are better able to sustain photosynthesis than are bryophytes.

2.  Explain why transpiration is best explained as a consequence of acquiring $CO_2$ from the atmosphere, as opposed to temperature regulation, nutrient transport, or utilization of water as a substrate in photosynthesis.

3.  Draw and label the features of a leaf that are necessary to maintain a well-hydrated interior while still allowing $CO_2$ uptake from the atmosphere.

4.  Diagram how the movement of solutes results in the opening and closing of stomata.

5.  Diagram how the enzyme PEP carboxylase allows CAM plants to reduce transpiration rates and $C_4$ plants to suppress photorespiration.

6.  Contrast how transport is generated in xylem and in phloem.

7.  Describe the structure of xylem conduits and sieve tubes in relation to their function.

8.  Explain why phloem is necessary to produce roots.

9.  List four adaptations of roots that enhance the uptake of nutrients from the soil.

Do you understand the chapter's Core Concepts? Log in to check your answers to the Self-Assessment questions, then practice what you've learned and reinforce this chapter's concepts by working through the problems and multimedia tutorials provided there.

🖳 **www.biologyhowlifeworks.com**

## CHAPTER 30

# PLANT REPRODUCTION

## Finding Mates and Dispersing Young

## Core Concepts

**30.1** The plant life cycle evolved in ways that enhance the ability to unite gametes and disperse offspring on land.

**30.2** Angiosperms (flowering plants) attract and reward animal pollinators, and they provide resources for seeds only after fertilization.

**30.3** Plants have sensory systems that control the timing of flowering and seed germination.

**30.4** Many plants also reproduce asexually.

As plants moved onto land, the challenges of carrying out photosynthesis in air were matched by the difficulties of completing their life cycle. The algal ancestors of land plants relied on water currents to carry sperm to egg and disperse their offspring. On land, the first plants were confronted with the challenge of moving gametes and offspring through air. Air is less buoyant than water, provides a poor buffer against changes in temperature and ultraviolet radiation, and increases the risk of drying out. As plants diversified on land, structures evolved that allow both gametes and offspring to survive being transported through air. For these structures to be useful, plant life cycles had to undergo radical change. The modification of land plant life cycles is a major theme in plant evolution.

The gametes and offspring of many land plants are carried passively by the movement of the air (or in a few cases, water—think coconuts). However, some plants evolved the capacity to harness animals as transport agents. In particular, the flowers and fruits of angiosperms influence animal behavior through their colors, scents, and food resources. Animals attracted by the food provided by flowers (or in a few cases, by the false promise of a mate) transport male gametes from one flower to another. Animals that eat fruits or inadvertently carry them attached to their fur are key agents of seed dispersal. As plants diversified on land, coevolution with animals emerged as a second major theme in plant evolution.

## 30.1 THE PLANT LIFE CYCLE AND EVOLUTION OF POLLEN AND SEEDS

Plants did not evolve the ability to reproduce on land all at once. Early diverging lineages evolved the capacity to disperse their offspring through the air, but they still required water for fertilization. How could these plants exist in both water and air? They couldn't—at least not at the same time. Instead, early land plants evolved a life cycle in which one generation, or phase of the life cycle, released sperm into a moist environment and the following generation dispersed offspring through the air. It was only with the evolution of pollen and seeds that plants no longer depended on external sources of water for fertilization. Yet the pattern of alternating generations remains a key feature of the plant life cycle.

**FIG. 30.1** The phylogeny of land plants, showing the evolutionary trajectory of the environmental setting for fertilization and dispersal.

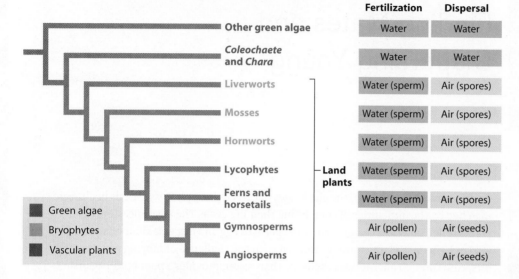

| | Fertilization | Dispersal |
|---|---|---|
| Other green algae | Water | Water |
| *Coleochaete* and *Chara* | Water | Water |
| Liverworts | Water (sperm) | Air (spores) |
| Mosses | Water (sperm) | Air (spores) |
| Hornworts | Water (sperm) | Air (spores) |
| Lycophytes | Water (sperm) | Air (spores) |
| Ferns and horsetails | Water (sperm) | Air (spores) |
| Gymnosperms | Air (pollen) | Air (seeds) |
| Angiosperms | Air (pollen) | Air (seeds) |

Land plants: Liverworts, Mosses, Hornworts, Lycophytes, Ferns and horsetails, Gymnosperms, Angiosperms

Legend:
- Green algae
- Bryophytes
- Vascular plants

### The algal sister groups of land plants have one multicellular generation in their life cycle.

Molecular sequence comparisons show that two groups of green algae are most closely related to land plants (**Fig. 30.1**): *Coleochaete*, an inconspicuous green pincushion that grows on aquatic plants or rocks in freshwater environments, and the erect stonewort *Chara* (and closely related genera) found along the margins of lakes and estuaries (**Fig. 30.2**). Like all sexually reproducing eukaryotes (Chapter 27), these algae alternate between a diploid ($2n$) phase and a haploid ($1n$) phase. The diploid phase is generated by the fusion of haploid gametes to form a zygote, and the haploid phase is generated from diploid cells by meiosis.

In animals, the multicellular body consists of diploid cells. *Coleochaete* and *Chara* have a multicellular body as well, shown in Fig. 30.2, but it consists entirely of haploid cells. Specialized cells of the haploid body produce haploid eggs and sperm by mitosis within multicellular reproductive organs. The egg is retained within the female reproductive organ and the sperm are released into the water. Fertilization results when one egg and one sperm fuse to form a diploid zygote. The zygote then undergoes meiosis to produce haploid cells that give rise to a new multicellular haploid generation (**Fig. 30.3**). In *Coleochaete*, the haploid products of meiosis disperse by swimming through the water, propelled by flagella. In *Chara*, the zygotes

disperse, carried by water currents until they settle and undergo meiosis.

In *Chara* and *Coleochaete*, both fertilization and dispersal take place in water. Thus, these algae are as dependent on water as fish are. Living along the margins of lakes and streams, however, *Chara* and *Coleochaete* are occasionally left high and dry by drought. Their photosynthetic bodies shrivel quickly, but these algae survive because their zygotes form a protective wall that allows the cell inside to tolerate exposure to air. When water once again covers the resting zygotes, they undergo meiosis and the resulting haploid cells escape from their protective coat. This is the condition from which the more complicated life cycles of land plants evolved.

### Bryophytes illustrate how the alternation of generations allows the dispersal of spores in the air.

As plants moved onto land, their life cycle pulled them in two directions. Fertilization requires that plants remain near the ground, where water is most likely to be found, but dispersal in air requires that they grow tall, away from that water. In Chapter 29, we saw that the bryophytes—mosses, hornworts, and liverworts—make up the earliest branches on the phylogenetic tree of land plants (see Fig. 30.1). Mosses, hornworts, and liverworts share many reproductive traits. Here, we use *Polytrichum commune*, a moss common in forests and the edges of fields, to illustrate how movement onto land was accompanied by the evolution of two multicellular generations: one specialized for fertilization and the other for dispersal.

**FIG. 30.2** *Coleochaete* (left) and *Chara* (right), the two green algae most closely related to land plants.

**FIG. 30.3   The life cycle of the green alga *Chara corallina*.** The only diploid cell is the zygote.

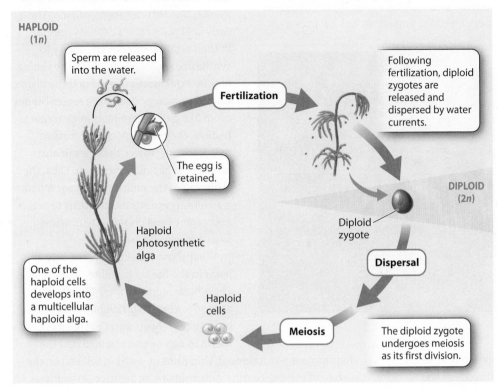

spores, it is called the **sporophyte**, and because the haploid multicellular generation gives rise to gametes, it is called the **gametophyte**. The resulting life cycle, in which a haploid gametophyte and a diploid sporophyte follow one after the other, is called **alternation of generations,** and it describes the basic life cycle of all land plants.

Because land plants are descended from green algae with only one haploid generation, we can infer that the diploid sporophyte of land plants is an evolutionary novelty on this branch of the phylogenetic tree. How might this new generation have made it possible for plants to colonize land?

To answer this question, we need to identify the sporophyte of *Polytrichum*. Late in the growing season, the green tufts of this moss sprout an extension—a small brown capsule at the end of a cylindrical stalk

Like *Chara* and *Coleochaete, Polytrichum* has a photosynthetic body—the familiar green tufts observed on the forest floor—that is made of haploid cells and that forms gametes by mitotic division of specialized cells. Sperm travel to eggs retained within reproductive organs, and the fusion of egg and sperm gives rise to a diploid zygote.

At this point, the life cycles of the green algae and *Polytrichum* diverge. In the moss, the zygote does not undergo meiosis and it does not disperse. Instead, the zygote is retained within the female reproductive organ, where it divides repeatedly by mitosis to produce a new multicellular generation—this one made of diploid cells (**Fig. 30.4**). Some cells within this diploid body eventually undergo meiosis, giving rise to haploid **spores,** cells that disperse and give rise to a new haploid generation. Because the diploid multicellular generation gives rise to

**FIG. 30.4   The life cycle of the moss *Polytrichum commune*.** The multicellular sporophyte generation is a major innovation of land plants.

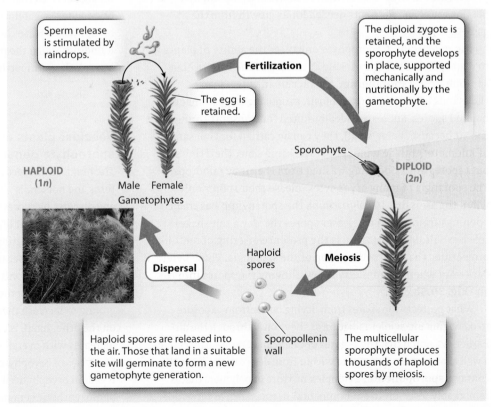

**FIG. 30.5**    **Moss sporangia.** (a) When spores of *Polytrichum commune* are mature, a cap falls off, allowing the salt shaker–like sporangium to release the spores. (b) The sporangia of *Ptychomitrium polyphyllum* have a ring of teethlike projections that curl back as they dry, flinging spores into the air.

a few centimeters high (**Fig. 30.5a**). This is the sporophyte, which originated from a fertilized egg. Because the fertilized egg was retained within the female reproductive organ, the sporophyte grows directly out of the gametophyte's body. Thus, the *Polytrichum* specimen pictured in Fig. 30.5a shows both the gametophyte (green) and the sporophyte (brown). The gametophyte is photosynthetically self-sufficient. In contrast, the sporophyte obtains water and nutrients needed for its growth from the gametophyte.

The multicellular sporophyte enhances the ability of plants to disperse on land. The capsule at the top of the *Polytrichum* sporophyte is a **sporangium**, a structure in which many thousands of cells undergo meiosis, producing large numbers of haploid spores. Spores are ideally suited for transport through the air because, being small, they can be carried for thousands of kilometers by the wind. But at the same time, their tiny size puts spores at risk of being washed from the air by raindrops. The sporangia of many bryophytes release their spores only when the air is dry. In *Polytrichum*, the sporangium has small openings from which it releases spores the way a salt shaker releases salt. More common is the presence of a ring of teethlike projections that curl over the top of the sporangia. These bend backward when air humidity is low, flinging the spores into the air (**Fig. 30.5b**).

What protects the spores from drying out or from exposure to damaging ultraviolet radiation as they travel through the air? Earlier, we noted that the zygotes of *Chara* and *Coleochaete* secrete a wall that protects the zygote within from desiccation. The wall contains **sporopollenin,** a complex mixture of polymers that is remarkably resistant to environmental stresses such as ultraviolet

radiation and desiccation. In mosses and other land plants, it is the spore, not the zygote, that secretes sporopollenin. Thus, early land plants had a capacity present in their algal ancestors—the ability to synthesize sporopollenin—but the timing of gene expression changed in these plants. Coated in sporopollenin, the spores cannot swim like meiotic products of *Coleochaete*. Instead, they are dispersed by wind to new locations, where they germinate to form a new haploid generation. Thus, the evolution of the multicellular sporophyte generation permitted land plants to make many spores and disperse them across large distances on land.

What about fertilization, the other point in the life cycle where *Chara* and *Coleochaete* require water? Like their algal ancestors, *Polytrichum* and other bryophytes rely on water to transport sperm to egg. Sperm released into the environment swim through thin films of water in soil and on the surface of the egg-bearing gametophyte. In addition, bryophytes tend to be small, so the distance between male and female gametes does not exceed the distance over which the sperm can swim, and they produce their gametes near the ground, where the likelihood of encountering a continuous film of water is greatest. A reliance on external water for fertilization restricts the growth of bryophytes to habitats that are, at least intermittently, wet. Many bryophytes release their sperm only when agitated by raindrops. Raindrops signal the presence of surface water, and their impact can splash sperm farther than they could swim on their own.

→   **Quick Check 1**  Give two reasons why spores are better suited for dispersal on land than are sperm.

## Vascular plants evolved a large photosynthetic sporophyte generation.

The first two lineages of vascular plants—the lycophytes and the ferns and horsetails—depend on swimming sperm for fertilization and disperse by spores that are released into the air (see Fig. 30.1). In these ways, their life cycle is similar to that of the bryophytes. However, the evolution of vascular systems had a major impact on their life cycle because it allowed them to grow tall. Xylem and phloem occur only in the sporophyte generation. This makes sense because gametophytes must remain small and close to the ground to increase the chances of fertilization. Spore dispersal, on the other hand, is enhanced by height, and spore production increases with overall size. An important difference between the life cycles of bryophytes and the spore-dispersing vascular plants is that in bryophytes it is the gametophyte that is the conspicuous photosynthetic generation, while in vascular plants it is the sporophyte generation that dominates in both physical size and photosynthetic output.

To illustrate how the evolution of vascular plants affected plant reproduction, let us examine the life cycle of the bracken fern *Pteridium aquilinum*, beginning with the largest component, the diploid sporophyte generation (**Fig. 30.6**). Bracken sporophytes appear to consist entirely of leaves because the stem grows underground. If you examine a bracken leaf closely, you may find tiny brown packets along its lower margin. These are sporangia, and each contains diploid cells that undergo meiosis to generate haploid spores. As in mosses, the spores become covered by a thick wall containing sporopollenin.

The structure of the sporangium aids in spore dispersal. The sporangium wall has a distinct ridge of asymmetrically thickened cells that, upon drying, produce a motion like that of a slingshot that hurls the spores away from the leaf surface to be carried off by air currents. Most spores travel only short distances, but a few may end up far from their parent. Bracken ferns can be found on every continent except Antarctica, as well as on isolated oceanic islands, testimony to the ability of spores to travel long distances.

Spores germinate to produce the haploid gametophyte generation. You have probably never seen a fern gametophyte— or never noticed one. Typical of most ferns, the bracken gametophyte is less than 2 cm long and only one to a few cells thick (Fig. 30.6). The small size of the fern gametophyte has led many to assume that the haploid generation is a weak link in the fern life cycle. In fact, because fern gametophytes tolerate drying out, they often live longer and withstand stressful environmental conditions better than the much larger sporophyte generation.

Most fern gametophytes are capable of producing both male and female gametes, although typically they produce only one type. Chemical signals released by gametophytes producing female gametes (eggs) stimulate gametophytes growing nearby to produce male gametes (sperm) that swim to the egg through a film of water. However, if fertilization does not occur, many fern gametophytes eventually produce both male and female gametes. The union of a male and a female gamete forms a diploid zygote, which is supported by the gametophyte as it begins to grow. Eventually, it forms leaves and roots that allow it to become a physiologically independent, diploid sporophyte.

Ferns, as well as the other spore-dispersing plants (lycophytes, horsetails, and bryophytes), release swimming sperm and thus are able to reproduce only when conditions are wet. What if a plant could eliminate that requirement? Seed plants did just that.

→ **Quick Check 2** In what ways is fern reproduction similar to moss reproduction? In what ways is fern reproduction different from moss reproduction?

---

**FIG. 30.6** **The life cycle of the bracken fern *Pteridium aquilinum*.** This life cycle illustrates the alternation between a small, free-living gametophyte generation and a taller, vascularized sporophyte generation.

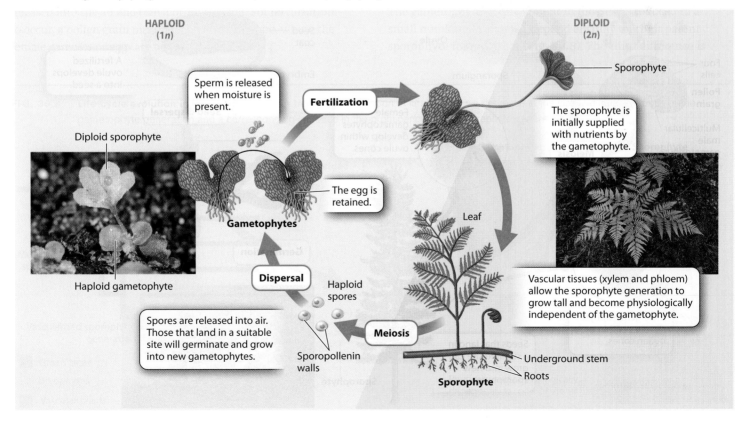

carpels and pollen-producing **stamens.** Here, we see an important difference between angiosperms and gymnosperms. Most flowers produce both pollen and ovules, whereas in gymnosperms, pollen and ovules are produced in separate structures. The significance of this arrangement in angiosperms is that because the sites of pollen and ovule formation are in proximity, a pollinator can deliver pollen to one plant, and take up pollen to carry to another in a single visit. Approximately 12% of flowering plant species produce flowers that form only pollen or ovules, but phylogenetic analysis makes it clear that all these species descend from ancestors whose flowers had both stamens and carpels. Such "unisexual" flowers are often found in plants that have reverted to wind pollination.

Let's look at the floral organs in more detail, starting with the carpels at the center of the flower (Fig. 30.12). Each flower produces several carpels (typically 3 to 5, but sometimes more than 20), but because the carpels are often fused, it may seem as though there is only one. Each carpel has a hollow **ovary** at the base in which one to many ovules develop. The ovary protects the ovules from being eaten or damaged by animals. The fact that the ovules are completely enclosed within the carpel is what gives rise to the name "angiosperm," which is from Greek words meaning "vessel" and "seed." In gymnosperms, the ovules are not enclosed in a vessel but exposed, and the name, again from the Greek, literally means "naked seeds."

The ovary makes it impossible for pollen to land directly on the ovule surface. Instead, carpels commonly have a cylindrical stalk at the top, called the **style.** The surface at the top of the style is called the **stigma.** Pollen is captured on the sticky or feathery stigma, where it germinates and extends its pollen tube down through the style to reach the egg (**Fig. 30.13**). In some plants the style can be more than 10 cm long—corn silks, for instance, are styles.

Immediately surrounding the carpels are the stamens. A stamen can have a leaflike structure bearing sporangia on its surface, but more commonly it forms a filament that supports a structure known as the **anther,** which contains several sporangia in which pollen is produced. In most flowers, the anther splits open, exposing the pollen grains. Once exposed, the pollen can come into contact with the body of a visiting pollinator or, in the case of a wind-pollinated species, be carried off by the wind. However, in some plants, for example tomato, small holes open at the top of anthers. To extract the pollen, bees land on the flower and vibrate at just the right frequency to shake loose the pollen inside. Orchids and milkweeds disperse their pollen in a completely different way. Instead of releasing individual pollen grains, they disperse their pollen all together in a package with a sticky tag that attaches to a visiting pollinator.

The outer whorls of the flower produce neither pollen nor ovules but instead contribute to reproductive success in other ways. Most flowers have two outer whorls, of which

**FIG. 30.13 Germination and growth of pollen tubes.** Pollen tubes of *Arabidopsis thaliana* can be seen because they have been stained with a fluorescent dye.

the outermost is made up of **sepals.** Sepals, which are often green, encase and protect the flower during its development. In contrast, **petals** are frequently brightly colored and distinctively shaped. Their role is to attract and orient animal pollinators. In addition to serving as visual cues, petals in many flowers produce volatile oils. These are the source of the distinctive odors, some pleasant, some decidedly not, that many flowers use to advertise their presence to pollinators.

→ **Quick Check 5** What is the name and function of the structures in each whorl of a flower?

## The diversity of floral morphology is related to modes of pollination.

The evolutionary histories of the angiosperms and their animal pollinators are closely intertwined. A rapid increase in angiosperm diversity occurred between 65 and 100 million years ago, about the same time as the diversification of bees (Hymenoptera) and butterflies (Lepidoptera). The evolution of flowers allowed

**FIG. 30.14 Flowers providing food or other rewards for their pollinators.** (a) A butterfly collecting pollen from flowers of a *Psiguria* vine; (b) a hummingbird visiting *Ravnia triflora* for nectar; and (c) a bat visiting *Agave palmeri* for nectar. (d) *Rhizanthes lowii* is pollinated by flies attracted to its smell of rotting flesh.

animals to specialize in new ways. At the same time, animal pollinators greatly facilitated the movement of pollen between plants within the same population.

So far, we have seen how flowers communicate their presence through both scent and color. For pollination to be reliable, however, it must be in the interest of the animal pollinator to move repeatedly (and with some constancy) between flowers of the same species. Flowers earn fidelity from their pollinators by providing rewards, frequently in the form of food (**Fig. 30.14**). Many insects consume some of the pollen itself, while many

flowers also produce nectar, a sugar-rich solution attractive to bats and birds as well as to insects.

Some species provide rewards other than food. For example, many orchids secrete chemicals that male bees need to make their own sexual pheromones. Other flowers provide an enclosed and sometimes heated chamber in which insects can aggregate. But some flowers do not provide rewards at all: Instead, they "trick" their pollinators into visiting. For example, flowers that look and smell like rotting flesh attract flies, and orchids that look like a female bee even emit similar pheromones to attract

**FIG. 30.15** *Ophrys ciliata*, **an orchid from Sicily, that mimics the shape and color of female bees.** These flowers induce males to attempt mating. Through repeated attempts on different flowers, the males transfer pollen.

male bees (**Fig. 30.15**). Male bees find these orchids irresistible and, as they attempt to copulate with the flower, deliver and receive pollen.

Producing rewards and attractants such as petals and scents requires resources. Thus, while animals can provide more efficient pollination than can the wind, the resources spent may be wasted if nonpollinating visitors "steal" pollen or nectar. For this reason, many flowers have mechanisms to protect their investments. For example, tubular cardinal flowers provide rewards to hummingbirds able to probe their depths, and snapdragons provide rewards only to insects that are strong enough to push apart their petals.

Many plants have evolved flowers that match rewards and attractants to the metabolic needs and sensory capabilities of their pollinators (see Fig. 30.14). For example, flowers pollinated by larger pollinators such as bats and birds produce copious amounts of nectar. Bat-pollinated flowers tend to be large, white or cream colored, and strongly scented, all traits appropriate to the size and nocturnal habits of their pollinators. The flowers are positioned among branches and leaves in such a way that they can be found by echolocation. In contrast, bird-pollinated flowers are often red, a color insects cannot see,

and have no scent. Because interactions between plants and pollinators affect gene flow, they can affect rates of speciation. A recent radiation in columbine occurred as the plants shifted from bumble bees to hummingbirds to hawkmoths as their primary pollinator (**Fig. 30.16**).

Angiosperms are thought to have evolved in tropical forests. Pollen cannot move freely through the air in the high density of foliage that is present year round in these forests, favoring the evolution of animal pollination. However, as angiosperms diversified and spread around the globe, some evolved wind pollination once again. About 20% of present-day angiosperm species release their pollen into the wind. The reversal to wind pollination is associated with habitats where pollinators are not reliably abundant, such as dry regions.

Wind pollination is most effective in species that are locally abundant. For example, many grasses rely on wind for pollination. Wind pollination is also common in temperate-zone trees, which produce flowers in the early spring, before leaves expand. Wind-pollinated species tend to produce large anthers and stigmas, the latter often branched like a feather duster, well adapted to capture pollen from the air. The sepals and petals of wind-pollinated plants have no attractive role and so tend to be small. Wind-pollinated species produce large amounts of pollen. Hay-fever is an allergic reaction to pollen, virtually all of which comes from wind-pollinated species.

### Angiosperms have mechanisms to increase outcrossing.

Plants do not have the elaborate courtship rituals that many animal species use to select an appropriate mate. Nevertheless, angiosperms have a wide range of mating systems. At one extreme are **self-compatible** species in which pollen and eggs produced by flowers on the same plant can unite and produce viable offspring. Self-compatible species are able to reproduce even when they are physically isolated from other individuals of the same species or when pollinators are rare. Many weedy species and the majority of crops are self-compatible.

At the other extreme are species in which pollen must be transferred between different plants. Approximately half of all of angiosperm species are **self-incompatible,** meaning that pollination by the same or a closely related individual does not lead to fertilization. Such self-recognition is based on the proteins produced from self-incompatibility genes, or S-genes. If the pollen's proteins match those of the carpel, the pollen either fails to germinate or germinates but the pollen tube grows slowly and eventually stops elongating. S-genes may have originated as a defense against fungal invaders, but subsequently evolved as a means of promoting outcrossing. Because there can be dozens of S-gene alleles in a population, only pollen transfers between closely related individuals are blocked. Apples are examples of self-incompatible plants, explaining why apple orchards always contain a mixture of different varieties.

**FIG. 30.16**

# Can pollinator shifts enhance rates of species formation?

**BACKGROUND** Columbines (*Aquilegia* species) are diverse and have visually striking flowers. Columbine flowers have nectar spurs, which are modified petals that form tubular outgrowths. Blue flowers with short nectar spurs are pollinated by bumblebees, red flowers with medium nectar spurs are pollinated by hummingbirds, and white or yellow flowers with long nectar spurs are pollinated by hawkmoths. The size of the nectar spur corresponds to the length of the tongue of the pollinator.

**HYPOTHESIS** The recent radiation of columbines corresponds with shifts from short-tongued pollinators (bumblebees) to ones with increasingly longer tongues (hummingbirds, hawkmoths).

**METHOD** Researchers mapped changes in flower color and nectar spur length onto a phylogenetic tree of columbines. To do this, they assigned each of 25 species examined to one of three groups based on length of tongue, flower color, and most common pollinator (bumblebees, hummingbirds, or hawkmoths). The group is indicated by the color circle at the tip of each branch in the phylogenetic tree in the figure. The researchers then did a statistical analysis that suggested the most likely pollinator type at each node representing an ancestor, also indicated by color circles in the figure. They used this information to infer the history of pollinator shifts during the radiation of this group.

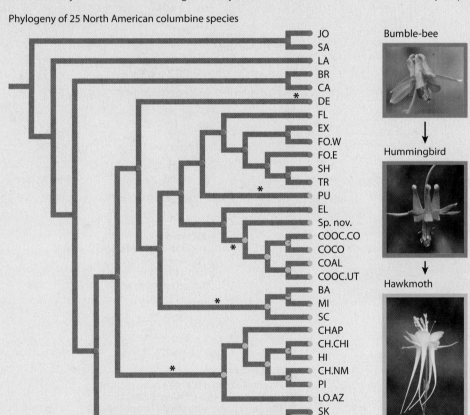

Phylogeny of 25 North American columbine species

Bumble-bee

Hummingbird

Hawkmoth

**RESULTS** Within the recent radiation of columbines, there are multiple instances of shifts from a shorter-tongued to a longer-tongued pollinator, and no reversals.

**CONCLUSION** The evolution of floral diversity and increasing nectar-spur length in columbines is associated with pollinator shifts. Large changes in nectar-spur length occur disproportionately at speciation events, indicating a key role in adaptive radiation.

**FOLLOW-UP WORK** Developmental studies show that the diversity of nectar spur length evolved as a result of changes in cell shape.

**SOURCES** Whittall, J. B. and S. A. Hodges. 2007. "Pollinator Shifts Drive Increasingly Long Nectar Spurs in Columbine Flowers." *Nature* 447:706–709; Puzey, J. R., S. J. Gerbode, S. A. Hodges, E. M. Kramer, and L. Mahadevan. 2011. "Evolution of Spur Length Diversity in *Aquilegia* Petals Is Achieved Solely Through Cell Shape Anisotropy." *Proceedings of the Royal Society, London, Series B,* 279:1640–1645.

● Short nectar spurs, blue petals, visited by bumble bees

● Medium-length nectar spurs, red petals, visited by hummingbirds

● Long nectar spurs, white or yellow petals, visited by hawkmoths

\* indicates a likely evolutionary shift in pollinator

**FIG. 30.17** **Reproduction in angiosperms.** (a) The angiosperm life cycle; (b) double fertilization, in which one sperm fuses with the egg to form a zygote, and another sperm fuses with the diploid gametophyte cell to form a triploid cell.

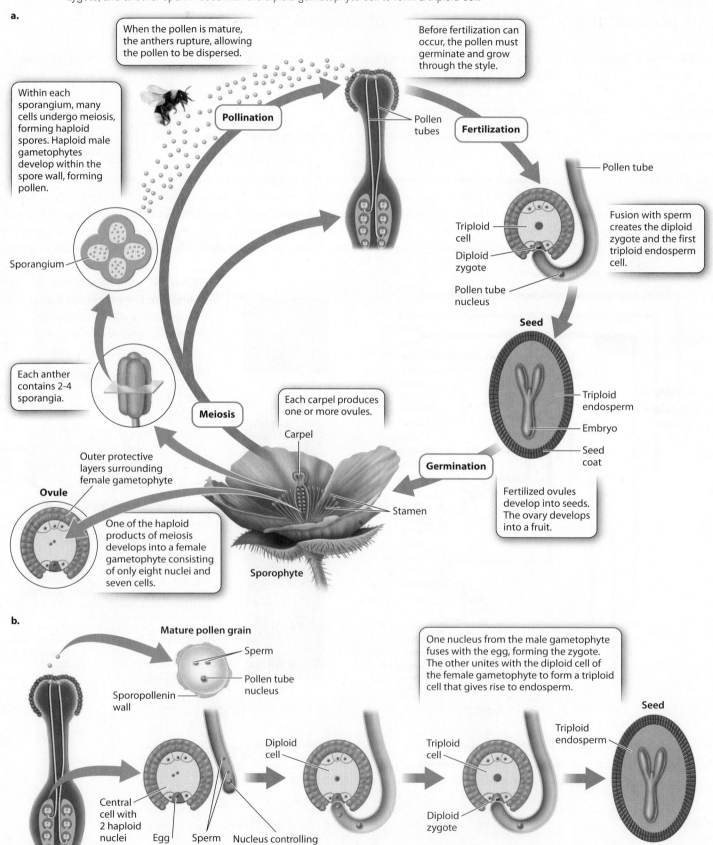

a.

When the pollen is mature, the anthers rupture, allowing the pollen to be dispersed.

Before fertilization can occur, the pollen must germinate and grow through the style.

Within each sporangium, many cells undergo meiosis, forming haploid spores. Haploid male gametophytes develop within the spore wall, forming pollen.

**Pollination**

Pollen tubes

**Fertilization**

Pollen tube

Sporangium

Triploid cell

Diploid zygote

Pollen tube nucleus

Fusion with sperm creates the diploid zygote and the first triploid endosperm cell.

Each anther contains 2-4 sporangia.

**Seed**

**Meiosis**

Each carpel produces one or more ovules.

Carpel

Triploid endosperm

Embryo

Seed coat

Outer protective layers surrounding female gametophyte

**Germination**

**Ovule**

One of the haploid products of meiosis develops into a female gametophyte consisting of only eight nuclei and seven cells.

Stamen

**Sporophyte**

Fertilized ovules develop into seeds. The ovary develops into a fruit.

b.

**Mature pollen grain**

Sperm

Pollen tube nucleus

Sporopollenin wall

One nucleus from the male gametophyte fuses with the egg, forming the zygote. The other unites with the diploid cell of the female gametophyte to form a triploid cell that gives rise to endosperm.

**Seed**

Diploid cell

Triploid cell

Triploid endosperm

Central cell with 2 haploid nuclei

Egg

Sperm

Nucleus controlling pollen tube growth

Diploid zygote

**FIG. 30.18 Endosperm.** Endosperm develops from the triploid cell produced by double fertilization.

In rice, the endosperm fills most of the seed and the embryo is small.

In a mature peanut seed, all of the endosperm's stored resources have already been transferred to the embryo.

Seed coat (2*n*, previous diploid generation)

Endosperm (3*n*)

Embryo (2*n*)

Female gametophyte tissue (1*n*) is too small to see, even at this scale.

Cotyledon (part of the embryo)

Shoot of embryo

Root of embryo

## Angiosperms delay provisioning their ovules until after fertilization.

A major trend in the evolution of plants is the decreasing size and independence of the gametophyte generation and the increasing prominence of the sporophyte generation, as we have seen. In angiosperms, this trend is taken even further (**Fig. 30.17a**). The male gametophyte has only three cells, one of which controls the growth of the pollen tube, while the other two are male gametes or sperm. In most species, the female gametophyte contains only eight nuclei, arranged as six haploid cells plus a central cell containing two nuclei. One of the haploid cells gives rise to the egg, and the two nuclei within the central cell fuse either before or during fertilization to form a diploid cell. The importance of this cell will become apparent. The female gametophyte of angiosperms is too small to support the

growth of the embryo, as it does in gymnosperms. How, then, do flowering plants provide nutrition for the embryo to grow?

This is where the diploid cell of the female gametophyte comes into play. During pollination, the two sperm travel down the pollen tube and enter the ovule (**Fig. 30.17b**). One of the sperm fuses with the egg to form a zygote, just as occurs in all plants. The other sperm, however, unites with the diploid gametophyte cell to form a triploid (3*n*) cell. This triploid cell undergoes many mitotic divisions, forming a new tissue called **endosperm.** In angiosperm seeds, it is the endosperm that supplies nutrition to the embryo. The process in which two sperm from a single pollen tube fuse with the egg and the diploid cell is called **double fertilization.**

What advantage is gained by double fertilization and endosperm formation? In pines and other gymnosperms, the female gametophyte provides food for the embryo. This provisioning occurs whether or not the egg is successfully fertilized, and so is potentially a significant waste of resources. In angiosperms, endosperm development occurs only when fertilization has occurred. In many flowering plants, including corn, wheat, and rice, the mature seed consists largely of endosperm (**Fig. 30.18**). The embryo itself is small. As the seeds germinate, the embryo draws resources from the endosperm to help it become established. It is no overstatement to suggest that the human population is sustained mostly by endosperm produced in the seeds of cereal crops.

In other angiosperms, growth of the embryo consumes all the endosperm before seed dispersal. For example, a peanut has virtually no endosperm left at the time it is dispersed. All the resources that were previously held within the endosperm have been transferred into two large embryonic leaves, called cotyledons, which form the two halves of the peanut we eat.

### Fruits enhance the dispersal of seeds.

The transformation of a flower into a **fruit** is as remarkable as the metamorphosis of a caterpillar into a butterfly (**Fig. 30.19**). As the fertilized egg develops to form an embryo and the endosperm

**FIG. 30.19 Developmental transition from flower to fruit.** Fruits, in this case an apple, develop from ovaries and provide a means of dispersing seed.

**FIG. 30.20 Fruits and seed dispersal.** (a) The seeds of the milkweed *Asclepias syriaca* are dispersed by wind; (b) a squirrel carries a fruit it will bury for later use; (c) a robin is eating berries of mountain ash (*Sorbus americana*); (d) cocklebur (*Xanthium strumarium*) fruits are hitching a ride on the fur of a mule deer.

In a number of species, including apples, bananas, and tomatoes, a gaseous hormone called ethylene triggers fruit ripening. It is the production of ethylene that explains why placing a ripe apple and an unripe banana together hastens the ripening of the banana. Tomatoes and bananas are usually picked in an immature state and are subsequently ripened by exposure to ethylene. The fruits can therefore be transported while they are still hard and less prone to damage. Because ethylene synthesis requires oxygen, the unripe fruits are typically stored in a low-oxygen environment. Without this precaution, the ethylene produced by the ripening of even a single fruit could set off a chain reaction in which all the stored fruit would ripen at once.

The tremendous morphological diversity of fruits reflects the many ways in which angiosperm seeds are dispersed (**Fig. 30.20**). The winged fruit of an ash tree is carried away on the wind, and a coconut can float on ocean currents for many months, protected from exposure to salt water by its thick outer husk. The burdock fruit has small hooks that attach to fur (or clothing)—this small hitchhiker was the inspiration for Velcro. Other fruits are not dispersed, but instead open to expose their seeds. For example, as milkweed fruits mature and dry out, they split apart to release hundreds of seeds, each with a tuft of silky hairs that floats them away on even a light breeze. Still others, such as *Impatiens* (jewelweed),

proliferates around it, the ovary wall develops into a fruit, stimulated by signaling molecules produced by the endosperm. In a tomato, for example, the fleshy fruit is a mature ovary that encloses the seeds. In some plants, other parts of the flower may also become incorporated into the fruit. For instance, the fleshy part of an apple is formed from the outer part of the ovary combined with the base of the petals and sepals. Other fruits, like pineapple, develop from multiple flowers.

Fruits serve two functions: They protect immature seeds from being preyed on by animals, and they enhance dispersal once the seeds are mature. In the case of fruits dispersed by animals, it is essential that the fruit not be consumed before the seeds are able to withstand a trip through their disperser's digestive tract. Thus, immature fleshy fruits are physically tough and their tissues highly astringent. As they ripen, the fruits are rapidly transformed in texture, palatability, and color. Ripening converts starches to sugars and loosens the connections between cell walls so that the fruit becomes softer.

produce fruits that forcibly eject their seeds.

Animals are probably the most important agents of seed transport, in many cases attracted by the nutritious flesh of the fruit. In fleshy fruits, the seeds are protected by a hard seed coat and pass unharmed through the animals' digestive tract. In some species, this passage actually enhances germination rates. In many cases, however, the seed is consumed, and successful dispersal occurs when the animals gather more seeds than they eat. Squirrels are a good example, storing many acorns that never get eaten. Humans represent an important dispersal agent, responsible for transporting seeds long distances, sometimes far outside their usual range. In some cases, accidentally introduced plant species can disrupt native communities.

→ **Quick Check 6** How does the fruit relate to structures found in the flower?

## 30.3 TIMING OF REPRODUCTIVE EVENTS

As plants moved onto land, they encountered environments with distinct seasonality. In temperate regions, temperatures vary from summer to winter; in most tropical habitats, wet seasons alternate with dry ones. Thus, in addition to overcoming the challenges of completing their life cycle on land, plants evolved mechanisms to control the timing of reproduction. This is true for all plants, which must produce gametes at a time when they can find suitable mates and which must disperse their young in a season suitable for their growth.

However, timing is particularly important for angiosperms. Not only must individuals of the same species flower at the same time, but they must do so when they can attract enough pollinators for efficient pollination. Furthermore, angiosperms are unique among plants in including species with life-spans of only one to two years. For these short-lived species, which are most commonly found in temperate regions, there is little margin for error in the timing of flower production and seed germination. In this section, we explore how plants gain information about their environment that enables them to regulate when flowers are produced and seeds germinate.

### Flowering time is affected by day length.

In the 1910s, Wightman Garner and Henry Allard, scientists at the U.S. Department of Agriculture, set out to understand why Maryland Mammoth, a variety of tobacco, does not produce flowers during the summer like other tobacco plants. Instead, these plants continue to grow throughout the summer, reaching heights several times that of a typical tobacco plant and producing many more leaves. Only in autumn does Maryland Mammoth finally produce flowers, too late in the year for seeds to mature.

One possible explanation for the observed delay in flowering time is that Maryland Mammoth must achieve a greater total size before it produces flowers. However, when Garner and Allard grew Maryland Mammoth plants in a heated greenhouse during the winter, the plants flowered at the same time and the same size as other tobacco plants. Therefore, size does not determine when Maryland Mammoth flowers. And because the heated greenhouse was set to the same temperatures as those found during the summer, temperature could also be eliminated as a factor. Garner and Allard were left with only one more environmental variable that was different between the summer and the winter: the length of the day. They proposed that plants can measure and respond to day length.

In a series of experiments in which day length was artificially shortened during summer and extended during winter, Garner and Allen showed that in Maryland Mammoth, as well as in many other plants, flowering is controlled by day length. They coined the term **photoperiodism** to describe the effect of the "photo period," or day length, on flowering. **Short-day plants,** like Maryland

Mammoth, flower only when the day length is less than a critical value. When the light period exceeds this threshold, a short-day plant continues to produce new leaves, but no flowers are formed. In contrast, **long-day plants,** which include radish, lettuce, and some varieties of wheat, flower only when the light period exceeds a certain length. Not all plants are photoperiodic. Flowering of **day-neutral plants** is independent of any change in day length.

Sensitivity to day length ensures that plants flower only when they are large enough to support the development of a large number of seeds *and* only when there is enough time for those seeds to mature before the onset of winter. A short-day plant that germinates in the spring will not flower until late summer, when the decreasing day length falls below the critical value for that species. This flowering strategy allows short-day plants to grow as large as possible and yet switch from producing leaves to producing flowers in time to complete seed development before winter. In contrast, many long-day plants germinate in the summer. Because they do not flower until a critical day length is exceeded, they do not flower in autumn, but instead during the following spring.

Understanding how photoperiod affects flowering has many commercial applications. For example, chrysanthemums are short-day plants and their natural flowering season is autumn. By artificially exposing plants to short day lengths, growers can induce chrysanthemums to flower at any time of the year. Photoperiodism can complicate the ability to grow crop species at latitudes different from the ones in which they evolved. For example, soybean is a short-day plant that normally germinates in the spring and flowers as the day length decreases in late summer. When planted in the tropics, where the day length is close to 12 hours throughout the year, soybean flowers when the plant is still quite small, reducing the number of seeds that can eventually be harvested. For this reason, crop species often become less sensitive to photoperiod as they are domesticated. For example, wild varieties of tobacco are photoperiodic, whereas cultivated tobacco, *Nicotiana tabaccum*, is day neutral.

### Photoreceptors enable plants to measure day length.

The first requirement for measuring day length is that a plant be able to sense when it is day and when it is night. To accomplish this, many plants produce **photoreceptors,** molecules whose chemical properties are altered when they absorb light. We saw in Chapter 9 how a signal can be conveyed to a cell by a change in the shape of a receptor molecule within the cell or on its surface. Photoreceptors in plants work similarly. The absorption of light by a photoreceptor produces a signal that triggers changes in the cell's metabolism or alters patterns of gene expression. Your eye relies on thousands of photoreceptors that work on the same principle as those in plants.

Although photoreceptors provide information about the presence of light, they do not, by themselves, explain how leaves are able to measure day length. An early clue to this ability came when scientists discovered that interrupting the dark period with a brief exposure to light was sufficient to alter the flowering response. In these experiments, both short-day plants and long-day plants were grown under a photoperiod that would normally cause the short-day plants to flower and the long-day plants to produce only leaves. A brief exposure to light during the dark period reversed this outcome. The short-day plants no longer produced flowers, but the long-day plants did. The reverse experiment, in which plants were placed under light-tight cover for a short period during the daytime, did not have any effect.

If no further experiments had been done, the scientists would have concluded— incorrectly—that plants were measuring the length of the dark period. More extensive experiments demonstrated that a plant's sensitivity to a "night interruption" varies according to a circadian—or 24–hour—rhythm. We now know that photoperiodism results from an interaction between photoreceptors that are activated by light and the product of a gene whose expression is controlled by the plant's circadian clock (**Fig. 30.21**). Circadian clocks are biochemical mechanisms found in all organisms that oscillate with a 24-hour period and are coordinated with the day-night cycle.

Other experiments demonstrated that day length is sensed by the leaves and not by the buds where the flowers will form. When conditions are right for flowering, a signal is transported from the leaves to the growing points of a plant. A flowering signal (named "florigen") had been hypothesized to exist for more than 70 years, but its nature had not been known, although widely sought. Only with the development of techniques for genetic analysis were scientists able to investigate the genetic basis of photoperiodism and determine the identity of this important signal.

### Vernalization prevents plants from flowering until winter has passed.

In a number of temperate-zone species, flowering can be induced only if the plant has experienced a prolonged period of cold temperatures, a process known as **vernalization.** Vernalization prevents plants from flowering prematurely in late summer or autumn. Plants are forced to wait until the following spring to flower, when the chances for successful pollination and seed development are greater.

An important feature of vernalization is that even parts of the plant that form after the cold stimulus is long past seem to "remember" that they received the required cold treatment. This "memory" of winter is important because many plants that overwinter grow rapidly in the spring. What is the basis of this "memory" of winter and how is it transmitted to new growth?

Plant species differ in the ways that vernalization affects the regulatory pathways that lead to flowering. These differences likely reflect the fact that the angiosperms evolved in tropical latitudes. As different groups migrated into temperate regions they independently evolved mechanisms for controlling flowering in highly seasonal environments.

One common mechanism acts through chromatin remodeling. The DNA of all eukaryotic organisms is bound with proteins to form chromatin. We saw in Chapter 19 that modifications to chromatin, such as DNA methylation, can change gene expression. By this means, a record of developmental events can be inscribed into a cell's DNA. In *Arabidopsis*, for example, vernalization results in chromatin remodeling that turns off a gene whose protein product represses flowering. Chromatin remodeling is stable through mitotic divisions, explaining why newly formed parts of a plant "remember" winter. However, the slate is wiped clean during meiosis, and the requirement for vernalization is reinstated with each generation.

### Dormant seeds can delay germination if they detect the presence of plants overhead.

Seed germination is a one-way street. Once a seed begins to germinate, it quickly reaches a point of no return and it either becomes established as a photosynthetically competent seedling or it dies. Dormant seeds may delay germination even though environmental conditions, notably temperature and moisture, are favorable. Dormancy increases the probability that some offspring will survive to maturity by extending the time over which the seeds germinate. From a farmer's perspective, however, dormancy is a problem because it means that some fraction of the seeds put into the ground will not germinate in that year. Therefore, agriculturalists have paid close attention to what environmental factors induce germination.

In some species, dormancy is enforced by a hard, tough seed coat. In such species, the seed coat must be weakened mechanically before the seed will germinate, either by passing through an animal's digestive system or by physical abrasion. In other plants, dormancy is controlled by the embryo. Two environmental variables are particularly effective at inducing such seeds to germinate. One is exposure to a prolonged period of cold temperatures, which signals that winter has past and thus prevents seeds from germinating on warm days in autumn. The second is that some seeds, particularly very small ones, will germinate only if they have been exposed to light.

When scientists first began to study the effects of light on seed germination they found that red light is particularly good at stimulating germination, while far-red light inhibits germination. A simple "flip-flop" experiment by H. A.

**FIG. 30.21**

# How do plants measure day length?

**BACKGROUND** *Arabidopsis thaliana*, a model organism for plant biology, is a long-day plant, flowering only when the light period exceeds about 16 hours. Genetic studies have shown that the *CONSTANS (CO)* gene is essential in allowing *Arabidopsis* to flower in response to day length. To determine the mechanism underlying photoperiodism, researchers set out to determine how this gene is affected by both the plant's circadian clock and the presence or absence of light.

**HYPOTHESIS** The ability to respond to day length in *Arabidopsis* results from a circadian (daily) increase and decrease in *CO* expression, as well as interactions with light.

**EXPERIMENT 1** The researchers determined how *CO* mRNA varies as a function of time of day and presence or absence of light.

**RESULT 1** *CO* mRNA levels follow a circadian up-and-down pattern. This pattern occurred independent of whether the plant was exposed to long or short days or even kept continually in light. The timing of these oscillations is such that high levels of *CO* mRNA coincide with the light period only when day-lengths are long.

Blue arrows indicate the amount of *CO* mRNA that would be present at the end of a short day (8-hour light period).

Red arrows indicate the amount of *CO* mRNA at the end of a long day (16-hour light period).

*CO* expression oscillates with a 24-hour (circadian) period, independent of light treatment.

**EXPERIMENT 2** Protein levels were measured following exposure to two days of light or dark. Protein extracts were first separated by gel electrophoresis and then probed with antibodies. Darker bands correspond to greater staining by protein-specific antibodies and thus provide visual information on the relative amount of the CO protein.

**RESULT 2** The much darker band after exposure to two days of light indicates that CO protein accumulates only in the light. In the dark, the CO protein does not accumulate.

After exposure to two days of light, CO levels are high, as indicated by the dark band at the left. After exposure to 2 days of dark, CO levels are low, as indicated by the light band at the right.

**CONCLUSION** Flowering in *Arabidopsis* is triggered by long days because there is overlap between high rates of CO expression and stabilization of the CO protein by light.

**FOLLOW-UP WORK** Studies with *Arabidopsis* mutants indicate a role for specific photoreceptors in stabilizing CO protein. Studies on other species indicate that photoperiod detection involves both the circadian clock and photoreceptors, but the way the CO protein is regulated by light varies.

**SOURCES** Suárez-López, P., K. Wheatley, F. Robson, H. Onouchi, F. Valverde, and G. Coupland. 2001. "*CONSTANS* Mediates Between the Circadian Clock and the Control of Flowering in *Arabidopsis*." *Nature* 410:1116–1120; Valverde, F., A. Mouradov, W. Soppel, D. Ravenscroft, A. Samach, and G. Coupland. 2004. "Photoreceptor Regulation of CONSTANS Protein in Photoperiodic Flowering." *Science* 303:1003–1006.

FIG. 30.22

# How do seeds detect the presence of plants growing overhead?

**BACKGROUND** To study the effect of light on seed germination, scientists exposed lettuce seeds that had been kept continually in the dark to different wavelengths of light and then counted what fraction of the seeds germinated. Red light had the greatest ability to stimulate germination, but surprisingly, far-red light inhibited germination such that fewer seeds germinated than in the control seeds, which were kept in darkness. In the rush to conduct more experiments, petri dishes with the light-treated seeds piled up by the sink until someone noticed that the seeds that had been experimentally treated with far-red light were now germinating.

**HYPOTHESIS** This observation suggested that the inhibitory effect of far-red light can be overcome by a subsequent exposure to red light.

**EXPERIMENT** The scientists exposed lettuce seeds to red and far-red light in an alternating pattern, ending with either red light or far-red light. They then placed the seeds in the dark for two days and afterward counted the number of seeds that had germinated.

**RESULTS** When the lettuce seeds were exposed to red light last,

nearly 100% of the seeds germinated. By contrast, when the last exposure of the seeds was to far-red light, the percentage of seeds that germinated was dramatically reduced.

**Lettuce Seeds Germinating After Exposure to Red and Far-red Light in Sequence**

| SEQUENCE OF LIGHT EXPOSURE | GERMINATION (%) |
|---|---|
| Dark (control) | 8.5 |
| Red | 98 |
| Red, Far-red | 54 |
| Red, Far-red, Red | 100 |
| Red, Far-red, Red, Far-red | 43 |
| Red, Far-red, Red, Far-red, Red | 99 |
| Red, Far-red, Red, Far-red, Red, Far-red | 54 |
| Red, Far-red, Red, Far-red, Red, Far-red, Red | 98 |

TIME →

**CONCLUSION** Seed germination in lettuce is triggered by exposure to red light and is inhibited by exposure to far-red light in a reversible fashion. As a result, plants are able to track changes in the relative amount of red and far-red light, which provides information on the presence or absence of plants overhead (see Fig. 30.23).

**FOLLOW-UP WORK** Additional studies showed that this result was due to a single pigment in a photoreceptor, now known as phytochrome, that is converted into an active form by red light and reversibly converted into an inactive form by far-red light.

**SOURCE** Borthwick, H. A., S. B. Hendricks, M. W. Parker, E. H. Toole, and V. K. Toole. 1952. "A Reversible Photoreaction Controlling Seed Germination." *Proceedings of the National Academy of Sciences* 38:662–666.

Red light chamber   Far-red light chamber

Ungerminated lettuce seeds

Borthwick and colleagues showed that the inhibitory effect of far-red light could be overcome by a subsequent exposure to red light (**Fig. 30.22**). Ultimately it was shown that only a single photoreversible pigment was needed to detect the exposure to both far-red and red light.

That pigment is present in **phytochrome,** a photoreceptor that switches back and forth between two stable forms depending on its exposure to light (**Fig. 30.23a**). Red light causes phytochrome to change into a form that absorbs primarily far-red light ($P_{fr}$); far-red light causes phytochrome to change back into the red-light-absorbing form ($P_r$). Because red light stimulates seed germination, we know that $P_{fr}$ is the active form of phytochrome.

In the dark, $P_{fr}$ slowly converts back into the inactive, $P_r$, form. The light-independent conversion of $P_{fr}$ to $P_r$ allows a seed that had been exposed to red light, but failed to germinate, to have its phytochrome reset so that it can again respond to red light.

Phytochrome allows dormant seeds to respond to the presence of plants overhead. For a small seed that has few reserves, germinating in the shade of another plant could be fatal. How does phytochrome allow plants to detect the presence of other plants? As sunlight passes through leaves, the red wavelengths are absorbed by chlorophyll but the far-red wavelengths are not. The ratio of red to far-red light is thus much higher under an open sky than it is in the shade of another plant (**Fig. 30.23b**). As a result, the proportion of phytochrome that is in the active, $P_{fr}$, form is much greater in open environments than in the forest understory. Furthermore, the ratio of red to far-red light is independent of light intensity. Thus, the ratio is a reliable signal, even for seeds covered by a thin layer of soil, that leaves are overhead.

Phytochrome plays other important roles in plant development. As we will discuss in Chapter 31, phytochrome allows plants to alter their growth in the presence of neighboring plants to compete more effectively for sunlight. Phytochrome also maintains the accuracy of the plant's circadian clock by resetting it each day at dawn. Finally, phytochrome is one of the photoreceptors that plants use to determine changes in day length.

? **CASE 6  Agriculture: Feeding a Growing Population**
**What is the basis for the spectacular increases in the yield of cereal grains during the Green Revolution?**

Mechanisms to control the timing of flower production and seed germination have evolved to optimize reproductive success under natural conditions. Agricultural conditions, however, are quite different. When Norman Borlaug, an American agronomist, led a project in the mid-twentieth century to make Mexico self-sufficient in bread wheat, his group encountered problems of flowering time and seed germination. Borlaug's aim was to breed wheat for enhanced resistance to infection by a devastating class of fungal pathogens known as rusts. To do this, he and his group hand-pollinated plants that exhibited resistance to rusts with ones that grew well or had seeds that produced good flour.

Under natural conditions, wheat usually self-pollinates. To introduce new sources of pollen required that the anthers of each flower be removed by forceps before the flowers opened. Each flower had to be covered to prevent unwanted pollen from coming into contact with the flower. Finally, when the stigma

was receptive, it had to be dusted with pollen collected from a plant with the desired characteristics. Under the hot sun, this is backbreaking work.

To speed the work, Borlaug proposed growing two crops of wheat each year. During summer, he conducted controlled crosses in the central highlands of Mexico, and then as winter approached he took his seeds north to the Yaqui Valley. There temperatures were warm enough for him to carry out another round of crosses during the winter.

**FIG. 30.23  Detection of shading by phytochrome.** Phytochrome detects the relative absence of red light under the canopy.

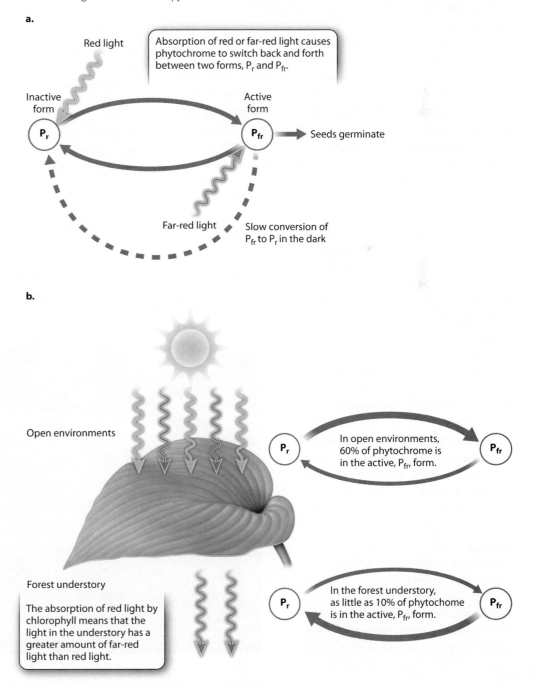

Wheat is a long-day plant that flowers in the long days of summer. By successively growing plants during the lengthening days of summer in the southern highlands and the shorter days of winter in the lowland north, Borlaug was unconsciously selecting for plants whose flowering was less sensitive to photoperiod. Although not the original goal of the project, the diminishing sensitivity to photoperiod had an important impact. The new varieties were day neutral, flowering when they achieved a certain size. As a result, Borlaug's varieties could be planted across a wide geographic range.

Before Borlaug's rust-resistant seeds would be ready to be grown throughout the world, however, one more problem had to be solved. Having succeeded in breeding wheat varieties that were disease resistant, Mexican farmers applied fertilizer in large amounts. Under these conditions, the new wheat varieties produced so many seeds that stalks became top heavy, falling over in the wind.

Borlaug knew that an answer to this problem was to cross his plants with wheat varieties that had shorter, sturdier stems. But given the selective advantage of height in natural populations, where was he to find such a plant? Japanese wheat breeders had identified and preserved a mutant dwarf plant that occurred spontaneously, and Borlaug obtained 80 seeds of this dwarf variety. However, in his haste to incorporate this new plant into his breeding program, he made a serious error. He forgot that the Japanese wheats, which were adapted to a more northerly climate, had to be vernalized before they would flower.

Because the Japanese wheats had not been exposed to a prolonged period of cold, they did not flower at the same time as the Mexican wheats. And when they did produce a few flowers, the blooms were of such low quality that no viable seeds were produced. As luck would have it, Borlaug had planted only 72 of the 80 seeds. Somehow 8 seeds had escaped his notice. Eventually, Borlaug found these at the bottom of the original envelope in which they had been sent. This time he took care to vernalize the seeds, and when the new plants were formed they flowered readily.

Much hard work and many thousands of hand pollinations lay ahead. But by 1963, more than 95% of Mexican wheat cultivation made use of Borlaug's high-yielding semidwarf wheat varieties, resulting in wheat yields six times greater than in 1944, the year he began work in Mexico. In 1965, the seeds were exported in large numbers, first to India and Pakistan and soon to the rest of the world. Used in combination with greater investments in fertilizer and irrigation, these varieties prevented the famines that had been predicted by many as a result of rising human populations. For his work in alleviating world hunger, Borlaug was awarded the Nobel Peace Prize in 1970.

## 30.4 VEGETATIVE REPRODUCTION

Eukaryotic organisms can reproduce asexually as well as sexually (Chapters 11 and 27). Asexual reproduction results in new individuals that are genetically identical to the parent. If the new individuals can disperse by walking, swimming, or floating away, asexual reproduction allows organisms to proliferate and spread.

In sexual reproduction, plants disperse by spores and seeds. What prevents asexually produced individuals from growing so close to the parent plant that they are in direct competition for resources? In mosses and liverworts, tiny plantlets form by mitosis at the base of shallow cups. Raindrops landing in this cup can dislodge the tiny plantlets, splashing some away from the parent plant.

Most plants that reproduce asexually do so by growing to a new location and only then producing a new plant. You need look no further than a front lawn to find evidence of such **vegetative reproduction.** Horizontal stems allow new upright grass shoots to be produced at a distance from the site where the parent plant originally germinated. Strawberries, bamboo, and spider plants (a common houseplant) also spread vegetatively by forming horizontal stems. In many cases, the connections that were needed initially to produce a new plant become severed. When

**FIG. 30.24 Vegetative growth of aspen trees.** Each distinct group of trees, distinguished by leaf color, is a group of genetically identical individuals produced by vegetative reproduction.

**FIG. 31.1  Shoot organization.** A grape vine (a) and barrel cacti (b) look very different, but their shoot systems are built from the same repeating units of nodes and internodes (c).

a.

b.

c.

Shoots are made up of repeating units of nodes and internodes; one or more leaves are attached at each node.

Leaf

Nodes

Internode

## 31.1 SHOOT GROWTH AND DEVELOPMENT

In tropical forests, vines weave their way upward through the branches of trees. Many of these vines have long, thin stems with widely spaced leaves (**Fig 31.1a**). In contrast, a barrel cactus, living in the desert, has thick stems and leaves modified to form sharp spines (**Fig. 31.1b**). Although they look nothing alike, these two plants are constructed in the same way. Plant shoots are modular, meaning that they are formed of repeating units. Each unit consists of a **node**, the point where one or more leaves are attached, and an **internode**, the segment between two nodes (**Fig. 31.1c**). In vines the internodes are long and the leaves large, whereas in cacti the internodes are short and the leaves thin and

sharp. The modular nature of plants helps us understand how the capacity for continued growth gives rise to the tremendous variation in plant form that we see around us.

### Shoots grow by adding new cells at their tips.

At the very tip of each branch, commonly hidden beneath an array of young leaves called **leaf primordia** (singular, primordium), lies a tiny dome of cells called the **shoot apical meristem** (**Fig. 31.2**). The shoot apical meristem is a group of totipotent cells that, like embryonic stem cells in animals, gives rise to new tissues. As we will see, plants have meristems in several parts of their bodies, but upward growth occurs exclusively in and just below the shoot apical meristem. This is where cell division occurs, generating all the new cells that serve

**FIG. 31.2  The shoot apical meristem.** The shoot apical meristem consists of undifferentiated cells that divide rapidly, giving rise to all the cells in stems and leaves. (a) The seedling is rowan (*Sorbus acuparia*). (b) This scanning electron micrograph shows the shoot apical meristem of tomato (*Solanum lycopersicum*).

a.

Shoot apical meristem

Leaf primordium

## CHAPTER 31

# PLANT GROWTH AND DEVELOPMENT

## Building the Plant Body

In the previous chapters, we have seen how two major innovations played key roles in enabling plants to thrive on land. The first is an internal vascular system capable of pulling water from the soil, allowing plants to elevate their leaves into the air to capture sunlight (Chapter 29). The second is a diploid sporophyte generation that allows plants to disperse by spores carried through the air (Chapter 30). These and other features of vascular plants depended on new ways of building the plant body.

The fossil record shows that the first vascular plants were little more than small photosynthetic stems with simple sporangia at their tips. Wood, leaves, and roots all came later. Today, most landscapes include trees with thick woody stems, extensive root systems, numerous leaves, and complex reproductive structures. To understand how these innovations evolved, we need to explore the ways that plants grow and develop.

To begin, let us remind ourselves briefly how our own bodies develop from fertilized eggs (Chapter 20). Humans and other mammals grow by the repeated division of cells throughout the body. Many cells migrate from one place to another, especially early in development. Embryonic stem cells, formed by the first divisions of the fertilized egg, have the potential to give rise to many different cell types. Cells with this ability are termed **totipotent**. As development proceeds, however, cells lose this totipotency and proceed down a genetically determined developmental pathway. Functionally distinct cell types differentiate as growth proceeds, forming tissues and organs. Once the body is mature, growth essentially stops.

In contrast, vascular plants grow according to entirely different principles, with cell division confined to discrete populations of totipotent cells called **meristems** and growth commonly lasting throughout the life of the plant. A bristlecone pine in western North America may be several thousand years old, but it continues to sprout new branches, new leaves, and new cones every year. The capacity for continued growth and development lies at the heart of how plants meet the challenges they face because they are rooted in one place.

Plants respond to the world around them not by moving about, but by modifying their size and shape. Plants acquire resources by growing to them, and they compete with neighboring plants by growing beyond them. Thus, growth fulfills many of the same roles in plants that behavior does in animals (Chapter 45). And although plants can't move out of the way of danger, they can rebuild after fire or frost damage, or after being struck by a falling tree. Even reproduction requires the continuing formation of new structures.

This chapter focuses on how vascular plants build their bodies. We begin by asking how the shoot system, consisting of leaves, stems, and reproductive structures, is formed. We then turn to the development of root systems, asking how plants are able to grow belowground. Finally, we consider how plants sense the environment and how they respond to what they sense by modifying their development.

Plants measure day length through an interaction between photoreceptors and a gene regulated by an internal clock. page 30-18

Vernalization is a requirement that a plant be exposed to prolonged cold temperatures before flowering can be induced. page 30-18

Dormant seeds must be triggered to germinate by appropriate environmental signals. page 30-18

Some seeds will not germinate unless they have been previously exposed to a prolonged cold period. page 30-18

Phytochrome is a protein photoreceptor that allows plants to determine the amount of red versus far-red light. Because light filtered through leaves has lower levels of red wavelengths, phytochrome allows plants to detect the presence of plants overhead. page 30-20

Plant reproduction is central to agriculture because it allows plant breeders to create new genotypes. page 30-21

**30.4 MANY PLANTS ALSO REPRODUCE ASEXUALLY.**

Vegetative reproduction produces plants that are genetically identical to the parent plant. page 30-22

Vegetative reproduction can occur when a plant spreads horizontally and develops new upright shoots. page 30-22

## Self-Assessment

1. Explain how the evolution of alternation of generations is an adaptation for reproduction on land.

2. Diagram the relationship between the sporophyte and gametophyte generations in bryophytes, ferns, gymnosperms, and angiosperms. Show the relative sizes and physical interactions (if any) of the two generations.

3. Explain how the organs in the different whorls of angiosperm flowers interact to promote fertilization.

4. Contrast the investments that angiosperms and gymnosperms make and the structures that they produce to enhance pollination.

5. Diagram the structure of a mature angiosperm and a mature gymnosperm seed, indicating the ploidy ($1n$, $2n$, $3n$) of each tissue and its role in seed development and germination.

6. Explain why a short-day plant that germinates in the spring will not flower until late summer and why a long-day plant that germinates at the end of summer will not flower until late the following spring.

7. Diagram how the amount of red versus far-red light allows plants to determine if they are in the open or in the forest understory.

8. Describe how vernalization can have an effect in cells that were not formed at the time of the cold treatment.

Do you understand the chapter's Core Concepts? Log in to check your answers to the Self-Assessment questions, then practice what you've learned and reinforce this chapter's concepts by working through the problems and multimedia tutorials provided there.

&#128421; **www.biologyhowlifeworks.com**

this happens, only genetic evidence can be used to determine if a plant is the result of sexual or asexual reproduction.

Vegetative reproduction also occurs in woody plants. For example, when a redwood tree falls over, new upright stems can form along the now-horizontal trunk. Perhaps the most impressive example of the ability to spread vegetatively is the quaking aspen (*Populus tremuloides*), a common tree of the Rocky Mountains that produces brilliant yellow foliage in autumn (**Fig. 30.24**). Aspens reproduce vegetatively by producing upright stems from a spreading root system. Although each shoot lives

for less than 200 years, a single individual can persist for tens of thousands of years by continuing to produce new stems. The current record holder is an aspen clone in Utah, consisting of nearly 50,000 stems, that is estimated to be 80,000 years old.

Plants are able to reproduce vegetatively because of how they build their bodies. As we will see in the next chapter, growth and development occur throughout a plant's life, allowing plants to produce new upright "individuals" from roots or horizontal stems.

## Core Concepts Summary

**30.1 THE PLANT LIFE CYCLE EVOLVED IN WAYS THAT ENHANCE THE ABILITY TO UNITE GAMETES AND DISPERSE OFFSPRING ON LAND.**

The life cycle of the algal sister groups of land plants has one multicellular haploid generation. page 30-2

The life cycle of land plants is characterized by an alternation of haploid and diploid generations, one generation specialized for fertilization and the other for dispersal. page 30-3

Gametes are haploid cells produced by mitosis from cells in the multicellular haploid generation, which is referred to as the gametophyte. page 30-3

Spores are haploid cells produced by meiosis from the multicellular diploid generation, which is referred to as the sporophyte generation. page 30-3

Bryophytes have a free-living haploid gametophyte that makes gametes and a dependent diploid sporophyte that makes spores that are released into the air. page 30-4

In spore-dispersing vascular plants, both the gametophyte generation and the sporophyte generation can survive on their own. page 30-4

Bryophytes and spore-dispersing vascular plants rely for fertilization on swimming sperm released into the environment. page 30-4

A trend in the evolution of plants is the progressive reduction in size and independence of the gametophyte and the increasing role of the sporophyte. page 30-7

Pollen is the male gametophyte that develops within the spore wall, and so can be transported through the air. page 30-7

The female gametophyte of seed plants is retained and nourished by the sporophyte. The female gametophyte, together with the surrounding layers of protective tissues,

is called an ovule. Ovules, when fertilized, develop into seeds. page 30-7

Pollination is the transport of pollen to the ovule, where fertilization takes place. page 30-7

Seeds contain the diploid embryo, a seed coat, and, in gymnosperms, nourishment in the form of the female gametophyte. page 30-8

**30.2 ANGIOSPERMS (FLOWERING PLANTS) ATTRACT AND REWARD ANIMAL POLLINATORS, AND THEY PROVIDE RESOURCES FOR SEEDS ONLY AFTER FERTILIZATION.**

The pollen and ovules of gymnosperms are produced in separate structures, whereas the pollen and ovules of angiosperms often form in one structure, the flower. page 30-10

Flowers consist of four whorls of organs: ovule-bearing carpels, pollen-producing staments, petals, and sepals. page 30-10

Many flowers attract animals because they provide a reward, such as food, shelter, or chemicals. Animals, in turn, transfer pollen. page 30-11

Flowering plants have genetic and structural mechanisms to prevent self-fertilization. page 30-12

Double fertilization, which is unique to angiosperms, is the formation of a diploid zygote and triploid endosperm. The endosperm nourishes the embryo that develops from the zygote. page 30-15

The fertilization of an ovule triggers the development of the ovary wall into a fruit, a structure that enhances seed dispersal. page 30-15

**30.3 PLANTS HAVE SENSORY SYSTEMS THAT CONTROL THE TIMING OF FLOWERING AND SEED GERMINATION.**

Many plants are photoperiodic, meaning that their flowering is controlled by the amount of daylight. page 30-17

**FIG. 31.3  The zones of cell division, elongation, and maturation.**
The photo to the right shows a longitudinal section of the shoot apical meristem of coleus (*Solenostemon scutellariodes*).

Shoot apical meristem

Elongating internodes

Fully expanded internodes

Time

Leaf primordium

Zone of cell division

Meristem cells appear dark in this section because the nuclei (stained pink) fill much of these actively dividing cells.

Zone of cell elongation

Growth

Within the zone of elongation, cells increase many times in length, and the vacuole comes to occupy over 90% of the interior volume of the cell.

Zone of cell maturation

Cell wall
Cytoplasm
Vacuole
Nucleus

to elongate stems. The shoot apical meristem also initiates leaves and produces new meristems, which allow plants to branch.

An important characteristic of the shoot meristem is that it maintains a constant size even though it is a region of active cell division. As new cells are added near the shoot tip, cells that are farther from the shoot tip cease to divide. This means that the cells of the meristem are constantly being lost and replaced. Cells near the shoot tip maintain their totipotency by the expression of **meristem identity genes,** which contribute to meristem stability and function. The expression of meristem identity genes is controlled by a network of chemical signals produced by cells at the very tip of the stem. Because these signals can be transmitted only over a limited distance, only the cells close to the shoot tip express meristem identity genes.

→  **Quick Check 1**  What are three functions of the shoot apical meristem?

**Stem elongation occurs primarily in a zone just below the apical meristem where new cells elongate.**
In animals, cell division and cell enlargement typically go hand in hand. In plants, most of the increase in cell size occurs after mitotic cell division is complete. Cells that are too far from the shoot tip to express meristem identity genes cease to divide, yet they continue to grow. This results in a zone of cell elongation located just beneath the shoot apical meristem (**Fig. 31.3**). It is here that most of the elongation of stems occurs. If you imagine all the cells in the stem of a 100-m redwood tree at the size they were when they left the zone of cell division—the tree would be less than 5 m tall.

**b.**
Developing leaves (leaf primordia)

200 µm

Shoot apical meristem

**FIG. 31.4** Diversity of leaf arrangements.

Alternate      Opposite      Whorled

In the zone of cell elongation, each cell grows many times more in length than in width. The reason is that the strong cellulose microfibrils wrapped around the cell make it difficult for the cell to expand in girth. The large central vacuole that characterizes mature plant cells forms when the cell is in the zone of cell elongation. In fact, most of the increase in cell volume is due to the uptake of water and solutes that fill the vacuole. This explains in part why plant growth is markedly reduced during periods of drought.

As we move even farther from the shoot tip, cells reach their final size and complete their differentiation into the mature cell types in leaves and stems. This organization into successive zones of cell division, cell elongation, and cell maturation allows stems to grow without any predetermined limit to their length. It also means that we can follow the time course of development by moving along the stem from the tip toward the base.

Once cells have matured, they no longer expand. Thus, during the upward growth of the shoot, it is the production and elongation of cells at the shoot tip that lift the meristem ever higher into the air. Imagine that, as a 10-year-old, you carved your initials into a tree. Twenty years later, your initials will be exactly the same distance from the ground as they were on the day you inscribed them. The letters, however, will be wider—a phenomenon we discuss later in this chapter.

## The shoot apical meristem controls the production and arrangement of leaves.

In most plants, leaves are the principal sites of photosynthesis. Because light is needed for photosynthesis, the arrangement of

the leaves along a stem has a major impact on their function. Each species has a characteristic number of leaves attached at each node along the stem. Some species have only a single leaf at each node, whereas others have two or more. How the leaves are positioned around the stem varies in a predictable fashion (**Fig. 31.4**). The regular placement of leaves reduces the shading of one leaf by another and thus enhances the ability of plants to obtain sunlight.

Leaves begin as small bumps, the leaf primordia, which form on the sides of the shoot apical meristem (see Fig. 31.2). The regular arrangement of leaves around the stem is controlled by the fact that each successive leaf primordium is located as far away as possible from all previously formed primordia. One hypothesis proposed to explain this developmental pattern is that the diffusion of chemical signals from developing leaf primordia creates regions where the growth of new primordia is inhibited. The result is an arrangement that prevents leaves from being produced one on top of another.

As noted previously, the earliest vascular plants were simple branching stems, and photosynthesis took place along the length of the shoot. As evolution proceeded, however, some branch systems became flattened, their axes growing in a plane that facilitated the capture of light. These planar branches lost the capacity for continued growth. By about 380 million years ago, the planar branches became modified into structures recognizable as leaves (**Fig. 31.5**).

The evolution of leaves required three developmental changes. First, the genetic program to produce a three-dimensional stem became modified to form flattened organs. Second, apical meristem identity genes were down-regulated and leaf identity genes up-regulated, resulting in a specialized organ incapable of continuing growth. And third, new meristems evolved, enabling leaves to expand into flattened photosynthetic structures that capture sunlight. In fern leaves, these meristems are located along the leaf margin. In pine needles, they occur at the base of each needle. In flowering plants, meristematic cells can be distributed throughout the developing leaf, making possible a diversity of leaf shape. In contrast to cells in the shoot apical meristems which can divide continuously throughout the lifetime of a plant, leaf meristematic cells divide only for a relatively short period of time. This explains why leaves grow to a final size.

When we think of leaves, it is the green photosynthetic ones that first come to mind. However, many plants produce leaves that are specialized for functions other than photosynthesis, including climbing, trapping insects, and attracting pollinators (**Fig. 31.6**). Plants that overwinter produce **bud scales** that protect shoot apical meristems from desiccation and damage

**FIG. 31.5**  **The evolution of leaves.** Fossils show that leaves evolved from branches that (1) lost the ability for continued growth, (2) developed in a plane, and (3) expanded the photosynthetic surface to fill the area between veins. The photo shows *Archaeopteris*, one of the earliest plants with flattened leaves.

Early vascular plants lacked leaves. Their shoots consisted of photosynthetic stems.

In some of these early vascular plants, side branches lost their apical meristems and became flattened into a single plane.

The development of new meristems allowed tissues to form between these side branches, leading to the development of leaves.

**FIG. 31.6**  Non-photosynthetic functions of leaves: (a) protection (bud scales); (b) climbing (tendrils); (c) trapping insects (spines); and (d) attracting pollinators (color).

Bud scales

Colored leaves

Flower

due to cold. Bud scales may not look like leaves, but they form from leaf primordia and are arranged in the same way around the stem as the green leaves produced in spring.

## The development of new apical meristems allows stems to branch.

Vascular plants evolved the ability to branch even before they evolved either roots or leaves. Branching was important to these first plants because it allowed them to produce more sporangia. Branching allows present-day plants to support greater numbers of both reproductive structures and leaves.

In lycophytes and in ferns and horsetails, branching occurs when the shoot apical meristem divides in two, giving rise to two stems. In seed plants, branches grow out from **axillary buds** (also

**FIG. 31.7** **Axillary buds.** In seed plants, branching results from axillary buds that are produced at nodes.

**FIG. 31.8** **Flower development.** The identity of floral organs is specified by the spatial expression of three classes of homeotic genes called A, B, and C.

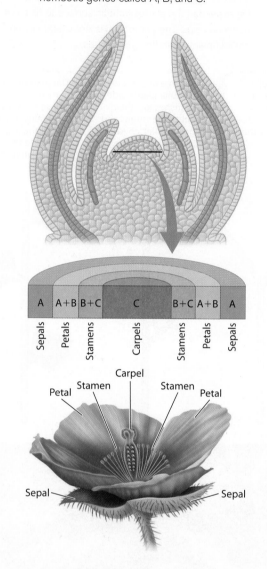

called lateral buds), which are meristems that form at the base of each leaf (**Fig. 31.7**). Axillary buds have the same structure and developmental potential as the apical meristem and express the same meristem identity genes. However, the axillary buds remain dormant until triggered to grow, remaining attached to the stem even after leaves are shed. Thus, axillary buds provide seed plants with many points along their stem where new branches can form.

**Flowers grow from and consume shoot meristems.**
Just as leaves and branches grow from meristems, so, too, do flowers. Flowers can be produced at the tip of a plant or shoot as a result of the conversion of the apical meristem into a floral meristem, or at the base of leaves as the products of axillary buds. As with leaves, floral meristems lose their capacity for continued growth. All the cells differentiate entirely during flower development.

As we saw in Chapter 30, flowering is triggered by florigen, a protein produced in leaves and transported through the phloem to apical meristems and axillary buds. Florigen triggers the transition to floral meristems by initiating the down-regulation of meristem identity genes and the up-regulation of genes that govern floral identity. The arrangement of primordia is altered to produce four whorls that give rise to the sepals, petals, stamens, and carpels. Once the meristem is launched along the trajectory for flower development, the identity of each whorl is controlled by the expression of homeotic genes (Chapter 20). Three classes of genes, referred to as A, B, and C, are expressed in overlapping rings around the meristem and serve as master controllers for the development of specific floral organs (**Fig. 31.8**).

## 31.2 PLANT HORMONES

A fundamental question in the development of multicellular organisms is what controls the differentiation of individual cells. In considering the formation of plant stems, we can ask: (1) What controls the arrangement of leaf primordia and the elongation of internodes? (2) What ensures that the differentiation of xylem and phloem cells produces a continuous vascular system extending from the roots to the leaves? (3) What determines whether an axillary bud remains dormant or develops into a new stem?

The answer, in each case, is that chemical signals guide the growth and differentiation of plant cells. Earlier, we discussed basic principles of cell signaling and communication (Chapter 9). Chemical signals play a central role in plant development because cell fate is determined by position. Some of the chemical signals important in plant development have highly localized effects. For example, signals that cause cells to express meristem identity genes are restricted to the region close to the shoot tip. Other signaling molecules, or their chemical precursors, enter the vascular system and are transported across the entire length of the plant. Because plants grow and develop in many places at once, signaling molecules that can move from one part of the plant to another play an important role in coordinating growth and development at the whole plant level.

## Hormones affect the growth and differentiation of plant cells.

**Hormones** are chemical signals that influence both physiology and development. They play a central role in establishing and maintaining basic patterns of plant development, as well as coordinating growth in different parts of the plant in response to both internal and external environmental factors. A key feature of hormones is that they are effective in extremely low concentrations. In plants, hormones can be synthesized throughout the plant body, although they are most often produced in meristems. In addition, plant hormones can affect both the cells in which they are produced as well as cells located in distant regions of the plant.

Plant hormones influence cell growth and differentiation by altering patterns of gene expression as well as rates of cell division and cell expansion. How a cell responds to a particular hormone depends both upon its developmental stage and its position within the plant. In addition, the effect of any one hormone is often affected by the presence of one or more other hormones. This observation explains how a small number of hormones is able to shape the ongoing development and growth of plants.

**TABLE 31.1    THE FIVE MAJOR HORMONES**

| HORMONE | SYNTHESIS AND TRANSPORT | EFFECTS |
|---|---|---|
| Auxin<br><br>Indole-3-acetic acid | Synthesized primarily in shoot apical meristems and young leaves.<br>Transported by polar transport and through phloem. | Induces cell wall extensibility.<br>Inhibits outgrowth of axillary meristems.<br>Stimulates formation of root meristems.<br>Stimulates differentiation of procambial cells.<br>Involved in phototropism and gravitropism.<br>Stimulates ethylene synthesis. |
| Gibberellic acid<br> | Synthesized in the growing regions of both roots and shoots. | Stimulates stem elongation and cell division.<br>Mobilizes seed resources for developing embryo.<br>Stimulates cytokinin synthesis. |
| Cytokinins<br><br>Zeatin | Synthesized in both root and shoot meristems.<br>Transported in both xylem and phloem. | Stimulates cell division.<br>Delays leaf senescence.<br>Acts synergistically with auxin. |
| Ethylene<br> | Synthesized in elongating regions of roots and stems, as well as in developing fruits.<br>Gaseous; precursors can be transported in xylem. | Reduces cell elongation.<br>Triggers fruit ripening.<br>Formation of air spaces in roots. |
| Abscisic acid<br> | Synthesized in root cap, developing seeds, and leaves.<br>Transported from roots to leaves in xylem. | Maintains seed dormancy.<br>Stimulates root elongation.<br>Triggers closing of stomata.<br>Antagonistic with ethylene. |

Traditionally, five major plant hormones have been recognized: auxin, gibberellic acid, cytokinins, ethylene, and abscisic acid (**Table 31.1**). One characteristic of these hormones is that each triggers a wide range of developmental effects. For example, ethylene can lead to fruit ripening (Chapter 30) and also slows the elongation of roots and stems. Several new chemicals with hormone-like properties have been recently discovered. Some of these have broad effects on plant development (e.g., strigolactone), whereas others appear to be more specific (e.g., florigen).

To illustrate how hormones influence plant growth and development, we consider three aspects of shoot development: the differentiation of vascular tissues, the elongation of internodes, and the formation of lateral branches.

### Auxin transport guides the development of vascular connections between leaves and stems.

During the initial growth of leaf primordia, the resources needed can be met by diffusion. As the young leaves increase in size, their demand for water and carbohydrates quickly exceeds what can be supplied by diffusion. If they are to grow to even a fraction of their full size, leaves must establish vascular connections with the xylem and phloem in the stem.

The plant hormone **auxin** was first identified by its ability to cause shoots to elongate. However, auxin has an even more remarkable property: It is transported preferentially from young leaves to the differentiated vascular tissues in the stem. In this way, the movement of auxin creates a template that guides vascular differentiation.

Auxin is synthesized and secreted by cells in young, expanding leaves. In the cell wall, auxin carries no net charge and thus can enter surrounding cells by diffusing across the plasma membrane. However, because the pH of the cytoplasm is higher than the pH of the cell wall, auxin loses a proton. As a result, auxin has a net negative charge when inside the cell, preventing it from moving easily across the plasma membrane. For auxin to leave the cell, a specific plasma membrane transport protein known as PIN is required. By restricting the placement of PIN proteins to only one face of the cell, plants can control the direction of auxin movement. Each cell in a plant's body has an apical end and a basal end that are maintained independent of gravity or orientation. PIN proteins typically are located only on the basal side of cells, the side farthest away from the shoot tip. As a result, the prevailing direction of auxin movement is in the apical-to-basal (top-to-bottom) direction. The coordinated movement of auxin across many cells is referred to as **polar transport** (**Fig. 31.9**).

---

**FIG. 31.9** **Polar transport of auxin.** The polar transport of auxin results in the movement of this hormone from the tip of each developing leaf to the base of the plant.

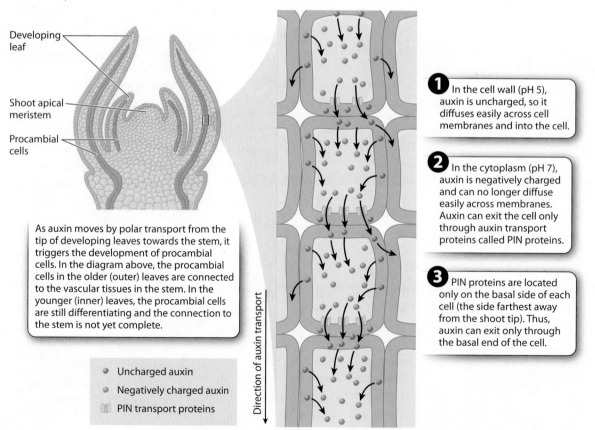

Developing leaf

Shoot apical meristem

Procambial cells

As auxin moves by polar transport from the tip of developing leaves towards the stem, it triggers the development of procambial cells. In the diagram above, the procambial cells in the older (outer) leaves are connected to the vascular tissues in the stem. In the younger (inner) leaves, the procambial cells are still differentiating and the connection to the stem is not yet complete.

Direction of auxin transport

- Uncharged auxin
- Negatively charged auxin
- PIN transport proteins

**1** In the cell wall (pH 5), auxin is uncharged, so it diffuses easily across cell membranes and into the cell.

**2** In the cytoplasm (pH 7), auxin is negatively charged and can no longer diffuse easily across membranes. Auxin can exit the cell only through auxin transport proteins called PIN proteins.

**3** PIN proteins are located only on the basal side of each cell (the side farthest away from the shoot tip). Thus, auxin can exit only through the basal end of the cell.

Auxin affects cells both by stimulating their elongation and increasing their production of PIN proteins. Thus, auxin transport results in linear rows of elongated cells as outflow from the basal side of one cell leads to enhanced auxin concentration in the cell immediately below. The elongated cells that carry auxin from the developing leaves to the stem differentiate to become **procambial cells.** Procambial cells retain the capacity for cell division and ultimately give rise to both xylem and phloem. The common origin of xylem and phloem from procambial cells explains why these two tissues are always found in close association throughout the length of the plant.

As auxin moves by polar transport out of developing leaves it is drawn towards the vascular tissues in the stem. The reason this happens is that auxin moves *between* cells by diffusion. Recall from Chapter 29 that molecules diffuse from regions of higher concentration to regions of lower concentration. The vascular tissues within the stem create a region of low auxin concentration because they transport auxin towards the roots faster than it can move through undifferentiated cells. In this way, the polar transport of auxin that originates in developing leaf primordia is responsible for establishing the vascular connections with the stem that allow leaves to grow and function.

The vascular system of an elongating stem is built from the successive addition of procambial strands that initiate in the developing leaves. The simplest arrangement, which is characteristic of the lycophytes, is a single cylinder of vascular tissues, with xylem at the center and phloem on the outside. The procambial strands that connect each leaf to the stem merge with this central cylinder.

In seed plants, the most common configuration is a ring of **vascular bundles** near the outside of the stem. The bundles have xylem conduits toward the center of the stem and phloem nearer the outside of the stem (**Fig. 31.10**). Typically, each leaf develops vascular connections with several of the stem's vascular bundles. The region between the epidermis and the vascular bundles is the **cortex,** and the region inside the ring of vascular bundles is the **pith.** The placement of the vascular bundles, which contain the stiff xylem conduits, near the outside of the stem increases the stem's ability to support its leaves without bending over.

**❓ CASE 6 Agriculture: Feeding a Growing Population
What is the developmental basis for the shorter stems of high-yielding rice and wheat?**
In natural ecosystems, competition for sunlight creates a strong selective advantage for growing tall. Compare the slender vine in Fig. 31.1, which grows in a dense forest, with the compact barrel cactus that grows where there is little danger of being shaded by a neighboring plant. Whereas the ancestors of today's crops competed for sunlight by growing taller, a key element of the high-yielding crop varieties developed during the Green Revolution is their shorter stems. These shorter-stemmed plants can support the much-larger seed mass induced by high rates of

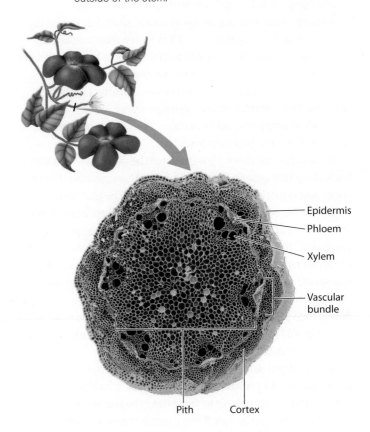

**FIG. 31.10    Arrangement of vascular bundles.** In most seed plants, the vascular bundles are arranged in a ring near the outside of the stem.

Epidermis
Phloem
Xylem
Vascular bundle
Pith        Cortex

fertilizer application (Chapter 29). Moreover, by investing less in stems, these plants can allocate a greater percentage of their resources to reproduction.

At the time that the semidwarf varieties were developed, the underlying basis for their reduced stem growth was not known. What happens inside plants of these high-yielding crop varieties to make the stems shorter? Understanding how plants control intermodal elongation provides some insight.

The extent to which stems elongate can vary even within a single plant. For example, a tree in the temperate zone produces widely spaced green leaves during spring, but as autumn approaches the internodes cease elongating. As a result, bud scales become tightly packed together. This dramatic change in internode elongation is controlled by hormones.

**Gibberellic acid** is a plant hormone that stimulates the elongation of stems. It was first identified as a chemical produced by fungal pathogens that caused the stems of young rice plants to grow to many times their normal length. Giberellic acid increases internode elongation by reducing the force needed to cause cell walls to expand.

The genes responsible for reduced stem growth have now been identified in both wheat and rice. The semidwarf wheats have reduced sensitivity to gibberellic acid, and the shorter rice varieties are deficient in the biosynthetic enzymes needed to produce this hormone. In both cases, the result is shorter but sturdier plants.

### Branching is affected by multiple hormones.

Branching allows plants to support more leaves, but producing branches too close together results in leaves that overlap and shade one another. What causes axillary buds to break dormancy and begin to grow? The answer to this question demonstrates how the interplay of different hormones can affect growth. In this case, the interplay of hormones produced in distinct parts of the plant communicates to axillary buds when conditions are right for the growth of new branches.

Removal of the shoot tip stimulates the outgrowth of axillary buds along that branch or stem (**Fig. 31.11**), suggesting that the shoot apical meristem somehow suppresses the growth of axillary buds. This suppression is referred to as **apical dominance.** If the shoot tip is cut off and a paste containing auxin is applied to the cut end of the stem, the axillary buds remain dormant. If instead the shoot tip is treated with a substance that inhibits auxin transport, branching is induced. Thus, branching is suppressed by the movement of auxin from the shoot tip toward the root. Apical dominance prevents branches from being formed too close to the shoot tip  and also allows new branches to form should the shoot meristem be damaged.

Auxin suppresses the growth of axillary buds by inhibiting the synthesis of **cytokinins,** hormones that stimulate the outgrowth of axillary buds by increasing the rate of cell division (Table 31.1). When the apical meristem is removed, cytokinin synthesis increases in the stem nodes, triggering the growth of axillary buds. Recent studies indicate that this top-down control is complemented by the action of another hormone, **strigolactone,** which is made in roots and transported upward in the xylem. Plants that are unable to synthesize strigolactone produce many more branches than do wild-type plants. This indicates that strigolactone inhibits the outgrowth of axillary buds.

Clearly, the development of axillary buds is affected by multiple hormones, although the details of how they interact and under what conditions they are synthesized in greatest amounts are not yet known. One hypothesis is that strigalactones are transported to the shoot when the root system is unable to supply the water and nutrients required by new branches.

## 31.3 SECONDARY GROWTH

Shoot apical meristems enable plants to grow in length and to branch. But a plant unable to grow in diameter would not become very tall before buckling under its own weight. And a plant unable to add vascular tissue could not supply its ever-increasing number of leaves with the water needed for photosynthesis. Growth in diameter thus serves two functions: It strengthens the stem, and it increases the capacity of the vascular system. Both are essential for supporting and supplying large numbers of leaves.

**Primary growth** is the increase in length made possible by apical meristems. An increase in diameter is referred to as **secondary growth** because it results from a new type of meristem, one that forms only after elongation below the growing tip is complete. The fossil record shows that the ability to grow in diameter evolved millions of years after primary growth, and it did so independently in several lineages, testimony to the benefits of being able to compete for sunlight by growing tall. Today, secondary growth occurs almost exclusively in seed plants, although this was not always the case in Earth history.

### Shoots produce two types of lateral meristem.

**Lateral meristems** are the source of the new cells that allow plants to grow in diameter. Lateral meristems are similar to shoot apical meristems in their ability to produce new cells that grow and differentiate. Lateral meristems, however, differ from their apical counterparts in several key ways. First, unlike shoot apical meristems, which are located at the shoot tip, lateral meristems form cylinders or, more properly, cones of cells that extend the entire length of the stem. Second, because lateral meristems form only after elongation is complete, the new cells they produce grow in diameter but not in length. Finally, as a stem becomes thicker, the number of meristem cells needed to encircle the stem also increases. Thus, lateral meristems become larger over time.

Plants produce two distinct lateral meristems that together result in secondary growth. One of these, the **vascular cambium** (plural, cambia), is the source of new xylem and

---

**FIG. 31.11  Apical dominance.**

Shoot apical meristem intact          Shoot apical meristem cut off

The growth of axillary buds is suppressed when the shoot apical meristem is intact.

When the shoot apical meristem is removed, growth of axillary buds along that stem is stimulated.

phloem. Secondary xylem is familiar to you: It's wood. Plants with secondary growth are often described as woody, whereas plants that lack secondary growth and so are produced entirely by apical meristems are described as herbaceous. In a large tree, most of the plant's mass consists of wood, generated from the vascular cambium.

As the activities of the vascular cambium increase the plant's diameter, epidermis formed during primary growth eventually ruptures. Thus, plants with secondary growth have a second lateral meristem, the **cork cambium,** which renews and maintains an outer layer that protects the stem against herbivores, mechanical damage, desiccation, and, in some species, fire. These tissues (along with secondary phloem) are commonly known as bark. The word "cambium" derives from the Latin word *cambiare,* meaning "to change." As we will see, the development of lateral meristems fundamentally changes the character of stems.

### The vascular cambium produces secondary xylem and phloem.

The vascular cambium forms at the interface between xylem and phloem. In stems, the vascular cambium derives from procambial cells within each vascular bundle and parenchyma cells in between the bundles. In forming vascular cambium, both types of cell take on a new identity, becoming meristem cells.

The vascular cambium produces new cells that differentiate on both its sides. Those to the inside of the vascular cambium become **secondary xylem,** and those on the outside develop as **secondary phloem (Fig. 31.12).** When you peel the bark off a stem, it separates from the wood at the vascular cambium. This is because the undifferentiated and actively dividing cells of the vascular cambium are weak and easily pulled apart. The vascular cambium forms a continuous layer, one to several cells thick, which surrounds the xylem and runs nearly the entire length of the plant.

The high rates of water loss by leaves associated with their uptake of $CO_2$ for photosynthesis (Chapter 29) means that plants must produce more secondary xylem than secondary phloem. As the vascular cambium forms new secondary xylem cells along its inner face, the newly formed cells push the vascular cambium outward, much the way the elongation of newly formed cells pushes the shoot apical meristem upward. The vascular cambium accommodates this outward movement by adding new cells that increase its circumference. In contrast, the secondary phloem, once formed, is unable to produce any additional cells. As the production of secondary xylem pushes the vascular cambium outward, the phloem becomes stretched and, eventually, damaged. Consequently, new secondary phloem must be continually produced.

Over time, xylem conduits lose their ability to transport water. Thus, in long-lived trees, the functional xylem is located in a **sapwood** layer adjacent to the vascular cambium, while

**FIG. 31.12 Lateral meristems.** The vascular cambium is the source of new xylem and phloem, while the cork cambium maintains an outer layer of bark.

| Pith | Vascular cambium | Cortex |
| Primary xylem | Secondary phloem | Cork cambium |
| Secondary xylem | Primary phloem | Cork |

Secondary growth involves the formation of a continuous vascular cambium that produces secondary xylem toward the center of the stem and secondary phloem toward the outside. The cork cambium maintains a protective outer layer.

the **heartwood,** in the center of the stem, does not conduct water. Heartwood is often darker in color than sapwood because it contains large amounts of chemicals such as resins. For this reason, heartwood is typically more resistant to decay than sapwood.

In many plants in seasonal climates, decreases in the size of secondary xylem cells at the end of the growing season result in **growth rings,** which make it possible to determine a tree's age. Growth rings also provide a window into the past because by measuring the width of each ring, it is possible to determine

**FIG. 31.13** **Annual growth.** In seasonal climates, tree rings provide a record of the annual accumulation of wood. The tree rings shown are from red pine (*Pinus resinosa*).

Wide growth rings indicate higher growth rates and thus suggest that the tree experienced more favorable conditions in recent years.

Narrow growth rings indicate that this tree grew slowly when it was young, perhaps because it was shaded by another tree.

how much wood was laid down in a given year (**Fig. 31.13**). Wood production is higher when the environmental conditions that enable photosynthesis are more favorable. Thus, annual growth rings provide a record of past climatic conditions such as temperature and moisture, as well as events such as fire.

### The cork cambium produces an outer protective layer.

The role of the cork cambium is to maintain a protective layer around a stem that is actively increasing in diameter (see Fig. 31.12). The cork cambium forms initially from cortex cells that dedifferentiate—that is, they regress to their earlier state—to become meristem cells. However, with time, this layer of actively dividing cells becomes increasingly cut off from its source of carbohydrates—the phloem. When the cork cambium becomes sufficiently cut off from its carbohydrate supply, a new cork cambium forms within the secondary phloem. Thus, each cork cambium is relatively short lived, and in many cases, new cork cambia form in patches rather than as continuous layers. As a result, the bark of many trees has a patchy and fissured appearance (**Fig. 31.14**).

The cells produced by the cork cambium are called cork. As cork cells mature, they become coated with **suberin,** a waxy compound that protects against mechanical damage and the entry of pathogens, while also forming a barrier to water loss. However, this layer of waxy, nonliving cells impedes the inward diffusion of oxygen needed to meet the respiratory demands of living cells. The outer bark cells are less tightly packed in regions called **lenticels,** allowing oxygen to diffuse into the stem (**Fig. 31.15**).

→ **Quick Check 2** Why do plants have two types of lateral meristem?

### Wood has both mechanical and transport functions.

We have seen that as a tree grows, its secondary xylem fulfills the two roles of secondary growth. The first is to provide trees with the strength and stability to stand upright and support large crowns. The second is to transport water and nutrients from the roots to the leaves. In most gymnosperms, a single cell type fulfills both functions. Cells on the xylem side of the vascular cambium differentiate to form tracheids (**Fig. 31.16a**). As discussed in Chapter 29, tracheids are single-celled xylem conduits that have thick lignified cell walls and are nonliving and empty at maturity. The wood of a pine tree, for example, is relatively homogeneous in structure, consisting almost entirely of tracheids. This type of wood does not vary dramatically from species to species.

Angiosperms, by contrast, exhibit a great diversity in wood structure, as the cells mature into several distinct types that separate the functions of water conduction and mechanical support (**Fig. 31.16b**). **Fibers** are narrow cells with extremely thick walls and almost no lumen; their primary function is support. In contrast, **vessel elements** can be extremely wide cells; their primary function is water transport (Chapter 29). The largest vessel elements are close to 0.5 mm in diameter and thus visible to the naked eye. Vessel elements can be much wider than tracheids because the fibers provide support. And because vessel elements develop open connections that link them end to end

**FIG. 31.14** **Outside (a) and inside (b and c) of the outer bark of a pine tree, showing how successive cork cambia are produced.** The growth of many, discontinuous cork cambia results in the patchy bark of pine trees.

**a.** Outside of tree, ponderosa pine (*Pinus ponderosa*)

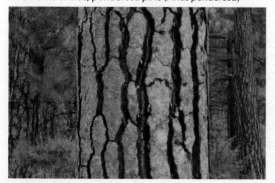

**b.** Cross-section of bark, white pine (*Pinus strobus*)

**c.**

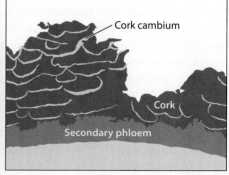

Cork cambium

Cork

Secondary phloem

**FIG. 31.15 Lenticels in (a) water birch (*Betula occidentalis*) and (b) basswood (*Tilia americana*).** Lenticels allow oxygen to diffuse through the outer bark to supply the respiratory needs of the cells within the stem.

with other vessel elements, creating multicellular xylem vessels, they allow angiosperms to produce wood that is more efficient at transporting water than the wood of gymnosperms.

The ability to form xylem vessels has a number of important consequences. One of these is that angiosperms can have higher rates of water flow through their stems than is possible with wood that contains only tracheids. As a result, angiosperms sustain higher rates of $CO_2$ uptake and thus higher growth rates.

In addition, because vessels can be much more efficient than tracheids at transporting water, angiosperms can devote a smaller fraction of their wood to water conduction, and more of their stem volume to other cell types. The wood of some angiosperms contains many fibers, in extreme cases resulting in wood that is so dense that it sinks. Such dense woods are extremely strong, allowing trees to create spreading crowns with outstretched branches. Such "hardwoods" are useful as flooring materials and furniture. The wood of other angiosperms contains a high percentage of living cells instead of fibers, and this wood has a very low density. Balsa trees, whose light wood is used in building model airplanes, are a good example.

## 31.4 ROOT GROWTH AND DEVELOPMENT

The roots that evolved in vascular plants enabled them to obtain water and nutrients from the soil and provided a mechanical anchor for plants capable of growing tall (Chapter 29). As in the case of leaves and wood, the fossil record shows that the first vascular plants lacked roots. These plants would have been limited to extremely wet environments in which they could absorb water through horizontal, and possibly submerged, stems. The evolution of roots, then, allowed vascular plants to expand into drier habitats and ensured the stability of taller stems.

The fossil record also indicates that roots evolved independently in the lycophytes and the lineage leading to the other vascular plants. Nevertheless, the roots of all vascular plants share many features. This suggests that the physical challenges of

**FIG. 31.16 Wood structure in (a) gymnosperms and (b) angiosperms.** These scanning electron micrographs show (a) Douglas fir (*Pseudotsuga menziesii*) and (b) English elm (*Ulmus procera*).

**a.** Gymnosperms

In gymnosperms, tracheids have both a mechanical and a transport function.

Tracheids

**b.** Angiosperms

In angiosperms, fibers provide mechanical support, allowing vessels to be specialized for water transport.

Vessels

Fiber

growing through a dense medium such as soil and the demands of obtaining water and nutrients from that medium place strong constraints on root structure. Here, we explore the growth and development of the plant body belowground, emphasizing how the elongation and branching of roots enables vascular plants to obtain the water and nutrients they need from the soil.

### Roots grow by producing new cells at their tips.

In Chapter 29, we saw that plants produce large numbers of thin roots to obtain sufficient nutrients from the soil. And to obtain water during periods without rain, plants produce roots that penetrate deep into the soil. Growing roots into the soil presents challenges different from those associated with elongating stems into the air. Although the dangers of drying out are much reduced belowground, deeper soil layers can have densities approaching that of concrete. How are the thin and permeable roots needed for water and nutrient uptake able to penetrate and grow through the soil?

At one level, roots grow in the same way as shoots, with new cells produced at their tip. The **root apical meristem** serves as the source of new cells. After cell division, these cells undergo elongation and complete their differentiation only after elongation has ceased (**Fig. 31.17**). Thus, as we move back from the elongating root tip, we encounter successive regions of cell division, elongation, and maturation, just as we saw in stems.

Roots, however, differ from stems in several important ways. One important difference is that roots are much thinner than stems. Plants with thin roots can more efficiently produce the large surface area needed for optimal uptake of water and nutrients from the soil. Thin roots can also more easily bend around soil particles and make use of channels created by invertebrate animals that burrow through the soil. A second major difference is that the root apical meristem is covered by a **root cap** that protects the meristem as it grows through the soil. As the root elongates through the soil, root cap cells are rubbed off or damaged. Thus, the root cap is continually supplied with new cells produced by the root apical meristem. A third difference between roots and stems, and one that can be seen without a microscope, is that the root apical meristem does not produce any lateral organs, whereas the shoot apical meristem makes leaves. Root hairs, the single-celled outgrowths of root epidermal cells that greatly increase the surface area for uptake (Chapter 29), are formed only after elongation has ceased.

Let's now look at how roots develop the specialized cells and tissues that they need to function. A cross section taken through the zone of cell maturation shows a series of concentric tissue layers, starting with the epidermis on the outside, followed by the root cortex, and then the endodermis (Fig. 31.17). Recall that the endodermis plays a key role in the selective uptake of nutrients (Chapter 29). The Casparian strip, a waxy band surrounding each endodermal cell, forces all materials that will enter the xylem to cross the plasma membranes of these cells.

At the center of the root lies a single vascular bundle in which the xylem is arranged as a fluted cylinder, with phloem nestled in each groove. In cross section, the xylem is arranged in a star shape, the number of points of the star differing in different plant species.

As in stems, the hormone auxin plays a central role in specifying which cells differentiate into xylem and phloem. Auxin produced by developing leaves is transported downward through the mature stem and into the roots by the phloem. At the end of the differentiated phloem, auxin moves by polar transport from cell to cell until it reaches the root apical meristem and the root cap. As auxin moves through the elongation zone, it triggers the formation of procambial cells that subsequently differentiate into phloem and xylem.

In this way, the production of auxin by developing leaves and the

---

**FIG. 31.17** The structure of a typical root, including the root apical meristem and root cap.

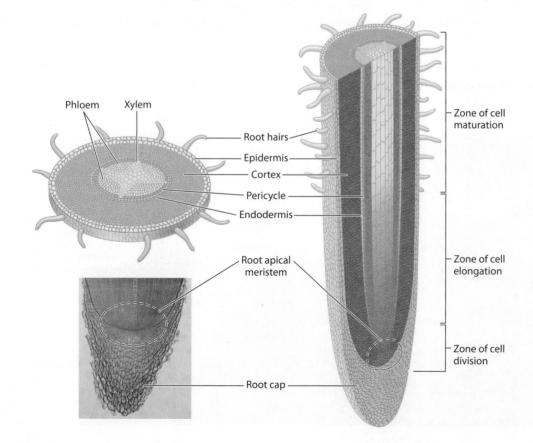

Phloem   Xylem

Root hairs
Epidermis
Cortex
Pericycle
Endodermis

Zone of cell maturation

Root apical meristem

Zone of cell elongation

Zone of cell division

Root cap

**FIG. 31.18 Root pericycle.** New root meristems develop from the pericycle and the developing lateral root grows through the root cortex before reaching the soil. The light micrographs show willow (*Salix*) roots.

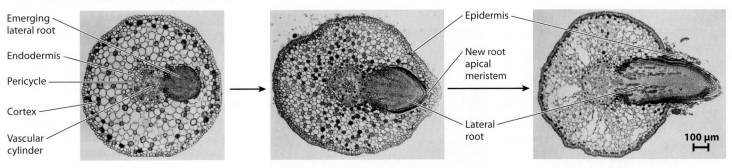

polar transport of auxin from cell to cell ensures that vascular continuity is established between roots and leaves. Phloem extends closer to the root tip than do fully lignified xylem conduits. This reflects the fact that the apical meristem can obtain water and nutrients directly from the soil but depends entirely on the leaves for carbohydrates needed for growth.

So far, we have described only the primary growth of roots. However, roots can undergo secondary growth as well. This allows roots to increase in diameter and to add new xylem and phloem. As in stems, the vascular cambium forms at the interface between the xylem and the phloem. Indeed, the vascular cambium must be continuous from the shoot to the roots to ensure that the secondary xylem and phloem provide an unbroken connection between the roots and the leaves. In older roots with extensive secondary growth, a cork cambium is also formed. As in stems, the cork cambium renews the outer layers of cells that protect the living tissues from damage and predation.

### The formation of new root apical meristems allows roots to branch.

Compared to stems, roots branch like crazy. Even a plant that has only a single stem aboveground will have many tens of thousands of elongating root tips. A major reason that roots branch so much is that the uptake of water and nutrients is concentrated in a zone near the root tip where root hairs are most abundant. Thus, branching is necessary to create enough root surface area to supply the water and nutrient needs of the leaves.

For a root to branch, a new root apical meristem must be formed. In roots, new meristems develop from the **pericycle,** a single layer of cells just to the inside of the endodermis (**Fig. 31.18**). The pericycle is immediately adjacent to the vascular bundle, so the new root is connected to the vascular system right from the beginning. But because the new root meristem develops internally, the new root must grow through the endodermis, cortex, and epidermis before reaching the soil.

One consequence of initiating new root meristems from the pericycle is that new roots can form anywhere along the root. As a result, plants have tremendous flexibility in the development

of their root systems and roots can proliferate locally in response to nutrient abundance. For example, nitrate ($NO_3^-$), the most common form of nitrogen in soil, triggers the formation of lateral roots by causing auxin to accumulate locally within the root tissues nearest to the nitrate-rich soil. Only a relatively small number of a plant's roots, however, continue to elongate and branch indefinitely. Most new roots are active for only a short time, after which they die and are shed. This turnover of fine roots allows the root system to respond dynamically to changes in the availability of water and nutrients but represents a major expenditure of carbon and energy for the plant.

Up to this point, we have described how new roots form from existing roots. New roots can also be produced by stems. This is important in plants that lack secondary growth, many of which produce horizontal stems that grow on or beneath the soil surface, for example many lycophytes and ferns. In these plants, new roots form at each successive node, so the uptake capacity of the root system is directly coupled to the elongation of the stem.

When the stem of a plant is severed—and if the shoot tip is not allowed to dehydrate—new roots often form at the cut end. The new root meristems are stimulated by auxin, which is produced by the young leaves and accumulates, by polar transport, at the cut end. The ability of cut stems to form new roots allows plants to survive and in some cases proliferate following mechanical damage. It is also the means by which many commercially important plants are propagated.

→ **Quick Check 3** How is root development similar to and how is it different from stem development?

### The structures and functions of root systems are diverse.

In the preceding sections, we focused on the growth and branching of individual roots. How a plant's roots are distributed in the soil has a tremendous impact on the ability of the root system to supply the shoot with water. Some plants produce very deep roots that provide them with access to a groundwater supply that is independent of the variability of rainfall. This can be advantageous, but it comes at the cost of producing and maintaining these

**FIG. 31.19  Root diversity.** (a) Climbing roots of poison ivy; (b) prop roots of a *Pandanus* tree; (c) storage roots of yams; and (d) breathing roots of black mangrove (shown at low tide).

non-photosynthetic structures. Roots can penetrate into mine shafts and caves more than 50 m below the soil surface.

At the other extreme are plants that produce shallow and spreading root systems. For example, barrel cacti produce roots that penetrate less than 10 cm into the soil but extend out several meters from the base of the plant. These roots cast a wide but shallow net to capture as much water as possible from intermittent rain showers.

Some plants produce distinctive roots whose principal role is other than the absorption of water and nutrients from the soil (**Fig. 31.19**). For example, climbing plants such as poison ivy produce roots along their stems that allow them to adhere to the sides of trees (Fig. 31.19a). The conspicuous prop roots produced by some tropical trees provide mechanical stability that allows these plants to grow tall despite their slender stems (Fig. 31.19b).

Many plants produce swollen roots in which resources such as starch can be stored (Fig. 31.19c). Belowground storage allows plants to produce new leaves and shoots quickly following either damage or a period in which drought or cold has caused the shoots to die. Cassava and sweet potato are examples of economically important species with storage roots.

Perhaps the most unusual rooting structures are the "breathing" roots produced by trees that grow in water. For example, black mangroves grow along the seashore in tropical regions and produce pencil-sized roots that extend vertically upward out of the sandy soil (Fig. 31.19d). Breathing roots do not actually breathe; instead, they contain internal air spaces that provide an easier pathway for oxygen to diffuse into the roots than the waterlogged soil. For this reason, breathing roots must extend above the surface of the water. Thus, the breathing roots of black mangrove provide an estimate of the water level at high tide.

## 31.5 THE ENVIRONMENTAL CONTEXT OF GROWTH AND DEVELOPMENT

We take it for granted that shoots grow up and roots grow down. But if you lay a plant on its side, its stems will turn upward and its roots will bend downward. Similarly, if you place a houseplant near a window, the plant will bend over and grow toward the light. Plants are able to modify their growth in response to both gravity and light. They can sense and respond to the world around them in other ways as well. For example, trees planted close together produce stems that are taller and thinner than those of trees planted far apart. Is this simply what happens when crowded plants compete for the resources that they need to grow? No. Experiments have shown that the taller, thinner stems result from individual plants' sensing the presence of their neighbors.

Plants gain information about the world around them in three principal ways. Photoreceptors, discussed in Chapter 30, provide information about the availability of light needed to drive photosynthesis. Receptors triggered by mechanical forces provide plants with information about physical influences such as gravity and wind. Chemical receptors allow plants to detect and respond to the presence of specific chemicals, as well as chemical gradients. When one of these sensory receptors is triggered, it sets in motion a chain of events that leads to a change in the way the plant develops and grows. Hormones play a key role in translating information gained by the plant's sensory receptors into an appropriate developmental response.

## Plants orient the growth of their stems and roots by light and gravity.

How do stems and roots control their orientation so that they grow in the right direction? In most cases, light provides the most useful signal to guide the growth of stems. Roots and seeds that germinate underground use several cues, including gravity. **Tropism** is the bending or turning of an organism in response to an external signal such as light or gravity. Plant stems are positively **phototropic,** meaning they bend toward the light. Plant stems are also negatively **gravitropic,** meaning they grow upward against the force of gravity. In contrast, plant roots grow down and away from the light. Thus, roots are positively gravitropic and negatively phototropic.

In 1880, Charles Darwin, working with his son Francis, carried out some of the earliest research on phototropism. Observing that grass seedlings bend toward light, Darwin conducted experiments to determine what part of the plant detects light. When he covered just the tip of the young grass plants, they no longer grew toward the light, but when he buried them in fine black sand so that only their tips were exposed, they did. This experiment established the shoot tip as the site where the light was perceived but left open the question of what kind of signal caused the plants to bend (**Fig. 31.20**).

In Chapter 30, we saw that plants produce photoreceptors that allow them to detect and respond to light. If a plant is exposed to

FIG. 31.20

# How do plants grow toward light?

**BACKGROUND** In 1926, Fritz Went, a Dutch graduate student, followed up on the experiments of Charles and Francis Darwin. In particular, Went sought to understand how the perception of light by the tip of the plant led to a growth response farther down the stem. He reasoned that if the signal from the shoot tip was a chemical, he could isolate it by cutting the tip off of a growing plant and putting it on a block of gelatin. The gelatin would act like a trap, capturing any chemical that moved out of the shoot tip.

**HYPOTHESIS** A chemical signal links the perception of light by the shoot tip with the growth of the stem toward light.

**EXPERIMENT AND RESULTS** Went cut off the tips of young oat plants (*Avena sativa*) and placed them on gelatin blocks for a period of 1 to 4 hours. Removing the tip of the plant caused the plants to stop elongating. When the gelatin block was placed on top of the cut plant (Experiment 1), growth resumed. This experiment demonstrated that a growth-inducing chemical was exported from the tip of the young oat plants.

Interestingly, Went found that if the block of gelatin was not centered on the tip but instead was a bit off center, the plant bent away from the side of the gelatin block (Experiment 2). In this case, the side of the plant exposed to the gelatin block grew faster, causing bending, because it received more of the chemical signal. This experiment demonstrates that increasing the amount of auxin on one side of a stem results in bending.

**CONCLUSION** A signal is produced in the tip of the plant and travels downward through the stem, toward the roots. This signal stimulates growth. Light causes the signal to move to the opposite side, stimulating growth on that side, which in turn causes the plant to bend toward the light. This signal is now known to be a plant hormone, called auxin from the Latin word *augere,* "to increase."

**FOLLOW-UP WORK** Went followed up his experiments by determining how the level of auxin relates to the degree of curvature. This assay became known as the Avena curvature test, after the experimental species. Other experiments showed that auxin moves in a directed fashion by polar transport. Later, auxin was shown to be indole-3-acetic acid (IAA).

**SOURCES** Darwin, C., and F. Darwin. 1880. *The Power of Movement in Plants.* London: John Murray; Went, F. W. 1926. "On Growth-Accelerating Substances in the Coleoptile of *Avena sativa.*" *Proceedings of the Royal Academy of Amsterdam* 30:10–19.

**FIG. 31.21    Phototropism in shoots, mediated by the hormone auxin.**

**a.**

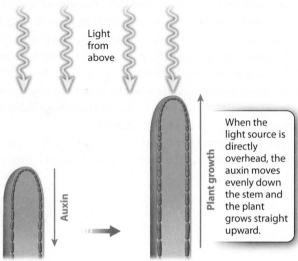

When the light source is directly overhead, the auxin moves evenly down the stem and the plant grows straight upward.

**b.**

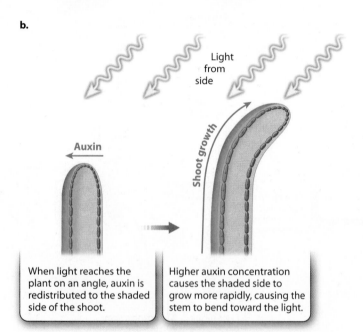

When light reaches the plant on an angle, auxin is redistributed to the shaded side of the shoot.

Higher auxin concentration causes the shaded side to grow more rapidly, causing the stem to bend toward the light.

light from only one side, only the photoreceptors on the illuminated side are activated. Phototropism in plants involves a photoreceptor that absorbs blue wavelengths of light. The absorption of blue light by the photoreceptor causes auxin to be transported preferentially to the shaded side of the stem. This results in high concentrations of auxin on the shaded side and lower auxin levels on the illuminated side. Because auxin stimulates cell expansion in stems, the shoot grows faster on the shaded side than it does on the illuminated side, causing the plant to bend toward the light (**Fig. 31.21**).

Phototropism in roots is similar, in that illuminating a root from the side causes auxin to be redistributed toward the shaded side. In this case, auxin is transported to the shaded side within the root cap, then moves upward into the zone of cell elongation. However, higher auxin concentrations *decrease* the rate of cell elongation in roots, causing roots to bend *away* from light. The reason that auxin

redistribution leads to such a different outcome in roots compared to shoots is that in roots high concentrations of auxin trigger the production of the hormone ethylene, which reduces cell elongation.

A similar mechanism orients plants with respect to gravity. Moving a plant to a horizontal position results in the transport of auxin to the lower side. The accumulation of auxin on the lower surface causes stems to bend up and roots to turn down. How is the plant able to detect which way is up, which way is down? Recall that plants store carbohydrates as starch. Starch is a large molecule that has a density greater than water. Starch-filled organelles thus sink to the bottom of the cell. The weight of these organelles pressing on the cytoskeleton or membranes is thought to provide plants with information regarding their orientation relative to gravity. In roots, for which gravity is a critical source of information regarding orientation, specialized gravity-sensing cells in the root cap contain large starch-filled organelles known as **statoliths** (**Fig. 31.22**). In stems, starch-filled organelles in bundle-sheath cells are thought to play a similar role.

**FIG. 31.22    Root orientation with respect to gravity.** Starch-filled statoliths settle to the bottom of cells, triggering a redistribution of auxin to the lower side of the root, where it slows elongation, thus causing the root to bend downward.

**FIG. 31.23** Seedlings grown in the light and the dark (left) or in full sun and shade (right).

Light

Dark

Plants adapted to full sun develop long, thin internodes, and fewer leaves, when grown in shade.

Full sun

Canopy shade

## Plants grow taller and branch less when light levels are low.

A newly germinated seedling must reach the soil surface before its stored resources are used up or it will die. Seeds that germinate in the dark produce completely white seedlings with thin, highly elongated internodes and suppressed leaf development (**Fig. 31.23**). By putting all their resources into elongation, seedlings grown in the dark maximize their chances of reaching the light. If and when they do, internodal elongation slows markedly and green leaves expand. These dramatic changes are triggered by photoreceptors, which signal the presence of light. The transformation from a dark-grown to a light-exposed seedling involves several photoreceptors, including one that absorbs blue wavelengths and one, phytochrome, that is activated by the absorption of far-red wavelengths.

Even after a plant emerges above the soil, the ability to sense and respond to light can enhance its chances for survival. Some species are adapted for shade, producing thin leaves that efficiently use low levels of light; others grow best in direct sunlight. When such sun-adapted species are grown in the shade of another plant, they direct their resources primarily into growing tall. These plant species produce stems that have long, thin internodes and fewer branches, and they invest less in root growth underground.

Interestingly, the effect is much less if the shade is produced by anything other than another plant. The reason for this is that

chlorophyll absorbs red wavelengths, but not far-red wavelengths of light. Thus, as light passes through leaves it becomes depleted in red wavelengths. As we saw in Chapter 30, phytochrome is a photoreversible photoreceptor that shuttles between a red-light and a far-red-light absorbing form. It allows plants to detect when they are underneath another plant. It also allows them to sense when they are surrounded by other plants that could grow to shade them in the future because light reflected off their neighbors' leaves is also depleted in red wavelengths of light. Thus, phytochrome provides an early-warning system of the presence of nearby competitors.

## Roots elongate more and branch less when water is scarce.

The developmental sensitivity of plant shoots is mirrored underground. When a plant experiences drought, it produces more roots, and these roots penetrate farther into the soil, in part because they produce fewer lateral branches. By producing deeper roots, a plant increases its chances of reaching moister soil.

The root cap appears to play an important role in allowing roots to sense and respond to the amount of water in the soil. Because the root cap is not well connected to the plant's vascular system, the water status of its cells provides a good indication of the moisture content of the soil. As soils dry, root cap cells produce a hormone called **abscisic acid,** commonly abbreviated as ABA (Table 31.1). Abscisic acid stimulates root elongation by suppressing ethylene synthesis. Ethylene affects the growth of cells by influencing the orientation of their cellulose microfibrils. Cells treated with ethylene exhibit an increase in diameter growth and lower rates of elongation. Thus, by inhibiting the production of ethylene, the production of ABA in drought-influenced roots leads to roots that penetrate deeper into the soil.

### Exposure to wind results in shorter and stronger stems.

In the 1980s, scientists studied the genetic model plant *Arabidopsis thaliana* to learn which genes are expressed in response to different hormones. The experimental treatment involved spraying *A. thaliana* plants with different solutions containing hormones; control plants were sprayed with water. The treatments were successful in that several genes were strongly up-regulated. However, the same genes were also turned on in the plants sprayed with water. Further experiments showed that what the plants were responding to was neither the hormones nor the water, but the fact that their stems were bent back and forth when the plants were sprayed. This led to the discovery of touch-sensitive genes in plants, which are activated by mechanical perturbation.

This discovery confirmed what foresters and horticulturists had long known. Plants exposed to wind produce stems that are shorter and wider than ones grown in more protected sites. In fact, commercial greenhouses often install fans, in part so that their plants will produce stems robust enough that the plants can thrive when moved outdoors. Flexing a stem back and forth triggers an increase in the synthesis of ethylene. Cells exposed to ethylene expand more in diameter and less in length, resulting in shorter and thicker stems.

### Plants use day length as a cue to prepare for winter.

In Chapter 30, we saw how plants can use day length to control when flowers are produced. Not surprisingly, day length also triggers other developmental events, particularly ones that allow plants that persist for more than a year to prepare for winter. One such developmental change is the formation of storage organs, for example in roots. Carbohydrates that accumulate in these structures can support the growth of new leaves and stems the following spring.

A second developmental change in response to day length is the formation of overwintering buds. As the days begin to shorten, plants stop producing photosynthetic leaves and begin forming bud scales (see Fig. 31.6a). Bud scales surround and protect the meristem from ice and water loss. The formation of bud scales accompanies a series of metabolic changes that allow meristems to remain in a dormant state throughout the winter. For example, plugs are produced that block the plasmodesmata between the meristem and the rest of the plant. These plugs prevent any growth-stimulating compounds from reaching the meristem.

The hormone abscisic acid contributes to the formation of overwintering buds, similarly to its role in promoting seed dormancy. Many temperate trees must experience sustained exposure to cold temperatures before they will break dormancy, thus preventing a warm period in the autumn from inappropriately triggering growth. In the spring, lengthening days and warmer temperatures reverse the metabolic changes that accompany dormancy and allow growth to resume.

This section has focused on the environmental context of plant growth and development—how plants alter their development to enhance the uptake of water, nutrients, and sunlight, and to cope with physical stresses such as winter. In the next chapter, we focus on the biological context of plant growth and development—how plants defend themselves from being eaten or parasitized.

## Core Concepts Summary

**31.1 IN PLANTS, UPWARD GROWTH BY STEMS OCCURS AT SHOOT APICAL MERISTEMS, POPULATIONS OF TOTIPOTENT CELLS THAT PRODUCE NEW CELLS FOR THE LIFETIME OF THE PLANT.**

Stems are built from repeating modules consisting of nodes, where one or more leaves are attached, and internodes, which are the regions of stem between each node. page 31-2

The shoot apical meristem, located at the tip of each stem, is the source of new cells from which the stem and the leaves are formed. page 31-2

Stems contain zones of cell division, elongation, and maturation. Most of the increase in size occurs in the zone of cell elongation. page 31-3

Leaves develop from leaf primordia produced by the shoot apical meristem. page 31-4

Leaves differ from stems in that they differentiate fully and thus have no capacity for continued growth. page 31-4

In seed plants, branching occurs at axillary buds at the base of each leaf. page 31-5

Shoot apical meristems and axillary buds can be transformed into floral meristems, at which point they lose their capacity for continued growth. page 31-6

**31.2 HORMONES ARE CHEMICAL SIGNALS THAT INFLUENCE THE GROWTH AND DIFFERENTIATION OF PLANT CELLS.**

Hormones are chemical signals that affect both physiology and development and are effective in extremely low concentrations. Traditionally five plant hormones have been identified: auxin, gibberellic acid, cytokinins, ethylene, and abscisic acid. page 31-7

The polar transport of auxin through PIN proteins guides the development of vascular channels (xylem and phloem) that connect leaves and stems. page 31-8

Xylem and phloem differentiate from procambial cells. page 31-9

In most seed plants, the stem's vascular bundles are organized as a ring. page 31-9

Gibberellic acid causes internodal elongation of shoots. page 31-9

Apical dominance is the suppression of the outgrowth of axillary buds caused by hormones produced by roots (strigolactone) and shoots (auxin). page 31-10

**31.3 LATERAL MERISTEMS ALLOW PLANTS TO GROW IN DIAMETER, INCREASING THEIR MECHANICAL STABILITY AND THE TRANSPORT CAPACITY OF THEIR VASCULAR SYSTEM.**

The formation of two lateral meristems—the vascular cambium and the cork cambium—allow plants to grow in diameter. page 31-10

The vascular cambium produces secondary xylem toward the center and phloem toward the outside. page 31-11

The cork cambium produces a protective outer layer. page 31-12

Gymnosperm tracheids function in both mechanical support and water transport. Angiosperm xylem produces fibers that influence the strength of wood and vessel elements that function in water transport. page 31-12

**31.4 THE ROOT APICAL MERISTEM PRODUCES NEW CELLS THAT ALLOW ROOTS TO GROW DOWNWARD INTO THE SOIL, ENABLING PLANTS TO OBTAIN WATER AND NUTRIENTS.**

The root apical meristem is the source of new cells that allow roots to elongate. page 31-14

The root apical meristem has a cap that protects it from damage as the root grows through the soil. page 31-14

The root's vascular tissues are located in the center of the root, surrounded by the pericycle, the endodermis, the cortex, and the epidermis. page 31-14

Lateral roots develop from new root apical meristems that form in the pericycle. page 31-15

Roots alter their development in response to conditions in the soil. page 31-15

**31.5 PLANTS RESPOND TO LIGHT, GRAVITY, AND WIND THROUGH CHANGES IN INTERNODE ELONGATION AND THE DEVELOPMENT OF LEAVES AND BRANCHES.**

Plant stems grow toward the light and away from gravity as a result of redistribution of auxin. page 31-18

Starch-filled statoliths allow roots to sense and respond to gravity. The weight of statoliths alters the lateral distribution of auxin, causing roots to bend downward. page 31-18

Seedlings germinating in the dark produce elongated internodes and suppress leaf expansion to increase their ability to reach the soil surface before their stored resources are depleted. page 31-19

Some plants respond to the presence of neighboring plants by elongating more rapidly and suppressing branching, which allow them to maximize height growth. page 31-19

Mechanical stress, such as the bending of stems in the wind, results in stems that are short and wide. page 31-20

Decreases in photoperiod trigger plants to prepare their meristems for winter by producing bud scales and reducing internode elongation. page 31-20

## Self-Assessment

1. Diagram the zones of cell division, elongation, and maturation, and explain why this organization allows stems to grow without a predetermined limit to their length.

2. Give two examples of how changes in internode elongation allow plants to alter the development of their stems.

3. Name one role of the plant hormone auxin and describe how auxin is transported within a plant.

4. Explain why a plant that has a vascular cambium also has a cork cambium.

5. Explain why the vascular cambium forms a continuous sheath that runs from near the tips of the branches to near the tips of the roots, whereas the cork cambium is discontinuous in both space and time.

6. List three structural differences between roots and shoots that allow roots to grow through the soil. Explain how you would tell whether an isolated piece of a plant came from the shoot or from the root.

7. Describe how lateral roots form and how lateral roots enhance the ability of plants to obtain nutrients and water from the soil.

8. Discuss the similarities and differences in the ways stems and roots respond to gravity and light.

9. Give an example of how a plant's ability to sense its environment improves the plant's chances for survival and reproduction.

Do you understand the chapter's Core Concepts? Log in to check your answers to the Self-Assessment questions, then practice what you've learned and reinforce this chapter's concepts by working through the problems and multimedia tutorials provided there.

🖥 **www.biologyhowlifeworks.com**

## CHAPTER 32

# PLANT DEFENSE

## Keeping the World Green

## Core Concepts

**32.1** Plants have evolved mechanisms to protect themselves from infection by pathogens, which include viruses, bacteria, fungi, worms, and even parasitic plants.

**32.2** Plants use chemical, mechanical, and ecological defenses to protect themselves from being eaten by herbivores.

**32.3** The production of defenses is costly, resulting in trade-offs between protection and growth.

**32.4** Interactions among plants, pathogens, and herbivores contribute to the origin and maintenance of plant diversity.

Plants have complicated interactions with animals. As we saw in Chapter 30, many plants rely on animals for pollination or seed dispersal. In return, plant seeds provide a rich source of food for animals, and leaves also furnish a vast, if lower-quality, food resource. The plant body also provides habitats for an array of viruses, bacteria, and protists, not to mention other plants, that grow on or within their tissues. How do plants defend themselves against the animals that would eat them? How do they defend against pathogens? The answers to these questions help us understand how plant populations are distributed in nature, and they help us solve practical problems as well. In the United States alone, crop damage by herbivores and pathogens costs farmers more than $2 billion every year.

The English poet Alfred, Lord Tennyson, created an indelible image: "Nature, red in tooth and claw." But that image does not recognize that some of the fiercest battles in nature are waged between plants and their predators.

Plants can't run away and hide, and so they lack a key defense employed by animals. Nonetheless, they can sense danger and are able to deploy a diverse and powerful defensive arsenal. Plant defenses include physical and chemical deterrents as well as aerial surveillance, armed guards, and sticky traps.

Of particular importance to humans, plants produce a wide array of chemicals that alter the metabolism and influence the behavior of animals that seek to consume them. These molecules also interact with our own biochemistry, and so compounds derived from plants have been used since antiquity as sources of medicine. Our early ancestors turned to the leaves and bark of willow trees in much the same way that we rely on aspirin—and for good reason. Willows produce a compound that deters herbivores and in humans relieves aches and pain; this compound in fact provides the chemical basis of aspirin. Compounds in rosemary improve the taste of food, and the compound quinine, derived from the tropical cinchona tree, is used in the treatment of malaria. Belladonna, a relative of the tomato, produces toxins that the ancient Romans used to eliminate rivals, but in very low doses it relieves gastrointestinal disorders. In fact, nearly 30% of all medicines are based on chemicals that plants produce to defend themselves.

## 32.1 PROTECTION AGAINST PATHOGENS

In September 1845, potato plants across Ireland began to die. Their leaves turned black and the entire plant began to rot. When the potatoes were dug up, those that were not already rotten soon shriveled and decayed into foul-smelling and inedible mush. The biological agent responsible for the failure of the potato crop looked like a fungus, but in fact was an oomycete called *Phytophthora infestans*, a type of protist (Chapter 27).

**FIG. 32.1**   A potato plant infected with *Phytophthora infestans,* an oomycete that causes blight.

*P. infestans* is one example of the diverse pathogens that infect plants. Plant pathogens include viruses, bacteria, fungi, nematode worms, and even other plants, but what they all have in common is the ability to grow on and in the tissues of plants, extracting resources to support their own growth and causing an array of disease symptoms. The outcome for the plant is often death (**Fig. 32.1**). The fact that most plants appear disease free is testimony to their ability to defend themselves. Plants have an innate immune system that allows them to recognize and respond to pathogens that would otherwise exploit them. In the Irish potato famine, which lasted from 1845–52, these defenses were breached, allowing *Phytophthora infestans* to spread and grow with devastating effects on the potato crop and, in consequence, the people of Ireland.

**Plant pathogens infect and exploit host plants by a variety of mechanisms.**

The epidermis, with its thick walls and waxy cuticle, is a plant's first line of defense against pathogens. Viruses and bacteria cannot penetrate the cuticle and so commonly enter through wounds or on the piercing mouthparts of insects and nematodes. Stomata form a natural entry point that enables bacteria, oomycetes, and fungi to gain access to leaves (**Fig. 32.2**). Some

bacteria and fungi even secrete chemicals that prevent stomata from closing, improving their ability to infect plant tissues. Other pathogens, notably parasitic plants and some species of fungi, secrete enzymes that weaken epidermal cell walls. These invaders then force their way through the weakened epidermis and gain access to the plant interior. Mechanical cultivation can damage crop plants, accelerating the spread of pathogens.

Once inside the plant, fungi can spread by growing through plant cell walls. Viruses move from one plant cell to another through plasmodesmata. Because the genomes of most plant viruses are single strands of RNA, they can travel by the plant's natural mechanism for moving messenger RNA between cells. The plant vascular system provides a highway that allows pathogens to move over longer distances. Viruses are transported within the phloem, whereas bacteria and fungi move through the xylem. During the 1950s, commercial banana production in Mesoamerica was nearly wiped out by *Fusarium oxysporum,* a fungal pathogen whose spores are carried in the xylem. More recently, the California wine industry has been threatened by Pierce's disease, which results from infection by *Xylella fastidiosa,* a bacterium that also moves through the xylem.

**Biotrophic pathogens** obtain resources from living cells, whereas **necrotrophic pathogens** kill cells before colonizing them. Defenses effective against one type are not necessarily effective against the other. Thus, plants respond differently

**FIG. 32.2**   **Entry points for pathogens.** A scanning electron micrograph shows *Phytophthora infestans* gaining access to the leaf interior by growing through stomata.

Stomata

*Phytophthora infestans*

**FIG. 32.3** **Parasitic plants.** (a) *Cuscuta*, or dodder, parasitizes the stems of many plants. It does not photosynthesize and instead receives everything it needs from its host. (b) *Rafflesia*, which produces the world's largest flower, also obtains everything it needs from its host. (c) Mistletoe carries out photosynthesis, but relies on its host plant for water and nutrients.

to these two classes of pathogens. Viruses are biotrophic pathogens; they must hijack a living cell's machinery to reproduce. Bacteria and fungal pathogens are present in both groups. For example, wheat leaf rust (*Puccinia* species) is a biotrophic fungus that significantly reduces wheat yields (Chapter 34). Rice blast (*Magnaporthe grisea*) is a necrotrophic fungus that each year reduces rice production by an amount that would feed 60 million people.

Parasitic plants obtain resources by infecting other plants (**Fig. 32.3**). Parasitic plants produce structures that penetrate into the stems or roots of other plants, eventually tapping into the host plant's vascular system. Some, like some mistletoes, are xylem parasites: They obtain water and nutrients from the xylem of the host plant without expending the costs of producing roots themselves. Other plant pathogens target both the xylem and the phloem. In the most extreme cases, the parasite receives all its carbohydrates from its host and carries out no photosynthesis of its own. *Rafflesia*, famous for producing the largest flower in the world, grows entirely within the tissues of its host, emerging only when it produces its giant flowers. The ability to parasitize other plants has evolved independently multiple times. Currently, more than 4000 species of parasitic plants have been identified.

How do parasitic plants find an appropriate host? Mistletoe fruits, which are eaten and dispersed by birds, have a sticky layer that adheres to the stems and branches of trees that the plant will parasitize. One of the most destructive plant pathogens is witchweed (*Striga* species), which infects the roots of a number of crop species, including corn, sorghum, and sugarcane. *Striga* seeds remain dormant in the soil until they detect chemicals released from the roots of suitable host species. The chemicals

that trigger the germination of *Striga* seeds are strigolactones (Chapter 31), a plant hormone also involved in establishing symbiotic interactions between plant roots and mychorrhizal fungi (Chapter 29) and inhibiting the outgrowth of axillary buds. *Striga* infects more than 40% of all agricultural areas in Africa, driving down crop yields, in some cases to zero.

### An innate immune system allows plants to detect and respond to pathogens.

Most plants are at risk of infection by only a small subset of the many plant pathogens known. **Host plants** of a particular pathogen are the species that can be infected by that pathogen. Not all infections have a significant effect on the host plant, however. **Virulent** pathogens are able to overcome the host plant's defenses and lead to disease. In contrast, **avirulent** pathogens damage only a small part of the plant because the host plant is able to contain the infection. Whether an interaction is virulent or avirulent depends on the genotypes of both the pathogen and its host plant.

In your own body, an innate immune system provides a generalized first line of defense against pathogens (Chapter 43). Plants, too, have an innate immune system that allows them to detect pathogens and mount an appropriate response. A key feature of the immune system is that plants, like animals, can recognize a cell or virus as foreign. How do they accomplish this?

Plant cells have protein receptors (Chapter 9) that bind to molecules produced by pathogens and recognize them as foreign to the plant body. Binding changes the conformation of the receptor, triggering a cascade of reactions that enhance the plant's ability to resist infection.

**FIG. 32.4    The plant immune system.** (a) In basal resistance, plasma membrane receptors recognize molecules produced by broad classes of pathogens. (b) Specific resistance depends on *R* genes that allow plant cells to identify and deactivate AVR proteins produced by specific pathogens.

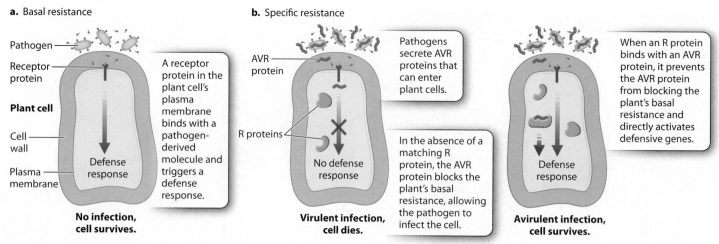

**a.** Basal resistance

Pathogen

Receptor protein

**Plant cell**

Cell wall

Plasma membrane

Defense response

A receptor protein in the plant cell's plasma membrane binds with a pathogen-derived molecule and triggers a defense response.

**No infection, cell survives.**

**b.** Specific resistance

AVR protein

R proteins

No defense response

Pathogens secrete AVR proteins that can enter plant cells.

In the absence of a matching R protein, the AVR protein blocks the plant's basal resistance, allowing the pathogen to infect the cell.

**Virulent infection, cell dies.**

Defense response

When an R protein binds with an AVR protein, it prevents the AVR protein from blocking the plant's basal resistance and directly activates defensive genes.

**Avirulent infection, cell survives.**

The plant immune system has two parts, one general, or basal, and the other specific (**Fig. 32.4**). The basal branch of the plant immune system consists of receptors located on the plasma membrane (Fig 32.4a). These receptors recognize highly conserved molecules generated by broad classes of pathogens. Examples of these molecules are flagellin, present in the flagella of bacteria, and chitin, a component of the fungal cell wall. When one of these molecules binds to the plant's receptor, an array of defense mechanisms are triggered that help protect the plant from infection.

The second branch of the plant immune system allows plants to resist specific pathogens (Fig. 32.4b). It also consists of receptors, in this case located inside the cell rather than on the plasma membrane. Plants harbor hundreds of these receptors, called **R proteins,** each expressed by a different gene —these genes are collectively called ***R* genes** (the "R" stands for "resistance"). Pathogens produce proteins called AVR proteins that enter into plant cells and facilitate infection ("AVR" stands for "avirulence"). Some AVR proteins block the basal branch of the immune system, while others cause the cell to secrete molecules that the pathogen needs to support its own growth. Each R protein recognizes a specific AVR protein. When an R protein binds with a pathogen-derived AVR protein, it both prevents the AVR protein from blocking the plant's defenses and activates additional defenses. Because this branch of the immune system depends on interactions between specific plant and pathogen genes, it is commonly called the gene-for-gene model of plant immunity.

The large number of *R* genes is the result of an evolutionary arms race that occurs as pathogens evolve ways to evade their host's detection systems. Pathogens improve their success rate when mutation results in an AVR protein that does not bind with any of the R proteins produced by its target. For example, *P. infestans,* the pathogen responsible for the Irish potato famine, produces

many more AVR proteins than related *Phytophthora* species. The large number of AVR proteins found in *P. infestans* increases the probability that at least one of these proteins is not detected by the infected plant. Conversely, when pathogens are successful, natural selection favors plant populations that produce new R proteins capable of binding to the AVR proteins of the invader.

→    **Quick Check 1** Can the specific (*R*-gene-mediated) branch of the plant immune system defend against both biotrophic and necrotrophic pathogens?

## Plants respond to infections by isolating infected regions.

Once a plant has detected a pathogen, how does it protect itself? One way is by reinforcing its natural barriers against pathogens. For example, plants commonly defend themselves by strengthening their cell walls, closing their stomata, and plugging their xylem. Or the plant can counterattack. Plants produce a variety of antimicrobial compounds, including ones that attack bacterial or fungal cell walls. A third type of defense is known as the **hypersensitive response.** In this case, uninfected cells surrounding the site of infection rapidly produce large numbers of reactive oxygen species, which trigger cell wall reinforcement and cause the cells to die (**Fig. 32.5**). The death of surrounding cells create a barrier of dead tissue that prevents the spread of biotrophic pathogens and slows the growth of necrotrophic ones.

Plants can defeat pathogens by isolating infected regions to a far greater degree than is possible in most animals. For example, plants respond to the presence of xylem-transported pathogens by plugging xylem conduits. If the infection is caught in time, the pathogen is prevented from spreading throughout the plant. But if containment fails and the pathogen spreads, the plant

**FIG. 32.5 Hypersensitive response.** A tobacco plant infected with tobacco mosaic virus responds by producing a region of dead tissue that surrounds the site of infection and prevents the virus from spreading to other parts of the plant.

Site of initial infection

plant were subsequently exposed to the virus, they suffered little or no damage (**Fig. 32.7**). Ross called this ability to resist future infections **systemic acquired resistance (SAR)**. Initially reported for viral infections, SAR occurs in response to a wide range of pathogens, although only when infection results in necrosis due to either a hypersensitive response or a necrotrophic pathogen.

How do uninfected leaves acquire resistance to pathogens? One hypothesis is that a chemical signal is transported from the infected region through the phloem. This chemical signal then triggers the expression of genes encoding many of the same proteins that defend against the pathogen in infected cells. The identity of the mobile signal has proved elusive, but experiments show that salicylic acid—the chemical basis of aspirin—is required for SAR. In experiments in which salicylic acid accumulation was blocked, plants exposed to pathogens failed to develop SAR. Furthermore, the experimental exposure of plants to salicylic acid induces SAR. The methylated form of salicylic acid, known familiarly as oil of wintergreen, is also a potent inducer of defense responses. Methyl salicylic acid is volatile and vaporizes readily, giving rise to the characteristic wintergreen smell. It has been proposed that the transport of methyl salicylic acid vapor through the air may play a role in signaling the presence of pathogens to more distant regions of the plant.

may contribute to its own demise by blocking its entire vascular system. Virulent pathogens that move through the xylem cause a variety of disorders collectively referred to as vascular wilt diseases (**Fig. 32.6**). The widespread death of native elm trees in both Europe and North America is the result of a vascular wilt disease caused by a fungus introduced from Asia, but first identified in the Netherlands and therefore referred to as Dutch elm disease.

### Mobile signals trigger defenses in uninfected tissues.

In 1961, American plant pathologist A. F. Ross reported that when individual tobacco leaves were infected with the tobacco mosaic virus (TMV), they developed necrotic patches characteristic of a hypersensitive response. However, when uninfected leaves of the same

**FIG. 32.6 Vascular wilt diseases.** Pathogens that are transmitted through the xylem can kill their host if plant defenses that prevent their spread also block water transport to the leaves. (a) Panama disease, which affects bananas, is caused by a fungal pathogen. (b) Pierce's disease, which affects grapes, is caused by a pathogenic bacterium.

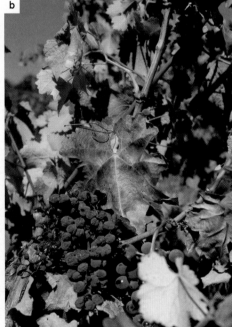

**FIG. 32.7**

# Can plants develop immunity to specific pathogens?

**BACKGROUND** Tobacco is susceptible to an infectious disease that causes the leaves to turn a mottled yellow, hence the name of the disease—tobacco mosaic disease. In the 1890s, experiments with filters showed that the infectious agent was smaller than a bacterium, a finding that led to the discovery of viruses. Some plants succumb to the disease, but others do not. Are the plants that survive the initial infection immune from further attacks?

**HYPOTHESIS** Plants that survive infection with tobacco mosaic virus (TMV) have acquired immunity and therefore resist further attack.

**EXPERIMENT** American plant pathologist A. F. Ross infected one leaf on a tobacco plant with TMV. One week later, he exposed another leaf on the same plant to the virus.

**RESULTS** Plants that had been previously exposed to TMV showed no signs of infection when exposed to the virus a second time.

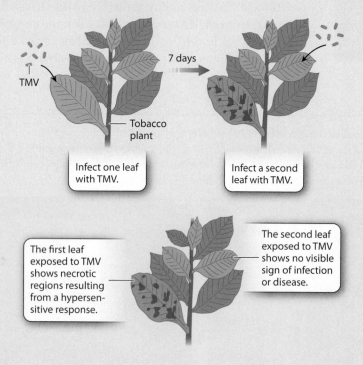

TMV

7 days

Tobacco plant

Infect one leaf with TMV.

Infect a second leaf with TMV.

The first leaf exposed to TMV shows necrotic regions resulting from a hypersensitive response.

The second leaf exposed to TMV shows no visible sign of infection or disease.

**CONCLUSION** Tobacco leaves that have never been exposed to TMV can acquire resistance to the pathogen if another leaf on the same plant has been previously exposed. This result indicates that a signal has been transmitted from the originally infected leaf to the undamaged parts of the plant, and that the transmitted signal subsequently triggers the development of an immune response that protects the plant from further infection.

**FOLLOW-UP WORK** Plants are now known to acquire immunity to a wide range of pathogens in addition to viruses. The genetic basis for the immune response is a system in which proteins encoded by plant *R* genes recognize specific AVR proteins produced by pathogens.

**SOURCE** Ross, A. F. 1961. "Systemic Acquired Resistance Induced by Localized Virus Infections in Plants." *Virology* 14:340–358.

Systemic acquired resistance increases the ability of uninfected tissues to resist infection. In general, these responses are not highly specific. Rather, SAR activates defenses effective against broad classes of pathogens. Only in the case of viral infections, discussed next, is the response targeted to an individual pathogen.

→ **Quick Check 2** How does the plant hypersensitive response differ from systemic acquired resistance (SAR)?

**Plants defend against viral infections by producing siRNA.**
Viruses can be potent infectious agents, but plants have evolved responses to viral infection—including a hypersensitive response—that are much like those mounted against other pathogens. In addition, plants have a form of defense that targets the virus itself (**Fig. 32.8**). Most plant viruses have genomes made of single-stranded RNA (ssRNA). During the replication of the viral genome, double-stranded RNA molecules (dsRNA) are formed. The production of double-stranded viral RNA opens the viral genome to counterattack by its host plant. Because plant cells do not normally make double stranded RNA, the replicating viral genomes are identified as foreign. Enzymes produced by the plant cell cleave the double-stranded RNA molecules into small pieces of 21 to 24 nucleotides, forming fragments called **small interfering RNA,** or **siRNA** (Chapter 3). These fragments play a role in targeting and destroying single-stranded RNA molecules that have a complementary sequence—that is, the viral genome (Chapter 19). The virus is unable to replicate and thus cannot spread. In this way, the plant cell uses RNA chemistry to eliminate viral infections.

When a cell is attacked by a virus, the siRNA molecules that are produced in response to the initial infection can also spread, allowing the plant to acquire immunity against specific viruses. The siRNA molecules move through plasmodesmata, enter the phloem, and from there spread throughout the plant.

**FIG. 32.8 Plant defense by small interfering RNAs (siRNAs).** siRNAs target complementary single-stranded RNA for destruction.

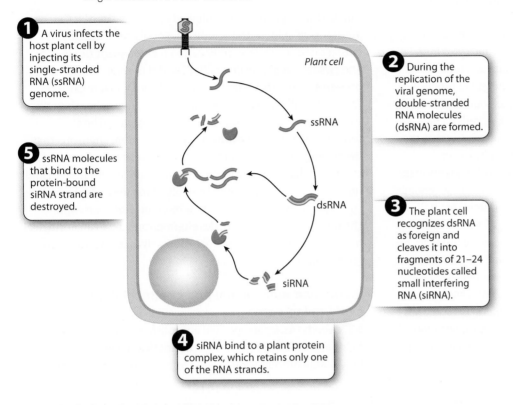

**①** A virus infects the host plant cell by injecting its single-stranded RNA (ssRNA) genome.

**②** During the replication of the viral genome, double-stranded RNA molecules (dsRNA) are formed.

**⑤** ssRNA molecules that bind to the protein-bound siRNA strand are destroyed.

**③** The plant cell recognizes dsRNA as foreign and cleaves it into fragments of 21–24 nucleotides called small interfering RNA (siRNA).

**④** siRNA bind to a plant protein complex, which retains only one of the RNA strands.

Thus, the systemic response to viral infection involves both general defense responses mediated by salicylic acid and the transport of highly specific molecules that can target and destroy viral genomes throughout the plant.

## A pathogenic bacterium provides a way to modify plant genomes.

Some pathogenic bacteria alter their host plant's biology by inserting their own genes into the host's genome. One such pathogen, *Rhizobium radiobacter* (formerly known as *Agrobacterium tumifaciens*), is beneficial to humans, despite the fact that it can cause widespread damage to crops such as apples and grapes. The reason? This bacterium's naturally evolved ability to incorporate its DNA into the genome of its host plant provides a surprisingly easy way for biotechnologists to engineer plants.

*R. radiobacter* is a soil bacterium with virulent strains that infect the roots and stems of many plants. Virulent bacteria enter plants through wounds and move through the plant's cell walls, propelled by flagella. A tumor, or gall, forms at the point of infection, often where the stem emerges from the soil, and for this reason *R. radiobacter* infection is called crown gall disease. *R. radiobacter* alters the growth and metabolism of infected cells by inserting some of its own genes into the chromosomal DNA of the host plant (**Fig. 32.9**).

**FIG. 32.9 The formation of a crown gall tumor.** *Rhizobium radiobacter*, a soil bacterium, can insert genes into the plant genome, altering the growth and metabolism of the plant in ways that enhance the growth of *R. radiobacter* bacteria.

Virulent strains of *R. radiobacter* contain a plasmid, called the **Ti plasmid**, which is a small (~200 kb) circular DNA molecule containing the genes that will be integrated into the host cell's genome, as well as all the genes needed to make this transfer. Some of the transferred genes—including ones that result in the synthesis of the two plant hormones auxin and cytokinin (Chapter 31)—cause infected cells to proliferate and form a gall. Other transferred genes induce cells in the gall to produce specialized compounds that the bacteria use as sources of carbon and nitrogen for growth. Note that because plant cells do not move (in contrast to animal cells), there is no possibility that the tumor-forming cells will spread to other parts of the plant.

Biologists interested in improving crops and other plants make use of *R. radiobacter*'s ability to insert genes into the host genome. The first step is to replace the genes in the Ti plasmid that will be inserted into the plant genome. Genes for gall formation are removed, and genes of interest are inserted. Genes of interest include those whose products increase the plant's nutritional value or confer resistance against disease. Researchers need some way to identify which cells have taken up the genes, and so a marker gene, for example one conferring antibiotic resistance, is typically included. In this way, cells that survive an application of the antibiotic are known to have successfully incorporated the Ti plasmid into their genome.

## 32.2 DEFENSE AGAINST HERBIVORES

The first land plants were followed closely in time by the ancestors of spiders and scorpions. Fossils show that these earliest land animals soon began to feed on plant fluids and tissues. As plants diversified, so did herbivores, and the pressure on plants to defend themselves against organisms from caterpillars to cows only increased. Given that they are unable to run away, plants seem like they would be the ideal food. But plants have evolved a diversity of mechanisms that deter would-be consumers. Because these mechanisms use resources that could otherwise be directed toward growth and reproduction, plants have also evolved means of deploying their defenses in a cost-effective manner. Here, we explore the mechanical, chemical, developmental, and even ecological means by which plants protect themselves in a world teeming with hungry herbivores.

### Plants use mechanical and chemical defenses to avoid being eaten.

Milkweeds (*Asclepias* species) are commonly found in open fields and roadsides across North America. They illustrate well how plants protect themselves from herbivorous animals.

**FIG. 32.10 How monarch caterpillars feed on milkweed.** Young monarch caterpillars disarm milkweed defenses by digging trenches, while older, larger caterpillars sever major veins. Both methods prevent the sticky latex from flowing into the feeding area.

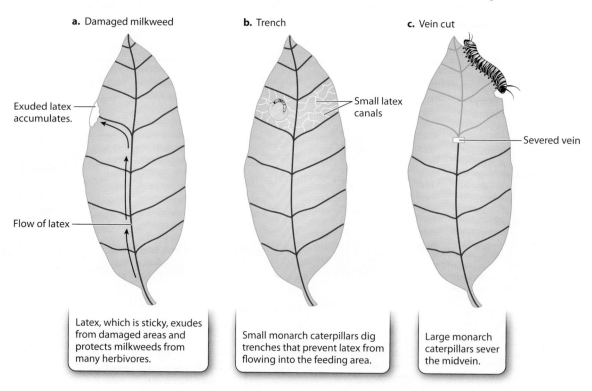

**a.** Damaged milkweed

Exuded latex accumulates.

Flow of latex

Latex, which is sticky, exudes from damaged areas and protects milkweeds from many herbivores.

**b.** Trench

Small latex canals

Small monarch caterpillars dig trenches that prevent latex from flowing into the feeding area.

**c.** Vein cut

Severed vein

Large monarch caterpillars sever the midvein.

In fact, most generalist herbivores, which eat a diversity of plants, do not use milkweeds as a food source. Only specialists, such as caterpillars of the monarch butterfly (*Danaus plexippus*), can feed on these well-defended plants.

In part, milkweed's defenses are mechanical. The leaves of many milkweed species are covered with dense hairs. An insect crawling on a leaf must make its way through as many as 3000 hairs per square centimeter, a journey likely more daunting to an insect than wading into a blackberry thicket is to a human. Monarch caterpillars may spend as much as an hour mowing off these hairs before they begin to feed. This long handling time represents a significant cost. To understand why monarch caterpillars make this investment, let's look at milkweed's other defenses.

A network of extracellular canals filled with a white sticky liquid called **latex** runs through the milkweed leaf, both within the major veins and extending into the regions between veins (**Fig. 32.10**). Latex canals are under pressure in the intact leaf. Thus, when the leaf is damaged, the latex flows out, sticking to everything it comes into contact with and rapidly congealing on exposure to air. A feeding insect is in danger of having its tiny mouthparts glued together, with potentially fatal consequences. Small monarch caterpillars cut trenches into the leaf that prevent latex from flowing onto the region where they are feeding. Larger caterpillars sever the midvein, relieving the pressure within the latex canals downstream of the cut. The caterpillars are then able to feed on the part of the leaf that lies beyond the severed midvein without risk.

Milkweed latex is not only sticky but also toxic. In particular, milkweed latex contains high concentrations of cardenolides, steroid compounds that cause heart arrest in animals. (In small doses, these compounds are used to treat irregular heart rhythms in humans.) Monarch caterpillars do not metabolize the cardenolides that they consume, but instead sequester them within specific regions of their bodies. Storing these compounds provides the monarchs with a chemical defense against their own predators. Because of the presence of these cardenolides, monarch caterpillars and even the adult

**FIG. 32.11 Mechanical defenses.** Spines on the beach palm (a) make it hard for animals to climb or penetrate its stem, and hairs on the leaves of nettle (b) and *Arabidopsis* (c) deter much smaller herbivores.

butterflies are highly toxic to most birds and other animals that would otherwise eat them.

Most plant species have mechanical or chemical defenses against herbivores, or even both. Hairs are common on leaves, and, as anyone who has ever rubbed against a stinging nettle knows, these hairs are sometimes armed with chemical irritants. Grasses and some other plants have a hard, mineral defense consisting of silica ($SiO_2$) plates formed within epidermal cells. Silica functions as a mechanical defense because it wears down insect mouthparts, so the insects feed less efficiently and grow more slowly. On a larger scale, some tropical trees have prickles, spines, or thorns (**Fig. 32.11**). Even something as simple as how tough a leaf is can have a large impact on the probability of the leaf's being eaten. In fact, leaves are most likely to be eaten while they are still growing and their tissues not fully hardened. Nonetheless, the most spectacular diversity in plant defenses lies in the realm of chemistry.

### Diverse chemical compounds deter herbivores.
The cardenolides produced by milkweeds are only one example of the vast chemical arsenal available to plants for protection.

Some plants produce **alkaloids,** nitrogen-bearing compounds that damage the nervous system of animals. Commonly bitter tasting, alkaloids include such well-known compounds as nicotine, caffeine, morphine, theobromine (found in chocolate), quinine (a treatment for malaria), strychnine, and atropine. Alkaloids are a costly defense because they are rich in nitrogen, an essential and often limiting element that plants need to build proteins for photosynthesis. However, they are specific in their action and therefore very small concentrations are effective. One of the first chemotherapy drugs (**Table 32.1**), vincristine, is an alkaloid extracted from the Madagascar periwinkle. It inhibits microtubule polymerization and thus prevents cell division (Chapter 11). Vincristine does not harm the plant's own cells because it is stored within an extracellular system of latex-containing canals.

A second group of defensive compounds, the **terpenes,** do not contain nitrogen. As a result, plants can produce them for defense without having to make use of nitrogen that could otherwise be used in protein synthesis. Small terpenes are volatile, vaporizing easily, and so make up many of the essential oils associated with plants. The distinctive smells of lemon peel, mint, sage, menthol, pine resins, and geranium leaves are all due to terpenes. While we find these odors pleasant, they are feeding deterrents to mammals such as squirrels and moose. In addition, terpenes obstruct the growth and metabolism of both fungi and insects. As a result, pyrethrin, a terpene derivative extracted from chrysanthemums, is marketed commercially as an insecticide.

Other compounds derived from terpenes interfere with insect development. For example, phytoecdysteroids are plant-produced

---

**TABLE 32.1    PLANT COMPOUNDS USED IN MEDICINE: A SAMPLING**

| CLASS | EXAMPLE | PLANT SOURCE | MEDICAL USE |
|---|---|---|---|
| Alkaloids | Atropine | Deadly nightshade | Dilate pupils; treat extremely low heart rate |
| | Vincristine | Madagascar periwinkle | Treat cancer |
| | Codeine | Opium poppy | Cough suppressant |
| | Morphine | Opium poppy | Painkiller |
| | Quinine | Cinchona tree | Treat malaria |
| | Ephedrine | *Ephedra* (Mormon tea) | Asthma relief |
| | Turocurare (curare) | *Strychnos toxifera* | Muscle relaxant |
| | Digitoxin | Foxglove | Treat congestive heart failure and arrhymthia |
| Morphine | Berberine | *Berberis* (barberry) | Antibiotic, anti-inflammatory |
| Terpenoids | Menthol | Peppermint | Topical analgesic; mouthwash |
| | Camphor | Camphor laurel | Anti-itch creams |
| | Eucalyptol | *Eucalyptus* species | Mouthwash; cough suppressant |
| | Taxol | Pacific yew | Treat cancer |
| | Artemisinin | *Artemisia annua* | Treat malaria |
| Menthol | | | |
| Phenols | Salicylic acid | Willow tree | Treat aches and pains |
| Salicylic acid | | | |

replicas of the hormones that induce molting in insects. Exposure to these compounds cause insects to molt prematurely and therefore keeps their populations in check. Taxadiene, precursor for the chemotherapy drug taxol, is a terpene that was first extracted from the bark of the Pacific yew tree.

**Phenols** form the third main class of defensive compounds, illustrated by the **tannins** found widely in plant tissues. Tannins bind with proteins, reducing their digestibility. Plants store tannins in cell vacuoles, and so these compounds come into contact with the protein-rich cytoplasm only when cells are damaged. Herbivores attempting to feed on tannin-producing plants obtain a poor reward for their efforts. For this reason, natural selection favors individuals that avoid tannin-rich plants. Many unripe fruits are high in tannins, and the unpleasant experience of biting into an unripe banana illustrates how tannins deter consumers. For thousands of years, humans have taken advantage of the protein-binding properties of tannins, using extracts from tree bark to process animal skins by "tanning" to produce leather.

Some of the chemical defenses used by plants are stored in a separate compartment from the enzymes that activate them. When you bite into a plant belonging to the cabbage family, the chemicals stored in the vacuole comingle with enzymes in the cytosol. The enzymes then catalyze the production of mustard oils from these compounds. Mustard oils give cabbage and its relatives their distinctive smell. They serve as feeding deterrents because they interfere with insect growth.

Cassava roots, an important food crop in Africa and South America, release the toxin hydrogen cyanide when their cells are damaged. Before cassava roots can be safely consumed, the roots must be ground and the cyanide-producing chemicals removed in running water. Although some cassava varieties, so-called sweet cassava, lack these chemicals and can be eaten without any special preparation, the bitter varieties are preferred crops in many cases because they often have higher yields.

A final chemical defense found in plants is protein based. Plants and animals use the same 20 amino acids to construct proteins, but some plants produce additional amino acids as well. Plants do not incorporate these additional amino acids into their proteins, but herbivores that ingest them do. The resulting proteins can no longer fulfill their function. Thus, insect herbivores that consume nonprotein amino acids grow slowly and often die early. A second protein-based defense is the production of antidigestive proteins called **protease inhibitors.** These proteins bind to the active site of enzymes that break down proteins in the herbivore's digestive system. This prevents proteins from being broken down into their individual amino acids and therefore reduces the nutritional value of the plant tissue. Insects that feed on plants that produce protease inhibitors have reduced growth rates.

## Some plants provide food and shelter for ants, which actively defend them.

While many plants have mechanical defenses against herbivores and nearly all have at least one form of chemical defense, a smaller number of species have evolved ecological defenses against herbivory. These plants "employ" animals as bodyguards in exchange for shelter or nourishment, much the way that many plants provide food or other rewards to animals that transfer pollen and disperse seeds (Chapter 30).

Nectar is typically associated with flowers. However, many plants produce nectar in glands located on their leaves (**Fig. 32.12**). These extrafloral nectaries, or nectar-producing regions, attract ants, which move actively throughout the plant in search of nectar. When the ants encounter either the eggs or larvae of other insects, they consume these as well. The value of this relationship to the plant can be demonstrated easily: When ants are prevented experimentally from patrolling certain branches, those branches suffer higher rates of herbivore damage.

A small number of plants, several hundred in total, have evolved a much closer relationship with ants. These so-called ant-plants provide both food and shelter for an entire colony of ants, which then defend their host. The bullhorn acacia (*Acacia cornigera*), which grows in dry areas of Mexico and Mesoamerica, produces hollow spines at the base of each leaf, as well as protein- and lipid-rich food bodies on the tips of its expanding leaves. Symbiotic ants (*Pseudomyrmex ferruginea*)

**FIG. 32.12  Ants attracted by an extrafloral nectary.** These ants get a food reward and also consume any insect eggs or larvae they encounter on the plant.

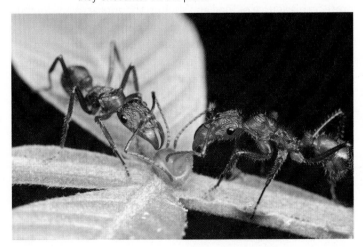

**FIG. 32.13　The symbiosis between ants and the bullhorn acacia.** (a) Ants live in the plant's thorns, and (b) the plant provides food (the yellow food bodies). The ants vigorously defend their home against animals and plants alike.

ants, an undefended plant would rapidly be consumed in nature because ant-plants lack the chemical defenses present in related species that do not host ants.

## Grasses can regrow quickly following grazing by mammals.

By far, the most important vertebrate herbivores are the bison, zebra, antelope, and their relatives that populate the great plains of East Africa and North America. Studies of herbivore damage in an East African savanna indicate that vertebrate herbivores annually consume approximately the same amount of leaf biomass as do herbivorous insects.

Grasses are well suited to cope with grazing mammals (**Fig. 32.14**). The vertically oriented leaves grow from stems with limited internodal elongation. This allows the shoot apical meristem to remain close to the ground, where it is less likely to be damaged. Grasses have persistent zones of cell division and elongation at the base of each leaf. Because grass leaves elongate from their base, grasses can replace leaf tissues removed by a grazing mammal, fire, or a lawn mower without having to add to the length of the stem. Only when grasses produce their flowering stalks does the apical meristem emerge above the leaves.

Grasses are thought to have evolved in forest environments. In fact, grass-dominated ecosystems have become widespread only

live in the spines and feed on the food bodies (**Fig. 32.13**). To protect this source of food and shelter, the ants actively defend the plant. When the ants detect an invader, they produce an alarm pheromone that causes the entire colony to swarm over the plant, attacking any insects or vertebrates (including scientists) that they encounter. The ants will attack the growing tip of a vine that might climb up the host and even destroy plants growing in the soil beneath the host *Acacia*.

This system works well until it comes time to reproduce. Without some mechanism to control the hypervigilant ants, any pollinator that tried to visit one of the host plant's flowers would be subject to the same treatment as a truly unwanted guest. To prevent pollinators from being turned away, flowers of the bullhorn acacia release a chemical that repels the resident ants, allowing pollinators to visit the flowers unharmed.

The interactions between ant-plants and their ants are highly coevolved. The ant species are found only in association with their host plants. And while ant-plants can be grown without

**FIG. 32.14　Adaptation for grazing.** The ability of grass leaves to elongate from the base allows grasses to regrow following grazing.

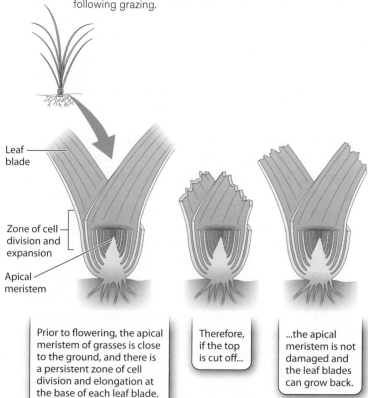

Leaf blade

Zone of cell division and expansion

Apical meristem

Prior to flowering, the apical meristem of grasses is close to the ground, and there is a persistent zone of cell division and elongation at the base of each leaf blade.

Therefore, if the top is cut off...

...the apical meristem is not damaged and the leaf blades can grow back.

during the past 25 million years. The fossil record shows that as grasslands expanded, grazing mammals evolved along with them. Horses, for example, originated as small browsers that fed on the leaves of small trees and shrubs. As grasslands expanded, larger horse species with teeth adapted to eating grass evolved. Why would specialized teeth be advantageous? The small plates of silica in grass cells appear to contribute to mechanical support and may provide mechanical defense against herbivores. Horses and other grazers that feed on these grasses wear down their tooth enamel grinding against the silica, so teeth with a thick enamel layer are favored by natural selection. It is not clear whether grasses evolved in response to grazing or to disturbances such as fire, but without question, their distinctive pattern of meristem activity minimizes the damage imparted by grazing mammals.

## 32.3 ALLOCATING RESOURCES TO DEFENSE

By now, it should be clear that plants invest substantial resources in defense, resources that might otherwise be used for growth and reproduction. Given the considerable cost of plant defenses, their benefits must be correspondingly large. Obviously, if a commonly encountered pathogen or herbivore is lethal, the resources invested in defense provide reproductive advantage. But what if the pathogen or herbivore usually damages only half of a plant's leaves? Or 10%? Or encounters the plant only infrequently?

In nature, the balance between cost and benefit is achieved in several different ways. When a threat is constant or generalized, plants mount **constitutive defenses**—defenses that are always active. When a threat is uncommon, however, there is an advantage to mounting a defense only when the threat is encountered. A defense that is mounted only when the plant senses the threat is an **inducible defense.**

### Plants can sense and respond to herbivores.

When a plant is attacked by herbivores, it begins to express a new set of genes. Many of these genes increase the production of chemical defenses, while others stimulate the mechanical reinforcement of cell walls. In addition, herbivore damage triggers the synthesis of **jasmonic acid,** a signal that is transmitted through the phloem. Exposure to jasmonic acid induces the transcription of defensive genes even in parts of the plant not directly affected by herbivores. Methyl-jasmonic acid is highly volatile and can move through the air to reach portions of the plant that are not actively importing carbohydrates through the phloem.

→ **Quick Check 3** What would happen to a plant in which jasmonic acid synthesis is blocked?

Another way in which plants can sense and respond to herbivores is illustrated by coyote tobacco (*Nicotiana attenuata*), a relative of tobacco native to dry open habitats of the American West

(**Fig. 32.15**). For more than 15 years, biologists have intensively studied the mechanisms by which N. *attenuata* defends itself against herbivores, mechanisms that provide a fascinating window into how plants protect themselves under natural conditions.

**FIG. 32.15** *Nicotiana attenuata*, **the coyote tobacco.** This species has been developed as a model organism for studying how plants protect themselves against herbivores. (a) *N. attenuata* in a Utah desert; (b) *Manduca sexta*, a specialist herbivore, feeding on *N. attenuata*.

*Nicotiana attenuata* is attacked by a wide variety of herbivores, both generalists and specialists. Which species cause the most damage, however, varies significantly from year to year. This means that *N. attenuata* must either produce costly defenses for herbivores that might not show up in any particular year or be able to identify and respond to specific threats.

A major component of *N. attenuata*'s defense against being eaten is the production of the alkaloid nicotine. This defense is metabolically expensive because it contains nitrogen. Nicotine is an effective defense against many generalist herbivores. However, *N. attenuata* is also attacked by a specialist herbivore, the tobacco hornworm caterpillar, *Manduca sexta*. These caterpillars, which can reach 7 cm in length, have evolved the ability to sequester and secrete nicotine. By this means, these caterpillars are able to consume leaves with high levels of nicotine. If nicotine is not an effective defense against this voracious herbivore, what can *N. attenuata* do to protect itself?

It turns out that *N. attenuata* is able to recognize specific chemicals in the saliva of *M. sexta*. When damaged by most herbivores—or a pair of scissors—*N. attenuata* responds by producing nicotine. But when attacked by *M. sexta*, *N. attenuata* does not waste time and resources synthesizing nicotine. Instead, it protects itself in an entirely different way, by attracting other insects that will attack its herbivores.

### Plants produce volatile signals that attract insects that prey upon herbivores.

How does *N. attenuata* attract its insect allies? A hint comes from the smell of a new-mown lawn. This distinctive smell originates from chemicals released from the cut blades of grass. Some of the chemicals released when plants are damaged represent nothing more than the exposure of cell interiors to the air. Others, however, are produced specifically in response to damage.

Studies first conducted in the 1980s showed that undamaged plants increase their synthesis of chemical defenses when neighboring plants are attacked by herbivores (**Fig. 32.16**). Although these plants were described in the popular press as "talking trees," the plants under attack are not altruistically warning their neighbors. Instead, subsequent studies suggest that these plants are sending signals directed at other insects. When *N. attenuata* is attacked by *M. sexta* caterpillars, it produces volatile signals that attract insects that prey on the eggs and larvae of *M. sexta*. The adage "The enemy of my enemy is my friend" seems to hold true when it comes to protecting plants from herbivore damage. Neighboring plants sense these chemical signals and respond by ramping up their own defenses.

**Nutrient-rich environments select for plants that allocate more resources to growth than to defense.** Plant species vary significantly in the extent to which they allocate resources to defense. Because defenses are costly to produce, you might think that plants growing in nutrient-rich habitats should be the best defended. However, consider the counterargument: Plants growing on nutrient-poor soils might invest heavily in defense because they cannot afford to replace tissues lost to herbivores. Such plants would grow relatively slowly because their resources were being used to build defenses rather than to produce new leaves and roots. By

---

HOW DO WE KNOW?

**FIG. 32.16**

# Can plants communicate?

**BACKGROUND** Plants respond to herbivore damage by producing chemicals that make their tissues less palatable. Field experiments in which caterpillars were added to target plants suggested that undamaged neighboring plants increased their production of defensive chemicals, whereas plants farther away did not.

**HYPOTHESIS** Volatile chemicals released from plants that have been attacked by herbivores elicit a defensive response in undamaged plants.

**EXPERIMENT** One set of plants was placed downwind of other plants in which two leaves were torn, simulating herbivore damage, and a set of control plants was placed downwind of an empty chamber. The concentrations of defensive compounds in all sets of plants were measured and compared.

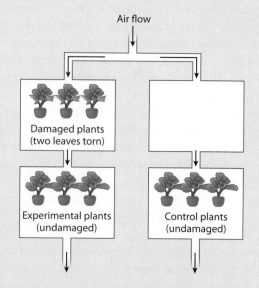

contrast, plants from nutrient-rich habitats would favor growth over defense because they could more readily replace lost or damaged tissues.

This situation is an example of a **trade-off:** Something is gained and, at the same time, something is lost. You can't have your cake and eat it, too—you have to make a choice. With plants, there can be a trade-off between growth and defense in which allocating resources for one means that those resources cannot be allocated for the other.

A recent study in the Amazon rain forest has focused on a possible trade-off between growth and defense. Researchers compared species found on nutrient-rich clay soils with closely related species found on nutrient-poor sandy soils. They found that species from the nutrient-rich clay soils invested less heavily in defenses than did species from the sandy sites. Because the clay-soil species divert fewer resources to defense, they grow faster in either soil type compared to sandy-soil species—as long as they are grown under a net that excludes herbivores (**Fig. 32.17**). However, when grown out in the open, each species grows best in the soil type on which it is naturally found.

The clay-soil species grow poorly on sandy soils when not protected by a net because, without ample soil nutrients

**RESULTS** When measured 52 hours after the initial damage, both the damaged plants and the undamaged experimental plants exhibited elevated levels of defensive compounds in their leaves compared to control plants.

**CONCLUSION** Volatile chemicals released from damaged plant tissues can trigger the production of defensive chemicals in undamaged plants.

**FOLLOW-UP WORK** The volatile chemicals released from damaged plants have been identified and shown to attract predatory animals that feed on the herbivores or parasitize their eggs or both. Thus, when plants respond to damaged neighbors they are "eavesdropping" on signals that likely evolved to attract predatory insects, rather than to warn neighboring plants.

**SOURCE** Baldwin, I. T., and J. C. Schulze. 1983. "Rapid Changes in Tree Leaf Chemistry Induced by Damage: Evidence for Communication Between Plants." *Science* 221:277–279.

**FIG. 32.17 A trade-off between growth and defense.** Plants from nutrient-rich clay habitats can overcome herbivory by growing fast, whereas plants from low-nutrient sandy habitats grow slowly and invest more in defense. The photos show that, without defenses, clay-soil plants are prone to being eaten by herbivores.

Clay soil plants growing in sandy soils

Unprotected from herbivores          Protected from herbivores

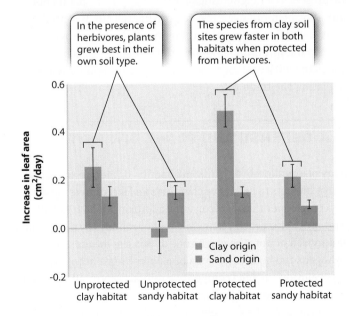

In the presence of herbivores, plants grew best in their own soil type.

The species from clay soil sites grew faster in both habitats when protected from herbivores.

available, they are unable to replace tissues consumed by herbivores. Conversely, when grown on the nutrient-rich soils, species from the sandy habitats are poor competitors because they allocate resources that could have been used for growth to defensive compounds and structures. What we see is a classic example of a trade-off.

→   **Quick Check 4** Does the trade-off between growth and defense favor a single species that dominates in all soil types or different species specialized for each habitat?

## Exposure to multiple threats can lead to trade-offs.

In the real world, plants are routinely confronted with more than one threat. A plant may be attacked simultaneously by both a pathogen and an herbivore, or it might be in danger of being shaded by a fast-growing neighbor at the same time that it must defend its leaves from being eaten. Given that the resources a plant can draw on are limited, how do plants respond to multiple threats?

Often, the response to multiple threats suggests a trade-off. For example, exposure to a pathogen can reduce the ability to respond to a subsequent herbivore attack. Tobacco plants in which systemic acquired resistance has been activated as a result of exposure to tobacco mosaic virus are more susceptible to the hornworm caterpillar, *Manduca sexta*. Studies indicate that cross talk between the signaling pathways for responding to pathogens versus herbivores is responsible for the reduced ability to respond subsequently to herbivores.

There can also be trade-offs between threats resulting from competition with neighboring plants and attack by herbivores. For example, when the phytochrome receptor in plants detects the presence of neighboring plants (Chapter 31), plants not only allocate more resources to growing tall but also produce fewer defensive chemicals in response to herbivore damage. Once again, interactions among the signaling molecules associated with each of these processes are responsible for mediating the trade-off.

## 32.4 DEFENSE AND PLANT DIVERSITY

Many plants allocate a significant fraction of their resource and energy budgets to defensive chemicals and structures. Despite this commitment, insects consume approximately 20% of all plant production each year, while pathogens destroy a substantial additional amount. The magnitude of this loss means that pathogens and herbivores must be a potent force in both ecology and evolution, shaping competitive interactions by allowing some species to prosper and limiting the success of others. In nature, plant fitness may be strongly influenced by the capacity to deter herbivores and resist pathogens.

Interactions between plants and their consumers are thought to have influenced patterns of plant diversification. Just as pollinating insects diversified along with angiosperms (the flowering plants), so, too, did plant-eating insects. The fact that angiosperms and plant-eating insects radiated at the same time is consistent with an important role for plant–insect coevolution in generating diversity. Further evidence that defense has influenced plant diversification comes from plant genes and secondary compounds. For example, the *R* genes that are instrumental in pathogen recognition form one of the largest gene families in plants, while chemical defenses against herbivores include about 6000 alkaloids and more than 10,000 terpenoid compounds. Such observations suggest that plants are locked into an evolutionary arms race with herbivores and pathogens. This arms race is very much in evidence in the efforts by farmers to limit crop losses by reducing the numbers of the pathogens and herbivores in fields and orchards.

## Pathogens, herbivores, and seed predators can increase plant diversity.

The lowland rain forests of eastern Ecuador have the highest diversity of tree species recorded anywhere: More than 1000 species of trees were identified in a 500 m × 500 m plot. What allows so many tree species to coexist is one of the great questions of modern biology. In the 1970s, two ecologists, Daniel Janzen and Joseph Connell, independently proposed that interactions with host-specific pathogens, herbivores, or seed predators play an important role.

The Janzen–Connell hypothesis, as it is now known, is based on the observation that seeds or seedlings often suffer high rates of mortality when growing directly beneath an adult tree of the same species (**Fig. 32.18**). Seedlings that grow farther away have a better chance of escaping pathogens that are common in the soil underneath an adult tree. By a similar argument, seeds that disperse far from their maternal plant may escape notice by seed predators, which often gather in higher densities near trees with mature fruit. Such density-dependent mortality provides rare species with an advantage over more common species. In other words, the offspring of a rare species is more likely to survive because its seeds are more likely to be dispersed into a location where there are no neighboring adult plants of the same species. By favoring the survival of less-common species, density-dependent mortality prevents one species from outcompeting all others.

In the tropical rain forests of Ecuador and elsewhere, only a small number of tree species are common; the majority of tree species are rare. The ability to persist at low population density allows rare species to escape herbivore and pathogen damage, but requires reliable ways of getting pollen from one plant to another. Thus, animal pollination, herbivory, and infection by pathogens are thought to play complementary roles in promoting angiosperm diversity.

**FIG. 32.18** **Density-dependent mortality promotes tree diversity.** Seedlings of rare species have a greater probability of surviving because they are more likely to escape infection by species-specific pathogens.

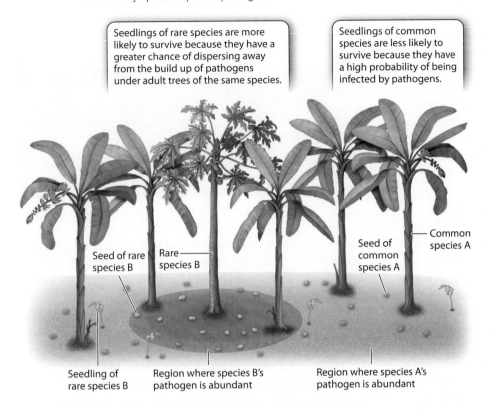

Seedlings of rare species are more likely to survive because they have a greater chance of dispersing away from the build up of pathogens under adult trees of the same species.

Seedlings of common species are less likely to survive because they have a high probability of being infected by pathogens.

Seed of rare species B

Rare species B

Seed of common species A

Common species A

Seedling of rare species B

Region where species B's pathogen is abundant

Region where species A's pathogen is abundant

### The evolution of new defenses may allow plants to diversify.

American scientists Paul Ehrlich and Peter Raven observed that closely related species of plants were fed upon by closely related species of butterflies. In 1964, they cited this observation as evidence that plants and their herbivores are caught up in a coevolutionary arms race: Plants evolve new forms of defense, and herbivores evolve mechanisms to overcome these defenses. In particular, Ehrlich and Raven hypothesized that the evolution of novel defenses has been an important force in the diversification of both plants and herbivorous insects.

An innovation in defense may allow a plant population to expand into new areas, increasing its potential to evolve into new species. Similarly, innovation in overcoming plant defenses may allow an insect or pathogen population access to new plant resources. Ehrlich and Raven's "escape and radiate" hypothesis thus predicts a burst of diversification following the evolution of a key innovation that allows plants to avoid being eaten. This hypothesis makes intuitive sense, but only with the advent of phylogenetic trees based on DNA sequence comparisons (Chapter 23) could it be tested. Biologists knew that latex or resin canals evolved independently in more than 40 different groups of plants. Reasoning that these features represent a novel defense, they asked whether the lineages that evolved latex or resin canals were more diverse than closely related groups that lacked this innovation. In 14 of the 16 groups examined, the lineages with the protective canals were significantly more diverse than their closest relatives that lacked this innovation. This pattern supports the hypothesis that the evolution of latex and resin canals provided plants with the freedom to expand into new habitats.

Escalation of defenses, however, is not the only possible evolutionary outcome. Phylogenetic research shows that, as milkweeds diversified, many of the newly evolved species produced *fewer* chemical defenses such as latex canals and cardenolides and instead replaced tissues by growing quickly following damage. The presence of specialized herbivores such as monarch caterpillars that can disarm milkweeds' defenses may help explain why these more recent plant species reduce investments in defense in favor of more rapid growth.

### ❓ CASE 6 Agriculture: Feeding a Growing Population
### Can modifying plants genetically protect crops from herbivores and pathogens?

The density-dependence of herbivore and pathogen damage is a particular challenge for agriculture as practiced in most countries. When acre after acre are covered by a single crop species, successful invaders will not be slowed by low target density. Large expanses of a single crop allow pest populations to build up to high levels, and the tendency to plant genetically uniform varieties increases the losses in yield. Not surprisingly, then, herbivores and pathogens exact a major toll on agricultural crops, and protecting crops from damage is a high priority for farmers.

The production of pesticides and herbicides increased dramatically in the second half of the twentieth century. Together with irrigation and fertilizers, these compounds are important contributors to the increased yields of the Green Revolution. However, large-scale application of pesticides is not without risk. In addition to the potential for toxic effects, widespread chemical treatments provide a strong selective force for the evolution of resistance. As pests become resistant to pesticides, one response is to apply more or stronger chemicals. The consequences of this approach, however, can spiral out of control as chemical applications indiscriminately kill off the natural predators of the pests themselves.

During the 1980s, farmers applied increasing amounts of pesticides to rice paddies in South Asia, yet crop losses continued to increase. This dangerous course was reversed by the introduction of a program first developed in the 1950s, in which farmers devote substantial time and effort determining what pests are actually present, whether or not they are at damaging levels, and whether or not there are natural predators of the pests. Only after understanding the ecology of the farm are pests removed, starting with mechanical means, then biological control, and, only as a last resort, industrial pesticides. This approach, called integrated pest management, allows farmers to minimize routine pesticide use and therefore prevent pest species from evolving resistance.

To further improve the ability of crops to defend themselves against herbivores and pathogens, crop breeders must emulate—on a faster timescale—the role played by evolution in selecting for resistance to newly evolving pests. For example, Norman Borlaug's original task was to develop wheat varieties for Mexico that were resistant to fungal pathogens. To do this, he obtained seeds of wheat varieties from around the world that showed evidence of pathogen resistance. He then crossed these disease-resistant varieties with ones adapted to the local conditions. Although it took thousands of hand-pollinations, Borlaug was successful at introducing resistance genes into wheat varieties that grew well in Mexico.

Crop breeding is a powerful way to alter the genetic makeup of cultivated species. However, because it relies on sexual reproduction, crop breeding is limited to the genetic diversity that exists within varieties that can interbreed. In contrast, *Rhizobium radiobacter* allows scientists to introduce specific genes found in distantly related organisms into cultivated species. One of the first commercial uses of this technology was to introduce chemical defenses from a bacterium into crop plants.

*Bacillus thuringiensis* (Bt), a soil-dwelling bacterium that produces proteins toxic to insects, has been used to control insect outbreaks in agriculture since the 1920s. At first, spores containing the toxins or the toxins themselves were applied directly to plants. In 1985, the genes that encode the toxins from *B. thuringiensis* were inserted into tobacco, where their expression conferred substantial protection against insects. Today, *Bt*-modified crops are some of the most widely planted genetically modified plants.

The introduction of *Bt* crops markedly reduced the application of pesticides, to the benefit of both farmworkers and the environment. However, the danger of planting *Bt*-expressing crops widely is that constant exposure to *Bt* toxins increases the probability that pest populations will evolve resistance. This possibility is of particular concern to organic farmers, who continue to use traditional application of *B. thuringiensis* spores to control insect outbreaks. To prevent the evolution of resistance, U.S. farmers are required by law to plant a fraction of their fields with non-*Bt*-expressing

**FIG. 32.19** *Bt*-crops co-planted with non-*Bt* varieties to prevent the development of resistance.

plants (**Fig. 32.19**). The proximity of non-*Bt*-expressing plants allows pests that are not resistant to *Bt* toxins to survive and interbreed with individuals feeding on the *Bt*-expressing plants, thus slowing the evolution of *Bt*-resistance. The efficacy of this practice, however, is called into question by reports of *Bt*-resistant pests and the need to apply increasing amounts of industrial pesticides to *Bt* crops.

If there is one practical lesson to be learned from our experience with *Bt*, it is that pathogens and herbivores will continue to evolve new ways of circumventing the defenses of crop plants. Agriculture will continue to require new forms of crop protection. While the use of chemical pesticides will probably remain widespread, crop protection will increasingly be based on the application of genetics, through both crop breeding and biotechnology. In either case, success will depend upon the availability of existing genes honed by natural selection. Without access to genes involved in pathogen recognition (R genes) or the production of chemical and mechanical defenses, we will have the tools, but not the raw materials, to protect our crops.

In some cases, crop protection will rely on genes from distantly related organisms, as illustrated by *Bt*. Commonly, however, defense strategies for crops will rely on genes from other plants, often ones that are closely related. For example, the specific R genes used for pathogen recognition typically function only within a single species or closely related species. Thus, recent efforts to develop potatoes that are resistant to *Phytophthora infestans* focus on transferring R genes from a wild relative of potato. The continuation of such efforts requires a commitment to safeguarding the genetic diversity found in natural populations of crop species and their close relatives. Only by tapping into these natural genetic resources can breeders produce crops that stay one step ahead of the constantly evolving threats of pathogens and herbivores.

# Core Concepts Summary

## 32.1 PLANTS HAVE EVOLVED MECHANISMS TO PROTECT THEMSELVES FROM INFECTION BY PATHOGENS, WHICH INCLUDE VIRUSES, BACTERIA, FUNGI, WORMS, AND EVEN PARASITIC PLANTS.

Pathogens enter plants through damaged tissue or stomata. page 32-2

Plant pathogens spread by growing or moving through the plant vascular system. Viruses move in the phloem, whereas bacteria and fungi move in the xylem. page 32-2

Biotrophic pathogens gain resources from living cells; necrotrophic pathogens kill cells before colonizing them. page 32-2

Plants have an innate immune system that allows them to detect and respond to pathogens. page 32-3

One part of the innate immune system acts generally and recognizes highly conserved molecules produced by broad classes of pathogens. page 32-4

A second part of the innate immune system acts specifically and consists of resistance or R genes that encode R proteins, which recognize pathogen-derived AVR proteins. page 32-4

Plants respond to pathogens by actively killing the cells surrounding the infection (the hypersensitive response) and by sending a signal to uninfected tissues so that they can mount a defense (systemic acquired resistance). page 32-5

The bacterium *Rhizobium radiobacter* infects plants by inserting some of its genes into the plant's genome, resulting in the formation of a tumor and providing a way to genetically engineer plants. page 32-7

## 32.2 PLANTS USE CHEMICAL, MECHANICAL, AND ECOLOGICAL DEFENSES TO PROTECT THEMSELVES FROM BEING EATEN BY HERBIVORES.

Plant defenses against herbivory include dense hairs, latex, and chemicals. page 32-9

Alkaloids are nitrogen-bearing compounds that damage the nervous system of animals. page 32-10

Terpenes are volatile compounds that give rise to many of the essential oils we associate with plants; they deter herbivores. page 32-10

Tannins bind with proteins, reducing their digestibility. page 32-11

Ant-plants, such as the bullhorn acacia, provide food and shelter for ants, which defend their host plant. page 32-11

## 32.3 THE PRODUCTION OF DEFENSES IS COSTLY, RESULTING IN TRADE-OFFS BETWEEN PROTECTION AND GROWTH.

Plants produce both constitutive defenses, which are produced whether or not a threat is present, as well as inducible defenses, which are triggered when a plant detects that it is being attacked. page 32-13

Plants produce volatile signals that attract insects that prey on herbivores. page 32-14

Experiments show that plants growing on nutrient-rich environments allocate more resources to growth than to defense. page 32-14

A trade-off is sometimes observed between plant growth and defense, in which plants favor one or the other, but cannot have both. page 32-15

## 32.4 INTERACTIONS AMONG PLANTS, PATHOGENS, AND HERBIVORES CONTRIBUTE TO THE ORIGIN AND MAINTENANCE OF PLANT DIVERSITY.

The Janzen–Connell hypothesis proposes that interactions with pathogens and herbivores increase plant diversity. page 32-16

The "escape and radiate" pattern of plant evolution suggests that plants undergo a burst of diversification following the evolution of a new form of defense against a pathogen or an herbivore. page 32-17

Herbivores and pathogens are a major concern for agriculture. page 32-17

Crop protection includes the use of chemical pesticides, integrated pest management, application of spores or toxins from *Bacillus thuringiensis* to plants, and inserting genes that encode for toxins from *B. thuringiensis* to make *Bt*-modified plants. page 32-18

# Self-Assessment

1. Describe how pathogens enter and move within the plant body.

2. Distinguish between biotrophic and necrotrophic plant pathogens.

3. Name and describe two components of the plant innate immune system.

4. Describe and contrast the hypersensitive response and systemic acquired resistance.

5. Describe a feature of *Rhizobium radiobacter* that makes it a useful tool in biotechnology.

6. Describe three ways that plants protect themselves from being eaten by herbivores.

7. Explain why there are often trade-offs between plant growth and plant defense.

8. Explain how pathogens and herbivores can increase plant biodiversity.

9. Draw a phylogenetic tree that illustrates an "escape and radiate" pattern of diversification for plants that evolve novel defenses.

10. Describe one benefit and one disadvantage of herbicide and pesticide use in agriculture.

Do you understand the chapter's Core Concepts? Log in to check your answers to the Self-Assessment questions, then practice what you've learned and reinforce this chapter's concepts by working through the problems and multimedia tutorials provided there.

⊡ **www.biologyhowlifeworks.com**

## CHAPTER 33

# PLANT DIVERSITY

## Core Concepts

**33.1** Plant diversity is dominated by angiosperms, which make up approximately 90% of all plant species found today.

**33.2** The bryophytes diverged before the evolution of vascular plants, and they grow in environments where the ability to obtain water from the soil does not provide an advantage.

**33.3** Spore-dispersing vascular plants are small, often epiphytic plants that usually grow in moist environments.

**33.4** Gymnosperms produce seeds and woody stems and are most common in seasonally cool or dry regions.

**33.5** Angiosperm diversity is partly explained by animal pollination.

In previous chapters, we explored how plants are built and how they function (Chapter 29); how these organisms, anchored to the ground, complete a sexual life cycle (Chapter 30); how the complex plant body develops from a fertilized egg (Chapter 31); and how plants defend themselves against predators and pathogens in terrestrial environments (Chapter 32). With this knowledge in hand, we are equipped to explore the diversity of land plants. How have evolutionary modifications of development and life cycle contributed to the patterns of plant diversity observed in forests, prairies, and tundra? Why are some groups of plants more diverse than others? How have changes in Earth's climate influenced the evolutionary history of plants? How did a small number of plant species come to support the nutritional needs of 7 billion humans?

## 33.1 PLANT DIVERSITY: AN EVOLUTIONARY OVERVIEW

In 1879, Charles Darwin wrote to botanist and long-time friend J. D. Hooker, "The rapid development as far as we can judge of all the higher plants within recent geological times is an abominable mystery." The mystery, according to Darwin, was the difficulty in explaining the relatively sudden and recent diversification of flowering plants, or angiosperms, in evolutionary history. If we tabulate the numbers of species within each major branch on the phylogenetic tree shown in **Fig. 33.1,** one fact stands out. Of the nearly 400,000 species of plants present today, approximately 90% are angiosperms. Thus, a central question in the study of plant evolution is: Why are angiosperms so much more diverse than any other group of plants?

Furthermore, angiosperms are relative newcomers. They appeared in the fossil record about 140 million years ago. The oldest known evidence of land plants is found in rocks approximately 465 million years old. Thus, for more than 300 million years, terrestrial vegetation was made up of plants other than angiosperms. During this period, lycophytes, ferns and horsetails, and gymnosperms, as well as many groups of now-extinct plants, dominated the land, forming forests and landscapes unlike those that surround us today.

Once the flowering plants gained an ecological foothold, however, the number of angiosperm species increased at an unprecedented rate. As the number of angiosperm species rose, the number of species of other groups fell. Yet the total number of plant species increased dramatically. Thus, angiosperms' hold on plant diversity is the result of both squeezing out other groups and increasing the number of species that can coexist (**Fig. 33.2**).

Angiosperms transformed the environments into which they diversified. With their multicellular xylem vessels, angiosperms can transport water through their stems at higher rates than plants that rely entirely on single-celled tracheids. Their uptake of $CO_2$ for photosynthesis increases as well because $CO_2$ uptake is directly linked to water lost to the atmosphere (Chapter 29). As a result, angiosperms can support higher rates of transpiration and photosynthesis than other groups of plants.

Recent climate simulations suggest that angiosperms may have been necessary for the formation of tropical rain forests as we know them today. Without the higher rates of

**FIG. 33.1   Phylogenetic tree of land plants.** The number of species is shown in parentheses. Note that 90% of all land plant species are angiosperms.

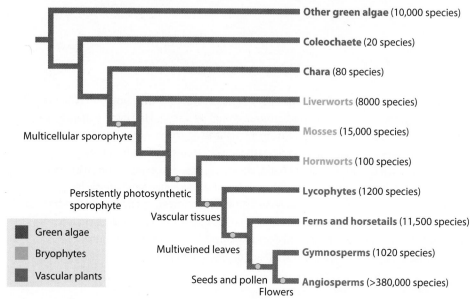

transpiration exhibited by angiosperms, many tropical regions would have higher temperatures and lower rainfall. The warmer, drier conditions would have been less conducive to the luxuriant plant growth found in tropical rain forests. The new tropical forests, with their dense shade and humid understories, provided new habitats into which both angiosperms and non-angiosperms could evolve. Thus, although early angiosperms outcompeted many older lineages for space and light, their evolution also stimulated continued evolution in at least some of the groups that preceded them.

## 33.2   BRYOPHYTES

Liverworts, mosses, and hornworts are referred to as **bryophytes**. They do not form a monophyletic group, but instead a paraphyletic group. Recall from Chapter 23 that a monophyletic group includes a common ancestor and *all* of its descendants, whereas a paraphyletic group includes a common ancestor and *some* of its descendants. Nevertheless, liverworts, mosses, and hornworts are discussed as a group and merit their own name because they share so many features despite having evolved independently for hundreds of millions of years.

As the living representatives of the first plant lineages to diverge after plants moved onto land, bryophytes provide us with insights into how plants gained a foothold on the terrestrial environment. This knowledge is particularly useful because the first plants are poorly represented in the fossil record. Only tiny spores and fragments of a cuticle-like covering record these early events.

Fig. 33.1 shows that liverworts, mosses, and hornworts diverged before the evolution of lignified xylem, about 425 million years ago. Thus, bryophytes add to our understanding of the evolutionary history of plants by showing how plants that lack xylem and phloem—key innovations in the evolution of plants—are able to survive in terrestrial environments alongside vascular plants. Of course, we must not lose sight of the fact that bryophytes have continued to evolve as the conditions for life on land have changed over the past 400+ million years. Where bryophytes grow and what they look like have been shaped by their long coexistence with vascular plants.

**FIG. 33.2   The diversity of fossil plant assemblages through time.** First appearing about 140 million years ago, angiosperms have come to dominate plant diversity.

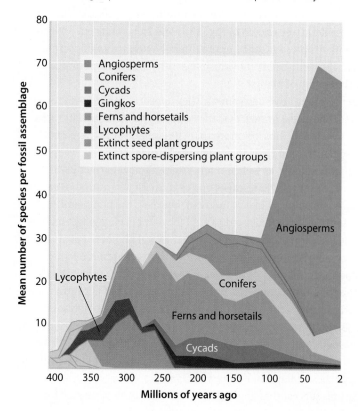

→   **Quick Check 1** Look at Fig. 33.1 and name the paraphyletic groups indicated by the tree.

## Bryophytes are small, simple, and tough.

Mosses are the most widely distributed of the bryophyte lineages and the most diverse, with about 15,000 species. You may have seen moss growing on a shady log or on top of rocks by a stream. However, mosses grow in all terrestrial environments, from deserts to tropical rain forests. Liverworts (about 8000 species) and hornworts (about 100 species) are less widespread and also less diverse. In considering the diversity of these three lineages, let's start by reviewing some of the features common to all bryophytes.

The most obvious feature that unites the bryophytes is that they are small, most of them only a couple of centimeters in height (**Fig. 33.3**). The major constraint on bryophyte size is thought to arise from their mode of fertilization. Bryophytes release their sperm into the environment. These sperm must swim through films of surface water or be transported by the splash of a raindrop to meet a female gamete. Sperm can travel in this manner only a relatively limited distance and only if water is present. A few mosses grow to be more than half a meter tall, but these giants of the bryophyte world are found only in the understory of very wet forests.

Bryophytes are not only small, they also have simple bodies. Some produce only a flattened photosynthetic structure called a **thallus.** Others consist of slender stalks and have a leafy appearance. However, these leaf-like structures are quite different from the leaves of vascular plants in that they are only one to several cells thick and lack internal air spaces or a water-conducting system. Both thalloid and leafy species are found in the liverworts, whereas all mosses are of the leafy type and all hornworts are of the thalloid type.

These conspicuous and persistently photosynthetic bryophyte bodies represent the haploid generation. Recall from Chapter 30 that a key innovation that appeared in the land plants is the alternation between a multicellular gamete-producing generation composed entirely of haploid cells and a multicellular spore-producing generation composed of diploid cells. The closest green algal relatives to the land plants have a life cycle in which there is only a single multicellular phase rather than two. The multicellular phase of the green algal relatives corresponds to the haploid, gamete-producing generation of land plants (the gametophyte generation) because both are formed entirely of haploid cells and produce gametes. Thus, the diploid, spore-producing generation (the sporophyte generation) represents a new component of the life cycle that evolved as plants moved onto land.

In bryophytes, the sporophyte remains physically attached to and, in varying degrees, nutritionally dependent upon the gametophyte. In a few liverworts, the sporophyte remains embedded within the thallus, releasing spores only when the gametophyte dies and its body decays. In this case, the multicellular sporophyte generation serves only to amplify the number of haploid spores. In most bryophytes, the sporophyte extends several centimeters above the gametophyte, greatly increasing the chances of spores being dispersed through the air. In a few cases, such as the liverwort *Marchantia* (Fig. 33.3b), tiny sporophytes grow from the top of upright structures produced by the gametophyte.

In mosses and liverworts, the sporophyte is short lived, drying out after the spores are dispersed. However, in hornworts, the sporophyte can live nearly as long as the gametophyte because it can produce new cells at its base, similar to the way grass blades elongate (Chapter 32). Thus, bryophytes illustrate an important trend in the evolution of

---

**FIG. 33.3**  **Bryophyte diversity.** Bryophytes include (a) mosses (*Polytricum commune* is shown); (b) liverworts (*Marchantia berteroanna*, a thalloid liverwort, is shown), and (c) hornworts (*Anthoceros* species).

plants, an increase in the persistence of the spore-producing generation.

Because bryophytes do not produce lignified xylem conduits, they cannot pull water from the soil. Instead, they absorb water and $CO_2$ through their surfaces, which have little or no waxy cuticle. Bryophytes can absorb enough water to remain metabolically active when the environment is wet, but they must be able to tolerate desiccation when the environment is dry. Thus, while the small size and lack of differentiated structures might suggest that bryophytes are delicate, nothing could be further from the truth. Bryophytes can be found from the equator to both the latitudinal and altitudinal limits of vegetation, and from swamps to deserts. They are the only plants that grow on the Antarctic continent. In fact, the entire flora of that continent consists of approximately 100 moss species and 25 species of liverworts.

Nevertheless, because bryophytes are so small, they are poor competitors for light and space. Instead, they thrive in local environments where roots do not provide an advantage—for example, on the surfaces of rocks. Many bryophytes live on the branches and trunks of trees rather than on the ground. Plants that grow on other plants are called **epiphytes** (from the Greek words *epi*, "on," and *phyton*, "plant"). Bryophytes are well suited to grow in this way because they are not dependent upon the soil as a source of water.

## Bryophytes exhibit several cases of convergent evolution with the vascular plants.

Bryophytes are interesting partly because they are so different from the more familiar vascular plants. However, given that they have long evolved in parallel with vascular plants, it is not surprising that bryophytes have evolved similar solutions to environmental challenges, a process referred to as convergent evolution (Chapter 23). For example, some mosses depend on insects to transport their spores (**Fig. 33.4a**). These mosses produce brightly colored sporangia, which emit volatile chemicals that mimic compounds released by rotting flesh or herbivore dung. These chemicals attract insects, and the insect legs become covered with spores when they land on the sporangia. When the insect subsequently lands on a real pile of dung, some of the spores fall off into what is, for them, an ideal site for supporting the growth of a new gametophyte.

Another example of convergent evolution is the presence in some mosses and liverworts of cells specialized for the transport of water and carbohydrates. It is clear from the structure of these cells that they evolved independently from the xylem and phloem of vascular plants. In particular, the water-conducting cells in bryophytes do not have lignified cell walls and thus are not sufficiently rigid to pull water from the soil. Nevertheless, water and carbohydrates can move more efficiently from one part of the plant to the other through these elongated cells. Not surprisingly, these internal transport cells are found in the largest of the bryophytes (**Fig. 33.4b**).

As discussed in Chapter 29, stomata are pores in the epidermis of vascular plants that can open and close, providing an active means of controlling water loss. Stomata also occur on the sporophytes of some mosses and hornworts. It is unknown whether the presence of these stomata is a case of convergent evolution. Answering that question will require further study of the structure and development of bryophyte stomata.

→ **Quick Check 2** If the stomata of bryophytes and vascular plants are not an example of convergent evolution, what is an alternative explanation?

---

**FIG. 33.4** **Moss specializations.** (a) The yellow moose-dung moss, *Splachnum luteum,* attracts insects that then transport its spores. (b) *Dawsonia superba* has internal transport tissues that allow it to grow more than 40 cm tall.

## Sphagnum moss plays an important role in the global carbon cycle.

In most ecosystems, bryophytes make only a small contribution to the total biomass. The one exception is **peat bogs**, wetlands in which dead organic matter accumulates. A major component of peat bogs is sphagnum moss, any of the several hundred species of the genus *Sphagnum*. These mosses play a key role in creating wet and acidic conditions that slow rates of decomposition. They have specialized cells that hold onto water, much like a sponge, and they secrete protons that acidify the surrounding water (**Fig. 33.5**). Unlike many plants with roots, sphagnum moss thrives under wet, acidic conditions. In addition, sphagnum mosses produce phenols, organic compounds that slow decomposition under waterlogged conditions.

Peat bogs occupy only 2% to 3% of the total land surface, but they store large amounts of organic carbon—on the order of 65 times the amount released each year from the combustion of fossil fuels. Thus, any increase in the decomposition rate of these naturally formed peats has the potential to increase significantly the $CO_2$ content of Earth's atmosphere.

What might cause an increase in the decomposition rate? Peat bogs are vulnerable to changes in the water table. As water levels fall, the peat becomes exposed to the air, allowing it to be broken down by microbes. Most peat bogs are located in northern latitudes, where the effects of climate change are predicted to be greatest. If rising summer temperatures result in higher rates of transpiration from surrounding forests, the result could be a lowering of the water table and a relatively rapid release of $CO_2$ from this moss-dominated ecosystem. Thus, peat bogs have the potential to accelerate rates of climate change over the next century.

---

## 33.3 SPORE-DISPERSING VASCULAR PLANTS

As noted in Chapters 29 and 30, vascular plants differ from bryophytes in several key aspects of form and life cycles. In vascular plants the diploid sporophyte is larger than the gametophyte and becomes capable of supplying all of its own needs. The sporophyte achieves greater height and independence with the help of vascular tissues that pull water from the soil and distribute carbohydrates throughout the plant body.

Vascular plants can be divided into two groups according to how they complete their life cycle. Lycophytes, as well as ferns and horsetails, disperse by spores and rely on swimming sperm for fertilization, just as occurs in bryophytes. Gymnosperms and angiosperms are seed plants: They are dispersed through the movement of seeds and fertilized through the movement of pollen. In this section, we focus on the spore-dispersing vascular plants: lycophytes, and ferns and horsetails. In both groups, the sporophyte dominates in terms of physical size; the gametophyte is only a few centimeters, thalloid in structure, and typically tolerant of desiccation. Thus, our discussion emphasizes the sporophytes of these plants. However, it is important to remember that because their gametophytes release sperm into the environment, they, like the bryophytes, are dependent on surface moisture for fertilization.

### Rhynie cherts provide a window into the early evolution of vascular plants.

Fossils recovered from a single site in Scotland document key stages in the evolution of vascular plants. Four hundred million years ago, near the present-day village of Rhynie, a wetland community thrived in a volcanic landscape. Mineral-laden springs similar to those found today in Yellowstone Park encased the newly deposited remains of local populations in chert, a mineral made of silica ($SiO_2$). These fossils preserve plants, fungi, protists, and small animals in remarkable detail. Rhynie cherts provide our

---

**FIG. 33.5** **Sphagnum moss.** Peat bogs have low rates of organic decomposition due to the wet and acidic conditions created by sphagnum moss.

Peat bog

Water-holding cell

Sphagnum moss

Photosynthetic cell

**FIG. 33.6**   **Rhynia, a fossil vascular plant from the 400-million-year-old Rhynie chert.** All vascular plants are descended from simple plants such as this.

best view of early vascular plants, which were small (up to about 15 cm tall) photosynthetic structures that branched repeatedly, forming sporangia at the tips of short side branches (**Fig. 33.6**). These plants had no leaves, and their only rooting structures were small hairlike extensions on the lower parts of stems that ran along the ground. The stems had a cuticle with stomata, and a tiny cylinder of vascular tissue ran through the center. Many of the gametophytes also had erect cylindrical stems, not at all like the reduced gametophytes of living vascular plants.

Rhynie cherts preserve another important glimpse of early plant evolution: Some fossils have a branched sporophyte, but no evidence of vascular tissue. Evidently, plants evolved the upright stature we associate with living vascular plants before they evolved the differentiated tissues of the vascular system (xylem and phloem). Once vascular tissues had evolved, however, they played a key role in shaping subsequent evolution of the plant sporophyte.

### Lycophytes are the sister group of all other vascular plants.

One of the fossil populations in Rhynie cherts is notably larger and more complex than the others. These fossils show leaf-like structures arranged on the stem in a spiral pattern. Through

the stem runs a thick-lobed cylinder of xylem (**Fig. 33.7**). These are among the earliest fossils of lycophytes, an early-branching group of vascular plants that can still be found from tropical rain forests to Arctic tundra. Because lycophytes are the earliest branching group of vascular plants, they are the sister group to all other vascular plants (see Fig. 33.1).

Fossils indicate that the leaves and roots of lycophytes evolved independently from those of ferns, horsetails, and seed plants. A distinctive feature of lycophyte leaves is that they have only a single vein, in contrast to the more complex leaf venation found in other vascular plants. About 1200 species of lycophytes can be found today. Plants assigned to the genus *Lycopodium* and its close relatives (**Fig. 33.8a**) grow low to the ground and produce branching stems covered by small leaves arranged in a helical pattern. Roots are produced from stems growing along the ground, while spores form in sporangia borne on leaves, either those arranged along the stems or those clustered near stem tips.

A second group of lycophytes, called *Selaginella,* have leaves that may be helically arranged around the stem or occur in four distinct rows with the leaves held in a single plane, giving these branched stems a distinctive appearance (**Fig. 33.8b**). Many species of

**FIG. 33.7**   **An early lycophyte preserved in the 400-million-year-old Rhynie chert.** The fossil has helically arranged leaflike structures, adventitious roots (that is, roots that arise from the shoot and not from the root system), and a central strand of xylem, features much like those of living lycophytes.

**FIG. 33.8** **Lycophyte diversity.** Living lycophytes include about 1200 species, in three major groups: (a) *Lycopodium annotinum*, showing yellow sporangia-bearing leaves at the shoot tips of this 5-to-30-cm tall plant; (b) *Selaginella wildenowii*, in a close-up showing the flattened arrangement of its 3-mm-long leaves; and (c) *Isoetes lacustris*, an aquatic lycophyte that grows from 8 to 20 cm tall.

*Selaginella* live as epiphytes in tropical forests. A few are vines and some, called resurrection plants, live in seasonally dry habitats where they tolerate desiccation, as more commonly do bryophytes.

The most distinctive living lycophytes are the quillworts (genus *Isoetes*), which are small plants that live along the margins of lakes and slow moving streams (**Fig. 33.8c**). A few species are desiccation tolerant and so can live in ponds that dry up in summer. Uniquely among living lycophytes, they have a vascular cambium. Seeing a quillwort today, you would never guess that its relatives once included enormous trees.

**Ancient lycophytes included giant trees that dominated coal swamps about 320 million years ago.**
Fossils show that ancient lycophytes evolved additional features convergently with seed plants, including a vascular cambium and cork cambium that enabled them to form trees up to 40 m tall and, in some cases, reproductive structures similar to seeds (**Fig. 33.9**). Indeed, swamps that formed widely about 320 million

years ago were dominated by tree-sized lycophytes. Unlike the vascular cambium of seed plants, the vascular cambium of these

**FIG. 33.9** **Giant lycophytes.** (a) Artist's vision of giant lycophytes. (b) A fossil grove of lycophyte trees showing rooting structures and thick stems. (c) Fossil showing the seedlike reproductive structures, about 1 to 2 cm long, evolved by these ancient trees.

lycophytes (**Fig. 33.10**) produced relatively little secondary xylem and no secondary phloem. Instead, the giant lycophytes relied on thick bark for mechanical support. Thus, tree-sized lycophytes were markedly different from the trees familiar to us today.

These forest giants were not outcompeted by seed plants, but persisted for millions of years in swampy environments alongside early seed plants. It was environmental change that spelled their doom. When changing climates dried out the swamps, the large

## HOW DO WE KNOW?

**FIG. 33.10**

# Did woody plants evolve more than once?

**OBSERVATIONS** Today, the vascular cambium occurs almost exclusively among seed plants, with only limited development of secondary xylem in some adder's tongue ferns and quillwort lycophytes (*Isoetes* species). The fossil record, however, shows that other plants, now extinct, also had a vascular cambium.

**HYPOTHESIS** The vascular cambium evolved convergently in several different groups of vascular plants.

**EXPERIMENT AND RESULTS** The vascular cambium is recorded in fossils by xylem cells in rows oriented radially in the stem or root. Thus, the giant tree lycophytes of Carboniferous coal swamps (the top photo at the lower right) had a vascular cambium, as did extinct tree-sized relatives of the horsetails (the bottom photo) and other extinct horsetail relatives. Phylogenetic trees generated from morphological features preserved in fossils unambiguously show

that woody lycophytes, woody horsetail relatives, and the group of seed plants and progymnosperms did not share a common ancestor that had a vascular cambium. Moreover, anatomical research shows that the vascular cambium of extinct woody lycophytes and the giant horsetails *Archaecalamites* and *Calamites* generated secondary xylem but not secondary phloem, unlike the vascular cambium of seed plants. Living horsetails do not have a vascular cambium, but fossils show that they are descended from ancestors that did make secondary xylem and have lost this trait through evolution.

**CONCLUSION** Fossils and phylogeny support the hypothesis that the vascular cambium and, hence, wood evolved more than once, reflecting a strong and persistent selection for tall sporophytes among vascular plants.

**SOURCE** Taylor, T. N., E. L. Taylor, and M. Krings. 2009. *Paleobotany.* Amsterdam: Elsevier.

**X** *Rhynia*
*Lycopodium*
*Selaginella*
*Isoetes*
**X** *Lepidodendron*
**X** *Psilophyton*
Adder's tongue ferns
Whisk ferns
Extant horsetails
**X** *Calamites*
**X** *Archaecalamites*
**X** *Sphenophyllum*
Ferns
**X** Progymnosperms
Seed plants

○ Vascular cambium present
● Vascular cambium lost
**X**  Extinct

**A small branch of *Lepidodendron*, an ancient giant lycophyte**

Cortex        Secondary        Pith      Vascular
              xylem                      cambium

2.5 mm

**A small stem of *Calamites*, an ancient giant horsetail**

Secondary xylem

Hollow center

Vascular cambium

5 mm

**FIG. 33.11 Diversity of ferns and horsetails.** (a) The adder's tongue fern *Ophioglossum*, (b) the whisk fern *Psilotum*, (c) the horsetail *Equisetum*, (d) a marattiod fern, (e) an aquatic fern (*Salvinia*), and (f) a polypod fern (in the middle is *Athyrium felix-femina*, the common lady-fern).

Lycophytes
**Ferns and horsetails**
Gymnosperms
Angiosperms

Adder's tongue ferns (114 species)

Whisk ferns (15 species)

Horsetails (15 species)

Distinct shoot organization

Marattiod ferns (240 species)

Leptosporangium

Earlier branching leptosporangiate ferns (1030 species)

The branching order of horsetails, marattiod ferns, and leptosporangiate ferns is unresolved.

Aquatic ferns (85 species)

Tree ferns (710 species)

Polypod ferns (9300 species)

lycophytes disappeared, along with their habitat. Little *Isoetes* was left as the only living reminder of a once-dominant group of plants. However, the other legacy of these ancient forests is the coal deposits we mine today. As trees died and fell over into the swamp, their bodies decomposed slowly. Over time, as material accumulated and became buried, the combined action of high temperature and pressure converted the dead organic matter first into peat and then into the carbon-rich material we call coal. Thus, a major source of energy used today by humans derives originally from photosynthesis carried out by these lycophyte trees.

### Ferns and horsetails are morphologically and ecologically diverse.

Ferns and horsetails form a monophyletic group that are the sister group to the seed plants (**Fig. 33.11**). The majority of species in this group are in fact ferns, which can be recognized by their distinctive leaves (Fig. 33.11) that uncoil during development from tightly wound "fiddleheads" (**Fig. 33.12**). This group also includes horsetails and whisk ferns, which traditionally were considered as distinct groups because of their unique body organizations. Whisk ferns have photosynthetic stems without leaves (see Fig. 33.11b) and they also do not form roots. They thus look similar to fossils of early vascular plants. Molecular sequence comparisons, however, support the hypothesis that whisk ferns are the simplified descendants of plants that produced both leaves and roots and that they are members of the same lineage as ferns.

The horsetails, represented by 15 living species of the genus *Equisetum*, produce tiny leaves arranged in whorls, giving the stem a jointed appearance (see Fig. 33.11c). Horsetail stems are hollow, and their cells accumulate high levels of silica, earning them the name "scouring rush." Today, horsetails are small plants,

**FIG. 33.12 A fern leaf uncoiling from a "fiddlehead."**

**FIG. 33.13**   **A tree fern and an aquatic fern.** (a) Silver tree ferns (*Cyathea dealbata*) grow 10 m tall or more. (b) *Azolla* ferns floating on the surface of still freshwater. Each fern is about 10 mm in diameter.

are by far the most diverse group of ferns (about 9300 species), and their distinctive sporangia, with their high potential for dispersal, may have played a role in their evolutionary success.

Although the fossil record shows that ferns originated more than 360 million years ago, the radiation of the polypod ferns occurred after the rise of the angiosperms. Thus, many of the fern species present today are likely to have evolved to occupy habitats newly created by the development of angiosperm forests. Approximately 40% of living ferns are epiphytes,

most less than a meter tall. However, tree-sized versions grew side by side with the ancient lycophyte trees (see Fig. 33.9).

Fern leaves can be large—in some cases up to 5 m long, although typically the photosynthetic surfaces are divided into smaller units called **pinnae.** Fern stems frequently grow underground, and only the leaves emerge into the air. Internode elongation can be very short, so that all the leaves are bunched together. Some species, such as bracken ferns, produce horizontal stems with widely spaced leaves. This arrangement allows bracken to spread rapidly. Because bracken is toxic to cattle and other livestock, its tendency to invade pastures is a serious problem.

Vascular tissue allows fern sporophytes to grow large. However, because ferns do not exhibit secondary growth, there are limits to how tall they can grow. Tree ferns can grow to more than 10 m by producing thick roots that descend from the leaves to the ground, adding both vascular capacity and mechanical support (**Fig. 31.13a**). Other ferns grow tall by producing leaves that twine around the stems of other plants, using the other plant for support. At the other extreme are tiny aquatic ferns such as *Salvinia* (see Fig. 33.11e) and *Azolla* (**Fig. 33.13b**) that float on the surface of the water.

Horsetails, whisk ferns, and a few other fern groups make large sporangia similar to those produced by early vascular plants. Most ferns, however, have a distinctive sporangium (called a leptosporangium) whose wall is only a single cell thick.  Polypod ferns (see Fig. 33.11f), in particular, have leptosporangia with a line of thick-walled cells that run along the sporangium surface. When the spores mature, the sporangium dries out and these cells contract, forcibly ejecting spores into the air. Polypod ferns

many of them growing in tropical rain forests dominated by angiosperm trees.

Epiphytic ferns are exposed to much more light than plants established on the ground. However, the canopy environment poses its own challenges, notably the danger of dislodgement and the risk of drying out. As a result, epiphytic ferns have adaptations that allow them to survive in the forest canopy. For example, staghorn ferns (**Fig. 33.14**) produce two leaf types: erect leaves that are photosynthetic and produce sporangia and basal leaves that grow close to the stem of the host tree, forming a pocket in which organic matter accumulates. The roots of staghorns can grow into the pocket of organic matter, where they are protected from damage and desiccation.

### An aquatic fern contributes to rice production.

*Azolla* is a free-floating aquatic fern that has been used for centuries in Asia as a biofertilizer of rice paddies (see Fig. 33.13b). Its success as a fertilizer is due to its symbiotic association with nitrogen-fixing cyanobacteria. *Azolla* is common worldwide on still water such as ponds and quiet streams. Its leaves contain an internal cavity in which colonies of nitrogen-fixing cyanobacteria become established. The cyanobacteria provide *Azolla* with a ready source of nitrogen, while the fern provides shelter and carbohydrates. Rates of nitrogen fixation by the cyanobacteria are reported to be greater when the cyanobacteria are housed within an *Azolla* leaf than when these organisms live freely in the environment.

When added to newly planted rice fields, this fast-growing fern suppresses weed growth by forming a dense, floating layer. At the same time, it provides organic nitrogen that eventually

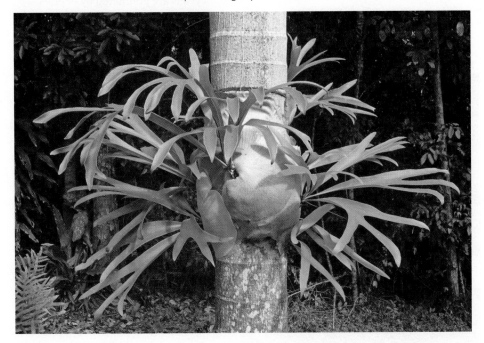

**FIG. 33.14 The staghorn fern.** This epiphyte produces two distinct types of leaf that enable it to thrive in the canopies of angiosperm trees.

more than a dozen additional groups, now extinct, were once present. These include a number of early seed plants that branched off before any living gymnosperm groups evolved. The floodplains adjacent to the coal swamps in which giant lycophytes thrived were populated by early seed plants with large fern-like leaves that produced large fleshy seeds. Gymnosperms probably reached their greatest diversity during the time of the dinosaurs. At that time, groups like cycads and conifers lived alongside a larger number of now-extinct groups.

**Cycads and ginkgos are the earliest diverging groups of living gymnosperms.**

The unobservant eye often confuses cycads with palm trees because both have unbranched stems and large leaves

finds its way into the developing rice plants. The floating fertilizer factories provided by *Azolla* may have made possible the long history of rice cultivation in Asia.

## 33.4 GYMNOSPERMS

The seed—and equally important, pollen—are evolutionary innovations that played important roles in the success of seed plants. Seed plants do not require external water for fertilization. Instead, the male gametophyte is transported through the air in a pollen grain. Successful pollination leads to fertilization and the formation of a seed. Seeds are multicellular structures that enhance dispersal success by transporting the embryo together with resources that it can draw on during germination. Before these innovations, the ancestors of the seed plants evolved the ability to grow large woody stems through the formation of both a vascular and a cork cambium. Therefore, woody stems and roots are the ancestral condition in seed plants.

The gymnosperms are one of two major groups of seed plants. The living gymnosperms include four groups: cycads, ginkgos, conifers, and gnetophytes (**Fig. 33.15**). However, the fossil record shows that

**FIG. 33.15 Seed plant diversity, showing a hypothesis of phylogenetic relationships among the gymnospermous plants.**

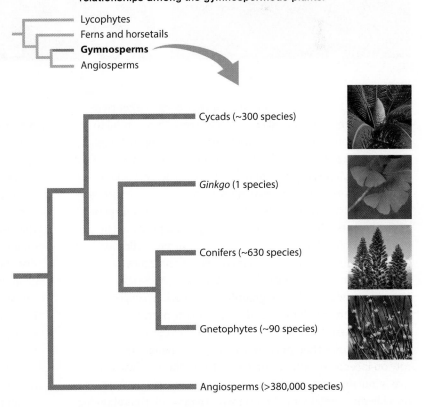

**FIG. 33.16 Cycads.** (a) Most living cycads have thick, unbranched stems that support large leaves and cones. Shown here is *Encephalartos transvenosus*. (b) Many cycads, such as *Encephalartos transvenosus,* have large cones. (c) Mature seeds are large (here showing those of *Cycas circinalis*), and contain a relatively large female gametophyte that provides nutrition for the embryo and germinating sporophyte.

(**Fig. 33.16**). However, the presence of large cones rather than flowers clearly marks cycads as gymnosperms. Cycads tend to have low photosynthetic rates and be slow growing. Although they have large stems, their vascular cambium produces little additional xylem, and most of their bulk is made up of a large pith and cortex. They form symbiotic associations with nitrogen-fixing cyanobacteria that allow them to grow in nutrient-poor environments. The apical meristem of cycads is protected by a dense layer of bud scales, enabling them to survive wildfires.

The approximately 300 species of cycads occur most commonly in tropical and subtropical regions. Most species have a limited and often fragmented distribution, suggesting that they were much more widespread in the past. Most cycad populations are small and often vulnerable to extinction. Cycads may owe their persistence in part to the fact that they rely on insects for pollination, as well as to their ability to live in nutrient-poor and fire-prone environments. Insects provide more efficient pollen transfer than wind, thus allowing species with small populations to reproduce successfully. Many cycads are pollinated by beetles, which are attracted by chemical signals produced by the cones. The insects use the large cones for shelter, and the pollen cones provide them with food. Following fertilization, brightly colored, fleshy seed coats attract a variety of birds and mammals that serve as dispersal agents. Because of their small numbers and fragmented distribution, two-thirds of all cycads are on the International Union for the Conservation of Nature's "red list" of threatened species, the highest percentage of any group of plants.

Cycads are an ancient lineage, with fossils at least 280 million years old. During the time of the dinosaurs, they were among the most common plants. Cycad diversity and abundance declined markedly during the Cretaceous Period, at the same time as the angiosperms rose to ecological prominence. A recent study shows, however, that all the present-day species within this group date from a burst of speciation that took place approximately 12 million years ago. At that time, changing continental positions and ocean circulation patterns resulted in increasingly cooler and more seasonal climates, providing new habitats for emerging species while dooming older species to extinction.

A second group of gymnosperms has only a single living representative: *Ginkgo biloba*. Like cycads, ginkgos date back about 270 million years, and during the time of the dinosaurs, they were common trees in temperate forests. Also like cycads, ginkgos declined in abundance, diversity, and geographic distribution over the past 100 million years, as a result of both changing climates and angiosperm diversification. *Ginkgo biloba* forms tall, branched trees, with fan-shaped leaves that turn a brilliant yellow in autumn (**Fig. 33.17**). Once again, like cycads, *Ginkgo* develops fleshy seeds, perhaps a feature inherited from their common seed plant ancestor.

Ginkgos have long been cultivated in China, especially near temples, and it is unclear whether any truly natural populations exist in the wild. Today, however, *Ginkgo* once again has a global

**FIG. 33.17** *Ginkgo biloba.* (a) The fan-shaped leaves of the ginkgo turn yellow in autumn and are then shed. (b) Its fleshy seeds enclose a large female gametophyte, as in cycads.

distribution, as its ability to tolerate pollution and other stresses associated with urban environments has made it a tree of choice for street plantings. The only drawback is that the fleshy seeds produce a chemical that smells like rancid butter. For this reason, individuals that produce pollen, as opposed to seeds, are strongly preferred.

### Conifers are forest giants that thrive in dry and cold climates.

The approximately 630 species of conifers are, for the most part, trees. Conifers include the tallest (over 100 m) and oldest (more than 5000 years) trees on Earth **(Fig. 33.18)**. Well-known conifers include pines, junipers, and redwoods. Conifer xylem consists almost entirely of tracheids, resulting in wood that is strong for its weight and has relatively uniform mechanical properties. Conifers have traditionally served as masts for sailing ships, and today are used for telephone poles. Conifers supply much of the world's timber, as well as the raw material for producing paper.

Most conifers are evergreen, meaning that they retain their often needle-like leaves throughout the year. In some species, the leaves are remarkably long lived, remaining green and photosynthetic for decades. Many conifers produce resin canals in their wood, bark, and leaves that deter insects and fungi. Conifer resins are harvested and distilled to produce turpentine and other solvents. In addition, conifer resins are the source of a number of important chemicals, including the drug taxol, which is used in the treatment of cancer.

In Chapter 30, we examined the life cycle of pine. Pollen is produced in small cones, and ovules develop in larger cones that mature slowly as the fertilized ovules develop into seeds. Conifers are wind pollinated, and most species also rely on

wind for seed dispersal. Some conifers, however, produce fleshy and often brightly colored tissues associated with their seeds that attract birds and other animals. In junipers, for example, the entire seed cone becomes fleshy. Juniper seeds are used as a seasoning; they are what gives gin its distinctive flavor.

Conifers dominate the vast boreal forests of Canada, Alaska, Siberia, and northern Europe. Conifers generally increase in abundance as elevation increases, and they are common in dry areas, such as the western parts of North America and much of Australia. Only a small number of conifer species are found in

**FIG. 33.18** **Conifers.** Giant sequoia (*Sequoiadendron giganteum*) can reach 85 meters in height and have stems that are 8 meters in diameter.

tropical latitudes, most commonly at higher elevations. Before the rise of the angiosperms, however, conifers were much more evenly distributed across the globe, including in the lowland tropics. One of the great questions in the evolutionary history of plants is why conifers were displaced from tropical latitudes by the angiosperms but continued to thrive in cold and dry environments.

An important difference between angiosperms and conifers is that angiosperms transport water in the multicellular xylem conduits called vessels, whereas conifers transport water in single-celled tracheids. In wet and warm environments, the xylem vessels of angiosperms can be both wide and long, increasing the efficiency of water transport through their stems. Conifer tracheids, which form from single cells, are much smaller. Thus, one hypothesis is that the evolution of highly conductive xylem vessels gave angiosperms a competitive advantage in many environments: They could support higher rates of photosynthesis and grow larger leaves. Once conifer populations began to dwindle, they could have disappeared completely because wind pollination, to be effective, requires large populations. In colder

or drier regions, the dangers of cavitation due to drought or freeze followed by thaw (Chapter 29) constrain the size of angiosperm vessels. This helps level the playing field in the competition between conifers and angiosperms and explains the greater coexistence of conifers and angiosperms in cold or dry climates.

### Gnetophytes are gymnosperms that have independently evolved xylem vessels and double fertilization.

Gnetophytes are a small group of gymnosperms (about 90 species) made up of three morphologically different genera (**Fig. 33.19**). *Welwitschia mirabilis* is found only in the deserts of southwestern Africa. It produces only two straplike leaves that elongate continuously from their base. *Gnetum* occur in tropical rain forests, where they form woody vines and small trees with large broad leaves. *Ephedra*, which is native to arid regions, produces shrubby plants with photosynthetic stems and small leaves.

Gnetophytes have long attracted the attention of plant biologists because their members exhibit a number of traits typically associated with angiosperms, notably the formation of multicellular xylem vessels and double fertilization, although the double fertilization in gnetophytes does not result in the formation of endosperm. Indeed, for years many botanists viewed gnetophytes as the sister group of angiosperms. DNA analysis, however, indicates that gnetophytes are not closely related to the angiosperms but are, instead, closely related to conifers. Thus, both vessels and double fertilization evolved independently and convergently in this group.

→ **Quick Check 3** Double fertilization evolved independently in gnetophytes and flowering plants. Name two other features usually associated with angiosperms that evolved independently in gymnosperms.

---

**FIG. 33.19 Gnetophytes.** The three genera in this group include some of the most distinctive seed plants: (a) *Welwitschia mirabilis*; (b) *Gnetum* species; and (c) *Ephedra* species.

---

## 33.5 ANGIOSPERMS

The evolution of the angiosperms completely transformed life on land. Not only did the total number of plant species increase significantly, but the productivity of terrestrial ecosystems increased as well. As we have seen, the evolution of angiosperms affected the abundance and distribution of other plant groups. In addition, the higher productivity of angiosperm-dominated environments increased the food available for insects, birds, and mammals. All these animal groups diversified along with the flowering plants.

### Angiosperm diversity remains a puzzle.

The question of why angiosperms are so diverse has occupied biologists since before Darwin's time. Despite their efforts, a definitive answer has so far not emerged. Diversity can result from high rates of species formation, and also from low rates of species loss. Interestingly, it may be that low rates of species loss have played an important role in the evolution of the angiosperms. Studies of the fossil record suggest that the rate at which new species appear is comparable for angiosperms and gymnosperms. However, angiosperm species disappear from the fossil record at a lower rate than gymnosperm species.

Flowers may contribute to angiosperm diversity by making it less likely that a rare species will become extinct. Wind-pollinated plants can reproduce only when their populations have a relatively high density. Since there is a high probability that no wind-carried pollen will fall on isolated individuals, species with low population density have a relatively high likelihood of extinction. But animal pollinators actively searching for rare species are much more likely to find them, and so animal-pollinated angiosperm populations can persist at low population densities.

A second feature of angiosperm reproduction that makes it likely that rare species will persist is the extreme reduction in the size of the female gametophyte (Chapter 30). Its small size is possible because the endosperm produced by double fertilization allows angiosperms to delay provisioning their offspring until after fertilization. The price in energy and resources that an angiosperm must pay to take a chance on being fertilized is much lower than in gymnosperms.

Tropical rain forests illustrate how the ability to reproduce at low population density contributes to angiosperm diversity. Rain forests may have several hundred different plant species growing on a single hectare of land, and most of these species are rare.

### Early diverging angiosperms have low diversity.

**Fig. 33.20** shows a phylogenetic tree of angiosperms. Note that while angiosperms as a whole are remarkably diverse, the early-branching groups of flowering plants are not. Early-branching angiosperms, however, provide important clues to the evolution of flowering plants.

The last common ancestor of angiosperms and living gymnosperms is thought to have lived more than 300 million years ago. In the absence of a closely related sister group, what tools can we use to understand the origin of the angiosperms? The first unambiguous fossils of angiosperms appear in rocks

**FIG. 33.20 Angiosperm phylogeny.** Most angiosperm species belong to one of two subgroups, the monocots and the eudicots.

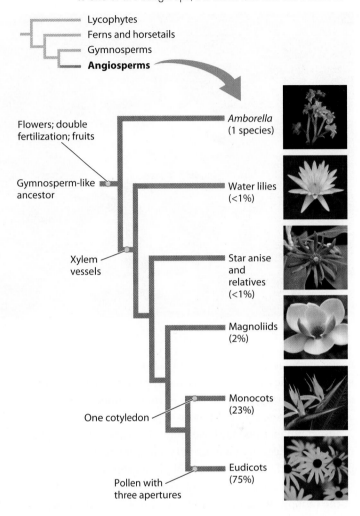

about 140 million years ago. These fossils belong to groups that branch early on the angiosperm phylogenetic tree, but they leave many questions unanswered regarding the ecological context of angiosperm evolution.

To gain further insights into the type of habitat in which the first angiosperms appeared, let's look at where living plants from the earliest-branching lineages of angiosperms are found today. Most occur in the understory of tropical rain forests; water lilies, which grow in shallow ponds, are one notable exception. This observation suggests that angiosperms may have evolved in wet and shady habitats. How could life in the understory of a tropical forest dominated by conifers have contributed to the evolution of flowers and xylem vessels?

Recall from Chapter 30 that flowers are a plant's way of enlisting animals to help them reproduce. Brightly colored petals and volatile scents advertise the presence of flowers, while rewards such as

**FIG. 33.21  Star anise, an early-branching angiosperm.** These flowers are pollinated by beetles.

nectar encourage pollinators to return repeatedly to individuals of the same species. Because many flowers produce pollen and ovules close to each other, the movement of animals from one flower to another provides an efficient mode of pollen transfer.

Insect pollination may have allowed the first angiosperms to exploit understory habitats of tropical rain forests. Wind provides an effective means of pollen transfer in open habitats or in plants that can produce reproductive structures at the top of the canopy. In the dense understory of a tropical forest, the likelihood that air movements alone would result in fertilization is close to zero. All of the living members of the earliest angiosperm lineages produce flowers that are visited by insects. **Fig. 33.21** shows the flower of star anise (*Illicium verum*), which is visited by beetles and whose fruit is the source of a popular seasoning used in Asian cooking.

Let's turn now to a second important feature of angiosperms, the formation of wood containing xylem vessels. As described in Chapter 31, most gymnosperms produce only tracheids, a single cell type that serves both the mechanical and the water transport roles of the stem. Angiosperms, in contrast, have two cell types, allowing them to separate the functions of mechanical support and water transport: Vessel elements stack to produce multicellular vessels through which water is transported, and thick-walled, elongate fibers provide mechanical support. This cellular specialization enables angiosperms to produce wood with larger diameter vessels

and thus much higher water transport capacity than found in tracheid-only plants. Faster water transport in turn allows greater stomatal opening and thus higher rates of photosynthesis.

In the humid understory of a tropical forest, where transpiration rates are extremely low, there would seem to be little advantage in producing wood with a greater capacity for water transport. In fact, the sole existing species of the earliest-diverging angiosperm lineage—*Amborella trichopoda*—produces only tracheids. Water lilies also lack vessels, but since they do not produce wood it is less clear whether this is an ancestral or a derived feature. Furthermore, early-diverging angiosperm lineages that do produce vessels do not conduct water at higher rates than those that do not. Thus, xylem vessels in angiosperms appear not to have evolved, at least initially, as a means of enhancing water transport.

Instead, the formation of separate cell types for water transport (vessel elements) and mechanical support (fibers) may have given the early angiosperms greater flexibility in form, allowing them to grow toward small patches of sunlight that filtered through the forest canopy. In the heavily shaded tropical rain forest understory, the more opportunistic growth forms exhibited by early-diverging angiosperm groups, including vines and sprawling shrubs, may have provided a distinct advantage.

The fossil record provides evidence that for the first 30 to 40 million years of their evolution, angiosperms were neither diverse nor ecologically dominant. In fact, fossil pollen indicates that between about 140 and 100 million years ago, gnetophytes diversified just as much as angiosperms. Later in the Cretaceous Period, however, beginning about 100 million years ago, angiosperm diversity began to increase at a much higher rate. Trees belonging to the magnoliids, the branch of the angiosperm tree that today includes magnolias, black pepper, and avocado, emerged as ecologically important members of forest canopies. More important, however, was the divergence of the two groups that would come to dominate both angiosperm diversity and the ecology of many terrestrial environments: the monocots and the eudicots. Together, monocots and eudicots make up more than 95% of angiosperm species.  What features may have contributed to their spectacular diversity?

## Monocots develop according to a novel body plan.
Monocots or monocotyledons make up nearly one quarter of all angiosperms. Monocots come in all shapes and sizes and are found in virtually every terrestrial habitat on Earth

(**Fig. 33.22**). Coconut palms, hanging Spanish moss (a relative of pineapple), and tiny floating duckweeds, less than 2 mm across, are all monocots, as are the sea grasses of tropical lagoons and the lilies and daffodils in gardens. Some monocots grow in arid regions—including agave, which is used to make tequila. Others, such as orchids, grow primarily as epiphytes in tropical forests. A few monocots form vines—for example, the rattan palms, whose stems are used to produce rattan furniture. Monocots are the most important group of plants in terms of what we eat. Grasses, for example corn, rice, wheat, and sugar cane, are staples in the diets of most people. Other monocots that make their way to our dinner table include banana, yams, ginger, asparagus, pineapple, and vanilla.

Monocots take their name from the fact that they have one embryonic seed leaf, or cotyledon, whereas all other angiosperms have two. However, monocots are distinct in form in so many other ways that one rarely has to count the number of cotyledons to identify a member of this large, monophyletic group. Monocots represent a major evolutionary departure in the way plants build their bodies, in that a vascular cambium is never formed. How has this group been so successful despite the loss of this key innovation? To understand, we first explore in more detail the novel body plan of the monocots.

In monocot leaves, the major veins are typically in parallel and the base of the leaf surrounds the stem, forming a continuous sheath. This type of leaf base means that only one leaf can be attached at any node, consistent with the formation of a single cotyledon. Although monocots do not form a vascular cambium, they can still produce stems that are quite large. For example, both corn and coconut palms are monocots. In monocots, all of the increase in stem diameter occurs in a narrow zone immediately below the apical meristem. As a result of this lateral expansion, the vascular bundles become distributed throughout the stem instead of being arranged in a ring as in all other seed plants.

The lack of a vascular cambium has a profound impact on the way monocots form roots. Because individual roots cannot increase their vascular capacity, monocots continuously initiate new roots from their stems. Thus, the root systems of monocots are more similar to those found in ferns and lycophytes than in other seed plants. Finally, monocot flowers typically produce organs in multiples of 3 (for example, 3, 6 or 9 stamens), whereas eudicot flower organs are most commonly in multiples of 4 or 5.

Monocots have a relatively poor fossil record, in part because they do not produce wood. What factors might have led to such a radical revision in how they build their bodies? The question requires that we look at where and how living monocots grow. One hypothesis is

---

**FIG. 33.22** **Monocot diversity.** This diverse group of angiosperms includes (a) *Agave shawii*, a desert succulent; (b) *Leucojum vernum*, a spring wildflower; (c) bamboo, a forest grass, and (d) *Costus* species, which grow in the rain forest understory.

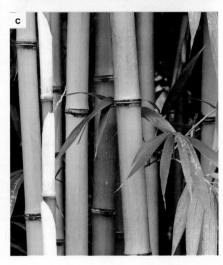

that monocots evolved from ancestors that produced creeping, horizontal stems as they grew along the shores of lakes and other wetlands. Many monocots today, including the earliest diverging groups of monocots, grow in such habitats, and many of the features of the monocot body plan are well suited to environments with loose substrates, flowing water, and fluctuating water levels. For example, their leaf base provides a firm attachment that prevents leaves from being pulled off by flowing water. Furthermore, many monocots produce strap-shaped leaves that elongate from a persistent zone of cell division and expansion located at the base of the leaf blade. By continually elongating from the base, monocot leaves can extend above fluctuating water levels. Finally, it is easier to imagine evolutionary changes that affected the vascular system occurring within an environment that makes only modest demands for water transport.

The environmental context surrounding the evolution of monocots is something we may never know for sure. We thus turn our attention to one of the most diverse groups within the monocots—the grasses. Many grasses produce stems that grow

## HOW DO WE KNOW?

**FIG. 33.23**

# When did grasslands expand over the land surface?

**BACKGROUND** Today, prairies, steppes, and other grasslands occur widely in the interiors of continents. Grasses, however, do not fossilize readily. How, then, can we understand how grasslands developed through time?

- ■ Forests
- ▫ $C_3$ grasslands
- ▪ $C_4$ grasslands
- ■ Crops

**HYPOTHESIS** Grasslands expanded as climate changed over the past 50 million years.

**OBSERVATIONS AND EXPERIMENTS** Grasses commonly make phytoliths, small structures of silica ($SiO_2$) in their cells. These preserve well, providing a direct record of grass expansion. Moreover, mammals that feed on grasses evolved high-crowned teeth, which also preserve well, giving us an indirect record of grassland history. Finally, $C_4$ grasses, which are adapted to hot sunny environments with limited rainfall, have a distinctive carbon isotopic composition imparted by the initial fixation of $CO_2$ by PEP carboxylase (Chapter 29). Measurements of $^{13}C$ and $^{12}C$ in mammal teeth, soil carbonate minerals, and more recently, tiny amounts of organic matter incorporated into phytoliths, allow scientists to track the $C_4$ grasslands through time.

horizontally and branch, allowing them to cover large areas. Grasses have linear leaves that elongate from the base, allowing them to survive grazing as well as fire and drought. Many grasses have evolved the ability to tolerate dry environments by producing roots that extend deep into the soil. In addition, $C_4$ photosynthesis has evolved within the grasses multiple times. As described in Chapter 29, $C_4$ photosynthesis allows plants to avoid photorespiration and thus photosynthesize with greater efficiency. Grasses are among the most successful group of plants, becoming widespread within the past 20 million years as climates changed (**Fig. 33.23**). Today, nearly 30% of terrestrial environments are grasslands.

→ **Quick Check 4** What are some of the distinctive features of monocots?

**Eudicots are the most diverse group of angiosperms.**
Eudicots first appear in the fossil record about 125 million years ago and, by 80 to 90 million years ago, most of the major groups present today can be distinguished. Today, there are estimated to be

**RESULTS** Studies of phytoliths, mammal tooth structure, and carbon isotopic composition of teeth enamel from the North American midcontinent clearly show that grasslands expanded 15 to 20 million years ago, and $C_4$ grasslands expanded, later, about 6 to 8 million years ago.

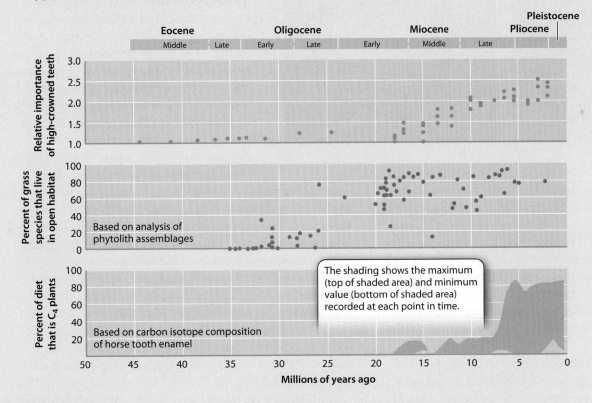

**CONCLUSION** In North America, grasslands expanded as atmospheric $CO_2$ levels declined and climates became drier. Other continents show evidence of a similar linkage of grassland expansion to climate change.

**FOLLOW-UP WORK** At present, atmospheric $CO_2$ levels are increasing rapidly, which may affect the competitive abilities of $C_3$ and $C_4$ grasses. Scientists are working to understand how global change will influence grasslands and other vegetation.

**SOURCES** Stömberg, C. A. E. 2011. "Evolution of Grasses and Grassland Ecosystems." *Annual Review of Earth and Planetary Sciences* 39:517-544. Map is from Edwards, E. J., C. O. Osborne, C. A. E. Stromberg, S. A. Smith, and the $C_4$ Grasses Consortium. 2010. "The Origins of $C_4$ Grasslands: Integrating Evolutionary and Ecosystem Science." *Science* 328:587-591.

**FIG. 33.24 Eudicot diversity.** Eudicots include (a) red oak trees (*Quercus rubra*), (b) tropical passionflower vines (*Passiflora caerulea*), (c) *Banksia* shrubs, native to Australia, and (d) beavertail cactus (*Opuntia basilaris*).

approximately 160,000 species of eudicots, nearly three-quarters of all angiosperm species (**Fig. 33.24**). Eudicots are well represented in the fossil record, in part because their pollen is easily distinguished. Each eudicot pollen grain has three openings from which the pollen tube can grow, whereas pollen in all other seed plants has only a single opening. Eudicots take their name from the fact that they produce two cotyledons, whereas monocots produce only one. But because the magnoliids and early angiosperm lineages also have two cotyledons this largest of all angiosperm groups is referred to as the "eu-" or "true" dicots.

Many eudicots produce highly conductive xylem. High rates of water transport, and thus high rates of photosynthesis, may explain why eudicot trees were able to replace magnoliid trees as the most ecologically important members of forest canopies. Today, tropical rain forest trees are an important component of the diversity of eudicots. Important eudicot trees in temperate regions include oaks, willows, and eucalyptus.

At the other end of the size spectrum are the herbaceous eudicots, which represent a substantial fraction of eudicot diversity. Herbaceous plants do not form woody stems. Instead, the aboveground shoot dies back each year rather than withstand

a period of drought or cold. At the extreme are annuals, herbaceous plants that complete their life cycle in less than a year, persisting during the unfavorable period as seeds. Annuals are unique to angiosperms and almost entirely occur within eudicots. Why might eudicots be successful as both herbs and trees? One factor is that the ability to produce highly conductive xylem may allow herbaceous eudicots to grow quickly and to produce inexpensive and thus easily replaced stems. Examples of herbaceous eudicots include violets, buttercups, and sunflowers.

The herbaceous growth form appears to have evolved many times within eudicots as tropical groups represented by woody plants expanded into temperate regions. For example, the common pea is an herbaceous plant that resides in many vegetable gardens; its relatives include many tropical rain forest trees. Peas are members of the legume, or bean, family, some of which can form symbiotic interactions with nitrogen-fixing bacteria (Chapter 29). The ability to trade carbon for nitrogen may explain why legumes are important in many habitats. Many of the trees of the African savannas, as well as trees in seasonally dry regions throughout the world, are legumes.

Eudicots are diverse in other ways as well. Most parasitic plants and virtually all carnivorous plants are eudicots, as are water-storing cacti that grow in deserts. Some, such as roses and blueberry, are woody shrubs. Others grow as epiphytes in the canopy of rain forests, and still others, such as grapes and honeysuckle, are vines. Perhaps the most unusual are the strangler fig trees. These begin as epiphytes and then produce roots that descend to the forest floor and fuse to form a solid cylinder that "strangles" their host tree. Finally, eudicots make a significant contribution to the diversity of the dinner table. Apples, carrots, pumpkins, and potatoes are all eudicots, as are coffee, cacao (the source of chocolate), and tea. Many plants with oil-rich seeds such as olives, walnuts, soybean, and canola are eudicots, as are buckwheat and quinoa.

The reproduction, growth, and physiology of angiosperms are summarized in **Fig. 33.26** on pages 33-22 and 33-23.

**? CASE 6 Agriculture: Feeding a Growing Population What can be done to protect the genetic diversity of crop species?**

Of the more than 400,000 species of plants on Earth today, we eat surprisingly few. Of those we do eat, nearly all are angiosperms. The only major exceptions are the female gametophyte of a small

**FIG. 33.25** "Centers of origin" where different crop species were domesticated.

Citrus, sugarcane, rice

Sugarcane, banana, yam, coconut

Barley, soybean, rice

Cabbage, lettuce, olives, oats, wheat

Wheat, barley

Grapes

Sorghum, wheat, coffee

Corn, tomato, bean, cotton

Cotton, potato, peanut

Potato

Rubber, chocolate

number of gymnosperms (pine nuts and ginkgos), as well as the young leaves of a few ferns. Most of the plants we eat come from species that are grown in cultivation, many of which have, as a consequence of artificial selection, lost the ability to survive and reproduce on their own. Of the approximately 200 such domesticated species, 12 account for over 80% of human caloric intake. Just three—wheat, rice, and corn—make up more than two-thirds.

Crop breeders select for varieties that grow well under cultivation and are easy to harvest, but a consequence of selective breeding is a decrease in genetic diversity. Thus, not only do we depend upon a small number of species for food, agriculture itself contributes to a narrowing of plant diversity. Because pathogens and pests continue to evolve, any loss in genetic diversity of crop species creates substantial risks. For example, in 1970, the fungal pathogen *Bipolaris maydis*, a disease-causing organism that had previously destroyed less than 1% of the U.S. corn crop annually, destroyed more than 15% of the corn crop in a single year. The severity of the epidemic was due to the genetic uniformity of the hybrid corn varieties planted at that time. In natural plant populations, genetic variation for pathogen resistance makes it highly unlikely that a newly evolved pathogen strain will be able to infect every plant.

Nicolai Vavilov, a Russian botanist and geneticist, was one of the first to recognize the importance of safeguarding the genetic diversity of both crop species and their wild relatives. In the early twentieth century, he mounted a series of expeditions to collect seeds from around the globe. Vavilov's observations of cultivated plants and their relatives led him to hypothesize that plants had been domesticated in specific locations (**Fig. 33.25**), from which they had subsequently expanded as a result of human migration and commerce. Vavilov called these regions "centers of origin," and he believed that they coincided with the centers of diversity for both the crop species and its wild relatives. In World War II, during the siege of Leningrad (1941–1944) in which more than 700,000 people perished, scientists at the Vavilov Institute sought to protect what was then the world's largest collection of seeds. Their dedication—at least one of the self-appointed caretakers died of starvation, despite being surrounded by vast quantities of edible seeds—illustrates the priceless nature of the genetic diversity on which our food supply rests.

Today, seed banks help preserve the genetic diversity of crop species and their wild relatives. However, seed banks can store only a fraction of the genetic diversity present in nature. To help make up the difference, some have suggested establishing protected areas that coincide with Vavilov's centers of origin. As we will see in the next chapter, new threats lie just over the horizon. Meeting these threats will require that every genetic resource be brought to bear.

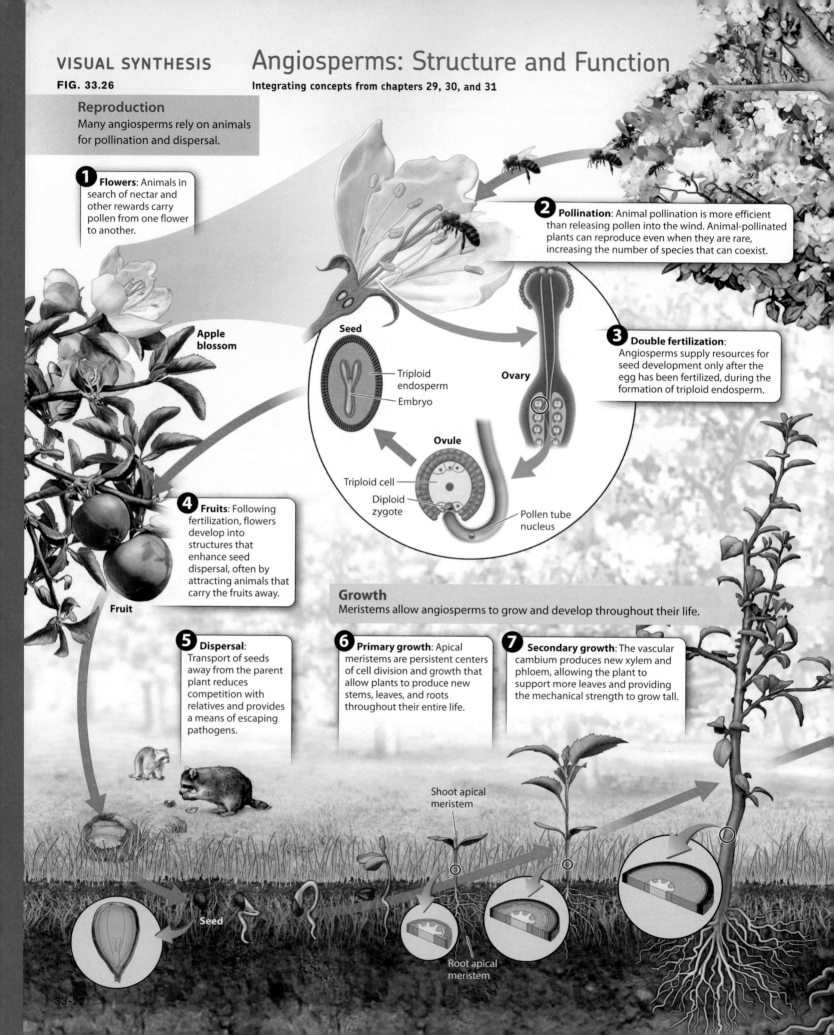

# Angiosperms: Structure and Function

**Integrating concepts from chapters 29, 30, and 31**

## Reproduction
Many angiosperms rely on animals for pollination and dispersal.

**1 Flowers:** Animals in search of nectar and other rewards carry pollen from one flower to another.

**2 Pollination:** Animal pollination is more efficient than releasing pollen into the wind. Animal-pollinated plants can reproduce even when they are rare, increasing the number of species that can coexist.

**3 Double fertilization:** Angiosperms supply resources for seed development only after the egg has been fertilized, during the formation of triploid endosperm.

**4 Fruits:** Following fertilization, flowers develop into structures that enhance seed dispersal, often by attracting animals that carry the fruits away.

**5 Dispersal:** Transport of seeds away from the parent plant reduces competition with relatives and provides a means of escaping pathogens.

**6 Primary growth:** Apical meristems are persistent centers of cell division and growth that allow plants to produce new stems, leaves, and roots throughout their entire life.

**7 Secondary growth:** The vascular cambium produces new xylem and phloem, allowing the plant to support more leaves and providing the mechanical strength to grow tall.

Apple blossom

Seed

Triploid endosperm

Embryo

Ovary

Ovule

Triploid cell

Diploid zygote

Pollen tube nucleus

Fruit

## Growth
Meristems allow angiosperms to grow and develop throughout their life.

Shoot apical meristem

Seed

Root apical meristem

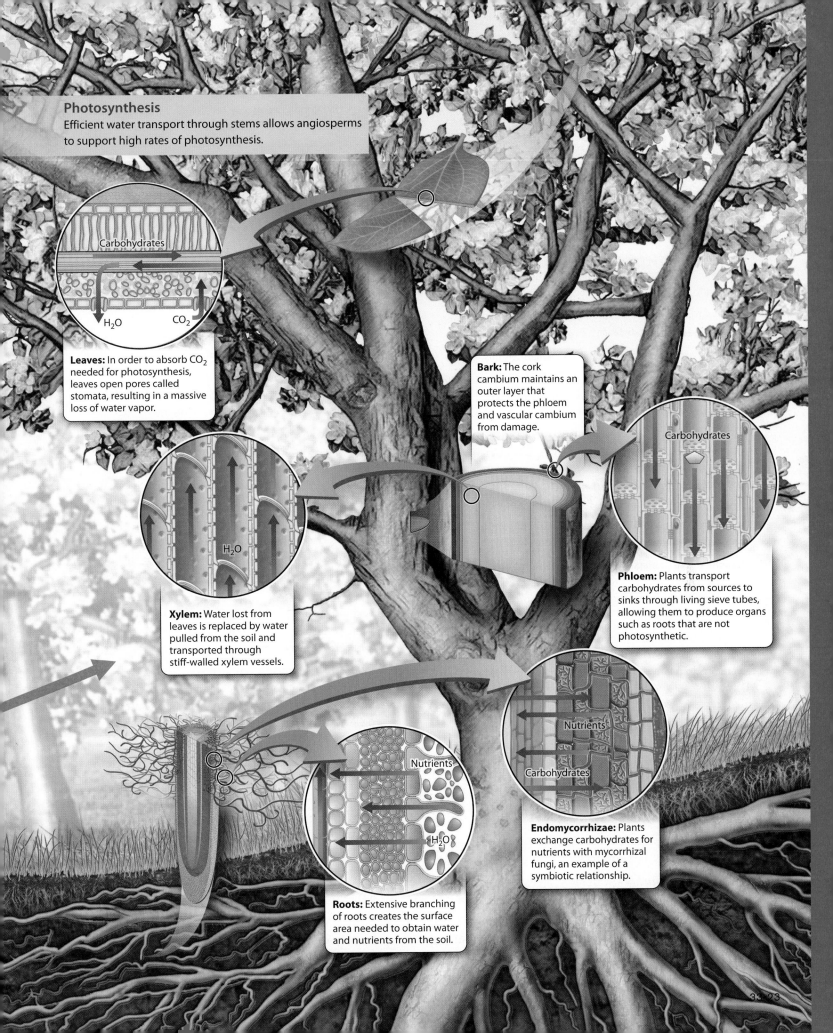

**Photosynthesis**
Efficient water transport through stems allows angiosperms to support high rates of photosynthesis.

Carbohydrates

$H_2O$    $CO_2$

**Leaves:** In order to absorb $CO_2$ needed for photosynthesis, leaves open pores called stomata, resulting in a massive loss of water vapor.

**Bark:** The cork cambium maintains an outer layer that protects the phloem and vascular cambium from damage.

Carbohydrates

$H_2O$

**Xylem:** Water lost from leaves is replaced by water pulled from the soil and transported through stiff-walled xylem vessels.

**Phloem:** Plants transport carbohydrates from sources to sinks through living sieve tubes, allowing them to produce organs such as roots that are not photosynthetic.

Nutrients

Nutrients

Carbohydrates

**Endomycorrhizae:** Plants exchange carbohydrates for nutrients with mycorrhizal fungi, an example of a symbiotic relationship.

$H_2O$

**Roots:** Extensive branching of roots creates the surface area needed to obtain water and nutrients from the soil.

# Core Concepts Summary

**33.1 PLANT DIVERSITY IS DOMINATED BY ANGIOSPERMS, WHICH MAKE UP APPROXIMATELY 90% OF ALL PLANT SPECIES FOUND TODAY.**

There are thought to be nearly 400,000 species of plants living today. Of these, approximately 90% are angiosperms. page 33-1

The other 10% of plant species is distributed among the other six major groups of plants. page 33-1

Angiosperms first appear in the fossil record about 140 million years ago, more than 300 million years after plants first moved onto land. page 33-1

The evolution of angiosperms resulted in a rapid and dramatic increase in total plant diversity. page 33-1

As angiosperm diversity increased, other plant groups, such as gymnosperms and ferns, declined in diversity. page 33-1

Moist tropical rain forests dominated by angiosperms provided new types of habitat into which other plants could evolve. page 33-2

**33.2 BRYOPHYTES DIVERGED BEFORE THE EVOLUTION OF VASCULAR PLANTS, AND THEY GROW IN ENVIRONMENTS WHERE THE ABILITY TO OBTAIN WATER FROM THE SOIL DOES NOT PROVIDE AN ADVANTAGE.**

Bryophytes constitute a paraphyletic group consisting of three groups of plants: mosses, liverworts, and hornworts. page 33-2

Bryophytes are small plants that produce one of two morphological types: a flattened thallus or an upright leafy type. They do not form roots, but instead absorb water through their surfaces. page 33-3

The persistent component of the life cycle is the gametophyte. Sporophytes range from tiny and non-photosynthetic in some liverworts to relatively long-lived and photosynthetic in hornworts. page 33-3

There are several examples of convergent evolution between bryophytes and vascular plants, including insect dispersal of spores in some mosses and the evolution of internal transport cells in some mosses and liverworts. page 33-4

Sphagnum moss is the dominant plant of peat bogs. page 33-5

Sphagnum moss produces water-holding cells that allow it to soak up water, and it acidifies the environment. Both characteristics help slow decomposition, so large amounts of organic carbon build up year after year. page 33-5

**33.3 SPORE-DISPERSING VASCULAR PLANTS ARE SMALL, OFTEN EPIPHYTIC PLANTS THAT USUALLY GROW IN MOIST ENVIRONMENTS.**

Only two groups of seedless vascular plants have living relatives today: lycophytes, and ferns and horsetails. page 33-5

Three hundred million years ago, lycophytes included large trees that dominated swamp forests. Today, lycophytes are small plants that grow either in the forest understory as epiphytes, or occur in shallow ponds. page 33-6

Ferns and horsetails are morphologically diverse. Ferns produce large leaves that uncoil as they grow; horsetails have tiny leaves and whisk ferns with no leaves at all. page 33-9

Although ferns have a long history, most present-day species are the result of a radiation that occurred after the rise of the angiosperms. page 33-10

**33.4 GYMNOSPERMS PRODUCE SEEDS AND WOODY STEMS AND ARE MOST COMMON IN COLD OR DRY REGIONS.**

Of the many groups of seed plants that have evolved, only two can be found today: the gymnosperms, with fewer than 1000 species, and the angiosperms, with more than 380,000 species. page 33-11

Gymnosperms are composed of four groups of woody plants: cycads, ginkgos, conifers, and the gnetophytes. page 33-11

Cycads produce large leaves on stout, unbranched stems. Although they once were widely distributed, they now occur in small, fragmented populations, primarily in the tropics and subtropics. Many cycads are insect pollinated, and all form symbiotic associations with nitrogen-fixing bacteria. page 33-12

*Ginkgo* is the single living species of a group that was distributed globally before the evolution of the angiosperms. *Ginkgo* is wind pollinated and produces tall, branched trees. page 33-12

Conifers include the tallest and longest-lived trees on Earth. Wind-pollinated and largely evergreen, they are found primarily in cool to cold environments. page 33-13

Before the angiosperms appeared, conifers were widespread. Their almost complete absence in the tropics and persistence in temperate regions may be due to their dependence upon wind pollination and xylem formed entirely of tracheids. page 33-14

The gnetophytes are a small group, containing only three genera and few species, that has independently evolved xylem vessels. page 33-14

**33.5 ANGIOSPERM DIVERSITY IS PARTLY EXPLAINED BY ANIMAL POLLINATION AND XYLEM VESSELS.**

Angiosperm diversity may result as much from low rates of extinction as from high rates of species formation. page 33-15

Plants having flowers can reproduce even if they are far apart, allowing rare species to persist and reproduce. page 33-15

Xylem vessels make it possible for angiosperms to have a diversity of form and to grow toward light. page 33-16

The angiosperm phylogeny indicates a major split between two diverse groups, monocots and eudicots. page 33-16

Monocots include grasses, coconut palms, bananas, ginger, and orchids. page 33-17

Monocots have a single cotyledon, or embryonic seed leaf, and they do not form a vascular cambium. page 33-17

Eudicots are diverse, including many familiar plants such as legumes, roses, cabbage, pumpkin, coffee, tea, and cacao, which yields chocolate. page 33-20

Eudicots are characterized by pollen grains with three openings through which the pollen tube can grow. page 33-20

Modern agriculture is based on just a few plant species with low genetic diversity, making it vulnerable to pests and pathogens. page 33-21

## Self-Assessment

1. Using information in Fig. 33.2, draw a pie chart that depicts proportional plant diversity just before the appearance of the angiosperms.

2. Imagine a world in which mosses, liverworts, and hornworts formed a monophyletic group. How would your ability to infer what the first land plants looked like be affected?

3. Describe three environments that allow bryophytes to coexist with vascular plants.

4. Describe the habitats in which lycophytes are found today.

5. List three ways that ferns, which lack secondary growth, are able to elevate their leaves and thus access more sunlight.

6. Describe how fern diversity has been affected by the evolution of the angiosperms.

7. Contrast the ways in which the evolution of angiosperms has affected the distribution of cycads and of conifers.

8. Explain how xylem produced by conifers differs from that of angiosperms and how that difference may have influenced the present-day distribution of conifers.

9. Compare the movement of pollen in an animal-pollinated angiosperm and a wind-pollinated conifer, noting what features of angiosperm reproduction increase the efficiency (or lower the costs) of pollen transfer.

10. Name several features that might account for the diversity and success of angiosperms and discuss possible advantages of these features.

Do you understand the chapter's Core Concepts? Log in to check your answers to the Self-Assessment questions, then practice what you've learned and reinforce this chapter's concepts by working through the problems and multimedia tutorials provided there.

📖 **www.biologyhowlifeworks.com**

**FIG. 34.4** **Decomposition of wood by fungi.** (a) A dead tree being decomposed by brown rot fungi. (b) These scanning electron microscope images show (left) normal cedar wood and (right) rotted cedar wood from the coffin in the tomb of King Midas in Gordion, Turkey.

decompose lignin, but in nature, fungi account for most of the decomposition of wood (**Fig. 34.4**).

To break down lignified plant cell walls, fungi require oxygen. However, oxygen concentrations are typically low in water-logged environments, and in that case decay proceeds slowly if at all. For this reason, woody biomass, once submerged in water, tends to accumulate, leading to the buildup of partially decayed plant material as peat. Over geologically long time intervals, peat becomes coal, an energy staple of the industrial world that owes its existence to the inhibition of fungal decay.

### Fungi are important plant and animal pathogens.

Fungi not only feed on dead plants and animals, but they are also well adapted to infect living tissues (**Fig. 34.5**). Plants, for example, are vulnerable to a diverse array of fungal pathogens—rusts, smuts, and molds—that cause huge losses in agricultural production, as we discuss in section 34.4.

Invasion is the key to success for plant pathogens (Chapter 32). Most plants have physical or chemical means of inhibiting

fungal invasion, so successful pathogens must be able to get past these defenses. In many cases, fungi infect plants through wounds, which provide a route around a plant's outer defenses. Some fungi enter through stomata, while others penetrate epidermal cells directly, degrading the wall with enzymes and then pushing their hyphae into the plant interior by turgor pressure. The extent of infection usually depends on the ability of hyphae to expand into new tissues. Vascular wilt fungi (discussed in Chapter 32) send their hyphae through xylem vessels, enabling them to spread throughout the entire plant. Old photographs of towns in the eastern United States show streets shaded by

**FIG. 34.5** **Fungal infection of living tissue.** (a) This stained section of heart tissue (pink) is infected with the fungus *Candida albicans* (purple threadlike structures). (b) A scanning electron micrograph shows hyphae of *Ceratocystis ulmi*, a vascular wilt fungus, growing in xylem vessels of an elm tree.

American elm trees, but today nearly all those trees are dead, victims of a vascular wilt fungus introduced into the country early in the twentieth century.

Aboveground, plant infections are usually transmitted by fungal spores, carried either by the wind or on the bodies of insects. For example, many fungi that attack fruits are transmitted by insects also seeking those fruits; yeasts that feed on nectar within flowers hitch rides on pollinators. Some fungi secrete sugar-rich droplets that attract insects, which then transport their spores to other sites.

Belowground, infection is typically transmitted by hyphae that penetrate the root. Hyphae can feed on organic matter in the soil until the root of a suitable host plant is encountered. The persistence of fungal hyphae in soils means that populations of pathogenic fungi can build up in the soil. This is why seedlings germinating close to an adult tree of the same species are at risk of infection, whereas seedlings of distantly related species may be spared. By favoring the establishment of seedlings of less-common species, fungal pathogens are thought to play an important role in maintaining high levels of species diversity in tropical rain forests (Chapter 32).

Insects and other invertebrates are also susceptible to fungal pathogens. Certain fungi exploit living animals in a unique way: by trapping them. Some fungal predators form sticky traps with their hyphae, whereas others actually lasso their prey (**Fig. 34.6**). The hyphae of these fungi form rings that can inflate within seconds, trapping anything within the ring. Nematodes, tiny worms common in soils, inadvertently pass through the rings, setting off hyphal inflation. The nematodes are then trapped, providing food for this fungus.

Fungal infection is relatively rare in vertebrates. Severe infections are more frequent among fish and amphibians than in mammals, perhaps because fungi grow poorly at mammalian body temperatures. An apparent exception is the fungal infection that has caused dramatic declines in North American bat populations, but the fungi infect the bats during their winter hibernation, when the bats have lowered body temperatures that conserve energy. In humans, most fungal infections are annoying rather than life threatening—athlete's foot and yeast infections are prime examples. However, in individuals with immune systems compromised by HIV or other illnesses, fungal infection can be fatal.

## Many fungi form symbiotic associations with plants and animals.

While some interactions between species benefit the fungus at the expense of its host, others benefit both partners. In Chapter 29, we saw that mycorrhizal fungi supply plant roots with nutrients such as phosphorus from the soil and in return receive carbohydrates from their host (**Fig. 34.7**). There are two main types of mycorrhizae. The hyphae of **ectomycorrhizal** fungi surround, but do not penetrate, root cells. In contrast, the hyphae of **endomycorrhizal** fungi penetrate into root cells, where they produce highly branched structures that provide a large surface area for nutrient exchange. As we will see in section 34.3, the endomycorrhizal fungi constitute a monophyletic group dating back over 400 million years. In contrast, ectomycorrhizae are present in multiple groups and appear to have evolved more than once.

**FIG. 34.7 Mutualistic relationship between fungi and plants.** Mycorrhizal fungi (white) surround roots (brown) and branch extensively in the soil.

**FIG. 34.6 A fungal lasso.** Hyphae of *Drechslerella doedycoides* form inflated rings that capture nematode worms moving through the soil.

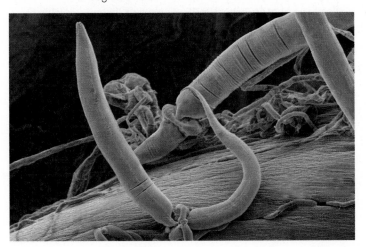

**FIG. 34.8** **Lichens.** A lichen is a mutualistic association between a fungus and a photosynthetic microorganism.

Other fungi, called **endophytes,** live within leaves. While endophytes are much less studied than mycorrhizae, they may be as widespread. Endophytic hyphae grow within cell walls and in the spaces between cells. Long thought to be harmless, endophytes are now recognized as beneficial. They may help the host plant by producing chemicals that deter pathogens and herbivorous insects.

Mutually beneficial associations between fungi and animals are much less common, but a few examples are known, none more striking than insects that actually grow fungi for food. The fungi benefit because the insects provide shelter, food, and protection from predators and pathogens. Insect–fungal agriculture has evolved at least three times: in leaf-cutter ants that live in tropical forests (Chapter 48), in a group of African termites, and in some wood-boring beetles. These beetles maintain fungal gardens within tunnels that they excavate in damaged or dead trees.

### Lichens are symbioses between a fungus and a green alga or a cyanobacterium.

Lichens are familiar sights in many environments, often forming colorful growths on rocks or tree trunks (**Fig. 34.8**). Lichens look, function, and even reproduce as single organisms, but they are actually stable associations between a fungus and a photosynthetic microorganism, usually a green alga but sometimes a cyanobacterium. Nearly 15% of all known fungal species grow as lichens.

The dual nature of lichens was first proposed by a Swiss botanist, Simon Schwendener, in 1867, but his hypothesis found little favor at the time. The existence of a composite "organism" challenged the idea that life could be divided into discrete categories such as animals and plants and raised the question of how an association of distinct species could function as an integrated whole. Given what we now understand about the widespread nature of mutualisms, the opposition to Schwendener's proposal may seem surprising. Yet it was only in 1939, when Eugen Thomas showed that lichens could be separated into their individual parts and then reassembled, that their dual nature became widely accepted.

Lichens consist mostly of fungal hyphae, with the photosynthetic algae or cyanobacteria forming a thin layer just under the surface (**Fig. 34.9**). The hyphae anchor the lichen to a rock or tree, aid in the uptake and retention of water and nutrients, and produce chemicals that protect against excess light and herbivorous animals. In turn, the photosynthetic partners provide a source of reduced carbon, such as carbohydrates. Cyanobacteria living in lichens also provide a source of fixed nitrogen, like ammonia. The two partners exchange nutrients through fungal hyphae that tightly encircle or even penetrate the walls of the photosynthetic cells. As a result, lichens are able to thrive at sites where neither partner could exist on its own.

It remains an open question whether either partner can exist independently in nature. In cases where the fungal partner can be grown in the laboratory in culture, it produces a relatively undifferentiated hyphal mass. Thus, while the bulk of the lichen structure comes from the fungus, chemical signals from the photosynthetic partner influence the form and shape of the fungus.

Surprisingly, these associations are not species specific. There are approximately 13,500 known lichens, but only about 100 participating photosynthetic species. This means that different lichens can have the same algae or cyanobacteria.

**FIG. 34.9** **Lichen anatomy.** The fungus provides structure, and photosynthetic algae form a thin layer under the surface. The scanning electron micrograph shows a section through *Xanthoria flammea.*

Fungal hyphae

Algal cells

Lichens consist mostly of fungal hyphae that take up water and nutrients from the soil.

Photosynthetic algal cells form a thin layer just under the surface of the lichen.

Conversely, a given fungal partner may associate with more than one photosynthetic species, without affecting the lichen's outward form. Because of the intimate association between lichen morphology and fungal species, the lichen and fungus are assigned the same scientific name.

Lichens spread asexually by fragmentation or through the formation of dispersal units consisting of a single photosynthetic cell surrounded by hyphae. Sexual reproduction, at least by the fungi, is common, and the photosynthetic cells reproduce asexually by mitotic cell division. Whether sexual or asexual reproduction is more important in allowing lichens to establish in new habitats, however, is not known.

Lichens are remarkable for their ability to grow on the surfaces of rocks and tree trunks. They are among the first colonizers of lava flows and the barren land left after glacial retreat. In some habitats, for example Arctic tundra and semiarid landscapes, lichens make a measurable contribution to net productivity. How do lichens obtain the resources they need in these barren environments? They obtain nutrients from rainfall or by secreting organic acids that help release some nutrients from even rocky surfaces. Not surprisingly, lichens in these harsh environments grow slowly. Some lichens grow on soil, for example reindeer "moss," which is eaten by caribou. Reindeer moss is thought to compete successfully for space by secreting chemicals that hinder the growth of plants.

Given their small size and exposure to the environment, lichens must be able to tolerate drying out from time to time. All lichens have a high tolerance for dessication, and can tolerate wide fluctuations in temperature and light. In contrast, lichens are quite sensitive to air pollution, particularly sulfur dioxide ($SO_2$).

For this reason, lichen growth is sometimes used as an indicator of industrial pollution.

## 34.2 REPRODUCTION

Like plants, fungi face two challenges in completing their life cycles. First, to maintain genetic diversity within populations, they must find other individuals with which to mate. Second, they must be able to disperse from one place to another. The ways that fungi generate genetic diversity are unique. However, fungal adaptations for dispersal at least broadly resemble those of plants. Fungi rely on wind, water, or animals to carry spores through the environment. Recall from Chapter 30 that spores are specialized cells well adapted for dispersal and long-term survival.

We most often come into contact with fungi through their reproductive structures. Fungal spores commonly cause respiratory illness, and mushrooms attract our attention because they are delicious or poisonous. Yet many aspects of fungal reproduction are not easily observed, including life-cycle features that distinguish fungi from other eukaryotes. Fungi have a diverse set of reproductive strategies. In this section, we emphasize the most general principles and patterns.

### Fungi proliferate and disperse using spores.

The tissues, living or dead, that support fungal growth are sometimes hard to come by and have a patchy distribution. For this reason, fungi must be able to disperse from one food source to another. The extensive networks of hyphae within soils or host organisms can spread locally but cannot disperse over great

**FIG. 34.10 Fungal spores dispersed by wind.**

**a.** Devil's snuffbox (*Lycoperdon perlatum*)

**b.** Black bread mold (*Rhizopus*)

Sporangium

Spores

distances, so fungi produce spores that can be carried by the wind (**Fig. 34.10a**), in water, or attached to (or within) animals. In early-diverging fungi that still live in aquatic environments, spores have flagella that allow them to swim. The great majority of fungi, however, live on land, and their spores have no flagella. Instead, the spores are encased in a thick wall that protects them as they are dispersed over habitats unsuitable for growth.

The probability that any given spore will come to rest in a favorable habitat is low, so fungi produce huge numbers of spores. Fungal spores remain viable, able to grow if provided with an appropriate environment, for periods ranging from only a few hours in some species to many years in others. Thus, spores allow fungi to use resources that are patchy in time as well as in space. In fact, a shortage of resources is one of the cues that triggers spore formation.

Spores can form by meiotic cell division as part of sexual reproduction, and they can also form asexually. Asexual spores are formed by mitotic cell division and therefore are genetically identical to their parent. Asexual spores allow fungi to proliferate and disperse to new environments. In many species, asexual spores are produced within sporangia that form at the ends of erect hyphae, facilitating the release of the spores into the air. A close look at a moldy piece of bread reveals that the surface is covered with hyphae carrying sporangia containing asexual spores (**Fig. 34.10b**).

Fungal species vary in the extent to which they rely on asexual as opposed to sexual spore production. In some

groups, such as the lineage that includes the most common mushrooms, sexual reproduction is the dominant means of spore production. On the other hand, in a small number of fungi that includes the endomycorrhizal species, spores produced by meiotic cell division have never been observed. Although sexual reproduction appears to be absent in some fungal species, asexual fungi have other mechanisms for producing genetic diversity, as discussed later in this section.

### Multicellular fruiting bodies facilitate the dispersal of sexually produced spores.

Fungi employ an astonishing array of mechanisms to enhance spore dispersal. Particularly conspicuous are the multicellular **fruiting bodies** produced by some fungi, which facilitate the dispersal of sexually produced spores. (Fungal fruiting bodies should not be confused with the fruits of flowering plants, which are unrelated and very different structures.) Mushrooms, stinkhorns, puffballs, bracket fungi, truffles, and many other well-known structures are fungal fruiting bodies.

Fruiting bodies are highly ordered and compact structures compared to the mycelia from which they grow, yet they are constructed entirely of hyphae (**Fig. 34.11**). In fact, in many cases their mechanisms of spore dispersal demand a high degree of structural precision. Yet we know little of the developmental processes that allow tip-growing hyphae to produce such complex and regular structures.

**FIG. 34.11 Fruiting bodies, complex multicellular structures built from hyphae.** Shown are fruiting bodies of *Hygrophorus miniatus*, known commonly as the vermillion waxcap.

The fruiting bodies of many fungi rise above the ground or grow from the trunks of dead trees, so the sexually produced spores are released high above the ground. But elevation by itself is not enough to ensure dispersal. Thus, many fungi forcibly eject their spores, achieving velocities of more than 1 m/s that allow the tiny spores to penetrate and travel beyond the layer of stagnant air that surrounds the fruiting body. Studies have shown that spores move efficiently through air (**Fig. 34.12**). Other fungi rely on external agents such as raindrops or animals to move their spores around. In section 34.3 we will encounter examples of the diverse dispersal mechanisms found in the fungi.

## HOW DO WE KNOW?

### FIG. 34.12

# What determines the shape of fungal spores that are ejected into the air?

**BACKGROUND** Many fungi disperse their spores by ejecting them forcibly into the air. For spores to be picked up by the wind, however, they must escape a layer of still air called the boundary layer that lies close to the ground. Escape from the boundary layer is easier if spores have a size and shape that minimize drag, the resistance of air to the movement of an object.

**HYPOTHESIS** In fungi that eject their spores into the atmosphere, natural selection favors spore shapes that minimize drag.

**EXPERIMENT** Working from computer models for minimizing drag on airplane wings, scientists modeled spore shapes that minimize drag. These models took into account spore size as well as the physical characteristics of the air through which spores travel. The scientists then measured spore shape for more than 100 species of spore-ejecting fungi.

Fungal spores with outlines (blue) of computer-modeled shapes that minimize drag.

**RESULTS** Nearly three-quarters of the examined fungi had spore shapes that came within 1% of the shape calculated to minimize drag. Related species that do not eject spores forcibly were less likely to have drag-minimizing shapes.

**CONCLUSION** Natural selection has acted on fungi to facilitate spore dispersal by wind.

**SOURCE** Roper, M., R. E. Pepper, M. P. Brenner, and A. Pringle. 2008. "Explosively Launched Spores of Ascomycete Fungi Have Drag-Minimizing Shapes." *Proceedings of the National Academy of Sciences, USA* 105:20583–20588.

## The fungal life cycle often includes a stage in which haploid cells fuse, but nuclei do not.

Like other sexually reproducing eukaryotes, fungi have life cycles that include haploid (1*n*) and diploid (2*n*) stages. The nuclei in fungal hyphae are haploid, and the fungal life cycle is therefore similar to haploid-dominant life cycles found among eukaryotic organisms (see Fig. 27.3a). In haploid-dominant organisms, asexual reproduction involves the production of haploid spores by mitosis, while sexual reproduction involves the fusion of haploid cells (often differentiated as gametes) to form a diploid zygote, which undergoes meiosis as its first division. However, sexual reproduction in fungi differs from all other haploid-dominant organisms in one important respect: In fungi, the fusion of haploid cells is not immediately followed by the fusion of their nuclei.

In most fungi, the sexual phase of the life cycle involves the fusion of hyphal tips rather than specialized reproductive cells, or gametes. For mating to occur, two hyphae grow together and release enzymes that digest their cell walls at the point of contact. This allows the cell contents of the two hyphal cells to merge, forming a single cell with two haploid nuclei. In most sexually reproducing organisms, when two gametes merge, their nuclei fuse almost instantly to form a diploid zygote. In fungi, however, the cytoplasmic union of two cells (**plasmogamy**) is not always followed immediately by the fusion of their nuclei (**karyogamy**). Instead, the haploid nuclei retain their independent identities, resulting in what is referred to as a **heterokaryotic** ("different nuclei") stage (**Fig. 34.13**). In the heterokaryotic stage, cells have nuclei from two parental hyphae, but the nuclei remain distinct. The heterokaryotic stage ends with karyogamy, which leads to the formation of a diploid zygote. The zygote divides by meiotic cell division, giving rise to sexually produced haploid spores.

The separation between plasmogamy and karyogamy is unique to the fungi, and as will be discussed in the next section, the elaboration of this difference is a major theme in the evolution of fungi. In the earliest diverging fungal lineage, karyogamy follows fast on the heels of plasmogamy and there is no heterokaryotic stage. In some groups, the heterokaryotic stage consists of a single cell with many haploid nuclei. However, in other groups,

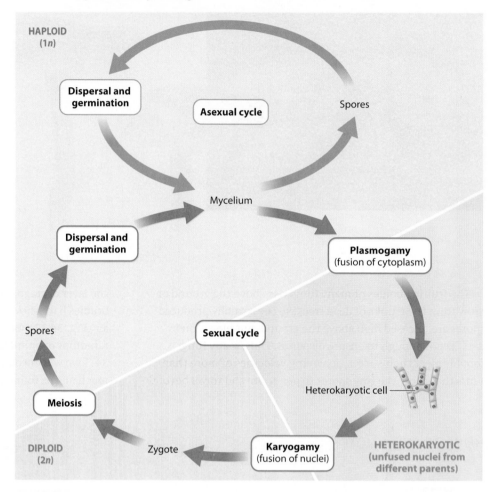

**FIG. 34.13 Generalized fungal life cycle.** In fungi, plasmogamy and karyogamy are separated by a heterokaryotic stage.

plasmogamy is followed by mitosis, which produces hyphae in which each cell contains two haploid nuclei, one from each parent. Heterokaryotic cells with just two genetically distinct haploid nuclei are referred to as **dikaryotic,** or ***n* + *n*,** cells. In fungi with dikaryotic cells, called dikaryotic fungi, there may be a small number of dikaryotic cells or extensive hyphal development made up of dikaryotic cells. For example, some edible mushrooms found on market shelves consist entirely of dikaryotic hyphae.

Dikaryotic fungi account for more than 98% of all known fungal species, suggesting that the separation of plasmogamy and karyogamy in time and space may confer an evolutionary advantage. One major advantage can be seen when we think about the environments where fungi live. Mating takes place principally within the soil or inside the trunks of rotting trees, sites where hyphae are most likely to come into contact. Dispersal, however, is most effective when spores can be released into the air. The separation of plasmogamy and karyogamy allows mating and spore production to occur where each is most effective. In some dikaryotic fungi, however, the separation of plasmogamy and karyogamy is neither distant in space nor long in time. In these fungi, the highly branched dikaryotic hyphae may serve primarily as a means of increasing the number of cells in which karyogamy will eventually take place, therefore leading to the production of more sexually produced spores.

At present, we really don't know why fungi proliferate $n + n$ cells rather than forming a multicellular diploid phase, and there may be no single answer to the question of why the dikaryotic fungi have this unique cell type.

→ **Quick Check 2** What is the difference between a diploid cell and a dikaryotic cell?

### Genetically distinct mating types promote outcrossing.

Under most conditions found in nature, genetically diverse populations persist better than those lacking diversity. Indeed, sexual reproduction is widely viewed as a means of promoting genetic diversity for long-term ecological success in variable environments (Chapters 11 and 42). With the exception of early-diverging aquatic lineages, fungi do not produce male and female gametes. How, then, is sex accomplished, and what prevents an individual from mating with itself?

Many individual fungi of a species look similar but nevertheless have different **mating types** that are genetically determined and prevent self-fertilization. The mating type of an individual is determined by a mating-type gene. Fertilization can take place only between individuals that have different alleles at the mating type gene. In some species, there are only two mating-type alleles. In this case, mating patterns are identical to those for species with male and female sexes. If the two mating-type alleles are spread evenly throughout the population, an individual should be able to mate with 50% of the general population.

Some fungi have more than two mating-type alleles. These fungi have a greater likelihood of encountering a compatible genotype in the general population. That likelihood is even higher for fungi in which mating type is determined by two different mating-type genes that each have multiple alleles. Fungi in the group that includes the common mushrooms can have as many as 20,000 different mating-type alleles. In that case, the odds of finding a compatible mate are close to 100%.

Genetic compatibility is a prerequisite for mating. How do fungi go about finding a suitable mate? The answer is that they secrete chemical signals that attract fungi with mating types different from their own. The chemical nature of these signals differs among groups.

### Fungi that lack sexual reproduction have other means of generating genetic diversity.

Asexual reproduction occurs in most groups of fungi. Indeed, about 20% of fungi appear to lack sexual reproduction altogether. Among these are such well-known groups as *Penicillium* (the source of the antibiotic penicillin), *Aspergillus* (the major industrial source of vitamin C), and all the endomycorrhizal fungi.

How are asexual species able to persist without a mechanism for generating genetic diversity? The short answer is, they don't. Fungal species that lack an observable sexual cycle have another mechanism of generating genetic diversity: the crossing over of DNA during mitosis. Such species are described as **parasexual.** The parasexual cycle (**Fig. 34.14**) is not in fact a coordinated cycle but rather a series of four events. First, two hyphae fuse, forming a heterokaryotic cell. Then, karyogamy produces a diploid nucleus from two genetically distinct haploid nuclei—so far, a familiar pattern. Next, however, during mitosis, crossing over may occur between the two sets of chromosomes. Crossing over is common

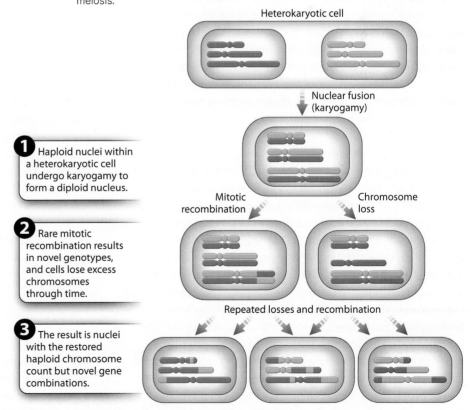

**FIG. 34.14** **The parasexual cycle in fungi.** This cycle can generate genetic diversity without meiosis.*

Heterokaryotic cell

Nuclear fusion (karyogamy)

❶ Haploid nuclei within a heterokaryotic cell undergo karyogamy to form a diploid nucleus.

Mitotic recombination

Chromosome loss

❷ Rare mitotic recombination results in novel genotypes, and cells lose excess chromosomes through time.

Repeated losses and recombination

❸ The result is nuclei with the restored haploid chromosome count but novel gene combinations.

*Adapted from http://what-when-how.com/molecular-biology/parasexual-cycle-molecular-biology/

during meiosis (Chapter 11) but rare during mitosis. The haploid chromosome number is restored not by meiotic division but by the progressive loss of chromosomes. Like sexual reproduction, these processes produce novel gene combinations. The key difference between parasexuality and sexual reproduction is that parasexual species do not undergo meiosis.

Entirely asexual species of fungi posed a problem for early taxonomists because these species lack many of the morphological characters such as fruiting bodies or sporangium-bearing stalks previously used to define different groups. With the advent of DNA sequence data, it has become increasingly clear that many asexual species are closely related to sexual forms, indicating that their evolutionary history as asexual species is not long. A major exception are the Glomeromycota, for which fossils as old as 450 million years have been reported.

## 34.3 DIVERSITY

When Aristotle classified the living world into animals and plants, he grouped fungi with plants because of their lack of motility. As discussed in Chapter 27, molecular sequence comparisons now make it clear that fungi are actually related more closely to animals, such that it would not be amiss to refer to mushrooms and molds as our "cousins."

Fungi share a number of features with animals: Their motile cells, when present, have a single flagellum attached to their posterior ends; fungi and many animals synthesize chitin; and fungi store energy as glycogen, as do animals. Nevertheless, the last common ancestor of fungi and animals was a single-celled microorganism that lived in aquatic environments about 1 billion years ago. Multicellularity evolved independently in fungi and animals (Chapter 28), and so the fungi have many features that are unique in the biological world.

### Fungi are highly diverse.

By any measure, fungi are diverse, although just how diverse is not known. About 75,000 species have been formally described, but estimates of true fungal diversity run as high as 5 million species. Among eukaryotes, the only organisms more diverse than fungi are the animals. The availability of DNA sequence data has greatly advanced our understanding of phylogenetic relationships within the fungi, which are shown in **Fig. 34.15.** The tree clearly shows how the characters present in familiar mushrooms accumulated through the course of evolution: first chitinous cell walls, then hyphae, then regularly placed septa, and finally the complex multicellular reproductive bodies we call mushrooms. Motile single cells are found only in the most ancient lineages, while dikaryotic cells and complexity of form are found only on the most recent branches.

The evolutionary history of fungi can be viewed as the transition from a flagellated ancestral form that lived in aquatic environments to soil-dwelling forms in which hyphae divided

**FIG. 34.15 A phylogeny of the fungi.** Approximate numbers of described species are given in parentheses.

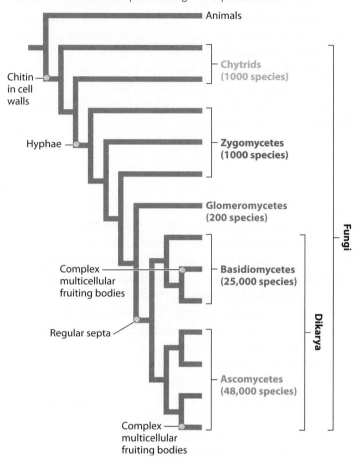

by septa developed complex morphologies visible to the naked eye. As we saw in the preceding section, another evolutionary trend is the progressive separation in time between plasmogamy and karyogamy in fungal life cycles. Many of the distinguishing features of fungi evolved as adaptations to the challenges faced by nonmotile, absorptive heterotrophs on land.

We also see that the numbers of species are not spread evenly across the phylogeny. Instead, more ancient lineages include less than 2% of known species, while the two dikaryotic groups include more than 98% of known fungi. Clearly, dikaryotic fungi are well adapted to many different habitats, including other organisms, both living and dead.

→   **Quick Check 3** From the phylogenetic tree shown in Fig. 34.15, would you say that complex multicellular fruiting bodies evolved once or twice (independently) in fungi?

### Fungi evolved from aquatic, unicellular, and flagellated ancestors.

Fungi are opisthokonts, members of the eukaryotic superkingdom that includes animals (Chapter 27). Among fungi, the Chytridiomycota (commonly referred to as **chytrids**)

**FIG. 34.16 Chytrids.** Chytrids are aquatic fungi that attach to decomposing organic matter by rhizoids, seen in the lower half of the image. This is *Chytriomyces hyalinus.*

20 μm

lie at the base of the phylogenetic tree. This is a small group of about 1000 species found in aquatic or moist environments (**Fig. 34.16**). Many chytrids are single cells with walls of chitin. They may also form short multinucleate structures, but they lack the well-defined hyphae characteristic of other fungi. Chytrids, therefore, do not form a true mycelium, although in some species elongated cellular outgrowths called rhizoids penetrate into organic substrates. Rhizoids anchor the organism in place and absorb food molecules. Chytrids generally disperse by flagellated spores. Sexual reproduction appears to be rare, but there is substantial life-cycle diversity within this group. Chytrids lack a heterokaryotic stage but form flagellated gametes that swim through their aqueous environment.

Most chytrids are decomposers, and a few of them are pathogenic. Notably, the chytrid *Batrachochytrium dendrobatidis* is associated with widespread mortality in amphibian populations (Chapter 48). Molecular clock analysis estimates that chytrids began to diversify 700 to 800 million years ago.

### Zygomycetes produce hyphae undivided by septa.

Moving through the phylogenetic tree, we next encounter several groups traditionally united as the Zygomycota, or **zygomycetes** (see Fig. 34.15), which share a number of traits.

Collectively, zygomycetes make up less than 1% of known fungal diversity. Some are decomposers, specializing on dead leaves, animal feces, and food—there are probably zygomycetes in your kitchen. Others live on and in plants, animals, and even other fungi. Zygomycetes have many typical fungal traits, including the growth of a mycelium and the production of aerial spores. These traits may have evolved as adaptations for finding food and dispersing on land. The loss of flagellated spores may also reflect adaptation for life on land. Unlike more complex fungi, zygomycetes do not form regular septa along their hyphae, nor do they produce multicellular fruiting bodies.

Earlier, we introduced the black bread mold *Rhizopus*, which is a zygomycete (see Fig. 34.10b). *Rhizopus* and its relatives are specialists on substrates containing abundant, easy-to-digest carbon compounds, such as bread, ripe fruits, and the dung of herbivorous animals. These fungi consume their substrates rapidly, and once their meal is finished, they release large numbers of aerial spores to locate another food source.

Sexual reproduction occurs when two compatible hyphal tips fuse to form a thick-walled structure containing many nuclei of each mating type. Karyogamy and meiosis are followed by germination of the haploid cell to form an elevated stalk. Each stalk develops a sporangium that contains spores produced asexually by mitotic cell division. If you look closely at a bread mold, you can see the white filamentous hyphae and small black sporangia on slender stalks. Each sporangium can produce as many as 100,000 spores that are dispersed by the wind, explaining how these organisms seem to get everywhere.

Although zygomycete fungi do not form multicellular fruiting bodies, some have evolved truly spectacular means of dispersal. *Pilobilus*, which consumes the dung of herbivorous animals, has a life cycle similar to *Rhizopus* except that instead of releasing individual spores, it forcibly ejects the entire sporangium (**Fig. 34.17**). Turgor pressure generated in the supporting stalk propels the sporangia as far as 2 m. Light-sensitive pigments in the stalk's hyphae control the

**FIG. 34.17 Spore dispersal in a zygomycete.** In *Pilobilus*, light-sensing pigments orient the turgor-propelled dispersal of the sporangia.

orientation of this water cannon, ensuring that the spores have the best chance of escaping the dung pile and landing on vegetation that is attractive to grazing herbivores. Feeding herbivores disperse the spores further, while supplying them with fresh dung for food. The remarkable dispersal of *Pilobilus* sporangia benefits other species, as well: Parasitic nematode worms hitch a ride on the sporangia to find a new host.

### Glomeromycetes form endomycorrhizae.

The Glomeromycota, or **glomeromycetes**, are a monophyletic group of apparently low diversity (there are about 200 described species) but tremendous ecological importance. All known glomeromycetes occur in association with plant roots. Most famous are those that form endomycorrhizae. Molecular clock estimates suggest that glomeromycetes diverged from other fungi 500 to 600 million years ago, before the evolution of vascular plants. Establishment of glomerophyte–root symbiosis changed the course of evolution for both participants. Today, most vascular plant species harbor endomycorrhizae, dramatically increasing nutrient uptake from the soil.

Glomeromycetes are hard to study because they cannot be grown independently of their plant partners, and so many unanswered questions remain. For example, no evidence of sexual reproduction has ever been found in this group. Yet genetic studies of glomeromycete populations show some genetic diversity, which is thought to result from parasexual processes (see Fig. 34.14). Glomerocytes produce extremely large (0.1 to 0.5 mm diameter) multinucleate spores asexually. These large spores can persist in soils until they come into contact with an uninfected root.

### The Dikarya produce regular septa during mitosis.

The final major node, or branching point, in the fungal phylogeny marks the origin of the **Dikarya,** a vast group that includes about 98% of all described fungal species. A key feature of the dikaryotic fungi is that every mitotic division is accompanied by the formation of a new septum. This innovation allows these fungi to control the number of nuclei within each cell and thus to proliferate dikaryotic cells. The Dikarya include all the edible mushrooms, the yeast species used in the production of beer, bread, and cheese, the major wood-rotting fungi, and pathogens of both crops and humans.

The Dikarya contain two monophyletic subgroups (see Fig. 34.15), each named for the shape of the cell in which the key reproductive processes of nuclear fusion (karyogamy) and meiosis take place. The Ascomycota, or **ascomycetes,** are sometimes called sac fungi because nuclear fusion and meiosis take place in an elongated saclike cell called an ascus (*askos* is Greek for "leather bag"). The Basidiomycota, or **basidiomycetes,** are popularly known as club fungi because the nuclear fusion

and meiosis take place in a club-shaped cell called a basidium. ("Basidium" is derived from the Greek *basis,* which means "base" or "pedestal" in Greek; the reference is to the position of sexually produced spores at the tips of this club-shaped body.)

Although ascomycetes and basidiomycetes share the same basic life cycle, they diverged from each other as much as 600 to 500 million years ago and evolved complex multicellular structures independently of each other. The morel in the supermarket is an ascomycete, whereas the shiitake mushroom next to it is a basidiomycete (**Fig. 34.18**). In addition to asci and basidia, the two groups differ in several other features. For example, the septal pores that connect adjacent cells within hyphae are distinct in the two groups, as are the ways that cells can plug the pores to minimize damage to a mycelium. Also, when ascomycetes reproduce, mating cells form and their nuclei fuse in relatively close succession. Thus, the heterokaryotic stage is brief. In contrast, in the basidiomycetes, the nuclei of two fused cells may remain separate for a long time. In this case, hyphae containing both types of nuclei (dikaryotic hyphae) can continue to absorb nutrients from the substrate for a while before nuclear fusion occurs.

### Ascomycetes are the most diverse group of fungi.

Ascomycetes make up 64% of all known fungal species. They include important wood-rotting fungi, many ectomychorrizal species, and significant pathogens of both animals and plants. Ascomycetes form the fungal partner in most lichens— approximately 40% of all ascomycete species occur in lichens. Ascomycetes loom large in human history, contributing the baker's and brewer's yeasts used to make bread and beer, the antibiotic penicillin, and model systems employed in laboratory investigations of eukaryotic cell biology and genetics. Ascomycetes are used to produce soy sauce, sake, rice vinegar, and

**FIG. 34.18  Two groups of dikaryotic fungi.** The morel (a) is an ascomycete, whereas the shiitake mushroom (b) is a basidiomycete.

miso, and to transform milk into Brie, Camembert, and Roquefort cheeses. Athlete's foot and other skin infections are caused by ascomycetes, as are many more serious fungal diseases.

Ascomycetes include both species that form multicellular fruiting bodies and ones that form unicellular yeasts. A remarkable feature of ascomycete fruiting bodies is that they contain both dikaryotic and haploid cells. The proliferation of dikaryotic cells leads to the production of many asci, while the growth of haploid cells contributes to the bulk of the fruiting body. Two ascomycete lineages consist largely of yeasts. The earliest branching members in these groups, however, produce hyphae. Thus, the unicellular nature of yeasts is a derived feature evolved through loss of hyphae and is not an ancestral condition.

**Fig. 34.19** illustrates the life cycle of a common ascomycete, the brown cup fungus. In ascomycetes, meiosis is followed by a single round of mitosis resulting in asci that contain eight haploid spores. In many ascomycetes, fruiting bodies elevate the asci on one or more cup-shaped surfaces, from where they are easily caught and carried away by wind. When mature, the spores are ejected from the top of the asci, expelled by turgor pressure.

In the fruiting bodies of some ascomycetes, the asci are completely enclosed by a layer of tissue and thus must be dispersed by other organisms rather than by the wind. The truffles prized in cooking provide an example. The edible truffle is the fruiting body of an ascomycete that grows as an ectomycorrhizal fungus on tree roots. Not only do truffles encase their spores in protective tissues,

**FIG. 34.19** The sexual life cycle of a common ascomycete, the brown cup fungus (*Peziza* species).

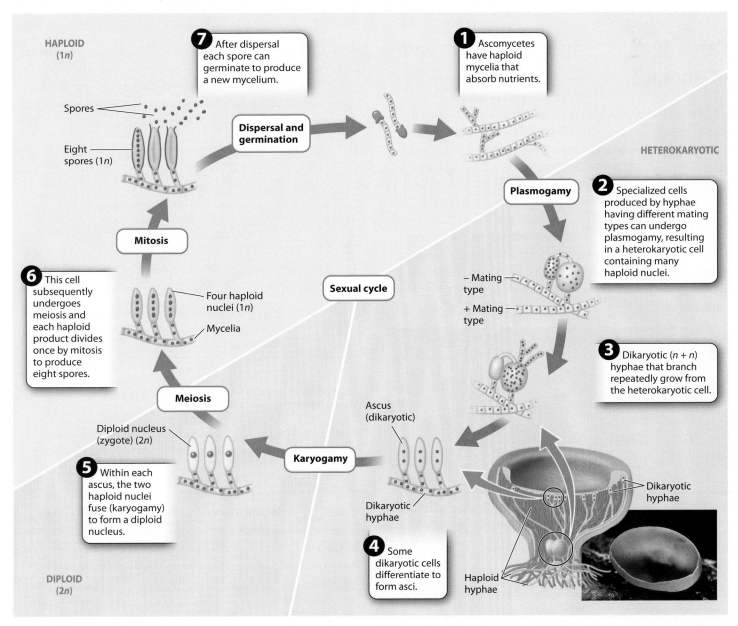

but they also develop underground. How, then, can their spores be dispersed? Developing truffles release androstenol, a hormone also produced by boars before mating. The hormone attracts female pigs, which unearth and consume the fruiting body. The spores pass through the pig's digestive tract without damage and are released into the environment in feces, thereby dispersing the spores.

Another ascomycete is thought to have played a role in the events that unfolded in Salem, Massachusetts, in 1692. That year, several girls came down with an unknown illness manifested by delirium, hallucinations, convulsions, and a crawling sensation on the skin. The girls were thought to be bewitched, and subsequent

accusations led to the infamous Salem witch trials. They were executed for witchcraft, but today many scholars prefer a simpler diagnosis for the illness: poisoning by the ascomycete fungus ergot. Ergot is a common pathogen of rye and related grasses, and ingestion of contaminated plants can lead to the symptoms reported in Salem. Like many defensive compounds produced by plants, alkaloid molecules produced by ergot and their derivatives are used in medicine in low doses, such as in the treatment of migraines.

Ascomycetes are even known for producing so-called zombie ants, a topic explored in **Fig. 34.20**.

## HOW DO WE KNOW?

### FIG. 34.20

# Can a fungus influence the behavior of an ant?

**BACKGROUND** A curious death ritual unfolds in the rain forest of Thailand. Spores of the ascomycete fungus *Ophiocordyceps* infect *Camponotus leonardi* ants, growing hyphae inside their bodies. The fungus eventually kills the ant, and fruiting bodies emerge from the dead ant's head to disperse spores that begin the life cycle anew. *C. leonardi* nests and forages for food high in the forest canopy. Infected ants, however, undergo convulsions that cause them to fall to the forest floor. The infected ants wander erratically and are unable to climb more than a few meters above the ground before convulsions make them fall again. In their final act, the ants bite into leaves and die. The fungus is then able to complete its life cycle within the humid forest understory. Is the ants' behavior induced by the fungi?

An infected ant attached to a leaf by having bitten into it in a "death grip."

**HYPOTHESIS** Parasitic fungi manipulate ant behavior to complete their own reproductive cycle.

**OBSERVATIONS** Infected and uninfected ants were observed for many hours and behavioral events were recorded.

**RESULTS** Infected ants have repeated convulsions, but uninfected ants show no such behavior. Transitions from erratic wandering to a "death grip," in which the ant bites into a leaf, occurred at about the same time. Dissections showed that the ants' death grip results from jaw-muscle wasting caused by the fungus.

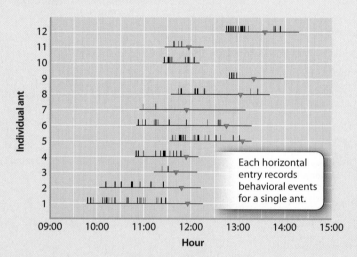

Observations of numerous infected ants (vertical bars indicate convulsions; red triangles indicate the time of ant biting).

**CONCLUSION** The fungi change the behavior of infected ants, inducing a stereotypical behavior that facilitates completion of the fungus' life cycle. The molecular basis of this manipulation is not yet known.

**FOLLOW-UP WORK** The zombie-like behavior induced by the fungus is an example of what has been called the extended phenotype: the idea that a phenotype should include the effects that an organism has on its environment—in this case, its effects on host behavior.

**SOURCE** Hughes, D. P., S. B. Andersen, N. L. Hywel-Jones, W. Himaman, J. Billen, and J. J. Boomsma. 2011. "Behavioral Mechanisms and Morphological Symptoms of Zombie Ants Dying from Fungal Infections." *BMC Ecology* 11:13–22.

**FIG. 34.21 Basidiomycetes.** (a) Corn smut (*Ustilago maydis*); (b) pseudoflowers that form when plants are infected by the rust *Puccinia monoica;* (c) a toadstool fungus—the highly toxic *Amanita.*

## Basidiomycetes include smuts, rusts, and mushrooms.

The basidiomycetes, which make up 34% of all described fungal species, include three major groups (**Fig. 34.21**). Two of these—the smuts and rusts—consist primarily of plant pathogens. Smut fungi, which infect the reproductive tissues of grasses and related plants, take their name from the black sooty spores that they produce. *Ustilago maydis,* the corn smut, turns developing corn kernels into soft gray masses that are a culinary delicacy in Mexico. Because smuts infect seeds, their spread through crops is magnified by the harvest, storage, and eventual sowing of seeds. Before the introduction of chemical seed treatments in the 1930s, infection by *Tilletia* (the stinking smut) commonly caused losses of up to 50% of the wheat harvest. Smuts remain a problem in parts of the world where untreated seeds are still planted.

Rusts can also have profound impacts on plant function and development. For example, the rust *Puccinia monoica* alters leaf development in its host plant *Arabis,* resulting in a pseudoflower that attracts insects by visual and olfactory cues as well as nectar rewards. These pseudopollinators transport spores, thus enhancing outcrossing as well as dispersal of the fungi but provide no service to the plant. We discuss the biology of rusts in greater detail in the next section when we examine Ug99, a highly virulent wheat rust first reported in Uganda in 1999.

The third basidiomycete group is characterized by the formation of multicellular fruiting bodies, including the iconic toadstools with their central stalk and umbrella-like cap. However, this group also includes the diverse shapes seen in stinkhorns, puffballs, and bracket fungi (**Fig. 34.22**).

**FIG. 34.22 Diverse fruiting bodies of basidiomycetes.** (a) Stinkhorn (*Dictyophora indusiata*); (b) puffball (*Calvatia gigantea*); (c) bracket fungi (*Laetiporus sulphureus*).

**FIG. 34.23** The sexual life cycle of a basidiomycete that forms multicellular fruiting bodies.

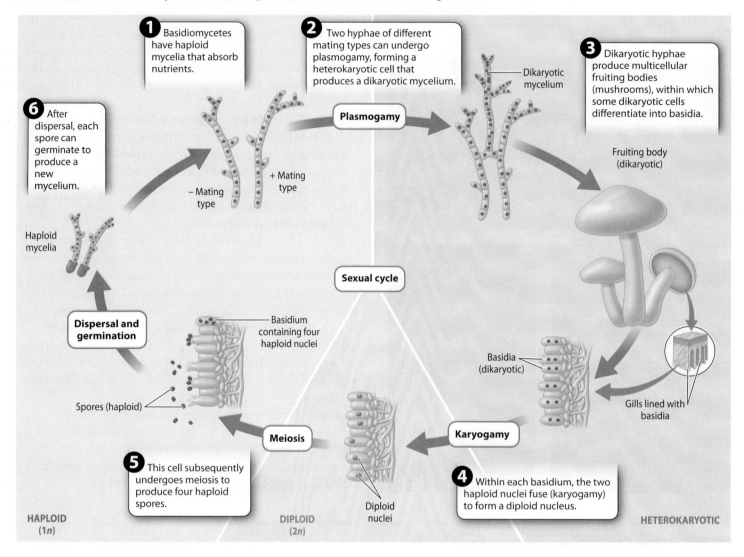

**1** Basidiomycetes have haploid mycelia that absorb nutrients.

**2** Two hyphae of different mating types can undergo plasmogamy, forming a heterokaryotic cell that produces a dikaryotic mycelium.

**3** Dikaryotic hyphae produce multicellular fruiting bodies (mushrooms), within which some dikaryotic cells differentiate into basidia.

**6** After dispersal, each spore can germinate to produce a new mycelium.

Dikaryotic mycelium

Plasmogamy

Fruiting body (dikaryotic)

– Mating type

+ Mating type

Haploid mycelia

Sexual cycle

**Dispersal and germination**

Basidium containing four haploid nuclei

Basidia (dikaryotic)

Gills lined with basidia

Spores (haploid)

Meiosis

Karyogamy

**5** This cell subsequently undergoes meiosis to produce four haploid spores.

Diploid nuclei

**4** Within each basidium, the two haploid nuclei fuse (karyogamy) to form a diploid nucleus.

**HAPLOID** (1n)

**DIPLOID** (2n)

**HETEROKARYOTIC**

**Fig. 34.23** shows the life cycle of a typical basidiomycete mushroom. The life cycle is similar to that of ascomycetes (see Fig. 34.19), except that the fruiting body is made up entirely of dikaryotic hyphae instead of a combination of dikaryotic and haploid hyphae as in ascomycetes. In addition, nuclear fusion (karyogamy) takes place in a specialized cell called a basidium rather than an ascus. Finally, the haploid products of meiosis (the spores) do not undergo mitosis, so four spores are produced from each basidium rather than eight spores from each ascus.

→ **Quick Check 4** When you eat a mushroom, what stage of its life cycle are you consuming?

Most species in this group live by forming ectomycorrhizal associations or by decomposing wood and other substrates. The fruiting bodies that we see elevated above a rotting log or the soil are connected to extensive mycelia that provide resources for fruiting-body development.

Elevation alone does not ensure dispersal because the spores must penetrate a boundary layer of air that surrounds the fruiting body. Many basidiomycetes depend on surface tension to catapult their spores through the air. The placement of their four spores on the ends of short stalks is the key to this mechanism. Both the spores and the supporting basidial cell actively secrete solutes that cause water to condense on their surfaces. At first, the droplets are independent, but as they grow they eventually come into contact with one another, and at this point the high surface tension of water causes them to merge into a single, smooth droplet. This change in shape shifts the center of mass of the water with such force that it catapults the spore into the air.

Basidiomycetes that form conspicuous fruiting bodies have diverse mechanisms for dispersing their spores. For example, puffballs function like bellows: The impact of raindrops forces the loose, dry spores out of a small hole at the tip of the ball (see Fig. 34.10a). Bird's nest fungi function like splash cups: In this case, raindrops displace groups of spores (the "eggs" in the nest) onto nearby vegetation. Herbivores consume the spores when they feed on the vegetation and transport them to a new dung pile. Stinkhorns produce spores in a stinky, sticky mass that attracts insects, which then carry off spores that stick to their legs and body.

## ? CASE 6 Agriculture: Feeding a Growing Population
## How do fungi threaten global wheat production?

Throughout much of human history, crops have been vulnerable to infection by *Puccinia graminis*, the black stem rust of wheat. *P. graminis* infects wheat leaves and stems through their stomata and then extends within the plant's living tissues to fuel its own growth. Rust-colored pustules erupt along the stem and then release a vast number of spores, leaving the host plant to wither and die. Crop losses from *P. graminis* can be devastating. In the past, both the United States and the Soviet Union tried to develop black stem rust as a biological weapon. The U.S. version was cluster bombs (since destroyed) containing turkey feathers smeared with rust spores, produced with the aim of disrupting crop yields in enemy countries.

Many rusts have a life history in which they alternate between different host species. For example, to complete its life cycle, *P. graminis* must infect wheat first and then barberry (*Berberis* species). After a particularly severe outbreak of wheat rust during World War I, in which crop losses equaled one-third of total U.S. wheat consumption, a major program of barberry eradication was started to bring stem rusts under control. This program reduced the threat, but did not totally eliminate it.

The Green Revolution seemed to give growers a decisive advantage over stem rust. During World War II, Norman Borlaug, whose original training was in plant pathology, was sent to Mexico by the Rockefeller Foundation to help combat crop losses due to stem rust. Borlaug's breeding efforts led to resistant wheat varieties. Much of this resistance could be attributed to variation in a small number of genes. Stem rusts were apparently vanquished—that is, until the appearance of Ug99, a wheat stem rust first identified in Uganda in 1999 (hence the name). Ug99 is capable of defeating all major resistance genes. Given that wheat accounts for 20% of daily global calorie consumption by humans and that 90% of cultivated wheat has little or no resistance to Ug99, the appearance of this resistant fungus is cause for major concern. At present, Ug99 shows every sign of spreading (**Fig. 34.24**)—not surprising given that a single hectare of infected wheat produces upward of a billion spores. It currently

**FIG. 34.24 The spread of Ug99.** Ug99 is a highly resistant stem rust of wheat and a major threat to global food production.

- ● Ug99 sites
- → Movement
- ---→ Possible spread
- High wheat production
- Low wheat production

devastates wheat production in Kenya and has been found in Yemen (2006), Iran (2007), and South Africa (2009). As you read this, Ug99 is poised to reach the major wheat growing regions of Turkey and South Asia. Even more rapid spread is possible if spores accidentally lodge on cargo or airline passengers.

Just before his death in 2009, Norman Borlaug urged the world to take this threat seriously. Today, plant breeders are actively searching ancestral wheat varieties for genetic sources of resistance to Ug99, while other scientists are trying to understand how plants defend themselves against rusts and what makes Ug99 so virulent. The resurgence of *P. graminis* as a serious threat to world food production underscores the importance of safeguarding the genetic diversity of agricultural species so that the inevitable evolutionary battles with pathogenic species may be waged successfully.

# Core Concepts Summary

## 34.1 FUNGI ARE HETEROTROPHIC EUKARYOTES THAT FEED BY ABSORPTION.

Fungi are heterotrophs that depend on preformed organic molecules for both carbon and energy. page 34-1

Fungi break down their food and then absorb it. page 34-1

Fungi use the process of growth to find and obtain food in their environment. page 34-2

Most fungi have numerous branched filaments called hyphae, which they use to absorb nutrients. page 34-2

When fungi encounter a rich source of food, their hyphae form a network called a mycelium. page 34-2

Fungal cell walls are made of chitin, the same compound found in the exoskeletons of insects. page 34-2

In early fungal groups, the hyphae form a continuous compartment, with many nuclei but with no cell walls, but in later evolving groups, nuclear divisions are accompanied by the formation of septa that divide the cytoplasm into separate cells. page 34-2

Some fungi, such as yeasts, do not produce hyphae. page 34-2

Most fungi feed on dead organic matter. page 34-3

Fungi are critical elements of the carbon cycle, converting dead organic matter back into carbon dioxide and water. page 34-3

Many fungi feed on living animals and plants, causing diseases. page 34-4

Fungi have repeatedly evolved mutually beneficial relationships with plants and animals. page 34-5

Lichens are stable associations between a fungus and a photosynthetic microorganism that look, function, and even reproduce as single organisms. page 34-6

## 34.2 FUNGI REPRODUCE BOTH SEXUALLY AND ASEXUALLY, AND DISPERSE BY SPORES.

Fungi produce haploid spores that are dispersed by wind, water, or animals. page 34-8

Spores can be produced asexually (by mitosis) or sexually (by cell fusion and meiosis). page 34-8

Many fungi produce spores within multicellular fruiting bodies that enhance spore dispersal. page 34-8

The fungal life cycle is similar to haploid-dominant life cycles found in many eukaryotic organisms, except that plasmogamy (cell fusion) and karyogamy (nuclear fusion) are often separated, resulting in a stage in which there are genetically distinct haploid nuclei in one cell, called a heterokaryotic stage. page 34-10

Genetically determined mating types prevent self-fertilization in many fungi. page 34-11

Asexual fungal species have mechanisms of generating genetic diversity, including a parasexual cycle, which involves crossing over in mitosis and chromosome loss. page 34-11

## 34.3 NEXT TO THE ANIMALS, FUNGI ARE THE MOST DIVERSE GROUP OF EUKARYOTIC ORGANISMS.

About 75,000 fungal species have been formally described, but total diversity may be as high as 5 million species. page 34-12

The chytrids, a small group of about 1000 species, are at the base of the fungal tree. page 34-12

Zygomycetes produce hyphae without septa and make up less than 1% of known fungal diversity. page 34-13

The glomeromycetes are a group of fungi of low diversity but tremendous ecological importance due to their association with plant roots. page 34-14

Most fungi belong to the Dikarya (ascomycetes and basidiomycetes), which form septa along hyphae. During the heterokaryotic stage, these fungi produce cells that contain two haploid nuclei, one from each parent. page 34-14

Ascomycetes include wood-rotting fungi, many ectomychorrizal species, plant and animal pathogens, the fungal partner in most lichens, and baker's and brewer's yeasts. page 34-14

Basidiomycetes include familiar toadstools and puffballs, as well as plant pathogens. page 34-17

# Self-Assessment

1. Describe two ways in which fungi differ from other heterotrophic organisms in how they obtain and digest their food.

2. Explain how hyphae and cell walls made of chitin allow fungi to obtain nutrients from their environment.

3. Describe how fungi contribute to the terrestrial (land-based) carbon cycle.

4. Name the two organisms that make up a lichen and describe how each partner benefits from the association.

5. Describe how fungi disperse.

6. Draw the life cycle of an ascomycete, and indicate the heterokaryotic stage.

7. Draw the life cycle of a basidiomycete, and indicate the heterokaryotic stage.

8. Name and describe several key innovations in the evolutionary history of fungi that allowed them to move from water to land.

9. Explain how the evolution of vascular plants has provided opportunities for fungal diversification.

Do you understand the chapter's Core Concepts? Log in to check your answers to the Self-Assessment questions, then practice what you've learned and reinforce this chapter's concepts by working through the problems and multimedia tutorials provided there.

⌨ www.biologyhowlifeworks.com

**FIG. 36.1** **Diverse sensory receptors.** (a) Many insects can detect ultraviolet light. (b) Rattlesnakes can see at night by detecting infrared radiation. (c) Dogs have a keen sense of smell.

share many similar properties, differences have also evolved. Consequently, animals perceive the world differently from one another. Many insects, for instance, are sensitive to ultraviolet light; nocturnal snakes can see at night by sensing infrared radiation; and dogs can distinguish odor compounds as much as 100 million times lower in concentration than humans can detect (**Fig. 36.1**).

## Specialized sensory receptors detect diverse stimuli.

Animals have evolved a diverse array of sensory receptors that respond to different stimuli, among them touch, light, and the oscillations in air pressure we call sound waves. How is a physical phenomenon in the world outside an organism transformed into a nerve impulse? The initial transformation takes place inside the sensory receptor (**Fig. 36.2**). A physical or chemical stimulus is converted into a change in the sensory receptor's membrane potential. Recall from Chapter 35 that a cell's membrane potential is the charge difference between the inside and the outside of the cell membrane, and that a depolarization in membrane potential is the first step in firing an action potential. Sensory receptors either fire action potentials themselves or synapse with other neurons that fire action potentials, which are then transmitted to the brain.

In many cases, the sound wave, touch, or other stimulus causes ion channels in the cell's plasma membrane to open. The influx of ions changes the membrane potential by altering the distribution of charged ions on either side of the membrane. If the influx of ions reduces the charge difference, then the cell is depolarized (Chapter 35).

The most ancient type of sensory detection is chemoreception. All organisms respond to chemical cues in their environment. Even the earliest branching groups of Bacteria and Archaea have protein receptors in their cell membranes that respond to molecules in the environment. In animals, **chemoreceptors** respond to molecules

that bind to specific protein receptors on the cell membrane of the sensory receptor (Fig. 36.2a). Many animals detect food in their environment by sensing key molecules such as oxygen ($O_2$), carbon dioxide ($CO_2$), glucose, and amino acids. Mosquitoes track $CO_2$ levels to detect prey for blood meals, and coral polyps respond to simple amino acids in the water, extending their bodies and tentacles toward areas of greater concentration to feed. Other arthropods, such as flies and crabs, have chemosensory hairs on their feet. These animals taste potential food sources by walking on them. Salmon rely on chemoreception to detect chemical traces of the home waters of the river where they hatched, and where they will return to mate and spawn.

Chemoreception also underlies the sense of smell and taste. In most cases, the binding of molecules to a protein receptor on a taste receptor causes the protein receptor to change conformation. That change in conformation in turn triggers the opening of $Na^+$ channels through G protein signal transduction pathways similar to those discussed in Chapter 9. The influx of $Na^+$ ions depolarizes the receptor cell membrane. No action potential fires, but the depolarization travels far enough down the receptor's short axon to trigger the release of neurotransmitters.

**Mechanoreceptors** respond to physical deformations of their membrane produced by touch, stretch, pressure, motion, and sound. Deformation of the receptor membrane opens sodium channels, causing a depolarization of the endings of the cell's dendrites (Fig. 36.2b). Early mechanoreceptors in bacteria sensed local physical forces and internal cell pressure. Other mechanoreceptors in roundworms and anemones are linked to externally projecting cilia at the animal's body surface.

One well-studied example of a mechanoreceptor is a touch receptor in the roundworm *Caenorhabditis elegans*. Deformation of the surface of the worm—its cuticle—exerts pressure on proteins connected to an ion channel. The mechanical force

## CHAPTER 36

# ANIMAL SENSORY SYSTEMS AND BRAIN FUNCTION

## Core Concepts

To see, your eyes detect light, which is a form of electromagnetic radiation; to hear, your ears detect sound waves; to smell, your nose detects odor molecules present in the air. In all these cases, your senses are detecting a physical or chemical stimulus in the environment. The first half of this chapter explores how specialized nerve cells detect these signals and code them as information that can be transmitted and processed by the nervous system. How is a dog able to distinguish thousands of different odors? How is a hawk able to see a small rodent far below? How are humans able to detect small differences in the pitch of a sound?

The second half of the chapter follows nervous system pathways from the senses into the brain, on the way exploring basic principles of brain function. Processing of sensory information in the brain takes place in regions specialized for each sense, and within many of these regions information is represented in the form of topographic maps.

## 36.1 ANIMAL SENSORY SYSTEMS

Animals are able to sense the physical properties of their environment, including light, chemicals, temperature, pressure, and sound, useful in finding mates and food and avoiding predators and noxious environments. Early in evolutionary history, organisms evolved specialized membrane **receptors** that functioned to detect these critical features in their environment. For example, before the evolution of nerve cells and a nervous system, bacteria evolved receptors located in the cell membrane to sense osmotic pressures that might otherwise rupture the membrane (Chapter 5). Bacteria and sponges also evolved membrane receptors that could detect chemicals—including nutrients—in their environment.

The senses of smell, taste, and sight in multicellular organisms with a nervous system also rely on membrane receptors. However, these receptors are embedded in specialized membranes of sensory neurons, forming a **sensory receptor.** In this case, a sensory receptor refers to the entire neuron, not just a membrane protein. In most multicellular animals, the sensory receptors are organized into specialized **sensory organs** that convert particular physical and chemical stimuli into nerve impulses that are processed by a nervous system and sent to a brain. Cnidarians (which include jellyfish, corals, and anemones) and roundworms evolved sensory receptors to sense physical contact, and cnidarians and flatworms were among the first multicellular animals to evolve simple light-sensing photoreceptive organs.

The conversion of physical or chemical stimuli into nerve impulses is called **sensory transduction.** For example, receptors located in the ear convert the energy of sound waves into nerve impulses that allow an animal to distinguish loud versus soft sounds and high- versus low-pitched sounds. Although the sense organs of different animals

3. Diagram and label the basic features of a neuron, indicating where information is received and where it is sent.

4. Graph an action potential, showing the change in electrical potential on the y-axis and time on the x-axis. Indicate on the graph the phases when voltage-gated $Na^+$ and $K^+$ ion channels are opened and when they are closed.

5. Explain what is meant by saying action potentials are "all-or-nothing."

6. Explain why action potentials propagate along an axon only in a single direction.

7. Briefly describe how myelinated axons increase the speed of signal transmission.

8. Diagram a chemical synapse, labeling the vesicles that contain neurotransmitter molecules and the receptors that bind the neurotransmitter to produce either an inhibitory or excitatory stimulus in the postsynaptic cell.

9. Briefly describe how neurotransmitter binding to receptors on a postsynaptic cell causes inhibition or excitation.

10. Describe how vertebrate nervous systems are organized into voluntary and involuntary components, listing which functions of an animal are controlled by each component.

11. Diagram a simple circuit that includes a sensory neuron that synapses with a motor neuron to produce a reflex. Indicate where in the nervous system this synapse is found.

Do you understand the chapter's Core Concepts? Log in to check your answers to the Self-Assessment questions, then practice what you've learned and reinforce this chapter's concepts by working through the problems and multimedia tutorials provided there.

📖 **www.biologyhowlifeworks.com**

# Core Concepts Summary

**35.1 ANIMAL NERVOUS SYSTEMS ALLOW ORGANISMS TO SENSE AND RESPOND TO THE ENVIRONMENT, COORDINATE MOVEMENT, AND REGULATE INTERNAL FUNCTIONS OF THE BODY.**

Nerve cells, or neurons, receive and send signals and are the functional unit of the nervous system. page 35-1

Animal nervous systems include three types of neuron: sensory neurons that respond to signals, interneurons that integrate and process sensory information, and motor neurons that produce a response. page 35-2

Ganglia are localized collections of nerve cell bodies that integrate and process information. page 35-2

Simply organized animals, such as cnidarians, have a nerve net to coordinate sensory and motor function. page 35-3

Forward locomotion led to the evolution of specialized sense organs in the head, along with concentrated groupings of nerve cells to form ganglia and a brain. page 35-3

**35.2 THE NEURON IS THE FUNCTIONAL UNIT OF THE NERVOUS SYSTEM. NEURONS HAVE DENDRITES THAT RECEIVE INFORMATION AND AXONS THAT TRANSMIT INFORMATION.**

Neurons share a common organization, with dendrites that receive inputs, a cell body that receives and sums the inputs, and axons that transmit signals to other nerve cells. page 35-4

Glial cells provide nutritional and physical support for neurons. page 35-6

In vertebrates, glial cells also produce the myelin sheath that insulates axons, increasing the speed of nerve impulses. page 35-6

**35.3 THE ELECTRICAL PROPERTIES OF NEURONS ALLOW THEM TO COMMUNICATE RAPIDLY WITH ONE ANOTHER.**

Neurons have electrically excitable membranes that code information by changes in membrane voltage and transmit information in the form of electrical signals called action potentials. page 35-6

Ion channels open and close in response to changes in membrane voltage, underlying the production of action potentials in nerve cells. page 35-8

Action potentials fire in an all-or-nothing fashion followed by a brief refractory period. page 35-8

Action potentials are conducted in a saltatory fashion in myelinated axons, firing at nodes of Ranvier where the axon membrane is exposed and not insulated by myelin. page 35-11

Most neurons communicate by chemical synapses formed between an axon terminal and a neighboring nerve or muscle cell. page 35-12

Communication across the synapse occurs when an arriving action potential triggers the release of neurotransmitters from vesicles within the axon terminal. page 35-12

Neurotransmitters released from presynaptic vesicles bind to receptors in the postsynaptic membrane, causing either an excitatory or an inhibitory stimulus. page 35-13

Excitatory stimuli depolarize the membrane, producing an excitatory postsynaptic potential (EPSP), whereas inhibitory stimuli hyperpolarize the membrane, producing an inhibitory postsynaptic potential (IPSP). page 35-13

**35.4 ANIMAL NERVOUS SYSTEMS CAN BE ORGANIZED INTO CENTRAL AND PERIPHERAL COMPONENTS.**

Animal nervous systems are organized into central and peripheral components called the central nervous system (CNS) and peripheral nervous system (PNS). page 35-15

The central nervous system includes the brain and one or more main trunks of nerve cells, such as the spinal cord. The peripheral nervous system is distributed throughout the animal's body and is composed of sensory and motor nerve cells. page 35-16

In many invertebrates and vertebrates, the peripheral nervous system is divided into voluntary and involuntary components. page 35-16

In vertebrates, the voluntary component is somatic, and the involuntary is autonomic. page 35-16

The autonomic system regulates body functions through opposing actions of the sympathetic and parasympathetic divisions. page 35-17

The nervous system helps to regulate physiological functions to actively maintain stable conditions inside a cell or an organism, a property known as homeostasis. page 35-17

Homeostasis is often achieved by negative feedback, in which the response inhibits the stimulus. page 35-17

Simple reflex circuits can involve as few as two neurons: a sensory neuron from the periphery that synapses with a motor neuron in the spinal cord that sends a signal to a muscle. page 35-18

# Self-Assessment

1. Diagram a simple nervous system of an animal that lacks cephalization and compare that system with the general organizational features of a nervous system that exhibits cephalization.

2. Name the three basic categories of neuron and describe their functions.

**FIG. 35.20 The knee-jerk reflex.** The knee-jerk reflex involves a sensory neuron, a single synapse, and a motor neuron.

**2** A stretch receptor in an extensor muscle responds by sending a signal along the sensory nerve.

Stretch receptor

Extensor muscle

**3** The sensory neuron synapses with a motor neuron in the spinal cord.

Spinal cord

Neuron from stretch receptor

Patella

Patellar tendon

Flexor muscle

**4** The motor neuron sends an excitatory signal to the same extensor muscle, which responds by contracting.

Motor neuron to extensor muscle originally stretched

Motor neuron to flexor muscle

Interneuron for reciprocal inhibition of flexor muscle

**5** An inhibitory interneuron inhibits contraction of the opposing flexor muscle.

**1** A physician strikes the patellar tendon with a reflex hammer.

beside the spinal cord. These cell bodies have axons that extend into the spinal cord. In response to a stretch, a signal is sent from the stretch receptor, through dendrite and cell body, to the axon. In the spinal cord, the axon of the sensory neuron synapses directly on motor neurons that travel from the spinal cord back to the muscle where the stretch originated. The signal from the muscle stretch receptor stimulates the motor neurons to increase the activation of the muscle, causing the muscle to contract and the leg to extend at the knee. This motion is the knee-jerk response.

This reflex arc does not include an interneuron: It is composed of just two neurons and one synapse. It provides a rapid change in muscle contraction because only one synapse is required to relay the sensory information back to the muscle. Neural circuits containing more neurons and synapses take longer to transmit nerve signals because transmission is delayed by communication at the synapses.

As well as being a useful test in examination, the knee-jerk reflex has a normal physiological role. In running or landing from a jump, the knee joint is flexed (that is, bent), stretching the quadriceps muscles that extend (that is, straighten) the knee joint. Knee flexion allows rapid adjustments in muscle force and knee position that help stabilize the body during running and jumping.

Because muscles can only contract, opposing movements at a joint such as flexion and extension require flexor and extensor muscles on either side of the joint (Chapter 37). These are often activated out of phase (that is, at different times) so that, when one set of muscles is activated (contracting), the other is inhibited (relaxed). The alternating movements of the limbs during walking and running provide an example. This pattern of joint and limb movement is achieved by **reciprocal inhibition:**

When stretch receptors of the knee extensor muscles are activated to stimulate these muscles to extend the knee, they also inhibit the activity of opposing muscles that flex the knee (Fig. 35.20).

Reciprocal inhibition of opposing sets of muscles occurs in the spinal cord. Axons of the stretch receptor neurons not only synapse with motor neurons of extensor muscles but also synapse with inhibitory interneurons that inhibit motor neuron stimulation of the opposing flexor muscles. This inhibitory reflex pathway contains two synapses, one between the sensory neuron and interneuron and the second between the interneuron and the motor neuron to the flexor muscle.

Reciprocal inhibition also operates between the right and left sides of the body. For example, reciprocal inhibition helps to control the movement of the right and left limbs. In this case, interneurons cross the spinal cord to control the timing of activity by flexor and extensor muscles of the opposite limb. Reciprocal inhibition is also involved in the Mauthner cell circuit of fish to ensure that only one side of the fish's body bends away from the stimulus to escape. The local spinal circuits that provide reciprocal inhibition are therefore fundamental to the alternating motion of the body and limbs that characterizes the movements of most animals.

In other cases, more complex circuits in the brain may act to coordinate sensory input and motor output. These circuits integrate other sources of sensory information, such as vision and balance, with conscious commands from the brain to voluntarily control motor behavior. In the next chapter, we look more closely at these sensory circuits.

(**Fig. 35.19**). In negative feedback, a stimulus acts on a sensor that communicates with an effector, which produces a response that opposes the initial stimulus. For example, negative feedback is used to maintain a constant temperature in a house. Cool temperature (the stimulus) is detected by a thermostat (the sensor). The thermostat sends a signal to the heater (the effector), producing heat (the response). In this way, a stable temperature is maintained (Fig. 35.19a).

In a similar fashion, humans and other mammals maintain a steady body temperature even as the temperature outside fluctuates. Nerve cells in the hypothalamus (located in the base of the brain) act as the body's thermostat (Fig. 35.19b). A drop in body temperature signals the hypothalamus to activate the somatic nervous system to induce shivering and metabolic heat, as discussed in Chapter 40. At the same time, the hypothalamus activates the autonomic nervous system, causing peripheral blood vessels to constrict. The reduction in blood flow near the body's surface in turn reduces heat loss to the surrounding air. An increase in temperature signals sweat glands and vasodilation of peripheral blood vessels to aid heat loss from the skin.

Homeostatic regulation, therefore, relies on negative feedback to maintain a set point, which in this case represents an animal's preferred body temperature. The ability to maintain a constant body temperature is known as thermoregulation, and it is just one of many parameters that the body actively maintains, as we discuss in subsequent chapters.

### Simple reflex circuits provide rapid responses to stimuli.

An animal that has perceived a predator has an advantage if it can move more quickly than the predator. Fast responses are made possible by simple reflex circuits that bypass the brain by directly connecting sensory neurons (detecting the presence of the predator) with motor neurons (the quick movement necessary to escape the predator). Reflex circuits are common in both the somatic and autonomic components of the nervous system.

Simple spinal reflex circuits connect sensory and motor neurons directly in the spinal cord so that a vertebrate can respond rapidly to sensory stimuli without the need for delay-causing conscious input from the brain. The large Mauthner neurons of fish are an example of such a circuit. Sensory neurons respond to threatening visual, tactile, or vibratory cues from the environment and transmit signals to the hindbrain of a fish, activating a large Mauthner neuron on the side opposite the stimulus. The Mauthner cell in turn rapidly activates motor neurons along the body, causing the fish to bend away from the threatening stimulus to initiate a rapid escape. At the same time, the Mauthner cell and motor nerves on the other side of the fish are inhibited. The giant axons of squid are part of a similar reflex circuit that allows them to swim quickly away from a predator.

The **knee-jerk reflex** in humans is an example of a simple nerve circuit that includes only a single synapse between two neurons—a sensory neuron and a motor neuron (**Fig. 35.20**). Physicians commonly use this reflex to evaluate peripheral nervous and muscular system function. Let's follow this reflex pathway. It starts with specialized "stretch" receptors in the extensor muscles of the leg, including the quadriceps muscle. These stretch receptors sense the stretch of the muscle that occurs during movement or in response to a physician's strike of a reflex hammer on the stretch receptors. The receptors are part of a sensory nerve with dendrites that extend from the stretch receptors to cell bodies in ganglia

---

**FIG. 35.19 Temperature regulation by negative feedback in (a) a house and (b) a mammal.** In negative feedback, a response (such as heat) opposes the stimulus (cold), leading to a stable state (a steady temperature).

awareness. The autonomic nervous system, in turn, is divided into two major subdivisions, a **sympathetic division** and a **parasympathetic division.** Both divisions continuously monitor and regulate internal functions of the body. Generally, sympathetic and parasympathetic nerves have opposite effects, but not always. For example, sympathetic neurons stimulate the heart to beat faster, whereas parasympathetic neurons cause the heart to beat slower.

The sympathetic nervous system generally results in arousal and increased activity. It is the pathway activated when animals are exposed to threatening conditions, resulting in what is often referred to as the fight-or-flight response. This response includes an increase in the heart and breathing rates, increase in glucose release by the liver, and inhibition of digestion. The parasympathetic nervous system, in contrast, slows the heart and stimulates digestion as well as metabolic processes that store energy—in other words, it enables the body to "rest and digest."

The sympathetic and the parasympathetic nerves have different anatomical distributions (Fig. 35.18) as well as different functions. Sympathetic nerves leave the CNS from the middle region of the spinal cord, forming ganglia along much of the length (thoracic and lumbar) of the spinal cord. Parasympathetic nerves leave from the brain by cranial nerves, as well as from lower levels (sacral) of the spinal cord.

**FIG. 35.18** **Function and organization of the human involuntary (autonomic) nervous system.** The autonomic system, which controls involuntary functions, is divided into sympathetic and parasympathetic systems.

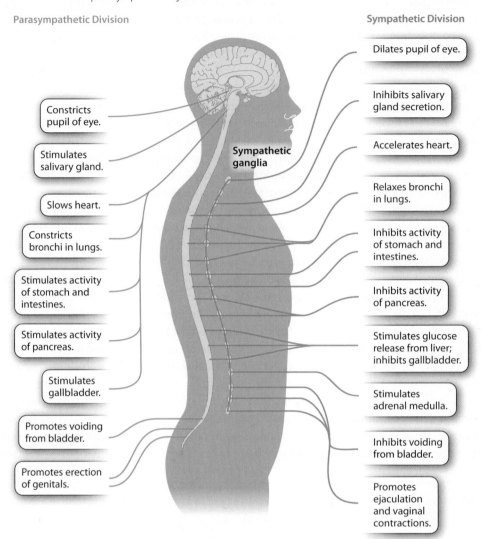

Parasympathetic Division

Sympathetic Division

Constricts pupil of eye.

Stimulates salivary gland.

Slows heart.

Constricts bronchi in lungs.

Stimulates activity of stomach and intestines.

Stimulates activity of pancreas.

Stimulates gallbladder.

Promotes voiding from bladder.

Promotes erection of genitals.

Sympathetic ganglia

Dilates pupil of eye.

Inihibits salivary gland secretion.

Accelerates heart.

Relaxes bronchi in lungs.

Inhibits activity of stomach and intestines.

Inhibits activity of pancreas.

Stimulates glucose release from liver; inhibits gallbladder.

Stimulates adrenal medulla.

Inhibits voiding from bladder.

Promotes ejaculation and vaginal contractions.

## The nervous system helps to maintain homeostasis.

The autonomic nervous system plays a key role in homeostasis, which is the ability to maintain a steady physiological state in the face of changing environmental conditions. Many physiological parameters are maintained in a narrow range of conditions, including pH, temperature, and ion concentrations. Homeostasis is maintained within individual cells as well as within the organism as a whole. Enzymes within a cell, for example, often work effectively only in narrow pH and temperature ranges. Ion concentrations also must be maintained within narrow ranges for normal cell function. As we have seen, the firing of action potentials by neurons requires particular ion concentrations on either side of the membrane. Similarly,

the water content of cells and the body as a whole is kept stable through the careful regulation of ions and other solutes, as discussed in Chapter 41.

The concept of homeostasis was first described as regulation of the body's "interior milieu" in the late 1800s by the French physiologist Claude Bernard, who is often credited with bringing the scientific method to the field of medicine. The term "homeostasis" was coined by the American physiologist Walter Cannon, whose book *The Wisdom of the Body* (first published in 1932) popularized the concept.

Maintaining steady and stable conditions takes work in the face of changing environmental conditions. That is, a cell (or organism) actively maintains homeostasis. How does the body, and in particular the nervous system, maintain homeostasis? Homeostatic regulation often depends on **negative feedback** surprisingly, your eyes, sense of touch, and other sensory organs are located on the surface of the body, where they can receive signals from the environment. In contrast, the brain, centrally located ganglia, and a main nerve cord are located in the interior.

How do structures in the interior and near the surface communicate? Nerves form the lines of communication between these nervous system structures. In general, neuron cell bodies are grouped compactly together in sensory organs, ganglia, and a main nerve cord that extends from the brain. In contrast, nerves are composed mainly of axons from many different nerve cells. For example, the optic nerve contains axons that travel from nerve cells in the eye to the brain.

Sensory and motor nerves make up the **peripheral nervous system (PNS).** These nerves communicate with the brain, a main nerve cord, or—in the case of organisms such as flatworms that lack a brain—with centralized information-processing ganglia. The brain, main nerve cord, and centralized ganglia together make up the **central nervous system (CNS).** Information gathered by sensory organs arrives mainly from the animal's periphery, at or near its body surface. After processing the information, the ganglia and brain send commands back to peripheral nerves that coordinate the activity of muscles and glands in different regions of the animal's body. This organization ensures that an animal achieves a coordinated behavioral and physiological response to the stimuli received from its environment. Only the diffuse nerve nets of cnidarians lack defined central and peripheral components.

Specialized sensory organs transmit information from the animal's periphery by **afferent neurons,** which send information toward the CNS. **Efferent neurons** send signals away from the CNS, communicating with muscles and other tissues and organs. The peripheral nervous system also includes interneurons and ganglia that integrate and process information in local regions of the animal's body. For example, vertebrates and invertebrates have ganglia that lie outside each segment of their primary nerve cord. However, the bulk of information processing, particularly in animals that exhibit a greater capacity for learning and memory, occurs in the CNS.

The central nervous system of vertebrates includes both the brain and **spinal cord** (Fig. 35.17). The spinal cord is a central tract of neurons that passes through the vertebrae to transmit information between the brain and the periphery of the body. The vertebrate spinal cord is divided into segments, each controlling body movement in a particular region along the animal's length. Each spinal cord segment contains axons from peripheral sensory neurons, a set of interneurons, and a set of motor neuron cell bodies. These are distinct from, but often associated with, segmental ganglia that lie outside the spinal cord.

In humans and other vertebrate animals, the peripheral nervous system is organized into left and right sets of **cranial nerves** located within the head and **spinal nerves** running

from the spinal cord to the periphery. Most of the cranial nerves and all of the spinal nerves contain axons of both sensory and motor neurons. Cranial nerves link specialized sensory organs (eyes, ears, tongue) to the brain. Cranial nerves also control eye movement, facial expression, speech, and feeding. Some cranial nerves, such as the olfactory and optic nerves, contain only sensory axons. Spinal nerves exit from the spinal cord through openings located between adjacent vertebrae to thread through the trunk and limbs of an animal's body. These nerves receive sensory information from receptors in nearby body regions along the length of the body and carry motor signals from the spinal cord back to those regions (Fig. 35.17).

## Nervous systems have voluntary and involuntary components.

As bodies with distinct internal organ systems evolved, two separate components of the nervous system emerged. When a gazelle senses a predator, some nerve circuits send a signal to run, an action that is under conscious control, while others signal the heart to beat faster and blood vessels supplying muscles to dilate, actions that occur unconsciously. Conscious reactions are under the control of the **voluntary** component of the nervous system, and unconscious ones are under the control of the **involuntary** component. Voluntary components mainly handle sensing and responding to external stimuli, whereas involuntary components typically regulate internal bodily functions. Both nervous system components are found in invertebrate and vertebrate animals.

Let's look at these components in insects and crustaceans. In these animals, a nerve circuit regulates foregut function together with nerve circuits that regulate other parts of the animal's digestive system. These circuits are an involuntary component of an animal's nervous system. Sensory structures such as the antennae and eyes transmit information to the animal's brain, which allows voluntary responses to stimuli received by these sensory organs. Other invertebrates such as mollusks also have involuntary components that regulate the function of internal organ systems that are distinct from their voluntary sensory–motor system.

In vertebrates, the peripheral nervous system is divided into **somatic** (voluntary) and **autonomic** (involuntary or visceral) components. The somatic nervous system is made up of sensory neurons that respond to external stimuli and motor neurons that synapse with voluntary muscles. This system is considered voluntary because it is under conscious control. However, many reflexes are controlled at lower spinal levels (discussed in the following section) or by the brainstem, independent of conscious control by the central nervous system.

The autonomic nervous system (**Fig. 35.18**) controls internal functions of the body such as heart rate, blood flow, digestion, excretion, and temperature. It includes both sensory and motor components, which usually act without our conscious

**FIG. 35.16 Vertebrate motor endplate.** The motor endplate is an excitatory synapse between a motor nerve and a muscle cell; acetylcholine is the neurotransmitter.

Motor neuron axon    Muscle cell

Acetylcholine binds to muscle membrane receptors, causing a depolarization of the muscle cell and contraction.

## 35.4 NERVOUS SYSTEM ORGANIZATION

The brain is considered the body's command center: It receives sensory information from the eyes, ears, and skin and issues instructions to the rest of the body. Up to this point, we have examined how individual neurons send signals to other nerves and muscles. How are these neurons organized in the body to sense stimuli, process them, and issue an appropriate response? How does the brain coordinate the movement of limbs necessary to walk or run, and adjust heart rate and breathing in response to exercise? In this section, we look at how the nervous system is organized to allow the brain and the body to communicate with each other.

### Nervous systems are organized into peripheral and central components.

As animals evolved the ability to sense and coordinate responses to increasingly complex stimuli, their nervous systems became organized into peripheral and central components (**Fig. 35.17**). Not

**FIG. 35.17 Human central (yellow) and peripheral (blue) nervous systems.**

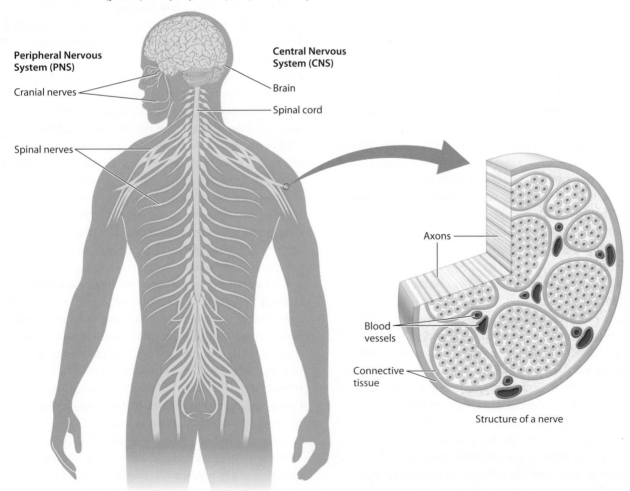

Peripheral Nervous System (PNS)

Cranial nerves

Spinal nerves

Central Nervous System (CNS)

Brain

Spinal cord

Axons

Blood vessels

Connective tissue

Structure of a nerve

**FIG. 35.15  Summation of excitatory and inhibitory postsynaptic potentials.** EPSPs and IPSPs can be summed in time (b) or space (c), or even cancel each other out (d). Widely spaced EPSPs (a) do not sum.

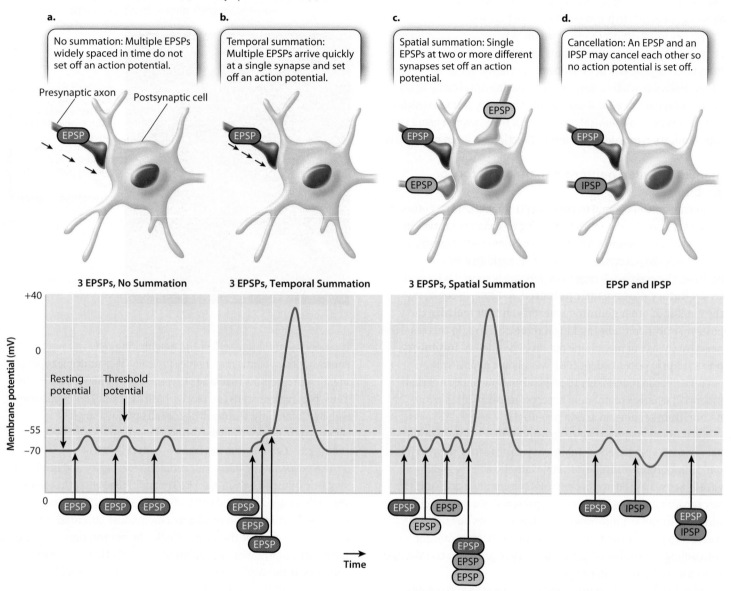

stimulus is responded to or ignored. The role of temporal and spatial summation is explored further in Chapter 36.

More than 25 neurotransmitters are now recognized, and more are likely to be discovered. The amino acids glutamate (excitatory), glycine (inhibitory), and GABA (inhibitory) are key neurotransmitters that operate at synapses in the brain. The related amino acid derivatives dopamine, norepinephrine, and serotonin are also neurotransmitters acting in the brain. Simple peptides serve as neurotransmitters for sensory neurons. More recently, two gases, nitrous oxide and carbon monoxide, have been discovered to function as messengers between nerve cells in various regions of the nervous system, even though they do not behave as standard neurotransmitters that bind to membrane receptors.

Acetylcholine is one of the key neurotransmitters produced in a variety of nerve cells. It is the excitatory neurotransmitter released by motor neurons to stimulate muscle fibers at a type of synapse called the motor endplate (**Fig. 35.16**). All vertebrate motor synapses rely on acetylcholine and are therefore excitatory only. The release of acetylcholine produces a depolarization of the muscle cell membrane, causing the muscle to contract.

→  **Quick Check 3** Acetylcholinesterase is the enzyme that breaks down and inactivates acetylcholine. Some nerve gases used as chemical weapons block acetylcholinesterase. What effect would such nerve gases have on muscle contraction?

diffuse rapidly across the cleft and bind to postsynaptic membrane receptors of the neighboring cell. The binding of neurotransmitters causes a change in the postsynaptic cell membrane potential, allowing the signal to propagate along the next neuron.

The neurotransmitter's effect on the postsynaptic membrane ends shortly after binding. Neurotransmitters become unbound from the receptor, and neurotransmitters in the synaptic cleft are either broken down by a deactivating enzyme or taken up again by the presynaptic cell and reassembled in newly formed vesicles. The structural and chemical properties of the synapse ensure that communication proceeds in a single direction between nerve cells.

### Signals between neurons can be excitatory or inhibitory.

Neurotransmitters bound to postsynaptic membrane receptors can either stimulate or inhibit the firing of action potentials in the postsynaptic neuron. The binding of some neurotransmitters depolarizes the postsynaptic membrane, making it more positive. The positive change in membrane potential is termed an **excitatory postsynaptic potential,** or **EPSP.** In contrast, the binding of other neurotransmitters can hyperpolarize the postsynaptic membrane, making it more negative. In this case, the negative change in membrane potential is called an **inhibitory postsynaptic potential,** or **IPSP.** Membrane depolarization makes a neuron more likely to respond and transmit an action potential, whereas membrane hyperpolarization inhibits the neuron from sending an action potential.

Excitatory synapses tend to transmit relevant information between nerve cells, and inhibitory synapses often serve to filter out unimportant information. For example, when a grazing gazelle senses a predator, its nervous system must transmit the key sensory information about the presence of a predator but filter out irrelevant information such as the flies buzzing over its back and ears. Inhibitory synapses are also important for controlling the timing of muscle activity required for coordinated movement, discussed in Chapter 37.

How does neurotransmitter binding change the postsynaptic membrane potential? A neurotransmitter binds to receptors that trigger the opening or closing of ligand-gated ion channels in the postsynaptic membrane. The binding of excitatory neurotransmitters causes $Na^+$ channels to open. The postsynaptic membrane becomes depolarized because the opening of $Na^+$ channels allows $Na^+$ ions to enter the cell. In contrast, the binding of inhibitory neurotransmitters causes $Cl^-$, or sometimes $K^+$, channels to open. $Cl^-$ ions diffuse into the cell or $K^+$ ions diffuse out of the cell through these channels, causing the membrane potential to become more negative, or hyperpolarized, than the outside.

Different types of nerve cell contain different neurotransmitters, but each type of nerve cell releases only one type of neurotransmitter. A particular type of nerve cell exerts either

**FIG. 35.14 Integration of information.** Multiple synapses—often hundreds of them—from communicating neurons enable a neuron to integrate diverse sources of information.

Each yellow-green dot represents an axon terminal from another neuron synapsing on the dendrites of this (green) postsynaptic cell.

excitatory or inhibitory effects on neighboring cells, but not both. In contrast, postsynaptic membranes typically have multiple types of membrane receptors that bind different kinds of neurotransmitters. Thus, postsynaptic cells can receive both excitatory and inhibitory signals. Commonly, the dendrites or cell body of a single postsynaptic nerve cell make connections to hundreds or even thousands of axons from other nerve cells (**Fig. 35.14**).

The postsynaptic nerve cell sums the excitatory and inhibitory synaptic inputs (the summed EPSPs and IPSPs) that it receives through its dendrites and cell body (**Fig. 35.15**). If the sum of synaptic stimulation results in a membrane potential that exceeds threshold at the axon hillock, the neuron fires an action potential, communicating an impulse to cells that it in turn contacts. If the sum of inputs does not exceed threshold, no action potential fires (Fig. 35.15a).

EPSPs and IPSPs can be summed over time and space. When summed over time, the frequency of synaptic stimuli determines whether the postsynaptic cell fires an action potential (Fig. 35.15b). This is referred to as **temporal summation.** When summed over space, the number of synaptic stimuli received from different regions of the postsynaptic cell's dendrites determines if the cell fires an action potential (Fig. 35.15c). This is referred to as **spatial summation.** Sometimes, excitatory and inhibitory signals may cancel each other out (Fig. 35.15d). Temporal and spatial summation of EPSPs and IPSPs are the fundamental forms of information processing carried out by the nervous system. They provide mechanisms for determining whether a particular sensory

Physiology or Medicine along with John Eccles, who worked out the membrane properties of synapses, which we explore next.

## Neurons communicate at synapses.

We have seen that nerve cells communicate at specialized junctions called synapses. There are two types of synapse, electrical and chemical. Electrical synapses provide direct electrical communication through gap junctions that form between neighboring cells (Chapter 10). They enable rapid communication but limit the ability to process and integrate information. Electrical synapses are found in the giant axons of squid, as well as in large motor nerves in fish for escape swimming. Electrical synapses can also be found in the mammalian brain. They likely help to speed up information

processing, but their role in the brain has not been well studied.

Chemical synapses are by far the more common type of synapse in animal nervous systems. The signals conveyed at chemical synapses are chemicals called neurotransmitters, which are contained within small vesicles in the knoblike axon terminal. When an action potential reaches the end of an axon (**Fig. 35.13**), the resulting depolarization induces voltage-gated $Ca^{2+}$ ion channels to open. These channels are found only in the axon terminal membrane. Because of their higher concentration outside the cell, $Ca^{2+}$ ions diffuse through these channels into the axon terminal. In response to the rise in $Ca^{2+}$ concentration, the vesicles fuse with the presynaptic membrane and release neurotransmitter molecules into the synaptic cleft by exocytosis. The neurotransmitters

**FIG. 35.13 A chemical synapse.** Neurons communicate with other neurons or with muscle cells by releasing neurotransmitters at a synapse.

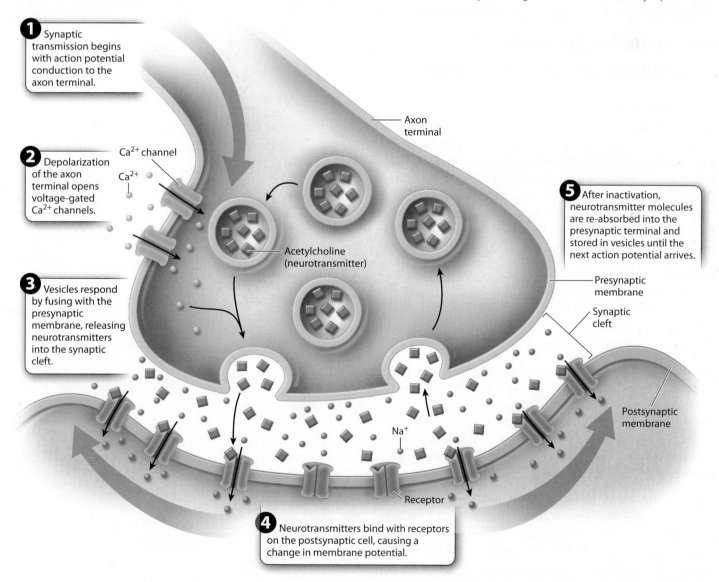

the absence of myelin. At regular intervals, the axon membrane is exposed at sites called **nodes of Ranvier** that lie between adjacent segments wrapped with myelin (**Fig. 35.11**). Voltage-gated Na⁺ and K⁺ channels are concentrated at these nodes. As a result, rather than being conducted in a continuous fashion, as is the case for unmyelinated axons, the action potentials in myelinated axons "jump" from node to node. This **saltatory propagation** (from the Latin *saltus*, "jump") of axon potentials by myelinated axons greatly increases the speed of signal transmission, enabling vertebrate animals to respond rapidly to stimuli in their environment.

→ **Quick Check 2** Multiple sclerosis is a disease in which the immune system attacks and destroys the myelin sheath surrounding nerve cells. What effect would you expect the loss of myelin to have on the speed of nerve impulses?

The British neurophysiologists Alan Hodgkin and Andrew Huxley first worked out the electrical properties of the axon plasma membrane by studying the giant axon in squid that stimulates contractions of their body muscles for swimming (**Fig. 35.12**). For this work, they shared the 1963 Nobel Prize in

## HOW DO WE KNOW?

**FIG. 35.12**

# What is the resting membrane potential and what changes in electrical activity occur during an action potential?

**BACKGROUND** In order to record the voltage of the inside of a nerve cell relative to the outside, a small electrical recording device, a microelectrode, was inserted into a neuron. The technique is more easily performed on large cells, such as the squid giant axon. This axon, as its name suggests, is quite large, measuring 0.5 mm in diameter (Fig. 35.12a). The large diameter of the axon allows electrical signals to be propagated to the muscles very quickly. Vertebrate neurons are much smaller; they rely on myelin sheaths rather than large size for rapid conduction of electrical signals.

**EXPERIMENT** In 1939, two British neurophysiologists, Alan Hodgkin and Andrew Huxley, inserted a microelectrode into a squid giant axon with a reference electrode on the outside (Fig. 35.12b). They then used a separate set of electrodes (not shown) to depolarize the cell to threshold, triggering an action potential.

**RESULTS** Fig. 35.12c is a trace from Hodgkin and Huxley's 1939 paper showing the resting membrane potential and the course of an action potential recorded by the electrode inside a giant squid axon. Note that the resting potential is negative on the inside of the axon relative to the outside, and that the action potential is a rapid spike in potential, with the inside of the cell quickly becoming positive, then negative again. The large size of the spike was a surprise.

**FOLLOW-UP WORK** This work was performed before the ion channels responsible for the changes in current across the membrane were identified. Subsequent work focused on identifying these channels and their roles in generating the action potential.

**SOURCES** Hodgkin, A. L., and A. F. Huxley. 1939. "Action Potentials Recorded from Inside a Nerve Fibre." *Nature* 144:710–711; R. D. Keynes. 1989. "The Role of Giant Axons in Studies of the Nerve Impulse." *Bioessays* 10:90–94.

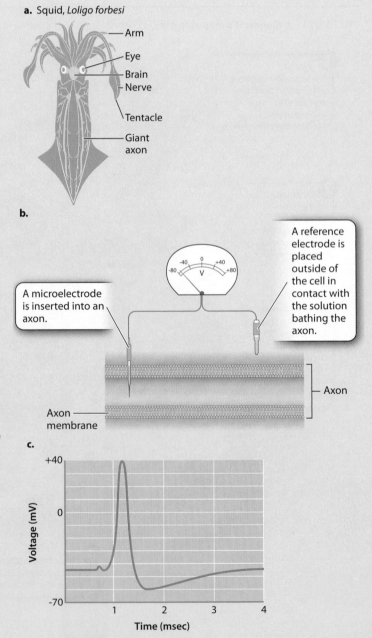

**a.** Squid, *Loligo forbesi*

Arm
Eye
Brain
Nerve
Tentacle
Giant axon

**b.**

A microelectrode is inserted into an axon.

A reference electrode is placed outside of the cell in contact with the solution bathing the axon.

Axon membrane

Axon

**c.**

**FIG. 35.10** **Propagation of action potentials.** (a) Local membrane potential depolarization triggers the opening of nearby voltage-gated Na⁺ channels, producing an action potential that spreads along the membrane. (b) The action potential moves in only one direction (shown here as left to right) because of the refractory period.

**a.**

**b.**

farther along the axon (**Fig. 35.10**). The inward sodium current depolarizes the membrane above threshold, triggering the opening of nearby voltage-gated Na⁺ channels still farther along the axon. By this means, the depolarization—the location of the action potential spike—spreads down the axon (Fig. 35.10a). Neighboring voltage-gated K⁺ channels subsequently also open and close to reestablish a resting membrane potential after an action potential has fired. Action potentials are thus **self-propagating.** Action potentials propagate only in one direction, normally from the cell body at the axon hillock to the axon terminal. The refractory period following an action potential prevents the membrane from reaching threshold and firing an action potential in the reverse direction (Fig. 35.10b).

The conduction speed of action potentials is limited by the neuron's membrane properties and the size of its axon. Yet fast transmission speeds are essential for enabling rapid responses to stimuli that may require communication over long distances within an animal's body. For example, when you mistakenly touch something very hot, your fingers sense and relay this information rapidly to ensure that you withdraw your hand quickly and avoid worse injury.

Vertebrate axons, as we have seen, are wrapped in myelin (see Fig. 35.7). The myelin sheath insulates the axon's membrane, spreading the charge from a local action potential over a much greater distance along the axon's length than would be possible in

**FIG. 35.11** **Saltatory conduction.** Myelination insulates axons, enabling faster conduction of action potentials.

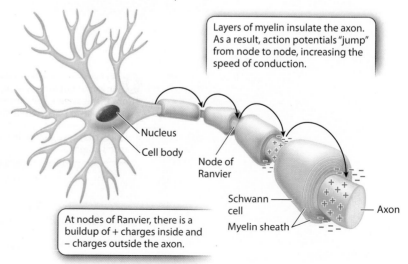

Layers of myelin insulate the axon. As a result, action potentials "jump" from node to node, increasing the speed of conduction.

At nodes of Ranvier, there is a buildup of + charges inside and – charges outside the axon.

**FIG. 35.9** **An action potential.** An action potential results from the opening and closing of voltage-gated ion channels and allows the membrane potential briefly to be positively charged.

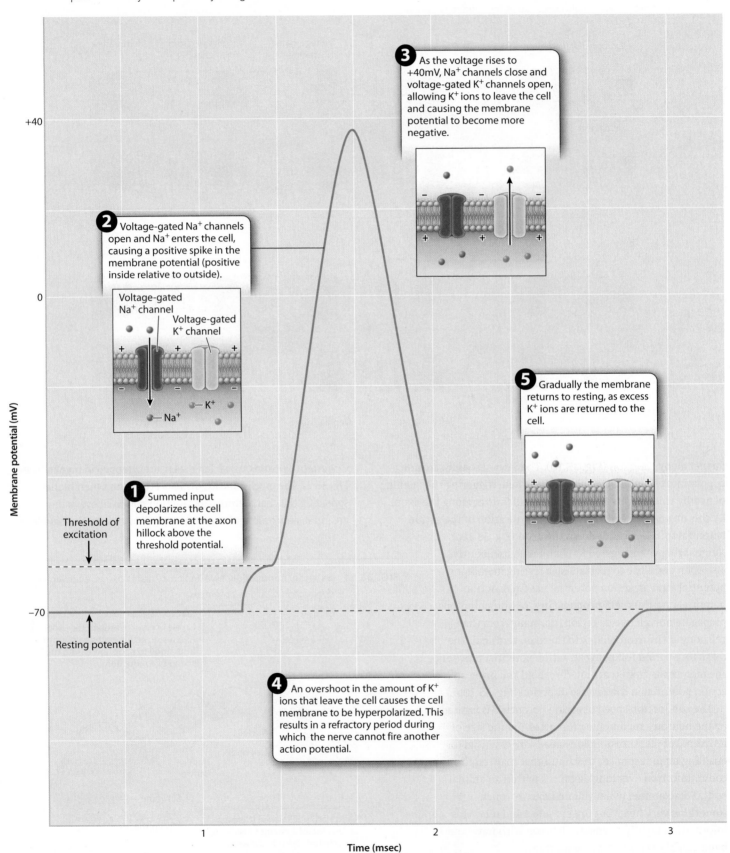

## Neurons are excitable cells that transmit information by action potentials.

When a nerve cell is excited, its membrane potential becomes less negative—that is, the inside becomes less negative, or more positive, than the outside of the cell. The increase in membrane potential is therefore referred to as a **depolarization** of the membrane. At the axon hillock of the cell body, this depolarization causes voltage-gated sodium channels to open, allowing $Na^+$ ions to enter the cell. Voltage-gated channels open and close in response to changes in membrane potential. Voltage-gated sodium channels are closed at the resting membrane potential but open rapidly in response to a depolarization of the cell membrane (**Fig. 35.9**).

The influx of $Na^+$ causes a flow of positive charge, like a current, along the inside of the cell, toward more negatively charged nearby regions. This flow of charge is reduced at longer distances and quickly dissipates so that the membrane potential returns to a resting state unless additional excitatory stimuli further depolarize the membrane.

If the excitatory signal is strong enough to depolarize the membrane of the nerve cell body to a voltage of approximately 15 mV above the resting membrane potential (about –50 mV), the nerve fires an **action potential** at the axon hillock. An action potential is a rapid, short-lasting rise and fall in membrane potential, as shown in Fig. 35.9. The critical depolarization voltage of –50 mV required for an action potential is the cell's **threshold potential.** Nerve cells in different animals have slightly different resting and threshold potentials, but all operate similarly with respect to firing an action potential. When the threshold potential is exceeded, the nerve fires an action potential in an all-or-nothing fashion. This means that once the threshold potential has been exceeded, the magnitude of the action potential is always the same and is independent of the strength of the stimulating input. At this point, a rapid spike in voltage to approximately +40 mV occurs over a very brief time period, 1 to 2 msec.

An action potential results from dramatic changes in the state of voltage-gated $Na^+$ and voltage-gated $K^+$ channels in the axon membrane. As the cell membrane crosses its threshold potential, a large number of voltage-gated $Na^+$ channels suddenly open, allowing $Na^+$ ions to rush inside the cell. The rise in voltage that results causes additional voltage-gated $Na^+$ channels to open. This is an example of **positive feedback,** in which a signal (depolarization) causes a response (open voltage-gated $Na^+$ channels) that leads to an enhancement of the signal (more depolarization) that leads to an even larger response (more open voltage-gated $Na^+$ channels). During the rising phase of the action potential, voltage-gated $K^+$ channels remain closed, keeping $K^+$ ions inside the cell.

As you can see from Fig. 35.9, the rising phase of the action potential (rapid depolarization) is followed by the falling phase (rapid repolarization). What causes this sudden reversal of the membrane potential? There are two important factors.

First, voltage-gated $Na^+$ channels close. These gated $Na^+$ channels do not close in response to the change in voltage, but instead close automatically after a brief period of time. Second, voltage-gated $K^+$ channels open. These channels, like voltage-gated $Na^+$ channels, open in response to the change in voltage, but they are slower to respond than the voltage-gated $Na^+$ channels. As a result of the closing of voltage-gated $Na^+$ channels and the opening of voltage-gated $K^+$ channels, the membrane voltage peaks and then rapidly falls as $K^+$ ions diffuse out of the axon through open voltage-gated $K^+$ channels.

The voltage inside the axon does not return immediately to the resting membrane potential. Instead, it briefly falls below the resting potential (in what is known as hyperpolarization or undershoot), then returns to the resting potential after another few milliseconds as $K^+$ channels close to restore the resting concentration of $Na^+$ and $K^+$ ions on either side of the cell membrane. The continuous action of the sodium-potassium pumps also helps to reestablish the resting membrane potential.

The period during which the inner membrane voltage falls below and then returns to the resting potential is the **refractory period** (Fig. 35.9). During the refractory period, a neuron cannot fire a second action potential. The refractory period results in part from the fact that when voltage-gated $Na^+$ channels close, they require a certain amount of time before they will open again in response to a new wave of depolarization. In addition, open voltage-gated $K^+$ channels make it difficult for the cell to reach the threshold potential. The duration of the refractory period varies for different kinds of nerve cells but limits a nerve cell's fastest firing frequency to less than 200 times per second. Most nerve cells fire at lower rates.

It might seem that many $Na^+$ and $K^+$ ions need to diffuse across the axon membrane to produce an action potential. However, only about 1 in 10 million available $Na^+$ and $K^+$ ions need to cross the membrane. Thus, enough ions are always available to generate repeated action potentials.

The nature of the action potential, which is both all-or-nothing and stereotyped, means that the action potential itself contains little information. Most neurons code information by changing the rate and temporal pattern of action potentials that they transmit. Generally, a higher firing frequency codes for a more intense stimulus, such as a brighter light or louder noise, or a stronger signal transmitted by the nerve cell to other cells it contacts.

## Neurons propagate action potentials along their axons by sequentially opening and closing adjacent $Na^+$ and $K^+$ ion channels.

Once initiated, an action potential propagates along the axon. The local membrane depolarization initiated at the axon hillock triggers the opening of neighboring voltage-gated $Na^+$ channels an electrical signal in a nerve cell is the movement of positively and negatively charged ions across the cell's membrane. This movement creates a change in electrical charge across the membrane that constitutes the electrical signal.

There is a difference in electrical charge across the cell membrane because unequal numbers of charged ions are located outside and inside the cell. The difference in electrical charge is the voltage and is measured in volts. The charge difference between the inside and outside of a neuron due to differences in charged ions is the cell's **membrane potential**. Other cells also have membrane potentials, but only nerve cells and muscle cells (Chapter 37) respond to changes in their membrane potential. Therefore, these cells are considered electrically excitable. Changes in membrane potential provide the basis for transmitting signals between neurons and from neurons to muscle cells.

At rest, when no signal is being received or sent, the cell's membrane voltage is negative on its inside relative to its outside (**Fig. 35.8**). The resting membrane potential of the cell is said to be **polarized.** This means that there is a buildup of negatively charged ions on the inside surface of the cell's plasma membrane and positively charged ions on its outer surface. The negative voltage across the membrane at rest is referred to as the cell's **resting membrane potential.** The resting membrane potential ranges from -40 to -85 millivolts (mV) depending on the type of nerve cell and most commonly is about -65 to -70 mV. The voltage of the cell's interior can be measured with respect to its outside voltage by small glass electrodes on the inside and outside of the cell.

At rest, nerve (and muscle) cells have a greater concentration of sodium (Na⁺) ions outside the cell and a greater concentration of potassium (K⁺) ions inside the cell. This distribution of ions results in part from the action of the sodium-potassium pump, discussed in Chapter 5. The sodium-potassium pump uses the energy of ATP to move three Na⁺ ions outside the cell for every two K⁺ ions moved in (Fig. 35.8). The action of the pump makes the inside of the cell less positive, and therefore more negative, than the outside of the cell.

The exact value of the resting membrane potential, however, depends on the movement of K⁺ ions back out of the cell by passive diffusion through potassium ion channels (Fig. 35.8). As discussed in Chapter 5, ion channels are protein pores embedded in the cell membrane's lipid bilayer. The most important ions that move across the cell membrane are sodium ($Na^+$), potassium ($K^+$), chloride ($Cl^-$), and calcium ($Ca^{2+}$). When a nerve cell is at rest, more K⁺ ion channels are open, giving K⁺ greater permeability compared with all other ions. As a result, K⁺ ions move out of the cell. The movement of K⁺ ions from the inside to the outside causes positive ions to build up on the outside of the cell, whereas negatively charged ions (largely Cl⁻ and proteins) remain on the inside of the cell, making the inside of the nerve cell more negative than its outside.

The relative proportion of ions does not by itself determine the resting membrane potential. This is because the number of charged ions that build up at the cell's surface is a tiny fraction of the total number of charged ions and proteins located inside and outside of the cell. It is the *movement* of K⁺ ions relative to other ions, particularly Na⁺ ions, that largely determines the resting membrane potential.

**FIG. 35.8  Neuron resting membrane potential.** The resting potential of a neuron is negative and results primarily from the movement of K⁺ ions out of the cell.

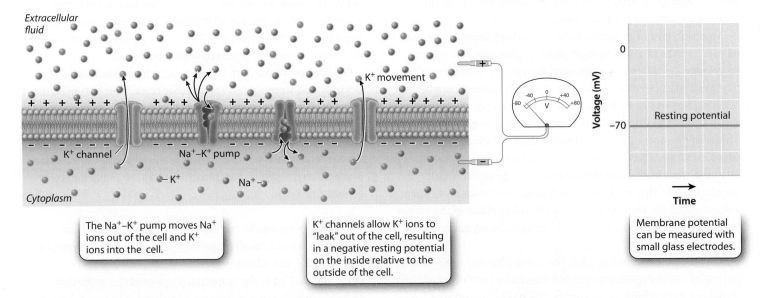

The Na⁺–K⁺ pump moves Na⁺ ions out of the cell and K⁺ ions into the cell.

K⁺ channels allow K⁺ ions to "leak" out of the cell, resulting in a negative resting potential on the inside relative to the outside of the cell.

Membrane potential can be measured with small glass electrodes.

in the mammalian brain, were elegantly drawn by the Spanish neuroanatomist Santiago Ramón y Cajal, who shared the 1906 Nobel Prize in Medicine or Physiology with the Italian scientist Camillo Golgi "in recognition of their work on the structure of the nervous system."

The degree of branching of a neuron's extensions can vary from one neuron to the next. A neuron's dendrites form a finely branched field that receives inputs from many other nerve cells or receptors. A neuron with more dendritic branches receives inputs from more sources than one with fewer branches. Some neurons receive information only from nearby regions through a few dendrites that branch close to the cell body. Other neurons receive information from many branching dendrites and transmit their output along a single axon branch or only a few axon branches. Pyramidal cells provide an example (Fig. 35.6). They have many highly branched dendrites that converge on the cell body, but just a single axon. Pyramidal cells illustrate a general organizational feature of nervous systems: Information arriving from many sources often converges on a smaller subset of neurons.

Axons typically have fewer branches than dendrites. Signals from all the dendrites, converging on the neuron cell body, determine the strength, or frequency, and timing of signals carried by the neuron's axon. The process of bringing together information gathered from different sources is referred to as the integration of information. The degree of branching and the number of synapses that a neuron makes with neighboring nerve cells, therefore, reflect how information is processed and integrated by the cell as part of a larger circuit.

### Neurons are supported by other types of cell.

Neurons in many body regions, and in particular within the brain, are supported by other types of cell that do not themselves transmit electrical signals. **Glial cells** are a major class of supporting cell. Indeed, the human brain has more glial cells than neurons. Glial cells surround neurons and provide them with nutrition and physical support. During development, glial cells help orient neurons as they develop their connections.

Star-shaped glial cells called **astrocytes** contribute to the blood–brain barrier, a set of structural adaptations of the blood vessels supplying the brain that prevent pathogens and toxic compounds in the blood from entering the brain. Astrocytes surround blood vessels in the brain, limiting the size of compounds that can diffuse from the blood into the brain. Because nerve cells have limited capacity to regenerate after being damaged, protection of the brain and spinal cord is critically important. Nevertheless, lipid-soluble compounds such as alcohol, certain anesthetics, and other drugs readily diffuse across the blood–brain barrier. As a result, these compounds affect the functioning of the brain and an animal's

**FIG. 35.7** **Schwann cell.** In vertebrates, Schwann cells insulate the axons of sensory and motor neurons.

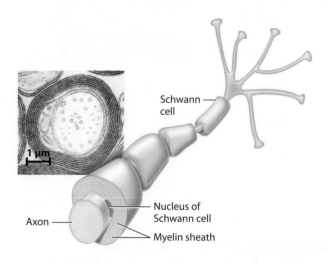

Schwann cell

1 µm

Axon

Nucleus of Schwann cell

Myelin sheath

mental state. They can also lead to damage and even destruction of neurons in the brain.

Glial cells also provide electrical insulation to vertebrate neurons. Glial cells form multiple lipid-rich layers or sheaths, called **myelin,** that wrap around the axons of neurons (**Fig. 35.7**). Myelin gives many nerves their glistening white appearance. Glial cells called **Schwann cells** insulate sensory and motor neurons, and glial cells called **oligodendrocytes** insulate cells in the brain and spinal cord. By insulating the axon, myelin greatly increases the speed of nerve signal transmission. As a result, the nervous system can respond rapidly to stimuli, as we will see after first discussing how nerve cells transmit electrical signals.

## 35.3 NEURON FUNCTION

In addition to sharing a common organization, neurons share another feature: They are electrically excitable cells. All neurons encode information by changes in their membrane voltage and transmit information in the form of electrical signals. Nerve signals travel at high speeds, ranging up to 200 m/s (450 mph). Although the speed of nerve signals is fast by biological standards, it is slow by comparison with the speed of electrical signals transmitted by a wire. The speed of signal transmission affects how nervous systems are organized and function.

### The resting membrane potential is negative and results in part from the movement of potassium ions.

How are the electrical signals transmitted by animal nervous systems generated within each neuron? The key to producing

**FIG. 35.5** **Variation in neuron shape.** Neurons display a diversity of shapes, reflecting their different functions.

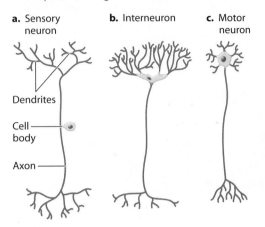

**a.** Sensory neuron

**b.** Interneuron

**c.** Motor neuron

Dendrites

Cell body

Axon

## Neurons share a common organization.

Neurons all have some basic features in common. These include a large cell body from which emerge two kinds of fiberlike extension, the **dendrites** and **axons** (**Fig. 35.4**). These extensions are the input and output ends of the nerve cell. Both types of cellular extension can be highly branched, enabling neurons to communicate over large distances with many other cells. A neuron's dendrites receive signals from other nerve cells or, in the case of sensory nerves, from specialized sensory endings. These signals travel along the dendrites to the neuron's cell body. At the junction of the cell body and its axon, the **axon hillock,** the signals are summed. If the sum of the signals is high enough, the neuron fires an **action potential,** or nerve impulse, that travels down the axon. An action potential is a brief electrical signal transmitted from the cell body along one or more axon branches.

Axons generally transmit signals away from the nerve's cell body. The end of each axon forms a swelling called the axon terminal. An axon terminal communicates with a neighboring cell through a junction called a **synapse.** A space, the **synaptic cleft,** separates the end of the axon of the presynaptic cell and the neighboring postsynaptic cell. The synaptic cleft is commonly only about 10 to 20 nm wide.

How does a signal cross the synaptic cleft? Molecules called **neurotransmitters** convey the signal from the end of the axon to the post-synaptic target cell. The arrival of a nerve signal at the axon terminal triggers the release of neurotransmitter molecules from vesicles located in the terminal. The

vesicles fuse with the axon's membrane, releasing neurotransmitter molecules into the synaptic cleft. The molecules diffuse across the synapse and bind to receptors on the plasma membrane of the target cell. The binding of neurotransmitters to these receptors causes a change in the electrical charge across the membrane of the receiving postsynaptic cell, continuing the signal in the second cell. Most neurons communicate by hundreds to thousands of synapses that they form with other cells. The massive number of interconnections enables the formation of complex nerve cell circuits and the transmission of information from one circuit to others in the nervous system.

Some neurons do not synapse with other neurons, but instead with other types of cell that produce some physiological response in the animal. Examples are muscle cells, which contract in response to nerve signals, and secretory cells in glands, which release hormones in response to nerve signals.

→ **Quick Check 1** How are signals transmitted from one end of a nerve cell to the other, and from one neuron to another?

## Neurons differ in size and shape.

In spite of their common organization, neurons show a remarkable diversity in size and shape. Some neurons, such as interneurons, are very short, whereas others, like the motor neuron that extends from your spinal cord to your big toe, are several feet long. The position of the cell body can also vary in different neurons (**Fig. 35.5**). Some sensory neurons involved in smell, sight, and taste have just two extensions coming off the cell body (Fig. 35.5a), whereas other neurons have many more (Fig. 35.5b and 35.5c).

The cell bodies of some neurons have distinctive shapes. For example, pyramidal neurons, as their name suggests, have triangular-shaped cell bodies (**Fig. 35.6**). These cells, located

**FIG. 35.6** **Pyramidal cell.** (a) These neurons, drawn by Santiago Ramón y Cajal (1852–1934), show the triangular cell body, highly branched dendrites, and single axon. (b) The triangular shape of the cell body is clear in the photomicrograph at right.

a

b

Although animal nervous systems differ in organization and complexity, their nerve cells have fundamentally similar molecular and cellular features and use fundamentally the same mechanisms to communicate with neighboring cells. These mechanisms have been retained over the course of evolution to code and reliably transmit information. The ability to receive sensory information from the environment, and transmit and process this information within a nervous system, is therefore shared across a wide diversity of animal life, contributing to the success of multicellular animals.

**? CASE 7 Predator–Prey: A Game of Life and Death**
**What body features arose as adaptations for successful predation?**

A notable feature of most animal nervous systems is that nervous system tissue, including specialized sense organs such as eyes, becomes concentrated at one end of the body. For example, the paired ganglia and eyespot of the flatworm are located at one end of its body, as are the brain and sense organs of the earthworm, squid, and insect, as shown in Fig. 35.3. The concentration of nervous system components at one end of the body, defined as the "front," is referred to as **cephalization.** Cephalization is a key feature of the body plan of most multicellular animals.

Cephalization is thought be an adaptation for forward locomotion because it allows animals to take in sensory information from the environment ahead of them as they move forward. As the quality and amount of sensory information increased, brain size and complexity increased. Cephalization is also considered to be an adaptation for predation, allowing animals to better detect and capture prey.

Cephalization evolved independently multiple times in different animal groups and is therefore thought to confer certain advantages. These advantages likely include the ability to sense environmental stimuli encountered by forward motion and to process this information quickly to enable a suitable behavioral response. Although cephalization is a feature of many animals, it has been particularly well studied in vertebrates. In vertebrates, the brain, many sense organs, and mouth are all located in the head region. Vertebrates also evolved several novel features, including a jaw, teeth, and tongue. These are all thought to be adaptations for successful predation, or more generally the acquisition and processing of food.

As a result of evolutionary selection for enhanced sensory perception and the ability to respond to important cues in the environment, the brain, sensory organs, and nervous system of many animals are complex in their organization. These are linked to more sophisticated decision-making abilities that allow for a broad range of behaviors, as well as to improved memory and learning. These abilities are critical to the success of both predators and prey and underlie the complex behavioral interactions that occur among members of a species when they mate, reproduce and disperse, and care for their young.

## 35.2 NEURON STRUCTURE

We have seen that nerve cells can be classified as sensory neurons, interneurons, or motor neurons. All share the same basic organization, adapted to their common function in receiving and transmitting information. Neurons have extensions that receive information and relay it to the cell body. Other extensions transmit information away from the cell body. Although all neurons share this basic plan, they differ remarkably in size and number of extensions, according to their specific function.

**FIG. 35.4** **Neuron organization.** Nerve cells typically have dendrites that receive signals and axons that transmit signals.

1 Stimuli are received by the dendrites and cell body.

2 Synaptic stimuli are summed at the axon hillock, where an action potential is initiated if the sum of the arriving signals is high enough.

3 Action potentials are conducted to the axon terminal, where they cause the release of neurotransmitter molecules. These bind to receptors on the postsynaptic cell membrane, creating a new signal in the postsynaptic neuron.

Axon hillock

Action potential

Nucleus

Cell body

Axon

Dendrites

Axon terminal

Presynaptic cell

Postsynaptic cell

Synapse

Synaptic cleft

Neurotransmitter

Receptor

Postsynaptic membrane

FIG. 35.2 **A simplified phylogenetic tree of multicellular animals.** All multicellular organisms, except sponges, have a nervous system.

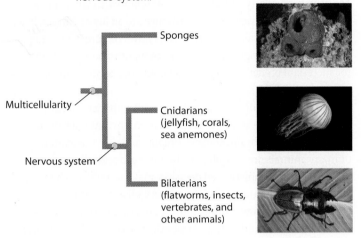

these animals has relatively few neurons and no ganglia or central brain (**Fig. 35.3a**). In these animals, the nerve cells are arranged like a net. The nerve net of a sea anemone is organized around its body cavity. At one end of its body is a mouth that opens into a central body cavity where food is digested. Sensory neurons sensitive to touch signal when one of the anemone's tentacles has captured food prey, then motor neurons react to coordinate the tentacle's movement toward the mouth for feeding and digestion. Sensory neurons responding to a noxious stimulus signal the sea anemone to retract or detach from its substrate.

More complex nerve connections develop in the head end of flatworms, segmented annelid worms, insects, and cephalopod mollusks, such as squid and octopus. These animals are all bilaterians, animals with symmetrical right and left sides and a distinct front and back end. Flatworms are representatives of one of the earliest branching groups of bilaterians, and their bodies are well adapted for forward locomotion. Sensory neurons are most

numerous in the animal's front end, so the flatworm can perceive the environment ahead of it as it moves forward. The flatworm's nervous system can compare inputs from different sensory neurons and then send signals to the appropriate muscles to move toward a target. Paired ganglia (one on each side of the animal) in the head end of flatworms handle this processing task (**Fig. 35.3b**).

In animals with an organized nervous system, information is transmitted to different regions along the length of the animal's body by distinct **nerve cords** and **nerves** (Fig. 35.3). These are bundles of the long fiberlike extensions from multiple nerve cells.

In earthworms, paired ganglia, each made up of many interneurons, are located in each of the body's segments (**Fig. 35.3c**). These segmental ganglia help to regulate the muscles within each body segment that move the earthworm's body. Paired ganglia near the head regulate the activity of motor neurons that control movements of the mouth for feeding. Ganglia serve to regulate key processes in local regions and organs of the animal's body. For example, local ganglia assist in processing information from the eyes (Chapter 36) or control the digestive state of an animal's gut (Chapter 40).

Centralized collections of neurons forming a brain are present in annelid worms, insects, and mollusks (**Figs. 35.3c–35.3e**). Sense organs also developed in the head region, from eyespots in flatworms that respond to light to more sophisticated image-forming eyes in octopus and squid. The evolution of a brain in these invertebrate animals enabled them to learn and perform complex behaviors. Many of these nervous system capabilities are shared with vertebrate animals.

The evolution of a brain is also a hallmark of vertebrate animals (**Fig. 35.3f**). It enabled them to evolve complex behaviors that rely on learning and memory and, in certain vertebrates, the ability to reason. At the same time, animal brains became increasingly complex in the number of nerve cells they contain (fruit fly, about 0.25 million; cockroach, about 1 million; mouse, about 75 million; humans, about 100 billion), in the number of connections between neurons, and in the variety of nerve cell features.

FIG. 35.3 **Animal nervous systems.** (a) Sea anemones have a nerve net; (b) flatworms have paired ganglia; (c) earthworms, (d) insects, (e) squid, and (f) frogs have a brain.

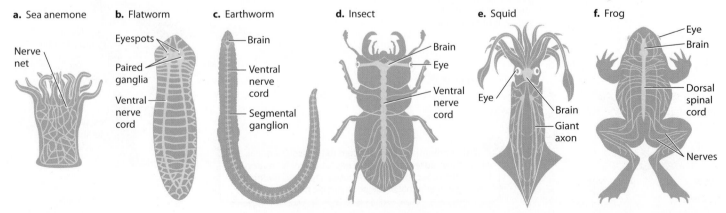

**FIG. 35.1**    **Predator and prey.** Interactions of predator and prey, such as this cheetah and gazelle, are mediated by the nervous system.

also made possible more sophisticated behaviors that relied on improved decision making, learning, and memory. These abilities, possessed by many invertebrate and vertebrate animals, increased their fitness by improving their ability to survive, find and select a mate, and reproduce.

### Animal nervous systems have three types of nerve cell.

All animal nervous systems are made up of three types of nerve cell: **sensory neurons, interneurons,** and **motor neurons.** Each type of neuron has a different function. Sensory neurons receive and transmit information about an animal's environment or its internal physiological state. These neurons respond to physical features such as temperature, light, and touch or to chemical signals such as odor and taste. Interneurons process the information received by sensory neurons and transmit it to different body regions, communicating with motor neurons at the end of the pathway to produce suitable responses. For example, a motor neuron may stimulate a muscle to contract to produce movement. Other motor neurons may adjust an animal's internal physiology, constricting blood vessels to adjust blood flow or causing wavelike contractions of the gut to aid digestion. As a result, nervous system function is fundamental to **homeostasis,** the ability of animals, organs, and cells to actively regulate and maintain a stable internal state.

Most nerve cells have fiberlike extensions that receive information and other extensions that transit information. The result is a network of interconnected nerve cells that form a circuit. These circuits allow information to be received, processed, and delivered. In most animals, interneurons form specialized circuits that may consist of a diverse array of nerve cells having different shapes, sizes, and chemical properties.

The ability to process information first evolved with the formation of **ganglia** (singular, ganglion). Ganglia are groups of nerve cell bodies that process sensory information received from a local, nearby region, resulting in a signal to motor neurons that control some physiological function of the animal. They can be thought of as relay stations or processing points in nerve cell circuits. Eventually, the centralized concentration of neurons that we call a **brain** evolved, and this organ was able to process increasingly complex sensory stimuli received from the environment or anywhere in the body. Neurons in the brain form complex circuits, allowing the brain to mediate a broad array of behaviors, as well as to learn and retain memories of past experiences.

### Nervous systems range from simple to complex.

Even single-celled animals have cell-surface receptors that enable them to sense and respond to their environment. By contrast, most multicellular animals have a nervous system. In general, the organization and complexity of an animal's nervous system reflects its lifestyle. Sessile, or immobile, animals, like sea anemones or clams, have relatively simple sense organs and nervous systems. In contrast, active animals such as arthropods (which include insects, spiders, and crustaceans), squid, and vertebrates have more sophisticated sense organs linked to a brain and more complex nervous systems.

The sponges are the only multicellular organisms that lack a nervous system (**Fig. 35.2**). Their simple body plan means that a nervous system is not required to coordinate the functions of different body regions. Instead, groups of cells within sponges respond to local chemical and physical cues.

Among multicellular organisms, the simplest nervous systems are found in cnidarians, which are radially symmetric animals such as jellyfish and sea anemones (Fig. 35.2). The nervous system of

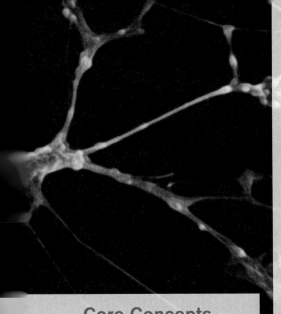

# ANIMAL NERVOUS SYSTEMS

## Core Concepts

**35.1** Animal nervous systems allow organisms to sense and respond to the environment, coordinate movement, and regulate internal functions of the body.

**35.2** The neuron is the basic functional unit of the nervous system. Neurons have dendrites that receive information and axons that transmit information.

**35.3** The electrical properties of neurons allow them to communicate rapidly with one another.

**35.4** Animal nervous systems can be organized into central and peripheral components.

Gazelle grazing on an African plain are alert to the threat of predators nearby. Members of the herd scan the distant landscape for visual cues that might reveal the stealthy approach of a cheetah (**Fig. 35.1**). Olfactory (smell) and auditory (sound) cues from the environment can also warn of the cheetah's approach. Herd members communicate the presence of the cheetah to one another by the directed attention of a dominant male or the pawing of a hoof. At what point is the threat real? When does each animal decide to stop grazing and turn to run? What direction should the escape take to keep the herd together as a group?

The cheetah scans the horizon, spotting the grazing gazelle with its keen vision. The cheetah's ability to recognize wind direction enables it to move downwind of the gazelle to minimize auditory and olfactory cues that might otherwise be carried by the air toward suspecting prey. At what point does the cheetah decide to sprint in pursuit of its quick-running prey?

The perceptions and reactions of these two animals depend on the activity of their nervous systems. In this and the following chapter, we explore the organization of the nervous system, including the brain and the nerve cells that compose it. The nervous system has the remarkable ability to process all kinds of information. It takes in information from the environment through the animal's sensory organs and processes that information through the activity of networks of nerve cells. The results of that processing guide an animal's motor responses. Such processes underlie the perceptions and interaction of cheetah and gazelle as predator and prey, as well as the complex emotional and cognitive responses triggered, for example, by seeing a friend or relative after a long separation.

---

## 35.1 NERVOUS SYSTEM FUNCTION AND EVOLUTION

Except for sponges, all multicellular animals possess a **nervous system,** a network of many interconnected nerve cells. This network allows an animal to sense and respond to the environment, coordinate the action of muscles, and control the internal function of its body. Nerve cells, or **neurons,** are the basic functional units of nervous systems.

Nervous systems first evolved in simple animals. These early nervous systems could sense basic features of the environment, such as light, temperature, chemical odors, and physical forces. Guided by these cues, early nervous systems could trigger movements useful for obtaining food, finding a mate, or choosing a suitable habitat. As animals evolved more complex bodies with increasingly specialized organ systems for respiration, circulation, digestion, and reproduction, their nervous systems also became more complex. These more complex nervous systems could regulate internal body functions in response to sensory cues received from the environment. They

In the modern world, our own species, *Homo sapiens*, occupies a unique position at the top of the food web. While humans are occasionally killed by animals such as bears or sharks, we are, for the most part, no longer constrained by the fear of predation by other species. But it wasn't always so. Some scientists have proposed that fear of predation among early humans led to the evolution of cooperative social behavior and large brains. Others have theorized that it was our hunting of other animals that led us to evolve big brains and the ability to work together. Perhaps both factors played a role. Either way, the importance of the predator–prey system can't be ignored. To understand our roots, it seems, we must take a good look at nature, red in tooth and claw.

## ? CASE 7 QUESTIONS

*Answers to Case 7 questions can be found in Chapters 35–41.*

1. What body features arose as adaptations for successful predation? *See page 35-4.*
2. How have sensory systems evolved in predators and prey? *See page 36-10.*
3. How do different types of muscle fiber affect the speed of predators and prey? *See page 37-12.*
4. How does the endocrine system influence predators and prey? *See page 38-15.*
5. How do hormones and nerves provide homeostatic regulation of blood flow as well as allow an animal to respond to stress? *See page 39-17.*
6. Does body temperature limit activity level in predators and prey? *See page 40-5.*
7. Can the loss of water and electrolytes in exercise be exploited as a strategy to hunt prey? *See page 41-6.*

**Wolves and moose on Isle Royale.** The populations of this predator and prey have been studied for more than 50 years and the two have achieved a delicate balance over time.

forms to help them hide from hungry carnivores. Still others have developed protective behaviors, such as living together in herds for security against predation.

Often, predator and prey evolve in lockstep, driving each other's adaptations. Clams may have evolved thicker shells to protect themselves from hungry crabs. In turn, the crabs evolved larger, stronger claws for cracking clamshells. This type of coevolution is often described as an evolutionary arms race. And it has happened time and time again.

The interactions between eaters and eaten have also influenced the evolution of physiological systems. The skills that a predator needs to hunt its prey and that its prey needs to escape depend in large part on adaptations of their sensory systems, musculoskeletal systems, nervous systems, and even their circulatory and respiratory systems.

Consider sensory systems. Animals rely on visual, auditory, tactile, and chemical stimuli to warn them of approaching predators—or to guide them to suitable prey.

A bat relies on sonar to locate insects. A spider responds to the flutter of silk when an insect becomes ensnared in its web. Gazelles are always attentive, watching and listening for signs of a cheetah or lion in the grass.

When predators are nearby, prey species experience fear and anxiety. Stress hormones produced by prey animals can influence an animal's biology in a number of ways. In snowshoe hares, levels of the stress hormone cortisol increase when predators such as lynx and coyote are plentiful. The hormones trigger behaviors, such as alertness and fearfulness, that help the hares avoid becoming a lynx's lunch. But the behavioral benefit comes at a cost. Research has shown that stressed hares give birth to fewer and smaller offspring.

In short, predation has left a physical imprint on both predator and prey, from nose to tail—their body shapes, their behaviors, their physiology, even their muscle fibers have been influenced over time by predator–prey interactions.

# Predator–Prey

## A GAME OF LIFE AND DEATH

In his 1850 poem *In Memoriam A.H.H.*, Alfred, Lord Tennyson wrote of "Nature, red in tooth and claw." The phrase has endured as a powerful description of the brutality of wild nature. Thanks to nature documentaries and our human fascination with wild animals, most of us have witnessed plenty of images of this red-toothed nature: a lion chasing down a zebra, a hawk scooping a mouse into its talons, a shark sinking its teeth into a sea lion.

Predator–prey interactions are important in every ecosystem on Earth. After all, every living animal can be classified as either "predator" or "prey"—and often, both labels apply. To a fly, a toad is a fearsome predator. To a snake, that same toad may be choice prey.

Very few species enjoy the luxury of having no natural predators; these are called top predators. For the vast majority of animals, the threat of predation is simply a fact of life. Not surprisingly, that threat has been a powerful evolutionary force. Predator–prey dynamics have influenced both the evolution of individual organisms and the shape of the ecosystems in which they live.

On Isle Royale, an island in Lake Superior, ecologists have been studying moose and wolves since 1958 in the longest-running study of a single predator–prey system in the world. The moose arrived on the island about 100 years ago, presumably by swimming about 15 miles from the nearest shoreline. They arrived to find a predator-free paradise, and the moose population quickly exploded. Then, about 1950, a pair of wolves crossed an ice bridge to Isle Royale. As the wolf population grew, the moose population declined, until a delicate balance was established.

As a predator population grows, the prey population shrinks—a seemingly logical relationship. But predator–prey interactions are complex and sometimes even counterintuitive. Removing a top predator can actually reduce biodiversity. Sea stars, for instance, prey on mussels in the rocky intertidal zone of the Pacific Northwest. When sea star numbers fall, mussel numbers increase. The mussels crowd out other species, such as barnacles and seaweed, that compete for space on the rocks. The result is a drop in the overall number of species in that habitat.

The same patterns have been found among birds of prey in the Italian Alps. These raptors—including the goshawk and four types of owl—are the top predators of their food chains. Spanish researchers found that locations where the birds were present had a greater diversity of trees, birds, and butterflies than did comparable control sites.

One reason that the predator and prey populations eventually achieve a delicate balance is that both are well adapted to their roles. Predation has exerted powerful evolutionary pressure that has influenced body plans over the long term. That pressure has shaped predators by giving them claws, teeth, venom, and powerful muscles for hunting prey.

Of course, predation pressure has also shaped prey. Some animals, including some insects and frogs, have evolved toxins to deter would-be predators. These toxic species often exhibit warning colors that tell the predators to steer clear. Other prey animals have evolved camouflage colors and

> Predator–prey dynamics have influenced both the evolution of individual organisms and the shape of the ecosystems in which they live.

exerted changes the shape of the ion channel, causing it to open and leading to a change in membrane potential.

A mechanoreceptor in humans and other mammals is the sensory receptors found in the skin that sense touch and pressure. The cell bodies of these neurons are located in ganglia near the spinal cord, with extensions going in two directions: to the skin and to the spinal cord. In the skin, the neuron has branched tips, which are the initial sensors of touch and pressure. If the stimulus is strong enough, local depolarization leads to the firing of an action potential that travels all the way to the spinal cord. Thus, in contrast to a chemoreceptor, this type of mechanoreceptor transmits an action potential. The axon is much too long for simple depolarization to spread to its other end.

Mechanoreceptors are also found at the base of whiskers that rodents, cats, dogs, and other mammals use to sense touch with their snouts. Stretch receptors found in muscles are also mechanoreceptors that influence a muscle's motor activation, helping to control its length and force. A very different group of specialized mechanoreceptors called hair cells are the sensory receptors for balance, gravity sensing, and hearing, discussed in section 36.3.

**Thermoreceptors** in the skin and in specialized regions of the central nervous system (CNS) respond to heat and cold. Thermoreceptors help to control an animal's metabolism and patterns of blood flow, regulating body temperature by affecting rates of heat gain and loss (Chapter 40). As a result, they help to maintain homeostasis. Many invertebrate animals rely on thermoreception when seeking environments that provide favorable temperatures. Thermoreceptors in the skin of vertebrate animals are believed to be simple dendrite endings that respond to changes in skin temperature by altering the firing rate of the cell's action potentials.

**Nociceptors**, or pain receptors, are another class of nerve cell with dendrites in the skin and connective tissues of the body. When these sensory receptors are exposed to an excessive mechanical, thermal, or chemical stimulus, they send action potentials to the brain or spinal cord, typically stimulating rapid withdrawal from the painful stimulus. Pain is a subjective sensation, and therefore difficult to assess and treat. Although it is clear that many animals experience pain, it is difficult to judge how different species associate the sensation of pain with underlying physical or chemical stimuli.

**FIG. 36.2   Sensory transduction, which converts an external stimulus into a change in membrane potential.** (a) Chemoreceptors are located in the antennae of the luna moth; (b) mechanoreceptors below the cutical of roundworms; and (c) photoreceptors in the eyes of squid.

Light-detecting **photoreceptors** are the sensory receptors in eyes (Fig. 36.2c). Photoreceptors respond to individual photons of light energy by closing Na⁺ channels, causing the cell membrane to become hyperpolarized rather than depolarized. Most receptors excite the neurons that they synapse with, but vertebrate photoreceptors are unusual in that they inhibit the firing rate of other neurons within the eye rather than exciting them. Photoreception in the vertebrate eye is discussed in section 36.4.

Some fish, such as catfish, contain specialized **electroreceptors** arranged in a lateral line along their bodies. Electroreceptors enable these fish to detect weak electrical signals emitted by all organisms. They likely evolved as an adaptation for locating prey or potential predators in poorly lit habitats where vision was less useful. Some specialized "weakly electric" fish actually generate an electromagnetic field by emitting pulses from an electric organ located in the tail. Disturbances in the field detected by electroreceptors of the lateral line system signal the location of nearby prey. These fish also inhabit rivers with poor visibility. The bill of the duckbilled platypus also contains electroreceptors that locate prey in dimly lit water.

→  **Quick Check 1** Aspirin and ibuprofen reduce pain by inhibiting the synthesis of chemicals called prostaglandins. What effect would you predict prostaglandins to have on the firing rate of nociceptors?

### Stimuli are transmitted by changes in the firing rate of action potentials.

Sensory reception depends on converting the energy of a physical or chemical stimulus into an action potential, either in the sensory receptor itself or in the neuron it synapses with. How do these nerve impulses convey information to the brain? In effect, action potentials can be considered a code that the brain deciphers.

To convert information from the environment into a code, the nerve impulses conveyed by sensory organs carry out the following functions:

1. Convey the strength of the incoming signal—the brightness of light, the loudness of a sound, or the intensity of an odor.

2. Carry information about even weak signals when necessary.

3. Convey the location of a signal's source—for example, where is the body being touched?

4. Filter out unimportant background signals.

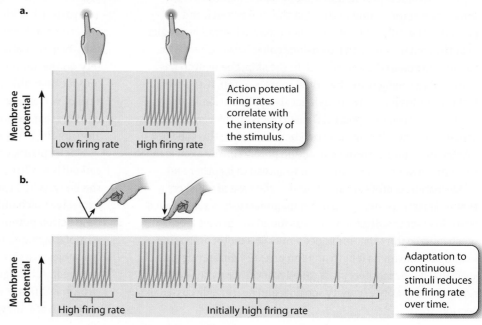

**FIG. 36.3** **Sensory receptor signals.** The firing rate of action potentials increases with (a) the intensity of the stimulus and (b) a novel compared to a continuous stimulus.

Action potential firing rates correlate with the intensity of the stimulus.

Adaptation to continuous stimuli reduces the firing rate over time.

Because nerve cells transmit action potentials in an all-or-nothing fashion, information about the strength of the signal is coded by the number of action potentials fired over a period of time, which is the nerve cell's **firing rate.** Most commonly, strong signals produce high firing rates and weak signals produce low or background firing rates (**Fig. 36.3a**). For example, the firing rate of mechanoreceptors increases as the tactile pressure they sense increases.

Several mechanisms increase a nerve cell's sensitivity to even weak signals. Following sensory transduction, the signal typically undergoes initial processing. As discussed in Chapter 35, sensory stimuli may be summed over space or time. Multiple receptors that receive a stimulus often converge onto a neighboring neuron that increases its firing rate proportionally to the number of signals received. This is an example of **spatial summation.** Combining the synaptic input increases the receptor organ's sensitivity to external stimuli. **Temporal summation** is the integration of sensory stimuli that are received repeatedly over time by the same sensory cell. When a sensory cell is stimulated more frequently, the excitatory postsynaptic potentials (EPSPs) sum, making it more likely that the cell is depolarized above threshold and fires an action potential.

Sensory receptors also distinguish between discrete (that is, intermittent) and continuous stimuli, providing a way to ignore background noise. Novel stimuli are generally the most important for an animal to respond to. Consequently, sensory receptors initially respond most strongly when a stimulus is first received (**Fig. 36.3b**). If the stimulus continues over a longer time period, sensory receptors typically reduce their firing rate through a process called **adaptation** (this is *not* the adaptation that results from natural selection). Adaptation of firing rate enables an animal's sensory receptors to focus on novel stimuli, which are those likely to be most important. You can experience adaptation by touching your skin lightly at a particular location. Initially, skin mechanoreceptors fire in response to the touch, but after a while their firing rate diminishes and the sensation of a touch becomes less noticeable. Adaptation also ensures that the clothes touching your skin do not cause an irritating and distracting long-term sensation.

In addition, sensory receptors locate a signal's source by **lateral inhibition** (**Fig. 36.4**). Lateral inhibition enhances the strength of a sensory signal locally but diminish it peripherally. In this process, receptors inhibit interneurons that receive sensory inputs from adjacent regions of the sensory organ, while sending excitatory signals to interneurons within their local region. This increases the contrast in signal strength between adjacent regions. For example, lateral inhibition in the retina of the eye enhances edge detection and sharpens image contrast. It also enables the sense of touch to be localized to a small region of the body surface. The inhibitory and excitatory signals are transmitted by synapses that have different neurotransmitters.

**FIG. 36.4** **Lateral inhibition.** The firing rate of action potentials is higher in the center of the sensory receptive field than in surrounding regions.

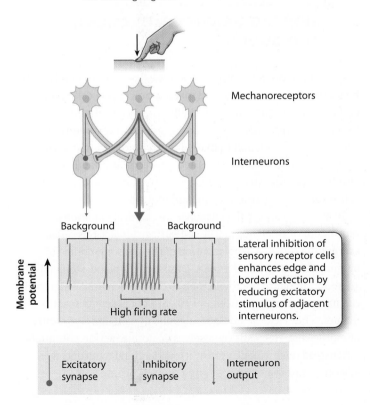

We have not yet described all the information in the code. How do action potentials convey pitch or color or a particular odor? This information is conveyed by the identity of the particular sensory receptors that are activated. Sensory organs are so constructed that, for example, tone A activates "tone A" sensory neurons in the ear, red light activates "red" sensory neurons in the eye, and the fragrance of a rose activates "rose" sensory neurons in the nose. The way in which the "correct" sensory neuron is excited varies for each of the senses. In the next three sections, we discuss how this happens for smell and taste, hearing, and vision.

## 36.2 SMELL AND TASTE

In many animals, specialized sensory organs for smell (**olfaction**) and taste (**gustation**) have evolved that detect a variety of chemical compounds. These organs contain distinct types of protein membrane receptor, and each of them binds to a specific class of molecule of complementary chemical structure. A large number of genes code for different chemical odor receptors, and the presence of so many receptors enables the detection of broad categories of smell and taste.

### Smell and taste depend on chemoreception of molecules carried in the environment and in food.

The sense of smell in mammals depends on specialized sensory neurons that extend into the nasal cavity. These neurons communicate directly with the brain by the olfactory nerve, one of the cranial nerves of the head (**Fig. 36.5**). Odor molecules are captured by nasal mucus during inhalation and sensed by the cells' long, thin hairlike extensions, which project into the mucus. Membrane receptors on these extensions bind odor molecules that are complementary in structure to the receptor. When bound, the odor molecules produce excitatory postsynaptic potentials (EPSPs). If enough odor molecules bind to the receptor cell, the EPSPs are summed and transmit an action potential to the brain.

Each sensory receptor has one type of membrane receptor, but there are typically many different sensory receptors for many different molecules. Humans have nearly 1000 genes that express particular odorant receptor proteins. Mice have about 1500 genes that express odorant receptor proteins, constituting about 15% of all their genes. The olfactory system can detect specific odor molecules in quite low concentrations.

Taste, or gustation, is achieved by clusters of chemosensory receptor cells located in specialized **taste buds,** the sensory organs for taste (Fig. 36.5). Some fish have taste buds in their skin that sense amino acids in the water, helping them to localize food. The taste buds of terrestrial vertebrates are in the mouth; most of the taste buds in mammals are on the tongue. In humans, the tongue has about 10,000 taste buds, which are contained within raised structures called papillae, giving the tongue its rough surface. Each taste bud has a pore through which extend fingerlike

**FIG. 36.5    The senses of smell and taste.** Olfactory receptors respond to chemical odors in the nasal passages and communicate directly with the brain. Taste cells are chemosensory receptor cells located in taste buds.

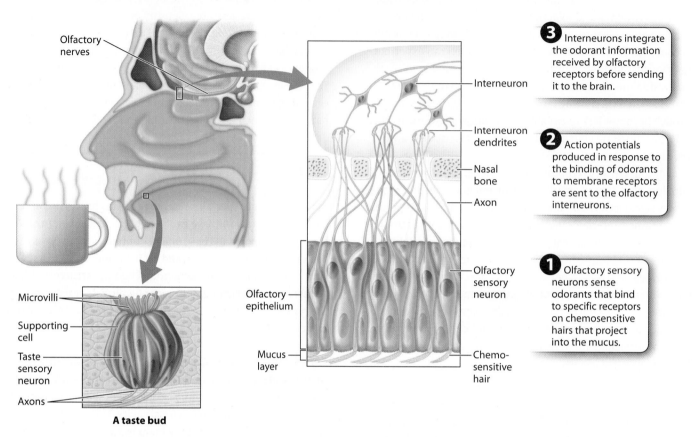

**3** Interneurons integrate the odorant information received by olfactory receptors before sending it to the brain.

**2** Action potentials produced in response to the binding of odorants to membrane receptors are sent to the olfactory interneurons.

**1** Olfactory sensory neurons sense odorants that bind to specific receptors on chemosensitive hairs that project into the mucus.

Olfactory nerves

Interneuron

Interneuron dendrites

Nasal bone

Axon

Olfactory sensory neuron

Chemo-sensitive hair

Olfactory epithelium

Mucus layer

Microvilli

Supporting cell

Taste sensory neuron

Axons

**A taste bud**

projections of sensory cells called microvilli that contact food items. The microvilli provide a large surface area that is rich in membrane receptors. These receptors bind to specific chemical compounds that give the food its taste. Taste bud sensory cells synapse on interneurons, which fire action potentials when a sufficient number of sensory cells are depolarized by a particular food item. In contrast to olfaction, taste generally requires higher concentrations of food chemicals coming into contact with taste bud receptors to stimulate receptor depolarization.

Human taste can be divided into five categories—sweet, sour, bitter, salty, and savory. The tongue contains only five types of taste receptor, one specialized for each of these categories. Variation in taste is achieved by combining the signals received by receptors in the five categories. Your sense of taste is substantially enhanced by airborne odors that emanate from a food or beverage and are sensed by chemoreceptors in the nose. Signals from taste bud and olfactory receptors are combined in the brain, eliciting the subtle sensation of taste and flavor of a particular food or drink.

→   **Quick Check 2** Why is your sense of smell and taste diminished when you have a cold?

## 36.3 SENSING GRAVITY, MOVEMENT, AND SOUND

The ability to detect motion, orient with respect to gravity, and hear all depend on specialized mechanoreceptors called **hair cells** that sense movement and vibration. They are found in fishes and amphibians, in which they detect movement of the surrounding water, and in many invertebrates, in which they sense gravity and other forces acting on the animal. They are also found in the ears of terrestrial vertebrates, in which they sense sound, body orientation, and motion.

In all these cases, hair cells sense mechanical vibrations. Movement of small nonmotile cell-surface projections called **stereocilia** causes a depolarization of the cell's membrane by opening or closing ion channels. Despite their name, stereocilia are more similar to microvilli than to cilia, which unlike stereocilia can move on their own (Chapter 10). Stereocilia contain actin filaments but lack the microtubules and dynein motor proteins of motile cilia. Hair cells themselves do not fire action potentials but instead release neurotransmitters that bind to receptors in adjacent sensory neurons, altering their firing rate.

## Hair cells sense gravity and motion.

Hair cells detect gravity and motion in many animals. Hair cells contained in the **lateral line system** of fish and sharks sense water vibrations that indicate nearby threats or prey, as well as their own motion (**Fig. 36.6a**). The lateral line is a sensory organ along both sides of the body that uses hair cells to detect movement of the surrounding water.

A sense of gravity helps animals to orient their body within the environment, providing a sense of "up" and "down." Even buoyant aquatic animals depend on a sense of gravity to keep oriented to the water's surface or within the water column. Gravity-sensing organs called **statocysts** are found in most invertebrates (**Fig. 36.6b**). The statocyst is made up of an internal chamber lined by hair cells with stereocilia that project into the chamber. Small granules of sand or other material form a

**FIG. 36.6** **Motion and gravity sensation.** (a) Hair cells within the lateral line system of fish sense water vibrations for motion and prey detection. (b) Gravity-sensing organs (statocysts) in lobsters provide cues about body orientation.

**a.** Lateral line

Lateral line

**b.** Statocyst

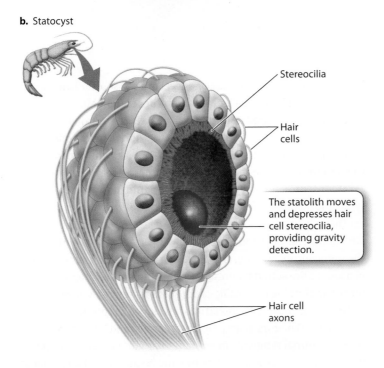

Stereocilia

Hair cells

The statolith moves and depresses hair cell stereocilia, providing gravity detection.

Hair cell axons

**FIG. 36.7** **The vestibular system.** The vestibular system senses angular rotations of the head from movements of fluid within three semicircular canals that deflect hair cells. It detects gravity and posture by means of statoliths in two statocyst chambers.

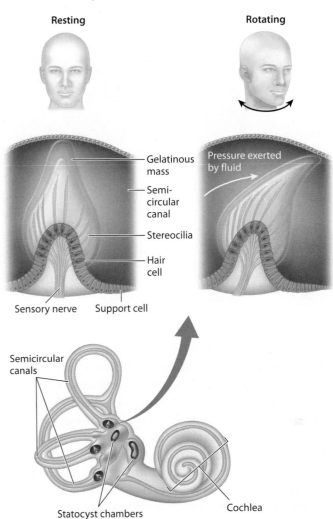

Resting

Rotating

Gelatinous mass

Semi-circular canal

Stereocilia

Hair cell

Sensory nerve    Support cell

Pressure exerted by fluid

Semicircular canals

Statocyst chambers

Cochlea

dense particle called a **statolith** that is free to move within the statocyst organ. By pressing down on hair cells at the "bottom" of the chamber, the statolith activates those cells, indicating the direction of gravity. Statoliths help anemones and jellyfish orient themselves in the water and direct their tentacles. When statoliths of lobster and crayfish are replaced by magnetic particles and subjected to an experimental magnetic field, the lobsters and crayfish swim upside down or on their sides in response to the experimental change in the direction of the magnetic field.

The mammalian inner ear contains organs that sense motions of the head and its orientation with respect to gravity. These organs make up the **vestibular system** (**Fig. 36.7**), which consists of two statocyst chambers and three **semicircular canals.** These statocyst chambers are similar to those of invertebrates, providing

**FIG. 36.8** **The human ear.** (a) The tympanic membrane and bones of the middle ear transmit external sound vibrations to the oval window of the inner ear. (b) Hair cells in the cochlea respond to the vibrations, sending signals to the brain that distinguish sound amplitude and pitch. (c) A scanning electron micrograph shows the V-shaped arrangement of stereocilia on hair cells.

a sense of gravity and of body orientation with respect to motion. Hair cells located within the semicircular canals sense angular motions of the head in three perpendicular planes, providing a sense of balance. When the head rotates, the motion of gelatinous fluid in the semicircular canals is accelerated, deflecting the stereocilia of the hair cells to activate sensory neurons. The brain interprets differences in the motion of the fluid with respect to the hair cells among the three semicircular canals to resolve angular motions of the head in any direction. As a result, mammals and birds react much faster to head motion than to visual cues, and that ability helps them to stabilize their gaze.

→ **Quick Check 3** After you spin in place or take a ride on a merry-go-round, why are you unstable when you try to walk?

### Hair cells detect the physical vibrations of sound.

What our ears interpret as sound are alternating waves of pressure traveling through the air. The frequency of the waves—the number of waves per second—determines the pitch of a sound, and the amplitude, or height, of the waves determines its loudness. Sound waves cause the stereocilia of hair cells to bend, and the hair cells become excited, depolarizing and releasing neurotransmitters. The process sounds simple, but how can the bending of stereocilia produce all the variations in pitch and loudness you hear in the world around you?

Insects hear by sensing airborne vibrations with small hairs on their antennae or other regions of their body. Hairs of different

length detect different frequencies because the shorter hairs are stiffer than long hairs and vibrate at higher frequencies. Frequency is measured in Hertz (Hz), or cycles per second. Male mosquitoes have small antennal hairs that vibrate in response to the 500-Hz hum of a female mosquito's flight, which attracts them to the female to mate. Many insects also have specialized ears that sense sound by means of a **tympanic membrane.** The tympanic membrane is a thin sheet of tissue at the surface of the ear that vibrates in response to sound waves, amplifying airborne vibrations. The vibration excites hair cells attached to the inside of the tympanic membrane, and these hair cells in turn excite other neurons to fire action potentials. Crickets and other singing insects sense the chirping or buzzing of potential mates by their ears.

Hearing is most widespread and developed in terrestrial vertebrates. The ears of amphibians, reptiles, and birds have a simple external tympanic membrane that vibrates when sound waves strike its surface. This membrane is similar to the tympanic membrane of insects but evolved independently and therefore is a case of convergent evolution. Terrestrial vertebrate ears can detect a broad range of sound frequencies. For humans, the range of audible frequencies is from 20 to 20,000 Hz. For dogs, the audible range is 70 to 44,000 Hz, and some rodents can hear up to 80,000 Hz. Birds generally hear in the same frequency range as mammals—one reason bird songs are pleasing to our ear. Frogs and reptiles generally hear at lower frequencies than mammals and birds. Vertebrate ears can also make fine distinctions between close frequencies and can detect softer sounds because external structures amplify the sounds before they reach the hair cells.

The ears of mammals have an external structure, the **pinna,** that enhances the reception of sound waves contacting the ear. Notice the heightened sensitivity that this structure provides for hearing when you cup your hands behind your ears. It is no coincidence that dogs and rabbits perk up their ears when alerted by a sound: This action orients them to the direction of the sound's source.

**Fig. 36.8** shows the structure of the human ear. The pinna is part of the **outer ear,** which also includes the ear canal and tympanic membrane, or mammalian **eardrum,** which transmits airborne sounds into the ear. The **middle ear** contains three small bones or ossicles, called the **malleus, incus,** and **stapes,** which amplify the waves that strike the tympanic membrane. The stapes connects to a thin membrane called the **oval window** of the cochlea in the **inner ear.** The **cochlea** (which means "snail" in Latin) is a coiled chamber within the skull that contains hair cells that convert pressure waves into an electrical impulse that is sent to the brain.

The mammalian eardrum evolved from the tympanic membrane of reptiles and earlier amphibians. The three middle-ear bones are also evolutionarily related to three bones in reptiles, but only the stapes functions in reptiles to transmit vibrations of the tympanic membrane to the inner ear. The malleus and incus help support jaw movements during feeding in reptiles. During the evolution of mammals from their reptilian ancestors, these two bones were incorporated together with the stapes to form the middle-ear bones that transmit sounds from the eardrum to the inner ear.

The process of hearing can be separated into three stages:

1. *Amplification.* Sound vibrations received by the outer ear are transmitted by the eardrum and amplified by the three bones in the middle ear (Fig. 36.8a). Through piston-like actions, the stapes transmits the energy of its movements to vibrations of the oval window of the cochlea in the inner ear. Vibrations transmitted from the eardrum through the middle ear to the oval window are amplified more than 30 times because of the larger size of the eardrum compared to the oval window and the lever-like action of the three bones.

2. *Transfer of sound vibration to fluid pressure waves.* The cochlea contains fluid and an upper and a lower canal separated by a **basilar membrane** and **cochlear duct** (Fig. 36.8b). Vibrations of the oval window cause fluid pressure waves in both canals at nearly the same time.

3. *Mechanoreception by hair cells within the cochlea.* The cochlear duct contains the **organ of Corti,** which has specialized hair cells with stereocilia supported by the basilar membrane. The stereocilia of mammalian hair cells form a "V" and project into a rigid **tectorial membrane** that does not move (Fig. 36.8c). The fluid vibrations within the cochlear canals induce motions of the basilar membrane relative to the tectorial membrane, bending the stereocilia of the hair cells back and forth in localized regions of the cochlea (**Fig. 36.9**). Bending of the stereocilia stimulates them to release excitatory neurotransmitters that cause postsynaptic neurons to fire action potentials. These

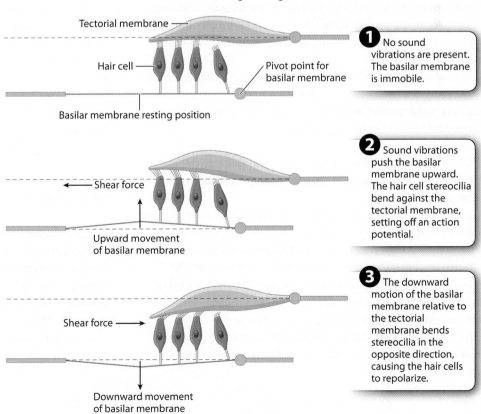

**FIG. 36.9** **Activation of hair cells.** Upward movement of the basilar membrane pushes hair cells into the tectorial membrane, causing bending of the stereocilia.

Tectorial membrane

Hair cell

Pivot point for basilar membrane

Basilar membrane resting position

**1** No sound vibrations are present. The basilar membrane is immobile.

Shear force

Upward movement of basilar membrane

**2** Sound vibrations push the basilar membrane upward. The hair cell stereocilia bend against the tectorial membrane, setting off an action potential.

Shear force

Downward movement of basilar membrane

**3** The downward motion of the basilar membrane relative to the tectorial membrane bends stereocilia in the opposite direction, causing the hair cells to repolarize.

are sensed as sound by neuronal networks in the **auditory cortex,** the area of the brain that processes sound.

Sound is characterized by amplitude (loudness) and frequency (pitch). Louder sounds produce larger fluid vibrations that cause the hair cell stereocilia to bend more, increasing their release of excitatory neurotransmitters. The release of more neurotransmitters increases the firing rate of the postsynaptic cell. The firing rate indicates intensity, or in the case of hearing, loudness.

Many people can tell pure tones apart that differ by only a fraction of a percentage point in frequency. The ability to discriminate different sound frequencies is largely the result of differences in the mechanical properties of the basilar membrane along its length. At the apex, the basilar membrane is widest but thinnest and most flexible. At the base, the basilar membrane is narrow but thick and stiff. Thus, the basilar membrane is mechanically tuned to respond in different regions to different frequencies. Low-frequency vibrations excite hair cells in the apex of the basilar membrane, midrange-frequency vibrations excite hair cells in the middle region of the basilar membrane, and high-frequency vibrations excite hair cells closest to the base.

Sound amplitude and pitch, together with vestibular sensing of gravity and head motion, are transmitted by the vestibulocochlear nerve (another cranial nerve of the head) to the brain's auditory cortex, discussed later in the chapter.

**FIG. 36.10**   A bat about to catch a moth by high-frequency auditory echolocation.

**? CASE 7 Predator–Prey: A Game of Life and Death**
**How have sensory systems evolved in predators and prey?**

Many insect-eating bats rely on **echolocation** to find and apprehend their flying prey (**Fig. 36.10**). As it flies, a bat emits short bursts of high-frequency sound. These sound pulses bounce off surrounding objects and are reflected back to the bat. The echoes are detected by the ears and processed by the bat's brain to locate the prey. Bats typically increase their call rate as they approach the prey to locate it more accurately and judge its flight trajectory and its speed. The sharpness of an image is measured by its resolution, the smallest distance between two features or objects that can be perceived. Bats can resolve prey and other objects at less than 1 mm, an ability that exceeds that of the most sophisticated sonar developed by human engineers.

Together with mosquitoes and other flying insects, nocturnal moths are one of the bat's primary food sources. In response to the evolution of echolocation by bats, several moths have evolved the ability to emit sounds that jam the bat's sonar signal. These moths are more likely to escape capture. The evolution of sophisticated sensory systems in both groups of animals, and the brain organization that goes with it, is the result of a kind of evolutionary arms race. It illustrates how animal sensory systems and brains can achieve impressive detection and information processing performance that improves an animal's chance of survival.

## 36.4 VISION

**Electromagnetic receptors** respond to electrical, magnetic, and light stimuli. Of these, light-detecting photoreceptors are the most common and diverse. Most animals sense light in their environment. Although animals have evolved different light-sensing organs, all rely on the same light-sensitive photopigment to convert light energy into a nervous signal. Their use of the same photopigment suggests a common evolutionary origin for the considerable diversity of animal eyes that dates from the

Cambrian Period (approximately 500 million years ago), when diverse forms of multicellular animal life first evolved.

## All animals use a similar photosensitive protein called opsin to detect light.

Most multicellular animals have cells called photoreceptors that respond to light. Animals with simple forms of photosensitivity can move toward or away from light sources. The photoreceptors of other animals are arranged to form eyes that produce an image. In some animals, detecting the motion of shapes and simple patterns of light is most important. Other animals, such as vertebrates and cephalopod mollusks like squid and octopus, perceive sharper images. Despite the diversity of light-sensitive organs and eyes that have evolved, all of them depend on a similar photosensitive protein known as **opsin** to convert the energy of light photons into electrical signals in the receptor cell.

Genes involved in the development of photoreceptors in flatworms are shared both with other invertebrates and with vertebrates, indicating that an early common ancestor of invertebrates and vertebrates first evolved the underlying biochemical machinery necessary for perceiving light. In these animals, a master regulatory gene, *Pax6*, is now known to function as a transcription factor regulating the development of the eye. In experiments performed by the Swiss developmental biologist Walter Gehring, *Pax6* was recently found to regulate eye development in fruit flies and mice. Gehring and his group were able to insert the mouse *Pax6* gene into fruit flies, leading to the formation of an additional small eye on the fly's antenna

(Chapter 20). *Pax 6* has also been found expressed in flatworm photoreceptors. Remarkably, all light-sensing and image-formation systems therefore ultimately derive from the ancestor of the flatworm's original photoreceptor.

Rhodopsin is the specific transmembrane protein found in the photosensitive cells of vertebrates. This protein is covalently bound to **retinal,** a derivative of vitamin A that absorbs light (**Fig. 36.11**). Opsin molecules are arranged in cylindrical groups in the plasma membrane of most photosensitive cells. These cells are modified neurons with leaky $Na^+$ channels. As a result, their resting membrane potential (in the dark) is less negative (about −35 mV) than that of other nerve cells. When retinal absorbs a light photon, it undergoes a conformational change from a *cis* to a *trans* configuration, and that conformational change causes $Na^+$ channels to close. Without $Na^+$ ions entering the cell, the cell membrane becomes hyperpolarized (rather than depolarized) and the cell's neurotransmitter release is reduced. Photoreceptors themselves do not fire action potentials, but when stimulated by light energy they inhibit the firing rate of other neurons within the eye. The decrease in firing rate provides information about the intensity and pattern of light received.

## Animals see the world through different types of eyes.

Although all eyes rely on similar opsin photopigments, the way the visual world is perceived—what it looks like to the organism—depends on the structure of the eye in which the receptors are embedded. Here, we look at three main types of

**FIG. 36.11 Rhodopsin, the light-sensitive pigment found in vertebrates.** Rhodopsin is made up of the transmembrane protein opsin bound to the light-absorbing retinal.

**FIG. 36.12 Three types of eye.**

**a.** Eyecup

**b.** Compound eye

**c.** Single-lens eye

Photoreceptors

Light

Pigment layer

The flatworm *Planaria* uses simple photoreceptors and a pigmented epithelium to sense the direction of light.

Light

Lens
Photoreceptors

The compound eye of insects, such as the common housefly, are composed of hundreds of ommatidia, each with a lens, that individually sense light.

Lens

Light

Photoreceptors

The single-lens eye of a squid focuses light on a retina and allows for a high degree of acuity.

eye structure shown in **Fig. 36.12: eyecups**, **compound eyes**, and **single-lens eyes**. Each has very different capabilities.

In flatworms, two simple eyecups on the dorsal, or back, surface of the head detect the direction and intensity of light sources (Fig. 36.12a). Each eyecup contains photoreceptors that point up and to the left or right. A pigmented epithelium blocks light from behind, so the photoreceptors receive light only from above and in front of the animal. In response to light, flatworms turn to move directly away from the light source, seeking a dark region that hides them from potential predators. To find a dark region, the flatworm compares the light intensity received by photoreceptors of the two eyecups, and then moves in the direction of lower light intensity. Many invertebrates use simple light-sensitive photoreceptors in a similar way.

Two major types of image-forming eye evolved in other invertebrates: compound eyes and single-lens eyes. Insects and crustaceans have compound eyes that consist of individual light-focusing elements called **ommatidia** (Fig. 36.12b). The number of ommatidia in the eye determines the resolution, or sharpness, of the image. The compound eyes of ants have only a few ommatidia, and those of fruit flies have about 800. In contrast, the eyes of dragonflies have 10,000 or more. Dragonflies are predators, and their high number of ommatidia is an adaptation for visually tracking their prey.

Within a single ommatidium, light is focused through a lens onto a central region formed from multiple overlapping photoreceptors. Each ommatidium is sensitive to a narrow angle of light (about 1° to 2°) in the animal's visual field. Compound eyes of insects and crustaceans provide a mosaic image because individual light regions are sensed by separate ommatidia. It is likely that the image is sharpened within the brain, but the resolution of images produced

by compound eyes is not nearly as good as those produced by single-lens eyes. Nevertheless, compound eyes are extremely good at detecting motion and rapid flashes of light, more than 300 per second compared to 50 per second for a human eye. Insect eyes also have good color vision, and many insects can perceive ultraviolet light. Many pollinating insects sense UV light to locate the flowers they prefer to visit for nectar (**Fig. 36.13**).

The single-lens eyes of vertebrates and cephalopod mollusks like squid and octopus work like a camera to produce a sharply defined image of the animal's visual field (see Fig. 36.12c). In spite of their similar outward appearance, there is clear evidence that single-lens eyes evolved independently in vertebrates and in some mollusks (**Fig. 36.14**). Other invertebrates, such as some spiders and annelid worms, also evolved single-lens eyes independently. An advantage of single-lens eyes compared to compound eyes is that the single

**FIG. 36.13 The same flower seen with visual and UV light.** Many pollinating insects see ultraviolet light, which can dramatically alter the image that attracts them to a particular flower. (Tormentil is shown here.)

Seen with visual light

Seen with UV light

**FIG. 36.14 Convergent evolution of single-lens eyes.** Single-lens eyes evolved independently in vertebrates (fish and mammals) and in some mollusks (octopus and squid).

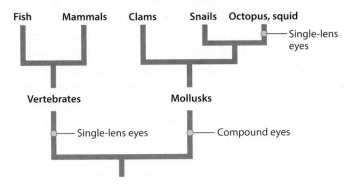

lens can focus light rays on a particular region of photoreceptors, improving both image quality and light sensitivity. As a result, animals with single-lens eyes can detect prey effectively by their motion and shape.

We focus much of our subsequent discussion on the human eye, but many of its organizational features and properties apply to vertebrates more generally.

### The structure and function of the vertebrate eye underlie image processing.

Vertebrate eyes are generally protected within the eye socket of the skull. The anatomy of the eye is shown in **Fig. 36.15**.

The eye is surrounded by the **sclera,** a tough, white outer layer. Below the sclera, a thinner layer, called the choroid, carries blood vessels to nourish the eye. Mucus secretions produced over the sclera keep the eye moist. The portion of the sclera in the front of the eye is transparent; this is the **cornea.** Light passes through the transparent cornea, enters the eye through an opening called the **pupil,** and then passes through a convex **lens.** Small muscles called ciliary muscles attach between the sclera and lens. Contraction or relaxation of these muscles adjusts the shape of the lens to focus light images.

At the front of the eye, surrounding the pupil, the choroid forms a donut-shaped **iris** that opens and closes to adjust the amount of light that enters through the pupil, in the manner of the diaphragm of a camera. In bright light, the iris constricts to make the pupil small. In dim light, the iris dilates to allow in more light. The iris also gives the eye its color.

The focusing muscles and lens separate the eye into two regions. The interior region in front of the lens is filled with a clear watery liquid, the **aqueous humor.** This liquid is produced continuously and drained by small ducts at the base of the eye. Blockage of the ducts results in increased pressure within the eye that if sustained can cause glaucoma, which sometimes leads to blindness. The large cavity behind the lens is filled with a gel-like substance, the **vitreous humor,** that makes up most of the eye's volume.

To form an image, the cornea and lens bend incoming light rays, focusing them on the **retina** (Fig. 36.15). In doing so, the light rays cross, inverting the image on the retina and changing up to down and left to right. The retina is a thin tissue located in the posterior of

**FIG. 36.15 Anatomy of the human eye.** The eyes of other mammals and of birds have a similar organization.

**FIG. 36.16 Focusing an image.** The eyes of humans and other mammals and of birds focus the light of an image on the retina by changing the shape of the lens.

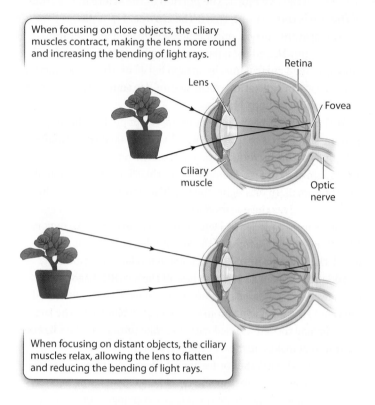

When focusing on close objects, the ciliary muscles contract, making the lens more round and increasing the bending of light rays.

When focusing on distant objects, the ciliary muscles relax, allowing the lens to flatten and reducing the bending of light rays.

the eye. It contains the photoreceptors and other nerve cells that sense and initially process light stimuli. Humans, other mammals, and birds focus light on the retina by changing the shape of the lens (**Fig. 36.16**). When we focus on objects at close range, as when we read a book, the ciliary muscles contract, making the lens more rounded to bend the light rays more. When we focus on distant objects, the ciliary muscles relax, allowing the lens to flatten, reducing the diffraction, or bending, of light. In contrast, the eyes of octopus and squid focus by moving the lens back and forth within the eye, similar to a camera, rather than by changing its shape.

Why do so many animals have two eyes? One reason is that two eyes provide a wider field of vision. Close one eye and you will see the visual field of the other eye. With both eyes open, the visual field is wider than the visual field of either eye alone. You'll also notice considerable overlap in the visual field of both eyes. This overlap exists in humans and other primates, as well as in other mammals, and makes possible binocular vision, the ability to combine images from both eyes to produce a single visual image. Binocular vision allows for depth and distance perception.

Prey animals, such as some birds, antelope, and rabbits, have eyes positioned more to the side, expanding their visual field up to 360°, but limiting or preventing binocular vision. Animals that require greater visual sharpness, such as predatory hawks, cats, and snakes, have eyes pointing forward, increasing their binocular field of view.

## Color vision detects different wavelengths of light.

Color vision is crucial for many invertebrate and vertebrate animals. It is achieved by photoreceptor cells that contain visual pigments sensitive to different wavelengths of light (**Fig. 36.17**). Color-sensitive pigments are found within the **cone cells** of vertebrates. Cone cells are distinguished by their shape from the more numerous **rod cells,** which are sensitive to light but not color. Rod cells detect shades ranging from white to shades of gray and black. The human retina contains about 6 million cone cells and about 125 million rod cells. Because of their greater number and sensitivity to light, rod cells enable animals to see in low light. Together, rod and cone cells constitute approximately 70% of all sensory receptor cells in the human body, highlighting the importance of vision to perceive our environment.

Vertebrate cone cells likely evolved from a rod cell precursor. Most vertebrates have cone cells with four photopigments, whereas the cone cells of most mammals have only two. It is likely that two photopigments were lost early in mammal evolution, when mammals were small, nocturnal, and burrowing. Old World primates, apes, and humans re-evolved trichromatic (that is, three-color) vision by means of gene duplication (some New World primates also regained trichromatic vision), likely in response to selection for better ability to locate fruit, a key part of their diet. Each human cone cell therefore has one of three photopigments, which absorb light at blue, green, or red wavelengths (**Fig. 36.18**). Stimulation of cone cells in varying combinations of these three photopigments allows humans and other primates to see a full range of color (violet to red). Fish, amphibians, reptiles, and birds also have good color vision, including for many the ability to see ultraviolet.

The low sensitivity of cone cells to light makes it hard to detect color at night. Cone cells are most concentrated within the **fovea** of the retina, the center of the visual field of most vertebrates (see Fig. 36.15). Cone cells provide the sharpest vision. Animals

**FIG. 36.17 Vertebrate photoreceptors: rods and cones.**

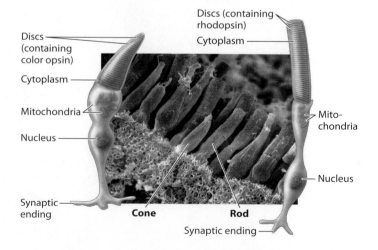

**FIG. 36.18 Absorption spectra of the three opsins in human cone cells.** Each cone cell expresses just one of the three opsins.

with particularly sharp vision, such as hawks and other birds of prey, have an extremely concentrated number of cone cells, approaching 1 million/mm² (compared to about 150,000/mm² in the human fovea). Birds of prey can see small prey on the ground from high in the air. Many birds also have eyes with two foveae, one projecting forward for binocular vision and one projecting to the side to enhance the sharpness of their side vision.

In contrast, rod cells are absent in the fovea but predominate in the periphery. Their positioning makes it easier to detect motion in the periphery, particularly at night. The eyes of many nocturnal predatory mammals, such as foxes and cats, contain mainly rod cells, enhancing nighttime vision. These animals also have a pigment within the retina that reflects light past the photoreceptors, enhancing the ability to sense light under low-light conditions. The reflection of light by this pigment creates the "eyeshine" that you see when a flashlight or a car's headlights shines on a nocturnal predator's eyes at night. Great white sharks are able to hunt at night because they also have a retinal reflective pigment.

### Local sensory processing of light determines basic features of shape and movement.

The rods and cones detect the intensity, color, and pattern of light entering the eye through the lens. However, it is their interaction with a highly ordered array of interneurons in the retina that begins the first steps of visual sensory processing (**Fig. 36.19**). The retina consists of five layers of neurons that form a signal-processing network just in front of a pigmented epithelium. The rods and cones are at the back of the retina, so that light must pass through several layers of neurons before reaching these light-sensitive cells. Because they are photoreceptors, rod and cone cells hyperpolarize in response to light but do not fire action potentials. They synapse on to **bipolar cells,** which also do not fire action potentials, but rather adjust their release of neurotransmitter in response to the input from multiple rod

and cone cells. The bipolar cells, in turn, synapse on to **ganglion cells** located on the front of the retina. If activated, ganglion cells transmit action potentials by the **optic nerve** (a cranial nerve) to the **visual cortex** in the brain, the part of the brain that processes images. The optic nerve begins at the front of the retina and exits at the back, creating an area without light-sensitive cells: This is the blind spot. In octopus and squid, photoreceptors are in the front of the retina, not in the back as they are in vertebrates. As a result, octopus and squid don't have a blind spot.

In addition to these three primary layers of neurons, two additional sets of nerve cells combine information laterally across the retina. **Horizontal cells** communicate between neighboring pairs of photoreceptors and bipolar cells, enhancing contrast through lateral inhibition to sharpen the image. **Amacrine cells** communicate between neighboring bipolar cells and ganglion cells, enhancing motion detection and adjusting for changes in illumination of the visual scene.

Photoreceptor cells, bipolar cells, ganglion cells, horizontal cells, and amacrine cells form a neural network that processes visual information before it is sent to the brain for further processing and interpretation. Visual signals from approximately 125 million photoreceptors converge to approximately 1 million

**FIG. 36.19 Cellular organization of the human retina.** The retina is made up of five layers of cells: rods and cones, bipolar cells, ganglion cells, horizontal cells, and amacrine cells.

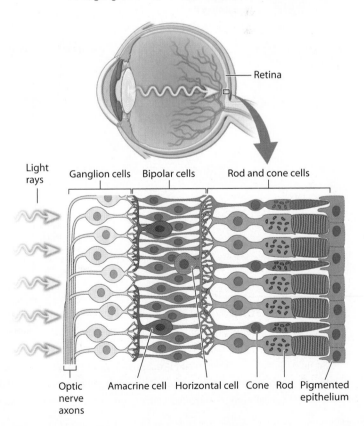

ganglion cell neurons that communicate with the brain. Properties of this network were first studied in the mammalian retina by the American neurophysiologists Stephen Kuffler, David Hubel, and Torsten Wiesel in the 1950s and 1960s (**Fig. 36.20**). Hubel and Wiesel shared the 1981 Nobel Prize in Physiology or Medicine for their later work on how visual information is processed in the brain.

→ **Quick Check 4** What is the primary role of rod cells, and what are the two roles of cone cells in the retina?

## HOW DO WE KNOW?

**FIG. 36.20**

# How does the retina process visual information?

**BACKGROUND** In the 1950s, the American neurophysiologist Stephen Kuffler was interested in understanding how the retina helps to process light information before it is sent to the brain. He focused on the activity of ganglion cells in the retina because they receive input from the photoreceptors and bipolar cells.

**EXPERIMENT** Kuffler stimulated different regions of the cat's retina with localized points of light while recording the action potentials produced by ganglion cells.

**RESULTS** Kuffler found that there are two types of ganglion cell: on-center and off-center cells. On-center ganglion cells fire more action potentials when light shines on the center of the cell's receptive field compared to the surrounding region, and off-center cells fire more when light is shown in the periphery and less on the center. These patterns are explained by lateral inhibition of input to the ganglion cells by the photoreceptors and bipolar neurons in the retina.

**FOLLOW-UP WORK** In the 1960s, Hubel and Wiesel found similar center–surround neural receptive fields, though with enhanced opposition, in part of the thalamus and in the visual cortex of the brain. Cells with these fields enable cats and other mammals to detect shapes of a given orientation moving through their visual field. Similar center–surround receptive fields have also been found in the somatosensory and auditory cortex, highlighting the use of lateral inhibition to enhance sensory acuity and edge detection. Other studies have found similar center–surround sensory processing in invertebrates and other vertebrates.

**SOURCES** Kuffler, S. W. 1953. "Discharge Patterns and Functional Organization of Mammalian Retina." *Journal of Neurophysiology* 16:37–68; Hubel, D. H. 1963. "The Visual Cortex of the Brain." *Scientific American* 209:54–62.

## 36.5 BRAIN ORGANIZATION AND FUNCTION

Up to this point, we have looked at how sensory information, such as smell, taste, hearing, and vision, is detected by sensory receptors. In each case, the sensory information is sent to the brain, where it is processed and integrated. Much of what we know about how the brain functions has been accumulated over many years by studies of patients who suffered brain injuries. A well-known example is Phineas Gage, a railroad worker who in 1848 survived severe damage to his brain's frontal lobe after a tamping iron he was using to set a dynamite charge caused the dynamite accidentally to explode, sending the iron through his skull and the frontal cortex of his brain (**Fig. 36.21**). Gage had been hard-working, responsible, and well liked; after the accident he was given to fitful arguments, lacked self-control, and showed reduced intellectual capacity. These changes provided some of the first evidence that particular brain regions control particular aspects of a person's behavior and cognitive function. The remainder of this chapter explores the functional organization of the vertebrate brain, highlighting examples that show how the brain integrates, organizes, and maps sensory information, and how memories of sights and sounds, for example, are formed that contribute to learning.

### The brain processes and integrates information received from different sensory systems.

With the evolution of greater cognitive ability, animal brains became larger and more complex. Different brain regions became specialized to carry out different functions. The vertebrate brain is organized into a **hindbrain, midbrain,** and **forebrain,** with a

**FIG. 36.21** **The skull of Phineas Gage.** Phineas Gage survived an accident in which a tamping iron passed through his skull, but his personality changed and his intellectual capacity diminished.

cerebral cortex formed from a portion of the forebrain (**Fig. 36.22a**). The hindbrain and midbrain control basic body functions and behaviors, and the forebrain, particularly the cerebral cortex, governs more-advanced cognitive functions. The cerebral cortex of mammals—particularly primates and humans—is greatly expanded relative to that of other vertebrates.

In adult vertebrates, the hindbrain develops into the **cerebellum** and a portion of the **brainstem** (Fig. 36.22a). The remainder of the

**FIG. 36.22** **Human brain development and organization.** (a) The brain develops from the forebrain, midbrain, and hindbrain. (b) Part of the forebrain develops into the limbic system, which controls drives and emotions.

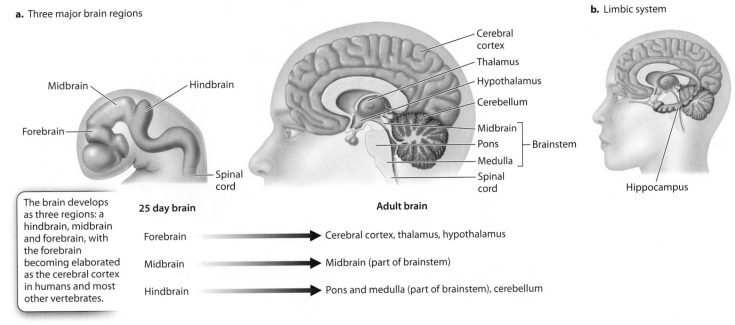

**a.** Three major brain regions

**b.** Limbic system

Midbrain

Hindbrain

Forebrain

Spinal cord

The brain develops as three regions: a hindbrain, midbrain and forebrain, with the forebrain becoming elaborated as the cerebral cortex in humans and most other vertebrates.

**25 day brain**

Cerebral cortex

Thalamus

Hypothalamus

Cerebellum

Midbrain
Pons — Brainstem
Medulla

Spinal cord

**Adult brain**

Hippocampus

Forebrain ⟶ Cerebral cortex, thalamus, hypothalamus

Midbrain ⟶ Midbrain (part of brainstem)

Hindbrain ⟶ Pons and medulla (part of brainstem), cerebellum brainstem develops from the midbrain. The cerebellum coordinates complex motor tasks, such as catching a ball or learning to write and talk. It integrates both motor and sensory information. The brainstem, consisting of the **medulla, pons,** and **midbrain,** initiates and regulates motor functions such as walking and controlling posture and coordinates breathing and swallowing. The brainstem activates the forebrain by relaying information from lower spinal levels. High levels of activity within the brainstem maintain a wakeful state; low levels enable sleep. If the midbrain is damaged, loss of consciousness and then coma result.

The forebrain consists of an inner brain region that forms the **thalamus** and the underlying **hypothalamus,** and a more anterior region that develops into the **cerebrum,** the outer left and right hemispheres of the cerebral cortex (Fig. 36.22a). The thalamus is a central relay station for sensory information sent to higher brain centers of the cerebrum. The hypothalamus interacts closely with the autonomic and endocrine systems to regulate the general physiological state of the body. In humans and most other primates, the cerebral cortex is the largest part of the brain, overseeing sensory perception, memory, and learning. Whereas the forebrain is essential to normal behavior in mammals, it appears less important for other vertebrates.

Other inner components of the forebrain constitute the **limbic system,** which controls physiological drives, instincts, emotions, and motivation and, through interactions with midbrain regions, the sense of reward (**Fig. 36.22b**). Stimulation of the limbic system can induce strong sensations of pleasure, pain, or rage, reinforcing

learning related to basic physiological drives. A posterior region of the limbic system, the **hippocampus,** is involved in long-term memory formation, discussed in the next section.

Sensory information reaches the cerebral cortex from the cranial nerves and nerves passing through the spinal cord. This information passes through the brainstem, and then through the thalamus, the central relay station for sensory information. From the thalamus, information from each of the senses goes to a different region of the brain specialized to further process that information, in a manner discussed next.

## The brain is divided into lobes with specialized functions.

The cerebral hemispheres are the largest structures of the mammalian brain (**Fig. 36.23**). They consist of a highly folded outer layer of **gray matter** about 4 mm thick, called the **cortex,** that is made up of densely packed neuron cell bodies and their dendrites (Fig. 36.23a). The folds greatly increase the surface area of the cortex and are the result of selection for an increased number of cortical neurons within the limited volume of the skull. Deep inside the cortex is the **white matter,** which contains the axons of cortical neurons. It is the fatty myelin produced by glial cells surrounding the axon that makes this region of the brain white. These axons are the means by which neurons in different regions of the gray matter communicate with neurons in other regions of the cortex and in deeper regions of the forebrain, midbrain, and hindbrain.

**FIG. 36.23　Human brain organization.** (a) The folding of the brain and (b) its lobes.

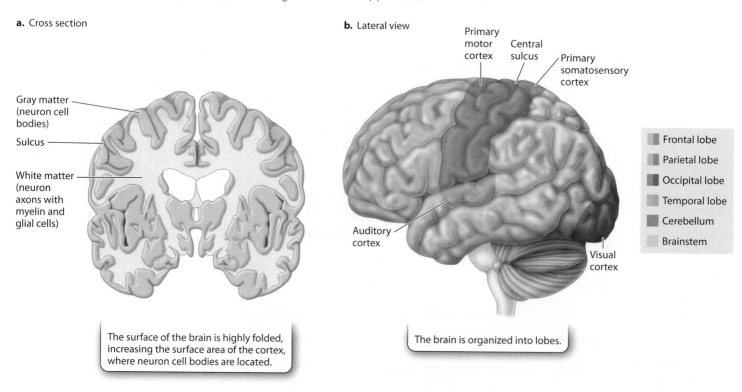

**a.** Cross section

Gray matter (neuron cell bodies)

Sulcus

White matter (neuron axons with myelin and glial cells)

The surface of the brain is highly folded, increasing the surface area of the cortex, where neuron cell bodies are located.

**b.** Lateral view

Primary motor cortex　Central sulcus　Primary somatosensory cortex

Auditory cortex

Visual cortex

Frontal lobe
Parietal lobe
Occipital lobe
Temporal lobe
Cerebellum
Brainstem

The brain is organized into lobes.

**FIG. 36.24** Somatosensory topographic map.

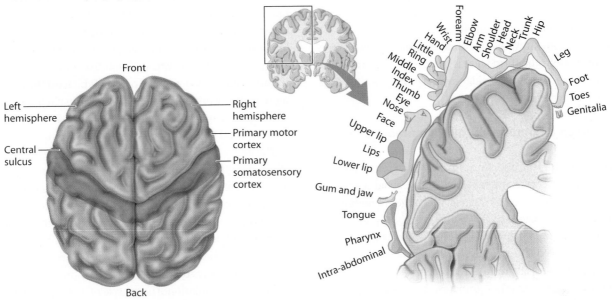

Major regions of the cerebral hemispheres are defined by clearly visible anatomical lobes separated in most cases by deep crevices called **sulci** (singular, sulcus) (Fig. 36.23b). The **frontal lobe** is located in the anterior region of the cerebral cortex (behind your forehead). It is important in decision making and planning. The two **parietal lobes** are located posterior to the frontal lobe. The parietal lobes are separated from the frontal lobe by the central sulcus. The parietal lobe controls body awareness and the ability to perform complex tasks, such as dressing or brushing one's hair. The two **temporal lobes** lie to the side, below the parietal lobes. They are involved in processing sound, as well as performing other functions discussed below. Located behind the parietal lobe, at the back of the brain, is the **occipital lobe,** which processes visual information from the eyes.

The central sulcus separates the **primary motor cortex** of the frontal lobe from the **primary somatosensory cortex** of the parietal lobe (**Fig. 36.24**). "Command" neurons in the primary motor cortex produce complex coordinated behaviors by controlling skeletal muscle movements. The primary somatosensory cortex integrates tactile information from specific body regions and relays this information to the motor cortex. Both cortices make connections to the opposite side of the body by sending axons that cross over in the brainstem and spinal cord. As a result, the right cortex controls the left side of the body and the left cortex controls the right side of the body.

### Information is topographically mapped into the vertebrate cerebral cortex.

The primary motor cortex and somatosensory cortex are organized as maps that represent different body regions, as shown in Fig. 36.24. The neurons that control movements of the foot and human lower limb are located near the mid-axis of the motor cortex. Those that control motions of the face, jaws, and lips map to the side and farther down in the primary motor cortex. There is a similar topographic map in the somatosensory cortex for pressure and touch sensation. Regions such as the fingers, hands, and face that are involved in fine motor movements and distinguishing fine sensations are represented by larger areas of the cortex.

The auditory cortex in the temporal lobe is similarly organized as a map. This region processes sound information transmitted from the cochlea of the ear. Neurons in the auditory cortex are organized by pitch: Neurons sensitive to low frequencies are located at one end and neurons sensitive to high frequencies at the other, helping to distinguish fine differences in pitch. With the evolution of language in humans, regions farther back in the temporal lobe developed into language and reading centers. These centers are linked by pathways to a specialized motor center controlling speech located at the base of the primary motor cortex of the frontal lobe. Functions of the temporal lobe include object identification and naming. Damage to the temporal lobe can lead to an individual's loss of facial recognition, even though others may still be recognized by voice or other body features.

The occipital lobe topographically maps visual information received from the optic nerves. The topographic mapping of information from retinal ganglion cells enables the brain to detect patterns and motion that are precisely related to different regions of the visual field. As a result, highly visual animals, such as humans, primates, and birds, are able to detect complex patterns and movements in their environment.

## 36.6 MEMORY AND COGNITION

The ability of the brain to process and integrate complex sources of information, interpret and remember past events, solve problems, reason, and form ideas is broadly referred to as **cognition.** In this section, we consider memory and learning, as examples of cognitive abilities.

### The brain serves an important role in memory and learning.

Memory is the basis for learning. An animal that can remember its encounter with a predator or noxious plant will avoid that danger in the future. A memory is formed by changes to neural circuits within specific regions of the brain—specifically, by changes in the synaptic connections between neurons in a neural circuit. These changes enable an animal, for example, to recognize other individuals in its social group or to recognize the threat of a predator. In humans and other primates, the hippocampus, a brain structure in the limbic region, plays a special role in memory formation. We can remember much of the information we receive for several minutes. However, unless reinforced by repetition or by paying particular attention, the information is typically forgotten. The hippocampus transforms reinforced short-term memories into long-term memories. Long-term memory is essential for learning.

The hippocampus forms long-term memories by repetitively relaying information to regions of the cerebral cortex. The molecular and cellular basis for memory and learning remains uncertain but depends on establishing particular neural circuits in the cortex. These circuits can be activated to recall a memory that is triggered by relevant stimuli. The formation of a memory circuit requires changes in synapses, which establish a particular set of connections between neurons in the circuit. Some synapses are weakened or removed altogether, and others are formed or strengthened to establish and refine the circuit.

The ability to adjust synaptic connections between neurons is called **synaptic plasticity** (**Fig. 36.25**). Long-term potentiation (LTP) is an example of synaptic plasticity that is believed to underlie memory and learning. In response to repeated excitation, neurons in excited circuits within the hippocampus release the neurotransmitter glutamate, opening $Na^+$ and $Ca^{2+}$ channels. The increased $Ca^{2+}$ stimulates protein synthesis and signaling pathways that make the cells more responsive to subsequent stimulation. The release of glutamate stimulates the placement of new receptors in the postsynaptic cell membrane as well as the formation of new dendrites, strengthening synaptic signaling between the two cells.

Through repeated activation of neurons within the hippocampus and associated regions of the cerebral cortex, neural circuits are established that may be reinforced by LTP to create a memory that can be retrieved over long periods of time. Although

**FIG. 36.25** **Synaptic plasticity.** Memory and learning depend on long-term potentiation (LTP) of synapses within neural circuits.

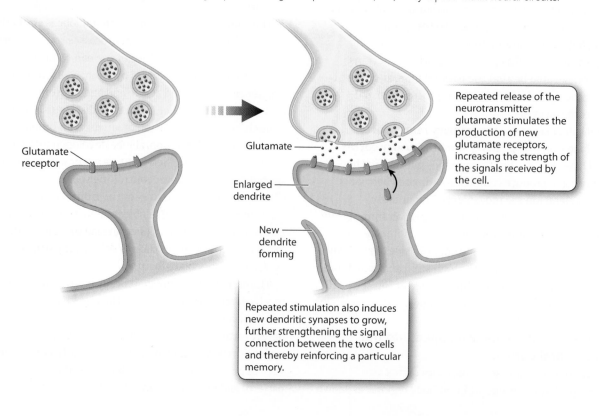

Glutamate receptor

Glutamate

Enlarged dendrite

New dendrite forming

Repeated release of the neurotransmitter glutamate stimulates the production of new glutamate receptors, increasing the strength of the signals received by the cell.

Repeated stimulation also induces new dendritic synapses to grow, further strengthening the signal connection between the two cells and thereby reinforcing a particular memory.

memories may be associated with certain regions of the cortex, they also rely on the interaction of nerve cells and networks within multiple brain regions. Consequently, when a local brain region suffers damage, memories that are lost as a result may sometimes be recovered by reinforcement of the memory stored within other regions.

Although the brain regions involved in memory and learning vary among different animals, the basic features of how memories are constructed by means of synaptic plasticity within neural circuits are likely broadly shared across animals capable of forming memories. For example, the ability of an octopus to learn and remember how to perform motor tasks rivals that of certain vertebrates. Even the roundworm *C. elegans* has been found to be capable of memory and learning: Individual roundworms learn to avoid pathogenic bacteria in favor of bacteria that are nutritious.

→ **Quick Check 5** Which three areas of the human brain are important for remembering and recognizing a face?

### Cognition involves brain information processing and decision making.

Cognitive brain function ultimately gives rise to a state of consciousness—an awareness of oneself and one's actions in relation to others. Consciousness allows judgments to be made about events and experiences in one's environment. When you suddenly see someone you recognize from long ago, rapid neural associations within the cortex identify that person by name, retrieve multiple memories of past events you shared with that person, and elicit conscious thoughts, including your opinion of that person. At the same time, extraneous sensory information is filtered and other thoughts recede, allowing you to focus your attention on the person you have just seen and recognized.

Mental processes are studied by relating changes in brain activity to changes in an animal's behavioral state. The study of the brain is made challenging by the complexity of the circuits involved. Nevertheless, considerable progress is being made through non-invasive neuroimaging techniques (**Fig. 36.26**). These techniques allow the interactions of different brain regions within human and other animal subjects to be investigated while the subjects perform a particular mental or motor task. The results of neuroimaging studies can be linked to studies of the molecular and cellular properties of nerve cells, to advance our understanding of how nerve cells form and maintain nerve cell circuits among different brain regions, and of how these circuits are linked to particular behaviors and cognitive processes.

**FIG. 36.26  Non-invasive imaging of brain function.** PET scan imaging reveals regions of metabolic brain cell activity. The reddish areas show the locations of (upper left) the visual area activated by sight, (upper right) the auditory area activating by hearing, (lower left) the tactile area activated by touching braille, and (lower right) areas of the frontal cortex activated by generating words.

Do animals other than humans have a conscious state of awareness? Although the conscious awareness of other animals is difficult to study, it seems clear that many animals exhibit conscious states of awareness that influence their experience and behavior. Donald Griffin, known for his co-discovery as an undergraduate of bat echolocation, advocated the view not only that animals are conscious, but also that animal consciousness is amenable to scientific study. Behavioral biologist Marion Dawkins summarized her review of studies on animal cognition and consciousness as follows:

> Our near-certainty about [human] shared experiences is based, amongst other things, on a mixture of the complexity of their behavior, their ability to "think" intelligently and on their being able to demonstrate to us that they have a point of view in which what happens to them matters to them. We now know that these three attributes—complexity, thinking, and minding about the world—are also present in other species. The conclusion that they, too, are consciously aware is therefore compelling.

# Core Concepts Summary

**36.1 ANIMAL SENSORY RECEPTORS DETECT PHYSICAL AND CHEMICAL STIMULI BY CHANGES IN MEMBRANE POTENTIAL.**

Sensory neurons include chemoreceptors, mechanoreceptors, thermoreceptors, pain receptors, and electromagnetic receptors. page 36-2

Action potential firing rate correlates with the strength of a stimulus. Generally, strong stimuli induce high firing rates and weak stimuli induce low firing rates. page 36-4

Animal sensory receptors increase their sensitivity to stimuli by temporal and spatial summation, enhance their acuity by lateral inhibition, and adapt to continuous stimuli. page 36-4

**36.2 SPECIALIZED CHEMORECEPTORS RELAY INFORMATION ABOUT SMELL AND TASTE.**

Odorant molecules bind to membrane receptors on the surface of olfactory chemoreceptors. page 36-5

In humans, five taste receptors—for sweet, bitter, sour, salty, and savory—are activated in combination to determine a specific taste. page 36-6

**36.3 HAIR CELLS CONVEY INFORMATION ABOUT MOVEMENT AND SOUND.**

Hair cells are specialized mechanoreceptors that detect motion and vibration. page 36-6

The statocysts of invertebrates are sensitive to gravity, orienting an organism to "up" and "down." page 36-7

The vestibular system of the vertebrate ear provides a sense of balance and gravity. page 36-7

The tympanic membrane in the outer ear transmits airborne sound waves to the middle ear, where the signal is amplified, and then to the cochlea in the inner ear, where sound waves are converted to fluid pressure waves and then sensed by hair cells that convert the signal to an electrical impulse. page 36-8

**36.4 THE ABILITY TO SENSE LIGHT AND FORM IMAGES DEPENDS ON PHOTOSENSITIVE CELLS WITH LIGHT-ABSORBING PROTEINS.**

The protein opsin present in photoreceptors converts light energy into chemical signals, altering the firing rate of neurons they inhibit or stimulate. page 36-11

The eyecups of flatworms detect light and dark. page 36-12

The compound eyes of arthropods sense light using individual light-focusing elements called ommatidia, providing low acuity but rapid motion detection. page 36-12

The single-lens eyes of vertebrates and cephalopod mollusks focus images on a retina, which contains photoreceptor cells and interneurons that process the light stimuli. page 36-12

Photoreceptor cells of the retina include rod cells, which detect light but not color, and cone cells, which detect color. page 36-14

Photoreceptor cells, bipolar cells, ganglion cells, horizontal cells, and amacrine cells form a network that processes visual information in the retina. page 36-15

**36.5 THE BRAIN PROCESSES AND INTEGRATES INFORMATION FROM MULTIPLE SENSORY SYSTEMS, WITH TACTILE, VISUAL, AND AUDITORY STIMULI MAPPED IN THE CEREBRAL CORTEX.**

The vertebrate brain is organized into a hindbrain, midbrain, and forebrain, with the forebrain elaborated into the cerebral cortex in birds and mammals. page 36-17

The brain is divided into frontal, parietal, temporal, and occipital lobes that are specialized for different functions. page 36-19

Somatosensory, motor, auditory, and visual information is topographically mapped to specific areas in the cerebral cortex. page 36-19

**36.6 COGNITION IS THE ABILITY OF THE BRAIN TO PROCESS AND INTEGRATE COMPLEX SOURCES OF INFORMATION, TO REMEMBER AND INTERPRET PAST EVENTS, TO SOLVE PROBLEMS, TO REASON, AND TO FORM IDEAS.**

The cerebral cortex integrates and processes information from diverse sources, giving rise to problem solving, reasoning, and decision-making. page 36-20

The brain stores memories of past experiences and enables learning through the creation of more permanent neural circuits. page 36-20

Consciousness is an awareness of one's own identity in relation to others. page 36-21

# Self-Assessment

1. Give an example of a chemosensory neuron, a mechanosensory neuron, and an electromagnetic sensory neuron.

2. Diagram a generalized receptor neuron, showing how it changes its firing rate in response to detected stimuli.

3. State a hypothesis that explains why all animals have chemoreceptors.

4. Describe the three main stages by which sound is detected and coded by the mammalian ear.

5. Compare and contrast the roles of rod cells and cone cells in the retina.

6. Describe the role of the cornea and lens in vertebrate eyes.

7. Describe three different types of eye in animals.

8. Draw the brain, label the lobes, and describe their primary functions.

9. Describe the importance of topographic mapping of sensory input to the cortex, using the primary somatosensory cortex as an example.

10. Explain how brain function can be understood by studying patients with brain injuries.

Do you understand the chapter's Core Concepts? Log in to check your answers to the Self-Assessment questions, then practice what you've learned and reinforce this chapter's concepts by working through the problems and multimedia tutorials provided there.

**www.biologyhowlifeworks.com**

## CHAPTER 37

# ANIMAL MOVEMENT

## Muscles and Skeletons

The ability to move is a defining feature of animal life, allowing animals to explore new environments and avoid inhospitable ones, escape predators, mate, feed, and play. An animal's motor and nervous systems (Chapters 35 and 36) work together to enable it to sense and respond to its environment. How are movements produced, and how are they controlled? What determines an athlete's performance? What are the mechanical requirements for movement and support? This chapter explores the organization and function of the muscles and skeletons that provide mechanical support and power the movements of larger multicellular animals (**Fig. 37.1**).

## Core Concepts

**37.1** Muscles are biological motors composed of actin and myosin that generate force and produce movement for body support, locomotion, and control of the body's internal physiological functions.

**37.2** Muscle force depends on muscle size, degree of actin–myosin overlap, contractile velocity, and stimulation rate.

**37.3** Hydrostatic skeletons, exoskeletons, or endoskeletons provide animals with mechanical support and protection.

**37.4** Vertebrate endoskeletons allow for growth and repair and transmit muscle forces across joints.

## 37.1 MUSCLES: BIOLOGICAL MOTORS THAT GENERATE FORCE AND PRODUCE MOVEMENT

We rely on our muscles for all kinds of movement. Muscles function as biological motors within the body because they generate force and produce movement. Some of these movements are obvious—running, climbing up stairs, raising your hand, playing the piano. We are less aware of others—moving food through the digestive tract, breathing in and out, pumping blood through the body. A muscle's ability to produce movement depends on a combination of electrically excitable muscle cells and contractile proteins within these cells that can be activated by the nervous system.

The contractile machinery of muscles is ancient. The proteins underlying muscle contraction are found in eukaryotes that lived more than 1 billion years ago, and the first muscle fibers common to all animals are found in cnidarians (jellyfish and sea anemones), which lived at least 600 million years ago. Thus, basic features of muscle organization and function are conserved across the vast diversity of eukaryotic life. As we will discuss, the geometry and organization of these proteins largely determines how muscles contract to produce force and movement.

### Muscles can be striated or smooth.

Muscle cells, or **fibers,** produce forces within an animal's body, as well as exert forces on the environment. Despite their diversity, all muscle cells produce force by a similar mechanism. Skeletal muscle fibers produce forces by shortening, thus pulling on bones of the skeleton. Muscles pull on the bones of your leg and foot to swing them forward when you kick a ball. Muscle fibers in the heart contract to pump blood. Still other muscle fibers control blood flow through arteries by contracting to change an artery's cross-sectional shape, or are used for movements involved in digestion and reproduction.

Muscles can shorten very quickly, and they can produce large forces for their weight. The forces they produce can be many times an animal's body weight, allowing

**FIG. 37.1    Muscle and bones.** Muscle and bones work together to produce movement.

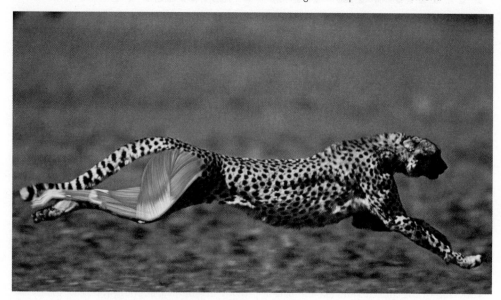

move the animal's limbs and torso, and **cardiac muscle** (Fig. 37.2b), which contracts to pump blood. In contrast, **smooth muscles** appear uniform under the light microscope (Fig. 37.2c). They are found in the walls of arteries to regulate blood flow, in the respiratory system to control airflow, and in the digestive and excretory systems to help transport food and waste products. Smooth muscles contract slowly compared to cardiac and skeletal muscles.

Striated and smooth muscles use the same sets of muscle proteins—**actin** and **myosin** (Chapter 10)—to contract and generate force (force is mass times acceleration, and in this case is the ability to accelerate the motion of the body). These proteins are organized into thin threads called **filaments** that interact with one another to cause muscles to shorten.

the animal to accelerate quickly. Some small muscles contract extremely quickly: The muscles of many flying insects contract 200 to 500 times per second. Even in vertebrate animals, muscles can contract as many as 50 to 100 times per second. Other muscles, such as the closing muscle of a clam, contract slowly, but do so for prolonged periods without tiring.

All muscles contain contractile proteins that enable them to shorten and produce force, but they are divided into two broad groups on the basis of their function and appearance under a light microscope (**Fig. 37.2**). As their name suggests, **striated muscles** appear striped under a light microscope. They include **skeletal muscles** (Fig. 37.2a), which connect to the body skeleton to

The actin and myosin filaments in striated muscles are arranged in a regularly repeating pattern, producing the striations that can be observed under the microscope. In contrast, the organization of actin and myosin filaments in smooth muscles is irregular, giving these muscles a smooth appearance when viewed under the light microscope.

## Skeletal muscle fibers are organized into repeating contractile units called sarcomeres.

The visible bands in skeletal muscles turn out to be an important clue to the mechanism used to produce muscle contractions.

**FIG. 37.2    Light micrographs of (a) skeletal muscle, (b) cardiac muscle, and (c) smooth muscle.** Skeletal and cardiac muscles have a striated, or striped, appearance; smooth muscle does not.

Whole muscles are made up of parallel bundles of individual muscle fibers (**Fig. 37.3**). Each muscle fiber is an elongated single cell that can be up to 20 cm long. It has several nuclei and therefore is described as **multinucleated.** The cell's multinucleated organization means that gene expression must be coordinated among all nuclei in the muscle fiber. Coordinated gene expression is necessary when endurance training increases the cell's rate of ATP production, so that the entire cell contracts uniformly. Each muscle fiber contains hundreds of long rodlike structures called **myofibrils** that contain parallel arrays of the actin and myosin filaments that cause a muscle to contract.

Let's consider the structural organization of myosin and actin filaments in more detail (**Fig. 37.4**). Each myosin molecule consists of two long polypeptide chains coiled together, each ending with a globular head. Consequently, the myosin molecule resembles a double-headed golf club. The myosin molecules are arranged in parallel to form a **thick filament,** with the

**FIG. 37.4** **Thin and thick filaments.** Thin filaments are two actin filaments wound around each other. Thick filaments are made up of numerous myosin molecules arranged in parallel.

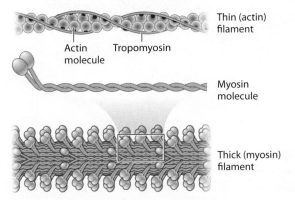

Thin (actin) filament

Actin molecule

Tropomyosin

Myosin molecule

Thick (myosin) filament

numerous myosin heads extending out from their flexible necks along the myosin filament. Two helically arranged actin filaments are twisted together to form a **thin filament.** The protein **tropomyosin** runs in the grooves formed by the actin helices.

The thin filaments are attached to protein backbones called **Z discs** that are regularly spaced along the length of the myofibril (**Fig. 37.5a**). The region from one Z disc to the next, called the **sarcomere,** is the basic contractile unit of a muscle. The sarcomeres are repeating contractile units arranged in series along the length of the muscle fiber. The shortening of thousands of sarcomeres along a fiber leads to shortening and contraction of a muscle.

Sets of actin thin filaments extend from both sides of the Z discs toward the midline of the sarcomere (Fig. 37.5a). In the middle of the sarcomere, not directly contacting the Z discs, are myosin thick filaments. The thin filaments overlap with the myosin thick filaments, forming two regions of overlap within a sarcomere. Toward either end of the sarcomere, as well as in the middle, are regions where the actin and myosin filaments do not overlap. These regions of non-overlap are lighter in color than the regions that overlap, which appear dark under an electron microscope. A third large protein, called **titin,** links the myosin filaments to the Z discs at the ends of the sarcomere. Titin is believed to help with assembly and to protect the sarcomeres from being overstretched, thus contributing to muscle elasticity.

As we have seen, the regular pattern of actin and myosin filaments within sarcomeres along the length of the fiber gives skeletal muscles their striated appearance (Fig. 37.5a). In cross section, thick and thin filaments are arranged in a hexagonal lattice that allows each thick filament to interact and slide with respect to six adjacent thin filaments (**Fig. 37.5b**). The precise geometry of thin and thick filaments is critical to their interaction, discussed on the following page.

**FIG. 37.3** **Muscle organization.** Muscles are made up of bundles of muscle fibers (cells), each of which contains myofibrils.

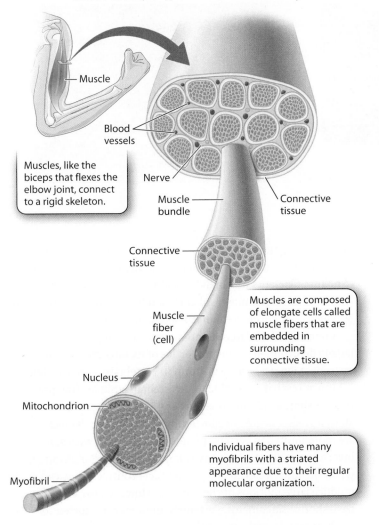

Muscle

Muscles, like the biceps that flexes the elbow joint, connect to a rigid skeleton.

Blood vessels

Nerve

Muscle bundle

Connective tissue

Connective tissue

Muscles are composed of elongate cells called muscle fibers that are embedded in surrounding connective tissue.

Muscle fiber (cell)

Nucleus

Mitochondrion

Individual fibers have many myofibrils with a striated appearance due to their regular molecular organization.

Myofibril

**FIG. 37.5 Muscle sarcomere.** (a) Longitudinal and (b) cross-sectional views of a sarcomere, including two transmission electron micrographs. The sarcomere is the region between Z discs. It is the contractile unit of a muscle.

Muscle fiber

Myofibril

Sarcomere

Z discs

**a.**

Titin    Z disc    Thin filament    Thick filament    Z disc

A longitudinal section shows the organization of thick (myosin) and thin (actin) filaments.

**b.**

Actin (thin)    Myosin (thick)

A cross section shows the regular hexagonal arrangement of thick and thin filaments.

Z disc    Z disc

Sarcomere

## Muscles contract by the sliding of myosin and actin filaments.

By using high-resolution light microscopy, two independent teams (physiologists Hugh Huxley and Jean Hanson in the United States and Andrew Huxley and Rolf Niedergerke in England) were able to quantify the changing banding pattern of sarcomeres when rabbit or frog myofibrils were stimulated to contract to different lengths. Their results led them to hypothesize that muscles produce force and change length by the sliding of actin filaments relative to myosin filaments. Their theory is known as the **sliding filament model** of muscle contraction.

The banding patterns of sarcomeres reveal the extent of overlap between actin and myosin filaments. When myofibrils contracted to short lengths, the sarcomeres were observed to have increased actin–myosin overlap. When myofibrils were stretched to longer lengths, actin–myosin overlap decreased (**Fig. 37.6**). However, the lengths of the myosin and actin filaments never varied.

Consequently, nearly all of the length change during a muscle contraction results from the sliding of actin filaments with respect to myosin filaments within individual sarcomeres. The length change of the whole muscle fiber is therefore a sum of the fractions by which each sarcomere shortens along the fiber's length.

A muscle's ability to generate force and change length is largely determined by the properties of its sarcomeres. Whereas sarcomere length is quite uniform in vertebrate animals (about 2.3 μm at rest), it is variable in invertebrate animals (ranging from as short as 1.3 μm to as long as 43 μm). Longer sarcomeres allow a greater degree of shortening. Thus, length changes are more limited in vertebrate muscles compared with some invertebrate muscles that have long sarcomeres. For example, when the muscles in an octopus tentacle contract, they can produce large motions of the tentacle. However, because they have shorter sarcomeres, vertebrate muscles can shorten to produce movements more quickly.

**FIG. 37.6 Sliding filament model.** Muscle contraction, or shortening, results from the sliding of actin thin filaments relative to myosin thick filaments. Filament sliding changes the sarcomere banding pattern between Z discs.

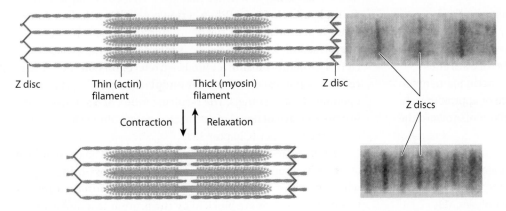

Z disc     Thin (actin) filament     Thick (myosin) filament     Z disc

Contraction ↓↑ Relaxation

Z discs

Interactions between the myosin and actin filaments are what cause a muscle fiber to shorten and produce force. Along the myosin filament, the two heads of myosin molecules bind to actin at specific sites to form **cross-bridges** between the myosin and actin filaments. The myosin filaments pull the actin filaments toward each other by means of the cross-bridges they form with actin filaments. The key to making the

filaments slide relative to each other is the ability of the myosin head to undergo a conformational change and therefore pivot back and forth. The myosin head attaches to the actin filament and pivots forward, sliding the actin filament toward the middle of the sarcomere. The myosin head then detaches from the actin filament, cocks back, reattaches, and the cycle repeats. The movement of the myosin head is powered by ATP. Together, these events describe the **cross-bridge cycle**.

The cross-bridge cycle is just that—a cycle (**Fig. 37.7**). Let's look at it in more detail:

1. The myosin head binds ATP. Binding of ATP allows the myosin head to detach from actin.

2. The myosin head hydrolyzes ATP to ADP and inorganic phosphate ($P_i$). Hydrolysis of ATP results in a conformational change in which the myosin head is cocked back. ADP and $P_i$ remain bound to the myosin head. Because ADP and $P_i$ are

**FIG. 37.7 Cross-bridge cycle.** The myosin head binds to actin and uses the energy of ATP to pull on the actin filament, causing muscle shortening.

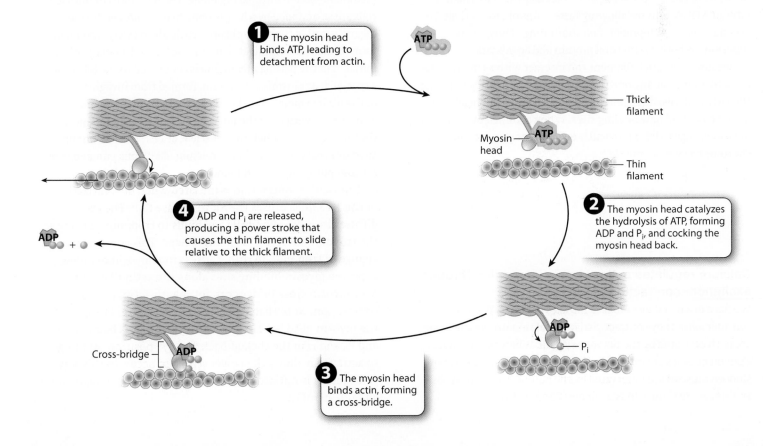

❶ The myosin head binds ATP, leading to detachment from actin.

ATP

Thick filament

Myosin head — ATP

Thin filament

❷ The myosin head catalyzes the hydrolysis of ATP, forming ADP and $P_i$, and cocking the myosin head back.

ADP
$P_i$

❹ ADP and $P_i$ are released, producing a power stroke that causes the thin filament to slide relative to the thick filament.

ADP +

Cross-bridge — ADP

❸ The myosin head binds actin, forming a cross-bridge.

bound rather than released, the myosin head is in a high-energy state.

3.  The myosin head then binds actin, forming a cross-bridge.

4.  When the myosin head binds actin, the myosin head releases ADP and $P_i$. The result is another conformational change in the myosin head, called the **power stroke.** During the power stroke, the myosin head pivots forward and generates a force, causing the myosin and actin filaments to slide relative to each other over a distance of approximately 7 nm. The power stroke pulls the actin filaments toward the sarcomere midline.

The myosin head then binds a new ATP molecule (step 1 again), allowing it to detach from actin, and the cycle repeats. After detachment, the myosin head again hydrolyzes ATP, allowing it to return to its original conformation and bind to a new site farther along the actin filament. With the release of ADP and $P_i$, the myosin head undergoes another step of force generation and movement.

Individual muscle contractions are the result of many successive cycles of cross-bridge formation and detachment, during which the chemical energy released by ATP hydrolysis is converted into mechanical work (the product of force and the distance moved—that is, the amount of shortening). Muscle fibers that contract especially quickly, such as those that power insect wing flapping or produce the sound of cicadas and rattlesnake tails, express myosin molecules that have higher rates of ATP hydrolysis, allowing faster rates of cross-bridge cycling, force development, and shortening.  Thus, myosin functions as both a structural protein and an enzyme.

Because each thick filament can interact with as many as six actin filaments, at any one time numerous cross-bridges anchor the lattice of myosin filaments, while other myosin heads are detached to find new binding sites. When summed over millions of cross-bridges, these molecular events generate the force that shortens the whole muscle.

→   **Quick Check 1** When an animal dies, its limbs and body become stiff because its muscles go into rigor mortis (literally, "stiffness of death"). Why would the loss of ATP following death cause this to happen?

## Calcium regulates actin–myosin interaction through excitation–contraction coupling.

We have discussed *what* makes muscles contract. We next consider *when* they contract. Striated and smooth muscle fibers are both activated by the nervous system. Whereas vertebrate skeletal muscles are innervated by the somatic nervous system, smooth muscles are innervated by the autonomic nervous system, as are cardiac muscle fibers (Chapter 35).

Like nerve cells, muscle fibers are electrically excitable. Skeletal muscle fibers are activated by impulses transmitted by motor nerves to synaptic junctions (**Fig. 37.8**). Motor neuron axons have branches that allow them to synapse with multiple muscle fibers. When action potentials traveling down a motor neuron arrive at the neuromuscular junction, the neurotransmitter acetylcholine is released into the synaptic cleft. The neurotransmitter binds with receptors on the muscle cell at a region called the **motor endplate,** triggering the opening of $Na^+$ channels. The resulting influx of $Na^+$ ions in turn initiates a wave of depolarization that passes along the length of the muscle fiber (Chapter 35).

How does membrane depolarization lead to the cross-bridge cycle? Actin and myosin filaments can form cross-bridges only when the myosin-binding sites on actin are exposed. At rest, these binding sites are blocked by the protein tropomyosin. The wave of depolarization in the muscle cell initiates a chain of events that leads to a change in the conformation of a second protein, called **troponin,** which in turn moves tropomyosin away from these binding sites, allowing cross-bridges between actin and myosin to form and the muscle to contract (Fig. 37.8).

The chain of events that concludes with the exposure of the myosin-binding sites begins at the motor endplate. After the initial membrane depolarization at the motor endplate, the wave of depolarization is conducted inward to deeper regions. You saw in Chapter 5 that eukaryotic cells contain several types of membrane-bound internal organelle. The myofibrils of muscle cells are surrounded by a highly branched membrane-bound organelle called the **sarcoplasmic reticulum (SR),** a modified form of the endoplasmic reticulum (Fig. 37.8). Depolarization initiated in the plasma membrane is conducted to the SR through a specialized transmission system, formed from invaginations of the plasma membrane, known as the **T-tubule system** that is in contact with both the muscle cell's plasma membrane and the SR (Fig. 37.8). When the muscle is at rest, the SR contains a large internal concentration of calcium ($Ca^{2+}$) ions pumped in by calcium pumps in its membranes.

A muscular contraction is initiated when depolarization of the T-tubules causes the SR to release $Ca^{2+}$. The $Ca^{2+}$ diffuses into the myofibrils and binds to troponin, causing the troponin molecule to change shape. This conformational change of troponin, in turn, causes tropomyosin to move, exposing myosin-binding sites along the actin filament. Now myosin cross-bridges can form with actin, producing a contraction. Note that at this stage in the cross-bridge cycle, the myosin head has already hydrolyzed ATP, is bound to ADP and $P_i$, and is in the cocked-back "ready" position. Binding to actin then allows the power stroke to occur. In this way, contraction is initiated immediately following depolarization and release of $Ca^{2+}$.

**FIG. 37.8   Excitation–contraction coupling.** Depolarization (excitation) leads to shortening (contraction) of the muscle.

**a.** Scanning electron micrograph of a motor nerve endplate

Motor nerve axon

Muscle fiber

**b.** Muscle excitation–contraction coupling

**1** An action potential from a motor neuron leads to release of acetylcholine and depolarization of the muscle cell.

Plasma membrane

Nerve

Action potential

Motor endplate

Nucleus

Muscle cell

T-tubule

Myofibrils

**2** The depolarization is conducted into the interior of the fiber by the T-tubules.

Sarcoplasmic reticulum

**3** Depolarization leads to the release of $Ca^{2+}$ from the sarcoplasmic reticulum.

Tropomyosin

$Ca^{2+}$

Myosin-binding sites on actin blocked

Actin   Troponin

Myosin-binding sites on actin exposed

**4** $Ca^{2+}$ binds to troponin, which causes movement of tropomyosin, exposure of myosin-binding sites on actin, and formation of cross-bridges to produce shortening of the muscle.

Thus, $Ca^{2+}$ release from the SR and binding to the troponin–tropomyosin complex on the actin filament results in **excitation–contraction coupling** (so called because excitation of the muscle cell is coupled to contraction of the muscle), producing force and movement. Together, these events are the molecular "switch" that causes a muscle to contract. Muscle relaxation follows the end of neural stimulation, when acetylcholine is broken down or reabsorbed and $Ca^{2+}$ is actively transported back into the sarcoplasmic reticulum, causing tropomyosin molecules to block myosin-binding sites along the actin filaments.

### Calmodulin regulates $Ca^{2+}$ activation and relaxation of smooth muscle.

How is the contraction of smooth muscle regulated? The smooth muscle that controls many internal organs can be activated by stimulation from the autonomic nervous system but also responds to stretch of the muscle, local hormones, or other local factors such as pH, oxygen, carbon dioxide, or nitric oxide. For example, smooth muscle in the walls of arteries contracts when stimulated by the release of epinephrine to reduce blood flow to a body region (Chapter 39), and smooth muscle within the mammalian uterus contracts when oxytocin is released to stimulate contractions (Chapter 42). As in skeletal muscle, these initiators trigger the release of $Ca^{2+}$ from the SR. However, unique to smooth muscle, $Ca^{2+}$ also enters through voltage-gated and stretch-receptor calcium channels in the cell's plasma membrane.

Smooth muscle lacks the troponin–tropomyosin mechanism for regulating contraction. Instead, the switch that activates smooth muscle cells is the binding of the protein **calmodulin** with $Ca^{2+}$ released from the SR or entering through the cell's membrane. The calmodulin–$Ca^{2+}$ complex activates the enzyme myosin kinase, which phosphorylates the smooth muscle myosin heads, causing them to bind actin and begin the cross-bridge cycle. A second enzyme dephosphorylates the myosin heads, disrupting their ability to bind to actin, and the muscle relaxes.

The SR of smooth muscle cells is less extensive and has many fewer calcium pumps than the SR of skeletal muscle cells. As a result, calcium is returned more gradually from the myofibrils back to the SR, or pumped back out through the cell's membrane, slowing the relaxation of smooth muscle and resulting in longer periods of contraction compared with skeletal muscle fibers.

→ **Quick Check 2** Curare is a paralyzing compound that blocks the action of the neurotransmitter acetylcholine at the muscle fiber's motor endplate. What effect do you think curare has on the release of calcium ions from the sarcoplasmic reticulum of the muscle cell?

## 37.2 MUSCLE CONTRACTILE PROPERTIES

Actin and myosin are evolutionarily conserved across all animal life, so all muscles have similar force-generating properties. The huge muscles of a weight lifter are evidence that larger muscles can exert more force. Why is that? The summed force produced by adjacent myofibrils of activated muscle fibers determines the force the whole muscle can exert. This means that larger muscles with more fibers produce greater forces than smaller muscles with fewer fibers. However, the force that a muscle produces also depends on its contraction length and the speed of contraction, as we discuss next.

### Muscle length affects actin-myosin overlap and generation of force.

Why is it difficult to jump while standing on your toes? A muscle's ability to generate force depends in part on how much it is stretched before contraction begins. The sliding filament model for muscle contraction, like any good model, makes specific predictions. In this case, it predicts that the amount of overlap between actin and myosin filaments within the sarcomere

## HOW DO WE KNOW?

**FIG. 37.9**

# How does filament overlap affect force generation in muscles?

**BACKGROUND** From the 1930s through the 1960s, British muscle physiologists studied the properties of striated frog muscle fibers. In the first set of experiments, A. V. Hill examined how a muscle's shortening velocity affects its force production. In the second set of studies, British physiologists Andrew Gordon and Andrew Huxley, together with American physiologist Fred Julian, studied how a muscle's contractile length affects its force-generating ability.

**HYPOTHESIS** They hypothesized that changes in overlap between myosin and actin affect the number of cross-bridges and thus the force the muscle can produce.

**EXPERIMENT** Gordon, Huxley, and Julian isolated a single muscle fiber and used optical microscopy to measure sarcomere length. They kept this length constant while the fiber was stimulated to produce force. Measurements of force were obtained at different sarcomere lengths.

**RESULTS** The top graph shows the data charted by Gordon, Huxley, and Julian, and the bottom graph reinterprets their data in terms of force and percent of muscle length. Their experiments showed that the muscle fiber produced maximal force at an intermediate sarcomere length (approximately 2.3 μm). At this length, the greatest number of cross-bridges were hypothesized to form, generating maximal force. As the sarcomere length was decreased or stretched, and then stimulated to contract, force declined. These results were consistent with a decrease in the number of cross-bridges that could form as a result of decreased actin–myosin overlap.

**FOLLOW-UP WORK** Other studies confirmed that changes in myosin cross-bridge formation are linked to changes in actin and myosin filament overlap at different sarcomere lengths, supporting the sliding filament model of muscle contraction.

**SOURCE** Gordon, A. M., A. F. Huxley, and F. J. Julian. 1966. "The Variation in Isometric Tension with Sarcomere Length in Vertebrate Muscle Fibres." *Journal of Physiology* 184:170–192.

determines the number of cross-bridges that can form. The more overlap, the more force that can be produced. Conversely, when a muscle fiber is pulled to long lengths and then contracts, it generates lower force because the overlap between myosin and actin filaments is reduced. When a muscle fiber contracts at short lengths, it also produces lower force because myosin filaments begin to run into the Z disc at the end of the sarcomere, disrupting the geometry of the myosin filament lattice and hindering cross-bridge formation. At intermediate lengths the muscle fiber generates maximum force because actin–myosin filament overlap is greatest. These predictions were tested by experiments (**Fig. 37.9**).

This influence of muscle fiber length on force generated is the reason that you can't jump well if you are on your toes and your ankles are fully extended: This position shortens your calf muscle fibers. Conversely, if your ankles are overly flexed, your calf muscles are lengthened, and again jumping force is limited. Through training, athletes and dancers learn techniques, such as bending the knees before jumping, so that their leg muscles contract at intermediate lengths to maximize muscle-force output.

The relationship between force and length of striated muscle also explains Starling's Law for the function of the heart as a pump (discussed in Chapter 39), which ensures that the heart contracts more strongly when it is filled by larger amounts of blood returning from the veins.

→ **Quick Check 3** Why can you lift a larger load when your elbow is slightly flexed and your biceps muscle is at an intermediate length than you can when your elbow is fully extended?

### Muscle force and shortening velocity are inversely related.

In experiments carried out in the 1930s, A. V. Hill (who had won the 1922 Nobel Prize in Physiology or Medicine for his work on muscle energy use) observed that a muscle shortens fastest when producing low forces (**Fig. 37.10**). To produce larger forces, the muscle must shorten at progressively slower velocities. For example, your arm moves faster when you throw a light ball than when you throw a heavy one.

The term "muscle contraction" suggests that muscles shorten when stimulated to generate force (in a shortening contraction), but in fact muscles may also exert force and remain a uniform length or even be lengthened. Consequently, the term "contraction" needs to be broadened beyond its familiar reference to muscle shortening to include those instances when a muscle exerts force while remaining the same length or being actively stretched.

When a muscle contracts but cannot shorten (for example, when you carry a heavy bucket of water), the muscle's shortening velocity is zero (Fig. 37.10), and it generates an **isometric** force (*iso-* means

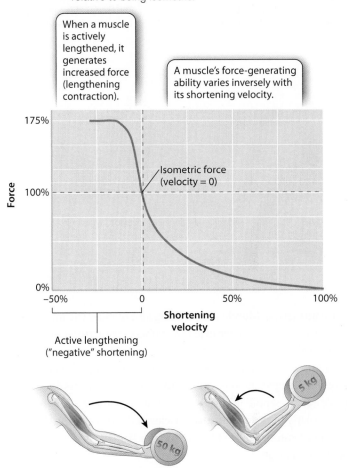

**FIG. 37.10 Force-velocity relationship of muscle.** Skeletal muscles exert greater force when they shorten more slowly and produce even greater force when they are stretched relative to being isometric.

When a muscle is actively lengthened, it generates increased force (lengthening contraction).

A muscle's force-generating ability varies inversely with its shortening velocity.

Isometric force (velocity = 0)

Force

Shortening velocity

Active lengthening ("negative" shortening)

"same" and *metric* refers to "measure"). That is, the muscle still generates force, even though it stays the same length and does not shorten. Isometric exercises use this type of muscle contraction: Holding a weight in a fixed position or exerting force against an immovable object helps maintain muscle strength and may be beneficial in rehabilitation after injury by avoiding joint motion.

Muscles can also be stretched when the external load against which they contract exceeds their force output. This action is known as a **lengthening contraction.** During a lengthening contraction, the muscle generates much more force (up to 75% more, Fig. 37.10) than when it remains isometric or shortens. This kind of contraction occurs when you lower a box from a shelf or when your leg muscles are stretched to support your body in landing from a jump. It also happens during every stride you take when running. A brief phase of muscle stretch at the onset of contraction is common in animal movement because it allows the muscles to generate more force than if they remained isometric or if they began shortening at the start of

**FIG. 37.11  Antagonist muscles.** Skeletal muscles can produce force only by pulling on a skeletal attachment, so they are arranged as antagonists to produce opposing motions at a joint.

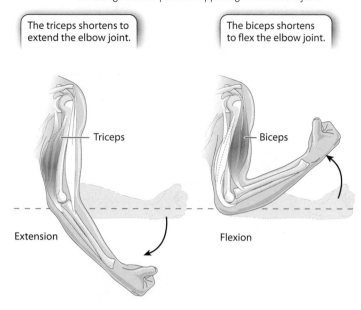

The triceps shortens to extend the elbow joint.

The biceps shortens to flex the elbow joint.

Triceps

Biceps

Extension

Flexion

the contraction. Muscle lengthening also reduces the muscle's energy consumption because fewer fibers are recruited to generate a given level of force.

Injury can result if a muscle is too rapidly or repetitively stretched. Rapid and repetitive stretching often occurs during prolonged downhill hiking, causing muscle damage and soreness, particularly in the quadriceps muscles in the front of the thigh. Suitable training and warm-up exercises help muscles to stretch without damage.

## Antagonist pairs of muscles produce reciprocal motions at a joint.

Because muscles can generate force only by pulling, not pushing, on the skeleton, to produce reciprocal joint and limb movements they must be arranged as pairs of **antagonist muscles** that pull in opposite directions. **Flexion** is the joint motion in which bone segments rotate closer together, and **extension** describes the motion when they move apart. For example, muscles at the elbow joint are organized as flexors (the biceps muscle) that contract to bring the arm toward the shoulder, and extensors (the triceps muscle) that contract to extend the arm (**Fig. 37.11**). Additional pairs of antagonist muscles must be present for movements in other directions. Muscles in the body wall of cnidarians (jellyfish and sea anemones) and segmented annelid worms, as well as the smooth muscles in the walls of intestines, are organized in longitudinal (lengthwise) and circumferential (circular) layers, allowing them to function as muscle antagonists that control movement and shape.

In contrast, when muscles combine to produce similar motions they are termed muscle **agonists**. Muscles arranged as agonists increase the strength and improve the control of joint motion. The three heads of the triceps are agonists of each other (Fig. 37.11).

## Muscle force is summed by an increase in stimulation frequency and the recruitment of motor units.

How are animals able to vary the amount of force exerted by a muscle? As we have seen, skeletal muscles are stimulated by motor nerves, which conduct action potentials to the neuromuscular junction. The force exerted by a muscle depends on the frequency of stimulation by the motor nerve because more-frequent stimulation increases the amount of calcium

**FIG. 37.12  (a) Twitch and (b) fused tetanus muscle contractions.**

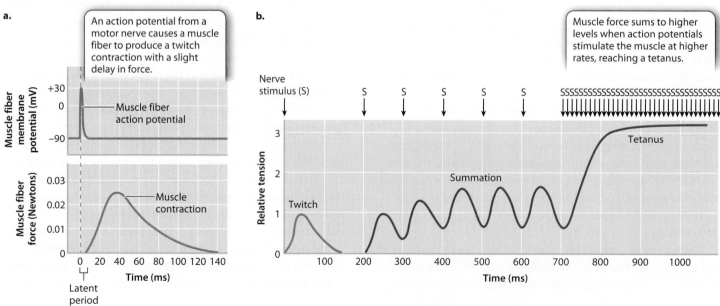

An action potential from a motor nerve causes a muscle fiber to produce a twitch contraction with a slight delay in force.

Muscle fiber action potential

Muscle contraction

Latent period

Muscle force sums to higher levels when action potentials stimulate the muscle at higher rates, reaching a tetanus.

Nerve stimulus (S)

Tetanus

Summation

Twitch

**FIG. 37.13 Motor units.** A motor unit is the motor neuron and the population of cells (fibers) it innervates.

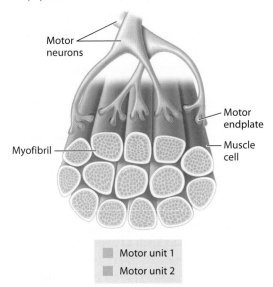

Motor neurons

Motor endplate

Muscle cell

Myofibril

☐ Motor unit 1
☐ Motor unit 2

that is released to activate cross-bridge formation in the muscle fibers. A single action potential results in a **twitch** contraction of a certain force (**Fig. 37.12a**). If sufficient time is allowed for the muscle to pump Ca²⁺ back into the sarcoplasmic reticulum, the contraction ends, force falls to zero, and a second stimulus can elicit a twitch contraction of the same force. However, if a second action potential arrives before the muscle has relaxed, a greater force is produced. With increasing frequency of stimulation, muscle force increases (**Fig. 37.12b**). This property of muscle contraction is called force summation.

At a sufficiently high stimulation frequency, muscle force reaches a plateau and remains steady as stimulation continues. This muscle contraction of sustained force is described as a **tetanus.** Measurements in animals indicate that skeletal muscles commonly produce tetanic contractions, which generate steady and large forces. Tetanus is a normal physiological response of a muscle, but it can also be induced by the bacterium *Clostridium tetani*, so named because it causes severe muscle spasms, particularly of the jaw, chest, back, and abdomen.

A motor neuron and the population of muscle fibers (cells) that it innervates is collectively termed a **motor unit** (**Fig. 37.13**). Motor units can include relatively few to several hundred muscle fibers. The number of muscle fibers innervated by a given motor nerve, and hence the size of the motor unit, affects how finely a muscle's force can be controlled. For example, finger muscles have small motor units allowing fine adjustments in muscle force, whereas leg muscles have large motor units providing larger adjustments in force. Together with stimulation frequency, a muscle's force output depends on the number of motor units (and therefore the number of muscle fibers) that are activated. Within each motor unit, muscle fibers generally share similar contractile and metabolic properties. Differences in these properties can give muscles very different capabilities, as discussed in the next section.

**Skeletal muscles have slow-twitch and fast-twitch fibers.**
The difference between a sprinter and a marathoner is not just training but also the properties of their muscles. There are different types of muscle fiber (**Fig. 37.14a**), allowing

**FIG. 37.14 Slow-twitch (darkly stained) and fast-twitch (lightly stained) skeletal muscle fibers.** These two types of fiber provide differences in endurance and the speed of contraction. (a) Mammalian limb muscle; (b) muscles of two different fish species.

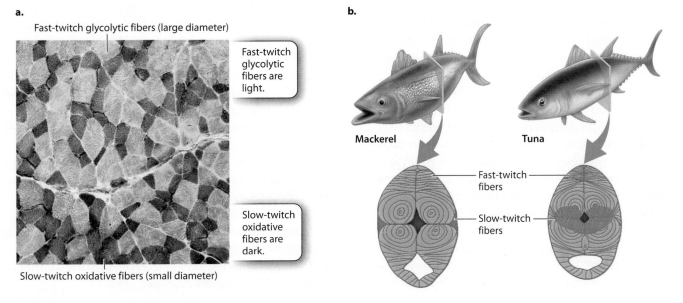

**a.**

Fast-twitch glycolytic fibers (large diameter)

Fast-twitch glycolytic fibers are light.

Slow-twitch oxidative fibers are dark.

Slow-twitch oxidative fibers (small diameter)

**b.**

Mackerel

Tuna

Fast-twitch fibers

Slow-twitch fibers

variation in the way muscles contract. In vertebrates, these are most simply classified as red **slow-twitch** fibers and white **fast-twitch** fibers. Slow-twitch fibers are found in muscles that contract slowly and consume less ATP to produce force. An animal uses slow-twitch fibers to control its posture or move slowly and economically. Fast-twitch fibers generate force quickly, producing rapid movements, but they consume more ATP. An animal depends on fast-twitch fibers to sprint or lunge at its prey. A key difference between the two types of fiber is that slow-twitch fibers obtain their energy through oxidative phosphorylation (aerobic respiration), and fast-twitch fibers obtain their energy mainly through glycolysis (Chapter 7).

Although slow-twitch fibers develop force more slowly, they have greater resistance to fatigue in response to repetitive stimulation—that is, their loss of force over time is less. Their greater endurance is explained by the ability of their mitochondria to supply ATP to muscle fibers by means of aerobic respiration. Slow-twitch fibers have many mitochondria and are well supplied by capillaries. They contain an abundance of **myoglobin,** an oxygen-binding protein related to hemoglobin that facilitates oxygen delivery to the mitochondria (Chapter 39). The iron in myoglobin gives muscles with these fibers a red appearance. Slow-twitch muscle fibers express a chemically "slow" form of myosin with a relatively low rate of ATP hydrolysis, limiting their speed of cross-bridge cycling and force development.

White fast-twitch muscle fibers have fewer mitochondria and capillaries, contain little myoglobin, and rely heavily on anaerobic glycolysis (lactic acid fermentation) to produce ATP. Fast-twitch fibers express a chemically "fast" form of myosin with a high rate of ATP hydrolysis, favoring rapid force development and movement. However, fast-twitch fibers fatigue quickly because of the drop in pH caused by the accumulation of lactic acid. Because white fibers are larger than red fibers, they generate more force. In general, most skeletal muscles are composed of mixed populations of red and white fiber types, providing a broad range of contractile function.

Athletes who have more oxidative slow-twitch fibers excel at endurance and long-distance competitions. In contrast, athletes who have larger numbers of fast-twitch fibers perform better in sprint races and weight-lifting competitions. Because the distribution and properties of muscle fiber types are strongly inherited, it is largely true that champions are born and not made. Nevertheless, aerobic training can significantly increase the energy output through oxidative phosphorylation of an individual's muscle fibers, just as weight lifting and speed training can enhance an individual's strength and sprinting ability. Contrary to popular belief that weight lifting builds muscle by straining the muscle to build new fibers, it actually builds muscle by increasing the size of existing muscle fibers through the synthesis of additional myosin and actin filaments. This synthesis increases the cross-sectional area of the fiber (cell) and thus its force-generating capacity.

**? CASE 7 Predator–Prey: A Game of Life and Death How do different types of muscle fiber affect the speed of predators and prey?**

We have just seen that slow-twitch muscle fibers provide endurance for sustained activity, whereas larger fast-twitch fibers produce greater speed and strength but fatigue quickly. The relative distribution of these fiber types in the muscles of different animals is influenced mainly by genetic heritage. This affects the locomotive behavior and capacities of both predator and prey animals. Whereas cheetahs have exceptional speed, their muscles tire quickly. As their prey, antelope have muscles that enable endurance long-distance running and contract rapidly for maneuvering.

Highly aerobic animals such as dogs and antelope have large fractions of slow-twitch fibers *and* specialized fast-twitch muscle fibers that also have a high oxidative capacity in addition to using glycolysis for rapid ATP synthesis. These muscle fibers allow the animals to move for extended periods of time at relatively fast speeds. In contrast, cats have mainly fast-twitch fibers, enabling them to sprint and pounce quickly. Nevertheless, cats and other animals use particular muscles with high concentrations of slow-twitch fibers (these are the dark-staining fibers shown in Fig. 37.14a) for slower postural movements and stealth. Animals such as sloths and lorises that move slowly have mostly oxidative slow-twitch fibers and few glycolytic fast-twitch fibers.

More generally, animals with high body temperatures and high metabolic rates (birds and mammals) have larger numbers of oxidative fibers compared with animals that have lower body temperatures and lower metabolic rates (reptiles). As a result, lizards use glycolytic fast-twitch fibers to sprint for brief periods but must then recover from the acid buildup produced by their muscles' anaerobic activity in longer periods of rest (Chapter 40). In contrast, birds and mammals can sustain activity over longer time periods because they have oxidative muscle fibers supported by aerobic ATP supply.

Most fish have a narrow band of slow-twitch red muscle fibers that runs beneath their skin, which they use for slow steady swimming. Tuna and some sharks have evolved deeper regions of red muscle that they can keep warm (**Fig. 37.14b**), enabling their muscles to contract longer and faster to enhance their swimming. When escaping from a predator or attacking prey, fish recruit their larger fast-twitch white trunk musculature for rapid swimming.

The evolutionary arms race between predators and prey has resulted in selection for particular muscle fiber types and other musculoskeletal specializations for rapid movement and maneuvering. Features of the skeleton that contribute to these specializations are discussed next.

## 37.3 ANIMAL SKELETONS

Most animal skeletons provide a rigid set of elements that articulate, that is, they meet at joints, to transmit muscle forces for movement and body support. They also enable many animals to manipulate their environment by digging burrows, building nests, or constructing webs to catch food. Animals have evolved three types of skeletal system: **hydrostatic skeletons, endoskeletons,** and **exoskeletons.** All three types of skeleton use rigid elements to resist the pull of antagonist sets of muscles.

### Hydrostatic skeletons support animals by muscles that act on a fluid-filled cavity.

Hydrostatic skeletons evolved early in multicellular animals with the first cnidarians (jellyfish and anemones) approximately 600 million years ago. They are found in nearly all multicellular animals as well as in many vascular plants. In animals that depend on a hydrostatic skeleton, fluid contained within a body cavity serves as the supportive component of the skeleton. Muscles exert pressure against the fluid to produce movement.

Two sets of muscles surround the fluid, controlling the width and length of the body cavity and, in many cases, of the whole animal (**Fig. 37.15a**). Circular muscles reduce the diameter of the body cavity, and longitudinal muscles reduce its length. A sea anemone can bend its body in various directions to feed or to resist water currents by closing its mouth to keep its fluid volume constant and contracting its circular and longitudinal muscles. Sea anemones extend themselves into the water column by contracting their circular muscles and relaxing their longitudinal muscles. Controlling the amount of fluid in their body cavity also allows sea anemones to adjust their shape. When threatened,

**FIG. 37.15 Diverse examples of hydrostatic skeletal support.** (a) Sea anemone; (b) earthworm; (c) intervertebral discs; (d) articular cartilage.

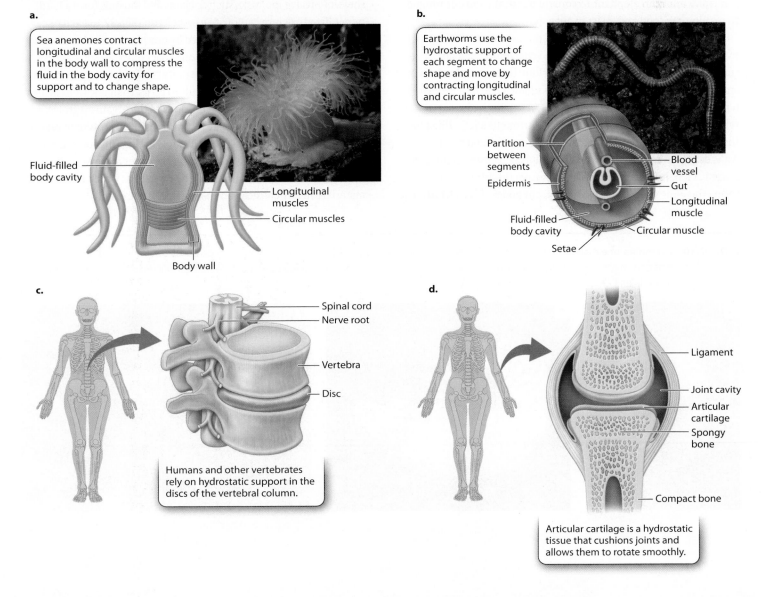

a.

Sea anemones contract longitudinal and circular muscles in the body wall to compress the fluid in the body cavity for support and to change shape.

Fluid-filled body cavity

Longitudinal muscles

Circular muscles

Body wall

b.

Earthworms use the hydrostatic support of each segment to change shape and move by contracting longitudinal and circular muscles.

Partition between segments

Epidermis

Blood vessel

Gut

Longitudinal muscle

Fluid-filled body cavity

Circular muscle

Setae

c.

Spinal cord

Nerve root

Vertebra

Disc

Humans and other vertebrates rely on hydrostatic support in the discs of the vertebral column.

d.

Ligament

Joint cavity

Articular cartilage

Spongy bone

Compact bone

Articular cartilage is a hydrostatic tissue that cushions joints and allows them to rotate smoothly.

a sea anemone rapidly retracts by contracting its longitudinal muscles and allowing water to escape from its body cavity.

Similarly, earthworms burrow through the soil by moving a series of hydrostatic segments along their body (**Fig. 37.15b**). By contracting longitudinal muscles in a few segments, they shorten those segments and also widen them to anchor those segments against the soil. They then extend intervening segments between anchor points by contracting circular muscles, causing the segments to lengthen and move the animal's body through the soil.

Because of their simple but effective organization, hydrostatic skeletons are well adapted for many uses. In groups such as squid and jellyfish, the hydrostatic skeleton has become specialized for jet-propelled locomotion. Large circular muscles contract to eject fluid from the body cavity, propelling the animal in the opposite direction. The muscles of other mollusks also employ a hydrostatic skeleton, allowing a clam, for example, to burrow into the sediment with a muscular "foot" and an octopus to adeptly manipulate its flexible tentacles. (Clams also have shells that form an exoskeleton to protect the body.) These same principles allow an elephant to control its trunk, and our tongue to manipulate food and assist in speech.

Even vertebrate animals with rigid endoskeletons have hydrostatic elements for flexibility and cushioning of loads transmitted by the skeleton. These include **intervertebral discs** and **cartilage.** Intervertebral discs are sandwiched between the bony vertebrae of the backbone (**Fig. 37.15c**). Each disc has a wall reinforced with connective tissue that surrounds a jelly-like fluid. Intervertebral discs enable the backbone to twist and bend. Disc walls damaged by repeated stress can rupture, causing significant lower back pain. Articular cartilage, another fluid-filled tissue, forms the joint surfaces between adjacent bones (**Fig. 37.15d**). The fluid within the cartilage resists the pressure between the ends of articulating bones, cushioning forces transmitted across the joint.

## Exoskeletons provide hard external support and protection.

An exoskeleton is a rigid skeleton that lies external to the animal's soft tissues. The first mineralized skeletons arose with sponges about 650 million years ago, but more successful exoskeletons subsequently evolved in other invertebrate groups adapted for life in aquatic and terrestrial environments (**Fig. 37.16**). Arthropod exoskeletons, which first evolved in aquatic crustaceans, protect animals from desiccation and physical insults. Consequently, exoskeletons were key to the success and diversification of insects in terrestrial environments about 450 million years ago. Because they are on the exterior of the animal, exoskeletons provide hard external support and protection, allowing muscles to attach from the inside. However, a main disadvantage is that exoskeletons limit growth.

Different invertebrates have different kinds of exoskeleton. The shells of bivalve mollusks such as clams and mussels form a hard, mineralized calcium carbonate exoskeleton (Fig. 37.16a and b). The calcium carbonate is reinforced by proteins and is therefore an example of a composite material, one that combines substances with different properties. Because of its composite nature, the shell is less brittle and more difficult to break than if it were made of just one substance. The shell expands as the organism grows, as epidermal cells deposit new layers of mineralized protein in regular geometric patterns. You can see the growth rings if you examine a mollusk shell. Although some marine bivalves achieve large size over many years, growth is limited by the rate at which new skeletal material can be deposited on the outside of the shell.

**FIG. 37.16 Examples of exoskeletons.** The shells of a bivalve mollusk (a) and nautilus (b), made of calcium carbonate, surround and protect internal soft body parts. The exoskeleton of arthropods, such as a grasshopper (c), is made of stiff cuticle that surrounds internal muscles and tendons and other internal body parts.

Arthropods, such as insects, have more complex exoskeletons formed of a **cuticle** that covers their entire body (Fig. 37.16c). The cuticle of insects is composed mainly of **chitin,** a nitrogen-containing polysaccharide. Marine crustaceans, such as crabs and lobsters, also incorporate calcium carbonate in their cuticle. The calcium carbonate makes the exoskeleton hard and stiff, helping to protect them from predators, much like the hardened cuticle of terrestrial insects. When initially formed, the cuticle of arthropods is soft, so the animal can grow before the cuticle is hardened. The cuticle remains flexible at joints, allowing motion between body segments. Mature cuticle consists of two layers. A thin outer, waxy layer minimizes water loss. Water loss is especially dangerous for arthropods, given their small size. The much thicker inner layer, whether flexible or stiff, is tough, making it hard to break.

Because a rigid exoskeleton restricts growth, arthropods shed their cuticle at intervals, a process termed **molting.** Molting allows arthropods to expand and grow before forming a new rigid exoskeleton. While offering several protective benefits, exoskeletons pose risks as well. Animals are vulnerable when the newly formed exoskeleton has not yet hardened. Exoskeletons are also hard to repair. If a skeleton is damaged while growing, the animal must produce an entirely new one. Because the exoskeleton is a thin-walled structure, it is prone to breaking if its surface area is very large. Consequently, the imperviousness of the monstrous insects depicted in science-fiction films is improbable—they would easily break from a blow to their exoskeleton.

### The rigid bones of vertebrate endoskeletons are jointed for motion and can be repaired if damaged.

Endoskeletons first evolved in vertebrate animals about 500 million years ago, during the Cambrian explosion. This type of skeleton lies internal to most of an animal's soft tissues. In contrast to exoskeletons, the bony endoskeletons of vertebrate animals can grow extensively and, when broken, can be repaired. Endoskeletons also provide protection for internal organs, such as the brain, lungs, and heart.

**FIG. 37.17** The vertebrate skeleton, showing axial and appendicular regions.

**a.** Human

**b.** Cat

Axial skeleton

Appendicular skeleton

The bones of vertebrate skeletons consist of a variety of tubular, rodlike, and platelike elements that form a scaffold to which the muscles attach. Because bones are rigid and hard, there are joints between adjacent bones to enable movement. Muscles attach to the skeleton by connective tissue and specialized **tendons** made of collagen. Tendons transmit muscle forces, allowing the forces to be redirected and transmitted over a wide range of joint motion. Tendons, like a spring, also store and recover elastic energy.

The vertebrate skeleton can be separated into **axial** and **appendicular** regions (**Fig. 37.17**). The axial skeleton consists of the skull and jaws of the head, the vertebrae of the spinal column, and the ribs. Bones of the skull protect the brain and sense organs of the head. The appendicular skeleton consists of the bones of the limbs, including the shoulder and pelvis. The axial and appendicular regions reflect the evolutionary ancestry of vertebrates. The axial skeleton formed first, to protect the head and provide support and movement of the animal's body by bending its body axis. As a result, the axial skeleton is the main skeletal component of fishes. The appendicular skeleton comprises the pectoral and pelvic fins of fishes, which were elaborated into limbs when vertebrate animals first evolved to live on land. In a special group of lobed-fin fishes, the fin bones became adapted for an amphibious lifestyle, evolving into the limb bones that terrestrial vertebrates use for support against gravity and for movement on land.

Skeletal tissues have relatively few cells for their volume, similar to invertebrate exoskeletons. These tissues consist mostly of an extensive **extracellular matrix** that is secreted by specialized cells, forming a connective tissue external to the cells (Chapter 10). Three main extracellular tissues are produced in the formation of the vertebrate skeleton: bone, tooth enamel, and cartilage.

Bone tissue–forming cells called **osteoblasts** synthesize and secrete calcium phosphate as **hydroxyapatite** mineral crystals in close association with the protein **collagen.** Generally, bone consists of two-thirds hydroxyapatite mineral and one-third type I collagen protein. The combination of mineral and protein makes bone a composite material, like exoskeletons. Bone is rigid (because of the properties of the mineral), but also hard to break because it absorbs much energy before fracture (because of the properties of the collagen). The genetic bone disorder osteogenesis imperfecta leads to bones that are brittle and fragile because the osteoblasts produce insufficient collagen in the extracellular matrix. Bone gradually becomes more mineralized and brittle as a normal process of aging. Sharks and other elasmobranchs (the cartilaginous fishes) evolved unusual skeletons of calcified cartilage rather than mineralized bone tissue to achieve rigid mechanical support.

Articular cartilage located at the joint surfaces of a bone forms a gel-like matrix that resists fluid being squeezed out when forces press on the cartilage during movement. The cartilage is reinforced by type II collagen fibers, allowing the cartilage to cushion and distribute loads as the joint moves. Articular cartilage consists of 70% water, 15% large molecules called proteoglycans, and the remaining 15% collagen. Cartilage may degenerate for a variety of reasons, including rheumatoid arthritis and repetitive physical injury, but in general it lasts over an individual's lifetime.

→   **Quick Check 4** Glass is strong and rigid but can be easily broken. In contrast, a bone is strong and rigid but hard to break. How can this be?

## 37.4 VERTEBRATE SKELETONS

Although both vertebrate skeletons and invertebrate exoskeletons are rigid and have flexible joints, certain features distinguish them. Here, we explore the biology of vertebrate skeletons, focusing on how bones develop and grow and how they are repaired and remodeled.

### Vertebrate bones form by intramembranous and endochondral ossification.

Vertebrate bones form and develop by two distinct processes. Bones of the skull and the ribs form when precursor cells differentiate into osteoblasts that immediately begin producing bone as a skeletal sheet or membrane. These bones are often referred to as membranous bones.

Most other bones of the body—the vertebrae, shoulder, pelvis, and limb bones—are formed as cartilage first (**Fig. 37.18**). The precursor cells initially become cartilage-producing cells, known as chondroblasts, that in the embryo grow into cartilage models of most bones of the developing body (Fig. 37.18a). For example, at approximately 9 weeks' development, a human fetal hand consists mainly of this cartilage model (Fig. 37.18b). This embryonic cartilage is the same as the articular cartilage

**FIG. 37.18 Endochondral ossification.** (a) Most bones other than the skull undergo endochondral ossification, which involves the initial formation of a cartilage model of the bone, followed by vascularization and the transformation into bone tissue by osteoblasts. (b) Cartilage model of a human fetal hand, showing initial sites of ossification at about 9 weeks of embryonic development.

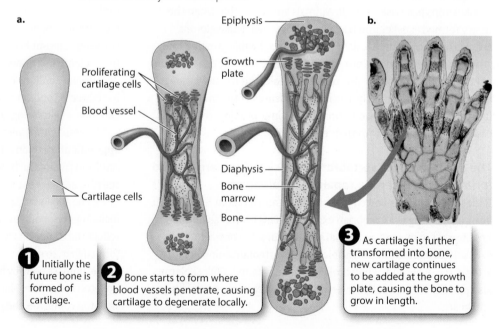

a.

**1** Initially the future bone is formed of cartilage.

**2** Bone starts to form where blood vessels penetrate, causing cartilage to degenerate locally.

**3** As cartilage is further transformed into bone, new cartilage continues to be added at the growth plate, causing the bone to grow in length.

Cartilage cells

Proliferating cartilage cells

Blood vessel

Epiphysis

Growth plate

Diaphysis

Bone marrow

Bone

b.

**FIG. 37.19  Compact and spongy bone.**

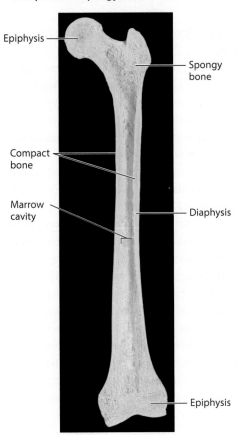

Epiphysis

Spongy bone

Compact bone

Marrow cavity

Diaphysis

Epiphysis

that remains at the ends of bones to form the joint surfaces in the growing and mature animal. Because cartilage is a pliable fluid-filled matrix, it grows by expansion within the structure as well as at the surface, enabling the fetal skeleton to grow rapidly. As the fetus grows and its skeleton matures, the cartilage model of the bone is invaded by blood vessels. The presence of blood vessels triggers the transformation of cartilage into hard, mineralized bone.

The cartilage template of a bone is replaced by one of two main types of bone tissue (**Fig. 37.19**). Both types contribute to the overall form of many bones in the body. **Compact bone** forms the walls of the bone's shaft. It consists mainly of dense mineralized bone tissue containing bone cells known as **osteocytes** and the network of blood vessels supplying them with oxygen and nutrients.

A second type of bone, called **spongy bone,** consists of small plates and rods known as **trabeculae** with spaces between them. Spongy bone is found in the ends of limb bones and within the vertebrae. It reduces the weight of bone in these regions, but it is also stiff so that forces are transmitted

effectively from the joint to the bone shaft. **Bone marrow,** a fatty tissue found between trabeculae and also within the bone's central cavity, contains many important cell populations, including blood-forming cells, other stem cells (Chapter 20), and immune system cells (Chapter 43), as well as fat cells.

Bones can grow in length as well as in width. Growth in length occurs at a **growth plate** (see Fig. 37.18)**,** a region of cartilage between a middle region, called the **diaphysis,** and the end, called the **epiphysis**. Blood vessels invade the central diaphysis region, as well as each epiphysis, triggering the transformation of cartilage into bone. The growth plate is left in between. The growth plate adds new cartilage toward the bone's diaphysis, enabling bone length to continue to increase after birth (see Fig. 37.18). At maturity, the growth plates fuse as the remaining cartilage is replaced with bone, preventing further growth in length. The fusion of growth plates is typical of most mammals and birds. In amphibians and reptiles, the growth plates often do not fuse and growth may continue at a slower pace over much of their lifetime.

Limb bones grow in diameter when osteoblasts deposit new bone on the bone's external surface. At the same time, bone is removed from the inner surface, expanding the marrow cavity. Bone removal is slower than bone growth, thickening the walls during growth. Bone is removed from the marrow cavity by a group of cells called **osteoclasts** that secrete digestive enzymes and acid to dissolve the calcium mineral and collagen. These dissolved compounds are reabsorbed and recycled for bone formation in other regions of the skeleton. Through this process, the vertebrate skeleton serves as an important store of calcium and phosphate ions. For example, in female birds, calcium removed from the skeleton is regularly used to form the eggshell.

A great advantage of endoskeletons compared with exoskeletons is that a vertebrate's skeleton can be repaired if damaged by osteoblasts and osteoclasts forming and removing mineralized tissue in particular regions of the bone. Generally, physically active younger adults have thicker bones than less-active individuals. However, in older humans, bone tissue is gradually lost as osteoclasts remove more bone than is produced by osteoblasts. Bone loss is particularly severe in women after menopause, in part because of hormone shifts, and can lead to osteoporosis, a condition that significantly increases the risk of bone fracture.

### Joint shape determines range of motion and skeletal muscle organization.

The shapes of the bone surfaces that meet at a joint determine the range of motion at that joint. Joints range from simple **hinge joints** that allow one axis of rotation to **ball-and-socket**

**FIG. 37.20 Hinge and ball-and-socket joints.** The elbow and knee are examples of hinge joints. The shoulder and hip are examples of ball-and-socket joints.

**Ball-and-socket**

Highly mobile ball-and-socket joints allow motion along all three axes.

**Hinge**

Hinge joints limit motion to one primary axis.

muscles is needed to control the movements of the arm at the shoulder joint or the leg at the hip joint.

### Muscles exert forces by skeletal levers to produce joint motion.

By serving as a rigid set of levers, the skeleton enables muscles to transmit forces that cause joint rotation. Analogous to the gears of a bicycle, muscles that attach farther from a joint's axis of rotation produce slower but stronger movements, whereas muscles that attach closer to a joint produce rapid but weaker movements. This is why you open a door by pulling on a doorknob located opposite from its hinges. The doorknob's location increases the lever distance, making it easier to pull the door open than if it were placed next to the hinge.

The strength of rotational movement, or torque, is determined by the product of a muscle's force ($F$) and perpendicular lever distance ($r$) from the axis of joint rotation: $F \times r$. Muscles produce joint torques to counteract torques that are applied by external forces, such as a weight that is carried by your hand. In **Fig. 37.21**, the axis of rotation is the elbow, and contraction of the biceps muscle causes the arm to flex at the elbow. The strength of elbow rotation is determined by the biceps muscle's force ($F$) and the lever distance between the elbow joint and the site of muscle attachment, where the muscle pulls on the arm bone ($r$). Muscles produce stronger rotational movements either by exerting a larger force or by having a longer lever distance between the joint

**joints** that allow rotation in three axes (**Fig. 37.20**). The human elbow joint and the ankle joint of a dog are examples of hinge joints. The shoulder and hip joints are examples of ball-and-socket joints, which allow the widest range of motion, as when you throw or kick a ball. The joints at the base of each finger are intermediate: They allow you to flex and extend your fingers as well as spread them laterally or move them together when making a fist or grasping different-sized objects.

Joints with a broader range of motion are generally less stable. The shoulder is the most mobile joint in the human body, but it is also the most often dislocated or injured. In contrast, the ankle joint of dogs, horses, and other animals is a stable hinge joint that is unlikely to be dislocated, but its range of motion is limited to flexion and extension. Because muscles are arranged as paired sets of antagonists to move a joint in opposing directions, hinge joints are controlled by as few as two antagonist muscles (generally referred to as a flexor and an extensor). In contrast, ball-and-socket joints minimally have three sets of muscle antagonists to control motion in three different planes. As a result, a more complex organization of

**FIG. 37.21 Torque.** The torque ($F \times r$) depends on the muscle's force ($F$) and the distance the force acts from the joint's axis of rotation ($r$), referred to as the muscle's lever arm. This torque must counteract the opposing torque of the weight and its lever arm ($L$).

$F$ ↑ 35 kg

Because the distance between the joint and the muscle attachment site is short, and the distance between the joint and the load is long, the arm can lift the weight quickly but needs more muscle force than if the muscle attached further from the joint.

Site of muscle attachment

Joint axis of rotation

Muscle lever arm

5 cm

$r$

**5 kg**

35 cm

5 kg

$L$

functional state of its body. Changes in light or temperature, the threat of a predator, or the presence of a potential mate stimulate responses of an animal's nervous system and endocrine system. Endocrine responses are commonly triggered by sensory signals received by the nervous system that are relayed to the endocrine system. The endocrine system, with its slower and more prolonged signaling, reinforces physiological changes in the animal's body that better suit the environmental cues received by its nervous system.

When a gazelle sees or smells a predator, its endocrine system helps to ready its body for rapid escape. When a female cardinal sees a bright red, singing male early in spring, endocrine signals transmitted within the female cardinal's brain initiate changes in its reproductive organs and increase its behavioral responsiveness, encouraging it to select the male as its mate. And when particular smells stimulate a sense of hunger and digestive function in animals, such as dogs and humans, the animals are attracted to food. In each of these cases, the endocrine system helps an organism respond appropriately to environmental cues.

### The endocrine system is involved in growth and development.

The endocrine system plays important roles in growth and development, which require broad changes in many different organ systems. The release of circulating hormones accomplishes these changes, regulating how animals develop and grow. For

example, we discuss in Chapter 42 the importance of the sex hormones estrogen and testosterone in determining female or male sexual characteristics during embryonic development and in sexual development during puberty.

As another example, consider the well-studied endocrine system of insects, which regulates their growth and development. Insects produce a rigid exoskeleton (Chapter 37), which must be periodically shed and a larger new one made to enable the insect to grow (**Fig. 38.1**). Shedding of the exoskeleton is referred to as **molting.** Each molting produces a new larval stage. In some insects, such as moths, the animal's body also changes dramatically, undergoing **metamorphosis** at key stages in development (Fig. 38.1a). In other insects, such as grasshoppers and crickets, the animal grows larger after each molting, but with little change in body form (Fig. 38.1b).

Molting and metamorphosis are regulated by hormones released from tissues in the insect's head. British physiologist Sir Vincent Wigglesworth studied their effects on the growth and development of the blood-sucking insect *Rhodnius*. Normally, *Rhodnius* goes through five juvenile stages before developing into its final adult body form. Like grasshoppers and crickets, it does not undergo metamorphosis. Each of the developmental stages is triggered by a blood meal. Wigglesworth showed that he could prevent molting if the head of the insect were removed shortly after a blood meal (**Fig. 38.2**). However, if the insect takes a blood meal and the head is removed a week later, the animal's body

**FIG. 38.1** Growth and development in (a) moths and caterpillars and (b) grasshoppers.

a.

Moths and caterpillars go through several larval stages before undergoing metamorphosis into an adult form.

b.

The rigid exoskeleton of many insects, such as grasshoppers, is shed periodically, allowing the animal to grow before a new exoskeleton is produced and hardened.

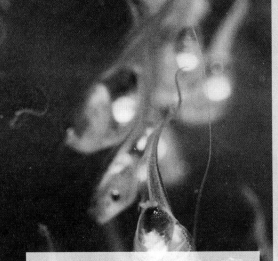

## CHAPTER 38

# ANIMAL ENDOCRINE SYSTEMS

## Core Concepts

**38.1** Animal endocrine systems release chemical signals called hormones into the bloodstream.These signals respond to the environment, regulate growth and development, and maintain homeostasis.

**38.2** Hormones achieve specificity by binding to receptors on or inside their target cells. Their signals are amplified to exert strong effects on their target cells.

**38.3** In vertebrate animals, the hypothalamus and the pituitary gland control and integrate diverse bodily functions and behaviors.

**38.4** Chemical communication can also occur locally between neighboring cells or, through the release of pheromones, between individuals, coordinating social interactions.

Animals, like all organisms, respond to their environment, regulating their metabolism to meet changes in energy required by different activities and maintaining stable internal bodily functions. We have seen how the nervous system allows an animal to sense and respond to its environment, coordinating the animal's movement and behavior (Chapters 35–37). The nervous system also works closely with the **endocrine system** to regulate an animal's internal physiological functions. This coordination is accomplished by signals sent from the nervous system to the endocrine system, which distributes chemical signals throughout the body.

The signals communicated by the nervous and endocrine systems differ greatly in their modes of transmission and the times over which they act. The nervous system sends signals rapidly by action potentials running along nerve axons, and communication between adjacent nerve cells occurs by means of neurotransmitters in proximity at synapses. The endocrine system, on the other hand, relies on cells and glands that secrete chemical signals called **hormones,** which are released into the bloodstream and circulate throughout the body. As a result, endocrine communication is generally slower and more prolonged than the rapid and brief signals transmitted by nerve cells. Hormones exert specific effects on particular target cells within the body by binding to receptors on or in target cells. The downstream influence of hormones is amplified in a series of steps that occur along a particular endocrine pathway so that a small amount of hormone can have dramatic effects in the body.

In this chapter, we explore the properties and actions of hormones that help to coordinate and maintain a broad set of physiological functions of the organism. In particular, we examine the mechanisms by which hormones trigger cellular responses in their target organs. Because most hormones are evolutionarily conserved, their chemical organization is similar across a diverse array of animals. However, their functions often evolve rapidly, enabling hormones to serve new and broader roles.

## 38.1 AN OVERVIEW OF ENDOCRINE FUNCTION

The word "hormone" often popularly connotes teenagers and the changes that occur to their bodies as they grow and mature. Hormones, as we will see, do have key functions in animal growth and development—but that is only one of their roles. Hormones play diverse physiological roles in the body, including regulating an organism's response to the environment and helping to maintain stable physiological conditions within cells or within the animal as a whole. We start by highlighting these functions before addressing the molecular mechanism of how hormones work.

### The endocrine system helps to regulate an organism's response to its environment.

Organisms face constant changes in their environment. These changes often present physiological challenges or stresses that the organism must respond to by altering the

## Self-Assessment

1. Diagram a sarcomere, showing the basic organization of thin (actin) and thick (myosin) filaments. Indicate on your diagram the regions where myosin cross-bridges form with actin.

2. Describe the sequence of events that occurs when an action potential arrives at a muscle fiber's motor endplate, causing the muscle to be depolarized.

3. Draw a graph of the isometric force–length relationship of striated muscle, indicating where maximal overlap between actin and myosin filaments occurs.

4. Draw a graph of the force–shortening velocity relationship of striated muscle.

5. Compare the force produced by a muscle over time when it is stimulated by a single twitch stimulus with the force produced by multiple stimuli at low frequency, and then with the force produced when the stimulation frequency is increased.

6. Name the two basic structural elements common to all animal skeletons.

7. Compare and contrast three features of an exoskeleton and an endoskeleton.

8. Identify the two primary components of bone tissue and explain how each of these contributes to a bone's strength, stiffness, and resistance to fracture.

9. Diagram a limb bone, such as the tibia, showing the regions of articular cartilage, spongy trabecular bone, compact bone tissue, and the marrow cavity.

Do you understand the chapter's Core Concepts? Log in to check your answers to the Self-Assessment questions, then practice what you've learned and reinforce this chapter's concepts by working through the problems and multimedia tutorials provided there.

**www.biologyhowlifeworks.com**

# Core Concepts Summary

**37.1 MUSCLES ARE BIOLOGICAL MOTORS COMPOSED OF ACTIN AND MYOSIN THAT GENERATE FORCE AND PRODUCE MOVEMENT FOR BODY SUPPORT, LOCOMOTION, AND CONTROL OF THE BODY'S INTERNAL PHYSIOLOGICAL FUNCTIONS.**

There are two main types of muscle: striated (skeletal and cardiac) and smooth. page 37-1

Skeletal and cardiac muscle fibers appear striated when viewed under the microscope because of the regular spacing of sarcomeres along their length. page 37-2

Smooth muscle fibers, which regulate airflow for breathing, blood flow through arteries, and the passage of food through the gut, lack regular sarcomere organization and appear smooth when viewed under the microscope. page 37-2

Skeletal muscle fibers are long thin cells composed of parallel sets of myofibrils built up from smaller parallel arrays of actin and myosin filaments. page 37-2

The sarcomere is the basic contractile unit of a skeletal muscle. Sarcomeres are arranged in series along the length of a myofibril. page 37-2

Muscles change length and produce force by the formation of actin–myosin cross-bridges, causing the sliding of myosin and actin filaments relative to each other. page 37-5

Muscles are stimulated by motor neurons at the fiber's motor endplate, leading to depolarization of the muscle cell that triggers the release of $Ca^{2+}$ ions from the sarcoplasmic reticulum. $Ca^{2+}$ binds troponin, which moves tropomyosin off the myosin-binding sites on actin, allowing myosin heads to form cross-bridges with actin. page 37-6

Excitation–contraction coupling is the process in which depolarization of the muscle cell leads to its shortening. page 37-6

**37.2 MUSCLE FORCE DEPENDS ON MUSCLE SIZE, DEGREE OF ACTIN–MYOSIN OVERLAP, CONTRACTILE VELOCITY, AND STIMULATION RATE.**

Muscles exert their highest force when actin–myosin filaments have maximal overlap, allowing the most cross-bridges to form. page 37-8

Muscles exert more force when they contract at slow velocities compared to when they contract at high velocities. page 37-9

Because muscles can transmit force only by pulling on the skeleton, they are arranged as antagonist pairs to produce reciprocal motions of a joint or limb. page 37-10

In vertebrates, a motor unit consists of a single motor neuron and the muscle fibers (cells) it innervates. page 37-11

Muscle force is increased by increasing the motor neuron firing rate and, in vertebrates, by the number of motor units that are activated. page 37-11

Vertebrate muscles have two types of fiber: red slow-twitch fibers that contract slowly over longer time periods and white fast-twitch fibers that contract rapidly but fatigue quickly. page 37-11

**37.3 HYDROSTATIC SKELETONS, EXOSKELETONS, OR ENDOSKELETONS PROVIDE ANIMALS WITH MECHANICAL SUPPORT AND PROTECTION.**

The rigid element of a hydrostatic skeleton is an incompressible fluid within a body cavity, used to support the body and change shape. page 37-13

Invertebrate exoskeletons form an external rigid support system, which protects the animal and limits water loss, but also limits growth and repair. page 37-14

Vertebrates have a bony endoskeleton that provides rigid support and protection of body organs, and that can grow and be repaired. page 37-15

The vertebrate endoskeleton is organized into axial (central) and appendicular (limb) components. page 37-15

Vertebrate endoskeletons consist of bone and cartilage. page 37-16

Bone is a composite tissue that consists of calcium phosphate mineral and type I collagen and that is rigid and hard to break. page 37-16

Cartilage is a fluid-based gel reinforced by type II collagen and proteoglycans, providing cushioning support at joint surfaces. page 37-16

**37.4 VERTEBRATE ENDOSKELETONS ALLOW FOR GROWTH AND REPAIR AND TRANSMIT MUSCLE FORCES ACROSS JOINTS.**

Vertebrate bones develop by one of two processes: either directly or by way of a cartilage model (endochondral ossification). page 37-16

Cartilage forms much of the embryonic skeleton and remains as growth plates within the bone to provide rapid growth during adolescence, as well as forming the bone's joint (articular) surfaces. page 37-16

Bone has two basic structures: Compact bone forms the solid walls of a bone's shaft, and spongy bone forms a mesh that supports the cartilage at the bone's ends. page 37-17

Bone formation by osteoblasts and removal by osteoclasts is a continuing process of growth, shape change, and repair. page 37-17

The shape of a bone's joint surfaces largely determines the range of motion and stability of a joint. page 37-17

Muscles produce joint movements by transmitting forces by means of rigid skeletal levers. page 37-18

**FIG. 37.22 A trade-off between force and velocity in animal skeletons.** The armadillo forelimb skeleton is adapted for digging, whereas the horse forelimb is adapted for speed and weight support.

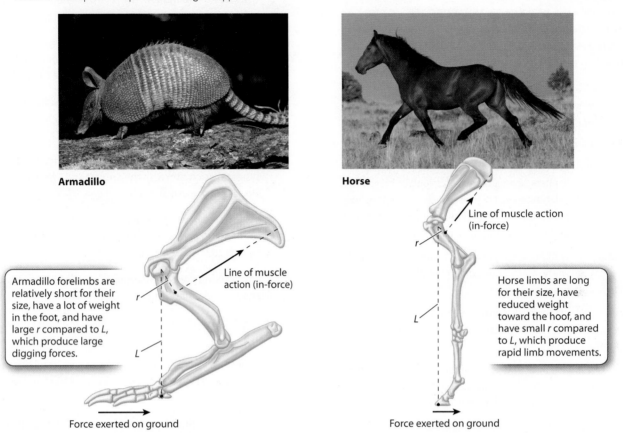

Armadillo

Armadillo forelimbs are relatively short for their size, have a lot of weight in the foot, and have large $r$ compared to $L$, which produce large digging forces.

Line of muscle action (in-force)

Force exerted on ground

Horse

Line of muscle action (in-force)

Horse limbs are long for their size, have reduced weight toward the hoof, and have small $r$ compared to $L$, which produce rapid limb movements.

Force exerted on ground

and muscle attachment site. Muscles that transmit force closer to the joint (smaller lever distance) result in less rotational strength but can produce large ranges of movement compared to muscles that attach farther from the joint.

As a result, the trade-off of force versus velocity that we observed earlier as a basic contractile property of a muscle also applies to the way muscles pull on the skeleton to cause joint movement. Muscles that transmit large forces for body movement and weight support, such as the calf muscles that extend your ankle joint, typically have large leverage for strong joint rotation. The heel of your foot gives the Achilles tendon of your calf muscles this greater leverage.

Animal joints located a long distance from the muscle attachment site (high muscle leverage) and a shorter distance from where the output force is exerted produce strong but slow movements. Moles, armadillos, and spade-foot toads are

examples of digging animals that have specialized muscle–joint arrangements like this (**Fig. 37.22**). In contrast, joints located a short distance from the muscle attachment site (low muscle leverage) and a longer distance from where the output force is exerted produce faster movements but with less force. These are found in animals adapted for high-speed running, such as antelope, cheetahs, and horses.

Larger, faster animals also increase their economy of movement by reducing the mass of muscles and bones toward the end of their limbs. This lowers the energy cost of swinging the limb, which antelope rely on for fast economical locomotion to escape predators. Predators generally have heavier limbs that require larger muscles to move but which make them stronger.

**FIG. 38.2**

# How are growth and development controlled in insects?

**BACKGROUND** During the 1930s, British physiologist Sir Vincent Wigglesworth studied how a blood meal taken by the bug *Rhodnius* triggered its molting and growth. This work pioneered the discovery of hormonal substances that stimulated its growth. *Rhodnius* goes through five successive larval stages (called nymphs) before becoming a winged adult, as shown in Figure 38.2a. Each developmental step is triggered by a blood meal.

**HYPOTHESIS** Wigglesworth hypothesized that a substance (specifically, a hormone) that diffuses from the head triggers molt in *Rhodnius*.

**EXPERIMENT** Wigglesworth decapitated juvenile bugs at different intervals of time after a blood meal and observed whether or not molting occurred. (Decapitation does not kill the insect.)

**RESULTS** Wigglesworth showed that if a bug is decapitated less than an hour after a blood meal, it fails to molt. If a bug is decapitated 1 week after the blood meal, it molts (Fig. 38.2b, Experiment 1). He also found that if he used a fluid tube to join the body of a bug decapitated immediately after a blood meal with the body of one that wasn't decapitated until a week after feeding, the bodies of both bugs molted (Fig. 38.2b, Experiment 2). This experiment demonstrated that the diffusing hormone could trigger molting in the bug that lacked the hormone because of immediate decapitation.

**a.**

**Development of *Rhodnius***

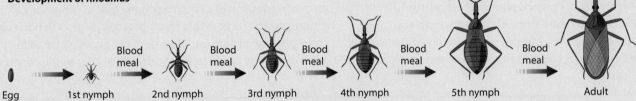

Egg — Blood meal — 1st nymph — Blood meal — 2nd nymph — Blood meal — 3rd nymph — Blood meal — 4th nymph — Blood meal — 5th nymph — Blood meal — Adult

**b.**

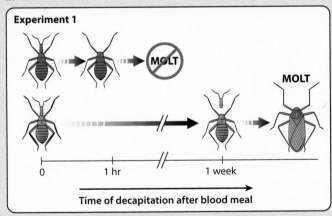

Experiment 1

MOLT (crossed out) · MOLT

0 — 1 hr — 1 week

Time of decapitation after blood meal

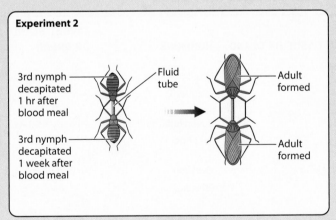

Experiment 2

3rd nymph decapitated 1 hr after blood meal · 3rd nymph decapitated 1 week after blood meal · Fluid tube · Adult formed · Adult formed

**CONCLUSION** A substance diffuses from the head in *Rhodnius* and triggers the molting process.

**FOLLOW-UP WORK** Subsequent studies by the German endocrinologist Alfred Kuhn confirmed the result in experiments performed on moth caterpillars. More recent molecular analyses performed by Japanese molecular endocrinologists Nagasawa and colleagues (1984) and Kawakami and colleagues (1990) isolated and sequenced the structure of the hormone from silkworm moths. It is now known as prothoracicotropic hormone, or PTTH, because it acts on the prothoracic gland of the insect to release the molting hormone, ecdysone.

**SOURCES** Wigglesworth, V.B. 1934. "The Physiology of Ecdysis in *Rhodnius prolixus* (Hemiptera). II. Factors Controlling Moulting and 'Metamorphosis.'"*Quarterly Journal of Microscopical Sciences* 77:191–223; Nagasawa, H., H. Kataoka, A. Isogai, S. Tamura, A. Suzuki, H. Ishizaki, A. Mizoguchi, Y. Fujiwara, A. Suzuki. 1984. "Amino-Terminal Amino Acid Sequence of the Silkworm Prothoracicotropic Hormone: Homology with Insulin." *Science* 226:1344–1345; Kawakami, A., H. Kataoka, T. Oka, A. Mizoguchi, M. Kimura-Kawakami, T. Adachi, M. Iwami, H. Nagasawa, A. Suzuki, H. Ishizaki. 1990. "Molecular cloning of the *Bombyx mori* Prothoracicotropic Hormone." *Science* 247:1333–1335.

undergoes a molt and grows into its adult form. Wigglesworth hypothesized that a diffusing substance from the animal's head controls its molt. In the first case, this substance does not have time to reach the body to trigger a molt, but in the second it does.

The German endocrinologist Alfred Kuhn found a similar result in experiments performed on moth caterpillars. Tying the caterpillar's head off from the rest of the body prevented the body from molting. Wigglesworth and Kuhn considered the diffusing substance to be a brain hormone. Subsequent work showed that cells in a specialized region of the insect brain secrete the peptide now known as prothoracicotropic hormone (PTTH) (**Table 38.1**). In *Rhodnius*, PTTH is released after a blood meal and acts to trigger the animal's molt by stimulating the release of ecdysone from the prothoracic gland.

Two decades later, the German endocrinologist Peter Karlson isolated and purified a second hormone, termed ecdysone. Whereas PTTH triggers molting, ecdysone coordinates the growth and reorganization of body tissues during a molt. Ecdysone is released from a gland when stimulated by PTTH (Table 38.1). Karlson was able to purify a small amount of ecdysone from large numbers of silk moth *Bombyx* caterpillars. He ultimately isolated a mere 25-mg sample from 500 kg of moth larvae! This tiny amount highlights the general principle that relatively few hormone molecules can have a large effect on an organism.

In his studies of *Rhodnius*, Wigglesworth also noted that, whatever the larval stage, the decapitated bug always molted into an adult after its blood meal. By removing just the front brain region of the head, Wigglesworth was able to show that a region in the head close to the brain releases a hormone that normally prevents the earlier larval stages from molting into the adult form (Table 38.1). This hormone, called juvenile hormone, is released in decreasing amounts during each successive larval molt. After the fifth and final stage, the level of juvenile hormone is so low that it no longer blocks maturation, allowing the final molt to undergo metamorphosis into the adult form (**Fig. 38.3**).

These studies reveal how small amounts of hormones released from key glands within the animal's body regulate major stages of growth and changes in body form during metamorphosis. When an animal's body requires coordinated changes in multiple organ systems, hormonal regulation by the endocrine system plays a critical role.

Hormones also regulate growth in humans and other vertebrate animals. Growth hormone controls the growth of the skeleton and many other tissues in the human body. Growth hormone is produced by the **pituitary gland,** which is located beneath the brain. Tumors of the pituitary that cause an overproduction of growth hormone lead to conditions of gigantism, and tumors that result in too little growth hormone cause pituitary dwarfism. The role of the pituitary gland in regulating body function will be described later in the chapter.

**TABLE 38.1**   Major Invertebrate Hormones

| SECRETING TISSUE OR GLAND | HORMONE | TARGET GLAND OR ORGAN | ACTION |
|---|---|---|---|
| Brain | Brain hormone (peptide) | Prothoracic gland | Stimulates release of ecdysone to trigger molt and metamorphosis |
| Corpora allata | Juvenile hormone (peptide) | All body tissues | Inhibits metamorphosis to adult stages; stimulates retention of juvenile characteristics |
| Prothoracic gland | Ecdysone (steroid) | All body tissues | Stimulates molt and, in absence or low levels of juvenile hormone, stimulates metamorphosis |
| Nervous system | Ecdysone (steroid) | All body tissues | Stimulates growth and regeneration in sea anemones, flat worms, nematodes, annelids, snails, and sea stars |
| Brain | Melanocyte-stimulating hormone (peptide) | Chromatophores | Stimulates pigmentation changes in cephalopods (octopus, squid, and cuttlefish) |
| Brain and eye stalks | Chromatotropins (peptides) | Chromatophores | Stimulates pigmentation changes in crustaceans |
| Reproductive gland | Androgen (peptide) | Reproductive tract | Regulates development of testes and male secondary sexual characteristics of crustaceans |

**FIG. 38.3** Hormonal control of growth and development in *Rhodnius*.

## The endocrine system underlies homeostasis.

In addition to helping an animal respond to the environment and regulating growth and development, hormones synchronize and coordinate multiple bodily processes over longer time periods. Consequently, endocrine control of internal body functions is central to **homeostasis,** the maintenance of a steady physiological state within a cell or an organism (Chapters 5 and 35).

Physiological mechanisms that maintain a stable internal environment are necessary because without them, changing environmental conditions would lead to dangerous shifts in an animal's physiological function. For example, an animal's body weight depends on the regulation of its energy intake relative to energy expenditure. Disruption of hormones that regulate appetite and food intake can lead to food disorders that promote obesity or lead to weakness and lethargy. Similarly, hormones that regulate the concentration of key ions in the body, such as $Na^+$ and $K^+$, are fundamental to healthy nerve and muscle function (Chapters 35 and 36) and to fluid balance within the body (Chapter 41).

How does the body, and in particular the endocrine system, maintain homeostasis? Hormones are released in response to stimuli and have an effect on distant organs. Homeostatic regulation, or maintenance of homeostasis, depends on **feedback** from the target organ sent back to the endocrine gland that secretes the hormone. In feedback, a stimulus evokes a response, which in turn affects the stimulus. Thus, feedback is the means by which a signal produced by the regulatory organ can modify its own subsequent production, either inhibiting or stimulating it. There are two general types of feedback, **negative feedback** and **positive feedback.**

Homeostasis typically depends on negative feedback for control. In negative feedback, a change in a system causes a response that brings the system back to the starting point. If we consider the components of such a system, we can define a stimulus (the change itself), a sensor (which detects the change), an effector (which leads to a response), and the response (which ultimately opposes the initial change), thereby feeding back information to the sensor that

limits a further response (**Fig. 38.4**). In Chapter 35, we saw how both the temperature in a house and the core body temperature of an animal are maintained at a constant level by this type of control.

In fact, many physiological parameters are maintained at steady levels by negative feedback. In the case of the endocrine system, a stimulus leads to release of a hormone, and the hormone causes a response in the body that opposes the initial stimulus so that secretion of the hormone is decreased and there is a limit to further change. For example, blood-glucose and calcium levels are kept within a narrow range by negative feedback. Although the parameter—temperature, glucose levels, calcium levels—does not change, constant feedback between the response and sensor maintains a set point. That is, the maintenance of homeostasis is an active process.

Let's examine an example in more detail. Like core body temperature, the amount of glucose in the blood of animals is

**FIG. 38.4** **Negative feedback.** Negative feedback results in steady conditions, or homeostasis.

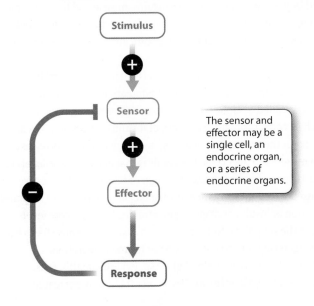

The sensor and effector may be a single cell, an endocrine organ, or a series of endocrine organs.

**FIG. 38.5**   Control of glucose levels in the blood by negative feedback.

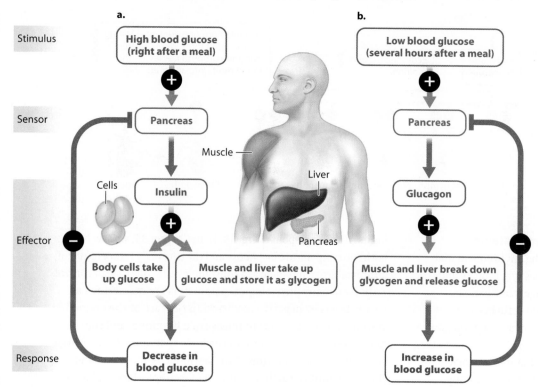

maintained at a steady level (**Fig. 38.5**). If glucose levels are too low, cells of the body do not have a ready source of energy. If glucose levels are too high for too long, they can damage organs.

Maintaining steady blood glucose levels is challenging because glucose is commonly absorbed into the bloodstream from the intestine after a meal, and glucose is readily taken up by cells to meet their energy needs. How then are constant levels maintained in the body? After a meal, when blood glucose rises, β (beta) cells of the pancreas secrete the hormone insulin, which circulates in the blood (Fig. 38.5a). In response to insulin, muscle and liver cells take up glucose from the blood and convert it to a storage form called glycogen (Chapter 7). In this way, insulin guards against high levels of glucose in the blood.

What happens if blood glucose levels fall too low several hours after a meal? In this case, a different population of cells in the pancreas, called α (alpha) cells, secrete the hormone glucagon, which has effects opposite to those of insulin (Fig. 38.5b). Glucagon stimulates the breakdown of glycogen into glucose and its release from muscle and liver cells. The result is that blood glucose levels rise. In both cases, the stimulus (either high or low blood glucose levels) is sensed by cells of the pancreas (β or α cells) and triggers a response (secretion of insulin or of glucagon, bringing blood glucose levels back to the set point). Note that in each case the response feeds back to the secreting cells to reduce further hormone secretion.

When the control of blood glucose levels by insulin fails, a disease called **diabetes mellitus** results. When untreated,

diabetic individuals excrete excess glucose in their urine because blood glucose levels are too high. Diabetes also causes cardiovascular and neurological problems, including loss of sensation in the extremities, particularly the feet.

In some instances, it is necessary to enhance the production of a hormone, accelerating the response of the target cells for a period of time. Positive feedback provides this enhancement (**Fig. 38.6**). In positive feedback, a stimulus causes a response in the same direction as the initial stimulus, which leads to a further response, and so on. In positive feedback in the endocrine system, a stimulus leads to secretion of a hormone that causes a response, and the response causes the release of more hormone. The result is an escalation of the response. A positive feedback loop reinforces itself until it is interrupted or broken by some sort of external signal.

Positive feedback occurs in mammals during childbirth (Chapter 42). In response to uterine contractions, the hormone oxytocin is released from the pituitary gland. Uterine

**FIG. 38.6**   **Positive feedback.** In positive feedback, a stimulus causes a response, and that response causes a further response in the same direction.

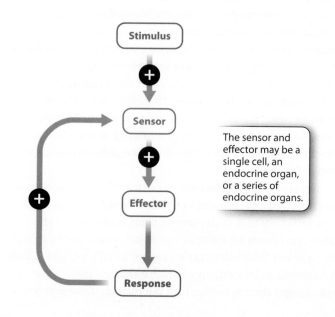

contractions in turn stimulate the pituitary gland to secrete even more oxytocin, causing the uterine muscles to contract more forcefully and more frequently.

→ **Quick Check 1** Diabetes mellitus is a disease characterized by high blood glucose levels. What two different physiological conditions can produce diabetes?

## 38.2 PROPERTIES OF HORMONES

Hormones play diverse roles in the body, enabling regulation and coordination of multiple bodily functions in response to environmental cues. Because hormones are released into the bloodstream, they have the potential to affect many organ systems. Before discussing other important hormonal regulatory pathways, we must first consider how hormones target specific tissues and how different hormones exert their effect on different target cell types. Receptors on target organs respond to blood-borne hormones by binding the hormone, allowing the hormone to influence the function of specific cell targets. Certain classes of hormone (amine and peptide hormones) bind to receptors on the cell membrane and influence cellular function by signaling pathways (Chapter 9). The other general class of hormone (steroid hormones) is lipid soluble, so these hormones can diffuse across the cell membrane. Steroid hormones bind to receptors in the cytoplasm or in the nucleus and regulate transcriptional control of protein synthesis of the target cell.

### Three main classes of hormone are peptide, amine, and steroid hormones.

Hormones are grouped into three general classes: hydrophilic **peptide hormones,** hydrophilic **amine hormones,** and hydrophobic **steroid hormones,** all of which are illustrated in **Fig. 38.7.** Peptide hormones and amine hormones are both derived from amino acids. Peptide hormones are short chains of amino acids, whereas amine hormones are derived from a single aromatic amino acid, such as tyrosine. All steroid hormones are derived from cholesterol. Whereas peptide hormones (which are sometimes large enough to be considered proteins) can evolve through changes in their amino acid sequence, steroid hormones cannot. Evolutionary changes in steroid hormone function depend instead on changes in the receptors they bind and the cellular responses they trigger.

**FIG. 38.7** (a) Peptide, (b) amine, and (c) steroid hormones.

**a.** Peptide hormones

Oxytocin

Antidiuretic hormone (ADH)

Oxytocin and ADH are both 9 amino acid peptide hormones released by the posterior pituitary gland that share a similar structure (only 2 amino acids differ).

**b.** Amine hormones

L-dopa

Dopamine

Norepinephrine

Epinephrine

Amine hormones are derived from aromatic amino acids such as tyrosine.

The steroid hormones all share a similar structure and are derived from cholesterol.

**c.** Steroid hormones

Progesterone

Testosterone

Cortisol

Cholesterol

Peptide and amine hormones are more abundant than steroid hormones and are more diverse in their actions. Because most peptide and amine hormones are hydrophilic, they cannot diffuse across the plasma membrane. Instead, nearly all peptide and amine hormones bind to membrane receptors on the surface of the cell, triggering the intracellular second messenger pathways discussed in Chapter 9 (**Fig. 38.8a**). Peptide and amine hormones alter the biochemical activity of the target cell by activating or inactivating enzymes within the cell. Typically, they affect protein kinases (which phosphorylate other proteins), initiating signaling cascades within the target cell. These signaling cascades can lead to changes in gene expression or metabolism, trigger a cell to grow, divide, or change shape, or lead to the release of other hormones. Peptide and amine hormones act on timescales of minutes to hours.

There are many examples of peptide hormones. In vertebrates, for instance, growth hormone stimulates protein synthesis and growth of many body tissues, particularly the musculoskeletal system (**Table 38.2**). As we saw, insulin and glucagon are peptide hormones released by the pancreas that regulate glucose metabolism by the liver, muscles, and other tissues. Gastrin and cholecystokinin are peptide hormones that regulate mammalian digestive function (Chapter 40). In addition to serving as a neurotransmitter released by the sympathetic nervous system (Chapter 35), epinephrine and norepinephrine (also known as adrenaline and noradrenaline) act as amine hormones, supporting an animal's fight-or-flight response to stress.

In contrast to peptide and amine hormones, steroid hormones are hydrophobic. Therefore, they diffuse freely across the cell membrane to bind with receptors in the cytoplasm or nucleus (**Fig. 38.8b**), forming a steroid hormone–receptor complex. Hormone–receptor complexes that form in the cytoplasm are transported into the nucleus of the cell. These complexes most commonly act as transcription factors, stimulating or repressing gene expression and, therefore, altering the proteins produced by the target cell. Consequently, steroid hormones typically exert profound and long-lasting effects on the cells and tissues they target, with timescales of days to months. The actions of steroid hormones, therefore, tend to have a broader impact on body functions than those exerted by amine and peptide hormones.

The sex hormones estrogen, progesterone, and testosterone (Table 38.2), which we discuss in Chapter 42, are steroid hormones that regulate the differentiation, maturation, and functional state of the vertebrate reproductive organs. Cortisol is a steroid hormone that mediates a vertebrate animal's response to stress and inhibits inflammation. Steroid hormones similar to cortisol are used medically to reduce the symptoms of certain inflammation disorders. However, they carry risks because they also depress the immune system, increasing the chance of infection and diminishing wound healing.

**FIG. 38.8** Mechanism of action of (a) peptide and (b) steroid hormones.

**a.** Cell-surface receptors

Peptide and amine hormones

Receptor

Plasma membrane of target cell

Peptide and amine hormones are hydrophilic and bind to cell-membrane receptors activating second messenger pathways, which change the functional state of the target cell.

**b.** Intracellular receptors

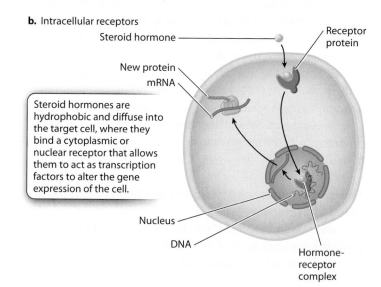

Steroid hormone

Receptor protein

New protein

mRNA

Steroid hormones are hydrophobic and diffuse into the target cell, where they bind a cytoplasmic or nuclear receptor that allows them to act as transcription factors to alter the gene expression of the cell.

Nucleus

DNA

Hormone-receptor complex

## Hormonal signals are typically amplified.

Hormones are typically released in small amounts. So how is it possible for them to have large effects on the overall physiology of an organism? The answer is that hormone signals are amplified. Amplification occurs through a series of signaling steps between different endocrine glands in a hormonal pathway, as well as signal transduction steps within the target cells after the hormone binds to a cell receptor. Signals are thus amplified at each step of the pathway, resulting in a large effect on the target organ system.

Hormonal signaling pathways between endocrine glands and tissues are often referred to as endocrine axes. For example, in vertebrates, the hypothalamic–pituitary axis involves hormonal signals that are amplified along a pathway from the hypothalamus

**TABLE 38.2** **Major Vertebrate Hormones**

| SECRETING TISSUE OR GLAND | HORMONE | TARGET GLAND OR ORGAN | ACTION |
|---|---|---|---|
| Hypothalamus | Releasing factors (peptides) | Anterior pituitary gland | Stimulate secretion of anterior pituitary hormones |
| Anterior pituitary gland | Thyroid-stimulating hormone (TSH) (glycoprotein) | Thyroid gland | Stimulates synthesis and secretion of thyroid hormones by the thyroid gland |
| | Follicle-stimulating hormone (FSH) (glycoprotein) | Gonads | Stimulates maturation of eggs in females; stimulates sperm production in males |
| | Luteinizing hormone (LH) (glycoprotein) | Gonads | Stimulates production and secretion of sex hormones in ovaries (estrogen and progesterone) and testes (testosterone) |
| | Adrenocorticotropic hormone (ACTH) (peptide) | Adrenal glands | Stimulates production and release of cortisol |
| | Growth hormone (GH) (protein) | Bones, muscles, liver | Stimulates protein synthesis and body growth |
| | Prolactin (protein) | Mammary glands | Stimulates milk production |
| | Melanocyte-stimulating hormone (peptide) | Melanocytes | Regulates skin (and scale) pigmentation |
| Posterior pituitary gland | Oxytocin (peptide) | Uterus, breast, brain | Stimulates uterine contraction and milk let-down; influences social behavior |
| | Antidiuretic hormone (ADH) (vasopressin) (peptide) | Kidneys, brain | Stimulates uptake of water from the kidneys; involved in pair bonding |
| Thyroid gland | Thyroid hormones (peptides) | Many tissues | Stimulate and maintain metabolism for development and growth |
| | Calcitonin (peptide) | Bone | Stimulates bone formation by osteoblasts |
| Ovaries | Estrogen (steroid) | Uterus, breast, other tissues | Stimulates development of female secondary sexual characteristics and regulates reproductive behavior |
| | Progesterone (steroid) | Uterus | Maintains female secondary sexual characteristics and sustains pregnancy |
| Testes | Testosterone (steroid) | Various tissues | Stimulates development of male secondary sexual characteristics; regulates male reproductive behavior and stimulates sperm production |
| Adrenal cortex | Cortisol (steroid) | Liver, muscles, immune system | Regulates response to stress by increasing blood glucose levels and reduces inflammation |

*(continues on page 38-10)*

**TABLE 38.2**     **Major Vertebrate Hormones** (continued)

| SECRETING TISSUE OR GLAND | HORMONE | TARGET GLAND OR ORGAN | ACTION |
|---|---|---|---|
| Adrenal medulla | (Nor-) Epinephrine (peptide) | Heart, blood vessels, liver | Stimulates heart rate, blood flow to muscles, and elevation of blood glucose level as part of fight-or-flight response |
| Parathyroid glands | Parathyroid hormone (PTH) (protein) | Bone | Stimulates bone resorption by osteoclasts to increase blood $Ca^{2+}$ levels |
| Pancreas | Insulin (protein) | Liver, muscles, fat, other tissues | Stimulates uptake of blood glucose and storage as glycogen |
| | Glucagon (protein) | Liver | Stimulates breakdown of glycogen and glucose release into blood |
| | Somatostatin (peptide) | Digestive tract | Inhibits insulin and glucagon release; decreases digestive activity (secretion, absorption, and motility) |
| Stomach | Gastrin (peptide) | Stomach | Stimulates protein digestion by secretion of digestive enzymes and acid; stimulates gastric motility |
| Small intestine | Cholecystokinin (peptide) | Pancreas, liver, gallbladder | Stimulates secretion of digestive enzymes and products from liver and gallbladder |
| | Secretin (peptide) | Pancreas | Stimulates bicarbonate secretion from pancreas |
| Pineal gland | Melatonin (peptide) | Brain, various organs | Regulates circadian rhythms |

to the anterior pituitary gland, and then from the anterior pituitary gland to target glands or tissues in the body.

The vertebrate hypothalamic–pituitary axis begins with trace amounts of peptide hormones called **releasing factors** that are initially released by the hypothalamus (**Fig. 38.9**). They signal to the anterior pituitary gland, leading to a much larger release of associated hormones from that organ. For example, corticotropin releasing factor secreted by the hypothalamus stimulates the release of a larger amount of adrenocorticotropic hormone (ACTH) by the anterior pituitary gland.

Hormones released by the anterior pituitary gland in turn bind cell receptors in the target organ. In this case, ACTH acts on cells of the adrenal cortex, stimulating their secretion of the hormone cortisol. Cortisol acts on many different cells and tissues in the body, causing what is known as an acute stress response.

Among its effects, it causes the liver to convert glucose to glycogen (although through effects on other cells it also increases blood glucose levels). This action yields 56,000x more glycogen molecules than the initial number of releasing factors secreted by the hypothalamus.

Similar signaling amplification cascades occur in invertebrates in the regulation of insect molting and metamorphosis (see Fig. 38.3). The amount of brain hormone that is produced is dwarfed by the amount of ecdysone or juvenile hormone that is released to regulate the growth and metamorphosis of the insect's body. As in vertebrates, second messenger signaling cascades within receptor cells of the body respond to the release of these hormones by amplifying the hormone signal in the cell. Thus, amplification by signal transduction pathways applies both to second messenger systems within a cell as well as the chemical

## FIG. 38.9 Amplification of a hormonal signal.

At each step, the hormonal signal is amplified, so that a small amount of corticotropin releasing factor leads to a large effect (in this case, the production of glycogen by the liver).

signals that are transmitted as hormones between glands and tissues in an organism's endocrine system.

→ **Quick Check 2** Why do steroid and peptide hormones bind different kinds of receptor and how does this difference affect the resulting signaling pathways in the target cell?

## Hormones act specifically on cells with receptors that bind the hormone.

Hormones are chemical signals that enable communication between different cells in an organism. Because hormones are secreted into the bloodstream, they achieve specificity by binding to receptors on the surface of or inside the target cells. Therefore, it is the presence or absence of a receptor for a given hormone that determines which cells respond and which ones do not. For example, when the hormone oxytocin is released into the bloodstream of mammals, it affects only cells that express a receptor on the cell surface capable of binding oxytocin

as it flows by in the bloodstream. These cells are uterine muscle cells and secretory cells in breast tissue. When oxytocin binds these cell-surface receptors, uterine muscle cells are stimulated to contract, and secretory cells in breast tissue release milk during breastfeeding. The effect of oxytocin also depends on cells of other organs *not* expressing the receptor so that the hormone can exert its effect on these specific tissues.

The binding of a hormone to its receptor leads to changes in the target cell, resulting in a cellular response. The specific action of a hormone depends on the kind of response that it triggers in the target cell. For example, the binding of a hormone can lead to changes in ion fluxes across the cell membrane, activation of an intracellular signal transduction cascade, or changes in gene and protein expression.

→ **Quick Check 3** What general features make a chemical compound a hormone, and how do hormones achieve specificity for certain kinds of target cells?

## Hormones are evolutionarily conserved molecules with diverse functions.

Most hormones have an ancient evolutionary history. The structures of many hormones are evolutionarily conserved. Some vertebrate hormones can also be found in many invertebrates, but typically they serve different functions. Since the first vertebrate animals diverged from invertebrates over 500 million years ago, this indicates that some vertebrate hormones are even older. And, in many cases, their roles in many invertebrate animals have yet to be discovered.

Thyroid-stimulating hormone (TSH), a hormone released by the anterior pituitary gland that targets the thyroid gland, regulates metabolism in vertebrate animals but triggers metamorphosis in amphibians and feather molt in birds. It has even been found in snails and other invertebrates that lack a thyroid gland. Its function in snails appears to be to stimulate the number of sperm or eggs produced.

Recent genomic analysis has also revealed that the receptors for many hormones evolved well before the hormones with which they now interact. Even though the structure of a hormone or its receptor is often largely unchanged across diverse groups of organisms, hormones and their receptors can readily be selected to take on new roles. An example is the shift in function of TSH to regulate metabolism in mammals and feather molt in birds from an earlier role stimulating sperm or egg production in snails. Thus, the biochemical evolution of hormones and their target cell receptors can be readily altered as organisms evolve new behaviors and exploit new environments.

Another intriguing finding is that many peptides originally identified as hormones in various tissues have also been found to function as neurotransmitters in the nervous system. For example, oxytocin, which stimulates uterine contractions and the release of milk, also serves as a neurotransmitter in the brain and is believed to influence social behavior, as well as stimulate sexual arousal in mammals. Similarly, another peptide hormone, antidiuretic hormone, which regulates water uptake in the kidneys (Chapter 41), also functions as a neurotransmitter in the brain influencing mammalian mating and pair-bonding behavior. Hence, the same compound, expressed in two different cell types, may have entirely different functions. The roles of hormones as chemical messengers are varied within an organism and easily changed over the course of evolution.

## 38.3 THE VERTEBRATE ENDOCRINE SYSTEM

The vertebrate endocrine system regulates changes in the animal's physiological and behavioral states in response to sensory cues received by its nervous system both from the environment and from internal body organs. These sensory signals are processed within the brain and transmitted to the endocrine system primarily by the hypothalamus, which is located in the forebrain.

The hypothalamus, in turn, relays the signals to the pituitary gland, which lies just below it. The pituitary gland is a central regulating gland of the vertebrate endocrine system, releasing hormones that coordinate the action of many other endocrine glands and tissues. These glands, illustrated in **Fig. 38.10**, control the growth and maturation of the body, regulate reproductive development functions of the animal, coordinate digestion and metabolism, and control water balance. These effects reflect the broad range of roles that the endocrine and nervous systems perform in maintaining homeostasis.

The pituitary gland communicates with many cells, tissues, and organs of the body. Some of these, like the thyroid and adrenal glands, secrete hormones in response to signals from the pituitary gland and therefore have exclusively endocrine functions. Others, like the lungs, kidneys, and digestive tract, harbor endocrine cells that secrete hormones and also have other physiological roles in the body. Consequently, the vertebrate endocrine system is not localized in one part of the body, but is present throughout.

**FIG. 38.10**   Human endocrine organs and the hormones they secrete.

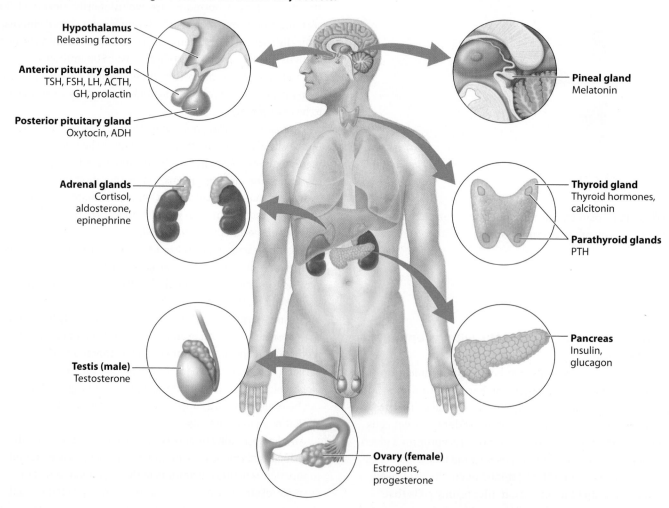

**FIG. 38.11** The hypothalamus and the pituitary gland.

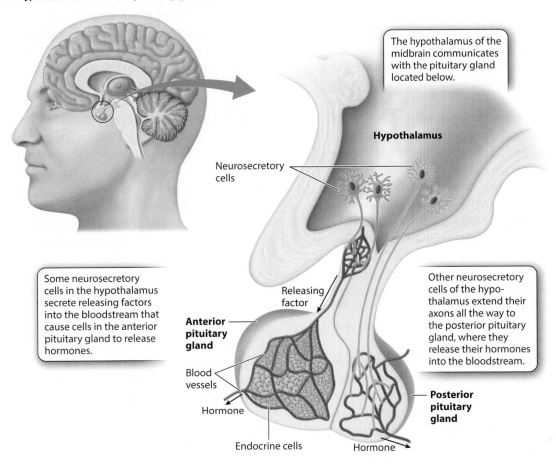

The hypothalamus of the midbrain communicates with the pituitary gland located below.

**Hypothalamus**

Neurosecretory cells

Some neurosecretory cells in the hypothalamus secrete releasing factors into the bloodstream that cause cells in the anterior pituitary gland to release hormones.

Releasing factor

**Anterior pituitary gland**

Blood vessels

Hormone

Endocrine cells

Other neurosecretory cells of the hypothalamus extend their axons all the way to the posterior pituitary gland, where they release their hormones into the bloodstream.

**Posterior pituitary gland**

Hormone

## The pituitary gland integrates diverse bodily functions by secreting hormones in response to signals from the hypothalamus.

The hypothalamus is the main route by which nervous system signals are transmitted to the vertebrate endocrine system. The function of the hypothalamus is to transmit these signals to the pituitary gland, the endocrine gland that acts as a control center for most other endocrine glands in the body.

The pituitary gland is divided into anterior and posterior regions (**Fig. 38.11**). This division is not arbitrary—the two regions have distinct functions, organizations, and embryonic origins. The **anterior pituitary gland** forms from epithelial cells that develop and push up from the roof of the mouth, whereas the **posterior pituitary gland** develops from neural tissue at the base of the brain. Both sets of ectodermal cells form pouches that develop into glands, which come to lie adjacent to each other. The anterior and posterior pituitary glands should therefore be thought of as two distinct glands, not one.

Because of these developmental differences, the anterior and posterior pituitary glands receive input from the hypothalamus in different ways. The hypothalamus contains

**neurosecretory cells.** These cells are located in the hypothalamus, which is part of the brain, and therefore they are neurons. However, instead of secreting neurotransmitters that bind to another neuron or to muscle, these cells release hormones into the bloodstream. Some of these neurosecretory cells communicate with the anterior pituitary gland. In this case, they secrete hormones called releasing factors into small blood vessels that travel to and supply the anterior pituitary gland (Fig. 38.11). In response, cells of the anterior pituitary gland release hormones into the bloodstream, in which they circulate throughout the body and bind to receptors on target cells, tissues, and organs.

In contrast, communication between the hypothalamus and the posterior pituitary gland does not involve hypothalamic releasing factors. Instead, the posterior pituitary gland contains the axons of neurosecretory cells whose cell bodies are located in the hypothalamus. These axons release hormones directly into the bloodstream, by which they are transported to distant sites (Fig. 38.11). Consequently, the posterior pituitary is part of the nervous system itself.

In response to signals from the hypothalamus, distinct hormones are secreted by the anterior and posterior pituitary glands (Table 38.2). The anterior pituitary gland hormones include thyroid stimulating hormone (TSH), which acts on the thyroid gland, the gonadotropic hormones follicle-stimulating hormone (FSH) and luteinizing hormone (LH), which act on the female and male gonads (the ovaries and testes), and adrenocorticotropic hormone (ACTH), which acts on the adrenal glands (see Fig. 38.10). In each of these cases, the anterior pituitary hormone acts on an endocrine gland to cause release of other hormones. Hormones that control the release of other hormones are called **tropic hormones.** In response to TSH, the thyroid gland releases thyroid hormones that regulate the metabolic state of the body. In response to FSH and LH, the ovaries release estrogen and progesterone and the testes release testosterone. In response to ACTH, the adrenal glands release cortisol, which has diverse effects that include stimulating glucose release into the bloodstream, maintaining blood pressure, and suppressing the immune system.

The anterior pituitary gland also secretes growth hormone (GH), which acts generally on the muscles, bones, and other body tissues to stimulate their growth, and prolactin, which stimulates milk production in the breast of female mammals in response to an infant's suckling at the mother's nipple.

The posterior pituitary gland hormones include **oxytocin** and **antidiuretic hormone** (**ADH;** also called **vasopressin**). These are two evolutionarily related peptide hormones, each consisting of nine amino acids. As we discussed earlier, oxytocin plays several roles related to female reproduction: It causes uterine contraction during labor and stimulates the release of milk during breastfeeding. Antidiuretic hormone acts on the kidneys (Chapter 41), regulating the concentration of urine that an animal excretes, which is critical to maintaining water and solute balance in the body.

In addition to their roles in reproduction and kidney function, oxytocin and vasopressin are also released by cells in the brain and may play roles in social behaviors. For instance, oxytocin is important in regulating maternal behavior toward infants. Recent research has suggested that oxytocin may also have a role in behaviors such as trust as a prelude to mating in vertebrate animals. These functions reflect the long evolutionary history of oxytocin, as it also plays a role in the social behavior of insects.

Recent evidence suggests that antidiuretic hormone plays a parallel role in regulating male social behavior and parental behavior in mammals. This is not surprising because oxytocin and antidiuretic hormone have very similar chemical structures (see Fig. 38.7). The integration of reproductive function with parental and social behavior for the care of young is likely favored strongly by natural selection. Receptors for these hormones are expressed within the brain. Thus, in addition to acting on reproductive and renal organs of the body, oxytocin and antidiuretic hormone affect mammalian behavior by acting directly on the brain.

## Many targets of pituitary hormones are endocrine tissues that also secrete hormones.

As we have seen, some of the hormones released by the anterior pituitary gland act on endocrine organs that themselves then release hormones. These tropic hormones are TSH, FSH and LH, and ACTH. Here, we look at their target organs in more detail.

TSH acts on the **thyroid gland,** which is located in the front of the neck (see Fig. 38.10), and leads to the release of two peptide hormones, thyroxine (T4) and triiodothyronine (T3). These hormones regulate cellular metabolism throughout the body. Overproduction of thyroid hormones (hyperthyroidism) or thyroid hormone deficiency (hypothyroidism) creates symptoms that reflect either an overly active metabolic state (increased appetite and weight loss) or one that is too low (fatigue and sluggishness). Both conditions are diagnosed by blood tests that monitor circulating levels of thyroid hormones, and both can be treated by medication.

Because thyroid hormones require iodine for their formation, individuals who do not acquire enough iodine from their diets produce too little thyroid hormone, resulting in metabolic problems. Additionally, iodine insufficiency stimulates increased production of TSH by the anterior pituitary gland because of the absence of negative feedback. Over time, the thyroid gland enlarges to form a goiter, which is observed as an enlargement of the throat, to compensate and produce more thyroid hormone. The introduction of iodized salt has eliminated goiter formation and hypothyroidism in many countries, but in many underdeveloped areas these remain significant public health problems.

The gonadotropic hormones are FSH and LH. They target the female and male gonads—the **ovary** and **testis** (see Fig. 38.10). In response to FSH and LH, the ovary and testis each secrete sex hormones that regulate their own development as well as the sexual differentiation and maturation of secondary sexual characteristics of the body. Female sex hormones include the steroids estrogen and progesterone. The principal male sex hormone is the androgen testosterone. These sex hormones are common to a vast majority of vertebrates, regulating sexual differentiation, gonadal maturation, and reproductive behavior. The role of these sex hormones in regulating reproductive function is discussed in greater detail in Chapter 42.

Testosterone is a naturally occurring anabolic steroid that stimulates the synthesis in the testes of proteins needed for sperm production and the development of male sexual features and body tissue growth, particularly in muscles. Anabolic steroids promote protein synthesis to build body tissues and anabolic metabolism to store energy within cells. A variety of synthetic anabolic steroids have been developed to treat muscle wasting due to loss of appetite and diseases such as cancer and AIDS. Anabolic steroids are used illegally by both male and female athletes to promote muscle strengthening, undermining fair competition. The risks to health of continued use of anabolic steroids are considerable: liver damage, heart disease, and high blood pressure. Tissue damage occurs because elevated circulating levels of testosterone produced by anabolic steroid use inhibits the normal synthesis of anterior pituitary tropic hormones.

ACTH released by the anterior pituitary gland is also a tropic hormone. It acts on the cortex (the outer portion) of the paired **adrenal glands** (see Fig. 38.10). The adrenal glands are located adjacent to the kidneys ("ad-" meaning "near" and "renal" referring to the kidneys). In humans, each small adrenal gland lies just above each kidney. During times of stress, such as starvation, fear, or intense physical exertion, ACTH stimulates adrenal cortex cells to secrete cortisol.

## Other endocrine organs have diverse functions.

Although many endocrine organs receive signals from the hypothalamic–pituitary axis, some respond instead to internal physiological states of the body. One of these, adjacent to the thyroid gland, is the **parathyroid gland** (see Fig. 38.10), which secretes parathyroid hormone (PTH). Calcitonin, which is secreted by the thyroid gland, and PTH together regulate the actions of bone cells (osteoblasts and osteoclasts, Chapter 37) that control bone formation and bone removal and therefore the levels of calcium in the blood.

When circulating levels of calcium fall too low, the parathyroid gland triggers release of PTH, stimulating osteoclasts to reabsorb bone mineral, halting bone formation and raising blood calcium levels. When calcium levels are too high, the production of PTH is inhibited and calcitonin is released to shift bone metabolism toward net bone formation, building bone that stores calcium in the skeleton. Note that, as with blood glucose levels, calcium levels are maintained in a narrow range by negative feedback: The response to the hormone (increasing or decreasing levels of calcium in the blood) is the opposite of the stimulus (low or high levels of calcium in the blood) so that a stable set point is maintained.

Several organs of the digestive system, including the pancreas, stomach, and duodenum, produce hormones that regulate digestion, many of which are discussed in Chapter 40.

The **pineal gland** is located in the thalamic region of the brain (see Fig. 38.10). It responds to autonomic nervous system input by secreting melatonin, a hormone that helps control an animal's state of wakefulness. When melatonin levels rise, animals sleep. In diurnal animals (those that are active during the day), increased levels occur at night; in nocturnal animals (those that are active during the night), increased

levels occur during the day. The pineal gland is sensitive to changes in light and darkness. It releases melatonin in response to daily (and seasonal) light cycles, helping to maintain circadian rhythms of the body. Circadian rhythms are cycles of about 24 hours (or longer) in which an animal's biochemical, physiological, and behavioral state shifts in response to changes in daily and seasonal environmental conditions. Exposure to sunlight inhibits the production of melatonin in diurnal animals, whereas darkness stimulates its release. To overcome jet lag, travelers are advised to seek sunlight and often to take melatonin as a medication before sleeping to help reset their 24-hour circadian rhythm. In other vertebrates, such as lampreys, fishes, and reptiles, the pineal gland functions as a "third eye," sensing light through an opening in the skull. Light shining on the pineal gland regulates the animal's general state of activity, influencing its behavioral response to changing light conditions.

In several animals, including ground squirrels, bats, bears, and rattlesnakes, the pineal gland regulates hibernation, controlling the animal's metabolic state over longer time periods. In many animals, release of melatonin also regulates seasonal breeding cycles.

**?** **CASE 7 Predator–Prey: A Game of Life and Death**
**How does the endocrine system influence predators and prey?**

The endocrine system works closely with the nervous system to enable animals to respond to external cues in the environment. In Chapters 35 and 36, we discussed how prey animals, such as warthogs, use visual and olfactory cues processed by the nervous system to detect the distant movement of a predator, such as a lion, which in some instances occurs too late for a successful escape (**Fig. 38.12**). We saw how the sympathetic nervous system,

**FIG. 38.12 Predator and prey.** Animals rely on rapid integration of sensory information with coordination of bodily functions to escape a predator or to catch their prey.

in the fight-or-flight response, can trigger broad physiological changes, such as increased heart and breathing rates and changes in metabolism that ready the predator to attack or the prey to respond.

These actions are coordinated by the endocrine system. The sympathetic nervous system sends axons to the **adrenal medulla,** the inner part of the adrenal gland. In response to stimulation by the sympathetic nervous system, cells of the adrenal medulla secrete two hormones, epinephrine and norepinephrine (also known as adrenaline and noradrenaline). These are hormones released by the adrenal gland with targets throughout the body, leading to the physiological changes of the fight-or-flight response. Interestingly, epinephrine and norepinephrine also act as neurotransmitters in the brain. In fact, cells of the adrenal medulla are modified nerve cells that have lost their axons and dendrites.

The same set of responses occurs when, walking along a deserted sidewalk, we note a shadowy movement or hear a footstep behind us. The perceived threat sets in motion a coordinated physiological reaction, mediated by the combined action of the nervous and endocrine systems. These changes make us alert and ready for action to avoid or resist a perceived threat. They also inhibit digestive functions and eliminate a sense of appetite.

Earlier, we considered the importance of homeostasis in maintaining a variety of parameters, such as temperature and glucose levels, at a steady level. In this case, there is an adjustment of this set point, which is critical in enabling an organism to respond to external cues and adjust its physiological response appropriately. When the real or perceived threat is gone, the body is able to return to its prior state by reestablishing the earlier set point.

## 38.4 OTHER FORMS OF CHEMICAL COMMUNICATION

As we have seen, hormones are chemical compounds secreted by glands that enter the circulation and act on distant cells, whose response is determined by the presence of receptors that bind the hormones. However, other chemical compounds act over short distances, and still others are released into the environment to signal other individuals of the same species. Here, we explore additional modes of chemical communication and the compounds that produce them.

### Local chemical signals regulate neighboring target cells.

Whereas hormones enter the bloodstream to be transmitted to more distant target cells (**Fig. 38.13a**), other chemical compounds act locally on neighboring cells (**Fig. 38.13b**). In

**FIG. 38.13 Modes of chemical signaling.** Chemical signaling can occur (a) over long-distances (endocrine), or (b) locally (paracrine and synaptic).

**a.** Long-distance signaling

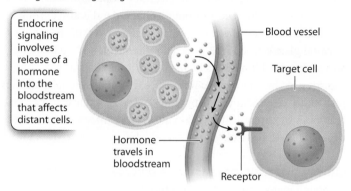

Endocrine signaling involves release of a hormone into the bloodstream that affects distant cells.

Blood vessel

Target cell

Hormone travels in bloodstream

Receptor

**b.** Local signaling

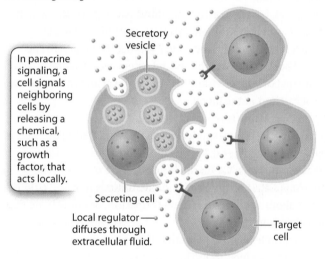

In paracrine signaling, a cell signals neighboring cells by releasing a chemical, such as a growth factor, that acts locally.

Secretory vesicle

Secreting cell

Local regulator diffuses through extracellular fluid.

Target cell

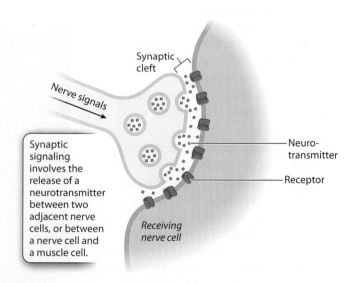

Nerve signals

Synaptic cleft

Synaptic signaling involves the release of a neurotransmitter between two adjacent nerve cells, or between a nerve cell and a muscle cell.

Neurotransmitter

Receptor

Receiving nerve cell

order to take up the chemical signal, these cells must have receptors for the compound to bind to before the signal degrades or diffuses into the bloodstream.

Chemical compounds that act locally are said to have **paracrine** function (Fig. 38.13b). If the compounds act on the secreting cell itself, they are described as having **autocrine** function. That is, the chemical signals may stimulate or inhibit their own secretion. Growth factors (also referred to as cytokines and distinct from the growth hormone released by the anterior pituitary gland) and histamine are examples of paracrine chemical agents released in small amounts that act locally on neighboring cells.

Growth factors (Chapter 9) enhance the differentiation and growth of particular kinds of tissue. For example, bone morphogenetic proteins (BMPs) promote the formation of bone in skeletal growth, and fibroblast growth factors (FGFs) stimulate the formation of connective tissue. Similarly, nerve growth factors (NGFs) promote the survival and growth of nerve cells. When local tissues are damaged, specialized cells release histamine, which dilates blood vessels, allowing blood proteins and white blood cells to move into the region to fight infection and induce repair (Chapter 43). The release of histamine causes the swelling, redness, and warmth that commonly surround a wound. Locally acting signals have immediate, short-term effects because they are quickly degraded.

Another form of local chemical signaling is the synaptic release of neurotransmitters that occurs between directly communicating nerve cells or between nerve cells and muscles at neuromuscular junctions (Fig. 38.13b). Synaptic signaling is extremely rapid and brief compared to the other modes of chemical communication.

## Pheromones are chemical compounds released into the environment to signal behavioral cues to other species members.

Animal endocrine systems generally transmit chemical signals that integrate the physiological functions of organ systems within the body. However, many animals release small chemical compounds into the environment to signal and influence the behavior of other members of their species. These water or airborne compounds are called **pheromones.** For example, female bees and wasps release pheromones to attract male mates. Even plants, such as orchids, release pheromones. Some plant pheromones can attract pollinating bees or wasps by mimicking the chemical signals released by the females of these species. There are many different pheromones, which serve very different functions, as shown in **Fig. 38.14**.

Female silk moths use pheromones for long-distance communication to attract males. The sex pheromone bombykol, released from a gland in the female's abdomen, signals her location to males looking for a mate. Sex pheromones play a key role in the mating behavior of many invertebrate and vertebrate species. In most mammals, amphibians, and reptiles, pheromones are detected by a vomeronasal organ with chemosensory neurons in the nasal region of the skull. The release of sex pheromones is one of the most common types of pheromone signaling to attract a mate and trigger ritualistic social and mating behaviors (Chapter 45). Most studies indicate that humans and nonhuman primates lack a vomeronasal organ and do not produce pheromones.

Other pheromones signal territorial boundaries or predatory threats. Dogs and other canids (this family also includes wolves and coyotes), as well as cats, secrete pheromones in their urine that mark their territories. Social seabirds, such as cormorants and boobies, mark their nesting sites by pheromones.

**FIG. 38.14 Pheromones.** (a) Pheromones are means of attracting mates, as in these ladybugs; (b) wolves secrete pheromones in their urine that mark their territory; (c) honeybees swarm to defend their nest after a bee that stings a threatening animal releases an alarm pheromone.

**FIG. 38.15** **Trail pheromones in ants.** Here, a chemical trail has been laid down by worker ants signaling a food source for the colony.

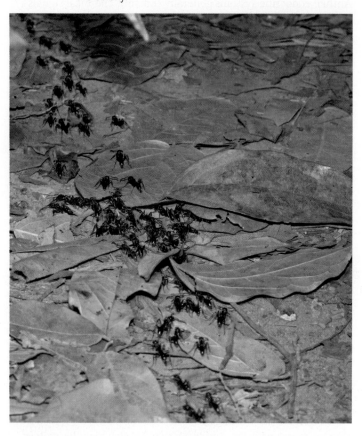

**FIG. 38.16** **A brightly colored male cichlid fish.** The behavior and markings of this fish are under hormonal control and triggered by social context.

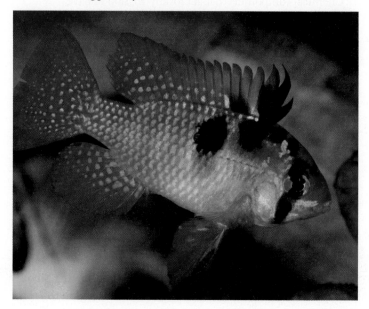

Ants and other social insects commonly deploy alarm pheromones to warn of an attack by an advancing army from another colony. The release of alarm pheromones from a stinging bee causes other bees to swarm and join the attack to defend their nest. In many other species, an animal under attack releases an alarm pheromone, warning its neighbors. Similarly, social mammals such as deer warn of an approaching predator by means of alarm pheromones.

Social insects, such as ants, also deploy trail pheromones that provide chemical cues that other worker ants can follow to a food source (**Fig. 38.15**). An ant returning to the colony from a food source releases trail pheromones from its abdomen onto the soil. Other worker ants use their antennae to detect and track the concentration of the trail pheromone guiding them along the trail. If you have watched ants marching in line across a sidewalk, a trail, or into and out of your house, you have seen worker ants using pheromone cues to signpost the way to food.

Aquatic species also depend on pheromone signaling. Fish, salamanders, and tadpoles release alarm pheromones into the water to warn of a threatening predator. Prey species that share the same habitat may even sense and respond to the release of another species' alarm pheromone.

Fish also release pheromones to coordinate mating and to regulate social interactions. In dense populations of cichlid fish living in Lake Tanganyika in central Africa, the ratio of brightly colored breeding males (**Fig. 38.16**) to females is highly regulated. Only about 10% of males are brightly colored and defend their territory to attract and mate with females. This restriction on the number of dominant breeding males is regulated by behavioral and chemical pheromone cues. When a dominant male dies, subordinate males fight for the vacated territory. When a new male assumes dominance, it becomes brightly colored to attract females and defend its newly won territory. The dominant breeding male reinforces its dominance by releasing pheromones in its urine that signal females and subordinate males in its breeding area. The color change is mediated by hormones released from the male fish's pituitary gland, which act on pigment cells called melanocytes that are located in the epidermis of its scales. Thus, the fish makes use of both endocrine and pheromone signals to change its behavior and the behavior of other members of its species.

→   **Quick Check 4** What distinguishes paracrine and pheromone signals from endocrine signals?

## Core Concepts Summary

### 38.1 ANIMAL ENDOCRINE SYSTEMS RELEASE CHEMICAL SIGNALS CALLED HORMONES INTO THE BLOODSTREAM. THESE SIGNALS RESPOND TO THE ENVIRONMENT, REGULATE GROWTH AND DEVELOPMENT, AND MAINTAIN HOMEOSTASIS.

The nervous system responds rapidly to sensory information, providing immediate physiological responses, whereas the endocrine system acts more slowly and over prolonged periods to affect broad changes in the physiological and behavioral state of an animal. page 38-1

The endocrine system helps an animal respond to its environment. page 38-1

The endocrine system is involved in growth and development. page 38-2

The nervous and endocrine systems regulate physiological functions to maintain homeostasis. page 38-5

Homeostasis is often achieved by negative feedback, in which the response opposes the stimulus. page 38-5

A stimulus can be amplified by positive feedback, in which the response amplifies the initial stimulus. page 38-6

### 38.2 HORMONES ACHIEVE SPECIFICITY BY BINDING TO RECEPTORS ON OR INSIDE THEIR TARGET CELLS. THEIR SIGNALS ARE AMPLIFIED TO EXERT STRONG EFFECTS ON THEIR TARGET CELLS.

Hormones are chemical compounds that are secreted and transported by the circulatory system, often to distant targets. page 38-7

Hormones bind to specific receptors located on the surface or inside the target cells that receive their signals. page 38-7

The three classes of hormone are peptide, amine, and steroid hormones. page 38-7

Peptide and amine hormones are hydrophilic and bind to cell membrane receptors, stimulating second messenger pathways to alter the biochemical function of target cells. page 38-8

Steroids are hydrophobic and bind to intracellular or nuclear receptors, which function as transcription factors to regulate gene expression and protein synthesis. page 38-8

Hormones communicate by signaling cascades to amplify the strength of their downstream effect on target cells. page 38-8

The hypothalamic–pituitary axis forms the central endocrine pathway by which vertebrates regulate and integrate diverse bodily functions. page 38-10

Hormones are evolutionarily conserved molecules common to diverse groups of animals, but which have evolved novel functions. page 38-11

### 38.3 IN VERTEBRATE ANIMALS, THE HYPOTHALAMUS AND THE PITUITARY GLAND CONTROL AND INTEGRATE DIVERSE BODILY FUNCTIONS AND BEHAVIORS.

The anterior pituitary gland secretes hormones in response to releasing factors produced and secreted by the hypothalamus. page 38-13

Hormones released by the anterior pituitary gland include thyroid-stimulating hormone (TSH), the gonadotropic hormones follicle-stimulating hormone (FSH) and luteinizing hormone (LH), adrenocorticotropic hormone (ACTH), growth hormone (GH), and prolactin. page 38-14

Tropic hormones are hormones that act on endocrine glands to produce additional hormones. TSH causes the thyroid gland to release thyroid hormones; the gonadotropic hormones FSH and LH act on the female ovaries to produce estrogen and progesterone and the male testes to produce testosterone; and ACTH acts on the adrenal gland to release cortisol. page 38-14

The posterior pituitary gland contains the axons of neurosecretory cells that release the peptide hormones oxytocin and antidiuretic hormone into the bloodstream. page 38-14

Some endocrine glands do not respond to signals from the pituitary gland. These include the parathyroid gland, which releases parathyroid hormone that helps to control calcium levels; the pancreas, which secretes insulin and glucagon to regulate blood glucose levels; and the pineal gland, which secretes melatonin in response to light and dark cycles. page 38-15

### 38.4 CHEMICAL COMMUNICATION CAN ALSO OCCUR LOCALLY BETWEEN NEIGHBORING CELLS OR, THROUGH THE RELEASE OF PHEROMONES, BETWEEN INDIVIDUALS, COORDINATING SOCIAL INTERACTIONS.

Paracrine signals act locally over short times to affect neighboring cells, compared to the more distant and longer-lasting signals transmitted by blood-borne hormones. page 38-17

Pheromones are chemical signals released into the air or water, signaling behavioral cues between individuals of a species. page 38-17

Sex pheromones attract mates and coordinate ritualized mating behavior; alarm pheromones signal predatory threats; trail pheromones signal the location of food sources; and territorial pheromones signal a male's dominance and breeding area. page 38-17

## Self-Assessment

1. Describe three general functions of the endocrine system and provide an example of each.

2.  Diagram an endocrine pathway, such as parathyroid regulation of blood calcium levels, that maintains homeostasis of a bodily function by means of negative feedback.

3.  Diagram an endocrine pathway, such as oxytocin regulation of female labor during birth, that relies on positive feedback.

4.  Identify and provide an example of each of the three general classes of hormone.

5.  Explain how hormones act specifically on only some cells or organs despite being transported by the bloodstream throughout the body.

6.  Describe how hormones exert substantial and broad effects on their target cells despite being released in small amounts by the initial signaling gland.

7.  Explain how peptide hormones affect the function of a target cell.

8.  Explain how steroid hormones affect the function of a target cell.

9.  Explain the difference between the way in which the hypothalamus controls the anterior pituitary gland and the way it controls the posterior pituitary gland.

10. Explain the difference in action between a hormone, a paracrine signal, a neurotransmitter, and a pheromone.

Do you understand the chapter's Core Concepts? Log in to check your answers to the Self-Assessment questions, then practice what you've learned and reinforce this chapter's concepts by working through the problems and multimedia tutorials provided there.

📖 **www.biologyhowlifeworks.com**

**CHAPTER 39**

# ANIMAL CARDIOVASCULAR AND RESPIRATORY SYSTEMS

## Core Concepts

**39.1** Respiration and circulation depend on diffusion over short distances and bulk flow over long distances.

**39.2** Respiration provides oxygen and eliminates carbon dioxide in support of cellular metabolism.

**39.3** Red blood cells produce hemoglobin, greatly increasing the amount of oxygen transported by the blood.

**39.4** Circulatory systems have different-sized vessels that facilitate bulk flow and diffusion.

**39.5** The evolution of animal hearts reflects selection for a high metabolic rate, achieved by increasing the delivery of oxygen to metabolically active cells.

As animals evolved to be larger and more complex, they faced the challenge of obtaining an adequate supply of oxygen ($O_2$) and nutrients and, at the same time, eliminating carbon dioxide ($CO_2$) and other waste products. In this chapter, we explore how animal respiratory and circulatory systems evolved to meet the metabolic needs of an animal's cells. In particular, we focus on the uptake of $O_2$ and the release of $CO_2$, key gases in cellular respiration (Chapter 7).

Gases, nutrients, and wastes are continually being exchanged between an organism and its environment. Because of their small size or thin bodies, single-celled organisms and simple multicellular animals exchange compounds with the environment by diffusion. Because diffusion works effectively only over very short distances, larger and more complex animals cannot rely on diffusion as the sole means of gas, nutrient, and waste exchange (Chapter 28). These animals evolved respiratory structures at the body's surface for taking up $O_2$ and releasing $CO_2$, and internal circulatory systems for long-distance transport to different body regions.

Because similar physical laws govern gas and fluid movement, animal respiratory and circulatory systems share many features. However, fish obtain $O_2$ from water, whereas land animals breathe air. Water and air have very different physical properties. Thus, aquatic and terrestrial environments pose distinct challenges to the animal in obtaining $O_2$. Because life first began in the water, the evolution of life on land required substantial changes in how animals exchange gases, take up nutrients, and eliminate wastes.

## 39.1 DELIVERY OF OXYGEN AND ELIMINATION OF CARBON DIOXIDE

Eukaryotic cells use $O_2$ to burn organic fuels—carbohydrate, lipid, and protein—for ATP production (Chapter 7). The increase in atmospheric $O_2$ with the evolution of land plants was key to the evolution of $O_2$-based cellular respiration, and that in turn contributed to the diversification of multicellular animal life during the Cambrian Period about 500 million years ago. In the process of burning fuel by aerobic respiration, $CO_2$ is produced. Animals have thus evolved mechanisms to acquire $O_2$ from their environment and, at the same time, to eliminate excess $CO_2$ from their body to the environment. These mechanisms make up the process known as animal respiration—or "breathing." The transport of $O_2$ and $CO_2$ between an animal and its environment is referred to as **gas exchange,** and is fundamental to all eukaryotic animals, as well as to photosynthetic plants.

## Diffusion governs gas exchange over short distances.

Single-celled organisms and simple animals and plants exchange gases by diffusion. **Diffusion** is the random movement of individual molecules (Chapter 5). If there is a difference in concentration of a substance between two regions, diffusion results in the net movement of molecules from regions of higher concentration to regions of lower concentration until the substance is evenly distributed.

Given a large enough surface, diffusion is an extremely effective way to exchange gases and other substances over short distances between two compartments, for example between the lung and the blood vessels supplying the lung. The rate of diffusion is directly proportional to the surface area over which exchange occurs and to the concentration difference, and inversely proportional to the distance over which the molecules move (or the thickness of the barrier). Diffusion is generally ineffective over distances exceeding about 100 μm, or 0.1 mm. Consequently, respiratory and circulatory structures have large surface areas and thin membranes.

Because effective transport by diffusion is limited to short distances, some of the earliest and simplest invertebrate animals

**FIG. 39.2**   **Partial pressure.** The partial pressure of a gas is its fractional concentration multiplied by the total atmospheric pressure.

Atmospheric pressure = 760 mmHg

The sum of the partial pressures of a mixture of gases equals the total atmospheric pressure.

N$_2$ (78%)

O$_2$ (21%)

CO$_2$ and trace gases (0.03%)

Partial pressure of N$_2$ ($p$N$_2$) = 760 x 78% mmHg = 593 mmHg

Partial pressure of O$_2$ ($p$O$_2$) = 760 x 21% mmHg = 160 mmHg

Partial pressure of CO$_2$ ($p$CO$_2$) = 760 x 0.03% mmHg = 0.2 mmHg

**FIG. 39.1**   **The thin body of a flatworm.** Gases can be exchanged by diffusion in flatworms because their bodies are only a few cell layers thick.

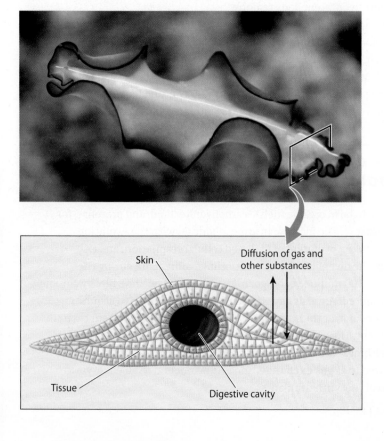

Skin

Diffusion of gas and other substances

Tissue

Digestive cavity

are composed of thin sheets of cells with a large surface area. For example, sponges and sea anemones have a thin body with a central cavity through which water circulates, and flatworms have a flattened body surface (**Fig. 39.1**).

Net movement of a substance results from diffusion when there is a concentration difference between two regions. When considering the diffusion of a gas, however, we usually refer to its partial pressure rather than its concentration. The **partial pressure ($p$)** of a gas is defined as its fractional concentration relative to other gases present, multiplied by the overall (that is, the atmospheric) pressure (**Fig. 39.2**). Pressure is measured in units of millimeters of mercury (mmHg). Oxygen makes up approximately 21% of the gases in air (nitrogen contributes about 78%, carbon dioxide about 0.03%, and the remainder is trace gases). At sea level, atmospheric pressure is approximately 760 mmHg, which means that O$_2$ in the atmosphere exerts a partial pressure ($p$O$_2$) of about 160 mmHg (0.21 × 760). For O$_2$ to diffuse from the air into cells, the partial pressure of O$_2$ inside the cells must be lower than the partial pressure of O$_2$ in the atmosphere.

→   **Quick Check 1** What two structural features favor diffusion?

## Bulk flow moves fluid over long distances.

Larger, more complex animals transport O$_2$ and CO$_2$ longer distances to cells within their body. These animals rely on **bulk flow** in addition to diffusion. Bulk flow is the physical movement

**FIG. 39.3** The four steps of gas transport and exchange.

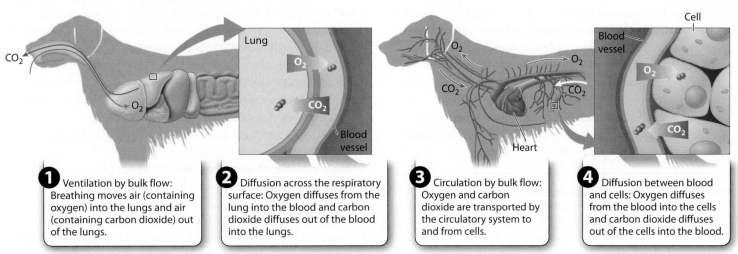

**1** Ventilation by bulk flow: Breathing moves air (containing oxygen) into the lungs and air (containing carbon dioxide) out of the lungs.

**2** Diffusion across the respiratory surface: Oxygen diffuses from the lung into the blood and carbon dioxide diffuses out of the blood into the lungs.

**3** Circulation by bulk flow: Oxygen and carbon dioxide are transported by the circulatory system to and from cells.

**4** Diffusion between blood and cells: Oxygen diffuses from the blood into the cells and carbon dioxide diffuses out of the cells into the blood.

of fluid, and the gases and compounds carried by the fluid, over a given distance. Specialized pumps, like the heart, generate the pressure that is required to move the fluid.

Bulk flow occurs in two steps to meet the gas exchange needs of cells in larger animals. The first is **ventilation.** Ventilation is the movement of the animal's respiratory medium—water or air—past a specialized respiratory surface. The second is **circulation.** Circulation is the movement of a specialized body fluid that carries $O_2$ and $CO_2$. The circulatory fluid is called **hemolymph** in invertebrates and **blood** in vertebrates. This fluid delivers $O_2$ to cells within different regions of the body and carries $CO_2$ back to the respiratory exchange surface.

Ventilation and circulation each require a pump to produce a pressure ($P$) that drives flow ($Q$) against the resistance ($R$) to flow. The rate of flow is governed by the following simple equation:

$$Q = P/R$$

Resistance to flow measures the difficulty of pumping the fluid through a network of chambers and vessels located within the respiratory and circulatory systems. If the resistance to flow doubles, the flow rate is halved. The longer the network of vessels and the narrower the vessels themselves, the greater the vessels' resistance to the fluid (whether air, water, or blood) moving through them. Consider how much harder it is to blow fluid out of a long thin tube than a short wide one.

In summary, we can view gas exchange in complex multicellular organisms as four steps that are linked in series (**Fig. 39.3**). (1) To deliver $O_2$ to the mitochondria within their cells, animals move fresh air or water past their respiratory exchange surface in the process of ventilation (bulk flow). Ventilation maximizes the concentration of $O_2$ in the air or water on the outside of the respiratory surface. (2) The buildup of $O_2$

favors the diffusion of $O_2$ into the animal across its respiratory surface. (3) Following diffusion into the blood, $O_2$ is transported by the circulation (bulk flow) to the tissues. Internal circulation again serves to maximize the concentration of $O_2$ outside cells. (4) Oxygen then diffuses from the blood across the cell membrane and into the mitochondria, where it burns fuels for ATP production.

The same four steps occur in reverse to remove $CO_2$ from the body: $CO_2$ moves from cells to the blood by diffusion, then is carried in the circulation to the respiratory surface by bulk flow, then moves across the respiratory surface by diffusion, and finally moves out of the animal by bulk flow. Because diffusion and bulk flow are coupled, it is important that each step of transport (whether for a gas, nutrient, or waste) has a similar capacity— that is, the same amount of $O_2$ enters the blood through diffusion as is transported by the circulation. This matching of coupled transport processes reflects an evolutionary principle of minimizing excess capacity, which is costly to the animal.

→ **Quick Check 2** Those who suffer an asthma attack have difficulty breathing. From the discussion of bulk flow, what do you think happens to these individuals' airways that makes it difficult for them to breathe?

## 39.2 RESPIRATORY GAS EXCHANGE

All animals obtain $O_2$ from the surrounding air or water. Aquatic animals like crabs, aquatic salamanders, and fish that take in $O_2$ from water breathe through **gills,** highly folded delicate structures that facilitate gas exchange with the surrounding water

(**Fig. 39.4a**). Some animals, such as frogs and salamanders, also breathe through their skin. In contrast, many terrestrial animals, such as reptiles, birds, and mammals, have internal **lungs** for gas exchange (**Fig. 39.4b**). Instead of lungs, terrestrial insects evolved a system of air tubes called **tracheae** that branch from openings along their abdominal surface into smaller airways. These smaller airways, termed tracheoles, supply air directly to the cells within their body.

For both aquatic and terrestrial animals, ventilation supplies $O_2$ to gas exchange surfaces, and then $O_2$ crosses the surface through diffusion. Recall that diffusion is much slower than the bulk flow of ventilation. How can the rate of diffusion through the gas exchange surface in lungs or gills keep up with the rate of $O_2$ supplied by ventilation? The gas exchange surface must have a large surface area and be extremely thin. The respiratory surfaces of more complex and active animals are highly folded, creating a large surface area within a small space. These surfaces are also only one or two cell layers thick, providing a diffusion distance of as little as 1 to 2 μm. In fact, this short distance is all that separates the air you breathe into your lungs from the bloodstream into which $O_2$ diffuses.

### Many aquatic animals breathe through gills.

Most large aquatic animals (with the notable exception of marine mammals like dolphins and whales) use gills to breathe $O_2$-containing water. Some sessile invertebrates, including tube worms and sea urchins, as well as nudibranch mollusks and aquatic amphibians, have external gills (Fig. 39.4a). However, because external gills are easily damaged or may be eaten by predators, most mollusks, crustaceans, and fishes have evolved internal gills within protective cavities. Whereas the external gills of invertebrates often rely on the natural motions of the water to move water past the gills, invertebrates with internal gills often have cilia that direct water over the gill's surface.

Fish need sufficient $O_2$ to meet the energy demands of swimming. As a result, they actively pump water through their mouth and over the gills, which are located in a chamber behind the mouth cavity (**Fig. 39.5**). The effort required to ventilate the gills could be high because of the density and viscosity of water, but fish greatly reduce the energy cost of moving water by maintaining a continuous, unidirectional flow of water past their gills. In bony fishes, a protective flap overlying the gills, called the operculum, expands laterally to draw water over the gills while the mouth is refilling before its next pumping cycle.

The gills of fishes typically consist of a series of gill arches located on either side of the animal behind the mouth cavity and, in bony fishes, beneath the operculum. Each gill arch consists of two stacked rows of flat leaf-shaped structures called gill filaments. Numerous **lamellae**, thin sheetlike structures, are evenly but tightly spaced along the length of each gill filament and extend upward from the filament's surface. A series of blood vessels brings $O_2$-poor blood from the heart to the lamellar surfaces. The lamellae are composed of flattened epithelial cells and are extremely thin, so the water passing outside and blood passing inside the lamellae are separated by only about 1 to 2 μm. The lamellae give the gills an enormous surface area for their size.

The lamellae are oriented so that the blood flowing through them in a capillary network moves in a direction opposite to the flow of water past the gills (Fig. 39.5). This type of organization, in which fluids with different properties move in opposite directions, is an efficient way to exchange properties between the two fluids. The property can be a chemical property, such as $O_2$ concentration, or a physical property, such as heat.

---

**FIG. 39.4** **Animal gas exchange organs.** Gas exchange organs are shaped and folded to provide large surfaces for gas exchange.

**a.** Aquatic animals

**Tube worm: External gills**

**Salamander: External gills**

**Fish: Internal gills**

**b.** Terrestrial animals

**Grasshopper: Tracheae**

**Dog: Internal lungs**

**FIG. 39.5** **Respiration by fish gills.** Fish gills extract $O_2$ from water efficiently through a unidirectional flow of water and countercurrent flow.

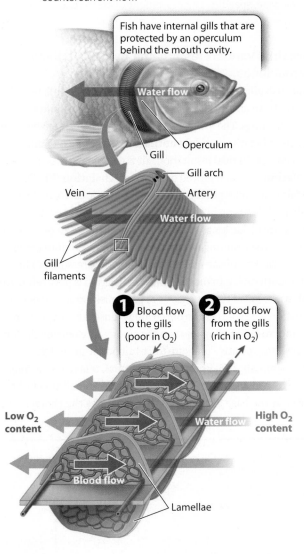

Fish have internal gills that are protected by an operculum behind the mouth cavity.

Water flow

Operculum

Gill

Gill arch

Vein

Artery

Water flow

Gill filaments

❶ Blood flow to the gills (poor in $O_2$)

❷ Blood flow from the gills (rich in $O_2$)

Low $O_2$ content

Water flow

High $O_2$ content

Blood flow

Lamellae

Consider for a moment two tubes right next to each other, one carrying hot water and the other carrying cold water (**Fig. 39.6**). If we want to make the hot water cold and the cold water hot, the efficiency of heat transfer between the two tubes depends on the direction of flow. When the water in the tubes flows in the same direction (concurrent flow), the hot water gets colder and the cold water gets hotter since heat is transferred between the two tubes (Fig. 39.6a). Given enough distance, the water will be at an intermediate (warm) temperature at the end of the tubes, reaching the same average temperature in both tubes.

Now consider what happens when the water in the two tubes flows in opposite directions (countercurrent flow). In this case, the cold water encounters increasingly hot temperatures and, given enough distance, becomes nearly as hot as the hot water entering the other tube (Fig 39.6b). Similarly, the hot water encounters increasingly cold temperatures, and becomes nearly as cold as the

cold water entering the other tube. With countercurrent flow, the two essentially exchange properties (heat, in this example).

This mechanism, known as **countercurrent exchange**, is used widely in nature. As a result of countercurrent exchange, fish gills can extract nearly all the $O_2$ in the water that passes over them (Fig. 39.6c). At the same time, $CO_2$ readily diffuses out of the blood vessels and into the water that leaves the gill chamber. A set of blood vessels carries the $O_2$-rich blood away from the gills to supply the fish's body.

### Insects breathe air through tracheae.

With the evolution of life on land, animals that breathe air were able to increase their uptake of $O_2$ and achieve higher rates of metabolism. Oxygen uptake increased for the following reasons:

1. The $O_2$ content of air is typically 50 times greater than that of a similar volume of water. Fresh air contains approximately

**FIG. 39.6** **Concurrent and countercurrent exchange.** Countercurrent exchange is used widely in nature, such as in fish gills, in cases where two fluids exchange properties (such as $O_2$, $CO_2$, or heat) efficiently.

**a.** Concurrent flow

Cold

Warm

Heat exchange

Hot

Warm

When hot and cold water move in the same direction, the cold and hot water both become warm, reaching the same temperature.

**b.** Countercurrent flow

Cold

Hot

Heat exchange

Cold

Hot

When hot and cold water move in opposite directions, the cold water becomes hot and the hot water becomes cold.

**c.** Countercurrent flow in fish gills

10%     Blood flow     90%

Low $O_2$

High $O_2$

$O_2$ exchange

20%     Water flow     100%

210 ml $O_2$ per liter of air (21%), whereas the concentration of $O_2$ in well-mixed fresh or salt water rarely exceeds 4 ml $O_2$ per liter of water (0.4%).

2. Oxygen diffuses about 8000 times faster in air than in water. For this reason, the $O_2$ content can be close to zero a few centimeters beneath the surface of a stagnant pond.

3. Water is 800 times denser and 50 times more viscous than air. Thus, it requires more energy to pump water than to pump air past a gas exchange surface.

Because of their small size, insects can employ a direct pathway of air transport that gets air right to their tissues (see Fig. 39.4b). There is no respiratory surface. In contrast to terrestrial vertebrates, insects rely on a two-step process of ventilation and diffusion to supply their cells with $O_2$ and eliminate $CO_2$. First, air enters an insect through openings, called **spiracles,** along either side of its abdomen. The spiracles can be opened or closed to limit water loss and regulate $O_2$ delivery, much like the leaf stomata of plants (Chapter 29). Inside the insect body, air is ventilated through a branching series of air tubes—the tracheae and tracheoles—directly to the cells. Second, diffusion occurs at the cell: $O_2$ supplied by the fine airways diffuses into the cells, and $CO_2$ diffuses out and is eliminated through the insect's tracheae.

The mitochondria of the flight muscles and other metabolically active tissues of insects are located within a few micrometers of tracheole airways. Because $O_2$ and $CO_2$ diffuse rapidly in air, the tracheal system of insects delivers $O_2$ to cellular mitochondria at high rates. Insects also have an air sac system connected to their tracheae. The air sacs act like bellows to pump air through the tracheae, speeding the movement of air to tissues so that gas exchange is faster.

### Terrestrial vertebrates breathe by tidal ventilation of internal lungs.

Unlike fish gills, the lungs of most land vertebrates inflate and deflate to move fresh air with $O_2$ into the lungs and expire stale air with $CO_2$ out of the lungs. The low density and viscosity of air enables these animals to breathe by **tidal ventilation** without expending too much energy. In tidal respiration, air is drawn into the lungs during **inhalation** and then moved out during **exhalation** (**Fig. 39.7a**).

Mammals and reptiles expand their thoracic cavity to draw air inside their lungs on inhalation. The expansion of the lungs causes the air pressure inside lungs to become lower than the air pressure outside the lungs. The resulting negative pressure draws air into the lungs. By contrast, amphibians inflate their lungs by pressure produced by their mouth cavity. In most animals, exhalation is passively driven by elastic recoil of tissues that were previously stretched during inhalation. The contraction of the lungs causes the air pressure inside the lungs to become higher than the air pressure outside the lungs. The resulting positive pressure forces air out of the lungs.

In mammals, inhalation during normal, relaxed breathing is driven by contraction of the **diaphragm**, a domed sheet of muscle located at the base of the lungs that separates the thoracic and

**FIG. 39.7** **Tidal ventilation of the lungs.** (a) The movement of the diaphragm causes lung volume to increase and decrease during breathing, drawing in air, then expelling it. (b) The change in lung volume is typically about 0.5 liter.

**a.**

Inhalation

Exhalation

Rib
Lung
Thoracic cavity
Diaphragm

Negative pressure draws air into lungs.

Positive pressure expels air from lungs.

Humans and other mammals ventilate their lungs by increasing their lung volume by expanding their thoracic cavity (outlined in yellow) to draw oxygen-rich air into the lungs (inhalation) and reducing their lung volume to expire oxygen-poor air from the lungs (exhalation).

**b.**

Maximum possible inhalation
Total lung capacity
Tidal volume
Maximum possible exhalation

Lung volume (mL)

Time (sec)

**FIG. 39.8**   **Human lung anatomy.** (a) Millions of alveoli provide a vast surface for gas exchange. (b) A scanning electron micrograph of the lung alveolar surface. (c) A transmission electron micrograph showing a cross section of the alveolar wall and lung capillary.

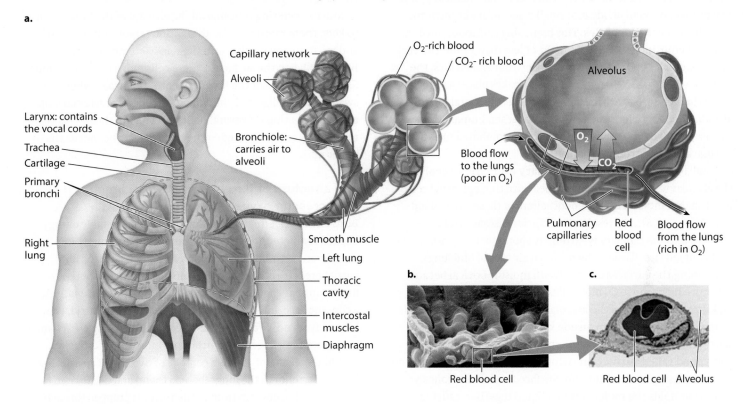

abdominal cavities (Fig. 39.7a). Exhalation occurs passively by elastic recoil of the lungs and chest wall. During exercise, other muscles come into play to assist with inhalation and exhalation. For example, **intercostal muscles**, which are attached to adjacent pairs of ribs, assist the diaphragm by elevating the ribs on inhalation and depressing them during exhalation. The action of the intercostal muscles helps to produce larger changes in the volume of the thoracic cavity, increasing the negative pressure that draws air into the lungs during inhalation and assisting elastic recoil of the lungs and chest wall to pump air out of the lungs during exhalation.

At rest, we inhale and exhale about 0.5 liter of air every cycle (**Fig. 39.7b**). This amount represents the **tidal volume** of the lungs. With a breathing frequency of 12 breaths per minute, the ventilation rate (breathing frequency × tidal volume) is 6 liters per minute. When more $O_2$ is needed during exercise, both breathing frequency and tidal volume increase to elevate ventilation rate.

The fresh air inhaled during tidal breathing mixes with $O_2$-depleted stale air that remains in the airways after exhalation. As a result, the $pO_2$ in the lung (approximately 100 mmHg in humans) is lower than the $pO_2$ of freshly inhaled air (approximately 160 mmHg at sea level). Consequently, the fraction of $O_2$ that can be extracted is lower than the fraction that can be extracted by the countercurrent flow of fish gills, which can reach 90% or more.

Typically, mammalian lungs extract less than 25% of the $O_2$ in the air, and reptile and amphibian lungs extract even less. Nevertheless, the disadvantages of tidal respiration are offset by the ease of ventilating the lung at high rates and the high $O_2$ content of air.

→   **Quick Check 3** What properties of air make it possible for humans and other mammals to breathe by tidal ventilation?

**Mammalian lungs are well adapted for gas exchange.**
Despite the limitations of tidal respiration, mammalian lungs are well adapted for breathing air. They have an enormous surface area and a short diffusion distance for gas exchange. Consequently, the lungs of mammals supply $O_2$ quickly enough to support high metabolic rates.

Except in the case of birds (discussed below), the lungs of air-breathing vertebrates are blind-ended sacs located within the thoracic cavity. Air is taken in through the mouth and nasal passages, and then passes through the **larynx,** within which the **vocal cords** are located. (The vocal cords enable speech, song, and sound production.) Air then enters the **trachea,** the central airway leading to the lungs (**Fig. 39.8**). The trachea divides into two airways, called **primary bronchi,** one of which supplies each lung. The trachea and bronchi are supported by rings of cartilage that prevent them from collapsing during respiration, ensuring that airflow meets with minimal resistance. You can feel some of these cartilage rings at the front of your throat below your larynx (the part of the larynx you can feel is your Adam's apple). The primary bronchi divide into smaller secondary bronchi and again into finer **bronchioles**. This branching continues until the terminal bronchioles have a diameter of less than 1 mm. The very fine bronchioles end with clusters of tiny thin-walled sacs, the **alveoli,** where gas exchange by diffusion takes place. As a result of this branching pattern, each lung consists of several million alveoli, providing a large surface area. Together, human lungs consist of 300 to 500 million alveoli with a combined surface area as large as a tennis court—about 100 m$^2$!

Small blood vessels, called **pulmonary capillaries**, supply the alveolar wall. Each alveolus is lined with thin epithelial cells in intimate contact with the endothelium of these small blood vessels (Fig. 39.8). As a result, the diffusion distance from the alveolus into the capillary is extremely short (about 2 μm), similar to that of gill lamellae and the terminal airways of bird lungs.

Keeping the surfaces of the alveoli moist is critical because moisture helps move $O_2$ molecules from the air into solution and thus diffuse across the alveolar wall. Mucus-secreting cells lining the alveoli keep the inside surface of the alveoli coated with a fluid film. Other alveolar epithelial cells produce a compound called **surfactant**, which acts like soap to reduce the surface tension of the fluid film. Surface tension is a cohesive force that holds the molecules of a liquid together, causing the surface to act like an elastic membrane. Because of surface tension, it requires greater pressure to inflate a small balloon than a large one. The surfactant allows the lungs to be inflated easily at low volumes when the alveoli are partially collapsed

and small. It also ensures that alveoli of different sizes inflate with similar ease, enabling more nearly uniform ventilation of the lung.

Mucus-secreting cells line all the airways of the lung not only to keep them moist but also to trap and remove foreign particles and microorganisms that an animal may breathe in with air. Beating cilia on the surface of these cells move the mucus and foreign debris out of the lungs and into the throat, where it is swallowed and digested. Smoking even one cigarette can stop the cilia's beating for several hours or even kill them. This is one reason why smokers cough—they must clear the mucus that builds up.

### The structure of bird lungs allow unidirectional airflow for increased oxygen uptake.

Birds extract $O_2$ efficiently to supply the considerable energy needed for flight. An efficient $O_2$ supply is particularly important for migratory birds that fly at high altitude where the air's partial pressure of $O_2$ is relatively low. The lungs of birds are unique in that they are rigid and do not inflate and deflate. As a consequence, birds do not use tidal breathing. Instead, they benefit from unidirectional airflow that enhances gas exchange by maintaining larger concentration gradients for diffusion.

A bird's respiratory system consists of two sets of air sacs: one located posterior to the lungs (behind them) and one located anterior to the lungs (in front of them). Air is pumped through these air sacs in a bellows-like action (**Fig. 39.9**). As the set of posterior air sacs expand, air is drawn from the bird's mouth into those sacs. As the sacs are then compressed by surrounding muscles acting on the skeleton, air is pumped through the bird's rigid lungs.

---

**FIG. 39.9**   **Respiration by bird lungs.** Anterior and posterior air sacs move air unidirectionally through a rigid lung, increasing the uptake of $O_2$ at low $pO_2$.

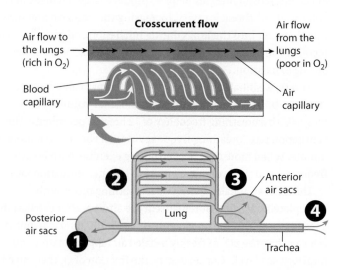

The lungs consist of a series of larger air channels that branch into fine air capillaries having a large surface area. The air capillaries direct the air so that it moves **crosscurrent** (at 90°) to the path of blood flow through the walls of the air capillaries. This crosscurrent flow is not as efficient as the countercurrent flow of fish gills, but it enables birds to extract more $O_2$ than mammals can.

After losing $O_2$ and gaining $CO_2$, the air leaves the lungs and enters the set of anterior air sacs. After fresh air is inhaled into the posterior air sacs, the stale air in the anterior air sacs is exhaled out of the bird's trachea and mouth. As a result, birds achieve a continuous supply of fresh air into the lung during both inhalation *and* exhalation. In contrast to other air-breathing vertebrates, therefore, birds move air through their respiratory system in two cycles of ventilation.

### Voluntary and involuntary mechanisms control breathing.

Because an animal's need for $O_2$ varies with activity level, animals adjust their respiratory rate to meet their cells' changing demand for $O_2$. Respiration is a unique physiological process in that it is controlled by both the voluntary and involuntary components of the nervous system (Chapter 35). In sleep, breathing is maintained at a resting rate by the involuntary part of the nervous system. Indeed, in most circumstances, breathing is controlled unconsciously.

The regulation of blood $O_2$ levels is a key example of homeostasis, as we saw in the case of core body temperature (Chapter 35) and blood glucose levels (Chapter 38). Recall that homeostasis often depends on sensors that monitor the levels of the chemical being regulated. In the case of breathing, the sensors are chemoreceptors located within the brainstem and in sensory structures called the **carotid** and **aortic bodies (Fig. 39.10).** The carotid bodies sense $O_2$ and proton ($H^+$) concentrations of the blood going to the brain, and the aortic bodies monitor their levels in blood moving to the body. In contrast, chemoreceptors in the brainstem sense $CO_2$ and $H^+$ concentrations. The most important factor in the control of breathing is the amount of $CO_2$ in the blood. If the concentration of $CO_2$ in the blood is too high, chemoreceptors in the brainstem stimulate motor neurons that activate the respiratory muscles to contract more strongly or more frequently. Stronger or faster breathing rids the blood of excess $CO_2$ and increases the supply of $O_2$ to the body.

Breathing can also be controlled voluntarily. It is a simple matter to voluntarily hold your breath. Holding the breath makes it possible for humans and marine mammals to dive under water; it is also critical to the production of speech, song, and sound for communication. Sound is produced by voluntarily adjusting the magnitude and rate of airflow over the vocal cords of mammals, the syrinx of songbirds, or the glottal folds of calling amphibians, such as some toads and frogs.

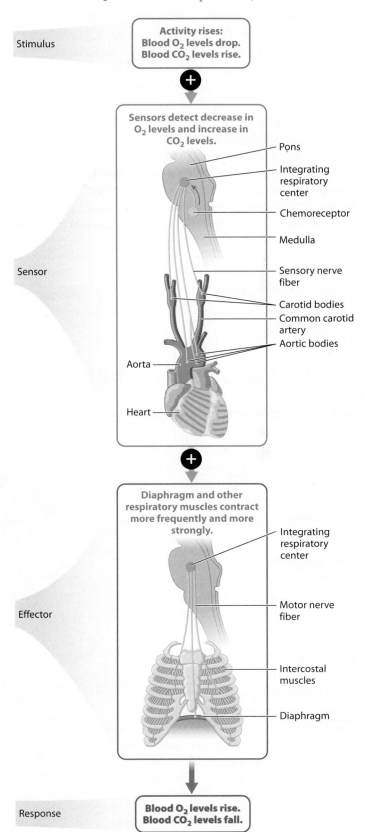

**FIG. 39.10  Homeostatic control of breathing.** Breathing becomes stronger or faster when $O_2$ levels drop.

## 39.3 OXYGEN TRANSPORT BY HEMOGLOBIN

Once $O_2$ is extracted from the environment and diffuses into the animal, it is transported to respiring cells. Oxygen is transported in the circulation by specialized body fluids—the blood of vertebrates or the hemolymph of invertebrates.

Blood plasma is the fluid portion of blood without the cells. It can hold only as much $O_2$ or $CO_2$ as can be dissolved in solution. How much $O_2$ goes into solution at a given partial pressure is a measure of its **solubility**. Because $O_2$ is about 30 times less soluble than $CO_2$, only about 0.2 ml of $O_2$ can be carried in 100 ml of blood. In both vertebrates and invertebrates, **hemoglobin** evolved as a specialized iron-containing, or heme, molecule for $O_2$ transport. By binding $O_2$ and removing it from solution, hemoglobin increases the amount of $O_2$ in the blood a hundredfold. Because of its greater

solubility, $CO_2$ is transported in solution within the blood. Rather than remaining dissolved in solution as a gas, most of the $CO_2$ (about 95%) is converted to carbonic acid, which dissociates to form bicarbonate ions ($HCO_3^-$) and protons (Chapter 6).

Whereas some invertebrate hemoglobins exist in solution in the hemolymph and others exist within cells carried by the hemolymph, all vertebrate hemoglobins are produced by and reside in red blood cells. Hemoglobin gives the cells and the blood their red appearance.

### Hemoglobin is an ancient molecule with diverse roles related to oxygen binding and transport.

Hemoglobin is an ancient molecule that has been found in organisms representing all kingdoms of life. In bacteria, archaeons, and fungi, similar forms of hemoglobin evolved as

---

## HOW DO WE KNOW?

**FIG. 39.11**

# What is the molecular structure of hemoglobin and myoglobin?

**BACKGROUND** In the 1950s, the Austrian Max Perutz worked with John Kendrew in the Cavendish Laboratory of the University of Cambridge to determine the molecular structure of globular proteins. Perutz and Kendrew were interested in understanding how the structures of hemoglobin and myoglobin enabled binding and transport of $O_2$.

**EXPERIMENT** Perutz and Kendrew developed and applied the new technique of X-ray crystallography to determine the three-dimensional molecular structures of hemoglobin and myoglobin.

**RESULTS** Perutz and Kendrew showed that adult hemoglobin consists of four subunits, two α (alpha) and two β (beta) subunits. Each subunit contains a heme group that contains iron, which is the site of $O_2$ binding. By contrast, myoglobin consists of only a single subunit with one heme group. These differences in molecular structure underlie the $O_2$-binding and dissociation properties of the two $O_2$ transport proteins.

Hemoglobin

Myoglobin

**FOLLOW-UP WORK** Work by Perutz in the 1960s showed how the molecular structure of hemoglobin changes when it binds $O_2$ versus when it is in an unbound state. Perutz also supervised Francis Crick as a doctoral student, who later worked with James Watson to resolve the structure of DNA. Perutz and Kendrew received the 1962 Nobel Prize in Chemistry for their work.

**SOURCES** Kendrew J. C., M. F. Perutz. 1948. "A Comparative X-Ray Study of Foetal and Adult Sheep Haemoglobins." *Proceedings of the Royal Society, London, Series A* 194:375–98; Perutz, M. F. 1964. "The Hemoglobin Molecule." *Scientific American* 211:64–76; Perutz, M. F. 1978. "Hemoglobin Structure and Respiratory Ttransport." *Scientific American* 239:92–125.

**FIG. 39.12** (a) Hemoglobin and (b) myoglobin dissociation curves.

**a.**

The curve is sigmoidal due to cooperative binding of oxygen by hemoglobin. In the middle of the curve, small increases (or decreases) in $pO_2$ result in large increases (or decreases) in hemoglobin saturation.

**b.**

At any given $pO_2$, myoglobin binds oxygen more readily than hemoglobin. As a result, hemoglobin delivers oxygen to myoglobin in muscle tissues.

$P_{50}$ is the $pO_2$ at which hemoglobin is 50% saturated with $O_2$.

a terminal electron acceptor in cellular respiration and may have also served a role in scavenging $O_2$ to avoid damage due to oxidative stress.

Plant hemoglobins are known that bind and transport $O_2$ within the cell. In soybeans, hemoglobin binds $O_2$ in root nodule cells to keep concentrations low enough to avoid inhibiting nitrogenase, the enzyme used by symbiotic rhizobial bacteria in the root nodules for nitrogen fixation (Chapter 29).

Invertebrate hemoglobins found in roundworms, annelid worms, and arthropods bind $O_2$ in the circulatory fluid, as discussed above. They share considerable gene and amino acid sequence similarity with vertebrate hemoglobins, indicating that a common ancestor to invertebrate and vertebrate animals evolved a shared form of hemoglobin more than 670 million years ago.

### Hemoglobin reversibly binds oxygen.

Hemoglobin exists in large concentrations within red blood cells. It is a globular protein that consists of four polypeptide units (**Fig. 39.11**). Each of these four units surrounds a heme group that contains iron, which reversibly binds one $O_2$ molecule. After $O_2$ diffuses into the blood, it diffuses into the red blood cells and binds to the heme groups in hemoglobin. Hemoglobin's binding of $O_2$ removes $O_2$ from solution, keeping the $pO_2$ of the red blood cell below that of the blood plasma so that $O_2$ continues to diffuse into the cell. The removal of $O_2$ from the plasma, in turn,

keeps the $pO_2$ of the plasma below that of the lung alveolus, so $O_2$ continues to diffuse from the lungs into the blood.

When more $O_2$ is present in blood plasma, we expect more $O_2$ to become bound to hemoglobin, and that is what happens. But there is an interesting wrinkle that enables hemoglobin to readily bind $O_2$ leaving the lung. If we plot blood $pO_2$ against the percentage of $O_2$ bound to hemoglobin (the relation defined as "hemoglobin saturation"), we get the curve shown in **Fig. 39.12a**. As blood $pO_2$ increases, hemoglobin saturation rises slowly at first, then more steeply, and then more slowly again until it levels out. At 100% saturation, all hemoglobin molecules bind four $O_2$ molecules. The curve is called the hemoglobin's **oxygen dissociation curve,** and it has a sigmoidal shape—that is, a shape like an "S."

The shape of this curve can be explained as a result of changes in the ability of hemoglobin to bind $O_2$ (a property called its binding affinity) at different $O_2$ partial pressures. At 25% saturation, each hemoglobin molecule binds on average one $O_2$ molecule. As $pO_2$ rises, hemoglobin binds $O_2$ with increasing binding affinity. This increase in affinity results from the interaction of adjacent heme groups in each hemoglobin molecule. After one heme group binds the first $O_2$ molecule, hemoglobin undergoes a conformational change that increases the binding affinity of the remaining heme groups for additional $O_2$. The increase in binding affinity with additional binding of $O_2$ is called **cooperative binding**, and it gives the $O_2$ dissociation curve for hemoglobin its sigmoidal shape.

The increasing affinity of hemoglobin for $O_2$ with increasing $O_2$ concentration has important physiological consequences. In the middle part of the $O_2$ dissociation curve, small increases in $O_2$ concentration lead to large increases in hemoglobin saturation. Therefore, under normal circumstances, hemoglobin in the blood leaving the lung is fully saturated with $O_2$—each hemoglobin molecule is bound to four $O_2$ molecules.

The shape of the $O_2$ dissociation curve also helps to explain how $O_2$ is delivered to respiring cells. When the hemoglobin in red blood cells reaches tissues needing $O_2$ to supply their mitochondria, the $O_2$ is released from the hemoglobin. As active cells consume $O_2$, they reduce the local $pO_2$ of the cell and surrounding tissues to 40 mmHg or less. At these lower $pO_2$ values, the hemoglobin's $O_2$ dissociation curve has a steep slope (Fig. 39.12a). The steepness of the slope indicates that for a relatively small decrease in $pO_2$, large amounts of $O_2$ can be released from hemoglobin to diffuse into the cell.

## Myoglobin stores oxygen, enhancing delivery to muscle mitochondria.

**Myoglobin** is a specialized $O_2$ carrier within the cells of vertebrate muscles. In contrast to hemoglobin, myoglobin is a monomer that contains only a single heme group. Because there are no interacting subunits, the $O_2$ dissociation curve for myoglobin has a different shape from that of hemoglobin (**Fig. 39.12b**). In fact, over the physiological range of $pO_2$, myoglobin has a greater affinity for $O_2$ than hemoglobin does

and binds $O_2$ more tightly. As a result, hemoglobin releases $O_2$ to exercising muscles. Once the exercising muscles consume the available $O_2$, the intracellular $pO_2$ drops, and the myoglobin releases its bound $O_2$ to the muscle cell's mitochondria.

Red muscle cells that depend mainly on aerobic respiration to produce ATP (Chapter 37) store large amounts of myoglobin. The myoglobin in these cells can release $O_2$ quickly at the onset of activity, before respiratory and circulatory systems have had time to increase the supply of $O_2$. Diving marine mammals, such as whales and seals, carry large amounts of myoglobin within their muscles. The myoglobin loads up with $O_2$ when the animals breathe at the surface before a dive. The $O_2$ bound to the myoglobin is then used to supply ATP during the dive, when the animal cannot breathe. These marine mammals can stay under water for 30 minutes or longer and dive to considerable depths (**Fig. 39.13**).

## Many factors affect hemoglobin–oxygen binding.

Obtaining enough $O_2$ is a special challenge in some circumstances. How does a mammalian fetus in its mother's uterus take up $O_2$ from its mother's blood? How do animals that live at high altitude obtain $O_2$ under conditions of low $pO_2$? In each case, the species adapts by evolving forms of hemoglobin with higher binding affinity.

In mammals, the mother's respiratory and cardiovascular systems must provide the fetus with $O_2$ and remove $CO_2$, as well as supply nutrients and remove waste. Maternal and fetal blood do not mix. Rather, they exchange gases across the placenta. Yet the $O_2$ concentration gradient at this exchange point does not favor the movement of $O_2$ from the mother's blood to the fetal blood. Fetal mammals solve the problem of extracting $O_2$ from their mother's hemoglobin by producing a form of hemoglobin before birth that has a higher affinity for $O_2$ than their mother's hemoglobin. The $O_2$ dissociation curve for fetal hemoglobin is shifted to the left of the curve for maternal hemoglobin (**Fig. 39.14a**), allowing the fetal hemoglobin to extract $O_2$ from its mother's circulation. At birth, when the newborn begins to breathe, its red blood cells rapidly shift to producing the adult form of hemoglobin, maintaining this form throughout life.

Many animals live at high altitude. For example, birds have been observed flying over the highest peaks (taller than 9000 m) despite the fact that air at this altitude has less than one-third the $O_2$ content of air at

**FIG. 39.13** **Myoglobin as an $O_2$ store for long-lasting dives.** The graph tracks the dives of a beaked whale (*Ziphius cavirostris*) fitted with a tracking device.

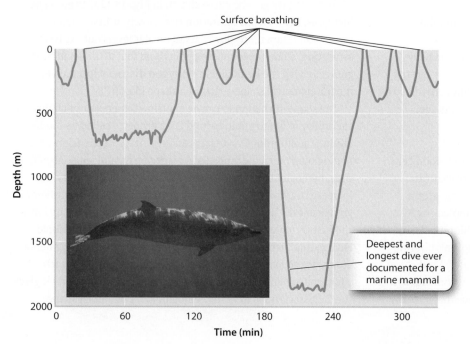

Surface breathing

Deepest and longest dive ever documented for a marine mammal

Depth (m)

Time (min)

**FIG. 39.14 Changes in hemoglobin's O$_2$ dissociation curve.**

a.

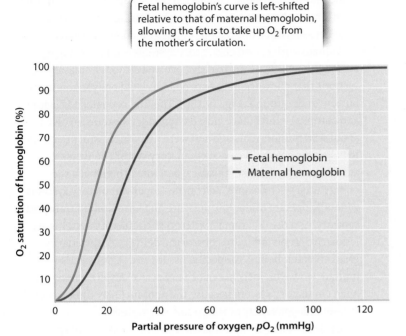

Fetal hemoglobin's curve is left-shifted relative to that of maternal hemoglobin, allowing the fetus to take up O$_2$ from the mother's circulation.

b.

A decrease in pH, often the result of increased cell activity, causes a right shift in hemoglobin's O$_2$ dissociation curve. Hemoglobin then releases more O$_2$ to the active cells.

sea level. Similarly, llamas inhabit the Andes at altitudes up to 5000 m, where at a $pO_2$ of approximately 80 mmHg the air has about one-half the O$_2$ content of air at sea level. Consequently, the lung $pO_2$ of a llama is only about 50 mmHg. How do high-flying birds and llamas obtain enough O$_2$? Their hemoglobin has a higher affinity for O$_2$ and binds O$_2$ more readily than that of mammals living at sea level. Consequently, their O$_2$ dissociation curve is shifted to the left, favoring the binding of more O$_2$ at a given $pO_2$.

Tissue and blood pH also has an important physiological effect on the O$_2$ affinity of hemoglobin. When a decrease in pH (that is, an increase in H$^+$ concentration) occurs during exercise, the affinity of hemoglobin for O$_2$ *decreases*, resulting in a rightward shift in the O$_2$ dissociation curve (**Fig. 39.14b**). This phenomenon is called the Bohr effect after Christian Bohr, the Danish physiologist who first described it in 1904. Decreases in pH occur when CO$_2$ is released from metabolizing cells, or when an inadequate supply of O$_2$ leads to the production of lactic acid (Chapter 7). Because hemoglobin's affinity for O$_2$ is reduced, more O$_2$ is released and supplied to the cells for aerobic ATP synthesis.

Carbon dioxide also reacts with the amine (NH$_2$) groups of hemoglobin, reducing hemoglobin's affinity for O$_2$. Thus, when released from respiring tissues, CO$_2$ promotes increased O$_2$ delivery both through its direct effect on hemoglobin and its contribution to a decrease in blood pH by the Bohr effect. While the production of CO$_2$ promotes O$_2$ release at the tissues, its

elimination at the lung increases hemoglobin's affinity for O$_2$, enhancing O$_2$ uptake.

→ **Quick Check 4** Sickle-cell anemia is a genetic disorder of individuals homozygous for a mutation of hemoglobin that causes their red blood cells to be sickle shaped and stiff under conditions of low $pO_2$. Why is this disease life threatening?

## 39.4 CIRCULATORY SYSTEMS

The O$_2$ carried by hemoglobin in red blood cells is transported to tissues throughout the body by the circulatory system. The **closed circulatory systems** of animals are made up of a set of internal vessels and a pump—the **heart**—to transport the blood to different regions of the body. Closed circulatory systems have two conflicting requirements: They must produce enough pressure to carry the circulating blood to all the tissues, yet once the blood reaches smaller vessels that supply the cells within the tissues the blood pressure and flow rate must not be too high. High pressure would push fluid through the walls of the smaller vessels and cause the blood to flow so quickly that there would not be enough time to exchange gases.

Generally, smaller animals, such as insects and many mollusks, have **open circulatory systems** that contain few blood vessels: Most of the circulating fluid, the hemolymph, is contained within

**FIG. 39.15** (a) Open and (b) closed invertebrate circulatory systems.

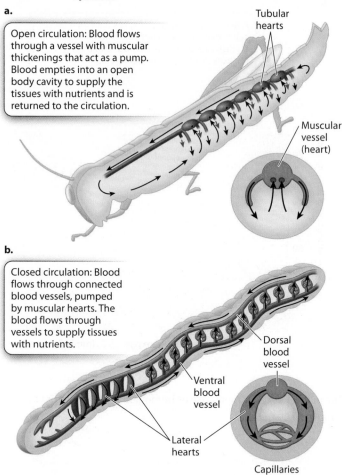

**a.**

Open circulation: Blood flows through a vessel with muscular thickenings that act as a pump. Blood empties into an open body cavity to supply the tissues with nutrients and is returned to the circulation.

Tubular hearts

Muscular vessel (heart)

**b.**

Closed circulation: Blood flows through connected blood vessels, pumped by muscular hearts. The blood flows through vessels to supply tissues with nutrients.

Dorsal blood vessel

Ventral blood vessel

Lateral hearts

Capillaries

the animal's body cavity. The hemolymph bathes the animal's tissues and organs (**Fig. 39.15a**). Open circulatory systems have limited control of where the fluid moves. Muscles that are active in locomotion can assist circulation of the hemolymph, and some invertebrates have simple hearts with openings that help pump fluid between different regions of the animal's body cavity.

Animals with closed circulatory systems tend to be larger, although segmented worms have evolved closed circulatory systems that allow them to expand individual body segments to burrow underground (**Fig. 39.15b**). The first cephalopods (squid and octopus), which have closed circulatory systems, evolved about 480 million years ago and were early successful predators in Paleozoic oceans. A closed circulatory system delivers $O_2$ at high rates to exercising tissues, providing them with the energy used to chase prey. In contrast, open circulatory systems generally operate under low pressure and have limited transport capacity. As a result, animals with open circulatory systems are less active. Their ability to control the delivery of respiratory gases and metabolites to specific tissues and regions is also limited. However, insects can be very active despite having an open circulatory system because they obtain $O_2$ by their tracheal system independently of the circulatory system.

Closed circulatory systems can control blood flow to specific regions of the body by varying the resistance to flow. For example, wading birds reduce blood flow to their legs when they are in cold water to reduce heat loss, and vertebrates can increase blood flow to their muscles to deliver $O_2$ and nutrients during exercise. In order to pump blood through a set of closed interconnected vessels, a muscular heart is needed to produce sufficient pressure to overcome the flow resistance of the vessels.

**FIG. 39.16** **Flow and vessel size.** (a) Vessel hierarchy of circulatory systems; (b) total vessel cross-sectional area versus flow velocity.

**a.**

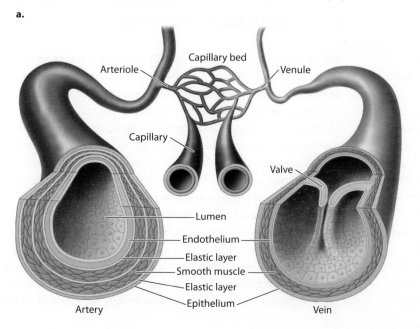

Capillary bed

Arteriole

Venule

Capillary

Valve

Lumen

Endothelium

Elastic layer

Smooth muscle

Elastic layer

Epithelium

Artery

Vein

**b.**

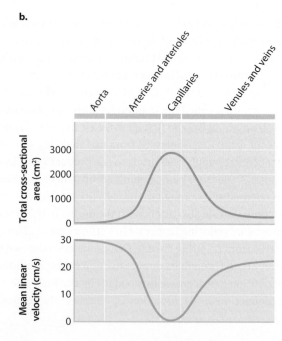

Aorta

Arteries and arterioles

Capillaries

Venules and veins

Total cross-sectional area (cm²)

3000

2000

1000

0

Mean linear velocity (cm/s)

30

20

10

0

## Circulatory systems have vessels of different sizes to facilitate bulk flow and diffusion.

For a fluid like blood to flow through a set of pipes, pressure (P) is required to overcome the resistance (R) to flow. The pressure developed by a heart when it contracts establishes a rate of blood flow that is governed by P/R. That is, the rate of blood flow increases with an increase in pressure and decreases with an increase in resistance. The resistance to flow is determined in part by the fluid's stickiness (its **viscosity**) and the vessel's length. Longer vessels of a given size impose greater resistance. However, the main factor governing resistance to flow is the vessel's radius (r). Resistance is proportional to $1/r^4$, which means that if a vessel's radius is reduced by half, its resistance to flow increases 16 times.

The high resistance of narrow vessels presents a challenge. To keep diffusion distances short, gas exchange between cells and blood requires narrow vessels. How does the circulatory system overcome the dramatic increase in resistance with narrower diameters? Animal circulatory systems are organized to have blood flow over longer distances in a relatively few large-diameter vessels with low resistance to flow (**Fig. 39.16a**).

**Arteries** are the large, high-pressure vessels that move blood flow away from the heart to the tissues. **Veins** are the large, low-pressure vessels that return blood to the heart. Arteries branch into blood vessels of progressively smaller diameter called **arterioles,** and these arterioles ultimately connect to finely branched networks of very small blood vessels called **capillaries.** It is at the capillaries that gases are exchanged by diffusion with the surrounding tissues. The number of smaller-diameter vessels at each branching point in the circulatory system greatly exceeds the number of larger diameter vessels. The reverse organization is found on the return side of circulation: Numerous capillaries drain into vessels of progressively larger diameter called **venules,** and the venules drain into a few larger veins that return blood to the heart.

This organization has two advantages. First, it maintains the same volume of blood flow at all levels within the circulatory system: The increased resistance to flow in the capillaries is offset by the large increase in the number of capillaries. Second, it enables the blood to flow more slowly in smaller vessels (**Fig. 39.16b**), providing time for gases and metabolites to diffuse into and out of the neighboring cells. Indeed, the diameter of a capillary is so small that red blood cells pass through one at a time, and must be flexible to do even that.

## Arteries are muscular, elastic vessels that carry blood away from the heart under high pressure.

When an animal becomes active after a period of rest and feeding, blood flow must be increased to its muscles and reduced to its digestive organs (**Fig. 39.17**). How do animals accomplish this? The answer comes from physics: the strong effect of vessel radius on flow resistance. Arterioles supplying regions of the body that need less blood shrink in diameter by contracting the circular smooth muscle fibers in their wall. The reduction in diameter increases the resistance to flow and thereby reduces the rate of blood flow. For example, when you put your hands or feet into cold water, the arterioles contract, reducing the loss of heat from blood flowing to your hands and feet.

Arterioles supplying regions of the body that need more blood expand in diameter by relaxing their smooth muscles. The larger vessels offer less resistance, so more blood can flow through them. What is the mechanism that causes the

**FIG. 39.17 Changes in blood flow from resting to exercise as measured in humans.** Numbers in parentheses indicate percent of total blood flow.

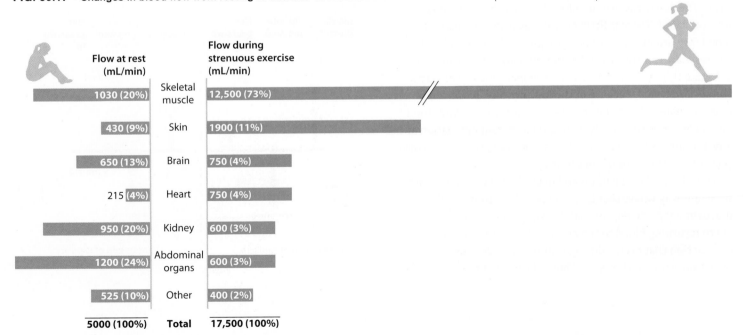

| | Flow at rest (mL/min) | | Flow during strenuous exercise (mL/min) |
|---|---|---|---|
| Skeletal muscle | 1030 (20%) | | 12,500 (73%) |
| Skin | 430 (9%) | | 1900 (11%) |
| Brain | 650 (13%) | | 750 (4%) |
| Heart | 215 (4%) | | 750 (4%) |
| Kidney | 950 (20%) | | 600 (3%) |
| Abdominal organs | 1200 (24%) | | 600 (3%) |
| Other | 525 (10%) | | 400 (2%) |
| **Total** | **5000 (100%)** | | **17,500 (100%)** |

smooth muscles to relax when cells in neighboring tissue become more active? As the rate of cellular respiration in these cells increases, they begin releasing more $CO_2$ as well as other metabolites, such as lactate and hydrogen ions, that are produced in respiration. The increase of these molecules in the bloodstream acts as a signal to the smooth muscle to begin relaxing. In this way, blood flow increases to the capillaries to match $O_2$ and nutrient delivery to the metabolic need of the cells.

A strong pulse of flow pushes against the arterial walls about every second in a human as the heart beats to pump blood through the arteries. This pulse of flow causes a momentary expansion of the arteries. Artery walls are able to withstand these repeated pressure pulses because they contain multiple elastic layers composed of two proteins: **collagen** and **elastin.** Collagen fibers are strong and resist the expansion of the arterial wall during these pressure pulses. Together with elastin fibers, the collagen provides an elastic rebound of the arterial wall once the pulse has passed, returning energy to help smooth out blood flow. The collagen and elastin resist overexpansion of the arterial wall at local sites. If an artery wall deteriorates, the collagen and elastin can become so thin that the artery bulges outward. The outward bulge, called an aneurysm, can lead to a life-threatening rupture.

→ **Quick Check 5** What change in vessel shape most affects resistance to blood flow?

### Veins are thin-walled vessels that return blood to the heart under low pressure.

Blood collected from local capillary networks returns to the heart through progressively larger veins that ultimately drain into the two largest veins, the **venae cavae** (singular, vena cava). The venae cavae drain blood from the head and body into the heart. There is little pressure available to push the blood forward in the veins because pressure has been lost to the resistance of the arterioles and capillaries. Consequently, veins are thin walled and have little smooth muscle or elastic connective tissue. Because of the low pressure, blood tends to accumulate within the veins: As much as 80% of your total blood volume resides in the venous side of your circulation at any one time. When you stand or sit for a long time, blood may pool within the veins of your limbs.

Veins located in the limbs and in the body below the heart have one-way valves that help prevent blood from pooling due to gravity (see Fig. 39.16a). However, the most important aid to returning blood to the heart is the voluntary muscle contraction that occurs during walking and exercise, which exerts pressure on the veins. Above the level of the heart, the

return of blood to the heart through the veins is assisted by gravity.

### Compounds and fluid move across capillary walls by diffusion, filtration, and osmosis.

We have seen that $O_2$ and $CO_2$ move between capillaries and tissues by diffusion. In addition, water, certain ions, and other small molecules, but not proteins, move from capillaries into the surrounding interstitial fluid (the extracellular fluid surrounding vessels) by filtration, forced by blood pressure.

Why doesn't the blood plasma lose all its water and ions over time? As water is filtered out of the capillaries, some ions, proteins, and other large compounds remain inside, increasing in concentration (**Fig. 39.18**). As a result, there is a tendency of water to flow back into the capillary by osmosis (Chapter 5). At the same time, blood pressure decreases because of the flow resistance imposed by the capillaries. When the pressure is no longer high enough to counter osmosis, water moves back into the capillaries.

In vertebrates, excess interstitial fluid is also returned to the bloodstream by means of the **lymphatic system,** a network of vessels distributed throughout the body. The fluid that enters the lymphatic system is called the **lymph.** Small lymphatic vessels merge with progressively larger thin-walled vessels, draining the lymph into lymphatic ducts, which empty into the venous system and then the heart. Lymphatic vessels have one-way valves, similar to those in the veins, that assist the return of lymph to the circulatory system. Local muscle contractions also

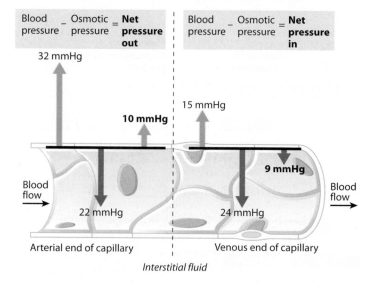

**FIG. 39.18** Pressures pushing fluid out of blood vessels (blood pressure) and into them (osmotic pressure).

| Blood pressure | − | Osmotic pressure | = | **Net pressure out** |
| --- | --- | --- | --- | --- |

32 mmHg

10 mmHg

22 mmHg

Blood flow

Arterial end of capillary

| Blood pressure | − | Osmotic pressure | = | **Net pressure in** |
| --- | --- | --- | --- | --- |

15 mmHg

9 mmHg

24 mmHg

Blood flow

Venous end of capillary

*Interstitial fluid*

help to pump the lymph toward the thoracic duct. In addition to its role in returning plasma to the bloodstream, the lymphatic system has important functions in the immune system (Chapter 43).

**? CASE 7 Predator–Prey: A Game of Life and Death**
**How do hormones and nerves provide homeostatic regulation of blood flow as well as allow an animal to respond to stress?**

An animal that is dehydrated or has lost blood following an injury may experience a prolonged drop in blood pressure. In such cases, the posterior pituitary gland releases antidiuretic hormone (ADH), also known as vasopressin (Chapter 38), into the circulation. ADH causes the arteries to contract, increasing their resistance to blood flow. The higher resistance elevates blood pressure throughout the body.

Even under less stressful conditions, blood pressure is continually being adjusted to remain within normal physiological limits, another example of homeostasis. The autonomic nervous system plays an important role in keeping blood pressure at healthy levels (Chapter 35). If blood pressure in the general circulation drops, animals reduce the supply of blood to the limbs by constricting arteries that supply the limbs. This response helps maintain blood pressure to the heart, brain, and kidneys. In this case, sympathetic neurons synapsing on the smooth muscles of arterioles in the limbs stimulate these muscles to contract. In contrast, when blood pressure is too high, sympathetic neurons that synapse on the smooth muscles are inhibited. The smooth muscles relax, reducing resistance in the arteries and increasing blood flow. These two responses are called **vasoconstriction** and **vasodilation**.

In some circumstances, it may be advantageous for an animal's blood pressure to rise higher than normal levels. One example is during the fight-or-flight response that occurs when animals are faced with a stressful situation, as when prey animals are alerted to the presence of a predator, and when predators locate a potential prey. The sympathetic nervous system stimulates the adrenal gland to release the hormone **adrenaline** (also called **epinephrine**). The surge of adrenaline into the bloodstream—the "adrenaline rush"—causes an animal to become alert and aroused. At the same time, the sympathetic nervous system acts to increase heart rate and respiratory rate and inhibit digestive processes. Arteries supplying limb muscles dilate to enhance blood flow to exercising muscles, while arteries supplying organs within the digestive system constrict. The increase in heart and breathing rates increases the rate of $O_2$ delivery to the exercising muscles to meet their increased demand for ATP. These physiological changes, along with a heightening of sensory acuity, prepare an animal for the more intensive activity needed to chase or flee.

In contrast, when an animal has recently eaten a meal, its sympathetic nervous system is inhibited and its parasympathetic nervous system is stimulated, increasing digestive secretions and the activity of gut smooth muscle (Chapter 40). Thus, local regulation of blood flow within different regions and organ systems of the body is integrated with more general feedback from the nervous and endocrine systems. The result is homeostatic regulation of blood pressure that monitors digestion and meets the need for $O_2$ delivery to all metabolizing cells of the animal. These shifts in blood flow are readily controlled by vasoconstriction or dilation of the arteries supplying the different organs.

## 39.5 THE EVOLUTION, STRUCTURE, AND FUNCTION OF THE HEART

The pressure required to drive blood through blood vessels is generated by the heart. Animal hearts are pumps made of muscle that rhythmically contract, producing this pressure. Early hearts were as simple as a muscular thickening of a small region of a blood vessel. Such hearts are still found in insects (see Fig. 39.15a). More complex hearts have chambers that expand and fill with deoxygenated blood returning from the animal's tissues. After filling, the heart muscle contracts, pumping deoxygenated blood to lungs or gills and newly oxygenated blood onward to the animal's tissues. The closed circulatory system of cephalopod mollusks has three hearts—two that pump deoxygenated blood through the gills and another that pumps oxygenated blood to the body. In contrast, vertebrates have a single heart with at least two chambers, one for receiving blood and the other for pumping blood to the body during each heartbeat cycle. Hearts also have valves to ensure that blood does not flow backward when the heart contracts.

A theme in the evolution of vertebrate circulatory systems is the progressive separation of circulation to the gas exchange organ from the circulation to the rest of the body. In fish, deoxygenated blood is first pumped to the gills to gain $O_2$ before flowing to the rest of the body and then returning to the heart. This single circulation path limits the rate of blood flow to metabolically active body tissues: The resistance imposed by the small vessels in the gills slows the blood flow considerably before it even reaches the tissues. By contrast, birds and mammals evolved a separate **pulmonary circulation** to the lungs and **systemic circulation** to the rest of the body. The evolution of separate pulmonary and systemic circulations was made possible by the evolution of a four-chambered heart. This organization has two advantages: It increases the supply of oxygenated blood to active tissues, and it increases the uptake of $O_2$ at the gas exchange surface.

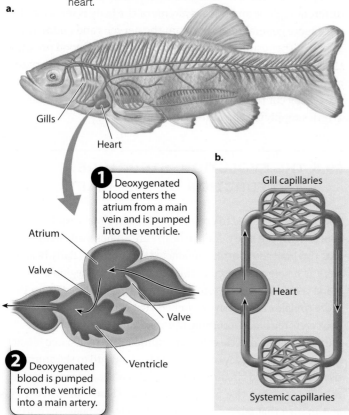

**FIG. 39.19** **Fish heart and circulatory system.** (a) Fish have a two-chambered heart that pumps deoxygenated blood. (b) Blood flows in a single circuit from the heart, through gills, to the tissues for gas exchange, and back to the heart.

**a.**

Gills

Heart

**b.**

Gill capillaries

Heart

Systemic capillaries

**1** Deoxygenated blood enters the atrium from a main vein and is pumped into the ventricle.

Atrium

Valve

Valve

**2** Deoxygenated blood is pumped from the ventricle into a main artery.

Ventricle

## Fish have two-chambered hearts and a single circulatory system.

Fish hearts have two chambers **(Fig. 39.19),** the **atrium** and **ventricle**. Deoxygenated blood returning from the fish's tissues enters the atrium, which fills and then contracts to move the blood into a thicker-walled ventricle. The muscular ventricle pumps the blood through a main artery to the gills for uptake of $O_2$ and elimination of $CO_2$. Oxygenated blood collected from the gills travels to the tissues through a large artery called the aorta.

The small gill capillaries impose a large resistance to flow. As a result, much of the blood pressure is lost in moving blood through the gills. This loss of pressure limits the flow of oxygenated blood to body tissues.

## Amphibians and reptiles have three-chambered hearts and partially divided circulations.

As animals moved onto land, the transition from breathing water to breathing air had important consequences for the organization of vertebrate circulatory systems. Land vertebrates evolved hearts that separated the circulation of deoxygenated blood pumped to their gas exchange organs from circulation of oxygenated blood delivered to their body tissues. Gas exchange became more efficient and $O_2$ delivery increased. Metabolic rates rose, and animals became capable of greater activity.

Reflecting their lifestyle, both aquatic and terrestrial, amphibians evolved a variety of ways to breathe. Some retain gills, while others use lungs or their skin, or some combination of these, to breathe. Amphibians evolved an additional atrium, dividing their heart into two separate atria and a single ventricle

**FIG. 39.20** **Amphibian heart and circulatory system.** (a) The frog heart has three chambers: two atria and one ventricle. (b) Mixed blood leaving the common ventricle goes either to the lung and back to the heart, or to the tissues and back to the heart.

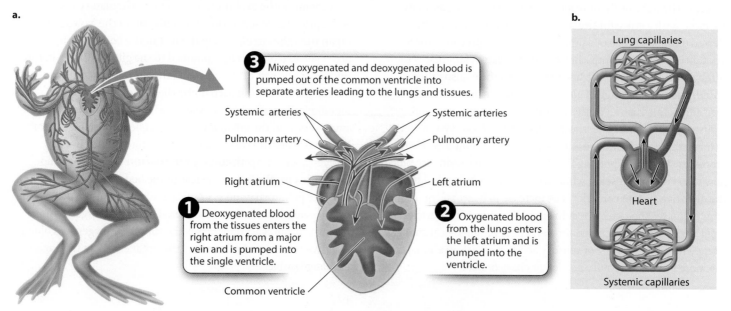

**a.**

**3** Mixed oxygenated and deoxygenated blood is pumped out of the common ventricle into separate arteries leading to the lungs and tissues.

Systemic arteries

Pulmonary artery

Right atrium

**1** Deoxygenated blood from the tissues enters the right atrium from a major vein and is pumped into the single ventricle.

Common ventricle

Systemic arteries

Pulmonary artery

Left atrium

**2** Oxygenated blood from the lungs enters the left atrium and is pumped into the ventricle.

**b.**

Lung capillaries

Heart

Systemic capillaries

**FIG. 39.21 The mammalian heart and circulatory system.** (a) Mammals have a fully divided four-chambered heart with two atria and two ventricles. (b) Mammals have separate pulmonary and systemic circulations.

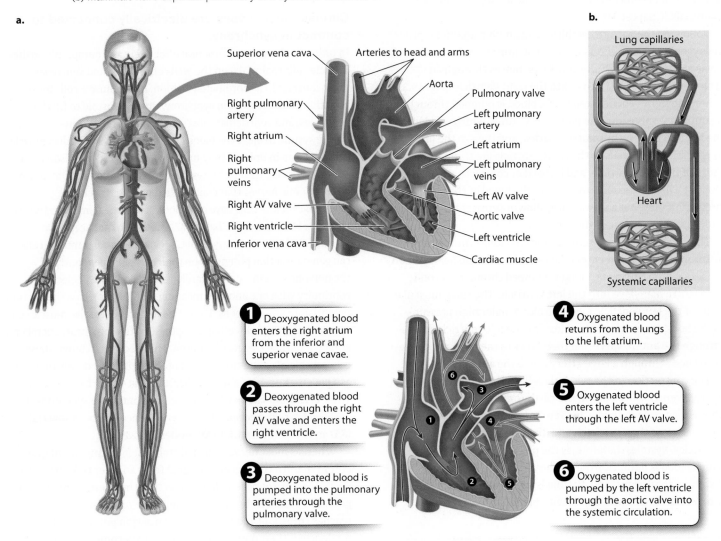

a.

Superior vena cava

Arteries to head and arms

Aorta

Right pulmonary artery

Pulmonary valve

Left pulmonary artery

Right atrium

Left atrium

Right pulmonary veins

Left pulmonary veins

Left AV valve

Right AV valve

Aortic valve

Right ventricle

Left ventricle

Inferior vena cava

Cardiac muscle

b.

Lung capillaries

Heart

Systemic capillaries

**1** Deoxygenated blood enters the right atrium from the inferior and superior venae cavae.

**2** Deoxygenated blood passes through the right AV valve and enters the right ventricle.

**3** Deoxygenated blood is pumped into the pulmonary arteries through the pulmonary valve.

**4** Oxygenated blood returns from the lungs to the left atrium.

**5** Oxygenated blood enters the left ventricle through the left AV valve.

**6** Oxygenated blood is pumped by the left ventricle through the aortic valve into the systemic circulation.

(**Fig. 39.20**). This arrangement partially separates the pulmonary and systemic circulations, so in amphibians freshly oxygenated blood can be pumped under higher pressure to the body than is the case in fishes. However, because amphibians have a single ventricle, oxygenated blood returning to the heart from the gills or lungs mixes with deoxygenated blood returning from the animal's body before being pumped from the ventricle. Anatomical features of the ventricle help to limit this mixing, but nonetheless mixing still hinders the supply of oxygenated blood to an amphibian's body.

Most reptiles also have a three-chambered heart, but their single ventricle has internal ridges to improve the separation of deoxygenated and oxygenated blood. As a result, reptiles are able to deliver more $O_2$ to their tissues than amphibians can. At the same time, the incompletely divided ventricle provides reptiles with versatile circulatory responses during exercise and when

diving underwater. For example, when they dive, the blood of turtles and sea snakes bypasses their lungs, which they cannot use underwater.

**Mammals and birds have four-chambered hearts and fully divided pulmonary and systemic circulations.**

Mammals and birds evolved four-chambered hearts with separate atria and separate ventricles (**Fig. 39.21**). This arrangement completely separates blood flow to the lungs from blood flow to the tissues. As a result, these animals can pump blood to their lungs under lower pressure, allowing increased uptake of $O_2$, while at the same time supplying blood gases and nutrients to their tissues at high pressure. Whereas the systemic blood pressure of fish is only a few mmHg, in a bird or mammal it is 100 mmHg or higher. Moreover, blood flowing to the tissues is fully oxygenated at the outset, delivering more $O_2$ to body tissues at a higher rate. This change in heart structure and circulatory pattern was linked to an increase in the metabolic rates and the evolution of endothermic lifestyles in birds and mammals (Chapter 40).

Following the pattern established in amphibians and reptiles, deoxygenated blood enters the right atrium from the venae cavae and, when the atrium contracts, moves through an **atrioventricular (AV) valve** into the right ventricle (Fig. 39.21). When the right ventricle contracts, blood is pumped through the **pulmonary valve** into the pulmonary trunk, which divides into the left and right **pulmonary arteries**, and then to the lungs for oxygenation. The walls of the right ventricle are thinner than those of the left ventricle, and its weaker contractions eject blood at a lower pressure. As a result, the blood of the pulmonary circulation moves at a slower rate, allowing greater time for gases to diffuse into and out of the lungs.

The oxygenated blood returns from the lungs through the **pulmonary veins** and enters the left atrium of the heart. When the left atrium contracts, blood is pumped through a second atrioventricular valve into the left ventricle. The thick muscular walls of the left ventricle eject the blood under high pressure to the body. Oxygenated blood leaving the left ventricle passes through the **aortic valve** and then flows to the head and the rest of the body through a large artery called the **aorta**.

The contraction of the two atria followed by contraction of the two ventricles makes up the **cardiac cycle (Fig. 39.22)**. The cardiac cycle is divided into two main phases. **Systole** is the contraction of the ventricles, and **diastole** is the relaxation of the ventricles. Systole, therefore, is the phase of the cardiac cycle in which blood is pumped from the heart into the pulmonary and systemic circulations, and diastole is the phase in which the atria contract and the ventricles fill with blood.

---

**FIG. 39.22   The four-chamber cardiac cycle.**

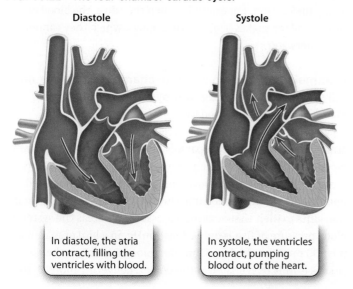

Diastole    Systole

In diastole, the atria contract, filling the ventricles with blood.

In systole, the ventricles contract, pumping blood out of the heart.

---

→    **Quick Check 6** What pressure difference between the ventricle and the atrium causes the atrioventricular valve to close?

## Cardiac muscle cells are electrically connected to contract in synchrony.

In order for the heart to function effectively as a pump, the **cardiac muscle** cells that make up the walls of the atria and ventricles must contract in a coordinated fashion. Atrial muscle cells must be activated to contract in synchrony during diastole to fill the ventricles, and ventricular muscle cells must contract in synchrony during systole to eject the blood from the heart. For cardiac muscle cells to pump in unison, they must be depolarized in unison by an action potential. In vertebrates, this action potential is transmitted by autonomic nerves that control the heart beat cycle.

Cardiac muscle cells have contractile properties similar to those of skeletal muscle cells (Chapter 37). However, cardiac muscle cells are distinct in two ways. First, specialized cardiac muscle cells can generate action potentials on their own, independently of the nervous system. Second, cardiac muscle cells are in electrical continuity with one another, meaning that they can pass their action potentials to adjacent cells. These two features of cardiac muscle cells ensure that all muscle cells in a region surrounding a heart chamber are activated and contract in unison. The specialized cardiac muscle cells capable of generating action potentials independently function as a **pacemaker** that causes the heart to beat with a basic rhythm. These cells stimulate neighboring cells to contract in synchrony. They are found in two specialized regions of the heart—the **sinoatrial (SA)** and **atrioventricular (AV) nodes (Fig. 39.23)**.

Whereas most nerve cells maintain their resting membrane potential until they receive a signal from another cell, the critical feature of pacemaker cells is that their resting membrane potential gradually becomes less negative on its own until it reaches threshold and the cell fires an action potential. The tendency of their membrane potential to become less negative is due to slow leakage of sodium ions into the cell and decreased flow of potassium out of the cell. When the cell reaches threshold, voltage-gated calcium ions open, causing more rapid depolarization. Like nerve cells, after firing an action potential, the pacemaker cells repolarize to reach their resting potential, keeping the heart relaxed between contractions. Also like nerve cells, pacemaker cells repolarize by opening voltage-gated potassium channels to allow potassium ions to leave the cell.

The heartbeat is initiated at the sinoatrial node, located at the junction of the vena cava and right atrium (Fig. 39.23). Because adjacent cardiac muscle cells are in electrical contact, action potentials initiated at the sinoatrial node spread rapidly from one cell to the next. Thus, when the pacemaker cells in the SA node fire an action potential, their depolarization spreads electrically throughout the right and left atria, causing them to contract in unison. Because there is no electrical contact between the atria and ventricles, the ventricles do not contract. Instead, the

**FIG. 39.23  Control of heart beat rhythm.** The pacemaker cells of the sinoatrial and atrioventricular nodes keep the heart beating with a basic rhythm. This rhythm can be assessed by recording the electrocardiogram (EKG) of the heart.

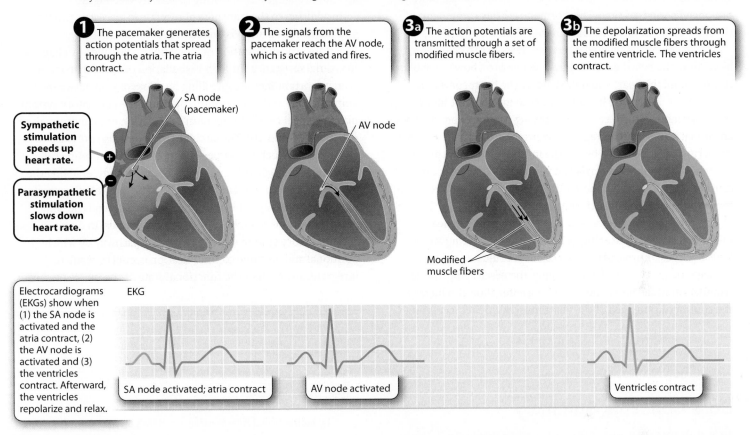

**1** The pacemaker generates action potentials that spread through the atria. The atria contract.

**2** The signals from the pacemaker reach the AV node, which is activated and fires.

**3a** The action potentials are transmitted through a set of modified muscle fibers.

**3b** The depolarization spreads from the modified muscle fibers through the entire ventricle. The ventricles contract.

SA node (pacemaker)

AV node

Modified muscle fibers

**Sympathetic stimulation speeds up heart rate.**

**Parasympathetic stimulation slows down heart rate.**

Electrocardiograms (EKGs) show when (1) the SA node is activated and the atria contract, (2) the AV node is activated and (3) the ventricles contract. Afterward, the ventricles repolarize and relax.

EKG

SA node activated; atria contract

AV node activated

Ventricles contract

depolarization in the atria reaches a second set of pacemaker cells, located in the atrioventricular (AV) node.

Activation of the AV node transmits the action potential to the ventricles. A modified set of cardiac muscle fibers transmits the action potential from the AV node to the base of the ventricles. From there, the depolarization spreads throughout the ventricle walls, causing the ventricles to contract in unison. The delay in transmission from the AV node and through the conducting fibers ensures that the ventricles do not contract until they are fully filled with blood from the atria.

Electrodes placed over the surface of the chest and other body regions can record the electrical currents produced by the depolarization of the heart while it beats. This type of recording is called an electrocardiogram (EKG, because "kardia" is Greek for heart, although ECG is also used; Figure 39.23). Cardiologists use EKGs to diagnose heart problems or to determine what region of the heart has been damaged following a heart attack.

### Cardiac output is regulated by the autonomic nervous system.

The volume of blood pumped by the heart over a given interval of time is its **cardiac output (CO)**. This quantity is the key measure of heart function. Cardiac output is determined by calculating the product of **heart rate (HR)** and the volume of blood pumped during each beat, the **stroke volume (SV)**: $CO = HR \times SV$. For example, if heart rate increases 50% and the heart's stroke volume increases 10%, the heart's cardiac output increases 65% $((1.5 \times 1.1) - 1 \times 100\%)$. Cardiac output rises or falls in response to the metabolic demand for $O_2$, which in turn depends on an animal's state of activity. When an animal sleeps, cardiac output is minimal, but when it runs, cardiac output increases to meet the demand for $O_2$ of its active muscles. Animals can increase their cardiac output by increasing heart rate or stroke volume or both. Whereas humans and other mammals rely most heavily on increases in heart rate to increase cardiac output, fish rely more on increases in stroke volume.

The nervous system controls the rate of the heartbeat. Nerve signals influence the rate at which the cardiac pacemaker cells depolarize in between action potentials. When there is no signal from the nervous system, the depolarization rate of the pacemaker cells sets the resting heart rate. Stimulation by sympathetic nerves causes the pacemaker cells of the sinoatrial node to depolarize more rapidly and the heartbeat to speed up. Stimulation by parasympathetic nerves causes these cells to depolarize more slowly and the heartbeat to slow down. By acting as a hormone (Chapter 38), adrenaline released by the adrenal gland into the circulation also elevates the heart rate for more prolonged periods of time. This hormone is a key component of the fight-or-flight response of many vertebrate animals.

Stroke volume is adjusted directly in response to changes in blood flow and heartbeat strength. It depends on how much blood returns to fill the heart and how strongly the heart contracts. These two processes are linked in order to match heart filling (that is, venous return) to heart emptying. During exercise, when an animal's limb muscles regularly contract, their action increases the return of blood to the heart. The influx of additional blood stretches the walls of the atria and ventricles to a greater extent, causing them to contract more forcefully. As a result, more blood is ejected during each heartbeat.

The correspondence of stroke volume to changes in heart filling is described by **Starling's Law.** It provides the means by which venous return and cardiac output are adjusted similarly to meet changing rates of blood supply. Increased sympathetic stimulation to the heart and circulating adrenaline also increase the contractile strength of the heart muscle, further increasing the heart's stroke volume. These changes in heart rate and stroke volume underlie the increase in cardiac output that occurs when an animal changes from resting to activity (see Fig. 39.17).

In addition to the transport of respiratory gases, nutrients, and wastes, circulation is linked to several other bodily functions. The distribution of blood flow, controlled by changes in arterial resistance through vasodilation or vasoconstriction, is an important factor controlling heat loss or gain between an animal's body and its environment. Thus, the circulatory system plays a central role in temperature regulation of many animals (Chapters 35 and 40). The circulatory system also provides the route by which hormones (Chapter 38) and immune cells (Chapter 43) are distributed throughout the body, enabling the endocrine system to maintain homeostatic control of diverse bodily functions and the immune system to defend the body against injurious microbes and toxic compounds. In doing so, the circulatory system serves as a general pathway by which hormonal and immune communication occurs to regulate, integrate, and protect the functional state of the whole organism.

## Core Concepts Summary

**39.1 RESPIRATION AND CIRCULATION DEPEND ON DIFFUSION OVER SHORT DISTANCES AND BULK FLOW OVER LONG DISTANCES.**

Diffusion is effective only over short distances and requires large exchange surface areas with thin barriers. page 39-2

Ventilation and circulation provide bulk transport over long distances. page 39-3

Oxygen is delivered to tissues in four steps: (1) bulk flow of water or air past the respiratory surface (gills and lungs); (2) diffusion of $O_2$ across the respiratory surface into the circulatory system; (3) bulk flow through the circulatory system; and (4) diffusion of $O_2$ into tissues and cells. page 39-3

**39.2 RESPIRATION PROVIDES OXYGEN AND ELIMINATES CARBON DIOXIDE IN SUPPORT OF CELLULAR METABOLISM.**

Many aquatic animals exchange respiratory gases with water through gills. page 39-4

Countercurrent flow of water relative to blood in gills enhances $O_2$ extraction from water. page 39-5

Terrestrial animals breathe air by means of internal tracheae or lungs. page 39-6

Terrestrial vertebrates inflate and deflate their lungs by bidirectional tidal ventilation driven by changes in pressure. page 39-6

A unidirectional flow of air maximizes $O_2$ uptake by bird lungs. page 39-8

Both the voluntary and involuntary nervous systems control breathing. page 39-9

**39.3 RED BLOOD CELLS PRODUCE HEMOGLOBIN, GREATLY INCREASING THE AMOUNT OF OXYGEN TRANSPORTED BY THE BLOOD.**

Hemoglobin greatly increases the transport capacity of $O_2$ in the blood. page 39-10

Red blood cells contain hemoglobin with iron-containing heme groups that reversibly bind and release $O_2$. page 39-11

A sigmoidal $O_2$-dissociation curve describes the change in the binding affinity of hemoglobin for $O_2$ with changes in partial pressure. The shape of the curve results from cooperative binding of $O_2$ by hemoglobin. page 39-11

Myoglobin binds and stores $O_2$ in muscle cells, increasing the delivery of $O_2$ to muscle mitochondria for activity in general and in marine mammals for diving. page 39-12

Fetal hemoglobin is expressed by the mammalian fetus to allow $O_2$ uptake from the mother's blood. page 39-12

The affinity of hemoglobin for $O_2$ is affected by pH and $CO_2$. page 39-13

## 39.4 CIRCULATORY SYSTEMS HAVE DIFFERENT-SIZED VESSELS THAT FACILITATE BULK FLOW AND DIFFUSION.

The fluid in a closed circulatory system travels through a set of internal vessels, moved by a pump, the heart. The fluid in an open circulatory system moves through only a few vessels and is mostly contained within the animal's body cavity. page 39-13

Resistance to blood flow is most strongly controlled by vessel diameter. page 39-15

Arteries are thick-walled vessels that contain elastic fibers for support against pressure. page 39-15

Arterioles control blood flow within the body by changing their resistance through contraction and relaxation of the smooth muscle in their walls. page 39-15

Capillaries are small-diameter, thin-walled vessels that facilitate diffusion of $O_2$ and $CO_2$. page 39-15

Veins are thin-walled vessels that return blood to the heart under low pressure. page 39-16

Fluids, gases, and other compounds move across capillary walls by diffusion, filtration, and osmosis. In vertebrates, some fluid is returned to the bloodstream by the lymphatic system. page 39-16

## 39.5 THE EVOLUTION OF ANIMAL HEARTS REFLECTS SELECTION FOR A HIGH METABOLIC RATE, ACHIEVED BY INCREASING THE DELIVERY OF OXYGEN TO METABOLICALLY ACTIVE CELLS.

Birds and mammals have a separate pulmonary circulation, from the heart to the lungs, and systemic circulation, from the heart to the rest of the body. page 39-17

Fish have two-chambered hearts and a single circulatory path. page 39-18

Amphibians and reptiles have three-chambered hearts and a partially divided circulatory path. page 39-18

Birds and mammals have four-chambered hearts and fully divided pulmonary and systemic circulations to enhance $O_2$ uptake and delivery to metabolizing tissues. page 39-19

One-way valves control blood flow through the heart. page 39-20

The cardiac cycle has discrete filling (diastole) and emptying (systole) phases. page 39-20

Cardiac muscle cells are electrically connected, allowing them to contract in synchrony as a unified heart beat. page 39-20

Contraction of the heart is controlled by signals from pacemaker cells of the sinoatrial and atrioventricular nodes, transmitted by electrical conducting fibers to the ventricles. page 39-20

The vertebrate autonomic nervous system regulates the heart's cardiac output by changes in heart rate and stroke volume. page 39-21

## Self-Assessment

1. Diagram the four basic steps of $O_2$ transport from an animal's respiratory medium (air or water) to its cells.

2. For an aquatic animal, describe what features of its gills favor diffusion of $O_2$ and $CO_2$.

3. Name two differences in the physical properties of water and air that affect gas exchange in aquatic versus terrestrial animals.

4. Explain how the tracheal respiratory system of insects enables high metabolic rates.

5. Explain how gill respiration is facilitated by unidirectional water flow, whereas lung ventilation can depend on bidirectional airflow.

6. Describe the pressure changes that are needed to draw air into the lungs and to expel air out of the lungs.

7. Explain how cooperative binding by hemoglobin facilitates $O_2$ uptake into blood.

8. Explain how the branching of larger arteries into many smaller vessels affects the rate of and resistance to blood flow in the smaller vessels.

9. Diagram the path of blood flow through the heart, lungs, and body of a fish and of a mammal or bird.

10. Describe two mechanisms by which the cardiac output of an animal's heart can be adjusted and indicate which is more important for increasing cardiac output in mammals.

---

Do you understand the chapter's Core Concepts? Log in to check your answers to the Self-Assessment questions, then practice what you've learned and reinforce this chapter's concepts by working through the problems and multimedia tutorials provided there.

📖 www.biologyhowlifeworks.com

CHAPTER 40

# ANIMAL METABOLISM, NUTRITION, AND DIGESTION

## Core Concepts

**40.1** Metabolic rate depends on level of activity, body size, and body temperature.

**40.2** An animal's diet supplies the energy it needs for homeostasis and essential nutrients it cannot synthesize on its own.

**40.3** Different animals have different adaptations for feeding.

**40.4** The digestive tract is a tubelike structure with regions specialized for different functions.

It is literally true that "you are what you eat." Not only do animals build their bodies and obtain energy from the food they eat, but also their very appearance often reflects what they eat. Consider a cat. We know it's a carnivore—an animal that eats other animals—just from looking at it. It has sharp teeth and powerful jaws specialized for tearing meat; strong legs, sharp claws, and keen eyesight for hunting prey; and a short gut because breaking down animal protein is easier than breaking down plant material. Humans are omnivores, adapted for eating both other animals and plants, and this is reflected in our anatomy. Our teeth include sharp canines for tearing meat and flat molars for grinding plant material. Finally, consider herbivores, such as cows, that eat plants. Cows have evolved a variety of adaptations, including a four-chambered stomach, that allow them to digest cellulose, which is tough but rich in energy.

Animals are heterotrophs, obtaining food from other organisms (Chapter 6). The energy animals obtain from food is essential for building bodies, moving, surviving, and reproducing. Much of an animal's activity therefore centers on obtaining food. In turn, the biology of an animal is in large part shaped by the type of food the animal consumes. Furthermore, an animal's metabolism, the chemical reactions by which it breaks down its food, is closely linked to the energy and nutrients contained in that food.

The food that an animal eats also affects its place in an ecosystem. Primary producers, such as plants and some algae, are food for a great diversity of animals. Animals that feed on plants tend to exist in large numbers. Large animals that feed on other, smaller animals exist in smaller numbers and commonly represent fewer species.

This chapter looks at how animals obtain and break down the food they eat. We begin at the cellular level, building on the biochemistry of metabolism discussed earlier (Chapter 7). Then, we examine the dietary needs and the organization and function of animal digestive systems: how an animal eats, digests, and absorbs food to supply its cells with the energy required for its function. The relationship between an animal's diet and its place in an ecosystem is reserved until Chapter 47.

## 40.1 PATTERNS OF ANIMAL METABOLISM

As discussed in Chapter 7, the biochemical pathways that make up animal metabolism are remarkably conserved among diverse groups of organisms. These pathways evolved early in the history of life and have been retained throughout the evolution and diversification of life. In Chapter 7, we examined the pathway of glucose breakdown by cellular respiration, which provides ATP to fuel the work of a cell. Here, we put this and related pathways in an organismal context. We explore how carbohydrates, proteins, and fats are metabolized for building body structures and supplying energy needs and how metabolic rate varies with an animal's level of activity, internal body temperature, and size.

## Animals rely on anaerobic and aerobic metabolism.

Nearly all animals depend on three main molecules as sources of energy and building blocks for growth and development: carbohydrates, fats, and proteins. The conversion of these food sources into biologically useful forms typically involves

**FIG. 40.1  Cellular metabolism, which comprises anaerobic and aerobic pathways.** Animals obtain energy to fuel their activities and development from the breakdown of carbohydrates, fats, and proteins.

Glycolysis (anaerobic metabolism) occurs in the cytosol and breaks down carbohydrates into pyruvate or, in the absence of oxygen, lactic acid.

When oxygen is available (aerobic metabolism), pyruvate enters the citric acid cycle in the mitochondria, generating NADH and $FADH_2$.

NADH and $FADH_2$ power ATP synthesis by electron transport in the inner mitochondrial membrane.

means that the average rate at which each gram of body tissue consumes energy is *less* in larger animals compared to smaller ones. Or, put another way, the larger the organism, the *lower* the metabolic rate per gram of body tissue. This scaling pattern of cellular energy metabolism is remarkable because it holds across a diverse size range of unicellular and multicellular organisms, as can be seen in Fig. 40.4.

In addition to basal metabolic rate, we can also examine how metabolic rate varies with respect to exercise in different-sized animals. In this case, for example, we are looking at how energy is used to allow animals to move at different speeds. The American comparative physiologist C. Richard Taylor has extensively studied how metabolic rate varies with running speed in different-sized terrestrial animals (**Fig. 40.5**). He found that metabolic rate increases linearly with speed, and that larger animals expend less energy per unit mass than smaller ones.

→ **Quick Check 2** Does a dog that is twice as heavy as a cat have twice the resting metabolic rate?

### Metabolic rate is linked to body temperature.

Metabolic rate is also affected by an animal's internal body temperature. Temperature affects the rates of chemical reactions, which in turn determine how fast fuel molecules can be mobilized and broken down to supply energy for a cell. Animals can be categorized by the sources of most of their heat. Animals that produce most of their own heat as by-products of metabolic reactions, including the breakdown of food, are **endotherms.** These animals usually, but not always, maintain a constant body temperature that is higher than that of their environment. As we saw earlier (Chapter 35), the maintenance of a constant core body temperature is a form of homeostasis and requires balancing heat production and heat loss. Mammals, including humans, and birds are endotherms.

By contrast, animals that obtain most of their heat from the environment are **ectotherms.** Ectotherms often regulate their body temperatures by behavioral means such as moving into or out of the sun. Think of a turtle or another reptile sunning itself on rock: It is raising its body temperature. The temperatures of these animals usually fluctuate with the outside environment. Most fish, amphibians, reptiles, and invertebrates are ectotherms.

Endotherms are sometimes referred to as "warm-blooded" and ectotherms as "cold-blooded," but these terms are misleading because, depending on environmental conditions, cold-blooded animals can have core body temperatures higher than warm-blooded animals. It should also be kept in mind that these

thermoregulatory mechanisms are not completely distinct, but instead represent extremes of a continuum.

Nevertheless, thermoregulatory mechanisms have profound implications for an animal's metabolic rate. Endotherms have a higher metabolic rate than ectotherms. As a result, they are able to be active over a broader range of external temperatures than ectotherms. The activity level and metabolic rate of ectotherms both increase with increasing body temperature. However, ectotherms have metabolic rates that are about 25% of endotherms of similar body mass and at similar body temperatures. Although they can achieve activity levels similar to those of endotherms when their body temperatures are similar, ectotherms cannot sustain prolonged activity. Therefore, ectotherms such as lizards or insects rely on brief bouts of activity, followed by longer periods of inactivity compared to endotherms.

Endotherms benefit from having an active lifestyle. However, their high metabolic rate requires long periods of foraging to acquire enough food to keep warm and be active. By contrast, although ectotherms are limited in their ability to sustain activity, they can survive much longer periods without food.

As we have discussed, the control of core body temperature, or thermoregulation, is a form of homeostasis and requires the coordinated activities of the nervous, muscular, endocrine, circulatory, and digestive systems. We summarize how these systems work together in endotherms and ectotherms in **Fig. 40.6** on the following page.

## 40.2 ANIMAL NUTRITION AND DIET

Running, jumping, moving, growing, reproducing, and all other activities performed by animals require energy. Autotrophs like plants and certain microbes are able to capture energy from the sun or from inorganic compounds (Chapters 8 and 26), but animals must acquire food in the form of plants or other animals to support their growth, reproduction, and other activities. Food does not consist just of carbohydrates, fats, and proteins, however. To enable various cell processes, animals must also obtain in their diet certain minerals and chemical compounds that they cannot synthesize on their own. In this section, we explore what animals must obtain from their food, and how much of it, in order to sustain their cells and tissues.

### Energy balance is a form of homeostasis.

Like core body temperature (Fig. 40.6), blood-glucose levels (Chapter 37), and blood pressure (Chapter 39), the **energy balance** of an organism is often maintained at a constant level.

# Homeostasis and Thermoregulation

Integrating concepts from chapters 35-40

## Thermoregulation in an endotherm

Endotherms regulate their body temperature primarily by physiological means, such as increasing or decreasing their metabolic activity, which generates heat as a by-product.

Temperature

Core body temperature

Outside temperature

Burrow temperature

4 am          8 am          12 pm

**Rock Squirrel**

Hypothalamus

Muscle

Blood vessel

---

**Cold nighttime temperature**

⊕

Hypothalamus

⊕

Motor cortex

⊕     ⊕

**Physiological:**
• Shivering
• Peripheral vasoconstriction
• Increase in metabolic rate

**Behavioral:**
• Burrowing
• Huddling

⊖

**Increased heat production
Decreased heat loss**

At night when it is cold, the rock squirrel enters its burrow, and physiological mechanisms generate heat to maintain a constant core body temperature.

---

**Cool daytime temperature**

⊕

Hypothalamus

⊕

Motor cortex

⊕     ⊕

**Physiological:**
• Peripheral vasoconstriction

**Behavioral:**
• Increasing activity

⊖

**Increased heat production
Decreased heat loss**

The squirrel is most active in the early morning and late afternoon, when it is cool.

---

**Hot daytime temperature**

⊕

Hypothalamus

⊕

Motor cortex

⊕     ⊕

**Physiological:**
• Panting
• Peripheral vasodilation
• Decrease in metabolic rate

**Behavioral:**
• Seeking shade
• Licking forelimbs

⊖

**Decreased heat production
Increased heat loss**

To avoid overheating during the day, the squirrel decreases its exposure to the sun, and physiological mechanisms cool the body.

## Thermoregulation in an ectotherm

Ectotherms gain heat from or lose heat to their environment, and so their primary means of regulating their body temperature is behavioral, increasing or decreasing their exposure to the sun and hot surfaces.

Temperature

Core body temperature

Outside temperature

Burrow temperature

4 am          8 am          12 pm

Hypothalamus
Blood vessel
Muscle
**Zebra-tailed Lizard**

**Cold nighttime temperature**
+
Hypothalamus
+
Motor cortex
+
**Behavioral:**
• Burrowing
−
**Decreased heat loss**

At night, lizards retreat to burrows in order to maintain a constant, intermediate temperature.

**Cool daytime temperature**
+
Hypothalamus
+
Motor cortex
+        +
**Physiological:**
• Peripheral vasodilation (heat gain from sun/rock)
**Behavioral:**
• Increasing exposure to sun
• Lying on hot surfaces
• Increasing activity
−
**Increased heat gain**

In the morning, the lizard basks in the heat of the sun. In the evening, it warms itself by heat radiated from rocks heated by the daytime sun.

**Hot daytime temperature**
+
Hypothalamus
+
Motor cortex
+
**Behavioral:**
• Seeking shade
• Lying on cool ground
• Decreasing activity
−
**Increased heat loss**

During the hottest parts of day, the lizard regulates its temperature by seeking shade.

An animal in energy balance takes in the same amount of calories of energy from food that it uses over time to meet its metabolic needs. Energy balance can be thought of as a form of homeostasis. We consider sources of energy, or **energy intake,** and ways in which energy is expended, or **energy use.**

For animals, the source of energy is the diet. In turn, energy is used to do work (maintain tissues, grow, move about, and the like). Energy required for basic life processes accounts for the majority (about 70%) of energy use. The amount of energy used also depends on the level of physical activity, with higher levels of activity using more energy, as we saw. Finally, digestion and absorption of food from the diet itself requires energy.

When energy intake does not equal energy used, there is an energy imbalance. If an animal eats more food than it requires, energy stores such as fat deposits grow over time. The result is that the body shifts its metabolism mostly to anabolic processes that build energy stores. Many animals achieve a net positive energy intake during the late summer and fall when food is plentiful, before it becomes scarce in winter. Other animals maintain a constant energy intake throughout the year by migrating to areas with more abundant food and avoiding colder temperatures that require increased energy expenditure to remain active and warm. Still other animals hibernate, or become less active, to conserve their energy use over the winter when food is scarce or unavailable.

Animals that cannot acquire enough food are in negative energy balance and become undernourished. During prolonged periods of an inadequate food supply or starvation, an animal consumes its own internal fuel reserves. Starvation forces animals to deplete their glycogen and fat reserves first, and then, if no food is found, to resort to protein stores, primarily in muscle tissue. This ultimately leads to muscle wasting. Undernourishment and starvation are particularly serious human health problems in many developing countries, especially those ravaged by war and internal strife.

Humans, like most other animals, store excess food calories as fat. This is because over much of our evolutionary history food was less abundant and was more unpredictable in its availability than it is today. With the development of agriculture and domestication of livestock, food supplies rapidly increased, allowing many human populations to grow and consume increasing amounts of food. With the rise of mechanized agricultural food production and highly processed foods developed by the modern food industry, excessive intake of food calories has led to an increasing and now critical public health problem: obesity.

Obesity is now an epidemic in many industrialized nations, including the United States, where about 36% of the adult population is considered obese. Obesity is a major public health concern because it increases the risk of diabetes, heart disease, and stroke and contributes to a shorter life-span. For most animals, however, acquiring food and storing its products efficiently in the body allow them to have a fuel reserve to meet seasonal energy requirements. This rationing of stored energy remains an essential part of their metabolic and digestive physiology.

### An animal's diet must supply nutrients that it cannot synthesize.

Animals' metabolic pathways enable them to obtain energy from the environment as well as to synthesize many of the compounds needed to sustain life. However, many nutrients necessary for life cannot be synthesized by an animal's metabolism and therefore must be acquired in the food that they eat.

Recall that amino acids are the basic units of proteins (Chapter 4). There are 20 different amino acids that are typically found in proteins (see Fig. 4.2). Although all of them are necessary for life, an **essential amino acid** is formally defined as one that cannot be synthesized by cellular biochemical pathways and instead must be ingested. Most animals can synthesize about half of their amino acids. Humans are unable to synthesize 8 of the 20 amino acids (**Table 40.1**). We have to obtain these eight essential amino acids in our diets. Infants require additional amino acids. The most reliable source of all eight essential amino acids is meat. This is not to say that only omnivorous humans can lead healthy lives, though. While most plant proteins lack at least one of the eight amino acids essential to humans, and diets heavy in a particular food, such as corn (which lacks lysine), can lead to protein deficiency, vegetarians can achieve a healthy intake of all essential amino acids by combining plant foods. For example, beans supply the lysine that corn lacks, and corn supplies the methionine that beans lack.

**Dietary minerals** are chemical elements other than carbon, hydrogen, oxygen, and nitrogen that are required in the diet and must be obtained in the food that an animal eats (**Fig. 40.7**).

**TABLE 40.1 Essential Amino Acids for Humans**

| ESSENTIAL AMINO ACID | COMMON SOURCES |
| --- | --- |
| Isoleucine | Meat, dairy, lentils, wheat |
| Leucine | Meat, cottage cheese, lentils, peanuts |
| Lysine | Meat, beans, lentils, spinach |
| Methionine | Meat, dairy, whole grains, corn |
| Phenylalanine | Meat, dairy, peanuts, seeds |
| Threonine | Meat, dairy, nuts, seeds |
| Tryptophan | Meat, dairy, oats |
| Valine | Meat, dairy, grains |

**FIG. 40.7** **Commonly recognized minerals required by humans.** Dietary minerals are chemical elements other than carbon, hydrogen, oxygen, and nitrogen that are required in the diet.

They include such elements as calcium, iron, phosphorus, potassium, zinc, and magnesium. Calcium is required for building skeletons (Chapter 37), and iron in hemoglobin binds oxygen and transports it in the blood (Chapter 39). Humans typically obtain the minerals sodium and chloride, which ionize to form salt crystals, through common table salt and other foods. Many wild animals seek exposed rock that they lick to obtain minerals and salts.

In addition to amino acids and minerals, animals must ingest essential **vitamins,** organic molecules that are required in very small amounts in the diet. Vitamins have diverse roles, some binding to and increasing the activity of particular enzymes, others acting as antioxidants or chemical signals. Different animals have different vitamin requirements, so a molecule that functions as a vitamin for one may not be a vitamin for another. Thirteen essential vitamins have been identified for humans (**Table 40.2**). Knowing which vitamins are required for an animal's health is critical to proper medical and veterinary care. Animals have evolved diets that ensure that their need for vitamins, as well as for essential amino acids and minerals, is met through the combination of foods that they eat.

In humans as well as in other animals, vitamin deficiency can have serious consequences. Whereas most mammals can synthesize ascorbic acid (vitamin C), which is necessary for building connective tissue, primates (including humans) cannot. Without ingesting sufficient vitamin C, humans develop scurvy, a disease characterized by bleeding gums, loss of teeth, and slow wound healing. Until it was discovered that green vegetables and fruit supply vitamin C, scurvy was a prevalent disease that could lead to death for sailors who were deprived of fresh produce for long periods of time while at sea.

Deficiencies of vitamins $B_1$, $B_6$, and $B_{12}$ can cause nervous system disorders and various forms of anemia (evidenced by low levels of red blood cells or hemoglobin in the blood). Vitamin D is essential for the absorption of calcium in the diet and thus to skeletal growth and health. With adequate exposure to ultraviolet solar radiation, skin cells synthesize enough vitamin D to sustain a growing body. However, fairer-skinned people inhabiting more northern regions of the world produce less vitamin D and therefore require more of it in their diet. A clear role for vitamin E remains uncertain, but its absence is often linked to anemia.

→ **Quick Check 3** In the context of nutrition, what does "essential" mean?

**TABLE 40.2** **Essential Vitamins for Humans**

| VITAMIN DESIGNATION | NAME | COMMON SOURCES |
|---|---|---|
| Vitamin A | Retinol | Green and orange vegetables, liver, milk |
| Vitamin $B_1$ | Thiamine | Pork, grains |
| Vitamin $B_2$ | Riboflavin | Dairy products, eggs, leafy green vegetables |
| Vitamin $B_3$ | Niacin | Meat, dairy products, eggs, vegetables |
| Vitamin $B_5$ | Pantothenic acid | Meat, grains |
| Vitamin $B_6$ | Pyridoxine | Meat, grains |
| Vitamin $B_7$ | Biotin | Meat, eggs |
| Vitamin $B_9$ | Folic acid | Leafy green vegetables, eggs |
| Vitamin $B_{12}$ | Cobalamin | Liver, eggs |
| Vitamin C | Ascorbic acid | Fruit, vegetables |
| Vitamin D | Calciferol | Fish, eggs |
| Vitamin E | Tocopherol | Fruit, vegetable oils |
| Vitamin K | Phylloquinone | Leafy green vegetables |

## 40.3 ADAPTATIONS FOR FEEDING

Animals have evolved a variety of adaptations to enhance their ability to acquire energy from the environment. Animals eat other organisms, whether they are other animals, plants, or fungi. Herbivores are adapted to eat plants; carnivores are adapted to eat animals; and omnivores are adapted to eat both. To acquire energy effectively from the food that they eat, animals have evolved a diversity of ways to capture food and mechanisms for breaking down that food before its digestion in the gut.

### Suspension filter feeding is common in many aquatic animals.

The most common form of food capture by animals is **suspension filter feeding,** in which water with food suspended in it is passed through a sievelike structure (**Fig. 40.8**). Suspension filter feeding is possible only in aquatic environments, and its ubiquity there speaks to the great diversity of animals found in oceans and freshwater bodies. This form of feeding evolved multiple times independently on many different branches of the tree of life, and so represents a form of convergent evolution (Chapter 23).

Many worms and bivalve mollusks, such as scallops, clams, and oysters, pump water over their gills to trap food particles suspended in the water (Fig. 40.8a). Cells of the gills produce mucus that acts as an adhesive to trap food particles, and the cells' beating cilia sweep food trapped in the mucus to the mouth and gut.

Other aquatic organisms move water with food particles suspended in it through filters in their oral cavity and then convey the food into the gut. Large baleen whales have long comblike blades of baleen (which is made of keratin, the same material that

**FIG. 40.9**    **Suction feeding.** This form of feeding involves a rapid expansion of the fish's mouth that draws in water and prey.

makes up hair) in their mouth to capture small shrimp called krill that are plentiful in ocean waters (Fig. 40.8b).

### Large aquatic animals apprehend prey by suction feeding and active swimming.

Instead of relying on suspension filter feeding, many large aquatic animals capture their prey in other ways. Many fish feed by suction. A rapid expansion of the fish's mouth cavity draws water and the desired prey into the mouth (**Fig. 40.9**). After the fish closes its mouth, the water is pumped out of the mouth cavity past the gills. The prey is trapped inside the mouth, moves into the pharynx (part of the throat), and is broken up by specialized pharyngeal jaws (a second set of jaws in the throat) before being

**FIG. 40.8**    Suspension filter feeding by (a) a scallop and (b) a baleen whale.

**FIG. 40.10 Feeding by active swimming.** Large marine predators, like this great white shark, can use their size and speed to capture prey.

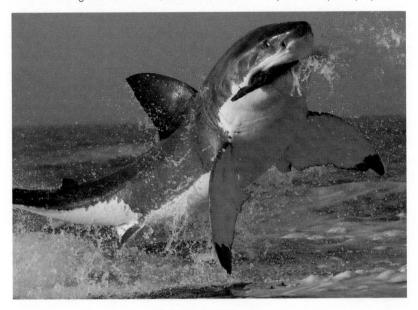

vertebrates, the jawed fishes, became dominant in their aquatic environment through the ability to swim and bite forcefully to obtain their food. **Fig. 40.11** shows a phylogenetic tree of vertebrates indicating the appearance of jaws. Jawed fish evolved from jawless ancestors, and jaws can be found in present-day fish, amphibians, reptiles, and mammals. As you can see from the number of jawed groups, jaws are key to the evolutionary success of vertebrates. They are thought to have evolved from cartilage that supported the gills, providing a spectacular example of an organ adapted for one function (gill support) changing over time to become adapted for an entirely different function (predation).

Among vertebrates, mammals evolved a specialized jaw joint, the **temporomandibular joint,** as well as a great diversity of specialized forms of teeth. The temporomandibular joint allowed the teeth of the lower and upper jaws to fit together precisely. This anatomy facilitated the specialization of teeth with

swallowed. Suction feeding is the most generalized form of prey capture among fish, as it allows them to be "sit-and-wait" predators, hiding within a coral reef or under a rock before rapidly striking to capture prey moving in front of them. Suction feeding is effectively used by a wide variety of fish, as well as by aquatic salamanders, and has contributed to their evolutionary diversification and success.

Many insects that bite to obtain a blood meal also rely on suction to draw the blood of their prey into their digestive system. Young mammals also feed by suckling milk from their mother's breast, using their tongue and also generating suction as fish do.

As top predators in marine and freshwater environments, larger fish and marine mammals like sharks, whales, and dolphins actively swim to capture their prey (**Fig. 40.10**). Their larger size and speed enables them to catch smaller or similar-sized prey. As discussed below, the evolution of jaws and teeth considerably enhanced the ability of these and other vertebrate animals to capture greater amounts of food, allowing them to lead a more active and energetically demanding lifestyle.

### Jaws and teeth provide specialized food capture and mechanical breakdown of food.

Jaws and teeth were an important evolutionary innovation for active predators. Some of the first

**FIG. 40.11 A phylogeny of vertebrates, showing the appearance of jaws.** Jaws, which evolved more than 400 million years ago from a group of jawless ancestors, contributed to the success of many groups of animals.

**FIG. 40.12  Jaws and teeth of carnivorous, herbivorous, and omnivorous mammals.**
Mammals evolved a diverse arrangement of teeth for diverse diets.

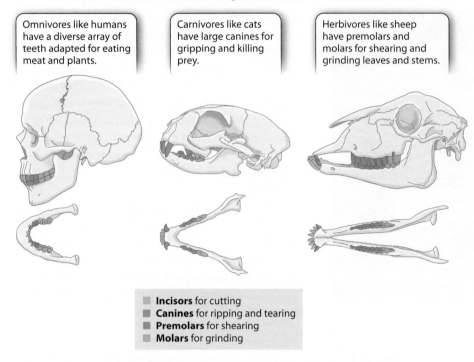

Omnivores like humans have a diverse array of teeth adapted for eating meat and plants.

Carnivores like cats have large canines for gripping and killing prey.

Herbivores like sheep have premolars and molars for shearing and grinding leaves and stems.

- **Incisors** for cutting
- **Canines** for ripping and tearing
- **Premolars** for shearing
- **Molars** for grinding

cutting and crushing surfaces, enabling mammals to break down a variety of foods mechanically before swallowing. **Fig. 40.12** illustrates the arrangement of specialized teeth in mammals with different diets. Teeth in the front of the mouth, called **incisors,** are specialized for biting. Others, such as the **canines** of dogs, cats, and other carnivores, are specialized for piercing the body of their prey. Saber-toothed cats, which lived as recently as 10,000 years ago, represent some of the most remarkable examples of animals with long canines (up to 16 inches or 40 cm!) specialized for killing large prey. **Molars** and **premolars** are teeth in the back of the mouth that are specialized for crushing and shredding tougher foods, such as meat and fibrous plant material.

Herbivorous animals, such as cattle, sheep, and horses, have specialized molars and premolars with prominent surface ridges. These teeth enable them to shred the tough plant material that they eat before it is swallowed and digested. Mammalian herbivores use their front incisors and canines to bite grasses and leaves, and then move the food to the rear of the mouth, where it is ground and crushed between their ridged premolars and molars.

## 40.4 DIGESTION AND ABSORPTION OF FOOD

After animals obtain food, they break it down through the process of digestion. Digestion requires that the food be isolated in a specialized compartment so that it can be broken down chemically without damage to other cellular organelles or structures of the body. Organisms on different branches of the tree of life have evolved very different means of isolating and digesting food. The process in single-celled protists, which obtain food particles by phagocytosis, is **intracellular digestion,** in which food is broken down within cells. Lysosomes containing hydrolytic enzymes fuse with vacuoles containing food particles, so the enzymes are mixed with the food (Chapter 5). Following their chemical breakdown, the food products of intracellular digestion are absorbed into the cell for use. In most animals, however, the process is **extracellular digestion,** in which food is isolated and broken down outside a cell, in a body compartment.

### The digestive tract has regional specializations.

Extracellular digestion involves a specialized cavity or body compartment in which food is broken down. Some animals, like sea anemones, have an internal cavity in which digestion occurs. Cells lining this cavity secrete digestive enzymes that break down the food, allowing the smaller chemical compounds to be absorbed by the cells.

Other animals have more elaborate digestive systems that allow the transport of food by a digestive tube that runs from an animal's mouth to its anus. Collectively, the passages that connect the mouth, digestive organs, and anus constitute an animal's **gut** or **digestive tract.** Because the food is moved in a single direction through a tubelike gut, particular regions of the digestive tract can be specialized for different functions. These functions include storage, chemical breakdown of different kinds of food, absorption of released nutrients, and elimination of waste products.

The digestive tracts of animals are commonly divided into a **foregut, midgut,** and **hindgut** (**Fig. 40.13**). Most animals' digestive systems have the same basic plan, though some animals have also evolved specialized structures. The foregut includes the **mouth, esophagus,** and **stomach** or **crop,** which serves as an initial storage and digestive chamber. Next is the midgut, which includes the **small intestine,** where the remainder of digestion and most nutrient absorption takes place. Here, specialized organs secrete enzymes and other chemicals that aid in the breakdown of particular macromolecules, such as fats and carbohydrates. Finally, the digested material reaches the hindgut, which includes the **large intestine** and **rectum.** Within the hindgut, water and minerals are reabsorbed, leaving

**FIG. 40.13** Organization of animal digestive tracts: foregut, midgut, and hindgut.

Most animals' digestive systems have the same basic plan, though some animals have evolved more specialized structures.

*Earthworm*

*Grasshopper*

Mouth
Esophagus

Stomach
Large intestine
Small intestine

*Human*

*Dog*

Foregut (mouth–stomach)
Midgut (small intestine)
Hindgut (large intestine–anus)

the waste products, or **feces,** which are stored in the rectum until being eliminated from the body. Elimination of the feces, or **defecation,** involves relaxation of an anal sphincter combined with muscular contraction of the rectum and body wall.

### Digestion begins in the mouth.

Most animals jump-start digestion by breaking down food mechanically. After ingesting food, many animals manipulate the food in the mouth or **buccal cavity.** As we saw in the previous section, many animals break down food with the aid of jaws and teeth. Insects, which lack jaws and teeth, manipulate and break down their food with mouthparts called mandibles before it passes into their gut. In some animals, such as mammals, food entering the mouth is mixed with salivary secretions that contain **amylase,** an enzyme that breaks down carbohydrates, to begin the digestion of sugars and starches.

Mammals and other land vertebrates have a muscular **tongue** that facilitates food manipulation and transport within the mouth cavity. The tongue moves the food into position within the mouth for effective cutting and maceration into smaller pieces. When a bolus of food has been chewed and is ready to be swallowed, the tongue moves it to the rear of the mouth cavity.

**Swallowing,** which is controlled by the autonomic nervous system, is a complex set of motor reflexes that involves several muscles and structures in the rear of the mouth and the **pharynx,** the region of the throat that connects the nasal and mouth cavities (**Fig. 40.14**). Once initiated, swallowing reflexes

**FIG. 40.14 Swallowing.** Swallowing moves food from the mouth to the stomach.

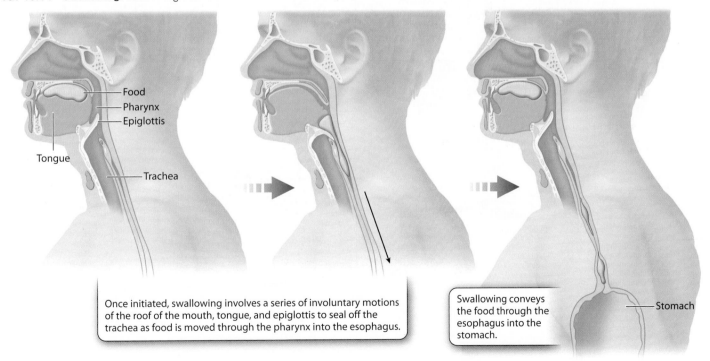

Food
Pharynx
Epiglottis

Tongue

Trachea

Once initiated, swallowing involves a series of involuntary motions of the roof of the mouth, tongue, and epiglottis to seal off the trachea as food is moved through the pharynx into the esophagus.

Swallowing conveys the food through the esophagus into the stomach.

Stomach

occur without voluntary control. During swallowing, food is pushed through the pharynx, over the **epiglottis,** a flap of tissue that protects food from entering the trachea and lungs, and into the esophagus, the tube that connects the pharynx to the stomach.

Birds, alligators, crocodiles, and earthworms break down food into smaller pieces further along their digestive tracts in the **gizzard,** a compartment with thick muscular walls. Birds and earthworms often ingest small rocks or sediment to help grind the food into smaller pieces within the muscular gizzard. Breaking food into smaller sizes aids the chemical digestion that follows.

→ **Quick Check 4** The rocks and sediments that break apart food mechanically play a role similar to what structures in vertebrates?

## The stomach is an initial storage and digestive chamber.

After being ingested and mechanically broken down, the food is transported into a stomach or crop (see Fig. 40.13). Stomachs and crops further fragment and mix the food with digestive enzymes, storing the food in a partially digested form before releasing it for further digestion and nutrient absorption, as needed by the animal.

The stomach is one of the main sites of protein breakdown. The sight, smell, and taste of food send signals to the brain that stimulate a sense of appetite, and to the stomach, stimulating the secretion of hydrochloric acid (HCl) and digestive enzymes that break down proteins into amino acids, which are then absorbed into the bloodstream. Cells secrete these digestive enzymes in an inactive form; otherwise, they would be digested themselves. After secretion, the inactive enzymes are activated by a change in

**FIG. 40.15   Control of digestion in the stomach.**

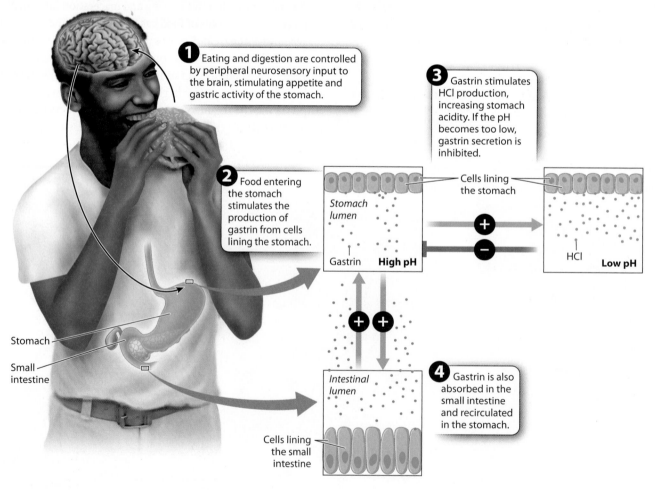

their chemical structure. The primary digestive enzyme produced in the stomach is **pepsin,** an enzyme that breaks down proteins. Cells lining the stomach release pepsin in an inactive form called pepsinogen that is activated by the low pH of the stomach, which is in the range of pH 1–3. The low pH also aids active pepsin in cleaving bonds between amino acids of the peptides. Finally, the acidic environment and digestive enzymes of the stomach kill most pathogens that may be ingested with the food.

In addition to digestive enzymes and HCl, the cells lining the stomach secrete a peptide hormone called **gastrin** in response to food in the stomach (**Fig. 40.15**). Gastrin stimulates the cells lining the stomach to increase their production of HCl. Gastrin secreted by the stomach is absorbed within the small intestine and recirculated to the cells lining the stomach wall. There, the recirculated gastrin stimulates the stomach lining cells to increase their production of gastrin, forming a positive feedback loop (Chapter 38). The feedback loop, starting with the secretion of pepsin to the amplified secretion of gastrin, ensures that protein digestion occurs in response to the start of a meal as food enters the stomach. If the pH of the stomach becomes too low, gastrin secretion is inhibited by negative feedback control.

Glands in the lining of the stomach secrete mucus to protect the stomach wall from the acid its cells secrete to digest food. Sometimes, this defense can break down, resulting in an **ulcer,** a break or erosion in the cells lining the stomach. Although stress and oversecretion of digestive juices can cause an ulcer, Australian pathologists Barry Marshall and Robin Warren made the remarkable discovery that most ulcers result from an infection by the bacterium *Heliobacter pylori,* which can survive in extremely acidic environments. Ulcers produced by *H. pylori* are worsened by the stomach's secretions of HCl and pepsin. The two researchers won the Nobel Prize in Physiology or Medicine in 2005 for their work.

The stomach walls contract to mix the contents of the stomach, aiding their digestion. Waves of muscular contraction, called **peristalsis,** move the food toward the base of the stomach. There, the **pyloric sphincter,** a band of muscle, opens and allows small amounts of digested food to enter the small intestine. Opening and closing of the pyloric sphincter regulates the rate at which the stomach empties, allowing time for digestion and absorption of the food products released into the small intestine.

Animals often eat large amounts of food in a short time, but it takes much longer to digest that food. The stomach serves as a storage compartment as well as a digestion compartment. In humans, it typically takes about four hours for the stomach to empty, allowing digestion and absorption of nutrients to occur between meals. In other animals, stomach emptying can take

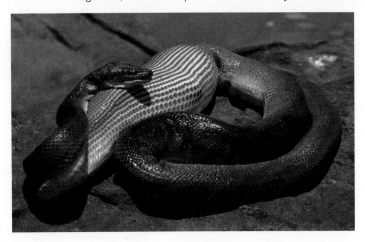

**FIG. 40.16 Python and prey.** Because the prey (here, a duck) is eaten whole, the python's metabolic rate is high during digestion, which takes place over several days.

much longer. Most carnivorous animals consume large and infrequent meals, resulting in long periods of digestion and nutrient absorption compared with humans and other animals that eat more regularly. Snakes such as pythons that engulf large prey whole (**Fig. 40.16**) elevate their metabolic rate to high levels during digestion and spend several days digesting and absorbing the nutrients from their meal. This requires an extensive remodeling of their gut to produce new secretory and absorptive cell surfaces to digest their single large meal. These extra cells are not retained between meals because of the high energy cost of maintaining them.

## The small intestine is specialized for nutrient absorption.

After the food enters the small intestine, protein and carbohydrate digestion continue, and fat digestion begins. The small intestine is so named because of its small diameter, but it is a long (about 20 feet or 6 m in an adult human) and therefore large organ. Because of its length, the small intestine is coiled to fit within the abdominal cavity. The small intestine is made up of three sections (**Fig. 40.17**). The initial section is the **duodenum,** into which the food enters from the stomach. Most of the remaining digestion takes place in the duodenum. The next two sections, the **jejunum** and **ileum,** carry out most nutrient absorption.

Two accessory organs, the **liver** and **pancreas,** aid in the digestion of proteins, carbohydrates, and fats in the duodenum of the small intestine. The liver produces **bile,** which is composed of bile salts, bile acids, and bicarbonate ions. Bile

**FIG. 40.17 The small intestine and associated organs.** Digestion in the small intestine is aided by secretions from the liver, gallbladder, and pancreas. The small intestine is the main site of nutrient absorption.

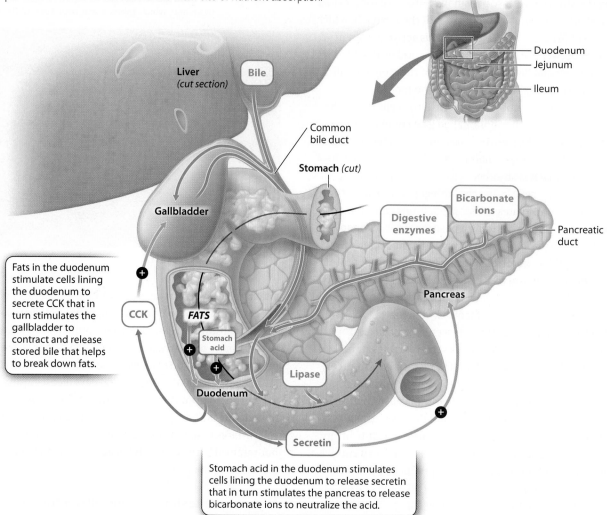

Fats in the duodenum stimulate cells lining the duodenum to secrete CCK that in turn stimulates the gallbladder to contract and release stored bile that helps to break down fats.

Stomach acid in the duodenum stimulates cells lining the duodenum to release secretin that in turn stimulates the pancreas to release bicarbonate ions to neutralize the acid.

aids in fat digestion by breaking large clusters of fats into smaller lipid droplets, a process called emulsification. Bile produced by the liver is stored in the **gallbladder** (Fig. 40.17). When fats enter the duodenum, cells lining the duodenum release a peptide hormone called **cholecystokinin (CCK),** which causes the gallbladder to contract, thus releasing the bile into the duodenum. Bile in turn breaks up the fats into smaller droplets, allowing **lipases,** enzymes that break apart lipids such as triacylglycerol, to digest the lipids more effectively, breaking them down into free fatty acids and glycerol that can be readily absorbed across the wall of the small intestine and eventually into the bloodstream.

The pancreas is a secretory gland that lies below the stomach, posterior to the duodenum (Fig. 40.17). It functions as both an endocrine gland, secreting hormones directly into the blood as discussed in Chapter 38, and an exocrine gland, secreting substances into ducts that connect to the duodenum. The pancreas produces a variety of digestive enzymes, including lipase, which breaks down fats, and **trypsin,** which further breaks down proteins. Like pepsin produced by the stomach, the pancreas produces trypsin in an inactive form called trypsinogen to avoid digesting itself.

The pancreas also secretes bicarbonate ions, which neutralize the acid produced by the stomach. Bicarbonate secretion by the pancreas is stimulated by the hormone **secretin** (the first hormone ever to be identified), which is released by cells lining the duodenum in response to the acidic pH of the stomach contents entering the small intestine. Enzymes that

**FIG. 40.18 Intestinal villi and microvilli.** These structures increase the surface area for absorption of nutrients.

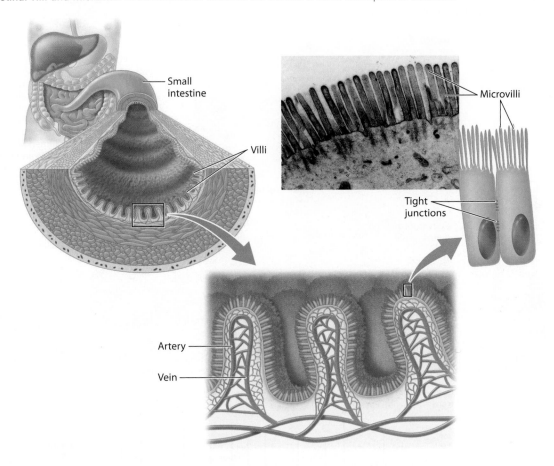

break down proteins work best in acidic environments like that of the stomach, but most proteins, like those that break down carbohydrates and fats, are denatured (that is, unfolded and thus made inactive) in acidic environments. In order to digest carbohydrates and fats, the stomach acid is neutralized by the secretion of bicarbonate ions by the pancreas, which then allows the enzymes that digest carbohydrates and fat to operate.

Final digestion of proteins and carbohydrates occurs in the jejunum and ileum. These two sections of the small intestine have highly folded inner surfaces, called **villi,** and the cells constituting the villi themselves have highly folded surfaces, called **microvilli (Fig. 40.18)**. Together, villi and microvilli greatly increase the surface area for the absorption of nutrients. Maximizing surface area by folding into villi and microvilli is much like the way the lung maximizes surface area for gas exchange with alveoli (Chapter 39). Cells that line the small intestine are connected by **tight junctions** (Fig. 40.17 and Chapter 10). These junctions force the products of digestion

to be absorbed across the microvilli surfaces, controlling their movement through the cell and into the bloodstream, rather than leaking between the cells.

The microvilli of the cells lining the small intestine secrete enzymes that cleave peptides into amino acids, which can then be absorbed across the plasma membrane. They also secrete enzymes that break sugars into their subunits, which can also be readily absorbed. Lactose intolerance can arise in humans if they stop producing lactase, an enzyme that normally breaks lactose into its component parts. This can be a critical problem for infants who fail to produce lactase, as milk is their principal source of nutrition. Lactose intolerance is readily treated by lactase supplements to the diet.

Nutrient molecules, such as glucose and amino acids, are often co-transported into cells lining the intestine with sodium ions from gut contents. The sodium ion binds to a transmembrane protein on the surface of cells lining the intestine that also binds the nutrient molecule. Absorption of the nutrient molecule into the cell is driven by the movement of the sodium ion down its

**FIG. 40.19 Glucose absorption by cells of the small intestine.** Absorption of glucose into the cell is driven by the movement of the sodium ions down a concentration gradient.

Glucose enters an intestinal cell along with Na$^+$ driven by the Na$^+$ concentration difference between the lumen of the intestine and the cytoplasm of the cell.

Na$^+$ concentration is kept low inside the cell by the action of the Na$^+$–K$^+$ ATPase.

Glucose exits the intestinal cell passively by a glucose transport protein.

concentration gradient (**Fig. 40.19**). The concentration of sodium ions inside the cell is kept low by the action of the sodium-potassium pump (Chapter 5). Molecules such as glucose are then moved from the intestinal cell into the blood by a pump on the basal side of the cell. Because the products of fat digestion—fatty acids and glycerol—are lipid soluble, they do not require a carrier protein for transport. Instead, they readily diffuse across the lipid plasma membrane, entering the mucosal cell, where they are further broken down and eventually transported to the bloodstream.

**The large intestine reabsorbs water and stores waste.**

After the absorption of nutrients in the small intestine, water and mineral ions are reabsorbed in the large intestine, or **colon,** of the hindgut (see Fig. 40.13). Peristalsis moves the contents of the small intestine into the colon at a pace that allows time

for nutrient absorption in the small intestine. By the time the gut contents reach the large intestine, the nutrients have been absorbed into the body, but water and inorganic ions remain. These are absorbed in the large intestine until the contents form semisolid feces. Excess water absorption can cause constipation, whereas too little water reabsorption results in diarrhea. Diarrhea and constipation also can be caused by toxins released by microorganisms , such as the bacterium *Vibrio cholera*, the cause of cholera, in the food an animal eats. Feces are stored in the final segment of the colon, the **rectum,** until they are periodically eliminated through the anus.

As we saw in Case 5: The Human Microbiome, digestion is not a solo endeavor. Large populations of bacteria reside in the small and large intestine and help extract nutrients that the animal's body cannot extract itself. The principle gut resident is *Escherichia coli*, the bacterium commonly used for research in molecular biology laboratories. In the process of nourishing themselves by aiding in the digestion of the host's gut contents, the bacteria provide nutrients and certain vitamins, such as biotin and vitamin K, that the animal cannot produce itself. Vitamin deficiency therefore can sometimes result from prolonged antibiotic medication that kills large numbers of gut bacteria. Another by-product of bacterial action in the gut is gases such as methane and hydrogen sulfide. These are expelled when foods contain compounds that the bacteria can digest but which cannot be digested and absorbed by the host animal, as when humans eat beans. The mutual benefits to the host animal and the bacteria ensure the success of their symbiotic relationship.

### The lining of the digestive tract is composed of distinct layers.

As we have seen, the digestive tract is not a passive tube. It secretes enzymes and other chemical compounds, absorbs nutrients, and actively moves food through the body. How does a series of tubes accomplish all of this? As shown in **Fig. 40.20,** the digestive tract is made up of several layers of tissue, each with a specialized function. The central space, or **lumen,** through which the gut contents travel is surrounded by an inner tissue layer, the **mucosa,** which has secretory and absorptive functions. The cells of the mucosa secrete mucus to protect the gut wall from digestive enzymes, and, in the stomach, hydrochloric acid. Surrounding the mucosa is the **submucosa,** a layer containing blood vessels, lymph vessels, and nerves.

Outside these layers are two smooth muscle layers (Fig. 40.20). An inner **circular muscle** layer contracts to reduce the size of the lumen. An outer **longitudinal muscle** layer contracts to shorten small sections of the gut. These two muscle layers contract alternately to mix gut contents, acting as a traveling wave during peristalsis to move the contents along the digestive tract from compartment to compartment. Between the two muscle layers are autonomic nerves that control the contractions of the two sets of smooth muscle.

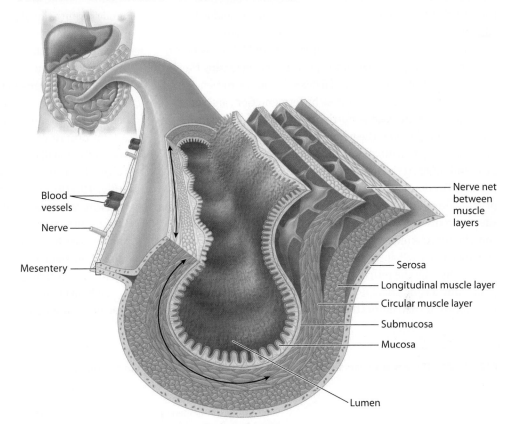

**FIG. 40.20 Cross section of the mammalian intestine.**

Blood vessels

Nerve

Mesentery

Nerve net between muscle layers

Serosa

Longitudinal muscle layer

Circular muscle layer

Submucosa

Mucosa

Lumen

An outer layer of cells and connective tissue called the **serosa** covers and protects the gut (Fig. 40.20). The gut is supported in the abdominal cavity by a membrane called the **mesentery,** through which blood vessels, nerves, and lymph travel to supply the gut.

### Plant-eating animals have specialized digestive tracts that reflect their diets.

Herbivorous (that is, plant-eating) animals evolved specialized digestive tracts that enhance their ability to digest cellulose and other plant compounds. Most herbivores, such as cows and termites, lack **cellulase,** the enzyme that breaks down cellulose. Instead, they have specialized compartments in their digestive tract that contain large bacterial populations, which do produce cellulase. The bacteria are able to digest the cellulose, a main constituent of plants. This association of herbivores and their gut microbes is another example of a symbiotic relationship.

Rather than the single stomach of other mammals, ruminants (cattle, sheep, and goats) have a four-chambered stomach that is highly specialized to enhance the ability of their gut bacteria to digest plants (**Fig. 40.21a**). The first two chambers—the **rumen** and the **reticulum**—harbor large populations of anaerobic bacteria that break down cellulose. To facilitate breakdown, the rumen contents are frequently regurgitated (in a process called rumination) for further chewing in the mouth. This repeated chewing decreases the size of the plant fibers, increasing the surface area available for bacterial digestion in the rumen. The bacteria break down cellulose by fermentation, and some of the products are used as nutrients by the host. (The site of fermentation provides the name "foregut fermenters" for these mammals.) Carbon dioxide and methane gas are also produced as a result of bacterial fermentation. The amount of methane produced by domesticated ruminants is second only to the production of industrial methane. Since methane is a greenhouse gas that traps heat from the sun, methane production by cattle contributes to atmospheric increases in methane and global climate change (Chapter 25).

After leaving the reticulum, the mixture of food and bacteria passes into a third chamber, the **omasum,** where water is reabsorbed. It then enters the **abomasum,** the last of the four

chambers, where protein digestion begins as it does in other mammals. The acid and protease secretions kill the bacteria, and the nutritional products are absorbed in the small intestine. The bacteria are an important protein source for the ruminant, yielding as much as 100 g of protein each day for a cow. This is important because plants are a poor source of protein but contain considerable nitrogen that the bacteria use to synthesize their own amino acids. The bacteria lost to digestion are balanced by the continual reproduction of bacteria in the rumen.

In contrast to the foregut fermentation of cows and sheep, other mammalian herbivores, such as koalas, rabbits, and horses, digest the plant material they eat by hindgut fermentation (**Fig. 40.21b**). Hindgut fermentation occurs in the colon and in the **cecum,** a chamber that branches off the large intestine. Because the fermentation products released from the cecum of a koala or horse have already passed through the small intestine, the main site of nutrient absorption, hindgut fermentation is less efficient than foregut fermentation in terms of nutrient extraction.

In humans and other animals such as lemurs and squirrels, the cecum is present but small and includes the **appendix,** a narrow,

**FIG. 40.21** Digestive systems of (a) a foregut fermenter and (b) a hindgut fermenter.

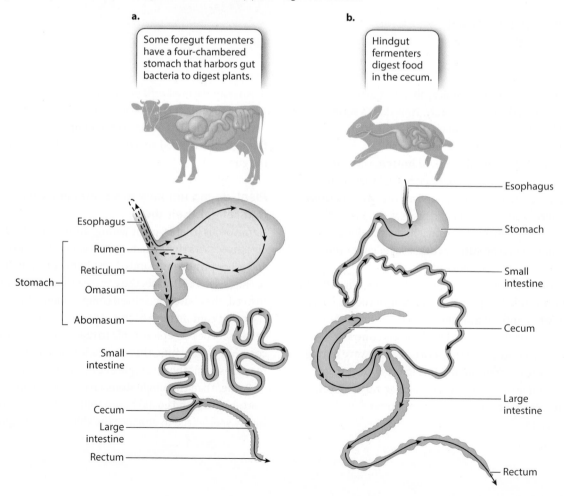

a.

Some foregut fermenters have a four-chambered stomach that harbors gut bacteria to digest plants.

b.

Hindgut fermenters digest food in the cecum.

Esophagus

Rumen

Reticulum

Stomach

Omasum

Abomasum

Small intestine

Cecum

Large intestine

Rectum

Esophagus

Stomach

Small intestine

Cecum

Large intestine

Rectum

tubelike structure that extends from the cecum. The appendix is an example of a **vestigial structure,** a structure that has lost its original function over time and is much reduced in size. Although the appendix has no clear function in nonherbivorous animals, evidence suggests it may play a role in the immune system. The appendix can cause damage to the organism if it becomes infected by partially digested material falling into it. Unless the infected appendix is removed, it can burst, releasing gut contents into the abdominal cavity, a condition that can be life threatening.

Charles Darwin put forth a hypothesis explaining how the appendix might have evolved to take on its current form in humans. He suggested that in the ancestors of modern humans the cecum once harbored bacteria that aided in the digestion of plant material. However, as their diet changed, they became less reliant on plant material such as cellulose to meet their nutritional needs, so the cecum became smaller and lost its function over time, becoming what we see today as the appendix. Vestigial structures, such as the appendix in humans, remnants of hind limbs in whales, small eyes in cave-dwelling animals, and wings in flightless birds, provide an illuminating window into the past and strong evidence for evolution.

→ **Quick Check 5** Some animals, such as rabbits, whose digestive systems include hindgut fermentation, re-ingest the products of digestion. What function do you think this serves?

## Core Concepts Summary

### 40.1 METABOLIC RATE DEPENDS ON LEVEL OF ACTIVITY, BODY SIZE, AND BODY TEMPERATURE.

Animals acquire carbohydrates, fats, and protein from their diet for energy to fuel cellular processes and activities. page 40-1

Carbohydrates are broken down rapidly in the cytosol by anaerobic glycolysis to produce a small amount of ATP in the absence of oxygen. page 40-2

Carbohydrates and fats produce large amounts of ATP by aerobic respiration through the citric acid cycle and oxidative phosphorylation. page 40-2

An animal's metabolic rate is its overall rate of energy use. page 40-3

Metabolic rate increases with increased physical activity. page 40-3

As the end product of anaerobic glycolysis, lactic acid decreases cellular pH and limits sustainable exercise. page 40-4

Larger animals consume more energy and have higher metabolic rates than smaller ones, but the increase in metabolic rate is not proportional to mass. page 40-5

Endothermic animals maintain elevated body temperature by metabolic heat production and have high metabolic rates. page 40-7

Ectothermic animals depend on external heat sources to warm their bodies. They have lower metabolic rates compared to endothermic animals. page 40-7

### 40.2 AN ANIMAL'S DIET SUPPLIES THE ENERGY IT NEEDS FOR HOMEOSTASIS AND ESSENTIAL NUTRIENTS IT CANNOT SYNTHESIZE ON ITS OWN.

An animal is in energy balance when the amount of energy that it takes in is equal to the amount it uses to sustain life and perform its functions. page 40-10

Animals must ingest essential amino acids, minerals, and vitamins, those that cannot be synthesized in their cells. page 40-10

### 40.3 DIFFERENT ANIMALS HAVE DIFFERENT ADAPTATIONS FOR FEEDING.

Aquatic animals take in food by suspension filter feeding, suction feeding, and active swimming. page 40-12

Vertebrates have specialized jaws and teeth for biting and mechanically processing food. page 40-13

Carnivorous animals have enlarged canines and slicing molars for catching prey and eating meat. page 40-13

Herbivorous animals have molars and premolars with surface ridges that enable them to grind plant matter. page 40-13

### 40.4 THE DIGESTIVE TRACT IS A TUBELIKE STRUCTURE WITH REGIONS SPECIALIZED FOR DIFFERENT FUNCTIONS.

The animal digestive tract consists of a series of compartments linked by a gut tube that allows for spatial and temporal control of digestion and absorption of nutrients. page 40-14

Most animals break down their food into small pieces in the mouth, making it easier to digest. page 40-14

Peristalsis mixes and moves food and digestion products through the gut by means of smooth muscle contractions. page 40-15

The stomach stores food and initiates protein digestion through secretions of hydrochloric acid and digestive enzymes, including pepsin. page 40-16

Carbohydrates and lipids are broken down in the duodenum of the vertebrate midgut, facilitated by the secretion of digestive enzymes, bile salts, and bicarbonate. page 40-17

Small intestine nutrient absorption is enhanced by an enlarged surface area of villi and smaller microvilli on the surface of cells. page 40-17

The colon of the hindgut reabsorbs water before waste products are eliminated as semisolid feces from the rectum through the anus. page 40-20

Herbivorous animals have specialized digestive chambers (rumen and cecum) that enable bacterial fermentation to break down plant cellulose. page 40-21

## Self-Assessment

1. Discuss how metabolic rate changes with levels of activity, body temperature regulation, and size.

2. Describe how ATP is produced for short and rapid activities versus long and sustained activities.

3. Explain how you would measure the metabolic rate of an animal.

4. Explain why the world record speed for a 10-m race is faster than that for a marathon.

5. Describe what happens in an endotherm and an ectotherm when outside temperature gets cold, and what happens when it gets hot.

6. Describe the order in which energy reserves are used in negative energy balance, such as starvation.

7. Name four ways in which animals capture prey and, for each, name one organism that uses it.

8. Draw the path of food through the vertebrate digestive tract, naming and describing the major function of each part.

9. Name the principal sites of digestion of proteins, carbohydrates, and fats, and indicate the principal enzymes involved in breaking down each.

10. Compare and contrast foregut and hindgut fermentation.

Do you understand the chapter's Core Concepts? Log in to check your answers to the Self-Assessment questions, then practice what you've learned and reinforce this chapter's concepts by working through the problems and multimedia tutorials provided there.

 www.biologyhowlifeworks.com

## CHAPTER 41

# ANIMAL RENAL SYSTEMS

## Water and Waste

### Core Concepts

**41.1** All animals regulate the water and electrolyte levels within their cells.

**41.2** Excretory organs eliminate nitrogenous wastes and regulate water and electrolyte levels.

**41.3** The mammalian kidney can produce urine that is more concentrated than blood as an adaptation for living on land.

In the 1960s, researchers at the University of Florida developed a new sports drink for players on the football team, the Gators. In addition to water and sugar, the drink contained electrolytes, or ions such as sodium ($Na^+$) and potassium ($K^+$) that give the solution electrical properties. The researchers who came up with the formula for the drink reasoned that it was important to replace both water and electrolytes lost in sweat. The drink—known as Gatorade—was a success.

Water and electrolytes are two key substances that allow cells and organisms to maintain their size and shape, as well as to carry out essential functions. Life originated and diversified in a watery environment. Consequently, the chemical functions of cells and organisms crucially depend on the properties of water (Chapter 2) and the relative amounts of water and electrolytes inside a cell. In general, the internal environment of a cell is maintained within a narrow window of conditions. The set of chemical reactions that constitute metabolism, for example, requires enzymes that work best at a particular pH (Chapter 6). **Homeostasis**—the maintenance of a stable internal environment in the face of a changing external environment—is a dynamic process.

Animals must also maintain homeostasis in terms of the molecules produced by normal activity. Metabolism generates wastes that are not needed and, in many cases, are actually toxic to the cell. These must be eliminated, or **excreted.** In most organisms, the excretion of wastes is closely tied to the maintenance of water and electrolyte balance.

This chapter explores the physiological and cellular mechanisms that underlie water and electrolyte balance, as well as waste elimination. It examines how animals meet the demands for water and electrolyte balance in different environments. It also examines how water and electrolyte balance is linked to waste elimination and explores the diversity of excretory organs animals have evolved to regulate these processes, which are critical to maintaining a stable internal body chemistry. In vertebrates, a **renal system** with paired **kidneys** serves these roles. The functional organization of mammalian kidneys is covered in the last section of the chapter.

## 41.1 WATER AND ELECTROLYTE BALANCE

A healthy person can live several weeks without food but cannot survive much more than a few days without water. A loss of more than 10% of body water threatens survival. Homeostasis of water and electrolyte concentration is critical to normal physiological function, not just in humans, but in all organisms.

## Osmosis governs the movement of water across cell membranes.

Water, like all molecules, moves by diffusion, or the random motion of molecules. When there is a difference in water concentration between two regions, there is a net movement of water molecules from regions of higher *water* concentration to regions of lower *water* concentration, a process called **osmosis** (Chapter 5). Water often contains other dissolved molecules, such as electrolytes, amino acids, and sugars, which are collectively called **solutes.** Because the concentration of a solute is typically known and the concentration of water is not, it is sometimes easier to think about water moving from regions of lower *solute* concentration (higher water concentration) to higher *solute* concentration (lower water concentration). Whether in terms of water concentration or solute concentration, the direction of net water movement is the same.

In organisms, the ability to control osmosis, and therefore to regulate levels of water and electrolyte in cells, depends on the properties of their plasma membranes and embedded proteins. When a barrier allows water or solutes to diffuse freely, it is said to be permeable. When it blocks the diffusion of water or solutes entirely, the membrane is considered impermeable. When it allows the movement of some molecules but not others, it is **semipermeable,** or selectively permeable. The membranes of cells are selectively permeable because of the combination of the lipid bilayer and transmembrane proteins, as we saw in Chapter 5.

Osmosis governs the movement of water across semipermeable membranes (**Fig. 41.1**). Plasma membranes and sheets of cells that are semipermeable allow water and smaller solutes to diffuse freely but prevent larger solutes, like proteins, from passing through them. Water is able to move directly through lipid bilayers by diffusion or through channels called **aquaporins** by facilitated diffusion (Chapter 5).

Consider then a situation in which there is a difference in solute concentration on the two sides of a semipermeable membrane, and the membrane is impermeable to the solute (Fig. 41.1a). Water moves by osmosis from regions of higher *water* concentration to regions of lower *water* concentration, or, from regions of lower *solute* concentration to regions of higher *solute* concentration (Fig. 41.1b).

→ **Quick Check 1** What happens when a cell is placed in a hypotonic solution (one with a lower solute concentration than that of the cell)? What happens when a cell is placed in a hypertonic solution (one with a higher solute concentration than that of a cell)?

The difference in concentration of solutes on the two sides of the membrane creates **osmotic pressure** that drives the movement of water. Osmotic pressure is the tendency of water to move from one solution into another by osmosis. The higher the solute concentration, the higher the osmotic pressure of that solution (that is, there is a greater tendency for water to move *into* that solution). Water diffuses into a region with a higher solute concentration (higher osmotic pressure) until it exerts a hydrostatic pressure (the pressure exerted by water itself due to gravity, not differences in solute concentration) that balances the osmotic pressure. When osmotic pressure equals hydrostatic pressure, the net movement of water stops and equilibrium is reached (Fig. 41.1c). At equilibrium, water

**FIG. 41.1    Osmosis.**

a.

Lid with pressure sensors

Time

Na⁺
Cl⁻

Semipermeable membrane

2% NaCl          10% NaCl

A semipermeable membrane allows movement of water but not sodium or chloride ions.

b.

Time

Water movement

4% NaCl          8% NaCl

Water moves by osmosis from a region of lower to higher solute concentration.

c.

Low pressure          High pressure

6% NaCl          6% NaCl

Net movement of water stops when osmotic pressure equals hydrostatic pressure.

molecules still move across the membrane in both directions by diffusion, but there is no net movement of water molecules.

Whether osmotic pressure drives water into or out of a cell depends on the solute concentration of the inside of the cell relative to the outside. Similarly, whether osmotic pressure drives movements of water into or out of an organism depends on solute concentration inside an animal relative to its environment. Note that osmosis is driven by differences in *total* solute concentration between the two sides of a cell membrane. That is, if a particular electrolyte such as potassium has a high concentration inside the cell and low concentration outside a cell, but another electrolyte, such as sodium, has exactly the opposite distribution, there will be no net water movement because the total concentrations of the solutes on the two sides of the membrane are the same. Also note that water moves across semipermeable membranes in response to concentration differences of *any* solute. That is, the solute can be an electrolyte, such as sodium or potassium, or any other dissolved molecule, such as glucose or an amino acid or even waste, as long as the membrane is impermeable to it.

→ **Quick Check 2** A solution of water and dissolved ions is separated by a semipermeable membrane that allows the passage of water but not small ions. The concentration of sodium is higher on one side of the membrane than on the other side. However, there is no net movement of water across the membrane. How is this possible?

## Osmoregulation is the control of osmotic pressure inside cells and organisms.

Animal life on Earth exists in a wide range of environments by adopting different means of **osmoregulation,** or the regulation of water and solute levels to control osmotic pressure. Osmoregulation, like energy balance (Chapter 40), is a form of homeostasis. High osmotic pressures inside a cell can damage the cell, sometimes even causing it to burst and thus disrupt some of the animal's functions. Although cells and tissues can tolerate dehydration, often as high as 50% or more, excessive dehydration impairs a cell's metabolic function, as many chemical reactions, such as the hydrolysis reactions that build proteins and nucleic acids from individual subunits, depend on the presence of water.

How do cells control their internal osmotic pressure? To regulate water and solute levels, and hence the osmotic pressure inside cells, the cell controls the solute concentration of the inside of the cell relative to the outside of the cell because water follows solutes across semipermeable cell membranes. At the level of the cell, then, osmoregulation is achieved by the movement of solutes, particularly electrolytes.

At the level of the organism, osmoregulation is achieved by balancing input and output of water and electrolytes. The water that animals need is gained and lost in different ways. **Fig. 41.2** shows sites of water gain and loss in a saltwater fish, which vary

**FIG. 41.2  Sites of water and electrolyte gain and loss in a saltwater fish.** Osmoregulation is achieved by balancing water and electrolyte input and output.

depending on the organism. In general, animals gain water by drinking water that is less concentrated in solutes (hypotonic) than their body fluids. Humans and many other vertebrates, both freshwater and terrestrial, gain most of their water this way. Water is also acquired through the food that animals eat. Additionally, animals produce water as a product of cellular respiration when two protons ($H^+$) combine with oxygen ($O_2$) to form water (Chapter 7). This is the primary source of water for some animals adapted to deserts and oceans, where drinkable water is not readily available. Animals lose water in their urine and feces. Terrestrial animals lose water through evaporation from their lungs, and, in the case of humans, by sweating. Freshwater animals gain water through their gills, whereas most marine animals lose water through their gills.

Similarly, electrolytes are taken in and lost in different ways. Animals gain electrolytes in the food they eat. If the water they drink has a higher solute concentration than their body fluids, animals can gain electrolytes by drinking. Marine aquatic animals gain electrolytes as hypertonic water moves across their gills and body surface. In contrast, freshwater aquatic animals lose electrolytes through diffusion across their gills and skin into the hypotonic watery environment. Humans lose electrolytes as well as water when they sweat. Drinking beverages high in electrolytes is helpful before demanding physical activities—including sports—because these activities can result in substantial water and electrolyte loss through sweating. Electrolytes are also lost by specialized glands (discussed below) and in urine and feces.

The ability of many organisms to regulate their water and electrolyte levels has enabled them to maintain homeostasis and a stable body chemistry in the face of changing environmental conditions, and to live in diverse environments. Most forms of life originated in salty marine environments, but many groups of organisms subsequently moved onto land and into freshwater

**FIG. 41.3**    **The salmon life cycle.** Salmon live in a range of environments at different stages of their life cycle, which require different osmoregulatory strategies.

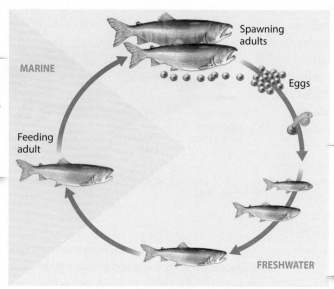

The marine environment has a high solute concentration, so the salmon must avoid excessive uptake of electrolytes and loss of water.

The freshwater environment has a low solute concentration, so the salmon needs to avoid losing electrolytes and gaining too much water.

## Osmoconformers match their internal solute concentration to that of the environment.

Animals regulate their internal osmotic pressure in different ways. Some animals match their internal osmotic pressure to that of their external environment. These animals are **osmoconformers.** They keep their internal fluids at the same osmotic pressure as the surrounding environment, which reduces the movement of water and solutes into or out of their bodies. Because the solute concentrations inside and outside osmoconformers are similar, osmoconformers don't have to spend a lot of energy regulating osmotic pressure. However, they have to adapt to the solute

lakes and rivers that present challenges for water and electrolyte balance. Living in freshwater environments requires animals to avoid excess water uptake into their tissues and control the loss of electrolytes from their body. Living on land requires animals to minimize their water loss in order to avoid desiccation, or drying out. Only insects and terrestrial vertebrates have achieved the ability to restrict water loss.

An extreme example of the ability of an animal to regulate water and electrolyte levels is provided by the life cycle of salmon. Mature salmon swim in open ocean saltwater environments but return upriver to spawn in freshwater environments (**Fig. 41.3**). As they do, they swim from high-salt ocean water through water of intermediate saltiness into freshwater rivers and streams that have little salt. After spawning, the young feed and grow in freshwater over a period of up to 3 years, eventually returning to the open ocean, where they live until they become reproductively mature and return to their home river system to spawn again.

These changes in aquatic habitat demand that the salmon make dramatic adjustments in the way they handle water and solutes. When in the ocean, the salmon must avoid excessive uptake of electrolytes and loss of water. However, when mature salmon migrate upriver to spawn and as young salmon continue to grow in freshwater streams and rivers, the challenge is reversed: They must avoid losing critical electrolytes and gaining too much water across their gills. In salmon, radical changes in osmoregulation are required over the course of an animal's lifetime to enable reproduction in freshwater streams and a return to the sea for further growth and maturation. Next, we explore how gills and kidneys meet these challenges.

concentration of their external environment, which is typically seawater. Osmoconformers tend to live in environments with stable solute concentrations, which provide an effective way of maintaining stable internal solute concentrations.

Although osmoconformers generally match the *overall* concentration of their tissues with their external environment, they often expend energy to regulate the concentrations of *particular* ions such as sodium, potassium, and chloride ($Cl^-$), and other solutes such as amino acids and glucose that are important to their function. For example, whereas the intracellular space of nearly all multicellular animals has a relatively high concentration of potassium ions and low concentration of sodium ions, the proportion in the extracellular space of most animals is just the opposite—low in potassium but high in sodium. This means that these cells are actively pumping sodium out and potassium in (Chapter 5).

Most marine invertebrates are osmoconformers, matching their total intracellular solute concentration to the solute concentration of seawater. These organisms maintain high concentrations of sodium and chloride in their cells to achieve an overall solute concentration close to that of seawater. Consequently, they have few specializations for osmoregulation beyond the need to regulate specific internal ion concentrations relative to their environment. Many marine vertebrates are also osmoconformers. Some, like hagfish and lampreys, match their total internal solute concentration to that of seawater but maintain high internal concentrations of particular electrolytes, just as marine invertebrates do.

Other marine vertebrates that are osmoconformers, like sharks, rays and coelocanths, match seawater's solute concentration by maintaining a high internal concentration of a compound called **urea.** Urea is a waste product of protein metabolism that many animals excrete, so it does not build up to high concentrations (section 41.2). By retaining this solute, these marine vertebrates are able to achieve osmotic equilibrium with the surrounding seawater. Even though marine osmoconformers seek to match the osmotic concentration of their environment, these animals also regulate specific ions for particular physiological functions within their body, as discussed in more detail below.

## Osmoregulators have internal solute concentrations that differ from that of their environment.

Many animals maintain internal solute concentrations that differ from that of their environment. These animals are **osmoregulators.** Osmoregulators expend considerable energy pumping ions across cell membranes in order to regulate the movement of water into or out of their bodies. Some freshwater and marine fishes that are osmoregulators are estimated to expend 50% of their resting metabolic energy on osmoregulation. However, the ability to regulate osmotic pressure allows osmoregulators to live in diverse environments—saltwater, freshwater, and land.

The external surface (such as the skin) of most osmoregulators prevents water and solutes from passing freely between the environment and the inside of the animal. Although the animal's skin maintains different total concentrations of solutes between the outside and inside of the animal, most of the internal cells do not maintain differences in total solute concentrations on the two sides of the plasma membrane. As a result, in general, there are usually no lasting differences in osmotic pressure between the inside and outside of internal cells of osmoregulators.

Teleosts, or bony fish (the largest group of marine vertebrates), as well as all freshwater and terrestrial animals, are osmoregulators. These animals have evolved a variety of specializations for water and electrolyte balance and waste excretion. Whereas marine teleosts maintain concentrations of electrolytes much lower than the surrounding seawater, freshwater fishes and amphibians maintain concentrations that are higher than those of their aquatic environment.

Aquatic animals use their gills to breathe as well as to regulate electrolyte levels. The thin gill filaments, with their extensive surface area, are well adapted for gas exchange and also facilitate electrolyte transport (**Fig. 41.4**). In marine bony fishes, specialized cells in the gills called **chloride cells** counter the ingestion and diffusion of excess electrolytes into the animal by pumping chloride ions into the surrounding seawater. By pumping negatively charged chloride ions out of the body, chloride cells create an electrical gradient that is balanced by positively charged sodium ions moving out of the body as well (Fig. 41.4a). In contrast, freshwater fishes have gill chloride cells with reversed polarity. Their chloride cells pump chloride ions (with sodium ions following) in the reverse direction, moving the chloride ions into the body to counter their ongoing loss from the gills into the surrounding freshwater (Fig. 41.4b). In both cases, the gills maintain a substantial difference in solute concentration between the fish's internal and external environment.

Sharks and rays excrete excess salt by a rectal gland, which eliminates salt with their digestive wastes. Similarly, many

**FIG. 41.4** **Osmoregulators.** Osmoregulators actively maintain osmotic pressure different from that of their environment.

**a.** Marine bony fish

Marine fish eliminate excess salt by active transport from gill chloride cells, which pump chloride ions out with sodium ions following.

Drinks seawater

Gill filaments

Gill chloride cells

Active ion elimination by chloride cells and osmotic $H_2O$ loss through gills

**b.** Freshwater bony fish

Freshwater fish use gill chloride cells to pump chloride ions in, with sodium ions following.

Freshwater fish do not drink water.

Active ion uptake by chloride cells and osmotic $H_2O$ uptake

→ Direction of electrolyte movement ($Na^+$, $K^+$, $Cl^-$)
→ Direction of water movement

**FIG. 41.5** **Salt excretion from glands.** Many marine birds have glands on the tops of the skull that actively excrete salt.

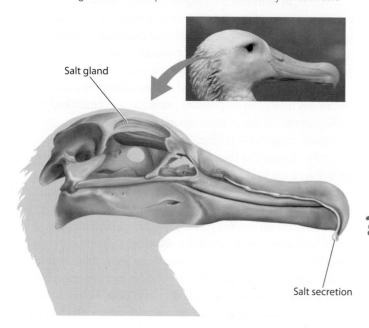

Salt gland

Salt secretion

marine birds have evolved specialized nasal salt glands to rid themselves of the high salt content of their diet (**Fig. 41.5**). These salt glands work by actively pumping ions from the blood into cells that make up the gland, followed by excretion of salt from the body through the gland and the nasal cavity. Their salt glands allow these animals to gain net water intake by drinking seawater. In contrast, humans and many other non-marine-adapted animals are incapable of deriving useful net water gain by drinking seawater because they cannot eliminate salt in high enough concentration. The high concentration of magnesium sulfate in seawater results in diarrhea if too much is consumed, exacerbating the net loss of water in the feces.

Some organisms that encounter a range of environments use different means at different times. For example, the brine shrimp *Artemia* usually acts as an osmoconformer but can act as an osmoregulator in extreme environments. *Artemia* lives in salt lakes such as the Great Salt Lake of Utah, where it usually matches its internal solute concentration to that of the outside environment. However, it sometimes encounters exceptionally high salt concentrations. It can survive in this high-salt environment by excreting excess salt ions from its gills to maintain an internal concentration that is hypotonic relative to its environment. Alternatively, in low saltwater environments (but not in freshwater, which they cannot tolerate), *Artemia* reverse their gill ion exchange and maintain a hypertonic internal body fluid concentration relative to the surrounding water.

In addition to respiratory surfaces such as gills and specialized salt glands, the other major regulator of water and electrolyte levels is the kidney, which we discuss in the next two sections.

**? CASE 7 Predator–Prey: A Game of Life and Death**
**Can the loss of water and electrolytes in exercise be exploited as a strategy to hunt prey?**
As we have seen, exercise can lead to loss of water and electrolytes, a response that is exploited by African bushmen as they hunt large prey. African bushmen are hunter-gatherers; their tribes live in hot, dry grasslands. Equipped only with wood spears, bows, and poisoned arrows, they have historically hunted kudus, antelope, buffalo, and other large grazing herbivores (**Fig. 41.6**). Bushmen hunters pursue and track down their prey, exploiting effects of the physiological stress that occurs when their prey run during the heat of the day.

Before hunting, bushmen drink as much water as possible, hydrating their bodies. They select a large prey animal, which

**FIG. 41.6** **Exploiting an animal's physiological stress.** (a) A kudu will be forced to run in the heat of the day by (b) African bushmen hunters. The animal will lose water and electrolytes, and will eventually become too dehydrated to escape.

a

b

they chase from a shaded refuge into the open sun when temperatures exceed the animal's core body temperature. The bushmen can deal with the high temperatures, cooling their bodies through sweating, but because they hydrated thoroughly before the hunt, their sweating does not dehydrate them. Not so for the prey. The bushmen hunters track and chase their prey until the prey becomes dehydrated and hyperthermic, thus rendered unable to escape.

Even if the bushmen hunters strike the animal successfully with a poisoned arrow, they still must track it for long distances and over long periods of time. Hunts typically last several hours, but sometimes as much as 1 or 2 days, and cover distances of 25 km (about 16 miles) or more. The loss of water in intensive heat disrupts the animal's water and electrolyte balance and thus contributes to its inability to move and defend itself.

## 41.2 EXCRETION OF WASTES IN RELATION TO ELECTROLYTE BALANCE

In most organisms, the regulation of water and solute levels is closely tied to waste excretion. Excretion is the elimination of waste products and toxic compounds from the body. Some of these wastes are generated as a by-product of metabolism, particularly protein and nucleic acid breakdown. Others were originally ingested, like the undigested remains of food, and need to be eliminated from the organism.

Most multicellular animals have evolved specialized excretory organs, such as the vertebrate kidney, to isolate, store, and eliminate toxic compounds and metabolic waste products. Because these compounds are eliminated as solutes that are dissolved and transported within the blood or other body fluids, their elimination from the body is intimately tied to water and electrolyte balance. The excretory organs of most animals therefore function to maintain water and electrolyte balance in addition to eliminating waste products.

Both the gastrointestinal tract (Chapter 40) and excretory organs like the vertebrate kidneys eliminate wastes from the body. However, the gastrointestinal tract is continuous with the outside world, and therefore its role in the elimination of wastes is quite different from that of excretory organs that separate toxic compounds from essential nutrients, electrolytes, and cells circulating in the blood.

### The excretion of nitrogenous wastes is linked to an animal's habitat and evolutionary history.

When proteins and nucleic acids are broken down by metabolism, one of the by-products is ammonia ($NH_3$). Ammonia contains nitrogen and is toxic to organisms, so it is a form of **nitrogenous waste** (**Fig. 41.7**). By far, most ammonia is produced from the breakdown of proteins. Ammonia can disturb the pH balance of

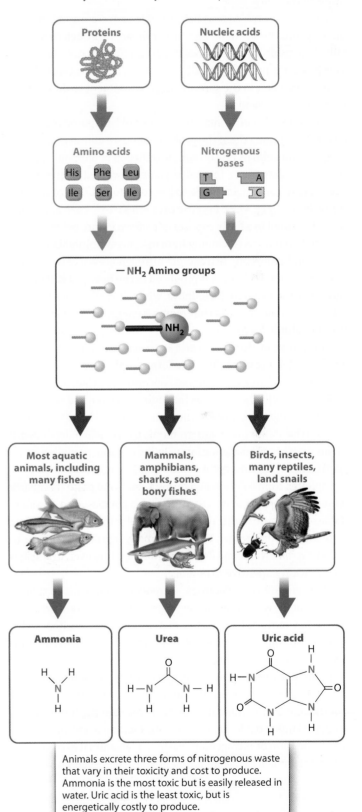

**FIG. 41.7 Nitrogenous waste.** Animals excrete nitrogenous waste in three major forms—ammonia, urea, and uric acid—which vary in their toxicity and cost to produce.

Animals excrete three forms of nitrogenous waste that vary in their toxicity and cost to produce. Ammonia is the most toxic but is easily released in water. Uric acid is the least toxic, but is energetically costly to produce.

cells and cause neuronal cell injury. At high levels, it is lethal to organisms. It is either eliminated from the body or converted into a less toxic form. In either case, ammonia is kept at very low levels in the body.

Some organisms, such as fish, are able to excrete ammonia directly into the surrounding water, primarily through their gills. As a result, ammonia does not accumulate to toxic levels. This process is not energetically expensive, but it requires high volumes of water in the surrounding environment to dilute the highly toxic ammonia.

The movement of animals from water to land posed many challenges, among them the problem of waste elimination. No longer could animals rely on simple diffusion of ammonia into the surrounding water. Instead, organisms evolved biochemical pathways to convert ammonia to less toxic forms. One of these forms is urea (Fig. 41.7). Because urea is less toxic than ammonia, it can be stored in a fairly concentrated form before being excreted. Mammals (including humans), many amphibians, sharks, and some bony fish excrete nitrogenous waste in the form of urea. In mammals, urea is produced in the liver and is then carried by the blood to the kidneys, where it is eliminated. Urea is less toxic than ammonia but requires energy to produce it and water to eliminate it.

Ammonia can also be converted into uric acid (Fig. 41.7). Uric acid is the least toxic of the three forms of nitrogenous waste, so animals can store it at a higher concentration than urea. Birds and many reptiles, arthropods (including insects), and land snails excrete nitrogenous waste in the form of uric acid. These groups evolved the ability to convert ammonia to uric acid independently of one another as an adaptation to living on land, providing a good example of convergent evolution (Chapter 23). Uric acid precipitates from solution and forms a semisolid paste (familiar as white bird droppings). Because uric acid is not dissolved in water, it does not exert osmotic pressure and is eliminated with minimal water loss for the animal. This allows many reptiles and insects to live in extremely hot and dry environments. However, uric acid is energetically expensive to produce.

Many amphibians, like frogs, excrete ammonia early in their development, when they live in water, but then excrete urea when they move onto land after metamorphosis. Alligators and turtles also adjust what they excrete: ammonia when in the water and uric acid when on land.

Ammonia, urea, and uric acid are the three major forms of nitrogenous wastes excreted by animals. However, they are not the only forms. Animals excrete nitrogen in the forms of trimethylamine oxide, creatine, creatinine, and even amino acids. The form of nitrogenous waste that an animal excretes is ultimately linked to its environment and evolutionary history.

→  **Quick Check 3** Why do mammals convert ammonia to urea rather than simply excreting it, as fish do?

## Excretory organs work by filtration, reabsorption and secretion.

Excreting wastes does not occur by simply pumping the wastes out of the body. Instead, animals filter or secrete wastes out of the blood along with water, electrolytes, and even essential nutrients, but then reabsorb the nutrients they need, excreting the waste and enough water and electrolytes to maintain appropriate levels of water and electrolytes for hydration and osmotic balance. While this process may seem inefficient, it has several advantages compared to simply pumping wastes out. First, as we will see, it allows organisms to excrete wastes and at the same time balance their water and electrolyte levels. Second, it allows any toxic compound that ends up in the blood to be excreted. If organisms relied on molecule-specific active transport of wastes, novel toxic compounds ingested by organisms could not be excreted.

The first step is to isolate the waste into an extracellular space. In the case of unicellular organisms such as *Paramecium* and simple multicellular animals such as sponges and cnidarians (jellyfish and sea anemones), waste compounds are isolated in a cellular compartment called a **contractile vacuole** (**Fig. 41.8**). Fusion of the contractile vacuole with the cell membrane eliminates the waste contents by exocytosis from the cell (Chapter 5).

Multicellular animals with pressurized circulatory systems (annelid worms, mollusks, and vertebrates) also isolate wastes

**FIG. 41.8**  Contractile vacuole of a *Paramecium* when (a) full and (b) after emptying.

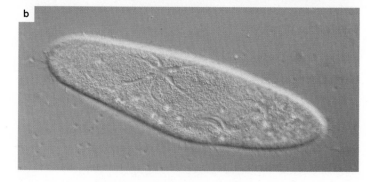

**FIG. 41.9 Filtration, reabsorption, and secretion.** These three processes eliminate wastes and provide osmoregulation in multicellular animals with pressurized circulatory systems.

Blood flow

Capillary

Flow of filtrate

Excretory tubule

Urine

**1** Filtration produces a filtrate of the blood.

**2** Reabsorption removes useful solutes from the filtrate and returns them to the blood.

**3** Secretion adds solutes to the filtrate.

## Animals have diverse excretory organs.

The processes of filtration, reabsorption, and secretion serve two important and complementary functions: They allow wastes to be isolated and subsequently removed from the body, and they allow the organism to adjust the amounts of water and electrolytes required to meet its osmoregulatory needs. These two functions are carried out by different excretory organs in different animals. For example, in flatworms, excretion involves isolating the fluid from the body cavity in excretory organs called **protonephridia** without filtering it first (**Fig. 41.10**). Cells with beating cilia move the fluid down the lumen of the tubule to an excretory pore. Freshwater flatworms, such as *Planaria*, have an extensive network of tubules that eliminate excess fluid along with waste products. As the fluid passes down the tubule, cells lining the tubule modify the fluid's contents through reabsorption and secretion. The urine leaving a freshwater flatworm contains nitrogenous waste and is less concentrated (that is, more dilute) than its body to balance the water it ingests as it feeds.

In segmented annelid worms such as earthworms, the body fluid is filtered through small capillaries into a pair of excretory organs

**FIG. 41.10 Excretory organ of a flatworm.**

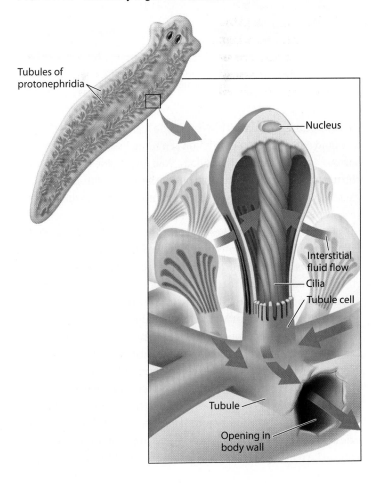

Tubules of protonephridia

Nucleus

Interstitial fluid flow

Cilia

Tubule cell

Tubule

Opening in body wall

from the blood, but they do so by **filtration** (**Fig. 41.9**). Circulatory pressure pushes fluid containing the wastes through specialized filters into an extracellular space. The filtered fluid, called the filtrate, contains waste products along with water, electrolytes, and other solutes, which drain into **excretory tubules** that connect to the outside. The excretory filter, like a cell's plasma membrane, is selectively permeable. It allows smaller ions and solutes to pass into the tubule lumen along with water but retains larger solutes such as proteins and cells.

Filtration is followed by **reabsorption** of key ions and solutes (Fig. 41.9). Reabsorption is an active process for some molecules and a passive one for others. In both cases, substances that are important for the animal to retain, such as electrolytes, amino acids, vitamins, and simple sugars, are taken up by cells of the excretory tubule and returned to the bloodstream. Water is reabsorbed into the cells by osmosis, driven by the osmotic pressure established by the active transport of solutes into the cell.

The final step in waste removal is the **secretion** of additional toxic compounds and excess ions (Fig. 41.9). Secretion is also an active process and involves eliminating substances that were not filtered from the blood earlier. Filtration, reabsorption, and secretion are the three steps that all animal excretory systems perform to eliminate toxic wastes, retain key electrolytes and solutes, and regulate water volume.

**FIG. 41.11   Excretory organ of an earthworm.**

Bladder   External   Collecting   Internal   Capillary
          opening    tubule       opening    network

Components of
a metanephridium

called **metanephridia** within each body segment (**Fig. 41.11**). The fluid enters a metanephridium at a funnel-shaped opening surrounded by cilia that drains body fluid into a series of tubes in the next body segment and then out of the animal through an excretory pore. Blood vessels and cells surrounding the tubules reabsorb and secrete additional compounds as urine is formed. Like freshwater flatworms, earthworms reabsorb electrolytes and eliminate dilute urine containing dissolved nitrogenous wastes.

In insects and other terrestrial arthropods, fluid passes from the main body cavity into a series of tubes called **Malpighian tubules** that empty into the hindgut (**Fig. 41.12**). (These structures are named for Marcello Malpighi, the seventeenth-century Italian physician who first described them.) An insect can have two Malpighian tubules or more than 100. The tubules actively secrete uric acid and excess electrolytes from the animal's tissues while retaining key solutes. Muscle fibers in the walls of the tubules help push the fluid to the hindgut. In the hindgut, the fluid becomes more acidic because of the buildup of uric acid, causing uric acid to precipitate from solution when it enters the rectum. Because uric acid is removed from the fluid,

it does not influence the movement of water. Water is therefore free to diffuse into the tissues with the reabsorbed electrolytes that are transported from the rectum before the uric acid and remaining dried digestive wastes are eliminated. Malpighian tubules represent one of the most successful adaptations for excreting nitrogenous wastes with minimal water loss, enabling insects to live in diverse and arid terrestrial environments.

### Vertebrates filter blood under pressure through paired kidneys.

The excretory organs of vertebrates are the kidneys. Like all excretory organs, they not only serve to eliminate wastes, but also play an important role in water and electrolyte balance. Vertebrate kidneys also reflect the evolutionary transition of vertebrates from an aquatic to a terrestrial environment, where there is a need to minimize water loss. Mammals and birds, for example, can excrete urine that is more concentrated than their blood and tissues.

The kidneys lie posterior to the abdominal cavity. Whereas the kidneys of humans and other mammals are bean-shaped, the kidneys of birds and other vertebrate animals are elongated and segmented (**Fig. 41.13**). Paired vertebrate kidneys receive blood supply from the heart by renal arteries that branch from the abdominal aorta.

Vertebrates filter their blood through specialized capillaries that have openings in their walls. These porous capillaries form a

**FIG. 41.12   Excretory organ of an insect.**

**FIG. 41.22 Hormonal control of urine concentration by ADH.** (a) In the presence of ADH, the collecting duct is more permeable to water than (b) in the absence of ADH.

**a.**

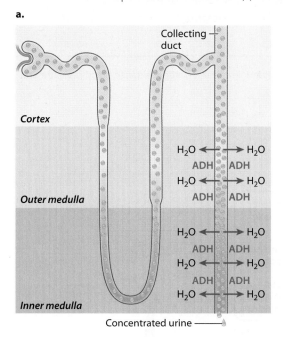

ADH released from the posterior pituitary gland increases the collecting ducts' permeability to water, allowing water to diffuse out of the filtrate with the gradient created by the loop of Henle.

**b.**

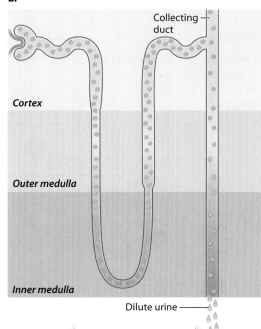

In the absence of ADH, the collecting duct is less permeable to water, resulting in dilute urine.

**RESULTS** Their results, shown in the table, are consistent with the predictions of Kuhn's hypothesis. That is, the solute concentrations of the fluid in the loop of Henle and in the collecting ducts are similar to each other and higher than that of the blood.

**CONCLUSION** These results are consistent with a model in which concentrated urine is produced by first generating a concentration gradient between the cortex and the medulla, and then allowing water to move out of the collecting ducts by osmosis to produce concentrated urine.

**FOLLOW-UP WORK** Similar measurements in other areas of the kidney and in other mammals and birds supported the model.

SOURCES Gottschalk, C. W., and M. Mylle. 1958. "Evidence That the Mammalian Nephron Functions as a Countercurrent Multiplier System." *Science* 128:594; Gottschalk, C. W., and M. Mylle. 1959. "Micropuncture Study of the mammalian urinary concentrating mechanism: evidence for the countercurrent hypothesis." *American Journal of Physiology* 196 (4):927–936.

## The distal convoluted tubule secretes additional wastes.

The filtrate entering the distal convoluted tubule, which is located within the cortex, is hypotonic relative to the interstitial space, having lost nearly all of its electrolytes. As a result, urea remains the principal solute. Other wastes from the bloodstream are also actively secreted into the distal convoluted tubule.

At this point, then, the filtrate is dilute, and the main solutes are urea and other wastes. The dilute filtrate enters the collecting ducts, which drain into the renal pelvis at the base of the kidney, passing from the cortex to the medulla. The concentration gradient established by the loop of Henle from the cortex to the medulla allows water to be reabsorbed as needed depending on the water levels of the organism, as we will see next.

## The final concentration of urine is determined in the collecting ducts and is under hormonal control.

The final segment of the nephron is the collecting duct. At this point, the key processes of filtration, reabsorption, and secretion have occurred. The filtrate is dilute but contains wastes and water. The collecting duct is the site where water levels are adjusted to meet the osmoregulatory needs of the organism and therefore maintain homeostasis. The concentration gradient from cortex to medulla established by the loop of Henle allows the collecting ducts to regulate water levels.

The permeability of the collecting ducts to water determines how dilute or concentrated the urine is. The water permeability of the collecting ducts is controlled by the peptide hormone **antidiuretic hormone (ADH),** also called **vasopressin,** which is secreted by the posterior pituitary gland (Chapter 38). In the presence of ADH, the collecting duct walls become permeable to water (**Fig. 41.22**). This change in permeability occurs because aquaporins insert in the membrane of cells of the collecting duct. The aquaporins allow water in the collecting duct to diffuse into the interstitial fluid, thus making the urine in the tubule more concentrated. When a person is dehydrated, the urine leaving the collecting ducts can become as concentrated in the interstitial space at the base of the kidney.

In the absence of ADH, the collecting duct walls are impermeable to water, preventing the water from leaving. This produces dilute urine. Some diuretic substances ("diuretic" is from a Greek word meaning "to flow through"),

**FIG. 41.23  Control of blood volume and blood pressure by the kidney.** Cells of the efferent arteriole of the glomerulus sense blood pressure and start a hormone cascade when blood pressure gets too low.

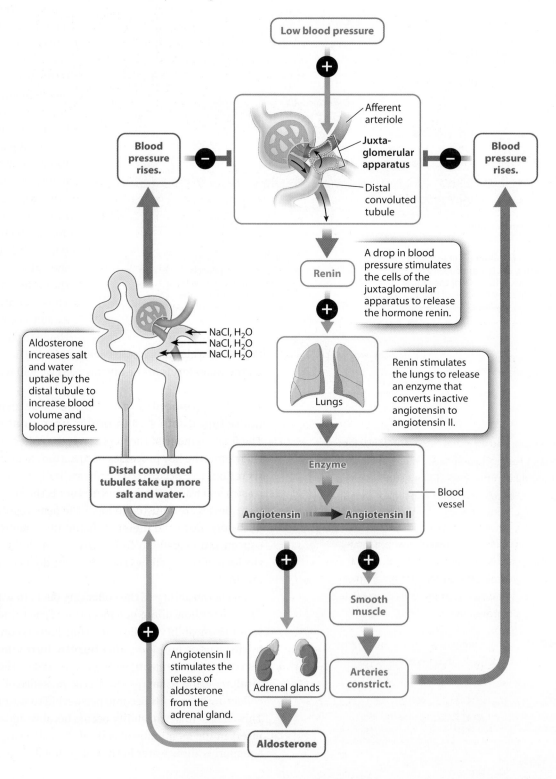

such as alcohol, promote the dilution of urine by interfering with ADH production and therefore making the collecting duct walls more impermeable to water. If hypotonic urine is produced, there is net water loss. Consequently, diuretic drinks often cause dehydration.

The regulation of water and solute reabsorption by the kidney is remarkably precise. The kidneys of a human filter 180 liters (L) of blood each day (125 ml/min). The total blood volume of a human is about 5 L, which means that the blood is filtered about 60 times per day. Urine production in humans averages 0.5 L per day, which means that under normal conditions, the kidneys reabsorb 99.9% of the water entering the renal tubules through glomerular filtration. In addition, the kidneys reabsorb all the glucose and amino acids that were filtered out of the blood in the glomerulus, together with 99.9% of the electrolytes in the blood. Human kidneys achieve this efficient transport of key ions, solutes, and water because of the renal tubules' extensive microvilli and the large number of mitochondria and membrane transporters in their epithelial cells.

The exceptional ability of the kidneys to filter and reabsorb key ions, solutes, and water also shows why kidney damage can have such serious consequences. People with kidney disease require that their blood be filtered by dialysis treatment as frequently as three times per week, each treatment lasting several hours. Otherwise, toxic wastes would build up to dangerous levels in the body. Renal failure resulting from kidney disease remains a substantial health problem in the world today, costing more than $34 billion in the United States alone.

**The kidneys help regulate blood pressure and blood volume.**

Because the kidneys receive a large fraction of the blood leaving the heart, they are well suited for monitoring changes in blood pressure and maintaining it at relatively constant levels, another example of homeostasis. Specialized cells of the efferent arteriole leaving the glomerulus of each nephron form the **juxtaglomerular apparatus** (**Fig. 41.23**). In response to a drop in blood pressure, these cells secrete the hormone **renin** into the bloodstream. Renin stimulates epithelial cells of the lung to release an enzyme that converts an inactive hormone (angiotensinogen) in the bloodstream into its active form, **angiotensin II.** Angiotensin II acts on the smooth muscles of arterioles throughout the body, causing them to constrict, thus increasing blood pressure and directing more blood back to the heart.

Angiotensin II also stimulates the release of **aldosterone,** a hormone produced by the adrenal glands that stimulates the distal convoluted tubule and collecting ducts to take up more salt and water. Increased salt and water uptake in turn leads to increased blood volume and therefore blood pressure. Thus, the kidneys participate in the control of blood pressure that is also monitored by baroreceptors (specialized mechanoreceptors that sense blood pressure) in the aorta and carotid bodies. These communicate with the autonomic nervous system to control heart rate and contractile force, as discussed in Chapter 39.

The kidneys, therefore, play many roles. In addition to the elimination of wastes and other toxic compounds, they serve important homeostatic functions, including water and electrolyte balance and the regulation of blood volume and blood pressure.

## Core Concepts Summary

### 41.1 ALL ANIMALS REGULATE THE WATER AND ELECTROLYTE LEVELS WITHIN THEIR CELLS.

Cell membranes and sheets of cells act as semipermeable membranes, allowing the passage of water but restricting or controlling the movement of many solutes. page 41-2

Water moves across semipermeable membranes by osmosis from regions of lower solute concentration to regions of higher solute concentration. page 41-2

Osmotic pressure is the pressure exerted by water as it moves from one solution to another by osmosis. page 41-2

Osmoregulation is the regulation of osmotic pressure inside cells or organisms. page 41-3

Osmoconformers maintain their internal solute concentration similar to that of the environment, whereas osmoregulators have internal solute concentrations different from that of the environment. page 41-4

Both osmoconformers and osmoregulators regulate the levels of particular solutes, especially sodium, potassium, and chloride. page 41-4

Osmoregulators that live in high-salt environments excrete excess electrolytes and minimize water loss. page 41-5

Osmoregulators that inhabit low-salt or freshwater environments excrete excess water and minimize electrolyte loss. page 41-5

Osmoregulators that live on land minimize water loss. page 41-5

## 41.2 EXCRETORY ORGANS ELIMINATE NITROGENOUS WASTES AND REGULATE WATER AND ELECTROLYTE LEVELS.

Osmoregulation and excretion are closely coordinated processes controlled by the excretory organs of animals.  page 41-7

Ammonia, urea, and uric acid are three major forms of nitrogenous waste that animals excrete, depending on their evolutionary history and habitat.  page 41-7

Ammonia is toxic but readily diffuses into water; urea is less toxic and can be stored before it is eliminated; uric acid is a semisolid and can be eliminated with minimal water loss as an adaptation for living on land.  page 41-7

All animal excretory systems first isolate or filter fluid into an extracellular space. After being isolated or filtered, key electrolytes, solutes, and water are reabsorbed from the filtrate. Additional wastes are then secreted into the remaining filtrate before it is eliminated from the body.  page 41-8

Examples of excretory organs are the protonephridia of flatworms, the metanephridia of segmented annelid worms, the Malpighian tubules of insects, and the kidneys of vertebrates.  page 41-9

The urine produced by the kidneys and other excretory organs is stored in a bladder or the cloaca until being eliminated from the body.  page 41-10

The functional unit of the vertebrate kidney is the nephron, which consists of a glomerulus, renal tubules, and collecting duct.  page 41-11

## 41.3 THE MAMMALIAN KIDNEY CAN PRODUCE URINE THAT IS MORE CONCENTRATED THAN BLOOD AS AN ADAPTATION FOR LIVING ON LAND.

The kidney has an outer cortex and an inner medulla.  page 41-12

The renal tubules have specialized regions, including the proximal convoluted tubule, the loop of Henle, and the distal convoluted tubule.  page 41-14

The loop of Henle of some nephrons extends all the way into the medulla before looping back to the cortex. These nephrons create a concentration gradient from the outer cortex to the deeper medulla of the kidney.  page 41-14

The ability to create a concentration gradient from the cortex to the medulla results from a countercurrent multiplier mechanism of the loops of Henle.  page 41-14

The final concentration of urine is regulated by the control of water permeability within the walls of the collecting ducts, which pass from the cortex to the medulla.  page 41-17

Release of antidiuretic hormone by the posterior pituitary gland increases the water permeability of the collecting ducts, concentrating the urine that empties from vertebrate kidneys.  page 41-17

The kidneys help to regulate blood volume and pressure by secreting the hormone renin, leading to the production of angiotensin II, which constricts blood vessels, and aldosterone, which increases reabsorption of salt and water by the kidneys.  page 41-19

## Self-Assessment

1. Given a semipermeable membrane that is permeable to water but not to a particular solute with different concentrations on the two sides of the membrane, show the direction of water movement and label the side with the higher osmotic pressure.

2. Describe how animals gain and lose water and electrolytes.

3. Name two animals that are osmoconformers and two that are osmoregulators. Explain the difference between the two types of animal.

4. List three forms of nitrogenous waste and describe how each is an adaptation for the environment in which the animal lives.

5. Describe the three steps in which organisms excrete wastes.

6. Draw a mammalian nephron, label and describe the primary function of each part, and show the direction of water and electrolyte in each part.

7. Explain how the loop of Henle creates a concentration gradient from the cortex to the medulla.

8. Describe the role of ADH in the regulation of urine concentration.

9. Explain how the kidneys help to regulate blood volume and blood pressure.

Do you understand the chapter's Core Concepts? Log in to check your answers to the Self-Assessment questions, then practice what you've learned and reinforce this chapter's concepts by working through the problems and multimedia tutorials provided there.

    www.biologyhowlifeworks.com

## CHAPTER 42

# ANIMAL REPRODUCTION AND DEVELOPMENT

## Core Concepts

Children are sometimes gently introduced to the subject of human reproduction by referring to "the birds and the bees." Since birds and bees reproduce, it is assumed that they can substitute for humans in discussions about reproduction. But how relevant is bird and bee reproduction to human reproduction? It might come as a surprise that most male birds don't have penises, and that male bees don't even have fathers. Perhaps, then, the lesson of the "birds and the bees" is not so much in what it teaches us about human reproduction, but, as we explore in this chapter, in showing us that reproductive strategies among organisms are spectacularly diverse.

Reproduction is a striking and conspicuous feature of the natural world. Flowers burst into bloom each spring to attract pollinators, male fireflies light up brilliantly on warm summer nights as they signal to females in the grass below, birds sing and display ornate plumage to attract mates. Most living organisms have the ability to reproduce—that is, to make new organisms like themselves. This remarkable ability is not unique to living organisms—after all, computer viruses and self-replicating molecules are capable of reproduction but are not themselves living. However, reproduction is a key attribute of life. All living things came about by reproduction, and most are capable in turn of reproduction. Of course, this statement does not address the question of how life originated in the first place, a topic considered in Case 1: The First Cell.

This chapter focuses on reproduction and development. We begin by considering reproduction broadly across the living world, then turn our attention to animal, and specifically vertebrate, reproduction and the surprise that even among vertebrates there is quite an array of reproductive strategies. In the last two sections, we focus on human reproduction, starting with reproductive anatomy and physiology and then following reproduction from gamete formation to fertilization, pregnancy, and birth.

## 42.1 THE EVOLUTIONARY HISTORY OF REPRODUCTION

While the ability to reproduce is shared across all domains of life, the particular way in which animals reproduce varies among species, from the familiar to what might strike us as truly bizarre. To take just one of many examples: After birth, a male anglerfish finds a female anglerfish, bites her skin, and fuses with her. His body atrophies, leaving the male reproductive organs physically attached to the female and allowing her eggs to be fertilized when she is ready. To begin to make sense of the diversity of reproductive strategies, we consider how different forms of reproduction evolved, what organisms use them, and their adaptive significance.

**FIG. 42.1    Asexually reproducing organisms.**  (a) Bacteria divide by binary fission. (b) The yeast *Saccharomyces cerevisiae* reproduces by budding. (c) Corals are propagated by fragmentation. (d) Komodo dragons can reproduce by parthenogenesis.

## Asexual reproduction produces clones.

Organisms have two potential ways to reproduce. One is the production of genetically identical cells or individuals called **clones.** This form of reproduction is called **asexual reproduction,** and it occurs in a wide variety of organisms (**Fig. 42.1**).

For example, bacteria (Fig. 42.1a) and archaeons reproduce by binary fission, a form of cell division in which the genome replicates and then the cell divides in two (Chapter 11). The result is two cells that are either identical to the parent cell or nearly so, differing only because of chance mutations.

Prokaryotes can nevertheless increase genetic variation in various ways (Chapter 26). For example, they can transfer DNA from one individual to another, usually of the same species, and much more rarely to an individual of another species, during conjugation. They can also obtain DNA directly from the environment and by viral infection, incorporating genes that might have evolved in even distantly related species. These processes allow prokaryotes to acquire novel genetic sequences that can affect fitness.

Eukaryotic cells divide by mitosis (Chapter 11), a process that evolved early and was present in the common ancestor of all living eukaryotes. Mitosis is a form of growth and development in all eukaryotes, and is the basis of asexual reproduction in many eukaryotes.

One form of asexual reproduction in fungi, plants, and some animals is **budding,** in which a bud, or protrusion, forms on an organism and eventually breaks off to form a new organism that is smaller than its parent. The budding yeast, *Saccharomyces cerevisiae*, which is used to make bread and beer, is so named because of its ability to produce new individuals by budding (Fig. 42.1b). Budding can also occur in animals such as the water-dwelling hydra, which can grow and subsequently break off part of itself to form a new, free-living individual. Whether used to describe reproduction in unicellular or multicellular organisms, budding involves an unequal division between a mother and daughter cell or organism. Budding is the result of mitosis and therefore, like binary fission, results in an offspring that is genetically identical to the parent.

Another form of asexual reproduction is **fragmentation,** in which new individuals arise asexually by the splitting of one organism into pieces, each of which develops into a new individual. Certain molds, algae, worms, sea stars, and corals reproduce by fragmentation. While fragmentation occurs naturally, it is also done intentionally to propagate corals in reef restoration projects (Fig. 42.1c). Once again, the new individual is a clone of the parent.

A final example is **parthenogenesis,** literally "virgin birth." In this form of asexual reproduction, females produce eggs that are not fertilized by males but divide by mitosis and develop into new individuals. Parthenogenesis occurs in invertebrates, such as bees, other insects, and crustaceans, as well as in vertebrates, including some fish and reptiles (Fig. 42.1d). The details of parthenogenesis differ depending on the organism. Egg cells are usually the product of meiosis and are haploid, having one set of chromosomes. In some organisms, this haploid egg cell divides by mitosis to become a haploid adult; in others, there is a doubling of the DNA content during development, so the adult is diploid, having two sets of chromosomes; and in still others, the egg cell does not undergo meiosis (it is diploid) and divides by mitosis to become a diploid adult. In all cases, parthenogenesis involves a single maternal parent.

### Sexual reproduction involves the formation and fusion of gametes.

The second way that organisms reproduce is by combining a complete set of genetic information from two individuals to make a new genetically unique individual. This form of reproduction is called **sexual reproduction.**

At its core are two basic biological processes—**meiosis** and **fertilization** (**Fig. 42.2**). Meiosis is a form of cell division that halves the number of chromosomes; fertilization involves fusion of specialized cells to restore the original chromosomal content. Meiosis evolved in the ancestors of modern eukaryotes, and so sexual reproduction occurs only in eukaryotes.

As we saw in Chapter 11, meiosis consists of one round of chromosome replication followed by two rounds of cell division, resulting in four cells each with half the number of chromosomes of the parent cell. So, in a diploid species with $2n$ chromosomes, the result of meiosis is four cells with $n$ chromosomes. The resulting cells are haploid and are called **gametes** or **spores.** Humans, for example, have 23 pairs of chromosomes ($2n$ = 46 chromosomes) and human gametes have half this number ($n$ = 23 chromosomes).

**FIG. 42.2 Sexual reproduction.** This form of reproduction involves the production of gametes by meiosis and their fusion by fertilization.

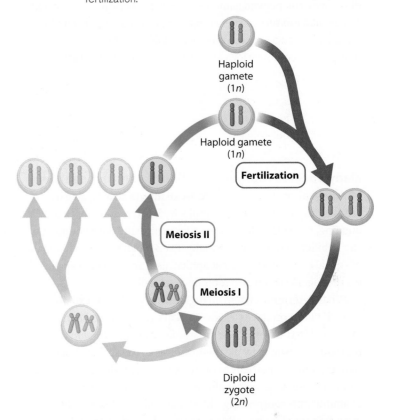

Haploid gamete ($1n$)

Haploid gamete ($1n$)

Fertilization

Meiosis II

Meiosis I

Diploid zygote ($2n$)

Many species produce two types of gametes that differ in shape and size. The smaller, male gametes are called **spermatozoa,** or **sperm,** and the larger, female gametes are called **ova,** or **eggs.** For other organisms, including green algae, diatoms, slime molds, and most fungi, the two types of gametes are the same size. In spite of their similar size and appearance, however, they come in two (or more) distinct mating types and cannot reproduce sexually with individuals of the same type.

Fertilization involves the fusion of two gametes, one from each parent (Fig. 42.2). The result is a **zygote.** If each gamete has $n$ chromosomes, the resulting zygote has $2n$ chromosomes. Fertilization results in a net reduction in cell number—two haploid gametes fuse to make one diploid zygote. For unicellular organisms with a prominent haploid phase, increase in population size occurs when the zygote undergoes meiosis and the four resulting cells divide by mitosis (Chapter 27). For diploid multicellular organisms, however, fertilization is the direct basis for reproduction. The zygote divides by mitosis and develops into an **embryo,** an early stage of multicellular development.

Meiosis and fertilization have significant genetic consequences. During meiosis, pairs of homologous chromosomes undergo recombination, shuffling the combination of alleles between the chromosomes. In addition, meiosis randomly sorts homologous chromosomes into different gametes. As a result, the products of meiosis, the haploid gametes, are each genetically unique and represent a mixture of the parental genetic makeup. Then, during fertilization, two of these unique haploid gametes fuse, creating new genetic combinations and a unique diploid individual.

→   **Quick Check 1** Sexual reproduction results in offspring that are genetically different from one another and from their parents. Describe four mechanisms that produce this genetic variation. Are any of these mechanisms shared with asexual reproduction?

### Many species reproduce both sexually and asexually.

Most organisms that reproduce asexually are also capable of reproducing sexually. For example, hydras reproduce asexually by budding, but this ability to reproduce asexually does not preclude their ability to reproduce sexually. Even budding yeast are capable of sexual reproduction. Most aphids, fish, and reptiles capable of parthenogenesis can also reproduce sexually.

What determines when an organism reproduces sexually and when it reproduces asexually? This question has been examined in the freshwater crustacean *Daphnia*. *Daphnia* reproduce asexually by parthenogenesis in the spring, when food is plentiful, perhaps because asexual reproduction is a rapid form of reproduction that allows the organism to take advantage of abundant resources (**Fig. 42.3**). Later in the season, when conditions are less favorable because of increased crowding and decreased food and temperature, *Daphnia* reproduce sexually, mating to produce a zygote that starts to form an embryo. At an early stage, however, development is arrested and metabolism slows dramatically. The embryo forms a thick-walled structure, called a cyst, which enables it to resist freezing and drying and therefore last through the winter.

This link between environmental conditions and mode of reproduction has been observed in many other species, from algae to slime molds. The life cycle of *Daphnia* hints that asexual and sexual reproduction have costs and benefits, a subject we turn to now.

### Exclusive asexuality is often an evolutionary dead end.

Asexual reproduction has some advantages compared to sexual reproduction. It does not involve finding and attracting a mate, which takes time and energy. Certain birds, for example, go to elaborate lengths to attract mates. Male birds of paradise invest a significant amount of energy in growing colorful plumage, and male bowerbirds take time to build elaborate homes for their potential mates (**Fig. 42.4**). For other organisms, finding a mate

**FIG. 42.3    Sexual and asexual reproduction in *Daphnia*.** *Daphnia* can reproduce sexually or asexually, depending on environmental conditions.

**Favorable conditions:**
• abundant food supply
• warm temperatures

**Less favorable conditions:**
• decreased food supply
• cool temperatures
• overcrowding
• accumulation of wastes

can be difficult because of very low population density—this is the case for anglerfish, and may help to explain their apparently unusual reproductive strategy, described earlier.

In addition, asexual reproduction is rapid, allowing organisms to increase their numbers quickly. Consider rates of population growth for organisms that reproduce asexually and those that reproduce sexually. In asexual organisms, all offspring can produce more offspring, allowing for exponential population growth if conditions are favorable. By contrast, in sexually reproducing organisms, only females produce new offspring. The consequence of this difference is that an asexual group can have a much greater rate of population growth compared to that of a sexually reproducing one. The British evolutionary biologist John Maynard Smith called this effect the **twofold cost of sex** (**Fig. 42.5**), and it is one of the key disadvantages of sexual reproduction.

In spite of the advantages of asexual reproduction, a survey of the natural world reveals a striking pattern: Most organisms reproduce sexually at least part of the time. And even prokaryotes, as we have seen, have mechanisms for genetic exchange. The observation that the vast majority of organisms reproduce

**FIG. 42.4 Time and energy costs of sexual reproduction.** (a) A male bird of paradise has colorful plumage to attract a mate. (b) A bowerbird builds an elaborate hut to entice a female.

sexually at some point suggests that sexual reproduction offers advantages over asexual reproduction. Asexual reproduction is quick, but it produces offspring that are genetically identical to one another and to the parent. The only source of genetic variation is chance mutations. By contrast, one of the defining features of sexual reproduction is the production of offspring that are genetically different from one another and from their parents.

Many hypotheses have been developed to explain the advantages of producing genetically distinct offspring in spite of the costs of sexual reproduction. One suggests that sexual reproduction allows organisms to adapt faster than asexual ones because rare beneficial mutations that arise in different organisms can be brought together, increasing the overall fitness of the population. Another set of hypotheses suggests that sexual reproduction allows a population to purge itself of harmful mutations more quickly than could a population of asexual individuals.

A third hypothesis—the Red Queen hypothesis—suggests that sexual reproduction is a mechanism of parasite defense. This hypothesis takes its name from Lewis Carroll's *Through the Looking-Glass*, in which the Red Queen tells Alice that she needs to keep running to stay in the same place. Similarly, in host–parasite coevolution, the host evolves defenses against the parasite, and the parasite evolves new ways to infect the host, so each evolves rapidly, or keeps moving, simply to maintain its same ecological position—that is, to stay in the same place. This kind of rapid

**FIG. 42.5 The twofold cost of sex.** Population size increases more rapidly in asexually reproducing organisms compared to sexually reproducing organisms.

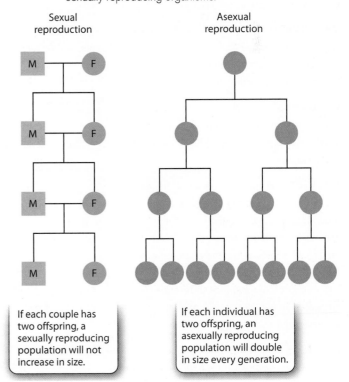

Sexual reproduction

Asexual reproduction

If each couple has two offspring, a sexually reproducing population will not increase in size.

If each individual has two offspring, an asexually reproducing population will double in size every generation.

evolution requires lots of genetic variation on which natural selection can act, and sexual reproduction increases genetic variation among individuals more than asexual reproduction does.

These different explanations for the prevalence of sexual reproduction are not mutually exclusive. It may be that several mechanisms work together, or that different ones are important for different organisms.

It is instructive to consider groups of eukaryotes that have lost the ability to reproduce sexually, and those that have lost the ability to reproduce asexually. While there are organisms such as the New Mexico whiptail lizard (*Cnemidophorus neomexicanus*) that reproduce only asexually, these species are not very old in evolutionary terms. In other words, there are very few, if any, *ancient* asexual groups. It appears that asexual reproduction on its own rarely allows a species to persist for very long in evolutionary terms.

A possible exception to this rule may be bdelloid rotifers, a group of microscopic freshwater invertebrates that are thought to have reproduced asexually for at least 35 to 40 million years (**Fig. 42.6**). The biology of this group is being intensely scrutinized for hints of its pattern of reproduction over time. If indeed bdelloid rotifers have persisted for tens of millions of years without sexual reproduction, then we have an interesting puzzle. We might predict that the rotifers have either an unusual way to escape the long-term

## HOW DO WE KNOW?

**FIG. 42.6**

# Do bdelloid rotifers only reproduce asexually?

**BACKGROUND**  Many groups of animals reproduce asexually, but most of these organisms reproduce sexually at least some of the time. The bdelloid rotifer, one of several species of rotifer, may be an exception.

**OBSERVATION**  All bdelloid rotifers found to date are females. No males or hermaphrodites (organisms with both male and female reproductive organs) have ever been observed. Meiosis has never been documented.

**HYPOTHESIS**  Bdelloid rotifers are asexual and reproduce solely by parthenogenesis, in which unfertilized eggs develop into adult females.

**PREDICTION**  In the absence of meiosis, the two copies (alleles) of a gene do not recombine and mutations that accumulate are not reshuffled between them. As a result, two alleles of a gene in an asexually reproducing organism will show a much higher degree of sequence difference compared to two alleles of a gene in a sexually reproducing organism.

**RESULTS**  Following is the sequence from the two copies of the *hsp82* gene in a sexually reproducing rotifer (Fig. 42.6a) and an asexually reproducing rotifer (Fig. 42.6b). As can be seen, the two copies are similar in Fig. 42.6a but very different in Fig. 42.6b.

**CONCLUSION**  All evidence to date suggests that bdelloid rotifers are exclusively asexual.

**SOURCE**  Welch, M. D., and M. S. Meselson, 2000. "Evidence for the Evolution of Bdelloid Rotifers Without Sexual Reproduction or Genetic Exchange." *Science* 288 (5469):1211–1215.

a. The two copies of *hsp82* found in a sexually reproducing rotifer.

The two copies of the *hsp82* gene are very similar in the sexually reproducing rotifer.

b. The two copies of *hsp82* found in an asexually reproducing bdelloid rotifer.

The two copies of the *hsp82* gene are very different in the asexually reproducing rotifer.

Asterisks (*) indicate bases that are identical in the two gene copies, and red shading indicates bases that are different.

evolutionary costs of genetic uniformity, or an as yet unknown mechanism for exchange of genetic material.

Mammals are unusual in the opposite way. Mammals have lost the ability to reproduce asexually. Parthenogenesis, as we have seen, occurs in some vertebrates, but does not occur in mammals. Although this form of reproduction may seem odd from our point of view, scientists have wondered just the opposite—why can't female mammals reproduce without a father? Asked another way, why do mammalian embryos need both a maternal and a paternal genome? Wouldn't two maternal genomes from a single parent do the trick? Evidently not. Scientists have known for some time that mammalian maternal and paternal genomes are not quite the same. There are some genes, called imprinted genes, that are expressed differently depending on their parent of origin. Imprinting ensures that mammals develop from embryos with one maternal and one paternal genome. When researchers manipulated key imprinted genes in a maternal genome, they were able to breed a parthenogenetic mouse—one with two copies of a maternal genome from a single parent.

We can speculate that early mammals gained fitness advantages in the evolution of imprinting of genes. However, a consequence was the loss of parthenogenesis in mammals. Parthenogenesis might offer greater fitness advantages than strictly sexual reproduction in some species of mammals living today, but that option is no longer available.

## 42.2 MOVEMENT ONTO LAND AND REPRODUCTIVE ADAPTATIONS

We have considered the two modes of reproduction observed across all living organisms. In this section, we narrow our focus to vertebrates: fish, amphibians, reptiles, birds, and mammals. Interestingly, even among vertebrates, there are many different reproductive strategies. Understanding these strategies not only gives us a picture of how these organisms reproduce but also sheds light on how they live and behave.

How vertebrates reproduce is also intimately connected to the transition of vertebrates from water to land 375 to 400 million years ago. Moving from an aquatic environment to a terrestrial one involved many adaptations. Among these were changes in the way that reproduction took place, so it is helpful to consider reproductive strategies in an evolutionary context.

### Fertilization can take place externally or internally.

Eggs and sperm require a wet environment to survive. In aquatic organisms like fish and amphibians, eggs and sperm can be released directly into the water in **external fertilization.**

**FIG. 42.7 External fertilization.** Tree frogs reproduce by external fertilization, with the male frog holding onto the female and fertilizing her eggs as she lays them.

External fertilization, as you might imagine, is a chance affair: Somehow, sperm must meet egg in a large environment. Animals that release gametes externally therefore have developed strategies to increase the probability of fertilization. One strategy involves releasing large numbers of gametes. In addition, many female fish, including salmon and perch, lay eggs on or near a substrate, such as a rock, and males release their sperm onto the eggs to increase the probability of fertilization. Externally fertilizing animals may also come close together physically to improve chances that the sperm will fertilize the eggs. Male frogs grasp females, even though fertilization occurs externally, releasing sperm as the female lays her eggs (**Fig. 42.7**).

A challenge for land-dwelling organisms is how to keep gametes from drying out. **Internal fertilization,** in which fertilization takes place inside the body of the female, is an adaptation for living on land. Reptiles, birds, and mammals all use internal fertilization. However, this mode of fertilization is not exclusive to land dwellers. Marine mammals such as whales and dolphins use internal fertilization, as did the land-dwelling ancestors from which they evolved. Even some fish, such as guppies, reproduce by internal fertilization.

### *r*-strategists and *K*-strategists differ in number of offspring and parental care.

While the terms "external" and "internal" fertilization focus our attention on where fertilization takes place, the two modes of reproduction are associated with a host of other behavioral and reproductive differences. One key difference is the number of offspring produced.

External fertilization is generally associated with the release of large numbers of gametes and the production of large numbers of offspring, each of which has a low probability of survival in part because the offspring typically receive little, if any, parental care. External fertilization is a game of numbers. In internal fertilization, however, there is much more control over fertilization and subsequent development of the embryo because both fertilization and embryonic development take place inside the female. As a result, animals that use internal fertilization typically produce far fewer offspring and invest considerable time and energy into raising those offspring.

These two strategies represent two ends of a continuum of reproductive strategies that was first described by the American ecologists Robert H. MacArthur and E. O. Wilson in 1967. Organisms that produce large numbers of offspring without a lot of parental investment are called ***r*-strategists,** and those that produce few offspring but put in a lot of parental investment are ***K*-strategists** (**Fig. 42.8**). Fish are *r*-strategists; humans are *K*-strategists.

According to this model, the environment places selective pressures on organisms to drive them in one or the other direction. In general, *r*-strategists evolve in unstable, changing, and unpredictable environments. In such an environment, there is an advantage to reproducing quickly and producing many offspring when conditions are favorable. By contrast, *K*-strategists often evolve in stable, unchanging, and predictable environments, where there tends to be more crowding and larger populations. Here, there is intense competition for limited resources, so traits such as increased parental care and few offspring are favored.

These two strategies represent ends of a spectrum, and there are notable exceptions. An octopus, for example, can lay up to 150,000 eggs, typical of *r*-strategists, but the female guards and takes care of them for a month or more, typical of *K*-strategists. However, the two strategies highlight different reproductive patterns across the animal kingdom and suggest important links between number of offspring, parental care, and the environment (Chapter 46).

### Oviparous animals lay eggs, and viviparous animals give birth to live young.

When we speak of parental care, it is natural to think of tending the young. However, parental care actually begins much earlier in development, in the provisioning of nutrients for the developing embryo. The eggs of animals with external fertilization have a substance called **yolk** that provides all the nutrients that the developing embryo needs until it hatches.

---

**FIG. 42.8** *r*-strategists and *K*-strategists. *r*-strategists, like fish, produce many offspring with low probability of survival, whereas *K*-strategists, like primates, produce few offspring and put a lot of care into raising their young.

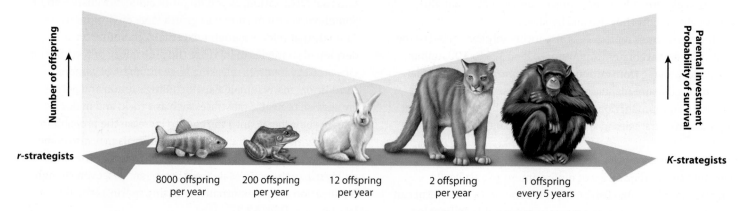

*r*-strategists     Number of offspring     Parental investment / Probability of survival     *K*-strategists

| 8000 offspring per year | 200 offspring per year | 12 offspring per year | 2 offspring per year | 1 offspring every 5 years |

Water and oxygen are obtained directly from the aquatic environment and metabolic wastes easily diffuse out.

Animals that are internally fertilized can either lay eggs or give birth to live young. These two strategies are called **oviparity** ("egg birth") and **viviparity** ("live birth"). Oviparity is used by fish and amphibians, as well as reptiles, birds, and most insects. Even some mammals—the monotremes, which include the spiny anteater and platypus—lay eggs. For oviparous animals there is very little, if any, embryonic development inside the mother, so all the nutrients for the developing embryo come from the yolk.

The movement onto land was associated with important changes in the anatomy of the egg. No longer could the developing embryo depend simply on diffusion of key substances into and out of a watery environment. A key innovation is the **amnion,** a membrane surrounding a fluid-filled cavity that allows the embryo to develop in a watery environment. Reptiles, birds, and mammals are known as amniotes to reflect the presence of this key membrane (**Fig. 42.9**). A second membrane, called the **allantois,** encloses a space where metabolic wastes collect. Finally, a third membrane, called the **chorion,** surrounds the entire embryo along with its yolk and allantoic sac. The embryo is further protected in many species by a hard shell that prevents drying out and facilitates gas exchange. The yolk sac, amnion, allantois, and chorion are **extraembryonic membranes,** sheets of cells that extend out from the developing embryo (**Fig. 42.10**).

**FIG. 42.9 The evolution of the amnion.** The amnion is a fluid-filled cavity surrounding the developing embryo that evolved during the transition from water to land.

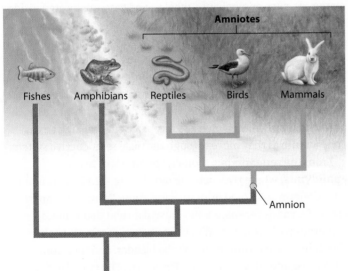

**FIG. 42.10 The amniotic egg.** The amniotic egg includes four extraembryonic membranes: the yolk sac, amnion, allantois, and chorion.

Viviparous animals give birth to live young. Most mammals are viviparous, but viviparity is not unique to mammals; it evolved independently in several different groups. For example, some sharks and snakes are capable of giving birth to live young. In viviparous animals, the embryo develops inside the mother. In placental mammals, such as humans, the chorion and allantois fuse to form the **placenta,** an organ that allows nutrients to be obtained directly from the mother.

## 42.3 HUMAN REPRODUCTIVE ANATOMY AND PHYSIOLOGY

In this and the next section, we continue to narrow our focus in order to consider reproduction in more detail. Here, we cover human reproduction and development. While many aspects of human reproduction are shared with other mammals, others are unique. Our focus on human reproduction allows for a detailed treatment of the topic in one organism and also gives us a chance to explore ways we intervene in the process to promote, prevent, and follow pregnancy. Humans, like all mammals, are amniotes. They reproduce sexually, use internal fertilization, and are viviparous.

**The male reproductive system is specialized for the production and delivery of sperm.**

Let's start by considering the human male reproductive system. Male gametes, called spermatozoa, or sperm, are produced in the

**FIG. 42.11  The male reproductive system**.  Haploid sperm are produced in the testes, then travel through the epididymis, vas deferens, ejaculatory duct, and urethra.

two male **gonads,** or **testes** (singular, testis; **Fig. 42.11**). Over the course of his life, a man is capable of making hundreds of billions of sperm. In fact, a single ejaculate typically contains several hundred million sperm. If it takes just one sperm to fertilize an egg, why do males make so many? A likely explanation involves sperm competition. In species like humans, where females can mate with more than one male, a male's reproductive success is at least partially correlated with the number of sperm—the more sperm, the better chance of fertilizing the egg. A second explanation relates to the female reproductive tract, which is hostile to sperm; many sperm are needed to ensure that at least one makes it to the egg. Of the several hundred million sperm released by a male during a single ejaculation, only several hundred sperm make it to the egg.

Sperm are haploid cells with a unique shape and properties adapted for their function of delivering genetic material to the egg (**Fig. 42.12**). They have a small head containing minimal cytoplasm and a very densely packed nucleus. The head is surrounded by a specialized organelle, the **acrosome,** that contains enzymes that are used by sperm to transverse the outer coating of the egg. Sperm have a long tail, or **flagellum,** which moves the cell by a whipping motion powered by the sperm's mitochondria.

The testes are located outside the abdominal cavity in a sac called the **scrotum** (see Fig. 42.11). This location is critical for the production of sperm, which require cooler temperatures than are found in the abdominal cavity. However, the testes of some mammals, such as elephants, whales, and bats, are located inside the abdominal cavity, indicating that there are species-specific differences in sperm temperature requirements.

Sperm are produced within the testes in a series of tubes called the **seminiferous** ("seed-bearing") **tubules** (**Fig. 42.13**). This is where diploid cells undergo meiosis to produce haploid sperm. When sperm are first produced, they are not fully mature: They are unable to move on their own and incapable of fertilizing an egg. They mature as they move through the male reproductive system.

From the seminiferous tubules, sperm travel to the **epididymis,** where they become motile (Fig. 42.13). This is also where sperm are stored prior to ejaculation. From there, sperm enter the **vas deferens,** a long, muscular tube that follows a circuitous path, starting off in the scrotum, passing into the abdominal cavity, running along the bladder, and then connecting with the **ejaculatory duct** (see Fig. 42.11). The two ejaculatory ducts merge at the **urethra,** just below the bladder. The urethra is used by both the urinary and reproductive systems in males.

**FIG. 42.12 Sperm.** Sperm are specialized cells with minimal cytoplasm, a densely packed nucleus, an acrosome with digestive enzymes, and a flagellum.

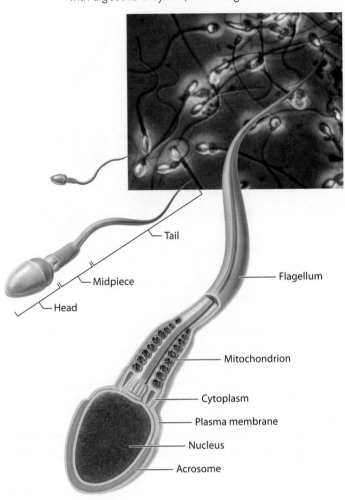

**FIG. 42.13 Seminiferous tubules, epididymis, and vas deferens.**

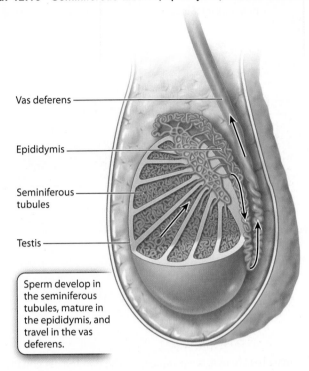

Vas deferens

Epididymis

Seminiferous tubules

Testis

Sperm develop in the seminiferous tubules, mature in the epididymis, and travel in the vas deferens.

Along this path, there are several exocrine glands that produce components of **semen,** the fluid that nourishes and sustains sperm as they travel in the male and then female reproductive tracts. The first is the **prostate gland,** located just beneath the bladder (see Fig. 42.11). The urethra and ejaculatory ducts pass directly through this gland. The prostate gland produces a thin, slightly alkaline fluid that helps maintain sperm motility and counteracts the acidity of the female reproductive tract. Two **seminal vesicles,** located at the junction of the vas deferens and the prostate gland, secrete a protein- and sugar-rich fluid that makes up most of the semen and provides energy for sperm motility. Finally, the **bulbourethral glands,** below the prostate gland, produce a clear fluid that lubricates the urethra for passage of sperm.

The urethra is enclosed in the **penis,** the male copulatory organ (see Fig. 42.11). When a male is sexually aroused, the penis becomes erect as a result of changes in blood flow—the arteries become dilated and the veins compressed. During ejaculation, semen is expelled from the penis.

The male reproductive system is well adapted for the production, storage, and delivery of sperm. Therefore, male contraceptive methods focus on blocking sperm. For example, a condom is a sheath of rubber or latex that covers the penis and blocks sperm from entering the female during sexual intercourse.

→ **Quick Check 2** What effect would cutting the vas deferens and tying off the ends have on male fertility?

The human penis is a relatively simple organ, consisting of a shaft and a head called the **glans penis.** In other animals, the penis can be quite elaborate, with spikes, hooks, and grooves, whose function is to stimulate the female, grip the female, or in some cases even remove sperm from a previous copulation. In marsupial mammals like kangaroos and koalas, males have forked penises, and in snakes and lizards, males have two penises, called hemipenes. In insects, the external genitalia are often distinct enough to be used in species identification. By contrast, most species of birds do not have a penis.

### The female reproductive system produces eggs and supports the developing embryo.

The female reproductive system, like the male reproductive system, produces gametes, but it is also specialized to support the developing embryo. Developing female gametes, called **oocytes,**

**FIG. 42.14   The female reproductive system.** Oocytes are produced in the ovaries and then travel through the fallopian tubes to the uterus, where they implant if they are fertilized; if they are not fertilized, they are shed during menstruation

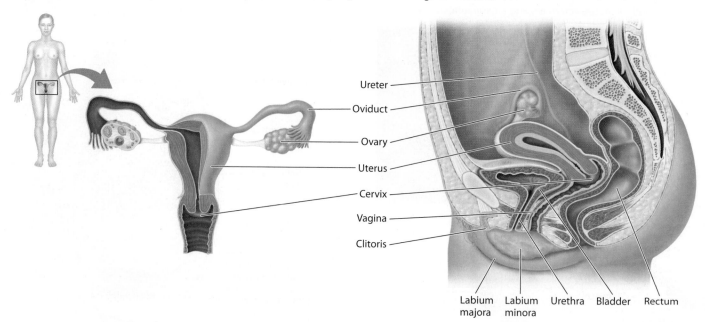

Ureter
Oviduct
Ovary
Uterus
Cervix
Vagina
Clitoris

Labium majora   Labium minora   Urethra   Bladder   Rectum

which mature into ova, are produced in the two female gonads, called **ovaries** (**Fig. 42.14**), and are released monthly in response to hormones (discussed in the next section).

The oocyte is a remarkable and unique cell (**Fig. 42.15**). In contrast to sperm, which are small and motile, the oocyte is the largest cell by volume in the human body and is visible to the naked eye; it is about the size of the period at the end of this sentence. Its large size reflects the presence of large amounts of cytoplasm containing molecules that direct the early cell divisions following fertilization and food supply in the form of yolk. While a male produces millions of sperm, a female usually produces just one oocyte each month, as she will nourish and support the developing embryo if the oocyte is fertilized.

Following release from the ovary, the oocyte travels through the **fallopian tube,** or **oviduct** (see Fig. 42.14). There are two fallopian tubes, one on each side. The fallopian tubes have featherlike projections on their ends that help channel the egg into the tube and not into the abdominal cavity.

The oocyte travels from the fallopian tube to the **uterus,** or womb (see Fig. 42.14). The uterus is a hollow organ with thick, muscular walls that is adapted to support the developing embryo if fertilization occurs and to deliver the baby during birth. In humans, the uterus is usually a single pear-shaped structure, but it can sometimes have two horns, which can interfere with pregnancy. Other mammals differ in the shape of the uterus, ranging from pear-shaped (humans and other primates), to two horns (pigs, horses), to more completely divided but still connected at the base (cats, dogs), to completely separate (rodents, rabbits).

The end of the uterus is called the **cervix,** or neck, of the uterus (see Fig. 42.14). Under the influence of hormones, the

cervix produces different kinds of mucus capable of either blocking or guiding sperm through the cervix.

The uterus is continuous with the **vagina,** or birth canal (see Fig. 42.14). The vagina is a tubular channel connecting the uterus to the exterior of the body. It is highly elastic, allowing it to stretch during sexual intercourse and birth.

The external genitalia of the female are collectively called the **vulva** (**Fig. 42.16**). The vulva includes two folds of skin, the **labia**

**FIG. 42.15   A female gamete.** Oocytes are produced in the ovaries, mature into ova (eggs), and are by volume the largest cells in the human body.

**FIG. 42.16** The female external genitalia.

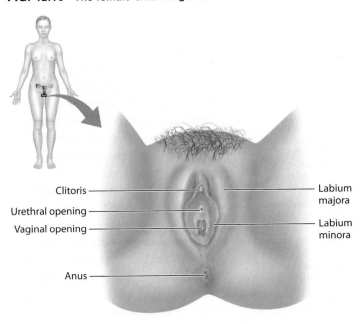

- Clitoris
- Urethral opening
- Vaginal opening
- Anus
- Labium majora
- Labium minora

**majora** and **labia minora.** The labia minora meet at the **clitoris,** the female homolog of the glans penis.

## Hormones regulate the human reproductive system.

The male testes and female ovaries not only are the site of gamete production, but also are part of the endocrine system (Chapter 38). That is, they respond to hormones and in turn secrete hormones. These hormones play key roles during embryonic development and puberty, and also in the adult. Hormonal control of reproduction is critically important, as it allows reproduction to take place only when, for example, a woman is able to support the developing embryo.

The master regulator of the endocrine system is the hypothalamus, which is located in the brain (Chapter 38). The hypothalamus secretes hormones that act on the pituitary gland, which in turn releases hormones that act on target organs (**Fig. 42.17**). In the case of the reproductive system, the hypothalamus releases **gonadotropin-releasing hormone (GnRH),** which stimulates the anterior pituitary gland to secrete **luteinizing hormone (LH)** and **follicle-stimulating hormone (FSH).** These hormones then act on the male and female gonads. In males, these hormones are secreted almost continually, whereas in females, they are secreted cyclically. In response, the testes secrete the hormone **testosterone** and the ovaries secrete the hormones **estrogen** and **progesterone.**

In males, LH acts on cells in the testes called **Leydig cells,** which secrete testosterone. Testosterone is a steroid hormone derived from cholesterol that plays key roles in male growth, development, and reproduction. Testosterone is important during

embryogenesis to direct development of the male reproductive organs. At puberty, levels of testosterone increase, leading to the development of male **secondary sexual characteristics,** traits that characterize and differentiate the two sexes but that do not relate directly to reproduction. For males, these include a deep voice, facial, body, and pubic hair, and increased muscle mass. Testosterone together with FSH act on another population of cells in the testes called **Sertoli cells** to stimulate sperm production. Sertoli cells, located in the seminiferous tubules, support the process of sperm production. In this way, hormones help to regulate the timing of sexual maturity.

In females, FSH acts on cells surrounding the oocyte called **follicle cells,** which support the developing oocyte and secrete estrogen. At puberty, an increase in the level of estrogen leads to the development of female secondary sexual characteristics, such as enlargement of the breasts, growth of body and pubic hair, and changes in the distribution of muscle and fat. Also at

**FIG. 42.17** Hormonal control of the reproductive system.

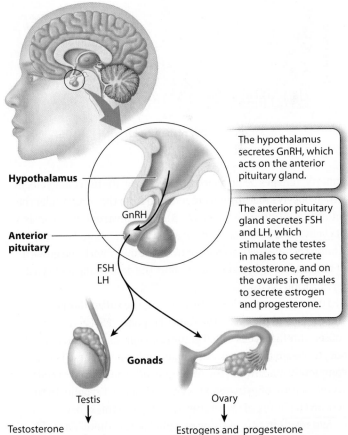

- Hypothalamus
- Anterior pituitary
- GnRH
- FSH LH
- Gonads
- Testis
- Ovary

The hypothalamus secretes GnRH, which acts on the anterior pituitary gland.

The anterior pituitary gland secretes FSH and LH, which stimulate the testes in males to secrete testosterone, and on the ovaries in females to secrete estrogen and progesterone.

Testosterone
- Development of male reproductive structures during embryogenesis
- Development of male secondary sexual characteristics during puberty
- Maintenance of health in adult

Estrogens and progesterone
- Development of female reproductive structures during embryogenesis
- Development of female secondary sexual characteristics during puberty
- Regulation of menstrual cycle
- Support of pregnancy
- Maintenance of health in adult

**FIG. 42.18 The menstrual cycle.** Note the cyclical and coordinated changes in the anterior pituitary hormones, ovary, ovarian hormones, and uterine lining.

The LH surge causes ovulation.

Follicle cells secrete estradiol and the corpus luteum secretes progesterone.

High levels of estradiol cause the uterine lining to thicken, and high levels of progesterone maintain the thickened uterine lining.

The changes in the uterine lining are coordinated with those of the developing oocyte.

appearance. The corpus luteum is a temporary endocrine structure that secretes the hormone progesterone, which maintains the thickened and vascularized uterine lining.

The oocyte, meanwhile, is swept into the fallopian tube and travels to the uterus. If it is fertilized, the developing embryo implants into the uterine lining. The corpus luteum continues to secrete progesterone, and is first maintained by LH and then by the hormone **human chorionic gonadotropin (hCG),** which is released by the developing embryo. Eventually, the placenta takes over estrogen and progesterone production to maintain the uterine lining and stimulate growth of the uterus. High levels of estrogen and progesterone during pregnancy also block ovulation because they suppress the release of GnRH, FSH, and LH.

If the oocyte is not fertilized, the corpus luteum degenerates, estrogen and progesterone levels drop, and the uterine lining is shed. The monthly shedding of the uterine lining is known as **menstruation.** Menstrual cycles usually start occurring around age 12 in the United States and continue until approximately age 45 to 55. The cessation of menstrual cycles, called **menopause,** results from decreasing production of estradiol and progesterone by the ovaries.

As we have seen, the brain, anterior pituitary, ovaries, uterus, and placenta pass hormonal messages back and forth to make sure that behavior, gamete release, the uterine lining, and embryo development are coordinated. If hormone levels are low or absent, as occurs normally before puberty or as the result of malnutrition or some diseases, ovulation does not occur. Similarly, too much hormone at the wrong time can interfere with fertility, which is the basis for oral contraceptive pills ("the pill"). Oral contraceptive pills contain various combinations of synthetic estrogens or progesterones or both.

→ **Quick Check 3** From what you know about hormone levels and ovulation during pregnancy, how do you think the pill works?

Humans and chimpanzees both have menstrual cycles. Other placental mammals have an **estrus cycle,** which is characterized by phases in which females are sexually receptive. In these mammals, the uterine lining is reabsorbed instead of shed if fertilization does not occur.

puberty, females begin a monthly cycle, the **menstrual cycle,** in which oocytes mature and are released from the ovary under the influence of hormones (**Fig. 42.18**). The menstrual cycle consists of cyclical and coordinated changes in the ovaries and uterus, timing the release of an oocyte from the ovary with the growth of the uterine lining so it can support the developing embryo if fertilization occurs.

The menstrual cycle has two phases, the **follicular phase** and the **luteal phase.** In the follicular phase, FSH acts on a subset of follicle cells called granulosa cells. As a result, several oocytes begin to mature, but usually only one of these becomes completely mature and the others die off. The granulosa cells secrete a form of estrogen known as estradiol. Estradiol in turn acts on the lining of the uterus, causing it to thicken.

As a result in part of rising estradiol levels, there is a rapid increase followed by a sharp decrease in the level of LH produced by the anterior pituitary gland. This surge causes ovulation, the release of the oocyte from the ovary. Ovulation marks the beginning of the luteal phase. The follicle cells, now devoid of the oocyte, are converted to a structure known as the **corpus luteum,** literally "yellow body," so called because of its

## 42.4 GAMETE FORMATION TO BIRTH IN HUMANS

With an understanding of male and female reproductive anatomy and endocrinology, we are ready to follow human reproduction and development, from the formation of sperm and eggs, to fertilization itself, to subsequent development of the embryo, and finally to birth. Along the way, we consider how these processes vary in different organisms. Our understanding of these processes is in large part due to studies in model organisms, including the sea urchin (*Strongylocentrotus purpuratus*), roundworm (*Caenorhabditis elegans*), fruit fly (*Drosophila melanogaster*), African clawed frog (*Xenopus laevis*), and house mouse (*Mus musculus*). These organisms have particular reproductive and developmental features that make them amenable to laboratory study.

### Male and female gametogenesis have shared and distinct features.

The formation of gametes is called **gametogenesis.** Gametogenesis follows the same steps in males and females,

but there are striking differences in timing (**Fig. 42.19**). **Spermatogenesis** is the formation of sperm. It occurs in the seminiferous tubules of the testes, begins at puberty, and continues throughout life. The process starts with spermatogonia, diploid stem cells that divide by mitosis to ensure a continuous population of precursor cells. These cells differentiate into diploid **primary spermatocytes,** which undergo the first meiotic division to produce two **secondary spermatocytes** and the second meiotic division to produce four haploid spermatids joined by cytoplasmic bridges. They subsequently separate from each other to become distinct haploid cells. The maturation process, in which spermatids become spermatozoa or sperm, begins in the testes and continues in the epididymis, where they acquire motility. Spermatogenesis takes about 2 to 3 months.

**Oogenesis** is the formation of ova or eggs. The diploid precursor cells are called oogonia. These cells form during embryonic development and divide by mitosis. Some oogonia differentiate into diploid **primary oocytes** during fetal development. These enter the first meiotic division but arrest immediately in prophase I, the first stage of meiosis. All these steps take place before the female is even born. At the time

**FIG. 42.19** Spermatogenesis in the male and oogenesis in the female.

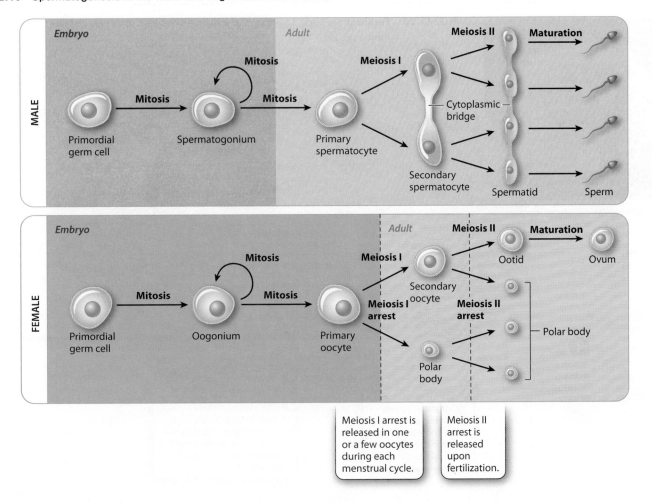

of birth, a female has about 1 to 2 million primary oocytes, all arrested in prophase I of meiosis. They do not resume the first meiotic division until a menstrual cycle. Therefore, remarkably, they all remain arrested for at least 12 years (from birth to the age of menarche) and some as long as 50 years (to the age of menopause).

In the first phase of the menstrual cycle, approximately 10 to 20 follicles begin to mature because of rising FSH levels. One of these follicles becomes dominant, allowing the primary oocyte to complete the first meiotic division. This division is asymmetric, leading to the formation of a large **secondary oocyte** and a much smaller **polar body.** The unequal division allows most of the cytoplasm to be partitioned into the secondary oocyte. The secondary oocyte is released from the ovary and immediately enters the second meiotic division, but again arrests, this time in metaphase II. The secondary oocyte remains arrested in metaphase II until fertilization takes place. Upon fertilization, the second meiotic division continues, producing an ootid and another polar body, again allowing most of the cytoplasm to end up in the developing egg. The ootid quickly develops into a mature ovum or egg.

→   **Quick Check 4** When are primary oocytes and spermatocytes produced, and how many gametes result from one diploid cell in females and males?

## Fertilization occurs when a sperm fuses with an oocyte.

Fertilization involves the fusion of gametes, restoring the diploid chromosome content (**Fig. 42.20**). In humans, male and female gametes are usually brought together by sexual intercourse. Sexual arousal leads to a host of physiological changes, including erection of the penis and erection of the clitoris and lubrication of the vagina. The erect penis is inserted into the vagina until orgasm occurs. Orgasm is characterized by feelings of pleasure and contractions of the pelvic muscles. In males, it is usually accompanied by ejaculation, in which semen, containing sperm, is delivered into the vagina.

Sperm travel through the cervix and into the uterus and fallopian tubes. Fertilization usually occurs in one of the fallopian tubes. As a result, many of the details of how sperm travel through the female reproductive tract are poorly understood. Sperm are motile, but from studies of the time it takes sperm to reach the fallopian tubes, it appears that the female reproductive

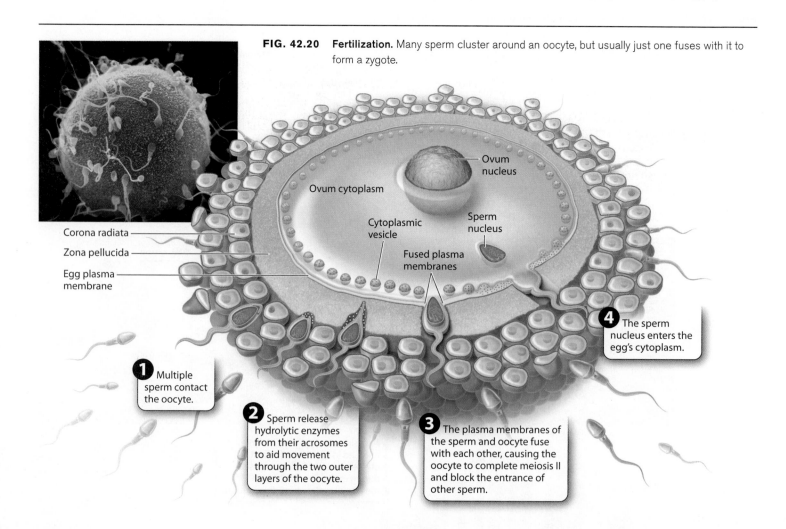

**FIG. 42.20** **Fertilization.** Many sperm cluster around an oocyte, but usually just one fuses with it to form a zygote.

Corona radiata

Zona pellucida

Egg plasma membrane

Ovum cytoplasm

Ovum nucleus

Cytoplasmic vesicle

Sperm nucleus

Fused plasma membranes

**❶** Multiple sperm contact the oocyte.

**❷** Sperm release hydrolytic enzymes from their acrosomes to aid movement through the two outer layers of the oocyte.

**❸** The plasma membranes of the sperm and oocyte fuse with each other, causing the oocyte to complete meiosis II and block the entrance of other sperm.

**❹** The sperm nucleus enters the egg's cytoplasm.

tract plays an active role in moving sperm to the oocyte, for example through muscular contractions.

In the female reproductive tract, sperm go through a key developmental step, called **capacitation,** a process that is unique to mammals. Capacitation is a series of physiological changes that allow the sperm to fertilize the egg. These changes include alterations in the fluidity of the plasma membrane, loss of some surface membrane proteins, and changes in the charge across the membrane, called the membrane potential. As a result, sperm show increased motility.

In many species, the oocyte releases soluble molecules, creating a concentration gradient that sperm use as a guide to find the oocyte. On reaching the oocyte, the sperm passes through two layers—an outer layer of follicle cells called the corona radiata and an inner matrix of glycoproteins called the zona pellucida—before fusing with the plasma membrane of the oocyte (Fig. 42.20). In sea urchins, by contrast, there is a single vitelline membrane surrounding the egg. To facilitate its passage, the head of the sperm releases enzymes from the acrosome. In addition, surface proteins on the gametes allow them to interact in a species-specific manner. For organisms with external fertilization, such as sea urchins, these proteins are especially important.

After passing through the layers surrounding the ooctye, the plasma membranes of the sperm and oocyte fuse (Fig. 42.20). The oocyte then completes meiosis II and changes occur that prevent **polyspermy,** or fertilization by more than one sperm. These mechanisms are classified according to how quickly they act and have been extensively studied in sea urchins. The first line of defense, or fast block to polyspermy, is caused by changes in the membrane potential of the egg, which is usually negative but quickly becomes positive on fertilization. The fast block is quick but temporary and does not occur in mammals.

There is also a second, more stable defense, or slow block to polyspermy. In this process, vesicles in the oocyte fuse with the plasma membrane and release their contents, leading to modification of the structure of the zona pellucida in mammals and the vitelline membrane in sea urchins. The fertilized, diploid egg is a zygote—the first cell in the development of a new organism.

Although sexual intercourse is one way that gametes are brought together, there are now a variety of other methods to achieve fertilization and pregnancy in humans. These are collectively called assisted reproductive technologies and are typically used to treat infertility. For example, **in vitro fertilization (IVF)** is a process in which eggs and sperm are brought together in a petri dish, where fertilization and cell divisions occur. The developing embryos are then inserted into the uterus. The first IVF or "test-tube" baby was born in 1978, and since then several million babies have been conceived in this way. Assisted reproductive technologies such as IVF have opened the door to pre-implantation genetic diagnosis, in which oocytes or embryos are screened for genetic diseases before implantation. This technique raises ethical concerns, as it can be used to determine and select the gender of the embryo and other traits not associated with disease.

### The first trimester includes cleavage, gastrulation, and organogenesis.

How does a single cell become a complex, multicellular organism? Development depends on a precisely orchestrated set of cell divisions, movements, and specializations coordinated by both maternal and zygotic genetic instructions (Chapter 20). In this section, we focus on three key processes in early embryonic development: cleavage, gastrulation, and organogenesis.

After fertilization, the single-celled zygote divides by mitosis, a process called **cleavage (Fig. 42.21).** This process divides the single large egg into many smaller cells. In most organisms, these mitotic divisions are rapid, occur at the same time, and are not accompanied by an overall increase in size of the embryo. Furthermore, they occur in the absence of expression of the zygote's genes. Instead, molecules in the egg's cytoplasm, including mRNA and protein encoded by maternal effect genes (Chapter 20) and stored in the egg, direct these early cell divisions.

Cleavage in mammals is unique in several respects. It is slow compared to cleavage in other organisms. In addition, in mammals the early divisions do not occur at the same time, so it is not unusual to find an odd number of cells in the developing embryo. Finally, zygotic transcription begins very early, as early as the first mitotic division, or two-cell stage.

The pattern of cleavage is different for different organisms and is determined in part by the amount of yolk present

**FIG. 42.21 Cleavage divisions.**

| Zygote | Two-cell stage | Four-cell stage | Morula |

FIG. 42.22  **Cleavage patterns.** (a) Partial cleavage, as shown in this zebrafish embryo. (b) Full cleavage, as shown in this sea urchin embryo.

Animal pole
(no yolk)

Vegetal pole
(yolk)

in the egg (**Fig. 42.22**). In animals with lots of yolk, such as insects, fish, and birds, the embryo has two poles—a vegetal pole, where the yolk is concentrated, and an animal pole, where it is less concentrated. The presence of yolk inhibits cell division. Therefore, in these animals, cell division does not extend through the entire egg (Fig. 42.22a). In animals with very little yolk,

such as mammals and sea urchins, cell division extends completely through the egg (Fig. 42.22b). In the absence of yolk, the developing embryo needs another source of nutrition, which in humans and other placental mammals is provided through the placenta.

Cleavage results in a solid ball of cells called a **morula** (see Fig. 42.21). Further cell divisions result in a fluid-filled ball of cells called a **blastula.** It is around this time, about 5 days after fertilization, that the developing embryo implants into the uterine lining, which is thickened and highly vascularized under the influence of progesterone from the corpus luteum.

Implantation marks the beginning of **pregnancy,** the carrying of one or more embryos in the uterus. It lasts approximately 38 weeks from fertilization. Because the time of fertilization is often not known, human pregnancy is typically measured as 40 weeks from the last menstrual period. Pregnancy is divided into three periods called **trimesters**, each lasting about 3 months (**Fig. 42.23**). The divisions between the three trimesters are arbitrary, but they provide a convenient way to follow the major changes that occur during pregnancy.

The first trimester is characterized by spectacular changes in the developing embryo under the influence of genes and developmental pathways that are largely conserved across vertebrates (Chapter 20). During this time, the embryo develops all its major organs and begins to be recognizably human. Toward the end of the first trimester, a human embryo is known as a **fetus.**

In mammals, such as humans, the blastula has an inner cell mass and an outer layer of cells, and is known as a blastocyst (**Fig. 42.24a**). The inner cell mass becomes the embryo itself, and the outer cell layer forms part of the placenta. The placenta is composed of both maternal and embryonic cells and is the site of gas exchange, nutrient uptake, and excretion of waste between maternal and fetal blood. The placenta also secretes the hormones estrogen and progesterone, which maintain the uterine lining during pregnancy. Meanwhile, the inner cell mass differentiates into a two-layered (bilaminar) structure consisting of the epiblast and hypoblast (**Fig. 42.24b**).

**Gastrulation** is the second key developmental process in embryogenesis. It is a highly coordinated set of cell movements that leads to a fundamental reorganization of the embryo. In humans, cells of the epiblast divide and migrate inward (**Fig. 42.24c**). These migrating epiblast cells form three layers or sheets of cells called

FIG. 42.23  **The stages of human pregnancy.**

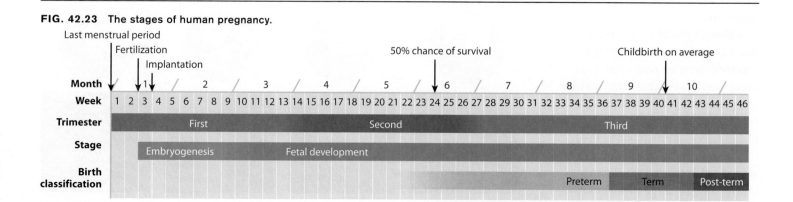

germ layers, which include the outer **ectoderm**, intermediate **mesoderm,** and inner **endoderm** (**Fig. 42.24d**). All of the adult tissues and organs are made from the three germ layers. At this stage, the embryo is known as a trilaminar (three-layered) embryo to reflect the three germ layers.

As with cleavage, the details of gastrulation vary in different organisms. One notable difference is the fate of the opening through which cells migrate during gastrulation. In protostomes ("mouth first"), such as worms, mollusks, and arthropods, it becomes the mouth, whereas in deuterostomes ("mouth second"), such as vertebrates and sea urchins, it becomes the anus and the mouth forms later (Chapter 44).

**Organogenesis,** the third key developmental process, is the transformation of the three germ layers into all the organ systems of the body. The ectoderm becomes the outer layer of the skin and nervous system, the mesoderm makes up the circulatory system, muscle, and bone, and the endoderm becomes the lining of the digestive tract and lungs (Fig. 42.24d). Because all the basic structures and organ systems develop during the first trimester of pregnancy, it is during this time that the developing fetus is most vulnerable to toxins, drugs, and infections.

It is also during the first trimester that the embryo begins to develop male or female characteristics. In humans, **sex determination** is achieved by the presence or absence of the Y chromosome. The Y chromosome contains a gene, *SRY*, that encodes a protein called testis-determining factor (TDF). TDF in turn acts on the developing gonad to cause it to differentiate into a testis. In the absence of TDF (in females), the immature gonad becomes an ovary. Once sex is determined, sexual differentiation follows under the influence of hormones produced by the gonads. In humans, then, sex determination is genetic, as is the case in other mammals and in birds. For other organisms, such as some reptiles and fish, sex determination depends instead on environmental factors, relying on such cues as temperature and crowding.

### The second and third trimesters are characterized by fetal growth.

The second trimester, from 13 to 27 weeks, is characterized by continued development and growth of the fetus. This is the time when fetal movements can first be felt by the mother. Babies born in the last few weeks of the second trimester are often able to survive, and are considered premature. Because of improvements in the intensive care of premature babies, babies born as early as 22 weeks after fertilization can survive.

The third trimester, from 28 to 40 weeks, is characterized by rapid growth, with the fetus doubling in size during the last two months. Weight gain is critical, as birth weight is a major factor in infant mortality. The fetus often moves less frequently during this trimester as it becomes increasingly compressed by limited space. Although 40 weeks (from the last menstrual period) is the average length of pregnancy, a baby born between 37 and 42 weeks is considered term. Human development from fertilization to the end of pregnancy is summarized in **Fig. 42.25.**

**FIG. 42.24  Early human embryonic development.**

**a.** Blastocyst

Inner cell mass

The inner cell mass differentiates into two layers of cells – the epiblast and hypoblast.

**b.** Bilaminar embryo

Epiblast

Hypoblast

Epiblast

Hypoblast

**c.** Gastrulation

The epiblast cells migrate inward, displace the hypoblast cells, and differentiate into all 3 germ layers – ectoderm, mesoderm, and endoderm.

**d.** Trilaminar embryo

Ectoderm

Mesoderm

Endoderm

- Outer layer of the skin
- Nervous system
- Posterior pituitary gland
- Cornea and lens of the eye

- Muscle
- Bone
- Connective tissue
- Circulatory system
- Kidneys
- Gonads

- Lining of the respiratory tract
- Lining of the digestive tract
- Liver
- Gallbladder
- Pancreas
- Thyroid gland

# Reproduction and Development

**Integrating concepts from chapters 10, 20, and 42**

## Early development

The fertilized egg is a totipotent stem cell because it can give rise to a complete organism. During development, cells become increasingly restricted in their developmental potential as a result of changes in gene expression. During gastrulation, cells further differentiate and move relative to one another until the embryo is organized in three germ layers.

Inner cell mass

Blastocyst cavity

Female and male nuclei fusing

Maternal transcription

Zygotic transcription

**1** **Fertilization.** The fusion of a haploid sperm with a haploid egg to form a diploid zygote is called fertilization. The most common site of fertilization is the fallopian tube.

Zygote

Two-cell stage

Four-cell stage

Morula

Blastocyst

Uterus

**2** **Cleavage.** The single-celled zygote divides rapidly by mitosis, resulting in a solid ball of cells called a morula. At this stage, the developing embryo approaches the uterus.

Meoitic spindle

Sperm

Zona pellucida

Polar body

Corona radiata

Bilaminar embryo

Uterine wall

Epiblast     Hypoblast

**3** **Implantation.** By the end of the first week, the developing embryo, now a fluid-filled blastocyst, implants into the uterine wall.

**4** **Bilaminar Embryo.** During the second week of development, cells of the inner cell mass differentiate into two cell layers, the epiblast and hypoblast.

Epiblast

Hypoblast

**5** **Gastrulation.** In the third week, cells of the epiblast migrate inward to form all three germ layers – the ectoderm, mesoderm, and endoderm.

Ectoderm

Mesoderm

Endoderm

Notochord

## Late development
The three germ layers form all of the organs of the fetus by organogenesis. The organs, such as the heart, lungs, and kidneys, are specialized for different functions but work in a coordinated fashion.

**7 Fetal Growth.** The third month to birth is characterized by growth and maturation of the fetus. Pregnancy lasts about 40 weeks from the last menstrual period or 38 weeks after fertilization.

Placenta

Uterus

Day 44 embryo

Day 25 embryo

Neural tube

Notocord

Somite

Yolk sac

Day 21 embryo

**6 Organogenesis.** From the third to the eighth week, the three germ layers give rise to the major organs. The ectoderm becomes the outer layer of the skin and nervous system, the mesoderm makes up the circulatory system, muscle, and bone, and the endoderm becomes the lining of the digestive tract and lungs.

**FIG. 42.26 Childbirth.** Childbirth involves changes to the cervix, delivery of the baby, and delivery of the placenta.

Delivery of the baby

Delivery of the placenta

→ **Quick Check 5** A multicellular organism starts off as a single cell, which divides by mitosis to produce many cells. Therefore, for the most part, all the cells in a multicellular organism are genetically identical—they are clones—yet in time different groups of cells look different and function quite differently from one another. How can this be?

## Childbirth is initiated by hormonal changes.

Childbirth marks the end of pregnancy. Although we tend to think of childbirth as the delivery of the baby, it actually consists of three events—changes in the cervix to allow passage of the baby, descent and delivery of the baby, and delivery of the placenta. These three events result from strong muscular contractions of the uterus mediated by the hormone oxytocin from the posterior pituitary gland. Oxytocin causes uterine contractions, which stimulate more oxytocin release, which in turn leads to further contractions, and so on.

During pregnancy, the cervix is a tube about 4 cm long closed with a mucus plug. To allow passage of the infant, the cervix shortens and widens. These changes are not enough to accommodate passage of the baby during delivery. The head and shoulders of the fetus are wide relative to the size of the pelvis. Humans have a narrow pelvis relative to that of other primates as an adaptation for bipedalism. How, then, is the baby born? The baby goes through a series of movements that facilitate its passage through the pelvis (**Fig. 42.26**). Babies are usually born head first with the face pointing down (toward the mother's back) when passing through the birth canal. Other positions make vaginal birth more difficult, sometimes impossible.

The third and final stage of childbirth is the delivery of the placenta, which usually occurs about 15 to 30 minutes after the baby is born.

Although birth normally occurs vaginally, in some cases, such as when the baby is too large or positioned feet first, it is necessary to remove the baby by making an incision in the abdomen and uterus in a surgical procedure known as a Caesarian section. Interventions such as Caesarian sections, the use of antibiotics, and pre- and postnatal care have greatly reduced maternal and fetal mortality associated with childbirth.

Whichever way the baby is delivered, when the umbilical cord is cut and the baby takes its first breath, it begins its life independent of the mother—anatomically independent, that is. Otherwise, human newborns are completely dependent on others. Humans are unusual among animals in having an extended period of helplessness and immaturity while their brains and bodies grow and develop. This time means that parents usually put considerable effort into raising and caring for their offspring, and it allows for extended periods of learning, the development of strong social bonds, and the transmission of culture to the next generation.

# Core Concepts Summary

## 42.1 REPRODUCTION IS A FUNDAMENTAL FEATURE OF LIVING ORGANISMS AND OCCURS ASEXUALLY AND SEXUALLY.

Asexual reproduction involves a single parent. Offspring are produced by binary fission or mitosis and are clones of the parent. page 42-2

Examples of asexual reproduction include budding, fragmentation, and parthenogenesis. page 42-2

Sexual reproduction involves the production of haploid gametes by meiosis and the fusion of gametes to make a diploid zygote. page 42-3

Many organisms reproduce both sexually and asexually. For some organisms, environmental conditions favor one mode or the other. page 42-4

Asexual reproduction allows for rapid population growth but produces clones and results in genetic uniformity. page 42-4

Sexual reproduction produces genetically unique offspring, a feature that is thought to explain why most eukaryotes reproduce sexually at least some of the time. page 42-4

## 42.2 THE MOVEMENT OF VERTEBRATES FROM WATER TO LAND INVOLVED CHANGES IN REPRODUCTIVE STRATEGIES, INCLUDING INTERNAL FERTILIZATION AND THE AMNIOTIC EGG.

Fertilization can occur outside or inside the body of the female. External fertilization can occur only in an aquatic environment. Internal fertilization is an adaptation to terrestrial living. page 42-7

External fertilization is usually associated with the production of many offspring, little or no parental care, and high mortality. These are characteristics of *r*-strategists. page 42-8

Internal fertilization is often associated with the production of fewer offspring, increased parental care, and low mortality. These are characteristics of *K*-strategists. page 42-8

A watertight sac, called the amnion, is an adaptation for reproduction on land. page 42-8

Oviparous animals lay eggs and development occurs outside the female; viviparous animals give birth to live young and development occurs inside the female. page 42-8

## 42.3 THE MALE REPRODUCTIVE SYSTEM IS ADAPTED FOR THE PRODUCTION AND DELIVERY OF SPERM, AND THE FEMALE REPRODUCTIVE SYSTEM IS ADAPTED FOR THE PRODUCTION OF EGGS AND, IN MANY CASES, SUPPORT OF THE DEVELOPING FETUS.

Male gametes, called sperm, develop in the seminiferous tubules of the testes. page 42-10

Sperm travel from the seminiferous tubules to the epididymis to the vas deferens to the ejaculatory duct and finally to the urethra before being released during ejaculation. page 42-10

Semen is a mixture of sperm and fluid from the prostate gland, seminal vesicles, and bulbourethral glands. page 42-10

Developing female gametes, called oocytes, are produced in the ovaries. Mature oocytes are called eggs, or ova. page 42-11

An oocyte is released from the ovary, travels through the fallopian tube, and implants in the uterus if it is fertilized. page 42-11

The reproductive system responds to and produces hormones. page 42-13

The hypothalamus releases GnRH, which stimulates the anterior pituitary gland to release FSH and LH, which in turn stimulate the ovaries to release estrogen and progesterone or the testes to secrete testosterone. page 42-13

In males, LH stimulates Leydig cells to secrete testosterone, and FSH and testosterone act on Sertoli cells to support sperm production. page 42-13

In females, monthly menstrual cycles are driven by the interplay of hypothalamic, anterior pituitary, and ovarian hormones, leading to ovulation, which is the maturation and release of an oocyte from the ovary, and to changes to the uterine lining. page 42-13

## 42.4 HUMAN REPRODUCTION INVOLVES THE FORMATION OF GAMETES, FERTILIZATION, AND GROWTH AND DEVELOPMENT.

Gametogenesis is called spermatogenesis in males and oogenesis in females. page 42-15

In males, spermatogonia differentiate into primary spermatocytes that undergo meiosis to make immature haploid spermatids and mature spermatozoa or sperm. page 42-15

Spermatogenesis begins at puberty and continues throughout life. page 42-15

In females, oogonia undergo differentiation into primary oocytes during fetal development, which begin meiosis but arrest in prophase I. During a menstrual cycle, primary oocytes are released from arrest, finish meiosis I, but then arrest in metaphase II. They remain arrested in metaphase II until they are fertilized. page 42-15

Fertilization is the fusion of the oocyte and sperm plasma membranes. It usually occurs in the fallopian tube and results in the formation of a diploid zygote. page 42-16

Cleavage is the division of the zygote by mitosis into smaller cells. At the blastocyst stage, it implants in the uterus. page 42-17

Gastrulation results in the formation of the three germ layers—ectoderm, mesoderm, and endoderm. page 42-17

Organogenesis is the formation of organs. page 42-17

Pregnancy lasts about 38 weeks from fertilization until birth and is divided into three trimesters. page 42-19

Childbirth, mediated by the posterior pituitary hormone oxytocin, involves changes in the cervix, delivery of the baby, and delivery of the placenta. page 42-22

## Self-Assessment

1. List similarities and differences between asexual and sexual reproduction.

2. Explain the roles of meiosis and fertilization in sexual reproduction.

3. Describe the costs and benefits of asexual and sexual reproduction.

4. Provide an explanation for the observation that there are very few, if any, ancient asexual organisms.

5. Name three adaptations that allow reproduction to take place on land.

6. Describe the pathway of sperm from its site of production to the urethra.

7. Describe the pathway of an oocyte (and the egg it develops into) from its site of production to the uterus.

8. Explain the relationships among changes in levels of anterior pituitary hormones, ovarian hormones, oocyte development, and the uterine lining during a menstrual cycle.

9. Describe similarities and differences between male and female gametogenesis.

10. Name and describe three key developmental steps in the development of a single-celled zygote to a multicellular individual.

Do you understand the chapter's Core Concepts? Log in to check your answers to the Self-Assessment questions, then practice what you've learned and reinforce this chapter's concepts by working through the problems and multimedia tutorials provided there.

📖 **www.biologyhowlifeworks.com**

## CHAPTER 43

# ANIMAL IMMUNE SYSTEMS

## Core Concepts

**43.1** The innate immune system acts against a diversity of foreign pathogens.

**43.2** The adaptive immune system includes B cells that produce antibodies against specific pathogens.

**43.3** The adaptive immune system includes helper and cytotoxic T cells that attack pathogens, foreign cells, and diseased cells.

**43.4** Some pathogens have evolved mechanisms that enable them to evade the immune system.

As our bodies' main line of defense, the immune system keeps watch against harmful pathogens. Some pathogens in turn have evolved a means of hiding within our own cells. In this way, they evade easy detection by our immune system, not unlike the Greek soldiers who hid inside a wooden horse during the Trojan War. As the story goes, the Trojans wheeled the magnificent horse within the protective gates of Troy. That evening, the stealthy Greek soldiers emerged and opened the gates, letting in the rest of the Greek army.

As pathogens evolve new strategies, the immune system in turn evolves new responses. The result of this evolutionary back and forth is an elaborate and sophisticated immune system. The immune system is remarkable in its ability in most cases to protect us from harmful pathogens, including viruses like influenza and HIV, bacteria that cause strep throat and tuberculosis, fungi such as ringworm, and single-celled eukaryotes like those that cause malaria. It has evolved defenses that act against a broad range of pathogens and defenses that target specific pathogens.

The activity of the immune system also must be carefully regulated. If it is overactive or acts when it is not supposed to, it may attack the cells of its own body, leading to allergies or debilitating autoimmune diseases. If it is underactive, it won't adequately protect an organism. A weak immune system is most likely to occur at the extremes of age or when infections or disease compromise its function.

We begin with an overview of general defenses that provide a first line of protection against diverse pathogens. We then turn to a powerful part of the immune system that is tailored to attack specific pathogens and that has the ability to learn and remember, attributes we usually associate with the nervous system (Chapters 35 and 36). Finally, we consider three infections that illustrate how pathogens have evolved ways to evade these defenses.

---

## 43.1 INNATE IMMUNITY

The immune system consists of two parts that work in different ways and interact with each other to protect against infection. The first, called **innate**, or **natural, immunity**, provides protection in a nonspecific manner against all kinds of infection. This form of immunity is present in plants, fungi, and animals and is an evolutionarily early form of immunity. It is considered *innate* because it does not depend on exposure to a pathogen. By contrast, the second form of immunity, called **adaptive**, or **acquired, immunity**, is specific to a given pathogen. This form of immunity "remembers" past infections, and subsequent encounters with the same pathogen generate a stronger response from the host. In other words, the immune system *adapts* over time, and immunity is *acquired* after an initial exposure. This form of immunity evolved later than the innate form

**TABLE 43.1** Features of the Innate and Adaptive Immune Systems

| | INNATE IMMUNE SYSTEM | ADAPTIVE IMMUNE SYSTEM |
|---|---|---|
| Targets | Diverse pathogens | Diverse pathogens |
| Ability to distinguish self from non-self? | Yes | Yes |
| Specificity? | No | Yes |
| Memory? | No | Yes |
| Organisms | All organisms | Vertebrates |

and is unique to vertebrates. **Table 43.1** summarizes the major differences between these two forms of immunity.

We begin by considering forms of defense in the innate immune system. These include tissues, such as the skin, that act as a barrier to infection, cells that act as sentries against pathogens, and proteins and other molecules circulating in the blood that signal the presence of infection. All these defenses are nonspecific, act immediately, and do not depend on past exposure to elicit a response.

### The skin and mucous membranes provide the first line of defense against infection.

Organisms have a barrier that surrounds them, like the walls of Troy, which can at times be breached. Ordinarily, the barrier around organisms, among its other functions, protects against infection. Examples include the cell wall of bacteria, the bark of trees, the cuticle of leaves, the exoskeleton of insects, the scales of fish, the shells of eggs, and the skin of mammals (**Fig. 43.1**). These structures act as physical barriers against infection.

To understand the importance of the barrier, consider what happens when it is breached. Burn victims who lose the outer layer of their skin are particularly vulnerable to infection. The bite of a dog or the sting of an insect can seed viruses, bacteria, or parasites directly into the bloodstream, bypassing the protective layer of skin. Even a small puncture wound can provide entry for the bacterium *Clostridium tetani*, the cause of tetanus.

In humans, this physical barrier is the skin (**Fig. 43.2**). It has two layers, the outer epidermis and the inner dermis (Chapter 10). The **epidermis** consists of several layers of cells covered with the protein keratin. The **dermis** consists of connective tissue, hair follicles, blood and lymphatic vessels, and glands. Together, these two layers provide a physical barrier against microorganisms.

Any surface that is in contact with the external environment is a potential entry site for pathogens. Like the skin, the mucous membranes of the respiratory, gastrointestinal, and genitourinary tracts act as physical barriers. However, they have additional protective mechanisms. For example, parts of the respiratory tract have **cilia,** hairlike projections that sweep mucus along with trapped particles and microorganisms out of the body. Saliva and tears contain enzymes that help to protect against bacteria. The stomach, with its low pH and enzymes that break down food, is inhospitable to many microorganisms.

In addition to acting as a physical and chemical barrier, the skin and mucous membranes provide a home to many nonpathogenic microorganisms. By one estimate, of the 100 trillion cells in the human body, only 10% are human—the rest are mostly bacteria. These microorganisms in some cases provide protection against harmful pathogens by competing with them for food and space. They may also maintain health in other ways, such as by aiding digestion. In this way, we live in a mutualistic relationship with these microorganisms (Case 5: The Human Microbiome). The Human Microbiome Project, sponsored by the National Institutes of Health, is an effort to describe and understand the ecology of these microorganisms.

→ **Quick Check 1** Certain diseases, as well as cigarette smoke, can damage cilia lining the respiratory tract. What effects would you expect from damage to respiratory cilia?

**FIG. 43.1** **Physical barriers against infection.** (a) Tree bark; (b) leaf cuticle; (c) insect exoskeleton; (d) egg shell.

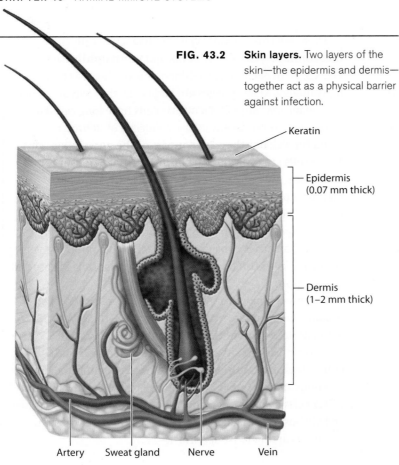

**FIG. 43.2** **Skin layers.** Two layers of the skin—the epidermis and dermis—together act as a physical barrier against infection.

Keratin

Epidermis (0.07 mm thick)

Dermis (1–2 mm thick)

Artery  Sweat gland  Nerve  Vein

## Some cells act broadly against diverse pathogens.

The innate immune system has numerous cells in addition to those in the skin and mucous membranes, found in different locations in the body and having different mechanisms of action. These cells are **white blood cells,** or **leukocytes,** and they arise by differentiation from stem cells in the bone marrow (**Fig. 43.3**). Many of these cells have important roles in the adaptive immune response as well. Here, we focus on their role in innate immunity.

**Phagocytes** (literally, "eating cells") are immune cells that engulf and destroy foreign cells or particles. The engulfing of a cell or particle by another cell is called **phagocytosis.** The Russian microbiologist Ilya Mechnikov first described this remarkable process, for which he shared the Nobel Prize in Physiology or Medicine with the German scientist Paul Ehrlich in 1908.

The process begins when a phagocyte encounters a cell or particle that it recognizes as foreign and binds to it. The phagocyte then extends its plasma membrane completely around the cell or particle until it is within a separate compartment inside

**FIG. 43.3** Development of white blood cells important in the immune system.

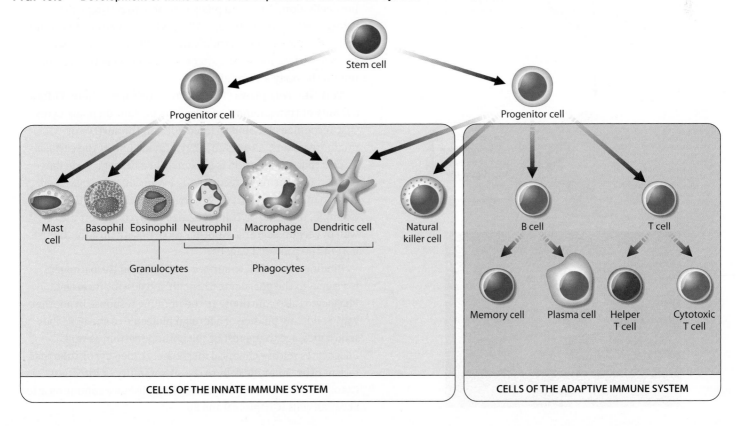

Stem cell

Progenitor cell

Progenitor cell

Mast cell  Basophil  Eosinophil  Neutrophil  Macrophage  Dendritic cell  Natural killer cell

Granulocytes  Phagocytes

B cell  T cell

Memory cell  Plasma cell  Helper T cell  Cytotoxic T cell

**CELLS OF THE INNATE IMMUNE SYSTEM**

**CELLS OF THE ADAPTIVE IMMUNE SYSTEM**

the phagocyte (**Fig. 43.4**). In some cases, the compartment merges with a lysosome, and enzymes within the lysosome digest the foreign cell or particle. Phagocytosis can also trigger a **respiratory burst,** a process that generates reactive oxygen species, such as superoxide radicals and hydrogen peroxide, and reactive nitrogen species, such as nitrogen oxide and nitrogen dioxide. These molecules have antimicrobial properties that directly damage pathogens.

---

**FIG. 43.4    Phagocytosis.** A phagocyte recognizes, engulfs, and digests foreign particles and cells. The scanning electron micrograph at the bottom shows a macrophage engulfing tuberculosis bacteria.

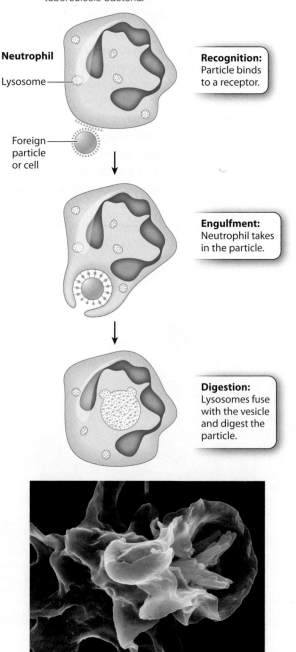

Neutrophil

Lysosome

**Recognition:** Particle binds to a receptor.

Foreign particle or cell

**Engulfment:** Neutrophil takes in the particle.

**Digestion:** Lysosomes fuse with the vesicle and digest the particle.

There are three major types of phagocytic cell: **macrophages, dendritic cells,** and **neutrophils** (see Fig. 43.3). In each case, their names give clues to their morphology. Macrophages are large cells that patrol the body (*macro* means "large"). Dendritic cells have long cellular projections reminiscent of the dendrites of neurons (Chapter 35), but they are not part of the nervous system. These cells are typically part of the natural defenses found in the skin and mucous membranes. Neutrophils are members of a group of cells called **granulocytes** because of the presence of granules in their cytoplasm. The granules of neutrophils take up both acidic and basic dyes, so their staining pattern is considered neutral. Neutrophils are very abundant in the blood and are often the first cells to respond to infection.

The innate immune system contains several other cells in addition to phagocytic cells (see Fig. 43.3). Two types of granulocyte, **eosinophils** and **basophils,** defend against parasitic infections but also contribute to allergies. **Mast cells** release **histamine,** an important contributor to allergic reactions and inflammation, discussed below. Finally, **natural killer cells** do not recognize foreign cells, but instead recognize and kill host cells that are infected by a virus or have become cancerous or otherwise abnormal.

### Phagocytes recognize foreign molecules and send signals to other cells.

Phagocytes attack foreign cells that they encounter but leave host cells alone. How is a phagocyte able to distinguish a foreign cell from a host cell? All cells have an array of proteins and other molecules on their surfaces. Phagocytes detect specific surface molecules that are present on pathogens but not on the body's own cells.

**Toll-like receptors (TLRs)** play a critical role here. TLRs are a family of transmembrane receptors present on phagocytes that recognize and bind to molecules on the surface of microorganisms. As a result, they provide one of the earliest signals that an infection is present. TLRs detect a wide range of microorganisms by recognizing evolutionarily conserved surface molecules shared by many microorganisms. For example, one type of TLR recognizes a glycolipid on the cell wall of certain bacteria. Another recognizes a protein that is part of the bacterial flagella.

Binding of the TLR to surface molecules on the pathogen is a signal to the phagocyte to engulf and destroy its target. Phagocytes also contribute to the immune response in another way. Following binding of a foreign molecule to the TLR, they send a message to the rest of the immune system as well. Phagocytes release chemical messengers called **cytokines** that recruit other immune cells to the site of injury or infection. Cytokines, like hormones, provide long-distance communication between cells (Chapters 9 and 38).

**FIG. 43.5** **Inflammation.** Inflammation involves increased blood flow, leaky blood vessels, and activation of the immune system. Thus, a mosquito bite causes redness and swelling (photograph at bottom).

Bacteria

Dendritic cell

Mast cell

Blood flow

Blood vessel

**1** Bacteria and other pathogens enter a wound.

**2** Mast cells release histamine that increases blood flow (causing redness and heat) and makes blood vessels leaky (causing swelling).

**3** Immune system cells in the tissue release chemical messengers called cytokines that bind to and recruit phagocytes in nearby blood vessels.

Neutrophil

**4** Phagocytes enter the infected site from the blood and remove pathogens by phagocytosis.

## Inflammation is a coordinated response to tissue injury.

While there are generally some immune cells already present at the site of tissue injury or infection, many more must be recruited to fight infection successfully. That recruitment happens as part of a process called **inflammation.** The redness and swelling you see following an injury, a break in the skin, or an infection are signs of inflammation. More precisely, inflammation is a physiological response of the body to injury that removes the inciting agent if present and begins the healing process. It works through the coordinated responses of the local tissue, the vascular system (blood and lymph), and the innate immune system. Inflammation therefore is a useful, beneficial process. However, left unchecked or allowed to continue for a long time, it can be debilitating.

Inflammation is characterized by four classic signs, described by their Latin names: *rubor* (redness), *calor* (heat), *dolor* (pain), and *tumor* (swelling). The process begins following tissue injury or infection (**Fig. 43.5**). Certain cells in the tissue, such as dendritic and mast cells, are activated and release cytokines and other chemical messengers. Some of these chemical messengers recruit additional white blood cells to the site of infection or injury. Others make it easier for white blood cells to reach this site quickly. For example, histamine, released by mast cells and basophils, acts directly on blood vessels to cause vasodilation, increasing blood flow to the site of infection or injury. Vasodilation causes the characteristic redness and heat of inflammation. In addition, histamine increases the permeability of the blood vessel wall. Fluid leaks out of the blood vessel, carrying white blood cells into the damaged tissue. The increased fluid in the tissue surrounding the blood vessels is visible as swelling. Some chemical messengers act directly on nerve fibers, causing pain.

Phagocytes that travel in the blood can move from a blood vessel to the site of infection. This process is called **extravasation** (**Fig. 43.6**). The phagocyte travels along the vessel wall in a rolling motion, grabbing weakly onto the wall as glycoproteins on the phagocyte surface bind transiently to proteins on the endothelial cells. Stronger interactions put a brake on this rolling motion and allow the phagocyte to adhere more firmly to the vessel wall. As it nears the site of infection, the phagocyte encounters and binds to cytokines. The phagocyte then changes shape and moves in between cells lining the blood vessel and into the surrounding tissue.

Several proteins play key roles in inflammation. In the early 1900s, researchers studied patients with pneumonia caused by the bacterium *Streptococcus pneumoniae*. They looked for and found a change in the level of proteins in the blood as a marker for the initial phase of the disease, which became known as the **acute phase response.** One of these proteins, C-reactive protein or CRP, recognizes and binds to molecules on the surface of some bacteria. These CRP-coated bacteria are more easily taken up by

**FIG. 43.6** **Extravasation.** In extravasation, an activated phagocyte moves from a blood vessel to the surrounding tissue.

Blood vessel lumen

**1** **Rolling:** The phagocyte travels along the inside of the vessel wall.

**2** **Adhesion:** The phagocyte binds to the vessel wall.

**3** **Extravasation:** The phagocyte changes shape, exits the vessel, and enters the surrounding tissue to travel to the site of damage.

Phagocyte

Endothelial cells

Basal lamina

Tissue outside blood vessel

Macrophage

Bacteria

phagocytes. The binding of a molecule like CRP to a pathogen to facilitate uptake by a phagocyte is called **opsonization.**

### The complement system participates in the innate and adaptive immune systems.

In addition to individual proteins, such as CRP, there is an entire system of proteins circulating in the blood that participates in innate immune function. Collectively, these proteins make up the **complement system.** First described by Paul Ehrlich in the late 1800s, this system is so named because it supplements or *complements* other parts of the immune system.

The complement system consists of more than 25 proteins that circulate in the blood in an inactive form (**Fig. 43.7**). The system is activated when these proteins bind to molecules specific to microorganisms or to antibodies (discussed below). Activation in turn sets off a biochemical cascade in which the product of one reaction is the enzyme that catalyzes the next. Such sequential activation leads to amplification of the response, as we saw earlier in cell signaling (Chapter 9) and in the endocrine system (Chapter 38), where a small stimulus can have a dramatic response.

Activation of the complement system has three effects. The most dramatic is breaking open cells, or lysis. Lysis results when complement proteins form a **membrane attack complex (MAC)** that makes holes in bacterial cells. Second, the complement system targets pathogens for phagocytosis by coating bacteria with a protein that phagocytes recognize (opsonization). Finally, activation of the complement system leads to the production of

**FIG. 43.7** **The complement system.** The proteins of the complement system contribute to the defense against pathogens by cell lysis, phagocytosis, and inflammation.

Microorganisms

Antibodies

Complement system

Bacterial cell    MAC

Bacterial cell

Mast cell

**Formation of the membrane attack complex (MAC)**

**Opsonization**

**Cell signaling**

— Histamine

**Cell lysis**

**Phagocytosis**

**Inflammation**

activated proteins that attract other components of the immune system. For example, some of these proteins stimulate mast cells, which release histamine and initiate the inflammatory response.

You can appreciate the importance of the complement system in the immune system by considering what occurs in individuals who lack key components. Complement deficiencies are a rare form of **immunodeficiency,** any disease in which part of the immune system does not function properly. Their effects depend on which component of the complement system is lacking. For example, one type of complement deficiency makes individuals vulnerable to bacterial infections because fewer pathogens undergo cell lysis and opsonization.

Because antibodies are an important signal activating the complement system, the complement system provides an important link between the innate, nonspecific arm of the immune system and the adaptive, specific arm, a topic we consider next.

## 43.2 ADAPTIVE IMMUNITY: B CELLS, ANTIBODIES, AND HUMORAL IMMUNITY

The innate immune system can react to diverse pathogens and has the capacity to distinguish self (host cells) from non-self (foreign cells). The adaptive immune system shares these features, and has two additional features not present in the innate immune system. The first is specificity: The adaptive immune system targets responses to specific pathogens. As a result, it is often more effective at eliminating a pathogen. The second is memory: The adaptive immune system remembers past infections and mounts a stronger response on re-exposure. How can the adaptive immune system respond to a single pathogen among the diversity of pathogens that it meets? How is immunological memory achieved? We explore answers to these questions in the next two sections.

While many cells and tissues play a role in adaptive immunity, two types of cell are particularly important: **B lymphocytes,** or **B cells,** and **T lymphocytes,** or **T cells** (see Fig. 43.3). B and T cells are named for their site of maturation. In birds, B cells mature in the bursa of Fabricius, an organ of the lymphatic system located off the digestive tract (**Fig. 43.8**). In mammals, B cells mature in the bone marrow. T cells mature in the thymus, an organ of the immune system located just behind the sternum, or breastbone. B and T cells circulate in blood and lymphatic vessels, and also can be found in the spleen, liver, and lymph nodes (Fig. 43.8). We begin with a discussion of the role of B cells in adaptive immunity.

### B cells produce antibodies.

One of the hallmarks of adaptive immunity is the ability to target specific pathogens. But there are a tremendous number of different viruses, bacteria, single-celled eukaryotes, fungi,

and worms that can lead to infection. How does the adaptive immune system recognize each one specifically? The specificity of the adaptive immune system is in part the result of **antibodies** produced by B cells.

An antibody is a large protein that carries sugar molecules attached to some amino acids (hence it is a glycoprotein). An antibody binds to foreign molecules that occur naturally on or in microorganisms and participate in normal cellular functions. Such a molecule is termed an **antigen,** which is a molecule that binds to and leads to the production of antibodies. (The word "antigen" is a contraction of "antibody" and "generator.") The great diversity of antigens on different microorganisms is matched by a great diversity of antibodies. It is estimated that humans produce approximately 10 billion different antibodies, each with the ability to recognize one or a few antigens on microorganisms.

Antibodies can be found on the surface of B cells or free in the blood or tissues. Collectively, antibodies make up a significant portion of the proteins present in the blood, typically about 20%. Because antibodies are present in the blood, tissue fluid, and secretions, which used to be called humors, this part of

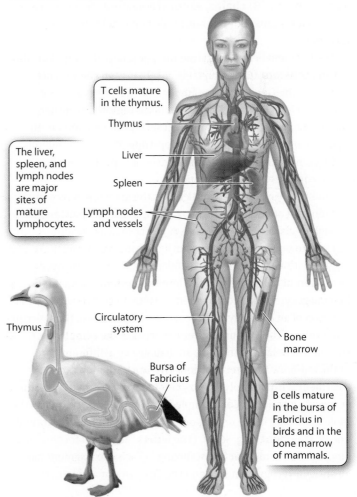

**FIG. 43.8** Major organs of the immune system.

T cells mature in the thymus.

Thymus

Liver

Spleen

Lymph nodes and vessels

The liver, spleen, and lymph nodes are major sites of mature lymphocytes.

Thymus

Circulatory system

Bone marrow

Bursa of Fabricius

B cells mature in the bursa of Fabricius in birds and in the bone marrow of mammals.

**FIG. 43.9   Antibody structure.** An antibody has variable regions that distinguish it from the billions of other antibodies and allow it to bind a specific antigen.

Heavy (H) chain
Light (L) chain

the adaptive immune system is sometimes called **humoral immunity.**

Antibodies have a distinctive structure that reflects their specificity for a given antigen. The simplest antibody molecule, shown in **Fig. 43.9,** is a Y-shaped protein made up of four polypeptide chains—two identical **light (L) chains** and two identical **heavy (H) chains,** so called because of their relative molecular weights. Covalent disulfide bonds hold the four chains together.

The light and heavy chains are further subdivided into **variable (V)** and **constant (C)** regions (Fig. 43.9). Within the variable regions of the L and H chains are regions that are even more variable, which are called **hypervariable regions.** Together, these hypervariable regions interact in a specific manner with a portion of the antigen called the **epitope.** Each of the billions of different antibodies has a distinct set of hypervariable regions that recognizes a unique epitope.

Binding of antibodies to antigens is the first step in recognition and removal of microorganisms. Binding alone is sometimes enough to disable a microorganism. In these cases, binding can lead to precipitation of the antibody–antigen complex (agglutination). More often, the microorganism is destroyed by a different component of the immune system, for example, by a phagocyte or the complement system. In this case, the function of an antibody is to recognize the pathogen, then recruit other cells of the immune system or activate the complement system. These latter functions are handled by a different part of the antibody from recognition. This can be demonstrated by treatment with proteolytic enzymes (enzymes that cut proteins at specific locations). For example, one type of enzyme breaks the molecule at the hinge region, producing two **Fab fragments** and one **Fc fragment** (Fig. 43.9). (The letters "ab" stand for "antigen-binding"; "c" stands for "crystallizing.") Each Fab fragment has an antigen-recognition site, and the Fc fragment activates the

complement system and binds to cell-surface receptors of other cells of the immune system.

## Mammals produce five classes of antibody with different biological functions.

Antibodies are members of a family of proteins with common structural features. As a group, they are called **immunoglobulins (Ig).** There are five classes of immunoglobulin—**IgG, IgM, IgA, IgD,** and **IgE**—each with a different function (**Fig. 43.10**). These classes are defined by their heavy chains, which differ in the amino acid sequences of their constant regions. There are also two types of light chain. These two types of light chain occur in all five classes of antibody, but any given antibody contains only one type of light chain.

IgG is by far the most abundant of the five antibody classes. It is the Y-shaped antibody depicted in Figs. 43.9 and 43.10. IgG circulates in the blood and is particularly effective against bacteria and viruses. In addition, IgG is the only class of antibody that can cross the placenta because of the presence of receptors on placental cells for the Fc portion of IgG. The ability of IgG to cross the placenta provides protection to the developing fetus.

IgM is a pentamer in mammals and a tetramer in fish. (A pentamer is a molecule made of five monomers; a tetramer is a molecule made of four monomers.) The individual units—the monomers—are linked by a joining chain (Fig. 43.10). IgM also exists as a monomer on the surface of B cells. IgM is particularly important in the early response to infection

**FIG. 43.10   Five classes of antibody.**

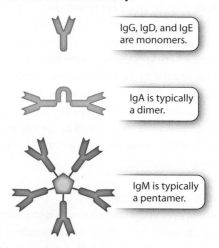

IgG, IgD, and IgE are monomers.

IgA is typically a dimer.

IgM is typically a pentamer.

(discussed below) and is very efficient in activating the complement system and stimulating an immune response.

IgA is usually a dimer consisting of two antibody molecules linked by a joining chain (Fig. 43.10). It is the major antibody on mucosal surfaces, such as those of the respiratory, gastrointestinal, and genitourinary tracts. It helps to protect mucous membranes from infection. It is also present in secretions, including tears, saliva, and breast milk.

IgD and IgE, like IgG, are monomers. IgD is typically found on the surface of B cells. This immunoglobulin helps initiate inflammation. IgE plays a central role in allergies, asthma, and other **immediate hypersensitivity reactions,** which are characterized by a heightened or an inappropriate immune response to common antigens. Immediate hypersensitivity reactions result from the binding of the Fc portion of IgE to receptors on the surface of mast cells and basophils. Binding leads to the release of histamine and cytokines, resulting in inflammation that can be life threatening.

The five classes of antibody are also called **isotypes.** B cells can change the class of antibody they make, a process known as **isotype** or **class switching,** discussed below.

### Clonal selection is the basis for antibody specificity.

A central question in immunology is how antibody specificity is achieved. There are seemingly limitless types of antigen that an organism might come across. How can B cells generate antibodies that interact only with specific antigens among this tremendous diversity?

Two alternative hypotheses were suggested to explain antibody specificity. In 1900, Paul Ehrlich suggested that there exists a large pool of B cells, each with a different antibody on its surface. The antigen binds to one or a few of these antibodies, stimulating the B cell to divide. In this model, specificity is achieved before any exposure to the antigen. Ehrlich's model drew on concepts from chemistry such as the binding of an enzyme to a substrate in the manner of a lock and key (Chapter 6). Similarly, the antigen and antibody interact in a specific way.

Later, in the early 1900s, a different hypothesis was suggested. This hypothesis proposed that the antigen instructs the antibody to fold in a particular way so that the two interact in a specific manner. Here, the antigen plays a more active role in the process and antibody specificity is determined only after interaction with the antibody.

Which of these two hypotheses is correct? Does the antigen *select* a B cell with a preexisting surface antibody with which it interacts specifically, as in the first hypothesis, or does it *instruct* the B cell to make a specific antibody, as in the second?

Experimental data on the genetic basis for antibody diversity (described below) provided strong evidence for the first hypothesis. Now called **clonal selection,** it is one of the central principles of immunology (**Fig. 43.11**). Every individual has a very large pool of B cells, each with a different antibody on its cell surface. Collectively, these antibodies recognize a great diversity of antigens, but each individual antibody can recognize only one antigen or a few antigens. An antigen interacts with the

**FIG. 43.11 Clonal selection.** Those B cells that recognize a foreign antigen are selected to divide and differentiate into plasma cells and memory cells.

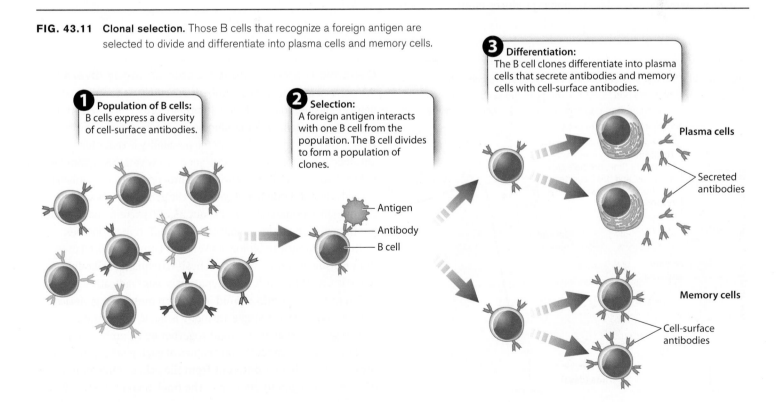

**❶ Population of B cells:** B cells express a diversity of cell-surface antibodies.

**❷ Selection:** A foreign antigen interacts with one B cell from the population. The B cell divides to form a population of clones.

**❸ Differentiation:** The B cell clones differentiate into plasma cells that secrete antibodies and memory cells with cell-surface antibodies.

Antigen
Antibody
B cell

Plasma cells
Secreted antibodies

Memory cells
Cell-surface antibodies

B cell, or a small set of B cells, that has the best fit between the antigen and the membrane-bound antibody. Binding leads the B cell to divide and differentiate into two types of cell. **Plasma,** or **effector, cells** secrete antibodies. **Memory cells** are long-lived cells that contain membrane-bound antibody having the same specificity as the parent cell. Since the resulting population of B cells arose from one B cell, they are clones of one another. Almost counterintuitively, antibody specificity is achieved before any exposure to the antigen.

### Clonal selection also explains immunological memory.

Clonal selection provides a mechanism for explaining antibody specificity. It also explains the second feature of the adaptive immune system: the ability to remember past infections—that is, to react more vigorously on re-exposure. The first encounter with an antigen leads to a **primary response** (**Fig. 43.12**). In this response, there is a short lag before antibody is produced. The lag is the time required for B cells to divide and form plasma cells. The plasma cells secrete antibodies, typically IgM. The level of IgM in the blood increases, peaks, and then declines.

On re-exposure to the same antigen, even months or years after the first exposure, there is a **secondary response,** which is quicker, stronger, and longer than the primary response (Fig. 43.12). During this response, the antibodies are released by memory B cells. These cells respond more quickly to antigen exposure than naïve B cells (B cells that haven't been previously exposed to antigen). In addition, the pool of memory B cells is larger than the pool of naïve B cells for a given antigen, explaining why the secondary response produces more

antibodies for a longer period of time. IgG is the most abundant antibody in the secondary response.

It is the presence of long-lived memory cells produced following the primary response that provides natural, long-lasting immunity after an infection such as chicken pox. We have learned to take advantage of immunological memory in **vaccination.** An antigen from a pathogen is deliberately given to a patient in a vaccine to induce a primary response but not the disease, thereby providing future protection from infection.

Until the early 1700s, it was common practice to inoculate people with smallpox in an effort to induce a mild disease and prevent a more severe, even lethal, infection. However, the English scientist Edward Jenner is usually credited with the first demonstration of the protective effects of vaccination. It was well known that milkmaids, exposed to the relatively benign cowpox virus from milking cows, were immune to the related but more deadly smallpox virus. In 1796, Jenner tested the consequences of exposure to cowpox by deliberately inoculating a young boy with cowpox and demonstrating that he became immune to smallpox. In fact, the modern word "vaccine" is derived from the Latin *vacca*, which means "cow."

Today, vaccines take many forms: They can be a protein or a part of a protein from the pathogen, a live but weakened form of the pathogen, or a killed pathogen. They are among the most effective public health measures ever developed.

→ **Quick Check 2** Look again at Fig. 43.12. Imagine that a different novel antigen is added on day 40 rather than a second injection of the same antigen. Would you predict that there would be a primary or a secondary response to this second antigen?

**FIG. 43.12** **Primary and secondary responses.** The immune system responds more strongly to the second exposure to an antigen than to the first exposure because of the presence of memory B cells.

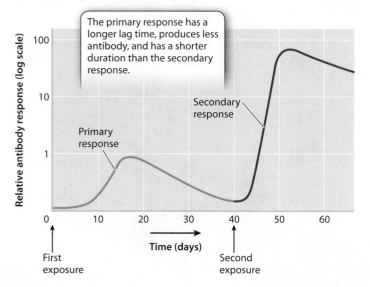

### Genomic rearrangement creates antibody diversity.

We pointed out earlier that antibody specificity is determined before exposure to the antigen. In other words, clonal selection requires a mechanism for producing a great diversity of antibodies. How is this diversity achieved? One possibility is that each antibody is encoded by a separate gene. However, this cannot be the case, as the number of antibodies produced by an organism far exceeds the total number of genes in the genome. For example, the human genome consists of about 25,000 protein-coding genes, but humans can produce about 10 billion different antibodies.

In 1965, American biologists William Dreyer and J. Claude Bennett suggested a novel hypothesis to explain antibody diversity. As they put it, this hypothesis was "radically different from anything found in modern molecular genetics." They proposed that a single antibody is made by separate gene segments that are brought together by recombination. According to this model, many copies of each gene segment are present, each slightly different from the others. Only one copy of each gene segment ends up in the final, recombined antibody gene (and hence in the antibody protein). Diversity is therefore achieved by combining different copies of each segment in different B cells as they mature.

At the time Dreyer and Bennett proposed their model, evidence for it was indirect. More than 10 years later, direct experimental evidence showed that their hypothesis is correct (**Fig. 43.13**). As a B cell differentiates, different gene segments are joined in a process called **genomic rearrangement** that produces a specific antibody.

## HOW DO WE KNOW?

**FIG. 43.13**

# How is antibody diversity generated?

**BACKGROUND** Humans produce an estimated 10 billion different antibodies. Antibodies are proteins, encoded by genes. However, there are only about 25,000 genes in the human genome. How then can humans (and other vertebrates) produce such a diversity of antibodies from a limited set of genes? This was one of the central problems in immunology until the 1960s and 1970s.

**HYPOTHESIS** In 1965, American biologists William Dreyer and J. Claude Bennett proposed a model in which there are many copies of two gene segments, which they called the V (variable) and C (constant) segments. An antibody gene is assembled by recombining single copies of each of these segments. At the time, there was no direct experimental evidence to support this hypothesis.

**EXPERIMENT** In 1976, Japanese immunologists Nobumichi Hozumi and Susumu Tonegawa tested the Dreyer and Bennett hypothesis. Using mice as a model system, they isolated DNA from embryonic cells (before recombination was thought to occur) and from adult B cell tumor lines that produce one type of antibody (after recombination was thought to occur). The procedure they followed was slightly different from what is described here but conceptually the same. DNA is cut by restriction enzymes, the fragments are separated on a gel and transferred to a filter paper, and light chain DNA (containing both V and C regions) is used as a radioactive probe to determine the size and positions of the V and C regions (Southern blot, Chapter 12).

**RESULTS** Hozumi and Tonegawa found that the V and C regions were far apart in the DNA of embryonic cells and close together in adult cells, suggesting that the gene segments were brought together during B cell differentiation.

**CONCLUSION** Antibody genes are assembled by recombination of individual gene segments, providing a mechanism for generating antibody diversity.

**FOLLOW-UP WORK** Further studies showed that there are multiple copies of several different gene segments, now called the V, D (diversity), J (joining), and C regions for heavy chains, and V, J, and C regions for light chains. These are recombined in unique patterns in different B cells, so each B cell produces one antibody and a population of B cells produces a diversity of antibodies.

Embryonic cell

B cell

DNA is extracted and cleaved into fragments by restriction enzymes.

DNA fragments

DNA fragments are separated by gel electrophoresis, transferred to a filter (blot), and probed with light-chain DNA containing both V and C regions.

Restriction fragments

$V_1$ to $V_n$        C        $V_2$   C

There are two restriction fragments from embryonic cells and just one restriction fragment in B cells, indicating that the genome was rearranged in the process of B cell maturation.

**SOURCES** Dreyer, W. J., and J. C. Bennett. 1965. "The Molecular Basis of Antibody Formation: A Paradox." *Proceedings of the National Academy of Sciences* 54:864–869; Hozumi, N. and S. Tonegawa. 1976. "Evidence for Somatic Rearrangement of Immunoglobulin Genes Coding for Variable and Constant Regions." *Proceedings of the National Academy of Sciences* 73:3628–3632.

**FIG. 43.14  Genomic rearrangement in mice.** In genomic rearrangement, different gene segments recombine to generate a functional gene encoding the parts of an antibody.

For H chains, there are multiple V (variable), D (diversity), J (joining), and C (constant) gene segments (**Fig. 43.14**). In mice H chains, for example, there are 100 different V segments, 30 D segments, 6 J segments, and 8 C segments. During B cell differentiation, the DNA in this region undergoes recombination so that just one of each of these segments makes part of a functional H chain gene, and the intervening DNA is deleted. The assembled VDJ segment encodes the variable portion, and the C segment encodes the constant region. L chain genes also contain multiple copies of gene segments, except they do not have a D gene segment, so the VJ segment encodes the variable portion and the C segment encodes the constant region. The result of this process is that each B cell encodes a single H chain and a single L chain, and each B cell expresses a unique antibody.

It is worth pausing here to consider the significance of this mechanism. We generally think of DNA as being a stable blueprint present in identical copies in all of the somatic cells in our bodies. B cells are an exception. As a result of genomic rearrangement, the DNA in each mature B cell is different from the DNA in every other mature B cell and different from the DNA in other cells in the body.

Genomic rearrangement is the key mechanism that creates antibody diversity, but other mechanisms contribute as well. One is simply the association of different L and H chains to make a functional antibody. In addition, recombination is sometimes imprecise, creating different sequences at the junctions between the different gene segments. There are also mechanisms for adding nucleotides at the junctions by a process called N-nucleotide addition, and replacing nucleotides within VJ or VDJ segments by a process called somatic hypermutation. Finally, alternative splicing of mRNA (Chapter 3) allows a single B cell to express both membrane-bound and secreted antibodies.

We have seen that B cells as a group produce an enormous diversity of antibodies, whereas a given B cell makes one and only one antibody. But B cells are diploid, containing maternal and paternal copies of the genes that encode for antibodies. Maternal and paternal alleles of the antibody genes are similar, but not identical. As a result, a B cell able to use both homologs would make two different antibodies, yet B cells always make one. Remarkably, once one allele undergoes genomic rearrangement, the other allele is prevented from doing so. As a result, B cells act as if they have a single allele for each antibody gene—they express the maternal or the paternal gene, but not both.

As we have seen, B cells are able to switch between different classes, making IgM early in the response to infection and IgG later, while maintaining specificity to the same antigen. How is isotype switching achieved? Isotype switching results from further genome rearrangement, in which a given V(D)J region can join to different C segments during B cell differentiation.

The mechanisms we have considered pertain to humans and mice. Whereas B cells mature in the bone marrow in humans and mice, they mature in gut-associated lymphoid tissue (GALT) in many other vertebrates. Furthermore, genomic rearrangement is less important in many vertebrates. Instead, somatic hypermutation appears to play a more important role.

→ **Quick Check 3** If a given B cell produces only one type of antibody, how do organisms produce a great diversity of antibodies?

## 43.3 ADAPTIVE IMMUNITY: T CELLS AND CELL-MEDIATED IMMUNITY

Antibodies are remarkable for their diversity and specificity. They provide protection against many kinds of bacteria, especially those encased in a polysaccharide capsule, such as the bacteria that cause pneumonia and meningitis. The B cells that generate them exhibit immunological memory, as we saw by contrasting the primary and secondary immune responses. However, B cells and their antibodies have limitations. For example, B cells on their own can make antibodies only against some antigens; for other antigens, they require the assistance of other cells. Furthermore, they are sometimes ineffective against pathogens that take up residence inside a cell, as do the bacteria that cause tuberculosis.

To handle these and other kinds of pathogens, there is another part of the adaptive immune system, which depends on the second type of lymphocyte—the T cell. T cells do not secrete antibodies. Instead, they participate in **cell-mediated immunity**, so named because cells, not antibodies, recognize and act against pathogens. The importance of cell-mediated immunity is most dramatically demonstrated by individuals taking immunosuppressive drugs or infected by HIV. Both of these suppress the activity of T cells. The result can be overwhelming infections and even tumors.

### T cells include helper and cytotoxic cells.

T cells, like B cells, originate in the bone marrow. However, unlike B cells, T cells mature in the thymus. A mature T cell

**TABLE 43.2** Comparison of Helper and Cytotoxic T Cells

| | HELPER T CELL | CYTOTOXIC T CELL |
|---|---|---|
| Functions | Activation of macrophages Activation of B cells Activation of cytotoxic T cells Secretion of cytokines | Cell killing |
| Surface molecule | CD4 | CD8 |
| MHC molecule recognized by the T cell | Class II | Class I |

is characterized by the presence of a **T cell receptor** (TCR) on the plasma membrane. The T cell receptor is a protein receptor that recognizes and binds to the antigen. In this way, the T cell receptor is similar to an antibody. However, whereas antibodies may be either on the cell surface of B cells or secreted, TCRs are always found on the cell surface of T cells.

There are two major subpopulations of T cells with different names, functions, and cell-surface markers—**helper T cells** and **cytotoxic T cells** (**Table 43.2;** see Fig. 43.3). Helper T cells do just that—they *help* other cells of the immune system by secreting cytokines. One of their key roles is to activate B cells to secrete antibodies. Although B cells can work on their own, particularly against encapsulated bacteria, in most cases they require the participation of helper T cells. Helper T cells also activate macrophages, cytotoxic T cells, and other cells of the immune system.

Cytotoxic T cells, as their name implies, are cytotoxic—they can *kill* other cells. Like B cells, they are activated by cytokines released from helper T cells. They are particularly effective against altered host cells, such as those infected with a virus or that have become cancerous.

These two types of T cell are distinguished by the presence of different glycoproteins on their surface—CD4 on helper T cells and CD8 on cytotoxic T cells. The ratio of these two classes of T cell can be used to assess immune function. In healthy individuals, the CD4:CD8 ratio is usually about 2. A decreased ratio, indicating the presence of fewer helper T cells relative to cytotoxic T cells, is typical of individuals infected with HIV, the virus that causes AIDS. HIV infects and kills helper T cells, disarming a key player in the immune system.

**FIG. 43.15  The T cell receptor (TCR).** The TCR is a protein on the surface of T cells that binds foreign antigens that are bound to MHC proteins, triggering the division of the T cell into clones.

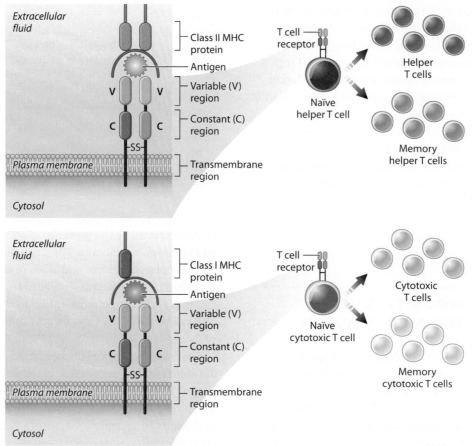

### T cells have T cell receptors on their surface.

T cells have TCRs on their surface that recognize and bind antigens (**Fig. 43.15**). In several respects, TCRs are similar to antibodies on the surface of B cells. For example, TCRs recognize antigens with a specific structure. In addition, there is a great diversity of TCRs that differ from one another, but each T cell has just one type of TCR on its surface. Binding of TCR to an antigen triggers the T cell to divide into clones, resulting in a pool of T cells that are each specific for a given antigen. Finally, the diversity of TCRs among different T cells results from genomic rearrangement of V, D, J, and C gene segments.

However, TCRs are different from antibodies in important ways. First, they are composed of two, rather than four, polypeptide chains. Second, they are not secreted like antibodies, but are always membrane bound on the T cell surface. Finally, the TCR does not recognize free antigen. It only recognizes antigen in association with proteins of the **major histocompatibility complex,** or **MHC,** proteins that appear on the surface of most mammalian cells (Fig. 43.15).

Once activated, T cells divide and form helper and cytotoxic T cells. Some cells of each type are memory cells that provide long-lasting immunity following an initial infection, as in the case of B cells (Fig. 43.15). Also like B cells, T cells can sometimes be activated too strongly. Earlier, we saw how IgE bound to mast cells and basophils can lead to immediate hypersensitivity reactions characteristic of allergies and asthma. The counterpart in T cells is a **delayed hypersensitivity reaction,** which, as its name suggests, does not begin right away. For example, if you touch poison ivy, your skin will turn red and start itching only after a delay of several hours or days. Delayed hypersensitivity reactions are initiated by helper T cells, which release cytokines that attract macrophages to the site of exposure, which is typically the skin.

### T cell activation requires the presence of antigen in association with MHC proteins.

Because of their central role in activating different components of the immune system, T cells themselves require not one signal but two signals to become activated. One of these signals is an antigen, which indicates the presence of foreign cells or particles. The second signal is MHC proteins. T cells interact only with antigens that are bound to molecules of the MHC on the surface of host cells. The requirement for antigen in association with MHC proteins explains why T cells target infected or cancerous host cells. MHCs were first discovered in transplantation biology because it is their presence that leads to acceptance or rejection of transplanted tissues.

The MHC is a cluster of genes present in all mammals that encode proteins on the surface of cells. The MHC is composed of many genes with a high rate of polymorphism, meaning that there is a lot of variability in the gene sequence (and consequently the protein sequence) among individuals. In humans and mice, the genes are divided into three classes: **class I** genes are expressed on the surface of all nucleated cells; **class II** genes are expressed on the surface of macrophages, dendritic cells, and B cells; and **class III** genes encode several proteins of the complement system and proteins involved in inflammation.

Let's consider the activation of helper T cells (**Fig. 43.16**). When an antigen enters the immune system, it may be recognized by an antibody directly or be taken up by **antigen-presenting cells.**

These cells, which include macrophages, dendritic cells, and B cells, take up the antigen and return portions of it to the cell surface bound to MHC class II proteins. Helper T cells recognize processed antigen along with MHC class II molecules by their T cell receptors. As a result of TCR binding to antigen and MHC class II proteins, the helper T cells release cytokines that activate other parts of the immune system, including macrophages, B cells, and cyotoxic T cells.

Cytotoxic T cells also recognize antigen displayed by host cells, but in this case the antigen is presented in association with MHC class I molecules (Fig. 43.16). Because class I molecules are present on virtually all cells, cytotoxic T cells recognize and kill any host cell that becomes abnormal in some way. For example, a virus-infected cell often expresses viral antigens and MHC class I molecules on its surface. Cytotoxic T cells recognize the antigen and MHC class I molecules and kill the cell. Tumor cells express novel antigens along with MHC class I proteins, and may also be eliminated in some cases by cytotoxic T cells.

→ **Quick Check 4** How does T cell activation differ from B cell activation?

---

**FIG. 43.16** **Three-way interactions among the T cell receptor, MHC molecule, and processed antigen.** Helper T cells recognize processed antigen in association with class II MHC protein, and cytotoxic cells recognize processed antigen in association with class I MHC protein.

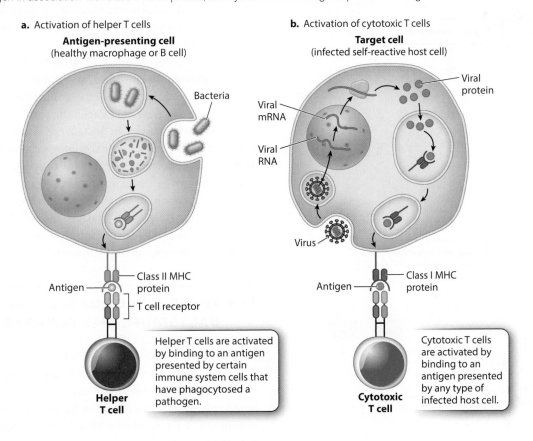

**The ability to distinguish between self and non-self is acquired during T cell maturation.**

Of the possible T cell receptors that are generated by genomic rearrangement, only some are useful: those that react with the host's own MHC molecules. In addition, useful T cell receptors must not react with molecules normally present in or on cells of the host. In other words, T cells must not be activated in response to an organism's own cells. Molecules normally present on a host cell that can bind antibodies are called self antigens.

A sorting process is therefore necessary so that only some T cells mature and others are eliminated (**Fig. 43.17**). As T cells

mature in the thymus, they interact with cells of the epithelium of the thymus. Those that recognize self MHC molecules on epithelium cells are **positively selected** and continue to mature. Those that react too strongly to self antigens in association with MHC are **negatively selected** and eliminated through cell death. Note that, in spite of their names, both processes involve elimination of some T cells. By far, the majority of T cells are eliminated through a combination of positive and negative selection.

The result of this sorting process is twofold. First, T cells become MHC restricted, so helper T cells interact with foreign antigen plus MHC class II proteins and cytotoxic T cells interact with foreign antigen plus MHC class I proteins. Second, T cells exhibit **tolerance**—in other words, they do not respond to self antigens, even though the immune system functions normally otherwise. Those antigens present as T cells mature are labeled as self and do not elicit a response; those not present are non-self and, if encountered, do elicit a response.

B cells are not MHC restricted. However, they, too, exhibit tolerance to self antigens because they go through a similar process of negative selection in the bone marrow. During that process, B cells that react strongly to self antigens are eliminated through cell death. Although most self-reactive T and B cells are eliminated, some T and B cells that react to self antigens escape this developmental check and end up in the circulation. As a result, additional mechanisms are in place to eliminate these self-reactive cells, either through the activity of T cells or through cell death.

The ability to distinguish self from non-self is critical. Failure leads to **autoimmune diseases,** in which tolerance is lost and the immune system becomes active against antigens of the host. Autoimmune diseases can be debilitating as T cells or antibodies attack cells and organs of the host. In rheumatoid arthritis, self-reactive T cells attack the joints, causing inflammation and tissue damage; in type I diabetes, attack is directed against insulin-producing cells of the pancreas; and in multiple sclerosis, the target is the myelin sheath surrounding nerves of the central nervous system. It is not fully understood how these self-reactive T cells evade checks on their development. Perhaps one clue is that some autoimmune diseases are associated with particular MHC alleles.

## 43.4 THREE INFECTIONS: A VIRUS, BACTERIUM, AND EUKARYOTE

Although we have presented the different components of the immune system one at a time, in life they work together to provide a coordinated response to pathogens. We have seen how B cells often require T cell help, how the complement system can be activated by antibodies produced by B cells, and how macrophages present antigens in association with MHC

**FIG. 43.17  Positive and negative selection of T cells.** Selection ensures that T cells react with self MHC molecules but do not react against antigens on the body's own cells.

T cell receptor

Immature T cells

Positive selection retains T cells that react with self MHC molecules and removes those that do not react with self MHC molecules.

**Positive selection**

Class II MHC protein

Class I MHC protein

Thymus cell

Negative selection removes cells that react too strongly with self antigens in association with MHC molecules.

**Negative selection**

Mature T cells

Thymus cell

molecules to T cells. In turn, pathogens have evolved means of evading the immune system. In this section, we consider the flu virus, the bacterium that causes tuberculosis, and the malaria parasite. These three pathogens illustrate how pathogens evolve and adapt in response to the immune system, and how the immune system in turn evolves and adapts in response to pathogens.

### The flu virus evades the immune system through antigenic drift and shift.

Everyone knows the symptoms: fever, chills, sore throat, cough, weakness, and fatigue, perhaps a headache and muscle pain. Many viruses cause these common flu-like symptoms. However, the common cold, stomach flu, and 24-hour flu are caused by unrelated viruses, not by the flu virus.

The flu—its full name is influenza—is caused by an RNA virus that infects mammals and birds. It is notable for causing seasonal outbreaks, or epidemics. Four times in the last century—the Spanish flu of 1918, the Asian flu of 1957–58, the Hong Kong flu of 1968–69, and the swine flu of 2009—the virus spread more widely and infected more people, and the disease reached pandemic levels. The flu is more than just a nuisance: It causes hundreds of thousands of deaths each year, and millions of deaths during pandemics.

Part of the success of the virus lies in its ability to spread easily. The flu virus can spread in the air by small droplets produced when an infected person coughs or sneezes. It can also spread by direct contact or by touching a contaminated surface, such as a doorknob. Hand washing, therefore, is one of the most effective means of protection.

Another characteristic of the flu is its ability to evade the immune system. Like the Greek soldiers in their wooden horse, the flu virus can change its outer coat so memory cells, produced from earlier infections, no longer recognize it. There are three major types of flu virus—called A, B, and C—and many different strains. For example, the swine flu pandemic of 2009 was caused by a type A flu virus of the strain H1N1. These strains differ in several respects, including the structure of a cell-surface glycoprotein called hemagglutinin (HA). HA binds to epithelial cells and controls viral entry into these cells. It can bind only to cells that display a complementary cell-surface protein. As a result, hemagglutinin determines in part which organisms the flu virus infects—humans, pigs, or birds—and which part of the body it infects—the nose, throat, or lungs.

Cells infected by viruses secrete cytokines that bring macrophages, T cells, and B cells to the site of infection. These cytokines, produced in abundance, lead to many of the symptoms commonly associated with the flu. For example, the cytokine interleukin-1 produces fever.

Once in the cell, the virus replicates its genome and makes more virus particles. However, replication is prone to error, so

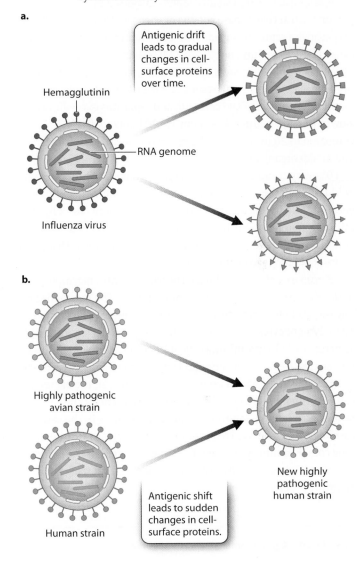

**FIG. 43.18** **(a) Antigenic drift and (b) antigenic shift.** Virus populations evolve over time and evade the immune system's memory cells.

a.

Antigenic drift leads to gradual changes in cell-surface proteins over time.

Hemagglutinin

RNA genome

Influenza virus

b.

Highly pathogenic avian strain

Human strain

Antigenic shift leads to sudden changes in cell-surface proteins.

New highly pathogenic human strain

there is a high rate of mutation that leads to changes in the amino acid sequences of antigens present on the viral surface, including HA. This process is called **antigenic drift (Fig. 43.18a).** It allows a population of viruses to evolve over time and evade memory T and B cells that remember past infections.

Antigenic drift leads to a gradual change of the virus over time. The flu virus is also capable of sudden changes by a process known as **antigenic shift (Fig. 43.18b).** The viral genome consists of eight linear RNA strands. If a single cell is co-infected with two or more different flu strains, the RNA strands can reassort to generate a new strain. H1N1, for example, has genetic elements from human, pig, and bird flu viruses. Antigenic drift and shift make it difficult to predict from year to year which strains will be most prevalent and therefore what vaccine will be most effective.

## Tuberculosis is caused by a slow-growing, intracellular bacterium.

Our second example is tuberculosis, also known as TB. Like the flu, it is spread through small droplets dispersed in the air when a person with active disease coughs or sneezes. Also like the flu, the TB pathogen replicates inside host cells. But unlike the flu, TB is caused by a bacterium, not a virus. TB results from infection with *Mycobacterium tuberculosis,* a bacterium with several unusual properties. It is very small with a compact genome, lacks a plasma membrane outside of the cell wall, and replicates exceedingly slowly. *M. tuberculosis* takes about 16–20 hours to divide, compared to 20 minutes for the common intestinal bacterium and model organism *E. coli.*

TB is very common. It is estimated that about a third of the world's population has TB. Symptoms are absent in most cases because the bacteria are in a dormant state, kept in check by the immune system. About 10% of cases are active, characterized by chronic cough, fever, night sweats, and the weight loss that gives TB its other name: consumption.

TB primarily affects the lungs. The bacteria enter macrophages in the alveoli of the lungs, where they replicate. The macrophages release cytokines, recruiting T and B cells to the site of infection. These lymphocytes surround the infected macrophages, forming a structure called a **granuloma (Fig. 43.19).** The granuloma helps to prevent the spread of the infection and aids in killing infected cells.

Many TB cases remain asymptomatic. However, the bacteria can sometimes overcome host defenses, especially if the immune system is compromised or suppressed, as in the case of co-infection with HIV. Treatment for active disease requires a long course of multiple antibiotics. In many areas of the world, however, TB has evolved antibiotic resistance, a growing and alarming problem.

Diagnosis poses challenges, especially in the case of asymptomatic infections. The tuberculin skin test is performed by injecting protein from the bacteria under the skin and observing whether or not a reaction, in the form of a hard, raised area, occurs. A positive skin test is an example of a delayed hypersensitivity reaction caused by the recruitment of memory T cells to the site of inoculation. Unfortunately, the test is not always accurate, and newer tests are being developed.

Developing an effective vaccine is challenging. The only one currently available is BCG (Bacillus Calmette-Guérin, named for Albert Calmette, a French bacteriologist, and Camille Guérin, a French veterinarian, who developed it early in the twentieth century). BCG is a live but weakened strain of a related bacterium that infects cows. In fact, this vaccine was developed following the success of using cowpox to immunize against smallpox. However, BCG is only sometimes effective, and a public health disaster resulting from contaminated vaccine occurred in 1930 in Germany. Consequently, it is not used for mass immunization in the United States.

## The malaria parasite uses antigenic variation to change surface molecules.

"Bad air." That's the literal translation of "malaria," a contraction of the Italian *mala* ("bad") and *aria* ("air"). But malaria is not caused by the air. Nor is it caused by a virus or bacterium. Malaria is caused by a single-celled eukaryote of the genus *Plasmodium* (Case 4: Malaria). Several species infect humans, but *P. falciparum* is the most common and most virulent.

Malaria is both devastating and common: There are about 500 million cases and 2 million deaths every year, mostly in sub-Saharan Africa, and mostly of children. Malaria causes cyclical fevers and chills and can lead to coma and even death.

**FIG. 43.19**  A granuloma in a tubercular lung.

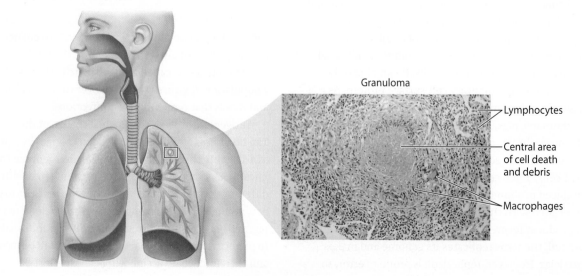

Granuloma

Lymphocytes

Central area of cell death and debris

Macrophages

**FIG. 43.20 Malaria transmission.** (a) The malaria parasite, a single-celled eukaryote, is transmitted by the bite of a mosquito. (b) Following infection, red blood cells adhere to blood vessel walls by interactions with a malaria-encoded cell-surface protein and receptors on the blood vessel wall.

a.

b.

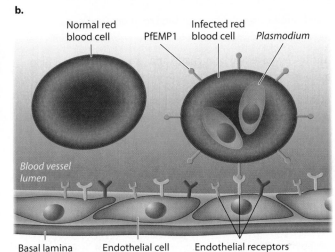

The malaria parasite is transmitted to humans by the bite of an infected mosquito (**Fig. 43.20a**). Using mosquitoes as a vector, the malaria parasite deftly bypasses the natural protective barrier provided by the skin. The malaria parasite spreads to the liver and then to red blood cells, where it completes its life cycle. The progeny can be taken up again by mosquitoes.

Malaria is an ancient parasite that has coevolved with humans (Case 4: Malaria). We have already seen how altered hemoglobin provides some protection against infection (Chapter 21). But the malaria parasite, too, has adapted—it is able to evade our immune system. It resides in liver cells and red blood cells, where it cannot be easily detected by macrophages, B cells, or T cells. Infected red blood cells can be removed by the spleen, but the malaria parasite has evolved a way to avoid this filtering action. It expresses an adhesive protein that inserts itself on the surface of red blood cells. This protein interacts with proteins on the surface of cells lining blood vessels, helping infected cells stick to blood vessel walls and

keeping them in the circulatory system and out of the spleen (**Fig. 43.20b**). The protein, PfEMP1 (*Plasmodium falciparum* erythrocyte membrane protein 1), could conceivably be a target for antibodies or TCRs. However, it is encoded by one of about 60 genes, each slightly different from the others. The parasite expresses just one of these genes at a time and can change which one is expressed, in a process called **antigenic variation.** The great and ever-changing diversity of this protein, both in a single parasite and in the population as a whole, makes it a moving target for the immune system.

Despite continuing research, there is no vaccine. Researchers hope that the sequence of the malaria genome, completed in 2002, will provide new targets for vaccine research. In the meantime, efforts to control malaria focus on measures to control mosquitoes, such as nets and insecticides, and on treatment with antibiotics. The evolution of antibiotic resistance, however, is an all too familiar problem.

# Core Concepts Summary

**43.1 THE INNATE IMMUNE SYSTEM ACTS AGAINST A DIVERSITY OF FOREIGN PATHOGENS.**

The skin is a physical barrier and contains chemicals, cells, and microorganisms that provide protection against infection. page 43-2

Phagocytes, including macrophages, dendritic cells, and neutrophils, engulf other cells and debris. page 43-3

Toll-like receptors on the surface of phagocytes play a central role in recognition of foreign molecules. page 43-4

Eosinophils, basophils, mast cells, and natural killer cells are key players in the innate immune system. page 43-4

Inflammation is the response of the body to infection or injury. It entails increases in blood flow and blood vessel permeability and recruitment of phagocytes. page 43-5

The complement system consists of circulating proteins that activate one another. Activation results in the formation of a membrane attack complex that lyses bacteria and other cells, production of proteins that enhance phagocytosis, and generation of cytokines. page 43-6

## 43.2 THE ADAPTIVE IMMUNE SYSTEM INCLUDES B CELLS THAT PRODUCE ANTIBODIES AGAINST SPECIFIC PATHOGENS.

B cells produce antibodies that bind to foreign antigens. page 43-7

Antibodies have two light chains and two heavy chains joined by covalent disulfide bonds. page 43-8

Light and heavy chains have constant and variable regions. page 43-8

Mammals produce five types of antibody: IgG, IgM, IgA, IgD, and IgE. IgG is the most abundant; IgM and IgD are found on the surface of B cells; IgA is secreted; and IgE is involved in allergic reactions. page 43-8

Antibody specificity is achieved by clonal selection, in which an antigen binds to one or a few B cells on the basis of the fit with its cell-surface antibody receptor, leading to B cell differentiation into plasma and memory cells. page 43-9

The primary immune response is characterized by a long lag period and a short response. The secondary immune response is quicker and more robust and provides the basis for immunity following infection or vaccination. page 43-10

Antibody diversity is generated by genomic rearrangement, where individual gene segments are joined to make functional genes. page 43-10

## 43.3 THE ADAPTIVE IMMUNE SYSTEM INCLUDES HELPER AND CYTOTOXIC T CELLS THAT ATTACK PATHOGENS, FOREIGN CELLS, AND DISEASED CELLS.

Two types of T cell are helper T cells and cytoxic T cells. page 43-13

Helper T cells activate other cells of the immune system, including macrophages, B cells, and cytotoxic T cells. page 43-13

Cytotoxic T cells kill altered host cells. page 43-13

T cells are activated by the binding of a T cell receptor to an antigen in association with an MHC protein. page 43-14

As T cells mature in the thymus, they undergo positive and negative selection, so only T cells that recognize self MHC, but not self antigens, survive. page 43-15

The ability to distinguish self from non-self antigens is critical for proper immune function; autoimmune diseases result when the immune system attacks cells of the host. page 43-15

## 43.4 SOME PATHOGENS HAVE EVOLVED MECHANISMS THAT ENABLE THEM TO EVADE THE IMMUNE SYSTEM.

The flu is caused by an RNA virus that can change its surface proteins through antigenic drift and antigenic shift. page 43-16

The bacterium that causes tuberculosis grows slowly and resides inside cells, particularly macrophages. page 43-17

Malaria is caused by a single-celled eukaryote with a complex life cycle having stages spent in mosquitoes and humans. The malaria parasite has evolved several mechanisms, including antigenic variation, that help it reproduce in humans without being eliminated by the immune system. page 43-18

## Self-Assessment

1. Compare and contrast features of the innate and adaptive immune systems.

2. List components of the innate immune system.

3. Describe how cells of the innate immune system recognize pathogens.

4. Explain how the four signs of inflammation relate to the changes that occur in the underlying tissues.

5. Draw a picture of an antibody molecule, labeling each of the parts. Include the light and heavy chains, the variable and constant regions, and the antigen-binding site.

6. Describe how B cells produce a great variety of different antibodies, each specific for a particular antigen.

7. Explain how immunological memory is achieved.

8. Compare and contrast the activation of T cells and B cells.

9. Describe how positive and negative selection leads to a population of T cells that react with foreign antigens but not self antigens, and what happens when tolerance to self antigens is lost.

10. Describe three mechanisms that allow pathogens to evade the immune system.

Do you understand the chapter's Core Concepts? Log in to check your answers to the Self-Assessment questions, then practice what you've learned and reinforce this chapter's concepts by working through the problems and multimedia tutorials provided there.

www.biologyhowlifeworks.com

# Biodiversity Hotspots

## RAIN FORESTS AND CORAL REEFS

When we think about mass extinctions, the demise of the dinosaurs is often the first example to spring to mind. But many conservationists believe we are in the midst of a great mass extinction at this very moment. The International Union for the Conservation of Nature has estimated that 800 plant and animal species have died out in the last 500 years, and more than 16,000 are currently at risk of extinction.

The cause of this mass die-off isn't an asteroid impact or a cataclysmic volcanic event. Instead, human activity is dramatically altering the planet and its biological diversity.

Despite our role in the extinctions, it's clearly not possible to save every single species threatened with extinction. Conservationists must make difficult choices, prioritizing which species to protect. To get the most bang for every conservation buck, ecologists have identified 25 priority areas, or biodiversity hotspots.

These hotspots are areas with disproportionately large concentrations of native species. Scientists estimate that as many as 44% of all vascular plants and 35% of vertebrate species, for example, are confined to these 25 hotspots. Those figures are impressive, considering that the identified hotspots make up just 1.4% of Earth's land surfaces.

The islands of the Caribbean—the West Indies—constitute one such biodiversity hotspot. The Caribbean Sea contains more than 7000 islands rich in biological diversity, including an estimated 6550 native plant species. All the Caribbean islands are flush with flora and fauna, but one island in particular epitomizes the concept of a biodiversity hotspot. Hispaniola—the second largest of the Caribbean islands—was the site of the first New World colonies, founded by Columbus. Today, Hispaniola is shared by two countries, Haiti to the west and the Dominican Republic to the east.

Hispaniola is only about the size of Ireland, yet it contains a striking assortment of habitats including desert, savanna, rain forests, dry forests, pine forests, coastal estuaries, and interior mountains. With such a rich abundance of habitats, Hispaniola is, in some ways, a microcosm of the planet's many ecosystems. And like the Earth's other natural places, Hispaniola faces many of the same challenges.

Of all the habitats on Hispaniola, the forests have the greatest variety of plants, animals, and other organisms. Ferns, orchids, and bromeliads—spiky relatives of the pineapple—grow in the treetops far above the forest floor. These plants, known as epiphytes, absorb water from dew, rainwater, and even the air. Their roots never reach the soil.

The forest is alive with activity. Ants march silently. Crickets and cicadas buzz and whine in the treetops. Vertebrate vocalists such as frogs and birds fill the air with chirps and whistles, making themselves known both to potential mates and rivals. Bats swoop through the darkened skies, using ultrasonic calls to locate insect prey. About 30,000 species of multicellular organisms are found on Hispaniola, and a third of those are endemic—they are found nowhere else in the world. One example is the Hispaniolan solenodon, a long-nosed mammal that resembles an overgrown shrew and secretes toxic saliva. Listed as endangered, the solenodon has flirted with extinction for decades.

> Conservationists must make difficult choices, prioritizing which species to protect. To get the most bang for every conservation buck, ecologists have identified 25 priority areas, or biodiversity hotspots.

How did all these species come to live on the island? Over tens of millions of years, organisms traveled to Hispaniola by air and by sea. Geographically separated from their ancestors on the mainland, many of these organisms evolved to become distinct species. As they adapted to fill vacant niches on the island, animals such as *Anolis* lizards, frogs, and crickets evolved an impressive diversity of forms.

In some cases, scientists have been able to explore Hispaniola's evolutionary history directly. They have found remains of insects and other species entombed in fossilized tree resin, or amber, 30–23 million years ago. Many of these preserved specimens resemble species still living on Hispaniola. Others are unlike any creatures found on the island today.

The stunning diversity of Hispaniola doesn't just exist on land. Conservationists consider the warm, shallow reefs of the Caribbean one of the world's marine biodiversity hotspots. Coral reefs in general are among the world's most biologically diverse ecosystems. Species from all recognized phyla of animals are found on reefs.

Around the world, reefs support hundreds of thousands of species, including snails, corals, lobsters, and fish. Many, but not all, of these creatures spend their entire life in this habitat. Reefs also act as vital nursery grounds for animals that live in other parts of the ocean, from sea turtles to sharks. Reef biodiversity is critical to maintaining fisheries that provide food and income for millions of people.

Hispaniola's rain forests and reefs—and indeed all the biodiversity hotspots identified by conservationists— aren't just notable for their biological richness. They have also been given priority status because they are facing serious threats.

Reefs around the world are being damaged by human activities. Overfishing and pollution take their toll on reef ecosystems. Climate change is increasing seawater temperature and altering ocean chemistry. Those changes are threatening the coral organisms that form the backbone of the reefs.

Ecosystems on land, too, face significant challenges. The Caribbean islands retain only about 11% of their original primary vegetation. Humans have cleared great swaths of land for agriculture and chopped trees to use as fuel. Humans have also introduced non-native, or invasive, species to the islands, both intentionally and accidentally. On Hispaniola, Europeans introduced the Javanese mongoose to control rats in sugarcane fields. The mongoose has devastated local populations of reptiles, amphibians, and birds.

If Hispaniola serves as a microcosm of the planet's ecosystems, it also serves as a cautionary tale. Two

**Biodiversity hotspots.** The areas highlighted in orange have particularly high species diversity.

**Coral reefs.** Coral reefs support many species but are threatened by ocean warming and acidification.

divergent paths have emerged on the single island. Haiti has historically suffered greater poverty than the Dominican Republic, and has much greater population density. Those factors put more pressure on the environment in Haiti, and have contributed to much greater deforestation on the western half of the island.

The case of Hispaniola illustrates the complex challenges facing conservationists. Solutions to environmental challenges must take into consideration the effects of poverty, culture, and other social pressures. Balancing the needs of humans with the needs of natural ecosystems is no easy task. But as a relatively small island, Hispaniola may in fact be in a better position to turn things around than larger, more politically complex countries. The world's biodiversity hotspots might yet lead the charge in finding ways for humans and the rest of the planet's inhabitants to peacefully coexist.

## ? CASE 8 QUESTIONS

*Answers to Case 8 questions can be found in Chapters 44–48.*

1. **How have coral reefs changed through time?** *See page 44-29.*
2. **How do islands promote species diversification?** *See page 46-15.*
3. **Why are tropical species so diverse?** *See page 47-21.*
4. **How has climate change affected coral reefs around the world?** *See page 48-7.*
5. **How has human activity affected biological diversity?** *See page 48-15.*

## CHAPTER 44

# ANIMAL DIVERSITY

## Core Concepts

**44.1** The animal tree of life consists of more than a million species.

**44.2** The earliest branches of the animal tree include sponges, cnidarians, ctenophores, and placozoans.

**44.3** Bilaterians, including protostomes and deuterostomes, have bilateral symmetry and develop from three germ layers.

**44.4** Vertebrates have a bony cranium and vertebral column, and include fish, reptiles, birds, and mammals.

**44.5** Animals first evolved more than 600 million years ago in the oceans, and by 500 million years ago, the major structural and functional body plans of animal phyla were in place.

Biologists have described about 1.8 million species of eukaryotic organisms from the world's forests, deserts, grasslands, and oceans. Of these, about 1.3 million species are animals. There is reason to believe that animal diversification began in the oceans, but today most animal species are found on land. The majority of all known animal species are insects.

In previous chapters, we saw how the organ systems of animals enable them to move, feed, and behave in different ways. Clearly, this wide range of functions has permitted animals to diversify to a degree unmatched by plants, fungi, algae, or protozoans. And just as clearly, species diversity is unevenly distributed among major animal groups. How can we bring phylogenetic order to animal diversity, and how, in turn, can we understand the biological traits that have given rise to such great diversity on a few branches of the animal tree of life?

## 44.1 A TREE OF LIFE FOR MORE THAN A MILLION ANIMAL SPECIES

As discussed in Chapter 23, it is relatively easy to understand how the visible features of humans, chimpanzees, and gorillas show them to be more closely related to one another than any of them is to other animal species. Furthermore, it isn't hard to see that humans, chimpanzees, and gorillas are more closely related to monkeys than they are to lemurs, and that humans, chimpanzees, gorillas, and monkeys are more closely related to lemurs than to horses. Anatomy and morphology reveal key evolutionary relationships among vertebrate animals, but how do we come to understand the place of vertebrates in a broader tree that includes all animals? More generally, how can we construct an animal phylogeny that includes organisms with body plans as different as those of sponges, sea stars, earthworms, and mussels?

### Phylogenetic trees propose an evolutionary history of animals.

The anatomy of animals ranges from simple (think of sponges) to remarkably complex (lobsters, for example, or humans). All animals are multicellular heterotrophs, gaining both food and energy from organic molecules (Chapter 6). Most fungi are also multicellular heterotrophs (Chapter 34), but fungi have cell walls and animals do not. Animals are also distinguished by embryos that include a gastrula stage (Chapter 42) and by the presence of collagen, a fiberlike protein that makes up connective tissues such as basal lamina, tendon, and cartilage.

While they share some general features, animals vary widely in many respects. For example, they differ in the ways that they gather food. Tiny animals called placozoans that live on the seafloor absorb organic compounds directly from the environment and take in particulate matter by phagocytosis. In contrast, bilaterally symmetrical animals from insects to eagles have highly adapted nervous and muscular systems that enable them to capture plants or other animals and digest them in a specialized gut.

**FIG. 44.1    A simple phylogenetic tree of animals.** The tree shows increasing complexity from the closest protistan relatives of animals, the choanoflagellates, to bilaterian animals such as insects and mammals.

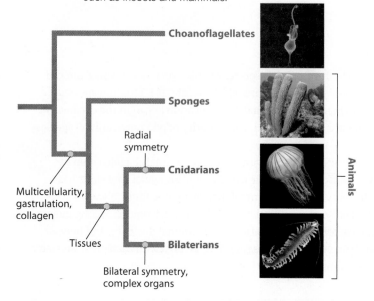

Choanoflagellates

Sponges

Radial symmetry

Cnidarians

Multicellularity, gastrulation, collagen

Tissues

Bilaterians

Bilateral symmetry, complex organs

Animals

Sponges have no nerve cells, and sea anemones have only simple networks of nerves, but humans have brains containing trillions of cells that permit us to make and communicate discoveries about sponges, sea anemones, and other organisms we encounter. How the diverse morphologies, anatomies, and behaviors that characterize animal species came to be is a question of evolution, and phylogenetic trees provide hypotheses about the evolution of animal diversity.

Early in the history of biology, taxonomists such as Linnaeus recognized that for all their diversity, animals can be sorted into a limited number of distinctive body plans, each group today formally called a **phylum** (plural, phyla). The problem for biologists has long been how to understand the evolutionary relationships among animal phyla, still an active field of research today.

**Fig. 44.1** shows a simple phylogenetic tree. Recall the basic logic of phylogenies, presented in Chapter 23. The fundamental mechanism by which biological diversity increases is speciation, the divergence of two populations from a common ancestor. Played out repeatedly through time, speciation gives rise to a treelike pattern of evolutionary relatedness, with more closely related taxa branching from points close to the tips of the tree and more distantly related taxa diverging from branch points nearer its base. Distinctive features of organisms, called characters, evolve throughout evolutionary history, and the time when they arose can be estimated from their shared presence in the descendants of the population in which the character first evolved. But a prominent feature of evolution is change, and in fact many characters change from one form to another and sometimes back again. Thus, mammals are descended from

egg-laying ancestors, but most living mammals do not lay eggs—that character has been lost, replaced by live birth.

A character can evolve more than once in separate groups by convergent evolution. Such characters, called analogies, include the single-lens eye in octopus and vertebrates, and wings in birds and bats. These characters are not helpful in reconstructing evolutionary relationships. Close studies of morphological and molecular characters can often distinguish analogies from similarities that result from common ancestry, called homologies, which are useful in understanding evolutionary relationships. A large number of studies have resolved many but not all of the evolutionary relationships among animals.

With these points in mind, Fig. 44.1 proposes that animals are closely related to protists called choanoflagellates (Chapter 27), but differ from them in the presence (among other features) of persistent multicellularity, the formation of a gastrula during early development, and the synthesis of collagen. **Poriferans** (the phylum that is made up of sponges) diverged from other animals early, and **cnidarians** (the phylum that contains jellyfish and sea anemones) diverged later from animals whose descendants have well-defined and complex organs. Note that this diagram does not tell us that sponges are older than more complex animals. Instead, it says that sponges and more complex animals diverged from a common ancestor but doesn't tell us what that common ancestor looked like. Similarly, cnidarians and more complex animals with bilateral symmetry diverged from their last common ancestor at a single point in time and so are equally old, but structurally distinct.

We'll use phylogeny as our framework for discussing the immense diversity of animals, but first, we need to consider how the tree itself has taken shape through more than a century of biological research.

### Nineteenth-century biologists grouped animals by anatomical and embryological features.

Since the earliest days of comparative anatomy, it has been clear that animals vary in their structural complexity. Sponges have only a few cell types, and those cells are not organized into tightly coordinated tissues or organs. Jellyfish and sea anemones have simple tissues but not complex organs. Crabs and alligators have complex organs that govern movement, behavior, digestion, and gas exchange.

Organisms also differ in their symmetry. Sponges are often very irregular in form. Jellyfish and sea anemones display **radial symmetry,** meaning that their bodies have an axis that runs from mouth to base with many planes of symmetry through this axis (**Fig. 44.2a**). This organization allows jellyfish to move up and down in the water column by flexing muscles around their bell-like bodies, and permits sea anemones and corals to wave their ring of food-gathering tentacles in all directions at once. Most simple animals with radial symmetry are grouped together on one branch of the animal tree, in the phylum Cnidaria.

Other organisms show **bilateral symmetry:** Their bodies have a distinct head and tail, marking front and back, with a

**FIG. 44.2** **Symmetry in animal form.** (a) Radial or (b) bilateral symmetry defines distinct groups of animals.

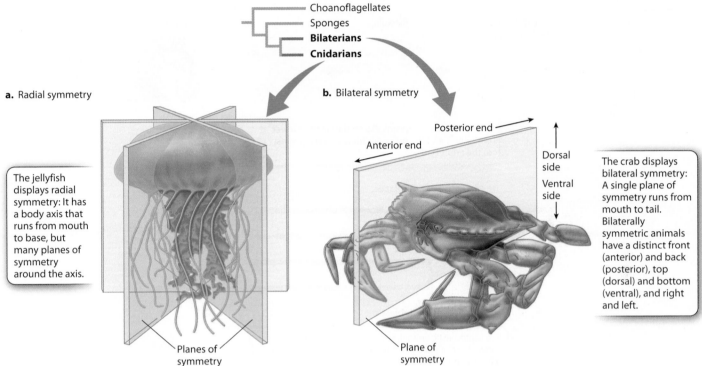

**a.** Radial symmetry

The jellyfish displays radial symmetry: It has a body axis that runs from mouth to base, but many planes of symmetry around the axis.

Planes of symmetry

**b.** Bilateral symmetry

Posterior end

Anterior end

Dorsal side

Ventral side

The crab displays bilateral symmetry: A single plane of symmetry runs from mouth to tail. Bilaterally symmetric animals have a distinct front (anterior) and back (posterior), top (dorsal) and bottom (ventral), and right and left.

Plane of symmetry

single plane of symmetry running between them at the midline (**Fig. 44.2b**). Bilateral symmetry enables animals to move in one horizontal direction to capture prey, find shelter, or escape from enemies. It also allows the development of specialized sensory organs at the front end for guidance, and specialized appendages along both sides for locomotion, sensing the environment, or defense. Animals with bilateral symmetry cluster together on one branch of the animal tree and are called **Bilateria.**

Biologists understood early that cnidarians are simpler than animals with bilateral symmetry. Bilaterian animals develop, to varying degrees, digestive and circulatory systems, nervous systems, and muscles. From these observations, early biologists concluded that sponges branched near the base of the animal tree, cnidarians in the middle, and bilaterians closer to the top.

Within the bilaterian group, however, understanding evolutionary relationships proved more difficult. Some biologists pointed to the presence or absence of a cavity, or coelom, in tissues surrounding the gut, dividing bilaterians into three groups: **acoelomates** (without a body cavity), **coelomates** (with a body cavity), and **pseudocoelomates** (with a body cavity that does not completely surround the internal organs). These groups are shown in **Fig. 44.3.** A fluid-filled body cavity cushions the internal organs against hard blows to the body and enables the body to turn without twisting these organs. A body cavity also allows internal organs like the stomach to expand, enhancing digestive function.

Early biologists also noted that some animals, notably insects and earthworms, show a striking pattern of body segmentation,

but others do not. These variations in anatomy were critically important in defining phyla within bilaterian animals, but in the end they did not provide a way to understand the details of evolutionary relationships among these phyla.

New insights became possible with the advent of microscopes that enabled the direct study of development from its earliest

**FIG. 44.3** **Major types of anatomical organization of bilaterian animals.** Bilaterians vary in the presence, absence, or structure of a body cavity.

Digestive tract

Cavity

Acoelomate

Pseudocoelomate

Coelomate

Acoelomates do not have a cavity outside the digestive tract.

Pseudocoelomates have a body cavity that does not completely surround the internal organs.

Coelomates have a body cavity that completely surrounds the internal organs.

**FIG. 44.4    Variation in embryonic development.** Early development of germ tissues in the embryo separates the diploblastic cnidarians from the triploblastic bilaterians.

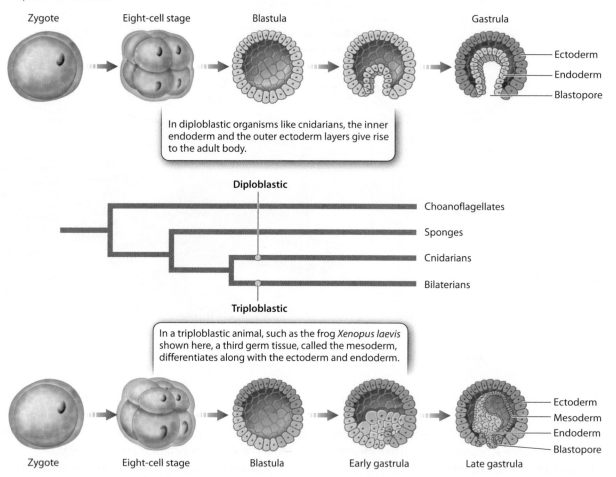

In diploblastic organisms like cnidarians, the inner endoderm and the outer ectoderm layers give rise to the adult body.

In a triploblastic animal, such as the frog *Xenopus laevis* shown here, a third germ tissue, called the mesoderm, differentiates along with the ectoderm and endoderm.

stages. One important observation was that some animals that look very different as adults share patterns of early embryological development. For example, adult sea stars and catfish have few morphological features in common, but their embryos show a number of key similarities, including the number of embryonic tissue layers they develop and the way that the early cell divisions occur. In the Cnidaria, the embryo has two germ layers, termed endoderm and ectoderm, from which the adult tissues develop. These animals are therefore called diploblastic (**Fig. 44.4**). The bilaterian animals are triploblastic, with a third germ layer, the mesoderm, lying between the endoderm and ectoderm, and developing into muscles and connective tissues (Fig. 44.4).

Comparative embryology also enabled biologists to divide bilaterian animals into the two groups shown in **Fig. 44.5,** called **protostomes** (from the Greek for "first mouth") and **deuterostomes** (from the Greek for "second mouth"). In protostomes, the first opening to the internal cavity of the developing embryo, called the blastopore, becomes the mouth. In deuterostomes, the blastopore becomes the anus. However, only in the age of molecular sequence comparisons have the relationships among phyla within each of these two groups become clear.

### Molecular sequence comparisons have confirmed some relationships and raised new questions.

Over the past two decades, comparisons among DNA, RNA, and amino acid sequences have greatly improved our understanding of evolutionary relationships among major groups of animals. Molecular comparisons support many of the conclusions reached earlier on the basis of comparative anatomy and embryonic development, including the early divergence of sponges. Other hypotheses have been rejected. For example, the traditional phylogenetic division of bilaterians into acoelomate, coelomate, and pseudocoelomate groups gets little support from molecular studies. Neither do molecular sequence comparisons support the once widespread view that segmented bodies indicate a close relationship between earthworms and lobsters.

Some of the biggest surprises provided by molecular data concern the relationships of phyla within the protostomes, showing that the protostome phyla can be divided into two subgroups, the **lophotrochozoans** and the **ecdysozoans** (Fig. 44.5). As we will see, biologists have identified developmental and anatomical characters that unite the phyla on each branch, but the phylogenetic significance of these features was recognized

## FIG. 44.5 A simple phylogenetic tree of bilaterian animals.

Embryological features and similarities in the molecular sequences of genes show that bilaterian animals can be divided into two major groups, the deuterostomes and protostomes. Protostomes can be further subdivided into lophotrochozoans and ecdysozoans.

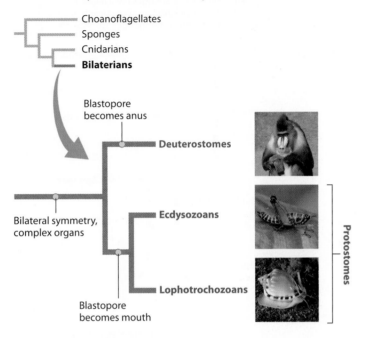

only after gene sequences revised the thinking about protostome animals. Molecular sequence comparisons have also overturned traditional views on relationships among the deuterostome phyla, discussed later in this chapter.

→ **Quick Check 1** Which animals are more closely related to sponges, Cnidaria or Bilateria?

## 44.2 THE SIMPLEST ANIMALS: SPONGES, CNIDARIANS, CTENOPHORES, AND PLACOZOANS

The human body is complex, our diverse functions made possible by sophisticated, interacting organ systems: from muscle pairs attached to skeletons that allow us to run, jump, and dance to lungs that exchange gases with the atmosphere and a brain and nervous system that enable us to sense our environment and respond to it. None of these organ systems is well developed in early-branching animal groups. How then do sponges, cnidarians, ctenophores, and placozoans function, and what are the consequences of their anatomical organization for diversity and ecology? On the other hand, these animals have survived a half billion years of Earth history. What can we learn from the diversity and continuity of their remarkable adaptations?

### Sponges are simple and widespread in the oceans.

As we discussed in Chapters 27 and 28, the unicellular ancestors of animals carried out all functions from metabolism to reproduction as individual cells. The cells making up sponges retain much of this independence, while also having some benefits of multicellularity. Coordination among cells means that there can be distinct kinds of cell specialized for different functions—cells that form a protective skin, for example, and others that secrete enzymes for digestion. Most dramatically, multicellularity permits sponges to extend above the seafloor, giving them access to food suspended in water currents.

The sponge body plan resembles a flower vase, with many small pores along its sides and a larger opening at the top (**Fig. 44.6**). The cells on the outer surface of the sponge are tough and act as the sponge's skin. The interior surface is lined by cells called **choanocytes** that have flagella and function in nutrition and gas exchange. Choanocyte cells have a collar of small cilia around their flagellum, much like the cells of our closest unicellular relatives, the choanoflagellates (Chapter 27). Between the interior and exterior cell layers lies a gelatinous mass called the **mesohyl** (Fig. 44.6). Mesohyl is mostly noncellular, but it contains some amoeba-like cells that function in skeleton formation and the dispersal of nutrients.

Sponges gain nutrition by intracellular digestion (Chapter 40). The choanocytes that surround the interior chamber of the sponge

### FIG. 44.6 Anatomy of a sponge.

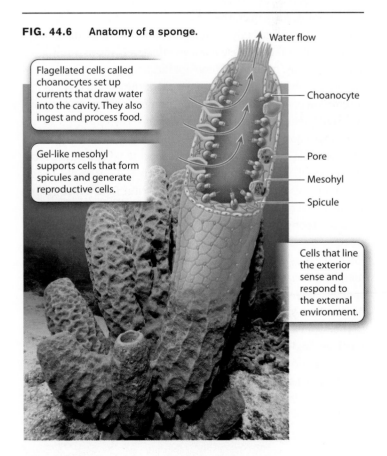

Flagellated cells called choanocytes set up currents that draw water into the cavity. They also ingest and process food.

Gel-like mesohyl supports cells that form spicules and generate reproductive cells.

Water flow

Choanocyte

Pore

Mesohyl

Spicule

Cells that line the exterior sense and respond to the external environment.

beat their flagella, creating a current that draws water from outside the body, through the pores in its walls and upward through the central cavity of the sponge, where it exits through the large opening at the top. The circulating water contains food particles and dissolved organic matter, which cells lining the cavity capture by endocytosis. The individual cells then metabolize the food in their interiors, much the way that protozoans feed.

From this observation, we might conclude that sponges are simply a group of uncoordinated cells, each working for itself, but that isn't the case. The choanocyte flagella beat in a coordinated pattern, helping to draw water into the body interior through the pore system and outward again through the vase opening. Moreover, the shape of the sponge body itself directs water movement across feeding cell surfaces. These features effectively move water across the cells, facilitating food uptake. Sponge cells require oxygen for respiration and must get rid of the carbon dioxide respiration generates. Gas exchange occurs by diffusion, aided by the movement of water through the sponge cavity.

Sponges don't have highly developed reproductive organs. Instead, cells recruited from the choanocyte layer migrate into the mesohyl, where they undergo meiosis and differentiate as sperm or eggs. Sperm released into the water fuse with eggs in the mesohyl of other sponges.

Many sponges build skeletons of simple structures called spicules. Some sponges precipitate spicules of glass-like silica ($SiO_2$). Others, however, make their spicules—and sometimes massive skeletons—of calcium carbonate ($CaCO_3$). Still other sponges have a skeleton made up of proteins (these are often used as bath sponges). So we see that sponges are intermediate in the sophistication of their bodies. Their cells function much as single-celled protozoa do, but they coordinate their activities and so are more efficient in extracting food and oxygen from seawater.

**FIG. 44.7**  **Two hypotheses for the phylogenetic relationships of sponges to other animals.** One group of sponges, the homoscleromorphs, has epithelium-like cell sheets. In the left-hand hypothesis, this character is interpreted as a convergence with eumetazoans; in the right-hand hypothesis, it is viewed as a shared derived character for a homoscleromorph sponge + eumetazoan group.

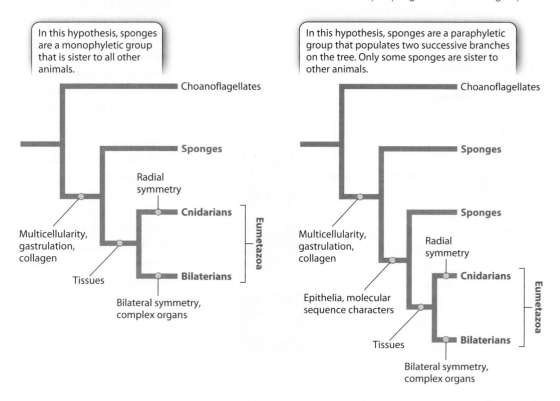

Despite their anatomic simplicity, sponges are major contributors to seafloor communities. Approximately 9000 species of sponge have been described, most of them ocean dwelling, but a few that live in freshwater lakes. Many sponges obtain at least part of their nutrition from symbiotic algae living within their bodies. In fact, all sponges are full of bacterial cells. Many of these are probably symbionts, but only a few experiments have demonstrated their function.

Until recently, most biologists regarded sponges as a monophyletic group, one that includes all the descendants of a common ancestor. Some molecular data, however, now suggest that sponges are paraphyletic—that is, that they contain some but not all the descendants of their common ancestor (Chapter 23).

**Fig. 44.**7 shows these two possibilities. If the tree on the left is correct, the two main limbs of the animal tree contain very different organisms—the simple sponges on one branch and all more complex animals on the other. As noted earlier, this figure indicates that sponges are distinct from other animals but leaves open the question of what ancestral animals were like. In contrast, the hypothesis illustrated by the tree on the right suggests that the last common ancestor of all living animals, in essence, *was*

a sponge. At present, biologists are divided on this issue; more research is needed to demonstrate how sponges relate to each other and to other animals, which are grouped together as the **Eumetazoa** (Fig. 44.7).

### Cnidarians are the architects of life's largest constructions: coral reefs.

Many of us have encountered at least a few cnidarians. Jellyfish flourish in marine environments from coastlines to the deep sea; sea anemones cling to the seafloor, extending their tentacles upward to gather food and deter predators; corals secrete massive skeletons of calcium carbonate, forming reefs that fringe continents and islands in tropical oceans.

Jellyfish and sea anemones look strikingly different, but they share a body plan common to all cnidarians (**Fig. 44.8**). At one end of the radially symmetrical body is a mouth surrounded by tentacles armed with stinging cells that subdue prey and defend against enemies. An anemone is like a jellyfish stuck upside down onto the seafloor with its tentacles and mouth facing upward. We call a free-floating jellyfish a medusa, and the sessile form of an anemone a polyp. Some cnidarians develop into one form or the other, but many have life cycles that alternate between the two.

The mouth of both medusa and polyp opens into a closed internal **gastric cavity,** the site of extracellular digestion and excretion. Rather than digesting food particles inside individual cells, cnidarians receive both food and digestive enzymes in the gastric cavity. (The digestive enzymes are secreted by cells that line the walls of the cavity.) This arrangement permits cnidarians to digest large food items, such as a whole fish (paralyzed by the stinging cells on the tentacles), whose dissolved nutrients can then be absorbed through the walls of the cavity. Cnidarians, then, can consume many types of food that are unavailable to sponges.

The cnidarian body develops from a diploblastic embryo (one with two germ layers). It has an outer layer, the **epidermis,** which develops from the ectoderm, and an inner lining, the **endodermis,** that derives from the endoderm. These tissues enclose a gelatinous mass called the **mesoglea** (the "jelly" of jellyfish). This organization sounds a bit like the sponge body plan, but there are important differences.

First, in cnidarians, the cells that form the epidermis and endodermis occur as closely packed layers of cells embedded in a protein-rich matrix, forming an epithelium (Chapter 10). Epithelia line compartments (like the gut) within animal bodies, but they do not occur in most sponges. These tissues are formed from an unusually tight layer of specialized cells, each connected to its neighbors with specialized junctions that regulate the passage of ions or other molecules. Epithelial layers often absorb or secrete substances into the compartments they line, making possible organs such as those of the digestive system (Chapter 40).

Second, cnidarians have a wider array of cell types than sponges do, permitting more sophisticated tissue function. Muscle cells allow jellyfish to swim through the ocean. A simple network of nerve cells permits cnidarians to sense their environment and respond to it by directional movement. No such cells occur in sponges. Cnidarians don't have a brain, but at least a few have light-sensitive cells that function as simple eyes.

**FIG. 44.8**  **Two cnidarians:** (a) jellyfish and (b) sea anemones. All cnidarians have a similar organization.

**a.** Jellyfish (medusa)

- Epidermis
- Mesoglea
- Endodermis
- Gastric cavity
- Mouth
- Tentacle

Note the position of the mouth in the two forms.

**b.** Sea anemone (polyp)

- Mouth
- Gastric cavity
- Tentacle
- Epidermis
- Mesoglea
- Endodermis

Moreover, whereas sponges filter water to gain food, cnidarians are predators, capturing prey with their tentacles and digesting it in the gastric cavity. Specialized cells on the tentacles contain a tiny harpoon-like organelle called a nematocyst, often tipped with a powerful neurotoxin that greatly aids prey capture and defense against other predators (**Fig. 44.9**). Other specialized cells lining the gastric cavity secrete digestive enzymes that break down ingested food into molecules that can be taken up by cells lining the cavity by endocytosis. There is no specialized passage for waste removal. Instead, waste is excreted back into the gastric cavity and leaves by way of the mouth. Oxygen uptake and carbon dioxide release occur by diffusion.

Many cnidarians reproduce asexually to form colonies. For example, corals form extensive rounded, fan-shaped, or hornlike colonies by budding, building potentially massive structures on the seafloor (**Fig. 44.10a**). In the Portuguese man-of-war, different individuals in the same colony develop distinct morphologies, some specialized for flotation, others for prey capture, and still others for reproduction (**Fig. 44.10b**). Among animals without complex organs, such colonies represent the height of morphological complexity.

About 9000 species of cnidarians live in the oceans. Reef corals are remarkable not only for the size of their skeletal colonies, but also for their mode of nutrition. Most have lost the ability to capture prey and instead gain nutrition from symbiotic algae in their surface tissues.

**FIG. 44.9    Nematocysts.** Nematocysts are harpoon-like organelles that capture prey.

Harpoon-like nematocysts are microscopic structures found in the epidermis of cnidarians.

Tentacle

**1** Contact triggers the ejection of nematocysts.

"Trigger"

Coiled thread

Nematocyst

**2** Nematocysts of some cnidarians inject toxins into predators or prey.

**FIG. 44.10    Colony formation in cnidarians.** (a) In soft corals, the individual members of the colony are similar in form and function. (b) In the Portuguese man-of-war, the members of the colony differentiate to serve distinct functions, such as floating, feeding, and reproduction.

a

b

## Ctenophores and placozoans represent the extremes of body organization among early branching animals.

**Ctenophores**, or comb-jellies, resemble cnidarians in body plan, and for many years most biologists thought that the two were close relatives. Like cnidarians, comb-jellies have radial symmetry, with an outer epithelium and an inner endodermis that enclose a gelatinous interior (**Fig. 44.11**). Comb-jellies also differentiate muscle cells and a simple nerve net, as well as rudimentary gonads. They are predators that feed by ingestion, digesting prey within their gastric cavity by enzymes secreted from the cells lining the gut cavity. Gas exchange occurs by diffusion. However, there are important differences between cnidarians and comb-jellies.

Comb-jellies propel themselves through the oceans by the coordinated beating of cilia that extend from epidermal cells. These cilia are usually arranged in comb-like groups,

## FIG. 44.11 Ctenophores, or comb-jellies.

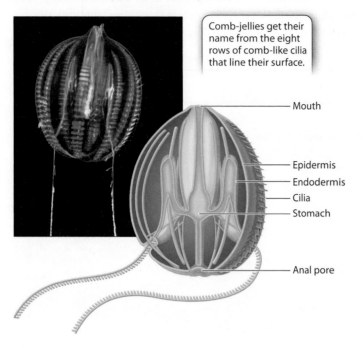

Comb-jellies get their name from the eight rows of comb-like cilia that line their surface.

Mouth

Epidermis

Endodermis

Cilia

Stomach

Anal pore

hence the common name for this phylum. Some but not all comb-jellies also have a pair of long tentacles that aid in feeding. Like cnidarians, comb-jellies have specialized cells on their tentacles, but in comb-jellies, the cells secrete adhesive molecules that entangle prey.

Unlike cnidarians, digestive wastes generated by comb-jellies move through a gut cavity for elimination through an anal pore opposite the mouth. This simple flow-through gut provides comb-jellies with an anterior–posterior axis that cnidarians do not have. This arrangement allows each section of the gut to be specialized

for a particular function, such as temporary storage (near the mouth), digestion (in the middle), and absorption and excretion (at the rear), increasing the overall efficiency in processing food. Cnidarians digest and absorb and store food in the same pouch, and so are much less efficient.

Also, the pattern of muscle cell development in comb-jellies suggests that these cells originate from a rudimentary mesoderm situated between the ectoderm and endoderm. Flow-through guts, mesoderm, and axial symmetry are characteristic of bilaterian animals, as is another feature of comb-jellies—the use of acetylcholine as a neurotransmitter (Chapter 35). This unique combination of traits makes placement on the tree of life difficult. About 100–150 species of comb-jelly have been named, and photographs from deep-sea submersibles make it clear that many more remain to be discovered in the oceans.

→ **Quick Check 2** Which would you expect to use available nutrients more effectively and reproduce more rapidly, a jellyfish or a sponge of the same size?

If comb-jellies are the most complex animals discussed so far, **placozoans** are the simplest. These tiny (millimeter-scale) animals each contain only a few thousand cells arranged into upper and lower epithelia that sandwich an interior fluid crisscrossed by a network of multinucleate fiber cells (**Fig. 44.12**). Placozoans can absorb dissolved organic molecules, but commonly feed by surrounding food particles and secreting digestive enzymes to break them down. Individual cells then bring food particles in by endocytosis. Placozoans have no specialized tissues and few differentiated cell types. Cilia on cell surfaces allow movement, and gas exchange occurs by diffusion. Placozoans reproduce asexually but can also form egg and sperm cells for sexual reproduction.

## FIG. 44.12 The placozoan, *Trichoplax adhaerens*.

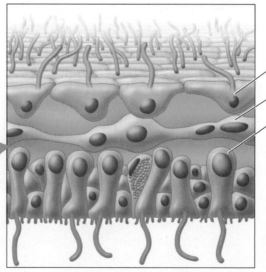

Upper epithelium

Multinucleated fiber cell

Lower epithelium

Placozoans may be the simplest of all animals, consisting of a few thousand cells organized into upper and lower epithelia, with multinucleated fiber cells in between.

Despite their morphological simplicity, placozoans have a genome that contains many of the genes for transcription factors and signaling molecules that are present in cnidarians and bilaterian animals. What roles these genes play in placozoan biology remains unclear. A single species of placozoan, *Trichoplax adhaerens*, is known. First observed adhering to the inner wall of a marine aquarium, *Trichoplax* occurs widely in tropical and subtropical oceans. A second species was described in 1896 but has not been

**TABLE 44.1   Animal Phyla.**

| PHYLUM | COMMON NAME | LIMB OF ANIMAL TREE | APPROXIMATE NUMBER OF DESCRIBED SPECIES |
|---|---|---|---|
| Acanthocephala | Thorny-headed worms | Lophotrochozoa | 1000 |
| Acoelomorpha | Acoel flatworms | Basal Bilateria (?) | 350 |
| Annelida | Segmented worms | Lophotrochozoa | 15,000 |
| Arthropoda | Arthropods | Ecdysozoa | 800,000–1,000,000 |
| Brachiopoda | Lamp shells | Lophotrochozoa | 350 |
| Bryozoa | Moss animals, sea mats | Lophotrochozoa | 5000 |
| Chaetognatha | Arrow worms | Protostomia | 100 |
| Chordata | Chordates | Deuterostomia | 50,000 |
| Cnidaria | Corals, jelly fish, etc. | Early branching Eumetazoa | 9000 |
| Ctenophora | Comb-jellies | Early branching Eumetazoa | 100–150 |
| Cycliophora | *Symbion* | Lophotrochozoa | At least 3 |
| Echinodermata | Echinoderms | Deuterostomia | 7000 |
| Echiura | Spoon worms | Lophotrochozoa | 150 |
| Entoprocta | Goblet worms | Lophotrochozoa | 150 |
| Gastrotricha | Hairy backs | Lophotrochozoa | 700 |
| Gnathostomulida | Jaw worms | Lophotrochozoa | 100 |
| Hemichordata | Acorn worms, pterobranchs | Deuterostomia | 90 |
| Kinorhyncha | Mud dragons | Ecdysozoa | 150 |
| Loricifera | Brush heads | Ecdysozoa | 120 |
| Micrognathozoa | Jaw animals | Lophotrochozoa | 1 |
| Mollusca | Mollusks | Lophotrochozoa | 80,000 |
| Nematoda | Roundworms | Ecdysozoa | 20,000 |
| Nematomorpha | Horsehair worms | Ecdysozoa | 320 |
| Nemertea | Ribbon worms | Lophotrochzoa | 1200 |
| Onychophora | Velvet worms | Ecdysozoa | 75 |
| Phoronida | Horseshoe worms | Lophotrochozoa | 20 |
| Placozoa | *Trichoplax adhaerens* | Early branching Eumetazoa | 1 |
| Platyhelminthes | Flatworms | Lophotrochozoa | 25,000 |
| Porifera | Sponges | Basal animals | 9000 |
| Priapulida | Priapulid worms | Ecdysozoa | 17 |
| Rhombozoa | — | Lophotrochozoa | 75 |
| Rotifera | Rotifers | Lophotrochozoa | 2000 |
| Sipuncula | Peanut worms | Lophotrochozoa | 350 |
| Tardigrada | Water bears | Ecdysozoa | More than 600 |
| Xenoturbellida | — | Deuterostomia | 2 |

**FIG. 44.13** **A phylogenetic tree of animals, with placozoans and ctenophores added.** The placements of sponge subgroups, placozoans, and ctenophores remain uncertain and should be treated as hypotheses to be tested.

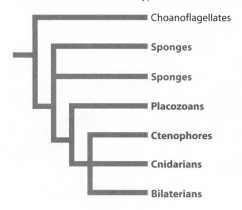

Choanoflagellates
Sponges
Sponges
Placozoans
Ctenophores
Cnidarians
Bilaterians

observed since that time. **Fig. 44.13** shows one current view of the phylogenetic relationships among sponges and eumetazoans. It should be regarded as a set of hypotheses to be tested against new data rather than an unambiguously "correct" tree. See **Table 44.1** for a list of all currently recognized animal phyla.

## 44.3 BILATERIAN ANIMALS

The phyla discussed in the preceding section include about 20,000 species, a tiny fraction of all animal diversity. The final stop in this broad tour is the bilaterian branch, on which most animal species reside. Earlier, we pointed out that basic patterns of embryo formation divide bilaterians into two major groups, the Protostomia and the Deuterostomia (see Fig. 44.5). Molecular sequence data support the protostome–deuterostome groupings of bilaterian animals and show that the protostome animals can be further divided into two groups: the Lophotrochozoa (which include mollusks and annelid worms) and the Ecdysozoa (which include insects and other arthropods).

As noted above, bilaterian animals have bilateral symmetry and complex organs that develop from a triploblastic (three-germ layer) embryo. The anatomical complexity of bilaterian animals makes possible types of locomotion, feeding, gas exchange, behavior, and reproduction that are unknown in earlier branching groups. These capabilities underpin the remarkable diversity of bilaterian animals.

**Lophotrochozoans make up nearly half of all animal phyla, including the diverse and ecologically important annelids and mollusks.**

The Lophotrochozoa are a group of animals first recognized on the basis of unique molecular characters and now known to include several very different types of animal. On the one hand, there are the brachiopods, Bryozoa, and related phyla that have

a tentacle-lined organ for filter feeding called a lophophore. Then, there are the mollusks, annelid worms, and their close relatives, characterized by a type of larva called a trochophore. The odd name "lophotrochozoan" is a contraction of the terms "lophophore" and "trochophore," with "zoan" added to indicate that these organisms are animals (*zoon* is Greek for "animal"). The majority of lophotrochozoan species also have a distinctive form of spiral cleavage early in development.

The Lophotrochozoa contains 17 phyla, mostly small marine animals of limited diversity, but it also includes the diverse and ecologically important **annelid worms** and **mollusks** (**Fig. 44.14**). We'll focus on these two phyla, which are closely

**FIG. 44.14** **A phylogenetic tree for lophotrochozoan animals, showing all 17 phyla in this group.** Phyla discussed in the text are shown in red. Uncertainties are shown by nodes with more than two branches.

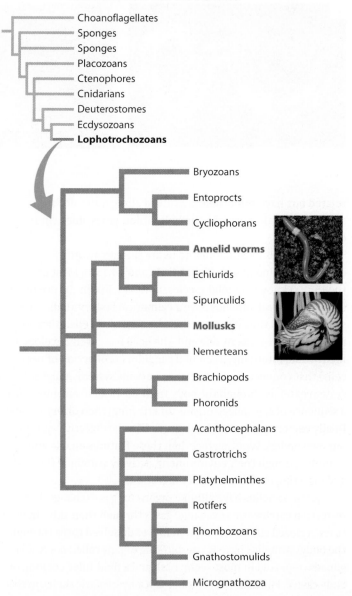

Choanoflagellates
Sponges
Sponges
Placozoans
Ctenophores
Cnidarians
Deuterostomes
Ecdysozoans
**Lophotrochozoans**

Bryozoans
Entoprocts
Cycliophorans
**Annelid worms**
Echiurids
Sipunculids
**Mollusks**
Nemerteans
Brachiopods
Phoronids
Acanthocephalans
Gastrotrichs
Platyhelminthes
Rotifers
Rhombozoans
Gnathostomulids
Micrognathozoa

**FIG. 44.15    Annelids.** (a and b) Earthworms typify the segmented body plan found throughout the Annelida. (c) The most diverse annelids are polychaete worms, which live in the oceans. (d) Leeches are specialized for ingesting blood meals from mammal hosts. (e) Vestimentiferan worms live along hydrothermal rift vents in the oceans, gaining nutrition from chemosynthetic bacteria in specialized organs (red in the image).

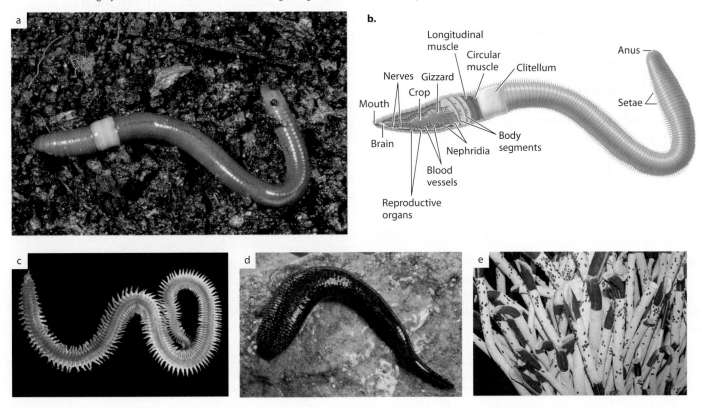

related but have nonetheless evolved along quite different anatomical trajectories in the half billion years since their origins.

The annelids most familiar to us are the earthworms, commonly found in soil (**Figs. 44.15a** and **44.15b**). Most of the 15,000 known annelid species, however, live in the oceans (**Fig. 44.15c**). All annelids have a cylindrical body with distinct segments, and they illustrate the advantages of a bilaterian body plan (see Fig. 44.2b). At one end, the head has a well-developed mouth and, internally, a cerebral ganglion (a collection of nerve cells) that connects to an extensive nervous system. A digestive system extends through the body from the head to an anus, with a sequence of specialized organs for crushing, then digesting, and finally excreting ingested food, much like the digestive system in our own bodies. Working together, these features enable annelids to move through their environment, actively searching for food and digesting it efficiently after ingestion.

Aquatic annelids have gill-like organs for gas exchange, but terrestrial earthworms exchange gases through their skin. In both cases, a closed circulatory system moves dissolved gases through the body. Annelids have waste-filtering organs called nephridia, gonads (repeated in most segments), and a fluid-filled coelom, or body cavity. Fluids in the coelom form a hydrostatic skeleton that

works in coordination with paired muscles in each segment to direct movement.

Many annelids are predators that capture and ingest prey, but some, like earthworms, ingest sediment, digesting the organic matter it contains and excreting mineral particles. A few marine annelids have evolved tentacles that enable them to filter food particles from water, while leeches, a specialized group of freshwater annelids, attach themselves to vertebrates to suck out a meal of blood (**Fig. 44.15d**). And, remarkably, the meter-long vestimentiferan worms that live around hydrothermal vents in the oceans have given up ingestion altogether (**Fig. 44.15e**). Without mouths, these enormous worms gain nutrition from chemosynthetic bacteria that live within a collar of specialized tissue. Discovered in the 1970s, vestimentiferan worms were originally placed in their own phylum, but molecular sequence comparisons indicate that they are annelids.

Mollusks are the second major phylum of the Lophotrochozoa. Mollusks develop a distinctive larva called a trocophore that has a tuft of cilia at its top and additional cilia bands around its middle. Marine annelids develop from trochophores as well, and the presence of this unique form of larva in both phyla suggests a close evolutionary relationship between the two. This hypothesis is supported by molecular sequence comparisons. Nonetheless,

**FIG. 44.16** A gastropod—here, a snail—showing a body plan typical of the phylum Mollusca.

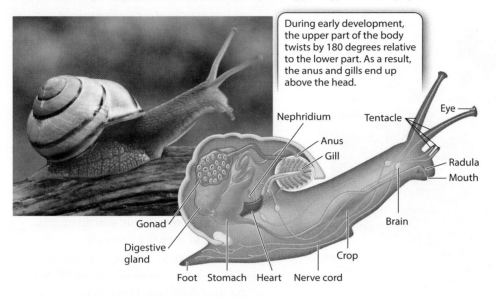

During early development, the upper part of the body twists by 180 degrees relative to the lower part. As a result, the anus and gills end up above the head.

Nephridium · Anus · Gill · Tentacle · Eye · Radula · Mouth · Gonad · Digestive gland · Foot · Stomach · Heart · Nerve cord · Crop · Brain

mollusks and annelids share very few features of the adult body plan. Mollusks are distinguished by a unique structure called the mantle that plays a major role in breathing and excretion, and which forms their shells when present. In fact, mollusks by themselves exhibit a remarkable variety of forms and functions. More than 80,000 species of mollusks have been described, most of them gastropods.

**Gastropods**—snails and slugs—provide an illuminating example of the mollusk body plan (**Fig. 44.16**). Gastropods have a head with a well-developed mouth that contains a toothlike structure called a radula for feeding. The mouth connects to a gut cavity that extends to an anus. Feather-like gills facilitate gas exchange, and a muscular foot is used for locomotion. A neural ganglion in the head coordinates a nervous system that extends through a body that also contains a well-developed circulatory system, gonads, and nephridia. Gastropods are coelomates, but the body cavity is generally reduced to small pouches that surround the heart and other organs. The outer surface of the body consists of the mantle. In many gastropods the mantle tissues secrete external skeletons of calcium carbonate, which form shells.

Some gastropods eat algae, but many are predators. About half occur in the ocean, and the other half are freshwater and terrestrial species that primarily feed on plants. In land snails and slugs, the only terrestrial mollusks, the gills have been lost, and gas exchange occurs in an internal cavity that has been modified to function as a lung.

The second great class of mollusks is the **cephalopods**—about 700 species of squid, cuttlefish, octopus, and the chambered nautilus (**Fig. 44.17**). Cephalopods share a number of features with gastropods, including much of the internal anatomy, feather-like gills, and mantle. But they also have distinctive adaptations for their unique modes of life. Most obviously, cephalopods have muscular tentacles that capture prey and sense the environment. Surprisingly, these leglike appendages are found on the head region rather than on the sides of the body as in other animals.

**FIG. 44.17** Cephalopods. *Nautilus* (a and b), octopus (c), and squid are ecologically important predators in the oceans.

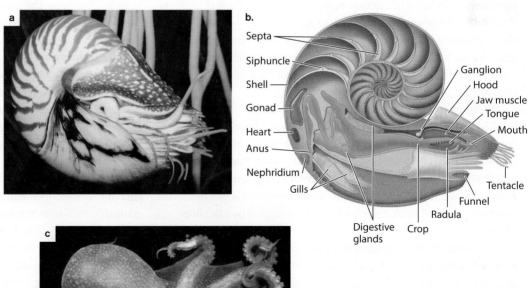

Septa · Siphuncle · Shell · Gonad · Heart · Anus · Nephridium · Gills · Digestive glands · Crop · Radula · Funnel · Tentacle · Mouth · Tongue · Jaw muscle · Hood · Ganglion

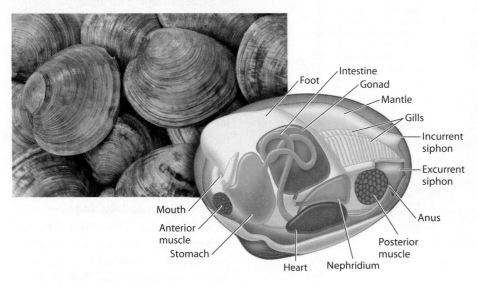

**Fig. 44.18** *Mercenaria mercenaria*, a bivalve mollusk commonly harvested from coastal waters of eastern North America.

Cephalopods are superb swimmers, able to dart through the water by means of a jet propulsion system that forces water through mantle tissue that is fused to form a siphon. It is this combination of tentacles and rapid locomotion that makes cephalopods important predators in the oceans. Moreover, cephalopods have exceptional eyesight and exhibit the most complex behavior of any invertebrate animals, able to learn visual patterns and solve puzzles to gain food. The chambered nautilus secretes a coiled shell of calcium carbonate (Figs. 44.17a and 44.17b), but in other cephalopods, like squid and octopus, the shell is reduced or absent (Fig. 44.17c). In oceans of the Mesozoic Era (252–65 million years ago), however, shelled cephalopods called ammonites were among the most abundant and diverse predators.

A third major group of mollusks, the **bivalves,** includes the familiar clams, oysters, and mussels. Once again, these animals have anatomical features that point to their close relationship to snails and squids, but also have distinctive features that set them apart (**Fig. 44.18**). Notably, bivalves have (literally) lost their heads. As their name implies, bivalves have evolved a skeleton in which two hard shells called valves are connected by a flexible hinge.

Most bivalves obtain food by filtering particles from seawater. However, many bivalves live within marine sediments, burrowing into sand or mud with their muscular foot. How can bivalves filter seawater within the sediment? Once again, this group of mollusks has a modified mantle that allows a new function. Flaps of mantle tissue have fused to form a pair of siphons that

extend upward from the bivalve's body to the surface of the seafloor above it. One siphon draws water containing food and oxygen into the body. The second siphon then returns water and waste materials to the environment. About 2000 species of bivalves have been described, mostly from the oceans, but also many from freshwater.

→   **Quick Check 3** From the shared traits of present-day snails, squids, and clams, what features do you think the common ancestor of all three had?

## Ecdysozoans include arthropods, the most diverse animals.

The second major group of protostome animals is the Ecdysozoa (**Fig. 44.19**), which get their name from the process of ecdysis, or molting. All ecdysozoans secrete a cuticle made of protein that covers their bodies. This tough cuticle is like a suit of flexible, lightweight armor, protecting bodies from injury and physical challenges of the environment such as drying. This "armor" has a special property: The hard cuticle can be used to form appendages that function as tools, weapons, or even wings. Like a suit of armor, the cuticle is always a perfect fit for the wearer but does not stretch, and so it must be exchanged episodically during growth for a larger size to fit a larger, growing body. Ecdysozoans, then, are animals that molt their external cuticle during growth, producing a very soft, larger replacement underneath that soon hardens into a new protective covering.

The Ecdysozoa include eight phyla, but one dwarfs the rest in anatomical complexity, diversity, and ecological importance. As noted earlier, the **Arthropoda,** including insects, contains more than half of all known animal species. What features have led to the remarkable evolutionary success of this branch?

Perhaps the most obvious feature of arthropods is the one for which they are named: jointed legs (*arthro* means "jointed," and *pod* means "foot" or "leg"). The jointed leg is the most versatile part of the arthropod body; it has been modified through evolution into structures that function as paddles, spears, stilts, pincers, needles, hammers, and more. The jointed leg allowed arthropods to colonize land by providing a multipoint, independent suspension—much like that of an all-terrain vehicle—able to support a relatively large body and move over the most uneven ground. So well adapted for locomotion is the arthropod leg that NASA engineers are using it as a model for planetary exploration vehicles. Researchers have conducted

**FIG. 44.19** **A phylogenetic tree for ecdysozoan animals, showing all eight phyla in this group.** Phyla discussed in the text are shown in red. Uncertainties are shown by nodes with more than two branches.

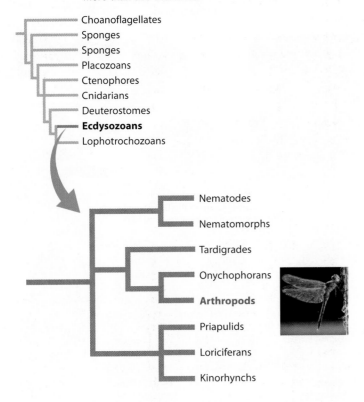

many studies to understand how this one type of limb could give rise to so many diverse structures with different functions (**Fig. 44.20**).

The arthropods' other defining characteristic is the material that forms their hard external skeleton: a strong, lightweight and nearly indestructible polysaccharide called **chitin.** Chitin, you may recall from Chapter 34, also forms the wall of fungal cells. Next to the cellulose that forms the wall of plant cells, chitin may be the most abundant biomolecule on Earth.

There are four main groups of arthropods (**Fig. 44.21**), of which the most diverse by far is the **insects.** The other three are **chelicerates,** including spiders, scorpions, and their relatives; **myriapods**, including centipedes and millipedes; and **crustaceans**, including lobsters, shrimp, crabs, and their relatives.

Chelicerates, named for the pincer-like claws called chelicerae, are the only arthropods that lack antennae. The chelicerates are mostly carnivores, except for plant-eating mites. Scorpions and spiders are known for their venoms, used to subdue prey. In spiders, the venom is injected through chelicerae modified to form fanglike appendages, while in scorpions the venom is

injected through the curved tail segment of their abdomen. Other familiar chelicerates include the harvestmen (often called daddy long-legs in the United States), ticks, and horseshoe crabs. Giant sea scorpions called eurypterids, up to 2 m long, were dominant predators in warm shallow seas 450–250 million years ago. The chelicerates were among the very first animals to invade land, about 400 million years ago, aided by their protective cuticle and supported by their arrays of jointed legs.

Myriapods are named for their many pairs of legs, which make them easy to recognize. Centipedes ("hundred legs") may have up to 300 pairs of legs and are fast-moving predators with front legs modified into venomous, fanglike organs. Millipedes ("thousand legs") may have several hundred pairs of legs (though never a thousand) and are slow-moving herbivores and detritivores (eaters of dead organisms). Millipedes often have glands lining their long bodies that produce a substance that contains cyanide, which they secrete to defend against predators.

Crustaceans also have distinctive appendages, notably their branched legs. They navigate their world with the aid of two pairs of highly sensory antennae, the larger of which has two branches. The unique crustacean larva, called a nauplius, swims with the aid of the smaller antenna and has a single eye that typically disappears in the adults. In the sea, these versatile arthropods fill many of the ecological roles that insects play on land, eating plants, other animals, or detritus. The tiny shrimplike plankton known as krill occur in vast numbers in polar seas, supporting ecosystems of fish, birds, and mammals such as whales. One Antarctic krill species is estimated to total over 500 million tons of biomass, more than the global total for humans. Lobsters and crabs are largely detritivores, scouring the ocean bottoms for dead and dying animals. Some crustaceans are even parasites, attaching themselves to the skin or inside the mouths of fish or sea mammals in order to feed.

Insects were the first animals to evolve wings, nearly 350 million years ago, and modern-day mayflies and dragonflies still bear structural similarities to these first fliers. The earliest branching members of this group, the silverfish and rockhoppers, are referred to as primitively wingless because they evolved before the origin of wings. Some other insects, like fleas, have lost wings that were present in their common ancestor, presumably because their parasitic lifestyles are better served by jumping than flying. Like all arthropods, insects repeatedly shed their exoskeleton as they grow into adulthood. No insect can molt wings, however, and so only adults can fly.

Within the insects, the main evolutionary dividing line is between those that undergo **metamorphosis,** a major change in form from one developmental stage to another, and those such as grasshoppers and water bugs that do not. In insects that do not

**FIG. 44.20**

# How did the diverse feeding appendages of arthropods arise?

**BACKGROUND** Arthropod species diversity is matched by their diversity of form and function, challenging scientists to recognize and relate the equivalent parts of the very different front ends of, for example, spiders, crabs, and centipedes.

**HYPOTHESIS** Researchers hypothesized that the diverse front appendages of the different classes of arthropods develop from similar limb buds clearly visible in developing embryos, much as the limb buds leading to bird wings, whale flippers, and human arms are present in vertebrate embryos.

**EXPERIMENT** The early stages of arthropod embryos have a definite head region and several body segments, each with a pair of appendages. *Hox* genes specify the identity of each body segment (Chapter 20). The expression patterns of different *Hox* genes can be visualized by means of labeled probes on whole embryos. In this way, the development of appendages on the same segment in different types of arthropod can be compared.

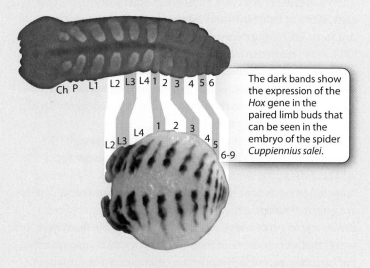

The dark bands show the expression of the *Hox* gene in the paired limb buds that can be seen in the embryo of the spider *Cuppiennius salei*.

**RESULTS** Study of the embryos revealed that, for example, the jawlike mandibles of insects and crustaceans form from appendages on the same segments of the developing body, whereas this segment on spiders and scorpions bears a pair of leglike feelers. Behind the mandibles are two pairs of feelers that are represented by legs in spiders and scorpions. On the other hand, the fangs of spiders and their counterparts in scorpions arise from a segment closer to the front end than the mandibles.

Feeding structures in the head regions of arthropods develop from paired appendages in the embryo.

Blue segments are in the head region, and the green regions bear walking legs.

| Crustaceans* | Insects | Myriapods | Chelicerates |
|---|---|---|---|
| pre-Ant | pre-Ant | pre-Ant | pre-Ant |
| Antenna | Antenna | Antenna | x |
| Antenna | | | Chelicerae |
| Mandibles | Mandibles | Mandibles | Secondary maxillae |
| Maxillae | Maxillae | Maxillae | Walking legs |
| Maxillae | Bottom of mouth | Maxillae | Walking legs |
| Walking legs | Walking legs | Walking legs | Walking legs |
| Walking legs | Walking legs | Walking legs | Walking legs |
| Walking legs | Walking legs | Walking legs | A1 |
| Walking legs | A1 | Walking legs | A2 |
| Walking legs | A2 | Walking legs | A3 |

*Crustaceans are paraphyletic.

**CONCLUSIONS** The diverse feeding structures of arthropods arise from limb buds in the first segments of the developing embryo. The mandibles of crustaceans and insects arise from the same body segment and indicate the common ancestry of these animals, while spiders, scorpions, and their relatives are more distantly related. Evolution has modified simple legs into the remarkable diversity of forms, enabling structures as specialized as a spider's fangs and the grinding mandibles of crabs.

**FOLLOW-UP WORK** There has been recent controversy over the front appendages of sea spiders and whether or not they are equivalent to the great appendage on some Cambrian arthropod fossils.

**SOURCE** Damen, W. G. M., M. Hausdorf, E.-A. Seyfarth, and D. Tautz. 1998. "A Conserved Mode of Head Segmentation in Arthropod Revealed by the Expression Pattern of HOX Genes in a Spider." *Proceedings of the National Academy of Sciences USA* 95:10665–10670.

**FIG. 44.21 Arthropods.** This diverse phylum includes four main groups: chelicerates (spiders, scorpions, and ticks), myriapods (centipedes and millipedes), crustaceans (including lobsters, shrimp, and crabs), and insects. Molecular data indicate that insects are derived from crustacean ancestors.

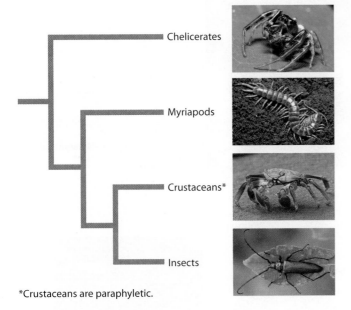

Chelicerates

Myriapods

Crustaceans*

Insects

*Crustaceans are paraphyletic.

undergo metamorphosis, the main change in body form involves the appearance of wings, though some, like dragonflies, may also change the form of their legs or eyes as they leave the aquatic environment.

However, for those insects undergoing metamorphosis, the body changes from a wormlike larva specialized for feeding to a stage called a **pupa** (**Fig. 44.22**). During the pupa stage, the body tissues undergo a remarkable transformation from the relatively simple larva to a very different looking adult such as a fly, butterfly, wasp, or beetle, usually specialized for reproduction. Insects undergo a nearly complete dissolution of their muscles and other body parts, and so require the immobile, coffin-like pupa in which to transform their bodies from something as simple as a maggot, which lacks even an obvious head or tail, into an animal as sophisticated as a fly, able to walk on water, hover motionless in front of a flower, or fly at 60 mph. This evolutionary step has little parallel elsewhere in life and is one of the keys to insect diversification.

Unlike other terrestrial arthropods, all insects, whether or not they undergo metamorphosis, have highly specialized eggshells that can withstand desiccation while still allowing gas exchange. This shell allows insect eggs to survive in dry settings such as the surfaces of leaves, and so they can occupy habitats that are difficult for other terrestrial arthropods to access.

Insects have evolved one other critical adaptation for life on land. Aquatic arthropods obtain oxygen through gills, which don't work well in air because their large surface area dries out. Some terrestrial chelicerates have evolved simple lungs, called book lungs: modified appendages filled with hemolymph (blood) that have a large surface area for gas exchange, but are kept moist inside a protective pocket. Some crabs are able to make extended forays on land by keeping their gills moist underneath their carapace. In contrast, insects exchange gases through small pores in their exoskeletons called **spiracles** (see Fig. 39.4b). The spiracles connect to an internal system of tubes, the **tracheae,** that direct oxygen to and remove carbon dioxide from respiring tissues, much like air exchange ducts in a building.

Insects are extraordinarily diverse, with approximately 1 million described species making up 80% or more of all known animal species. Their success may be a function of the three adaptations that allow them to live in diverse habitats: dessication-resistant eggs, wings, and metamorphosis. The ability of insect eggs to resist drying out means that they can be placed on virtually any surface. Their wings allow them to reach habitats and feed in ways other animals cannot. Metamorphosis allows them to have two bodies in their life cycles: a larval form specialized for feeding and an adult form specialized for mating and reproduction. These three adaptations enabled insects to use flowering plants directly as food. In turn, many insects pollinate plants, and so pollinating insects are in part responsible for flowering plant diversity (Chapter 33).

**FIG. 44.22 Development in butterflies, a group of insects that undergoes metamorphosis.**

Complete metamorphosis involves a pupal stage (center) in which the tissues of the larva (caterpillar on left) are reorganized into a very different adult form (butterfly on right).

## Deuterostomes include humans and other chordates, but also acorn worms and sea stars.

In a classification recognized long ago by shared features of larval development, and strongly supported by molecular sequence data, the Deuterostomia (**Fig. 44.23**) include three major phyla: **Chordata** (vertebrates and closely related invertebrate animals such as sea squirts), **Hemichordata** (acorn worms), and **Echinodermata** (sea urchins and sea stars).

**FIG. 44.23** A phylogenetic tree for deuterostome animals, showing all phyla (and the major subdivisions of the chordates) in this group. Phyla discussed in the text are shown in red.

**FIG. 44.24** Hemichordates, including (a) acorn worms, with gill slits, and (b) pterobranchs, with their filter-feeding tentacles.

Let's begin with the deuterostomes most distantly related to the vertebrates, the hemichordates and echinoderms. Hemichordates include acorn worms, about 75 species of wormlike animals that move through seafloor sediments in search of food particles. Pterobranchs, also hemichordates, consist of about a dozen species of animals that attach to the seafloor and use tentacles to filter food from seawater (**Fig. 44.24**). Hemichordates all have a mouth on an elongate protuberance called a proboscis that connects to the digestive tract by a tube called the **pharynx,** which contains a number of vertical openings called **pharyngeal slits** separated by stiff rods of protein. They also have a **dorsal nerve cord.** Additional

**FIG. 44.25 Echinoderms.** (a) Sea stars are familiar sights on the shallow seafloor; (b) sea urchins are motile echinoderms found on most seafloors.

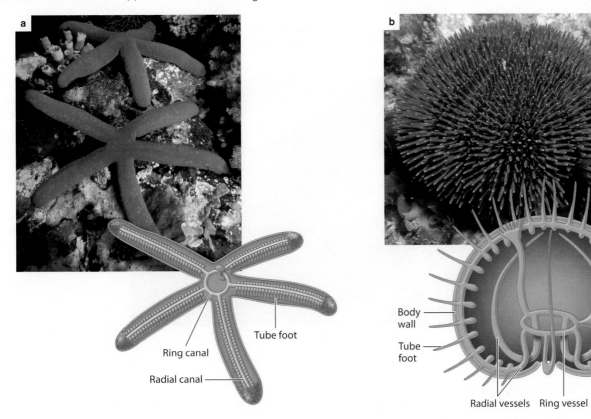

characters that relate hemichordates to other deuterostomes are apparent in embryological and larval development. Nonetheless, living hemichordates have body plans quite distinct from those of chordates and echinoderms.

Another major group of deuterostomes is the echinoderms, among the most distinctive of all animal phyla. Echinoderms include sea stars and sea urchins (including sand dollars), as well as brittle stars, sea cucumbers, and sea lilies—all of which share a unique fivefold symmetry on top of their basic bilaterian organization (**Fig. 44.25**). Echinoderms also form distinctive skeletons made of interlocking plates of porous calcite, a form of calcium carbonate. The third unique feature of echinoderms is the **water vascular system,** a series of fluid-filled canals that permits bulk transport of oxygen and nutrients. **Tube feet,** small projections of the water vascular system that extend outward from the body surface, facilitate locomotion, sensory perception, food capture, and gas exchange. About 7000 echinoderm species reside in present-day oceans, but the fossil record suggests that this phylum was more diverse in the past.

Zoologists have long known that echinoderm and hemichordate larvae share many features. Molecular sequence data now confirm that echinoderms are the closest relatives of hemichordates, with chordates a sister group to the group containing echinoderms and hemichordates (see Fig. 44.23).

## Chordates include vertebrates, cephalochordates, and tunicates.

The other great branch of the deuterostome tree is the chordates, the phylum that includes vertebrate animals. Within the phylum Chordata, there are three subphyla: the **cephalochordates,** the **tunicates,** and the **vertebrates** (also called **craniates**). Like hemichordates, chordates all have a pharynx with pharyngeal slits (well illustrated by the invertebrate chordate amphioxus in **Fig. 44.26**). In fish, these pharyngeal slits form the gills, but in terrestrial animals like humans, these slits can be seen only in developing embryos. The **notochord,** a stiff rod of collagen and other proteins, runs along the back, providing support for the axis in some chordates. In vertebrates, the notochord is apparent

**FIG. 44.26  Amphioxus, a cephalochordate.** Amphioxus clearly has many features in common with vertebrates, but it lacks a well-differentiated head and has no cranium or vertebrae.

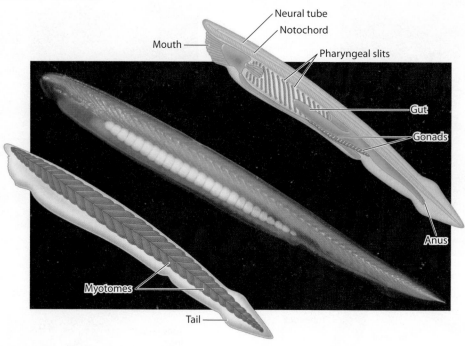

**FIG. 44.27  Tunicates.** (a and b) Adult form; (c) larval form.

a.

b.

c.

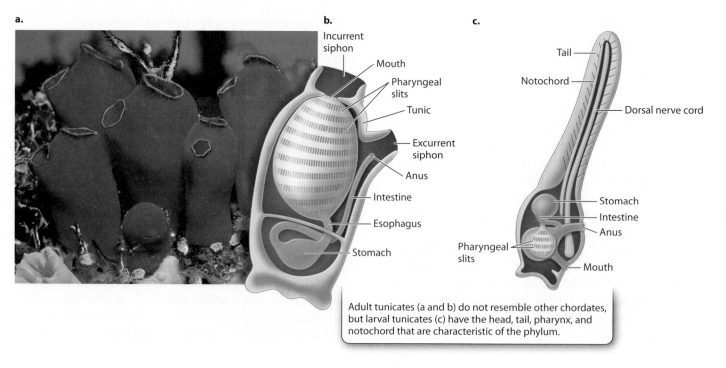

Adult tunicates (a and b) do not resemble other chordates, but larval tunicates (c) have the head, tail, pharynx, and notochord that are characteristic of the phylum.

only during early embryogenesis and is replaced functionally by the development of a **vertebral column.** Also forming during early development is the **neural tube,** a cylinder of embryological tissue that develops into a dorsal nerve cord. Body musculature is organized into a series of segments called **myotomes.** Chordates have a tail (reduced but still internally present in humans and other apes) that extends posterior to the anus and muscularized appendages that include the fins of fish and legs of terrestrial vertebrates.

Amphioxus (Fig. 44.26) belongs to the cephalochordates. These animals share key features of body organization with vertebrates but lack a well-developed brain and eyes, have no lateral appendages, and do not have a mineralized skeleton. Nonetheless, their many similarities to vertebrates suggest they are closely related, and molecular sequence comparisons confirm this hypothesis.

The tunicates also share key features of the chordate body plan during early development, but they have a unique adult form (**Fig. 44.27**). Tunicates include about 3000 species of filter-feeding marine animals, such as sea squirts anchored to the seafloor and salps that float in the sea. The adult tunicate body is a basket-like structure that is highly modified for filter-feeding. It might be mistaken for a sponge, but its complex internal structure shows that it lies far from sponges on the animal tree. The tunicate's body wall, or tunic, has a siphon-like mouth at one end that draws water through an expanded pharynx that captures food particles and exchanges gases. Water and wastes are expelled through an anal siphon. In the adult tunicate, the only obvious similarities to other chordates are the pharynx and its pharyngeal slits. Larval tunicates, however, resemble other chordates like fish or tadpoles with a more typical chordate body plan, including a notochord, neural tube, and a long tail with muscles arranged in myotomes.

For many years, zoologists considered cephalochordates to be the closest relatives of vertebrate animals, but molecular data now suggest that, in fact, the tunicates are our closest invertebrate relatives. This means that the common ancestor of tunicates, amphioxus, and vertebrate animals was a small animal much like amphioxus in organization. After the divergence of the three major groups of chordates, however, the tunicates' adaptation for filter-feeding led to the unique anatomy they have today.

→ **Quick Check 4** You do not have a notochord, a tail, or gill slits. How do we know that you are a chordate?

## 44.4 VERTEBRATE DIVERSITY

In many ways, the last stop on our tour of animal diversity is the most familiar because it includes our own species, *Homo sapiens.* The animals known as vertebrates are named for their jointed skeleton that runs along the main axis of the body, forming a series of hard segments collectively termed **vertebrae** (singular, vertebra). In addition to features shared with other chordates, vertebrate animals are distinguished by a cranium that protects a well-developed brain, a pair of eyes, a distinctive mouth for food capture and ingestion, and an internal skeleton commonly mineralized by calcium phosphate (**Fig. 44.28**). The vertebrates also have a coelom in which the organs are suspended, and a closed circulatory system. Like the other chordates, vertebrates have pharyngeal slits in at least the early embryonic stages.

Many of the features that separate vertebrates from invertebrate chordates can be found in the head, including a bony cranium that protects the brain, joined in most vertebrates by the mandible (or jawbone) to form a skull. This protective helmet for the delicate neuronal tissues permits the development of a larger brain than would otherwise be possible. The mandible permits eating foods such as other living animals, themselves often protected by hard shells, and plants containing hard lignin. Because the bottom scavenger hagfish have a cranium but not vertebrae (or a mandible), some biologists use the name "Craniata" for this group, rather than the traditional "Vertebrata."

**FIG. 44.28 Vertebrate characters.** Anatomical features of vertebrates that are characteristic of all chordates are shown in red, and those that are unique to vertebrates are shown in blue.

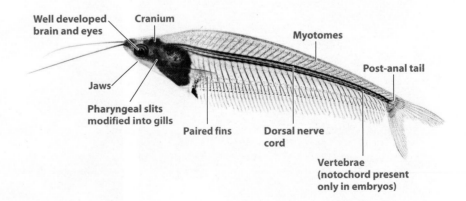

Well developed brain and eyes

Cranium

Myotomes

Post-anal tail

Jaws

Pharyngeal slits modified into gills

Paired fins

Dorsal nerve cord

Vertebrae (notochord present only in embryos)

**FIG. 44.29** **A phylogeny of the craniates.** This phylogeny is favored by recent molecular data, but, like all phylogenies, is a hypothesis. The placements of hagfish relative to other vertebrates and turtles relative to other sauropsids have provoked debate for many years.

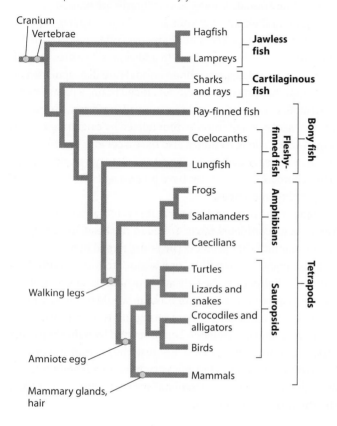

## Fish are the earliest-branching and most diverse vertebrate animals.

We sometimes think of fish as one group, but the aquatic animals we commonly call "fish" include four distinct groups of aquatic vertebrates, not all of them monophyletic (**Fig. 44.29**).

The earliest branching craniates are the **hagfish** and **lampreys** (**Fig. 44.30**). These animals have a cranium built of cartilage but lack jaws. Lampreys also have a vertebral column built of cartilage. Hagfish lack vertebrae, though current research suggests that this trait existed in a common ancestor of hagfish and lampreys, but was lost in hagfish. These eel-like organisms feed on soft foods without the aid of jaws because they diverged before jaws evolved from pharyngeal slits. Hagfish feed on marine worms and on dead and dying sea animals, while lampreys live parasitically, sucking body fluids from fish prey. Both hagfish and lampreys have a series of gill slits through which water enters to bring oxygen to the gills.

Biologists have long debated the phylogenetic relationships among hagfish, lampreys, and other vertebrate animals. The presence of a vertebral column in lampreys convinces many biologists that lampreys are the sister group to other vertebrates and hagfish the sister group to all vertebrates, including lampreys. The debate continues, but several lines of molecular data now favor the view shown in Fig. 44.29, that hagfish and lampreys together form the sister group to all other vertebrates. About 65 species of lampreys and hagfish live in freshwater and coastal oceans. Fossils record the existence of many extinct species of jawless fishes. When we include fossils, the jawless

**FIG. 44.30** **Hagfish and lampreys.** Sometimes grouped together as jawless fish, (a) hagfish and (b) lampreys are the earliest branching craniates alive today.

FIG. 44.31 **Cartilaginous fish.** The skeletons and jaws of sharks are made of cartilage.

fishes as a whole are paraphyletic because one of their groups gave rise to fish with jaws.

**Chondrichthyes,** or cartilaginous fish, form the next deepest branch on the vertebrate tree. This monophyletic group includes about 800 species of sharks, rays, and chimaeras, all of which have jaws and a skeleton made of cartilage (**Fig. 44.31**). These fish deposit calcium phosphate minerals only in their teeth and in small toothlike structures called denticles embedded in the skin. The best-known cartilaginous fish are the sharks, many of which are predators that put those mineralized teeth to good use. On the other hand, the group also includes whale sharks, filter-feeding animals that, at a length up to 40 feet (12 m), are the gentle giants of the sea. Rays are closely related to sharks but have a flattened body adapted for life on the seafloor, where they feed on clams and crabs. The cartilaginous fish can retain levels of urea in their tissues that would poison other vertebrates, a physiological adaptation that helps them to maintain their salt balance without well-developed kidneys (Chapter 41). Nearly all Chondrichthyes occur in the oceans, but Bull Sharks (*Carcharhinus leucas*) commonly swim into rivers—they've been recorded well up the Mississippi—and there are several freshwater species of rays.

Moving up one more branch in the vertebrate tree, we encounter the **Osteichthyes,** or bony fish (see Fig. 44.29). Bony fish have a cranium, jaws, and bones mineralized by calcium phosphate. Numbering about 20,000–25,000 fresh and seawater species, these are the fish that we most commonly encounter (**Fig. 44.32**). Bony fish are by far the most diverse group of vertebrates, possessing several unique features that facilitate their occupation of diverse

FIG. 44.32 **Bony fish.** The char, well known to those who fish in freshwater, illustrates the principal anatomical features of bony fish: a cranium, jaws, and mineralized bones.

**FIG. 44.33** **Fleshy-finned fish.** The fish most closely related to tetrapods are the fleshy-finned fish: (a) the coelacanth and, even closer, (b) the lungfish. These fish have paired fins with an anatomical structure similar to that of tetrapod legs.

niches. First, they have a system of movable elements in their jaws that allows them to specialize and diversify their feeding on many different types of food. Second, they possess a unique gas-filled sac called a swim bladder that permits exquisite control over their position in the water column through changes in buoyancy. Third, bony fish have kidneys that allow them to regulate water balance and so occupy waters over a wide range of salinity.

The swim bladder plays a special role in vertebrate evolution. Early vertebrates evolved a gut sac that enabled them to gulp air to obtain additional oxygen. In some fish, this sac evolved into the swim bladder, but in one group the sac was modified to become a lung. Charles Darwin was among the first to recognize that the swim bladder and the vertebrate lung are homologous organs.

### The common ancestor of tetrapods had four limbs.

Most animal phyla occur in the oceans, and many are exclusively marine. A subset of animal phyla has successfully colonized freshwater, and a smaller subset has radiated onto the land. Eleven groups on the animal phylogenetic tree contain both aquatic *and* terrestrial species: nematodes, flatworms, annelids (earthworms and leeches), snails, tardigrades (microscopic animals with eight legs, sometimes called water bears), onychophorans (velvet worms), four groups of arthropods (millipedes and centipedes, scorpions and spiders, land crabs, and insects), and vertebrates. In each of these groups, the aquatic species occupy the earliest branches, and the terrestrial species occupy later branches. No two of these groups share a last common ancestor that lived on land; thus, they all made the transition independently. Earlier, we outlined the extraordinary diversity of terrestrial arthropods. Here, we focus on vertebrates as the other great animal colonists of the land.

Most bony fish have fins that are supported by a raylike array of thin bones. About a dozen closely related species, however, have paired pectoral and pelvic fins. These fish are called **fleshy-finned fish** and include the **coelacanth** and **lungfish** (**Fig. 44.33**). Although these animals resemble other fish, the coelacanth and lungfish are the nearest relatives of tetrapods, four-legged

**FIG. 44.34** **The amphibian life cycle.**

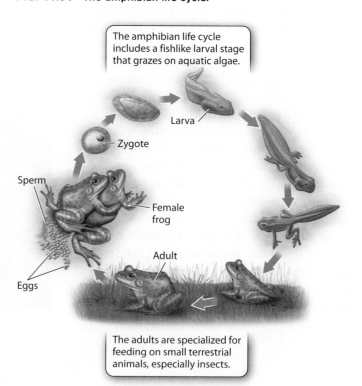

The amphibian life cycle includes a fishlike larval stage that grazes on aquatic algae.

Larva

Zygote

Sperm

Female frog

Adult

Eggs

The adults are specialized for feeding on small terrestrial animals, especially insects.

animals. Lungfish can survive periods when their watery habitat dries by burying themselves in moist mud and breathing air. Coelacanths first appeared over 400 million years ago and were thought to have been extinct for over 80 million years until a specimen was discovered in a fish market in South Africa in 1938. Hence, they are sometimes referred to as living fossils. There are two living species of coelacanths and three species of lungfish, one each in South America, Africa, and Australia. Because the fleshy-finned fish are the closest relatives of tetrapods, there is no monophyletic group that corresponds to our popular notion of "fish." To be monophyletic, such a group must include the tetrapods.

Fossils document the anatomical transition in vertebrate animals as they moved from water to land, including changes in the form of the limbs, rib cage, and skull (see Fig. 23.20). Living tetrapods include amphibians, such as frogs and salamanders; lizards, turtles, crocodilians, and birds; and mammals. The last common ancestor of all these animals had four limbs, hence the name **Tetrapoda** ("four legs"). Some, however, like snakes and a few amphibians and lizards, have lost their legs in the course of evolution.

There are over 4500 species of **Amphibia** ("double life"), ranging in size from tiny frogs a few millimeters in length to the Chinese Giant Salamander, which is over 1 m long. Their name reflects their distinctive life cycle (**Fig. 44.34**). Most species have an aquatic larval form with gills that permit breathing under water and an adult form that is terrestrial and usually has lungs for breathing air. Amphibians must reproduce in the water, and so are not completely terrestrial. Whereas many amphibian larvae graze on algae, the adults are predators and often have a muscular tongue for capturing prey. Many are protected by toxins they secrete from glands in their skin, sometimes advertised by their brilliant color patterns. Most amphibian adults require moist skin for breathing and consequently have small lungs, with toads and red efts being notable exceptions.

### Amniotes evolved terrestrial eggs.

Like insects, some vertebrates evolved an egg adapted to tolerate dry conditions that accompany life on land. The amniotic egg has a desiccation-resistant shell and four membranes that permit gas exchange and management of waste products produced

by the embryo (see Fig. 42.10). These eggs must be fertilized internally before the eggshell is produced by the female because sperm cannot penetrate the shell. Amniotic eggs permit long development times, with the embryo's nutrition supported by a large yolk or the placenta. They keep wastes separated from the embryo so they do not poison it, giving the embryo time to build more complex bodies than otherwise would be possible.

The **amniotic egg** can exchange gases while retaining water, and so permits the group of vertebrates known as **amniotes** to live in dry terrestrial habitats that amphibian eggs cannot tolerate. Amniotes include lizards, snakes, turtles, and crocodilians (about 6000 species of scaly animals commonly referred to as "reptiles"), as well as the birds and mammals (**Fig. 44.35**). Most mammals, many lizards, and some snakes have evolved live birth rather than laying eggs. Nonetheless, they retain the specialized membranes characteristic of all amniotes. Instead of being wrapped around the

**FIG. 44.35** Amniotes, including (a) turtles, (b) lizards, (c) mammals, and (d) birds.

embryo inside a tough egg shell, these membranes surround and protect the embryo inside the womb.

As shown in Fig. 44.29, amniotes form a monophyletic group with two major branches. One is made up of mammals, and the other contains turtles, birds, and the amniotes traditionally grouped together as reptiles. Like fish, however, reptiles are not monophyletic—birds are descended from reptiles, although you have to look closely to see the similarities between the two groups. Birds lack teeth, most of their scales have been modified into feathers (except on the legs and toes), and they are generally adapted for flight. Adaptations for flight include hollow bones that are light but remain strong, and a method of breathing that extracts much oxygen from each breath, permitting their high-performance flight (Chapter 39).

One consequence of this streamlined, lightweight body is a lack of the heavy jaws needed for grinding plant leaves or other low-nutrition foods. Instead, birds have their grinding organ, the gizzard (also found in other reptiles), inside their bodies near their center of gravity, where it interferes less with flight. Only one kind of bird, the weak-flying Amazonian Hoatzin, can actually feed on leaves. As discussed in Chapter 23, fossils show that birds diverged from dinosaur ancestors, and molecular sequence comparisons indicate that crocodiles are the closest living relatives of birds.

→ **Quick Check 5** Why do bird feet have scales like those found on snakes and lizards?

All **mammals** are covered with hair and feed their young milk from the mammary glands for which the class Mammalia is named. Like many other groups of animals and plants, the early-branching mammals show intermediate stages in the evolution of the body plans that dominate on Earth today. The first fossils that clearly show hair appeared about 200 million years ago, during the Triassic Period. Like feathers in birds, hair permits retention of body heat. The earliest-branching living mammals, the monotremes, lay eggs, like birds or lizards (**Fig. 44.36**), but their hatched young drink milk secreted from pores in the skin of the mother's belly. Among these are the Australian platypus and the echidna, or spiny anteater.

The first mammals that gave birth to live young appeared in the Jurassic Period, about 125 million years ago. These animals gave rise to the two major groups of living mammals, marsupial and placental mammals. **Marsupials** include kangaroos, koalas, and related groups native to Australia, as well as the opossums found in the Americas. Their young are born at an early stage of development, and the tiny babies must crawl to a pouch where mammary glands equipped with nipples provide them with milk. Only later do fully formed juveniles emerge from the pouch and begin life on the ground.

**FIG. 44.36** A phylogenetic tree for mammals, illustrating the three major groups: monotremes, marsupials, and placental mammals.

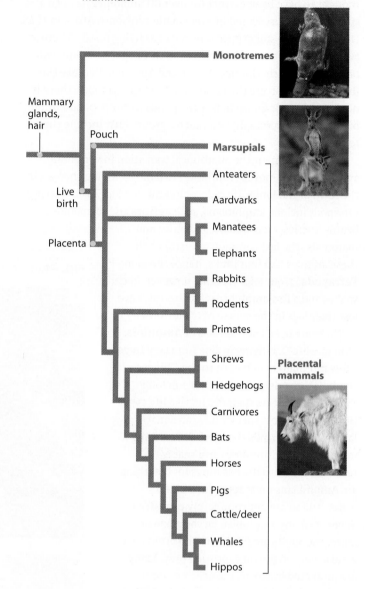

The **placental mammals** are named for a temporary organ called the **placenta** that develops in the uterus along with the embryo, providing nutrition that enables the offspring to be larger and more quickly independent when born (Chapter 42). Most living mammals fall into this group. Placental mammals include the carnivores, such as lions and weasels; the primates, including monkeys, apes, and humans; and the hooved mammals, which include cattle, pigs, deer, and—perhaps surprisingly—their marine off-shoot, the whales.

All together, the familiar mammals mentioned so far number about 1000 species. The most diverse mammals, however, belong to two groups of mostly small animals, the bats and the rodents. Bats, which evolved flight and a form of sonar called echolocation to find food at night, include more than 1000 species. The largely plant-feeding rodents are as diverse as the large mammals and bats combined, numbering more than 2000 species. Beavers, gophers, rats, mice, hamsters, and squirrels are all rodents. Their evolutionary success is often attributed to innovations in tooth development. Their teeth are unusual in growing continually throughout life. This constant growth permits constant wear, and so allows these animals to gnaw through hard protective coverings to obtain nutrition from a remarkably wide variety of otherwise hard-to-access sources like the insides of seeds.

## 44.5 THE EVOLUTIONARY HISTORY OF ANIMALS

Because many animals form mineralized skeletons, sedimentary rocks deposited during the past 542 million years contain a rich fossil record of shells and bones. Animals also leave a sedimentary calling card in their tracks, trails, and burrows, again widespread in sedimentary rocks formed after 542 million years ago. Not all animal phyla are well represented in the fossil record—fossil annelids, for example, are rare and flatworms are essentially unknown as fossils because they lack mineralized body parts (Chapter 23). In contrast, the fossilized shells and bones of bivalve mollusks, brachiopods, echinoderms, and mammals preserve an excellent record of evolutionary history within these groups. If we step back and look at the big picture, what do fossils tell us about the evolutionary history of the animal kingdom?

**Fossils and phylogeny show that animal forms were initially simple but rapidly evolved complexity.**
Phylogeny suggests that animals are relative latecomers in evolutionary history, and the fossil record confirms this hypothesis. Life originated more than 3.5 billion years ago, but, as discussed in previous chapters, microorganisms dominated ecosystems for most of our planet's history. As noted in Chapter 28, macroscopic fossils of organisms thought to be animals first appear in rocks deposited only 575 million years ago. Called Ediacaran fossils after the Ediacara Hills of South Australia where they were discovered, these fossils have simple shapes that are not easily classified among living animal groups (**Fig. 44.37**; see also Fig. 28.13).

**FIG. 44.37** *Dickinsonia*, an Ediacaran fossil. *Dickinsonia* has been interpreted as an early animal that acquired food by phagocytosis and by absorbing organic molecules, and exchanged gases by diffusion. The structural simplicity of *Dickinsonia* and similar fossils suggests that they record early branches of the animal tree, perhaps as placed in the diagram.

Phylogeny suggests that we should look for sponges, cnidarians, and other diploblastic animals among the oldest animal fossils, but sponges, at least, are rare among Ediacaran fossils. Instead, a majority of Ediacaran fossils show simple, fluid-filled tubes, without identifiable mouths or other organs. Many may have formed colonies, gaining complexity through colonial growth and differentiation, as some living cnidarians do. These early animals had epithelia and are thought to have obtained food by taking in dissolved organic matter while exchanging gases by diffusion, somewhat like living placozoans. Most Ediacaran fossils probably form an early branch or branches on the animal tree, falling among the limbs whose survivors include cnidarians, ctenophores, and placozoans (Fig. 44.37).

Why did the first animal radiation occur so late in the evolutionary day? Scientists continue to debate this question, but part of the answer appears to lie in Earth's environmental history. Geochemical data suggest that only during the Ediacaran Period did the atmosphere and oceans come to contain oxygen in quantities sufficient to support the metabolism of large, active animals.

**The animal body plans we see today emerged during the Cambrian Period.**
Ediacaran fossils differ markedly from the shapes of living animals, but in the next interval of geologic history, the Cambrian

**FIG. 44.38  Fossils of the Cambrian Period.** Cambrian rocks preserve early representatives of present-day phyla, including (a) early arthropods called trilobites, (b) annelid worms, (c) early deuterostomes with a pharynx, and even (d) primitive vertebrates.

Period (542–489 million years ago), we begin to see the remains of animals with familiar body plans (**Fig. 44.38**). Cambrian fossils commonly include skeletons made of silica, calcium carbonate, and calcium phosphate minerals, and these record the presence of arthropods, echinoderms, mollusks, brachiopods, and other bilaterian animals in the oceans. Spicules made by sponges are common in Cambrian rocks as well. The rocks also preserve complex tracks and burrows made by organisms with hydrostatic skeletons, muscular feet, and the jointed legs of arthropods. And in a few places, notably at Chengjiang in China and the Burgess Shale in Canada, unusual environmental conditions have preserved a treasure trove of animals that did not form mineralized skeletons (Chapter 23).

These exceptional windows on early animal evolution show that, during the first 40 million years of the Cambrian Period, the body plans characteristic of most bilaterian phyla took shape in a

transition sometimes called the **Cambrian explosion.** Sponges and cnidarians radiated as well, producing through time the biomass- and diversity-enhancing habitats of reefs and imparting an ecological structure to life in the sea broadly similar to what we see today.

Scientists sometimes argue about whether the name "Cambrian explosion" is apt. The fossil record makes it clear that bilaterian body plans did not suddenly appear fully formed, so the event was not truly "explosive." Rather, fossils demonstrate a huge accumulation of new characters in a relatively short period of time, during which the key attributes of modern animal phyla emerged. For example, living arthropods combine segmented bodies with a protective cuticle, jointed legs, other appendages specialized for feeding or sensing the environment, and compound eyes with many lenses. Cambrian fossils include the remains of organisms with some but not all of the major features combined today in arthropods. In short, the first 40 million years of Cambrian animal evolution ushered in a world utterly distinct from anything known in the preceding 3 billion years.

Despite the burst of evolution recorded by Cambrian fossils, there were still relatively few species of marine animals at the end of the Cambrian Period. The ensuing Ordovician Period (489–444 million years ago) was a time of renewed animal diversification, especially the evolution of heavily skeletonized animals in the world's oceans. The Ordovician radiation established a marine ecosystem that persisted for more than 200 million years. Interestingly, however, if you had walked along an Ordovician beach, the shells washing about your feet would have been far different from the ones you see today. The dominant shells were those of brachiopods, not clams. Broken corals in the surf were the skeletons of now-extinct cnidarians only distantly related to modern reef-forming corals. And arthropod shells, molted during growth, were those of now-extinct trilobites, not lobsters or crabs. This ecosystem was devastated by a mass extinction 252 million years ago (Chapter 23).

### Animals began to colonize the land 420 million years ago.

The colonization of land by animals began only after plants had established themselves there. Land plants evolved about 460 million years ago. Arthropods followed by about 420 million years ago, especially chelicerates that included the ancestors of spiders, mites, and scorpions. Insects, descended from crustacean ancestors that independently gained access to the land surface, appeared at about the same time. The radiation of the major groups of insects, however, began 360 million years ago with a marked diversification of dragonflies and the ancestors of cockroaches and grasshoppers. Some of the dragonfly-like insects that darted among the plants of Carboniferous coal swamps had bodies the size of a lobster and 75-cm wing spans! One might well wonder what prevents insects from attaining this size today. These immense insects flew through an atmosphere that contained more oxygen than is present today, and this higher level of atmospheric oxygen is thought to have been a necessary condition for insect gigantism. After some time, levels of atmospheric oxygen declined to levels closer to those seen today, and the giant insects disappeared. Flies, beetles, bees, wasps, butterflies, and moths radiated later, beginning in the early Mesozoic Era. Their rise in diversity parallels that of the flowering plants and reflects coevolution of these pollinators and flowering plants.

Tetrapod fossils first appear in sedimentary rocks deposited near the end of the Devonian Period, about the same time as the early insect radiations 360 million years ago. As discussed in Chapter 23, the fossil record documents in some detail the shifts in skull, trunk, and limb morphology that allowed vertebrate animals to colonize land. These include the evolution of muscled, articulated legs from fins, together with a set of strong, articulated digits where the limbs meet the ground, internal lungs rather than gills, and an erect, elevated head with eyes oriented for forward vision. The appearance of amphibians and then reptiles follows the predictions of phylogenies based on comparative biology.

The most remarkable animals among the reptiles are surely the dinosaurs. Dinosaurs first evolved about 210 million years ago from small, bipedal ancestors that radiated following the mass extinction 252 million years ago. Once established, the dinosaurs radiated to produce many hundreds of species, dominating terrestrial ecosystems until 65 million years ago, when nearly all dinosaur species disappeared, the most familiar victims of an enormous and Earth-changing asteroid impact (see Fig. 1.3). We say "nearly all" because the fossil record documents the divergence of the ancestors of modern birds from a specific subgroup of dinosaurs, about 150 million years ago (see Fig. 23.20).

Mammals have been the dominant vertebrates in most terrestrial ecosystems since the extinction of dinosaurs, but the group originated much earlier, at least 210 million years ago. During the age of dinosaurs, most mammals were small nocturnal or tree-dwelling animals that stayed out of the way of large dinosaurs, although fossils from China show clearly that the largest mammals of this interval ate small dinosaurs and their eggs.

We summarize the biological diversity of all organisms, including animals, along with a timeline of Earth's history, in **Fig. 44.39**.

## ? CASE 8  Biodiversity Hotspots: Rain Forests and Coral Reefs

### How have coral reefs changed through time?

On land, tropical rain forests are hotspots of diversity, and coral reefs support unusually large numbers of species in the oceans. Reefs have existed through most of Earth's recorded history, and rain forests have existed since the origin of trees more than 370 million years ago. Through time these have been Earth's most diverse communities in the sea and on land. But the taxonomic composition of these communities has changed through time because of both evolutionary innovation and mass extinction.

Bacteria built the first reefs, layered mounds of limestone that rose above the seafloor. With the Cambrian diversification of animals, calcified sponges rose to ecological prominence as participants in reef construction, but these communities were short lived, as mass extinction about 515 million years ago essentially wiped out the reef-building sponges. Subsequent renewed diversification of sponges, bryozoans, and corals with massive calcium carbonate skeletons resulted in new and different reef communities beginning in the Ordovician Period. Eventually these communities were also destroyed by mass extinction during the Devonian Period. Reefs were widespread again by the Permian Period (300–252 million years ago), dominated this time by different groups of sponges and bryozoans, along with skeleton-forming green algae.

The devastating extinction at the end of the Permian Period removed all corals and calcifying sponges, and for several million years following this event, reefs were built only by microorganisms, much as they were before the evolution of animals. Then, the evolution of new skeleton-forming corals (from sea anemone-like ancestors) and a resurgence of sponges capable of massive calcification introduced new reef systems into the ocean. Given the iconic status of coral reefs as examples of marine biodiversity, it is important to note that reefs like those of the present day have developed only over 40 million years of Earth history.

As discussed in Chapter 48, we may be nearing another great change in reef biology as global climate change threatens reef communities in many parts of the world. Today's coral species are at their tolerance limits of ocean temperature and pH, and both may be changing more rapidly than species can adapt. From the perspective of Earth history, the loss of coral reefs would simply be one more in a long series of extinction episodes. For humans, however, the commercial and ecological consequences would be substantial and permanent.

# Diversity through Time

Integrating concepts from Chapters 26, 27, 33, 34, and 44

## Diversity over time

If we stand back and look at evolution over time, we see a pattern of increasing complexity, which has enabled increasingly specialized function and therefore greater species diversity. We also see a pattern of accumulation through time. Bacteria and Archaea were not replaced by later evolving eukaryotes. Prokaryotic microorganisms remain at the heart of functioning ecosystems, with complex interactions of protists, plants, animals, and fungi built around them.

## Oxygen

Early in Earth's history, the atmosphere and oceans had little or no oxygen, and metabolism was entirely anaerobic. All of the oxygen we breathe today is the result of photosynthesis that produces $O_2$ as a by-product. Oxygen levels first increased about 2.4 billion years ago. As oxygen levels in the atmosphere and oceans increased, new metabolisms evolved, expanding possibilities for both cellular respiration and chemoautotrophy.

## Prokaryotic diversity

The two great prokaryotic domains, Bacteria and Archaea, originated 3.5 - 4 billion years ago. They diversified greatly with the rise in oxygen levels in the atmosphere and oceans about 2.4 billion years ago, and again with the diversification of animals during the Cambrian explosion, reflecting co-evolution between eukaryotes and prokaryotes.

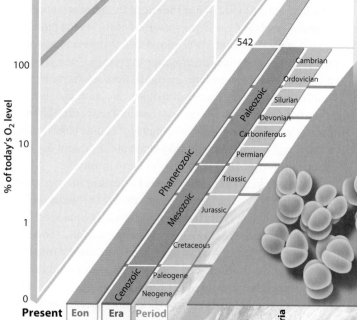

Animals and plants have provided bacteria and, to an extent, archaeons with new habitats for diversification.

Millions of years ago

4567

Archean

2500

Proterozoic

Eukaryotes

542

Phanerozoic

Paleozoic
Cambrian
Ordovician
Silurian
Devonian
Carboniferous
Permian

Mesozoic
Triassic
Jurassic
Cretaceous

Cenozoic
Paleogene
Neogene

% of today's $O_2$ level

100

10

1

0

**Present**  Eon  Era  Period

Bacteria

Archaea

Excavates

Rhizarians

Alveolates

Stramenopiles

**Glaciation**

Some aspects of Earth's environment change directionally through time– oxygen (on the left) is a good example. In contrast, Earth's climate has shifted back and forth through time between "icehouses" when continental glaciers covered significant parts of the land and "greenhouses" when glacial ice was limited or absent. Large ice ages have occurred intermittently through Earth history; most ice sheets expanded from the poles to mid-latitudes, but at least three times (shown in solid purple lines) ice covered the planet, or nearly so. These "Snowball Earth" events must have presented challenges for life at the beginning and near the end of the Proteozoic Eon.

▪▪▪▪ Regional glaciation
▬▬▬ Global glaciation

4567

2500

**Eukaryotic origins**

Eukaryotes, which may have originated by symbiosis between an archaeon and a proteobacterium, began to diversify after oxygen started to accumulate in the atmosphere and surface oceans.

**Eukaryotic diversity**

Eukaryotic diversity continued to rise as oxygen levels increased further. Especially notable were the radiations of marine animals and, later, land animals (particularly insects), land plants, and fungi.

Archaeoplastids

Opisthokonts

*Millions of years ago*

**Mass extinction**

Over the past 542 million years, five major environmental perturbations have reduced species diversity by half or more. These mass extinctions devastated ecosystems, but provided new evolutionary possibilities for the survivors. Some scientists think we may experience a sixth mass extinction as a result of human activities.

▬▬ Mass extinction event

Simple animals diversified during the Ediacaran Period.

Bilaterian animals diversified during the Cambrian Period.

Tetrapods have diversified markedly on land, but fish remain the most diverse vertebrates.

542

Tetrapods

Seed plant diversity increased markedly with the evolution of flowering plants.

Dinosaurs

Green algae

Mosses

Lycophytes

Ferns

Seed plants

Amoebozoans

Fungi

Sponges

Cnidarians

Arthropods

Mollusks

Jawless fish

Ray-finned fish

Lungfish

Amphibians

Lizards and Snakes

Birds

Mammals

Arthropods have been ecologically dominant animals in the oceans since the Cambrian Period, but much of their diversity reflects the late Paleozoic radiation of insects and their subsequent co-evolution with flowering plants.

# Core Concepts Summary

## 44.1 THE ANIMAL TREE OF LIFE CONSISTS OF MORE THAN A MILLION SPECIES.

Sponges lie on the deepest branch of the animal tree of life. page 44-1

All animals other than sponges are eumetazoans and fall into four major groups: cnidarians, ctenophores, placozoans, and bilaterians. page 44-1

Early biologists grouped animals on the basis of shared features of adult bodies, such as symmetry and type of body cavity. page 44-2

Bilaterians are characterized by bilateral symmetry, three germ layers, and well-developed organ systems. They include insects, worms, and mammals, and can be grouped as protostomes and deuterostomes. page 44-3

Embryology provided additional information on animal phylogeny, and, in the age of molecular biology, molecular sequence comparisons have provided still more information. page 44-4

## 44.2 THE EARLY BRANCHES OF THE ANIMAL TREE INCLUDE SPONGES, CNIDARIANS, CTENOPHORES, AND PLACOZOANS.

Sponges have a simple anatomical organization, feed by drawing water containing dissolved food particles into their interiors, and are widespread in the oceans. page 44-5

Cnidarians form true epithelial tissue, differentiate muscle and nerve cells, and act as predators by means of specialized cells called nematocysts. page 44-7

Ctenophores, or comb-jellies, resemble cnidarians but move by the beating of cilia, have an anal pore for waste excretion, and have a rudimentary mesoderm germ layer. They remain a difficult limb to locate on the animal tree of life. page 44-8

Placozoans have a very simple organization, but genomic data support the hypothesis that they are the sister group to all other eumetazoans. page 44-9

## 44.3 BILATERIANS, INCLUDING PROTOSTOMES AND DEUTEROSTOMES, HAVE BILATERAL SYMMETRY AND DEVELOP FROM THREE GERM LAYERS.

Protostomes are divided into lophotrochozoans and ecdysozoans. page 44-11

Lophotrochozoans consist of 17 phyla, including annelid worms and mollusks. page 44-11

Annelid worms include earthworms and leeches, but most species of annelids live in the oceans. page 44-11

Mollusks include gastropods (snails, slugs), cephalopods (squid, octopus), and bivalve mollusks (clams, oysters). page 44-11

Ecdysozoans include several phyla that molt their cuticle, notably the arthropods. page 44-14

Arthropods can be divided into four main groups—insects, chelicerates (spiders, scorpions), myriapods (centipedes, millipedes), and crustaceans (lobsters, shrimp). page 44-14

Deuterostomes include three main phyla—Chordata, Hemichordata, and Echinodermata. page 44-18

Well-known echinoderms are sea stars and sea lilies, which have outward fivefold symmetry. page 44-18

Chordates include vertebrates, cephalochordates, and tunicates. page 44-19

## 44.4 VERTEBRATES HAVE A CRANIUM AND VERTEBRAL COLUMN, AND INCLUDE FISH, REPTILES, BIRDS, AND MAMMALS.

The group commonly known as "fish" consists of four distinct groups of aquatic vertebrates: hagfish and lampreys, cartilaginous fish, bony fish, and fleshy-finned fish (lungfish). page 44-22

Lungfish are the closest relatives of tetrapods, which include amphibians, lizards, turtles, crocodilians, birds, and mammals. page 44-24

Amphibians have an aquatic larval form and a terrestrial adult form. page 44-24

Amniotes, such as lizards, snakes, crocodiles, birds, and mammals, have an amniotic egg, permitting movement into dry, terrestrial habitats. page 44-25

Mammals are covered with hair and feed their young milk from mammary glands. page 44-25

## 44.5 ANIMALS EVOLVED MORE THAN 575 MILLION YEARS AGO IN THE OCEANS, AND BY 500 MILLION YEARS AGO THE MAJOR STRUCTURAL AND FUNCTIONAL BODY PLANS OF ANIMAL PHYLA WERE IN PLACE.

Ediacaran fossils from 575 million years ago provide evidence of early animals. page 44-27

The Cambrian explosion was an interval of rapid diversification beginning 542 million years ago, during which time most of the animal body plans we see today first evolved. page 44-27

Chelicerates and insects were the first animals to colonize the land, sometime after 420 million years ago. page 44-28

Tetrapods first appear in the fossil record about 360 million years ago. page 44-28

Mammals originated at least 210 million years ago, but became dominant only after the extinction of the non-avian dinosaurs 65 million years ago. page 44-28

The fossil record of reef ecosystems reflects both evolutionary innovation and mass extinction. page 44-29

## Self-Assessment

1. Draw a simplified animal tree of life, indicating the relationships among sponges and the four other major groups of animals.

2. Describe what kinds of trait are used to classify animals.

3. Compare and contrast the anatomical organization and feeding patterns of sponges, cnidarians, comb-jellies, and placozoans.

4. Describe feeding habits that do not require a head.

5. Name three features that account at least in part for the success of insects.

6. Draw a phylogeny of deuterostomes, indicating the most likely relationships among chordates, hemichordates, and echinoderms, with common examples of each group.

7. List features common to fish, amphibians, amniotes, and mammals.

8. Describe the evolutionary significance of the Cambrian explosion.

Do you understand the chapter's Core Concepts? Log in to check your answers to the Self-Assessment questions, then practice what you've learned and reinforce this chapter's concepts by working through the problems and multimedia tutorials provided there.

📖 **www.biologyhowlifeworks.com**

# ANIMAL BEHAVIOR

## Core Concepts

There is no Nobel Prize for biology. It is the Prize for Physiology or Medicine that is awarded for biological research. Biologists specializing in other areas, such as behavior and ecology, are not even in the running for science's ultimate prize—except, that is, in 1973, when the Nobel Committee bent its own rules and awarded the Prize for Physiology or Medicine to three scientists with no physiological or medical credentials. The three were Niko Tinbergen, Konrad Lorenz, and Karl von Frisch. The award was made in recognition of their roles in pioneering the study of animal behavior under natural conditions.

Why "under natural conditions"? In the early 1900s, Carl von Hess, a prominent visual physiologist, claimed that bees are color blind, a conclusion based on experiments in laboratory settings. However, by training bees to go to food sources associated with differently colored cards in a natural context, the Austrian behavioral biologist Karl von Frisch demonstrated that, in fact, bees have excellent color vision. He emphasized this point at a scientific meeting where he was showing bees that had been trained to associate blue cards with food. By chance, the audience was wearing blue name tags, which proved highly popular with the blue-fixated bees! Von Frisch showed that behavior cannot be studied in isolation, but can only be fully understood when studied in a natural context—in this case, that of foraging for food.

In this chapter, we explore animal behavior—how animals learn, communicate with one another, form social groups, and choose mates. In each case, animal behavior can be considered at the individual level; individual behavior is determined in part by a complex interplay of the nervous and endocrine systems. Behavior is also influenced by the environment, as we saw with von Frisch's bees. Finally, behaviors, like all traits, are shaped by natural selection.

## 45.1 TINBERGEN'S QUESTIONS

We begin our exploration of animal behavior by asking a simple question: Why does an animal exhibit a particular behavior? The Dutch behavioral biologist Niko Tinbergen divided this over-arching question into four separate questions, each one focusing on a different aspect of the behavior. Let's consider an example—Why does a bird sing? While apparently straightforward, this question can be broken down into more specific questions and then answered in a number of different ways:

1. *Causation.* What physiological mechanisms cause the behavior? This question can have multiple answers. A bird sings because its hormone levels have changed in response to changes in day length. Or, more immediately, a bird sings because air passing through its specialized singing organ, the syrinx, causes membranes to vibrate rhythmically.

2. *Development.* How did the behavior develop? Here the focus is on the role of genes and the environment in shaping the development of the behavior. In birds, typically it's the male that sings, and he has learned the song from his father.

3. *Adaptive function.* How does the behavior promote the individual's ability to survive and reproduce? In this case, the answer to the question might be that a male bird sings in order to attract a mate and then reproduce.

4. *Evolutionary history.* How did the behavior evolve over time? Complex bird songs may have evolved from vocalizations made by ancestors that were reinforced and became increasingly stereotyped or ritualized over time, so much so that we can often identify a bird species simply by hearing it sing. A behavior may have originated to fulfill a function different from the one it currently serves. For example, the song of a particular species may have first evolved to claim a territory but now is used to attract mates.

Tinbergen's analysis allows us to see that different answers to the question "Why does a bird sing?" can all be correct. The first two, causation and development, are mechanistic explanations for behavior, whereas the second two, adaptive function and evolutionary history, provide evolutionary explanations of how natural selection has shaped a behavior over time. Tinbergen's four questions are complementary ways of looking at the same problem, and are a good starting point for the analysis of behavior.

## 45.2 GENES AND BEHAVIOR

The answers to Tinbergen's questions rely on an interplay between genes and the environment. The influence of genes is especially clear in **innate** behaviors, those that are instinctive and carried out regardless of earlier experience. **Fig. 45.1** shows that a male silkworm moth of the genus *Bombyx* flies upwind toward the source of a female-produced pheromone, an airborne chemical signal that communicates with members of the same species, in this case to attract the opposite sex. Pheromones released from a female's abdominal gland are sensed by small hairs on the male's antennae. In the presence of pheromones, the antennal sensory hairs fire action potentials (Chapter 35). When about 200 hairs are activated per second, a male flies upwind distances of a kilometer or more, tracking the increasing pheromone concentration until he finds the female. The male does not need to learn this behavior; he performs it spontaneously. His genes encode molecular receptors to which the pheromone binds, and this binding triggers a cascade of events that result in the moth heading up the pheromone's concentration gradient.

A particular behavior may also depend on an individual's experience. In this case, the behavior has been **learned.** As we will see, even neurologically simple organisms have a considerable capacity to learn. Fruit flies, for example, learn to avoid certain locations or substances if they associate them with an unpleasant experience.

We can consider, then, the extent to which a particular behavior is genetically encoded (part of the animal's **nature**) and the extent to which it is conditioned by the environment in which the animal develops (part of the animal's **nurture**). However, as we discussed in Chapter 18, nature and nurture are inextricably linked. For example, most human behaviors are the product of an interaction between nature and nurture. Consider language in the context of Tinbergen's second question, concerning the development of behavior: We have an innate ability for acquiring language, but the specific language that we acquire depends entirely on our nurture. A baby in Italy learns Italian, and one in Finland learns Finnish.

### The fixed action pattern is a stereotyped behavior.

Behavior can be complex and therefore influenced by a large number of factors. For that reason, complex behaviors can sometimes defy analysis. Scientists studying behavior tend to focus on simple behaviors because they are more easily studied and understood.

Some of the first behaviors to be carefully analyzed were **displays,** which are patterns of behavior that are species specific and tend to be highly repeatable and similar from one individual to the next. Presumably, natural selection has favored display behaviors that are unmistakable in their function as signals. Consider, for example, courtship displays that are often precursors to mating in birds (**Fig. 45.2**). Experiments have shown that a bird raised in complete isolation from other members of its species will still perform a courtship display with great precision.

Displays are an example of a **fixed action pattern (FAP).** A FAP is a sequence of behaviors that, once triggered, is followed through to completion. A classic example, originally studied by Tinbergen, is the response of a goose to an egg that has fallen from its nest (**Fig. 45.3**). The stimulus that initiates the behavior is the **key stimulus** and, in this case, it is the sight of the misplaced egg. This sight provokes in the goose an egg-retrieval

**FIG. 45.1** **An example of innate behavior.** A male moth flies toward the source of a female-produced pheromone.

Female moth

Pheromone

Male moth

**FIG. 45.2 Courtship display in the Raggiana Bird-of-paradise.** The male (left) is performing a highly repeated set of behaviors to entice the female to mate with him.

FAP, which consists of rolling the egg back to the nest with the underside of its beak (Fig. 45.3a). This response cannot be broken down into smaller subunits and it is always carried out to the end, even if it is interrupted. In fact, the goose will persist in its efforts even if the researcher ties a string around the misplaced egg and removes it while the goose is in mid-action. The goose will still continue the task of rolling the now-absent egg (Fig. 45.3b).

It is possible to understand this behavior by varying attributes of the key stimulus (in this case, the misplaced egg). The researcher can make models that vary in only one attribute to investigate how sensitive the goose is to aspects such as color or shape. A remarkable finding is that many birds respond most strongly to the largest round object provided, even a soccer ball, as illustrated in Fig. 45.3c. A soccer ball in fact not only elicits the egg-retrieval FAP, but it does so even more strongly than does a normal egg. The soccer ball is considered a **supernormal stimulus** because it is larger than any egg the goose would naturally encounter and elicits an exaggerated response. Natural selection has likely favored geese that recognize and respond to large eggs that have rolled outside the nest, but since eggs never grow to the size of soccer balls in normal circumstances, selection has not shaped an appropriate response to unrealistically large egg-shaped objects.

## The nervous system processes stimuli and evokes behaviors.

How do animals recognize the stimulus that leads to a particular behavior? As we have seen, a goose attempts to retrieve an egg that is nearby but outside its nest. How does it recognize that the object in question is an egg? At its root, this is an extraordinarily difficult problem. The world is full of stimuli and there are many

**FIG. 45.3 A fixed action pattern.** A goose displays and completes the behavior to retrieve an egg (a), even if the behavior is interrupted (b) or the stimulus altered (c).

**a.**

Normal egg retrieval by a goose is an example of a fixed action pattern, a stereotyped behavior in response to a stimulus.

**b.**

This behavior persists even if the egg is tied with a string, preventing it from being brought to the nest.

**c.**

A soccer ball elicits an even stronger response.

things that look like an egg. Thus, the challenge for the animal is to filter out the correct signal (the egg) from the surrounding noise (everything else). An animal has to process many different kinds of information in three dimensions.

We know that stimulus recognition is often carried out by **feature detectors,** specialized sensory receptors or groups of sensory receptors that respond to important signals in the environment. Frogs provide a good example. Male frogs call to attract females and warn off other males. A single pond may contain many species of frog, each with its specific advertisement vocalization, or call. A heavily populated frog pond is positively cacophonous, but it is critical to an individual frog that it be able to distinguish the call of other members of its own species from those of other species. Further complicating the matter is that different species of frog can have similar calls, often virtually indistinguishable to the human ear.

One approach to the problem of understanding how frogs recognize their species-specific call has been to experiment with recordings of frog calls. By recording frog calls, altering aspects of the call in specific ways, playing them back, and then monitoring the responses of frogs, we can identify the impact of different components—pitch, duration, and pulse frequency, for

example—of the call. These kinds of study have demonstrated that the frog's auditory nerves can act as a feature detector, each one "tuned" to a specific component of the call, as shown in **Fig. 45.4.** The combination of different feature detectors stimulated by different components adds up to the recognition by the frog of a specific call. Once the sound has been correctly identified, the appropriate behavior follows.

## Hormones can trigger certain behaviors.

As we have seen, stimuli are detected and processed by sensory receptors. In some cases, the response includes the release of a hormone. The power of a hormone lies in its ability to affect multiple cells in target organs simultaneously (Chapter 38). A neuron, on the other hand, is typically much more specific in its action, connecting directly to another neuron or neurons, or to a muscle (Chapter 35).

Hormones can affect behavior. The hormone testosterone, for example, is critical in sex determination in mammals, including humans (Chapter 42). In addition, in many respects males and females behave differently, and testosterone is at least partially responsible for many sex-specific behaviors.

*Anolis* lizards demonstrate the important role of hormones in sexual behavior, showing both how social stimuli can affect the release of hormones and how hormones can affect behavior (**Fig. 45.5**). Females of *Anolis carolinensis* collected in the spring and housed in the laboratory under a spring light-dark cycle prepare for reproduction, and about 80% of individuals have active egg follicles in their ovaries (Fig. 45.5a). If one male is added to a group of females, however, that figure increases to 100% (Fig. 45.5b). The courting behavior of the male lizard stimulates the females to produce hormones that cause the full development of the ovaries, making the females reproductively active. If a group of males instead of a single male is added to an all-female population, on the other hand, only about 40% of the females undergo ovarian development (Fig. 45.5c). This unexpected result seems to occur because the males interact and fight among themselves rather than court the females, and therefore the courtship stimulus is lacking. Castrated males (which do not produce testosterone) that are added to a group of females have no effect on rates of ovarian development, which remains about 80% (Fig. 45.5d). They fail to court because the behavior is

**FIG. 45.4    A feature detector.** The frog has a feature detector tuned to its species-specific call, allowing the frog to distinguish the call of its species (red arrow) from the calls of other frogs (blue arrows).

**Sonograms  representing frog calls**

Sound frequency

Call of species 1

Species 2

Species 3

Species 4

An individual of species 3

Species 5

Species 6

Time

**FIG. 45.5** Role of hormones in behavior, demonstrated by groups of *Anolis* lizards.

**a.** + 0 males

In a group of all female *Anolis carolinensis* collected in the spring, 80% of the females are prepared for reproduction with egg follicles in their ovaries.

**b.** + 1 male

A male displaying courtship behavior added to the all-female group causes 100% of females to become reproductively active.

**c.** + more than one male

More than one male added to a group of females fight with each other instead of courting females. Now only 40% of the females are reproductively active.

**d.** + castrated male

Castrated males do not display courtship behavior, and so do not affect the females' reproductive readiness.

**e.** + castrated male injected with testosterone

An injection of testosterone causes a castrated male to display courtship behavior, affecting the group of females in the same way as a non-castrated male.

■ Reproductive  ■ Nonreproductive

**FIG. 45.6** **Artificial selection.** Artificial selection can alter behavior, as shown by the remarkable differences in behavior we see among breeds of dogs: (a) a Pointer and (b) a Border Collie.

testosterone-mediated. However, a castrated male that is injected with testosterone does display courting behavior and has the same effect as the presence of a single male, inducing all the females to undergo ovarian development (Fig. 45.5e). These simple laboratory experiments demonstrate the complex interplay between hormones and behavior.

## Breeding experiments can help determine the degree to which a behavior is genetic.

The nervous and endocrine systems that in part shape behaviors are the product of genetic instructions. If you take an extreme view, then, all behaviors have a genetic component. After all, it is the instructions in the genome that build the nervous system, muscles, and glands that lead to behavior. On the other hand, all behavior can be considered environmental since, without the appropriate environment, an organism would not develop the ability to perform the behavior. On a very basic level, an organism in an environment without appropriate nutrients would lack the necessary energy to perform a given behavior. How, then, do we study the genetic basis of behavior? Modern approaches harness the power of molecular genetics, but more traditional analyses are also highly informative.

Artificial selection, in which humans breed animals and plants for particular traits, provides strong evidence of the role of genetics in influencing behavior. Dogs were domesticated from wolves about 10,000 years ago. Today, there is great diversity among dogs, although they are all the same species. Obviously, there has been extensive selection on physical traits, with the result that a Dachshund, for example, looks quite different from a German Shepherd or Great Dane. Selection has also been applied to behavior: A Pointer has extraordinary "pointing" behavior that specifies the location of a hunter's prey, and a Border Collie is an excellent herder (**Fig. 45.6**).

In the late 1950s, William Dilger studied the nest-building behavior of lovebirds in the genus *Agapornis*. Some species transport their nesting material in their beaks, whereas others tuck pieces into their tail feathers. Is there a genetic basis to the nesting behavior of these birds? To answer this question, Dilger set up crosses between species with different nest-building techniques to produce hybrids. He then observed how the hybrids built their nests. Interestingly, the hybrid offspring showed nest-building behavior intermediate between that of the parents: They tried to tuck material under their feathers, but were not successful. These experiments suggest that this behavior has a genetic basis.

It was not possible to do a true Mendelian dissection of the lovebirds' behavior (Chapter 16). Such an analysis would require additional crosses, which were not possible because the hybrids were sterile. Like most behaviors, the lovebirds' nest-building behavior is probably controlled by a large number of genes, each with a relatively small effect (Chapter 18). Therefore, it is difficult to identify which genes are responsible for the behavior. Even with molecular methods, the task of mapping and identifying the underlying genes when a large number of different genes govern a particular trait is daunting.

## Molecular techniques provide new ways of testing the role of genes in behavior.

Molecular biology is changing the way we approach behavioral genetics. Some studies have identified complex behaviors that are strongly influenced by a single gene. One well understood instance of a single gene's effect on behavior comes from the fruit fly *Drosophila*.

In *Drosophila*, there are two different alleles of the *foraging* (*for*) gene—*for*$^s$ and *for*$^R$—that are common in populations. These two alleles have different effects on the behavior of *Drosophila* larvae. The gene encodes a cGMP-dependent protein kinase, an enzyme expressed in the brain that affects neuronal activity and alters behavior. In the absence of food, both "sitter" (*for*$^s$) and "rover" (*for*$^R$) larvae move about in search of food. In the presence of food, however, sitters barely move, feeding on the patch on which they find themselves, whereas rovers move extensively both within a patch of food and between patches (**Fig 45.7a**).

Both forms are present in natural populations (**Fig. 45.7b**); typically, 70% are rovers, 30% are sitters. This proportion suggests that both strategies are adaptive. Furthermore, studies have shown that rovers are selected for in crowded environments in which there is an advantage to seeking new food sources, and sitters are selected for in less-crowded

**FIG. 45.7** **The effect of genotype at the foraging locus on the feeding behavior of *Drosophila melanogaster* larvae.** (a) When food is present, rover larvae travel farther than sitter larvae. (b) Both types of larvae exist in natural populations.

a.

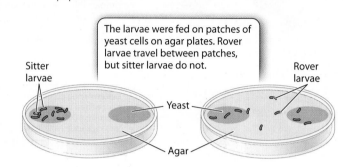

The larvae were fed on patches of yeast cells on agar plates. Rover larvae travel between patches, but sitter larvae do not.

Sitter larvae

Rover larvae

Yeast

Agar

b.

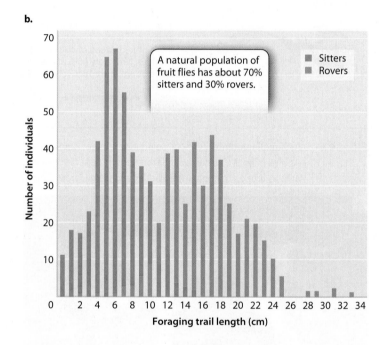

A natural population of fruit flies has about 70% sitters and 30% rovers.

■ Sitters
■ Rovers

ones because they take maximum advantage of their current food source. In this case, variation at a single gene affects a complex behavior in fruit flies.

The *foraging* gene is present in other insects as well, including honeybees, suggesting that it has been evolutionary conserved. Honeybees with low levels of *for* expression in their brain tend to stay in the hive, whereas those with high levels of *for* expression are likely to be foragers (**Fig. 45.8**).

→ **Quick Check 1** The *for* gene affects foraging behavior in fruit flies and honeybees. In what way does it act differently in the two insects?

FIG. 45.8

# Can genes influence behavior?

**BACKGROUND** In fruit flies (*Drosophila melanogaster*), variation at the *foraging* (*for*) gene affects feeding behavior of larvae. Two alleles of the *for* gene—*for$^s$* and *for$^R$*—exist in natural populations; *for$^s$* larvae tend to stay on a patch of food ("sitters") and *for$^R$* larvae tend to move from patch to patch ("rovers"). Honeybees (*Apis mellifera*) also forage for food, but, in contrast to fruit flies, this behavior changes with age. Young honeybees stay at the hive ("nurses"), and older ones forage for nectar ("foragers"). The *for* gene is present in honeybees, raising the possibility that this gene is involved in the developmental change in foraging in honeybees.

**HYPOTHESIS** Foraging in honeybees is influenced by the expression of the *for* gene, and nurses have lower expression levels than foragers.

**EXPERIMENT 1** Levels of *for* mRNA were measured in honeybee nurses and foragers.

**RESULTS** Levels of *for* mRNA are significantly higher in foragers than in nurses (Fig. 45.8a). However, foragers are older than nurses, so it was unclear whether the increased expression in foragers compared to nurses is associated with differences in age or differences in behavior. Therefore, the experiment was repeated with honeybees that forage at a young age. These young foragers also have higher *for* mRNA levels than do nurses (Fig. 45.8b).

**EXPERIMENT 2** The *for* gene encodes a cGMP-dependent kinase that phosphorylates other proteins (Chapter 9). Honeybee nurses were treated with cGMP to activate the kinase and their subsequent behavior was monitored. A related compound, cAMP, with a chemical structure similar to cGMP but which does not affect the kinase, was used as a control, to ensure that any effect observed was specific for the pathway and not the result of the treatment.

**RESULTS** Treatment with cGMP changed the behavior of nurses, causing them to forage (Fig. 45.8c). Furthermore, cGMP acted in a dose-dependent fashion: The higher the dose, the more foraging behavior was observed. No effect on foraging was seen in the control treatment (Fig. 45.8d).

**CONCLUSION** The same gene that is involved in two different behavioral phenotypes in fruit flies (sitters vs. rovers) is also involved in a developmental behavioral change in honeybees (nurses vs. foragers). This finding suggests that a gene can have related but different functions in different organisms.

**FOLLOW-UP WORK** Researchers have looked at the expression profiles of many other genes in the honeybee genome to learn which genes are associated with the change of foraging behavior of honeybees.

**SOURCE** Ben-Shahar, Y., A. Robichon, M. B. Sokolowski, and G. E. Robinson. 2002. "Influence of Gene Action Across Different Time Scales on Behavior." *Science* 296:741–744.

**FIG. 45.9**  **Differing mating behavior in voles.**  The different behaviors in (a) monogamous prairie voles and (b) promiscuous montane voles results from variation at a single gene.

Studies of the North American vole, in the genus *Microtus,* have also revealed how genes influence behavior (**Fig. 45.9**). The prairie vole is monogamous: Couples pair bond for life, meaning that they mate only with each other. By contrast, the prairie vole's close relative, the montane vole, is promiscuous: A male mates and moves on, and a female, over her lifetime, produces litters by several different males. What differences underlie these different behaviors?

Hormones and their receptors provide an answer. In all mammals, antidiuretic hormone (ADH) controls the concentration of urine (Chapter 41). The receptor for ADH is not only expressed in the kidney, but also in the brain, where it affects behavior. Both the monogamous prairie vole and the promiscuous montane vole produce ADH, but the hormone receptors in the two species are different. The American neuroscientist Thomas Insel analyzed the genes that encode the ADH receptor in the two vole species. He found a difference in the region of the gene that determines under what conditions and in what cells it is expressed. Because of this difference, the distribution of ADH receptors in the brain is different in the prairie vole and in the montane vole.

Does the difference in gene expression explain the different sexual behaviors of the two species? The answer seems to be yes. Insel and his colleague Larry Young inserted the prairie vole ADH receptor gene, complete with its adjacent regulatory region, into a laboratory mouse (a promiscuous species, like the montane vole). Interestingly, Insel and Young observed a marked change in the mouse's behavior. Rather than mating with a female and then moving on to another mate, the transgenic male mouse pair bonded with the female. In short, the addition of a single gene affected a complex behavior such as pair bonding, making the otherwise promiscuous mouse more likely to pair bond.

## 45.3 LEARNING

Genes are not the only determinants of behavior. We all know that experience often leads to changes in behavior. We call this process **learning.** Humans are extraordinary learners, but even many relatively simple organisms respond to experience. Researchers have identified several types of learning. The categories are not strict, and there is some overlap among them, but they are a useful way to organize our thinking and give us deeper insights into the subject.

### Non-associative learning occurs without linking two events.

**Non-associative learning** is learning that occurs in the absence of any particular outcome, such as a reward or punishment. One type of non-associative learning is **habituation,** which is the reduction or elimination of a behavioral response to a repeatedly presented stimulus. Chicks presented with silhouettes flying overhead provide an example. Initially, any overhead silhouette provokes a defensive, crouching posture, but eventually, chicks habituate to overhead silhouettes that have proved not to be threatening. In the absence of any consequences, chicks no longer crouch in response to a harmless silhouette passing repeatedly overhead.

**Sensitization** is another form of non-associative learning. Sensitization is the enhancement of a response to a stimulus that is achieved by presenting a strong or novel stimulus *first.* This pre-stimulus makes the animal more alert and responsive to the next stimulus. The sea slug, *Aplysia,* a model organism among scientists interested in the neuronal basis of behavior, exhibits sensitization. *Aplysia* withdraws its gills in response to a touch on the siphon, the tube through which it draws water over its gills. Interestingly, if the animal is given a weak electric shock first, its response to a touch on its siphon is much more rapid, suggesting that the shock has made the slug more sensitive to later stimuli.

### Associative learning occurs when two events are linked.

**Associative learning** (also called **conditioning**) occurs when an animal learns to link (or associate) two events. Perhaps the most famous example of conditioning is Ivan Pavlov's dogs. Pavlov first presented the dogs with meat powder, and they salivated in response. He then presented the dogs with an additional cue, a ringing bell, whenever he presented the dogs with meat powder. After repeated experience of the two stimuli together, the dogs salivated at the sound of the bell alone, in expectation of the meat reward. This form of conditioning, in which two stimuli are paired, is called **classical conditioning.** In this case, a stimulus that leads to a behavior (the meat powder) was paired with a neutral stimulus (the bell) that initially had nothing to do with salivation. Eventually, a novel association is made between the sound of the bell and food, and the sound of the bell alone elicits salivation.

A second form of associative learning involves linking a behavior with a reward or punishment. If the behavior is rewarded (positively reinforced), it is more likely to occur the next time around. It is not just out of the goodness of her heart that the trainer at Sea World gives a sea lion or dolphin a fish reward at the end of each trick that the animal performs. However, if the behavior is punished (negatively reinforced), the response becomes less likely. This form of associative learning is called **operant conditioning.**

In classical conditioning, an association is made between a stimulus and a behavior, whereas in operant conditioning, an association is made between a behavior and a response. In addition, operant conditioning involves a novel, undirected behavior that becomes more or less likely over time. Consider a rat in a cage with some levers in it, each a different color. The rat has no impulse to press any of the levers, but by chance, it presses the red one. A food reward tumbles into the cage. Eventually the

rat learns to associate pressing the red lever with the food reward; it has been operantly conditioned. It is thought that play in young animals provides them with the opportunity to explore their environment much in the same way as the rat does when first placed in the cage. Like the rat, they may find that some behaviors are more rewarding or more punishing than others.

→ **Quick Check 2** Both classical and operant conditioning involve learning to associate two events. What then is the difference between the two kinds of conditioning?

### Learning takes many forms.

The capacity to learn often has adaptive functions. Much learning, including human learning, is based on **imitation:** One individual copies another. Learning by imitation was famously observed in the days when people in Britain had milk deliveries left outside on the doorstep: A number of birds, including Great Tits and Blue Tits, learned to open foil-topped milk bottles, and this behavior spread rapidly through populations of these birds as naïve birds copied the successful birds' approach. Individuals can also learn by imitating individuals of a different species. An octopus can learn to open a jar with a reward inside by watching a human or another octopus do it.

The capacity to learn a particular task seems in many instances to be innate. Female digger wasps are foragers. Tinbergen showed that a digger wasp learns the landmarks around her nest and then uses this information to find her way back to it from hunting. In his experiment, illustrated in **Fig. 45.10,** Tinbergen placed a ring of pine cones around a wasp's nest and, after she had flown out and returned a few times, moved the circle of pine cones to the side of the nest. The wasp flew past her own nest and went directly to the center of the ring. Such innate responses improve the survival or reproductive success of the individual.

This adaptive aspect of animal learning is also revealed by taste aversion experiments, in which an animal typically learns to avoid certain flavors associated with a negative outcome. Rats learn to avoid flavored water if consuming it is associated with nausea. However, they do *not* learn to avoid flavored water if it is associated with a different kind of negative reinforcement, such as a mild electric shock. Rats, it turns out, can make some associations, but not others. The ones they can learn are the biologically meaningful ones, those that favor survival. In the course of evolution, the rat's ancestors encountered poisoned food that resulted in nausea and the aversion response evolved. Until humans started doing experiments on them, rats had never encountered a bad meal that resulted in an electric shock. It is not surprising, then, that the ability to pair these two phenomena never evolved. These experiments also show that the specific

**FIG. 45.10**

# To what extent are insects capable of learning?

**BACKGROUND** European digger wasps, *Philanthus triangulum,* live in the sand. These wasps are sometimes called "bee wolves" because they specialize on hunting for honeybees to feed their developing young. After mating, each female digs a long burrow with a few chambers at the end where she lays her eggs. She then forages for honeybees that she brings back to these chambers for her larvae to eat. The wasp faces a navigational challenge: Having captured her prey, how can she find her way back to and recognize her nest? Niko Tinbergen noticed that wasps lingered briefly in the vicinity of a new nest before heading off to hunt, and thought that they were learning local landmarks associated with the nest.

**HYPOTHESIS** The wasp learns visual cues around her nest to help her locate it upon her return.

**METHOD** Tinbergen recognized that a good test of the learning abilities of an insect should take place in its natural environment. He combined his skills as a naturalist and as an experimentalist to devise an elegant demonstration of the way in which female wasps learn landmarks for navigation. Tinbergen's approach was simple. He placed a ring of pine cones around the nest of a wasp, and then, once she had left to hunt, he shifted them to a new location away from the nest entrance. If visual cues are key, the wasp would return to the displaced pine cone ring. On the other hand, if cues are, for example, olfactory, she would return directly to the nest.

**RESULTS** Females carried out a brief landmark-learning flight on departure from the nest (Fig. 45.10a). When the landmarks were displaced, the females returned to the wrong location (Fig. 45.10b).

**CONCLUSION** Female digger wasps learn and then use local landmarks, such as a ring of pine cones, as cues to the nest location.

**SOURCE** Tinbergen, N. 1958. *Curious Naturalists.* New York: Basic Books.

**a.**

**b.**

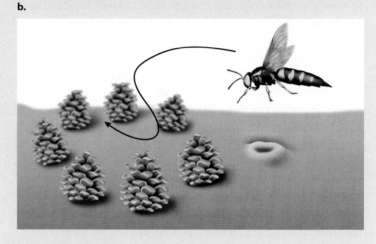

---

evolutionary history of the species in question matters when analyzing animal behavior.

In addition to adaptive predispositions for *what* can be learned and not learned, many species exhibit predispositions for *when* learning takes place. This is particularly evident in **imprinting,** a form of learning typically seen in young animals in which they acquire a certain behavior in response to key experiences during a critical period of development. Konrad Lorenz made imprinting famous by exploiting the observation that newly hatched goslings and ducklings rapidly learn to treat any animal they happen to see shortly after hatching as their mother. Lorenz found that,

if he was the first person the hatchlings saw, they would follow him as though he were their mother (**Fig. 45.11**). This behavior is adaptive because the first being that a hatchling normally sees is a parent. This type of imprinting is called **filial imprinting,** and it is most common in species whose offspring leave the nest and walk around while still young, like chicks and ducklings. Filial imprinting is rare in species of birds whose young stay in the nest until they are able to fly away.

Experiments have shown that filial imprinting typically occurs during a specific, sensitive period in the animal's life and that the results are usually irreversible. After Lorenz's baby ducks had

**FIG. 45.11** Imprinting on (a) Konrad Lorenz or (b) an ultralight aircraft. By manipulating newly hatched birds' first sight of a moving being, it is possible to induce powerful attachments in the birds.

imprinted on him, they would not change their minds about who their parent was even when presented with their real mother duck. The timing of the sensitive period varies from species to species. In some cliff-nesting sea birds, it has been shown that imprinting on auditory stimuli (like the call of the parents) begins while the chick is still in the egg.

## 45.4 ORIENTATION, NAVIGATION, AND BIOLOGICAL CLOCKS

Learning is a form of information processing, in which experience shapes behavior. At a mechanistic level, it typically involves a change in the strength of connections between neurons. The central nervous system of even relatively simple animals is capable of extraordinary feats of information processing. The ways in which animal nervous and endocrine systems and the input of experience are integrated to generate adaptive behaviors are remarkable. Some of these adaptive behaviors include ways of moving, navigating, and keeping time. Here, we look at how organisms integrate environmental stimuli to produce these behaviors.

### Orientation involves a directed response to a stimulus.

Even the simplest bacteria and protozoa are capable of moving in response to stimuli. A *Paramecium* that finds itself in an unfavorable environment, such as water that is too warm or too salty, increases its speed and begins to make random turns. When it finds favorable conditions, such as cooler water, it slows and reduces its turning rate. These random, undirected movements are termed **kineses.**

In contrast, **taxes** (singular, taxis) are movements in a specific direction in response to a stimulus. An interesting example of a taxis is movement oriented to a magnetic field, a behavior called magnetotaxis—the term was first used by the American microbiologist Richard Blakemore in 1975. Blakemore had found that anaerobic bacteria in the genus *Aquaspirillum,* which swim by means of flagella, tend to swim toward magnetic north. These bacteria can be attracted to the side of a dish with a bar magnet. Little bits of magnetized iron oxide, arranged in a row inside the bacterial cell, allow the bacteria to sense the magnetic field. Blakemore hypothesized that the bacteria swim north in order to swim deeper. In the northern hemisphere, the north magnetic pole is inclined downward. At Woods Hole, Massachusetts, where these bacteria were found, the magnetic pole is at about a 70-degree incline. Since the bacteria are anaerobic, they must stay buried in sediment, and they remain buried by moving downward along the magnetic gradient.

To test his hypothesis, Blakemore looked for and found bacteria in New Zealand that exhibited similar behavior to the Woods Hole bacteria except that they swam toward the *south* magnetic pole, bringing them downward in the southern hemisphere. Finally, he took New Zealand cultures back to Woods Hole, where the bacteria swam up into the oxygen and died. These experiments demonstrated that bacteria are able to sense a magnetic field and move in a directed fashion relative to that field.

### Navigation is illustrated by the remarkable ability of homing in birds.

Many animals use environmental cues to migrate long distances. The navigational achievements of homing pigeons are legendary. They can home—that is, find their way back to the place where they are housed and fed—over extremely long distances, even more than a thousand miles. It seems that they use a wealth of cues when homing. A compass may tell you which way is north, but you must also have **map information**—you must know where you are with respect to your goal—if the compass is to be useful for finding a particular location. So pigeons must have both compass and map senses. That map sense is presumably based on landmarks. Given the long distances the pigeons travel, these landmarks must vary in their type and probably include olfactory as well as visual cues.

The pigeon compass relies on different kinds of information. For example, pigeons can navigate during the day using the sun as a compass, and at night they use the stars. And, like *Aquaspirillum*, they can detect the Earth's magnetic field. The magnetic navigation system is presumably important to pigeons on cloudy days. Researchers have performed experiments with pigeons wearing magnetic helmets, which disrupt the pigeons' ability to detect the Earth's magnetic field. On cloudy days, the pigeons wearing the helmets are unable to home. On sunny days, however, the pigeons' sun compass overrides the erroneous magnetic information of the helmets.

The sun compass requires information about time as well. Every hour, the sun moves 15 degrees through the sky. To determine where north is, a pigeon needs information from the sun as well as some way to keep time. It turns out that pigeons have a clock—a biological one.

### Biological clocks provide important time cues for many behaviors.

Like us, other animals live in space *and* time. So far, we have focused on their behavior in space. For many species, time is a life-and-death matter. When to migrate or mate is a critical decision in a seasonal environment. Researchers are beginning to unravel the neural and genetic underpinnings of the clocks in some species. In model organisms like *Drosophila*, for example, researchers have identified a number of genes that, when mutated, cause the clock to run slow or fast.

A biological clock is a set of molecular mechanisms that cycles on its own and therefore keeps a regular rhythm. It helps to control physiological or behavioral aspects of an animal. Different clocks work on different timescales. Daily cycles are governed by a **circadian clock.** Circadian clocks regulate many daily rhythms in animals, such as feeding, sleeping, hormone production, and core body temperature. Some species, especially seacoast species living in habitats where the tides are important, time activities by a **lunar** (moon-based) **clock.** There are also **annual** (yearly) **clocks**. For example, periodical cicadas are insects in the genus *Magicicada* that have a generation time of either 13 or 17 years: In a given location, there will be a cicada outbreak every 13 or 17 years.

Circadian clocks are observed in many organisms, including plants, fungi, and animals. Humans have their own circadian clocks, as jetlag never fails to remind us. The circadian clock is based on a set of clock genes. The protein products of these genes oscillate through a series of feedback loops in a roughly 24-hour cycle. Thus, when animals that, like us, are active during the day and inactive at night are placed in artificial conditions that are always lit, they continue to follow a basic day–night, active–inactive cycle. The clocks are not perfect, though: As the period

spent in constant light is prolonged, the circadian clock drifts slowly until the animal is eventually no longer synchronized with the true day–night cycle.

Biological clocks remain synchronized with the day–night cycle because they are often reset, or entrained, by external inputs. For the circadian clock, light is the primary input, so the natural light–dark cycle keeps the clock from drifting. For clocks related to the seasons, day length (known as **photoperiod**) is the critical input because it is a good indicator of the time of the year (Chapter 30). For example, many mammals produce their offspring in the spring so that they can grow over the

---

## HOW DO WE KNOW?

**FIG. 45.12**

# Does a biological clock play a role in birds' ability to orient?

**BACKGROUND** One suggestion for how pigeons home is that they use a sun compass. If you are in the northern hemisphere and you know the time is 12 noon, then the sun is due south of you. Orienting yourself by this method is possible only if you know the time, so the question arises whether homing birds have the ability to tell the time. One way to answer this question is to "clockshift" the birds. Researchers raise birds in an artificial day–night cycle that is out of sync with the actual one. When released into a sunlit environment, these birds' sense of time is shifted by a set number of hours:

| event | dawn | dusk |
|---|---|---|
| actual time | 6 am | 6 pm |
| clockshifted time | 12 noon | 12 midnight |

**HYPOTHESIS** If a bird's ability to home is dependent on an internal clock, clockshifting should affect the bird's homing ability in a predictable way. Given that the sun travels 360 degrees in 24 hours, a 6-hour clockshift will result in a 90-degree error in homing direction because $360/(24/6) = 90$.

summer, when resources are most abundant. These mammals have been selected over many generations to synchronize their reproduction at a time when the young have the greatest chance of survival. Photoperiod determines the timing of many kinds of behaviors, including migration, development, and reproduction.

The importance of the sun compass coupled with the biological clock in homing pigeons can be demonstrated by experimentally disturbing the birds' clock, as discussed in **Fig. 45.12.**

→ **Quick Check 3** Why is photoperiod such a widely used time cue?

## 45.5 COMMUNICATION

Up to this point, we have focused on individual behaviors. Behaviors often depend on environmental cues, as we saw with the examples of navigation and keeping time. In addition, behaviors are often shaped by interactions with other individuals. Learning sometimes depends on communication between a teacher and student, and, as anyone knows who has ever watched a mother duck tending her ducklings, communication is central to filial imprinting.

**EXPERIMENT** Birds were clockshifted by raising them in a chamber under an artificial light. Birds from a home loft in Ithaca, New York, were released on a sunny day at Marathon, New York, about 30 km east of Ithaca. Release on a sunny day made it possible for the birds to use the sun's position to navigate.

**RESULTS** As expected, the control birds (those that were not clockshifted) were usually good at picking the direction of their home loft, heading approximately westward toward Ithaca. The results for the clockshifted birds were very different: They miscalculated the appropriate direction. These birds headed approximately northward, as shown by the positions of the red triangles on the compass in the figure.

**INTERPRETATION** Assume the birds are released at 12 noon, when the sun is due south. The control birds know to fly in a direction 90 degrees clockwise from the direction of the sun, but the clockshifted birds "think" it is 6 p.m., so they expect the sun to be in the west. Their 90-degree clockwise correction, then, has them flying due north.

**CONCLUSION** The clear difference between control and clockshifted birds in the experiment shows that an internal time-based sun compass is an important component of the birds' homing abilities. However, the scatter of points (for both experiment and control) suggests that other factors are also important. This conclusion is reinforced by the observation that birds home well on cloudy days, when they cannot use a sun compass, suggesting that birds use multiple cues and navigational systems when they are homing.

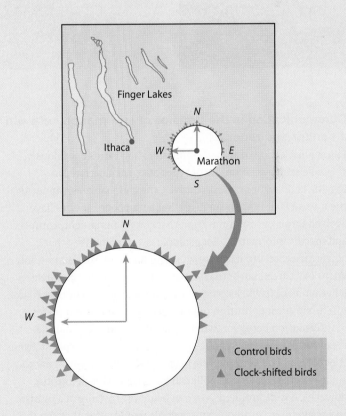

**SOURCE** Keeton, W. T. 1969. "Orientation by Pigeons: Is the Sun Necessary?" *Science* 165:922–928

**FIG. 45.13 Types of signal in different forms of communications.** Signals can be (a) visual, as in the male frigatebird's display, (b) auditory, such as a male frog's call, (c) electrical, as in African electric fish, which generate species-specific electric pulses, (d) chemical, such as the pheromones with which a queen ant controls her colony, and (e) mechanical, as when spiders "twang" webs to interact.

## Communication is the transfer of information between a sender and receiver.

The sophistication of animal communication varies enormously, as can be seen by the variety of types of signal illustrated in **Fig. 45.13.** Even our closest relatives, the chimpanzees, cannot rival the human facility with language, but a number of species have evolved forms of communication that convey remarkably complex and specific information. For example, a vervet monkey that perceives a threat to its group will utter an alarm call that not only warns of a threat, but actually specifies the nature of the threat—whether it is a hawk, leopard, snake, or some other type of predator.

The simplest definition of **communication** is the transfer of information between two individuals, the **sender** and the **receiver.** The sender supplies a signal that elicits a response from the receiver. For example, the bright petals of a flower signal to an insect that nectar and pollen are available. This definition, however, has its problems. An owl hears the rustling of a mouse and responds accordingly. Most people would agree that the owl is not really communicating with the mouse (nor the mouse with the owl). For this reason, some biologists prefer to define communication as attempts by the sender to manipulate in some way the behavior of the receiver.

How has communication evolved? It is thought that communication has often evolved through co-opting and modifying behaviors used in another context. This process

is called **ritualization,** and it involves (1) increasing the conspicuousness of the behavior; (2) reducing the amount of variation in the behavior so that it can be immediately recognized; and (3) increasing its separation from the original function. The scent markings with which mammals mark their territory provide an example. The original function was simply the elimination of waste, but this function has been modified through evolution: Strategically placed marks communicate with other individuals, indicating, for example, the extent of an individual's territory. The communication advantage is that an intruder detecting a territorial scent may be less likely to invade, thereby avoiding what could be an expensive fight.

Forms of communication have frequently evolved that prevent animals from coming to harm in what has sometimes been called limited-war strategies. Thus, two males, instead of battling it out, may engage in elaborate displays to size each other up, either literally by standing side by side or through displays of physical prowess such as roaring, in order to determine who is dominant. Fighting would be disadvantageous to both since even the winner might be seriously injured. In this case, communication may have evolved as a way for individuals to assess each other.

In some cases, communication in the natural world can be deceitful. A male may attempt to convince other males (or a female) that he is bigger (and stronger) than he really is. Possibly one reason that dogs circling in a fight raise the hair along their back is to inflate their apparent size. A potential prey may attempt to convince a predator that it is not, in fact, prey. This form of deceit is especially evident in some species of butterfly that mimic leaves or bark when their wings are at rest. Alternatively, predators may emit deceitful signals to entice their prey. Females of one species of firefly, for instance, mimic the flashes of the females of another species in order to lure males of the second species, which they then eat.

### Some forms of communication are complex and learned during a sensitive period.

Bird song is one of the best known and richest forms of animal communication. Because of the clear connection between sensory perception and motor output involved in this form of communication, songbirds have served as a model system in the study of learning and communication. These songs are complex sequences of sounds, often repeated over and over. Like cricket and frog calls, bird songs are **advertisement displays,** behaviors by which individuals draw attention to their status (for example, that they are sexually available or are holding territory). Bird songs are typically produced in the breeding season and usually only by males, although in some species females also sing and perform "duets" with their mates.

In many species of birds that sing, some or all of the song is learned, often during a specific, sensitive period. Detailed studies pioneered by British-American neurobiologist and ethologist Peter Marler of the White-Crowned Sparrow, shown in **Fig. 45.14,** have yielded the following general picture of the process.

Song in White-Crowned Sparrows is learned by imprinting: The young male hears adult song during a sensitive period, 10–50 days after hatching. Then, shortly afterward, young male White-Crowned Sparrows produce unstructured twittering sounds, known as a subsong, comparable to the babbling of human babies. If deprived of hearing adult song, for example by isolation during the sensitive period, the bird sings for the rest of his life a song not much different from unstructured twittering, even if he hears his father sing both before and after isolation.

Between about 250 and 300 days, the male sings an imperfect copy of his father's song (known as plastic song), and by about 300 to 350 days after hatching, song acquisition is complete (this song is known as structured song). At this point, even if the bird is deafened in the lab, he will still sing correctly. The song he produces is a precise copy of the one he learned, typically by hearing his father's song during the sensitive period, complete with most of its individual as well as species-specific characteristics.

So, White-Crowned Sparrows have a programmed predisposition for *when* song learning takes place. Even more interestingly, *what* can be learned is similarly constrained. If a tape of another species' song is played during the sensitive period, even the song of closely related Swamp Sparrows, the White-Crowned male cannot learn that song; his adult song ends up being not much different from his unstructured twittering. However, if he is provided with a live tutor, such as a live male Swamp Sparrow rather than a tape, his ability to learn the other bird's songs is greatly improved. Furthermore, if he hears tapes of both the song of his own species and that of closely related species played together, he can pick out the correct elements and sing a perfect White-Crowned Sparrow song. Thus, birds of this species preferentially learn their species-specific song, but they cannot sing a normal song without learning it.

### Other forms of communication convey specific information.

Honeybees have an elaborate means of communication that is quite different from the sound-based communication of birds. To collect food for a honeybee hive, about 10,000 foragers leave the nest each day, making several million foraging trips in the course of a summer, each of which may take them up to 10 km from the nest. The hive can be considered a kind of giant organism, fixed in one spot but sending out foragers in all directions. As social insect expert and evolutionary biologist E. O. Wilson points out, if honeybee workers were as large as people, they would cover an area the size of Texas in their daily search for food. Given the

---

**FIG. 45.14 Song acquisition in the White-Crowned Sparrow.** Sonograms show differences between the unstructured twittering of early life and the final, structured song.

vastness of this task, it is not surprising that returning workers use a variety of cues to inform outgoing foragers about the location and nature of the food resources they have found.

Returning workers pass samples of the collected food to other workers that gather around the forager when she returns to the hive. The returning foragers then begin a series of movements in specific patterns that encode information about the direction and distance of food sources. This is the dance language of honeybees, originally discovered by Karl von Frisch.

If the food source is within about 50 m of the hive, a simple dance called the round dance—rapid movement in little circles, in both clockwise and counterclockwise directions—is sufficient. Outgoing foragers then know that the source is close and they search at random in the vicinity of the hive until they find the source.

If the food source is farther away, the incoming forager does a more complex dance, the waggle dance. The steps of the dance are illustrated in **Fig. 45.15**. The forager moves quickly back and forth ("waggles") as it moves forward, circles back and repeats the waggle, then circles back the other way, and so on. This dance usually goes on for several minutes and may

**FIG. 45.15   The waggle dance of honeybees.** This form of communication provides information to other foragers about the distance and direction of a food source.

| The distance to the food source is communicated by the length of time it takes to move up the middle of the circle (the "waggle"). | The direction of the food source relative to the sun is communicated by the angle of the waggle run relative to vertical. |

take much longer. The distance to the food source is conveyed by the length of time it takes to waggle up the middle of the circle. The direction, with reference to the position of the sun, is conveyed by the angle of the line of the waggle run: If the food source is in a direct line with the sun, the dancer's waggle line points straight up; if food is at an angle of 45 degrees to the left of the sun, the dancer's line is 45 degrees to the left of vertical. Short of human language, the bees' dance language is among the natural world's most extraordinary and effective methods of communication.

## 45.6 SOCIAL BEHAVIOR

Returning to Tinbergen's questions, we can consider further the adaptive function and evolutionary history of animal behavior. The evolutionary advantages of most behaviors are readily apparent. The imprinting of Lorenz's goslings on the first moving object they encountered makes sense when you consider that that object is usually their parent (not Konrad Lorenz!) and that sticking close to their parent is an excellent survival strategy in a world filled with predators.

Consider cooperative hunting seen in killer whales. Killer whales participate in a behavior called wave washing, in which a pod of whales collectively creates waves to wash a prey species—typically a seal—off an iceberg (**Fig. 45.16**). This behavior is collaborative, and it is clear why each individual participates. By itself, a single killer whale is incapable of generating a big enough wave, but, through cooperation, an individual can contribute to prey capture and gain some portion of the prey.

Sometimes, however, how a behavior enhances survival and reproduction is less clear. Perhaps the most prominent examples of this kind of behavior are sacrifices made by individuals apparently for the good of others. Formally, we describe an act of self-sacrifice as **altruistic**. Such acts seem to fly in the face of the natural selection, which posits that every individual competes for resources and mates to maximize its genetic contribution to the next generation. This contradiction is particularly stark in the case of species like honeybees: Honeybee workers, which are females, do not reproduce but instead assist the reproduction of the queen. Darwin confessed that such behaviors caused him sleepless nights. How, Darwin wondered, could altruism arise?

### Group selection is a weak explanation of altruistic behavior.

Darwin himself suggested a solution to the altruism problem, a solution that today we call **group selection.** The idea is that natural selection operating on individuals is a less powerful force than another form of selection that operates on groups. Consider two groups, one with individuals that are altruistic, helping

**FIG. 45.16 Mutually beneficial cooperation.** Killer whales cooperate to create a wave to wash prey off an iceberg.

one another out, and another with individuals that act entirely selfishly. Darwin argued that, in a conflict between the two groups, the first group would beat the second because the first group works better as a unit. Here, then, we see that selection in favor of altruism that benefits the group has trumped selection in favor of selfish behavior.

Although an attractive idea, group selection is probably not important in the evolution of altruism. Altruism under group selection is typically not an **evolutionarily stable strategy,** meaning that this kind of behavior can be readily driven to extinction by an alternative strategy. It used to be thought that lemmings, a species of Scandinavian rodent, would collectively kill themselves by plunging into the ocean when the population became too large to sustain. It was claimed that the lemmings were altruistic: Their self-sacrificial act would ensure that the population did not exhaust available resources, preventing a population crash.

This is a nice story, but untrue. Lemmings do not kill themselves for the good of the species. An individual that "cheats" by not performing the altruistic act will survive and breed, and its altruistic companions will not. Now consider that this behavior—to kill oneself or not—is genetically encoded. A mutation that arises in a population that causes the individual to "cheat" and not kill itself will spread rapidly through the population by natural selection until, after many generations, it has completely replaced the altruistic allele that causes the individual to kill itself. The lemming behavior, then, is not an evolutionarily stable strategy because natural selection favors an alternative strategy.

This description is oversimplified because there is unlikely to be a simple genetic basis to traits such as these, but the point stands: Group selection is in general not an evolutionarily stable strategy because it can readily be overthrown by "selfish" strategies.

→ **Quick Check 4** It's sometimes said that the reason an animal does something is that it's "for the good of the species." Why is this argument incorrect?

### Reciprocal altruism is one way that altruism can evolve.

If group selection does not provide a strong explanation for the evolution of altruism, how can we explain this behavior? **Reciprocal altruism,** whereby individuals exchange favors, is one way that altruism can evolve. This idea is perhaps best summarized as "You scratch my back, and I'll scratch yours." For this kind of behavior to work, individuals have to be able to recognize each other and remember previous interactions. If I did you a favor, I need to be able to recall who you are so that I can be sure that I get the return I expect.

A famous example of reciprocal altruism comes from an unlikely source—vampire bats. Adult female vampire bats live in groups of 8 to 12. Individuals have been recorded as living as long as 18 years, and in one study, two tagged females shared the same roost for 12 years. As is the case for most mammals, young males disperse from their natal group, but females remain. Thus, the female groups are composed partly of kin, although not entirely, since they are occasionally joined by unrelated females. In addition, because they have long-term associations, there is opportunity for reciprocal interactions.

Feeding for these bats is a "boom or bust" activity—either a bat has found and successfully fed on a blood source (boom), or a bat has failed to feed (bust). As a result, bats are often either overfed or underfed. Underfed bats, in turn, are at risk of dying by starvation. Bats returning from a successful feeding expedition often regurgitate blood to unsuccessful ones, on the expectation that the favor will be returned at a future date.

**Kin selection is based on the idea that it is possible to contribute genetically to future generations by helping close relatives.**

Only in the 1960s did we come to understand that the relatedness of participating individuals has played an important role in the evolution of altruism. This understanding lay in the insight that there are at least two ways to contribute genetically to the next generation: Either an individual can produce its own offspring, or it can produce no offspring and help close relatives reproduce instead. The latter strategy can become established in a population through a process called **kin selection,** a form of natural selection that favors the spread of alleles that promote behaviors that help close relatives, or kin.

What relationships qualify as kin? Consider an animal (gray shading in the figure) that has an allele *A* at a particular locus (**Fig. 45.17**). If the animal reproduces, the probability of each offspring inheriting an *A* allele is 0.5. But what if the animal does not reproduce? According to Mendel's laws, there is a 50% chance that the animal's parents passed on the same *A* allele to the animal's sibling (Chapter 16). If the sibling does, in fact, inherit allele *A*, the probability that it passes allele *A* to each of its offspring is also 0.5. Given the two probabilities (that the sibling has an *A*, and that the allele passed on to an offspring), the probability that the animal's niece or nephew has *A* is 0.25. In other words, a diploid animal (such as humans) is twice as closely related to its own offspring than to its nephew or niece. Therefore, if the animal does not reproduce in order to help its sibling have two offspring, the animal will make the same net genetic contribution to the next generation as it would having one of its own offspring.

→  **Quick Check 5** How many cousins are "genetically equivalent" to a single offspring?

**FIG. 45.17 Computing relatedness.** One's relatedness (*r*) to another individual is the probability that one shares, through inheritance, a given gene with that individual. Here, relatedness is calculated with reference to the individual indicated by the gray background.

Generation 1

OR

Generation 2

The probability that a sibling has *A* is 0.5.

Sibling          Sibling

Generation 3          OR          OR

Offspring   Offspring          Niece      Niece

The probability that an offspring has *A* is 0.5.

The probability that a niece has *A* is the probability that the sibling has *A* (0.5) multiplied by the probability that the sibling's offspring has *A* (0.5), or 0.25.

This logic was formalized in 1964 in a famous paper by British evolutionary biologist William D. Hamilton in terms of three key variables: *B*, the benefit of the behavior to the recipient (how many individuals does the behavior save?); *C*, the cost of the behavior to the donor (how many offspring are lost?); and *r*, the degree of relatedness between the recipient and donor. Hamilton asserted that as long as *rB* exceeds *C*, altruism can evolve.

How in practice does kin selection work? Let's start with an extreme case of relatedness: clonal organisms that are genetically identical to each other (meaning that the probability of a sibling having allele *A* is 1). Many aphid species have a phase of their life cycle in which a "mother" aphid reproduces parthenogenetically to produce a small colony of clones of herself (Chapter 42). Any aphid in the colony is capable of reproducing in this way, but in a case of reproductive altruism, some individuals do not reproduce at all. They are "soldier" aphids, who defend the colony against predators, typically other insects. As specialized defenders, they do not have the ability to reproduce. In this case, the Hamilton calculation is simple. By helping their clone-mates survive, the sterile soldier aphids are ensuring that their own genes survive to be passed on to the next generation.

The most famous cases of reproductive altruism occur in the insect group Hymenoptera, which includes ants, bees, and wasps. Many species of Hymenoptera are **eusocial,** meaning that they have overlapping generations in a nest, cooperative care of the young, and clear and consistent division of labor between reproducers (the queen of a honeybee colony) and nonreproducers (the workers). Often called social insects, these species are one of evolution's most extraordinary success stories.

What is it about Hymenoptera that predisposed them to eusociality? E. O. Wilson argues that natural selection has strongly favored group behavior in these insects. They typically receive group benefits from living in and defending a nest. For example, they can easily communicate with one another about the location of food resources such as nectar.

William Hamilton also noted that most Hymenoptera have an unusual mode of sex determination: Females (the queen and workers) are diploid, whereas males (drones) are haploid, produced from unfertilized eggs. As a result, the degree of relatedness, $r$, is higher for the sister–sister relationship (0.75) than for the mother–daughter one (0.5), as shown in **Fig. 45.18.** Put another way, a female is more closely related to her sister than she is to her daughter. Therefore, it makes more evolutionary sense, in terms of genetic representation in the next generation, for a worker to help her mother, the queen, to produce more sisters than for her to produce her own offspring.

This form of social organization can result in a huge colony of related individuals all working collectively, much as different cells work collectively within a body. Often, we see specialization into castes: Some ants are soldiers, defending the colony, others are foragers, and so on. If we extend the analogy of individuals acting like cells in a body, the multiple castes are equivalent to organs, all coordinating and working together. The term "superorganism" is sometimes used to describe a single social insect colony.

**FIG. 45.18** **Relatedness in haplodiploid insects, such as ants.** Sisters are more related to each other (0.75) than to their offspring (0.5).

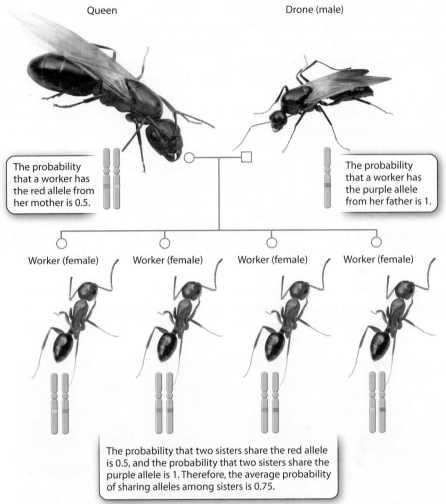

## 45.7 BEHAVIOR AND SEXUAL SELECTION

Throughout this chapter, we have discussed the development of certain behaviors in terms of natural selection. In many species, behavior is also affected by **sexual selection,** in which traits evolve that increase the probability of finding and attracting mates (Chapter 21). Traits favored by sexual selection sometimes appear to run counter to those favored by natural selection. For example, a peacock's tail is a compromise between sexual selection, which acts to increase the size and color of the display, and natural selection, which acts to minimize the amount a male invests in its cumbersome and metabolically expensive tail.

**Patterns of sexual selection are governed by differences between the sexes in their investment in offspring.**

In peacocks, it is the male that has the showy tail, not the female. Male birds often have bright colors, complex songs, and elaborate displays, and females usually do not. Differences in color or size between males and females are examples of **sexual dimorphisms,** or phenotypic differences between the sexes (**Fig. 45.19**).

Why do we see such differences between males and females? The key is the level of investment that an individual makes in an offspring. Consider a typical rodent species in which a male mates with a female and then moves on in search of additional mates. The female invests time, energy, and resources in the offspring, both during gestation and, after birth, during lactation. The male, on the other hand, has invested one ejaculate, no more.

In many species, eggs are fertilized externally and neither males nor females care for offspring. But, even in these species, there are differences in parental investment between males and females. Eggs are larger than sperm, and it therefore takes more energy and resources to make eggs than to make sperm. As a result, females still invest more in offspring than males do in these species.

These differences in parental investment lead to different selection pressures on females and on males. The reproductive success of females (that is, her number of offspring) is limited by the number of eggs she can produce or the number of pregnancies she can carry. By contrast, the reproductive success of males is limited by the number of mates. As a result, any trait that increases a male's ability to attract a female is favored by sexual selection and will evolve over time.

Therefore, sexual selection tends to act more strongly on males than on females, leading to the evolution of the bright colors and striking displays we often associate with male birds. Tellingly, in species in which this typical pattern of parental investment is reversed (the male invests more in the offspring than the female), we see a corresponding reversal of sexual selection (females that are more colorful than males). For example, in phalaropes and sandpipers, males incubate the eggs and females are brightly colored.

---

**FIG. 45.19  Sexual dimorphism, or phenotypic differences between the sexes.** Males (on the left in each photograph) look different from females, as shown by (a) lions, (b) cardinals, and (c) stag beetles.

**Sexual selection can be intrasexual or intersexual.**

In *On the Origin of Species,* Darwin recognized two forms of sexual selection (**Fig. 45.20**), and later, in *The Descent of Man* and *Selection in Relation to Sex,* he developed his ideas more fully. In one form, members of one sex (usually the males) compete with one another for access to the other sex (usually the females). This form is called **intrasexual selection** since it focuses on interactions between individuals of one sex (Fig. 45.20a). Because competition typically occurs among males, it is in males that we see physical traits such as large size and horns and other elaborate weaponry, as well as behaviors such as fighting. Larger, more powerful males tend to win more fights, hold larger territories, and have access to more females.

Darwin also recognized a second form of sexual selection. Here, males (typically) do not fight with one another, but instead compete for the attention of the female with bright colors or advertisement displays. In this case, females choose their mates. This form of selection is called **intersexual selection,** since it focuses on interactions between females and males (Fig 45.20b). The peacock's tail is thought to be the product of intersexual selection: Its evolution has been driven by a female preference for ever-showier tails.

**FIG. 45.20 Sexual selection.** (a) Intrasexual selection often involves competition between males, as shown by this battle between two male Elk. (b) Intersexual selection often involves bright colors and displays by males to attract females, as shown by this male and female Japanese Red-crowned Crane.

Often, these eye-catching ornaments are associated with elaborate behaviors on the part of the male. Male sage grouse, for example, congregate and put on displays in the mating season so that females can select their mates. Male bowerbirds accumulate shiny, bright objects in carefully tended displays to attract females (see Fig. 42.4b).

In general, therefore, females have evolved behaviors to ensure the quality of their mate (they tend to be choosy), whereas males have evolved behaviors to maximize the number of matings they have (they tend to be competitive). However, as we saw earlier,

when the male invests more in the offspring than do the females, there is a reversal of these patterns so that males are choosy and females competitive. In seahorses, for example, males brood the offspring in a specialized pouch and behave in the choosy way we see in females of other species. It is also important to keep in mind that other factors can influence and modify these general behavioral patterns. Nevertheless, sexually selected traits remind us of the power of selection in shaping all kinds of behaviors.

# Core Concepts Summary

**45.1 FOR ANY BEHAVIOR, WE CAN ASK WHAT CAUSES IT, HOW IT DEVELOPS, WHAT ADAPTIVE FUNCTION IT SERVES, AND HOW IT EVOLVED.**

Tinbergen pointed out that questions about animal behavior can be asked at four different levels of analysis, focusing on cause, development, function, and evolution. page 45-1

The four questions have different answers, and all may be correct explanations of the behavior. page 45-2

**45.2 ANIMAL BEHAVIOR IS SHAPED IN PART BY GENES ACTING THROUGH THE NERVOUS AND ENDOCRINE SYSTEMS.**

Behavior can be innate, meaning that it is carried out in the absence of experience, or learned, meaning that it depends on experience. page 45-2

A fixed action pattern is a stereotyped sequence of behaviors that, once initiated by a key stimulus, always goes to completion. page 45-2

Stimuli are recognized by feature detectors, which are specialized sensory receptors of the nervous system. page 45-4

Hormones can affect behavior. page 45-4

The extent to which genes influence a behavior can be determined by crossing closely related species having different behaviors and analyzing the behavior of the offspring. page 45-5

Most behaviors are influenced by many genes, but a few are strongly influenced by a single gene. page 45-6

Molecular studies are providing new ways to understand the role of genes in behavior in laboratory animals. page 45-6

**45.3 LEARNING IS A CHANGE OF BEHAVIOR AS A RESULT OF EXPERIENCE.**

Non-associative learning is a form of learning that occurs without linking two events. page 45-8

Habituation, a form of non-associative learning, is the reduction or elimination of a response to a repeated stimulus. page 45-8

Sensitization, another form of non-associative learning, is the enhancement of a response to a stimulus that is achieved by first presenting a novel stimulus. page 45-9

Associative learning occurs when an animal links two separate events. page 45-9

Classical conditioning is a form of associative learning that occurs when two stimuli are paired, and a novel association is made between a formerly neutral stimulus and a behavior. page 45-9

Operant conditioning is a form of associative learning that occurs when a behavior is rewarded or punished, making the behavior more or less likely to occur, respectively. page 45-9

Imitation and imprinting are also forms of learning. page 45-9

**45.4 ORIENTATION, NAVIGATION, AND BIOLOGICAL CLOCKS ALL REQUIRE INFORMATION PROCESSING.**

Kineses are undirected movements, whereas taxes are directed movements. page 45-11

Long-distance navigation may require the processing of many external cues, including the position of the sun, the Earth's magnetic field, and information about time. page 45-11

Biological clocks control some physiological and behavioral aspects of animals. page 45-12

**45.5 COMMUNICATION INVOLVES AN INTERACTION BETWEEN A SENDER AND A RECEIVER.**

Communication is an interaction between two individuals, the sender and the receiver, and often involves attempts by the sender to manipulate the behavior of the receiver. page 45-14

Communication can evolve by the modification of noncommunicative aspects of an animal's behavior. page 45-14

Bird song is a rich form of communication that is often learned during a sensitive period. Acquisition of bird song proceeds through distinct stages. page 45-15

The dance language of honeybees communicates information about the location and direction of a food source. page 45-15

**45.6 SOCIAL BEHAVIOR IS SHAPED BY NATURAL SELECTION.**

One of the most perplexing behaviors from the standpoint of natural selection is altruism, in which an organism reduces its own fitness by acts that benefit others. page 45-16

Group selection is the idea that altruistic behaviors can evolve if they benefit the group, not the individual, but this idea is not widely accepted. page 45-16

Reciprocal altruism requires that individuals interact regularly and remember each other so that they can "pay back" past favors. page 45-17

Kin selection is the concept that altruistic behaviors can evolve if they benefit close relatives. page 45-18

**45.7 SOME BEHAVIORS ARE INFLUENCED BY SEXUAL SELECTION.**

In most animal species, females invest more in offspring than do males. page 45-20

Differences in parental investment lead to different selection pressures on males and females. Typically, the reproductive success of males depends on the number of mates, so sexual selection favors traits in males that increase the probability of finding and attracting females. page 45-20

Intrasexual selection involves competition among individuals of one sex (usually the males) for access to the other. page 45-20

Intersexual selection involves bright colors, ornaments, songs, and other advertisements usually by males to attract females to mate, and accounts for many of the spectacular displays we see among animals. page 45-20

## Self-Assessment

1.  Choose a behavior described in this chapter and ask the four kinds of questions that Tinbergen might have asked about it.

2.  Provide an example of an innate and a learned behavior.

3.  Describe two approaches used to determine the extent to which genes influence a particular behavior.

4.  Give two examples of non-associative learning.

5.  Give two examples of associative learning.

6.  Differentiate between a kinesis and a taxis.

7.  Explain why many biologists would not consider any transfer of information between a sender and receiver to be a form of communication.

8.  Describe how an altruistic behavior might evolve by kin selection.

9.  Differentiate between intra- and intersexual selection, and give one example of each.

Do you understand the chapter's Core Concepts? Log in to check your answers to the Self-Assessment questions, then practice what you've learned and reinforce this chapter's concepts by working through the problems and multimedia tutorials provided there.

www.biologyhowlifeworks.com

## CHAPTER 46

# POPULATION ECOLOGY

## Core Concepts

**46.1** A population consists of all the individuals of a given species that live and reproduce in a particular place and is characterized by its size, range, and density.

**46.2** The age structure of a population helps ecologists understand past and predict future changes in population size.

**46.3** The dynamics of populations are influenced by the colonization and extinction of smaller, interconnected populations that make up a metapopulation.

In 2004, undergraduates on a field trip to the Dominican Republic, on the island of Hispaniola, swept up an unfamiliar black-and-white butterfly in their nets. That butterfly was a Lime Swallowtail (*Papilio demoleus*), the very first of its species, native to Asia, to be captured in the New World.

Soon, Lime Swallowtails would spread to all corners of Hispaniola, a serious concern because their caterpillars strip the leaves off young lime and orange trees. Within 3 years, the Lime Swallowtail had also established itself in nearby Puerto Rico and Jamaica, and it has just reached Cuba. From there, it is just a short flight to Florida and the multibillion-dollar citrus industry of the United States. Can we enlist basic principles of biology to help understand this butterfly's Caribbean expansion and predict its future as it continues to spread?

**Ecology** is the study of the relationships of organisms to one another and to the environment. Ecological relationships sustain a flow of energy and materials from the sun, Earth, and atmosphere through organisms, with carbon and other elements eventually returning to the environment (Chapter 25). And, in the end, it is these interactions that determine the shifting ranges of Lime Swallowtails and other species on Earth.

In previous chapters, we discussed the anatomical, physiological, reproductive, and behavioral characteristics of organisms. These attributes are the products of evolution, and they determine the ways in which individual organisms interact with their physical and biological environment. Abiotic factors that influence evolution include climate and nutrient availability; biotic factors include competitors, predators, parasites, and prey, as well as organisms that provide shelter or food. Ecology and evolution, then, are closely intertwined.

## 46.1 POPULATIONS AND THEIR PROPERTIES

When the first United States census was completed in 1790, the British scholar Thomas Malthus supposed that the remarkable recent growth of the American population, which had doubled in the previous 25 years, was a consequence of ample food supply and the active encouragement of marriages, neither of which characterized Europe at the time. In his 1798 volume *An Essay on the Principle of Population,* Malthus argued that a similar periodic doubling of population in the British Isles, on the other hand, would outstrip the more slowly increasing food supply in 50 years and lead to starvation. That is, increasing the number of births per year would eventually result in an increased number of deaths because the amount of food the land could produce was limited. This view of the human condition would influence the thinking of Charles Darwin, who as a young, inquisitive student read the most exciting works of the day. Darwin realized that the tension between population growth and resource limitation applies to all species, leading to a struggle for existence that results in adaptations that enhance survival (Chapter 21).

In Chapter 21, we introduced the concept of population as a fundamental unit of evolution—it is populations, and not individuals, that evolve. Here, we return to populations, this time as equally fundamental units of ecology. A **population** consists of all the individuals of a given species that live and reproduce in a particular place. Natural selection enables populations to adapt to the physical and biological components of their environment. It is the traits evolved under natural selection that determine whether a population will succeed or fail under a given set of physical and biological conditions—whether it will grow and expand its range or shrink, perhaps disappearing entirely from a local environment.

What controls changes in population size? We first look at what happens when there are more births than deaths. We then examine what happens when resources such as food start to run out. We consider how individuals of different ages contribute to population growth, and finally, in the next chapter, we introduce interactions with other species and see how they shape the growth of a population. All these factors directly influence the flow of energy and materials through producers and consumers and, ultimately, the flow of genes through a population.

### Three key features of a population are its size, range, and density.

Populations are defined by three key features: size, range, and density (**Fig. 46.1**). **Population size** is simply the number of individuals of all ages alive at a particular time in a particular place. For example, the population of krill in the cold waters around Antarctica is estimated to be 800 trillion, making them by mass the most abundant animals on Earth (Fig. 46.1a).

The size of a population may increase, decrease, or stay the same over time, and trends in population size are a key focus of ecological research and its application to conservation biology. A decrease in size through time, for example, might signal declining health in a local population—a matter of practical concern when the population in question is a commercially harvested fish or an endangered species. In the 1990s, catastrophic declines in cod populations led to a ban on commercial cod fishing on the Grand Banks of Newfoundland. The decline affected both cod and cod fishers, and it also led to changes in Grand Banks ecosystems— crab and lobster populations have increased as predation pressure by cod has declined.

A second key feature of a population is its **geographic range:** how widely a population is spread out. Fig. 46.1b, for example, shows that the krill population ranges through an area of 19 million square kilometers around Antarctica. The geographic range of any species determines how much variation in climate a population can tolerate and how many other species the population encounters. The geographic range of a plant population, for example, influences how many kinds of insect eat it, and the size of the range of an insect determines how many kinds of plant it encounters.

Taken together, population size and range contribute to a third key attribute of a population, its density: how crowded or dispersed the individuals are that make up the population. **Population density** is defined as a population's size divided by

its range. For krill, 800 trillion animals are spread over 19 million square kilometers—a density of 42 million animals per square kilometer (Fig. 46.1c). As we will see, population density can affect population size.

---

**FIG. 46.1**    **(a) Size, (b) range, and (c) density of a population.** The places where populations occur can be mapped, and the numbers of individuals within those populations can be estimated. This information helps us understand what factors limit population distributions.

**a.** Size

The size of the worldwide population of krill is estimated to be 800 trillion animals.

**b.** Range

Their range is approximately 19 million square kilometers in the Southern Ocean.

**c.** Density

800 trillion krill over 19 million square kilometers of the Southern Ocean means the population density is 42 million animals per square kilometer on average.

Mean no. krill/km²

| | |
|---|---|
| <2 | 32–64 |
| 2–4 | 64–128 |
| 4–8 | 128–256 |
| 8–16 | 256–512 |
| 16–32 | >512 |

**FIG. 46.2** **Distribution of individuals within populations.** A population may be (a) randomly distributed, (b) clumped, or (c) overdispersed.

**a.**

The distributions of individual trees or other organisms can appear to be random, with no clear pattern to where they occur.

**b.**

If resources are clustered or spatial proximity to other individuals enhances fitness, populations may be clustered.

**c.**

When resources are limited or predators target a single species, an individual might be better off if it is as far from others as possible, producing an over-dispersed pattern of distribution.

This way of characterizing population density, as average density across the range of a species, is often useful, but it can be misleading if organisms occur unevenly within their range. In nature, the individuals within a population may be distributed randomly—that is, a new individual has an equal chance of occupying any position within the range, and the location of one individual has no influence on where the next will occur (**Fig. 46.2a**). Sometimes, however, resources are distributed patchily within the range, or the chances of a new individual's surviving are enhanced by the presence of other individuals. In these circumstances, the population may be

clumped, with individuals clustered together (**Fig. 46.2b**). In some cases, one individual of a population might prevent another from settling nearby, for example if all individuals depend on the same set of resources. Or, predators, having located one prey individual, may search for other members of the same species nearby. Ecologists call such populations overdispersed, meaning that they are distributed more nearly uniformly than would be predicted by chance (**Fig. 46.2c**). Also, habitats suitable for growth may themselves occur in isolated patches. We discuss the implications of habitat patchiness in greater detail later in the chapter.

**Population size can increase or decrease over time.**
Many factors affect population size. An obvious process that contributes to population increase is birth. Another is immigration: Individuals may arrive in a population from elsewhere. In contrast, mortality (death) and emigration, the departure of individuals from a population, both act to decrease population size (**Fig. 46.3**).

Population size changes continually as individuals are born or immigrate into the population, and as they die or emigrate away from a population. We can describe the changes in a population's size mathematically, defining population size as N and the change in population time over a given time interval as $\Delta N$, where the Greek letter $\Delta$ (delta) means "change in." $\Delta N$ is the number of individuals at a given time (time 2) minus the number of individuals at an earlier time (time 1), which is symbolized as $N_2 - N_1$. As the processes leading to a change in population size through time include births (B), deaths (D), immigration (I), and emigration (E), we can quantify $\Delta N$ as:

$$\Delta N = N_2 - N_1 = (B - D) + (I - E)$$

**FIG. 46.3** **Factors affecting the size of a population: birth, mortality, immigration, and emigration.**

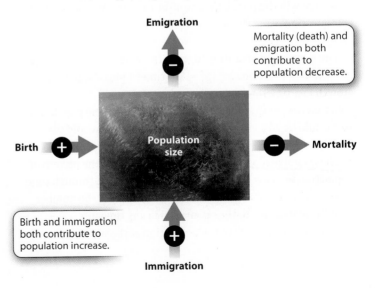

Mortality (death) and emigration both contribute to population decrease.

Birth and immigration both contribute to population increase.

Commonly, ecologists want to know not just whether population size is increasing or decreasing, but also the rate at which population size is changing—that is, the change in population size ($\Delta N$) in a given period of time ($\Delta t$), or $\Delta N/\Delta t$. Let's say a population starts with 80 individuals and after 2 years has grown to 120 individuals. The population, then, has gained 40 individuals in 2 years, for a rate of $40/2 = 20$ individuals per year.

The importance we place on such a number strongly depends on the actual population size. An increase of 20 individuals in a year is a big deal if the starting population had 80 individuals, but it is small if the starting population numbered 10,000. Usually, therefore, we are most interested in the proportional increase or decrease over time, especially the growth rate per individual, or the per capita growth rate (*capita* in Latin means "head"). The per capita growth rate is commonly symbolized by $r$ and is calculated as the change in population size per unit of time divided by the number of individuals at the start (time 1):

$$r = (\Delta N/\Delta t)/N_1$$

In our example above, $r$ equals 20 individuals per year divided by 80 individuals, for a per capita growth rate of 0.25 per year.

It is important to realize that while rates are important in ecology, the data we usually have in hand actually consist of the numbers of individuals that we have counted, and we often do not know exactly how large a natural population really is and so must rely on estimates. We consider how we make such population size estimates later in this chapter, but for the rest of this discussion we focus on the per capita increase (or decrease) in a population over time and assume we know the population sizes for species of interest.

→ **Quick Check 1** If a population triples in size in a year, what is the per capita growth rate?

In the example discussed in the preceding paragraphs, an initial population of 80 individuals increased by 20 per year, numbering 120 after 2 years, for a per capita rate of increase of 0.25 per year. Should this per capita rate of increase continue, how large will the population be after 10 years? We can answer this question using the following formula:

$$N_t = N_1(1 + r)^t$$

The population size after 10 years ($N_t$, where $t = 10$) will equal the initial population size ($N_1 = 80$) times 1 plus the per capita growth rate (0.25) raised to the power $t$ (10 years). Solving the equation for our example shows that in 10 years the population size would be 745. After 15 years, the population would swell to 2273.

This example illustrates **exponential growth,** the pattern of population increase that results when $r$ is constant through time. A prominent feature of exponential growth is that the number of individuals added to the population in any time interval is proportional to the size of the population at the start of the

interval. That is, while the per capita growth rate remains constant, the actual number of individuals added per time unit increases as the population grows. This is because the new individuals added also reproduce themselves, so through time more and more individuals contribute to the growing population. **Fig. 46.4** shows the characteristic shape of an exponential growth curve.

A second example brings home the remarkable consequences of exponential growth. A chessboard has 8 rows of 8 squares each, a total of 64 squares. Suppose we place a grain of wheat on the first square of the chessboard, two on the second square, four on the third, and keep doubling the number at each successive square. By the time we reach the 64th square, the total number of grains on the board is 18,446,744,073,709,551,615, more than 80 times what the Earth could produce in a year if wheat were planted on every square inch of farmland, and about a million times the number of stars in the Milky Way galaxy.

→ **Quick Check 2** Would you rather have a million dollars today, or start with one cent and double the amount each day for a month? Why?

The per capita growth rate, $r$, is commonly called the **intrinsic growth rate** of a population, the maximum rate of growth when no environmental factors limit population increase. Intrinsic rates of growth in mammals range from 82% per year in muskrats to 50% in snowshoe hares and 10% in moose. Malthus recognized that exponential growth cannot continue for long in nature because the number of individuals will eventually outstrip the resources available to support the population. Darwin, in turn, recognized that when the number of young produced at birth exceeds the number

---

**FIG. 46.4** **Exponential growth.** When per capita growth rate, $r$, is constant, the curve shows ever-larger changes in population size over time.

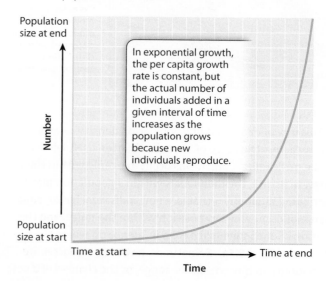

In exponential growth, the per capita growth rate is constant, but the actual number of individuals added in a given interval of time increases as the population grows because new individuals reproduce.

of adults that can be supported by available resources, individuals within the population will compete to obtain the resources needed for growth and reproduction. This **intraspecific** (within-species) **competition** results in natural selection (Chapter 21), linking the ecological and evolutionary aspects of populations. Individuals from different species can also compete for resources. Such **interspecific** (between-species) **competition** can result in an increase or decrease in population size. We focus for the most part on intraspecific competition in this chapter, reserving a detailed discussion of the effects of interspecific competition until Chapter 47.

### Carrying capacity is the maximum number of individuals a habitat can support.

As a population grows, crowding commonly results in a decrease in the availability or quality of resources required for continuing growth. As a consequence, growth rate declines. On Hispaniola, for example, Lime Swallowtails have probably reached the limit of available citrus groves. Individual females have to search longer to find suitable plants that do not have eggs already laid by other females. With that delay, and the pressure of eggs that are continuing to mature inside their bodies, females may end up laying too many eggs in one place or on poor-quality food, so there is not enough food for all larvae to survive. All of these factors lead to fewer births than occurred immediately after the Lime Swallowtails first arrived in Hispaniola, when resources were not limiting.

For many organisms, crowding affects the death rate as well as the birth rate. For example, as the fish population of a pond increases, there is less food available for each individual, and oxygen may be used up by microbes that proliferate because of nutrients in the fish waste.

In other words, birth and death rates are themselves affected by population density, which we defined earlier as the number of individuals per unit of area or (for aquatic species) volume of habitat. Increasing density is not always detrimental to a population. For example, eggs of a salmon or sturgeon are more likely to be fertilized in rivers or streams swarming with their species than when there are few reproducing adults. Generally, however, at some point increasing population density causes the environment to deteriorate because of a scarcity of resources. When the density of a growing population reaches this point, the rate of growth slows and eventually levels off as the death rate increases to equal the birth rate.

We call the maximum number of individuals any habitat can support its **carrying capacity (K).** In a sense, K represents the interplay among the functional requirements of individuals for growth and reproduction and the environmental resources such as food and space available to support these needs.

Many factors can keep a population below K. Predation and parasitism are two of them. For example, increasing numbers of Lime Swallowtails on Hispaniola have no doubt attracted predators and parasites that feed on them. Similarly, a high population density of fish in a pond increases the likelihood of infection by viruses, bacteria, and fungi because these agents are transmitted from fish to fish. Predation and parasitism can reduce the population size below K, but they do not change K itself.

As noted, when a population is small, resources are not limiting and the population can grow at or near its intrinsic growth rate, r. As the population increases, however, drawing closer to K, the worsening conditions due to crowding slow the rate of growth. For any given population at any given moment, such as deer today in New England forests or striped bass in Chesapeake Bay, we may wish to know what percentage of the carrying capacity is available for further population growth. The answer, expressed mathematically, is $(K - N)/K$.

The effects on population growth as a habitat fills up with individuals can be seen in **Fig. 46.5.** Growth is exponential at first and then slows down as the population size approaches its maximum sustainable size, K. The resulting pattern of population growth, with its characteristic S-shaped curve, is termed **logistic growth.** Population size at any time $t$ ($N_t$) can be calculated using the equation:

$$N_t = \frac{KN_1 e^{rt}}{K} + N_1(e^{rt} - 1)$$

Note that $N_t$ depends on both K and r ($e$ is Euler's number, a constant used in natural logarithms and approximately equal to 2.718).

---

**FIG. 46.5 Logistic growth in a population.** The graph shows the characteristic S-shaped growth curve, which is determined by r and K.

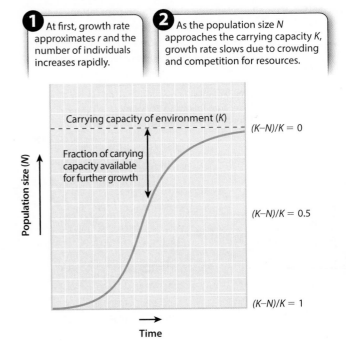

**1** At first, growth rate approximates r and the number of individuals increases rapidly.

**2** As the population size N approaches the carrying capacity K, growth rate slows due to crowding and competition for resources.

Carrying capacity of environment (K)  $(K{-}N)/K = 0$

Fraction of carrying capacity available for further growth

$(K{-}N)/K = 0.5$

$(K{-}N)/K = 1$

Population size (N)

Time

Logistic growth curves, then, are S shaped because they actually reflect two sets of processes, each related to population size. At first, growth reflects only the rate at which individuals can reproduce themselves, so the curve rises steeply. The second part of the curve, where it starts to level off, reflects the onset of additional factors that become important above a certain population size, such as decreasing availability of food and space.

Virtually all natural populations grow this way. Logistic curves also apply to many other phenomena in biology, as well as in chemistry and such diverse fields as sociology, political science, and economics. Logistic growth therefore is a mathematical concept worth knowing. For example, a logistic curve describes the rate of binding of oxygen to a population of hemoglobin molecules, as seen in Chapter 39. Similarly, the number of fertilized salmon eggs goes up with the amount of sperm (which reflects the number of males in the population) deposited in a stream by males but slows again as the percentage of unfertilized eggs available for the sperm to fertilize drops.

→ **Quick Check 3** Why is the growth curve for most populations S shaped?

### Factors that influence population growth can be dependent on or independent of its density.

In the preceding discussion, the factors that limit population size depended on the density of the population. At low density, food and other resources do not limit growth, but as density

increases, they exert more and more influence on population growth, spurring competition for available resources. Similarly, at high population density, individuals may be more vulnerable to predation or to infection. Factors such as resources and predation are called **density-dependent** factors.

Not all changes in population growth rates result from population density, however. **Density-independent** factors influence population size without regard for the population's density. They include events like severe drought or a prolonged cold period, either of which can cause widespread mortality independent of population density. A graph of density-independent effects on populations would show instantaneous, rather than gradual, drops in population size, and these changes would occur at different starting densities.

For example, population growth in Song Sparrows is a function of the number of nesting territories established by males each year, and of their young's success in those territories. **Fig. 46.6** shows that the number of territories generally increases for several years until an unusually dry spring occurs, which greatly decreases the amount of vegetation available to support the insects on which the sparrows feed. The number of territories then plummets, regardless of territory numbers the previous year.

### Ecologists estimate population size and density by sampling.

Imagine trying to calculate the growth rate of a population of Lime Swallowtails by counting every individual in the Dominican Republic. The task is hopeless. Like that of the Lime Swallowtail, the populations of most species are either too large or too widely dispersed for every individual to be counted. Instead, biologists estimate population size by taking repeated samples of a population and from these samples estimating the total number of individuals and their density. Because of the importance of population size in ecology, biologists have developed a number of sophisticated sampling strategies.

For sessile, or sedentary, organisms like milkweeds in a field or barnacles on a rocky beach, ecologists commonly lay a rope across the area of interest and count every individual that touches it, or they throw a hoop and count the individuals inside it. Each such count gives one sample. In practice, ecologists take a number of samples and use them to estimate the number and distribution of individuals in the population. For example, suppose 1 m² hoops are placed in three different spots in a field, encircling respectively 4, 5, and 3 individual milkweeds. We can multiply the average number of individuals (in this case, 4) per 1 m² sample by the total square meters in the area of interest (in this case, the size of the entire field) to give an estimate of population size.

It is relatively easy to count sedentary organisms along a straight line or within a grid area, but many species move around.

**FIG. 46.6   Density-independent effects on population size.** Independent of population density, extreme weather (such as drought) lowers the quality of habitat for Song Sparrows, reducing the number of territories occupied.

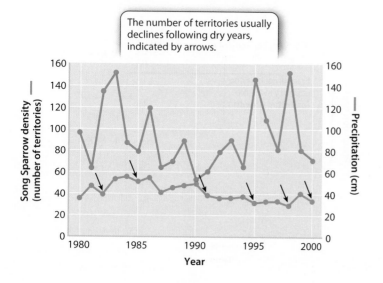

The number of territories usually declines following dry years, indicated by arrows.

How do ecologists estimate population size for turtles in a pond, or mice in a forest? In these cases, a method called **mark-and-recapture** is useful. Biologists capture individuals, mark them in a way that doesn't affect their function or behavior, release them back into the wild, and then capture a second set of individuals, some of which were previously marked.

In **Fig. 46.7,** for example, an ecologist has marked a butterfly with a painted number, which does not hinder its activity in any way. Later, ecologists return to the area and capture another set of samples from the population. This second sample will most likely contain a few individuals who were captured the first time around, identifiable by the paint mark; these are the recaptures. The rest are newly caught individuals that have no paint mark.

If a large number of butterflies are caught, marked, released, and then recaptured, it is possible to make an estimate of the total population by extrapolating from the proportion of the second sample that is marked.

These methods are simple, and they make a number of assumptions about the population. For example, we assume that the population has not changed in size between our first and second samples. Nonetheless, mark-and-recapture methods have proved effective in estimating the sizes of populations in nature. More sophisticated analyses are available for more complicated situations, for example when more than two samples are caught, or when immigration or emigration occurs during the time between samples.

## HOW DO WE KNOW?

**FIG. 46.7**

# How many butterflies are there in a given population?

**BACKGROUND** Individuals in many populations of butterflies (and other animals) most often live and die within the patch of habitat where they were born. Population sizes of such animals can be estimated by a technique called mark-and-recapture, in which organisms are captured and released on two successive days. We assume that the population size is about the same on the second day as the first day. We also assume that the captured and released animals mixed with the population and do not avoid recapture on the second day.

**METHOD** Butterflies, such as the Karner Blue Butterfly, are captured in an area, marked with a spot or identification number on the wing as shown here, and then released. The number of marked butterflies is recorded. The next day, butterflies are caught and the number of marked and unmarked butterflies is recorded. Butterflies with marks are the recaptures.

**RESULTS** To find the population size (*N*), we take the total number of marked butterflies and unmarked butterflies caught on the second day (*C*) and divide by the number of recaptures (*R*). We multiply this number by the number of butterflies marked on the first day (*M*), as follows:

$$N = C/R \times M$$

**CONCLUSION** Let's say we capture 100 butterflies on the first day, mark them, and release them. On the second day, we capture

120 butterflies, 30 of which are marked and the rest unmarked. We conclude that the population size is (120/30) x 100, or 400 butterflies total in the population sampled.

**RELATED METHODS** Mark-and-recapture techniques have proved extremely useful in estimating population sizes of organisms from polar bears to disease victims. There are refinements to the mark-and-recapture approach that take into account the movement of individuals between populations; these methods involve taking multiple samples and adding variables to the equation. Many animals have distinctive color markings, such as the blotches on humpback whale tailfins and frog skin-spotting, and so can be recorded without having to be marked.

**SOURCE** Lincoln, F.C. 1930. "Calculating waterfowl abundance on the basis of banding returns." *Cir. U.S. Department of Agriculture* 118:1–4.

## 46.2 AGE-STRUCTURED POPULATION GROWTH

Direct counting and mark-and-recapture methods permit ecologists to assess the size and density of populations in a particular place. However, not all organisms in a population contribute equally to population growth because some reproduce more than others. Therefore, it is also useful to take into account individuals that have differing reproductive capacities, which most often depend on differences in age. Biologists interested in the potential for population growth seek information on the **age structure** of a population, the number of individuals within each age group of the population studied. A population in which most individuals are past their age of greatest reproduction (because for some reason young are not surviving) or a population mostly composed of individuals younger than this age (because older individuals are not surviving) will not increase as rapidly as one dominated by individuals capable of greatest reproductive output.

### Birth and death rates vary with age and environment.

To understand past and predict future changes in populations, ecologists use birth rates, expected longevity of individuals, and the proportion of individuals in a population that is able to reproduce. Insights into these parameters come from the age structure of the population.

Commonly, researchers divide the population, or a sample of the population, into age classes, for example individuals born within a single year or within an interval of several years. Age can often be determined by measuring an aspect of the organism's anatomy (for example, counting growth rings in fish scales or a core drilled into a tree trunk) or estimated by overall size (for example, length of a fish, or tree height).

Estimating the number of individuals of differing ages enables ecologists to predict whether the size of a population will increase or decrease. This method applies to human populations as well as those of other species. A growing population usually shows a pyramid-shaped age distribution, with the youngest classes much more abundant than older classes, whereas a stable population shows a more even distribution of age classes. **Fig. 46.8** shows the difference in age structure between France, a country with a stable population, and India, a country with a growing population.

In humans, such differences reflect, in part, socioeconomic variation among countries, but differences in the age structure of other species can reveal threats from various sources. For example, the age structure of fish populations can be affected by the preferential capture of the largest fish, a recurring problem in commercial fisheries. Because the larger fish produce most of the eggs each year, this pattern can eventually lead to a rapid decrease in the population.

**Fig. 46.9** shows the effects of overfishing on the age structure of a species of long-lived Pacific rockfish, determined by comparing the ages of fish sampled in unfished waters (Fig. 46.9a) with those in heavily fished waters (Fig. 46.9b). In the unfished population, the curve shows many individuals close to and past the age of 10 years, when reproductive ability starts to climb. In the fished populations, the curve shows a sharp decline of individuals older than 6 years because mostly older and larger fish are caught. Because these older fish produce most of the eggs and young fish, the annual reproductive effort of fished populations is severely affected by their removal.

**FIG. 46.8   Age structure of two populations, with data from 2010.** The data are depicted as histograms, bar graphs in which the length of each bar indicates the number of individuals in the group (here, age group).

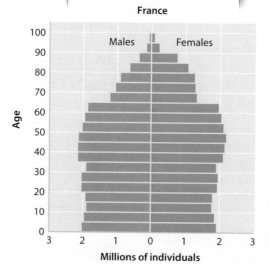

The number of males and females in France is uniform from childhood through the reproductive years, showing that younger individuals replace older individuals through time and so ensure the stability of the population.

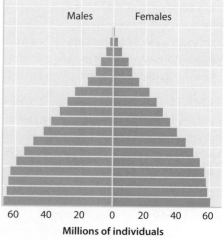

The number of males and females in India is much greater in younger age classes than in older classes. The population will continue to grow as individuals in the younger age classes mature and reproduce.

**FIG. 46.9   Age structure in Pacific rockfish populations in fished and unfished waters.** Harvesting large individuals reduces the population's capacity for growth, jeopardizing commercial stocks.

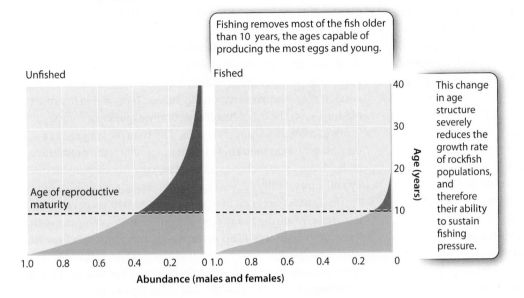

Unfished

Fished

Fishing removes most of the fish older than 10 years, the ages capable of producing the most eggs and young.

Age of reproductive maturity

This change in age structure severely reduces the growth rate of rockfish populations, and therefore their ability to sustain fishing pressure.

Age (years)

Abundance (males and females)

## Survivorship curves record changes in survival probability over an organism's life-span.

The age structure of a population reflects death rates as well as births. **Demography** is the study of the size, structure, and distribution of populations over time and includes changes in response to birth, aging, migration, and death. The demography of the invasive Lime Swallowtail butterfly has been studied in Southeast Asia, where it is native, and this investigation may be helpful in predicting the population dynamics of this species in coming years as it spreads from Hispaniola across the Americas. Like many other insects, Lime Swallowtails have a pupal stage in which they metamorphose from a caterpillar larva specialized for feeding on leaves into a winged adult that can search for other plants on which to lay eggs (Chapter 44). Neither the larvae nor the pupae reproduce, but the adults that emerge from the pupae do.

The factors that cause insect mortality commonly differ from one life stage to the next, and knowing their effects can help us understand the physical and biological influences on population growth. **Fig. 46.10** shows the different dangers to each stage of the Lime Swallowtail's life. These butterflies lay 30 or more eggs, but only a fraction of their eggs survive because of parasitism, drying out, falling from leaves, or predation by ants or other insects. The larvae have many defenses against predators, among them camouflage and chemical toxins, but they still fall prey to wasps, as well as to lizards and other animals, so the faster they develop into adults, the more likely they are to escape discovery by their enemies. The pupae, which are large and immobile, are perhaps the most vulnerable stage in the Lime

**FIG. 46.10   Lime Swallowtail life cycle and natural enemies.** The factors that cause insect mortality commonly differ from one life stage to the next.

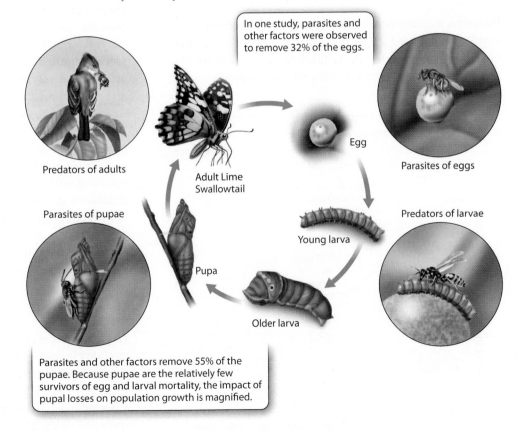

In one study, parasites and other factors were observed to remove 32% of the eggs.

Predators of adults

Adult Lime Swallowtail

Egg

Parasites of eggs

Parasites of pupae

Pupa

Young larva

Predators of larvae

Older larva

Parasites and other factors remove 55% of the pupae. Because pupae are the relatively few survivors of egg and larval mortality, the impact of pupal losses on population growth is magnified.

Swallowtail life cycle, and they are often highly parasitized by a particular kind of wasp.

**Fig. 46.11** presents data on the duration of each stage in the Lime Swallowtail life cycle and the numbers killed in that stage. We can consider the individual eggs that begin our study to form a **cohort,** a group defined as the individuals born at a given time, and we can trace their survival as they age. The proportion of individuals from the initial cohort that survive to each successive stage of the life cycle is called **survivorship.** When survivorship curves are plotted through time, they show an ever-declining fraction of the starting number of eggs (Fig. 46.11).

As expected from the preceding discussion, in this example, survival is lowest among eggs (68% survive to the next stage) and pupae (only 45% survive to the next stage), which of course cannot move to escape harm. If we divide the number lost at each stage by the number of days at that stage, we see that the egg stage loses 16 per day, the pupal stage loses 6 per day, but the larval stage loses only 0.6 per day. There is therefore strong selective pressure to decrease the length of each stage, especially those stages most vulnerable to mortality. Nevertheless, if we remember that each adult female can lay 30 or more eggs, only a very few eggs need to survive to the adult egg-laying stage to ensure the growth of the population. In warm climates like the Caribbean, these butterflies may reproduce eight or nine times in a year. Most of their eggs will not survive to become adults; if they did, available resources would soon be depleted.

### Patterns of survivorship vary among organisms.

Populations differ in the ages at which they experience the most mortality. In some species, many offspring are produced, but high mortality early in life means that few survive to reproductive age.

---

**FIG. 46.11  Survivorship curve.** A survivorship curve plots the proportion of the original cohort alive at the end of each stage or time period.

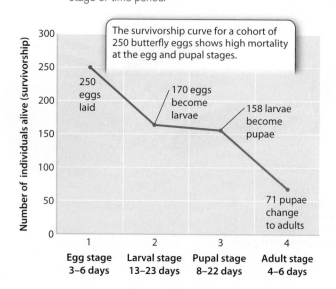

In other species, mortality is highest in older individuals, while in still other species, mortality occurs with equal frequency at all ages. **Fig. 46.12** illustrates the three principal types of survivorship curve—Type I, Type II, and Type III— which correspond to the pattern of mortality: whether most mortality occurs late in life, is even throughout life, or is concentrated early in life, respectively.

Human populations in affluent countries, which experience most mortality late in life, typically show a Type I survivorship curve (Fig. 46.12a). Other large animals, such as elephants and whales, show similar survivorship curves. Birds and many small mammals, on the other hand, commonly experience consistent levels of mortality throughout life, which can be graphed as a declining straight line. These animals, for which the probability of death does not change much through the life cycle, exhibit a Type II survivorship curve (Fig. 46.12b). Finally, organisms like Turtle Grass reflect high mortality at the earliest stages of their life cycles, resulting in a sharply declining survivorship curve that levels off (Fig. 46.12c). This pattern is characteristic of a Type III survivorship curve, which is typical of small herbaceous plants, small animals such as mice, insects like the Lime Swallowtail, and other organisms that grow fast and reproduce early.

### Reproductive patterns reflect the predictability of a species' environment.

In the preceding sections, we discussed population size and growth curves, and we outlined age structure and survivorship. These topics converge when we consider patterns of reproduction. Some species produce large numbers of offspring but provide few resources for their support. In Chapter 42, such species were described as **r-strategists.** In contrast, some species produce relatively few young but invest considerable resources into their support. These are termed **K-strategists.** How do these concepts relate to $r$ and $K$ in population growth?

In general, species that live in unpredictable environments produce many young. For example, when the environment is full of predators watching for vulnerable prey, such as gulls that watch for young turtles on a beach, or when resources are patchily distributed and abundant for only a short time, the chances of a young organism surviving to adulthood are low. Laying an egg, then, is like buying a lottery ticket—there is a limited chance of success. It is more advantageous to buy many tickets cheaply (that is, to have many offspring and make little parental investment in any of them) than to pay a lot for one. Thus, salmon and other fish produce many thousands of young, each with an extremely small chance of survival. As we saw earlier, Type III strategists like insects and plants also experience high mortality in the first stages of life. Because their reproductive success depends on the small chance that a few of their young will survive all the hazards they encounter, parents produce many offspring and invest relatively little in any of them. We refer to species like most fish as $r$-strategists because they reproduce at rates approaching $r$, the intrinsic rate of growth.

**FIG. 46.12** Three different types of survivorship curve.

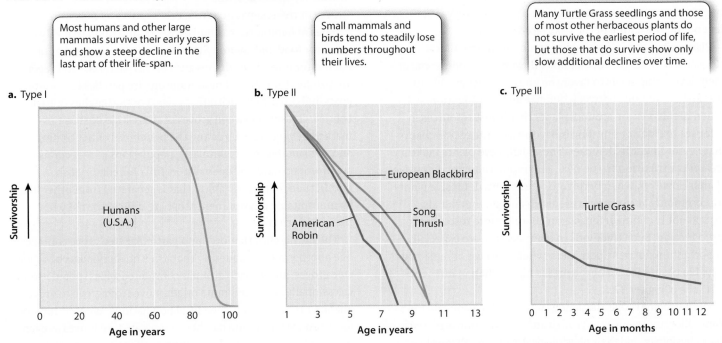

a. Type I

Most humans and other large mammals survive their early years and show a steep decline in the last part of their life-span.

b. Type II

Small mammals and birds tend to steadily lose numbers throughout their lives.

c. Type III

Many Turtle Grass seedlings and those of most other herbaceous plants do not survive the earliest period of life, but those that do survive show only slow additional declines over time.

In contrast, species living where resources are predictable often produce a lower number of offspring and invest their reproductive resources in just a few, often larger and better-cared-for, offspring. These offspring are therefore better protected from predators, able to compete with others when they are young, and more likely to survive periods of bad weather. Large animals, such as eagles and bears, produce only two or three young, each of which is much more likely to survive to reproductive age than the offspring of small animals like mice, which have many predators and are easily killed by cold weather. Animals like eagles and bear are called *K*-strategists because their population densities commonly lie near the carrying capacity *K*, when resources become scarce.

These two strategies reflect an evolutionary trade-off. A reproductive individual has access to a finite amount of resources and can invest either in many inexpensive young or few well-provisioned offspring. Depending on the combination of physiology and environment, natural selection in a given set of circumstances tends to favor one strategy over the other. In fact, *r*- and *K*-strategists represent the opposite ends of a continuum in reproductive pattern. Many species lie between these extremes.

### The life history of an organism shows trade-offs among physiological functions.

We've noted that population growth can be limited by resource availability. How an organism makes use of the resources that are available can also affect population growth. Each individual must devote some of its available food and energy intake to growth, some to the maintenance of cells and tissues, and some to reproduction. This allocation is very much like a

household budget, in which incoming resources are divided among the day-to-day expenses of acquiring food and paying for housing and transportation, meeting present and future medical and educational expenses, and savings held in reserve for emergencies. When one category goes up in cost—it needs more money, or more resources—another must come down. In nature, individual organisms strike a balance in how they allocate resources to the basic needs of growth, maintenance, and reproduction, a balance that has evolved to minimize waste while retaining the flexibility to use resources in different ways depending on environmental circumstances. These kinds of trade-offs evolve through natural selection and are the rule in ecology, explaining in part why species tend to produce either many small offspring or a few larger ones.

The typical pattern of resource investment in each stage of a given species' lifetime is called its **life history.** The amount of energy available for reproduction varies over an individual's lifetime, and the degree of control an individual has over resource allocation also varies from species to species. Lizards, for example, show considerable flexibility in allocating resources to reproduction. Many lizards show a physiological trade-off from year to year between the number of eggs produced and the average size of each egg. There is a finite amount of resources to put into offspring, and lizards adjust the number and size of eggs depending on conditions in the environment. Larger eggs require more resources to build, leaving less for additional eggs, and larger eggs hatch into larger offspring that are better able to fend for themselves. But if the environment carries a high predation risk, for example, then laying more eggs scattered in different places may be a better strategy than laying a few larger eggs in one place. Does this flexibility make these lizards *K*-strategists or *r*-strategists? The answer is neither. Which strategy is optimal depends on the nature of the risks faced by young.

Plants also show physiological trade-offs in the allocation of resources to seeds. For example, flowers may abort some of their developing seeds to favor the growth of others. Plants also make trade-offs between defense and reproduction. When herbivores are around, plants that invest resources in defenses like poisons or thorns have an advantage over plants that do not invest in defenses. These defensive plants make a greater contribution to the next generation than nondefensive plants. However, when herbivores are absent, defensive plants are commonly at a competitive disadvantage. Individuals that do not invest in defenses can allocate more resources to reproduction. Such trade-offs in plant defense are especially important on nutrient-poor soils, where the cost of replacing valuable tissues lost to herbivores is higher than in nutrient-rich soils (Chapter 32).

The key point is that the various ways in which animals and plants allocate resources to growth and reproduction is rooted in their physiology, and their physiological traits are, themselves, evolved traits. Thus, as stated at the beginning of this chapter, ecology both reflects and determines how organisms function in nature.

→ **Quick Check 4** What ecological conditions require a plant or animal to experience trade-offs among growth, reproduction, defenses, and other adaptive traits?

## 46.3 METAPOPULATION DYNAMICS

In nature, the populations of many, if not most, species are patchily distributed across the landscape. Deer occur only where there are forest patches of suitable size, milkweeds occur only in open fields, and ducks need to be near water. Local populations are therefore usually separated from one another in space and are connected only occasionally by individuals that migrate between habitat patches. Moreover, some habitat areas that could, in principle, support populations of a given species are unoccupied. How does this kind of distribution influence the dynamics—that is, the change in abundances—of populations in nature?

### A metapopulation is a group of populations linked by immigrants.

Groves of trees, separated from other groves by fields, are patches of habitat for the birds, insects, and fungi that rely on them for food and shelter. Similarly, humans are patches of habitat for infectious viruses and certain bacteria, coral reefs are patches for fish, and mountaintops are patches for mountain goats and other high altitude species. For nuthatches or bark beetles, crossing between groves of trees is a risky business, marked by predatory birds and the possibility of not finding another suitable grove. Similarly, populations of reef corals are often separated by open waters full of predatory fish. A **patch,** then, is a bit of habitat that is separated from other bits by inhospitable environments that are difficult or risky for individuals to cross.

The nuthatches that occupy a particular tree grove contact one another more frequently than they do individuals of other tree groves, and so each group is considered a local population. Occasionally, a few migrants might move from grove to grove, enabling gene flow between local populations. We can think of local populations of nuthatches in separate tree groves as each part of a larger, more inclusive group called a metapopulation. A **metapopulation** is a large population made up of smaller populations linked by migration (**Fig. 46.13**).

In order to understand the dynamics of populations, then, we must consider not only the birth and death of individuals, but also

**FIG. 46.13 Metapopulations.** A metapopulation is a group of independent populations connected by occasional immigrants, illustrated here by populations of nuthatches in groves of trees.

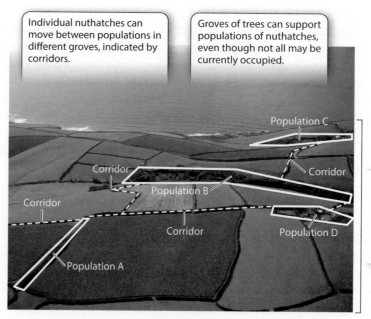

Individual nuthatches can move between populations in different groves, indicated by corridors.

Groves of trees can support populations of nuthatches, even though not all may be currently occupied.

Population C

Corridor

Corridor

Population B

Corridor

Corridor

Population D

Population A

The independent populations, connected by migrants, constitute a metapopulation of nuthatches.

FIG. 46.14 **Experimental patches of habitat.** This study investigated the effects of patch size and connections in moss and the rates of colonization and extinction of the tiny arthropods that live there.

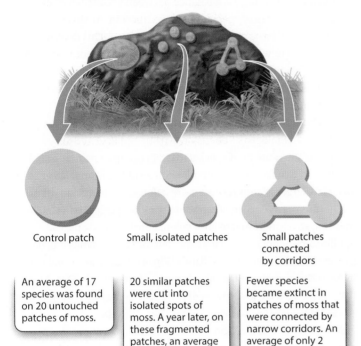

Control patch

Small, isolated patches

Small patches connected by corridors

An average of 17 species was found on 20 untouched patches of moss.

20 similar patches were cut into isolated spots of moss. A year later, on these fragmented patches, an average of 7 species had become extinct.

Fewer species became extinct in patches of moss that were connected by narrow corridors. An average of only 2 species became extinct.

Experiments with patches of moss growing on boulders showed that the size and connectedness of the patches of moss affected the probability that species of arthropods inhabiting the moss would become extinct (**Fig. 46.14**). Researchers cut large patches of moss into several small fragments and sampled the number of insect species in each patch over time. As expected, isolated patches showed decreases in species diversity, up to 40% in 1 year; some species became extinct. However, when nearly isolated moss patches were left with a narrow corridor of moss connecting them, the rate of extinction was lower. Even narrow corridors of suitable habitat can allow species to move between patches of habitat and prevent extinction, or serve to reintroduce species that become locally extinct.

In a world increasingly marked by habitat fragmentation due to human activities, studies of metapopulations are playing an important role in conservation biology (**Fig. 46.15**). Species survival depends not only on the health of local populations, but also on the ability of individuals to colonize new habitat patches. As the simple experiment with mosses shows, corridors that connect endangered populations can dramatically increase the probability that a species will survive. However, corridors are important only to species that use them to cross barriers between patches. For species like ferns that produce spores able to float between patches on breezes, corridors may not be as important to patch colonization. Dispersal ability differs among organisms, strongly influencing their capacity to colonize distant patches.

the colonization and extinction of local populations. Local populations within a metapopulation have independent fates and commonly become extinct because of their relatively small sizes. In general, small populations become extinct more often than large populations because they are more susceptible to random factors that increase mortality, such as weather, natural disasters, or predation. For example, a nest of predators can eliminate a local prey species; or, if population density falls too low in a small local population because of disease or lack of resources, survivors may not be able to find each other for mating. However, as long as some local populations remain—if they do not all become extinct at the same time—these populations can recolonize old habitats or find new habitat patches, thus ensuring survival of the metapopulation.

FIG. 46.15 **The Biological Dynamics of Forest Fragments Project.** Researchers have monitored nine forest patches near Manaus, Brazil, for more than 30 years and have found that fragmentation strongly affects forest biodiversity, providing clues to the design of effective refuges.

## Island biogeography explains species diversity on habitat islands.

Islands represent extremes in habitat patchiness for colonizing species—organisms that live on a mainland must successfully cross a barrier of water to establish a new population on an island. Ecologists commonly expand the use of the term "island" to include any habitat patch that is surrounded by a substantial expanse of inhospitable environment. While this use certainly includes parcels of land surrounded by water, it can also apply to a body of water surrounded by land, an expanse of forest surrounded by grasslands, or a mountaintop alpine field surrounded by forests at lower elevation. Where do colonists of these habitat islands come from, and what determines how many succeed in establishing a new population?

Species diversity reflects both the rate at which new species arrive on the island, and the rate at which species already on the island become extinct (**Fig. 46.16**). As species arrive on an initially uninhabited island, the colonization rate is high because there are no predators yet and the small number of colonists means that competition for resources is limited. Inevitably, as more and more species successfully establish themselves on the island, the rate of colonization goes down. This occurs both because competition

and predation increase, making successful colonization more difficult, and because of the diminishing number of species left in the pool of potential colonists. In contrast, extinction rate goes up as newly arrived species compete.

At some point, there is a balance between the arrival of new species and the loss of species due to extinction. At this point, diversity is said to be at equilibrium (Fig. 46.16). This does not mean that no new species will colonize the island, but that each new colonist species will be accompanied by the extinction of one already in place. There is generally turnover in species composition through time.

The **theory of island biogeography,** articulated in 1967 by the American biologists Robert MacArthur and E. O. Wilson, states that the number of species that can occupy a habitat island depends on two factors. The first is the size of the island. Because of their size, larger islands receive more colonists than smaller ones. Furthermore, larger islands can support more species than smaller ones can, so extinction rate goes down. For these reasons, the equilibrium number of species expected for larger islands will be greater than that for smaller islands (**Fig. 46.17a**). The second key factor is the distance of the island from a source of colonists. More distant islands have lower rates of colonization, moving equilibrium species diversity toward lower values (**Fig. 46. 17b**).

A number of observations and experiments have provided support for the theory of island biogeography. Wilson and his student Daniel Simberloff tested the theory of island biogeography in a novel way. Working among small mangrove islands off the Florida Keys, they used short-term insecticides to remove all the insects living on individual islands. Over the next few years, new insect populations established themselves on the mangrove islands. The researchers found that the species diversity of islands differing in size and distance from shore corresponded to those predicted by the theory. Larger islands had more successful colonists, whereas smaller and more distant islands had fewer.

At a larger scale, the number of animal species counted on tropical islands generally fits the predictions of island biographic theory. **Fig. 46.18** shows not only that larger islands support more species than smaller ones, but also that there is a quantitative relationship between island size and species number. Similar relationships between island size and equilibrium species diversity have been observed repeatedly. The relationship can be described mathematically as a **species–area relationship:**

$$S = cA^x$$

where $S$ is the number of species at equilibrium, $c$ is a mathematical constant, $A$ is the habitat area, and $x$ is an experimentally determined exponent, relating species number to

**FIG. 46.16** Island species diversity, a balance of colonization and extinction.

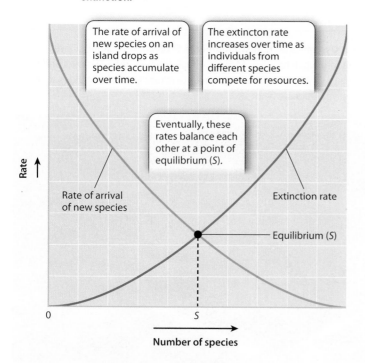

The rate of arrival of new species on an island drops as species accumulate over time.

The extincton rate increases over time as individuals from different species compete for resources.

Eventually, these rates balance each other at a point of equilibrium (S).

Rate of arrival of new species

Extinction rate

Equilibrium (S)

Rate

0    S

Number of species

**FIG. 46.17** Effect of (a) island size and (b) distance from source populations on the equilibrium number of species on an island.

**a.**

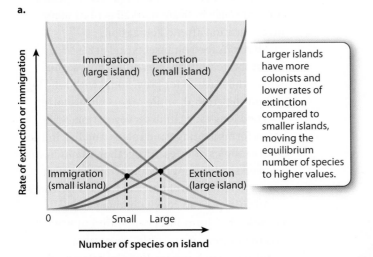

Larger islands have more colonists and lower rates of extinction compared to smaller islands, moving the equilibrium number of species to higher values.

**b.**

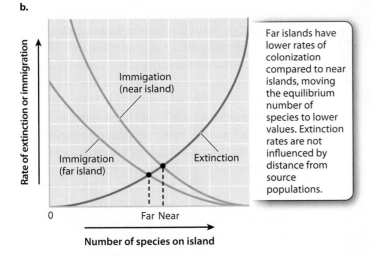

Far islands have lower rates of colonization compared to near islands, moving the equilibrium number of species to lower values. Extinction rates are not influenced by distance from source populations.

habitat area. In many experiments, $x$ has fallen within the narrow range of 0.1 to 0.4.

A useful back-of-the-envelope calculation is that a tenfold increase in island size results in a doubling of species diversity. Perhaps more relevant for issues of conservation, a 90% reduction of habitat area results in a 50% loss of species diversity if no other factors are taken into account. This relationship has many applications in conservation biology, among them the determination of how large a protected area must be to preserve biodiversity and how close reserves must be to one another to ensure the movement of at least some individuals between them.

Island biogeography theory has been widely applied, but it has some limitations in representing island colonizations over long evolutionary time frames. For example, the theory treats all potential colonists as equally able to colonize a new island, when in fact they are not all equally able to arrive or form an association with a species already in place. These abilities are often determined by a species' particular evolutionary history. For example, flying animals like bats and birds more frequently arrive on oceanic islands than nonflying animals like rodents, lizards, and frogs. In contrast, the nonflying animals that colonize successfully more often evolve into multiple new species on islands.

Among the species that have colonized the island of Hispaniola, for example, there are many birds, bats, and strong flying insects like butterflies and dragonflies. Nonflying animals such as frogs, lizards, and fish became established by relatively few colonizations, but successful colonists diversified, evolving a larger number of closely related species now found nowhere else in the world; these are called endemic species.

→ **Quick Check 5** Two islands lie off the coast of Florida. One is larger, the other is closer to the mainland. What determines which one will have more species?

**?** **CASE 8 Biodiversity Hotspots: Rain Forests and Coral Reefs**
**How do islands promote species diversification?**
*Anolis* lizards provide a classic and dramatic example of evolutionary diversification on the island of Hispaniola,

**FIG. 46.18 Island size and number of species.** The diversity of amphibian and reptile species on Caribbean islands fits the predictions of island biogeography theory.

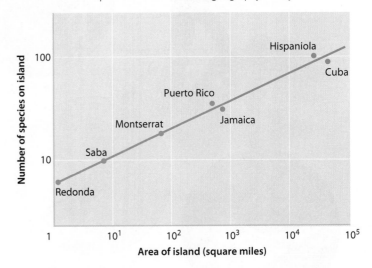

**FIG. 46.19** *Anolis* **lizards in Hispaniola.** These lizards arrived as colonists from elsewhere in the Caribbean and subsequently diversified to produce a variety of species reflecting specialization in their insect-hunting in different parts of local vegetation.

Crown-giant anole

Trunk-crown anole

Tree crown

Crown of tree trunk

Twig anole

Twig

Trunk

Bottom of tree trunk

Grass and bush

Trunk anole

Trunk-ground anole

Grass-bush anole

where many dozens of species of closely related species coexist. Phylogenies suggest that *Anolis* lizards colonized Hispaniola in five or six separate events, but the dozens of species present today indicate marked evolutionary diversification after colonization.

All *Anolis* feed on insects and other invertebrates, but species differ in how and where they search for prey. Some with long legs run on the ground, while others with shorter legs and long tails climb on vegetation. Still other forms are large and robust and live in the high canopies of trees. Altogether, *Anolis* have evolved six different feeding strategies with corresponding morphological traits (**Fig. 46.19**).

All six types can coexist in one place because they hunt insects in different ways and in different parts of the vegetation. Careful research is revealing the pattern of population expansion and fragmentation, followed by natural selection within different subpopulations, that has resulted in the diversity observed on Hispaniola today. Through evolution and complex ecological interactions, then, habitat diversity can depart from the simple predictions of island biogeography. Nonetheless, the theory remains a powerful way of thinking about species diversity and an important consideration in conservation biology.

## Species coexistence depends on habitat diversity.

Species diversity on islands depends not only on the size of the island and its distance from the mainland, but also on interactions among species. Early experiments by the Russian biologist Georgii Gause showed that a simple system with one predator and one prey population is inherently unstable. The predator overexploits the prey, driving it to extinction, and then becomes extinct itself.

In 1958, the American scientist Carl Huffaker demonstrated that if the prey had refuges where some individuals could escape from predators, they could persist while predator populations declined. Through time, the prey population would recover and rise in population density to a point where predators would again expand and cause prey abundance to decline—and then decline again

## FIG. 46.20

# Can predators and prey coexist stably in certain environments?

**BACKGROUND** In the 1950s, it was not clear whether a simple system of predators and prey could be stable and coexist indefinitely or whether it would become extinct as predators consumed all available prey. The results of experiments up to that time were equivocal, and ecologists knew that predators introduced to islands could hunt their prey to extinction.

**HYPOTHESIS** Ecologist Carl Huffaker hypothesized that predators and prey could stably coexist if temporary refuges were available for the prey.

**EXPERIMENT** Huffaker studied two kinds of mites: One feeds on the surface of oranges, and the other preys on the orange-eating mite. Huffaker put two sets of oranges on a table, separating them with barriers of petroleum jelly, essentially making each orange its own habitat patch. He added toothpicks to one set of oranges so that the orange-eating mites could let out silk and float over the petroleum jelly barriers and, at least temporarily, escape predation.

**RESULTS** In the setup in which prey could not escape, a single cycle of increase and decline was observed: The population size of prey increased but was closely tracked by rising populations of their predators, and eventually both declined to extinction.

In experiments in which toothpicks were provided, the orange-feeding mites climbed the toothpicks and dispersed on silken threads they produced, moving away from predators and toward other oranges without mites. These populations of predators and prey went through three cycles of increase and decline, and cycles would probably have continued if oranges continued to be supplied.

**CONCLUSION** Predator–prey systems can be stable if there are sufficient areas available where prey can escape predators, at least temporarily.

Experiments without prey refuges    Experiments with prey refuges

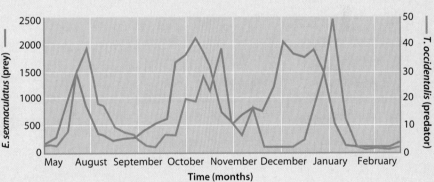

**SOURCE** Huffaker, C. B. 1958. "Experimental Studies on Predation: Dispersion Factors and Predator–Prey Oscillations." *Hilgardia: A Journal of Agricultural Science* 27:795–834.

themselves (**Fig. 46.20**). Huffaker showed that predators and prey cycle repeatedly through periods of increasing and then decreasing density, as predators track their prey and some prey escape predation. In other words, a long-term,

stable oscillation pattern can be achieved that links the population densities of predator and prey. We consider predator–prey interactions in more detail in the next chapter.

# Core Concepts Summary

**46.1 A POPULATION CONSISTS OF ALL THE INDIVIDUALS OF A GIVEN SPECIES THAT LIVE AND REPRODUCE IN A PARTICULAR PLACE AND IS CHARACTERIZED BY ITS SIZE, RANGE, AND DENSITY.**

Population density is calculated by the population size divided by its range. page 46-2

Population size is influenced by births, deaths, immigration, and emigration. page 46-3

The population growth rate is the change in population size, commonly expressed as the per capita rate of growth—the rate of growth (or decline) per individual in the initial population. page 46-3

Initially, growth rates are often exponential, so that the number of individuals added to the population increases over time. page 46-3

Population density often limits population growth because of decreasing resources. This limit defines the carrying capacity of the environment. page 46-5

The growth curve of populations is commonly logistic, an S-shaped curve in which growth is exponential at first but tapers off as population size approaches the carrying capacity of the environment. page 46-5

Population density can be limited by density-dependent factors, such as resources and predation, and by density-independent factors, such as drought and fire. page 46-6

Population size can be estimated by sampling and mark-and-recapture techniques. page 46-6

**46.2 THE AGE STRUCTURE OF A POPULATION HELPS ECOLOGISTS UNDERSTAND PAST AND PREDICT FUTURE CHANGES IN POPULATION SIZE.**

The age structure of a population is the number of individuals of each age group and is often depicted as a histogram. page 46-8

A growing population shows a pyramid-shaped age distribution, with the youngest classes more abundant than older classes, whereas a stable population shows an even distribution of age classes. page 46-8

Survivorship is the proportion of individuals from an initial cohort that survives to each successive age or stage of the life cycle. page 46-9

Type I, Type II, and Type III survivorship curves correspond to mortality that occurs predominantly late in life, is even throughout life, or is concentrated early in life, respectively. page 46-10

*r*-strategists are organisms that live in unpredictable environments, producing many young but investing relatively little in each offspring. page 46-10

*K*-strategists live where resources are predictable and produce a lower number of offspring but invest more in each. page 46-10

*r*- and *K*- strategists represent extremes of a continuum of reproductive strategies. page 46-11

Organisms often exhibit trade-offs between reproduction and other physiological functions. page 46-11

The life history of an organism is the evolved pattern of resource investment in each stage of a given species' lifetime. page 46-11

**46.3 THE DYNAMICS OF POPULATIONS ARE INFLUENCED BY THE COLONIZATION AND EXTINCTION OF SMALLER, INTERCONNECTED POPULATIONS THAT TOGETHER MAKE UP A METAPOPULATION.**

The size of local populations (patches) and the degree of connectedness between patches determines in part whether a given patch will persist or not. page 46-12

An ecological island may be a true island surrounded by water or any habitat surrounded by uninhabitable areas. page 46-12

The theory of island biogeography proposes that the number of species that can occupy an island depends on the size of the island and its distance from the mainland. page 46-14

Species diversity on habitat islands reflects both the rate at which new species arrive and the rate at which colonist species become extinct. page 46-14

Species diversity on islands depends not only on size of the island and distance from the mainland, but also on interactions within and between species. page 46-16

# Self-Assessment

1. Name the four factors that affect population size.

2. Draw an exponential and a logistic growth curve, and explain what accounts for their different shapes.

3. Name two density-dependent and two density-independent factors that can limit the size of a population.

4. Draw a graph showing the age structure of a population that is growing rapidly and a graph of one that is not.

5. Plot a survivorship curve for a species with high rates of predation early in life and one for a species with high mortality late in life. Name the type of survivorship this species displays.

6. Explain how *r* and *K* strategies relate to the predictability of the environment and in what kinds of environment each strategist would be more successful.

7.  Describe what is meant by a trade-off in physiological functions and give an example.

8.  Give three examples of a habitat island and explain what makes them islands.

9.  Name factors that determine the diversity of species on a habitat island and explain the relevance of these factors in managing the habitat of an endangered species.

Do you understand the chapter's Core Concepts? Log in to check your answers to the Self-Assessment questions, then practice what you've learned and reinforce this chapter's concepts by working through the problems and multimedia tutorials provided there.

**www.biologyhowlifeworks.com**

**CHAPTER 47**

# SPECIES INTERACTIONS, COMMUNITIES, AND ECOSYSTEMS

## Core Concepts

**47.1** The niche is the ecological role played by a species.

**47.2** Competition, predation, and parasitism are antagonistic interactions in which one species benefits at the expense of another.

**47.3** In mutualisms, the benefits to both species outweigh the costs of participation.

**47.4** Communities are shaped by the evolved features of organisms, which shape the interactions with other organisms and with the physical environment.

**47.5** The cycling of materials and the flow of energy through ecosystems reflect the ways in which organisms interact in communities.

**47.6** Biomes are broad, ecologically uniform areas distinguished by their climate, soil, and characteristic species.

The chocolate tree, *Theobroma cacao*, known simply as cacao, is native to tropical America. Cacao grows where the environment is consistently warm and humid. Physical environment alone, however, does not determine where cacao can live. Cacao reproduction, and ultimately distribution, is limited by the requirement that tiny flies called midges pollinate them. The midges' eggs and young need deep shade and a thick layer of moist rotting vegetation, whether from the natural fall of leaves onto the forest floor or from banana plants left to decay in cacao plantations. The midges' needs mean that cacao cannot reproduce in areas with full sun or on clean ground that otherwise would support their growth—at least not without human help. The natural distribution of cacao plants, therefore, depends on interactions with both the physical environment and other species.

The reliance of chocolate trees on midges is not unusual. Every species interacts with other species in ways that affect growth and reproduction. These other species may be the insects that help a plant pollinate its flowers or ones that eat its leaves. They may be a fungus that helps a tree gain nutrients or another that attacks it. They may be other species that compete for resources, predators and pathogens that reduce population size, prey species that provide a specialized food resource, or microbes that provide or take nutrients while living inside their hosts (Case 5: The Human Microbiome). Associations among species, whether positive, negative, or neutral, form a complex web of interactions in which each species influences the population size and distribution of other species, whether they are partners, enemies, or food.

In this chapter, we explore interactions that shape the distributions of populations and, therefore, the composition, diversity, and flow of material and energy in communities. We also discuss how adaptations to the physical and biological aspects of habitats result in the stunning variety of ecosystems seen around the world, from the poles to the tropics.

## 47.1 THE NICHE

As discussed in Chapter 46, the habitat occupied by a species is sometimes patchy in its distribution, like islands in a sea, and each patch can contain a number of different species. Distinct species can coexist within shared habitat patches because they use different resources or are active at different times. Nearly a century ago, the American ecologist Joseph Grinnell defined the niche as the sum of the habitat requirements needed for a species' survival and reproduction. Later, ecologist Charles Elton redefined the niche as the role a species plays in a community, switching emphasis from the habitat to the population itself. By the middle of the twentieth century, ecologist G. Evelyn Hutchinson had combined these ideas, popularizing the concept of the **niche** as a multidimensional habitat that allows a species to practice its way of

**FIG. 47.1** **Adaptations of *Anolis* lizards for feeding.** (a) *Anolis* species that feed on twigs have short limbs and a prehensile tail that they use to support themselves when they sleep; (b) those that feed on trunks have long limbs.

life. The niche is therefore determined by both physical (abiotic) parameters such as climate and soil chemistry and biological (biotic) factors based on interactions with other species.

### The niche is the ecological role played by a species in its community.

The niche, then, reflects the interactions between populations and the physical and biological components of their surroundings. These interactions lead to adaptations, the products of natural selection. Adaptations in turn enable different species to coexist by exploiting different combinations of resources.

*Anolis* lizards, for example, hunt small insects in humid forests on Hispaniola and other islands in the Caribbean (see Fig. 46.19). The lizards have features of anatomy and physiology that determine, in part, where they can live. They cannot, for example, endure freezing temperatures. Within the forest, different *Anolis* species eat insects found in distinct parts of forest trees. That is, over time, they have evolved different adaptations for feeding as a result of competition for space and resources (**Fig. 47.1**). *Anolis* lizards, then, show that species whose niches overlap may diverge to minimize the overlap, a pattern called **resource partitioning**.

Interactions with predators such as lizard cuckoos and curly-tailed lizards

further influence, and limit, the distributions of different species. Moreover, there is reason to believe that the ability (or inability) to disperse contributes to the distributions of individual species. For example, when *Anolis* species are introduced by accident or deliberately to places beyond their dispersal capabilities, they commonly thrive. Therefore, competition, predation, and dispersal are all factors that influence the distribution of species.

**FIG. 47.2** **Fundamental and realized niches.** The fundamental niche includes all the habitat potentially available to a population, whereas the realized niche is the subset of that habitat actually occupied by the population.

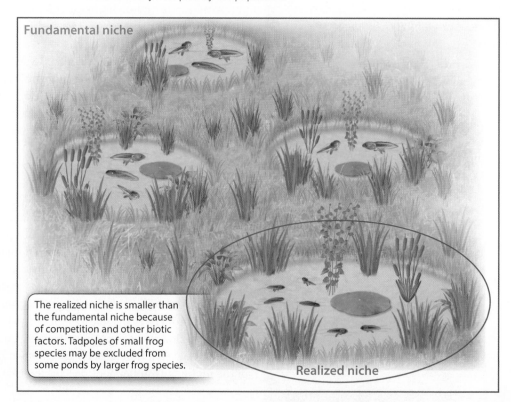

Fundamental niche

The realized niche is smaller than the fundamental niche because of competition and other biotic factors. Tadpoles of small frog species may be excluded from some ponds by larger frog species.

Realized niche

Species are commonly associated with a specific habitat, but "niche" and "habitat" are not interchangeable, as the niche refers to the ways that organisms respond to the resources and other species found within the habitat. American ecologist Eugene Odum summarized the distinction succinctly: The habitat is the organism's "address" and the niche its "profession." This distinction illustrates the dual nature of niches: They reflect where organisms occur and what they do there. The profession of *Anolis* lizards is insect-hunting; their addresses are the trees and shrubs of humid tropical America.

**FIG. 47.3** **A mutualistic interaction, in which both participants benefit.** Development of seed-rich cacao fruits (a), whether for the plant's reproduction or to provide raw material for chocolate, depends on pollination by midges (b).

### The realized niche of a species is more restricted than its fundamental niche.

The **fundamental niche** comprises the full range of climate conditions and food resources that permits the individuals in a species to live. In nature, however, many species do not occupy all the habitats permitted by their anatomy and physiology. That is because other species compete for available resources, prey on the organisms in question, or influence their growth and reproduction, reducing the range actually occupied. This actual range of habitats occupied by a species is called its **realized niche (Fig. 47.2).**

For example, tree frogs can reproduce in small ponds near forests throughout eastern North America, their fundamental niche. However, tree frogs are excluded from ponds that also have larger frogs because the larger frogs eat most of the algal food available, limiting the tree frogs' realized niche to ponds without larger frogs. Therefore, competition with other species—that is, **interspecific competition**—can determine the size of the realized niche.

→ **Quick Check 1** Which is larger—the fundamental or the realized niche?

## 47.2 ANTAGONISTIC INTERACTIONS BETWEEN SPECIES

The realized niche of tree frogs compared with their fundamental niche makes it clear that interactions with other species play key roles in the distribution and abundance of populations. Close interactions between species that have evolved over long periods of time are called **symbioses** (Chapter 26). Many of these interactions enhance the reproduction and population growth of both ecological partners, like chocolate trees and midges (**Fig. 47.3**). These mutually beneficial interactions are called

**mutualisms**. Many other interactions are one sided: One organism actually consumes another or competes with it for food or some other resource (**Fig. 47.4**). Such one-sided interactions, in which at least one participant loses more than it gains, are called **antagonisms**.

All interactions combine costs (often in physical resources) and benefits (usually increases in reproduction), and the ratio of costs to benefits for each of the participants reveals whether an overall

**FIG. 47.4** **An antagonistic interaction, in which one participant benefits and one participant loses.**

interaction is mutualistic or antagonistic. Some interactions are direct, where species physically interact (such as plants and their pollinators). Others are indirect, where they influence one another through competition for a shared resource, such as food. Virtually every organism is involved in both mutualisms and antagonisms with other species—think of plants and their pollinators or predators and their prey. These interactions often narrow the geographic distributions of species within the larger boundaries of their physical requirements, and so help shape their realized niches.

### Limited resources foster competition.

Whenever there are fewer resources than individuals seeking them, there can be competition. **Competition** is an interaction in which the use of a mutually needed resource by one individual or group of individuals lowers the availability of the resource for another individual or group. Different species of sea anemones compete for space in rocky shallows, swallows compete with one another for nesting holes in tree or cliffs, and many different species of dung beetles compete for animal droppings (some even cling to monkeys, waiting for feces to emerge). Male birds, mammals, insects, and other animals commonly compete for female mates.

In Chapter 46, we saw that in many populations the tension between intrinsic rate of reproduction and environmental carrying capacity results in intraspecific competition—competition within a species. The increase in competition between individuals of a species that accompanies increased population density is a main reason why population growth slows as the species' environment approaches its carrying capacity. As noted earlier in this chapter, competition can also occur between individuals of different species, in which case it is interspecific competition. Whether competition occurs between individuals of one species or two different ones, it is a lose–lose situation: Both sides spend energy they would not in the absence of the other. Indeed, Darwin recognized that the "struggle for existence," as he called it, is a primary driver for natural selection, leading to the evolution of traits that aid survival and reproduction by reducing competition in one way or another.

### Competition promotes niche divergence.

In general, when two co-occurring species overlap strongly in resource use, one will either become extinct in that place or change its resource use. For example, coyotes and foxes eat the same small prey, mice and rabbits, but when they meet, a coyote will waste energy by chasing the fox and of course the fox loses energy by running away from the coyote. In regions where foxes and coyotes occur together, the foxes avoid interactions with the larger coyotes by leaving the area, placing their dens in locations where coyotes do not occur. This result of such an antagonistic interaction, in which one species is prevented from occupying a particular habitat or niche, is called **competitive exclusion**. This is another way that a fundamental niche can be reduced to a smaller realized niche. Over time, competitive exclusion leads to resource partitioning, as we saw with the *Anolis* lizards (Chapter 46).

Species can compete with each other directly or indirectly. For example, some trees compete with other trees directly by producing toxins in the soil around them that the seedlings of other species cannot tolerate, and corals may compete by stinging overlapping colonies. Some species compete with others indirectly by interfering with the resources used by a different species. For example, diatom algae make skeletons of silica

---

**FIG. 47.5  Competition for space.** (a) Herons, egrets, and cormorants compete for nest sites in wetland trees. (b) In the sea, corals compete for space on reef surfaces.

(SiO$_2$), depleting this substance in the surface waters of the world's oceans. Radiolaria also make skeletons of silica, and for the past 65 million years, radiolarians have evolved silica-saving modifications of their skeletons in response to the increasing sequestration of SiO$_2$ by diatoms.

### Species compete for resources other than food.

Food is the resource we most often think of as the limiting factor for population growth and reproduction, but many species just require a place to stand, so to speak, or suitable conditions for a nest. Barnacle and mussel larvae compete for a place to attach on submerged rocks. Their food, tiny plankton, is almost always abundant and therefore is not as likely to limit population growth as attachment space is.

Similarly, plants compete for space in fields and forests so they can collect sunlight and obtain water and nutrients from the soil. Sunlight cannot itself be exhausted by use, but room to collect sunlight on the ground below can be filled. Nesting sites for birds that lay their eggs in tree hollows or on cliff faces can also be strongly limited by the availability of space, as can the settling of corals on reef surfaces (**Fig. 47.5**).

### Predators and parasites can limit prey population size, minimizing competition.

Predation is another form of antagonism, as the predator benefits at the expense of the prey. Predators can limit the population sizes of their prey, preventing prey populations from increasing to the level where competitive exclusion occurs. We have seen that when species overlap in resource use and their populations increase to the point where they compete for resources, they must be separated in space or time or risk local extinction through competitive exclusion when resources become limited.

If, however, populations do not rise to densities at which resources are limiting, competition is reduced and species can overlap in niches without excluding one another. For example, in ponds throughout the southern United States, tadpoles of the large Southern Toad, the Eastern Spadefoot, and the tiny Spring Peeper all use the same food source. In the presence of the two larger species, Peeper tadpoles compete poorly for food (**Fig. 47.6**). As a result, they have low survival rates, and those individuals that do make it to maturity are often undersized

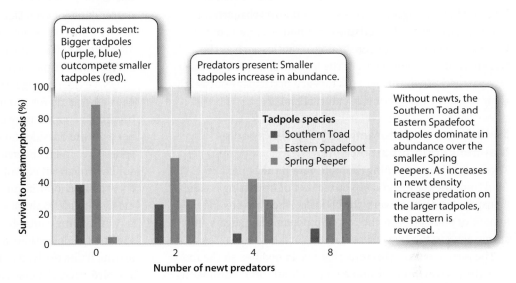

**FIG. 47.6** **Effect of predators on the outcome of competition.** Newt predators in ponds prefer to eat bigger tadpoles that compete well for food, enabling smaller tadpoles that are poorer competitors to survive to adulthood.

Predators absent: Bigger tadpoles (purple, blue) outcompete smaller tadpoles (red).

Predators present: Smaller tadpoles increase in abundance.

Without newts, the Southern Toad and Eastern Spadefoot tadpoles dominate in abundance over the smaller Spring Peepers. As increases in newt density increase predation on the larger tadpoles, the pattern is reversed.

**Tadpole species**
- Southern Toad
- Eastern Spadefoot
- Spring Peeper

*y-axis:* Survival to metamorphosis (%)
*x-axis:* Number of newt predators

because as tadpoles they did not have access to adequate nutrition. In the presence of a predatory Red-Spotted Newt, however, the tables are turned. The newt prefers to eat the larger tadpole species, and so prevents them from becoming sufficiently abundant to outcompete the smaller Peeper tadpoles. The Peeper tadpoles, then, have the highest survival rates. With competition for food reduced, these small tadpoles grow to their maximal body size, and become the most abundant tadpole species in the ponds.

Parasites can also limit the population size of their host, keeping numbers well below the carrying capacity of the environment. An extreme example is the fungus that infects the American chestnut, a tree once dominant in eastern North American forests. The fungus attacks the vascular system of young chestnut trees, which usually succumb when they reach 2 to 3 m in height. Today, there are only a few isolated patches of large chestnut trees in Vermont and a handful of other places that have escaped the fungus. As chestnut populations have declined, populations of oak, beech, and other trees that grow in the same forest have increased.

## 47.3 MUTUALISTIC INTERACTIONS BETWEEN SPECIES

Interactions between species do not necessarily pair a gain for one participant with a loss for the other. As the example of cacao and midges shows, some interactions benefit both participants. Typical benefits include access to nutrients, shelter from enemies or weather, and direct help in reproduction, but all benefits are ultimately measured by natural selection in terms of reproductive output. Some typical costs include the proteins, fats, and carbohydrates that are invested in building structures such as flowers that attract pollinators, or specialized tissues that house bacteria or algae. Costs can also be energy-consuming activities such as the transport of pollen or seeds, or the loss of food resources used by a partner. When the benefits for each participant outweigh their costs, the interaction is a mutualism.

### Mutualisms are interactions between species that benefit both participants.

The midges in our opening example obtain the benefit of food from the cacao blossoms at the cost of unwitting pollen transport between flowers; the chocolate tree obtains the benefit of pollination at the cost of producing the sugars and amino acids in the flowers' nectar.

The nitrogen-fixing bacteria that live in nodules on the roots of soybeans (Chapters 26 and 29) provide another illuminating example. The bacteria occur naturally in the soil and are attracted to the plant roots, where they take up residence in nodules produced by the plant. The bacteria provide their host plant with nitrogen in a biologically useful form, and the plant provides the bacteria with food and a stable environment. In response to signals from the bacteria, the plant builds the bacteria's home—the nodules—out of root tissue, a cost outweighed by the benefit of enhanced access to nitrogen, a benefit measured in greater growth and reproductive output. In the interaction between plant and bacteria, both sides win.

Despite the observation that both partners in a mutualism benefit, it is important to remember that each side is acting in its own self-interest and bears costs that are weighed against benefits in terms of growth and reproduction. Associations that, overall, are beneficial to one partner are often also beneficial to the other partner and, therefore, more often represented in the next generation. In other words, mutualisms are subject to natural selection, just like any other adaptation (Chapter 21).

### Mutualisms may evolve increasing interdependence.

One of the best-studied mutualisms involves aphids (and their insect relatives) and closely associated bacteria. The insects suck plant sap for food, and the bacteria live in tissues that develop in the insects adjacent to their digestive system. The insects provide a home for the bacteria at some physiological cost but gain an important benefit: Plant sap contains relatively few nutrients, and the bacteria provide their hosts with essential amino acids. The bacteria, in turn, benefit from a stable and favorable environment at the cost of some loss of nutrients supplied to the aphids. The aphids pass the bacteria from mother to daughter in egg cells, guaranteeing that

offspring will have bacteria as effective as those associated with earlier aphid generations. When the interaction between species drives reciprocal adaptations in both participants, this opens the possibility for long-term coevolution because the descendants of each side will also be associated with each other, just as were past generations. Any mutations that arise can affect both sides of the partnership because neither aphids nor bacteria are found without the other.

DNA evidence suggests that aphids and their bacteria have been associated for 100 million years or more, and that the bacterial genomes have been gradually losing genes they would need for life outside the insects (**Fig. 47.7**). Some bacterial genes have in fact been acquired by the insects. We saw a similar process of coevolution in Chapter 26. As we discussed there, mitochondria and chloroplasts are probably nature's most dramatic examples of coevolved mutualism.

While some mutualisms involve close interactions between specific species, others are less particular. For example, sweet-tasting fruits evolved in flowering plants, attracting mammals and birds that disperse their seeds, but not in response to any particular bird or mammal species. Similarly, flowers evolved in response to insect pollinators such as bees and flies, whose own adaptations for visiting flowers evolved in response to flower availability.

### Mutualisms may be obligate or facultative.

When one or both sides of a mutualism cannot survive without the other, the association is said to be **obligate**. For example, the association of aphids and bacteria is obligatory for both sides: Bacteria cannot live without the shelter the aphids provide, and aphids cannot live without the nutrients provided by the bacteria.

Many associations, however, are not so tightly intertwined, and one or both participants can survive without the other. These interactions are called **facultative**. For the midges that pollinate cacao, the association is facultative because the midges have other sources of food, including other flowers, in the wet forests they inhabit. For cacao, however, the relationship is obligate because bees and other insects do not usually pollinate them, and so their reproduction more or less relies on visits by midges.

Many obligate mutualisms are thought to have begun as facultative relationships that became reinforced over time by natural selection. Thus, the mutation that originally made a cacao flower sprout from the tree trunk close to the ground rather than high out on a branch made the flower easier to pollinate by low-flying midges while decreasing the chances of discovery and pollination by bees. If midges were consistently more reliable pollinators than bees, the mutation would have been selected and then spread throughout the population of cacao (Chapters 21 and 22).

FIG. 47.7

# Have aphids and their symbiotic bacteria coevolved?

**BACKGROUND** Aphids are small insects that are common pests of garden plants. Specialized cells in the aphids harbor populations of the bacterium *Buchnera* that provide their host with amino acids essential for growth. Most, but not all, aphid reproduction occurs without males, with mothers producing daughters. Mother aphids pass the bacteria through their eggs to their daughters.

**HYPOTHESIS** Aphids and *Buchnera* bacteria have coevolved for millions of years.

**EXPERIMENT** Evolutionary ecologist Nancy Moran sequenced the DNA of bacteria and the aphids they come from to establish the phylogenetic relationships of aphid species and their symbionts. She reconstructed the phylogeny of both groups and compared their relationships. If the two phylogenies matched each other, the hypothesis of coevolution would be supported.

**RESULTS** The phylogenetic trees of aphids and their associated bacteria matched perfectly, just as if the bacteria were a gene of the aphids instead of a separate lineage. This matching of phylogenetic trees was seen both among aphid species in different genera and for aphid populations within a single species. A DNA-based molecular clock further showed that aphids, and the insects related to them, have all coevolved with these bacteria over nearly 200 million years.

**CONCLUSION** A single lineage of *Buchnera* has been passed down through the generations and the diversification of its aphid hosts for nearly 200 million years. Host-symbiont systems can be stable over long time intervals and show long-term coevolution.

**FOLLOW-UP WORK** Researchers are studying the changes in the genomes of bacteria living inside aphids to see how they lose genes necessary for living independently of their aphid hosts.

**SOURCE** Moran, N. A., M. A. Munson, P. Baumann, and H. Ishikawa. 1993. "A Molecular Clock in Endosymbiotic Bacteria Is Calibrated Using the Insect Hosts." *Proceedings of the Royal Society. London, Series B*, 253:167–171.

**TABLE 47.1**  Types of Species Interactions

| INTERACTION | EFFECT ON EACH SPECIES | EXAMPLE |
|---|---|---|
| Mutualism | +/+ | Flowers and bees: Flowers gain pollination; bees gain nectar and some pollen. |
| Antagonism | +/− or −/− | |
|    Competition | −/− | A grass and a wildflower: Each species loses the water, nutrients, and access to sunlight that the other takes. |
|    Predation | +/− | Arctic foxes and lemmings: Foxes benefit from eating lemmings; lemmings lose opportunities to reproduce. |
|    Herbivory | +/− | Bison and grass: Bison benefit from eating grass; grass loses biomass that is eaten. |
|    Parasitism | +/− | Tapeworms and humans: Tapeworms benefit from absorbing nutrients in human intestine; humans lose nutrients. |
| Commensalism | +/0 | Egrets and cattle: Egrets benefit from insects stirred up by cattle; cattle are unaffected by egrets. |

### The costs and benefits of species interactions can change over time.

In addition to interactions that benefit both partners (mutualism) and those in which at least one partner is harmed (antagonism), there are interactions in which one partner benefits with no apparent effect on the other (**commensalism**). For example, Grey Whales in the Pacific Ocean are commonly festooned with barnacles. The barnacles benefit from the association, obtaining both a substrate for growth and a free ride through waters rich in planktonic food, but the presence or absence of barnacles doesn't seem to affect the whales one way or the other.

Some ecologists also recognize a fourth class of interaction, called amensalism, in which one partner is harmed with no apparent effect on the other. A commonly cited example is the Black Walnut, a forest tree that produces compounds that inhibit the growth of other plant species in its immediate vicinity. A fair criticism of this and many other cited examples of amensalism is that the Black Walnut actually benefits by limiting competition for nutrients in the soil. The spectrum of costs and benefits associated with interactions among organisms is summarized in **Table 47.1**.

Associations are not fixed—they can change over time. A mutualism can in some cases become antagonistic if one of the partners "cheats" by imposing a larger cost than benefit on the other. In fact, mutualisms that are loose associations among changing partners can become one sided rather quickly. For example, many plants have evolved tubular flowers that guide bees past their anthers or stigma on the way to the nectar at the base, taking just

**FIG. 47.8  Commensalism and mutualism.** Egrets feed on insects stirred up by water buffalo, which are usually not affected by the egrets—a commensalism. If, however, the egrets spot lions and the poor-sighted buffalo take warning, the relationship becomes a mutualism.

a little more time from bees who try to visit as many flowers as possible. Some bees have short-circuited this plant mechanism by nipping the flower base from the outside and then drinking the nectar without pollinating the flower. Most plants have enough successful pollination to ensure seed production, but losses from cheating can still be costly.

Some interactions begin as a benefit to one species, with no benefit or cost to the other. For example, cattle egrets follow water buffalo to pick up the insects stirred up by the buffalo passing. The egrets gain food, but the buffalo are not affected either way—an example of commensalism (**Fig. 47.8**). However, cattle–egret commensalism can become a mutualism over time if the egrets give early warnings of nearby predators such as lions or eat insects like tsetse flies that carry buffalo diseases.

→   **Quick Check 2** What are some of the costs and benefits to an apple tree and to the honeybee that pollinates it?

## 47.4 ECOLOGICAL COMMUNITIES

We have been discussing how species interact and how these interactions can shape the distributions of populations, limiting distributions in one way or another. Now we look collectively at all the species that occur together in the same place. How, and to what extent, do they interact? And what are the properties of these spatial aggregations of species?

### Species that live in the same place make up communities.

A **community** is the set of all populations found in a given place. That definition seems straightforward, but it raises some large questions about the nature of communities. Two different views were proposed nearly a century ago, and ecologists have debated their merits ever since. Frederic Clements, a prominent American plant ecologist of the early twentieth century, likened communities to a "superorganism" in which species interact strongly and predictably, like the organs within a body. In contrast, his contemporary, Henry Gleason viewed communities as simply the products of species acting individually in time and space. Not surprisingly, research inspired by their disagreement suggests that most communities lie somewhere between these extremes.

Populations in a community are tied together by the various interactions that secure their spot in a food web (Chapter 25), as well as by their physical location. But when we look at the details of where particular species occur, it turns out that almost no two species have exactly the same geographic distributions. This is partly a result of competitive exclusion

(unless they are obligate associates of each other, like aphids and their bacteria). The Black Spruce and Bog Club Moss that inhabit freshwater bogs in northern New England overlap in their mutual preference for water-saturated, acidic, sandy soils found in cold temperate bogs, but they do not occur in the same places over most of their ranges.

Even within a local habitat like a meadow, species have different distributions that may reflect differences in soil moisture or patterns of sunlight exposure. For example, two kinds of milkweed occur in meadows, but one prefers damper sites and the other drier areas. For milkweeds and other plants, interactions also vary from place to place as the local set of species changes. Because a given species' nearest neighbors differ over its geographic range, natural selection may also differ from place to place as particular competitors, partners, predators, or prey drop out or enter its distribution. For example, in the northern part of their range, Least Weasels prey only on lemmings, and so are under selection to coordinate their reproduction with that of the lemmings. But in the southern part of their range, lemmings are not present and so the weasels hunt mice and rabbits, placing different selection pressures on them.

The activities of different populations in a community also vary in time. For example, flies are active during the day, pollinating flowers that open in the morning and falling prey to dragonflies, birds, and other predators. In contrast, moths are important pollinators at night, when spiders and bats that prey on them are also active. Similarly, in coastal marine communities, oysters, crabs, and many other animals follow the daily rise and fall of tides. And, of course, the activities of many species change seasonally in responses to sunlight, temperature, and rainfall. Protein-rich insects that birds prey on to feed their young are most abundant in the spring, and fruits and seeds rich in the oils that migratory birds need to supply fat reserves are mostly available in late summer and fall.

### A single herbivore species can affect other herbivores and their predators.

Predators and prey interact in settings as different as a tropical rain forest or Arctic tundra. In habitats with many different species of predators and prey, identifying the way each species affects others can be challenging. For this reason, much research on how predators and prey influence community structure has been carried out in the Arctic, where communities tend to have relatively few species.

One well-studied community is found on Bylot Island in the Canadian territory of Nunavut. There, two species of lemming, the main herbivores in tundra communities, affect the populations of other herbivores and predators. Snow Geese are herbivores that live alongside the lemmings, and both geese

**FIG. 47.9**  **Predator–prey relationships in the Arctic grassland community on Bylot Island, Canada.** Species have either a positive effect (+) or a negative effect (−) on other species.

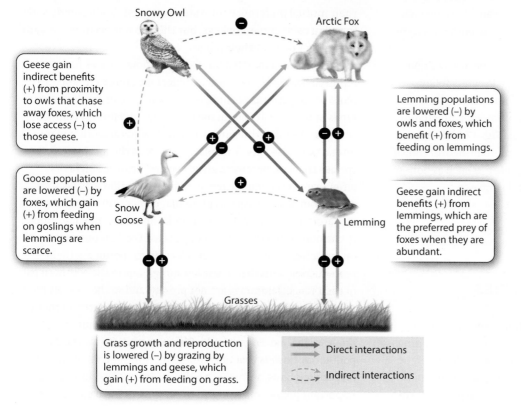

Geese gain indirect benefits (+) from proximity to owls that chase away foxes, which lose access (−) to those geese.

Lemming populations are lowered (−) by owls and foxes, which benefit (+) from feeding on lemmings.

Goose populations are lowered (−) by foxes, which gain (+) from feeding on goslings when lemmings are scarce.

Geese gain indirect benefits (+) from lemmings, which are the preferred prey of foxes when they are abundant.

Grass growth and reproduction is lowered (−) by grazing by lemmings and geese, which gain (+) from feeding on grass.

Direct interactions

Indirect interactions

## Keystone species have disproportionate effects on communities.

While all or nearly all species in a given place have some effect on one another, the integrity of a community may depend on a single species. This situation can occur when one species influences the transfer of a large proportion of biomass and energy from one trophic level to another or when the species modifies its physical environment, as beavers do when they make ponds. Such pivotal populations are called **keystone species** because they support a community in much the way a keystone supports an arch. Though all species contribute to the structure of a community, a keystone species affects other members of the community in ways that are disproportionate to its abundance or biomass.

A classic experiment by American zoologist Robert Paine gave birth to the concept of keystone species. Along the northwestern coast of the United States, mussels adhere to rocks in the intertidal zone, where they coexist with about two dozen other species of invertebrate animals and seaweeds. Mussel populations are regulated by their principal predator, a sea star called *Pisaster ochraceus*. When Paine removed the sea star from experimental plots, mussel populations exploded, outcompeting other animals and seaweeds for space and so eliminating them from the rocky shoreline. Control plots in which the sea star remained showed no such change. From these results, Paine concluded that the activities of the predatory sea star had a singularly strong influence on the diversity and distribution of other populations in the rocky intertidal community. *Pisaster ochraceus* is a keystone species.

Another experiment, this time unintended, cements the point. Along the west coast of the United States and Canada, giant kelp grow more than 50 m tall, providing a home for a remarkable diversity of fish and invertebrates (**Fig. 47.10**). The kelp community depends strongly on a single predator: the sea otter. Sea otters are the predators of sea urchins, which in turn graze on the rootlike holdfasts that anchor kelp to the seafloor. When the otters were removed by overhunting

and lemmings eat the grasses and other low plants that grow on the island. The main predators of lemmings and Snow Geese are Arctic Foxes and Snowy Owls. Lemmings, foxes, Snowy Owls, geese, and plants, then, are linked by a set of mutualisms and antagonisms, shown in **Fig. 47.9**. Changes in one species can affect all of the others.

Lemming populations rise and fall every 4 years or so. When lemming populations decline, foxes turn to other prey, including the eggs of Snow Geese (while the now harder-to-find lemmings escape and their populations begin to increase again). Snow Geese, therefore, are indirectly influenced by lemming population densities. The population density of foxes also reflects that of lemmings—as lemming populations increase to the point at which they can again become a significant part of fox diets, the foxes have access to enough resources to have more pups, increasing the fox population. In time, the additional predators again cause the lemming population to decline, and fox reproduction drops as a result. The population density of lemmings also influences grass abundance through its indirect effect on whether or not foxes eat Snow Goose eggs; young geese are heavy grazers in years when goose reproduction is high.

**FIG. 47.10  Keystone species in Pacific kelp forests.** The activities of keystone species strongly influence the communities in which they live.

Sea otter predation is key to the health of a kelp forest. Notice the many fish beneath the kelp canopy.

When otters are absent, their sea urchin prey rise in abundance and consume most of the kelp holdfasts. Without kelp forests, many animals decline in abundance.

decades ago, sea urchin abundance soared and the kelp nearly disappeared, along with many other populations that live in kelp forests. Now, where otters are protected they once again keep urchin populations in check, enabling the kelp to return, and along with it, the diversity of fish and other animals for which kelp provides a habitat.

Keystone species called ecosystem engineers actively shape the physical environment, creating habitat for others. The classic example is beavers, which produce ponds by damming streams in forests, creating habitat that would not otherwise exist (**Fig. 47.11**).

→  **Quick Check 3** Lemmings affect many other species. Why are lemmings not considered a keystone species?

### Disturbance can modify community composition.

We have seen how predation, herbivory, and parasitism can reduce the effects of competition. Severe climatic events, such as unusual rainfall or extreme temperatures, can also dramatically lower the abundances of some species, for example when an early frost kills mosquitoes and other insects. Severe physical impacts on a habitat, such as those caused by

storms, earthquakes, or road building, are known as **disturbances**, and they have effects on populations of interacting species that are independent of their densities, as we saw in Chapter 46. Disturbances often affect multiple species in the same community and so can exert a strong influence over community composition. We usually think of disturbances as abiotic, but the line is not so clear with road-building by humans, for example, or dam-building by beavers. The main point is that catastrophic changes in a habitat affect populations in ways independent of their densities. Such changes are not uncommon, but are usually unpredictable.

Severe weather in forests demonstrates how disturbance can affect a community. When fire or a hurricane takes down tall trees in a forest, it also takes away the shade that had prevented some species from establishing new seedlings. Fires are frequent sources of disturbance in regions with a seasonally dry climate, such as parts of western North America and southwestern Australia. Like storms, fires remove incumbent plants and animals, opening up habitat for recolonization. Some species have, in fact, adapted to frequent fires. For example, several pine species have evolved thick bark and high branches, traits that help protect them from fire. Others, which have evolved

**FIG. 47.11  A beaver dam visible from space.** This NASA satellite image shows a beaver dam in northern Alberta, Canada.

Beaver lodges block the flow of water from the surrounding wetlands, making a pond that drowns all the trees downstream of the lodges, creating homes for cavity-nesting birds, and food for insects and fungi.

Beaver lodges

Direction of ground water flow

166 m

Google earth

cones that open only in the wake of fire, use the habitat space opened up by the fire and the nutrients that dead plant matter releases into the soil.

Disturbances allow species to enter or remain in the community as long as they can establish their young in the newly opened habitat patch. Noting the influence of storms on community composition in forests and coral reefs, American ecologist Joseph Connell proposed that the diversity of species in a given community reflects, at least in part, the frequency and intensity of disturbance. If disturbance is intense, few species can tolerate the physical conditions of the habitat and diversity will be low. On the other hand, if disturbance is rare or weak, competition may take over and only a few, stronger, competitors may remain. Species diversity tends to be highest when disturbance is frequent or intense enough to inhibit competition, but not so strong as to limit the number of species that can tolerate the environment.

### Succession describes the community response to new habitats or disturbance.

Following disturbance or the appearance of a new habitat, a community goes through a series of changes. These changes involve a predictable sequence of species that colonize and then transform the community, in what can appear to be a linear process of maturation. For example, following a large storm

or fire, newly vacated habitat is commonly colonized by plant species adapted for rapid growth and strong sun. Eventually, these species are replaced by others able to grow in increasing shade. The pace of this transformation may be rapid, occurring over a few decades, or it may be much slower, depending on the environment.

This process of species replacing each other in time is called **succession** (**Fig. 47.12**). Clements viewed the pattern of species replacement in succession as strong evidence in support of his view that communities are highly integrated. Gleason, in contrast, explained succession purely in terms of the habitat preferences of individual populations.

The pin cherry, a resident of New England forests, illustrates how the life histories of individual species can contribute to succession. If you walk through a forest in New Hampshire, you will see oak trees and beeches, pine and hemlock. A mature forest may contain few or no pin cherry trees, but if you examine seeds lying dormant within the forest soil, you will see numerous pin cherry seeds. How did they get there?

Years earlier, disturbance from a storm or fire opened up the forest floor and pin cherries established themselves in the sunlit soils, growing from seeds left in the droppings of birds that had consumed the cherries elsewhere. The seedlings of pin cherries grow well in full sunlight, so the trees grew rapidly to reproductive maturity. A forest canopy was eventually

---

**FIG. 47.12 Succession.** In new habitats or following disturbance, species colonize and transform the habitat in a predictable sequence that reflects the adaptations of participating species.

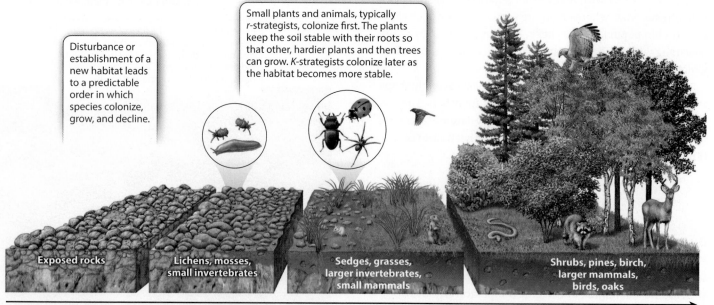

Disturbance or establishment of a new habitat leads to a predictable order in which species colonize, grow, and decline.

Small plants and animals, typically *r*-strategists, colonize first. The plants keep the soil stable with their roots so that other, hardier plants and then trees can grow. *K*-strategists colonize later as the habitat becomes more stable.

Exposed rocks

Lichens, mosses, small invertebrates

Sedges, grasses, larger invertebrates, small mammals

Shrubs, pines, birch, larger mammals, birds, oaks

Time

reestablished, but pin cherry seeds do not germinate in the shade. In contrast, oak seedlings prosper in shade, so through time oaks and other species came to dominate this patch of forest. They will continue to dominate it until the next disturbance, when the removal of mature trees will once more expose the forest floor to bright sunlight, literally giving pin cherries another day in the sun.

The succession of species reflects the adaptations of each to the changes in sunlight, soil, water, and space that occur as the first organisms arrive, mature, and die. These organisms' bodies influence their habitat in both life and death, whether aquatic or terrestrial. Succession over thousands of years can turn bare rock into forest. Lichens are often the first organisms to colonize bare rock newly formed by cooling lava or exposed by landslides. The lichens may start off with only a few mites and other tiny invertebrates living on them. As the stone weathers, and microorganisms feed on the lichen and one another, they form particles of soil that accumulate in cracks and crevices. Eventually, small plants, such as sedges and grasses, take root, and the growing plant community begins to gain height and diversity as other plants establish themselves.

Soon, larger insects, spiders, and other invertebrates join the community, as well as the first small mammals and birds. Many of these early colonizing species are *r*-strategists (Chapter 46), growing and reproducing quickly, and spreading their young far and wide. Birds carry the seeds of shrubs and other plants in their droppings, and the gradually deepening soil eventually permits the establishment of shrubs that in turn provide habitat for larger mammals and birds, and perhaps snakes and salamanders.

The first sun-loving trees, such as birches and pines, appear next, providing shade for the seedlings of later colonizing trees like oaks and maples. In many forests, these stable, long-lived *K*-strategists (Chapter 46) make up a final stage in the succession, forming a mature assembly commonly called a **climax community.** A climax community is one in which there is little further change in species composition.

## 47.5 ECOSYSTEMS

A community of organisms and the physical environment it occupies together form an **ecosystem (Fig. 47.13).** The physical and biological components of ecosystems are linked by the processes that cycle nutrients and transfer energy through the system. In Chapter 25, we introduced the concepts of the **food web** and the **trophic pyramid** as depictions of carbon cycling and energy flow through ecosystems, respectively. Now we can revisit these concepts, understanding them as reflections of the functional properties of individual species and the mutualisms, antagonisms, and other interactions among species within a community.

**Species interactions result in food webs that cycle carbon and other elements through ecosystems.**
Of all the interactions observed among species, none are more prominent than predation and herbivory. Heterotrophic organisms, from bacteria to blue whales, obtain carbon and other elements needed for growth from other organisms. Thus, when a sea star preys on a mussel, we can view this interaction both as an antagonism between two species and as a link in the carbon

**FIG. 47.13  A savanna ecosystem in Africa.** The ecosystem includes the grasses, trees, animals, and microorganisms in the community, as well as the climate, soil chemistry, and other features that define its physical environment.

cycle. Heterotrophic organisms commonly consume other heterotrophs, but, ultimately, ecosystem function depends on autotrophs—photosynthetic or chemosynthetic organisms that can form organic molecules by the reduction of carbon dioxide available in the physical environment (Chapter 6).

In ecology, photosynthetic and chemosynthetic organisms are known as **primary producers**, organisms that take up inorganic carbon, nitrogen, phosphorus, and other compounds from the environment and convert them biochemically into proteins, nucleic acids, lipids, and more (Chapter 25). **Consumers**, heterotrophic organisms of all kinds, depend on primary production, directly consuming primary producers or consuming those that do. Predators hunt and eat prey, herbivores eat plants, and parasites infect their hosts.

The various species present in an ecosystem can therefore be placed in an order that describes how one organism feeds on another, moving carbon through the system. This order is commonly depicted as linear, and it is called a food chain. In nature, however, every species is connected to many others. Many ecologists also stress that most communities contain species that are omnivorous, eating both plants and animals. For these reasons, the movement of carbon through an ecosystem is more realistically thought of as a **food web** (**Fig. 47.14**).

An organism's typical place in a food web is its trophic level (Fig. 47.14). The first trophic level consists of primary producers; consumers occupy the levels above the first trophic level. At the second trophic level are primary consumers; these are the herbivores, organisms that eat plants or other primary producers. Predators, organisms that eat other consumers, are secondary consumers, at the third trophic level, and potentially higher levels as well. Top or apex predators, which have few, if any, natural predators, occupy the highest trophic level. Parasites exploit all levels of the system, obtaining carbon and other nutrients from primary producers and consumers alike. Finally, there are the all-important decomposers and detritivores. Decomposers feed on the dead cells or bodies of other organisms, and detritivores do much the same, consuming partially decomposed plants or animals in soil or sediment, as well as organic compounds in feces. These organisms return carbon dioxide and other inorganic compounds to the environment, completing the cycle of elements through the ecosystem.

**FIG. 47.14  A food web.** A food web diagrams the cycling of carbon and other elements from the environment through a succession of organisms and back to the environment.

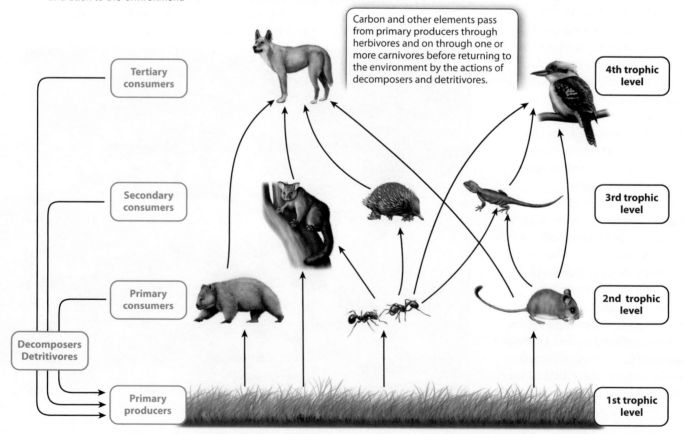

Carbon and other elements pass from primary producers through herbivores and on through one or more carnivores before returning to the environment by the actions of decomposers and detritivores.

Tertiary consumers

Secondary consumers

Primary consumers

Decomposers Detritivores

Primary producers

4th trophic level

3rd trophic level

2nd trophic level

1st trophic level

## Species interactions form trophic pyramids that transfer energy through ecosystems.

The interactions among species transfer energy as well as carbon and other elements through ecosystems. Unlike carbon, energy does not cycle through an ecosystem, and so new energy must continually be harvested from the environment to sustain the community. In essentially all ecosystems where sunlight is available, photosynthesis lies at the heart of ecosystem function. Plants, algae, and photosynthetic bacteria capture energy from the sun and use it to synthesize organic molecules. Where sunlight is absent, especially in the vast depths of ocean, Bacteria and Archaea fuel primary production by chemical reactions.

Some of the energy harvested by primary producers is stored as chemical bonds in organic molecules such as carbohydrates (Chapter 8) and so will be available to organisms at the next trophic level. Some, however, will be dissipated as heat or used to do work. Because building organic molecules is not 100% energy efficient, and because organisms at one trophic level rarely consume all the resources in the level below them, biomass and the energy it represents generally decrease from one trophic level to the next. The result is a **trophic pyramid**.

The broad base of biomass available in primary producers decreases steadily with each ascending step in trophic level (**Fig. 47.15**). In general, about 10% of the energy and biomass available at one trophic level is passed to the next. So 1000 kg of grass biomass will support about 100 kg of lemmings (about 400 animals), and that, in turn, will support 10 kg of small predators like foxes. At the apex of the pyramid, we find a single kilogram of eagle tissue. In this way, primary production exerts a powerful influence on the rest of the community. To get more foxes and eagles, we would need more photosynthesis.

### Light, water, nutrients, and diversity all influence rates of primary production.

If most ecosystems rest on a base of primary production, what controls rates of photosynthesis? Sunlight plays a role. Although all ecosystems, from polar deserts to equatorial forests, receive the same total number of hours of sunlight annually, the intensity of solar radiation declines from the equator to the poles because of Earth's curvature. Seasonality also becomes more pronounced toward the poles, influencing the ability of primary producers to sustain growth throughout the year at higher latitudes.

In addition to sunlight, in most ecosystems water and nutrients govern rates of primary production. In Chapter 29, we saw that plants perform a physiological balancing act, opening their stomata to allow carbon dioxide into leaves but losing water vapor as a result. Because photosynthesis requires both carbon dioxide and water, photosynthetic rate—the amount of $CO_2$ fixed into organic matter in a given time interval—is closely tied to water

**FIG. 47.15 A trophic pyramid.** A trophic pyramid graphically shows the transfer of energy through an ecosystem.

Trophic level

Predator (tertiary consumer): 1 kg of eagle — 4

Predator (secondary consumer): 10 kg of fox — 3

Herbivore (primary consumer): 100 kg of lemmings — 2

Primary producer: 1000 kg of grasses — 1

From the base of plant productivity, about 10% of the biomass of each trophic level is passed on to the next-higher level.

availability. Lack of regular access to water limits the productivity of photosynthetic organisms in deserts and seasonally arid landscapes, explaining why irrigation increases crop yields in places like southern California, northern Africa, and the Middle East.

Even where water is not limiting, rates of photosynthesis are constrained by the availability of nutrients. This is why fertilizer generally increases crop yields (Chapter 29). Plants, algae, and photosynthetic bacteria all need nitrogen, phosphorus, sulfur, iron, and an array of trace elements to grow, and any one of these can limit primary production. In forests, nitrogen is commonly the soil nutrient that is least abundant relative to the needs of plants. In the open ocean, nitrogen can also be limiting, but commonly it is a scarcity of phosphorus that limits photosynthesis. And in some parts of the ocean, nitrogen and phosphorus are both present in abundance, but primary production is still limited—in this case by iron, needed for many enzymes and electron transport molecules. The idea that primary production is limited by the nutrient that is least available relative to its use by primary producers is called **Liebig's Law of the Minimum**, after Justus von Liebig, a nineteenth-century German scientist who first popularized the concept.

Biological diversity in a given region is, in some ways, a result of high productivity. We have seen that the population sizes of predators depend on prey population size, which depends on primary production, and greater primary production also means a greater variety of plants of different sizes and kinds that can support a greater variety of herbivorous animals. While this

FIG. 47.16

# Does species diversity promote primary productivity?

**BACKGROUND** The different plant species found in communities have combinations of leaves, stems, and roots that tap different sources of water and nutrients in the soil and capture light at different levels of sun and shade. Do the differences among plant species result in higher levels of primary production than would be possible with fewer species?

**HYPOTHESIS** Plant diversity promotes primary production within a community.

**EXPERIMENT** Ecologist David Tilman conducted a long-term experiment in Minnesota, seeding each of 11 plots with 1, 2, 4, or 16 grassland species, and controlling each plot's amount of nutrients, water, and carbon dioxide, as well as manipulating the plots' exposure to herbivory and disturbance by fire. The primary production of each plot was measured annually for up to 23 years.

**RESULTS** The graph shows that the addition of nitrogen fertilizer [quantified as kg of N per hectare (ha)] increased primary production significantly, and additional water, higher $CO_2$, herbivory, and disturbance also resulted in at least a modest increase in productivity. The difference in primary production between high-diversity (16 species) and low-diversity (1, 2, or 4 species) plots was at least as great as the largest additions of fertilizer—and much greater than any of the other treatments.

**CONCLUSION** The hypothesis was supported: In the grassland community under study, species diversity promotes primary production.

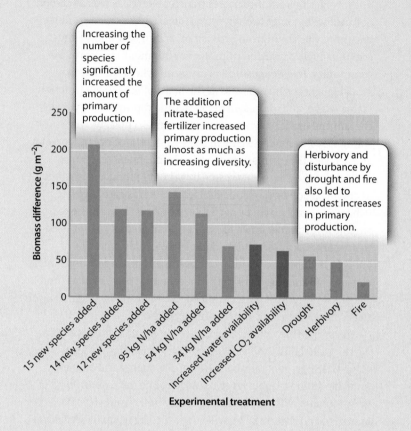

**FOLLOW-UP WORK** Researchers continue to study the ecosystem effects of species diversity, steadily accumulating data that show the importance of conserving the species diversity found in nature.

**SOURCE** Tilman, D., P. B. Reich, and F. Isbell. 2012. "Biodiversity Impacts Ecosystem Productivity as Much as Resources, Disturbance, or Herbivory." *Proceedings of the National Academy of Sciences USA*, 109:10394–10397.

suggests that high productivity can foster high diversity, recent experiments show that the reverse may also be true: Biological diversity may itself play an important role in ecosystem productivity (**Fig. 47.16**).

→ **Quick Check 4** What is the difference between a community and an ecosystem?

## 47.6 BIOMES AND DIVERSITY GRADIENTS

Populations are dynamic, continually increasing and decreasing in response to local physical and biological influences.

Communities and ecosystems are also dynamic, with species coming and going as a result of disturbance and succession. Nevertheless, the assemblages of species found over broad regions of Earth are both distinctive and stable. Such broad, ecologically uniform areas are called **biomes**. On land, different biomes are recognized by their characteristic vegetation, which reflects the evolutionary adaptation of form and physiology to climate. Terrestrial biomes range from the low-growing tundra found at high latitudes to tropical rain forests, and from grass-dominated steppes and deserts to coastal estuaries (**Fig. 47.17**). Biomes occur in the oceans as well as on land, and the largest and least-understood biome of all is the vast expanse of cold dark water in the deep sea.

**FIG. 47.17 Some common land biomes.**

### Tundra

Tundra occurs close to the North Pole, above 65° N. The South Pole is largely surrounded by the ice and seas of Antarctica, and so there is very little area with plants. Tundra is the coldest biome, and precipitation and evaporation are minimal. Even with little precipitation, the lack of evaporation and drainage means the ground is usually waterlogged, and permanent ice occurs below a few centimeters of soil. The plants are mostly mosses, lichens, herbs, and low shrubs. Grasses and sedges occur in drier places, as do other flowering plants. Plant diversity is low, and most plants are small.

### Alpine

The Alpine biome is similar to tundra but lacks permanent ice below the soil, and the temperatures vary more widely. Alpine areas occur throughout the world, often at about 10,000 feet (3000 m) at lower latitudes, but always just below the snow line. Because of their altitude, these are windy, cold places. The thin atmosphere provides only limited protection from UV radiation. Many Alpine plants are therefore low and slow growing.

### Taiga

These cool, moist forests occur from 50° to 65° N. The short summer brings rain, and most of the plants are conifers like spruce, fir, larch, and pine, with an understory dominated by shrubs in the blueberry and rose families. The soils are deep with accumulated organic matter because the low temperatures result in slow decomposition, but they are acidic and poor in nutrients.

### Temperate coniferous forest

Two broad areas of temperate coniferous forests occur below 50° N in North America, northern Japan, and parts of Europe and continental Asia. Along the Pacific coast of the United States, abundant precipitation permits growth of enormous conifers such as Douglas-fir, Redcedar, Sitka Spruce, and redwoods. Much of the undergrowth is ferns and members of the blueberry family. In the interior of North America, much less precipitation and colder winter temperatures support drought-resistant conifers such as Ponderosa and Lodgepole Pines and Englemann Spruce.

*continues on page 47-18*

**FIG. 47.17  Some common land biomes.** *(continued from page 47-17)*

### Deciduous forest

A moderate climate and dominance of hardwood deciduous trees occurs across much of North America, Europe, and Asia. Much of this biome has been subjected to human disturbance for agriculture and urban development. There are usually 15 to 25 species of trees, including maples, oaks, poplars, and birches. Springtime sun passes through seasonally leafless trees to reach a diverse understory flora. Soils are rich in nutrients from annual leaf fall, and the moderate temperatures and precipitation promote decomposition, while the cool winters promote accumulation of organic materials.

### Temperate grassland

Before settlement, this biome occupied most of the midwestern United States, where it is dominated by blue-stem and buffalo grasses. Fire helps to maintain grass populations in this biome. Lack of precipitation also prevents many species of tree from growing, and those trees that grow are usually in low areas with some moisture. Where there is enough moisture to support decomposition, the soils accumulate nutrients, providing some of the most productive agricultural lands in the world.

### Desert

Desert occurs in continental interiors around the world north and south of the equator from 25° to 35°. Wind patterns prevent this biome from receiving more than a few centimeters of precipitation annually. The deep-rooted plants are adapted to store water, like cactus and euphorbia. Primary production is low, and soils are poor in nutrients but may have high surface salt due to evaporation.

### Chaparral

Like deserts, the distribution of chaparral reflects a narrow range of climate conditions and occurs on the western edge of continents from 32° to 40° north and south of the equator. Precipitation ranges from 30 to 75 cm per year, usually falling in 2 to 4 months. Typical plants are annual herbs, evergreen shrubs, and small trees. Olives, eucalyptus, acacia, and oaks are typical woody species, always drought resistant and often adapted to withstand fire. Limited precipitation means soils are not rich in organic materials.

### Savanna

Tall, perennial grasses dominate this biome, which occurs in eastern Africa, southern South America, and Australia. Rain is seasonal and ranges from 75 to 150 cm per year. Scattered trees and shrubs usually drop their leaves in the dry season to conserve moisture. Animal diversity can be high and includes the large mammals well known in Africa.

### Rain forest

This moist, highly diverse forest extends north and south of the equator from 10° N to 10° S. Annual precipitation is commonly more than 250 cm, and tree diversity alone often exceeds 300 species per hectare. Trees grow tall, and many have buttressed roots for support. Lianas and other epiphytic plants are common. Most leaves are evergreen and leathery and many have long pointed tips that facilitate drainage of excess moisture. Because of the high temperatures and heavy rains, decomposition is very rapid, preventing the accumulation of organic materials in clay-rich or sandy soils.

## Biomes reflect the interaction of Earth and life.

Globally, the distribution of biomes corresponds closely to climate (**Fig. 47.18**). As noted earlier, this correspondence reflects the need for plants to balance $CO_2$ intake against water loss, a balance commonly struck by means of adaptations in the shape and physiology of leaves, stems, and roots (Chapter 29). To explore this further, we need to discuss a new concept: **evapotranspiration ratio**.

The intensity of solar radiation varies with latitude and altitude, as does the amount of annual precipitation and wind. These factors determine the amount of evaporation from Earth's surface to the atmosphere, including the amount transpired by plants. The

**FIG. 47.18 Climate regions.** Distributions of biomes around the world reflect regional climates.

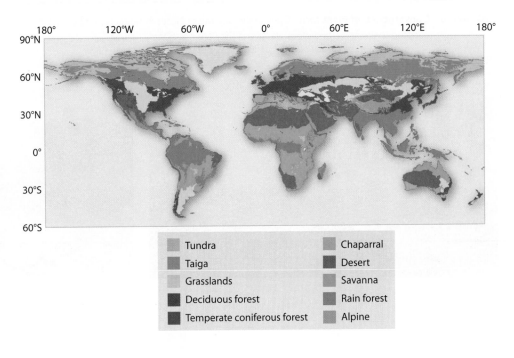

Tundra

Taiga

Grasslands

Deciduous forest

Temperate coniferous forest

Chaparral

Desert

Savanna

Rain forest

Alpine

**FIG. 47.19    Evaporation and transpiration.** The amount of evaporation and transpiration changes with latitude and altitude.

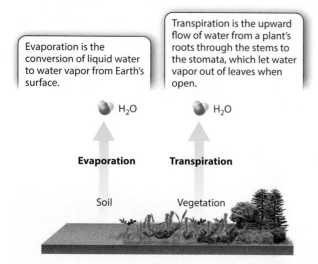

Evaporation is the conversion of liquid water to water vapor from Earth's surface.

Transpiration is the upward flow of water from a plant's roots through the stems to the stomata, which let water vapor out of leaves when open.

evapotranspiration ratio is the ratio of the amount of water evaporated from the Earth's surface directly from ponds, rivers, and soils, to the amount of water vapor transpired by plants (**Fig. 47.19**).

The ratio of evaporation to transpiration effectively tells us what type of vegetation can be supported in a given region. Deserts, with low annual precipitation, have a high evapotranspiration ratio: Soil moisture evaporates rapidly, and plants must keep transpiration (and therefore photosynthetic rate) to a minimum. At the other extreme, rain

forests have low evapotranspiration ratios: High precipitation permits nearly unlimited transpiration (and therefore high photosynthetic rates, leading to luxuriant growth), enabling the development of wet forests with high biomass. Intermediate evapotranspiration ratios occur in the temperate zones that encompass much of the United States, Europe, and parts of eastern Asia, as well as at mid-altitudes, below the highest mountain peaks but well above the coastline. Both latitude and altitude influence the evapotranspiration ratio through their effects on temperature and precipitation, and the combined effects of these factors are reflected in the distribution of biomes around the world.

In similar biomes on different continents, dominant plants generally look similar but are not closely related. That is, plants have independently evolved structures by convergent evolution as adaptations to particular climatic regimes. For example, the cacti that grow in deserts in western North America look very similar to plants in drylands of Africa, but in Africa, these plants belong to a different family entirely (**Fig. 47.20**). Convergent evolution is also seen in rain forest trees that commonly have leaves with long tips that allow excess water to drip away, a reddish color in new leaves that protects them from intense sunlight, and roots with high buttresses that support tall trees atop thin soil.

### Tropical biomes usually have more species than temperate biomes.

On a half-hour walk through a temperate forest—for example, in the United States or southern Canada—we can expect to see maybe a dozen different kinds of tree: two or three

**FIG. 47.20    Convergence of plant form.** Cacti in the deserts of North America (a) and euphorb plants in Africa (b) look similar because they are both adapted to conserve and store water, but they are not closely related.

**FIG. 47.21 Latitudinal diversity gradient in mammals.** The number of mammal species decreases from the equator to the poles.

There are few mammal species near the poles, mostly rodents and their carnivore predators, plus seals and whales in the sea.

Tropical regions have a great diversity of different kinds of mammals, representing all of the major groups from anteaters to apes.

0  22  44  66  88  110  132  154  172  198  220  242  267  No data

**Number of mammal species**

kinds of oak perhaps, a couple of maple species, birch, beech, hornbeam, ironwood, one or two pine species, and possibly a sycamore, poplar, or spruce, depending on just where we are. Standing in front of a tree and turning around, we can almost always see several more trees of the same kind. In contrast, on a similar walk through a forest in the Amazon, or in Borneo, we would pass 300 or 400 different tree species, with almost no near neighbors of the same kind. The great Victorian naturalist Alfred Russel Wallace took note of the low population density of trees in tropical forests and observed, "If the traveller notices a particular species and wishes to find more like it, he may often turn his eyes in vain in every direction."

It isn't just trees. Mammals, birds, insects, and many other kinds of animals show similar patterns of species diversity (**Fig. 47.21**). One survey found more than 700 species of leaf beetles in one small set of trees in the Amazon, a number equal to the total diversity of leaf beetles in all of eastern North America. A single tree in this same forest harbored 43 species of ants, exceeding the total for all of Great Britain.

When we compare biomes across the continents, the pattern just described stands out clearly: Species diversity generally peaks near the equator and declines toward the poles. What are causes of this **latitudinal diversity gradient,** and what are its consequences? Does this global pattern reflect the ecological processes that characterize populations?

**? CASE 8 Biodiversity Hotspots: Rain Forests and Coral Reefs Why are tropical species so diverse?**

Theories abound on why the tropics contain so many species. One hypothesis, for which there is broad support, looks to the relative ages of tropical and higher-latitude biomes. Tropical biomes tend to be older, having evolved over tens of millions of years, whereas biomes at higher latitudes have been shaped by climatic change, and ultimately an ice age, within the past few million years. Because tropical habitats have existed much longer than temperate ones, biologists hypothesize that there has been more time for species to evolve and diversify at low latitudes. Several studies of diversification rate, in frogs for example, show comparable rates of diversification in temperate and tropical biomes, supporting the hypothesis that the higher tropical diversity reflects the longer interval through which diversification has taken place.

Time, however, is just one of several factors that influence diversity. Another hypothesis is that temperate communities have fewer species because it is relatively difficult to adapt successfully to cold, dry winters and the range of weather typically experienced at higher latitudes. A tree growing in a temperate area like New England experiences a wide range of temperatures over a year (and sometimes even over a single day) as winds blow and snow falls, or heat waves move through the area. Because species living at high latitudes are adapted to a greater range of conditions on average than those nearer the equator, they tend to have larger geographic ranges than tropical species, which are adapted to the narrow range of environmental variation experienced in any particular place.

It has also been suggested that the many insects and fungi specialized to attack particular kinds of trees are generally more abundant in wet tropical forests, and so trees of a given species tend to be as far apart as they can be, thus reducing the chance of being found by their enemies. This means that there would be more room for different species in any single hectare of land. It has also been noted that there is more land surface area at low latitudes, so the species-area curve (Chapter 46) plays a role in the latitudinal diversity gradient.

Very likely, the latitudinal diversity gradient reflects all these influences and more not yet appreciated. What is clear, however, is that where there are more species of trees, there are also more species of animals and other consumers. Tropical temperatures and high humidity have permitted a wide range of plant species to evolve over the past 100 million years, in turn creating many niches for the animals that depend on them.

The high diversity—and therefore the low population density—of many tropical forests prompts another question. What determines how far apart individual trees can be and still maintain a viable population? One important consideration goes back to mutualisms: Animal pollination ensures that pollen from one tree can reach another individual of the same species if it is within the foraging range of a pollinator carrying its pollen. Of course, if trees are farther apart than any single pollinator can visit within a foraging excursion, they will not be pollinated. Animal pollination, therefore, contributes to the diversity differences between tropical trees and higher-latitude species, which are more likely to be pollinated by wind.

→   **Quick Check 5** From what you know about how species diversity changes with latitude, predict how species diversity changes with altitude.

### Evolutionary and ecological history underpins diversity.

Earth's biomes are the historical outcome of natural selection and environmental change through time. For this reason, the fossil record can shed light on how current patterns of diversity came to be. In general, fossils only tell us a little about the origins of tropical forests, but on the island of Hispaniola, there is a

**FIG. 47.22  Fossils in 30–23-million-year-old Dominican amber.** These document the evolutionary history of (a) orchid-pollinating stingless bees, now extinct on Hispaniola, and (b) *Anolis* lizards that, like their modern descendants, hunted insects in trees.

remarkable exception: nuggets of amber that provide a time capsule of ancient rain forest life. Dominican amber provides an extraordinary window on the diversity of life in the rain forest of Hispaniola 30 to 23 million years ago.

Amber is fossilized tree resin, and in Hispaniola, it comes from the tree *Hymenea*, a member of the pea family. *Hymenea* trees grew in the Hispaniola rain forest 30 to 23 million years ago, much as they do today. Lizards, tree frogs, insects, plants, and fungi were all captured in their sticky sap, which hardened into amber, preserving a record of forest diversity and ecology (**Fig. 47.22**). Studies of these fossils reveal ecological relationships similar to some observed on Hispaniola today. Not surprisingly, some of the most abundant animals in the amber are insects whose modern counterparts have ecological associations with *Hymenea*. Today, ants commonly patrol the bark of *Hymenea* and other tropical trees in search of food, and beetles often feed on trees that secrete resin as a defense mechanism, while stingless bees gather resins for their nests. These insects are preserved intact in the amber as if they were just caught. Indeed, many of the insects seen in this amber, perhaps even the majority, are practically indistinguishable from species that continue to exist today on Hispaniola.

Dominican amber also reveals species that are no longer found on this island. For example, many of the more specialized kinds of ants found in the amber no longer occur on Hispaniola. A kind of beetle that feeds on palm trees in South America is found in amber but not on the island today, even though there are plenty of palms. There are no longer stingless bees on Hispaniola, although the flowers they visit persist, and the bees occur on nearby Jamaica.

Vertebrate fossils are rare in the Dominican amber, but they include *Anolis* lizards closely related to anoles specialized for life in the canopies and high on the trunks of forest trees—resource partitioning began long ago in the Hispaniola rain forest. More vertebrate fossils occur in other settings, and these show that most of the birds and mammals found 25 million years ago are quite different from those in modern forests.

Such fossils allow us to add a deeper dimension of time to our understanding of how ecosystems have come to be. Whether forest, reef, or grassland, every ecosystem contains the present-day survivors of groups representing a broad range of origins in evolutionary time and geographic space. The amber fossils remind us that what G. Evelyn Hutchinson called "the ecological theater and the evolutionary play" have been intertwined throughout our planet's history, producing the diversity of species we see today.

## Core Concepts Summary

### 47.1 THE NICHE IS THE ECOLOGICAL ROLE PLAYED BY A SPECIES.

The niche reflects the biological and nonbiological factors that determine where an organism lives and how it functions. page 47-1

The fundamental niche is the full range of physical conditions and food resources that permit an organism to live. page 47-3

The realized niche is the actual range of habitats where an organism lives, and it is often smaller than the fundamental niche because of competition, predation, and other interactions between species. page 47-3

### 47.2 COMPETITION, PREDATION, AND PARASITISM ARE ANTAGONISTIC INTERACTIONS IN WHICH ONE SPECIES BENEFITS AT THE EXPENSE OF ANOTHER.

Competition is a form of antagonism that occurs when two individuals of the same or different species use the same limited resource. page 47-4

Species can compete for food, space, mates, or other resources. page 47-4

Competition promotes resource partitioning, which occurs when two co-occurring species overlap in resource use and as a result one changes its resource use. page 47-4

Predators and parasites can reduce competition by limiting the population sizes of successful competitors. page 47-5

### 47.3 IN MUTUALISMS, THE BENEFITS TO BOTH SPECIES OUTWEIGH THE COSTS OF PARTICIPATION.

Mutualisms can be obligate, meaning that they are required for survival or reproduction, or facultative, meaning that they are not required. page 47-6

Interactions in which one species benefits and the other is unaffected are called commensalisms. page 47-8

Species interactions can change over time. For example, partners can become parasites. page 47-8

### 47.4 COMMUNITIES ARE SHAPED BY THE EVOLVED FEATURES OF ORGANISMS, WHICH SHAPE THE INTERACTIONS WITH OTHER ORGANISMS AND WITH THE PHYSICAL ENVIRONMENT.

An ecological community consists of all the organisms living in one place. page 47-9

Species that have a large influence on the composition of a community disproportionate to their numbers are known as keystone species. page 47-10

Physical disturbances, such as a storm or drought, can reshape communities. page 47-11

Succession is the predictable order of species colonization and replacement in a new or newly disturbed patch of habitat. page 47-12

### 47.5 THE CYCLING OF MATERIALS AND THE FLOW OF ENERGY THROUGH ECOSYSTEMS REFLECT WAYS IN WHICH ORGANISMS INTERACT IN COMMUNITIES.

An ecosystem is the community and physical habitat in which the organisms live. page 47-13

Carbon cycles through ecosystems, but energy is used and dissipated. page 47-13

A food web describes how carbon cycles through ecosystems, and a trophic pyramid describes how energy flows through ecosystems. page 47-15

Because of energy lost to heat, work, and the inefficiencies of consumers, each trophic level in an ecosystem generally has about 10% of the biomass found at the next-lower level. page 47-15

In most terrestrial ecosystems, water and nutrients limit rates of primary production. In the sea, primary production is commonly limited by nutrients. page 47-15

### 47.6 BIOMES ARE BROAD, ECOLOGICALLY UNIFORM AREAS CHARACTERIZED BY THEIR CLIMATE, SOIL, AND PLANT SPECIES.

The distribution of plants reflects in large part abiotic factors such as climate and soil. page 47-19

Annual precipitation and sun intensity affect the amount of evaporation and plant transpiration. page 47-19

Earth's biomes are the historical outcome of environmental change through time and natural selection. page 47-20

Species diversity generally declines from the equator toward the poles, a pattern known as a latitudinal diversity gradient. page 47-20

## Self-Assessment

1. Choose an organism, such as an oak tree, and define its niche.

2. Give an example of an antagonism and a mutualism, and in each case, describe the benefits and costs to the participants.

3. Name three factors that help determine the species composition of a community.

4. Describe how a physical disturbance, such as a drought, can affect community composition.

5. Explain what is meant by "ecological succession" and give an example.

6. Describe how herbivores can affect the abundances of organisms at higher and lower trophic levels.

7. Choose five biomes, describe their climate and vegetation, and explain why they differ from one another.

8. Describe the general pattern of diversity from the equator to the poles. Provide two hypotheses to explain this pattern of diversity.

Do you understand the chapter's Core Concepts? Log in to check your answers to the Self-Assessment questions, then practice what you've learned and reinforce this chapter's concepts by working through the problems and multimedia tutorials provided there.

⌨ www.biologyhowlifeworks.com

## CHAPTER 48

# THE ANTHROPOCENE

## Humans as a Planetary Force

## Core Concepts

**48.1** Some scientists call the time in which we are living the Anthropocene Period to reflect our significant impact on the planet.

**48.2** Humans have a major impact on the carbon cycle, primarily through the burning of fossil fuels, which returns carbon dioxide to the atmosphere.

**48.3** Humans have an important impact on the nitrogen and phosphorus cycles, primarily through the use of fertilizer in agriculture.

**48.4** The impact of humans on the environment is changing the stage on which evolution acts.

**48.5** In the 21st century, biologists, doctors, engineers, teachers, and informed citizens have vital roles to play in understanding our changing planet and making wise choices for our future.

By almost any measure, you were born into humanity's golden age. Beginning with the invention of the steam engine, humans have learned to harness Earth's store of chemical energy, dramatically extending the power of muscles. You probably don't find it remarkable to drive across a continent in less than a week or fly across an ocean overnight, but your great-great-grandparents would have been astonished. At the same time, advances in antibiotics and public health have drastically reduced infectious disease in many parts of the world. Smallpox, long one of life's deadliest realities, has been eliminated, and polio may eventually follow suit. And the Green Revolution of the last 50 years provides unprecedented amounts of food for a hungry planet (Case 6: Agriculture).

One consequence of this success is an increasing human population: We recently crossed the 7-billion mark. Another is that our actions affect the world around us, directly or indirectly. In this chapter, we look at some of these impacts, and what we can do about them. Can we sustain or control current patterns of population growth and resource use, and if not, will we approach the end of humanity's golden age? As discussed in the following sections, there is reason both for concern and for optimism. Together, scientific insight, technical ingenuity, and a wise citizenship provide our best hope of sustaining hard-won gains in food, energy, and health through the 21st century and beyond.

---

## 48.1 THE ANTHROPOCENE PERIOD

Increasingly, scientists refer to the modern era as the **Anthropocene Period**, emphasizing the dominant (and, in part, conscious) impact of humans on the Earth. Our impact is partly a consequence of our sheer numbers, which in turn drive our energy and land use, but it also reflects the individual and societal choices we make.

### Humans are a major force on the planet.

There are more humans than ever before, and our numbers are climbing rapidly. A world that had about 500 million inhabitants when Christopher Columbus set sail in 1492 now supports more than 7 billion people (**Fig. 48.1**). By the time you retire, our numbers will probably exceed 9 billion.

→ **Quick Check 1** Take a look at Fig. 48.1. How long did it take our numbers to double, from 1 billion to 2 billion? How about from 2 billion to 4 billion? What does this say about the rate of population growth?

Of course, population size doesn't tell the whole story. Whether driving a car, living in a heated apartment, plugging in a computer or television, or eating fruit

FIG. 48.1    **Growth through time of the human population.**

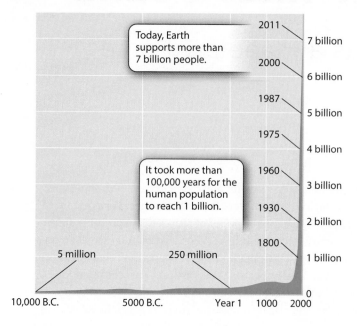

shipped from another hemisphere, you use a great deal more energy than your ancestors did. To give just one example, it has been suggested that the electrical outlets in our homes let us accomplish about the same amount of work performed by 30 servants in a Roman villa. **Fig. 48.2** shows annual per capita energy use in selected countries. Energy use by individuals varies greatly from one part of the world to another. Americans, Canadians, Europeans, and Australians live energy-intensive lives, whereas populations in rural Africa and parts of Asia consume relatively little energy per person.

Our direct use of energy, however, reflects only part of our impact on the planet. Nearly everything we use requires both land and energy. All of us consume food grown in fields, fertilized and harvested through energy, and shipped to a local store. We live in houses built of wood, brick, and stone, and use clothing, furniture, and appliances fashioned from plant materials grown in fields and forests or from minerals extracted from the Earth by mining. It takes land to grow plants, and energy to sow seeds, make and distribute fertilizer, harvest useful materials, and manufacture, ship, and clean the finished products.

The concept of the **ecological footprint** is an attempt to quantify our individual claims on global resources by adding up all the energy, food, materials, and services we use and estimating how much land is required to provide those resources. **Fig. 48.3** plots the average ecological footprint per person against a ranking of the Human Development Index, a measure of standard of living maintained by the United Nations, for many countries. Not surprisingly, developed countries tend to have large ecological footprints—it takes about 7 hectares of land to support an average American, more than 6 to support an average Australian. In contrast, in many parts of Asia and Africa, only a single hectare supports the average citizen.

In many countries, living standards have risen over the past 50 years, nearly always increasing the mean ecological footprint of their citizens. By some estimates, the total ecological footprint of humanity now exceeds the actual surface area of the Earth. Taken together, then, trends in population size and the ecological footprint highlight both the remarkable technological successes of the past century and the challenges we face as citizens, policy makers, scientists, and engineers in the decades to come.

FIG. 48.2    **Energy use by individuals among countries.** Energy use is distributed unevenly among countries.

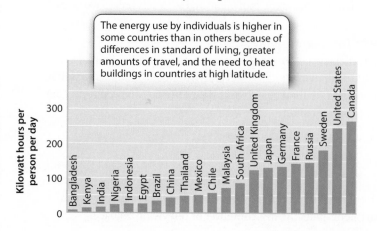

FIG. 48.3    **Our ecological footprint.** The graph shows the number of hectares needed to support an average person compared to a ranking of the Human Development Index, a measure of standard of living.

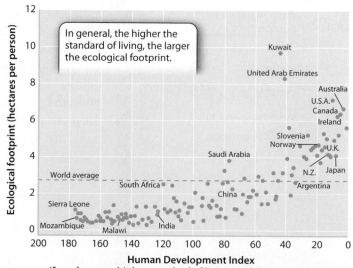

## 48.2 HUMAN INFLUENCE ON THE CARBON CYCLE

In Chapter 25, we discussed the carbon cycle, explaining how photosynthesis removes carbon dioxide from the atmosphere and respiration replaces it. The remarkable data plotted in the Keeling curve (see Fig. 25.1) show that atmospheric $CO_2$ levels have increased annually over the past 60 years, and analyses of air bubbles trapped in glacial ice show that this pattern of persistent increase began earlier, with the onset of the Industrial Revolution in the nineteenth century (see Fig. 25.3).

There is no longer any doubt that humans have become major contributors to the carbon cycle, particularly the part of the cycle in which ancient organic matter, or fossil fuel, is oxidized to $CO_2$, returning $CO_2$ to the atmosphere (**Fig. 48.4**). As discussed

**FIG. 48.5** Global warming.

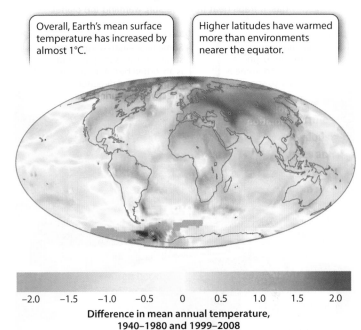

Overall, Earth's mean surface temperature has increased by almost 1°C.

Higher latitudes have warmed more than environments nearer the equator.

-2.0   -1.5   -1.0   -0.5   0   0.5   1.0   1.5   2.0

**Difference in mean annual temperature, 1940–1980 and 1999–2008**

**FIG. 48.4** **Human-generated sources of carbon dioxide.** Every year, humans convert organic matter to carbon dioxide at rates far higher than those found in nature, predominantly by (a) burning fossil fuels and also through (b) biomass burning, as seen in this Amazon forest.

a

b

in Chapter 25, burning fossil fuel adds $CO_2$ to the atmosphere at rates about 100 times greater than all Earth's volcanoes together. As we saw in Chapter 25, measurements over the past several decades of the relative abundances of $^{12}C$, $^{13}C$, and $^{14}C$ in atmospheric $CO_2$ make it clear that the $CO_2$ now being added to our air comes primarily from the combustion of fossil fuels. No other source of $CO_2$ has the right isotopic composition to explain the measurements (see Fig. 25.4). Clearing forests for agriculture generates still more carbon dioxide (Fig. 48.4). When land is cleared for agriculture, the existing vegetation is generally burned off, converting much of the organic carbon in biomass and soil organic matter to $CO_2$. No counteracting process removes $CO_2$ at comparable rates.

**As atmospheric carbon dioxide levels have increased, so has mean surface temperature.**

For the past century, scientists, sailors, and interested citizens have monitored temperature at weather stations around the world. More recently, satellites have enabled us to measure temperature in places as remote as the high Arctic and the middle of the ocean. The results are clear: In most parts of the world, mean annual temperature during the decade 1999–2008 was warmer than 1940–1980 averages (**Fig. 48.5**). In some places, the temperature change has been slight, but in others, especially at high latitudes, the increase has been as much as 2°C.

We can measure $CO_2$ levels in the atmosphere, and they are increasing. We can measure global temperature, and it is increasing. Is increasing $CO_2$ responsible for observed

**FIG. 48.14** Alternative energy sources, including (a) solar and (b) wind energy.

## What can be done?

From the preceding discussion, it is clear that rising $CO_2$ has both direct effects on the physiology of land and sea organisms and indirect effects related to climate change. In order to ensure our environmental future, scientists must understand the climate system and biological responses to global change much better than they do now. Because the principal source of rising $CO_2$ is the burning of fossil fuels, the solutions we develop as citizens, however, will clearly focus on energy—how we obtain it and how we use it. As the global population continues to increase and people aspire to higher standards of living, human impact on the carbon cycle will continue to grow, at least if we continue business as usual. Business as usual, however, may not be a viable option for the long term because population growth may simply outstrip resource availability, especially for petroleum. Ultimately, our success in both slowing rates of global warming and sustaining energy supplies will depend on our ability to use energy more efficiently and generate new forms of power, most likely from the sun, wind, and nuclear sources.

Half a century ago, geologist M. King Hubbard predicted from patterns of oil use and the rate at which new oil fields are discovered that petroleum production in the United States would peak in the 1970s and decline thereafter. Global production, he suggested, would peak later, in the early decades of the 21st century. King's predictions proved accurate for the United States: Domestic petroleum production peaked in 1972, although petroleum use continues to climb. In addition, global petroleum production also seems on course to fulfill Hubbard's predictions, although the timescale remains a contentious question. Coal,

natural gas, and unconventional reserves such as the oil sands found widely in western Canada can quench the world's thirst for fossil fuels for another century or more, and biofuels— fuels generated from plant biomass—may also help conserve petroleum, but the energy produced will be expensive and global change will continue.

Like fossil fuels, alternative energy sources—wind, the sun, tidal energy, and nuclear power (**Fig. 48.14**)—all have important pros and cons, not the least of which, in the case of nuclear power, is long-term storage of radioactive waste. On the other hand, all these sources have the potential to benefit humanity at least three ways: by reducing the amount of $CO_2$ emitted per kilowatt of energy generated, by conserving finite natural resources, and by helping to contain the cost of energy production. The development of more fuel-efficient cars, airplanes, and buildings will also help us to maintain energy-intensive lifestyles while decreasing their environmental consequences. As individuals, we can choose energy-efficient transportation and appliances and develop habits that decrease energy waste. When one person turns off the light in an empty room, the energy saved is tiny, but when millions of people do it, the benefits add up.

Wise choices by citizens and consumers can slow the growth of atmospheric $CO_2$, but the $CO_2$ already there isn't going to go away quickly. As noted earlier, we have dramatically increased the rate at which $CO_2$ is added to the atmosphere, but we have not changed the rate at which it is removed. The geologic processes that control atmospheric $CO_2$ work on longer timescales. Therefore, even if we could cut $CO_2$ emissions by 90% tomorrow, $CO_2$ would not return to nineteenth-century levels for hundreds or even thousands of years.

Another possibility is to actively remove $CO_2$ from the air. Reforestation of previously cleared landscapes removes carbon from the atmosphere since plants build biomass from $CO_2$ during photosynthesis (Chapter 8). It is also possible to capture $CO_2$ as it rises upward through smokestacks, although it is expensive to do so currently. We do not yet know how to scrub industrial exhaust of $CO_2$, but if $CO_2$ could be removed from the air in this way at reasonable costs, it could apply a significant brake to global warming.

Clearly, forestry and 21st-century technology have potentially important roles to fill in undoing the consequences of the past hundred years. For the most part, however, gaining control over our environmental future will entail wise personal and governmental choices about how we obtain and use energy.

## 48.3 HUMAN INFLUENCE ON THE NITROGEN AND PHOSPHORUS CYCLES

Human activities influence the environment not because we operate outside the carbon cycle, but because we have become an important part of it. Our interest in improved crop production also results in unprecedented human impact on the nitrogen and phosphorus cycles. Once again, environmental changes tied to human nitrogen and phosphorus use are altering the ecological landscape for species in lakes and the sea.

### Nitrogen fertilizer transported to lakes and the sea causes eutrophication.

As discussed in Chapter 26, all organisms require nitrogen to synthesize proteins, nucleic acids, and other molecules. Although nitrogen is plentiful in the atmosphere as $N_2$ gas, this form is not biologically available to animals and plants. As we saw in Chapter 26, bacteria and archaeons are able to convert $N_2$ to biologically available forms of nitrogen, called fixed nitrogen, such as ammonia ($NH_3$) and nitrate ($NO_3^-$). Plants then obtain nitrogen from ammonia and nitrate in the soil, and animals get the nitrogen needed for growth from the food they eat. Crops, harvested year after year in the same fields, deplete available nitrogen, and so high and sustained yields require that farmers add biologically available nitrogen to the soil as fertilizer.

Through the use of fertilizer and other activities, humans add about 150 million tons of fixed nitrogen to the biosphere each year, an amount comparable to that contributed by microbial nitrogen fixation and lightning combined. The majority of this fixed nitrogen (125 million tons per year) is created industrially by reacting nitrogen gas and hydrogen in the presence of a catalyst to produce ammonia. Most commercially produced ammonia is further processed to nitrate that will be spread across fields as fertilizer to increase crop production. Success in feeding the world's people owes much to fertilization with nitrates, but in fact, only about 10% of the nitrogen added to croplands ends up in food. Much of the nitrate fertilizer leaves fields as surface runoff and travels by river to lakes or the sea. Denitrifying bacteria also use some of the added nitrate for respiration, returning $N_2$ to the atmosphere—and also generating $N_2O$, yet another greenhouse gas.

In lakes and oceans, the fertilizing effects of agricultural runoff are finally realized, but in unintended ways. The added nutrients lead to a great increase in the populations of algae and cyanobacteria in a process called **eutrophication.** These algal and cyanobacterial masses eventually sink to the bottom, where heterotrophic organisms, mostly bacteria, feed on them, fueling high rates of aerobic respiration. The bacteria's demand for oxygen can completely deplete $O_2$ in bottom waters, with potentially catastrophic consequences for animal life on the sea or lake floor.

In the Gulf of Mexico, nutrients leached from croplands and transported down the Mississippi River result in an ominously named dead zone of oxygen-depleted bottom water. This dead zone expands seasonally from the mouth of the river west toward Texas, covering an area about the size of New Jersey (**Fig. 48.15**). Within the dead zone, fish and seafloor invertebrates die in masses. Several hundred dead zones have been identified in coastal oceans around the world, and most are expanding year by year. Those in the Gulf of Mexico and the Baltic Sea are among the largest.

Nitrogen applied as fertilizer but spread by runoff and streamflow can impact the full range of ecological interactions discussed in chapters 46 and 47 (**Fig. 48.16**). We benefit tremendously from increased crop production but pay a price in the eutrophication of lakes and oceans. We pay another price as well—as much as 1% to 2% of all fossil fuel consumption goes to make nitrogen fertilizer.

**FIG. 48.15** A dead zone, visible in this NASA satellite image as billows of green.

# Succession: Ecology in Microcosm
### Integrating concepts from Chapters 46 through 48

**Human impact**
Agricultural runoff can greatly speed up pond succession by adding nutrients.

**Population growth**
When organisms first colonize a pond, their populations increase exponentially until competition for food or space slows further growth. Zooplankton and algae increase exponentially, as do the aquatic insect larvae and tadpoles that feed on them.

*Time*

Runoff

Excess deposition of N and P

Zooplankton (*Daphnia*)

Water beetles

Fly larvae

Tiny insects

Natural deposition of C, N, and P

Algae

Bacteria

Colonizing plants, animals, and bacteria arrive, carried by the wind or on the legs of birds. Some follow cues of light reflecting from water surfaces as well as breezes carrying humid air.

At first, populations grow exponentially, unchecked by competition, predation, or dwindling resources.

Carrying capacity of environment (*K*)

Population size (*N*)

Time

Algae, bacteria, and zooplankton

Aquatic insects and other arthropods
Amphibians and fish
Aquatic plants

Grasses and sedges
Herbaceous plants

## Species interactions

With increasing diversity comes an increase in the numbers of connections between species as predators, prey, or symbionts. Interactions help to control population sizes.

## Succession

As forest communities mature and stabilize, shade-tolerant species may only be disrupted by fire or other disturbances. This can set communities back to earlier successional stages. As soils deepen and store the seeds of sun-loving species, they establish themselves in any openings that are created.

**Time**

## Human impact

Logging, starting fires (or suppressing natural fires), and introducing invasive species like Purple Loosestrife can alter the composition of a forest or field community.

**Increased accumulation of dead organic material**

A succession of plant communities and associated animals increases the biomass dependent on the water source, but eventually sediment runoff from surrounding land fills in the pond, and a terrestrial ecosystem becomes established.

The dominant plants change the physical habitat by shading and spurring organic decay, excluding some animals and plants while permitting others to establish.

Terrestrial animals

Shrubs and trees

## Phosphate fertilizer is also used in agriculture, but has finite sources.

The other major nutrient added to fields as fertilizer is phosphate $(PO_4{}^3)$, and here the problem is a different one. Like nitrate, much of the phosphate added to fields as fertilizer leaves as runoff, and phosphate runoff can be an important contributor to eutrophication. The Everglades of south Florida provide a good example. This ecosystem is built of plants that have adapted to low phosphate levels. However, phosphate levels in parts of the Everglades have increased up to a hundredfold. Chemical tracers show that these increased phosphate levels stem from phosphate fertilizer used on agricultural lands to the north. As a consequence, phosphate-loving cattails and other introduced plants are expanding at the expense of native plants, decreasing the diversity of this remarkable ecosystem.

There's no question that overuse of phosphates has detrimental effects on certain natural ecosystems, and yet we need to be concerned that we will eventually run out of phosphates to use as fertilizers. Unlike for nitrate, there is no industrial process to produce phosphate. Thus, the only source of phosphate fertilizer is mining. Only a few places in the world have sedimentary rock with phosphates in high enough concentration to be worth mining, and these stores are being depleted.

Globally, phosphate production peaked about 1990, raising a serious question of how the world's need for phosphate fertilizer will be satisfied a century from now. In many ways, the problem of phosphate availability parallels that of petroleum production. As abundant sources become depleted, we will have to mine poorer and poorer resources, raising production costs and ultimately calling into question our capacity to sustain agricultural production.

Human encroachment on the nitrogen cycle and long-term phosphate availability, then, converge on a single question: Can we manage future agricultural production in ways that feed the growing human population in a sustainable fashion?

## What can be done?

The Green Revolution in agriculture began in the 1940s with the introduction of high-yield, disease-resistant varieties of wheat to Mexico (Case 6: Agriculture). In subsequent decades, new strains of corn, wheat, and rice, widespread application of industrial fertilizer, and the spread of farm machinery increased global crop production to levels that could scarcely have been imagined at the beginning of the twentieth century. As population growth has accelerated, however, the need to grow ever more food remains strong. And other factors, notably rising meat consumption in many countries and the increasing use of corn to generate biofuel, now place still greater demands on global agriculture. It is

estimated that over the next decade alone global corn and wheat production will have to increase by 15% to meet demand.

How can we meet the needs of a growing population? Conventionally, agronomists see two options: We can devote more land to crops, or we can increase the yield of fields already in place. **Fig. 48.17** shows that globally both the amount of land devoted to growing corn and corn yield have increased steadily over the past 50 years. It is possible to increase yield further, as the yield per hectare in the United States, Canada, China, and western Europe far exceeds that of most other countries. Increasing yields, however, will require increasing the use of fertilizer and the use of fossil fuels to power farm machinery, with all the attendant problems of climate change and eutrophication noted earlier. Increasing crop area also comes at a cost, as forests and natural grasslands go under the plow, decreasing the biological storage of carbon and threatening biological diversity, discussed more fully below.

For more than a century, biologists have developed improved varieties of crop plants by breeding, the highly effective practice that Darwin called "variation under domestication" and used to frame his arguments about natural selection. Today, we have another tool available: genetic engineering. Biologists can now introduce genes for desirable traits into crop plants, improving their yield, disease resistance, or nutritional value (Chapter 32). The potential benefits of genetically modified crops are obvious, but some biologists see risks as well, many of which are detailed in Chapter 12.

Conventional breeding and genetic modification may improve crop yields, but in a world of changing climate and evolving pests, these gains may have something of a "Red Queen" flavor to them (Chapter 42). In Lewis Carroll's *Through the Looking-Glass*, the Red Queen tells Alice, "Now, here, you see, it takes all the running you can do, to keep in the same place. If you want to get somewhere else, you must run at least twice as fast as that!" Much of evolution appears to work this way: As the physical or biological environment changes, continued adaptation is needed just to maintain fitness. Strains of wheat, modified to improve drought tolerance, may simply maintain high yields in a changing environment, not increase them. Increases in yields through genetic modification may be possible, but limited. Thus, massive fertilization by nitrate and phosphate fertilizers will continue to be a key component of agriculture.

Another type of biological intervention is possible, however. We are not the only species with a nutritional interest in croplands. Fungi, bacteria, protists, nematodes, insects, birds, and mice all target cornfields, sharply reducing yield. In the United States alone, such unwanted guests cause about $1.5 billion worth of damage per year. Conventional breeding and genetic modification have produced improved resistance to pests, but

**FIG. 48.17  Corn yield and area in the United States and globally from 1961 to 2007.** Global corn yield and area have increased at about the same rate, whereas U.S. corn yield has risen much more quickly than corn area because of energy-intensive agriculture and use of fertilizer.

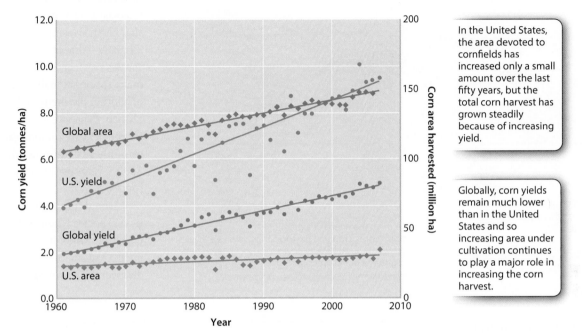

In the United States, the area devoted to cornfields has increased only a small amount over the last fifty years, but the total corn harvest has grown steadily because of increasing yield.

Globally, corn yields remain much lower than in the United States and so increasing area under cultivation continues to play a major role in increasing the corn harvest.

again, the Red Queen looms large. Fungus-resistant plants provide strong selective pressure for fungi that can circumvent the mechanisms of resistance, gaining access to otherwise unavailable food.

Food spoilage after harvest also reduces the effective yield of agricultural lands. It is estimated that one-third of the food grown in the United States is never eaten, and figures for Europe are similar. Fungi and bacteria spoil food during storage and transportation, on supermarket shelves, and in refrigerators. The problem also exists in less-developed countries. The United Nations estimates that in some parts of Africa a quarter of all crops spoil before they ever reach the market. More efficient storage, transportation, and packaging could sharply increase the effective yield of fields and pastures.

As with energy, a sustainable future for agriculture will require creative biologists, ingenious engineers, and wise citizens. Geneticists and physiologists can help to develop plant breeds that improve crop yield while using less fertilizer. Working with microbiologists, engineers can discover new ways of storing and transporting food that increase the likelihood that it will be eaten before it spoils. As individuals, we make daily choices about the food we eat. We can choose to eat more plant products, for example. The basic ecological logic of the trophic pyramid (Chapter 47) tells us that a hectare of grain fed directly to humans feeds more people than the meat of cattle

fed on the same crops. We can also choose to eat locally grown products where possible, saving energy while decreasing the likelihood of spoilage.

## 48.4 HUMAN INFLUENCE ON EVOLUTION

It has sometimes been claimed that as humans exert increasing influence on nature, evolution will grind to a halt. Nothing could be further from the truth. As examples in the preceding sections show, what is changing, and what will continue to change as the human footprint grows, is the selective landscape within which evolution operates. Some species, for example rats and cockroaches, thrive in the urban environments we have constructed for our own benefit. Others will expand or decline as we replace prairies and forests with crop and pasture land. What are our conscious and inadvertent effects on evolution and biodiversity?

**?  CASE 8 Biodiversity Hotspots:
Rain Forests and Coral Reefs
How has human activity affected biological diversity?**
As noted in Chapter 1, it has been estimated that humans make use of nearly 25% of the entire photosynthetic output on land. That is, through the crops and livestock we harvest, the biomass we burn, and primary production we eliminate when we clear-cut

**FIG. 48.18 Global agriculture land.** Through time, land under cultivation has expanded, diminishing natural landscapes and the species diversity they support.

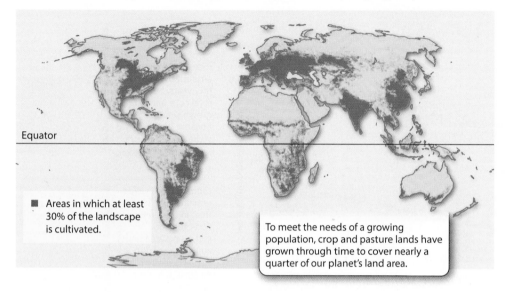

Equator

■ Areas in which at least 30% of the landscape is cultivated.

To meet the needs of a growing population, crop and pasture lands have grown through time to cover nearly a quarter of our planet's land area.

As the forest disappears, so, too, will untold numbers of insect, mite, and other animal species. The renowned biologist E. O. Wilson has estimated that Earth is losing 0.25% of its species annually—that is, 5000 to 25,000 species per year, depending on how many species actually exist. In the words of another distinguished biologist, Paul Ehrlich, "The fate of biological diversity for the next 10 million years will almost certainly be determined during the next 50–100 years by the activities of a single species."

Without question, there will be fewer species when the 21st century ends than when it began. Lost species means lost opportunities for the discovery of novel compounds for

forests, we use one in every four grams of organic matter produced by land plants. Ecologists also estimate that crops cover 10% to 15% of Earth's land surface, and pasture land another 6% to 8% (**Fig. 48.18**).

The populations of a few species have expanded dramatically during the Anthropocene Period. For example, 10,000 years ago, corn (*Zea mays*) was an inconspicuous herb in southwestern Mexico; today, corn fields cover nearly 150 million hectares (ha) across the globe. As crop and pasture lands expand, along with urban and suburban environments, indigenous populations become displaced and their natural habitats disrupted.

Habitat loss in tropical rain forests poses one of the most important threats to biological diversity (**Fig. 48.19**). These forests represent the greatest hotspots of biodiversity on land, but clear-cutting for grazing and croplands has destroyed more than half of their pre-industrial area. It has been estimated that deforestation over the next century could eliminate 40% to 50% of all tree species in the Amazon forest.

**FIG. 48.19 Forests under threat.** (a) The map shows the extent of deforestation worldwide. (b) This desolate scene in Amazonia was once an expanse of tropical rain forest. Nearly 20% of the Amazon rain forest has been cleared for crops, pasture, or mining.

a.

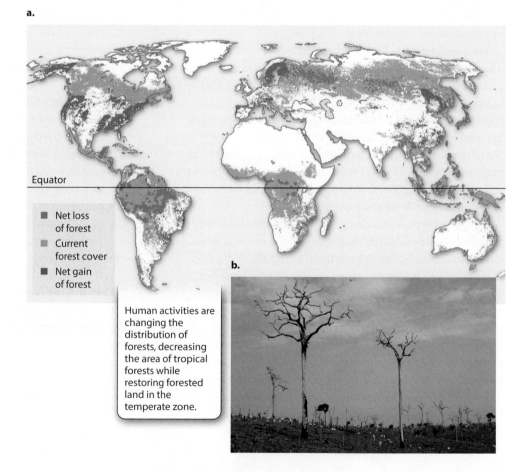

Equator

■ Net loss of forest
■ Current forest cover
■ Net gain of forest

b.

Human activities are changing the distribution of forests, decreasing the area of tropical forests while restoring forested land in the temperate zone.

medical research. Diminished diversity may make communities less productive (Chapter 47) and less resilient to fires, hurricanes, or other environmental events. As discussed in the preceding section, there is a pressing need to feed the world, and there are also many good reasons to conserve the biological diversity that has evolved over 4 billion years. Balancing these two imperatives will entail creative approaches to land policy, including thoughtful ways of using land in multiple ways that serve both people and conservation. Reforestation of former crop and pasture lands will help, as will the preservation or establishment of ecological corridors to facilitate migration of populations.

### Humans play an important role in the dispersal of species.

Biodiversity loss has many causes. In addition to habitat destruction, one of the most insidious is the introduction of non-native species, often called **invasive species**. Removed from natural constraints on population growth, invasive species can expand dramatically when introduced into new areas, often with negative consequences for native species and ecosystems.

Some species can be transported long distances by natural processes. For example, the Gulf Stream and North Atlantic Drift occasionally carry coconuts from the Caribbean to the coast of Ireland. Migratory birds can transport microorganisms across continents. In the Anthropocene Period, however, dispersal has increased dramatically at the hands of humans. Ships fill their ballast tanks with seawater in Indonesia and empty it into Los Angeles Harbor, carrying organisms across the Pacific Ocean. Insects in fruit from South America wind up on a dock in London. Humans have become major agents of dispersal, adding new complexity to 21st-century ecology.

It has been estimated that 49,000 invasive species have been introduced into the United States (**Fig. 48.20**). Many of these provide food or pleasure (for example, ornamental plants), and only a few have escaped into natural ecosystems. Some of the invasive species that have become established in their new surroundings do little beyond increasing community diversity, but others put strong pressure on native species.

Kudzu, for example, a plant imported from Japan to retard soil erosion, now covers more than 3 million hectares in the southeastern United States, displacing native plants through competition for space and resources (Fig. 48.20a). The Zebra Mussel, a European bivalve that hitchhiked to the Great Lakes in the ballast water of cargo ships, has also multiplied dramatically in its new habitat, displacing native species and clogging intake pipes to power plants (Fig. 48.20b). Nearly 500 introduced plant species have become weeds, contributing disproportionately to crops lost each year to competition from invasive plants.

Invasive species have particularly devastating effects on islands because island species have commonly evolved with relatively few competitors or predators. Bones in Hawaiian lava caves record more than 40 bird species that existed 1000 years ago but are now extinct, eliminated by rats and pigs introduced by Polynesian colonists, as well as by landscape alteration. Introduced predators have had a similar impact on the island of Guam. Brown tree snakes introduced from Australasia since World War II have reduced native bird and reptile diversity by 75% (Fig. 48.20c).

**FIG. 48.20** **Invasive species.** (a) Kudzu (*Pueraria lobata*), a Japanese vine, was introduced to the United States in 1876 and now covers more than 3 million hectares. (b) The Zebra Mussel (*Dreissena polymorpha*), introduced into the Great Lakes from Europe in 1985, disrupts lake ecosystems, threatening native biological diversity. (c) The brown tree snake (*Boiga irregularis*), introduced into Guam from Asia and Australia shortly after World War II, has decimated indigenous bird populations across the island.

Such introductions contribute to the changing diversity and ecology of 21st-century landscapes, but it may be the spread of modern diseases that should impress (and concern) us most. In the world of airplanes and oceangoing ships, bacteria and viruses spread much more rapidly than they did in our pre-industrial past. The H1N1 (swine) flu, a global concern in 2009, is a recent example. The first diagnosed case was identified on April 13 in Mexico, and by mid-May this flu had been reported on five continents.

→  **Quick Check 4**  How do invasive species affect species diversity of communities and ecosystems?

### Humans have altered the selective landscape for many pathogens.

From the Black Death of the fourteenth century to the flu pandemic of 1918, disease has played a major role in human history. In the Anthropocene Period, we have unprecedented opportunity to control or even eradicate some of the great diseases of the past and present. At the same time, however, the strong selective pressures placed on pathogens by antibiotics pose a key challenge for 21st-century medicine.

Malaria, discussed in Case 4, provides an illustrative case. In any given year, malaria infects an estimated 500 million people, causing a million or more deaths. Infection is particularly severe in parts of Africa, where the direct and indirect effects of malaria account for 40% to 49% of all childhood mortality. Insecticide-soaked bed nets have proven highly effective in affected areas, reducing transmission by as much as 90%. Natural selection, however, has begun to chip away at success in controlling the disease: Mosquito strains have begun to emerge that are resistant to DDT and another family of insecticides called pyrethroids.

Quinine, a chemical extracted from the bark of the cinchona tree, was the first effective treatment for malaria, and remained in widespread use until the 1940s. Since that time, a small arsenal of antibiotics has been developed, including chloroquine, mefloquine, primaquine, and doxycycline. Their usefulness, however, has declined through time as resistance to each of these drugs has evolved in populations of *Plasmodium*, the infectious agent of malaria. More recently, artemisinins,

antibiotics based on a traditional Chinese remedy, have proved effective in treatment. Again, however, resistant strains of malaria have been identified.

→  **Quick Check 5**  The interplay of pathogens and antibiotics again brings up the problem of the Red Queen. How does the Red Queen hypothesis apply to these kinds of interaction?

The arms race between antibiotics and pathogens is not limited to malaria. It characterizes nearly the whole range of diseases caused by viruses, bacteria, protists, and fungi that infect humans. For example, tuberculosis, a lung infection spread by inhalation of the bacterium *Mycobacterium tuberculosis*, causes 1.8 million deaths annually, mostly in sub-Saharan Africa, where many people are also infected with HIV (Chapter 43). Antibiotic treatment is effective, but many who are infected do not or are unable to comply with the strict regimen of daily treatment for 6 months, hastening the evolution of drug-resistant *M. tuberculosis* strains. This occurs because an incomplete course of antibiotics does not eradicate all the bacteria, selecting for those bacteria that are least susceptible to the antibiotic. In 2007, about half a million cases of antibiotic-resistant tuberculosis were reported. More troubling, a small percentage of these infections have been characterized as "extremely drug resistant," meaning that the pathogens resist both standard treatments and alternative antibiotics developed over the past 20 years (**Fig. 48.21**). In effect, extremely drug-resistant tuberculosis is untreatable.

Because bacteria can pass genes horizontally (Chapter 26), antibiotic resistance evolved in one species can jump to another.

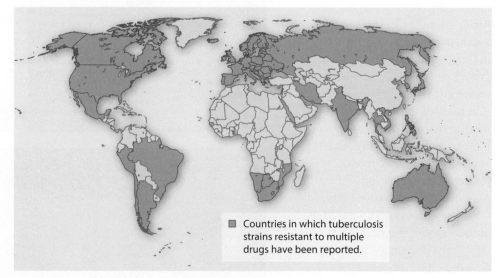

**FIG. 48.21  Distribution of extremely drug-resistant tuberculosis.** By 2008, confirmed cases had been reported in nearly 50 countries.

Countries in which tuberculosis strains resistant to multiple drugs have been reported.

**FIG. 48.22 Amphibians under threat.** The Harlequin Toad (*Atelopus varius*), from Costa Rica and Panama, has been classified as critically endangered.

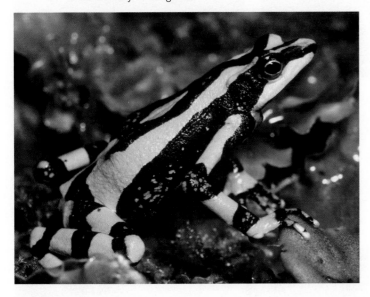

Thus, the bacterial incubators for drug resistance can arise anywhere and spread rapidly.

Clearly, despite the remarkable gains of the twentieth century, public health will remain a key issue in the decades ahead. Improved sanitation—clean drinking water and safe food—can do much to reduce transmission of Earth's great diseases, and the search is on for vaccines that can further reduce or even eliminate some historic scourges. But the ongoing evolution of drug resistance will require the continuing efforts of scientists to stay one step ahead of pathogens.

**Are amphibians ecology's "canary in the coal mine"?**
In the nineteenth and well into the twentieth century, British, American, and Canadian coal miners sometimes took canaries with them down long mine shafts. Canaries are more sensitive than humans to carbon monoxide and other toxic gases that can accumulate in mines. When the canaries passed out, the miners knew it was time to leave. Because many amphibian species have narrow environmental tolerances and exchange gases through their skin, they can be especially sensitive to environmental disruption. For this reason, some biologists see them as "canaries" for the global environment.

If amphibians are indeed our early-warning system, we should be concerned. Fully one-third of all amphibian species are threatened with extinction, and more than 40% have undergone significant population declines over the past decade (**Fig. 48.22**). What accounts for this pattern? Just about all

the environmental issues discussed in this chapter come into play. Amphibian habitat is being destroyed in many parts of the world, especially old-growth forests that provide damp, food-rich environments in abundance. Pesticides and other toxins of human manufacture take their toll as well. For example, the common herbicide atrazine has been shown experimentally to impair development in frogs. Studies on placental cells grown in the laboratory suggest that atrazine in our food may also affect human reproduction. When atrazine was applied to human placental cells, it caused overexpression of genes associated with abnormal fetal development. Because of concerns about public health, atrazine has been banned in several European countries.

Recently, infection has also been implicated in amphibian decline. The fungus *Batrachochytrium dendrobatidis,* or *Bd,* infects the skin of frogs and impairs many of the skin's functions. Recent studies of *Bd* show the classic features of epidemics—rapid geographic spread and high mortality—along the expanding front of fungal populations. How *Bd* spreads and whether it acts in concert with other facets of global change remain controversial, but there can be no doubt that disease has joined habitat destruction and pollution as a major cause of amphibian decline.

## 48.5 SCIENTISTS AND CITIZENS IN THE 21ST CENTURY

The decades ahead will be a time of great opportunity, but they will also present increasing challenges as the human footprint continues to grow around the world. How we will meet those challenges is, in large measure, up to you.

If you become a biologist, you will be able to help improve our planet's health in many different ways (**Fig. 48.23**). Field biologists are needed to document Earth's true biological diversity, even as much of it comes under threat. Physiologists are needed to understand how species respond to environmental change. And ecologists and population geneticists are needed to discover the principles that govern species interactions in communities, so that we can understand how physiological responses to global change in one species will affect the fortunes of others. Only then will we be able to design effective ways to preserve the world's biodiversity. Molecular biologists will help us to combat disease in new ways, and geneticists will develop crops that yield more food, even as temperature and rainfall patterns change. Perhaps most important, biologists will work with economists, lawmakers, and others to ensure that government policies reflect what we know about the natural world on which we all depend.

**FIG. 48.23  Human impact.** Scientists, doctors, teachers, and informed citizens can make a difference.

If you become a doctor, you will have many opportunities to improve the health of people in countries around the world (Fig. 48.23). Through laboratory research physicians will discover new ways to treat cancer, malaria, HIV, and a host of other diseases. And by treating patients in new ways, doctors will reduce the threat of antibiotic-resistant pathogens. Working with government and private agencies, doctors will help to ensure clean water and adequate food for people in parts of Africa, Asia, and wherever else the basic needs of life pose an ongoing challenge.

If you become a teacher, you will have a particularly important role in making sure that our children have the knowledge they need to meet the challenges of life in the Anthropocene Period (Fig. 48.23). And, perhaps most important, by choosing to be a responsible citizen, you can help to ensure that your family, your workplace, and your government make wise choices. As individuals, we all have a responsibility to make wise choices in our lives, and in an age of environmental change, knowledge of Earth and life provides a key framework for personal choice. As citizens, our votes shape governments, and biological education can help to ensure wise leadership.

We cannot change the world we've been given, but we can and will influence the one our children inherit. It is easy to despair about the future, but as the novelist Philip Pullman put it, "Hope is not the name of a temperament, or of an emotion, but of a virtue." Hope inspires action, and our collective capacity for positive change is great. Knowledge of biology will not by itself ensure that humanity's golden age can be sustained in coming years, but it will surely be an important part of our solutions.

# Core Concepts Summary

**48.1 SOME SCIENTISTS CALL THE TIME IN WHICH WE ARE LIVING THE ANTHROPOCENE PERIOD TO REFLECT OUR SIGNIFICANT IMPACT ON THE PLANET.**

The human population recently passed the 7-billion mark. page 48-1

The ecological footprint is the amount of land required to support an individual at an average standard of living. page 48-1

Our ecological footprint reflects all the activities that support our daily life, from growing and transporting the foods that we eat to supplying the materials and energy for our homes and the vehicles that provide our transportation. page 48-1

**48.2 HUMANS HAVE A MAJOR IMPACT ON THE CARBON CYCLE, PRIMARILY THROUGH THE BURNING OF FOSSIL FUELS, WHICH RETURNS CARBON DIOXIDE TO THE ATMOSPHERE.**

Carbon dioxide is a greenhouse gas, trapping heat and thereby keeping the surface of the Earth warm. page 48-3

Greenhouse gases play an important role in maintaining temperatures on Earth compatible with life, but increasing levels of greenhouse gases are resulting in higher mean surface temperatures. page 48-3

Changing climate has already affected the distribution of species and the composition of communities, and will continue to do so in the 21st century. page 48-5

Increasing levels of $CO_2$ are increasing the temperature and decreasing the pH of the oceans, affecting marine organisms, including corals. page 48-7

We can reduce $CO_2$ emissions, conserve finite natural resources, and contain the cost of energy by developing energy sources other than fossil fuels and improving energy efficiency. page 48-10

**48.3 HUMANS HAVE AN IMPORTANT IMPACT ON THE NITROGEN AND PHOSPHORUS CYCLES, PRIMARILY THROUGH THE USE OF FERTILIZER IN AGRICULTURE.**

Humans add large amounts of industrially fixed nitrogen to the environment, which enters the biosphere as fertilizer to increase food production. page 48-11

Only a small percentage of fixed nitrogen is taken up by crops; the rest is returned to the atmosphere by soil microorganisms or ends up in lakes and oceans, where it results in eutrophication, a boost in primary production that depletes oxygen and leads to the death of fish and other aquatic organisms. page 48-11

A second major nutrient in fertilizer is phosphorus, which is mined rather than produced industrially. The finite supply of phosphorus raises questions about our ability to maintain agricultural production in the future. page 48-14

We have increased agricultural yields through the use of fertilizer, high-yield and disease-resistant plants, and farm machinery. page 48-14

To feed a growing population, we can devote more land to agriculture or increase productivity further, but each option brings with it costs. page 48-14

**48.4 THE IMPACT OF HUMANS ON THE ENVIRONMENT IS CHANGING THE STAGE ON WHICH EVOLUTION ACTS.**

Humans have altered the landscape in dramatic ways, leading to the expansion of some species, like corn and wheat, and the decrease and extinction of others. page 48-15

Habitat loss may cause extinction in some of Earth's most diverse ecosystems, such as tropical rain forests. page 48-15

Invasive species, spread through human activity, are major players in biodiversity loss. page 48-17

The use of antibiotics is one of the success stories of twentieth-century medicine, but has also led to the spread of antibiotic resistance, a major challenge for 21st-century biologists. page 48-18

Amphibian populations have declined dramatically because of a combination of habitat loss, pesticide use, and the spread of infection. page 48-19

**48.5 IN THE 21ST CENTURY, BIOLOGISTS, DOCTORS, ENGINEERS, TEACHERS, AND INFORMED CITIZENS HAVE VITAL ROLES TO PLAY IN UNDERSTANDING OUR CHANGING PLANET AND MAKING WISE CHOICES FOR OUR FUTURE.**

Human impact can be positive as well as negative. page 48-19

# Self-Assessment

1. Describe what is meant by the term "ecological footprint."

2. Name three sources of atmospheric $CO_2$.

3. Explain the relationship between atmospheric $CO_2$ levels and mean temperature.

4. Provide evidence that indicates that human activities are responsible for increases in atmospheric $CO_2$ over the last century.

5. Describe two ways in which organisms respond to changes in $CO_2$ levels and temperatures.

6. Describe several possible solutions to the problem of increased atmospheric $CO_2$.

7. Describe the causes and consequences of eutrophication.

8. Describe several possible solutions to the problem of feeding a growing human population.

9. Give three examples of species that have benefited from human activity and three examples of species that have been harmed by human activity.

10. Explain why amphibians are sometimes called "canaries in a coal mine."

Do you understand the chapter's Core Concepts? Log in to check your answers to the Self-Assessment questions, then practice what you've learned and reinforce this chapter's concepts by working through the problems and multimedia tutorials provided there.

🖵 **www.biologyhowlifeworks.com**

# QUICK CHECK ANSWERS

## Chapter 1

1. One hypothesis is that their tan fur protects the mice from predators by allowing them to blend in with their surroundings. To test this hypothesis, you could design an experiment in which dark and tan mice (or, to be more humane, artificial mice) are exposed to a predator, such as a hawk, on a variety of backgrounds. Do hawks preferentially attack dark mice on light sands and light mice on dark backgrounds?

2. One test would be to take two large populations of laboratory mice and introduce cigarette smoke into the cages of one population but not the other. Do mice subjected to smoke develop cancer at rates that are significantly higher than those of the control group? We can also make observations of human populations: Do smokers develop lung cancer at rates significantly higher than nonsmokers?

3. DNA directs the formation of proteins that do the cell's work. Mutations can result in proteins with altered structure and, therefore, different functions. Changes in protein function can cause the cell to work improperly, or fail altogether, resulting in some cases in disease.

4. The use of antibiotics selects for antibiotic resistance among bacteria. Bacteria do not become antibiotic resistant out of "need" or because resistance would be advantageous for them. Instead, before the application of antibiotics, bacteria with antibiotic resistance exist in low numbers. The application of antibiotics allows these bacteria to grow and reproduce more successfully than those that are susceptible to antibiotics. Mutations, such as those responsible for antibiotic resistance, are not influenced by whether or not the organism is in an environment in which that mutation would be advantageous.

## Chapter 2

1. One conclusion is that atoms consist mainly of empty space, and hence most negatively charged particles passing through the gold foil do not come close enough to any positive charge to be deflected. Another conclusion is that the positively charged protons in the nucleus must be small and densely packed.

2. Hydrogen and lithium are in the same column, or group, in the periodic table. They both have one valence electron in their outer orbital. As a result, one atom of lithium combines with one atom of hydrogen to make lithium hydride, with a full complement of two electrons in the single molecular orbital.

3. Ice is less dense than liquid water. As a result, when water freezes, it expands in volume and can burst closed containers, such as cans of soda or water pipes in houses. This property is unusual. For most substances, the solid phase is more dense than the liquid phase.

4. Glucose and galactose differ only in the orientation of the –OH and –H groups attached to carbon number 4.

## Chapter 3

1. Because R = A or G, then %R = %A + %G; and because %A = %T and %G = %C, we can write %R = %A + %G = %T + %C. But T or C is Y, and so %T + %C = %Y. It follows that %R = %Y.

2. The RNA transcript has the sequence 5'–AUCGCUGAAAGU–3'.

3. The incorporation of a nucleotide with a 3' H rather than a 3' OH will stop subsequent elongation because the 3' OH is necessary to attack the high energy phosphate bond of the incoming nucleoside triphosphate. The incorporation of a nucleotide with a 2' H rather than a 2' OH will have no effect on elongation, as this group is not involved in the polymerization reaction.

4. The eukaryotic DNA sequence contains introns, which the bacterial cell cannot splice out properly, and so the correct protein is not produced from the information in the bacterial RNA transcript.

## Chapter 4

1. The sequence of amino acids in a protein determines how a protein folds, so a change in even a single amino acid can affect the way the protein folds and can disrupt its function.

2. The three reading frames are

   UUU GGG UUU GGG..., which codes for repeating Phe–Gly–Phe–Gly...

   UUG GUU UUG GGU..., which codes for repeating Leu–Gly–Leu–Gly...

   UGG GUU UGG GUU..., which codes for repeating Trp–Val–Trp–Val...

3. With proper eukaryotic processing, the RNA transcript from the bacterial DNA will be capped at the 5' end. The initiation complex will form at the 5' cap and move along the mRNA until the first AUG codon is encountered, and then translation begins. When one of the termination codons is encountered, the polypeptide is released. Translation of the downstream polypeptides cannot take place because the Shine–Dalgarno sequences preceding them are not recognized by the eukaryotic translational machinery.

4. The eukaryotic DNA sequence contains introns, which the bacterial cell cannot splice out properly, and so the correct protein is not produced from the information in the bacterial RNA transcript.

## Chapter 5

1. Saturated fatty acids have reduced membrane mobility compared to that of unsaturated fatty acids. As a result, fatty acids tend to be solid at room temperature, while unsaturated fatty acids tend to be liquid. Margarine and animal fats contain saturated fatty acids and are solid, while many plant and fish oils, containing unsaturated fatty acids, are liquid at room temperature.

2. Water molecules move in both directions, but the *net* movement of water molecules is from side A to side B. This process is called

osmosis: Water moves from regions of higher water concentration to regions of lower water concentration. Likewise, sodium and chloride ions move in both directions, but the *net* movement of sodium and chloride ions is from side B to side A. Movement of water and ions results from diffusion, the random motion of substances. Even when the concentration of all molecules is the same on the two sides, diffusion still occurs, but there is no net movement of water molecules or ions.

3. If the sodium-potassium pump is poisoned, cells tend to swell and even burst, as the intracellular fluid becomes hypertonic relative to the outside of the cell and water moves into the cell by osmosis.

## Chapter 6

1. No, because the second law of thermodynamics applies to the universe as a whole. This means that we have to consider not just the air in the room but the heat released to the outdoors as well. An air conditioner produces more hot air than cold air, and therefore total entropy increases, as expected from the second law of thermodynamics.

2. Increasing the temperature increases the value of $T\Delta S$, which decreases $\Delta G$, since $\Delta G = \Delta H - T\Delta S$. As a result, an increase in temperature makes it more likely that a reaction will proceed without a net input of energy.

3. Enzymes increase the reaction rate and decrease the activation energy. The other parameters are not changed by enzymes.

## Chapter 7

1. The higher-energy forms are ATP, NADH, $FADH_2$, $C_6H_{12}O_6$, and reduced molecules; the lower-energy forms are ADP, $NAD^+$, FAD, $CO_2$, and corresponding oxidized molecules.

2. At the end of glycolysis, the energy in the original glucose molecule is contained in pyruvate, ATP, and NADH.

3. At the end of the citric acid cycle, the energy in the original glucose molecule is held in ATP, NADH, and $FADH_2$.

4. Oxygen is consumed in cellular respiration. Oxygen is the final electron acceptor and is converted to water in the electron transport chain.

5. Uncoupling agents decrease the proton gradient and therefore decrease the levels of ATP. The energy of the proton gradient is not used for oxidative phosphorylation, but instead is dissipated as heat. Uncoupling agents are found naturally in certain tissues, such as fat, for heat generation. They can also act as poisons.

6. Yeast cells are eukaryotes. In breadmaking, yeast can use sugar as a food source for ethanol fermentation. The carbon dioxide produced in the process causes the bread to rise. The ethanol is removed in the baking process.

## Chapter 8

1. You should label the oxygen in $CO_2$ (that is, inject $C^{18}O_2$) because the entire $CO_2$ molecule is combined with RuBP by rubisco, whereas

$H_2O$ is the electron donor needed for the reduction step of the Calvin cycle. The extraction of electrons from water releases $O_2$ as a by-product.

2. NADPH supplies the major input of energy that is used to synthesize carbohydrates in the Calvin cycle.

3. Antenna chlorophylls transfer absorbed light energy from one antenna chlorophyll molecule to another, and ultimately to the reaction center. Reaction center chlorophylls transfer absorbed light energy and a high-energy electron to an electron acceptor, resulting in the oxidation of the reaction center chlorophyll molecule.

4. The energy difference between the oxidation of water and the reduction of $NADP^+$ is greater than the amount of energy that can be harvested when a single photosystem absorbs light energy.

5. Rubisco faces a fundamental trade-off between selectivity and speed because it can use both $CO_2$ and $O_2$ as substrates. High selectivity of $CO_2$ over $O_2$ requires that the reaction have a high energy barrier, leading to a lower catalytic rate.

6. Like cellular respiration, photorespiration releases $CO_2$. Unlike photorespiration, it consumes rather than produces ATP.

## Chapter 9

1. Only cells that have receptors for the hormone respond to the signal. Therefore, signaling can be specific for particular cells.

2. No. A G protein-coupled receptor is a transmembrane receptor protein, which interacts with another protein called a G protein located inside the cell on the cytoplasmic side of the plasma membrane.

3. The signal to increase heart rate carried by adrenaline can be reversed in at least four ways: (1) Decreasing the amounts of the signaling molecule available to bind and activate the G protein-coupled receptor; (2) inactivating the G protein; (3) decreasing the amount of the second messenger cAMP; and (4) dephosphorylating the target proteins that cause the increased rate of contraction of the muscle cells.

## Chapter 10

1. A defect in dynein would cause the melanin granules to remain dispersed because dynein transports the granules back toward the minus end of the microtubule during granule aggregation.

2. Adherens junctions form a belt around cells, whereas desmosomes are button-like attachments. In addition, adherens junctions connect to actin microfilaments, whereas desmosomes connect to intermediate filaments.

3. Tight junctions prevent the passage of materials through the space between cells. Gap junctions in animal cells and plasmodesmata in plant cells permit the exchange of materials between the cytoplasm of adjacent cells.

4. Integrins are responsible for the production of milk proteins by mammary cells in response to the extracellular matrix because integrins bind extracellular matrix proteins. By contrast, cadherins bind to other cadherins when cells adhere to other cells.

## Chapter 11

1. A mutation that disrupts the function of the FtsZ protein will block cell division.

2. Sister chromatids are the result of DNA replication during S phase, and so they have identical DNA sequences (with the exception of a few changes due to rare mutations). The two homologous chromosomes are inherited from two different parents. The DNA sequences of these chromosomes are therefore similar, but not identical.

3. A cell that undergoes mitosis but not cytokinesis will become a single cell with two nuclei (and therefore twice the normal amount of DNA); this type of cell is called a multinucleate cell.

4. In human cells at the end of prophase I, there are 92 chromatids, 46 centromeres, and 23 bivalents.

5. In meiosis I, homologous chromosomes pair, undergo crossing over, and segregate from each other. These events do not occur in mitosis. In mitosis, centromeres divide and sister chromatids separate, events that do not take place in meiosis I.

6. The products of meiosis are different from each other as a result of two key processes: (1) crossing over, which occurs at essentially random positions along the chromosomes and creates unique combinations of genetic differences that may be present in the maternal and paternal chromosomes, and (2) random orientation of the homologous chromosomes on the spindle in metaphase I, so each nucleus receives a random combination of maternal and paternal homologs.

7. The function of the p53 protein can be disrupted by a mutation in the *p53* gene. Alternatively, certain viral proteins, such as the E6 protein of HPV discussed in Case 2: Cancer, can interfere with the function of p53.

8. An oncogene causes cancer by producing an excess of protein activity that pushes the cell to divide. A tumor suppressor like p53 functions oppositely. Its normal function is to prevent cell division and its absence is what allows the cell to divide uncontrollably.

## Chapter 12

1. After one round of replication, you would predict only $^{14}N/^{15}N$ hybrid DNA, which has a density of 1.708 gm/cm³. After two rounds of replication, you would predict half the molecules to be $^{14}N/^{15}N$ hybrid DNA (density 1.708 gm/cm³) and half to be $^{15}N/^{15}N$ heavy DNA (density 1.722 gm/cm³).

2. There are many possible reasons that might explain the absence of a band following PCR and gel electrophoresis. The DNA polymerase might be defective or have low activity, or you may have inadvertently left out one of the components of the reaction, or the primer sequences might have been incorrect so that there was not a complementary sequence for them to anneal to, or you may have connected the positive and negative leads to the wrong ends of the gel box so that the products ran off the top of the gel.

3. The sequence of the template strand is antiparallel to the synthesized strand and inferred from complementary base pairing as 5'-ACTCGGTAGT-3'.

4. The value of sticky ends is that they give the researcher greater control over which restriction fragments can come together and be attached. Sticky ends can pair only with other sticky ends that have complementary 3' and 5' overhangs. Hence, restriction fragments produced by *Bam*HI can combine only with other fragments produced by *Bam*HI, and not with fragments produced by *Hind*III, for example. However, any blunt end can be attached to any other blunt end, for example, a *Hpa*I end to a *Sma*I end.

## Chapter 13

1. DNA sequencing technology is limited to DNA molecules that are less than about 1000 nucleotides, so the challenge of genome sequencing is the size of many genomes, which can far exceed 1000 nucleotides. Sequencing whole genomes requires a method of piecing together smaller sequenced DNA fragments.

2. Even when short repeated sequences can be sequenced, there is an assembly problem similar to that for longer repeats in which the researcher has no way of knowing where in the repeat any particular sequenced fragment should be assigned. The result is that the total number of repeats remains unresolved. It could be in the hundreds, thousands, tens of thousands, hundreds of thousands, or even millions.

3. The observation that closely related lentiviruses have closely related host organisms suggests that each type of lentivirus can infect only one or a small number of closely related organisms.

4. You can't tell. The C-value paradox means that you cannot predict genome size from the complexity of the organism. In fact, the genome of an amoeba is much larger than the genome of a human.

## Chapter 14

1. Mutant genes that are harmful or neutral are much less likely to persist in a population than ones that result in increased survival and reproduction because of greater fitness.

2. The number of nucleotides that are inserted or deleted is almost always an exact multiple of 3 because each codon in the genetic code consists of three nucleotides. Any insertion or deletion that includes a number of nucleotides that is not an exact multiple of 3 shifts the reading frame, and the resulting sequencing will most likely code for a nonfunctional protein.

3. Mutations are random; they are not directed by the environment. This does not mean that the environment cannot affect the *rate* of mutation. Mutagens increase the rate of the mutation, but they cannot induce specific mutations that would be beneficial to the organism in response to the environment.

## Chapter 15

1. Mutations that cause antibiotic resistance are clearly beneficial to bacteria when antibiotic is present, but are neutral or even harmful in the absence of antibiotic. The effect of a mutation on an organism often depends on the environment.

2. Only a small fraction of the human genome codes for proteins or other functional elements (Chapter 13), so most mutations are neutral.

3. If the suspect is the actual source of a sample, then the DNA "fingerprints" must match exactly, and any mismatch rules out the suspect as the source of the sample. When there is a match, however, there is always a small chance (typically very small) that the sample came from another person with the same DNA "fingerprint."

4. In a VNTR, the restriction fragments are different lengths because of a variable number of repeats between restriction sites (one, two, three, four, or more). In a RFLP, the restriction fragments are different lengths because a restriction site is removed in one DNA sequence and not in another, making the distances between flanking sites different.

5. A point mutation is a change in a single nucleotide in an individual cell, such as C to G. A SNP results from a point mutation that occurred at some time in the past, so that now in the population there are two or more different single nucleotides at a given position. For example, some people might have C at a certain position and others might have G at that position.

6. Both Y chromosomes in the XYY baby must come from the father, so nondisjunction took place in the father. Consider now normal meiosis, shown in Fig. 15.12a, and suppose we let dark blue represent the X chromosome and light blue the Y chromosome. The first meiotic division separates the X chromosome from the Y chromosome, and the second meiotic division separates the sister chromatids of the X chromosome and the sister chromatids of the Y chromosome. For the Y chromosomes to remain together and be included in the same sperm, nondisjunction must take place in the second meiotic division (Fig. 15.12c).

## Chapter 16

1. For the true-breeding plants with yellow seeds, the phenotype is yellow seeds, and the genotype is $AA$. For the true-breeding plants with green seeds, the phenotype is green seeds, and the genotype is $aa$.

2. Yes in both cases. When an allele is recessive, the homozygous dominant and heterozygous individuals have the same phenotype but different genotypes, as in $AA$ and $Aa$ plants. When a trait is influenced by the environment, individuals with the same genotype can have different phenotypes because of the environment. An example discussed in Chapter 15 is how smoking tobacco increases the severity of emphysema resulting from the $PiZ$ allele of alpha-1 antitrypsin.

3. Half the progeny will have the $AA$ genotype and half will have the $Aa$ genotype. All the plants will have yellow seeds.

4. The probability that the first pea is green is ¼; the probability that the second pea is green is also ¼; the probability that the third pea is yellow is ¾; *and* the probability that the fourth pea is yellow is ¾. We use the multiplication rule (because these are mutually exclusive events) to figure out that the probability of this configuration is ¼ × ¼ × ¾ × ¾ = 9/256. There are six different ways that a pod of four seeds can have two green seeds and two yellow seeds (the green seeds can be in positions 1 and 2; 1 and 3; 1 and 4; 2 and 3; 2 and 4; or 3 and 4). Each of these configurations occurs with a probability of 9/256 (as previously described). The probability of *any* of these occurring (*either* one configuration *or* another *or* another) is given by the addition rule, or 9/256 + 9/256 + 9/256 + 9/256 + 9/256 + 9/256 = (9/256) × 6 = 54/256, or about 21%.

5. The $F_1$ chickens have white feathers because the dominant inhibitor allele $I$ inhibits expression of the pigment allele $C$.

6. In the population as a whole, there are many copies of each chromosome, so any gene can have multiple alleles present in the different copies. Any one individual can have only two copies of any chromosome, and so any individual can have no more than two different alleles.

## Chapter 17

1. A woman whose father is color blind must be heterozygous for the mutant allele. If she has children with a man who is color blind, then half of the female offspring are expected to be homozygous mutant and therefore color blind.

2. Even when one (or more) crossover occurs between the genes, only two of the four products of meiosis are recombinant because crossing over takes place at the four-strand stage of meiosis (Fig. 17.10a). With two strands that are recombinant and two that are nonrecombinant, the frequency of recombination is 2/4 = 50%, and so this is the maximum.

3. Independent assortment means that a doubly heterozygous genotype like $AB\ ab$ produces gametes in the ratio 1/4 $AB$, 1/4 $Ab$, 1/4 $aB$, 1/4 $ab$. The first and last are nonrecombinant gametes, and the second and third are recombinant gametes. The frequency of recombination is therefore 1/4 + 1/4 = 1/2, or 50%. This means that independent assortment is observed for genes that are far apart in the same chromosome as well as for genes in different chromosomes.

## Chapter 18

1. No. Regression toward the mean simply indicates that, in any population, parents with extreme phenotypes (very tall or very short, for example) will tend to have offspring that are closer to the average height of the population.

2. Let's consider some possibilities, some true, some false.

   a. In the development of fingerprint ridges in the embryo, genes play a predominant role.

b. In the development of fingerprint ridges in the embryo, the environment plays only a minor role.

c. Differences in the number of fingerprint ridges among individuals are determined largely by differences in genotype.

d. Differences in the number of fingerprint ridges among individuals are almost unaffected by differences in the environment.

e. The average number of fingerprint ridges in the offspring is not much different from the average number of fingerprint ridges in the parents.

f. The average number of fingerprint ridges in the offspring is almost unrelated to the average number of fingerprint ridges in the parents.

Statements a and b are false no matter what the heritability because heritability refers to differences (variation) among individuals and not to the phenotype of any one individual. High heritability implies that statements c, d, and e are all correct. Statement f is false because such a trait would have a low heritability.

3. The easier genes to detect are the few genes with large effects. The magnitude of gene effects that can be detected depends on the number of individuals studied. The larger the number, the smaller the effects that can be detected. Whatever the size of the study, however, the genes with the largest effects are always easier to detect, even though there may be few of them.

## Chapter 19

1. Combinatorial control requires far fewer genes because there is not a one-to-one correspondence between the genes that are regulated and those that do the regulating. The products of a relatively small number of regulatory genes can be used in combination to control the expression of a far larger number of target genes. As an analogy, consider the virtually limitless combinations of sounds that can be played on a standard piano that has only 88 keys (36 black and 52 white).

2. Small regulatory RNAs are not translated into proteins (Chapter 3). As a result, they are often called noncoding RNAs since they do not encode for proteins. Their function is often to regulate the expression of other genes. By contrast, mRNAs are translated into proteins.

3. An activator turns on the expression of a gene directly by binding to DNA. An inducer binds a repressor, preventing the repressor from turning off the expression of a gene.

4. A mutation in the repressor gene that does not allow the repressor protein to bind allactose will lead to a cell that is not able to produce β-galactosidase, in the presence or absence of lactose. In other words, the lactose operon would not be inducible.

## Chapter 20

1. A cell from the inner cell mass of a blastocyst has more developmental potential than a cell from the ectoderm, which is one of the three germ layers, because the blastocyst occurs earlier in development than the formation of the three germ layers.

2. In Chapter 19, we saw that many common traits result from an interaction of genes and the environment, and so even clones can look different from each other because of environmental influences. Furthermore, if the clone is produced by nuclear transplantation, as in the cases of Dolly and CopyCat, the nuclear genomes of the clone and parent are identical, but their mitochondrial genomes are different.

3. Yes. The embryo would develop normally, even though the embryo does not have normal *bicoid* function, because *bicoid* is a maternal-effect gene, in which the mother's genotype affects the phenotype of the offspring.

4. The gap genes control the expression of the pair-rule genes, which in turn control the expression of segment-polarity genes. Therefore, if gap-gene expression is not normal, you would predict that segment-polarity gene expression would also not be normal.

5. Sexually reproducing organisms carry two copies of most genes, one from the mother and the other from the father. If either of these is knocked out by a loss-of-function mutation, the other is still present and compensates for the mutant. Hence, a loss-of-function mutation is expected to be recessive. In a gain-of-function mutation, one of the gene copies is expressed in the wrong amount, or the wrong tissue, or at the wrong time, and if such a gene turns on a developmental pathway, expression of only one copy of the gene is sufficient to turn the pathway on. Hence, a gain-of-function mutation in a gene that controls a developmental pathway is expected to be dominant.

6. Because A and B together result in petals, and B and C together result in stamens, a flower expressing B activity in all four whorls would have, from outside to inside, petals, petals, stamens, and stamens.

## Chapter 21

1. The allele frequency of *a* was calculated as follows:

$$\text{frequency}(a) = [2 \times (\text{number } aa) + 1 \times (\text{number } Aa)]/$$
$$[2 \times (\text{total number})]$$

This equation can be rewritten as

$$\text{frequency}(a) = [(\text{number } aa) + \tfrac{1}{2} \times (\text{number } Aa)] /$$
$$(\text{total number})$$

Note that

$$\text{number } aa \,/\, \text{total number} = \text{freqency}(aa)$$

and

$$\tfrac{1}{2} \times (\text{number } Aa) \,/\, \text{total number} = \tfrac{1}{2}\,\text{frequency}(Aa)$$

Therefore,

$$\text{frequency}(a) = \text{frequency}(aa) + \tfrac{1}{2}\,\text{frequency}(Aa)$$

Stated in words, the frequency of allele *a* equals the frequency of *aa* homozygotes plus half the frequency of *Aa* heterozygotes. Because we were given genotype frequencies (50% *aa*, 25% *Aa*, and 25% *AA*), we can simply substitute into the preceding equation to solve for the frequency of *a*:

$$0.50 + ½(0.25) = 0.625, \text{ or } 62.5\%$$

By similar logic,

$$\text{frequency}(A) = \text{frequency}(AA) + ½ \text{ frequency}(Aa)$$

which equals

$$0.25 + ½(0.25) = 0.375, \text{ or } 37.5\%$$

These equations are very useful for determining allele frequencies directly from genotype frequencies.

2. There is no next step. DNA sequence provides *full* genetic resolution.

3. We can conclude that the population is evolving. What we cannot tell is what mechanism—selection, migration, mutation, genetic drift, or non-random mating—is causing it to evolve. To determine what mechanism(s) is (are) driving the process requires more detailed population genetics analysis.

4. Adaptation is the fit between an organism and its environment. Of all the evolutionary mechanisms, only selection causes allele frequencies to change based on how they contribute to the success of an individual in terms of survival and reproduction. This means that allele frequencies in the next generation are ultimately governed by the environment. Because phenotype is in part determined by genotype, organisms become adapted to their environment under the influence of selection over time.

## Chapter 22

1. Species change over time, making a single definition that can be applied in all cases difficult. The biological species concept (BSC) works well in many cases and is useful in that it gives us a specific phenomenon—reproductive isolation—to study. In addition, it focuses our attention on the flow of genes between individuals of a single species. However, there are plenty of cases, including asexual organisms and organisms known only from fossils, in which it is not applicable.

2. Volcanic island chains offer many examples of adaptive radiation because they are dependent on the serendipitous process of colonization. This means that only some plants and animals are able to get there. In addition, because of the absence of competitors on these islands, there are often many available ecological opportunities for the colonizers. For example, there may be no insect-eating birds on the islands, providing an ecological niche to be filled.

3. If populations of fish became separated from other populations of fish in different ponds during dry periods, this scenario suggests that speciation was more likely to be allopatric rather than sympatric. The allopatrically evolved species became sympatric only when the lake flooded again, combining all the separate ponds into a single body of water.

## Chapter 23

1. No. The two trees are equivalent. The node shared by humans and mice is simply rotated in one tree compared to the other tree. In both trees, the closest relative of humans is mice.

2. A group called "fish" is paraphyletic because it includes some, but not all, of the descendants of a common ancestor. The descendants of the common ancestor of fish also include amphibians, sauropsids, and mammals.

3. The traits are analogous, the result of convergent evolution. Both animals are adapted for swimming in water and converged independently on similar traits, including a streamlined body and fins. However, they are only distantly related, as one is a fish and the other is a mammal.

4. Character traits can be measured for fossils as well as living organisms, so you can examine the features of the skeleton and then construct a phylogenetic tree that takes into account the skeletal traits they share with known vertebrates.

## Chapter 24

1. No. Modern humans and modern chimpanzees share a common ancestor. We are not descended from chimpanzees because chimpanzees are living today.

2. The multiregional hypothesis suggests that the most recent common ancestor of all living humans lived about 2 million years ago, when *H. ergaster* first migrated from Africa to colonize Europe and Asia. The out-of-Africa hypothesis suggests that *H. sapiens* evolved much more recently in Africa and then moved out of Africa, replacing the remnants of populations originally established by *H. ergaster* when it left Africa. The out-of-Africa hypothesis suggests a much more recent common ancestor for all present-day humans, one that lived about 200,000 years ago.

3. Studies of mtDNA, which is maternally transmitted, indicated that there was no input of Neanderthal mtDNA into the modern human population. In contrast, studies of Neanderthal genomic DNA showed that 1% to 4% of non-Africans' genomes are composed of Neanderthal DNA, implying that there was interbreeding between the ancestors of non-Africans and Neanderthals, presumably when the *H. sapiens* population first left Africa. If there is evidence of genetic input but no evidence of a female-based genetic contribution, we can hypothesize that male Neanderthals were the key contributors.

4. It is inaccurate to label *FOXP2* the "language gene" for several reasons. First, the gene is expressed in many tissues, so its effects are not limited to speech and language. Second, many genes are required for language, not just *FOXP2*. Nevertheless, its association with vocal communication in multiple species is striking.

## Chapter 25

1. The coincidence in timing between the Industrial Revolution and an increase in $CO_2$ levels—following a millennium of little change—suggests that human activities played a role in recent

changes in atmospheric composition. Alternative hypotheses would focus on other processes that add $CO_2$ to the atmosphere. Perhaps, for example, increased volcanic activity added more $CO_2$ to the atmosphere, or perhaps warming induced by changes in solar radiation caused thawing of permafrost, facilitating increased respiration of soil organic matter at high latitudes. These hypotheses can be tested against the historical record of volcanic activity and measurements of solar radiation over the past century.

2. All else being equal, increasing the elevation of mountains should increase rates of chemical weathering and erosion. As chemical weathering of continental rocks consumes $CO_2$, atmospheric $CO_2$ levels should decline.

3. Antelopes are primary consumers—that is, animals that eat vegetation. Lions are secondary consumers that eat antelopes. Because the transfer of energy from one level in the food web to the next higher level is inefficient, a given abundance of antelopes can support only a much smaller biomass of lions.

## Chapter 26

1. In prokaryotic organisms, horizontal gene transfer transports genes from one organism to another, facilitating the generation of new gene combinations.

2. Archaeal cell and genome organization are similar to those of Bacteria. It was only through the use of molecular-sequence comparisons that Archaea were shown to occupy a distinct branch on the tree of life.

3. Primary production could be accomplished by anoxygenic photosynthesis, which does not generate oxygen. Oxidation of organic matter to carbon dioxide could be accomplished by anaerobic respiration (and fermentation), which does not use oxygen.

4. It wouldn't. Bacteria and archaeons cycle nitrogen between the atmosphere (nitrogen gas) and biologically useful forms. In the absence of these prokaryotes, life as we know it would not be possible.

5. Not necessarily. Today, hyperthermophiles live in hot springs and hydrothermal ridges on the ocean floor. These might have been the environment in which the last common ancestors of these organisms thrived.

## Chapter 27

1. In general, eukaryotes employ a subset of the metabolic pathways used by bacteria. Most eukaryotes are capable of aerobic respiration; most can also gain at least some energy by fermentation; and some can also photosynthesize. In many ways, then, carbon cycling by eukaryotes is much like carbon cycling by aerobic prokaryotes. The novel contribution of eukaryotes is the ability to capture and ingest other cells, thus introducing predation into the carbon cycle.

2. No. Plants form one branch of the green algal tree, whose early branches are all occupied by unicellular forms. Similarly, the branch containing animals has choanoflagellates and other unicellular forms in its lower branches. Thus, animals and plants evolved

multicellularity independently of each other. (Note that genetic evidence further supports this phylogenetic hypothesis. Plants and animals use different sets of genes to regulate multicellular development; see Chapter 28.)

3. Molecular sequences for genes in red and green algae show the close evolutionary relationship of these organelles to free-living cyanobacteria. This supports the hypothesis that eukaryotes initially gained the ability to photosynthesize by the symbiotic incorporation of cyanobacterial cells. Gene sequences in the chloroplasts of other photosynthetic eukaryotes (brown algae, diatoms, euglenids, and other algae) are very similar to those of red and green algal chloroplasts, indicating that photosynthesis spread throughout the eukaryotic tree by the symbiotic incorporation of green or red algal cells into other eukaryotes.

## Chapter 28

1. In simple multicellular organisms, all or nearly all cells are in direct contact with the environment. In complex multicellular organisms, most cells are completely surrounded by other cells.

2. Without mechanisms for bulk transport, the movement of oxygen, nutrients, and molecular signals through organisms is limited by diffusion. Rates of diffusion, in turn, limit the size and shape that an organism can achieve. Bulk transport allows key molecules to be transported over distances much greater than those possible by diffusion alone, making larger organisms possible.

3. All multicellular organisms require molecules that promote adhesion between cells, communication between cells, and differentiation into different cell types within the body. Plants and animals have largely distinct sets of molecules for these functions. Many of the differences between plants and animals reflect the fact that plant cells have cell walls, whereas animal cells do not.

## Chapter 29

1. Plants transpire to obtain $CO_2$ from the atmosphere. Plants cannot completely seal themselves off against water loss because that would prevent them from obtaining the $CO_2$ needed for photosynthesis. As $CO_2$ diffuses into a leaf to reach the photosynthetic cells, water vapor evaporates from surfaces within the leaf and diffuses out of the leaf. Thus, transpiration is an unavoidable byproduct of acquiring $CO_2$ by diffusion from the atmosphere.

2. CAM plants open their stomata at night when rates of evaporation are low and close them during the day to conserve water. During the night, $CO_2$ that diffuses into the leaf becomes incorporated into 4-carbon organic acids that are stored in the vacuole. During the day, these organic acids are retrieved from the vacuole and broken apart, and $CO_2$ is released into the cytoplasm. Because the stomata are closed, this $CO_2$ does not diffuse out of the leaf but instead can be used in the Calvin cycle to synthesize carbohydrates. $C_4$ plants produce 4-carbon organic acids in mesophyll cells, which then diffuse through plasmodesmata into bundle-sheath cells. In the

bundle-sheath cells, each 4-carbon acid is broken up into a 3-carbon compound and $CO_2$. The $CO_2$ is used in the Calvin cycle to produce carbohydrates, while the 3-carbon compound diffuses back to the mesophyll cells, completing the $C_4$ cycle. Because the $C_4$ cycle operates much faster than the Calvin cycle, the concentration of $CO_2$ in the bundle-sheath cells is greatly increased. It is this ability to concentrate $CO_2$ that allows $C_4$ plants to suppress photorespiration.

3. The statement is correct in that the force that pulls water from the soil results from the evaporation of water from leaf cells, the energy for which comes from the heating of the leaf by the sun. It is incorrect in that there are structures within the plant that make this possible, notably the xylem conduits, and also the roots and the stem. The plant must expend energy to produce all these structures.

4. The movement of carbohydrates in phloem is from source to sink. Sources can be actively photosynthesizing leaves, or any organ with a large amount of stored carbohydrates. Carrots are examples of roots with large amounts of stored carbohydrates. In contrast, sinks are organs that require an import of carbohydrates to meet their growth and respiratory needs. Examples of sinks include young leaves, developing fruits, and most roots. Typically, the shoot is the source and the roots are a sink. At the end of winter or following a severe drought, however, a plant may regrow its shoot by drawing upon carbohydrates that had been stored in the roots. In this case, the movement of carbohydrates is from roots to shoot.

## Chapter 30

1. Spores can be dispersed by the wind. In addition, their walls contain sporopollenin, a tough, resistant covering that allows spores to withstand environmental stresses such as ultraviolet radiation and desiccation.

2. Both mosses and ferns, like all plants, exhibit an alternation of a haploid (gametophyte) generation and a diploid (sporophyte) generation. However, in mosses, the sporophyte is dependent on the gametophyte, whereas in ferns the gametophyte and sporophyte are independent of each other.

3. Pollen is the male gametophyte of seed plants that develops within the spore wall. A seed develops from an ovule, which is the female gametophyte enclosed in layers of the sporophyte.

4. Compared to spores, seeds can store more resources, slow down their metabolism, and exhibit dormancy, all of which aid their dispersal.

5. A flower consists of four whorls of organs: From outside in, they are the sepals, petals, stamens, and carpels. The sepals generally protect developing flowers; the petals commonly serve to attract pollinators; the stamens produce pollen; and the carpel protects the ovules inside that develop into seeds following fertilization.

6. A fruit forms from the ovary that encloses the seeds and, in some plants, other parts of the flower, including the petals and sepals.

## Chapter 31

1. The shoot apical meristem is the site of cell division and therefore of the growth of stems. It controls the pattern of nodes and internodes

and therefore determines the patterns of leaves. Finally, it produces new apical meristems, allowing stems to branch.

2. One of a plant's two lateral meristems produces new vascular tissues, and the other maintains an outer protective layer.

3. Roots and stems both grow from apical meristems, have regions of cell division, elongation, and differentiation, and respond to light and gravity. However, root apical meristems are covered by a root cap, have a single vascular bundle, and initiate branching from the pericycle; all these features are different from what is seen in stems.

## Chapter 32

1. No. The specific branch of the plant immune system can defend only against biotrophic pathogens, not necrotrophic ones, because this branch of the immune system is based on receptor molecules within living cells. Necrotrophic pathogens kill cells before colonizing them.

2. In the hypersensitive response, plants block off an infected area. In SAR, plants send a signal to uninfected tissue, enabling it to mount a defense and prevent infection.

3. Preventing jasmonic acid synthesis results in plants that are unable to mount a defense throughout the plant body and as a result suffer more damage from herbivores.

4. The trade-off between growth and defense creates a diversifying force that favors different species that are specialized for each habitat, rather than a single species that dominates across all soil types.

## Chapter 33

1. The green algae are paraphyletic because the last common ancestor of all green algae also gave rise to land plants; the bryophytes are paraphyletic because the last common ancestor of mosses, liverworts, and hornworts also gave rise to vascular plants. Spore-bearing vascular plants are paraphyletic because the last common ancestor of lycophytes, ferns, and horsetails also gave rise to seed plants. You can't tell by the figure, but when fossils are considered, gymnosperms are also paraphyletic because the last common ancestor of all gymnosperms also gave rise to the angiopserms.

2. An alternative explanation is that stomata evolved once, in the common ancestor of bryophytes and vascular plants.

3. Xylem vessels evolved convergently in gnetophytes and angiosperms; insect pollination evolved convergently in cycads and angiosperms; fleshy tissues to facilitate seed dispersal evolved convergently in early seed plants (reflected today by ginkgo and cycad seeds), conifers such as juniper and yew, and flowering plants.

4. Monocots produce only a single cotyledon; monocot roots are produced directly from the stem; many monocots produce creeping stems; and the strap-shaped leaves of many monocots elongate from the base and the many parallel veins connect to the stem. Because the vascular bundles of monocot stems are distributed throughout the cross section, as opposed to arranged in a ring, they do not form a vascular cambium.

## Chapter 34

1. Hyphae enable fungi to seek out new food resources; their mechanical strength allows them to penetrate large food bodies such as rotting logs or animal carcasses. In addition, hyphae transport nutrients from one part of the fungus to another.

2. A diploid cell has a single nucleus with two complete sets of chromosomes; a dikaryotic cell has two genetically distinct haploid nuclei.

3. Complex multicellular fruiting bodies evolved independently on two branches, the Basidiomycetes and the Ascomycetes.

4. You are eating the multicellular fruiting body built from dikaryotic hyphae.

## Chapter 35

1. Nerve signals are transmitted electrically (in the form of an action potential) from one end of a neuron to the other. They are transmitted chemically (by neurotransmitters) across a synapse from one neuron to another.

2. Nerve impulses are slowed or completely stopped in response to loss of myelin surrounding nerve cells.

3. Nerve gases that block acetylcholinesterase, the enzyme that inactivates acetylcholine, cause violent muscle contractions, leading to paralysis of the diaphragm (the main muscle involved in breathing) and death by asphyxiation.

## Chapter 36

1. Prostaglandins depolarize nociceptors and increase their firing rate.

2. Excess mucus secretion blocks airborne odor molecules from reaching the chemosensory hairs of olfactory neurons, reducing their ability to bind odorants and therefore reducing the sense of smell. Taste is also affected because much of the sense of taste depends on smell.

3. Spinning causes the hair cells in the semicircular canals to accommodate to the new motion. When you stop spinning, the hair cells detect a change in angular motion of the head (relative to the motion when your head was spinning), which leads to a sense of imbalance when you stand or try to walk.

4. Rod cells are extremely sensitive to light, allowing an animal to see in dimly lit conditions. Cone cells provide color vision through photopigments with different light wavelength absorption peaks, and they provide high spatial acuity, and thus sharp vision.

5. The three areas of the brain that are used to remember and recognize a face are the occipital cortex for processing visual information, the hippocampus for forming long-term memory of a face, and the temporal lobe for facial recognition.

## Chapter 37

1. Without newly synthesized ATP, myosin cross-bridges formed with actin cannot detach from their actin binding sites, so they remain in the bound state and make the muscle stiff.

2. Because acetylcholine is blocked and cannot depolarize the muscle cell membrane, calcium is not released from the sarcoplasmic reticulum, and the muscle is unable to contract.

3. By contracting at an intermediate length, actin and myosin filament overlap of the biceps fibers is maximized, allowing the greatest number of cross-bridges to form and produce force to lift a weight.

4. Glass is composed of a single material (silica), which is rigid but can crack easily. Bone is a composite material, built of crystalline mineral reinforced by the protein collagen, enabling bone to absorb much more energy before breaking.

## Chapter 38

1. Diabetes can result either from decreased insulin production by the pancreas (type 1 diabetes) or decreased effect of insulin on target cells (type 2 diabetes). Type 1 diabetes is an autoimmune disease in which the insulin-producing cells in the pancreas are attacked by the immune system. The result is inadequate insulin secretion. Daily injections of insulin can maintain stable normal circulating levels of blood glucose. In individuals with type 2 diabetes, insulin production is unaffected, but cells are not able to respond to normal circulating levels of insulin. Type 2 diabetes is commonly linked to obesity and is most effectively treated by proper diet and exercise.

2. Peptide hormones bind cell membrane receptors because they are hydrophilic and cannot diffuse across the cell membrane, activating intracellular second messengers that stimulate signaling cascades within the target cell. Steroid hormones diffuse across the cell membrane and bind intracellular cytoplasmic or nuclear receptors, leading to changes in gene expression and protein synthesis within the target cell.

3. Hormones are secreted by a cell or gland and transported in the bloodstream to distant sites where they exert their effect. Hormones achieve specificity by binding to receptors, which are present only on their target cells.

4. Paracrine signals are chemicals secreted by a cell that influence the activity of neighboring cells without entering the bloodstream. Pheromones are chemicals secreted into the environment and transmitted to other species members, influencing their behavior. Hormones are released into the bloodstream and have effects on distant cells.

## Chapter 39

1. Diffusion is increased by a large surface area for exchange and a short diffusion distance.

2. In individuals with asthma, resistance to airflow increases, thereby decreasing the flow of air. Resistance increases because the diameter of the airways decreases as a result of smooth muscle contraction in the airway walls, as well as mucus secretion.

3. Air has a low density and viscosity, and a high $O_2$ content relative to water.

4. Sickle-cell anemia is life threatening because transport of the deformed sickle-shaped red blood cells is limited in small blood vessels. Therefore, the supply of $O_2$ to tissues is reduced and cellular metabolism in the tissues supplied by the blood vessels is disrupted, possibly leading to organ failure.

5. Change in a vessel's diameter (or radius) affects blood flow most. For example, a twofold reduction in radius increases resistance 16 times.

6. The pressure in the ventricle exceeds the pressure in the atrium.

## Chapter 40

1. Glycogen stores are used first, then fat, and finally muscle protein.

2. No. Metabolic rate increases with increasing mass, but the relationship between the two is not linear. Instead, metabolic rate increases with mass raised to the ¾ power. This means that a dog that is twice as heavy as a cat, for example, has a metabolic rate that is less than twice as much as the cat.

3. "Essential," when used to describe dietary amino acids, minerals, or vitamins, does not mean that other nutrients are less important or useful. "Essential" in this context means that these nutrients must be obtained in the diet—the organism cannot synthesize them on its own. In other words, it is "essential" that these nutrients are consumed.

4. Rocks and sediments in gizzards break apart food, as teeth do in vertebrates.

5. By eating the products of the cecum, rabbits can pass the partially digested food by the small intestine a second time, where nutrients are then absorbed.

## Chapter 41

1. When a cell is placed in a hypotonic solution, water moves into the cell by osmosis and the cell swells or bursts. When a cell is placed in a hypertonic solution, water leaves the cell by osmosis and the cell shrinks.

2. Osmosis depends on the difference in *total* solute concentration on the two sides of a membrane. There is no net movement of water molecules across a semipermeable membrane with a high concentration of a *particular* solute on one side of a membrane if this concentration is matched by a similar concentration of some other solute on the other side of the membrane. For example, the concentration of sodium ions may be high on one side of a semipermeable membrane and low on the other, whereas the distribution of potassium ions may be just the reverse, so the total solute concentration is the same on the two sides of the membrane. The result is no net water movement.

3. Ammonia excreted by fish is rapidly diluted to nontoxic levels in the surrounding water. Mammals convert ammonia to urea, which is much less toxic than ammonia and therefore can be concentrated and stored before being eliminated.

4. Although the filtrate is dilute at the start and end of the loop of Henle, the loop creates a concentration gradient from the cortex to the medulla, which is important for subsequent water reabsorption from the collecting ducts. The loops also leave urea as the main solute.

## Chapter 42

1. Sexual reproduction generates genetically unique offspring by (1) chance mutation; (2) recombination between homologous chromosomes during meiosis I; (3) random segregation of homologous chromosomes during meiosis I; and (4) new combinations of chromosomes in the process of fertilization. Asexual reproduction, like sexual reproduction, is also subject to chance mutations.

2. The vas deferens carries sperm from the epididymis to the ejaculatory duct. Cutting the vas deferens on both sides prevents sperm from entering the ejaculatory duct and leads to male sterility. This procedure, called a vasectomy, is a method of male contraception.

3. Estrogens or progesterones (or a combination of these) in the pill suppress GnRH, FSH, and LH and therefore block oocyte development and ovulation.

4. In females, oogonia differentiate into primary oocytes before birth and therefore a female is born with all the primary oocytes she will have. In males, by contrast, spermatogonia do not differentiate into primary spermatocytes until puberty, but sperm can be produced throughout a male's lifetime. In addition, only one of the four products of meiosis becomes a gamete in females, whereas all four products of meiosis become sperm in males.

5. In the process of development, cells differentiate into different cell types. The process of differentiation involves the turning on and turning off of specific genes. So, for example, genes that are expressed in a skin cell are different from those that are expressed in a liver or muscle cell. These changes in gene expression, once established, are for the most part stable through mitotic cell divisions.

## Chapter 43

1. Damage to cilia in the respiratory tract increases the probability of lung infection by impairing the ability of the respiratory tract to sweep out microorganisms.

2. A second novel antigen injected on day 40 will produce a primary response, not a secondary response, because it is a novel antigen that has not been encountered before by the immune system.

3. Although a given B cell produces only one antibody, the entire population of B cells produces many different antibodies. Each B cell produces a different antibody as the result of a unique pattern of genomic rearrangement.

4. T cells are activated when their surface T cell receptors bind to an antigen in association with an MHC molecule. By contrast, B cells are activated when their surface antibodies bind to a free antigen.

## Chapter 44

1. Neither Cnidaria nor Bilateria is more closely related to sponges. They share a common ancestor that is the closest relative of sponges.

2. A jellyfish would use more available nutrients than a sponge of the same size because it can move toward nutrients and capture and digest larger food items.

3. The shared features among snails, squids, and clams suggest that all mollusks descended from a common ancestor that had a head, radula, foot, mantle, and well-differentiated organ systems.

4. Young humans and other animals often have features that they lose as they age. Our first "baby" teeth are an example. If you could look at your own photograph when you were a one-month-old, inch-long embryo, you would see a notochord, gill slits, and a tail.

5. Scales are found on all of these animals due to their presence in a common ancestor. Even feathers are modified scales.

## Chapter 45

1. In fruit flies, there are two alleles of the *for* gene that are expressed in larvae, leading to two different phenotypes in the population. As a result, larvae are either sitters or rovers. By contrast, in honeybees, the *for* gene is up-regulated during development, so a single individual begins by staying in the hive and then becomes a forager.

2. In classical conditioning, one stimulus is exchanged for another (the smell of food is exchanged for the sound of a bell in the case of Pavlov's dogs). As a result, a novel stimulus evokes a behavior. In operant conditioning, a behavior is associated with a reward or punishment, so the behavior becomes more likely (with a reward) or less likely (with a punishment). Therefore, in the case of classical conditioning, an association is made between a stimulus and a behavior, and in operant conditioning, an association is made between a behavior and a response.

3. Photoperiod is a good marker for time of year, and therefore for season. Many species have season-specific behaviors—hibernation in bears, migration in birds—that must be timed appropriately.

4. Organisms do not act "for the good of the species" because natural selection operates on individuals: It is the individual that lives or dies, reproduces or fails to reproduce. Traits that are disadvantageous to the individual are therefore selected against by natural selection even if they are beneficial to the species as a whole.

5. With reference to Fig. 45.17, the probability of each offspring inheriting an $A$ allele is 0.5. The probability of a cousin sharing an $A$ allele is 0.25, so the probability of each of the animal's cousin's offspring having an $A$ is $0.25 \times 0.5 = 0.125$. In this case, the animal would have to substitute four of its cousin's offspring for each of its own.

## Chapter 46

1. The per capita growth rate, $r$, equals $r = (\Delta N/\Delta t)/N_1$. In this case, the change in time $\Delta t$ is 1 year, so $r = \Delta N/N_1$. If the starting population is $x$, at triple the size it is $3x$. Therefore, $\Delta N$ is $3x - x$, or $2x$, and $r = 2x/x$, or 2. In general, when a population triples in size in a year, it grows by 200%, so the per capita growth rate is 2.

2. One cent per day, doubled each day (a rate of increase of 100%) for 30 days exceeds $10,000,000.

3. Logistic curves are S shaped because exponential increase during early stages of population growth causes the first half of the curve to rise steeply, but then growth becomes limited by resource availability and so slows as population size approaches the carrying capacity of the environment.

4. Every organism invests all available energy in the ways that make most sense for the current environment. Increase in one area means decrease in another. To increase investments in all these areas, more resources in terms of food or energy would have to be acquired.

5. The closer island will initially receive more colonists than the farther one, but the larger island can eventually support more kinds of species.

## Chapter 47

1. The fundamental niche is larger than the realized niche because the fundamental niche represents all the habitats where a species could possibly live, but competition among different species usually restricts a species to a smaller area, which is the realized niche.

2. The cost to the apple tree is resources in the form of nectar, which takes energy to produce, and the benefit is pollination. The cost to the bee is the energy required to search for food and transfer pollen, and the benefit is food.

3. Although they can affect many other species in the Arctic, lemmings are not a keystone species because their effects are a function of their great abundance. Keystone species have effects on communities that are disproportionate to their numbers because of the particular roles they play, often as predators.

4. A community is a group of organisms that share a given place. A community together with the physical environment in which the organisms live constitutes an ecosystem.

5. A diversity gradient can be seen with altitude that is similar to the one seen with latitude: Typically, the higher the altitude, the fewer the number of species.

## Chapter 48

1. In the 130 years from 1800 to 1930, our numbers doubled from 1 to 2 billion, and they doubled again in just the 45 years from 1930 to 1975. This pattern indicates that the *rate* of human population growth increased in this time.

2. Global warming is the measured increase in Earth surface temperatures over the past 50 years. The greenhouse effect describes a process by which global warming can occur. The greenhouse effect is the result of the capacity of some molecules—especially carbon dioxide, methane, and water vapor—to absorb heat energy and then emit it in all directions. As a result, the energy incoming from solar radiation exceeds the amount radiated back into space as heat. Without the greenhouse effect, Earth would not be habitable. However, in recent decades, increasing

amounts of greenhouse gases in the atmosphere have resulted in global warming.

3. As ecosystems warm, plants able to open flowers earlier appear to have higher fitness, enabling them to maintain or increase their geographic ranges.

4. Invasive species may outcompete native species for available resources, diminishing the population size and, in time, the diversity of native species. Introduced animals may prey on native species, reducing their numbers.

5. The Red Queen hypothesis suggests that organisms need to keep evolving (running) just to stay in the same ecological niche (place). As new antibiotics are developed, pathogens like malaria frequently evolve resistance, so researchers must continually strive to develop new medicines as natural selection erodes the effectiveness of the old.

# GLOSSARY

**10-nm chromatin fiber**  A relaxed 30-nm chromatin fiber, the state of the chromatin fiber in regions of the nucleus where transcription is currently taking place.

**3′ end**  The end of a nucleic acid strand that carries a free 3′ hydroxyl.

**30-nm chromatin fiber**  A chromosomal conformation created by the folding of the nucleosome fiber of DNA and histones.

**3-phosphoglycerate (3-PGA)**  A 3-carbon molecule; two molecules of 3-PGA are the first stable products of the Calvin cycle.

**5′ cap**  The modification of the 5′ end of the primary transcript by the addition of a special nucleotide attached in an unusual chemical linkage.

**5′ end**  The end of a nucleic acid strand containing a free 5′ phosphate group.

**ABC model**  A model of floral development that invokes three activities, A, B, and C, each of which represents the function of a protein or proteins hypothesized to be present in the cells of each whorl.

**abomasum**  The fourth chamber in the stomach of ruminants, where protein digestion begins.

**abscisic acid**  A hormone that stimulates root elongation by suppressing ethylene synthesis.

**absolute temperature (T)**  Temperature measured on the Kelvin scale.

**accessory pigment**  A pigment other than chlorophyll in the thylakoid membrane; carotenoids are important accessory pigments.

**acidic**  Describes a solution in which the concentration of protons is higher than that of hydroxide ions (the pH is lower than 7).

**acoelomate**  A bilaterian without a body cavity.

**acrosome**  An organelle that surrounds the head of the sperm containing enzymes that enable sperm to transverse the outer coating of the egg.

**actin**  A protein subunit that makes up microfilaments; used by both striated and smooth muscles to contract and generate force.

**action potential**  A brief electrical signal transmitted from the nerve cell body along one or more axon branches.

**activated**  The state of the receptor after binding the signaling molecule; the activated receptor transmits the information through the cytoplasm of the cell.

**activation energy ($E_A$)**  The energy input necessary to reach the transition state.

**activator**  A synthesized compound that increases the activity of an enzyme.

**active site**  The portion of the enzyme that binds substrate and converts it to product.

**active transport**  The "uphill" movement of substances against a concentration gradient.

**acute phase response**  A change in the level of proteins in the blood, which is a marker for the initial phase of a disease.

**adaptation**  In sensory reception, the process in which sensory receptors reduce their firing rate when a stimulus continues over a period of time.

**adapted**  Better able to survive and reproduce in a given environment.

**adaptive (acquired) immunity**  The part of the immune system that is specific to given pathogens.

**adaptive radiation**  A bout of unusually rapid evolutionary diversification in which natural selection accelerates the rates of both speciation and adaptation.

**addition rule**  The principle that the probability of either of two mutually exclusive outcomes occurring is given by the sum of their individual probabilities.

**adenine (A)**  A purine base.

**adenosine triphosphate (ATP)**  The molecule that provides energy in a form that all cells can readily use to perform the work of the cell. ATP is the universal energy currency for all cells.

**adherens junction**  A beltlike junctional complex composed of cadherins that attaches a band of actin to the plasma membrane.

**adrenal glands**  Paired glands located adjacent to the kidneys that secrete cortisol in times of stress.

**adrenal medulla**  The inner part of the adrenal gland, which is stimulated by the sympathetic nervous system.

**adrenaline (epinephrine)**  A hormone released by the adrenal gland that causes alertness and arousal.

**advantageous**  Describes mutations that improve their carriers' chances of survival or reproduction.

**advertisement display**  Behavior by which individuals draw attention to their status.

**aerobic**  Utilizing oxygen.

**aerobic metabolism**  Metabolism in the presence of oxygen in the mitochondria of eukaryotic animals that provides a steady supply of ATP for long-term, sustainable activity.

**afferent neuron**  A neuron that sends information toward the central nervous system.

**age structure**  The number of individuals within each age group of the population studied.

**agonist muscles**  Muscle pairs that combine to produce similar motions.

**aldose**  A monosaccharide with an aldehyde group.

**aldosterone**  A hormone produced by the adrenal glands that stimulates the distal convoluted tubule and collecting ducts to take up more salt and water.

**alga (plural, algae)**  A photosynthetic protist.

**alkaloid**    Any one of a group of nitrogen-bearing compounds that damages the nervous system of animals, produced by some plants as a defensive mechanism.

**allantois**    In the amniotic egg, a membrane that encloses a space where metabolic wastes collect.

**allele frequency**    The rate of occurrence of an allele in populations.

**alleles**    The different forms of a gene, corresponding to different DNA sequences in each different form.

**allopatric**    Describes populations that are geographically separated from each other.

**allosteric effect**    A change in the activity or affinity of a protein as the result of binding of a molecule to a site other than the active site.

**allosteric enzyme**    An enzyme whose activity is affected by binding a molecule at a site other than the active site. Typically, allosteric enzymes change their shape on binding an activator or inhibitor.

**alpha (α) carbon**    The central carbon atom of each amino acid.

**alpha (α) helix**    One of the two principal types of secondary structure found in proteins.

**alternation of generations**    The life cycle in which a haploid gametophyte and a diploid sporophyte follow one after the other.

**alternative splicing**    A process in which primary transcripts from the same gene can be spliced in different ways to yield different mRNAs and therefore different protein products.

**altruistic**    Describes an act of self-sacrifice.

**Alveolata**    A eukaryotic superkingdom, defined by the presence of cortical alveoli, small vesicles that, in some species, store calcium ions.

**alveolus** (plural, **alveoli**)    A cluster of tiny thin-walled sacs where gas exchange by diffusion takes place, found at the ends of bronchioles.

**amacrine cell**    A type of interneuron in the retina that communicates between neighboring bipolar cells and ganglion cells, enhancing motion detection and adjusting for changes in illumination of the visual scene.

**amine hormone**    A hormones that is derived from a single aromatic amino acid, such as tyrosine.

**amino acid replacement**    A change in the identity of an amino acid at a particular site in a protein resulting from a mutation in the gene.

**amino acid**    An organic molecule containing a central carbon atom, a carboxyl group, an amino group, a hydrogen atom, and a side chain. Amino acids are the building blocks of proteins.

**amino end**    The end of a polypeptide chain that has a free amino group.

**amino group**    $NH_2$; a nitrogen atom bonded to two hydrogen atoms, covalently linked to the central carbon atom of an amino acid.

**aminoacyl (A) site**    One of three binding sites for tRNA on the large subunit of a ribosome.

**aminoacyl tRNA synthetase**    An enzyme that attaches a specific amino acid to a specific tRNA molecule.

**amnion**    In the amniotic egg, a membrane surrounding a fluid-filled cavity that allows the embryo to develop in a watery environment.

**amniotes**    The group of vertebrate animals that produces amniotic eggs; this group includes lizards, snakes, turtles, and crocodilians.

**amniotic egg**    An egg that can exchange gases while retaining water, permitting amniotes to live in dry terrestrial habitats that amphibian eggs cannot tolerate.

**Amoebozoa**    A superkingdom of eukaryotes with amoeba-like cells that move and gather food by means of pseudopodia.

**Amphibia**    Species of vertebrates, including frogs and salamanders, with an aquatic larval form with gills and an adult terrestrial form that usually has lungs.

**amphipathic**    Having both hydrophilic and hydrophobic regions.

**amplified**    In PCR technology, an alternative term for "replicated."

**amylase**    An enzyme in salivary secretions that breaks down carbohydrates to begin the digestion of sugars and starches.

**anabolism**    The set of chemical reactions that build molecules from smaller units utilizing an input of energy, usually in the form of ATP. Anabolic reactions result in net energy storage within cells and the organism.

**anaerobic metabolism**    Metabolism in the absence of oxygen, which produces ATP that provides rapid but short-term energy to the cell and the organism.

**analogous**    Describes similar characters that evolved independently in different organisms as a result of adaptation to similar environments.

**anammox**    Anaerobic ammonia oxidation; the oxidation of ammonium ion to nitrogen gas and the reduction of nitrite to nitrogen gas, with water produced as a by-product.

**anaphase**    The stage of mitosis in which the sister chromatids separate.

**anaphase I**    The stage of meiosis I in which the two homologous chromosomes of each bivalent separate as they are pulled in opposite directions, but the sister chromatids remained joined at the centromere.

**anaphase II**    The stage of meiosis II in which the centromere of each chromosome splits and the separated chromatids are pulled toward opposite poles of the spindle.

**anchor**    A membrane protein that attaches to other proteins and helps to maintain cell structure and shape.

**angiosperms**    The flowering plants; angiosperms are one of the two monophyletic groups of vascular plants.

**angiotensin II**    The active form of the hormone angiotensinogen, which causes the smooth muscles of arterioles throughout the body to constrict, thus increasing blood pressure and directing more blood back to the heart.

**annealing**    The coming together of complementary strands of single-stranded nucleic acids by base pairing.

**annelid worms**    A phylum of worms that have a cylindrical body with distinct segments and a bilaterian body plan.

**annual clock**    A year-based biological clock.

**anoxygenic** Not producing oxygen; anoxygenic photosynthetic bacteria do not gain electrons from water and so do not generate oxygen gas.

**antagonism** A one-sided interaction between species in which at least one participant loses more than it gains.

**antagonist muscles** Muscle pairs that pull in opposite directions.

**anterior pituitary gland** The region of the pituitary gland that forms from epithelial cells that develop and push up from the roof of the mouth; it receives hormones from the hypothalamus that stimulate it to release hormones in turn.

**anther** A structure supported by a filament from the stamen that contains several sporangia in which pollen is produced.

**Anthropocene Period** The modern era, so named to reflect the dominant impact of humans on Earth.

**antibody** A large protein that carries sugar molecules attached to some amino acids that binds to foreign molecules occurring naturally on or in microorganisms that participate in normal cellular functions.

**antidiuretic hormone (ADH)** A posterior pituitary gland hormone that acts on the kidneys and controls the water permeability of the collecting ducts, thus regulating the concentration of urine that an animal excretes; also called *vasopressin*.

**antigen** A foreign molecule that binds to an antibody and leads to the production of more antibodies.

**antigenic drift** The gradual process in which a high rate of mutation leads to changes in the amino acid sequences of antigens, thus allowing a population of viruses to evolve over time and evade memory T and B cells.

**antigenic shift** Reassortment of RNA strands in the viral genome, leading to sudden changes in cell-surface proteins, thus making it difficult to predict from year to year which virus strains will be most prevalent and therefore what vaccine will be most effective.

**antigenic variation** The encoding of a protein at different times by any one of a number of different genes.

**antigen-presenting cell** A type of cell (including macrophages, dendritic cells, and B cells) that takes up an antigen and returns portions of it to the cell surface bound to MHC class II proteins.

**antiparallel** Oriented in opposite directions; the strands in a DNA duplex are antiparallel.

**aorta** A large artery through which oxygenated blood flows from the left ventricle to the head and rest of the body.

**aortic body** A sensory structure that monitors the levels of oxygen and protons moving to the body.

**aortic valve** A valve beween the left ventricle and the aorta.

**apical dominance** The suppression of growth of axillary buds by the shoot apical meristem.

**appendicular** Describes the part of the vertebrate skeleton that consists of the bones of the limbs, including the shoulder and pelvis.

**appendix** A narrow, tubelike structure that extends from the cecum. The appendix is a vestigial structure that has no clear function in nonherbivorous animals.

**aquaporin** A protein channel that allows water to flow through the plasma membrane more readily by facilitated diffusion.

**aqueous** Watery.

**aqueous humor** A clear watery liquid that fills the interior region in front of the lens of the vertebrate eye.

**Archaea** One of the three domains of life, consisting of single-celled organisms without true chromosomes or a nucleus that divide by binary fission, differing from bacteria in many aspects of their cell and molecular biology.

**Archaeplastida** A eukaryotic superkingdom of photosynthetic organisms; includes the land plants.

**Ardi** A specimen of *Ardipethecus ramidus,* an important early hominin, dating from about 4.4 million years ago.

**arteriole** A small branch of an artery.

**artery** A large, high-pressure vessel that moves blood flow away from the heart to the tissues.

**Arthropoda** An animal group that includes insects and contains more than half of all known animal species; distinguished by their jointed legs.

**artificial selection** A form of directional selection analogous to natural selection, but without the competitive element; successful genotypes are determined by the breeder, not by competition.

**ascomycetes** A monophyletic fungal subgroup of the Dikarya, making up 64% of all fungal species, in which nuclear fusion and meiosis take place in an elongated saclike cell called an ascus; also called *sac fungi.*

**asexual reproduction** The reproduction of cells or single-celled organisms by cell division; offspring are clones of the parent.

**assimilation** The process in which plants take up sulfate ions from the soil and reduce them within their cells to hydrogen sulfide that can be incorporated into cysteine and other biomolecules.

**associative learning (conditioning)** Learning that two events are correlated.

**astrocyte** A type of star-shaped glial cell that contributes to the blood–brain barrier by surrounding blood vessels in the brain and thus limiting the size of compounds that can diffuse from the blood into the brain.

**atom** The basic unit of matter.

**atomic mass** The mass of the atom determined by the number of protons and neutrons.

**ATP synthase** An enzyme that couples the movement of protons through the enzyme with the synthesis of ATP.

**atrioventricular (AV) node** A specialized region of the heart containing pacemaker cells that transmit action potentials from the sinoatrial nodes to the ventricles of the heart.

**atrioventricular (AV) valve** A valve between the right atrium and the right ventricle and between the left ventricle and left atrium.

**atrium**   A heart chamber that receives blood from the lungs or the rest of the body.

**auditory cortex**   The area of the brain that processes sound.

**autocrine signaling**   Signaling between different parts of a cell; the signaling cell and the responding cell are one and the same.

**autoimmune disease**   A disease in which tolerance is lost and the immune system becomes active against antigens of the host.

**autonomic nervous system**   The involuntary component of the peripheral nervous system, which controls internal functions of the body such as heart rate, blood flow, digestion, excretion, and temperature.

**autosome**   Any chromosome other than the sex chromosomes.

**autotroph**   Any organism that is able to convert carbon dioxide into glucose, thus making its own organic source of carbon.

**auxin**   The plant hormone that causes shoots to elongate and guides vascular differentiation.

**avirulent**   Describes pathogens that damage only a small part of the plant because the host plant is able to contain the infection.

**axial**   Describes the part of the vertebrate skeleton that consists of the skull and jaws of the head, the vertebrae of the spinal column, and the ribs.

**axillary bud**   A meristem that forms at the base of each leaf.

**axon**   The fiberlike extension from the cell body of a neuron that transmits signals away from the nerve's cell body; the output end of a nerve cell.

**axon hillock**   The junction of the nerve cell body and its axon.

**B lymphocyte (B cell)**   A cell type that matures in the bone marrow of humans and produces antibodies.

**Bacteria**   One of the three domains of life, consisting of single-celled organisms without true chromosomes or a nucleus that divide by binary fission, differing from archaeons in many aspects of their cell and molecular biology.

**bacteriochlorophyll**   A light-harvesting pigment closely related to the chlorophyll found in plants, algae, and cyanobacteria.

**bacteriophage**   Virus that infects bacterial cells.

**balancing selection**   Natural selection that acts to maintain two or more alleles of a given gene in a population.

**ball-and-socket joint**   A joint that allows rotation in three axes, like the hip and shoulder.

**Baltimore system**   A classification of viruses into seven major groups, I–VII, by type of nucleic acid and method of mRNA synthesis; developed by David Baltimore.

**band**   A crosswise striation in a chromosome, or a horizontal stripe in an electrophoresis gel.

**basal lamina**   A specialized form of extracellular matrix that underlies and supports all epithelial tissues.

**base**   A nitrogen-containing compound that makes up part of a nucleotide.

**base excision pair**   A specialized repair system in which an improper DNA base and its sugar are both removed and the resulting gap is repaired.

**base stacking**   Stabilizing interactions between bases in the same strand of DNA.

**basic**   Describes a solution in which the concentration of protons is lower than that of hydroxide ions (the pH is higher than 7).

**basidiomycetes**   A monophyletic fungal subgroup of the Dikarya, including smuts, rusts, and mushrooms, in which nuclear fusion and meiosis take place in a club-shaped cell called a basidium; also called club fungi.

**basilar membrane**   The membrane that, with the cochlear duct, separates the upper and lower canal of the cochlea.

**basophil**   A type of granulocyte that, along with eosinphils, defends against parasitic infections but also contributes to allergies.

**behaviorally isolated**   Describes individuals that only mate with other individuals on the basis of specific courtship rituals, songs, and other behaviors.

**beta (β) sheet**   One of the two principal types of secondary structure found in proteins.

**beta-(β-)oxidation**   The process of shortening fatty acids by a series of reactions that sequentially remove two carbon units from their ends.

**bilateral symmetry**   Symmetry on both sides of a midline; animals with bilateral symmetry have a distinct head and tail, marking front and back, with a single plane of symmetry running between them at the midline.

**bilateria**   The branch of the animal tree that includes animals with bilateral symmetry.

**bilayer**   A two-layered structure of the cell membrane with hydrophilic "heads" pointing outward toward the aqueous environment and hydrophobic "tails" oriented inward, away from water.

**bile**   A fluid produced by the liver that aids in fat digestion by breaking large clusters of fats into smaller lipid droplets.

**binary fission**   The process by which cells of bacteria or archaeons divide.

**binding affinity**   The tightness of the binding between the receptor and the signaling molecule.

**biological species concept (BSC)**   As described by Ernst Mayr, the concept that "[s]pecies are groups of actually or potentially interbreeding populations that are reproductively isolated from other such groups." The BSC is the most widely used and accepted definition of a species.

**biologist**   A scientist who studies life.

**biology**   The science of life and how it works.

**biomass**   The total mass of organisms in a given area.

**biome**   The distinctive and stable assemblage of species found over a broad region of Earth; terrestrial biomes are recognized by their vegetation.

**biomineralization**    The precipitation of minerals by organisms, as in the formation of skeletons.

**biotrophic pathogen**    A plant pathogen that obtains resources from living cells.

**biparental inheritance**    A type of inheritance in which the organelles in the offspring cells derive from those in both parents.

**bipedal**    Moving by two feet and habitually walking upright.

**bipolar cell**    A type of interneuron in the retina that adjusts its release of neurotransmitter in response to input from rod and cone cells.

**bivalent**    The four-stranded structure consisting of two pairs of sister chromatids aligned along their length and held together by chiasmata.

**bivalves**    A group of mollusks that includes clams, oysters, and mussels; they have evolved a skeleton in which two hard shells are connected by a flexible hinge.

**bladder**    A hollow organ in mammals and fishes for the storage and elimination of urine.

**blastocyst**    A hollow sphere produced by cells in the morula that move in relation to one another, pushing against and expanding the membrane that encloses them. A blastocyst forms from the blastula, has an inner cell mass, and occurs only in mammals.

**blastula**    A fluid-filled ball of undifferentiated cells formed after the fertilized egg has undergone several rounds of mitotic cell division following the morula stage.

**blending inheritance**    The now-discredited model in which heredity factors transmitted by the parents become intermingled in the offspring instead of retaining their individual genetic identities.

**blood**    The circulatory fluid in vertebrates.

**bone marrow**    A fatty tissue between trabeculae and within the central cavity of a bone that contains many important cell populations.

**bottleneck**    An extreme case of genetic drift that occurs when a population falls to just a few individuals.

**Bowman's capsule**    A membranous sac that encases the glomerulus.

**brain**    The centralized concentration of neurons in an organ that processes complex sensory stimuli from the environment or from anywhere in the body.

**brainstem**    Part of the vertebrate brain, formed from the midbrain, which activates the forebrain by relaying information from lower spinal levels.

**bronchiole**    Any one of the fine branches of secondary bronchi.

**bryophytes**    A paraphyletic group of plants that includes the mosses, liverworts, and hornworts.

**buccal cavity**    The mouth.

**bud scale**    One of many like structures formed from leaf primordia that, together, protect shoot apical meristems from desiccation and damage due to cold.

**budding**    A form of asexual reproduction in fungi, plants, and some animals in which a bud forms on the organism and eventually breaks off to form a new organism that is smaller than its parent.

**bulbourethral glands**    Glands below the prostate gland that produce a clear fluid that lubricates the urethra for passage of the sperm.

**bulk flow**    The movement of molecules through organisms in the active circulation of fluids at rates beyond those possible by diffusion across a concentration gradient; also known as *bulk transport*.

**bulk transport**    The movement of molecules through organisms in the active circulation of fluids at rates beyond those possible by diffusion across a concentration gradient; also known as *bulk flow*.

**bundle sheath**    A cylinder of cells that surrounds each vein in $C_4$ plants in which carbon dioxide is concentrated, suppressing photorespiration.

**Burgess Shale**    A sedimentary rock formation on the seafloor covering what is now British Columbia, Canada, that preserves a remarkable sampling of marine life during the initial diversification of animals.

**$C_3$ plant**    A plant that does not use 4-carbon organic acids to supply the Calvin cycle with carbon dioxide.

**$C_4$ plant**    A plant that suppresses photorespiration by increasing the concentration of carbon dioxide in the immediate vicinity of rubisco.

**cadherin**    A calcium-dependent adherence protein, important in the adhesion of cells to other cells.

**calmodulin**    A protein that binds with $Ca^{2+}$ and activates the enzyme myosin kinase.

**Calvin cycle**    The process in which carbon dioxide is reduced to synthesize carbohydrates, with ATP and NADPH as the energy sources.

**Cambrian explosion**    A transition period in geologic time during which the body plans characteristic of most bilaterian phyla developed.

**cancer**    A group of diseases characterized by uncontrolled cell division.

**canine**    One of the teeth in carnivores specialized for piercing the body of prey.

**capacitation**    A series of physiological changes that allow the sperm to fertilize the egg.

**capillary**    A very small blood vessel, arranged in finely branched networks connected to arterioles or venules, where gases are exchanged by diffusion with surrounding tissues.

**capsid**    The protein coat that surrounds the nucleic acid of a virus.

**carbohydrate**    An organic molecule containing C, H, and O atoms that provides a source of energy for metabolism and that makes up the cell wall in bacteria, plants, and algae.

**carbon cycle**    The intricately linked network of biological and physical processes that shuttles carbon among rocks, soil, oceans, air, and organisms.

**carboxyl end**    The end of a polypeptide chain that has a free carboxyl group.

**carboxyl group**    COOH; a carbon atom with a double bond to oxygen and a single bond to a hydroxyl group.

**carboxylation**    The first step of the Calvin cycle, in which carbon dioxide absorbed from the air is added to a 5-carbon molecule.

**cardiac cycle**   The contraction of the two atria of the heart followed by contraction of the two ventricles.

**cardiac muscle**   Muscle cells that make up the walls of the atria and ventricles and contract to pump blood through the heart.

**cardiac output (CO)**   The volume of blood pumped by the heart over a given interval of time, the key measure of heart function.

**caroid body**   A sensory structure that senses oxygen and proton concentrations of the blood moving to the brain.

**carpel**   An ovule-producing floral organ in the center whorl.

**carrier**   A transporter that facilitates movement of molecules.

**carrying capacity (K)**   The maximum number of individuals a habitat can support.

**cartilage**   Fluid-filled tissue that forms the surfaces between adjacent bones, cushioning forces transmitted across the joint.

**Casparian strip**   A thin band of hydrophobic material that encircles each cell of the endodermis of a root, controlling which materials enter the xylem.

**catabolism**   The set of chemical reactions that break down molecules into smaller units and, in the process, produces ATP to meet the energy needs of the cell.

**causation**   A relationship in which one event leads to another.

**cavitation**   The abrupt replacement of the water in a conduit by water vapor, which blocks water flow in xylem.

**cecum**   A chamber that branches off the large intestine; along with the colon, the site of hindgut fermentation.

**cell**   The simplest self-replicating entity that can exist as an independent unit of life.

**cell adhesion molecule**   A cell-surface protein that attaches cells to one another and to the extracellular matrix.

**cell cycle**   The collective name for the steps that make up eukaryotic cell division.

**cell division**   The process by which cells make more cells.

**cell plate**   In dividing plant cells, a new cell wall formed in the middle of the cell from the fusing of vesicles during late anaphase and telophase.

**cell theory**   The theory that the cell is the fundamental unit of life in all organisms and that cells come only from preexisting cells.

**cell wall**   A defining boundary in many organism, external to the cell membrane, that helps maintain the shape and internal composition of the cell.

**cell-mediated immunity**   The ability of T cells, which do not secrete antibodies, to recognize and act against pathogens directly.

**cellular blastoderm**   In *Drosophila* development, the structure formed by the nuclei in the single-cell embryo when they migrate to the periphery of the embryo and each nucleus becomes enclosed in its own cell membrane.

**cellular junction**   A region in the plasma membrane, consisting of cell adhesion molecules and other cytosolic proteins, where a cell makes contact with and adheres to another cell or the extracellular matrix.

**cellular respiration**   A series of chemical reactions that convert the energy stored in nutrients into a chemical form that can be readily used by cells.

**cellulase**   The enzyme that breaks down cellulose.

**central dogma**   The theory that information transfer in a cell usually goes from DNA to RNA to protein.

**central nervous system (CNS)**   In vertebrates, the brain and spinal cord; in invertebrates, centralized information-processing ganglia.

**centromere**   A constriction that physically holds sister chromatids together; the site of the attachment of the spindle fibers that move the chromosome in cell division.

**centrosome**   A compact structure that is the microtubule organizing center for animal cells.

**cephalization**   The concentration of nervous system components at one end of the body.

**cephalochordates**   A subphylum of Chordata that shares key features of body organization with vertebrates but lacks a well-developed brain and eyes, has no lateral appendages, and does not have a mineralized skeleton.

**cephalopods**   A group of mollusks, including squid, cuttlefish, octopus, and chambered nautilus, with distinctive adaptations such as muscular tentacles that capture prey and sense the environment.

**cerebellum**   Part of the vertebrate brain, formed from the hindbrain, which coordinates complex motor tasks by integrating motor and sensory information.

**cerebral cortex**   Part of the vertebrate brain formed from a portion of the forebrain that is greatly expanded in mammals, particularly primates.

**cerebrum**   The outer left and right hemispheres of the cerebral cortex.

**cervix**   The end, or neck, of the uterus.

**chain terminator**   A term for a dideoxynucleotide, which if incorporated into a growing daughter strand stops strand growth because there is no hydroxyl group to attack the incoming nucleotide.

**channel**   A transporter with a passage that allows the movement of molecules through it.

**chaperone**   A protein that helps shield a slow-folding protein until it can attain its proper three-dimensional structure.

**character**   One of the anatomical, physiological, or molecular features that make up organisms' bodies.

**character state**   The observed condition of a character, such as presence or absence of lungs or arrangement of petals.

**checkpoint**   One of multiple regulatory mechanism that coordinate the temporal sequence of events in the cell cycle.

**chelicerates**   One of the four main groups of arthropods, including spiders and scorpions, chelicerates have pincer-like claws and are the only arthropods that lack antennae.

**chemical bond**   Any form of attraction between atoms that holds them together.

**chemical reaction** The process by which molecules are transformed into different molecules.

**chemoautotroph** A microorganism that obtains energy from chemical compounds, not from sunlight.

**chemoreceptor** A receptor that responds to molecules that bind to specific protein receptors on the cell membrane of the sensory receptor.

**chemotroph** An organism that derives its energy directly from organic molecules such as glucose.

**chiasma** (plural, **chiasmata**) A crosslike structure within a bivalent constituting a physical manifestation of crossing over.

**chitin** A modified polysaccharide containing nitrogen that makes up the cell walls of fungi and the hard exoskeletons of arthropods.

**chloride cell** A type of specialized cell in the gills of marine bony fishes that counters the ingestion and diffusion of excess electrolytes into the animal by pumping chloride ions into the surrounding seawater; chloride cells in freshwater fishes have opposite polarity.

**chlorophyll** The major photosynthetic pigment contained in the thylakoid membrane; it plays a key role in the chloroplast's ability to capture energy from sunlight. Chlorophyll appears green because it is poor at absorbing green wavelengths.

**chloroplast** An organelle that converts energy of sunlight into chemical energy by synthesizing simple sugars.

**chloroplast genome** In plant eykaryotic cells, the genome of the chloroplast.

**choanocyte** A type of cell that lines the interior surface of a sponge; choanocytes have flagella and function in nutrition and gas exchange.

**choanoflagellate** One of a group of mostly unicellular protists characterized by a ring of microvilli around the cell's single flagellum.

**cholecystokinin (CCK)** A peptide hormone that causes the gallbladder to contract and thus release bile into the duodenum.

**cholesterol** An amphipathic lipid that is a major component of animal cell membranes.

**Chondrichthyes** Cartilaginous fish, a monophyletic group that includes about 800 species of sharks, rays, and chimaeras.

**Chordata** One of the three major phyla of deuterosomes, this group includes vertebrates and closely related invertebrate animals such as sea squirts.

**chorion** In the amniotic egg, a membrane that surrounds the entire embryo along with its yolk and allantoic sac.

**chromatin** A complex of DNA, RNA, and proteins that gives chromosomes their structure; chromatin fibers are either 30 nm in diameter or, in a relaxed state, 10 nm.

**chromatin remodeling** The process in which the nucleosomes are repositioned to expose different stretches of DNA to the nuclear environment.

**chromosome** In eukaryotes, the physical structure in which DNA in the nucleus is packaged; used more loosely to refer to the DNA in bacterial cells or archaeons.

**chromosome condensation** The progressive coiling of the chromatin fiber, an active, energy-consuming process requiring the participation of several types of proteins.

**chytrid** A single-celled aquatic fungus with chitin walls that attaches to decomposing organic matter.

**cilium** (plural, **cilia**) A hairlike organelle that propels the movement of cells or of substances within cells or out of the body; shorter than a flagellum.

**circadian clock** A biological clock, on a daily cycle, that regulates many daily rhythms in animals, such as feeding, sleeping, hormone production, and core body temperature.

**circular muscle** Smooth muscle that encircles the body or an organ; in the digestive tract, a circular muscle layer contracts to reduce the size of the lumen. A circular muscle layer contracts alternately with longitudinal muscle to move contents through the digestive tract and to enable locomotion in animals with hydrostatic skeletons.

**circulation** The movement of a specialized body fluid that carries oxygen and carbon dioxide.

**cis-regulatory element** A short DNA sequence adjacent to a gene, usually at the 5′ end, that interacts with transcription factors.

**cisternae** The series of flattened membrane sacs that make up the Golgi apparatus.

**citric acid cycle** The third stage of cellular respiration, in which acetyl-CoA is broken down and more carbon dioxide is released.

**cladistics** Phylogenetic reconstruction on the basis of synapomorphies.

**class** A group of closely related orders.

**classical conditioning** Associative learning in which two stimuli are paired.

**cleavage** The successive mitotic divisions of the zygote after fertilization, in which the single large egg is divided into many smaller cells.

**climax community** A mature assembly, a final stage in succession, in which there is little further change in species composition.

**clitoris** The female homolog of the glans penis.

**cloaca** A hollow organ in amphibians, reptiles, and birds for the storage and elimination of urine.

**clonal selection** A hypothesis proposing that the antigen instructs the antibody to fold in a particular way so that the two interact in a specific manner; now a central principle of immunology.

**clone** An individual that carries an exact copy of the nuclear genome of another individual; clones are genetically identical cells or individuals.

**closed circulatory system** A circulatory system made up of a set of internal vessels and a heart that functions as a pump to move blood to different regions of the body.

**cochlea** A coiled chamber within the skull containing hair cells that convert pressure waves into an electrical impulse that is sent to the brain.

**cochlear duct**   A fluid-filled cavity in the cochlea, next to the upper canal, that houses the organ of Corti.

**codon**   A group of three adjacent nucleotides that specifies an amino acid in a protein or that terminates polypeptide synthesis.

**coelacanth**   A species of fleshy-finned fish thought to have been extinct for 80 million years; along with lungfish, the nearest relative of tetrapods.

**coelomate**   A bilaterian with a body cavity.

**coenocytic**   Containing many nuclei within one giant cell; the nucleus divides multiple times, but the nuclei are not partitioned into individual cells.

**coenzyme Q**   The final electron acceptor from both complexes I and II in the electron transport chain.

**coevolution**   The process in which species evolve together, each responding to selective pressures from the other.

**cofactor**   A substance that associates with an enzyme and plays a key role in its function.

**cognition**   The ability of the brain to process and integrate complex sources of information, interpret and remember past events, solve problems, reason, and form ideas.

**cohort**   A group of the individuals born at a given time.

**collagen**   A strong protein fiber found in bone and artery walls.

**collecting duct**   A type of duct in the vertebrate kidney where urine collects.

**colon**   Part of the hindgut and the site of reabsorption of water and minerals; also known as the *large intestine.*

**combinatorial control**   Regulation of gene transcription by means of multiple transcription factors acting together.

**commensalism**   An interaction between species in which one partner benefits with no apparent effect on the other.

**communication**   The transfer of information between two individuals, the sender and the receiver.

**community**   The set of all populations found in a given place.

**compact bone**   Dense, mineralized bone tissue that forms the walls of a bone's shaft.

**companion cell**   A cell adjacent to a sieve element, it carries out cellular functions such as protein synthesis.

**comparative genomics**   The analysis of the similarities and differences in protein-coding genes and other types of sequence in the genomes of different species.

**competition**   An interaction in which the use of a mutually needed resource by one individual or group of individuals lowers the availability of the resource for another individual or group.

**competitive exclusion**   The result of an antagonistic interaction in which one species is prevented from occupying a particular habitat or niche.

**competitive inhibitor**   A reversible inhibitor that reduces the affinity of the substrate for the active site of an enzyme but can be overcome by excess substrate, so the maximum velocity of the reaction is not changed. Competitive inhibitors usually have a structure similar to that of the substrate and therefore bind to the active site of the enzyme.

**complement system**   The collective name for certain proteins circulating in the blood that participate in innate immune function and thus complement other parts of the immune system.

**complementary**   Describes the relationship of purine and pyrimidine bases, in which the base A pairs only with T and G pairs only with C.

**complex carbohydrate**   A long, branched chain of monosaccharides.

**complex multicellular organism**   An organism composed of specialized cells for specific functions, so the organism as a whole can perform a broad range of tasks but individual cells for the most part cannot.

**complex trait**   A trait that is influenced by multiple genes as well as by the environment.

**compound eye**   An eye structure found in insects and crustaceans that consists of a number of ommatidia, individual light-focusing elements.

**concordance**   The percentage of cases in which both members of a pair of twins show the trait when it is known that at least one member shows it.

**cone cell**   A type of photoreceptor cell on the retina that detects color.

**conjugation**   The direct cell-to-cell transfer of DNA, usually in the form of a plasmid.

**connective tissue**   A type of tissue characterized by a few cells and substantial amounts of extracellular matrix; a major component of the dermis.

**conserved**   Describes sequences that are similar in different organisms.

**constant (C)**   Describes an unchanging region of the H and L chains.

**constitutive**   Describes expression of a gene that occurs continuously.

**constitutive defense**   A defense that is always active.

**consumer**   An organism that obtains the carbon it needs for growth and reproduction from the foods it eats and gains energy by respiring food molecules; heterotrophic organisms of all kinds that directly consume primary producers or consume those that do.

**continuous**   Describes variation that occurs across a spectrum.

**contractile ring**   In animal cells, a ring of actin filaments that forms at the equator of the cell perpendicular to the axis of what was the spindle at the beginning of cytokinesis.

**contractile vacuole**   A type of cellular compartment that takes up excess water and waste products from inside the cell and expels them into the external environment.

**cooperative binding**   The increase in binding affinity with additional binding of $O_2$.

**copy-number variation (CNV)**   Differences among individuals in the number of copies of a region of the genome.

**cork cambium**   Lateral meristem that renews and maintains an outer layer that protects the stem against herbivores, mechanical damage, desiccation, and fire.

**cornea**   The transparent portion of the sclera in the front of the vertebrate eye.

**corpus luteum**   A temporary endocrine structure that secretes progesterone.

**correlation**   The co-occurrence of two events or processes; correlation does not imply causation.

**cortex**   In a stem, the region between the epidermis and the vascular bundles, composed of parenchyma cells. In the mammalian brain, the highly folded outer layer of gray matter, about 4 mm thick, made up of densely packed neuron cell bodies and their dendrites. In the mammalian renal system, the outer layer of the kidney.

**co-speciation**   A process in which two groups of organisms speciate in response to each other and at the same time.

**countercurrent exchange**   A mechanism in which two fluids flow in opposite directions, exchanging properties.

**countercurrent multiplier**   A system that generates a concentration gradient as two fluids move in parallel but in opposite directions.

**covalent bond**   A chemical bond formed by a shared pair of electrons holding two different atoms together.

**CpG island**   A cluster of CpG sites on a DNA strand where cytosine (C) is adjacent to guanosine (G); the "p" represents the phosphate in the backbone.

**cranial nerve**   In vertebrates, a nerve that links specialized sensory organs to the brain; most contain axons of both sensory and motor neurons.

**Craniates**   A subphylum of Chordata, distinguished by a bony cranium that protects the brain.

**crassulacean acid metabolism (CAM)**   The mechanism in plants that helps balance carbon dioxide gain and water loss.

**Crenarchaeota**   One of the three major divisions of Archaea; includes acid-loving microorganisms.

**crisscross inheritance**   A pattern in which an X chromosome present in a male in one generation is transmitted to a female in the next generation, and in the generation after that can be transmitted back to a male.

**Cro-Magnon**   The first population of *Homo sapiens* to arrive in Europe, named for the site in France where specimens were first described.

**cross-bridge**   The binding of the head of a myosin molecule to actin at a specific site between the myosin and actin filaments.

**cross-bridge cycle**   Repeated sequential interactions between myosin and actin filaments at cross-bridges that cause a muscle fiber to contract.

**crosscurrent**   Running across another current at 90 degrees.

**crossover**   The physical breakage, exchange of parts, and reunion between non-sister chromatids.

**crustaceans**   One of the four main groups of arthropods, including lobsters, shrimp, and crabs; their most notable distinction is their branched legs.

**cuticle**   In leaves, a protective layer of a waxy substance secreted by epidermal cells that limits water loss; also, an exoskeleton that covers the bodies of invertebrates such as nematodes and arthropods.

**C-value paradox**   The disconnect between genome size and organismal complexity (the C-value is the amount of DNA in a reproductive cell).

**cyanobacteria**   All bacteria capable of oxygenic photosynthesis.

**cyclic electron transport**   An alternative pathway for electrons during the Calvin cycle that increases the production of ATP.

**cyclin-dependent kinase (CDK)**   A kinase that is always present within the cell but active only when bound to the appropriate cyclin.

**cyclin**   A regulatory protein whose levels rise and fall with each round of the cell cycle.

**cytochrome $b_6f$ complex**   Part of the photosynthetic electron transport chain, through which electrons pass between photosystem II and photosystem I.

**cytochrome c**   The enzyme to which electrons are transferred in complex III of the electron transport chain.

**cytokine**   A chemical messenger released by phagocytes that recruits other immune cells to the site of injury or infection.

**cytokinesis**   In eukaryotic cells, the division of the cytoplasm into two separate cells.

**cytokinin**   A hormone that stimulates the outgrowth of auxillary buds by increasing the rate of cell division.

**cytoplasm**   The contents of the cell other than the nucleus.

**cytosine (C)**   A pyrimidine base.

**cytoskeleton**   In eukaryotes, an internal protein scaffold that helps cells to maintain their shape and serves as a network of tracks for the movement of substances within cells.

**cytosol**   The region of the cell inside the plasma membrane but outside the organelles; the jelly-like internal environment that surrounds the organelles.

**cytotoxic T cell**   One of a subpopulation of T cells, activated by cytokines released from helper T cells, that can kill other cells.

**daughter strand**   In DNA replication, the strand synthesized from a parental template strand.

**day-neutral plant**   A plant that flowers independently of any change in day length.

**decomposer**   An organism that breaks down dead tissues, feeding on the dead cells or bodies of other organisms.

**defecation**   The elimination of feces.

**delayed hypersensitivity reaction**   Reactions initiated by helper T cells, which release cytokines that attract macrophages to the site of exposure, which is typically the skin.

**deleterious**   Describes mutations that are harmful to an organism.

**deletion**   A missing region of a gene or chromosome.

**endergonic**   Describes reactions with a positive $\Delta G$ that are not spontaneous and so require an input of energy.

**endocrine signaling**   Signaling by molecules that travel through the bloodstream.

**endocrine system**   A system of cells and glands that secretes hormones and works with the nervous system to regulate an animal's internal physiological functions.

**endocytosis**   The process in which a vesicle buds off from the plasma membrane, bringing material from outside the cell into that vesicle, which can then fuse with other organelles.

**endoderm**   The germ layer that differentiates into cells of the lining of the digestive tract and lung, liver cells, pancreas cells, and gallbladder cells.

**endodermis**   In plants, a layer of cells surrounding the xylem and phloem at the center of the root that controls the movement of nutrients into the xylem. Also, the inner lining of the cnidarians body.

**endomembrane system**   A cellular system that includes the nuclear envelope, the endoplasmic reticulum, the Golgi apparatus, lysosomes, the plasma membrane, and the vesicles that move between them.

**endomycorrhizae**   One of the two main types of mycorrhizae; endomycorrhizal hyphae penetrate into root cells, where they produce highly branched structures (arbuscules) that provide a large surface area for nutrient exchange.

**endophyte**   A fungus that lives within leaves.

**endoplasmic reticulum (ER)**   The organelle involved in the synthesis of proteins and lipids.

**endoskeleton**   The bony skeletal system of vertebrate animals, which lies internal to most of the animal's soft tissues.

**endosperm**   A tissue formed by many mitotic divisions of a triploid cell, it supplies nutrition to the angiosperm embryo.

**endosymbiosis**   A symbiosis in which one partner lives within the other.

**endotherm**   An animal that produces most of its own heat as by-products of metabolic reactions.

**energetic coupling**   The driving of a non-spontaneous reaction by a spontaneous reaction.

**energy balance**   A form of homeostasis in which the amount of energy calories from food taken in equals the amount of calories used over time to meet metabolic needs.

**energy intake**   Sources of energy.

**energy use**   The ways in which energy is expended.

**enhancer**   A specific DNA sequence necessary for transcription.

**enthalpy (H)**   The total amount of energy in a system.

**entropy (S)**   The degree of disorder in a system.

**envelope**   A lipid structure that surrounds the capsids of some viruses.

**environmental risk factor**   A characteristic in a person's surroundings that increases the likelihood of developing a particular disease.

**environmental variation**   Variation among individuals that is due to differences in the environment.

**enzyme**   A protein that functions as a catalyst to accelerate the rate of a chemical reaction; enzymes are critical in determining which chemical reactions take place in a cell.

**eosinophil**   A type of granulocyte that, along with basophils, defends against parasitic infections but also contributes to allergies.

**epidermis**   In mammals, the outer layer of skin, which serves as a water-resistant, protective barrier. In plants, sheets of cells that line the leaf's upper and lower surfaces and constitute the outer cell layer of roots. Also, the outer layer of the cnidarian body.

**epididymis**   An organ that lies above the testes where sperm become motile and are stored prior to ejaculation.

**epigenetic**   Describes effects on gene expression due to differences in DNA packaging, such as modifications in histones or chromatin structure.

**epiglottis**   A flap of tissue at the bottom of the pharynx that prevents food from entering the trachea and lungs.

**epiphysis**   The end region of a bone; blood vessels invading the epiphysis and diaphysis trigger the transformation of cartilage into bone.

**epiphyte**   A plant that grows high in the canopy of other plants, or on branches or trunks of trees, without contact with the soil.

**epistasis**   Interaction between genes that modifies the phenotypic expression of genotypes.

**epithelial tissue**   A type of animal tissue, made up of epithelial cells, that covers the outside of the body and lines many internal structures.

**epitope**   The portion of an antigen that interacts in a specific manner with a hypervariable region of an antibody and binds to it.

**equational division**   Another name for meiosis II because cells in meiosis II have the same number of chromosomes at the beginning and at the end of the process.

**esophagus**   Part of the foregut; the passage from the mouth to the stomach.

**essential amino acid**   An amino acid that cannot be synthesized by cellular biochemical pathways and instead must be ingested.

**estrogen**   A hormone secreted by the ovaries that stimulates the development of female secondary sexual characteristics.

**estrus cycle**   A cycle in placental mammals other than humans and chimpanzees characterized by phases in which females are sexually receptive.

**ethanol fermentation**   The fermentation pathway in plants and fungi during which pyruvate releases carbon dioxide to form acetaldehyde and electrons from NADH are transferred to acetaldehyde to produce ethanol and $NAD^+$.

**Eukarya**   One of the three domains of life, consisting of cells with a true nucleus containing chromosomes that divide by mitosis; cells of Eukarya are eukaryotes.

**eukaryote**   A cell with a true nucleus containing chromosomes that divide by mitosis.

**Eumetazoa**   Animal groups other than sponges.

**Euryarchaeota**   One of the three major divisions of Archaea; includes acid-loving, heat-loving, methane-producing, and salt-loving microorganisms.

**eusocial**   Describes behavior of many species of Hymenoptera insects, in which they have overlapping generations in a nest, cooperative care of the young, and clear and consistent division of labor between reproducers (the queen of a honeybee colony) and nonreproducers (the workers).

**eutrophication**   The process in which added nutrients lead to a great increase in the populations of algae and cyanobacteria.

**evaporation**   The amount of water evaporated from the Earth's surface, including ponds, rivers, and soil.

**evapotranspiration ratio**   The amount of water evaporated from the Earth's surface divided by the amount of water transpired by leaves of plants.

**evo-devo**   The short name for evolutionary-developmental biology, a field of study that compares the genetic programs for growth and development in species on different branches of phylogenetic reconstructions.

**evolution**   Changes in the genetic make-up of populations over time, resulting in some cases in adaptation to the environment and the origin of new species.

**evolutionarily conserved**   Little changed through evolution and therefore similar from one organism to the next.

**evolutionarily stable strategy**   A type of behavior that cannot readily be driven to extinction by an alternative strategy.

**evolutionary species concept (EvSC)**   The concept that members of a species all share a common ancestry and a common fate.

**excitation–contraction coupling**   The process that produces muscle force and movement, by excitation of the muscle cell coupled to contraction of the muscle.

**excitatory postsynaptic potential (EPSP)**   A positive change in the postsynaptic membrane potential.

**excreted**   Eliminated (referring to waste generated by metabolism).

**excretory tubule**   In renal systems, a type of tube that drains waste products and connects to the outside of the body.

**exergonic**   Describes reactions with a negative $\Delta G$ that release energy and proceed spontaneously.

**exhalation**   The expelling of oxygen-poor air by the elastic recoil of the lungs and chest wall.

**exit (E) site**   One of three binding sites for tRNA on the large subunit of a ribosome.

**exocytosis**   The process in which a vesicle fuses with the plasma membrane and empties its contents into the extracellular space or delivers proteins to the plasma membrane.

**exon**   A sequence that is left intact in mRNA after RNA splicing.

**exoskeleton**   A rigid skeletal system that lies external to the animal's soft tissues.

**experimentation**   A disciplined and controlled way of learning about the world and testing hypotheses in an unbiased manner.

**exponential growth**   The pattern of population increase that results when $r$ (the per capita growth rate) is constant through time.

**expressed**   Turned on or activated, as a gene or protein.

**extension (PCR)**   A step in the polymerase chain reaction (PCR) for producing new DNA fragments in which the reaction mixture is heated to the optimal temperature for DNA polymerase, and each primer is elongated by means of deoxynucleoside triphophosphates.

**extension (joint)**   The joint motion in which bone segments move apart.

**external fertilization**   Fertilization that takes place outside the body of the female; in aquatic organisms, for example, eggs and sperm are released into the water.

**extracellular digestion**   The process in most animals in which food is isolated and broken down outside a cell, in a body compartment.

**extracellular matrix**   A meshwork of proteins and polysaccharides outside the cell; the main constituent of connective tissue.

**extraembryonic membrane**   In the amniotic egg, one of several sheets of cells that extend out from the developing embryo and form the yolk sac, amnion, allantois, and chorion.

**extravasation**   The process in which phagocytes that travel in the blood move from a blood vessel to the site of infection.

**eyecup**   An eye structure found in flatworms that contains photoreceptors that point up and to the left or right.

**$F_1$ generation**   The first filial, or offspring, generation.

**$F_2$ generation**   The second filial generation; the offspring of the $F_1$ generation.

**Fab fragment**   The part of an antigen molecule broken by an enzyme that has an antigen-recognition site.

**facilitated diffusion**   Diffusion through a membrane protein, bypassing the lipid bilayer.

**facultative**   Describes a mutualism in which one or both sides can survive without the other.

**fallopian tube (oviduct)**   A tube from each ovary, through one of which a released oocyte passes.

**family**   A group of closely related genera.

**fast-twitch**   Describes muscle fibers that generate force quickly, producing rapid movements, but consume much more ATP than do slow-twitch fibers.

**fatty acid**   A long chain of carbons attached to a carboxyl group; three fatty acid chains attached to glycerol form a triacylglycerol, a lipid used for energy storage.

**Fc fragment**   The part of an antigen molecule broken by an enzyme that activates the complement system and binds to cell-surface receptors on other cells of the immune system.

**feature detector**   A specialized sensory receptor or group of sensory receptors that respond to important signals in the environment.

**feces**    Waste products that are eliminated from the body.

**feedback**    A response evoked by a stimulus that in turn affects that stimulus; feedback is the means by which a signal produced by a regulatory organ can modify its own subsequent production, either inhibiting or stimulating it.

**fermentation**    A process of breaking down pyruvate through a wide variety of metabolic pathways that extract energy from fuel molecules such as glucose; the partial oxidation of complex carbon molecules to molecules that are less oxidized than carbon dioxide.

**ferns and horsetails**    A monophyletic group of vascular plants.

**fertilization**    The union of gametes to produce a zygote that restores the original chromosomal content of the parental organisms.

**fetus**    In humans, the embryo toward the end of the first trimester.

**fiber**    In angiosperm plants, a narrow cell with thick walls that provides mechanical support in wood, allowing vessels to be specialized for water transport. In animals, a term for a muscle cell, which produces forces within an animal's body and exerts forces on the environment.

**filament**    A thin thread of proteins that interacts with other filaments to cause muscles to shorten.

**filial imprinting**    Imprinting in which newborn offspring rapidly learn to treat any animal they see shortly after birth as their mother.

**filtration**    The separation of solids from fluids, as when circulatory pressure pushes fluid containing wastes through specialized filters into an extracellular space.

**firing rate**    The number of action potentials fired over a given period of time.

**first law of thermodynamics**    The law of conservation of energy: Energy can neither be created nor destroyed—it can only be transformed from one form into another.

**first-division nondisjunction**    Failure of chromosome separation in meiosis I.

**fitness**    A measure of the extent to which an individual's genotype is represented in the next generation

**fixation**    The point at which a given allele has a frequency of 1; it has become the only allele of that gene.

**fixed**    Describes a population that exhibits only one allele at a particular gene.

**fixed action pattern (FAP)**    A sequence of behaviors that, once triggered, is followed through to completion.

**flagellum (plural, flagella)**    An organelle that propels the movement of cells or of substances within cells; longer than a cilium.

**fleshy-finned fish**    Species of fish with paired pectoral and pelvic fins.

**flexion**    The joint motion in which bone segments rotate closer together.

**fluid**    Describes lipids that are able to move in the plane of the cell membrane.

**fluid mosaic model**    A model that proposes that the lipid bilayer is a fluid structure that allows molecules to move laterally within the membrane and is a mosaic of two types of molecules, lipids and proteins.

**flux**    The rate at which a substance, for example carbon, flows from one reservoir to another.

**folding domain**    A region of a protein that folds in a similar way across a protein family relatively independently of the rest of the protein.

**follicle cell**    A type of cell surrounding the oocyte that supports the developing oocyte and secretes estrogen.

**follicle-stimulating hormone (FSH)**    A hormone secreted by the anterior pituitary gland that stimulates the male and female gonads to secrete testosterone in males and estrogen and progesterone in females.

**follicular phase**    The phase of the menstrual cycle during which FSH acts on granulosa cells, resulting in the maturation of several oocytes, of which, usually, only one becomes completely mature.

**food chain**    The linear transfer of carbon from one organism to another.

**food web**    An interaction among organisms within the carbon cycle; the movement of carbon through an ecosystem.

**forebrain**    The region of the vertebrate brain that governs cognitive functions.

**foregut**    The first part of an animal's digestive tract, including the mouth, esophagus, and stomach.

**fossil**    The remains of a once-living organism, preserved through time in sedimentary rocks.

**founder event**    A type of bottleneck that occurs when only a few individuals establish a new population.

**fovea**    The center of the visual field of most vertebrates, where cone cells are most concentrated; the region of greatest acuity.

**fragmentation**    A form of asexual reproduction in which new individuals arise by the splitting of one organism into pieces, each of which develops into a new individual.

**frameshift mutation**    A mutation in which the insertion or deletion of a single nucleotide causes a one-nucleotide shift in the reading frame of the mRNA, changing all following codons.

**fraternal (dizygotic) twins**    Twins that arise when two separate eggs, produced by double ovulation, are fertilized by two different sperm.

**frequency of recombination**    The proportion of recombinant chromosomes among the total number of chromosomes observed.

**frontal lobe**    The region of the brain located in the anterior region of the cerebral cortex, important in decision making and planning.

**fruit**    The structure that develops from the ovary and serves to protect immature seeds and enhance dispersal once the seeds are mature.

**fruiting body**    A multicellular structure in some fungi that facilitates the dispersal of sexually produced spores.

**fundamental niche**    The full range of climate conditions and food resources that permit the individuals in a species to live.

**fungi**    An abundant and diverse group of heterotrophic eukaryotic organisms, principally responsible for the decomposition of plant and animal tissues.

**G protein-coupled receptor**   A receptor that couples to G proteins, which bind to the guanine nucleotides GTP and GDP.

**G protein**   A protein that binds to the guanine nucleotides GTP and GDP.

**$G_0$ phase**   The gap phase in which cells pause in the cell cycle between M phase and S phase; may last for periods ranging from days to more than a year.

**$G_1$ phase**   The gap phase in which the size and protein content of the cell increase and specific regulatory proteins are made and activated in preparation for S-phase DNA synthesis.

**$G_2$ phase**   The gap phase in which the size and protein content of the cell increase in preparation for M-phase mitosis and cytokinesis.

**gain-of-function mutation**   Any mutation in which a gene is expressed in the wrong place or at the wrong time.

**gallbladder**   The organ in which bile produced by the liver is stored.

**gamete**   A reproductive haploid cell resulting from meiotic cell division (in some species gametes are called spores). In many species, there are two types of gametes: eggs in females, sperm in males.

**gametogenesis**   The formation of gametes.

**gametophyte**   Describes the haploid multicellular generation that gives rise to gametes.

**ganglion** (plural, **ganglia**)   A group of nerve cell bodies that processes sensory information received from a local, nearby region, resulting in a signal to motor neurons that control some physiological function of the animal.

**ganglion cell**   A type of interneuron in the retina that synapses with bipolar cells and, if activated, transmits action potentials along the optic nerve to the visual cortex in the brain.

**gap junction**   A type of connection between the plasma membranes of adjacent animal cells that permits materials to pass directly from the cytoplasm of one cell to the cytoplasm of another.

**gas exchange**   The transport of oxygen and carbon dioxide between an animal and its environment.

**gastric cavity**   In cnidarians, a closed internal site where extracellular digestion and excretion take place.

**gastrin**   A peptide hormone produced in the stomach that stimulates cells lining the stomach to increase their production of HCl.

**gastropods**   A group of mollusks consisting of snails and slugs.

**gastrula**   A layered structure formed when the inner cell mass cells of the blastocyst migrate and reorganize.

**gastrulation**   A highly coordinated set of cell movements in which the cells of the blastoderm migrate inward, creating germ layers of cells within the embryo.

**gel electrophoresis**   A procedure to determine the size of a DNA fragment, in which DNA samples are inserted into slots or wells in a gel and a current passed through. Fragments move toward the positive pole according to size.

**gene**   The unit of heredity; the stretch of DNA that affects one or more traits in an organism, usually through an encoded protein or noncoding RNA.

**gene expression**   The production of a functional gene product.

**gene family**   A group of genes with related functions, usually resulting from multiple rounds of duplication and divergence.

**gene flow**   The movement of alleles from one population to another.

**gene pool**   All the alleles present in all individuals in a species.

**gene regulation**   The various ways in which cells control gene expression.

**general transcription factors**   A set of proteins that bind to the promoter of a gene whose combined action is necessary for transcription.

**genetic code**   The correspondence between codons and amino acids, in which 20 amino acids are specified by 64 codons.

**genetic drift**   A change in the frequency of an allele due to the random effects of small population size.

**genetic incompatibility**   Genetic dissimilarity between two organisms, such as different numbers of chromosomes, that is sufficient to act as a pre-zygotic isolating factor.

**genetic information**   Information carried in DNA, organized in the form of genes.

**genetic map**   A diagram showing the relative positions of genes along a chromosome.

**genetic risk factor**   Any mutation that increases the risk of a given disease in an individual.

**genetic test**   A method of identifying the genotype of an individual.

**genetic variation**   Differences in genotype among individuals in a population.

**genetically modified organism (GMO)**   An organism that has been genetically engineered, such as modified viruses and bacteria, laboratory organisms, agricultural crops, and domestic animals. Also known as a *transgenic organism*.

**genome**   The genetic material transmitted from a parental cell or organism to its offspring.

**genome annotation**   The process by which researchers identify the various types of sequence present in genomes.

**genomic rearrangement**   The process of joining different gene segments as a B cell differentiates to produce a specific antibody.

**genotype**   The genetic makeup of a cell or organism; the particular combination of alleles present in an individual.

**genotype-by-environment interaction**   Variation in the effects of the environment on different genotypes, resulting in different phenotypes.

**genus**   A group of closely related species.

**geographic range**   How widely a population is spread and the factors that determine its distribution.

**geologic timescale**   The series of time divisions that mark Earth's long history.

**germ cells**   The reproductive cells that produce sperm or eggs and the cells that give rise to them.

**germ layers**   Three sheets of cells, the ectoderm, mesoderm, and endoderm, formed by migrating cells of the gastrula that differentiate further into specialized cells.

**germ-line mutation**   A mutation that occurs in eggs and sperm or in the cells that give rise to these reproductive cells and therefore is passed on to the next generation.

**gibberellic acid**   A plant hormone that stimulates the elongation of stems.

**Gibbs free energy (G)**   The amount of energy available to do work.

**gills**   Highly folded delicate structures in aquatic animals that facilitate gas exchange with the surrounding water.

**gizzard**   In birds, alligators, crocodiles, and earthworms, a compartment with thick muscular walls in the digestive tract where food mixed with ingested rock or sediment is broken down into smaller pieces.

**glans penis**   The head of the human penis.

**glial cell**   A type of cell that surrounds neurons and provides them with nutrition and physical support.

**glomeromycetes**   A monophyletic fungal group of apparently low diversity but tremendous ecological importance that occurs in association with plant roots.

**glomerulus**   A tufted loop of porous capillaries in the vertebrate kidney that filters blood.

**glycerol**   A 3-carbon molecule with OH groups attached to each carbon.

**glycogen**   The form in which glucose is stored in animals.

**glycolysis**   The breakdown of glucose to pyruvate; the first stage of cellular respiration. (The second of the four stages is the conversion of pyruvate to acetyl-CoA and the release of $CO_2$.)

**glycosidic bond**   A covalent bond that attaches one monosaccharide to another.

**Golgi apparatus**   The organelle that modifies proteins and lipids produced by the ER and acts as a sorting station as they move to their final destinations.

**gonad**   The part of the reproductive system where haploid gametes are produced. Male gonads are testes, where sperm are produced. Female gonads are ovaries, where eggs are produced.

**gonadotropin-releasing hormone (GnRH)**   A hormone released by the hypothalamus that stimulates the anterior pituitary gland to secrete luteinizing hormone and follicle-stimulating hormone.

**gram-positive bacteria**   Bacteria that retain, in their thick peptidoglycan walls, the diagnostic dye developed by Hans Christian Gram. (Bacteria with thin walls, which do not retain the dye, are said to be gram negative.)

**grana**   Interlinked structures that form the thylakoid membrane.

**granulocyte**   A type of phagocytic cell that contains granules in its cytoplasm.

**granuloma**   A structure formed by lymphocytes surrounding infected macrophages that helps to prevent the spread of an infection and aids in killing infected cells.

**gravitropic**   Bending in response to gravity. A negative gravitropic response, as in stems, is growth upward against the force of gravity; a positive gravitropic response, as in roots, is with the force of gravity.

**gray matter**   Densely packed neuron cell bodies and dendrites that make up the cortex, a highly folded outer layer of the mammalian brain about 4 mm thick.

**greenhouse gas**   A gas that absorbs heat energy from incoming solar radiation and allows it to reach Earth's surface but traps heat that is reemitted from land and sea.

**group selection**   A form of natural selection that operates on groups rather than on individuals.

**growth factor**   Any one of a group of small, soluble molecules, usually the signal in paracrine signaling, that affect cell growth, cell division, and changes in gene expression.

**growth plate**   A region of cartilage near the end of a bone where growth in length occurs.

**growth ring**   One of the many rings apparent in the cross section of the trunk of a tree, produced by decreases in the size of secondary xylem cells at the end of the growing season, that make it possible to determine the tree's age.

**guanine (G)**   A purine base.

**guard cell**   One of two cells surrounding the central pore of a stoma.

**gustation**   The sense of taste.

**gut**   Collectively, the passages that connect the mouth, digestive organs, and anus; also known as the *digestive tract*.

**gymnosperms**   One of the two groups of vascular plants; gymnosperms include pine trees and other conifers.

**habituation**   The reduction or elimination of a behavioral response to a repeatedly presented stimulus.

**hagfish**   One of the earliest-branching craniates, with a cranium built of cartilage but no jaws; hagfish feed on marine worms and dead and dying sea animals.

**hair cell**   A specialized mechanoreceptor that senses movement and vibration.

**hairpin** (structures)   Stems-and-loops formed in self-complementary, single-stranded nucleic acid molecules, stabilized by base pairing in the stem.

**half-life**   The time it takes for half of the atoms in a given sample of a substance to decay.

**haploid**   Describes a cell with one complete set of chromosomes.

**haplotype**   A haploid genotype, as the particular combination of alleles present in any particular region of a chromosome.

**Hardy–Weinberg equilibrium**   A state in which allele and genotype frequencies do not change over time, suggesting that evolutionary mechanisms are not acting on the population being studied. It also specifies a mathematical relationship between allele frequencies and genotype frequencies.

**heart** The pump of the circulatory system, which moves blood to different regions of the body.

**Heart rate (HR)** The number of heartbeats per unit time.

**heartwood** The center of the stem in long-lived trees, which does not conduct water.

**heavy (H) chains** Two of the four polypeptide chains that make up the simplest antibody molecule.

**helicase** A protein that unwinds the parental double helix at the replication fork.

**helper T cell** One of a subpopulation of T cells that help other cells of the immune system by secreting cytokines, thus activating B cells to secrete antibodies.

**Hemichordata** One of the three major phyla of deuterosomes, this group includes acorn worms and pterobranchs.

**hemidesmosome** A type of desmosome in which integrins are the prominent cell adhesion molecules.

**hemoglobin** An iron-containing molecule specialized for oxygen transport.

**hemolymph** The circulatory fluid in invertebrates.

**hemophilia** A trait characterized by excessive bleeding that results from a recessive mutation in a gene encoding a protein necessary for blood clotting.

**heterokaryotic** Describes a stage in in the life cycle of some fungi, following plasmogamy, in which cells have two haploid nuclei, one from each parent, and the nuclei have not yet fused in karyogamy to produce a diploid zygote.

**heterotroph** An organism that obtains its carbon from organic molecules synthesized by other organisms.

**heterozygote advantage** A form of balancing selection in which the heterozygote's fitness is higher than that of either of the homozygotes, resulting in selection that ensures that both alleles remain in the population at intermediate frequencies.

**heterozygous** Describes an individual who inherits different types of alleles from the parents, or genotypes in which the two alleles for a given gene are different.

**hierarchical** Describes gene regulation during development, in which the genes expressed at each stage in the process control the expression of genes that act later.

**high-energy phosphate bond** In nucleic acid polymerization, the bond connecting the innermost phosphate to the next. The cleaved bond provides the energy to drive the reaction that creates the phosphodiester bond attaching the incoming nucleotide to the 3′ end of the growing chain.

**highly repetitive DNA** A type of noncoding DNA consting of sequences present in many thousands of copies per genome.

**hindbrain** Along with the midbrain, the region of the vertebrate brain that controls basic body functions and behaviors.

**hindgut** The last part of an animal's digestive tract, including the large intestine and rectum.

**hinge joint** A simple joint that allows one axis of rotation, like the elbow and knee.

**hippocampus** A posterior region of the limbic system involved in long-term memory formation.

**histamine** A chemical messenger released by mast cells and basophils; an important contributor to allergic reactions and inflammation.

**histone** A protein found in all eukaryotes that interacts with DNA to form chromatin.

**histone code** The pattern of modifications of the histone tails that affects the chromatin structure and gene transcription.

**histone tail** A string of amino acids that protrudes from a histone protein in the nucleosome.

**homeobox** A DNA sequence within homeotic genes, which function in development, that specifies the homeodomain.

**homeodomain** The DNA-binding domain in homeotic proteins, a sequence of 60 amino acids whose sequences are very similar from one homeotic protein to the next.

**homeostasis** The active regulation and maintenance, in animals, organs, or cells, of a stable internal physiological state in the face of a changing external environment.

**homeotic gene** A gene that specifies the identity of a body part or segment during embryonic development; also known as a *Hox gene*.

**hominins** A member of one of the different species in the group leading to humans.

**homologous** Describes characters that are similar in different species because of descent from a common ancestor.

**homologous chromosomes** Pairs of chromosomes, matching in size and appearance, that carry the same set of genes; one of each pair was received from the mother, the other from the father.

**homozygous** Describes an individual who inherits an allele of the same type from each parent, or a genotype in which both alleles for a given gene are the same.

**horizontal cell** A type of interneuron in the retina that communicates between neighboring pairs of photoreceptors and bipolar cells, enhancing contrast through lateral inhibition to sharpen the image.

**horizontal gene transfer** The transfer of genetic material between organisms that are not parent and offspring.

**hormone** A chemical signal that influences physiology and development in both plants and animals; in animals hormones are released into the bloodstream and circulate throughout the body.

**host cell** A cell in which viral reproduction occurs.

**host plant** A plant species that can be infected by a given pathogen.

**hotspot** A site in the genome that is especially mutable.

**housekeeping gene** A gene that is transcribed continually because its product is needed at all times and in all cells.

***Hox* gene** A gene that species the identity of a body part or segment during embryonic development; also known as a *homeotic gene*.

**human chorionic gonadotropin (hCG)**   A hormone released by the developing embryo that maintains the corpus luteum.

**humoral immunity**   The part of the adaptive immune system consisting of blood, tissue fluid, and secretions, where antibodies are found.

**hybridization**   Interbreeding between two different varieties or species of an organism.

**hydrogen bond**   A weak bond between a hydrogen atom in one molecule and an electronegative atom in another molecule.

**hydrophilic**   "Water loving"; describes a class of molecules with which water can undergo hydrogen bonding.

**hydrophobic**   "Water fearing"; describes a class of molecules poorly able to undergo hydrogen bonding with water.

**hydrophobic effect**   The exclusion of nonpolar molecules by polar molecules, which drives biological processes such as the formation of cell membranes and the folding of proteins.

**hydrostatic skeleton**   A skeletal system in which fluid contained within a body cavity is the supporting element; found in nearly all multicellular animals and many vascular plants.

**hydroxyapatite**   Mineral crystals of calcium phosphate, a major component of bone.

**hypersensitive response**   A type of plant defense against infection in which uninfected cells surrounding the site of infection rapidly produce large numbers of reactive oxygen species, triggering cell wall reinforcement and causing the cells to die, thus creating a barrier of dead tissue.

**hyperthermophile**   An organism that requires an environment with high temperature.

**hypervariable region**   In antibodies, an even more variable region within the variable regions of the heavy and light chains.

**hyphae**   In fungi, highly branched filaments that provide a large surface area for absorbing nutrients.

**hypothalamus**   The underlying brain region of the forebrain, which interacts with the autonomic and endocrine systems to regulate the general physiological state of the body.

**hypothesis**   A tentative explanation for one or more observations that makes predictions that can be tested by experiments or additional observations.

**identical (monozygotic) twins**   Twins that arise from a single fertilized egg, which after several rounds of cell division, separates into two distinct, but genetically identical, embryos.

**IgA**   One of the five antibody classes, IgA is typically a dimer and the major antibody on mucosal surfaces.

**IgD**   One of the five antibody classes, IgD is a monomer and is typically found on the surface of B cells; it helps initiate inflammation.

**IgE**   One of the five antibody classes, IgE is a monomer that plays a central role in allergies, asthma, and other immediate hypersensitivity reactions.

**IgG**   The most abundant of the five antibody classes, IgG is a monomer that circulates in the blood and is particularly effective against bacteria and viruses.

**IgM**   One of the five antibody classes, IgM is a pentamer in mammals and a tetramer in fish and is particularly important in the early response to infection, activating the complement system and stimulating an immune response.

**ileum**   A section of the small intestine that, with the jejunum, carries out most nutrient absorption.

**imitation**   Observing and copying the behavior of another.

**immediate hypersensitivity reaction**   A reaction characterized by a heightened or an inappropriate immune response to common antigens.

**immunodeficiency**   Any disease in which part of the immune system does not function properly.

**immunoglobulins (Ig)**   A family of proteins with common structural features that includes antibodies.

**imprinting**   A form of learning typically seen in young animals in which a certain behavior is acquired in response to key experiences during a critical period of development.

**in vitro fertilization (IVF)**   A process in which eggs and sperm are brought together in a petri dish, where fertilization and early cell divisions occur.

**inbred line**   A true-breeding, homozygous strain.

**incisor**   One of the teeth in the front of the mouth, used for biting.

**incomplete dominance**   Describes inheritance in which the phenotype of the heterozygous genotype is intermediate between those of homozygous genotypes.

**incomplete penetrance**   The phenomenon in which some individuals with a genotype corresponding to a trait do not show the phenotype, either because of environmental effects or because of interactions with other genes.

**incus**   A small bone in the middle ear that helps amplify the waves that strike the tympanic membrane.

**induced pluripotent cell (iPS cell)**   A cell that has been reprogrammed to become pluripotent by activation of certain genes, most of them encoding transcription factors or chromatin proteins.

**inducer**   A small molecule that elicits gene expression.

**inducible defense**   A defense that is mounted only when a threat is sensed.

**inflammation**   A physiological response of the body to injury that removes the inciting agent if present and begins the healing process.

**inhalation**   The drawing of oxygen-rich blood into the lungs by the expansion of the thoracic cavity.

**inhibitor**   A synthesized compound that decreases the activity of an enzyme.

**inhibitory postsynaptic potential (IPSP)**   A negative change in the postsynaptic membrane potential.

**initiation**   The stage of translation in which methionine is established as the first amino acid in a new polypeptide chain.

**initiation factor**   A protein that binds to mRNA to initiate translation.

**innate**   Describes behaviors that are instinctive and carried out regardless of earlier experience.

**innate (natural) immunity**   The part of the immune system that provides protection in a nonspecific manner against all kinds of infection; it does not depend on exposure to a pathogen.

**inner cell mass**   A mass of cells in one region of the inner wall of the blastocyst from which the body of the embryo develops.

**inner ear**   The part of the inner ear that includes the cochlea and semicircular canals.

**insects**   The most diverse of the four main groups of arthropods.

**instantaneous speciation**   Speciation that occurs in a single generation.

**integral membrane protein**   A protein that is permanently associated with the cell membrane and cannot be separated from the membrane experimentally without destroying the membrane itself.

**integrin**   A transmembrane protein, present on the surface of virtually every animal cell, that enables cells to adhere to the extracellular matrix.

**intercostal muscles**   Muscles attached to adjacent pairs of ribs that assist the diaphragm by elevating the ribs on inhalation and depressing them during exhalation.

**intermediate filament**   A polymer of proteins, which vary according to cell type, that combine to form strong, cable-like filaments that provide animal cells with mechanical strength.

**intermembrane space**   The space between the inner and outer mitochondrial membranes.

**internal fertilization**   Fertilization that takes place inside the body of the female.

**interneuron**   A neuron that processes information received by sensory neurons and transmits it to motor neurons in different body regions.

**internode**   The segment between two nodes on a shoot.

**interphase**   The time between two successive M phases.

**intersexual selection**   A form of sexual selection that focuses on interaction between males and females, as when males compete for the attention of the female, who chooses her mate.

**interspecific competition**   Competition between individuals of different species.

**intervertebral disc**   A fluid-filled support structure found between the bony vertebrae of the backbone that enables flexibility and provides cushioning of loads.

**intracellular digestion**   The process in single-celled protists in which food is broken down within cells.

**intrasexual selection**   A form of sexual selection that focuses on interactions between individuals of one sex, as when members of one sex compete with one another for access to the other sex.

**intraspecific competition**   Competition within species.

**intrinsic growth rate**   The per capita growth rate; the maximum rate of growth when no environmental factors limit population increase.

**intron**   A sequence that is excised from the primary transcript and degraded during RNA splicing.

**invasive species**   Non-native species; since they are removed from natural constraints on population growth, invasive species can expand dramatically when introduced into new areas, often with devastating consequences for native species and ecosystems.

**inversion**   The reversal of the normal order of a block of genes.

**involuntary**   Describes the component of the nervous system that regulates internal bodily functions.

**ion**   An electrically charged atom or molecule.

**ionic bond**   The association of two atoms resulting from the attraction of opposite charges.

**iris**   A structure found at the front of the vertebrate eye, surrounding the pupil, that opens and closes to adjust the amount of light that enters the eye.

**irreversible inhibitor**   Any one of a class of inhibitors that usually forms covalent bonds with enzymes and irreversibly inactivates them.

**island population**   A distant, isolated population.

**isomers**   Molecules that have the same chemical formula but different structures.

**isometric**   Describes the generation of force without muscle movement.

**isotopes**   Atoms of the same element that have different numbers of neutrons.

**isotype**   In immunology, any one of the five classes of antibody.

**isotype (class) switching**   The ability of B cells to change the class of antibody they make.

**jasmonic acid**   A chemical signal that is triggered by herbivore damage; exposure to jasmonic acid induces the transcription of defensive genes.

**jejunum**   A section of the small intestine that, with the ileum, carries out most nutrient absorption.

**juxtacrine signaling**   Signaling by direct physical contact of one cell with another, with no chemical signal that diffuses or circulates through an external medium.

**juxtaglomerular apparatus**   The structure formed by specialized cells of the efferent arteriole leaving the glomerulus of each nephron, which secretes rennin into the bloodstream.

**karyogamy**   The fusion of two nuclei following plasmogamy.

**karyotype**   A standard arrangement of chromosomes, showing the number and shapes of the chromosomes representative of a species.

**ketose**   A monosaccharide with a ketone group.

**key stimulus**   A stimulus that initiates a fixed action pattern.

**keystone species**   Pivotal populations that affect other members of the community in ways that are disproportionate to their abundance or biomass.

**kidneys**   In vertebrates, paired organs of the renal system that remove waste products and excess fluid; their action contributes to homeostasis.

**kin selection**   A form of natural selection that favors the spread of alleles that promote behaviors that help close relatives.

**kinesin**   A motor protein, similar in structure to myosin, that transports cargo toward the plus end of microtubules.

**kinesis** (plural, **kineses**)   A random, undirected movement in response to a stimulus.

**kinetic energy**   The energy of motion.

**kinetochore**   The protein complexes on a chromatid where spindle fibers attach.

**kingdom**   A group of closely related phyla.

**Klinefelter syndrome**   A sex-chromosomal abnormality in which an individual has 47 chromosomes, including two X chromosomes and one Y chromosome.

**knee-jerk reflex**   A reflex commonly tested by physicians to evaluate peripheral nervous and muscular system function.

*K*-**strategist**   A species that produces relatively few young but invests considerable resources into their support.

**labia majora**   Outer folds of skin in the vulva.

**labia minora**   Inner folds of skin in the vulva that meet at the clitoris.

**lactic acid fermentation**   The fermentation pathway in animals and bacteria during which electrons from NADH are transferred to pyruvate to produce lactic acid and $NAD^+$.

**lagging strand**   A daughter strand that has its 5′ end pointed toward the replication fork, so as the parental double helix unwinds, a new DNA piece is initiated at intervals, and each new piece is elongated at its 3′ end until it reaches the piece in front of it.

**lamella** (plural, **lamellae**)   One of many thin, sheetlike structures spread along the length of each gill filament, giving gills an enormous surface area relative to their size.

**lampreys**   One of the earliest-branching craniates, with a cranium and vertebral column built of cartilage but no jaws; lampreys live parasitically, sucking body fluids from fish prey.

**large intestine**   Part of the hindgut and the site of reabsorption of water and minerals; also known as the *colon*.

**lariat**   A loop and tail of RNA formed after RNA splicing.

**larynx**   The structure, above the trachea, that contains the vocal cords.

**lateral inhibition**   Inhibition of a process in cells adjacent to the cell receiving a signal inducing that process, enhancing the strength of a signal locally but diminishing it peripherally.

**lateral line system**   In fish and sharks, a sensory organ along both sides of the body that uses hair cells to detect movement of the surrounding water.

**lateral meristem**   The source of new cells that allows plants to grow in diameter.

**latex**   A white sticky liquid produced in some plants.

**latitudinal diversity gradient**   The increase in species diversity from the poles to the equator.

**leading strand**   A daughter strand that has its 3′ end pointed toward the replication fork, so as the parental double helix unwinds, this daughter strand can be synthesized as one long, continuous polymer.

**leaf**   The principal site of photosynthesis in vascular plants.

**leaf primordia**   Young leaves.

**learned**   Describes a behavior that depends on an individual's experience.

**learning**   The process in which experience leads to changes in behavior.

**lengthening contraction**   The contraction of a muscle against a load greater than the muscle's force output, leading to a lengthening of the muscle.

**lens**   A flexible structure in the vertebrate eye through which light passes after entering through the pupil; it is controlled by ciliary muscles that contract or relax to the adjust the shape of the lens to focus light images.

**lenticel**   A region of less tightly packed cells in the outer bark that allows oxygen to diffuse into the stem.

**Leydig cell**   A type of cell in the testes that secretes testosterone.

**Liebig's Law of the Minimum**   The principle that primary production is limited by the nutrient that is least available relative to its use by primary producers.

**life history**   The typical pattern of resource investment in each stage of a given species' lifetime.

**ligand**   Alternative term for a signaling molecule that binds with a receptor, usually a protein.

**ligand-binding site**   The specific location on the receptor protein where a signaling molecule binds.

**ligand-gated ion channel**   A receptor that alters the flow of ions across the plasma membrane when bound by its ligand.

**light (L) chains**   Two of the four polypeptide chains that make up the simplest antibody molecule.

**limbic system**   Inner components of the forebrain that control physiological drives, instincts, emotions, motivation, and the sense of reward.

**LINE**   Long interspersed nuclear element of about 1000 base pairs present in multiple copies in a genome owing to transposition.

**linked**   Describes genes that are sufficiently close together in the same chromosome that they do not assort independently.

**lipase**   A type of enzyme produced by the pancreas that breaks apart lipids, thus enabling their more effective digestion.

**lipid**   An organic molecule that stores energy, acts as a signaling molecule, and is a component of cell membranes.

**lipid raft**   Lipids assembled in a defined patch in the cell membrane.

**liposome**   An enclosed bilayer structure spontaneously formed by phospholipids in environments with neutral pH, like water.

**liver**   An organ that aids in the digestion of proteins, carbohydrates, and fats in the duodenum by producing bile, which breaks down fat.

**lock and key**   In mating, describes a system that requires both components, whether physical or biochemical, to match for a successful interaction to take place.

**logistic growth**   The pattern of population growth that results as growth potential slows down as the population size approaches K, its maximum sustainable size.

**long-day plant**   A plant that flowers only when the light period exceeds a critical value.

**longitudinal muscle**   Smooth muscle that runs lengthwise along a body or organ; in the digestive tract, a longitudinal muscle layer contracts to shorten small sections of the gut. A longitudinal muscle layer contracts alternately with circular muscle to move contents through the digestive tract and to enable locomotion in animals with hydrostatic skeletons.

**loop of Henle**   The middle portion of the renal tubule, which creates a concentration gradient that allows water passing through the collecting duct to be reabsorbed.

**Lophotrochozoa**   A group of animals that contains 17 phyla, mostly small marine animals of limited diversity, but also including the annelid worms and mollusks.

**loss-of-function mutation**   A mutation that inactivates the normal function of a gene.

**LTR element**   A type of transposable element characterized by long repeated sequences, called long terminal repeats, at its ends, and that transposes by means of an RNA intermediate.

**Lucy**   An unusually complete early hominin fossil, found in 1974 in Ethiopia.

**lumen**   In eukaryotes, the continuous interior of the endoplasmic reticulum; in plants, a fluid-filled compartment enclosed by the thylakoid membrane; generally, the interior of any tubelike structure.

**lunar clock**   A moon-based biological clock that times activities in some species, especially those living in habitats where tides are important.

**lungfish**   A species of fleshy-finned fish that can survive periods when their watery habitat dries by burying themselves in moist mud and breathing air; along with coelacanths, the nearest relative of tetrapods.

**lungs**   The internal organs for gas exchange in many terrestrial animals.

**luteal phase**   The phase of the menstrual cycle beginning with ovulation.

**luteinizing hormone (LH)**   A hormone secreted by the anterior pituitary gland that stimulates the male and female gonads to secrete testosterone in males and estrogen and progesterone in females.

**lycophytes**   A major vascular plant group, the spore-dispersing lycophytes are the sister group to all other vascular plants.

**lymph**   The fluid in the lymphatic system in which T and B cells ciculate.

**lymphatic system**   A network of vessels distributed through the body with important functions in the immune system.

**lysogenic pathway**   The alternative to the lytic pathway; in the lysogenic pathway, a virus integrates its DNA into the host cell's DNA, which is then transmitted to offspring cells.

**lysosome**   A vesicle derived from the Golgi apparatus that contains enzymes that break down macromolecules such as proteins, nucleic acids, lipids, and complex carbohydrates.

**lytic pathway**   The alternative to the lysogenic pathway; in the lytic pathway, a virus bursts, or lyses, the cell it infects, releasing new virus particles.

**M phase**   The stage of the cell cycle consisting of mitosis and cytokinesis, in which the parent cell divides into two daughter cells.

**macrophage**   A type of large phagocytic cell that patrols the body.

**mainland population**   The central population of a species.

**major groove**   The larger of two uneven grooves on the outside of a DNA duplex.

**major histocompatibility complex (MHC)**   A group of proteins that appear on the surface of most mammalian cells; only the antigen associated with MHC proteins is recognized by T cell receptors.

**malleus**   A small bone in the middle ear that helps amplify the waves that strike the tympanic membrane.

**Malpighian tubule**   One of the tubes in the main body cavity of insects and other terrestrial arthropods through which fluid passes and which empties into the hindgut.

**mammals**   A class of vertebrates distinguished by body hair and mammary glands from which they feed their young.

**map information**   The knowledge of where an individual is in respect to the goal.

**MAP kinase pathway**   A series of kinases that are triggered by activated GTP-bound Ras; the final kinase enters the nucleus, where it phosphorylates its target proteins.

**map unit**   A unit of distance in a genetic map equal to the distance between genes resulting in 1% recombination.

**mark-and-recapture**   A method in which individuals are captured, marked in way that doesn't affect their function or behavior, and then released.

**marsupials**   A group of mammals that includes kangaroos, koalas, and opposums; their young are born at an early stage of development and must crawl to a pouch where mammary glands equipped with nipples provide them with milk.

**mass extinction**   A catastrophic drop in recorded diversity, which has occurred several times in the past 542 million years.

**mast cell**   A cell that releases histamine.

**maternal inheritance**   A type of inheritance in which the organelles in the offspring cells derive from those in the mother.

**maternal-effect gene**   A gene that is expressed by the mother that affects the phenotype of the offspring, typically through the composition or organization of the oocyte.

**mating types**   Genetically distinct forms of individuals of a fungus species that, by enabling fertilization only between different types, prevent self-fertilization.

**mechanoreceptor**   A sensory receptor that responds to physical deformations of its membrane produced by touch, stretch, pressure, motion, and sound.

**Neanderthal** *Homo neanderthalensis,* a species similar to humans, but with thicker bones and flatter heads that contained brains about the same size as humans'; appeared in the fossil record 60,000–30,000 years ago.

**necrotrophic pathogen** A plant pathogen that kills cells before colonizing them.

**negative feedback** Describes the effect in which the final product of a biochemical pathway inhibits the first step; the process in which a stimulus acts on a sensor that communicates with an effector, producing a response that opposes the initial stimulus. Negative feedback is used to maintain steady conditions, or homeostasis.

**negative regulation** The process in which a regulatory molecule must bind to the DNA at a site near the gene to prevent transcription.

**negative selection** Natural selection that reduces the frequency of a deleterious allele.

**negatively selected** Describes T cells that react too strongly to self antigens in association with MHC and are eliminated through cell death.

**neoteny** The long-term evolutionary process in which the timing of development is altered so that a sexually mature organism retains the physical characteristics of the juvenile form.

**nephron** The functional unit of the kidney, consisting of the glomerulus, capsule, renal tubules, and collecting ducts.

**nerve** A bundle of long fiberlike extensions from multiple nerve cells.

**nerve cord** A bundle of long fiberlike extensions from multiple nerve cells that serves as the central nervous system of invertebrates such as flatworms and earthworms.

**nervous system** A network of many interconnected nerve cells.

**neural tube** In chordates, a cylinder of embryological tissue that develops into a dorsal nerve cord.

**neuron** Nerve cell; the basic fundamental unit of nervous systems.

**neurosecretory cell** A neuron in the hypothalamus that secretes hormones into the bloodstream.

**neurotransmitter** A molecule that conveys a signal from the end of the axon to the postsynaptic target cell.

**neutral** Describes mutations that have no effect or negligible effects on the organism, or whose effects are not associated with differences in survival or reproduction.

**neutron** An electrically neutral particle in the atomic nucleus.

**neutrophil** A type of phagocytic cell that is very abundant in the blood and is often one of the first cells to respond to infection.

**niche** A multidimensional habitat that allows a species to practice its way of life.

**nicotinamide adenine dinucleotide phosphate (NADPH)** An important cofactor in many biosynthetic reactions; the reducing agent used in the Calvin cycle.

**nitrification** The process by which chemoautotrophic bacteria oxidize ammonia ($NH_3$) to nitrate ($NO_3^-$).

**nitrogen fixation** The process in which nitrogen gas ($N_2$) is converted into ammonia ($NH_3$), a form biologically useful to primary producers.

**nitrogenous waste** Waste in the form of ammonia, urea, and uric acid, which are toxic to organisms in varying degrees.

**nociceptor** A type of nerve cell with dendrites in the skin and connective tissues of the body that responds to excessive mechanical, thermal, or chemical stimuli by withdrawal from the stimulus and by the sensation of pain.

**node** In phylogenetic trees, the point where a branch splits, representing the common ancestor from which the descendant species diverged. In plants, the point on a shoot where one or more leaves are attached.

**nodes of Ranvier** Sites on an axon that lie between adjacent myelin-wrapped segments, where the axon membrane is exposed.

**non-associative learning** Learning that occurs in the absence of any particular outcome, such as a reward or punishment.

**non-competitve inhibitor** A reversible inhibitor that reduces the maximum velocity of the reaction, but does not affect the affinity of the substrate for the active site of the enzyme. It usually has a structure very different from that of the substrate and binds to the enzyme at a site different from the active site.

**nondisjunction** The failure of a pair of chromosomes to separate normally during anaphase of cell division.

**non-random mating** Mate selection in which genotype is a factor.

**nonrecombinants** Progeny in which the alleles are present in the same combination as that present in a parent.

**nonsense mutation** A mutation that creates a stop codon, terminating translation.

**non-sister chromatids** Chromatids of differerent members of a pair of homologous chromosomes; although they carry the same complement of genes, they are not genetically identical.

**nonsynonymous (missense) mutation** A point mutation (nucleotide substitution) that causes an amino acid replacement.

**nontemplate strand** The untranscribed partner of the template strand of DNA used in transcription.

**norm of reaction** A graphical depiction of the change in phenotype across a range of environments.

**normal distribution** A distribution whose plot is a bell-shaped curve.

**notochord** In some chordates, a stiff rod of collagen and other proteins that runs along the back and provides support for the axis of the body.

**nuclear envelope** The cell structure, composed of two membranes, inner and outer, that defines the boundary of the nucleus.

**nuclear genome** In eukaryotic cells, the DNA in the chromosomes.

**nuclear localization signal** The signal sequence for the nucleus that enables proteins to move through pores in the nuclear envelope.

**nuclear pore** One of many protein channels in the nuclear envelope that act as gateways that allow molecules to move into and out of the nucleus and are thus essential for the nucleus to communicate with the rest of the cell.

**nuclear transfer**   A procedure in which a hollow glass needle is used to insert the nucleus of a cell into the cytoplasm of an egg whose own nucleus has been destroyed or removed.

**nucleic acid**   A polymer of nucleotides that encodes and transmits genetic information.

**nucleoid**   In prokaryotes, a cell structure with multiple loops formed from supercoils of DNA.

**nucleoside**   A molecule consisting of a 5-carbon sugar and a base.

**nucleosome**   A beadlike repeating unit of histone proteins wrapped with DNA making up the 10-nm chromatin fiber.

**nucleotide**   The subunit of nucleic acids, consisting of a 5-carbon sugar, a nitrogen-containing base, and one or more phosphate groups.

**nucleotide excision repair**   The repair of multiple mismatched or damaged bases across a region; a process similar to mismatch repair, but over a much longer piece of DNA, sometimes thousands of nucleotides.

**nucleotide substitution**   A mutation in which a base pair is replaced by a different base pair; this is the most frequent type of mutation. Also known as a *point mutation*.

**nucleus** (of a cell)   The compartment of the cell that houses the DNA in chromosomes.

**nucleus** (of an atom)   The dense central part of an atom containing protons and neutrons.

**nurture**   Conditioning of behavior by the environment or personal experience.

**obligate**   Describes a mutualism in which one or both sides cannot survive without the other.

**observation**   The act of viewing the world around us.

**occipital lobe**   The region of the brain that processes visual information from the eyes.

**ocean acidification**   An increase in the abundance of carbon dioxide in the oceans that causes the pH of seawater to go down.

**Okazaki fragment**   Any one of the many short DNA pieces in the lagging strand.

**olfaction**   The sense of smell.

**oligodendrocyte**   A type of glial cell that insulates cells in the brain and spinal cord by forming a myelin sheath.

**oligonucleotide**   A short (typically 20 to 30 nucleotides), single-stranded molecule of known sequence produced by chemical synthesis; oligonucleotides are often used as primer sequences in the polymerase chain reaction.

**omasum**   The third chamber in the stomach of ruminants, into which the mixture of food and bacteria passes and where water is reabsorbed.

**ommatidia**   Individual light-focusing elements that make up the compound eyes of insects and crustaceans; the number of ommatidia determines the resolution of the image.

**oncogene**   A cancer-causing gene.

**oocyte**   The unfertilized egg cell produced by the mother; the developing female gamete.

**oogenesis**   The formation of ova or eggs.

**open circulatory system**   A circulatory system found in many smaller animals that contains few blood vessels and in which most of the circulating fluid is contained within the animal's body cavity.

**open reading frame (ORF)**   A stretch of DNA or RNA consisting of codons for amino acids uninterrupted by a stop codon. In genome annotation, this sequence motif identifies the region as potentially protein coding.

**operant conditioning**   Associative learning in which a novel behavior that was initially undirected has become paired with a particular stimulus through reinforcement.

**operator**   The binding site for a repressor protein.

**operon**   A group of functionally related genes located in tandem along the DNA and transcribed as a single unit from one promoter; the region of DNA consisting of the promoter, the operator, and the coding sequence for the structural genes.

**opsin**   A photosensitive protein that converts the energy of light photons into electrical signals in the receptor cell.

**optic nerve**   A cranial nerve that transmits action potentials from ganglion cells in the retina to the visual cortex of the brain.

**orbital**   A region in space where an electron is present most of the time.

**order**   A group of closely related families.

**organ**   Two or more tissues that combine and function together.

**organ of Corti**   A structure in the cochlear duct, supported by the basilar membrane, with specialized hair cells with stereocilia, that functions to convert mechanical vibrations to electrical impulses.

**organelle**   Any one of several compartments in eukaryotes that divide the cell contents into smaller spaces specialized for different functions.

**organic molecule**   A carbon-containing molecule.

**organogenesis**   The transformation of the three germ layers into all the organ systems of the body.

**origin of replication**   Each point on a DNA molecule at which DNA synthesis is initiated.

**osmoconformer**   An animal that matches its internal osmotic pressure to that of its external environment.

**osmoregulation**   The regulation of water and solute levels to control osmotic pressure.

**osmoregulator**   An animal that maintains internal solute concentrations that differ from that of its environment.

**osmosis**   The net movement of water molecules from regions of higher water concentration to regions of lower water concentration.

**osmotic pressure**   The tendency of water to move from one solution into another by osmosis.

**osponization**   The binding of a molecule to a pathogen to facilitate uptake by a phagocyte.

**Osteichthyes**   Bony fish that have a cranium, jaws, and mineralized bones; there are about 20,000–25,000 species.

**osteoblast**   A type of cell that forms bone tissue.

**osteoclast**   A type of cell that secretes digestive enzymes and acid that dissolves the calcium mineral and collagen in bone.

**osteocyte**   Bone cell.

**outer ear**   The part of the human ear that includes the pinna, the ear canal, and tympanic membrane.

**Out-of-Africa hypothesis**   The idea that modern humans arose from *Homo ergaster* descendants in Africa about 200,000 years ago.

**oval window**   The thin membrane between the stapes of the middle ear and the cochlea of the inner ear.

**ovary**   In plants, a hollow structure at the base of the carpel in which the ovules develop and which protects the ovules from being eaten or damaged by animals; generally, the female gonad where eggs are produced.

**oviparity**   Laying eggs.

**ovule**   The female reproductive structure in plants that contains the sporangium, now filled with a female gametophyte.

**ovule cone**   A type of reproductive structure in pines that occurs on upper branches and produces female gametophytes and female gametes.

**ovum (egg)** (plural, **ova**)   The larger, female gamete.

**oxidation reaction**   A reaction in which a molecule loses electrons and releases energy.

**oxidation–reduction reaction**   A reaction involving the loss and gain of electrons between reactants. In biological systems these reactions are often used to store or release chemical energy.

**oxidative phosphorylation**   The fourth stage of cellular respiration, in which electron carriers generated in stages 1–3 donate their high-energy electrons to an electron transport chain.

**oxidizing agent**   An electron acceptor.

**oxygen dissociation curve**   The curve that results when blood $pO_2$ is plotted against the percentage of $O_2$ bound to hemoglobin.

**oxygenic**   Producing oxygen.

**oxytocin**   A posterior pituitary gland hormone that causes uterine contraction during labor and stimulates the release of milk during breastfeeding.

**$P_1$ generation**   The parental generation in a series of crosses.

**pacemaker**   Describes cardiac muscle cells that function as a regulator of heart rhythm.

**palindromic**   Describes sequence identity in the paired strands of a duplex DNA molecule; a symmetry typical of restriction sites.

**pancreas**   A secretory gland that has both endocrine function, secreting hormones, including insulin, directly into the blood, and exocrine function, aiding the digestion of proteins, carbohydrates, and fats by secreting digestive enzymes into ducts that connect to the duodenum.

**paracrine signaling**   Signaling by a molecule that travels a short distance to the nearest neighboring cell to bind its receptor and deliver its message.

**paraphyletic**   Describes groupings that include some, but not all, the descendants of a common ancestor.

**parasexual**   Describes asexual species that generate genetic diversity by the crossing over of DNA during mitosis.

**parasympathetic division**   The division of the autonomic nervous system that slows the heart and stimulates digestion and metabolic processes that store energy, enabling the body to "rest and digest."

**parathyroid gland**   A gland adjacent to the thyroid gland that secretes parathyroid hormone (PTH), which, with calcitonin, regulates the actions of bone cells.

**parenchyma**   Describes thin-walled, undifferentiated cells in the stem of a vascular plant.

**parietal lobe**   The brain region, posterior to the frontal lobe, that controls body awareness and the ability to perform complex tasks.

**parsimony**   Choosing the simplest hypothesis to account for a given set of observations.

**parthenogenesis**   A form of asexual reproduction in which females produce eggs that are not fertilized by males but divide by mitosis and develop into new individuals.

**partial pressure (*p*)**   The fractional concentration of a gas relative to other gases present multiplied by the atmospheric pressure exerted on the gases.

**partially reproductively isolated**   Describes populations that have not yet diverged as separate species but whose genetic differences are extensive enough that the hybrid offspring they produce have reduced fertility or viability compared to offspring produced by crosses between individuals within each population.

**patch**   A bit of habitat that is separated from other bits by inhospitable environments that are difficult or risky for individuals to cross.

**paternal inheritance**   A type of inheritance in which the organelles in the offspring cells derive from those in the father.

**peat bog**   Wetland in which dead organic matter accumulates.

**pedigree**   A diagram of family history that summarizes the record of the ancestral relationships among individuals.

**penis**   The male copulatory organ.

**pepsin**   An enzyme produced in the stomach that breaks down proteins.

**peptide bond**   A covalent bond that links the carbon atom in the carboxyl group of one amino acid to the nitrogen atom in the amino group of another amino acid.

**peptide hormone**   A hormone that is a short chain of linked amino acids.

**peptidoglycan**   A complex polymer of sugars and amino acids that makes up the cell wall.

**peptidyl (P) site**   One of three binding sites for tRNA on the large subunit of a ribosome.

**pericycle**   In roots, a single layer of cells just to the inside of the endodermis from which new root apical meristem is formed.

**periodic selection**   The episodic loss of diversity as a successful variant outcompetes others.

**periodic table of the elements**   The arangement of the chemical elements in tabular form, organized by their chemical properties.

**peripatric speciation**   A specific kind of allopatric speciation in which a few individuals from a mainland population disperse to a new location remote from the original population and evolve separately.

**peripheral membrane protein**   A protein that is temporarily associated with the lipid bilayer or with integral membrane proteins through weak noncovalent interactions.

**peripheral nervous system (PNS)**   Collectively, the sensory and motor nerves, including the cranial and spinal nerves, and interneurons and ganglia.

**peristalsis**   Waves of muscular contraction that move food toward the base of the stomach.

**personalized medicine**   An approach in which the treatment is matched to the patient, not the disease; examination of an individual's genome sequence, by revealing his or her disease susceptibilities and drug sensitivities, allows treatments to be tailored to that individual.

**petal**   A structure, often brightly colored and distinctively shaped, that forms the next-to-outermost whorl of a flower with other petals; petals attract and orient animal pollinators.

**phagocyte**   A type of immune cell that engulfs and destroys foreign cells or particles.

**phagocytosis**   The engulfing of a cell or particle by another cell.

**pharyngeal slit**   A vertical opening separated from other slits by stiff rods of protein in the pharynx of hemichordates.

**pharynx**   The region of the throat that connects the nasal and mouth cavities; in hemichordates, a tube that connects the mouth and the digestive tract.

**phenol**   Any one of a class of compounds, produced by some plants as a defensive mechanism.

**phenotype**   The expression of a physical, behavioral, or biochemical trait; an individual's observable phenotypes include height, weight, eye color, and so forth.

**pheromone**   A water- or airborne chemical compound released by animals into the environment that signals and influences the behavior of other members of their species.

**phloem**   The outer tissue of a stem, which transports carbohydrates from leaves to the rest of the plant body.

**phloem sap**   The sugar-rich solution that flows through both the lumen of the sieve tubes and the sieve plate pores.

**phosphatase**   An enzyme that removes a phosphate group from another molecule.

**phosphate group**   A chemical group consisting of a phosphorus atom bonded to four oxygen atoms.

**phosphodiester bond**   A bond that forms when a phosphate group in one nucleotide is covalently joined to the sugar unit in another nucleotide. Phosophodiester bonds are relatively stable and form the backbone of a DNA strand.

**phospholipid**   A type of lipid and a major component of the cell membrane.

**photic zone**   The surface layer of the ocean through which enough sunlight penetrates to enable photosynthesis.

**photoheterotroph**   An organism that uses the energy from sunlight to make ATP and relies on organic molecules obtained from the environment as the source of carbon for growth and other vital functions.

**photoperiod**   Day length.

**photoperiodism**   The effect of the photoperiod, or day length, on flowering.

**photoreceptor**   A molecule whose chemical properties are altered when it absorbs light; also, photoreceptors are the sensory receptors in the eye.

**photorespiration**   The process in which ATP is used to drive the reactions that convert a portion of the carbon atoms in 2-phosphoglycolate into 3-phosphoglyceric acid, which can reenter the Calvin cycle.

**photosynthesis**   The biochemical process in which carbohydrates are built from carbon dioxide and the energy of sunlight; oxygen is released as a waste product.

**photosynthetic electron transport chain**   A series of redox reactions in which electrons are passed from one compound to another.

**photosystem**   A protein-pigment complex that absorbs light energy to drive redox reactions and thereby sets the photosynthetic electron transport chain in motion.

**photosystem I**   The photosystem that energizes electrons with a second input of light energy so they have enough energy to reduce $NADP^+$.

**photosystem II**   The photosystem that supplies electrons to the beginning of the electron transport chain. When photosystem II loses an electron it can pull electrons from water.

**phototroph**   An organism that captures energy from sunlight.

**phototropic**   Bending in response to light. A positive phototropic response, as in stems, is toward light; a negative phototropic response, as in roots, is away from light.

**phragmoplast**   In dividing plant cells, a structure formed by overlapping microtubules that guide vesicles containing cell wall components to the middle of the cell.

**phylogenetic tree**   A diagrammed hypothesis about the evolutionary history, or phylogeny, of a species.

**phylogeny**   The history of descent with modification and the accumulation of change over time.

**phylum**   A group of closely related classes, defined by having one of a number of distinctive body plans.

**phytochrome**   A photoreceptor that switches back and forth between two stable forms, active and inactive, depending on its exposure to light.

**pilus** (plural, **pili**)   A threadlike, hollow structure through which plasmids are transferred between bacteria.

**pineal gland**   A gland located in the thalamic region of the brain that responds to autonomic nervous system input by secreting melatonin, which controls wakefulness.

**pinna** (plural, **pinnae**)   Small unit of the photosynthetic surface on fern leaves; also, the external structure of mammalian ears that enhance the reception of sound waves contacting the ear.

**pit**   A circular or ovoid region in the walls of xylem cells where the lignified cell wall layer is not produced.

**pith**   In a stem, the region inside the ring of vascular bundles.

**pituitary gland**   A gland beneath the brain that produces a number of hormones, including growth hormone.

**placenta**   In placental mammals, an organ formed by the fusion of the chorion and allantois that allows the embryo to obtain nutrients directly from the mother.

**placental mammal**   A mammal that provides nutrition to the embryo through the placenta, a temporary organ that develops in the uterus; placental mammals include carnivores, primates, hooved mammals, and whales.

**plasma (effector) cell**   A cell that secretes antibodies.

**plasma membrane**   The membrane that defines the space of the cell, separating the living material within the cell from the nonliving environment around it.

**plasmid**   In bacteria, a small circular molecule of DNA carrying a small number of genes that can replicate independently of the bacterial genomic DNA.

**plasmodesmata**   Connections in plant cells between the plasma membranes of adjacent cells that permit materials to pass directly from the cytoplasm of one cell to the cytoplasm of another.

**plasmogamy**   The cytoplasmic union of two cells.

**plate tectonics**   The dynamic movement of Earth's crust, the outer layer of Earth.

**pleiotropy**   The phenomenon in which a single gene has multiple effects on seemingly unrelated traits.

**pluripotent**   Describes embryonic stem cells (cells of the inner mass), which can give rise to any of the three germ layers and therefore to any cell of the body.

**podocytes**   Cells with footlike processes that make up one of the three layers of the filtration barrier of the glomerulus.

**point mutation**   A mutation in which a base pair is replaced by a different base pair; this is the most frequent type of mutation. Also known as a *nucleotide substitution*.

**Pol II**   The RNA polymerase complex responsible for transcription of protein-coding genes.

**polar (molecule)**   A molecule that has regions of positive and negative charge.

**polar body**   A small cell produced by the asymmetric first meiotic division of the primary oocyte.

**polar covalent bond**   Bonds that do not share electrons equally.

**polar transport**   The coordinated movement of auxin across many cells.

**polarity**   An asymmetry such that one end of a structure differs from the other.

**polarized**   Having opposite properties in opposite parts; describes a resting membrane potential in which there is a buildup of negatively charged ions on the inside surface of the cell's plasma membrane and positively charged ions on its outer surface.

**pollen cone**   A type of reproductive structure in pines that occurs near the tips of lower branches and produces male gametophytes and male gametes.

**pollen tube**   A structure produced by the male gametophyte that grows outward through an opening in the sporopollenin coat.

**pollination**   The process in which pollen is carried to an ovule.

**poly(A) tail**   The nucleotides added to the 3′ end of the primary transcript by polyadenylation.

**polyadenylation**   The addition of a long string of consecutive A-bearing ribonucelotideas to the 3′ end of the primary transcript.

**polycistronic mRNA**   A single molecule of messenger RNA that is formed by the transcription of a group of functionally related genes located next to one another along the bacterial DNA.

**polymer**   A complex organic molecule made up of repeated simpler units connected by covalent bonds.

**polymerase chain reaction (PCR)**   A selective and highly sensitive method for making copies of a piece of DNA, which allows a targeted region of a DNA molecule to be replicated into as many copies as desired.

**polymorphism**   Any genetic difference among individuals sufficiently common that it is likely to be present in a group of 50 randomly chosen individuals.

**polypeptide**   A polymer of amino acids connected by peptide bonds.

**polyphyletic**   Describes groupings that do not include the last common ancestor of all members.

**polyploidy**   The condition of having more than two complete sets of chromosomes in the genome.

**polysaccharide**   A polymer of simple sugars. Polysaccharides provide long-term energy storage or structural support.

**polyspermy**   Fertilization by more than one sperm.

**pons**   A part of the brainstem.

**population**   All the individuals of a given species that live and reproduce in a particular place; one of several interbreeding groups of organisms of the same species living in the same geographical area.

**population density**   The size of a population divided by its range.

**population size**   The number of individuals of all ages alive at a particular time in a particular place.

**positive feedback**   In the nervous system, the type of feedback in which a stimulus causes a response that leads to an enhancement of the original stimulus that leads to a larger response. In the endocrine system, the type of feedback in which a stimulus causes a response, and that response causes a further response in the same direction. In both cases, the process reinforces itself until interrupted.

**positive regulation**   The process in which a regulatory molecule must bind to the DNA at a site near the gene in order for transcription to take place.

**positive selection**   Natural selection that increases the frequency of a favorable allele.

**positively selected**   Describes T cells that recognize self MHC molecules on epithelium cells and continue to mature.

**posterior pituitary gland**   The region of the pituitary gland that develops from neural tissue at the base of the brain and into which neurosecretory cells of the hypothalamus extend that secrete releasing factors.

**posttranslational modification**   The modification, after translation, of proteins in ways that regulate their structure and function.

**post-zygotic**   Describes factors that cause the failure of the fertilized egg to develop into a fertile individual, thus causing reproductive isolation.

**potential energy**   Stored energy that is released by a change in an object's structure or position.

**power stroke**   The stage in the cross-bridge cycle in which the myosin head pivots forward and generates a force, causing the myosin and actin filaments to slide relative to each other.

**prediction**   An informed guess about the outcome of an experiment or observation based on a hypothesis.

**pregnancy**   The carrying of one or more embryos in the uterus.

**premolar**   One of the teeth between the canines and molars that are specialized for shearing tough foods.

**pre-zygotic**   Describes factors that that prevent the fertilization of an egg, thus causing reproductive isolation.

**primary active transport**   Active transport that uses the energy of ATP directly.

**primary bronchi**   The two divided airways from the trachea, supported by cartilage rings, each airway leading to a lung.

**primary growth**   The increase in plant length made possible by apical meristems.

**primary motor cortex**   The part of the frontal lobe of the brain that produces complex coordinated behaviors by controlling skeletal muscle movements.

**primary oocyte**   A diploid cell formed by mitotic division of oogonia during fetal development.

**primary producer**   An organism that takes up inorganic carbon, nitrogen, phosphorus, and other compounds from the environment and converts them into organic compounds that will provide food for other organisms in the local environment.

**primary response**   The response to the first encounter with an antigen, during which there is a short lag before antibody is produced.

**primary somatosensory cortex**   The part of the parietal lobe that integrates tactile information from specific body regions and relays it to the motor cortex.

**primary spermatocyte**   A diploid cell formed by mitotic division of spermatogonia at the beginning of spermatogenesis.

**primary structure**   The sequence of amino acids in a protein.

**primary transcript**   The initial RNA transcript that comes off the template DNA strand.

**primate**   A member of an order of mammals that share a number of general features that distinguish them from other mammals, including nails rather than claws, front-facing eyes, and an opposable thumb.

**primer**   A short stretch of RNA at the beginning of each new DNA strand that serves as a starter for DNA synthesis; an oligonucleotide that serves as a starter in the polymerase chain reaction.

**principle of independent assortment**   The principle that segregation of one set of alleles of a gene pair is independent of the segregation of another set of alleles of a different gene pair.

**principle of segregation**   The principle by which half the gametes receive one allele of a gene and half receive the other allele.

**probability**   Among a very large number of observations, the expected proportion of observations that are of a specified type.

**probe**   A labeled DNA fragment that can be tracked in a procedure such as a Southern blot.

**procambial cell**   A cell that retains the capacity for cell division and gives rise to both xylem and phloem.

**product**   Any one of the transformed molecules that result from a chemical reaction.

**progesterone**   A hormone secreted by the ovaries that maintains the thickened and vascularized uterine lining.

**prokaryote**   A unicellular organism without a nucleus. Often used to refer collectively to archaeons and bacteria.

**prometaphase**   The mitosis in which the nuclear envelope breaks down and the microtubules of the mitotic spindle attach to chromosomes.

**promoter**   A regulatory region where RNA polymerase and associated proteins bind to the DNA duplex.

**proofreading**   The process in which DNA polymerases correct their own errors by excising and replacing a mismatched base.

**prophase**   The stage of mitosis characterized by the appearance of visible chromosomes.

**prophase I**   The beginning of meiosis I, marked by the visible manifestation of chromosome condensation.

**prophase II**   The stage of meiosis II in which the chromosomes in the now-haploid nuclei recondense to their maximum extent.

**prostate gland**   An exocrine gland that produces a thin, slightly alkaline fluid that helps maintain sperm motility and counteracts the acidity of the female reproductive tract.

**reversible inhibitor** Any one of the class of inhibitors that form weak bonds with enzymes and easily dissociate from them.

**rhizosphere** The soil layer that surrounds actively growing roots.

**ribonucleic acid (RNA)** A molecule closely related to DNA that is synthesized by proteins from a DNA template.

**ribose** The sugar in RNA.

**ribosomal RNA (rRNA)** Noncoding RNA found in all ribosomes that aid in translation.

**ribosome** A complex structure of RNA and protein, bound to the cytosolic face of the RER in the cytoplasm, on which proteins are synthesized.

**ribulose bisphosphate carboxylase oxygenase (rubisco)** The enzyme that catalyzes a carboxylation reaction.

**ribulose-1,5-bisphosphate (RuBP)** The 5-carbon sugar to which carbon dioxide is added in carboxylation.

**ring species** Species that contain populations that are reproductively isolated from each other but can exchange genetic material through other, linking populations.

**RISC (RNA-induced silencing complex)** A protein complex that is targeted to specific mRNA molecules by base pairing with short regions on the target mRNA, inhibiting translation or degrading the RNA.

**ritualization** The process of co-opting and modifying begaviors used in another context by increasing the conspicuousness of the behavior, reducing the amount of variation in the behavior so that it can be immediately recognized, and increasing its separation from the original function.

**RNA editing** The process in which some RNA molecules become a substrate for enzymes that modify particular bases in the RNA, thereby changing its sequence and what it codes for.

**RNA polymerase** The enzyme that carries out polymerization of ribonucleoside triphosphates from a DNA template to produce an RNA transcript.

**RNA polymerase complex** Aggregate of proteins that synthesize the RNA transcript complementary to the template strand of DNA.

**RNA primase** An RNA polymerase that synthesizes a short piece of RNA complementary to the DNA template and does not require a primer.

**RNA processing** Chemical modification that converts the primary transcript into finished mRNA, enabling the RNA molecule to be transported to the cytoplasm and recognized by the translational machinery.

**RNA splicing** The process of intron removal from the primary transcript.

**RNA transcript** The RNA sequence synthesized from a DNA template.

**RNA world hypothesis** The belief that RNA, not DNA, was the original information-storage molecule in the earliest forms of life on Earth.

**rod cell** A type of photoreceptor cell on the retina that detects light and shades ranging from white to shades of gray and black, but not color.

**root apical meristem** A group of totipotent cells near the tip of a root that is the source of new root cells.

**root cap** A structure that covers and protects the root apical meristem as it grows through the soil.

**root hair** A slender outgrowth produced by epidermal cells in active areas of the root, which greatly increases the surface area of the root.

**root nodule** A structure, formed by dividing root cells, into which nitrogen-fixing bacteria enter.

**roots** A major organ system of vascular plants, generally belowground.

**rough endoplasmic reticulum (RER)** The part of the endoplasmic reticulum with attached ribosomes.

**r-strategist** A species that produces large numbers of offspring but provides few resources for their support.

**rumen** The first chamber in the stomach of ruminants, which, along with the reticulum, harbors large populations of anaerobic bacteria that break down cellulose.

**S phase** The phase of interphase in which the entire DNA content of the nucleus is replicated.

**saccharide** The simplest carbohydrate molecule, also called a *sugar*; saccharides store energy in their bonds.

**saltatory propagation** The movement of an action potential along a myelinated axon, "jumping" from node to node.

**Sanger sequencing** A procedure in which the terminated daughter strands help in determining the DNA sequence.

**sapwood** In long-lived trees, the layer adjacent to the vascular cambium that contains the functional xylem.

**sarcomere** The region from one Z disc to the next, the basic contractile unit of a muscle.

**sarcoplasmic reticulum (SR)** A modified form of the endoplasmic reticulum surrounding the myofibrils of muscle cells.

**saturated** Describes fatty acids that do not contain double bonds; the maximum number of hydrogen atoms is attached to each carbon atom, "saturating" the carbons with hydrogen atoms.

**scaffold** A supporting protein structure in a metaphase chromosome.

**Schwann cell** A type of glial cell that insulates sensory and motor neurons by forming a myelin sheath.

**scientific method** A deliberate, careful, and unbiased way of learning about the natural world.

**sclera** A tough, white outer layer surrounding the vertebrate eye.

**scrotum** A sac outside the abdominal cavity of the male that holds the testes.

**second law of thermodynamics** The principle that the transformation of energy is associated with an increase in the degree of disorder in the universe.

**second messenger** An intermediate cytosolic signaling molecule that transmits signals from a receptor to a target within the cell. (First messengers transmit signals from outside the cell to a receptor.)

**secondary active transport**   Active transport that uses the energy of an electrochemical gradient to drive the movement of molecules.

**secondary growth**   The increase in plant diameter, which results from meristem that forms only after elongation below the growing tip is complete.

**secondary oocyte**   A large cell produced by the asymmetric first meiotic division of the primary oocyte.

**secondary phloem**   New phloem cells produced by the vascular cambium, which appear to the outside of the vascular cambium.

**secondary response**   The response to re-exposure to an antigen, which is quicker, stronger, and longer than the primary response.

**secondary sexual characteristic**   A trait that characterizes and differentiates the two sexes but that does not relate directly to reproduction.

**secondary spermatocyte**   A diploid cell formed during the first meiotic division of the primary spermatocyte.

**secondary structure**   The structure formed by interactions between stretches of amino acids in a protein.

**secondary xylem**   New xylem cells produced by vascular cambium, which appear to the inside of the vascular cambium.

**second-division nondisjunction**   Disjunction in the second meiotic division.

**secretin**   A hormone released by cells lining the duodenum in response to the acidic pH of the stomach contents entering the small intestine and that stimulates the pancreas to secrete bicarbonate ions.

**secretion**   In renal systems, an active process that eliminates substances that were not previously filtered from the blood.

**seed**   A fertilized ovule; seeds are multicellular structures that allow offspring to disperse away from the parent plant.

**seed coat**   A protective outer structure surrounding the seed formed from tissues that surround the sporangial wall.

**segmentation**   The formation of discrete parts or segments in the insect embryo.

**segregate**   Separate; applies to chromosomes or members of a gene pair moving into different gametes.

**selection**   The retention or elimination of random mutations in a population of organisms.

**selective barrier**   Describes the plasma membrane, which lets some molecules in and out freely, lets others in and out only under certain conditions, and prevents other molecules from passing through at all.

**self-compatible**   Describes species in which pollen and eggs produced by flowers on the same plant can unite and produce viable offspring.

**self-incompatible**   Describes species in which pollination by the same or a closely related individual does not lead to fertilization.

**self-propagating**   Continuing without input from an outside source; action potentials are self-propagating in that they move along axons by sequentially opening and closing adjacent ion channels.

**semen**   A fluid that nourishes and sustains sperm as they travel in the male and then the female reproductive tracts.

**semicircular canal**   One of three connected fluid-filled tubes in the mammalian inner ear that contains hair cells that sense angular motions of the head in three perpendicular planes.

**semiconservative replication**   The mechanism of DNA replication in which each strand of a parental DNA duplex serves as a template for the synthesis of a new daughter strand.

**seminal vesicles**   Two glands at the junction of the vas deferens and the prostate gland that secrete a protein- and sugar-rich fluid that makes up most of the semen and provides energy for sperm motility.

**seminiferous tubules**   A series of tubes in the testes where sperm are produced.

**semipermeable**   Selectively permeable; a semipermeable cell membrane allows the movement of some molecules but not others.

**sender**   The indivdual who, during communication, supplies a signal that elicits a response from the receiver.

**sensitization**   The enhancement of a response to a stimulus that is achieved by first presenting a strong or novel stimulus.

**sensory neuron**   A neuron that receives and transmits information about an animal's environment or its internal physiological state.

**sensory organ**   A group of sensory receptors that converts particular physical and chemical stimuli into nerve impulses that are processed by a nervous system and sent to a brain.

**sensory receptor**   A sensory neuron with specialized membranes in which receptor proteins are embedded.

**sensory transduction**   The conversion of physical or chemical stimuli into nerve impulses.

**sepal**   A structure, often green, that forms the outermost whorl of a flower with other sepals and encases and protects the flower during its development.

**septum** (plural, **septa**)   In fungi, a wall that partially divides the cytoplasm into separate cells in hyphae.

**sequence assembly**   The process in which short nucleotide sequences of a long DNA molecule are arranged in the correct order to generate the complete sequence.

**sequence motif**   Any of a number of sequences or sequence arrangements that indicate the likely function of a segment of DNA.

**serosa**   An outer layer of cells and connective tissue that covers and protects the gut.

**Sertoli cell**   A type of cell in the seminiferous tubules that supports sperm production.

**sex chromosome**   One of the chromosomes associated with sex, in most animals denoted the X and Y chromosomes.

**sex determination**   The development of male or female characteristics, resulting from the presence or absence of the Y chromosome.

**sexual dimorphism**   A phenotypic difference between the sexes.

**sexual reproduction**   The process of producing offspring that receive genetic material from two parents; in eukaryotes, the process occurs through meiosis and fertilization.

**sexual selection**   A form of selection that promotes traits that increase an individual's access to reproductive opportunities.

**shell** (of an atom)   An energy level.

**shoot**   The collective name for the leaves, stems, and reproductive organs, the major aboveground organ systems of vascular plants.

**shoot apical meristem**   A group of totipotent cells near the tip of a branch that gives rise to new shoot tissues in plants.

**short-day plant**   A plant that flowers only when the day length is less than a critical value.

**shotgun sequencing**   DNA sequencing method in which the sequenced fragments do not originate from a particular gene or region but from sites scattered randomly across the molecule.

**sickle-cell anemia**   A condition in which hemoglobin molecules tend to crystallize when exposed to lower-than-normal levels of oxygen, causing the red blood cells to collapse and block capillary blood vessels.

**side chain**   A chemical group attached to the central carbon atom of an amino acid, whose structure and composition determine the identity of the amino acid; also known as an *R group*.

**sieve element**   A highly modified cell, lacking most intracellular structure, that is connected end to end with other sieve elements to make up a sieve tube.

**sieve plate**   A modified end wall with large pores that links sieve elements.

**sieve tube**   A multicellular unit composed of sieve elements that are connected end to end, through which phloem transport takes place.

**sigma factor**   A protein that associates with RNA polymerase that facilitates its binding to specific promoters.

**signal sequence**   An amino acid sequence that directs a protein to its proper cellular compartment.

**signal transduction**   The process in which an extracellular molecule acts as a signal to activate a receptor, which transmits information through the cytoplasm.

**signaling cell**   The source of the signaling molecule.

**signaling molecule**   The carrier of information transmitted when the signaling molecule binds to a receptor; also referred to as a *ligand*.

**signal-recognition particle (SRP)**   An RNA–protein complex that binds with part of a polypeptide chain and marks the molecule for incorporation into the endoplasmic reticulum (eukaryotes) or the plasma membrane (prokaryotes).

**simple multicellularity**   Multicellularity, achieved by adhesive molecules, in which there is relatively little communication or transfer of resources between cells and little differentiation of specialized cell types.

**SINE**   Any one of many short interspersed nuclear elements of about 300 base pairs present in multiple copies in a genome owing to transposition.

**single-gene trait**   A trait determined by Mendelian alleles of a single gene without much influence from the environment.

**single-lens eye**   An eye structure found in vertebrates and cephalopod mollusks that works like a camera to produce a sharply defined image of the animal's visual field.

**single-nucleotide polymorphism (SNP)**   A site in the genome where a base pair differs among individuals in a population.

**single-stranded binding protein**   A protein that binds single-stranded nucleic acids.

**sink**   In plants, any portion of the plant that needs carbohydrates to fuel growth and respiration, such as a root, young leaf, or developing fruit.

**sinoatrial (SA) node**   A specialized region of the heart containing pacemaker cells where the heartbeat is initiated.

**sister chromatids**   The two identical copies of chromosomes produced by DNA replication.

**sister groups**   Groups that are more closely related to each other than either of them is to any other group.

**skeletal muscle**   Muscle that connects to the body skeleton to move an animal's limbs and torso.

**sliding filament model**   The hypothesis that muscles produce force and change length by the sliding of actin filaments relative to myosin filaments.

**slow-twitch**   Describes muscle fibers that contract slowly and consume less ATP than do fast-twitch fibers to produce force.

**small interfering RNA (siRNA)**   A type of small double-stranded regulatory RNA that becomes part of a complex able to cleave and destroy single-stranded RNA with a complementary sequence.

**small intestine**   Part of the midgut; the site of the last part of digestion and most nutrient absorption.

**small nuclear RNA (snRNA)**   Noncoding RNA found in eukaryotes and involved in splicing, polyadenylation, and other processes in the nucleus.

**small regulatory RNA**   A short RNA molecule that works primarily by blocking transcription or translation.

**smooth endoplasmic reticulum (SER)**   The portion of the endoplasmic reticulum that lacks ribosomes.

**smooth muscle**   The muscle in the walls of arteries, the respiratory system, and the digestive and excretory systems; smooth muscle appears uniform under the light microscope.

**solubility**   The ability of a substance to dissolve.

**solute**   A dissolved molecule such as the electrolytes, amino acids, and sugars often found in water, a solvent.

**solvent**   A liquid capable of dissolving a substance.

**somatic cell**   A nonreproductive cell, the most common type of cell in body.

**somatic mutation**   A mutation that occurs in somatic cells.

**somatic nervous system**   The voluntary component of the peripheral nervous system, which is made up of sensory neurons that respond to external stimuli and motor neurons that synapse with voluntary muscles.

**source**   In plants, a region that produces or stores carbohydrates.

**Southern blot**   A method for determining the size and number of copies of a DNA sequence of interest by means of a labeled probe.

**spatial summation**   The converging of multiple receptors onto a neighboring neuron, increasing its firing rate proportionally to the number of signals received.

**speciation**   The process that produces new and distinct forms of life.

**species**   A group of individuals that can exchange genetic material through interbreeding to produce fertile offspring.

**species–area relationship**   The relationship between island size and equilibrium species diversity.

**spermatogenesis**   The formation of sperm.

**spermatozoa (sperm)**   The smaller, male gametes.

**spinal cord**   In vertebrates, a central tract of neurons that passes through the vertebrae to transmit information between the brain and the periphery of the body.

**spinal nerve**   In vertebrates, a nerve running from the spinal cord to the periphery containing axons of both sensory and motor neurons.

**spindle apparatus**   The organelle formed by microtubules that separates replicated chromosomes during eukaryotic cell division.

**spiracle**   An opening in the exoskeleton on either side of an insect's abdomen through which gases are exchanged.

**spliceosome**   A complex of RNA and protein that catalyzes RNA splicing.

**spongy bone**   Bone tissue consisting of trabeculae, and thus lighter than compact bone, found in the ends of limb bones and within vertebrae.

**spontaneous**   Occurring in the absence of any assignable cause; most mutations are spontaneous.

**sporangium**   A plant structure in which thousands of cells undergo meiosis, producing large numbers of haploid spores.

**spore**   In plants, the haploid cells resulting from meiotic cell division that disperse and give rise to a new haploid generation. In animals, reproductive haploid cells resulting from meiotic cell division (in some species called gametes).

**sporophyte**   Describes the diploid multicellular generation that gives rise to spores.

**sporopollenin**   A complex mixture of polymers that is remarkably resistant to environmental stresses such as ultraviolet radiation and desiccation; a wall containing sporopollenin protects the plant zygote.

**stabilizing selection**   A form of selection that maintains the status quo and selects against extremes.

**stamen**   A pollen-producing floral organ in the center whorl.

**stapes**   A small bone in the middle ear that helps amplify the waves that strike the tympanic membrane; the stapes connects to the oval window of the cochlea.

**starch**   The form in which glucose is stored in plants.

**Starling's Law**   The correspondence between change in stroke volume and change in the volume of blood filling the heart.

**statocyst**   A type of gravity-sensing organ found in most invertebrates.

**statolith**   In plants, a large starch-filled organelle in the root cap that senses gravity; in animals, a dense particle that moves freely within a statocyst, enabling it to sense gravity.

**stem cell**   An undifferentiated cell that can undergo an unlimited number of mitotic divisions and differentiate into any of a large number of specialized cell types.

**stereocilia**   Nonmotile cell-surface projections on hair cells whose movement causes a depolarization of the cell's membrane.

**steroid**   A type of lipid.

**steroid hormone**   A hormone that is derived from cholesterol.

**stigma**   The surface of the top of the style, to which pollen adheres.

**stomach (crop)**   The last part of the foregut, which serves as an initial storage and digestive chamber.

**stomata**   Pores in the epidermis of a leaf that regulate the diffusion of gases between the interior of the leaf and the atmosphere.

**Stramenopila**   A eukaryotic superkingdom including unicellular organisms, giant kelps, algae, protozoa, free-living cells, and parasites; distinguished by a flagellum with two rows of stiff hairs and, usually, a second, smooth flagellum.

**striated muscle**   Skeletal muscle and cardiac muscle, which appear striped under a light microscope.

**strigolactone**   A hormone, produced in roots and transported upward in the xylem, that inhibits the outgrowth of axillary buds.

**stroke volume (SV)**   The volume of blood pumped during each beat.

**stroma**   The region surrounding the thylakoid, where carbohydrate synthesis takes place.

**stromatolite**   A layered structure that records sediment accumulation by microbial communities.

**structural gene**   A gene that codes for the sequence of amino acids in a polypeptide chain.

**style**   A cylindrical stalk at the top of a carpel.

**suberin**   A waxy compound coating cork cells that protects against mechanical damage, the entry of pathogens, and water loss.

**submucosa**   A tissue layer surrounding the mucosa that contains blood vessels, lymph vessels, and nerves.

**subspecies**   Allopatric populations that have yet to evolve even partial reproductive isolation but which have accumulated a few population-specific traits.

**substrate (S)**   A molecule acted upon by an enzyme.

**substrate-level phosphorylation**   A way of generating ATP in which a phosphate group is transferred to ADP from an organic molecule, which acts as a phosphate donor or substrate.

**succession**   The replacement of species by other species over time.

**sugar** The simplest carbohydrate molecule; also called a *saccharide*.

**sulci** Deep crevices in the brain that separate the lobes of the cerebral hemispheres.

**supercoil** A coil of coils; a circular molecule of DNA can coil upon itself to form a supercoil.

**superkingdom** One of seven major groups of eukaryotic organisms, classified by molecular sequence comparisons.

**supernormal stimulus** An exaggerated stimulus that elicits a response more strongly than the normal stimulus.

**surfactant** A compound that reduces the surface tension of a fluid film.

**survivorship** The proportion of individuals from an initial cohort that survive to each successive stage of the life cycle.

**suspension filter feeding** The most common form of food capture by animals, in which water with food suspended in it passes through a sievelike structure.

**swallowing** A complex set of motor reflexes that conveys food through the pharynx and into the esophagus and stomach.

**symbiont** An organism that lives in closely evolved association with another species.

**symbiosis** (plural, **symbioses**) A close interaction between species that have evolved over long periods of time.

**sympathetic division** The division of the autonomic nervous system that generally produces arousal and increased activity; active in the fight-or-flight response.

**sympatric** Describes populations that are in the same geographic location.

**synapomorphy** A shared derived character; a homology shared by some, but not all, members of a group.

**synapse** A junction through which the axon terminal communicates with a neighboring cell.

**synapsis** The gene-for-gene pairing of homologous chromosomes in prophase I of meiosis.

**synaptic cleft** The space between the axon of the presynaptic cell and the neighboring postsynaptic cell.

**synaptic plasticity** The ability to adjust synaptic connections between neurons.

**synonymous (silent) mutation** A mutation in a codon that does not alter the corresponding amino acid in the polypeptide.

**systemic acquired resistance (SAR)** The ability of a plant to resist future infections, occurring in response to a wide range of pathogens.

**systemic circulation** Circulation of the blood to the body, excluding the lungs.

**systole** The contraction of the ventricles.

**T cell receptor (TCR)** A protein receptor on a mature T cell that recognizes and binds to an antigen.

**T lymphocyte (T cell)** A cell type that matures in the thymus and includes helper and cytotoxic cells.

**tannin** Any one of a group of phenols found widely in plant tissues that bind with proteins and reduce their digestibility.

**taste bud** One of the sensory organs for taste.

**TATA box** A DNA sequence present in many promoters in eukaryotes and archaeons that serves as a protein-binding site for a key general transcription factor.

**taxis** (plural, **taxes**) Movement in a specific direction in response to a stimulus.

**taxon** (plural, **taxa**) All the species in a taxonomic entity such as family or genus.

**tectorial membrane** A rigid membrane in the cochlear duct, against which the stereocilia of hair cells in the organ of Corti bend when stimulated by vibration, setting off an action potential.

**telomerase** An enzyme that synthesizes telomere repeats.

**telomere** A repeating sequence at each end of a eukaryotic chromosome.

**telophase** The stage of mitosis in which the nuclei of the daughter cells are formed and the chromosomes uncoil to their original state.

**telophase I** The stage of meiosis I in which the chromosomes uncoil slightly, a nuclear envelope briefly reappears, and in many species the cytoplasm divides, producing two separate cells.

**telophase II** The stage of meiosis II in which the chromosomes uncoil and become diffuse, a nuclear envelope forms around each set of chromosomes, and the cytoplasm divides by cytokinesis.

**template** A strand of DNA or RNA whose squence of nucleotides is used to sythesis a complementary strand.

**template strand** In DNA replication, the parental strand whose sequence is used to synthesize a complementary daughter strand.

**temporal lobe** The region of the brain involved in the processing of sound, language and reading, and object identification and naming.

**temporal separation** Separation in time, as the pre-zygotic isolation of both plants and animals in time.

**temporal summation** The frequency of synaptic stimuli; the integration of sensory stimuli that are received repeatedly over time by the same sensory cell.

**temporomandibular joint** A specialized jaw joint in mammals that allows the teeth of the lower and upper jaws to fit together precisely.

**tendon** A collagen structure that attaches muscles to the skeleton and transmits muscle forces over a wide range of joint motion.

**termination** In protein translation, the stage in which the addition of amino acids stops and the completed polypeptide chain is released from the ribosome. In cell communication, the stopping of a signal.

**terminator** A DNA sequence at which transcription stops and the transcript is released.

**terpene** Any one of a group of compounds that do not contain nitrogen and produced by some plants as a defensive mechanism.

**tertiary structure**   The overall three-dimensional shape of a protein, formed by interactions between secondary structures.

**test** (of a hypothesis)   An experiment or observation to determine whether a prediction made by the hypothesis holds true.

**test** (of a protist)   A "house" constructed of organic molecules that shelters a protist.

**testcross**   Any cross of an unknown genotype with a homozygous recessive genotype.

**testis** (plural, **testes**)   The male gonad, where sperm are produced.

**testosterone**   A steroid hormone, secreted by the testes, that plays key roles in male growth, development, and reproduction.

**tetanus**   A muscle contraction of sustained force.

**tetraploid**   A cell or organism with four complete sets of chromosomes; a double diploid.

**Tetrapoda**   Animals whose last common ancestor had four limbs; this group includes amphibians, lizards, turtles, crocodilians, birds, and mammals (some tetrapods, like snakes, have lost their legs in the course of evolution).

**thalamus**   The inner brain region of the forebrain, which acts as a relay station for sensory information sent to the cerebrum.

**thallus**   A flattened photosynthetic structure produced by some bryophytes.

**Thaumarchaeota**   One of the three major divisions of Archaea; thaumarchaeota are chemotrophs, deriving energy from the oxidation of ammonia.

**theory**   A general explanation of a natural phenomenon supported by a large body of experiments and observations.

**theory of island biogeography**   A theory that states that the number of species that can occupy a habitat island depends on two factors: the size of the island and the distance of the island from a source of colonists.

**thermoreceptor**   A sensory receptor in the skin and in specialized regions of the central nervous system that responds to heat and cold.

**thick filament**   A parallel grouping of myosin molecules that makes up the myosin filament.

**thin filament**   Two helically arranged actin filaments twisted together that make up the actin filament.

**threshold potential**   The critical depolarization voltage of −50 mV required for an action potential.

**thylakoid**   The internal membrane-bound compartment in the center of chloroplasts, consisting of the highly folded thylakoid membrane, which contains light-collecting pigments and is the site of the photosynthetic electron transport chain and the interior lumen.

**thylakoid membrane**   A highly folded membrane in the center of the chloroplast that contains light-collecting pigments and that is the site of the photosynthetic electron transport chain.

**thymine (T)**   A pyrimidine base.

**thyroid gland**   A gland located in the front of the neck that leads to the release of two peptide hormones, thyroxine and triiodothyorine.

**Ti plasmid**   A small circular DNA molecule in virulent strains of *R. radiobacter* containing genes that can be integrated into the host cell's genome, as well as the genes needed to make this transfer.

**tidal ventilation**   A breathing technique in most land vertebrates in which air is drawn into the lungs during inhalation and moved out during exhalation.

**tidal volume**   The amount of air inhaled and exhaled in a cycle; in humans, tidal volume is 0.5 liter.

**tight junction**   A junctional complex that establishes a seal between cells so that the only way a substance can travel from one side of a sheet of epithelial cells to the other is by moving through the cells by a cellular transport mechanism.

**tissue**   A collection of cells that work together to perform a specific function.

**titin**   A large protein that links myosin filaments to Z discs at the ends of a sarcomere.

**tolerance**   The ability of T cells not to respond to self antigens even though the immune system functions normally otherwise.

**toll-like receptors (TLRs)**   A family of transmembrane receptors on phagocytes that recognize and bind to molecules on the surface of microorganisms, providing an early signal that an infection is present.

**tongue**   A muscle in mammals and other land vertebrates that facilitates food manipulation and transport within the mouth.

**topoisomerase**   Any one of a class of enzymes that regulates the supercoiling of DNA by cleaving one or both strands of the DNA double helix, and later repairing the break.

**topoisomerase II**   An enzyme that breaks a DNA double helix, rotates the ends, and seals the break.

**totipotent**   Describes cells that have the potential to give rise to a complete organism; a fertilized egg is a totipotent cell.

**trabeculae**   Small plates and rods with spaces between them, found in spongy bone.

**trace fossil**   A track or trail, such as a dinosaur track or the feeding trails of snails and trilobites, left by an animal as it moves about or burrows into sediments.

**trachea**   The central airway leading to the lungs, supported by cartilage rings.

**tracheae**   An internal system of tubes in insects that branch from openings along the abdominal surface into smaller airways, directing oxygen to and removing carbon dioxide from respiring tissues.

**tracheid**   A unicellular xylem conduit.

**trade-off**   An exchange in which something is gained at the expense of something lost.

**trait**   A characteristic of an individual.

**transcription**   The synthesis of RNA from a DNA template.

**transcriptional activator protein**   A protein that binds to an enhancer to enable transcription to begin.

**transcriptional regulation**   The mechanisms that collectively regulate whether or not transcription occurs.

**transduction**   Horizontal gene transfer by means of viruses.

**transfer RNA (tRNA)**   Noncoding RNA that carries individual amino acids for use in translation.

**transformation**   The conversion of cells from one state to another, as from nonvirulent to virulent, when DNA released to the environment by cell breakdown is taken up by recipient cells. In recombinant DNA technology, the introduction of recombinant DNA into a recipient cell.

**transgenic organisms**   An alternative term for genetically modified organisms.

**transition state**   The brief time in a chemical reaction in which chemical bonds in the reactants are broken and new bonds in the product are formed.

**translation**   Synthesis of a polypeptide chain corresponding to the coding sequence present in a molecule of messanger RNA.

**transmembrane proteins**   Proteins that span the entire lipid bilayer; most integral membrane proteins are transmembrane proteins

**transmission genetics**   The discipline that deals with the manner in which genetic differences among individuals are passed from generation to generation.

**transpiration**   The loss of water vapor from leaves.

**transporters**   Membrane proteins that move ions or other molecules across the cell membrane.

**transposable element**   A DNA sequence that can replicate and move from one location to another in a DNA molecule.

**transposition**   The movement of a transposable element.

**tree of life**   The full set of evolutionary relationships among all organisms.

**triacylglycerol**   A lipid that stores energy.

**trimesters**   The three periods of pregnancy, each lasting about 3 months.

**triose phosphate**   A 3-carbon carbohydrate molecule; triose phosphates are the true products of the Calvin cycle because they are the molecules exported from the chloroplast.

**triploid**   A cell or organism with three complete sets of chromosomes.

**trisomy 21**   A condition resulting from the presence of three, rather than two, copies of chromosome 21; also known as *Down syndrome*.

**trophic pyramid**   A diagram that traces the flow of energy through communities, showing the amount of energy available at each level to feed the next. The pyramid shape results because biomass and the energy it represents generally decrease from one trophic level to the next.

**tropic hormone**   A hormone that controls the release of other hormones.

**tropism**   The bending or turning of an organism in response to an external signal such as light or gravity.

**tropomyosin**   A protein that runs in the grooves formed by the actin helices and blocks the myosin-binding sites.

**troponin**   A protein that moves tropomyosin away from myosin-binding sites, allowing cross-bridges between actin and myosin to form and the muscle to contract.

**true breeding**   Describes a trait whose physical appearance in each successive generation is identical to that in the previous one.

**trypsin**   A digestive enzyme produced by the pancreas that breaks down proteins.

**T-tubule system**   A system for the conduction of depolarization of a muscle cell through tubules formed from invaginations of the plasma membrane.

**tube feet**   Small projections of the water vascular system that extend outward from the body surface and facilitate locomotion, sensory perception, food capture, and gas exchange in echinoderms.

**tubulin**   Dimers (composed of an a tubulin and a b tubulin) that assemble into microfilaments.

**tumor suppressors**   A family of genes that encode proteins whose normal activities inhibit cell division.

**tunicates**   A subphylum of Chordata that includes about 3000 species of filter-feeding marine animals, such as sea squirts and salps.

**turgor pressure**   The pressure exerted by water against an object, which provides structural support for many plants, fungi, and bacteria.

**Turner syndrome**   A sex-chromosomal abnormality in which an individual has 45 chromosomes, including only one X chromosome.

**twitch**   A muscle contraction that results from a single action potential.

**twofold cost of sex**   Population size can increase more rapidly in asexually reproducing organisms than in sexually reproducing organisms because only female produce offspring, and females have only half the fitness of asexual parents.

**tympanic membrane**   A thin sheet of tissue at the surface of the ear that vibrates in response to sound waves, amplifying airborne vibrations; in mammals, also known as the *eardrum*.

**ulcer**   A break or erosion in the cells lining the stomach.

**unbalanced translocation**   Tranlocation in which only part of a reciprocal translocation (and one of the nontranslocated chromosomes) is inherited from one of the parents.

**unsaturated**   Describes fatty acids that contain carbon–carbon double bonds.

**uracil (U)**   A pyrimidine base in RNA, where it replaces the thymine found in DNA.

**urea**   A waste product of protein metabolism that many animals excrete.

**ureter**   A large tube in the vertebrate kidney that brings urine from the kidneys to the bladder.

**urethra**   A tube from the bladder that in males carries semen as well as urine from the body.

**uterus**   A hollow organ with thick, muscular walls that is adapted to support the developing embryo if fertilization occurs and to deliver the baby during birth.

**vaccination**   The deliberate delivery in a vaccine of an antigen from a pathogen to induce a primary response but not the disease, thereby providing future protection from infection.

**vacuole**   A cell structure that absorbs water and contributes to turgor pressure.

**vagina**   A tubular channel connecting the uterus to the exterior of the body; also known as the birth canal.

**valence electrons**   The electrons farthest from the nucleus, which are at the highest energy level.

**van der Waals forces**   The binding of temporarily polarized molecules because of the attraction of opposite charges.

**variable** (in experimentation)   The feature of an experiment that is changed by the experimenter from one treatment to the next.

**variable (V) region**   A region of the heavy (H) and light (L) chains of an antibody; a change in the variable region distinguishes a given antibody from all others.

**variable expressivity**   The phenomenon in which a particular phenotype is expressed with a different degree of severity in different individuals.

**variable number tandem repeat (VNTR)**   A genetic difference in which the number of short repeated sequences of DNA differs from one chromosome to the next.

**vas deferens**   A long, muscular tube from the scrotum, through the abdominal cavity, along the bladder, and connecting with the ejaculatory duct.

**vasa recta**   The blood vessels in the kidneys.

**vascular bundle**   A bundle of xylem conduits and phloem; multiple vascular bundles are arranged in a ring near the outside of the stem.

**vascular cambium**   Lateral meristem that is the source of new xylem and phloem.

**vascular plant**   A photosynthetic land plant that can draw water from soil and limit water loss from their leaves.

**vasoconstriction**   The process in the supply of blood to the limbs is reduced by constriction of the arteries that supply the limbs.

**vasodilation**   The process in which resistance in the arteries is decreased and blood flow increased following relaxation of the smooth arterial muscles.

**vasopressin**   See *antidiuretic hormone (ADH)*.

**vector**   In recombinant DNA, a carrier of the donor fragment, usually a plasmid.

**vegetative reproduction**   The production of upright shoots from horizontal stems, permitting new plants to be produced at a distance from the site where the parent plant originally germinated.

**veins**   In plants, the system of vascular conduits that connects the leaf to the rest of the plant; in animals, the large, low-pressure vessels that return blood to the heart.

**vena cava** (plural **venae cavae**)   One of the two largest veins in the body, which drain blood from the head and body into the heart.

**ventilation**   The movement of an animal's respiratory medium—water or air—past a specialized respiratory surface.

**ventricle**   A heart chamber that pumps blood to the lungs or the rest of the body.

**venule**   A blood vessel into which capillaries drain as blood is returned to the heart.

**vernalization**   A prolonged period of exposure to cold temperatures necessary to induce flowering.

**vertebrae**   The series of hard segments making up the jointed skeleton that runs along the main axis of the body in vertebrates.

**vertebral column**   A structure in vertebrates that functionally replaces the embryonic notochord.

**vesicle**   A small membrane-enclosed sac that transports substances within the cell.

**vessel**   A multicellular xylem conduit.

**vessel element**   An individual cell that is part of a vessel for water transport in plants; in contrast to fibers, vessel elements can be quite wide.

**vestibular system**   A system in the mammalian inner ear made up of two statocyst chambers and three semicircular canals.

**vestigial structure**   A structure that has lost its original function over time and is much reduced in size.

**vicariance**   The process in which a geographic barrier arises within a single population, separating it into two or more isolated populations.

**villi**   Highly folded inner surfaces of the jejunum and ileum of the small intestine.

**virulent**   Describes pathogens that are able to overcome the host plant's defenses and lead to disease.

**virus**   A small infectious agent that contains a nucleic acid genome packaged inside a protein coat called a capsid.

**viscosity**   The stickiness of a fluid.

**visible light**   The portion of the electromagnetic spectrum apparent to our eyes.

**visual cortex**   The part of the brain that processes images.

**vitamin**   An organic molecule that is required in very small amounts in the diet.

**vitreous humor**   A gel-like substance filling the large cavity behind the lens that makes up most of the volume of the vertebrate eye.

**viviparity**   Giving birth to live young.

**vocal cords**   Twin organs in the larynx that vibrate as air passes over them, enabling speech, song, and sound production.

**voluntary**   Describes the component of the nervous system that handles sensing and responding to external stimuli.

**vulva**   The external genitalia of the female.

**water vascular system**   A series of fluid-filled canals that permit bulk transport of oxygen and nutrients in echinoderms.

**white blood cell (leukocyte)** A type of cell in the immune system that arises by differentiation from stem cells in the bone marrow.

**white matter** Collectively, the axons of cortical neurons in the interior of the brain; it is the fatty myelin produced by glial cells surrounding the axons that makes this region of the brain white.

**wild type** The most common allele, genotype, or phenotype present in a population; nonmutant.

**X chromosome** One of the two sex chromosomes; a normal human female has two copies of the X chromosome, a normal male has one X and one Y chromosome.

**xanthophyll** Any one of several yellow-orange pigments that slow the formation of reactive oxygen species by reducing excess light energy; these pigments accept absorbed light energy directly from chlorophyll and convert this energy to heat.

**X-inactivation** The process in mammals in which dosage compensation occurs through the inactivation of one X chromosome in each cell in females.

**X-linked gene** A gene in the X chromosome.

**xylem** The inner tissue of a stem, which transports water and nutrients from the roots to the leaves.

**Y chromosome** One of the two sex chromosomes; a normal human male has one X and one Y chromosome.

**yeast** A single-celled fungus found in moist, nutrient-rich environments.

**Y-linked gene** A gene that is present in the region of the Y chromosome that shares no homology with the X chromosome.

**yolk** A substance in the eggs of animals with external fertilization that provides all the nutrients that the developing embryo needs until it hatches.

**Z disc** A protein backbone found regularly spaced along the length of a myofibril.

**Z scheme** Another name for the photosynthetic electron transport chain, so called because the overall energy trajectory resembles a "Z."

**zygomycetes** Fungi groups that produce hyphae undivided by septa and do not form multicellular fruiting bodies; they make up less than 1% of known fungal diversity.

**zygote** The diploid fertilized egg cell formed by the fusion of two haploid gametes.

# CREDITS/SOURCES

**Part 1** Visuals Unlimited, Inc./Gregory Basco/Getty Images

## Chapter 1

*Opening photo* Frans Lanting/www.lanting.com. *Fig. 1.1* Charles J. Smith. *Fig. 1.3 (T)* Kirk Johnson, Denver Museum of Nature & Science, *(B, L)* Dr. David Kring/Photo Researchers, Inc., *(B, R)* Image courtesy of V. L. Sharpton/Lunar and Planetary Institute. *Fig. 1.4* Scott Hailstone/iStockphoto. *Fig. 1.10 (a & b)* Steve Gschmeissner/Photo Researchers, Inc., *(c)* Arthur Morris/Visuals Unlimited, Inc., *(d)* Sven-Olof Lindblad/ Photo Researchers, Inc., *(e)* Megapress/Alamy. *Fig. 1.11 (a)* Biophoto Associates/Photo Researchers, Inc., *(b)* Dr. Jonathan Clarke/Wellcome Images, *(c)* Hemis/Alamy. *Fig. 1.14* Dr. Gopal Murti/Visuals Unlimited, Inc. *Fig. 1.15* © Rob Brodman 2011. *Fig. 1.16* John Giustina/Photoshot. *Fig. 1.18* Neerja Hajela, Michigan State University. *Fig. 1.19* Donald Specker/Animals Animals–Earth Scenes. *Fig. 1.20 (a)* Frans Lanting/www.lanting.com, *(b)* Ken Lucas/Visuals Unlimited, Inc. *Fig. 1.21 (a)* Fred Dimmick/iStockphoto, *(b)* Arndt Sven-Erik/age fotostock, *(c)* Nigel Cattlin/Visuals Unlimited, Inc. *Fig. 1.22 (a)* Photo Researchers, Inc., *(b)* G. I. Bernard/Photo Researchers, Inc., *(c)* Look and Learn/The Bridgeman Art Library, *(d)* P.W. Sykes/U.S. Fish and Wildlife Service, *(e)* James Warwick/Photo Researchers, Inc.

## Case 1

*Snottites* Kenneth Ingham Photography. *Hydrothermal vent* Image courtesy of New Zealand American Submarine Ring of Fire 2007 Exploration, NOAA Vents Program, the Institute of Geological & Nuclear Sciences, and NOAA-OE.

## Chapter 2

*Opening photo* Science Photo Library/Alamy. *Fig. 2.30* Ray L. Frost, Professor of Physical Chemistry, Queensland University of Technology, Australia.

## Chapter 3

*Opening photo* Comstock/Getty Images. *Fig. 3.13* Rendered by Dale Muzzey.

## Chapter 4

*Opening photo* Pasieka/SPL/Getty Images. *Fig. 4.5* Courtesy of William E. Royer, University of Massachusetts Medical School, and Vukica Srajer, BioCARS, Center for Advanced Radiation Sources, The University of Chicago. *Fig. 4.8, Fig. 4.9, Fig. 4.10, Fig. 4.21* Rendered by Dale Muzzey.

## Chapter 5

*Opening photo* Dr. George Chapman/Visuals Unlimited, Inc. *Fig. 5.1 (L)* Science Museum/SSPL/The Image Works, *(R)* Ted Kinsman/Photo Researchers, Inc. *Fig. 5.8* FRAP of cytoplasmic EGFP in living HeLa cells, performed using an UltraVIEW® spinning disk confocal system (PerkinElmer Inc.). HeLa cells were transfected with pEGFP-C1 (Clontech Laboratories, Inc.) using GeneJuice® transfection reagent (Novagen®). A region of interest in the cytoplasm was photobleached using the UltraVIEW® photokinesis unit, and the recovery of fluorescence in this region was observed. (Fluorescence Recovery After Photobleaching (FRAP) using the UltraVIEW PhotoKinesis accessory, PerkinElmer Technical Note). *Fig. 5.15* Biophoto Associates/Photo Researchers, Inc. *Fig. 5.19* Don W. Fawcett/Photo Researchers, Inc. *Fig. 5.20 (T)* David M. Phillips/Photo Researchers, Inc., *(B)* Dr. Donald Fawcett/Visuals Unlimited, Inc. *Fig. 5.21* Biophoto Associates/Photo Researchers, Inc. *Fig. 5.22* Dr. Donald Fawcett/Visuals Unlimited, Inc. *Fig. 5.27* Dr. Donald Fawcett, K. Porter/Visuals Unlimited, Inc. *Fig. 5.28* Dr. Jeremy Burgess/Photo Researchers, Inc.

## Chapter 6

*Opening photo* Purestock/Getty Images. *Fig. 6.1 (T, L–R)* Dr. Tony Brain/Photo Researchers, Inc., Science PR/Getty Images, copyright 1997 Microbial Diversity, Rolf Schauder, Dr. Martin Oeggerli/Visuals Unlimited, Inc., *(B, L–R)* DNY59/iStockphoto, Science VU/Visuals Unlimited, Inc., EM image by Manfred Rohde, Helmholtz Centre for Infection Research, Braunschweig, Germany, Anna Omelchenko/Dreamstime.com. *Fig. 6.16* Rendered by Dale Muzzey.

## Chapter 7

*Opening photo* David Madison/Getty Images.

## Chapter 8

*Opening photo* Imagewerks Japan/Getty Images. *Fig. 8.2* Sinclair Stammers/Photo Researchers, Inc. *Fig. 8.3 (a)* Photo by Zoe G. Cardon from cover of *Plant, Cell, & Environment* 30(10), Wiley-Blackwell Publishers, *(b)* Olivier Grunewald, *(c)* Shattil & Rozinski/Naturepl.com. *Fig. 8.5* Manfred Kage. *Fig. 8.8* Dr. Howard Berg/Visuals Unlimited, Inc.

## Chapter 9

*Opening photo* Steve Gschmeissner/Photo Researchers, Inc. *Fig. 9.12 (T, L)* Mark Boulton/Photo Researchers, Inc., *(T, R)* Mark Smith/Photo Researchers, Inc., *(B, L)* Werner Bollmann/Oxford Scientific/Getty Images, *(B, R)* Phillip Colla.

## Chapter 10

*Opening photo* Dr. Torsten Wittmann/Photo Researchers, Inc. *Fig. 10.1 (a)* Cheryl Power/Photo Researchers, Inc., *(b)* Professors P. Motta & T. Naguro/Photo Researchers, Inc., *(c)* Innerspace Imaging/Photo Researchers, Inc., *(d)* Biophoto Associates/Photo Researchers, Inc., *(e)* Don W. Fawcett/Photo Researchers, Inc. Colorization by Mary Martin. *Fig. 10.2* Jose Luis Pelaez Inc/Blend Images/Getty Images. *Fig. 10.4 (L–R)* Dennis Kunkel Microscopy, Inc., Eye of Science/Photo Researchers, Inc., Andrew Syred/Photo Researchers, Inc., Dennis Kunkel Microscopy, Inc./Visuals Unlimited, Inc. *Fig. 10.5* N. Hirokawa, L. G. Tilney, K. Fujiwara, and J. E. Heuser, 1982, "Organization of Actin, Myosin, and Intermediate Filaments in the Brush Border of Intestinal Epithelial Cells," *Journal of Cell Biology* 94:425–443. *Fig. 10.6* Courtesy of R. D. Goldman. *Fig. 10.7* Helen Osler/Northscot/Rex USA. *Fig. 10.12 (T)* Maryann Haldi, Adam Amsterdam, and Nancy Hopkins, *(B)* Courtesy of Darren Logan. *Fig. 10.13 (L–R)* Dennis Kunkel Microscopy, Inc., Dennis Kunkel Microsopy, Inc./Visuals Unlimited, Inc., Dr. Gopal Murti/Photo Researchers, Inc. *Fig. 10.14* Rachel Fink, Mount Holyoke College. *Fig. 10.15 (T)* Borut Furlan/WaterFrame/age fotostock. *(B)* Franco Banfi/WaterFrame/age fotostock. *Fig. 10.18* Biophoto Associates/Photo Researchers, Inc. *Fig. 10.19* Biophoto Associates/Photo Researchers, Inc. *Fig. 10.20 (T)* Tom Grundy/Alamy, *(M)* Egon Bömsch/age fotostock, *(B)* Eye of Science/Photo Researchers, Inc. *Fig. 10.21* Dr. Alvin Telser/Visuals Unlimited, Inc. *Fig. 10.23* F. Tao, S. Chaudry, B. Tolloczko, J. G. Martin, and S. M. Kelly, 2003, "Modulation of Smooth Muscle Phenotype in Vitro by Homologous Cell Substrate," *American Journal of Physiology/Cell Physiology* 284(6):C1531–C1541. *Fig. 10.24* Courtesy of Motoyoshi Nomizu. *Fig. 10.25* Eye of Science/Photo Researchers, Inc.

## Chapter 11

*Opening photo* Jennifer C. Waters/Photo Researchers, Inc. *Fig. 11.3* ISM/Phototake. *Fig. 11.5* Conly Rieder and Alexey Khodjakov/Visuals Unlimited, Inc. *Fig. 11.7 (a)* Dr. Paul Andrews, University of Dundee/Photo Researchers,

Inc., *(b)* Dr. Robert Calentine/Visuals Unlimited, Inc. *Fig. 11.14 (L)* SeaPics.com, *(R)* Biology Pics/Photo Researchers, Inc. *Fig. 11.21* Jose Luis Pelaez Inc./Blend Images/Getty Images.

## Case 3

*Claudia Gilmore* Dominic Gutierrez, courtesy of Claudia Gilmore. *Karyotypes (L–R)* L. Willatt/Photo Researchers, Inc., ISM/Phototake, ISM/Phototake.

## Chapter 12

*Opening photo* Dr. Gopal Murti/Photo Researchers, Inc. *Fig. 12.3* Daniel Hartl. *Fig.12.14* Guy Tear/Wellcome Images. *Fig. 12.16* Reproduced with permission from A. A. Da'dara and R. D. Walter, 1998, "Molecular and Biochemical Characterization of S-Adenosylmethionine Decarboxylase from the Free-Living Nematode *Caenorhabditis elegans*," *Biochemical Journal* 336(Pt 3):545–550. © The Biochemical Society. *Fig. 12.20 (a)* Adam Hart-Davis/Photo Researchers, Inc., *(b)* Steve Percival/Photo Researchers, Inc., *(c)* RIA Novosti/Photo Researchers, Inc., *(d)* Murdo Macleod/Polaris, *(e)* Barrett & MacKay Photo– barretmacay.com, *(f)* Steven Morse/University of Missouri/Polaris.

## Chapter 13

*Opening Photo* SSPL/Getty Images. *Fig. 13.6* Data from A. Katzourakis, M. Tristem, O. G. Pybus, and R. J. Gifford, 2007, "Discovery and Analysis of the First Endogenous Lentivirus," *Proceedings of the National Academy of Sciences USA* 104(15):6261–6265. *Table 13.1, Table 13.2* Data from "Sequence Composition of the Human Genome" in International Human Genome Sequencing Consortium, 2001, "Initial Sequencing and Analysis of the Human Genome" (PDF), *Nature* 409(6822):860–921, doi:10.1038/35057062, PMID 11237011. *Fig. 13.8* After Fig. 1, p. 186, in T. R. Gregory, 2004, "Macroevolution, Hierarchy Theory, and the C-Value Enigma," *Paleobiology* 30:179–202. *Fig. 13.9* Chromosome sets courtesy of M. Ørgaard, N. Jacobsen, and J. S. Heslop-Harrison, 1995, "The Hybrid Origin of Two Cultivars of Crocus (Iridaceae) Analysed by Molecular Cytogenetics Including Genomic Southern and *in situ* Hybridization," *Annals of Botany* 76:253–262 (photo of crocuses courtesy of the authors). *Fig. 13.11* Dr. Klaus Boller/Photo Researchers, Inc. *Fig. 13.13* Courtesy of Ulrich Laemmli. *Fig. 13.14* Y. Garini, M. Macville, S. du Manoir, R. A. Buckwald, M. Lavi, N. Katzir, D. Wine, I. Bar-Am, E. Schröck, D. Cabib, and T. Ried, 1996, "Spectral Karyotyping," *Bioimaging* 4:65–72. *Fig. 13.15* National Institutes of Health, National Human Genome Research Institute. *Fig. 13.16* Courtesy of H. D. Shew, American Phytopathological Society. *Fig. 13.18 (a)* Dr. Harold Fisher/Visuals Unlimited, Inc., *(b)* Dr. Timothy Baker/Visuals Unlimited, Inc., *(c)* BSIP/Photo Researchers, Inc.

## Chapter 14

*Opening photo* Tetra Images/Getty Images. *Fig. 14.1, Fig. 14.2* After J. W. Drake, B. Charlesworth, D. Charlesworth, J. F. Crow, 1998, "Rates of Spontaneous Mutation," *Genetics* 148:1667–1686. *Fig. 14.3* Courtesy of Atsushi Hoshino, National Institute for Basic Biology, Japan. *Fig. 14.4* Kathleen R. Cho, University of Michigan Medical School. *Fig. 14.11 (a)* B. Runk/S. Schoenburger/Grant Heilman Photography, *(b)* Rob Martienssen, Cold Spring Harbor Laboratory. *Fig. 14.13* After Y. A. Trusov and P. H. Dear, 1996, "A Molecular Clock Based on the Expansion of Gene Families," *Nucleic Acids Research* 24(6):995–999.

## Chapter 15

*Opening photo* Tastyart Ltd Rob White/Getty Images. *Fig. 15.7* Photodisc, Stockbroker xtra/age fotostock. *Fig. 15.10* Data from G. H. Perry, N. J. Dominy, K. G. Claw, A. S. Lee, H. Fiegler, R. Redon, J. Werner, F. A. Villanea, J. L. Mountain, R. Misra, N. P. Carter, C. Lee, and A. C. Stone, 2007, "Diet and the Evolution of Human Amylase Gene Copy Number Variation," *Nature Genetics*

39:1256–1260. *Fig. 15.13* After D. L. Hartl and M. Ruvolo, 2012, *Genetics: Analysis of Genes and Genomes,* 8th ed., Burlington, MA: Jones & Bartlett, Table 8.2, p. 268. *Fig. 15.14* Rex Features via AP Images, Biophoto Associates/Photo Researchers, Inc. Colorization by Mary Martin. *Fig. 15.16* After D. L. Hartl and M. Ruvolo, 2012, *Genetics: Analysis of Genes and Genomes,* 8th ed., Burlington, MA: Jones & Bartlett, Table 8.2, p. 268.

## Chapter 16

*Opening photo* Koichi Saito/A.collection/Getty Images. *Fig. 16.2* Juliette Wade/Getty Images, *(inset)* Photo by Authenticated News/Getty Images. *Fig. 16.16* Reprinted by permission from Macmillan Publishers Ltd: B. Gao, J. Guo, C. She, A. Shu, M. Yang, Z. Tan, X. Yang, S. Guo, G. Feng, and L. He, 2001, "Mutations in *IHH,* Encoding Indian hedgehog, Cause Brachydactyly Type A-1," *Nature Genetics* 28:386–388. Copyright 2001. Image Courtesy of Bo Gao. *Fig. 16.19* National Anthropological Archives, Smithsonian Institution (NAA INV 06404400).

## Chapter 17

*Opening photo* SSPL/Getty Images. *Fig. 17.1* Science Photo Library/Photo Researchers, Inc. *Fig. 17.7* Dorling Kindersley/Getty Images. *Fig. 17.15* After M. A. Jobling and C. Tyler-Smith, 2003, "The Human Y Chromosome: An Evolutionary Marker Comes of Age," *Nature Reviews Genetics* 4:598–612. *Fig. 17.16* Courtesy of Dr. Kurenai Tanji, Columbia University Medical Center, New York, NY.

## Chapter 18

*Opening photo* Photononstop/SuperStock. *Fig. 18.1 (a)* Randy Faris/Corbis, *(b)* Ashley Cooper/Visuals Unlimited, Inc./Getty Images, *(c)* Heidi & Hans-Juergen Koch/Minden Pictures, *(d)* Radius Images/Getty Images. *Fig. 18.2 (a)* Burwell and Burwell Photography/iStockphoto, *(b)* Ocean/Corbis, *(c)* Junophoto/Getty Images, *(d)* Imago/ZUMApress.com. *Fig. 18.3* Bruce Leighty/Getty Images. *Fig. 18.5* Data courtesy of J. W. Dudley. *Fig. 18.6* Phil Jones/GHSU. *Fig. 18.7* Data from F. Galton, 1888, "Co-Relations and Their Measurement, Chiefly from Anthropometric Data," *Proceedings of the Royal Society, London* 45:135–145, reprinted in K. Pearson, 1920, "Notes on the History of Correlation," *Biometrika* 13:25–45. *Table 18.1* Data from M. McGue and T. J. Bouchard, Jr., 1998, "Genetic and Environmental Influences on Human Behavioral Differences," *Annual Review of Neuroscience* 21:1–24. *Fig. 18.10 (a)* AP Photo/Macomb Daily, Craig Gaffield, *(b)* Bruce Roberts/Photo Researchers, Inc. *Fig. 18.12* Data from Centers for Disease Control and Prevention; National Newborn Screening and Genetics Resource Center; March of Dimes. *Fig. 18.13* Data from X. Wang and B. Paigen, 2005, "Genetics of Variation in HDL Cholesterol in Humans and Mice," *Circulation Research* 96:27–42.

## Chapter 19

*Opening photo* Science Photo Library/Photo Researchers, Inc. *Fig. 19.4* Eyal Nahmias/Alamy.

## Chapter 20

*Opening photo* Stephen Dalton/Minden Pictures. *Fig. 20.3* J. B. Gurdon and D. A. Melton, "Nuclear Reprogramming in Cells," 2008, *Science* 322:1811–1815. *Fig. 20.4* AP Photo/John Chadwick, AP Photo/Pat Sullivan. *Fig. 20.9, Fig. 20.10, Fig. 20.11* James Langeland, Steve Paddock and Sean Carroll, HHMI, University of Wisconsin–Madison. *Fig. 20.12 (a)* F. Rudolf Turner, Ph.D., Indiana University, *(b)* Thomas Deerinck, NCMIR/Science Photo Library/Photo Researchers, Inc., David Scharf/Science Faction/Getty Images. *Fig. 20.16 (a)* Tom Adams/Visuals Unlimited, Inc., *(b)* Norbert Wu/SuperStock, *(c)* Reinhard Dirscherl/age fotostock, *(d)* Andrey Armyagov/iStockphoto, *(e)* Julian Brooks/age fotostock, *(f)* Walter

Past 400,000 Years," from Vital Climate Graphics, UNEP/GRID–Arendal Maps & Graphics Library, JPG file, 2005, http://www.grida.no/publications/vg/climate/page/3057.aspx. Based on J. R. Petit et al., 1999, "Climate and Atmospheric History of the Past 420,000 Years from the Vostok Ice Core, Antarctica," *Nature* 399:429–436. *Fig. 25.11* Plate tectonic maps and continental drift animations by C. R. Scotese, PALEOMAP Project (www.scotese.com). *Fig. 25.12* After Fig 3.4 in K. Wallmann and G. Aloisi, 2012, "The Global Carbon Cycle: Geological Processes," in *Fundamentals of Geobiology*, A. H. Knoll, D. E. Canfield, and K. O. Konhauser (eds.), Chichester, England: Wiley-Blackwell, pp. 20–35. *Fig. 25.15* Albert Aanensen/Nature Picture Library.

## Case 5

*Seated woman* Mimi Haddon/Getty Images. *Pie charts* Data from I. Chung and M. J. Blaser, 2012, "The human microbiome: at the interface of health and disease." *Nature Reviews Genetics* 13:260–270, doi: 10.1038/nrg3182. *Bobtail squid* Gary Bell/OceanwideImages.com.

## Chapter 26

*Opening photo* Steve Gschmeissner/SPL/Getty Images. *Fig. 26.2 (a)* Eye of Science/Photo Researchers, Inc., *(b)* Dennis Kunkel Microscopy, Inc./Phototake, *(c)* Courtesy Mike Dyall-Smith, *(d)* David Scharf/Photo Researchers, Inc., *(e)* © ANIMA RES. *Fig. 26.3* Heide Schulz-Vogt, MPI Bremen. *Fig. 26.4* Dennis Kunkel Microscopy, Inc./Phototake. *Fig. 26.6* lecates via Flickr. *Fig. 26.7 (a)* Andrew Knoll, Harvard University, *(b)* With kind permission from Springer Science+Business Media. Jörg Overmann, 2006, *The Prokaryotes*, Springer eBook, Fig. 7. *Photo* courtesy of Jörg Overmann. *Fig. 26.9* Woods Hole Oceanographic Institution/Visuals Unlimited, Inc. *Fig. 26.12* Wally Eberhart/Visuals Unlimited, Inc. *Fig. 26.16 (a)* Dr. Ralf Wagner, *(b)* Wim van Egmond/Visuals Unlimited, Inc., *(c)* Tom Adams/Visuals Unlimited, Inc., *(d)* Jason Oyadomari. *Fig. 26.14* Adapted from W. F. Doolittle, 2000, "Uprooting the Tree of Life,"*Scientific American* 282:90–95. *Fig. 26.17* Adapted from M. Groussin and M. Gouy, 2011, "Adaptation to Environmental Temperature Is a Major Determinant of Molecular Evolutionary Rates in Archaea," *Molecular Biology and Evolution* 28(9):2667, doi: 10.1093/molbev/msr098. © The Author 2011. Published by Oxford University Press on behalf of the Society for Molecular Biology and Evolution. *Fig. 26.18* Andrew Knoll, Harvard University. *Fig. 26.20 (a–c)* Andrew Knoll, Harvard University. *Fig. 26.21 (T)* Andrew Knoll, Harvard University, *(B)* François Gohier/Photo Researchers, Inc. *Fig. 26.22* Image by Zoe Veneti and Kostas Bourtzis. *Fig. 26.23* Data from J. Peterson et al., 2009, "The NIH Human Microbiome Project," *Genome Research* 19:2317–2323; originally published online October 9, 2009. *Fig. 26.24* Data from C. De Filippo, D. Cavalieri, M. Di Paola, M. Ramazzotti, J. B. Poullet, S. M., S. Collini, G. Pieraccini, and P. Lionetti, 2010, "Impact of Diet in Shaping Gut Microbiota Revealed by a Comparative Study in Children from Europe and Rural Africa," *Proceedings of the National Academy of Sciences USA* 107(33):14693, doi: 10.1073/pnas.1005963107.

## Chapter 27

*Opening photo* Biophoto Associates/Photo Researchers, Inc. *Fig. 27.2* V. Mercanti, S. J. Charette, N. Bennett, J. J. Ryckewaert, F. Letourneur, and P. Cosson, 2006, "Selective Membrane Exclusion in Phagocytic and Macropinocytic Cups, *Journal of Cell Science* 119:4079–4067. *Fig. 27.4 (a, T–B)* Dr. Kari Lounatmaa/Photo Researchers, Inc., Dennis Kunkel Microscopy, Inc., Dr. Jeremy Burgess/Photo Researchers, Inc. *Fig. 27.5* Courtesy of Hwan Su Yoon. *Fig. 27.6* Reprinted by permission from Macmillan Publishers Ltd: B. Boxma, et al., 2005, "An Anaerobic Mitochondrion That Produces Hydrogen, *Nature* 434:74–79. Copyright 2005. *Fig. 27.8* Reprinted by permission from Macmillan Publishers Ltd: V. P. Edgcomb, S. A. Breglia, N. Yubuki, D. Beaudoin, D. J. Patterson, B. S. Leander, and J. M. Bernhard, 2011, "Identity of Epibiotic Bacteria on Symbiontid

Euglenozoans in $O_2$-Depleted Marine Sediments: Evidence for Symbiont and Host Co-evolution, *The ISME Journal* 5:231–243. Copyright 2010. Courtesy of Naoji Yubuki. *Fig. 27.10* Nicole King, Steve Paddock and Sean Carroll, HHMI, University of Wisconsin. *Fig. 27.11 (b)* Wim van Egmond/Visuals Unlimited, Inc., *(c)* Carolina Biological Supply Co./Visuals Unlimited, Inc. *Fig. 27.12 (a)* Matt Meadows/Photo Researchers, Inc., *(b)* Masana Izawa/Minden Pictures. *Fig. 27.13* Carolina Biological Supply Co./Visuals Unlimited, Inc. *Fig. 27.14 (b)* Jerome Pickett-Heaps/Photo Researchers, Inc., *(c)* Wim van Egmond/Visuals Unlimited, Inc., *(d)* David Patterson and Bob Andersen, image used under license to MBL (micro*scope). *Fig. 27.15 (a)* Dr. Peter Siver/Visuals Unlimited, Inc./Corbis, *(b)* Wim van Egmond/Visuals Unlimited, Inc., *(c)* Gerd Guenther/Photo Researchers, Inc., *(d)* Frank Fox/Photo Researchers, Inc., *(e)* Linda Sims/Visuals Unlimited, Inc. *Fig. 27.16* D. J. Patterson, image used under license to MBL (micro*scope). *Fig. 27.17* Mark Conlin/Alamy. *Fig. 27.18 (a & b)* Steve Gschmeissner/Photo Researchers, Inc., *(c)* Andrew Syred/Photo Researchers, Inc., *(d)* Wim van Egmond/Visuals Unlimited, Inc. *Fig. 27.19 (a)* Electron Microscopy Unit, Australian Antarctic Division © Commonwealth of Australia, *(b)* Andrew Syred/Photo Researchers, Inc., *(c)* Gert Hansen–SCCAP, University of Copenhagen. *Fig. 27.23 (a)* Andrew Knoll, Harvard University, *(b)* Nicholas J. Butterfield, University of Cambridge, *(c)* Susannah Porter, University of California, Santa Barbara. *Fig. 27.24 (a)* Astrid & Hanns-Frieder Michler/Photo Researchers, Inc., *(b)* Visuals Unlimited, Inc./Corbis. *Fig. 27.25 (a)* Copyright The Natural History Museum, London, *(b)* Provided by the SeaWiFS Project, NASA/Goddard Space Flight Center, and ORBIMAGE.

## Chapter 28

*Opening photo* irawansubingarphotography/Getty Images. *Fig. 28.1 (a)* Jason Oyadomari, *(b)* D. J. Patterson and Aimlee Ladermann, image used under license to MBL (micro*scope), *(c)* Fabio Rindi. *Fig. 28.2 (a)* Marevision/age fotostock, *(b)* Wolfgang Poelzer/WaterFrame/age fotostock. *Fig. 28.3 (a)* Ingo Arndt/Minden Pictures, *(b)* Richard Herrmann/Visuals Unlimited, Inc., *(c)* Filmfoto/Dreamstime.com, *(d)* Jack Milchanowski/age fotostock, *(e)* David M. Cobb/Visuals Unlimited, Inc. *Fig. 28.5 (a)* Andrew J. Martinez/SeaPics.com, *(b)* D. R. Schrichte/SeaPics.com. *Fig. 28.7* Reprinted from M. J. Dayel, R. Alegado, S. Fairclough, T. Levin, S. A. Nichols, K. L. McDonald, and N. King, 2011, "Cell Differentiation and Morphgenesis in the Colony-Forming Choanoflagellate *Salpingoeca rosetta*," *Developmental Biology* 357:73–82. Copyright 2011, with permission from Elsevier. *Fig. 28.9* Biophoto Associates/Photo Researchers, Inc. *Fig. 28.10* Biodisc/Visuals Unlimited, Inc. *Fig. 28.13 (a)* Guy Narbonne, *(b)* Mikhahil Fedonkin, *(c)* Andrew Knoll, Harvard University. *Fig. 28.15 (a)* Reprinted from F. Hueber, 2001, "Rotted Wood–Alga–Fungus: The History and Life of Prototaxites Dawson 1959," *Review of Paleobotany and Palynology* 116:123–158, p. 146, Smithsonian Institution, copyright 2001, with permission from Elsevier, *(b)* L. E. Graham, M. E. Cook, D. T. Hanson, K. B. Pigg, and J. M. Graham, 2010, "Structural, Physiological, and Stable Carbon Isotopic Evidence That the Enigmatic Paleozoic Fossil Prototaxites Formed from Rolled Liverwort Mats," *American Journal of Botany* 97(2):268–275, Botanical Society of America; Brooklyn Botanic Garden, copyright 2010. Reproduced with permission of Botanical Society of America, Inc., via Copyright Clearance Center. Image courtesy of Martha E. Cook. *Fig. 28.16* Reprinted by permission from Macmillan Publishers Ltd: P. M. Brakefield, J. Gates, D. Keyes, F. Kesbeke, P. J. Wijngaarden, A. Monteiro, V. French, and S. B. Carroll, 1996, "Development, Plasticity and Evolution of Butterfly Eyespot Patterns," *Nature* 384:236–242. Copyright 1996.

## Case 6

*Wild wheat* James King-Holmes/Photo Researchers, Inc. *Domesticated wheat* Oliver Brandt/Getty Images. *Comparison of wild wheat and domesticated wheat* Biophoto Associates/Photo Researchers, Inc.

## Chapter 29

*Opening photo* Frank Krahmer/Getty Images. *Fig. 29.2* Brent Mishler and Michael Hamilton. *Fig. 29.5* Dr. Jeremy Burgess/Photo Researchers, Inc. *Fig. 29.7* Clément Philippe/age fotostock. *Fig. 29.8* Carolina Biological Supply Co./Visuals Unlimited, Inc. *Fig. 29.10* Carolina Biological Supply Co./Visuals Unlimited, Inc. *Fig. 29.17 (a)* Topic Photo Agency/age fotostock, *(b)* Photoshot/SuperStock. *Fig. 29.20* Inga Spence/Photo Researchers, Inc.

## Chapter 30

*Opening photo* Frans Lanting/www.lanting.com. *Fig. 30.2 (L)* Dr. Ralf Wagner. email&lt;mikroskopie@dr-ralf-wagner.de&gt;/http://www.dr-ralf-wagner.de, *(R)* Show_ryu/http://commons.wikimedia.org/wiki/File:Chara_braunii_1.JPG. *Fig. 30.4* Malcolm Storey, 2010, www.bioimages.org.uk. *Fig. 30.5 (a)* Keith Burdett/age fotostock, *(b)* Malcolm Storey, 2010, www.bioimages.org.uk. *Fig. 30.6 (L)* Biophoto Associates/Photo Researchers, Inc., *(R)* Masana Izawa/Nature Production/Minden Pictures. *Fig. 30.7* Charles O. Cecil/Alamy. *Fig. 30.9* Jim French. *Fig. 30.10 (T)* Perennou Nuridsany/Photo Researchers, Inc., *(B)* Kevin O´Hara/age fotostock. *Fig. 30.11 (a)* Jim Brandenburg/Minden Pictures, *(b)* J S Sira/Gap Photo/Visuals Unlimited, Inc., *(c)* Rob Whitworth/Gap Photo/Visuals Unlimited, Inc., *(d)* Thomas Marent/Visuals Unlimited, Inc. *Fig. 30.12* Cora Niele/Getty Images. *Fig. 30.13* K. Chae and E. M. Lord, 2011, "Pollen Tube Growth and Guidance: Roles of Small, Secreted Proteins," *Annals of Botany* 108(4):627–636. By permission of Oxford University Press. *Fig. 30.14 (a & b)* Michael & Patricia Fogden/Minden Pictures, *(c)* Charles Melton/Visuals Unlimited, Inc., *(d)* Ch'ien Lee/Minden Pictures. *Fig. 30.15* Hans Christoph Kappel/NPL/Minden Pictures. *Fig. 30.16 (T)* Paul Oomen/Getty Images, *(M)* Kevin Schafer/Getty Images, *(B)* Daniela Duncan/Getty Images. *Fig. 30.18* Greg Lawler/ Alamy. *Fig. 30.19* Dwight Kuhn. *Fig. 30.20 (a)* Don Johnston/age fotostock, *(b)* David Norton/Alamy, *(c)* Don Johnston/age fotostock, *(d)* David Cavagnaro/Visuals Unlimited, Inc. *Fig. 30.21* F. Valverde, A. Mouradov, W. Soppe, D. Ravenscroft, A. Samach, and G. Coupland, 2004, "Photoreceptor Regulation of CONSTANS Protein in Photoperiodic Flowering," *Science* 303:1003–1006. Reprinted with permission from AAAS. *Fig. 30.24* Scott Smith/Corbis.

## Chapter 31

*Opening photo* Dietrich Rose/Getty Images. *Fig. 31.1 (a)* Elena Elisseeva/Dreamstime.com, *(b)* LianeM/Alamy. *Fig. 31.2 (a, L–R)* Ian Gowland/Photo Researchers, Inc., M. I. Walker/Photo Researchers, Inc., *(b)* Siobhan Braybrook. *Fig. 31.3* M. I. Walker/Photo Researchers, Inc. *Fig. 31.4 (L–R)* Jonathan Buckley/Gap Photo/Visuals Unlimited, Inc., Jerome Wexler/Visuals Unlimited, Inc., Christian Hütter/imagebroker/age fotostock. *Fig. 31.5* Topham/The Image Works. *Fig. 31.6 (a)* Dr. Keith Wheeler/Photo Researchers, Inc. *(b)* Marga Werner/age fotostock, *(c)* J Garden/Alamy, *(d)* Adisa/Dreamstime.com. *Fig. 31.7* Jacqueline S. Wong/ViewFinder Exis, LLC. *Fig. 31.10* Steve Gschmeissner/Photo Researchers, Inc. *Fig. 31.11* Nigel Cattlin/Visuals Unlimited, Inc. *Fig. 31.13* Wally Eberhart/Visuals Unlimited, Inc. *Fig. 31.14 (a)* Alan Majchrowicz/age fotostock, *(b)* Wally Eberhart/Visuals Unlimited, Inc. *Fig. 31.15* Robert & Jean Pollock/Visuals Unlimited, Inc., Scientifica/Visuals Unlimited, Inc. *Fig. 31.16 (a)* Scanning electron micrograph courtesy of the N. Brown Center for Ultrastructure Studies, State University of New York College of Environmental Science and Forestry, Syracuse, NY, *(b)* Andrew Syred/Photo Researchers, Inc. *Fig. 31.17* Carolina Biological Supply Co./Visuals Unlimited, Inc./Corbis. *Fig. 31.18* Science Source/Photo Researchers, Inc. *Fig. 31.19 (a)* Jack Dermid/Visuals Unlimited, Inc., *(b)* A Jagel/age fotostock, *(c)* Jane Grushow/Grant Heilman Photography, *(d)* Patrick Lynch/Alamy. *Fig. 31.22* C. Jourdan, N. Michaux-Ferrière, and G. Perbal, 2000, "Root System Architecture and Gravitropism in the Oil Palm," *Annals of Botany* 85(6):861–868, Fig. 3E. By permission of Oxford University Press. Courtesy of Dr. Christophe Jordan. *Fig. 31.23* G. R. "Dick" Roberts/NSIL/Visuals Unlimited, Inc.

## Chapter 32

*Opening photo* Don Johnston/Getty Images. *Fig. 32.1* Nigel Cattlin/Visuals Unlimited, Inc. *Fig. 32. 2* Clouds Hill Imaging Ltd/Photo Researchers, Inc. *Fig. 32.3 (a)* Clive Varlack/clivevbugs/AGPix, *(b)* Paul Kennedy/Gettty Images, *(c)* Mark Boulton/NHPA. *Fig. 32.5* Norm Thomas/Photo Researchers, Inc. *Fig. 32.6 (a)* Guy Blomme/Bioversity International, *(b)* Bruce Fleming/Cephas Picture Library. *Fig. 32.11 (a)* Fletcher & Baylis/Photo Researchers, Inc., *(b)* Dr. Keith Wheeler/Photo Researchers, Inc., *(c)* Dr. Stanley Flegler/Visuals Unlimited, Inc. *Fig. 32.12* Alex Wild/alexanderwild.com. *Fig. 32.13* Dan L. Perlman/EcoLibrary.org. *Fig. 32.15 (a)* Danny Kessler, Max Planck Institute for Chemical Ecology, Jena, Germany *(b)* Celia Diezel, Max Planck Institute for Chemical Ecology, Jena, Germany. *Fig. 32.17* Data from P. V. A. Fine, I. Mesones, and P. D. Coley, 2004, "Herbivores Promote Habitat Specialization by Trees in Amazonian Forests," *Science* 305:663. *Photos*: Paul Fine. *Fig. 32.19* John L. Obermeyer, IPM Specialist, Department of Entomology, Purdue University.

## Chapter 33

*Opening photo* Frans Lanting/www.lanting.com. *Fig. 33.2* Adapted from A. H. Knoll and K. J. Niklas, 1987, "Adaptation, Plant Evolution, and the Fossil Record," *Review of Palaeobotany and Palynology*, 50:127–149. *Fig. 33.3 (a)* Laurie Campbell /NHPA, *(b)* Densey Clyne/Minden Pictures, Inc., *(c)* Daniel Vega/age fotostock. *Fig. 33.4 (a)* Biopix: A Neumann. *(b)* Piotr Naskrecki/Minden Pictures. *Fig. 33.5 (T)* Science Photo Library/Alamy, *(M)* All Canada Photos/Alamy, *(B)* Daniel Vega/age fotostock. *Fig. 33.6 (T)* Courtesy of Hans Kerp © Palaeobotanical Research Group, University of Münster, *(M & B)* Hans Steur. *Fig. 33.7 (T)* Photo by William Chaloner, Crown Copyright, *(B)* Photo by John Hall © Botanical Society of America. *Fig. 33.8 (a)* Philippe Clement/Nature Picture Library, *(b)* Thomas Marent/Visuals Unlimited, Inc., *(c)* Biopix: JC Schou. *Fig. 33.9 (a)* John Gurche, *(b)* Biophoto Associates/Photo Researchers, Inc., *(c)* Tom Phillips, University of Illinois. *Fig. 33.10 (T & B)* Andrew Knoll, Harvard University. *Fig. 33.11 (a)* Ron Watts/First Light/Corbis, *(b)* David Sieren/Visuals Unlimited, Inc., *(c)* Jonathan Buckley/age fotostock, *(d)* Courtesy of Gary Higgins, *(e)* Doug Sokell/Visuals Unlimited, Inc., *(f)* Biophoto Associates/Photo Researchers, Inc. *Fig. 33.12* Jean-Paul Ferrero/Auscape/Minden Pictures. *Fig. 33.13 (a)* John Shaw/Auscape/Minden Pictures, *(b)* Gerry Bishop/Visuals Unlimited, Inc. *Fig. 33.14* Clay Perry/Corbis. *Fig. 33.15 (T–B)* Pete Ryan/Getty Images, Michael P. Gadomski/Photo Researchers, Inc., Gerald & Buff Corsi/Visuals Unlimited, Inc., Robert Davis/age fotostock. *Fig. 33.16 (a)* Piotr Naskrecki/Minden Pictures, *(b & c)* Andrew Knoll, Harvard University. *Fig. 33.17 (a)* JTB/Photoshot, *(b)* Kathy Merrifield/Photo Researchers, Inc. *Fig. 33.18* Andrew Knoll, Harvard University. *Fig. 33.19 (a)* M. Philip Kahl/Photo Researchers, Inc., *(b)* Ahmad Fuad Morad [Flickr page] 072112, *(c)* Bob Gibbons/Photo Researchers, Inc. *Fig. 33.20 (T–B)* Sangtae Kim, Frans Lanting/National Geographic Stock, Rob & Ann Simpson/Visuals Unlimited, Inc., Jo Whitworth/age fotostock, Tropicals JR Mau/PhotoResourceHawaii.com, Image Source/Getty Images. *Fig. 33.21* Keith Rushforth/FLPA/Minden Pictures. *Fig. 33.22 (a)* Robert Gustafson/Visuals Unlimited, Inc., *(b)* Kerstin Hinze/naturepl.com/NaturePL, *(c)* Axle71/Dreamstime.com, *(d)* Tropicals JR Mau/PhotoResourceHawaii.com. *Fig. 33.24 (a)* Michael P Gadomski/Getty Images, *(b)* Sergio Hayashi/Dreamstime.com, *(c)* Smileyjoanne/Dreamstime.com, *(d)* Danita Delimont/Getty Images.

## Chapter 34

*Opening photo* Chris Ted/Getty Images. *Fig. 34.1* Dennis Kunkel Microscopy, Inc./Visuals Unlimited, Inc. *Fig. 34.3* Simko/Visuals Unlimited, Inc. *Fig. 34.4 (a)* Science Photo Library/Alamy, *(b)* T. R. Filley, R. A. Blanchette, E. Simpson, and M. L. Fogel, 20001, "Nitrogen Cycling by Wood Decomposing Soft-Rot Fungi in the 'King Midas tomb,' Gordion, Turkey," *Proceedings of the National Academy of Sciences USA* 98(23):13346–13350. Copyright 2001

National Academy of Sciences, U.S.A. *Photo*s courtesy of Robert A. Blanchette, University of Minnesota. *Fig. 34.5 (a)* Science Source/Photo Researchers, Inc., *(b)* Biophoto Associates/Photo Researchers, Inc. *Fig. 34.6* Reprinted from J. K. Pirri and M. J. Alkema, 2011, "The Neuroethology of *C. elegans* Escape," *Current Opinion in Neurobiology* 22:187–193, Fig. 3. Copyright 2011, with permission from Elsevier. Reproduced with permission from Mark Alkema. *Fig. 34.7* Dr. Jeremy Burgess/Photo Researchers, Inc. *Fig. 34.8 (L–R)* Wallace Garrison/Getty Images, Stephen Sharnoff/National Geographic Stock, Stephen Sharnoff/National Geographic Stock. *Fig. 34.9* Eye of Science/Photo Researchers, Inc. *Fig. 34.10 (a)* Satoshi Kuribayashi/Minden Pictures, *(b, T)* E. R. Degginger/Photo Researchers, Inc., *(b, B)* Garry DeLong/Getty Images. *Fig. 34.11* Matt Meadow/Getty Images. *Fig. 34.12 (L)* Reprinted from R. Dulymamode, P. F. Cannon, and A. Peerally, 1998, "Fungi from Mauritius *Anthostomella* Species on *Pandanus*," *Mycological Research* 102(11): 1319–1324. Copyright 1998, with permission from Elsevier, *(M)* Reprinted from N. L. Glass, R. L. Metzenberg, and N. B. Raju, 1990, "Homothallic Sordariaceae from Nature: The Absence of Strains Containing Only the a Mating Type Sequence," *Experimenal Mycology* 14(3):274–289. Copyright 1990, with permission from Elsevier, *(R)* I. Schmitt, H. T. Lumbsch, and C. Bratt, 2006, "Two New Brown-Spored Species of *Pertusaria* from South-Western North America," *The Lichenologist* 38(5):411–416, Cambridge University Press. *Fig. 34.16* John Taylor/Visuals Unlimited, Inc. *Fig. 34.17* Carolina Biological Supply Co./Visuals Unlimited, Inc. *Fig. 34.18 (a)* Topic Photo Agency/age fotostock, *(b)* Matt Meadows/Getty Images. *Fig. 34.19* Ed Reschke/Getty Images. *Fig. 34.20* David Hughes. *Fig. 34.21 (a)* Inga Spence/Visuals Unlimited, Inc., *(b)* Bruce Perrault, *(c)* David Clapp/Getty Images. *Fig. 34.22 (a)* Albert Lleal/Minden Pictures, *(b)* Wally Eberhart/Getty Images, *(c)* Matthias Breiter/Minden Pictures. *Fig. 34.24* After "The Spread of Wheat Stem Rust UG99 Lineage," Food and Agriculture Organization of the United Nations, JPEG image, 2010, http://www. fao.org/agriculture/crops/rust/stem/rust-report/stem-ug99racettksk/en/. *Photo and inset photo:* Yue Jin, USDA-ARS, Cereal Disease Laboratory.

### Case 7

*Wolf* Samuel R. Maglione/Photo Researchers, Inc. *Moose* Ron Erwin/age fotostock. *Graph* Data from J. A. Vucetich and R. O. Peterson, 2012, "The population biology of Isle Royale wolves and moose: an overview," www.isleroyalewolf.org.

### Chapter 35

*Opening photo* Dr. Michael Delannoy/Getty Images. *Fig. 35.1* Winfried Wisniewski/Corbis. *Fig. 35.2 (T)* Andrew J. Martinez/Photo Researchers, Inc., *(M)* UMI NO KAZE/a.collectionRF /Getty Images, *(B)* Robert Pickett/Visuals Unlimited, Inc. *Fig. 35.6 (L)* Cajal Legacy, Instituto Cajal (CSIC), Madrid, Spain, *(R)* Bob Jacobs, Ph.D., Laboratory of Quantitative Neuromorphology, Department of Psychology, Colorado College. *Fig. 35.7* Science VU/C. Raine/Visuals Unlimited, Inc. *Fig. 35.14* Mary B. Kennedy, Division of Biology, Caltech. *Fig. 35.16* David B. Fankhauser, Ph.D., Professor of Biology and Chemistry, University of Cincinnati, Clermont College.

### Chapter 36

*Opening photo* Custom Medical Stock Photo/Getty Images. *Fig. 36.1 (a)* Gerry Bishop/Visuals Unlimited, Inc., *(b)* Joe McDonald/Visuals Unlimited, Inc., *(c)* Monika Ondru ová/Dreamstime.com. *Fig. 36.2 (a)* Gary Meszaros/Visuals Unlimited, Inc., *(b)* Sinclair Stammers/Photo Researchers, Inc., *(c)* Danté Fenolio/Photo Researchers, Inc. *Fig. 36.8* SPL/Photo Researchers, Inc. *Fig. 36.10* Michael Durham/Minden Pictures. *Fig. 36.11* Rendered by Dale Muzzey. *Fig. 36.12 (a)* Tom Adams/Visuals Unlimited, Inc., *(b)* Eye of Science/Photo Researchers, Inc., *(c)* Danté Fenolio/Photo Researchers, Inc. *Fig. 36.13* Bjorn Rorslett/Photo Researchers, Inc. *Fig. 36.17* Science Source/Photo Researchers, Inc. *Fig. 36.21*

The National Library of Medicine. *Fig. 36.26* WDCN/University College London/ Photo Researchers, Inc.

### Chapter 37

*Opening photo* Gustoimages/Science Photo Library/Photo Researchers, Inc. *Fig. 37.1* Tom Brakefield/Corbis. *Fig. 37.2* Dr. Gladden Willis/Visuals Unlimited, Inc. *Fig. 37.5 (a)* Don W. Fawcett/Photo Researchers, Inc. *(b)* Biophoto Associates/ Photo Researchers, Inc. *Fig. 37.6* Reprinted by permission from Macmillan Publishers Ltd: H. Huxley and J. Hanson, "Changes in the Cross-Striations of Muscle During Contraction and Stretch and Their Structural Interpretation," 1954, *Nature* 173:973–976. Copyright 1954. *Fig. 37.8* Don W. Fawcett/Photo Researchers, Inc. *Fig. 37.12* Adapted from E. P. Widmaier, H. Raff, and K. T. Strang, 2011, *Vander's Human Physiology*, New York: McGraw-Hill. *Fig. 37.14* Dr. Gladden Willis/Visuals Unlimited, Inc. *Fig. 37.15 (a)* Jan Van Arkel/ Minden Pictures, *(b)* Nigel Cattlin/Visuals Unlimited, Inc. *Fig. 37.16 (a)* Rita van den Broek/age fotostock, *(b)* Reinhard Dirscherl/Visuals Unlimited, Inc., *(c)* Thomas Marent/Visuals Unlimited, Inc. *Fig. 37.18* Science Photo Library/Photo Researchers, Inc. *Fig. 37.19* Ralph Hutchings/Visuals Unlimited, Inc. *Fig. 37.22* Based on M. J. Smith and R. J. G. Savage, 1956, "Some Locomotory Adaptations in Mammals," *Zoological Journal of the Linnean Society* 42:603–622. *Photo*: Norbert Wu/Minden Pictures, J. L. Klein & M. L. Hubert/Photo Researchers, Inc.

### Chapter 38

*Opening photo* Heidi & Hans-Jurgen Koch/Minden Pictures/Getty Images. *Fig. 38.12* Peter Blackwell/Minden Pictures. *Fig. 38.14 (L–R)* Jef Meul/ Minden Pictures, Jorg & Petra Werner/Animals Animals–Earth Scenes, Steven Smith/Alamy. *Fig. 38.15* Piotr Naskrecki/Minden Pictures/Corbis. *Fig. 38.16* blickwinkel/Alamy.

### Chapter 39

*Opening photo* Science Photo Library/Photo Researchers, Inc. *Fig. 39.1* Amar & Isabelle Guillen/SeaPics.com. *Fig. 39.4 (T)* José B. Ruiz/Minden Pictures, *(B)* Stephen Dalton/Animals Animals. *Fig. 39.8* Reprinted from E. Weibel, B. Sapoval, and M. Filoche, 2005, "Structure and Function in the Periphery of the Lung," *Respiratory Physiology & Neurobiology* 148:3–21. Copyright 2005, with permission from Elsevier. *Fig. 39.11* Rendered by Dale Muzzey. *Fig. 39.13* Based on Fig. 3A in P. L. Tyack, M. Johnson, N. Aguilar de Soto, A. Sturlese, and P. T. Madsen, 2006, "Extreme Diving Behaviour of Beaked Whale Species Known to Strand in Conjunction with Use of Military Sonars, *Journal of Experimental Biology* 209:4238–4253. *Photo*: Bill Curtsinger/Getty Images. *Fig. 39.17* Data from C. B. Chapman and J. H. Mitchell, 1965, "The Physiology of Exercise," *Scientific American* 212:88–96.

### Chapter 40

*Opening photo* Science Source/Photo Researchers, Inc. *Fig. 40.2* age fotostock/ SuperStock. *Fig. 40.4* Adapted from A. M. Hemingsen, 1960, "Energy metabolism as related to body size and respiratory surfaces, and its evolution." *Re. Steno Memorial Hospital Nordisk Insulinlaboratorium.* 9:1–110. *Fig. 40.5* Cary Wolinsky/Getty Images. *Fig. 40.8 (a)* Fred Bavendam/Minden Pictures, *(b)* © 2012 Brandon Cole www.norbertwu.com. *Fig. 40.9* Marevision/Getty Images. *Fig. 40.10* C & M Fallows/SeaPics.com. *Fig. 40.16* Frank Woerle/ Minden Pictures. *Fig. 40.18* Biophoto Associates/Photo Researchers, Inc.

### Chapter 41

*Opening photo* Steve Allen/Getty Images. *Fig. 41.5* David Wall/Alamy. *Fig. 41.6 (a)* Peter Betts/iStockphoto, *(b)* Roger De La Harpe; Gallo Images/Corbis. *Fig. 41.8 (a)* Michael Abbey/Photo Researchers, Inc., *(b)* Michael Abbey/Photo Researchers, Inc. *Fig. 41.15* Minden Pictures/SuperStock.

J. E. Kaldy and K. H. Dunton. "Ontogenetic Photosynthetic Changes, Dispersal and Survival of *Thalassia testudinum* (Turtlegrass) Seedlings in a Sub-tropical Lagoon," *Journal of Experimental Marine Biology and Ecology* 240(2):193–212. *Fig. 46.13* Charles O'Rear/Corbis. *Fig. 46.15* Mark Moffett/Minden Pictures. *Fig. 46.18* After R. H. MacArthur and E. O. Wilson, 1967, *The Theory of Island Biogeography*, Princeton, NJ: Princeton University Press. *Fig. 46.19 (T–B)* Eladio M. Fernandez, Eladio M. Fernandez, Eladio M. Fernandez, © 2013 Jake Scott, Edward Myles/FLPA/Minden Pictures, Eladio M. Fernandez.

## Chapter 47

*Opening photo* Charles J. Smith. *Fig. 47.1* Eladio M. Fernandez. *Fig. 47.3 (a)* Michael & Patricia Fogden/Minden Pictures, *(b)* Mark Moffett/Minden Pictures. *Fig. 47.4* Charles J. Smith. *Fig. 47.5* Courtesy of Jim Maddox, Galloway Twp., NJ; picture-newsletter.com. *Fig. 47.6* Data from P. J. Morin, 1981, "Predatory Salamanders Reverse Outcome of Competition Among Three Species of Anuran Tadpoles," *Science* 212:1284–1286. *Fig. 47.7* Gerry Bishop/ Visuals Unlimited, Inc. *Fig. 47.8* Biosphoto /Gérard Lacz. *Fig. 47.10* Carl Gwinn, blackcormorant.net. *Fig. 47.11* © 2012 DigitalGlobe © 2012 Google. *Fig. 47.13* Dan L. Perlman/EcoLibrary.org. *Fig. 47.17 (Tundra)* Helge Schulz /Picture Press/ Getty Images, *(Alpine)* Christian Ziegler/Minden Pictures, *(Taiga)* Gerry Ellis/ Minden Pictures, *(Temperate coniferous forest)* Charlie Ott/Photo Researchers, Inc., *(Deciduous forest)* Tim Fitzharris/Minden Pictures, *(Temperate grassland)* Robert and Jean Pollock/Photo Researchers, Inc., *(Desert)* Yva Momatiuk & John Eastcott/Minden Pictures, *(Chaparral)* Richard Herrmann/Visuals Unlimited, Inc., *(Savanna)* Richard Du Toit/Minden Pictures, *(Rain forest)* Charles J. Smith. *Fig. 47.18* After D. M. Olson et al., 2001, "Terrestrial Ecoregions of the World: New Map of Life on Earth," *Bioscience* 51:933–938. *Fig. 47.21* After D. M. Kaufman, 1995, "Diversity of New World Mammals: Universality of the Latitudinal Gradients of Species and Bauplans," *Journal of Mammalogy* 76:322–334. *Fig. 47.22* Courtesy of George Poinar.

## Chapter 48

*Opening photo* Michael Utech/Getty Images. *Fig. 48.1* Data from "Historical Estimates of World Population," United States Census Bureau International Data Base, last modified June 2012, http://www.census.gov/population/international/ data/worldpop/table_history.php. *Fig. 48.2* Data from *United Nations* Department of Economic and Social Affairs, 2009, *World Economic and Social Survey 2009: Promoting Development, Saving the Planet*, New York: United Nations, p. 44, http://www.un.org/en/development/desa/news/policy/wess-2009.shtml. *Fig. 48.3* Data from *Global Footprint Network, 2006; United Nations Development Programme, 2006. Fig. 48.4* Tim McCaig/iStockphoto, Stockbyte /Getty Images. *Fig. 48.5* Data from R. A. Rohde, Global Warming Art, http://www. globalwarmingart.com/. Based on data from the Hadley Centre Coupled Model version 3 (HadCM3), http://badc.nerc.ac.uk/view/badc.nerc.ac.uk__ATOM__ dpt_1162913571289262. *Fig. 48.6* Fotosearch Stock Images. *Fig. 48.7* Adapted from R. A. Rohde, "Global Warming Predictions," from Global Warming Art, JPEG image, http://www.globalwarmingart.com/wiki/File:Global_Warming_ Predictions_Map_jpg. Based on data from the Hadley Centre Coupled Model version 3 (HadCM3), http://badc.nerc.ac.uk/view/badc.nerc.ac.uk__ATOM__ dpt_1162913571289262. *Fig. 48.8* Data from A. J. Miller-Rushing and R. B. Primack, 2008, "Global Warming and Flowering Times in Thoreau's Concord: A Community Perspective," *Ecology* 89(2):332–341, doi: 10.1890/07-0068.1. *Fig. 48.9* After M. B. Davis, 1976, "Pleistocene Biogeography of Temperate Deciduous Forests," *Geoscience and Man* 13:13–26. *Fig. 48.11* Fred Bavendam/ Minden Pictures. *Fig. 48.13* Reprinted by permission from Macmillan Publishers Ltd: U. Riebesell, I. Zondervan, B. Rost, P. D. Tortell, R. E. Zeebe, and F. M. M. Morel, 2000, "Reduced Calcification of Marine Plankton in Response to Increased Atmospheric CO$_2$," *Nature* 407:364–367. Copyright 2000. *Fig. 48.14* euroluftbild. de/dpa/Corbis, Travel Pix/Robert Harding. *Fig. 48.15* Image courtesy of Liam Gumley, Space Science and Engineering Center, University of Wisconsin– Madison and the MODIS Science Team. *Fig. 48.17* Data from M. D. Edgerton, 2009, "Increasing Crop Productivity to Meet Global Needs for Feed, Food, and Fuel," *Plant Physiology* 149:9, http://www.plantphysiol.org/content/149/1/7. *Fig. 48.18, Fig. 48.18 map* After Millennium Ecosystem Assessment, 2005, *Ecosystems and Human Well-being: Synthesis*, Washington, DC: Island Press). http://www.millenniumassessment.org/documents/document.356.aspx.pdf. *Fig. 48.19* AP Photo/Andre Penner. *Fig. 48.20* James K. York/Dreamstime. com, AP Photo/U.S. Department of Agriculture, Ch'ien Lee/Minden Pictures. *Fig. 48.22* Michael Fogden/Animals Animals. *Fig. 48.23 (Doctor)* AP Photo/ Sebastian John, *(Science teacher)* WA/Dann Tardif/Blend Images/Getty Images, *(Scientist in the field)* Alexis Rosenfeld/Photo Researchers, Inc., *(Scientists testifying before congressional subcommittee)* Matthew Cavanaugh/EPA/ Newscom, Chas Metivier/ZUMA Press/Newscom.

# INDEX

**Bold face** indicates a definition, *italics* indicates a figure, 1-1*t* indicates a table, C1-1 indicates a Case.

Due to a printing error, the opposite page is incorrect.

Please see the back of this page for a corrected version of page I-19.